Transients in Electrical Systems

Analysis, Recognition, and Mitigation

J. C. Das

McGraw Hill

New York Chicago San Francisco Lisbon London Madrid
Mexico City Milan New Delhi San Juan Seoul
Singapore Sydney Toronto

The McGraw·Hill Companies

Library of Congress Cataloging-in-Publication Data

Das, J. C., date.
Transients in electrical systems : analysis, recognition, and mitigation / J.C. Das.
p. cm.
Includes bibliographical references.
ISBN 978-0-07-162248-6 (alk. paper)
1. Transients (Electricity) I. Title.
TK3226.D3165 2010
621.319'21—dc22
2010011147

McGraw-Hill books are available at special quantity discounts to use as premiums and sales promotions, or for use in corporate training programs. To contact a representative please e-mail us at bulksales@mcgraw-hill.com.

Transients in Electrical Systems

1 2 3 4 5 6 7 8 9 0 CCW/CCW 1 6 5 4 3 2 1 0

ISBN 978-0-07-162248-6
MHID 0-07-162248-9

This book is printed on acid-free paper.

Sponsoring Editor
Michael Penn

Acquisitions Coordinator
Michael Mulcahy

Editorial Supervisor
David E. Fogarty

Project Manager
Vastavikta Sharma, Glyph International

Copy Editor
Ragini Pandey, Glyph International

Proofreader
Connie Blazewicz

Production Supervisor
Richard C. Ruzycka

Composition
Glyph International

Art Director, Cover
Jeff Weeks

To My Parents

About the Author

J. C. Das is currently Staff Consultant, Electrical Power Systems, AMEC Inc., Tucker, Georgia, USA. He has varied experience in the utility industry, industrial establishments, hydroelectric generation, and atomic energy. He is responsible for power system studies, including short-circuit, load flow, harmonics, stability, arc-flash hazard, grounding, switching transients, and also, protective relaying. He conducts courses for continuing education in power systems and has authored or coauthored about 60 technical publications. He is author of the book *Power System Analysis, Short-Circuit, Load Flow and Harmonics* (New York, Marcel Dekker, 2002); its second edition is forthcoming. His interests include power system transients, EMTP simulations, harmonics, power quality, protection, and relaying. He has published 185 electrical power systems study reports for his clients.

He is a Life Fellow of the Institute of Electrical and Electronics Engineers, IEEE (USA), a member of the IEEE Industry Applications and IEEE Power Engineering societies, a Fellow of Institution of Engineering Technology (UK), a Life Fellow of the Institution of Engineers (India), a member of the Federation of European Engineers (France), and a member of CIGRE (France). He is a registered Professional Engineer in the states of Georgia and Oklahoma, a Chartered Engineer (C. Eng.) in the UK, and a European Engineer (Eur. Ing.).

He received a MSEE degree from Tulsa University, Tulsa, Oklahoma in 1982 and BA (mathematics) and BEE degrees in India.

Contents

Appendix D Sequence Impedances of Transmission Lines and Cables

Appendix E Energy Functions and Stability

Appendix F Statistics and Probability

Appendix G Numerical Techniques

PREFACE

The book aims to serve as a textbook for upper undergraduate and graduate level students in the universities, a practical and analytical guide for practicing engineers, and a standard reference book on transients. At the undergraduate level, the subject of transients is covered under circuit theory, which does not go very far for understanding the nature and impact of transients. The transient analyses must account for special modeling and frequency-dependent behavior and are important in the context of modern power systems of increasing complexity.

Often, it is difficult to predict intuitively that a transient problem exists in a certain section of the system. Dynamic modeling in the planning stage of the systems may not be fully investigated. The book addresses analyses, recognition, and mitigation. Chapters on surge protection, TVSS (transient voltage surge suppression), and insulation coordination are included to meet this objective.

The book is a harmonious combination of theory and practice. The theory must lead to solutions of practical importance and real world situations.

A specialist or a beginner will find the book equally engrossing and interesting because, starting from the fundamentals, gradually, the subjects are developed to a higher level of understanding. In this process, enough material is provided to sustain a reader's interest and motivate him to explore further and deeper into an aspect of his/her liking.

The comprehensive nature of the book is its foremost asset. All the transient frequencies, in the frequency range from 0.1 Hz to 50 MHz, which are classified into four groups: (1) low frequency oscillations, (2) slow front surges, (3) fast front surges, and (4) very fast front surges, are discussed. Transients that affect power system stability and transients in transmission lines, transformers, rotating machines, electronic equipment, FACTs, bus transfer schemes, grounding systems, gas insulated substations, and dc systems are covered. A review of the contents will provide further details of the subject matter covered and the organization of the book.

An aspect of importance is the practical and real world "feel" of the transients. Computer and EMTP simulations provide a vivid visual impact. Many illustrative examples at each stage of the development of a subject provide deeper understanding.

The author is thankful to Taisuke Soda of McGraw-Hill for his help and suggestions in the preparation of the manuscript and subsequent printing.

J. C. Das

CHAPTER 1

INTRODUCTION TO TRANSIENTS

Electrical power systems are highly nonlinear and dynamic in nature: circuit breakers are closing and opening, faults are being cleared, generation is varying in response to load demand, and the power systems are subjected to atmospheric disturbances, that is, lightning. Assuming a given steady state, the system must settle to a new *acceptable* steady state in a short duration. Thus, the electromagnetic and electromechanical energy is constantly being redistributed in the power systems, among the system components. These energy exchanges cannot take place instantaneously, but take some time period which brings about the transient state. The energy statuses of the sources can also undergo changes and may subject the system to higher stresses resulting from increased currents and voltages.

The analysis of these excursions, for example, currents, voltages, speeds, frequency, torques, in the electrical systems is the main objective of transient analysis and simulation of transients in power systems.

1-1 CLASSIFICATION OF TRANSIENTS

Broadly, the transients are studied in two categories, based upon their origin:

1. Of atmospheric origin, that is, lightning

2. Of switching origin, that is, all switching operations, load rejection, and faults

Another classification can be done based upon the mode of generation of transients:

1. *Electromagnetic transients*. Generated predominantly by the interaction between the electrical fields of capacitance and magnetic fields of inductances in the power systems. The electromagnetic phenomena may appear as traveling waves on transmission lines, cables, bus sections, and oscillations between inductance and capacitance.

2. *Electromechanical transients*. Interaction between the electrical energy stored in the system and the mechanical energy stored in the inertia of the rotating machines, that is, generators and motors.

As an example, in transient stability analysis, both these effects are present. The term transient, synonymous with surges, is used loosely to describe a wide range of frequencies and magnitudes. Table 1-1 shows the power system transients with respect to the time duration of the phenomena.

1-2 CLASSIFICATION WITH RESPECT TO FREQUENCY GROUPS

The study of transients in power systems involves frequency range from dc to about 50 MHz and in specific cases even more. Table 1-2 gives the origin of transients and most common frequency ranges. Usually, transients above power frequency involve electromagnetic phenomena. Below power frequency, electromechanical transients in rotating machines occur.

Table 1-3 shows the division into four groups, and also the phenomena giving rise to transients in a certain group is indicated. This classification is more appropriate from system modeling considerations and is proposed by CIGRE Working Group 33.02.[1]

Transients in the frequency range of 100 kHz to 50 MHz are termed very fast transients (VFT), also called very fast front transients. These belong to the highest range of transients in power systems. According to IEC 60071-1,[2] the shape of a very fast transient is usually unidirectional with time to peak less than 0.1 μs, total duration less than 3 ms, and with superimposed oscillation at a frequency of 30 kHz $< f <$ 100 MHz. Generally, the term is applied to transients of frequencies above 1 MHz. These transients can originate in gas-insulated substations (GIS), by switching of motors and transformers with short connections to the switchgear, by certain lightning conditions, as per IEC 60071-2.[2]

Lightning is the fastest disturbance, from nanoseconds to microseconds. The peak currents can approach 100 kA in the first stroke and even higher in the subsequent strokes.

Nonpermanent departures form the normal line voltage, and frequency can be classified as power system disturbances. These deviations can be in wave shape, frequency, phase relationship, voltage unbalance, outages and interruptions, *surges and sags,* and *impulses and noise.* The phenomena shown in italics may loosely be called transients.[3] A stricter definition is that a transient is a subcycle disturbance in the ac waveform that is evidenced by a sharp, brief discontinuity in the waveform, which may be additive or subtractive

TABLE 1-1 Time Duration of Transient Phenomena in Electrical Systems

Nature of the Transient Phenomena	Time Duration
Lightning	0.1 μs–1.0 ms
Switching	10 μs to less than a second
Subsynchronous resonance	0.1 ms–5 s
Transient stability	1 ms–10 s
Dynamic stability, long-term dynamics	0.5–1000 s
Tie line regulation	10–1000 s
Daily load management, operator actions	Up to 24 h

TABLE 1-2 Frequency Ranges of Transients

Origin of Transient	Frequency Range
Restrikes on disconnectors and faults in GIS	100 kHz–50 MHz
Lightning surges	10 kHz–3 MHz
Multiple restrikes in circuit breakers	10 kHz–1 MHz
Transient recovery voltage:	
Terminal faults	50/60 Hz–20 kHz
Short-line faults	50/60 kHz–100 kHz
Fault clearing	50/60 Hz–3 kHz
Fault initiation	50/60 Hz–20 kHz
Fault energization	50/60 Hz–3 kHz
Load rejection	0.1 Hz–3 kHz
Transformer energization	(dc) 0.1 Hz–1 kHz
Ferroresonance	(dc) 0.1 Hz–1 kHz

TABLE 1-3 Classification of Frequency Ranges[1]

Group	Frequency Range for Representation	Shape Designation	Representation Mainly for
I	0.1 Hz–3 kHz	Low-frequency oscillations	Temporary overvoltages
II	50/60 Hz–20 kHz	Slow front surges	Switching overvoltages
III	10 kHz–3 MHz	Fast front surges	Lightning overvoltages
IV	100 kHz–50 MHz	Very fast front surges	Restrike overvoltages, GIS

from the original waveform. Yet, in common use, the term transients embraces overvoltages of various origins, transients in the control systems, transient and dynamic stability of power systems, and dynamics of the power system on short circuits, starting of motors, operation of current limiting fuses, grounding systems, and the like.

The switching and fault events give rise to overvoltages, up to three times the rated voltage for phase-to-ground transients, and up to four times for phase-to-phase transients. The rise time varies from 50 μs to some thousands of microseconds. The simulation time may be in several cycles, if system recovery from disturbance is required to be investigated.

The physical characteristics of a specific network element, which affect a certain transient phenomena, must receive detailed considerations. Specimen examples are:

- The saturation characteristics of transformers and reactors can be of importance in case of fault clearing, transformer energization, and if significant temporary overvoltages are expected. Temporary overvoltages originate from transformer energization, fault overvoltages, and overvoltages due to load rejection and resonance.

- On transmission line switching, not only the characteristics of the line itself, but also of the feeding and terminating networks will be of interest. If details of initial rate of rise of overvoltages are of importance, the substation details, capacitances of measuring transformers, and the number of outgoing lines and their surge impedances also become equally important.

- When studying phenomena above 1 MHz, for example, in GIS caused by a disconnector strike, the small capacitances and inductances of each section of GIS become important.

These are some representative statements. The system configuration under study and the component models of the system are of major importance. Therefore, the importance of frequency-dependent models cannot be overstated. Referring to Table 1-3, note that the groups assigned are not hard and fast with respect to the phenomena described, that is, faults of switching origin may also create steep fronted surges in the local vicinity.

1-3 FREQUENCY-DEPENDENT MODELING

The power system components have frequency-dependent behavior, and the development of models that are accurate enough for a wide range of frequencies is a difficult task. The mathematical representation of each power system component can, generally, be developed for a specific frequency range. This means that one model cannot be applied to every type of transient study. This can lead to considerable errors and results far removed from the real-world situations. This leads to the importance of correct modeling for each specific study type, which is not so straightforward in every case.[4]

1-3-1 SoFT

SoFT™ (Swiss Technology Award, 2006) is a new approach that measures the true and full frequency-dependent behavior of the electrical equipment. This reveals the interplay between the three phases of an ac system, equipment interaction, and system resonances to achieve the most accurate frequency-dependent models of electrical components. The three-step process is:

1. On-site measurements
2. Determining the frequency dependent models
3. Simulation and modeling

The modeling fits a state-space model to the measured data, based upon vector fitting techniques. Five frequency-independent matrices representing the state-space are generated, and in the frequency domain, the matrix techniques are used to eliminate the state vector $\dot{\mathbf{x}}$. An admittance matrix is then generated. The matrices of state-space can be directly imported into programs like EMTP-RV. Thus, a highly accurate simulation can be performed.

Apart from the reference here, this book does not discuss the field measurement techniques for ascertaining system data for modeling.

1-4 OTHER SOURCES OF TRANSIENTS

Detonation of nuclear devices at high altitudes, 40 km and higher, gives rise to transients called high-altitude electromagnetic pulse (HEMP). These are not discussed in this book.

Strong geomagnetic storms are caused by sunspot activity every 11 years or so, and this can induce dc currents in the transmission lines and magnetize the cores of the transformers connected to the end of transmission lines. This can result in much saturation of the iron core. In 1989 a large blackout was reported in the U.S. and Canadian electrical utilities due to geomagnetic storms (Chap. 14).

Extremely low magnetic fields (ELF), with a frequency of 60 Hz with higher harmonics up to 300 Hz and lower harmonics up to 5 Hz, are created by alternating current, and associations have been made between various cancers and leukemia in some epidemiological studies (Chap. 22).

1-5 STUDY OF TRANSIENTS

The transients can be studied from the following angles:

- Recognition
- Prediction
- Mitigation

This study of transients is fairly involved, as it must consider the behavior of the equipment and amplification or attenuation of the transient in the equipment itself. The transient voltage excitation can produce equipment responses that may not be easy to decipher intuitively or at first glance.

Again coming back to the models of the power system equipment, these can be generated on two precepts: (1) based on lumped parameters, that is, motors, capacitors, and reactors (though wave propagation can be applied to transient studies in motor windings) and (2) based on distributed system parameters, that is, overhead lines and underground cables (though simplifying techniques and lumped parameters can be used with certain assumptions). It is important that transient simulations and models must reproduce frequency variations, nonlinearity, magnetic saturation, surge-arresters characteristics, circuit breaker, and power fuse operation.

The transient waveforms may contain one or more oscillatory components and can be characterized by the natural frequencies of these oscillations, which are dependent upon the nature of the power system equipment.

Transients are generated due to phenomena internal to the equipment, or of atmospheric origin. Therefore, the transients are inherent in the electrical systems. Mitigation through surge arresters, transient voltage surge suppressors (TVSS), active and passive filters, chokes, coils and capacitors, snubber and damping circuits requires knowledge of the characteristics of these devices for appropriate analyses. Standards establish the surge performance of the electrical equipment by application of a number of test wave shapes and rigorous testing, yet to apply proper strategies and devices for a certain design configuration of a large system, for example, high-voltage transmission networks, detailed modeling and analysis are required. Thus, for mitigation of transients we get back to analysis and recognition of the transient problem.

This shows that all the three aspects, analysis, recognition, and mitigation, are interdependent, the share of analysis being more than 75 percent. After all, a mitigation strategy must again be analyzed and its effectiveness be proven by modeling before implementation.

It should not, however, be construed that we need to start from the very beginning every time. Much work has been done. Over the past 100 years at least 1000 papers have been written on the subject and ANSI/IEEE and IEC standards provide guidelines.

1-6 TNAs—ANALOG COMPUTERS

The term TNA stands for transient network analyzer. The power system can be modeled by discrete scaled down components of the power system and their interconnections. Low voltage and current levels are used. The analog computer basically solves differential equations, with several units for specific functions, like adders, integrators, multipliers, CRT displays, and the like. The TNAs work in real time; many runs can be performed quickly and the system data changed, though the setting up of the base system model may be fairly time-consuming. The behavior of actual control hardware can be studied and validated before field applications. The advancement in digital computation and simulation is somewhat overshadowing the TNA models, yet these remain a powerful analog research tool. It is obvious that these simulators could be relied upon to solve relatively simple problems. The digital computers provide more accurate and general solutions for large complex systems.

1-7 DIGITAL SIMULATIONS, EMTP/ATP, AND SIMILAR PROGRAMS

The electrical power systems parameters and variables to be studied are continuous functions, while digital simulation, by its nature, is discrete. Therefore, the development of algorithms to solve digitally the differential and algebraic equations of the power system was the starting point. H.W. Dommel of Bonneville Power Administration (BPA) published a paper in 1969,[5] enumerating digital solution of power system electromagnetic transients based on difference equations (App. C). The method was called Electromagnetic Transients Program (EMTP). It immediately became an industrial standard all over the world. Many research projects and the Electrical Power Research Institute (EPRI) contributed to it. EMTP was made available to the worldwide community as the Alternate Transient Program (ATP), developed with W. S. Meyer of BPA as the coordinator.[6] A major contribution, Transient Analysis of Control Systems (TACS), was added by L. Dubé in 1976.

A mention of the state variable method seems appropriate here. It is a popular technique for numerical integration of differential equations that will not give rise to a numerical instability problem inherent in numerical integration (App. G). This can be circumvented by proper modeling techniques.

The versatility of EMTP lies in the component models, which can be freely assembled to represent a system under study. Nonlinear resistances, inductances, time-varying resistances, switches, lumped elements (single or three-phase), two or three winding transformers, transposed and untransposed transmission lines, detailed generator models according to Park's transformation, converter circuits, and surge arresters can be modeled. The insulation coordination, transient stability, fault currents, overvoltages due to switching surges, circuit breaker operations, transient behavior of power system under electronic control subsynchronous resonance, and ferroresonance phenomena can all be studied.

Electromagnetic Transients Program for DC (EMTDC) was designed by D. A. Woodford of Manitoba Hydro and A. Gole and R. Menzies in 1970. The original program ran on mainframe computers. The EMTDC version for PC use was released in the 1980s. Manitoba HVDC Research Center developed a comprehensive graphic user interface called Power System Computer Aided Design (PSCAD), and PSCAD/EMTDC version 2 was released in the 1990s for UNIX work stations, followed by a Windows/PC-based version in 1998. EMTP-RV is the restructured version of EMTP.[7,8]

Other EMTP type programs are: MicroTran by Micro Tran Power System Analysis Corporation, Transient Program Analyzer (TPA) based upon MATLAB; NETOMAC by Siemens; SABER by Avant.[8]

It seems that in a large number of cases dynamic analyses are carried out occasionally in the planning stage and in some situations dynamic analysis is not carried at all.[9] The reasons of lack of analysis were identified as:

- A resource problem
- Lack of data
- Lack of experience

Further, the following problems were identified as the most crucial, in the order of priority:

- Lack of models for wind farms
- Lack of models for new network equipment
- Lack of models for dispersed generation (equivalent dynamic models for transmission studies)
- Lack of verified models (specially dynamic models) and data for loads
- Lack of field verifications and manufacturer's data to ensure that generator parameters are correct
- Lack of open cycle and combined cycle (CC) gas turbine models in some cases

Thus, data gathering and verifications of the correct data is of great importance for dynamic analysis.

EMTP/ATP is used to simulate illustrative examples of transient phenomena discussed in this book. In all simulations it is necessary that the system has a ground node. Consider, for example, the delta winding of a transformer or a three-phase ungrounded capacitor bank. These do have some capacitance to ground. This may not have been shown in the circuit diagrams of configurations for simplicity, but the ground node is always implied in all simulations using EMTP. This book also uses both SI and FPS units, the latter being still in practical use in the United States.

REFERENCES

1. CIGRE joint WG 33.02, Guidelines for Representation of Networks Elements when Calculating Transients, CIGRE Brochure, 1990.
2. IEC 60071-1, ed. 8, Insulation Coordination, Definitions, Principles, and Rules, 2006; IEC 60071-2, 3rd ed., Application Guide, 1996.
3. ANSI/IEEE Std. 446, IEEE Recommended Practice for Emergency and Standby Power Systems for Industrial and Commercial Applications, 1987.
4. IEEE, Modeling and Analysis of System Transients Using Digital Programs, Document TP-133-0, 1998. (This document provides 985 further references).
5. H. W. Dommel, "Digital Computer Solution of Electromagnetic Transients in Single and Multiphase Networks," *IEEE Trans. Power Apparatus and Systems*, vol. PAS-88, no. 4, pp. 388–399, Apr. 1969.
6. ATP Rule Book, ATP User Group, Portland, OR, 1992.
7. J. Mahseredjian, S. Dennetiere, L. Dubé, B. Khodabakhehian, and L. Gerin-Lajoie, "A New Approach for the Simulation of Transients in Power Systems," *International Conference on Power System Transients*, Montreal, Canada, June 2005.
8. EMTP, www.emtp.org; NETOMAC, www.ev.siemens.de/en/pages; EMTAP, www.edsa.com; TPA, www.mpr.com; PSCAD/EMTDC, www.hvdc.ca; EMTP-RV, www.emtp.com.
9. CIGRE WG C1.04, "Application and Required Developments of Dynamic Models to Support Practical Planning," *Electra,* no. 230, pp. 18–32, Feb. 2007.

FURTHER READING

A. Clerici, "Analogue and Digital Simulation for Transient Voltage Determinations," *Electra*, no. 22, pp. 111–138, 1972.

H. W. Dommel and W. S. Meyer, "Computation of Electromagnetic Transients," *IEE Proc.* no. 62, pp. 983–993, 1974.

L. Dube and H. W. Dommel, "Simulation of Control Systems in An Electromagnetic Transients Program with TACS," *Proc. IEEE PICA*, pp. 266–271, 1977.

M. Erche, "Network Analyzer for Study of Electromagnetic Transients in High-Voltage Networks," *Siemens Power Engineering and Automation*, no. 7, pp. 285–290, 1985.

B. Gustavsen and A. Semlyen, "Rational Approximation of Frequency Domain Responses by Vector Fitting," *IEEE Trans. PD*, vol. 14, no. 3, pp. 1052–1061, July 1999.

B. Gustavsen and A. Semlyen, "Enforcing Passivity of Admittance Matrices Approximated by Rational Functions," *IEEE Trans. PS*, vol. 16, no. 1, pp. 97–104, Feb 2001.

M. Zitnik, "Numerical Modeling of Transients in Electrical Systems," Uppsal Dissertations from the Faculty of Science and Technology (35), Elanders Gutab, Stockholm, 2001.

CHAPTER 2

TRANSIENTS IN LUMPED CIRCUITS

In this chapter the transients in lumped, passive, linear circuits are studied. Complex electrical systems can be modeled with certain constraints and interconnections of passive system components, which can be excited from a variety of sources. A familiarity with basic circuit concepts, circuit theorems, and matrices is assumed. A reader may like to pursue the synopsis of differential equations, Laplace transform, and z-transform in Apps. A, B, and C, respectively, before proceeding with this chapter. Fourier transform can also be used for transient analysis; while Laplace transform converts a time domain function into complex frequency ($s = \sigma + \omega$), the Fourier transform converts it into imaginary frequency of $j\omega$. We will confine our discussion to Laplace transform in this chapter.

2-1 LUMPED AND DISTRIBUTED PARAMETERS

A lumped parameter system is that in which the disturbance originating at one point of the system is propagated instantaneously to every other point in the system. The assumption is valid if the largest physical dimension of the system is small compared to the wavelength of the highest significant frequency. These systems can be modeled by ordinary differential equations.

In a distributed parameter system, it takes a finite time for a disturbance at one point to be transmitted to the other point. Thus, we deal with space variable in addition to independent time variable. The equations describing distributed parameter systems are partial differential equations.

All systems are in fact, to an extent, distributed parameter systems. The power transmission line models are an example. Each elemental section of the line has resistance, inductance, shunt conductance, and shunt capacitance. For short lines we ignore shunt capacitance and conductance all together, for medium long lines we approximate with lumped T and Π models, and for long lines we use distributed parameter models (see Chap. 4).

2-2 TIME INVARIANCE

When the characteristics of the system do not change with time it is said to be a time invariant or stationary system.

Mathematically, if the state of the system at $t = t_0$ is $x(t)$ and for a delayed input it is $w(t)$ then the system changes its state in the stationary or time invariant manner if:

$$w(t+\tau) = x(t) \qquad \qquad (2\text{-}1)$$
$$w(t) = x(t-\tau)$$

This is shown in Fig. 2-1. A shift in waveform by τ will have no effect on the waveform of the state variables except for a shift by τ. This suggests that in time invariant systems the time origin t_0 is not important. The reference time for a time invariant system can be chosen as zero, Therefore:

$$\mathbf{x}(t) = \phi[\mathbf{x}(0), \mathbf{r}(0, t)] \qquad \qquad (2\text{-}2)$$

To some extent physical systems do vary with time, for example, due to aging and tolerances in component values. A time invariant system is, thus, an idealization of a practical system or, in other words, we say that the changes are very slow with respect to the input.

2-3 LINEAR AND NONLINEAR SYSTEMS

Linearity implies two conditions:

1. Homogeneity
2. Superposition

Consider the state of a system defined by (see Sec. 2-14 on state equations):

$$\dot{\mathbf{x}} = \mathbf{f}[\mathbf{x}(t), \mathbf{r}(t), t] \qquad \qquad (2\text{-}3)$$

If $\mathbf{x}(t)$ is the solution to this differential equation with initial conditions $\mathbf{x}(t_0)$ at $t = t_0$ and input $\mathbf{r}(t)$, $t > t_0$:

$$\mathbf{x}(t) = \phi[\mathbf{x}(t_0), \mathbf{r}(t)] \qquad \qquad (2\text{-}4)$$

Then homogeneity implies that:

$$\boldsymbol{\phi}[\mathbf{x}(t_0), \alpha\mathbf{r}(t)] = \alpha\boldsymbol{\phi}[\mathbf{x}(t_0), \mathbf{r}(t)] \qquad \qquad (2\text{-}5)$$

where α is a scalar constant. This means that $\mathbf{x}(t)$ with input $\alpha\mathbf{r}(t)$ is equal to α times $\mathbf{x}(t)$ with input $\mathbf{r}(t)$ for any scalar α.

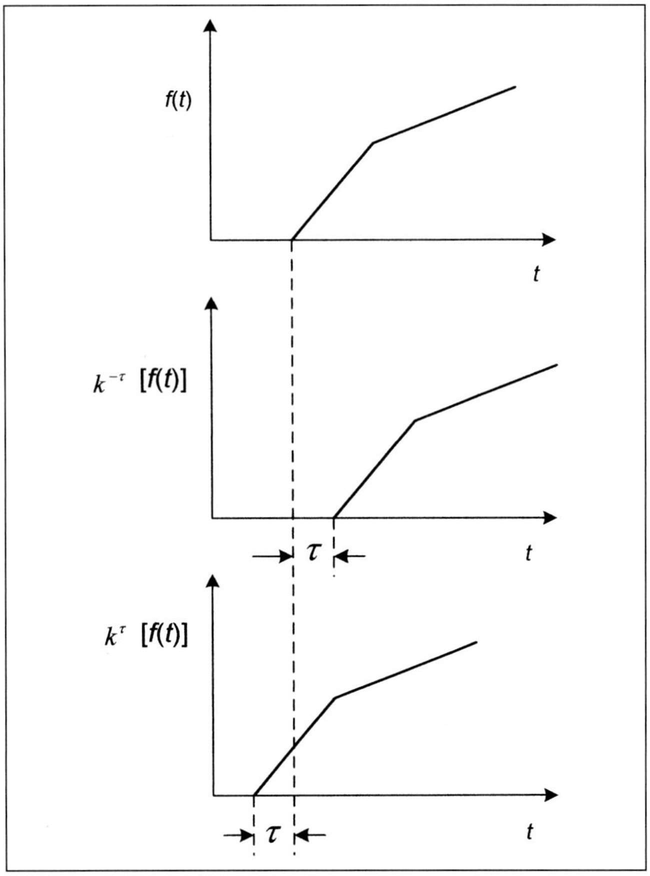

FIGURE 2-1 A time invariant system, effect of shifted input by τ.

Superposition implies that:

$$\phi[\mathbf{x}(t_0),\mathbf{r}_1(t)+\mathbf{r}_2(t)]=\phi[\mathbf{x}(t_0),\mathbf{r}_1(t)]+\phi[\mathbf{x}(t_0),\mathbf{r}_2(t)] \qquad (2\text{-}6)$$

That is, $\mathbf{x}(t)$ with inputs $\mathbf{r}_1(t) + \mathbf{r}_2(t)$ is = sum of $\mathbf{x}(t)$ with input $\mathbf{r}_1(t)$ and $\mathbf{x}(t)$ with input $\mathbf{r}_2(t)$. Thus linearity is superimposition plus homogeneity.

2-4 PROPERTY OF DECOMPOSITION

A system is said to be linear if it satisfies the decomposition property and the decomposed components are linear.

If $\mathbf{x}'(t)$ is solution of Eq. (2-3) when system is in zero state for all inputs $\mathbf{r}(t)$, that is:

$$\mathbf{x}'(t)=\phi[0,\mathbf{r}(t)] \qquad (2\text{-}7)$$

And $\mathbf{x}''(t)$ is the solution when for all states $\mathbf{x}(t_0)$, the input $\mathbf{r}(t)$ is zero, that is:

$$\mathbf{x}''(t)=\phi[\mathbf{x}(t_0),0] \qquad (2\text{-}8)$$

Then, the system is said to have the decomposition property if:

$$\mathbf{x}(t)=\mathbf{x}'(t)+\mathbf{x}''(t) \qquad (2\text{-}9)$$

The zero input response and zero state response satisfy the properties of homogeneity and superimposition with respect to initial states and initial inputs, respectively. If this is not true, then the system is nonlinear.

Electrical power systems are perhaps the most *nonlinear* systems in the physical world. For nonlinear systems, general methods of solutions are not available and each system must be studied specifically. Yet, we apply linear techniques of solution to nonlinear systems over a certain time interval. Perhaps the system is not changing so fast, and for a certain range of applications linearity can be applied. Thus, the linear system analysis forms the very fundamental aspect of the study.

2-5 TIME DOMAIN ANALYSIS OF LINEAR SYSTEMS

We can study the behavior of an electrical system in the time domain. A linear system can be described by a set of linear differential or *difference* equations (App. C). The output of the system for some given inputs can be studied. If the behavior of the system at all points in the system is to be studied, then a mathematical description of the system can be obtained in state variable form.

A *transform* of the time signals in another form can often express the problem in a more simplified way. Examples of transform techniques are Laplace transform, Fourier transform, *z*-transform, and integral transform, which are powerful analytical tools. There are inherently three steps in applying a transform:

1. The original problem is transformed into a simpler form for solution using a transform.

2. The problem is solved, and possibly the transformed form is mathematically easy to manipulate and solve.

3. Inverse transform is applied to get to the original solution.

As we will see, all three steps may not be necessary, and sometimes a direct solution can be more easily obtained.

2-6 STATIC AND DYNAMIC SYSTEMS

Consider a time-invariant, linear resistor element across a voltage source. The output, that is, the voltage across the resistor, is solely dependent upon the input voltage at that instant. We may say that the resistor does not have a *memory, and is a static system.* On the other hand the voltage across a capacitor depends not only upon the input, but also upon its initial charge, that is, the past history of current flow. We say that the capacitor has a *memory* and is a dynamic system.

The state of the system with memory is determined by state variables that vary with time. The state transition from $x(t_1)$ at time t_1 to $x(t_2)$ at time t_2 is a dynamic process that can be described by differential equations. For a capacitor connected to a voltage source, the dynamics of the state variable $x(t) = e(t)$ can be described by:

$$\dot{x}=\frac{1}{C}r(t) \qquad r(t)=i(t) \qquad (2\text{-}10)$$

We will revert to the state variable form in Sec. 2-14. By this definition, practically all electrical systems are dynamic in nature.

2-7 FUNDAMENTAL CONCEPTS

Some basic concepts are outlined for the solution of transients, which are discussed in many texts.

2-7-1 Representation of Sources

We will represent the independent and dependent current and voltage sources as shown in Fig. 2-2*a*, *b*, and *c*. Recall that in a dependent controlled source the controlling physical parameter may be current, voltage, light intensity, temperature, and the like. An ideal voltage source will have a Thévenin impedance of zero, that is, any amount of current can be taken from the source without altering the source voltage. An ideal current source (Norton equivalent)

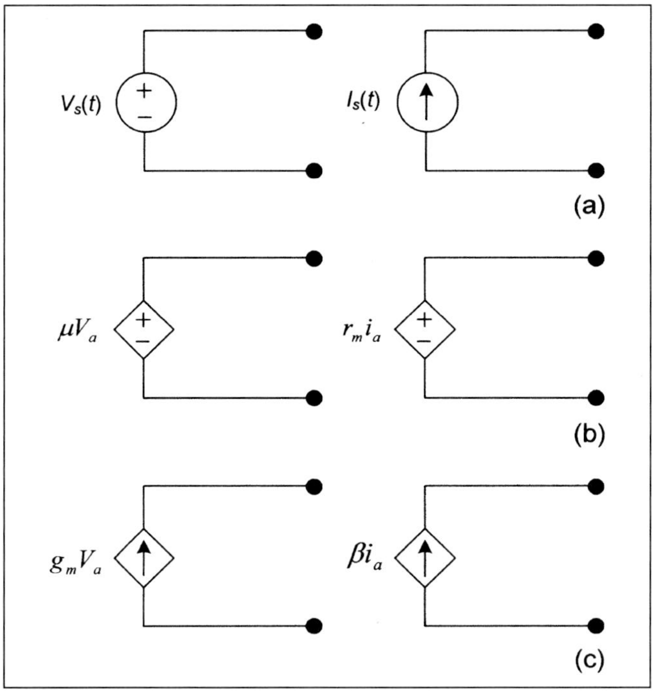

FIGURE 2-2 (*a*) Independent voltage and current source. (*b*) Voltage and current controlled voltage source. (*c*) Voltage and current controlled current source.

will have an infinite admittance across it. In practice a large generator approximates to an ideal current and voltage source. Sometimes utility systems are modeled as ideal sources, but this can lead to appreciable errors depending upon the problem under study.

2-7-2 Inductance and Capacitance Excited by DC Source

Consider that an ideal dc voltage source is connected through a switch, normally open ($t = 0^-$) to a parallel combination of a capacitor and resistors shown in Fig. 2-3*a*. Further consider that there is no prior charge on the capacitor. When the switch is suddenly closed at $t = 0^+$, the capacitor acts like a short circuit across the resistor and, the charging current assuming no source resistance and ignoring resistance of connections, will be theoretically infinite. Practically some resistance in the circuit, shown dotted as R_1, will limit the current. Note that the symbol $t = 0^+$ signifies the time after the switch is closed:

$$i_C(0^+) = \frac{V_s}{R_1}$$

As the current in the capacitor is given by:

$$i_C = C\frac{dv}{dt}$$

We can write:

$$\frac{dv}{dt} = \frac{V_s}{R_1 C}$$

When the capacitor is fully charged, dv/dt and i_C are zero and the current through R_2 is given by:

$$i_2 = \frac{V_s}{R_1 + R_2}$$

And the voltage across the capacitor as well as the resistor R_2 is:

$$v_C(\infty) = i_2 R_2 = \frac{V_s R_2}{R_1 + R_2}$$

We have not calculated the time-charging current transient profile and have arrived at the initial and final value results by elementary circuit conditions.

Now consider that the capacitor is replaced by an inductor as shown in Fig. 2-3*b*. Again consider that there is no stored energy in the reactor prior to closing the switch. Inductance acts like an open circuit on closing the switch, therefore all the current flows through R_2.

Thus, the voltage across resistor or inductor is:

$$v_L = \frac{V_s R_2}{R_1 + R_2} = L\frac{di_L}{dt}$$

$$\frac{di_L}{dt} = \frac{V_s R_2}{L(R_1 + R_2)}$$

In steady state $di_L/dt = 0$, there is no voltage drop across the inductor. It acts like a short circuit across R_2, and the current will be limited only by R_1. It is equal to V/R_1.

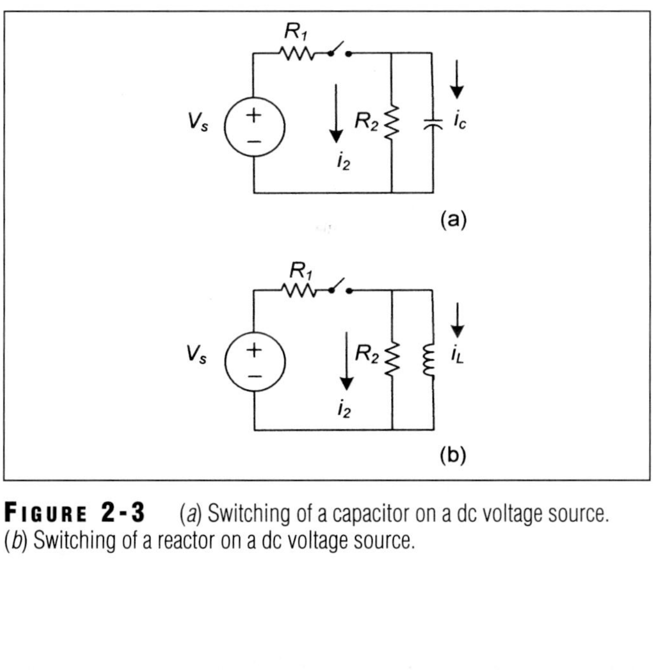

FIGURE 2-3 (*a*) Switching of a capacitor on a dc voltage source. (*b*) Switching of a reactor on a dc voltage source.

2-7-3 Coupled Coils

Two coupled coils are shown in Fig. 2-4. We can write the following equations relating current and voltages:

$$v_1 = L_1\frac{di_1}{dt} + M\frac{di_2}{dt}$$
$$v_2 = M\frac{di_1}{dt} + L_2\frac{di_2}{dt} \qquad (2\text{-}11)$$

where M is given by the coefficient of coupling:

$$M = k\sqrt{L_1 L_2} \qquad (2\text{-}12)$$

In an ideal transformer, $k = 1$. These equations can be treated as two loop equations; the voltage generated in loop 1 is due to current in loop 2 and vice versa. This is the example of a bilateral

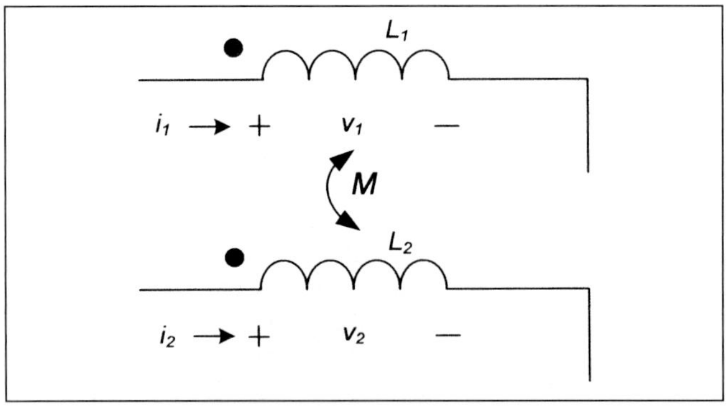

FIGURE 2-4 To illustrate mutually coupled coils.

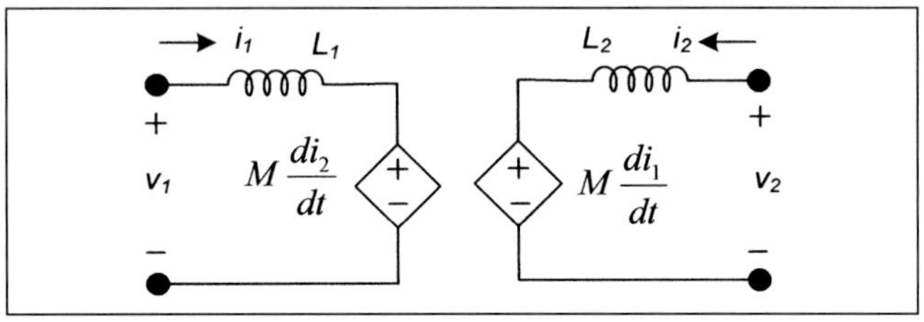

FIGURE 2-5 Equivalent circuit model of coupled coils using controlled sources.

circuit, which can be represented by an equivalent circuit of controlled sources (Fig. 2-5).

In a three-terminal device, with voltages measured to common third terminal (Fig. 2-6), a matrix equation of the following form can be written:

$$\begin{vmatrix} v_1 \\ v_2 \end{vmatrix} = \begin{vmatrix} z_i & z_r \\ z_f & z_0 \end{vmatrix} \begin{vmatrix} i_1 \\ i_2 \end{vmatrix} \tag{2-13}$$

where z_i is input impedance, z_r is reverse impedance, z_f is forward impedance, and z_0 is output impedance. The z parameters result in current-controlled voltage sources in series with impedances. These can be converted to voltage-controlled current sources in parallel with admittances—y-parameter formation.

For example, consider a three-terminal device, with following y parameters:

$$y_i = 0.012 \qquad y_r = -0.001 \qquad y_f = 0.0067 \qquad y_0 = 0.002$$

All the above values are in mhos. Then the following equations can be written:

$$i_1 = 0.012v_1 - 0.001v_2$$

$$i_2 = 0.0067v_1 + 0.002v_2$$

Each of these equations describes connection to one node, and the voltages are measured with respect to the reference node. These can be represented by the equivalent circuit shown in Fig. 2-7.

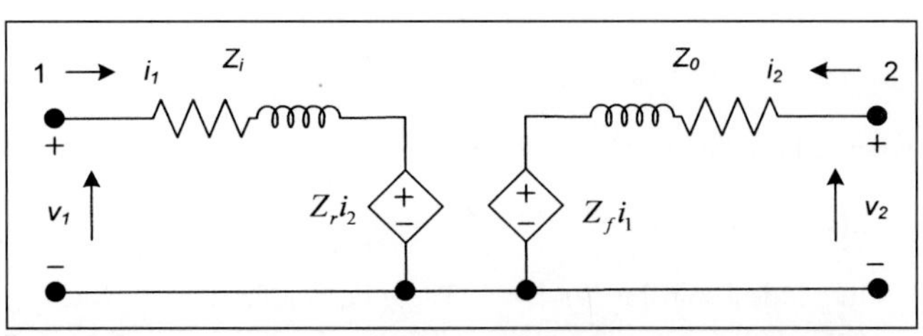

FIGURE 2-6 Equivalent circuit, voltage of controlled sources.

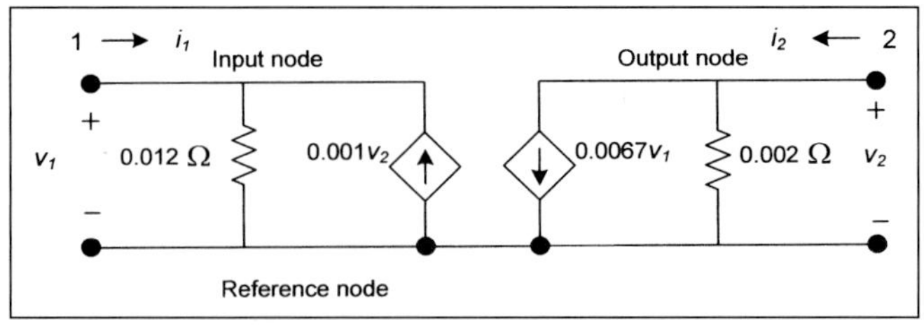

FIGURE 2-7 Equivalent circuit of admittances, y-parameter representation.

2-7-4 Two-Port Networks

Two-port networks such as transformers, transistors, and transmission lines may be three- or four-terminal devices. They are assumed to be linear. A representation of such a network is shown in Fig. 2-8, with four variables which are related with the following matrix equation:

$$\begin{vmatrix} v_1 \\ v_2 \end{vmatrix} = \begin{vmatrix} z_{11} & z_{12} \\ z_{21} & z_{22} \end{vmatrix} \begin{vmatrix} i_1 \\ i_2 \end{vmatrix} \tag{2-14}$$

Note the convention used for the current flow and the voltage polarity. The subscript 1 pertains to input port and the input terminals; the subscript 2 indicates output port and output terminals. The four port variables can be dependent or independent, that is, the independent variables may be currents and the dependent variables may be voltages.

By choosing voltage as the independent variable, y parameters are obtained, and by choosing the input current and the output voltage as independent variable h, parameters are obtained.

2-7-5 Network Reductions

Circuit reductions; loop and mesh equations; and Thévenin, Norton, Miller, maximum power transfer, and superposition theorems, which are fundamental to circuit concepts, are not discussed, and a knowledge of these basic concepts is assumed. A network for study of transients can be simplified using these theorems. The following simple example illustrates this.

Example 2-1 Consider the circuit configuration shown in Fig. 2-9*a*. It is required to write the differential equation for the voltage across the capacitor.

We could write three loop equations and then solve these simultaneous equations for the current in the capacitor. However, this can be much simplified, using a basic circuit transformation. It is seen from Fig. 2-9*a* that the capacitor and inductor are neither in a series or a parallel configuration. A step-wise reduction of the system is shown in Figs. 2-9*b*, *c*, and *d*. In Fig. 2-9*b* the voltage source is converted to a current source, and 20 Ω and 40 Ω resistances in parallel are combined. In Fig. 2-9*c*, the current source is converted back to the voltage source. Finally, we can write the following differential equation:

$$0.028v_s = \frac{v_c}{23.33} + C\left(\frac{dv_c}{dt}\right) + i_L$$

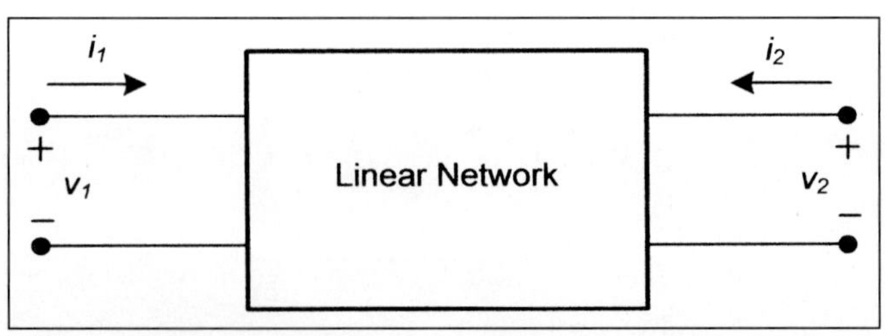

FIGURE 2-8 Two-port network, showing defined directions of currents and polarity of voltages.

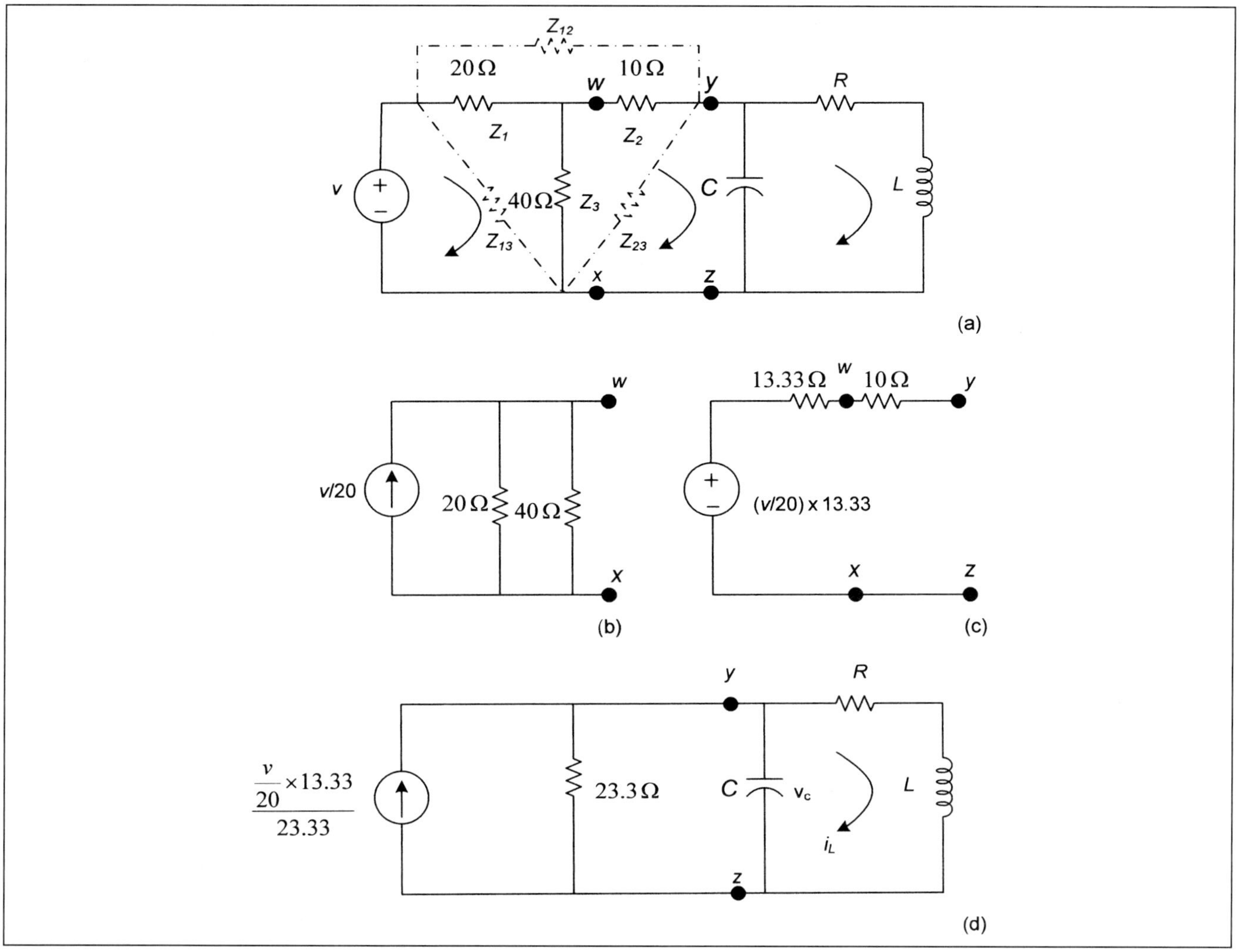

FIGURE 2-9 (*a*), (*b*), (*c*), and (*d*) Progressive reduction of a network by source transformations/Dotted lines in Fig. 2-9*a* show wye-delta and delta-wye impedance transformations.

A simpler reduction could be obtained by wye-delta transformation. Consider the impedances shown dotted in Fig. 2-9*a*, then:

$$Z_{12} = \frac{Z_1Z_2 + Z_1Z_3 + Z_2Z_3}{Z_3}$$

$$Z_{13} = \frac{Z_1Z_2 + Z_1Z_3 + Z_2Z_3}{Z_2} \qquad (2\text{-}15)$$

$$Z_{23} = \frac{Z_1Z_2 + Z_1Z_3 + Z_2Z_3}{Z_1}$$

Conversely:

$$Z_1 = \frac{Z_{12}Z_{13}}{Z_{12} + Z_{13} + Z_{23}}$$

$$Z_2 = \frac{Z_{12}Z_{23}}{Z_{12} + Z_{13} + Z_{23}} \qquad (2\text{-}16)$$

$$Z_3 = \frac{Z_{13}Z_{23}}{Z_{12} + Z_{13} + Z_{23}}$$

2-7-6 Impedance Forms

For transient and stability analysis, the following impedance forms of simple combination of circuit elements are useful:

Inductance

$$v = L\frac{di}{dt} = sLi \qquad z_L = sL$$

Capacitance

$$i = C\frac{dv}{dt} = sCv \qquad z_C = \frac{1}{sC}$$

Series RL

$$z = R\left(1 + \frac{sL}{R}\right) \qquad (2\text{-}17)$$

Series RC

$$z = \frac{1 + sCR}{sC} \qquad (2\text{-}18)$$

Parallel RL

$$z = \frac{sL}{1+sL/R} \tag{2-19}$$

Parallel RC

$$z = \frac{R}{1+sCR} \tag{2-20}$$

Parallel LC

$$z = \frac{sL}{1+s^2LC} \tag{2-21}$$

Series LC

$$z = \frac{1+s^2LC}{SC} \tag{2-22}$$

Series RLC

$$z = \frac{1+sCR+s^2LC}{sC} \tag{2-23}$$

Parallel RLC

$$z = \frac{sL}{1+(sL/R)+s^2LC} \tag{2-24}$$

Response to the application of a voltage V and the resulting current flow can simply be found by the expression:

$$i(s) = \frac{V}{s}\frac{1}{z} \tag{2-25}$$

Equation (2-25) is applicable if there is no initial charge on the capacitors and there is no prior stored energy in the reactors. The capacitance voltage does not change instantaneously, and therefore, $v_C(0^+) = v_C(0)$. The capacitance voltage and current are transformed according to the equations:

$$\begin{aligned} &\mathscr{L}[v_C(t)] = V_C(s) \\ &\mathscr{L}[i_C(t)] = \mathscr{L}\left[C\frac{dv_C(t)}{dt}\right] = sCV_C(s) - CV_C(0) \end{aligned} \tag{2-26}$$

Figure 2-10*a* shows the equivalent capacitor circuit. Note that the two-source current model is transformed into an impedance and current source model.

In an inductance the transformation is:

$$\begin{aligned} &\mathscr{L}[i_L(t)] = I_L(s) \\ &\mathscr{L}[v_L(t)] = \mathscr{L}\left(L\frac{di_L(t)}{dt}\right) = sLI_L(s) - Ii_L(0) \end{aligned} \tag{2-27}$$

This two-source model and the equivalent impedance and source model are shown in Fig. 2-10*b*.

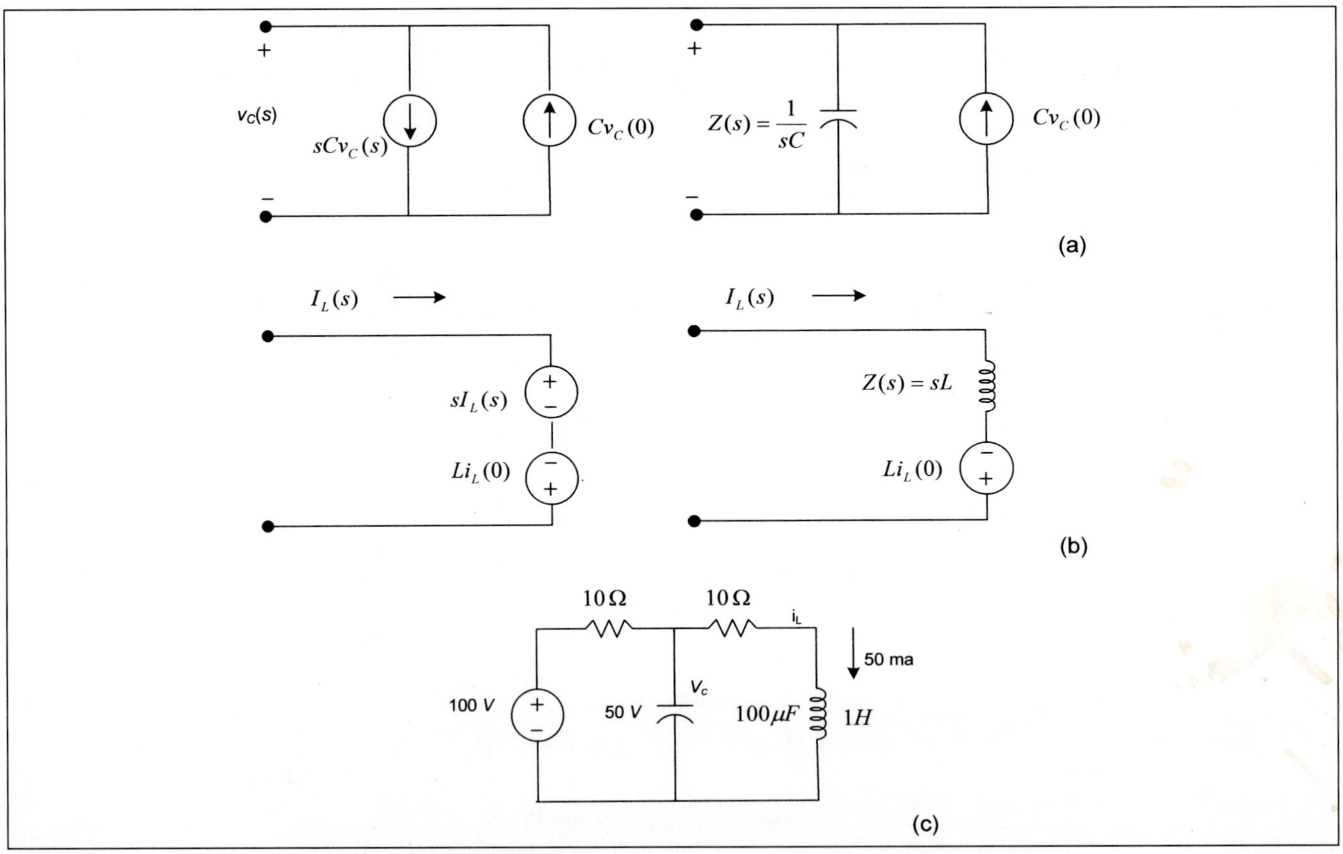

FIGURE 2-10 (*a*) Transformed equivalent circuit for the initial conditions of voltage on a capacitor. (*b*) Transformed equivalent circuit for the current in an inductor. (*c*) Circuit diagram for Example 2-2.

Example 2-2 Consider a circuit of Fig. 2-10*c*. Initially the state variables are 50 V across the capacitor and a current of 500 mA flows in the inductor. It is required to find the voltage across the capacitor for $t > 0$.

We can write two differential equations for the left-hand and right-hand loops, using Kirchoff's voltage laws:

$$100 = 10\left(i_L + 10^{-4}\frac{dV_c}{dt}\right) + V_c$$

$$V_c = 10i_L + 1.0\frac{di_L}{dt}$$

Now apply Laplace transform:

$$\frac{100}{s} = 10i_L + 10^{-3}[s(V_c) - V_c(0)] + V_c$$

$$V_c = 10i_L + si_L - I_L(0)$$

$$I_L(0) = 0.5\text{ A} \qquad V_c(0) = 50\text{ V}$$

These values can be substituted and the equations solved for capacitor voltage (App. B).

2-8 FIRST-ORDER TRANSIENTS

The energy storage elements in electrical circuits are inductors and capacitors. The first-order transients occur when the circuit contains only one energy storage element, that is, inductance or capacitance. This results in a first-order differential equation which can be easily solved. The circuit may be excited by:

- DC source, giving rise to dc transients
- AC source, giving rise to ac transients.

When switching devices operate, they change the topology of the circuit. Some parts of the system may be connected or disconnected. Hence these may be called switching transients. On the other hand, pulse transients do not change the topology of the circuit, as only the current or voltage waveforms of a source are changed.

Example 2-3 Consider Fig. 2-11*a*, with the following values.

$R_1 = 1\ \Omega \qquad R_2 = 10\ \Omega \qquad L = 0.15\text{ H} \qquad V = 10\text{ V}$

With the given parameters and the switch closed, we reduce the circuit to an equivalent Thévenin voltage $V_{th} = 9.09$ V and series Thévenin resistance $R_{th} = 0.909\ \Omega$ (Fig. 2-11*b*). Therefore we can write the following differential equation:

$$9.09 = 0.909i + 0.15\frac{di}{dt}$$

When steady-state condition is reached, $di/dt = 0$, and the steady-state current is 10 A, the reactor acts as an open circuit.

When the switch is opened, $V_{th} = 0$ V and $R_{th} = 10\ \Omega$. Therefore:

$$0 = 10i + 0.15\frac{di}{dt}$$

$di/dt = 0$ in steady state and therefore, current = 0.

Example 2-4 In Example 2-3, we replace the inductor with a capacitor of 1 μF and solve for the capacitor current, with resistances and voltage remaining unchanged. The Thévenin voltage and impedance on closing the switch are the same as calculated in Example 2-3. Therefore we can write the following differential equation:

$$v_c + iR_{th} = v_{th}$$

The current in the capacitor is given by:

$$i_C = C\frac{dv_C}{dt}$$

Thus, we can write:

$$9.09 = 0.909\times10^{-6}\frac{dv_C}{dt} + v_c$$

For steady state, $dv_C/dt = 0$ and the capacitor is charged to 9.09 V. When the switch is opened:

$$0 = 10^{-5}\frac{dv_C}{dt} + v_C$$

Again for steady state, dv_C/dt is zero and the capacitor voltage is zero. The above examples show the reduction of the system to a simple RL or RC in series excited by a dc source. The differential equation for series RL circuit is:

$$Ri + L\frac{di}{dt} = V \tag{2-28}$$

This is a first-order linear differential equation, and from App. A its solution is:

$$ie^{\frac{R}{L}t} = \frac{V}{L}\frac{L}{R}e^{\frac{R}{L}t} + A$$

$$i = \frac{V}{R} + Ae^{-\frac{R}{L}t}$$

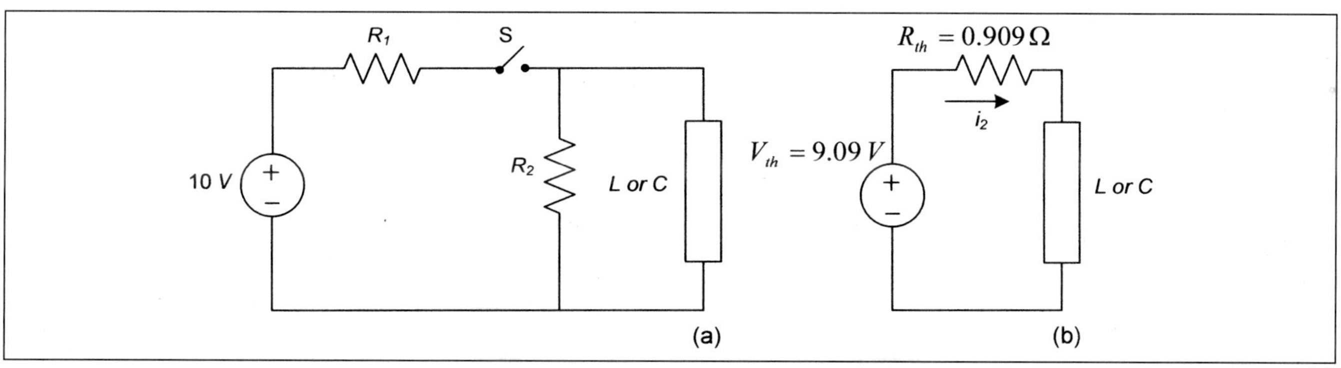

FIGURE 2-11 (*a*) Circuit with switch open, Example 2-3. (*b*) Circuit with switch closed, Example 2-3.

The coefficient A can be found from the initial conditions, at $t=0$ $i=0$. Thus, $A=-V/R$:

$$i=\frac{V}{R}\left[1-e^{-\frac{R}{L}t}\right] \tag{2-29}$$

The current increases with time and attains a maximum value of V/R, as before. The voltage across the reactor is $L\,di/dt$:

$$V_L=Ve^{-\frac{R}{L}t} \tag{2-30}$$

That is, the inductor is an open circuit at the instant of switching and is a short circuit ultimately.

We could also arrive at the same results by Laplace transform. Taking Laplace transform of Eq. (2-28):

$$Ri(s)+Lsi(s)-LI(0)=\frac{V}{s}$$

$$i(s)=\frac{V}{L}\frac{1}{s(s+R/L)}$$

The inverse Laplace transform gives the same result as Eq. (2-29) (see App. B). For solution of differential equations, it is not always necessary to solve using Laplace transform. A direct solution can, sometimes, be straightforward.

For the series RC circuit excited by a dc voltage, we can write the following general equation:

$$\frac{di}{dt}+\frac{i}{RC}=\frac{V-V_C(0)}{R}$$

The solution of this equation gives:

$$i=\frac{V-V_C(0)}{R}e^{-t/RC} \tag{2-31}$$

$V_C(0)$ is the initial voltage on the capacitor.

2-8-1 RL Series Circuit Excited by an AC Sinusoidal Source

Short circuit of a passive reactor and resistor in series excited by a sinusoidal source is rather an important transient. This depicts the basics of short circuit of a synchronous generator, except that the reactances of a synchronous generator are not time-invariant:

$$L\frac{di}{dt}+Ri=V\sin(\omega t+\theta)$$

The integrating factor is:

$$e^{\int(R/L)dt}=e^{Rt/L}$$

The solution is:

$$ie^{\frac{R}{L}t}=\frac{V}{L}\int e^{\frac{R}{L}t}\sin(\omega t+\theta)dt+A$$

$$=\frac{V}{L}\frac{e^{\frac{R}{L}t}}{\sqrt{\frac{R^2}{L^2}+\omega^2}}\sin\left(\omega t+\theta-\tan^{-1}\frac{\omega L}{R}\right)+A$$

Thus:

$$i=\frac{V}{\sqrt{R^2+\omega^2L^2}}\sin(\omega t+\theta-\phi)+Ae^{-\frac{R}{L}t}$$

where:

$$\phi=\tan^{-1}\frac{\omega L}{R}$$

The constant A can be found from the initial conditions at $t=0$, $i=0$.

This gives:

$$i=\frac{V}{\sqrt{R^2+(\omega L)^2}}[\sin(\omega t+\theta-\phi)-\sin(\theta-\phi)e^{-Rt/L}]$$

$$=I_m\sin(\omega t+\theta-\phi)-I_m\sin(\theta-\phi)e^{-Rt/L} \tag{2-32}$$

where I_m is the peak value of the steady-state current (V is the peak value of the applied sinusoidal voltage).

The same result can be obtained using Laplace transform though more steps are required. Taking Lapalce transform:

$$Ri(s)+Lsi(s)-LI(0)=V(\sin\omega t\cos\theta+\cos\omega t\sin\theta)$$

$$=V\left(\frac{\omega\cos\theta}{s^2+\omega^2}+\frac{s\sin\theta}{s^2+\omega^2}\right)$$

$$i(s)=\frac{V}{L}\frac{1}{s+(R/L)}\left[\frac{\omega\cos\theta}{s^2+\omega^2}+\frac{s\sin\theta}{s^2+\omega^2}\right]$$

This can be resolved into partial fractions:

$$\frac{1}{(s+\alpha)(s^2+\omega^2)}=\frac{1}{\alpha^2+\omega^2}\left[\frac{1}{s+\alpha}-\frac{s}{s^2+\omega^2}+\frac{\alpha}{s^2+\omega^2}\right] \tag{2-33}$$

where $\alpha=R/L$, therefore:

$$\begin{aligned}i(s)=&\frac{V\omega\cos\theta}{L(\alpha^2+\omega^2)}\left[\frac{1}{s+R/L}-\frac{s}{s^2+\omega^2}+\frac{\alpha}{s^2+\omega^2}\right]\\&+\frac{V\sin\theta}{L(\alpha^2+\omega^2)}\left[\frac{1}{s+R/L}-\frac{s}{s^2+\omega^2}+\frac{\alpha}{s^2+\omega^2}\right]\end{aligned} \tag{2-34}$$

The inverse Laplace transform of function in Eq. (2-33) is:

$$\mathcal{L}^{-1}\frac{1}{(s+\alpha)(s^2+\omega^2)}=\frac{1}{(\alpha^2+\omega^2)}\left[e^{-\alpha t}-\cos\omega t+\frac{\alpha}{\omega}\sin\omega t\right]$$

Therefore:

$$\mathcal{L}^{-1}\frac{s}{(s+\alpha)(s^2+\omega^2)}=\frac{1}{(\alpha^2+\omega^2)}[-\alpha e^{-\alpha t}+\omega\sin\omega t+\cos\omega t]$$

Substituting these results in Eq. (2-34):

$$\begin{aligned}i=&\frac{V}{L(\alpha^2+\omega^2)}\left[\omega\cos\theta\left(e^{-\alpha t}-\cos\omega t+\frac{\alpha}{\omega}\sin\omega t\right)\right.\\&\left.+\sin\theta(\alpha\cos\omega t+\omega\sin\omega t-\alpha e^{-\alpha t})\right]\\=&\frac{V}{L(\alpha^2+\omega^2)}[(\omega\cos\theta-\alpha\sin\theta)e^{-\alpha t}\\&-(\omega\cos\theta-\alpha\sin\theta)\cos\omega t+(\alpha\cos\theta+\omega\sin\theta)\sin\omega t]\end{aligned} \tag{2-35}$$

Remembering that:

$$\tan\phi = \omega L/R = \omega\alpha \quad \sin\phi = \omega/(\alpha^2+\omega^2)^{1/2} \quad \cos\phi = \alpha/(\alpha^2+\omega^2)^{1/2}$$

$$i = \frac{V}{(R^2+\omega^2L^2)^{1/2}}[\sin(\omega t+\theta-\phi)-\sin(\theta-\phi)e^{-\alpha t}] \qquad (2\text{-}36)$$

This is the same result as obtained in Eq. (2-32). Again, the derivation shows that direct solution of differential equation is simpler.

In power systems *X/R* ratio is high. A 100-MVA, 0.85 power factor generator may have an *X/R* of 110 and a transformer of the same rating, an *X/R* = 45. The *X/R* ratios for low-voltage systems may be of the order of 2–8. Consider that $\phi \approx 90°$.

If the short circuit occurs when the switch is closed at an instant at $t = 0$, $\theta = 0$, that is, when the voltage wave is crossing through zero amplitude on *x* axis, the instantaneous value of the short circuit current from Eq. (2-36) is 2 I_m. This is sometimes called the *doubling effect*.

If the short circuit occurs when the switch is closed at an instant at $t=0$, $\theta=\pi/2$, that is, when the voltage wave peaks, the second term in Eq. (2-36) is zero and there is no transient component. This is illustrated in Fig. 2-12*a* and *b*.

A physical explanation of the dc transient is that the inductance component of the impedance is high. If the short circuit occurs at the peak of the voltage, the current is zero. No dc component is required to satisfy the physical law that the current in an inductance cannot change suddenly. When the fault occurs at an instant when $\theta - \phi = 0$, there has to be a transient current whose value is equal and opposite to the instantaneous value of the ac short-circuit current. This transient current can be called a dc component and it decays at an exponential rate. Equation (2-36) can be simply written as:

$$i = I_m \sin\omega t + I_{dc}e^{-Rt/L} \qquad (2\text{-}37)$$

This circuit transient is important in power systems from short-circuit considerations. The following inferences are of interest.

There are two distinct components of the short-circuit current: (1) an ac component and (2) a decaying dc component at an

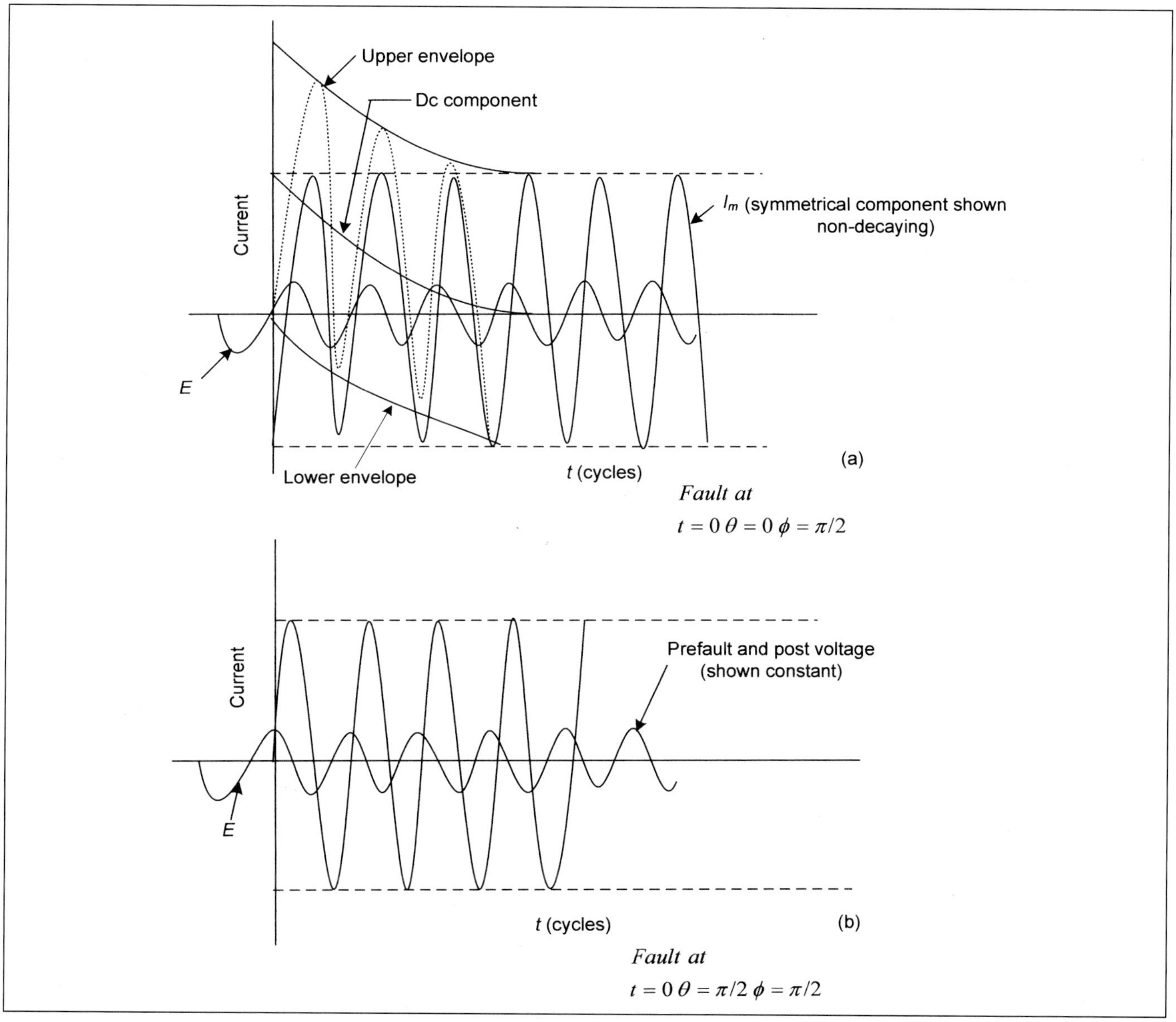

FIGURE 2-12 (*a*) Short circuit of a passive RL circuit on ac source, switch closed at zero crossing of the voltage wave. (*b*) Short circuit of a passive RL circuit on ac source, switch closed at the crest of the voltage wave.

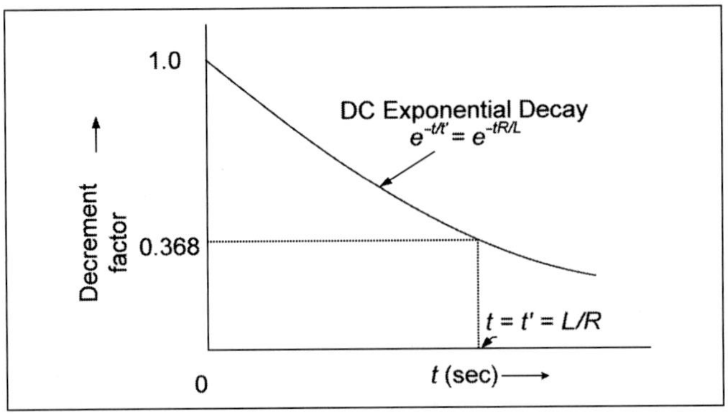

FIGURE 2-13 To illustrate concept of time constant of an RL circuit.

exponential rate, the initial value of the dc component being equal to the maximum value of the ac component.

Factor L/R can be termed as a time constant. The exponential then becomes $I_{dc}\, e^{-t/t'}$, where $t' = L/R$. In this equation making $t = t'$ will result in approximately 62.3 percent decay from its original value, that is, the transient current is reduced to a value of 0.368 per unit after a elapsed time equal to the time constant (Fig. 2-13). The dc component always decays to zero in a short time. Consider a modest X/R of 15; the dc component decays to 88 percent of its initial value in five cycles. This phenomenon is important for the rating structure of circuit breakers. Higher is the X/R ratio, slower is the decay.

The presence of the dc component makes the total short-circuit current wave asymmetrical about zero line; Fig. 2-12*a* clearly shows this. The total asymmetrical rms current is given by:

$$i_t(\text{rms,asym}) = \sqrt{(\text{ac component})^2 + (\text{dc component})^2} \quad (2\text{-}38)$$

In a three-phase system, the phases are displaced from each other by 120°. If a fault occurs when the dc component in phase *a* is zero, the phase *b* component will be positive and the phase component will be equal in magnitude but negative. As the fault is symmetrical, the identity, $Ia + Ib + Ic = 0$, is approximately maintained.

Example 2-5 We will illustrate the transient in an RL circuit with EMTP simulation. Consider $R = 3.76\ \Omega$ and $X = 37.6\ \Omega$, the source voltage is 13.8 kV, three phase, 60 Hz. The simulated transients in the three phases are shown in Fig. 2-14. The switch is closed when the voltage wave in phase *a* crests. The steady-state short-circuit current is 211A rms < 84.28°. This figure clearly shows the asymmetry in phases *b* and *c* and in phase *a* there is no transient. In phases *b* and *c* the current does not reach double the steady-state value, here the angle $\phi = 84.28°$. The lower is the resistance, higher is the asymmetry.

The short circuit of synchronous machines is more complex and is discussed in Chaps. 10 and 11.

2-8-2 First-Order Pulse Transients

Pulse transients are not caused by switching, but by the pulses generated in the sources. Pulses are represented by unit steps (App. B). The unit step is defined as:

$$\begin{aligned} u(t-t_1) &= 0 \qquad t < t_1 \\ &= 1 \qquad t > t_1 \end{aligned} \qquad (2\text{–}39)$$

Any pulse train can be constructed from a series of unit steps and the response determined by superimposition (Fig. 2-15). As the network is linear, the step responses are all the same, only shifted in time by the appropriate amount.

Consider the response of inductance and capacitance to a pulse transient. The response to a unit step function is determined by the state variable of the energy storage device. When the unit step

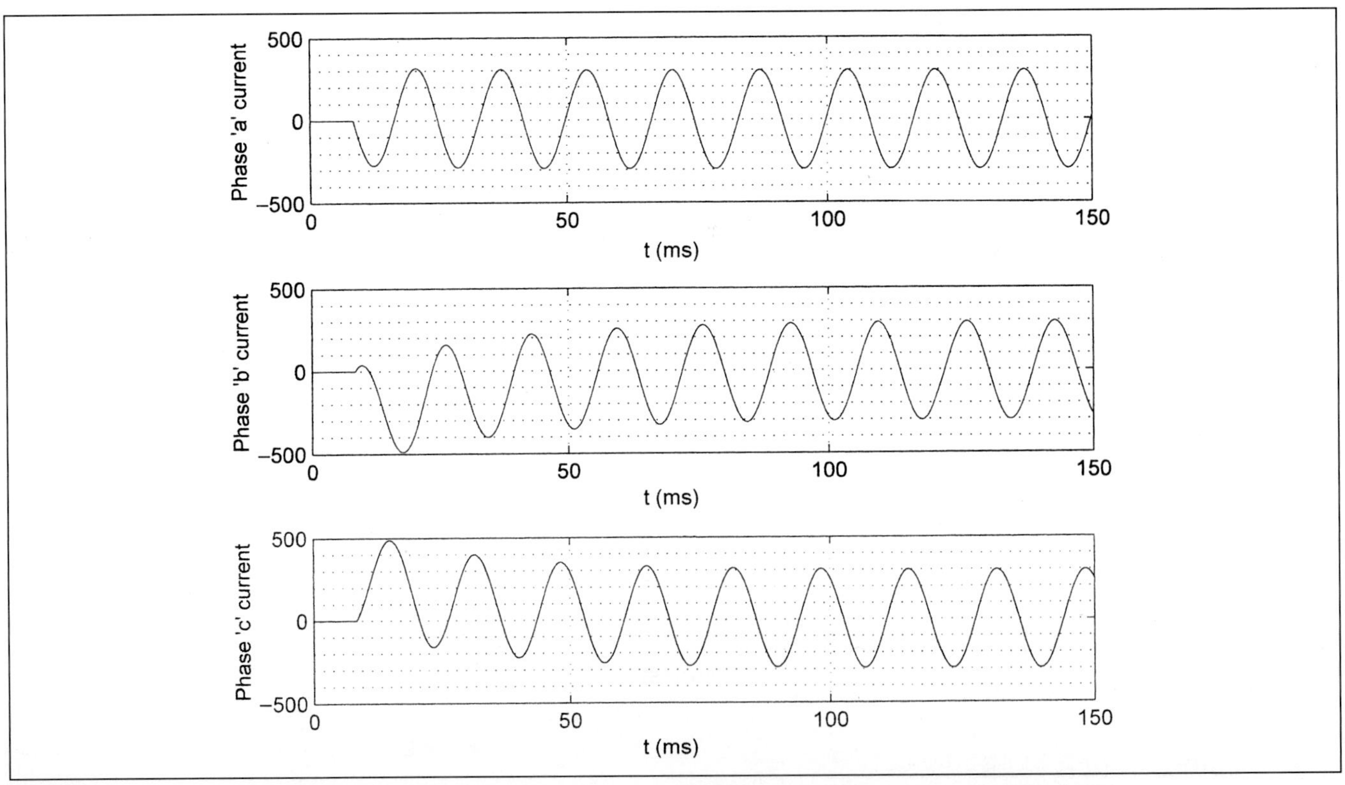

FIGURE 2-14 EMTP simulation of transients in passive RL circuit on three-phase, 13.8 kV, 60 Hz source, transient initiated at peak of the voltage in phase *a*.

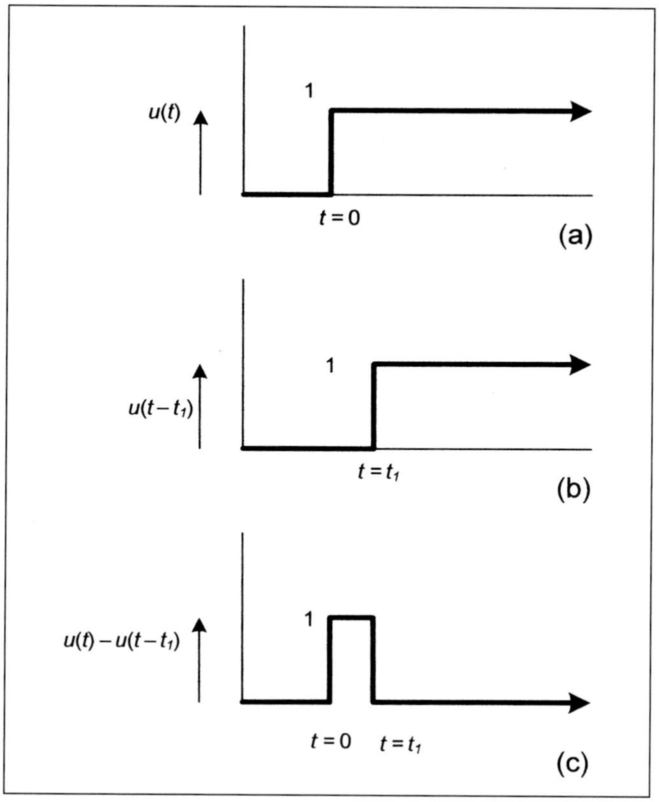

FIGURE 2-15 (*a*) Unit step $u(t)$. (*b*) Unit step $u(t-t_1)$. (*c*) Pulse $u(t)-u(t-t_1)$.

function is zero, there is no voltage on the capacitor and no current in the inductance. At $t = t_1$:

$$v_C(t) = u(t-t_1)(1-e^{-(t-t_1)/\tau})$$
$$i_L(t) = \frac{1}{R_{th}}u(t-t_1)(1-e^{-(t-t_1)/\tau}) \tag{2-40}$$

where $\tau = R_{th}C$ for the capacitance and L/R_{th} for the inductance and R_{th} is the Thévenin resistance.

Example 2-6 Consider a pulse source, which puts out two pulses of 5 V, 5 ms in length spaced by 10 ms. Find the voltage on a capacitor of 100 μF, R_{th} = 1000 Ω.

The voltage source can be written as:

$$v = 5u(t) - 5u(t-0.05) + 5u(t-0.015) - 5u(t-0.02)$$

The time constant is given by:

$$\tau = 10^3 \times 100 \times 10^{-6} = 10^{-1}$$

Thus, the voltage on the capacitor is:

$$5u(t-e^{-5t}) - 5u(t-0.05)(1-e^{-5(t-0.05)})$$
$$+5u(t-0.015)(1-e^{-5(t-0.015)}) - 5u(t-0.02)(1-e^{-5(t-0.02)})$$

2-9 SECOND-ORDER TRANSIENTS

There are two energy storage elements, and a second-order differential equation describes these systems.

2-9-1 DC Excitation

We will first consider dc excitation. Consider resistance, inductance, and capacitance in series excited by a dc source. There is no charge on the capacitor and no energy stored in the inductance prior to closing the switch.

In App. A we studied the second-order differential equation of the form:

$$f(t) = a\frac{d^2q}{dt^2} + b\frac{dq}{dt} + cq \tag{2-41}$$

The solution of the complementary function (CF) depends upon whether the roots are equal, imaginary or real, and different. In a series RLC circuit, the current equation is written as:

$$L\frac{di}{dt} + \frac{\int i\,dt}{C} + Ri = f(t) \tag{2-42}$$

Differentiating:

$$\frac{d^2i}{dt^2} + \frac{R}{L}\frac{di}{dt} + \frac{i}{LC} = f'(t) \tag{2-43}$$

Case 1 (Overdamped)

$$\text{If } R > \sqrt{4L/C} \tag{2-44}$$

The system is overdamped. The solution can be written as:

$$q(t) = Ae^{s_1t} + Be^{s_2t} + q_{ss} \tag{2-45}$$

where q_{ss} is the steady-state solution and s_1 and s_2 are given by:

$$s_1 = \frac{-b+\sqrt{b^2-4ac}}{2a}$$
$$s_2 = \frac{-b-\sqrt{b^2-4ac}}{2a} \tag{2-46}$$

The constants A and B can be found from the initial conditions. Consider that the transient is initiated at $t = 0^+$, the initial values of q and q' are found:

$$q(0^+) = A + B + q_{ss}$$
$$q'(0^+) = s_1A + s_2B + q'_{ss} \tag{2-47}$$

These equations give the values of A and B.

Case 2 (Critically Damped)
If both the roots are equal:

$$b^2 = 4ac \quad (R = \sqrt{4L/C}) \tag{2-48}$$

the system is critically damped. The solution is given by:

$$q(t) = Ae^{st} + Bte^{st} + q_{ss} \tag{2-49}$$

where s is given by:

$$s = \frac{-b}{2a} \tag{2-50}$$

The constants A and B are found from the initial conditions.

Case 3 (Underdamped)
If the roots are imaginary, that is,

$$b^2 < 4ac \qquad (R < \sqrt{4L/C}) \tag{2-51}$$

The system is underdamped and the response will be oscillatory. This is the most common situation in the electrical systems as the resistance component of the impedance is small. The solution is given by:

$$q(t) = Ae^{-\alpha t}\cos\beta t + Be^{-\alpha t}\sin\beta t + q_{ss} \tag{2-52}$$

where:

$$\alpha = \frac{b}{2a}$$
$$\beta = \frac{\sqrt{4ac-b^2}}{2a} \tag{2-53}$$

Again, the constants A and B are found from the initial conditions. The role of resistance in the switching circuit is obvious; it will damp the transients. This principle is employed in resistance switching as we will examine further in the chapters to follow.

Example 2-7 A series RLC circuit with $R = 1\ \Omega$, $L = 0.2$ H, and $C = 1.25$ F, excited by a 10 V dc source. The initial conditions are that at $t = 0$, $i = 0$, and $di/dt = 0$. In the steady state, di/dt and d^2i/dt^2 must both be zero, irrespective of the initial conditions. These values are rather hypothetical for the purpose of the example and are not representative of a practical power system. We can write the differential equation:

$$\frac{d^2i}{dt^2} + 5\frac{di}{dt} + 4i = 10$$

Taking Laplace transform:

$$s^2i(s) - si(0) - I'(0) + 5[si(s) - I(0)] + 4i(s) = \frac{10}{s}$$

$$s^2i(s) + 5si(s) + 4i(s) = \frac{10}{s}$$

$$i(s) = \frac{10}{s(s^2+5s+4)}$$

Resolve into partial fractions:

$$i(s) = \frac{2.5}{s} - \frac{2}{(s+1)} - \frac{0.5}{(s+4)}$$

Taking inverse transform, the solution is:

$$i = -2e^{-t} - 0.5e^{-4t} + 2.5$$

This is an overdamped circuit. Alternatively, the solution can be found as follows:

$$i_{ss} = \frac{10}{4} = 2.5$$
$$i'_{ss} = 0$$
$$s_1 = \frac{-5+\sqrt{25-16}}{2} = -1$$
$$s_2 = \frac{-5-\sqrt{25-16}}{2} = -4$$

Therefore the solution is:

$$i = Ae^{-t} + Be^{-4t} + 2.5$$

A and B can be found from the initial conditions:

$$i(0^+) = A + B + 2.5 = 0$$
$$i'(0^+) = (-1)A + (-4)B + 0 = 0$$

Given the initial conditions that $i(0^+)$ and $i'(0^+)$ are both zero, $A = -2$, $B = -1/2$ and the complete solution is:

$$i = -2e^{-t} - 0.5e^{-4t} + 2.5$$

It may not be obvious that the solution represents a damped response. The equation can be evaluated at small time intervals and the results plotted.

Example 2-8 Consider now the same circuit, but let us change the inductance to 2.5 H. This gives the equation:

$$\frac{d^2i}{dt^2} + 4\frac{di}{dt} + 4 = 10$$

Here the roots are equal and the system is critically damped.

$$i_{ss} = \frac{10}{4} = 2.5$$
$$i'_{ss} = 0$$
$$s = -2$$

Therefore the solution is:

$$i = Ae^{-2t} + Bte^{-2t} + 2.5$$

From initial conditions:

$$i(0^+) = A + 0 + 2.5 = 0$$
$$i'(0^+) = (-2)A + B + i'_{ss}$$

Thus, $A = -2.5$, $B = -5$

The solution is:

$$i = 2.5(1 - e^{-2t}) - 5te^{-2t}$$

Example 2-9 Next consider an underdamped circuit. $R = 1\ \Omega$, $L = 0.2$ H, and $C = 0.5$ F, again excited by 10 V dc source; initial conditions are the same. This gives the equation:

$$\frac{d^2i}{dt^2} + 5\frac{di}{dt} + 10i = 10$$

Therefore from Eq. (2-53):

$$\alpha = \frac{b}{2a} = 2.5$$
$$\beta = \frac{\sqrt{4ac-b^2}}{2a} = 1.94$$

Therefore the solution can be written as:

$$i = Ae^{-2.5t}\cos 1.94t + Be^{-2.5t}\sin 1.94t + i_{ss}$$
$$i_{ss} = 1$$
$$i(0^+) = A + i_{ss} = 0 \quad A = -1$$
$$i'(0^+) = -\alpha A + \beta B + i'_{ss} = -(2.5)(-1) + 1.94B = 0 \quad B = -1.29$$

Therefore, from Eq. (2-52) the solution is:

$$i = -e^{-2.5t}\cos 1.94t - 1.29e^{-2.5t}\sin 1.94t + 1$$

Alternatively, the solution can be written as:

$$i = Ce^{-\alpha t}\cos(\beta t - \phi) + i_{ss}$$

where:

$$C = \sqrt{A^2 + B^2}$$

$$\tan\phi = \frac{B}{A}$$

This gives:

$$i = 1.63e^{-2.5t}\cos(1.94t - 52.2°) + 1$$

2-9-2 RLC Circuit on AC Sinusoidal Excitation

We studied the response with a dc forcing function. These examples will be repeated with a sinusoidal function, $10\cos 2t$.

Example 2-10: Overdamped Circuit The differential equation is:

$$\frac{d^2i}{dt^2} + 5\frac{di}{dt} + 4i = 10\cos 2t$$

Given the same initial conditions that $i = 0$, and $di/dt = 0$ at $t = 0$, as before, we can write the Particular integral (PI) as:

$$\text{PI} = \frac{1}{D^2 + 5D + 4}10\cos 2t = 10\frac{1}{(-2^2) + 5D + 4}\cos 2t$$

$$= 10\frac{1}{5D}\cos 2t = 10\frac{D\cos 2t}{5(-2^2)} = \sin 2t$$

We can write the CF of the equation as before in Example 2-7:

$$\text{CF} = Be^{-t} + C^{-4t}$$

The complete solution is:

$$i = Be^{-t} + Ce^{-4t} + \sin 2t$$

The constants B and C are found from the initial conditions:

$$i(0^+) = B + C + 0 = 0$$

$$i'(0^+) = -B + 4C + 2$$

This gives $B = 2/3$ and $C = -2/3$. The solution is:

$$i = \frac{2}{3}e^{-t} - \frac{2}{3}e^{-4t} + \sin 2t$$

Alternatively, we can arrive at the same result by using Laplace transform:

$$[s^2 i(s) - si(0) - I'(0)] + 5[si(s) - I(0)] + 4i(s) = \frac{20}{s^2 + 4}$$

This gives:

$$i(s) = \frac{20}{(s^2 + 4)(s + 4)(s + 1)}$$

$$= \frac{2}{s^2 + 4} - \frac{2}{3(s + 4)} + \frac{2}{3(s + 1)}$$

Taking the inverse Laplace transform, we get the same result.

Example 2-11: Critically Damped Circuit The differential equation is:

$$\frac{d^2i}{dt^2} + 4\frac{di}{dt} + 4 = 10\cos 2t$$

The PI can be calculated as $1.25\sin 2t$ (see App. A). The complete solution is:

$$i = Be^{-2t} + Cte^{-2t} + 1.25\sin 2t$$

Again the constants B and C are found from the initial conditions:

$$i(0^+) = B = 0$$

$$i'(0^+) = (-2)B + C + 2.5 = 0$$

Therefore the solution is:

$$i = 1.25\sin 2t - 2.5te^{-2t}$$

Example 2-12 Lastly, we will consider the underdamped circuit, excited by an ac source. As before the differential equation is:

$$\frac{d^2i}{dt^2} + 5\frac{di}{dt} = 10i = 10\cos 2t$$

The PI, that is, the steady-state solution is:

$$\text{PI} = 10\frac{1}{D^2 + 5D + 10}\cos 2t = \frac{10\cos 2t}{(5D + 6)}$$

$$= \frac{10(5D - 6)\cos 2t}{25(-2^2) - 36}$$

$$i_{ss} = 0.44\cos 2t + 0.74\sin 2t$$

The complete solution is:

$$i = Ae^{-2.5t}\cos 1.94t + Be^{-2.5t}\sin 1.94t$$
$$+ 0.44\cos 2t + 0.74\sin 2t$$

The constants A and B are found as before by differentiating and substituting the initial values, $A = -0.44$, $B = 1.22$. The solution is therefore:

$$i = 1.22e^{-2.5t}\cos 1.94t + 0.44e^{-2.5t}\sin 1.94t$$
$$+ 0.44\cos 2t + 0.74\sin 2t$$

In general, we can write the roots of the characteristic equation of a series RLC circuit as:

$$s_1, s_2 = -\alpha \pm \sqrt{\alpha^2 - \omega_d^2}$$
$$= \alpha \pm \omega_n^2 \quad (2\text{-}54)$$

where:

$$\alpha = R/2L \qquad \omega_d = 1/\sqrt{LC} \quad (2\text{-}55)$$

We can term α as the *exponential damping coefficient*, ω_d as the *resonant frequency* of the circuit and ω_n is the *natural frequency*. Figure 2-16*a*, *b*, and *c* shows underdamped, critically damped, and overdamped responses, respectively.

We can write the natural response of RLC circuit in the general form:

$$i = I_m e^{-\alpha t}\sin(\omega_n t + \beta) \quad (2\text{-}56)$$

The voltage across the capacitor is:

$$v_C = \frac{1}{C}\int i\,dt = I_m\sqrt{\frac{L}{C}}e^{-\alpha t}\sin[\omega_n t + \beta - (90^0 + \delta)] \quad (2\text{-}57)$$

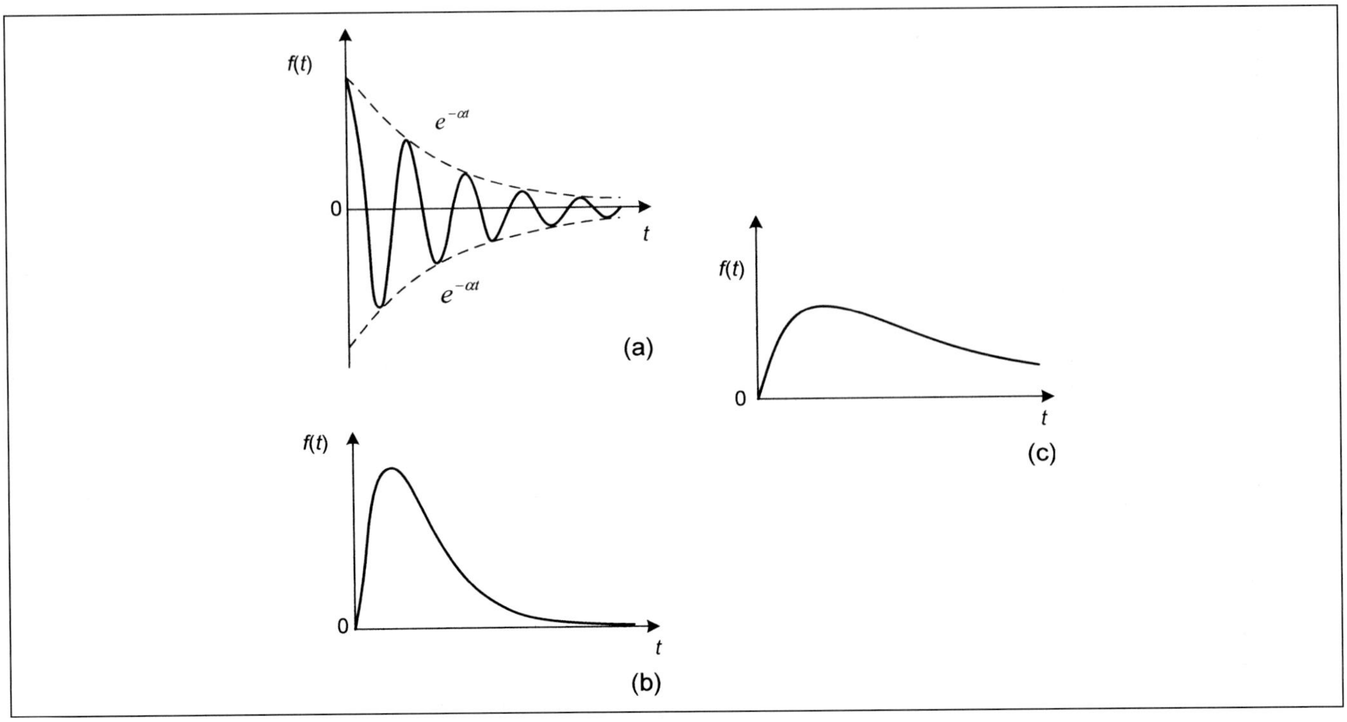

FIGURE 2-16 (*a*) Underdamped response. (*b*) Critically damped response. (*c*) Overdamped response patterns.

where:

$$\delta = \tan^{-1}\left(\frac{\alpha}{\omega_n}\right) \quad \text{and} \quad (\alpha^2+\omega_n^2) = \omega_d^2 + \frac{1}{LC} \tag{2-58}$$

The voltage across the inductor is:

$$v_L = L\frac{di}{dt} = I_m\sqrt{\frac{L}{C}}e^{-\alpha t}\sin[\omega_n t + \beta + (90^0 + \delta)] \tag{2-59}$$

Note that voltage across the capacitor is lagging more than 90° with respect to the current. Also the voltage across the inductor is leading slightly more than 90° with respect to current. In the steady-state solution, these are exactly in phase quadrature. The difference is expressed by angle δ, due to exponential damping. This angle denotes the displacement of the deviation of the displacement angle. As the resistance is generally small, we can write:

$$\omega_n \approx \frac{1}{\sqrt{LC}} \quad \text{and} \quad \tan\delta \approx \frac{R}{2\sqrt{LC}} \tag{2-60}$$

In most power systems, δ can be neglected.

In the preceding examples we have calculated the constants of integration from the initial conditions. In general, to find n constants, we need:

- The transient response $f(0)$ and its $(n-1)$ derivatives
- The initial value of the forced response $ff(0)$ and its $n-1$ derivatives.

Consider for example a second-order system:

$$i_n(t) = A_1e^{s_1t} + A_2e^{s_2t}$$
$$i'_n(t) = s_1A_1e^{s_1t} + s_2A_2e^{s_2t}$$

Then

$$\begin{aligned} i_n(0) &= i(0) + i_f(0) = A_1 + A_2 \\ i'_n(0) &= i'(0) + i'_f(0) = s_1A_1 + s_2A_2 \end{aligned} \tag{2-61}$$

From Eq. (2.61) the constants A_1 and A_2 can be calculated.

From the above examples we observe that the end results do not immediately indicate what type of response can be expected. The response can be calculated in small increments of time and plotted to show a graphical representation.

Example 2-13 A 500-kvar capacitor bank is switched through an impedance of $0.0069 + j0.067\ \Omega$, representing the impedance of cable circuit and bus work. The supply system voltage is 480 V, three phase, 60 Hz, and the supply source has an available three-phase short-circuit current of 35 kA at < 80.5°. The switch is closed when the phase *a* voltage peaks in the negative direction.

The result of EMTP simulated transients of inrush current and voltage for 50 ms are shown in Fig. 2-17*a*. This shows that the maximum switching current is 2866 A peak (= 2027 A rms), and that the voltage transient, shown in Fig. 2-17*b* results in approximately twice the normal system voltage. The calculated steady-state current is 507 A rms. This example is a practical case of capacitor switching transients in low-voltage systems, and we will further revert to this subject in greater detail in the following chapters.

To continue with this example, the resistance is changed to correspond to critical damping circuit. The response is shown in Fig. 2-17*c*; the current transient is eliminated. The voltage transients also disappear, which is not shown.

2-10 PARALLEL RLC CIRCUIT

The differential equation governing the parallel RLC circuit is similar to the series circuit. It can be written as:

$$\frac{d^2\varphi}{dt^2} + \frac{1}{RC}\frac{d\varphi}{dt} + \frac{\varphi}{LC} = f(t) \tag{2-62}$$

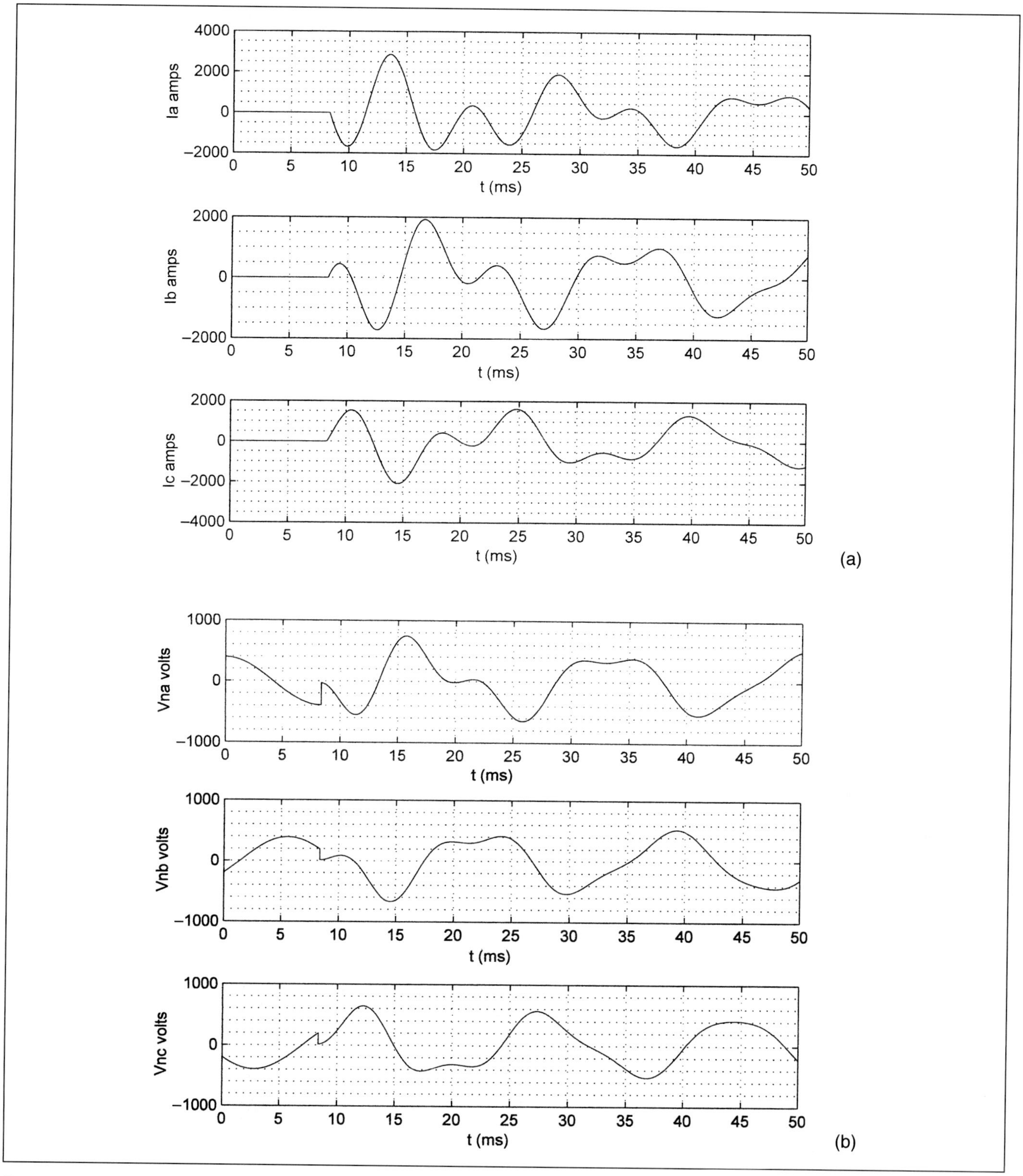

FIGURE 2-17 EMTP simulation of switching of a 500 kvar capacitor on 480-V, three-phase, 60-Hz source of 35-kA short-circuit current, switch closed at the crest of voltage in phase *a*, underdamped circuit. (*a*) Current transients. (*b*) Voltage transients. (*c*) Current transients, critically damped circuit.

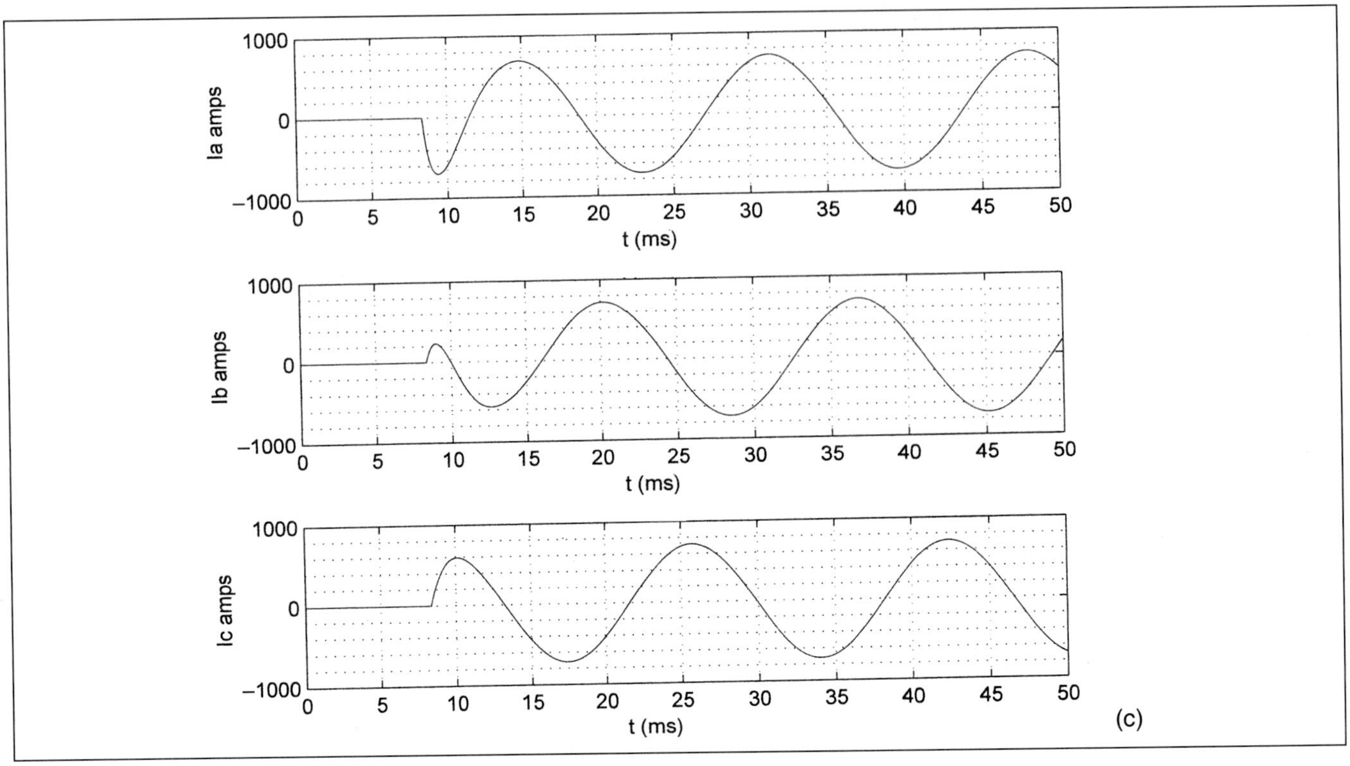

FIGURE 2-17 (*Continued*)

where $f(t)$ is the forcing function and φ can be current or voltage. Comparing with Eq. (2-43) for the series circuit we note that parallel circuit time constant RC is akin to the series circuit time constant L/R. The product of these two time constants gives LC. If we define a parameter:

$$\eta = R/Z_0 = R/\sqrt{L/C} \tag{2-63}$$

Then $RC/(L/R)$, that is, ratio of the time constants or parallel and series circuits is η^2. This leads *to duality in the analysis of the series and parallel circuits*. Note that $Z_0 = \sqrt{L/C}$ is called the characteristic impedance or surge impedance, which is of much significance in transmission line analysis (Chap. 4).

Consider in a parallel RLC circuit, the capacitor is charged and suddenly connected to the RL circuit in parallel, and there is no external excitation. When the switch is closed, the current from the charged capacitor divides between the resistance and inductance:

$$I_C = I_R + I_L$$

$$-C\frac{dV_C}{dt} = I_L + \frac{V_C}{R} = I_L + \frac{L}{R}\frac{dI_L}{dt}$$

Eliminate V_C from the above equations:

$$\frac{d^2I_L}{dt^2} + \frac{1}{RC}\frac{dI_L}{dt} + \frac{I_L}{LC} = 0 \tag{2-64}$$

This equation can be solved like a series circuit differential equation, as before. The equation can be written as:

$$\frac{d^2I_L}{dt^2} + \frac{1}{\tau_p}\frac{dI_L}{dt} + \frac{I_L}{\tau^2} = 0 \tag{2-65}$$

The current in the inductor is given by:

$$\left(s^2 + \frac{s}{\tau_p} + \frac{1}{\tau^2}\right) i_L(s) = \left(s + \frac{1}{\tau_p}\right) I_L(0) + I_L'(0) \tag{2-66}$$

However, the initial current is zero and $I_L'(0) = V_C(0)/L$. Therefore:

$$i_L(s) = \frac{V_C(0)}{L}\frac{1}{\left(s^2 + \frac{s}{\tau_p} + \frac{1}{\tau^2}\right)} \tag{2-67}$$

Again the form of solution will depend upon the values of the roots of the quadratic equation:

$$s_1, s_2 = -\frac{1}{2\tau_p} \pm \frac{1}{2}\sqrt{\frac{1}{\tau_p^2} - \frac{4}{\tau^2}} \tag{2-68}$$

If we want to find the capacitor voltage, then we can write the equation:

$$\frac{d^2V_C}{dt^2} + \frac{1}{RC}\frac{dV_C}{dt} + \frac{V_C}{LC} = 0 \tag{2-69}$$

And in terms of Laplace transform, the voltage across the capacitor is:

$$v_c(s) = \frac{V_C(0)}{\left(s^2 + \frac{s}{\tau_p} + \frac{1}{\tau^2}\right)} \tag{2-70}$$

Note the similarity between Eq. (2-62) and Eq. (2-43). These differential equations can be solved as before and again the response can be damped, critically damped, or oscillatory.

2-11 SECOND-ORDER STEP RESPONSE

The step response of second-order systems for zero initial conditions is suitable in analysis of pulse transients. Consider zero initial conditions. The final output with a step input is a constant:

$$q_{ss} = \frac{f(t)}{c} \tag{2-71}$$

Here c is same as in the general differential equation [Eq. (2-41)]. The response will be overdamped, critically damped, or underdamped depending upon the roots of the quadratic equation, as before.

Overdamped

$b^2 > 4ac$

$$q(t) = -\frac{1}{2c}\left(1+\frac{b}{\sqrt{b^2-4ac}}\right)e^{s_1 t} - \frac{1}{2c}\left(1-\frac{b}{\sqrt{b^2-4ac}}\right)e^{s_2 t} + \frac{1}{c} \tag{2-72}$$

Critically damped

$b^2 = 4ac$

$$q(t) = -\frac{1}{c}e^{-st} - \frac{b}{2ac}te^{-st} + \frac{1}{c} \tag{2-73}$$

Underdamped

$b^2 < 4ac$

$$q(t) = \frac{1}{c} - \frac{1}{c}e^{-\alpha t}\cos\beta t - \frac{\alpha}{\beta c}e^{-\alpha t}\sin\beta t \tag{2-74}$$

Example 2-14 Consider a network given by the following equation:

$$f(t) = 10\frac{d^2 t}{dt^2} + 5\frac{di}{dt} = 5i$$

This represents an underdamped system. The solution is:

$$i(t) = \frac{1}{c} - \frac{1}{c}e^{-\alpha t}\cos\beta t - \frac{\alpha}{\beta c}e^{\alpha t}\sin\beta t$$

Here

$$\alpha = \frac{b}{2a} = \frac{5}{20} = 0.25$$

$$\beta = \frac{\sqrt{4ac-b^2}}{2a} = 0.66$$

$$c = 5$$

Therefore:

$$i(t) = \frac{1}{5}(1 - e^{-0.25t}\cos 0.66t - 0.378e^{-0.25t}\sin 0.66t)u(t)$$

Consider a pulse of:

$$2u(t) - 2u(t-0.1)$$

Response to first step is:

$$i_1(t) = \frac{2}{5}(1 - e^{-0.25t}\cos 0.66t - 0.378e^{-0.25t}\sin 0.66t)u(t)$$

And the response to the second step is:

$$i_2(t) = \frac{1}{5}[1 - e^{-0.25(t-0.1)}\cos 0.66(t-0.1) - 0.378e^{-0.25(t-0.1)}\sin 0.66(t-0.1)]\,u(t)$$

The total current is:

$$i_1(t) + i_2(t)$$

2-12 RESONANCE IN SERIES AND PARALLEL RLC CIRCUITS

Amplification of the voltage and currents can occur in RLC resonant circuits. The natural frequency of a power distribution system is much higher than any of the load-generated harmonics. However, application of capacitors, in presence of harmonic producing loads, can bring about a resonant condition (Chap. 6).

2-12-1 Series RLC Circuit

The impedance of a series circuit (Fig. 2–18a) is:

$$z = R + j\left(\omega L - \frac{1}{\omega C}\right)$$

At a certain frequency, say f_0, z is minimum when:

$$\omega L - \frac{1}{\omega C} = 0 \quad \text{or} \quad \omega = \omega_0 = \frac{1}{\sqrt{LC}} \tag{2-75}$$

Figure 2-18b shows the frequency response. The capacitive reactance inversely proportional to frequency is higher at low frequencies, while the inductive reactance directly proportional to frequency is higher at higher frequencies. Thus, the reactance is capacitive and angle of z is negative below f_0, and above f_0 the circuit is inductive and angle z is positive (Fig. 2-18c). The voltage transfer function, $H_v = V2/V1 = R/Z$, is shown in Fig. 2-18d. This curve is reciprocal of Fig. 2-18b. The half-power frequencies are expressed as:

$$\omega_h = \frac{R}{2L} + \sqrt{\left(\frac{R}{2L}\right)^2 + \frac{1}{LC}} = \omega_0\left(\sqrt{1+\frac{1}{4Q_0^2}} + \frac{1}{2Q_0}\right)$$
$$\omega_l = -\frac{R}{2L} + \sqrt{\left(\frac{R}{2L}\right)^2 + \frac{1}{LC}} = \omega_0\left(\sqrt{1+\frac{1}{4Q_0^2}} - \frac{1}{2Q_0}\right) \tag{2-76}$$

The bandwidth β shown in Fig. 2.18d is given by:

$$\beta = \frac{R}{L} = \frac{\omega_0}{Q_0}$$

or

$$Q_0 = \frac{\omega_0 L}{R} \tag{2-77}$$

The quality factor Q is defined as:

$$Q_0 = 2\pi\left(\frac{\text{maximum energy stored}}{\text{energy dissipated per cycle}}\right) \tag{2-78}$$

2-12-2 Parallel RLC Circuit

In a parallel RLC circuit (Fig. 2-19a), the impedance is high at resonance:

$$y = \frac{1}{R} + \frac{1}{j\omega L} + j\omega C$$

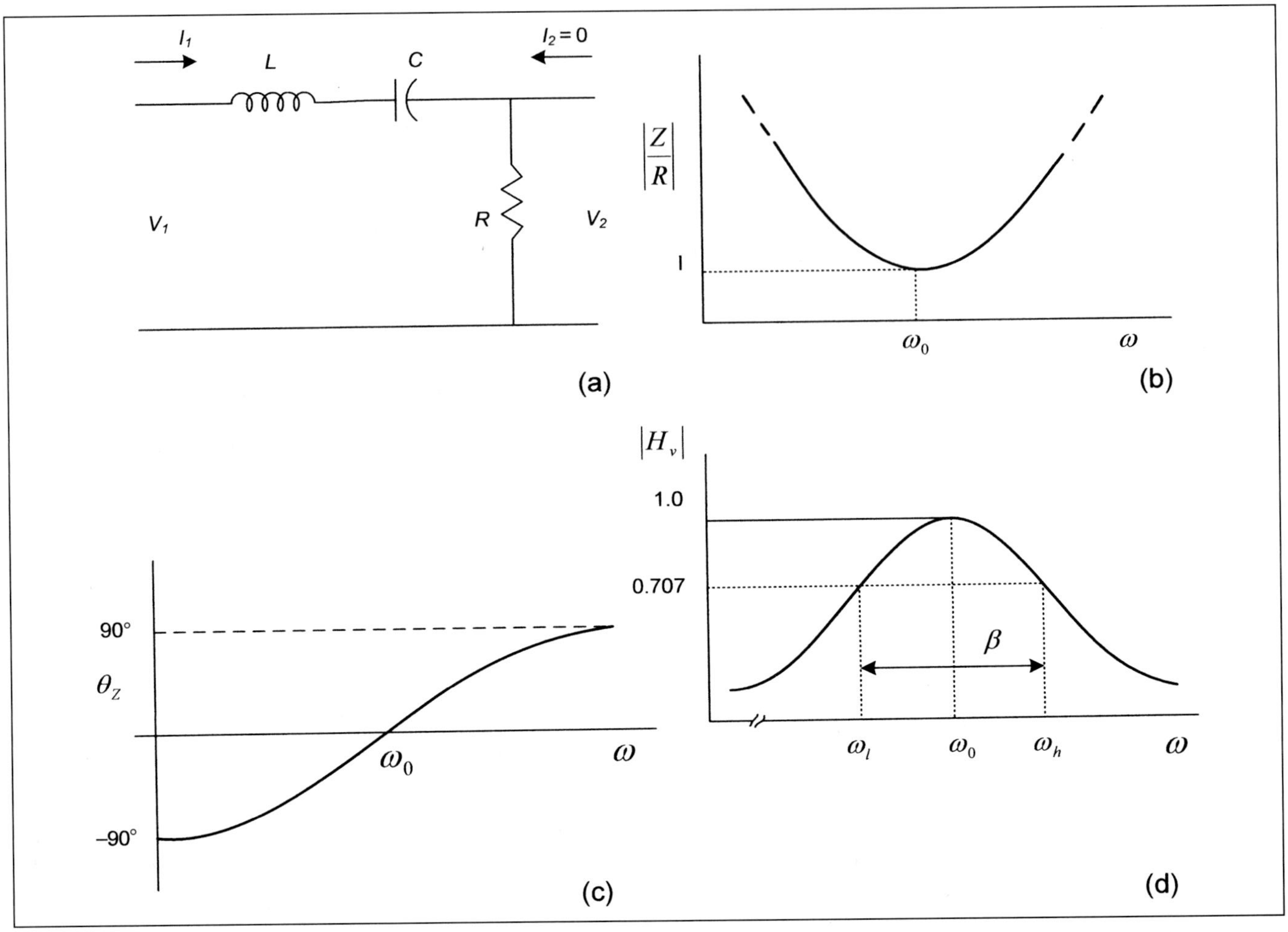

FIGURE 2-18 (*a*). Series RLC resonant circuit. (*b*) Impedance amplitude *Z/R* versus angular frequency. (*c*) Impedance angle versus angular frequency. (*d*) Voltage transfer function.

Thus, the resonant condition is:

$$-\frac{1}{\omega L}+\omega C=0 \quad \text{or} \quad \omega=\omega_a=\frac{1}{\sqrt{LC}} \tag{2-79}$$

Compare Eq. (2-79) with Eq. (2-75) for the series circuit.

The magnitude *Z/R* is plotted in Fig. 2-19*b*. Half-power frequencies are indicated in the plot. The bandwidth is given by:

$$\beta=\frac{\omega_a}{Q_a} \tag{2-80}$$

where the quality factor for the parallel circuit is given by:

$$Q_a=\frac{R}{\omega_a L}=\omega_a RC=R\sqrt{\frac{C}{L}} \tag{2-81}$$

Example 2-15 This example is a simulation of the current in a series resonant circuit excited by a three-phase, 480-V, and 60-Hz source. The components in the series circuits are chosen so that the resonant frequency is close to the source frequency of 60 Hz (L = 5 mH, C = 0.00141 F) and the resistance in the circuit is 0.1 Ω. The transients are initiated by closing the switch at peak of the phase *a* voltage in the negative direction and these will reach their maximum after a period of three to five times the time constant of the exponential term. As the resistance is low, the maximums may be reached only after a number of cycles. Figure 2-20*a* shows the current transient in phase *a* only.

A simulation of the transient in the same phase when the resonant frequency of the circuit is slightly different from the source frequency is shown in Fig. 2-20*b*. As the natural and system frequencies are different, we cannot combine the natural and steady-state harmonic functions, and these will be displaced in time on switching. The subtraction and addition of the two components occur periodically and beats of total current/voltage appear. These beats will diminish gradually and will decay in three to five time constants τ. The circuit behavior after a short time of switching can be represented by:

$$i=2I_m\sin\frac{\omega-\omega_n}{2}t\times\cos\frac{\omega-\omega_n}{2}t \tag{2-82}$$

2-12-3 Normalized Damping Curves

Consider the parallel RLC circuit. The roots of the auxiliary equation can be written as:

$$s_1, s_2=\frac{1}{2RC}\pm\frac{1}{2RC}\sqrt{1-4Q_a^2} \tag{2-83}$$

The response to a step input of voltage and current can be expressed as a family of damping curves for the series and parallel circuits. To

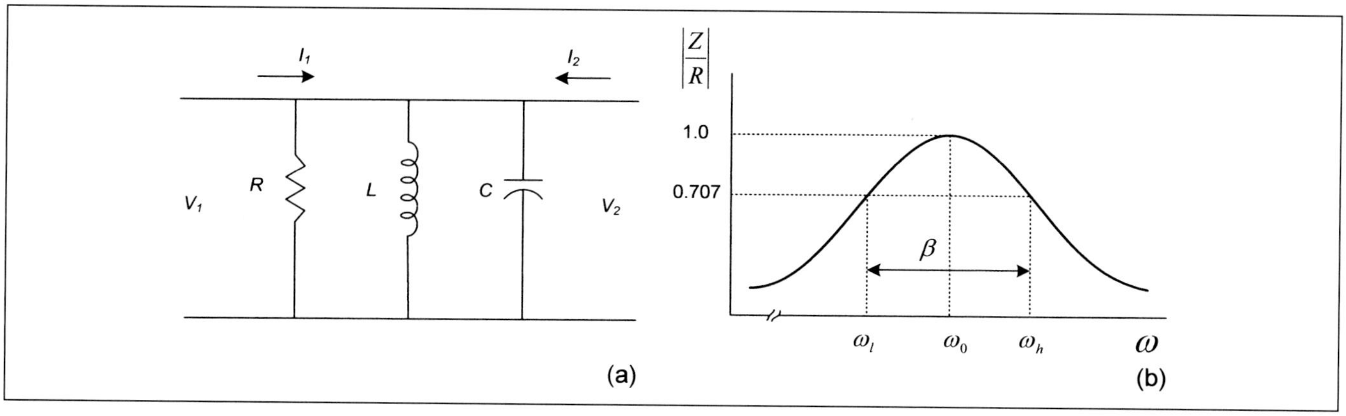

FIGURE 2-19 (*a*) Parallel RLC resonant circuit. (*b*) Impedance amplitude *Z*/*R* versus angular frequency.

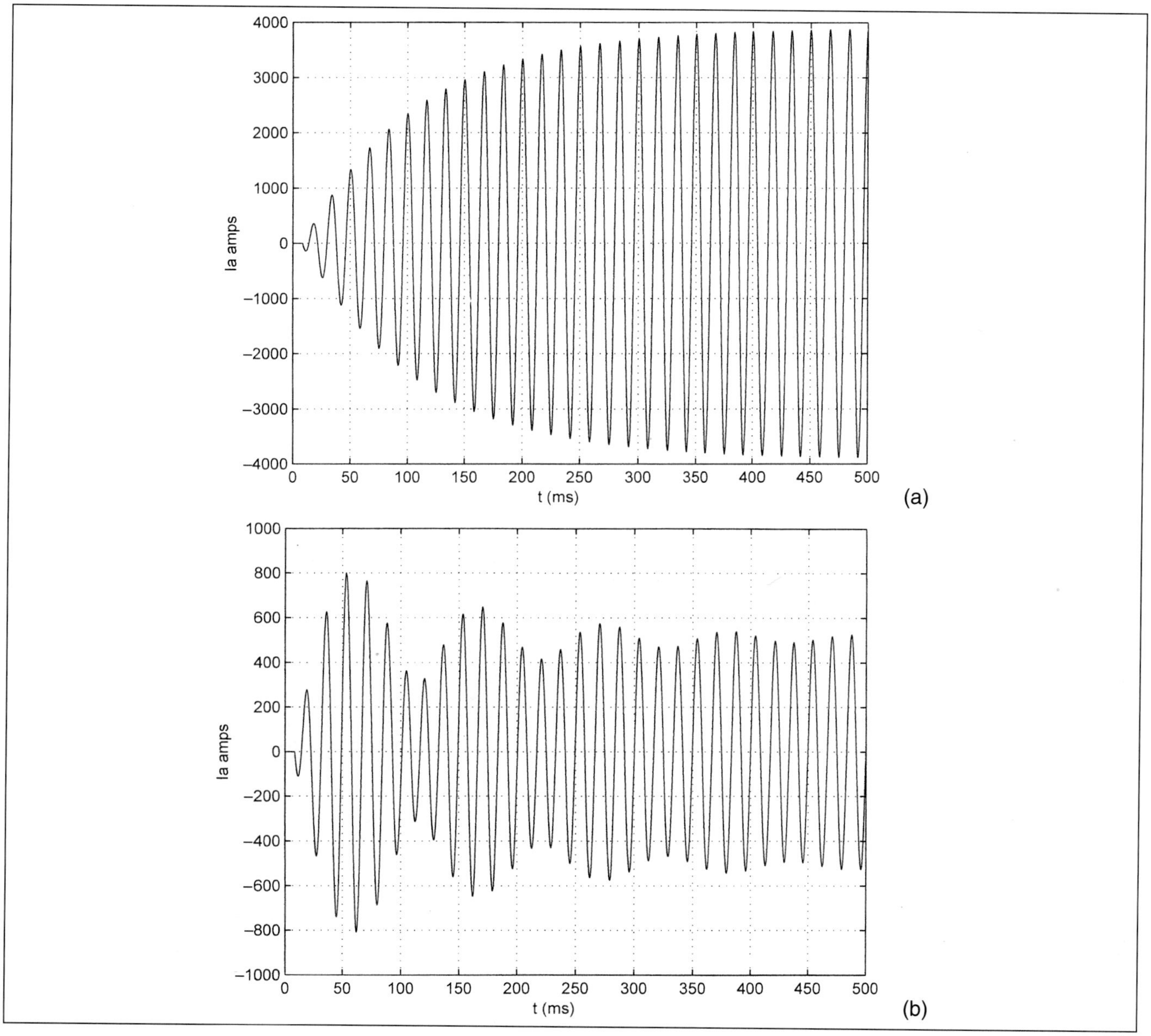

FIGURE 2-20 (*a*) EMTP simulation of RLC series resonance in 480-V, three-phase, 60-Hz system; the resonant frequency coincides with the supply source frequency. Current transient in phase *a*, switch closed at the crest of the voltage wave in phase *a*. (*b*) Transient in phase *a*, the resonant frequency close to the source frequency.

unitize the solutions, we use response of a parallel LC circuit. For the voltage and current:

$$v(t)=\frac{1}{\omega_a C}\sin(\omega_a t) \tag{2-84}$$

$$i(t)=\frac{1}{\omega_a L}\sin(\omega_a t) \tag{2-85}$$

The maximum voltage or current occurs at an angular displacement of $\omega_a t=\pi/2$. If we set $\omega_a t=\theta$, then $t/(2RC)$ can be replaced with $\theta/2Q_a$.

Then the solution for *underdamped* circuit:

$$f(Q_a,\theta)=\frac{2Q_a e^{-\theta/2Q_a}}{\sqrt{1-4Q_a^2}}\left[\sinh\left(\frac{\theta\sqrt{1-4Q_a^2}}{2Q_a}\right)\right] \tag{2-86}$$

for critically damped:

$$f(Q_a,\theta)=\theta e^{-\theta/2Q_a} \tag{2-87}$$

and for the overdamped circuit:

$$f(Q_a,\theta)=\frac{2Q_a e^{-\theta/2Q_a}}{\sqrt{1-4Q_a^2}}\left[\sin\left(\frac{\theta\sqrt{1-4Q_a^2}}{2Q_a}\right)\right] \tag{2-88}$$

Figure 2-21 shows the normalized curves from which the response can be ascertained. For the series RLC circuit, similar curves can be drawn by replacing Q_a with Q_0. This is shown in Fig. 2-22. These normalized curves are based upon the Figures in Reference 1.

2-13 LOOP AND NODAL MATRIX METHODS FOR TRANSIENT ANALYSIS

This section provides an overview of matrix methods for transient analysis, which are used extensively for the steady-state solutions of networks. Network equations that can be formed in bus (nodal), loop (mesh), or branch frame of reference, that is, in the bus frame the performance is described by $n-1$ linear independent equations for n number of nodes. The reference node which is at ground potential is always neglected. In the admittance form:

$$\overline{I}_B=\overline{Y}_B\overline{V}_B \tag{2-89}$$

where $\overline{I}_B$ is the vector of injection of bus currents. The usual convention for the flow of the current is that it is positive when flowing toward the bus and negative when flowing away from it. $\overline{V}_B$ is the vector of the bus or nodal voltages measured from the reference node:

$$\begin{vmatrix} I_1 \\ I_2 \\ \cdot \\ I_{(n-1)} \end{vmatrix}=\begin{vmatrix} Y_{11} & Y_{12} & \cdot & Y_{1,n-1} \\ Y_{21} & Y_{22} & \cdot & Y_{2,n-1} \\ \cdot & \cdot & \cdot & \cdot \\ Y_{n-1,1} & Y_{n-2,2} & \cdot & Y_{n-1,n-1} \end{vmatrix}\begin{vmatrix} V_1 \\ V_2 \\ \cdot \\ V_n \end{vmatrix} \tag{2-90}$$

$\overline{Y}_B$ is nonsingular square matrix of order $(n-1)(n-1)$. It has an inverse:

$$\overline{Y}_B^{-1}=\overline{Z}_B \tag{2-91}$$

where $\overline{Z}_B$ is the bus impedance matrix, also of order $(n-1)$ $(n-1)$.

In this book we will denote a matrix Z by a top bar, that is, $\overline{Z}$, the transpose of the matrix by $\overline{Z}^t$ the inverse of the matrix by $\overline{Z}^{-1}$, the conjugate of a matrix by $\overline{Z}^*$, the adjoint of a matrix by $\overline{Z}_{adj}$. $\overline{Y}_{ii}$ ($i=1, 2, 3, 4\ldots$) is the self-admittance or driving point admittance of node i, given by the diagonal elements, and it is equal to the algebraic sum of all admittances terminating in that node. $\overline{Y}_{ik}$ (i, $k=1, 2, 3, 4\ldots$) is the mutual admittance between nodes i and k, and it is equal to negative of the sum of all admittances directly connected between these nodes.

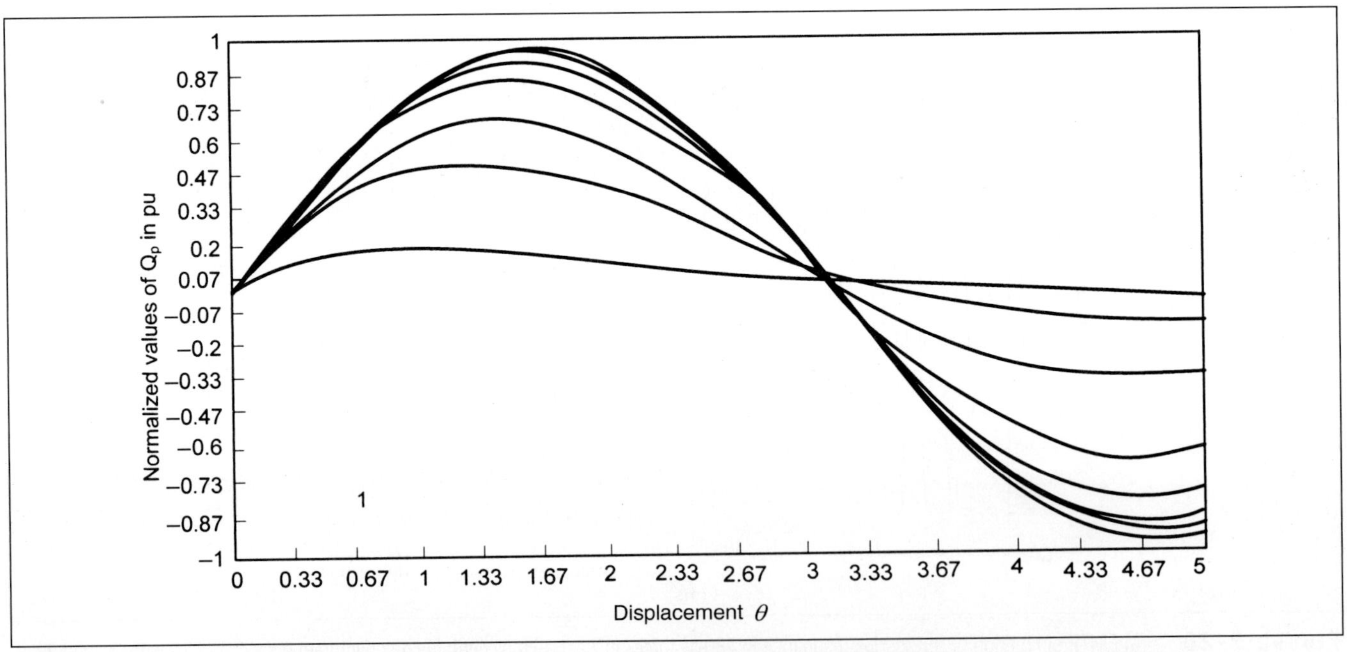

FIGURE 2-21 Normalized damping curves, parallel RLC circuit, $0.50 \le Q_p \le 75.0$, 0.50, 1.0, 2.0, 5.0, 10.0, 15.0, 30.0, and 75.0, with $Q_P(75)=1.00$ *pu*.

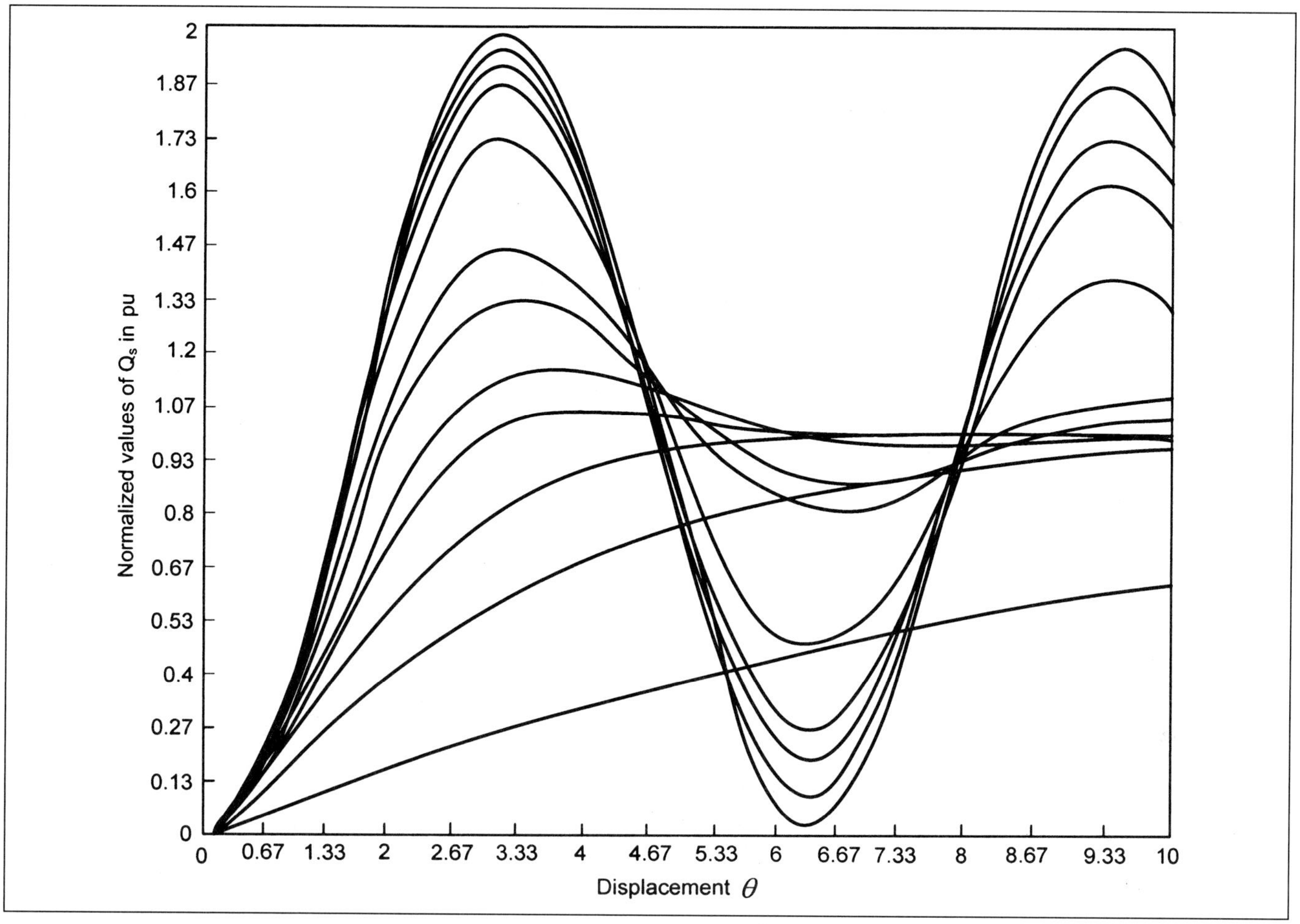

FIGURE 2-22 Normalized damping curves, series RLC circuit, $0.10 \le Q_S \le 100.0$, 0.10, 0.30, 0.50, 0.75, 1.0, 1.5, 2.0, 5.0, 10.0, 15.0, 30.0, and 100.0, with $Q_S(100) = 1.99$.

In the loop frame of reference:

$$\bar{V}_L = \bar{Z}_L \bar{I}_L \tag{2-92}$$

where $\bar{V}_L$ is the vector of loop voltages, $\bar{I}_L$ is the vector of unknown loop currents, and $\bar{Z}_L$ is the loop impedance matrix of the order of $l \times l$. The loop impedance matrix is derived from basic loop impedance equations, and it is based upon Kirchoff's voltage law. It can be constructed without writing the loop equations, just like the admittance matrix in the bus frame of reference.

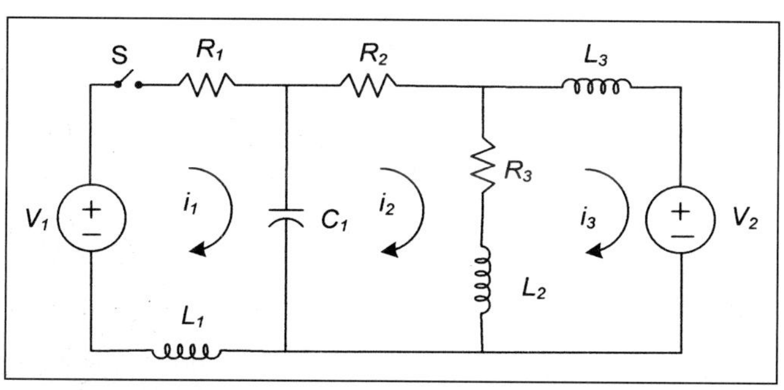

FIGURE 2-23 Circuit for loop matrix development with three loop currents, i_1, i_2, and i_3.

Consider the circuit of Fig. 2-23. By inspection, we can write the following equations:

$$\begin{vmatrix} R_1 + L_1 p + \dfrac{1}{C_1 p} & -\dfrac{1}{C_1 p} & \cdot \\ -\dfrac{1}{C_1 p} & R_2 + R_3 + L_2 p + \dfrac{1}{C_1 p} & -(R_3 + L_2 p) \\ 0 & -(R_3 + L_2 p) & L_2 p + L_3 p + R_3 + \dfrac{1}{C_1 p} \end{vmatrix} \begin{vmatrix} i_1 \\ i_2 \\ i_3 \end{vmatrix} = \begin{vmatrix} V_1 \\ 0 \\ V_2 \end{vmatrix} \tag{2-93}$$

where the operator p is defined as d/dt, and $1/p$ is written in place of the integrator. Circuit theorems and reduction techniques applied in steady state can be adapted to transient state.

2-14 STATE VARIABLE REPRESENTATION

The differential equations of a system can be written as first-order differential equations. A state variable representation of nth order can be arranged in n first-order differential equations:

$$\begin{aligned} \dot{x}_1 &= f_1(x_1, x_2, \ldots, x_n; r_1, r_2, \ldots, r_n) \\ &\vdots \\ \dot{x}_n &= f_n(x_1, x_2, \ldots, x_n; r_1, r_2, \ldots, r_n) \end{aligned} \tag{2-94}$$

We write this in the following form:

$$\dot{\mathbf{x}}(t)=\mathbf{f}[\mathbf{x}(t),\mathbf{r}(t),t] \tag{2-95}$$

where $\mathbf{x}$ is n-dimensional vector of unknown variables, called the state vector, $\mathbf{r}$ is m-dimensional input vector, $\mathbf{f}$ is the n-dimensional function vector, and t is the time variable. The output can be expressed in terms of state $\mathbf{x}(t)$ and input $\mathbf{r}(t)$ and can be written as:

$$\mathbf{y}(t)=\mathbf{g}[\mathbf{x}(t),\mathbf{r}(t),t] \tag{2-96}$$

The state and output equations constitute the state of the system.

In a model of a linear time-invariant system, the derivative of each state is a linear combination of the system states and inputs:

$$\begin{aligned}\dot{x}_1 &= a_{11}x_1+a_{12}x_2+\cdots+a_{1n}x_n+b_{11}r_1+b_{12}r_2+\cdots+b_{1m}r_m\\ &\vdots\\ \dot{x}_n &= a_{n1}x_1+a_{n2}x_2+\cdots+a_{nn}x_n+b_{n1}r_1+b_{n2}r_2+\cdots+b_{nm}r_n\end{aligned} \tag{2-97}$$

This can be written in the matrix form:

$$\begin{vmatrix}\dot{x}_1\\ \dot{x}_2\\ \cdot\\ \dot{x}_n\end{vmatrix}=\begin{vmatrix}a_{11} & a_{12} & \cdot & a_{1n}\\ a_{21} & a_{22} & \cdot & a_{2n}\\ \cdot & \cdot & \cdot & \cdot\\ a_{n1} & a_{n2} & \cdot & a_{nn}\end{vmatrix}\begin{vmatrix}x_1\\ x_2\\ \cdot\\ x_n\end{vmatrix}+\begin{vmatrix}b_{11} & b_{12} & \cdot & b_{1m}\\ b_{21} & b_{22} & \cdot & b_{2m}\\ \cdot & \cdot & \cdot & \cdot\\ b_{n1} & b_{n2} & \cdot & b_{nn}\end{vmatrix}\begin{vmatrix}r_1\\ r_2\\ \cdot\\ r_n\end{vmatrix} \tag{2-98}$$

In the abbreviated form we can write:

$$\dot{\mathbf{x}}=\bar{\mathbf{A}}\mathbf{x}+\bar{\mathbf{B}}\mathbf{r} \tag{2-99}$$

Similarly the output is a linear combination of system states and inputs, and the output equations can be written as:

$$\begin{aligned}y_1 &= c_{11}x_1+c_{12}x_2+\cdots+c_{1n}x_n+d_{11}r_1+d_{12}r_2+\cdots+d_{1m}r_m\\ y_2 &= c_{21}x_1+c_{22}x_2+\cdots+c_{2n}x_n+d_{21}r_1+d_{22}r_2+\cdots+d_{2m}r_m\\ &\vdots\\ y_p &= c_{p1}x_1+c_{p2}x_2+\cdots+c_{p1n}x_n+d_{p1}r_1+d_{p2}r_2+\cdots+d_{pm}r_m\end{aligned} \tag{2-100}$$

or in the matrix form:

$$\begin{vmatrix}y_1\\ y_2\\ \vdots\\ y_p\end{vmatrix}=\begin{vmatrix}c_{11} & c_{12} & \cdots & c_{1n}\\ c_{21} & c_{22} & \cdots & c_{2n}\\ \vdots & \vdots & \ddots & \vdots\\ c_{p1} & c_{p2} & \cdots & c_{pn}\end{vmatrix}\begin{vmatrix}x_1\\ x_2\\ \vdots\\ x_n\end{vmatrix}+\begin{vmatrix}d_{11} & d_{12} & \cdots & d_{1m}\\ d_{12} & d_{22} & \cdots & d_{2m}\\ \vdots & \vdots & \ddots & \vdots\\ d_{p1} & d_{p2} & \cdots & d_{pm}\end{vmatrix}\begin{vmatrix}r_1\\ r_2\\ \vdots\\ r_m\end{vmatrix} \tag{2-101}$$

Thus, we can write the two equations concisely as:

$$\begin{aligned}\dot{\mathbf{x}} &= \bar{\mathbf{A}}\mathbf{x}+\bar{\mathbf{B}}\mathbf{r} \quad \text{state equation}\\ \mathbf{y} &= \bar{\mathbf{C}}\mathbf{x}+\bar{\mathbf{D}}\mathbf{r} \quad \text{output equation}\end{aligned} \tag{2-102}$$

where $\mathbf{x}$ is n-dimensional state vector, $\mathbf{r}$ is m-dimensional input vector, $\mathbf{y}$ is p-dimensional output vector, $\bar{\mathbf{A}}$ is $n \times n$ square matrix, $\bar{\mathbf{B}}$ is $n \times m$ matrix, $\bar{\mathbf{C}}$ is $p \times n$ matrix, and $\bar{\mathbf{D}}$ is $p \times m$ matrix.

In a linear time-varying system, the elements of matrices $\bar{A},\bar{B},\bar{C},\bar{D}$ are functions of time.

The advantages of the state model are:

- The state variable model makes all the states visible for all time, besides providing the output information. In an input-output relation only, the intermediate states are missed. If at any time t_0, the state is known, then the state equations unequally determine the state at any time $t > t_0$ for any given input. Thus, given the state of the circuit at $t = 0$, and all inputs, the behavior of the circuit is completely determined for $t > t_0$.
- A state variable model is more amenable to simulation on a digital computer.
- It permits compact description of complex systems.
- The state variable method lends itself easily to optimization techniques.
- It is easy to apply the state variable method to time-varying and nonlinear systems.

Example 2-16 Consider the circuit of Fig. 2-24. The system contains two energy storage elements. The state of the system can be described in terms of set of state variables, which are somewhat arbitrary, say x_1 and x_2. Let x_1 be the capacitor voltage and x_2, the inductor current. Then:

$$\mathbf{x}(t)=\begin{bmatrix}v_C(t)\\ i_L(t)\end{bmatrix}$$

is a vector of state variables.

$$\begin{aligned}i_c &= C\frac{dv_c}{dt}=i(t)-i_L\\ L\frac{di_L}{dt} &= -Ri_L+v_C\end{aligned} \tag{2-103}$$

And the output of the system is given by:

$$Ri_L(t)=v_0$$

We can write Eqs. (2-103) in the state variables x_1 and x_2:

$$\begin{aligned}\frac{dx_1}{dt} &= -\frac{1}{C}x_2+\frac{1}{C}i(t)\\ \frac{dx_2}{dt} &= \frac{1}{L}x_1-\frac{R}{L}x_2\end{aligned}$$

The output equation is:

$$y_1(t)=v_0(t)=Rx_2$$

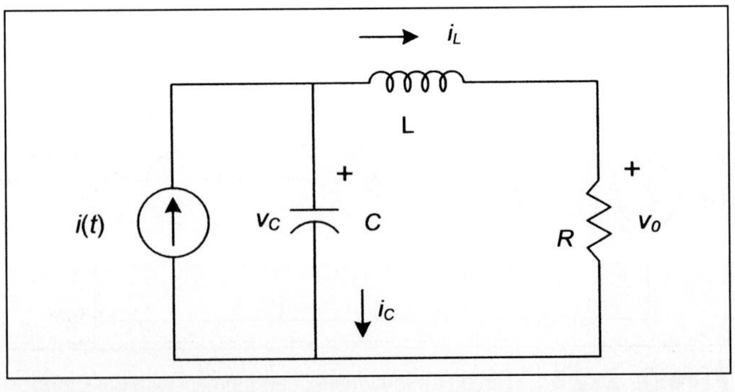

FIGURE 2-24 An RLC circuit excited by a current source, Example 2-16.

These equations can be further written as:

$$\dot{\mathbf{x}} = \begin{vmatrix} 0 & \dfrac{-1}{C} \\ \dfrac{1}{L} & \dfrac{-R}{L} \end{vmatrix} \mathbf{x} + \begin{vmatrix} \dfrac{1}{C} \\ 0 \end{vmatrix} i(t)$$

$$= \bar{\mathbf{A}}x + \bar{\mathbf{B}}i(t) \tag{2-104}$$

Thus matrix $\bar{A}$ is constant 2×2 matrix and $\bar{B}$ is a constant vector. And output equation is:

$$y = \begin{vmatrix} 0 & R \end{vmatrix} \mathbf{x} \tag{2-105}$$

For solving Eq. (2-104) the initial conditions of the capacitor voltage and inductor current are required. Thus the pair $v_C(0) = V_0$ and $i_L(0) = I_0$ is called the initial state:

$$\mathbf{x}_0 = \begin{bmatrix} V_0 \\ I_0 \end{bmatrix}$$

The zero input response is given by:

$$\frac{d\mathbf{x}(t)}{dt} = \bar{\mathbf{A}}\mathbf{x}(t)$$

and is determined by the initial state equation $\mathbf{x}_0$. If we consider the plot of $i_L(t)$, $v_C(t)$ as t increases from 0 to infinity. A curve is traced called the *state-space trajectory* and the plane i_L-v_C is called the *state-space* of the circuit. This trajectory will start at the initial point V_0, I_0 and end at the origin (0, 0), when $t = \infty$. The velocity of the trajectory $(di_L/dt, dv_C/dt)$ can be obtained from Eq. (2-103). Thus, the trajectory of a space vector in a two-dimensional space characterizes the behavior of a second-order circuit.

Figures 2-25*a* and 2-25*b* are the overdamped and underdamped responses, while Fig. 2.25*c* shows three kinds of trajectories for (a) overdamped, (b) underdamped, and (c) loss-free circuits. In the underdamped case the trajectory is a shrinking spiral and in the loss-free cases it is an ellipse, which shows continuous oscillations. In overdamped case (a), segment starts at (0.7, 0.9). The semiaxes of the ellipse depend upon L and C and the initial state $i_L(0)$ and $v_C(0)$.

For various initial states in the i_L-v_C plane, usually uniformly spaced points, we obtain a family of trajectories called the *phase portrait*, as shown in Fig. 2-26.

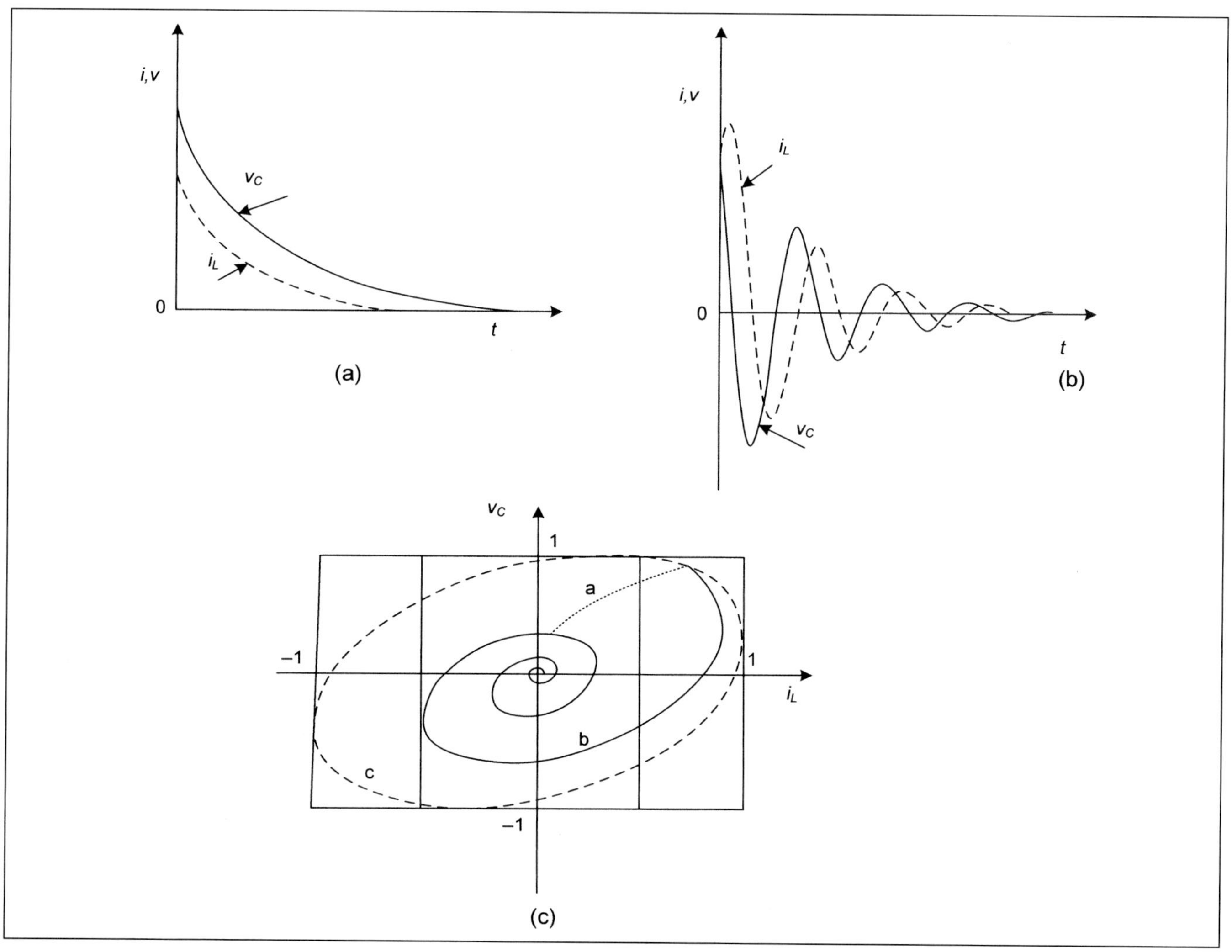

FIGURE 2-25 (*a*) Overdamped response of i_L and V_c. (*b*) Underdamped response of i_L and V_c. (*c*) State-space trajectory of the response. The dotted ellipse shows response of a loss-less circuit.

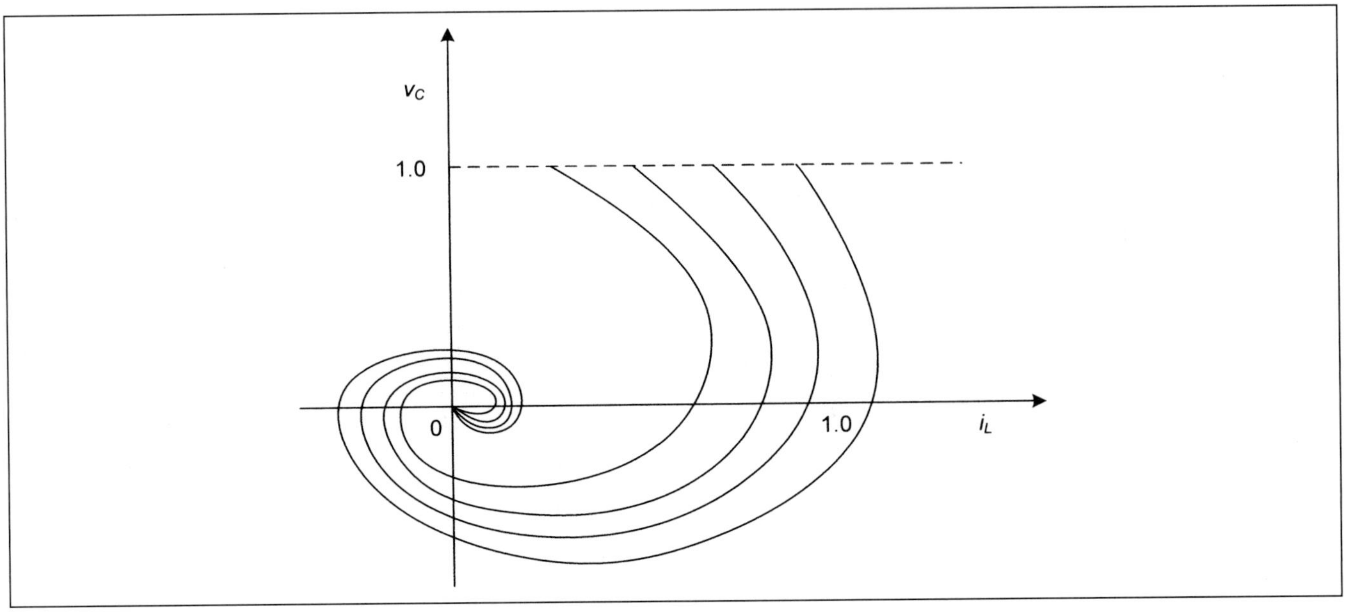

FIGURE 2-26 A phase portrait of the trajectories.

A trajectory can be plotted by starting at the initial state:

$$\mathbf{x}_0 \left| v_C(0), i_L(0) \right|^t$$

And finding an estimate of new state x at increased small time interval by Euler's method:

$$\mathbf{x}_{\text{new}} = \mathbf{x}_{\text{old}} + \Delta t \left(\frac{d\mathbf{x}}{dt} \right)_{\text{old}} \tag{2-106}$$

where Δt is the step length. Essentially during this small interval, it is assumed that dx/dt is approximately constant. Thus the straight line segment is approximated by:

$$\Delta \mathbf{x} = \left(\frac{d\mathbf{x}}{dt} \right)_{\text{const}} \Delta t \tag{2-107}$$

2-15 DISCRETE-TIME SYSTEMS

In discrete-time systems, the essential signals are discrete-time in nature. Continuous-time signals can be analyzed numerically on the digital computer by approximating the continuous-time model by a discrete-time model.

In digital data-processing systems, the continuous-time signal is converted into discrete-time signal, which is fed into the discrete-time system. The discrete-time system operates on this signal to perform the required data processing. Figure 2-27 shows the interface between the discrete- and continuous-time signals.

Consider the integral:

$$y(t) = \int_0^{t_n} r(\tau)\,d\tau$$

or

$$y(kT) = \int_0^{kT} r(\tau)\,d\tau \tag{2-108}$$

The value of $y(t)$ is the area of curve under $r(\tau)$ in the time interval 0 to t_n, shown in Fig. 2-28. Divide the curve into n small strips, kT, $k = 0, 1, 2, 3, \ldots n$. Above equation can be written as:

$$y(kT) = \int_0^{kT-T} r(\tau)\,d\tau + \int_{kT-T}^{kT} r(\tau)\,d\tau$$

$$\approx y(kT - T) + Tr(kT) \tag{2-109}$$

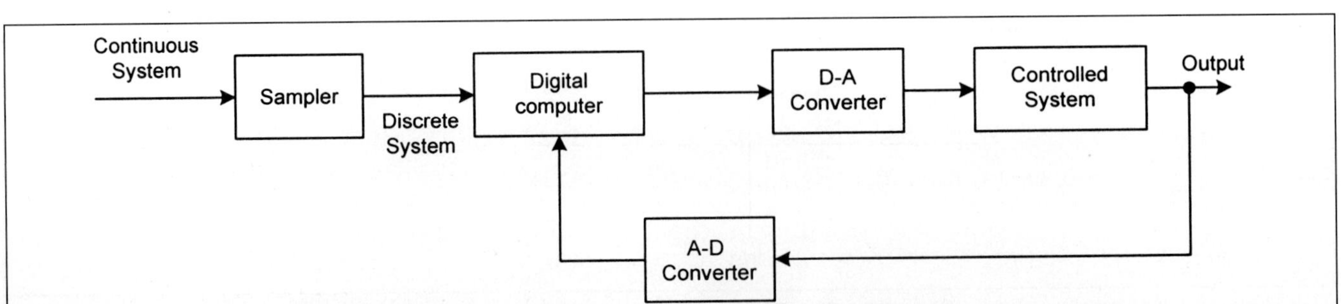

FIGURE 2-27 Interface between continuous-time and discrete-time systems with output in continuous time.

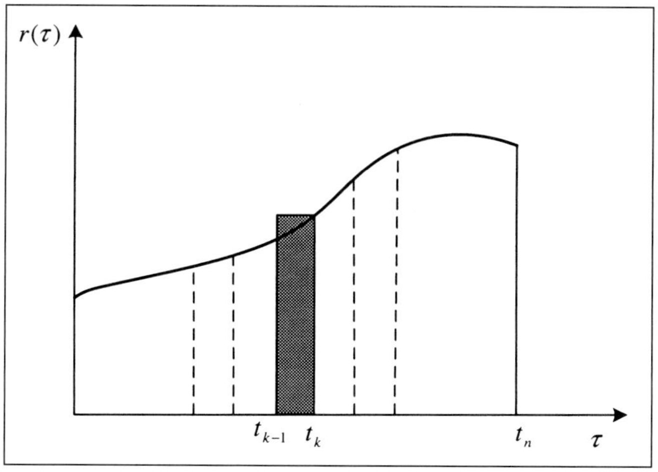

FIGURE 2-28 Rectangular estimation of an integral.

The area under the integral:

$$\int_{kT-T}^{kT} r(x)dx$$

is equal to $Tr(kT)$, shown shaded if T is sufficiently small. The Eq. (2-109) can be written as *difference equation*:

$$y(k) = y(k-1) + Tr(k) \qquad k = 1, 2, 3, \ldots$$

or (2-110)

$$y(k+1) - y(k) = Tr(k+1) \qquad k = 0, 1, 2, \ldots$$

The integral of a continuous function can be converted into a difference equation, which can be solved numerically.

2-15-1 First-Order Difference Equation

Consider a first-order differential equation given by:

$$\frac{dy(t)}{dt} + ay(t) = x(t)y(0) = y_0 \tag{2-111}$$

This can be written as:

$$\frac{dy(kT)}{dt} + ay(kT) = x(kT) \qquad k = 0, 1, 2, 3, \ldots \tag{2-112}$$

The derivative of $y(t)$ at $t = kT$ is the slope of curve $y(t)$ at $t = kT$ and can be approximated by the equation:

$$\frac{dy(kT)}{dt} = \frac{y(kT+T) - y(kT)}{T} \tag{2-113}$$

By making T smaller and smaller, the accuracy of calculation improves. Thus Eq. (2-113) becomes:

$$y(k+1) + (aT-1)y(k) = Tx(k) \qquad k = 0,1,2,\ldots \tag{2-114}$$

This is a first-order difference equation with the initial condition that $y(0) = y_0$

2-15-2 Second-Order Difference Equation

Now consider second-order differential equation given by:

$$\frac{d^2y(t)}{dt^2} + a_1\frac{dy(t)}{dt} + a_0y(t) = x(t) \tag{2-115}$$

Intial conditions:

$$y(0) = y_0 \quad \left.\frac{dy(t)}{dt}\right|_{t=0} = \dot{y}_0$$

The equation can be written as:

$$\frac{d^2y(kT)}{dt^2} + a_1\frac{dy(kT)}{dt} + a_0y(kT) = x(kT) \qquad k = 0, 1, 2, \ldots \tag{2-116}$$

The second derivative can be approximated as:

$$\begin{aligned}\frac{d^2y(kT)}{dt^2} &= \frac{\left.\frac{dy}{dt}\right|_{t=kT+T} - \left.\frac{dy}{dt}\right|_{t=kT}}{T}\\ &= \frac{\frac{y(kT+T)-y(kT)}{T} - \frac{y(kT)-y(kT-T)}{T}}{T}\\ &= \frac{y(kT+T)-2y(kT)+y(kT-T)}{T^2}\end{aligned} \tag{2-117}$$

Thus, the second-order equation can be approximated with the following difference equation:

$$y(k+1) + \frac{a_0T^2 - a_1T - 2}{1+a_1T}y(k) + \frac{1}{1+a_1T}y(k-1) = \frac{T^2}{1+a_1T}x(k)$$

or

$$\begin{aligned}&y(k+2) + \frac{a_0T^2 - a_1T - 2}{1+a_1T}y(k+1) + \frac{1}{1+a_1T}y(k)\\ &= \frac{T^2}{1+a_1T}x(k+1) \qquad k = 0, 1, 2, \ldots\end{aligned} \tag{2-118}$$

These concepts can be generalized. An nth-order differential equation

$$\begin{aligned}&\frac{d^ny(t)}{dt^n} + a_{n-1}\frac{d^{n-1}y(t)}{dt^{n-1}} + \cdots + a_1\frac{dy(t)}{dt} + a_0y(t)\\ &= b_m\frac{d^mx(t)}{dt^m} + \cdots + b_0x(t)\end{aligned} \tag{2-119}$$

can be approximated with the following nth-order difference equation,

$$\begin{aligned}&y(k+n) + \alpha_{n-1}y(k+n-1) + \cdots + \alpha_1 y(k+1) + \alpha_0y(k)\\ &= \beta_m x(k+m) + \cdots + \beta_0 x(k) \qquad k = 0, 1, 2, \ldots\end{aligned} \tag{2.120}$$

where α's and β's are appropriate constants.

As we stated in Chap. 1, the EMTP algorithms are based upon difference equations.

2-16 STATE VARIABLE MODEL OF A DISCRETE SYSTEM

Similar to continuous-time case the state variable model of a discrete system is given by:

$$\begin{aligned}\mathbf{x}(k+1) &= \bar{\mathbf{A}}\mathbf{x}(k)+\bar{\mathbf{B}}\mathbf{r}(k) \quad \text{state equation}\\ \mathbf{y}(k) &= \bar{\mathbf{C}}\mathbf{x}(k)+\bar{\mathbf{D}}\mathbf{r}(k) \quad \text{output equation}\end{aligned} \tag{2-121}$$

where $\mathbf{x}(k)$ is n-dimensional state vector, $\mathbf{r}(k)$ is m-dimensional input vector, $\mathbf{y}(k)$ is p-dimensional output vector, $\bar{\mathbf{A}} = (n \times n)$ matrix of constants, $\bar{\mathbf{B}} = (n \times m)$ matrix of constants, $\bar{\mathbf{C}} = (p \times n)$ matrix of constants, and $\bar{\mathbf{D}} = (p \times m)$ matrix of constants. The state equation is a set of n (order of the system) first-order difference equations.

2-17 LINEAR APPROXIMATION

Electrical systems are highly nonlinear in nature. Even a linear element is linear under constraints. Its characteristics undergo a change with temperature and current. Linear systems can be analyzed easily and powerful analytical tools are available. The behavior of many systems can be modeled with linear mathematical models to a high accuracy provided the range of variables is limited.

First consider a static system, with input $r(t)$ and output $y(t)$, the input-output relation shown in Fig. 2-29. Though the relationship is nonlinear, it is continuous.

Consider the operating point given by r_0, y_0. A Taylor series around the operating point gives:

$$y=\varphi(r)=\varphi(r_0)+\left(\frac{d\varphi}{dr}\right)_0\frac{r-r_0}{1!}+\left(\frac{d^2\varphi}{dr^2}\right)_0\frac{(r-r_0)^2}{2!}+\cdots \tag{2-122}$$

For small variations of input around the operating point, only the first term of the series may be considered, giving a linear approximation:

$$y=y_0+m(r-r_0) \tag{2-123}$$

where m is the slope of $\varphi(r)$ at the operating point. We can write:

$$y-y_0=m(r-r_0)$$

or

$$\tilde{y}=m\tilde{r} \tag{2-124}$$

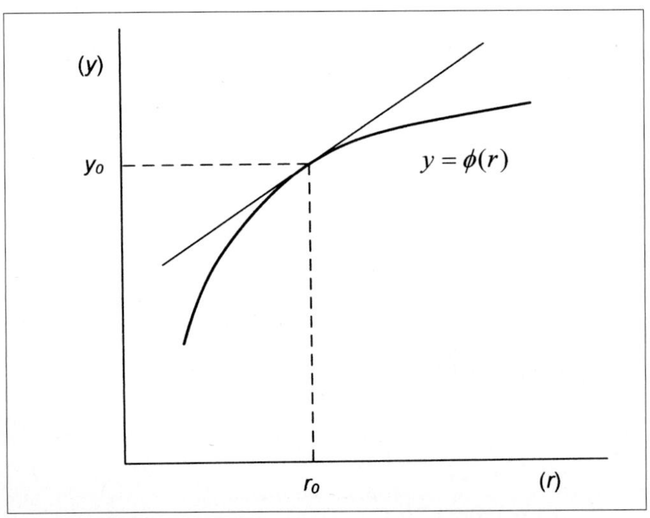

FIGURE 2-29 Linearization of a non-linear system.

2-17-1 Linearization of a Dynamic System

Consider a nonlinear time-invariant dynamic system, whose state equations is given by:

$$\mathbf{x}(t)=\mathbf{f}[\mathbf{x}(t),\mathbf{r}(t)] \tag{2-125}$$

$$\mathbf{x}(k+1)=\mathbf{f}[\mathbf{x}(k),\mathbf{r}(k)] \tag{2-126}$$

Equation (2-125) is for a continuous-time system, while (2-126) is for a discrete-time system. It is desired to approximate the system with a linear time-invariant model, given by:

$$\begin{aligned}\dot{\tilde{\mathbf{x}}}(t)&=\bar{\mathbf{A}}\tilde{\mathbf{x}}(t)+\bar{\mathbf{B}}\tilde{\mathbf{r}}(t)\\ \tilde{\mathbf{x}}(k+1)&=\bar{\mathbf{A}}\tilde{\mathbf{x}}(k)+\bar{\mathbf{B}}\tilde{\mathbf{r}}(k)\end{aligned} \tag{2-127}$$

For a system with n state variables and m inputs, the state equation is written as:

$$\begin{aligned}\dot{x}_1 &= f_1(x_1,\ldots,x_n,r_1,\ldots,r_m)\\ &\cdots\cdots\\ \dot{x}_n &= f_n(x_1,\ldots,x_n,r_1,\ldots,r_m)\end{aligned} \tag{2-128}$$

The Taylor theorem can be used to expand these equations about the equilibrium point (x_0, r_0), retaining only the first-order terms:

$$\begin{aligned}\dot{x}_1=\dot{\tilde{x}}_1&=\left(\frac{\partial f_1}{\partial x_1}\right)_0\tilde{x}_1+\cdots+\left(\frac{\partial f_1}{\partial x_n}\right)_0\tilde{x}_n+\left(\frac{\partial f_1}{\partial r_1}\right)_0\tilde{r}_1+\cdots+\left(\frac{\partial f_1}{\partial r_m}\right)_0\tilde{r}_m\\ &\cdots\\ \dot{x}_n=\dot{\tilde{x}}_n&=\left(\frac{\partial f_n}{\partial x_1}\right)_0\tilde{x}_1+\cdots+\left(\frac{\partial f_n}{\partial x_n}\right)_0\tilde{x}_n+\left(\frac{\partial f_n}{\partial r_1}\right)_0\tilde{r}_1+\cdots+\left(\frac{\partial f_n}{\partial r_m}\right)_0\tilde{r}_m\end{aligned} \tag{2-129}$$

These equations can be written as:

$$\dot{\tilde{\mathbf{x}}}=\bar{\mathbf{A}}\tilde{\mathbf{x}}+\bar{\mathbf{B}}\tilde{\mathbf{r}} \tag{2-130}$$

where:

$$\bar{\mathbf{A}}=\left|\begin{matrix}\dfrac{\partial f_1}{\partial x_1} & \cdot & \dfrac{\partial f_1}{\partial x_n}\\ \cdot & \cdot & \cdot\\ \dfrac{\partial f_n}{\partial x_1} & \cdot & \dfrac{\partial f_n}{\partial x_n}\end{matrix}\right|_0 \tag{2-131}$$

$$\bar{\mathbf{B}}=\left|\begin{matrix}\dfrac{\partial f_1}{\partial r_1} & \cdot & \dfrac{\partial f_n}{\partial r_m}\\ \cdot & \cdot & \cdot\\ \dfrac{\partial f_n}{\partial r_1} & \cdot & \dfrac{\partial f_n}{\partial r_m}\end{matrix}\right|_0 \tag{2-132}$$

The partial derivatives of f are evaluated at the equilibrium points, which must be first found out. At the equilibrium points the inputs reach steady-state values, which we may denote by: $r_{10}, r_{20}, \ldots, r_{m0}$. Then for a linear or nonlinear system, x_0 is an equilibrium point, if it does not change with constant input:

$$\begin{aligned}\frac{d\mathbf{x}(t)}{dt}&=\mathbf{f}[\mathbf{x}(t),\mathbf{r}_0]=0\\ x(k+1)-x(k)&=f[x(k),r_0]=0\end{aligned} \tag{2-133}$$

In this chapter we studied simple transients in lumped circuits. We did not attach physical meaning to the various studies, except the short circuit of an inductor with resistance connected to an ac source and capacitor switching. However, many circuits studied can be related to the practical phenomena, that is, switching of capacitors, recovery voltages on current interruption in ac circuits, harmonic resonance, lumped parameters of transmission lines, damping of transients and the like. We also formed a conceptual base of the space-state equations, difference equations, and numerical solutions of the differential equations, linearity, time-invariance, and the like. These will be further discussed in appropriate chapters in this book. We used Laplace method of analysis; however, this has limited applications for modeling of large networks on digital computers; also saturation and nonlinearity cannot be accounted for. See App. G for continuing discussion.

PROBLEMS

1. A system is described by:

$$10 \sin 3t = \frac{d^2 i}{dt^2} + 4 \sin \frac{di}{dt} = 4i = 0$$

Given that:

$$i(0) = 1 \qquad \frac{di(0)}{dt} = 4$$

find

$i(t)$ for $t > 0$

2. In Fig. 2-10c, the initial voltage on capacitor = 40 V, $t = 0$, initial current in inductor = 1A, $t = 0$. Find the capacitor voltage for $t > 0$.

3. Consider the circuit of Fig. 2-P1. Write the equations of the system in state variable form.

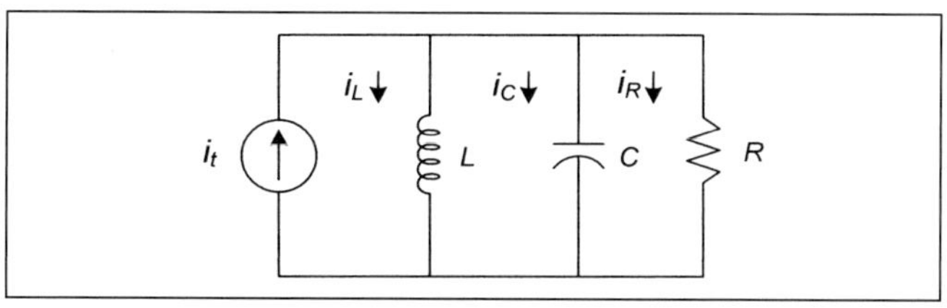

FIGURE 2-P1 Circuit for Prob. 3.

4. Consider the circuit of Fig. 2-P2. Write a differential equation relating $e_1(t)$ to $e(t)$.

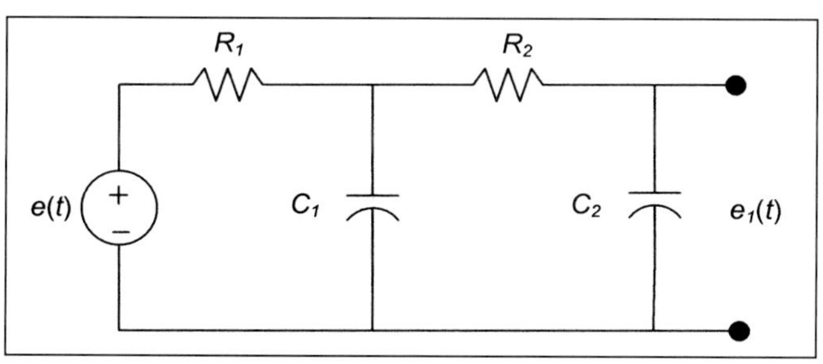

FIGURE 2-P2 Circuit for Prob. 4.

5. Consider an ideal generator of 13.8 kV, with a series impedance of $0.01 + j0.10\ \Omega$. Calculate the first peak of the short-circuit current if a fault occurs at a time when the voltage (rms) is decreasing and is at 5.0 kV.

6. Consider a series RLC circuit, $R = 4\ \Omega$, $L = 2.0$ H, $C = 0.05\ \mu$F. Find the charge and current at any time $t > 0$ in the capacitor, if the switch is closed at $t = 0$. Consider a source voltage of 100 V, and repeat for a voltage of 100 sin 4t. Solve using differential equations.

7. Solve Problem 6 using Laplace transform.

8. In Fig. 2-P3, find the pulse response for a pulse duration of 2 ms.

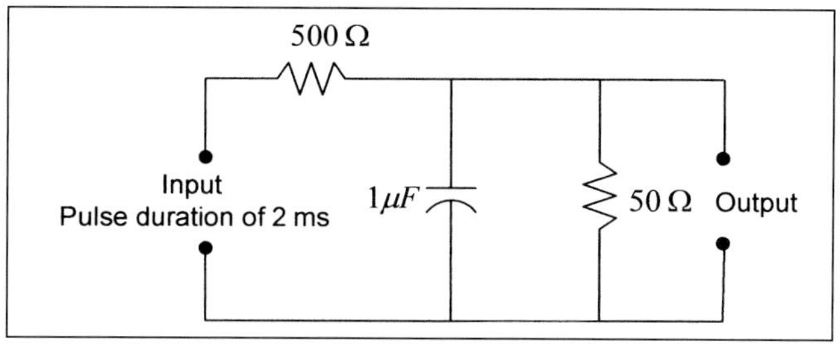

FIGURE 2-P3 Circuit for Prob. 8.

9. Find Laplace transform of a half-wave rectifier

$$0 \le t \le 2\pi \qquad f(t) = \sin t \quad 0 \le t \le \pi$$
$$= 0 \quad \pi \le t \le 2\pi$$

10. Consider the circuit of Fig. 2-P4. Write a matrix equation of the form:

$$\bar{Z}\bar{I} = \bar{V}$$

and calculate the currents i_1, i_2 and i_3.

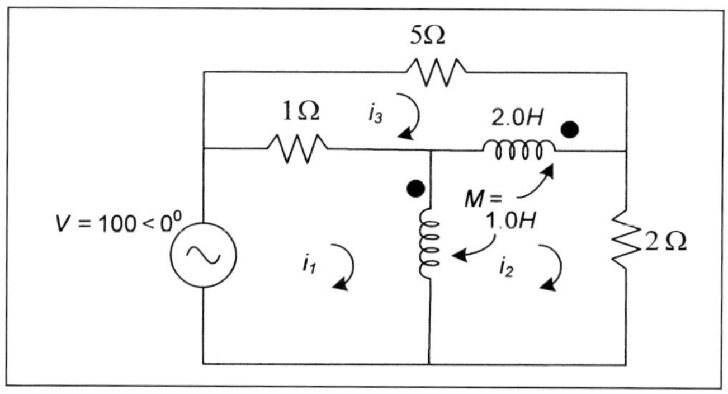

FIGURE 2-P4 Circuit for Prob. 10.

11. What is the value of R_L for maximum power transfer from the source in Fig. 2-P5? What is the value of the maximum power transferred?

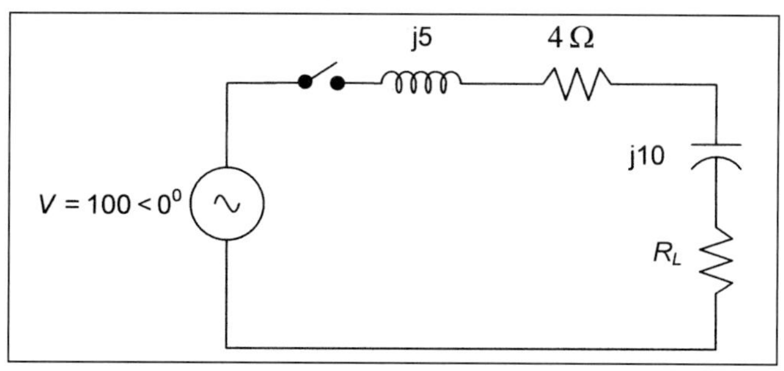

FIGURE 2-P5 Circuit for Probs. 11 and 14.

12. Design a parallel tuned circuit for resonance, consisting of an inductor and capacitor in parallel, at 660 Hz, with bandwidth of 60 Hz. If the resistance of the coil is 10 Ω, what is the impedance of the circuit at resonance?

13. A capacitor C_1 of 1 μF is charged to 100 V. It is suddenly connected to another discharged capacitor C_2 of 2μF with a series resistor of R Ω. Find the energy stored in each capacitor when the switch is closed, energy stored after a long time, and the energy dissipated in the resistance R.

14. In Fig. 2-P5, the switch is closed at $t = 0$, after it has been open for a long time. Obtain the Laplace transform of the current, $t > 0$.

15. Select the correct option for the arrangement of resistor, inductor, and capacitor: (1) all are linear, (2) all are nonlinear, (3) resistance is linear, while the capacitor and inductor can be nonlinear because the reactance varies with the frequency, (4) all can be linear or nonlinear.

16. Write a difference equation for the second-order differential equation:

$$\frac{d^2y}{dt^2}+5\frac{dy}{dt}=10y=100$$

$$y(0)=10 \qquad \left.\frac{dy}{dt}\right|_{t=0}=-2$$

17. Compare a space-state model of a system with a model derived through differential equations and Laplace transform.

18. Solve the following simultaneous differential equations relating to a transmission line:

$$-\frac{dV}{dx}=Ri \qquad -\frac{di}{dx}=GV$$

Given the conditions that $V = 0$, at $x = l$, and $V = V_0$ at $x = 0$

19. Write a step-by-step proof of Eqs. (2-85), (2-86), and (2-87) in the text of this chapter.

REFERENCE

1. IEEE Std. 399, Brown Book, Power Systems Analysis, Chapter 11, 1997.

FURTHER READING

D. L. Beeman, ed., *Industrial Power System Handbook*, McGraw-Hill, New York, 1955.

J. O. Bird, *Electrical Circuit Theory and Technology*, Butterworth, Heinemann, Oxford, UK, 1997.

J. A. Cadzow, *Discrete Time Systems*, Prentice Hall, Englewood Cliffs, NJ, 1973.

D. K. Cheng, *Analysis of Linear Systems*, Addison-Wesley Publishing Co., New York, 1959.

J. C. Das, *Power System Analysis*, Marcel and Dekker, New York, 2002.

J. A. Edminster, *Theory and Problems of Electrical Circuits*, 2d ed., Schaum's Outline Series, McGraw-Hill, New York, 1983.

O. I. Elgord, *Electrical Energy System Theory—An Introduction*, McGraw-Hill, New York, 1971.

S. Fisch, *Transient Analysis in Electrical Engineering*, Prentice Hall, New York, 1971.

A. Greenwood, *Electrical Transients in Power Systems,* 2d ed., John Wiley, New York, 1991.

W. R. LePage and S. Scely, *General Network Analysis*, McGraw-Hill, New York, 1952.

W. E. Lewis and D. G. Pryce, *The Application of Matrix Theory to Electrical Engineering*, E and FN Spon, London, 1965.

A. H. Peterson, *Transients in Power Systems*, John Wiley, New York, 1951.

M. B. Reed, *Alternating Current Theory*, 2d ed., Harper and Row, New York, 1956.

L. A. Zedeh and C. A. Desoer, *Linear System Theory*, McGraw-Hill, New York, 1963.

CHAPTER 3

CONTROL SYSTEMS

The control systems interface with electrical power systems and may have a profound effect on the transient behavior of power systems. One example is the excitation systems of synchronous generators and their impact on transient stability. A badly tuned and applied control system interface with a power system can result in oscillations and instability. Control engineering is based upon feedback theory, linear system analysis, and network theory. It is not exclusive to electrical engineering and has far-reaching applications into automation, robotics, artificial intelligence, man-machine interface (MMI), and intelligent systems. We will form a basic understanding of the transfer functions, damping and stability, feedback control systems, root locus, Bode plots, block diagrams, and frequency response methods.

We studied the overdamped, critically damped, and underdamped response of an RLC circuit in Chap. 2. Consider a second-order, linear, constant coefficient differential equation of the form:

$$\frac{d^2y}{dt^2}+2\xi\omega_n\frac{dy}{dt}+\omega_n^2 y=\omega_n^2 u \tag{3-1}$$

We can write $y(s)$ as:

$$y(s)=\frac{(s+2\xi\omega_n)y_0}{s^2+2\xi\omega_n s+\omega_n^2}=\frac{p(s)}{q(s)} \tag{3-2}$$

The denominator polynomial $q(s)$ when set to zero is called the *characteristic equation,* because the roots of this equation determine the character of the time response. The roots of this equation are also called the *poles* of the system.

The roots of the numerator polynomial $p(s)$ are called the *zeros* of the system. Poles and zeros are critical frequencies. At the poles the function $y(s)$ becomes infinite, whereas at zeros it becomes zero. The complex frequency s-plane plot of poles and zeros graphically portrays the character of the transient response. The poles and zeros are the complex numbers determined by two real variables, one representing the real part and the other representing the imaginary part. In the s plane, the abscissa is called the σ axis, and the ordinate is called the $j\omega$ axis. In the z plane, the abscissa is called the μ axis, and the ordinate is called the $j\upsilon$ axis. From Eq. (3.2), the characteristic equation is:

$$s^2+2\xi\omega_n s+\omega_n^2=0 \tag{3-3}$$

In Eq. (3.1), ξ is called the *damping ratio,* and ω_n is called the *undamped natural frequency*. The roots of the denominator are:

$$\begin{aligned} s_1, s_2 &= -\xi\omega_n \pm j\omega_n\sqrt{1-\xi^2} \\ &= -\alpha \pm j\omega_d \end{aligned} \tag{3-4}$$

where $\alpha = \xi\omega_n$ is called the *damping coefficient*, ω_d is called the *damped natural frequency,* and α is the inverse of the time constant of the system, that is, $\alpha = 1/\tau$. The weighting function of Eq. (3.1) is:

$$w(t)=\frac{1}{\omega_d}e^{-\alpha t}\sin\omega_d t \tag{3-5}$$

The unit step response is given by:

$$y_1(t)=\int_0^t w(t-\tau)\omega_n^2\,d\tau=1-\frac{\omega_n e^{-\alpha t}}{\omega_d}\sin(\omega_d t+\phi) \tag{3-6}$$

where $\phi=\tan^{-1}(\omega_d/\alpha)$ and w = weighting function

The response in Fig. 3-1 is a parametric representation of the unit step function, normalized with respect to $\omega_n t$. This can be compared with Fig. 2-21.

From Chap. 2 we know that if $\xi > 1$, the roots are real, when $\xi < 1$, the roots are complex and conjugate, and when $\xi = 1$, the roots are repeated and real, corresponding to critical damping.

The s-plane plot of poles and zeros is shown in Fig. 3-2*a*. As ξ varies with ω_n *constant*, the complex conjugate roots follow a circular locus (Fig. 3-2*b*). The transient response is increasingly oscillatory as the roots approach the imaginary axis and ξ approaches zero.

3-1 TRANSFER FUNCTION

A transfer function is the Laplace transform of the ratio of variables, which can be current ratio, voltage ratio, admittance, or impedance. Transfer functions assume that the initial conditions are zero.

Example 3-1 Consider a simple RC network as shown in Fig. 3-3. The transfer function is:

$$\frac{\text{Output}}{\text{Input}}=G(s)=\frac{v_2(s)}{v_1(s)}=\frac{1}{RCs+1}=\frac{(1/\tau)}{s+1/\tau} \tag{3-7}$$

where τ is the time constant of the network.

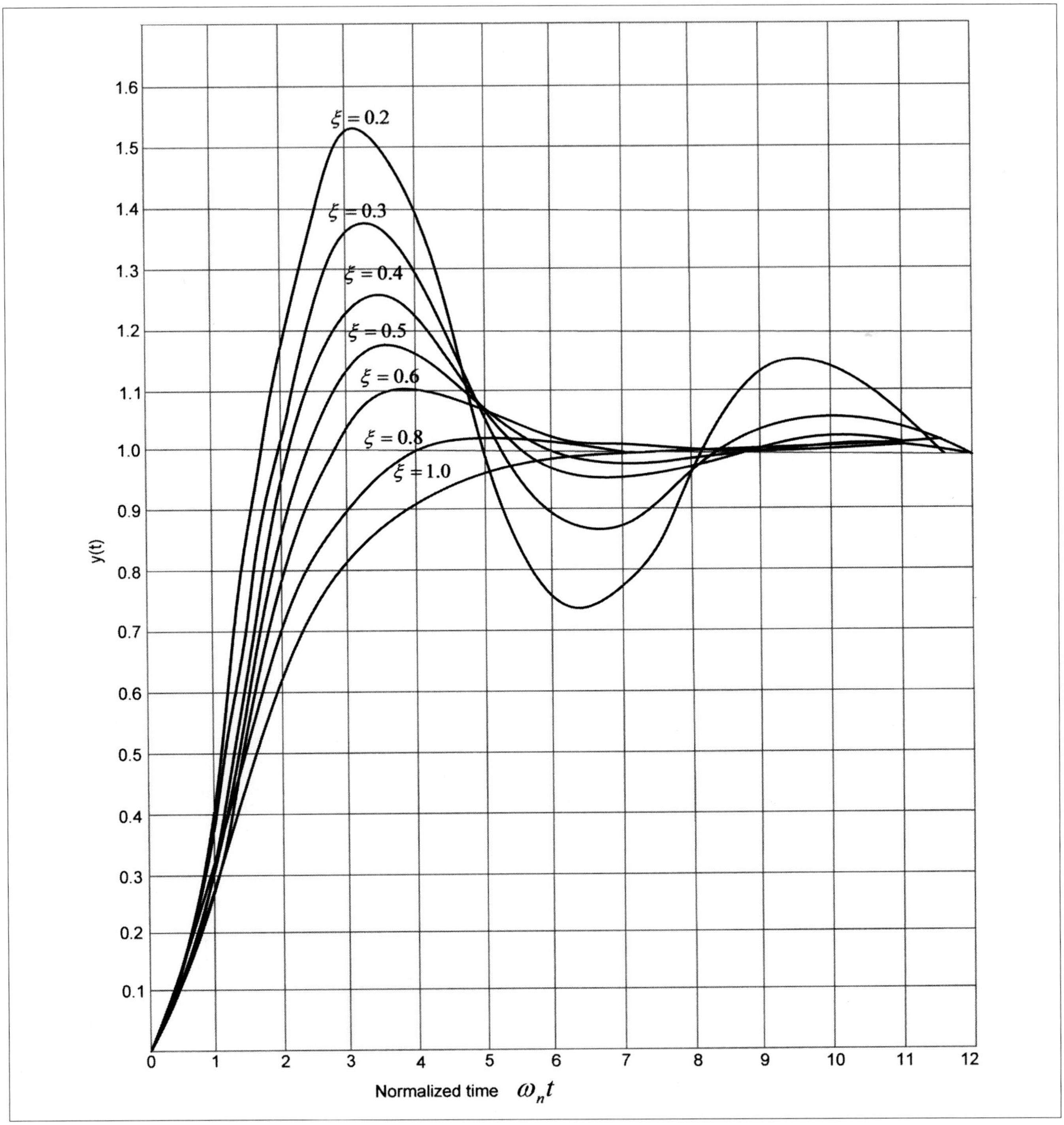

FIGURE 3-1 Normalized damping curves, second-order systems, unit step response.

A multiloop electrical circuit will result in a number of simultaneous equations in Laplace variable. It is convenient to solve the simultaneous equations using matrix techniques. Consider a dynamic system represented by the differential equation:

$$\frac{d^n y}{dt^n}+q_{n-1}\frac{d^{n-1}y}{dt^{n-1}}+\cdots+q_0 y=p_{n-1}\frac{d^{n-1}r}{dt^{n-1}}+p_{n-2}\frac{d^{n-2}r}{dt^{n-2}}+\cdots+p_0 r \tag{3-8}$$

If the initial conditions are all zero, that is, the system is at rest, the transfer function is:

$$y(s)=G(s)R(s)=\frac{p(s)}{q(s)}R(s)$$

$$=\frac{p_{n-1}s^{n-1}+p_{n-2}s^{n-2}+\cdots+p_0}{(s^n+q_{n-1}s^{n-1}+\cdots+q_0)}R(s) \tag{3-9}$$

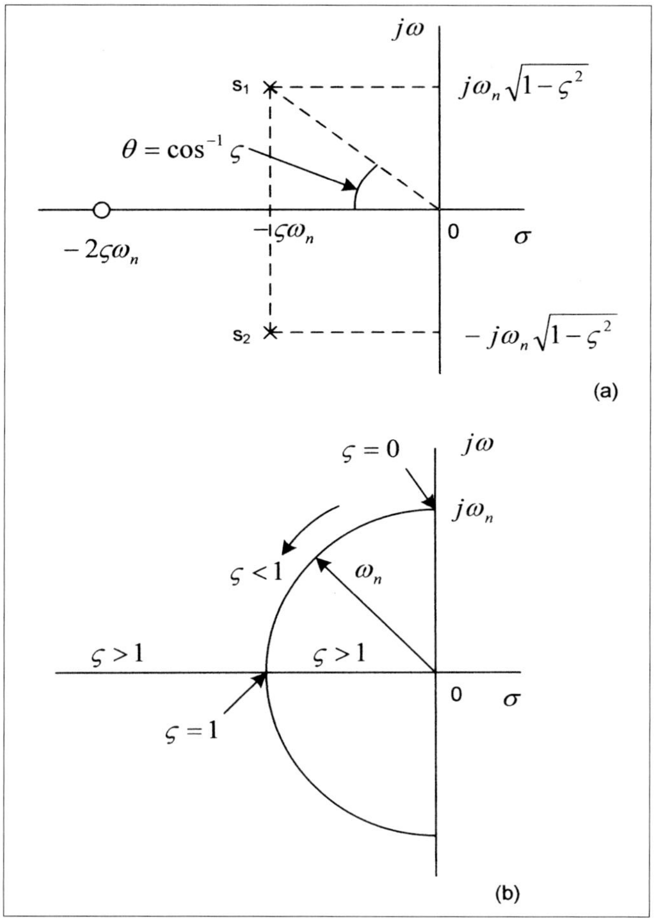

FIGURE 3-2 Second-order system, location of poles in the *s*-plane.

The output response consists of a natural response and a forced response, determined by the initial conditions and the input. If the input has the rational form:

$$R(s)=\frac{n(s)}{d(s)} \tag{3-10}$$

We can write:

$$y(s)=\frac{m(s)}{q(s)}+\frac{p(s)}{q(s)}\frac{n(s)}{d(s)}$$
$$=Y_1(s)+Y_2(s)+Y_3(s) \tag{3-11}$$

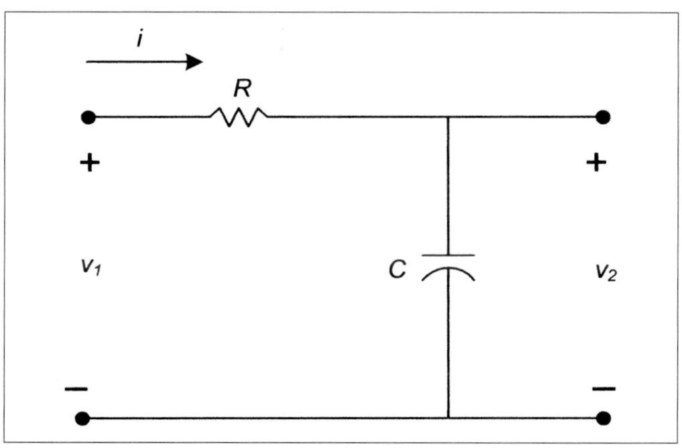

FIGURE 3-3 Example 3-1, transfer function of a *RC* circuit.

where $q(s) = 0$ is the characteristic equation, $Y_1(s)$ is the partial fraction expansion of the natural response, $Y_2(s)$ is the term involving factors of $q(s)$, and $Y_3(s)$ is the term involving partial fraction expansion of the terms involving $d(s)$. Taking inverse Laplace transform:

$$y(t)=y_1(t)+y_2(t)+y_3(t) \tag{3-12}$$

$y_1(t) + y_2(t)$ is the transient response and $y_3(t)$ is the steady-state response.

Not all transfer functions are rational algebraic expressions. The transfer function of a system having time delays contains terms of the form e^{-st} (Chap. 2).

3-2 GENERAL FEEDBACK THEORY

Linear control systems are represented in block diagram forms, and Fig. 3-4 shows such a block representation. *C* is the controlled variable, *R* is the reference (input variable), *E* is the error. The forward transfer function, expressed as *GcGp*, consists of a controlled plant and a compensator to improve stability or reduce the steady-state error. It is also called the direct transfer function. *H* is the feedback transfer function, and the primary feedback signal *B* is a function of the output. *GcGp H* is called the loop transfer function. This is equivalent to open-loop transfer function. The negative feedback means that summing point is a subtractor. The closed-loop transfer function can be written as:

$$\frac{C}{R}=\frac{G_cG_p}{1\pm H(G_cG_p)} \tag{3-13}$$

The ratio *E/R* is called the actuating signal ratio or the error ratio, and *B/R* is the primary feedback ratio. Ratio *C/R* is also called control ratio.

$$\frac{E}{R}=\frac{1}{1\pm G_cG_pH}$$
$$\frac{B}{R}=\frac{G_cG_pH}{1\pm G_cG_pH} \tag{3-14}$$

The negative feedback means that the summing junction is a subtractor, that is, $E = R - B$. Conversely for the positive feedback, it is an adder, that is, $E = R + B$.

The term controller in a feedback system is associated with the elements of the forward path, between error signal *E* and the control variable *U*. Sometimes it may include the summing point, the feedback elements, or both. The terms controller and compensator are sometimes used synonymously.

A proportional controller (*P*) gives an output *U* proportional to its input, that is, $U = K_pE$; a derivative controller (*D*) provides an output *E* that is derivative of its input; an integrator (*I*) provides an output which is integral of its input *E*. Thus, a PID has an output of the form:

$$U_{PID}=K_pE+K_D\frac{dE}{dt}+K_I\int E(t)dt \tag{3-15}$$

The transfer function of a PID controller is:

$$P_{PID}(s)=K_P+K_Ds+\frac{K_I}{s}$$
$$=\frac{K_Ds^2+K_Ps+K_I}{s} \tag{3-16}$$

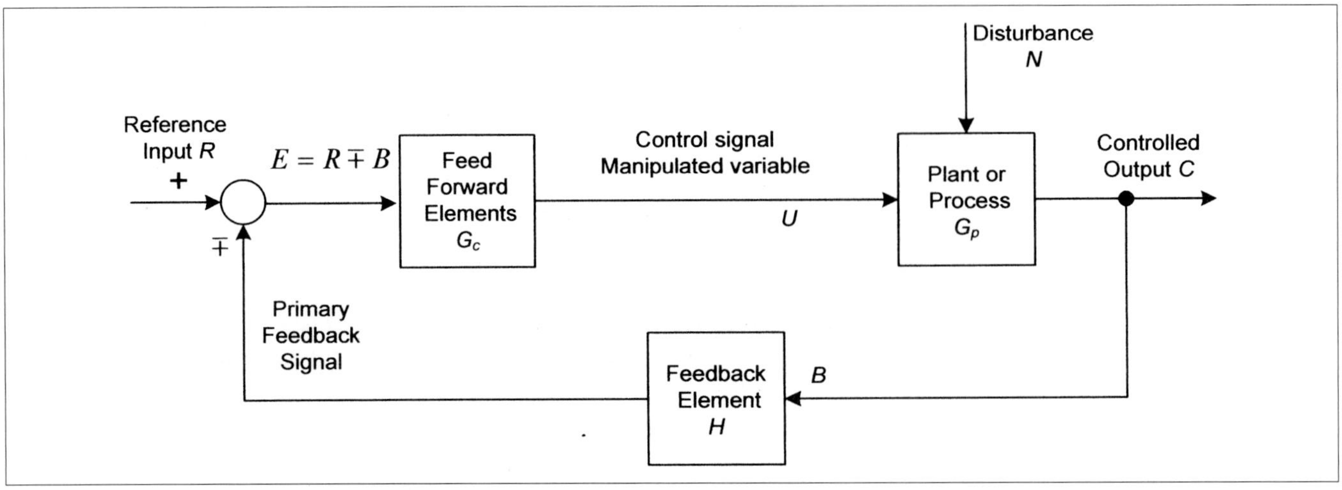

FIGURE 3-4 To illustrate elements of a closed-loop control system.

3-2-1 Compensators

Compensation networks are introduced to improve the performance of a control system. These can be introduced in the forward path or the feedback path and the compensation can be passive or active.

The three compensators using passive RC networks are illustrated in Fig. 3-5*a*, *b*, and *c*.

Lead Compensator (Fig. 3-5a)

$$P_{\text{Lead}}(s) = \frac{s+a}{s+b}$$

$$\text{Zero} = -a = -\left[\frac{1}{R_1 C}\right]$$

$$\text{Pole} = -b = -\left[\frac{1}{R_1 C} + \frac{1}{R_2 C}\right] \tag{3-17}$$

$$b > a$$

Lag Compensator (Fig. 3-5b)

$$P_{\text{Lag}}(s) = \frac{a(s+b)}{b(s+a)}$$

$$\text{Zero} = -b = -\left[\frac{1}{R_2 C}\right]$$

$$\text{Pole} = -a = -\left[\frac{1}{(R_1 + R_2)C}\right] \tag{3-18}$$

$$b > a$$

Lag-Lead Compensator (Fig. 3-5c)

The transfer function is:

$$P_{\text{LL}}(s) = \frac{(s+a_1)(s+b_2)}{(s+b_1)(s+a_2)} \tag{3-19}$$

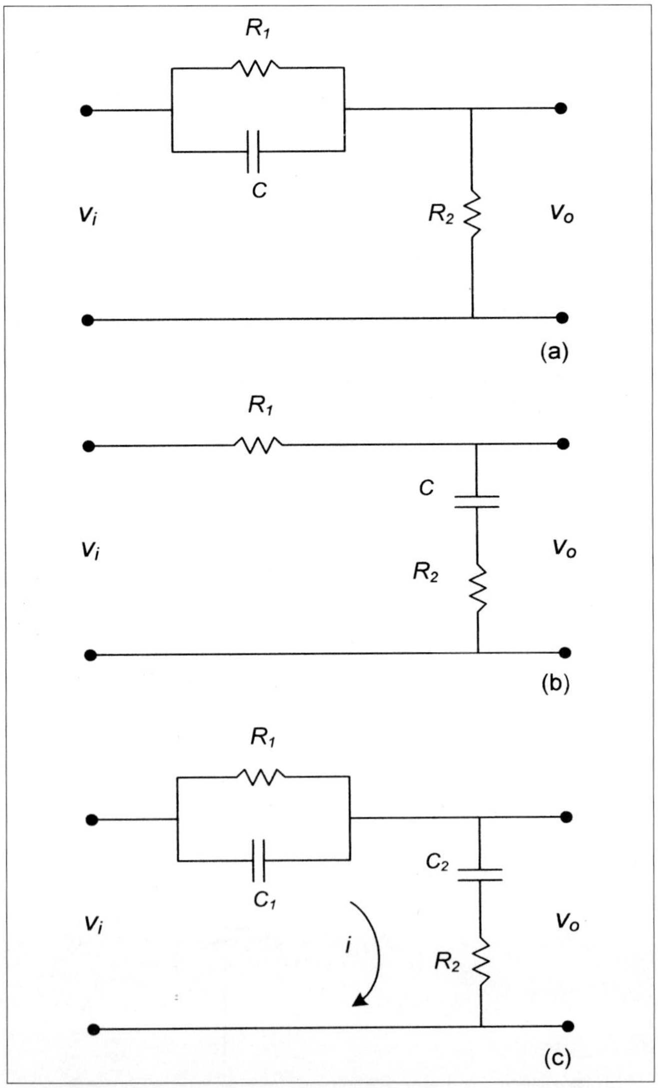

FIGURE 3-5 (*a*) Passive lead compensator. (*b*) Lag compensator. (*c*) Lead-lag compensator.

Equation (3.19) can be derived using techniques discussed in Chap. 2. In Fig. 3-5c, the current i in the loop is:

$$\frac{V_i - V_o}{R_1} + C_1 \frac{d}{dt}(V_i - V_o) = i$$

$$V_0 = \frac{1}{C_2}\int_0^t i\,dt + iR_2$$

Taking Laplace transform with zero initial conditions and eliminating $i(s)$,

$$\left(\frac{1}{R_1} + C_1 s\right)[V_i(s) - V_0(s)] = \frac{V_0(s)}{(1/sC_2) + R_2}$$

The transfer function is:

$$P_{LL} = \frac{V_0(s)}{V_i(s)} = \frac{\left(s + \frac{1}{R_1C_1}\right)\left(s + \frac{1}{R_2C_2}\right)}{s^2 + \left(\frac{1}{R_2C_2} + \frac{1}{R_2C_1} + \frac{1}{R_1C_1}\right)s + \frac{1}{R_1C_1R_2C_2}}$$

$$= \frac{(s+a_1)(s+b_2)}{(s+b_1)(s+a_2)}$$

Simplifying, it can be shown that:

$$a_1 = \frac{1}{R_1C_1} \qquad b_2 = \frac{1}{R_2C_2}$$
$$b_1a_2 = a_1b_2 \qquad b_1 + a_2 = a_1 + b_2 + \frac{1}{R_2C_1} \tag{3-20}$$

Example 3-2 Consider a control system given by the block diagram in Fig. 3-6a. It is desired to modify the system so that it has a damping ratio ξ of 0.707 and an undamped natural frequency of 4. The transfer function is:

$$\frac{0.4K/[s(s+0.4)]}{1+0.4K/[s(s+0.4)]} = \frac{0.4K}{s^2+0.4s+0.4K}$$

Comparing this with Eq. (3-2), the natural frequency should be 4. Thus:

$$\omega_n = 4 = \sqrt{0.4K}$$
$$K = 40$$

The damping ratio is:

$$2\xi\omega_n = 0.4s$$
$$\xi = 0.4/2\omega_n = 0.05$$

We need to change it to 0.7. Modify the control system by adding a compensator, as shown in Fig. 3-6b. The transfer function is:

$$\frac{\frac{(0.4K)\,(s+\omega_1)}{s(s+0.4)\,(s+\omega_2)}}{1+\frac{(0.4K)\,(s+\omega_1)}{s(s+0.4)\,(s+\omega_2)}} = \frac{0.4K(s+\omega_1)}{s(s+0.4)\,(s+\omega_2)+0.4K(s+\omega_1)}$$

We can cancel out $(s+\omega_1)$ and $(s+0.4)$ by making $\omega_1 = 0.4$. Then the transfer function reduces to:

$$\frac{0.4K}{s(s+\omega_2)+0.4K} = \frac{0.4K}{s^2+\omega_2 s+0.4K}$$

To get the required damping ratio:

$$\xi = 0.7 = \frac{\omega_2}{2\omega_n} \qquad \omega_2 = 5.6\ rad/\sec$$

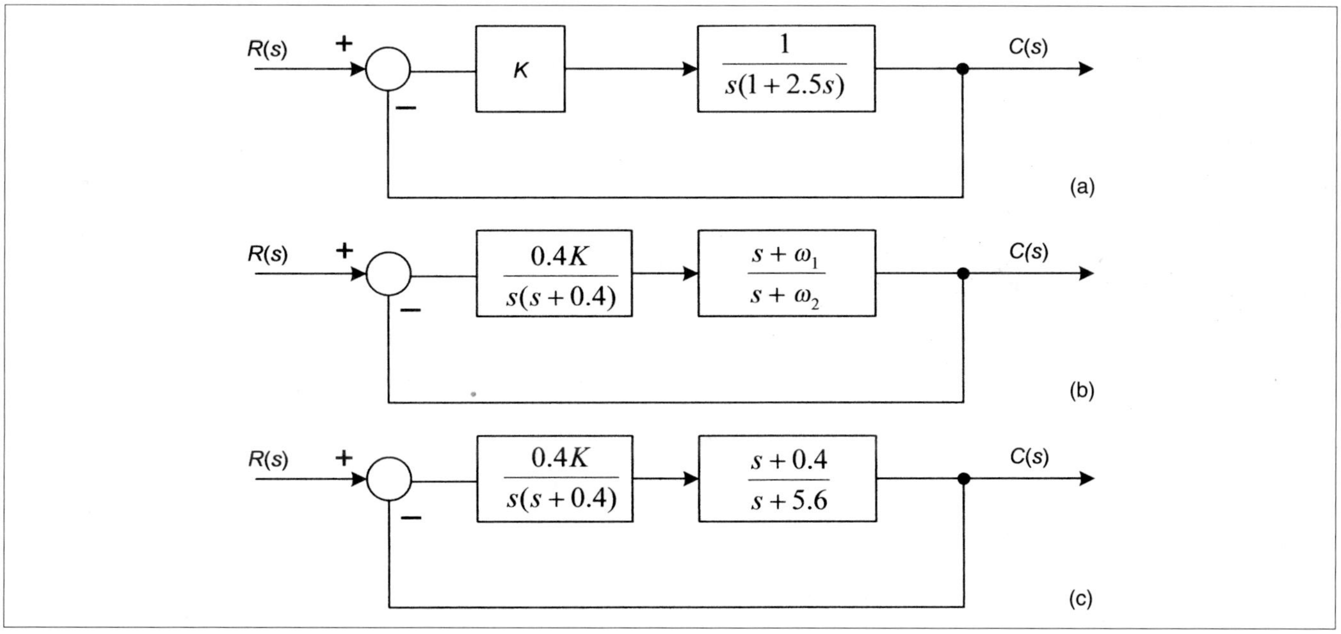

FIGURE 3-6 Example 3-2, altering the response of a control system by adding a compensator.

The control system is now drawn as shown in Fig. 3-6*c*. The gain of the compensated system is given by:

$$G(s)=\frac{s+0.4}{s+5.6}\times 16\times\frac{1}{s(s+0.4)}=\frac{16}{s(s+5.6)}=2.86$$

3-3 CONTINUOUS SYSTEM FREQUENCY RESPONSE

The steady-state response of a stable system to an input $u = A\sin\omega t$ is given by:

$$y_{ss}=A\left|P(j\omega)\right|\sin(\omega t+\phi) \tag{3-21}$$

where ϕ is arg $P(j\omega)$. The complex number $P(j\omega)$ is obtained by replacing s with $j\omega$ in $P(s)$. The magnitude and the angle for all ω's define the system frequency response. It is generally represented by two graphs: one for the magnitude of $P(j\omega)$, called the *gain* of the system, and the other for the arg $P(j\omega)$. It can be determined graphically in the s plane from a pole-zero map of $P(s)$.

For a step-function input of magnitude A, often called *dc input*, the Laplace transform of the output is:

$$Y(s)=P(s)\frac{A}{s} \tag{3-22}$$

If system is stable then the steady-state response is a step function of $AP(0)$, that is, the amplitude of the input signal is multiplied by $P(0)$ to determine amplitude of output. $P(0)$ is, therefore, called the *dc gain* of the system.

Example 3-3 Consider a system with transfer function:

$$P(s)=\frac{1}{(s+1)(s+3)}$$

We will compute the magnitude and angles for $j\omega = 1$. Then from Fig. 3-7*a*:

$$\left|P(j1)\right|=\frac{1}{\sqrt{10}\sqrt{2}}=0.223$$

$$\arg\left|P(j1)\right|=-18.4°-45°=63.43°$$

The values of $|P(j\omega)|$ and arg $|P(j\omega)|$ at various values of $j\omega$ can be calculated and a graph plotted, as shown in Fig. 3-7*b* and *c*.

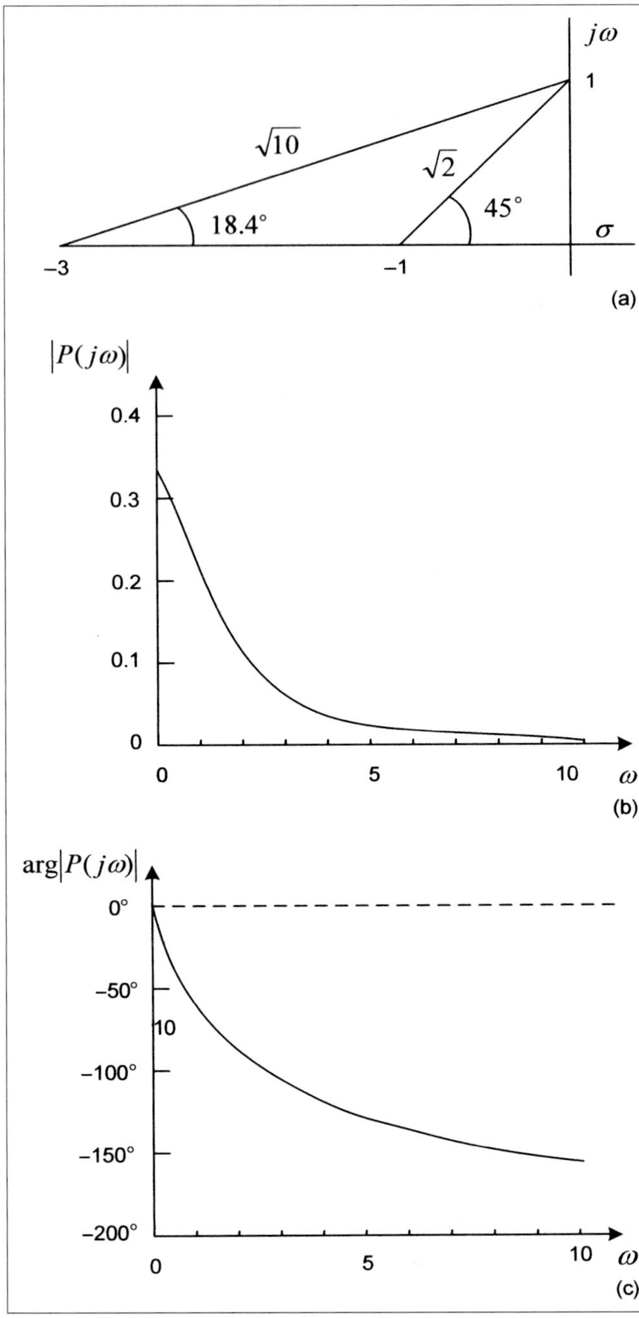

FIGURE 3-7 Example 3-3, computation of the magnitude and angle of a transfer function.

3-4 TRANSFER FUNCTION OF A DISCRETE-TIME SYSTEM

If all terms due to initial condition are zero, then for an input $U(z)$:

$$P(z)=\frac{Y(z)}{U(z)}=\frac{K(z+z_1)(z+z_2)+\cdots+(z+z_m)}{(z+p_1)(z+p_2)+\cdots+(z+p_n)} \tag{3-23}$$

that is, $P(z)$ is specified by system poles and zeros and a gain factor K.

The system poles and zeros can be represented in a pole-zero map in the z plane. The pole-zero map of the output $y(z)$ can be constructed from pole-zero map of $P(z)$ by including poles and zeros of the input $U(z)$.

The order of the denominator polynomial of a transfer function of a casual discrete-time system must be greater than or equal to the order of the numerator polynomial. The steady-state response to a unit-step input is called the dc gain and is given by the final value theorem (Chap. 2).

Example 3-4 The transfer functions of a discrete-time system described by the equation:

$$y(k+2)+1.1y(k+1)+0.3y(k)=u(k+2)+0.2u(k+1)$$

can be found by taking z-transform:

$$(z^2+1.1z+0.3)Y(z)=(z^2+0.2z)U(z)$$

Therefore:

$$P(z)=\frac{z(z+0.2)}{z^2+1.1z+0.3}=\frac{z(z+0.2)}{(z+0.5)(z+0.6)}$$

The system has zeros at –0.2 and two poles at –0.5 and –0.6. The system is stable as the poles are within the unit circle and the dc gain is:

$$P(1)=\frac{1.2}{1.5\times 1.6}=0.5$$

3-4-1 Discrete-Time System Frequency Response

The frequency response of a stable discrete system to an input sequence $u(k) = A \sin \omega kT$, where $k = 0, 1, 2,\ldots$, which has a transfer function $P(z)$ given by:

$$y_{ss}=A\left|Pe^{j\omega t}\right|\sin(\omega kT+\phi) \qquad k=0,1,2,\ldots \tag{3-24}$$

The complex function $P(e^{j\omega T})$ is determined from $P(z)$ by replacing z with $e^{j\omega T}$. The system response is a sequence of samples of the sinusoidal with the same frequency as the input sinusoidal. The output sequence is obtained by multiplying A of input with $\left|Pe^{j\omega T}\right|$, and shifting the phase angle of the input by arg $\left|Pe^{j\omega T}\right|$. The magnitude $\left|Pe^{j\omega T}\right|$, gain, and arg $\left|Pe^{j\omega T}\right|$ for all ω's define discrete-system frequency function.

The discrete-time system frequency response function can be determined in the z plane from a pole-zero map of $P(z)$ in the same manner as calculation of residues (App. C).

3-5 STABILITY

We may characterize a continuous or discrete system as stable, if every bounded input produced a bounded output. A necessary and sufficient condition for the system to be stable is that real parts of the roots of the characteristic equation have negative real parts. The impulse response will then decay exponentially with time. If the roots have real parts equal to zero, the system is marginally stable, the impulse response does not decay to zero, and some other inputs may produce unbounded outputs. *Thus, marginally stable systems are unstable.* For example, a system with roots, –4, –2, and 0 is marginally stable. The system with roots $-1+j$ and $-1-j$ is stable as the real part is negative. A system which has poles at –1, –4, and zeros at –2 and 1 is stable, though the system has a zero with positive real part.

3-5-1 Routh Stability Criteria

The stability of a system with nth-order characteristic equation of the form:

$$a_n s^n + a_{n-1}s^{n-1}+\cdots+a_1 s+a_0=0 \tag{3-25}$$

can be determined by constructing a Routh table as follows:

$$\begin{array}{c|ccccc} s^n & a_n & a_{n-2} & a_{n-4} & \cdots \\ s^{n-1} & a_{n-1} & a_{n-3} & a_{n-5} & \cdots \\ s^{n-2} & b_1 & b_2 & b_3 & \cdots \\ \cdot & c_1 & c_2 & c_3 & \cdots \\ \cdot & \cdot & \cdot & \cdot & \cdot \end{array} \tag{3-26}$$

where:

$$\begin{aligned} b_1 &= \frac{a_{n-1}a_{n-2}-a_n a_{n-3}}{a_{n-1}} \\ b_2 &= \frac{a_{n-1}a_{n-4}-a_n a_{n-5}}{a_{n-1}} \\ &\ldots \\ c_1 &= \frac{b_1 a_{n-3}-a_{n-1}b_2}{b_1} \\ c_2 &= \frac{b_1 a_{n-5}-a_{n-1}b_2}{b_1} \\ &\ldots \end{aligned} \tag{3-27}$$

The table is continued horizontally and vertically till only zeros are obtained. All the roots of the characteristic equation have negative real parts if and only if the elements of the first column of the table have the same sign. If not, then the number of roots with the positive real parts is equal to the number of changes of sign.

Example 3-5 Consider a system as shown in the block diagram of Fig. 3-8 and find the value of K for stability:

$$GH=\frac{K}{s(s+3)(s^2+7s+12)}$$

The characteristic equation is:

$$s^4+10s^3+33s^2+36s+K=0$$

The Routh table is:

$$\begin{array}{cccc} s^4 & 1 & 33 & K \\ s^3 & 10 & 36 & 0 \\ s^2 & 29.4 & K & 0 \\ s^1 & 36-0.34K & 0 & 0 \\ s^0 & K & & \end{array}$$

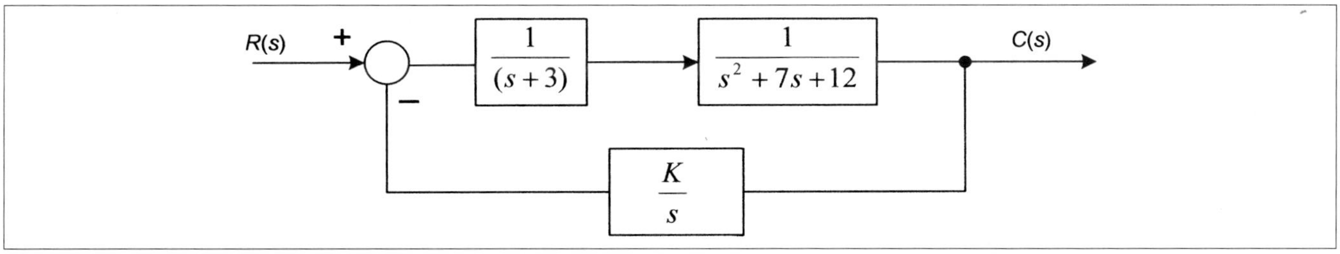

FIGURE 3-8 Example 3-5, block circuit diagram of a control system, value of K for stability.

Therefore:

$$K > 0$$

$$36 - 0.34K > 0$$

$$K = 105.9$$

For stability:

$$0 < K < 105.9$$

If rows are zero for s^1 row of the table, they form an auxiliary equation:

$$As^2 + B = 0 \tag{3-28}$$

where A and B are the first and second elements of the row s^2. Replace the zeros of row s^1 with the differential of the auxiliary equation, that is, $2A$ and zero in this case.

3-5-2 Hurwitz Stability Criteria

The criteria are applied using the determinants formed from the coefficients of the characteristic equation in the matrix:

$$\Delta_n = \begin{vmatrix} a_{n-1} & a_{n-3} & \cdot & \begin{bmatrix} a_0 (n \text{ odd}) \\ a_1 (n \text{ even}) \end{bmatrix} & 0 & \cdot & 0 \\ a_n & a_{n-2} & \cdot & \begin{bmatrix} a_1 (n \text{ odd}) \\ a_0 (n \text{ even}) \end{bmatrix} & 0 & \cdot & 0 \\ 0 & a_{n-1} & a_{n-3} & \cdot & \cdot & \cdot & 0 \\ 0 & a_n & a_{n-2} & \cdot & \cdot & \cdot & 0 \\ \cdot & \cdot & \cdot & \cdot & \cdot & \cdot & \cdot \\ 0 & \cdot & \cdot & \cdot & \cdot & \cdot & a_0 \end{vmatrix} \tag{3-29}$$

The determinants are formed as follows:

$$\Delta_1 = a_{n-1}$$

$$\Delta_2 = \begin{vmatrix} a_{n-1} & a_{n-3} \\ a_n & a_{n-2} \end{vmatrix} \tag{3-30}$$

$$\Delta_3 = \begin{vmatrix} a_{n-1} & a_{n-3} & a_{n-5} \\ a_n & a_{n-2} & a_{n-4} \\ 0 & a_{n-1} & a_{n-3} \end{vmatrix}$$

All the roots have negative real parts if and only if

$$\Delta_i > 0 \qquad i = 1, 2, \ldots, n \tag{3-31}$$

Note that the coefficient a_n is positive.

Example 3-6 Apply Hurwitz criteria to ascertain whether the system represented by:

$$s^3 + 6s^2 + 10s + 20 = 0$$

is a stable or unstable system. Calculate the determinants as follows:

$$\Delta_3 = \begin{vmatrix} a_{n-1} & a_{n-3} & a_{n-5} \\ a_n & a_{n-2} & a_{n-4} \\ 0 & a_{n-1} & a_{n-3} \end{vmatrix} = \begin{vmatrix} 6 & 20 & 0 \\ 1 & 10 & 0 \\ 0 & 6 & 20 \end{vmatrix}$$

$$\Delta_2 = \begin{vmatrix} a_{n-1} & a_{n-3} \\ a_n & a_{n-2} \end{vmatrix} = \begin{vmatrix} 6 & 20 \\ 1 & 10 \end{vmatrix}$$

$$\Delta_1 = a_{n-1} = 6$$

As all the determinants are positive, the system is stable.

3-5-3 Stability of Discrete-Time Systems

Routh criteria can also be applied to discrete systems. The stability is determined from the characteristic equation:

$$Q(z) = a_n z^n + a_{n-1} z^{n-1} + \cdots + a_1 z + a_0 = 0 \tag{3-32}$$

The stability region is defined by a unit circle $|z| = 1$ in the z plane. The necessary and sufficient condition of the stability is that all the roots of the characteristic equation have a magnitude less than 1, that is, these are within the unit circle. The stability criteria similar to discrete systems are called the *jury tests.* The coefficients are first arranged in a *jury array*:

$$\begin{array}{c|cccccc} 1 & a_0 & a_1 & a_2 & \cdot & a_{n-1} & a_n \\ 2 & a_n & a_{n-1} & a_{n-2} & \cdot & a_1 & a_0 \\ 3 & b_0 & b_1 & b_2 & \cdot & b_{n-1} & \\ 4 & b_{n-1} & b_{n-2} & b_{n-3} & \cdot & b_0 & \\ 5 & c_0 & c_1 & c_2 & \cdot & c_{n-2} & \\ 6 & c_{n-2} & c_{n-3} & c_{n-4} & \cdot & c_0 & \\ \cdot & \cdot & \cdot & \cdot & \cdot & \cdot & \\ 2n-5 & r_0 & r_1 & r_2 & r_3 & & \\ 2n-4 & r_3 & r_2 & r_1 & r_0 & & \\ 2n-3 & s_0 & s_1 & s_3 & & & \end{array} \tag{3-33}$$

where:

$$b_k = \begin{vmatrix} a_0 & a_{n-k} \\ a_n & a_k \end{vmatrix}$$

$$c_k = \begin{vmatrix} b_0 & b_{n-1-k} \\ b_{n-1} & b_k \end{vmatrix} \tag{3-34}$$

$$s_0 = \begin{vmatrix} r_0 & r_3 \\ r_3 & r_0 \end{vmatrix} \qquad s_1 = \begin{vmatrix} r_0 & r_2 \\ r_3 & r_1 \end{vmatrix} \qquad s_3 = \begin{vmatrix} r_0 & r_1 \\ r_3 & r_2 \end{vmatrix}$$

The first two rows are constructed using coefficients of the characteristic equation; next two rows are computed using determinant relations shown above. This is continued with each succeeding pair of rows having one less column than the previous pair, until row $2n - 3$ is computed, which has only three entries. The array is then terminated.

The necessary and sufficient conditions of stability are:

$$Q(1) > 0$$

$$Q(-1) > 0 \qquad (n \text{ even})$$

$$Q(-1) < 0 \qquad (n \text{ odd})$$

$$
\begin{aligned}
&|a_0| < a_n \\
&|b_0| > |b_{n-1}| \\
&|c_0| > |c_{n-2}| \\
&\vdots \\
&|r_0| > |r_3| \\
&|s_0| > |s_2|
\end{aligned}
\tag{3-35}
$$

The array may not be constructed if some of the initial conditions of stability, that is $Q(1)$, $Q(-1)$, a_0, and a_n are not met, and then the system will be unstable.

Example 3-7 A continuous system is given by:

$$Q(z) = 3z^4 + 2z^3 + z^2 + 2z + 1 = 0$$

Ascertain whether the system is stable:

$$Q(1) = 3+2+1+2+1 = 9 > 0$$

$$Q(-1) = 3-2+1-2+1 = 1 > 0$$

Thus, the jury array can be constructed:

1	1	2	1	2	3
2	3	2	1	2	1
3	−8	−4	−2	−4	
4	−4	−2	−4	−8	
5	80	24	0		

Examine the remaining conditions:

$$|a_0| = 1 < 3 = a_n$$

$$|b_0| = |-8| > |-4| = |b_{n-1}|$$

$$|c_0| = 80 > 0 = |c_{n-2}|$$

Thus, the system is stable.

3-6 BLOCK DIAGRAMS

A block diagram is a convenient way of representing a physical system. We used block diagrams in the examples above to calculate transfer functions and simple combinations of forward blocks. Block diagram transformations are shown in Fig. 3-9*a*. A given control system can be simplified by using these transformations.

Example 3-8 Consider the block diagram of control system in Fig. 3-9*b*. This can be simplified using the transformations shown in Fig. 3-9*a*, and the simplified control system is progressively shown in Fig. 3-9*c* and 3-9*d*.

3-7 SIGNAL-FLOW GRAPHS

Though block diagrams are used extensively for the representation of control systems, signal-flow graphs are yet another representation. A signal-flow graph is a pictorial representation of the simultaneous equations describing a system. As a simple example Fig. 3-10*a* is a representation of Ohm's law, $E = RI$. Note the designations of nodes and branches.

Addition Rule (Fig. 3-10b)

$$X_i = \sum_{j=1}^{n} A_{ij} X_j \tag{3-36}$$

Multiplication Rule (Fig. 3-10c)

$$X_n = A_{21} \cdot A_{32} \ldots A_{n(n-1)} X_1 \tag{3-37}$$

Transmission Rule (Fig. 3-10d)

$$
\begin{aligned}
&X_i = A_{ik} X_k \\
&i = 1, 2, \ldots n,\ k
\end{aligned}
\tag{3-38}
$$

A loop is a closed path that originates and terminates on the same node, *and along the path no node is met twice.* Two loops are said to be nontouching if they do not have a common node. Two touching loops share one or more common nodes.

To construct a signal-flow graph write the equations in the form:

$$
\begin{aligned}
&X_1 = A_{11}X_1 + A_{12}X_2 + \cdots + A_{1n}X_n \\
&X_2 = A_{21}X_1 + A_{22}X_2 + \cdots + A_{2n}X_n \\
&\vdots \\
&X_m = A_{m1}X_1 + A_{m2}X_2 + \cdots + A_{mn}X_n
\end{aligned}
\tag{3-39}
$$

- An equation in X_1 is not required if X_1 is an input node.
- Arrange larger nodes from left to right.
- Connect the nodes by appropriate branches A_{11}, A_{12} . . .
- If desired output node has outgoing branches, add a dummy node and unity gain branch.

Example 3-9 Consider the circuit shown in Fig. 3-11*a*. The following equations can be written:

$$i_1 = \frac{1}{R_1}v_1 - \frac{1}{R_1}v_2$$

$$v_2 = R_3 i_1 - R_3 i_2$$

$$i_2 = \frac{1}{R_2}v_2 - \frac{1}{R_2}v_3$$

$$v_3 = R_4 i_2$$

The signal-flow graph is shown in Fig. 3-11*b*.

3-7-1 Mason's Signal-Flow Gain Formula

The relationship between an input variable and output variable of a signal-flow graph is given by the net gain between input and output nodes and is called the overall gain of the system. Mason's gain formula is:[1]

$$T_{ij} = \frac{\sum_k P_{ijk}\Delta_{ijk}}{\Delta} \tag{3-40}$$

where:

P_{ijk} = kth path from variable x_i to variable x_j
Δ = determinant of the path
Δ_{ijk} = cofactor of the path P_{ijk}

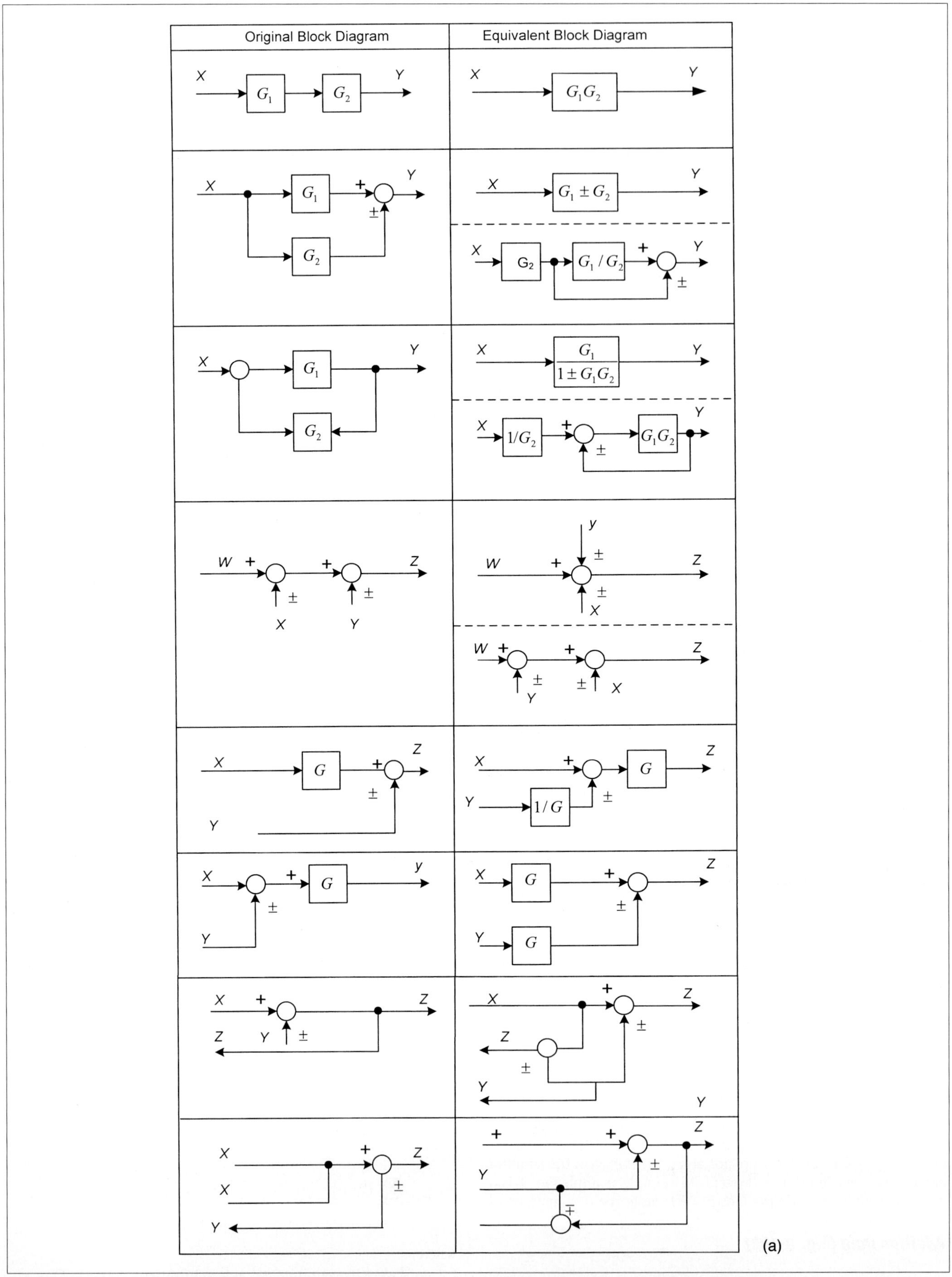

FIGURE 3-9 (*a*) Transformation of control circuit diagrams. (*b*) Original block diagram of a control system. (*c*) and (*d*) Progressive simplification of the control system block diagram. (*Continued*)

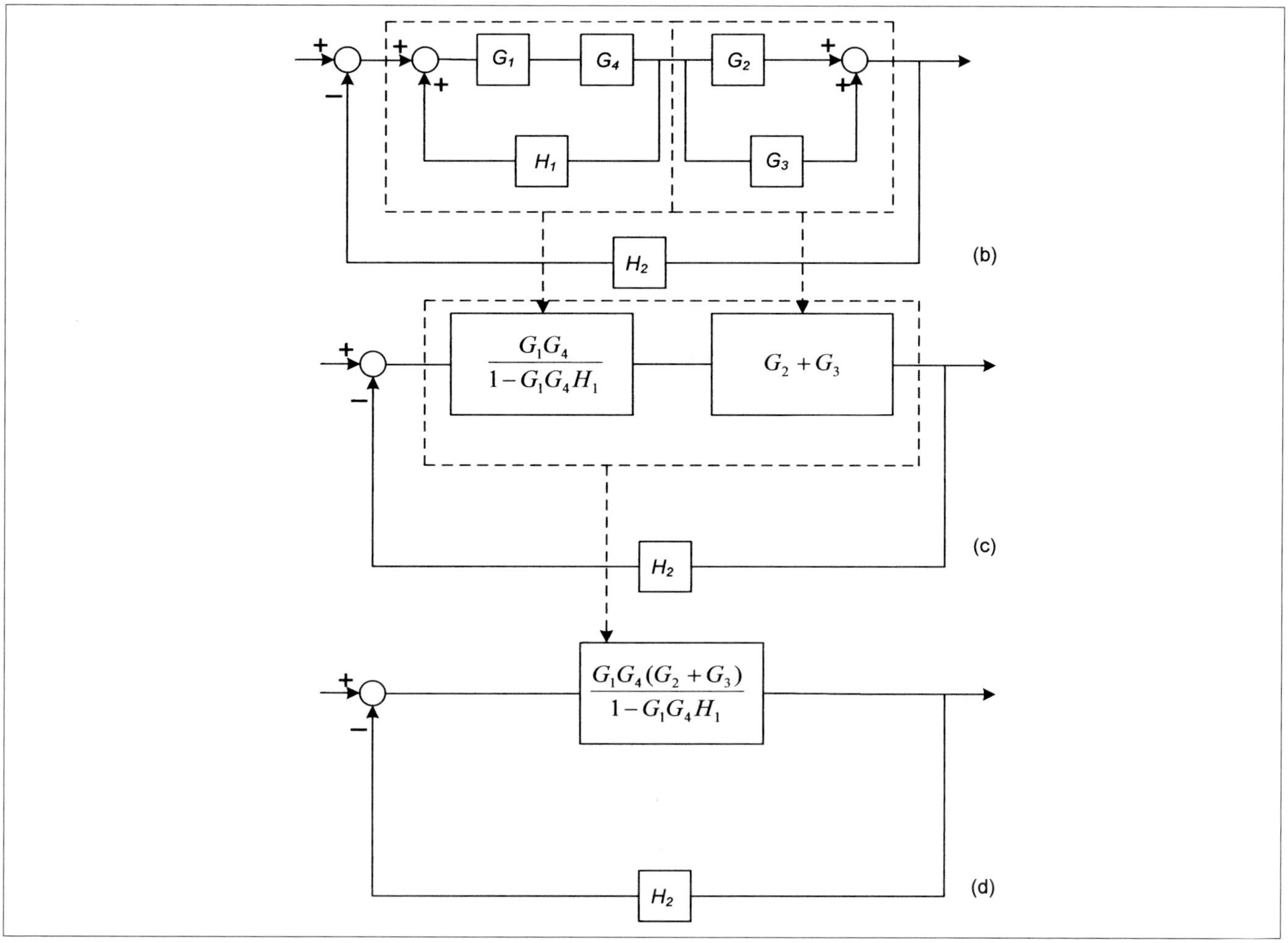

FIGURE 3-9 (*Continued*)

The summation is taken for all possible paths from x_i to x_j. The cofactor Δ_{ijk} is the determinant with the loops touching the removed kth path. The determinant Δ is

$$\Delta = 1 - \sum_{n=1}^{N} L_n + \sum_{m=1,q=1}^{M,Q} L_m L_q - \sum L_r L_s L_t + \cdots \tag{3-41}$$

where L_q equals the qth loop transmittance. Therefore, the rule of evaluating Δ is:

Δ = 1– (Sum of all different loop gains)

+ Sum of gain products of all combinations of two nontouching loops)

– (Sum of gains products of all combinations of three nontouching loops)

+ · · · (3-42)

Example 3-10 Consider the signal-flow graph shown in Fig. 3-12. The paths connecting input to output are:

$$\text{Path 1}: P_1 = G_1 G_2 G_3 G_4$$

$$\text{Path 2}: P_2 = G_5 G_6 G_7$$

$$\text{Loop 1}: L_1 = G_2 H_2$$

$$\text{Loop 2}: L_2 = G_3 H_3$$

$$\text{Loop 3}: L_3 = G_6 H_4$$

Loops L_1 and L_2 do not touch L_3. Therefore the determinant is:

$$\Delta = 1 - (L_1 + L_2 + L_3) + (L_1 L_3 + L_2 L_3)$$

Cofactor of the determinant along path 1 is evaluated by removing the loops that touch path 1 from Δ. Therefore:

$$L_1 = L_2 = 0 \qquad \Delta_1 = 1 - L_3$$

The cofactor for path 2 is:

$$\Delta_2 = 1 - L_1 - L_2$$

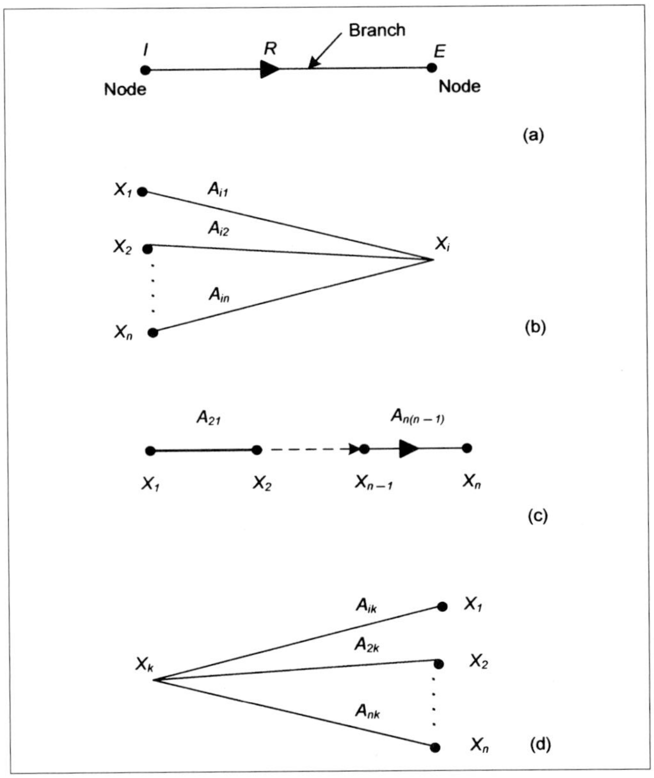

FIGURE 3-10 (*a*) Signal-flow graph of ohms law. (*b*) Rule of addition. (*c*) Rule of multiplication. (*d*) Rule of transmission.

Then:

$$\frac{Y(s)}{R(s)} = T(s) = \frac{P_1\Delta_1 + P_2\Delta_2}{\Delta}$$

$$= \frac{G_1G_2G_3G_4(1-L_3)+G_5G_6G_7(1-L_1-L_2)}{1-L_1-L_2-L_3+L_1L_3+L_2L_3}$$

Example 3-11 Draw the signal graph of the control system shown in Fig. 3-13*a* and find overall gain. Using the procedure outlined above, the signal-flow graph is constructed and is shown in Fig. 3-13*b*. There are two forward paths:

$$\text{Path 1}: P_1 = G_1G_2G_4$$

$$\text{Path 2}: P_2 = G_1G_3G_4$$

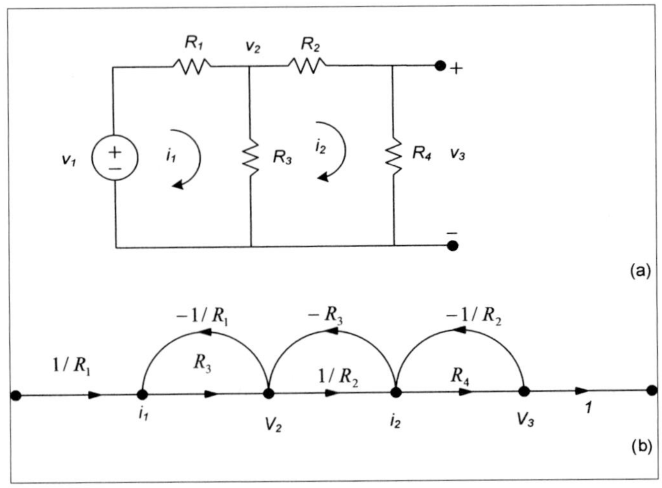

FIGURE 3-11 Example 3-9, (*a*) Original circuit diagram. (*b*) Signal-flow graph.

There are three loops:

$$L_1 = G_1G_4H_1$$

$$L_2 = -G_1G_2G_4H_2$$

$$L_3 = -G_1G_3G_4H_2$$

There are no nontouching loops. Thus:

$$\Delta = 1-(L_1+L_2+L_3)$$

$$\Delta_1 = \Delta_2 = 1$$

Thus:

$$T(s) = \frac{P_1+P_2}{\Delta}$$

3-8 BLOCK DIAGRAMS OF STATE MODELS

In Chap. 2 we observed that in the state model the dynamics of the system can be represented by first-order differential equations or by the state differential equations. It is useful to develop a state flow graph model and use this model to relate state variable concept with the transfer function representation.

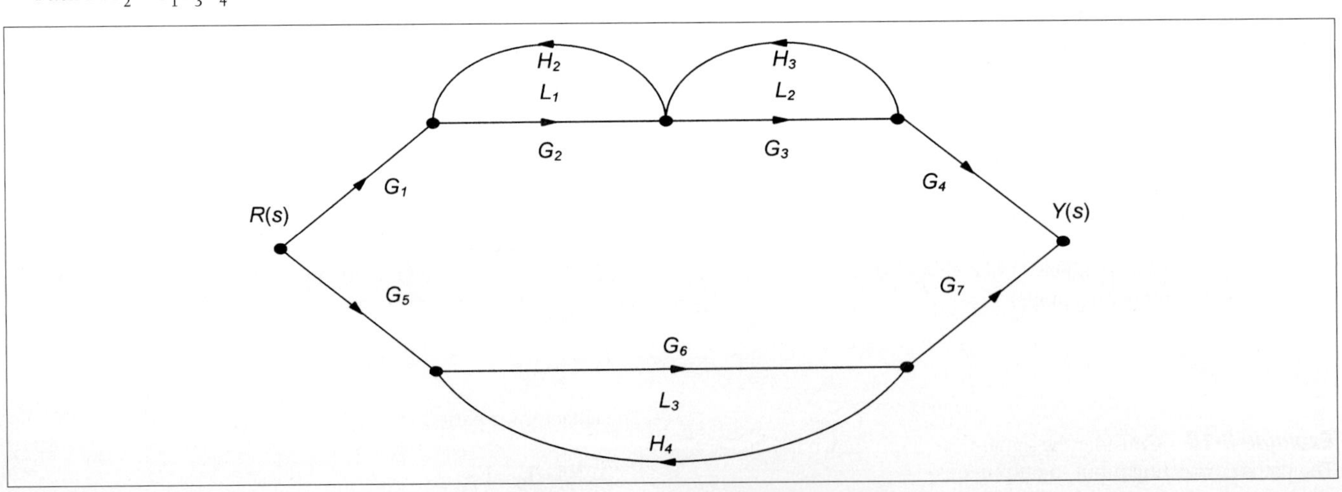

FIGURE 3-12 Example 3-10, signal-flow graph.

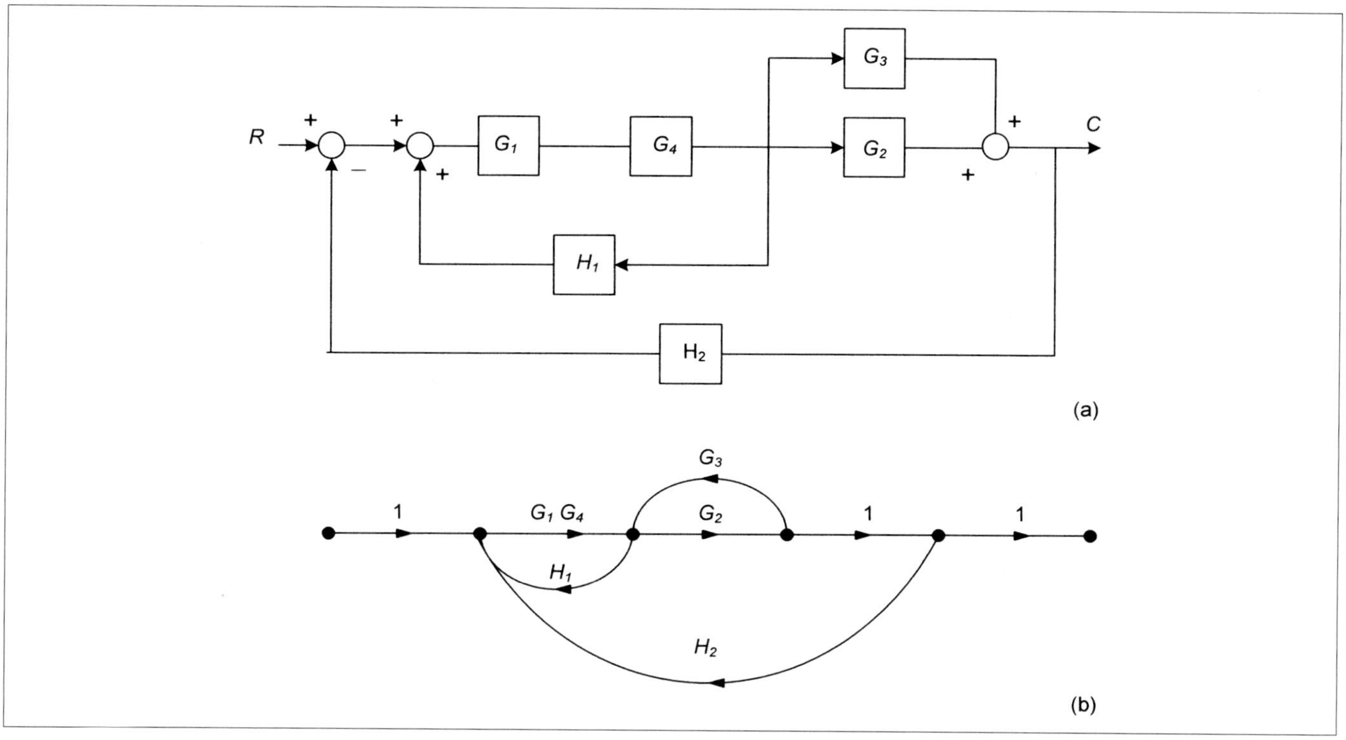

FIGURE 3-13 Example 3-11, (*a*) Block circuit diagram of a control system. (*b*) Signal-flow graph of the control system in (*a*).

From Example 2-16, we had following equations:

$$\dot{x}_1 = -\frac{1}{C}x_2 + \frac{1}{C}u(t)$$
$$\dot{x}_2 = \frac{1}{L}x_1 - \frac{R}{L}x_2 \tag{3-43}$$
$$v_0 = Rx_2$$

The flow graph is shown in Fig. 13-14. Consider a first-order differential equation:

$$\dot{x}(t) = ax(t) + br(t) \qquad x(0) = x_0 \tag{3-44}$$

The state diagram for this differential equation is shown in Fig. 3-15. Start with the signals $x(t)$ and $r(t)$ and form the right-hand side to give $\dot{x}(t)$. The initial condition $x(0) = x_0$ is added to the integrator.

Now consider a system with single input $r(t)$ and output $y(t)$ given by two state variables:

$$\dot{x}_1 = a_{11}x_1 + a_{12}x_2 + b_1 r \qquad x_1(0) = x_{10}$$
$$\dot{x}_2 = a_{21}x_1 + a_{22}x_2 + b_2 r \qquad x_2(0) = x_{20} \tag{3-45}$$
$$y = c_1x_1 + c_2x_2 + dr$$

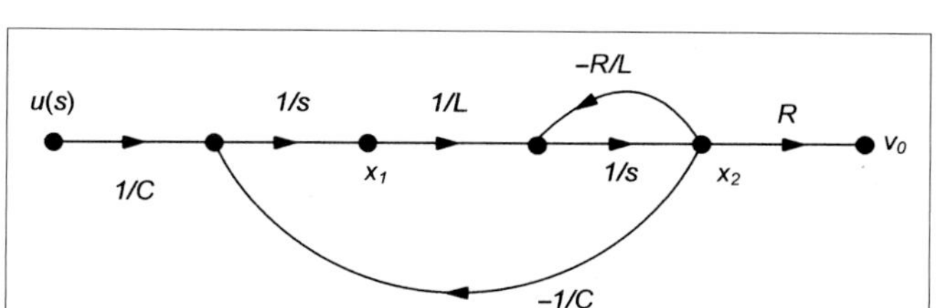

FIGURE 3-14 Example 2-16, flow graph of the state space model.

The state diagram using two integrators is shown in Fig. 3-16. This can be generalized for n integrators.

3-9 STATE DIAGRAMS OF DIFFERENTIAL EQUATIONS

The state diagram for a differential equation is not unique and a number of state models are possible for a given differential equation. Consider a differential equation:

$$\frac{d^n y}{dt^n} + a_{n-1}\frac{d^{n-1}y}{dt^{n-1}} + \cdots + a_1\frac{dy}{dt} + a_0 y(t) = r(t)$$
$$y(0), \dot{y}(0), \ldots, \frac{d^{n-1}y(0)}{dt^{n-1}} \qquad \text{initial conditions} \tag{3-46}$$

To arrive at a state model define:

$$x_1 = y$$
$$x_2 = \dot{y}$$
$$\vdots \tag{3-47}$$
$$x_n = \frac{d^{n-1}y}{dt^{n-1}}$$

Then the given equation of nth order is reduced to n first-order differential equations:

$$\dot{x}_1 = x_2$$
$$\dot{x}_2 = x_3$$
$$\vdots \tag{3-48}$$
$$\dot{x}_{n-1} = x_n$$
$$\dot{x}_n = -a_0x_1 - a_1x_2 - \cdots - a_{n-1}x_n + r$$

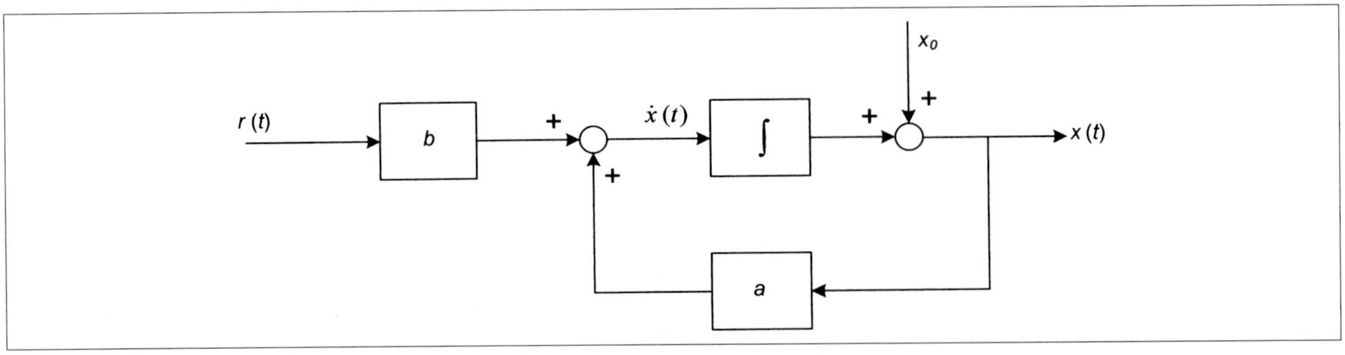

FIGURE 3-15 State diagram for first-order differential equation, one integrator.

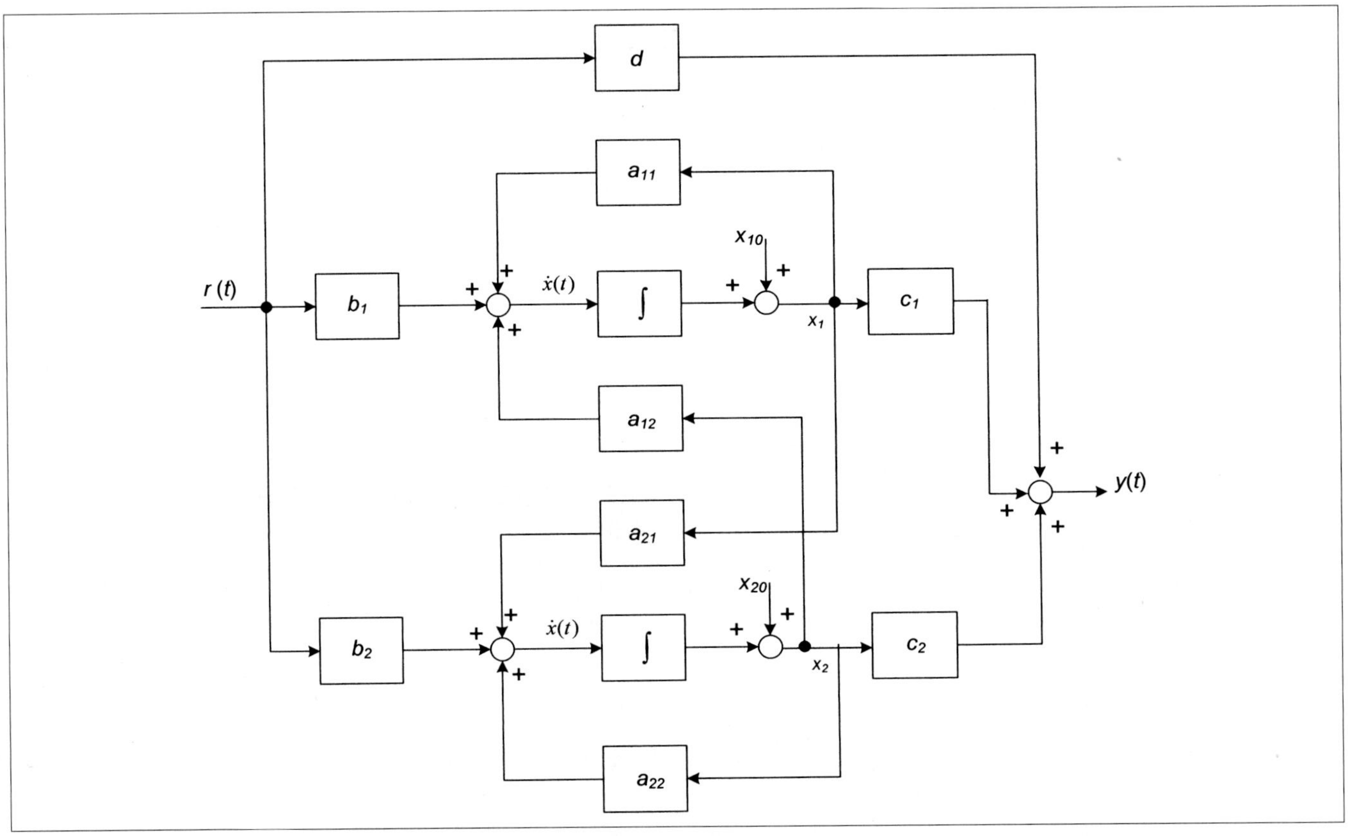

FIGURE 3-16 State diagram with two state variables and two integrators.

These form the state model:

$$\begin{vmatrix} \dot{x}_1 \\ \dot{x}_2 \\ \cdot \\ \dot{x}_{n-1} \\ \dot{x}_n \end{vmatrix} = \begin{vmatrix} 0 & 1 & 0 & \cdot & 0 \\ 0 & 0 & 1 & \cdot & 0 \\ \cdot & \cdot & \cdot & \cdot & \cdot \\ 0 & 0 & 0 & \cdot & 1 \\ -a_0 & -a_1 & -a_2 & \cdot & -a_{n-1} \end{vmatrix} \begin{vmatrix} x_1 \\ x_2 \\ \cdot \\ x_{n-1} \\ x_n \end{vmatrix} + \begin{vmatrix} 0 \\ 0 \\ \cdot \\ 0 \\ 1 \end{vmatrix} r$$

$$y = \begin{vmatrix} 1 & 0 & 0 & \cdot & 0 \end{vmatrix} \begin{vmatrix} x_1 \\ x_2 \\ \cdot \\ x_{n-1} \\ x_n \end{vmatrix} \tag{3-49}$$

The state diagram of the nth-order differential equation is shown in Fig. 3-17. Now consider a transfer function defined by:

$$\frac{C}{R} = \frac{b_2 s^2 + b_1 s + b_0}{s^3 + a_2 s + a_1 s} \tag{3-50}$$

We can write

$$\begin{aligned} r(t) &= \dddot{x}_1 + a_2 \ddot{x}_1 + a_1 \dot{x}_1 \\ y(t) &= c(t) = b_2 \ddot{x}_1 + b_1 \dot{x}_1 + b_0 x_1 \end{aligned} \tag{3-51}$$

or in the matrix form:

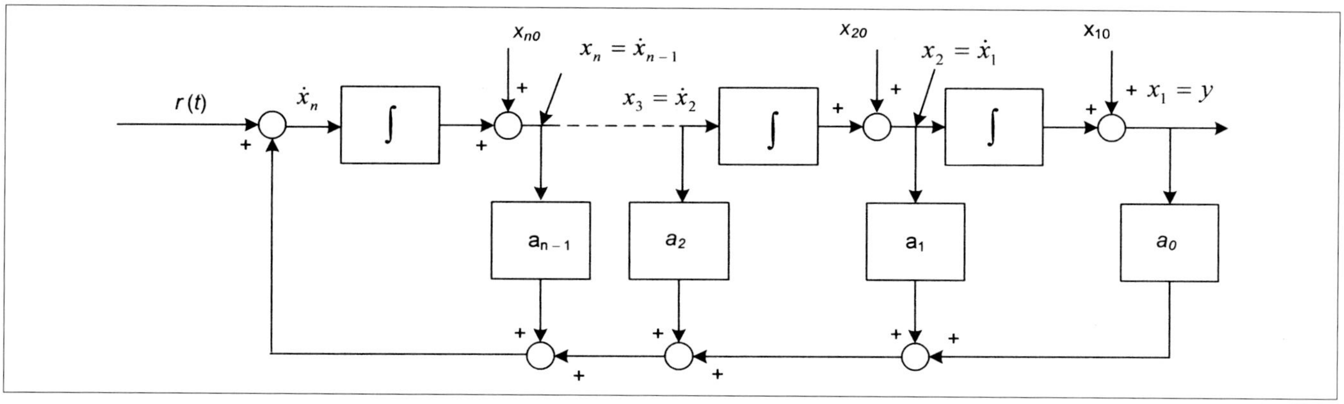

FIGURE 3-17 State diagram for *n*-th order differential equation.

$$\begin{vmatrix} \dot{x}_1 \\ \dot{x}_2 \\ \dot{x}_3 \end{vmatrix} = \begin{vmatrix} 0 & 1 & 0 \\ 0 & 0 & 1 \\ 0 & -a_1 & -a_2 \end{vmatrix} \begin{vmatrix} x_1 \\ x_2 \\ x_3 \end{vmatrix} + \begin{vmatrix} 0 \\ 0 \\ 1 \end{vmatrix} r(t)$$

$$y(t) = \begin{vmatrix} b_0 & b_1 & b_2 \end{vmatrix} \begin{vmatrix} x_1 \\ x_2 \\ x_3 \end{vmatrix} \quad (3\text{-}52)$$

Figure 13-18 shows the development of the block diagram.

Example 3-12 Consider the state variable block diagram given in Fig. 13-19. What is the transfer function?

By reversing the process, we can find the required transfer function, which is:

$$\frac{Y(s)}{R(s)} = \frac{3s+1}{(s^3+s^2+3s+7)}$$

3-10 STEADY-STATE ERRORS

If open-loop transfer function and input r are expressed as a function of s, then error as a function of s is:

$$E(s) = R(s)\frac{1}{1+GH(s)} \quad (3\text{-}53)$$

Three types of inputs for which the errors are well defined are:

- Step input: position error
- Ramp or velocity input: error in velocity
- Parabolic input: error in acceleration

The response is determined by the number of integrators in denominator term of GH of the form s^n, where n is the number of integrators. A type-one system has one integrator, a type-two system has two integrators, and a type-zero system has no integrators in the open-loop transfer function.

3-10-1 Position Error

Laplace transform of unit step is $1/s$, so the error for unit step is:

$$E(s) = \frac{R}{s}\frac{1}{1+GH(s)} \quad (3\text{-}54)$$

Using Laplace final value theorem:

$$e_{ss}(t) = \lim_{s\to 0} \frac{R}{1+GH(s)} \quad (3\text{-}55)$$

The term GH(s) can be expressed as:

$$GH(s) = \frac{K\left(1+\frac{s}{z_1}\right)\left(1+\frac{s}{z_2}\right)\cdots\left(1+\frac{s}{z_k}\right)}{s^n\left(1+\frac{s}{p_1}\right)\left(1+\frac{s}{p_2}\right)\cdots\left(1+\frac{s}{p_m}\right)}$$

Therefore:

$$\lim_{s\to 0} GH(s) = \lim_{s\to 0} \frac{K}{s^n} \quad (3\text{-}56)$$

The steady-state error is:

$$e_{ss}(t) = \lim_{s\to 0} \frac{Rs^n}{s^n+K} \quad (3\text{-}57)$$

It is $R/(1 + K)$ for type-zero system and 0 for type-one and type-two systems.

3-10-2 Velocity Error

The velocity is the term used for a ramp input, $r(t) = R'(t)$. The Laplace transform is R'/s^2. R' is the slope of the ramp.

$$E(s) = \frac{R'}{s^2[1+GH(s)]}$$

$$e_{ss}(t) = \lim_{s\to 0} s\frac{R'}{s^2[1+GH(s)]} = \lim_{s\to 0} \frac{\frac{R'}{s}}{1+\frac{K}{s^n}} \quad (3\text{-}58)$$

$$= \lim_{s\to 0} \frac{R's^{n-1}}{s^n+K} = \lim_{s\to 0} \frac{R's^{n-1}}{K}$$

For a type-zero system, the error is infinite; for type-one system, it is R'/K; and for type-two system, it is zero.

3-10-3 Acceleration Error

It can be similarly proved that for type-zero and type-one systems the steady-state error will be infinite and for type-two system it will be finite, $2R''/K$.

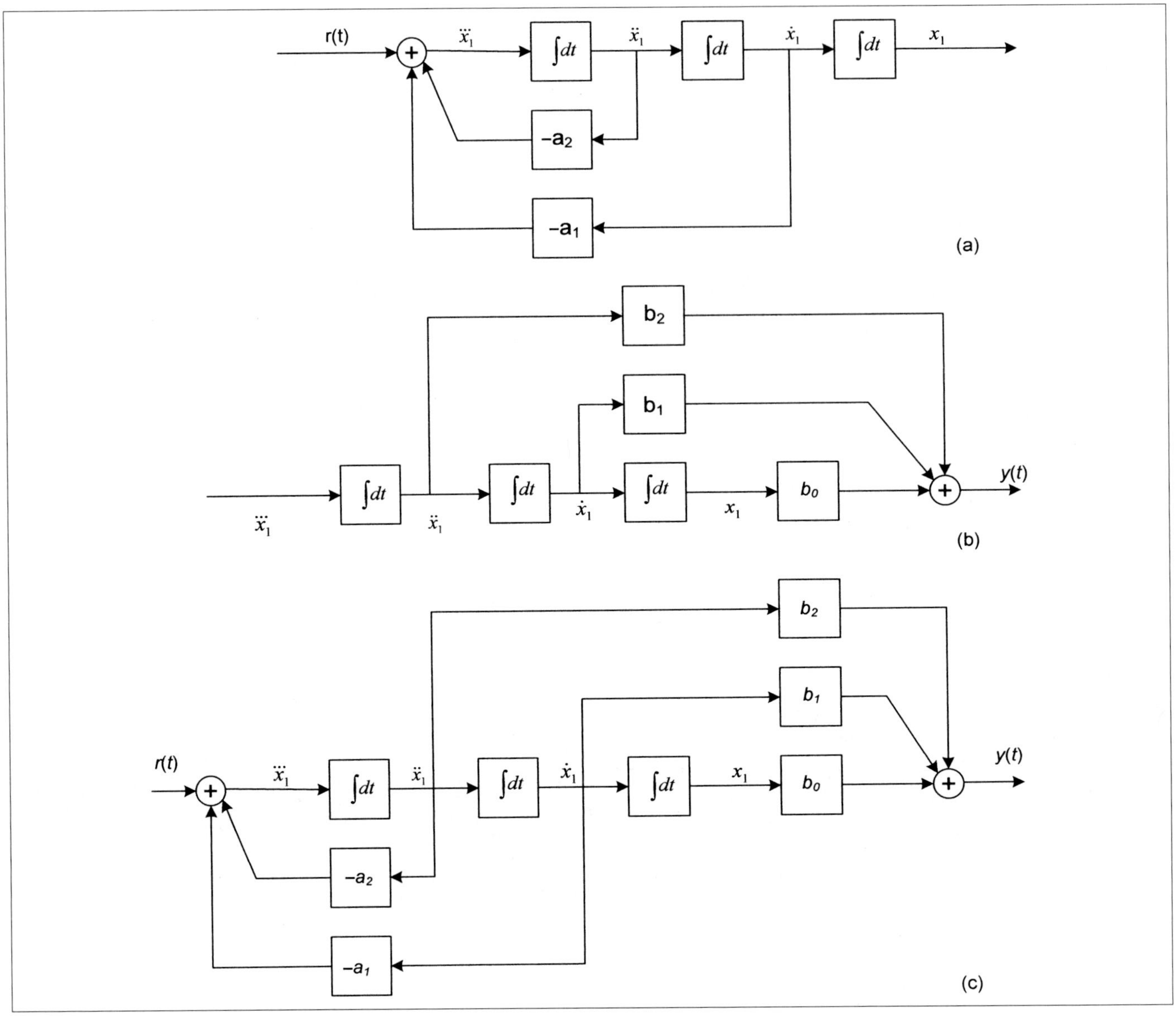

FIGURE 3-18 Development of the block diagram of a second-order differential equation.

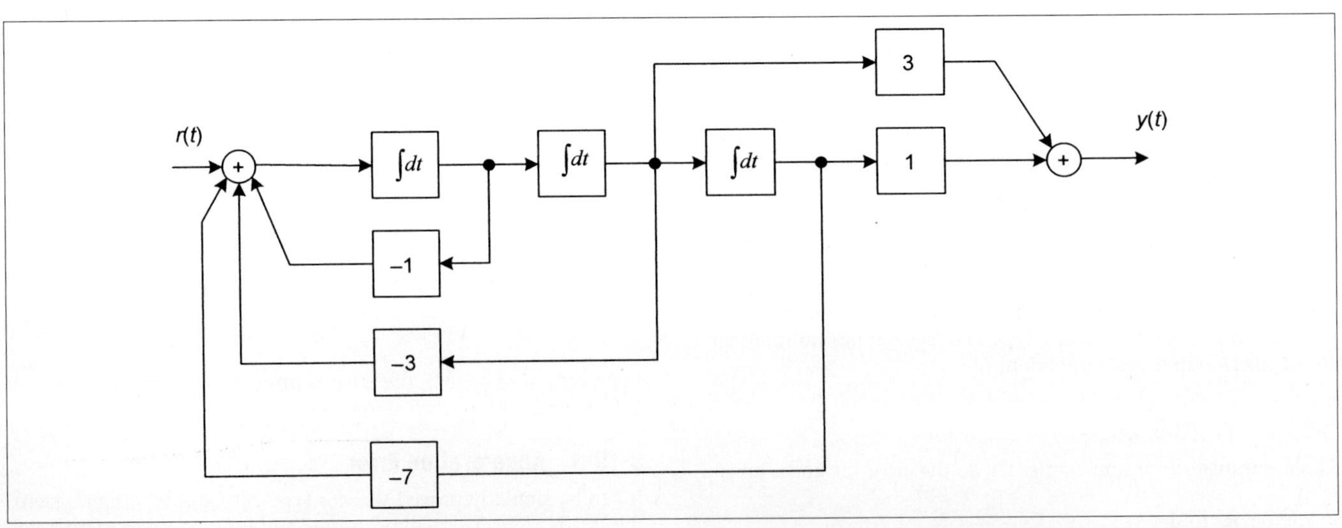

FIGURE 3-19 Example 3-12, calculation of transfer function from the block circuit diagram.

We can summarize that the system with zero integrator in the open loop has a finite position error and infinite velocity and acceleration errors. An infinite steady-state error indicates that the system cannot track the particular type of function which causes the error.

Given enough time, a system with one integrator in the open loop will track position perfectly and will follow a ramp function with finite error but cannot track an acceleration input at all.

A system with two integrators in the open-loop function can track both position and velocity with no long-run errors, but will have constant error with respect to acceleration input. Systems with more than two integrators in the open loop will have closed-loop performance able to track all derivates equal to one less than the number of integrators. However, systems with even two integrators have difficulty in damping transients. Systems with more than two integrators are very difficult to stabilize.

3-11 FREQUENCY-DOMAIN RESPONSE SPECIFICATIONS

The frequency-domain specifications form a design criteria and are stated in the following parameters:

Gain margin Gain margin is a measure of relative stability and defined as reciprocal of open-loop transfer function at a frequency at which phase angle is –180°.

$$\text{Gain margin} \equiv \frac{1}{|\mathrm{GH}(\omega_\pi)|} \tag{3-59}$$

where arg $\mathrm{GH}(\omega_\pi) = -180°$, and ω_π is called the phase crossover frequency

Phase Margin Phase margin is given by:

$$\phi_{\mathrm{PM}} = [180° + \arg \mathrm{GH}(\omega_1)] \tag{3-60}$$

where $|\mathrm{GH}(\omega_1)| = 1$.

that is, it is 180° plus the phase angle of open-loop transfer function at unity gain. ω_1 is called the gain crossover frequency. Figure 3-20 shows phase and gain margins of typical continuous-time control system.

Delay Time Delay time, as a frequency-domain specification, is interpreted as a measure of the speed of response:

$$T_d(\omega) = -\frac{d\gamma}{d\omega} \tag{3-61}$$

where $\gamma = \arg(C/R)$.

Average value of $T_d(\omega)$ over frequencies of interest is specified.

Bandwidth, Cutoff Frequencies It is generally defined as range of frequencies over which the values do not differ by more than –3 dB, (Fig. 3-21). Decibel, abbreviated as dB, is defined as:

$$\mathrm{dB} = 20\log_{10}(\textit{any magnitude ratio}) \tag{3-62}$$

The magnitude ratio is 0.707, approximately –3dB at the cutoff frequencies shown in Fig. 3-21

Resonant Peak and Resonant Frequency The resonant peak is maximum value of the magnitude of closed-loop frequency response:

$$M_p \equiv \max\left|\frac{C}{R}\right| \tag{3-63}$$

and the frequency at which it occurs is called the *resonant frequency*. It is a measure of the stability. The resonant peak M_p and the resonant frequency ω_p for an underdamped second-order system are illustrated in Fig. 3-22.

3-12 TIME-DOMAIN RESPONSE SPECIFICATIONS

Similar specifications for the time-domain response can be written from the response characteristics shown in Fig. 3-23. Transient response is described in terms of unit-step function response.

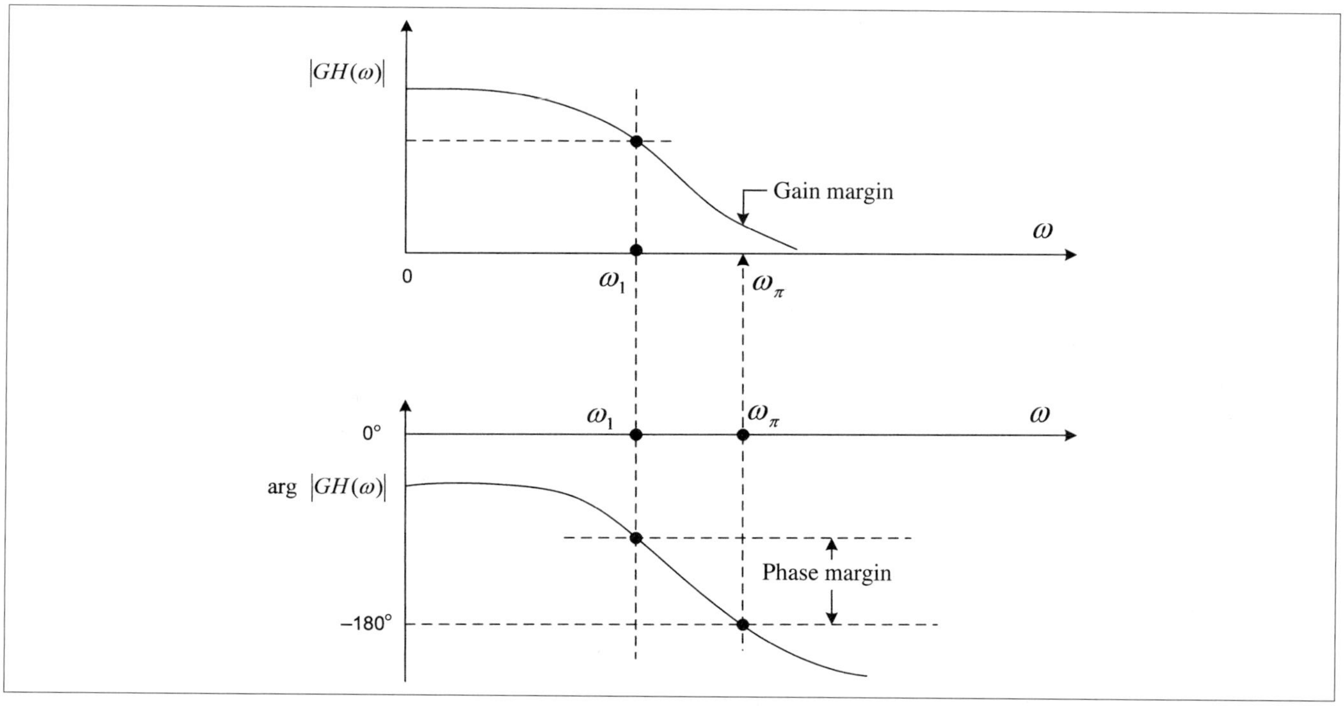

FIGURE 3-20 Phase and gain margins for stability—continuous control system.

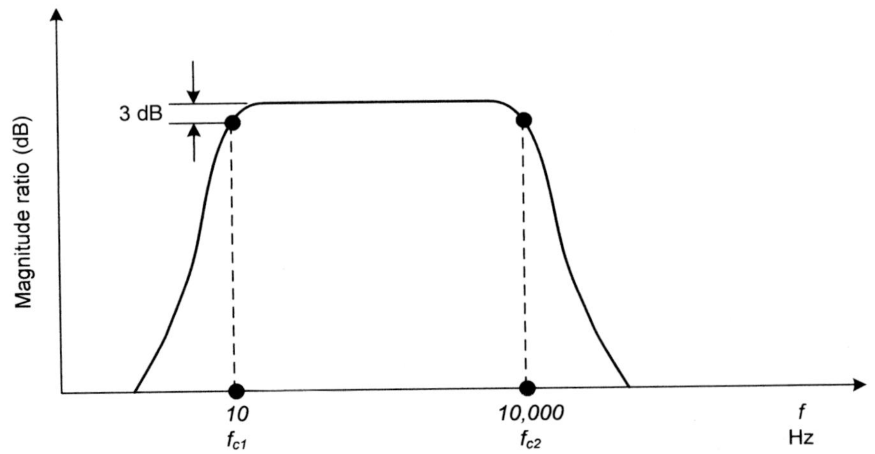

FIGURE 3-21 Cut-off frequencies, continuous control system.

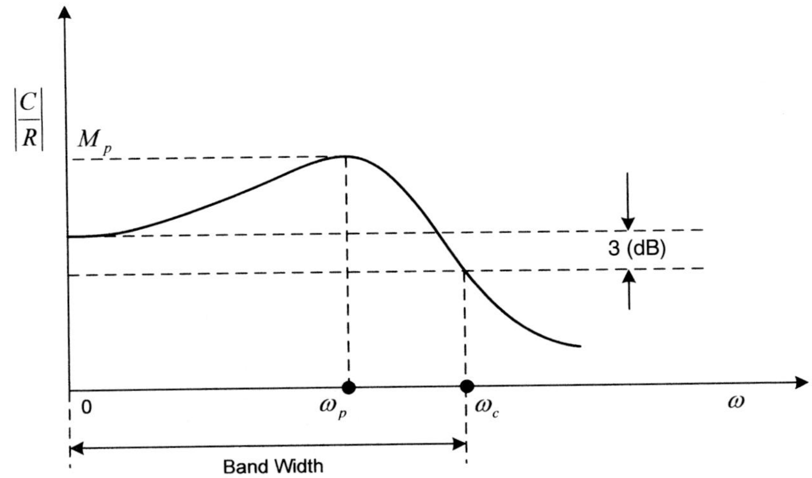

FIGURE 3-22 The resonant peak and bandwidth of a continuous control system.

Overshoot The maximum difference between the transient and steady-state solution for a unit-step input.

Time Delay T_d The time required for the response to a unit-step function to reach 50 percent of its initial value.

Rise Time T_r The time required for the unit-step input to rise from 10 to 90 percent of the final value.

Settling Time T_s The time required to reach and remain within a specified value, generally 2 to 5 percent of its final value.

Time Constant T The time when the exponential reaches 63 percent of its initial value as shown in the exponential envelope of Fig. 3-23. For continuous feedback control systems of the order higher than two, the exponential is given by:

$$\tau \leq \frac{1}{\xi \omega_n} \tag{3-64}$$

3-13 ROOT-LOCUS ANALYSIS

The root locus is helpful in determining the relative stability and transient response of a closed-loop system, as these are directly related to the location of the closed-loop roots of the characteristic equation. The movement of these roots in the complex plane can be examined as the system parameters vary. This graphical method can be used to evaluate the initial design and subsequent modifications. A computer simulation is generally used.

In order to evaluate the relative stability and transient performance of a closed-loop system, the roots of the denominator of the closed-loop transfer function are determined. The open-loop transfer function GH can be represented by:

$$\mathrm{GH} \equiv \frac{KN}{D} \tag{3-65}$$

where N and D are polynomials in complex variables s or z, and K is the open-loop gain. The closed-loop transfer function can be written as:

$$\frac{C}{R} = \frac{G}{1+GH} = \frac{GD}{D+KN} \tag{3-66}$$

The closed-loop poles are the roots of the characteristic equation:

$$D + KN = 0 \tag{3-67}$$

Thus, the location of the roots in s or z plane changes as the open loop-gain factor K changes. A locus of these roots plotted in s or z plane as a function of K is called a *root locus*.

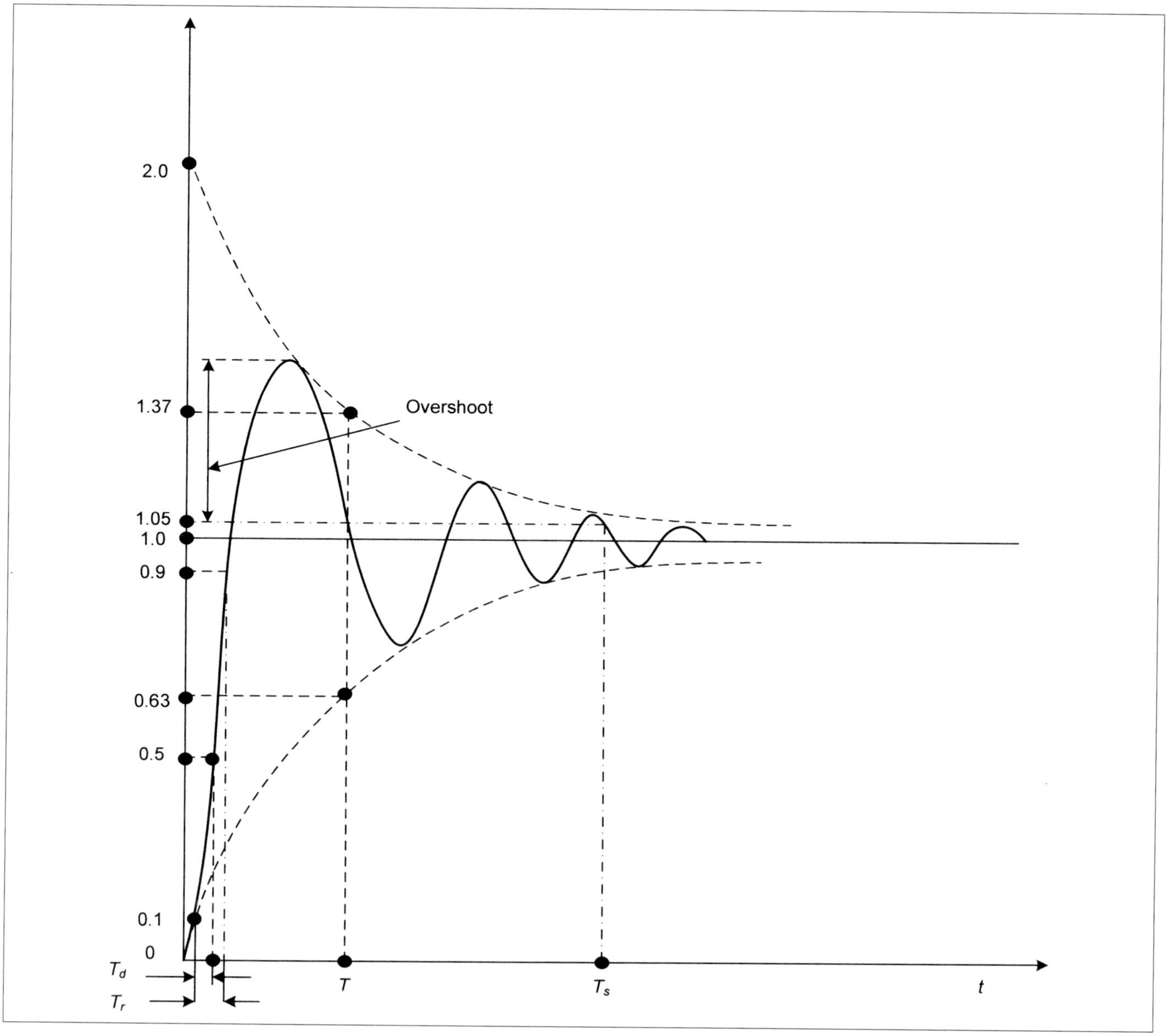

FIGURE 3-23 Time-domain transient response, step input function.

Consider $K = 0$. Then the roots of Eq. (3-67) are roots of the polynomial D, which are the same as the poles of the open-loop transfer function GH. Consider K very large; the roots approach those of polynomial N, that is, the open-loop zeros. Thus as K is increased from zero to infinity, the close-loop poles originate from open-loop poles and terminate at open-loop zeros. The closed-loop transfer function C/R is determined from the root-locus plot, for a specified value of K. For a unity feedback ($H = 1$), we can write:

$$G=\frac{KN}{D}=\frac{K(s+z_1)(s+z_2)\cdots(s+z_m)}{(s+p_1)(s+p_2)\cdots(s+p_n)} \tag{3-68}$$

Therefore:

$$\frac{C}{R}=\frac{KN}{D+KN}=\frac{KN}{(s+\alpha_1)(s+\alpha_2)\cdots(s+\alpha_n)} \tag{3-69}$$

Thus, C/R and G have the same zeros, but not the same poles, except when $K = 0$.

Example 3-13 A system with one integrator and variable gain has an open-loop transfer function GH = K/s. It is required to place the closed-loop pole at $s = -4$. There are no poles or zeros to the right of the origin in s plane. Thus, the positive axis is not part of the root locus, and the entire negative axis is the root locus. GH = 0, at s = infinity. To find the value of K for closed-loop pole at $s = -4$, we write GH = −1, and insert desired value of s to solve for K, which gives $K = 4$.

3-13-1 Angle and Magnitude

For root locus to pass through a point p_1 in complex plane, it should be a root of the characteristic equation (3-67) for some real value of K.

$$D(p_1)+KN(p_1)=0$$

or

$$\text{GH}=\frac{KN(p_1)}{D(p_1)}=-1$$

Complex number $GH(p_1)$ must have phase angles of $180°+360n°$, therefore:

$$\arg GH(p_1) = 180° + 360n°$$
$$= (2n+1)\pi \text{ radians}$$

or

$$\arg\left[\frac{N(p_1)}{D(p_1)}\right] = (2n+1)\pi \qquad K>0 \tag{3-70}$$
$$= 2n\pi \qquad K<0$$

where

$$n = 0, \pm 1, \pm 2, \ldots$$

Also the magnitude criteria should be satisfied. Thus K must have the particular value that satisfies:

$$|K| = \left|\frac{D(p_1)}{N(p_1)}\right| \tag{3-71}$$

3-13-2 The Number of Loci

The number of loci, that is, the branches of the root locus are equal to the number of poles of the open-loop transfer function GH.

For $K > 0$. Points of the root locus on the real axis lie to the *left of the odd number of finite poles and zeros.*

For $K < 0$. Points of the root locus on the real axis lie to the *left of an even number of finite poles and zeros.*

If no points on the real axis lie to the left of an odd number of finite poles and zeros, then no portion of the root locus for $K > 0$ lies on the real axis. A similar statement is true for $K < 0$

3-13-3 Asymptotes

The asymptotes originate from a point on the real axis called the center of asymptotes, which is given by:

$$\sigma_c = -\frac{\sum_{i=1}^{i=n} p_i - \sum_{i=1}^{i=m} z_i}{n-m} \tag{3-72}$$

where n is the number of poles and m is the number of zeros, and $-p_i$ are the poles and $-z_i$ are the zeros of GH.

The angles between the asymptotes and the real axis are given by:

$$\beta = \begin{vmatrix} \dfrac{(2l+1)180}{n-m} & \text{degrees} & K>0 \\ \dfrac{(2l)180}{n-m} & \text{degrees} & K<0 \end{vmatrix} \tag{3-73}$$

where l is an arbitrary integer.

3-13-4 Breakaway Points

A breakaway point is defined where two or more branches of the root locus arrive or depart at the real axis. The location is given by:

$$\sum_{i=1}^{n} \frac{1}{(\sigma_b + p_i)} = \sum_{i=1}^{m} \frac{1}{(\sigma_b + z_i)} \tag{3-74}$$

where p_i and z_i are the poles and zeros of GH, and σ_b is the breakaway point.

3-13-5 Departure and Arrival Angles

The departure and arrival angles, θ_D, θ_A, of the root locus from complex poles are given by:

$$\theta_D = 180° + \arg GH'$$
$$\theta_A = 180° - \arg GH'' \tag{3-75}$$

where GH′ is the phase angle of GH computed at complex pole, ignoring the contribution of that particular pole, and GH″ is the phase angle of GH at complex zero, ignoring the effect of that zero.

3-13-6 Procedure for Constructing Root Locus

The procedure for constructing root locus can be summarized as follows:

- Write characteristic equation
- Factor polynomial if required in the form of poles and zeros
- Locate poles and zeros on s plane
- Locate segments on real axis that are root loci
- Determine number of separate loci
- Root locus must be symmetrical with respect to real axis
- Calculate asymptotes and departure angles
- Calculate breakaway points
- Determine the points at which the locus crosses the imaginary axis
- Determine the angle of locus departure from complex poles and angle of arrival at complex zeros, using phase criteria:

$$< P(s) = 180° \pm q360° \quad \text{at} \quad s = p_j \quad \text{or} \quad z_i \tag{3-76}$$

- Determine the root locations that satisfy the criteria:

$$< P(s) = 180° \pm q360° \quad \text{at root } s_x \tag{3-77}$$

- Determine the parameter K_x at root location s_x:

$$K_x = \frac{\prod_{j=1}^{n_p} |s+p_j|}{\prod_{i=1}^{n_z} |s+z_i|} \tag{3-78}$$

We will illustrate the root-locus analysis and procedure through some solved examples.

Example 3-14 Consider an open-loop transfer function:

$$GH = \frac{K(s+1)}{s^2(s+4)}$$

Find the center of the asymptotes and their angles. At the real axis, there are three poles and one zero; therefore $n - m = 2$. From Eq. (3-72):

$$\sigma_c = -\frac{4-1}{2} = -1.5$$

There are two asymptotes. Their angles with the real axis are given by Eq. (3-73). These are 90° and 270°, respectively.

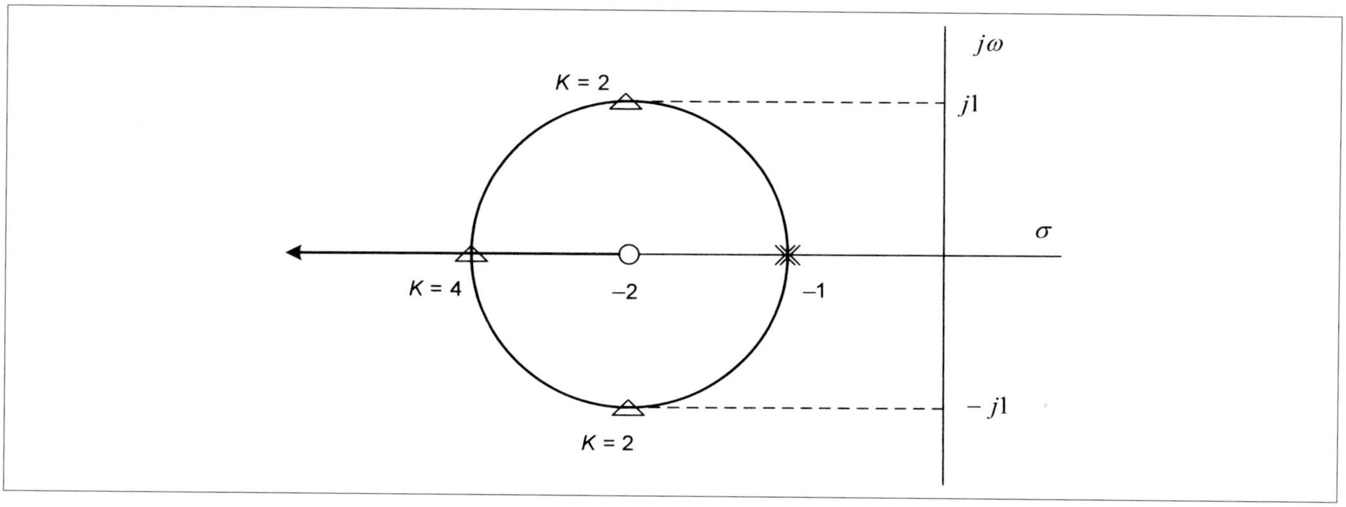

FIGURE 3-24 Example 3-16, root-locus diagram.

Example 3-15 Determine the breakaway points of:

$$GH = \frac{K}{s(s+1)(s+3)}$$

From Eq. (3-74), the following equations must be solved:

$$\frac{1}{\sigma_b} + \frac{1}{\sigma_b + 1} + \frac{1}{\sigma_b + 3} = 0$$

$$(\sigma_b + 1)(\sigma_b + 3) + \sigma_b(\sigma_b + 3) + \sigma_b(\sigma_b + 1) = 0$$

$$3\sigma_b^2 + 8\sigma_b + 3 = 0$$

The roots are −0.45 and −2.22. Thus for $K > 0$, there are branches between 0 and −1, and the breakaway point is at −0.45. For $K < 0$, the breakaway point is −2.22, since the portion of the real axis between −1 and −3 is on the root locus for $K < 0$.

Example 3-16 Consider a continuous-system, open-loop function defined by:

$$G = \frac{K(s+2)}{(s+1)^2}$$

It is required to construct the root locus and find a closed-loop pole at $K = 2$. When we have two adjacent poles with a zero, the root locus is a circle, centered at zero and having a radius which is a geometric mean of the distances from zeros to the poles. The radius of the circle is:

$$[(z - p_1)(z - p_2)]^{1/2} \tag{3-79}$$

The root locus is as shown in Fig. 3-24. We know that *C/R* and *G* have the same zeros, but not the same poles unless $K = 0$. Closed-loop poles may be determined from the root locus for a given *K*, but closed-loop zeros are not equal to the open-loop zeros. Several values of *K* are shown in Fig. 3-24 by small triangles. These are the closed-loop poles corresponding to the specified values of *K*. As an example for $K = 2$, the closed-loop poles are:

$$\frac{2(s+2)}{(s+2+j)(s+2-j)}$$

Example 3-17 A system with unity feedback has an open-loop transfer function given by:

$$\frac{K(s+9)}{s(s+3)}$$

Find K for damping ratio $\xi = 0.707$. Also find closed-loop response to a unit-step function. Here, we have two adjacent poles and a zero. Thus, the root locus is a circle, and the radius is given by:

$$r = \sqrt{(z - p_1)(z - p_2)} = \sqrt{-9 \times -6} = 7.35$$

$$\xi = 0.707 \qquad \cos^{-1}\xi = 45°$$

From Fig. 3-25, the closed loop-poles can be calculated graphically. Alternatively, the equation of the circle for $\xi = 0.707$ and $\omega = -\sigma$

$$(\sigma + 9)^2 + \omega^2 = 54$$

$$\sigma^2 + 18\sigma + 81 + \sigma^2 = 54$$

This gives:

$$\sigma = -1.902, -7.098$$

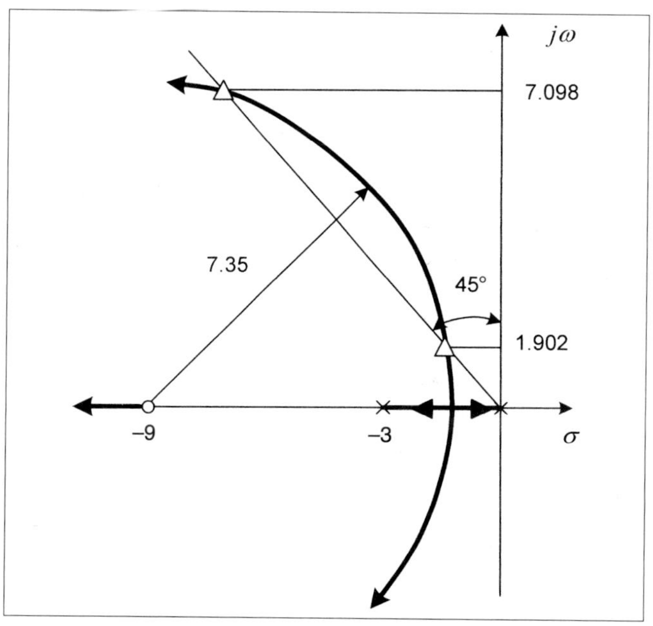

FIGURE 3-25 Example 3-17, root-locus diagram.

For the first set of poles:

$$\sigma = -1.902 \pm j1.902$$

Characteristic equation is:

$$1 + GH = 0$$

$$1 + \frac{K(s+9)}{s(s+3)} = 0$$

$$s^2 + (3+K)s + 9K = 0$$

Also:

$$(s + 1.902 + j1.902)(s + 1.902 - j1.902) = 0$$

$$s^2 + 2 \times 1.902s + 2 \times 1.902^2 = 0$$

Comparing:

$$3 + K = 2 \times 1.902$$

$$9K = 2 \times 1.902^2$$

Both of these relations give $K = 0.8038$. Thus, the closed-loop transfer function is:

$$\frac{(0.8038)(s+9)}{s^2 + 3.8038s + 7.235}$$

Step response is given by:

$$c(s) = \frac{1}{s}\left[\frac{G}{1+GH}\right] = \frac{1}{s}\frac{(0.8038)(s+9)}{s^2 + 3.8038s + 7.235}$$

Resolve into partial fractions:

$$c(s) = \frac{1}{s} - \frac{s + 1.902}{(s+1.902)^2 + (1.902)^2} - \frac{0.577(1.902)}{(s+1.902)^2 + (1.902)^2}$$

Thus:

$$C(t) = 1 - e^{-1.902t}[\cos 1.902t + 0.577 \sin 1.902t]$$

This can be repeated for the second set of poles at $-7.098 \pm j7.098$

Example 3-18 Construct the root locus for closed-loop continuous system for $K > 0$, given the open-loop transfer function:

$$GH = \frac{K}{s(s+2)(s+4)}$$

Mark the poles 0, 2, and 4 on the real axis, as shown in Fig. 3-26. The root locus on the real axis is straightforward. Next, find asymptotes:

$$\sigma_c = -\left[\frac{(2+4) - 0}{3 - 0}\right] = -2$$

This gives the center of the asymptotes; there are three asymptotes. These angles are calculated from Eq. (3-73) and are $\beta = 60°, 180°$, and $300°$. There is a breakaway point between 0 and –2 on the real axis. Using Eq. (3-74) this breakaway point is at –0.845. The value of K at crossing of the root locus on imaginary axis can be found from the characteristic equation:

$$s(s+2)(s+4) + K = 0$$

$s = j\omega = 0$ at the imaginary axis. This gives $K = 48$ and $\omega = 2.82$. The values of K can be calculated similarly for other points on the axis. The gain factor K required to give a specific damping ratio ξ is determined from:

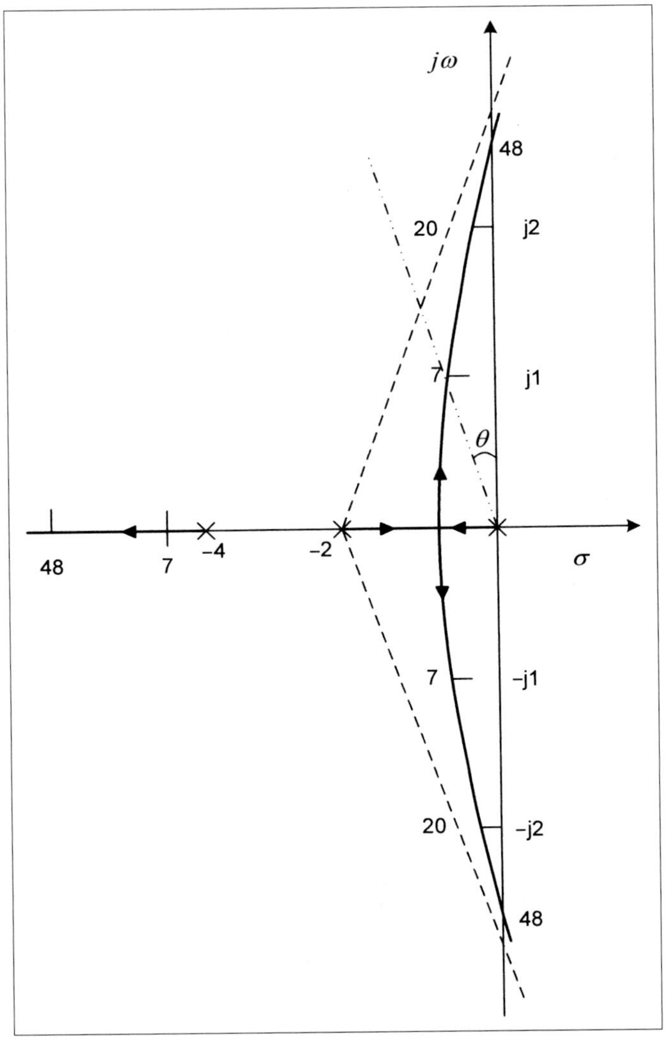

FIGURE 3-26 Example 3-18, root locus diagram.

$$\theta = \cos^{-1} \xi \tag{3-80}$$

The procedure can be applied to any pair of complex conjugate poles for systems of second order or higher.

Let us find the gain which gives a damping ratio of 0.55. Here, $\cos^{-1} 0.55 = 56.6°$. This intersects the root locus at $K = 7$.

Example 3-19 Consider the characteristic equation of a fourth-order system as K varies, $K > 0$:

$$1 + \frac{K}{s^4 + 9s^3 + 36s^2 + 54s} = 0$$

This can be written as:

$$1 + \frac{K}{s(s+3)(s+3+3j)(s+3-3j)}$$

Thus, as K varies from zero to infinity, the system has no finite zeros. Locate the poles in the s plane as shown in Fig. 3-27a.

Root locus between $s = 0$ and $s = -3$ can be easily drawn. For the fourth-order system, we should have four separate loci and the

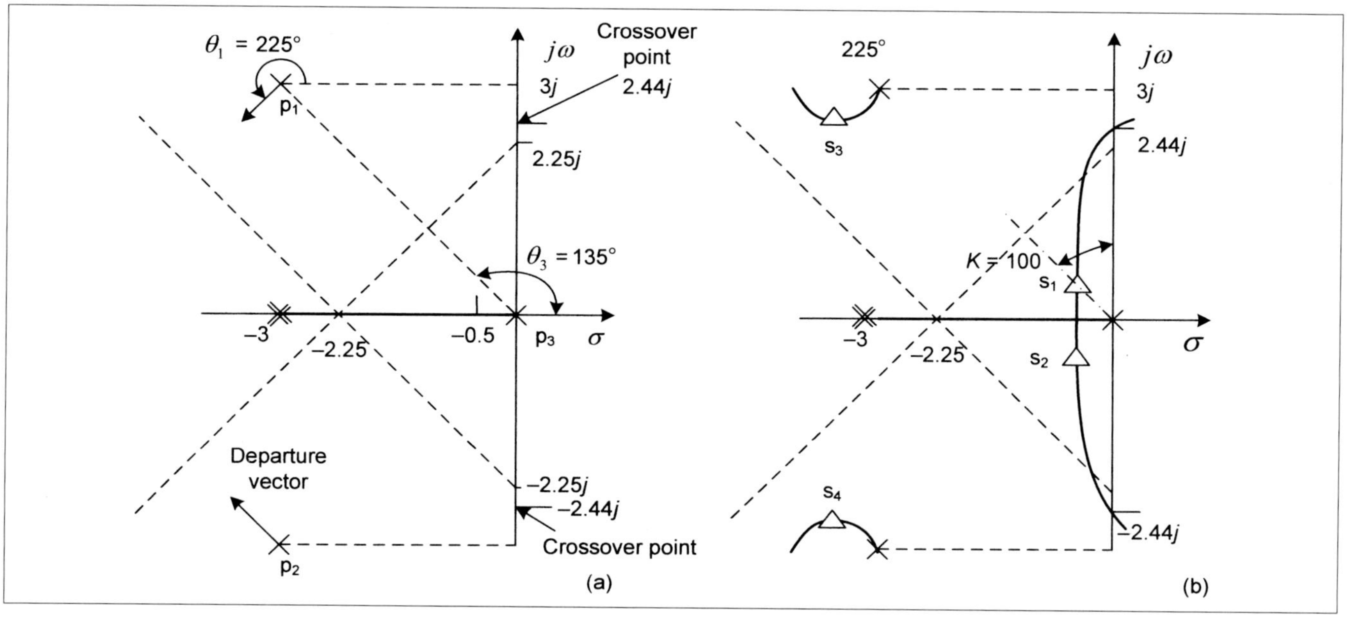

FIGURE 3-27 Example 3-19, root locus diagram.

root locus is symmetrical with respect to real axis. The angles of the asymptotes are found from Eq. (3-73) and are:

$$\beta = \frac{(2l+1)180^\circ}{4} = 45^\circ, 135^\circ, 225^\circ, 315^\circ$$

The center of the asymptotes is calculated from Eq. (3-72), which gives:

$$\sigma_c = \frac{-3-3-3}{4} = 2.25$$

The asymptotes can be drawn, as shown in Fig. 3-27*a*. The characteristic equation is:

$$s(s+3)(s^2+6s+18)+K = s^4+9s^3+36s^2+54s+K = 0$$

Apply Routh's criteria of stability:

s^4	1	36	K
s^3	9	54	
s^2	30	K	
s^1	$54-0.3K$		
s^0	K		

The limiting value of K for stability is:

$$54-0.3K=0 \qquad K=180$$

The auxiliary equation is:

$$30s^2+180=0$$

$$30(s^2+6)=0$$

$$30(s+j2.44)(s-j2.44)$$

The points where the root locus crosses the imaginary axis are shown in Fig. 3-27*a*. Estimate the breakaway point between $s = 0$ and $s = -3$.

$$\frac{1}{\sigma_b}+\frac{1}{\sigma_b+3}+\frac{1}{\sigma_b+3+3j}+\frac{1}{\sigma_b+3-3j}=0$$

The solution can be found by hit and trial between 0 and −3. An estimated value of $-s = -0.5$ satisfies this equation. We can also find the break point by maximizing the polynomial for s:

$$-s(s+3)(s+3+j3)(s+3-j3)$$

The angle of departure at the complex pole p_1 can be found by utilizing the angle criteria:

$$\theta_1+90^\circ+90^\circ+\theta_3=180^\circ$$

$$\theta_3=135^\circ$$

Therefore:

$$\theta_1=-135^\circ=+225^\circ$$

Note that θ_3 is the angle of the vector from the pole at p_3.

The completed root locus plot is shown in Fig. 3-27*b*. Consider a damping ratio, $\xi=0.707$. Then we can graphically find the gain K to the root location at s_1 as follows:

$$K=|s_1||s_1+3||s_1-p_1||s_1-p_2|\approx 100 \qquad \text{approximately}$$

Roots s_3 and s_4 occur when $K = 100$. The effect of these roots on transient response will be negligible compared to roots s_1 and s_2. The complex conjugate roots near the origin of the s plane relative to other roots are labeled dominant roots of the system because they represent dominant transient response.

3-14 BODE PLOT

The Bode plot or Bode diagram is a frequency plot of the system, where the amplitude in decibels and the phase angle in degrees are plotted against the logarithm of the frequency. Thus, the Bode plot consists of two graphs. The representation of the open-loop frequency response function GH(ω) may refer to a discrete-time or continuous-time system. Bode plot clearly illustrates the stability of the system, and often the gain and phase margins are defined in terms of Bode plots (Chap. 13).

The Bode form of the function can be written by replacing s with $j\omega$, that is

$$\begin{aligned} GH(j\omega) &= \frac{K(j\omega+z_1)(j\omega+z_2)\cdots(j\omega+z_m)}{(j\omega)^l(j\omega+p_1)(j\omega+p_2)\cdots(j\omega+p_n)} \\ &= \frac{K_B(1+j\omega/z_1)(1+j\omega/z_2)\cdots(1+j\omega/z_m)}{(j\omega)^l(1+j\omega/p_1)(1+j\omega/p_2)\cdots(1+j\omega/p_n)} \end{aligned} \tag{3-81}$$

where l is any nonnegative integer, and K_B is defined as the Bode gain given by:

$$K_B = \frac{K\prod_{i=1}^{m} z_i}{\prod_{i=1}^{n} p_i} \tag{3-82}$$

Taking logs:

$$\begin{aligned} 20\log|GH(j\omega)| &= 20\log|K_B| + 20\log\left|1+\frac{j\omega}{z_1}\right| + \cdots + 20\log\left|1+\frac{j\omega}{z_m}\right| \\ &- 20\log|(j\omega)^l| - 20\log\left|1+\frac{j\omega}{p_1}\right| - \cdots - 20\log\left|1+\frac{j\omega}{p_n}\right| \end{aligned} \tag{3-83}$$

and:

$$\begin{aligned} \arg GH(j\omega) &= \arg K_B + \arg\left(1+\frac{j\omega}{z_1}\right) + \cdots + \arg\left(1+\frac{j\omega}{z_m}\right) \\ &+ \arg\left(\frac{1}{(j\omega)^l}\right) + \arg\left(\frac{1}{1+j\omega/p_1}\right) + \cdots + \arg\left(\frac{1}{1+j\omega/p_n}\right) \end{aligned} \tag{3-84}$$

Here, we denote log as the log to the base 10

Each part of GH in the factored form can be plotted separately and the results added. The Bode plots are drawn on semilog paper.

3-14-1 Bode Plots of Simple Functions

The magnitude plot of $20\log_{10}|K_B|$ is a straight line and can be easily plotted. K_B has a phase angle of zero if positive, and −180° if negative. The magnitude and phase angle plots are shown in Fig. 3–28*a* and *b*.

The frequency response functions of the pole at origin, that is, $1/(j\omega)^l$, $20\log s^{-l}$ are straight lines, and have a slope of 20 dB per unit of log ω (Fig. 3-29*a*). Args of $1/(j\omega)^l$ are straight lines, as shown in Fig. 3-29*b*. Since a unit for a logarithmic scale is a change by a factor of 10 (log 1000 − log 100 = 3 − 2 = 1), the logarithmic unit is a decade of frequency, 20 dB per decade for $l = 1$. For $l = 2$, it will be 40 dB per decade (Fig. 3-29*a*).

Similarly, the frequency response and args of 20 log $(j\omega)^l$ are shown in Fig. 3-30*a* and *b*. *The slope is positive for zero and negative for a pole.* The plots for the single-pole transfer functions $1/(1 + j\omega/p)$ are shown in Fig. 3-31*a* and *b*. The asymptotic approximations are arrived as follows:

For $\omega/p \ll 1$ or $\omega \ll p$

$$20\log\left|\frac{1}{1+j\omega/p}\right| \cong 20\log 1 = 0\text{ dB} \tag{3-85}$$

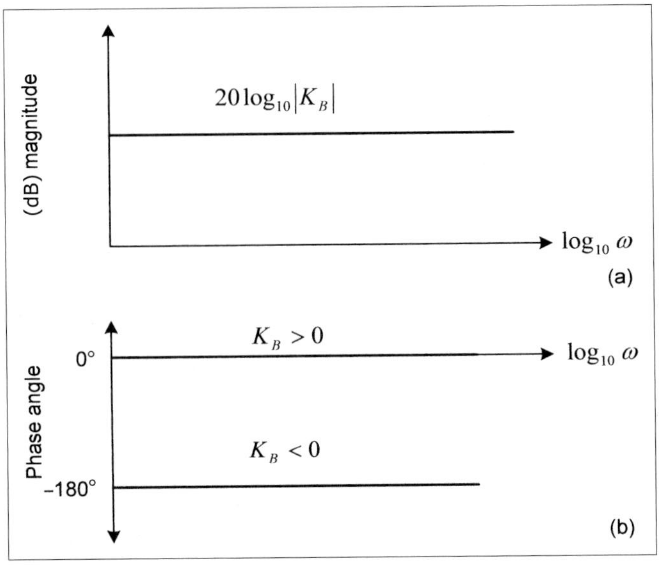

FIGURE 3-28 (*a*) Bode plot magnitude. (*b*) Phase angle for K_B.

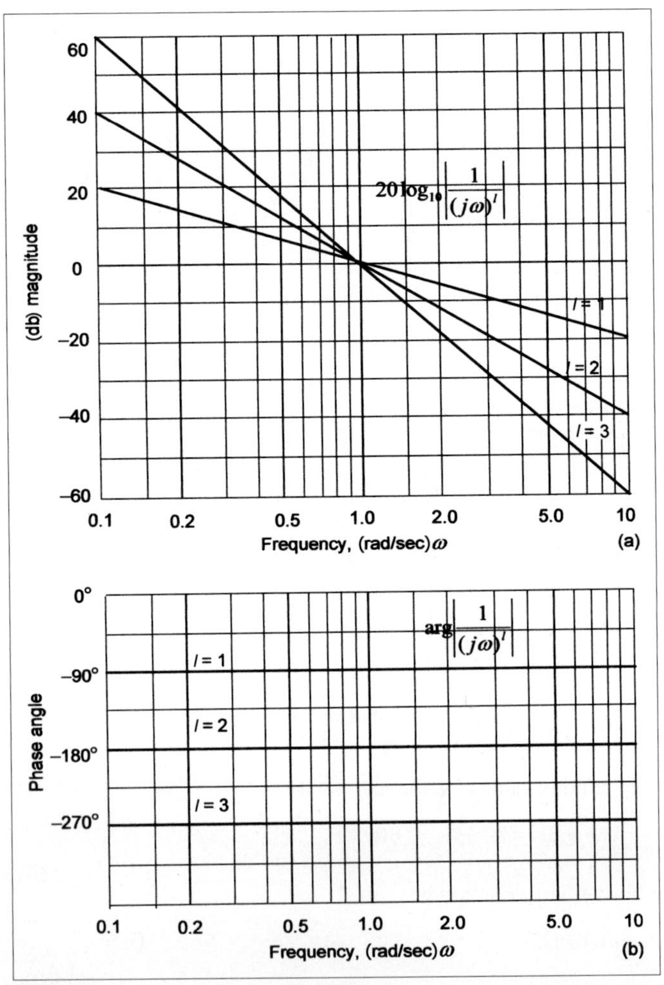

FIGURE 3-29 (*a*) Bode plot, magnitude. (*b*) Phase angle for pole of order *l* at the origin.

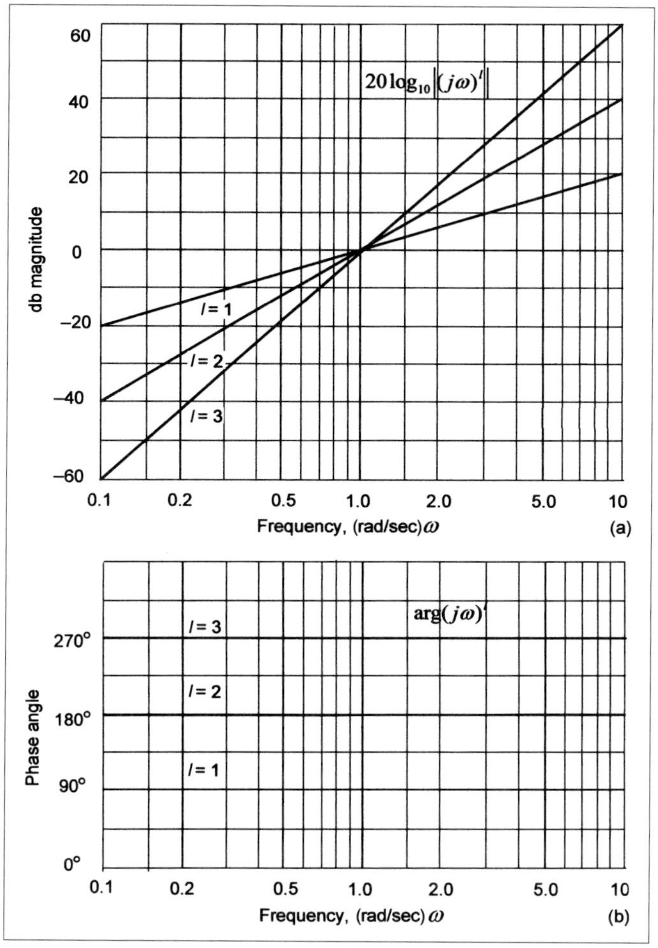

FIGURE 3-30 (a) Bode plot, magnitude. (b) Phase angle for zero of order l at the origin.

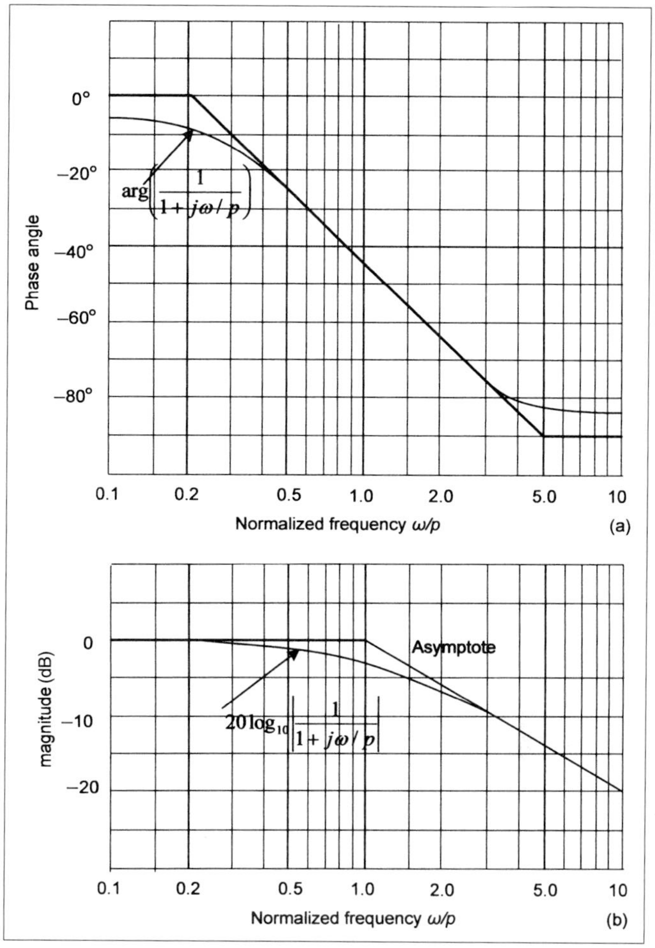

FIGURE 3-31 (a) Bode plot, phase angle. (b) Magnitude for single-pole.

For $\omega/p >> 1$ or $\omega >> p$

$$20\log_{10}\left|\frac{1}{1+j\omega/p}\right| \cong -20\log(\omega/p) \qquad (3\text{-}86)$$

The Bode magnitude plot asymptotically approaches a straight line horizontally at 0 dB as ω/p approaches zero and $-20\log_{10}(\omega/p)$ as ω/p approaches infinity. Note that the horizontal axis is normalized to ω/p. The high-frequency asymptote is a straight line with a slope of –20 dB/decade or –6 dB/octave, when plotted on a logarithmic frequency scale as shown. The two asymptotes intersect at the *corner frequency*, where $\omega = p$ rad/s.

For $\omega/p << 1$ or $\omega << p$

$$\arg\left(\frac{1}{1+j\omega/p}\right) = -\tan^{-1}(\omega/p)\Big|_{\omega \ll p} \cong 0° \qquad (3\text{-}87)$$

For $\omega/p >> 1$ or $\omega >>p$

$$\arg\left(\frac{1}{1+j\omega/p}\right) = -\tan^{-1}(\omega/p)\Big|_{\omega \gg p} \cong -90° \qquad (3\text{-}88)$$

Bode plot angle asymptotically approaches 0° as ω/p approaches zero and –90° when ω/p approaches infinity. Note the straight line portion (asymptote) between $\omega = p/5$ and $\omega = 5p$. It is tangent to the exact curve at $\omega = p$.

The magnitude error when $\omega = p$ is –3 dB maximum and the phase angle error is zero. At $\omega = 5p$ and $5/p$, the phase angle error is +11.3° and –11.3°, respectively, and the magnitude error is –0.17dB.

Similarly, the Bode plots and their asymptotic approximations for single zero frequency response function, $1 + j\omega/z$, are shown in Figs. 3-32*a* and *b*.

3-14-2 Bode Plot for Second-Order Frequency Response

The Bode plot and their asymptotic approximations for the second-order frequency response function with complex poles:

$$\frac{1}{1+j2\xi\omega/\omega_n-(\omega/\omega_n)^2} \qquad 0 \le \xi \le 1 \qquad (3\text{-}89)$$

are constructed as shown in Fig. 3-33*a* and *b*. The magnitude asymptote has a corner frequency at $\omega = \omega_n$ and a high-frequency asymptote twice that of single-pole case, that is, 40 dB. The phase angle asymptote is similar to that of a single pole, except that the high-frequency portion is at –180°, instead of –90° for the single-pole case. The point of inflection is at –90°. Note the effect of damping ratio ξ.

The Bode plots for a *complex* pair of zeros will be the reflections about 0 dB and 0° lines of those for the complex poles.

3-14-3 Construction of Bode Plots

The Bode plots can be constructed based upon the concepts outlined above. This will be illustrated by an example.

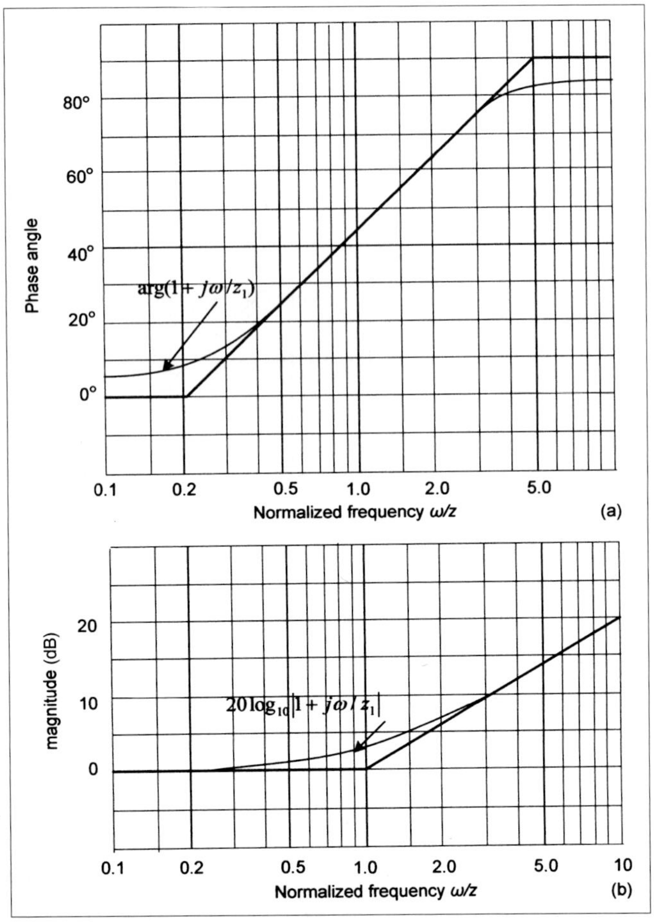

FIGURE 3-32 (*a*) Bode plot, phase angle. (*b*) Magnitude for single-zero.

Example 3-20 Construct the asymptotic Bode plots for the frequency response function:

$$GH(j\omega) = \frac{10(1+j\omega)}{(j\omega)^2[1+j\omega/3-(\omega/3)^2]}$$

The asymptotes are constructed using the equations already derived above. Taking logs,

$$20\log_{10}|GH(\omega)| = 20\log 10 + 20\log|1+j\omega|$$

$$+20\log\left|\frac{1}{(j\omega)^2}\right| + 20\log\left|\frac{1}{1+j\omega/3-(\omega/3)^2}\right|$$

The construction is shown in Fig. 3-34*a* and *b*. Note the construction of asymptote for each component using the techniques as graphically illustrated before. The complex poles are constructed as follows:

> Magnitude asymptote is $\omega = \omega_n = 3$, which gives the corner frequency. Draw a –40 dB slope between 3 and 30 rad/s.
>
> Phase angle asymptote $5p = 15$ and $0.2p = 0.6$. Draw an asymptote spanning 180°.
>
> The asymptotic Bode plot is obtained by summation of all the asymptotes, shown in bold. The Bode plots can be constructed around these asymptotes. A computer simulation is commonly used.

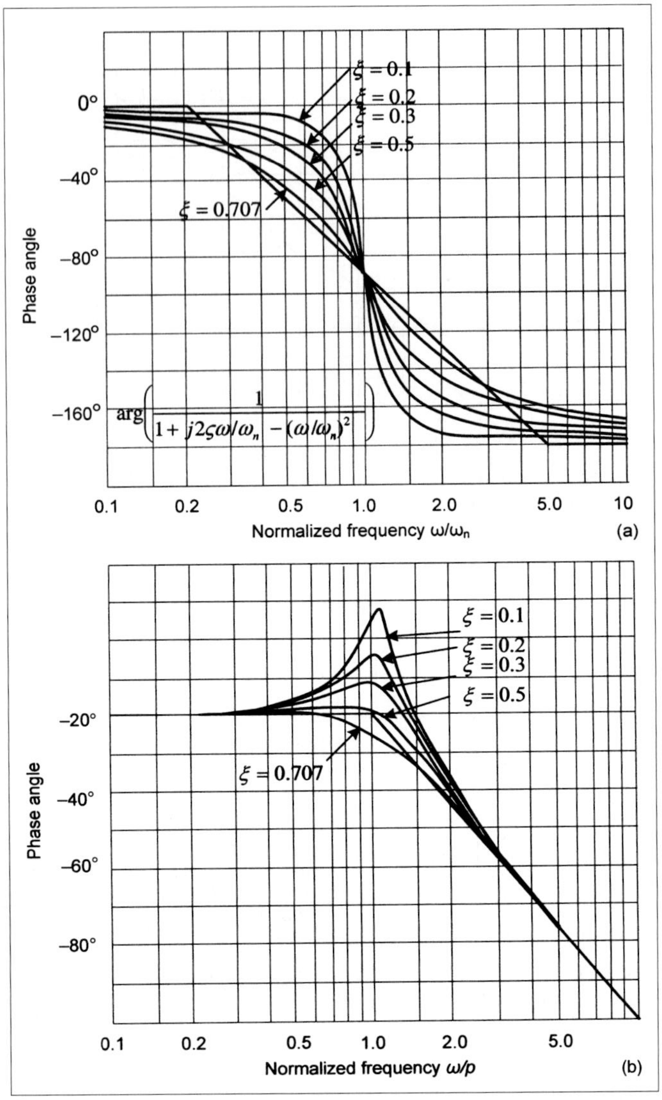

FIGURE 3-33 (*a*) Bode plot, phase angle. (*b*) Magnitude for second-order frequency function with complex poles.

3-15 RELATIVE STABILITY

The criteria of relative stability in terms of *system open-loop frequency response* are phase and gain margins, for both discrete-time and continuous-time systems. These are easily determined from Bode plots (Fig. 3-35).

0 dB corresponds to a magnitude of 1. The gain margin is the number of decibels the magnitude of GH is below 0 dB, at the phase crossover frequency:

$$[\arg GH(\omega)] = -180° = \omega_\pi \qquad (3\text{-}90)$$

The phase margin is the number of degrees arg (GH) is above –180° at the gain crossover frequency ω_1

$$|GH(\omega)| = 1 \qquad (3\text{-}91)$$

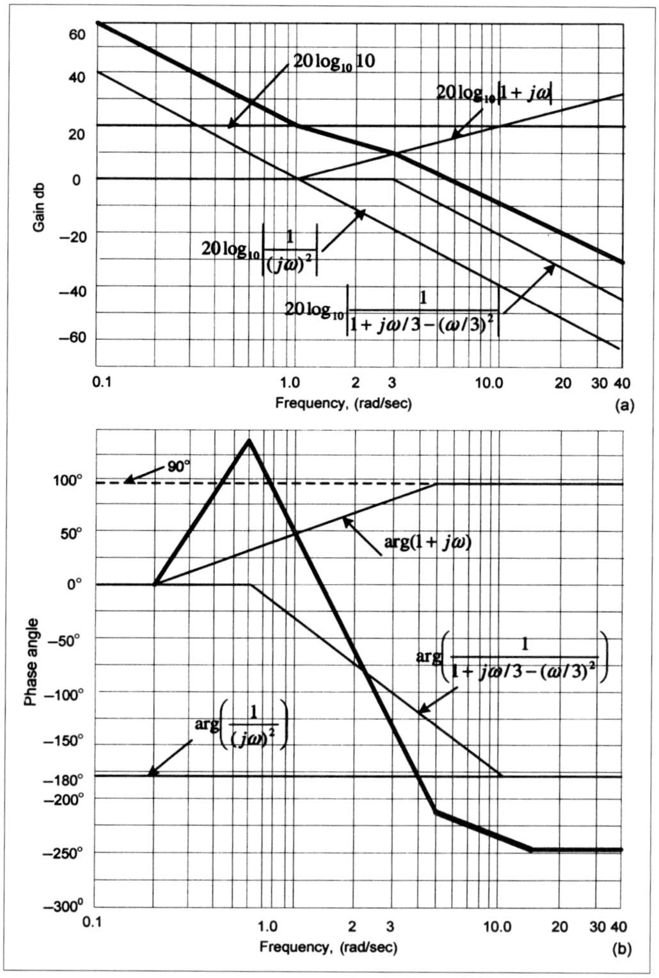

FIGURE 3-34 Example 3-20, asymptotic Bode plots.

3-15-1 Closed-Loop Frequency Response

In most cases the phase margin and the gain margin, shown in Eqs. (3-90) and (3-91) will ensure relative stability. The absolute stability can be ascertained by Routh's array or Nyquist stability plots. There is no straightforward method of plotting the closed-loop response from the Bode plots. The closed-loop frequency response can be approximated as follows:

$$|GH(\omega)| >> 1$$

$$\left.\frac{C}{R}(\omega)\right|_{|GH(\omega)|>>1} \approx \frac{G(\omega)}{GH(\omega)} = \frac{1}{H(\omega)} \quad (3\text{-}92)$$

and

$$|GH(\omega)| << 1$$

$$\left.\frac{C}{R}(\omega)\right|_{|GH(\omega)|<<1} \approx G(\omega) \quad (3\text{-}93)$$

The open-loop frequency response for most systems shows high gain, which decreases as the frequency is increased, because the poles predominate over zeros. The closed-loop frequency response for unity feedback systems is approximated by magnitude of 1 (0 dB) and phase angle of 0° for frequency below gain crossover frequency of ω_1. For higher frequencies, above ω_1, the close-loop frequency response may be approximated by magnitude and phase angle of $G(\omega)$.

Example 3-21 Consider the *given* Bode plot of the open-loop frequency response function of the form in Fig. 3-36:

$$GH(s) = \frac{K}{s(s+\omega_1)(s+\omega_2)}$$

Both magnitude and phase angle are given. Find ω_1, ω_2, K and phase margin. Adjust K to give a phase margin of 45°. ω_1 and ω_2 can be easily found from the Bode plot by observing the two break points at 10 and 100 rad/s. Therefore:

$$T_1 = 1/\omega_1 = 0.1\text{s} \qquad T_2 = 1/\omega_2 = 0.01\text{s}$$

$$GH(s) = \frac{K'}{s(0.1s+1)(0.01s+1)} = \frac{K}{s(s+10)(s+100)}$$

$$GH(j\omega) = \frac{K}{j\omega(j\omega+10)(j\omega+100)}$$

At 0.1 rad/s, the gain in dB = 60. Substituting:

$$|GH(j\omega)| = 60\text{ dB} = 1000 = \left|\frac{K}{(0.1j)(0.1j+10)(0.1j+100)}\right|$$

$$K = 10{,}0000$$

From Fig. 3-36 at 0 dB, the phase angle is –175°. Thus, the phase margin is 5°, and the system is unstable.

For a phase margin of 45°, the phase plot should cross at an angle of –135° at 0 dB. Again referring to Fig. 3-36, at –135°, our gain is 20 dB at a frequency of 10 rad/s. Thus, we can reduce our K by 20 dB, that is, by a factor of 10 ($K = 10^4$).

Example 3-22 Consider two open-loop functions in cascade represented by the Bode magnitude plot in Fig. 3-37. Find the open-loop transfer function. If the two functions are in a forward loop with unity feedback, is the system stable?

Consider the plot of the function $G_1(s)$. The break point is clearly at 50 rad/s, and if we examine the plot from 0.5 through 50 rad/s, it has a roll of 20 dB and 40 dB past 50 rad/s. Therefore, the function is:

$$G_1(s) = \frac{K_1}{s(0.2s+1)}$$

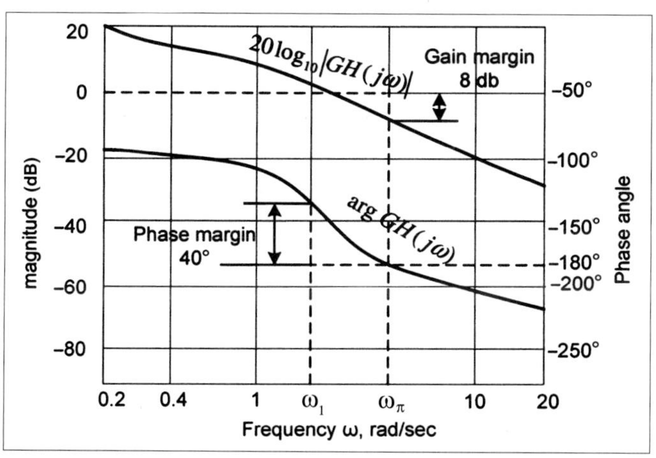

FIGURE 3-35 Relative stability, phase and gain margins, system open-loop frequency response.

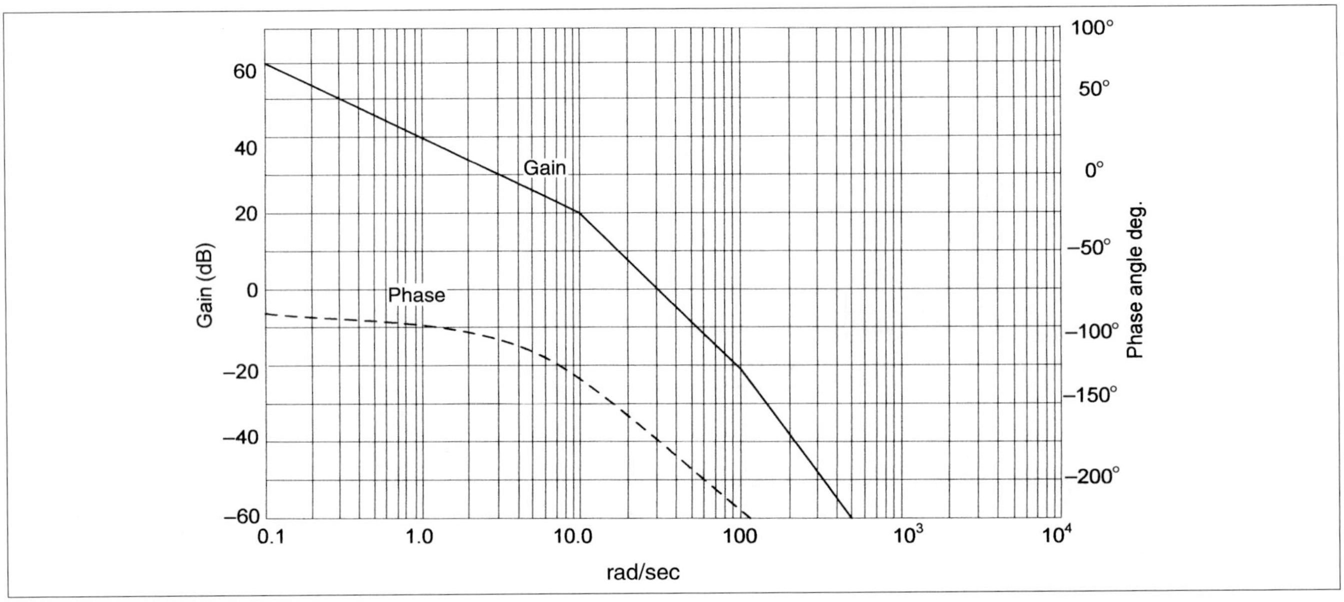

FIGURE 3-36 Example 3-21, Bode plot, magnitude and phase angle.

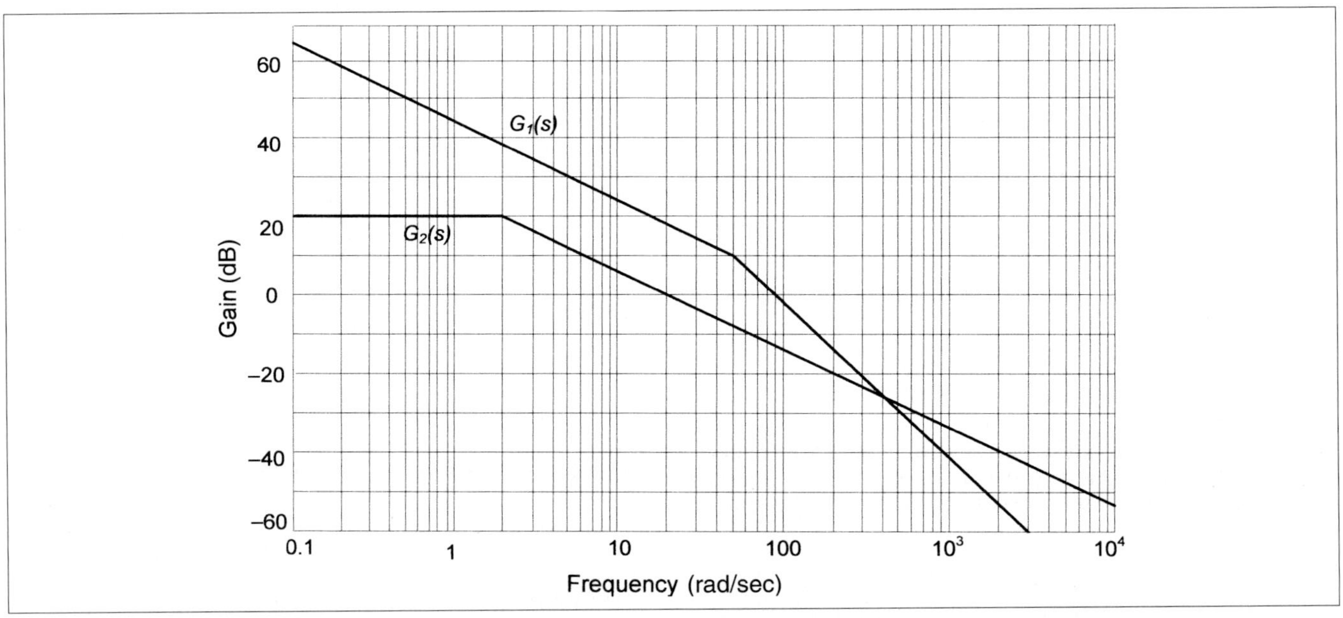

FIGURE 3-37 Example 3-22, Bode plot, gain magnitude of two functions.

At $j(0.5)$, the gain is 50 dB. Pick up a point as far away from the break point as possible. By substitution $K_1 = 158$. The function $G_2(s)$ has a single pole given by:

$$G_2(s) = \frac{K_2}{0.5s+1}$$

K_2 is obviously the dc gain when $s = 0$ and is 20 dB, $K = 10$. The open-loop transfer function is:

$$G(s) = \frac{1580}{s(0.02s+1)(0.5s+1)}$$

By examining Fig. 3-37, the Bode plot of the overall function will cross over 0 dB at 50 rad/s. Substituting the values:

$$G(j50) = 0.89 < -223°$$

It has negative phase margin at a gain of 0.893, and the system is definitely unstable.

3-16 THE NYQUIST DIAGRAM

In the Nyquist diagram, the amplitude of GH is plotted against phase angle in a polar form. The points are found by ascertaining the magnitude as well as the phase angles at each frequency.

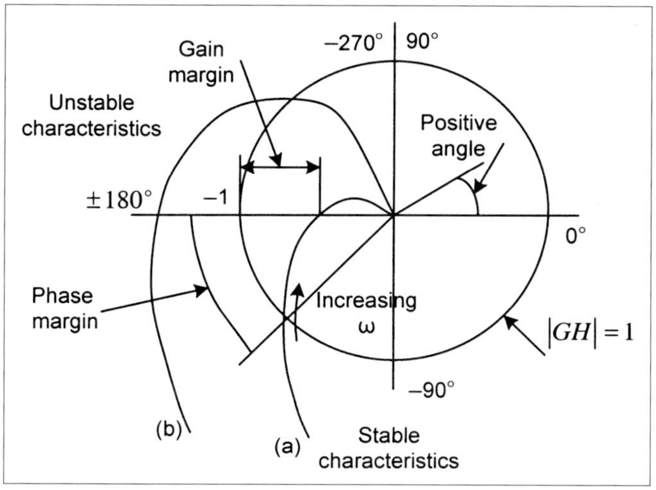

FIGURE 3-38 Nyquist diagram to illustrate stability concepts.

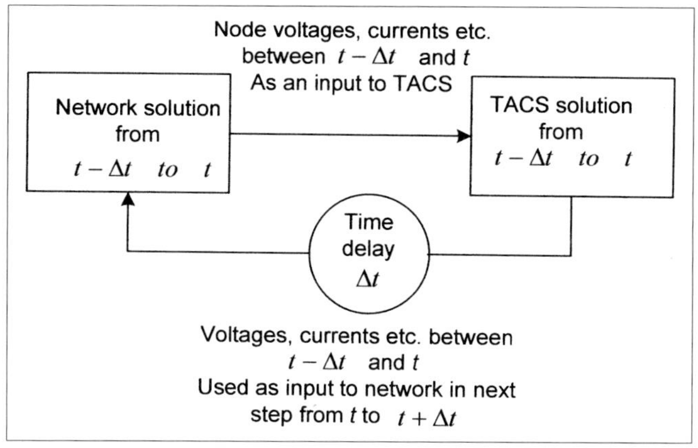

FIGURE 3-39 Data flow between power system and TACS in EMTP.

The absolute stability of a closed-loop system can be determined from the behavior of the curve near the −1 point. Figure 3-38 shows the curves plotted with unit circle which is the magnitude of GH. The system will be absolutely stable if the curve of $|GH|$ versus < GH does not encircle the −1 point on the Nyquist diagram.

The behavior of the closed-loop system will be determined mainly by the closeness of the curve to the −1 point. The gain and phase margins are available on the curve. The stable and unstable curves shown in the figure are for a single integrator. As the gain is increased the system becomes unstable.

The Nyquist diagram is more difficult to construct as compared to the Bode plot. While Bode plots and root-locus methods are common, the Nyquist diagram is occasionally used. This book does not provide a detailed analysis of the Nyquist diagram.

3-17 TACS IN EMTP

Transient analysis of control systems (TACS) was added to EMTP by L. Dubé. In 1983–1984 Ma Ren Ming made major revisions to it.[2] It was originally written to simulate operation of high-voltage DC (HVDC) converters, but soon had wider applications, for example, excitation systems of synchronous machines, arcing phenomena in circuit breakers, and surge arrester modeling.

The control system in TACS has been solved separately. Output quantities from network are used as input quantities in TACS over the same time step. Output quantities from TACS become the input quantities to the network solution *only over the next time step*. TACS accepts as input network voltages, currents, switch currents, status of switches, and internal variables, and the network solution accepts output signals from TACS (Fig. 3-39). In many cases the error is due to time delay; Δt may not be of significance, but not always. With continuous voltage and current source functions and fast varying data coming out of TACS, the time delay can become critical and can cause numerical instability.

The solution can be straightforward, if all devices in the control system are linear. Consider a transfer function of the form:

$$\frac{K(N_0+N_1s+\cdots+N_ms^m)}{D_0+D_1s+\cdots D_ns^n};\ m<n \tag{3-94}$$

Input is $u(s)$ and output $x(s)$. In changing nth-order differential equation to n differential equations of first order, we have an algebraic equation:

$$D_0x+D_1x_1+\cdots+D_nx_n=K(N_0u+N_1u_1+\cdots+N_mu_m) \tag{3-95}$$

The differential equations can be converted into difference equations with trapezoidal rule of integration (see App. G):

$$x_i(t)=\frac{2}{\Delta t}x_{i-1}(t)-\left\{x_i(t-\Delta t)+\frac{2}{\Delta t}x_{i-1}(t-\Delta t)\right\}\quad i=1,\ldots,n \tag{3-96}$$

$$u_j(t)=\frac{2}{\Delta t}u_{j-1}(t)-\left\{u_j(t-\Delta t)+\frac{2}{\Delta t}u_{j-1}(t-\Delta t)\right\}\quad j=1,\ldots,n \tag{3-97}$$

Expressing x_n as a function of x_{n-1} and then expressing x_{n-1} as a function of x_{n-2} and so on, and then using the same procedure for u, we have a single output-input relation:

$$cx(t)=Kdu(t)+\text{hist}(t-\Delta t) \tag{3-98}$$

where "hist" is the history term after the solution at each time step, n history terms must be updated to obtain single term "hist" for the next time step. If recursive formulas are used:

$$\text{hist}_n(t)=Kd_nu(t)-c_nx(t) \tag{3-99}$$

Equation (3-98) is used in the transient solution of the control system. Fast varying components may create instabilities and end up in wrong operating region. More discussion on improvements of TACS is available in Refs. 3 to 6.

A new approach for eliminating delays in the solution of control systems is presented in Ref. 7. The complete system is formulated using a jacobian matrix and solved through an iterative process. The iterative methods, noniterative method, and ordering issues are discussed with some test results.

PROBLEMS

1. A unity feedback control system has a forward transfer function:

$$G=\frac{20}{s(s+10)}$$

A tachometer feedback is applied, with a transfer function ks. Determine k for a damping ratio of 0.707.

2. A control system has the following open-loop transfer function:

$$GH = \frac{10^6}{(s^2+10s+100)(s+100)}$$

Is the system stable? Add a lag compensator of the form $1/(1+Cs)$, and select C to provide 45° phase margin.

3. A closed-loop control transfer function is given by:

$$\frac{K}{s^3+2s^2+4s+K}$$

For positive values of K, sketch the root locus and find (a) breakaway points on real axis, (b) frequency and gain at any point where instability begins, and (c) complex frequency and gain for a damping ratio of 0.707.

4. The characteristic equation of system is given by:

$$s^4+5s^3+10s^2+6s+K=0$$

Find the value of K so that the system is stable.

5. Use Hurwitz stability criteria to find out if the control system given by the following equation is stable or unstable:

$$s^3+10s^2+13s+20=0$$

6. Find if the discrete system given by the following equation is stable:

$$z^4+2z^3+4z^2+z+1=0$$

7. Find the transfer function of the differential equation represented by the following equation:

$$\frac{d^2y}{dt^2}+6\frac{dy}{dt}+5y=2u+\frac{du}{dt}$$

8. Find the transfer function of the electrical network shown in Fig. 3-P1*a* and *b*

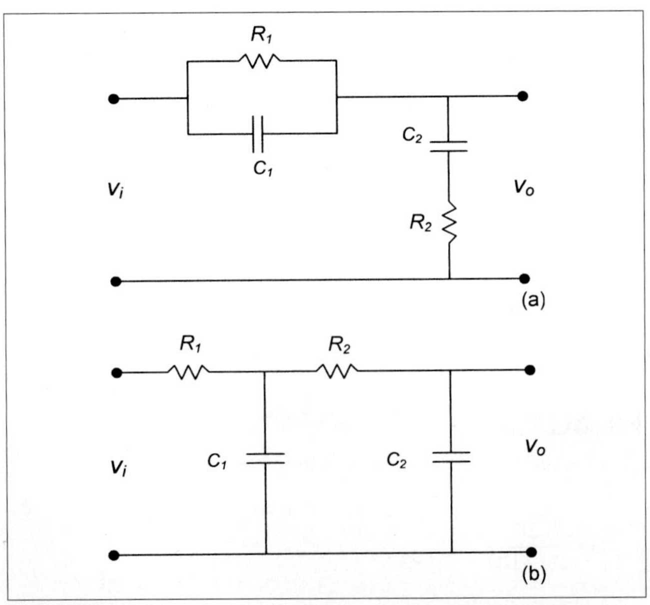

FIGURE 3-P1 (*a*) and (*b*). Calculation of transfer functions, Prob. 8.

9. Reduce the block diagram of Fig. 3-P2, and eliminate feedback loop H_2 and the summing junction 1.

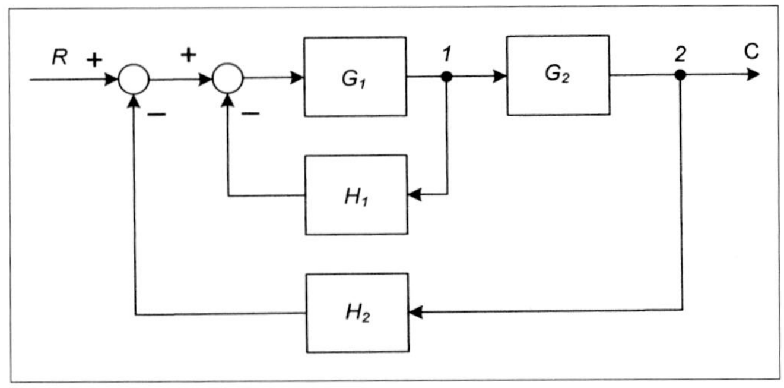

FIGURE 3-P2 Block diagram of a control system, Prob. 9.

10. Consider the signal-flow graph as shown in Fig. 3-P3. Find forward paths, feedback paths, nontouching loops, touching loops, loop gain of feedback loops, and path gains of the forward loops.

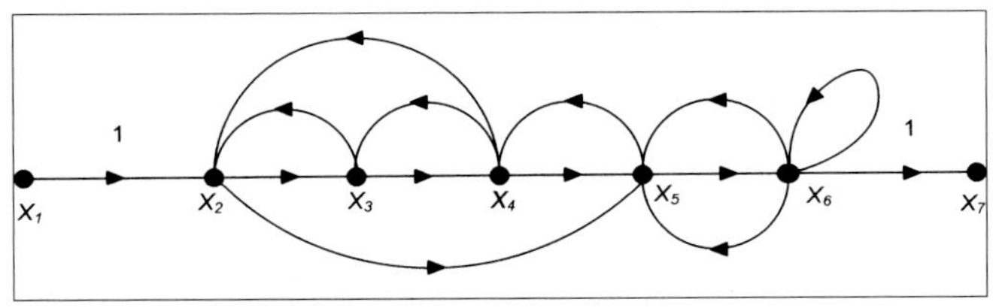

FIGURE 3-P3 Signal-flow graph, Prob. 10.

11. Draw a signal-flow graph for the electrical network shown in Fig. 3-P4.

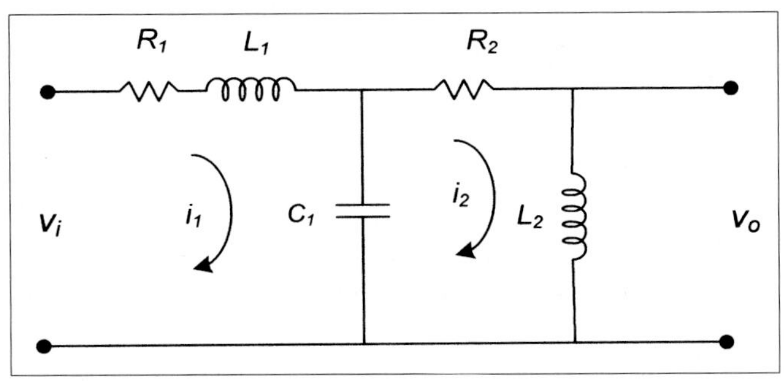

FIGURE 3-P4 Circuit for drawing signal-flow graph, Prob. 11.

12. The system defined by the Fig. 3-P5 is a (1) type-one system, (2) type-two system (3) type-zero system? Describe the position, velocity, and acceleration errors of type-zero, type-one and type-two systems. What is the major problem of type-three systems?

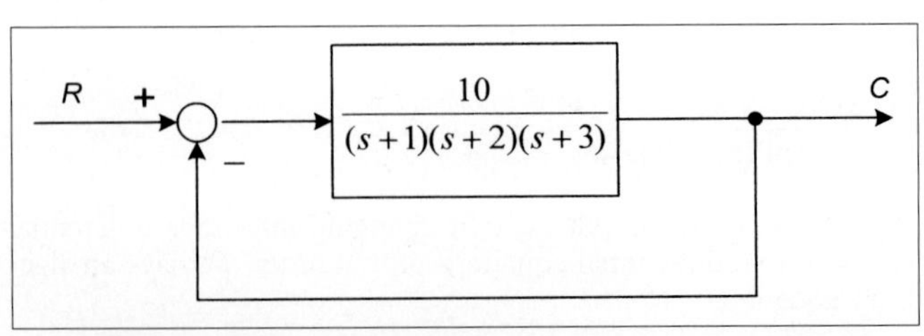

FIGURE 3-P5 Block diagram of a control system, Prob. 12.

13. Find the breakaway point for:

$$GH = \frac{K(s+1)}{(s+1+j\sqrt{2})(s+1-j\sqrt{2})}$$

and draw the root locus.

14. Construct the root locus for:

$$\frac{K}{(s+2)(s+2-j)(s+2+j)} \qquad K>0$$

15. Construct the Bode plot for the open-loop frequency response function:

$$GH(j\omega) = \frac{1+0.33j\omega-(0.33\omega)^2}{j\omega(1+0.25j\omega)(1+0.5j\omega)}$$

16. Obtain the state diagram for the following equations:

$$\ddot{y}(t)+3\dot{y}(t)+5y(t)=r(t)$$

$$y(k+2)+ay(k+1)+by(k)=r(k)$$

17. Consider the system shown in Fig. 3-P6. Find a compensator so that closed-loop transfer function has damping ratio of 0.707 and natural frequency of 0.15.

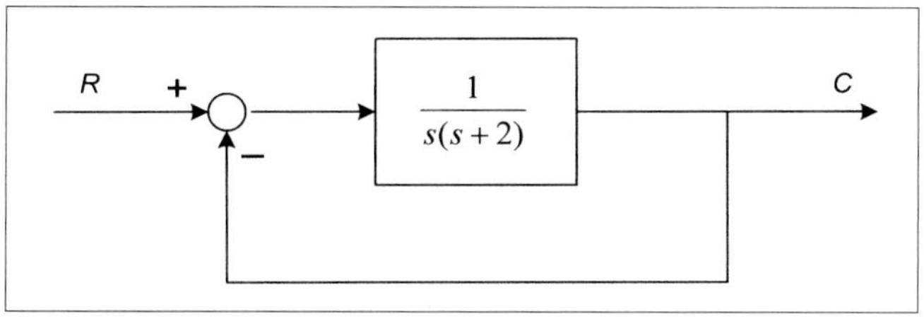

FIGURE 3-P6 Control circuit for designing a compensator, Prob. 17.

18. Draw the state diagram of a system represented by the following differential equation:

$$a\frac{d^2y}{dt^2}+b\frac{dy}{dt}+cy=u(t)$$

REFERENCES

1. G. Newton, L. Gould, and J. Kaiser, *Analytical Design of Feedback Control*, John Wiley, New York, 1957.
2. M. Ren Ming, "The Challenge of Better EMTP-TACS Variable Ordering," *EMTP Newsletter*, vol. 4, no. 4, pp.1–6, Aug. 1984.
3. R. H. Lasseter and J. Zhou, "TACS Enhancements for the Electromagnetic Transient Program," *IEEE Trans. PS*, vol. 9, no. 2, pp. 736–742, May 1994.
4. S. Lefebvre and J. Mahseredjian, "Improved Control System Simulation in the EMTP Through Compensation," *IEEE Trans. PS*, vol. 9, no. 3, pp. 1654–1662, July 1994.
5. X. Cao, A. Kurita, T. Yamanaka, Y. Tada, and H. Mitsuma, "Supression of Numerical Oscillation Caused by the EMTP-TACS Interface using Filter Incorporation," *IEEE Trans. PD*, vol. 11, no. 4, pp. 2049–2055, Oct. 1996.
6. A. E. A. Araujo, H. W. Dommel, and J. R. Marti, "Simultaneous Solution of Power and Control System Equations," *IEEE Trans. PS*, vol. 8, no. 4, pp. 1483–1489, Nov. 1993.
7. J. Mahseredjian, L. Dube, M. Zou, S. Dennetiere, and G. Joos, "Simultaneous Solution of Control System Equations in EMTP," *International Conference on Power System Transients*, Montreal, June 2005.

FURTHER READING

B. K. Bose, *Power Electronics and Variable Frequency Drives*, IEEE Press, Piscataway, NJ, 1997.

J. A. Cadzow, *Discrete Time Systems*, Prentice Hall, Englewood Cliffs, NJ, 1973.

C., Chi-Tsing, *Linear System Theory and Design*, Holt, Rinehart, and Winston, New York, 1984.

R. C. Dorf, *Electrical Engineering Handbook*, 2d ed., CRC Press, Boca Raton, FL, 1998.

R. C. Dorf and R. H. Bishop, *Modern Control Systems*, 8th ed., Addison-Wesley, Berkeley, CA, 1999.

W. R. Evans, *Control System Dynamics*, McGraw Hill, New York, 1954.

M. M. Gupta, *Intelligent Control*, IEEE Press, Piscataway, NJ, 1995.

R. R. Kadiyala, "A Tool Box for Approximate Linearization of Non-Linear Systems," *IEEE Control Systems*, pp. 47–56, April 1993.

E. W. Kamen and B. S. Heck, *Fundamentals of Signals and Systems Using MALAB*, Prentice Hall, Englewood Cliffs, NJ, 1997.

B. C. Kuo. *Automatic Control Systems*, 5th ed., Prentice Hall, Englewood Cliffs, NJ, 1996.

W. S. Levine. *The Control Handbook*, CRC Press, Boca Raton, FL, 1996.

G. Norris, "Boeing Seventh Wonder," *IEEE Spectrum*, pp. 20–23, October 1995.

CHAPTER 4

MODELING OF TRANSMISSION LINES AND CABLES FOR TRANSIENT STUDIES

Electrical power systems can be classified into generation, transmission, sub-transmission, and distribution systems. Individual power systems are organized in the form of electrically connected areas of *regional* grids, which are interconnected through transmission lines to form *national* grids and also international grids. Each area is interconnected to another area in terms of some contracted parameters like generation and scheduling and tie line power flow, contingency operations. The irreplaceable sources of power generation are petroleum, natural gas, oil and nuclear fuels. The fission of heavy atomic weight elements like uranium and thorium and fusion of lightweight elements, such as deuterium, offer almost limitless reserves. Replaceable sources are elevated water, pumped storage systems, solar, geothermal, wind and fuel cells, which in recent times have received much attention. Single shaft steam units of 1500 MW are in operation, and superconducting single units of 5000 MW or more are a possibility. On the other hand, dispersed generating units integrated with grid may produce only a few kilowatts of power.

The generation voltage is a low 13.8–25 kV, the transmission voltages have risen to 765 kV or higher, and many HVDC links around the world are in operation. While, maintaining acceptable voltage profile and load frequency control is one issue, the transient phenomena on transmission lines due to lightning and switching and mitigation of these transients through shielding, surge arresters, and control of switching overvoltages are equally important and covered in this book. Synchronous condensers, shunt capacitors, static var compensators, and FACTS devices are employed to improve power system stability and enhance power handling capability of transmission lines and also impact the transient behavior. The sub-transmission voltage levels are 23 kV to approximately 69 kV, though for large industrial consumers voltages of 230 kV and 138 kV are also used. Sub-transmission systems connect high voltage substations to local distribution substations. The voltage is further educed to 12.47 kV, and several distribution lines and cables emanate from distribution stations. The refinements in modeling and analytical techniques continue, as we will discuss in this chapter.

The modeling of transmission lines for transient analysis should be based on the frequency. For load flow and power transmission, the models are based on the length of the line, though the same load flow models (in modified form) are used for the transient studies. Conventionally, lines exceeding approximately 150 mi are considered long lines and we consider wave propagation, that is, the voltage and currents propagate on a conductor with finite velocity, which are reflected or refracted at the impedance discontinuities. As discussed in Chap. 2, partial differential equations are used to derive the wave equations, the capacitance of the transmission lines playing an important role. For short lines of less than 50 mi, the line is modeled based on conductor series resistance and inductance only. For the transmission lines in the range of 50 to 150 mi, there are two models in use, the nominal Π and nominal T-circuit models. In the T-circuit model, the shunt admittance is connected at the midpoint of the line, while in the nominal Π model, it is equally divided at the sending end and the receiving end.

Table 4-1 shows the models with respect to frequency and transmission line parameters. This table is based on CIGRE guidelines.[1] Depending on the nature of specific transient study, additional details of shielding, tower footing resistance, towers themselves, and soil resistivity data will be required to derive an appropriate model. The length of the line modeled depends on the frequency involved. For high-frequency studies, it will be appropriate to model a couple of spans of the transmission line, and for low-frequency transients, the whole line length may be included in the studies. For VFT, even for short lines the load flow model is not proper, and distributed parameter model should be used (Table 4-1).

While deriving appropriate line models in EMTP-like programs, more data of the lines will be required than are available in the line constant tabulations in most handbooks. Appendix D is devoted to the estimation of line constants, and the reader may like to study it before proceeding with this chapter.

We will revert to the models for the transient studies in a later section of this chapter.

4-1 *ABCD* PARAMETERS

A transmission line of any length can be represented by a four-terminal network, Fig. 4-1*a*. In terms of *A, B, C*, and *D* parameters, the relation

TABLE 4-1 Modeling of Transmission Lines

Topic	Low Frequency Transients Group I	Slow Front Transients Group II	Fast Front Transients Group III	Very Fast Front Transients Group IV
Representation of transposed lines	Lumped parameters; multiphase Π circuits	Distributed parameter; multiphase model; multiphase Π circuits possible	Distributed parameter; multiphase model	Distributed parameter; single-phase model
Line asymmetry	Important	Capacitive and inductive asymmetry important; for statistical study inductive asymmetry not important	Negligible for single-phase simulations, otherwise important	Negligible
Frequency-dependent parameters	Important	Important	Important	Important
Corona effect	Important if phase conductor voltage can exceed the corona inception voltage	Negligible	Very important	Important

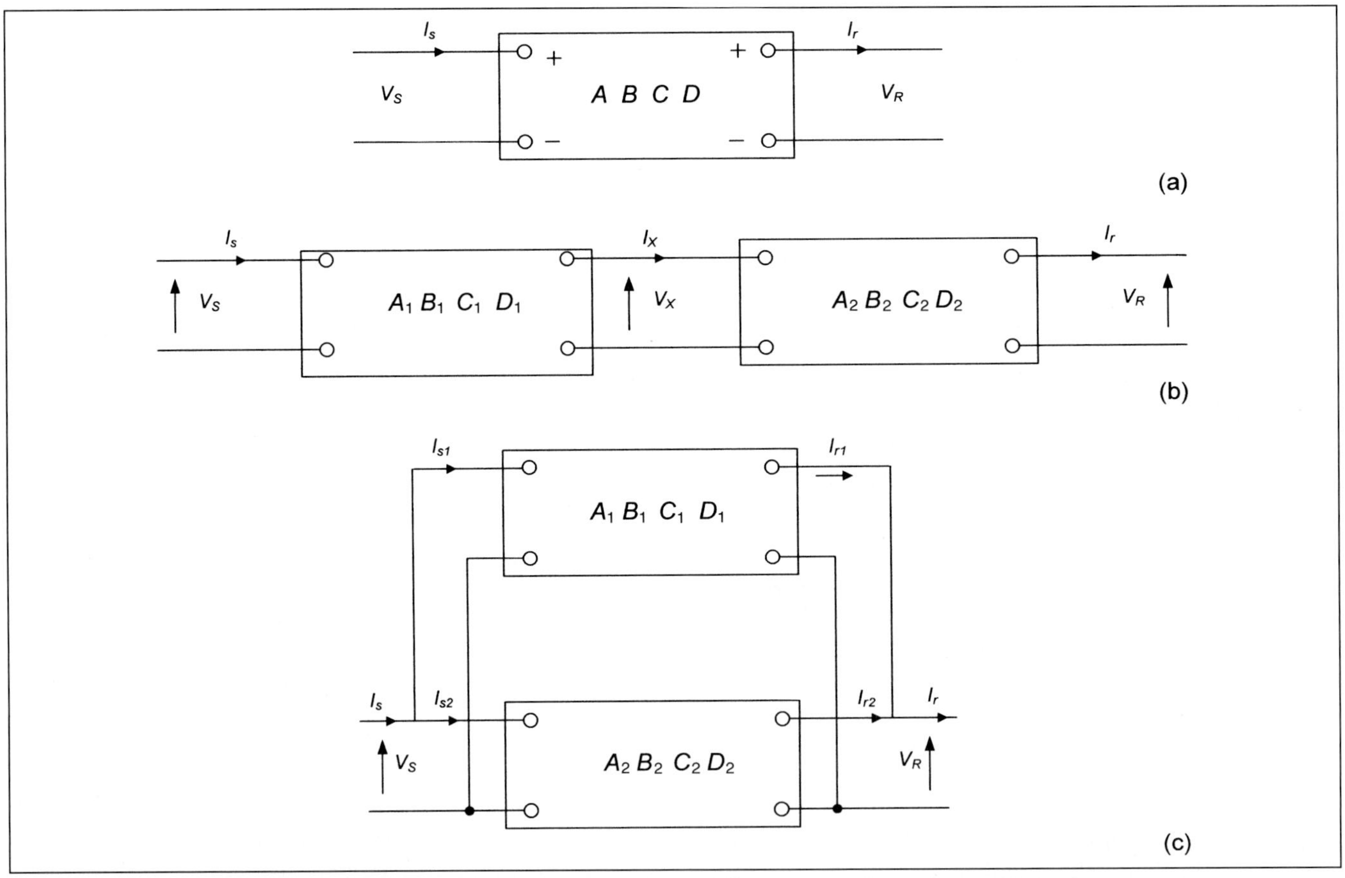

FIGURE 4-1 (*a*) Schematic representation of a two-terminal network using ABCD constants; (*b*) two networks in series; (*c*) two networks in parallel.

between sending- and receiving-end voltages and currents can be expressed as:

$$\begin{vmatrix} V_s \\ I_s \end{vmatrix} = \begin{vmatrix} A & B \\ C & D \end{vmatrix} \begin{vmatrix} V_r \\ I_r \end{vmatrix} \tag{4-1}$$

where V_s and I_s are the sending-end voltage and current and V_r and I_r are the receiving-end voltage and current, respectively.

In case sending-end voltages and currents are known, the receiving-end voltage and current can be found by:

$$\begin{vmatrix} V_r \\ I_r \end{vmatrix} = \begin{vmatrix} D & -B \\ -C & A \end{vmatrix} \begin{vmatrix} V_s \\ I_s \end{vmatrix} \tag{4-2}$$

The significance of *A*, *B*, *C*, and *D* parameters can be stated as follows:

$A = V_s/V_r$, when I_r is 0, that is, the receiving end is open-circuited. It is the ratio of two voltages and is, thus, dimensionless.

$B = V_r/I_r$, when V_r is 0, that is, the receiving end is short-circuited. It has the dimensions of impedance and is specified in ohms.

$C = I_s/V_r$, when the receiving end is open-circuited and I_r is 0. It has the dimensions of admittance.

$D = I_s/I_r$, when V_r is 0, that is, the receiving end is short-circuited. It is the ratio of two currents and is, thus, dimensionless.

Two *ABCD* networks in series (Fig. 4-1*b*) can be reduced to a single equivalent network as follows:

$$\begin{vmatrix} V_s \\ I_s \end{vmatrix} = \begin{vmatrix} A_1 & B_1 \\ C_1 & D_1 \end{vmatrix} \begin{vmatrix} A_2 & B_2 \\ C_2 & D_2 \end{vmatrix} \begin{vmatrix} V_r \\ I_r \end{vmatrix} = \begin{vmatrix} A_1A_2+B_1C_2 & A_1B_2+B_1D_2 \\ C_1A_2+D_1C_2 & C_1B_2+D_1D_2 \end{vmatrix} \begin{vmatrix} V_r \\ I_r \end{vmatrix} \tag{4-3}$$

For parallel *ABCD* networks (Fig. 4-1*c*) the combined *ABCD* parameters are:

$$\begin{aligned} A &= (A_1B_2 + A_2B_1)/(B_1+B_2) \\ B &= (B_1B_2)/(B_1+B_2) \\ C &= (C_1+C_2) \\ D &= (B_2D_1 + B_1D_2)/(B_1+B_2) \end{aligned} \tag{4-4}$$

4-2 *ABCD* PARAMETERS OF TRANSMISSION LINE MODELS

Table 4-2 gives the *ABCD* parameters of transmission line models—T or Π representation parameters and distributed parameters model. For the medium long lines, the Π representation is more commonly used, compared to the T representation. A reader can derive T and Π models by considering the terminal current and voltage conditions, as shown in Fig. 4-2*a*, *b*, *c*, and *d*.

4-3 LONG TRANSMISSION LINE MODEL-WAVE EQUATION

Lumping together the shunt admittance of the lines at sending and receiving ends is an approximation, and for line lengths over 150 mi (240 km), distributed parameter representation of a line is used. Each elemental section of line has series impedance and shunt admittance associated with it. The operation of a long line can be examined by considering an elemental section of impedance z per unit length, $z = r_{sc} + jx_{sc} = r_{sc} + j\omega L$ and admittance y per unit length. Here, we have denoted r_{sc} and x_{sc} as the series resistance and reactance of the elemental section. The impedance for the elemental section of length dx is $z\,dx$ and the admittance is $y\,dx$, $Z = zl$, where l is the length of the line and $Y = yl$. Referring to Fig. 4-3, by Kirchoff's voltage law, the voltage drop in the elemental length dx due to resistance and inductance is:

$$\begin{aligned} dV &= \frac{\partial V}{\partial x}dx = ir_{sc}dx + L\frac{\partial}{\partial t}(idx) \\ &= \left(r_{sc} + L\frac{\partial}{\partial t}\right)idx \end{aligned} \tag{4-5}$$

If we ignore the series resistance, the equation can be written as:

$$\frac{\partial V}{\partial x} = L\frac{\partial i}{\partial t} \tag{4-6}$$

The change in the flux with respect to time produces a voltage along the line given by the above expression. The term $L\,\partial i/\partial t$ by definition is equal to the rate of change in the flux.

Define the following elements, to distinguish from line series elements:

g_{sh} = shunt conductance, b_{sh} = shunt susceptance, x_{sh} = shunt capacitive reactance, and C = capacitance per unit length. Shunt conductance is small and ignoring it gives $y = j/x_{sh}$.

The shunt current through the conductance and capacitance is:

$$\begin{aligned} \partial I &= \frac{\partial I}{\partial x}dx = Vg_{sh}dx + \frac{\partial}{\partial t}(\Delta\varphi) \\ &= Vg_{sh}dx + \frac{\partial}{\partial t}(VCdx) \\ &= \left(g_{sh} + C\frac{\partial}{\partial t}\right)Vdx \end{aligned} \tag{4-7}$$

where $\Delta\varphi$ is the change in the electrostatic flux between the conductors. Ignoring shunt conductance, this equation can be written as:

$$\frac{\partial i}{\partial t} = C\frac{\partial V}{\partial t} \tag{4-8}$$

A change of current with time produces a change of voltage with position along the line, and a change of voltage with time produces a change of current with position along the line.

Taking Laplace transform with respect to time variable t, the equations can be written as:

$$\begin{aligned} \frac{\partial V}{\partial x} &= (r_{sc} + Ls)\,i = zi \\ \frac{\partial i}{\partial x} &= (g_{sh} + Cs)\,i = yi \end{aligned} \tag{4-9}$$

TABLE 4-2 *ABCD* Parameters

Line Length	Equivalent Circuit	*A*	*B*	*C*	*D*
Short	Series impedance only	1	Z	0	1
Medium	Nominal Π	$1+\frac{1}{2}YZ$	Z	$Z\left[1+\frac{1}{4}YZ\right]$	$1+\frac{1}{2}YZ$
Medium	Nominal T	$1+\frac{1}{2}YZ$	$Z\left[1+\frac{1}{4}YZ\right]$	Y	$1+\frac{1}{2}YZ$
Long	Distributed parameters	$\cosh\gamma l$	$Z_0\sinh\gamma l$	$\frac{\sinh\gamma l}{Z_0}$	$\cosh\gamma l$

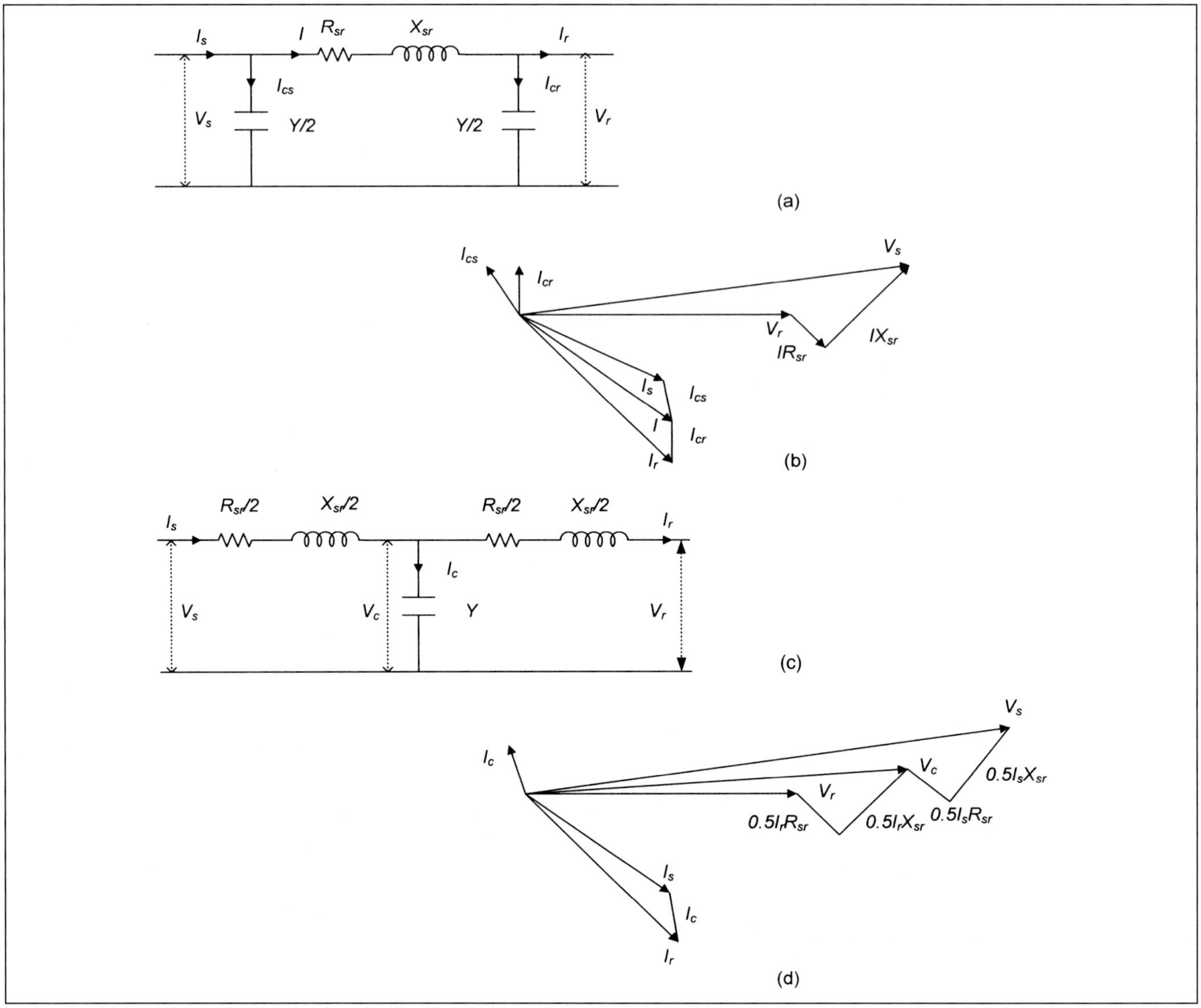

FIGURE 4-2 (*a*) Π representation of a transmission line; (*b*) phasor diagram of the Π circuit; (*c*) T-representation of a transmission line; (*d*) phasor diagram of T-circuit.

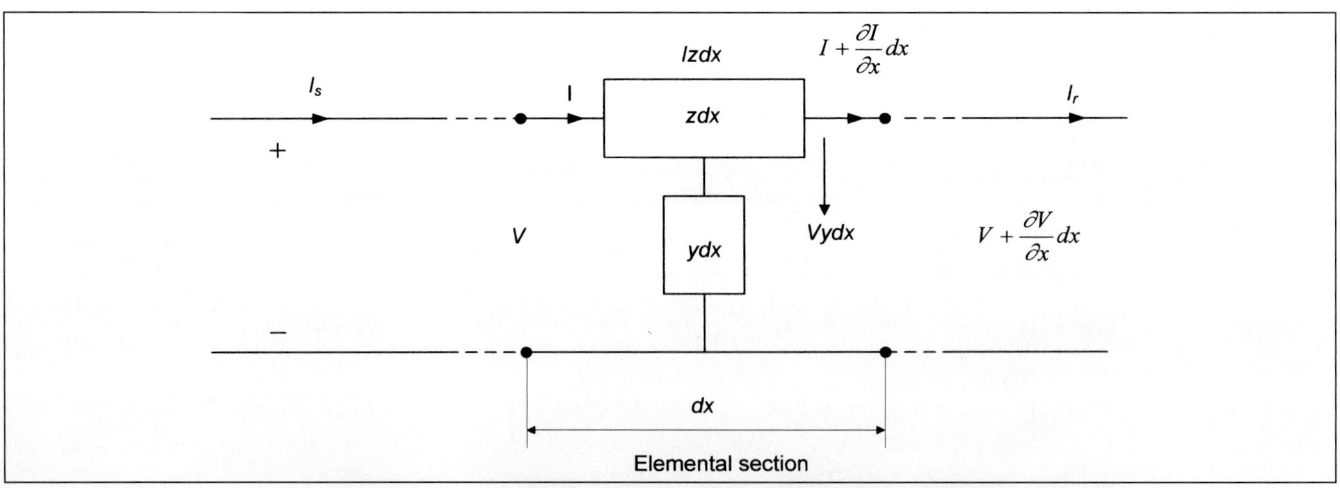

FIGURE 4-3 Modal of a small section of a long transmission line.

Differentiating these equations gives:

$$\frac{\partial^2 V}{\partial x^2} = yzV$$
$$\frac{\partial^2 I}{\partial x^2} = yzI \qquad (4\text{-}10)$$

These differential equations have solutions of the form:

$$V = V_1 e^{\gamma x} + V_2 e^{-\gamma x} \qquad (4\text{-}11)$$

Or in a more general form:

$$f_1(t)e^{\gamma x} + f_2(t)e^{-\gamma x} \qquad (4\text{-}12)$$

where γ is the *propagation constant* and is defined as:

$$\begin{aligned}\gamma &= \sqrt{yz}\\ &= [r_{sc}g_{sh} + (r_{sc}C + Lg_{sh})s + LCs^2]^{1/2}\\ &= [(r_{sc} + Ls)(g_{sh} + Cs)]^{1/2}\\ &= \frac{1}{\sqrt{LC}}\left(s + \frac{r_{sc}}{L}\right)^{1/2}\left(s + \frac{g_{sh}}{C}\right)^{1/2}\end{aligned} \qquad (4\text{-}13)$$

If we ignore the shunt conductance:

$$\begin{aligned}y &= jb_{sh}\\ yz &= -b_{sh}x_{sc} + jr_{sc}b_{sh}\\ \gamma &= \sqrt{b_{sh}}(r_{sc}^2 + x_{sc}^2)^{1/4}\end{aligned} \qquad (4\text{-}14)$$

The complex propagation constant can be written as:

$$\begin{aligned}\gamma &= \alpha + j\beta\\ &= \frac{1}{2}\left(\frac{r_{sc}}{L} + \frac{g_{sh}}{C}\right) + j\frac{1}{2}\left(\frac{r_{sc}}{L} - \frac{g_{sh}}{C}\right)\end{aligned} \qquad (4\text{-}15)$$

where α is defined as *attenuation constant*. Common units are nepers per mile or per kilometer. β is the *phase constant*. Common units are radians per mile. Again ignoring shunt conductance, the equations are simplified to the following form:

$$\begin{aligned}\alpha &= |\gamma|\cos\left[\frac{1}{2}\tan^{-1}\left(\frac{-r_{sc}}{x_{sc}}\right)\right]\\ \beta &= |\gamma|\sin\left[\frac{1}{2}\tan^{-1}\left(\frac{-r_{sc}}{x_{sc}}\right)\right]\end{aligned} \qquad (4\text{-}16)$$

The expressions for α and β can be written as:

$$\begin{aligned}\alpha &= \sqrt{\frac{r_{sc}g_{sh} - \omega^2 LC + \sqrt{(r_{sc}g_{sh} - \omega^2 LC)^2 + \omega^2(Lg_{sh} + Cr_{sc})^2}}{2}}\\ \beta &= \sqrt{\frac{-r_{sc}g_{sh} + \omega^2 LC + \sqrt{(r_{sc}g_{sh} - \omega^2 LC)^2 + \omega^2(Lg_{sh} + Cr_{sc})^2}}{2}}\end{aligned} \qquad (4\text{-}17)$$

The *characteristic impedance*, also called surge impedance, is:

$$\begin{aligned}Z_0 &= \sqrt{\frac{z}{y}}\\ &= \sqrt{\frac{L}{C}}\left[\left(\frac{s + \alpha + \beta}{s + \alpha - \beta}\right)\right]^{1/2}\end{aligned} \qquad (4\text{-}18)$$

Again neglecting g_{sh}:

$$\begin{aligned}\frac{z}{y} &= \frac{r_{sc} + jx_{sc}}{jb_{sh}}\\ Z_0 &= \sqrt{x_{sh}}(r_{sc}^2 + x_{sc}^2)^{1/4}\\ < Z_0 &= \frac{1}{2}\tan^{-1}\left(\frac{-r_{sc}}{x_{sc}}\right)\end{aligned} \qquad (4\text{-}19)$$

If we neglect the line series resistance also:

$$Z_0 = \sqrt{\frac{L}{C}} \qquad (4\text{-}20)$$

Equation (4-20) is of much practical importance in the transient behavior of the transmission lines.

The voltage at any distance x can be written as:

$$V_x = \left|\frac{V_r + Z_0 I_r}{2}\right| e^{\alpha x + j\beta x} + \left|\frac{V_r + Z_0 I_r}{2}\right| e^{-\alpha x - j\beta x} \qquad (4\text{-}21)$$

These equations represent traveling waves. The solution consists of two terms, each of which is a function of two variables, time and distance. At any instant of time the first term, the incident wave, is distributed sinusoidally along the line, with amplitude increasing exponentially from the receiving end. After a time interval Δt, the distribution advances in phase by $\omega\Delta t/\beta$, and the wave is traveling toward the receiving end. The second term is the reflected wave, and after time interval Δt, the distribution retards in distance phase by $\omega\Delta t/\beta$, the wave traveling from the receiving end to the sending end.

Similar explanation holds true for the current. It is given by:

$$I_x = \left|\frac{V_r/Z_0 + I_r}{2}\right| e^{\alpha x + j\beta x} - \left|\frac{V_r/Z_0 - I_r}{2}\right| e^{-\alpha x - j\beta x} \qquad (4\text{-}22)$$

These equations can be written as:

$$\begin{aligned}V_x &= V_r\left(\frac{e^{\gamma x} + e^{-\gamma x}}{2}\right) + I_r Z_0\left(\frac{e^{\gamma x} - e^{-\gamma x}}{2}\right)\\ I_x &= \frac{V_r}{Z_0}\left(\frac{e^{\gamma x} - e^{-\gamma x}}{2}\right) + I_r\left(\frac{e^{\gamma x} + e^{-\gamma x}}{2}\right)\end{aligned} \qquad (4\text{-}23)$$

Or in matrix form as:

$$\begin{vmatrix} V_s \\ I_s \end{vmatrix} = \begin{vmatrix} \cosh\gamma l & Z_0 \sinh\gamma l \\ \dfrac{1}{Z_0}\sinh\gamma l & \cosh\gamma l \end{vmatrix} \qquad (4\text{-}24)$$

This result is also shown in Table 4-2 for the ABCD constants of the distributed parameter lines.

Wavelength A complete voltage or current cycle along the line, corresponding to a change of 2π radians in angular argument of β is defined as the wavelength λ. If β is expressed in radians per mile or radians per meter:

$$\lambda = \frac{2\pi}{\beta} \qquad (4\text{-}25)$$

For a lossless line, that is, a line with zero series resistance and shunt conductance:

$$\beta = \omega\sqrt{LC} \qquad (4\text{-}26)$$

Therefore:

$$\lambda = \frac{1}{f\sqrt{LC}} \tag{4-27}$$

and the velocity of propagation of the wave:

$$v = f\lambda = \frac{1}{\sqrt{LC}} \tag{4-28}$$

Substituting the values of L and C for a two-conductor line from App. D:

$$v = \frac{1}{\sqrt{\mu_0 k_0 \left[\frac{1}{4\ln(d/a)} + 1\right]}}$$

$$\approx \frac{1}{\sqrt{\mu_0 k_0}} \tag{4-29}$$

where μ_0 is $4\pi \times 10^{-7}$ and k_0 is 8.854×10^{-12}. Therefore, $1/(\sqrt{\mu_0 k_0})$ = 3×10^{10} cm/s or 186,000 mi/s is velocity of light.

The actual velocity of the propagation of wave along the line is somewhat less than the speed of light because of reactance, resistance, and leakage of the line. The value of attenuation constant is determined from Eq. (4-16) and is a function of the frequency. Thus, all the frequencies will not be attenuated equally. Also, phase constant is a function of frequency, and as the velocity of propagation is inversely proportional to phase constant and directly to ω, which also involves frequency, the velocity will be some function of frequency. All frequencies applied to the transmission line will not have the same time of transmission, some frequencies being delayed more than the others, giving rise to phase distortion. This is an important concept.

4-4 REFLECTION AND TRANSMISSION AT TRANSITION POINTS

When there is a change in the parameters of a transmission line, that is, open circuit, short circuit, terminations through a cable, or other impedances, the traveling wave goes through a transition—part of the wave is reflected and sent back and only part of the wave is transmitted.

Consider a general junction point, as shown in Fig. 4-4, with n lines originating from the junction point and also ground impedance connected to the junction point. An example of ground impedance is a capacitor bank. The n lines have surge impedances $Z_1 \ldots Zn$, and impedances $Z_1(s)$, $Z_2(s)$, ... are lumped impedances expressed in Laplace transform. Let V_1, I_1 be the voltage and current related to incident wave, V_1', I_1' related to reflected wave, and V_k'', I_k'' related to transmitted wave. The total impedance of the line is:

$$Z_t(s) = Z_1(s) + \frac{1}{\left[\frac{1}{Z_g(s)} + \sum_{k=2}^{k=n} \frac{1}{(Z_k(s) + Z_k)}\right]} \tag{4-30}$$

We can write the following relations from Kirchoff's laws:

$$\begin{aligned} V_1 &= Z_1 I_1 \\ V_1' &= -Z_1 I_1' \\ V_k'' &= Z_k I_k'' \end{aligned} \tag{4-31}$$

The total voltage and current at the transition point t are:

$$V_t = V_1 + V_1' = Z_t(s) I_t = Z_t(s)(I_1 + I_2) \tag{4-32}$$

Thus, the following relations can be written:

$$\begin{aligned} V_1' &= \frac{Z_t(s) - Z_1}{Z_t(s) + Z_1} V_1 \\ V_t &= \frac{2Z_t(s)}{Z_t(s) + Z_1} V_1 \\ I_1' &= \frac{Z_t(s) - Z_1}{Z_t(s) + Z_1} I_1 \\ I_t &= \frac{2Z_t(s)}{Z_t(s) + Z_1} I_1 \end{aligned} \tag{4-33}$$

The ground current is:

$$I_g = \frac{V_1}{Z_g(s)} = \frac{2Z(s)}{Z_g(s)[Z_t(s) + Z_1]} V_1 \tag{4-34}$$

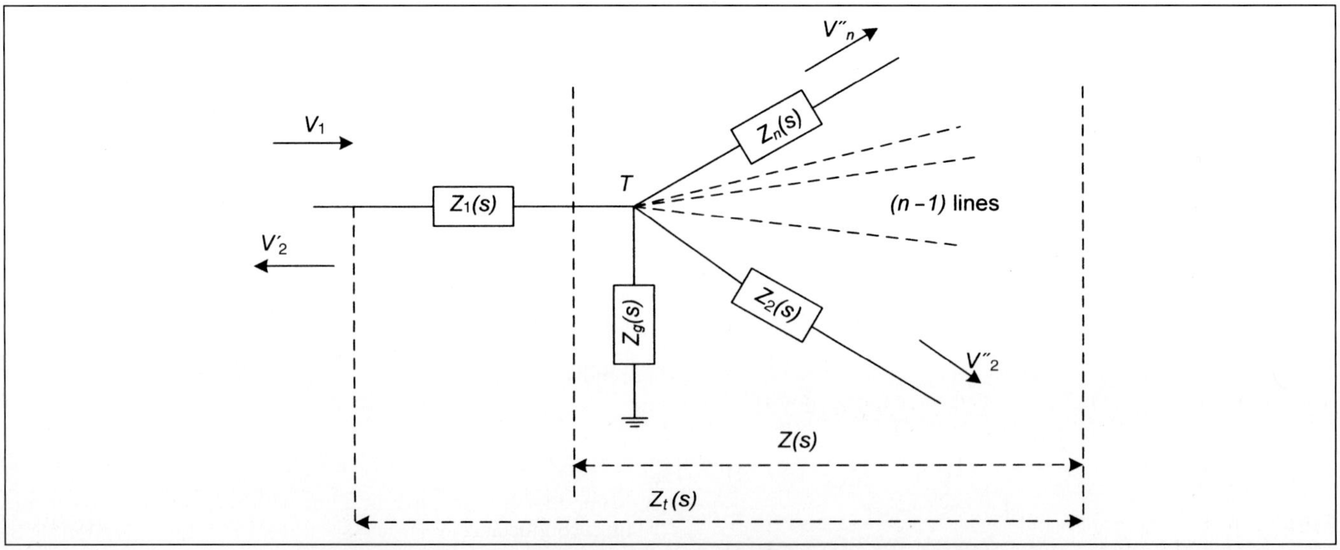

FIGURE 4-4 Multiple line terminations on a junction T, propagation of waves.

The transmitted voltage and current are:

$$I_k'' = \frac{V_g}{Z_k(s)+Z_k} = \frac{2Z(s)}{[Z_t(s)+Z_1][Z_k(s)+Z_k]}V_1$$
$$V_k'' = Z_k I_k'' = \frac{2Z(s)Z_k}{[Z_t(s)+Z_1][Z_k(s)+Z_k]}V_1 \tag{4-35}$$

The voltage drop across the lumped series impedances is:

$$V_k = Z_k(s)I_k'' = \frac{2Z(s)Z_k(s)}{[Z_t(s)+Z_1][Z_k(s)+Z_k]}V_1 \tag{4-36}$$

These are functions of s. The inverse transform gives the desired time functions.

When the junction consists of two impedances only, the reflection and transmission coefficients are simpler. The reflection coefficient is defined at the load as the ratio of the amplitudes of the backward and forward traveling waves. For a line terminated in a load impedance Z_2:

$$V_1' = \left(\frac{Z_2 - Z_1}{Z_2 + Z_1}\right)V_1 \tag{4-37}$$

Therefore, the reflection coefficient at the load end is:

$$\rho_L = \left(\frac{Z_2 - Z_1}{Z_2 + Z_1}\right) \tag{4-38}$$

The current reflection coefficient is negative of the voltage reflection coefficient. The reflected wave at an impedance discontinuity is a mirror image of the incident wave moving in the opposite direction. Every point on the wave is the corresponding point on the incident wave multiplied by the reflection coefficient, but a mirror image. Figure 4-5*a*, *b*, and *c* clearly shows these phenomena. The source reflection coefficient akin to the load reflection coefficient can be defined as:

$$\rho_s = \left(\frac{Z_s - Z_1}{Z_s + Z_1}\right) \tag{4-39}$$

The transmission coefficient is:

$$V_1'' = (1+\rho)V_1 = \frac{2Z_2}{Z_2 + Z_1}$$
$$I_1' = (1-\rho)I_1 = \frac{2Z_1}{Z_2 + Z_1} \tag{4-40}$$

4-5 LATTICE DIAGRAMS

The forward and backward traveling waves can be shown on a lattice diagram. The horizontal axis is the distance of the line from the source and the vertical axis is labeled in time increments, each increment being the time required for the wave to travel the line in one direction, that is, from the source to the load. Consider a point P on the pulse shape of Fig. 4-6*a* at time t'. The time to travel in one direction is l/u, where u is close to the velocity of light. The point P then reaches the load end at $t' + l/u$ seconds, and is reflected back. The corresponding point on the reflected wave is $P\rho_L$. At the sending end, it is reflected as $P\rho_L\rho_s$. This is shown in Fig. 4-6*b*.

Example 4-1 A lossless line has a surge impedance of 300 Ω. It is terminated in a resistance of 600 Ω. A 120-V dc source is applied to the line at $t = 0$ at the sending end. The voltage profile along the line for several time periods is drawn considering that the transient time in one direction is t seconds.

The reflection coefficient of voltage at the receiving end (load end) is 1/3 and at the source end is –1. (The source impedance is zero.) The lattice diagram is in Fig. 4-7*a* and the receiving end voltage is shown in Fig. 4-7*b*.

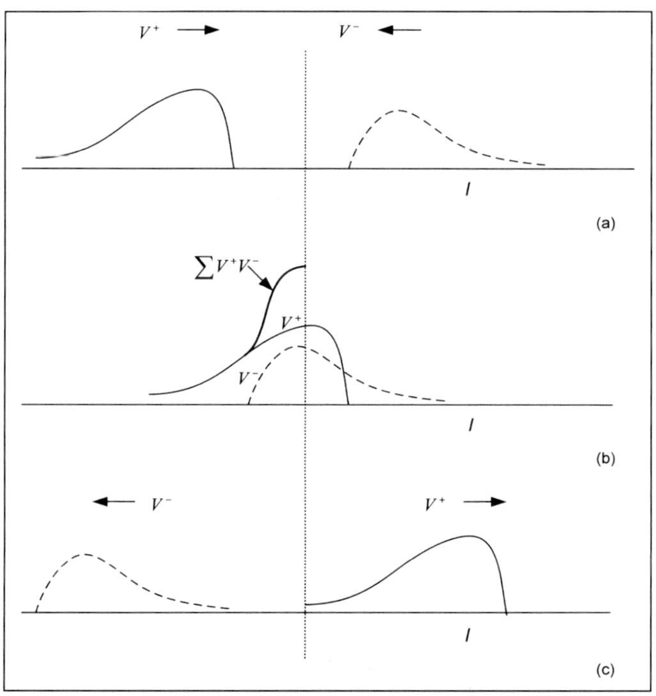

FIGURE 4-5 (*a*) Incident and reflected waves at an impedance discontinuity; (*b*) reinforcement of incident and reflected waves; (*c*) incident and reflected waves crossing each other.

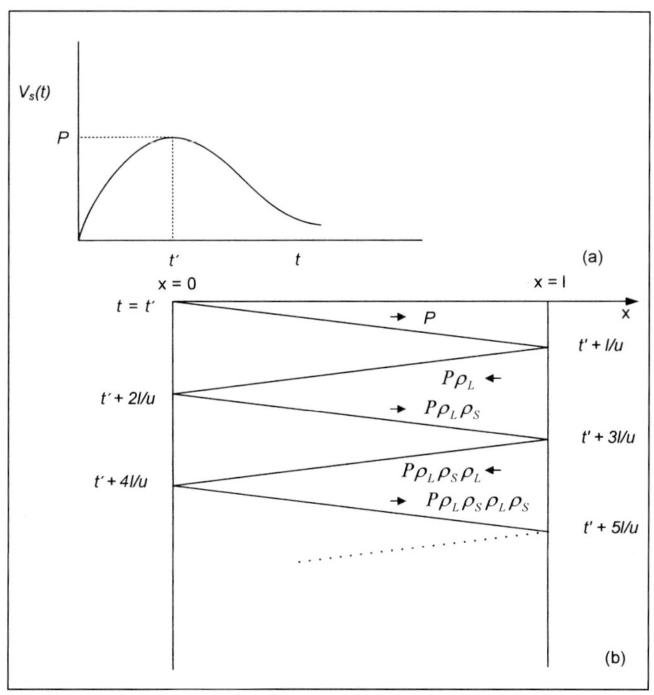

FIGURE 4-6 (*a*) A point P at time t' on a pulse signal applied to the sending end of a line; (*b*) lattice diagram.

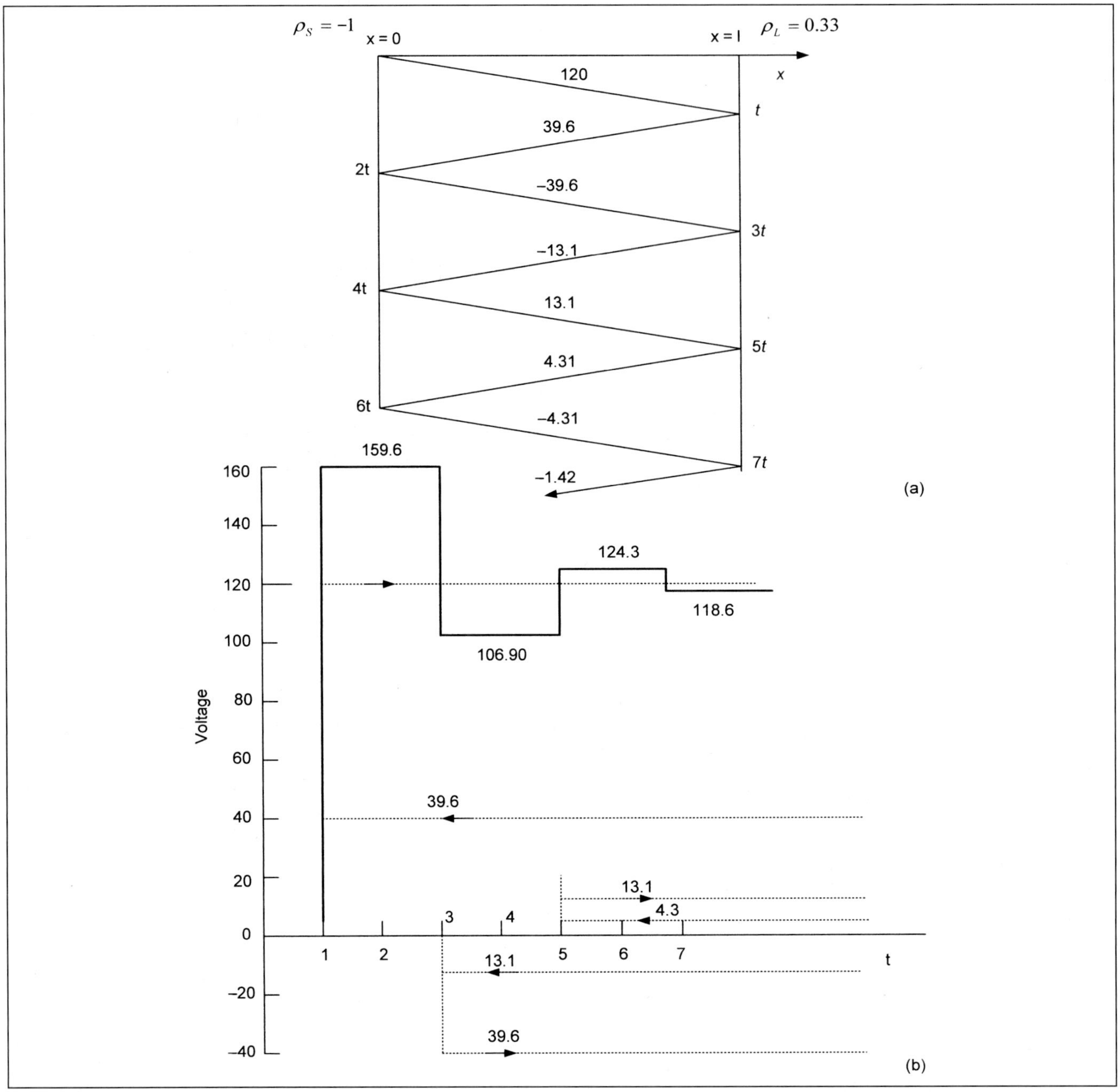

FIGURE 4-7 (*a*) Lattice diagram of Example 4-1; (*b*) voltage at receiving end of the line.

4-6 BEHAVIOR WITH UNIT STEP FUNCTIONS AT TRANSITION POINTS

Some typical cases can be studied as follows:

Line Open at Far End As the line is open, $Z_2 = \infty$. Therefore, the reflection coefficient is:

$$\rho_L = \left(\frac{Z_2 - Z_1}{Z_2 + Z_1}\right)$$

$$= \left(\frac{1 - Z_1/Z_L}{1 + Z_1/Z_L}\right) = 1 \tag{4-41}$$

The reflected wave is equal to the incident wave and the transmitted wave is:

$$V'' = (1 + \rho_L)V = 2V \tag{4-42}$$

The voltage will be doubled and the current is zero. If the line is terminated in impedance, which is much higher than its surge impedance, the voltage will be almost doubled. The highest overvoltages result when a circuit consists of overhead lines and cables, particularly if the length of one is smaller than the other.

Line Short-Circuited at the Far End As the line is short-circuited, $Z_2 = 0$; therefore, from Eq. (4-38) the reflection coefficient = −1. The reflected voltage wave is, therefore, reverse of the incident voltage wave and the transmitted voltage wave = 0, the current will be doubled. A negative reflected voltage wave starts traveling the line canceling the incident wave, while a positive current wave travels the line increasing the current.

Line Terminated in a Reactor Consider that the line is terminated with a reactor, and a step voltage is applied. We can calculate the expressions for transmitted and reflected voltages as follows:

We can write $Z_2 = Ls$ in Laplace transform and let $Z_1 = Z$. Then the reflection coefficient is:

$$\rho_L = \frac{(s - Z/L)}{(s + Z/L)} \tag{4-43}$$

The reflected wave is:

$$V'(s) = \frac{(s - Z/L)}{(s + Z/L)} \frac{V}{s} = \left[-\frac{1}{s} + \frac{2}{s + Z/L} \right] \tag{4-44}$$

Taking inverse transform:

$$V' = [-1 + 2e^{-(Z/L)t}]Vu(t), \text{ where } u(t) \text{ is a step function.} \tag{4-45}$$

The voltage of the transmitted wave is:

$$V''(s) = (1 + \rho_L)V(s) = \frac{2V}{(s + Z/L)} \tag{4-46}$$

Thus:

$$V'' = 2Ve^{-(Z/L)t} \tag{4-47}$$

The voltage across the inductor rises initially to double the incident voltage and decays exponentially.

Line Terminated in a Capacitor Here we write $Z_2 = 1/Cs$ The reflection coefficient is:

$$\rho_L = \frac{\left(\dfrac{1}{Cs} - Z\right)}{\left(\dfrac{1}{Cs} + Z\right)} = \frac{1 - CZs}{1 + CZs} \tag{4-48}$$

The reflected voltage wave is:

$$V'(s) = \frac{1 - CZs}{1 + CZs} \frac{V}{s} = \left[\frac{1}{s} - \frac{2CZs}{1 + CZs} \right] V \tag{4-49}$$

$$V' = [1 - 2e^{-t/cz}]Vu(t)$$

And the transmitted voltage wave is:

$$V'' = 2[1 - e^{-t/cz}]Vu(t) \tag{4-50}$$

From the expression of V'' in Eq. (4-50), the steepness of the front is reduced and the voltage rises slowly in an exponential manner. The capacitor behaves like a short circuit at $t = 0$, and will be charged through line impedance Z. The voltage at junction will finally rise to two times the incident voltage.

Two Parallel Conductors Connected to a Bus The voltage at the bus will be the same, though the currents in the conductors will differ. In the matrix form:

$$\bar{V} = \bar{\bar{Z}}\bar{I} \tag{4-51}$$

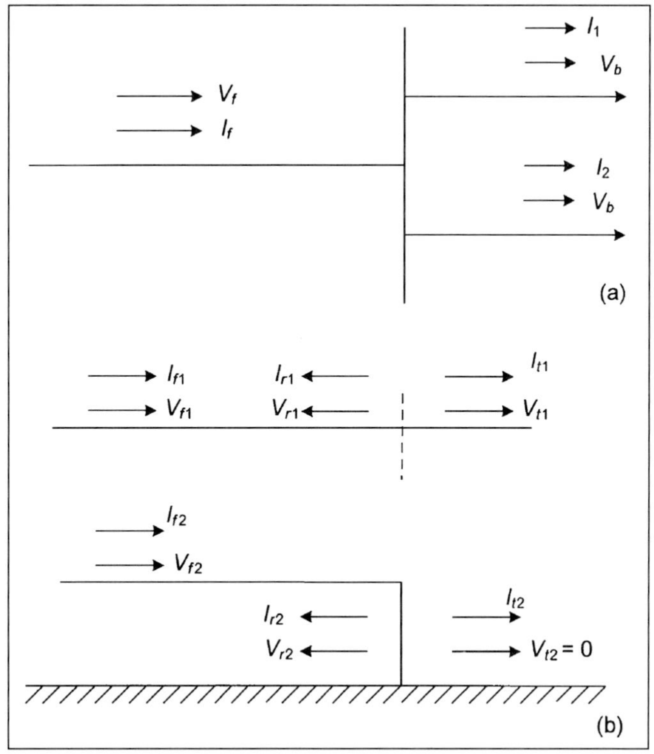

FIGURE 4-8 (*a*) Two parallel conductors connected to a bus, propagation of waves; (*b*) two parallel conductors, one shorted to ground, other continuous; propagation of waves.

Therefore, with reference to Fig. 4-8*a*, if V_b is the bus voltage:

$$I_1 = V_b \frac{\begin{vmatrix} 1 & Z_{12} \\ 1 & Z_{22} \end{vmatrix}}{\bar{\bar{Z}}} \qquad I_2 = V_b \frac{\begin{vmatrix} Z_{11} & 1 \\ Z_{12} & 1 \end{vmatrix}}{\bar{\bar{Z}}} \tag{4-52}$$

Now consider that one of the conductors is grounded, for example, a shield wire or sky wire on a transmission line. Then, the voltage on the grounded conductor is zero.

$$V_{f2} = 0 \quad I_{f2} = -\frac{Z_{12}}{Z_{22}} I_{f1} \quad V_{f1} = Z_{11}I_{f1} + Z_{12}I_{f2} = I_{f1}\left(Z_{11} - \frac{Z_{12}^2}{Z_{22}}\right) \tag{4-53}$$

One Conductor Shorter Than the Other and Grounded This situation is shown in Fig. 4-8*b*. We know that the following relations are applicable:

$$\begin{aligned} V_{t1} &= V_{f1} + V_{r1} = Z_{11}I_{t1} \\ V_{t2} &= V_{f2} + V_{r2} = 0 \\ I_{t1} &= I_{f1} + I_{r1} \end{aligned} \tag{4-54}$$

Then, it can be shown that:

$$V_{r1} = \frac{Z_{12}^2}{2Z_{11}Z_{22} - Z_{12}^2} V_{f1} - \frac{Z_{12}^2}{2Z_{11}Z_{22} - Z_{12}^2} V_{f2} \tag{4-55}$$

$$V_{t1} = \frac{2Z_{11}Z_{22}}{2Z_{11}Z_{22} - Z_{12}^2} V_{f1} - \frac{2Z_{11}Z_{12}}{2Z_{11}Z_{22} - Z_{12}^2} V_f \tag{4-56}$$

and

$$I_{r2} = \frac{\begin{vmatrix} Z_{11} & -V_{r1} \\ Z_{12} & -V_{r2} \end{vmatrix}}{\bar{Z}} \tag{4-57}$$

4-7 INFINITE LINE

When the line is terminated in its characteristic load impedance, that is, $R = Z_0$, the reflected wave is zero and the transmitted wave is equal to the incident wave. Such a line is called an *infinite line*, and the incident wave cannot distinguish between the termination and the continuation of the line.

The *characteristic impedance*, also called surge *impedance*, is approximately 400 Ω for overhead transmission lines, and its phase angle may vary from 0° to 15°. For underground cables, the surge impedance is much lower, approximately 1/10 of the overhead lines.

4-7-1 Surge Impedance Loading

The *surge impedance loading* (SIL) of the line is defined as the power delivered to a purely resistive load equal in value to the surge impedance of the line.

$$\text{SIL} = \frac{V_r^2}{Z_0} \tag{4-58}$$

For a 400-Ω surge impedance, the SIL in kW is 2.5 multiplied by the square of the receiving-end voltage in kV. The surge impedance is a real number, and therefore the power factor along the line is unity, that is, no reactive power compensation is required. The SIL loading is also called the *natural loading of the transmission line*. Practically, this loading is not achieved.

4-8 TUNED POWER LINE

In the long transmission line model, if the shunt conductance and series resistance are neglected, then:

$$\begin{aligned} \gamma &= \sqrt{YZ} = j\omega\sqrt{LC} \\ \text{Cosh}\gamma l &= \cosh j\omega l\sqrt{LC} = \cos\omega l\sqrt{LC} \\ \text{Sinh}\gamma l &= \sinh j\omega l\sqrt{LC} = j\sin\omega l\sqrt{LC} \end{aligned} \tag{4-59}$$

where l is the length of the line. This simplifies *ABCD* parameters and the following relation results:

$$\begin{vmatrix} V_s \\ I_s \end{vmatrix} = \begin{vmatrix} \cos\omega l\sqrt{LC} & jZ_0\omega l\sqrt{LC} \\ \dfrac{j}{Z_0}\sin\omega l\sqrt{LC} & \cos\omega l\sqrt{LC} \end{vmatrix} \begin{vmatrix} V_r \\ I_r \end{vmatrix} \tag{4-60}$$

If:

$$\omega l\sqrt{LC} = n\pi \qquad (n = 1, 2, 3, \ldots) \tag{4-61}$$

Then:

$$|V_s| = |V_r| \qquad \text{and} \qquad |I_s| = |I_r| \tag{4-62}$$

This will be an ideal situation to operate a transmission line; the receiving-end voltage and currents are equal to the sending-end voltage and current. The line has a flat voltage profile.

$$\beta = \frac{2\pi f}{v} = \frac{2\pi}{\lambda} \tag{4-63}$$

As $1/\sqrt{LC}$ is equal to the velocity of light, the line length λ is 3100, 6200, . . . mi or $\beta = 0.116°$ per mile. The quantity βl = *electrical length* of the line. The line length calculated earlier is too long to avail this ideal property. This suggests that power lines can be tuned with series capacitors to cancel the effect of inductance and shunt inductors to neutralize the effect of line capacitance. This compensation may be done by sectionalizing the line. For power lines, series and shunt capacitors for heavy load conditions and shunt reactors under light load conditions are used to improve power transfer and line regulation.

Also, see Chap. 15 for flexible ac transmission system (FACTS) controllers, which improve the performance of transmission lines based on the concepts outlined here.

4-9 FERRANTI EFFECT

As the transmission line length increases, the receiving-end voltage rises above the sending-end voltage due to line capacitance. This is called Ferranti effect. In a long line model, at no load ($I_R = 0$), the sending-end voltage is:

$$V_s = \frac{V_r}{2}e^{\alpha l + j\beta l} + \frac{V_r}{2}e^{-\alpha l - j\beta l} \tag{4-64}$$

At $l = 0$, both incident and reflected waves are equal to $V_R/2$. As l increases, the incident wave increases exponentially, while the reflected wave decreases. Thus, the receiving-end voltage rises. Another explanation of this phenomenon is provided by considering the line capacitance lumped at the receiving end. On open circuit, the sending-end current is:

$$I_s = \frac{V_s}{\left(j\omega Ll - \dfrac{1}{j\omega Cl}\right)} \tag{4-65}$$

In Eq. (4-65) C is small in comparison with L. Thus, ωLl can be neglected. We can write the receiving-end voltage considering only the voltage drop in the series inductance of the transmission line:

$$\begin{aligned} V_r &= V_s - I_s(j\omega Ll) \\ &= V_s + V_s\omega^2CLl^2 \\ &= V_s(1 + \omega^2CLl^2) \end{aligned} \tag{4-66}$$

This gives a voltage rise at receiving end of:

$$|V_s|\omega^2CLl^2 = |V_s|\omega^2l^2/v^2 \tag{4-67}$$

where v is the velocity of propagation. Considering that the velocity of propagation is constant, the receiving-end voltage rises as the line length is increased.

Also from Eq. (4-60) the voltage at any distance x in terms of sending-end voltage, with the line open circuited and resistance neglected is:

$$V_x = V_s\frac{\cos\beta(l - x)}{\cos\beta l} \tag{4-68}$$

and the current is:

$$I_x = j\frac{V_s}{Z_0}\frac{\sin\beta(l - x)}{\cos\beta l} \tag{4-69}$$

Example 4-2 A 230-kV three-phase transmission line has 795 kcmil aluminum conductor steel reinforced (ACSR) conductors, one per phase, neglecting resistance, $z = j0.8$ Ω/mi, and $y = j5.4 \times 10^{-6}$ siemens per mile. Calculate the voltage rise at the receiving end for a 400-mi long line.

Using these expressions, $Z_0 = 385\ \Omega$, $\beta = 2.078 \times 10^{-3}$ rad/mi = 0.119° per mile. $\beta l = 0.119 \times 400 = 47.6°$, the receiving-end voltage *rise* from Eq. (4-68) is 48.3 percent and at 756 mi, one-quarter wavelength it will be infinite.

Even a voltage rise of 10 percent at the receiving end is not acceptable as it may give rise to insulation stresses and affect the consumer apparatus. Practically, the voltage rise will be more than calculated here. As the load is thrown off, the sending-end voltage will rise before the generator voltage regulators and excitation systems act to reduce the voltage, further increasing the voltage on the line. It is imperative that the transmission lines are compensated to control voltage excursions.

In the above example, the sending-end charging current is 1.18 per unit and falls to zero at the receiving end. This means that the capacitive charging current flowing in the line is 118 percent of the line natural load current.

4-10 SYMMETRICAL LINE AT NO LOAD

If we consider a symmetrical line at no load, with the sending-end and receiving-end voltages maintained at the same level, these voltages have to be in phase as no power is transferred. Half the charging current is supplied from each end, and the line is equivalent to two equal half-sections connected back-to-back. The voltage rises at the midpoint, where the charging current falls to zero and reverses direction. The synchronous machines at the sending end absorb leading reactive power, while the synchronous machines at the receiving end generate lagging reactive power (Fig. 4-9*a*, *b*, and *c*).

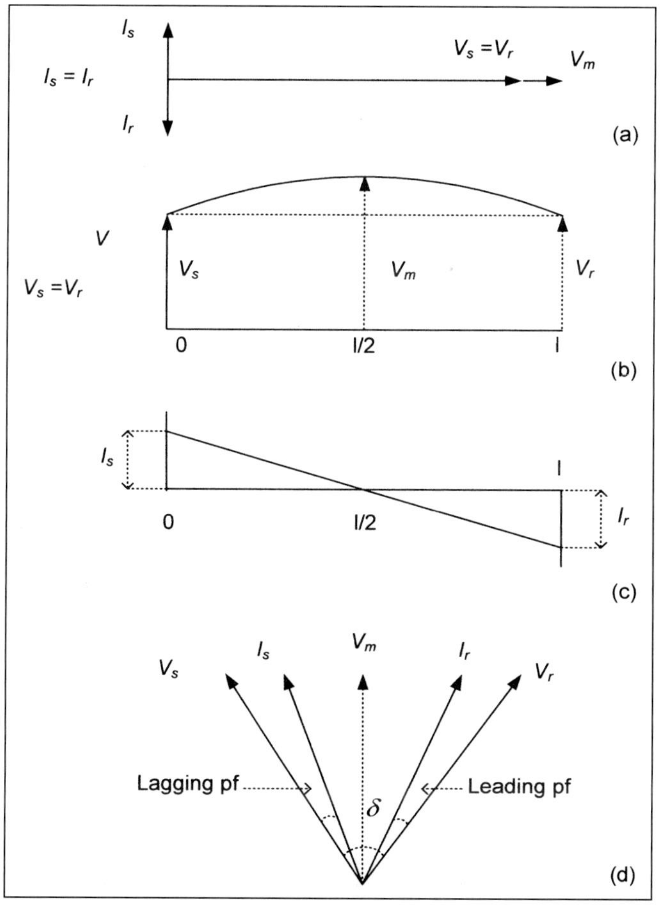

FIGURE 4-9 (*a*) Phasor diagram of a symmetrical line at no load; (*b*) the voltage profile of a symmetrical line at no load; (*c*) charging current profile; (*d*) phasor diagram of the symmetrical line with reference to midpoint, at no load.

The midpoint voltage is therefore equal to the voltage as if the line was half the length.

On loading, the vector diagram shown in Fig. 4-9*d* is applicable. By symmetry, the midpoint voltage vector exactly bisects the sending- and receiving-end voltage vectors. The power factor angle at both ends are equal but of the opposite sign. Therefore, receiving-end voltage on a symmetric line of length $2l$ is the same as that of line of length l at unity power factor load. From Eq. (4-60), the equations for sending-end voltage and current for a symmetrical line can be written with βl replaced with $\beta l/2 = \theta/2$.

$$\begin{aligned} V_s &= V_m \cos\left(\frac{\theta}{2}\right) + jZ_0 I_m \sin\left(\frac{\theta}{2}\right) \\ I_s &= j\frac{V_m}{Z_0}\sin\left(\frac{\theta}{2}\right) + I_m \cos\left(\frac{\theta}{2}\right) \end{aligned} \tag{4-70}$$

At the midpoint:

$$P_m + jQ_m = V_m I_m^* = P$$

$$Q_s = \text{Img } V_s I_s^* = j\frac{\sin\theta}{2}\left[Z_0 I_m^2 - \frac{V_m^2}{Z_0}\right] \tag{4-71}$$

where P is the transmitted power. No reactive power flows past the midpoint and it is supplied by the sending end.

Example 4-3 Consider a line of medium length represented by a nominal Π circuit. The impedances and shunt susceptances are shown in Fig. 4-10 in per unit on a 100-MVA base. Based on the given data, calculate *ABCD* parameters. Consider that 1 per unit current at 0.9 power factor lagging is required to be supplied to the receiving end. Calculate sending-end voltage, currents, power, and losses.

$$\begin{aligned} D = A &= 1 + \frac{1}{2}YZ \\ &= 1 + 0.5[j0.0538][0.0746 + j0.394] \\ &= 0.989 + j0.002 \end{aligned}$$

$$\begin{aligned} C &= Y\left(1 + \frac{1}{4}YZ\right) \\ &= (j0.0538)[1 + (-0.0053 + j0.9947)] \\ &= -0.000054 + j0.0535 \end{aligned}$$

$$B = Z = 0.0746 + j0.394$$

Voltage at receiving bus is $1 < 0°$. The receiving-end power is:

$$V_2 I_2^* = (1 < 0°)(I_2 < 25.8°) = 0.9 + j0.436$$

The sending-end voltage is:

$$\begin{aligned} V_1 &= AV_2 + BI_2 \\ &= (0.989 + j0.002)(1 < 0°) + (0.0746 \\ &\quad + j0.394)(1 < 25.8°) \\ &= 1.227 + j0.234 \\ |V_1| &= 1.269 < 14.79° \end{aligned}$$

The sending-end current is given by:

$$\begin{aligned} I_1 &= CV_2 + DI_2 \\ &= (-0.000054 + j0.0535) + (0.989 + j0.0021)(1 < 25.8°) \\ &= 0.8903 - j0.3769 \\ &= 0.9668 < -22.944° \end{aligned}$$

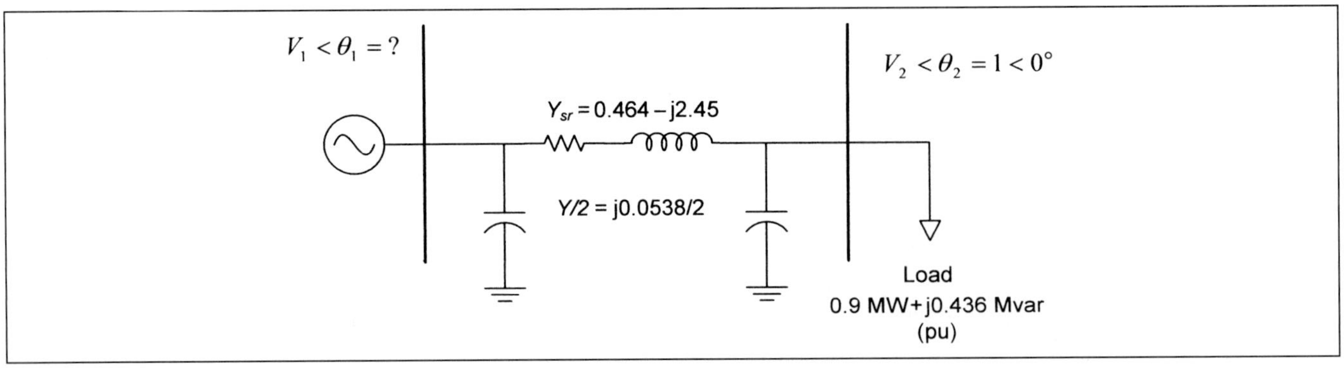

FIGURE 4-10 Transmission line and load parameters for Examples 4-3 and 4-4.

The sending-end power is, therefore:

$$V_1 I_1^* = (1.269 < 14.79°)(0.9668 < 22.944°)$$
$$= 0.971 + j0.75$$

The active power loss is 0.071 per unit and the reactive power loss is 0.314 per unit. The reactive power loss increases as the load power factor becomes more lagging.

The power supplied, based on known sending-end and receiving-end voltages, can also be calculated from the equations:

$$[V_1^2 - V_1V_2\cos(\theta_1 - \theta_2)g_{12} - [V_1V_2\sin(\theta_1 - \theta_2]b_{12}$$
$$= [1.269^2 - 1.269\cos(14.79°)0.464$$
$$- [1.269\ \sin\ (14.79°)](-2.450)$$
$$= 0.971 \qquad (4\text{-}72)$$

This gives the same result as calculated before. A dilemma of the calculation is that for a given load, neither the sending-end voltage nor the receiving-end current is known.

The sending-end reactive power is:

$$Q_{12} = [-V_1V_2\sin(\theta_1 - \theta_2)]g_{12} - [V_1^2 - V_1V_2\cos(\theta_1 - \theta_2]b_{12} - \frac{Y}{2}V_1^2$$
$$= [11.269\sin 14.79°]\,0.464 - [1.269^2 - 1.269\cos 14.79°]$$
$$\times(-2.450) - (0.0269)(1.269^2)$$
$$= 0.75 \qquad (4\text{-}73)$$

The receiving-end active power is:

$$P_{21} = [-V_2V_1\sin(\theta_2 - \theta_1)]\,b_{12} + [V_2^2 - V_2V_1\cos(\theta_2 - \theta_1)]\,g_{12}$$
$$= 0.9 \qquad (4\text{-}74)$$

Also, receiving-end reactive power is:

$$Q_{21} = [-V_2V_1\sin(\theta_2 - \theta_1)]\,g_{12} - [V_2^2 - V_2V_1\cos(\theta_2 - \theta_1)]\,b_{12} - \frac{Y}{2}V_2^2$$
$$= [-1.269\sin(-14.79°)]\,0.464 - [1 - 1.269\cos(-14.79°)]$$
$$\times(-2.450) - (0.0269)(1)$$
$$= 0.436 \qquad (4\text{-}75)$$

Example 4-4 Example 4-3 is repeated with long line model and the results are compared.

We can write:

$$Z = 0.0746 + j0.394$$
$$Y = 0.0538$$
$$ZY = 0.021571 < 169.32°$$

From Eq. (4.17):

$$\alpha = 0.0136687\ \text{Np}$$
$$\beta = 0.146235\ \text{rad}$$

The hyperbolic functions cosh γ and sinh γ are given by:

$$\cosh\gamma = \cosh\alpha\cos\beta + j\sinh\alpha\sin\beta$$
$$= (1.00009)(0.987362) + j(0.013668)(0.14571)$$
$$= 0.990519 < 0.1152°$$

$$\text{Sinh}\ \gamma = \sinh\alpha\cos\beta + j\cosh\alpha\sin\beta$$
$$= (0.013668)(0.989326) + j(1.00009)(0.14571)$$
$$= 0.146349 < 84.698°$$

The sending-end voltage is:

$$V_1 = \cosh\gamma V_2 + Z_0\sinh\gamma I_2$$
$$= (0.990519 < 0.1152°)(1 < 0) + (2.73 < -5.361°)$$
$$\times(0.1463449 < 84.698°)(1 < -25.842°)$$
$$= 1.269 < 14.73°$$

The result is fairly close to the one calculated with Π model.

The sending-end current is:

$$I_2 = \left(\frac{1}{Z_0}\right)(\sinh\gamma)V_2 + (\cosh\gamma)I_2$$
$$= (0.3663 < 5.361°)(0.146349 < 84.698°)$$
$$+ (0.990519 < 0.1152°)(1 < -25.84°)$$
$$= 0.968 < -22.865°$$

Again this result is close to the one calculated with Π model. The parameters of the transmission line shown in this example are for a 138-kV line, length of approximately 120 mi. For longer lines, the difference between the calculations using two models will diverge.

These two examples, though on power transmission, do underline some voltage problems, reactive power supply, and loss, which play a role in the transient analysis.

4-11 LOSSLESS LINE

To eliminate frequency and delay distortion in a line, α and velocity of propagation should be independent of frequency. As the velocity of propagation is given by:

$$v = \frac{\omega}{\beta} \tag{4-76}$$

β should be made a direct function of frequency.

If we make:

$$Lg_{sh} = Cr_{sc} \tag{4-77}$$

then, from Eq. (4-17)

$$\beta = \omega\sqrt{LC} \tag{4-78}$$

and the velocity of propagation is:

$$v = \frac{1}{\sqrt{LC}} \tag{4-79}$$

This will be the same for all frequencies. Also, from Eq. (4-17):

$$\alpha = \sqrt{r_{sc}g_{sh}} \tag{4-80}$$

which is also independent of frequency.

Such a hypothetical line is not practical. From Eq. (4-77), it requires a very large value of inductance L as shunt conductance g_{sh} is small. If g_{sh} is intentionally increased, then α increases, that is, the attenuation increases. To reduce r_{sc}, the size and cost of conductors increase beyond economical limits, so the hypothetical results cannot be achieved in practice.

4-12 GENERALIZED WAVE EQUATIONS

It is recommended that the reader studies App. D on transmission line constants before continuing with the rest of this chapter.

A lossless line can be represented by its self-inductance, mutual inductance, capacitance to ground, and mutual capacitances. We had arrived at the inductance and capacitance matrices of three-phase transmission lines in App. D. The following is the alternative approach.

The voltage drop across a section of the line can be written as:

$$\Delta V_1 = \Delta x\left[L_{11}\frac{\partial I_1}{\partial t} + L_{12}\frac{\partial I_2}{\partial t} + \cdots + L_{1n}\frac{\partial I_n}{\partial t}\right]a$$

$$\ldots$$

$$\Delta V_n = \Delta x\left[L_{n1}\frac{\partial I_1}{\partial t} + L_{n2}\frac{\partial I_2}{\partial t} + \cdots + L_{nn}\frac{\partial I_n}{\partial t}\right] \tag{4-81}$$

or in the matrix form as:

$$\frac{\partial \bar{V}}{\partial x} = -\bar{L}\frac{\partial \bar{I}}{\partial t} \tag{4-82}$$

The matrix L for n-phase line is:

$$\bar{L} = \begin{vmatrix} L_{11} & L_{12} & \cdot & L_{1n} \\ L_{21} & L_{22} & \cdot & L_{2n} \\ \cdot & \cdot & \cdot & \cdot \\ L_{n1} & L_{n2} & \cdot & L_{nn} \end{vmatrix} \tag{4-83}$$

Similarly, we can write the current equation in the matrix form:

$$\frac{\partial I_1}{\partial x} = \left[C_{11}\frac{\partial V_1}{\partial t} + C_{12}\frac{\partial V_2}{\partial t} + \cdots + C_{1n}\frac{\partial V_n}{\partial t}\right]$$

$$\frac{\partial I_n}{\partial x} = \left[C_{n1}\frac{\partial V_1}{\partial t} + C_{n2}\frac{\partial V_2}{\partial t} + \cdots + C_{nn}\frac{\partial V_n}{\partial t}\right] \tag{4-84}$$

or in the matrix form:

$$\frac{\partial \bar{I}}{\partial x} = -\bar{C}\frac{\partial V}{\partial t} \tag{4-85}$$

The capacitance matrix for a three-phase line can be written as (App. D):

$$\bar{C}_{123} = \begin{vmatrix} C_{11} & -C_{12} & -C_{13} \\ -C_{21} & C_{22} & -C_{23} \\ -C_{31} & -C_{32} & C_{33} \end{vmatrix} \tag{4-86}$$

From Eqs. (4-85) and (4-86):

$$\frac{\partial^2 \bar{V}}{\partial x^2} = \bar{L}\bar{C}\frac{\partial^2 \bar{V}}{\partial t^2} \tag{4-87}$$

For a perfect earth, we can write:

$$\bar{L}\bar{C} = \frac{\bar{I}}{v^2} = \bar{M} \tag{4-88}$$

For kth conductor:

$$\frac{\partial^2 V_k}{\partial x^2} = \frac{1}{v^2}\frac{\partial^2 V_k}{\partial t^2} \tag{4-89}$$

This shows that the wave equation of each conductor is independent of the mutual influence of other conductors in the system—a result which is only partially true. However, for lightning surges and high frequencies, the depth of the current image almost coincides with the perfect earth. This may not be true for switching surges. The M matrix in Eq. (4-88) is not a diagonal matrix. We will use decoupling to diagonalize the matrix (App. D). With respect to transmission lines, this is called *modal analysis* as it gives different modes of propagation.

4-13 MODAL ANALYSIS

Consider a matrix of modal voltages V_m and a transformation matrix T, transforming conductor voltage matrix V:

$$\bar{V} = \bar{T}\bar{V}_m \tag{4-90}$$

Then, the wave equation can be decoupled and written as:

$$\begin{aligned} \frac{\partial^2 \bar{V}_m}{\partial x^2} &= \bar{T}^{-1}[\bar{L}\bar{C}]\bar{T}\frac{\partial^2 \bar{V}_m}{\partial t^2} \\ &= \bar{T}^{-1}\bar{M}^{-1}\bar{T}\frac{\partial^2 \bar{V}_m}{\partial t^2} \\ &= \bar{\lambda}\frac{\partial^2 \bar{V}_m}{\partial t^2} \end{aligned} \tag{4-91}$$

For decoupling matrix, $\bar{\lambda} = \bar{T}^{-1}\bar{M}^{-1}\bar{T}$ must be diagonal. This is done by finding the eigenvalues of $\bar{M}$ from the solution of its characteristic equation. The significance of this analysis is that for n conductor system, the matrices are of order n, and n number of

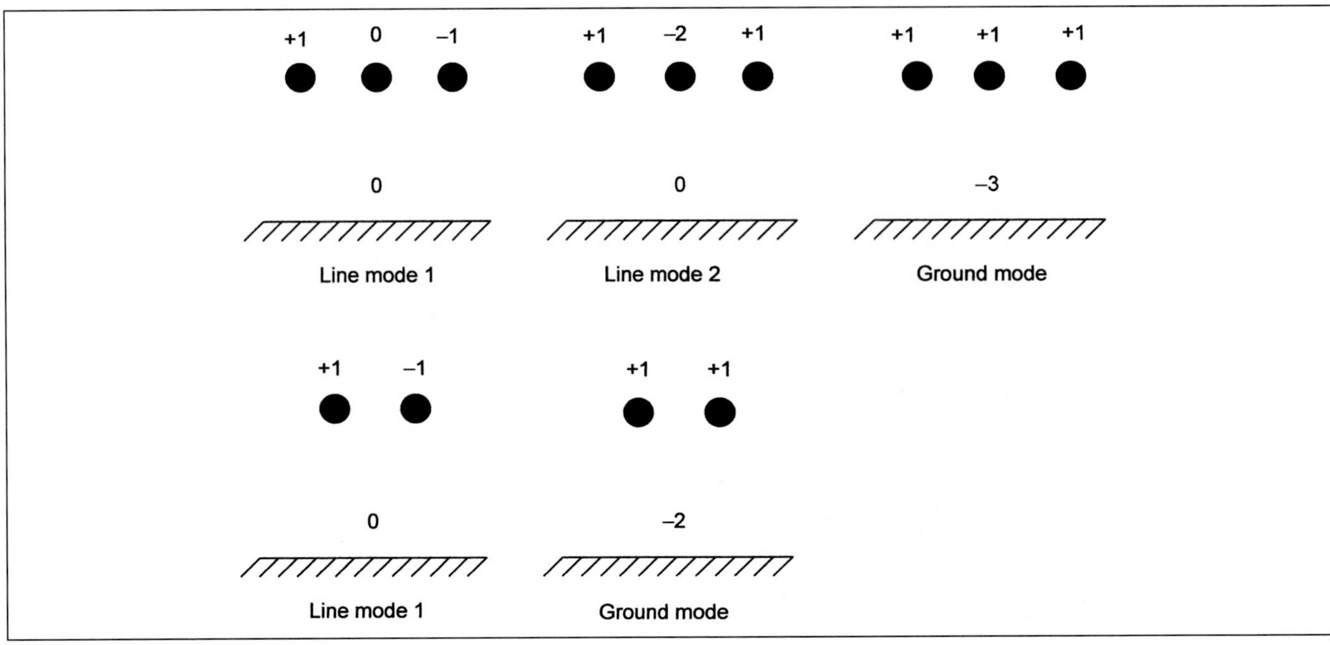

FIGURE 4-11 Modes of propagation for three-conductor and two-conductor lines.

modal voltages are generated which are independent of each other. Each wave travels with a velocity:

$$v_n = \frac{1}{\lambda_k} \qquad k = 1, 2, \ldots, n \tag{4-92}$$

and the actual voltage on the conductors is given by Eq. (4-90).

For a three-phase line:

$$\bar{T} = \begin{vmatrix} 1 & 1 & 1 \\ -1 & 0 & 1 \\ 0 & -1 & 1 \end{vmatrix} \tag{4-93}$$

Figure 4-11 shows modes for two-conductor and three-conductor lines. For a two-conductor line, there are two modes. In the line mode, the voltage and current travel over one conductor returning through the other, none flowing through the ground. In the ground mode, modal quantities travel over both the conductors and return through the ground. For a three-conductor line, there are two line modes and one ground mode.

The line-to-line modes of propagation are close to the speed of light and encounter less attenuation and distortion compared to ground modes. The ground mode has more resistance, and hence more attenuation and distortion. The resistance of the conductors and earth resistivity plays an important role.

4-13-1 Distortion Caused by Mode Propagation

Mode propagation gives rise to further distortion of multiconductor lines. For voltages below corona inception voltage (Sec. 4-15), the velocity of propagation of the line components is close to the velocity of light and the distortion is negligible. For ground currents, the charges induced are near the surface, while the return current is well below the surface, at a depth depending on the frequency and earth's resistivity. For a perfectly conducting earth, the current will flow at the earth's surface at a velocity equal to the velocity of light.

$$\begin{aligned} V_1 &= V_{m1} + V_{m2} + V_g \\ V_2 &= -V_{m1} + 0 + V_g \\ V_3 &= 0 - V_{m2} + V_g \end{aligned} \tag{4-94}$$

The ground component V_g of a three-phase line carrying surge voltages to ground of V_1, V_2, V_3 is defined as:

$$V_g = \frac{1}{3}(V_1 + V_2 + V_3) \tag{4-95}$$

Thus, all three components of the model voltages are present on the struck conductor, and one-line mode voltage is missing on conductors 2 and 3. This shows that the charge on phase a is balanced by the charges on phases b and c. Figure 4-12 shows the resolution into components and the distortion produced after the components have traveled some distance.

The calculation of the transmission line parameters for mode propagation becomes important, based on the collected transmission line data. We will illustrate these calculations in Example 4-6. Further analysis of the transition points in multiconductor systems can be made using modal analysis.

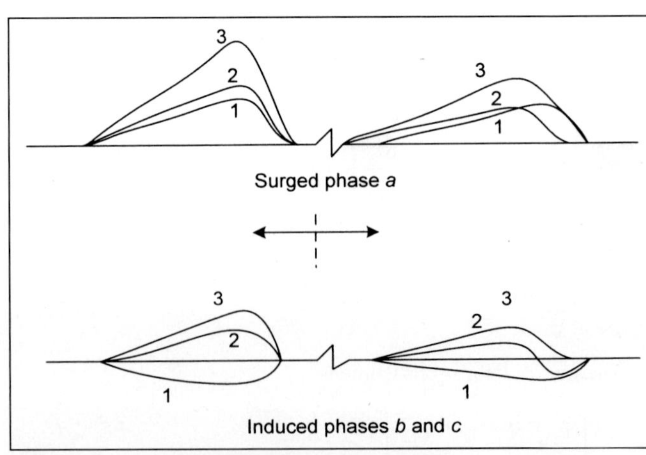

FIGURE 4-12 Mode propagation and distortion on a balanced three-phase line: 1. Line components; 2. ground components; 3. total voltage wave.

4-13-2 Approximate Long-Line Parameters

Regardless of voltage, conductor size, or spacing of a line, the series reactance is approximately 0.8 Ω/mi and the shunt-capacitive reactance is 0.2 MΩ/mi. This gives β of 1.998×10^{-3} per mile or 0.1145° per mile. These may be considered for the line mode and will vary over large values with respect to the mode of propagation. It is necessary to calculate parameters accurately for the study of transients with respect to each mode of propagation. The numbers stated here provide "ball-park" figures.

4-14 DAMPING AND ATTENUATION

We alluded to the phase and frequency distortion and that an ideal lossless line with $LG = CR$ is not practical. The decrease in the magnitude of the wave as it propagates is called *attenuation*. The current and voltage wave shapes become dissimilar, though these may be the same initially. Attenuation is caused by the energy loss in the line and the distortion is caused by change in the inductance and capacitance of the line. The constant surge impedance assumes that the wave shapes are not altered. The conductor resistance is modified by skin effect and the changes in the ground resistance. The changes in the inductance occur because of nonuniform distribution of currents and proximity to steel structures, that is, line supports. The capacitance can change due to change in the insulation nearest to the line supports.

The energy stored in the traveling wave is the sum of the electrostatic and electromagnetic energies:

$$E = \frac{1}{2}CV^2 + \frac{1}{2}LI^2 \tag{4-96}$$

As the two components of the energy are equal:

$$\frac{1}{2}CV^2 = \frac{1}{2}LI^2$$
$$\frac{V}{I} = \sqrt{\frac{L}{C}} = Z_0 \tag{4-97}$$

where Z_0 is the surge impedance. The total energy content is then estimated by integration over the length of the line:

$$E_{\text{total}} = \int E dx = C\int V^2 dx = L\int I^2 dx \tag{4-98}$$

The power of the pulses that pass through a given section is given by the product of energy content and the velocity of propagation:

$$W = E_{\text{total}} v = I^2 L \times \frac{1}{\sqrt{LC}} = I^2 Z_0 = \frac{V^2}{Z_0} \tag{4-99}$$

For the same voltage, the power is much higher in cables than in overhead lines, as the cables have a much lower surge impedance (50 Ω approximately) compared to overheads lines (400 Ω approximately). The surge impedance of paper-insulated high-voltage (HV) cables is higher than 50 Ω, as indicated earlier.

These equations have not considered the resistance and insulation resistance (G), and thus the waves are propagated without loss. If we consider R = series resistance of the line and G = the conductance of the line, then the heat loss (energy loss) is given by:

$$dW = I^2 R dx + V^2 G dx$$
$$= I^2(R + Z_0^2 G)dx \tag{4-100}$$

From Eqs. (4-100) and (4-101):

$$\frac{di}{i} = \frac{1}{2}\left(\frac{R}{Z_0} + Z_0 G\right)dx \tag{4-101}$$

Integration gives the change of current along the line:

$$I = I_0 \exp\left[-\frac{1}{2}\left(\frac{R}{Z_0} + Z_0 G\right)x\right] \tag{4-102}$$

where I_0 is the amplitude of the current pulse at any selected arbitrary origin or starting point. For the voltage pulse, the required change along the line is:

$$V = V_0 \exp\left[-\frac{1}{2}\left(\frac{R}{Z_0} + Z_0 G\right)x\right] \tag{4-103}$$

The G of the overhead lines is small, that is, the transmission lines have high insulation resistance, and if we ignore G:

$$\frac{V}{V_0} = \frac{I}{I_0} = e^{-Rx/2Z_0} \tag{4-104}$$

V_0 and I_0 are the voltage and current pulses at an arbitrary selected origin. Equation (4-103) shows that damping is directly proportional to the resistance and inversely proportional to the surge impedance. Therefore, it is much higher in cables than in overhead lines.

For the rapidly propagating pulses, the resistance of the conductors is increased above dc resistance values due to skin effect (App. D).

Example 4-5 Calculate the attenuation in an overhead line with a resistance of 0.5 Ω/km, and also in cables of the same cross section, length in each case being 200 km. The surge impedance of the overhead line is 400 Ω and that of the cable is 50 Ω.

For the overhead line, the attenuation is:

$$e^{-0.5 \times 200/400} = 0.778$$

For the cable, it is:

$$e^{-0.5 \times 200/50} = 0.135$$

4-15 CORONA

Corona discharges follow an avalanche process and an ionization process is set in. As the voltage increases, and the surrounding air is ionized, the energy is supplied until the critical field intensity for the air is reached. Once this level is reached, the field intensity cannot increase anymore with the rise in voltage. After the crest of the voltage is reached and the wave starts trailing, the space charge does not diminish or vanish. It is not the intention to go into the streamer theory, corona modes, except to consider some aspects of corona with reference to HV transmission lines. The impact of corona discharges has been recognized since the early days of the electric power transmissions. These effects can be summarized as:

Corona Noise A pulsating sine wave is generated from the discharge due to increase in pressure of the air and is audible in the immediate vicinity of the HV lines. Analyses to predict levels of audible noise consider A-weighted sound level during rain, L50, which means that the level is exceeded 50 percent of the time during rain; and L5, which is the level exceeded 5 percent of the time during rain. A comparison of the audible noise formulas that have been developed in various countries is included in the IEEE Task Force Report of 1982.[2]

Modal analysis provides a convenient approach to evaluate noise currents on line conductors. First, the noise currents are transposed into their modal components, which propagate without distortion along the line conductors at their own velocity.

$$\bar{I}_m = \frac{1}{\sqrt{\alpha}}\bar{M}^{-1}\bar{I}_L \tag{4-105}$$

where α is the real part of the propagation constant, as defined in Eq. (4-15). From the model currents, the line currents are obtained by reverse transformation. The magnetic and electrical fields produced by noise currents in the conductors can then be calculated. Moreau and Gary obtained good correlation between the calculated and experimental results.[3] They used Clarke's transformation for the model matrix *M*. (Clarke's transformation is discussed in (Sec. 4-16-2.)

Radio and Television Noise The noise is caused by complete electrical discharges across small gaps (insulators, line hardware, which should be corona-free), and by partial discharges. Gap discharges may generate sharp pulses with nanosecond rise time and may interfere with TV reception.

Ground-Level Electric Fields The electric fields in the vicinity of a transmission line are the superimposition of three-phase conductors and the conducting earth, represented by the image charges below the earth equal to the height of the conductors. The effects on humans is due to discharges from the objects typically insulated from ground, like vehicles, buildings, and fences, which become electrically charged due to induction from the line.

Ground-Level Magnetic Fields Magnetic field couplings affect objects which parallel the line for a distance. The parallelism of pipelines is a major concern, which is discussed in Chapter 21.

4-15-1 Corona Loss

Corona loss is described in terms of energy loss per unit length of the line. These may be negligible under fair-weather conditions, but may reach values of several hundred kW/km of line length during foul-weather conditions. The foul-weather losses are evaluated in test cages under artificial rain conditions. Peek's formula for corona loss per unit length of the line is:

$$P_{loss} = \frac{K}{n}(V - V_c)^2 \tag{4-106}$$

where K is a constant, which relates crest voltage in volts to energy loss in Joules per foot per half cycle. For HV transmission lines, it is of the order of 4×10^{-12}. The factor n may be considered = 2.4 for the positive waves and $n = 4$ for the negative waves.

Corona will reduce the crest of the voltage wave limiting it to the critical corona voltage. The voltage above the corona voltage will cause power loss by ionizing the surrounding air. This increases the effective diameter of the conductor, which in turn increases the capacitance of the conductor and velocity of propagation.

The corona inception voltage is given by:

$$V_c = \frac{2\pi r\varepsilon_0 E_c}{C\times 10^6} \tag{4-107}$$

where r is the radius of conductor in m, ε_0 is the permittivity of free space, E_c is the corona inception electric field in kV/m on the conductor surface, and C is the capacitance of the overhead line in μF/m. The corona onset voltage is, generally, 30 to 40 percent above the rated operating voltage.

Using Peek's equation, V_c can be expressed as:

$$V_c = \frac{1.66\times10^{-3}\, nmk_d^{2/3}\, r\left(1+\dfrac{0.3}{\sqrt{r}}\right)}{k_h C} \tag{4-108}$$

where n is number of subconductors in a bundle, m is conductor surface factor which varies from 0.7 to 0.8, k_d is relative air density, r is conductor radius in cm, C is capacitance as above in μF/m, and k_b is ratio of maximum-to-average surface gradient for bundle conductors. This is given by the expression:

$$k_b = 1 + 2(n-1)\sin\left(\frac{\pi}{n}\right)\left(\frac{r}{A}\right) \tag{4-109}$$

where A is the distance between adjacent subconductors in cm.

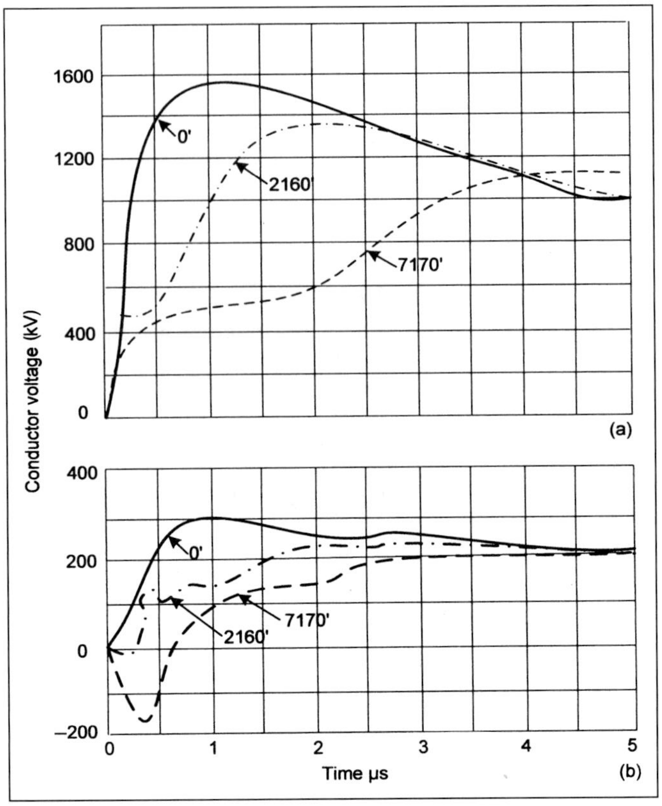

FIGURE 4-13 (*a*) Voltage waveshapes on the surged conductor at various distances from the origin; (*b*) voltage waveshape on parallel unsurged conductor. Positive polarity wave applied to a 2.0-in diameter ACSR.

As the surge voltage is much higher than the corona inception voltage, the corona discharge adds to the energy losses and retards the portion of the voltage wave front above V_c. A sort of shearing back of the voltage wave occurs, as shown in Fig. 4-13.

A voltage wave can be divided into sections of voltage levels of increasing magnitude, each section corresponding to a different velocity of propagation. Prior to ionization, the velocity of propagation is given by, say, C_n, and above inception of corona voltage by C_d, a variable capacitance depending on the voltage and corona inception level. As C_d increases with voltage, the voltage elements above the corona voltage will be retarded behind the voltage elements of lower voltage level. This will flatten the wave. Figure 4-13*a* shows the voltages on the surged conductor and Fig. 4-13*b* shows the voltages on parallel unsurged conductors. A point above the corona inception voltage will be retarded by:

$$t_d = l\left(\frac{1}{v_d} - \frac{1}{v_c}\right) \tag{4-110}$$

where v_c is the velocity of propagation with capacitance C_n and v_d is the velocity of propagation with C_d.[4] While C_n is calculated by conventional expressions, C_d is much dependent on the polarity of the wave, the effect of positive polarity wave being significantly higher than that of negative polarity.

Skilling and Dykes[5] give the following expression for retardation and the distance traveled:

$$\frac{\Delta t}{D} = \frac{k}{v_c C_n}\left(1 - \frac{V_c}{V_d}\right) \tag{4-111}$$

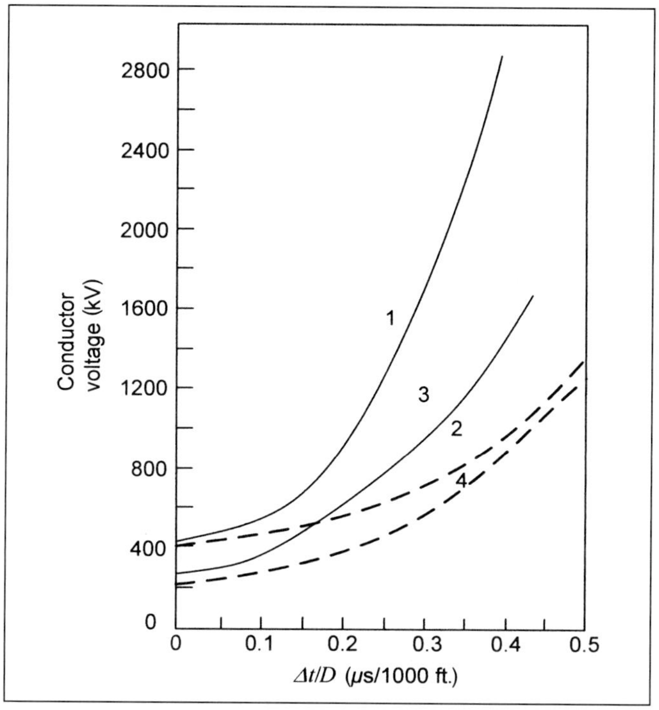

FIGURE 4-14 Slope reduction of traveling waves. 1, 2: 2-in diameter SCA (steel-cored aluminum conductor). 3, 4: 0.927 SCA. 1, 3: negative polarity surges. 2, 4: positive polarity surges.

where Δt is the retardation, D is distance traveled, V_d is voltage above corona voltage V_c, v_c is propagation velocity below V_c in m/s, C_n is capacitance of conductor in μF/m, below corona voltage V_c, and k is empirical constant determined by field tests.

Figure 4-14 gives the function $\Delta t/D$ versus conductor voltage without calculating the rest of Eq. (4-111). The following approximations can be applied with respect to front time prolongation and crest attenuation.

Front Time Prolongation Below V_c: 3 percent per km for first 5 km; 1.2 to 2 percent per km for the next 30 km.

Above V_c: 0.6 μs/km for negative surges and 1.2 μs/km for positive surges.

Crest Attenuation Below V_c: 3 percent per km for the first 5 km; 1.2 to 2 percent per km for the next 30 km.

Above V_c: Foust and Menger formula[6] can be used:

$$V_{\text{dcrest}} = \frac{V_{\text{0crest}}}{(k'DV_{\text{0crest}} + 1)} \tag{4-112}$$

where V_{dcrest} and V_{0crest} are the crest voltages at the origin and after travel of distance D in km, respectively. k' is a constant which increases from 0.0001 for waves of approximately 20 μs half-time to 0.0002 for waves of about 5 μs half-time. For chopped waves, it can be higher up to 0.00045.

4-16 TRANSMISSION LINE MODELS FOR TRANSIENT ANALYSIS

With the earlier background, we return to Table 4-1. CIGRE guidelines describe models for TNA analysis and digital computer simulations; the TNA models are not discussed here. TNA models, however, cannot be used for group III and group IV surges (fast front and very fast front surges). For digital computer simulation, mathematical transformations that allow different modes of propagation in the low-frequency range (groups I and II) are required. This is so because the propagation speed of electromagnetic waves in the ground mode of overhead lines (0.2 to 0.25 km/μs) is lower than the propagation speed of approximately 0.3 km/μs in the line mode. Except for lightning overvoltages, the earth wires can be removed and the impedance matrix is reduced to a 3 × 3 matrix.

For group III, lightning surge studies, earth wires must be included and the transformations to decouple n-conductors to n-equivalent single-conductor system are necessary; see App. D.

For group IV transients, it is adequate to represent an overhead line connected to gas-insulated substations (GIS) through its surge impedances. Within GIS, each small section is modeled; see Chap. 18.

Length of Line A line can be modeled with n Π sections, n depending on the frequency of transient oscillation. The highest frequency that can be attained by a Π section is the natural frequency of one individual element representing the total length: $l_{\text{section}} = l_{\text{total}}/n$. With $L = nL'$ and $C = nC'$, this gives:

$$f_{\max} = \frac{1}{2\pi n\sqrt{L'C'/2}} = \frac{v}{4.44n} \tag{4-113}$$

Or, in other words, if a maximum frequency $f_{\max}$ is to be represented, the maximum length of the section should be:

$$n \le \frac{v}{4.44 f_{\max}} \tag{4-114}$$

The considerations of rise time, attenuation, and phase shift also apply. In case the first capacitance of the Π section is switched in parallel with system capacitance, spurious oscillations may arise and a series resistance equal to the surge impedance is connected in series.

4-16-1 EMTP Models

EMTP permits a number of transmission line models for the transient studies. The Π model can be used for steady-state or frequency scan solutions and is not valid for time-domain solutions. The model data produced is in terms of a Y-matrix representation that includes series and shunt branches.

The constant parameter (CP) model is a frequency-independent transmission line model for the wave equation of the distributed parameter line. It is not accurate for zero sequence currents or high-frequency phenomena. It can be successfully used for problem analysis with limited frequency dispersion.

J. Marti's frequency-dependent model[7] is more accurate than the CP model, though computationally slower. It takes into account the frequency dependence of series resistance and inductance of the line. This model is also based on modal decomposition techniques.

4-16-2 Clarke's Transformations

EMTP uses the following transformation matrices. The Clarke's $\alpha\beta 0$ transformation matrix for m-phase balanced line is:

$$\bar{T}_i = \begin{vmatrix} \frac{1}{\sqrt{m}} & \frac{1}{\sqrt{2}} & \frac{1}{\sqrt{6}} & \cdot & \frac{1}{\sqrt{j(j-1)}} & \cdot & \frac{1}{\sqrt{m(m-1)}} \\ \frac{1}{\sqrt{m}} & -\frac{1}{\sqrt{2}} & \frac{1}{\sqrt{6}} & \cdot & \frac{1}{\sqrt{j(j-1)}} & \cdot & \frac{1}{\sqrt{m(m-1)}} \\ \frac{1}{\sqrt{m}} & 0 & -\frac{2}{\sqrt{6}} & \cdot & \cdot & \cdot & \cdot \\ \cdot & \cdot & 0 & \cdot & \frac{-(j-1)}{\sqrt{j(j-1)}} & \cdot & \cdot \\ \cdot & \cdot & \cdot & \cdot & 0 & \cdot & \cdot \\ \cdot & \cdot & \cdot & \cdot & \cdot & \cdot & \cdot \\ \frac{1}{\sqrt{m}} & 0 & 0 & \cdot & 0 & \cdot & \frac{-(m-1)}{\sqrt{m(m-1)}} \end{vmatrix} \tag{4-115}$$

Applying this m-phase transformation to the matrices of m-phase balanced lines will produce a diagonal matrix of the form:

$$\begin{vmatrix} Z_{g-m} & & & \\ & Z_{L-m} & & \\ & & \cdot & \\ & & & Z_{L-m} \end{vmatrix} \quad (4\text{-}116)$$

Z_{g-m} is the ground mode matrix and Z_{l-m} is the line mode matrix. The solution becomes simpler if m-phase transmission line equations (M-coupled equations) can be transformed into M-decoupled equations. Many transposed and even untransposed lines can be diagonalized with transformations to modal parameters based on eigenvalue/eigenvector theory. The phase differential equation:

$$\frac{d^2V_{\text{ph}}}{dx^2} = \bar{Z}_{\text{ph}}\bar{Y}_{\text{ph}}V_{\text{ph}} \quad (4\text{-}117)$$

becomes:

$$\frac{d^2V_{\text{mode}}}{dx^2} = \bar{\Lambda}V_{\text{mode}} \quad (4\text{-}118)$$

where:

$$\bar{\Lambda} = T_v^{-1}\bar{Z}_{\text{ph}}\bar{Y}_{\text{ph}}\bar{T}_v \quad (4\text{-}119)$$

The diagonal elements of $\bar{\Lambda}$ are eigenvalues of matrix product $\bar{Z}_{\text{ph}}\bar{Y}_{\text{ph}}$ and $\bar{T}_v$ is the matrix of eignvectors or modal matrix of that matrix product. Some methods of finding eigenvectors and eigenvalues are QR transformation and iteration schemes[8]; see App. G.

Similarly, for current:

$$\bar{I}_{\text{mode}} = \bar{T}_i\bar{I}_{\text{ph}} \quad (4\text{-}120)$$

$$\frac{d^2V_{\text{mode}}}{dx^2} = \bar{\Lambda}V_{\text{mode}} \quad (4\text{-}121)$$

$$\bar{T}_i = [\bar{T}_v^t]^{-1} \quad (4\text{-}122)$$

For a three-phase line:

$$\bar{T}_i = \begin{vmatrix} \frac{1}{\sqrt{3}} & \frac{1}{\sqrt{2}} & \frac{1}{\sqrt{6}} \\ \frac{1}{\sqrt{3}} & -\frac{1}{\sqrt{2}} & \frac{1}{\sqrt{6}} \\ \frac{1}{\sqrt{3}} & 0 & -\frac{2}{\sqrt{6}} \end{vmatrix} \quad (4\text{-}123)$$

We can write:

$$\bar{Z}_{012} = \bar{T}_i^{-1}\bar{Z}_{abc}\bar{T}_i \quad (4\text{-}124)$$

Then:

$$\bar{Z}_{123} = \begin{vmatrix} \frac{1}{\sqrt{3}} & \frac{1}{\sqrt{2}} & -\frac{1}{\sqrt{6}} \\ \frac{1}{\sqrt{3}} & -\frac{1}{\sqrt{2}} & \frac{1}{\sqrt{6}} \\ \frac{1}{\sqrt{3}} & 0 & -\frac{2}{\sqrt{6}} \end{vmatrix}^{-1} \begin{vmatrix} Z_s & M_u & M_u \\ M_u & Z_s & M_u \\ M_u & M_u & Z_s \end{vmatrix} \begin{vmatrix} \frac{1}{\sqrt{3}} & \frac{1}{\sqrt{2}} & \frac{1}{\sqrt{6}} \\ \frac{1}{\sqrt{3}} & -\frac{1}{\sqrt{2}} & \frac{1}{\sqrt{6}} \\ \frac{1}{\sqrt{3}} & 0 & -\frac{2}{\sqrt{6}} \end{vmatrix}$$

$$= \begin{vmatrix} Z_s + 2M_u & 0 & 0 \\ 0 & Z_s - M_u & 0 \\ 0 & 0 & Z_s - M_u \end{vmatrix} \quad (4\text{-}125)$$

In equation 4-125, Z_s is self impedance and M_u is neutral impedance. In App. D, we had the same result using symmetrical components. Similarly for the shunt elements:

$$\hat{\bar{Y}} = \bar{T}_v^{-1}\bar{Y}\bar{T}_v \quad (4\text{-}126)$$

Table 4-1 qualifies the representation of transposed lines. For this case, modal decoupling is still possible; however, transition matrices T_i and T_v are different for each line configuration and are a function of frequency.

Example 4-6 Consider the following parameters of a 400-kV transmission line:

Phase conductors: ACSR, 30 strands, 500 kcmil, resistance at 25°C = 0.187 Ω/mi, resistance at 50°C = 0.206 Ω/mi, X_a = 0.421 Ω/mi, outside diameter = 0.904 in, GMR = 0.0311 ft.

Ground wires: 7#8, 115.6 kcmil, 7 strands, resistance at 25°C = 2.44 Ω/mi, resistance at 50°C = 3.06 Ω/mi, X_a = 0.749 Ω/mi, outside diameter = 0.385 in, GMR = 0.00209 ft.

Line configuration: Phase conductors, flat formation, height above ground = 108 ft. Two ground wires, height above ground = 143 ft. Spacing between the ground wires = 50 ft.

Other data: Soil resistivity = 100 Ω/m, tower footing resistance = 20 Ω, span = 800 ft.

The calculated modal parameters for a CP, balanced line using EMTP routine are shown in Fig. 4-15. Note that this table shows two sets of parameters, one at 3543 Hz and the other at 60 Hz. The CP model will be fairly accurate up to 3543 Hz.

4-16-3 Frequency-Dependent Model, FD

The propagation constant can be defined as the ratio of the receiving-end voltage to the source voltage for an open-ended line if the line is fed through its characteristic impedance (Fig. 4-16*a*). Then there will be no reflection from the far end. In this case, $V_m + Z_cI_{mk} = V_s$. We can write the receiving-end voltage at k as:

$$V_k = V_sA(\omega)$$
$$\omega = \exp(-\gamma l) \quad (4\text{-}127)$$

If a unit voltage from dc to all frequencies is applied at the source end, then its time response will be unit impulse, infinitely narrow (area = 1.0), and integral of voltage = unit step (Chap. 2). We can write time response, a(t), to a unit impulse as inverse Fourier transform of $A(\omega)[A(\omega) = e^{-\gamma l}]$. This will not be attenuated and no longer infinitely narrow. The impulse response for a lossless line is unit impulse at $t = \tau$, with area 1.0. Setting $V_{\text{source}} = 1.0$ in Eq. (4-127) means that $A(\omega)$ transformed into time domain must be an impulse, which arrives at the other end k if the source is a unit impulse.

The history term $V_m/Z_c + I_{mk}$ at $t - \tau$ is picked up and weighted with $a(t)$ (App. G.) This weighting at the other end of line is done with convolution integral:

$$\text{hist}_{\text{propagation}} = \int_{\tau=\min}^{\tau=\max} i_{m-\text{total}}(t-u)a(u)dt \quad (4\text{-}128)$$

which can be evaluated with recursive convolution. $I_{m\text{-total}}$ is the sum of the line current I_{mk} and a current which will flow through characteristic impedance if a voltage V_m is applied to it. The approximation of Z_c, a frequency-dependent impedance, is done with a Foster-I-R-C network, Fig. 4-16*b*. Applying trapezoidal rule of integration, App. G, each RC block is a current source in parallel with

```
===========================================================================

       MODAL PARAMETERS FOR BALANCED-LINE TRANSFORMATION MATRICES

  FREQUENCY = 6.0000E+01  HZ                      LENGTH = 3.2187E+02  KM

     R           L           C          ZC        PH(ZC)   ATTENUATION  VELOCITY
 (OHMS/KM)    (MH/KM)   (MICROF/KM)   (OHMS)    (DEGREES) (E**-GAM*L)  (KM/SEC)
3.1424E-01 2.5878E+00 5.6182E-03 6.9563E+02-8.9244E+00 9.2903E-01 2.5901E+05
1.1937E-01 1.4559E+00 7.7265E-03 4.3913E+02-6.1331E+00 9.5694E-01 2.9643E+05
1.1937E-01 1.4559E+00 7.7265E-03 4.3913E+02-6.1331E+00 9.5694E-01 2.9643E+05

MODAL SHUNT CONDUCTANCE (MHOS/KM):

 2.0000E-10  2.0000E-10  2.0000E-10

===========================================================================

       MODAL PARAMETERS FOR BALANCED-LINE TRANSFORMATION MATRICES

  FREQUENCY = 3.5434E+03  HZ                      LENGTH = 3.2187E+02  KM

     R           L           C          ZC        PH(ZC)   ATTENUATION  VELOCITY
 (OHMS/KM)    (MH/KM)   (MICROF/KM)   (OHMS)    (DEGREES) (E**-GAM*L)  (KM/SEC)
2.6237E+00 2.2236E+00 5.6182E-03 6.2956E+02-1.5168E+00 5.1123E-01 2.8283E+05
1.4563E-01 1.4504E+00 7.7265E-03 4.3327E+02-1.2916E-01 9.4733E-01 2.9872E+05
1.4563E-01 1.4504E+00 7.7265E-03 4.3327E+02-1.2916E-01 9.4733E-01 2.9872E+05

MODAL SHUNT CONDUCTANCE (MHOS/KM):

 2.0000E-10  2.0000E-10  2.0000E-10

===========================================================================
```

FIGURE 4-15 Calculated modal parameters of the transmission line described in Example 4-6, EMTP calculation routine.

an equivalent resistance. Summing these up, a frequency dependent line is represented as shown in Fig. 4-16*c*. J. Marti[9] shows that it is best to sum up $A(\omega)$ and $Z(\omega)$ in the frequency domain.

$$A(s)=e^{-s\tau_{\min}}k\frac{(s+z_1)(s+z_2)\cdots(s+z_n)}{(s+p_1)(s+p_2)\cdots(s+p_n)} \tag{4-129}$$

From Chap. 3, partial fractions can be used and the time response is:

$$a(t)=[k_1e^{-p_1(t-\tau_{\min})}+k_2e^{-p_2(t-\tau_{\min})}+\cdots+k_me^{-p_m(t-\tau_{\min})}] \quad \text{for} \quad t\ge\tau_{\min}$$
$$=0 \quad \text{for} \quad t<\tau_{\min} \tag{4-130}$$

The weighting factor is used to calculate the history term in each time step. Similar expression of a transfer function of poles and zeros for Z_c is applicable.

$$Z_c(s)=k\frac{(s+z_1)(s+z_2)\cdots(s+z_n)}{(s+p_1)(s+p_2)\cdots(s+p_n)} \tag{4-131}$$

The success depends on the quality of approximation for $A(\omega)$ and $Zc(\omega)$. J. Marti used Bode's procedure for approximating the magnitudes of the functions. The M-phase lines, any of the M-modes can be specified as frequency dependent or with lumped resistances or distortionless. Field tests have verified the accuracy of calculations.[10]

Figure 4-17 shows the comparative simulation results of a 320-mi line, a dc voltage of 10 V connected to the sending end of the line, receiving end terminated with a shunt inductance of 100 mH, R = 0.0376 Ω/mi, L = 1.52 Ω/mi, and C = 14.3 nF/mi. The simulation results with Π sections (8 and 32) and distributed parameters are shown. Note the spurious oscillations due to lumpiness.

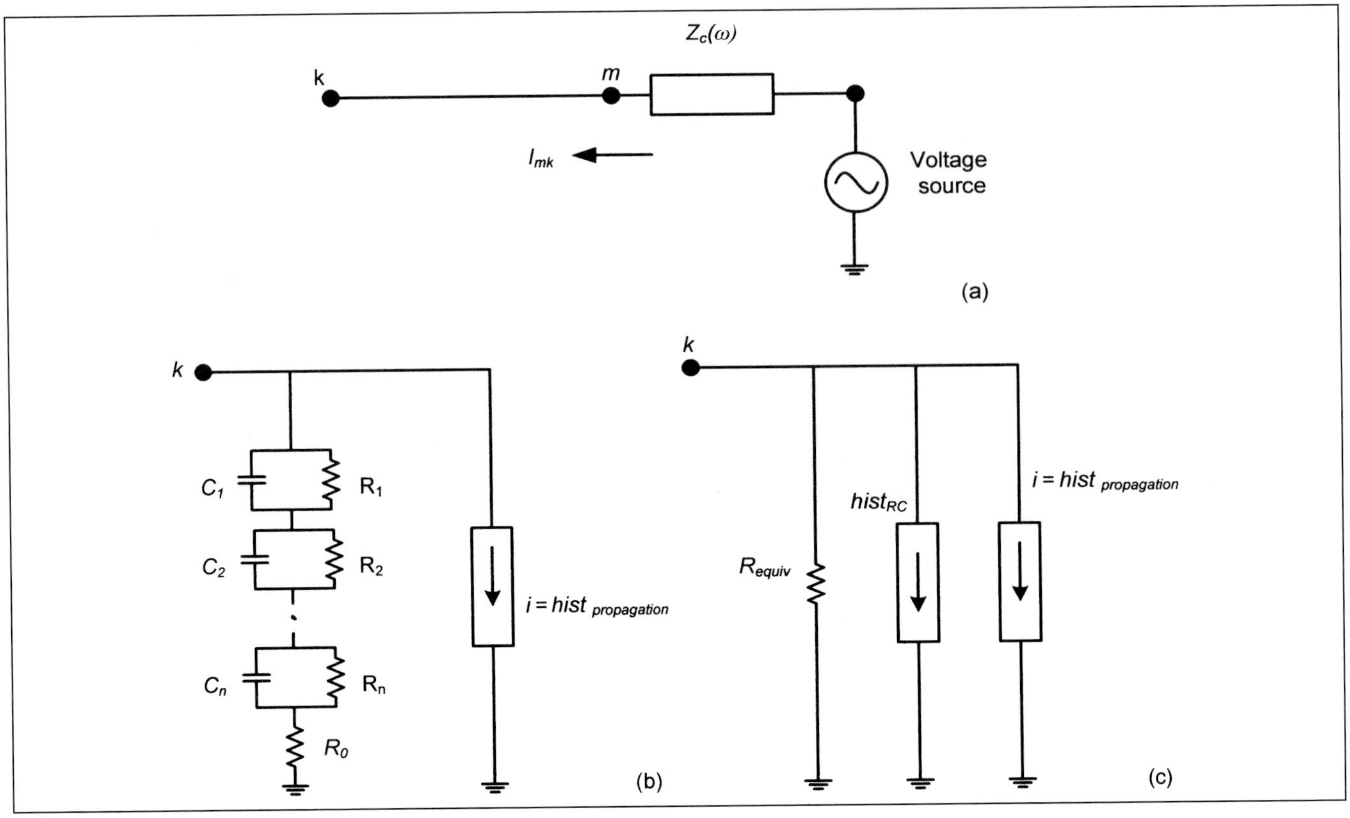

FIGURE 4-16 To explain J. Marti's frequency-dependent line model (FD) in EMTP: (*a*) a voltage source connected through matching impedance to node m; (*b*) RC network; (*c*) circuit with equivalent resistance after applying implicit integration.

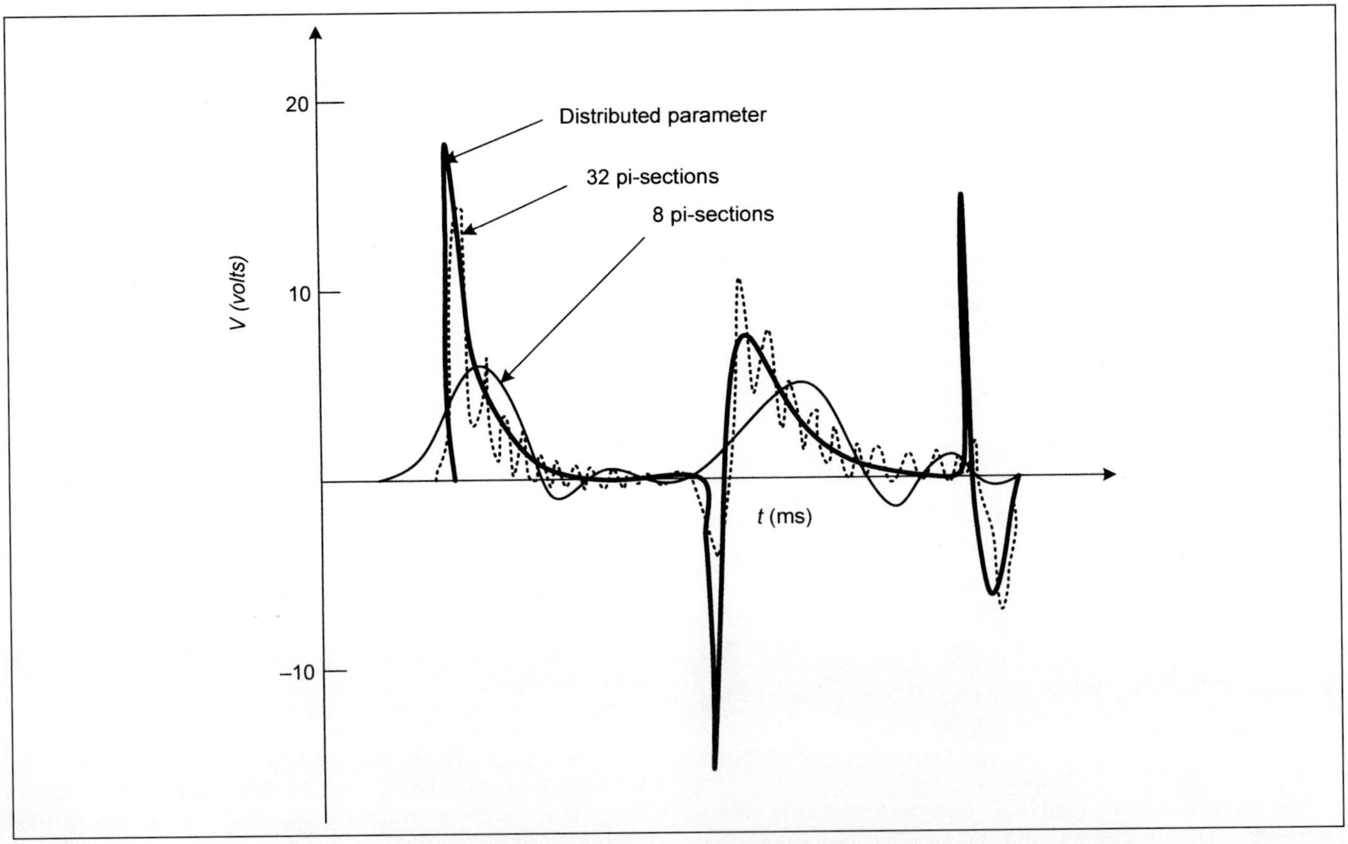

FIGURE 4-17 Comparative response of a transmission line with Π sections (8 and 32) and CP model. See text.

Example 4-7 Consider a three-phase 400-kV, 200-mi long line model, as arrived in Example 4-6, energized from a three-phase source ($Z^{+} = 0.3856 + j5.776\,\Omega$, $Z_0 = 0.5245 + j6.805\,\Omega$), by closing an ideal switch at $t = 0$, at the peak of phase *a* voltage. The receiving end is open-circuited. The EMTP simulation of voltage profile at the sending and receiving end is shown in Fig. 4-18*a* and *b*, respectively. The receiving-end voltage approximately doubles.

Example 4-8 This example models a lightning surge of 1.2/50 μs, 500 kV, which impacts phase *a* of a 400-kV transmission line at 1000' away from the receiving-end substation and the receiving end is open. The 400-kV line has two ground wires and a CP model is used. A description of 1.2/50 μs waveshape is provided in Chap. 5. Figure 4-19*a* shows the surge voltages at the receiving end of the line. The phase *a* voltage rises to approximately 920 kV peak. Voltages on coupled phases *b* and *c* are also shown. Figure 4-19*b* shows the surge voltages on the ground wires at the receiving end.

4-17 CABLE TYPES

Extruded dielectric (XD), also called solid dielectric—principally cross-linked polyethylene (XLPE) cables—have been used up to 500 kV. These are replacing pipe insulated cables and oil-filled paper-insulated cables because of lower costs and less maintenance. Extruded cables include ethylene-propylene rubber (EPR) and low-density polyethylene (LLDPE), though XLPE is the most common insulation system used for transmission cables. The XD cables also have a lower capacitance compared to the paper-insulted cables. In the United States, though XD cables are popular for the new installations, approximately 70 percent of circuit miles in service are pipe-type cables. Triple extrusion (inner shield, insulation, and outer conducting shield) processed in one sealed operation is responsible for the development of XLPE cables for higher voltages. Also, a dry vulcanization process with no steam or water is used in the modern manufacturing technology.

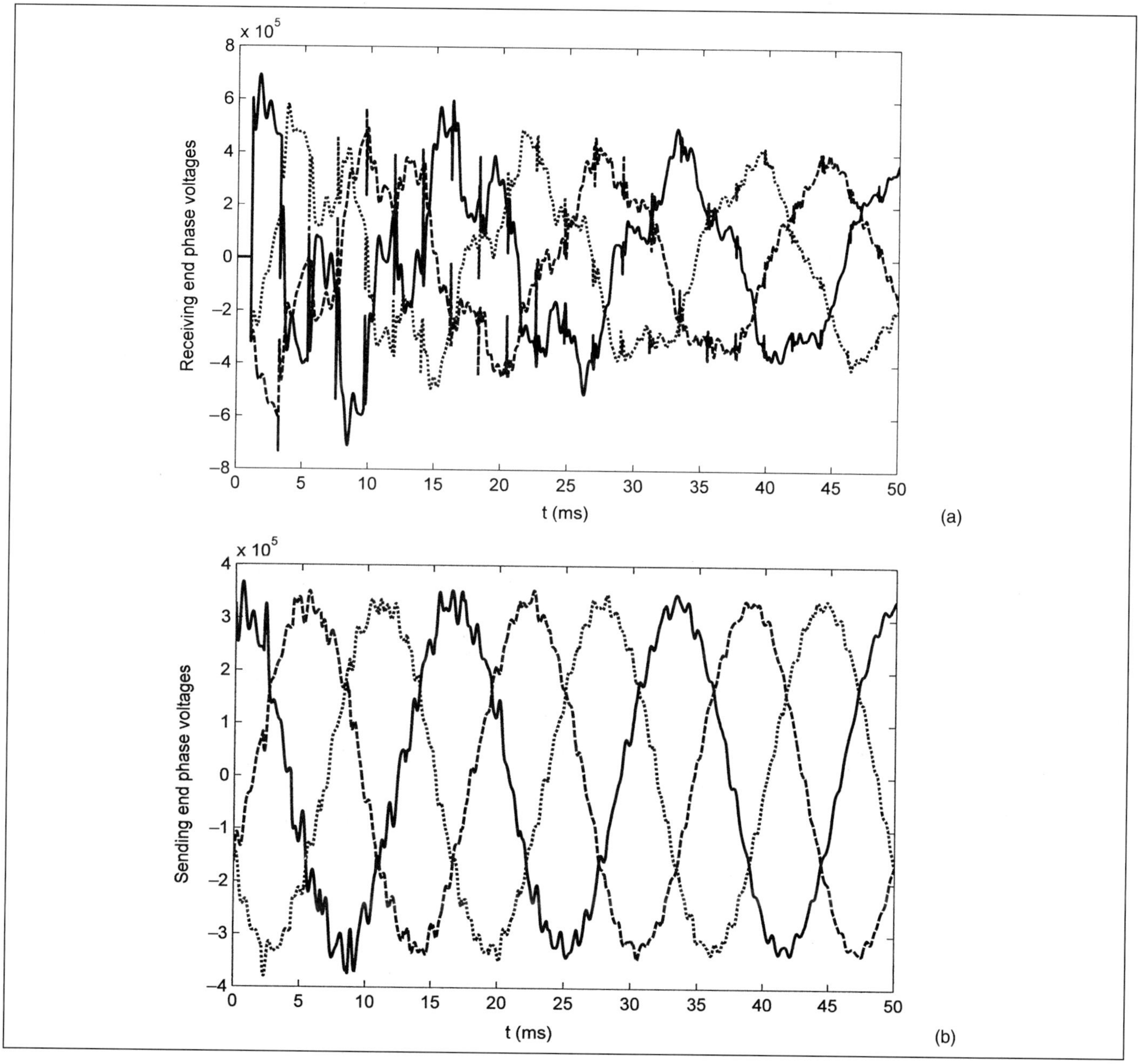

FIGURE 4-18 (*a*) Simulated three-phase receiving-end voltage transients, Example 4-7; (*b*) three-phase sending-end voltage transients, Example 4-7.

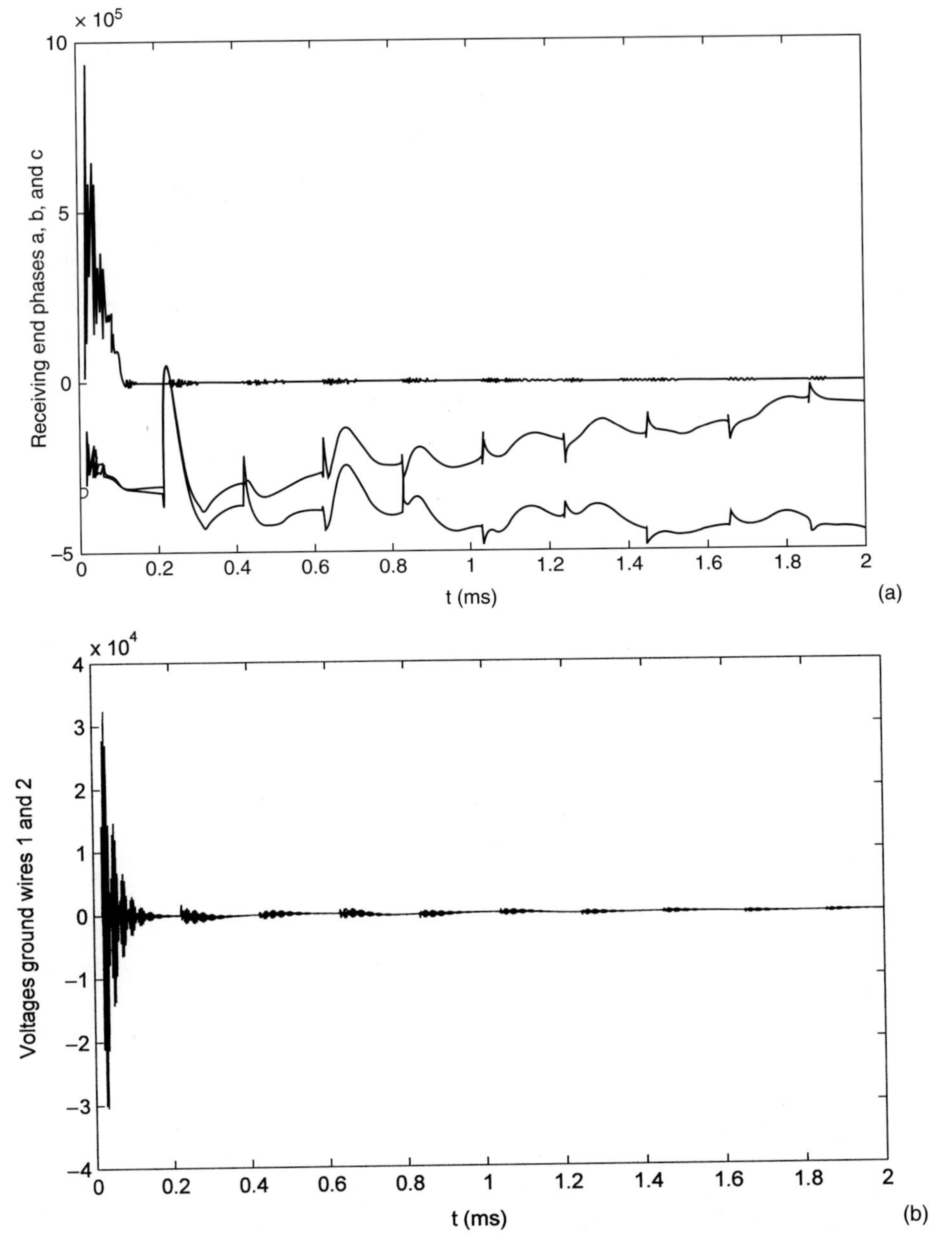

FIGURE 4-19 (*a*) Simulated surge voltage transients at the receiving end, Example 4-8; (*b*) voltage transients on the ground wires.

An EHV XLPE construction from inside to outside follows:

- Copper or aluminum conductor which may be sector shaped
- Carbon paper, and inner extruded semiconducting shield
- XLPE insulation
- Outer semiconducting extruded screen
- Semiconducting tapes
- Copper screen with counter helix
- Layer of swelling material, longitudinal water barrier; the XD cables are proven to be sensitive to moisture, so moisture barriers are used
- Semiconducting tapes
- Laminated aluminum tape (water vapor diffusion barrier)
- Outer polyethylene sheath

Over the course of years, the reliability of XLPE cables has increased, costs have decreased, and electrical stresses have increased, allowing construction of XLPE cables to be applied to high voltages.

Low-pressure oil-filled cables, also called self-contained liquid-filled systems (SCLF), were developed in 1920 and were extensively used worldwide until 1980, while in the United States, pipe-type cables (described next) were popular. There are many miles in operation at 525 kV, both land and submarine cables. The stranded conductors are formed to form a hollow duct in the center of the conductor. These may be wrapped with carbon paper, nonmagnetic steel tape, and conductor screen followed by many layers of oil-filled paper insulation. Insulation screen, lead sheath, bedding, reinforcement tapes, and outer polymeric sheath follow next. When the cable heats up during operation, the oil flows through the hollow duct in an axial direction into oil reservoirs connected with the sealing ends. The oil reservoirs are equipped with gas-filled flexible-wall metal cells. The oil flow causes the cells to compress, thus increasing the pressure. This pressure forces the oil back into the cable.

High-pressure fluid-filled (HPFF) cables have been a U.S. standard. The system is fairly rugged. A welded cathodically protected steel pipe, typically 8.625 or 10.75 in optical density, is pressure and vacuum tested, and three mass-impregnated cables are pulled into the pipe. The cable consists of copper or aluminum conductors, conductor shield, paper insulation, insulation shield, outer shielding, and skid wire for pulling cables into the pipe. The pipe is pressurized to approximately 200 psig with dielectric liquid to suppress ionization. The expansion and contraction of the liquid requires large reservoir tanks and sophisticated pressurizing systems. Nitrogen gas at 200 psi may be used to pressurize the cable at voltages up to 138 kV.

To state other constructions, mass-impregnated nondraining cables (MIND) are used for HVDC circuits. Superconductivity using liquid helium at 4 K has been known and, in the last couple of years, has produced high-temperature superconductors (HTS). These use liquid nitrogen temperatures (80 K).

Comparing with power lines, some advantages are:

- Environmentally little or no visual impact, no electric fields or low magnetic fields, high safety, no external corona discharges
- Not effected by weather conditions, like storm, snow, fog, or dust
- Low maintenance, lower power losses, less right of the way compared to power lines
- Fewer faults, higher short-term overload capability, and reliability

Conversely, a variety of problems are associated with long underground cables. The charging current of the cables is high due to higher capacitance. At high voltages, this becomes important, and reactive power compensation should be carefully considered. The charging current is given by $2\pi fCV_{lg}$, where V_{lg} is the line-to-ground voltage. A relatively high voltage can occur at the open end when a cable is connected to a relatively weak electrical system. Another potential problem can be self-excitation of a synchronous machine connected to the far end.

Temporary overvoltages can occur when a cable is switched with a transformer. These are all related to large cable capacitance. Shunt reactors are commonly used to compensate for cable capacitance. These need to be properly sized to prevent steady-state and transient overvoltages. The risk of overvoltages will be more under light load conditions. Table 4-3 compares the electrical characteristics of cables versus that of 230-kV overhead lines.[11]

4-17-1 Cable Models

Appendix D contains calculation routines for cable constants. The cable models for transient analysis depend on the cable construction and also on the geometry of installation, which become all the more important for submarine and pipe-type cables. A cable can be represented in a transient analysis by its resistance, characteristic impedance, attenuation, or velocity of propagation, much like a transmission line. A cable length can be considered short if the surge traveling time is lower than about 30 percent of the time constant of the main voltage rise time in the system. If a rectangular pulse travels on an overhead line and then a short length of cable, then the cable termination is seen more like a lumped capacitance. If the cable is modeled accurately as a lossless distributed parameter line, the buildup of the voltage has a staircase shape.

The Π model can be used for steady-state or frequency scans and is not valid for time-domain simulations. Spurious oscillations may occur due to lumped models, though a number of sections may be used (Fig. 4-2*a*).

The CP model assumes that R, L, C are constants, and they are calculated at a certain frequency. The model considers R and L to be distributed, while R is lumped at three places, at cable ends and cable middle. The conductance is neglected. The model is a lumped-impedance representation of the cable. The cross bonding of sheaths of the cables presents yet other considerations, and each subsection of the cable model has sheath bonding and grounding connections. To model a cross-bonded cable accurately, it is divided into sections, analogous to modeling a transposed transmission line. Sheath bonding and sheath connections are made using EMTP nodes. This modeling of short sections of 400 m or so requires a very small time step, and a number of these major sections must be connected to represent the entire cable length. This becomes computationally intensive.

Similar to transmission lines, the models for cable systems can be type FDQ, frequency-dependent, which provide an accurate representation of the distributed nature of all cable parameters, series and shunt elements, and their frequency dependence of transition matrices T_i and

TABLE 4-3 Typical Electrical Characteristics, 230-kV OH Lines versus Underground Cables

PARAMETER	OVERHEAD LINE	UNDERGROUND XLPE	UNDERGROUND HPFF
Shunt capacitance, μF/mi	0.015	0.30	0.61
Series inductance, mH/mi	2.0	0.95	0.59
Series reactance, Ω/mi	0.77	0.36	0.22
Charging current, A/mi	1.4	15.2	30.3
Dielectric loss, kW/mi	0+	0.2	2.9
Reactive charging power, MVA/mi	0.3	6.1	12.1
Capacitive energy, kJ/mi	0.26	2.3	7.6
Surge impedance, Ω	375	26.8	14.6
Surge impedance loading limit, MW	141	1975	3623

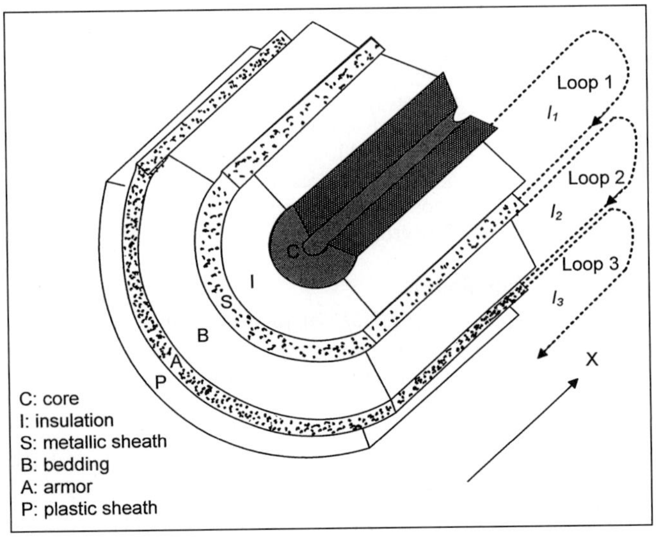

FIGURE 4-20 Cross section of a single core self-contained cable, with current loops. (Conductive layers are not shown.)

T_v in modal quantities. To create data for this model, it is necessary to approximate with rational functions the characteristic admittance Y_c and the propagation function $A = e^{-\gamma l}$ for each mode of the cable and the modal transformation matrix Q in the frequency domain.[12] Chapter 7 has an example of FDQ model for a 400-kV cable.

The cable parameters of coaxial arrangements, as shown in Fig. 4-20, are derived from coaxial loops.[13] Loop 1 is formed by core conductor and sheath as return, loop 2 by metallic sheath and armor, and loop 3 by armor and earth or sea water. The series impedance can be written as:

$$-\begin{vmatrix} \dfrac{dV_1}{dx} \\ \dfrac{dV_2}{dx} \\ \dfrac{dV_3}{dx} \end{vmatrix} = \begin{vmatrix} Z'_{11} & Z'_{12} & 0 \\ Z'_{21} & Z'_{22} & Z'_{23} \\ 0 & Z'_{32} & Z'_{33} \end{vmatrix} \begin{vmatrix} I_1 \\ I_2 \\ I_3 \end{vmatrix} \tag{4-132}$$

The self-impedance of loop 1 consists of three parts:

$$Z'_{11} = Z'_{\text{core-out}} + Z'_{\text{core/sheath-insulation}} + Z'_{\text{sheath-in}} \tag{4-133}$$

With

Z' core-out = internal impedance of tubular core conductor with return path outside tube, through sheath

Z' core/sheath-insulation = impedance of insulation between core and sheath

Z' sheath-in = internal impedance of the tubular sheath with return path inside the tube

$$\begin{aligned} Z'_{22} &= Z'_{\text{sheath-out}} + Z'_{\text{sheath/armor-insulation}} + Z'_{\text{armor-in}} \\ Z'_{33} &= Z'_{\text{armor-out}} + Z'_{\text{armor/earthinsulation}} + Z'_{\text{earth}} \end{aligned} \tag{4-134}$$

Analogous definitions apply. For shunt admittances:

$$\begin{aligned} -\frac{dI_1}{dx} &= (G'_1 + j\omega C'_1)V_1 \\ -\frac{dI_2}{dx} &= (G'_2 + j\omega C'_2)V_2 \\ -\frac{dI_3}{dx} &= (G'_3 + j\omega C'_3)V_3 \end{aligned} \tag{4-135}$$

The calculation of all the parameters becomes involved. Equation (4-132) is not yet supported in EMTP routine.

Example 4-9 A cable 5-km long, with a characteristic impedance of 50 Ω, propagation velocity 1E05 km/s is connected to a 1-km long line, with characteristic impedance 400 Ω, propagation velocity 3E05 km/s, and open at the receiving end. The sending end is impacted with a step of 10 kV, duration 10 μs. Figure 4-21 shows the voltages at the junction of the cable and transmission line and also at the open end of the transmission line. Due to different speeds of propagation on cables and lines, multiple reflections occur at the junction point; voltage escalation of 3.7 times occurs at the open end of the transmission line.

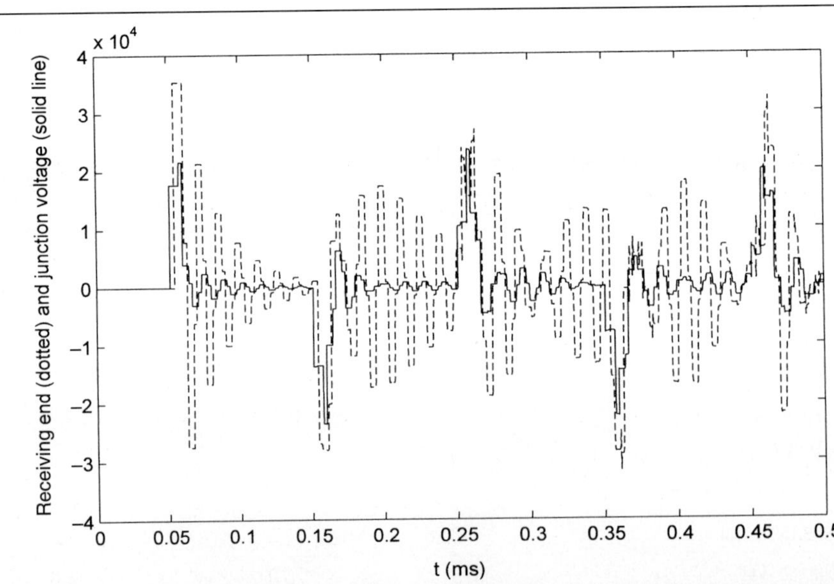

FIGURE 4-21 Transients at the open end of a transmission line connected to a cable and also at the junction of cable and transmission line, Example 4-9.

PROBLEMS

1. Write an equation relating *R*, *L*, *G*, and *C* for a lossless line. Why can such a line not be constructed? What is an infinite line? Why line compensation in the form of capacitors or reactors is required?

2. A rectangular voltage pulse of 50 V, 6 μs duration is applied to the sending end of a cable having a source impedance of 100 Ω, short-circuited at the far end. The cable parameters are: $C = 100$ pF/m, $L = 0.16$ μH/m. The cable is 400 m long. Draw a lattice diagram and plot the sending-end voltage for 15 μs.

3. Derive the ABCD constants for a line having resistance of 0.1 Ω/mi, reactance 0.86 Ω/mi, and capacitance 0.04 Ω/mi using long line model. What is the electrical length of the line?

4. In Problem 3, calculate the rise in voltage at the receiving end. Use a long line model and consider 400 mi of line length, the sending-end voltage being 230 kV.

5. In Problem 3, what is the SIL loading of the line?

6. Consider the parameters of Example 4-9. A voltage of 10 kV (continuous ramp) is applied to the sending end of the cable at $t = 0$. Draw a lattice diagram and plot the voltage profiles: (1) at the junction point, (2) at the open end of the transmission line, and (3) at the sending end.

7. A line of surge impedance 400 Ω is terminated in a resistor in parallel with a capacitor. Write an expression for the voltage across the resistor.

8. Draw the current and voltage profile of a symmetrical 230-kV, 300-mi long line, at no load and when supplying 150 MW of power at 0.8 power factor. Consider $L = 2$ mH/mi, $C = 0.12$ μF/mi.

9. An infinite rectangular wave on a line having a surge impedance of 400 Ω strikes a transmission line terminated in a capacitor of 0.004 μF. How much is the wave front retarded?

10. A long transmission line is energized by a unit step function at the sending end and is open-circuited at the receiving end. Construct Bewely lattice diagram and obtain the value of the voltage at the receiving end after a long time. Consider an attenuation factor of 0.8.

11. Derive the relations shown in Eq. (4-17).

12. To study transient behavior up to a maximum frequency of 3000 Hz, how many Π sections should be modeled for a line length of 150 mi?

REFERENCES

1. CIGRE WG 33.02, "Guidelines for Representation of Network Elements when Calculating Transients," CIGRE Brochure 39, 1990.
2. IEEE Task Force Report, "A Comparison of Methods for Calculating Audible Noise of High Voltage Transmission Lines," *IEEE Trans. Power Apparatus and Systems*, vol. PAS101, no. 10, p. 4290, Oct. 1982.
3. M. R. Moreau and C. H. Gary, "Predetermination of Radio-Interference Level of High-Voltage Transmission Lines-1: Predetermination of Excitation Function," *IEEE Trans. Power Apparatus and Systems,* vol. PAS91, p. 284, 1972.
4. P. Chowdhri, *Electromagnetic Transients in Power Systems*, Research Study Press, Somerset, England, 1996.
5. H. H. Skilling and P. Dykes, "Distortion of Traveling Waves by Corona," *AIEEE Trans*, vol. 56, pp. 850–875, 1937.
6. *Transmission and Distribution Handbook*, Westinghouse Electric Corporation, Pittsburgh, PA, 1964.
7. J. Marti, "Accurate Modeling of Frequency Dependent Transmission Lines in Electromagnetic Transient Simulations," *IEEE Trans. Power Apparatus and Systems*, vol. PAS101, pp. 147–157, 1982.
8. J. H. Wilkinson, *The Algebraic Evaluation Problem*, Oxford University Press, London, 1965.
9. J. R. Marti, "The Problem of Frequency Dependence in Transmission Line Modeling," PhD thesis, The University of British Columbia, Vancouver, Canada, April 1981.
10. W. S. Meyer and H. W. Dommel, "Numerical Modeling of Frequency Dependent Transmission Line Parameters in EMTP," *IEEE Trans. Power Apparatus and Systems*, vol. PAS93, pp. 1401–1409, Sept./Oct. 1974.
11. H. W. Beaty and D. G. Fink (eds.), *Standard Handbook for Electrical Engineers*, 15th ed., McGraw-Hill, New York, 2007.
12. L. Marti, "Simulation of Transients in Underground Cables with Frequency-Dependent Modal Transformation Matrices," *IEEE Trans. Power Delivery*, vol. 3, no. 3, pp. 1099–1110, 1988.
13. L. M. Wedepohl and D. J. Wilcox, "Transient Analysis of Underground Power Transmission Systems: System Model and Wave Propagation Characteristics," *Proc. IEE*, vol. 120, pp. 252–259, Feb. 1973.

FURTHER READING

G. W. Alexander and H. R. Armstrong, "Electrical Design of a Double Circuit Transmission Line, Including the Effects of Contamination," *IEEE Trans. Power Apparatus and Systems*, vol. 85, pp. 656–665, 1966.

J. G. Anderson, *Transmission Reference Book*, Edison Electric Company, New York, 1968.

J. Aubin, D. T. McGillis, and J. Parent, "Composite Insulation Strength of Hydro-Quebec 735 kV Towers," *IEEE Trans. Power Apparatus and Systems*, vol. 85, pp. 633–648, 1966.

L. V. Beweley, *Traveling Waves on Transmission Systems*, 2d ed., John Wiley & Sons, New York, 1951.

C. Gary, D. Cristescu, and G. Dragon, "Distortion and Attenuation of Traveling Waves Caused by Transient Corona," CIGRE Study Committee 33 Report, 1989.

IEEE Task Force Report, "Review of Technical Considerations on Limits of Interference from Power Lines and Stations," *IEEE Trans. Power Apparatus and Systems*, vol. PAS99, no. 1, p. 365, Jan./Feb. 1980.

L. B. Loeb, *Electrical Coronas: Their Basic Physical Mechanisms*, University of California Press, USA, 1965.

M. R. Moreau, and C. H. Gary, "Predetermination of Radio-Interference Level of High-Voltage Transmission Lines-II: Field Calculating Method," *IEEE Trans. Power Apparatus and Systems,* vol. PAS91, p. 292, 1972.

M. S. Naidu and V. Kamaraju, *High Voltage Engineering*, 2d ed., McGraw-Hill, New York, 1999.

R. Rudenberg, *Transient Performance of Electrical Systems*, McGraw-Hill, New York, 1950.

Transmission Line Reference Book, 345 kV and Above, EPR1, Palo Alto, CA, 1975.

C. F. Wagner, "A New Approach to the Calculation of Lightning Performance of Transmission Lines, Part I," *AIEE Trans.* vol. 75 Pt. III, pp. 1233–1256, 1956.

C. F. Wagner and A. R. Hileman, "A New Approach to the Calculation of Lightning Performance of Transmission Lines, Part II," *AIEE Trans.* vol. 78 Pt. III, pp. 996–1081, 1959.

C. F. Wagner and A. R. Hileman, "A New Approach to the Calculation of Lightning Performance of Transmission Lines, Part III," *AIEE Trans.* vol. 79 Pt. III, pp. 588–603, 1960.

C. F. Wagner and B. L. Lloyd, "Effects of Corona on Traveling Waves," *AIEE Trans.*, vol. 74, Pt. III, pp. 858–872, 1955.

CHAPTER 5

LIGHTNING STROKES, SHIELDING, AND BACKFLASHOVERS

The scientific study of lightning started in the summer of 1752, when Benjamin Franklin attached a metal key to a kite string and then flew the kite during a thunderstorm. Thus, more than 250 years ago he proved that lightning was an electrical discharge. Modern research began in the twentieth century with the work of C.T.R. Wilson, who was the first to infer the charge structure of the thunderclouds and the magnitude of charge involved in lightning. In 1930, the research was motivated primarily to protect power systems and understand lightning phenomena. In 1960, there was a renewed interest because of vulnerability of solid-state electronic devices to surges. The research was further accelerated in 1969 when the Apollo/Saturn crew survived two lightning flashes after takeoff. Research continues in this field, for example, rocket-triggered lightning tests (RTL) and linear finite difference time-domain (FDTD) electromagnetic solvers. On transmission lines, lightning is the main cause of unscheduled interruptions. In recent years much knowledge has been gained on very widely varying phenomena of statistical and nonlinear nature as evidenced by a spate of publications on the subject.

5-1 FORMATION OF CLOUDS

The clouds are composed of water droplets and ice crystals. The altitude at which supersaturation occurs in the rising air streams and clouds begin to form is called the lifted condensation level (LCL), which is within 1000 m of the earth's surface. A great majority of clouds form at temperatures above freezing and consist entirely of liquid droplets. Many observations disclose that a cloud must extend 2 to 3 km in the subfreezing portion of atmosphere before the first lighting is observed. A large variety of meteorological conditions are favorable for lightning to occur.

Figure 5-1 describes the tripole structure of the thundercloud, and Fig. 5-2, which shows three regions, is the classic Simpson's model. Below region A, the air currents travel above 800 cm/s and no raindrops fall through. In region A, the velocity of air currents is high enough to break falling raindrops, causing positive charge spray in the cloud and negative charges in the air. The spray is blown upward, but as the velocity decreases, the positively charged water drops combine with larger drops and fall again. Region A becomes positively charged, while region B becomes negatively charged due to air currents. In the upper regions, the temperature is below the freezing point and only ice crystals exist.

Though the theory has been modified, the three charge regions are confirmed by modern measurements. The main negative charge in the central portion is in the temperature zone between –10° and –20° C, often less than 1 km in vertical extent. The upper and lower regions in the cloud are separated by a quasi-neutral zone. The main charge regions in the cloud are confined vertically rather than horizontally. The lower positive charge is typically smaller in magnitude than the main negative charge. The top positive charge is more diffused. As the freezing of droplets progresses from outside to inside, the outer negatively charged shells fall off, and the remaining positively charged fragments are swept upward in the convection current. The exact mechanism of thundercloud electrification is still not clearly understood, although a number of theories attempt to account for the electrification mechanism. Some examples are

- Inductive mechanism and convective mechanism
- Selective ion capture theory
- Drop breakup theory
- Thermoelectric effects
- Surface potential theories

We will not go into these theories and Ref. 1 provides further reading. For lightning to occur, some conditions to be met are:

1. Cloud depth must be greater than 3 to 4 km.

2. Strong electrifications are not observed unless the cloud extends above freezing level.

3. Highly charged regions almost always coincide with coexistence of ice and supercooled water.

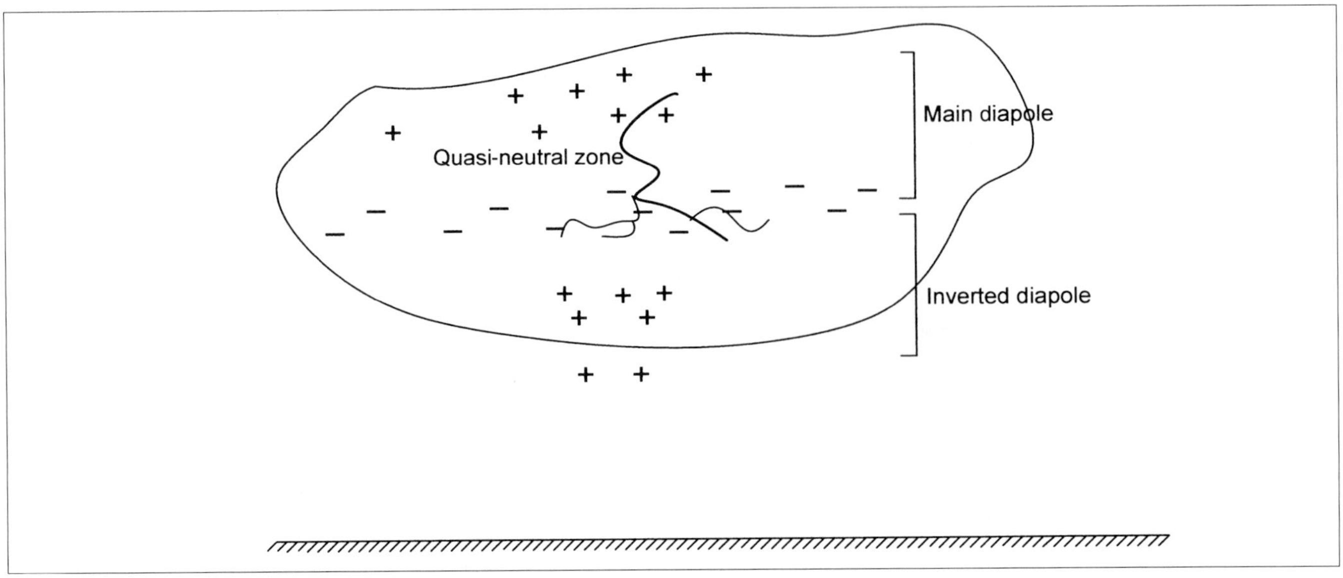

FIGURE 5-1 The tripole structure of thunderclouds.

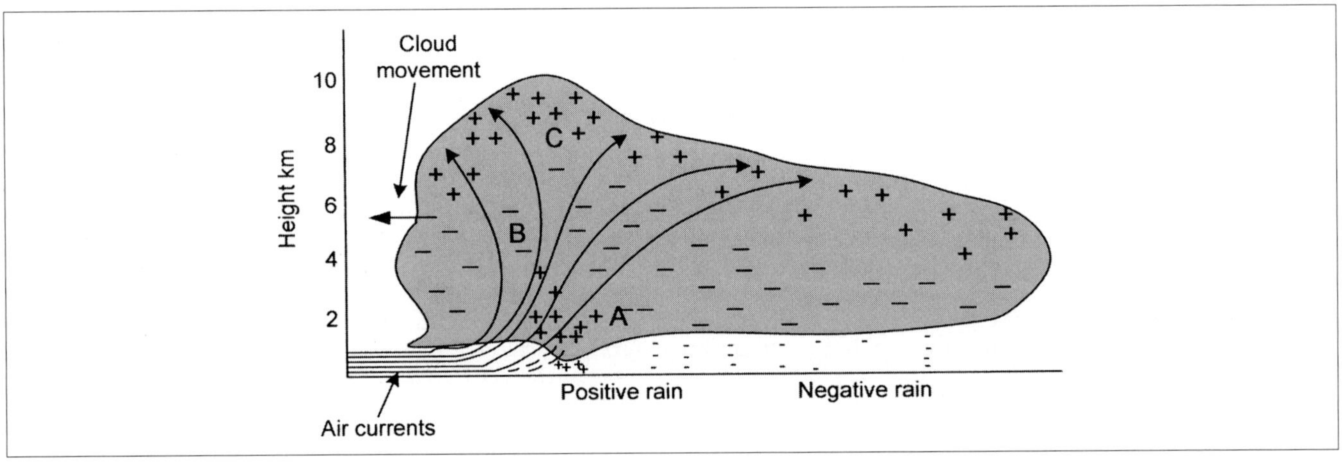

FIGURE 5-2 The classic charge structure of a thundercloud.

4. Strong electrification occurs when the cloud exhibits strong convective activity with rapid vertical development. The initial rate of electrification has a time constant of about 2 min. In the mature stage, electrical fields as high as 400 kV/m may be produced.

5-2 LIGHTNING DISCHARGE TYPES

The main lightning discharge types are (Fig. 5-3):

- *Intercloud flash*. The most common lightning discharge type is *intercloud flash*. This is a discharge between upper positive and main negative charge region of the cloud.

- *Cloud to ground flash*. The ground flash transfers negative charge (generally) from the main negative region of the cloud to ground; however, an initial discharge between the upper negative and the lower positive charge is an essential process. Also positive-charge transfers to ground can occur, though not so common.

- *Air discharge*. An air discharge is a discharge in the air; it does not touch the ground.

5-3 THE GROUND FLASH

The ground flash occurs between charge centers of the cloud and ground. A negative ground flash brings negative charge to ground and is most common, and a positive ground flash brings positive charge to ground. From the power systems point of view, this is of much concern and we will study its mechanism in the following sections.

5-3-1 Stepped Leader

The ground flash is initiated by electrical breakdown in the cloud, called the primary breakdown. This creates a column of charge called a *stepped leader* that travels from the cloud to ground in a stepped manner. On its way to ground, it may give rise to several branches. Periodically, the streamer reaches too far from the charge in the cloud, and the electrical field at the tip is too weak to propagate it further toward the ground. This slight time delay allows more charge to be transferred from the cloud to the tip of the streamer, and it moves further toward the ground, hence the name stepped leader. The propagation of each step is random, as the charge density at the tip of the streamer is dependent on local

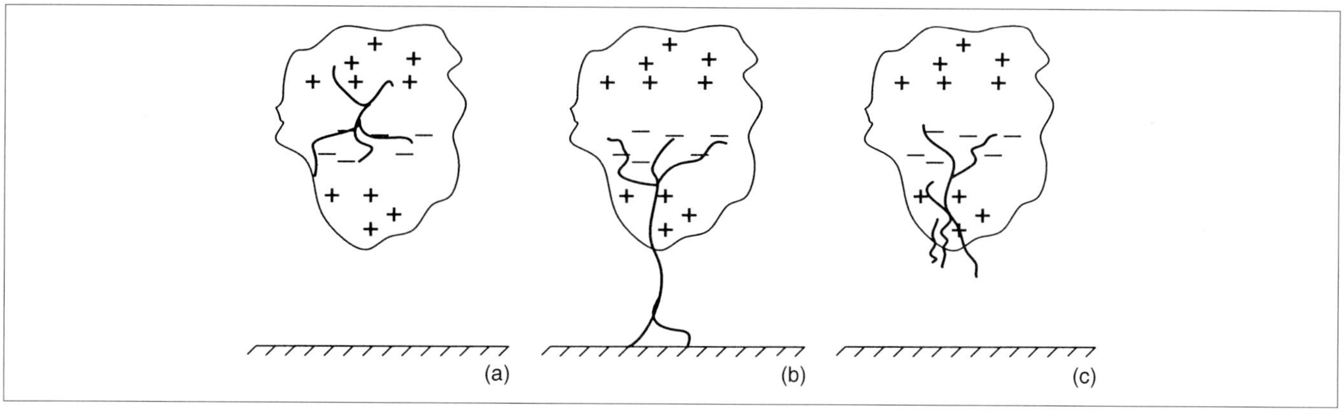

FIGURE 5-3 Predominant lightning types: (*a*) Intercloud flash; (*b*) cloud to ground flash; (*c*) air discharge.

conditions. Thus, most stepped leaders follow a rather twisted path to the ground. Figure 5-4 shows that several streamers can originate from the same charge cluster, and these become secondary leaders.

5-3-2 Return stroke

As the leader approaches the ground, the electrical field at ground level increases. Elevated objects, such as trees, power lines, structures, and buildings, are more vulnerable. Around an induced electrical field level of about 3 MV/m, a corona streamer of opposite polarity rises from the object to meet the downward leader. These discharges from the grounded buildings and structures are called *connecting leaders* and they travel toward the stepped leader. One of the connecting leaders may successfully bridge the gap between the ground and the stepped leader. This separation between the object struck and the tip of the stepped leader at the inception of the connecting leader is called the *striking distance*, which is of importance in shielding and lightning protection, as discussed in Sec. 5-6. This completes the current path between the cloud and the ground, and high discharge current flows to neutralize the charge, positive from ground to negative in the cloud. This is called the return stroke; it is highly luminous and strongly branched. Currents of the order of 100 kA and more are possible.

In most cases, the return stroke prevents any further propagation of the leader, as the current path is neutralized. However, two branches of the leader can reach the ground simultaneously, resulting in two return strokes.

An upward moving stroke may encounter a branch end and there is an immediate luminosity of the channel; such events are called branch components. In certain cases, the return stroke current may not go to zero quickly and continue to flow for tens to a few hundreds of milliseconds. Such long-duration currents are called continuing currents.

5-3-3 Dart leader and multiple flashes

The lightning flash may not end in the first flash. The depletion of charge in the original cloud cluster creates cloud to cloud (CC) discharges and connects the adjacent charge clusters to the original charge cluster, which may get sufficiently charged to create subsequent flashes. Because the discharge path from the original flash is still ionized, it takes less charge to start a flash. The preionized

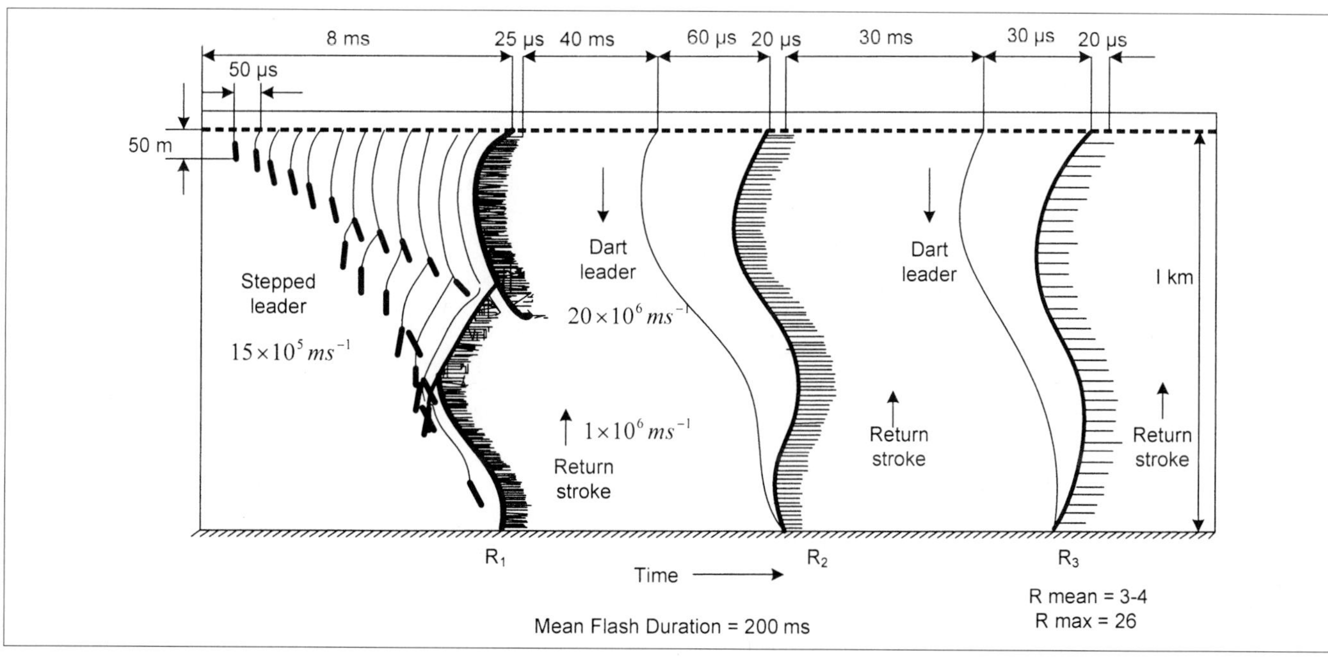

FIGURE 5-4 Typical parameters of lightning flash, U.S. Nuclear Regulatory Commission.[2]

channel results in subsequent strokes with about one-third the current and shorter rise time in comparison with the original stroke. Sometimes discharges originate several kilometers away from the end of the return stroke channel and travel toward it. These may die out before reaching the end of the return stroke point and are called the "K" changes. If these discharges do make contact with the return-stroke channel and the channel is carrying continuous current, it results in a discharge that travels toward the ground and is called the "M" component. If the return stroke channel is in partially conducting stage, it may initiate a dart leader that travels toward the ground.

Where ionization has decayed, it will prevent continuous propagation of the dart leader, and it may start propagating to ground as a stepped leader, called *dart-stepped leader*. If these leaders are successful in traveling all the way to ground, then subsequent return strokes occur. It is not uncommon for the dart leader to take a different path than the first stroke. A ground flash may last up to 0.5 s with a mean number of strokes ranging from 4 to 5. The separation between subsequent channels was observed to be a few kilometers on average.

The positive leaders propagate approximately in a similar manner as described for the negative strokes.

5-4 LIGHTNING PARAMETERS

Figure 5-5[3,4,5] shows cumulative distribution of lightning peak currents; Fig. 5-6[4,6] shows time to peak of lightning stroke currents; Fig. 5-7[7] shows rate of rise of lightning stroke current; and it also shows the front of a negative return stroke from Ref. 8. It is characterized by:

- Steep front rise
- Broad peak area with several minor peaks
- Slow decay to a low current.

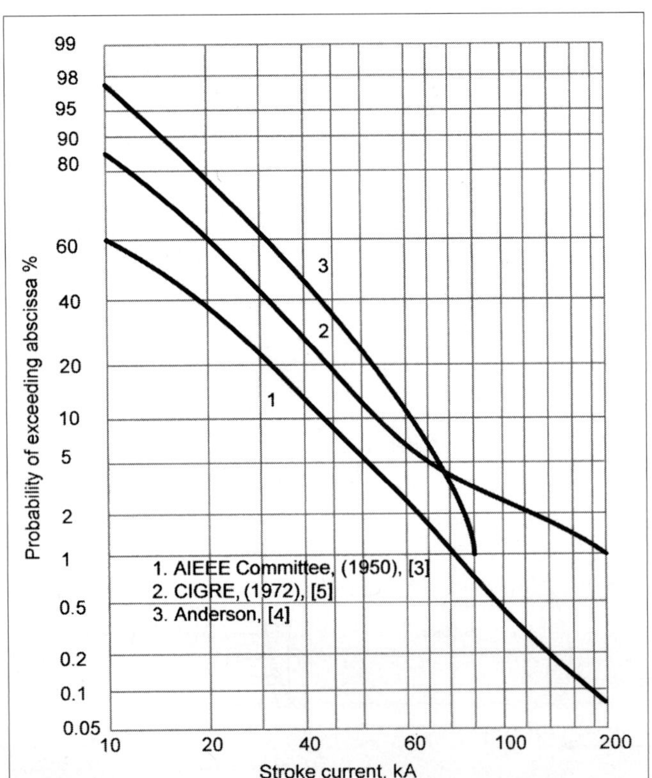

FIGURE 5-5 Cumulative distribution of lightning peak current distributions.[3,4,5]

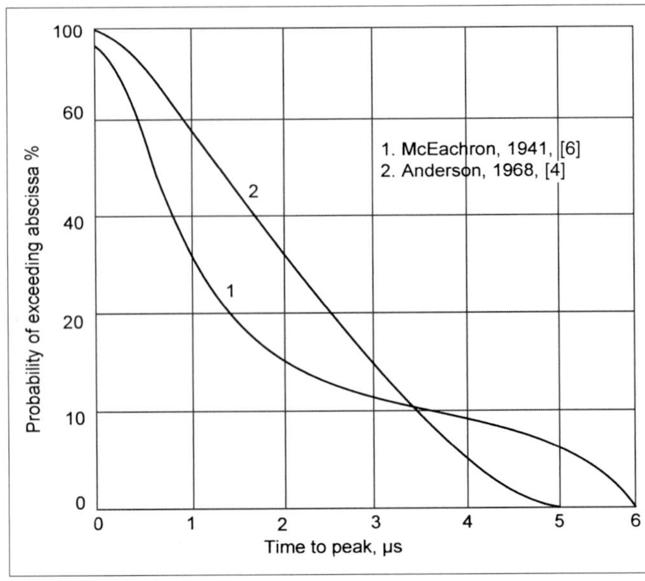

FIGURE 5-6 Time to peak of lightning stroke currents.[4,6]

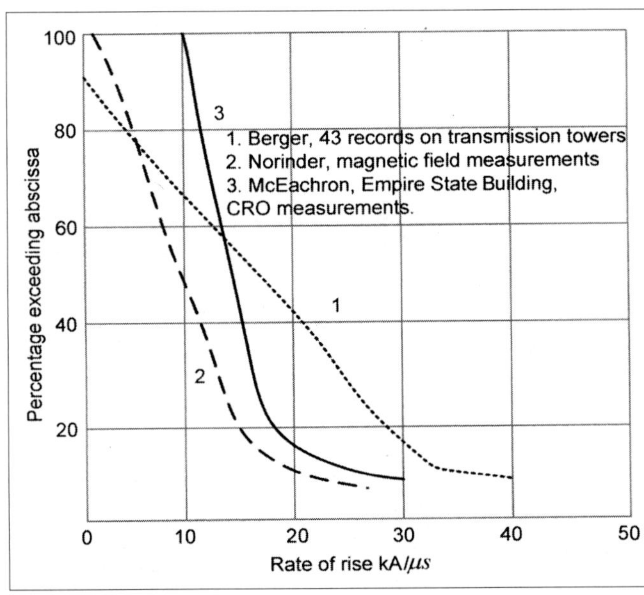

FIGURE 5-7 Rate of rise of lightning stroke current.

The critical part of the curve (Fig. 5-8) is the initial rise and the various parameters, which are used to define the range of lightning stroke waveshape. Table 5-1 provides typical values of these parameters. Actual parameters are spread randomly and their probability distribution is described mathematically as a "log-normal" function, when $\ln I_p$ has a normal distribution. The probability density function $P(I_p)$ of I_p can then be expressed as:

$$P(I_p) = \frac{1}{2\pi I_p \sigma_{\ln I_p}} \exp\left[-0.5\left(\frac{\ln I_p - \ln I_{pm}}{\sigma_{\ln I_p}}\right)^2\right] \tag{5-1}$$

where $\sigma_{\ln I_p}$ is standard deviation of $\ln I_p$, and I_{pm} is maximum value of return stroke current.

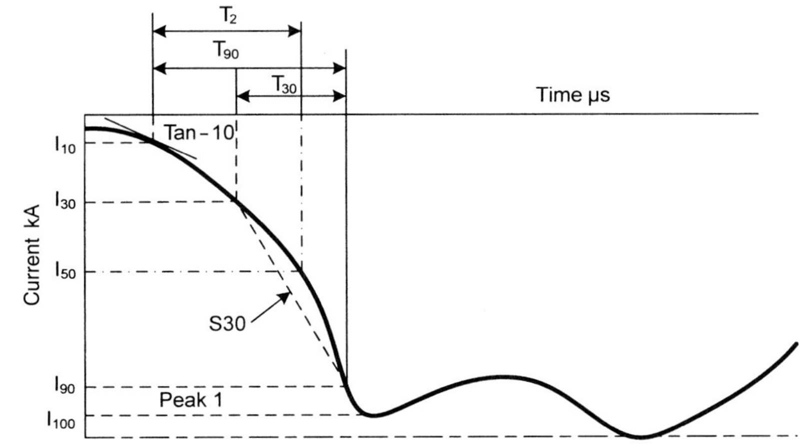

FIGURE 5-8 Parameters of the front of return stroke current wave.[8]

TABLE 5-1 Median Values of Stroke Parameters

PARAMETER	DESCRIPTION	FIRST STROKE	FOLLOWING STROKES
Peak (kA)	Peak value	31.1	12.3
Tan-10 (kA/μs)	Tangent at 10%	2.6	18.9
S-30 (kA/μs)	Slope from 30 to 90%	7.2	20.1
T-90 (μs)	Time from 10 to 90%	4.5	0.6
T_2 (μs)	Time to 50% on tail	80	32

The cumulative probability is given by:

$$P_c(I_p) = \frac{1}{\pi}\int_0^{\infty} e^{-u^2}\,du \tag{5-2}$$

where:

$$u = \frac{\ln I_p - \ln I_{pm}}{\sqrt{2}\sigma_{\ln I_p}} \tag{5-3}$$

The median value represents the 50 percent mark on the cumulative distribution curve, that is, 50 percent of the observed parameters are above and 50 percent below the mark. The cumulative distributions in Table 5-1 are shown in Figs. 5-9, 5-10, 5-11, and 5-12.[8,9]

The parameter S30/90 is the steepness of the front measured by a line drawn through 30 percent and 90 percent points. The parameter T30 is time to crest. For a wide range of peak current values, the field data distribution can be expressed by a much simpler equation by Anderson:[10]

$$P_c = \frac{100}{1+(I_p/31)^{2.6}} \tag{5-4}$$

where P_c is the probability of peak current exceeding I_p, the peak first stroke current. Consider a lightning voltage of 100 kV that includes all but 20 percent of the strokes:

- Peak current from Fig. 5-9 gives the first stroke current = 48 kA, and subsequent stroke current = 19 kA.
- Rise time from Fig. 5-10 gives first stroke current T90 = 7.5 μs, and rise time = 1.25 × 7.5 = 9.4 μs; following stroke current, T90 = 1.5 μs and rise time = 1.25 × 1.5 = 1.9 μs.
- Steepness from Fig. 5-11 gives first stroke current Tan-10 = 5.8 kA/μs, S-30 = 13 kA/μs; following stroke Tan-10 = 59 kA/μs, S-30 = 45 kA/μs.

Figure 5-13 is CIGRE/IEEE standard probability curves of first stroke, negative downward flash.[11]

The lightning flash includes several individual strokes and the total transfer of electric charge can only be determined by considering the entire succession of waveforms. Research shows that lightning stroke produces more energy than has been considered previously possible. A typical lightning stroke carries approximately 30 × 10^8 KW at approximately 125 MV and an average current of more than 20 kA.

To summarize, lightning currents can be very high, of the order of 200 kA. These currents rise at a very fast rate which varies from less than a microsecond to 10 to 20 ms and then decay to small values, a low-current component persisting even for a few seconds. A lightning stroke is a random phenomenon, and statistical data has been accumulated over the years. As the lightning current is so high the voltage developed depends upon the impedance of the system, and it can be of the order of several megavolts. Table 5-2 gives variations of lighting stroke parameters.

In power systems, most severe voltage surges are caused by lightning. As the system operating voltages increase, the switching surge transference becomes as important as lightning surges (Chap. 7). For application of surge arresters, the important parameters are overvoltages at the equipment terminals and the energy dissipation through the surge arrester. The effects of the lightening stroke currents can be summarized as:

- The crest current is responsible for ohmic voltage drops, especially in the ground resistance.
- The steepness of lightning current determines the inductive voltage drop, induced voltage, and magnetic couplings.
- The electric charge of the lightning current is a measure of the energy transferred by the lightning arc to metallic objects, causing melting and diffusion of the transient electromagnetic field through metallic screens.
- The integral of current squared multiplied by time is fundamental to mechanical effects and impulse heating.
- The lightning parameters vary over large values. The current and wave shape data are defined statistically. The frequency of lightning strokes is an important issue in determining the surge arrester duty.

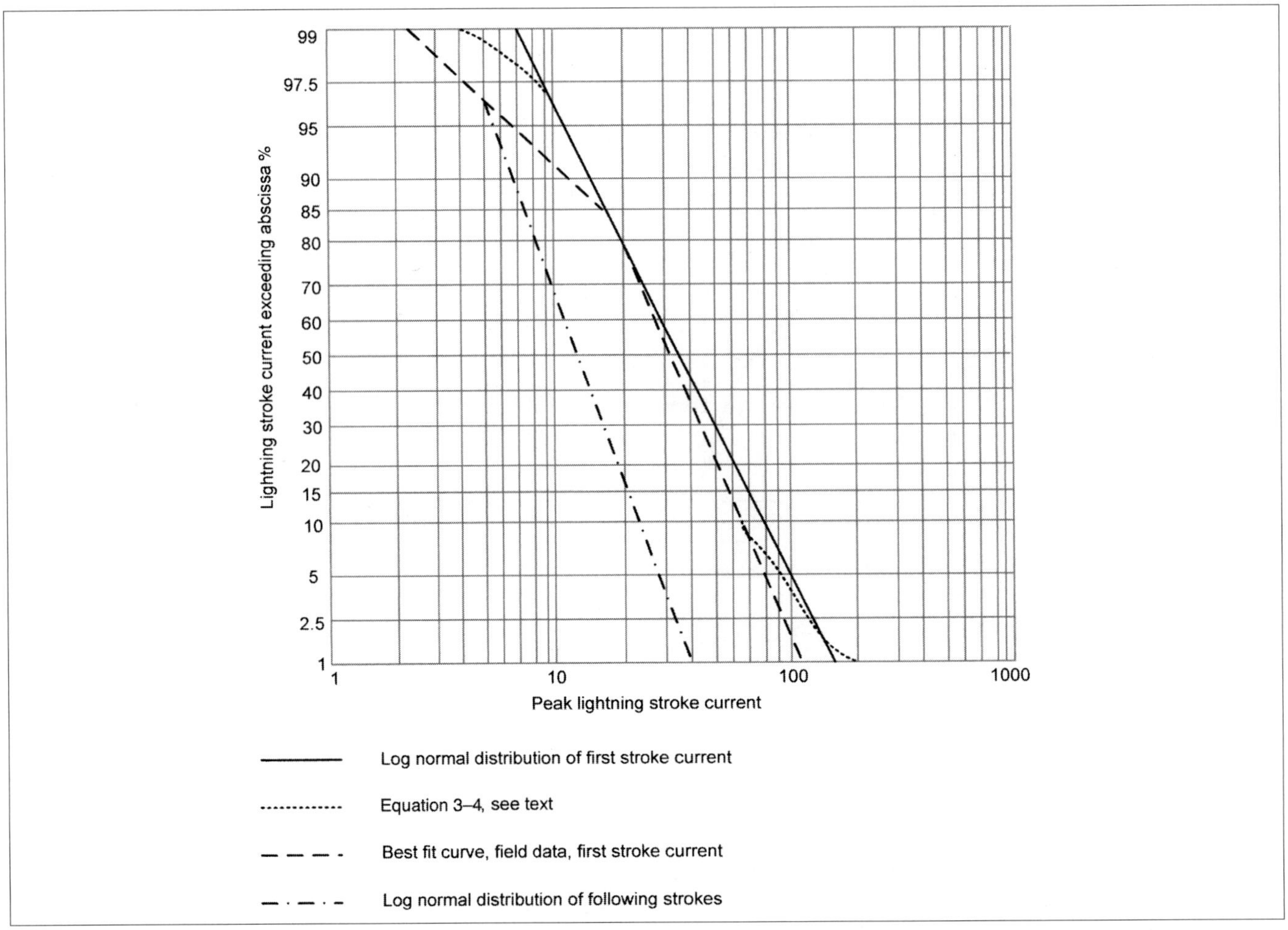

FIGURE 5-9 Cumulative distributions of peak stroke current, showing negative polarity.

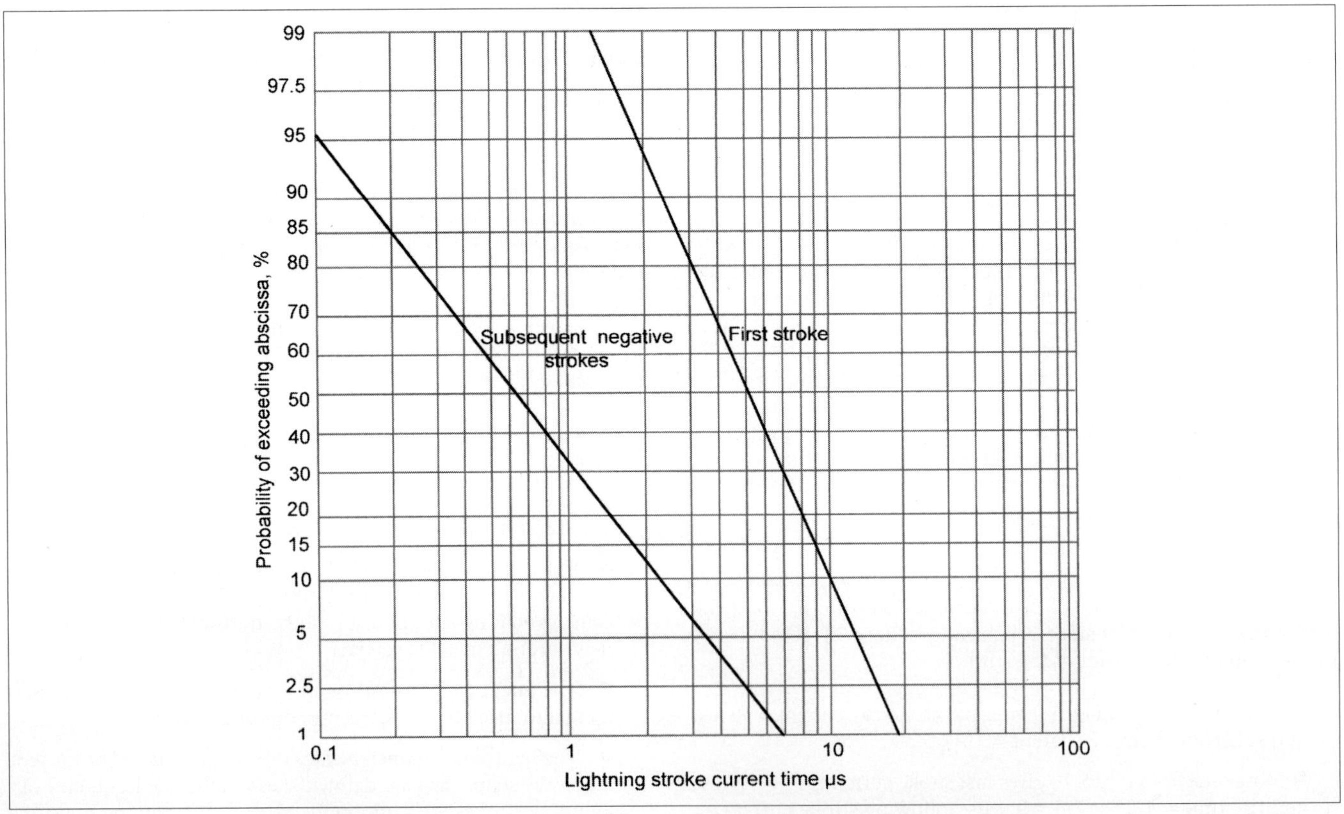

FIGURE 5-10 Cumulative distribution of T90 data for return stroke currents.[8]

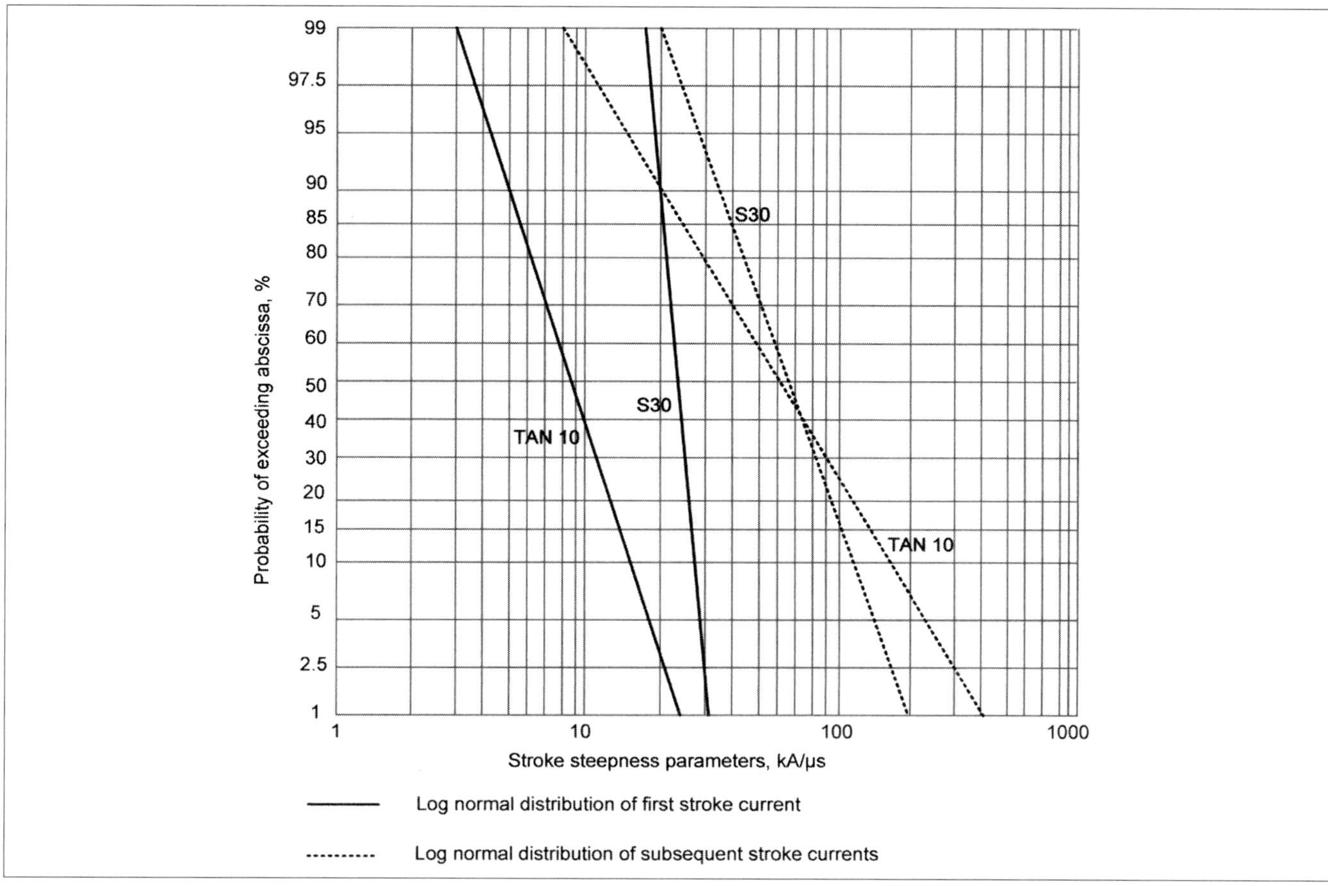

FIGURE 5-11 Cumulative distribution of Tan-10 and S-30 data for return stroke currents.[8]

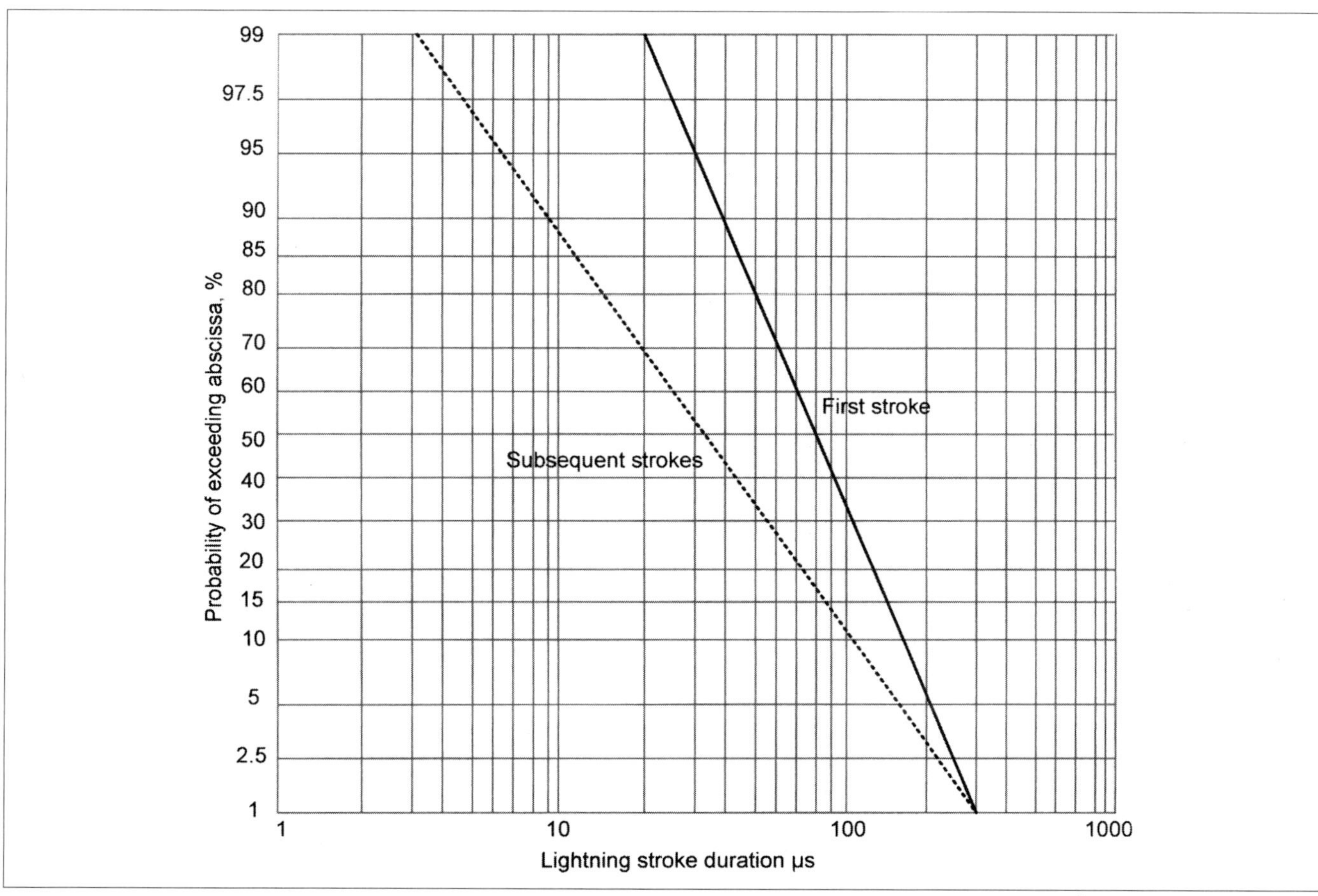

FIGURE 5-12 Cumulative distribution of stroke distribution T_2, data.[9]

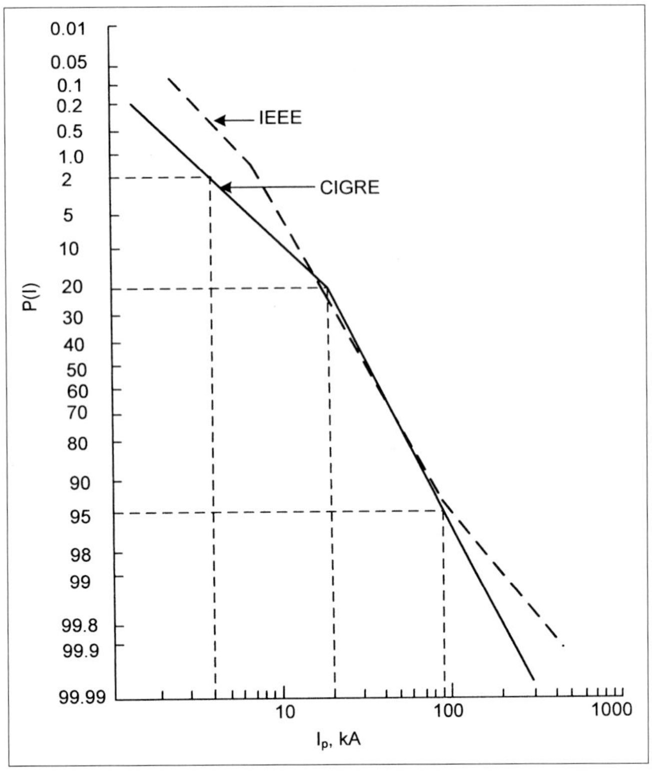

FIGURE 5-13 IEEE and CIGRE probability distributions of first stroke, negative downward flash.

5-5 GROUND FLASH DENSITY AND KERAUNIC LEVEL

The ground flash density is the number of lightning flashes to a specific area in a given time. It is measured in terms of flashes per square kilometer per year. This data may not be readily available. However, meteorological available data is in the form of counters depicting number of days (24 hour period) per year on which thunder is heard. This is also known as *keraunic level*. Keraunic level

TABLE 5-2 Variations of Lightning Stroke Parameters

PARAMETER	RANGE
Total charge transferred	2–300 C
Peak currents achieved	200–40 kA
Time duration to half value	10–250 μs/stroke
Current rise at 90% of its peak value	A few ns to 30 μs
Velocity of propagation	1–21 m/s
Time between strokes in one flash	3–100 ms
Number of strokes per flash	1–26
Energy per strike	$> 10^{10}$ J
RFI range	1 kHz–100 MHz

is rather a coarse measurement and its calculation requires simple observations. The keraunic-level data is available for a number of years all over the world. Figures 5-14 and 5-15[12] show the isokeraunic levels of the world and the United States, respectively.

An equation representing the relation between keraunic level and ground flash density is:

$$N_g = 0.14T_D^{1.03} \tag{5-5}$$

where N_g is the ground flash density in flashes per square kilometer per year and T_D is the keraunic level. Another relation used in a Canadian research is:

$$N_g = 0.054T_H^{1.1} \tag{5-6}$$

where T_H is the number of hours per year during which thunder is heard at a certain location. Figure 5-16 shows the variation of ground flash density with T_D. The following equation has been accepted by CIGRE and IEEE:

$$N_g = 0.04T_D^{1.25} \tag{5-7}$$

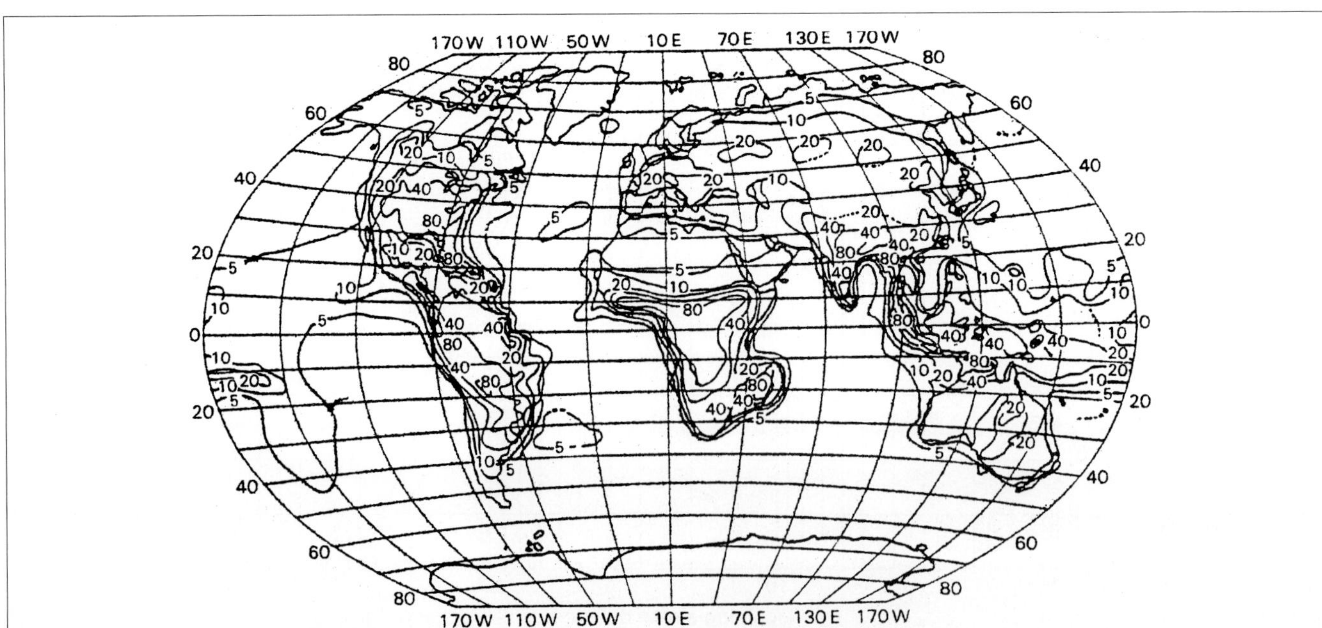

FIGURE 5-14 Isokeraunic map of the world.

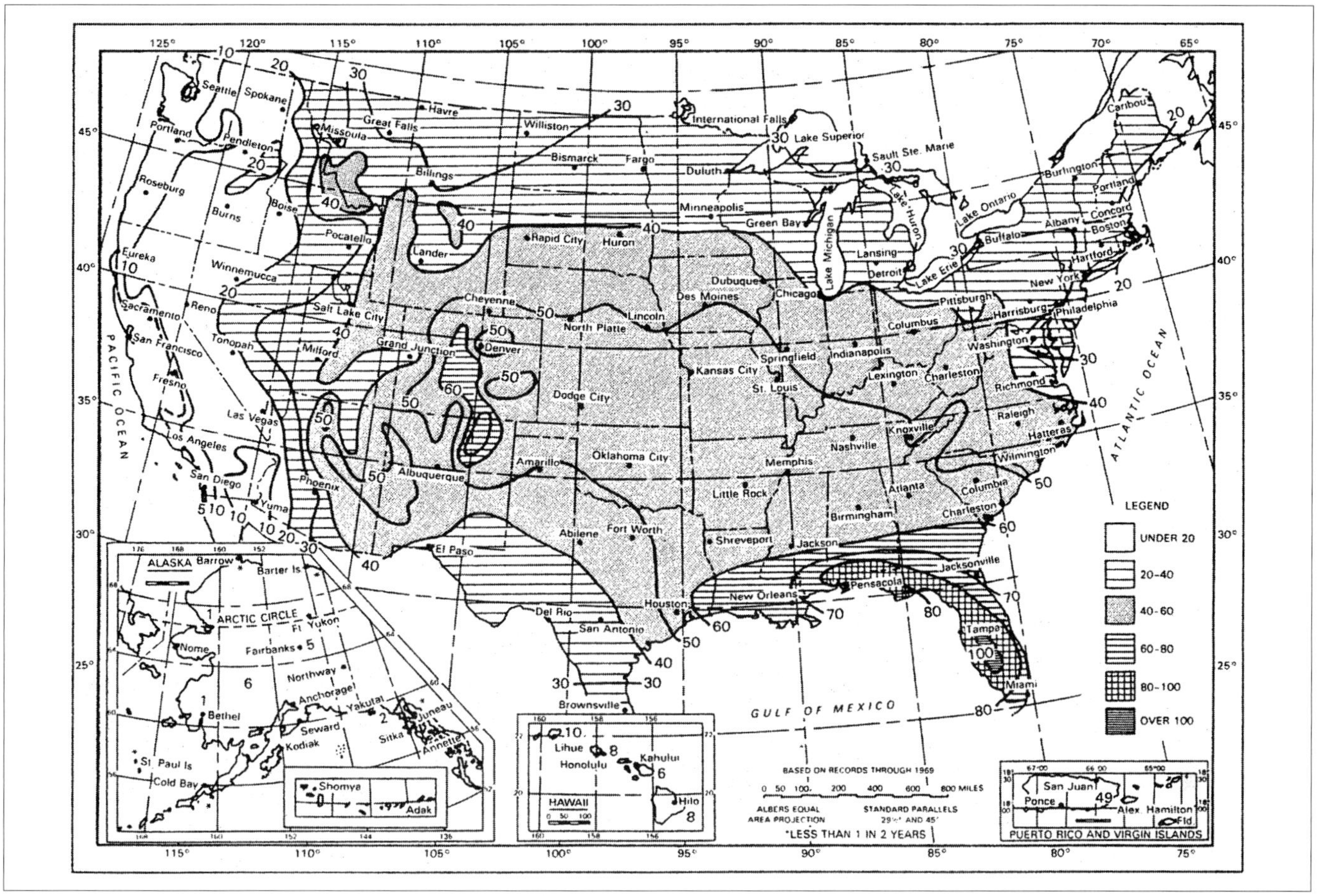

FIGURE 5-15 Isokeraunic map of the United States.

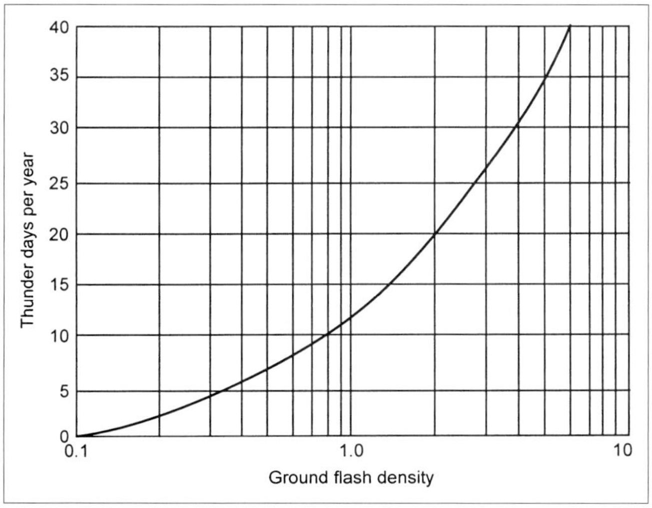

FIGURE 5-16 Ground flash density versus thunder days per year.

5-6 LIGHTNING STRIKES ON OVERHEAD LINES

The effect of lightning on the transmission lines is important in power system engineering. It gives rise to a number of related phenomena—surge protection, insulation coordination, shielding, safety of equipment and personnel, and concerns about continuity of power supply.

As a first step it is necessary to estimate the area and the number of lightning surges on a power system. The fundamentals of lightning mechanism described above are of help in this analysis.

5-6-1 Striking Distance

As the stepped leader descends downward and approaches an object, the intensity of the electric field is determined by the geometry of the object (for tall slender objects like transmission line towers and chimneys, the height above ground is the predominant factor). A corona steamer is initiated from the structure when the electric field intensity reaches a certain level (3 MV/m), and this ascending corona steamer creates a highly ionized channel. The distance between the tip of the leader and the structure at which it happens is called the *striking distance*. (Another definition of the striking distance is provided in Sec. 5-3-2) For transmission lines, it is the semicylindrical envelope around the conductor[8] given by:

$$r_s = (0.4 + 0.01h_s)I_s^{(1.4-0.001h_s)} \tag{5-8}$$

where r_s is the striking distance in meters, I_s is the peak lightning stroke current, h_s is the height of the structure above ground in meters. For a lightning current of 40 kA and a tower of 10 m, the striking distance is 84.6 m

A number of formulas for calculating the striking distance by various authors/researches are shown in Table 5-3. The most commonly used expression, which is in current use based upon IEEE committee report, is:[13]

$$r_s = 8I_p^{0.65} \tag{5-9}$$

where I_p, the return stroke current, varies from 1 to 200 kA.

TABLE 5-3 Expressions for the Striking Distance in Meters

AUTHOR/RESEARCHER	STRIKING DISTANCE EXPRESSION, r_s TO PHASE CONDUCTORS AND GROUND WIRE (IN METERS)	STRIKING DISTANCE EXPRESSION, r_g TO EARTH OR GROUND WIRES (IN METERS)
Darveniza, 1979	$2I_s + 30(1 - e^{-0.147I})$	—
Love, 1973	$10I_s^{0.65}$	$10I_s^{0.65}$
Brown-Whitehead, 1977	$7.1I_s^{0.75}$	$6.4I_s^{0.75}$
Suzuki, 1981	$3.3I_s^{0.78}$	—
Erickson, 1982	$0.67h^{0.6}I_s^{0.74}$	$0.67h^{0.6}I_s^{0.74}$
IEEE-1995 Subcommittee	$8h^{0.6}I_s^{0.65}$	$8h^{0.6}I_s^{0.65}$

I_s, the first return stroke current.

The upward leader from the structure may not contact the stepped leader; this may happen if the stepped leader crosses the striking distance envelope far away from the vertical axis. In this case the leader may proceed directly to the ground, without hitting the object. The loci of the stepped leader and the upward leader form a parabola above the structure, as shown in Fig. 5-17.

5-6-2 The Attractive Radius

Figure 5-17 shows the attractive radius, as the horizontal coordinate of the interception of the upward leader and the downward stepped leader. This means that all lightning strokes of current I_s will strike it, according to Eq. (5-10):

$$r_a = 0.84 I_s^{0.74} h_s^{0.6} \tag{5-10}$$

where r_a is in meters. For a 10 m tower and a lightning current of 40 kA, the attractive radius is 51.26 m. A relation independent of the stroke current gives the average attractive radius, solely dependent upon the height of the object:[14]

$$r_{av} = 14 h_s^{0.6} \tag{5-11}$$

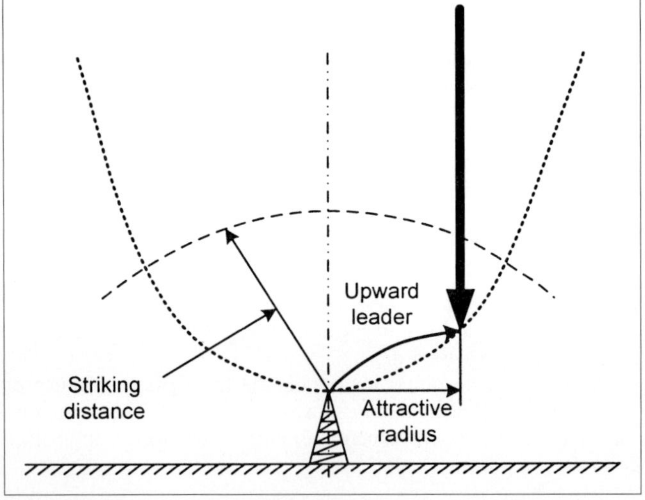

FIGURE 5-17 Loci of the interception of stepped leader by the upward leader from power line structure.[22]

5-7 BIL/CFO OF ELECTRICAL EQUIPMENT

For electrical equipment to withstand surges of atmospheric origin (lightning) and switching, test standards and test waveshapes have been established. Basic insulation level (BIL) shows the rating of the equipment to which it is impulse tested. Most electrical equipment, that is, line insulators, transformers, reactors, circuit breakers, switches, surge arresters, and the like have a BIL rating, which is an important parameter in insulation coordination and lightning protection of the equipment. The BIL may be statistical or conventional. The statistical BIL applies to self-restoring insulation, that is, insulation which completely recovers insulating properties after a flashover, for example, external insulators. The conventional BIL applies to non-self-restoring insulation, that is, insulation that does not recover after a disruptive discharge, for example, transformer windings. IEC terminology is "lightning impulse withstands voltage." We will often return to BIL in the chapters to follow.

Statistical and conventional BSL are similarly defined for switching test impulse. For every application of an impulse having standard waveshape, whose crest is equal to BIL or BSL, the probability of flashover or failure is 10 percent. The insulation strength can be represented by gaussian distribution (see App. F).

For self-restoring external insulation, for example, insulators and CFO, the critical flashover voltage is used. The CFO is defined as the impulse voltage level at which the probability of flashover is 50 percent, that is, half the impulse flashover. Considering the BIL and BSL at 10 percent point results in the definition that these are 1.28 standard deviation, σ_f, below CFO:

$$\begin{aligned} \text{BIL} &= \text{CFO}\left(1 - 1.28\frac{\sigma_f}{\text{CFO}}\right) \\ \text{BSL} &= \text{CFO}\left(1 - 1.28\frac{\sigma_f}{\text{CFO}}\right) \end{aligned} \tag{5-12}$$

σ_f in per unit of CFO is called the *coefficient of variation*. Thus, σ of 5 percent is a standard deviation of 5 percent of the CFO. For lightning the standard deviation is 2 to 3 percent, while for switching impulse it varies say 5 percent for tower insulation to 7 percent for station-type insulators. We will revert to these probability concepts in greater detail in Chap. 17 on insulation coordination.

5-7-1 1.2/50-μs Test Wave

Figure 5-18 shows the waveshape of the lightning current for laboratory BIL testing. The BIL tests are type tests to establish the integrity of the equipment to withstand steep-fronted lightning surges. The waveshape shown in Fig. 5-18 is according to ANSI/IEEE standards. It is called a 1.2/50-μs wave and the important parameters are as shown. The front time for the voltage wave is defined as:

$$1.67(t_{90} - t_{30}) = 1.2\ \mu\text{s} \pm 0.36\ \mu\text{s} \tag{5-13}$$

The duration is defined as the time between virtual origin and the time of 50 percent point on the tail. The virtual origin is a point where a straight line between 30 percent and 90 percent points on the leading edge of the waveform intersects $V = 0$ line.

$$\text{Duration} = 50\ \mu\text{s} \pm 10\ \mu\text{s} \tag{5-14}$$

The lightning impulse waveshape can be described by two exponentials:

$$v = V_p(e^{-\alpha t} - e^{\beta t}) \tag{5-15}$$

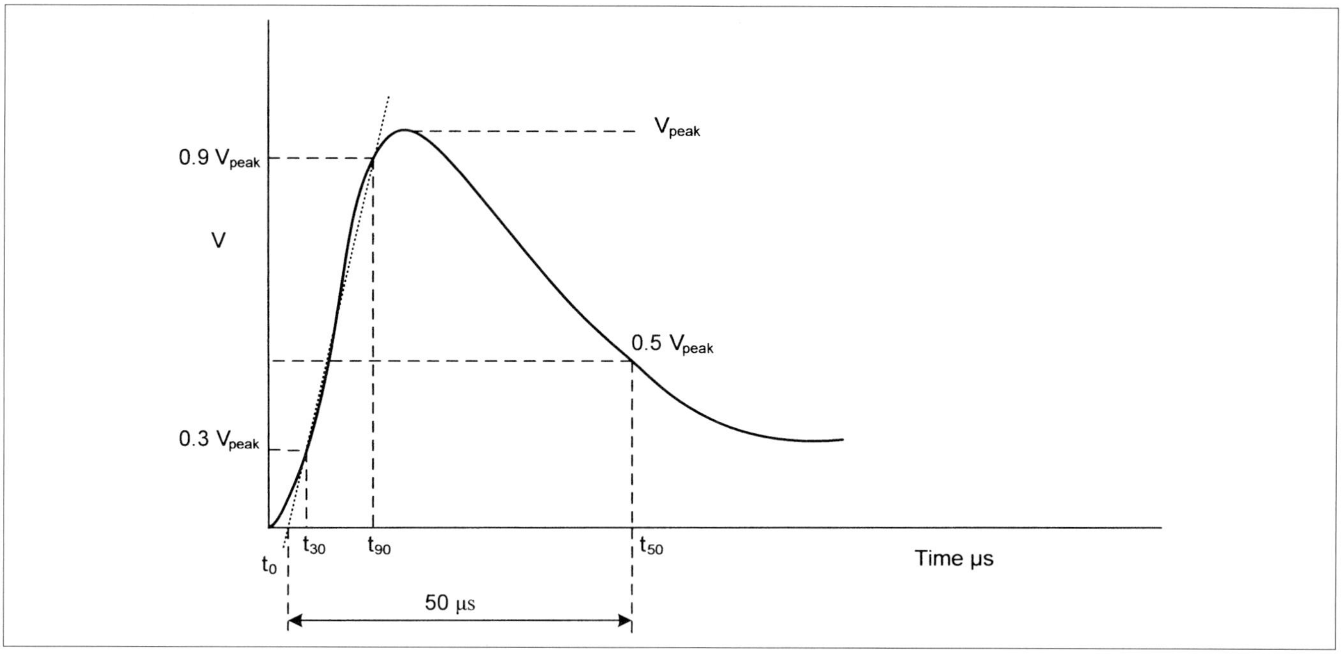

FIGURE 5-18 1.2/50-µs voltage test wave.

Selection of $\alpha = 1.4E4$ and $\beta = 4.5E6$ gives the 1.2/50-µs wave. The waveshape can also be represented by the equation:

$$V(t) = AV_p\left(1 - \exp\frac{-t}{\tau_1}\right)\exp\left(\frac{-t}{\tau_2}\right) \tag{5-16}$$

where $\tau_1 = 0.4074$ µs, $\tau_2 = 68.22$ µs and $A = 1.037$. Figure 5-18 shows a full impulse wave. This is the shape seen at some distance from the lightning impulse. The wave can get chopped on the tail or front. These waveshapes and volt-time characteristics of the insulation are described in Chap. 17.

5-7-2 8/20-µs Test Wave

Another test wave of interest is 8/20-µs current wave shape shown in Fig. 5-19. Here the front time is defined as:

$$1.25(t_{90} - t_{10}) \tag{5-17}$$

where t_{90} and t_{10} are the times for 90 percent and 10 percent amplitude points on the leading edge of the waveform. The duration is defined as the time between virtual origin and the time of the 50 percent amplitude point on the tail. Here the virtual origin is obtained by a straight line between 10 percent and 90 percent amplitude points on the leading edge of the waveform where it intersects $I = 0$ line (Fig. 5-19).

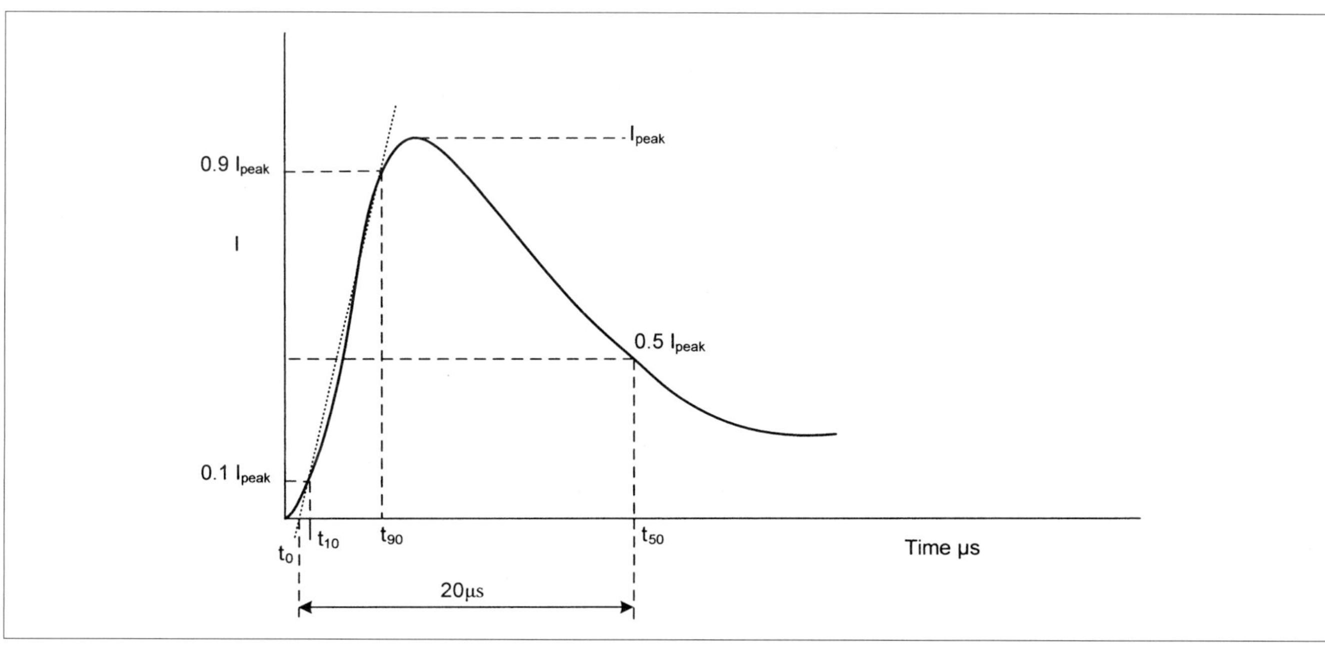

FIGURE 5-19 8/20-µs current test wave.

The wave can also be represented by double exponential equation [Eq. (5-15)] with appropriate choice of α and β. Another equation is:

$$I(t) = AI_p t^3 \exp\left(\frac{-t}{\tau}\right) \tag{5-18}$$

where $\tau = 3.911$ μs and $A = 0.01234$ (μs)$^{-3}$

These waveshapes are described as "impulse" in IEC. Both waves together are described as "combination waves" because these can be delivered by a generator, on open circuit, 1.2/50-μs wave and an 8/20-μs wave on short circuit. These can be readily generated in many laboratories and are commonly used for the study of lightning strikes. The earlier IEEE standard for the voltage wave was 1.5/40 μs. The other test waves are discussed in Chap. 19.

5-8 FREQUENCY OF DIRECT STROKES TO TRANSMISSION LINES

A direct stroke can occur on phase conductors, towers, or ground wires. This will give rise to overvoltages which may exceed CFO. Thus, direct strokes are important concerns for protection of the transmission lines. Indirect strokes may not induce high enough voltage to exceed CFO on high-voltage lines. Indirect lightning strokes can induce 50 to 60 kV or even higher overvoltages for a lightning stroke in the vicinity, with a relatively low-peak discharge current. This may exceed the insulation strength of distribution networks.

Once N_g is known, the frequency can be directly calculated from:

$$N_d = N_g A_e \tag{5-19}$$

where N_g is the ground flash density in flashes per square kilometer per year, and N_d is the number of direct strikes to the line. The problem is to calculate the area of exposure and various models have been in use. Some of these are described below:

(a) *Based upon attractive radius and height.* Referring to Fig. 5-20*a*, the area exposed to lightning is given by:

$$A_e = 0.001(2r_{av} + w)l \tag{5-20}$$

where l is the length of the line in kilometers, A_e is the exposed area in square kilometers and w is horizontal line width. It is zero for conductors placed vertically on the structure, that is, when there are no cross-arms, r_{av} is the average striking distance. This may be calculated from the conductor height less half the conductor sag. This equation applies to all lines that have shield wires. Tall objects in the vicinity of the line may divert lightning flashes. This shielding effect can be calculated in the flash frequency calculations.[15]

(b) *Protective shadow of a structure.* A vertical structure attracts lightning from a distance twice its height. If b is the spacing between the shield wires and H is the height, then the width of shadow W is:

$$W = b + 4h \text{ meters} \tag{5-21}$$

where h should be adjusted for the sag in the shield wires.

(c) *Strokes to shield wire, one shield wire.* The geometric construction is shown in Fig. 5-20*b*. This shows two striking distances, r_s and r_g. The breakdown gradient on top of tower differs from the gradient to cause flashover to ground. The numbers of strokes terminating on ground wire are:

$$N(G) = A_e N_g = (2D_g l)N_g \tag{5-22}$$

Area A_e that collects the strokes is $2D_g l$. The probability that this stroke current will occur is:

$$dN(G) = A_e N_g f(I)dI \tag{5-23}$$

and total number of strokes is:

$$N(G) = 2N_g l \int_3^\infty D_g f(I)dI \tag{5-24}$$

(d) *Two overhead wires.* The geometric construction is shown in Fig. 5-20*c*. Here:

$$N(G) = 2N_g l \int_3^\infty D_g f(I)dI + N_g l S_g \tag{5-25}$$

(e) *Eriksson's model.* There exists no striking distance to ground or earth (Fig. 5-20*d*):

$$D_g = r_s = 0.67h^{0.6}I^{0.74} \tag{5-26}$$

(f) *Electrogeometric model.* This is shown in Fig. 5-20*e*. Curve *a* enclosing line conductor is locus of stroke tip positions, which decide whether the leader will strike the conductor or the earth. However, even if the stroke tip is within the curved boundary, it will strike the earth if its distance from the conductor is greater than r_s.

$$r_g = \beta r_s \tag{5-27}$$

for β more than or equal to 1, the curve is a parabola. The width a_1b_1 is:

$$a_1b_1 = 2\sqrt{h(2r_{s1} - h)} \tag{5-28}$$

and for $r_{s1} \le h$

$$a_1b_1 = 2r_{s1} \tag{5-29}$$

If the probability of return stroke current exceeding I is P, then the number of strokes exceeding I and striking the line is:

$$N = A_e N_g P \tag{5-30}$$

where P can be calculated from (5-4).

Example 5-1 A transmission line runs through an area with thunderstorm activity T_D = 10 thunderstorm hours per year. The height of the top conductor = 11 m, the sag = 2 m, the horizontal line width = 3 m, and the line length = 20 km.

Calculate N_g using Eq. (5-5). This gives 1.5 flashes per square kilometer per year. Calculate attractive radius based upon a height of 10 m (11 m less 50 percent sag):

$$r_{av} = 14h_s^{0.6} = 14(10)^{0.6} = 56 \text{ m}$$

The area A_e from Eq. (5-20) is:

$$A_e = (0.001)(20)(2 \times 56 + 3) = 2.3 \text{ km}^2$$

Therefore from Eq. (5-19), N_d = 3.45 direct strikes per year. To use Eq. (5-21), assume b = 1.5 m. Then width of shadow is 40.18 m. This is calculated by adjusting h from the sag, that is, h = height of structure −0.667. The result is 1.2 strokes per year for 20 km of line length.

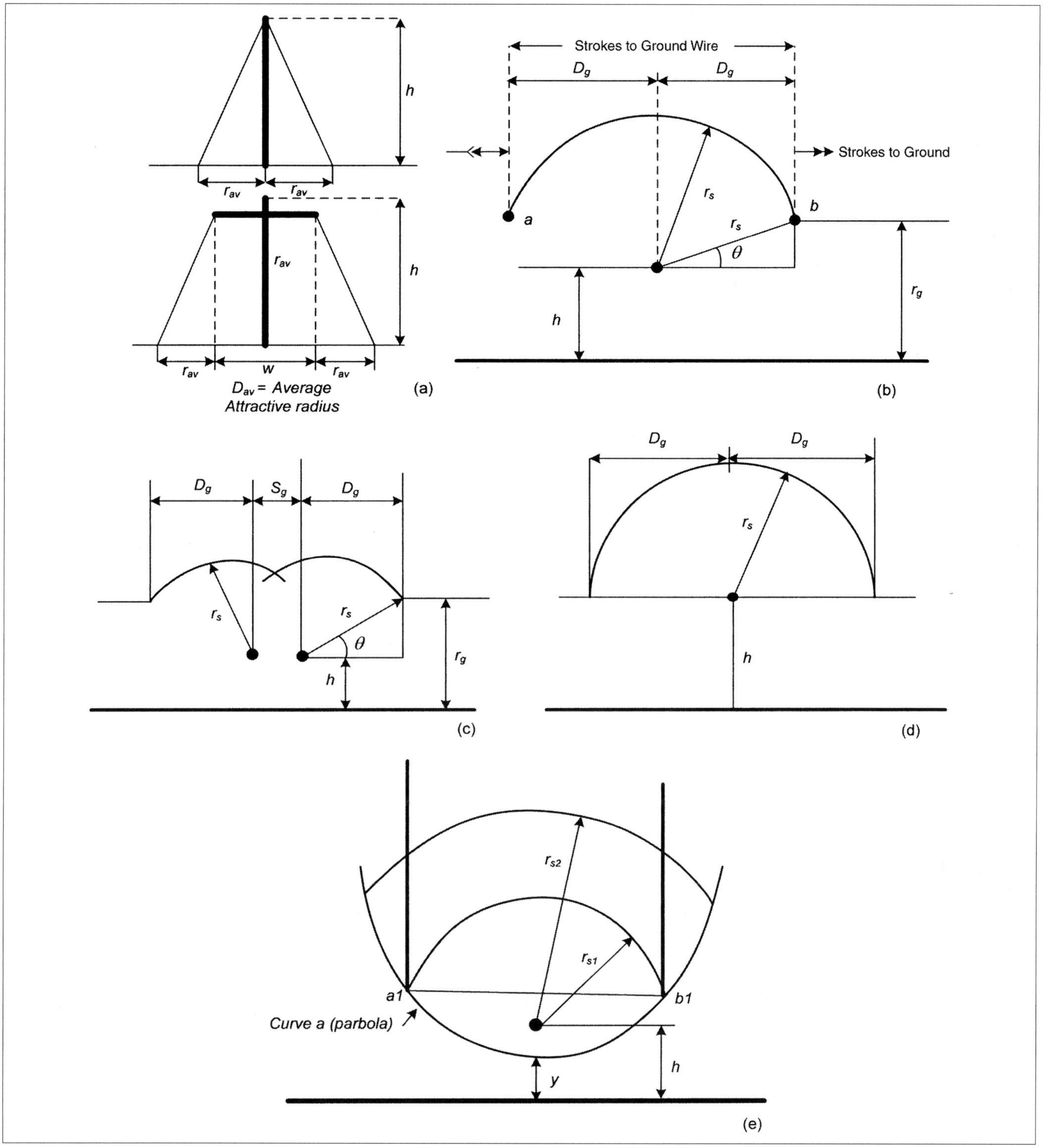

FIGURE 5-20 Models for calculating incident of lightning strokes: (*a*) model based upon average attractive radius; (*b*) and (*c*) geometric models for single and double ground wires, respectively; (*d*) Eriksson's geometric model; and (*e*) electro-geometric model.

To calculate r_s, assume a peak current I_p = 40 kA. Then from Eq. (5-26), r_s = 43.2 m. Then from Eriksson's model equation, number of strokes = 2.6 for 20 km of line per year. The electrogeometric model gives the same results as $r_{s1} \leq h$ and $a_1b_1 = 2r_s$ = 86.4 m and number of strokes = 2.6 for 20 km per year. The probability from Eq. (5-4) for 40-kA current = 0.3401. Then from Eq. (5-30), the number of strokes exceeding I_p = 0.88. There are variations in the results calculated based upon the model used. Also various models do not correlate so well with the field measurements.

To relate long-range data with short-range statistics, a probability distribution function (Poisson distribution) as follows can be used:

$$P_n = \frac{100(N_d t)e^{n-N_d t}}{n!} \tag{5-31}$$

where P_n is the probability of occurrence of exactly n flashes, n is the number of flashes striking the system, t is the time duration of interest in years.

Example 5-2 Find the probability of having more than five strikes per year in Example 5-1. Substituting the values in Eq. (5-31):

$$P_n = \frac{100(3.45)(1)e^{n-(3.45)(1)}}{n!}$$

Probability for

0 strike = 3.17 percent
1 strike = 10.90 percent
2 strikes = 18.86 percent
3 strikes = 21.60 percent
4 strikes = 18.71 percent
5 strikes = 12.90 percent
Total = 86.14 percent.

Thus, probability of having more than five direct strikes is 13.86 percent

5-9 DIRECT LIGHTNING STROKES

Direct lightning strokes to conductors, towers, and even shield or ground wires produce high voltages. The strokes to ground in the vicinity of a transmission line produce the lowest overvoltages. Consider a direct stroke to a conductor. As the stroke current will flow in either direction, the voltage induced V is approximately given by:

$$V = 0.5IZ_0 \tag{5-32}$$

where Z_0 is the surge impedance of the line. Thus, even a stroke current of 20 kA and a surge impedance of 400 Ω will give a voltage of 4000 kV. Thus, most strokes to phase conductors will result in a flashover. The voltage induced on other phases will be given by KV, where K is the coupling factor, and the voltage between two phase conductors will be $(1 - K)V$. A flashover between two conductors can occur, depending upon the separation between them. The voltages on the conductors will travel in either direction and a flashover will occur if CFO is exceeded. For a midspan stroke, the critical stroke current magnitude that will cause flashover is given by:

$$I_c = 2\text{CFO}/Z_0 \tag{5-33}$$

This follows from the definition of CFO. It is a standard practice to prevent direct strokes to conductors by arranging one or more shield (ground) wires, which are discussed in Sec. 5-13.

If Z is the impedance of the power system at the point of lightning strike, and Z_s the surge impedance of the stroke channel, then the stroke current I_0 (to zero earth resistance, infinite conductivity) will divide into Z_s and Z in parallel, and the voltage across Z, the stricken object, is given by:

$$V = \frac{I_0 Z Z_s}{(Z+Z_s)} \tag{5-34}$$

The problem lies in accurately calculating the surge channel impedance Z_s.

5-10 LIGHTNING STROKES TO TOWERS

If a lightning stroke occurs to a transmission line tower, the lightning currents in the phase conductors, ground wires, and towers are at right angles. The mutual effects of these currents do not obey conventional traveling wave theory which applies to parallel conductors. For steep-front lightning currents, a 30-m high tower behaves like a surge impedance of approximately 50 to 100 Ω for a 0.2-μs return traveling time. For lightning currents with wave fronts exceeding 0.5 μs, the tower behaves like an inductance of about 0.2 to 0.4 μH/m. The tower surge impedance has been computed and measured in the field and on scale models.[16,17] It varies along the tower and considerable attenuation exists. A constant value can, however, be derived for practical purposes. For conventional double circuit towers, the surge impedance is given by Refs. 16 and 17.

$$Z_t = 30\ln 2\left(1+\frac{h^2}{r^2}\right) \tag{5-35}$$

where h is the height of a cone of the same height as the tower and r is its base radius (mean periphery divided by 2π). For cylindrical towers the following expression can be used:

$$Z_t = 60\ln\left(\frac{h}{r}\right) + 90\left(\frac{r}{h}\right) - 60 \tag{5-36}$$

This equation can also be used for down leads with r representing the wire radius. For towers which cannot be approximated with cones or cylinders, scale models and testing are required to determine the effective surge impedance.

5-10-1 Tower Footing Resistance

Another parameter to be considered is the tower footing resistance, which will be much different from the resistance at power frequency currents. At the power frequency level the resistance is mainly a function of soil resistivity. Under currents of high intensity, the soil becomes ionized and may partially breakdown. The impulse value of the resistance is, therefore, lower than the power frequency resistance (Chap. 21).

The footing resistance at high impulse current can be determined from the following equation:

$$R_i = \frac{R_0}{\sqrt{1+\frac{I_R}{I_g}}} \tag{5-37}$$

where R_i is the footing resistance at impulse current, R_0 is the resistance measured at low current, and I_R is the lightning current through the footing resistance. I_g is the current required to produce a soil gradient, E_0, at which soil breakdown occurs. This current is given by:

$$I_g = \frac{1}{2\pi}\frac{\rho E_0}{R_0^2} \tag{5-38}$$

where ρ is the soil resistivity in ohms-meter, E_0 is the breakdown gradient, which is assumed = 400 kV/m. As the current increases, streamers are generated that evaporate the soil moisture, and arcs are produced. Within arcing and streamer zones, the resistivity decreases; we can imagine that the grounding rod becomes a hemisphere at high currents.

A single counterpoise behaves first with its surge impedance, which then falls to power frequency resistance. Consider a counterpoise of 100-m length; the counterpoise will have the surge impedance of buried wire, say equal to 150 Ω. A counterpoise may be represented by a series of leakage resistance (to duplicate distributed leakage current to ground) in parallel with R_i. Voltage and current waves travel down the tower impinge on this combination, and the waves travel at about one-third the speed of light. At a time equal to twice the travel time, the impedance is reduced to power frequency resistance, which is equal to total leakage impedance of the counterpoise.

For towers without shield wires, the voltage drop from the top of the tower to earth is:

$$\Delta V = L\frac{di}{dt} + R_f i \tag{5-39}$$

where R_f is the footing resistance of the tower and i is the surge current. For each frequency and current, the footing resistance varies. Neglecting induced voltages on phase conductors, ΔV appears across line insulators of the struck tower; the total voltage is, thus, the service voltage of the conductors plus ΔV. If this total voltage exceeds the insulator flashover voltage, a flash will occur. This is called *backflashover*. When this occurs, half of the surge impedance of the phase conductors acts in each direction, and the primary traveling waves on the struck conductor in each direction give rise to induced (coupled) traveling waves on the other two phases. This may reduce the induced voltage by 30 to 40 percent on the insulators of phases which were not struck by lightning and prevent further flashovers. Thus, the backflashovers remain on the phase originally struck by the lightning.

This is rather a simplified description of complex phenomena. The lightning current in the stroke channel and towers flows at right angles and does not observe the conventional traveling wave theory. The field theory concepts can be used, which get fairly involved. For practical use Wagner and Hileman[17] derived the voltage across the insulation by superimposing two voltage components: (1) the current injected into the tower and ground wire system and (2) the charge over the tower. The effect of (1) predominates for short-wave fronts up to 1 μs, while the effect of (2) increases for longer wave fronts above 2 μs.

To further complicate this picture, the voltage stresses on tower insulation can be reduced by the streamer projecting from the tower to meet the opposite polarity leader head. However, streamer length cannot be so precisely calculated. Most practical methods ignore this effect and consider only the injected current component.

For a tower with ground wires, the terminating impedance can be written as:

$$Z = \frac{Z_t Z_g}{Z_g + 2Z_t} \tag{5-40}$$

This expression shows that the total impedance is the tower surge impedance Z_t in parallel with the ground wire surge impedance Z_g. Considering ground wire extension in both directions, the effective surge impedance is 0.5 Z_g.

The tower top potential can now be calculated from (5-40), that is:

$$V_t = \frac{I_0(Z_t Z_g)Z_s}{Z_t Z_g + Z_s(Z_g + 2Z_t)} \tag{5-41}$$

The ground wire potential couples with the phase conductors and the voltage stressing the tower insulation is the difference in the potential between tower top potential and the voltage induced in the phase conductor. The induced voltage is given by:

$$V_i = (1-K)V_t \tag{5-42}$$

where, the coupling coefficient K varies from 0.15 to 0.30. The coupling coefficient is:

$$K = Z_{gp}/Z_g \tag{5-43}$$

where Z_{gp} is the mutual surge impedance between the shield wire and the phase conductor.

Greater the distance between the conductors, lower is the value of K. Two ground wires have greater coupling to the phase conductors than a single ground wire. A continuous counterpoise also increases the coupling factor. V_i is also the voltage developed across the insulator string. If the cross arm is not near the top of the tower, its voltage can be determined from the lattice diagram.

The corona increases the conductor's effective radius, which increases the coupling factor and reduces the insulation stress (see Chap. 4). The corona streamers carry significant quantities of charge away from the conductors and tower members. This effect creates nonlinear changes in the capacitance of conductors to ground and between phases (Chap. 4) and distorts the voltage waveshape appearing across the insulators. Corona moving outward is equivalent to a uniform extension of conductor diameter, until the gradient at the conductor drops to some critical extinction voltage, E_0—its value has been accepted to be 1500 kV/m. Corona diameters are plotted against V/E_0. For backflashovers, the voltage of the shield wire is taken as 1.8 times the CFO of insulators for 2 μs to crest. The corona does not change the inductance and this increases the no-corona surge impedance of the conductor.

The tower footing resistance is much less than the tower surge impedance. The incident wave traveling down the tower is reflected at the foot of the tower in an upward direction, up the tower, with opposite sign (Chap. 4). This may occur in a fraction of a microsecond, and this reflected wave is superimposed on the existing tower potential. Generally the wave front of the current surge will be much longer than the time taken by the reflected wave to move up the tower. This will reduce the rate of rise because the tower top potential has not yet reached its peak. A lower tower footing resistance will be effective in reducing the tower top potential. Considering a zero footing resistance, the incident wave will be reflected with the sign reversed multiplied by coupling factor K. The longer the surge front, the more effective will be the slope reduction.

The effect of the adjacent tower will be that the voltage waves traveling away from the struck tower on the ground conductors in both directions will be reflected back from the adjacent towers with opposite signs. If the time to crest is longer than the return time of the reflected wave, they will act like reflected waves from the base of the tower, further reducing the tower top voltage. If these arrive on the voltage tail, they will still have some effect on shortening the tail and reducing the stresses.

During the front of the wave, the impulse voltage induced across line insulation consists of two components: (1) due to surge response of tower and (2) resistive voltage rise of footing. During wave tail, the voltage across the tower reduces. If the flashovers do not occur during rise time or before return reflections from the adjacent towers, these are unlikely to occur subsequently.

Example 5-3 Consider a lightning stroke ramp rising at 40 kA/μs, with a wave front of 0.5 μs, and infinite tail. Tower surge impedance Z_t = 100 Ω. The ground wire impedance Z_g = 300 Ω. Tower height = 30 m. Velocity down the tower = 240 m/μs, velocity along the ground (shield) wire = 300 m/μs. The tower footing resistance = 10 Ω. Coupling factor K = 0.3, span = 300 m. Calculate the tower top voltage, insulation stress, voltage at base of the tower for the duration of the lightning wave front, that is, 0.5 μs. Consider a surge channel impedance of 1500 Ω.

The equivalent impedance of half the ground wire and tower surge impedance is:

$$Z = \frac{150 \times 100}{150 + 100} = 60\ \Omega$$

The rate of rise of the lightning current = 40 kA/μs. Therefore after 0.25 μs, the tower top potential is:

$$V_t = \frac{60}{(1 + 60/1500)} \times 40 \times 0.25 = 576\ \text{kV}$$

Ignoring the lightning channel impedance, it will be 600 kV.

The voltage stressing the insulation is:

$$V_i = (1-K)600 = 0.7 \times 600 = 430 \text{ kV}$$

Considering reflections from the tower base, the reflection coefficient at the tower base is:

$$\rho_{gr} = \frac{R_{tf} - Z_t}{R_{tf} + Z_t} = \frac{10-100}{110} = -0.818$$

With the given speed, the travel time of the wave to the base of the tower is 30/240 = 0.125 μs. Thus, the wave reflected from the base at ground level will reach the tower top in 0.25 μs. The reflection coefficient at the tower top is:

$$\rho_{tr} = \frac{Z_g - 2Z_t}{Z_g + 2Z_t} = \frac{300-200}{300+200} = 0.2$$

as

$$\rho_{tt} = 1 + \rho_{tr} = \frac{Z_g}{(0.5Z_g) + Z_t} = \frac{300}{150+100} = 1.2$$

After 0.25 μs, the voltage buildup on the tower top is reduced by:

$$(-0.818)(1.2)(40) = -39.2 \text{ kA/μs}$$

This reverses the voltage buildup. As per original ramp of 40 kA/μs, the voltage after 0.5 μs would have been 1200 KV. Considering the negative ramp of –39.2 kV/μs starting at 0.25 μs, the voltage at 0.5 μs will be 600 + (600 – 588) = 612 kV. The next reflected wave from the bottom of the tower will arrive in 0.5 μs. The reflection coefficient of the wave returning to top is 0.2, as calculated above. Thus the tower-top voltage under goes a change at 0.5 μs:

$$(0.2)(-0.818) = -0.1636 \text{ times the first}$$
$$= -6.413 \text{ kA/μs}$$

Thus, at 0.75 μs, the tower-top potential = 612 + (40 – 6.413 × 0.98) 60 × 0.25 = 1121 kV. The ladder diagram and the voltage buildup are as shown in Fig. 5-21. A curve can be drawn through the tower-top stepped potential profile.

Consider the reflections from the adjacent towers. The surge impedance of the towers will be:

$$Z_{ta} = \frac{Z_t Z_g}{Z_t + Z_g} = 75\ \Omega$$

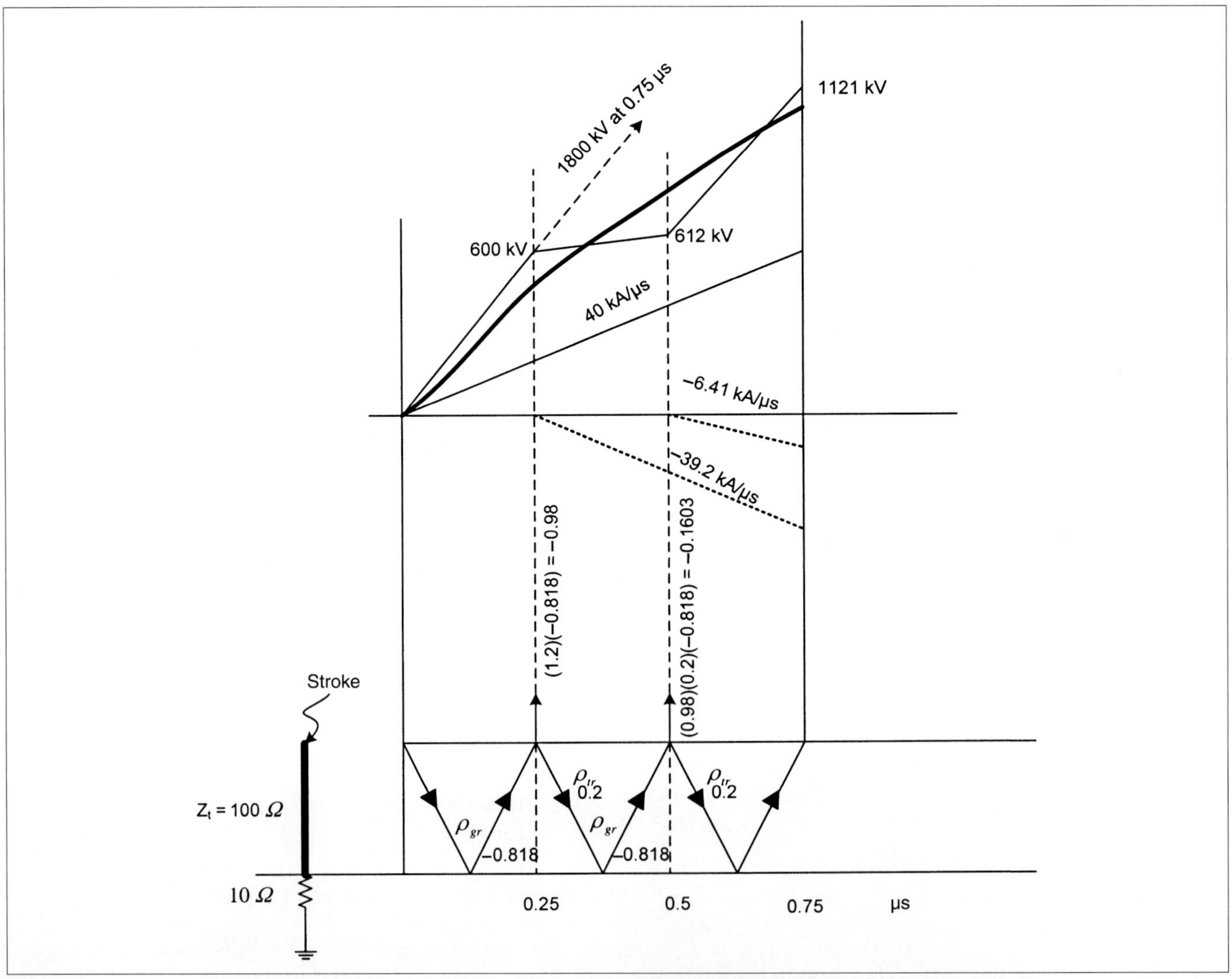

FIGURE 5-21 Calculation of tower-top potential, ladder diagram.

The reflection coefficient is:

$$\frac{Z_{ta}-Z_g}{Z_{ta}+Z_g}=-0.6$$

Therefore, the coefficient of the surge transmitted back to the struck tower = 1 – 0.6 = 0.4. Waves will arrive simultaneously from both the directions in 2 μs (as the span is 300 m). The associated reduction ramp is:

$$(-0.6)(0.4)40=-9.6 \text{ kA}/\mu\text{s}$$

The lighting surge may be at its tail by this time, and a backflashover might have already occurred. Thus, the reflections from the adjacent towers are not so significant and the reflections from the base of the tower dictate the reduction in the rate of rise. This shows the importance of having a lower tower footing resistance.

The voltage at the base of the tower can be calculated similarly. The stroke current passing down the tower is:

$$\frac{0.5Z_g}{0.5Z_g+Z_t}=0.6$$

Therefore, after the first travel time, the voltage at the base of the tower is:

$$0.6\times 40\times R_f=240 \text{ kV}/\mu\text{s}$$

Subsequently, the current transmitted to ground is:

$$\rho_{tt}=1+\rho_{tr}=\frac{2R_f}{Z_t+R_f} \tag{5-44}$$

Backflashover is an important parameter for insulation coordination, and two parameters to control it are: (1) increase the insulation strength, and (2) reduce the tower footing resistance, especially some distance from the substation. Figure 5-22 shows that the pattern of insulator strings voltage for a surge on a tower top, shielded line. Note the effect of impulse wave risk time on the insulator voltage. Other factors are tower height, footing resistance, tower surge impedance, surge channel impedance, width of cross arms, length of suspension strings, operating voltage, and span length, as negative reflections will take place from the adjacent towers. In Fig. 5-22, apart from stroke impulse current, no power frequency voltage is present.

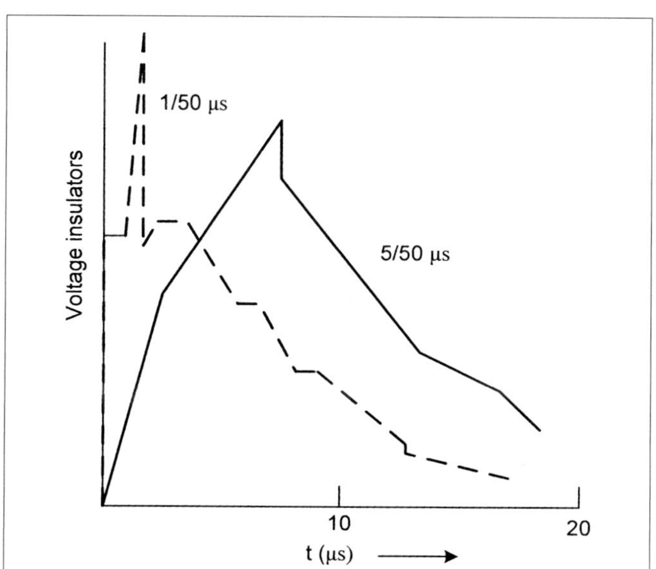

FIGURE 5-22 Effect of surge front on tower-top potential.

5-11 LIGHTNING STROKE TO GROUND WIRE

If a stroke occurs to a ground wire somewhere away from the tower, say at midspan, then the terminating impedance is half of the surge impedance of the ground wire, and the voltage on the ground wire will be:

$$V_g=\frac{I_0Z_gZ_s}{(2Z_s+Z_g)} \tag{5-45}$$

Voltage on the nearest phase conductor will be KV_g. The air clearances must withstand the difference $(1-K)V_g$. This voltage is much higher than any voltage across the insulator strings caused by a stroke of equal intensity to either the tower or the ground wire. Also the return time of the negative reflected wave from the adjacent towers is much longer because of higher effective terminating surge impedance at the midspan. For this reason, sometimes the midspan clearances between phase and ground conductors are increased. However, there is a voltage stress reduction due to predischarge current that flows between the ground wire and the phase conductor—a phenomena described by Wagner and Hileman.[18] This may delay the breakdown sufficiently to allow the negative reflected waves from the adjacent towers to prevent a breakdown. Generally, midspan flashes are rare. Assuming that no flashover occurs at the midspan, the voltage waves on the ground wires and phase conductors travel to the adjacent towers where these will be modified by reflections. The tower top voltage in this case is:

$$V_t'=V_gZ_t/(Z_t+0.5Z_g) \tag{5-46}$$

The voltage stressing the insulation is:

$$V_i'=\frac{(1-K)V_g}{(1+Z_g/2Z_t)} \tag{5-47}$$

Assuming that there are no midspan flashes, the traveling waves on ground and phase wires arrive at the adjacent towers, where they are modified by reflections, and tower flashovers can occur due to strokes on midspan, yet midspan flashovers are rare.

5-12 STROKES TO GROUND IN VICINITY OF TRANSMISSION LINES

The charges in the lower part of the clouds induce charges of opposite polarity on the ground. The classic picture is that an induced charge slowly builds up on the phase conductors of a transmission line, which are grounded through transformers, potential transformers, or neutral resistors. This induced voltage is given by $V = Q/C$. A flash to ground in the vicinity of the transmission line discharges the cloud and the charge on the transmission line is simultaneously released, initiating traveling waves in both direction of half the amplitude.

The theory was significantly advanced by Wagner in 1942.[19] He emphasized the predominating effect of the field of the lightning channel in comparison to the field of the thundercloud. The problem was further studied by many authors, and a mathematical solution was given by Rusck in 1958.[20]

These studies basically considered (1) the electric charge Q which is deposited in the leader channel and (2) the evacuation of this charge in the return stroke. It was concluded that the magnetic effect of the return-stroke current is much less and the electrostatic effects predominate, and that the voltage on the line will be a traveling wave in character, unipolar in nature because of electrostatic effect, and of a polarity opposite to that of the return stroke current. Chowdhuri and Gross[21] based on much observed data have drawn attention to the fact that there is also an electromagnetic induction from the return stroke current and magnetic effects cannot be ignored. The induced voltage can be bipolar in character depending upon the wave shape of the return stroke current, and the induced voltage is not entirely a traveling wave phenomena.

As the induced current is almost at right angles to the conductors, basic field theory is applied for the calculations. The induced voltage can be much higher then that will occur due to electrostatic effect only. These voltages can be sufficiently high to cause flashover of the insulation of lines in higher voltage range.

The mechanism of induction can be summarized as:

- Charged cloud induces charges on the line.
- When cloud discharges, the voltage and current traveling waves occur.
- The stepped leader further induces charges on the conductors.
- When stepped leader is neutralized by return stroke, the charges on conductors are released, producing traveling waves similar to cloud discharge.
- The residual charges in return stroke induce an electrostatic field in vicinity of line and hence induce a voltage.
- The *di/dt* of the return stroke produces a magnetically induced voltage on the line.

The parameters affecting these phenomena are many: height of the line, peak magnitude of the return-stroke current, cloud height, wave shape of return-stroke current, perpendicular distance of the struck point from line, horizontal distance along ground, return stroke velocity, and presence of shield wires (the induced voltages may be reduced from 10 to 30 percent). A lossless line and perfectly conducting ground is assumed.

Whether the line or the ground will be struck can be ascertained from the simple geometric relations in Fig. 5-23. If the perpendicular distance of the leader stroke is greater than a critical distance, the leader will strike the ground, otherwise the line. The least distance of ground strike can be calculated from the following expression:

$$\begin{aligned} y &= \sqrt{r_s^2-(r_s-h)^2} \qquad r_s>h \\ &= r_s \qquad\qquad\qquad\quad r_s\le h \end{aligned} \tag{5-48}$$

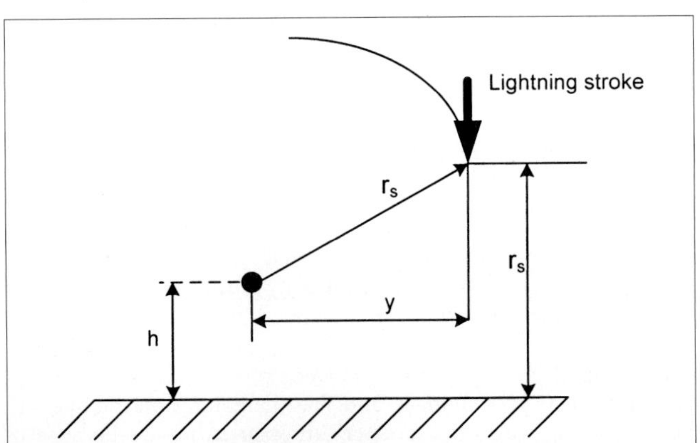

FIGURE 5-23 Direct and indirect lightning strokes.

Referring to Fig. 5-23, r_s is the striking distance, h is the height of the conductor above ground, and y = horizontal distance to the strike point.

5-13 SHIELDING

A shielding angle of 30° was popular for many years, until in the 1950s the failure rate on double circuit lines on towers up to 45-m high increased. This led to further study of shielding practices. It is well known that taller structures attract more lightning strokes. We defined attractive radius, the effect of return stroke current, and height of the structure in Eq. (5-10). Eriksson's equation for phase conductors and shield wires is:

$$r_a = 0.67h^{0.6}(I_p)^{0.74} \tag{5-49}$$

Comparing with Eq. (5-10), Eriksson assumed that the attractive radius for shield wires and phase conductors will be reduced by 80 percent.[22]

It is realized that several other parameters should also influence striking distance, that is, the charge distribution along leader stroke, return stroke velocity, presence of other conductors, and polarity of the leader stroke.

For each value of stroke current, the striking distance r_s and r_g define a surface ABCD, of which, the portion BC is the exposed boundary (Fig. 5-24). All strokes crossing BC are assumed to land on the phase conductors. The arc BC shrinks with increasing stroke currents, that is, increasing r_s until it becomes zero for a certain r_{s2}. For smaller currents the exposure increases, but we need to consider r_{s1} for the critical stroke current ($I = 2V/Z$). A flashover cannot occur if:

$$r_{s1} \ge r_{s2} \tag{5-50}$$

This clearly shows that the shielding angle should be reduced as the height increases. For partially shielded lines, the average number of strokes crossing the arc BC can be determined from geometric considerations. The breakdown gradient to the ground plane may differ from that of the shielding and the conductors. This is accounted for in a factor K_{sg}, R_{sg} in Fig. 5-24, $R_{sg} = r_s K_{sg}$. In Fig. 5-24, the number of strokes that terminate on phase conductor, shielding failure rate (SFR), is given by the area formed by D_c, length of line and ground flash density:

$$\text{SFR} = 2N_g D_c l \tag{5-51}$$

where l is the length of the line.

The probability of occurrence of this current is $f(I)\,dI$, so that SFR for all currents is:

$$\text{SFR} = 2N_g l \int_3^{I_m} D_c f(I)dI \tag{5-52}$$

The minimum value of current = 3 kA is from CIGRE, and at the maximum current I_m, the three arcs coincide and $D_c = 0$. All strokes that terminate on phase conductors will not result in flashover, therefore the equation for shielding failure flashover rate (SFFOR) becomes:

$$\text{SFFOR} = 2N_g l \int_{Ic}^{I_m} D_c f(I)dI \tag{5-53}$$

I_c is defined in Eq. 5-33. Anderson suggested that the average value of D_c over the interval from I_c to I_m is half of the value of D_c at $I = I_c$, and is a constant. Thus:

$$\text{SSFOR} = N_g l D_c P(I_m \ge I \ge I_c) \tag{5-54}$$

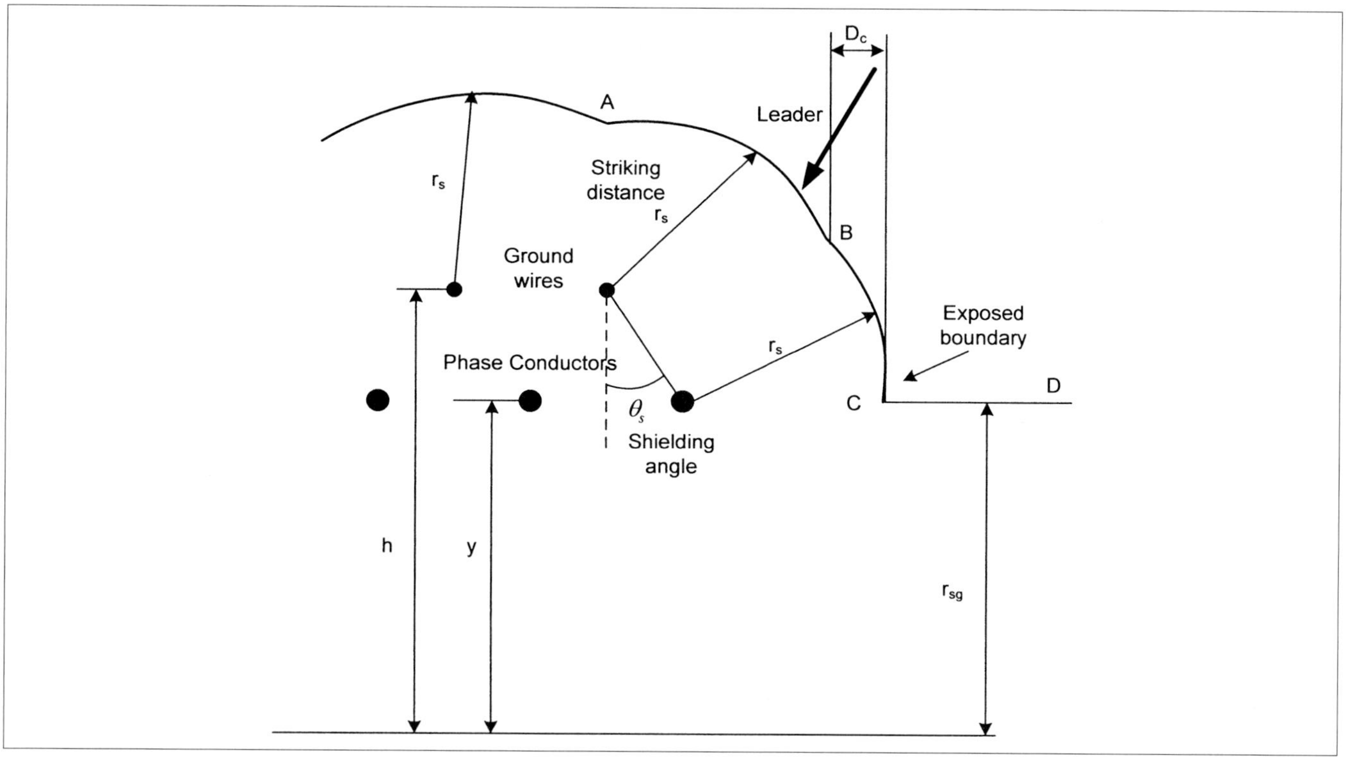

FIGURE 5-24 The geometric model for shielding of transmission lines.

We use log normal distribution to solve this equation, which can be written as:

$$N_g l D_c [Q(I_c) - Q(I_m)] \quad \text{where: } Q(I) = 1 - F(I) \tag{5-55}$$

To illustrate the geometric relations, a section of Fig. 5-24 is enlarged as shown in Fig. 5-25*a*. Figure 5-25*b* is the Eriksson model. From this geometric construction:

$$D_c = r_s[\cos\theta - \cos(\alpha+\beta)] \qquad D_g = r_s\cos(\alpha-\beta) \tag{5-56}$$

where:

$$\beta = \frac{1}{2}\sin^{-1}\frac{(h-y)\sqrt{1+\tan^2\alpha}}{2r_s} \qquad \theta = \sin^{-1}\frac{r_g - y}{r_s} \qquad \alpha = \tan^{-1}\frac{a}{h-y} \tag{5-57}$$

For I_m, where all striking distances coincide at a single point:

$$a = \sqrt{r_s^2 - (r_g - h)^2} - \sqrt{r_s^2 - (r_g - y)^2} \qquad r_{gm} = \frac{(h+y)/2}{1-(r_s/r_g)\sin\alpha} \tag{5-58}$$

$$I_m = \left[\frac{r_{gm}}{A}\right]^{1/b} \quad (r_g = A I_s^b) \text{ (Table 5-3)}$$

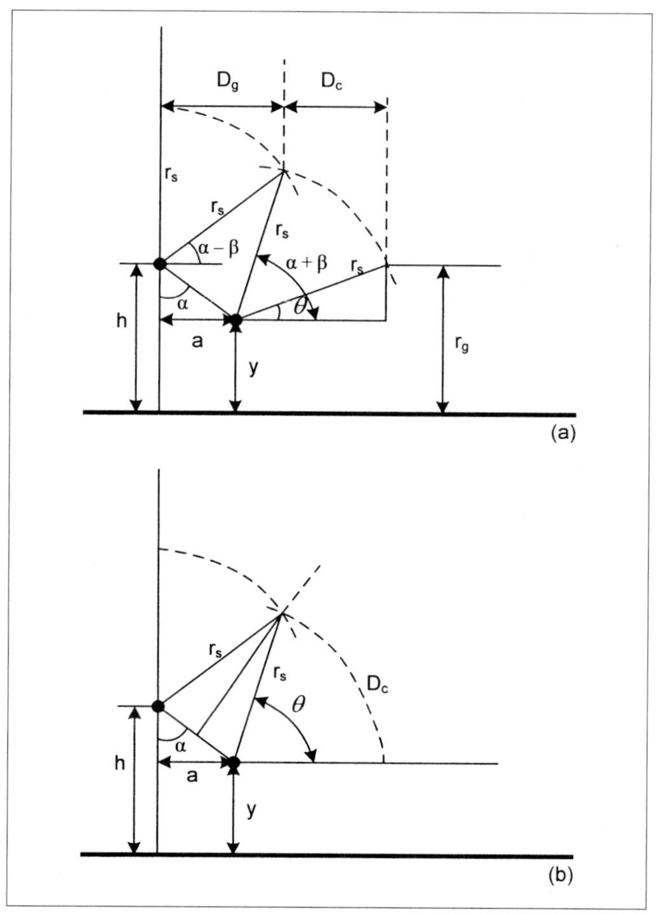

FIGURE 5-25 (*a*) Details of Fig. 5-24. (*b*) Eriksson's modifications.

TABLE 5-4 Summary of Pathfinder Operations Edison Electric Institute Research Project No. RP-50

DESCRIPTION OF DATA	No.
Total Pathfinder instrument operations	167
Number of separate lightning strokes	111
Number of strokes causing tripouts	94
Number of strokes causing shielding failure	51
TYPE OF FAULT FROM SHIELDING FAILURES	
Top single conductor tripouts	39
Middle single conductor tripouts	2
Multiple conductor tripouts	3
Double circuit tripouts	1
No tripout	6
Number of strokes causing backflashovers	52
TYPE OF FAULT FROM BACKFLASHOVERS	
Top single conductor tripouts	21
Multiple conductor tripouts	5
Double circuit tripouts	20
No tripout	6
Indicated number of negative polarity strokes	103
Indicated percentage negative polarity strokes	93
Instrument failures	8

For perfect shielding the angle is given by:

$$\alpha_p = \sin^{-1}\frac{r_g-(h+y)/2}{r_s} \approx \frac{r_g}{r_s}-\frac{1}{r_s}\left(\frac{h+y}{2}\right) \tag{5-59}$$

An extensive study for HV and EHV lines is "Pathfinder Project" conducted by Edison Electric Institute and reported in Refs. 23 and 24.

The Pathfinder Project reports results of 640 km of 110 to 220 and 345-kV lines, statistical survey covering 84,000 km/year.

Table 5-4 indicates the importance of backflashes and shielding failures. The majority of shielding failures are top single-conductor trip outs (39 out of 45). The backflashovers involving both circuits of double circuit lines are 20 out of 46 trip outs.

Another similar study is the CIGRE study, results of which are summarized in Table 5-5 and Fig. 5-26.[24,27] These clearly indicate that shielding angle and the height of the tower are two major factors in reducing the standard trip out rates (STR). The STR is defined as the number of outages caused by lightning per 100 km/year.

5-14 SHIELDING DESIGNS

We will discuss two methods of shielding designs. One method is based upon the work of Whitehead (Fig. 5-27) and the concepts discussed above. This figure gives the critical mean shielding angle based upon the conductor height h_p, ($h_p = y$ in Fig. 5-27) and conductor to ground wire spacing c. These quantities are normalized in per unit of striking distance r_s. The field results of Pathfinder Project correlate well with this method.

The second method is a simple geometric construction for placing the shield wires based upon the electromagnetic model of shielding.[25] The procedure can be discussed with reference to Fig. 5-28*a*. First find the critical shielding current. It is generally increased by 10 percent. Based upon this critical current, the striking distance r_s is known.

To locate the center C for describing an arc with radius equal to the striking distance r_s, the ordinate is given by r_{sg} and the abscissa is:

$$x_c = -d_p + \sqrt{r_s^2-(r_{sg}-h_p)^2} \qquad r_{sg} \geq h_p \tag{5-60}$$

Note that for factor $K_{sg} = 1$, $R_{sg} = r_s$. Draw an arc with center C and radius r_s passing through the outermost conductor. It may cut the vertical line through the middle conductor at some point S. A ground (shield) wire placed at this point will protect all three-phase conductors. Analytically the height of the shield wire is:

$$h_s = r_{sg} - \sqrt{r_s^2 - x_c^2} \tag{5-61}$$

A qualification applies that the minimum clearance required between the phase conductor and shield wire should be observed. If this clearance is S_{ps} and h_s is less than S_{ps}, then make $S_{ps} = h_s$.

TABLE 5-5 Lightning Outages for 100 km-yr and 40 Thunder Days per year (STR = 40) for Five HV and EHV Lines Operating in Comparable Environments except for the Heights and the Shielding Angles.

CIGRE LINE NO.	KM-YR	U (KV)	U (50%) (KV)	T_D (YR^{-1})	R_g (Ω)	H (M)	Y (M)	θ (DEG)	SC (M)	STR-40
32*	1760	345	1600	40	5	43	34	31	23.4§	5.70
31†	1570	345	1600	43	5	43	25	22	12.6	3.44
30‡	5900	230	1500	40	5	28	16	15	–4.8	0.24
39‡	3140	345	1600	40	5	29	16	–15	–27.7	0.19
40‡	4460	220	1580	32	12	27	19	14	–4.6	0.09

*Double-circuit vertical conductor configuration, one shield wire.
†Double-circuit vertical conductor configuration with two shield wires. Middle conductor data given; 67 percent of outages on this phase.
‡Single-circuit horizontal conductor configuration with two shield wires.
§Positive values of exposure arc S_c are an index of the shielding wire inefficiency.

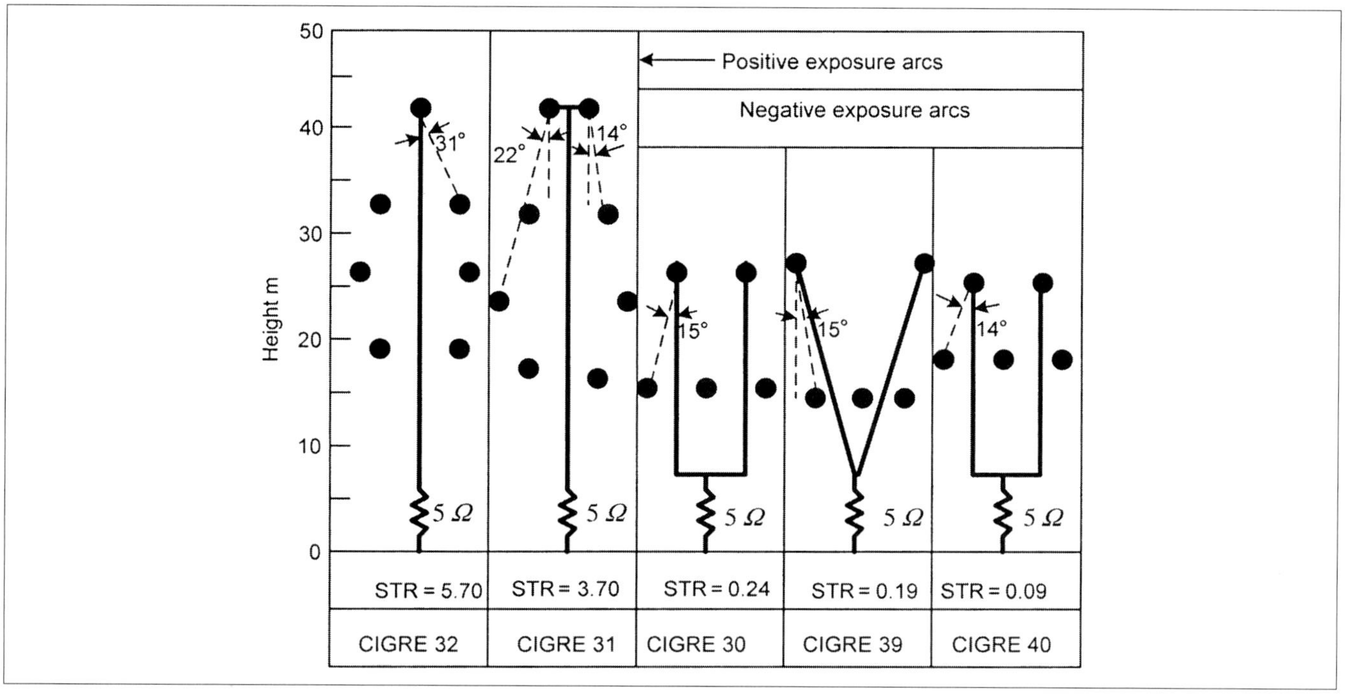

FIGURE 5-26 Standard trip out rates (STR) for different types of tower construction.[4,24]

The arc thus drawn in Fig. 5-28*a* may not cross the vertical through the middle conductor. Two shield wires are then required. Draw another arc with P_3 as center and S_{ps} as the radius (Fig. 5-28*b*). The intersection of these two arcs gives the location of the shield wire. The second shield wire is symmetrically placed.

Example 5-4 Calculate the SFFOR for a transmission line with two shield wires, located at 30 m, conductor height = 25 m, shielding angle = 30°, I_c = 15 kA, N_g = 4.

Let us use IEEE equations for r_s and r_g, from Table 5-3. These give $r_s = r_g = 46.5$ m, $r_c/r_g = 1$

$$r_{gm} = \frac{(30+25)/2}{1-\sin 30^\circ} = 55 \text{ m}$$

$$I_m = (55/8)^{1/0.65} = 19.4 \text{ kA} \qquad \beta = 3.56^\circ \qquad \theta = 27.5^\circ$$

This gives $D_c = 2.51$ m.

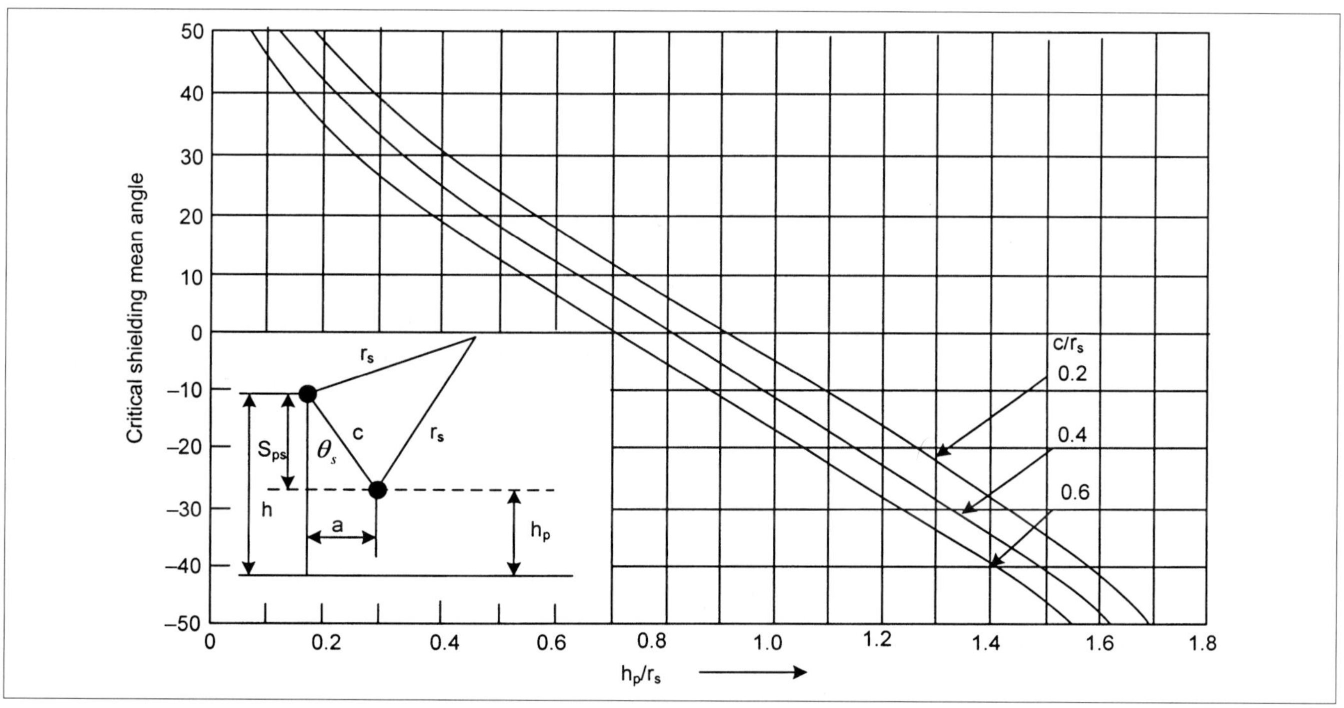

FIGURE 5-27 Critical shielding angle in terms of normalized geometry.[24]

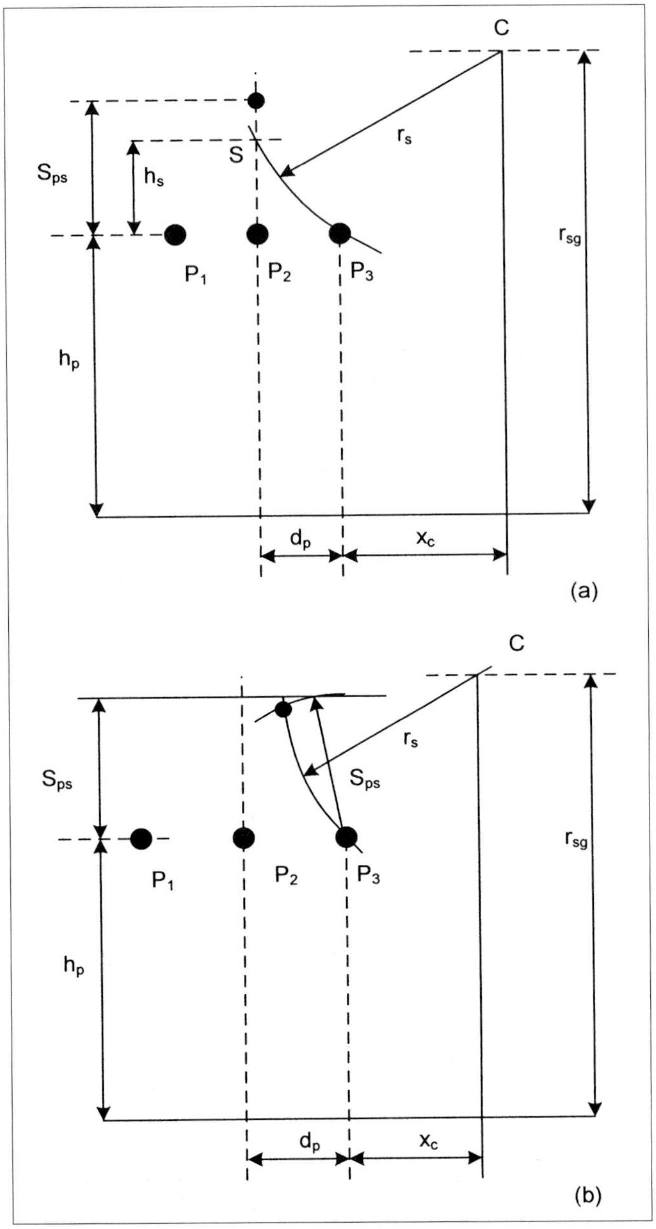

FIGURES 5-28 Geometric constructions for placement of shield wires: (*a*) one shield wire adequate; (*b*) two shield wires required.

An approximation to CIGRE cumulative distribution is shown in Table 5-6, and median and log standard deviations are shown in Table 5-7. Using the values from these tables first calculate:

$$z = \frac{\ln I - \ln M_1}{\beta_1}$$

TABLE 5-6 CIGRE Cumulative Distribution Approximations

RANGE OF CURRENT *I* (KA)	APPROXIMATE EQUATION OF *Q*
3–20	$1-0.31e^{-z^2/1.6}$
20–60	0.50–0.35Z
60–200	$0.278e^{-z^2/1.7}$

TABLE 5-7 Median and Log Standard Deviations for CIGRE Distribution

CURRENT RANGE (KA)	MEDIAN M_1	BETA β_1
3–20	61.1	1.33
>20	33.3	0.605

and then the values of Q:

$$z_c = \frac{\ln(15/61.1)}{1.33} = -1.05$$

$$Q_c = 1 - 0.31e^{-1.1025/1.6} = 0.844$$

$$z_m = \frac{\ln(19.4/61.1)}{1.33} = -0.863$$

$$Q_m = 1 - 0.31e^{-0.863/1.6} = 0.819$$

Therefore:

$$\text{SFFOR} = 4\times 100\times 2.51\times 0.025\times 10^{-3} = 0.0251/100 \text{ km}-\text{years.}$$

Table 5-8 shows required shielding angle for SFFOR of 0.05/100 km yr.[26]

Example 5-5 Consider an overhead line with the configuration as shown in Fig. 5-29. The spacing between conductors = 5 m, the minimum separation of shield wire as shown = 5 m, tower height = 27 m. T_D (keraunic level) = 27. The insulator strings have a CFO = 1500 kV, the surge impedance = 500 Ω. Estimate the shielding flashover rate. Redesign the shielding for better protection, using the two methods of calculations as described above. Is one shield wire adequate?

The shielding design is calculated based upon Figs. 5-27 and 5-28.

(*a*) *Based upon Fig. 5-27.* Multiply the critical current by 1.1, which gives 6.6 kA. Then r_s = 30.3 m. Let K_{sg} = 1.

Let height of the conductor h_p = 22 m. Normalize with respect to r_s: h_p = 22/30.3 = 0.726

In Fig. 5-27, we may consider c approximately equal to 5 m. Thus c/r_s = 5/30.3 = 0.166.

As a first approximation, we can use the curve marked c/r_s = 0.2. A shielding angle of 12° can be read. Check back to c/r_s estimate used in the calculation. From the geometry shown in Fig. 5-27,

$$c = s_{ps}/\cos\theta_s = 5.1 \text{ m}$$

TABLE 5-8 Required Shielding Angle for SFFOR of 0.05/100 km-yr.[26]

GROUND FLASH DENSITY (FLASHES km² YEAR)	GROUND WIRE HEIGHT (m)	PHASE CONDUCTOR HEIGHT (m)	SHIELDING ANGLE IEEE (DEG.)	SHIELDING ANGLE CIGRE, (DEG.)
0.5	30	22	38	42
	45	37	12	28
3.0	30	22	33.5	36
	45	37		19
5.0	30	33	32	35
	45	37	4	17

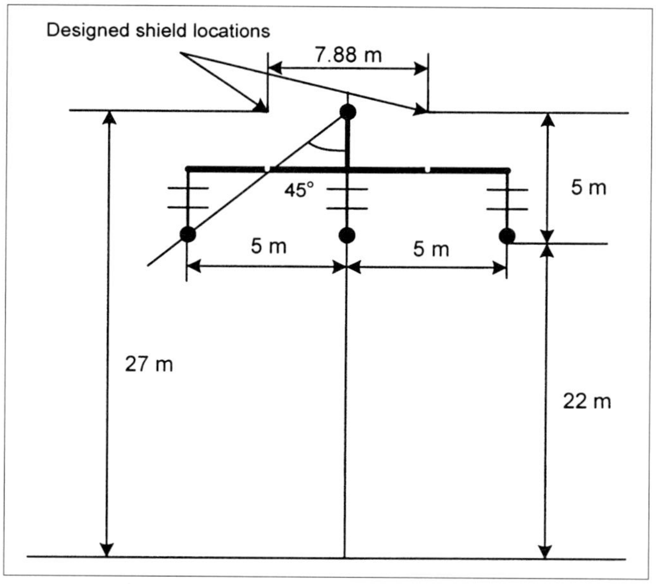

FIGURE 5-29 Placement of shield wires.

Thus, the initial estimate of c/r_s is acceptable. The horizontal spacing between the phase conductor and ground wire is:

$$h_s \tan 12^\circ = 5 \times 0.212 = 1.06 \text{ m}$$

Two ground wires with spacing between them = 7.88 m are required. See Fig. 5-29.

(b) *Based upon Fig. 5-28.* The center of the circle is located at C. The abscissa from Eq. (5-60) is:

$$x_c = -5 + \sqrt{30.3^2 - (30.3 - 22)^2} = 29.14 \text{ m}$$

And the ordinate is 30.3 m. A graphical construction gives approximately the same spacing as calculated before.

5-15 BACKFLASHOVERS

An outage can occur from backflash, which occurs if lightning strikes a shield wire or tower. The potential rise of the tower top due to flow of lightning current may exceed the insulation strength of the insulators. An arc between the ground wire and phase conductor will inject some portion of the lightning current into the phase conductors. Figure 5-30 shows a backflash. It is an important mechanism for surge transfer and breakdown and is discussed further.

There are two methods in use: CIGRE method and IEEE method. The calculations are performed using computer programs because of the iterative nature of the calculations. Anderson estimating method using a constant 2-μs front with minor modifications is adopted by IEEE working group.[27] The basic concepts of a backflashover are explained in Sec. 5-10, for example, the voltage across the insulation has to be greater than the CFO, for a certain critical current I_c. Thus, for a certain stroke current, the probability of flashover can be expressed as:

$$P(I \geq I_c) = \int_{I_c}^{\infty} f(I)dI \tag{5-62}$$

The BFR is then:

$$\text{BFR} = 0.6 N_L P(I_c) \tag{5-63}$$

where N_L are the number of strokes to transmission line, estimated from:

$$N_L = N_g \frac{(28h^{0.6} + S_g)}{10} \tag{5-64}$$

The S_g is shown in Fig. 5-20*c*.

Note the factor 0.6. As stated before, the flashovers within the span can be neglected, and the voltages at the tower for strokes

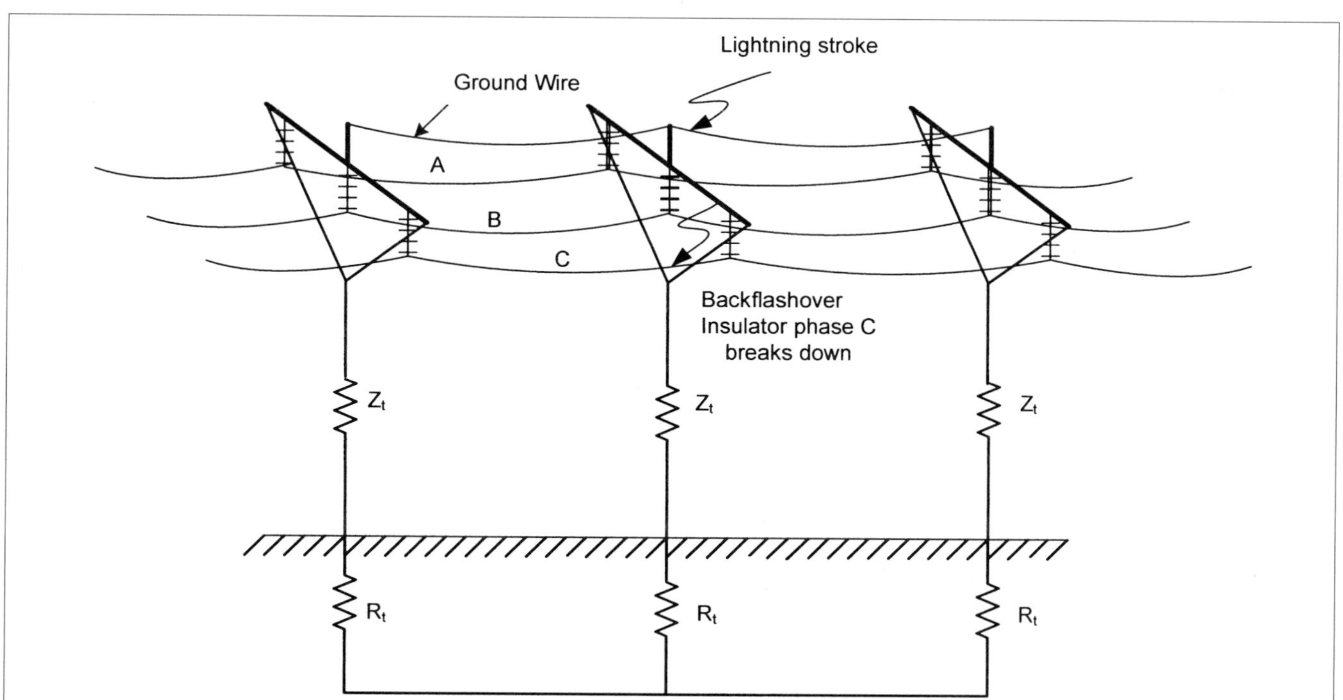

FIGURE 5-30 Illustration of a backflashover.

within the span are much lower compared to voltages for strokes on the tower itself. To account for the strokes terminating on the tower, adjustments are made to the BFR.

The voltage across insulation increases as the time to crest decreases. Thus, Eq. (5-63) should be modified to consider various fronts:

$$\mathrm{BFR} = \int_0^{\infty} (\mathrm{BFR}) f(t_f)\, dt_f \tag{5-65}$$

Above the corona inception voltage, the ground wire surge impedance decreases and the coupling factor K increases. The critical current increases as $(1 - K)$ decreases. Corona does not significantly alter BFR.

The CIGRE method basically starts with selection of crest time. For 115 to 230 kV lines, a front of 2.5 μs is appropriate, and for 345 kV and above, a front of 4.0 μs is appropriate. Assume a value of R_i equal to 50 percent of R_0, and solve for I_c:

$$I_c = \frac{\mathrm{CFO_{NS}} - V_{\mathrm{PF}}}{R_e(1-K)}, \quad R_e = \frac{R_i Z_g}{Z_g + 2R_i} \tag{5-66}$$

where $\mathrm{CFO_{NS}}$ is the nonstandard surge critical flashover, given by:

$$\mathrm{CFO_{NS}} = \left(0.977 + \frac{2.82}{\tau}\right)\left(1 - 0.2\frac{V_{\mathrm{PF}}}{\mathrm{CFO}}\right)\mathrm{CFO} \tag{5-67}$$

$$V_{\mathrm{PF}} = K_{\mathrm{PF}} V_{\mathrm{LN}}$$

$$\tau = \frac{Z_g}{R_i} T_s \tag{5-68}$$

where K_{PF} is a factor which depends upon the phase configuration, V_{LN} is the peak line to neutral voltage, Z_g is the surge impedance of the ground wire, τ is time constant of the tail, much akin to R/L, $L = Z_g/v$ (span length) $= Z_g T_s$, and v is velocity of light.

Calculate I_R, iterate on R_i, and when it is satisfactory, calculate median front for I_c; if this does not match assumed front, start over. CIGRE computer program BFR simplifies the calculation. It is obvious that BFR will vary with CFO and line construction.

Here we will use EMTP simulation and appropriate modeling, taking into account all the variable factors. The results of the simulation should be more accurate than hand calculations and other iterative methods, though no test verification is provided.

Example 5-6 This example is an EMTP simulation of a lightning stroke to the top of a transmission line tower. The following system data applies:

(i) *Towers*. Framed steel structure for 345-kV, three-phase, 60-Hz transmission line, the configuration of the phase conductors, shield wires, height of the tower, and other dimensions are shown in Fig. 5-31. There are two shield wires, and each phase consists of two bundle conductors spaced 18 in apart, 95400 circular mils (equivalent copper = 600000 circular mils), Aluminium Cable Steel Reinforced (ACSR). The span is 250 m (820 ft) and the sag = 4 m.

(ii) *System configuration*. Figure 5-32 shows the system configuration. 250-m spans between the last three transmission line towers T1, T2, and T3 are modeled with a FD model as discussed in Chap. 4. Similarly the FD model for the 40-km long line is derived. The line is terminated in a 200-MVA, wye-delta connected, 345 to 138 kV transformer at no-load. The transformer is modeled with bushing, winding, and interwinding capacitances (see Chap. 14).

(iii) *Tower surge impedance and footing resistance*. Each tower has a surge impedance of 50 Ω and a footing resistance of 15 Ω, controlled by an equivalent Norton current source to account for change in the resistance with surge current. There are no counterpoises.

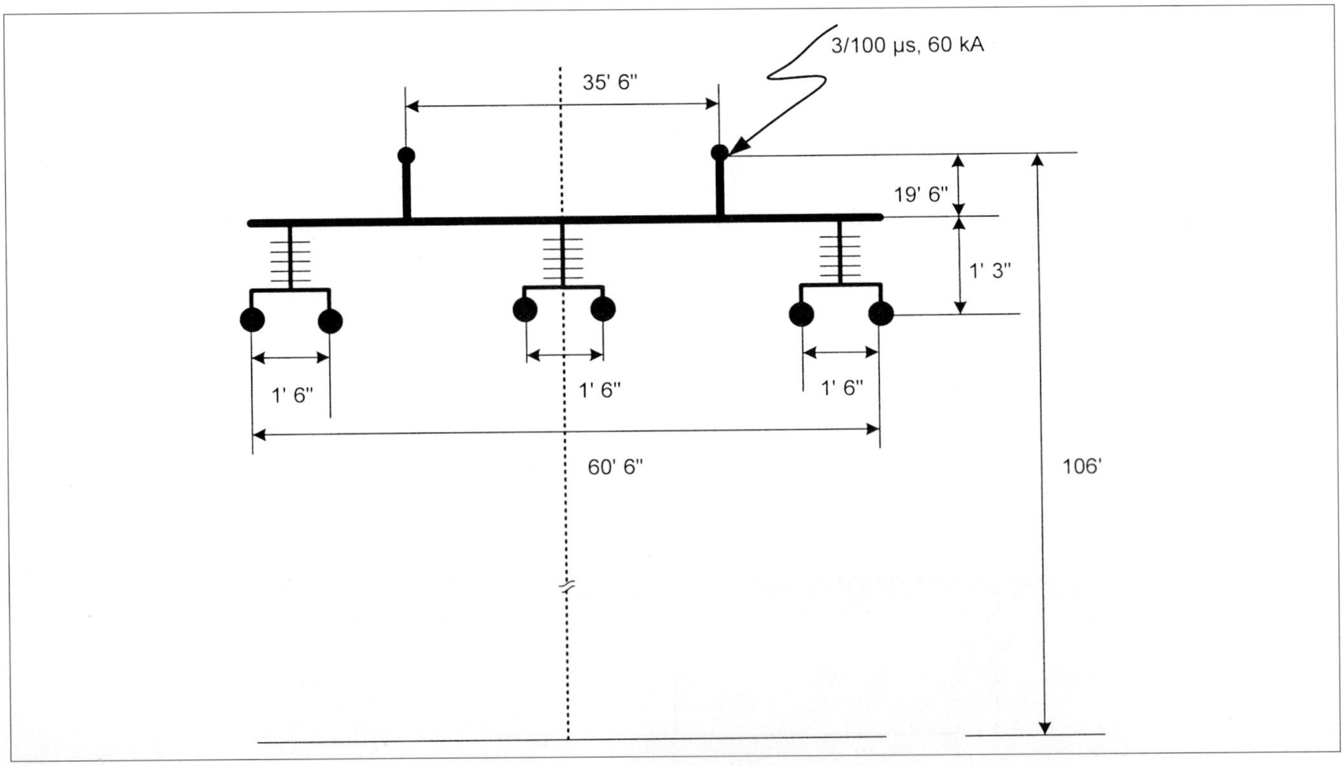

FIGURE 5-31 Tower configuration for Example 5-6.

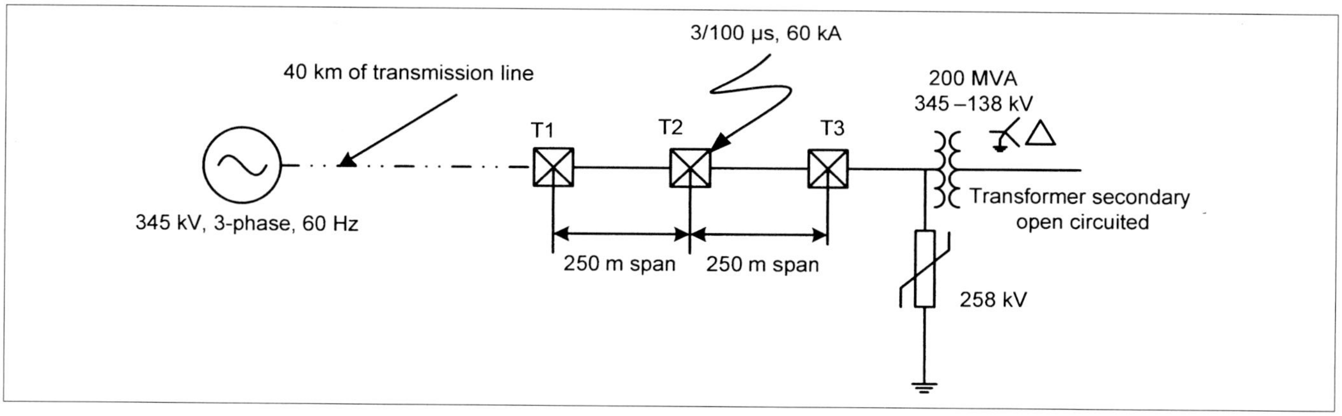

FIGURE 5-32 System configuration for study of backflashover.

(iv) Insulators—CFO. A CFO of 1210 kV is considered. These can be represented by voltage-dependent switches in parallel with capacitors connected between phases and tower. The capacitors simulate the coupling effect of conductors to the tower structure. Typical capacitance values for suspension insulators are 80 pF/unit. Ten insulators in a string will have a capacitance of 10 pF/string. An expression for the backflashover of insulators[28] considers a simplified model with an ideal short-circuit representation:

$$V_{vt} = K_1 + \frac{K_2}{t^{0.75}} \tag{5-69}$$

where:

K_1 = 400L
K_2 = 710L
V_{vt} = Flashover voltage, kV
L = length of insulator, m
t = Elapsed tie after lightning stroke, μs

(v) Lightning surge. A surge of 60 kA, 3/100 μs impacts the tower *T2*, as shown in Fig. 5-32.

(vi) Surge arrester. The simulation is repeated with a 258-kV, metal-oxide, gapless surge arrester at 200-MVA transformer terminals which will be invariably provided. Also a surge arrester on the transformer secondary (low-voltage side) terminals is required (Chaps. 14 and 20).

The results of the simulation are shown in Figs. 5-33 through 5-37, without 258-kV surge arrester.

Figure 5-33*a*, *b*, and *c* shows that the insulators in phases *b* and *c* of towers *T2* and *T3* flash over. None of the insulators in phase *a* on any three towers flash over. The insulators on phase *b* on *T1* also do not flash over.

Figure 5-34 shows the current in the tower footing resistance. Note that the maximum flashover current occurs in *T3*, a tower ahead of the one that is struck. The reflection coefficient will be different because of termination of the line in a transformer.

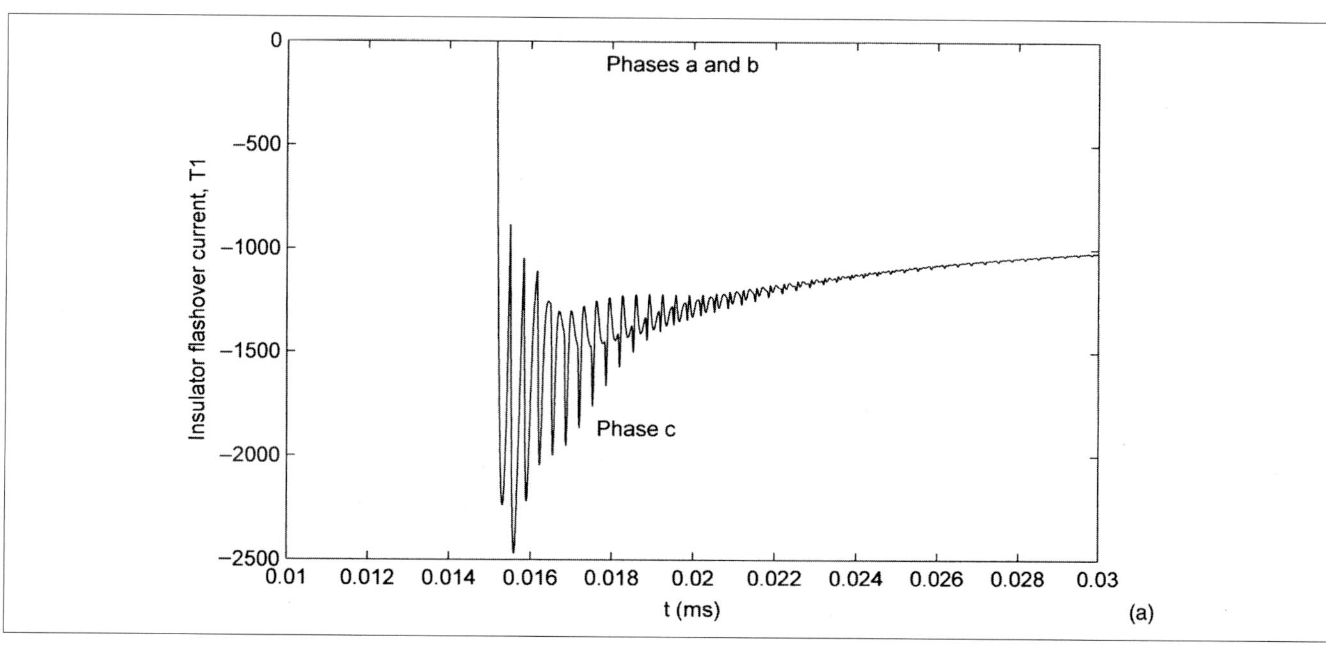

FIGURES 5-33 Insulator flashover currents of towers T1, T2, and T3. (*Continued*)

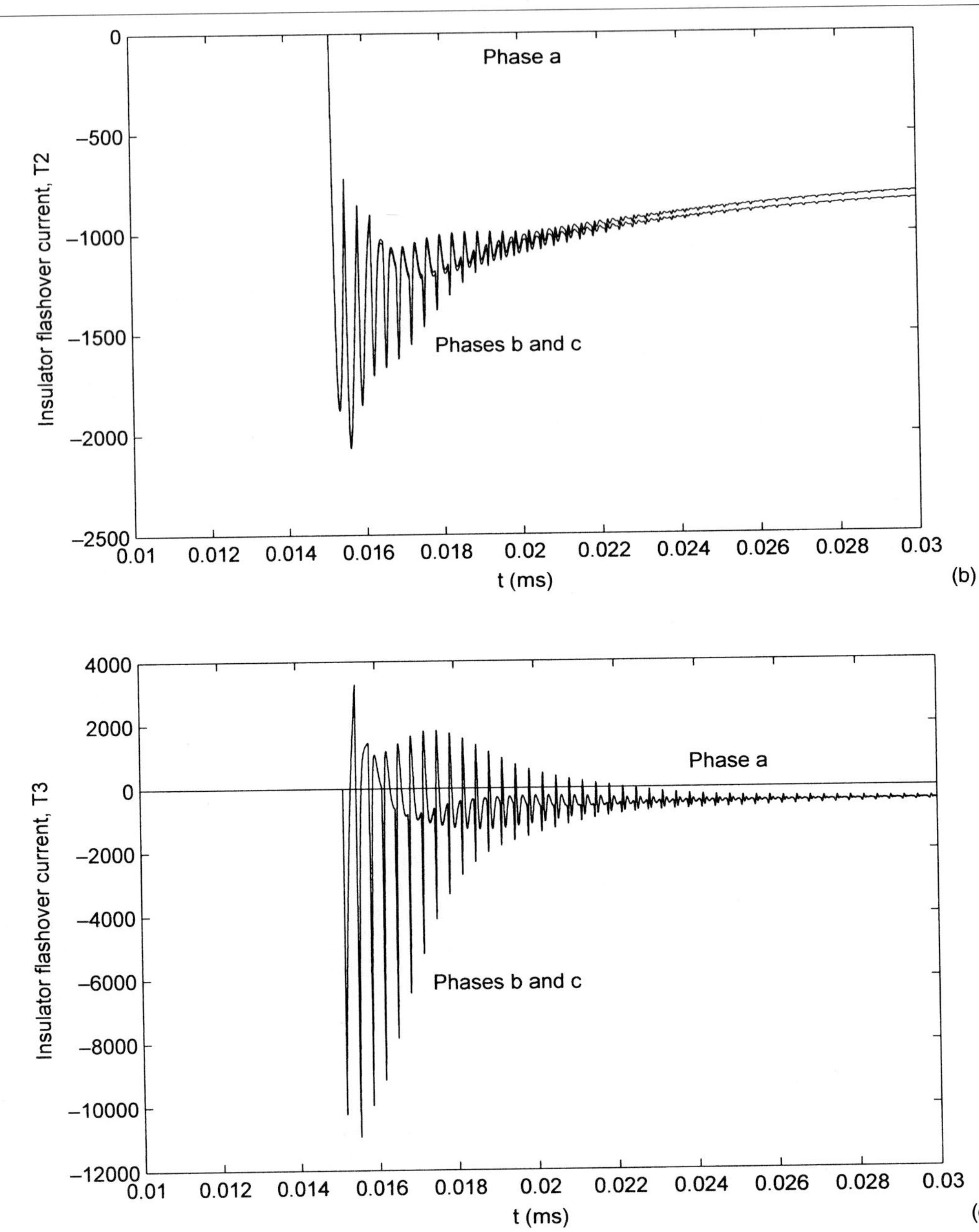

FIGURES 5-33 *(Continued)*

Figure 5-35 shows tower-top potential of the struck tower, *T*2. Figure 5-36*a* shows the gap voltages (voltages across the insulator strings) in phases *a*, *b*, and *c* of tower *T*2. Figure 5-36*b* is the picture of insulator flashovers towers *T*1, *T*2, and *T*3. Figure 5-37*a* shows the voltages across the 200-MVA transformer windings.

When a surge arrester is applied to the transformer primary windings, Fig. 5-37*b* depicts the voltage on the transformer primary windings. Figure 5-37*c* shows current through the surge arrester. Figure 5-38 illustrates that no flashover of insulators occurs on towers *T*1 and *T*2, and only insulator strings of phases *b* and *c* of Tower *T*3 flashover.

The effect of tower footing resistance may not be so obvious in this example, but it impacts the tower top potential and overvoltage for backflashover, as stated before. Figure 5-39 illustrates the effect of tower grounding on the tower top potential. For the first 0.2 μs, the tower top potential is not affected by the tower grounding. CIGRE has accepted the following volt–time curve for line-insulator flashover based upon the work in Ref. 29.

$$V_f = 0.4W + \frac{0.71W}{t^{0.75}} \tag{5-70}$$

where:

W = line insulator length, m
t = time to breakdown, s
V_f = flashover voltage for negative surges, MV

Counterpoise is used in the areas of high resistivity. If this is not sufficient to achieve the required backflash performance, the insulator string length can be increased, in other words, CFO can be increased as demonstrated in this example.

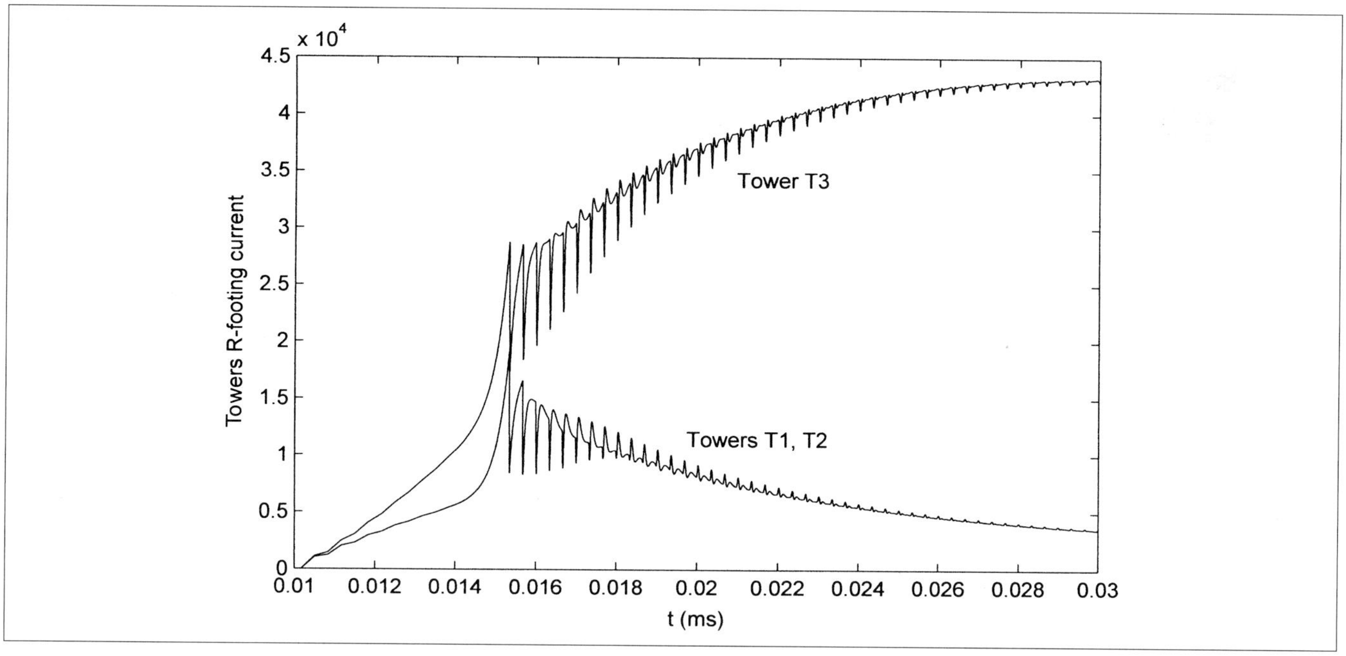

FIGURE 5-34 Tower footing currents.

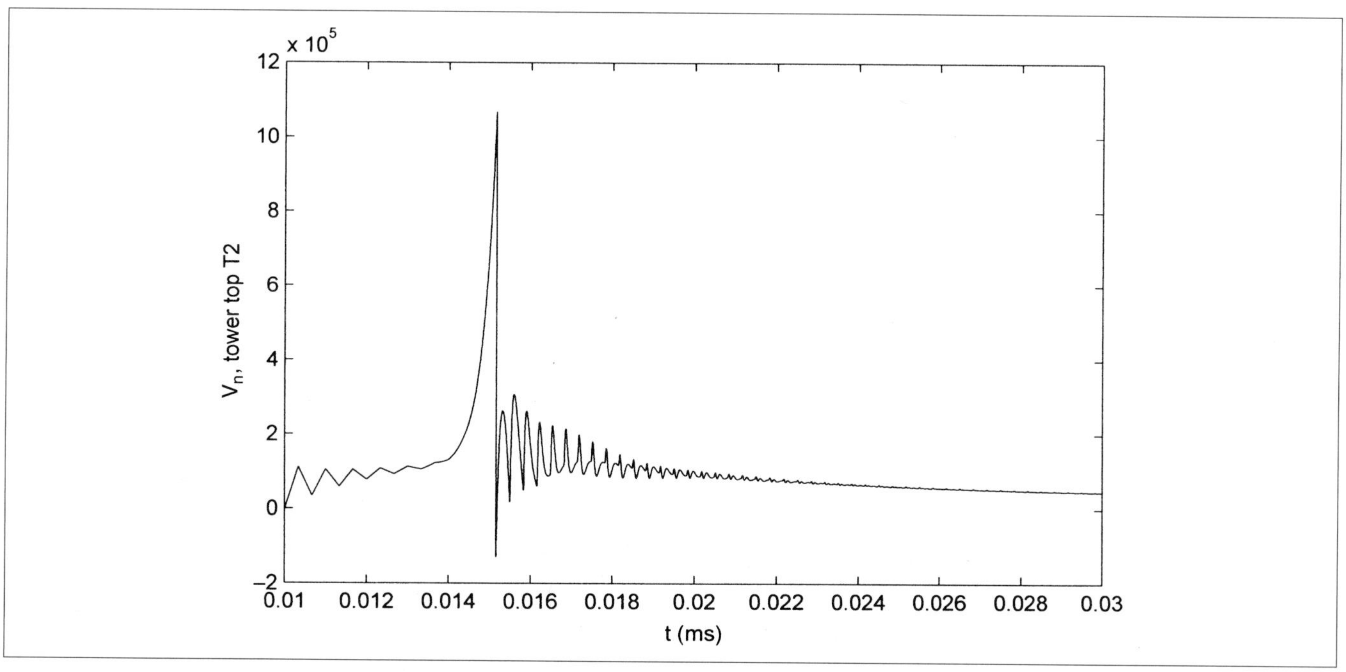

FIGURE 5-35 Tower top potential for tower *T*2.

It will be prohibitively expensive to design a system to cater to the worst lightning conditions, and a statistical approach is taken in the chapter on insulation coordination (Chap. 17).

PROBLEMS

1. What are K changes and M discharges? What is the typical stepped leader duration before a connection with the ground is made? Describe the significance of striking distance. Write three different expressions given in this chapter, explaining the difference between these.

2. Given $T_D = 30$, $h = 30$ m, $b = 6$ m, calculate the number of lightning strokes 100 km/year with the various analytical expressions given in this chapter. Is it necessary to calculate the attractive radius?

3. Which state in the United States experiences the highest lightning activity?

4. What is 98 percent probability of a peak current in a lightning stroke? What can be the maximum peak?

5. Calculate the tower-top and tower-base potential for a period of a lighting strike, impulse 60 kA/μs occurs at top of a

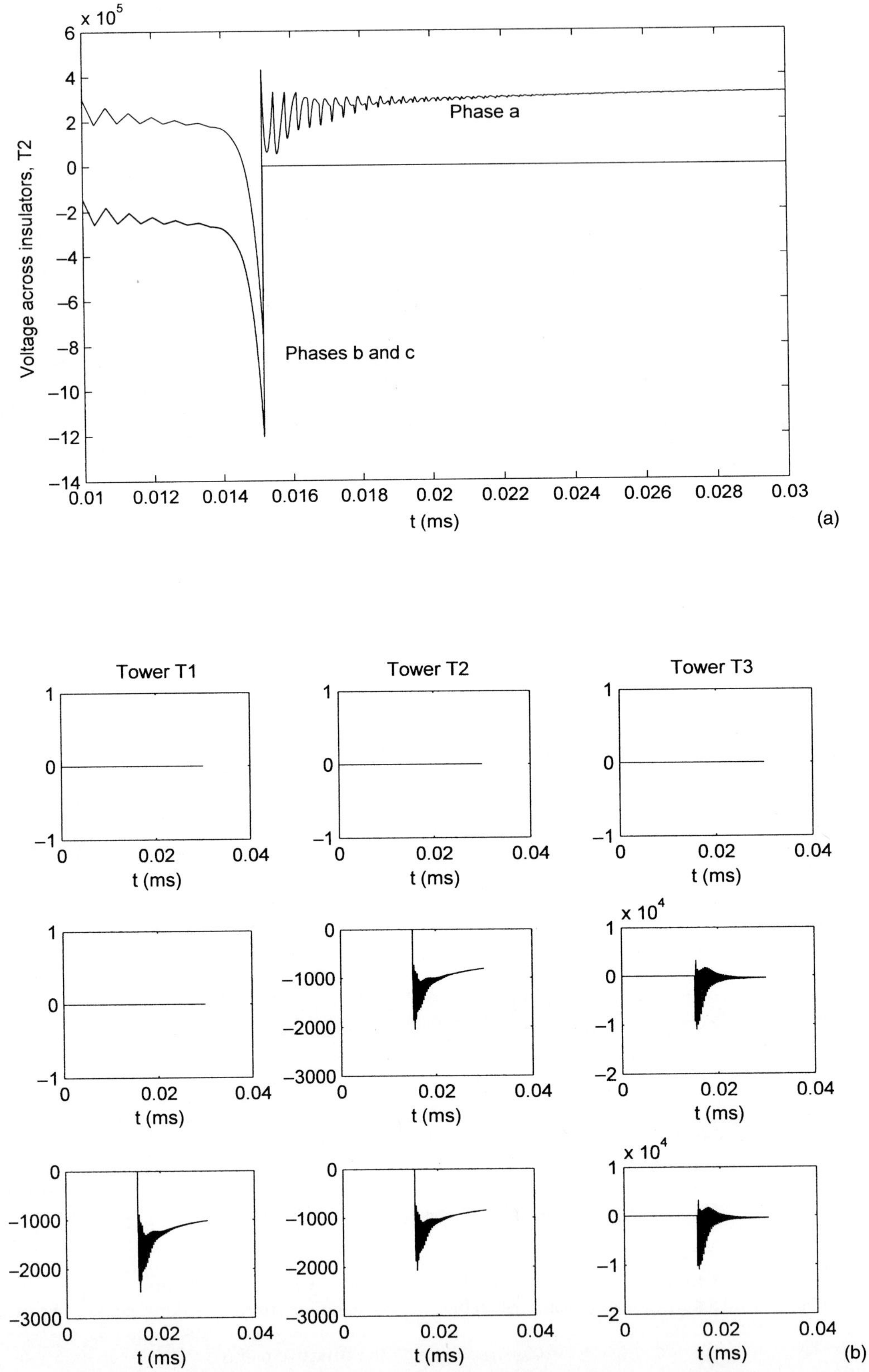

FIGURE 5-36 (*a*) Voltage across insulator strings of tower *T*2. (*b*) Tower insulator flashovers without surge arresters.

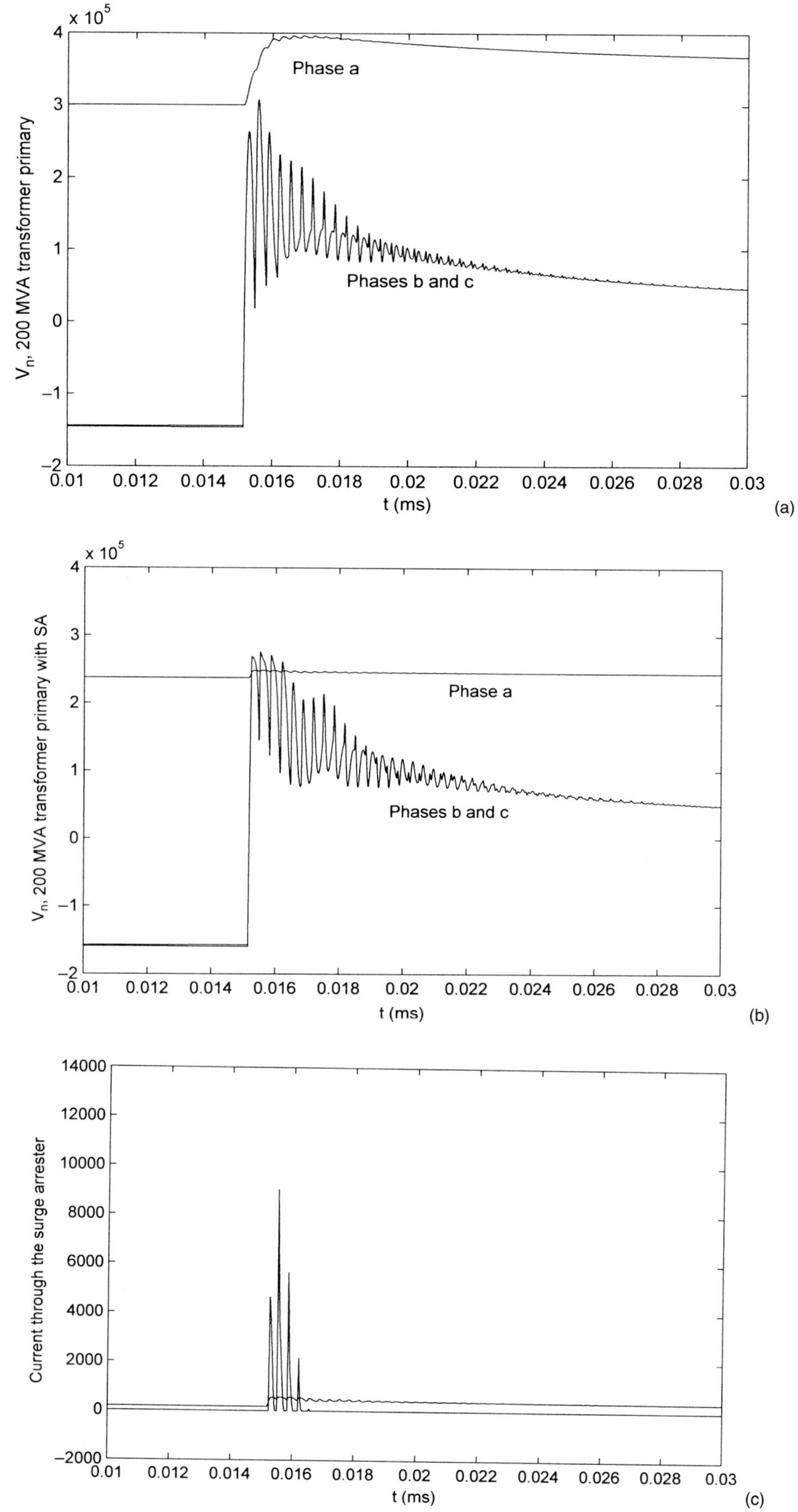

FIGURE 5-37 200-MVA transformer primary voltage (*a*) without and (*b*) with surge arrester, and (*c*) current through the surge arrester.

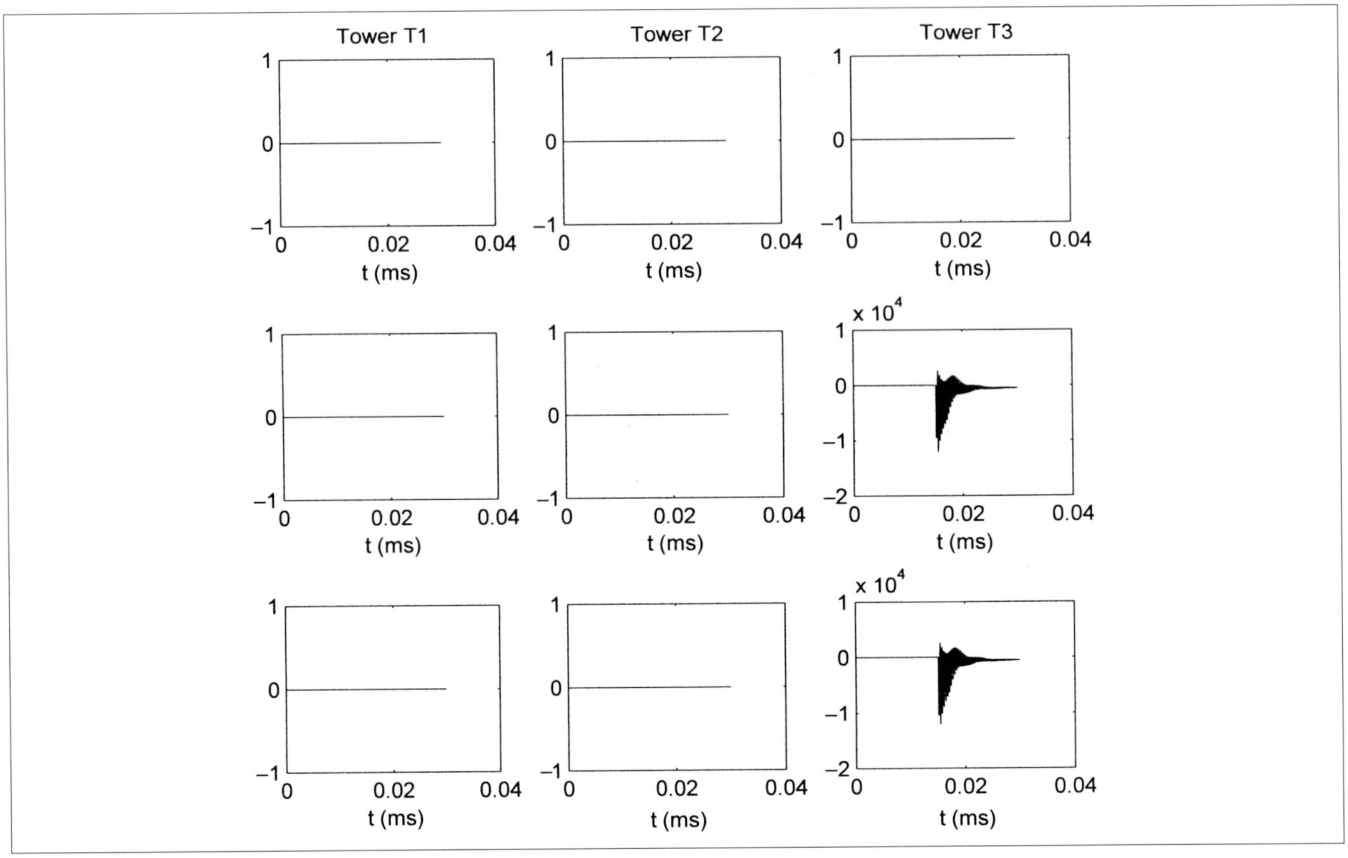

FIGURE 5-38 Tower insulator flashovers with a 258-kV metal-oxide surge arrester applied at 200-MVA transformer primary windings.

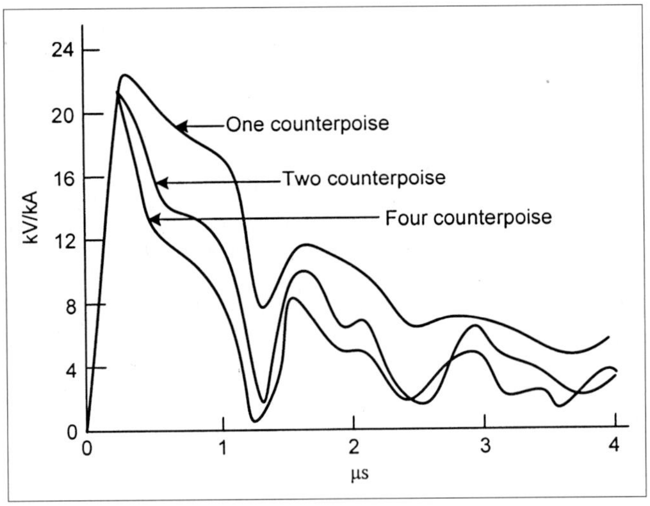

FIGURE 5-39 Effect of counterpoises on tower-top potential.

transmission tower for one μs. Consider tower footing resistance of 30 Ω, tower surge impedance of 150 Ω, neglect lightning channel impedance. The tower has one ground wire of surge impedance of 400 Ω. Consider a speed of 240 m/μs for the surge to travel down the tower, and a tower-height of 36 m. What is the maximum tower-top potential reached in one μs?

6. In Prob. 5, if the adjacent towers are 300 m apart, calculate the reflection from the adjacent towers and its travel time to reach the struck tower. How does it affect the tower-top potential?

7. In Prob. 5, consider a coupling factor $K = 30$ percent, what is the voltage stress on the insulation? If the insulators have a CFO of 1600 and 1000 kV, will a back flashover occur?

8. The line in Prob. 5 is located in an area with keraunic level of 60. Calculate shielding flashover rate (SFO) in flashovers/100 km years, using the expressions in this chapter.

9. Calculate the critical stroke current and the striking distance. Consider a surge impedance of 400 Ω.

10. Consider that the configuration of the transmission line in Example 5-6, as shown in Fig. 5-31. Ignoring the shield wire system shown in this figure, redesign using expressions in this chapter.

11. In Prob. 5, the lightning strike occurs at the ground wire 100 m away from a tower. Consider the same stroke current as specified in Prob. 5. Determine the voltage of the closet tower top after μs of the strike.

12. Repeat Prob. 5 with a tower footing resistance of 0 and 5 Ω. What is the maximum tower-top voltage in each case?

13. Describe CIGRE iterative method of BFR calculations. Why is CFO_{NS} different from conventional CFO? Relate it to a BIL of 1000 kV.

14. How does shielding of substations differ from that of transmission lines? Substation shielding is also provided by high masts. Comment on their efficacy and shielding angle.

REFERENCES

1. E. R. Williams, "Triple Pole Structure of Thunderstorms," *J. Geophysics Research*, vol. 94, pp. 13151–13167, 1989.
2. M. W. Maier, A. G. Boulangie, and R. I. Sax, "An Initial Assessment of Flash Density and Peak Current Characteristics of Lightning Flashes to Ground in South Florida," U.S. Nuclear Regulatory Commission, Washington, DC, NUREG/CR-102, 1979.
3. AIEE Committee Report, "A Method of Estimating Lightning Performance of Transmission Lines," *AIEEE Trans. Part III*, vol. 69, pp. 1187–1196, 1950.
4. J. G. Anderson, *EHV Transmission Reference Book*, Edison Electric Company, New York, 1968.
5. CIGRE Report, *Electra*, no. 22, pp. 139–147, 1972.
6. K. B. McEachron, "Lightning Stroke on Empire State Building," *AIEEE Trans.* vol. 60, pp. 885–962, 1941.
7. *Westinghouse Transmission and Distribution Reference Book*, East Pittsburg, PA, 1964.
8. R. B. Anderson and A. J. Eriksson, "A Summary of Lightning Parameters for Engineering Applications," *CIGRE Electra*, no. 69, pp. 65–102, March 1980.
9. K. Berger, R. B. Anderson, and H. Kroninger, "Parameters of Lightning Flashes," *CIGRE Electra*, no. 41, pp. 919–932, July 1975.
10. J. G. Anderson, "Lightning Performance of Transmission Lines," Chapter 12 of *Transmission Line Reference Book, 345 kV and Above*, 2d ed., EPRI, Palo Alto, CA, 1987.
11. CIGRE Working Group 33.01, "Guide to Procedures for Estimating the Lightning Performance of Transmission Lines," Brochure no. 63, Oct. 1991.
12. ANSI/IEEE Std. C62.42 IEEE Guide for the Application of Gas Tube Arresters-Low-Voltage Surge Protective Devices, 1987.
13. IEEE Working Group Report, "Estimating Lightning Performance of Transmission Lines II-Updates to Analytical Models," *IEEE Trans. PD*, vol. 8, pp. 1257–1267, 1993.
14. A. J. Eriksson, "The Incident of Lightning Strikes to Power Lines," *PWRD-2*, vol. 3, pp. 859–870, July 1987.
15. A. K. Mousa and K. D. Srivastva, "Effect of Shielding by Trees on Frequency of Lightning Strokes to Power Lines," *IEEE Trans. PWRD*, vol. 3, pp. 724–732, 1988.
16. M. Kawai, "Studies of Surge Response on a Transmission Tower," *IEEE Trans. PAS*, vol. 83, pp. 30–34, 1964.
17 C. F. Wagner and A. R. Hileman, "A New Approach to the Calculation of Lightning Performance of Transmission Lines, Part III," *AIEEE Trans. Part III*, pp. 589–603, 1960.
18. C. F. Wagner and A. R. Hileman, "Effect of Pre-Discharge Currents on Line Performance," *IEEE Trans. PAS*, vol. 82, pp. 117–131, 1963.
19. C. F. Wagner and G. D. McCain, "Induced Voltages on Transmission Lines," *AIEEE Trans.* no. 61, pp. 916–930, 1942.
20. R. Rusck, "Induced Lightning Overvoltages on Power Transmission Lines with Special Reference to Overvoltage Protection of Low-Voltage Networks," *Trans. Royal Institute of Tech.*, Stockholm, Sweden, no. 120, 1958.
21. P. Chowdhuri and E. T. B. Gross, "Voltages Induced on Overhead Multi-conductor Lines by Lightning Strokes," *Proc. IEE*, vol. 116, pp. 561–565, 1969.
22. A. J. Eriksson, "An Improved Electromagnetic Model for Transmission Line Shielding Analysis," *PWRD-2*, vol. 3, pp. 871–886, July 1987.
23. D. W. Gilman and E. R. Whitehead, "The Mechanism of Lightning Flashover on HV and EHV Lines," *Electra*, 27, pp. 69–89, 1973.
24. E. R. Whitehead, "CIGRE Survey of the Lightning Performance of Extra High Voltage Transmission Lines," *Electra*, vol. 33, pp. 63–89, 1974.
25. P. Chowdhuri, *Electromagnetic Transients in Power Systems*, Research Studies Press Ltd., Somerset, England, 1996.
26. ANSI/IEEE Standard 1313.2. IEEE Guide for Application of Insulation Coordination, 1999.
27. IEEE Working Group on Lightning Performance of Transmission Lines, "A Simplified Method of Estimating Lightning Performance of Transmission Lines," *IEEE Trans. PAS*, vol. 104, pp. 919–932, April 1985.
28. IEEE Report, "Modeling Guidelines for Fast Front Transients," *IEEE Trans. PD*, vol. 11, pp. 493–506, Jan. 1996.
29. E. R. Whitehead, "Protection of Transmission Lines," in *Lightning*, vol. 2, R. H. Golde, ed. Academic Press, New York, pp. 697–745, 1981.

FURTHER READING

H. R. Armstrong and E. R. Whitehead, "Field and Analytical Studies of Transmission Line Shielding," *IEEE Trans. PAS*, vol. 87, pp. 270–271, 1968.

K. Berger, "Observations on Lightning Discharges," *J. Franklin Institute*, No. 283, pp. 478–482, 1967.

G. W. Brown, and E. R. Whitehead, "Field and Analytical Studies of Transmission Line Shielding-II," *IEEE Trans. PAS*, vol. 88. pp. 617–626, 1968.

V. V. Burgsdrof, "Lightning Protection of Overhead Lines and Operating Experience in USSR," *Proc. CIGRE*, Report 326, 1958.

P. Chowdhuri, "Lightning Induced Overvoltages on Multi-Conductor Overhead Lines," *IEEE Trans. PWRD*, vol. 5, pp. 658–667, 1990.

M. Darveniza, F. Popolansky and E. R. Whitehead, "Lightning Protection of EHV Transmission Lines," *Electra*, vol. 41, pp. 36–39, July 1975.

EPRI, *Transmission Line Reference Book 345 kV and Above*, 2d ed., Chapter 12, Palo Alto, CA, 1987.

R. H. Golde, "Lightning Surges on Overhead Distribution Lines Caused by Indirect and Direct Lightning Strokes," *AIEE Trans.*, vol. 73, Part III, pp. 437–446, 1954.

R. H. Golde, *Lightning Protection*, Edward Arnold Publishers, London, 1973.

P. R. Krehbiel, M. Brook, and R. A NcRory, "An Analysis of the Charge Structure of Lightning Discharges to Ground," *J. Geophysics Research*, vol. 84, pp. 2432–2451, 1979.

A. C. Lewis and S. C. Mar, "Extension of the Chowdhari-Gross Model for Lightning Induced Voltage on Overhead Lines," *IEEE Trans. PWRD*, vol. 1, pp. 240–247, 1986.

D. R. MacGorman, M. W. Maier, and W. D. Rust, "Lightning Strike Density for the Contiguous United States from Thunderstorm Duration Records," U.S. Nuclear Regulatory Commission, NUREG/CR-3579, 1984.

NASA, Technical Memorandum 82473, Section 11, Atmospheric Electricity.

NASA Technical Memorandum 473, Section II, Atmospheric Electricity, 1982.

M. A. Sargent and M. Darvenzia, "Tower Surge Impedance," *IEEE Trans. PAS*, vol. 88, no. 5, pp. 680–687, 1969.

S. Yokoyama, "Advanced Observations of Lightning Induced Voltages on Power Distribution Lines-I," *IEEE Trans. PWRD*, vol. 1, pp. 129–139, 1986.

S. Yokoyama, "Advanced Observations of Lightning Induced Voltages on Power Distribution Lines-II," *IEEE Trans. PWRD*, vol. 4, pp. 2196–2203, 1989.

CHAPTER 6

TRANSIENTS OF SHUNT CAPACITOR BANKS

In Chap. 2, capacitor switching transients in lumped circuits were studied. These transients are of importance, as large capacitor banks are finding applications and acceptability in industrial distribution and utility systems. Shunt power capacitors have practically replaced rotating synchronous condensers, and are used on power transmission systems at voltage levels up to 500 kV, bank sizes ranging from a few Mvar to 300 Mvar. The size and location is based on the load flow and stability studies of the transmission network. These give rise to current and voltage transients, stress the switching devices and insulation systems, and can be detrimental to the sensitive loads. The switching transients of capacitor banks are at a frequency higher than that of the power frequency (Chap. 2). Historically, capacitor-switching transients have caused problems that have been studied in the existing literature. During the period from the late 1970s to 1980s, switching of capacitor banks in transmission systems caused high phase-to-phase voltages on transformers and magnification of transients at consumer-end distribution capacitors. Problems with switchgear restrikes caused even higher transients. Problems were common in industrial distributions with capacitors and dc drive systems, and the advent of pulse width modulation (PWM) inverters created a whole new concern of capacitor switching. These concerns and also the phenomena of secondary resonance, when a large utility capacitor bank is switched, while a capacitor bank at the downstream low-voltage distribution system remains energized, are investigated in this chapter; discussions are confined to shunt capacitor banks. The series capacitors and static var compensators (SVCs) used to enhance the power system stability limits are discussed in Chap. 15.

6-1 ORIGIN OF SWITCHING TRANSIENTS

The switching transients originate from:

1. Switching of a shunt capacitor bank, which may include switching on to a fault

2. Back-to-back switching, that is, switching of a second capacitor bank on the same bus in the presence of an already energized bank

3. Tripping or de-energizing a bank under normal operation and under fault conditions

4. Possible secondary resonance when the capacitors are applied at multivoltage level (i.e., at 13.8-kV level as well as at 480-V level) in a distribution system

5. Restrikes and prestrikes in the switching devices

6. Autoclosing with precharge on the capacitors

A transient, from its point of origin, will be propagated in either direction in the distribution system and will be transferred through the transformer inductive/capacitive couplings to other voltage levels. Transformer part-winding resonance can occur.[1]

The application of shunt capacitors can lead to the following additional side effects:

- Bring about severe harmonic distortion and resonance with load-generated harmonics

- Increase the transient inrush current of power transformers in the system, create overvoltages, and prolong its decay rate[1]

- Stress the capacitors themselves due to switching transients

- Increase the duty on switching devices, which may be stressed beyond the specified ratings in ANSI/IEEE standards[2,3]

- Discharge into an external fault, and produce damaging overvoltages across current transformer (CT) secondary terminals

- Impact the sensitive loads, that is, drive systems and bring about a shutdown

6-2 TRANSIENTS ON ENERGIZING A SINGLE CAPACITOR BANK

Consider the transients on energizing a single shunt capacitor bank. This assumes that there are no other capacitor banks in the immediate vicinity that will impact the transient behavior, though practically this will not be the case. On connecting to a power source, a capacitor is a sudden short circuit because the voltage across the capacitor cannot change suddenly (Chap. 2). The voltage of the bus to which the capacitor is connected will dip severely. This voltage dip and the transient step change is a function of the source impedance behind the bus. The voltage will then recover through a high-frequency oscillation. In the initial oscillation, the transient voltage can approach 2 per unit of the bus voltage. The initial step change and the subsequent oscillations are important. As these are propagated in the distribution system, they can couple across transformers and can be magnified. Transformer failures have been documented.[4,5] Surge arresters and surge capacitors can limit these transferred overvoltages and also reduce their frequency.[6] In a part-winding resonance, the predominant frequency of the transient can coincide with a natural frequency of the transformer. Secondary resonance, described further in this chapter, is a potential problem, and the sensitive loads connected in the distribution system may trip.

Example 6-1 In Chap. 2, switching transient of a series RLC circuit with sinusoidal excitation was discussed. A practical application is illustrated in Fig. 6-1. To improve the load power factor, a 6-Mvar capacitor bank is switched on 13.8-kV bus by closing its circuit breaker CB1. The term "bank" by definition means an assembly with all switching accessories, protective equipment, and controls required for a complete operating installation. The distribution can be reduced to Thévenin impedance, as seen from 13.8-kV bus. For switching transient studies, this is not permissible and complete distribution system should be modeled in all its details. This is so because much like lightning transients, the switching transients will be transmitted and reflected from the impedance discontinuities and will be transferred through the transformers windings too, altering the time-domain profile of the calculated transient. With this explicit qualification, this example illustrates the nature of capacitor switching transients. With this simplification, the equivalent impedance, as seen from the 13.8-kV bus, is $Z = (0.0115 + j0.1056$, per unit 100-MVA base). Thus, the transient analysis is reduced to excitation of a series RLC circuit. The reacrance in series with capacitor bank shown in this figure is not considered.

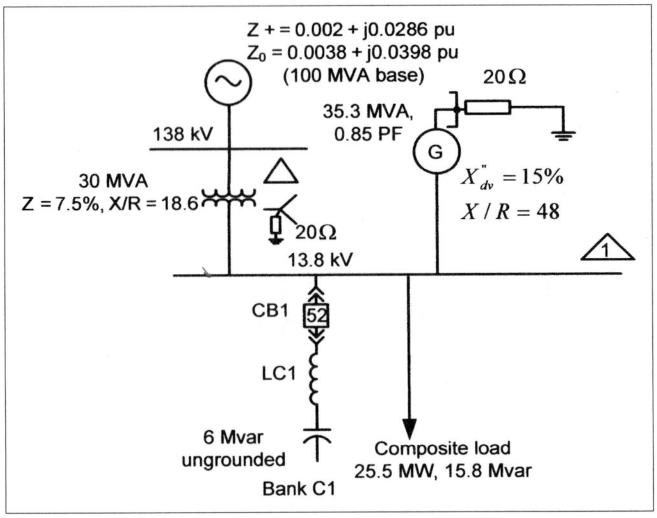

FIGURE 6-1 Circuit diagram for the study of switching transients of a 6-Mvar capacitor bank.

Figures 6-2 and 6-3 show EMTP simulation of current and voltage transients in phase *a*. With the given parameters and looking

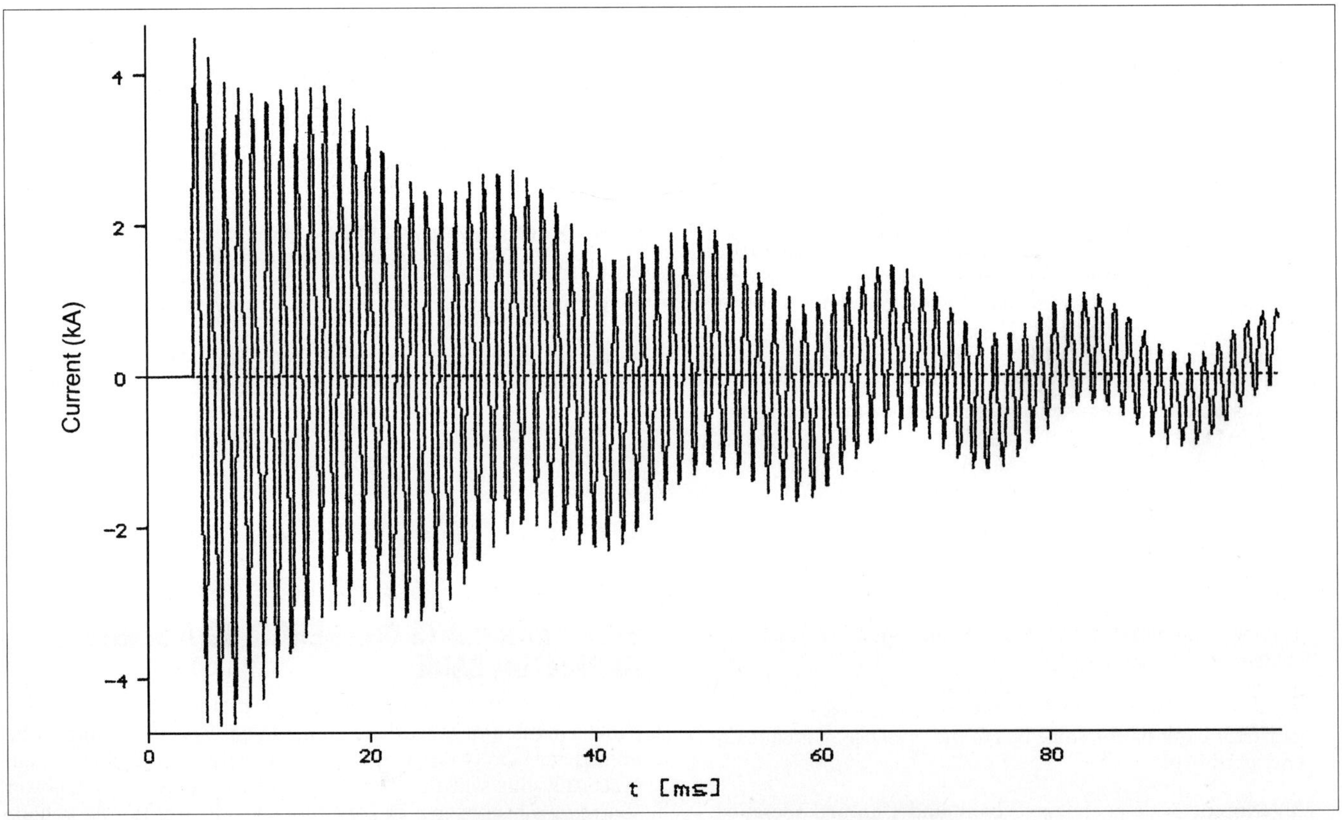

FIGURE 6-2 Switching current transient of 6-Mvar capacitor bank, Fig. 6-1.

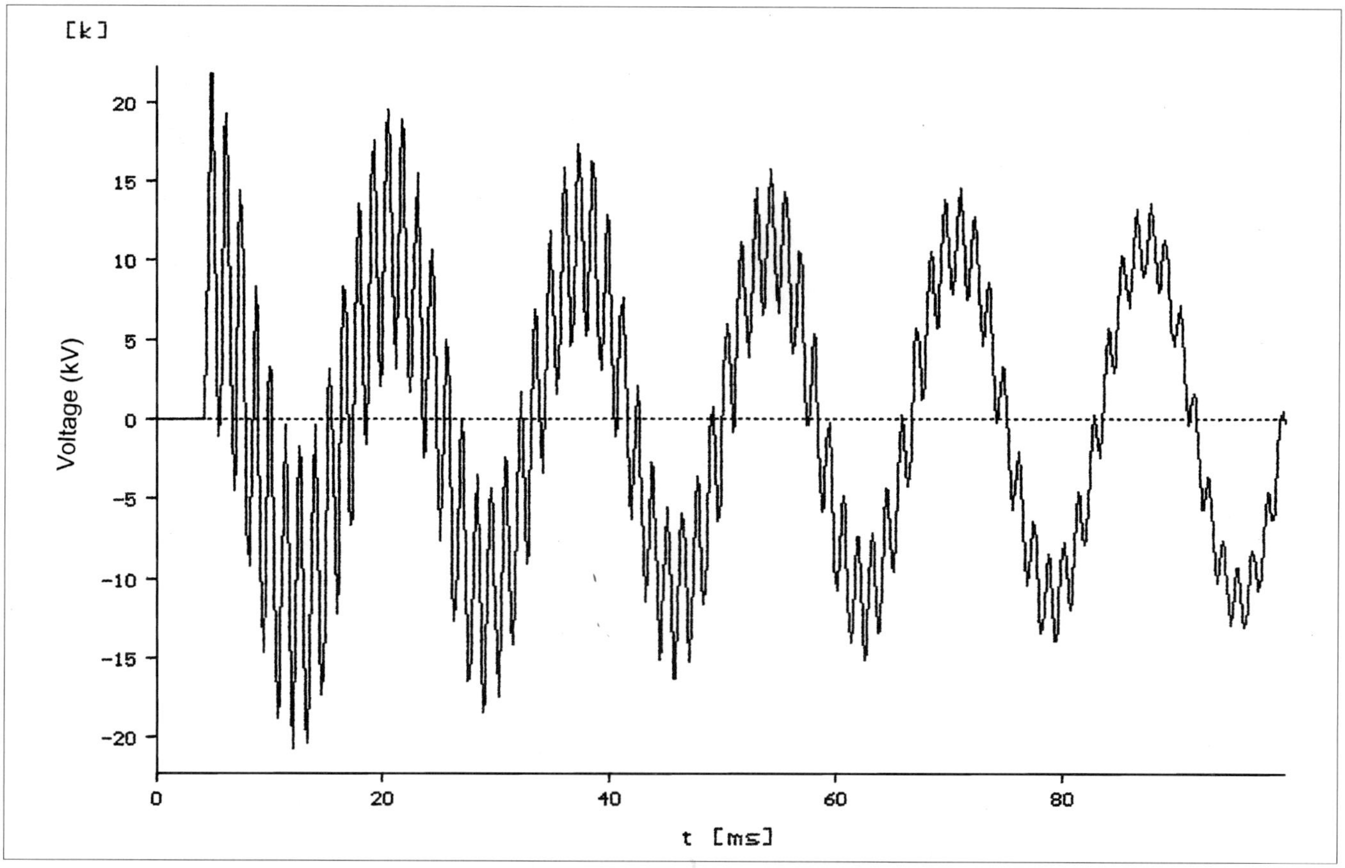

FIGURE 6-3 Voltage transient, 13.8-kV bus 1, on switching of 6-Mvar capacitor bank, Fig. 6-1.

at Thévenin impedance of 13.8-kV bus, the following differential equation for the inrush current can be written as:

$$5.08 \times 10^{-4}\frac{di}{dt} + 2.19 \times 10^{-2} i + \frac{\int i dt}{83.6 \times 10^{-6}} = 11.26 \times 10^{3} \sin \omega t \tag{6-1}$$

If damping in a series LC circuit is neglected and the circuit is excited through a step function V, the following equation can be written:

$$L\frac{di}{dt} + \frac{\int i dt}{C} = V \tag{6-2}$$

Taking Laplace transform:

$$s^2 i(s) - sI(0) - I'(0) + \frac{i(s)}{T^2} = 0 \tag{6-3}$$

The initial current is zero, and from Eq. (6-2):

$$\frac{di}{dt} = \frac{V}{L} = I'(0) \tag{6-4}$$

Therefore:

$$i(s) = \frac{V}{L}\frac{1}{(s^2 + 1/T^2)} \tag{6-5}$$

Taking inverse transform:

$$i = \frac{V}{L}\frac{1}{1/T}\sin(1/T)t = \frac{V}{L}\sqrt{LC}\sin\frac{1}{\sqrt{LC}}t$$

$$= V\sqrt{\frac{C}{L}}\sin\frac{1}{\sqrt{LC}}t \tag{6-6}$$

Thus, the peak inrush current can be written as:

$$i_{\text{max,peak}} = \frac{\sqrt{2}E_{LL}}{\sqrt{3}}\sqrt{\frac{C_{eq}}{L_{eq}}} \tag{6-7}$$

where $i_{\text{max,peak}}$ is the peak inrush current in amperes without damping, E_{LL} is the line-to-line voltage in volts, C_{eq} is the equivalent capacitance in farads, L_{eq} is the equivalent inductance in henries. (All inductances in the switching circuit, including that of cables and buses, must be considered.) The frequency of the inrush switching current f is given by:

$$f = \frac{1}{2\pi\sqrt{L_{eq}C_{eq}}} \tag{6-8}$$

The voltage across the capacitor, which will also be the bus voltage, is given by:

$$V_c(s) = \frac{i(s)}{sC} = \frac{V}{LC}\left[\frac{1}{s(s^2 + 1/T^2)}\right] \tag{6-9}$$

Resolving into partial fractions:

$$V_c(s)=\frac{V}{LC(1/T^2)}\left[\frac{1}{s}-\frac{s}{(s^2+1/T^2)}\right] \tag{6-10}$$

Taking inverse transform:

$$V_c=V\left(1-\cos\frac{1}{\sqrt{LC}}t\right)=V(1-\cos\omega_0 t) \tag{6-11}$$

Thus, the maximum voltage occurs at:

$$\omega_0 t=\pi \tag{6-12}$$

These equations give a peak current of 4.58 kA and a frequency of 770 Hz, which corresponds well with the simulation results shown in Fig. 6-2 and 6-3. The high-frequency oscillation is clearly visible in these figures. It takes about six cycles before the oscillations appreciably damp out. The damping is a function of the resistance and losses in the system (rigorously some resistance of the capacitors themselves should also be modeled). Figure 6-3 shows a peak voltage excursion of approximately 22 kV, that is, two times the bus-rated voltage of 13.8 kV rms. The inrush current, as read from Fig. 6-2, is 4.6 kA, and its frequency can be approximately determined around 700 Hz.

The damping due to the presence of the resistor has not been accounted for, yet the calculated results are close to the simulation. These can be further examined. Calculate Q_0:

$$Q_0=\frac{X}{R}\approx 10 \tag{6-13}$$

Thus, it does not have appreciable effect on damping. From Eq. (6-8)

$$f_0=\frac{1}{2\pi\sqrt{LC}}=770\ \text{Hz} \tag{6-14}$$

Thus, inrush current frequency is 770 Hz. (6-15)

6-2-1 Prior Charge on the Capacitors

A prior trapped charge on the capacitors will prolong the transients. As per ANSI/IEEE specifications,[7] the capacitors are provided with an internal discharge device that will reduce the residual charge to 50 V or less within 1 min for capacitors of 600 V or less and 5 min for capacitors over 600 V.

Thus, rapid switching in and switching out of the capacitors should be avoided to limit inrush current transients. Any switching arrangement should block reconnecting the capacitor banks to the power supply system without the required time delay. Up to twice the normal inrush currents are possible when a circuit breaker is employed to reswitch capacitive loads. When the bank is interrupted at or near current zero, the voltage trapped on the bank may be near the peak value. Reclosing will produce high inrush currents. When a capacitor bank is connected on the load side of a feeder, the high inrush currents can be avoided by isolating the capacitor bank before reclosing takes place or reclosing must be sufficiently time-delayed.

6-3 APPLICATION OF POWER CAPACITORS WITH NONLINEAR LOADS

Nonlinear loads like variable speed drives, wind power generation, switched mode power supplies, arc furnaces, and so on can give rise to a variety of line harmonic spectrums. It is not the intention to go into the much-involved subject of harmonics, except to relate it to switching transients and distortions that can be caused by the application of power capacitors with nonlinear loads. For the purpose of this analysis, we will confine the discussion to six-pulse current source converters, which produce characteristic harmonics of the order of 5th, 7th, 11th, 13th, . . . (Chap. 15).

Harmonics generated by nonlinear loads may be described as (1) having a Fourier series with fundamental frequency equal to the power system frequency, and a periodic steady state exists—this is the most common case; (2) a distorted waveform having submultiples of power system frequency, and a periodic steady state exists—certain types of pulsed loads and cycloconverters produce these types of waveforms; (3) the waveform is aperiodic, but perhaps almost periodic, and a trigonometric series expansion may still exist, examples being arcing devices, arc furnaces, and sodium vapor lighting; (4) the components in a Fourier series that are not a multiple of power frequency are called noninteger harmonics. Harmonics generating loads in the power systems are always on an increase due to the proliferation of power electronics. The effects of harmonics on power system operation are manifold. The presence of capacitors in a system does not generate harmonics in itself, but can accentuate these, create more distortion, and bring about resonant conditions.[8]

6-3-1 Harmonic Resonance

Series and parallel resonant circuit concepts of Chap. 2 are applied to harmonic resonance that occurs when power capacitors are applied in distribution systems which have nonlinear loads.

The shunt power capacitors act in parallel with the power system. Thévenin impedance, as seen from the point of application of the capacitor, ignoring resistance, is the impedance of the parallel combination:

$$\frac{j\omega L(1/j\omega C)}{(j\omega L+1/j\omega C)} \tag{6-16}$$

Assuming that L and C remain invariant with frequency, resonance will occur when the inductive and capacitive inductance of the denominator in Eq. (6-16) is equal and the denominator is zero. This means that the impedance of the combination is infinite for a lossless system. The impedance angle will change abruptly as the resonant frequency is crossed. Thus, at the resonant frequency f_n:

$$j2\pi f_n L=\frac{1}{j2\pi f_n C} \tag{6-17}$$

Without the presence of power capacitors, the natural resonant frequency of the power system is fairly high, much above any load-generated harmonic. As the frequency is increased, the capacitive reactance decreases and the inductive reactance increases. It may so happen that at a load-generated harmonic, say 5th, (f_n = 300 Hz), Eq. (6-17) is satisfied. The power capacitor acts like one branch of a parallel-tuned circuit while the rest of the system acts likes the other parallel branch. Such a circuit, when excited at the resonant frequency (5th harmonic, in this case), will result in the magnification of the harmonic current, which may even exceed the fundamental frequency current. This will overload the capacitors and all the system components with deleterious results. Resonance with one of the load-generated harmonics is a major concern in power systems, and this condition must be avoided in any application of the power capacitors. Equation (6-17) can be simply written as:

$$h=\frac{f_n}{f}=\sqrt{\frac{\text{kVA}_{sc}}{\text{kvar}_c}} \tag{6-18}$$

where h is the order of the harmonic, f_n is the resonant frequency, f is the supply system frequency, kVA_{sc} is the short-circuit kVA at the point of application of the capacitor, and kvar_c is the shunt capacitor rating. For a short-circuit level of 500 MVA, resonance at 5th, 7th, 11th, and 13th harmonics will occur for a capacitor bank size of 20, 10.20, 4.13, and 2.95 Mvar, respectively. The smaller the size of the power capacitor, the higher is the resonant frequency.

Sometimes, this simple artifice is used to size capacitor banks for a given distribution system to avoid resonance. A capacitor bank size is selected so that the resonant frequency does not coincide with any load-generated harmonics. However, the short-circuit level in a power system is not a constant parameter. It will vary with the switching conditions; for example, a generator or a tie-line circuit may be out of service or part of the motor loads may have been shut down, which will lower the short-circuit level. Thus, it can be concluded that:

- The resonant frequency will *float around* in the system, depending on the switching conditions.
- An expansion or reorganization of the system may bring about a resonant condition where none exited before.

6-3-2 Frequency Scan

An elementary tool to accurately ascertain the resonant frequency of the system in the presence of capacitors is to run a frequency scan on a digital computer. The frequency is applied in incremental steps, say 2 Hz, for the range of harmonics to be studied, say from the fundamental frequency to 2400 Hz. (This means 1200 calculations for the entire range.) The procedure is equally valid, whether there are one or more than one harmonic-producing loads in the system, so long as the principle of superimposition is held valid. Then, for the unit current injection, the calculated voltages give the driving point and transfer impedances, both modulus and phase angle. The Y_{bus} matrix contains only linear elements for each incremental frequency. Thus, an impedance plot can be made with varying frequency, which gives resonant frequencies. System component models used for power frequency applications, for example, transformers, generators, reactors, and motors, are modified for the higher frequencies, which are not discussed in this book.

Example 6-2 For this example, the circuit of Fig. 6-1 is modified and a 5-MVA transformer, which supplies a six-pulse drive system (three-phase fully controlled bridge circuit, Chap. 15) load of 5 MVA, is added; see Fig. 6-4. The harmonic spectrum is shown in Table 6-1, which is also a function of the source impedance (Chap. 15). A frequency scan of bus 2, both impedance modulus and angle, is shown in Figs. 6-5 and 6-6, respectively. A resonance occurs close to the 11th harmonic, and Fig. 6-6 shows that the angle abruptly changes at the resonant frequency. The calculated resonant frequency is 672 Hz. The calculation is made in an incremental step of 2 Hz. To capture the resonant frequency close to the actual, the incremental step should be small. Suppose an incremental step of 10 Hz is selected, the actual resonant frequency may be at a variation of ± 5 Hz.

Figure 6-7 shows the voltage spectrum of 13.8-kV bus 1 and 4.16-kV bus 2. This shows that the 11th harmonic voltage at 4.16-kV bus 2 is 17.2 percent, while at 13.8-kV bus 1, it is 11.8 percent. Figure 6-8 shows distorted waveform of the voltage at these two buses for one cycle. Figure 6-9 depicts the spectrum of current flow through the capacitor; the 11th harmonic current is 130 percent of the fundamental current = 326 A at 13.8 kV (the fundamental frequency current is not shown). Note that the injected 11th harmonic current, a current at 4.16-kV bus 1B, is 63 A based on the harmonic spectrum shown in Table 6-1. Thus, there is a magnification of approximately 17 times. This is the characteristic of a parallel-tuned circuit, that while the exciting current can be small, the capacitor forms a resonant tank circuit with the source impedance, resulting in much magnification of the injected current.

A harmonic source can be considered as a harmonic generator. The harmonic current injected at the source will flow to the 13.8-kV

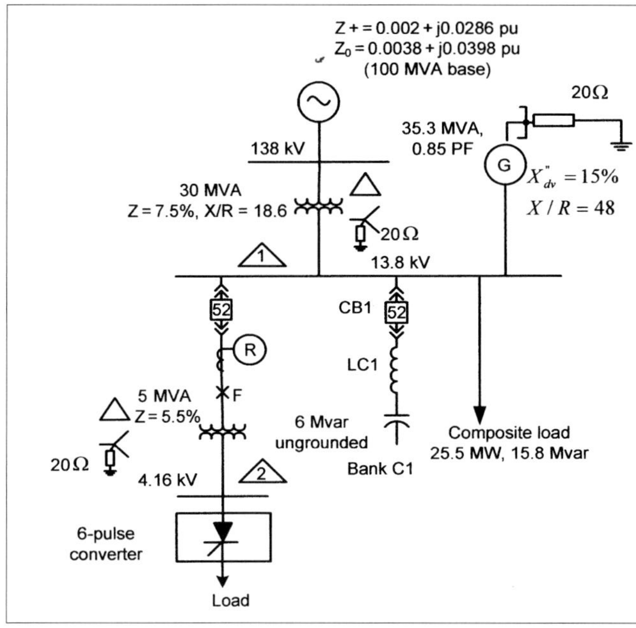

FIGURE 6-4 Circuit diagram to illustrate harmonic resonance.

TABLE 6-1 Harmonic Spectrum of Six-Pulse Load

h	5	7	11	13	17	19	23	25	29	31
%	18	13	6.5	4.8	2.8	1.5	0.5	0.4	0.3	0.2

bus through 5 MVA coupling transformer and divide into other system components connected to this bus, depending upon their harmonic impedances. Harmonic load flow is akin to the fundamental frequency load flow in this respect. Apart from overloading of the capacitors, the harmonic currents in the distribution systems seriously derate the transformers, produce additional losses, result in negative sequence overloading of the generators, give rise to transient torques and torsional oscillations in rotating machinery, and impact the protective relaying, to name a few effects.

Generators have a limited I_2^2t rating, where I_2 is the negative sequence current (Chap. 10). Harmonics of the order of $h = 6m+1$, where m is any integer, are forward going, that is, harmonics of the order of 7th, 13th, 19th, . . ., and these rotate at $1/h$ of the speed. Harmonics of the order $h = 6m-1$ are reverse going, that is, harmonics of the order of 5th, 11th, 17th, . . . In a synchronous machine, the frequency induced in the rotor is the net rotational difference between fundamental frequency and harmonic frequency. The 5th harmonic rotates reverse with respect to stator, and with respect to rotor, the induced frequency is that of 6th harmonic. Similarly, the forward-going 7th harmonic produces a frequency of 6th harmonic in the rotor. The harmonic pairs 11th and 13th will produce a frequency of 12th harmonic in the rotor. If the frequency of mechanical resonance happens to be close to these harmonics during starting, large mechanical forces can occur.

A similar phenomenon occurs in induction motors. Here, positive-sequence harmonics of the order of h = 1, 4, 7, 10, 13, . . . produce a torque of $(h-1+s)\omega$ in the direction of rotation, and negative-sequence harmonics, h = 2, 5, 8, 11, 14, . . ., produce a torque of $-(h+1-s)\omega$ opposite to the direction of rotation. Again, torque amplifications can occur.

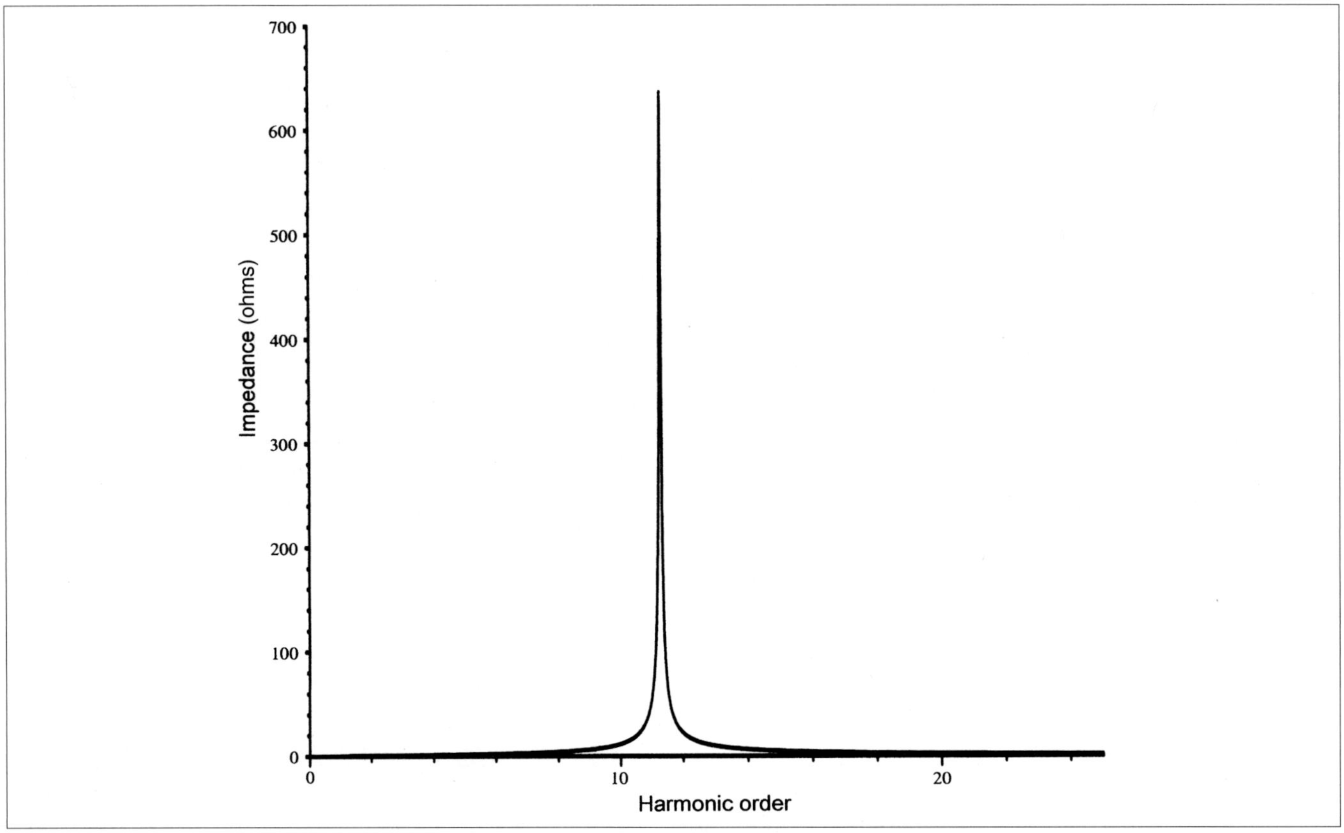

FIGURE 6-5 Impedance modulus, 4.16-kV bus 2, Fig. 6-4.

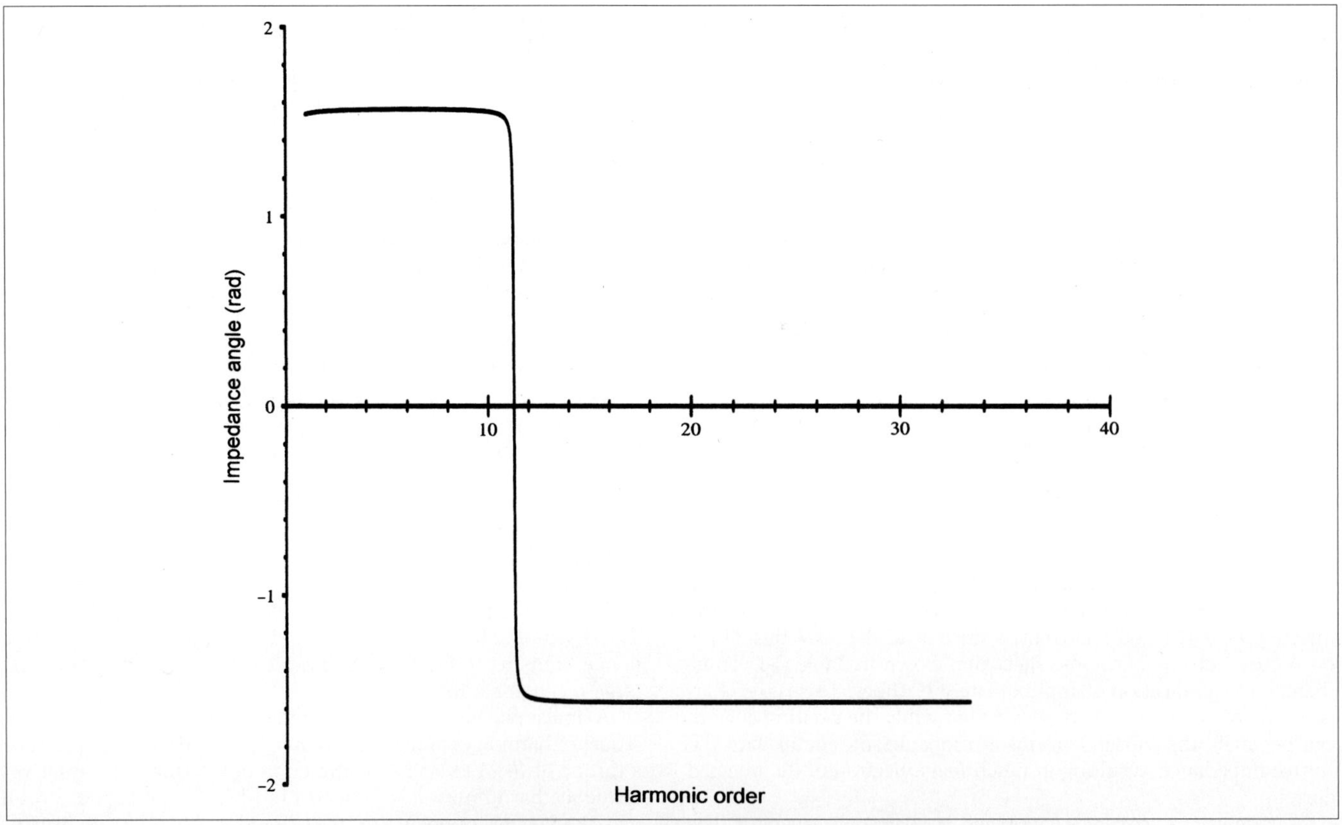

FIGURE 6-6 Impedance angle, 4.16-kV bus 2, Fig. 6-4

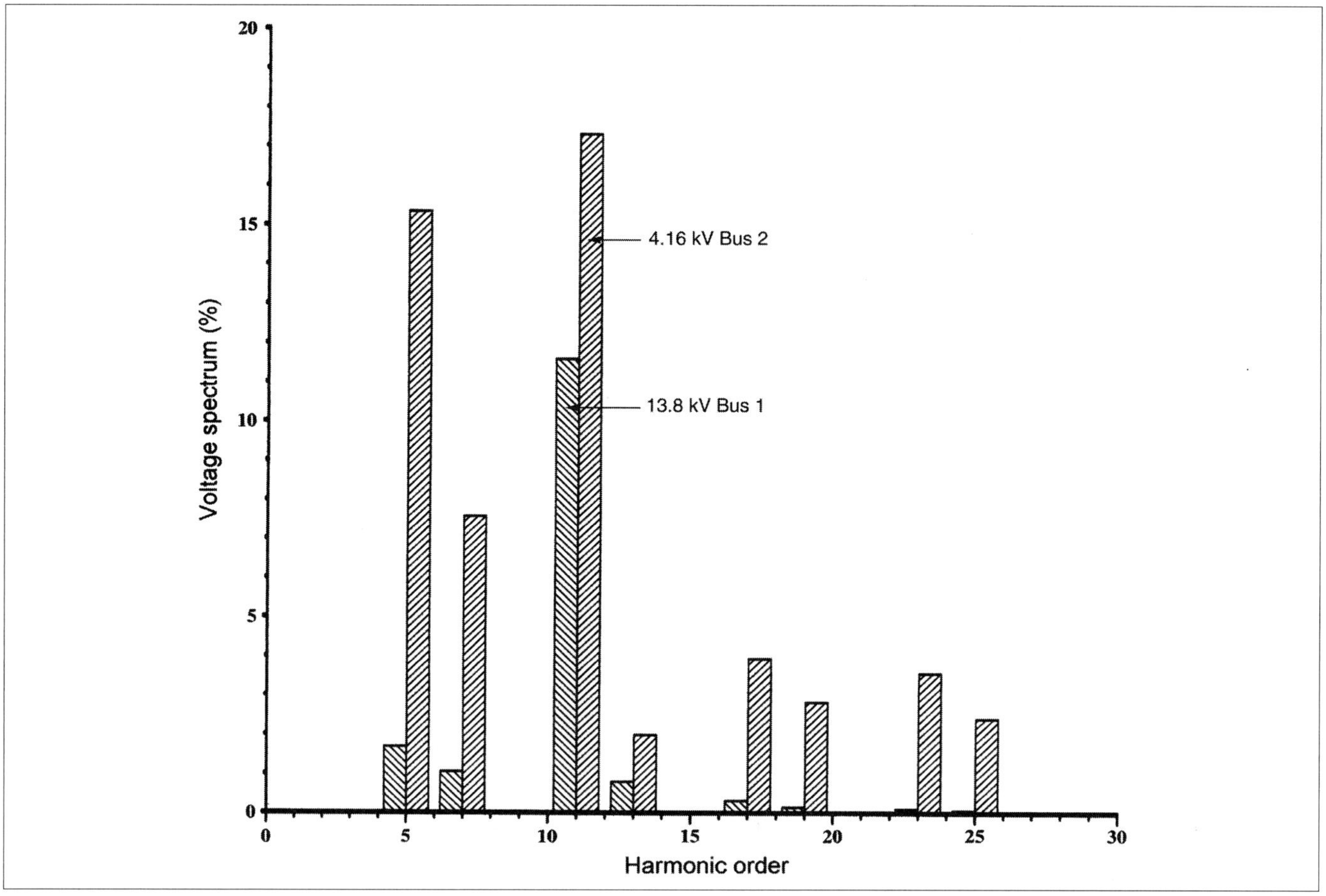

FIGURE 6-7 Voltage spectrum, 13.8-kV bus 1 and 4.16-kV bus 2, Fig. 6-4.

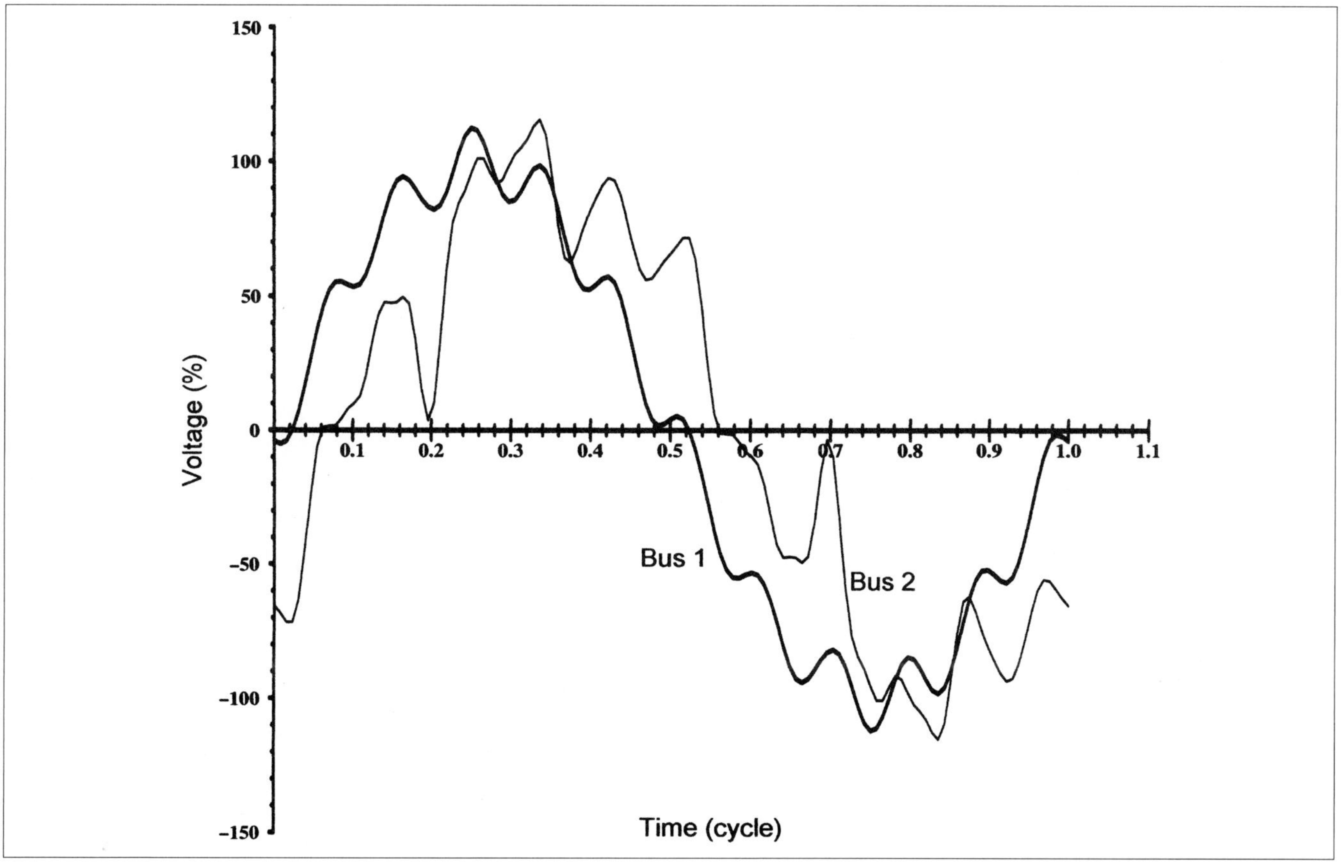

FIGURE 6-8 Distorted voltage waveforms, 13.8-kV bus 1 and 4.16-kV bus 2, due to harmonic resonance, Fig. 6-4.

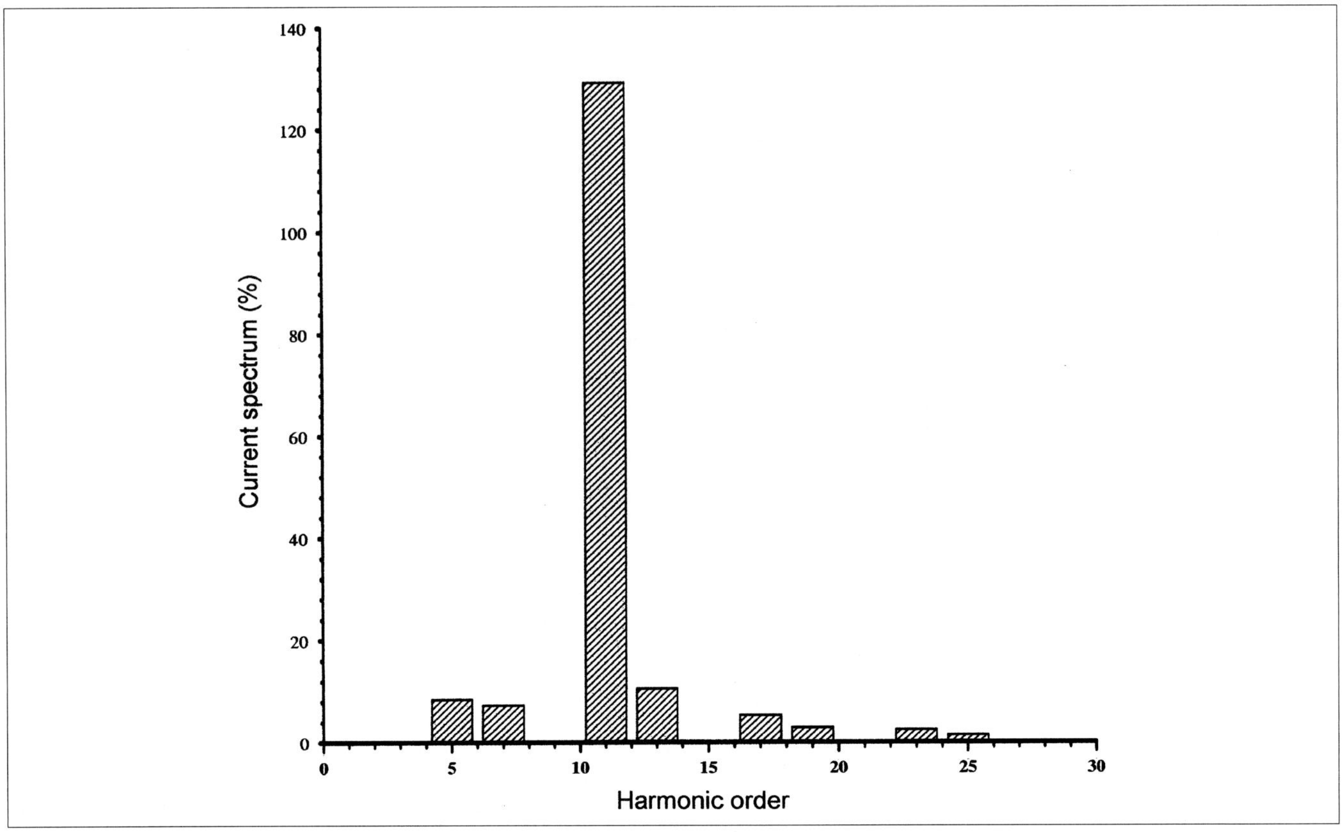

FIGURE 6-9 Harmonic current spectrum through 6-Mvar capacitor bank, Fig. 6-4.

6-3-3 Harmonic Propagation and Mitigation

Example 6-2 depicts a rather simple distribution system with respect to harmonic amplification. When capacitors are located close to a plant with significant harmonic-producing loads, harmonic resonance becomes a potential possibility, and a nearby distribution system, which does not have any nonlinear loads of its own, may be subjected to harmonic pollution. This means that the impact can be propagated through interconnections and impact remotely. There have been two approaches:

1. Consider capacitor placement from reactive power compensation point of view and then study harmonic effects.

2. Study fundamental frequency, voltage, and harmonic effects simultaneously.

For the power capacitors, the limitations due to harmonic loading are well defined.[7] Per unit kvar should not exceed 1.35:

$$\text{kvar}_{(pu)} \leq 1.35 = \sum_{h=1}^{h=h_{\max}} (V_h I_h) \tag{6-19}$$

The rms current should not exceed 135% of nominal current based upon rated kvar and rated voltage, including fundamental and harmonic currents.

$$I_{\text{rms}} \leq 1.35 = \left[\sum_{h=1}^{h=h_{\max}} I_h^2\right]^{1/2} \tag{6-20}$$

The limits on rms voltage are given by:

$$V_{\text{rms}} \leq 1.1 = \left[\sum_{h=1}^{h=h_{\max}} V_h^2\right]^{1/2} \tag{6-21}$$

The crest shall not exceed $1.2\times\sqrt{2}$ times rated rms voltage, including harmonics but excluding transients.

$$V_{(\text{rest})} \leq 1.2\sqrt{2} = \sum_{h=1}^{h=h_{\max}} V_h \tag{6-22}$$

Unbalances within a capacitor bank, due to capacitor element failures and/or individual fuse operations result in overvoltages on other capacitor units. In a filter bank, failure of a capacitor unit will cause detuning. The limitations of the capacitor bank loadings become of importance in the design of capacitor filters. A capacitor tested according to IEEE Std 18[20] will withstand a combined total of 300 applications of power frequency terminal-to-terminal overvoltages without superimposed transients or harmonic content. The capacitor unit is also expected to withstand transient currents inherent in the operation of power systems, which include infrequent high lightning currents and discharge currents due to nearby faults. Ref. 21 provides curves of voltage and transient currents withstand for shunt capacitors.

The impact of capacitor applications at transmission and subtransmission systems becomes fairly involved and is not the main subject of discussion in this book. Figure 6-10 shows the frequency scan of a 400-kV line with bundle conductors. IEEE Std. 519 defines the limits of harmonic indices, current, and voltage at the point of common coupling (PCC).[9] This can be the point of metering

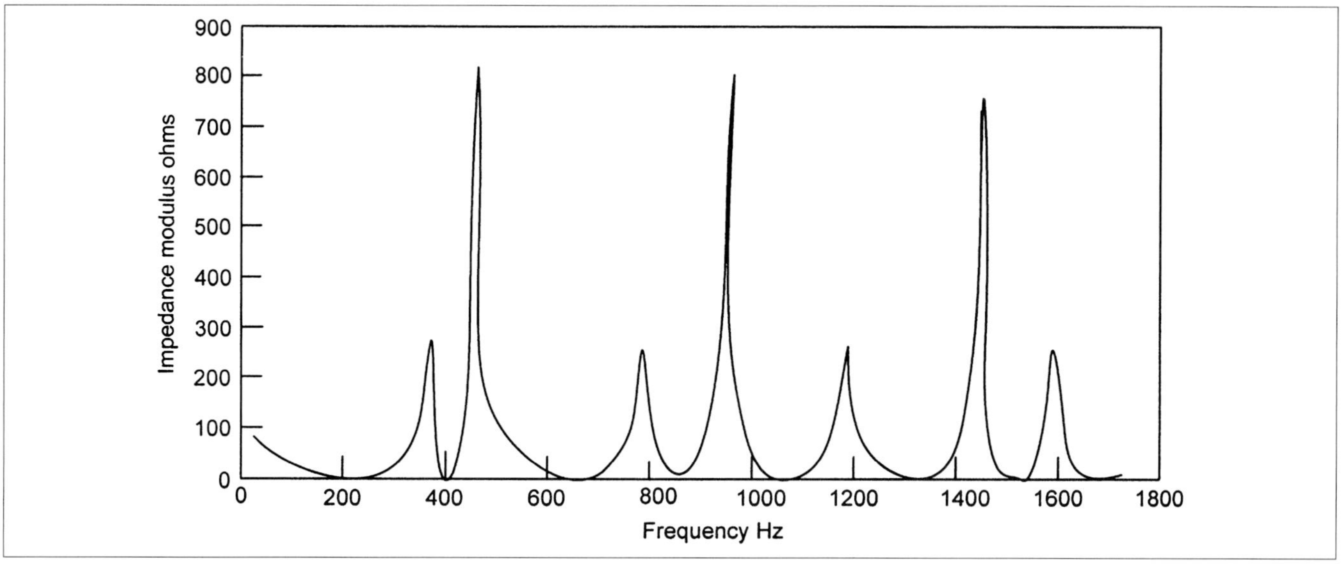

FIGURE 6-10 Impedance modulus of a 400-kV line.

or the point of connection of consumer apparatus with the utility supply company. Within a plant, PCC is the point between the nonlinear load and the other loads. The recommended current distortion limits in IEEE Standard 519 are concerned about total demand distortion,[9] which is defined as:

$$\mathrm{TDD} = \frac{\sqrt{\sum_{h=2}^{h=h_{max}} I_h^2}}{I_L} \tag{6-23}$$

where I_L is the load current demand for 15 or 30 min. The limits of the TDD are specified at each harmonic, and also total permissible TDD, as a function of I_{sc}/I_L, where I_{sc} is the short-circuit current.

The harmonics may be limited by (1) phase multiplication (Chap. 15), (2) use of passive or active harmonic filters, and (3) modern power electronics technology that limits the harmonics at the source.

The application of a single-tuned harmonic filter, also called a band pass filter, is briefly discussed, with reference to Fig. 6-11. A band pass filter consists of merely a reactor in series with a capacitor, and referring to Chap. 2, at its resonant frequency it offers a low impedance path to the harmonic of interest that is required to be shunted through the filter to minimize its propagation through the system. Two or three ST filters can be used, each appropriately tuned to a different harmonic that needs to be controlled and bypassed. Referring to the equivalent circuit, the injected harmonic current divides between the harmonic filter and the source impedance, as seen from the point of application of the capacitor bank. The harmonic injected current flows partly into the capacitor and partly into the source.

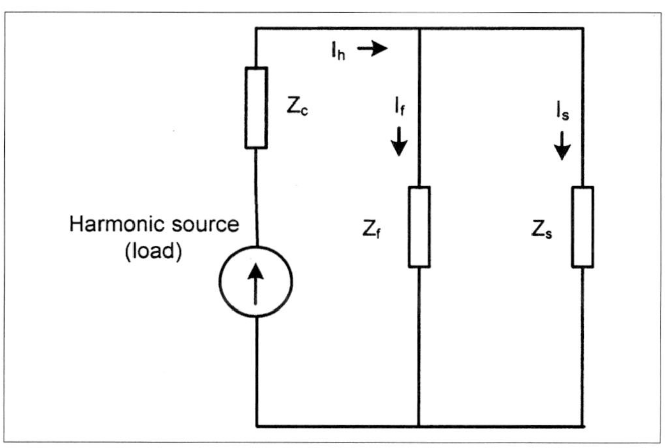

FIGURE 6-11 Equivalent harmonic injection circuit, distribution system represented by single impedance, *Zs*.

$$I_h = I_f + I_s \tag{6-24}$$

Also, voltage across the combination of reactor and capacitor should be equal to the voltage across *Zs*.

$$I_f Z_f = I_s Z_s \tag{6-25}$$

Then:

$$I_f = \left[\frac{Z_s}{Z_f + Z_s}\right] I_h = \rho_f I_h$$
$$I_s = \left[\frac{Z_f}{Z_f + Z_s}\right] I_h = \rho_s I_h \tag{6-26}$$

In harmonic flow analysis and harmonic filter design, the complex quantities ρ_s and ρ_f are of interest. To control the current injection into the utility system, these factors should be controlled, the system impedance playing an important role. As the utility source impedance may be considered fixed for a certain electrical distribution system, more current can be made to flow through the filter, by increasing the size of the capacitor bank. The effectiveness of harmonic filtering vis-à-vis source impedance is apparent from Eq. (6-26). The lower the source impedance, the greater is the harmonic current flowing into it, for the same filter size.

Generally, in the presence of load-generated harmonics, the capacitors are applied as harmonic filters, and the series tuning reactor serves the dual purpose of forming a low-impedance tuned circuit as well as limiting the inrush current and its frequency on switching a capacitor bank.

Example 6-3 The capacitor bank C1 shown in Fig. 6-4 is turned into a ST filter, the tuned frequency being 4.67 times the fundamental. The new frequency scan, impedance modulus, and angle are now depicted in Figs. 6-12(*a*) and 6-12(*b*), respectively. The application of a band pass filter has not eliminated the resonance, but now it occurs at a frequency below the tuned frequency that can be placed at a point so that the load-generated harmonics, and transformer inrush current harmonics are away from it. There will

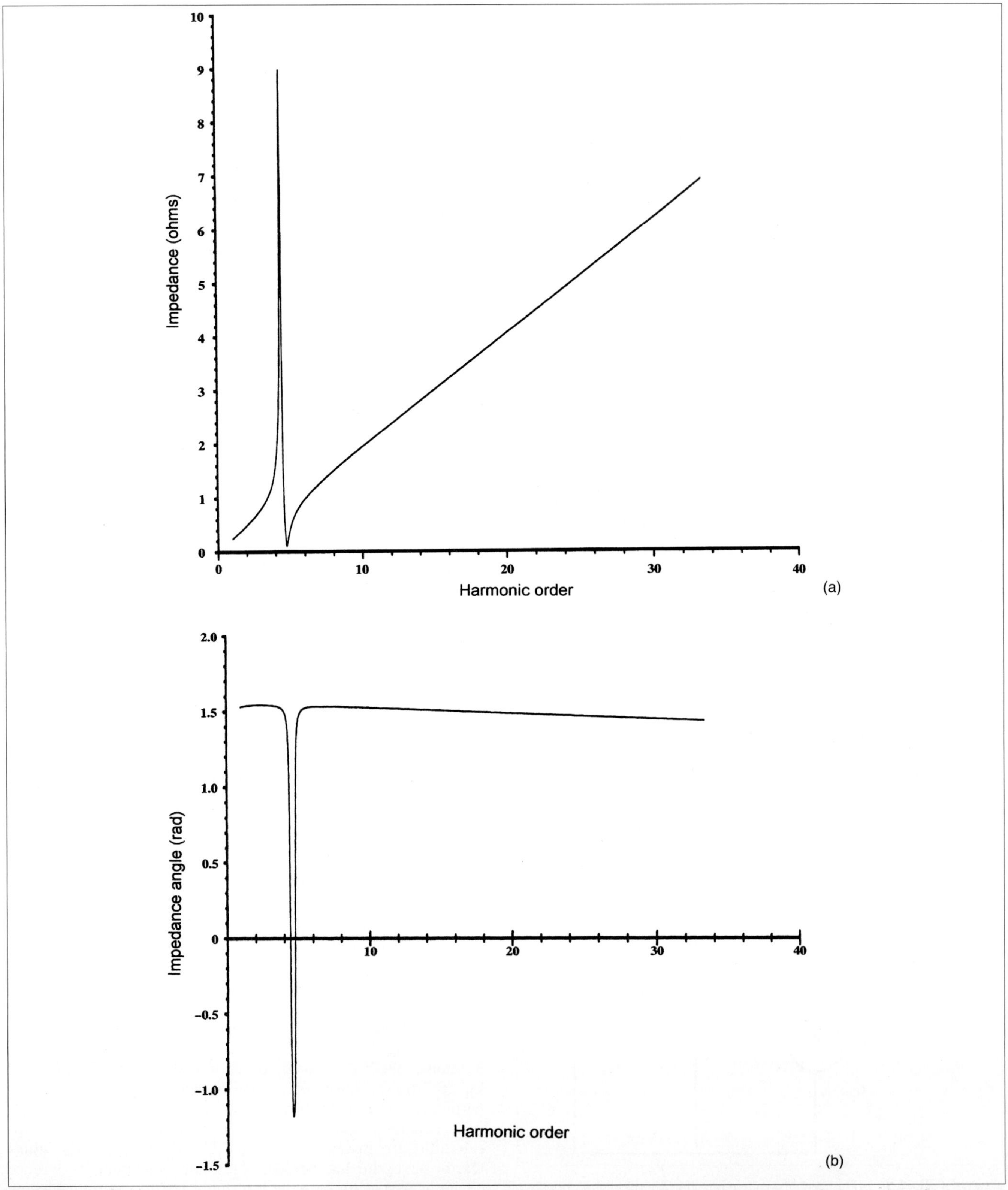

FIGURE 6-12 (*a*) Impedance modulus with 6-Mvar capacitor bank turned into a single tuned band-pass filter, tuning frequency 4.67th of fundamental, Fig. 6-4 (*b*) Impedance angle.

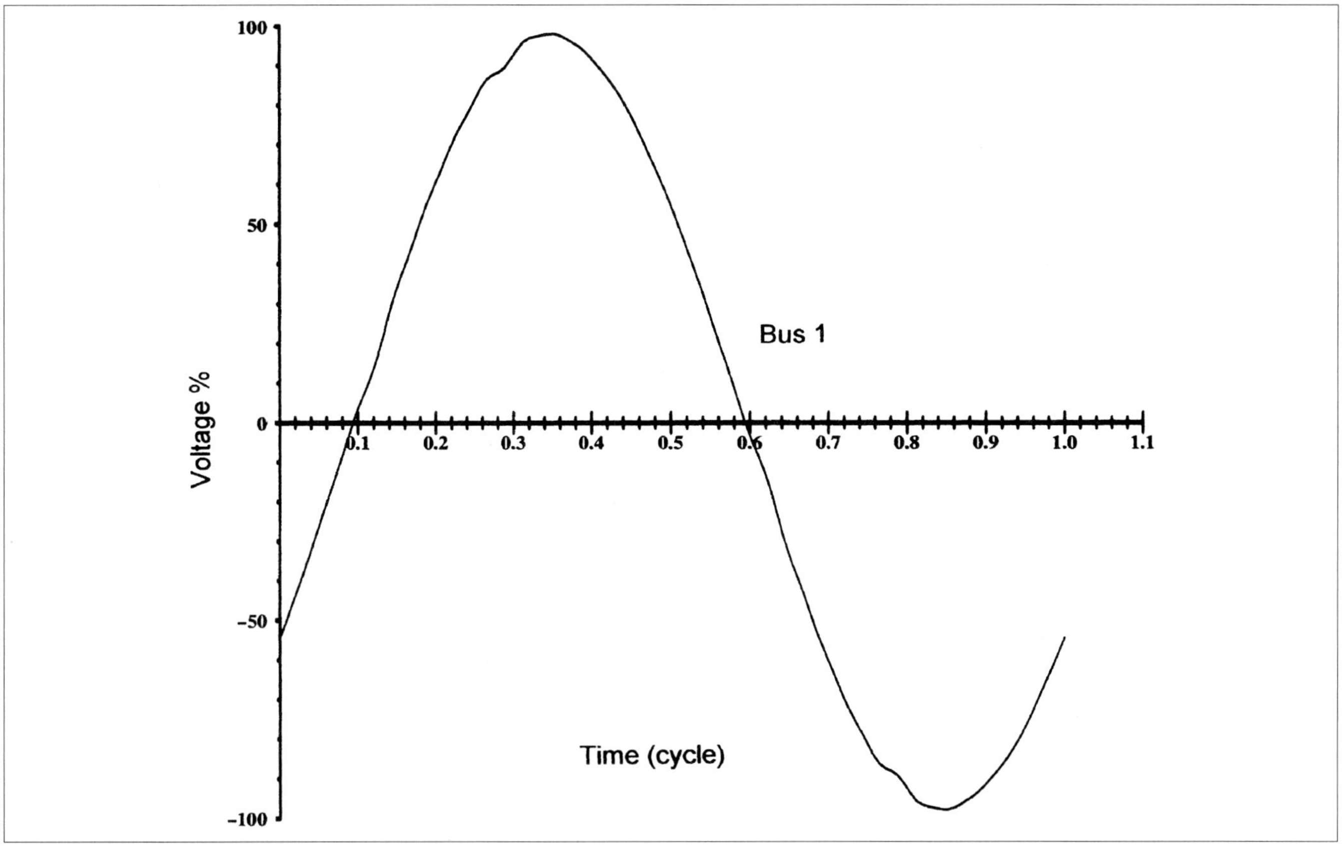

FIGURE 6-13 Voltage waveform, 13.8-kV bus 1, with 6-Mvar capacitor bank turned into a single-tuned band-pass filter, tuning frequency 4.67th of fundamental, Fig. 6-4.

be some swings in the tuned and resonant frequency with varied system switching conditions, and in a practical filter design, these must be considered with filter loading in each case.[8] Figure 6-13 shows 13.8-kV bus voltage with the applied filter; this can be compared with Fig. 6-8.

6-4 BACK-TO-BACK SWITCHING

The back-to-back switching involves energizing a capacitor bank on the same bus when another energized bank is present. The inrush transient in this case mainly consists of interchange of currents between the two banks, and the current supplied from the supply system can be ignored. This will not be true if the supply system impedance is comparable to the impedance between the banks being switched back to back.

The switching inrush current and frequency can be calculated from:

$$i_{\text{max,peak}} = \frac{\sqrt{2}E_{LL}}{\sqrt{3}}\sqrt{\frac{C_1C_2}{(C_1+C_2)L_m}} \tag{6-27}$$

And the frequency of the inrush current is:

$$f = \frac{1}{2\pi\sqrt{L_m\dfrac{C_1C_2}{(C_1+C_2)}}} \tag{6-28}$$

where C_1 and C_2 are the sizes of the capacitors in farad and L_m is the inductance between them in henries. Apart from stresses on the circuit breaker, the high-frequency transient can stress the other equipment too.

Example 6-4 The circuit of Fig. 6-1 is modified with the addition of another capacitor bank C2, similar to bank C1 of 6 Mvar, as shown in Fig. 6-14. C1 is already energized and its circuit breaker CB1 is closed. Bank C2 is switched on the same bus by closing its circuit breaker CB2. The equivalent switching circuit is shown in Fig. 6-15. To calculate the switching transients, it is necessary to calculate the inductance between the banks accurately.

- Inductance of cables C_1 and C_2 between the banks = 2-3/C 500 KCMIL, 11 m, $L_{c1} + L_{c2} = 4.6\ \mu H$

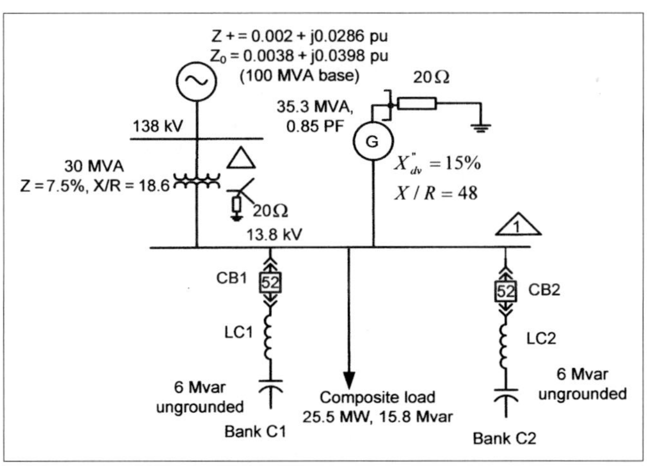

FIGURE 6-14 A circuit diagram for the study of back-to-back switching of capacitor banks.

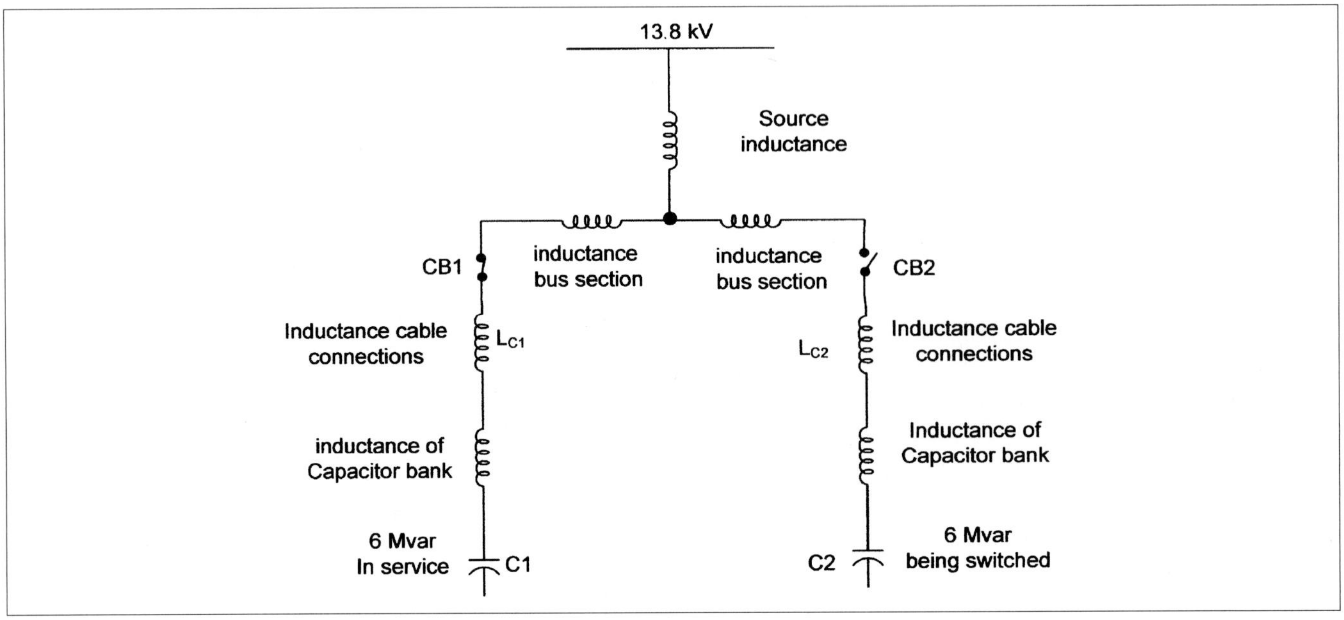

FIGURE 6-15 Calculation of inductance for back-to-back switching of capacitor banks.

- Inductance of bus duct between the circuit breakers, 13.8 kV, 1200 A, 2 m = 1.4 μH
- Inductance of each bank itself = 5 μH
- Total inductance = 16 μH

The source inductance of our example, as calculated earlier, is 588 μH. Thus, the inrush current can be calculated by ignoring the source inductance. Using these equations, the inrush current is 435 kA and its frequency is 6144 Hz, which are very high.

6-5 SWITCHING DEVICES FOR CAPACITOR BANKS

The derating of switching devices is required for capacitor switching duties, whether these are contactors, circuit breakers, or load break switches. The current requirement for rating the switching device should consider the effect of overvoltages (generally 10 percent), capacitor tolerances (105 to 115 percent), and harmonic components. For example, review the application of an indoor oil-less metal-clad 15-kV circuit breaker. The rated current of a 6-Mvar bank at 13.8 kV is 251 A. IEEE Standard C37.06-1987 specifies that a "general-purpose" 1200-A circuit breaker has a capacitor switching current rating of 250 A only.[2] A definite-purpose circuit breaker for capacitor switching, rated 1200 A, and 40-kA rms short-circuit rating, has a rated capacitance switching current of 630 A for isolated capacitor bank or for back-to-back switching. Furthermore, the inrush current and its frequency should not exceed 15 kA and 2000 Hz, respectively. This means that for the proper application of even a definite-purpose breaker, the inrush current and its frequency should be reduced for back-to-back switching applications. For high-voltage outdoor circuit breakers, similar derating applies. For example, a general-purpose 362-kV breaker, rated for 2000 or 3000 A and short-circuit rating of 65-kA rms symmetrical has an overhead line or isolated capacitor bank switching current of 250 A, and a definite-purpose circuit breaker must limit the inrush current to 25 kA, frequency 4250 Hz, on back-to-back switching. Gas circuit breakers, SF_6 or air-blast type, have a much better capability of capacitor switching and are now used exclusively for high-voltage networks. These can cope with high recovery voltages without restrikes. Installing closing resistors is one way to reduce switching transients; see Sec. 6-13-1. The vacuum switchgear, which is commonly used at medium voltage, cannot be provided with resistor switching, though two breakers with a separately mounted resistor can be used. The interruption of capacitor currents and effects on recovery voltages are covered in Chap. 8.

A bank is considered "isolated" if the rate of change of the transient inrush current does not exceed the maximum rate of change of symmetrical interrupting capability of the circuit breaker at the applied voltage.

Again, for a 15-kV indoor oil-less breaker, rated short-circuit current of 40 kA, K factor is 1.0. Then maximum rate of change is given by:

$$\left(\frac{di}{dt}\right)_{\max} = \sqrt{2}\omega(40.0\times10^{3})\times10^{-6} \qquad (6\text{-}29)$$

$$= 21.32\ \text{A}/\mu\text{s}$$

An inrush current of 4.58 kA (peak) and a frequency of 770 Hz was calculated when switching 6-Mvar bank in Example 6-1. This gives a rate of change of:

$$2\pi(770)(4580)\times10^{-6} = 22.16\ A/\mu\text{s} \qquad (6\text{-}30)$$

Therefore, switching of even a single 6-Mvar capacitor bank in Fig. 6-1 cannot be considered isolated bank switching. It requires a breaker of higher interrupting rating or, alternatively, the size of the capacitor bank should be reduced. Another option will be to provide an inrush current limiting reactor.

The year 2000 revision of ANSI/IEEE Standard C37.06-1987 has made K factor of oil-less indoor circuit breakers = 1. (This is an attempt to harmonize ANSI standards with IEC standards.) This reference may be seen for new rating tables of oil-less indoor circuit breakers.[2]

6-6 INRUSH CURRENT LIMITING REACTORS

When applying an inrush current limiting reactor, the following considerations apply:

1. An inrush current-limiting series reactor will tune the capacitor to a certain frequency. When load-generated harmonics are present, caution is required that the series reactor does not tune the capacitor close to a load-generated harmonic, unless the intention is to turn the capacitors into single-tuned filters, which requires detailed considerations. This is to avoid overloading of the capacitors, which will act as filters and create a low-impedance path to a load-generated harmonic.

2. A series reactor does reduce the capacitive reactance, yet, the net leading kvar output from the combination increases rather than decreases. This is so because the voltage drop across the reactor is additive to the capacitor voltage and the terminal voltage of the capacitor rises. As the reactive output of a capacitor changes in proportion to the square of the voltage, there is a net increase in the leading reactive kvar from the combination. This brings another consideration that the capacitor-rated voltage may have to be increased. Voltage at the junction of the capacitor and reactor will be:

$$V_c = \frac{n^2}{n^2 - 1} \qquad (6\text{-}31)$$

where $n = f_n/f$, f_n is the frequency in presence of the reactor. Also, the reactive power of the capacitor bank will be:

$$S_f = \frac{n^2}{n^2 - 1} \times \text{reactive power without reactor} \qquad (6\text{-}32)$$

Example 6-5 The limitation of switching inrush current and its frequency for isolated bank switching is demonstrated in this example by providing an inrush current-limiting reactor. A series reactor of 3.8 mH, with an $X/R = 50$ is provided in the switching circuit of 6-Mvar capacitor bank in Fig. 6-1. Then from Eqs. (6-7) and (6-8), the inrush current is reduced to 1633 A, and its frequency to 264 Hz. An EMTP simulation is shown in Fig. 6-16, which shows that in the first positive 1/2 cycle, the simulated value is close to the calculated value. The current increases in the negative 1/2 cycle, and this can be attributed to the exchange of energy between two energy storage elements, that is, the inductance and the capacitance. Also, it is noted that the decay of the current is prolonged due to added reactor. The transients are much reduced.

To reduce the inrush current on the back-to-back switching to be within the circuit breaker capability, its magnitude should be no more than 15 kA and frequency should be equal to or less than 2000 Hz.

Equation (6-7) can be used to calculate the minimum inductance between the banks to limit the transients. Both, inrush current limitation as well its frequency limitation should be considered. This calculation shows that minimum inductance should be 14.32 mH to reduce the frequency to 2000 Hz, and the inrush current will fall to 5.089 kA.

The ohmic losses in reactors can be substantial. A reactor of 4.3 mH, as calculated earlier and a Q factor of 40, has resistance of 0.04 Ω. The full-load current of 6-Mvar bank is 251 A. This gives an energy loss of 7.56 kWh, which, on a yearly basis, translates into 66 MWh of energy loss. For indoor reactors, the heat dissipation and ventilation for large reactors becomes a major consideration, and proper ventilation and heat dissipation measures should be considered.

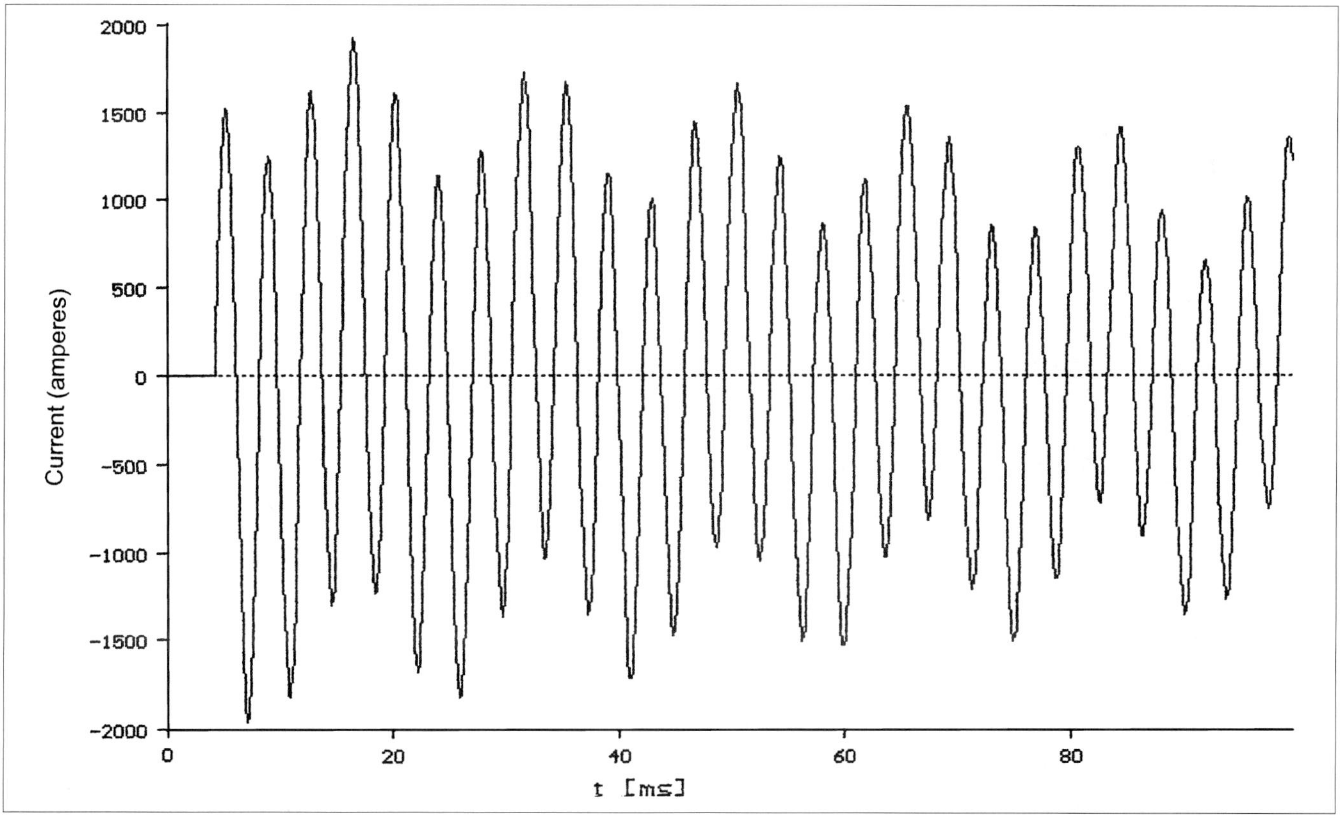

FIGURE 6-16 Reduction in current transient of 6-Mvar capacitor bank, Fig. 6-1, with a series inductor of 3.8 mH introduced in the capacitor bank.

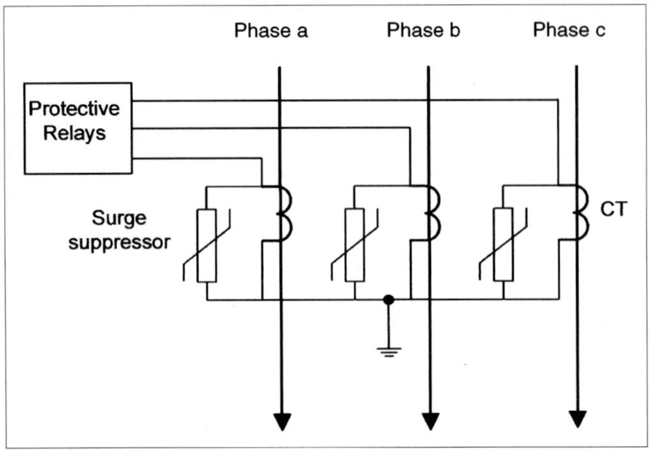

FIGURE 6-17 Surge suppressors across CT secondary windings to limit the voltages on discharge of capacitor currents on a fault.

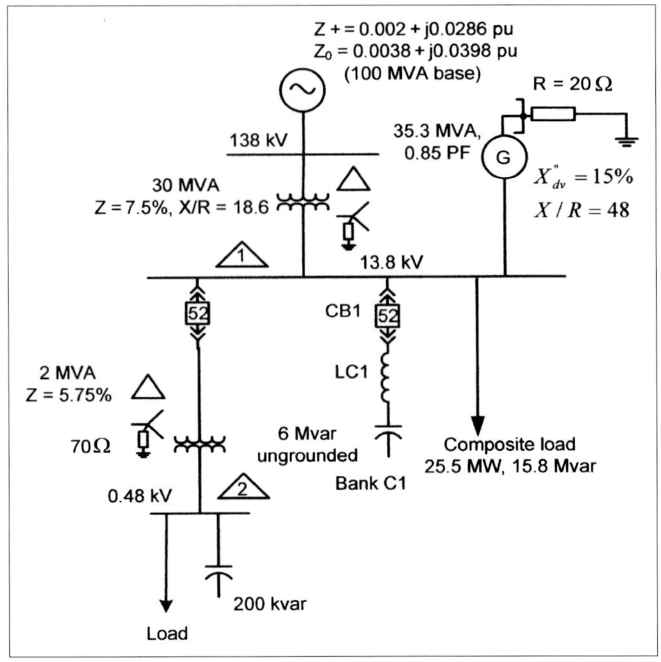

FIGURE 6-18 A circuit diagram for the study of secondary resonance.

6-7 DISCHARGE CURRENTS THROUGH PARALLEL BANKS

In Fig. 6-4, assume a fault occurs at *F*. The charged capacitors in the system will discharge into the fault. If the capacitors are ungrounded (the usual case for the industrial distributions), the maximum discharge current will occur for a three-phase fault. If the capacitors are grounded, the maximum discharge current will occur for a three-phase to ground or single-phase to ground fault.

The total discharge current will be the sum of the discharge currents from individual banks given by Eq. (6-7). Again the capability of the breaker to withstand this discharge current should not be exceeded. The effect on bushing CTs and linear couplers should be calculated.

$$V_{\text{cts}} = \frac{I_{\text{transient}}}{\text{CT ratio}} \times VA_{\text{ohms}} \times \frac{f_t}{f} \tag{6-33}$$

where V_{cts} is secondary voltage developed across BCT, VA_{ohms} is CT burden in ohms, f_t is transient frequency, and f is supply system frequency.

The voltage should be limited to safe levels. Sometimes, the provision of surge suppressors across CT terminals becomes necessary (Fig. 6-17).

For example, for a total discharge current of 20 kA, frequency of 2000 Hz, relay burden of 0.5 Ω and CT ratio of 1000/5, the secondary voltage across CT from Eq. (6-33) is 1.67 kV, which will impact all the relays and instruments connected across the secondary of the CT.

6-8 SECONDARY RESONANCE

Figure 6-1 is now redrawn as shown in Fig. 6-18, and a 2-MVA transformer with secondary load and a 200-kvar capacitor bank are added. Such a circuit gives rise to the possibility of secondary resonance. The switching of a large high-voltage capacitor can cause a much higher per unit transient voltage at the location of a smaller low-voltage capacitor bank. This occurs in the secondary circuits, which have their resonant frequencies close to the natural frequency of the switched capacitor bank. The initial surge can trigger oscillations in the secondary circuits, which are much greater than in the switched circuit:

$$\frac{f_c}{f_m} = \sqrt{\frac{L_m C_m}{L_s C_s}} \tag{6-34}$$

where f_c is coupled frequency, f_m is the main switching frequency, L_s and C_s are inductance and capacitance in the secondary circuit, and L_m and C_m are the inductance and capacitance in the main circuit. The lower the ratio f_c/f_m, the greater is the magnitude of the coupled transient. High-transient voltages of the order of five times the rated voltage are possible, as shown in Fig. 6-19.[10]

These transient voltages can result in failures and fuse operations in low-voltage power factor correcting capacitors and nuisance trips of power electronic-based devices, such as adjustable speed drives.

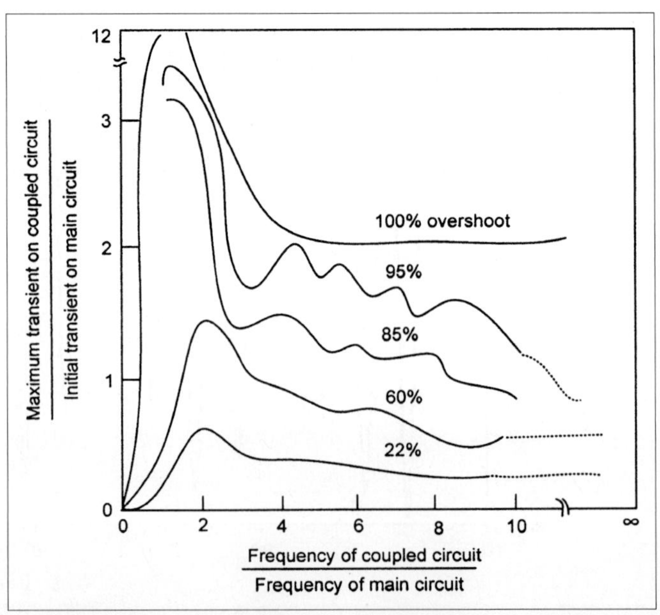

FIGURE 6-19 Overvoltages due to secondary resonance.

The studies show that:

1. Transients will predominate when the size of the switched capacitor bank is much larger than the capacitance of the low-voltage capacitor and cable, which is generally the case.

2. Highest transients will occur when the energizing frequency is close to the series-resonant frequency formed by the step-down transformer and low-voltage capacitance.

3. Problems predominate when there is little damping, which is very common in industrial distributions as the resistance is low.

Example 6-6 The 200-kvar capacitor connected to the 2-MVA, 13.8 to 0.48-kV transformer in Fig. 6-18 remains in service, while 6-Mvar capacitor at 13.8-kV bus is switched. The simulation in Fig. 6-20 shows that a peak voltage of 1220 V line-to-neutral is developed across the secondary of 480-V transformer. This overvoltage is 3.1 times the rated system voltage. The calculations show that $f_c/f_m \approx 3.8$. Figure 6-21 shows the transient current that flows through the 200-kvar capacitor; its peak value is 2200 A, approximately 6.5 times the full-load current of 200-kvar capacitor.

When capacitors are applied at multivoltage level in a distribution system, apart from the overvoltages due to the resonance illustrated earlier, these can also prolong the decay of the transients.

It is, therefore, best to apply capacitors at one voltage level only. If these are applied at multivoltage levels, a rigorous switching transient analysis to predetermine the resonance points and eliminate them is required. This requires lots of rigorous modeling not generally undertaken when placing capacitors in the distribution systems.

Example 6-7 This example is an EMTP simulation of switching transients of a transformer. The schematic single-line diagram is shown in Fig. 6-22. The EMTP transformer models are described in Chap. 14. The purpose of this example is to demonstrate that the transformer inrush transients escalate in amplitude and their decay is prolonged when a transformer and capacitor bank are energized together.

Figure 6-23 shows the switching inrush current transient on energizing the transformer when the 5-Mvar capacitor bank shown on the secondary is disconnected.

Figure 6-24 depicts the transient escalation in magnitude and time when the transformer and 5-Mvar capacitor bank on the secondary of the transformer are simultaneously energized. The transient profile shows some periodic escalations due to some higher-frequency resonance. This may not be a very practical situation. A load dependent or some other form of switching control of the capacitor bank is adopted so that leading reactive power is not supplied into the system and capacitor is switched off on loss of voltage.

Switching of a large capacitor bank when the system is unloaded gives rise to steady-state overvoltages, which can be calculated from the following expression:

$$\%\Delta V = \frac{\text{kvar}_c \times \%Z}{\text{kVA}_T} \tag{6-35}$$

where %Z is the transformer percentage impedance and kVA_T is the transformer kVA rating. The operating voltages may change by 10 percent for short periods, while adjustments are being made to the new operating conditions with the switched capacitors. This becomes important for the application of surge arresters (Chap. 20).

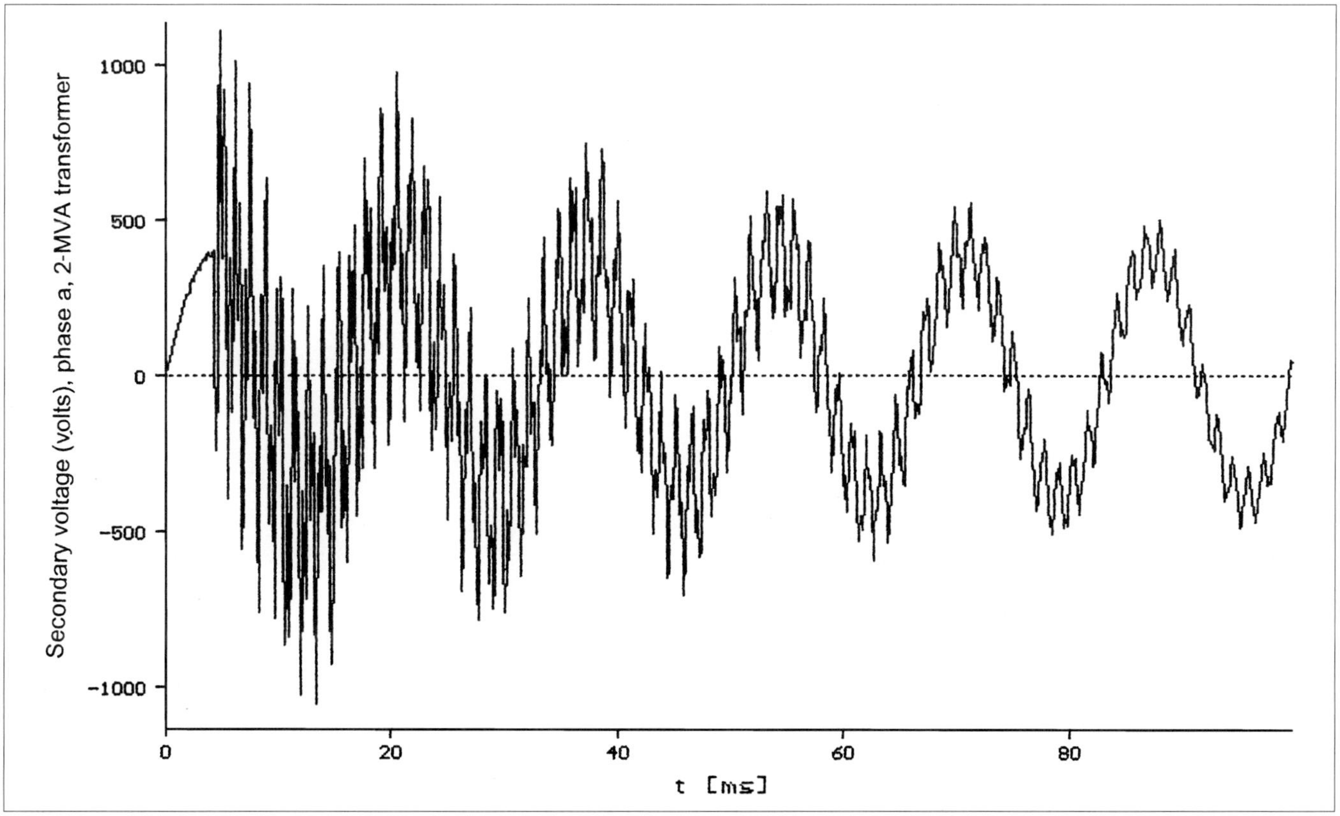

FIGURE 6-20 Voltage transient on 480-V bus 2, phase *a*, on switching of 6-Mvar capacitor bank, 200-kvar capacitor bank already energized, Fig. 6-18.

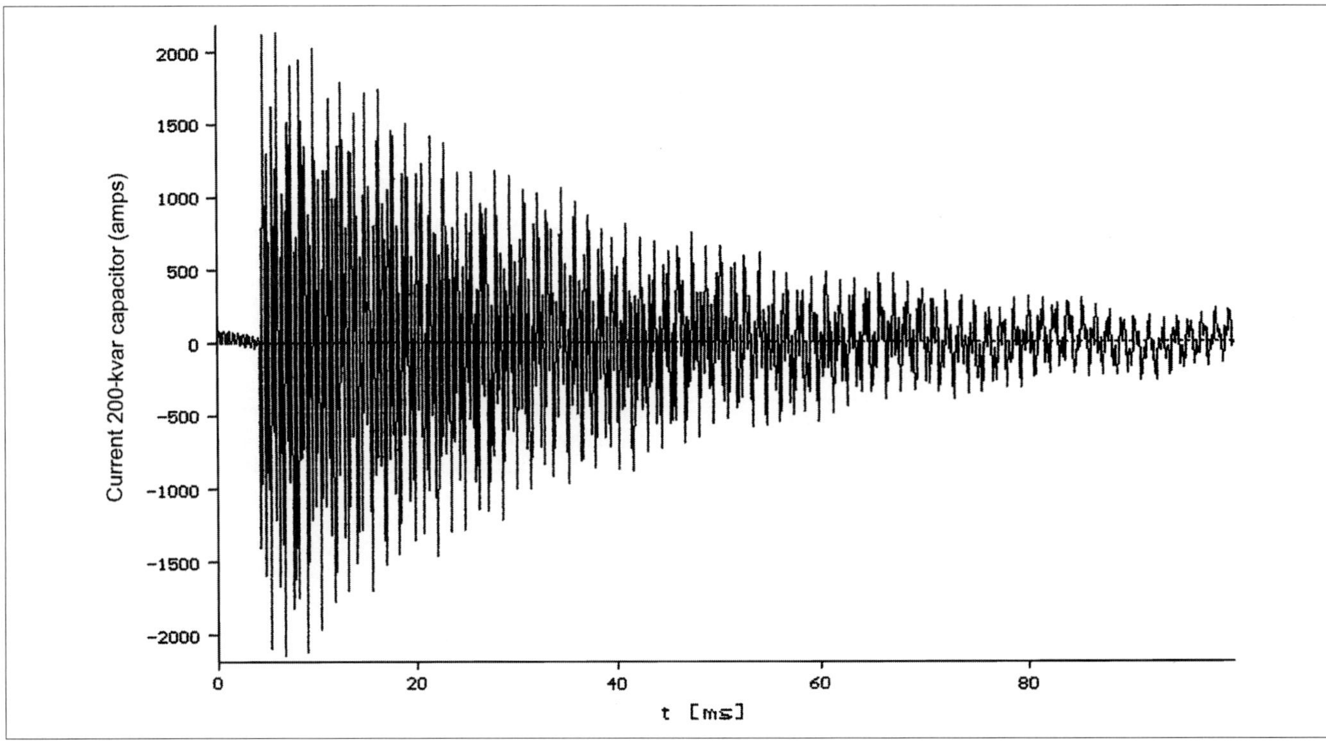

FIGURE 6-21 Current transient through 200-kvar capacitor bank.

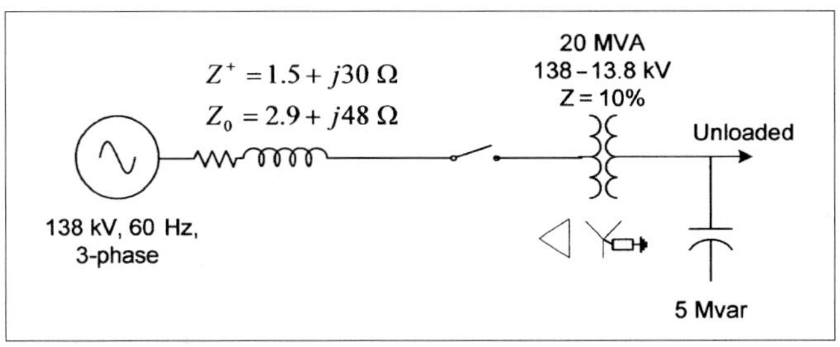

FIGURE 6-22 A simple circuit for switching transient of a transformer, with and without 5-Mvar capacitor bank.

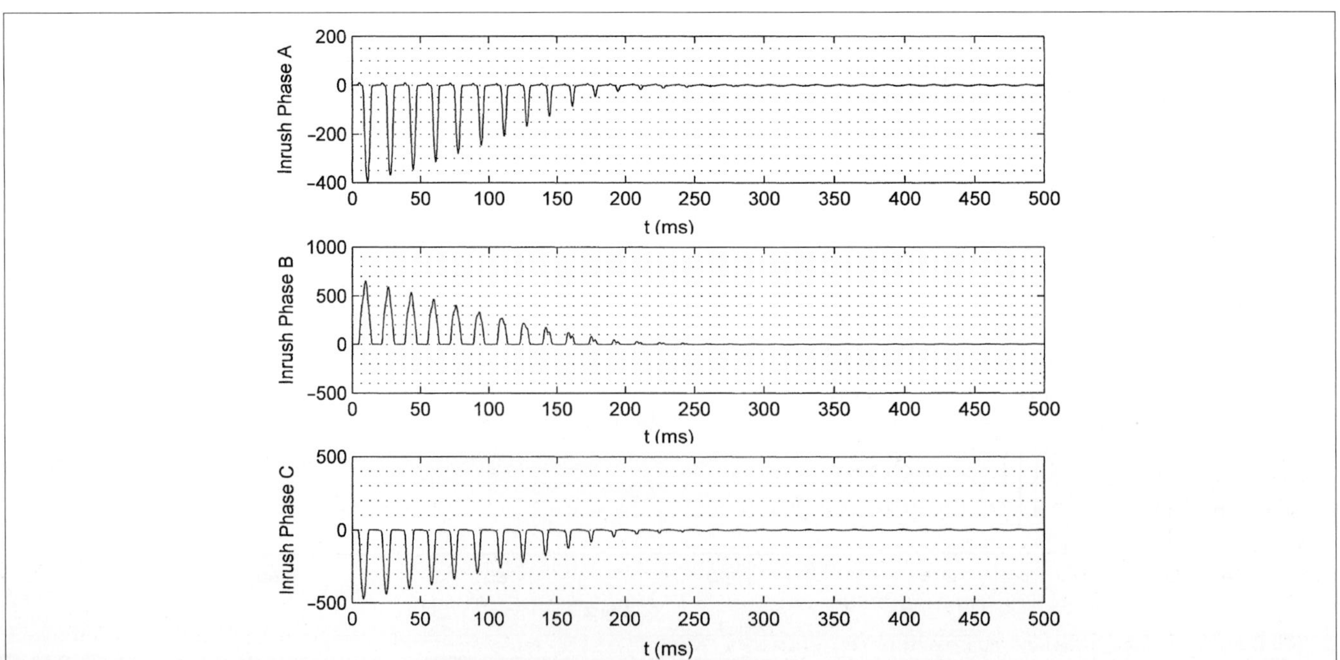

FIGURE 6-23 A 20-MVA transformer switching current transient without secondary 5-Mvar capacitor bank.

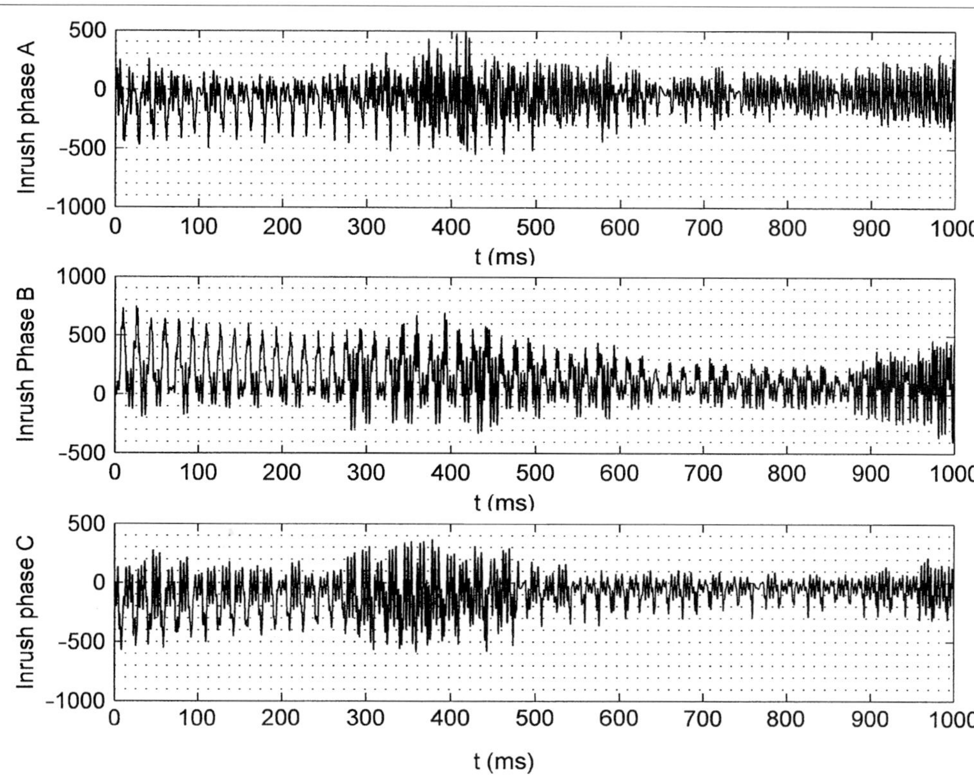

FIGURE 6-24 Current transients 20-MVA transformer and 5-Mvar bank switched together.

6-9 PHASE-TO-PHASE OVERVOLTAGES

The switching of capacitors can produce phase-to-phase overvoltages at a remote location. Consider a capacitor bank switched on a bus connected through a transmission line to a transformer. The surge will double at the transformer and the surges produced on energizing the capacitors are of opposite polarity on two phases; the transformer at the remote end will experience phase-to-phase surges. The surge arresters, normally connected phase-to-ground, will limit the phase-to-ground components, depending on their protective level, but the phase-to-phase overvoltages will be twice this value.[11] Restrikes in switching devices controlling ungrounded wye-connected banks can also generate high phase-to-phase voltages (Chap. 20).

Example 6-8 An electrical power system is shown in Fig. 6-25 for the simulation of capacitor bank switching transients. The following models are used:

Source impedance at 230 kV: This is simply a passive coupled RL branch, modeled with positive- and zero-sequence source impedance.

230-kV, 80-km line: A CP model is used, as discussed in Chap. 4.

40-MVA, 230- to 13.8-kV transformer: Transformer bushings, windings, and interwinding capacities are modeled; see Chap. 14, for transient models of transformers. The wye windings of the transformer are grounded through a resistance of 20 Ω to limit the ground fault current to 400 A.

Surge arresters: The surge arresters considered are gapless zinc-oxide surge arresters, rated voltages as shown in Fig. 6-21. Their models are discussed in Chap. 20.

400 ft of 3/C 500 KCMIL cable between the 13.8-kV bus and 2.5-MVA 13.8- to 0.48-kV transformer: This is a model with characteristic impedance, length, and attenuation (Chap. 4).

7.5-Mvar capacitor bank: The bank is ungrounded and switched at an instant when the voltage in phase *a* peaks. There is no inrush current-limiting reactor.

Loads at 13.8-kV bus and low-voltage transformer secondary: Passive load models consisting of appropriate RL elements and their connections.

2.5-MVA transformer: It is the same type of model as for main 40-MVA transformers. This transformer is high-resistance grounded; grounding resistance = 70 Ω.

The results of EMTP simulations are shown in Figs. 6-26 through 6-30.

Figure 6-26 shows the 13.8-kV three-phase voltages at bus V3 on capacitor bank switching. A sudden voltage dip and subsequent recovery profile can be seen from this figure. The peak excursion is approximately 18 kV from phase to ground.

Figure 6-27 illustrates phase *a* voltage on the primary of 2.5-MVA low-voltage transformer at V4 for about 4 ms from the instant of switching. The thick band is due to multiple reflections in the 400-ft cable. The voltage escalates and has high-frequency superimposed oscillations due to multiple reflections at the cable ends. To capture these high-frequency oscillations, a time step of 1 μs is used in EMTP simulations.

Figure 6-28 illustrated the secondary voltage V5 of 2.5-MVA transformer, simulation for 30 ms only. The voltage has high-frequency oscillations and rises to 2900 V peak from line to ground. Some resonant phenomena and amplification is occurring inside the transformer.

Figure 6-29 shows the 7.5-Mvar capacitor three-phase inrush currents which have high-frequency components, and finally, Fig. 6-30 depicts the 230-kV line current.

This is continued further in Chap. 20 to illustrate the control of high voltage on the secondary of 2.5-MVA transformer.

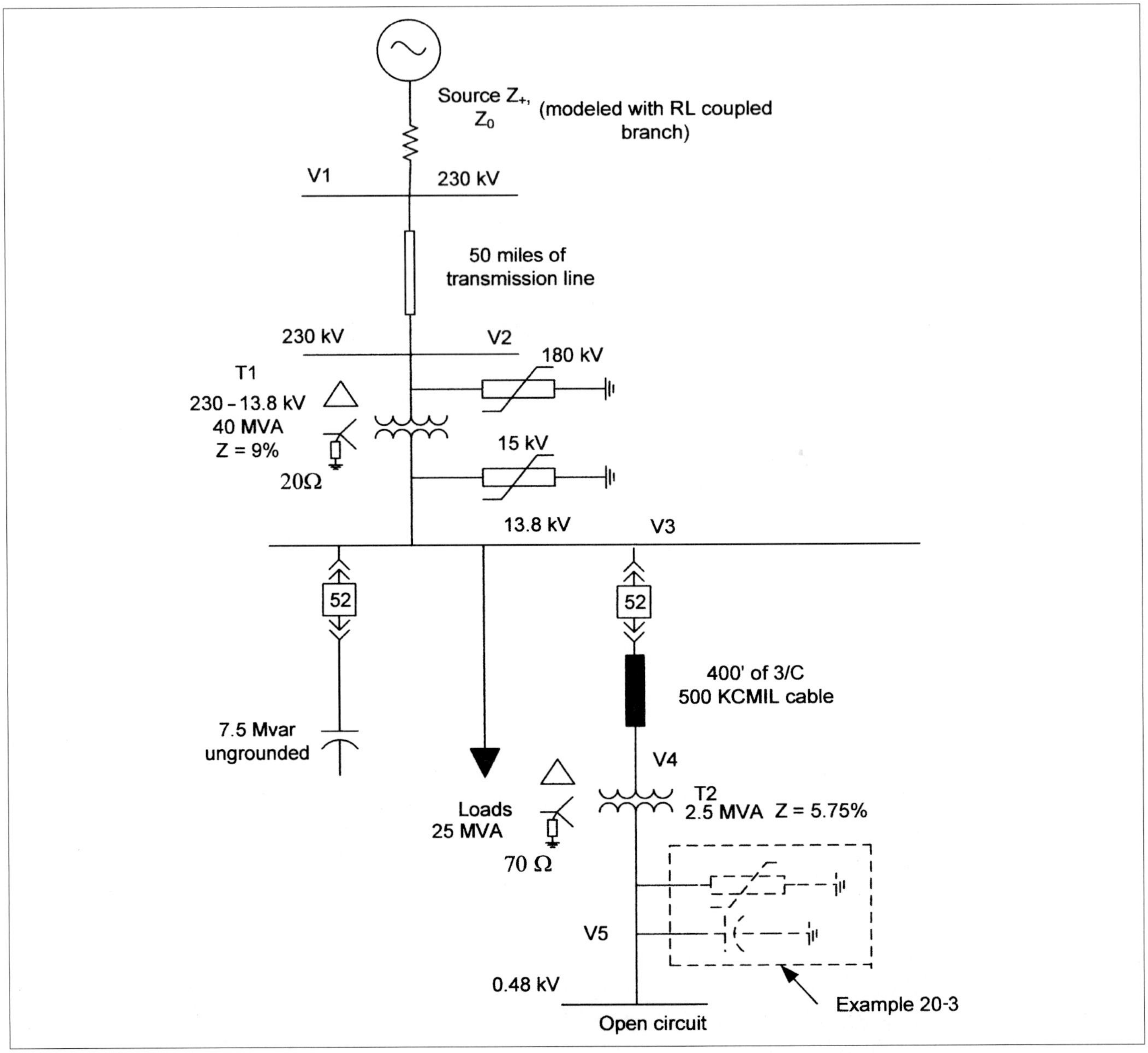

FIGURE 6-25 A power system configuration for the study of switching transients.

6-10 CAPACITOR SWITCHING IMPACT ON DRIVE SYSTEMS

An investigation of the failure of a drive system due to frequent cycling of a capacitor bank in the utility system, which raised the dc link voltage of a drive system to trip, is discussed in Ref. 12. This has been a major concern in the industry.

Example 6-9 This example is an EMTP simulation of escalation of the dc link voltage of a six-pulse converter feeding a dc motor load, Fig. 6-31, when a 6-Mvar capacitor bank is switched. The simulation in Fig. 6-32 shows that the voltage rises to approximately 2 pu. This can be controlled by all the means that reduce the capacitor inrush transients; see Sec. 6-13.

6-11 SWITCHING OF CAPACITORS WITH MOTORS

Capacitors are, generally, applied for the power factor correction of induction motors. The loaded induction motors operate at a power factor of 80 to 90 percent approximately. The power factor decreases for low-speed motors due to leakage reactance of stator overhang windings.

Generally, the induction motors do not operate at full load, which further lowers the operating power factor. Even though the power factor of the motor varies significantly with load, its reactive power requirement does not change much. Thus, with the application of power factor improvement capacitor, the motor power factor from no-load to full load will not vary much. Figure 6-33

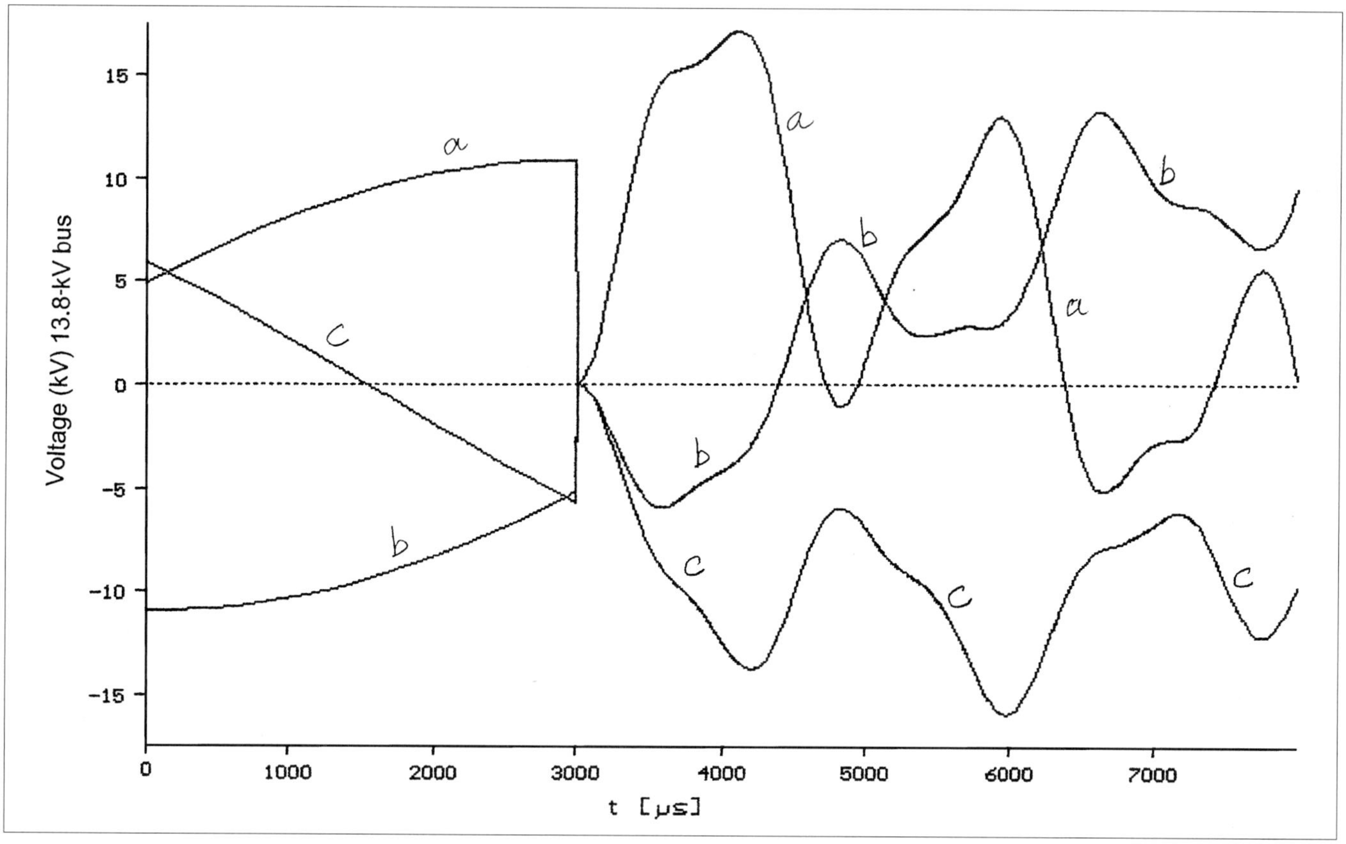

FIGURE 6-26 Voltage transients, 13.8-kV bus V3.

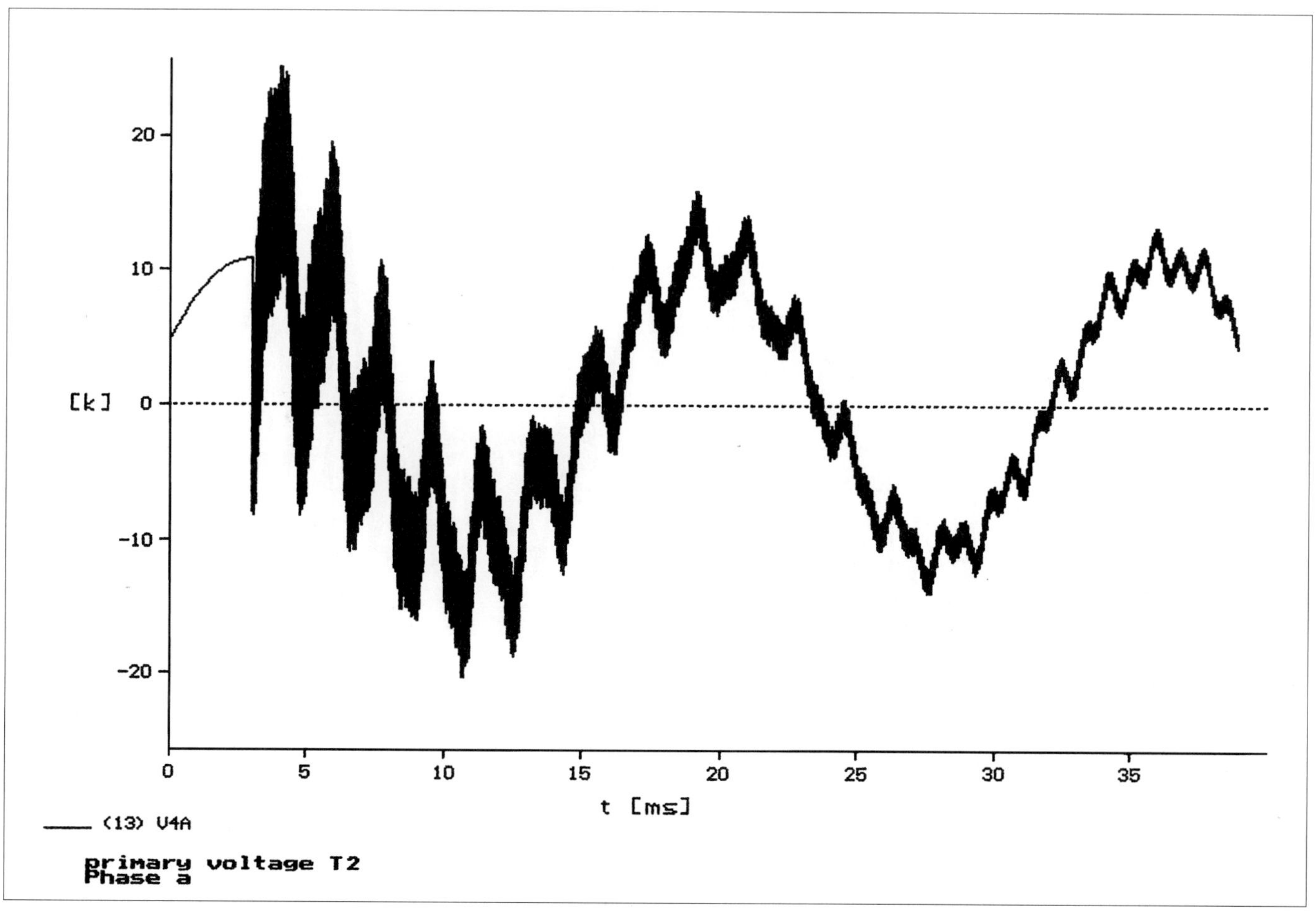

FIGURE 6-27 Voltage transient primary of transformer T2, V4.

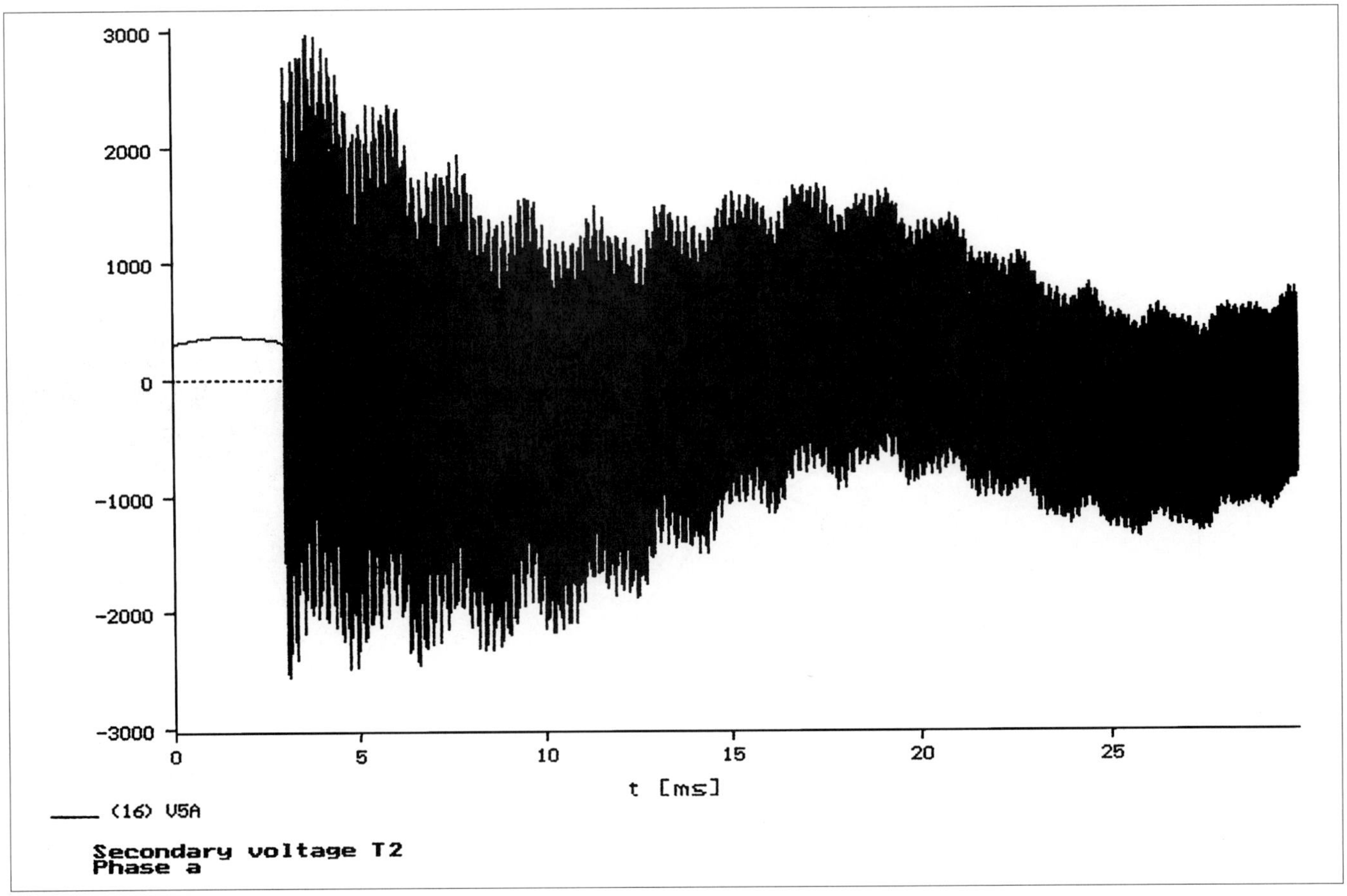

FIGURE 6-28 Voltage transient on secondary of T2, V5.

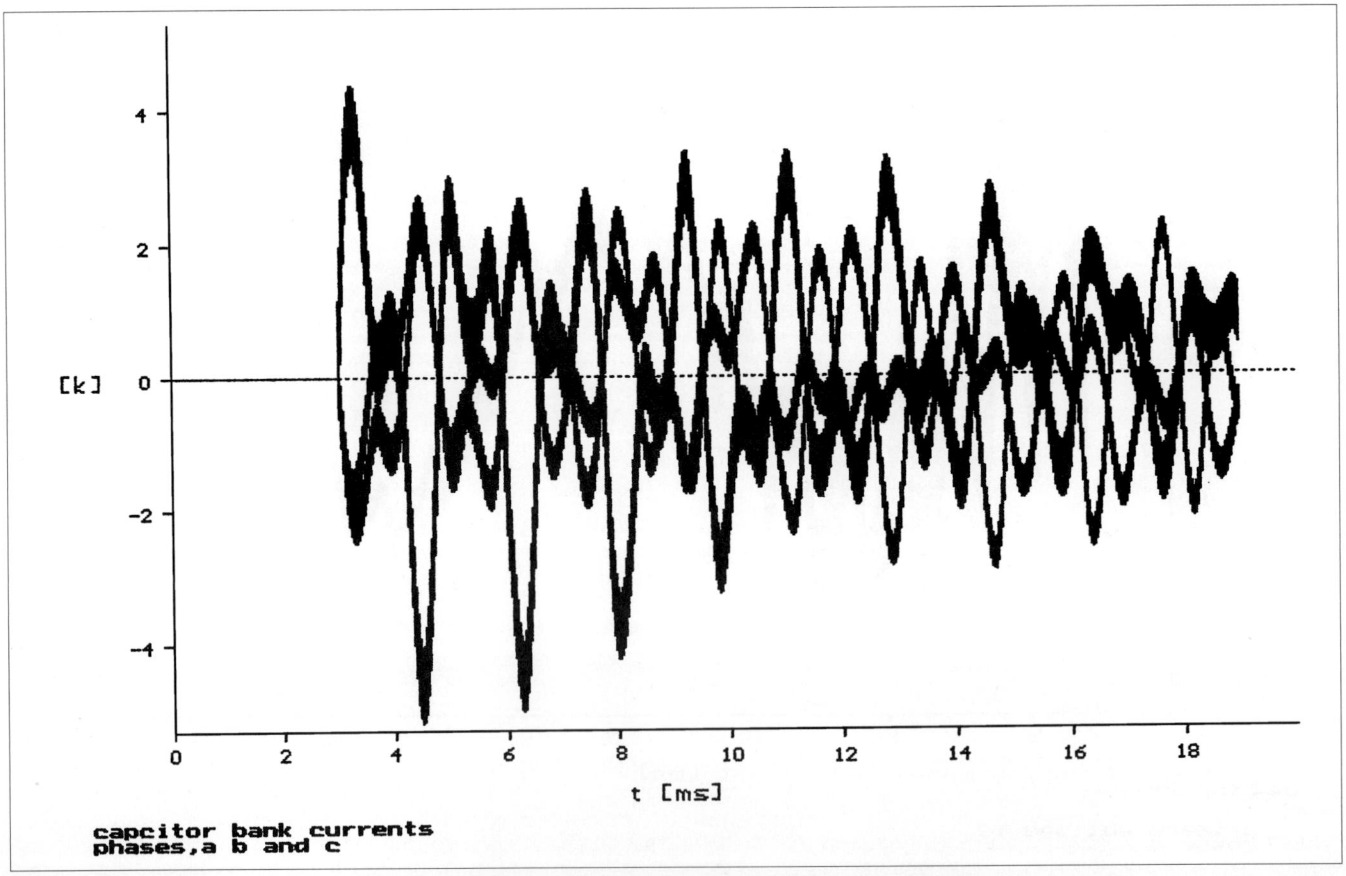

FIGURE 6-29 Current transients through 7.5-Mvar capacitor bank.

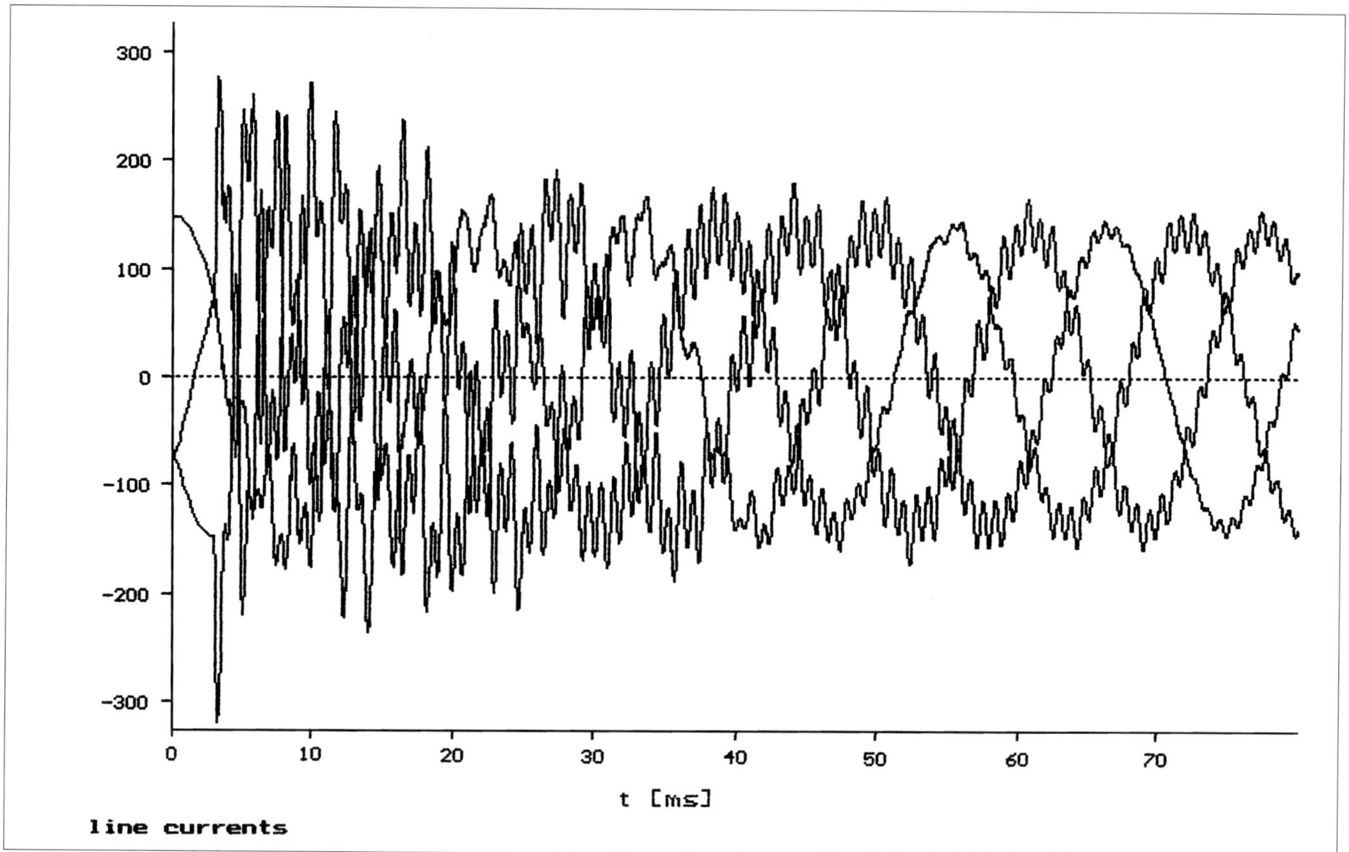

FIGURE 6-30 Line current through 230-kV line.

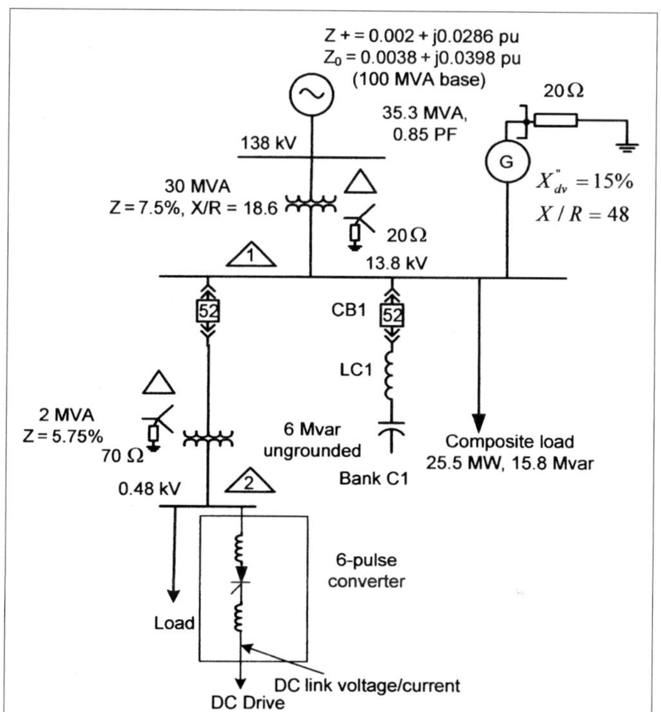

FIGURE 6-31 A circuit diagram for the study of escalation of dc link voltage in a drive system on switching of 6-Mvar capacitor bank.

shows three methods of location of the capacitors. (a) The capacitors are connected directly to the motor terminals; (b) the capacitor and motor are switched as a unit, and (c) the capacitor is switched independently of the motor contactor through a separate switching breaker interlocked with the motor starting breaker.

The switching of the capacitor and motor as a unit (second method) can result in problems due to:

- Presence of harmonic currents
- Overvoltages due to self-excitation
- Excessive transient torques and inrush currents due to out-of-phase closing

Overvoltages due to self-excitation are important for the proper application of capacitors with induction motors.

The magnetizing current of the induction motors varies with the motor design. Premium high-efficiency motors operate less saturated than the previous U- or T-frame designs.[13] The motor and capacitor combination in parallel will circulate a current between motor and capacitor corresponding to their terminal voltage. In this manner, the network is said to *self-excite.*

This is shown in Fig. 6-34*a* and *b*. The same size of capacitor applied to a standard and high-efficiency motor has different results because of the motor magnetizing characteristics. In case of high-efficiency design motor, it raises the terminal voltage to 680 V.

A capacitor size can be selected for the power factor improvement based on the following equation:

$$\text{kvar}_c \leq \sqrt{3}I_0 \sin\phi_0 \tag{6-36}$$

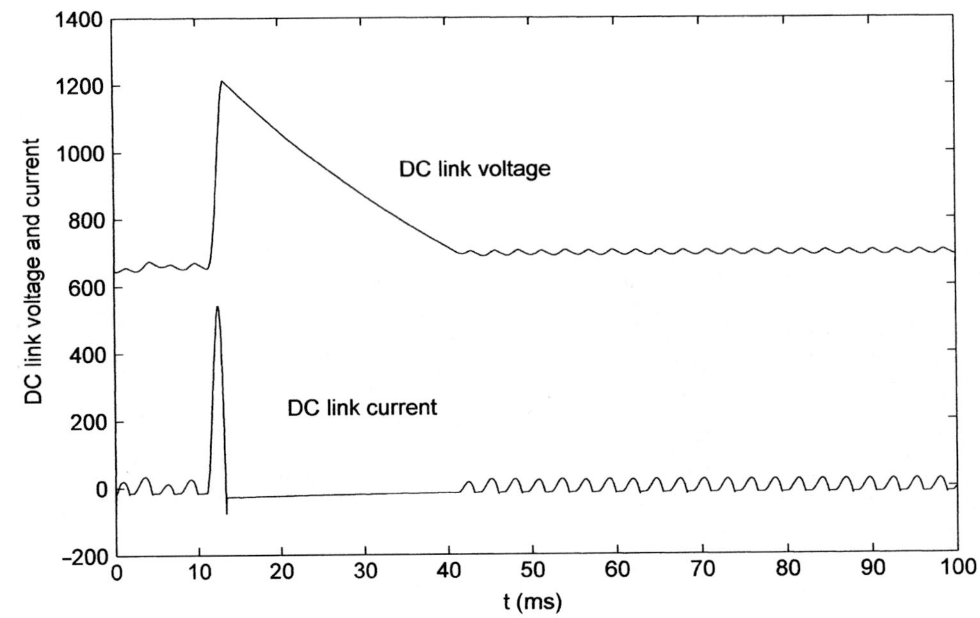

FIGURE 6-32 DC link voltage and current transients, Fig. 6-31.

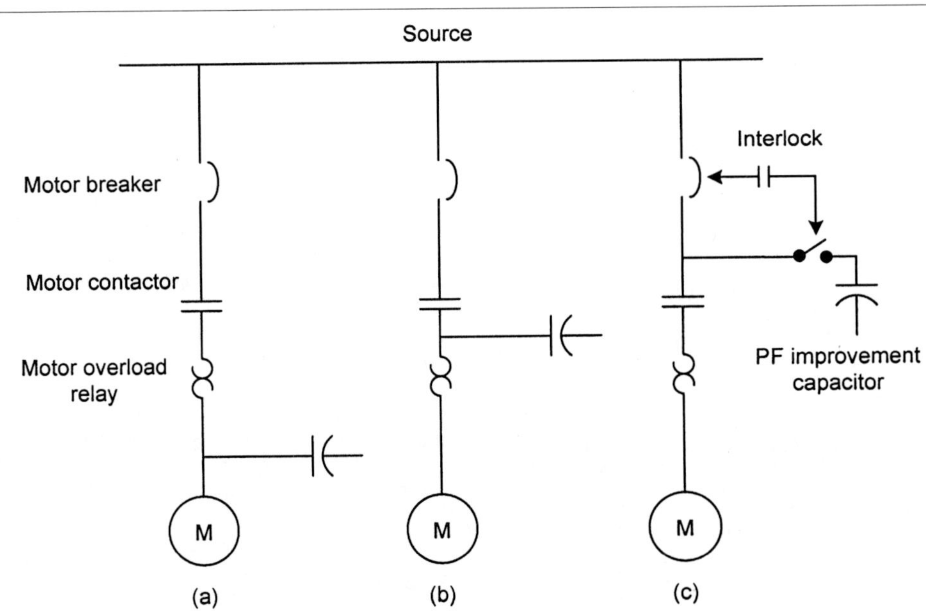

FIGURE 6-33 Various methods of connection of capacitors with motors.

where kvar_c is the maximum capacitor that can be applied, I_0 is the motor no-load current, and ϕ_0 is the no-load current power factor angle. This means that the capacitor size does not exceed the motor no-load reactive kvar. However, this no-load excitation motor data is not readily available and the size of capacitor application for a specific motor design should be based on the manufacturer's recommendations. The capacitors should not be directly connected to the motor terminals when:

- Solid-state starters are used
- Open transition starting methods are applied; see Chap. 11
- The motor is meant for repeated switching, jogging, inching, or plugging
- A reversing or multispeed motor is used
- A high inertia load may drive the motor, that is, a broken belt of a conveyor

6-12 INTERRUPTIONS OF CAPACITANCE CURRENTS

A breaker may be used for line dropping and interrupt charging currents of cables open at the far end or shunt capacitor currents. These duties impose voltage stresses on the breaker. A single-phase

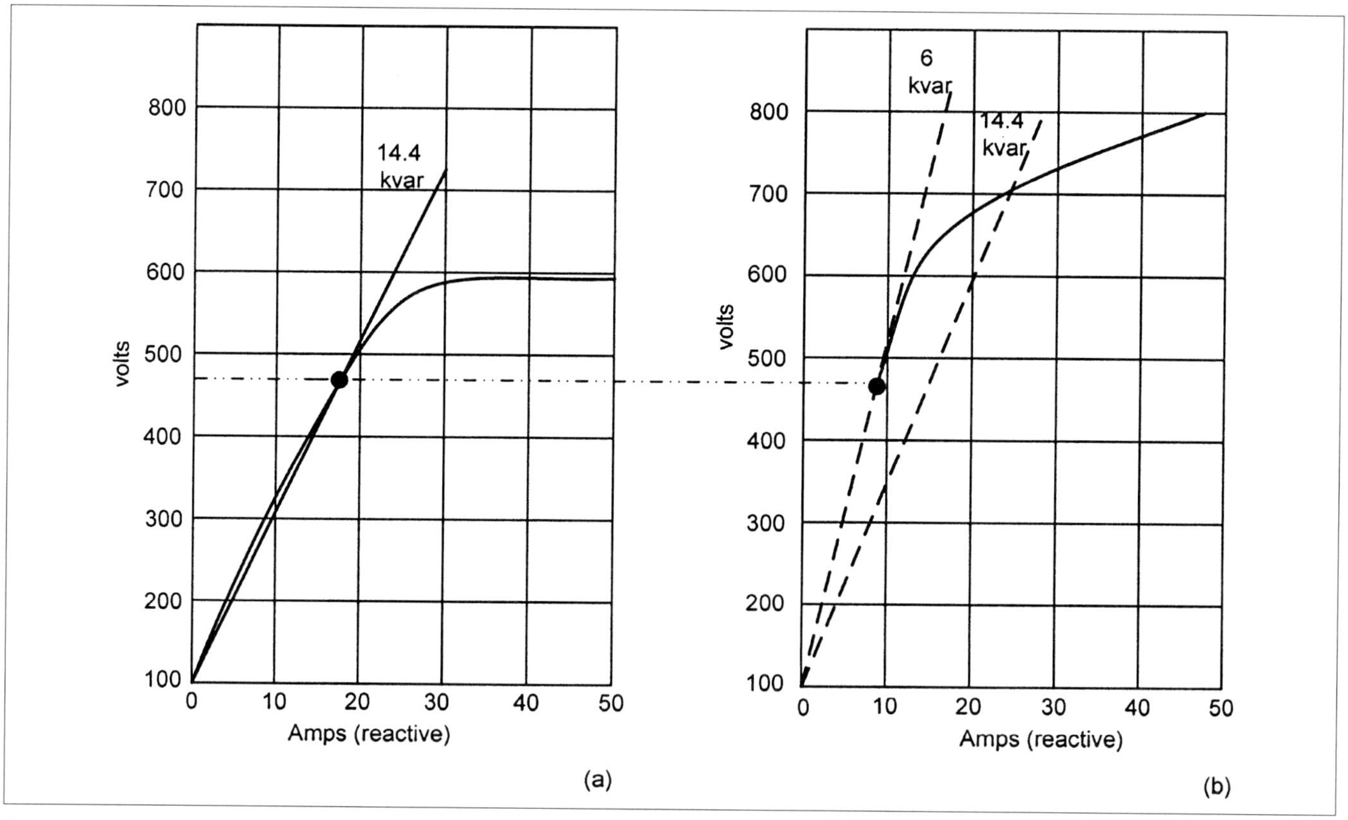

Figure 6-34 Self-excitation of motors with capacitors—typical motor saturation characteristics, (*a*) standard design, (*b*) high-efficiency design.

circuit for capacitive current interruption is in Fig. 6-35*a*. The distributed line capacitance is represented by a lumped capacitance C_2, or C_2 may be a power capacitor. C_1 is the source capacitance. The current and voltage waveforms of capacitance current interruption in a single pole of a circuit breaker under following three conditions are shown in Figs. 6-35*b*, *c*, and *d*.[14]

- Without restrike
- With restrike
- With restrike and current chopping

After interruption of the capacitive current, the voltage across the capacitance C_2 remains at the peak value of the power frequency voltage:

$$u_2 = \frac{\sqrt{2}u_n}{\sqrt{3}} \tag{6-37}$$

The voltage at the supply side oscillates at a frequency given by supply side C_1 and L_1, about the driving voltage u_n. The difference of these two voltages appears at the breaker pole. This can be more than double the rated voltage, with no prior charge on the capacitors. If the gap across poles of a circuit breaker has not recovered enough dielectric strength, restrike may occur. As the arc bridges the parting contacts, the capacitor being disconnected is again reconnected to the supply system. This results in a frequency higher than that of the natural frequency of the source side system being superimposed on the 60-Hz system voltage. The current may be interrupted at a zero crossing in reignition process. Thus, the high-frequency voltage at its crest is trapped on the capacitors. Therefore, after half a cycle following the restrike, the voltage across the breaker poles is the difference between the supply side and the disconnected side which is at peak voltage of the equalizing process and a second restrike may occur. Multiple restrikes can occur, pumping the capacitor voltage to three, five, and seven . . . times the system voltage at each restrike. The multiple restrikes can terminate in two ways: (1) these may cease as the breaker parting contacts increase the dielectric strength; and (2) these may continue for a number of cycles, until these are damped out.

A distinction should be made between reignitions in less than 5 ms of current zero and reignitions at 60-Hz power frequency. Reignitions in less than 5 ms have a small voltage across the circuit breaker gap and do not lead to overvoltages. Most breakers exhibit current chopping; see Chap. 8. It is usually caused by the instability of arc at low magnitude of currents due to high arc voltages. The chopping currents of modern breakers have been reduced, but it is clear that the capacitor charge will not be at peak if chopping occurs. This will *lower* the recovery voltage peak across the breaker gap. See Chap. 8 for recovery voltage profiles.

In Fig. 6-35*d*, the current is chopped before it reaches its natural zero. The voltage on the disconnected side does not remain at the service frequency peak voltage, but at a lower momentary value u_a of the driving voltage. Depending on the supply source impedance, the chopping current i_a creates an oscillation with an overvoltage, and the voltage u_s across the breaker contacts can rise to:

$$u_s = u_a + u_0 + \lambda\sqrt{u_a^2 + \frac{L_1}{C_1}i_a^2} \tag{6-38}$$

where λ is the damping coefficient <1.

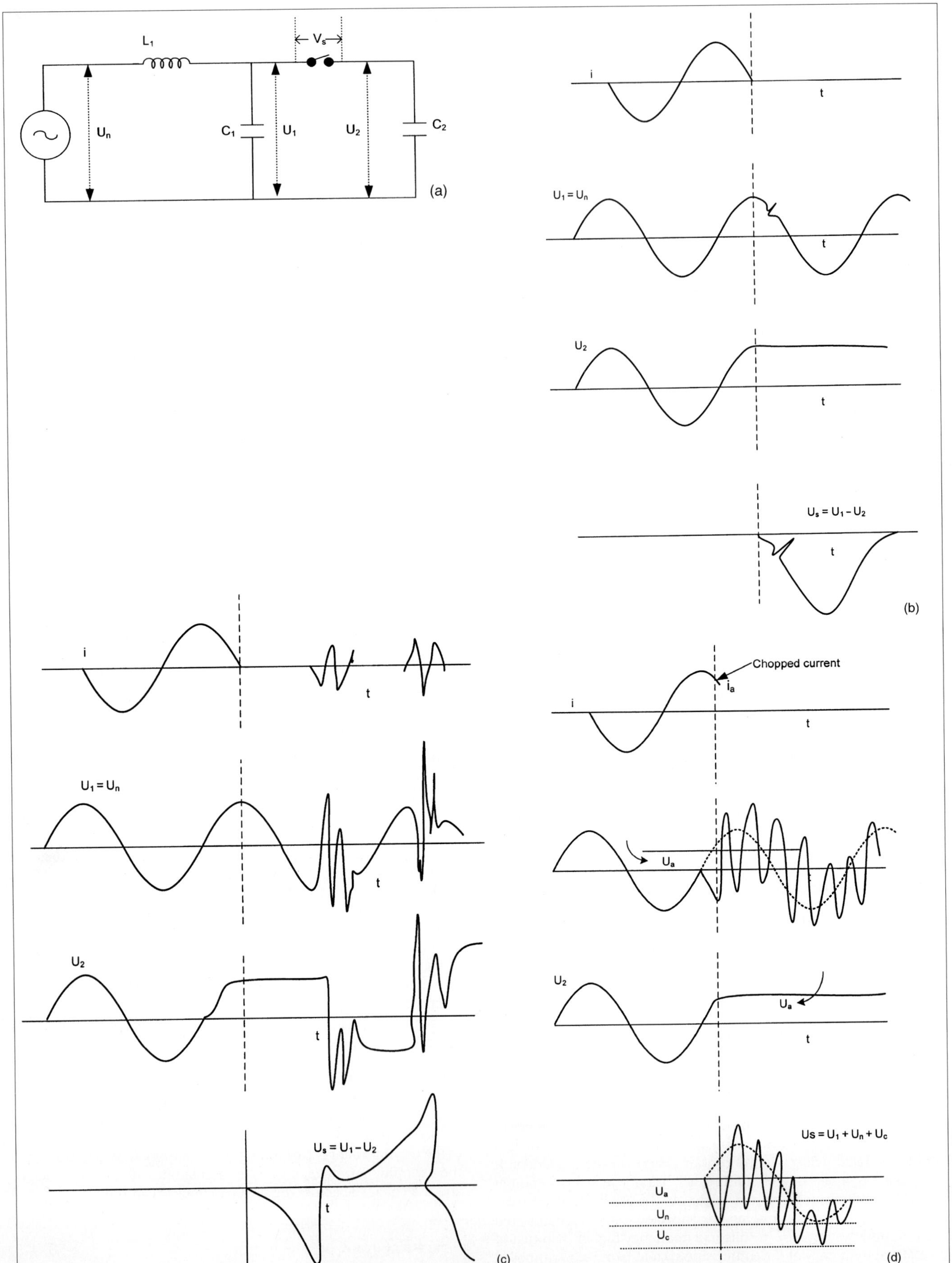

FIGURE 6-35 (*a*) A single phase circuit for capacitive current interruption. (*b*), (*c*), and (*d*) Interruption with no restrike, with restrike, and with current chopping, respectively.

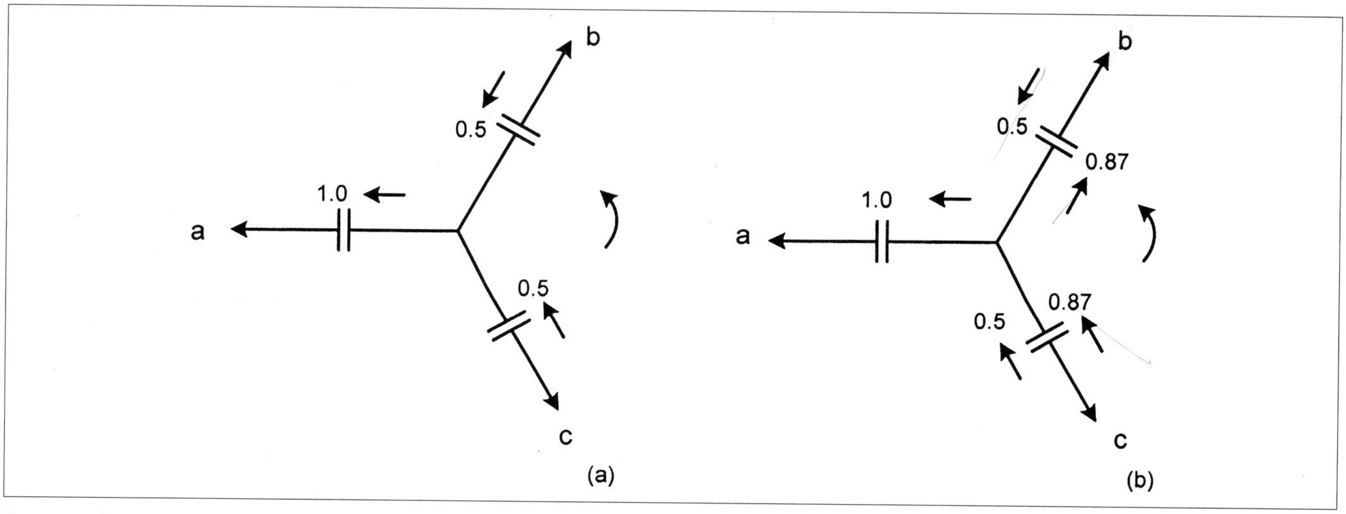

FIGURE 6-36 Sequence of creating trapped charges in an ungrounded three-phase capacitor bank. (*a*) Phase *a* clears first. (*b*) Phases *b* and *c* clear in series.

6-12-1 Disconnecting A Three-Phase Bank

Disconnecting a three-phase capacitor circuit is more complex. The instant of current interruption and trapped charge level depends on the circuit configuration. In an ungrounded three-phase wye-connected bank, commonly applied at medium- and high-voltage levels, let phase *a* current be first interrupted. This will occur when the voltage of phase *a* is at its peak. Figure 6-36*a* shows that phase *a* is interrupted first. The charge trapped in phase *a* is 1 per unit and that trapped in phases *b* and *c* is 0.5 per unit.

The interruption of phase *a* changes the circuit configuration and connects the capacitors in phases *b* and *c* in series. These capacitors are charged with equal and opposite polarities. The current in phases *b* and *c* will interrupt simultaneously as soon as phase-to-phase current becomes zero. This will occur at 90° after the current interruption in phase *a* at the crest of the phase-to-phase voltage so that an additional charge of $\sqrt{3}/2$ is stored in the capacitors, as shown in Fig. 6-36*b*. These charges will add to those already trapped on the capacitors in Fig. 6-36*a*, and thus voltages across the capacitor terminals are:

$E_a = 1.0$ per unit

$E_b = 0.37$ per unit

$E_c = 1.37$ per unit

Further escalation of voltages occurs if the phases *b* and *c* are not interrupted after 90° of current interruption in phase *a*. The voltage across the interrupting breaker pole will be the sum of supply system voltage and the charge trapped, maximum occurring in phase *c*, equal to 2.37 per unit. The capacitor bank is considered ungrounded and the stray capacitance to ground is ignored in the above analysis. It is hardly possible to take into account all modes of three-phase interruptions with restrikes.[14]

Advancements in the contact materials in vacuum interruption technology has reduced the restrikes and prestrikes of initial years, and so also the chopped currents.[15] However, it cannot be said that the restrikes are totally eliminated. The application of surge arresters provides a safeguard against these overvoltages and consequent insulation failures.[16]

6-13 CONTROL OF SWITCHING TRANSIENTS

The capacitor-switching transients can be controlled by:

- Using series inrush current limiting reactors.
- Using resistance switching.
- Using point-of-wave switching (synchronous breakers).
- Implementing the application of surge arresters.
- Dividing the capacitor bank into smaller size banks. The smaller the size of the capacitor bank being switched, the lesser is the transient.
- Avoiding application of capacitors at multivoltage levels to eliminate possibilities of secondary resonance. Metal-oxide varisitors (MOVs) can be applied at lower voltage buses.
- Converting the capacitor banks to capacitor filters; this is a must when harmonic-generating loads are present. This is one effective way to mitigate transients, eliminate harmonic resonance, and control harmonic distortion. Active filters can be used, depending on the size of the capacitive compensation and harmonic mitigation required for a particular system.
- Providing current-limiting reactors and chokes, which is a must for back-to-back switching and to have acceptable capacitor switching duties on circuit breakers and switching devices.
- Considering steady-state voltage rise due to the application of capacitors. The transformer taps may have to be adjusted.

6-13-1 Resistance Switching

In ac current interruption technology, the use of switching resistors in high-voltage breakers is well implemented to reduce the overvoltages and frequency of transient recovery voltage (TRV); see Chap. 8. In medium voltage cubical-type or metal-clad circuit breakers, as the switching resistors are not integral to the breakers, two breakers can be used, as shown in Fig. 6-37, to preinsert the resistor for a short duration.

Figure 6-38 shows a basic circuit of resistance switching. A resistor *r* is provided in parallel with the breaker pole and *R*, *L*, and *C* are the system parameters on the source side of the break. Consider the current loops in this figure. The following equations can be written:

$$u_n = iR + L\frac{di}{dt} + \frac{1}{C}\int i_c\,dt \tag{6-39}$$

$$\frac{1}{C}\int i_c\,dt = i_r r \tag{6-40}$$

$$i = i_r + i_c \tag{6-41}$$

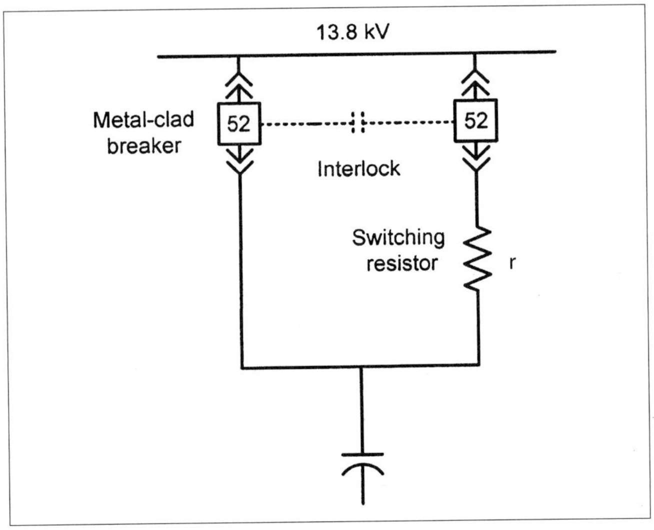

FIGURE 6-37 Two metal-clad indoor breakers organized for resistance switching.

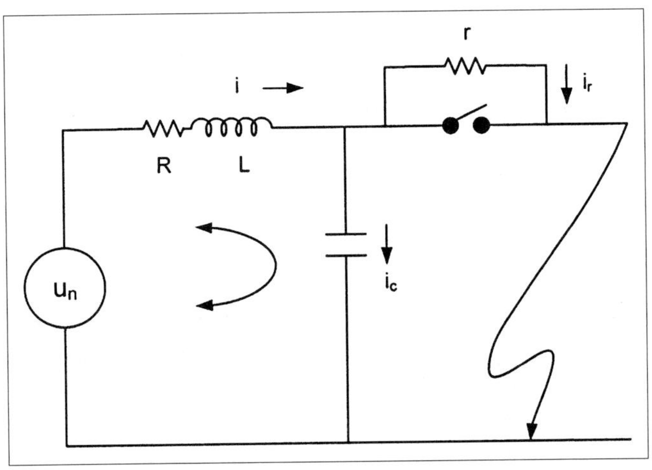

FIGURE 6-38 Diagram to explain resistance switching.

This gives:

$$\frac{d^2i_r}{dt^2}+\left(\frac{R}{L}+\frac{1}{rC}\right)\frac{di_r}{dt}+\left(\frac{1}{LC}+\frac{R}{rLC}\right)i_r=0 \tag{6-42}$$

The frequency of the transient is given by:

$$f_n=\frac{1}{2\pi}\sqrt{\frac{1}{LC}-\frac{1}{4}\left(\frac{R}{L}-\frac{1}{rC}\right)^2} \tag{6-43}$$

In power systems, R is << L. If a parallel resistor across the contacts of value $r<\frac{1}{2}\sqrt{L/C}$ is provided, the frequency reduces to zero. The value of r at which frequency reduces to zero is called the critical damping resistor. The critical resistance can be evaluated in terms of the system short-circuit current, I_{sc}:

$$r=\frac{1}{2}\sqrt{\frac{u_n}{I_{sc}\omega C}} \tag{6-44}$$

The same result could have been arrived at directly from Chap. 2.

Again reverting to the switching of 6-Mvar capacitor bank, and substituting the values, the transient disappears for a resistance of 1.32 Ω.

The simulation is shown in Figure 6-39*a* and *b* for current and voltage, respectively. A 1.32-Ω resistor is inserted for four cycles only. This clearly shows the effect of resistance switching on damping out the switching inrush transients. The current and voltage waveforms are at fundamental frequency; only minor excursions are visible.

6-13-2 Point-of-Wave Switching or Synchronous Operation

In all the examples, the transients are calculated with the switch closed at the peak of the voltage wave. A breaker can be designed to open or close with reference to the system voltage sensing and zero crossing.[17,18] An example of switching a transmission line with synchronous closing is discussed in Chap. 7. The switching device must have enough dielectric strength to withstand system voltage until its contacts touch on a closing operation. The consistency of closing within ± 0.5 ms is possible. Grounded capacitor banks are closed with three successive phase-to-ground voltages reaching zero, for example, 60° separations. Ungrounded banks are controlled by closing the first two phases at a phase-to-phase voltage of zero, and then delaying the third phase 90°, when phase-to-ground voltage is zero. The results obtained approximate resistance switching. See Chap. 7 for an EMTP simulation of synchronous closing.

6-13-3 Surge Arresters

Chapter 20 is devoted to surge arresters. Here it can be said that gap-type surge arresters in series with nonlinear resistors can be exposed to high stresses when protecting capacitor banks. If a transient triggers the arrester, the capacitors will discharge totally and the energy is dissipated in the arresters. Conversely, gapless metal oxide arresters will not discharge the capacitors below the system-rated voltage, as there is smooth transition from conducting to insulating condition.

The possibility of overvoltages due to lightning, switching surges, and temporary overvoltages requires a detailed evaluation before the application of surge arresters close to a capacitor bank.[19]

On an incoming surge, the capacitors will absorb the charge, not much dependent on the rate of rise of the incoming voltage. Shunt capacitor banks have low surge impedance and to an extent may be self-protecting in case there are other surge arresters properly rated for lightning surge duty. Depending on the substation configurations, additional surge arresters may be necessary.

The overvoltage protection should be considered at the following locations:

- On the switching circuit breakers to control the TRV when the shunt capacitors are switched out. See Chap. 8.
- At the primary of transformer to limit phase-to-phase voltage due to capacitor switching. See Chap. 14. Also, on the secondary of the transformer windings close to the switched capacitors, an example of which has been discussed in this chapter.
- At the end of transformer-terminated lines to limit the phase-to-phase overvoltages due to capacitor switching.
- On neutrals of ungrounded capacitor banks.
- On coupled low-voltage systems; Example 6-6 given in this chapter.

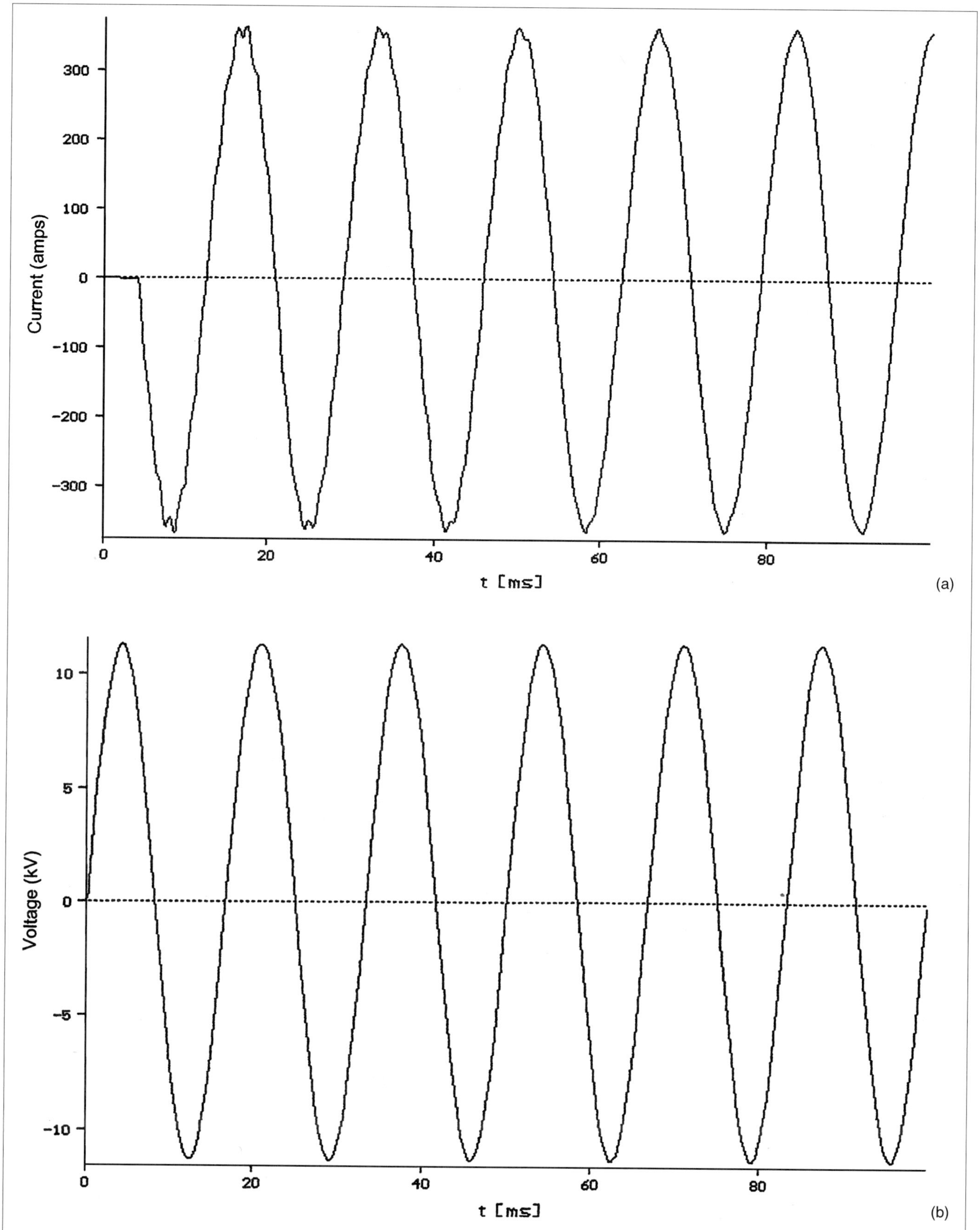

FIGURE 6-39 (*a*) Capacitor switching current with resistor switching, Fig. 6-1. (*b*) 13.8-kV bus voltage with resistance switching.

6-14 SHUNT CAPACITOR BANK ARRANGEMENTS

Shunt capacitor banks are formed from standard unit capacitor sizes available in certain kvar and voltage ratings. The unit sizes are generally limited to 100, 200, 300, and 400 kvar, and their voltage ratings are: 21.6, 19.92, 14.4, 13.8, 13.28, 12.47, 9.96, 9.54, 8.32, 7.96, 7.62, 7.2, 6.64, and so on, all in kV. To form a capacitor bank at a certain system voltage, a number of units are required to be connected in series and parallel combination, as shown in Fig. 6-40.

A capacitor bank for 500 kV can be formed with the following alternative capacitor units:

- 14 series strings of 21.6-kV capacitor units are required.
- 20 series strings of 14.4-kV capacitor units are required.
- 30 series strings of 9.54-kV capacitor units are required.
- 38 series strings of 7.62-kV capacitor units are required.

There are imitations on the formation of series strings and the number of capacitor units in a parallel string. Figure 6-40 shows that each capacitor unit is protected by a fuse. A minimum number of capacitors in parallel should be placed per series group to limit the overvoltages on the remaining units to 110 percent, in case one capacitor goes out, say due to operation of its fuse. Table 6-2 shows the minimum number of units in parallel per series group to limit voltage on remaining units to 110 percent with one unit out. Table 6-3 provides analytical expressions for the calculations of fault current and voltage on each remaining group in series with one group faulted.

Individual capacitor fusing is selected to protect the rupture/current rating withstand capability of the capacitor can, and the fuse operates to prevent a rupture of the capacitor can. The maximum number of capacitors that can be connected in parallel in a series group is dictated by the energy liberated and fed into a fault when a fuse operates. Figure 6-41 depicts that for a fault in one of the parallel units, the stored energy in the parallel capacitor units is fed into the fault and the protecting fuse must be able to handle this energy, depending on its characteristics. This energy release may be approximately calculated from the following equations:

$$E = 2.64\ \text{J/kvar}_c \text{ rated voltage} \tag{6-45}$$

$$E = 2.64(1.10)^2\ \text{J/kvar}_c\ 110\% \text{ voltage} \tag{6-46}$$

$$E = 2.64(1.20)^2\ \text{J/kvar}_c\ 120\% \text{ voltage} \tag{6-47}$$

Normally with expulsion-type fuses, commonly used for outdoor rack-mounted capacitor banks, 3100 kvar of capacitors in

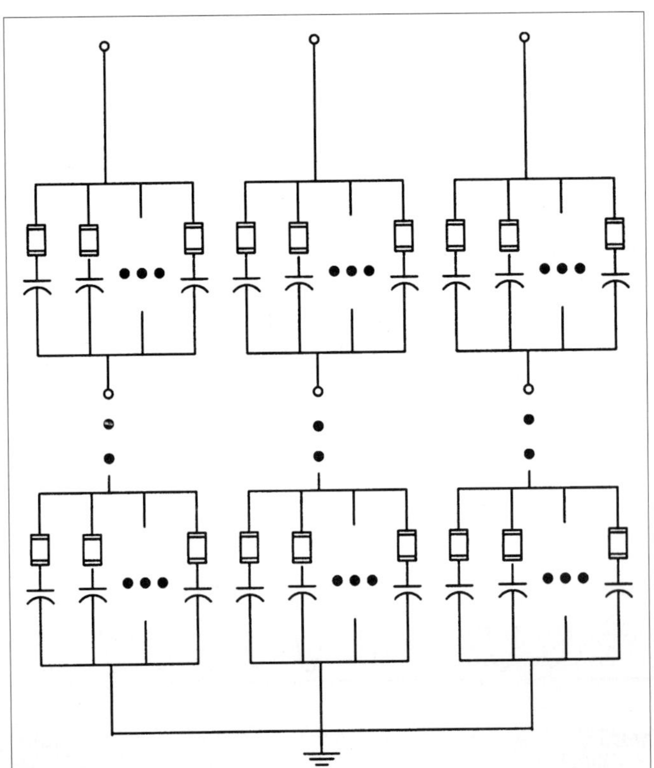

FIGURE 6-40 Typical arrangement of the formation of a high-voltage capacitor bank.

TABLE 6-2 Minimum Number of Units in Parallel per Series Group to Limit Voltage on the Remaining Units to 110% with One Unit Out

Number of Series Groups	Grounded Wye or Delta	Ungrounded Wye	Double-Wye, Equal Sections
1	—	4	2
2	6	8	7
3	8	9	8
4	9	10	9
5	9	10	10
6	10	10	10
7	10	10	10
8	10	11	10
9	10	11	11
10	10	11	11
11	10	11	11
12 and above	11	11	11

TABLE 6-3 60-Hz Fault Current and Voltages with One Unit Shorted

Bank Configuration	Fault Current	Voltage on Each Remaining Group in Series with Faulted Group
Wye, grounded	$\frac{S}{S-1}I_\phi{}^*$	$\frac{V_{LG}}{S-1}$
Single, ungrounded	$\frac{3S}{3S-2}I_\phi$	$\frac{3V_{LG}}{3S-2}$
Double, ungrounded	$\frac{6S}{6S-5}I_\phi$	$\frac{6V_{LG}}{6S-5}$

S = Number of series groups.
I_ϕ = Nominal phase current, A.
*For $S = 1$, the current is the system line-to-ground fault current.
V_{LG} = Line-to-ground voltage V. (Maximum value is used where appropriate).

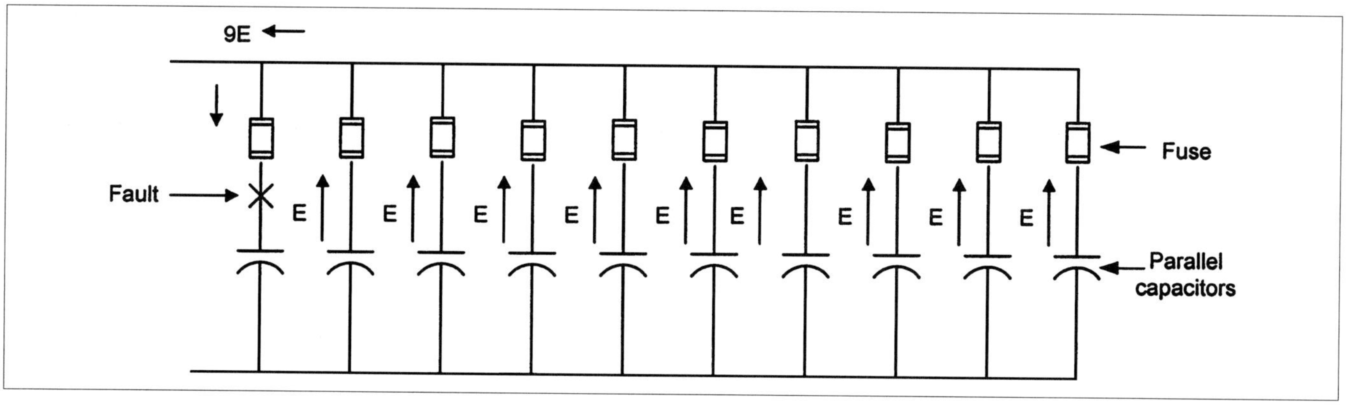

FIGURE 6-41 Limitation of number of capacitor units in parallel with expulsion-type fuses.

total can be connected in a parallel group, that is, a maximum of ten 300 kvar units. When current-limiting fuses are used, this limit can be increased.

6-14-1 Connections and Grounding of Capacitor Banks

The capacitor banks can be connected in a variety of three-phase connections, depending on their size, optimum utilization of the capacitor units, fusing, and protection. The following are the common connection methods:

- Multiple series groups, grounded wye connection
- Ungrounded wye connections
- Grounded double-wye neutrals
- Delta connections

These connections are shown in Fig. 6-42. Delta connection is common for the low-voltage systems, with one series group rated for line-to-line voltage. A wye-ungrounded group can be formed with one series group per phase when the required operating voltage corresponds to standard capacitor unit rating. For example, for 13.8-kV service voltage, one series group is adequate with capacitors rated for 7.96 kV ($\sqrt{3} \times 7.96 = 13.8$ kV). Grounded wye neutrals and multiple series groups are required for higher voltages above 35 kV. As a rule of thumb, when the system is effectively grounded, the capacitor bank neutral is grounded. The grounded capacitor banks have the following characteristics:

- These provide a low-impedance path to ground for the harmonic currents. The resulting harmonic currents may cause interference with the communication systems if the power lines are paralleled with the communication circuits.
- An open phase produces zero sequence currents which may cause ground fault relay operation.
- Third-harmonic resonance could be a problem.

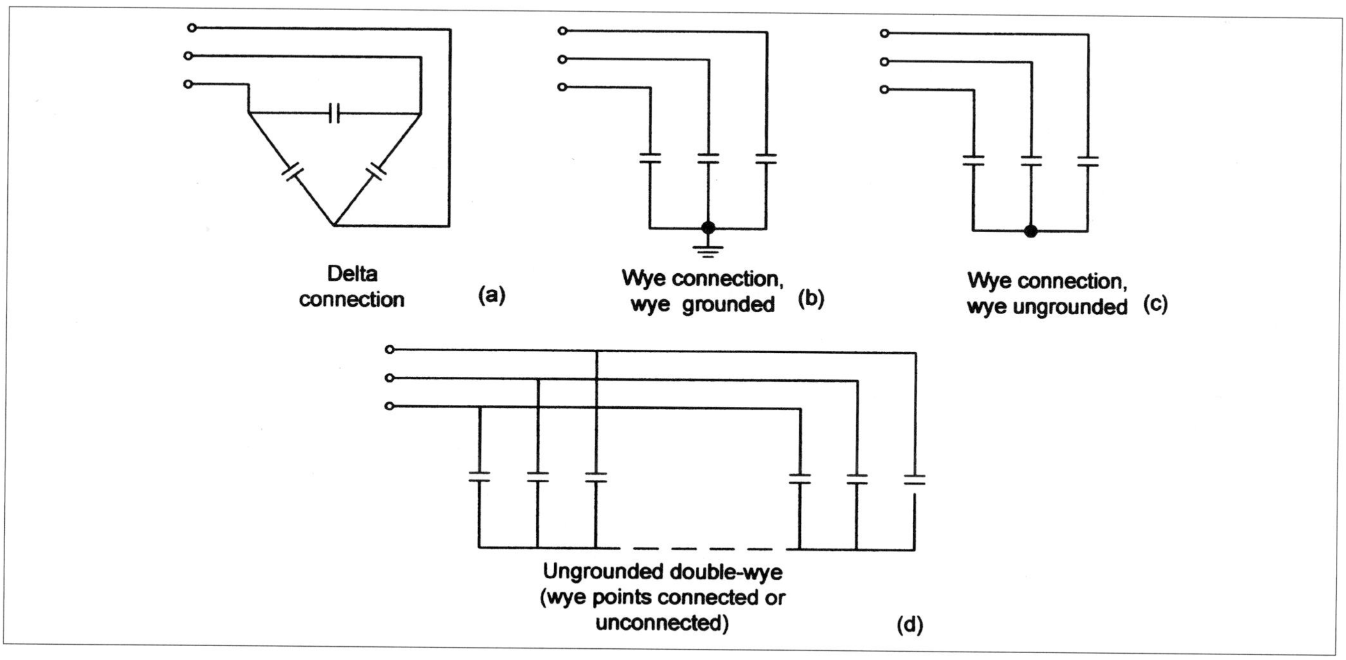

FIGURE 6-42 Grounding arrangements of three-phase capacitor banks.

- As the neutral is grounded, the recovery voltages on current interruption are reduced.

- A low-impedance path is provided to lightning surge currents and gives some protection from the surge voltages. Sometimes, the bank can be operated without surge arresters.

On the other hand, the ungrounded banks have the following characteristics:

- Banks do not provide any surge voltage protection and provide no path to ground for third-harmonic currents.

- The entire bank, including the neutral, must be insulated for line-to-line voltage.

When a bank is too large to meet the requirements of 3100 kvar per group as a maximum for the application of expulsion-type fuses or large enough to meet the minimum series group requirements, the bank can be split into two wye sections. The two neutrals are usually ungrounded.

PROBLEMS

1. A distribution system is shown in Fig. 6-P1. Calculate the voltage and current transients on sudden switching of 10-Mvar capacitor bank, when the bus section circuit breaker is closed. Ignore the nonlinear load shown in dotted lines.

2. Repeat Problem 1 with bus section circuit breaker open.

3. In Fig. 6-P1, the 10-Mvar capacitor bank is divided into two banks of 5 Mvar each. If the total inductance between the banks is 15 μH, calculate the inrush current and its frequency on energizing a 5.0-Mvar bank when the other 5.0-Mvar bank is in service.

4. In Problem 3, calculate the inductance to drop the inrush current and frequency of inrush current to 15 kA and 2000 Hz, respectively.

5. Find the value of R for resistance switching in Problem 1 with bus section circuit breaker open, and then repeat with the bus section circuit breaker closed.

6. Ignoring the nonlinear load, what is the Thévenin impedance of the system, as seen from 13.8-kV bus A? Use complex $R + jX$ calculation to combine the component impedances.

7. Consider that the nonlinear load shown within dotted lines has a harmonic current spectrum, as shown in Table 6-1. Find the current in the capacitor bank at each of the harmonics 5th, 7th, and 11th.

8. Calculate the current in the Thévenin impedance at each of the harmonics 5th, 7th, and 11th. From this calculation, find the harmonic current injected into 138-kV utility system at each of the harmonics.

9. Why can the power factor of an induction motor not be corrected to unity? Explain self-excitation and what gives rise to it. How it can be avoided?

10. Form a 58-Mvar, 34.5-kV, neutral grounded, wye-connected capacitor bank. Select a number of units in a parallel in each group and total number of series units in parallel, using the guidelines given in this chapter.

11. In Problem 10, calculate the fault current when one unit is shorted. Calculate the voltage in each remaining group in series with faulted group.

12. Figure 6-13 does not show the voltage waveform of 4.16-kV bus. Will it be the same as in Fig. 6-8? Will it be much improved because of filter?

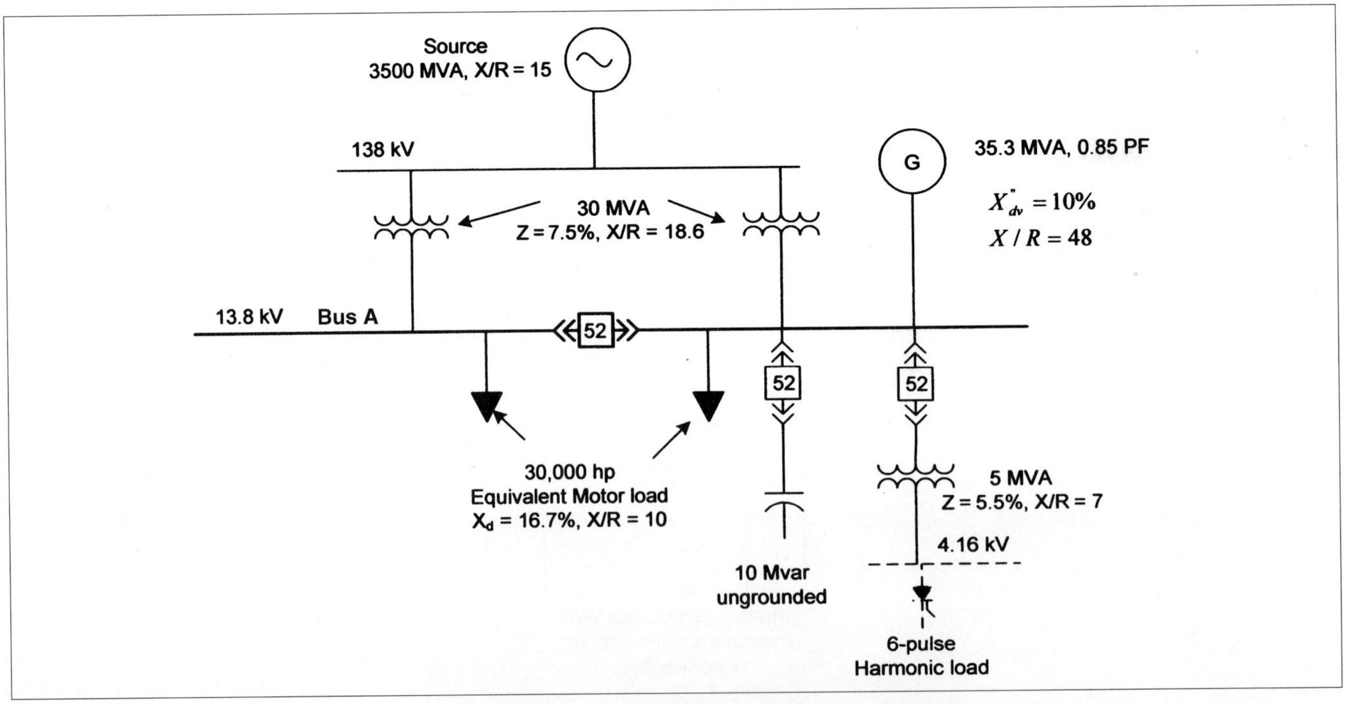

FIGURE 6-P1 A power system configuration for problems.

REFERENCES

1. J. F. Witte, F. P. DeCesaro, and S. R. Mendis, "Damaging Long Term Overvoltages on Industrial Capacitor Banks Due to Transformer Energization Inrush Currents," *IEEE Trans. Industry Applications*, vol. 30, no. 4, pp. 1107–1115, July/Aug. 1994.
2. ANSI/IEEE Std. C37.06-1987 (R2000), AC High Voltage Circuit Breakers Rated on a Symmetrical Current Basis—Preferred Ratings and Related Required Capabilities.
3 ANSI/IEEE Std. C37.012-1979 (R2000), IEEE Application Guide for Capacitance Current Switching for AC High Voltage Circuit Breakers Rated on a Symmetrical Current Basis.
4. H. M. Pflanz and G. N. Lester, "Control of Overvoltages on Energizing Capacitor Banks," *IEEE Trans. PAS*, vol. 92, no. 3, pp. 907–915, May/June 1973.
5. R. S. Bayless, J. D. Selmen, D. E. Traux, and W. E. Reid, "Capacitor Switching and Transformer Transients," *IEEE Trans. PWRD*, vol. 3, no. 1, pp. 349–357, Jan. 1988.
6. IEEE Report by Working Group 3.4.17, "Surge Protection of High Voltage Shunt Capacitor Banks on AC Power Systems—Survey Results and Application Considerations," *IEEE Trans. PWRD*, vol. 6, no. 3, pp. 1065–1072, July 1991.
7. IEEE Std. 1036-1992, IEEE Guide for Application of Shunt Power Capacitors, 1992.
8. J. C. Das, "Passive Filters—Potentialities and Limitations," *IEEE Trans. Industry Applications*, vol. 40, no.1, pp. 232–241, Jan./Feb. 2004.
9. IEEE Std. 519, IEEE Recommended Practice and Requirements for Harmonic Control in Electrical Power Systems, 1992.
10. Zabrosky and J. W. Rittenhouse, "Fundamental Aspects of Some Switching Overvoltages in Power Systems," *AIEE*, pp. 822–830, Feb 1963.
11. R. A. Jones and H. S. Fortson, Jr., "Considerations of Phase-to-Phase Surges in the Application of Capacitor Banks," *IEEE Trans. PWRD*, vol. PWRD-1, no. 3, pp. 240–244, July 1986.
12. V. E. Wagner, J. P. Staniak, and T. L. Oreloff, "Utility Capacitor Switching and Adjustable Speed Drives," *IEEE Trans. Industry Applications*, vol. 27, no. 4, pp. 645–652, July/Aug. 1991.
13. NEMA MG-1, Generators and Motors, 2003.
14. J. C. Das, "Analysis and Control of Large Shunt Capacitor Bank Switching Transients," *IEEE Trans. Industry Applications*, vol. 41, no. 6, pp. 1444–1451, Nov./Dec. 2005.
15. M. Gelinkowski, A. Greenwood, J. Hill, R. Mauro, and V. Varneckes, "Capacitance Switching with Vacuum Circuit Breaker—A Comparative Evaluation," *IEEE Trans. PWRD*, vol. 6, no.3, pp. 1088–1094, July 1991.
16. S. F. Farag and R. G. Bartheld, "Guidelines on the Application of Vacuum Contactors," *IEEE Trans. Industry Applications*, vol. IA-22, no. 1, pp. 102–108, Jan./Feb. 1986.
17 R. W. Alexander, "Synchronous Closing Control for Shunt Capacitor Banks," *IEEE Trans. PAS*, vol. 104, no. 9, pp. 2619–2626, Sep. 1985.
18 J. H. Brunke and G. G. Schockelt, "Synchronous Energization of Shunt Capacitors at 230 kV," *IEEE Trans. PAS*, vol. 97, no. 4, p. 1009, July/Aug. 1978.
19. M. P. McGranaghan, W. E. Reid, S. W. Law, D. W. Gresham, "Overvoltage Protection of Capacitor Banks," *IEEE Trans. PAS*, vol. 103, no. 8, pp. 2326–2336, Aug. 1984.
20. IEEE. Std-18, IEEE Standard for Shunt Power Capacitors, 2002.
21. IEEE. Plo36/D 13a, Draft Guide for Application of Shunt Power Capacitors, Sept. 2006.

FURTHER READING

E. W. Boehne and S. S. Low, "Shunt Capacitor Energization with Vacuum Interrupters—a Possible Source of Overvoltage," *IEEE Trans. PAS*, vol. 88, no. 3, pp. 1424–1443, Sep. 1969.

J. C. Das, "Effects of Medium Voltage Capacitor Bank Switching Surges in an Industrial Distribution System," in Conf. Record, *IEEE IC & PS Conf.* Pittsburgh, pp. 57–64, 1992.

A. A. Girgis, C. M. Fallon, J. C. P. Robino, and R. C. Catoe, Jr., "Harmonics and TRV Due to Capacitor Switching," *IEEE Trans. Industry Applications*, vol. 29, no. 6. pp. 1184-1189. Nov./Dec. 1993.

A. Greenwood, *Electrical Transients in Power Systems*, John Wiley & Sons, New York, 1994.

IEEE Working Group on Capacitance Switching, "Bibliography on Switching of Capacitance Circuits Exclusive of Series Capacitors," *IEEE Trans. PAS*, vol. 89, no. 6, pp. 1203–1207, July/Aug. 1970.

M. McGranagham, W. E. Reid, S. Law, and D. Gresham, "Overvoltage Protection of Shunt Capacitor Banks Using MOV Arresters," *IEEE Trans. PAS*, vol. 104, no. 8, pp. 2326–2336, Aug. 1984.

M. F. McGranaghan, R. M. Zavadil, G. Hensley, T. Singh, and M. Samotyj, "Impact of Utility Switched Capacitors on Customer Systems—Magnification at Low-Voltage Capacitors," *IEEE Trans. PWRD*, vol. 7, no. 2, pp. 862–868, April 1992.

R. P. O' Leary and R. H. Harner, "Evaluation of Methods for Controlling the Overvoltage Produced by Energization of Shunt Capacitor Banks," CIGRE 1988, Session Paper No. 13-05.

CHAPTER 7

SWITCHING TRANSIENTS AND TEMPORARY OVERVOLTAGES

In Chap. 4, the reflection, refraction, and transmission factors for various line terminations are calculated. Also, Ferranti effect, line compensation, and some EMTP simulations of surge voltages on lines and cables are examined. This investigation is continued in this chapter. Surges occur because of sudden interaction of the trapped charge with the system voltage. Sometimes the switching surges are superimposed upon short-term power frequency overvoltage. The charge is trapped in the line capacitance which is distributed along the length of the line. This introduces 168complexity in determining the shape and peak magnitude of the switching surges due to reflections, similar to lightning surges. Underground cables have a higher capacitance than overhead lines and the switching surges of overhead lines terminated in cables or vice versa become important. Akin to the switching of capacitors, most severe surges occur from restrikes and reclosing when the supply waveforms have reversed. An important factor of distributed capacitance and inductance of the overhead lines is that surges originating at a point take time to travel along a line and get reflected and re-reflected at the impedance discontinuities. Consequently, the wave shapes of the line-switching surges are quite different from the switching of lumped elements.

To calculate the maximum switching surge voltage, it is necessary to determine the magnitude of the initial surge and then follow it with appropriate delays and reflections at each discontinuity. The analysis can be simplified by EMTP and other computer programs, like TLTA.[1]

7-1 CLASSIFICATION OF VOLTAGE STRESSES

The voltage stresses in the electrical systems can be classified as follows:

- Continuous power frequency voltages vary in the power systems, and the limits are defined in the standards. For insulation coordination, maximum system voltages are used (Chap. 17).
- Lightning (fast front) overvoltages, caused by lightning (Chap. 5).
- Switching (slow front) overvoltages, caused by switching, fault initiation, or remote lightning strikes (this chapter and Chap. 8).
- Temporary overvoltages, caused by faults, load rejection, line energizing, resonant conditions, ferroresonance (this chapter and Chaps. 8, 9, and 14).
- Very fast front overvoltages usually associated with high-voltage disconnect switch operation and GIS, and cable connected motors (Chap. 18).

A diagrammatic representation of this classification is shown in Fig. 7-1 with relative amplitudes of the voltages. This figure does not show the time duration for which the overvoltage condition can last. Also the shape of waveforms in each case can vary. Thus, the figure is a simplified diagrammatic representation of the relative magnitudes. Overvoltages sustained for a longer period are called temporary overvoltages. This has some technical basis, in the sense that the breakdown of insulation is not only a function of the electrical stress to which it is subjected, but also of the time duration for which this condition lasts. A surge arrester will limit the transient overvoltage to its protective level (Chap. 20), but may not impact the temporary overvoltages of lower magnitude, though resonant conditions must be considered. Surge arresters can be applied to counteract the resonant conditions.

7-2 MAXIMUM SYSTEM VOLTAGE

The voltages in a power system and to the consumers must be maintained within certain narrow limits. The electrical systems generally operate at 5 to 10 percent below the maximum system voltage for which the system components are designed. The system operating voltage is regulated by the utilities, and a 5 percent overvoltage limit seems appropriate. This may be exceeded in some cases, that is, close to a generating station.

The *maximum system voltage* is defined as the highest system voltage for which equipment and other components are designed for satisfactory continuous operation without any derating. *Nominal system voltage* is defined as the voltage by which a portion of the system is designated and to which certain operating characteristics of the system are related. Each nominal system voltage pertains to a portion of the system bounded by transformers or utilization equipment.

The term low voltage embraces voltages up to 1000 V, medium voltage from 1000 V to 100 kV, and high voltage 100 to 230 kV.

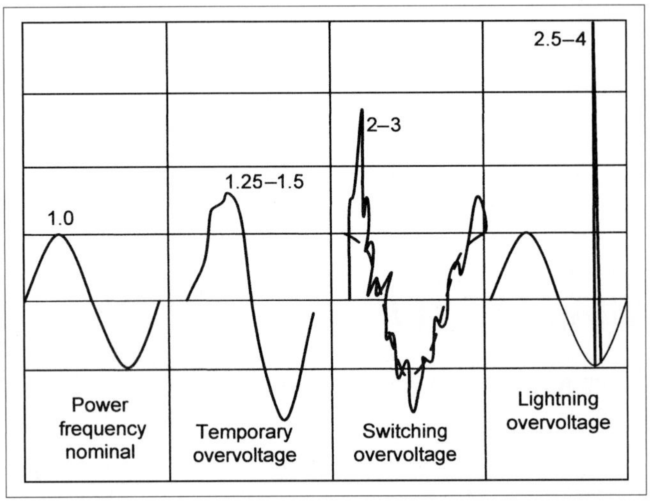

FIGURE 7-1 Relative magnitudes of temporary, switching, and lightning overvoltages—a generalization.

Extra high voltage (EHV) is 345 to 765 kV and ultrahigh voltage above 765 kV.

ANSI standard[2] defines two sets of tolerance limits: range A, which specifies the limits under most operating conditions, and range B, which specifies minor excursions outside range A. Corrective action should be taken when the voltage is in range B to restore it to range A. For transmission voltages over 345 kV, only the nominal and maximum voltages are specified, that is, 362 kV, which is higher by approximately 5 percent. As an example for 13.8-kV nominal voltage, range A = 14.19 to 12.46 kV, and range B = 14.5 to 13.11 kV. The control and regulation of the voltages is important for the operations of the electrical equipment, and the analysis of the steady-state voltage profiles in the power system are analyzed by load-flow methods. For the application of the surge arresters, maximum system voltages are considered.

7-3 TEMPORARY OVERVOLTAGES

The overvoltages can be classified with respect to the system operation. The overvoltages produced by the following operations may be called temporary overvoltages:

1. Fault overvoltages, generally produced due to ground faults. The system grounding plays an important role (Chaps. 9 and 21). The significance of the coefficient of grounding and earthling factor are discussed in Chap. 9 with a rigorous calculation of coefficient of grounding (COG) using the method of symmetrical components. In a four-wire system, with neutral grounded at multiple points, the fault current will divide between neutral conductor and earth; the accuracy of a symmetrical component method for calculations of overvoltages in multiple-grounded systems is discussed in Chap. 19.

2. Load rejection overvoltages, which originate when a loaded system is suddenly unloaded, say, due to load rejection. A system with relatively short lines and high short-circuit power will have low overvoltages, while a system with long lines and low short-circuit power will have high overvoltages. On a sudden load rejection, the generators will speed up and the voltage will rise. The speed governors and automatic voltage regulators will act and tend to limit this voltage rise. A rigorous calculation will require modeling of generator, governor, regulator, and transmission-line parameters. For turbo generators, the maximum speed rise on full load rejection is approximately 10 percent, and it occurs in less than 1 s. In water wheel generators the maximum speed rise on full load rejection can be as high as 60 percent, and it may take 10 s to reach it.

3. When a generator is suddenly off-loaded, say due to isolation of a load, the voltage will rise. A rigorous study with a transient generator model is included in Chap.12. For an approximate calculation it can be assumed that voltage behind subtransient reactance of the generator remains unchanged, but after a few cycles, the voltage E_d' behind transient reactance becomes the driving voltage. This gives the following relation:

$$V_1 = \left(\frac{f}{f_0}\right)\frac{E_d'}{\left[1-\frac{fX_s}{f_0X_c}\right]} \tag{7-1}$$

where X_s is the reactance between E'_d and V_1, the sending end voltage, X_c, is the capacitive input reactance of the open-circuited line at the increased frequency, and f/f_0 is the ratio of the instantaneous frequency at the time maximum voltage is reached and the supply system frequency. The problem occurs in estimating f.

4. Resonance and ferroresonance overvoltages arise because of interaction of capacitive elements and inductive elements having nonlinear characteristics, for example, transformers, Chapter 14.

5. Overvoltages due to transformer energizing (Chap. 14).

6. Some other causes of overvoltages are: ferranti effect, discussed in Chap. 4 (also see Example 4-2), backfeed through interconnected transformer windings, harmonic overvoltages. The overvoltages usually precede the transient overvoltage-causing event, for example, a switching operation.

Another classification of the temporary overvoltages is based upon the frequency characteristics. In Table 1-3, the temporary overvoltages have been classified in to Group 1 (0.1 to 3 kHz, low-frequency oscillations) and switching overvoltages into Group II (50/60 to 20 kHz, slow front surges). Also consider:

- Overvoltages with a frequency of oscillation equal to the power frequency. Here the overvoltage is sinusoid or almost a sinusoid, devoid of harmonics. Both types of overvoltages, appearing phase-to-phase and phase-to-ground are considered. Temporary overvoltages arising under linear conditions belong to this category.

- Overvoltages with a frequency of oscillation higher than the power frequency. Overvoltages due to harmonics can only arise if there is nonlinearity present in the system. This nonlinearity can result from power electronics, saturated magnetic characteristics of transformers, shunt reactors, and measurement transformers.

- Overvoltages with a frequency of oscillation lower than the power frequency (Chap. 12.)

IEC[3] classifies slow front overvoltages as the voltages arising from:

- Line energization and re-energization

- Faults and fault clearing
- Load rejection
- Switching of capacitive and inductive circuits
- Distant lightning strokes to the conductor of overhead lines

The probability distribution of the overvoltages without surge arresters is characterized by its 2 percent value, its deviation, and truncation value. This can be approximated with gaussian distribution with a 50 percent value at truncation, above which no values are assumed to exist (Chap. 17).

7-4 SWITCHING SURGES

Ignoring the distinction between temporary overvoltages and switching overvoltages, some mechanisms of generating these are:

- Charging an unloaded line, which has not been charged before, and open at far end, as studied in Chap. 4.
- Switching lines terminated in transformers.
- Series capacitor compensated lines, static condenser (STATCON), SVCs, and other flexible ac transmission system (FACT) devices, which are discussed in Chap. 15.
- Auto-reclosing.
- Load shedding.
- VHF overvoltages in GIS (Chap. 15).
- Disconnection of large inductances, unloaded transformers, reactors, and the like.
- Flashovers of longitudinal insulation configuration. This is an insulation configuration between terminals belonging to the same phase, but temporarily separated into two independently energized parts, for example, open switching devices with voltages on source and load sides.
- Short circuit and fault clearance, that is, when phase-to-phase or phase-to-ground faults are cleared.
- Single-pole closing of circuit breakers.

The surges produced by all these operations vary in severity and magnitude and are not equally dangerous to the system insulation. Compared to interrupting the low magnetizing currents of the transformers and disconnections of shunt reactors by circuit breakers, the closing operations cause the highest overvoltages.

Overvoltages of the same peak value can be of different significance. Autoclosing of a line with trapped charge can cause high overvoltage and precipitate a new fault. This can be avoided by reclosing with a longer pause or using appropriate switching devices. Faults and switching operations not only cause transients of a higher frequency, but also a change in the power frequency. Both the transient and the power frequency overvoltage define a *switching transient*. Switching overvoltages are damped by leakage resistance and conductivity. Generally 5 percent damping seems appropriate during the first half wave of the transient frequency, though a detailed simulation may give different results, that is, resonance is a distinct possibility.

7-5 SWITCHING SURGES AND SYSTEM VOLTAGE

The lightning surges are responsible for approximately 10 percent of all short circuits in substations and almost 50 percent of short circuits on lines and systems above 300 kV. Less than 1 percent short circuits are caused by switching surges.[4] Switching overvoltages do not cause flashovers to the same extent as caused by lightning. The frequency, point of occurrence, amplitude, and characteristics govern the selection of equipment, insulation level, and economical design. The switching surges gain more importance as the system voltage rises, and in EHV networks, it is the switching surges which are of primary importance in insulation coordination. The switching surges have, generally, a power frequency and a transient component, which bear certain relation to each other and may be of equal amplitude in the absence of damping. The nonsimultaneous closing of breaker poles can increase the transient component. The EHV installations are primarily concerned with the stresses imposed on the insulation by switching surges and the coordination of the insulation is based upon these values. To lower the costs, it is desirable to reduce the insulation levels at high voltages. On the transmission line towers, the flashover voltage cannot be allowed to increase along with service voltage, as beyond a certain point the electrical strength of air gaps can no longer be economically increased by increasing the clearances. Thus, while the external lightning voltages are limited, the switching overvoltages become a primary concern.

Table 7-1 from ANSI/IEEE standard[5] shows the switching surge levels versus the BIL for power transformers. BIL levels commonly used are shown in bold.

7-5-1 Line insulation with respect to switching surges

The clearance between the tower and conductors influences the switching surge withstand capability (Fig. 7-2).[6] For a particular insulator suspension arrangement, that is, V-formation or vertical, there is an upper limit of the surge-withstand voltage, which cannot be much varied by increasing the conductor clearance from the tower. This occurs at 1600 kV.[7,8] The horizontal suspension arrangement provides the best conditions (Fig. 7-2). In this figure k = ratio of the maximum amplitude of switching overvoltage to the amplitude of the service voltage, the symbol (a) is for insulator strings suspended vertically, (b) for insulator strings in V-formation with 45° to vertical, and symbol (c) is for horizontal strings of insulators.

Figure 7-2 also shows rapid increase in the number of insulator elements at higher voltages. To adhere to the limit of 1600 kV, for 800-kV systems, the switching surge should be limited to two times the peak value of the maximum voltage. The values shown in Table 7-2 have been found satisfactory for systems above 362 kV. In the design of an economical tower, it is necessary to consider proximity effect of the grounded metal which accentuates the saturation characteristics of the air gap. This is illustrated in Fig. 7-3 for the center phase of a 735-kV tower; for outer phases it will be less severe.[9] Coordination must be achieved between string length and clearances to tower metal. It is usually based upon a dry positive switching impulse, with a wavefront giving maximum CFO.

Thus, measures which reduce the switching surge, that is, resistor switching and synchronous closing, are very important at higher voltages. The role of surge arrestors becomes equally important, (Chap. 20). In order to keep the withstand levels as low as possible, the surge arresters should be selected so that they operate at the maximum switching surge and discharge the lines without damage to themselves.

All equipment for operating voltages above 300 kV is tested for switching impulses. No switching surge values are assigned to transformers 69 kV and below (Table 7-1).

Table 7-3,[10] shows rated line-closing switching surge factors for circuit breakers specifically designed to control line-switching closing surge. Note that the parameters are applicable for testing with standard reference transmission lines, as shown in this table.

TABLE 7-1 Dielectric Insulation Levels for Class II Transformers

NOMINAL SYSTEM VOLTAGE (KV)	BIL (KV CREST)	CHOPPED-WAVE LEVEL (KV CREST)	SWITCHING-IMPULSE LEVEL (KV CREST)	INDUCED VOLTAGE TEST (PHASE-TO-GROUND) 1-hr LEVEL (KV rms)	ENHANCEMENT LEVEL (KV rms)	APPLIED VOLTAGE TEST LEVEL (KV rms)
1	2	3	4	5	6	7
15 and below	110	120				34
25	150	165				50
34.5	200	220				70
46	250	275				95
69	250	275				95
	300	385				140
115	350	385	280	105	120	140
	450	495	375	105	120	185
	550	605	460	105	120	230
138	450	495	375	125	145	185
	550	605	460	125	145	230
	650	715	540	125	145	275
161	550	605	460	145	170	230
	650	715	540	145	170	275
	750	825	620	145	170	325
230	650	715	540	210	240	275
	750	825	620	210	240	325
	825	905	685	210	240	360
	900	990	745	210	240	395
345	900	990	745	315	360	395
	1050	1155	870	315	360	460
	1175	1290	975	315	360	520
500	1300	1430	1080	475	550	
	1425	1570	1180	475	550	
	1550	1705	1290	475	550	
	1675	1845	1390	475	550	
765	1800	1980	1500	690	800	
	1925	2120	1600	690	800	
	2050	2255	1700	690	800	

a. For chopped wave tests, the minimum time to flashover shall be 3.0 μs, except for 110-kV BIL, in which case, the minimum time to flashover shall be 2.0 μs.
b. Although Col. 4 establishes phase-to-ground switching impulse levels, it is not always possible to test these levels on low-voltage windings.
c. Columns 5 and 6 provide phase-to-ground test levels, which would normally be applicable to wye windings. When the test voltage is to be measured phase to phase, as is normally the case with delta windings, then the levels in Col. 5 are multiplied by 1.732 to obtain the requird phase-to-phase induced voltage test level.
d. The applied voltage test is not applicable to wye-windings terminals, unless they have been specified to be suitable for ungrounded systems.
e. The BIL levels in bold are commonly used.
Source: ANSI/IEEE Standard.[5]

7-6 CLOSING AND RECLOSING OF TRANSMISSION LINES

The transient phenomena will vary according to the system configuration, the source type, line length, and terminations. For a line open at the far end, three types of transients on switching are shown in Fig. 7-4.

Figure 7-4*a* shows a transient which can be expected when the source is mainly inductive, for example, a single line connected to a transformer. It is single-frequency transient, and if switching takes place at maximum voltage in a phase, the transient oscillates to almost twice the value of the system voltage across the entire line length. The losses in the system will dampen the transient by some percentage.

Figure 7-4*b* shows a high-frequency transient which can be expected with infinite source impedance. This means that the system from which the line originates has a number of cables or lines connected to it, and the line being switched is not longer than the incoming lines. Many line terminations and connections at the point of origination of the line being switched means that these terminations present an overall low characteristic impedance compared to the line being switched. As a result the transient occurs at its natural frequency.

Figure 7-4*c* shows the pattern of switching transient with complex source impedance, consisting of inductance of transformers

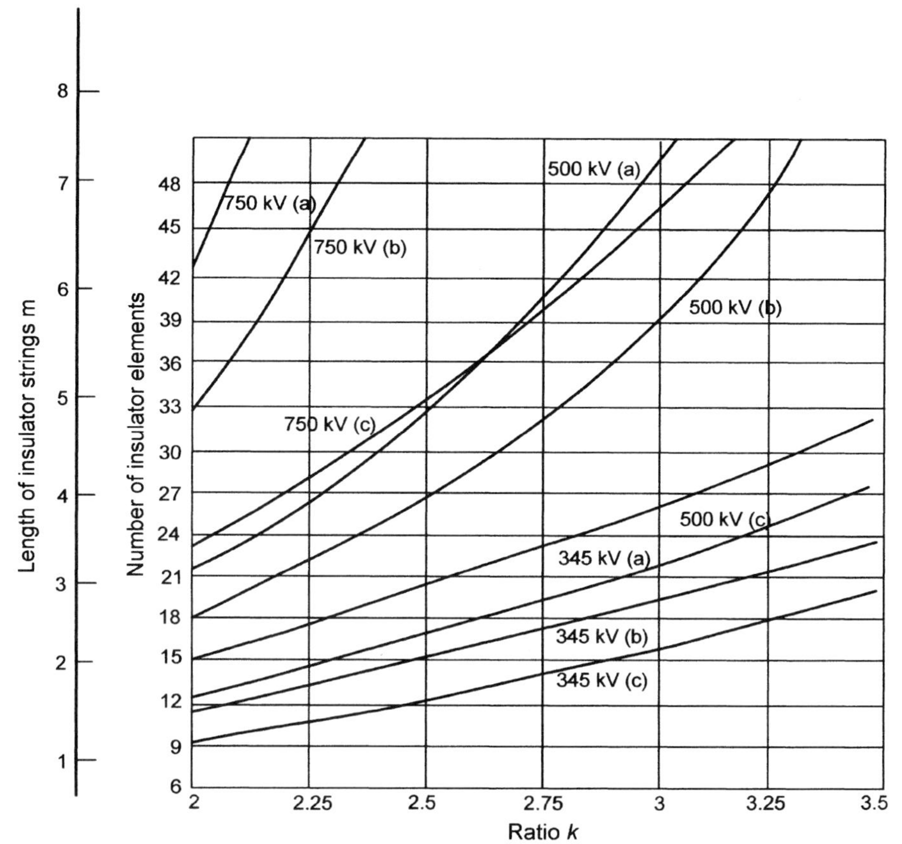

FIGURE 7-2 Relationship between the number of insulators in a string, their method of suspensions, and switching overvoltages at various voltage levels.

TABLE 7-2 Highest Permissible Switching Overvoltages with Respect to Highest System Voltage

Highest system voltage →	420	525	765	1150
Switching surge overvoltage = highest system voltage multiplied by factor →	2.5	2.25	2	1.8 or 1.9

and the surge impedance of other lines and cables feeding the system. The transient overvoltage occurs at a number of frequencies.

Highest overvoltages occur when unloaded high-voltage transmission lines are energized and reenergized and this imposes voltage stresses on circuit breakers. Figure 7-5*a* shows closing of a line of surge impedance Z_0 and length l, open at the far end. Before the breaker is closed, the voltage on the supply side of the breaker terminal is equal to the power system voltage, while the line voltage is zero. At the moment of closing, the voltage at the sending end must rise from zero to the power frequency voltage. This takes place in the form of a traveling wave on the line with its peak at u_m interacting with supply system parameters. As an unloaded line has capacitive impedance, the steady-state voltage at the supply end is higher than the system voltage, and due to Ferranti effect, the receiving-end voltage is higher than the sending end. Overvoltage factor can be defined as follows:

$$OV_{\text{total}} = \frac{u_m}{u_n} \tag{7-2}$$

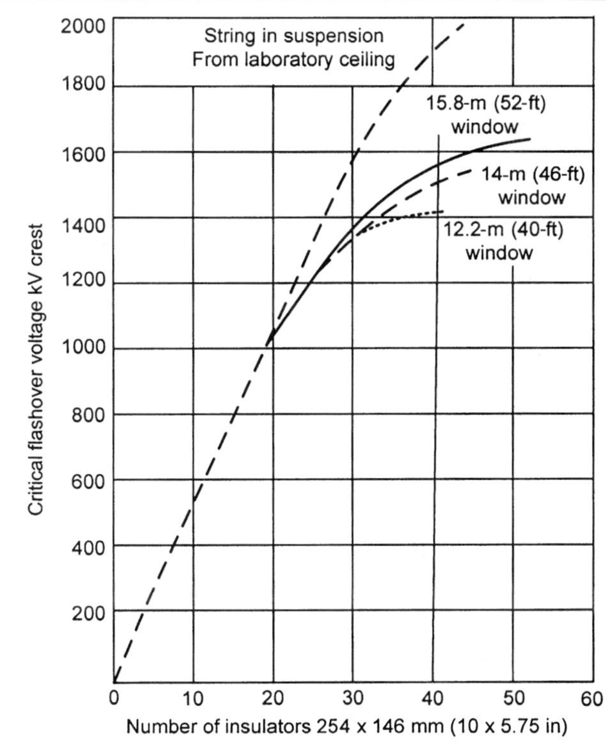

FIGURE 7-3 Proximity effect of tower window on insulation strength of single 900-V strings.[9]

TABLE 7-3 Rated Line-Closing Switching Surge Factors for Circuit Breakers, Specifically Designed to Control Line-Closing Switching Surge

Rated Maximum Voltage (kV rms)	Rated Line-Closing Switching Surge Factor	Line Length (mi)	Percentage Shunt Capacitance Divided Equally at Line Ends	L_1	L_0/L_1	R_1	R_0	C_1	C_1/C_0
362	2.4	150	0	1.6	3	0.05	0.5	0.02	1.5
500	2.2	200	0	1.6	3	0.03	0.5	0.02	1.5
800	2.0	200	60	1.4	3	0.02	0.5	0.02	1.5

Maximum voltage and parameters of standard reference transmission lines.
L_1 = Positive and negative sequence inductance in mH per mile; L_0 = zero sequence inductance in mH per mile; R_1 = positive and negative sequence resistance in ohms per mile; R_0 = zero sequence resistance in ohms per mile; C_1 = positive and negative sequence capacitance in μF per mile; C_0 = zero sequence capacitance in μF per mile.
Source: ANSI Standard.[10]

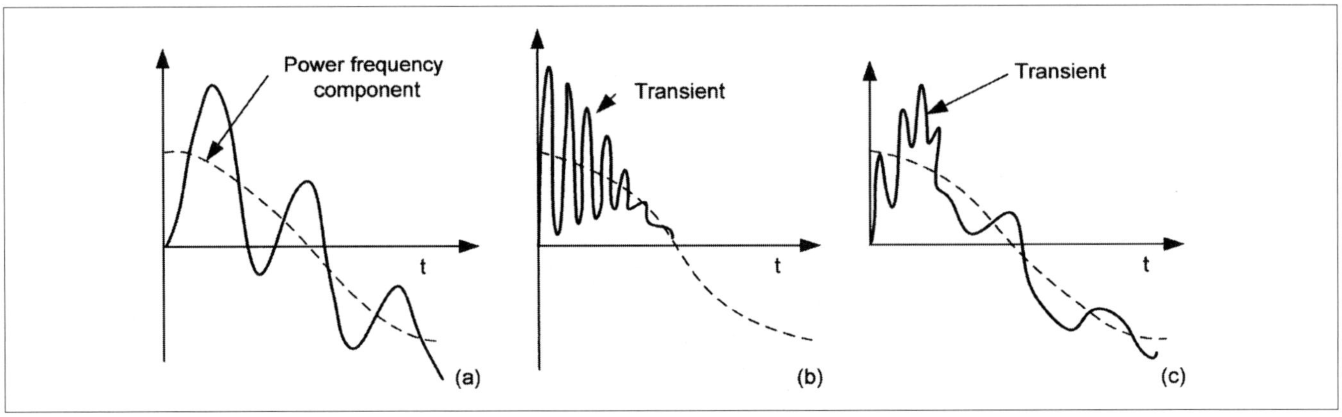

FIGURES 7-4 Profiles of switching surges on closing a line, dependent upon power system configuration.

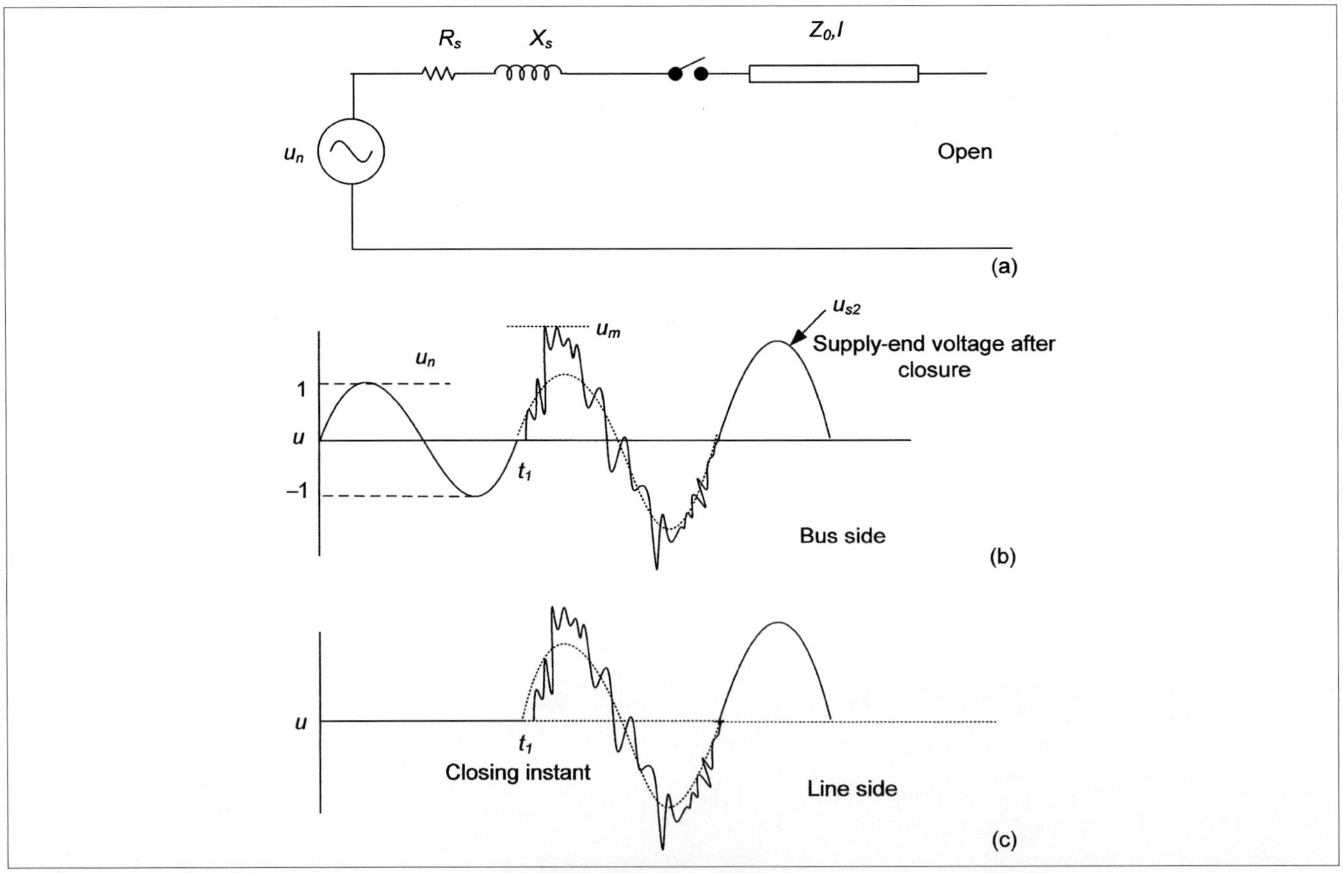

FIGURES 7-5 (*a*) Switching a line open circuited at the far end. (*b*) and (*c*) Voltage profiles on bus and line side, respectively.

where u_m is the highest peak voltage at a given point and u_n is the power-frequency voltage on the supply side of the breaker before switching. The power-frequency overvoltage factor is given by:

$$\mathrm{OV_{pf}} = \frac{u_{\mathrm{pf}}}{u_n} \tag{7-3}$$

This is the ratio of the power-frequency voltage u_{pf} after closure at a point and power-frequency voltage u_n on the supply side before closing. The transient overvoltage factor is:

$$\mathrm{OV_{tr}} = \frac{u_m}{u_{\mathrm{pf}}} \tag{7-4}$$

The power-frequency overvoltage factor can be calculated by known line parameters. This is given by:

$$\frac{1}{\cos \alpha l - X_s / Z_0 \sin \alpha l} \tag{7-5}$$

where the surge impedance and α are given by:

$$Z_0 = \sqrt{\frac{L_1}{C_1}}$$

$$\alpha = 2\pi f \sqrt{L_1 C_1} \tag{7-6}$$

The relation between sending- and receiving-end voltage is $1/\cos \alpha l$, as in Chap. 4.

This shows that the increase in power-frequency voltage depends considerably on the line length. The transient voltage is not so simple to determine and depends upon the phase angle at the closing instant, (Fig. 7-5). At instant $t = t_1$, maximum superposition of the transient and power-frequency voltages occurs.

7-6-1 Impact of trapped charges

Trapped charges occur on the transmission lines in three-pole autoclosure operations. Contact-making of three poles of a circuit breaker can be nonsimultaneous. Consider breakers at sending and receiving end of a line and a transient ground fault, which needs to be cleared by an auto-reclosure operation. The opening of the breakers at the sending and receiving ends can be nonsimultaneous. The breaker which opens later must clear two line phases at no load. These two phases can, therefore, remain charged at the peak of the power-frequency voltage, which is still present when the closure takes place. This means that after the dead time, one breaker has to close with two phases still charged. If the closing instant happens to be such that the trapped charge and the power frequency voltage are of opposite polarity, maximum transient overvoltage will occur.

Figure 7-6 shows the closing and reclosing transient of a 400 kV line, no surge arresters and no compensation. Figure 7-6*a* shows closing without a prior charge, while Fig. 7-6*b* shows closing with a trapped charge. Note that the overvoltage factor is 3.5 with line precharged versus 2 when there is no charge on the line. The voltage on the line can, after reflection of the wave, theoretically rise up to three times the system voltage. When three poles of a breaker close nonsimultaneously, even higher voltages can occur.

Utilities reclose to restore power for temporary faults. In areas of low lightning incidence, it may be single-shot reclosure, as majority of faults may be permanent. In lightning-prone regions, it is common to clear the fault as many as four times. This may consist of one fast and three delayed operations, or two fast and two delayed operations. First reclosure interval may be instantaneous or 12 to 30 cycles delayed. An instantaneous reclosure may create some conflicts with distributed generation trip times, as these will not tolerate a return of voltage which may be in phase-opposition to the generation voltage. Also single-pole reclosing is done to enhance the stability limits (Chap. 12).

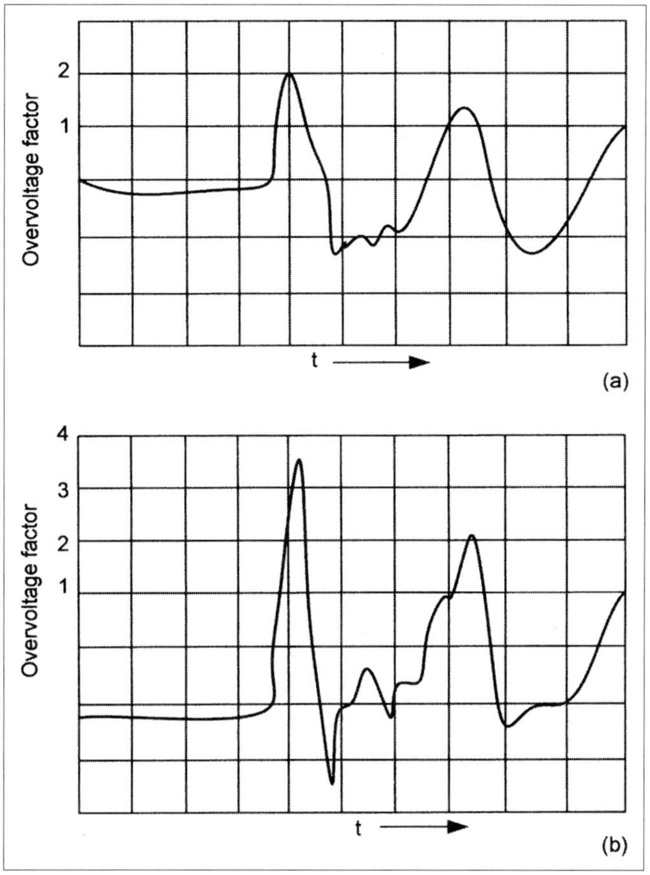

FIGURE 7-6 (*a*) Receiving-end overvoltages on a 400-kV, 400-km line, open at the far end without prior charge. (*b*) Overvoltages with prior charge.

Example 7-1 This example is an EMTP simulation of switching of a 50-mi long 138-kV line, configuration and switching options as shown in Fig. 7-7.

Figure 7-7*a* is the system configuration, the line is fed from a 230 to 138-kV step down transformer of 200 MVA, which is wye-wye connected, with wye windings solidly grounded. The percentage impedance = 10 percent and *X/R* ratio = 50. The source is modeled with positive- and zero-sequence impedances. The transformer is modeled as an ideal transformer. There are no surge arresters. A CP model for the 138-kV line is used. The calculated parameters of 138-kV line are given in Table 7-4. There are three methods of switching, which are as follows:

- Three-pole simultaneous closure, Fig. 7-7*b*
- Closure on point of wave, synchronous breakers, Fig. 7-7*c*
- Resistance switching, Fig. 7-7*d*

The switching transients in each of these closing methods are examined. First consider a three-phase closure at the peak of the voltage in phase *a*. The receiving-end and sending-end voltage transients in phase *a* are shown in Fig.7-8*a* and *b*, respectively. The receiving-end voltage rises to a maximum peak of 280 kV, phase to neutral, an overvoltage of 2.48 per unit (pu). It has superimposed transient frequency on the power frequency which decays considerably in the first four cycles. The sending-end voltage peak is 162 kV, phase to neutral, an overvoltage of 1.42 pu .The charging current is shown in Fig. 7-8*c*, peak of approximately 400 A.

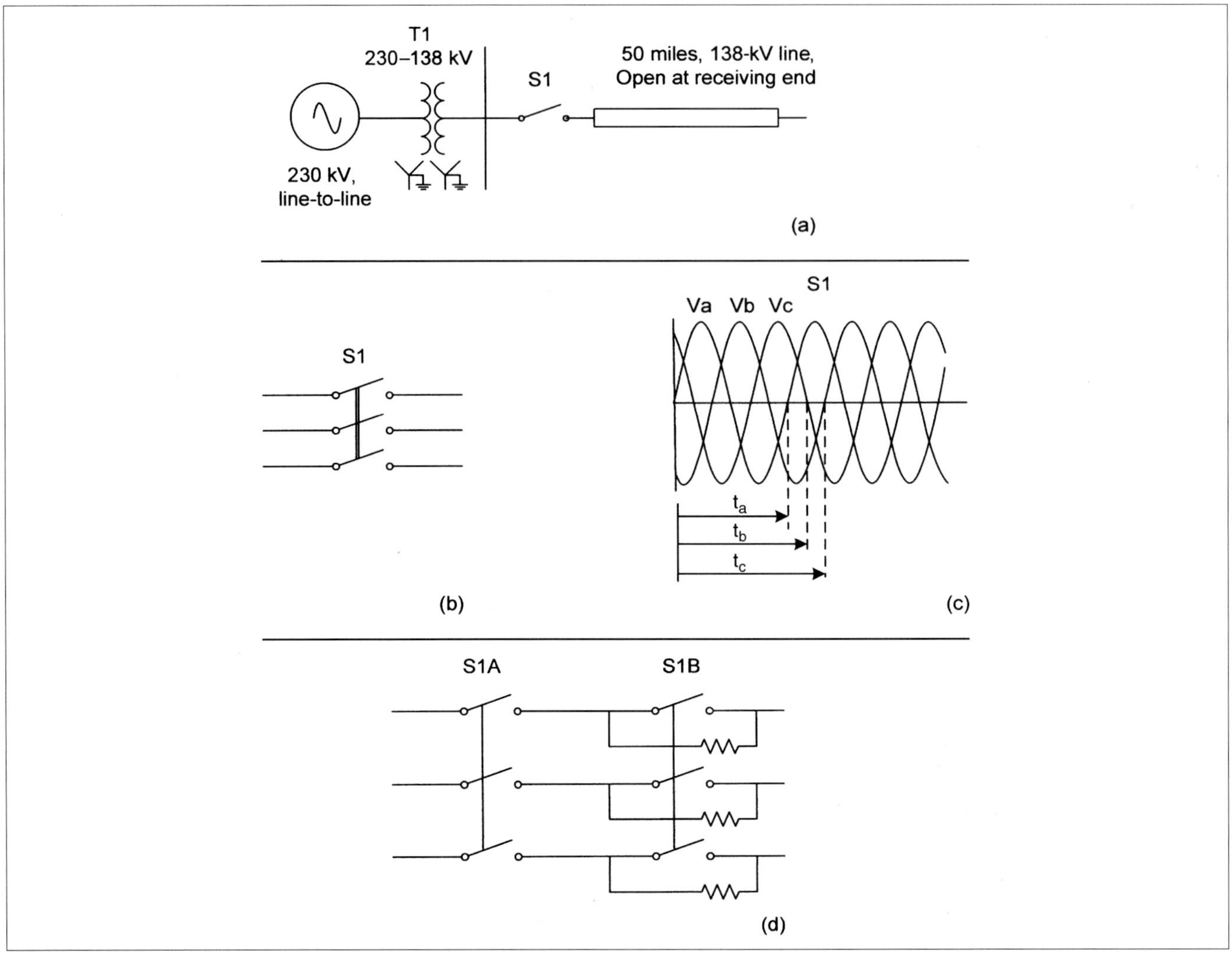

FIGURE 7-7 (*a*) Power system configuration for EMTP simulation of transients on energizing a line. (*b*) Three-phase simultaneous closure. (*c*) Synchronous closing. (*d*) Resistance closing.

The simulations with other switching methods will follow in the subsequent sections.

7-6-2 Phase-to-phase and phase-to-ground overvoltages

Figure 7-9 from Ref. 3 shows the range of 2 percent overvoltage values in per unit of system peak line to ground voltage, which can be expected at a location without the application of surge arresters. U_{e2} denotes the value of phase-to-earth voltages having a 2 percent probability of being exceeded. This figure is based upon a number of field results and studies and includes the effect of most of the factors determining the overvoltages.

Three-phase reenergization may generate high slow-front overvoltages due to trapped charges. In normal systems, single-phase reenergization does not generate overvoltages higher than those from energization. However, for lines where resonance or Ferranti effects may be significant, single-phase reclosing can result in higher voltages than three-phase energization.

TABLE 7-4 138-kV Line Parameters for Examples 7-1, 7-4, and 7-5

MODE	RΩ/mi	X/mi	S/mi	Z, REAL	Z, IMAGINARY	Z LOSSLESS	VELOCITY LOSSLESS (mi/s)	VELOCITY ACTUAL mi/s	ATTENUATION (np/mi)
1	5.326E–01	2.589E–0	3.207E–06	8.983E+02	–9.056E+01	8.940E+02	1.318E+05	1.312E+05	2.901E–04
2	1.163E–01	7.897E–01	5.489E–06	3.780E+02	–2.7887E+02	3.762E+02	1.829E+05	1.824E+05	1.527E–04
3	1.150E–01	6.568E–01	6.876E–06	3.248E+02	–2.756E+02	3.236E+02	1.828E+05	1.824E+05	1.756E–04

Note: EMTP provides results up to 10 floating decimal points. The numbers are rounded off in the above table.

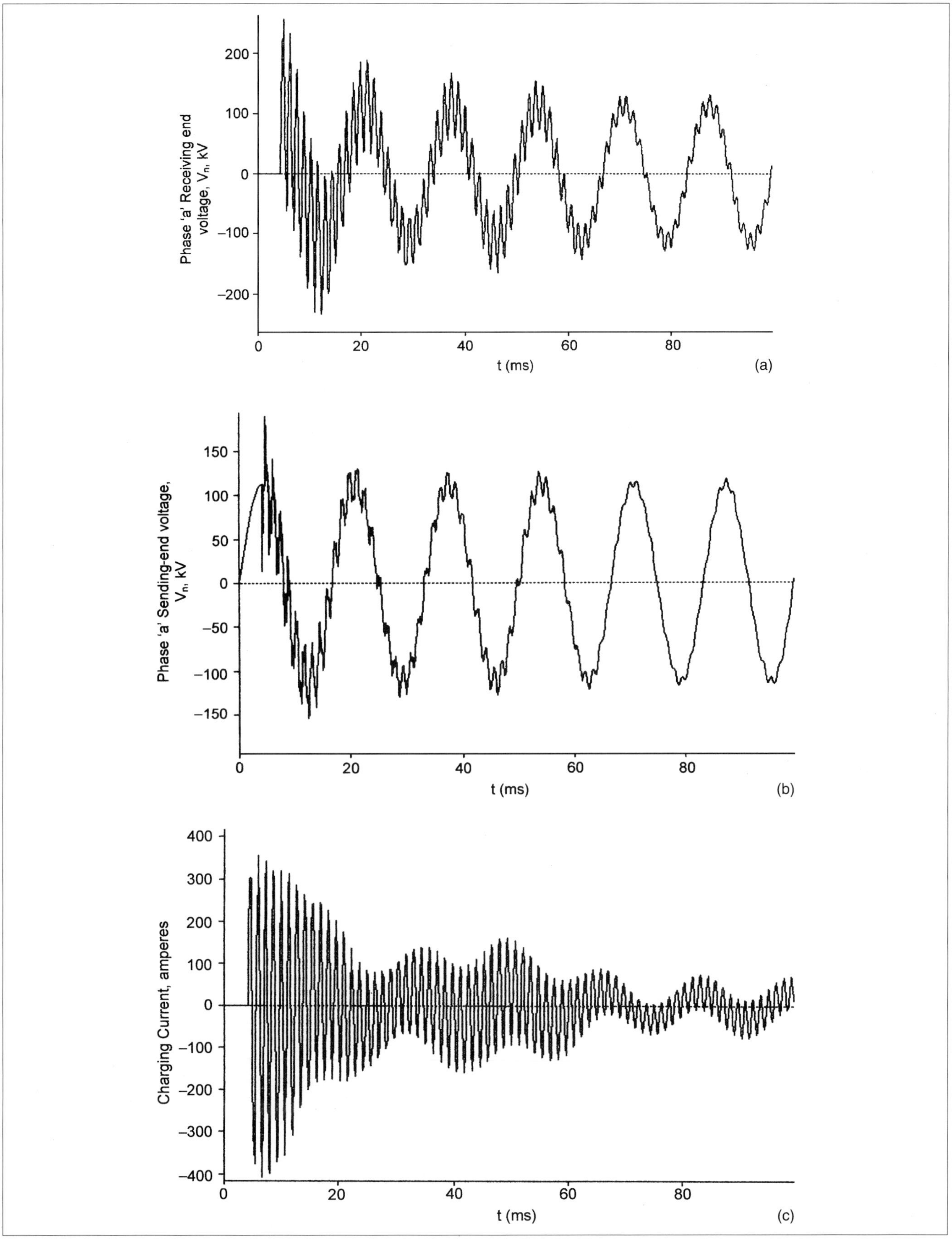

Figure 7-8 (*a*) Receiving-end voltage of phase *a* on three-phase simultaneous closure. (*b*) Sending-end voltage, phase *a*. (*c*) Line charging current.

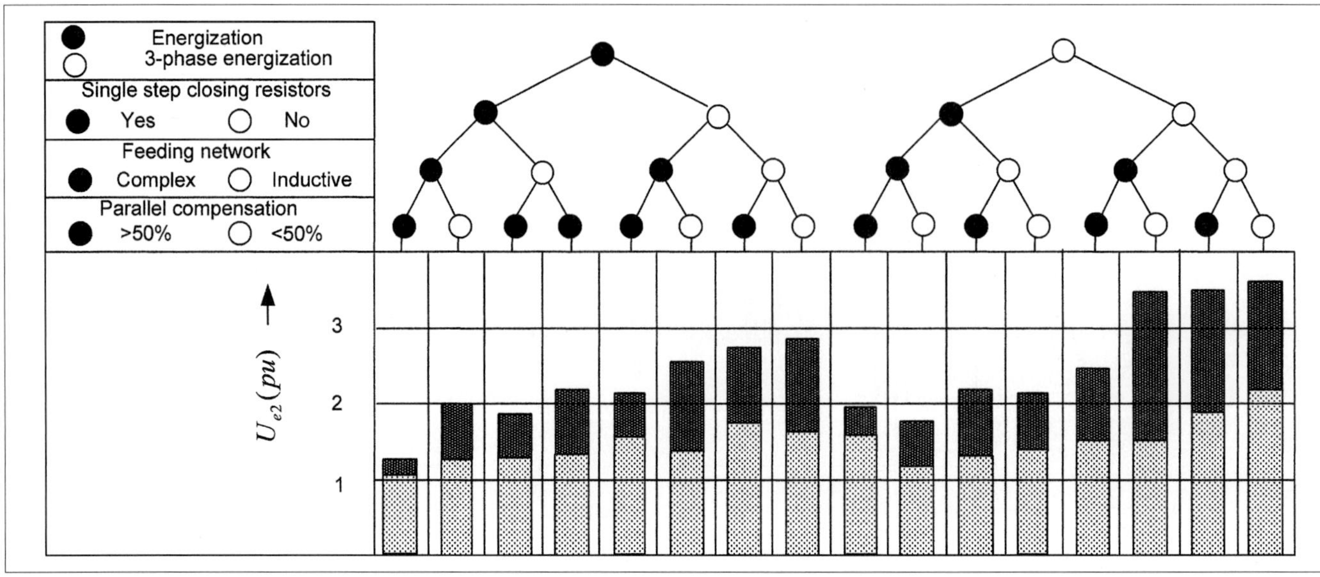

FIGURE 7-9 Range of 2 percent slow front overvoltages at the receiving end of the line on energization and re-energization. *Source:* IEC standard.[3]

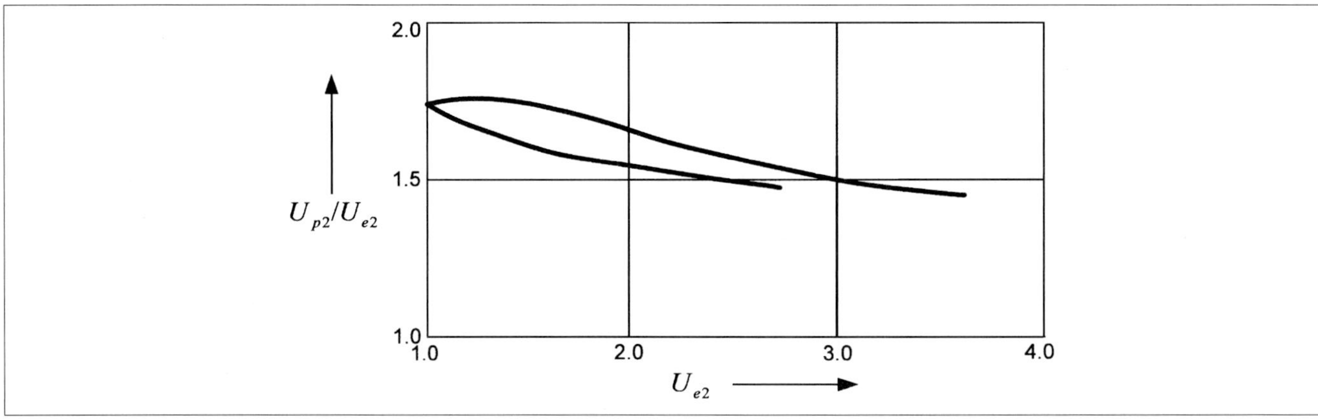

FIGURE 7-10 Ratio between 2 percent slow front overvoltages phase-to-phase and phase-to-earth. *Source:* IEC standard.[3]

The 2 percent phase-to-phase overvoltages can be determined from Figure 7-10.[3] This figure shows the range of possible ratios between 2 percent phase-to-phase and phase-to-earth voltages. U_{p2} denotes the value of phase-to-phase voltages having a 2 percent probability of being exceeded. The upper limit applies to fast three-phase reenergization overvoltages, the lower limit to three-phase energization voltages.

The instant of phase-to-phase overvoltage peak gives the highest phase-to-phase overvoltage value, typically insulation between windings and short air clearances. The phase-to-phase overvoltage at the instant of phase-to-earth voltage peak gives lower overvoltages than the instant of phase-to-phase overvoltage peak, but it may be more severe for some insulation configurations. Typical situations are large air clearances for which the instant of positive phase-to-earth peak is more severe, or gas-insulated substations for which negative peak is more severe[3].

7-7 OVERVOLTAGES DUE TO RESONANCE

The switching inrush current of transformers contains odd and even harmonics (Chap. 14). When a line and transformer are energized together, resonance overvoltages can occur. Figure 7-11 shows that a 765-kV line, 70-mi long is fed from a 400-kV system and has 800-MVA transformers at the sending and receiving ends. The system has a resonant frequency around third harmonic. An EMTP simulation of the receiving end voltage at transformer T2 primary (765 kV) is in Fig. 7-12. The saturation of a transformer reduces the receiving-end voltage, but the resonant phenomenon counteracts it. It is seen that the receiving-end voltage is 1470-KV peak, an overvoltage factor of 2.35. The effect of transformer inrush current harmonics is clearly visible in this figure. The nonlinearity can be supposed to give rise to harmonics, which, akin to fundamental frequency, can be modeled as two-port networks.

Switching charging currents through transformers can be less onerous than switching the same current directly. The capacitive currents will oscillate with the transformer inductance, which may also saturate. Switching through a transformer results in less recovery voltage and a lower possibility of a restrike. The switching overvoltage will also depend on the instant at which the breaker is closed on the voltage wave. Closing at the voltage peak will produce the maximum overvoltages.

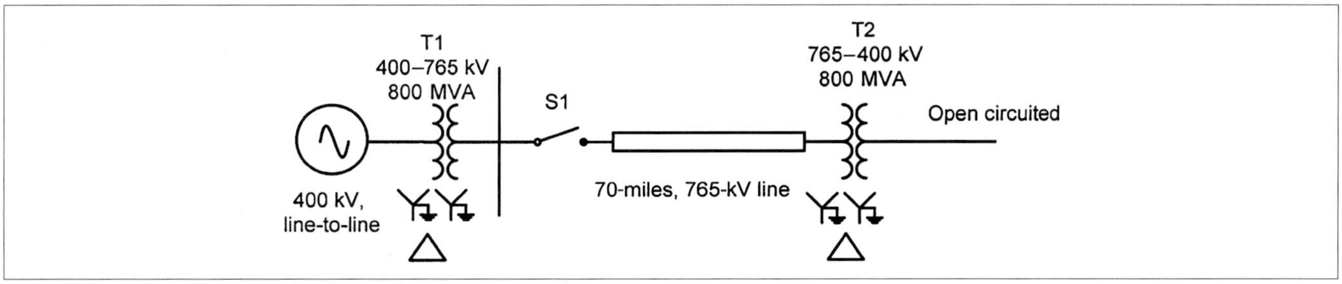

FIGURE 7-11 Configuration of 70-mi, 765-kV line terminated in a transformer, secondary open circuited, for EMTP simulation of receiving-end voltage on energization by closing switch *S*1 at the sending end.

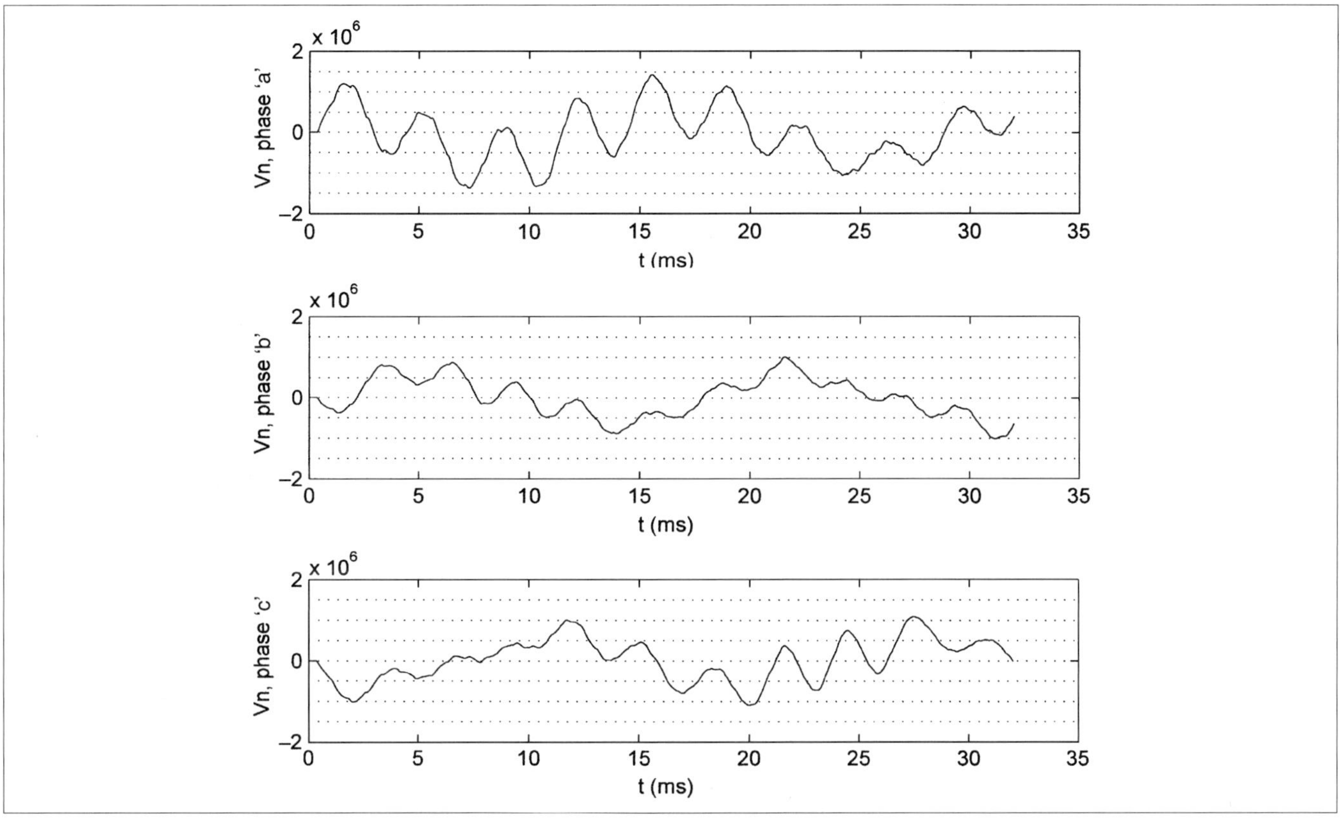

FIGURE 7-12 Receiving-end voltages in phases *a, b,* and *c* in system configuration of Fig. 7-11.

7-8 SWITCHING OVERVOLTAGES OF OVERHEAD LINES AND UNDERGROUND CABLES

When a combination of overhead line and underground cable is switched together, typically, the maximum surge voltage does not exceed 3.0 pu of line-to-ground voltage.

This is applicable to a line or cable switched from the substation, a line switched from a line, or a cable switched from a cable. When the cable is switched through a line or a line is switched through a cable the maximum switching voltage can escalate to 4.0 pu. For EHV systems, the switching surge voltages must be limited as stated before. Table 7-5 shows the maximum switching overvoltages normally encountered by switching of lines and underground cables. The highest overvoltages occur when it is mixed overhead and underground construction and particularly if the proportion of one (i.e., line or cable) is small compared to the other. The procedure of tracking and accumulating reflections using ladder diagrams is fairly complex and time consuming. Computer programs like EMTP and TLTA simplify the calculations.

The cable charging currents depend upon the length of cable, cable construction, system voltage, and insulation dielectric constants. Appendix D gives the calculations of cable capacitances. Prior to energizing, a cable is generally at ground potential or it may have a prior charge. Much akin to switching devices for capacitor banks, a cable is considered isolated if the maximum rate of change with respect to time of transient inrush current on energizing an

TABLE 7-5 Maximum Switching Overvoltages on Switching of Lines and Underground Cables

SEVERITY	DESCRIPTION	MAXIMUM VOLTAGE PER UNIT OF PHASE-TO-NEUTRAL VOLTAGE
Typical	Line or cable switched from station	3.0
	Line switched from line	3.0
	Cable switched from cable	3.0
Moderate	Cable switched from line	4.0
	Line switched from cable	4.0
	Cable trough line switched from station	4.0
Extreme	Line trough cable switched from the station	6.0

uncharged cable does not exceed the rate of change associated with maximum symmetrical interrupting current of the switching device. When switching a cable, the transient inrush current depends on:

- Cable surge impedance, which is much lower than that of transmission lines. In Chap. 4, we estimated that it is of the order of 50 Ω.
- Cable capacitive reactance, system reactance, switching voltage, and whether any damping devices, such as switching resistors, are present in the circuit.
- The cables may be switched back to back, much akin to capacitor banks.

Transient currents of high magnitude and initial high rate of change flow between cables when the switching circuit breaker is closed or on restrikes on opening. For isolated cable switching, we can use the following expression:

$$i = \frac{u_m - u_t}{Z}\left[1 - \exp\left(-\frac{Z}{L}t\right)\right] \tag{7-7}$$

where u_m is the supply system voltage, u_t is the trapped voltage on the cable being switched, Z is the cable surge impedance, and L is the source inductance. From Eq. (7-7) the maximum inrush current is:

$$i_p = \frac{u_m - u_t}{Z} \tag{7-8}$$

Equation (7-8) can be modified for back-to-back switching

$$i = \frac{u_m - u_t}{Z_1 + Z_2}\left[1 - \exp\left(-\frac{Z_1 + Z_2}{L}t\right)\right] \tag{7-9}$$

where Z_1 and Z_2 are the surge impedances of cable1 and cable 2, respectively. Again, akin to back-to-back switching of capacitors, the source reactance can be ignored.

The peak current is:

$$i_p = \frac{u_m - u_t}{Z_1 + Z_2} \tag{7-10}$$

and its frequency is given by:

$$f_{eq} = f\left[\frac{u_m - u_t}{\omega(L_1 + L_2)I_{ir}}\right] \tag{7-11}$$

where f_{eq} is the inrush frequency, I_{ir} is the charging current of one cable, L_1 and L_2 are the inductances of cables 1 and 2, respectively. The switching of single cable forms a series circuit and the transient will be oscillatory, damped or critically damped. Based upon Chap. 2:

$$I_m = \frac{\mu_m - \mu_t}{\sqrt{z^2 - 4L/c}}[e^{-\alpha t_m} - e^{-\beta t_m}] \quad \text{for} \quad Z > \sqrt{4L/c}$$

where

$$t_m = \frac{\ln \alpha/\beta}{\alpha - \beta} \tag{7-12}$$

$$\alpha = \frac{Z}{L} - \sqrt{\frac{Z^2}{L^2} - \frac{4}{LC}} \qquad \beta = \frac{Z}{L} + \sqrt{\frac{Z^2}{L^2} - \frac{4}{LC}} \tag{7-13}$$

The magnetic fields due to high inrush currents during back-to-back switching can induce voltage in control cables by capacitive and magnetic couplings. This is minimized by shielding the control cables.

7-9 CABLE MODELS

The cable models for transient analysis depend upon the cable construction and are discussed in Chap. 4. Consider a single-core shielded 400-kV XPLE cable for modeling in EMTP; the manufacturer's data for the cable are:

Conductor cross section = 1000 mm^2
Diameter of conductor (copper) = 32.9 mm
Insulation thickness = 29.00 mm
Diameter over the insulation = 98.1 mm
Cross section of the screen = 185 mm^2
Outside diameter = 114 mm
Inductance = 0.55 mH/km
Capacitance = 0.16 μF/km
Charging current per phase at 60 Hz = 13.56 A
Surge impedance = 34.5 Ω
Soil resistivity = 100 Ω-m

EMTP model requires the cable characteristics data and the geometric layout of the cables must be modeled. Resistivity of the conductor (ohm-meter), relative permeability and permittivity, loss factor of the surrounding insulation system, and the like are required to be modeled. For the physical configuration of the cable shown in Fig. 7-13 and the cable data as above, the results of the simulation are in Fig. 7-14.

Example 7-2 A 2-km cable length, FD model as derived in Fig. 7-14 is impacted with a surge of peak voltage of 400 kV, rise time 1 μs, tail half value at 50 μs, (arbitrary). There is no power-frequency voltage present at the sending end. The resulting receiving-end surge voltage profile with an EMTP simulation is shown in Fig. 7-15. Note that there are six modes of propagation and the phase-to-ground voltages are not identical in all the phases. The receiving-end voltage is 2.25 pu.

Figure 7-16 illustrates the receiving-end voltage when the cable length is 10 km, all other parameters remaining unchaged. Now, the receiving-end voltage escalates to 3.75 pu. The voltage escalation

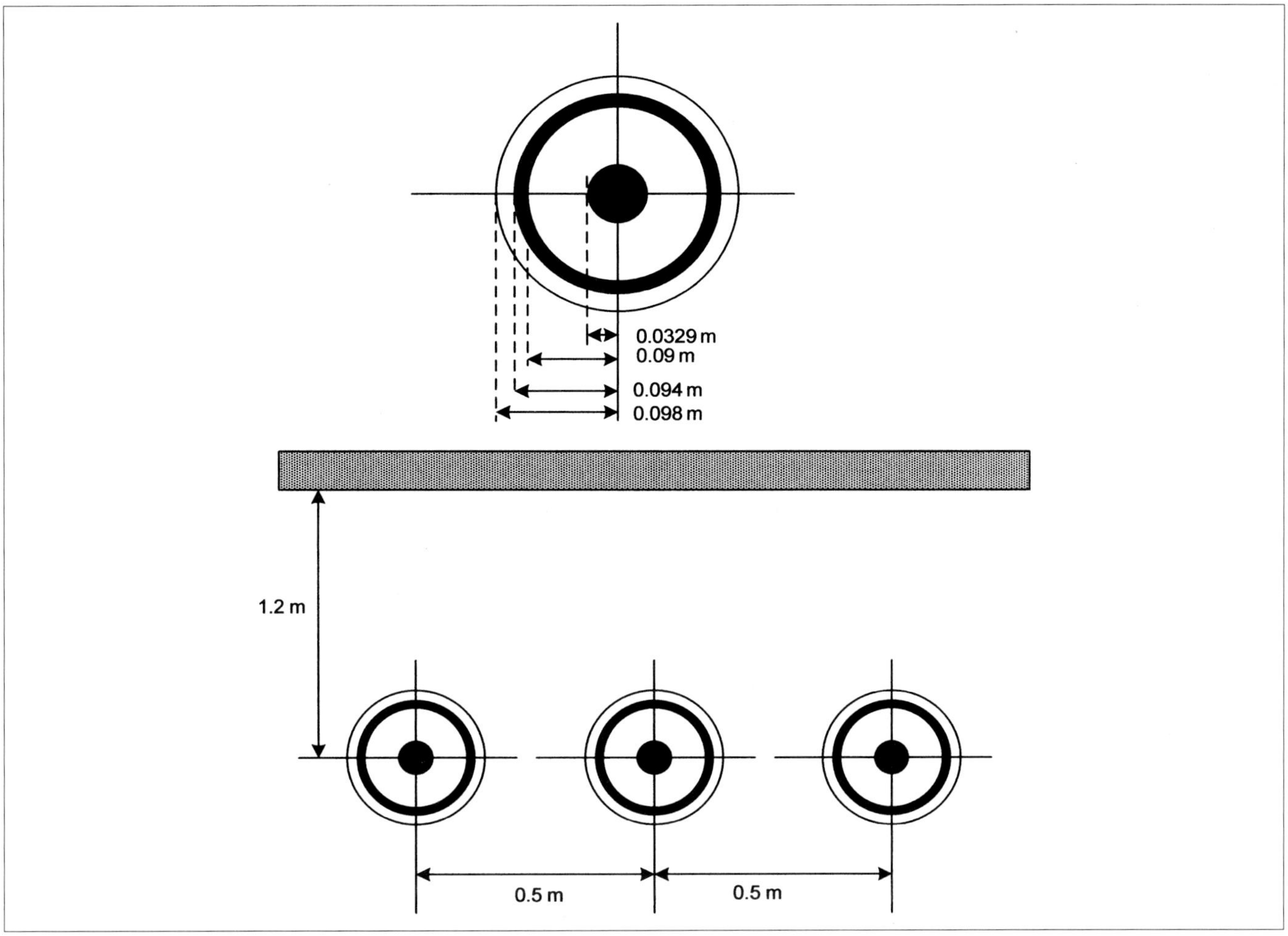

FIGURE 7-13 Physical layout and parameters of 400-kV cable for EMTP simulation.

MODE/ PHASE	FROM BUS	TO BUS	RESISTANCE OHMS/KM	CHAR.IMP. ZC OHMS	TRAVEL TIME SECONDS
1			5.04450E-02	8.60254E-01	8.07465E-05
2			7.39054E-02	2.09627E+01	1.42712E-04
3			5.12137E-02	8.80319E-01	7.88431E-05
4			3.07998E+00	7.22547E+01	5.32254E-04
5			5.09144E-02	8.77004E-01	7.99608E-05
6			8.33080E-02	2.61284E+01	1.84874E-04

<<< REAL TRANSFORMATION MATRIX Q >>>

	1	2	3	4	5	6
1	0.290901E+00	0.891101E-01	0.402989E+00	-0.210304E-04	-0.497050E+00	-0.499565
2	-0.300629E+00	0.348733E+00	-0.402987E+00	0.569608E+00	0.502691E+00	-0.673291
3	-0.560115E+00	-0.173031E+00	0.418569E+00	-0.880282E-04	0.105574E-13	-0.793855
4	0.579004E+00	-0.670012E+00	-0.418558E+00	0.591695E+00	-0.140056E-13	0.203973
5	0.290901E+00	0.891101E-01	0.402989E+00	-0.210304E-04	0.497050E+00	0.499565
6	-0.300629E+00	0.348733E+00	-0.402987E+00	0.569608E+00	-0.502691E+00	0.673291

FIGURE 7-14 EMTP simulation results of cable configuration in Fig. 7-13.

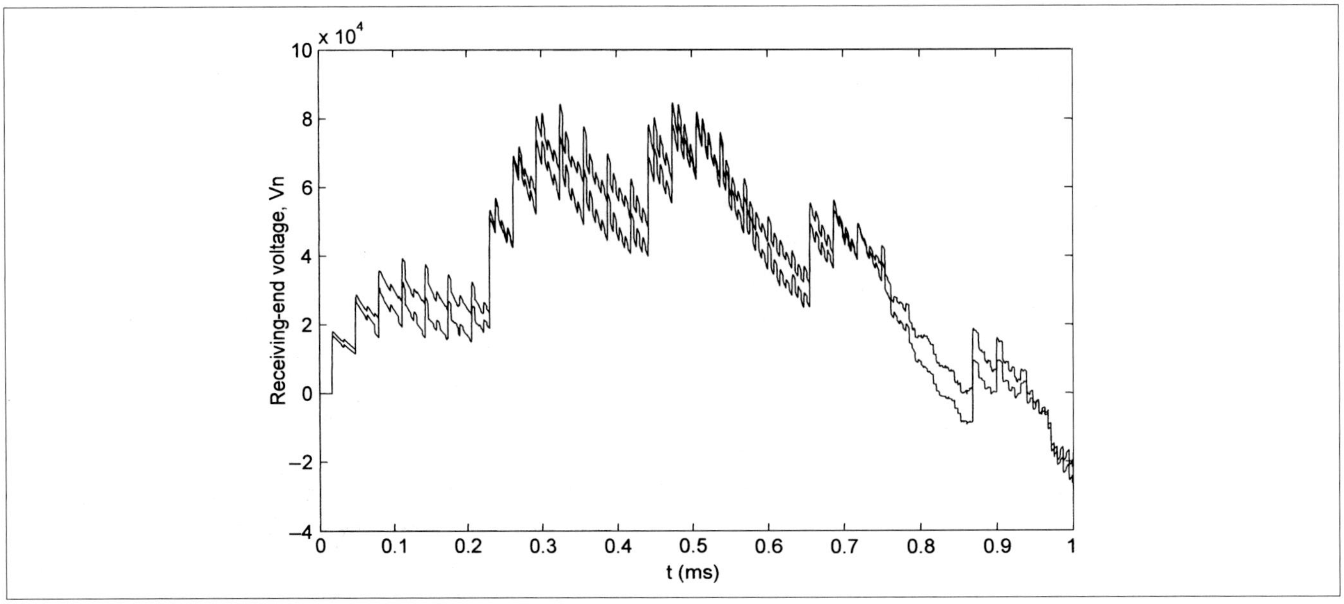

FIGURE 7-15 Receiving-end voltages of a 2-km-long cable, with input surge of 1/50 μs; model as developed in Fig. 7-14.

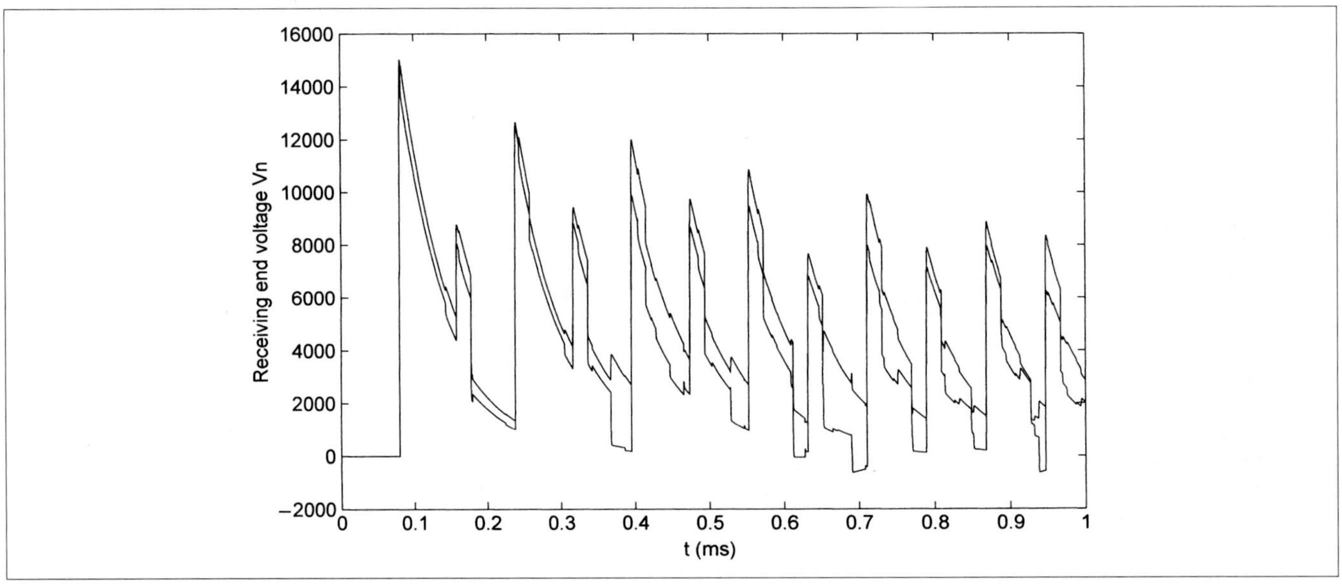

FIGURE 7-16 Receiving-end voltages of a 10-km-long cable, with input surge of 1/50 μs; model as developed in Fig. 7-14.

is the direct function of the cable length, the surge type, and dissipation of the wave energy will be dependent upon response to transients and the distance the wave travels.[11] Attenuation, distortion, and phase shift of the wave occur as demonstrated.

7-10 OVERVOLTAGES DUE TO LOAD REJECTION

Load flow in mainly reactive circuits requires a potential difference between the sending and receiving ends. When a load with an inductive component at the end of a line is suddenly thrown off, it results in the increase in power-frequency voltage at the end of the line. An electromagnetic transient is superimposed upon it and causes voltage surges in the system. The system can be simply modeled by reducing it to the short-circuit impedance at the point of load connection. The short-circuit impedance is mostly inductive and the load has also an inductive component, depending upon its power factor. Load rejection results in cancellation of the voltage drop in the short-circuit impedance, and the generated voltage appears on the system side of the breaker contacts. This voltage rise can be approximately calculated from the expression:

$$\frac{V_1}{V_2} = 1 + \frac{Z_{sc}}{Z_L} = 1 + \frac{P_L}{P_{sc}} \tag{7-14}$$

where V_1, V_2, Z_s, and Z_L are shown in Fig. 7-17. P_L is the load rejected and P_{sc} is the short-circuit power. The type of load and its ratio to the short-circuit power are the decisive factors for the increase in the power-frequency voltage on load rejection. When a number of lines and transformers are used in parallel and the load is subdivided, the 100 percent load rejection becomes a remote possibility.

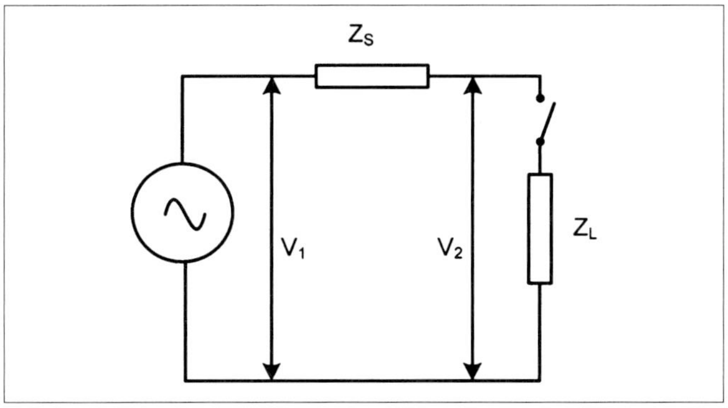

FIGURE 7-17 A schematic representation of load shed through an impedance Z_s.

Example 7-3 A 400-kV, 110-mi long transmission line is fed from a step-up transformer of 138 to 400 kV, which is wye-wye connected with grounded neutrals and tertiary delta windings, with a rating of 500 MVA. The connected load at the end of the transmission line is 350 MW at a power factor of 0.8 (Fig. 7-18). If the load is suddenly switched off, at the crest of the voltage in phase a at 10 ms, by opening switch S1, then the simulation of receiving-end voltage at L2 is shown in Fig. 7-19*a*. The rise in receiving-end voltage is 1.5 times the voltage prior to load shedding. The voltage after load shedding shows harmonic content, illustrating resonance with a natural frequency of the system. The occurrence of resonance phenomena depends upon whether the line and transformer reactances are tuned to one of higher harmonics of the inrush current. The sending-end line current at *L*1 is depicted in Fig. 7-19*b*.

7-11 FERRORESONANCE

In the presence of system capacitance, certain transformer and reactor combinations can give rise to ferroresonance phenomena, due to nonlinearity and saturation of the reactance. This is characterized by peaky current surges of short duration which generate overvoltages. These overvoltages can be predominantly harmonic or subharmonic and reach amplitude of 2 pu. The phenomena may be initiated by some system changes or disturbances, and the response may be stable or unstable, which may persist indefinitely or cease after a few seconds. The ferroresonance is documented in some cases, which are as follows:

- Transformer feeders on double circuits becoming energized through the mutual capacitance between lines and going into ferroresonance when one transformer feeder is switched out
- Transformers losing a phase, say due to operation of a current limiting fuse on ungrounded systems
- Grading capacitors of high-voltage breakers (provided across multibreaks per phase in high-voltage (HV) breakers for voltage distribution) remaining in service when the breaker opens
- Possibility of ferroresonance with a capacitive voltage transformer (CVT) or an electromagnetic potential transformer (PT) under certain operating conditions

Figure 7-20*a* depicts the basic circuit of ferroresonance, a nonlinear inductor, which may be PT or transformer windings, is energized through a resistor and capacitor. The phasor diagram is shown in Fig. 7-20*b*, *as long as the reactor operates below its saturation level*. In Fig. 7-20*c*, points *A* and *C* are stable operating points, while point *B* is not. Point *C* is characterized by large magnetizing current and voltage. This overvoltage appears between line and ground. The changeover from point *A* to point *C* is initiated by some transient in the system, and can be a nuisance due to its random nature.

Figure 7-20*d* shows two lines with mutual capacitance couplings. The transformer forms a series ferroresonance circuit with the line mutual capacitance. The ferroresonance can occur at lower frequency than the fundamental frequency and can overheat the transformers. Switching transients will be superimposed upon the ferroresonance voltages. The discussions are continued in Chap.14, and Example 14-10 is an EMTP simulation of the ferroresonance overvoltages.

7-12 COMPENSATION OF TRANSMISSION LINES

The compensation of transmission lines involves steady-state and dynamic/transient stability considerations. The line is compensated by lumped series and shunt elements at the mid point, terminals or in sections. Figure 7-21*a* illustrates midpoint compensation.[12] Each half-section of the line on either side of the compensating device is represented by a π equivalent model. The circuit of Fig. 7-21*a* can be redrawn as shown in Fig. 7-21*b*, and the phasor diagram is in Fig. 7-21*c*. The degree of compensation for the center held at a voltage equal to the sending-end or receiving end voltage is:

$$k_m = \frac{b_{\text{shcomp}}}{0.5b_{\text{sh}}} \tag{7-15}$$

For equal sending- and receiving-end voltages:

$$P = \frac{V^2}{X_{\text{sr}}(1-s)}\sin\delta \tag{7-16}$$

where:

$$s = \frac{X_{\text{sr}}}{2}\frac{b_{\text{sh}}}{4}(1-k_{\text{m}}) \tag{7-17}$$

The midpoint voltage is:

$$V_m = \frac{V\cos(\delta/2)}{1-s} \tag{7-18}$$

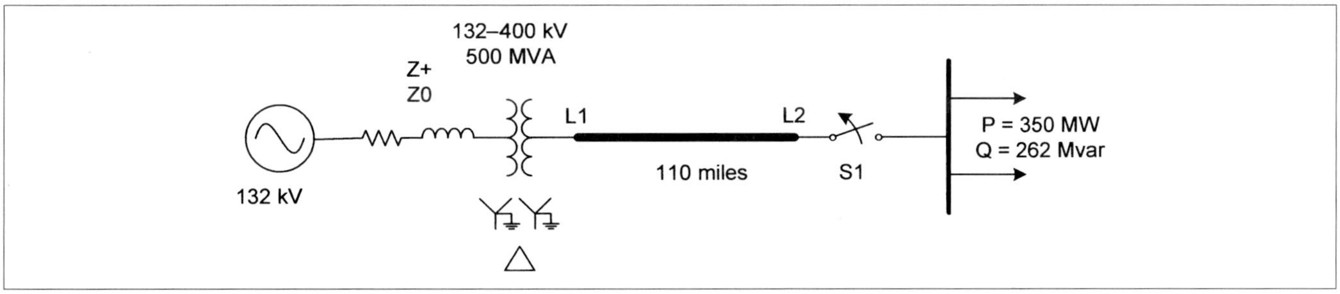

FIGURE 7-18 A power system configuration for EMTP simulation of 350-MW load shed.

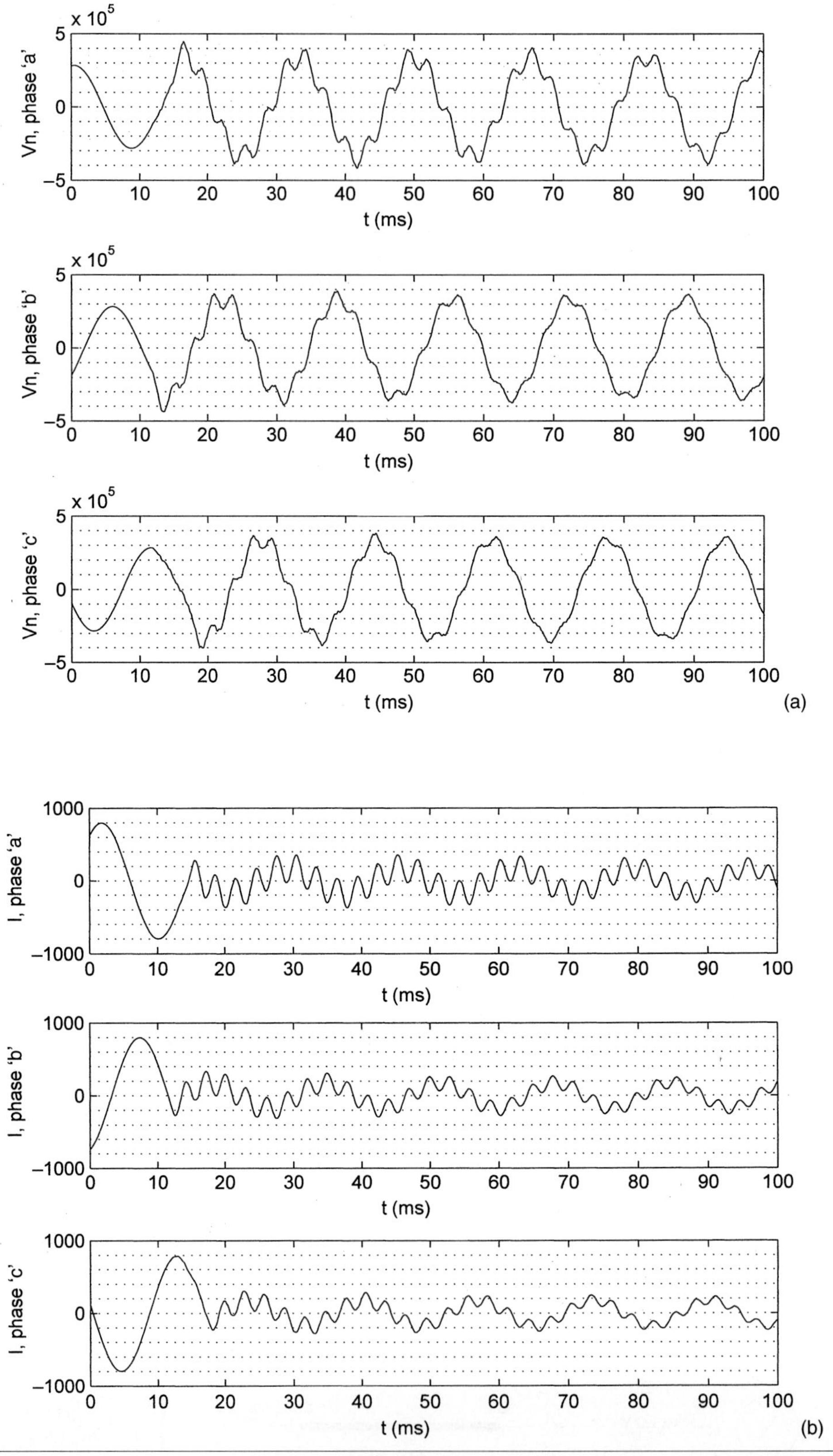

FIGURE 7-19 (*a*) Receiving-end voltage on load shed; opening of switch *S*1 in Fig. 7-18. (*b*) Sending-end current.

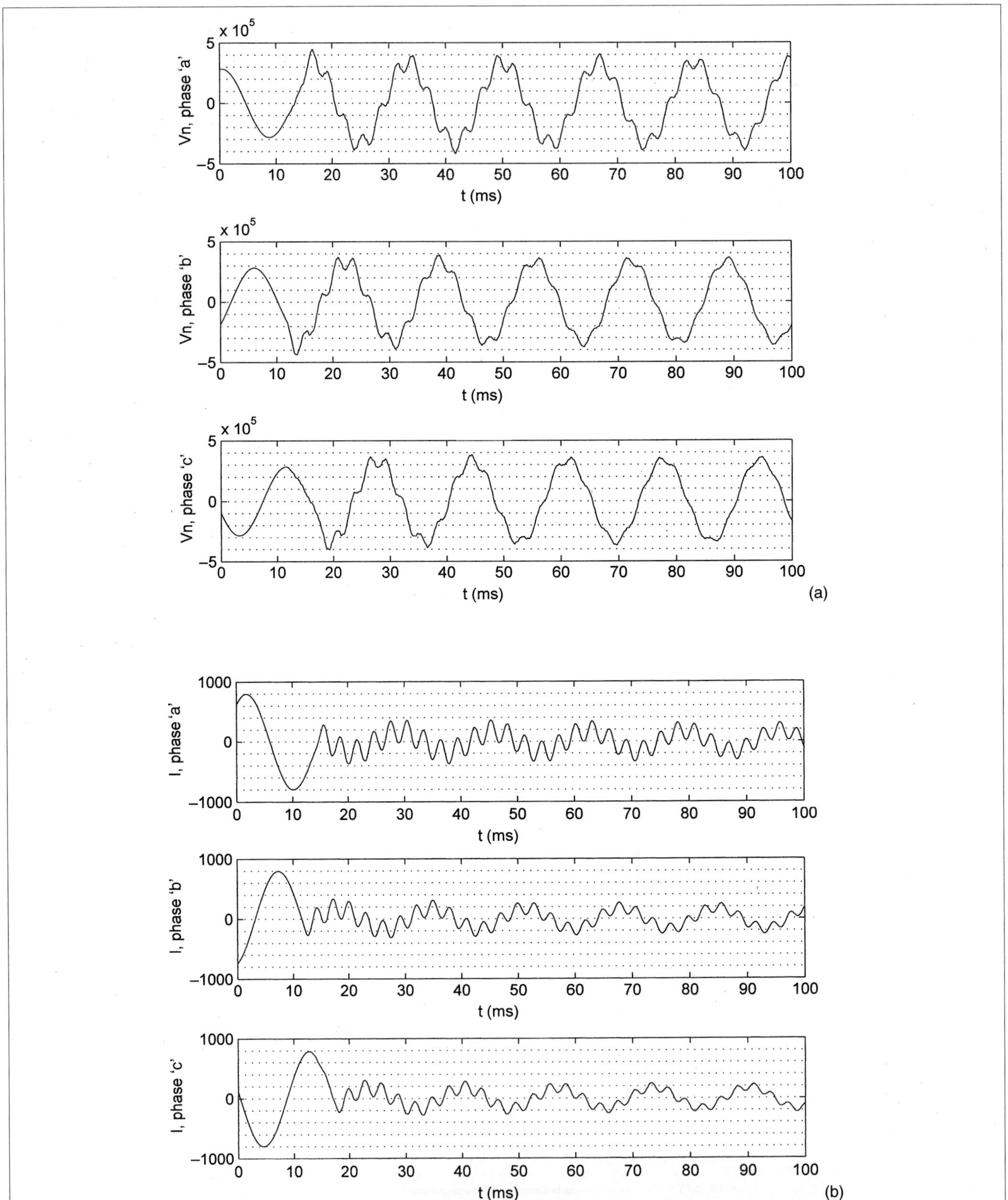

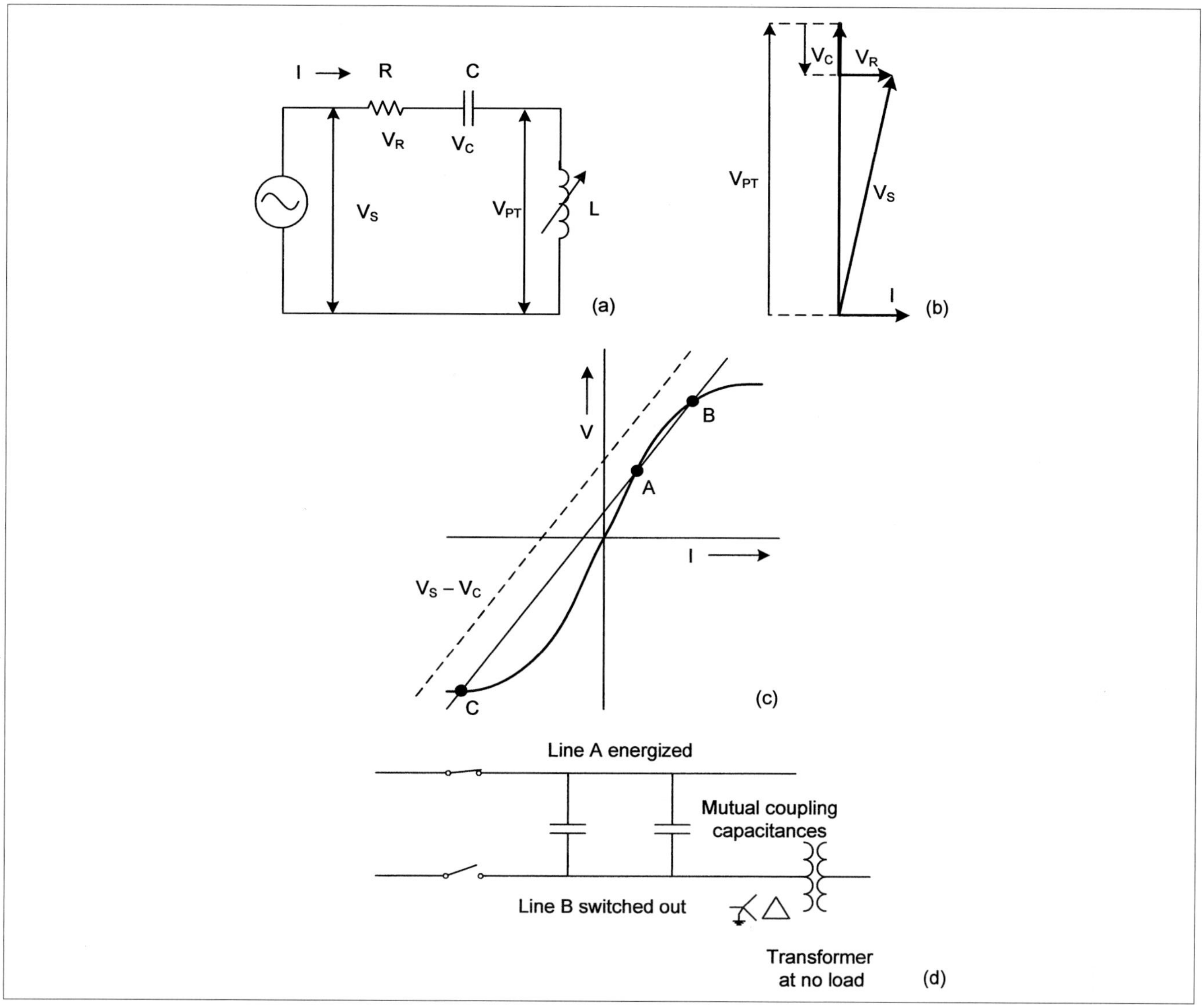

FIGURE 7-20 (*a*) A circuit connection with nonlinear resistor and capacitor that can give rise to ferroresonance. (*b*) Phasor diagram. (*c*) Showing stable operating points *A* and *C*, and unstable operating point *B*. (*d*) A possible power system configuration for ferroresonance.

The equivalent circuit of the line with compensation is shown in Fig. 7-21*d*. For $s = 1$, the midpoint compensation increases the midpoint voltage, which tends to offset the series voltage drop in the line. If the midpoint voltage is controlled by varying the susceptance of the compensating device, then the following relation for the power transfer can be written:

$$P = \frac{V^2}{X_{sr}(1-s)}\sin\delta = \frac{V_m V}{X_{sr}\cos(\delta/2)}\sin\delta$$

$$= 2\frac{V_m V}{X_{sr}}\sin\frac{\delta}{2} \tag{7-19}$$

If midpoint voltage is equal to the sending- and receiving-end voltage, that is, all three voltages are of the same magnitude:

$$P = 2\frac{V^2}{X_{sr}}\sin\frac{\delta}{2} \tag{7-20}$$

The power-transfer characteristics are sinusoids whose amplitude varies as *s* varies, (Fig. 7-21*e*). Each sinusoid of increasing value means higher power transfer: for $P > \sqrt{2}P_{max}$, angle $\delta > \pi/2$. As the transmission angle increases, the compensator responds by changing the susceptance to satisfy Eq. (7-20). The economic limit of a compensator to provide effective capacitive susceptance may be much lower than the maximum power transfer. When the compensator maximum limit is reached, it acts like a fixed capacitor and cannot maintain a constant voltage.

The shunt compensation to satisfy Eq. (7-18) can be calculated from the following equation:

$$b_{shcomp} = -\frac{4}{X_{sr}}\left[1 - \frac{V}{V_m}\cos\frac{\delta}{2}\right] + \frac{b_{sh}}{2} \tag{7-21}$$

The characteristics of series and shunt compensation can be summarized as follows:

- Capacitive shunt compensation increases β, reduces surge impedance, and increases power transfer. Inductive shunt compensation has the opposite effect. It decreases β, decreases power transfer, and increases surge impedance. A 100 percent capacitive shunt compensation will, theoretically, increase the surge impedance to infinity.

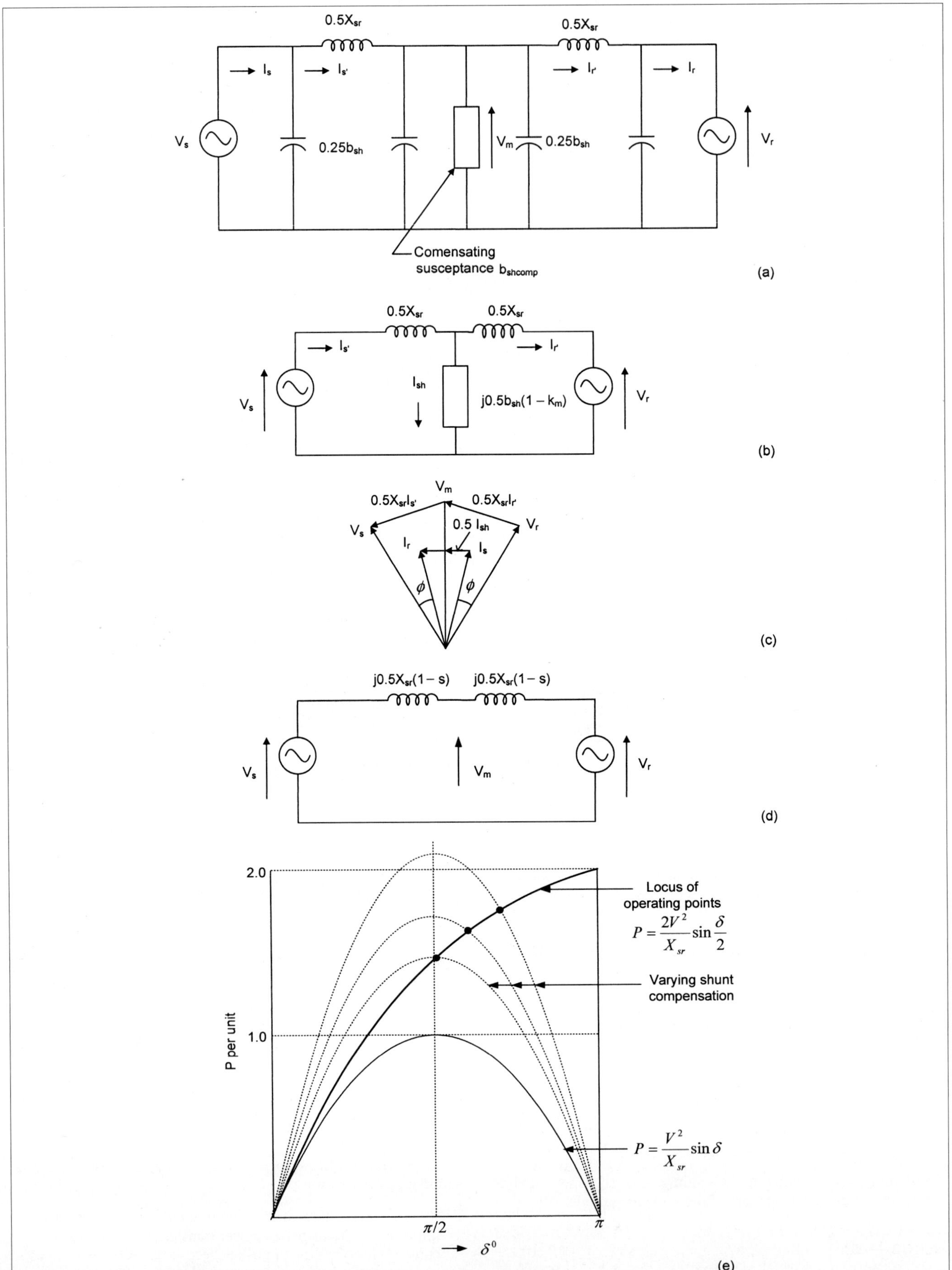

FIGURE 7-21 (*a*) A circuit diagram for midpoint compensation of a transmission line. (*b*). Reduced equivalent circuit. (*c*) Phasor diagram with series and shunt compensation. (*d*) Simplified equivalent circuit. (*e*) Power transfer characteristics.

$$P = \frac{2V^2}{X_{sr}}\sin\frac{\delta}{2}$$

$$P = \frac{V^2}{X_{sr}}\sin\delta$$

- Series capacitive compensation decreases surge impedance and β and increases power-transfer characteristics. Series compensation is applied from steady-state and transient considerations, rather than for power factor improvement. It provides better load division between parallel circuits, reduces voltage drop, and provides better adjustments of line loadings. Shunt compensation directly improves the load power factor. Both types of compensations improve the voltages and, thus, the power transfer capability.

7-13 OUT-OF-PHASE CLOSING

Out-of-phase closing places much stress on the system as well as on the circuit breaker. Out-of-phase closing can occur, due to human error, that is, two phases can be crossed over in a tie connection having voltage sources at either end. This will result in a 120° phase difference between two phases when closing the circuit breaker, with voltages on the load and source side. Figure 7-22 shows two interconnected systems which are totally out of phase. In Fig. 7-22*a* the two poles of a circuit breaker are shown arcing and about to open, while the third pole is open. A voltage equal to three times the system peak voltage appears across the breaker pole, while in Fig. 7-22*b*, a ground fault exists on two different phases at sending end and receiving end (rather an unusual condition). The maximum voltage across a breaker pole is $2 \times \sqrt{3}$ times the normal system peak voltage.

The present day high-speed relaying has reduced the tripping time and, thus, the divergence of generator rotors on fast closing is reduced. Simultaneously, the high-speed auto-reclosing to restore service and remove faults increase the possibility of out-of-phase closing, especially under fault conditions. The effect of the increased recovery voltage when the two systems have drifted apart can be stated in terms of the short-circuit power that needs to be interrupted. If the interrupting capacity of a circuit breaker remains unimpaired up to double the rated voltage, it will perform all events satisfactorily as a tie-line breaker when the two sections of the system are completely out of synchronism. The short-circuit power to be interrupted under out-of-step conditions is approximately equal to the total short-circuit power of the entire system, but reaches this level only if the two systems which are out of phase have the same capacity.

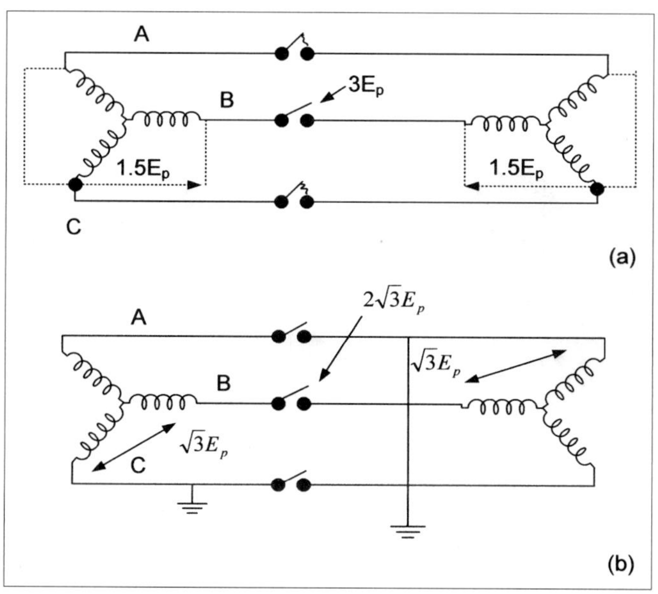

FIGURE 7-22 (*a*) Overvoltages due to two systems completely out of phase, unfaulted condition; the maximum voltage is equal to three times the peak system voltage. (*b*) Ground fault on different phases on source and load sides; the maximum voltage is equal to $2 \times \sqrt{3}$ times the peak system voltage.

If a breaker has an out-of phase switching rating, it is only 25 percent of the maximum short-circuit current in kiloamperes. The ANSI/IEEE standard[10] specifies the duty cycles. The conditions for out-of-phase switching are:

- Opening and closing operations in conformity with manufacturer's instructions, and closing angle limited to maximum out-of-phase angle of 90°
- Grounding conditions of the system corresponding to those for which the circuit breaker is tested
- Frequency within ± 20 percent of the rated frequency of the breaker
- Absence of fault on either side of the breaker, that is, the situation depicted in Fig. 7-22*b* is not applicable

Where frequent out-of-phase closing is expected, the recovery voltage should be evaluated. A special circuit breaker or the one rated for higher voltage may be required. The severity of out-of-phase closing can be reduced by using protective devices with coordinated impedance-sensitive elements to control the tripping instant, so that the interruption will occur substantially after or substantially before the instant when the phase angle is 180°. Polarity sensing and synchronous breakers can be employed.

7-14 OVERVOLTAGE CONTROL

The overvoltages in the system can be controlled by:

- Resistance switching
- Synchronous closing
- Line compensation
- Application of surge arresters
- Application of surge capacitors
- Protective relaying

We have discussed line compensation; the power frequency component of the overvoltage is controlled by connecting high-voltage reactors from line to ground at the sending and receiving ends of the transmission lines. In other words, the electrical length of the line is reduced. The effect of the trapped charge on the line can be eliminated if the closing takes place during that half cycle of the power-frequency voltage, which has the same polarity as the trapped charge. The high-voltage circuit breakers may be fitted with devices for polarity-dependent closing. Controlling overvoltages with switching resistors is yet another method.

Lines with trapped charge and no compensation and no switching resistors in breakers may have overvoltages greater than three times the rated voltage. Without a trapped charge, this overvoltage will be reduced to 2.0 to 2.8 times the rated voltage. With single stage closing resistors and compensated line, overvoltages are reduced to less than 2 times the rated voltage. With two-stage closing resistors or compensated lines with optimized closing resistors, the overvoltage factor is further reduced to 1.5.

7-14-1 Synchronous Operation

A breaker for synchronous operation opens and closes with fixed reference to the phase position of the system voltage or current oscillations. An electronic device records the system voltage, and the zero crossings of the voltage forms the reference point. Based upon the operating time of the circuit breaker, the tripping pulse is controlled so that the contacts touch or the contact separation takes place at the desired voltage phase position, depending upon the desired operation. In synchronous closing, the breaker is controlled so that the contacts touch at voltage zero or are at maximum system voltage. The contacts should close at voltage zero when energizing capacitor banks. In synchronous opening, the breaker is controlled so that current zero occurs at a definite contact gap. The maximums and zeros in different phases occur at different times.

The breaker is, therefore, either operated or controlled with individual mechanism for each pole, or its individual poles are coupled with each other such that contacts touch and separation occurs at different times. Independent pole operation is a standard feature for circuit breakers of 550 kV, but can be made applicable to circuit breakers of 138 KV. Figure 7-23 shows two effects on the resulting receiving-end overvoltage: (1) the angle of closure from source end voltage zero and (2) the effect of source impedance. The overvoltage will be minimum for zero-closing angle and infinite source impedance. Therefore, the problem of overvoltage control in long lines with relatively low short-circuits power. This figure also depicts that even a small deviation from the ideal closing angle can make much difference in the resulting overvoltage factor. In a practical implementation, the closing time of the breaker must be considered and there can be variations in each pole.

Operations for which synchronous switching can be effective are:

- Switching of capacitor banks
- Unloaded transformers
- Shunt reactors, that is, compensated transmission lines
- Switching to mitigate effects of residual charge on transmission lines.

Each of these operations has a different strategy as the switching transients vary. When closing is undertaken to be at the same polarity as the source and line side, to counteract the effect of trapped charge, the impact of the closing angle is not so critical. The success is dependent upon how the circuit breaker characteristics are matched with that of the operation under consideration. The circuit breaker parameters are the mechanical and electrical properties, temperature, contact velocity, buildup of dielectric strength in the contact gap, and wear and aging—which may be somewhat unpredictable. Note that neither the velocity of contact separation nor the closing and opening times are constant, nor can they vary evenly with the variations in the control voltage of the breaker.

Capacitance Switching An energization transient can be completely eliminated if there is no voltage across the breaker poles on the load and source side, which may not be practical. It seems that energization overvoltages can be well controlled if the closing occurs within 1 ms before and after current zero.[13,14,15] When capacitors are ungrounded, the first pole can be closed at random, as there is no current flow with only one pole closed. The second and third poles can then be closed at their respective current zeros. Alternatively, two poles can be closed simultaneously at current zeros

De-energizing a capacitor gap requires short arcing times (Chap. 8), and the arc can be established when there is not enough contact separation giving rise to restrikes. The contacts can be made to separate sufficiently ahead of a current zero, say about one-fourth of a cycle. With modern high-voltage circuit breakers of high speed, the contacts separate fast and there does not seem to be a need for synchronous de-energization.

Reactor Switching The operation may be considered reverse of capacitor switching. Generally on energization, the breakers may experience a prestrike (Chap. 8); however, this voltage is limited in magnitude and energization is not a concern, except during energization of a shunt-compensated line. In such cases, the frequency across the breaker contacts has a beat frequency, and it will be desirable to sense the polarity of the trapped charge and close when the voltage across the contacts is minimum. The synchronous operation on opening can reduce the overvoltages that may be generated due to current chopping, or reignitions that may occur on opening operation. The contacts are separated at a time which is larger than the minimum arcing time required for operation of that particular breaker, so that the contact gap is able to withstand the recovery voltage. This may not be so easy to predetermine and incorporate in the switching logic.

Example 7-4 Continuing with Example 7-1 with synchronous closing of the poles, all other parameters remaining unchanged, the switches in three phases are closed as shown in Fig. 7-7*c*.

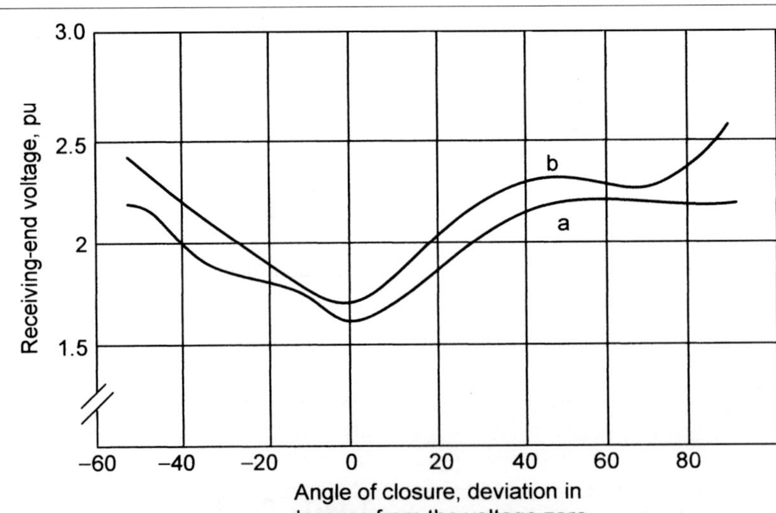

FIGURE 7-23 Effect of angle of closure and source impedance on the receiving-end voltage of a 400-kV line, 400 km long. Curve *a* for infinite source, curve *b* for 0.1-H source at 400 kV.

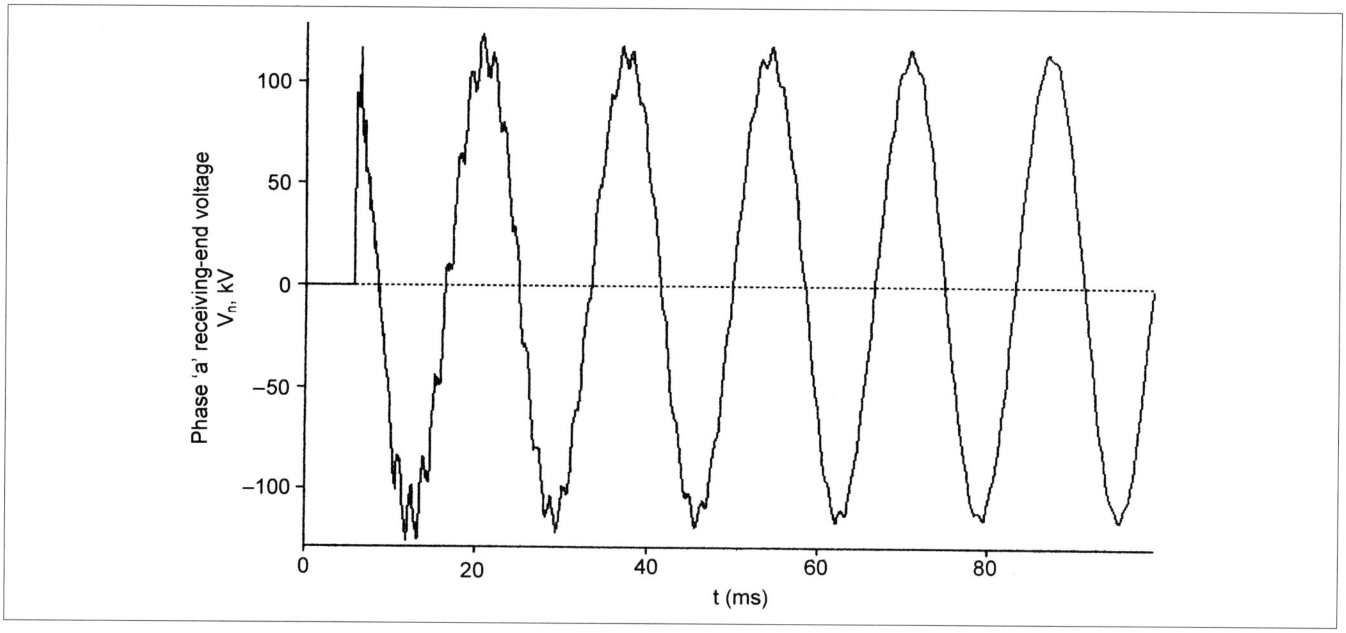

FIGURE 7-24 Receiving-end voltage of phase *a* synchronous closing, system configuration as in Fig. 7-7*c*.

The resulting receiving-end voltage, phase *a*, is illustrated in Fig. 7-24. Comparing with Fig. 7-8*b*, it is seen that:

- The amplitude is reduced from 2.48 to 1.24 pu.
- The high-frequency transients are reduced.

7-14-2 Resistance Switching

Resistance switching is discussed in Chap. 6 in connection with shunt capacitor bank switching. Circuit breaker resistors can be designed and arranged for resistance switching to meet the following objectives:

- To reduce switching surges and overvoltages
- For potential control across multibreaks per phase in the high-voltage breakers,
- To reduce natural frequency effects and breaker recovery voltage (see Chap. 8).

Opening resistors are also called switching resistors and are in parallel with the main break and in series with an auxiliary resistance break switch. On an opening operation, the resistor switch remains closed and opens with a certain delay after the main contacts have opened. The resistance switch may be formed by the moving parts in the interrupter or striking of an arc, dependent upon the circuit breaker design.

Figure 7-25 shows the sequence of opening and closing in a circuit breaker provided with both opening and closing resistors. The closing resistors control the switching overvoltage on energization of, say, long lines. An interrupting and closing operation is shown. The main break is S_B, the opening resistor is R_B, the closing resistor is R_C, the auxiliary breaker contact is S_C, and the resistor closing contact is S_R. On an opening operation, as the main contact start arcing, the current is diverted through the resistor R_B, which is interrupted by contacts S_C. In Fig. 7-25*d* the breaker is in open position. Figures 7-25*e* and *f* show the closing operation. Finally, the closed breaker is shown in Fig. 7-25*a*. The flow of main current and diverted current and the circuit not carrying any current are clearly depicted in this figure with different line weights.

Synchronous switching versus resistor switching is always a point of discussion. While both technologies try to achieve the overvoltage control, in synchronous switching the variations of contact scatter and opening and closing times may considerably alter the performance. In resistor switching the resistors are prefixed and for the tested conditions a uniform performance can be expected. In synchronous switching it is possible to program and adjust the contact operation.

Example 7-5 Example 7-1 is repeated with resistance switching to examine the effect on the sending- and receiving-end voltages. An auxiliary switch and a resistor equal to the surge impedance of the line is used (Figure 7-7*d*). The auxiliary switch contacts close in resistors at first voltage zero. Then the main contacts close in one-quarter cycles after preinsertion resistors. The receiving-end voltages are shown in Fig. 7-26.

Comparing Figs. 7-24 and 7-26, synchronous switching and resistance switching, respectively, the results are almost identical.

7-15 STATISTICAL STUDIES

The surge phenomena are statistical in nature. Statistical analysis is random data case, which is solved in the time domain. Many time-domain studies can be carried out on EMTP, as a parameter of interest varies, and a maximum and minimum values are obtained within certain assigned variations.

A random function can be used to generate the data. Number of simulations, maximum multiples of standard deviations, type of random algorithm, for example, gaussian distribution, with mean and standard deviations can be specified.

Example 7-6 Figure 7-12 is a simulation of the receiving-end voltage in the system configuration depicted in Fig. 7-11. Let the switch S1, all three poles, be modeled with Gaussian distribution, mean deviation $\sigma = 4$ ms and μ (mean) $= 28$ ms. The statistical simulation plots are shown in the following figures:

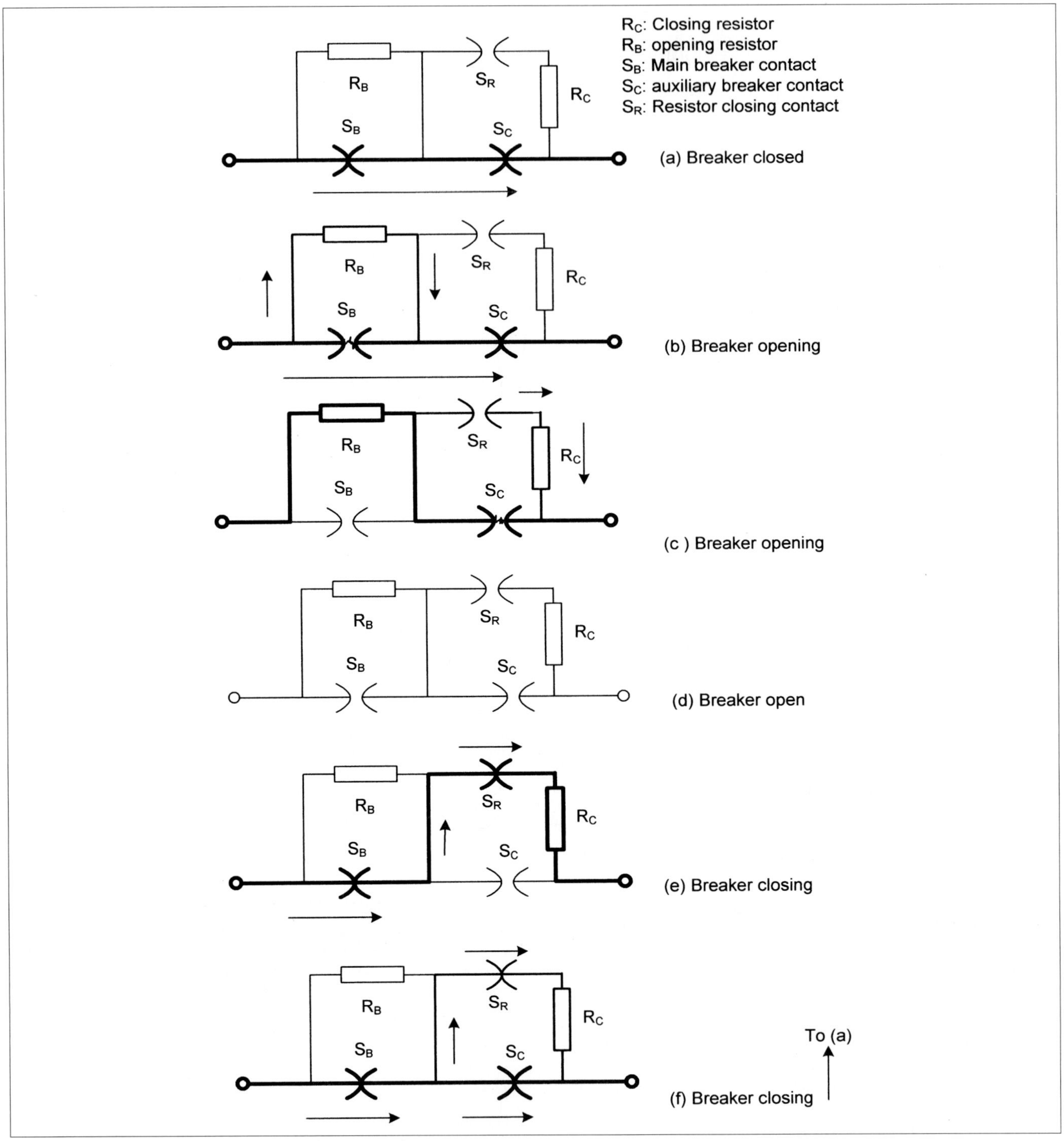

FIGURE 7-25 Schematic diagram of a high-voltage circuit breaker fitted with opening and closing resistors: (*a*) breaker closed, (*b*) and (*c*) breaker opening, (*d*) breaker open, (*e*) and (*f*) breaker closing.

Figure 7-27. The maximum and minimum receiving-end voltages on phase *a*, for 100 simulations.

Figure 7-28. The maximum and minimum currents through the closing switch *S*1, phase *a*.

Figure 7-29. A bar graph of the maximum and minimum receiving-end voltages, phase a.

Figure 7-30. Cumulative distribution plot, maximum receiving-end voltage, phase *a*. The calculated mean and standard deviation are shown on the top of the graph.

Figure 7-31. Histogram of receiving-end phase *a* voltage.

The statistical studies are powerful analytical tools and when properly applied can give valuable information of the system

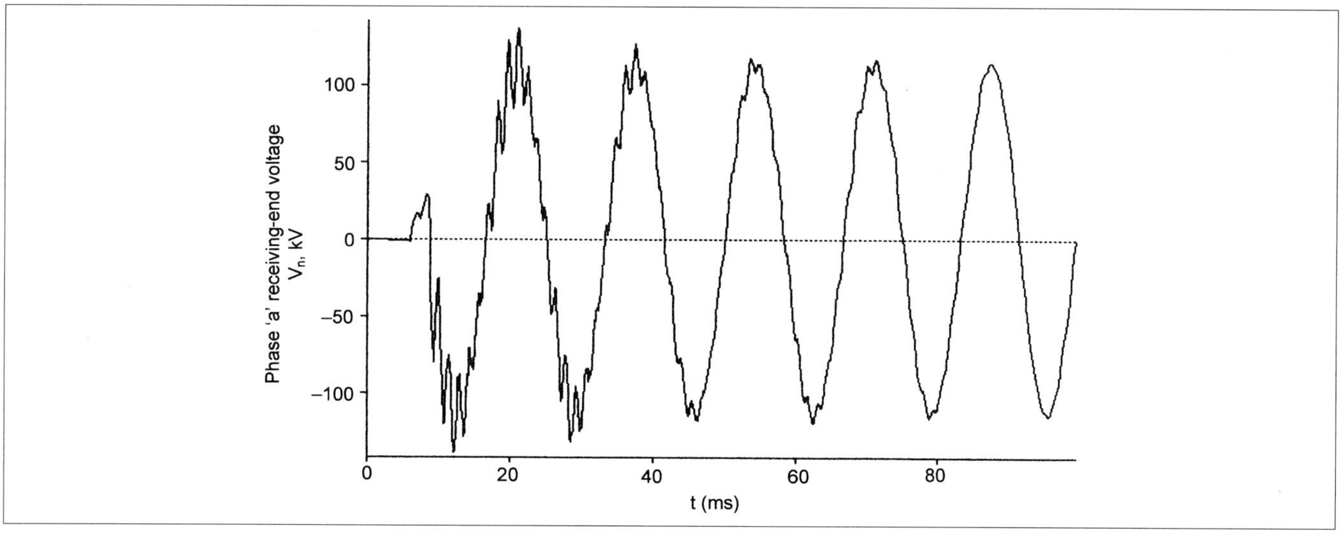

FIGURE 7-26 Receiving-end voltage, phase *a* resistance closing, system configuration as in Fig. 7-7*d*.

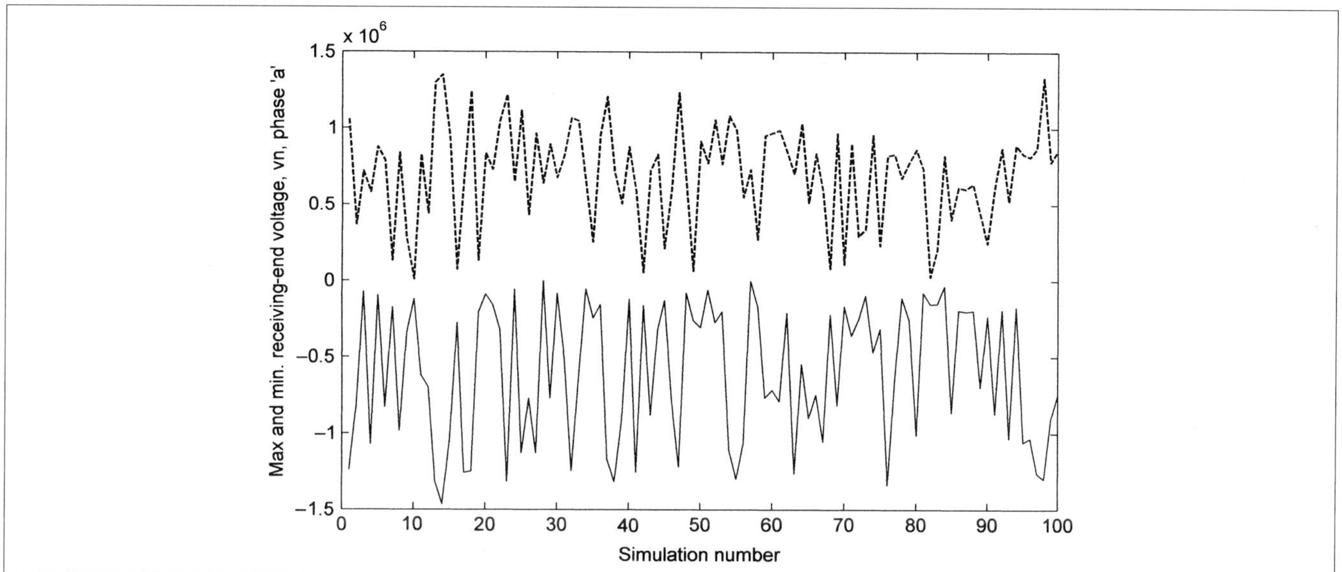

FIGURE 7-27 Statistical studies of system configuration in Fig. 7-11. Maximum and minimum receiving-end voltages for phase *a*; 100 simulations.

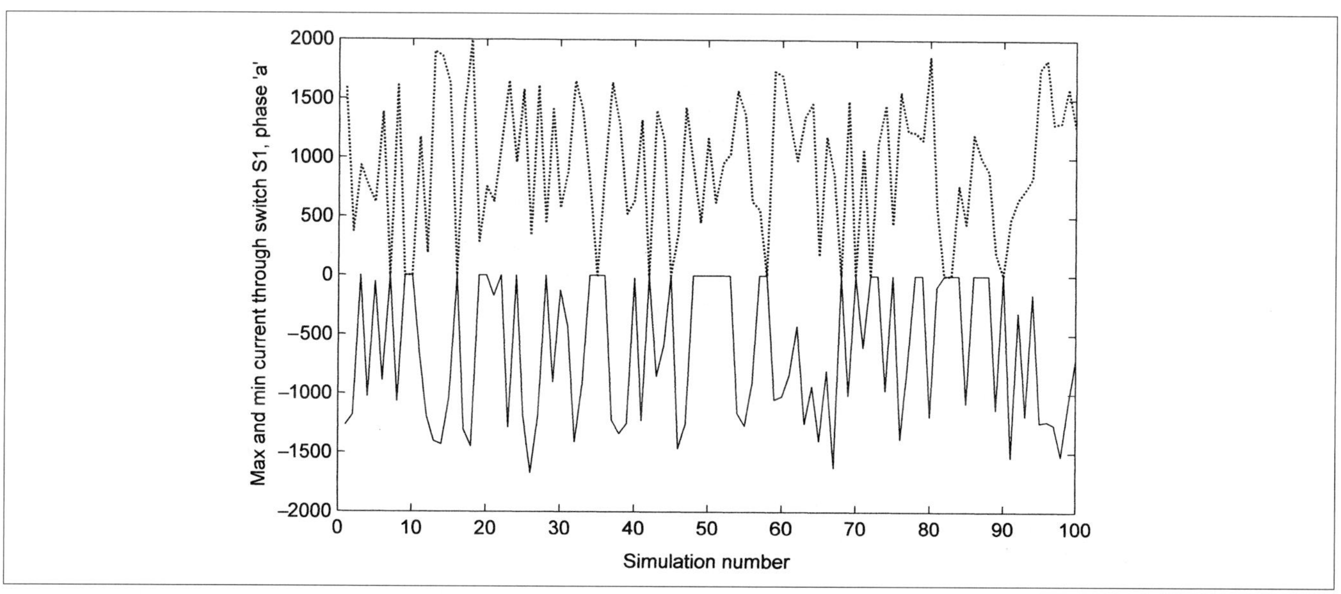

FIGURE 7-28 Statistical studies of system configuration in Fig. 7-11. Maximum and minimum current through switch *S*1, for phase *a*; 100 simulations.

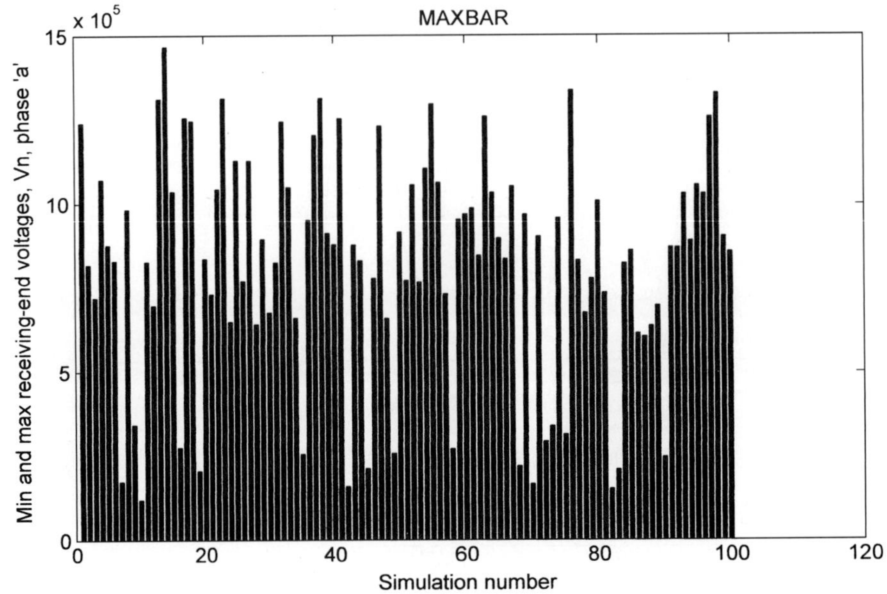

FIGURE 7-29 Statistical studies of system configuration in Fig. 7-11. Bar graph of maximum and minimum receiving-end voltages for phase *a*; 100 simulations.

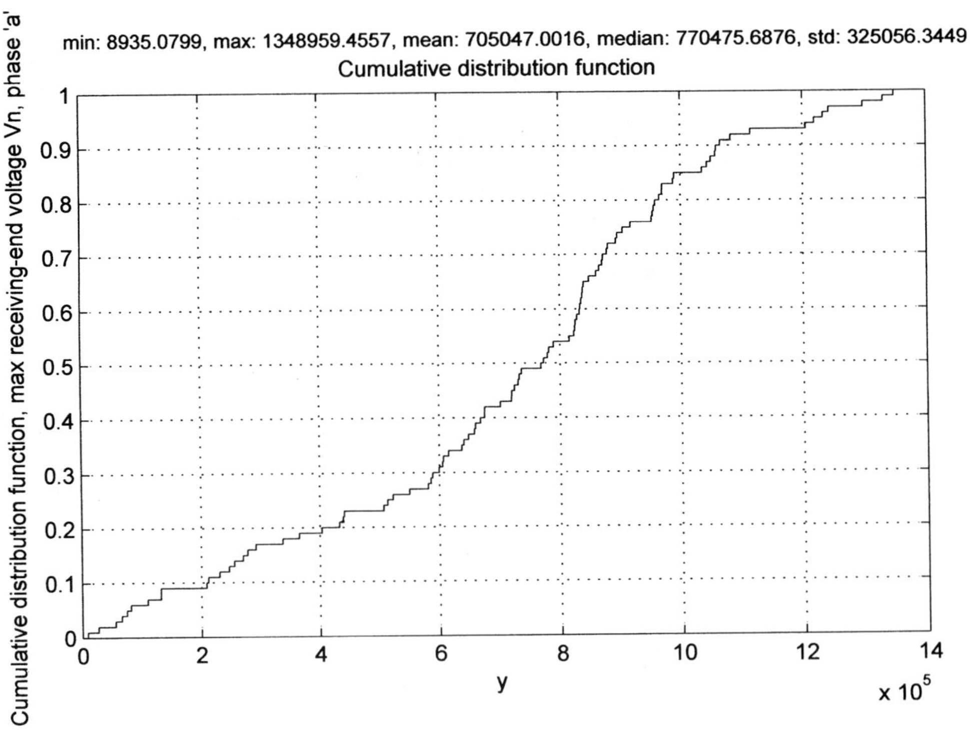

FIGURE 7-30 Statistical studies of system configuration in Fig. 7-11. Cumulative distribution function receiving-end voltages for phase *a*; 100 simulations.

behavior without hundreds of simulations in the time domain. The statistical data could be applied to each pole of the switch, and the closing and opening operation can be made dependent upon the operation of some other devices in the system, that is, a fault simulation switch.

Generally the lines are modeled as perfectly transposed lines. Asymmetry will introduce errors, which will vary. The pole span, that is, the time difference between the first pole and the last pole to close can be of the order of 5 ms, and that will introduce additional errors in the modeled results with simultaneous closing of poles.

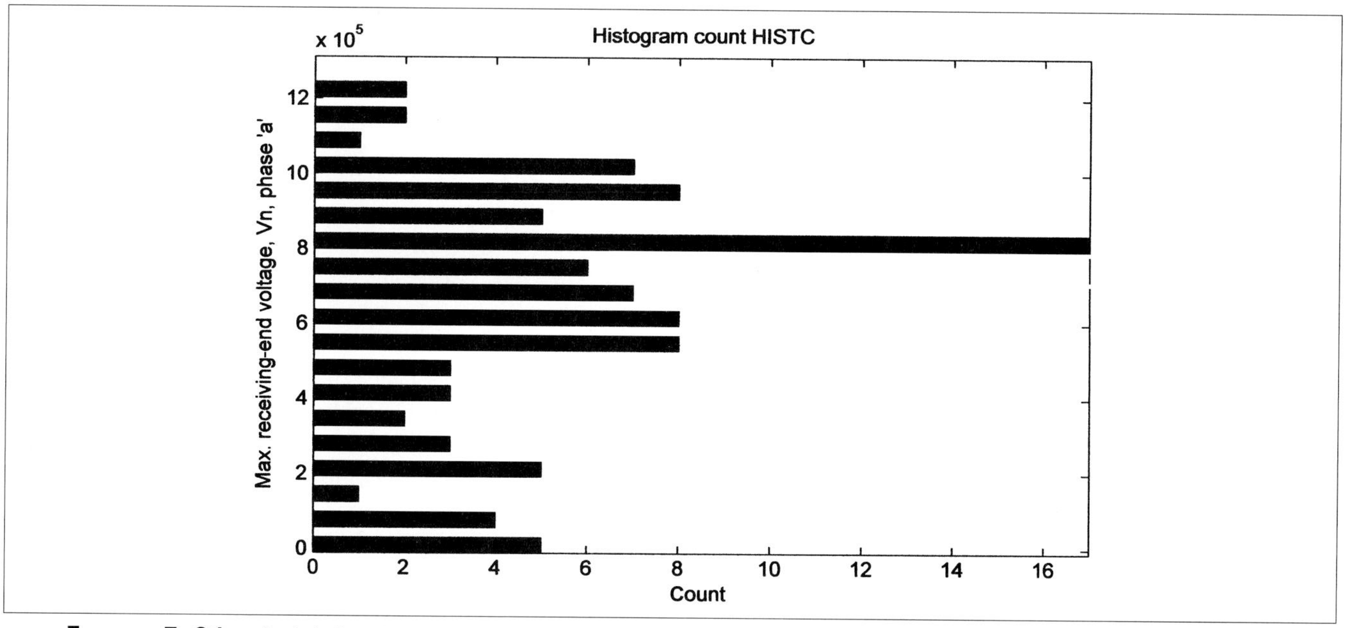

FIGURE 7-31 Statistical studies of system configuration in Fig. 7-11. Histogram of receiving-end voltages of phase *a*; 100 simulations.

The trapped charge oscillation is quite different on a balanced and on an untransposed line. A comparison of the statistical distributions of the overvoltages is one possibility to take into account the effect of asymmetry. EMTP allows modeling of unsymmetrical lines.

In this chapter, various modes and switching operations that result in overvoltages on the transmission lines and cables, and how these overvoltages can be controlled, are studied. The importance of switching surge overvoltages as the system voltage increases with respect to line insulation stresses is examined. At and above 500 kV, the switching transients gain importance over the lightning surges. We also discussed out-of-phase closing, pole scatter and control of switching transients through closing resistors, and synchronous closing and line compensation. The transients on symmetrical and unsymmetrical faults are carried over to Chap. 9.

PROBLEMS

1. A 200-mi-long transmission line is connected to a 400-kV system having a three-phase short-circuit current of 30 kA rms symmetrical, $X/R = 23$. The parameters of the transmission line are resistance = 0.06321 Ω/mi, inductance = 0.90325 Ω/mi, and shunt capacitance reactance = 0.2 M Ω/mile. Calculate the optimum value of the resistor to control the receiving-end voltage. What is the sending-end voltage without the switching resistance? The line is open at the receiving end.
2. Calculate the inrush current of a 400-kV cable, 10 mil long; the cable parameters are: resistance = 0.08 Ω/mi, reactance = 1.2 Ω/mi, $Z = 28$ Ω. The cable is connected to a source of L =.0.015mH, and $R = 0.022$ Ω.
3. The cable in Prob. 2 is paralleled with another cable of the same specifications. The inductance between cables is 0.002 mH. What is the peak back-to-back switching current? What is the frequency of this current?
4. Compare methods of switching overvoltage control described in this chapter with respect to their applicability and limitations.
5. Consider a 400-mi-long transmission line, with $\beta = 0.116°$/mi. What is the maximum power that can be transmitted on this line as function of the natural line load? The line is compensated with 60 percent series compensation. How does the maximum power transfer change? What is the new electrical length of the line?
6. Consider a system as shown in Fig. 7-P1. Calculate the maximum and minimum current through the breaker *S*1 on closing.

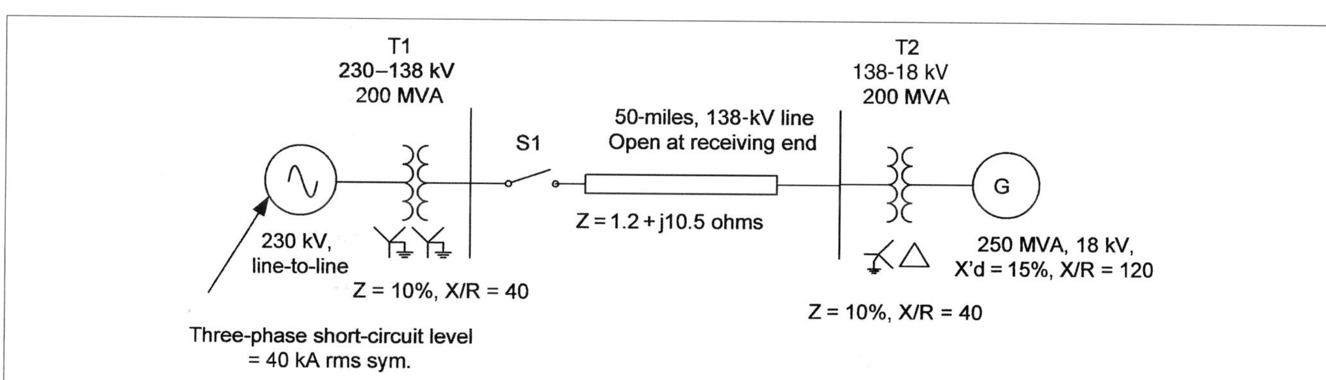

FIGURE 7-P1 Power system configuration for Prob 6.

7. The maximum switching overvoltage for 765 kV should be limited to 2, 2.15, 2.5, or 3 times the system voltage. What is the correct answer? What is the maximum permissible switching overvoltage for 400-kV systems?
8. A single conductor shielded solid dielectric cable has a conductor diameter of 25 mm, insulation thickness 10 mm, dielectric constant of insulation (permittivity) = 4.2. Calculate its surge impedance, electric length, and velocity of propagation.
9. The maximum overvoltage will be obtained when:
   A cable is connected to a short transmission line
   A transmission line is connected to a short cable length.
   Explain why it will be higher in one of the above two cases?
10. Explain how the overvoltages on reswitching of lines and cables, open circuited at the receiving end, which have trapped charges due to first switching operation can be avoided.
11. Compare resistance switching and point of wave switching. Which is better and why? Show how two sets of resistors for HV circuit breakers, one for closing and the other for interrupting, are used.
12. Explain why the switching overvoltage transients on an OH line are higher at the receiving end as compared to the sending end. Can these be higher at the sending end in any operation?
13. Write a one-page note without mathematical equations on BIL versus BSL as the system voltage increases.
14. List all the possible areas of transient simulations where statistical analyses can be recommended.

REFERENCES

1. N. Fujimoto, Transmission Line Transient Analysis Program, TLTA, Ontario Hydro Research Division, March 1985.
2. ANSI Standard C84.1, Voltage Ratings of Electrical Power Systems and Equipment (60 Hz.), 1989.
3. IEC 60071-2, *Insulation Coordination, Part 2-Application Guide*, 1996.
4. BBC Research Center—Surges in High-Voltage Networks, *BBC Symp. on Surges in High-Voltage Networks*, Beden, Switzerland 1979.
5. ANSI/IEEE Standard C57.12, IEEE Standard for Standard General Requirements for Liquid Immersed Distribution, Power, and Regulating Transformers, 2006.
6. H. Glavitsch, "Problems Associated with Switching Surges in EHV Networks," *BBC Review*, vol. 53, pp. 267–277, April/May 1966.
7. A. Pigini, L. Thione, K. H. Week, C. Menemenlis, G. N. Alexandrov and Y. A. Gersimov, CIGRE Working Group 33.03, "Part II-Switching Impulse Strength of Phase-to-Phase External Insulations," *ELECTRA*, pp. 138–157, May 1979.
8. E. W. Boehne and G. Carrara, "Switching Surge Insulation Strength of EHV Line and Station Insulation Structures," CIGRE Report 415, 1964.
9. J. Aubin, D. T. McGillis, and J. Parent, "Composite Insulation Strength of 735 kV Hydro-Quebec 735 kV Towers," *IEEE Trans. PAS*, vol. 85, pp. 633–648, 1966.
10. ANSI Standard C37.06, AC High Voltage Circuit Breakers Rated on Symmetrical Current Basis—Preferred Ratings and Related Capabilities, 2000.
11. E. W. Greenfield, "Transient Behavior of Short and Long Cables," *IEEE Trans. PAS*, vol. 103, no. 11, pp. 3193–3203, Nov. 1984.
12. T. J. E. Miller, (ed.) *Reactive Power Control in Electrical Power Systems*, John Wiley, New York, 1982. (Out of print, used copies can be purchased on Web site www.amazon.com)
13. B. J. Ware, J. G. Reckleff, G. Mauthe, and G. Schett, "Synchronous Switching of Power Systems," CIGRE Session, Report No. 13-205,1990.
14. G. Moraw, W. Richter, H. Hutegger, and J. Wogerbauer, "Point of Wave Switching, Controlled Switching of High-Voltage Circuit Breakers," CIGRE 13-02, 1988 session.
15. R. W. Alexander, "Synchronous Closing Control for Shunt Capacitors," *IEEE Trans. PAS*, vol. 104, no. 9, pp. 2619–2625, Sept. 1985.

FURTHER READING

G. W. Alexander and H. R. Armstrong, "Electrical Design of a 345 kV Double Circuit Transmission Line Including Influence of Contamination," *IEEE Trans. PAS*, vol. 85, pp. 656–665, 1966.

CIGRE Working Group 13-02, Switching Surge Phenomena in EHV Systems, "Switching Overvoltages in Transmission Lines with Special Reference to Closing and Reclosing Transmission Lines," *Electra*, vol. 30, pp. 70–122, 1973.

CIGRE Working Group 33-02, *Internal Overvoltages*, Document 33-78, 1978.

J. K. Dillard, J. M. Clayton, and L. A. Kilar, "Controlling Switching Surges on 1100-kV Transmission Systems," *IEEE Trans. PAS*, vol. 89, No. 8, pp. 1752–1759, 1970.

J. K. Dillard and A. R. Hileman, "Switching Surge Performance of Transmission Systems," *Proc. CIGRE*, Report 33-07, 1970.

E. W. Kimbark and A. C. Legate, "Fault Surge Verses Switching Surge: A Study of Transient Overvoltages Caused by Line-to-Ground Faults," *IEEE Trans. PAS*, PAS-87, pp. 1762–1769, 1968.

V. Koschik, S. R. Lambert, R. G. Rocamora, C. E. Wood, and G. Worner, "Long Line Single Phase Switching Transients and their Effect on Station Equipment," *IEEE Trans. PAS*, vol. 97, pp. 857–964, 1978.

J. F Perkins, "Evaluation of Switching Surge Overvoltage on Medium-Voltage Power Systems," *IEEE Trans. PAS*, vol. 101, no. 6, pp. 1727–1734, 1982.

L. Thione, "Evaluation of Switching Impulse Strength of External Insulation," *Electra*, pp. 77–95, May 1984.

Transmission Line Reference Book, 345 kV and Above, 2d ed., EPRI Palo, Alto, CA, 1982.

J. A. Williams, "Overhead Versus Underground Analysis," *Proc. Trans. Distri. World Expo*, Atlanta, Nov. 1997.

CHAPTER 8

CURRENT INTERRUPTION IN AC CIRCUITS

Current interruption in high-voltage ac networks has been intensively researched. Fundamental electrical phenomena occurring in the electrical network and the physical aspects of arc interruption process need to be considered simultaneously. Current interruption characteristics have a different effect under different conditions, and care must be exercised in applying generalizations.

The basic phenomena of interruption of transient currents are derived from the electrical system configurations and the modifying effect of the current interruption devices, that is, circuit breakers and fuses. Current limiting fuses are discussed in Chap. 20.

The two basic principles of ac interruption are: (1) high-resistance interruption or an ideal rheostatic circuit breaker; dc circuit breakers also employ high-resistance arc interruption, as discussed in Chap. 23; (2) low-resistance or zero-point arc extinction, which is discussed in this chapter.

8-1 ARC INTERRUPTION

The ionization of gaseous mediums and the contact space of circuit breakers depends on:

- Thermal ionization.
- Ionization by collision.
- Thermal emission from contact surfaces.
- Secondary emission from contact surface.
- Photoemission and field emission.
- The arc models in circuit breakers are further influenced by the contact geometry, turbulence, gas pressure, and arc extinguishing devices.[1]

Deionization can take place by recombination, diffusion, and conduction heat. The arc behaves like a nonlinear resistor with arc voltage in phase with the arc current. At low temperature, the arc has falling volt-ampere characteristics. At higher currents, voltage gradient levels out, practically becoming independent of current. Figure 8-1 shows radial temperature distribution between low- and high-current arcs. At high currents, arc temperature of the order of 20000 K occurs and tends to be limited by radiant heat transfer. The heat transfer and electrical conductivity have nearly constant values within the arc column.

It is rather perplexing that when the arc is cooled, the temperature increases. This happens due to the reduction in the diameter of the core, which results in higher current density (of the order of several thousand amperes/cm^2).

High- and low-pressure arcs can be distinguished. High-pressure arcs exist at pressures above atmosphere and appear as a bright column, characterized by a small burning core at high temperatures of the order of 20000 K. In low-pressure (vacuum arcs), the arc voltages are low of the order of 40 V, and the positive column of the arc is solely influenced by the electrode material, while that of high-pressure arcs is made up of ionized gases from arc's surrounding medium.

The central core is surrounded by a column of hot gases. A first attempt to relate the volt-ampere characteristic of a *steady* arc is given by the following expression, still in use:

$$V_{\text{arc}} = A + Bd + \frac{C + Dd}{I_{\text{arc}}} \tag{8-1}$$

where I_{arc} is the arc current; V_{arc} is the voltage across the arc; d is the length of the arc; and A, B, C, and D are constants. For small arc lengths, the voltage across arc length can be neglected, and we can write:

$$V_{\text{arc}} = A + \frac{D}{I_{\text{arc}}} \tag{8-2}$$

Voltage across the arc reduces as the current increases. Energy dissipated in the arc is:

$$E_{\text{arc}} = V_{\text{arc}} I_{\text{arc}} t \tag{8-3}$$

where t is the duration of arc in seconds. For a high-voltage current zero circuit breaker with an interrupting time of 2 cycles, the arcing time is 1 cycle. The approximate variation of arc resistance, r with time t, is obtained for different parameters of the arc by experimentation and theoretical analysis.

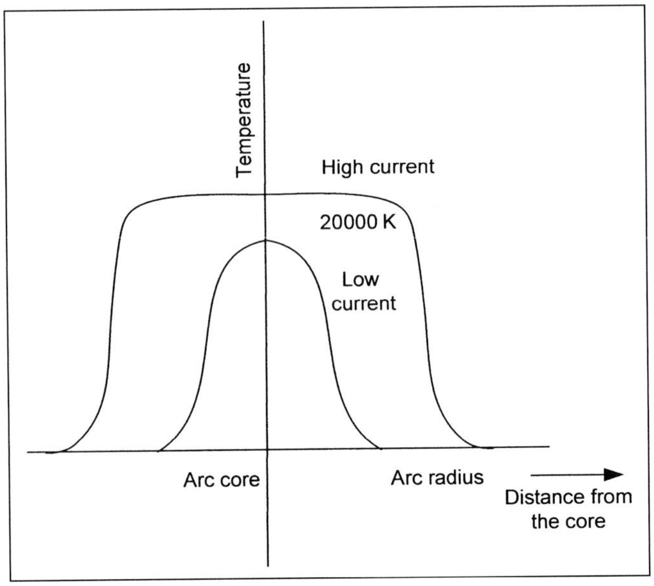

FIGURE 8-1 Arc temperature versus arc radius, low and high currents.

8-2 ARC INTERRUPTION THEORIES

8-2-1 *Slepian Theory*

It was proposed in 1928, and states that arc extension process is a race between the dielectric strength and restriking voltage. After current zero, there is a hot residual column of ionized gases. If the dielectric strength builds up quickly, so that it is always greater than the restriking voltage, the arc does not restrike. If the dielectric strength is less, the arc restrikes. The theory assumes that the restriking voltage and the buildup of dielectric strength are comparable quantities, which is not correct. Secondly, the theory does not consider energy functions in the arc extinction process.

8-2-2 *Cassie's Theory*

A differential equation describing the behavior of arc was presented by A. M. Cassie in 1939.[2] It assumes a constant temperature across the arc diameter. As the current varies, so does the arc cross section, but not the temperature inside the arc column. Under given assumptions, the conductance G of the model is proportional to current, so that the steady-state voltage gradient E_0 is fixed. The time constant:

$$\theta = \frac{Q}{N} \qquad \text{(8-4)}$$

where Q is energy storage capability and N is the finite rate of energy loss which defines the time lag due to energy storage and finite rate of energy loss. The following is the simplified Cassie's equation given in terms of instantaneous current:

$$\frac{d}{dt}(G^2) = \frac{2}{\theta}\left(\frac{I}{E_0}\right)^2 \qquad \text{(8-5)}$$

For high-current region, there is a good agreement with the model, but for current-zero region, agreement is good only for high rates of current decay.

8-2-3 *Mayr's Theory*

O. Mayr considered an arc column where arc diameter is constant and where arc temperature varies with time and radial dimensions. The decay of temperature of arc is assumed due to thermal conduction, and the electrical conductivity of arc is dependent on temperature:

$$\frac{1}{G}\frac{dG}{dt} = \frac{1}{\theta}\left(\frac{EI}{N_0} - 1\right) \qquad \text{(8-6)}$$

where:

$$\theta = \frac{Q_0}{N_0} \qquad \text{(8-7)}$$

The validity of the theory during current zero periods is acknowledged.

8-2-4 *Cassie Mayr Theory*

Mayr assumed arc temperature of 6000 K, but it is recognized to be in excess of 20000 K. At these high temperatures, there is a linear increase in gas conductivity. Assuming that before current zero, the current is defined by driving circuit, and that after current zero, the voltage across the gap is determined by arc circuit, we write the following two equations:

1. Cassie's period prior to current zero:

$$\frac{d}{dt}\left(\frac{1}{R^2}\right) + \frac{2}{\theta}\left(\frac{1}{R^2}\right) = \frac{2}{\theta}\left(\frac{1}{E_0}\right)^2 \qquad \text{(8-8)}$$

2. Mayr's period around current zero:

$$\frac{dR}{dt} - \frac{R}{\theta} = -\frac{e^2}{\theta N_0} \qquad \text{(8-9)}$$

This has been extensively used in circuit breaker designs, however, constant θ must be deduced from experimental results.

8-3 CURRENT-ZERO BREAKER

A circuit breaker does not operate instantaneously on a tripping or closing signal, nor do the contacts part and close simultaneously. There is a finite time before the switch/circuit breaker can interrupt a transient current. In a current-zero circuit breaker, the interruption takes place during the passage of current through zero, though there is some chopping of current because of instability of arc close to current zero. Before interruption occurs, the arc may persist for a number of cycles, depending on the design of the circuit breaker. At some initial current zeros, the arc reignites because the contact gap has not developed enough dielectric strength. We say that the *restriking* voltage is higher than the electrical strength of the gap. The dielectric strength of the arc gap is primarily a function of the interrupting devices, while voltage appearing across parting contacts is a function of system constants.

An arc can keep burning and will not extinguish itself even at some low amperes. The circuit breakers are called upon to interrupt currents of the order of tens of kilo-amperes or even more. Generator circuit breakers having an interrupting capability of 250 kA rms symmetrical are available. At the first current zero, the electrical conductivity of the arc is fairly high, and since the current-zero period is very short, the reignition cannot be prevented. The time lag between current and temperature is commonly called *arc hysteresis*.

When the current passes through zero, the arc voltage jumps to a value equal to the instantaneous peak value of extinguishing voltage from previous current loop, plus peak value of reignition value of next loop, associated with the reversal of current. The most favorable condition for interruption is that in which the applied voltage reaches zero when current is zero, that is, interrupting a resistive current.

As the contact separation increases, the transient voltage of the arc does not succeed in igniting the arc because the electrical strength of the break gap increases. The gap can now withstand the recovery voltage stress. All high-voltage breakers, whatever may be the arc-quenching medium (oil, air, or gas), use current-zero interruption. In an ideal circuit breaker, with no voltage drop before interruption, arc energy is zero. Modern circuit breakers approach this ideal on account of short arc duration and low arc voltage.

The various arc-quenching mediums in circuit breakers for arc interruption have different densities, thermal conductivities, dielectric strengths, arc time constants, and so on. We will briefly discuss the arc interruption in SF_6 in sec. 8-13.

Figure 8-2*b* shows a typical short-circuit current waveform in an inductive circuit of Fig. 8-2*a*, short circuit applied at $t = 0$. Short-circuit current is limited by the impedance $R + j\,\omega L$ and is interrupted by breaker *B*. The waveform shows asymmetry, as

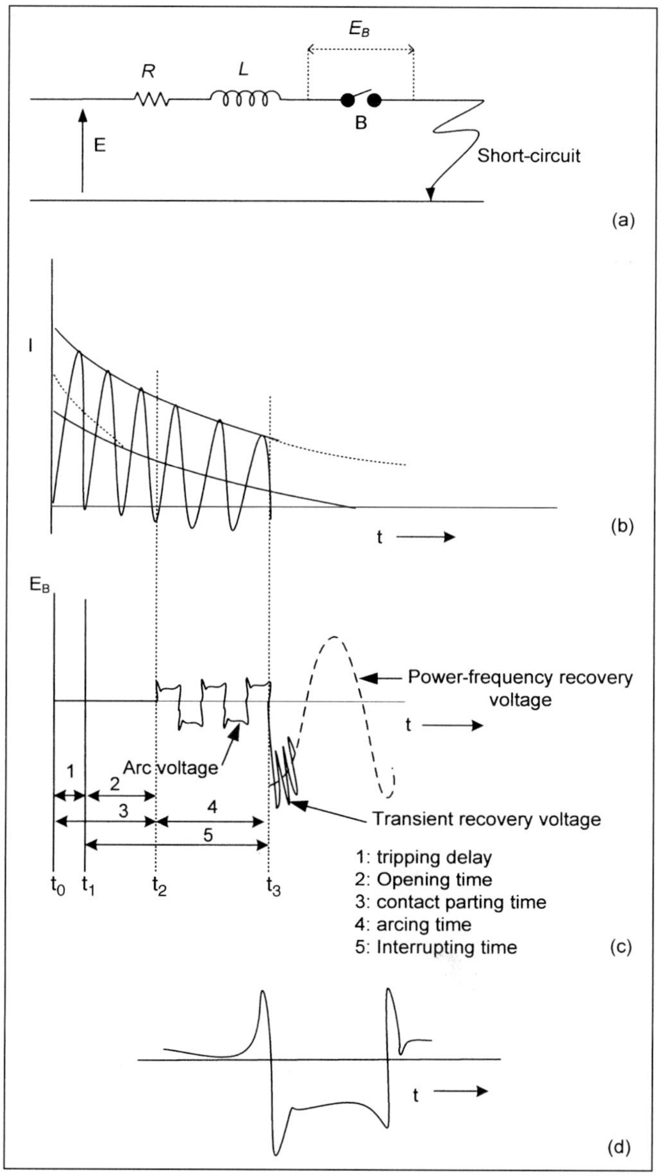

FIGURE 8-2 (*a*) Short circuit of an inductive circuit. (*b*) Asymmetrical short-circuit current profile. (*c*) Short-circuit current interruption. (*d*) Arc voltage profile.

discussed in Chap. 2. At t_2, the contacts start parting. The time $t_2 - t_0$ is termed *contact parting time* (IEC minimum trip delay), and consists of tripping delay, interval $t_1 - t_0$, and opening time interval $t_2 - t_1$. It takes some finite time for the breaker operating mechanism to set in motion and the protective relaying to signal a fault condition. In ANSI/IEEE and IEC standards for breakers,[3–5] this tripping delay is considered equal to 1/2 cycle.

As the contacts start parting, an arc is drawn, which is intensely cooled by the quenching medium (air, SF_6, or oil). The arc current varies for short duration and may have some oscillations. This voltage drop across the arc during arcing period is shown exaggeratedly in Fig. 8-2*d* for clarity. The arc is mostly resistive and the voltage in the arc is in-phase with the current. The contacts part at high speed to prevent continuation of the arc. The peculiar shape of arc voltage, shown in Fig. 8-2*d*, is the result of volt-ampere characteristic of the arc, and at a particular current zero, the dielectric strength of the arc space is sufficiently high. When the dielectric strength builds and a current zero occurs, the arc is interrupted. In Fig. 8-2*c*, it occurs at t_3, and the interval $t_3 - t_2$ is the arcing time. The interval $t_2 - t$ is the total interrupting time. With modern high-voltage breakers, it is as low as 2 cycles or even lower, based on the system power frequency.

It can be seen that the interruption process is much dependent on the current being interrupted, resistive, inductive, or capacitive. The breakers and switching devices are tested at a low power factor because the system inductance predominates and the short-circuit currents which impose the maximum switching duties on the circuit breakers are mostly reactive. ANSI standard defines the rated interrupting time of a circuit breaker as the time between trip circuit energization and power arc interruption on an opening operation, and it is used to classify breakers of different speeds. The rated interrupting time may be exceeded at low values of current and for close-open operations; also, the time for interruption of resistor current for interrupters equipped with resistors may exceed the rated interrupting time. The increase in interrupting time on close-open operation may be important from the standpoint of line damage and possible instability.

The significance of interrupting time on breaker interrupting duty can be seen from Fig. 8-2*b*. As the short-circuit current consists of decaying ac and dc components, the faster the breaker, the greater the asymmetry and the interrupted current.

Interrupting current at final arc extinction is asymmetrical in nature, consisting of an ac component and a dc component. The rms value of the asymmetrical current is given by:

$$I_{\text{rms,asym,total}} = \sqrt{I_{\text{acrms}}^2 + I_{\text{dc}}^2} \tag{8-10}$$

Table 18-1 shows the breaker interrupting times, contact parting time, and the asymmetrical current duties for breakers rated according to ANSI/IEEE standards.

8-4 TRANSIENT RECOVERY VOLTAGE

The essential problem of current interruption involves rapidly establishing an adequate electric strength across the break after current zero, so that restrikes are eliminated. Whatever may be the breaker design, it can be said that it is achieved in most interrupting mediums, that is, oil, air-blast, or SF_6 by an intense blast of gas. The flow velocities are always governed by aerodynamic laws. However, there are other factors that determine the rate of recovery of the dielectric medium: nature of quenching gases, mode of interaction of pressure and velocity of the arc, arc control devices, contact shape, number of breaks, and the circuit in which the breaker is connected.

At the final arc interruption, a high-frequency oscillation, superimposed upon the power-frequency voltage, appears across the breaker contacts. A short-circuit current loop is mainly inductive,

TABLE 8-1 Interrelation of Circuit Breaker Interrupting Time, Contact Parting Time, and Asymmetrical Ratings (ANSI/IEEE, Circuit Breakers Rated on Symmetrical Current Basis)

Breaker Interrupting Time Cycles	Tripping Delay (Cycles, 60-Hz Basis)	Breaker Contact Parting Time (Cycles, 60-Hz Basis)	Breaker Opening Time (Cycles, 60-Hz Basis)	Breaker Arcing Time (Cycles, 60-Hz Basis)	Required DC Component (%) Calculated at Contact Porting Time
2	0.5	1.5	1	1	58
3	0.5	2	1.5	1.5	49
5	0.5	3	2.5	2.5	32

Primary contact parting time is considered equal to the sum of 1/2 cycle (tripping delay) and the *lesser of*:

- Actual opening time of the circuit breaker, or
- 1.0, 1.5, 2.5 for breakers having an interrupting time of 2, 3, and 5 cycles, respectively.

and the power-frequency voltage has its peak at the current zero; however, a sudden voltage rise across the contacts is prevented by inherent capacitance of the system, and in simplest cases, a transient of the order of some hundred to 10000 c/s can occur. It is termed the *natural frequency* of the circuit. Figure 8-3 shows the recovery-voltage profile after final current extinction. The two components of the recovery voltage, (1) a high-frequency damped oscillation and (2) the power-frequency recovery voltage, are shown. The high-frequency component is called the *transient recovery voltage* (TRV) and sometimes the *restriking voltage*. Its frequency is given by:

$$f_n = \frac{1}{2\pi\sqrt{LC}} \tag{8-11}$$

where f_n is the natural frequency, and L and C are equivalent inductance and capacitance of the circuit, respectively.

If the contact space breaks down within a period of 1/4 cycle of initial arc extinction, the phenomena are called *reignitions*, and if the breakdown occurs after 1/4 of a cycle, the phenomena are called *restrikes*.

The transient oscillatory component subsides in a few microseconds and the power- frequency component continues. As an example, an inductor of 2 mH, with small effective capacitance of 100 pF, has its natural frequency of 356 kHz, a period of approximately 2.8 μs; see Eq. (8-11). On a 138-kV system, neglecting damping, the voltage will swing to twice the system voltage in half period, and the mean rate of rise of the recovery voltage is 160 kV per μs, which is very high and will cause restrikes in a circuit breaker. Capacitors and resistors are used in high-voltage circuit breakers to damp out the recovery voltage. The breakers are rated to withstand a certain recovery-voltage profile.

The calculation of the initial rate of rise is important and is not so simple because of traveling-wave phenomena and multiple reflections as a surge is impressed upon the two systems being disconnected by the breaker. The recovery-voltage profiles can be considered for fault currents, maximum and minimum, and the transformer-limited fault currents. Also nonfault currents can be resistive, inductive, and capacitive, and the circuit configurations play an important role, for example, terminal faults and short-line faults. The recovery-voltage profile varies in all these cases; the TRV profiles can be:

- Mostly supply frequency with little buzz when large capacitance currents are interrupted
- Overdamped
- Oscillatory; occur when switching a shunt reactor, unloaded transformer, or clearing a fault limited by transformer or series reactor

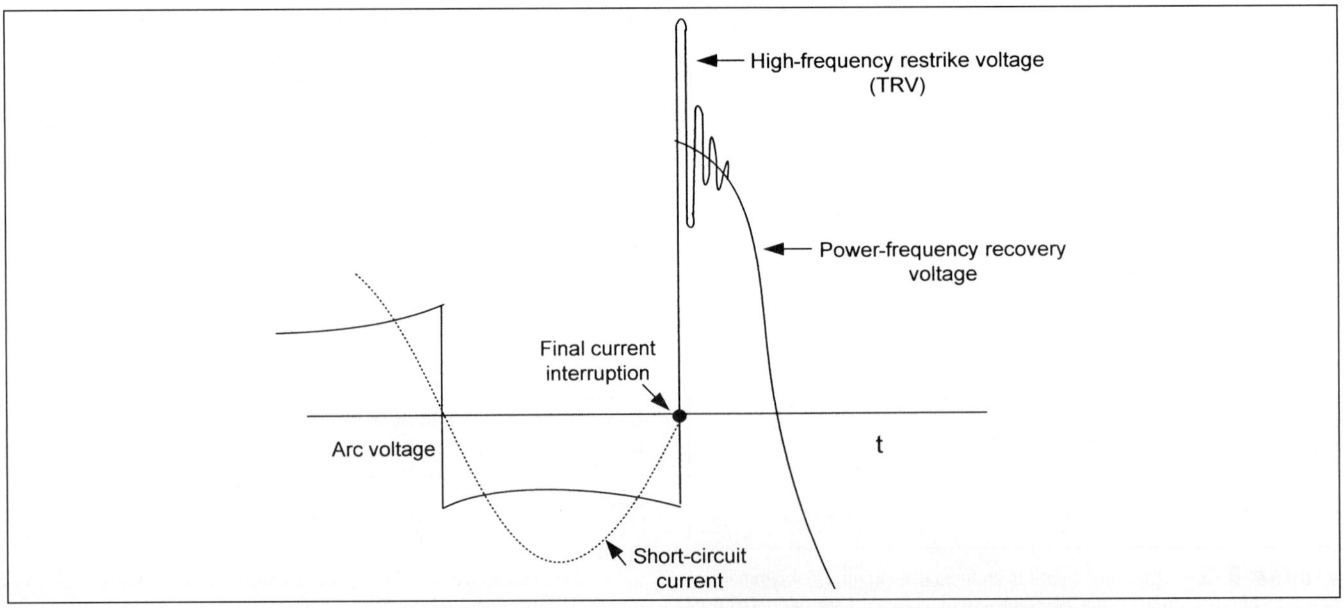

FIGURE 8-3 TRV and power-frequency recovery voltage components after current interruption.

- Triangular; occur in short-line faults
- Entirely different problems and waveforms, with escalation of voltage; occur if restrikes take place during the process

Consider that the two symmetrical three-phase networks are interconnected through a circuit breaker. Let the positive-, negative-, and zero-sequence impedances of left-hand and right-hand side networks be W', W'', W^0 and Z', Z'', Z^0, respectively. Let UW1, UW2, UW3 and UZ1, UZ2, UZ3 be the momentary values of the phase voltages across the open poles of the breaker and similarly $iW1$, $iZ1$, ... be the phase currents. Then:

$$t^k = \frac{2l}{v^k} \tag{8-12}$$

where l is the first point of reflection and superscript k denotes positive-, negative-, and zero-sequence modes. The shortest of these times then determines the RRRV (rate of rise of recovery voltage). Using symmetrical component transformations, for each system W', W'', W^0, and Z', Z'', and Z^0:

$$\begin{vmatrix} i_1 \\ i_2 \\ i_3 \end{vmatrix} = \frac{1}{3} \begin{vmatrix} \alpha_{Z,W} & \beta_{Z,W} & \gamma_{Z,W} \\ \gamma_{Z,W} & \alpha_{Z,W} & \beta_{Z,W} \\ \beta_{Z,W} & \gamma_{Z,W} & \alpha_{Z,W} \end{vmatrix} \tag{8-13}$$

where:

$$\alpha_Z = \frac{1}{Z'} + \frac{1}{Z''} + \frac{1}{Z^0}$$

$$\beta_Z = \frac{a}{Z'} + \frac{a^2}{Z''} + \frac{1}{Z^0} \tag{8-14}$$

$$\gamma_Z = \frac{a^2}{Z'} + \frac{a}{Z''} + \frac{1}{Z^0}$$

The expressions for initial recovery voltage are provided in Table 18-2.

8-4-1 First Pole to Clear Factor

TRV refers to the voltage across the first pole to clear because it is generally higher than the voltage across the other two poles of the breaker which clear later. Consider a three-phase ungrounded fault. The voltage across the breaker phase, first to clear, will be 1.5 times the phase voltage (Fig. 8-4). The arc interruption in three phases is not simultaneous, as the three phases are mutually 120° displaced. Thus, theoretically, the power-frequency voltage of the first pole to clear is 1.5 times the phase voltage. It may vary from 1.5 to 2, rarely exceeding 3 and can be calculated using symmetrical components. The first pole to clear factor is defined as the ratio of rms voltage between faulted phase and unfaulted phase and phase to neutral voltage with the fault removed. Figure 8-4 shows first pole to clear factors for three-phase terminal faults. The first pole to clear factor for three-phase fault with ground contact is calculated as:

$$1.5\frac{2X_0/X_1}{1+[2X_0/X_1]} \tag{8-15}$$

without ground contact, the following expression can be written:

$$1.5\left[\frac{2(X_0+Y_0)X_1}{1+[2(X_0+Y_0)/X_1]}\right] \tag{8-16}$$

where X_1 and X_0 are the positive- and zero-sequence reactances of the source side. Also, in Fig. 8-4, Y_1 and Y_0 are the sequence admittances of the load side. Currently, first pole to clear factor is 1.3 for circuit breaker for solidly grounded systems above 100 kV. It varies with type of fault. Say for a 550 kV breaker, it is 1.3 for terminal fault, 1.0 for short-line fault, and 2.0 for out-of-phase fault.[6]

Figure 8-5 illustrates the slopes of tangents to three TRV waveforms of different frequencies. As the natural frequency rises, the RRRV increases. Therefore, it can be concluded that:

1. Voltage across breaker contacts rises slowly as RRRV decreases.
2. There is a swing beyond recovery voltage value, the amplitude of which is determined by the circuit breaker and damping.
3. The higher the natural frequency of the circuit being interrupted, the lower the breaker interrupting rating.
4. Interrupting capacity/frequency characteristics of the breaker should not fall below that of the system.

The TRV is affected by many factors, amongst which the power factor of the current being interrupted is important. At zero power factors, maximum voltage is impressed across the gap at the instant of current zero, which tends to reignite the arc in the hot arc medium.

TRV can be defined by specifying the crest and the time to reach the crest, and alternatively, by defining the segments of lines which envelope TRV waveform.

The steepest rates of the rise in a system are due to short circuits beyond transformers and reactors which are connected to a system of high short-circuit power. In these cases, the capacitance which retards the voltage rise is small; however, the breaking capacity of the breaker for such faults need only be small compared to

TABLE 8-2 Initial Transient Recovery Voltage

PHASE TO CLEAR	RECOVERY VOLTAGE ACROSS BREAKER UP TO ARRIVAL OF FIRST REFLECTION
1	$U_{s_1} = i_1(t)\dfrac{3(W^0+Z^0)(W'+Z')(W''+Z'')}{(W^0+Z^0)(W'+Z')+(W^0+Z^0)(W''+Z'')+(W'+Z')(W''+Z'')}$
2	$U_{s_2} = i_2(t)\dfrac{(W^0+Z^0)(W'+Z')+(W^0+Z^0)(W''+Z'')+(W'+Z')(W''+Z'')}{W^0+W'+W''+Z^0+Z'+Z''}$
3	$U_{s_3} = i_3(t)\dfrac{1}{3}(W^0+W'+W''+Z^0+Z'+Z'')$

$i_1(t)$, $i_2(t)$, and $i_3(t)$ are the currents injected into first, second, and third phases to clear. Us_1, Us_2, Us_3 are initial TRVs.

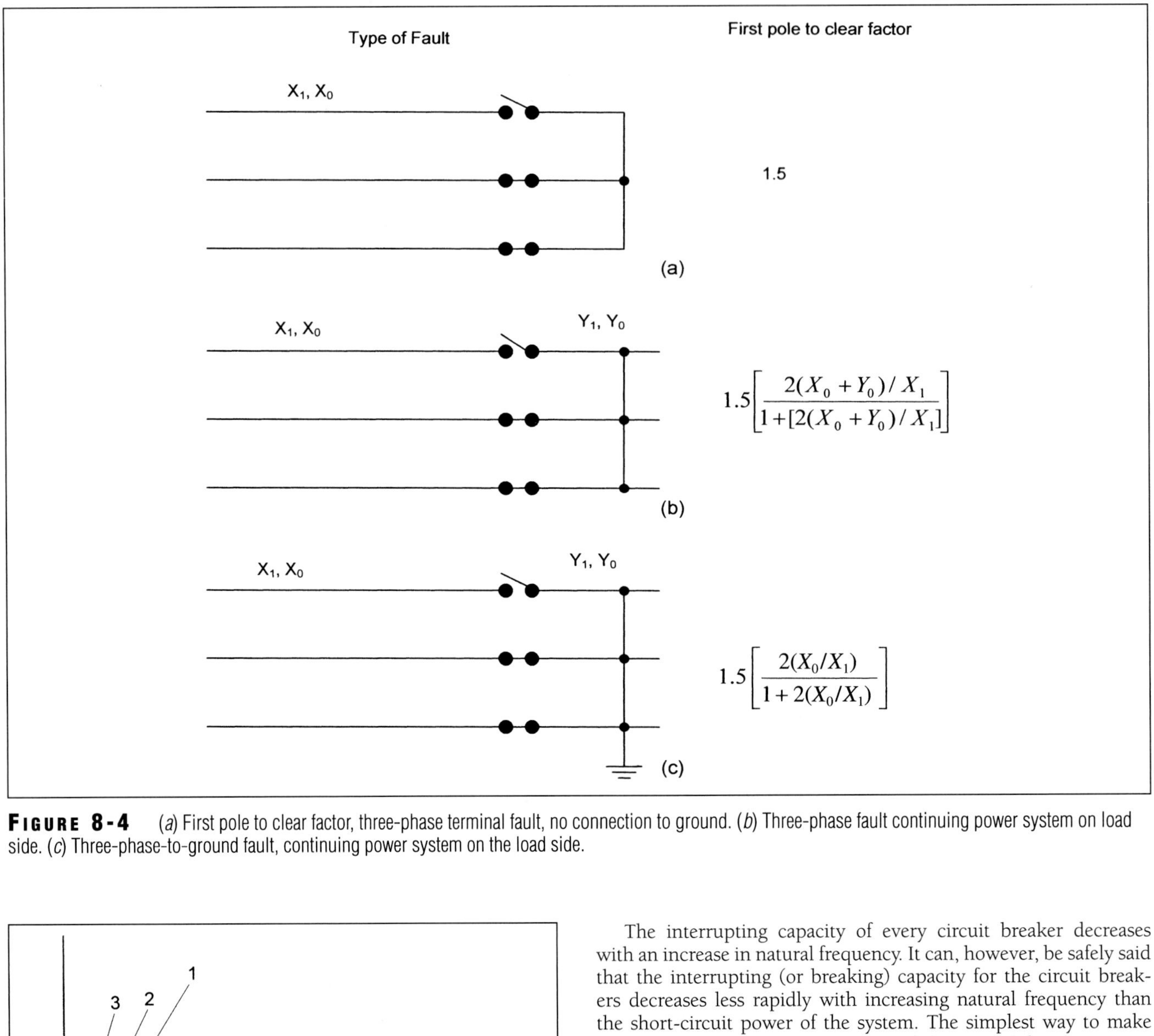

FIGURE 8-4 (*a*) First pole to clear factor, three-phase terminal fault, no connection to ground. (*b*) Three-phase fault continuing power system on load side. (*c*) Three-phase-to-ground fault, continuing power system on the load side.

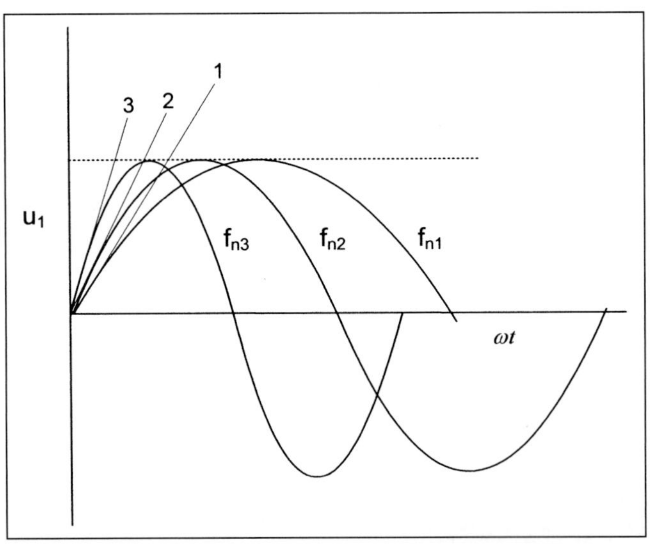

FIGURE 8-5 Effect of frequency of TRV on the RRRV (rate of rise of recovery voltage).

short-circuit power of the system, as the current is greatly reduced by the reactance of the transformers and reactors. It means that in most systems, high short-circuit levels of current and high natural frequencies may not occur simultaneously.

The interrupting capacity of every circuit breaker decreases with an increase in natural frequency. It can, however, be safely said that the interrupting (or breaking) capacity for the circuit breakers decreases less rapidly with increasing natural frequency than the short-circuit power of the system. The simplest way to make the breaking capacity independent of the natural frequency is to influence the rate of rise of recovery voltage across breaker contacts by resistors, which is discussed further. Yet, there may be special situations where the interrupting rating of a breaker may have to be reduced or a breaker of higher interrupting capacity is required. TRV is an important parameter for the specifications of circuit breakers, and ANSI/IEEE and IEC standards specify TRV parameters in the rating structure of the breakers.

8-5 SINGLE-FREQUENCY TRV TERMINAL FAULT

A circuit of a single-frequency transient occurs for a terminal fault in a power system composed of distributed capacitances and inductances. A *terminal fault* is defined as a fault close to the circuit breaker, and the reactance between the fault and the circuit breaker is negligible. TRV can vary from low to high values in the range of 20 to 10000 Hz.

Figure 8-6*a* shows power system constants, that is, resistance, inductance, and capacitance. The circuit shown may well represent the Π model of a transmission line (see Chap. 4). Figure 8-6*b* shows the basic parameters of the recovery-voltage transient for a terminal fault in simplified network of Fig. 8-6*a*. It shows behavior of recovery voltage, transient component, and power-frequency component. The amplitude of the power-frequency component is given by:

$$\alpha\sqrt{2}u_0 \tag{8-17}$$

where α depends on the type of fault and the network, and u_0 is the rated system rms voltage. The rate of rise of recovery voltage (RRRV = S) is the tangent to the transient recovery voltage starting from the zero point of the unaffected or inherent transient recovery voltage (ITRV). This requires some explanation. The TRV can change by the circuit breaker design and operation. The TRV measured across terminals of two circuit breakers can be different. The power system characteristics are calculated ignoring the effect of the breakers. This means that an ideal circuit breaker has zero terminal impedance when carrying its rated current, and when interrupting short-circuit current, its terminal impedance changes from zero to infinity instantaneously. The TRV is then called *inherent transient recovery voltage*.

Figure 8-6c shows an enlarged view of the slope. Under a few microseconds, the voltage behavior may be described by time

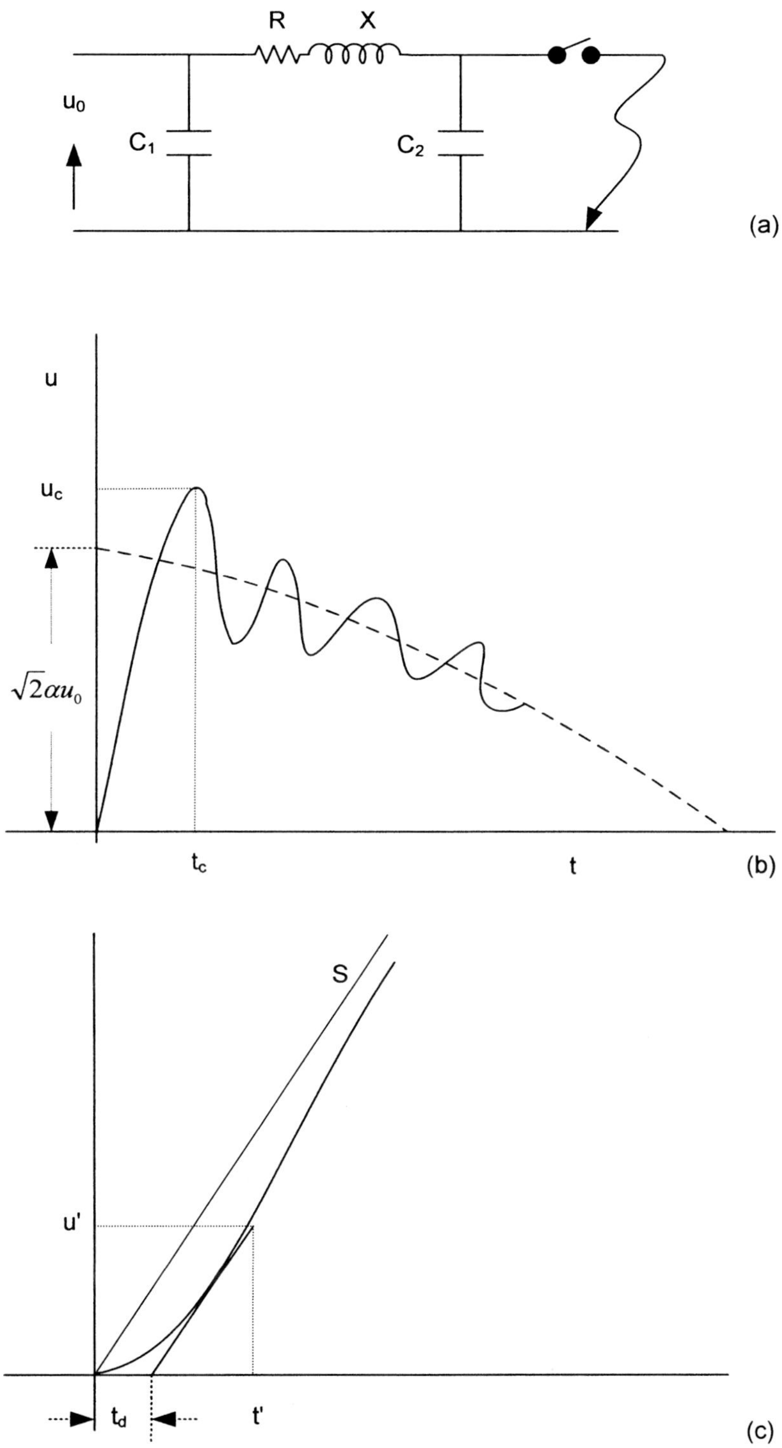

FIGURE 8-6 Recovery-voltage profile on a terminal fault. (*a*) System configuration. (*b*) Recovery-voltage profile. (*c*) Initial TRV curve, delay line and RRRV shown as *S*.

delay t_d, which is dependent on the ground capacitance of the circuit. The time delay t_d in Fig. 8-6c is approximated by:

$$t_d = CZ_0 \tag{8-18}$$

where C is the ground capacitance and Z_0 is the surge impedance. Measurements show that often a superimposed high-frequency oscillation appears. IEC specifications recommend a linear voltage rise with a surge impedance of 450 Ω and no time delay when the faulted phase is the last to be cleared in a single line-to-ground fault. This gives the maximum TRV. It is practical to distinguish between terminal faults and short-line faults for these phenomena.

8-5-1 TRV in Capacitive and Inductive Circuits

The TRV on interruption of capacitive and inductive circuits is illustrated in Fig. 8-7. In a capacitive circuit when the current passes through zero, Fig. 8-7*a*, the system voltage is trapped on the capacitors; also see Fig. 6-35*b*. The recovery voltage, the difference between the source side and load side of the breaker, reaches a maximum of 2.0 per unit after 1/2 cycle of current interruption, Fig. 8-7*b* and *c*. The TRV oscillations are practically absent as large capacitance suppresses the oscillatory frequency [Eq. (8-11)] and the rate of rise of the TRV is low. This may prompt circuit breaker contacts to interrupt, when there is not enough separation between them and precipitate restrikes.

This is not the case when disconnecting an inductive load (Fig. 8-7*d*, *e*, and *f*). The capacitance on the disconnected side is low and the frequency of the isolated circuit is high. The TRV, is, therefore, oscillatory. The rate of rise of TRV after disconnection is fairly high.

Above are examples of single-frequency transients which occur when the electrical energy is distributed between capacitive and inductive elements and no transmission line remains connected at the bus after the short circuit. Surge voltage is neglected; there is no initial charge on the capacitor or initial flow of current. Then the voltage at the capacitor, which can be equated to TRV, from Chap. 2, is:

$$\text{TRV} = \frac{V}{LC}\left(\frac{s}{s^2+\omega^2}\right)\left(\frac{1}{s^2+1/LC}\right) \tag{8-19}$$

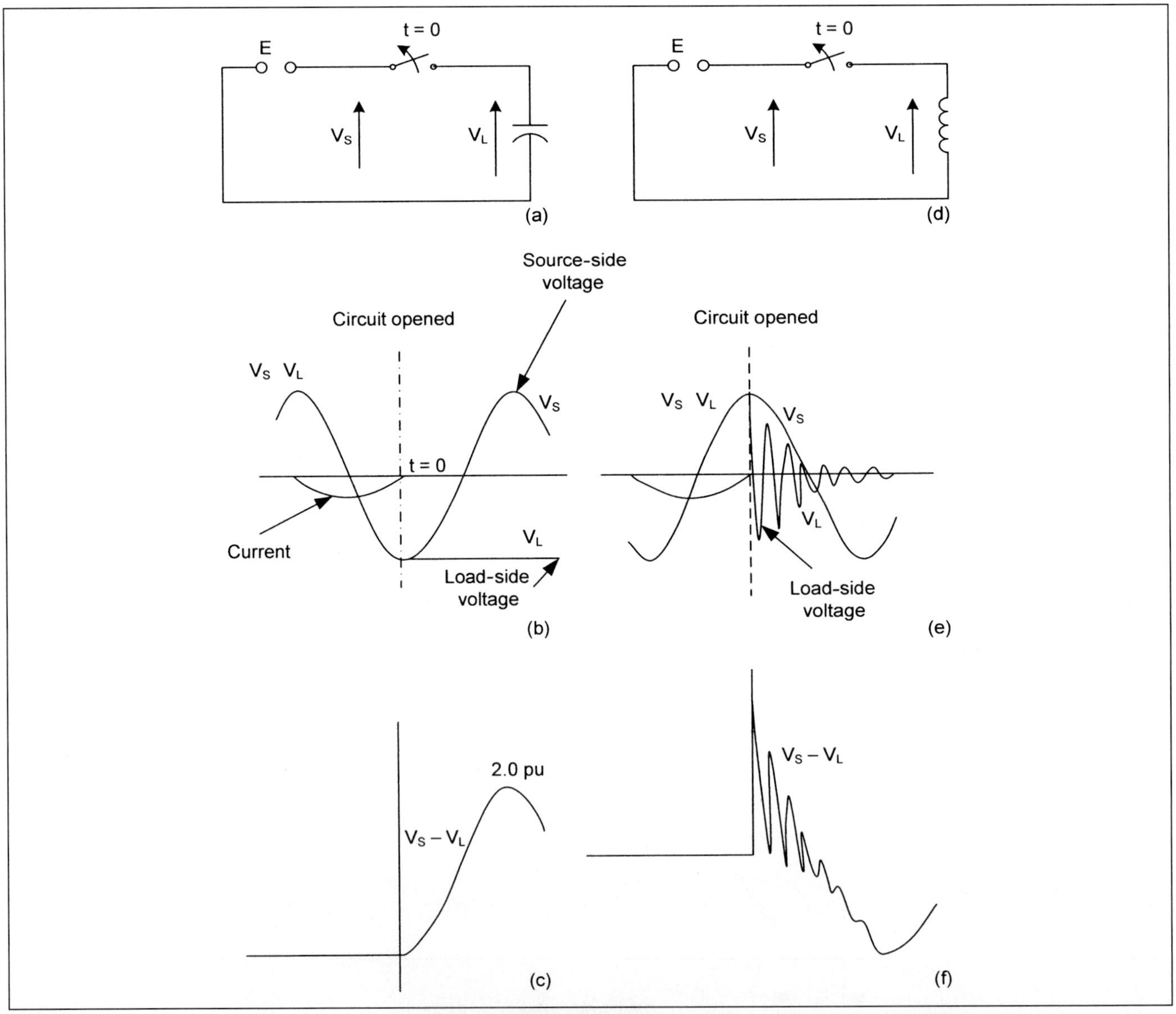

FIGURE 8-7 (*a*), (*b*), and (*c*) Interruption of capacitance current, circuit diagram, load- and source-side voltages, and TRV, respectively. (*d*), (*e*), and (*f*). Interruption of inductive current, circuit diagram, load- and source-side voltages, and TRV, respectively.

By substituting:

$$\omega_0 = \sqrt{1/LC}$$

$$\text{TRV} = \frac{V}{LC}\left[\frac{\cos\omega t - \cos\omega_0 t}{\omega_0^2 - \omega^2}\right] \qquad (8\text{-}20)$$

$$= V(1-\cos\omega_0 t) \qquad \text{for} \qquad \omega << \omega_0$$

Here V is generally taken as 1.88 times the system rms voltage based on a terminal fault. We discussed first pole to clear factor [Eqs. (8-15) and (8-16)], and stated that the power frequency overvoltage can be 1.5 to 3.0 per unit, but it rarely exceeds 2.0.

This is the so called "1-minus-cosine" curve of ANSI specifications.

Reverting to TRV on disconnection of a capacitance, IEEE standard [C37.04] specifies that voltage across capacitor neutral, not ground, during an opening operation should not exceed 3.0 per unit for a general-purpose breaker.[4] For the definite-purpose breaker rated 72.5 kV and lower, this limit is 2.5 per unit. For breakers rated 121 kV and above, capacitor banks are normally grounded (Chap. 6), and the voltage is limited to 2.0 per unit. It was observed that when interrupting capacitance current 2.0 per unit voltage occurs at small contact separation, this can lead to restrikes. With no damping, the voltage will reach 3.0 per unit (Fig. 6-35*c*). This suggests that one restrike is acceptable for general-purpose breakers, but not for definite-purpose breakers.

Some voltage-controlled methods are resistance switching and surge arresters. The resistance switching was discussed in Chap. 6 (Fig. 6-39*a*), which shows that with 1.32-Ω resistor inserted for 4 cycles, the switching transients almost disappear. The effect on voltage recovery is shown in the following example.

Example 8-1 This is an EMTP simulation of resistance switching to control recovery voltage. Again, consider 6-Mvar capacitor bank, 1.32 Ω switching resistance. The system is in steady state and the capacitor is energized. First open the switch at 4 ms to insert 1.32-Ω resistance in series with the break, and then at 8 ms open the resistor circuit. The resulting source and load-side voltages and TRV are shown in Fig. 8-8. This shows a TRV of approximately 2 per unit and no oscillations are observed. The tripping and closing resistors have conflicting requirements, and in HV breakers there may be separate resistors for these functions or the resistance value is optimized (Chap. 7).

8-6 DOUBLE-FREQUENCY TRV

A circuit with inductance and capacitance on both sides of the circuit breaker gives rise to double-frequency TRV. After fault interruption, both circuits oscillate at their own frequencies and a composite double-frequency transient appears across the circuit-breaker contacts. This can occur for a short-line fault. Recovery may feature traveling waves, depending on the faulted components in the network.

The circuit breakers in systems above 121 kV are usually applied in composite circuits where the faults can be fed from the transmission lines and transformers. Consider Fig. 8-9*a* where *n* transmission lines emanate from a bus; note that a transformer terminated line is included. The bus is fed from a transformer and the source reactance can be added to the transformer reactance; a three-phase ungrounded fault occurs on the bus, as shown in the figure.

Figure 8-9*b* shows the transient network when the first pole (in phase *a*) interrupts the fault, while the other two poles of the breaker are still closed. The phase *a* voltage to ground is 1.0 per unit and the phases *b* and *c* voltages are equal to –0.5 per unit. If the voltage on the load side of the breaker pole is denoted as V_L and on the source side as V_B, then the recovery voltage is the difference of these two voltages.

To derive at an equivalent circuit, as shown in Fig. 8-9*c*, the surge impedance of *n* lines in parallel is Z/n. The faulted line also has its surge impedance Z, and C is the capacitance of the transformer, bus, and breaker, all lumped together. On the bus side, the voltage V_B is determined by the voltage drop across transformer reactance L_s, bus capacitance C, and equivalent surge impedance of *n* lines in parallel. On the fault side of the breaker pole, the path is through phases *b* and *c* in parallel. This explains the bottom-half of Fig. 8-9*c*.

The circuit is solved by current injection method. In current injection method, a current equal and opposite to the short-circuit

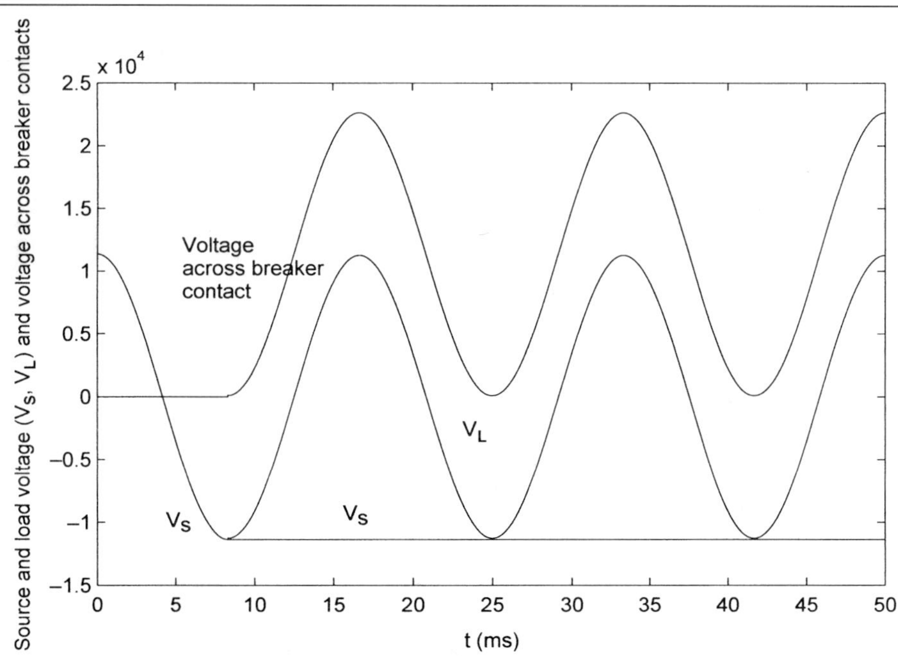

FIGURE 8-8 EMTP simulation of TRV, resistance switching, interruption of capacitive currents, Example 18-1.

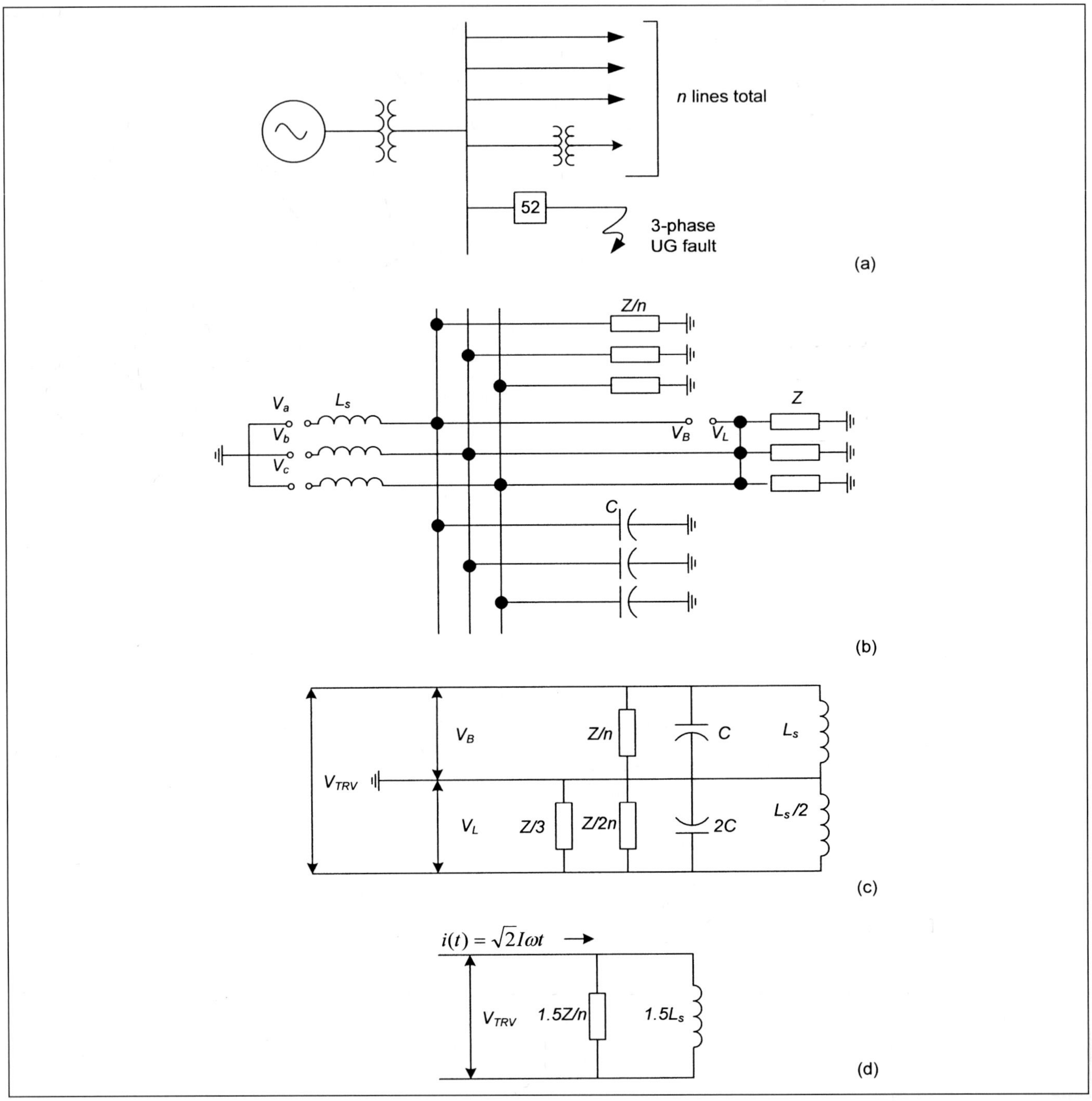

FIGURE 8-9 (*a*) A power system configuration. (*b*) Equivalent circuit drawn with first pole to clear; the other two poles still connected. (*c*) Development of circuit for TRV. (*d*) Simplification of the circuit in (*c*) for TRV calculations.

current that would have continued to flow in the event that interruption does not occur is flowing at the instant of current zero when the current interruption takes place. All bus voltage sources are equal to zero and a current pulse in injected into the breaker pole, which is equal and opposite to the current that would have flown if the breaker pole would have remained closed. Thus, 1/2 cycle of the three-phase fault current is injected into the open phase of the breaker.

A further simplification is shown in Fig. 8-9*d*. This ignores the capacitance, and the surge impedance of the faulted line which will be relatively high. Ignoring capacitance is justified on the basis that the surge impedance of the lines provides enough damping.

From Chap. 2, the operational impedance of the parallel circuit is given by:

$$Z_s = \frac{1.5sL_sZ_n}{Z_n + sL_s} \tag{8-21}$$

where $Z_n = Z/n$.

Therefore:

$$V_{TRV} = 1.5\sqrt{2}I\omega\left[\frac{Z_n}{s(s+Zn/L)}\right] \tag{8-22}$$

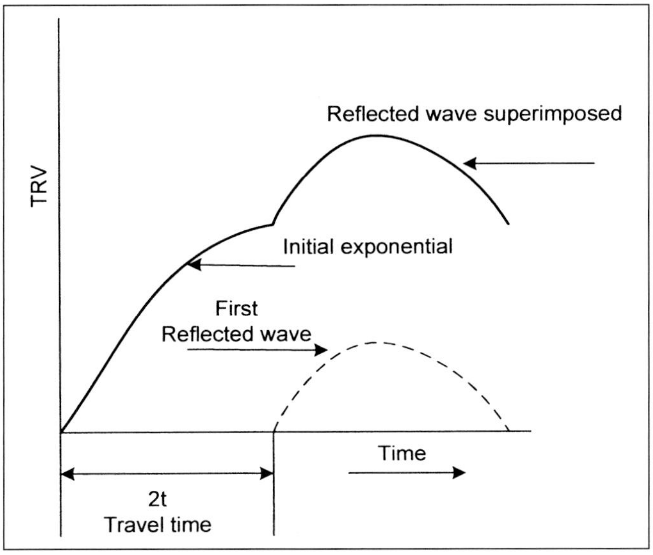

FIGURE 8-10 Profile of TRV, equivalent waveform.

Or in the time domain:

$$V_{TRV} = 1.5\sqrt{2}I\omega L_s(1-e^{-\alpha t}) \tag{8-23}$$

where $\alpha = Z_n/L_s$. (8-24)

We can also write:

$$V_{TRV} = 1.5\sqrt{\frac{2}{3}}V_{rated}(1-e^{-\alpha t}) \tag{8-25}$$

where V_{rated} is the maximum rated voltage of the device.

This voltage, as a traveling wave, will be reflected at the impedance discontinuity, and when it arrives at the point of the fault, its effective value is the initial magnitude, multiplied by reflected and refracted coefficients (Chap. 4). The TRV can then be found by first plotting the initial exponential wave given by Eq. (8-25) and then adding a time equal to approximately 30 km/s, a voltage equal to the reflected and refracted coefficients (Fig. 8-10).

The initial rate of rise is obtained by taking a derivative of Eq. (8-25), and at $t = 0$, this gives:

$$R_0 = 1.5\sqrt{2}I\omega Z_n \tag{8-26}$$

This means that initial TRV is directly proportional to fault current interrupted and inversely proportional to the number of transmission lines that remain connected to the bus.

8-7 ANSI/IEEE STANDARDS FOR TRV

According to ANSI/IEEE standards, for circuit breakers rated 100 kV and below, the rated transient voltage is defined as the envelope formed by 1-cosine wave shape using values of E_2 and T_2, as defined in the standards. E_2 is the peak of the TRV which is taken as 1.88 times the maximum rated voltage and T_2 is in microseconds to reach this peak, and is variable, depending on the circuit-breaker type, short-circuit type, and voltage rating. This is shown in Fig. 8-11*a*; for first 1/2 cycle, the power-frequency component is considered constant and is represented by a straight line E_2. The curve of Fig. 8-11*a* is called 1-minus cosine curve. The TRV is defined by the envelope formed by the curve.

For breakers rated 100 kV and above, the rated TRV is defined by higher of an exponential waveform and 1-cosine waveform, Fig. 8-11*b*, which closely represents the waveform arrived at in Fig. 8-10. For systems above 100 kV, most, if not all, systems are grounded and a first pole to clear factor of 1.3 is considered. E_1 = 1.06 V and E_2 = 1.49 V. Envelope formed by the exponential curve is obtained by reading the rated values E_1, R, T_1, E_2, and T_2 from standards and applying these values at the rated short-circuit current of the breaker. R is defined as the rated TRV rate ignoring the effect of bus-side capacitance, at which recovery voltage rises across the terminals of a first pole to interrupt for a three-phase, ungrounded load-side fault under the specified rated conditions. The rate is a close approximation of the *de*/*dt* in the rated envelope, but slightly higher because the bus-side capacitance is ignored. The equations are written as:

$$\begin{aligned} e_1 &= E_1(1-e^{-t/\tau}) \\ \tau &= E_1/R \\ e_2 &= \frac{E_2}{2}[1-\cos(\pi t/T_2)] \end{aligned} \tag{8-27}$$

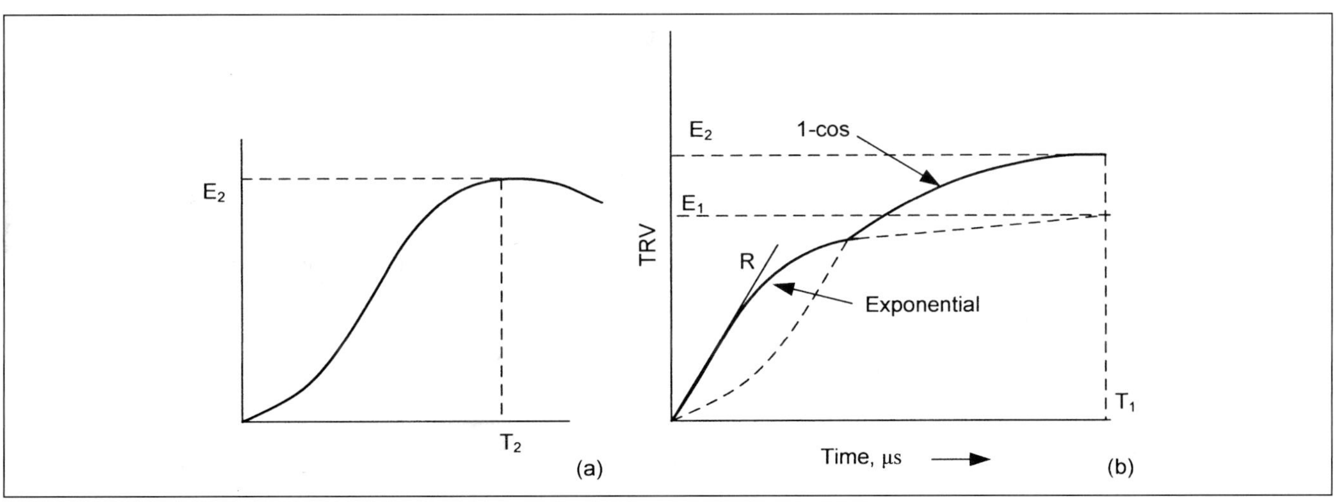

FIGURE 8-11 (*a*) ANSI/IEEE 1-minus-cosine curve for breakers rated 100 kV and below. (*b*) ANSI/IEEE TRV profile for breakers rated 100 kV and above.

There has been an attempt to harmonize ANSI/IEEE standards with IEC. Refer to IEEE unapproved draft standard for ac high-voltage circuit breakers rated on symmetrical current basis—preferred ratings and required related capabilities for voltage above 1000 V.[7] A joint task force group was established to harmonize requirements of TRV with that in IEC62271-100, including amendments 1 and 2. This draft standard also publishes new rating tables and completely revises the ratings and the nomenclature of the circuit breakers which has been in use in the United States for many years:

- Class S1 and S2 is used to denote traditional terms as "indoor" and "outdoor." The class S1 circuit breaker is for cable systems; indoor circuit breakers are predominantly used with cable distribution systems. Class S2 is for overhead line systems.
- The term "peak" is used—the term "crest" has been dropped from the usage. All tables show "prospective" or "inherent" characteristics of the current and voltages. The word prospective is used in conformance to international standard.
- Two and four parameter of representations of TRV is adopted in line with IEC standards.
- For class C0, general-purpose circuit breakers, no ratings are assigned for back-to-back capacitor switching. For class C0, exposed to transient currents for nearby capacitor banks during fault conditions, the capacitance transient current on closing shall not exceed the lesser of either 1.41 times rated short-circuit current or 50 kA peak. The product of transient inrush current peak and frequency shall not exceed 20 kAkHz. Definite-purpose circuit breakers are now identified as Class C1 and C2. Here the manufacturer shall specify the inrush current and frequency at which Class C1 or C2 performance is met.

Example 8-2 Plot the TRV characteristics of 550-kV breaker based on the ANSI/IEEE standards. To start with, consult ANSI standard C37.06[3] and the following ratings are specified:

K factor = 1

Consider breaker current rating of 2 kA.

Rated short-circuit current = 40 kA. The TRV should be plotted for the actual available short-circuit current. For this example, we will plot the curve for 40 kA.

$T_2 = 1{,}325\ \mu s$

R, rate of rise of recovery voltage = 2 kV/μs

Rated time delay $T_1 = 2\ \mu s$

$$E_1 = 1.06 \times \text{rated maximum voltage} = 583 \text{ kV}$$

$$E_2 = 1.49E_{max} = 819.5 \text{ kV}$$

$$\tau = E_1/R = 583/2 = 291.5\ \mu s$$

Then from Eq. (8-27):

$$e_1 = 583(1 - e^{-t/291.5})$$

$$e_2 = 409.8(1 - \cos 0.135t^0)$$

The calculated TRV profile is shown in Fig. 8-12*a*.

A circuit breaker should be capable of interrupting a short-circuit current which is less than the rated current. This requires withstanding a TRV envelope where E_2 value is higher and T_2 value is reduced. This is illustrated in Example 8-3.

Example 8-3 Calculate the TRV profile for the circuit breaker in Example 8-2 at 75 percent of the rated short-circuit current.

This requires the calculation of multiplying factors for adjustment based on the curves in ANSI standard C37.06, 1997. Note that the standard was revised in 1999, but this revised standard did not contain the curves included in the 1979 issue and merely referred to these curves. From these curves:

K_r = rate of rise multiplying factor = 1.625

$K_1 = E_2$ multiplying factor = 1.044

$K_t = T_2$ dividing factor = 1.625

This gives:

$E_1 = 583$ kV

$E_2 = (819.5)(1.044) = 855.5$ kV

$T_2 = (1{,}325)/1.625 = 815.4\ \mu s$

$T_1 = 2\ \mu s$ (unchanged)

$R = (1.6)(1.625) = 2.60$ kV/μs

$$\tau = E_1/R = (583)/3.25 = 179.4\ \mu s$$

The TRV for 75 percent fault current is also plotted in the Fig. 8-12*a* and it is higher than the TRV for 100 percent fault current.

8-7-1 Definite-Purpose Circuit Breakers TRV

Example 8-3 suggests that there can be situations when the TRV of a breaker may be exceeded. Consider, for example, a configuration, as shown in Fig. 8-13*a*. A three-phase *ungrounded* fault occurs on the low side of the transformer, as shown, and the transformer secondary winding is grounded. If we neglect the source impedance and its capacitance, the voltage on the transformer side of the first pole to clear of the breaker is 1.0 per unit, while on the fault side it is –0.5 per unit. Thus, voltage across the breaker pole is 1.50 per unit. (These values can be found by circuit reduction, first pole to open, as shown in Fig. 8-9.) The voltages on the transformer and fault side will oscillate at a frequency determined by the inductance and capacitance of the transformer Fig. 8-13*b*. The source-side voltage may reach a peak of 2 times and the fault side a peak of 1.0 per unit. Thus, the TRV may hit a peak of 3.0 per unit without damping.

For a three-phase grounded fault, delta-wye grounded transformer (Fig. 8-13*c*), the voltage on the fault side is 0, and the voltage on the transformer side of circuit breaker is 1.0 per unit. With no damping, this voltage overshoots and oscillated about E, and is 2.0 per unit, and with damping, somewhat less (Fig. 8-13*d*).

The TRV profile in Fig. 8-13*e* shows the fault on a reactor that will have a higher oscillation frequency because of lower capacitance. The effect of the source impedance and capacitance will be to reduce the transients.

ANSI standard C37.06.1[8] is for Definite Purpose Circuit Breakers for Fast Transient Recovery Voltages Rise Times. This is somewhat akin to the specifications of definite-purpose breakers for capacitor switching. Table 8-3 shows some comparisons of the TRV specifications from the standard breakers and definite-purpose breakers. The standard qualifies that:

1. No fast T_2 values or tests are proposed for fault currents >30 percent of the rated short-circuit current.

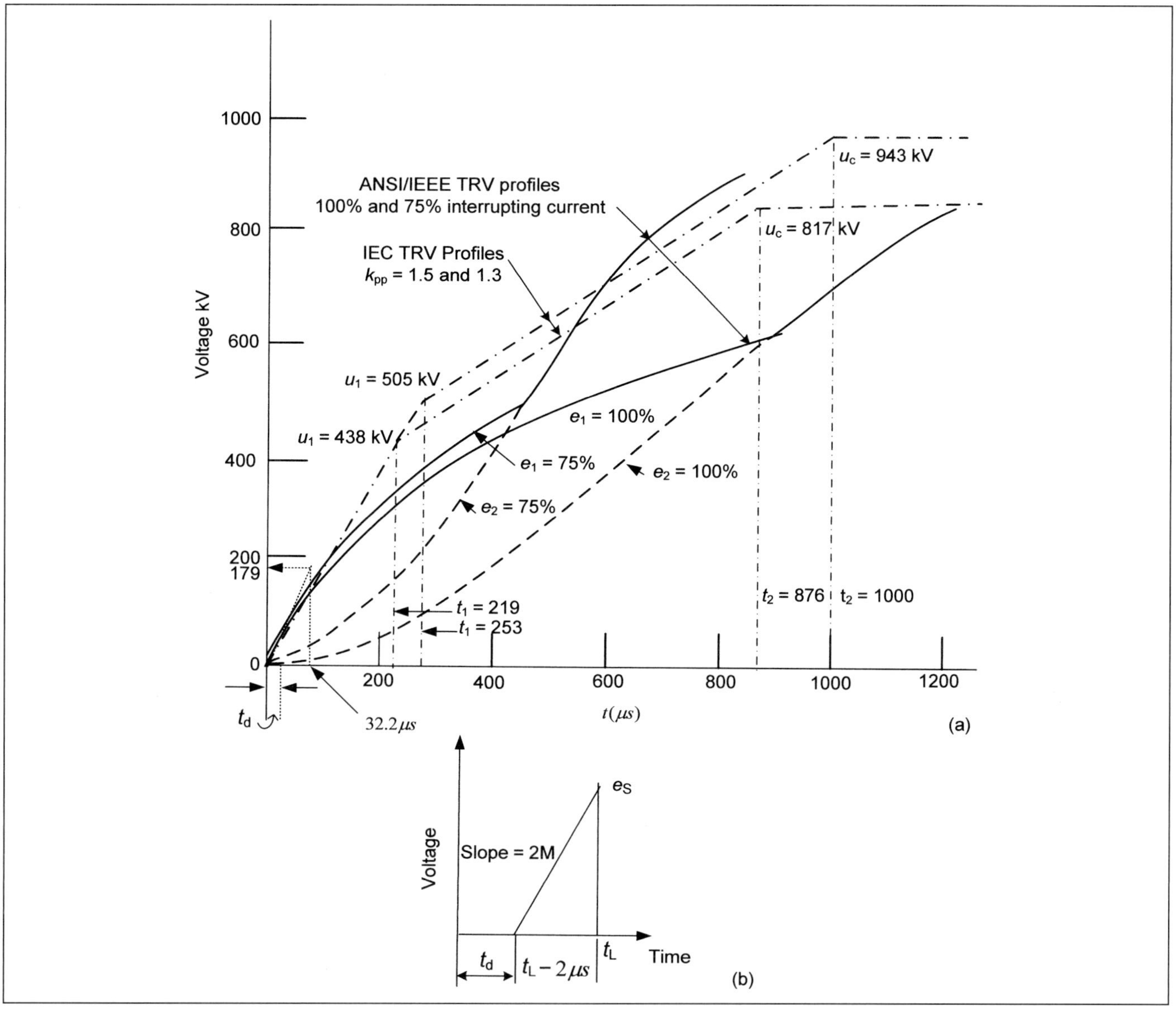

FIGURE 8-12 (*a*) Calculations of TRV profiles, Examples 8-2, 8-3, and 8-4, with superimposed IEC profiles (*b*) Initial rise, short-line fault.

2. The proposed T_2 values are chosen to meet 90 percent of the known TRV circuits, but even these fast values do not meet the requirements of all fast TRV applications.

3. A circuit breaker that meets the requirements of definite purpose for fast TRV may or may not meet the requirements of definite-purpose circuit breakers for capacitor switching (Chap. 6).

Table 8-3 shows some specimen values for fast TRV definite-purpose circuit breakers.

8-8 IEC TRV PROFILES

Two and four parameter methods of representation of TRV are used in IEC 62271–100[9], which is also adopted in IEEE draft standard.[7]

8-8-1 Four-Parameter Method

Figure 8-14 shows the representation of TRV wave by four-parameter method. Standards assume that for systems above 72.5 kV, clearing terminal faults higher than 30 percent of rating will result in TRV characteristics that have a four-parameter envelope. The TRV wave has an initial period of high rise, followed by a low rate of rise. Such waveforms can be represented by a four-parameter method:

u_1 = first-reference voltage in kV

t_1 = time to reach u_1, in μs

u_c = second-reference voltage, peak value of TRV

t_2 = time to reach u_c, in μs

IEC specifies values of u_1, uc, t_1, and t_2 for the circuit breakers. These values are also specified in Ref. 7. The amplitude factor k_{af} is given for various test duties, *T*100, *T*60, *T*30, *T*10, for example. The interrupting current tests are carried out on circuit breakers with specified TRVs. The segments can be plotted, as shown in Fig. 8-14, based on the parameters specified in standards. A table

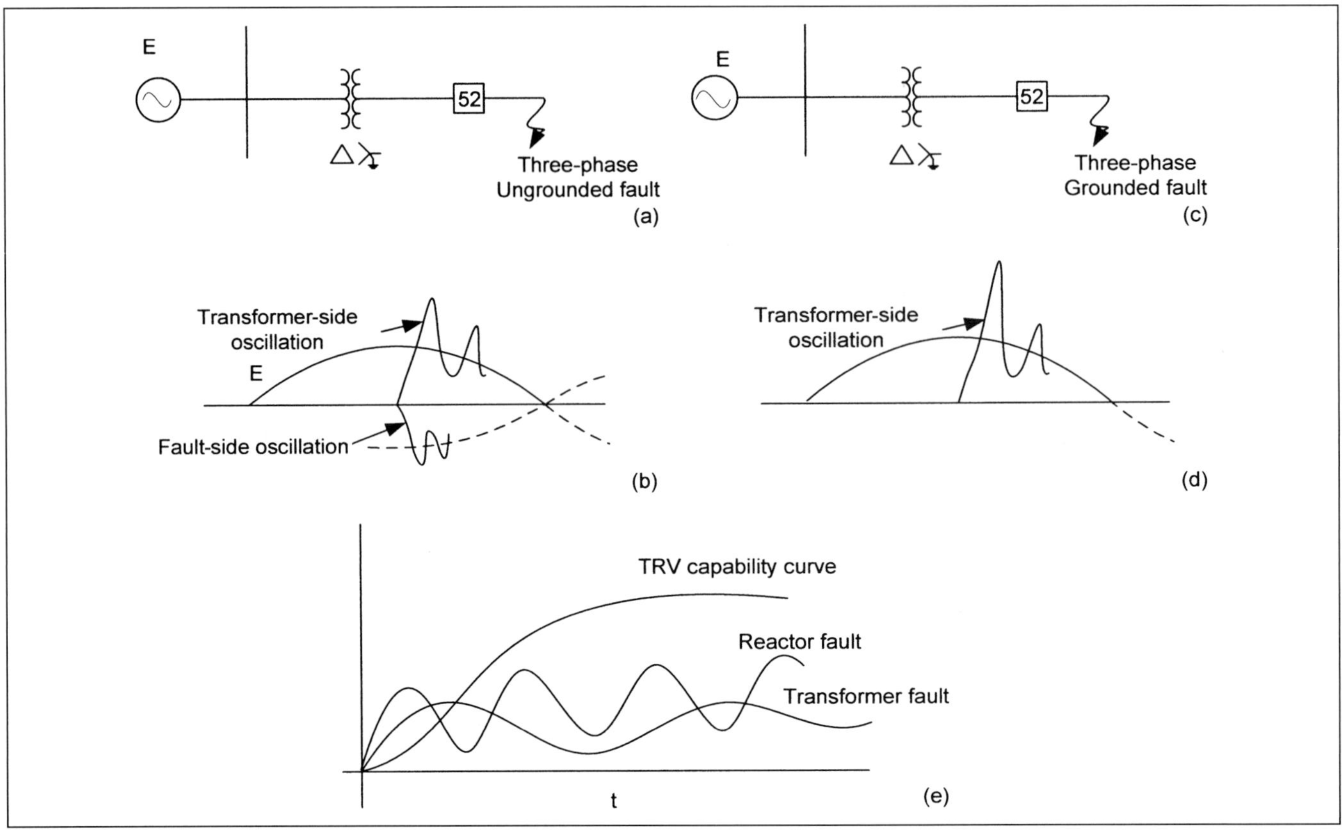

FIGURE 8-13 (*a*) and (*b*) Three-phase-ungrounded fault on the secondary side of a delta-wye-grounded transformer and TRV profile, respectively. (*c*) and (*d*) Three-phase-grounded fault on the secondary side of a delta-wye-grounded transformer and TRV profile, respectively. (*e*) Situations (transformer and reactor faults) where TRV can exceed the TRV capability curve of the breakers.

TABLE 8-3 Specimen ANSI/IEEE Recovery Voltage Parameters of Definite-Purpose Circuit Breakers (123 kV and Above)

		Rated Short-Circuit and	Rated Recovery Voltage		Definite-Purpose TRV Parameters at 30% of Rated Short-Circuit Current			Definite-Purpose TRV Parameters at 7% of Rated Short-Circuit Current		
Rated Maximum Voltage (kV)	K Factor	Short-Time Current (kA rms)	Peak Voltage, E_2 (kV), Peak	Time to Peak, T_2 (μs)	Current (kA rms)	Peak Voltage (kV), Peak	Time to Peak (μs)	Current (kA rms)	Peak Voltage (kV)	Time to Peak (μs)
245	1	63	431	520	19	487	30.3	4.4	505	43.8
362	1	40	637	775	12	720	40.7	2.8	745	63.2
362	1	63	637	775	19	720	37.1	4.4	745	55.7
550	1	40	968	1325	12	1094	49.0	2.8	1133	76.1
550	1	63	968	1325	19	1094	44.7	4.4	1133	63.9

in IEEE std. C37.011 provides factors for calculating rated TRV profiles for terminal faults.[6]

$$u_c = k_{af} \times k_{pp} \times \sqrt{2/3} \times U_r \tag{8-28}$$

where k_{pp} is the first pole to clear factor. For $T10$ (10% of interrupting current):

$$u_c(T10) = u_c(T100) \times Ku_c \tag{8-29}$$

where U_r is the breaker-rated voltage.

8-8-2 Two-Parameter Representation

For terminal faults between 10 and 30 percent for systems above 72.5 kV and for all terminal fault currents for systems 72.5 kV and below,[7] the standard assumes a TRV profile defined by two parameters, u_c and t_3. Figure 8-15 shows the representation of TRV wave by two-parameter method. This waveform occurs in systems less than 100 kV or locations where short-circuit current is low compared to the maximum short-circuit current in the system. TRV can be approximately represented by a single-frequency transient.

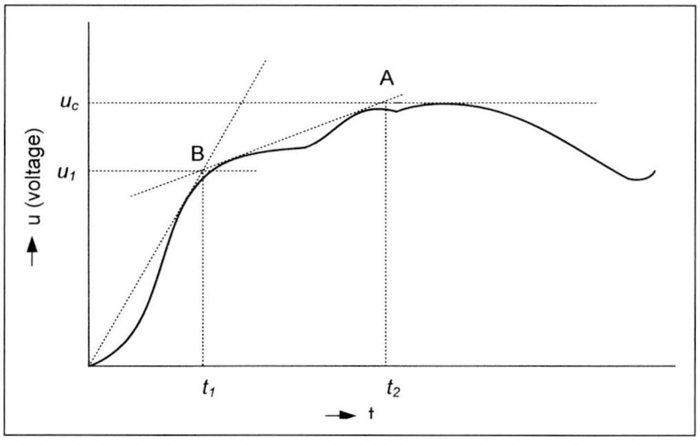

FIGURE 8-14 IEC, four-parameter representation of TRV.

u_c = peak of TRV wave, kV

t_3 = time to reach peak, μs

The initial rate of rise of TRV is contained within segments drawn according to two- or four-parameter methods by specifying the delay line, as shown in Fig. 8-15. Thus, there can be many varying TRV profiles depending upon (1) type of fault, (2) the system configuration and location where the fault occurs, and (3) modifying behavior of the circuit breaker interruption process. For a terminal fault and a typical switchgear layout of overhead lines, local transformers, and generators, the rate of rise of TRV can be estimated from:

$$S = 2\pi f \sqrt{2} I_k Z_r \tag{8-30}$$

where I_k is short-circuit current and Z_r is the resultant surge impedance. Z_r can be found from sequence impedances. With n equal outgoing lines:

$$Z_r = 1.5(Z_1/n)\frac{2Z_0/Z_1}{1+2Z_0/Z_1} \tag{8-31}$$

where Z_1, Z_0 are the surge impedances in positive and zero sequence of individual lines and n is the number of lines emanating from the substation. Factors 1.5 can be 1.3 in (Eq. 8.31). For plotting the TRV profile, traveling wave phenomena must be considered. For single-frequency transients, with concentrated components, IEC tangent (rate of rise) can be estimated from:

$$S = \frac{2\sqrt{2} f_n k a_f u_0}{0.85} \tag{8-32}$$

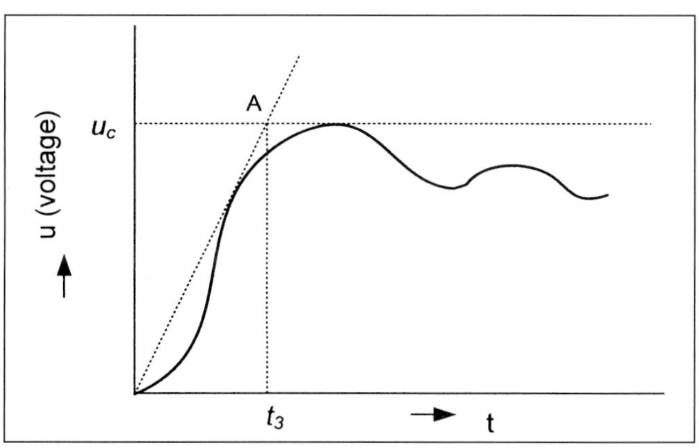

FIGURE 8-15 IEC, two- parameter representation of TRV.

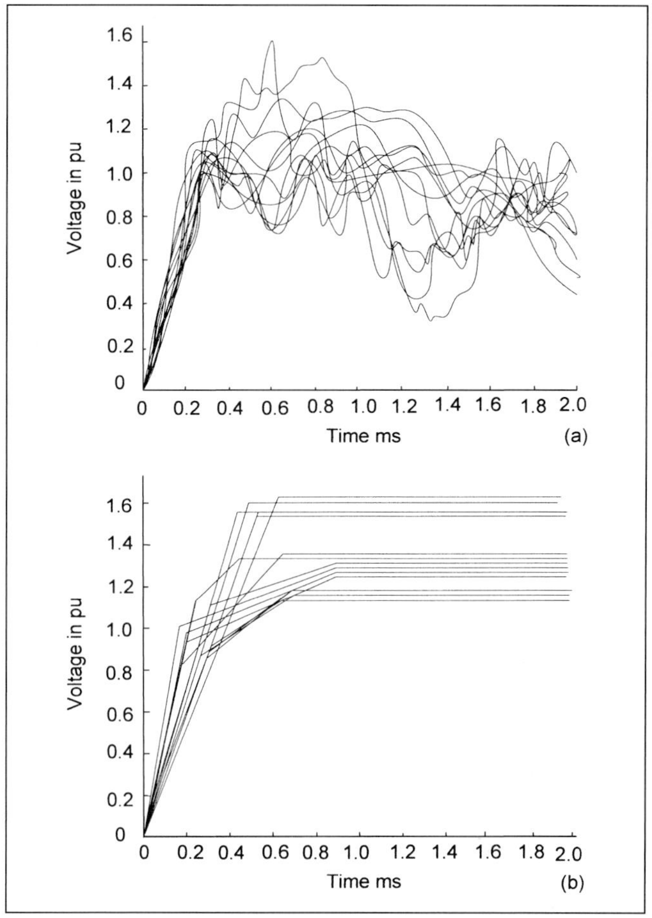

FIGURE 8-16 (*a*) TRV measurements in a 400-kV system. (*b*) Representation by IEC methods.

where f_n is the natural frequency and ka_f is the amplitude factor. The peak value u_c in a four-parameter method cannot be found easily due to many variations in the system configurations. The traveling waves are partially or totally reflected at the points of discontinuity of the surge impedance. A superimposition of all forward and reflected traveling waves gives the overall waveform.

Figure 8-16*a* shows the TRV profiles in a 400-kV system, with a three-phase ground fault, and Fig. 8-16*b* shows a four-parameter representation of the TRV according to IEC. The per unit voltage is defined as $(\sqrt{2} \times 400)/\sqrt{3}$. The variations in profiles of TRV are noteworthy.

8-9 SHORT-LINE FAULT

Faults occurring between a few kilometers to some hundreds of kilometers from the breaker are termed *short-line faults* (SLF), (Fig. 8-17*a*). The single-phase fault, the faulty phase being last to be disconnected, is taken as the basis, as this gives the largest rate of rise at the beginning of the recovery voltage. The line to the fault point is represented by its surge impedance. After the short-circuit current is interrupted, the breaker terminal at the line end assumes a sawtooth oscillation shape, as shown in Fig. 8-17*c*. Note the small time-delay t_d in this figure; this is because of the capacitance of the line-side apparatus. The rate of rise of voltage is directly proportional to the effective surge impedance (that can vary between 35 and 450 Ω, the lower value being applicable to cables) and the rate of rise of current at current zero. As the terminating impedance is zero, the reflection coefficient is –1, and a negative wave is produced traveling toward the breaker. The component on the supply

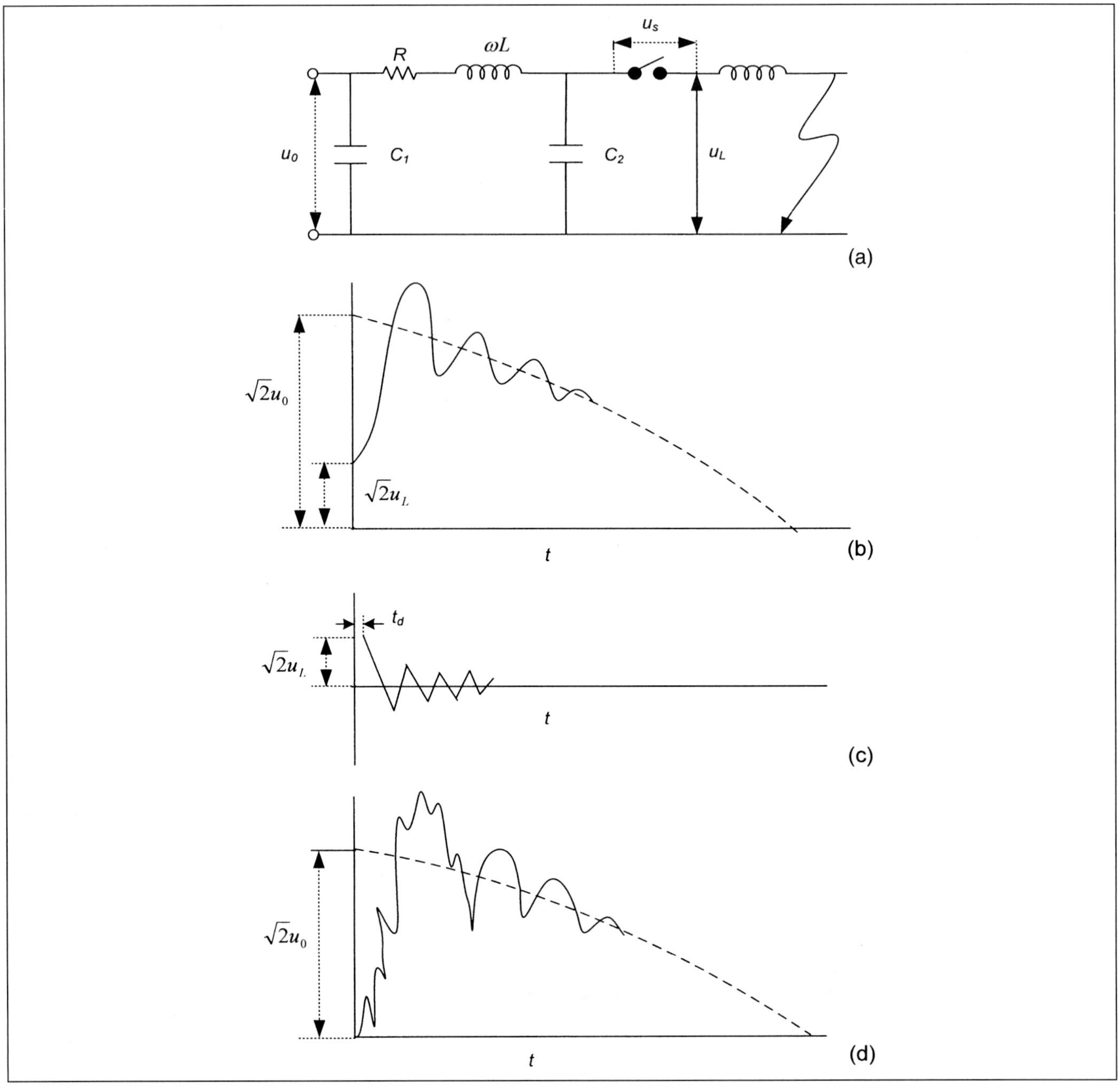

FIGURE 8-17 TRV for a short-line fault. (*a*) System equivalent circuit. (*b*) Recovery voltage on the source side. (*c*) Recovery voltage on the load side. (*d*) Voltage across the breaker contacts.

side exhibits the same waveform as for a terminal fault (Fig. 8-17*b*). The circuit breaker is stressed by the difference between these two voltages (Fig. 8-17*d*). Because of the high-frequency oscillation of the line-side terminal, the TRV has a very steep initial rate of rise. In many breaker designs, the short-line fault may become a limiting factor of the current-interrupting capability of the breaker.

IEEE Std. C37.04[4] specifies the TRV envelope and the time to first peak, depending on the breaker rating. The circuit breaker shall be capable of interrupting single-phase line faults at any distance from the circuit breaker on a system in which:

1. The voltage in the first ramp of the sawtooth wave is equal to or less than that in an ideal system in which the surge impedance and amplitude constants are as follows:

242 kV and below, single-conductor line: $Z = 450$, $d = 1.8$
362 kV and above, bundle conductors: $Z = 360$, $d = 1$

2. The amplitude constant d is the peak ratio of the sawtooth component that will appear across the circuit breaker terminals at the instant of interruption. The SLF TRV capability up to first peak of TRV is: $e = e_L + e_S$, where e_L is the line-side contribution and e_S is the source-side contribution. The triangular wave is defined as:

$$e_L = d(1-M)\sqrt{\frac{2}{3}}E_{\max} \tag{8-33}$$

$$e_s = 2M(t_L - t_d) \tag{8-34}$$

Also:

$$R_L = \sqrt{2}\omega MIZ \times 10^{-6}\ \text{kV}/\mu\text{s}$$
$$t_L = e/R_L\ \mu\text{s} \tag{8-35}$$

where R_L is the rate of rise, t_L is the time to peak, M is the ratio of the fault current to rated short-circuit current, I is the rated short-circuit current in kA, V is the rated voltage, Z is the surge impedance, and e is the peak voltage in kV.

We could write the frequency as:

$$f = \frac{10^6}{2(e/R_L)} \tag{8-36}$$

There is a delay of 0.5 μs for circuit breakers rated 245 kV and above and 0.2 μs for circuit breakers rated below 245 kV. It is not necessary to calculate SLF TRV as long as terminal-fault TRV is within rating and transmission-line parameter is within those specified in the standards.

Example 8-4 In Examples 8-2 and 8-3, 550-kV breaker TRV profiles have been plotted in Fig. 8-12. Plot the short-line capability for 75 percent short-circuit current and a surge impedance of 360 Ω.

Based on the given data, $M = 0.75$, $I = 40$ kA, $V = 550$ kV, $Z = 360$ Ω. Thus, e is:

$$e = 1.6(0.75)(550)\sqrt{\frac{2}{3}} = 179.6\ \text{kV}$$

$$R_L = \sqrt{2} \times 377 \times 0.75 \times 40 \times 360 \times 10^{-6} = 5.75\ \text{kV}/\mu\text{s}$$

Therefore, $t_L = 31.2$ μs.

The TRV profile is superimposed in Fig. 8-12*a*, and an enlarged profile is shown in Fig. 8-12*b*.

8-9-1 Control of Short-Line TRV

It is possible to reduce the voltage stresses of TRV by incorporating resistances or capacitances in parallel with the contact break and by synchronous switching (Chap. 7). Figure 8-18 shows some control methods. In (*a*) resistor switching is used. The main contacts open leaving the resistor in the circuit, which is opened after about 1 cycle. The current has a parallel path through the resistor which decreases the rate of rise. In (*b*) a shunt capacitor is provided. In (*c*) a capacitor is applied across the open poles of the circuit breaker. Most multibreak high-voltage circuit breakers (more than one breaker per phase) use grading capacitors to ensure voltage division across the multibreaks. The capacitor acts the same way as a shunt capacitor. The sawtooth wave is delayed by the time constant ZC.

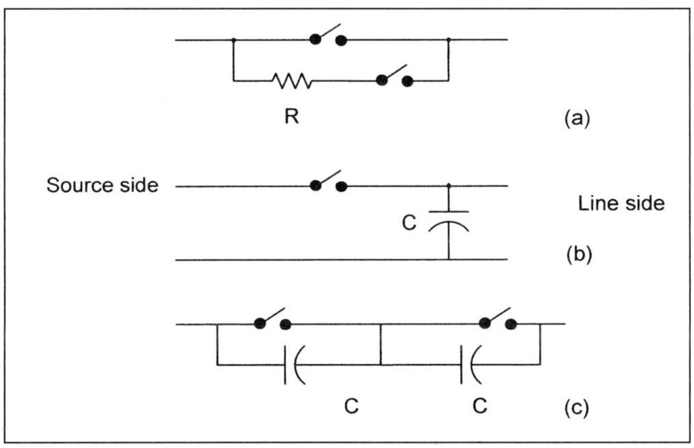

Figure 8-18 Control of TRV on short-line faults. (*a*) Resistors across circuit breaker main contacts inserted for a short duration during opening. (*b*) Shunt capacitors. (*c*) Series grading capacitors across multibreaks per phase in circuit breakers.

Sometimes, the grading capacitors can cause ferroresonance with the potential transformer (PT) reactance on the same bus (Fig. 14-43*a* and *b*). The grading capacitor and the bus capacitance act as a voltage divider, and the voltage on the bus is:

$$V_b = \frac{C_0}{C_0 + C_B} \tag{8-37}$$

where C_0 is the equivalent capacitance across the breaker gaps. This means that there will be some voltage imparted to a dead bus. With a PT connected to the bus, the bus voltage is:

$$V_b = \frac{X_m}{X_{C_0} - X_m} \tag{8-38}$$

X_m is nonlinear and the bus voltage is determined by the intersection of transformer saturation curve and the capacitance of the line. Resistance can be introduced in the secondary circuit of the PT to mitigate a ferroresonance problem.

8-9-2 Initial TRV

Circuit beakers rated 100 kV and above and short-circuit rating of 31.5 kA and above will have an initial TRV capability for phase-to-ground fault. This rises linearly from origin to first-peak voltage E_i and time T_i given in Ref. 4. E_i is given by the expression:

$$E_i = \omega \times \sqrt{2} \times I \times Z_b \times T_i \times 10^{-6}\text{kV} \tag{8-39}$$

where Z_b is the bus surge impedance = 450 Ω for outdoor substations, phase-to-ground faults and I is in the fault current in kA. The term initial TRV refers to the conditions during the first microsecond or so, following a current interruption, when the recovery voltage on the source side of the circuit breaker is influenced by proximity of buses, capacitors, isolators, and so on. Akin to short-line fault, voltage oscillation is produced, but this oscillation has a lower voltage-peak magnitude. The traveling wave will move down the bus where the first discontinuity occurs. The initial slope depends on the surge impedance and *di*/*dt*, and the peak of ITRV appears at a time equal to twice the traveling wave time. There can be as many variations of ITRV as the station layouts. T_i in μs, time to crest, is given in IEEE Std. C37.04 with respect to maximum system voltage: For 121, 145, 169, 362, 350, and 500 V, it is 0.3, 0.4, 0.5, 0.6, 0.8, 1.0, and 1.1, respectively.

The TRV capability envelope of a 550-kV breaker at 100 percent of its interrupting rating, as per Fig. 8 of IEEE Std. C37.011 is superimposed upon the calculated curves in Fig. 8*a*. The two profiles shown relate to first pole to clear factor k_{pp} of 1.3 and 1.5, respectively. The parameters shown are from IEEE Std. C37.011. This standard contains an example of TRV calculations in a practical electrical system, which an interested reader may refer to.

8-10 INTERRUPTION OF LOW INDUCTIVE CURRENTS

A circuit breaker is required to interrupt low inductive currents of transformers at no-load, high-voltage reactors, or locked rotor currents of motors. Because of arc instability in a low-current region, *current chopping* can occur, irrespective of the breaker interrupting medium, though some mediums may be more prone to current chopping, that is, vacuum technology compared to SF_6 breakers. In a low-current region, the characteristics of the arc lead to a negative resistance which lowers the damping of the circuit. This sets up a high-frequency oscillation, depending on the LC of the circuit.

Figure 8-19*a* shows the circuit diagram for interruption of low inductive currents. The inductance L_2 and capacitance C_2 on

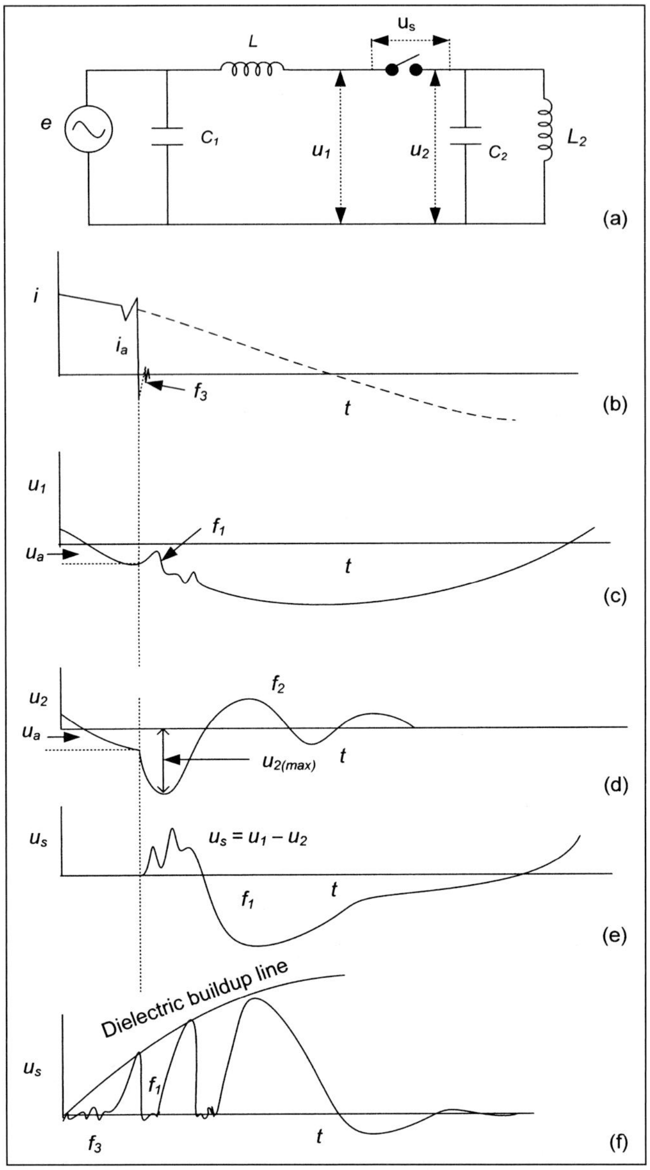

FIGURE 8-19 Interruption of low inductive currents. (*a*) The equivalent circuit diagram. (*b*) Chopped current i_a at u_a. (*c*) Source-side voltage. (*d*) Load-side voltage. (*e*) Voltage across breaker contacts. (*f*) Phenomena of repeated restrikes.

the load side can represent transformers and motors. As the arc becomes unstable at low currents, the capacitances C_1 and C_2 partake in an oscillatory process of frequency:

$$f_3 = \frac{1}{2\pi\sqrt{L\dfrac{C_1C_2}{C_1+C_2}}} \tag{8-40}$$

Practically, no current flows through the load inductance L_2. This forces a current zero, before the natural current zero, and the current is interrupted (Fig. 8-19*b*). This interrupted current i_a is the chopped current at voltage u_a (Fig. 8-19*c*). Thus, the chopped current is not only affected by the circuit breaker, but also the properties of the circuit. The energy stored in the load at this moment is:

$$i_a^2\frac{L_2}{2} + u_a^2\frac{C_2}{2} \tag{8-41}$$

which oscillates at the natural frequency of the disconnected circuit:

$$f_2 = \frac{1}{2\pi\sqrt{L_2C_2}} \tag{8-42}$$

This frequency may vary between 200 and 400 Hz for a transformer. The maximum voltage on the load side occurs when all the inductive energy is converted into capacitive energy:

$$u_{2\max}^2\frac{C_2}{2} = u_2^2\frac{C_2}{2} + i_a^2\frac{L_2}{2} \tag{8-43}$$

The source-side voltage builds up with the frequency:

$$f_1 = \frac{1}{2\pi\sqrt{L_1C_1}} \tag{8-44}$$

The frequency f_1 lies between 1 and 5 kHz. This is shown in Fig. 8.19*c*.The linear expression of magnetic energy at the time of current chopping is not strictly valid for transformers and should be replaced with:

$$\text{Volume of transformer core} \times \int_0^{B_m} H\,dB \tag{8-45}$$

The load-side voltage decays to zero due to system losses. The maximum load-side overvoltage is of concern (Fig. 8-19*d*) and from simplified relation Eq. (8-43), it is given by:

$$u_{2\max} = \sqrt{u_a^2 + \eta_m i_a^2\frac{L_2}{C_2}} \tag{8-46}$$

Here we have added a factor η_m to account for saturation. The factor is approximately unity for air-core reactors and is of the order of 0.3 to 0.5 for transformers at no load. Eq. (8-46) ignores the effect of circuit breaker behavior and is conservative. For low magnitude of chopping current at the peak of the supply voltage:

$$u_a = u_n\sqrt{\frac{2}{3}} \tag{8-47}$$

Then, the overvoltage factor k is given by:

$$k = u_2/u_a = \sqrt{1 + \frac{3}{2}\frac{i_a^2}{u_n^2}\eta_m\left(\sqrt{\frac{L_2}{C_2}}\right)^2} \tag{8-48}$$

If the current is chopped at the peak value, $u_a = 0$, and:

$$u_{2\max} = i_a\sqrt{\eta_m}\sqrt{\frac{L_2}{C_2}} \tag{8-49}$$

Thus, the overvoltage is dependent on the chopped current. The chopped currents in circuit breakers have been reduced with better designs and arc control. The voltage across the supply side of the breaker, neglecting arc voltage drop, is u_s and it oscillates at the frequency given by L and C_1. The voltage across the breaker contacts is $u_s = u_1 - u_2$ (Fig. 8-19*e*). The supply-side frequency is generally between 1000 and 5000 Hz.

If the circuit-breaker voltage intersects the dielectric recovery characteristics of the breaker, reignition occurs and the process is repeated anew (Fig. 8-19*f*). With every reignition, the energy stored is reduced until the dielectric strength is large enough and further reignitions are prevented. Overvoltages of the order of two to four times may be produced on the disconnection of inductive loads.

The disconnection of high-voltage reactors may involve interrupting reactive currents of the order of 100 to 200 A. The maximum overvoltage factor can be calculated from Eq. (8-49). Note that factor $\sqrt{L/C}$ for the reactors is high, of the order of 50 kΩ. In most practical cases, the overvoltage factor will be less than 2.5. Surge arresters can be used to limit the overvoltages (Chap. 20). In the previous equations for calculations of the overvoltage factors, the influence of the breaker itself is ignored; however, the equations seem to be conservative, as the breaker arc effects may lower the overvoltages given by the above equations.

The parallel compensated lines should be mentioned as the recovery voltages can be high. The voltage on the disconnected side oscillates with slight damping near the power frequency. The voltage across the breaker contacts can be high, and restrike free breakers are needed.

Example 8-5 An EMTP simulation of the recovery voltage of a 230-kV transmission line grounded at the far end, surge impedance of 300 Ω, is shown in Fig. 8-20. This shows the line-side and source-side voltages and also the voltage across the breaker contacts. The high-frequency recovery voltage is evident and the maximum magnitude is 2.13 per unit.

Figure 8-21 shows the recovery profile when a 138-kV transmission line at no load is opened, arcing time ignored, ideal switch. The maximum voltage is 2.52 per unit.

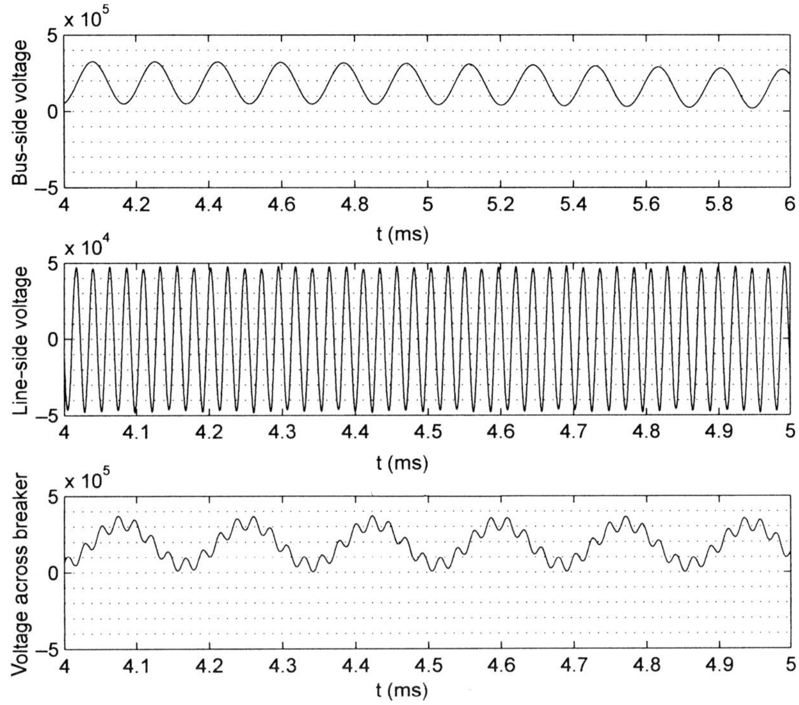

FIGURE 8-20 EMTP simulation of interruption of a line ground fault current. Voltages on the source and load side and the recovery voltage across breaker contacts showing high-frequency transients, Example 8-5.

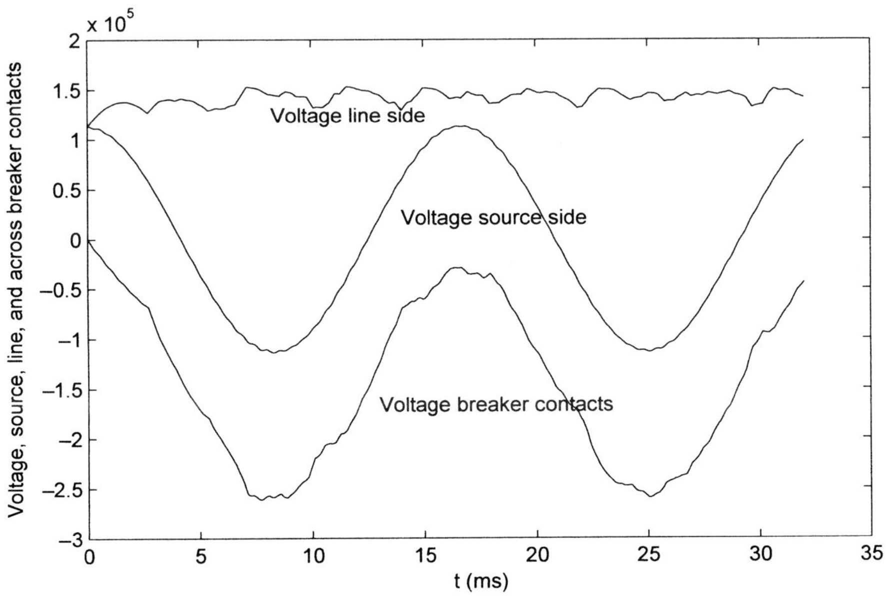

FIGURE 8-21 Simulation of the recovery voltage on interrupting a line at no load, Example 8-5.

8-11 INTERRUPTION OF CAPACITIVE CURRENTS

We discussed interruption of capacitor currents in Chap. 6 and 7. The rate of rise of the recovery voltage is low, and a circuit breaker will try to interrupt the current even at the first-current zero, when the contact gap is small and still ionized and has not recovered the dielectric strength. However, 1/2 cycle later, when the voltage reaches 2.0 per unit, a reignition may occur. The shunt capacitor being disconnected then discharges through the circuit breaker and the amplitude of this current depends on the instantaneous value of the capacitor charge. The high frequency of the current is caused by the inductance between the capacitor bank and breaker. This high-frequency current is superimposed upon the power-frequency current and creates additional current zeros (Fig. 6-35*c* and *d*). Some types of circuit breakers, vacuum and SF_6, will operate on these high-frequency current zeros.

Series reactors are employed to reduce the switching inrush currents of the capacitors (Chap. 6), but the discharge currents on the capacitors can be even higher than back-to-back switching currents. Again these can be reduced by reactors. Though, the rate of rise of the recovery voltage is low, its peak increases. The recovery voltage depends on the capacitor bank connections as follows:

1. A three-phase grounded capacitor bank can be represented by a single-phase circuit. The TRV across switching device depends on the ratio of positive-sequence capacitance to zero-sequence capacitance which is one for grounded banks. The maximum-recovery voltage is 2.0 per unit.

2. An ungrounded capacitor bank has C_1/C_0 = infinity, and the recovery voltage is 3.0 per unit. In Chap. 6, we discussed the current interruption in a three-phase capacitor bank. For overhead transmission lines, C_1/C_0 may be 1.5 to 2.0 per unit, and TRV will be between 2 and 3 per unit. For grounded or ungrounded capacitor banks, the recovery voltage profile is given by ANSI/IEEE 1-cosine curve. The following observations are of interest with respect to interruption of capacitive currents:

- For grounded capacitor banks, a nonsimultaneous closing of poles of the switching device by less than 1/4 cycle will result in recovery voltages across any pole of no more than 2.5 per unit. For 1/4 to 2/3 of a cycle of nonsimultaneous operation, the recovery voltage on a pole can be as high as 3.0 to 4.0 per unit.

- For ungrounded capacitor banks, the highest peak values of recovery voltages, though higher than those when switching grounded banks, are much lower than that in ungrounded banks, regardless of the degree of nonsimultaneity between poles.

- For a ratio of $C_1/C_0 = 2$ (for systems of 69 kV and lower), maximum-recovery voltage in any one pole can be expected to be 2.22 per unit, nonsimultaneity less than 1/6 cycle.

- For systems higher than 69 kV, C_1/C_0 will be less than 1.6 due to larger phase spacing, and recovery voltages with 1/6 cycle nonsimultaneity will be 2.15 per unit of the peak recovery voltage.

8-12 PRESTRIKES IN CIRCUIT BREAKERS

A prestrike may occur on closing of a circuit breaker, establishing the current flow before the contracts physically close. A prestrike occurs in a current flow at a frequency given by the inductances and capacitances of the supply circuit and the circuit being closed. In Fig. 8-22, this high-frequency current is interrupted at $t = t_1$. Assuming no trapped charge on the capacitors, the voltage rises to approximately 2 per unit. A high-frequency voltage given by source reactance and stray capacitance is superimposed upon the recovering bus voltage. If a second prestrike occurs at $t = t_2$, a further escalation of the bus voltage occurs. Thus, the transient conditions are similar as for restrikes; however, the voltage tends to decrease as the contacts come closer in a closing operation. In Fig. 8-22, u_m is the maximum system voltage, u_r is the recovery voltage, and u_s is the voltage across the breaker contacts.

8-13 BREAKDOWN IN GASES

The dielectric gases are (1) simple gases, for example, air, nitrogen, helium; (2) oxide gases, for example, carbon dioxide; (3) hydrocarbon gases, for example, methane; and (4) electronegative gases, for example, sulphur hexafluoride, SF_6, and freon. If the dielectric strength of helium is tested in a uniform electric field, under 13.6 atm (1 atm = 101325 Pa), which is taken as unity, the dielectric strength of nitrogen is 7.2, carbon dioxide 8, and for SF_6 it is 16. The chemical stability, corrosion, dielectric loss, toxicity, thermal conductivity, behavior under arcing, fire, and explosion hazards are some other considerations when applied to electrical equipment. SF_6 has superior properties under effects of electrical field and arcing.

Townsend (1900–1901) suggested that phenomena leading to rupture of an insulating gas can be represented by:

$$\gamma[e^{\alpha d}-1]=1 \tag{8-50}$$

where γ is the number of charged particles formed as a result of positive ion collision, also known as second Townsend coefficient; α is number of charged particles formed by collision of negative-charged ions, also designated as first Townsend coefficient; and d is the spacing of electrodes.

He studied that at low values of x/P, where x is the voltage gradient and P is the pressure of gas, the current increases as a linear function of the gap distance:

$$i = i_0 e^{\alpha x} \tag{8-51}$$

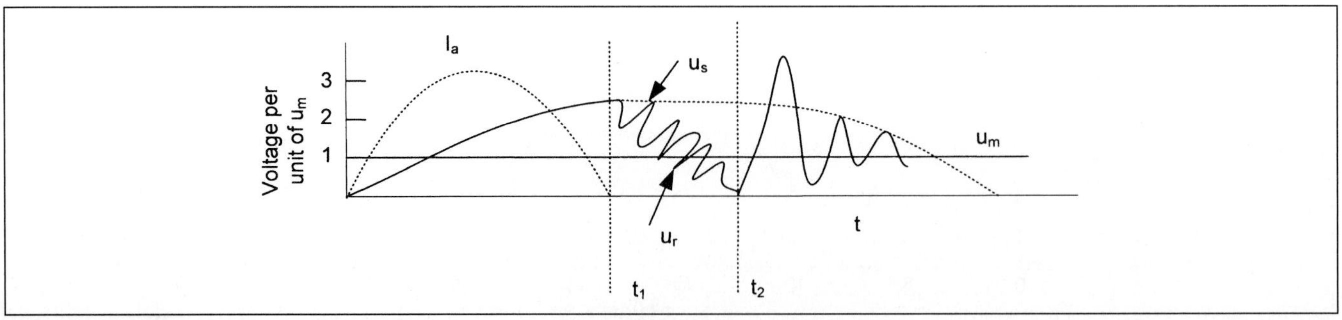

FIGURE 8-22 Prestrikes in circuit breakers.

where i_0 is the current at zero plate separation; α also represents number of new negative-charged particles created per centimeter of path along direction of electric field, that is, slope of log (i_0/i) and x and $1/\alpha$, thus represents average distance traveled by an electron to produce a new ion pair. α varies in a linear relationship to pressure up to 10 atm. For both air and nitrogen and in high-pressure range, α decreases rapidly[1] with increasing pressure and becomes negligible at 30 atm. To account for breakdown and deviation from Pachen's law at high pressure, Townsend suggested that at higher values of x/P, positive ions also ionize neutral atoms and molecules by impact.

The theory has been further modified by Loeb[10] and conducting particles can be created due to secondary liberation of electrons, photoelectric action at cathode, and field emission at contact surfaces.

The advantage of SF_6 is that it is an electronegative gas. The molecular structure of SF_6 has to be studied to understand the electronegativity. SF_6 is octahedral, with six fluorine atoms arranged symmetrically around sulfur atom. The molecular shapes can be explained in terms of mutual repulsion of covalent bonds. When two atoms are bonded by a covalent bond, both of them share a pair of electrons. The attraction that one of the atoms exerts on this shared pair of electrons is termed *electronegativity*. Atoms with nearly filled shells of electrons (halogens) tend to have higher electronegativity than those with sparsely filled shells.

The electronegativity is the important characteristic in electrical breakdown. Its ability to seize an electron and form a negative ion is reverse of that of the ionization process. Thus, if the attachment is equal to ionization, no breakdown is possible. The dissociation equation of SF_6 is:

$$SF_6 \rightarrow SF_6^+ + F^- \qquad (8\text{-}52)$$

The decomposition potential is 15.7 eV.

The molecular weight also plays a role. A gas of higher molecular weight ionizes at a higher potential and exhibits a greater attachment rate. SF_6 has a large collision diameter of the order of 4.77 Å. Energy can be stored in viberational levels of SF_6, forming ions of low mobility. These ions can reduce positive space charge around an electrode requiring a higher voltage to produce an arc across the gap. The Townsend criterion of breakdown of an electronegative gas can be written as:

$$\frac{\gamma\alpha}{\alpha-\eta}[e^{(\alpha-\eta)d}-1]=1 \qquad (8\text{-}53)$$

where η is the attachment coefficient.

The attachment coefficient varies with v/P.

Another breakdown criterion is known as *steamer criterion*:[11]

$$e^{(\alpha-\eta)d} \geq \text{NC} \qquad (8\text{-}54)$$

where NC = $3.10^8 - 10^9$, that is, breakdown will occur when a certain number of charge carriers in an avalanche head are exceeded.

The criterion may be extended to apply to nonuniform fields:

$$\exp\int_0^d (\alpha-\eta)dn \geq \text{NC}$$

or

$$\int_0^d [\alpha(x)-\eta(x)]dn \geq \ln \text{NC} \qquad (8\text{-}55)$$

The breakdown characteristics in nonuniform fields will be different because ionization may be locally sustained due to regions of high stress. This may be due to sharp corners, rough surfaces, and semiconducting particles left during manufacture. This is especially important for GIS, discussed in Chap. 18, where these particles can cause a lowering of the dielectric strength and breakdown. Extreme care is required in the field assembly of GIS.

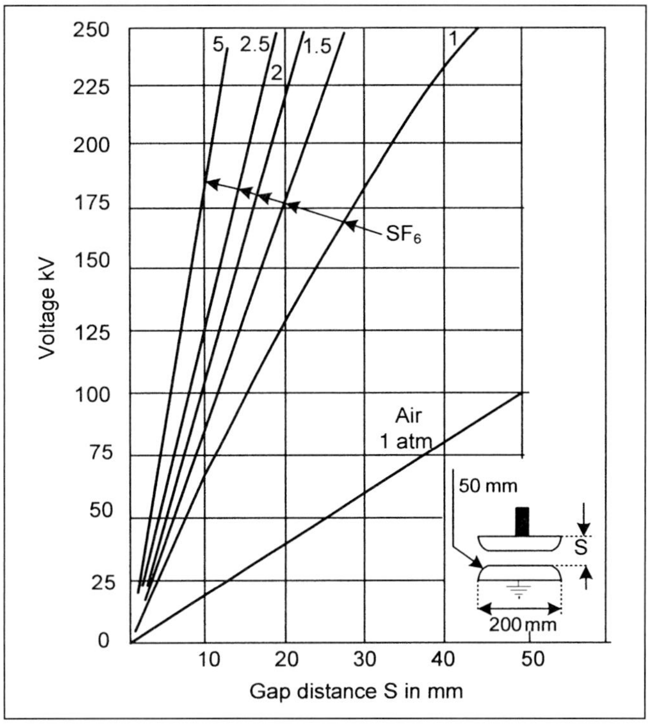

FIGURE 8-23 50-Hz breakdown voltage in SF_6, homogeneous field, in relation to distance between electrodes, at various pressures.

Figures 8-23, 8-24, and 8-25 show the superior characteristics of SF_6.[12] The following explanations apply:

1. Figure 8-23 shows 50-Hz breakdown voltage in SF_6 homogeneous field in relation to distance between electrodes and pressure.

2. Figure 8-24*a* depicts heat-transfer coefficients of SF_6 and transformer oil under natural convection. Tan δ, loss factor of SF_6 is low; less than 2×10^{-7} and SF_6 has excellent heat transfer characteristics. The heat absorption depends on the product of specific heat and specific weight; while specific heat is lower than that of air, its density is several times higher. Therefore, as a heat-absorbing medium, SF_6 is three times better than air. Figure 8-24*b* compares breakdown strength of air, SF_6, and transformer oil, and Fig. 8-24*c* is corona onset voltage related to pressure for air and SF_6.

3. Figure 8-25*a* illustrates interrupting capability of air, air and SF_6 mixture, and SF_6 versus pressure, and Fig. 8-25*b* is the arc radius of SF_6 and air versus temperature in K.

In ac circuit breakers, the phenomena of arc interruption are complex. Arc plasma temperatures of the order of 25000 to 5000 K are involved with conductivity changing billion times as fast as temperature in the critical range associated with thermal ionization. Supersonic turbulent flow occurs in changing flow and contact geometry at speeds from 100 to 1000 m/s in the arc. The contact system should accelerate from a stationary condition to high speeds in a very short duration.

With parabolic pressure distribution in the contact zone of a double-nozzle configuration, a cylindrical arc with arc at temperatures nearing 25000 K exists. Due to low density of gas at high temperatures, the plasma is strongly accelerated by axial pressure

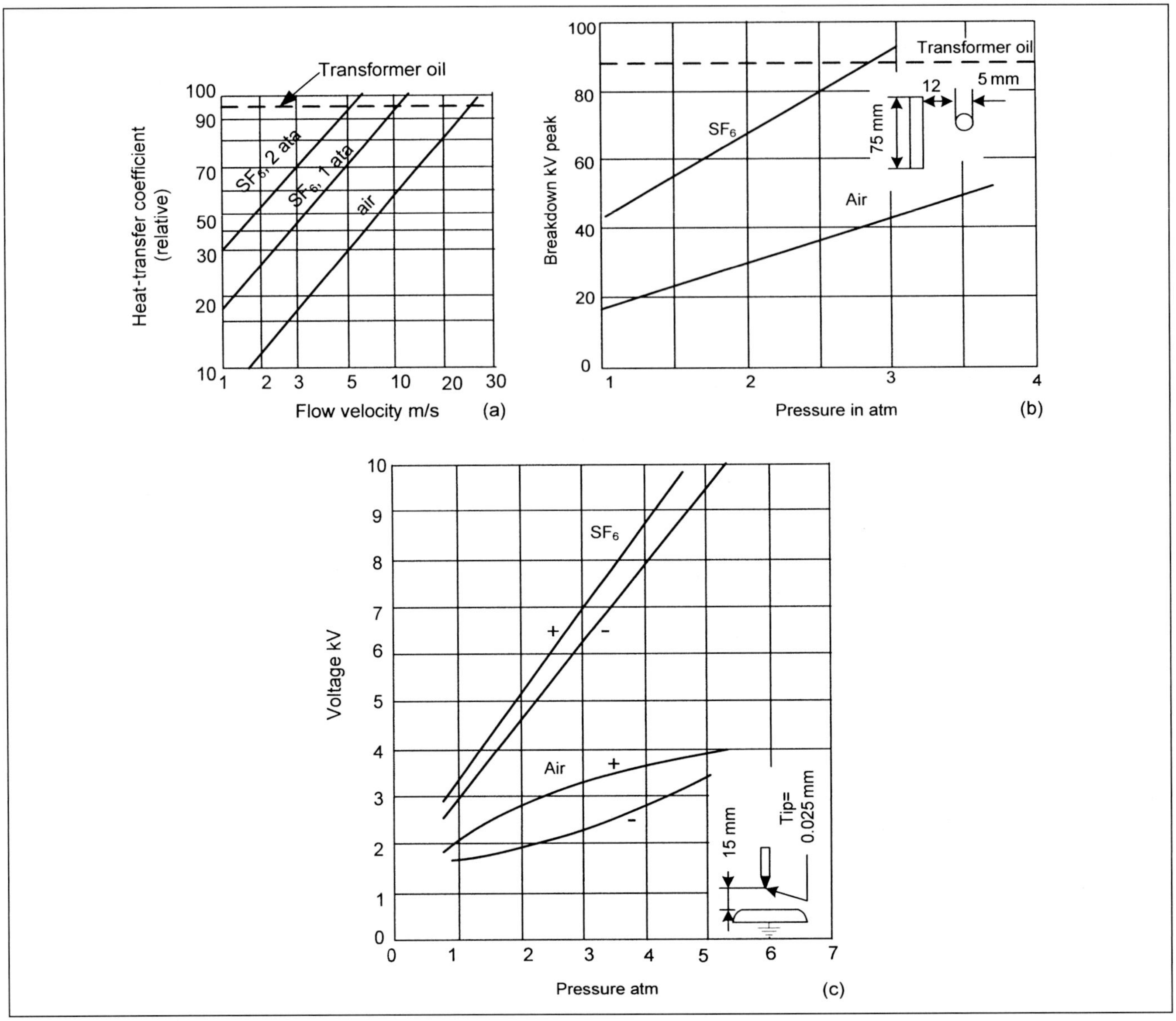

FIGURE 8-24 (*a*) Heat-transfer coefficient of SF_6 and transformer oil under natural convection. (*b*) Breakdown strength of air, SF_6, and transformer mineral oil. (*c*) Corona onset voltage related to pressure for SF_6 and air.

gradient. A so-called thermal arc boundary at temperature 300 to 2000 K exists. Arc downstream expands into nozzles, and in this region, the boundary layer between arc and gas is turbulent with the formation of vortices.

Two types of failures can occur: (1) dielectric failure which is usually coupled with a terminal fault and (2) thermal failure which is associated with a short-line fault. If after a current zero, RRRV is greater than a critical value, the decaying arc channel is reestablished by ohmic heating. This period, which is controlled by energy balance in the arc, is called the *thermal interruption mode*. Figure 8-26*a* and *b* show successful thermal interruption and a thermal failure, respectively. Within 2 ms after interruption, the voltage deviates from TRV. It decreases and approaches arc voltage.

Following thermal mode, a hot channel exists at a temperature from 300 to 5000 K, and the gas zone adjacent to arc diminishes at a slow rate. Recovering system voltage distorts and sets the dielectric limits. After successful thermal interruption, if the TRV can reach such a high-peak value that circuit breaker gap fails through, it is called a *dielectric failure mode*. This is shown in Fig. 8-27*a* and *b*. Figure 8-27*a* shows successful interruption, and 8-27*b* shows dielectric failure at the peak of the recovery voltage and rapid voltage decay.

The limit curves for circuit breakers can be plotted on log u and log I basis, as shown in Fig. 8-28. In this figure, u is the system voltage and I, the short-circuit current. The portion in thick lines shows dielectric limits, while the vertical portion in thin lines shows thermal limits. In thermal mode, (a) metal vapor production from contact surfaces, (b) *di/dt*, that is, the rate of decrease of the current at current zero, (c) arc constrictions in the nozzle due to finite velocity, (d) nozzle configurations, (e) presence of parallel capacitors and resistors, and (f) type of quenching medium and its pressure, are of importance. In the dielectric mode, the generation of electrons in electric field is governed by Townsend's equation:

$$\frac{dn_e}{dt} = \frac{-dn_e v_e}{dd} + (\alpha - \eta) n_e v_e \tag{8-56}$$

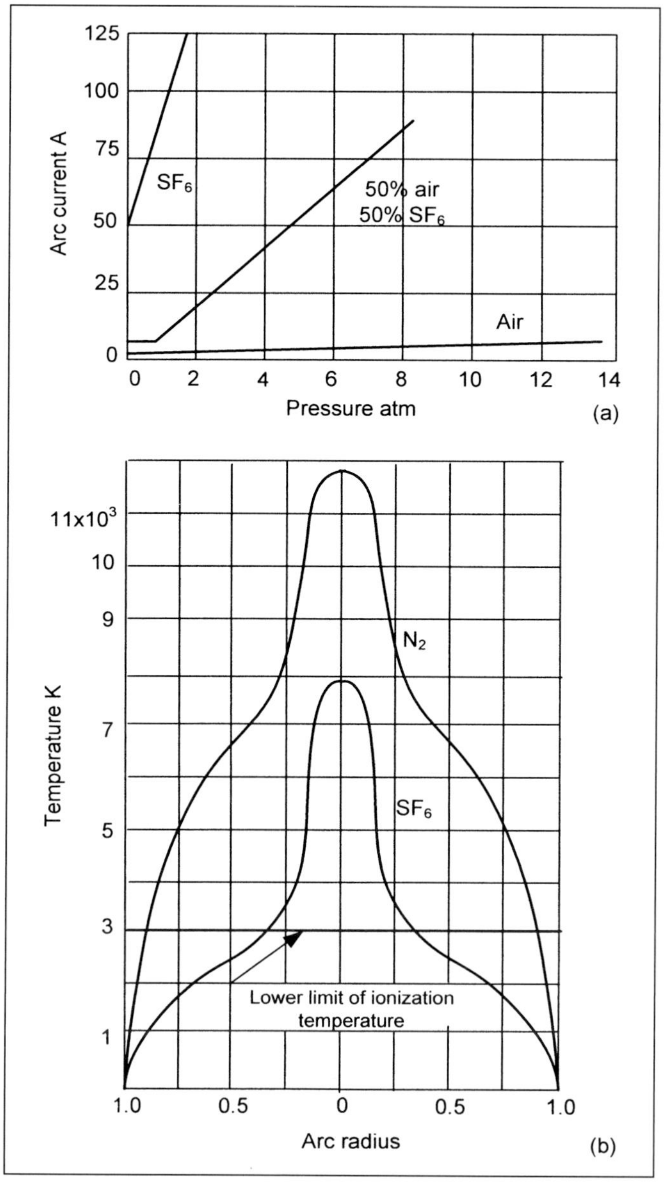

FIGURE 8-25 (*a*) Interrupting capacity of air, air and SF_6 mixture, and SF_6. (*b*) Arc radius of nitrogen and SF_6, and temperature.

where n_e is number of electrons, v_e is the electron drift velocity, and other symbols are as described before.

SF_6 breakers have an advantage as an electrical discharge causes the gas to decompose to an extent proportional to the applied energy. Part of the gas disintegrates in its atomic constituents:

$$SF_6\Delta^E \Leftrightarrow S+6F \tag{8-57}$$

where Δ^E is energy of the arc. When the temperature is lowered, the reaction proceeds in the reverse direction, resulting in the reformation of SF_6. The insulating properties are restored, provided there are no secondary reactions and vaporized electrode material. Arc in SF_6 has a fine core and its time constant may be expressed as:

$$\theta = c\pi r_0^2 \tag{8-58}$$

The cross section of arc decreases as the current is lowered. The arc in SF_6 is slender and well sustained even at lower currents (thus reducing the possibility of current chopping). Advantage lies in smaller thermal conductivity of arc core plasma for an equivalent electrical conductivity, which means a lower energy is transferred to the medium and a lower arc voltage is developed in the circuit. As current goes zero, the arc becomes gradually narrow before disappearing. The time constant featuring the speed at which arc loses its electrical conductivity is related to small volume occupied by the plasma. Figure 8-25*a* shows superior arc-interrupting performance and Fig. 8-24*a* shows high heat-transfer capacity and low ionization temperature of SF_6 compared to nitrogen. The time in which arc is quenched is approximately 100 times smaller than that in air.

It is necessary to point out the decomposition products and environmental effects. Metal fluorides such as CuF_2 and sulfur fluorides such as S_2F_2 may be formed. These primary decomposition products are good insulators so that the deposits on insulating surfaces do not impair operational efficiency. This is true so long as there is no moisture. In presence of moisture, secondary products are formed:

$$\begin{aligned}
&CuF_2 + H_2O \rightarrow CuO + 2HF \\
&SF_4 + H_2O \rightarrow SOF_2 + 2HF \\
&SOF_2 + H_2O \rightarrow SO_2 + 2HF \\
&SO_2 + 4HF \rightarrow SF_4 + 2H_2O
\end{aligned} \tag{8-59}$$

Note that moisture can be produced by the arcing process itself. HF products seriously attack materials containing silicone (glass

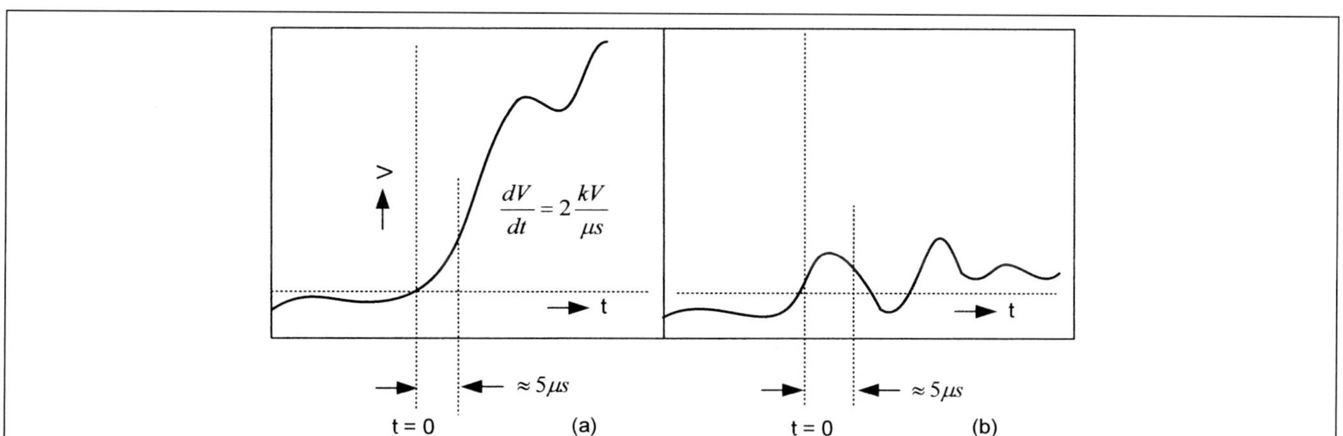

FIGURE 8-26 Thermal failure mode of a circuit breaker. (*a*) Successful interruption. (*b*) Failure in the thermal mode.

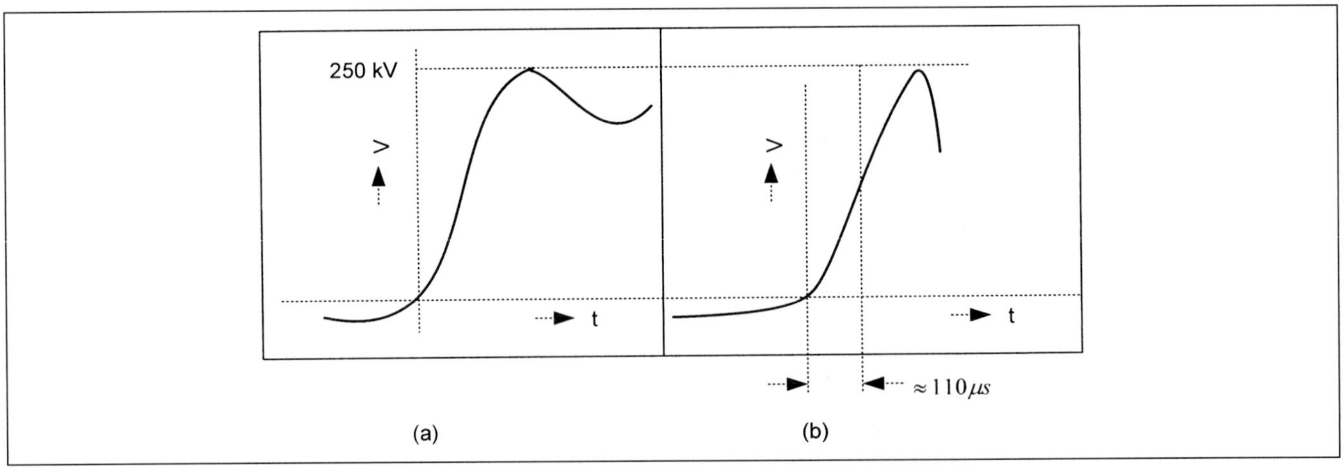

FIGURE 8-27 Dielectric failure mode of circuit breakers. (*a*) Successful interruption. (*b*) Failure at peak of TRV.

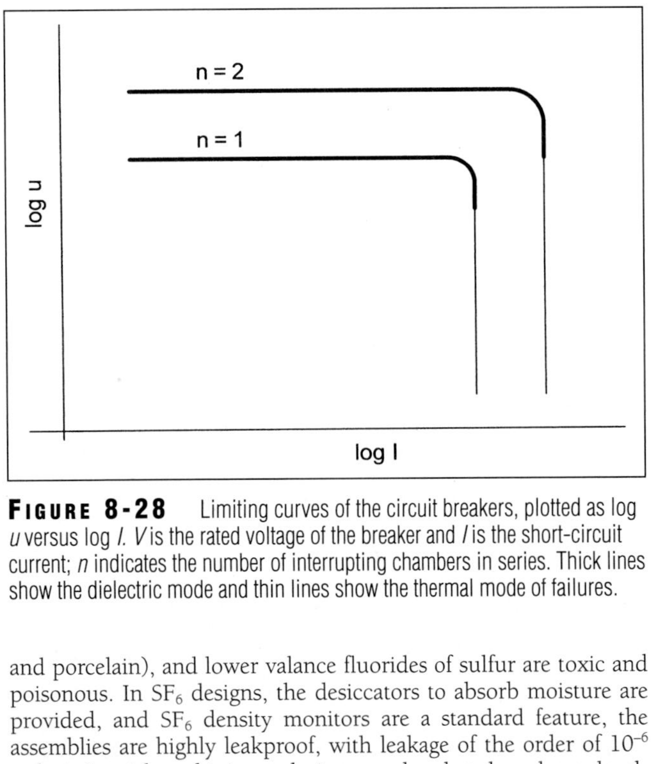

FIGURE 8-28 Limiting curves of the circuit breakers, plotted as log *u* versus log *I*. *V* is the rated voltage of the breaker and *I* is the short-circuit current; *n* indicates the number of interrupting chambers in series. Thick lines show the dielectric mode and thin lines show the thermal mode of failures.

and porcelain), and lower valance fluorides of sulfur are toxic and poisonous. In SF_6 designs, the desiccators to absorb moisture are provided, and SF_6 density monitors are a standard feature, the assemblies are highly leakproof, with leakage of the order of 10^{-6} or less. Special machining techniques and gaskets have been developed. A surface roughness of less than 0.8 microns is achieved.

The earlier designs of double-pressure breakers are obsolete, and "puffer-type" designs were introduced in 1967. The gas is sealed in the interrupting chamber at a low pressure and the contact movement itself generates a high pressure—part of the moving contact forms a cylinder and it moves against a stationary piston, thus the terminology puffer type—meaning blowing out the arc with a puff.

The last 20 years have seen the development of "self-blast technology" applicable to 800 kV. A valve was introduced between the expansion and compression volume. When the interrupting current is low, the valve opens under effect of overpressure and the interruption phenomena is similar to puffer design. At high-current interruption, the arc energy produces a high overpressure in the expansion volume and the valve remains closed. The overpressure for breaking is obtained by optimal use of thermal effect and nozzle clogging effect. The cross section of the arc significantly reduces the exhaust of gas in the nozzle. The self-blast technology is further optimized by double motion principle; this consists of displacing the two arcing contacts in opposite directions. This leads to further reduction in the operating energy. In a thermal blast chamber with arc-assisted opening, arc energy is used to generate the blast by thermal expansion and also to accelerate the moving part of the circuit breaker when interrupting high currents.[11,13]

Further descriptions of constructional features, interruptions in other mediums, for example, current interruption in vacuum, which has an entirely different conceptual base, are not discussed. It is sufficient to highlight that the interruption of currents in ac circuits is not an independent phenomenon; depending on the circuit conditions alone, it is seriously modified by the physical and electrical properties of the interrupting mediums in circuit breakers and the type of current being interrupted.

8-14 STRESSES IN CIRCUIT BREAKERS

The stresses in a circuit breaker under various operating conditions are summarized in Fig. 8-29. These stresses are shown in terms of three parameters, current, voltage, and *du/dt*, in a three-dimensional plane. Let the current stress be represented along *x*-axis, the *du/dt* stress along *y*-axis, and the voltage stress along *z*-axis. We see that a short-line fault (A_1, A_2, and A_3) gives the maximum RRRV stress, though the voltage stress is low. A terminal fault (B_1, B_2, B_3) results in the maximum interrupting current, while capacitor switching (*C*) and out-of-phase switching (*D*) give the maximum voltage stresses. All the stresses do not occur simultaneously in an interrupting process or in an electrical power system.

In this chapter, we continued the insight into current interruption and overvoltages gained in the previous chapters by addressing the postrecovery overvoltages after disconnection of various types of electrical circuits, modified by the circuit-breaker behavior. We noticed that arc characteristics of circuit breakers are important while disconnecting inductive currents, and on the other hand the dielectric characteristics gain importance for capacitive circuits. The recovery-voltage profiles in a number of cases, their impact on circuit breaker ratings, and the need for careful analysis depending on the applications are highlighted. The problems of interruption of low levels of inductive currents, restrike, reignitions, and prestrikes in breakers are the related aspects. Also, superior properties of SF_6 as an interrupting medium, which has overtaken all other interrupting mediums in high-voltage circuit breaker applications, are discussed with some failure modes. Yet, this chapter is not a guide for the application of circuit breakers where a number of other characteristics come into consideration. It is interesting to note that the failure rate of circuit breakers all over the world is

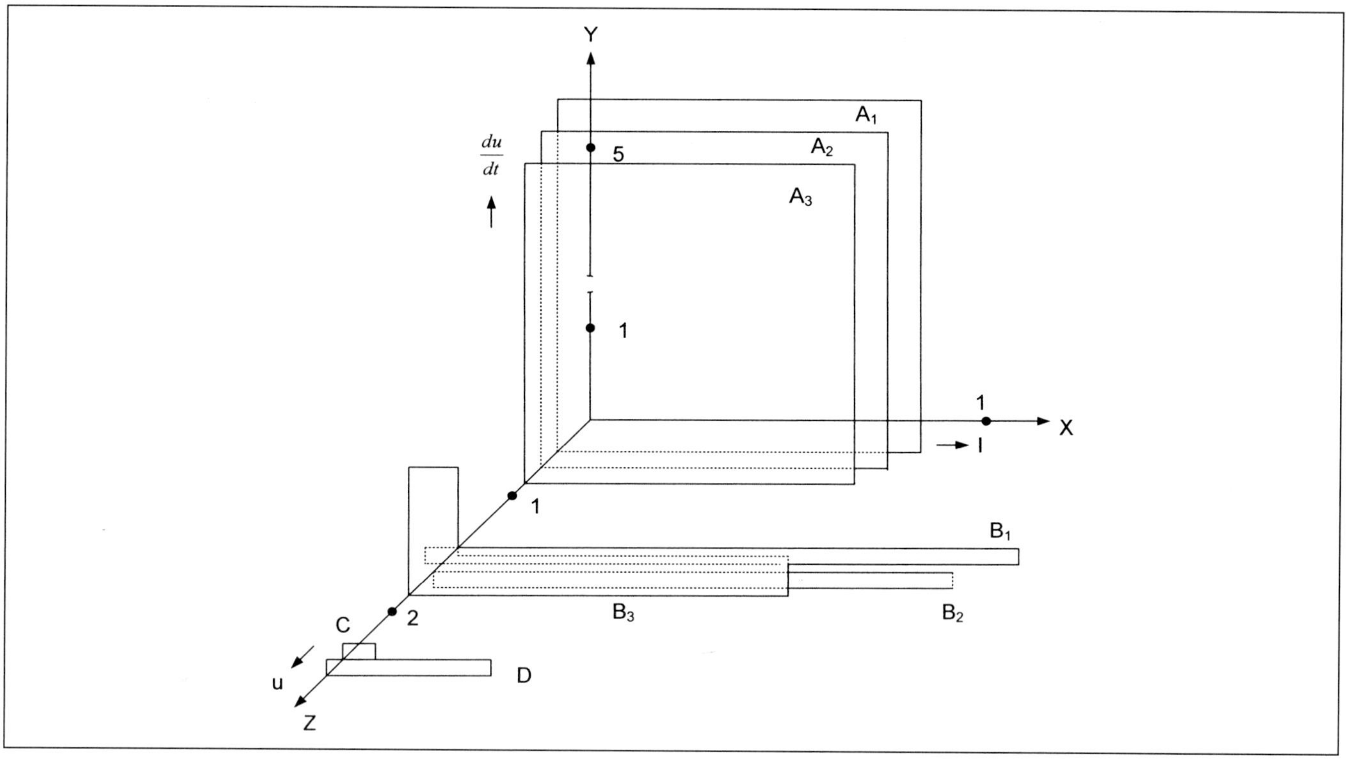

FIGURE 8-29 Stresses in high-voltage circuit breakers in terms of short-circuit current, RRRV, and maximum overvoltage (see text).

decreasing because of better designs and applications. The failure survey of circuit breaker control systems[14] attributes most failures to environmental effects.

PROBLEMS

1. Distinguish between reignition, restrike, prestrikes, and current chopping in high-voltage circuit breakers.

2. What is a delay line in TRV representation by two-parameter and four-parameter representation? Describe the parameters on which it depends and its calculation.

3. Describe two accepted failure modes of circuit breakers. Categorize the fault types that can lead to each of these two modes.

4. Find the recovery voltage across the breaker contacts while interrupting 4 A (peak) magnetizing current of 138-kV, 20-MVA transformer. Assume a capacitance of 4000 pF to ground and an inductance of 0.25 H.

5. What is the value of a switching resistor to eliminate the restriking transient in Prob. 4?

6. On the source side of a generator breaker, $L = 1.5$ mH and $C = 0.005$ μF. The breaker interrupts a current of 20 kA. Find (a) RRRV, (b) time to reach peak-recovery voltage, and (c) frequency of oscillation.

7. Explain the influence of power factor and first pole to clear on TRV. What is the effect of frequency of TRV and load current on interrupting duty of a circuit breaker?

8. A synchronous breaker is required to control a large shunt capacitor bank. Overvoltages can be reduced by closing the breaker at (1) peak of the voltage or (2) at zero crossing of the voltage. Which of the two statements is correct? Assume that the capacitors do not have a residual charge.

9. Comment on the correctness of these statements. (1) Interrupting an asymmetrical current gives rise to higher TRV than interrupting a symmetrical current. (2) As the current to be interrupted reduces, so does the initial RRRV. (3) Thermal mode of failure of breaker is excited when interrupting capacitor current due to higher TRV. (4) An oscillatory TRV occurs for a fault on a transformer connected to a transmission line. (5) Selecting a breaker of higher interrupting rating is an assurance that, in general, its TRV capability is better.

10. Describe a simple circuit element to control the RRRV when interrupting a high-magnetizing current.

11. Why is the TRV of a circuit breaker higher for interruption of short-circuit currents lower than its rated short-circuit current?

12. Compare the TRV profiles of ANSI/IEEE and IEC standards.

13. Draw a circuit diagram for TRV, first pole to clear, for the transformer secondary faults shown in Fig. 8-13.

14. An 18-kV, 200-MVA generator is connected through a step-up transformer of 18 to 230 kV, and serves a load of 150 MVA on the high-voltage side of the step-up transformer. The system runs in synchronism with the utility source. What is the expected waveform of TRV when the generator load is suddenly thrown off?

15. A breaker chops an inductive current of 5 A at its peak (thus, the source-side voltage is zero). Calculate the approximate maximum overvoltage that can be generated.

16. Referring to Table 8-3, plot the TRV profile for a breaker of maximum voltage of 362 kV, rated short-circuit current = 65 kA, when interrupting full short-circuit current, and when interrupting 30 percent short-circuit current, and when interrupting 7 percent short-circuit current.

17. Plot the short-line fault profile of the TRV for the breaker in Prob. 16 at 50 percent short-circuit current, surge impedance = 400 Ω. What is the frequency of the sawtooth wave?

18. Define tripping delay, opening time, arcing time, interrupting time, and contact parting time, as per ANSI/IEEE standards. What is the arcing time of a 2-cycle breaker?

19. Explain the superior properties of SF_6 as a dielectric medium in HV circuit breakers, the nature of corrosive arcing products, and the safeguards provided in the designs of the circuit breakers.

20. Write a brief note on the development of operating mechanisms of SF_6 circuit breakers.

REFERENCES

1. K. Ragaller, *Current Interruption in High Voltage Networks*, Plenum Press, New York, 1978.
2. A. M. Cassie, "Arc Rupture and Circuit Theory," CIGRE Report no. 102, 1939.
3. ANSI Std. C37.06, AC High Voltage Circuit Breakers Rated on Symmetrical Current Basis—Preferred Ratings and Related Required Capabilities, 1979 and 2000.
4. IEEE Std. C37.04, IEEE Standard Rating Structure for AC High-Voltage Circuit Breakers, 1999 (revision of 1979).
5. IEC Std. 60909-0, Short-Circuit Currents in Three-Phase AC Systems, Calculation of Currents, 2001-07.
6. IEEE Std. C37.011, IEEE Application Guide for Transient Recovery Voltage for AC High Voltage Circuit Breakers, 2005.
7. IEEE PC37.06, Draft: Standard AC High Voltage Circuit Breakers Rated on Symmetrical Current Basis—Preferred Ratings and Related Required Capabilities for Voltages above 1000 Volts, 2008.
8. ANSI C37.06.1, Guide for High Voltage Circuit Breakers Rated on a Symmetrical Current Basis Designated, Definite Purpose for Fast Transient Recovery Voltage Rise Times, 2000.
9. IEC Std. 62271-100, High Voltage Alternating Current Circuit Breakers, 2001.
10. L. B. Loeb, *Fundamental Processes of Electrical Breakdown in Gases*, John Wiley, New York, 1975.
11. D. Dufournet, "Generator Circuit Breakers: SF6 Breaking Chamber-Interruption of Current with Non-Zero Passage. Influence of Cable Connections on TRV of System-Fed Faults," *CIGRE*, pp. 13–101, Aug. 2002.
12. J. C. Das, "SF6 in High-Voltage Outdoor Switchgear," *Proc. IE*, vol. 61, pt. EL2, pp. 1–7, Oct. 1980.
13. D. Dufournet, F. Sciullo, J. Ozil, and A. Ludwig, "New Interrupting and Drive Techniques to Increase High Voltage Circuit Breakers Performance and Reliability," *CIGRE Session 1998*, pp. 13–104.
14. CIGRE WG A3.12, "Failure Survey on Circuit Breaker Controls Systems," *Electra*, no. 251, pp. 17–31, April 2007.

FURTHER READING

A. Braun, A. Eidinger, and E. Rouss, "Interruption of Short-Circuit Currents in High-Voltage AC Networks," *Brown Boveri Review*, vol. 66, pp. 240–245, April 1979.

CIGRE Working Group 13-02, "Switching Overvoltages in EHV and UHV Systems with Special Reference to Closing and Reclosing Transmission Lines," *Electra*, vol. 30, pp. 70–122, Oct. 1973.

J. C. Das, *Power System Analysis*, Chapter 5, Marcel Dekker, New York, 2002.

A. Greenwood, *Electrical Transients in Power Systems*, Wiley Interscience, New York, 1991.

W. Hermann and K. Ragaller, "Theoretical Description of Current Interruption in Gas Blast Circuit Breakers," *IEEE Trans. PAS*, vol. 96, pp. 1546–1555, 1977.

W. Hermann, K. Ragelschatz, L. Niemeyer, K. Ragaller, and E. Schade, "Investigations on Physical Phenomena Around Current Zero in HV Gas Blast Circuit Breakers," *IEEE Trans. PAS*, vol. 95, no. 4, pp. 1165–1178, 1976.

IEEE Std. C37.09, IEEE Test Procedure for AC High Voltage Circuit Breakers Rated on a Symmetrical Current Basis 1999 (R2007).

J. Kopainsky and E. Ruoss, "Interruption of Low Inductive and Capacitive Currents in High Voltage Systems," *BBC Review*, vol. 66, pp. 255–261, 1979.

NEMA SG-4, Alternating Current High Voltage Circuit Breakers, 1995.

L. V. D. Sluis and A. L. J. Jansen, "Clearing Faults Near Shunt Capacitor Banks," *IEEE Trans. PD*, vol. 5, no. 3, pp. 1346–1354, July 1990.

B. W. Swanson, "Theoretical Models for the Arc in Current Zero Regimes" in *Current Interruption in High Voltages Networks*, Plenum Press, New York, 1978.

H. Toda, Y. Ozaki, and I. Miwa, "Development of 800 kV Gas Insulated Switchgear," *IEEE Trans. PD*, vol. 7, no. 1, pp. 316–322, Jan. 1992.

C. L. Wagner and H. M. Smith, "Analysis of Transient Recovery Voltage Rating Concepts," *IEEE Trans. PAS*, vol. 103, pp. 3354–3364, 1984.

CHAPTER 9

SYMMETRICAL AND UNSYMMETRICAL SHORT-CIRCUIT CURRENTS

Short circuits in power systems cause decaying current transients, generally much above the load currents. The synchronous generators are the major source of short-circuit currents and we will discuss short-circuit of a synchronous generator in Chap. 10. The other sources of short-circuit currents in the power systems are:

- Asynchronous generators and motors.
- Synchronous motors.
- According to IEC standards the static converters in rolling mills contribute to the initial symmetrical short-circuit current and peak current (ANSI/IEEE close and latch current, expressed in peak kiloamperes), but not to the breaking (interrupting) current; exception is when the rotational masses of the motors and static equipment provide reverse transfer of energy for deceleration (a transient inverter operation). Hitherto ANSI/IEEE standards ignore any contributions to the short-circuit currents contributed by drive systems fed through static converters. The nonrotating loads and capacitors in parallel or series do not contribute to the short-circuit currents.

There are many facets of the short-circuit studies and calculations, for example, the switching devices should be able to carry and interrupt the short-circuit currents, and the equipments should be designed to have adequate short-circuit withstand capabilities according to relevant standards. With respect to transient analysis, the stability of interconnected systems to remain in synchronism until the faulty section of the system is isolated is important.

We discussed short circuit of a simple RL circuit in Chap. 2, origin of the dc component, its decay, and asymmetrical nature of the short-circuit currents. We also alluded to the fact that the reactances of a generator from the instant of short circuit until steady state are not constant and vary over a period of short time giving rise to ac decay. The ac decay will vary based on how far removed from the generator the fault is and on the intervening impedance between the generator and the fault point. The short-circuit currents of induction and synchronous motors are also decaying transients.

Rarely a simulation of short-circuit current in time domain is undertaken. The short-circuit currents for rating the switching devices are calculated using empirical methods according to prevalent standards.[1–3] The dynamic simulation, even for a relatively small power system, demands extensive system modeling and computing resources. Though such a study, for example, using EMTP, can validate the results obtained from an empirical calculation.

We will not get into details of the empirical short-circuit calculations, which have the primary objective of properly selecting the ratings of switching devices and providing input to the protective relaying.

9-1 SYMMETRICAL AND UNSYMMETRICAL FAULTS

In a three-phase system when all the phases are equally involved, it is a symmetrical fault. The term *bolted fault* is in common use and it implies that the three phases are connected together with links of zero impedance, that is, the fault current is only limited by the system and the machine impedances and the resistance to fault is zero. Bolted three-phase faults are rather uncommon. These generally (not always) give the maximum short-circuit currents in the system and form the basis of short-circuit ratings of the equipment.

Faults involving one phase, or more than one phase, and ground are called unsymmetrical faults. Under certain conditions a single line-to-ground or double line-to-ground fault current can exceed the three-phase symmetrical fault current. Unsymmetrical faults are more common as compared to three-phase symmetrical faults. The most common type is a line-to-ground fault. Approximately 70 percent of the faults in power systems are single line-to-ground faults.

The magnitude of fault currents in a system is related to the relative values of positive, negative, and zero sequence impedances as seen from the fault point. This is shown in Fig. 9-1.[2] To study unsymmetrical fault currents a basic knowledge of symmetrical component theory and matrices is required.[4–9] The section below provides basic concepts of the symmetrical component theory.

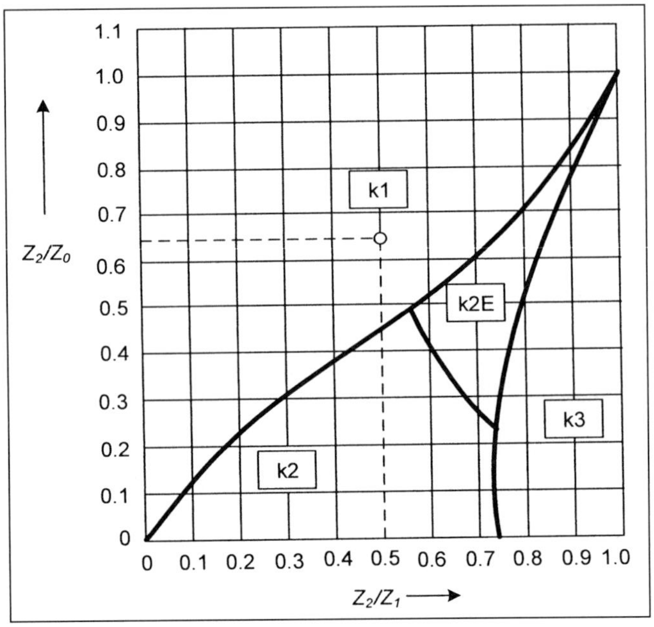

FIGURE 9-1 Magnitude of short-circuit currents depending upon the sequence impedances. $k1$: single phase-to-ground; $k2$: phase-to-phase, $k2E$: double phase-to-ground, $k3$: three-phase fault currents. For $Z_1/Z_2 \le 0.5$, $Z_2/Z_0 \le 0.65$, single line-to-ground current is the maximum.[2]

9-2 SYMMETRICAL COMPONENTS

The method of symmetrical components was originally presented by C.L. Fortescue in 1918[10] and has been popular ever since—it provided new dimensions to the electrical system modeling. It has been widely used in the analysis of unbalanced three-phase systems, unsymmetrical short-circuit currents, and rotating electrodynamic machinery.

Unbalance occurs in three-phase power systems due to faults, single-phase loads, untransposed transmission lines, or nonequilateral conductor spacing. In a three-phase balanced system, it is sufficient to determine the currents and voltages in one phase, and the currents and voltages in the other two phases are simply phase displaced. In an unbalanced system the simplicity of modeling a three-phase system as a single-phase system is not valid. A convenient way of analyzing unbalanced operation is through symmetrical components. Even the so-called balanced systems are not perfectly balanced. The three-phase voltages and currents, which may be unbalanced are transformed into three sets of balanced voltages and currents, called *symmetrical components*. The impedances presented by various power system components, that is, transformers, generators, transmission lines are *decoupled* from each other, resulting in independent networks for each component. These form a balanced set. This results in simplicity of calculations.

The basic theory of symmetrical components can be stated as a mathematical concept. A system of three coplanar vectors is completely defined by six parameters; the system can be said to possess six degrees of freedom. A point in a straight line being constrained to lie on the line possesses but one degree of freedom, and by the same analogy, a point in space has three degrees of freedom. A coplanar vector is defined by its terminal and length and therefore possesses two degrees of freedom. A system of coplanar vectors having six degrees of freedom, that is, three-phase unbalanced current or voltage vectors, can be represented by three symmetrical systems of vectors each having two degrees of freedom. In general, a system of n numbers can be resolved into n sets of component numbers each having n components, that is, a total of n^2 components. Fortescue demonstrated that an unbalanced set on n phasors can be resolved into n–1 balanced phase systems of different phase sequence and one zero sequence system, in which all phasors are of equal magnitude and cophasial:

$$\begin{aligned} V_a &= V_{a1} + V_{a2} + V_{a3} + \cdots + V_{an} \\ V_b &= V_{b1} + V_{b2} + V_{b3} + \cdots + V_{bn} \\ &\vdots \\ V_n &= V_{n1} + V_{n2} + V_{n3} + \cdots + V_{nn} \end{aligned} \tag{9-1}$$

where $V_a, V_b, \ldots, V_n$ are original n unbalanced voltage phasors. $V_{a1}, V_{b1}, \ldots, V_{n1}$ are the first set of n *balanced* phasors, at an angle of $2\pi/n$ between them. $V_{a2}, V_{b2}, \ldots, V_{n2}$ are the second set of n balanced phasors at an angle $4\pi/n$. And the final set $V_{an}, V_{bn}, \ldots, V_{nn}$ is the zero sequence set, all phasors at $n(2\pi/n) = 2\pi$, that is, cophasial.

In a symmetrical three-phase balanced system, the generators produce balanced voltages which are displaced from each other by $2\pi/3 = 120°$. These voltages can be called positive sequence voltages. If a vector operator a is defined which rotates a unit vector through 120° in a counterclockwise direction, then $a = -0.5 + j0.866$, $a^2 = -0.5 - j0.866$, $a^3 = 1$, $1 + a^2 + a = 0$. Considering a three-phase system, Eq. (9-1) reduces to:

$$\begin{aligned} V_a &= V_{a0} + V_{a1} + V_{a2} \\ V_b &= V_{b0} + V_{b1} + V_{b2} \\ V_c &= V_{c0} + V_{c1} + V_{c2} \end{aligned} \tag{9-2}$$

We can define the set consisting of V_{a0}, V_{b0}, and V_{c0} as the zero sequence set, the set V_{a1}, V_{b1}, and V_{c1}, as the positive sequence set, and the set V_{a2}, V_{b2}, and V_{c2} as the negative sequence set of voltages. The three original unbalanced voltage vectors give rise to nine voltage vectors, which must have constraints of freedom and are not totally independent. By definition of positive sequence, V_{a1}, V_{b1}, and V_{c1} should be related as follows, as in a normal balanced system:

$$V_{b1} = a^2 V_{a1},\ V_{c1} = aV_{a1}.$$

Note that V_{a1} phasor is taken as the reference vector. The negative sequence set can be similarly defined, but of opposite phase sequence:

$$V_{b2} = aV_{a2},\ V_{c2} = a^2 V_{a2}.$$

Also $V_{a0} = V_{b0} = V_{c0}$. With these relations defined, Eq. (9-2) can be written as:

$$\begin{vmatrix} V_a \\ V_b \\ V_c \end{vmatrix} = \begin{vmatrix} 1 & 1 & 1 \\ 1 & a^2 & a \\ 1 & a & a^2 \end{vmatrix} \begin{vmatrix} V_{a0} \\ V_{a1} \\ V_{a2} \end{vmatrix} \tag{9-3}$$

or in the abbreviated form:

$$\bar{V}_{abc} = \bar{T}_s \bar{V}_{012} \tag{9-4}$$

where the matrix T_s is the transformation matrix. Its inverse will give the reverse transformation:

$$\bar{V}_{012} = \bar{T}_s^{-1} \bar{V}_{abc} \tag{9-5}$$

Similar expressions apply for currents. The matrix $\bar{T}_s^{-1}$ is:

$$\bar{T}_s^{-1} = \frac{1}{3} \begin{vmatrix} 1 & 1 & 1 \\ 1 & a & a^2 \\ 1 & a^2 & a \end{vmatrix} \tag{9-6}$$

The impedance transformation is given by:

$$\bar{Z}_{abc} = \bar{T}_s \bar{Z}_{012} \bar{T}_s^{-1}$$
$$\bar{Z}_{012} = \bar{T}_s^{-1} \bar{Z}_{abc} \bar{T}_s \tag{9-7}$$

While this simple explanation may be adequate, a better insight into the symmetrical component theory can be gained through matrix concepts of similarity transformation, diagonalization, eigenvalues, and eigenvectors. It can be shown that:

- Eigenvectors giving rise to symmetrical component transformation are the same though the eigenvalues differ. Thus, these vectors are not unique.
- The Clarke component transformation (Chap. 4) is based upon the same eigenvectors, but different eigenvalues.[11]
- The symmetrical component transformation does not uncouple an initially unbalanced three-phase system. Prima facie, this is a contradiction of what we said earlier, that the main advantage of symmetrical components lies in decoupling unbalanced systems, which could then be represented much akin to three-phase balanced systems (see App. D, where the application to a three-phase system with unequal phase impedances shows that decoupling is not possible). In application for the fault analysis it is assumed that the system was perfectly symmetrical prior to the fault, and asymmetry occurs only at the fault point. In other words the unbalance part of the network is connected to a balanced system at the point of fault.

9-2-1 Characteristics of Symmetrical Components

Matrix equations [Eqs. (9-4) and (9-5)] are written in the expanded form:

$$V_a = V_0 + V_1 + V_2$$
$$V_b = V_0 + a^2 V_1 + aV_2 \tag{9-8}$$
$$V_c = V_0 + aV_1 + a^2 V_2$$

and

$$V_0 = \frac{1}{3}(V_a + V_b + V_c)$$
$$V_1 = \frac{1}{3}(V_a + aV_b + a^2 V_c) \tag{9-9}$$
$$V_2 = \frac{1}{3}(V_a + a^2 V_b + aV_c)$$

V_0, V_1, and V_2 are defined as follows:

- V_0 is the zero sequence voltage. It is of equal magnitude in all the three-phases and is cophasial.
- V_1 is the system of balanced positive sequence voltages, *of the same phase sequence* as the original unbalanced system of voltages. It is of equal magnitude in each phase, but displaced by 120°, component of phase *b* lagging the component of phase *a* by 120° and component of phase *c* leading the component of phase *a* by 120°.
- V_2 is the system of balanced negative sequence voltages. It is of equal magnitude in each phase, and there is 120° phase displacement between the voltages, the component of phase *c* lagging the component of phase *a*, and component of phase *b* leading the component of phase *a*.

Therefore, the positive and negative sequence voltages (or currents) can be defined as the order in which the three-phases attain a maximum value. For positive sequence, the order is "*abca*," while for the negative sequence, it is "*acba*." We can also define positive and negative sequence by the order in which the phasors pass a *fixed point* on the vector plot. Note that the rotation is counterclockwise for all three sets of sequence components, as was assumed for the original unbalanced vectors (Fig. 9-2). Sometimes, this is confused and negative sequence rotation is said to be *reverse of* positive sequence. The negative sequence vectors do not rotate in a direction opposite to the positive sequence vectors, though the negative phase sequence is opposite to the positive phase sequence.

In a symmetrical system of three phases, the resolution of voltages or currents into a system of zero, positive, and negative components is equivalent to three separate systems. Sequence voltages act in isolation and produce zero, positive, and negative sequence currents and the theorem of superposition applies. The following generalizations of symmetrical components can be made:

1. In a three-phase unfaulted system in which all the loads are balanced and generators produce positive sequence voltages, only positive sequence currents flow, resulting in balanced voltage drops of the same sequence. There are no negative sequence or zero sequence voltage drops.

2. In symmetrical systems, the currents and voltages of different sequences do not affect each other, that is, positive sequence currents produce only positive sequence voltage drops. By the same analogy, the negative sequence currents produce only negative sequence drops, and zero sequence currents produce only zero sequence drops.

3. Negative and zero sequence currents are set up in circuits of unbalanced impedances only, that is, a set of unbalanced impedances in a symmetrical system may be regarded as a source of negative and zero sequence current. Positive sequence currents flowing in an unbalanced system produce positive, negative, and possibly zero sequence voltage drops. The negative sequence currents flowing in an unbalanced system produce voltage drops of all three sequences. The same is true about zero sequence currents.

4. In a three-phase three-wire system, no zero sequence currents appear in the line conductors. This is so because $I_0 = (1/3)(I_a + I_b + I_c)$, and therefore there is no path for the zero sequence current to flow. In a three-phase four-wire system with a neutral return, the neutral carries out of the balance current, that is, $I_n = (I_a + I_b + I_c)$. Therefore, it follows that $I_n = 3I_0$. At the grounded neutral of three-phase wye system, positive and negative sequence voltages are zero. The neutral voltage is equal to the zero sequence voltage or product of zero sequence current and three times the neutral impedance, Z_n.

5. From what has been said in point 4 above, phase conductors emanating from ungrounded wye- or delta-connected transformer windings cannot have zero sequence current. In a delta winding, zero sequence currents, if present, set up circulating currents in the delta windings itself. This is because the delta winding forms a closed path of low impedance for the zero sequence currents; each phase zero sequence voltage is absorbed by its own phase voltage drop and there are no zero sequence components at the terminals.

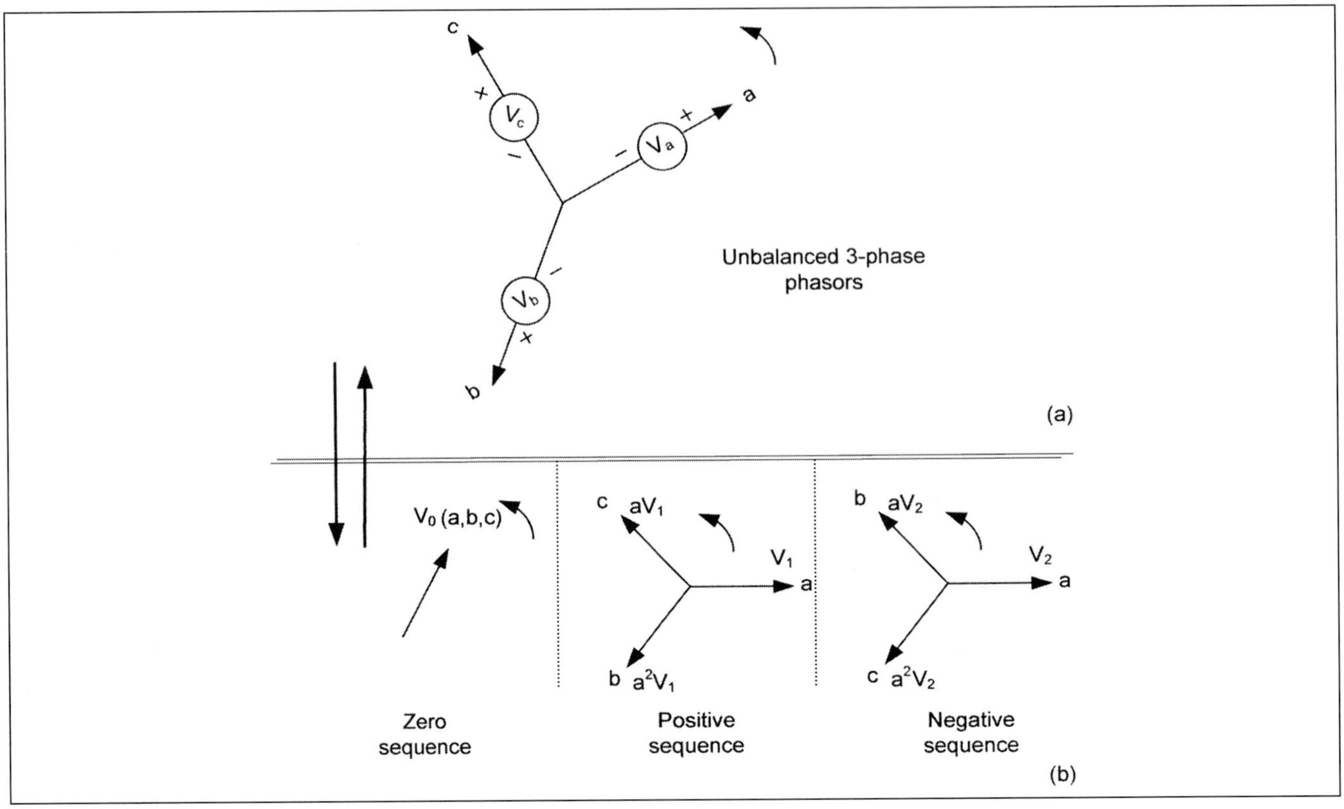

FIGURE 9-2 Symmetrical component transformations.

9-2-2 Power Invariance

The symmetrical component transformation is power invariant. The power is:

$$S = V_a I_a^* + V_b I_b^* + V_c I_c^* \tag{9-10}$$

where I_a^* is conjugate of I_a. This can be shown to be equal to:

$$S = 3(V_1 I_1^* + V_2 I_2^* + V_0 I_0^*) \tag{9-11}$$

9-3 SEQUENCE IMPEDANCE OF NETWORK COMPONENTS

The impedance encountered by the symmetrical components depends upon the type of power system equipment, that is, a generator, a transformer, or a transmission line. The sequence impedances are required for component modeling and analysis. Zero sequence impedance of overhead lines depends upon presence of the ground wires, tower footing resistance, and the grounding. It may vary between two and six times the positive sequence impedance. The line capacitance of overhead lines is ignored in short-circuit calculations. Appendix D details three-phase matrix models of transmission lines, bundle conductors, and cables and their transformation to symmetrical components. While estimating sequence impedances of power system components is one problem, constructing the zero, positive, and negative sequence impedance networks is the first step for unsymmetrical fault current calculations.

9-3-1 Construction of Sequence Networks

A sequence network shows how the sequence currents will flow in a system, if these are present. Connections between sequence component networks are necessary to achieve this objective. The sequence networks are constructed as viewed from the *fault point*, which can be defined as the point at which the unbalance occurs in a system, that is, a fault or load unbalance.

The voltages for the sequence networks are taken as line-to-neutral voltages. The only active network containing the voltage source is the positive sequence network. Phase *a* voltage is taken as the reference voltage and the voltages of the other two phases are expressed with reference to phase *a* voltage.

The sequence networks for positive, negative, and zero sequence will have per phase impedance values which may differ. Normally, the sequence impedance networks are constructed based upon per unit values on a common MVA base, and a base MVA of 100 is in common use. For nonrotating equipment like transformers, the impedance to negative sequence currents will be the same as for positive sequence currents. The impedance to negative sequence currents of rotating equipment will be different from the positive sequence impedance and, in general, for all apparatuses, the impedance to zero sequence currents will be different from the positive or negative sequence impedances. For a study involving sequence components, the sequence impedance data can be:

1. Calculated by using subroutine computer programs
2. Obtained from manufacturer's data
3. Calculated by long-hand calculations or
4. Estimated from tables in published references

The positive directions of current flow in each sequence network are outward at the faulted or unbalance point. This means that the sequence currents flow in the same direction in all three sequence networks.

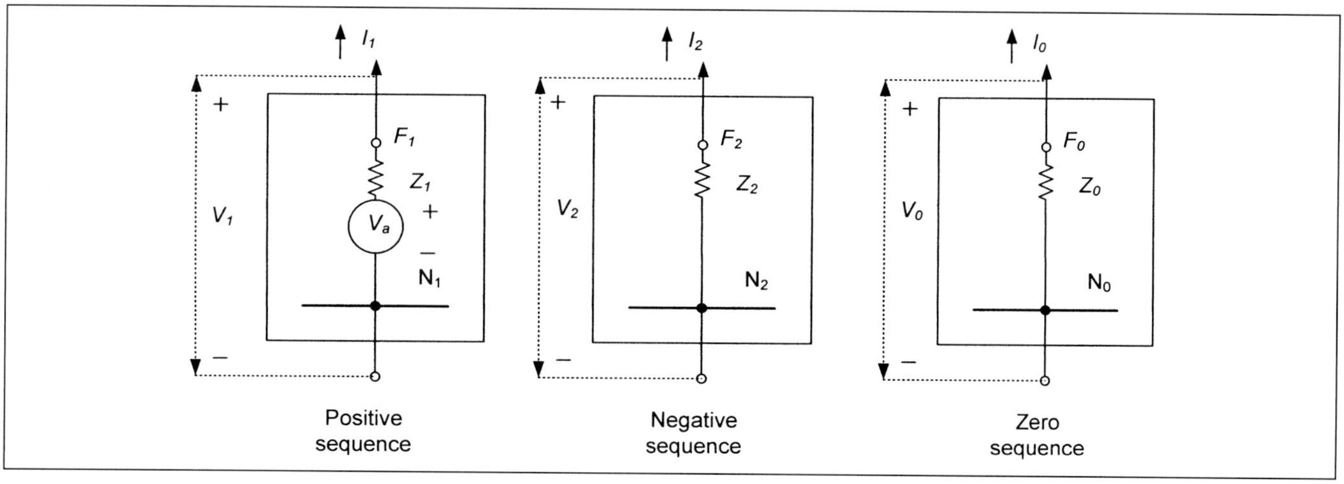

FIGURE 9-3 Representation of positive, negative, and zero sequence networks.

Sequence networks are shown schematically in boxes in which the fault points from which the sequence currents flow outward are marked as F_1, F_2, and F_0 and the neutral buses are designated as N_1, N_2, and N_0, respectively, for the positive, negative, and zero sequence impedance networks. Each network forms a two-port network with Thévenin sequence voltages across sequence impedances. Figure 9-3 illustrates this basic formation. Note the direction of currents. The voltage across the sequence impedance rises from N to F. As stated before, only the positive sequence network has a voltage source, which is the Thévenin equivalent. With this convention, appropriate signs must be allocated to the sequence voltages:

$$\begin{vmatrix} V_0 \\ V_1 \\ V_2 \end{vmatrix} = \begin{vmatrix} 0 \\ V_a \\ 0 \end{vmatrix} - \begin{vmatrix} Z_0 & 0 & 0 \\ 0 & Z_1 & 0 \\ 0 & 0 & Z_2 \end{vmatrix} \begin{vmatrix} I_0 \\ I_1 \\ I_2 \end{vmatrix} \tag{9-12}$$

Based on the discussions so far, we can graphically represent the sequence impedances of various system configurations, though some practice is required. This will be illustrated by construction of sequence impedances for unsymmetrical faults.

9-3-2 Zero Sequence Impedance of Transformers

The positive and negative sequence impedances of a transformer can be taken equal to its leakage impedance. As the transformer is a static device, the positive or negative sequence impedances do not change with phase sequence of balanced applied voltages. The zero sequence impedance can, however, vary from an open circuit to a low value, depending upon the transformer winding connection, method of neutral grounding, and transformer construction, that is, core or shell type. Figure 9-4 shows the sequence impedances of two-winding transformers, and Figure 9-5 for three-winding transformers of various winding arrangements and grounding.

9-4 FAULT ANALYSIS USING SYMMETRICAL COMPONENTS

While applying the symmetrical component method to fault analysis, we will ignore the load currents. This makes the positive sequence voltages of all the generators in the system identical and equal to prefault voltage.

In the analysis to follow, Z_1, Z_2, and Z_0 are the positive, negative, and zero sequence impedances as seen from the fault point. V_a, V_b, V_c are the phase-to-ground voltages at the fault point, prior to fault, that is, *if the fault does not exist,* and V_1, V_2, and V_0 are the corresponding sequence component voltages. Similarly, I_a, I_b, and I_c are the line currents, and I_1, I_2, and I_0 are their sequence components. A fault impedance of Z_f is assumed in every case. For a bolted fault, $Z_f = 0$

9-4-1 Line-To-Ground Fault

Figure 9-6*a* shows that phase *a* of a three-phase system goes to ground through an impedance Z_f. The flow of ground fault current depends upon the method of system grounding. A solidly grounded system with zero ground resistance is assumed. There will be some impedance to flow of the fault current in the form of impedance of the return ground conductor or the grounding grid resistance. A ground resistance can be added in series with the fault impedance Z_f. The ground fault current must have a return path through the grounded neutrals of generators or transformers. If there is no return path for the ground current, $Z_0 = \infty$, and the ground fault current is zero. This is an obvious conclusion.

Phase *a* is faulted in Fig. 9-6*a*. As the load current is neglected, currents in phases *b* and *c* are zero, and the voltage at the fault point $V_a = I_a Z_f$. The sequence components of the currents are given by:

$$\begin{vmatrix} I_0 \\ I_1 \\ I_2 \end{vmatrix} = \frac{1}{3} \begin{vmatrix} 1 & 1 & 1 \\ 1 & a & a^2 \\ 1 & a^2 & a \end{vmatrix} \begin{vmatrix} I_a \\ 0 \\ 0 \end{vmatrix} = \frac{1}{3} \begin{vmatrix} I_a \\ I_a \\ I_a \end{vmatrix} \tag{9-13}$$

Therefore:

$$I_0 = I_1 = I_2 = \frac{1}{3} I_a \tag{9-14}$$

and

$$3I_0 Z_f = V_0 + V_1 + V_2 = -I_0 Z_0 + (V_a - I_1 Z_1) - I_2 Z_2 \tag{9-15}$$

which gives:

$$I_0 = \frac{V_a}{Z_0 + Z_1 + Z_2 + 3Z_f} \tag{9-16}$$

The fault current I_a is:

$$I_a = 3I_0 = \frac{3V_a}{Z_0 + Z_1 + Z_2 + 3Z_f} \tag{9-17}$$

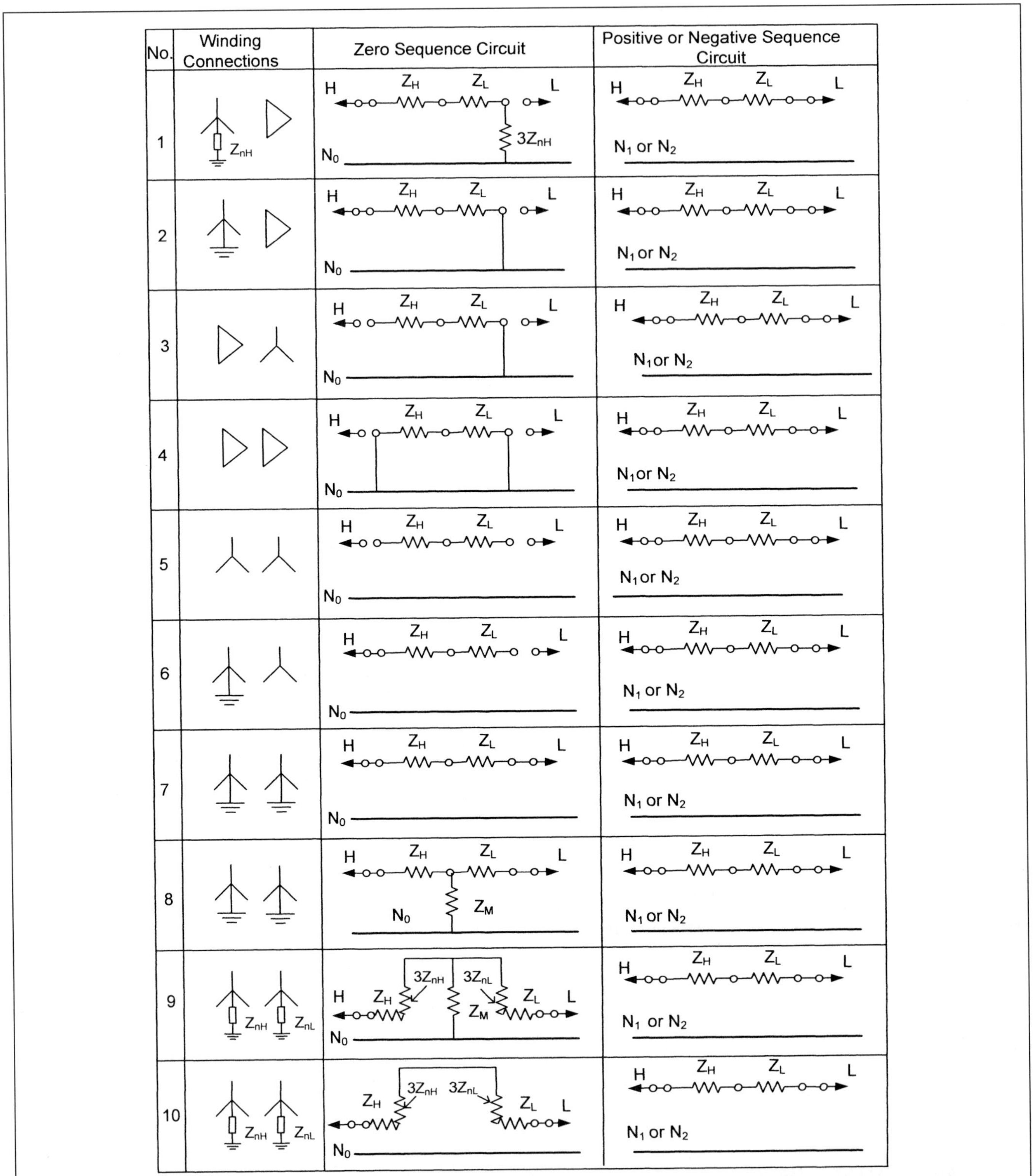

FIGURE 9-4 Positive, negative, and zero sequence circuits of two-winding transformers. Connections 8 and 9 are for core type transformers and connections 7 and 10 are for shell type transformers.

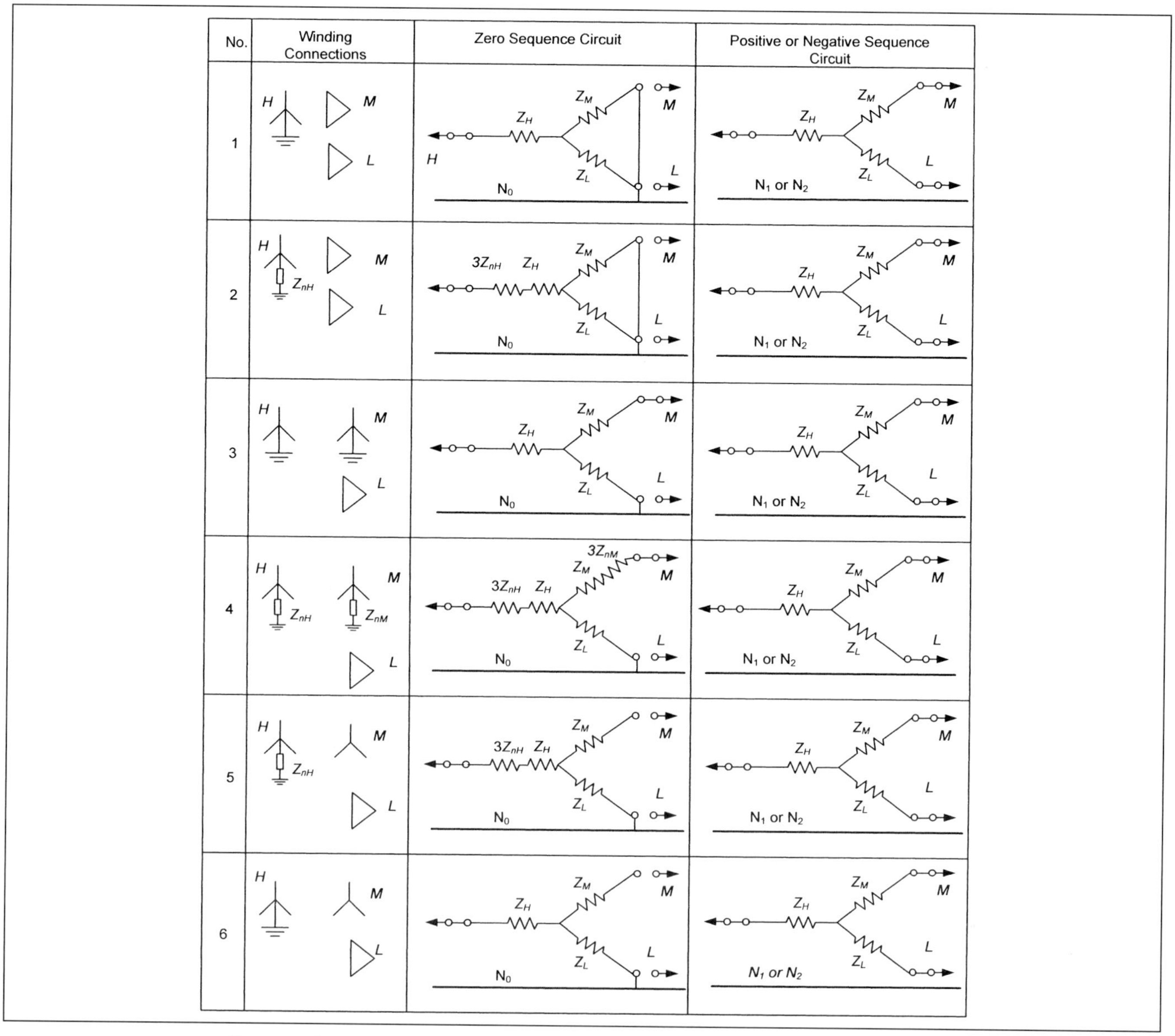

FIGURE 9-5 Positive, negative, and zero sequence circuits of three-winding transformers.

This shows that the equivalent fault circuit using sequence impedances can be constructed as shown in Fig. 9-6*b*. In terms of sequence impedances network blocks, the connections are shown in Fig. 9-6*c*. Voltage of phase *b* to ground under fault conditions is:

$$V_b = a^2V_1 + aV_2 + V_0$$

$$= V_a \frac{3a^2Z_f + Z_2(a^2 - a) + Z_0(a^2 - 1)}{(Z_1 + Z_2 + Z_0) + 3Z_f} \tag{9-18}$$

Similarly, the voltage of phase *c* can be calculated. An expression for the ground fault current for use in grounding grid designs and system grounding is as follows:

$$I_a = \frac{3V_a}{(R_0 + R_1 + R_2 + 3R_f + 3R_G) + j(X_0 + X_1 + X_2)} \tag{9-19}$$

where R_f is the fault resistance, and R_G is the resistance of the grounding grid. R_0, R_1, and R_2 are the sequence resistances and X_0, X_1, and X_2 are sequence reactances.

9-4-2 Line-To-Line Fault

Figure 9-7*a* shows a line-to-line fault. A short circuit occurs between phases *b* and *c*, through a fault impedance Z_f. The fault current circulates between phases *b* and *c*, flowing back to source through phase *b* and returning through phase *c*. $I_a = 0$ and $I_b = -I_c$. The sequence components of the currents are:

$$\begin{vmatrix} I_0 \\ I_1 \\ I_2 \end{vmatrix} = \frac{1}{3}\begin{vmatrix} 1 & 1 & 1 \\ 1 & a & a^2 \\ 1 & a^2 & a \end{vmatrix}\begin{vmatrix} 0 \\ -I_c \\ I_c \end{vmatrix} = \frac{1}{3}\begin{vmatrix} 0 \\ -a + a^2 \\ -a^2 + a \end{vmatrix} \tag{9-20}$$

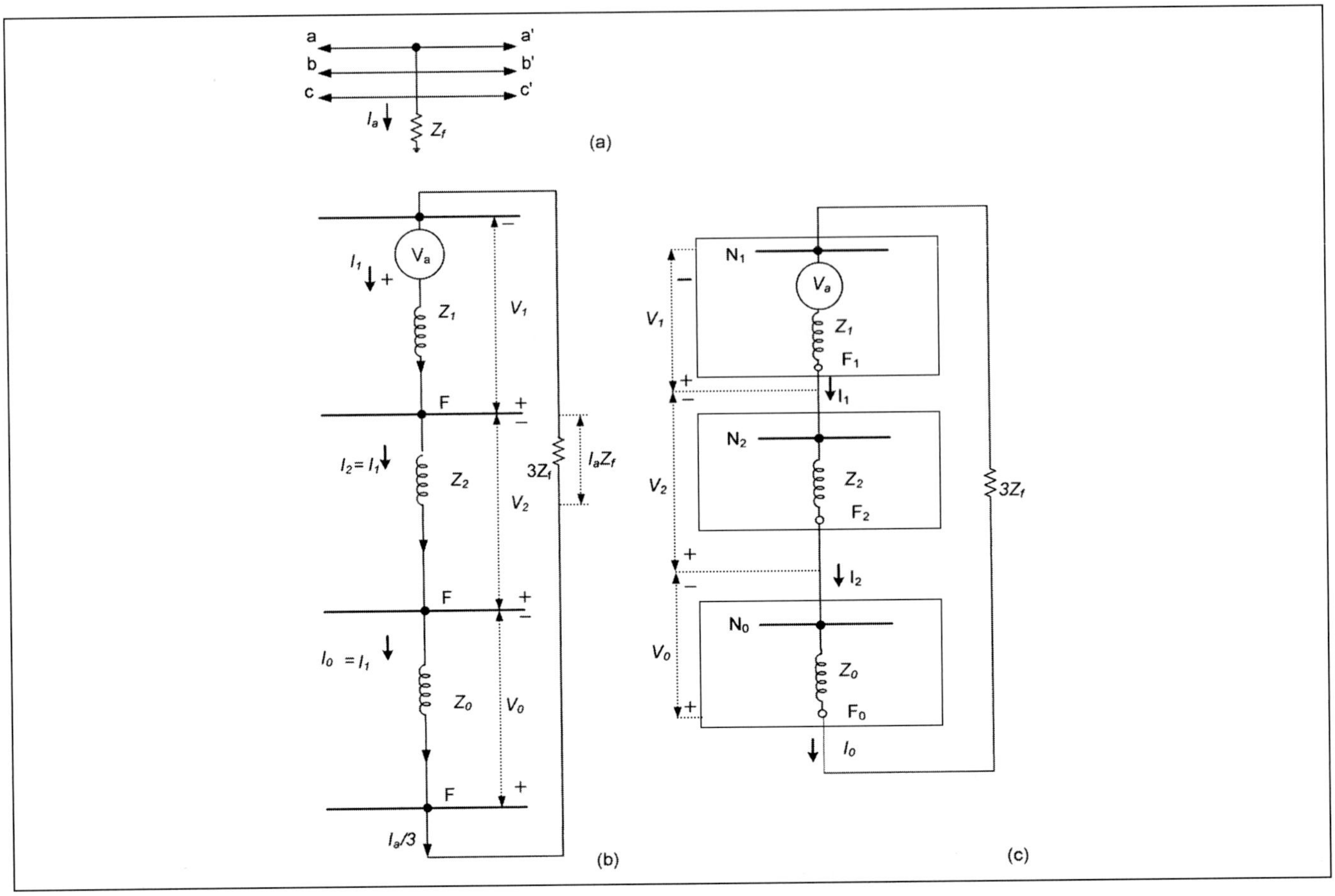

FIGURE 9-6 (*a*) Single line-to-ground fault. (*b*) Line-to-ground fault equivalent circuit connections. (*c*) Connections of sequence networks.

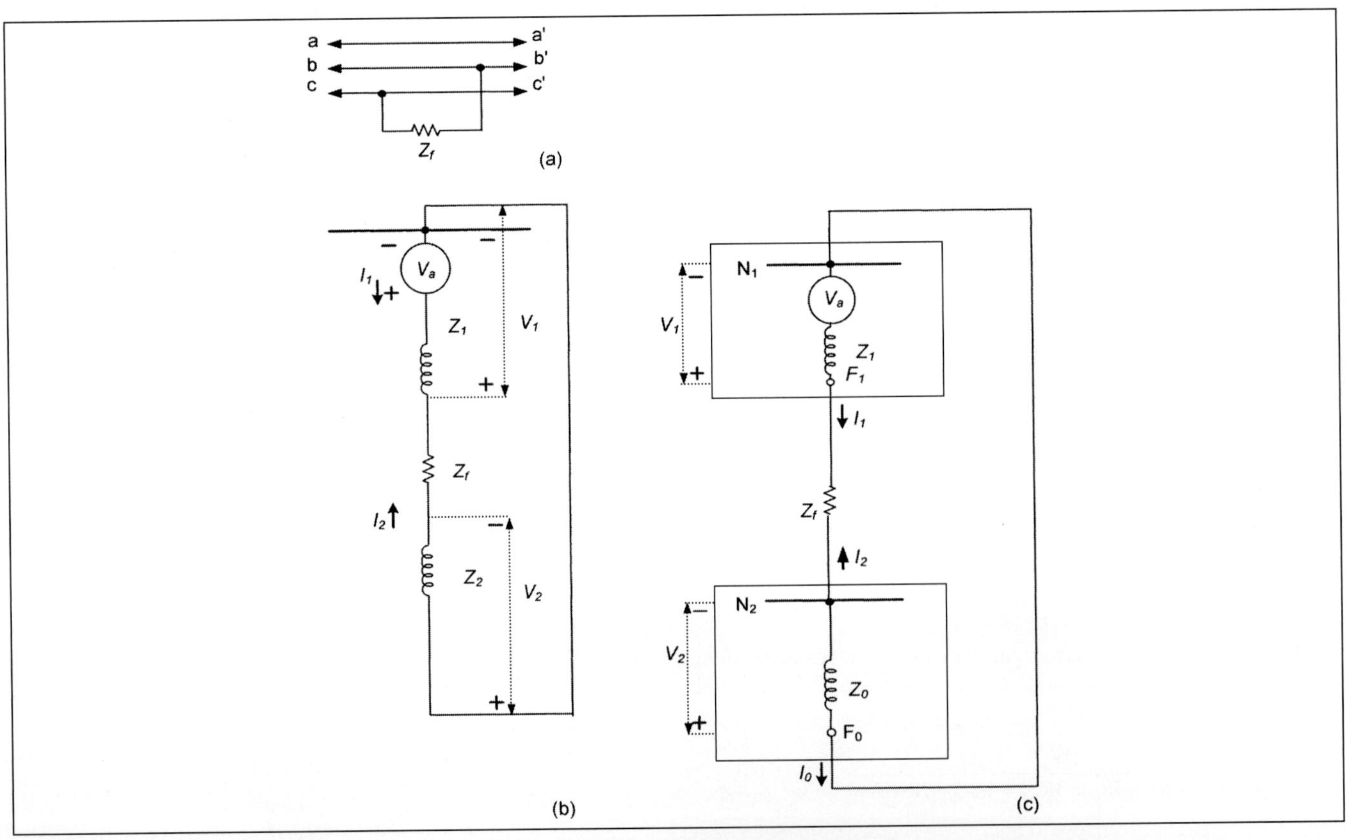

FIGURE 9-7 (*a*) Line-to-line fault. (*b*) Line-to-line fault equivalent circuit connections. (*c*) Connections of sequence networks.

From Eq. (9-20), $I_0 = 0$ and $I_1 = -I_2$. Therefore:

$$V_b - V_c = \begin{vmatrix} 0 & 1 & -1 \end{vmatrix} \begin{vmatrix} V_a \\ V_b \\ V_c \end{vmatrix} = \begin{vmatrix} 0 & 1 & -1 \end{vmatrix} \begin{vmatrix} 1 & 1 & 1 \\ 1 & a & a^2 \\ 1 & a^2 & a \end{vmatrix} \begin{vmatrix} V_0 \\ V_1 \\ V_2 \end{vmatrix}$$

$$= \begin{vmatrix} 0 & a^2 - a & a - a^2 \end{vmatrix} \begin{vmatrix} V_0 \\ V_1 \\ V_2 \end{vmatrix} \quad (9\text{-}21)$$

This gives:

$$\begin{aligned} V_b - V_c &= (a^2 - a)(V_1 - V_2) \\ &= (a^2 I_1 + a I_2) Z_f \\ &= (a^2 - a) I_1 Z_f \end{aligned} \quad (9\text{-}22)$$

Therefore:

$$(V_1 - V_2) = I_1 Z_f \quad (9\text{-}23)$$

The equivalent circuits are shown in Fig. 9-7*b* and *c*. Also:

$$I_b = (a^2 - a) I_1 = -j\sqrt{3} I_1 \quad (9\text{-}24)$$

and:

$$I_1 = \frac{V_a}{Z_1 + Z_2 + Z_f} \quad (9\text{-}25)$$

The fault current is:

$$I_b = -I_c = \frac{-j\sqrt{3} V_a}{Z_1 + Z_2 + Z_f} \quad (9\text{-}26)$$

9-4-3 Double Line-To-Ground Fault

A double line-to-ground fault is shown in Fig. 9-8*a*. Phases *b* and *c* go to ground through a fault impedance Z_f. Current in ungrounded phase is zero, that is, $I_a = 0$. Therefore, $I_1 + I_2 + I_0 = 0$.
So

$$V_b = V_c = (I_b + I_c) Z_f \quad (9\text{-}27)$$

Thus:

$$\begin{vmatrix} V_0 \\ V_1 \\ V_2 \end{vmatrix} = \frac{1}{3} \begin{vmatrix} 1 & 1 & 1 \\ 1 & a & a^2 \\ 1 & a^2 & a \end{vmatrix} \begin{vmatrix} V_a \\ V_b \\ V_c \end{vmatrix} = \frac{1}{3} \begin{vmatrix} V_a + 2V_b \\ V_a + (a + a^2) V_b \\ V_a + (a + a^2) V_b \end{vmatrix} \quad (9\text{-}28)$$

which gives $V_1 = V_2$ and

$$\begin{aligned} V_0 &= \frac{1}{3}(V_a + 2V_b) \\ &= \frac{1}{3}[(V_0 + V_1 + V_2) + 2(I_b + I_c) Z_f] \\ &= \frac{1}{3}[(V_0 + 2V_1) + 2(3I_0) Z_f] \\ &= V_1 + 3Z_f I_0 \end{aligned} \quad (9\text{-}29)$$

This gives the equivalent circuits of Fig. 9-8*b* and *c*. The fault current is:

$$\begin{aligned} I_1 &= \frac{V_a}{Z_1 + [Z_2 \,||\, (Z_0 + 3Z_f)]} \\ &= \frac{V_a}{Z_1 + \dfrac{Z_2 (Z_0 + 3Z_f)}{Z_2 + Z_0 + 3Z_f}} \end{aligned} \quad (9\text{-}30)$$

9-4-4 Three-Phase Fault

The three phases are short circuited through equal fault impedances Z_f (Fig. 9-9*a*.) The vector sum of fault currents is zero, as a symmetrical fault is considered, and there is no path to ground:

$$\begin{aligned} I_a + I_b + I_c &= 0 \\ I_0 &= 0 \end{aligned} \quad (9\text{-}31)$$

As the fault is symmetrical:

$$\begin{vmatrix} V_a \\ V_b \\ V_c \end{vmatrix} = \begin{vmatrix} Z_f & 0 & 0 \\ 0 & Z_f & 0 \\ 0 & 0 & Z_f \end{vmatrix} \begin{vmatrix} I_a \\ I_b \\ I_c \end{vmatrix} \quad (9\text{-}32)$$

The sequence voltages are given by:

$$\begin{vmatrix} V_0 \\ V_1 \\ V_2 \end{vmatrix} = \bar{T}_s^{-1} \begin{vmatrix} Z_f & 0 & 0 \\ 0 & Z_f & 0 \\ 0 & 0 & Z_f \end{vmatrix} \bar{T}_s \begin{vmatrix} I_0 \\ I_1 \\ I_2 \end{vmatrix} = \begin{vmatrix} Z_f & 0 & 0 \\ 0 & Z_f & 0 \\ 0 & 0 & Z_f \end{vmatrix} \begin{vmatrix} I_0 \\ I_1 \\ I_2 \end{vmatrix} \quad (9\text{-}33)$$

This gives equivalent circuits of Fig. 9-9*b* and *c*.

$$\begin{aligned} I_a &= I_1 = \frac{V_a}{Z_1 + Z_f} \\ I_b &= a^2 I_1 \qquad I_c = a I_1 \end{aligned} \quad (9\text{-}34)$$

Example 9-1 The calculations using symmetrical components can best be illustrated with an example. Consider a subtransmission system as shown in Fig. 9-10. A 13.8-kV generator G_1 voltage is stepped up to 138 kV. At the consumer end, the voltage is stepped down to 13.8 kV and generator G_2 operates in synchronism with the supply system. Bus *B* has a 10000 hp motor load. A line-to-ground fault occurs at bus *B*. It is required to calculate the fault current distribution throughout the system and also the fault voltages. The resistance of the system components is ignored in the calculations.

Impedance Data The reactance data of the system components is shown in Table 9-1, and all resistances are ignored for simplicity of calculations. Generators G_1 and G_2 are shown solidly grounded, which will not be the case in a practical installation. A high impedance grounding system is used by utilities for grounding generators in step-up transformer configurations. Generators in industrial facilities, directly connected to the load buses are low resistance grounded, and the ground fault currents are limited to 200–400 A. The simplifying assumptions in the example are not applicable to a practical installation, but clearly illustrate the procedure of calculations.

First step is to examine the given reactance data. Generator-saturated subtransient reactance is used in the short-circuit calculations and this is termed *positive sequence reactance*. A 138-kV transmission line reactance is calculated from the given data of conductor size and equivalent conductor spacing. The zero sequence

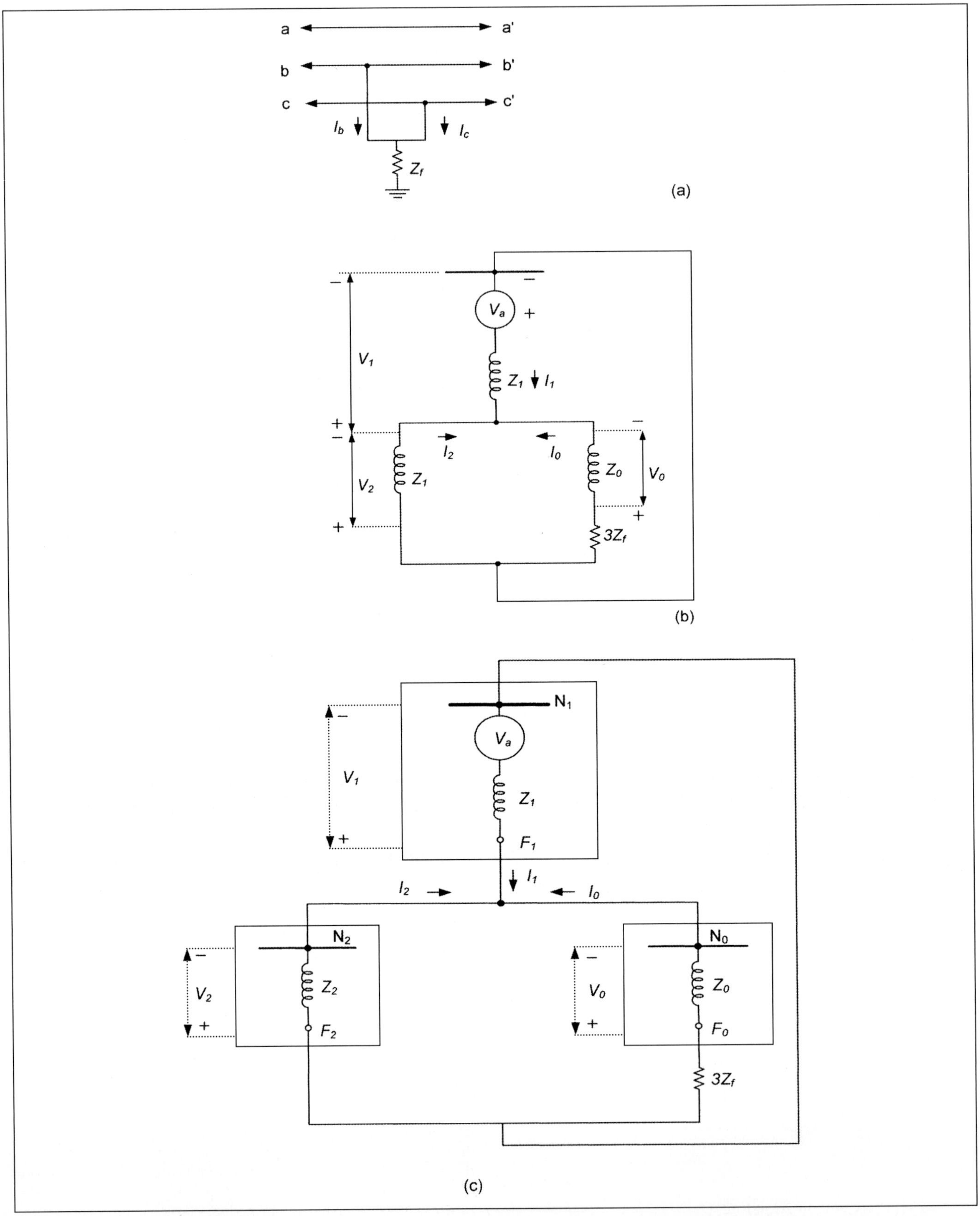

FIGURE 9-8 (*a*) Double line-to-ground fault. (*b*) Double line-to-ground fault equivalent circuit connections. (*c*) Connections of sequence networks.

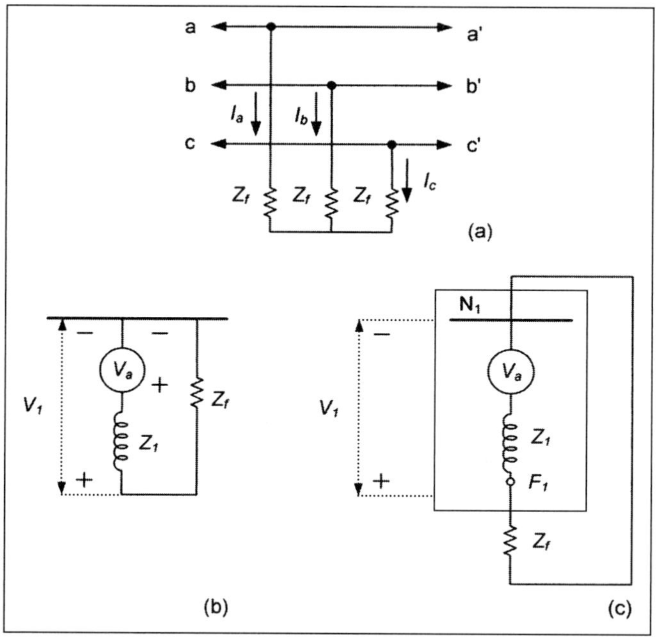

FIGURE 9-9 (*a*) Three-phase fault. (*b*) Three-phase fault equivalent circuit connections. (*c*) Connections of sequence networks.

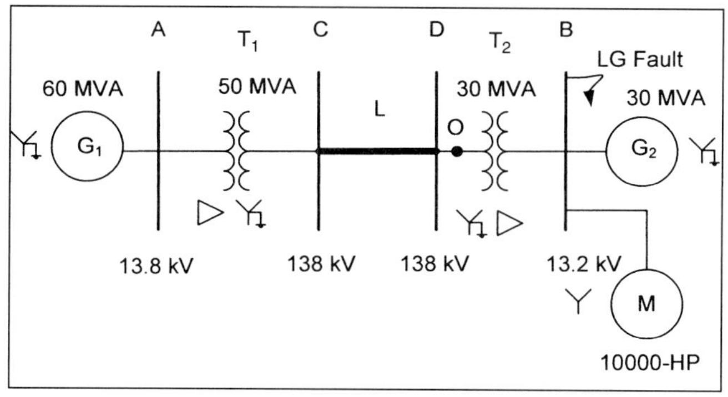

FIGURE 9-10 A power system for calculations of single line-to-ground fault for Example 9-1.

reactance of the transmission line cannot be completely calculated from the given data and is estimated based upon certain assumptions, that is, a soil resistivity of 100 Ω-m.

Compiling the impedance data of the system under study from the given parameters, from manufacturers' data or by calculation and estimation, can be time consuming. Most computer-based analysis programs have extensive data libraries and companion programs for calculation of system impedance data and line constants, which have partially removed the onerous task of generating the data from step-by-step analytical calculations. Appendix D provides models of line constants of coupled transmission lines, bundle conductors, and line capacitances.

Next, the impedance data is converted to a common MVA base. The voltage transformation ratio of transformer T_2 is 138 to 13.2 kV, while a bus voltage of 13.8 kV is specified, which should be considered in transforming impedance data on a common MVA base. Table 9-1 shows raw impedance data and its conversion to sequence impedances.

For a single line-to-ground fault at bus *B*, the sequence impedance network connections are shown in Fig. 9-11, with reactance data of components clearly marked. This figure is based upon the fault equivalent circuit shown in Fig. 9-6*b*, with fault impedance $Z_f = 0$. The calculation is carried out in per unit, and the units are not stated in every step of calculation.

The positive sequence reactance to the fault point is:

$$Z_1 = \frac{j(0.25+0.18+0.04+0.24)\times\dfrac{j0.37\times j1.67}{j(0.37+1.67)}}{j(0.25+0.18+0.04+0.24)+\dfrac{j0.37\times j1.67}{j(0.37+j1.67)}}$$

$$= j0.212$$

TABLE 9-1 Impedance Data for Example 9-1

EQUIPMENT	DESCRIPTION	IMPEDANCE DATA	PU IMPEDANCE 100-MVA BASE
G_1	13.8-kV, 60-MVA, 0.85 power factor generator	Subtransient reactance = 15% Transient reactance = 20% Zero sequence reactance = 8% Negative sequence reactance = 16.8%	$X_1 = 0.25$ $X_2 = 0.28$ $X_0 = 0.133$
T_1	13.8–138 kV step-up transformer, 50/84 MVA, delta-wye connected, wye neutral solidly grounded	$Z = 9\%$ on 50-MVA base	$X_1 = X_2 = X_0 = 0.18$
L_1	Transmission line, 5 mi long, 266.8 KCMIL, ACSR	Conductors at 15 ft (4.57 m) equivalent spacing	$X_1 = X_2 = 0.04$ $X_0 = 0.15$
T_2	138–13.2 kV, 30-MVA step-down transformer, wye-delta connected, high-voltage wye neutral solidly grounded	$Z = 8\%$	$X_1 = X_2 = X_0 = 0.24$
G_2	13.8-kV, 30-MVA, 0.85 power factor generator	Subtransient reactance = 11% Transient reactance = 15% Zero sequence reactance = 6% Negative sequence reactance = 16.5%	$X_1 = 0.37$ $X_2 = 0.55$ $X_0 = 0.20$
M	10000-hp induction motor load	Locked rotor reactance = 16.7% on motor base kVA (consider 1 hp, 1 kVA)	$X_1 = 1.67$ $X_2 = 1.80$ $X_0 = \infty$

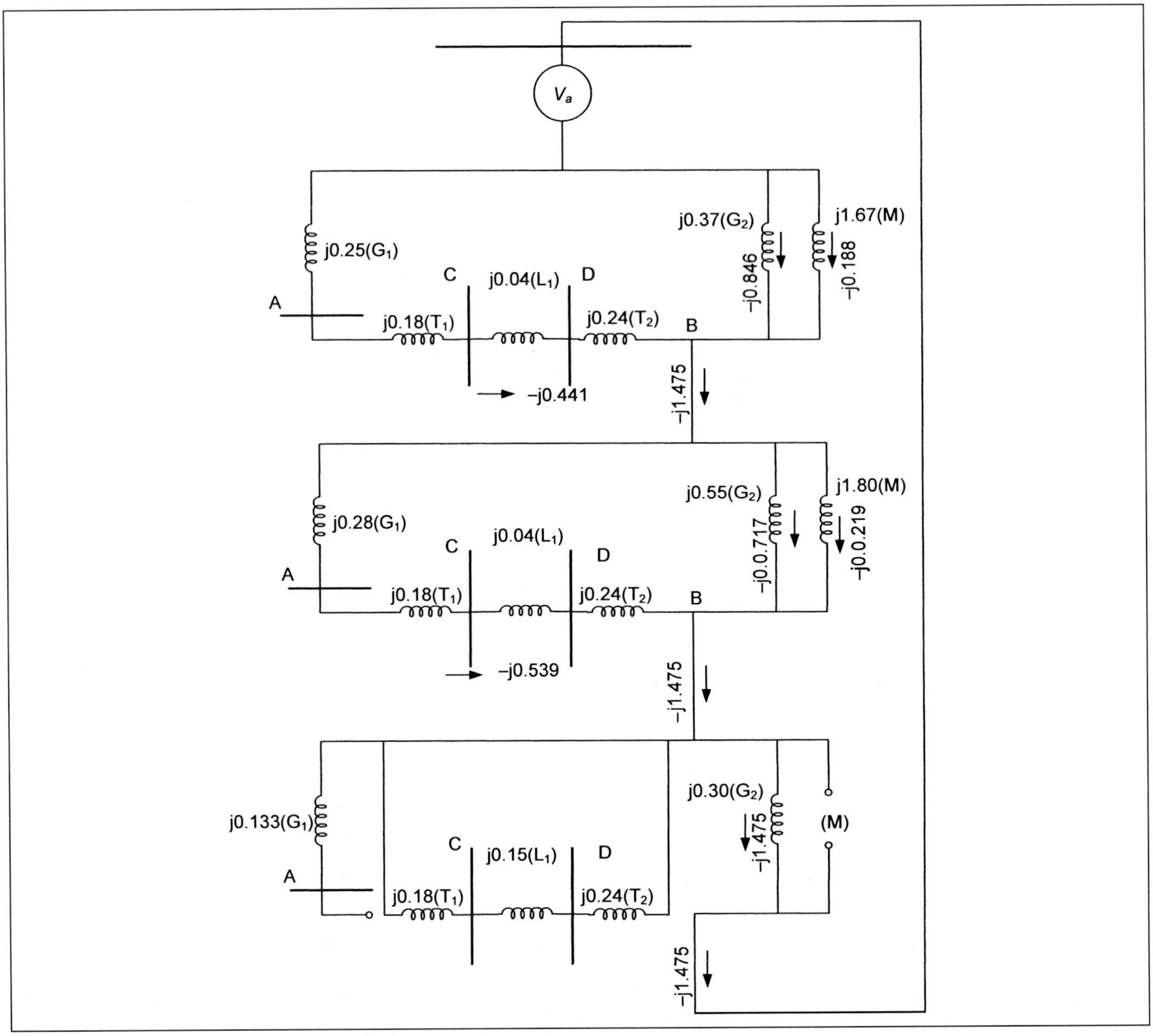

FIGURE 9-11 Circuit connections for single line-to-ground fault.

Z_2 can be similarly calculated to the fault point and this gives $X_2 = j0.266$ and $Z_0 = j0.2$. Therefore:

$$I_1 = \frac{E}{Z_1 + Z_2 + Z_0} = -j1.475 \text{ pu}$$

$$I_1 = I_2 = I_0 = -j1.475, \qquad I_a = -j4.425$$

As a base of 100 MVA is selected at 13.8 kV, 1 pu = 4.184 kA. Therefore, the single-phase line-to-ground fault current is 18.51 kA. The three-phase bolted fault current is given by the positive sequence reactance alone = $j4.717$ pu = 19.73 kA. Fault currents in phases b and c are zero:

$$I_b = I_c = 0$$

The sequence voltage to the fault point can therefore be calculated by:

$$V_0 = -I_0 Z_0 = -0.295$$

$$V_2 = -I_2 Z_2 = -0.392$$

$$V_1 = E - I_1 Z_1 = I_1(Z_0 + Z_2)$$

$$= 1 - (-j1.475 \times j0.212) = 0.687 \text{ pu}$$

The results can be checked. At the fault point $V_a = 0$:

$$V_a = V_0 + V_1 + V_2 = 0$$

The voltages of phases b and c at the fault point can be calculated as:

$$V_b = V_0 + aV_1 + a^2V_2 = -0.4425 - j0.9344$$

$$|V_b| = 1.034 \text{ pu}$$

$$V_c = V_0 - 0.5(V_1 + V_2) + j0.866(V_1 - V_2)$$

$$= -0.4425 + j0.9344$$

$$|V_c| = 1.034 \text{ pu}$$

Calculations of Current Flows The distribution of the sequence currents in the network is calculated from the known sequence impedances. The positive sequence current contributed from the right side of the fault, that is, by G_2 and motor M is:

$$-j1.475 \frac{j(0.25+0.18+0.04+0.24)}{j(0.25+0.18+0.04+0.24) + \dfrac{j0.37 \times j1.67}{j(0.37+1.67)}}$$

This gives $-j1.0338$. This current is composed of two components: one from the generator G_2 and the other from the motor M. The generator component is:

$$(-j1.0388)\frac{j1.67}{j(0.37+1.67)} = -j0.8463$$

The motor component is similarly calculated and is equal to $-j0.1875$. The positive sequence current from the left side of bus B is:

$$-j1.475 \frac{\dfrac{j0.37 \times j1.67}{j(0.37+1.67)}}{j(0.25+0.18+0.04+0.24) + \dfrac{j0.37 \times j1.67}{j(0.37+1.67)}}$$

This gives $-j0.441$. The currents from the right side and the left side should sum to $-j1.475$. This checks the calculation accuracy. The negative sequence currents are calculated likewise and are as follows:

In generator G_2	$= -j0.7172$
In motor M	$= -j0.2191$
From left side, bus B	$= -j0.5387$
From right side	$= -j0.9363$

The results are shown in Fig. 9-11. Again verify that the vector summation at the junctions confirms the accuracy of calculations.

Currents in Generator G_2

$$I_a(G_2) = I_1(G_2) + I_2(G_2) + I_0(G_2)$$

$$= -j0.8463 - j0.7172 - j1.475$$

$$= -j3.0385 = |3.0385| \text{ pu}$$

$$I_b(G_2) = I_0 - 0.5(I_1 + I_2) - j0.866(I_1 - I_2)$$

$$= -0.1118 - j0.6933 = |0.7023| \text{ pu}$$

$$I_c(G_2) = I_0 - 0.5(I_1 + I_2) + j0.866(I_1 - I_2)$$

$$= 0.1118 - j0.6933 = |0.7023| \text{ pu}$$

This large unbalance is noteworthy. It gives rise to increased thermal effects due to negative sequence currents and results in overheating of the generator rotor. A generator will be tripped quickly on negative sequence currents.

Currents in Motor M The zero sequence current in the motor is zero. In the United States the wye winding point of the motors is not grounded, irrespective of the size of the motor, thus:

$$I_a(M) = I_1(M) + I_2(M)$$

$$= -j0.1875 - j0.2191$$

$$= -j0.4066 = |0.4066| \text{ pu}$$

$$I_b(M) = -0.5(-j0.4066) - j0.866(0.0316)$$

$$= 0.0274 + j0.2033$$

$$I_c(M) = -0.0274 + j0.2033$$

$$|I_b(M)| = |I_c(M)| = 0.2051 \text{ pu}$$

The summation of the line currents in the motor M and generator G_2 are:

$$I_a(G_2) + I_a(M) = -j3.4451$$

$$I_b(G_2) + I_b(M) = -0.084 - j0.490$$

$$I_c(G_2) + I_c(M) = 0.084 - j0.490$$

Currents from the left side of the bus B are:

$$I_a = -j0.441 - j0.5387 = -j0.98$$

$$I_b = -0.5(-j0.441 - j0.5387)$$

$$- j0.866(-j0.441 + j0.5387) = 0.084 + j0.490$$

$$I_c = -0.084 + j0.490$$

These results are consistent as the sum of currents in phases b and c at the fault point from the right and left side is zero, and the summation of phase a currents gives the total ground fault current at $b = -j4.425$. The distribution of the currents is shown in three-line diagram, Fig. 9-13

Continuing with the example, the calculations of currents and voltages in the transformer T_2 windings must consider the phase shifts introduced in the primary and secondary vectors. We will not get into the phase shifts introduced in various types of transformer winding connections, and Refs. 12 to 14 provide further reading. This phase shift depends on:

- Transformer winding connections
- Sequence of the applied vectors.

Figure 9-12 shows the phase shifts for wye-delta and delta-wye connections according to ANSI/IEEE standards.[13] The low-voltage side vectors, whether in wye or delta connection, have a phase shift of 30° lagging with respect to high-voltage side phase to neutral voltage vectors.

The negative sequence currents and voltages *undergo a phase shift which is reverse of positive sequence currents and voltages*. We will correctly apply the phase shifts for positive and negative sequence components when passing from delta secondary to wye primary of the transformer. The positive and negative sequence current on wye side of transformer T2 are:

$$I_{1(p)} = I_1 < 30° = -j0.441 < 30° = 0.2205 - j0.382$$

$$I_{2(p)} = I_2 < -30° = -j0.539 < -30° = -0.2695 - j0.4668$$

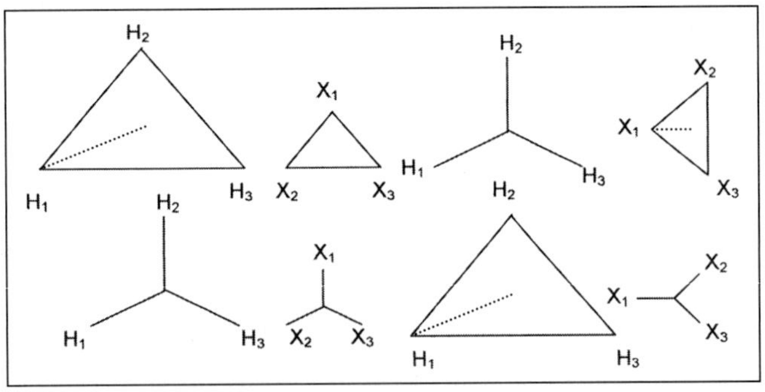

FIGURE 9-12 Phase shift in three-phase transformer winding connections, according to ANSI/IEEE standard.[13]

Also, the zero sequence current is zero. The primary currents are:

$$I_{a(p)} = I_0 + I_{1(p)} + I_{2(p)} = -0.049 - j0.8487$$

$$I_{b(p)} = a^2 I_{1(p)} + a I_{2(p)} = -0.0979$$

$$I_{c(p)} = a I_{1(p)} + a^2 I_{2(p)} = -0.049 - j0.8487$$

Currents in the lines on the delta side of the transformer *T*1 are similarly calculated. The positive sequence component, which underwent a 30° positive shift from delta to wye in transformer *T*2, undergoes a –30° phase shift; as for an ANSI connected transformer, it is the low-voltage vectors which lag the high-voltage side vectors. Similarly, the negative sequence component undergoes a positive phase shift. The currents on the delta side of transformer *T*1 and *T*2 are identical in amplitude and phase. Figure 9-13 shows the distribution of currents throughout the distribution system.

The voltage on the primary side of transformer *T*2 can be calculated. The voltages undergo the same phase shifts as the currents. Positive sequence voltage is the base fault positive sequence voltage, phase shifted by 30° (positive) minus the voltage drop in transformer reactance due to positive sequence current:

$$\begin{aligned} V_{1(p)} &= 1.0 < 30° - jI_{1(p)} X_{1t} \\ &= 1.0 < 30° - (j0.441 < 30°)(-j0.24) \\ &= 0.958 + j0.553 \end{aligned}$$

$$\begin{aligned} V_{2(p)} &= 0 - I_{2(p)} X_{2t} \\ &= -(0.539 < -30°)(0.24 < 270°) \\ &= 0.112 - j0.0647 \end{aligned}$$

Thus:

$$\begin{aligned} V_{a(p)} &= 0.958 + j0.553 + 0.112 - j0.0647 \\ &= 1.0697 + j0.4883 = 1.17 < 24.5° \end{aligned}$$

$$\begin{aligned} V_{b(p)} &= -0.5(V_{1(p)} + V_{2(p)}) - j0.866(V_{1(p)} - V_{2(p)}) \\ &= -j0.9763 \end{aligned}$$

$$\begin{aligned} V_{c(p)} &= -0.5(V_{1(p)} + V_{2(p)}) - j0.866(V_{2(p)} - V_{1(p)}) \\ &= 1.17 < 155.5° \end{aligned}$$

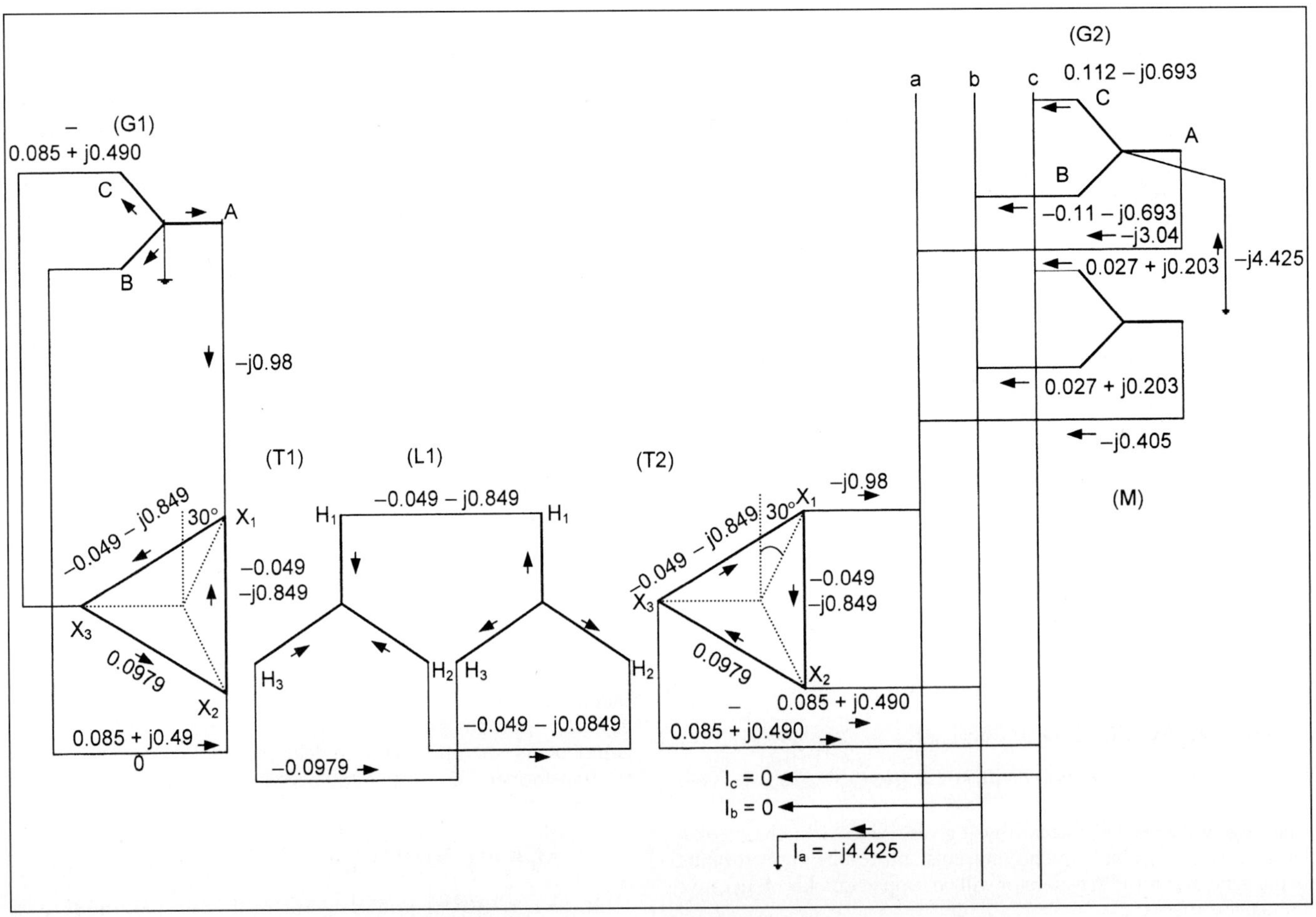

FIGURE 9-13 Distribution of currents for a single line-to-ground fault.

Note the voltage unbalance caused by the fault. Figure 9-13 shows the calculation of the currents throughout the distribution. It can be observed that short-circuit current transients result in associated voltage transients. We observed that while the voltage of the faulted phase is zero, that of the other two phases rises by 3.4 percent. This voltage rise is a function of the system grounding (Chap. 21).

9-5 MATRIX METHODS OF SHORT-CIRCUIT CURRENT CALCULATIONS

The step-by-step sequence component method illustrated above, though academically instructive, is rarely used in practice. Even for a simple system, the calculations are tedious and lengthy. For large networks consisting of thousands of nodes and branches these are impractical and matrix methods are used. The commercial computer programs use Z impedance matrix methods.

The network equations can be written in the bus (or nodal) frame of reference, in the loop frame of reference, and in the branch frame of reference. The bus frame of reference is important. For n-node system, the performance is described by $n - 1$ linear independent equations, and the reference node is at ground potential. In Chap. 2. (Sec. 2-13), we wrote the Y matrix by mere inspection. The impedance matrix cannot be written by mere inspection. It is formulated by:

- Inversion of Y matrix which is rarely done in computer applications.
- By open-circuit testing
- From graph theory[15]
- By step-by-step formulation

While the Y matrix is a sparse matrix and has a large number of zero terms, as a bus is not connected to every other bus in the system, a Z matrix is fully populated. Some matrix techniques are: triangulation and factorization—Crout's method, solution by forward-backward substitution, and sparsity and optimal ordering. These are not discussed here.

If we denote the sequence matrices by $\bar{Z}^1_{ss}, \bar{Z}^2_{ss}$, *and* $\bar{Z}^0_{ss}$ then: *For a line-to-ground fault:*

$$\bar{I}^0_{ss} = \bar{I}^1_{ss} = \bar{I}^2_{ss} = \frac{1}{\bar{Z}^1_{ss} + \bar{Z}^2_{ss} + \bar{Z}^0_{ss} + 3Z_f} \tag{9-35}$$

For a two-phase fault:

$$\bar{I}^1_{ss} = -\bar{I}^2_{ss} = \frac{1}{\bar{Z}^1_{ss} + \bar{Z}^2_{ss} + Z_f} \tag{9-36}$$

and for a double line-to-ground fault:

$$\bar{I}^1_s = \frac{1}{\bar{Z}^1_{ss} + \dfrac{\bar{Z}^2_{ss}(\bar{Z}^0_{ss} + 3Z_f)}{\bar{Z}^2_{ss} + (\bar{Z}^0_{ss} + 3Z_f)}}$$

$$\bar{I}^0_s = -\frac{\bar{Z}^2_{ss}}{\bar{Z}^2_{ss} + (\bar{Z}^0_{ss} + 3Z_f)} I^1_s \tag{9-37}$$

$$\bar{I}^2_s = \frac{(-\bar{Z}^0_{ss} + 3Z_f)}{\bar{Z}^2_{ss} + (\bar{Z}^0_{ss} + 3Z_f)} I^1_s$$

The phase currents are calculated by:

$$\bar{I}^{abc}_s = \bar{T}_s \bar{I}^{012}_s \tag{9-38}$$

The voltage at bus j of the system is given by:

$$\begin{vmatrix} \bar{V}^0_j \\ \bar{V}^1_j \\ \bar{V}^2_j \end{vmatrix} = \begin{vmatrix} 0 \\ 1 \\ 0 \end{vmatrix} - \begin{vmatrix} \bar{Z}^0_{js} & 0 & 0 \\ 0 & \bar{Z}^1_{js} & 0 \\ 0 & 0 & \bar{Z}^2_{js} \end{vmatrix} \begin{vmatrix} \bar{I}^0_s \\ \bar{I}^1_s \\ \bar{I}^2_s \end{vmatrix} \tag{9-39}$$

where $j = 1, 2, \ldots s, \ldots . m$ The fault currents from bus x to y are given by:

$$\begin{vmatrix} \bar{I}^0_{xy} \\ \bar{I}^1_{xy} \\ \bar{I}^2_{xy} \end{vmatrix} = \begin{vmatrix} \bar{Y}^0_{xy} & 0 & 0 \\ 0 & \bar{Y}^1_{xy} & 0 \\ 0 & 0 & \bar{Y}^2_{xy} \end{vmatrix} \begin{vmatrix} \bar{V}^0_x - \bar{V}^0_y \\ \bar{V}^1_x - \bar{V}^1_y \\ \bar{V}^2_x - \bar{V}^2_y \end{vmatrix} \tag{9-40}$$

where:

$$\bar{I}^0_{xy} = \begin{vmatrix} I^0_{12} \\ I^0_{13} \\ . \\ I^0_{mm} \end{vmatrix} \tag{9-41}$$

and

$$\bar{Y}^0_{xy} = \begin{vmatrix} \bar{Y}^0_{12,12} & \bar{Y}^0_{12,13} & \bar{Y}^0_{12,mn} \\ \bar{Y}^0_{13,12} & \bar{Y}^0_{13.13} & \bar{Y}^0_{13,mn} \\ \bar{Y}^0_{mn,12} & \bar{Y}^0_{mn,13} & \bar{Y}^0_{mn,mn} \end{vmatrix} \tag{9-42}$$

where $\bar{Y}^0_{xy}$ is the inverse of primitive matrix of the system. Similar expressions apply to positive sequence and negative sequence currents and voltages.

Example 9-2 We will illustrate the application of matrix techniques to Example 9-1, same system data. Figure 9-14*a* can be drawn for the positive sequence network and converted to Fig. 9.14*b* in terms of admittances. By inspection the positive sequence bus admittance matrix can be written as:

$$\bar{Y}^1_{bus} = \begin{vmatrix} Y^1_{aa} & Y^1_{ac} & Y^1_{ad} & Y^1_{ab} \\ Y^1_{ca} & Y^1_{cc} & Y^1_{cd} & Y^1_{cb} \\ Y^1_{da} & Y^1_{dc} & Y^1_{dd} & Y^1_{db} \\ Y^1_{ba} & Y^1_{bc} & Y^1_{bd} & Y^1_{bb} \end{vmatrix}$$

=	A	C	D	B
A	$-j9.55$	$j5.55$	0	0
C	$j5.55$	$-j30.55$	$j25$	0
D	0	$j25$	$-j29.17$	$j4.17$
B	0	0	$j4.17$	$-j7.47$

The inversion of this matrix gives the positive sequence bus matrix:

$\bar{Z}^1_{bus}$ =	A	C	D	B
A	$j0.188$	$j0.144$	$j0.134$	$j0.075$
C	$j0.144$	$j0.248$	$j0.231$	$j0.129$
D	$j0.134$	$j0.231$	$j0.252$	$j0.141$
B	$j0.075$	$j0.129$	$j0.141$	$j0.212$

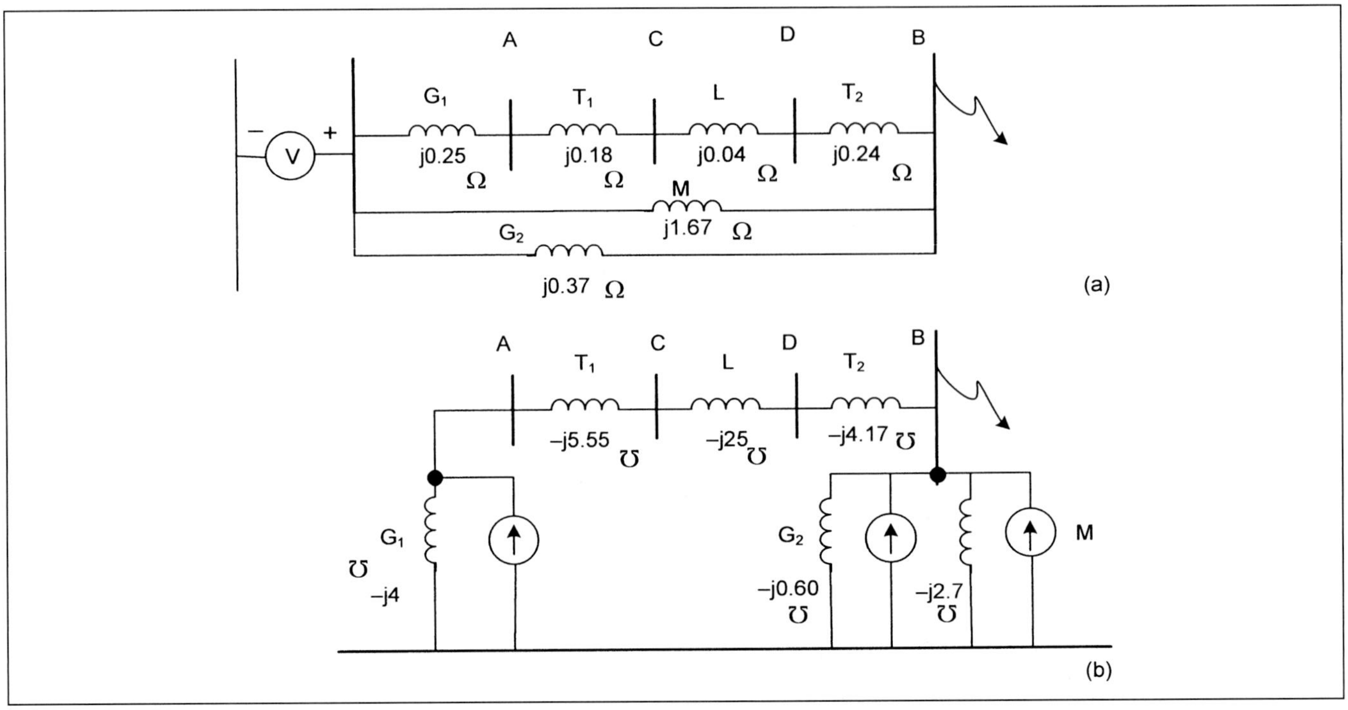

FIGURE 9-14 (*a*) Positive sequence network for a single line-to-ground fault. (*b*) Circuit of Fig. 9-14*a* reduced to admittance network.

All the elements are populated, there are no negative terms and the matrix is symmetric. Similarly, the negative sequence network can be drawn and the negative sequence Y^2_{bus} matrix is:

$$\bar{Y}^2_{bus} = \begin{array}{c|cccc|} & A & C & D & B \\ A & -j9.12 & j5.55 & 0 & 0 \\ C & j5.55 & -j30.55 & j25 & 0 \\ D & 0 & j25 & -29.17 & j4.17 \\ B & 0 & 0 & j4.17 & -j6.55 \end{array}$$

The inversion of this matrix gives:

$$\bar{Z}^2_{bus} = \begin{array}{c|cccc|} & A & C & D & B \\ A & j0.212 & j0.169 & j0.159 & j0.101 \\ C & j0.169 & j0.278 & j0.262 & j0.167 \\ D & j0.159 & j0.262 & j0.285 & j0.181 \\ B & j0.101 & j0.167 & j0.181 & j0.268 \end{array}$$

The formation of the zero sequence impedance is not so straightforward. At nodes *A* and *B*, the delta connection of the transformer results in an open circuit and thus no zero sequence current can flow from these nodes into the rest of the system (Fig. 9-15).

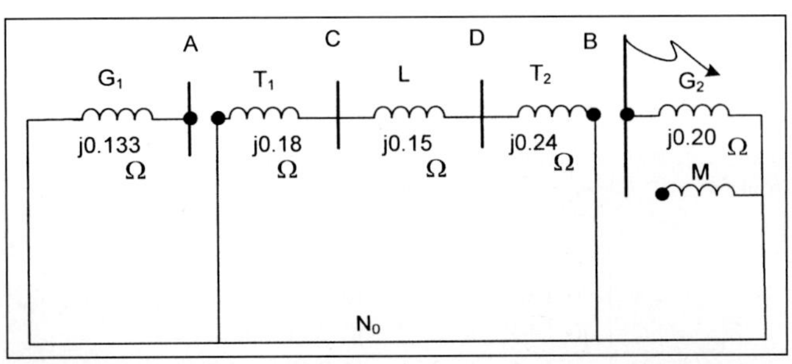

FIGURE 9-15 Zero sequence network for a single line-to-ground fault, showing discontinuities.

Thus, the zero sequence impedance at buses *A* and *B* is that of generators *G*1 and *G*2, respectively, that is, 0.133 pu and 0.20 pu, respectively.

A fault at buses C or D results in circulation of zero sequence currents through the wye-connected neutrals of the transformers. The zero sequence impedance of nodes *C* and *D* can be calculated similarly or by inversion of Y matrix. For buses *C* and *D*, this gives:

$$\bar{Z}^0_{bus,\,cd} = \begin{array}{c|cc|} & C & D \\ C & j0.123 & j0.076 \\ D & j0.076 & j0.139 \end{array}$$

We can therefore write the overall matrix as follows:

$$\bar{Z}^0_{bus} = \begin{array}{c|cccc|} & A & C & D & B \\ A & j0.133 & \text{open} & \text{open} & \text{open} \\ C & \text{open} & j0.123 & j0.076 & \text{open} \\ D & \text{open} & j0.076 & j0.139 & \text{open} \\ B & \text{open} & \text{open} & \text{open} & j0.20 \end{array}$$

The computer-based solutions address this discontinuity by maintaining the integrity of nodes. A fictitious bus *R* is created, and the delta-wye connected transformer can be modeled as illustrated in Fig. 9-16. In Fig. 9-16*a*, a delta-wye transformer is shown, and its zero sequence impedance when viewed from bus 1 side is an open circuit. Two possible approaches for computer solutions are shown in Fig. 9-16*b* and *c*. A fictitious bus *R* is created and the positive sequence impedance modified by dividing the positive sequence impedance into two parts, for high-and low-voltage windings. Infinite impedance at the junction of these impedance to bus *R* is connected (in computer calculations this is simulated by a large number, e.g., 999 + *j*9999) on a per unit basis. The zero sequence impedance is treated in a similar manner. Figure 9-16*c* shows another approach for maintaining the integrity of nodes.

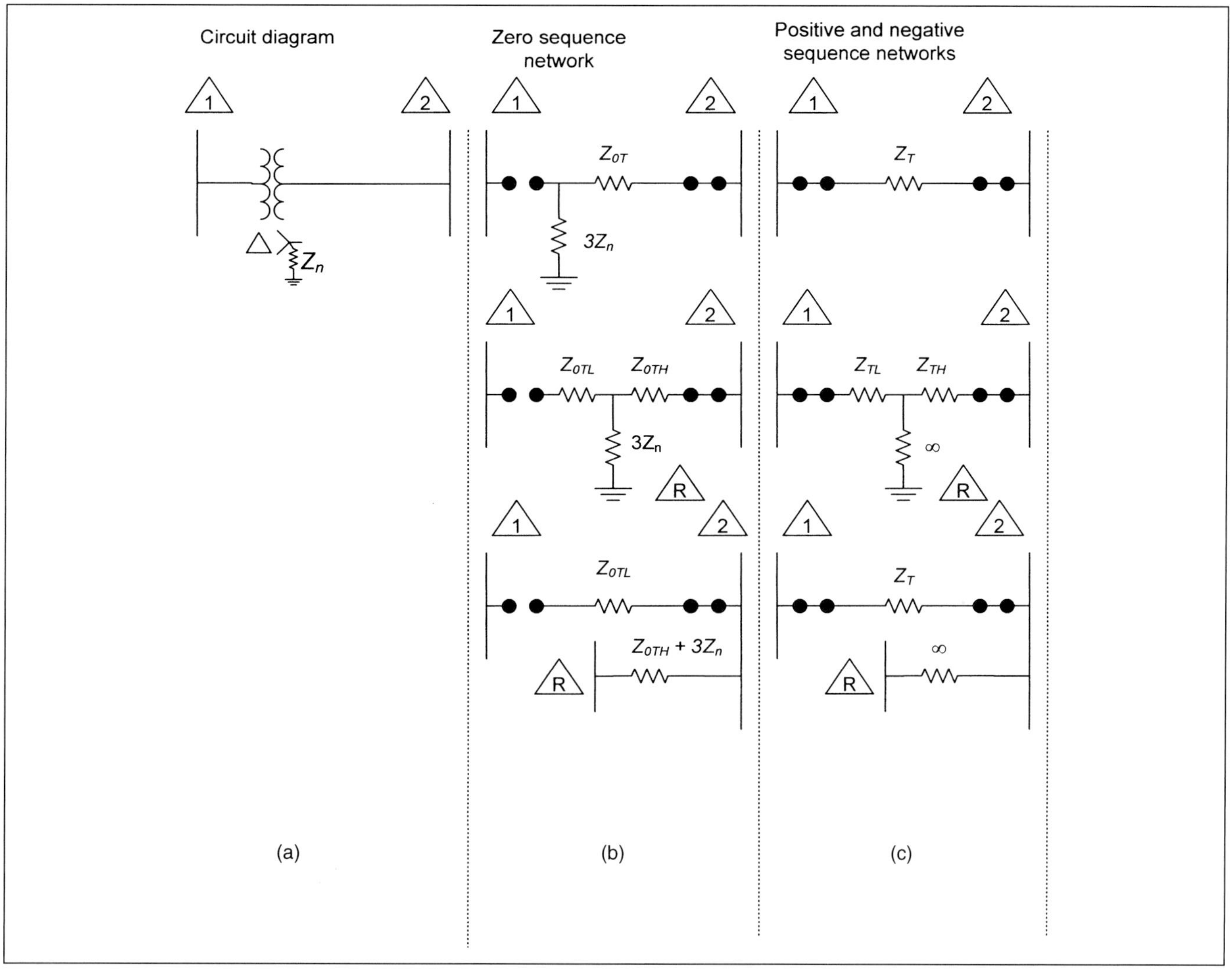

FIGURE 9-16 Computer methods to restore a node for an impedance discontinuity due to modeling of zero sequence impedance circuits of transformers.

Having generated the bus impedance matrices, it is easy to calculate the short-circuit currents using the equations derived above. A review of the matrices shows that the sequence impedances of bus B are:

$$Z_1^{BB} = j0.212$$

$$Z_2^{BB} = j0.268$$

$$Z_0^{BB} = j0.20$$

Therefore:

$$I_B^0 = I_B^1 = I_B^2 = \frac{1}{j0.212 + j0.268 + j0.20} = -j1.475 \text{ pu}$$

The line currents are given by:

$$\bar{I}_B^{abc} = \bar{T}_s \bar{I}_B^{012}$$

$$\begin{vmatrix} I_B^a \\ I_B^b \\ I_B^c \end{vmatrix} = \begin{vmatrix} 1 & 1 & 1 \\ 1 & a^2 & a \\ 1 & a & a^2 \end{vmatrix} \begin{vmatrix} -j1.425 \\ -j1.425 \\ -j1.425 \end{vmatrix} = \begin{vmatrix} -j4.425 \\ 0 \\ 0 \end{vmatrix}$$

The line current in phase a is $-j4.425$, as calculated before, and line currents in phases b and c are zero. Sequence voltages at bus B are:

$$\begin{vmatrix} V_B^0 \\ V_B^1 \\ V_B^2 \end{vmatrix} = \begin{vmatrix} 0 \\ 1 \\ 0 \end{vmatrix} - \begin{vmatrix} Z_{BS}^0 & 0 & 0 \\ 0 & Z_{BS}^1 & 0 \\ 0 & 0 & Z_{BS}^2 \end{vmatrix} \begin{vmatrix} I_B^0 \\ I_B^1 \\ I_B^2 \end{vmatrix}$$

$$= \begin{vmatrix} 0 \\ 1 \\ 0 \end{vmatrix} - \begin{vmatrix} j0.20 & 0 & 0 \\ 0 & j0.212 & 0 \\ 0 & 0 & j0.268 \end{vmatrix} \begin{vmatrix} -j1.475 \\ -j1.475 \\ -j1.475 \end{vmatrix}$$

$$= \begin{vmatrix} -0.295 \\ 0.687 \\ -0.392 \end{vmatrix}$$

Then the line voltages are:

$$\begin{vmatrix} V_B^a \\ V_B^b \\ V_B^c \end{vmatrix} = \begin{vmatrix} 1 & 1 & 1 \\ 1 & a^2 & a \\ 1 & a & a^2 \end{vmatrix} \begin{vmatrix} -0.295 \\ 0.687 \\ -0.392 \end{vmatrix} = \begin{vmatrix} 0 \\ -0.4425 - j0.9344 \\ -0.4425 + j0.9344 \end{vmatrix}$$

TABLE 9-2 Single Line-to-Ground Fault at Bus B—Computer Simulation, for Example 9-3

CONTRIBUTION		LINE-TO-GROUND FAULT					POSITIVE AND ZERO SEQUENCE IMPEDANCES			
FROM BUS	TO BUS	VOLTAGE (%)			KA SYM. rms		IMPEDANCE 100 MVA BASE (%)			
		V_a	V_b	V_c	I_a	$3I_0$	R_1	X_1	R_0	X_0
Bus B	Total	0.0	102.91	103.29	18.334	18.334	7.02E − 001	2.16E + 001	6.67E − 001	2.00E + 001
Bus D	Bus 4	70.40	105.50	70.92	3.932	0	2.40E + 000	7.36E + 001		
G2	Bus 4	100.0	100.0	100.0	12.69	18.334	1.22E + 000	3.67E + 001	6.67E − 001	2.00E + 001
M	Bus 4	104.55	104.55	104.55	1.712	0	5.17E + 000	1.81E + 002		

TABLE 9-3 Calculated Bus Sequence Impedances

BUS ID	POSITIVE SEQUENCE (Ω)	NEGATIVE SEQUENCE (Ω)	ZERO SEQUENCE (Ω)
A	0.00991 + *j*0.36179	0.00991 + *j*0.40671	0.00635 + *j*0.25392
B	0.41059 + *j*0.41081	0.01372 + *j*0.51163	0.01270 + *j*0.38088
C	1.51051 + *j*48.06746	1.50439 + *j*53.44850	0.78716 + *j*23.92366
D	1.55162 + *j*49.12745	1.53554 + *j*54.87076	0.94393 + *j*28.07056

This is the same result as in Example 9-1. In matrix techniques, it is easy to calculate the symmetrical and asymmetrical fault currents on any of the buses, using the calculated sequence impedances.

9-6 COMPUTER-BASED CALCULATIONS

The hand calculations of short-circuit currents, whether using symmetrical component methods or matrix methods, is a tedious exercise but establishes the basis of underlying methodology. Practically, the electrical systems are large, may consist of thousands of buses, generators, transformers, and a variety of load characteristics, and digital computers are invariably applied to such calculations. The basis of calculations are *Z* impedance matrices and matrix reduction techniques as stated earlier.

Example 9-3 A commercial short-circuit program is used to calculate the short-circuit currents and the results of calculation are in Table 9-2. We ignored the resistance component of all impedances in the hand calculations. A computer program will not accept a zero entry for the resistance, it has to be finite number, otherwise, it results in an indeterminate matrix due to division by zero. This table shows the flow of current from the adjacent bus *D* and also the currents contributed by generator *G*2 and motor *M*. Note that the zero sequence current contributed by the motor (windings not grounded) and from bus *D* (delta-wye transformer) are zero.

The calculated sequence impedances in ohms are shown in Table 9-3. The results shown closely agree with the matrices developed in Example 9-2. *These calculations do not show any transients.* This is because the calculations are based on the fixed values of the impedances, using algebraic equations. In order to simulate transients, the dynamic models of generators, motors, and loads are required.

9-7 OVERVOLTAGES DUE TO GROUND FAULTS

We discussed the temporary overvoltages in previous chapters and remarked that the overvoltages due to ground faults are an important parameter for electrical system analysis and application of surge arresters. These overvoltages depend on:

- Sequence impedance to faults, which also dictate the ground fault currents
- The fault resistance
- The method of system grounding, that is, solidly grounded, resistance grounded, high resistance grounded or ungrounded systems (Chap. 21).

The duration of overvoltages is dependent upon the fault clearance times; resonances can occur due to capacitive inductive couplings in ungrounded systems, and for application of surge arresters (Chap. 20), an estimation of the duration and magnitude of these overvoltages is required. In Example 9-1, the voltages on the unfaulted phases at the fault point rise by 3.4 percent.

9-7-1 Coefficient of Grounding

A measure of this overvoltage is the coefficient of grounding, defined as a ratio of E_{Lg}/E_{LL} in percentage, where E_{Lg} is the highest rms voltage on an unfaulted phase, at a selected location, during a fault effecting one or more phases to ground, and E_{LL} is the rms phase-to-phase power frequency voltage obtained at the location with the fault removed. In a solidly grounded system, no intentional impedance is introduced between system neutral and ground. These systems, generally, meet the definition of "effectively grounded systems" in which the ratio X_0/X_1 is positive and less than 3.0 and the ratio R_0/X_0 is positive and less than 1.0, where X_1, X_0, and R_0 are the positive sequence reactance, zero sequence reactance, and zero sequence resistance, respectively. These systems are, generally, characterized by COG of 80 percent. Approximately, a surge arrester with its rated voltage calculated on the basis of the system voltage multiplied by 0.8 can be applied (Chap. 20).

The single line-to-ground fault is the most important cause of overvoltages in the power systems. This type of fault produces the maximum fault voltages for all values of R_0/X_1. Sometimes we define EFF (IEC standards, earth fault factor). It is simply:

$$\text{EFF} = \sqrt{3}\text{COG} \tag{9-43}$$

In an ungrounded system, there is no intentional connection to ground except through potential transformers and metering devices of high impedance. Therefore, in reality, the ungrounded system is coupled to ground through the distributed phase capacitances. It is difficult to assign X_0/X_1 and R_0/X_0 values for ungrounded systems. The ratio X_0/X_1 is negative and may vary from low to high values. The COG may approach 120 percent. For values of X_0/X_1 between 0 and –40, the possibility of resonance with consequent generation of high voltages exits.

Figure 9-17 from Ref. 16 shows the overvoltages with respect to sequence impedances. The curves are applicable for $X_1 = X_2$. Also the curves include the effect of fault resistance. For each point, a fault resistance is chosen so that it produces maximum COG. In general, fault resistance will reduce COG, except in low-resistance systems. The numbers on the curves indicate coefficient of grounding for *any type of fault* in percent of unfaulted line-to-line voltage for the area bonded by the curve and the axes. All impedances must be on the same MVA base.

Figure 9-18 from Ref. 16 shows the maximum line-to-ground voltage during fault for isolated neutral systems as a function of X_0/X_1. COG can be calculated by the equations described below and more rigorously by the sequence component matrix methods, as illustrated above.

Single line-to-ground fault:

$$\mathrm{COG(phase}\ b) = -\frac{1}{2}\left(\frac{\sqrt{3}k}{2+k} + j1\right)$$

$$\mathrm{COG(phase}\ c) = -\frac{1}{2}\left(\frac{\sqrt{3}k}{2+k} - j1\right) \tag{9-44}$$

Double line-to-ground fault:

$$\mathrm{COG(phase}\ a) = \frac{\sqrt{3}k}{1+2k} \tag{9-45}$$

where k is given by:

$$k = \frac{Z_0}{Z_1} \tag{9-46}$$

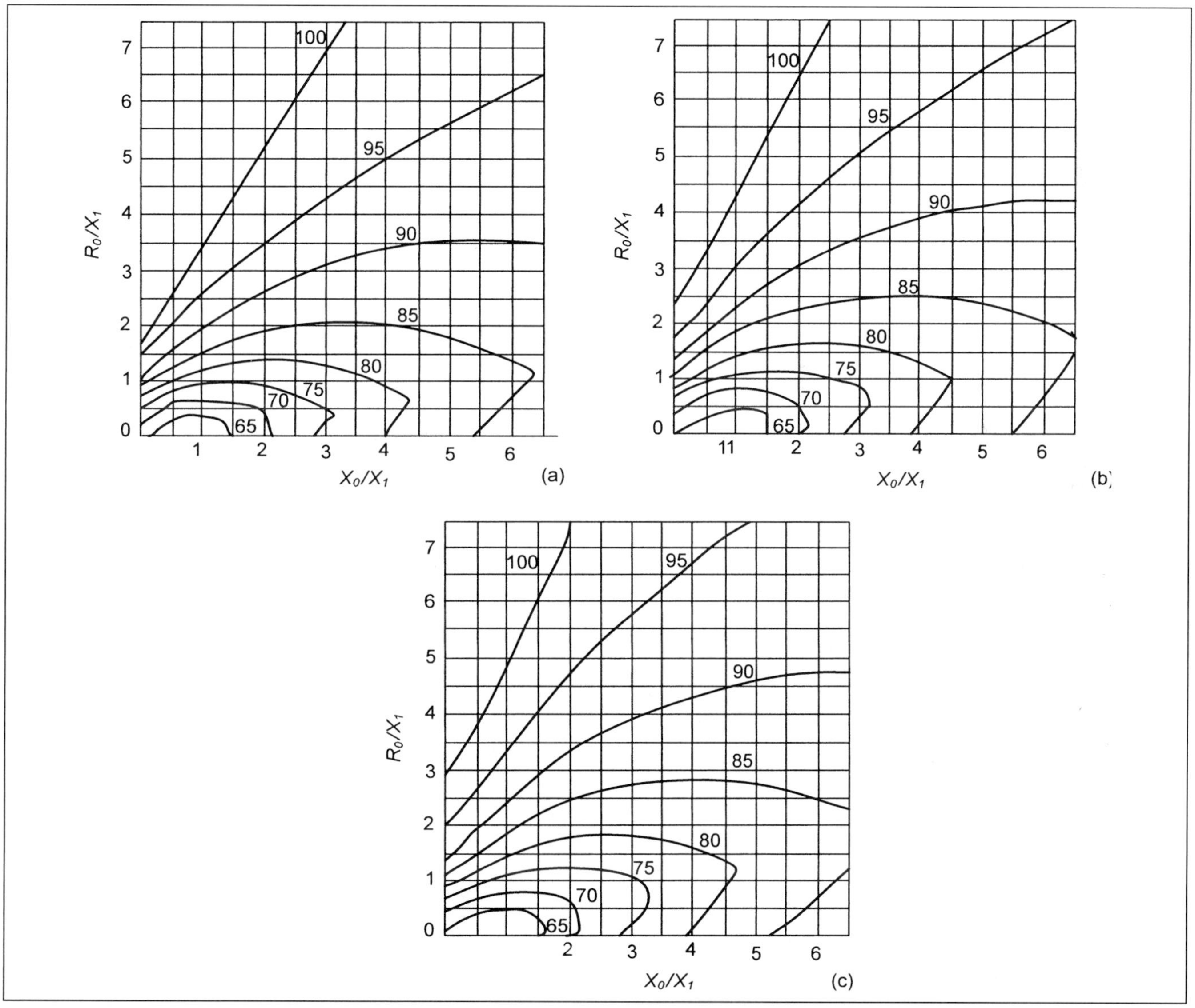

FIGURE 9-17 Maximum line-to-ground voltage at fault locations for grounded neutral systems; for *any fault* condition: (*a*) voltage conditions neglecting positive and negative sequence resistance; (*b*) voltage conditions for $R_1 = R_2 = 0.1X_1$; (*c*) voltage conditions for $R_1 = R_2 = 0.2X_1$.[16]

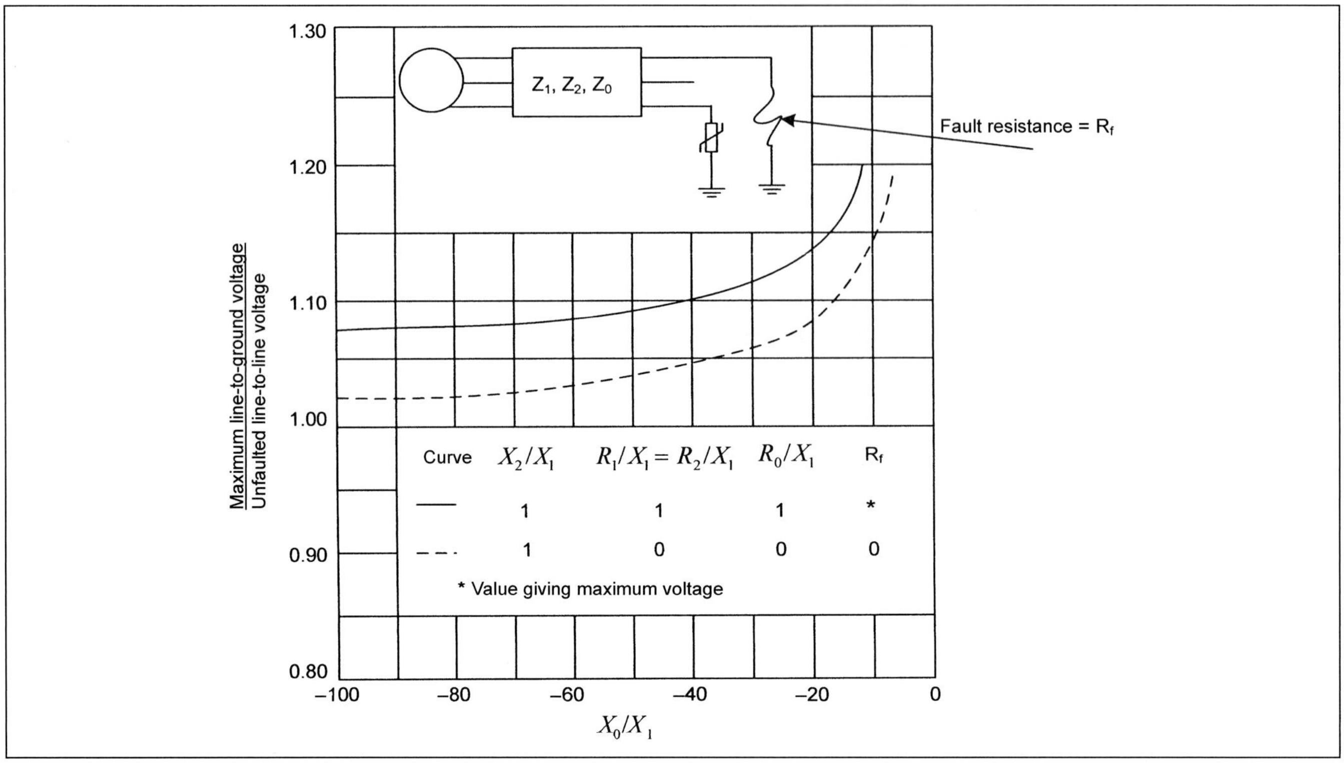

FIGURE 9-18 Maximum line-to-ground voltage at fault location for isolated neutral systems during fault.[16]

To take into account the fault resistance, k is modified as follows:
Single line-to-ground fault:

$$k = (R_0 + R_f + jX_0)/(R_1 + R_f + jX_1) \quad (9\text{-}47)$$

For double line-to-ground fault:

$$k = (R_0 + 2R_f + jX_0)/(R_1 + 2R_f + jX_1) \quad (9\text{-}48)$$

If R_0 and R_1 are zero, then the above equations reduce to:
For single line-to-ground fault:

$$\text{COG} = \frac{\sqrt{k^2 + k + 1}}{k + 2} \quad (9\text{-}49)$$

For double line-to-ground fault:

$$\text{COG} = \frac{k}{2k + 1} \quad (9\text{-}50)$$

$$\text{where } k = X_0/X_1 \quad (9\text{-}51)$$

Example 9-4 Calculate the COG at the faulted bus B in Example 9-1. If generator $G2$ is grounded through a 400-A resistor, what is the COG?

In Example 9-1, all resistances are ignored. From Fig. 9-17 the COG cannot be clearly read. But using Eq. (9-49), it is 0.57 (k = 0.9434). Also a voltage of 1.034 pu was calculated on the unfaulted phases, which gives a COG of 0.597. This is more accurate.

If the generator is grounded through 400-A resistor, then R_0 = 19.19 Ω, the positive sequence reactance is 0.4 Ω, and the zero sequence reactance is 0.38 Ω, which is much smaller than R_0. In fact, in a resistance grounded or high resistance grounded system, the sequence components are relatively small and the ground fault current can be calculated based upon the grounding resistor alone. The total ground fault current at bus 4 will reduce to approximately 400 A. This gives a COG of approximately 100 percent. This means that phase-to-ground voltage on unfaulted phases will be equal to line-to-line voltage.

Example 9-5 Though the dynamic models of generators and motors are covered in the chapters to follow, this example illustrates short-circuit transients in a power system (Fig. 9-19). The ratings of the major components are shown in this figure, the component models are discussed in other chapters.

Transients for a three-phase fault and then a single line-to-ground fault are simulated for a fault at 115-kV bus 1. Either fault occurs at 0.5 s and is removed at 1.5 s; fault duration of 1s. For a three-phase fault at bus 1, the results of EMTP simulations are:

Figure 9-20*a* shows voltage at the fault point. During the three-phase fault the voltage is not reduced to zero, because of some fault resistance considered in the model. The recovery of voltage after the fault is cleared at 1 s shows superimposed frequency beats (phenomena as shown in Fig. 2-20*b*). This is because of cyclic transients in the torque pulsations of generator and synchronous motor after the fault is cleared.

Figure 9-20*b* depicts the current transients in generator *G*1. The fault current rises during the fault duration, but shows the same behavior as the voltage to fault point, after the fault is cleared.

Figure 9-20*c* is the current transients, phases *a*, *b*, and *c* in the induction motor IM. At the instant of fault, the motor contributes high short-circuit current (maximum peak approximately 10 kA = nearly 12 times the motor full load current), which quickly decays. The transients show similar pattern as in Fig. 9.20*a* and *b*.

Figure 9-20*d* is the electrical torque of the generator and synchronous motor load. As the torque pulsates, the machines draw power and supply power into the system at a certain frequency, which explains the nature of the transients. The load

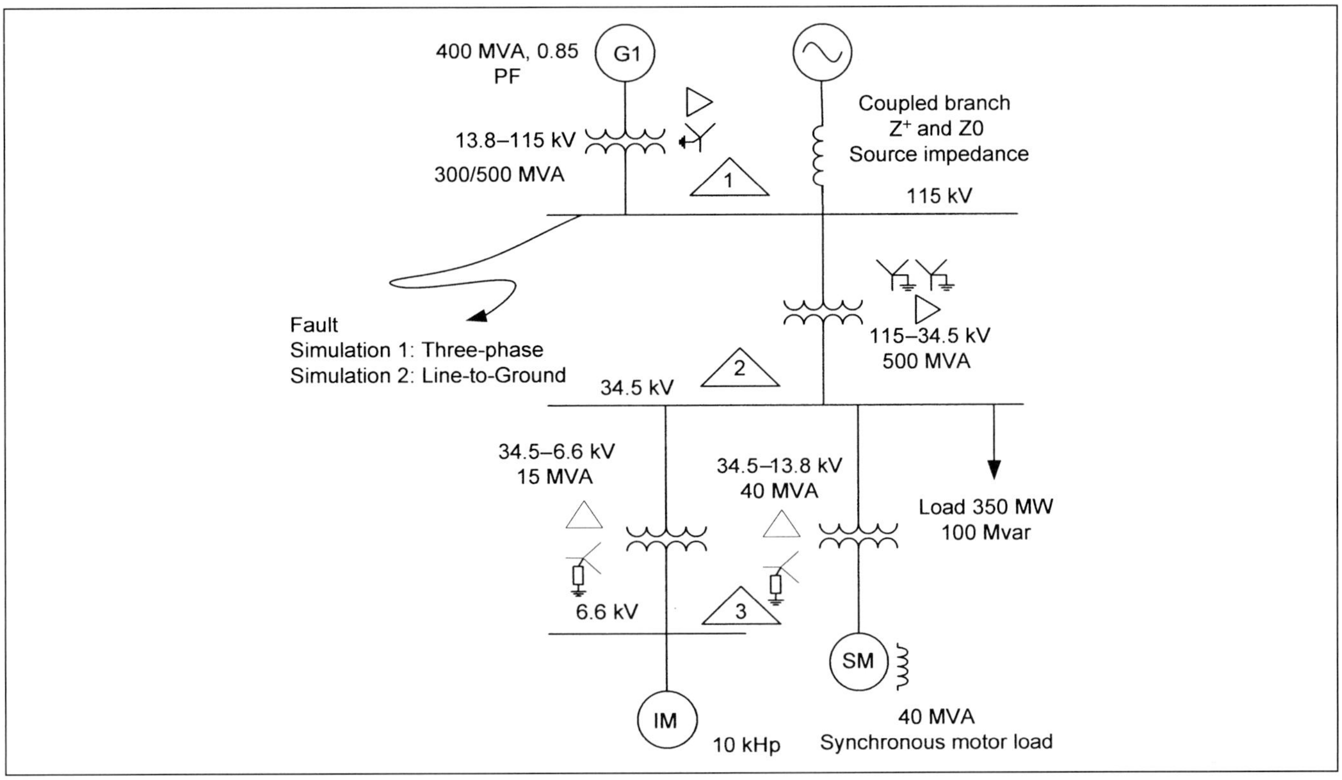

FIGURE 9-19 A power system for EMTP study of short-circuit transients.

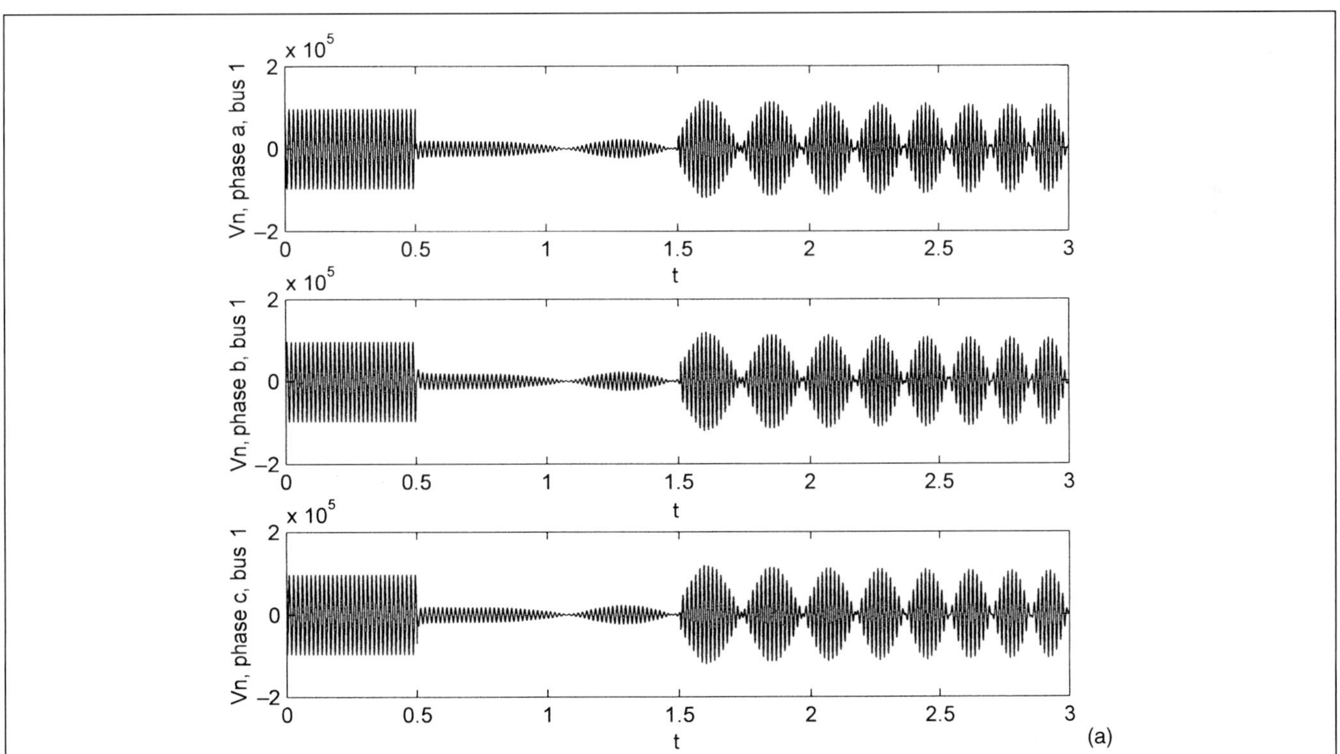

FIGURE 9-20 Transients for a three-phase fault on bus 1, which occurs at 0.5 s and is cleared at 1.5 s, fault duration of 1 s: (*a*) voltage at bus 1, phases *a*, *b*, and *c*; (*b*) generator *G*1 current (pu), phases *a*, *b*, and *c*; (*c*) induction motor IM currents. (*d*) electrical torque of generator *G*1 and synchronous motor load (pu); (*e*) power angle of generator *G*1 and synchronous motor load; (*f*) slip of the induction motor IM.

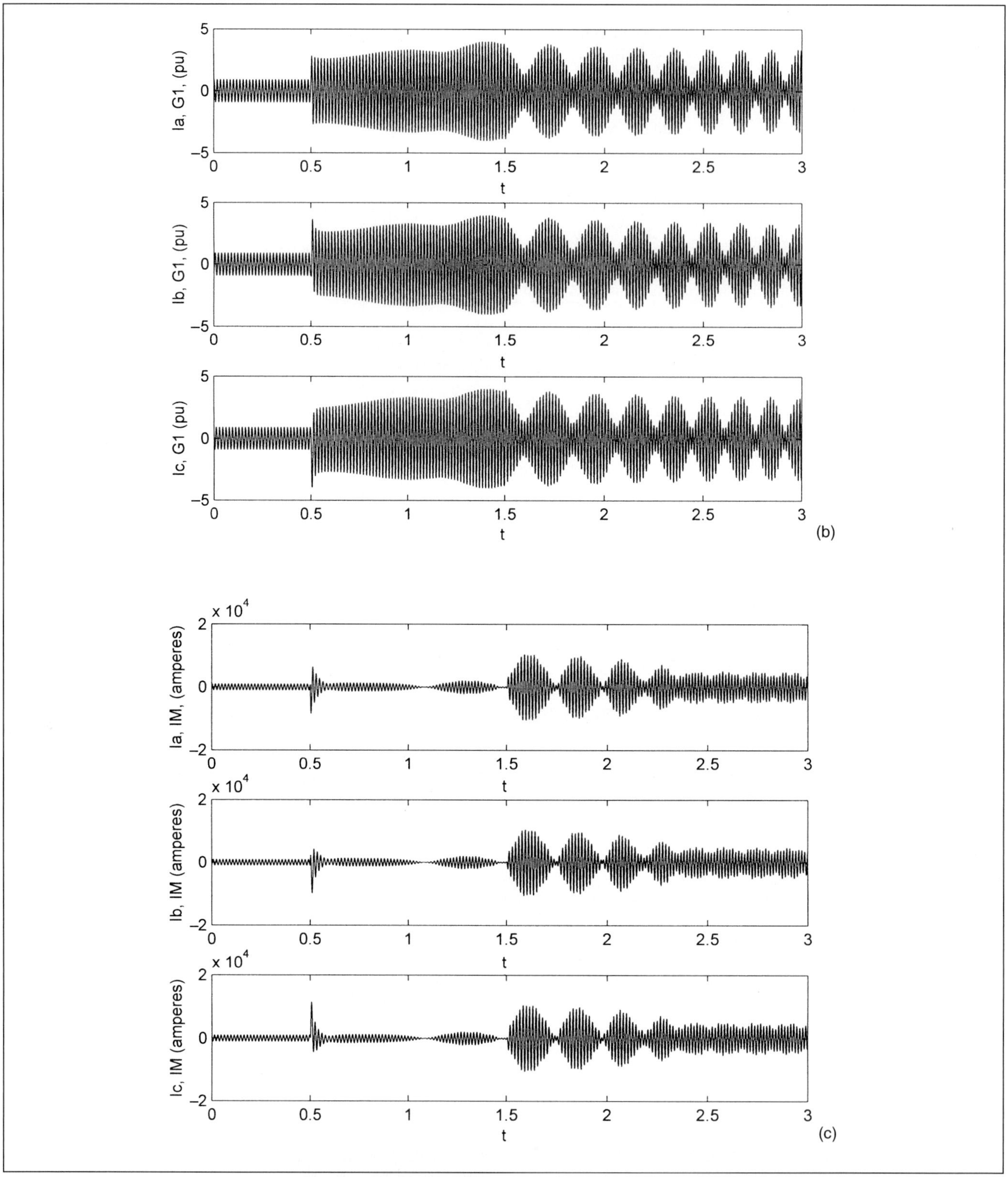

FIGURE 9-20 *(Continued)*

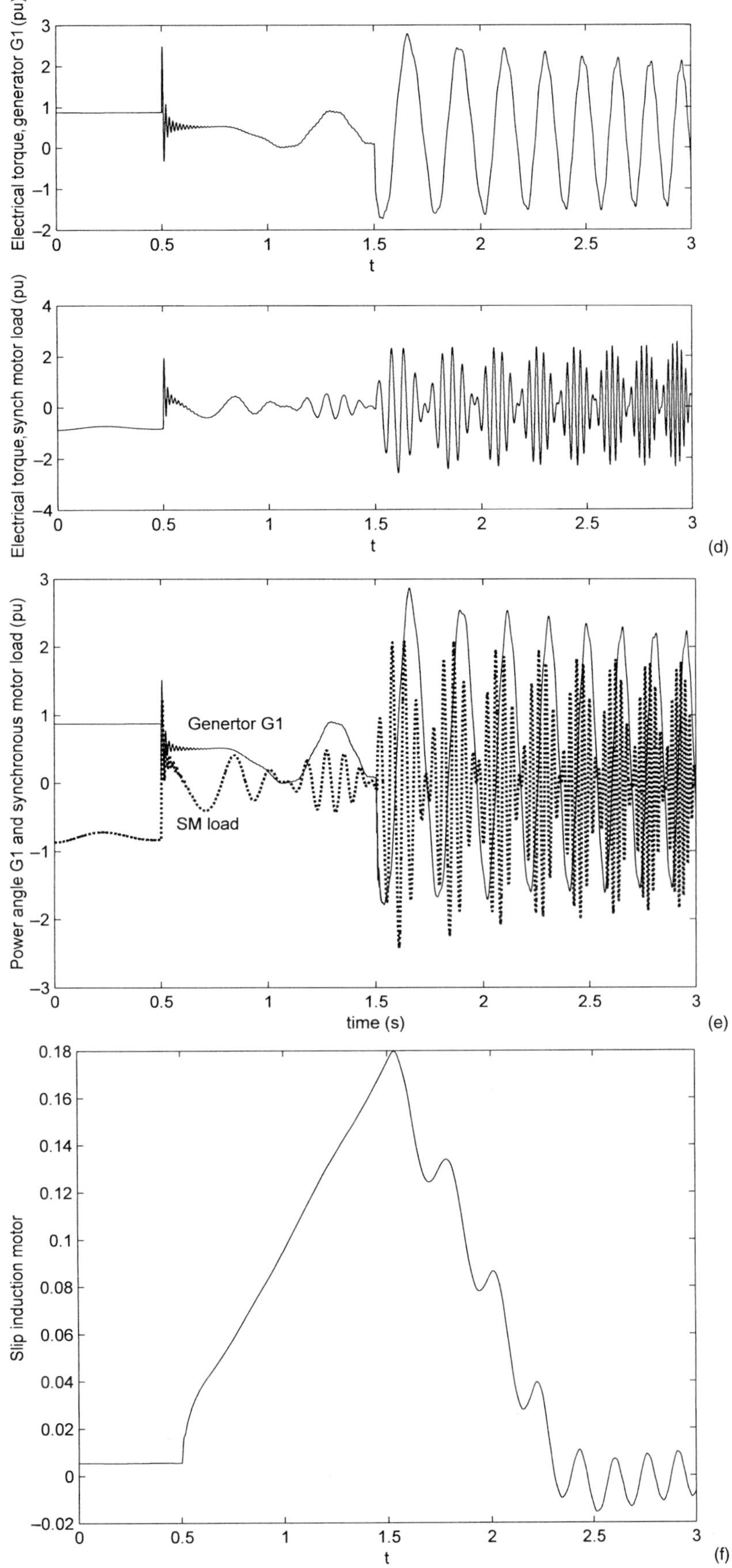

Figure 9-20 *(Continued)*

current cycles with respect to the bus voltage and it contributes to the accentuation of transients. Figure 9-20*e* is the power angle curve of generator *G*1. This figure and Fig. 9-20*d* depict that generator *G*1 and synchronous motor loads are unstable.

Figure 9-20*f* depicts the induction motor slip transient. Note the speed oscillations as the motor tries to speed up after the fault is removed.

Now consider a single phase-to-ground fault, phase *a*, at the same location. The transients are shown in Fig. 9-21 and are described as follows:

Figure 9-21*a* illustrates the voltage at the fault point in phases *a*, *b*, and *c*. Here the pattern after the fault clearance is entirely different because the torque pulsations of synchronous machines damp out after the fault is cleared (Figure 9-21*c*).

Figure 9-21*b* depicts the current transients in generator *G*1. Comparing this with Fig. 9-20*b*, there are no continued beat frequency transients.

Figure 9-21*c* illustrates the torque transients in the generator and synchronous motor load. These decay quickly after the fault is removed, and Fig. 9-21*d* shows the slip of the induction motor.

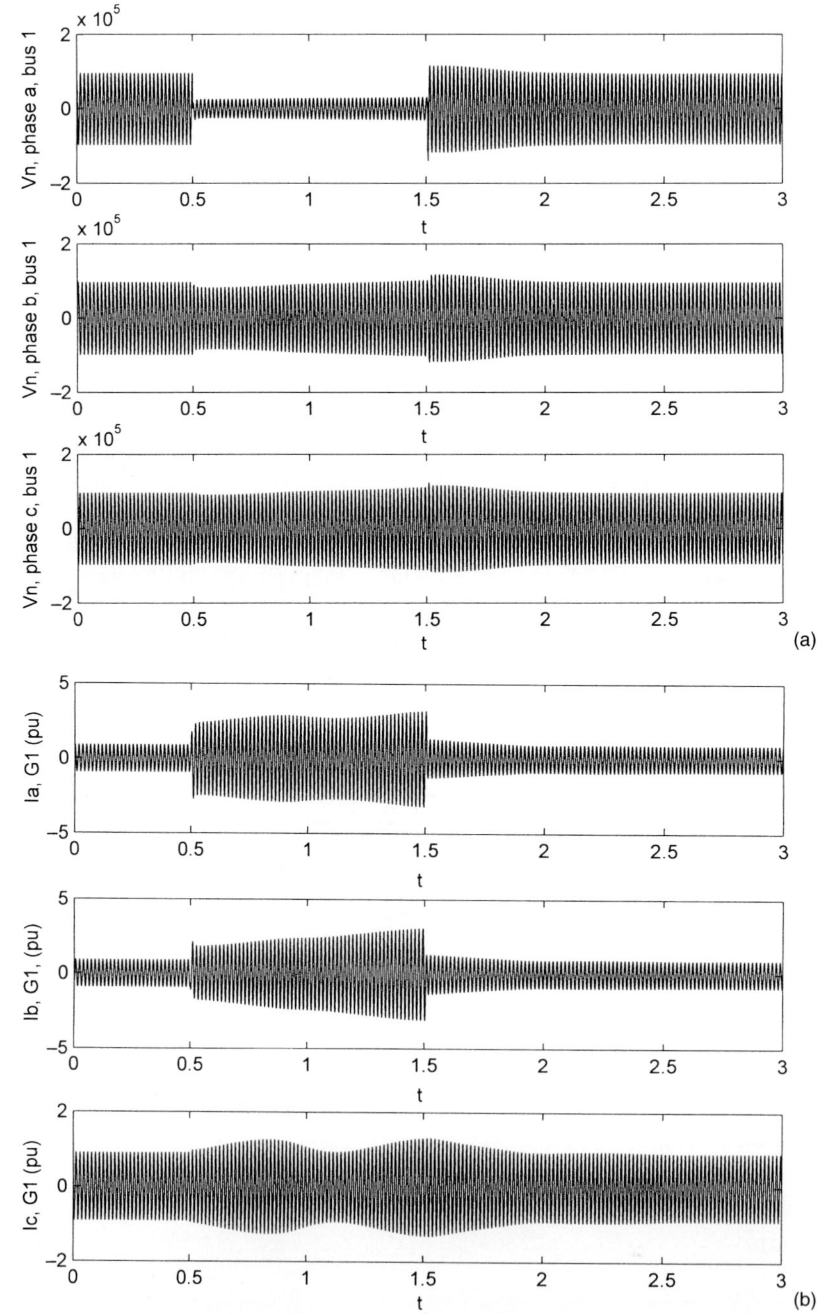

FIGURE 9-21 Transients for a single line-to-ground fault, phase *a* on bus 1, which occurs at 0.5 s and is removed at 1.5 s, fault duration of 1 s; (*a*) voltage at bus 1, phases *a*, *b*, and *c*; (*b*) generator *G*1 current, phases *a*, *b*, and *c*; (*c*) electrical torque of generator *G*1 and synchronous motor load (pu); (*d*) slip of the induction motor IM.

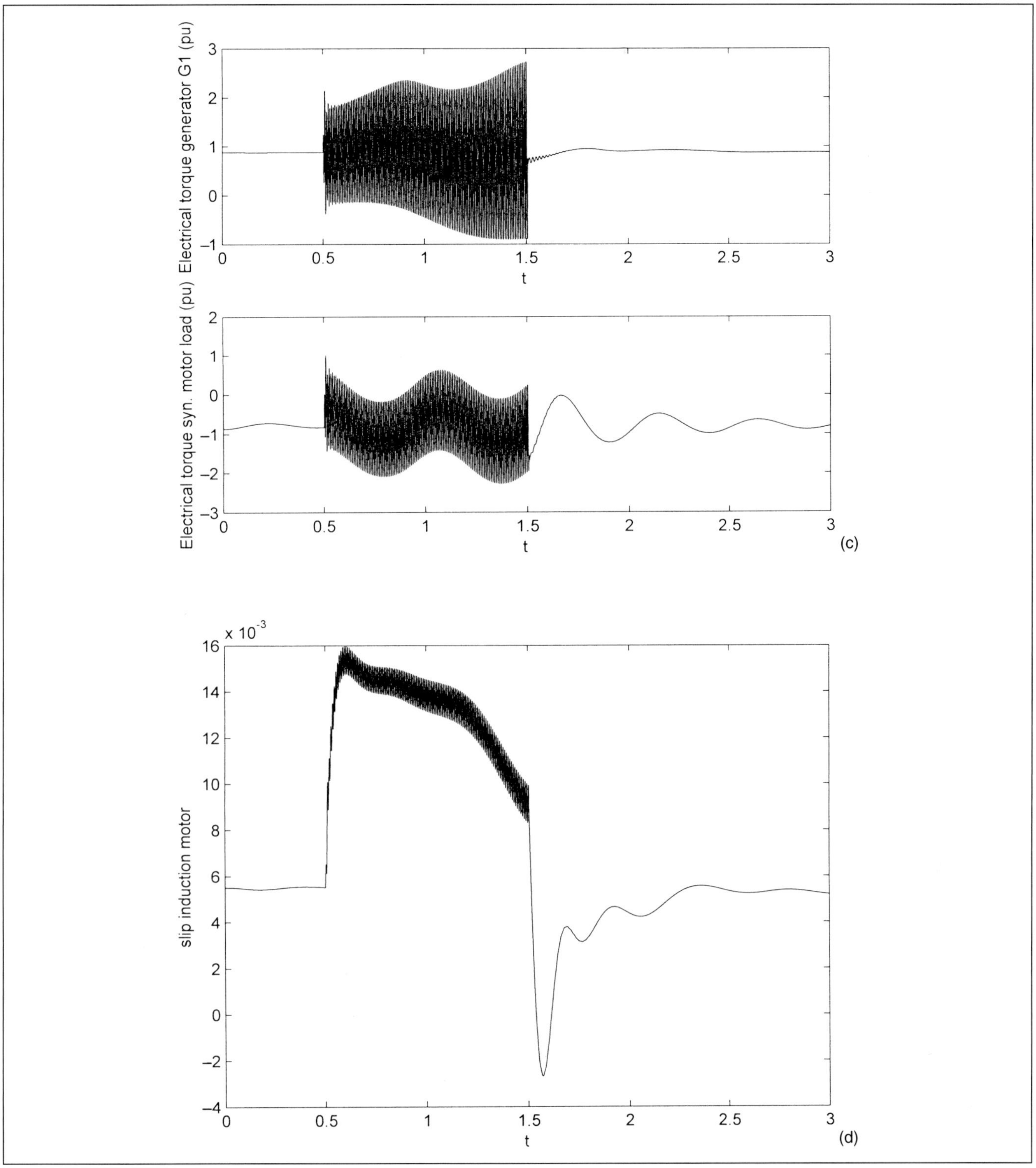

FIGURE 9-21 (*Continued*)

Fault duration of 1 s for three-phase fault considered in the simulation is rather large. Practically, the three-phase faults in a system as illustrated will be covered in differential zone of protection, operating time of the protective relaying of the order of one cycle or less. Add to it the breaker operating time. Figure 9-22 shows the simulation, voltages at the fault point, for three-phase fault duration of 0.3 s (18 cycles). This shows normal recovery voltage after the fault, and the swings in generator *G*1 recover quickly after the fault is cleared. Transient stability is of major consideration while evaluating the fault transients, which constitute severe disturbance to the system. This is followed up in the subsequent chapters.

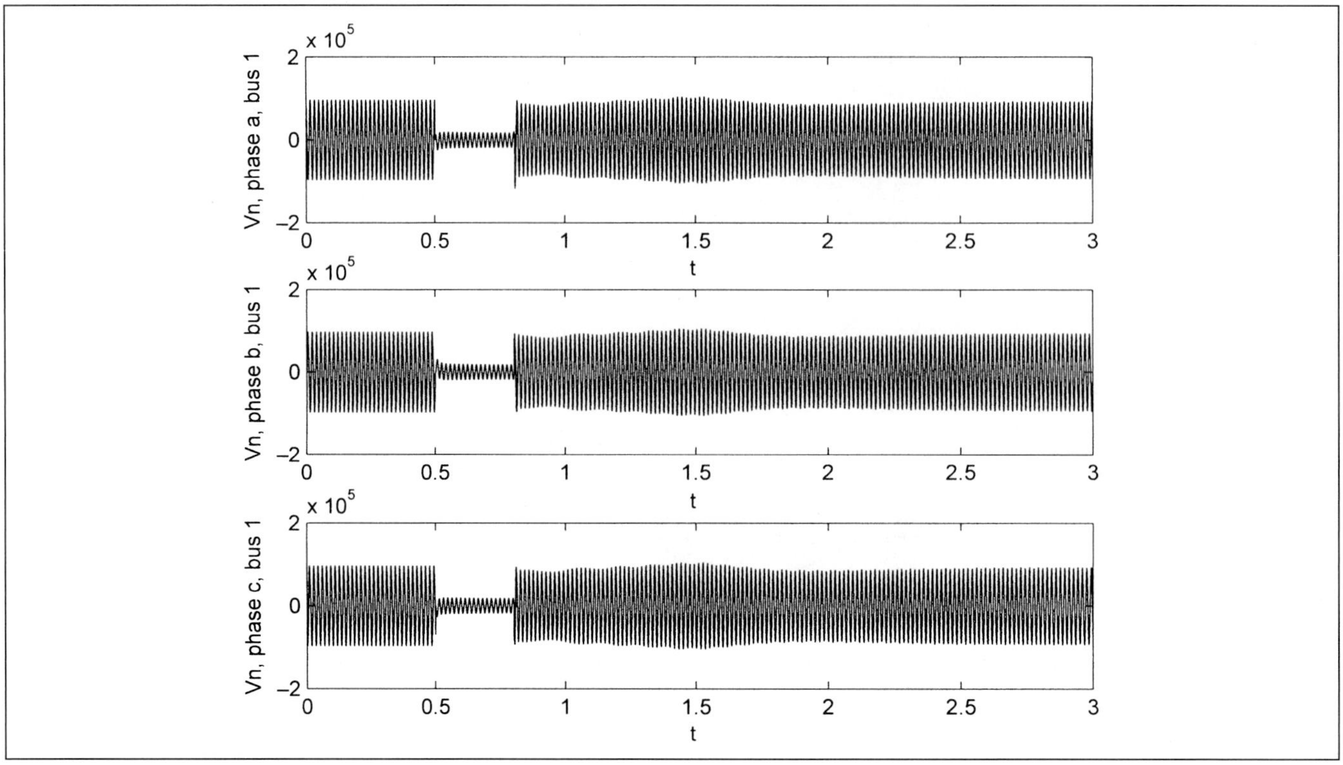

FIGURE 9-22 Transients for a three-phase fault on bus 1, that occurs at 0.5 s and is removed at 0.8 s, fault duration of 18 cycles.

PROBLEMS

1. A wye-wye connected step-down transformer has a primary voltage rating of 138 kV and the secondary voltage rating of 13.8 kV. The primary wye-connected winding neutral is ungrounded, while the secondary 13.8 kV neutral is solidly grounded. A single line-to-ground fault occurs on the 13.8 kV side. Does zero sequence current flow in the grounded neutral?

2. A wye-wye connected transformer with tertiary delta has a single line-to-ground fault on the secondary windings. Both wye-connected neutrals are grounded. Draw a zero sequence circuit of the transformer and show the current flows in all the windings of the transformer as well as in the lines connected to the transformer.

3. The unbalance voltages in a three-phase system are 0.8 pu, phase *a*, and 1.0 pu in phases *b* and *c*. Find the sequence components of the voltages and show the original voltages as well as zero sequence voltages in a sketch. Convert back from sequence voltages to original unbalance voltages.

4. A 138 to 13.8 kV, wye-wye connected transformer, with tertiary delta, 138-kV wye winding is solidly grounded and the 13.8-kV winding is ungrounded. A 15-Ω resistor is connected between phases *a* and *b* of the 13.8-kV windings. Show the flow of all currents in the transformer.

5. A simple distribution system is shown in Fig. 9-P1. The major equipment ratings are shown, but not the impedances. Assume practical values of impedances for all power system elements. Calculate three-phase, single line-to-ground and double line-to-ground faults at locations *F*1, *F*2, and *F*3, using method of symmetrical components and then the matrix methods. Calculate the voltages on all phases.

6. What are the COG factors in Prob. 5 for each case of calculation?

7. Repeat Probs. 5 and 6 when the transformer *T*2 is grounded through a resistance of 1 Ω.

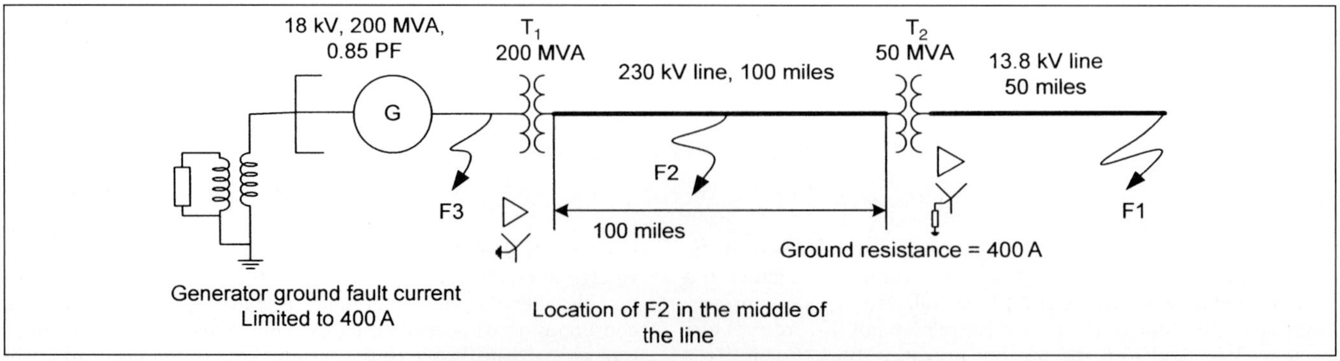

FIGURE 9-P1 Power system connections for Probs. 5, 6, and 7.

REFERENCES

1. ANSI/IEEE Std. C37.010, Guide for AC High Voltage Circuit Breakers Rated on Symmetrical Current Basis, 1999.
2. IEC 60909-0, Short-Circuit Currents in Three-Phase AC Systems, Part-0, Calculation of Currents, 2001; also IEC 60909-1, Factors for Calculation of Short-Circuit Currents in Three-Phase AC Systems According to IEC 60909-0, 1991.
3. IEEE Std. 551 (Violet Book), Calculating Short-Circuit Currents in Industrial and Commercial Power Systems, 2006.
4. W. E. Lewis and D. G. Pryce, *The Application of Matrix Theory to Electrical Engineering*, E&FN Spon, London, 1965.
5. L. J. Myatt, *Symmetrical Components*, Pergamon Press, Oxford, London, 1968.
6. J. L. Blackburn, *Symmetrical Components for Power System Engineering*, Marcel Dekker, New York, 1993.
7. H. E. Brown, *Solution of Large Networks by Matrix Methods*, Wiley Interscience, New York, 1975.
8. G. O. Calabrase, *Symmetrical Components Applied to Electric Power Networks*, Ronald Press Group, New York, 1959.
9. C. F. Wagnor and R. D. Evans, *Symmetrical Components*, McGraw Hill, New York, 1933.
10. C. L. Fortescue, "Method of Symmetrical Components Applied to the Solution of Polyphase Networks," *AIEE Trans*, vol. 37, pp. 1027–1140, 1918.
11. E. Clarke, *Circuit Analysis of Alternating Current Power Systems*, vol. 1, Wiley, New York, 1943.
12. Transformer Connections (including Auto-transformer Connections), General Electric, Publication No. GET-2H, Pittsfield, MA, 1967.
13. ANSI/IEEE Std. C57.12.00, General Requirements of Liquid Immersed Distribution, Power and Regulating Transformers, 2006.
14. ANSI Std. C57.12.70, Terminal Markings and Connections for Distribution and Power Transformers, 2000.
15. J. C. Das, *Power System Analysis Short-Circuit, Load Flow, and Harmonics*, Marcel Dekker, New York, 2002.
16. WESTINGHOUSE *Electrical Transmission and Distribution Reference Book*, 4th ed., Westinghouse Electric Corporation, East Pittsburgh, PA, 1964.

FURTHER READING

C. A. Gross, *Power System Analysis*, John Wiley, New York, 1979.

W. D. Stevenson, *Elements of Power System Analysis*, 4th ed., McGraw-Hill, New York, 1982.

CHAPTER 10

Transient Behavior of Synchronous Generators

The study of transients in generators has been carried out ever since the earlier nineteenth century. As the complexity of transmission and grid systems grew, so did the synchronous generator models to represent their behavior under steady-state and transient conditions. Synchronous generators are the prime source of power production and single shaft units of 1500 MVA, and more are in operation, while research continues on the development of larger superconducting units. In the utility systems, the synchronous generators are connected directly through a step-up transformer to the transmission systems, Fig. 10-1. In this figure, the ratings of generators and transformers are not shown for generality. Note that generators 1 and 4 do not have a generator breaker and the generator and transformer are protected as a unit with overlapping zones of differential protection (not discussed here). Generators 2 and 3 have generator breakers, which allow generator step-up transformers to be used as step-up down transformers during cold start-up. The power to the generation of auxiliary loads is duplicated with auto switching of a bus section breaker, and there is also a third standby power source. The auto switching of motor loads is discussed in Chap. 16.

Figure 10-2 shows a generator of 81.82 MVA, 12.47 kV, 0.85 power factor directly connected to a 12.47-kV bus, also powered by a 30/40/50 MVA, 115 to 12.47-kV utility transformer. The two sources are run in synchronism, and the plant running load is 45 MVA; the excess generated power is supplied into the utility system. The size of a generator that can be bus connected in an industrial distribution, primary distribution voltage of 13.8 kV, is approximately limited to 100 MVA, as an acceptable level of short circuit should be maintained at the medium voltage switchgear and the downstream distributions.

A utility or industrial generator is subjected to load regulation, near to the generator and far from generator faults, switching transients, and a worst-case scenario of terminal fault.

Under any perturbation from the steady-state operation, the currents in stator, rotor, field, and damper currents undergo transients. Swings occur in generator's active and reactive power output—electromagnetic torque, torque angle, frequency, and speed vary. A number of transient models to address these transients can be developed. The principles of the operation of the synchronous generators in the steady state form the background for development of the transient models; conversely, the steady-state models can be derived from the transient models. This chapter assumes that a reader has basic knowledge of the synchronous machine theory.

10-1 THREE-PHASE TERMINAL FAULT

A three-phase short circuit on the terminals of a generator is the largest perturbation to the operation of a generator and its survival to continue in operation after the protective relays clear the fault (assuming that the fault is not in the generator itself), and this transient is of practical importance too. We will study this transient as it leads to definitions of synchronous generator reactances. Short circuit has twofold effects. (1) Large disruptive forces are brought into play in the generator itself, and the generator should be designed to withstand these forces. (2) Short circuits should be removed quickly to limit fault damage and improve stability of the interconnected systems. The circuit breakers for generator application see a fault current of high asymmetry and must be rated to successfully interrupt such short-circuit currents, where a current zero may not occur at the contact parting time.

According to NEMA standards,[1] a synchronous generator shall be capable of withstanding, without injury, a 30-s, three-phase short circuit at its terminals when operating at rated kVA and power factor at 5-percent overvoltage, with fixed excitation. With a voltage regulator in service, the allowable duration t in seconds is determined from the following equation, where the regulator is designed to provide ceiling voltage continuously during a short circuit:

$$t=\left(\frac{\text{Nominal field voltage}}{\text{Exciter ceiling voltage}}\right)^2\times 30\ \text{s} \tag{10-1}$$

The generator should also be capable of withstanding without injury any other short circuits at its terminals for 30 s, provided (1) the following values of I_2^2t specified for the type of construction:

$$\begin{aligned} I_2^2t &= 40 \quad \text{for salient pole machines} \\ I_2^2t &= 30 \quad \text{for air-cooled cylindrical rotor machines} \end{aligned} \tag{10-2}$$

are not exceeded, and (2) the maximum current is limited by external means so as not to exceed the three-phase fault. I_2 is the negative-sequence current due to unsymmetrical faults.

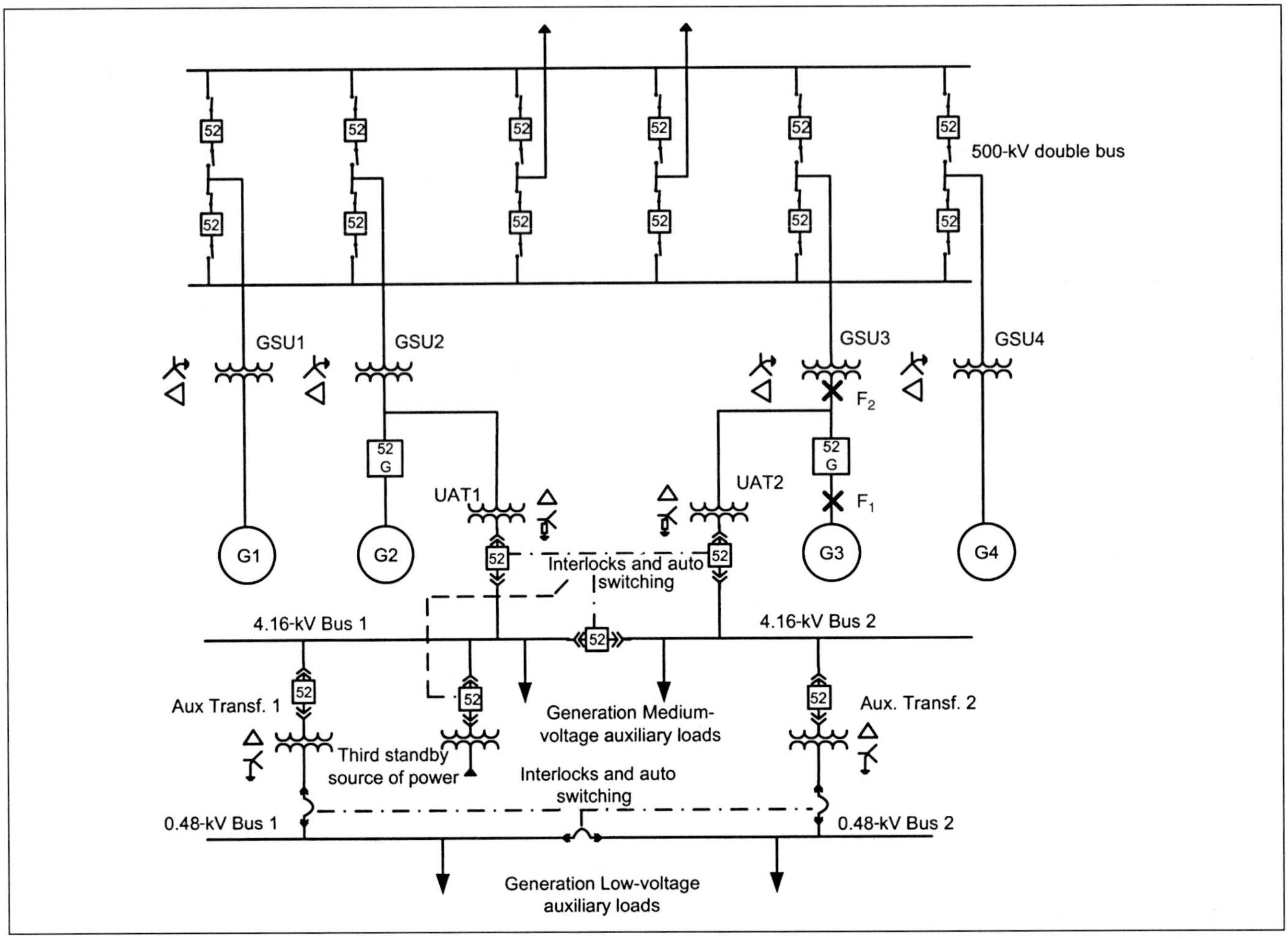

FIGURE 10-1 A single line diagram of connections of large synchronous generators in a utility generating station.

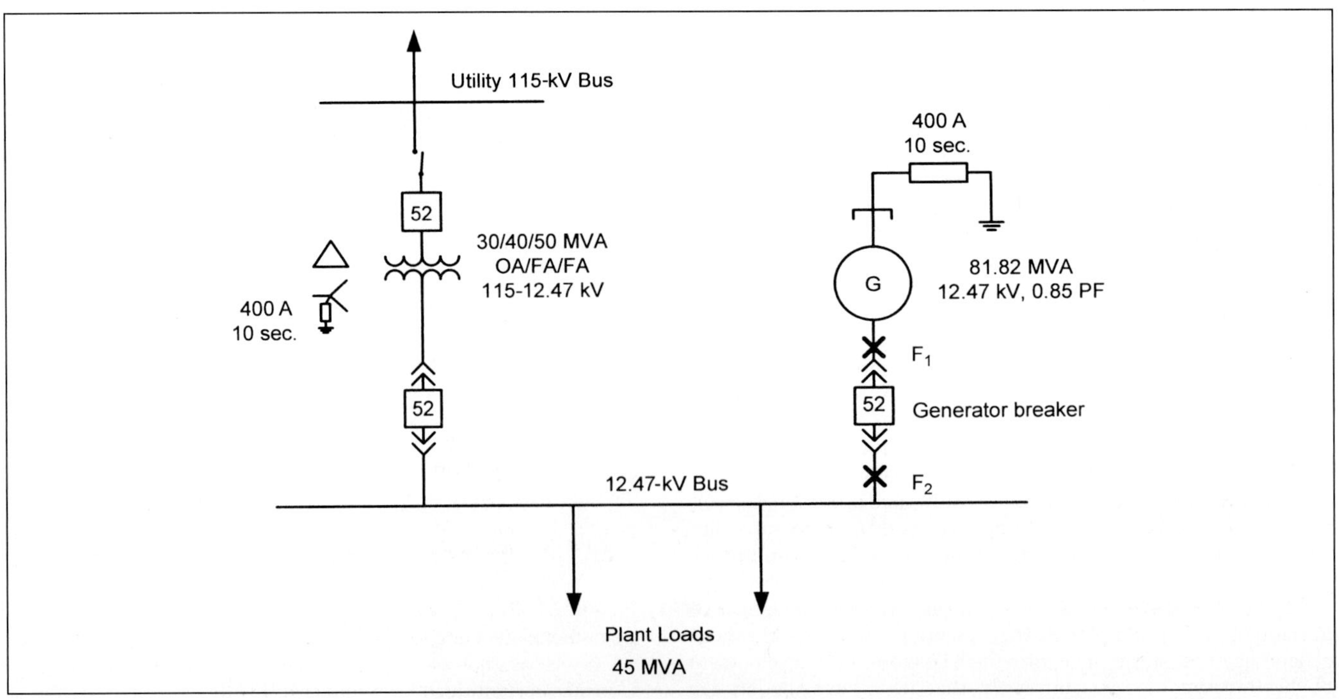

FIGURE 10-2 System configuration of a bus-connected industrial generator in cogeneration mode.

Synchronous generators are major sources of short-circuit currents in power systems. The fault current depends on:

1. The instant at which the short circuit occurs
2. The load and excitation of the generator immediately before the short circuit
3. The type of short circuits, that is, whether three phases or one or more than one phase and ground are involved
4. Constructional features of the generator, especially leakage and damping
5. The interconnecting impedances between generators

An insight into the physical behavior of the generator during short circuit can be made considering the theorem of constant flux linkages. For a closed circuit with resistance r and inductance L, $ri + Ldi/dt$ must be zero. If resistance is neglected, $Ldi/dt = 0$, that is, flux linkage Li must remain constant. In a generator, the resistance is small in comparison with inductance, the field winding is closed on the exciter, and the stator winding is closed due to short circuit. During the initial couple of cycles following short circuit, the flux linkages with these two windings must remain constant. On a terminal fault, the generated EMF acts on a closed circuit of stator windings and is analogous to an EMF being suddenly applied to an inductive circuit. Dynamically, the situation is more complex, that is, the lagging stator current has a demagnetizing effect on the field flux, and there are time constants associated with the penetration of the stator flux and decay of short-circuit currents.

10-2 REACTANCES OF A SYNCHRONOUS GENERATOR

The following definitions are applicable:

10-2-1 Leakage Reactance, X_l

The leakage reactance can be defined but cannot be tested. It is the reactance due to flux set up by armature windings, but not crossing the air gap. It can be divided into end-winding leakage and slot leakage. A convenient way of picturing the reactances is to view these in terms of permeances of various magnetic paths in the machine, which are functions of dimensions of iron and copper circuits and independent of the flux density or the current loading. The permeances, thus calculated, can be multiplied by a factor to consider the flux density and current. For example, the leakage reactance is mainly given by the slot permeance and the end-coil permeance.

10-2-2 Subtransient Reactance, X''_d

Subtransient reactance equals the leakage reactance plus the reactance due to the flux set up by stator currents crossing the air gap and penetrating the rotor as far as the damper windings in a laminated pole machine or as far as the surface damping currents in a solid pole machine. The subtransient conditions last for 1 to 5 cycles on a 60-Hz basis.

10-2-3 Transient Reactance, X'_d

Transient reactance is the reactance after all damping currents in the rotor surface or amortisseur windings have decayed, but before the damping currents in the field winding have decayed. The transient reactance equals the leakage reactance plus the reactance due to flux set up by the armature which penetrates the rotor to the field windings. Transient conditions last for 5–200 cycles on a 60-Hz basis.

10-2-4 Synchronous Reactance, X_d

Synchronous reactance is the steady-state reactance after all damping currents in the field windings have decayed. It is the sum of leakage reactance and a fictitious armature reaction reactance, which is much larger than the leakage reactance. Ignoring resistance, the per unit synchronous reactance is the ratio of per unit voltage on an open circuit divided by per unit armature current on short circuit for a given field excitation. This gives *saturated* synchronous reactance. The unsaturated value of the synchronous reactance is given by the per unit voltage on air-gap open circuit line divided by per unit armature current on short circuit. If 0.5 per unit field excitation produces full-load armature current on short circuit, the saturated synchronous reactance is 2.0 per unit. The saturated value may be only 70 to 90 percent of the unsaturated value.

10-2-5 Quadrature Axis Reactances, X''_q, X'_q, X_q

Quadrature axis reactances are similar to direct axis reactances, except that they involve the rotor permeance encountered when the stator flux enters one pole tip, crosses the pole, and leaves the other pole tip. The direct axis permeance is encountered by the flux crossing the air gap to the center of one pole, then crossing from one pole to the other pole and entering the stator from that pole. Figure 10-3 shows the armature reaction components. The total armature reaction F_a can be divided into two components, F_{ad} and F_{aq}. F_{ad} is directed across the direct axis and F_{aq}, across the quadrature axis. As these MMFs act on circuits of different permeances, the flux produced varies. This is the basis of two-reaction synchronous machine theory discussed in many texts.[2] If damper windings across pole faces are connected, X''_q is nearly equal to X''_d.

10-2-6 Negative-Sequence Reactance, X_2

The negative-sequence reactance is the reactance encountered by a voltage of reverse phase sequence applied to the stator, with the machine running. Negative-sequence flux revolves opposite to the rotor and is at twice the system frequency. Negative-sequence reactance is practically equal to the subtransient reactance as the damping currents in the damper windings or solid pole rotor surface prevent the flux from penetrating farther. The negative-sequence reactance is generally taken as the average of subtransient direct axis and quadrature axis reactances:

$$X_2 = \frac{X''_d + X''_q}{2} \tag{10-3}$$

10-2-7 Zero-Sequence Reactance, X_0

The zero-sequence reactance is the reactance effective when rated frequency currents enter all three terminals of the machine simultaneously and leave at the neutral of the machine. It is approximately equal to the leakage reactance of the machine with full-pitch coils. With two-thirds-pitch stator coils, the zero-sequence reactance will be a fraction of the leakage reactance.

10-2-8 Potier Reactance, X_p

Potier reactance is a reactance with numerical value between transient and subtransient reactances. It is used for the calculation of field current when open circuit and zero power factor curves are available. Figure 10-4 shows the saturation characteristics of a synchronous generator. Triangle XGS is the Potier triangle. Point S on the x-axis is field excitation to produce full-load current on short circuit, which is approximately 90° lagging. OS = $F_a + F_x$, where F_a represents direct demagnetizing armature reaction and F_x represents the field amperes to circulate current through the leakage reactance, which gives a voltage drop of iX_l. As we move up on the open circuit and zero power factor curves, S′G′ is not exactly

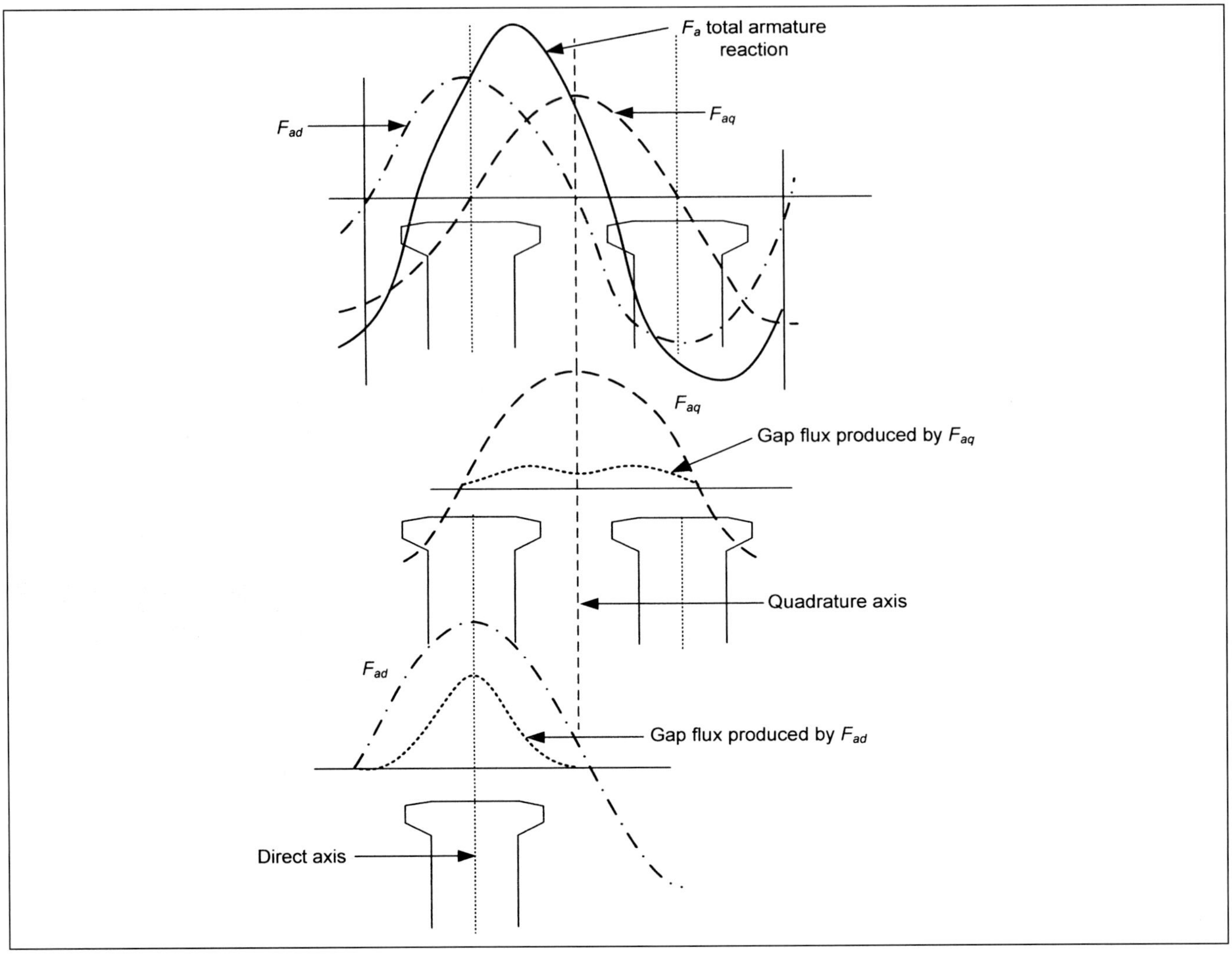

FIGURE 10-3 Resolution of total armature reaction in a salient-pole generator into direct-axis and quadrature-axis components.

equal to SG and also GX and G′X′ are not exactly identical due to different saturation of the curves. The value of X_l obtained from it is called the Potier reactance, and it exceeds the value based on pure armature leakage flux. We will return to saturation characteristics of the generator in Chap. 13. Due to different slopes of open circuit and zero power factor curves, G′X′ in Fig. 10-4 is slightly larger than GX and the value of reactance obtained from it is known as Potier reactance.

10-3 SATURATION OF REACTANCES

Saturation varies with voltage, current, and power factor. The saturation factor is usually applied to transient and synchronous reactances; in fact, all other reactances change, though slightly, with saturation. The saturated and unsaturated synchronous reactances are already defined above. In a typical generator, transient reactances may reduce from 5 to 25 percent on saturation. Saturated reactance is sometimes called the *rated voltage reactance* and is denoted by subscript *v* added to *d* and *q* axes, that is, X''_{dv} and X''_{qv} denote saturated subtransient reactances in direct and quadrature axes, respectively.

10-4 TIME CONSTANTS OF SYNCHRONOUS GENERATORS

10-4-1 Transient Open-Circuit Time Constant, T'_{do}

Transient open-circuit time constant expresses the rate of decay or build up of field current when the stator is open-circuited and there is zero resistance in the field circuit.

10-4-2 Subtransient Short-Circuit Time Constant, T''_d

Subtransient short-circuit time constant expresses the rate of decay of the subtransient component of current under a bolted (zero resistance), three-phase short circuit at the generator terminals.

10-4-3 Transient Short-Circuit Time Constant, T'_d

Transient short-circuit time constant expresses the rate of decay of the transient component of the current under a bolted (zero resistance), three-phase short circuit at the generator terminals. Manufacturers supply transient time constants for line-to-line and

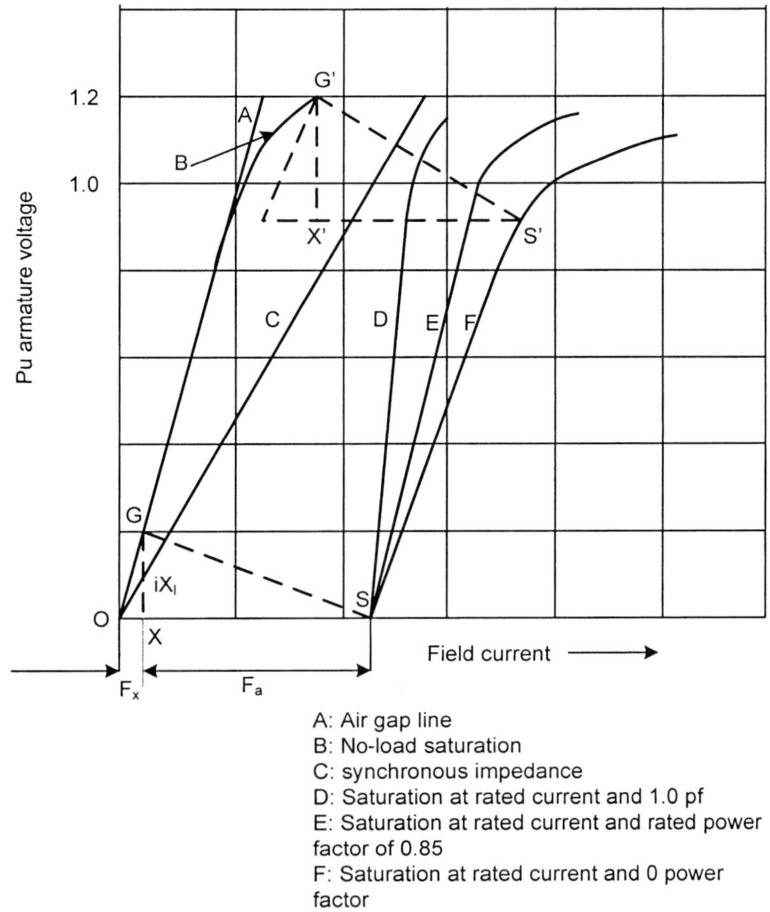

FIGURE 10-4 Potier reactance and other saturation characteristics of a generator.

line-to-neutral short-circuit currents also. To distinguish these time constants, a number is added to the subscript, that is, T'_{d3} denoting three-phase short-circuit transient time constant, T'_{d2}, denoting line-to-line short-circuit transient time constant, and T'_{d1} denoting line-to-neutral short-circuit time constant.

10-4-4 Armature Time Constant, T_a

Armature time constant expresses the rate of decay of the dc component of the short-circuit current under the same conditions. Again this is specified for three-phase, two-phase, and line-to-neutral faults. Table 10-1 shows electrical data, reactances, and time constants of a 13.8-kV, 112.1-MVA, and 0.85 power factor generator. In this table, all reactances are in per unit on generator MVA base and all time constants are in seconds.

10-5 SYNCHRONOUS GENERATOR BEHAVIOR ON TERMINAL SHORT-CIRCUIT

The time immediately after short circuit can be divided into three distinct periods:

- The subtransient period lasting from 1 to 5 cycles
- The transient period which may last up to 100 cycles or more
- The final or steady-state period follows after the transients have decayed and the terminal short circuit is limited by the synchronous reactance of the generator. Normally, the generator will be removed from service by protective relaying, much before the steady-state period is reached.

In the subtransient period, the conditions can be represented by the flux linking the stator and rotor windings. Any sudden change in the load or power factor of a generator produces changes in the MMF, both in direct and quadrature axes. At the moment of short circuit, the flux linking the stator from the rotor is trapped in the stator, giving a stationary replica of the main-pole flux. The rotor poles may be in a position of maximum or minimum flux linkage, and as these rotate, the flux linkages tend to change. This is counteracted by a current in the stator windings. The short-circuit current is, therefore, dependent on rotor angle. As the energy stored can be considered a function of armature and field linkages, the torque fluctuates and reverses cyclically. The dc component giving rise to asymmetry is caused by the flux trapped in the stator windings at the instant of short circuit, which sets up a dc transient in the armature circuit. This dc component establishes a component field in the air gap which is stationary in space, and which, therefore, induces a fundamental frequency voltage and current in the synchronously revolving rotor circuits. Thus, an increase in the stator current is followed by an increase in the field current. The field flux has superimposed on it a new flux pulsating with respect

TABLE 10-1 Manufacturer's Data of a 112.1-MVA Generator

DESCRIPTION	SYMBOL	DATA
Generator		
112.1 MVA, 2-pole, 13.8 kV, 0.85 pF, 95.286 MW, 4690 A, 235 V, field voltage, wye-connected		
SCR = 0.56		
Per unit reactance data, direct axis		
Saturated synchronous	X_{dv}	1.949
Unsaturated synchronous	X_d	1.949
Saturated transient	X'_{dv}	0.207
Unsaturated transient	X'_d	0.278
Saturated subtransient	X''_{dv}	0.164
Unsaturated subtransient	X''_d	0.193
Saturated negative sequence	X_{2v}	0.137
Unsaturated negative sequence	X_{2I}	0.185
Saturated zero sequence	X_{0v}	0.092
Unsaturated zero sequence	X_{0I}	0.111
Leakage reactance, overexcited	$X_{LM,OXE}$	0.164
Leakage reactance, underexcited	$X_{LM,UEX}$	0.164
Per unit reactance data, quadrature axis		
Saturated synchronous	X_{qv}	1.858
Unsaturated synchronous	X_q	1.858
Unsaturated transient	X'_q	0.434
Saturated subtransient	X''_{qv}	0.140
Unsaturated subtransient	X''_q	0.192
Field time constant data, direct axis		
Open circuit	T'_{do}	5.615
Three-phase short-circuit transient	T'_{d3}	0.597
Line-to-line short-circuit transient	T'_{d2}	0.927
Line-to-neutral short-circuit transient	T'_{d1}	1.124
Short-circuit subtransient	T''_d	0.015
Open-circuit subtransient	T''_{do}	0.022
Field time constant data, quadrature axis		
Open circuit	T'_{qo}	0.451
Three-phase short-circuit transient	T'_q	0.451
Short-circuit subtransient	T''_q	0.015
Open-circuit subtransient	T''_{qo}	0.046
Armature dc component time constant data		
Three-phase short-circuit	T_{a3}	0.330
Line-to-line short-circuit	T_{a2}	0.330
Line-to-neutral short-circuit	T_{a1}	0.294

All impedances in per unit on generator MVA base. All time constants in seconds.

to field windings at normal machine frequency. The single-phase-induced current in field can be resolved into two components, one stationary with respect to stator which counteracts the dc component of the stator current, and the other component travels at twice the synchronous speed with respect to stator and induces a second harmonic in it.

The armature and field are linked through the magnetic circuit, and the ac component of lagging current creates a demagnetizing effect. However, some time must elapse before it starts becoming effective in reducing the field current and the steady-state current is reached. The protective relays will normally operate to open the generator breaker and simultaneously the field circuit for suppression of generated voltage.

The above is rather an oversimplification of the transient phenomena in the generator on short circuit. In practicality, a generator will be connected in an interconnected system. The generator terminal voltage, rotor angle, and frequency all change, depending on the location of fault in the network, the network impedance, and the generator parameters. The generator output power will be affected by the change in the rotor winding EMF and the rotor position in addition to any changes in the impedance seen at the generator terminals. For a terminal fault, the voltage at the generator terminals is zero and, therefore, power supplied by the generator to load reduces to zero, while the prime mover output cannot change suddenly. Thus, the generator accelerates. In a multi-machine system with interconnecting impedances, the speeds of all machines change, so that these generate their share of synchronizing power in the overall impact, as these strive to reach a mean retardation through oscillations due to stored energy in the rotating masses.

In a dynamic simulation of short-circuit and transient stability calculations (Chaps. 12 and 13), the following are generally considered:

- Network before, during, and after the short circuit
- Dynamic modeling of induction motors, with zero excitation
- Dynamic modeling of synchronous machine, considering saturation
- Modeling of excitation systems
- Turbine and governor models

10-5-1 Equivalent Circuits During Fault

Figure 10-5 shows an envelope of decaying ac component of the short-circuit current wave, neglecting the dc component. The extrapolation of the current envelope to zero time gives the peak current. Note that immediately after the fault, the current decays fast and then more slowly. Only the upper-half of the transient is shown; the lower-half is symmetrical.

Transformer equivalent circuits of a salient-pole synchronous generator in the direct and quadrature axes at the instant of short circuit and during subsequent time delays help to derive the short-circuit current equations and explain the decaying ac component of Fig. 10-5. Based on the earlier discussions, these circuits are shown in Fig. 10-6, in the subtransient, transient, and steady-state periods. As the flux penetrates into the rotor, and the currents die down, it is equivalent to opening a circuit element; that is, from subtransient to transient state, the damper circuits are opened.

The direct-axis subtransient reactance is given by:

$$X''_d = X_l + \frac{1}{1/X_{\text{ad}} + 1/X_f + 1/X_{\text{kD}}} \tag{10-4}$$

where X_{ad} is the reactance corresponding to the fundamental space wave of armature in the direct axis, X_f is the reactance of the field windings, X_{kd} is the reactance of the damper windings in the direct axis, and X_l is the leakage reactance. X_f and X_{kD} are also akin to leakage reactances. The inductance giving rise to X_f is l_f, and to X_{kD} is l_{kD}. We write:

$$L_F = l_f + L_{\text{ad}}$$

$$X_F = X_f + X_{\text{ad}}$$

$$L_D = l_{\text{kD}} + L_{\text{ad}}$$

Therefore, Eq. (10-4) can also be written as:

$$L''_d = l_l + \frac{1}{1/L_{\text{ad}} + 1/l_f + 1/l_{\text{kD}}}$$

Similarly, the quadrature-axis subtransient reactance is given by:

$$X''_q = X_l + \frac{1}{1/X_{\text{aq}} + 1/X_{\text{kQ}}} \tag{10-5}$$

where X_{aq} is the reactance corresponding to fundamental space wave in the quadrature axis, and X_{kQ} is the reactance of the damper winding in the quadrature axis, akin to the leakage reactance. X_l is considered identical in the direct and quadrature axes, and the leakage reactance giving rise to X_l is l_l. That is:

$$l_l = l_d = l_q$$

The quadrature-axis rotor circuit does not carry a field winding, and this circuit is composed of damper bars or rotor iron in the inter-polar axis of a cylindrical rotor machine.

The direct-axis and quadrature-axis short-circuit time constants associated with decay of the subtransient component of the current are:

$$T''_d = \frac{1}{\omega r_D}\left[\frac{X_{\text{ad}} X_f X_l}{X_{\text{ad}} X_f + X_f X_l + X_{\text{ad}} X_l} + X_{\text{kD}}\right] \tag{10-6}$$

or

$$T''_d = \frac{1}{r_D}\left[\frac{L_{\text{ad}} l_f l_l}{L_{\text{ad}} l_f + l_f l_l + L_{\text{ad}} l_l} + l_{\text{kD}}\right]$$

$$T''_q = \frac{1}{\omega r_Q}\left[\frac{X_{\text{aq}} X_l}{X_{\text{aq}} + X_l} + X_{\text{kQ}}\right] \tag{10-7}$$

or

$$T''_q = \frac{1}{r_Q}\left[\frac{L_{\text{aq}} l_l}{L_{\text{aq}} + l_l} + l_{\text{kQ}}\right]$$

where r_D and r_Q are resistances of the damper windings in the direct and quadrature axes, respectively.

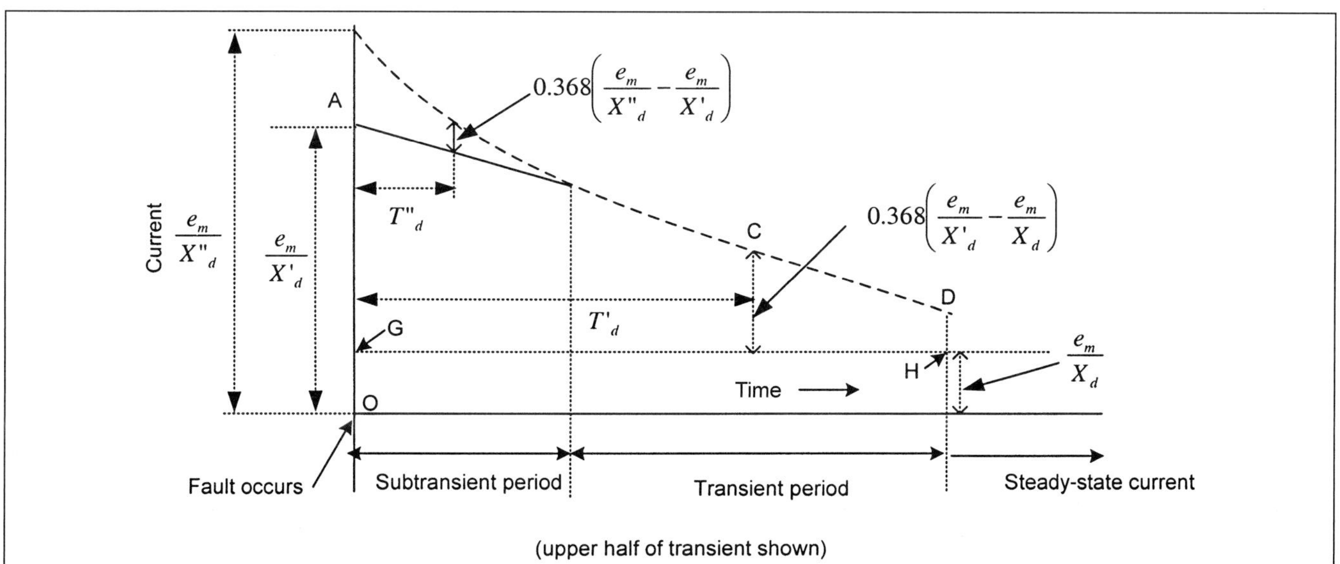

FIGURE 10-5 Decaying ac component of the short-circuit current from the instance of occurrence of fault to steady state. The subtransient, transient, and steady-state periods and decay characteristics of short-circuit current are shown.

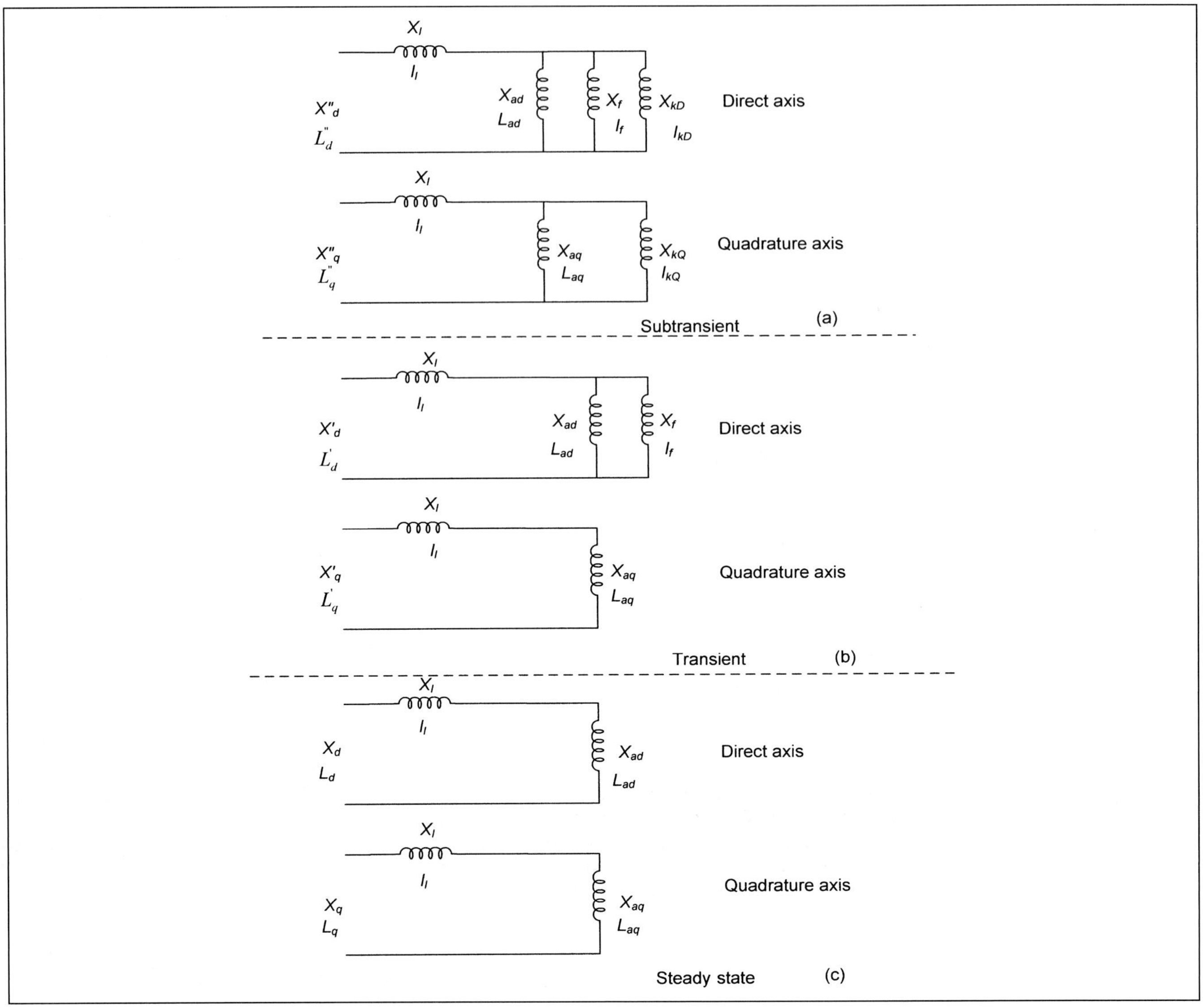

FIGURE 10-6 Equivalent circuits of a synchronous generator during subtransient, transient, and steady-state periods after a terminal fault.

When the flux has penetrated the air gap, the effect of the eddy currents in the pole face cease after a few cycles given by short-circuit subtransient time constants. The resistance of damper circuit is much higher than that of the field windings. This amounts to opening of the damper winding circuits, and the direct-axis and quadrature-axis transient reactances are given by:

$$X'_d = X_l + \frac{1}{1/X_{ad} + 1/X_f} = \left[\frac{X_{ad}X_f}{X_f + X_{ad}} + X_l\right] \tag{10-8}$$

$$X'_q = X_l + X_{aq} \tag{10-9}$$

The direct-axis transient time constant associated with this decay is:

$$T'_d = \frac{1}{\omega r_F}\left[\frac{X_{ad}X_l}{X_{ad} + X_l} + X_f\right] \tag{10-10}$$

where r_F is the resistance of the field windings.

Finally, when the currents in the field winding have also died down, given by transient short-circuit time constant, the steady-state short-circuit current is given by the synchronous reactance:

$$X_d = X_l + X_{ad}$$
$$X_q = X_l + X_{aq} \tag{10-11}$$

X_l, the leakage reactance in the direct and quadrature axes, is known to be nearly equal, as stated before. The reactances X_{ad} and X_{aq} are the mutual reactances between the stator and rotor, and X_d and X_q are total reactances in the two axes. Equations (10-9) and (10-11) show that X'_q is equal to X_q. The relative values of X''_q, X'_q, and X_q depend on generator construction. For cylindrical rotor generators, X_q is >> X'_q. Sometimes, one or two damper windings are modeled in the q-axis.

Reverting to Fig. 10-5, the direct-axis transient reactance determines the initial value OA of the symmetrical transient envelope ACD and the direct-axis transient short-circuit time constant T'_d determines the decay of this envelope. The direct-axis transient time constant $T'd$ is the time required by the transient envelope to decay to a point where the difference between it and the steady-state envelope GH is $1/\varepsilon$ (= 0.368) of the initial difference GA. Similar explanation applies to decay of the subtransient component in Fig. 10-5.

10-5-2 Fault-Decrement Curve

Based on Fig. 10-5, the expression for decaying ac component of the short-circuit current of a generator can be written as:

$$i_{ac} = \text{Decaying subtransient component} + \text{decaying transient component} + \text{steady-state component} \quad (10\text{-}12)$$

$$= (i''_d - i'_d)\varepsilon^{-t/T''_d} + (i'_d - i_d)\varepsilon^{-t/T'_d} + i_d$$

The subtransient current is given by:

$$i''_d = \frac{E''}{X''_d} \quad (10\text{-}13)$$

where E'' is the generator internal voltage behind subtransient reactance.

$$E'' = V_a + X''_d \sin\phi \quad (10\text{-}14)$$

where V_a is the generator terminal voltage, and ϕ is the load power factor angle, prior to fault.

Similarly, the transient component of the current is given by:

$$i'_d = \frac{E'}{X'_d} \quad (10\text{-}15)$$

where E' is the generator internal voltage behind transient reactance.

$$E' = V_a + X'_d \sin\phi \quad (10\text{-}16)$$

The steady-state component is given by:

$$i_d = \frac{V_a}{X_d}\left(\frac{i_F}{i_{Fg}}\right) \quad (10\text{-}17)$$

where i_F is the field current at given load conditions (when regulator action is taken into account) and i_{Fg} is the field current at no-load-rated voltage.

The dc component is given by:

$$i_{dc} = \sqrt{2}\, i''_d \varepsilon^{-t/T_a} \quad (10\text{-}18)$$

where T_a is the armature short-circuit time constant, given by:

$$T_a = \frac{1}{\omega r}\left[\frac{2X''_d X''_q}{X''_d + X''_q}\right] \quad (10\text{-}19)$$

where r is the stator resistance.

The open-circuit time constant describes the decay of the field transient when the field circuit is closed and the armature circuit is open. The direct-axis subtransient open circuit time constant is:

$$T''_{do} = \frac{1}{\omega r_D}\left[\frac{X_{ad}X_f}{X_{ad} + X_f} + X_{kD}\right] \quad (10\text{-}20)$$

And the quadrature-axis subtransient open-circuit time constant is:

$$T''_{qo} = \frac{1}{\omega r_Q}(X_{aq} + X_{kQ}) \quad (10\text{-}21)$$

The open-circuit direct-axis transient time constant is:

$$T'_{do} = \frac{1}{\omega r_F}[X_{ad} + X_f] \quad (10\text{-}22)$$

The short-circuit direct-axis transient time constant can be expressed as:

$$T'_d = T'_{do}\left[\frac{X'_d}{(X_{ad} + X_l)}\right] = T'_{do}\frac{X'_d}{X_d} \quad (10\text{-}23)$$

It may be observed that the resistances have been neglected in the previous expressions for current, but these can be included, that is, the subtransient current is:

$$i''_d = \frac{E''}{r_D + X''_d} \quad (10\text{-}24)$$

where r_D is defined as the resistance of the amortisseur windings on salient-pole generators and analogous body of cylindrical rotor generators. Similarly, the transient current is:

$$i'_d = \frac{E'}{r_f + X'_d} \quad (10\text{-}25)$$

As the resistance is much smaller compared to the reactances, the impact on the magnitude of the calculated current is negligible.

Example 10-1 Consider a 13.8-kV, 100-MVA, 0.85 power factor generator. Its rated full-load current is 4184 A. Other data are:

Saturated subtransient reactance, X''_{dv}	= 0.15 per unit
Saturated transient reactance, X'_{dv}	= 0.2 per unit
Synchronous reactance, X_d	= 2.0 per unit
Field current at rated load, i_f	= 3 per unit
Field current at no-load-rated voltage, i_{fg}	= 1 per unit
Subtransient short-circuit time constant, T''_d	= 0.012 s
Transient short-circuit time constant, T'_d	= 0.35 s
Armature short-circuit time constant, T_a	= 0.15 s
Effective resistance*	= 0.0012 per unit
Quadrature-axis synchronous reactance*	= 1.8 per unit

A three-phase short circuit occurs at the terminals of the generator when it is operating at its rated load and power factor. It is required to construct a fault decrement curve of the generator for (a) ac component, (b) dc component, and (c) total current. Data marked with an asterisk are intended for Example 10-4.

From Eq. (10-14), the voltage behind subtransient reactance at the generator-rated voltage, load, and power factor is:

$$E'' = V + X''_d \sin\phi = 1 + (0.15)(0.527) = 1.079$$

From Eq. (10-13), the subtransient component of the current is:

$$i''_d = \frac{E''}{X''_{dv}} = \frac{1.079}{0.15}\text{pu} = 30.10 \text{ kA}$$

Similarly from Eq. (10-16), E', the voltage behind transient reactance is 1.1054 per unit, and from Eq. (10-15), the transient component of the current is 23.12 kA.

From Eq. (10-17), current i_d at constant excitation is 2.09 kA rms. For a ratio of $i_f/i_{Fg} = 3$, current $i_d = 6.28$ kA rms. Therefore,

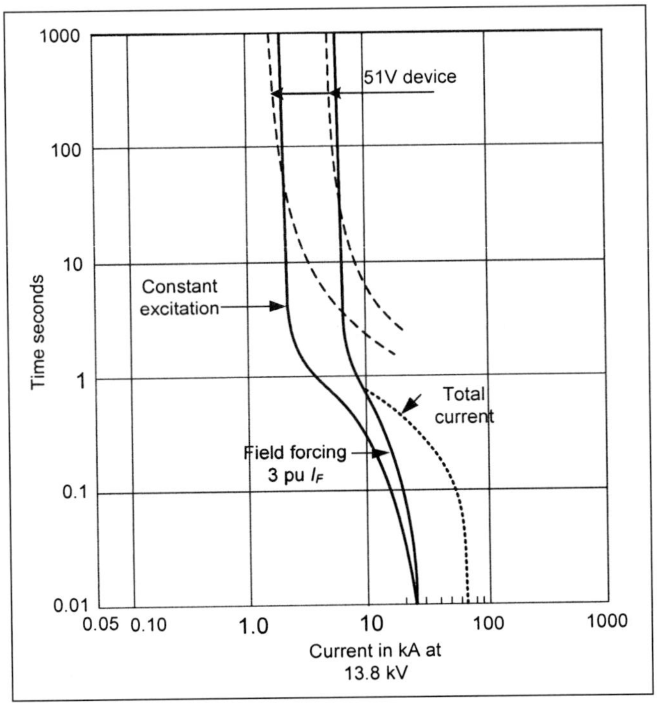

FIGURE 10-7 Calculated fault decrement curves of generator, Example 10-1.

the following equation can be written for the ac component of the current:

$$i_{ac} = 6.98e^{-t/0.012} + 20.03e^{-t/0.35} + 2.09 \text{ kA}$$

With full-load excitation:

$$i_{ac} = 6.98e^{-t/0.012} + 16.84e^{-t/0.35} + 6.28 \text{ kA}$$

The ac decaying component of the current can be plotted from these two equations, with lowest value of $t = 0.01$ s to $t = 1{,}000$ s. The dc component is given by Eq. (10-18):

$$i_{dc} = \sqrt{2}i''_d e^{-t/T_a} = 42.57e^{-t/0.15} \text{ kA}$$

At any instant, the total current is:

$$\sqrt{i_{ac}^2 + i_{dc}^2} \text{ kA rms}$$

The fault decrement curves are shown in Fig. 10-7. Short-circuit current with constant excitation is 50 percent of the generator full-load current. This can occur for a stuck voltage regulator condition. Though this current is lower than the generator full-load current, it cannot be allowed to be sustained. Voltage restraint or voltage-controlled overcurrent generator backup relays (ANSI/IEEE device number 51V) or distance relays (device 21) are set to pick up on this current. The generator fault decrement curve is often required for the appropriate setting and coordination of these relays with the system relays.

10-6 CIRCUIT EQUATIONS OF UNIT MACHINES

The behavior of machines can be analyzed in terms of circuit theory, which makes it useful not only for steady-state performance but also for transients. The circuit of machines can be simplified in terms of coils on the stationary (stator) and rotating (rotor) parts, and these coils interact with each other according to fundamental electromagnetic laws. The circuit of a unit machine can be derived

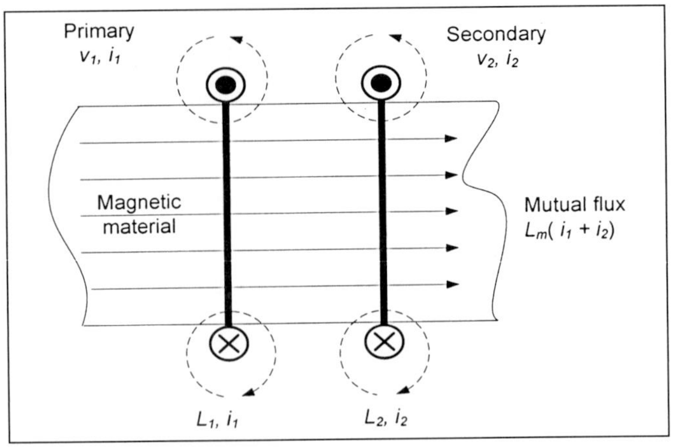

FIGURE 10-8 Representation of magnetically coupled coils; a two-winding transformer.

from the consideration of generation of EMF in coupled coils due to (a) transformer EMF, also called pulsation EMF, and (b) the EMF of rotation.

Consider two magnetically coupled, stationary, coaxial coils, as shown in Fig. 10-8. Let the applied voltages be v_1 and v_2 and currents i_1 and i_2, respectively. This is, in fact, the circuit of a two-winding transformer, the primary and secondary being represented by single-turn coils. The current in primary coil (any coil can be called a primary coil) sets up a total flux linkage Φ_{11}. The change of current in this coil induces an EMF given by:

$$e_{11} = -\frac{d\Phi_{11}}{dt} = -\frac{d\Phi_{11}}{di_1}\frac{di_1}{dt} = -L_{11}\frac{di_1}{dt} = -L_{11}pi \tag{10-26}$$

where $L_{11} = -d\Phi_{11}/di_1$ is the total primary self-inductance and the operator $p = d/dt$.

If Φ_1 is the leakage flux and Φ_{12} is the flux linking with secondary coil, then the variation of current in the primary coil induces in the secondary coil an EMF:

$$e_{12} = -\frac{d\Phi_{12}}{dt} = -\frac{d\Phi_{12}}{di_1}\cdot\frac{di_1}{dt} = -L_{12}\frac{di_i}{dt} = -L_{12}pi_1 \tag{10-27}$$

where $L_{12} = -d\Phi_{12}/di_1$ is the mutual inductance of the primary coil winding with the secondary coil winding.

Similar equations apply for the secondary coil winding. All the flux produced by the primary coil winding do not link with the secondary. The leakage inductance associated with the windings can be accounted for as:

$$\begin{aligned} L_{11} &= L_{12} + L_1 \\ L_{22} &= L_{21} + L_2 \end{aligned} \tag{10-28}$$

The mutual inductance between coils can be written as:

$$L_{12} = L_{21} = L_m = \sqrt{(L_{11} - L_1)(L_{22} - L_2)} = k\sqrt{L_{11}L_{22}} \tag{10-29}$$

Thus, the equations of a unit transformer are:

$$\begin{aligned} v_1 &= r_1 i_1 + (L_m + L_1)pi_1 + L_m pi_2 \\ v_2 &= r_2 i_2 + (L_m + L_2)pi_2 + L_m pi_1 \end{aligned} \tag{10-30}$$

Or, in the matrix form:

$$\begin{vmatrix} v_a \\ v_b \end{vmatrix} = \begin{vmatrix} r_1 + (L_1 + L_m)p & L_m p \\ L_m p & r_2 + (L_2 + L_m)p \end{vmatrix} \begin{vmatrix} i_1 \\ i_2 \end{vmatrix} \tag{10-31}$$

If the magnetic axes of the coupled coils are at right angles, no mutually induced pulsation or transformer EMF can be produced by variation of currents in either of the windings. However, if the coils are free to move, the coils with magnetic axes at right angles have an EMF of rotation, e_r, induced when the winding it represents rotates:

$$e_r = \omega_r \Phi \tag{10-32}$$

where ω_r is the angular speed of rotation and Φ is the flux. This EMF is maximum when the two coils are at right angles to each other and zero when these are cophasial.

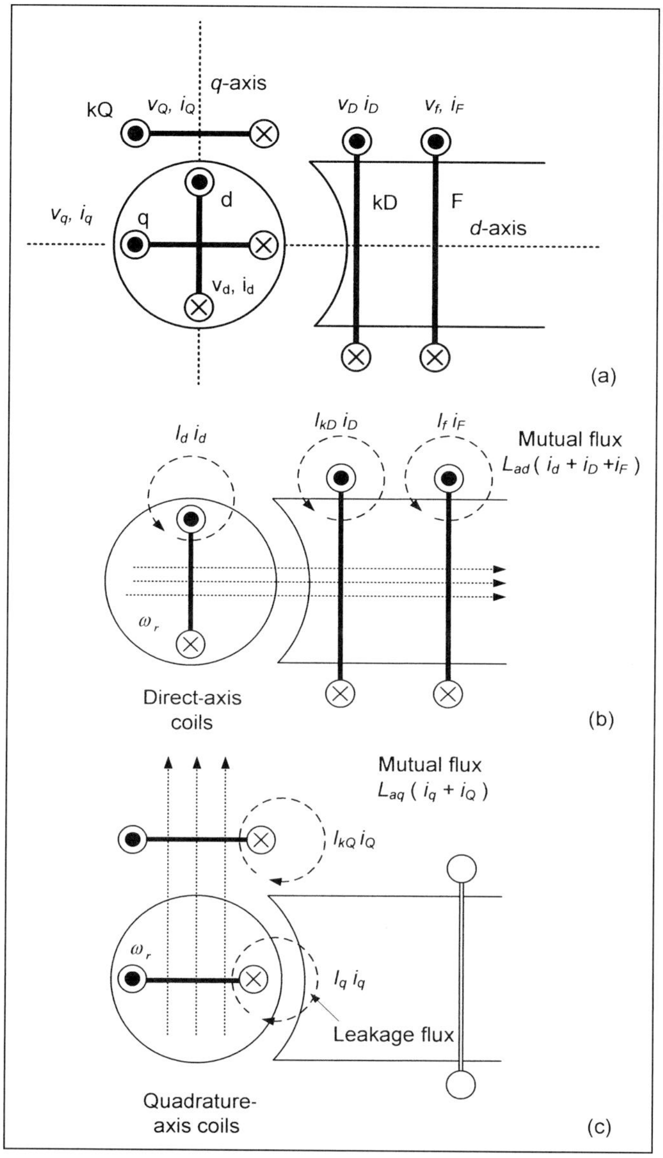

FIGURE 10-9 (*a*) Development of circuit of a synchronous generator; general machine theory and concepts of unit machine. (*b*), (*c*) Resolution of circuit of Fig. 10-9*a* into direct and quadrature axes.

To summarize, a pulsation EMF is developed in two coaxial coils and there is no rotational EMF. Conversely, a rotational EMF is developed in two coils at right angles, but no pulsation EMF. If the relative motion is at an angle θ, the EMF of rotation is multiplied by sin θ.

$$e_r = \omega_r \Phi \sin\theta \tag{10-33}$$

The equations of a unit machine may be constructed based on the simple derivation of EMF in the coils. Consider a generator with direct- and quadrature-axes coils, as shown in Fig 10-9*a*, and its resolution into Figs. 10-9*b* and *c*. Note that the armature is shown rotating, and it has two coils, D and Q, at right angles to each other in *dq* axes. The field winding F and the damper winding kD are shown stationary in the direct axis. All coils are single-turn coils. In the direct axis, there are three mutual inductances, that is, of D with kD, kD with F, and F with D. A simplification is to consider that these are all equal; inductance L_{ad}. Each coil has a leakage inductance of its own. Consequently, the total inductances of the coils are:

Coil D: $(l_d + L_{\text{ad}})$

Coil kD: $(l_{\text{kD}} + L_{\text{ad}})$

Coil F: $(l_f + L_{\text{ad}})$ (10-34)

The mutual linkage with armature coil D when all three coils carry currents is:

$$\lambda_d = L_{\text{ad}}(i_F + i_D + i_d) \tag{10-35}$$

where i_F, i_D, and i_d are the currents in the field, damper, and direct-axis coils, respectively. Similarly, in the quadrature axis:

$$\lambda_q = L_{\text{aq}}(i_q + i_Q) \tag{10-36}$$

The EMF equations in each of the coils can be written based on these observations. Field coil: No *q*-axis circuit will affect its flux, nor do any rotational voltages appear. The applied voltage v_f is:

$$v_f = r_F i_F + (L_{\text{ad}} + l_f)pi_F + L_{\text{ad}}pi_D + L_{\text{ad}}pi_d \tag{10-37}$$

Stator coil kD is similarly located as coil F:

$$v_D = r_D i_D + (L_{\text{ad}} + l_{\text{kD}})pi_D + L_{\text{ad}}pi_F + L_{\text{ad}}pi_d \tag{10-38}$$

Coil kQ has no rotational EMF, but will be affected magnetically by any current i_Q in coil Q:

$$v_Q = r_Q i_Q + (L_{\text{aq}} + l_{\text{kQ}})pi_Q + L_{\text{aq}}pi_q \tag{10-39}$$

Armature coils D and Q have the additional property of inducing rotational EMF:

$$\begin{aligned} v_d &= r_d i_d + (L_{\text{ad}} + l_d)pi_d + L_{\text{ad}}pi_F + L_{\text{ad}}pi_D \\ &\quad + L_{\text{aq}}\omega_r i_Q + (L_{\text{aq}} + l_q)\omega_r i_q \\ v_q &= r_q i_q + (L_{aq} + l_q)pi_q + L_{\text{aq}}pi_Q - L_{\text{ad}}\omega_r i_F \\ &\quad - L_{\text{ad}}\omega_r i_D - (L_{\text{ad}} + l_d)\omega_r i_d \end{aligned} \tag{10-40}$$

These equations can be written in a matrix form:

$$\begin{vmatrix} v_f \\ v_D \\ v_Q \\ v_d \\ v_q \end{vmatrix} = \begin{vmatrix} r_F + (L_{\text{ad}} + l_f)p & L_{\text{ad}}p & \cdot & L_{\text{ad}}p & \cdot \\ L_{\text{ad}}p & r_D + (L_{\text{ad}} + l_{\text{kD}})p & \cdot & L_{\text{ad}}p & \cdot \\ \cdot & \cdot & r_Q + (L_{\text{aq}} + l_{\text{kQ}})p & \cdot & L_{\text{aq}}p \\ L_{\text{ad}}p & L_{\text{ad}}p & L_{\text{aq}}\omega_r & r_d + (L_{\text{ad}} + l_d)p & (L_{\text{aq}} + l_q)\omega_r \\ -L_{\text{ad}}\omega_r & -L_{\text{ad}}\omega_r & L_{\text{aq}}p & -(L_{\text{ad}} + l_d)\omega_r & r_q + (L_{\text{aq}} + l_q)p \end{vmatrix} \begin{vmatrix} i_F \\ i_D \\ i_Q \\ i_d \\ i_q \end{vmatrix} \tag{10-41}$$

10-7 PARK'S TRANSFORMATION

Park's transformation greatly simplifies the mathematical model of synchronous machines. It describes a new set of variables, such as currents, voltages, and flux linkages, obtained by transformation of the actual (stator) variables in three axes: 0, d, and q. The d and q axes are already defined, the 0 axis is a stationary axis.

10-7-1 Reactance Matrix of a Synchronous Generator

Consider the normal construction of a three-phase synchronous generator, with three-phase stationary ac windings on the stator and the field and damper windings on the rotor, Fig. 10-10. The stator inductances vary, depending on the relative position of the stator and rotor. Consider that the field winding is cophasial with the direct axis, and also the direct axis carries a damper winding. The q-axis has also a damper winding. The phase windings are distributed, but are represented by single-turn coils aa, bb, and cc in Fig. 10-10. The field flux is directed along d-axis, and therefore the machine-generated voltage is at right angles to it, along q-axis. For generator action, the generated voltage vector **E** leads the terminal voltage vector **V** by an angle, and from basic machine theory we know that it is the torque angle. At $t = 0$, the voltage vector **V** is located along the axis of phase a, which is the reference axis in Fig. 10-10. The q-axis is at an angle δ, and the d-axis is at an angle $\delta+\pi/2$. For $t > 0$, the reference axis is at an angle $\omega_r t$ with respect to the axis of phase a. The d-axis of the rotor is therefore at:

$$\theta=\omega_r t+\delta+\pi/2 \text{ rad} \tag{10-42}$$

For synchronous operation, $\omega_r = \omega_0 =$ constant.

Consider phase a inductance. It is a combination of its own self-inductance L_{aa} and its mutual inductances L_{ab} and L_{bc} with phases b and c. All three inductances vary with the relative position of the rotor with respect to stator because of saliency of air gap. When the axis of phase a coincides with direct axis, that is, $\theta=0$ or π, the resulting flux of coil aa is maximum in the horizontal direction and its self-inductance is maximum. When at right angles to d-axis, $\theta=\pi/2$ or $3\pi/2$, its inductance is a minimum. Thus, L_{aa} fluctuates twice per revolution and can be expressed as:

$$L_{aa}=L_s+L_m\cos\theta \tag{10-43}$$

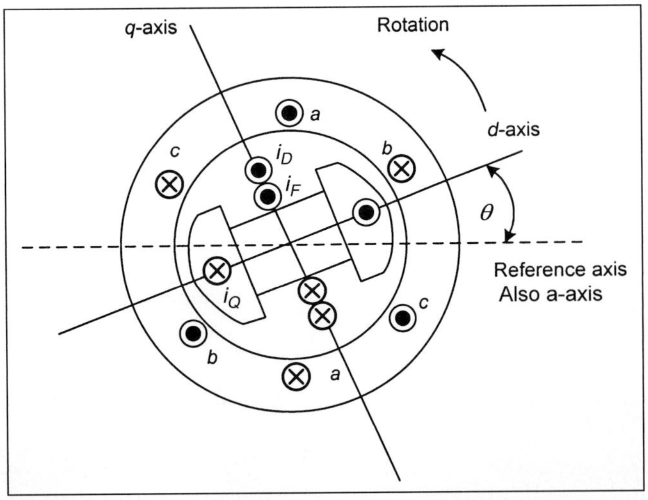

FIGURE 10-10 Representation of dq axes, windings, and reference axis of a synchronous generator.

Similarly, self-inductance of phase b is maximum at $\theta=2\pi/3$ and of phase c at $\theta=-2\pi/3$:

$$\begin{aligned} L_{bb}&=L_s+L_m\cos 2\left(\theta-2\frac{\pi}{3}\right)\\ L_{cc}&=L_s+L_m\cos 2\left(\theta+2\frac{\pi}{3}\right)\end{aligned} \tag{10-44}$$

Phase-to-phase mutual inductances are also a function of θ. L_{ab} is negative and is maximum at $\theta=-\pi/6$. This can be explained as follows: For the direction of currents shown in coils aa and bb, L_{ab} is negative. When the angle is $-\pi/3$, the current in phase b coil generates the maximum flux, but the linkage with phase a coil is better when the angle is zero degrees. However, at this angle the flux is reduced. The maximum flux linkage can be considered to take place at an angle which is the average of these two angles, that is, $\pi/6$.

$$\begin{aligned} L_{ab}&=-\left[M_s+L_m\cos 2\left(\theta+\frac{\pi}{6}\right)\right]\\ L_{bc}&=-\left[M_s+L_m\cos 2\left(\theta-\frac{\pi}{2}\right)\right]\\ L_{ca}&=-\left[M_s+L_m\cos 2\left(\theta+5\frac{\pi}{6}\right)\right]\end{aligned} \tag{10-45}$$

Stator-to-rotor mutual inductances are the inductances between stator windings and field windings, between stator windings and direct-axis damper windings, and between stator windings and quadrature-axis damper windings. These reactances are:

From stator phase windings to field windings:

$$\begin{aligned} L_{aF}&=M_F\cos\theta\\ L_{bF}&=M_F\cos(\theta-2\pi/3)\\ L_{cF}&=M_F\cos(\theta+2\pi/3)\end{aligned} \tag{10-46}$$

From stator phase windings to direct-axis damper windings:

$$\begin{aligned} L_{aD}&=M_D\cos\theta\\ L_{bD}&=M_D\cos(\theta-2\pi/3)\\ L_{cD}&=M_D\cos(\theta+2\pi/3)\end{aligned} \tag{10-47}$$

And from stator phase windings to damper windings in the quadrature axis:

$$\begin{aligned} L_{aQ}&=M_Q\sin\theta\\ L_{bQ}&=M_Q\sin(\theta-2\pi/3)\\ L_{cQ}&=M_Q\sin(\theta+2\pi/3)\end{aligned} \tag{10-48}$$

The rotor self-inductances are L_F, L_D, and L_Q. The mutual inductances are:

$$L_{DF}=M_R \qquad L_{FQ}=0 \qquad L_{DQ}=0 \tag{10-49}$$

The mutual inductance between field windings and direct-axis damper windings are constant and do not vary. Also d and q axes are 90° displaced and the mutual inductances between the field and direct-axis damper windings and quadrature-axis damper windings are zero.

The inductance matrix can therefore be written as:

$$\bar{L}=\begin{vmatrix}\bar{L}_{aa} & \bar{L}_{aR}\\ \bar{L}_{Ra} & \bar{L}_{RR}\end{vmatrix} \tag{10-50}$$

where $\bar{L}_{aa}$ is a stator-to-stator inductance matrix.

$$\bar{L}_{aa} = \begin{vmatrix} L_s + L_m \cos 2\theta & -M_s - L_m \cos 2(\theta + \pi/6) & -M_s - L_m \cos 2(\theta + 5\pi/6) \\ -M_s - L_m \cos 2(\theta + \pi/6) & L_s + L_m \cos 2(\theta - 2\pi/3) & -M_s - L_m \cos 2(\theta - \pi/2) \\ -M_s - L_m \cos 2(\theta + 5\pi/6) & -M_s - L_m \cos 2(\theta - \pi/2) & L_s + L_m \cos 2(\theta + 2\pi/3) \end{vmatrix} \tag{10-51}$$

$$\bar{L}_{aR} = \bar{L}_{Ra}^t = \begin{vmatrix} M_F \cos\theta & M_D \cos\theta & M_Q \sin\theta \\ M_F \cos(\theta - 2\pi/3) & M_D \cos(\theta - 2\pi/3) & M_Q \sin(\theta - 2\pi/3) \\ M_F \cos(\theta + 2\pi/3) & M_D \cos(\theta + 2\pi/3) & M_Q \sin(\theta + 2\pi/3) \end{vmatrix} \tag{10-52}$$

$\bar{L}_{RR}$ rotor-to-rotor inductance's matrix is:

$$\bar{L}_{RR} = \begin{vmatrix} L_F & M_R & 0 \\ M_R & L_D & 0 \\ 0 & 0 & L_Q \end{vmatrix} \tag{10-53}$$

The inductance matrix of Eq. (10-51) shows that the inductances vary with the angle θ. By referring the stator quantities to rotating rotor dq axes through Park's transformation, this dependence on θ is removed and a constant reactance matrix emerges.

10-7-2 Transformation of Reactance Matrix

Park's transformation describes a new set of variables, such as currents, voltages, and flux linkages in $0dq$ axes. The stator parameters are transferred to the rotor parameters. For the currents, this transformation is:

$$\begin{vmatrix} i_0 \\ i_d \\ i_q \end{vmatrix} = \sqrt{\frac{2}{3}} \begin{vmatrix} \frac{1}{\sqrt{2}} & \frac{1}{\sqrt{2}} & \frac{1}{\sqrt{2}} \\ \cos\theta & \cos\left(\theta - \frac{2\pi}{3}\right) & \cos\left(\theta + \frac{2\pi}{3}\right) \\ \sin\theta & \sin\left(\theta - \frac{2\pi}{3}\right) & \sin\left(\theta + \frac{2\pi}{3}\right) \end{vmatrix} \begin{vmatrix} i_a \\ i_b \\ i_c \end{vmatrix} \tag{10-54}$$

Using matrix notation:

$$\bar{i}_{0dq} = \bar{P}\,\bar{i}_{abc} \tag{10-55}$$

Similarly:

$$\bar{v}_{0dq} = \bar{P}\bar{v}_{abc}$$

$$\bar{\lambda}_{0dq} = \bar{P}\bar{\lambda}_{abc} \tag{10-56}$$

where $\bar{\lambda}$ is the flux linkage vector. The abc currents in the stator windings produce a synchronously rotating field, stationary with respect to rotor. This rotating field can be produced by constant currents in the fictitious rotating coils in dq axes. P is nonsingular and $\bar{P}^{-1} = \bar{P}^t$:

$$\bar{P}^{-1} = \bar{P}^t = \sqrt{\frac{2}{3}} \begin{vmatrix} \frac{1}{\sqrt{2}} & \cos\theta & \sin\theta \\ \frac{1}{\sqrt{2}} & \cos\left(\theta - \frac{2\pi}{3}\right) & \sin\left(\theta - \frac{2\pi}{3}\right) \\ \frac{1}{\sqrt{2}} & \cos\left(\theta + \frac{2\pi}{3}\right) & \sin\left(\theta + \frac{2\pi}{3}\right) \end{vmatrix} \tag{10-57}$$

To transform the stator-based variables into rotor-based variables, define a matrix as follows:

$$\begin{vmatrix} i_0 \\ i_d \\ i_q \\ i_F \\ i_D \\ i_Q \end{vmatrix} = \begin{vmatrix} \bar{P} & \bar{0} \\ \bar{0} & \bar{1} \end{vmatrix} \begin{vmatrix} i_a \\ i_b \\ i_c \\ i_F \\ i_D \\ i_Q \end{vmatrix} = \bar{B}\,\bar{i} \tag{10-58}$$

where $\bar{1}$ is 3×3 unity matrix and $\bar{0}$ is 3×3 zero matrix. The original rotor quantities are left unchanged. The time-varying inductances can be simplified by referring all quantities to rotor frame of reference:

$$\begin{vmatrix} \bar{\lambda}_{0dq} \\ \bar{\lambda}_{FDQ} \end{vmatrix} = \begin{vmatrix} \bar{P} & \bar{0} \\ \bar{0} & \bar{1} \end{vmatrix} \begin{vmatrix} \bar{\lambda}_{abc} \\ \bar{\lambda}_{FDQ} \end{vmatrix}$$

$$= \begin{vmatrix} \bar{P} & \bar{0} \\ \bar{0} & \bar{1} \end{vmatrix} \begin{vmatrix} \bar{L}_{aa} & \bar{L}_{aR} \\ \bar{L}_{Ra} & \bar{L}_{RR} \end{vmatrix} \begin{vmatrix} \bar{P}^{-1} & \bar{0} \\ \bar{0} & \bar{1} \end{vmatrix} \begin{vmatrix} \bar{P} & \bar{0} \\ \bar{0} & \bar{1} \end{vmatrix} \begin{vmatrix} \bar{i}_{abc} \\ \bar{i}_{FDQ} \end{vmatrix} \tag{10-59}$$

This transformation gives:

$$\begin{vmatrix} \lambda_0 \\ \lambda_d \\ \lambda_q \\ \lambda_F \\ \lambda_D \\ \lambda_Q \end{vmatrix} = \begin{vmatrix} L_0 & 0 & 0 & 0 & 0 & 0 \\ 0 & L_d & 0 & kM_F & kM_D & 0 \\ 0 & 0 & L_q & 0 & 0 & kM_q \\ 0 & kM_F & 0 & L_F & M_R & 0 \\ 0 & kM_D & 0 & M_R & L_D & 0 \\ 0 & 0 & kM_Q & 0 & 0 & L_Q \end{vmatrix} \begin{vmatrix} i_0 \\ i_d \\ i_q \\ i_F \\ i_D \\ i_Q \end{vmatrix} \tag{10-60}$$

Define:

$$\begin{aligned} L_d &= L_s + M_s + \frac{3}{2}L_m \\ L_q &= L_s + M_s - \frac{3}{2}L_m \\ L_0 &= L_s - 2M_s \\ k &= \sqrt{\frac{3}{2}} \end{aligned} \tag{10-61}$$

The inductance matrix is sparse, symmetric, and constant. It decouples the $0dq$ axes, as will be illustrated further.

10-8 PARK'S VOLTAGE EQUATION

The voltage equation in terms of current and flux linkages is:

$$\bar{v} = -\bar{R}\,\bar{i} - \frac{d\bar{\lambda}}{dt} \tag{10-62}$$

or

$$\begin{vmatrix} v_a \\ v_b \\ v_c \\ v_F \\ v_D \\ v_Q \end{vmatrix} = \begin{vmatrix} r & 0 & 0 & 0 & 0 & 0 \\ 0 & r & 0 & 0 & 0 & 0 \\ 0 & 0 & r & 0 & 0 & 0 \\ 0 & 0 & 0 & r_F & 0 & 0 \\ 0 & 0 & 0 & 0 & r_D & 0 \\ 0 & 0 & 0 & 0 & 0 & r_Q \end{vmatrix} \begin{vmatrix} i_a \\ i_b \\ i_c \\ i_F \\ i_D \\ i_Q \end{vmatrix} - \begin{vmatrix} \frac{di}{dt} \end{vmatrix} \begin{vmatrix} \lambda_a \\ \lambda_b \\ \lambda_c \\ \lambda_F \\ \lambda_D \\ \lambda_Q \end{vmatrix} \tag{10-63}$$

This can be partitioned as:

$$\begin{vmatrix} \bar{v}_{abc} \\ \bar{v}_{DFQ} \end{vmatrix} = -\begin{vmatrix} \bar{r} & \\ & \bar{r}_{FDQ} \end{vmatrix}\begin{vmatrix} \bar{i}_{abc} \\ \bar{i}_{FDQ} \end{vmatrix} - \frac{d}{dt}\begin{vmatrix} \bar{\lambda}_{abc} \\ \bar{\lambda}_{FDQ} \end{vmatrix} \tag{10-64}$$

The transformation is given by:

$$\bar{B}^{-1}\bar{v}_B = -\bar{R}\bar{B}^{-1}\bar{i}_B - \frac{d}{dt}(\bar{B}^{-1}\bar{\lambda}_B) \tag{10-65}$$

where:

$$\begin{vmatrix} \bar{P} & \bar{0} \\ \bar{0} & \bar{1} \end{vmatrix} = \bar{B} \qquad \bar{B}\begin{vmatrix} \bar{i}_{abc} \\ \bar{i}_{FDQ} \end{vmatrix} = \begin{vmatrix} \bar{i}_{0dq} \\ \bar{i}_{FDQ} \end{vmatrix} = \bar{i}_B$$

$$\bar{B}\begin{vmatrix} \bar{\lambda}_{abc} \\ \bar{\lambda}_{FDQ} \end{vmatrix} = \begin{vmatrix} \bar{\lambda}_{0dq} \\ \bar{\lambda}_{FDQ} \end{vmatrix} = \bar{\lambda}_B$$

$$\bar{B}\begin{vmatrix} \bar{v}_{abc} \\ \bar{v}_{FDQ} \end{vmatrix} = \begin{vmatrix} \bar{v}_{0dq} \\ \bar{v}_{FDQ} \end{vmatrix} = \bar{v}_B \tag{10-66}$$

Equation (10-65) can be written as:

$$\bar{v}_B = -\bar{B}R\bar{B}^{-1} - \bar{B}\frac{d}{dt}(\bar{B}^{-1}\bar{\lambda}_B) \tag{10-67}$$

First evaluate:

$$\bar{B}\frac{d\bar{B}^{-1}}{d\theta} = \begin{vmatrix} \bar{P} & \bar{0} \\ \bar{0} & \bar{1} \end{vmatrix}\begin{vmatrix} \frac{d\bar{P}^{-1}}{d\theta} & \bar{0} \\ \bar{0} & \bar{0} \end{vmatrix} = \begin{vmatrix} \bar{P}\frac{d\bar{P}^{-1}}{d\theta} & \bar{0} \\ \bar{0} & \bar{0} \end{vmatrix} \tag{10-68}$$

It can be shown that:

$$\bar{P}\frac{d\bar{P}^{-1}}{d\theta} = \begin{vmatrix} 0 & 0 & 0 \\ 0 & 0 & 1 \\ 0 & -1 & 0 \end{vmatrix} \tag{10-69}$$

We can write:

$$\frac{d\bar{B}^{-1}}{dt} = \frac{d\bar{B}^{-1}}{d\theta}\frac{d\theta}{dt} \tag{10-70}$$

$$\bar{B}\frac{d\bar{B}^{-1}}{d\theta} = \begin{vmatrix} 0 & 0 & 0 & 0 & 0 & 0 \\ 0 & 0 & 1 & 0 & 0 & 0 \\ 0 & -1 & 0 & 0 & 0 & 0 \\ 0 & 0 & 0 & 0 & 0 & 0 \\ 0 & 0 & 0 & 0 & 0 & 0 \\ 0 & 0 & 0 & 0 & 0 & 0 \end{vmatrix} \tag{10-71}$$

We write:

$$\bar{B}\left(\frac{d\bar{B}^{-1}}{dt}\right) = \bar{B}\left(\frac{d\bar{B}^{-1}}{d\theta}\right)\left(\frac{d\theta}{dt}\right) \tag{10-72}$$

Substituting in Eq. (10-67), the voltage equation becomes:

$$\bar{v}_B = \bar{R}\,\bar{i}_B - \frac{d\theta}{dt}\begin{vmatrix} 0 \\ \lambda_q \\ -\lambda_d \\ 0 \\ 0 \\ 0 \end{vmatrix} - \frac{d\bar{\lambda}_B}{dt} \tag{10-73}$$

When the shaft rotation is uniform, $d\theta/dt$ is a constant, and Eq. (10-73) is linear and time invariant.

The torque equation can be written as:

$$T_E = [i_0, i_d, i_q, i_F, i_D, i_Q]\begin{vmatrix} 0 \\ \lambda_q \\ -\lambda_d \\ 0 \\ 0 \\ 0 \end{vmatrix} = i_d\lambda_d - i_q\lambda_q \tag{10-74}$$

10-9 CIRCUIT MODEL OF SYNCHRONOUS GENERATORS

From the previous treatment, the following decoupled voltage equations can be written as follows:
Zero sequence:

$$v_0 = ri_0 - \frac{d\lambda_0}{dt} \tag{10-75}$$

Direct axis:

$$\begin{aligned} v_d &= -ri_d - \frac{d\theta}{dt}\lambda_q - \frac{d\lambda_d}{dt} \\ v_F &= r_F i_F + \frac{d\lambda_F}{dt} \\ v_D &= r_D i_D + \frac{d\lambda_D}{dt} = 0 \end{aligned} \tag{10-76}$$

Quadrature axis:

$$\begin{aligned} v_q &= -ri_q + \frac{d\theta}{dt}\lambda_d - \frac{d\lambda_q}{dt} \\ v_Q &= r_Q i_Q + \frac{d\lambda_Q}{dt} = 0 \end{aligned} \tag{10-77}$$

The decoupled equations relating to flux linkages and currents are:
Zero sequence:

$$\lambda_0 = L_0 i_0 \tag{10-78}$$

Direct axis:

$$\begin{vmatrix} \lambda_d \\ \lambda_F \\ \lambda_D \end{vmatrix} = \begin{vmatrix} L_d & kM_F & kM_D \\ kM_F & L_F & M_R \\ kM_D & M_R & L_D \end{vmatrix}\begin{vmatrix} i_d \\ i_F \\ i_D \end{vmatrix} \tag{10-79}$$

Quadrature axis:

$$\begin{vmatrix} \lambda_q \\ \lambda_Q \end{vmatrix} = \begin{vmatrix} L_q & kM_Q \\ kM_Q & L_Q \end{vmatrix}\begin{vmatrix} i_q \\ i_Q \end{vmatrix} \tag{10-80}$$

This decoupling is shown in equivalent circuits in Fig. 10-11. It can be shown that:
Direct axis:

$$\begin{aligned} v_d &= -ri_d - l_d i_d - L_{ad}(i_d + i_F + i_D) \\ -v_F &= r_F i_F - l_f i_F - L_{ad}(i_d + i_F + i_D) \\ v_D &= -r_D i_D - l_{kD} i_D - L_{ad}(i_d + i_F + i_D) \end{aligned} \tag{10-81}$$

where l_d, l_f, l_D = self-inductance in armature field and damper circuits and, L_{ad} is defined as:

$$L_{ad} = L_D - l_D = L_d - l_d = L_F - l_f = kM_F = kM_D = M_R \tag{10-82}$$

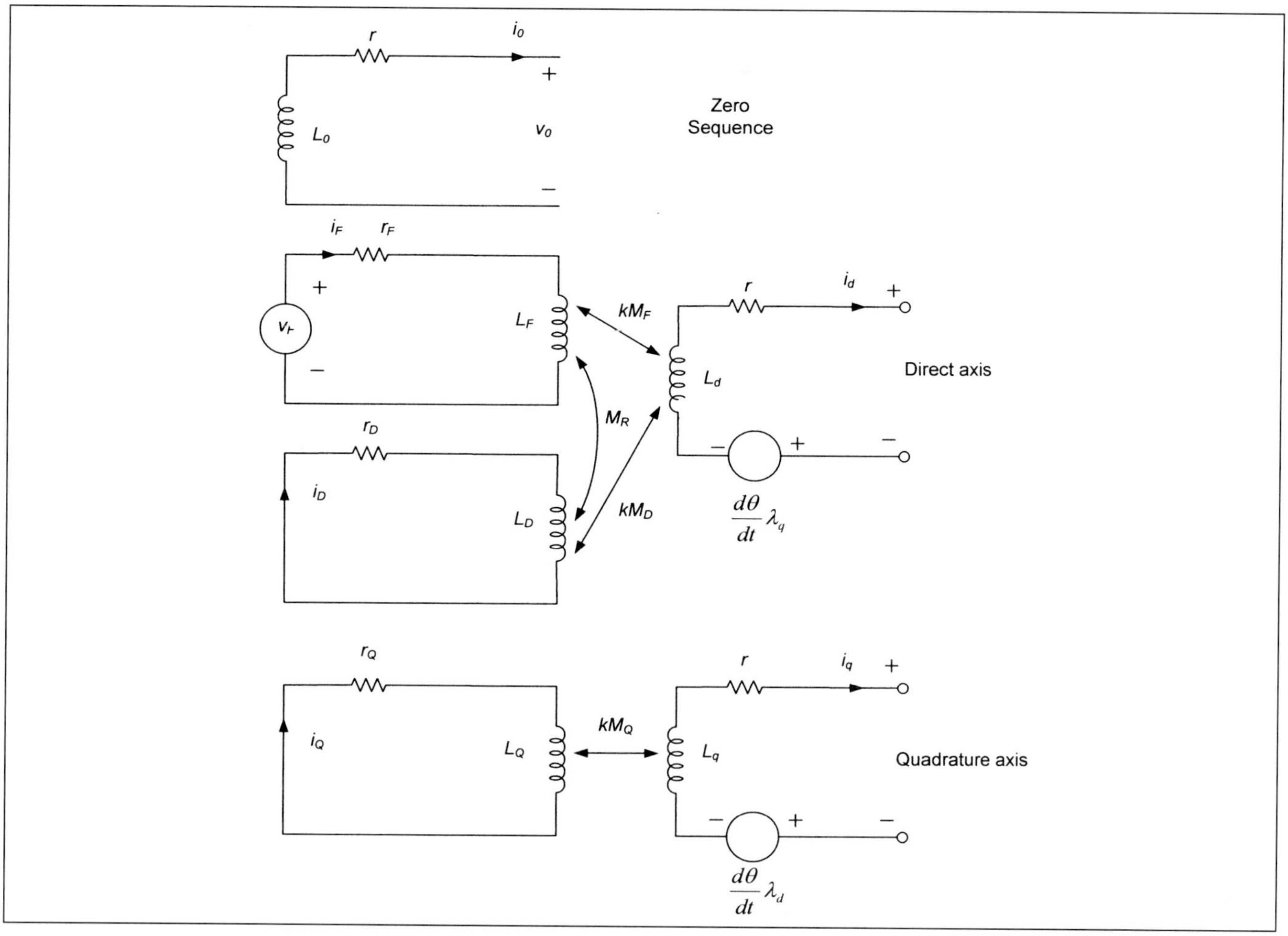

FIGURE 10-11 Decoupled circuits of a synchronous generator in $0dq$ axes.

Quadrature axis:

$$v_q = -r_q i_q - l_q \dot{i}_q - L_{aq}(\dot{i}_q + \dot{i}_Q) + \omega\lambda_d$$
$$v_Q = 0 = -r_Q i_Q - l_{kQ} \dot{i}_Q - L_{aq}(\dot{i}_q + \dot{i}_Q) \quad (10\text{-}83)$$

where:

$$L_{aq} = L_q - l_q = L_Q - l_{kQ} = kM_Q \quad (10\text{-}84)$$

Figure 10-12*a* and *b* depict *d* and *q* equivalent circuits. Figure 10-12*c* and *d* are commonly used simplified circuits, which do not show voltage-speed terms. These are adequate to derive λ_d and λ_q, including their derivatives.

10-10 CALCULATION PROCEDURE AND EXAMPLES

There are three steps involved:

1. The problem is normally defined in stator parameters, which are of interest. These are transformed into $0dq$ axes variables.

2. The problem is solved in $0dq$ axes parameters, using Park's transform or other means.

3. The results are transformed back to abc variables of interest, using inverse Park's transformation.

These three steps are inherent in any calculation using transformations. For simpler problems, it may be advantageous to solve directly in stator parameters. Figure 10-13 illustrates two-way transformations.

Example 10-2 Calculate time variation of direct-axis, quadrature-axis voltages and field current, when a step function of field voltage is suddenly applied to a generator at no load. Neglect damper circuits.

As the generator is operating without load, $i_{abc} = i_{0dq} = 0$. Therefore, from Eqs. (10-78)–(10-80):

$$\lambda_0 = 0 \qquad \lambda_d = kM_F i_F \qquad \lambda_F = L_F i_F \qquad \lambda_q = 0$$

From Eqs. (10-75)–(10-77):

$$v_0 = 0$$
$$v_d = -\frac{d\lambda_d}{dt} = -kM_F\frac{di_F}{dt}$$
$$v_F = r_F i_F + L_F\frac{di_F}{dt}$$
$$v_q = \omega_0\lambda_d = \omega_0 kM_F i_F$$

Therefore, as expected, the time variation of field current is:

$$i_F = \frac{v_F}{r_F}(1 - e^{(-r_F/L_F)t})$$

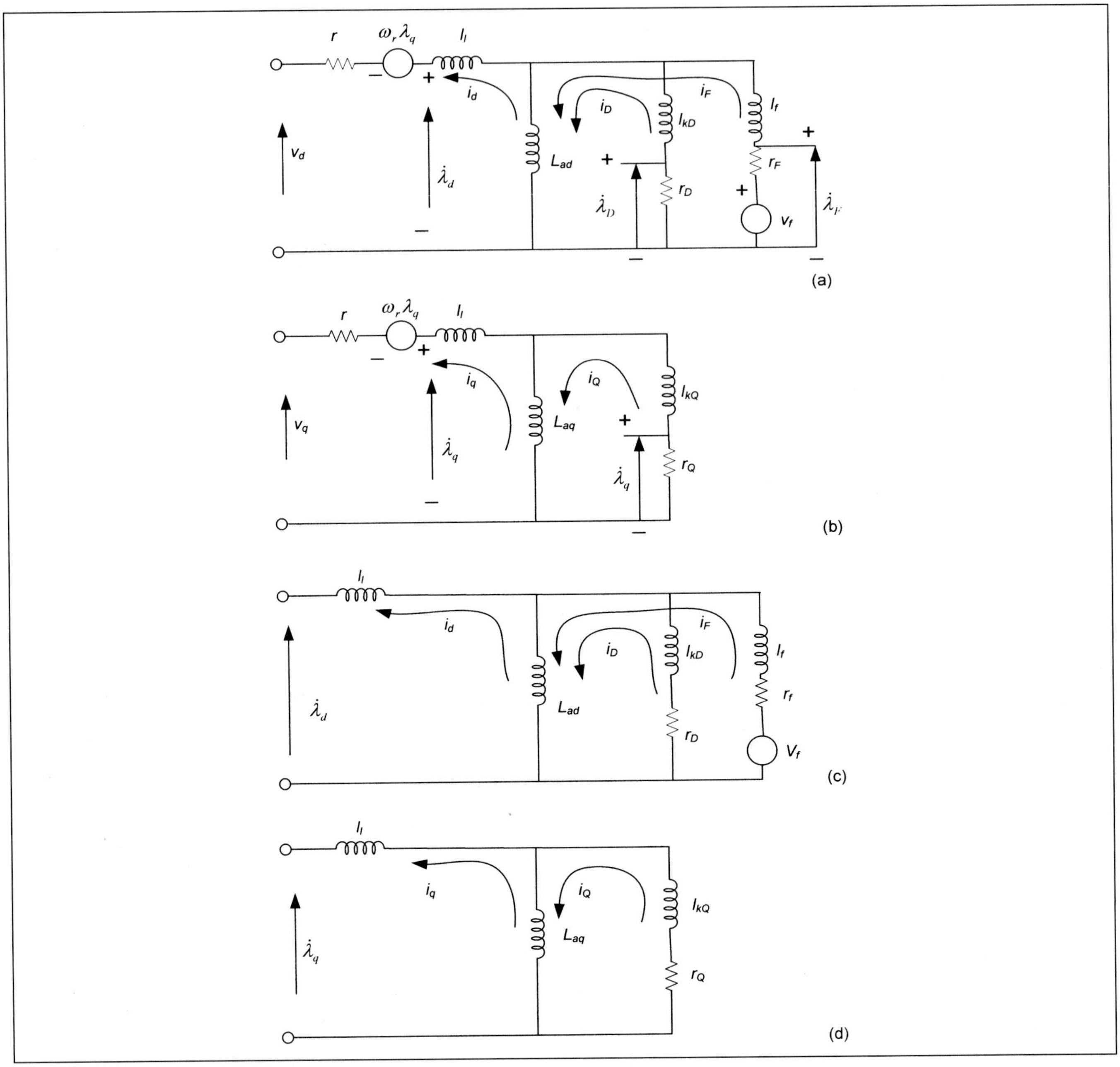

FIGURE 10-12 (*a*) and (*b*) Commonly used *dq* axes circuits of a synchronous generator, terminal voltage relations. (*c*) and (*d*) Flux relations.

The direct-axis and quadrature-axis voltages are given by:

$$v_d = -\frac{kM_F v_F}{L_F} e^{-(r_F/L_F)t}$$

$$v_q = -\frac{\omega k M_F v_F}{r_f}(1 - e^{-(r_F/L_F)t})$$

The phase voltages can be calculated using Park's transformation.

Example 10-3 A generator is operating with balanced positive sequence voltage of $v_a = \sqrt{2}\,|V|\cos(\omega_0 t + <V)$ across its terminals. The generator rotor is described by:

$$\theta = \omega_1 t + \frac{\pi}{2} + \delta$$

Find v_0, v_d, and v_q.

This is a simple case of transformation using Eq. (10-57):

$$\begin{vmatrix} v_0 \\ v_d \\ v_q \end{vmatrix} = \frac{2|V|}{\sqrt{3}} \begin{Vmatrix} \frac{1}{\sqrt{2}} & \frac{1}{\sqrt{2}} & \frac{1}{\sqrt{2}} \\ \cos\theta & \cos\left(\theta - \frac{2\pi}{3}\right) & \cos\left(\theta + \frac{2\pi}{3}\right) \\ \sin\theta & \sin\left(\theta - \frac{2\pi}{3}\right) & \sin\left(\theta + \frac{2\pi}{3}\right) \end{Vmatrix}$$

$$\begin{vmatrix} \cos(\omega_0 t + <V) \\ \cos\left(\omega_0 t + <V - \frac{2\pi}{3}\right) \\ \cos\left(\omega_0 t + <V - \frac{4\pi}{3}\right) \end{vmatrix}$$

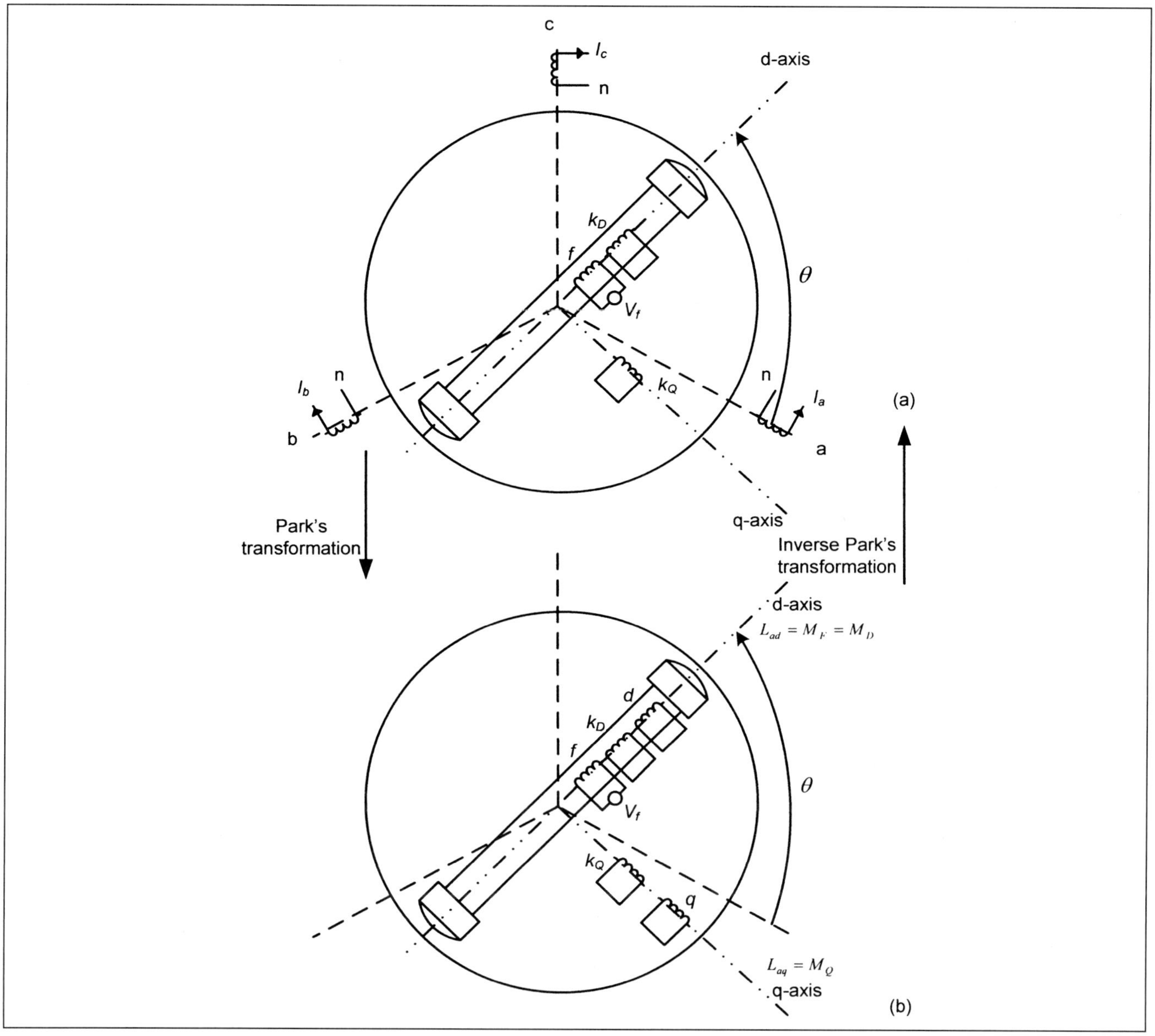

FIGURE 10-13 Park' s transformation and inverse Park's transformation of circuits of a synchronous generator.

A solution of this equation gives:

$$v_d = \sqrt{3}\,|V|\sin[(\omega_0 - \omega_1)t + \angle V - \delta]$$
$$v_q = \sqrt{3}\,|V|\cos[(\omega_0 - \omega_1)t + \angle V - \delta] \tag{10-85}$$

These relations apply equally well to derivation of i_d, i_q, λ_d, and λ_q. For synchronous operation, $\omega_1 = \omega_0$, and the equations reduce to:

$$v_d = \sqrt{3}\,|V|\sin[\angle V - \delta]$$
$$v_q = \sqrt{3}\,|V|\cos[\angle V - \delta] \tag{10-86}$$

Note that v_q and v_d are now constant and do not have the slip frequency term $\omega_1 - \omega_0$.

We can write:

$$v_q + jv_d = \sqrt{3}\,|V|\,e^{j(\angle V - \delta)} = \sqrt{3}V_a e^{-j\delta} \tag{10-87}$$

Therefore, V_a can be written as:

$$V_a = \left(\frac{v_q}{\sqrt{3}} + j\frac{v_d}{\sqrt{3}}\right)e^{j\delta} = (V_q + jV_d)\,e^{j\delta} \tag{10-88}$$

where:

$$V_q = v_q/\sqrt{3}, \qquad V_d = v_d/\sqrt{3} \tag{10-89}$$

This is shown in the phasor diagram of Fig. 10-14. We can write these equations in the following form:

$$\begin{vmatrix} \operatorname{Re} V_a \\ \operatorname{Im} V_a \end{vmatrix} = \begin{vmatrix} \cos\delta & -\sin\delta \\ \sin\delta & \cos\delta \end{vmatrix} \begin{vmatrix} V_q \\ V_d \end{vmatrix} \tag{10-90}$$

$$\begin{vmatrix} V_q \\ V_d \end{vmatrix} = \begin{vmatrix} \cos\delta & \sin\delta \\ -\sin\delta & \cos\delta \end{vmatrix} \begin{vmatrix} \operatorname{Re} V_a \\ \operatorname{Im} V_a \end{vmatrix} \tag{10-91}$$

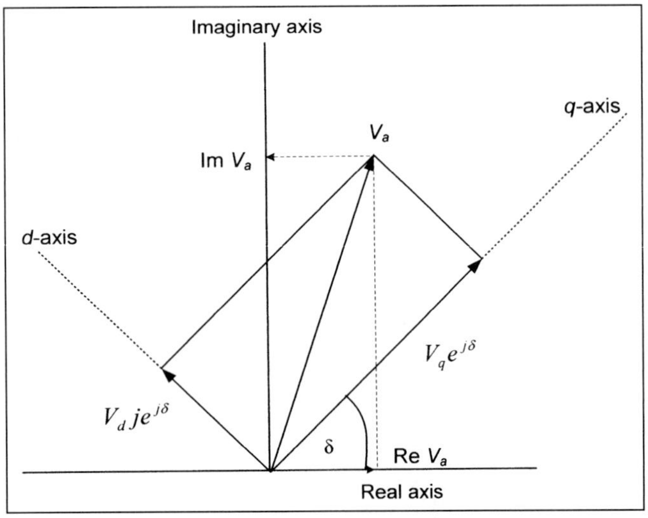

FIGURE 10-14 Phasor diagram illustrates relationship of *dq* axes voltage to terminal voltage.

10-11 STEADY-STATE MODEL OF SYNCHRONOUS GENERATOR

Based on previous equations, we can derive a steady-state model of a synchronous generator and its phasor diagram.

In the steady state, all the currents and flux linkages are constant. Also $i_0 = 0$ and rotor damper currents are zero. The Eqs. (10-78)–(10-80) reduce to:

$$v_d = -ri_d - \omega_0\lambda_q$$
$$v_q = -ri_q + \omega_0\lambda_d \tag{10-92}$$
$$v_F = r_F i_F$$

where:

$$\lambda_d = L_d i_d + kM_F i_F$$
$$\lambda_F = kM_F i_d + L_F i_F \tag{10-93}$$
$$\lambda_q = L_q i_q$$

Substitute values of λ_d and λ_q, and then from Eqs. (10-88) and (10-89), we can write the following equation:

$$V_a = -r(I_q + jI_d)e^{j\delta} + \omega_0 L_d I_d e^{j\delta} - j\omega_0 L_q I_q e^{j\delta} + \frac{1}{\sqrt{2}}\omega_0 M_F i_F e^{j\delta} \tag{10-94}$$

where:

$$i_d = \sqrt{3}I_d, \quad i_q = \sqrt{3}I_q \tag{10-95}$$

Define:

$$\sqrt{2}E = \omega_0 M_F i_F e^{j\delta} \tag{10-96}$$

This is the no-load voltage or the open-circuit voltage with generator current = 0. Then, we can write:

$$E = V_a + rI_a + jX_d I_d e^{j\delta} + jX_q I_q e^{j\delta} \tag{10-97}$$

The phasor diagram is shown in Fig. 10-15*a*. The open-circuit voltage on no load is a *q*-axis quantity and is equal to the terminal voltage.

To facilitate the location of *q*-axis, Fig. 10-15*b* is drawn. The terminal voltage, current, and power factor of the generator will be normally known. As the components I_d and I_q are not initially known, the phasor diagram is constructed by first laying out the terminal voltage V_a and line current I_a at correct phase angle ϕ, then adding the resistance drop and reactance drop $I_a X_q$. At the end of this vector, quadrature axis is located. Now the current is resolved into direct-axis and quadrature-axis components. This enables construction of vectors $I_q X_q$ and $I_d X_d$.

Figure 10-15*c* is further elucidation of the construction. The relations of triangle BCD in Fig. 10-15*b* and its construction are clearly shown in this figure.

Example 10-4 Consider generator data of Example 10-1. Calculate the direct- and quadrature-axis components of the currents and voltages and machine voltage *E* when the generator is delivering its full-load-rated current at its rated voltage. Also calculate all the angles shown in phasor diagram Fig. 10-15*a*. If this generator is connected to an infinite bus through an impedance of 0.01 + *j*0.1 per unit (100-MVA base), what is the voltage of the infinite bus?

The generator operates at a power factor of 0.85 at its rated voltage of 1.0 per unit. Therefore, $\phi = 31.8°$. The generator full-load current is 4183.8 A = 1.0 per unit. The terminal voltage vector can be drawn to scale, *Ir* drop (= 0.0012 per unit) is added and vector $I_a X_q$ = 1.8 per unit drawn to locate the *q*-axis. Current *I* can be resolved into direct-axis and quadrature-axis components and the phasor diagram completed, as shown in Fig. 10-16, and the values of V_d, I_d, V_q, I_q, and *E* can be read from it. This requires that the phasor diagram is constructed to scale. The analytical solution is as follows:

The load current is resolved into active and reactive components, $I_r = 0.85$ per unit and $I_x = 0.527$ per unit, respectively.

Then, from the geometric construction shown in Fig. 10-16:

$$(\delta - \beta) = \tan^{-1}\left(\frac{X_q I_r + rI_x}{V_a + rI_r - X_q I_x}\right)$$
$$= \tan^{-1}\left(\frac{(1.8)(0.85)+(0.0012)(0.527)}{1+(0.0012)(0.85)+(1.8)(0.527)}\right) = 38.14°$$

Note that the resistance can even be ignored from the above calculation without an appreciable error. Thus, $\delta - \beta + \phi = 69.93°$. This is the angle of current vector with *q*-axis. Therefore:

$$I_q = I_a\cos(\delta - \beta - \phi) = 0.343 \text{ pu}, \quad i_q = 0.594 \text{ pu}$$
$$I_d = -I_a\sin(\delta - \beta - \phi) = -0.939 \text{ pu}, \quad i_d = -1.626 \text{ pu}$$

And

$$V_q = V_a\cos(\delta - \beta) = 0.786 \text{ pu}, \quad v_q = 1.361 \text{ pu}$$
$$V_d = -V_a\sin(\delta - \beta) = -0.618 \text{ pu}, \quad v_d = 1.070 \text{ pu}$$

The machine-generated voltage is:

$$E = V_q + rI_q - X_d I_d = 2.66 \text{ pu}$$

The infinite bus voltage is simply the generator terminal voltage less the *IZ* drop subtracted as a vector:

$$V_\infty = V_a\angle 0° - I_a\angle 31.8° Z\angle 84.3° = 0.94\angle -4.8°$$

The infinite bus voltage lags the generator voltage by 4.8°. Practically, the infinite bus voltage will be held constant and the generator voltage changes, depending on the system impedance and generator output.

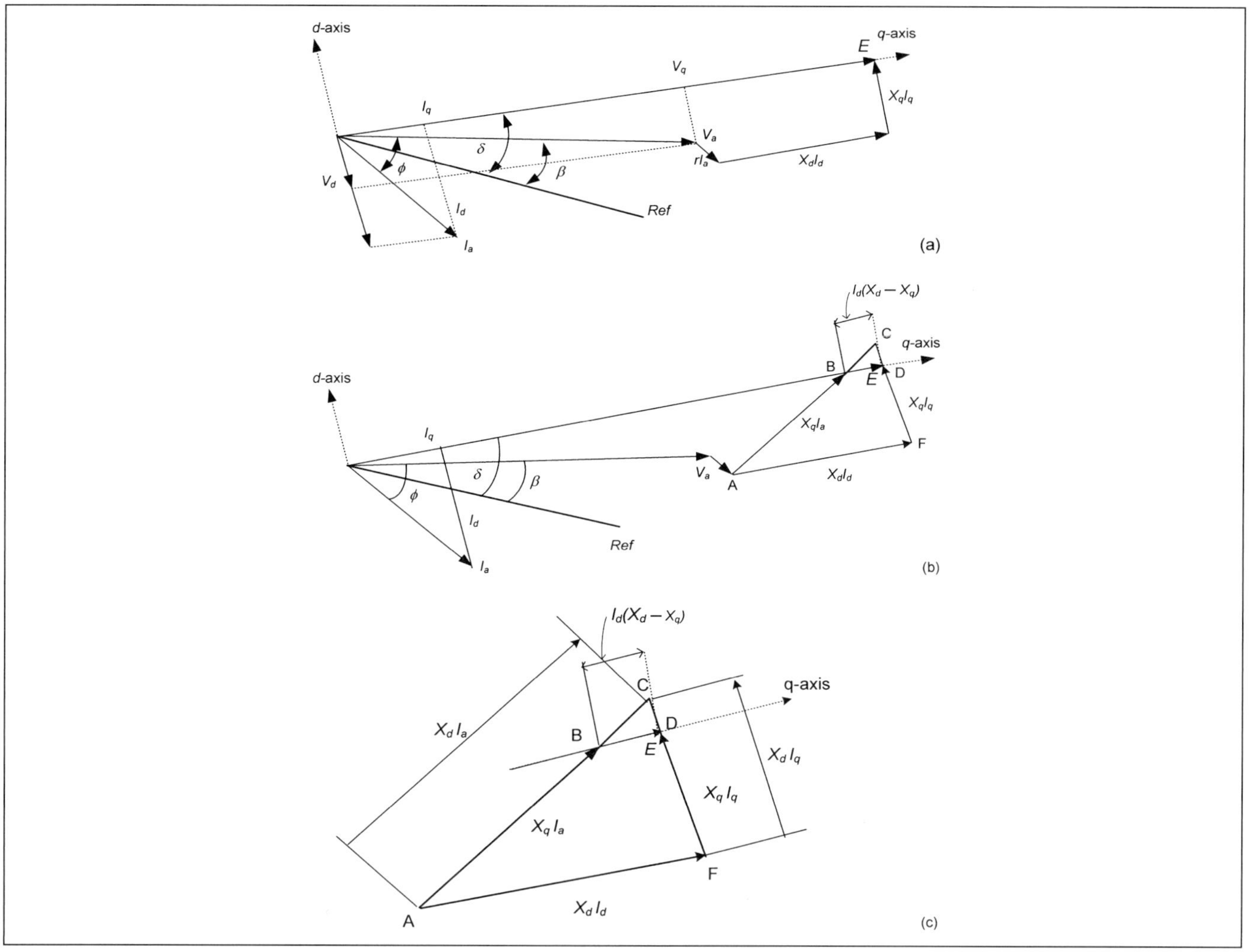

FIGURE 10-15 (*a*) Phasor diagram of a synchronous generator in steady state. (*b*) An alternate form of the phasor diagram. (*c*) Details of construction, and location of *q*-axis in phasor diagram of Fig. 10-15*b*.

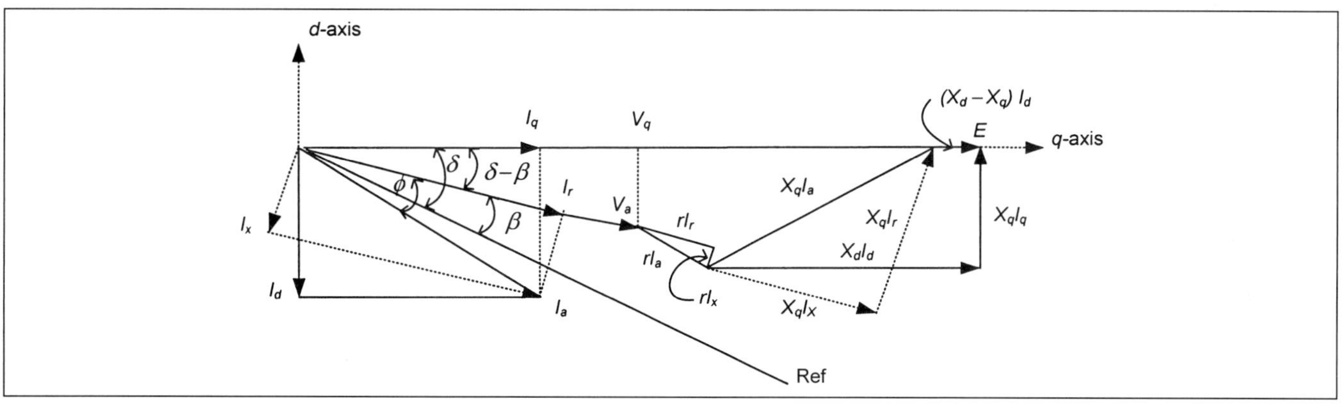

FIGURE 10-16 Phasor diagram for solution of Example 10-4.

10-12 SYMMETRICAL SHORT CIRCUIT OF A GENERATOR AT NO LOAD

The equations for the terminal three-phase short circuit of a generator at no load are important, in the sense that ANSI/IEEE calculations in [Ref. 3] are based on it. The ANSI/IEEE calculations for calculating of short-circuit duties on switching devices do not consider prior loading. We will ignore damper circuit and resistances and neglect the change in speed during short circuit.

Ignoring the damper circuit means that the subtransient effects are ignored. As the generator is operating at no load, prior to the fault, $i_{abc} = i_{0dq} = 0$.

Subsequent to fault instant, $v_{abc} = v_{0dq} = 0$. We will also ignore all resistances.

From Eq. (10-76) for $t \geq 0$.

Zero sequence:

$$v_0 = -L_0 \frac{di_0}{dt} = 0$$

Direct axis:

$$v_d = -\omega_0 \lambda_q - \frac{d\lambda_d}{dt} = 0$$

$$v_f = \frac{d\lambda_F}{dt}$$

Quadrature axis:

$$v_q = \omega_0 \lambda_d - \frac{d\lambda_q}{dt} = 0$$

The flux linkages can be expressed in terms of currents by using Eqs. (10-79) and (10-80):

$$\begin{aligned} &\omega_0 L_q i_q + L_d \frac{di_d}{dt} + kM_F \frac{di_F}{dt} = 0 \\ &v_f = kM_F \frac{di_d}{dt} + L_F \frac{di_F}{dt} \\ &-\omega_0 L_d i_d - \omega_0 kM_F i_F + L_q \frac{di_q}{dt} = 0 \end{aligned} \tag{10-98}$$

These equations can be solved using Laplace Transform. The initial conditions are:

$$i_d(0^-) = i_q(0^-) = 0$$

$$i_F(0^-) = i_F^0$$

Also from Eq. (10-96) we demonstrated that at no load, prior to fault, the terminal voltage is equal to the generated voltage, and this voltage is:

$$\sqrt{2}E_a = \omega_0 M_F i_F e^{j\delta}$$

and this is a quadrature axis quantity, also:

$$kM_F = L_{ad} = X_{ad}/\omega$$

$$L_F = (X_f + X_{ad})/\omega$$

$$\begin{vmatrix} sL_d & skM_F & \omega_0 L_q \\ skM_F & sL_F & 0 \\ -\omega_0 L_d & -\omega_0 kM_F & sL_q \end{vmatrix} \begin{vmatrix} i_d \\ i_F \\ i_q \end{vmatrix} = \begin{vmatrix} kM_F \\ L_F \\ 0 \end{vmatrix} i_F^0 \tag{10-99}$$

Solution of these equations, after simplification, will give:

$$i_d(t) = \frac{\sqrt{3}E_a(0)}{X'_d}(\cos\omega_0 t - 1) \tag{10-100}$$

$$i_q(t) = \frac{\sqrt{3}E_a(0)}{X_q}\sin\omega_0 t \tag{10-101}$$

$$i_F(t) = \left[\frac{X_d}{X'_d} - \left(\frac{X_d}{X'_d} - 1\right)\cos\omega_0 t\right] i_F^0 \tag{10-102}$$

Note that $k = \sqrt{3/2}$. Apply:

$$\bar{i}_{abc} = \bar{P}\bar{i}_{0dq}$$

With $\theta = \omega_0 t + \pi/2 + \delta$, the short-circuit current in phase a is:

$$i_a = \sqrt{2}\left|E\right|\left[\left(\frac{1}{X'_d}\right)\sin(\omega_0 t + \delta) - \frac{X_q - X'_d}{2X'_d X_q}\sin\delta - \frac{X_q - X'_d}{2X'_d X_q}\sin(2\omega t + \delta)\right] \tag{10-103}$$

The first term is normal-frequency short-circuit current, the second term is dc or unidirectional term, and the third term is double-frequency short-circuit current. The 120-Hz component imparts a nonsinusoidal characteristic to short-circuit current waveform. It rapidly decays to zero and is ignored in the calculation of short-circuit currents.

When damper winding circuit is considered, the short-circuit current can be expressed as:

$$i_a = \sqrt{2}E\begin{bmatrix} \left(\frac{1}{X_d}\right)\sin(\omega t + \delta) + \left(\frac{1}{X'_d} - \frac{1}{X_d}\right)e^{-t/T'_d}\sin(\omega t + \delta) \\ + \left(\frac{1}{X''_d} - \frac{1}{X'_d}\right)e^{-t/T''_d}\sin(\omega t + \delta) - \frac{(X''_d + X''_q)}{2X''_d X''_q}e^{-t/T_a}\sin\delta \\ - \frac{(X''_d - X''_q)}{2X''_d X''_q}e^{-t/T_a}\sin(2\omega t + \delta) \end{bmatrix} \tag{10-104}$$

- The first term is final steady-state short-circuit current.
- The second term is normal-frequency decaying transient current.
- The third term is normal-frequency decaying subtransient current.
- The forth term is asymmetric decaying dc current.
- The fifth term is double-frequency decaying current.

Example 10-5 Calculate the components of short-circuit currents at the instant of three-phase terminal short circuit of the generator, particulars as shown in Table 10-1. Assume that phase a is aligned with the field at the instant of short circuit, maximum asymmetry, that is $\delta = 0$. The generator is operating at no load prior to short circuit.

The calculations are performed by substituting the required numerical data from Table 10-1 in Eq. (10-104).

Steady-state current = 2.41 kA rms

Decaying transient current = 20.24 kA rms

Decaying subtransient current = 5.95 kA rms

Decaying dc component = 43.95 kA

Decaying second-harmonic component = 2.35 kA rms

Note that second-harmonic component is zero if the direct-axis and quadrature-axis subtransient reactances are equal. Also, dc component in this case is equal to 40.44 kA. These are the current values at the instant of short circuit.

10-13 MANUFACTURER'S DATA

The relation between the various inductances and the data commonly supplied by a manufacturer for a synchronous generator is not obvious. Following relations hold:

$$L_{ad} = L_d - l_l = kM_F = kM_D = M_R$$
$$L_{aq} = L_q - l_l = kM_Q \tag{10-105}$$

Here, l_l is the leakage inductance corresponding to X_l in Eq. (10-4). L_{ad} is the mutual inductance between the armature and rotor, which is = mutual inductance between field and rotor, which is = mutual inductance between damper and rotor. Similar relations are assumed in the q-axis. In some texts, different symbols are used. Field leakage reactance l_f is:

$$l_f = \frac{L_{ad}(L'_d - l_l)}{(L_d - L'_d)} \tag{10-106}$$

and

$$L_F = l_f + L_{ad} \tag{10-107}$$

The damper leakage reactance in the direct axis is:

$$L_{kD} = \frac{L_{ad} l_f (L''_d - l_l)}{L_{ad} l_f - L_F (L''_d - l_l)} \tag{10-108}$$

and

$$L_D = l_{kD} + L_{ad}$$

$$L_{kQ} = \frac{L_{aq}(L''_q - l_l)}{(L_q - L''_q)} \tag{10-109}$$

In the quadrature axis, the damper leakage reactance is:

$$L_Q = l_{kQ} + L_{aq} \tag{10-110}$$

The field resistance is:

$$r_F = \frac{L_F}{T'_{do}} \tag{10-111}$$

The damper resistances in direct axis can be obtained from:

$$T''_d = \frac{(L_D L_F - L^2_{ad})}{r_D L_F}\left(\frac{L''_d}{L'_d}\right) \tag{10-112}$$

and in the quadrature axis:

$$T''_q = \left(\frac{L''_q}{L_q}\right)\left(\frac{L_Q}{r_Q}\right) \tag{10-113}$$

In some texts, different symbols are used:

L_{ffd} = total field inductance = L_{fd} (field leakage inductance) + L_{afd} (mutual field inductance to d-axis).

This is identical to Eq. (10-107). Similarly, in the damper circuit ($L_{kkd} = L_{kd} + L_{akd}$; $L_{kkq} = L_{kq} + L_{akq}$), these are equivalent to Eqs. (10-109) and (10-110).

Example 10-6 Using the manufacturer's data given in Table 10-1, calculate the generator parameters in d-q axes.

Applying the equations derived above:

$L_{ad} = X_d - X_l = 1.949 - 0.164 = 1.785$ per unit $= kM_F = kM_D = M_R$

$L_{aq} = X_q - X_l = 1.858 - 0.164 = 1.694$ per unit $= kM_Q$

$l_f = (1.785)(0.278 - 0.164)/(1.964 - 0.278) = 0.121$ per unit

$L_F = 0.121 + 1.785 = 1.906$ per unit

$l_{kD} = (1.785)(0.121)(0.193 - 0.164)/\{(1.785)(0.164) - 1.096(0.193 - 0.164)\} = 0.026$ per unit

$L_D = 0.026 + 1.785 = 1.811$ per unit

$l_{kQ} = (1.694)(0.192 - 0.164)/(1.858 - 0.192) = 0.028$ per unit

$L_Q = 0.028 + 1.694 = 1.722$ per unit

$T'_{do} = 5.615$ s $= 2116.85$ rad

$r_F = 1.906/2116.85 = 1.005 \times 10^{-5}$ per unit

$$r_D = \frac{(1.811)(1.906) - 1.785^2}{(0.015)(377)(1.906)}\left(\frac{0.193}{0.278}\right) = 0.0131 \text{ per unit}$$

$$r_Q = \left(\frac{0.192}{1.858}\right)\left(\frac{1.722}{0.015 \times 377}\right) = 0.031 \text{ per unit}$$

Per unit system for synchronous generators is not straightforward, and variations in literature exist.[4]

10-14 INTERRUPTION OF CURRENTS WITH DELAYED CURRENT ZEROS

Large generators have high effective X/R ratios, and combined with other generator time constants and parameters, a current zero may not be obtained at contact parting of the circuit breaker, that is the dc component of the short-circuit current at the contact parting time is higher than the peak ac component. This is well documented in Refs. 5–7.

ANSI/IEEE Std. C37.010[7] cautions that the longer dc time constants can cause a problem with SF_6- type puffer circuit breakers. The interrupting window, which is the time difference between the minimum and maximum arcing times, may be exceeded because of delayed current zero, and arc energy and interruption window are of concern. The calculation methods described in this standard are qualified that the E/Z method of calculation with adjustments of ac and dc decrements can be used, provided the X/R does not exceed 45 at 60 Hz, that is, the dc time constant is not more than 120 ms, yet this qualification is mostly ignored in the industry.

IEEE Std. C37.013[8] for generator circuit breakers specifies that the the degree of asymmetry from generator source does not exceed 110 percent. It states that at the time of current interruption, the arc fault resistance will add to the generator armature resistance. This reduces the time constant of the dc component and forces it to decay faster:

$$T_a = \frac{X''_d}{2\pi f(R_a + R_{add})} \tag{10-114}$$

where T_a is the armature time constant, R_a is the armature resistance, R_{add} is the added arc resistance, X''_d is the subtransient reactance, and f is the system frequency. Figure 10-17 shows this effect on decay of the dc component and the current zero obtained at the contact parting time. However, the performance with arc fault resistance is difficult to simulate and demonstrate even in a test station.

When no current zero is obtained, the current interruption in this mode will be equivalent to that of interrupting a dc current without current zero crossing. The high-voltage circuit breakers have limited interrupting capability in this mode of operation, unless specifically designed to introduce resistance in the arc fault

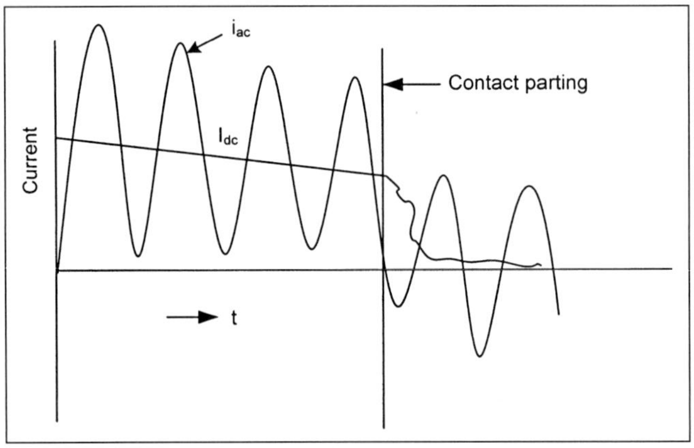

FIGURE 10-17 Illustrates that a current zero may be brought by arc resistance added at the time of current interruption, forcing a current zero.

path at current zero. Generator circuit breakers capable of interrupting with 130 percent asymmetry at the contact parting time are commercially available. Current technologies in some SF_6 breaker designs use arc rotation techniques to force a current zero. The vacuum interruption technology may also achieve the same results. The available continuous current rating and the interrupting symmetrical rating of generator breakers at the upper end is 50 kA and 250 kA, respectively.

IEC standards[9] do not discuss the asymmetry at the contact parting time of the breaker. The IEC standard showing the examples of short-circuit calculations, part 4, is yet to be published. IEC may adopt IEEE Std. C37.013[8] for the generator breakers.

Consider faults at locations F_1 and F_2 in Fig. 10-1. For a fault at F_2, there are three contributions of the short-circuit currents: (1) from the utility source, (2) from the auxiliary distribution system rotating loads through UAT (unit auxiliary transformer), (3) from the generator itself. However, the generator breaker sees only the component (3). Similarly for fault at F_1, the generator breaker sees the sum of the utility source and auxiliary distribution system short-circuit current contributions, but not the contribution from the generator itself. While selecting a generator breaker, the higher of these two fault currents at F_1 and F_2 is considered. Generally, the generator contribution for fault at F_2 gives rise to higher asymmetry than the fault at F_1 because large generators have a higher X/R ratio compared to X/R ratios in the utility systems.

Example 10-7 This example demonstrates the calculations and EMTP simulations of a terminal fault of a generator data, as shown in Table 10-2. Using these data and considering a 5-cycle breaker, with contact parting time of 3 cycles, consisting of $^1/_2$ cycle tripping delay and 2.5 cycles opening time, the calculated short-circuit currents, from Eq. (10-104) are:

- Close and latch = 112.2 kA peak
- Generator source ac symmetrical interrupting current = 30.9 kA rms
- Dc component = 59.22 kA
- Total rms asymmetrical interrupting current at contact parting = 66.80 kA
- Asymmetry factor = 135.5 percent and the current zero is not obtained.

TABLE 10-2 Generator Data for Example 10-7

DESCRIPTION	SYMBOL	DATA
Generator		
234 MVA, 2-pole, 18 kV, 0.85 pF, 198.9 MW, 7505 A, 60 Hz, 0.56 SCR, 350 field V, wye-connected, high impedance grounded through a distribution transformer, grounding current at 18 kV limited to 9A		
Per unit reactance data, direct axis		
Synchronous	X_{dv}	2.120
Transient	X'_{dv}	0.230
Subtransient	X''_{dv}	0.150
Saturated negative sequence	X_{2v}	0.150
Leakage reactance, overexcited	$X_{LM,OXE}$	0.135
Leakage reactance, underexcited	$X_{LM,UEX}$	0.150
Per unit reactance data, quadrature axis		
Synchronous	X_{qv}	1.858
Transient	X'_{qv}	0.434
Subtransient	X''_{qv}	0.140
Generator effective X/R	X/R	125
Field time constant data, direct axis		
Open circuit	T'_{do}	5.615
Three-phase short-circuit transient	T'_d	0.597
Short-circuit subtransient	T''_d	0.015
Open-circuit subtransient	T''_{do}	0.022
Field time constant data quadrature axis		
Open-circuit	T'_{qo}	0.451
Open-circuit subtransient	T''_{qo}	0.046

The asymmetry factor α is given by:

$$\alpha = \frac{\text{dc component}}{\sqrt{2}\ \text{symmetrical interrupting current}} \quad (10\text{-}115)$$

And the total asymmetrical interrupting current is given by:

$$I_{\text{total,asym}} = \sqrt{(\text{ac}_{\text{sym}})^2 + (\text{dc})^2} \quad (10\text{-}116)$$

An important parameter of calculation is the X/R ratio. The effective resistance of the generator used in the short-circuit calculations is calculated from the following expression:[3]

$$R_G = \frac{X_{2v}}{2\pi f T_a} \quad (10\text{-}117)$$

where R_G is the generator effective resistance, and all the symbols have been described in Table 10-2. Using appropriate values from Table 10-2, this gives an X/R of 125, which correlates with the data supplied by the manufacturer.

There are analytical and conceptual differences between the ANSI/IEEE methods of short circuit calculations and IEC,[10–13] which are not discussed here.

EMTP uses Park's transformations and calculates the transformed parameters based on the input manufacturer's data. It can also accept the transformed parameters in $0dq$ axes directly. Also, the system inertia constant and mechanical damping have been modeled.

The three-phase short-circuit current profile is shown in Fig. 10-18, for phases *a*, *b*, and *c*. It is seen that in phase *c*, current zero is not obtained for a number of cycles. The comparative results of calculations are shown in Table 10-3.

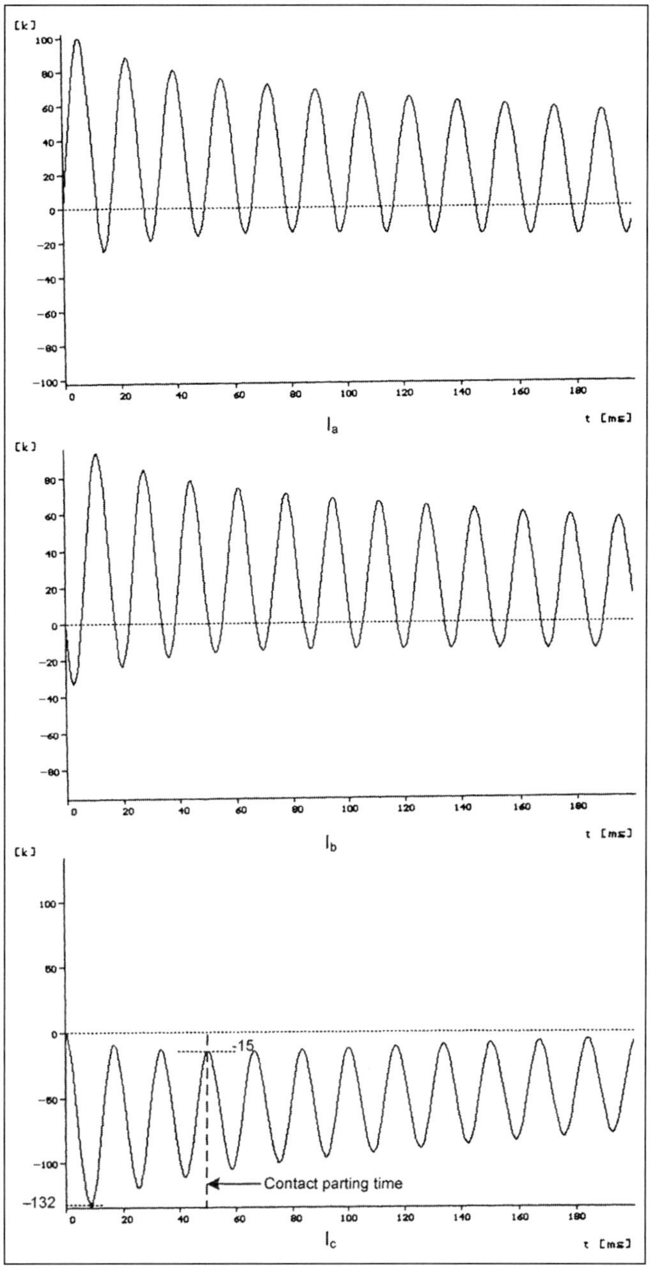

FIGURE 10-18 EMTP simulation of transients for a terminal fault of generator, Example 10-7. Current zero does not occur for a number of cycles in phase *c* as current interruption.

10-14-1 The Effect of Power Factor

The load power factor (lagging) in IEC equations (not discussed here) does not change the asymmetry at the contact parting time. However, the asymmetry *does* change with the power factor and prior load. This is clearly shown in Fig. 10-19*b*, with generator absorbing reactive power, that is, operating at leading power factor of 0.29. Generator load prior to short circuit = 28 MW, 92.4 Mvar. It is seen that the current zeros are further delayed, as compared to generator operating at no load. Fig. 10-19*a* shows the short-circuit current profile in phase *c* at no load, while Fig. 10-19*b* shows it with leading load. The prior leading power factor loading further delays the occurrence of current zeros.

Thus, there are differences in the calculations using the same data.

The example of calculation demonstrates that asymmetry at contact parting time can be even 130 percent or more. The delayed current zeros can also occur on short circuits in large industrial systems, with cogeneration facilities, as shown in Fig. 10-2. In this figure, the generator parameters are:

$$X_d'' = 0.162,\ X_d' = 0.223,\ X_d = 2.01,\ X_q''$$
$$= 0.159,\ T_d'' = 0.015,\ T_d' = 0.638,\ T_a = 0.476$$

All reactances are in per unit on generator MVA base of 81.82 and all time constants are in seconds. This gives an asymmetry factor of 132 percent at the contact parting time, 5-cycle symmetrical rated breaker.

The opening of a circuit breaker can be delayed till the current zeros occur. This can also lower the short-circuit duty on the circuit breaker.[14] However, stability and increased fault damage is of concern, as discussed in Chaps. 12 and 13. A circuit breaker capable of interrupting with higher asymmetry should normally be applied, and generator circuit breakers to meet this requirement are commercially available.

See also Example 16-4 which shows delayed current zeros due to out-of-phase synchronization.

10-15 SYNCHRONOUS GENERATOR ON INFINITE BUS

Figure 10-20 shows a synchronous generator with load connected to an infinite bus, with local load represented as a shunt and finite-series impedance. If we ignore damper windings and amortisseur effect, the model is simplified, but it cannot be used for transient stability analysis, following a large disturbance in the system. We consider only small perturbations around the original operating point; the system parameters are assumed unchanged and the system can be linearized.

TABLE 10-3 Comparison of Calculations Using IEEE/IEC Standards and EMTP-Simulations

CALCULATED PARAMETER	IEEE	IEC	EMTP
Close and latch, kA peak (IEC peak short-circuit current)	112.2	131.60	132.05
Generator source interrupting kA sym. RMS (IEC symmetrical breaking current i_{bsym}.)	30.90	38.50	33.59
DC component, kA	59.22	60.73	62.50
Total asymmetrical, kA RMS (IEC i_{basym})	66.80	71.90	70.90
Asymmetry factor	135%	112%	131%

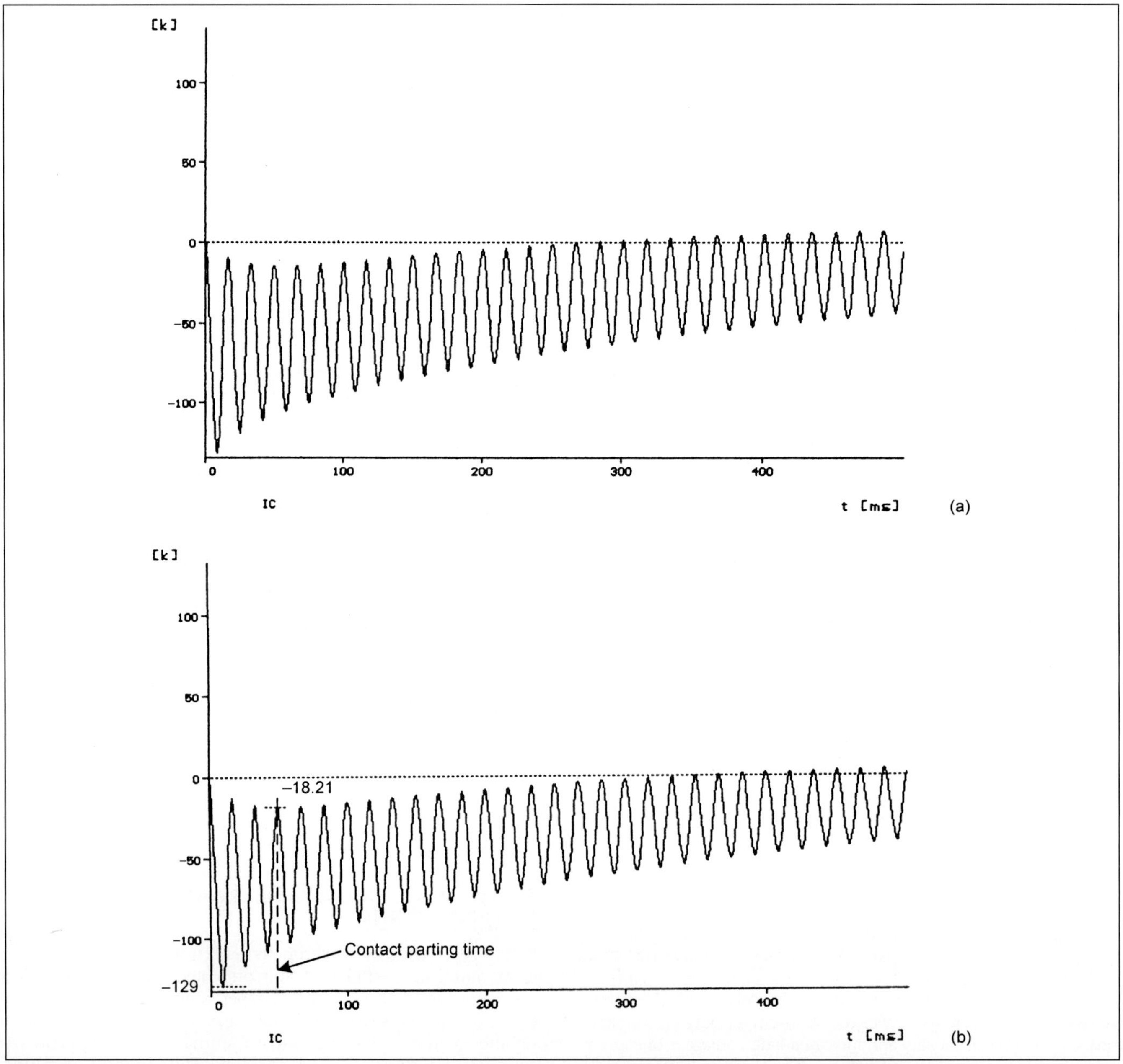

FIGURE 10-19 Comparison of the simulated transient in (*a*) and (*b*) shows that when a generator is operating at leading power factor (*b*), the current zeros are further delayed.

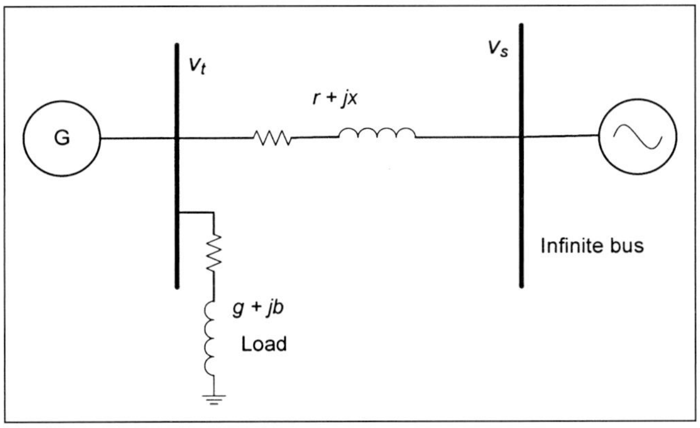

FIGURE 10-20 A generator connected to an infinite bus through an impedance and also supplying load connected to its terminals.

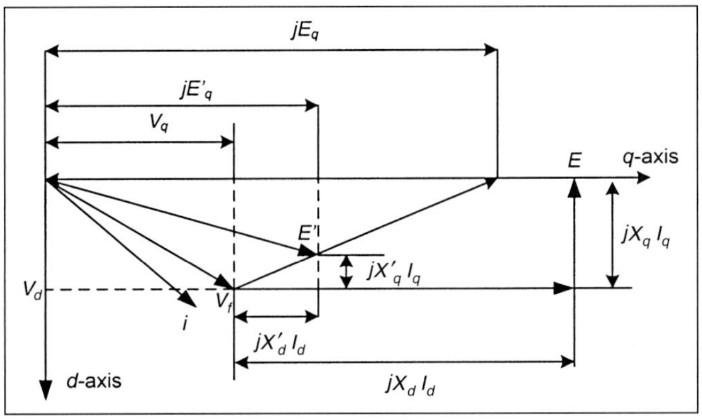

FIGURE 10-21 A simplified phasor diagram for development of small-signal model of a generator.

From Fig. 10-20, and the phasor diagram (Fig. 10-21) of the generator connected to infinite bus, we have:

$$I = YV_t + (V_t - V_s)/Z$$
$$ZI = ZYV_t + (V_t - V_s) \tag{10-118}$$

or

$$(R + jX)(I_d + jI_q) = (R + jX)(G + jB)(V_d + jV_q) - |V_s|(\sin\delta + j\cos\delta) \tag{10-119}$$

This can be written as:

$$\begin{vmatrix} R & -X \\ X & R \end{vmatrix}\begin{vmatrix} I_d \\ I_q \end{vmatrix} = \begin{vmatrix} 1+GR-BX & -(GX+BR) \\ GX+BR & 1+GR-BX \end{vmatrix}\begin{vmatrix} V_d \\ V_q \end{vmatrix} - |V_s|\begin{vmatrix} \sin\delta \\ \cos\delta \end{vmatrix} \tag{10-120}$$

We know that:

$$V_d = X_q I_q \qquad V_q = E'_q - X'_d I_d \tag{10-121}$$

Or in matrix form:

$$\begin{vmatrix} V_d \\ V_q \end{vmatrix} = \begin{vmatrix} 0 & X_q \\ -X'_d & 0 \end{vmatrix}\begin{vmatrix} I_d \\ I_q \end{vmatrix} + \begin{vmatrix} 0 \\ 1 \end{vmatrix}|E'_q| \tag{10-122}$$

Eliminating V_d and V_q, we get:

$$\begin{vmatrix} I_d \\ I_q \end{vmatrix} = \frac{1}{R_1R_2 - X_1X_2}\begin{vmatrix} R_2 & -X_2 \\ -X_1 & R_1 \end{vmatrix} \left\{\begin{vmatrix} -(GX+BR) \\ 1+RG-BX \end{vmatrix}|E'_q| - |V_s|\begin{vmatrix} \sin\delta \\ \cos\delta \end{vmatrix}\right\} \tag{10-123}$$

where:

$$\begin{aligned} R_1 &= R - (GX+BR)X'_d \\ R_2 &= R - (GX+BR)X_q \\ X_1 &= X + (1+RG-BX)X'_d \\ X_2 &= X + (1+RG-BX)X_q \end{aligned} \tag{10-124}$$

Because we are considering small disturbances, the equations can be linearized:

$$\begin{aligned} I_d &= I_{do} + \Delta I_d \qquad I_q = I_{qo} + \Delta I_q \\ E'_q &= E'_{qo} + \Delta E'_q \qquad \delta = \delta_o + \Delta\delta \end{aligned} \tag{10-125}$$

After some manipulations the equations are reduced to:

$$\begin{vmatrix} \Delta I_d \\ \Delta I_q \end{vmatrix} = \begin{vmatrix} Y_d \\ Y_q \end{vmatrix}|\Delta E'_q| + \begin{vmatrix} F_d \\ F_q \end{vmatrix}|\Delta\delta| \tag{10-126}$$

where:

$$\begin{aligned} Y_d &\cong [-R_2(BR+GX) - X_2(1+GR-BX)]/(R_1R_2 - X_1X_2) \\ Y_q &\cong [X_1(BR+GX) - R_1(1+GR-BX)]/(R_1R_2 - X_1X_2) \\ F_d &\cong -|V_s|(R_2\cos\delta_0 + X_2\sin\delta_0)/(R_1R_2 - X_1X_2) \\ F_q &\cong -|V_s|(X_1\cos\delta_0 + R_1\sin\delta_0)/(R_1R_2 - X_1X_2) \end{aligned} \tag{10-127}$$

Field circuit, neglecting dampers:

$$\begin{aligned} v_f &= r_F i_F + \frac{d\lambda_F}{dt} \\ \lambda_F &= L_F i_F + kM_F i_d \end{aligned} \tag{10-128}$$

Then, the v_f can be written as:

$$v_f = r_F\left[\left(\lambda_f/L_F + (L_F/r_F)\left(\frac{d\lambda_F}{dt}/L_F\right)\right) - kM_F i_d/L_F\right] \tag{10-129}$$

Also:

$$\sqrt{3}E_{FD} \cong \sqrt{3}\omega_0 M_F v_f/\sqrt{2}r_F$$

and (10-130)

$$\sqrt{3}E'_q \cong \sqrt{\frac{3}{2}}\omega_0 M_F \lambda_F/L_F$$

From these equations:

$$E_{FD} = [E'_q(1 + sT'_{do})] + (X_d - X'_d)I_d \tag{10-131}$$

where:

$$T'_{do} = \frac{L_F}{r_f} \qquad L'_d = L_d - \frac{L^2_{ad}}{L_F} \qquad \omega_0 L_d = X_d \qquad \omega_0 L'_d = X'_d \tag{10-132}$$

Linearizing Eq. (10-131):

$$\Delta E_{\mathrm{FD}} = \Delta E_q'(1+sT_{do}') + (X_d - X_d')\Delta I_d$$
$$= [1+(X_d - X_d')Y_d + sT_{do}']\Delta E_q' + (X_d - X_d')F_d\Delta\delta \quad (10\text{-}133)$$

This can be written as:

$$\Delta E_q' = (\Delta E_{\mathrm{FD}} - K_4\Delta\delta)K_3/(1+sT_{do}'K_3) \quad (10\text{-}134)$$

where:

$$K_3 = 1/[1+(X_d - X_d')Y_d]$$
$$K_4 = (X_d - X_d')F_d \quad (10\text{-}135)$$

Voltage equation:

$$|V_t|^2 = |V_d|^2 + |V_q|^2 \quad (10\text{-}136)$$

By linearizing:

$$2|V_{t0}|\Delta V_t = 2V_{d0}\Delta V_d + 2V_{q0}\Delta V_q \quad (10\text{-}137)$$

We know that:

$$\Delta V_d = X_q\Delta I_q = X_qY_q\Delta E_q + X_qF_q\Delta\delta$$
$$\Delta V_q = \Delta E_q' - X_d'\Delta I_d = \Delta E_q' - X_d'(Y_d\Delta E_q' + F_d\Delta\delta) \quad (10\text{-}138)$$

Substituting these equations into Eq. (10-137):

$$\Delta V_t = \frac{V_{d0}[X_q(Y_q\Delta E_q' + F_q\Delta\delta)] + v_{q0}[\Delta E_q' - X_d'(Y_d\Delta E_q' + F_d\Delta\delta)]}{|V_{t0}|} \quad (10\text{-}139)$$

This can be written as:

$$\Delta V_t = K_5\Delta\delta + K_6\Delta E_q' \quad (10\text{-}140)$$

where:

$$K_5 = \frac{V_{d0}X_qF_q - V_{q0}X_d'F_d}{|V_{t0}|}$$
$$K_6 = \frac{V_{d0}X_qY_q + V_{q0}(1-X_d'Y_d)}{|V_{t0}|} \quad (10\text{-}141)$$

Torque equation:

$$T_e = (V_dI_d + V_qI_q)/\omega_0 \quad (\omega_0 = 1 \text{ per unit})$$
$$= X_qI_qI_d + (E_q' - X_d'I_d)I_q \quad (10\text{-}142)$$

By linearizing:

$$\Delta T_e = X_q(I_{q0}\Delta I_d + I_{d0}\Delta I_q)$$
$$+ \Delta I_q(E_{do}' - X_d'I_{do}) + I_{q0}(\Delta E_q' - X_d'\Delta I_d) \quad (10\text{-}143)$$

By some manipulations, this can be written as:

$$\Delta T_e = K_1\Delta\delta + K_2\Delta E_q' \quad (10\text{-}144)$$

where:

$$K_1 = (X_q - X_d')(I_{q0}F_d + I_{d0}F_q) + E_{q0}'F_q$$
$$K_2 = I_{q0} + (X_q - X_d')(I_{q0}Y_d + I_{d0}Y_q) + E_{q0}'Y_q \quad (10\text{-}145)$$

Based on the above equations, the block diagram of synchronous machine is shown in Fig. 10-22. Consider IEEE type 1

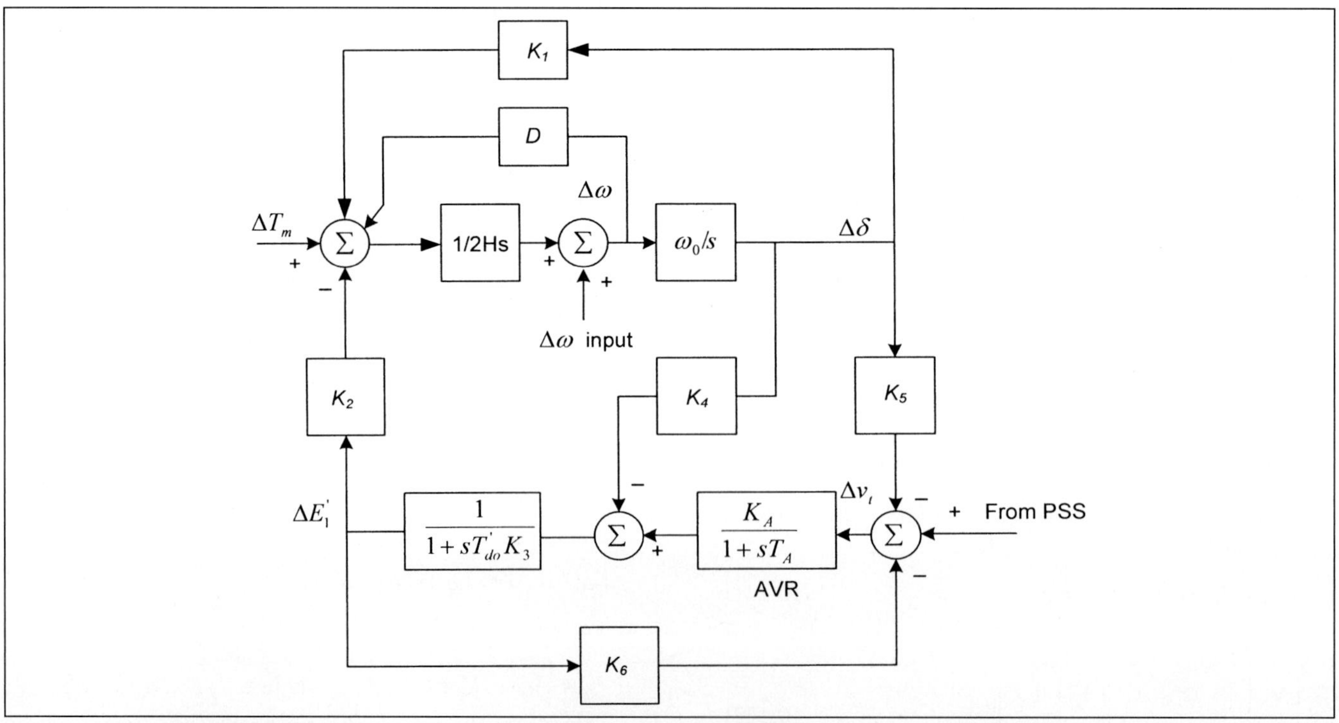

FIGURE 10-22 Control circuit block diagram of a generator, small-signal perturbation model, with AVR and input from PSS.

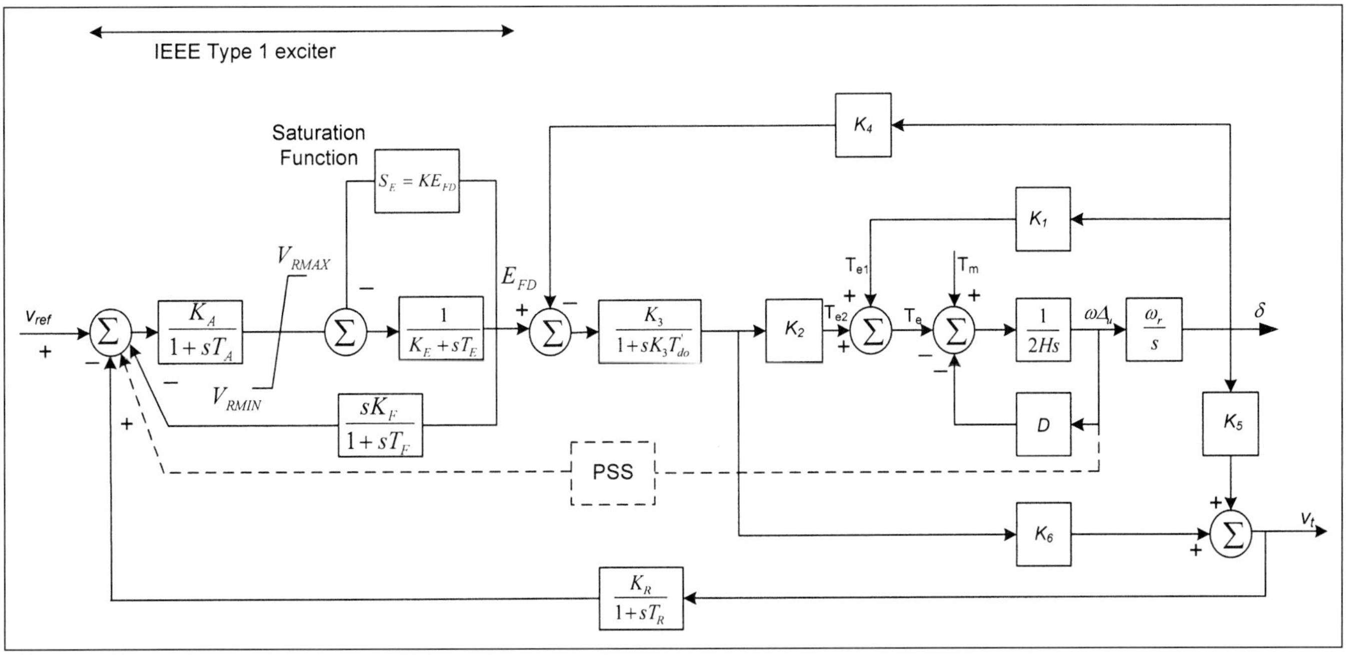

FIGURE 10-23 Control circuit block diagram, based on Fig. 10-22, with added control circuit block diagram of excitation system.

excitation system[15] and power system stabilizer (Chaps. 12 and 13), and incorporation of these makes the total block diagram of Fig. 10-23.

Example 10-8 A generator of specifications as in Table 10-1 is connected through a step-up transformer, 13.8–138 kV, delta-wye connected, of 10 percent impedance and operated in isolation to supply a 90 MW load at 0.85 power factor.

The load is suddenly dropped at 10 ms. The resulting transients are shown in Figs. 10-24*a* through *e*. These show generator terminal voltage, phases abc currents, currents in *dq* axes, generator active power transients, and currents in the damper circuits, respectively. The simulation time is 200 ms, and all transients have decaying trends.

No voltage regulator or excitation system is modeled. Fast excitation systems can even accentuate the initial transients (Chap. 13).

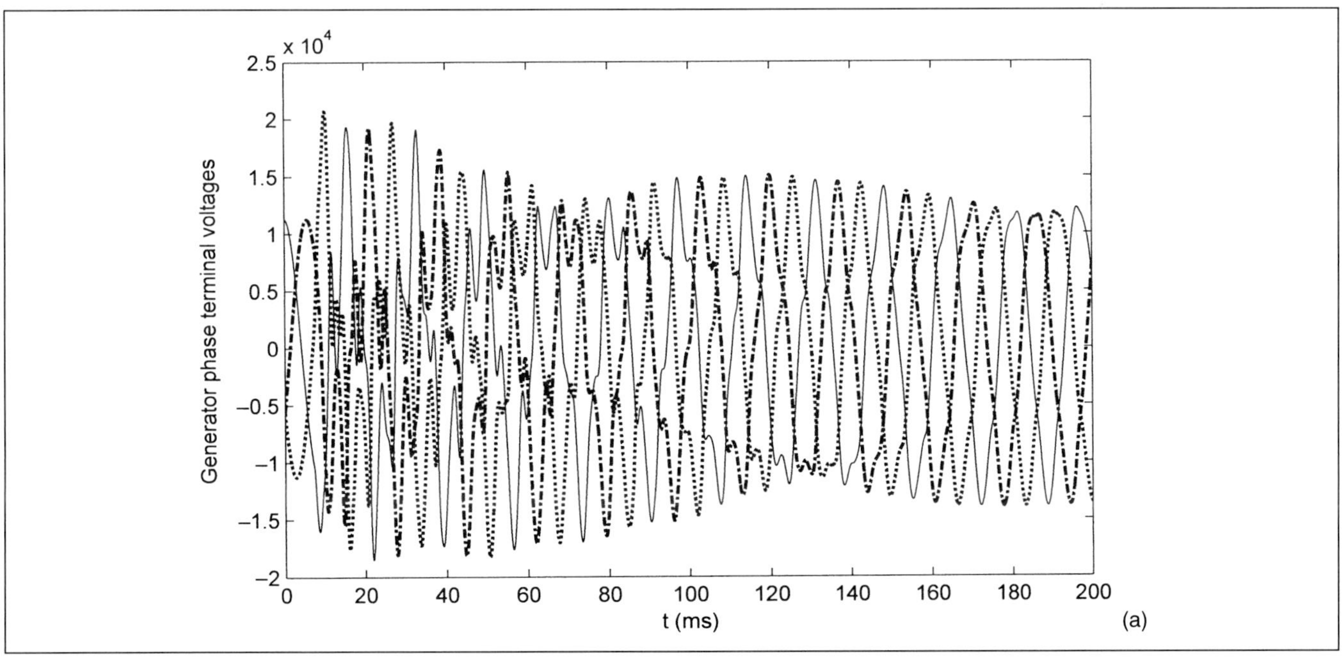

FIGURE 10-24 Transients on sudden unloading of a 112.1 MVA generator data in Table 10-1. (*a*) Generator terminal voltage, (*b*) phase currents, (*c*) currents in *dq* axes, (*d*) active power, and (*e*) damper currents, Example 10-8.

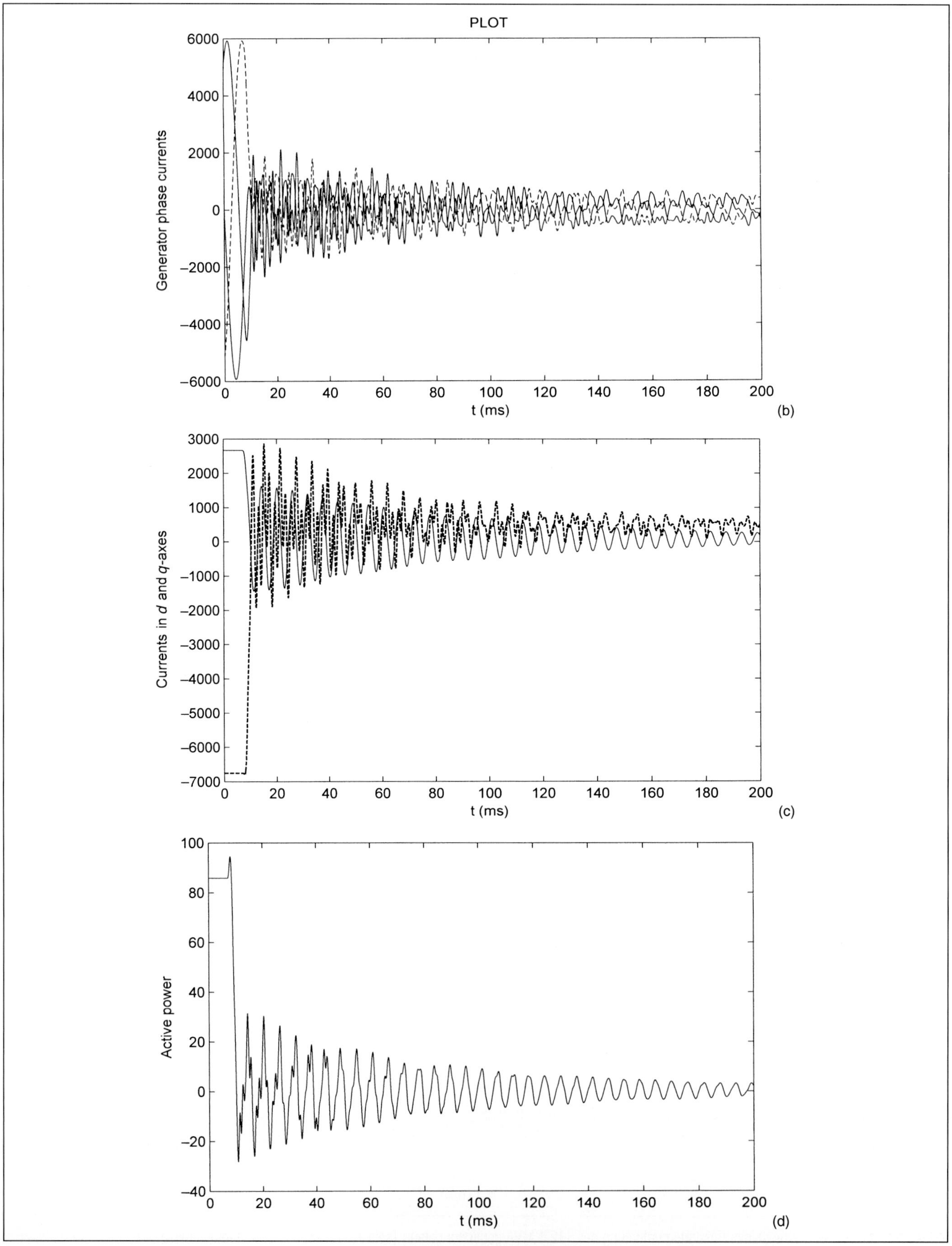

FIGURE 10-24 *(Continued)*

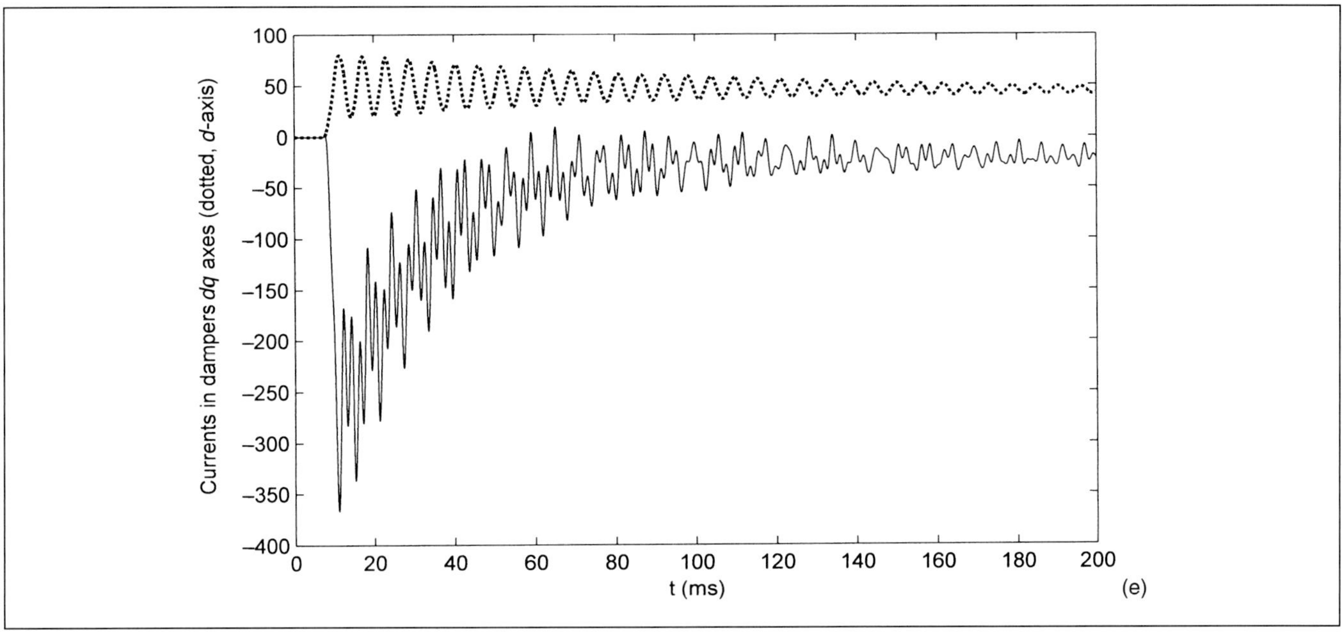

FIGURE 10-24 (*Continued*)

PROBLEMS

1. Calculate the fault decrement curves of the generator using the data given in Table 10-1. Calculate (a) ac decaying component, (b) dc component, and (c) total current. Plot the results similar to Fig. 10-7.

2. Consider a generator supplying power through an impedance, Fig. 10-P1, and the following generator data:

$$X''_{dv} = 0.16, X''_d = 0.18, X'_d = 0.20, X_d = 2.0, X_l = 0.15$$

$$X_q = 1.90, X'_q = 0.26, X''_q = 0.178, r_D = 0.015$$

$$r_Q = 0.025, T'_{do} = 4.8\text{s}, X/R = 110$$

All above reactances are in per unit on generator MVA base. Calculate (a) prefault voltages behind reactances X_d, X'_d and X''_d for faults at *G* and *F*, and (b) largest possible dc component for faults at *G* and *F*.

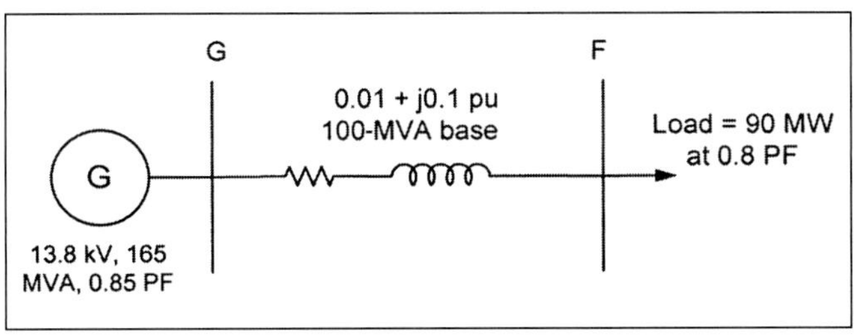

FIGURE 10-P1 Generator supplying load through an impedance circuit for Probs. 2 through 6.

3. Calculate field current in Prob. 2 on application of short circuit.

4. Calculate three-phase short-circuit subtransient and transient time constants in Prob. 2 for fault at *F*.

5. Write a numerical expression for the decaying ac component of current for fault at *G* and *F* in Prob. 2. What is the ac component of the fault current at 0.05 s and 0.10 s?

6. Transform the calculated direct-axis and quadrature-axis voltages derived in Example 10-2 into stator voltages using Park's transformation.

7. Draw a general steady-state phasor diagram of a synchronous generator operating at (a) leading power factor and (b) lagging power factor.

8. Construct a simplified dynamic phasor diagram (ignoring damper circuit and resistances) of a synchronous generator using Park's transformations. How does it differ from the steady-state phasor diagram?

9. Show that first column of P' is an eigenvector of L_{11} corresponding to eigenvalue $L_d = L_s - 2M_s$.

10. Prove that the transient inductance $L'_d = L_d - L^2_{ad}/L_F$.

11. Form an equivalent circuit similar to Fig. 10-11 with numerical values using the generator data from Table 10-1.

12. Given the parameters of the generator in Table 10-1 and considering that the generator is connected to an infinite bus through an impedance of 0.01 + 020 per unit and supplies its rated power at rated power factor in the system configuration of the figure, calculate all K constants in Fig. 10-22

13. The generator data shown in Table 10-1 is connected to a transformer of 120 MVA, Z_1 = 10 percent, X/R = 40, 13.8–138 kV, delta–wye grounded, in step-up configuration. The three-phase symmetrical short-circuit current of the 138-kV system is 30 kA, X/R = 15. Calculate total three-phase short-circuit currents when (a) a three-phase fault occurs on the 138 kV side of the transformer and (b) a fault occurs on 13.8 kV side of the transformer. Plot short-circuit current profiles: ac symmetrical, dc component, and total current, till the steady state is reached.

14. Table 10-2 gives the manufacturer's data of generator for Example 10-7. Convert it into *dq*0 axes.

REFERENCES

1. NEMA MG-1, Large Machines–Synchronous Generators, Part 22, 2003.
2. A. E. Fitzgerald, S. D. Umans, and C. Kingsley, *Electric Machinery*, McGraw-Hill Higher Education, New York, 2002.
3. ANSI/IEEE Std. C37.010, Guide for AC High Voltage Circuit Breakers Rated on Symmetrical Current Basis, 1999.
4. P. M. Anderson and A. Fouad, *Power System Control and Stability*, IEEE Press, New York, NY, 1991.
5. J. C. Das, "Study of Generator Source Short-Circuit Currents with Respect to Interrupting Duty of Generator Circuit Breakers, EMTP Simulation, ANSI/IEEE and IEC Methods," *Int. J. Emerging Elect. Power Syst.*, vol. 9, no. 3, article 3, 2008.
6. I. M. Canay and L.Warren, "Interrupting Sudden Asymmetrical Short-Circuit Currents Without Zero Transition," *BBC Rev.* vol. 56, pp. 484–493, 1969.
7. D. Dufournet, J. M. Willieme, and G. F. Montillet, "Design and Implementation of a SF_6 Interrupting Chamber Applied to Low Range Generator Breakers Suitable for Interrupting Currents Having a Non-Zero Passage," *IEEE Trans. Power Delivery*, vol. 17, pp. 963–969, Oct. 2002.
8. IEEE Std. C37.013, IEEE Standard for Generator Circuit Breakers Rated on Symmetrical Current Basis, 1997; and IEEE Std. C37.013a, Amendment 1, Supplement for Use With Generators Rated 10-100 MVA, 2007.
9. IEC 60909-0, Short-Circuit Currents in Three-Phase AC Systems, Calculation of Currents, 2001–07; and IEC 60909-1, Factors for Calculation of Short-Circuit Currents in Three-Phase AC Systems According to IEC 60909-0, 1991.
10. J. C. Das, "Short-Circuit Calculations—ANSI/IEEE & IEC Methods: Similarities and Differences," *Proc 8th International Symposium on Short-Circuit Currents in Power Systems*, Brussels, pp. 25–31, 1988.
11. J. C. Das, *Power System Analysis*, Chapter 8, Short-Circuit Calculations According to IEC Standards, Marcel Dekker, New York, 2002.
12. G. Knight and H. Sieling, "Comparison of ANSI and IEC 909 Short-Circuit Current Calculation Procedures," *IEEE Trans. Industry Applications*, vol. 29, no. 3, pp. 625–630, May/June 1993.
13. A. Berizzi, S. Massucco, A. Silvestri, and D. Zanin, "Short-Circuit Current Calculations: A Comparison Between Methods of IEC and ANSI Standards Using Dynamic Simulation as Reference," *IEEE Trans. Industry Applications*, vol. 30, no. 4, pp. 1099–1106, July/Aug. 1994.
14. J. C. Das, "Reducing Interrupting Duties of High-Voltage Circuit Breakers by Increasing Contact Parting Time," *IEEE Trans. Industry Applications*, vol. 44, no. 4, pp. 1027–1033, July/August 2008.
15. IEEE Std. 421.2, IEEE Guide for Identification and Evaluation of Dynamic Response of Excitation Control Systems, 1990.

FURTHER READING

ATP Rule Book, Canadian/American EMTP User Group, Portland, OR, 1987–92.

B. Adkins, *The General Theory of Electrical Machines*, Chapman and Hall, London, 1964.

P. M. Anderson, *Analysis of Faulted Power Systems*, Chapter 6, Iowa State University Press, Ames, 1973.

ANSI/IEEE Std. C37.09, IEEE Standard Test Procedure for AC High-Voltage Circuit Breakers Rated on a Symmetrical Current Basis, 1999.

I. Boldea, *Synchronous Generators*, CRC Press, Boca Raton, FL, 2005.

A. Braun, A. Edinger, and E. Rouss, "Interruption of Short-Circuit Currents in High Voltage AC Networks," *BBC Rev*, vol. 66, pp. 321–332, April 1979.

I. M. Canay, "Comparison of Generator Circuit Breaker Stresses in Test Laboratory and Real Service Condition," *IEEE Trans. Power Delivery*, vol. 16, pp. 415–421, 2001.

C. Concordia, *Synchronous Machines*, Wiley, New York, 1951.

J. R. Dunki-Jacobs, B. P. Lam, and R. P. Stratford, "A Comparison of ANSI-Based and Dynamically Rigorous Short-Circuit Current Calculation Procedures," *Trans. IEEE, Industry Applications*, vol. 24, no. 6, pp. 1180–1194, Nov./Dec. 1988.

N. N. Hancock, *Matrix Analysis of Electrical Machinery*, Pergamon Press, Oxford, 1964.

IEEE Committee Report, "Recommended Phasor Diagram for Synchronous Machines," *IEEE Trans. PAS*, vol. 88, pp. 1593–1610, 1963.

A. T. Morgan, *General Theory of Electrical Machines*, Heyden & Sons, London, 1979.

R. H. Park, "Two Reaction Theory of Synchronous Machines, Part 1," *AIEE Trans.*, vol. 48, pp. 716–730, 1929.

R. H. Park, "Two Reaction Theory of Synchronous Machines, Part II," *AIEE Trans.*, vol. 52, pp. 352–355, 1933.

CHAPTER 11

TRANSIENT BEHAVIOR OF INDUCTION AND SYNCHRONOUS MOTORS

In this chapter, we will study transient behavior of induction and synchronous motors, terminal short-circuits, and starting and switching transients. The voltage stability of the motors is an important criterion, addressed in Chap. 12. Also some other transients associated with the motors, during autobus transfer of motor loads, harmonics, cogging and crawling, and torsional vibrations are addressed in Chap. 16. This chapter may be considered parallel to Chap. 10 on synchronous generators, in the sense that the same basic approach is used in developing the transient models and analysis. A synchronous machine, whether it is a generator or motor, has the same theoretical and practical basis of analysis, and can be similarly modeled. The starting of synchronous motors must be addressed separately; while a synchronous generator is driven to the rated speed, a synchronous motor must accelerate and synchronize. A synchronous motor is started like an induction motor, but there are differences in the starting characteristics.

11-1 TRANSIENT AND STEADY-STATE MODELS OF INDUCTION MACHINES

We can study a rotating machine in steady state, depending upon its specific type, that is, synchronous, induction, dc, and the like, and arrive at its behavior using equations and models specific to the machine type. The generalized machine theory attempts to unify the piecemeal treatment of rotating machines, led by Park's two-axis equations for synchronous machines (Chap. 10). These ideas were further developed by Kron and others,[1,2] and treatment of electrical machines using tensors or matrices and linear transformation is the basis of generalized machine theory. Transformation here means that an old set of variables is transformed into a new set of variables and vice versa. One set of equations is good to describe the steady state and transient behavior. Yet, the theory has its limitations:

- It cannot model saturation, brush contact resistance, commutation effects, surge phenomena, skew, eddy current losses, stray load losses, and mechanical features such as vibrations, torsion, noise, critical speeds, and the like.
- It cannot be applied to machines with *both* rotor and stator of salient pole construction, such as an inductor alternator, as *dq* axis cannot be fixed with respect to stator or rotor. Another limitation is that the nonsalient pole part of the machine must have balanced windings. A single-phase generator has unbalanced windings, and the theory cannot be applied.

New approaches to machine analysis are: (a) energy state functions and (b) field equations. Much alike a synchronous machine in Chap. 10, here, an induction machine model is derived in terms of *dq* axes. Both the stator and rotor are cylindrical and symmetrical. The *d*-axis is chosen arbitrarily as the axis of phase *a*. The three-phase winding is converted into a two-phase winding so that the axis of the second phase becomes the *q*-axis.

Consider the transformation of a three-phase machine (Fig. 11-1*a*) to a two-phase machine (Fig. 11-1*b*). For the three-phase machine we can write:

$$\begin{aligned} i_A &= I_m \cos \omega t \\ i_B &= I_m \cos(\omega t - 120°) \\ i_C &= I_m \cos(\omega t + 120°) \end{aligned} \tag{11-1}$$

For a two-phase machine:

$$\begin{aligned} i_{\alpha s} &= I_m \cos \omega t \\ i_{\beta s} &= I_m \cos(\omega t - 90°) \end{aligned} \tag{11-2}$$

The rotary and stationary coils are denoted with subscripts *r* and *s*, respectively. While deriving equivalence, we have three choices:

- Keep effective turns on the coils the same and use multiplying factors for power and impedance.
- Keep current the same.
- Keep impedance the same.

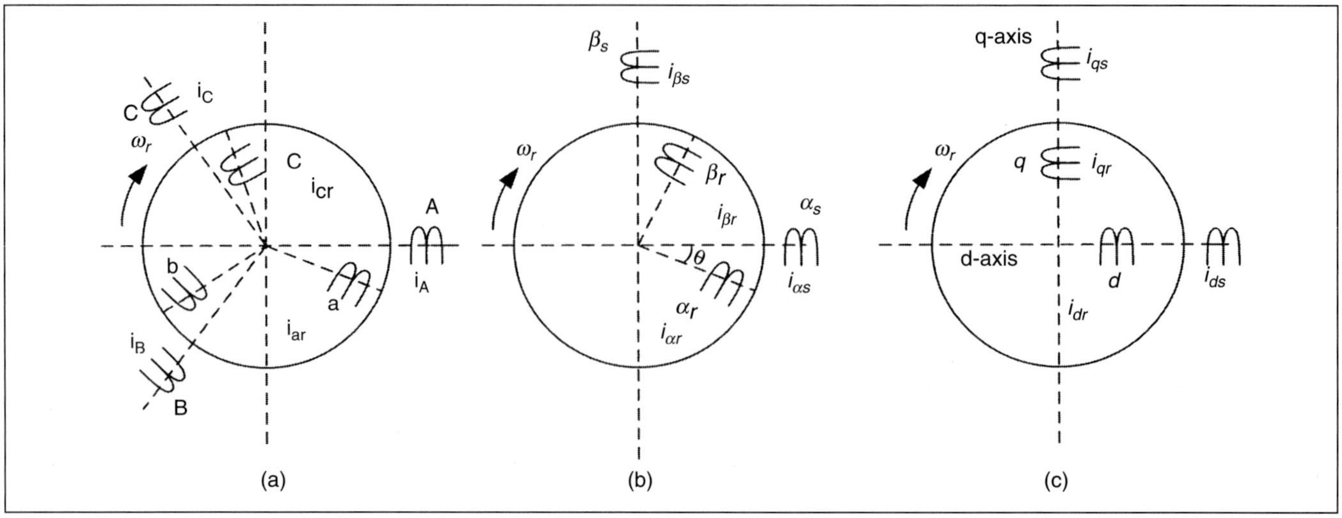

FIGURE 11-1 (*a*) Three-phase induction machine, stator, and rotor coils. (*b*) Transformation to a two-phase machine, α and β axes. (*c*) Transformation to *d* and *q* axes.

The parameters of equivalence in three cases are shown in Table 11-1. Case 3 is normally used. Referring to Fig. 11-1*a*, *b* and *c* for the stator, $\alpha\beta$ axes and dq axes coincide. Phase A-axis of three-phase machine coincides with phase-α axis of the two-phase machine. Thus, *d*-axis quantities apply to α-axis of two-phase machine or A-axis of three-phase machine. We define a transformation matrix $\bar{A}$, so that:

$$\begin{gathered}\left[i_{\alpha s}, i_{\beta s}, i_{0s}\right]=\left[\bar{A}\right]\left[i_A, i_B, i_C\right]\\ \left[i_A, i_B, i_C\right]=\left[\bar{A}^{-1}\right]\left[i_{\alpha s}, i_{\beta s}, i_{0s}\right]\end{gathered} \tag{11-3}$$

Note that we add another parameter:

$$i_0 = i_a + i_b + i_c \tag{11-4}$$

This will be zero for a three-phase balanced system. The matrix *A* is given by:

$$\bar{A}=\sqrt{\frac{2}{3}}\begin{vmatrix}\cos 0^\circ & \cos\frac{2\pi}{3} & \cos\frac{4\pi}{3}\\ \sin 0^\circ & \sin\frac{2\pi}{3} & \sin\frac{4\pi}{3}\\ \frac{1}{\sqrt{2}} & \frac{1}{\sqrt{2}} & \frac{1}{\sqrt{2}}\end{vmatrix} \tag{11-5}$$

For power invariance in the two systems, it can be shown that:

$$\begin{gathered}\bar{A}^t\bar{A}=\bar{1}\\ \bar{A}^{-1}=\bar{A}^t\end{gathered} \tag{11-6}$$

TABLE 11-1 Transformation from Three-phase to Two-phase—Option of Parameters for Equivalence

	Two-Phase Parameters		
Three-Phase Parameters	**Case 1 (*N* Same)**	**Case 2 (*I* Same)**	**Case3 (*Z* Same)**
Current *I*	3/2 *I*	*I*	$\sqrt{3}/2I$
Turns *N*	*N*	3/2*N*	$\sqrt{3}/2\ N$
Flux ϕ	ϕ	ϕ	ϕ
Voltage per phase *V*	*V*	3/2*V*	$\sqrt{3}/2V$
Power per phase *VI*	3/2*VI*	3/2*VI*	3/2*VI*
Total power 3*VI*	3*VI*	3*VI*	3*VI*
Impedance *Z*	2/3*Z*	3/2*Z*	*Z*

Now we will transform from rotating $\alpha\beta 0$ axes to pseudostationary axis $dq0$ (see Fig. 11-1*c*). In the dq axis:

$$\begin{vmatrix} i_{dr} \\ i_{qr} \\ i_{0r} \end{vmatrix} = \begin{vmatrix} \cos\theta & \sin\theta & 0 \\ -\sin\theta & \cos\theta & 0 \\ 0 & 0 & 1 \end{vmatrix} \begin{vmatrix} i_{\alpha r} \\ i_{\beta r} \\ i_{0r} \end{vmatrix} \tag{11-7}$$

Some other useful transformations to relate old current, voltage, and impedance to the new transformation axes are necessary. We define a connection matrix *C*, so that:

$$\begin{aligned} \bar{i} &= \bar{C}\bar{i}' \\ \bar{v}' &= \bar{C}^{*t} v \end{aligned} \tag{11-8}$$

and:

$$\bar{z}' = \bar{C}^{*t} z \bar{C} \tag{11-9}$$

Where the * shows a conjugate. For proofs of these transformations, Ref. 2 provides further reading. We will use these transformations to derive a model of induction machine. For the stator:

$$3-\phi\big|_{abc} \Leftrightarrow 2-\phi\big|_{\alpha\beta 0} \Leftrightarrow 2-\phi\big|_{dq0} \tag{11-10}$$

$$\begin{vmatrix} i_{\alpha s} \\ i_{\beta s} \\ i_{0s} \end{vmatrix} = \sqrt{\frac{2}{3}} \begin{vmatrix} 1 & -\frac{1}{2} & -\frac{1}{2} \\ 0 & \sqrt{\frac{3}{2}} & -\sqrt{\frac{3}{2}} \\ \frac{1}{\sqrt{2}} & \frac{1}{\sqrt{2}} & \frac{1}{\sqrt{2}} \end{vmatrix} \begin{vmatrix} i_A \\ i_B \\ i_C \end{vmatrix} \tag{11-11}$$

and:

$$\begin{vmatrix} i_{ds} \\ i_{qs} \\ i_{0s} \end{vmatrix} = \begin{vmatrix} 1 & 0 & 0 \\ 0 & 1 & 0 \\ 0 & 0 & 1 \end{vmatrix} \begin{vmatrix} i_{\alpha s} \\ i_{\beta s} \\ i_{0s} \end{vmatrix} \tag{11-12}$$

For the rotor:

$$3-\phi\big|_{abc} \Leftrightarrow 2-\phi\big|_{\alpha\beta 0} \Leftrightarrow 2-\phi\big|dqo \tag{11-13}$$

Thus we can write:

$$\begin{vmatrix} i_{\alpha r} \\ i_{\beta r} \\ i_{0r} \end{vmatrix} = \bar{A} \begin{vmatrix} i_{ar} \\ i_{br} \\ i_{cr} \end{vmatrix} \tag{11-14}$$

and Eq. (11-7) gives transformation to dq rotor axis.

From the basic concepts of coils on the stator and rotor, we can write the following matrix:

$$\begin{vmatrix} v_{ds} \\ v_{qs} \\ v_{dr} \\ v_{qr} \end{vmatrix} = \begin{vmatrix} r_{ds}+L_{ds}p & & M_d p & \\ & r_{qs}+L_{qs}p & & M_q p \\ M_d p & -M_q\omega_r & r_{dr}+L_{dr}p & -L_{qr}\omega_r \\ M_d\omega_r & M_q p & L_{dr}\omega_r & r_{qr}+L_{qr}p \end{vmatrix} \begin{vmatrix} i_{ds} \\ i_{qs} \\ i_{dr} \\ i_{qr} \end{vmatrix} \tag{11-15}$$

where p is the differential operator d/dt.

This can be simplified. As the coils are balanced and identical:

$$\begin{aligned} r_{ds} &= r_{qs} = r_s \\ r_{dr} &= r_{qr} = r_r \\ M_d &= M_q = M \end{aligned} \tag{11-16}$$

The air gap is uniform, therefore:

$$\begin{aligned} L_{ds} &= L_{qs} = L_s \\ L_{dr} &= L_{qr} = L_r \end{aligned} \tag{11-17}$$

Rotor coils are short-circuited:

$$\begin{aligned} v_{dr} &= 0 = v_{\alpha r}\cos\theta + v_{\beta r}\sin\theta \\ v_{qr} &= 0 = v_{\alpha r}\sin\theta + v_{\beta r}\cos\theta \end{aligned} \tag{11-18}$$

Thus, i_{dr} and i_{qr} are reversed. With change of signs, the following matrix can be written:

$$\begin{vmatrix} v_{ds} \\ v_{qs} \\ 0 \\ 0 \end{vmatrix} = \begin{vmatrix} r_s+L_s p & & -Mp & \\ & r_s+L_s p & & -Mp \\ Mp & -M\omega_r & -(r_r+L_r p) & L_r\omega_r \\ M\omega_r & Mp & -L_r\omega_r & -(r_r+L_r p) \end{vmatrix} \begin{vmatrix} i_{ds} \\ i_{qs} \\ i_{dr} \\ i_{qr} \end{vmatrix} \tag{11-19}$$

11-1-1 Steady-State Operation

For the steady-state operation:

$$\begin{aligned} \omega_r &= (1-s)\omega \\ p &= j\omega \end{aligned} \tag{11-20}$$

where s is the motor slip, ω_r is the actual operating speed of the motor in radians, and ω is 2πf. This gives:

$$\begin{vmatrix} v_{ds} \\ v_{qs} \\ 0 \\ 0 \end{vmatrix} = \begin{vmatrix} r_s+X_s & & -jX_m & \\ & r_s+jX_s & & -jX_m \\ jX_m & -(1-s)X_m & -(r_r+jX_r) & (1-s)X_r \\ (1-s)X_m & jX_m & -(1-s)X_r & -(r_r+jX_r) \end{vmatrix} \begin{vmatrix} i_{ds} \\ i_{qs} \\ i_{dr} \\ i_{qr} \end{vmatrix} \tag{11-21}$$

There is no difference in d and q axes except of time, therefore:

$$\begin{aligned} i_{qs} &= j i_{ds} \\ i_{qr} &= j i_{dr} \end{aligned} \tag{11-22}$$

Write the following transformation matrix:

$$\begin{vmatrix} i_{ds} \\ i_{qs} \\ i_{dr} \\ i_{qr} \end{vmatrix} = \begin{vmatrix} 1 & 0 \\ j & 0 \\ 0 & 1 \\ 0 & j \end{vmatrix} \begin{vmatrix} i_s \\ i_r \end{vmatrix} \tag{11-23}$$

The transformation of impedance is given by Eq. (11-9). Thus:

$$\bar{z}' = \bar{C}^t z \bar{C}$$

$$\bar{z}' = \begin{vmatrix} 1 & -j & 0 & 0 \\ 0 & 0 & 1 & -j \end{vmatrix} \bar{z} \begin{vmatrix} 1 & 0 \\ j & 0 \\ 0 & 1 \\ 0 & j \end{vmatrix} \quad (11\text{-}24)$$

After simplification and manipulation of the matrices, this transformation gives:

$$\bar{z}' = 2\begin{vmatrix} r_s + jX_s & -jX_m \\ jsX_m & -r_r - jsX_r \end{vmatrix} \quad (11\text{-}25)$$

The transformation of voltage is given by:

$$\bar{v}' = \bar{C}^{*t}\bar{v}$$

$$= \begin{vmatrix} 1 & -j & 0 & 0 \\ 0 & 0 & 1 & -j \end{vmatrix} \begin{vmatrix} v_{ds} \\ v_{qs} \\ 0 \\ 0 \end{vmatrix} \quad (11\text{-}26)$$

$$= \begin{vmatrix} v_{ds} - jv_{qs} \\ 0 \end{vmatrix} = \begin{vmatrix} 2v_s \\ 0 \end{vmatrix}$$

because

$$v_{ds} = v_s,\ v_{qs} = jv_s,\ \text{and}\ -jv_{qs} = +v_s. \quad (11\text{-}27)$$

We also know that the transformed voltages and currents are related with transformed impedance:

$$\bar{v}' = \bar{z}'\bar{i}' \quad (11\text{-}28)$$

Therefore:

$$\begin{vmatrix} v_s \\ 0 \end{vmatrix} = \begin{vmatrix} r_s + jX_s & -jX_m \\ jsX_m & -r_r - jsX_r \end{vmatrix} \begin{vmatrix} i_s \\ i_r \end{vmatrix} \quad (11\text{-}29)$$

This nonsymmetric matrix can be made symmetric:

$$\begin{vmatrix} v_s \\ 0 \end{vmatrix} = \begin{vmatrix} r_s + jX_s & -jX_m \\ jX_m & -r_r / s - jX_r \end{vmatrix} \begin{vmatrix} i_s \\ i_r \end{vmatrix} \quad (11\text{-}30)$$

This gives the equivalent circuit of the induction motor, shown in Fig.11-2*a* and *b*.

The leakage reactances of the stator and rotor are:

$$X_s = (L_s - M)$$
$$X_r = (L_r - M) \quad (11\text{-}31)$$

The Torque Equation The torque equation in general can be written as:

$$T_e = \frac{P}{\omega} = \bar{i}^t \bar{G} \bar{i} \quad (11\text{-}32)$$

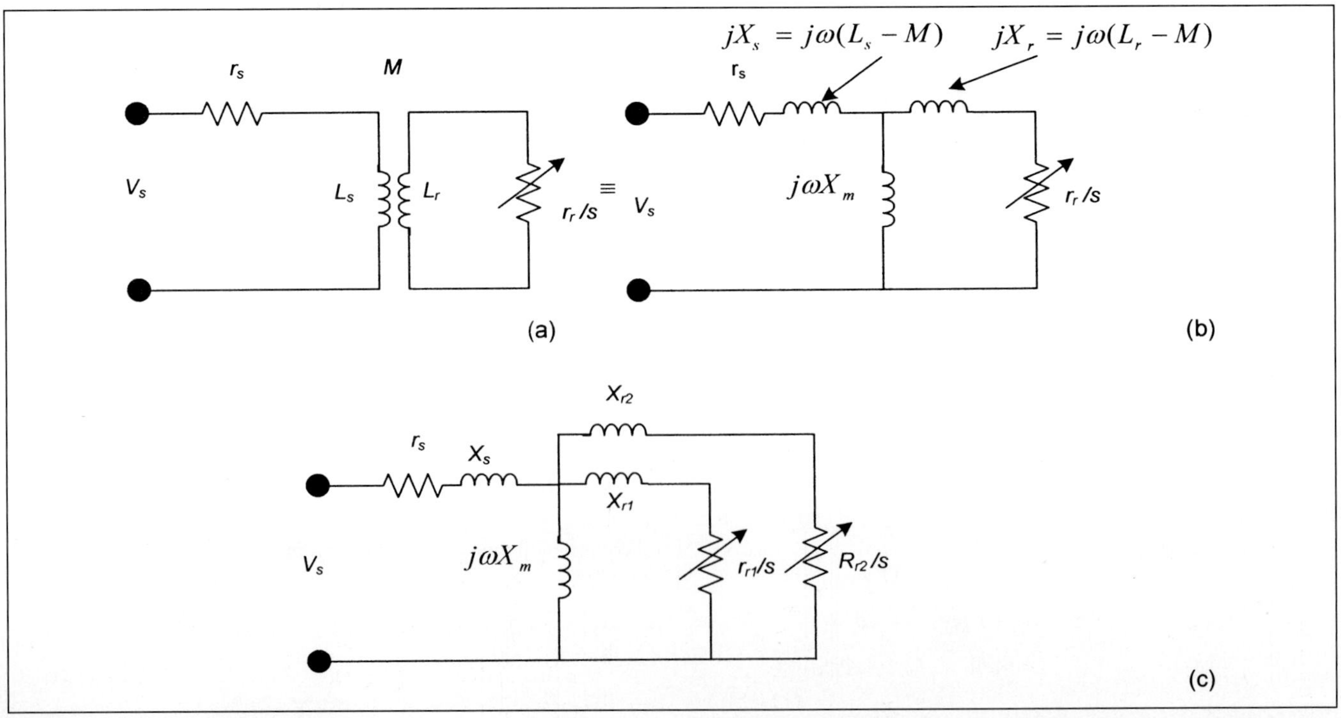

FIGURE 11-2 (*a*) and (*b*). Equivalent circuit of a single-cage induction motor in steady state. (c) Equivalent circuit of a double-cage rotor induction motor.

where $\bar{G}$ is the torque matrix and is a rotational inductance matrix.[2] For an induction motor:

$$\bar{G} = \begin{vmatrix} 0 & 0 & 0 & 0 \\ 0 & 0 & 0 & 0 \\ 0 & -M_q & 0 & -L_{qr} \\ M_d & 0 & L_{dr} & 0 \end{vmatrix} \tag{11-33}$$

The steady-state torque is:

$$P_e = \text{Re}[\bar{i}^{*t}\bar{G}\bar{i}]$$
$$= \text{Re}\begin{vmatrix} i^*_{ds} & i^*_{qs} & i^*_{dr} & i^*_{qr} \end{vmatrix}\bar{G}\begin{vmatrix} i_{ds} \\ i_{qs} \\ i_{dr} \\ i_{qr} \end{vmatrix} \tag{11-34}$$

The solution of Eq. (11-34) gives:

$$T_e = \text{Re}[j2Mi^*_r i_s] \tag{11-35}$$

also:

$$\begin{vmatrix} i_s \\ i_r \end{vmatrix} = \frac{1}{\Delta}\begin{vmatrix} \frac{r_r}{s} + jX_r & jX_m \\ jX_m & r_s + jX_s \end{vmatrix}\begin{vmatrix} v_s \\ 0 \end{vmatrix} \tag{11-36}$$

where Δ is determinant of $\bar{Z}$. This gives:

$$i_s = \frac{v_s}{\Delta}\left(\frac{r_r}{s} + jX_r\right)$$
$$i_r = \frac{v_s}{\Delta}(jX_m) \tag{11-37}$$
$$i^*_r = \frac{v^*_s}{\Delta^*}(jX_m)$$

Substituting these values in Eq. (11-35)

$$T_e = \text{Re}\left[j2M\frac{v^*_s}{\Delta^*}(-jX_m)\frac{v_s}{\Delta}\left(\frac{r_r}{s} + jX_r\right)\right]$$
$$= \text{Re}\left[\frac{2X^2_m}{\omega}\frac{v_s}{\Delta}\frac{v^*_s}{\Delta^*}\frac{r_r}{s}\right] \tag{11-38}$$
$$= \text{Re}\frac{2}{\omega}\left[i_r i^*_r \frac{r_r}{s}\right] = \frac{2}{\omega}\left|i^2_r\frac{r_r}{s}\right|$$

For a three-phase machine

$$T_e = \frac{3}{\omega}\left[i^2_r\frac{r_r}{s}\right]$$

Mechanical power developed is the power across the air gap minus copper loss in the rotor, that is,

$$(1-s)P_g \tag{11-39}$$

Thus, the motor torque T in newton-meters can be written as:

$$T = \frac{1}{\omega_s}i^2_r\frac{r_r}{s} \approx \frac{1}{\omega_s}\frac{v^2_s\left(\frac{r_r}{s}\right)}{\left(r_s + \frac{r_r}{s}\right)^2 + (X_s + X_r)^2} \tag{11-40}$$

The slip for maximum torque is obtained from above equation by $dT/ds = 0$, which gives:

$$s = \pm\frac{r_r}{\sqrt{r^2_s + (X_s + X_r)^2}} \approx \frac{r_r}{(X_s + X_r)} \tag{11-41}$$

The maximum torque is obtained by inserting the value of s in Eq. (11-40)

$$T_m = \frac{v^2_s}{2\left[\sqrt{\left(r^2_s + (X_s + X_r)^2\right)} \pm r_s\right]} \tag{11-42}$$

At small slip r_r/s is comparatively large compared to other terms in Eq. (11-40). Therefore, we can write:

$$T = sK \tag{11-43}$$

where K may be considered a constant. The negative sign in Eq. (11-42) is related to the operation as induction generator.

11-1-2 Double Cage Rotors

From Eq. (11-40) the maximum torque is proportional to the square of the applied voltage; it is reduced by stator resistance and leakage reactance, but is independent of rotor resistance. The slip at the maximum torque is directly proportional to the rotor resistance [Eq. (11-41)]. In many applications it is desirable to produce higher starting torque at the time of starting. External resistances can be introduced in wound rotor motors which are short-circuited when the motor speeds up. In squirrel-cage designs, deep rotor bars or double-cage rotors are used; a typical slot design is shown in Fig. 11-3. The top bar is of lower cross section and has higher resistance than the bottom bar of larger cross section. The flux patterns shown in Fig. 11-3 make the leakage resistance of the top cage negligible compared to the lower cage. At starting, the slip frequency is equal to the supply system frequency and most of the starting current flows in the top cage, and the motor produces high starting

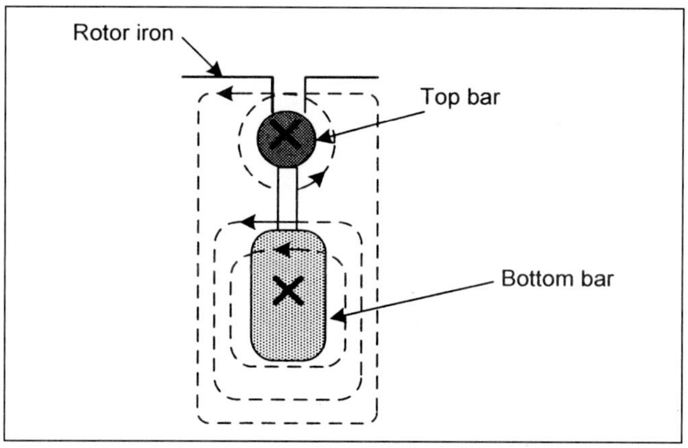

FIGURE 11-3 Flux patterns in a double-cage induction motor rotor construction.

torque. As the motor accelerates, the slip frequency decreases and the lower cage takes more current because of its low leakage reactance compared to the resistance. Following the same procedure as in the development of a single-cage induction motor, the equivalent circuit of a double-cage rotor is shown in Fig. 11-2*c*.

11-2 INDUCTION MACHINE MODEL WITH SATURATION

A double-cage machine model in *dq*0 axis can be represented as shown in Fig. 11-4*a*. We can write the following voltage equation in the direct axis:

$$\begin{vmatrix} V_{ds} \\ 0 \\ 0 \end{vmatrix} = -\begin{vmatrix} r_s & 0 & 0 \\ 0 & r_{r1} & 0 \\ 0 & 0 & r_{r2} \end{vmatrix}\begin{vmatrix} i_{ds} \\ i_{D1} \\ i_{D2} \end{vmatrix} - \begin{vmatrix} \omega\phi_q \\ (\omega-\omega_r)\phi_{Q1} \\ (\omega-\omega_r)\phi_{Q2} \end{vmatrix} - \begin{vmatrix} p\phi_d \\ p\phi_{D1} \\ p\phi_{D2} \end{vmatrix} \tag{11-44}$$

also:

$$\begin{vmatrix} \phi_d \\ \phi_{D1} \\ \phi_{D2} \end{vmatrix} = \begin{vmatrix} L_{md}+L_s & L_{md} & L_{md} \\ L_{md} & L_{md}+L_{r1} & L_{md} \\ L_{md} & L_{md} & L_{md}+L_{r2} \end{vmatrix} \tag{11-45}$$

where L_{md} = *d*-axis linkage reactance, L_s, L_{r1}, and L_{r2} are direct axis stator, rotor first-cage, and rotor second-cage leakage reactances, respectively. From Eqs. (11-44) and (11-45):

$$\begin{aligned} V_{ds} &= -r_s i_d - L_s p i_d - L_{md} p(i_{D1}+i_{D2}+i_d) - \omega\phi_q \\ 0 &= -r_{r1} i_{D1} - (\omega-\omega_r)\phi_{Q1} - L_{r1} p i_{D1} - L_{md} p(i_{D1}+i_{D2}+i_d) \\ 0 &= -r_{r2} i_{D2} - (\omega-\omega_r)\phi_{Q2} - L_{r1} p i_{D2} - L_{md} p(i_{D1}+i_{D2}+i_d) \end{aligned} \tag{11-46}$$

The circuit in the direct axis is in Fig. 11-4*b*.

The saturation is a function of current in the circuits. A piecewise linear approximation can be used to take account of the saturation. The magnetizing characteristics can be defined by a number of segments, and the monotonically increasing characteristics account for saturation of the magnetic branch. The unsaturated mutual flux is:

$$\phi_m = \sqrt{\phi_{md}^2 + \phi_{mq}^2} \tag{11-47}$$

The flux saturates to account for cross-magnetization, that is, the two axes affect each other. The leakage reactance becomes high when the motor takes starting currents. The skin effects in the deep bars change with the motor slip. A deep bar factor DF is defined as:

$$\text{DF} = d\sqrt{2\omega\mu_0\sigma} \tag{11-48}$$

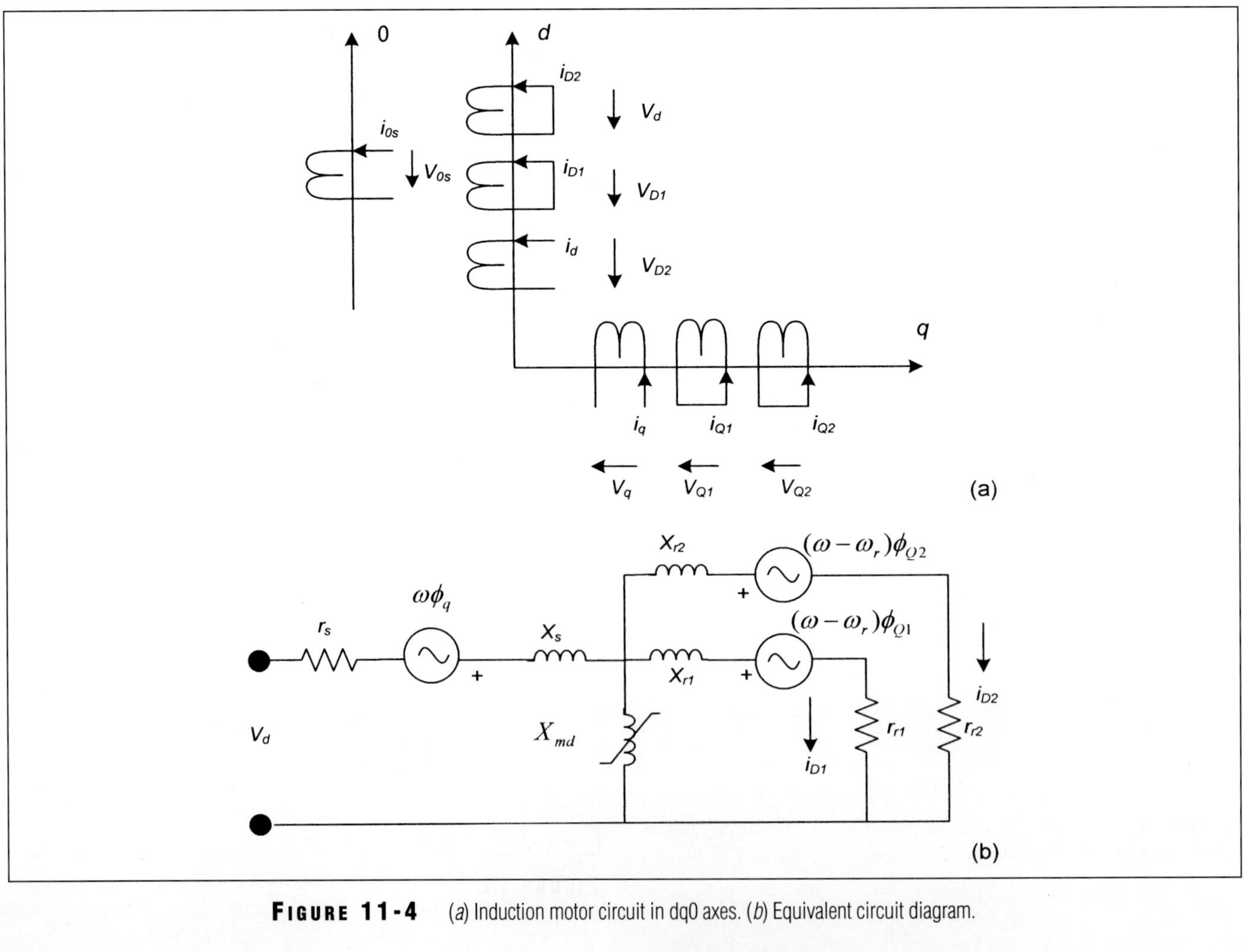

FIGURE 11-4 (*a*) Induction motor circuit in dq0 axes. (*b*) Equivalent circuit diagram.

where d is the thickness of deep bars and σ is the conductivity. The effective rotor impedance as a function of slip is given by:

$$r_r + jX_r = \frac{r_{r(\mathrm{slip}=0)}}{2}\theta\frac{\sinh\theta + \sin\theta + j(\sinh\theta - \sin\theta)}{\cosh\theta - \cos\theta} \tag{11-49}$$

where $\theta = \sqrt{\mathrm{slip}DF}$. EMTP uses above equations for modeling saturation.[3]

11-3 INDUCTION GENERATOR

If we plot torque slip characteristics of an induction machine for a negative slip, Fig. 11-5 results. An induction motor will act as an induction generator with the negative slip. At $s = 0$, the induction motor torque is zero, and if it is driven above its synchronous speed, the slip becomes negative and generator operation results. The negative r_s in Eq. (11-42) represents the maximum torque required to drive the machine as generator. The maximum torque is independent of the rotor resistance. For subsynchronous operation, the rotor resistance does affect the slip at which the maximum torque occurs. For maximum torque at starting, $s = 1$ and

$$r_r \approx \sqrt{r_s^2 + (X_s + X_r)^2} \tag{11-50}$$

For supersynchronous operation (generator operation), the maximum torque is independent of r_r, same as for the motor operation, but increases with the reduction of both the stator and the rotor reactances. Therefore, we can write:

$$\frac{T_{m,\mathrm{gen}}(\mathrm{supersyn})}{T_{m,\mathrm{motor}}(\mathrm{subsyn})} = \frac{\sqrt{r_s^2 + (X_s + X_r)^2} + r_s}{\sqrt{r_s^2 + (X_s + X_r)^2} - r_s} \approx \frac{X_s + X_r + r_s}{X_s + X_r - r_s} \tag{11-51}$$

The approximation holds as long as $r_s << X_s$. The torque-speed characteristics of the machine above synchronism are similar to that for running as an induction motor. If the prime mover develops a greater driving torque than the maximum counter torque, the speed rises into an unstable region and the slip increases. At some high value of the slip, the generating effect ceases and the machine becomes a brake.

Induction generators do not need synchronizing and can run in parallel without hunting and at any frequency; the speed variations of the prime mover are relatively unimportant. Thus, these machines are applied for wind power generation (Chap. 24).

An induction generator must draw its excitation from the supply system, which is mostly reactive power requirement. On a sudden short circuit the excitation fails, and with it the generator output; so in a way the generator is self-protecting.

As the rotor speed rises above synchronous speed, the rotor EMF becomes in phase opposition to its subsynchronous position because the rotor conductors are moving faster than the stator rotating field. This reverses the rotor current also and the stator component reverses. The rotor current locus is a complete circle. The stator current is clearly a leading current of definite phase angle. The output cannot be made to supply a lagging load.

An induction generator can be self-excited through a capacitor bank, without external dc source, but the frequency and generated voltage will be affected by speed, load, and capacitor rating. For an inductive load the magnetic energy circulation must be dealt with by the capacitor bank as induction generator cannot do so.

11-4 STABILITY OF INDUCTION MOTORS ON VOLTAGE DIPS

From Eq. (11-40) it can be inferred that the motor torque varies approximately as the square of the voltage. If the load torque remains constant and the voltage dips, there has to be an increase in the current to meet the load requirements. For the induction motors loads, the assumption that a balanced decrease in the terminal voltage results in a corresponding increase in the line current is conservative for stability studies, and is generally adopted in the load models.

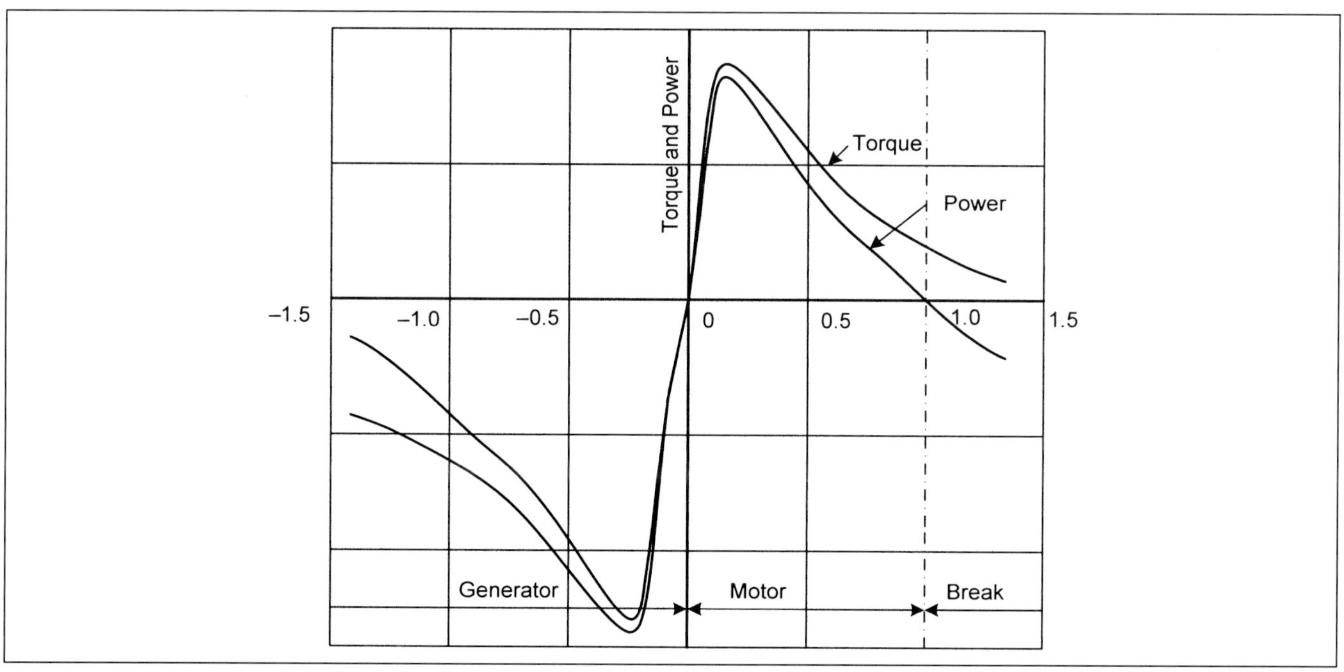

FIGURE 11-5 Torque-slip characteristics of induction machines. The characteristics in negative slip region relates to operation as an induction generator.

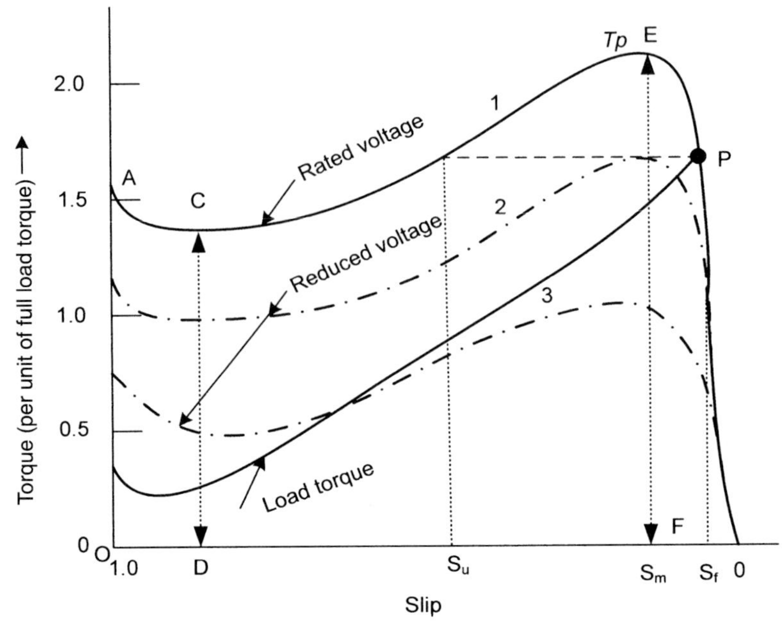

FIGURE 11-6 Torque-slip characteristics of an induction motor at rated voltage and reduced voltages.

Figure 11-6 shows typical torque-slip characteristics of an induction motor at rated voltages (curve 1), and reduced voltages (curves 2 and 3). The definitions of locked rotor torque, minimum accelerating torque, and breakdown torque during the starting cycle can be provided based upon this figure:

- OA is locked rotor torque or starting torque.
- CD is breakaway torque. There is a cusp or reverse curvature in accelerating torque curve, which gives the minimum accelerating torque or breakaway torque.
- EF is maximum torque, also called the breakdown torque. It occurs at slip s_m, the maximum slip at the breakdown torque.
- P is normal operating full-load point at full-load slip s_f. The starting load characteristics are shown for a fan or blower, and it varies widely depending upon the type of load to be accelerated. The operating point and the full-load slip vary with the change in the load.

Assume that the load torque remains constant, for example, in a conveyor motor, when a voltage dip occurs. The slip increases and the motor torque will be reduced. It should not fall below the load torque to prevent a stall. Considering a motor break down torque of 200 percent, the maximum voltage dip to prevent stalling is 29.3 percent.

Figure 11-7 is a comparison of the torque-speed characteristics for NEMA design motors, *A*, *B*, *C*, *D*, and *F*. A proper choice of motor type with respect to load is exercised; this is not discussed further.

The effect of voltage variations on induction motor's *steady-state* performance is shown in Table 11-2. We are not much concerned with the effect of frequency variation in load flow analysis, though this becomes important in harmonic analysis. The emf of a three-phase ac winding is given by:

$$E_{ph} = 4.44 K_w f T_{ph} \Phi \quad (V) \tag{11-52}$$

where E_{ph} is the phase emf, K_w is the winding factor, f is the system frequency, T_{ph} are the turns per phase, and $\mathbf{\Phi}$ is the flux. Maintaining the voltage constant, a variation in frequency results in an inverse variation in the flux. Thus, a lower frequency results in overfluxing the motor and its consequent derating. In variable frequency drive systems *V/f* is kept constant to maintain a constant flux relation.

From the transient analysis considerations, a motor can be stable even under complete collapse of voltage provided it is returned to normal soon enough. Consider the situation shown in Fig. 11-6. Assume that the load torque remains constant under a voltage dip. A voltage dip occurs that reduces the breakdown torque of the motor equal to load torque T'. Then we can write:

$$T'/T = (V/V_r)^2 \tag{11-53}$$

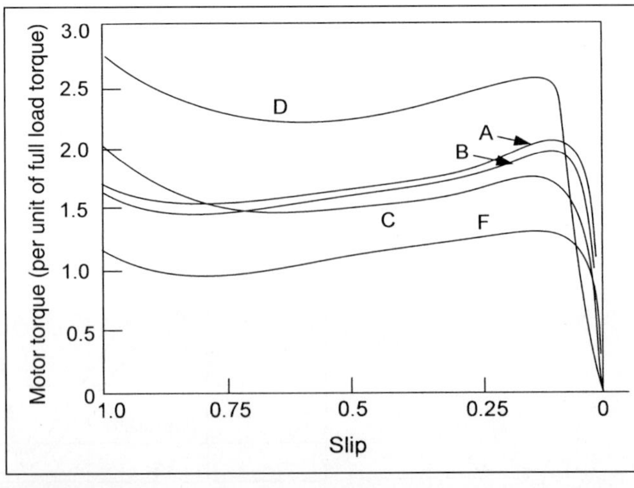

FIGURE 11-7 Torque-speed characteristics of NEMA design *A*, *B*, *C*, *D*, and *F* induction motors.

TABLE 11-2 Effects of Voltage Variations on Operation of Induction Motors

		Performance at Rated Voltage (1.0 pu) and Other Than Rated Voltage				
Characteristics of Induction Motor	**Variation with V**	**0.8**	**0.95**	**1.0**	**1.05**	**1.10**
Torque	$= V^2$	0.64	0.90	1.0	1.10	1.21
Full-load slip	$= 1/V^2$	1.56	1.11	1.0	0.91	0.83
Full-load current	$\approx 1/V$	1.28	1.04	1.0	0.956	0.935
Full-load efficiency		0.88	0.915	0.92	0.925	0.92
Full-load power factor		0.90	0.89	0.88	0.87	0.86
Starting current	$= V$	0.80	0.95	1.0	1.05	1.10
No load losses (W)	$= V^2$	0.016	0.023	0.025	0.028	0.030
No load losses (vars)	$= V^2$	0.16	0.226	0.25	0.276	0.303

where V_r is the reduced voltage and T' is torque at reduced voltage. A relation between the torque and the maximum slip S_m is:[4,5]

$$T/T_p = \frac{2}{(s/s_m)+(s_m/s)} \tag{11-54}$$

Then from Eqs. (11-53) and (11-54)

$$T'/T_l = \frac{2k}{(s/s_m)+(s_m/s)} \tag{11-55}$$

where:

$$k = (T_p/T_l)(V/V_r)^2 \tag{11-56}$$

For the voltage dip that reduces $T_p = T_l$, $k = 1$. The time to decelerate from the full-load slip s_f to s_u is given by:

$$\begin{aligned} t &= 2H\int_{s_f}^{s_u} \frac{-ds}{1-(T'/T_l)} \\ &= 2s_m H\int_{s_f}^{s_u} \frac{s/s_m + s_m/s}{(s/s_m + s_m/s) - 2k} d(s/s_m) \\ &= 2s_m HT \end{aligned} \tag{11-57}$$

where H is the inertia constant, defined as the kinetic energy at rated speed in kilowatts-seconds per kilovolt-ampere. It is given by:

$$H = \frac{(0.231)(WR^2)(r/\text{min})^2 \times 10^{-6}}{kVA} \text{s} \tag{11-58}$$

where WR^2 is the motor and load inertia in lb-ft^2.

$$T = \left\{ s/s_m + k\ln[s/s_m(s/s_m + s_m/s - 2k)] + \frac{2k^2}{\sqrt{1-k^2}}\tan^{-1}\left(\frac{s/s_m - k}{\sqrt{1-k^2}}\right)\right\}_{s_f}^{s_u} \tag{11-59}$$

The limits s_f and s_u are given by:

$$s_f,\ s_u = s_m\left(T_p/T_l \pm \sqrt{(T_p/T_l)^2 - 1}\right) \tag{11-60}$$

Example 11-1 Consider a 2000-hp, six-pole, 60-Hz load inertia = 16780 lb-ft^2; total inertia including motor inertia = 20000 lb-ft^2, breakdown torque T_p = 150 percent, and critical slip s_m = 10 percent. Calculated H = 3.6 s, s_f = 3.8 percent, and s_u = 26.2 percent.

The calculated stability on voltage dips using Eqs. (11-57) and (11-59) is shown in Fig. 11-8.[5] This figure also shows the effect of H, and the curve for H = 7.2 s is included for comparison. A higher inertia makes the machine more stable on voltage dips, due to stored energy in the rotating masses and the speed falls more slowly.

11-4-1 Negative-Sequence Characteristics

Figure 11-9 depicts the negative-sequence-equivalent circuit of an induction motor. When a negative-sequence voltage is applied, the mmf wave in the air gap rotates backward at a slip of 2.0 pu. The

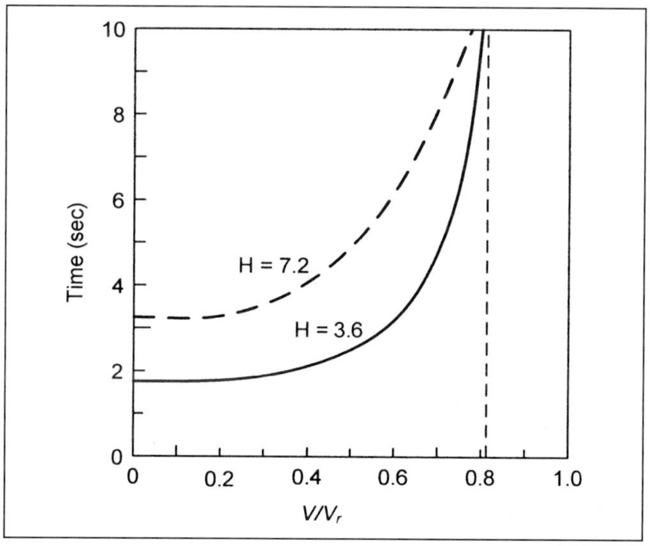

FIGURE 11-8 Calculated stability limits of a 2000-hp induction motor on voltage dips.

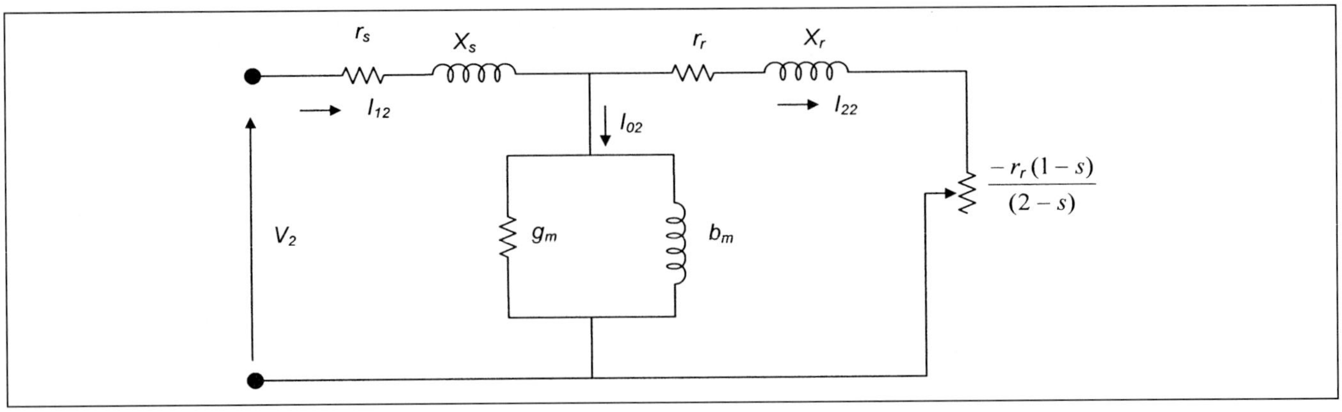

FIGURE 11-9 Equivalent circuit of an induction motor for negative slip.

slip of the rotor with respect to the backward rotating field is 2 s. This results in a retarding torque component, and the net motor torque reduces to:

$$T = \frac{r_r}{\omega_s}\left(\frac{I_r^2}{s} - \frac{I_{22}^2}{2-s}\right) \tag{11-61}$$

where I_{22} is the current in the negative-sequence circuit. From equivalent circuits of Figs. 11-2 and 11-9, we can write the approximate positive- and negative-sequence impedances of the motor as:

$$\begin{aligned} Z_1 &= [(r_s + r_r/s)^2 + (X_s + X_r)^2]^{1/2} \\ Z_2 &= [(r_s + r_r/(2-s))^2 + (X_s + X_r)^2]^{1/2} \end{aligned} \tag{11-62}$$

Therefore, approximately, the ratio $Z_1/Z_2 = I_s/I_f$, where I_s is the starting current or the locked rotor current of the motor and I_f is the full load current (s =1 at starting). For an induction motor with locked rotor current is generally six times the full load current, and the negative sequence impedance is one-sixth of the positive sequence impedance. A 5 percent negative sequence component in the supply system will produce 30 percent negative sequence current in the motor, which gives rise to additional heating and losses. Equations (11-62) are a simplification; the rotor resistance will change with respect to high rotor frequency and rotor losses are much higher than the stator losses. A 5 percent voltage unbalance may give rise to 38 percent negative sequence current with 50 percent increase in losses and 40° C higher temperature rise as compared to operation on a balanced voltage with zero negative sequence component. Also, the voltage unbalance should not be confused with negative sequence component. NEMA definition of percent voltage unbalance is maximum voltage deviation from average voltage divided by average voltage as a percentage. Operation above 5 percent unbalance is not recommended.[6]

The zero sequence impedance of motors, whether the windings are connected in wye or delta, is infinite. The motor windings are left ungrounded, as per industry practices in the United States.

11-5 SHORT-CIRCUIT TRANSIENTS OF AN INDUCTION MOTOR

Following the analogy of synchronous machines, the transient reactance of an induction machine is defined as:

$$X' = X_s + \frac{X_m X_r}{X_m + X_r} \tag{11-63}$$

This is also the motor-locked rotor reactance. The open-circuit transient time constant is:

$$T_0' = \frac{X_r + X_m}{\omega r_r} \tag{11-64}$$

Short-circuit transient time constant is:

$$\begin{aligned} T' &= T_0' \frac{X'}{X_s + X_m} \\ &\approx \frac{X'}{\omega r_r} \end{aligned} \tag{11-65}$$

The time constant for the decay of dc component is:

$$T_{\mathrm{dc}} = \frac{X'}{\omega r_s} \tag{11-66}$$

The ac symmetrical short-circuit component is:

$$i_{\mathrm{ac}} = \frac{E}{X'} e^{-t/T'} \tag{11-67}$$

and the dc component is:

$$i_{\mathrm{dc}} = \sqrt{2}\frac{E}{X'} e^{-t/T_{\mathrm{dc}}} \tag{11-68}$$

Example 11-2 Consider a 900-hp, 2.3-kV, 4-pole motor with the following data: full-load efficiency = 94.7 percent, full-load power factor = 88 percent, full-load current = 202A, stator resistance 0.08 Ω, rotor resistance = 0.09 Ω, stator leakage reactance = 0.32 Ω, rotor leakage reactance at locked rotor = 0.45 Ω, magnetizing reactance, both direct and quadrature axis = 16 Ω, magnetizing resistance = 100.00 pu, full load slip = 0.15 percent, H = 1 s.

The motor is connected in a system configuration as depicted in Fig. 11-10*a* and is running in the steady state; switch *S*2 is closed for a long time. A three-phase short-circuit is created by closing the switch *S*1 at 10 ms. (Figure 11-10*b* pertains to Example 11-3.)

Figure 11-11 shows the EMTP simulation of three-phase motor currents. The current at the instant of short circuit is approximately 14 times the motor's full-load current.

11-6 STARTING METHODS

From the equivalent circuit of an induction motor in Fig. 11-2, and neglecting the magnetizing and eddy current loss circuit, the starting

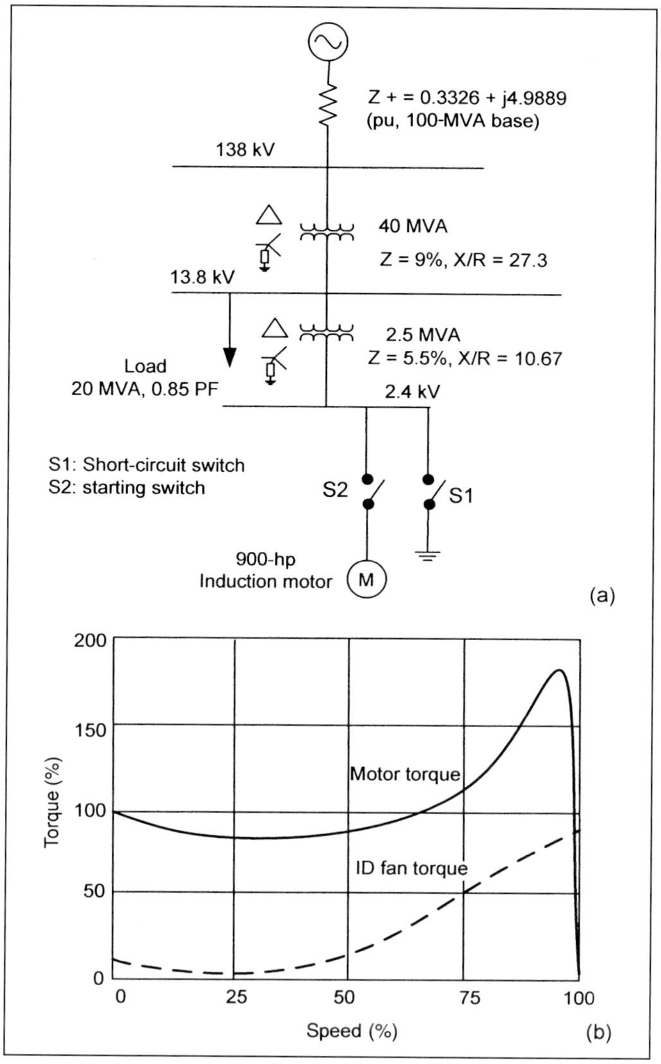

FIGURE 11-10 (*a*) System configuration for starting and terminal short circuit of a 900-hp motor. (*b*) Motor and load speed-torque characteristics.

current or the locked rotor current of the motor is:

$$I_s = \frac{V}{(r_s + r_r) + j(X_s + X_r)} \tag{11-69}$$

The locked rotor current of squirrel-cage induction motors is, generally, six times the full-load current on across the line starting, that is, full-rated voltage applied across the motor terminals with the motor at standstill. Higher or lower values are possible depending upon motor design. Wound rotor motors may be started with an external resistance in the rotor circuit to reduce the starting current and increase the motor-starting torque. The resistance may be cut out in steps as the motor speeds up. Synchronous motors are asynchronously started and their starting current is generally 3 to 4.5 times the rated full-load current on across the line starting. The starting currents are at a low power factor and may give rise to unacceptable voltage drops in the system and at motor terminals. On large voltage dips, the stability of running motors in the same system may be jeopardized, the motors may stall or the magnetic contactors may dropout, and other sensitive loads like adjustable speed drives (ASDs) may shutdown.

As the system impedances or motor reactances cannot be changed, impedance may be introduced in the motor circuit to reduce the starting voltage, and therefore, the starting current. This starting impedance is removed from the starting circuit as soon as the motor has accelerated to approximately 90 percent of its rated speed. Within certain limits, motors of lower locked rotor currents can be specified in the design stage of the project.

Table 11-3 shows a summary of the various starting methods. System network stiffness, acceptable voltage drops, starting reactive power limitations, motor and load parameters need to be considered when deciding upon a starting method. The characteristics of starting methods (Table 11-3), and the associated diagram of starting connections (Fig. 11-12) are summarized as follows:

Full Voltage Starting (Fig. 11-12a) This is the simplest method for required starting equipment and controls and is the most economical. It requires a stiff supply system because of mainly reactive high starting currents. The motor terminal voltage during starting and, therefore, the starting torque will be reduced, depending upon the voltage drop in the impedance of the supply system. It is the preferred method, especially when high inertia loads requiring high breakaway torques are required

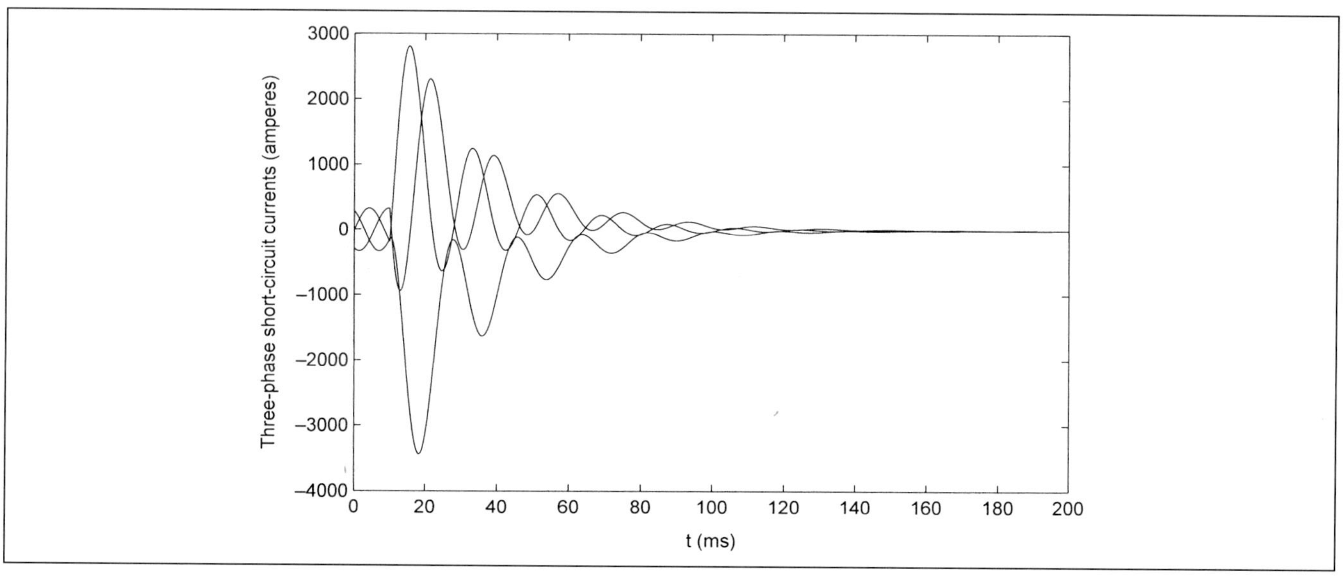

FIGURE 11-11 EMTP simulation of the terminal short circuit of a 900-hp induction motor.

TABLE 11-3 Starting Method of Motors

STARTING METHOD	REFERENCE FIGURE NO.	*I* STARTING	*T* STARTING	COST RATIO	QUALIFICATIONS
Full-voltage starting	Fig. 11-12*a*	1	1	1	Simplest starting method giving highest starting efficiency, provided voltage drops due to inrush currents are acceptable
Reactor starting	Fig. 11-12*b*	α	α^2	2.5	Simple switching closed transition, torque per kVA is lower as compared to autotransformer starting. A single reactor can be used to start a number of motors.
Krondrofer starting	Fig. 11-12*c*	α^2	α^2	3.5	Closed-circuit transition requires complex switching: (a) close Y, then S; (b) open Y; (c) close R; and (d) open S. Applicable to weak electrical systems, where the reduced starting torque of motor can still accelerate the loads.
Part-winding starting	Fig. 11-12*d*	α	α^2	Varies	Inrush current depends upon design of starting winding. Closed-circuit transition by switching the parallel stator winding by closing *R*. The starting torque cannot be varied and fixed at the design stage. Start: close *S*. Run: close *R*.
Shunt capacitor starting	Fig. 11-12*e*	Varies	Varies	3	May create harmonic pollution (harmonic filters are required). Capacitors switched off by voltage/current control when the speed approaches approximately 95 percent of the full-load speed (bus voltage rises as the motor current falls). Start: close *S*1, *S*2. Run: open *S*2.
Shunt capacitor in conjunction with other reduced voltage starting	Fig. 11-12*f*	Varies	Varies	Varies	A reactor starting with shunt capacitors has twofold reduction in starting current, due to reactor and capacitor. The motor terminal voltage and torque is increased as compared to reactor start.
Low-frequency starting	Fig. 11-12*g*	Varies (150–200% of load current)	Slightly more than load torque	6–7	May create harmonic pollution, not generally used due to high cost. Gives a smooth acceleration. The current from supply system can be reduced to 150–200% of the full load current. One low-frequency starting equipment can be used to start a number of motors in succession by appropriate switching.
Pony motor	Fig. 11-12*h*	-	-	Varies	Not generally used. The starting motor is high-torque intermittent-rated machine. Applicable when load torque during acceleration is small. Synchronizing required at *R*.

to be accelerated. Conversely, a very fast run up of low inertia loads can subject the drive shaft and coupling to high mechanical torsional stresses.

Reactor Starting (Fig. 11-12b) A starting reactor tapped at 40, 65, and 80 percent and rated on an intermittent duty basis is generally used. It allows a smooth start with an almost unobservable disturbance in transferring from the reduced to full voltage (when the bypass breaker is closed). The reactor starting equipment and controls are simple, though the torque efficiency is poor. Reactance of the reactor for a certain tap can be approximately calculated by the expression: $X = (1 - R)/R$, where R is the tap ratio or the voltage ratio in per unit of the rated voltage. Thus, the starting current at a particular tap is given by:

$$I_s = \frac{I}{Z_s + Z_m + Z_L} \tag{11-70}$$

It is in per unit based upon the motor rated voltage. Z_s, Z_m, and Z_L are the system, motor, and reactor impedances, reduced to a common base, say motor starting kVA base, and are vector quantities; respectively, they also give the starting power factor. As these impedances act as a potential divider, the starting voltage at the motor terminals is simply $I_s Z_m$.

The motor starting impedance can be calculated by $(V/\sqrt{3} \times I_s)$, where V is the motor-rated line-to-line voltage and I_s is the starting current at the motor at starting power factor.

Krondroffer Starting (Fig. 11-12c) An advantage of this method of starting is high torque efficiency, at the expense of additional control equipment and three switching devices per starter. Current from the supply system is further reduced by factor α, for the same starting torque obtained with a reactor start. Though close transition reduces the transient current and torque during transition, a quantitative evaluation is seldom attempted. The motor

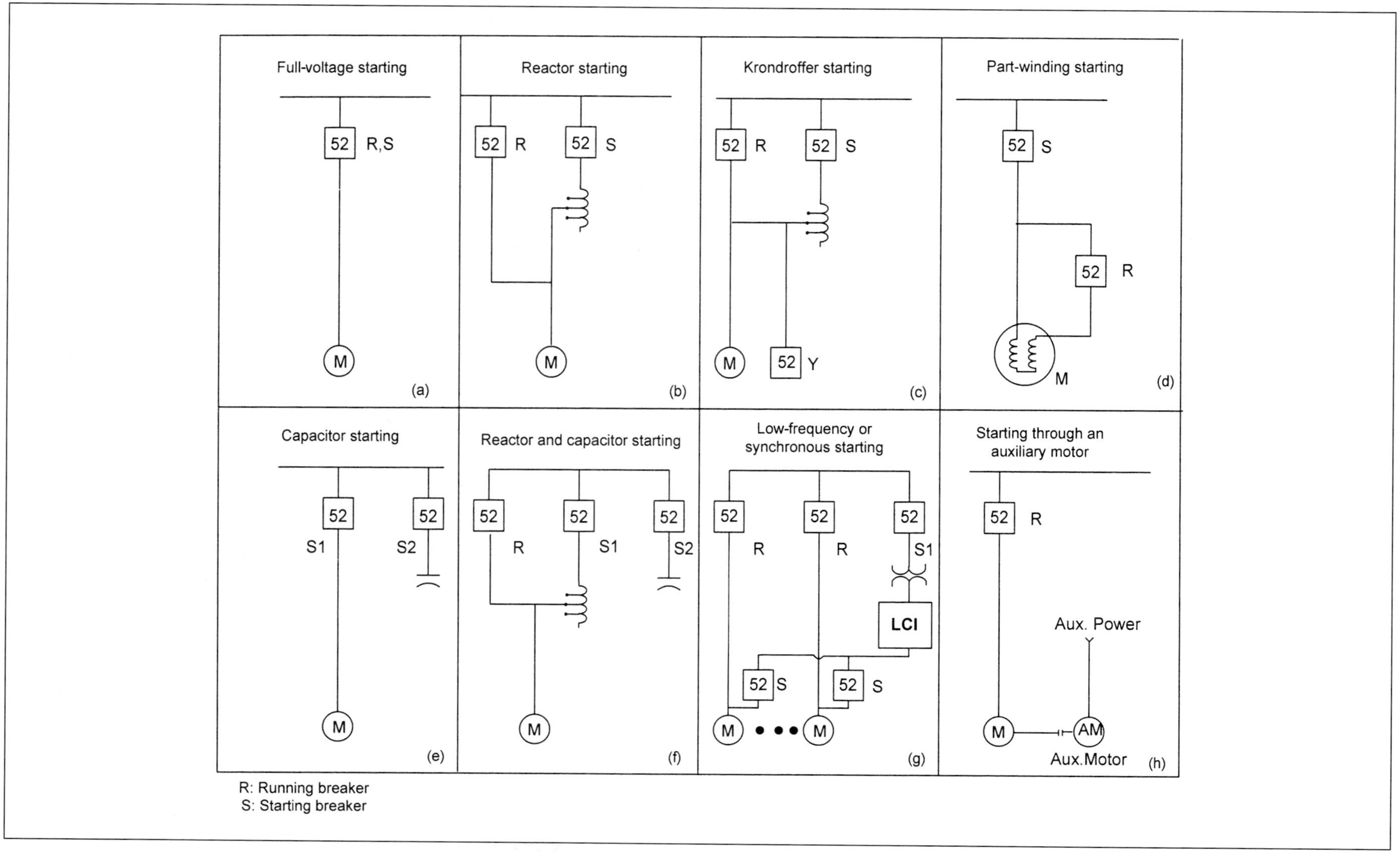

FIGURE 11-12 Starting methods of motors.

voltage during transition may drop significantly. For weak supply systems, this method of starting at 30 to 45 percent of the voltage can be considered, provided the reduced asynchronous torque is adequate to accelerate the load.

Part-Winding Starting (Fig. 11-12d) The method is applicable to large synchronous motors of thousands of horsepower designed for part-winding starting. These have at least two parallel circuits in the stator winding. This may add 5 to 10 percent to the motor cost. The two windings cannot be exactly symmetrical in fractional slot designs, and the motor design becomes specialized regarding winding pitch, number of slots, and the coil groupings. The starting winding may be designed for a higher temperature rise during starting. Proper sharing of the current between paralleled windings, limiting temperature rises, and avoiding hot spots become a design consideration. Though no external reduced voltage starting devices are required and the controls are inherently simple; the starting characteristics are fixed and cannot be altered. Part-winding starting has been applied to large thermo-mechanical pulping (TMP) synchronous motors of 10000 hp and above, yet some failures have been known to occur.

Capacitor Starting (Fig. 11-12e and f) The power factor of the starting current of even large motors is low, rarely exceeding 0.25. Starting voltage dip is dictated by the flow of starting reactive power over mainly inductive system impedances. Shunt-connected power capacitors can be sized to meet a part of the starting-reactive kvar, reducing the reactive power demand from the supply system. The voltage at the motor terminals improves and, thus, the available asynchronous torque. The size of the capacitors selected should ensure a certain starting voltage across the motor terminals, considering the starting characteristics and the system impedances. As the motor accelerates and the current starts falling, the voltage will increase and the capacitors are switched off at 100 to 104 percent of the normal voltage, sensed through a voltage relay. A redundant current switching is also provided. For infrequent starting, shunt capacitors rated at 60 percent of the motor voltage are acceptable. When harmonic resonance is of concern, shunt capacitor filters can be used.

Capacitor and reactor starting can be used in combination (Fig. 11-12*f*). A reactor reduces the starting inrush current and a capacitor compensates part of the lagging starting kvar requirements. These two effects in combination further reduce the starting voltage dip.[7]

Low-Frequency Starting or Synchronous Starting (Fig. 11-12g) Cycloconverters and load-commutated inverters (LCI) have also been used for motor starting.[8,9] During starting and at low speeds, the motor does not have enough back EMF, so to commute the inverter, thyristors and auxiliary means must be provided. A synchronous motor runs synchronized at low speed, with excitation applied, and accelerates smoothly as LCI frequency increases. The current from the supply system can be reduced to 150 to 200 percent of the full-load current, and the starting torque need only be slightly higher than the load torque. The disadvantages are cost, complexity, and large dimensions of the starter. Tuned capacitor filters can be incorporated to control harmonic distortion and possible resonance problems with the load-generated harmonics. Large motors require a coordinated starting equipment design.

Figure 11-12g shows that one starter can be switched to start a number of similar rated motors. In fact, any of the reduced-voltage systems can be used to start a number of similar motors, though additional switching complexity is involved.

Starting Through an Auxiliary Motor (Fig. 11-12h) It is an uncommon choice, yet a sound one in some circumstances involving a large machine in relation to power supply system capabilities. No asynchronous torque is required for acceleration and the motor design can be simplified. The starting motor provides an economical solution only when the load torque during acceleration is small. Disadvantages are increased shaft length and the rotational losses of the starting motor, after the main motor is synchronized.

11-6-1 Starting Considerations

When deciding upon a starting method, considerations include load characteristics, frequency of starting, motor and power system characteristics, and maximum acceptable voltage dip. When the motors are started on utility systems with no nearby generators, the starting voltage dip lasts approximately for the entire duration of starting. When there is a generator running in synchronism with a utility source, it will share a part of the starting kvar requirement, and the voltage recovery will be faster. The voltage recovery profiles in these two cases, therefore, differ widely.

Acceptable Voltage Dips A large starting voltage dip may have the following adverse effects:

- It may reduce the net accelerating torque of the motor below the load torque during acceleration and result in lockout and unsuccessful starting or synchronizing.
- DC motor contactors in NEMA E1 and E2 medium-voltage motor controllers[10] may tolerate a voltage dip of 30 percent or more; however, the auxiliary control relays may drop out earlier. The low-voltage motor contactors will be more susceptible to voltage dips and may drop out in the first cycle at a lower voltage dip. The exact drop out value is difficult to estimate because of residual flux trapped in the contactor coil. Drop out between 20 and 70 percent of the voltage can occur.
- Induction motors experience a higher slip, and the torque angle of synchronous motors increases. On restoration of the voltage, the induction motors will reaccelerate, increasing the current demand from the supply system, which will result in more voltage drops. This phenomenon is cumulative and the resulting large inrush currents on reacceleration may cause a shutdown.

AC motors can tolerate a much higher voltage dip for a longer duration, as compared to electronic systems. The *lower* tolerance limits to abort a process interruption will be set by these systems, rather than the stability of ac motors on voltage dips.

Limitations of the Supply System The utilities may impose restrictions on the reactive power demand and acceptable voltage dips in their systems during starting of large motors.

Motor Characteristics Only a limited choice can be exercised in a particular design to alter the starting characteristics, that is, a reduction in starting current will affect the complete starting torque-speed characteristics. In a synchronous motor, the starting winding should have a high resistance to develop a higher starting torque, and the same winding acts as a damper winding, after the motor is synchronized. To rapidly damp out the torque angle pulsations on a disturbance or fluctuating load, this winding should have lower resistance, which conflicts with higher torque starting requirements. Compromises are, therefore, necessary. A solid-pole synchronous motor, generally, has a higher starting power factor as compared to the laminated-pole motor, which reduces the reactive kvar demand per kVA of starting impact.

Number of Starts and Load Inertia NEMA[6] specifies two starts in succession with the motor at ambient temperature and for the WK^2 of load and the starting method for which the motor was designed, and one start with the motor initially at a temperature not exceeding its rated temperature when loaded. If the motor is required to be subjected to more frequent starts, it should be so designed. The starting time is approximately given by the expression:

$$t = \frac{2.74\sum WK^2 N_s^2 10^5}{P_r} \int \frac{dn}{T_a - T_l} \tag{11-71}$$

where t is the accelerating time in seconds, P_r is the rated output, and the other terms used have already been described before. As per Eq. (11-71), accelerating time is solely dependent upon the load inertia and inversely proportional to the accelerating torque. The normal inertia loads of the synchronous and induction motors are tabulated in Ref. 6. For a synchronous machine it is defined by the following equation:

$$\mathrm{WK}^2 = \frac{0.375(\text{hp rating})^{1.15}}{(\text{speed rpm}/1000)^2} \quad \text{lb-ft}^2 \qquad (11\text{-}72)$$

The heat produced during starting is given by the following expression:

$$h = 2.74\sum \mathrm{WK}^2 N_s^2 10^{-6} \int \frac{T_a}{T_a - T_l} s ds \qquad (11\text{-}73)$$

where h is the heat produced in kW-sec, and, again, it depends on load inertia. The importance of load inertia on the starting time and accelerating characteristics of the motor is demonstrated in Eq. (11-73).

11-7 STUDY OF STARTING TRANSIENTS

The methodology for evaluating the starting performance and transients of motors varies, depending on the specific needs of the system study. This requires simple to complex modeling, depending upon the nature of the study. Following is the industry practice:

1. *Snapshot study.* Many times the initial voltage dip on starting is of only interest. This can be simply done by a load-flow algorithm. While no idea of the transients or starting time and profiles of torque, slip, or current can be made, the load-flow algorithms give a steady-state picture for the specified loading conditions. The starting impact load of the motor is modeled and the system voltage dips for various switching conditions can be calculated, after the base case is established. This is also called the static motor starting study and is not of interest here.

2. *Dynamic motor starting.* The dynamic motor starting will plot out the starting motor torque, current, slip, accelerating time, and the like. A more detailed model of the load and motor torque-slip characteristics and inertia is required.

3. *Transient stability–type programs.* Transient stability–type programs (Chaps. 12 and 13) can be used. These will analyze the transients in the generators, load buses, and running motors on starting impact load of a motor (Example 11-5).

4. *EMTP-type programs.* EMTP-type programs give the real nature of transient behavior of the motors during starting in time or frequency domain.

Example 11-3 Dynamic starting of a 900-hp motor of specifications of Example 11-2 is simulated. The motor drives a boiler induced draft (ID) fan; H of the motor and load = 3.37 s. The system configuration is shown in Fig. 11-10*a*, and the motor and load starting torque profile is in Fig. 11-10*b*. Switch *S2* is closed to start the motor, and switch *S1* is open.

The starting transients are plotted in Fig. 11-13. These show motor kW and kvar demand and kW output; motor and load torque and the accelerating torque; the slip; the starting current; and the bus voltage. The starting time is 16 s. Some boiler ID fans may take 50 to 60 s to accelerate due to high load inertia.

Example 11-4 Example 11-3 is repeated with EMTP simulation. The motor is started unloaded to reduce the starting time, $H = 1$ s. The purpose is to illustrate the nature of transients and the difference with respect to conventional dynamic starting study. Figure 11-14*a*, *b*, and *c* shows the motor current in phase *a*, electromagnetic torque in Newton-meter (N-m), and the motor slip; respectively. Generally, EMTP simulation is not used due to complexity.

11-8 SYNCHRONOUS MOTORS

Synchronous motors of the revolving-field-type can be divided into two major types:

- *Salient solid-pole synchronous motors.* These are provided with concentrated field windings on the pole bodies

- *Cylindrical rotor synchronous motors.* These have distributed phase windings over the periphery of the rotor which is used for starting and excitation.

The salient-pole synchronous motor can again be classified into two types. The solid-pole type has poles in one-piece of solid mass, made of cast steel, forged steel, or alloy steel. The starting torque is produced by eddy currents induced in the pole surface. Sometimes, slots are provided in the pole surface to absorb thermal stresses. A solid-pole construction offers high thermal and mechanical reliability, and its large starting torque makes this construction suitable for starting high inertia loads, for example blowers, compressors, and sintering machines.

Salient-laminated-pole type has poles of punched sheet steel, laminated, compressed, and formed. The pole head may be equipped with starting windings of the deep-slot squirrel-cage type or double squirrel-cage type. A low-resistance starting winding will give low starting torque but high pullout torque, and vice versa. A double-cage winding can give high starting torque, while limiting the inrush currents, and when synchronous speed is reached, the low-resistance cage functions to produce large pull-in torque.

The performance characteristics can be depicted by V curves, extended into generator region to show the reversible nature of synchronous machines (Fig. 11-15). The effect of excitation on stability is apparent, and we will return to this topic in Chap. 13. A synchronous motor is more stable when operating at a leading power factor; conversely, a generator is more stable when operating at a lagging power factor, that is, supplying reactive power into the system.

A brief reference is made to synchronous induction motors, which can again be of a salient-pole or cylindrical rotor type. The rotor is constructed similar to a wound rotor induction motor, though the air gap is larger than that of an induction motor to increase the pullout torque. The rotor winding is of low resistance and can be connected to external resistors for starting, permitting an easy and smooth start even with heavy loads and giving highest torque efficiency. These motors can be designed to operate as induction motors for a short period when these fall out of step, that is, due to a sudden reduction in system voltage or excessive load torque and resynchronize automatically on reduction of load torque or restoration of voltage. The motor can operate as an induction motor for 10 to 15 s, when the system voltage has dipped to 45 percent of the rated voltage. Automatic resynchronizing on restoration of voltage to 90 percent of the rated value is possible. Some disadvantages are small thermal capacity of the rotor, special low-voltage excitation systems, and added starting, excitation, and control equipment costs. These have been used in rolling mills and mine ventilating fans; but are getting out of favor due to complexity of starting equipment and controls.

The starting of synchronous motors should consider similar factors as discussed for the induction motors, with the addition of synchronization, pulling out of step, and resynchronization.

The synchronous motor design permits lower starting currents for a given starting torque as compared with a squirrel-cage induction motor, typically, in ratio of 1:1.75. This may become an important system design consideration, when the starting voltage dips in the power system have to be limited to acceptable levels.

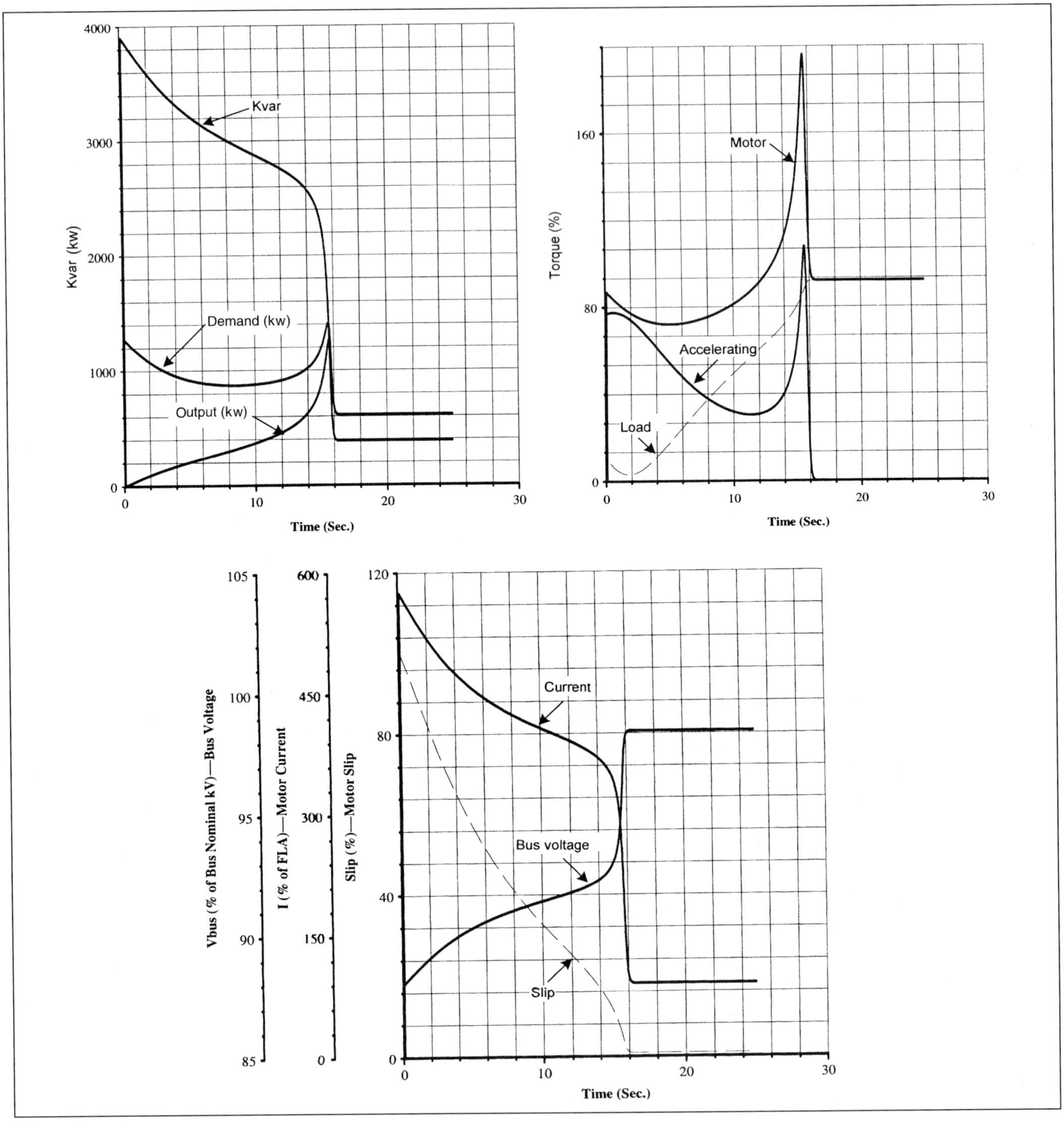

FIGURE 11-13 Dynamic starting study of 900-hp motor starting transients. FLA, full load current.

Based upon the motor rating and speed, a general application of synchronous and induction motors is depicted in Fig. 11-16. For low-speed applications, motor efficiency becomes of consideration—the induction motor efficiency drops because of leakage reactance of the stator winding overhangs. For a relative comparison, the efficiency of a 16-pole induction motor and synchronous motor may be 92 percent versus 96.5 percent.

11-8-1 Starting Characteristics

Figure 11-17 depicts the starting characteristics of two basic types of synchronous motors: salient pole and laminated pole. Salient-pole motors have higher starting torque (curve *a*) and pulsating or oscillating torque (curve *b*). The cylindrical rotor machines have starting torque characteristics akin to induction motors (curve *c*) and smaller oscillating torque (curve *d*). Salient-laminated-pole construction with damper windings may produce a characteristic somewhere in-between the solid-pole and cylindrical rotor designs.

Referring to Fig. 11-17, T_a is the asynchronous torque of the motor analogous to the induction motor starting torque, as a synchronous motor is started asynchronously. The rotor saliency causes another torque T_p, which pulsates at a frequency equal to twice the slip frequency. Considering a supply system frequency of f Hz, and a rotor slip of s, the magnetic field in the rotor has a frequency of sf. Due to saliency it can be divided into two components:

(a) a forward-revolving field at frequency sf in the same direction as the rotor, and (b) a field revolving in reverse direction at sf. Since the rotor revolves at a speed $(1 - f)s$, the forward field revolves at $(1 - s)f + sf$, that is, at fundamental frequency with respect to stator, in synchronism with the rotating field produced by the stator three-phase windings. The interaction of this field with the stator field produces the torque T_a in a fixed direction. Negative sequence field revolves at: $(1 - f)s - sf = (1 - 2f)s$ This is as viewed from the stator and, thus, it has a slip of $f - (1 - 2s)f = 2sf$ with respect to the field produced by the exciting current. Figure 11-18 shows these relative rotations of stator and rotor fields in a synchronous motor during starting. The total torque produced by the synchronous motor is therefore:

$$T_m = T_a + T_p \cos 2s\omega t \tag{11-74}$$

The currents corresponding to the positive and negative fields are I_a and I_p, respectively. The current I_a is at fundamental frequency and I_p is the pulsating current at $(1 - 2s)f$ Hz. The total current is given by the expression:

$$I = I_a \cos(\omega t - \phi) + I_p \cos(1 - 2s)\omega t \tag{11-75}$$

where ϕ is the power factor angle at starting.

The above treatment of the asynchronous torque during starting of the synchronous motors with solid poles or laminated poles and damper windings is not exhaustive. The asynchronous torque can be considered to have three components, generated by direct axis and quadrature axis circuit, and the single-axis motor-exciting winding also contributes to the starting torque. Due to electrical

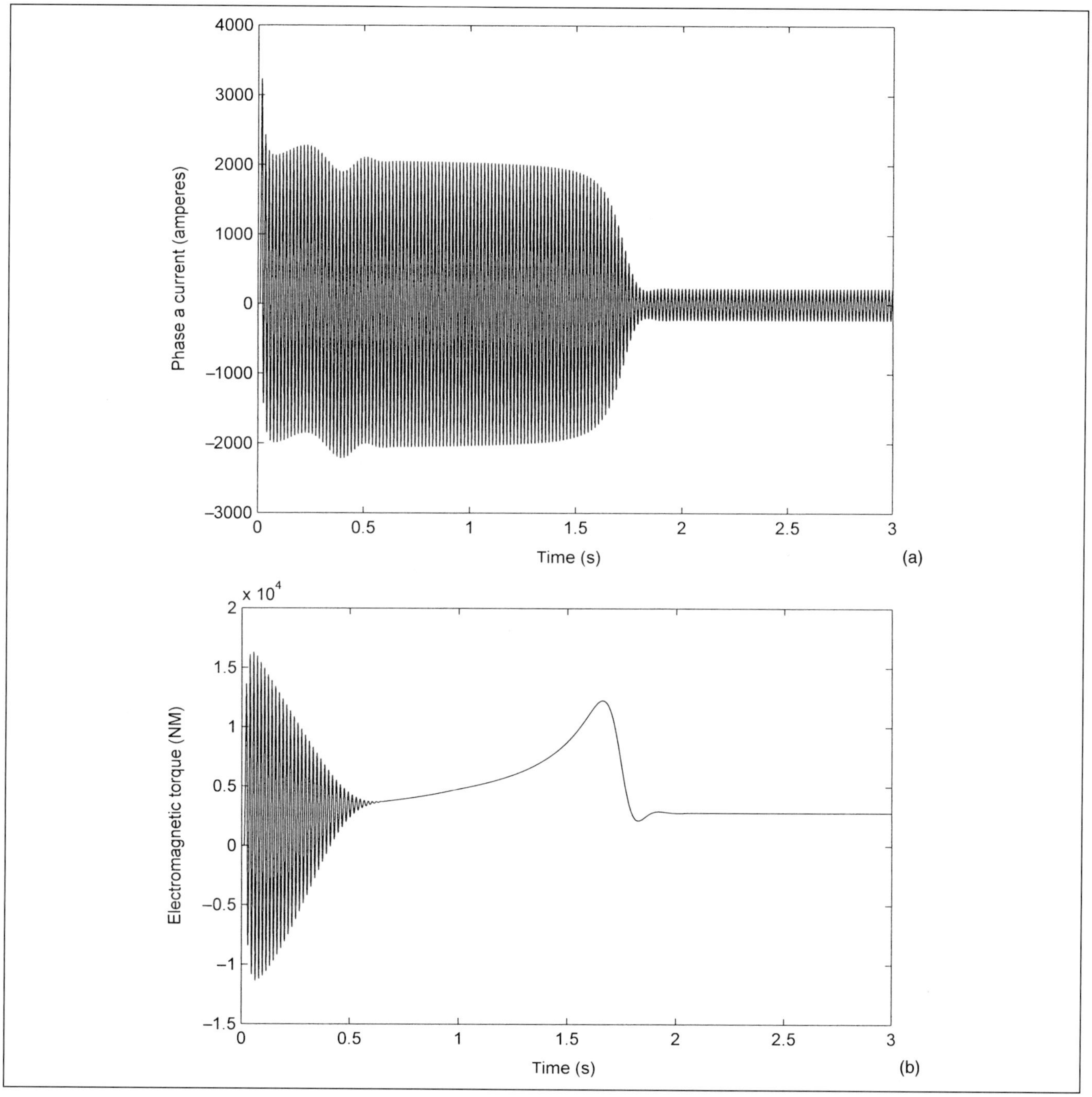

FIGURE 11-14 (*a*) Phase *a* current, (*b*) electromagnetic torque, and (*c*) slip of 900-hp motor starting transients.

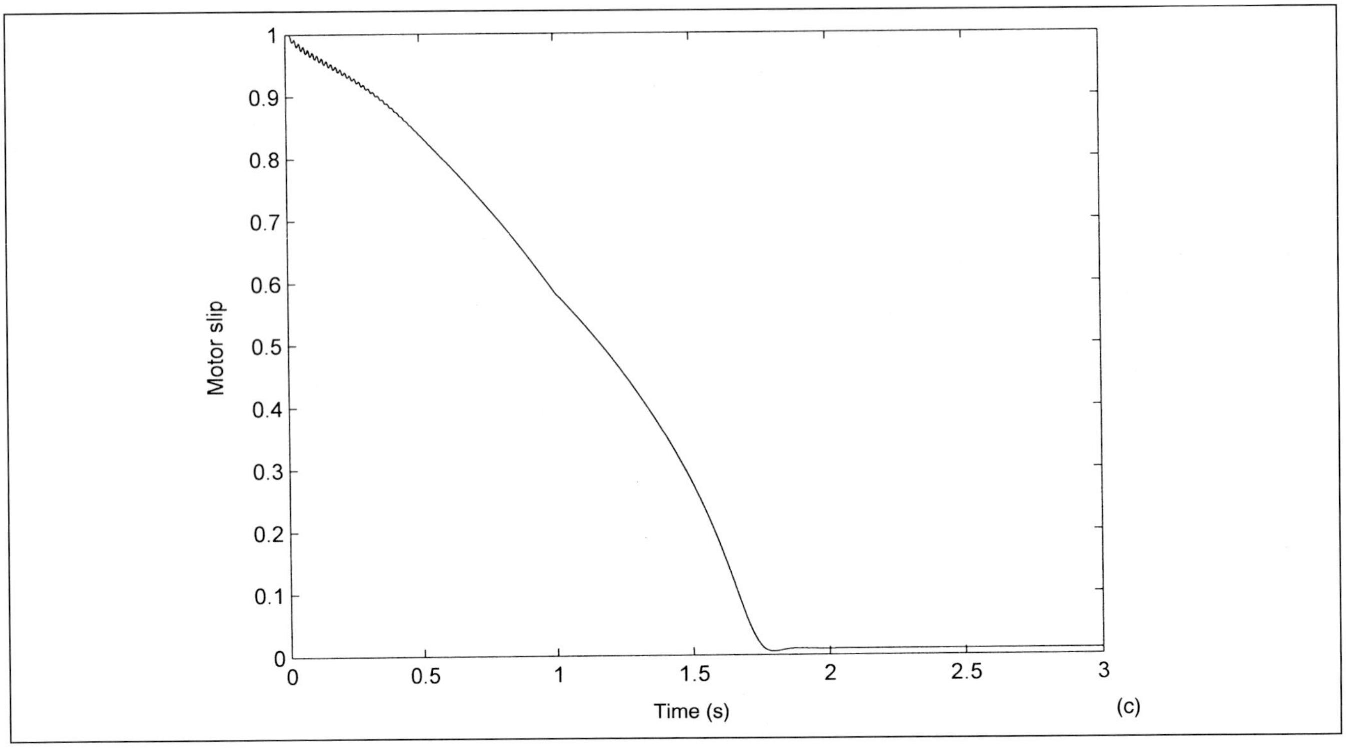

FIGURE 11-14 (*Continued*)

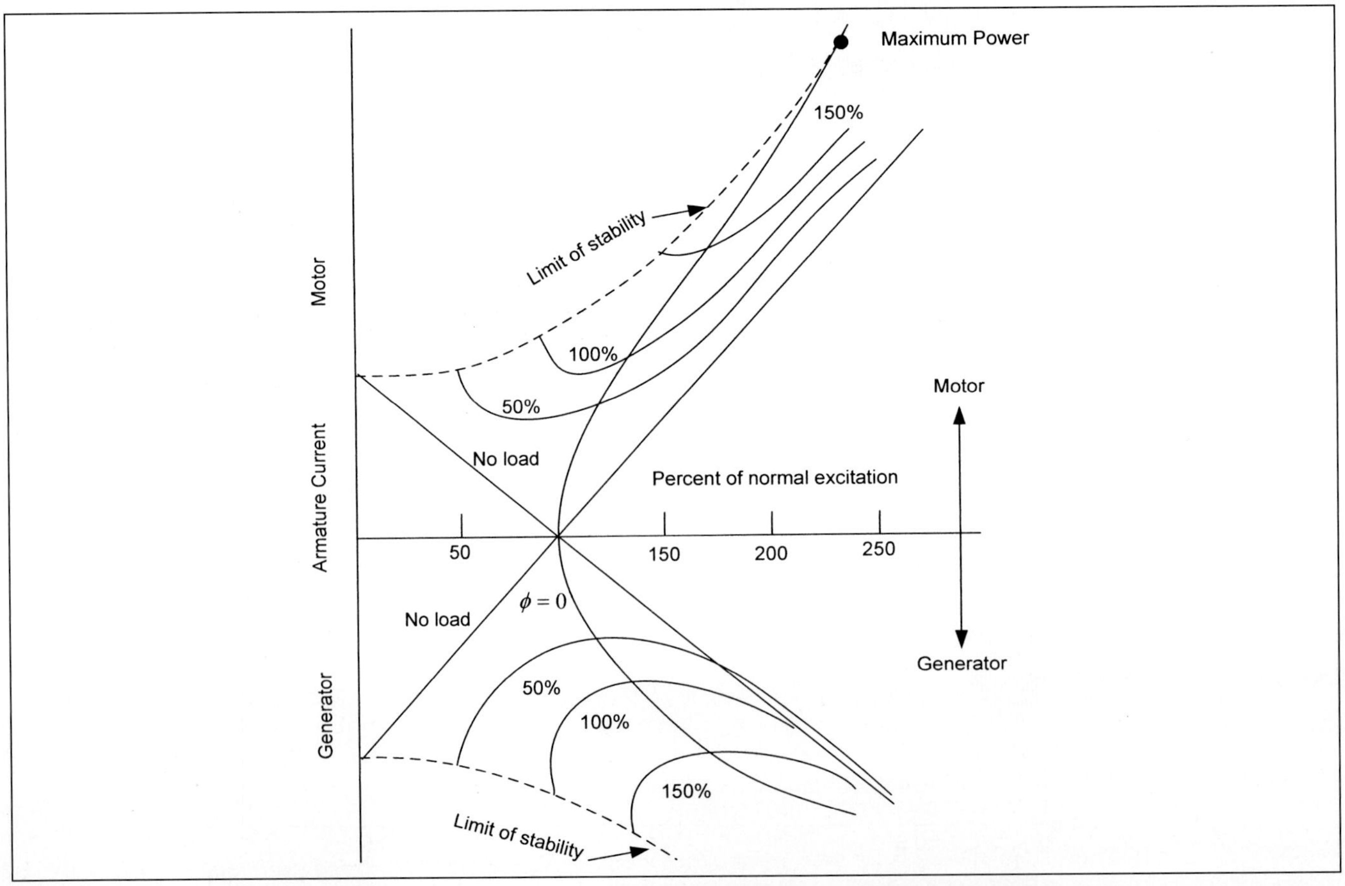

FIGURE 11-15 V curves showing the reversible nature of synchronous machines.

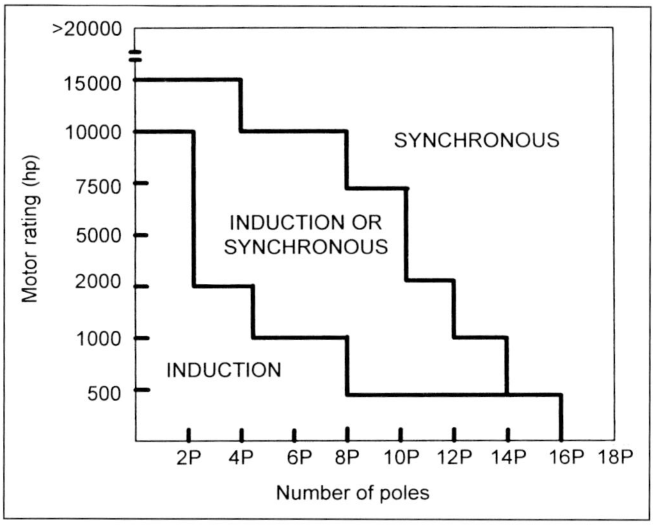

FIGURE 11-16 General application guidelines of synchronous and induction motors based upon motor rating and number of poles.

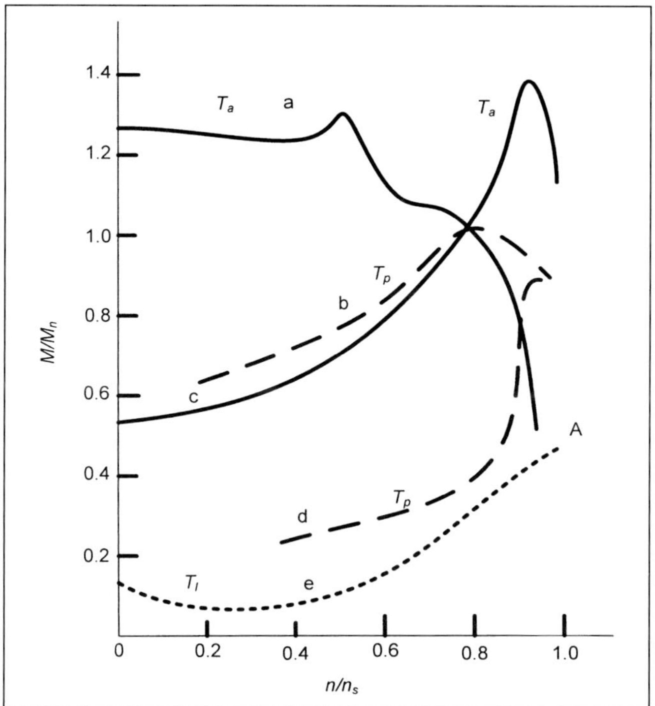

FIGURE 11-17 Torque-speed characteristics of synchronous motors. Curves *a* and *b* starting and pulsating torque of solid-pole synchronous motors, curves *c* and *d* starting and pulsating torque of cylindrical rotor synchronous motors, and curve *e* load torque curve.

and magnetic asymmetry, T_a develops a sudden change near 50 percent speed, resulting in a singular point. This is called the Gorges phenomenon.[11] The shape and size of this dip are governed by the machine data and the power system. If resistance is zero, the saddle disappears. Solid-poles motor without damper windings and slip-rings will give a more pronounced saddle, while the saddle is a minimum in cylindrical rotor and laminated-pole constructions with damper windings. The external resistances during asynchronous starting impact all T_a, T_p, and the saddle. At slip $s = 0.5$, the reverse rotating field comes to a halt as viewed from the stator, and torque T_p becomes zero, so also the pulsating current I_p. For slip

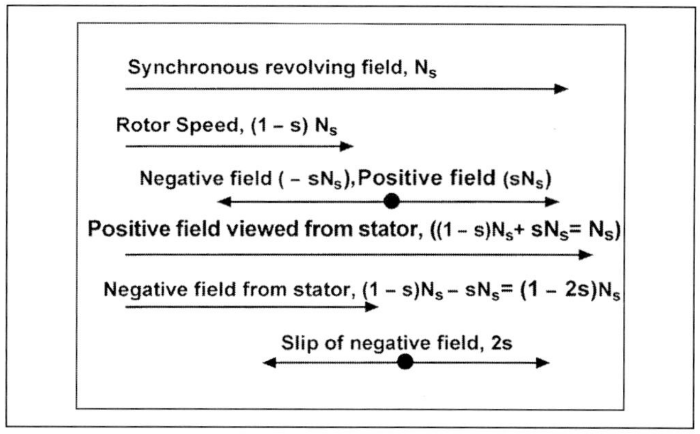

FIGURE 11-18 Relations of rotating fields in a synchronous motor, during starting.

less than 0.5, the direction of the torque changes; this produces a characteristic saddle. The average starting torque and the starting current can be calculated based upon the direct axis and quadrature axis equivalent circuits of the synchronous machine.[1,2]

Comparison of the starting torque speed characteristics of the cylindrical rotor versus solid-pole motors in Fig. 11-17 indicates that a solid-pole motor produces highest torque at the starting instant. The load curves follow mostly square law characteristics. The cylindrical rotor design has the following advantages as compared to solid-pole designs:[12,13]

1. At speeds close to the synchronizing speeds, the torque of a salient-pole motor reduces; this may make the synchronizing difficult.

2. The solid-pole motor has a considerably large pulsating torque as compared to a cylindrical rotor machine, as the magnetic asymmetry in the later type is minimized. These higher pulsating torques stress the shaft, motor, and drive system.

3. The first natural frequency of a motor-driven load lies between 5 and 25 Hz. Thus, higher torsional stresses will occur in solid-pole motors, every time these are started.

4. A cylindrical rotor machine, in principle, is suitable for asynchronous running, and depending upon the design of the starting winding, the motor can be operated at 50 to 90 percent of its rated output. The salient-pole motors have no continuous asynchronous running ratings, as the saliency causes torque pulsations, and slip-proportional losses lead to overheating of pole surfaces and the starting windings.

Conversely the solid-pole motors give a higher breakaway torque, and lesser heat energy is produced in the solid-pole rotors during starting. Solid-pole motor designs have been successfully applied in large four-pole motors of 25000 hp or higher.

11-8-2 Pullin Torque of a Synchronous Motor

Figure 11-17 also shows the load torque T_l during starting (curve *e*). The accelerating torque is $T_a - T_l$ and the motor will accelerate till point *A*, close to the synchronous speed. At this moment, pullin torque refers to the maximum load torque under which the motor will pull into synchronism while overcoming inertia of the load and the motor when the excitation is applied. The torque of the motor during acceleration as induction motor at 5 percent slip at rated voltage and frequency is called *nominal pullin torque*. Pullin and pullout torques of salient-pole synchronous motors for normal values of load inertia are specified in NEMA standards.[6] For motors of 1250 hp and above, speeds 500 to 1800 rpm, the pullin torque should not be less than 60 percent of the full-load torque. The maximum slip at pullin can be approximately determined from the following inequality:

$$s < \frac{242}{N_s}\sqrt{\frac{P_m}{\sum WK^2 f}} \tag{11-76}$$

where N_s is the synchronous speed in rpm, f is the system frequency, P_m is the maximum output of the synchronous motor with excitation applied, when pulling into synchronism, and WK^2 is the inertia of the motor and driven load in kg-m^2.

It is necessary to ascertain the load torque and operating requirements before selecting a synchronous-motor application. For example, wood chippers in the paper industry have large pullout torque requirements of 250 to 300 percent of the full-load torque. Banbury mixers require starting, pullin, and pullout torques of 125 percent, 125 percent, and 250 percent, respectively. Thus, the load types are carefully considered in applications. Load inertia impacts the heat produced during starting and the starting time. NEMA[6] gives the starting requirements of pumps and compressors and inertia ratios.

11-8-3 Synchronization Transients

A synchronous motor is synchronized during the starting cycle by application of the field excitation at (a) proper rotor speed and (b) proper rotor angle. The rotor speed is sensed by the frequency of induced currents in the field windings during starting, when the rotor has accelerated close to 92 to 97 percent of the rated speed. The rotor angle is sensed by the point where induced current passes through zero going from negative to positive and the excitation is applied with correct polarity so that maximum torque is obtained at the pullin. Application at wrong polarity will result in swinging. The successful synchronization is dependent upon:

1. Slip
2. Opposing torque
3. Field current applied
4. Machine data

A lightly loaded motor can synchronize without external excitation due to reluctance torque. Excitation system should also provide a discharge path for the current induced in the field during starting.

Excitation can be boosted during shock loads and system voltage depressions to counteract the possibility of the motor pulling out of step. If the motor does pull out of step, the excitation should be removed.

Earlier excitation systems were of the rotary type consisting of a special dc generator with a shaft-mounted commutator, and the dc voltage was applied to the motor through brushes and slip rings. These were replaced with single-phase or three-phase diode bridges, and the excitation could only be varied through taps on the rectifier transformer. Phase-controlled rectifiers and brushless excitation systems were subsequent developments.

The excitation system of a synchronous motor must provide a discharge path for the currents induced in the field windings of the motor during starting and open the discharge resistor when the excitation is applied, reinserting it in the circuit when the motor pulls out and the excitation is removed. At standstill, field windings are open circuited, and high voltages of the order of 10 to 40 kV can be induced in the field windings, which act akin to the open circuited secondary of a transformer.

The discharge resistor has also an effect upon the torque near the synchronous speed; this effect is more pronounced in salient-pole construction. A compromise in the resistor value is made depending upon the motor design.

11-8-4 Brushless Excitation Systems

Figure 11-19 shows a brushless excitation system which meets the above criteria. The ac output of the rotating armature exciter is converted into dc, but the control circuit blocks SCR1 from firing until the induced frequency in the rotor circuit is low. When the field terminal MF1 is positive, with respect to MF2, the diode D conducts, and on the reverse cycle SCR2 conducts during starting and connects the discharge resistor FD across the motor field windings MF. This protects the motor windings and electronics from high induced voltages. As the induced frequency of the current in the field windings decreases, the frequency part of the control circuit blocks SCR2, and simultaneously SCR1 conducts to apply the field excitation. A zero slip circuit is included to apply excitation if the motor synchronizes on reluctance torque.

On first half cycle after pullout, the induced field voltage across SCR1 switches it off, removing the motor excitation, and SCR2 conducts to reconnect the discharge resistor across the field windings. If the motor does not synchronize, it will slip a pole; the induced field voltage opposes the exciter voltage causing current to go to zero, turning off SCR1. The SCR2 is turned

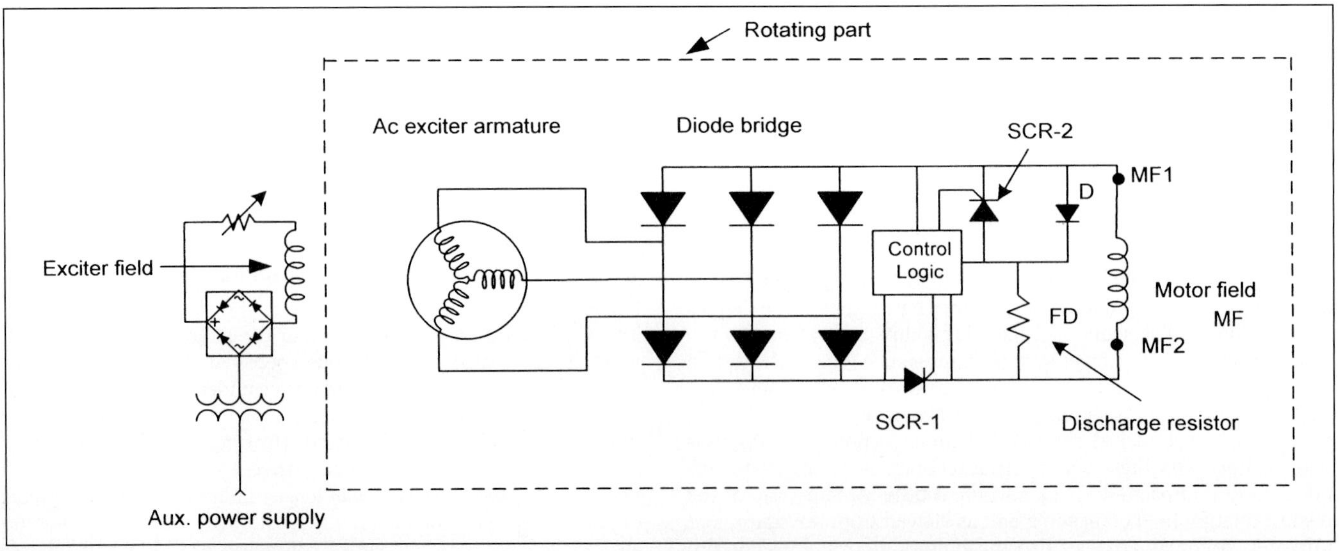

FIGURE 11-19 Brushless excitation system for a synchronous motor.

on only at a voltage higher than the exciter voltage and will not be conducting when SCR1 is conducting. There is a positive interlocking between SCR1 and SCR2. Figure 11-20 shows voltages during one cycle and Figs. 11-21 and 11-22 depict the field current and field voltage transients during starting. The circuit of Fig. 11-20 can be applied to brush type as well as to brushless type of excitation systems.

11-9 STABILITY OF SYNCHRONOUS MOTORS

The simplified current and power diagram of a synchronous motor is depicted in Fig. 11-23*a*. The derivation follows from Fig. 11-15, and is also arrived analytically in many texts.[4] The reactive power can be varied by control of excitation. The variation range of the excitation current I_e is limited by maximum operating temperature of the stator and rotor and by the stability limit of the motor. In this figure the stator temperature limit is shown by a dotted arc, and the rotor temperature limit is shown by a solid arc. The underexcited stability limit is given by dotted line. The horizontal lines parallel to *x*-axis

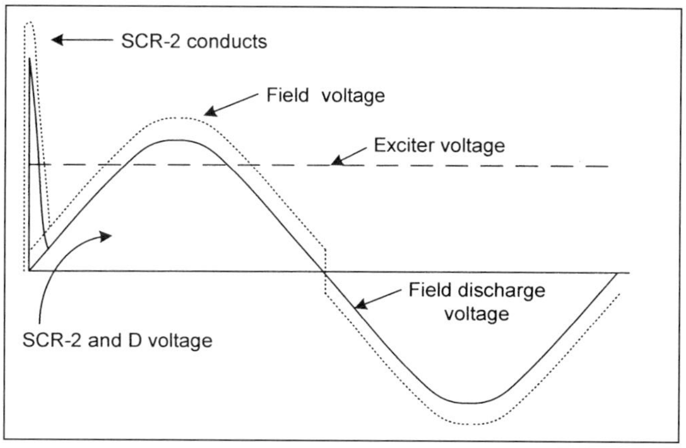

FIGURE 11-20 Relations during one cycle in the brushless excitation system of Fig. 11-19.

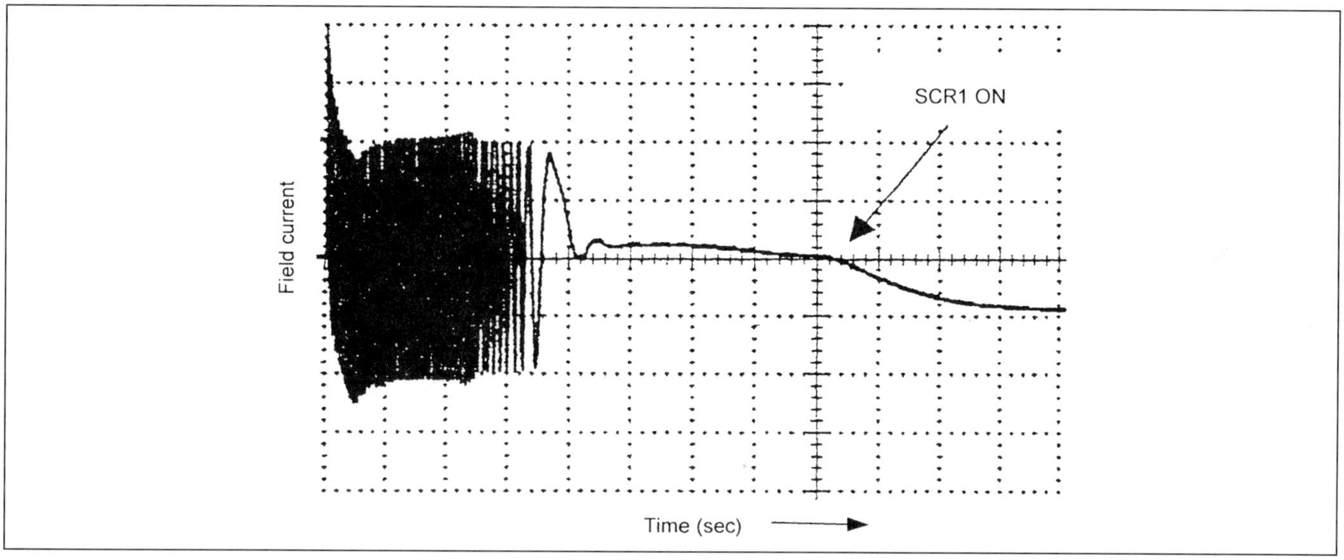

FIGURE 11-21 Field current from brushless excitation system of Fig. 11-19, during starting.

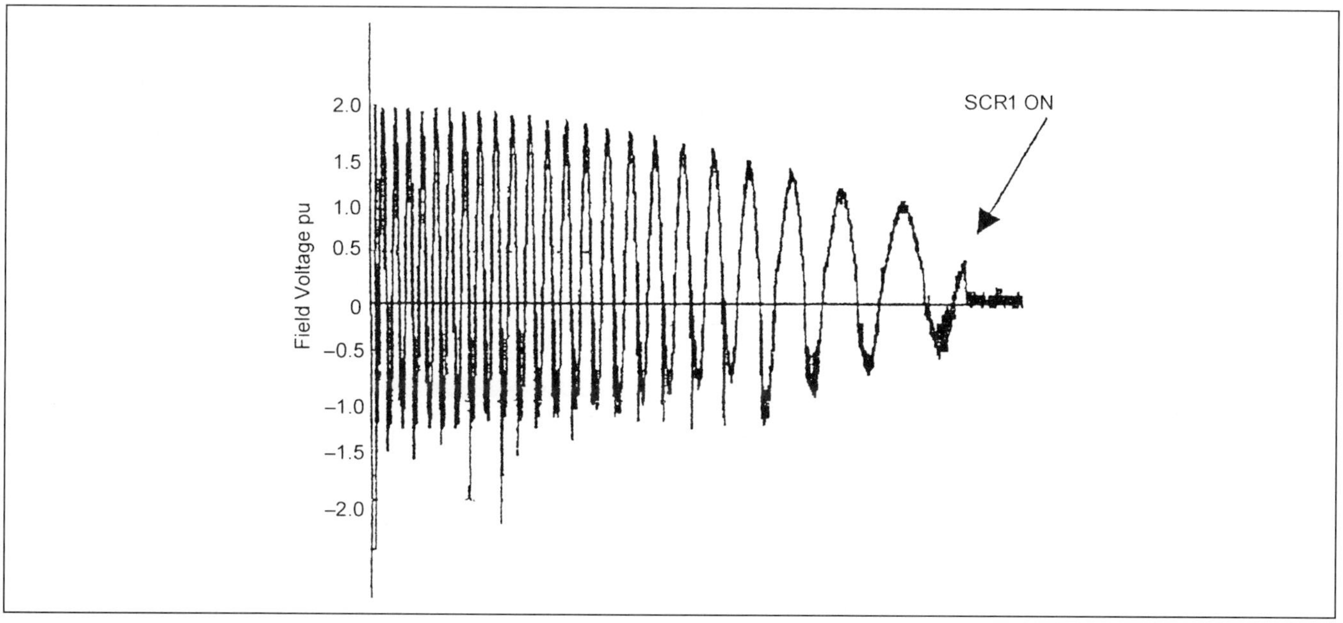

FIGURE 11-22 Field voltage from brushless excitation system of Fig. 11-19, during starting.

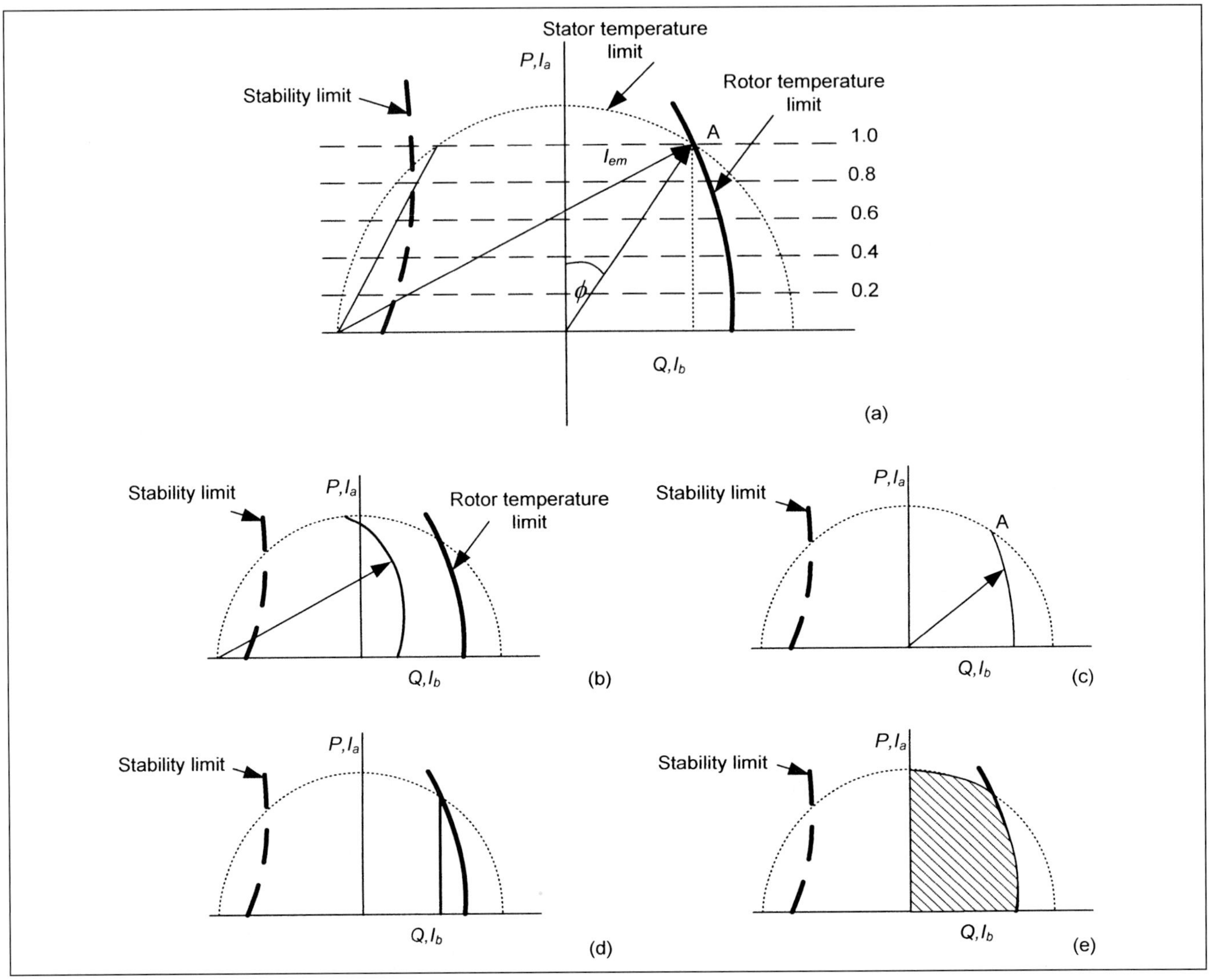

FIGURE 11-23 (*a*) Simplified current and power diagram of a synchronous motor. (*b*) excitation controllers, constant current excitation controller, (*c*) power factor excitation controller, (*d*) and (*e*) reactive power and voltage excitation controllers, respectively.

are constant-power output lines. At point *A* the motor operates at the maximum excitation current at its leading power factor ϕ. The stator current can be resolved into active and reactive components. The active power drawn from the system depends upon the load torque required, and the reactive power can be varied by control of excitation.

The unregulated controllers are the simplest in terms of hardware and are brush type. A fixed rectifier transformer with off-load tap adjustments supplies excitation power through a field contactor. The reactive power output at no load will exceed that at the full load. The system voltage can fluctuate, depending upon the size of the motor and the Thévenin impedance as seen from the bus to which the motor is connected. For fluctuating loads and large motors, this can give rise to instability on overloads, that is, a crusher motor connected to a weak electrical system. Four major types of regulated excitation systems are:

- *Constant field current.* This considers the variation of the field winding resistance as the motor is loaded; due to temperature rise, the resistance rises and the current falls. A constant field current is maintained by adjusting the conduction angle of SCRs (Fig. 11-23*b*).
- *Constant power factor.* The kvar output rises as the motor is loaded, and this will tend to raise the motor bus voltage. An impact load will cause power factor to dip and the controller will respond with a boosted output. This will enhance the stability limit. In brushless excitation systems, the time constants associated with the pilot exciter and the motor-field windings may not allow the excitation to respond quickly. Oscillations may occur on swinging loads and the stability may be jeopardized. The power factor controllers can be used to regulate the system power factor, rather than the individual motor (Fig. 11-23*c*).
- *Constant reactive power output.* The motor bus voltage will be maintained constant from no-load to full load. If a number of motors are connected to the same bus, the constant reactive power outputs of the running motor reduce the bus voltage dip when a motor is started (Fig. 11-23*d*).
- *Constant voltage controller.* The characteristics are similar to that of var controllers Fig. 11-23*e*.

Figure 11-24 shows the capability of a synchronous motor to ride through voltage dips under two conditions: (a) with excitation held constant, dotted lines for $H = 5$ and 1, and (b) with excitation

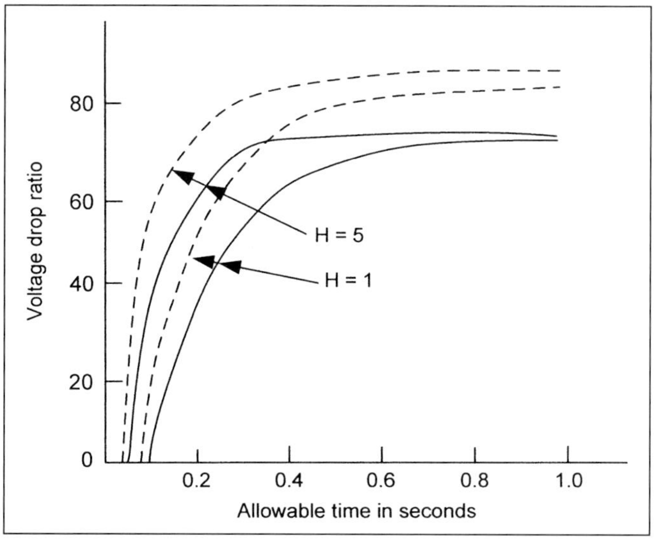

FIGURE 11-24 Stability of a synchronous motor on voltage dips, with excitation held constant and with excitation varying in direct proportion to supply system voltage dip. Curves for inertia constants of $H = 5$ s and 1 s are shown.

TABLE 11-4 Typical Time Constants of Excitation System

SYMBOL	RANGE OF VALUES
K_A	400–500
T_A	0.003–0.08
K_F	0.005–0.054
T_{F1}	0.22–0.65
T_{F2}	0.04–0.1
T_{F3}	0–0.004
K_E	1
T_E	0.55
$S_{E,\max}$	1.2–2.2
$S_{E,0.75\max}$	1.0–2.0

reducing in direct proportion to the supply system voltage, full lines for $H = 5$ and 1. On sudden reduction of voltage, the motor internal emf will not change instantaneously, and the motor tends to maintain a constant output by increasing the line current. The power factor swings to more leading before it goes lagging. The motor pullout torque can reduce below the load torque and the motor can still be stable provided the voltage is restored before the torque angle swings to the critical limit. In the transient stability studies, the synchronous motors are sometimes modeled with fixed excitation, but it can give erroneous results. The block circuit diagram of IEEE type AC5A exciter is shown in Fig. 11-25.[14] Typical time constants are in Table 11-4.[15]

Example 11-5 This example applies transient stability study to motor starting. Consider the configuration in Fig. 11-26. There are three synchronous generators, two utility tie transformers, and loads consisting of induction and synchronous motors. Total installed capacity is 164.7 MVA and the load demand is 109 MVA. The induction motor loads are lumped and connected through equivalent impedance representing transformers and cables. A 20000-hp synchronous motor is started on 13.8-kV bus *M*1 (as an induction motor, synchronizing transients not considered), while all the generators and motor loads *L*1, *L*2, *L*3, and *L*4 are in operation. The impact of starting a large 20000-hp motor on the stability of smaller 2000-hp, 4-kV synchronous motor, connected to 3.75-MVA transformer *T*3, as well as on the stability of running induction motor loads is examined. All motors are modeled with their inertia constants and load torque characteristics. All generators use subtransient models, and IEEE type ST1 excitation systems (Chap. 13).

This system configuration is for the purpose of an illustrative example. When such large motors are to be started, these are connected to dedicated buses, so that the starting impacts are not passed on to the smaller loads on the same buses. It implies that small system loads and large motor loads are connected to separate buses for better stability limits.

The starting impact of the 20000-hp motor is 15.2 MW of active and 72 Mvar of reactive power. The transients are shown in Fig. 11-27*a* through *h*.

1. Active and reactive power swings in running 2000-hp synchronous motor SM2 are depicted in Fig. 11-27*a*. These transients are of larger magnitude toward the lapse of starting period than at the instant of starting impact. Mechanical load is assumed constant.

2. Figure 11-27*b* shows the voltage at bus *M*1 and 2000-hp motor terminals SM2, which dips by 16 percent below rated voltage. Generator *G*1 terminal voltage dips by 13.5 percent. The starting time is approximately 5 s, and the voltage recovers to 95 percent within 1 s after starting impact. This is because of the generators excitation system response and generators taking a considerable impact of the starting demand. Note the overshoot in the voltage recovery profile, which is very typical.

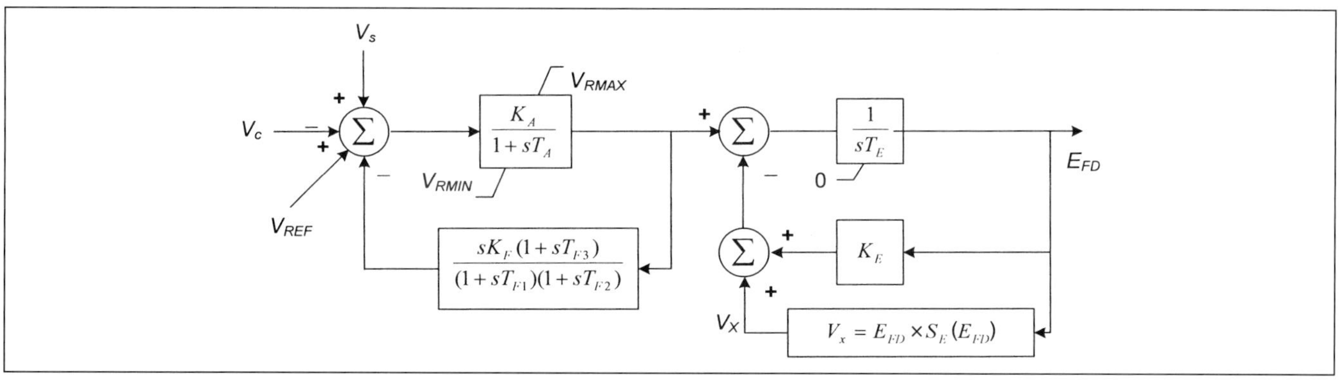

FIGURE 11-25 Control circuit block diagram of IEEE type AC5A excitation system.

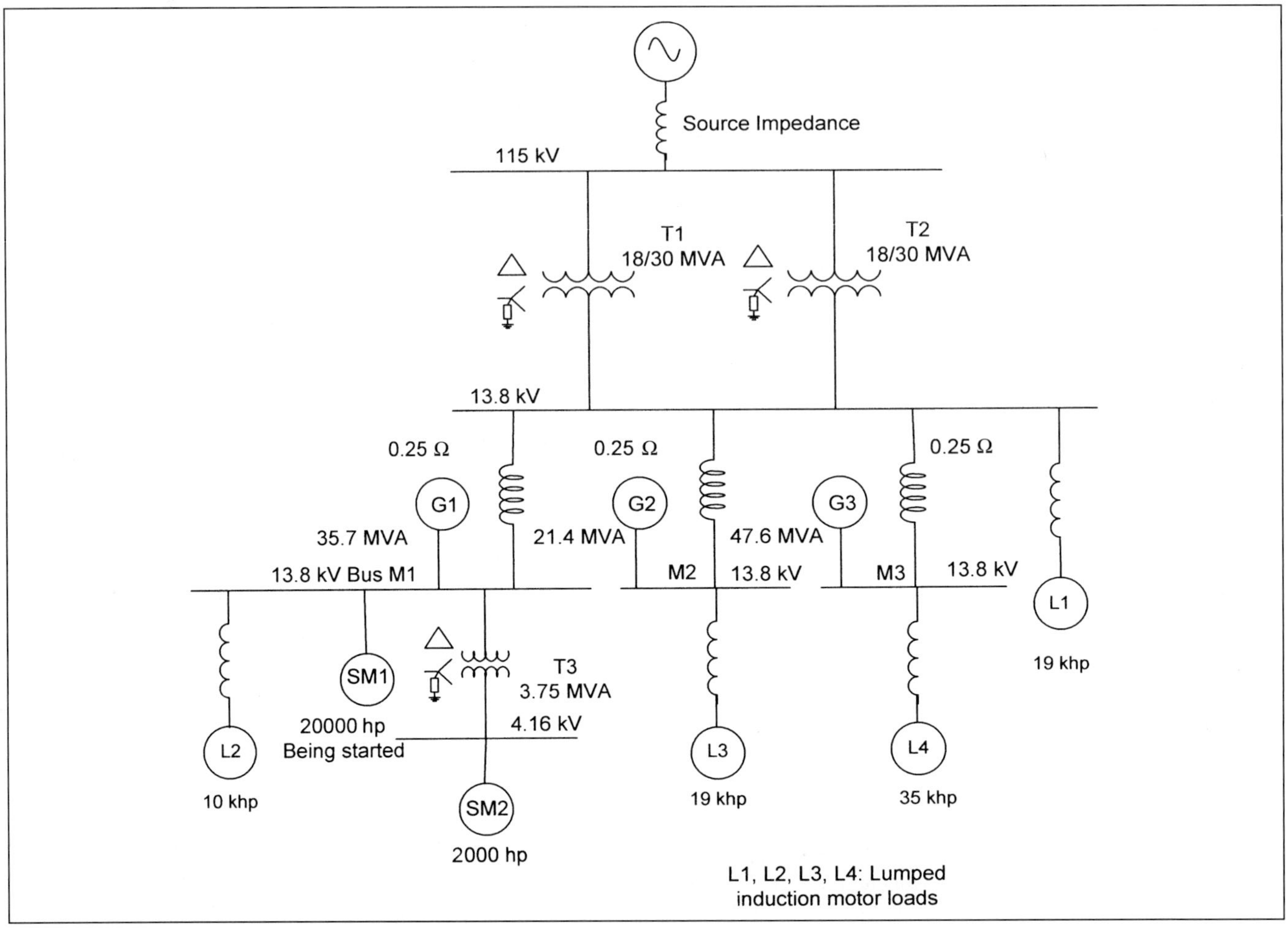

FIGURE 11-26 Circuit diagram for study of starting impact of a 20000-hp synchronous motor.

3. Figure 11-27*c* shows torque angle swings of 20000-hp synchronous motor and generator *G*1; the torque angle recovers quickly, showing stability.

4. Swings in the active and reactive power output of *G*1 can be noted in Fig. 11-27*d*. The reactive power of the generator jumps to approximately 60 Mvar in response to the starting reactive power demand. Generators *G*2 and *G*3 also experience similar transients, not shown.

5. Figure 11-27*e* illustrates that field current of the generator jumps to 6.2 times, and similar excursion in the field voltage, 5.2 pu, occurs. The other two generators' excitation systems experience similar transients to a lesser extent, not shown.

6. Active and reactive power oscillations in the 115 kV utility are depicted in Fig. 11-27*f*; note the post starting transients.

7. Figure 11-27*g* illustrates that speed of induction motors lumped on bus *M*1, which dips on the starting impact, recovers through an upward excursion of the speed transient.

8. Finally, Figure 11-27*h* shows active and reactive power oscillations of the induction motor loads connected to bus *M*2.

The simulation illustrates that the motor can be started; however, the bus *M*1 experiences a voltage dip of approximately 14 percent, which may not be acceptable for the stability of some of the loads connected to that bus.

PROBLEMS

1. Derive the equivalent circuit of a double-cage induction motor shown in Fig. 11-2*c*, following the mathematical model developed for single-cage rotor in the text.

2. The standstill impedances of the inner cage and outer cage of a double-cage induction motor are: inner cage = $0.15 + j1.6\ \Omega$, outer cage = $0.5 + j5\ \Omega$. Compare the relative currents and torques of these two cages at standstill and at a 4 percent slip.

3. A 1000-hp, 4-pole, 60-Hz induction motor has a breakdown torque of 175 percent, critical slip $s_m = 10$ percent, full load slip = 2 percent, and $H = 3.3$ s. Assume that the motor and load speed-torque characteristics are as shown in Fig. 11-10*b*. Calculate the stability curve under voltage dips similar to the one shown in Fig. 11-8.

4. A three-phase squirrel cage induction motor takes a starting current six times the full-load current and has a full-load slip of 3 percent. Calculate: (1) starting torque in terms of full-load torque, (2) slip at the maximum torque, and (3) the maximum torque in terms of full-load torque.

5. Consider the induction motor of Example 11-2. Calculate short-circuit current profiles, ac symmetrical component, dc component, and the total current over a period of eight cycles from the instant of short circuit, and compare the results with that of EMTP simulation shown in Fig. 11-11. Why do these differ?

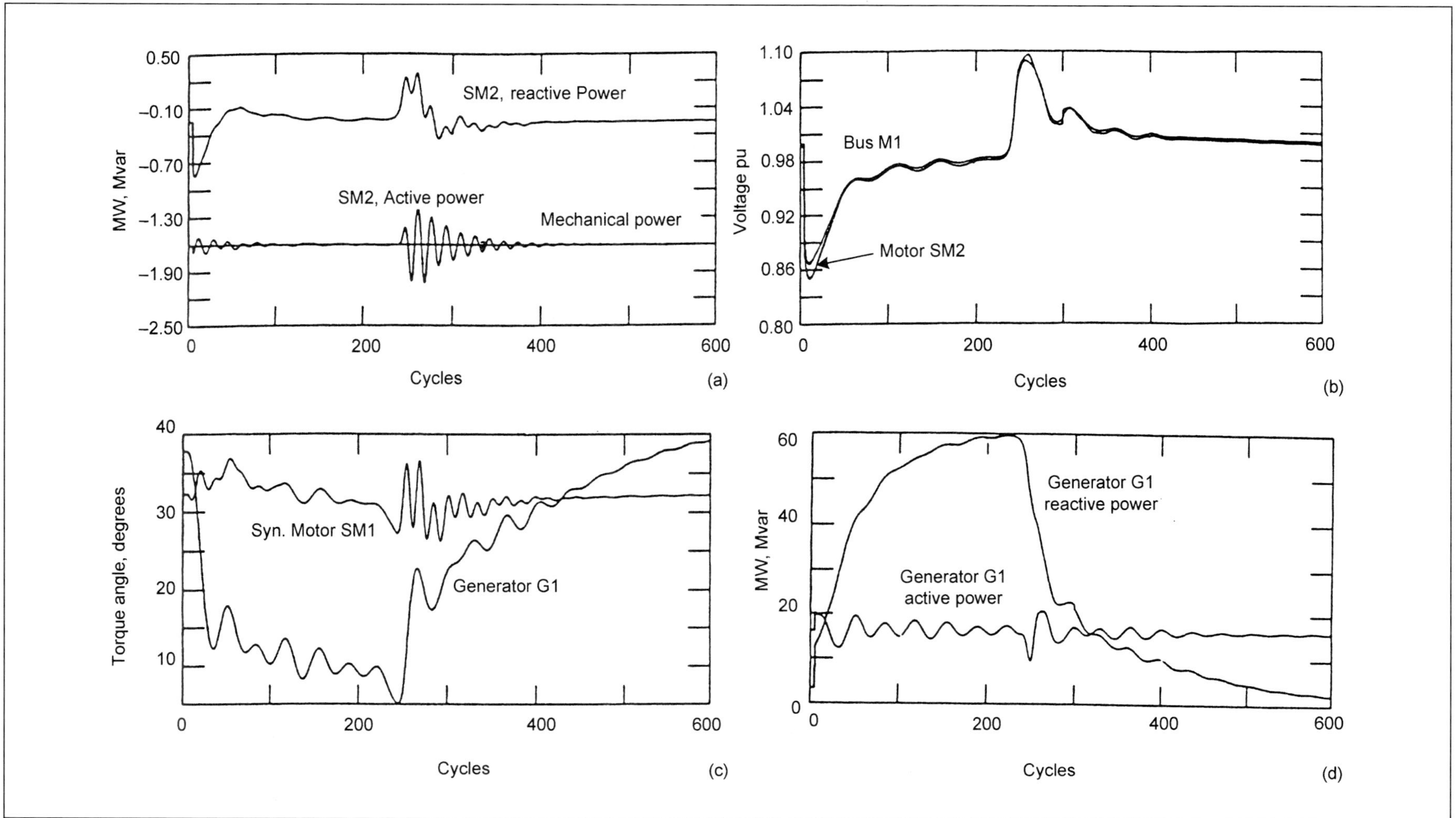

FIGURE 11-27 (*a*) Active and reactive power transients in synchronous motor SM2. (*b*) Bus *M*1 voltage and the voltage at the terminals of synchronous motor SM2. (*c*) Torque angle transients, synchronous motor SM1 and generator *G*1. (*d*) Active and reactive power transients, generator *G*1. (*e*) Field current and voltage generator *G*1. (*f*) Active and reactive power flow from the utility source. (*g*) Speed transient of induction motor loads *L*2. (*h*) Active and reactive power transients, induction motor loads *L*3. (*Continued*)

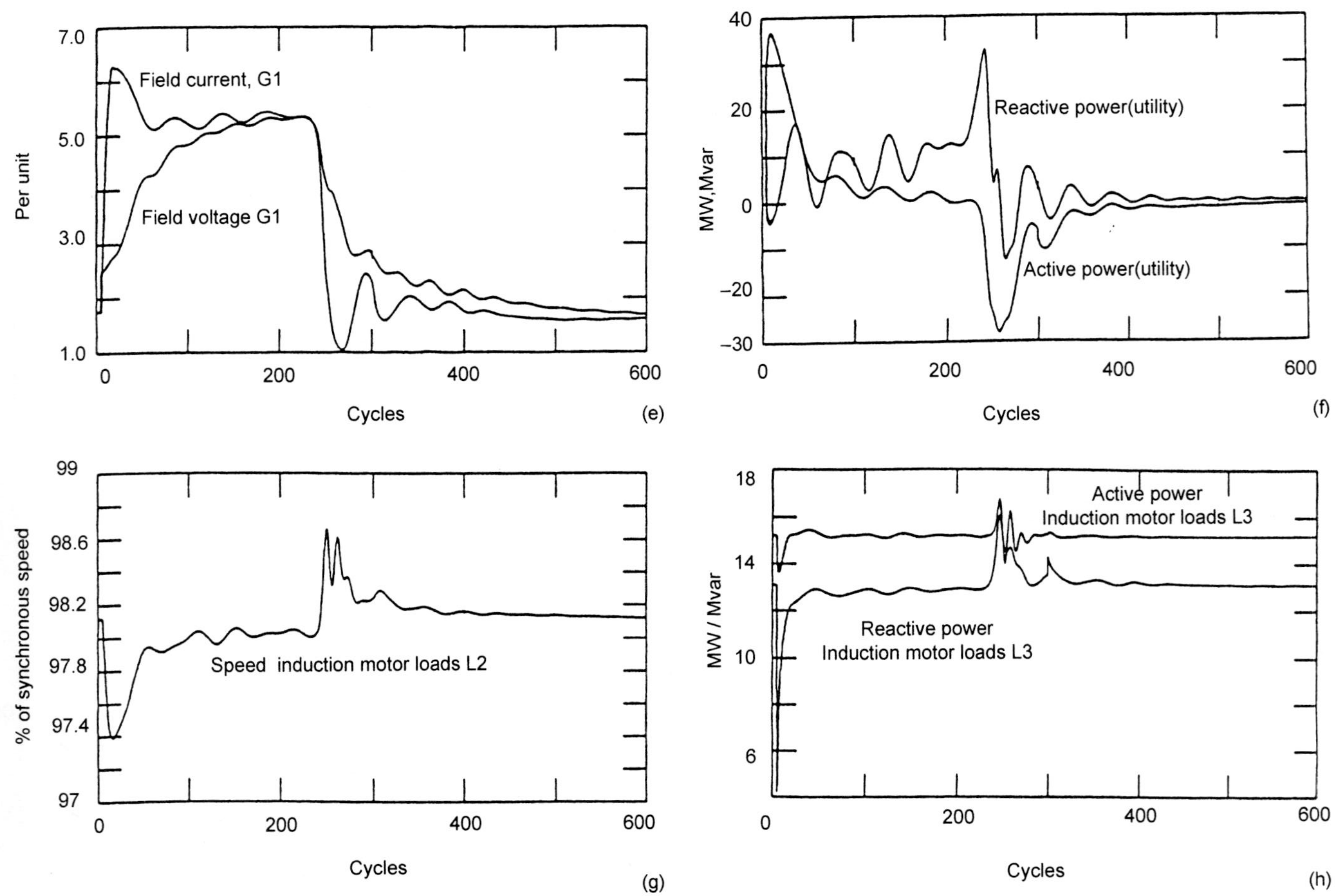

FIGURE 11-27 *(Continued)*

6. Compare the starting characteristics of cylindrical rotor and salient pole synchronous motors. What are the relative advantages and disadvantages? What is Gorges phenomena and what causes it?

7. A 2000-hp, three-phase, 2.3-kV, 4-pole, 60-Hz induction motor is connected to the secondary of a 2.5-MVA, 13.8 to 2.4-kV transformer; transformer percentage impedance = 5.5 percent, and $X/R = 10$. The transformer is served from a 13.8-kV system; source positive sequence impedance = $0.03 + j\,0.3$ pu on a 100-MVA base. The motor starting current is 2600 A at a power factor of 0.10. The total load and motor $H = 3.64$ at the motor speed. Full-load slip = 1.5 percent. Consider that the load torque and motor torque-speed characteristics are given by the curves in Fig. 11-10*b*.

Plot the starting current, bus voltage, and motor slip for the following methods of starting:

Across the line starting

Reactor starting at 0.5 tap

Autotransformer starting at 0.5 tap

Capacitor starting, with a shunt capacitor of 5 Mvar. Consider that the capacitor is disconnected when the motor has reached 95 percent speed.

8. A three-phase short circuit occurs at the terminals of an induction motor. Will the short-circuit current vary with the system impedance to which the motor is connected?

9. A synchronous motor connected to a weak electrical system drives a fluctuating load. What type of excitation controller will be appropriate? Will it be desirable to resort to field forcing as the load torque cycles?

10. In Fig. 11-27*h*, why does the reactive power of the running induction motor dip severely, and then recover rapidly? Explain the upward excursion in transients at the termination of the starting time.

REFERENCES

1. B. Adkins, *General Theory of Electrical Machines*, Chapman and Hall, London, 1964.
2. A. T. Morgan, *General Theory of Electrical Machines*, Heyden & Sons Ltd., London, 1979.
3. G. J. Rogers and D. Shirmphammadi, "Induction Machine Modeling for Electromagnetic Transient Program," *IEEE Trans. on Energy Conversion*, vol. EC-2, no. 4, Dec. 1987.
4. A. E. Fitzgerald, S. D. Umans, Jr., and C. Kingsley, *Electrical Machinery*, McGraw Hill Higher Education, New York, 2002.
5. J. C. Das, "Effects of Momentary Voltage Dips on Operation of Induction and Synchronous Motors," *IEEE Trans. Industry Applications*, vol. 36, no. 4, July/Aug. 1990.
6. NEMA MG-1, Motors and Generators, Large Machines, Parts 20 and 21, 2003.
7. G. S. Sangha, "Capacitor–Reactor Start of Large Synchronous Motor on a Limited Capacity Network," *IEEE Trans. Industry Applications*, vol. IA-20, no. 5, Sept./Oct. 1984.
8. J. Langer, "Static Frequency Changer Supply System for Synchronous Motors Driving Tube Mills," *Brown Boveri Review*, vol. 57, pp. 112–119, March 1970.
9. C. P. LeMone, "Large MV Motor Starting Using AC Inverters," in *ENTELEC Conf. Record*, TX. May 1984.
10. NEMA ICS 2-324, AC General Purpose HV Contactors and Class E Controllers, 50 Hz and 60 Hz, 1974.
11. G. L. Godwin, "The Nature of AC Machine Torques," *IEEE Trans. PAS*, vol. PAS-95, pp. 145–154, Jan./Feb. 1976.
12. J. Bredthauer and H. Tretzack, "HV Synchronous Motors with Cylindrical Rotors," *Siemens Power Engineering*, vol. V, no. 5, pp. 241–245, Sep./Oct. 1983.
13. H. E. Albright, "Applications of Large High Speed Synchronous Motors," *IEEE Trans. Industry Applications*, vol. IA 16, no. 1, pp. 134–143, Jan./Feb. 1980.
14. IEEE Std 421.5, "IEEE Recommended Practice for Excitation System Models for Power System Stability Studies," 1992.
15. J. C. Das and J. Casey, "Effects of Excitation Controls on Operation of Large Synchronous Motors," in *Conf. Record, IEEE I&CPS Technical Conference*, Sparks, Nevada, May, 1999.

FURTHER READING

K. Albrecht and K. P. Wever, "A Synchronous Motor with Brushless Excitation and Reactive Power Control," *Siemens Review*, no. 11, pp. 577–580, 1970.

M. Canny, "Methods of Starting Synchronous Motors," *Brown Boveri Review*, vol. 54, pp. 618–629, Sept. 1967.

C. Concordia, *Synchronous Machines—Theory and Performance*, John Wiley, New York, 1951.

J. C. Das and J. Casey, "Characteristics and Analysis of Starting of Large Synchronous Motors," in *Conf. Record, IEEE I&CPS Technical Conference*, Sparks, Nevada, May, 1999.

J. C. Das, "Application of Synchronous Motors in Pulp and Paper Industry," *TAPPI Proc., Joint Conf. Process Control, Electrical and Information*, Vancouver, BC, Canada, 1998.

J. R. Linders, "Effect of Power Supply Variations on AC Motor Characteristics," *IEEE Trans. Industry Applications*, vol. IA-8, pp. 383–400, July/Aug. 1972.

O. P. Malik and B. J. Croy, "Automatic Resynchronization of Synchronous Machines," *Proc. IEE*, vol. 113, pp. 1973–1976, Dec. 1966.

G. E. Publication GEH-5201, *Synchronous Motor Control with CR192 Microprocessor Based Starting and Protection Module*, General Electric, Mebane, NC, 1985.

CHAPTER 12

Power System Stability

In this chapter, the stability of power systems is discussed and analyzed. The texts on transients consider power system stability a separate subject, yet inclusion of this subject in this book provides an insight into an important aspect of transient behavior of power systems.

A power system is highly nonlinear and continuously experiences disturbances. From the stability point of view, these can be classified into two categories:

1. Contingency disturbances due to lightning, short circuits, insulation breakdowns, and incorrect relay operations. These can be called *large perturbations* or *event disturbances*.

2. Load disturbances because of random variations in the load demand.

There are many definitions of the power system stability in the literature, however, with respect to fault disturbances and an initial (prefault) steady-state equilibrium point (s.e.p), it explores whether the postfault trajectory will settle down to a new equilibrium point in an acceptable steady state. The definition of the final steady state is important. For example, if the postfault state has periodic oscillations, this will not be acceptable. Even small fluctuations in the voltage and frequency will be undesirable. Nor subsynchronous resonance, which may occur due to conversion of mechanical energy into electrical energy associated with subharmonic mode and electromagnetic oscillations (due to interactions between magnetic fields), can be tolerated.

12-1 CLASSIFICATION OF POWER SYSTEM STABILITY

The classification of the stability is necessary in view of:

1. The size of disturbance

2. Correct modeling and analysis of the specific disturbance

3. The time span for which the disturbance lasts

4. The system parameter which is most affected

The power system stability is classified into the following categories:

12-1-1 Rotor Angle Stability

This is large disturbance stability and concerned with the ability of interconnected synchronous machines to remain in synchronism after being subjected to a perturbation. This can be further subdivided into two categories:

1. Small-signal or small disturbance stability (sometimes termed dynamic stability)

2. Large disturbance rotor angle stability

As the nomenclature suggests, the small signal stability considers that the disturbances are small and the system equations can be linearized. Sometimes, the small rotor angle stability is termed dynamic stability. In modern systems, it is largely a problem of insufficient damping of the systems. The study is conducted for 10 to 20 s and sometimes even for a longer period after a disturbance.

Conditions for oscillations between major subsystems may depend on many variable factors, that is, loading of generators, load levels, and system voltages, which are difficult to predict with precision. The subharmonic frequencies can be reduced with the application of power system stabilizers (PSS). When a PSS is not applied, a fast-acting or wider-band excitation system has greater chance of destabilizing local-machine system oscillations. The National Electric Regulatory Council (NERC) recommends PSS for generators that exceed 30 MVA or a group of generators that exceeds 75 MVA.

The large rotor angle stability, or transient stability, as it is normally termed, is concerned when the power systems are subjected to large perturbations, that is, system faults. The resulting system response involves large excursions of rotor angle swings, and the system equations are nonlinear.

Classically, it was thought that due to insufficient damping torque, the instability occurs during the first swing (Fig. 12-1, curve *a*). In large power systems, transient instability may not occur during the first swing. It can be a result of superimposition of several oscillation modes, which may result in *larger* deviation of rotor angle in subsequent swings, Fig. 12-1, curve *b*. The curve *c* of Fig. 12-1 shows the transient stability as the rotor angle swings damp out. The time frame of interest may be 1 to 3 s.

The transient mode can sometimes lead to dynamic mode. The rotor angle may continue to swing at lower frequencies, that is, inter-area oscillations. Figure 12-2 shows dynamic instability as the oscillations increase with time. With the application of a PSS, these can be damped, as shown in the figure. The application of PSS is discussed in Chap. 13.

12-1-2 Voltage Instability

Voltage instability can again be categorized into the following two categories:

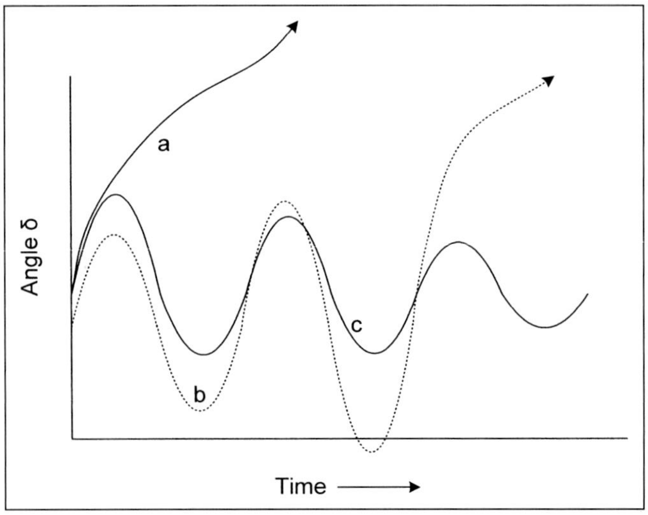

FIGURE 12-1 Rotor angle stability under different power system transients. Curve *a*, instability during first swing; curve *b*, instability due to larger rotor angle swings after a number of swings; curve *c*, decaying rotor angle transient depicting stability.

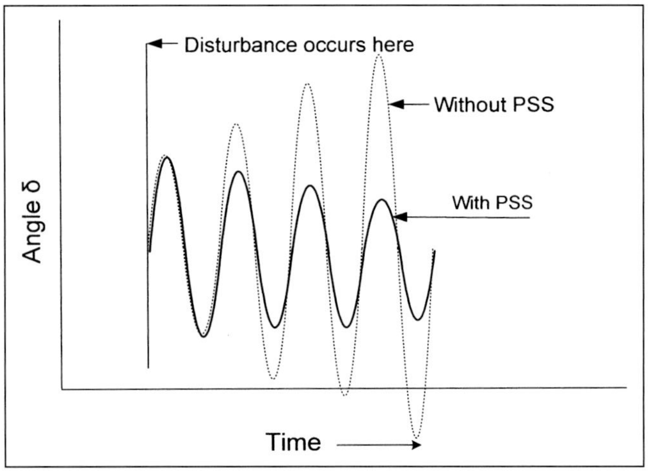

FIGURE 12-2 Dynamic instability, swings controlled by application of a PSS (power system stabilizer).

Large Disturbance Instability The voltages in the system should be controlled following large disturbances, such as faults, loss of generation, or circuit contingencies, which may force the power flow through alternate routes of higher impedances. The generators may hit their reactive power capability limit. The loads have certain voltage tolerance limits for successful ride-through capability of the disturbance before these will fall out of operation.

The analysis requires nonlinear dynamic performance of the system over a period of time that can extend to minutes. The rotor angle instability can lead to voltage instability. The gradual loss of synchronism, as the rotor angles between the machines depart more than 180°, would result in low voltages in the network close to the electrical center.

Small Disturbance Voltage Instability The instability is concerned with system ability to control voltages following small perturbations, that is, incremental changes in the system load. This is essentially of a steady-state nature. The reactive power

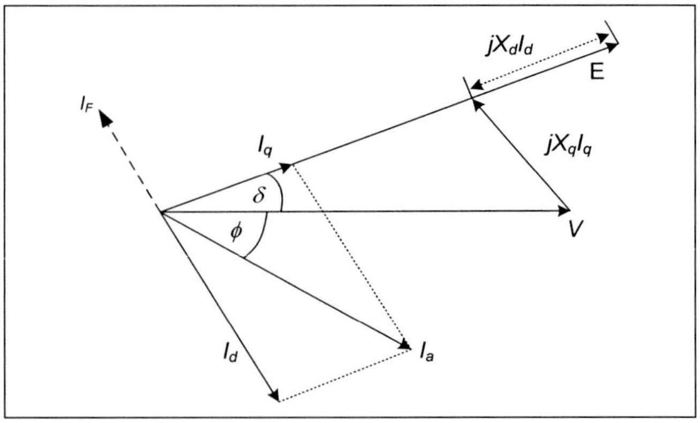

FIGURE 12-3 Simplified phasor diagram of a generator, resistances neglected.

flow in a mainly reactive tie circuit requires a difference of voltage at the tie ends, ΔV. A system is said to be stable if *V-Q* sensitivity is positive for every bus and unstable if *V-Q* sensitivity is negative for at least one bus. This *V-Q* relation is further discussed in Sec. 12-12.

12-1-3 Static Stability

The term *static stability* may seem a misnomer for the dynamic nature of the electrical systems, yet it can be assumed that the response of a system to a gradually occurring change is without an oscillation. Practically, this is not correct, but it forms the basis of some fundamental concepts.

The power angle characteristic of a synchronous generator can be derived from the basic phasor diagram shown in Fig. 12-3. All resistances and losses are neglected, and all parameters are as defined in Chap. 10. It can be shown that:

$$P = \frac{EV\sin\delta}{X_d} + \frac{1}{2}V^2\left(\frac{1}{X_q} - \frac{1}{X_d}\right)\sin 2\delta$$
$$Q = \frac{EV\cos\delta}{X_d} - V^2\left(\frac{\cos^2\delta}{X_d} + \frac{\sin^2\delta}{X_q}\right) \tag{12-1}$$

The second terms in these equations disappear if saliency is neglected and $X_d = X_q$. The saliency gives some element of stability even when the excitation of the generator is removed.

This shows that for a gradual increase in the power output of the generator, the torque angle should increase, with phase of *V* fixed. In this way, the energy balance is maintained. We say that the generator is statically stable if:

$$\frac{\partial P}{\partial \delta} > 0 \tag{12-2}$$

In fact, this ratio is called the *synchronizing power* and develops with every small variation of shaft power output, which results in a change in δ. From Eq. (12-1), the synchronizing power is:

$$P_s = \frac{EV\cos\delta}{x_d} - V^2\left(\frac{1}{X_d} - \frac{1}{X_q}\right)\cos 2\delta \tag{12-3}$$

Whenever there is any perturbation to the steady-state operation of a synchronous machine, the synchronizing power is brought into play, tending to counteract the disturbance and bring the system to a new stable point.

12-2 EQUAL AREA CONCEPT OF STABILITY

Equation (12-1) for the power output can be plotted as a power-angle relation, as shown in Fig. 12-4. The angle δ is called the *torque angle*. The second term in the power output relation has a sin 2δ term, which makes the power-angle curve of the generator peaky in the first half cycle. Referring to Fig. 12-4, if the output shaft power demands an increase, from P_1 to P_2, the torque angle increases from δ_1 to δ_2. The limit of δ_2 is reached at the peak of the torque angle curve, δp, then at δp synchronizing power is zero. If the load is increased very gradually and there are no oscillations due to a step change, then the maximum load that can be carried is given by torque angle δp. This maximum limit is much higher than the continuous rating of the generator and the operation at this point will be very unstable—a small excursion of load will result in instability.

It is obvious that excitation plays an important role [Eq. (12-1)]. The higher the E, the greater is the power output. On a simplistic basis, torque angle curves of a synchronous machine can be drawn with varying E. Neglecting second frequency term, these will be half sinusoids with varying maximum peak.

Practically, the situation will not be as depicted in Fig. 12-4. On a step variation of the shaft power (increase), the torque angle will overshoot to a point δ_3 (Fig. 12-5), which may pass the peak of the stability curve, δp. It will settle down to δ_2 only after a series of oscillations. If these oscillations damp out, we say that the stability will be achieved; if these oscillations diverge, then the stability will be lost. Note that:

1. The inertia of the synchronous generators and prime movers is fairly large.

2. The speed and power-angle transients are much slower than electrical current, and voltage transients.

In Fig. 12-5, the area of triangle ABC translates into the acceleration area due to variation of the kinetic energy of the rotating masses:

$$A_{\text{accelerating}} = \text{ABC} = \int_{\delta_1}^{\delta_2} (T_{\text{shaft}} - T_e)\,d\delta \qquad (12\text{-}4)$$

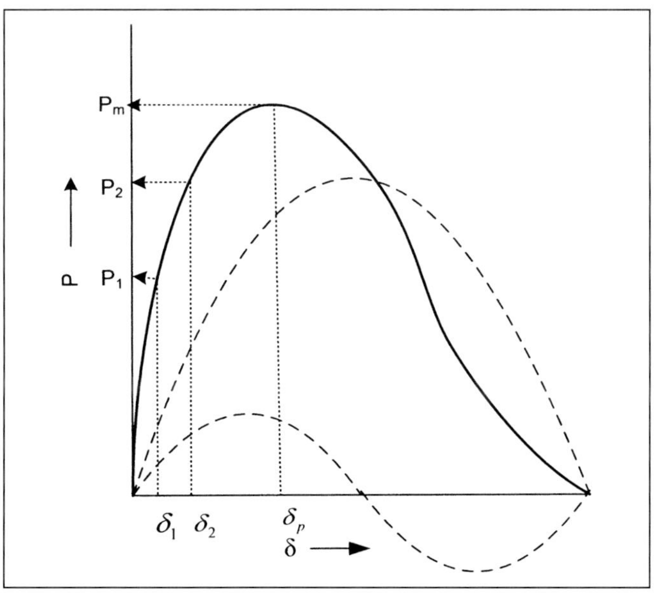

FIGURE 12-4 Power angle characteristics of a salient-pole generator.

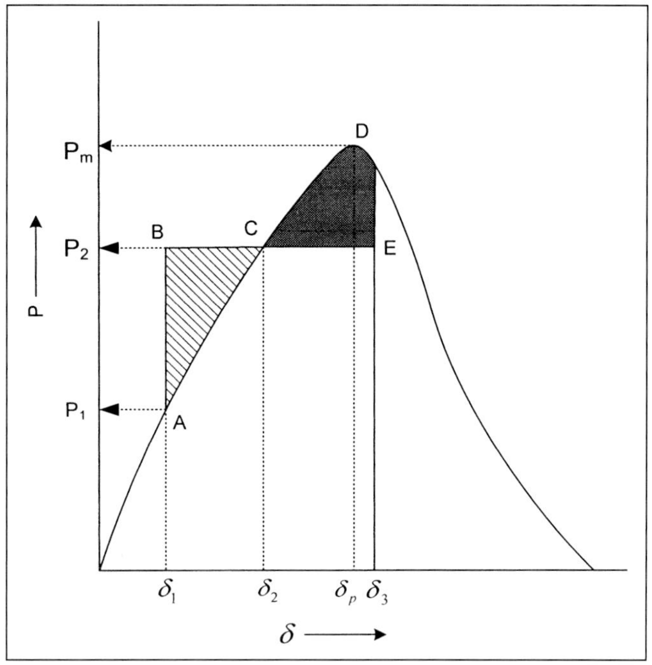

FIGURE 12-5 Torque angle swings on a sudden change of generator load, equal area criterion of stability.

The area CDE is the deacceleration area:

$$B_{\text{deaccelerating}} = \text{CDE} = \int_{\delta_2}^{\delta_3} (T_{\text{shaft}} - T_e)\,d\delta \qquad (12\text{-}5)$$

In accelerating and decelerating, the machine passes the equilibrium point given by δ_2, giving rise to oscillations that will either increase or decrease. If the initial impact is large enough, the machine will be unstable in the very first swing.

If:

$$A_{\text{accelerating}} = B_{\text{deaccelerating}} \qquad (12\text{-}6)$$

Then there are chances of the generator remaining in synchronism. The asynchronous torque produced by the dampers has been neglected in this analysis; also the synchronizing power is assumed to remain constant during the disturbance. It is clear that at point C, the accelerating power is zero, and assuming a generator connected to an infinite bus, the speed continues to increase. It is more than the speed of the infinite bus, and at point E the relative speed is zero and the torque angle ceases to increase, but the output is more than the input and the torque angle starts decreasing; rotor decelerates. But for damping, these oscillations can continue. This is the concept of *equal area criterion of stability*.

Figure 12-6 shows low-frequency oscillations in a multi-machine system, which has not completely died out even after 14 s. Such prolonged oscillations are not acceptable.

12-2-1 Critical Clearing Angle

From the equal area criterion, it is easy to infer that the disturbance, say a fault, should be removed quickly enough for the machine stability. The *critical clearing angle* is defined as the maximum angle at which the faulty section must be isolated to maintain stability of the system. Figure 12-7 shows three P/δ curves, which are, under normal operation, during fault, and after fault clearance; curves marked A, C and B, respectively. Note that during

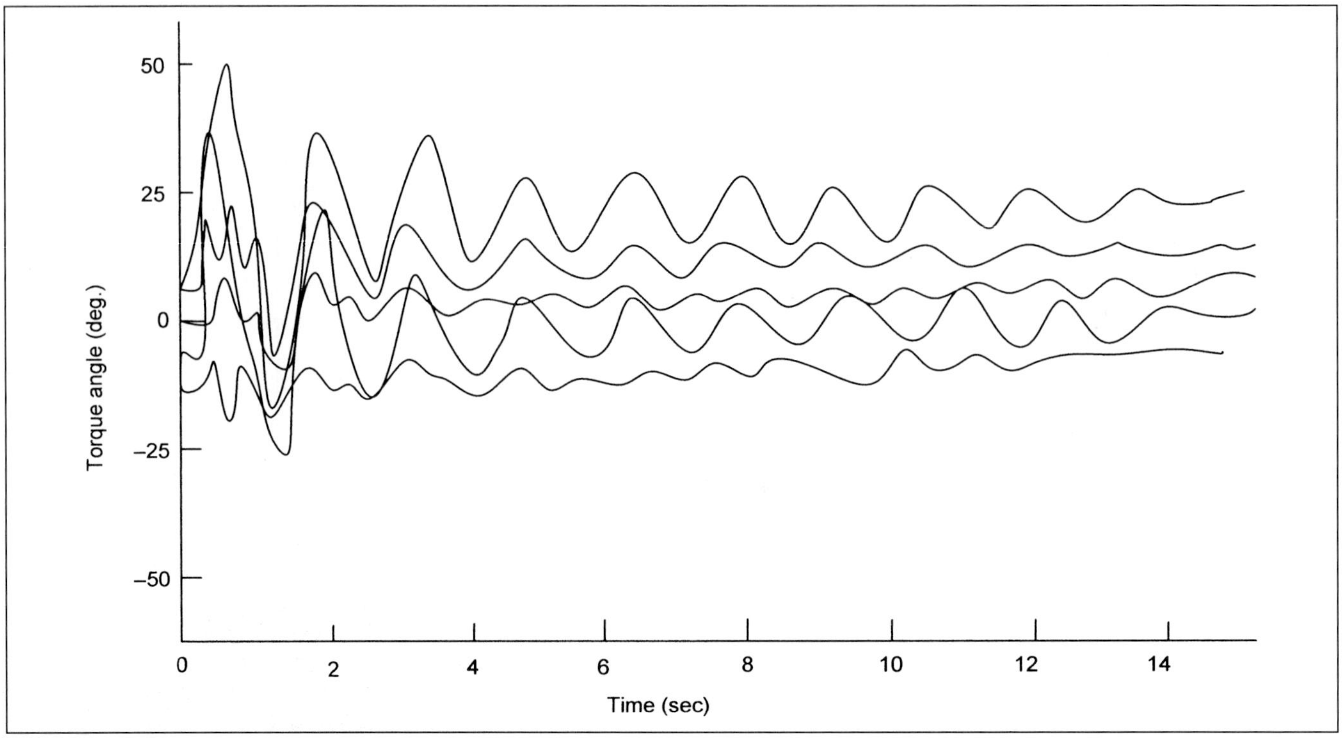

FIGURE 12-6 Oscillations, dynamic stability, in a multi-machine system.

fault, some synchronizing power can be transmitted, depending on the nature of fault and the system configuration. For the condition that area A1 = A2 (same results if we make rectangle abcd = dfghbc, because abcd = A1 + kebc and dfghbc = A2 + kebc), this gives the following equation:

$$(\delta_m - \delta_0)P_s = \int_{\delta_0}^{\delta_c} r_1 P_m \sin\delta d\delta + \int_{\delta_c}^{\delta_m} r_2 P_m \sin\delta d\delta \tag{12-7}$$

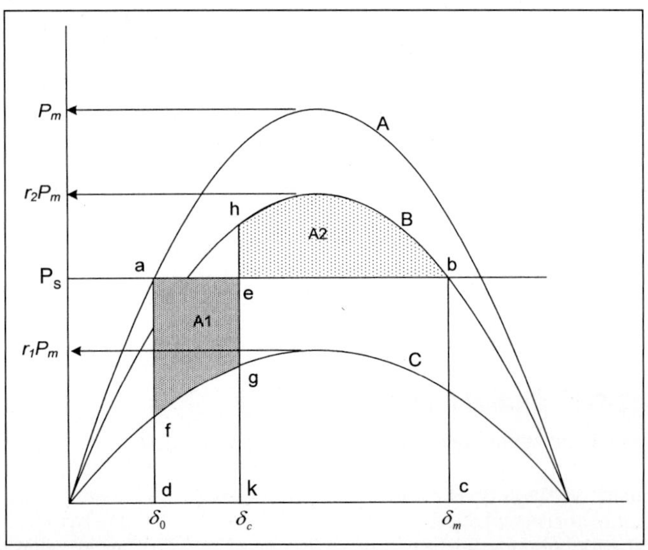

FIGURE 12-7 To illustrate critical clearing angle under fault conditions, and application of equal area criterion of stability.

Substituting:

$$P_s = P_m \sin\delta \tag{12-8}$$

and referring to Fig. 12-7, during fault, maximum power P_m is reduced by a factor r_1, and after fault it is r_2P_m. This gives:

$$\cos\delta_c = \frac{(\delta_m - \delta_0)\sin\delta_0 - r_1\cos\delta_0 + r_2\cos\delta_m}{r_2 - r_1} \tag{12-9}$$

where δ_c is the critical clearing angle. If the actual clearing angle is greater than the critical clearing angle, the system will be unstable. This brings an important factor in transient stability, that is, fast protective relaying and circuit breakers with lower interrupting time. The clearing time of the fault will be the relay operating time plus the breaker interrupting time.

From Fig. 12-7:

$$P_s = P_m \sin\delta_0 = r_2 P_m \sin\delta_m = r_2 P_m \sin(\pi - \delta_m)$$

Thus:

$$\sin\delta_0 = r_2 \sin(\pi - \delta_m)$$
$$\delta_m = \pi - \sin^{-1}\left(\frac{\sin\delta_0}{r_2}\right) \tag{12-10}$$

Note that P_s, the synchronizing power, is assumed constant during the disturbance and after the disturbance.

Example 12-1 Figure 12-8*a* illustrates a system configuration and *b* is the power-angle diagram for single-phase auto reclosing. Many line-to-ground faults are of intermittent nature, and utilities resort to autoclosing to clear such faults. Two or more autoclosing attempts may be made before the reclosures (shown as 52R

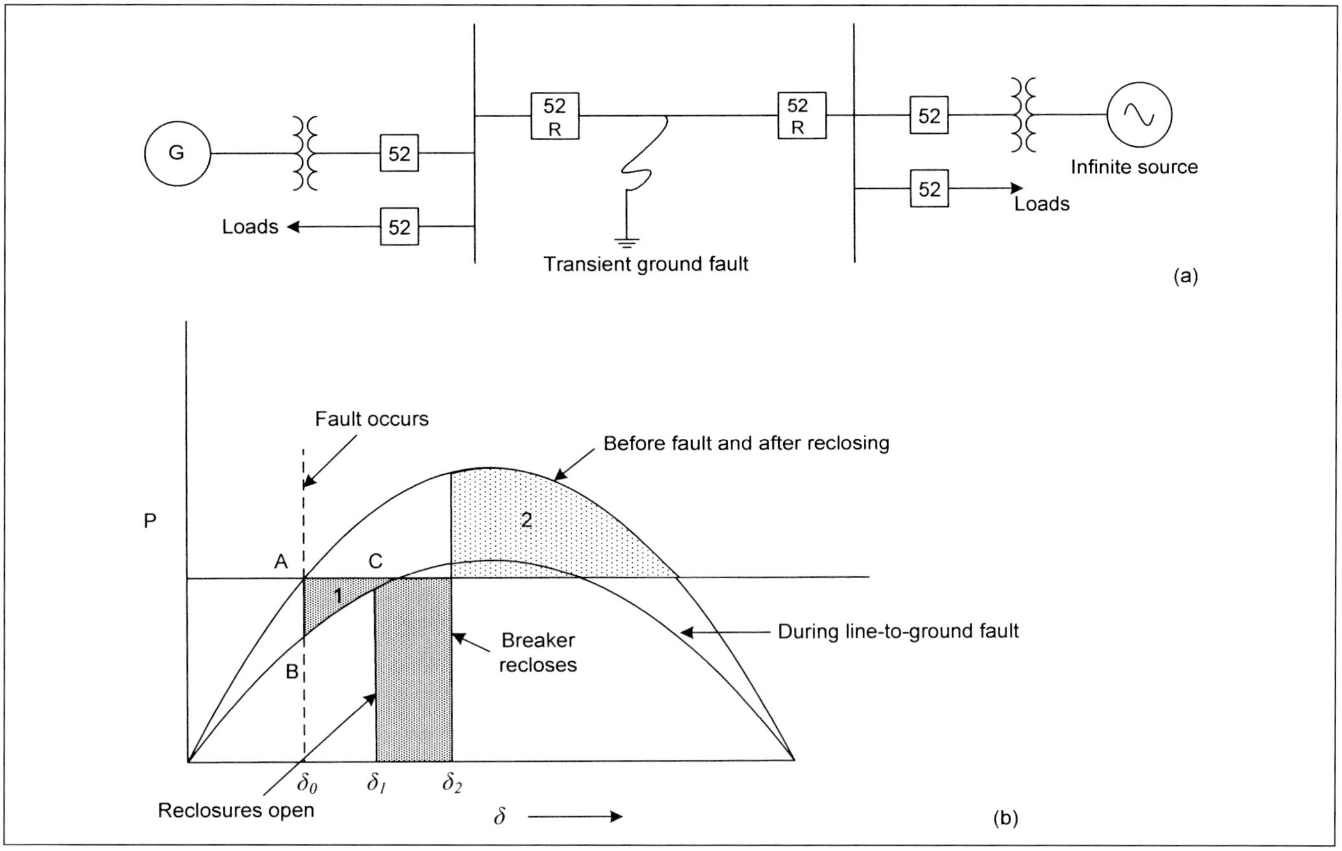

FIGURE 12-8 (*a*) A tie line transient fault, and (*b*) equal area criterion of stability, Example 12-1.

at either end of the line) lock out for a permanent fault. Consider that a transient single line-to-ground fault occurs on the transmission line interconnecting two stations as shown. The tie line reclosures open and then close after a short time delay; this time delay is called the *dead time* of the circuit breaker. Some synchronizing power flows through two unfaulted phases during a single line-to-ground fault, and no power flows during the dead time. The dead time of a circuit breaker on fast reclosing implies a time interval sufficient for the arc fault path to become deionized. In Fig. 12-8*b*, $\delta_2 - \delta_1$ is the dead time. Applying equal area criterion of stability, if shaded area 1 is equal to shaded area 2, stability is possible.

12-3 FACTORS AFFECTING STABILITY

The factors affecting stability are dependent on the type of stability problem. The considerations for large rotor angle instability, voltage instability, interarea oscillations, and turbine generator torsional problems are quite different. The following is a general list:

1. Short-circuit ratio of the generator (see Chap. 13).

2. Inertia constant H, as defined in Chap. 11.

3. Transient reactance of the generator, though all other generator parameters, as discussed in Chap. 10, impact to some extent. For example, single and double dampers.

4. Interconnecting system impedances between machines in a multi-machine system; the postfault system reactance, as seen from the generator.

5. The prior loading on the generator, which determines the initial torque angle, prior to disturbance. The higher the loading, the closer will be the generator to its P_m.

6. The type of fault or perturbation; a three-phase fault at the terminals of the machines will be a worst-case scenario for the large rotor angle stability of the synchronous generator. The transient stability on such faults is sacrificed due to economic reasons. The fault types in decreasing order of severity will be: two-phase-to-ground, phase-to-phase, and phase-to-ground.

7. The nature of the loads served; the presence of rotating motor loads may profoundly impact the results.

8. Speed of protective relaying.

9. Interrupting time of the circuit breakers; 1-cycle synchronous breakers have been developed.

10. Excitation system and voltage regulator types and response; redundant voltage regulators with bumpless transfer on failure, PV, or constant kvar controls.

11. Fast load shedding (frequency dependent with undervoltage settings).

12. Control system response of turbines, prime movers, governing systems, boilers, power system stabilizers, and voltage regulators. Power system stabilizers play a major role in the small signal stability.

13. The system interconnections, the parameters of transmission lines, and characteristics and ratings of other machines, relative to the stability of machine/machines under consideration.

14. Auto reclosing; single-pole switching has been used to enhance transient stability (Chap. 7). Only the faulted phase is tripped and the unfaulted phases support transfer of synchronizing power. This works well as approximately 70 percent of the faults in the transmission lines are of single phase-to-ground type. This may, however, impose negative-sequence loading on the generators in operation; it can excite 120-Hz torque oscillations and result in secondary arcing phenomena and overvoltages[1] (Chap. 7).

15. Series and shunt compensation of transmission lines, SVCs, and flexible ac transimission systems (FACTs) (Chap. 15).

16. Steam turbine fast valving (applicable to thermal power plants) involves rapid opening and closing of the steam valves in a predetermined manner to reduce generator accelerating power, following a severe transmission line fault. The reheat intercept valves in a turbine may control up to 70 percent of unit power, and rapid opening and closing of these valves can significantly reduce the turbine output.[2,3]

17. Fast tripping of a generator after it pulls out of step, as keeping it on line, results in avoidable stresses to the generator itself and the system as it draws power from and supplies power into the system, Example 12-5. While the unit is tripped from the system, the turbine is not tripped; turbine controls limit the overspeed, and the unit can be again brought on line quickly.

18. System separation and islanding the faulty section can prevent a major cascade and prevent propagating the disturbance to the rest of the system. Modern relays with adaptive controls have been applied. Much power is generated in dispersed generation and cogeneration facilities that run in synchronism with the utilities, and fast isolation is of importance.

19. Application of NH-SSR (Narain Hingorani subsynchronous resonance suppressor) (Chap. 15).

12-4 SWING EQUATION OF A GENERATOR

The swing equation relates the motion of the rotor, which we can write as:

$$J\frac{d^2\omega_r}{dt^2} = T_m - T_e \qquad (12\text{-}11)$$

where J is the total moment of inertia in kg-m^2, ω_r is angular displacement of rotor, mechanical in rad, and T_m and T_e are the mechanical and electrical torques in N-m.

This can be converted into a per unit form by noting the relation between H, the inertia constant, and J, the moment of inertia:

$$2H\frac{d^2\omega_{ru}}{dt^2} = T_{mu} - T_{eu} \qquad (12\text{-}12)$$

where:

$$H = \frac{J\omega_{om}^2}{2\text{VA}_{\text{base}}} \text{ and } \omega_{ru} = \omega_r/\omega_0 \qquad (12\text{-}13)$$

The subscript u indicates per unit values. H is the inertia constant, as defined in Eq. (11-58). It is repeated here in terms of units prevalent in the United States:

$$H = \frac{2.31 \times \text{RPM}^2 \times \text{WR}^2(\text{lb/ft}^2) \times 10^{-10}}{\text{MVA rating of machine}} \text{ MW-s/MVA} \qquad (12\text{-}14)$$

H does not vary over a large limit for various machines. Typical data for steam and hydro units of various sizes are provided in Ref. 4.

We are more interested in writing the swing equation in terms of angular position of the rotor in electrical degrees, δ, with reference to a synchronously rotating reference δ_0, at $t = 0$. Angle δ is easily interpreted from the phasor diagrams of the machine:

$$\begin{aligned} \delta &= \omega_r t - \omega_0 t + \delta_0 \\ \dot{\delta} &= \omega_r - \omega_0 = \Delta\omega_r \\ \ddot{\delta} &= \omega_0 \dot{\omega}_{ru} \end{aligned} \qquad (12\text{-}15)$$

Thus, we can write:

$$\frac{2H}{\omega_0}\frac{d^2\delta}{dt^2} = T_{mu} - T_{eu} \qquad (12\text{-}16)$$

We add another term to this equation to account for damping, proportional to the speed. Then the equation becomes:

$$\frac{2H}{\omega_0}\frac{d^2\delta}{dt^2} = T_{mu} - T_{eu} - K_D\omega_{ru} \qquad (12\text{-}17)$$

Equation (12-17) is called the *swing equation*. It can be represented in terms of two differential equations of the first order:

$$\begin{aligned} \frac{d\omega_r}{dt} &= \frac{1}{2H}(T_m - T_e - K_D\Delta\omega_r) \\ \frac{d\delta}{dt} &= \omega_0\Delta\omega_r \end{aligned} \qquad (12\text{-}18)$$

The additional subscript u has been dropped and the equation is understood to be in per unit.

Linearizing:

$$\begin{aligned} \Delta\dot{\omega}_r &= \frac{1}{2H}[\Delta T_m - K_1\Delta\delta - K_D\Delta\omega_r] \\ \Delta\dot{\delta} &= \omega_0\Delta\omega_r \end{aligned} \qquad (12\text{-}19)$$

K_1 (per unit ΔP/rad) = synchronizing coefficient.

The term $K_1\Delta\delta$ may be called *synchronizing power*, which acts to accelerate and decelerate the inertia to bring it to the stable operating point, if one exists. For small deviations, K_1 is the slope of the transient power angle curve, at the particular steady-state operating point, as shown in Fig. 12-9:

$$K_1 = \left.\frac{dP}{d\delta}\right|_{\delta_0} = \frac{VE'_q}{X'_d + X_e}\cos\delta_0 \qquad (12\text{-}20)$$

Referring to Fig. 12-10 of a synchronous generator connected to an infinite bus through a reactance X_e, and ignoring saliency, E'_q is internal voltage behind transient reactance, E_q is internal voltage behind synchronous reactance, V is infinite bus voltage, V_t is generator terminal voltage, and δ is the angle, as shown, between E'_q and V.

Equation (12-17) governs the dynamic response, having an oscillation frequency of approximately:

$$\omega_n \approx \sqrt{\frac{K_1\omega_s}{2H}} \text{ rad/s} \qquad (12\text{-}21)$$

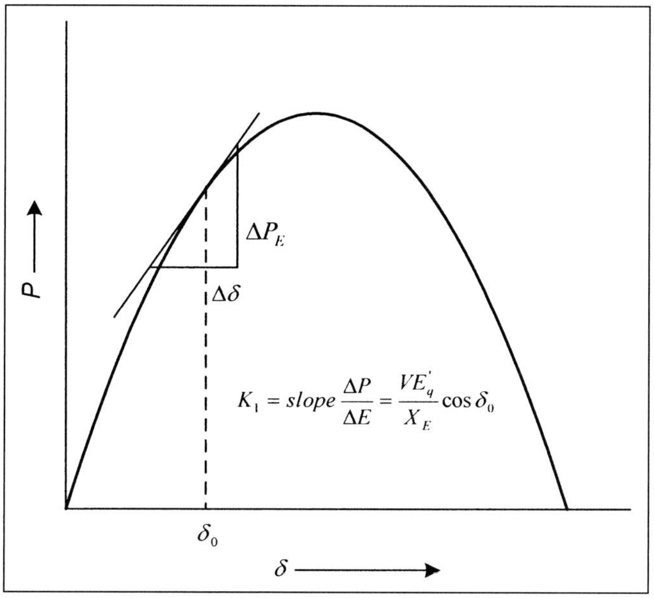

FIGURE 12-9 Power angle curve showing derivation of synchronizing coefficient K_1.

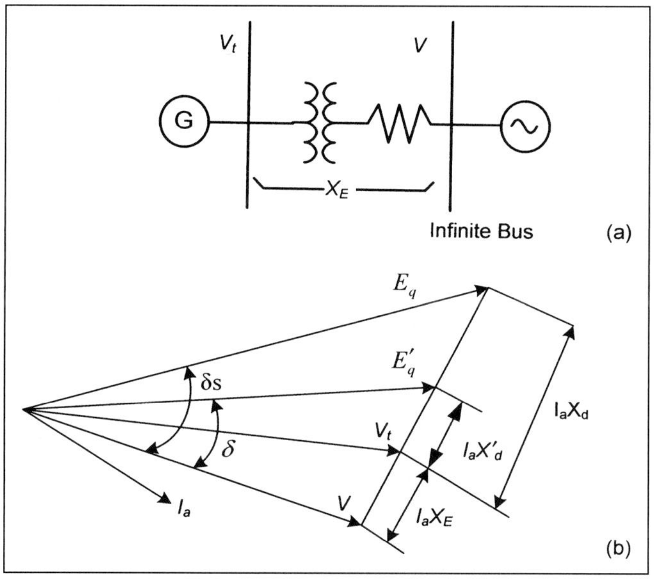

FIGURE 12-10 (*a*) A generator connected to an infinite bus through external impedance. (*b*) Phasor diagram of generator on an infinite bus.

12-5 CLASSICAL STABILITY MODEL

In the classical treatment of stability, the following assumptions are made:

1. The generator model is type 1, with all its assumptions represented by an internal generator voltage behind a transient reactance (Sec. 12-9).

2. The loads are represented as constant impedance loads.

3. The excitation systems, governing systems are not modeled. The mechanical power remains constant.

4. Damping or asynchronous power is neglected.

5. The mechanical rotor angle of the machine coincides with the angle of voltage behind transient reactance.

These assumptions may not be always valid, but they are illustrative of the methodology of simple stability calculations.

The swing equation reduces to:

$$\frac{2H}{\omega_s}\frac{d^2\delta}{dt^2} = P_m - P_e \tag{12-22}$$

The general network equations are:

$$\begin{vmatrix} \bar{I}_g \\ 0 \end{vmatrix} = \begin{vmatrix} \bar{Y}_{11} & \bar{Y}_{1j} \\ \bar{Y}_{j1} & \bar{Y}_{jj} \end{vmatrix} \begin{vmatrix} \bar{E} \\ \bar{V} \end{vmatrix} \tag{12-23}$$

where $\bar{I}_g$ is the vector of generator currents, $\bar{E}$ is the vector of internal generator voltages, and $\bar{V}$ the vector of bus voltages; $\bar{Y}_{11}$ is $n \times n$ matrix, $\bar{Y}_{1j}$ is $n \times m$ matrix, $\bar{Y}_{j1}$ is $m \times n$ matrix, and $\bar{Y}_{jj}$ is $m \times m$ matrix.

We eliminate all load nodes and are interested only in the generator currents.

From Eq. (12-23):

$$\begin{aligned} \bar{0} &= \bar{Y}_{j1}\bar{E} + \bar{Y}_{jj}\bar{V} \\ &\text{or} \\ \bar{V} &= -\bar{Y}_{jj}^{-1}\bar{Y}_{j1}\bar{E} \end{aligned} \tag{12-24}$$

Also:

$$\begin{aligned} \bar{I}_g &= \bar{Y}_{11}\bar{E} + \bar{Y}_{1j}\bar{V} \\ &= (\bar{Y}_{11} - \bar{Y}_{1j}\bar{Y}_{jj}^{-1}\bar{Y}_{j1})\bar{E} \\ &= \bar{Y}\bar{E} \end{aligned} \tag{12-25}$$

The $\bar{Y}$ matrix changes with the system conditions. In the prefault state, the load flow study is conducted to determine the $\bar{E}$ and the angles δ. At the instant of fault, the system configuration changes and the matrix is modified and again modified a second time after the fault is cleared. Thus, there are three distinct conditions in the solution.

Example 12-2 A system configuration, as illustrated in Fig. 12-11*a*, is considered. Two generators are connected to duplicate feeders; their transient reactances are as shown. Also, the feeder's reactances are shown. The resistances are neglected.

Figure 12-11*b* converts reactances to susceptances, and the full admittance matrix is:

$$\begin{vmatrix} Y_{11} & Y_{12} & Y_{13} & Y_{14} \\ Y_{21} & Y_{22} & Y_{23} & Y_{24} \\ Y_{31} & Y_{32} & Y_{33} & Y_{34} \\ Y_{41} & Y_{42} & Y_{43} & Y_{44} \end{vmatrix} = \begin{vmatrix} -j5 & 0 & j5 & 0 \\ 0 & -j6.7 & 0 & j6.7 \\ j5 & 0 & -j10 & j5 \\ 0 & j6.7 & j5 & -j11.7 \end{vmatrix}$$

Here:

$$\bar{Y}_{11} = \begin{vmatrix} -j5 & j0 \\ j0 & -j6.7 \end{vmatrix} \quad \bar{Y}_{1j} = \begin{vmatrix} j5 & 0 \\ 0 & j6.7 \end{vmatrix} = \bar{Y}_{j1} \quad \bar{Y}_{jj} = \begin{vmatrix} -j10 & j5 \\ j5 & -j11.7 \end{vmatrix}$$

$$\bar{Y}_{1j}\bar{Y}_{jj}^{-1}\bar{Y}_{j1} = \begin{vmatrix} -j3.179 & -j1.821 \\ -j1.821 & -j4.879 \end{vmatrix}$$

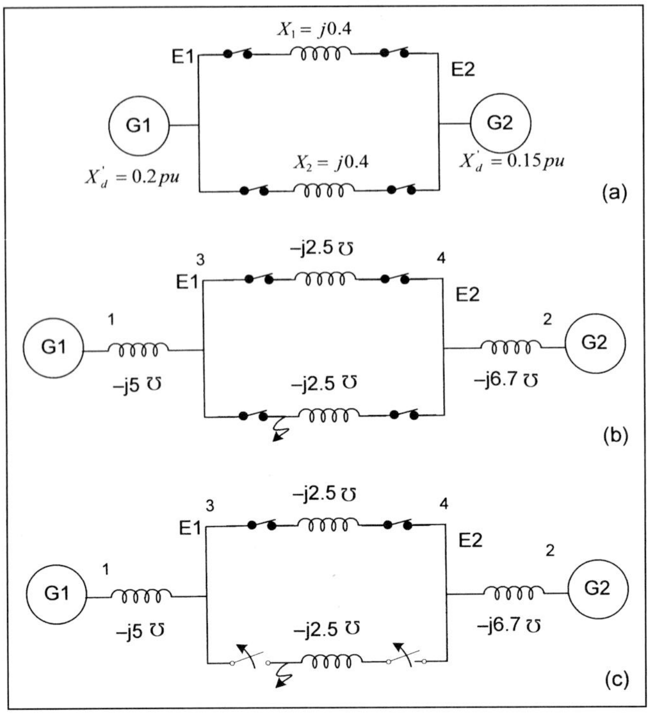

FIGURE 12-11 Circuit for Example 12-2. (*a*) Prior to fault, (*b*) during fault, (*c*) after fault clearance, one tie line opened.

Therefore, prefault Y is:

$$\overline{Y} = \begin{vmatrix} -j1.821 & j1.82 \\ j1.82 & -j1.821 \end{vmatrix}$$

The power transferred between nodes 1 and 2 or 2 and 1 is:

$$P_e = E_1 E_2 \overline{Y}_{12} \sin\delta$$

If we assume that the voltages at both ends are maintained the same (which is not practical for reactive power transfer), then:

$$P_e = 1.821 \sin\delta$$

During a three-phase bolted fault close to bus E1, on one of the tie lines, the breakers at both ends open (Fig. 12-11*b*). The bus voltage goes to zero during fault, and no power can be transferred. This can be illustrated by forming $\overline{Y}$ matrix. As node 3 voltage goes to zero, this amounts to eliminating third row and column from the original matrix as follows:

$$\begin{vmatrix} -j5 & 0 & 0 \\ 0 & -j6.7 & j6.7 \\ 0 & j6.7 & -j11.7 \end{vmatrix}$$

Now the reduced matrix Y following the same procedure is:

$$\overline{Y} = \begin{vmatrix} -j5 & 0 \\ 0 & -j6.7 \end{vmatrix} - \begin{vmatrix} 0 \\ j6.7 \end{vmatrix} \begin{vmatrix} -j11.7 \end{vmatrix}^{-1} \begin{vmatrix} 0 & j6.7 \end{vmatrix} = \begin{vmatrix} -j5 & 0 \\ 0 & -j2.88 \end{vmatrix}$$

As the matrix $Y_{12} = 0$, the power transmitted is zero.

Postfault condition

Consider that the faulty line is isolated by tripping the circuit breakers at either end (Fig. 12-11*c*). Then, one line remains in service and the full impedance matrix is:

$$\overline{Y} = \begin{vmatrix} -j5 & 0 & j5 & 0 \\ 0 & -j6.7 & 0 & j6.7 \\ j5 & 0 & -j7.5 & j2.5 \\ 0 & j6.7 & j2.5 & -j9.2 \end{vmatrix}$$

Following the same procedure as before:

$$\overline{Y} = \begin{vmatrix} -j1.335 & j1.335 \\ j1.335 & -j1.335 \end{vmatrix}$$

Thus:

$$P_e = 1.335 \sin\delta$$

12-5-1 Limitation of the Classical Method

The modern power systems are complex with lots of interconnections. The effects of modeling of excitation systems, governing systems, generator saturation, and prime mover, that is, steam or hydraulic governing systems are well documented.[5,6]

In a multi-machine system, the situation is complex. An incident impact will be unevenly distributed amongst various synchronous machines. Each machine will produce synchronizing power, and the total synchronizing power brought about by the impact may be considered a sum of the individual synchronizing powers. The rigid machines will take a higher share of impact, while the softer machines will share a smaller portion of the total impact. Under this impact, every machine is retarded or accelerated:

$$\alpha = \frac{v^2}{\omega} \Delta\delta \tag{12-26}$$

where v is the natural frequency of oscillation and α is the initial retardation.

The synchronizing forces brought into play cause all machines to strive for some mean retardation (Fig. 12-12). Each machine undergoes further oscillations to accomplish this. After decay of oscillations, the smaller machines are more retarded than they were under the first impact and are loaded to a higher degree. The larger machines are less retarded than initially, and are partially relieved of their overload. The final torque angle may be smaller or larger than the initial impact.

Figure 12-13 depicts this behavior. The initial angle δ_0 swings to δ_1, given by the initial impact angle $\Delta\delta$, and the machine produces its share of synchronizing power in the overall impact. All masses strive for some mean retardation which changes the angle to δ_2. This ultimate value is achieved only after some damped oscillations, in which torque angle swings to δ_3.

Figure 12-14 is a simplified picture of oscillations in a multi-machine system. The behavior depends on the rating of the machine, natural frequency, and the system stiffness. This figure shows that a smaller machine, which took a rather small swing under the initial impact because it produced a relatively small synchronizing power, may become unstable and fall out of step.

This is of importance in current impetus for "green energy" and smaller generating units (dispersed generation), when these operate in synchronism with the utility. The survival of these smaller units under system disturbances becomes a major consideration. Fast isolation from the utility systems on system disturbances becomes important. Generally, the following protective relay applications are used for system isolation:

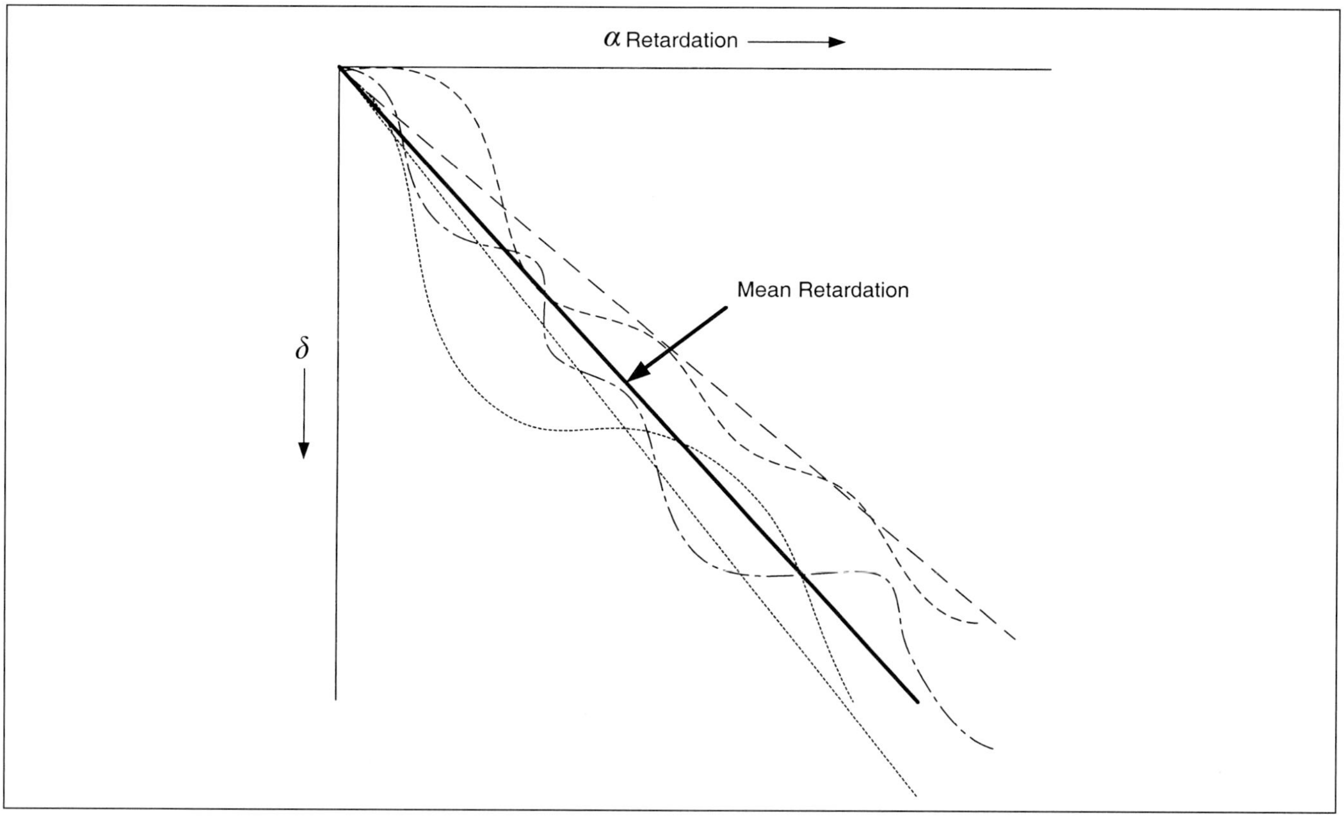

FIGURE 12-12 Illustration of relative swings of machines in a multi-machine system and mean retardation.

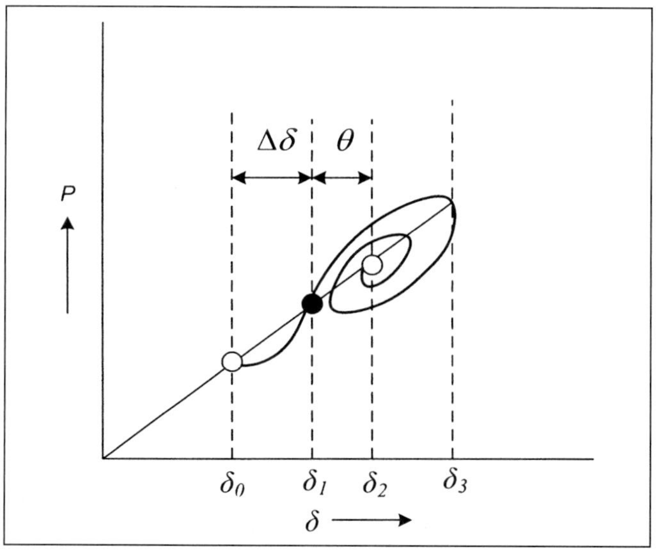

FIGURE 12-13 Damped oscillations around new stable position δ_1 on an impact.

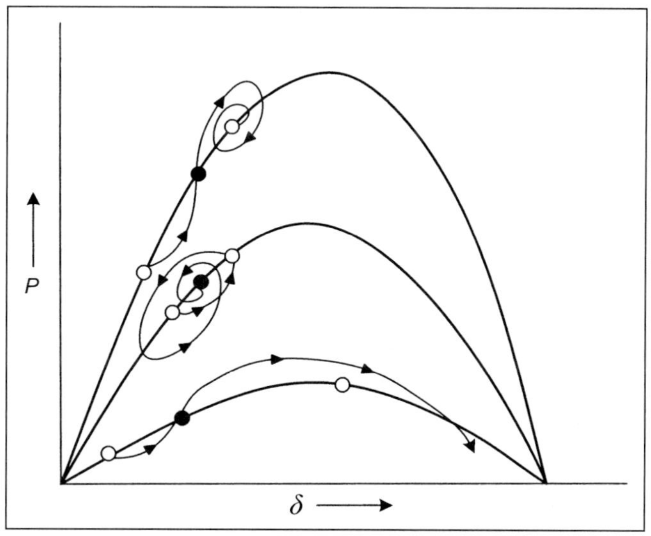

FIGURE 12-14 Relative swings in a multi-machine system, and a soft machine with relatively smaller rating and synchronizing power falling out of step. The impedance between the oscillating machines is considered minimal.

- Frequency relays, device 80
- Voltage relays, device 27/59
- Reverse current directional relays, device 67,67N
- Intertripping through fiber optic interfaces
- Differential current relays, device 87

12-6 DATA REQUIRED TO RUN A TRANSIENT STABILITY STUDY

A practical stability study starts with a converged load flow case. This will establish the prefault voltages, the torque angles of generators, and the active and reactive power flows through various branches and nodes throughout the power system.

In the second step, the disturbance to be studied is simulated. This may be a fault in the system, opening of a tie line, loss of generation, or sudden load increase, or any other transient condition. Any perturbation in the system will result in a dynamic response through oscillations (Example 10-8).

Next, the postfault or disturbance condition can be simulated. The data required for conducting a transient stability is rather extensive, and it can be categorized as follows:

12-6-1 System Data

1. Impedance data of all significant components, generators, transformers, transmission lines, cables, reactors, and so on. All the sequence impedances will be required, depending on the type of study, that is, a line-to-ground fault requires positive-, negative-, and zero-sequence data.

2. kVA rating, nominal voltage ratios, winding connections, tap settings on the transformers, regulating equipment, and auto transformers.

3. Short-circuit capabilities, Mvar ratings of shunt and series capacitors, SVCs, STATCOM data, and their control parameters for modeling.

4. Various switching data, normal and alternate operations.

12-6-2 Rotating Machine Data

1. The generator models can be simple to complex. All transient reactances, reactive capability curves, inertia constants, saturation information, Potier reactance, excitation system controls, voltage regulator, governing system, and prime mover data and control circuit block diagrams are required. Data for PSS and their control circuits are required, as applicable.

2. Large induction and synchronous motors are modeled dynamically; their ratings, inertia constants, torque speed characteristics, load data, method of starting, and so on. We discussed the transient models of rotating machines in the previous chapters.

12-6-3 Disturbance Data

1. The disturbance data to be studied, that is, fault type, protective relaying operating times, breaker interrupting times, limits of acceptable voltage, frequency, and power swings.

12-6-4 Study Parameters

1. The required study parameters are duration of study, integrating interval, and required data output.

12-7 STATE EQUATIONS

The swing equation of p-th generator is:

$$\begin{aligned}\frac{d^2\delta_p}{dt^2} &= \frac{\omega_s}{2H}(P_s - P_p) \\ &= \frac{\omega_s}{2H}\left(P_s - \sum_{q=1}^{q=n} E_p E_q Y_{pq}\cos(\delta_{pq} + \delta_p - \delta_q)\right)\end{aligned} \tag{12-27}$$

These are n-coupled nonlinear differential equations of the second order. In the state form, let:

$$x_1 \approx \delta \quad x_2 = \frac{d\delta}{dt} = \dot{\delta} \tag{12-28}$$

$$\begin{aligned}\dot{x}_1 &= x_2 = \frac{d\delta}{dt} = \omega - \omega_0 \\ \dot{x}_2 &= \frac{d^2\delta}{dt^2} = \frac{w_s}{2H}(P_s - P_p)\end{aligned} \tag{12-29}$$

The state vector:

$$|\,\omega_1, \delta_1 \omega_2, \delta_2 \ldots \omega_n, \delta_n\,|^t \tag{12-30}$$

is a vector of $2n \times 1$ dimensions, where n is the number of generators. The generator power is a function of δ. Thus:

$$\begin{aligned}\dot{x}_1 &= f_1(x_1, x_2) \\ \dot{x}_2 &= f_2(x_1, x_2)\end{aligned} \tag{12-31}$$

12-8 NUMERICAL TECHNIQUES

The state Equations (12-31) can be solved using Euler's method, modified Euler method, Runge–Kutta (R-K) method, trapezoidal rule, and so on. Consider a first-order differential equation:

$$\frac{dx}{dt} = f(x,t) \tag{12-32}$$

Referring to Fig. 12-15, we can approximate the curve representing the true solution by its tangent having a slope (see also Chap. 2):

$$\begin{aligned}\left.\frac{dx}{dt}\right|_{x=0} &= f(x_0, t_0) \\ \Delta x &= \left.\frac{dx}{dt}\right|_{x=0} \Delta t\end{aligned} \tag{12-33}$$

Then, the value of x at $t = t_1 = t_0 + \Delta t$ is given by:

$$x_1 = x_0 + \Delta x = x_0 + \left.\frac{dx}{dt}\right|_{x=0} \Delta t \tag{12-34}$$

The method is equivalent to using the first two terms of the Taylor series. To give sufficient accuracy, the time step must be small.

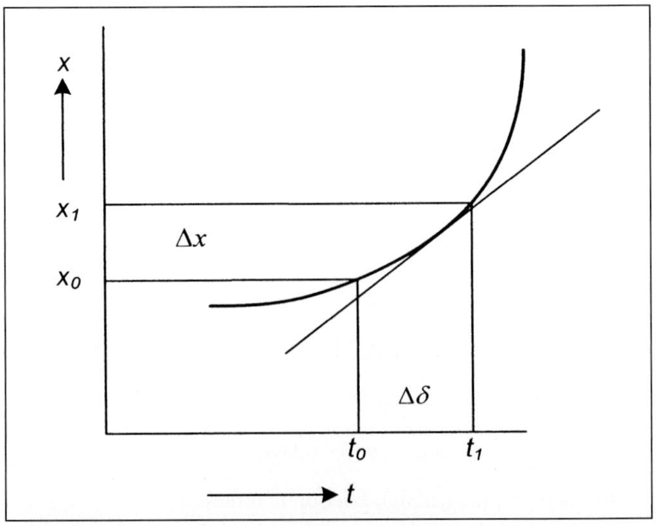

FIGURE 12-15 Estimation of a function using Taylor's series.

12-8-1 Modified Euler's Method

In the Euler's method, inaccuracies may result, as the derivative at the beginning of the interval is applied throughout the interval. The modified Euler's method uses derivatives at the beginning and end of each period and averages these in the following steps.

An initial condition X^0 is assumed. The derivative $\dot{X}^k = f(X^k)$ is computed.

The first estimate of the state vector is:

$$X^{k\pm1} = X^k + \dot{X}^k \Delta t \tag{12-35}$$

The second estimate of the state vector is:

$$\dot{X}^{k+1} = f(X^{k+1}) \tag{12-36}$$

The average of the state derivatives is:

$$\dot{X}^k_{\text{avg}} = \frac{1}{2}(\dot{X}^k + \dot{X}^{k+1}) \tag{12-37}$$

Evaluate the second estimate of the state vector:

$$X^{k+1} = X^k + \dot{X}^k_{\text{avg}}\, \Delta t \tag{12-38}$$

Repeat within the time t. The flowchart of the modified Euler's method is shown in Fig. 12-16.

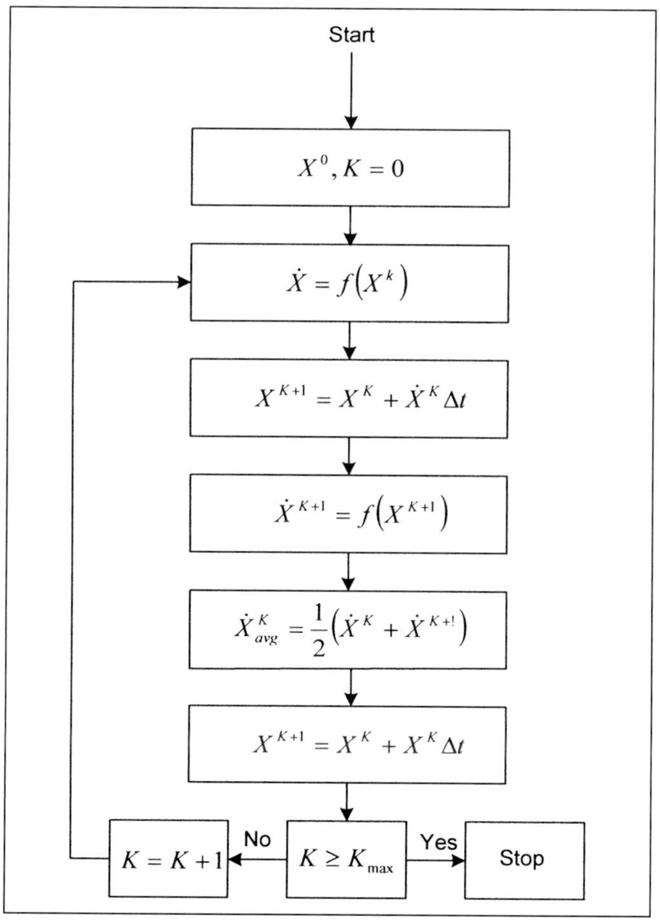

FIGURE 12-16 Flowchart of Euler's modified method, numerical solution.

12-8-2 Runge-Kutta Method

The R-K method approximates the Taylor series, but does not require explicit solution of higher-order derivates, other than the first. The effect of higher-order derivatives is included by several evaluations of the first-order derivative. There are R-K methods of different orders. A fourth-order R-K method has an error of the order of Δt.[5]

An R-K method calculates the following eight constants and can be applied to multi-machine system:

$$\begin{aligned}
K_1^k &= f_1(x_1^k, x_2^k)\Delta t \\
l_1^k &= f_2(x_1^k, x_2^k)\Delta t \\
K_2^k &= f_1\left(x_1^k + \frac{1}{2}K_1^k, x_2^k + \frac{1}{2}l_1^k\right)\Delta t \\
l_2^k &= f_2\left(x_1^k + \frac{1}{2}K_1^k, x_2^k + \frac{1}{2}l_1^k\right)\Delta t \\
K_3^k &= f_1\left(x_1^k + \frac{1}{2}K_2^k, x_2^k + \frac{1}{2}l_2^k\right)\Delta t \\
l_3^k &= f_2\left(x_1^k + \frac{1}{2}K_2^k, x_2^k + \frac{1}{2}l_2^k\right)\Delta t
\end{aligned} \tag{12-39}$$

Similarly, K_4^k, l_4^k.

The change in state vector is calculated as:

$$\begin{aligned}
\Delta \dot{x}_1^k &= \frac{1}{6}\left(K_1^k + 2K_2^k + 2K_3^k + K_4^k\right) \\
\Delta \dot{x}_2^k &= \frac{1}{6}\left(l_1^k + 2l_2^k + 2l_3^k + l_4^k\right)
\end{aligned} \tag{12-40}$$

The new state vector is evaluated as:

$$\begin{aligned}
\dot{x}_1^{k+1} &= \dot{x}_1^k + \Delta \dot{x}_1^k \\
\dot{x}_2^{k+1} &= \dot{x}_2^k + \Delta \dot{x}_2^k
\end{aligned} \tag{12-41}$$

The time constant is advanced to $k + 1$ and check if $k > k_{\max}$. If not, then the step 2 of calculating eight constants is repeated.

Example 12-3 A machine with $P_m = 3.0$ per unit, under fault $r_1 P_m = 1.5$, is considered, with $H = 4.0$ and a time step $\Delta t = 0.02$ s. The generator is operating at $P_e = 1.0$ per unit. The rotor angle and angular frequency at the end of 0.02 s are determined as follows:

Euler's Method

$$\delta_0 = \sin^{-1}\frac{1}{3} = 19.47° = 0.34 \text{ rad}$$

Initial angular velocity $= \omega_0 = \delta = 0$:

$$\frac{d\delta_p}{dt} = \omega_p - 2\pi f$$

$$\frac{d\omega_p}{dt} = \frac{\pi f}{H}(P_s - P_e) = \frac{\pi f}{4}(1 - 1.5\sin\delta) = 47.1(1 - 1.5\sin\delta)$$

Determine derivatives at $t = 0$:

$$\dot{\delta} = 0$$

$$\dot{\omega} = 47.1(1 - 1.5\sin 19.47°) = 23.55$$

Then:

$$\delta_1^1 = \delta_1^0 + \frac{d\delta t}{dt}\Delta t = 0.34 + 0 = 0.34 \text{ rad}$$

and

$$\omega_1^1 = 23.55\Delta t = 23.55 \times 0.02 = 0.471$$

There is no change in δ and the power generated remains the same. Again evaluate derivate of δ and angular velocity:

$$\delta = 0.471 \quad \dot{\omega} = 23.55$$
$$\delta_{avg} = \frac{0+0.47}{2} = 0.235$$
$$\dot{\omega}_{avg} = \frac{23.55+23.55}{2} = 23.55$$

New estimate of variables is:

$$\delta_1^1 = \delta_1^0 + \delta_{avg}\Delta t = 0.34 + 0.235 \times 0.02 = 0.3447$$
$$\omega_1^1 = \omega_1^0 + \omega_{avg}\Delta t = 0 + 23.55 \times 0.02 = 0.471$$

Runge-Kutta method

We can write:

$$K_1^k = f_1(\delta^k, \omega^k)\Delta t$$
$$l_1^k = f_2(\delta^k, \omega^k)\Delta t$$
$$K_2^k = f_1\left(\delta^k + \frac{1}{2}K_1^k, \omega^k + \frac{1}{2}l_1^k\right)$$

Therefore:

$$K_1^0 = 0$$
$$l_1^0 = 47.1(1 - 1.5\sin 0.34) \text{ rad}$$
$$\Delta t = 0.471$$
$$K_2^0 = (377 + 0.2355 - 377)0.02 = 0.00471$$
$$l_2^0 = 0.471$$
$$K_3^0 = 0.00471$$
$$l_3^0 = 47.1[1 - 1.5\sin(0.34 + 0.00236) \text{ rad}]0.02 = 0.4679$$
$$K_4^0 = 0.4679 \times 0.02 = 0.00936$$

$$l_4^0 = 47.1[1 - 1.5\sin(0.34 + 0.00236 + 0.00236) \text{ rad}] \times 0.02$$
$$= 0.4647$$

$$\Delta X_1^k = \frac{1}{6}\left(K_1^k + 2K_2^k + 2K_3^k + K_4^k\right)$$
$$= \frac{1}{6}(0 + 4 \times 0.00471 + 0.00936) = 0.0034$$

Therefore:

$$\delta^1 = 0.34 + 0.0034 = 0.3434$$

$$\Delta X_2^k = \frac{1}{6}\left(l_1^k + 2l_2^k + 2l_3^k + l_4^k\right)$$
$$= \frac{1}{6}(0.471 + 2 \times 0.471 + 2 \times 0.4679 + 0.4647) = 0.4689$$

Therefore:

$$\omega^1 = \omega_0 + \Delta\omega = 0.4689$$

These results can be reasonably compared with modified Euler's method, as calculated earlier. Here we have used only a four-segment method.

12-9 SYNCHRONOUS GENERATOR MODELS FOR STABILITY

These can be directly derived from the treatment of synchronous generators discussed in Chap. 10. The simplest generator model represents a machine with a transient reactance behind a voltage source. This is sometimes called the generator model 1.

12-9-1 E'_q Model

The reader may refer to the flux and voltage equations of the synchronous generator in Chap. 10. If we assume that:

$$\dot{\lambda}_d \text{ and } \dot{\lambda}_q \text{ are} << \omega_0\lambda_d \text{ and } \omega_0\lambda_q \tag{12-42}$$

And still neglect the damper circuits:

$$\begin{aligned} v_d &= -r\,i_d - \dot{\theta}\lambda_d \\ v_q &= -ri_q + \dot{\theta}\lambda_d \\ v_F &= r_F i_F + \frac{d\lambda_F}{dt} \end{aligned} \tag{12-43}$$

We see that except for the third equation, we need not calculate the rotor-based terms and can work with the stator quantities. We can express λ_F in stator terms as follows:

$$\lambda_F = kM_F i_d + L_F i_F \tag{12-44}$$

Define a stator voltage E', given by:

$$E'_q = \frac{\omega_0 M_F}{\sqrt{2}L_F}e^{j\delta}\lambda_F \tag{12-45}$$

Using Eq. (12-43), this can be written as:

$$E'_q = \frac{\omega_0 kM_F^2}{\sqrt{2}L_F}e^{j\delta}i_d + \frac{\omega_0 M_F}{\sqrt{2}}e^{j\delta}i_F \tag{12-46}$$

This can be reduced by substitution, as follows (Chap. 10):

$$\begin{aligned} &i_d = \sqrt{3}I_d, \quad k = \sqrt{3/2}, \quad E = \omega_0 M_F e^{j\delta} i_F/\sqrt{2}, \\ &L'_d = L_d - (kM_F)^2/L_F \end{aligned} \tag{12-47}$$

We have the following expression for E':

$$\begin{aligned} E'_q &= \omega_0(L_d - L'_d)I_d e^{j\delta} + E \\ &= j(X'_d - X_d)I_d e^{j\delta} + E \end{aligned}$$

Or:

$$\begin{aligned} E'_q &= V_a + rI_a + jX'_d jI_d e^{j\delta} + jX_q I_q e^{j\delta} \\ &= V_a + rI_a + jX'_d I_d + jX_q I_q \end{aligned} \tag{12-48}$$

Now from differential Equation (12-43), and multiplying v_F throughout $\frac{\omega_0 M_F}{\sqrt{2}r_F}$, we get:

$$\frac{\omega_0 M_F}{\sqrt{2}r_F}v_F = \frac{\omega_0 M_F}{\sqrt{2}r_F}r_F i_F + \frac{\omega_0 M_F}{\sqrt{2}r_F}\frac{d\lambda_F}{dt} \quad (12\text{-}49)$$

Define:

$$\frac{\omega_0 M_F}{\sqrt{2}r_F}v_F = E_{fd} \quad (12\text{-}50)$$

Then:

$$E_{fd} = E + \frac{L_F}{r_F}\frac{dE'_d}{dt} = E + T'_{do}\frac{dE'_q}{dt} \quad (12\text{-}51)$$

Rearranging:

$$\frac{dE'_q}{dt} = \frac{1}{T'_{do}}(E_{fd} - E) \quad (12\text{-}52)$$

E_{fd} may be considered as v_F in equivalent stator terms. The number of differential equations increased by one compared to model 1. The phasor diagram is shown in Fig. 12-17*a*. The equivalent circuit is obtained by ignoring dampers in circuits of Fig. 10-12.

12-9-2 *E″* model

We neglected the damper circuits and *q*-axis rotor circuits. In *E″* model, these are considered. Terms λ'_d and λ'_q are neglected, and $L''_d = L''_q$, $\lambda''_d = \lambda_d - L''_d i_d$ and in the quadrature axis, $\lambda'' = \lambda_q - L''_q i_q$. These flux linkages produce EMFs $e''_d = \omega\lambda''_d$, $e''_q = \omega\lambda''_q$. It can be show that:[5]

$$e''_d = -(X_q - X''_q)i_q - e_d$$

where:

$$e_d = \omega L_{aq} i_Q$$

The model equations are:

$$E''_d = \frac{-(X_q - X''_q)I_q}{(1 + T''_{qo}s)} \quad (12\text{-}53)$$

$$\lambda_D = \frac{\sqrt{3}[E'_q + (X'_d - X_l)]I_d}{(1 + T''_{do}s)} \quad (12\text{-}54)$$

$$E''_q = \left(\frac{X''_d - X_l}{X'_d - X_l}\right)E'_q + \frac{1}{\sqrt{3}}\left(1 - \frac{X''_d - X_l}{X'_d - X_l}\right)\lambda_D \quad (12\text{-}55)$$

The differential equations cannot be shown in the phasor models of Fig. 12-17*b*.

Figure 12-18 from IEEE Std. 1110[7] shows the various synchronous generator models of varying complexity. In this drawing, all symbols are as defined in Chap. 10. Note that L_{f1D} and L_{f12D}, the reactance between the field and dampers, is approximately equal to L_{ad}.

Example 12-4 Based on the previous discussions, this example is a study to demonstrate stability of a simple system under fault conditions and the effect of fast load shedding. It is demonstrated that without fast load shedding, the generator is unstable; however, with fast load shedding, the stability can be maintained.

The system configuration is shown in Fig. 12-19. A generator of 30 MW and a utility transformer of 30/50 MVA run in synchronism

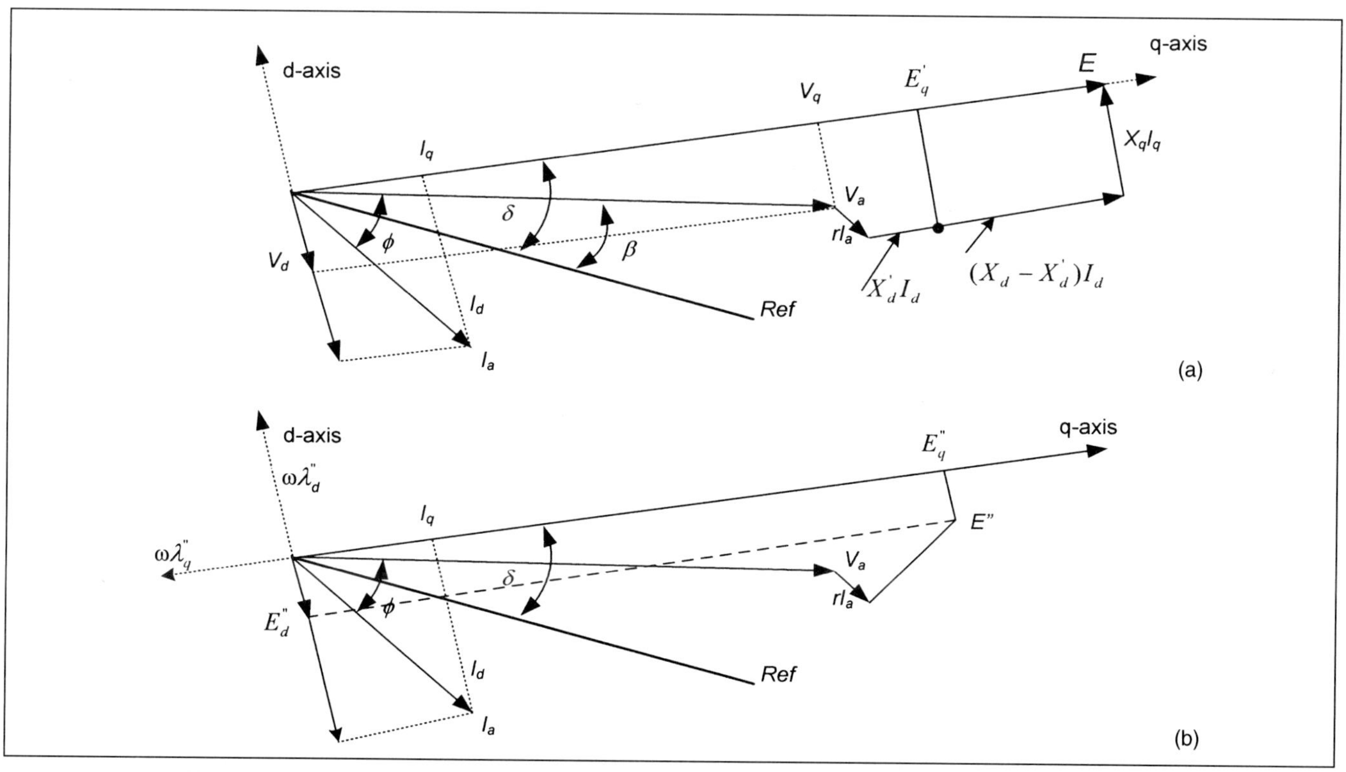

FIGURE 12-17 (*a*) Phasor diagram of a generator E'_q model. (*b*) Phasor diagram of a generator E'' model.

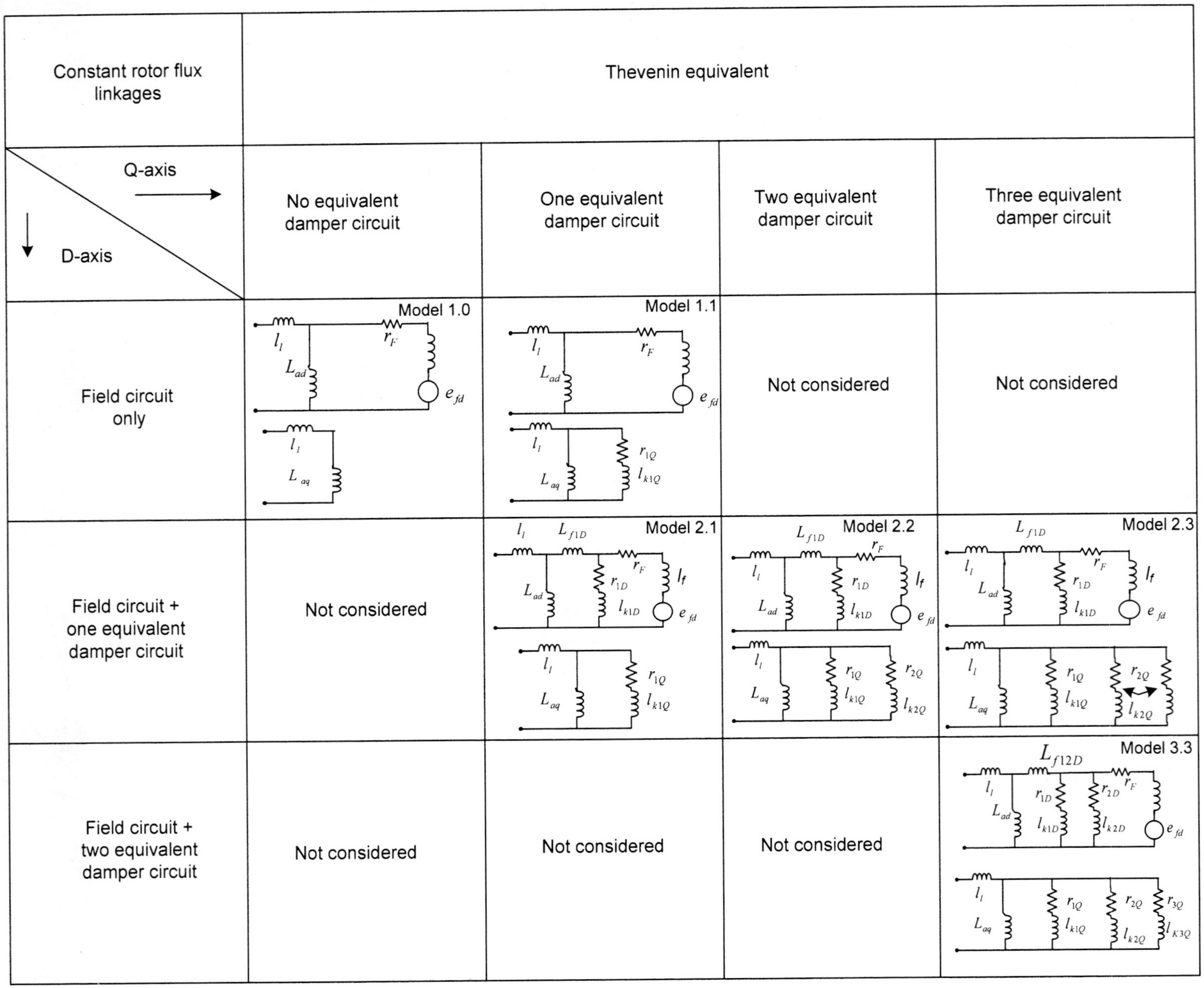

FIGURE 12-18 Synchronous generator models for stability studies, adapted from ANSI/IEEE Std. 1110.

on a 13.8-kV bus, and supply a composite load of 50 MVA. Following data are applicable:

1. 138 kV, bus 1 three-phase short-circuit level 11 kA rms symmetrical, $X/R = 14$, line-to-ground fault = 8.5 kA rms symmetrical, $X/R = 10.6$.

2. Transformer: 30/50 MVA (OA/FA/FA–50 MVA is maximum 65°C fan-cooled rating), 138-13.8 kV, primary windings delta-connected, secondary wye-connected, resistance grounded, $Z =$ 9 percent, $X/R = 24$. Transformer is provided with ±2.5 percent and ±5 percent taps on the 138-kV winding. The transformer taps are at −2.5 percent to provide secondary voltage boost to maintain bus 2 voltage close to the rated voltage. The operating bus 2 voltage is 99.84 percent, close to the rated voltage of 13.8 kV. A prior load flow is conducted to establish operating conditions prior to the fault.

3. Generator: 30 MW, 13.8 kV, 0.85 power factor and $X''_d =$ 12 percent, $X'_d = 23$ percent, $X_2 = X_0 = 11.5$ percent, $X/R = 49$, $X_d = 110$ percent, $X_q = 108$ percent, $T'_{do} = 5.5$ s, $X_1 = 11$ percent. Generator saturation is modeled (see Chap. 13), $H = 5.5$ (including prime mover and coupling), and no exciter, governor, or prime mover controls are modeled.

4. Load: 50 MVA, composite, 80 percent constant kVA, 20 percent constant Z, no dynamic models of the load, which includes essential load served from breaker 2.

5. Disturbance data: A three-phase fault occurs, as shown on a bus tie breaker load side, at 0.05 s, which is cleared in 0.4 s. (Practically the fault clearing time will be much less. Assume that the first level of protection does not operate.)

Consider:

Case 1: Study for 3 s, no other switching action takes place, except fault placement at 0.05 s and fault clearance at 0.45 s; fault remains active for 0.4 s.

Case 2: As case 1, except that fast load shedding occurs at 0.1 s, dumping a load of 37.75 MW by opening breaker 1 (Fig. 12-19). 4.71 MW of load, representing the generator auxiliary load and essential service load, connected through breaker 2 remains in service.

Figure 12-19 also shows the results of load flow prior to opening breaker 1 at $t = 0^-$. The generator supplies 30 MW and 18 Mvar to the load; the transformer load is 12.5 MW and 9.0 Mvar. Approximately 0.7 Mvar losses occur through the transformer.

The results of simulation are plotted in Figures 12-20 and 12-21.

- Figure 12-20*a* shows transients in generator active and reactive power, terminal current, rpm, and generator relative power angle. By the time the fault is cleared, the generator power angle has already swung to 178° and the speed goes on increasing with undulations. The active and reactive power and current swing violently. Examination of power angle (torque angle) swing of the machine gives a clear idea of instability.

- Figure 12-20*b* shows corresponding swings in generator bus voltage, bus 2, which again does not show any trend of recovery. It oscillates as the generator supplies power and draws power from the system.

- Figure 12-21*a* depicts similar transients as Fig. 12-20*a* but with major load shed at 0.1 s. Note that the power angle swing has been arrested (152° in the first swing) and the transients have a decaying trend, though slowly. These show that for the same fault clearance time, stability is obtained with fast load shedding.

- Figure 12-21*b* depicts the transients in the generator bus voltage, which is recovering to normal operating voltage through oscillations.

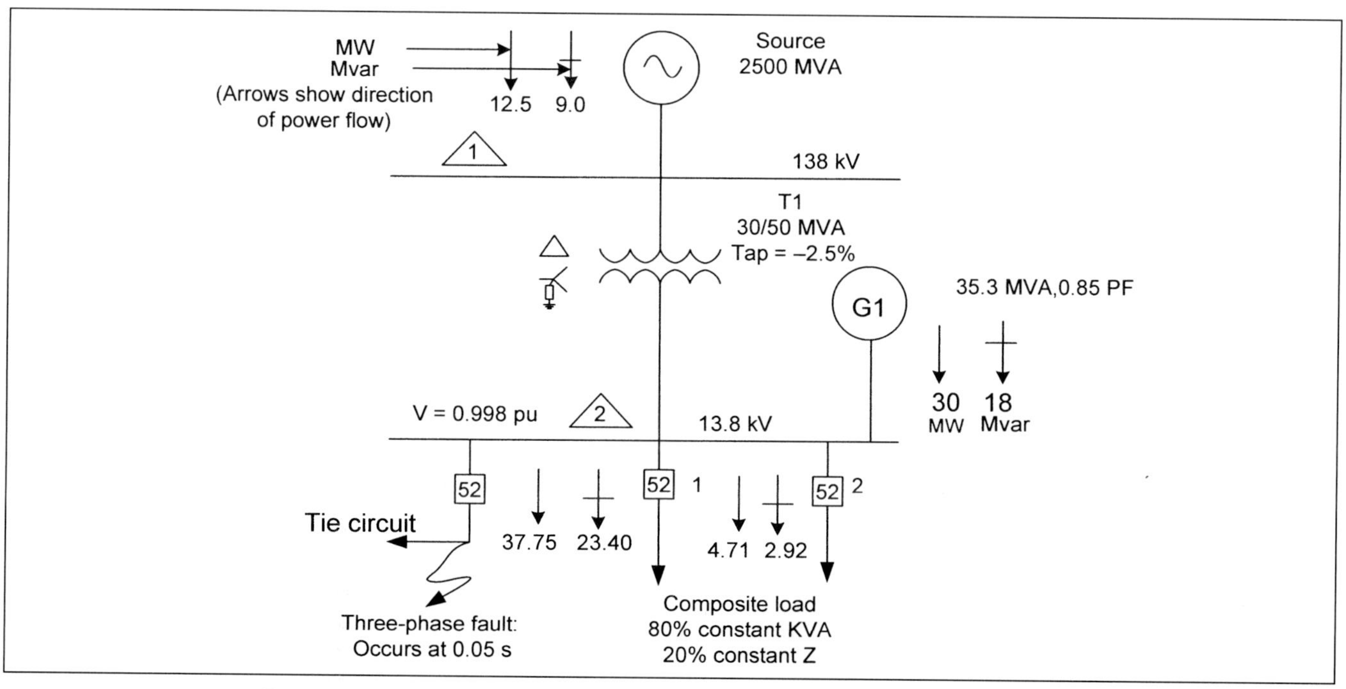

FIGURE 12-19 A power system configuration for transient stability study, Example 12-4.

The following explanation of the criterion for stability are valid:

1. A disturbance can be related to a power unbalance. A three-phase fault is, generally, the most severe condition; the fault current is mostly reactive; the voltage at the fault point falls to zero, while at other points in the system it will depend on the relative impedances and the current flows to the fault point. Due to severe voltage dips, the loads may drop out totally or rotating loads may take a proportionally higher current, for example, induction motors, which for a balanced voltage reduction, demand proportionally higher currents (Chap.11). These higher currents will aggravate the stability situation.

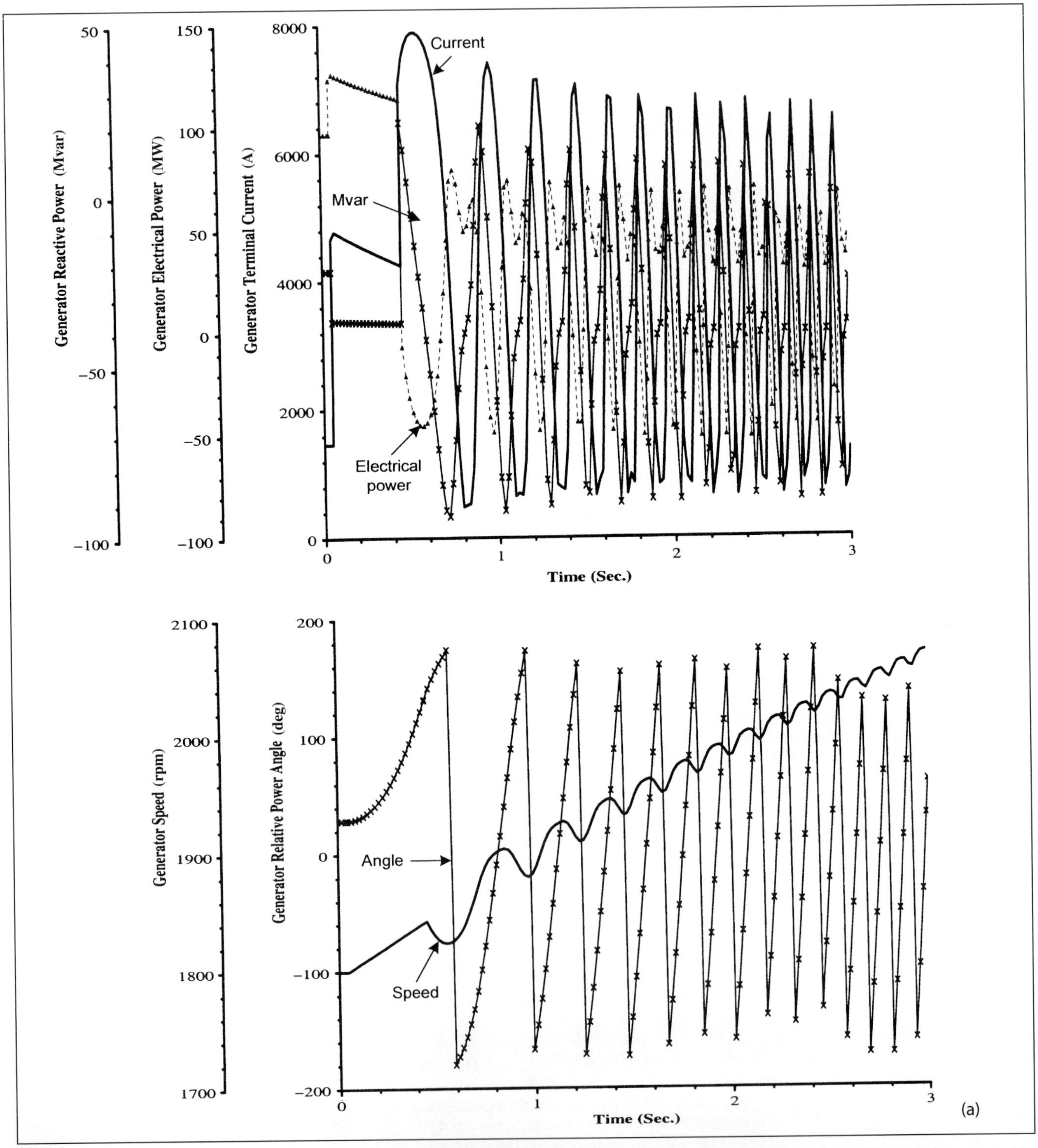

Figure 12-20 Transients for simulation of fault condition, without load shed, Example 12-4, when fault is cleared in 0.2 s. (*a*) Transients in generator G1. (*b*) Recovery transient of the bus voltage. The transients show instability of the system. (*Continued*)

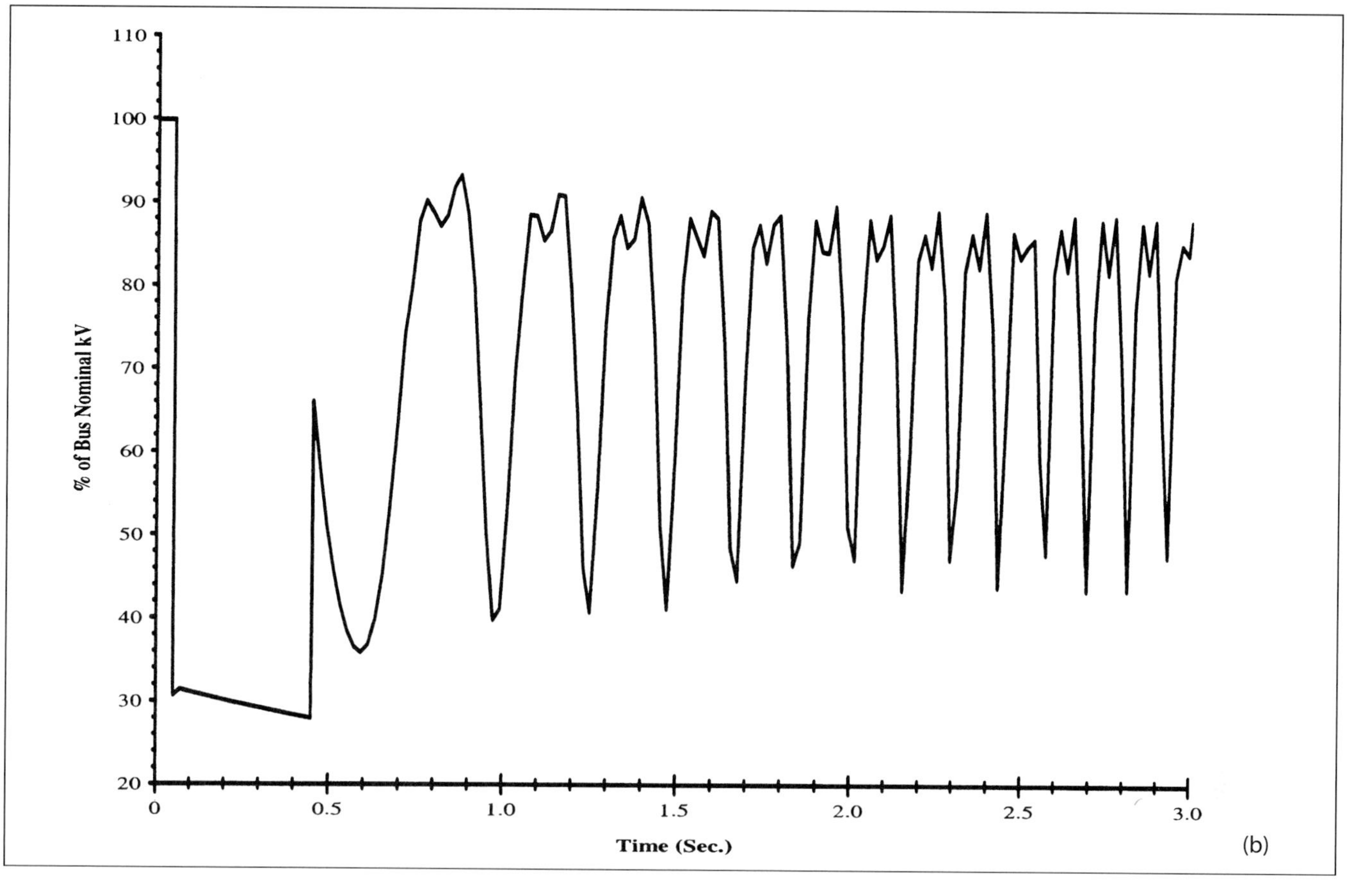

FIGURE 12-20 *(Continued)*

2. The frequency will swing, and this will create further oscillations in load current. As an approximation, a load decrease proportional to frequency may be considered.

3. Rotors of the synchronous machines swing, as accelerating and deaccelerating torques are exerted, and a machine may "slip a pole."

4. Loss of synchronism can happen in stages. A primary disturbance may give rise to a secondary disturbance; for example, a tie line in a transmission system carrying considerable amount of power can be interrupted, giving rise to superimposed transients, while the first set of transients may not have decayed.

5. The angle difference (not the absolute angle) between the swinging machines is the criterion of instability. Two machines of similar characteristics, on the same bus, may swing together. The larger the impedance between the swinging machines, the greater could be the divergence of angle between them.

Without load shedding, the generator torque angle keeps oscillating after the fault has been cleared. The reactive and active power output of the generator, bus voltage, and current shows violent cyclic swings. Once a generator falls out of synchronism, it is subjected to cyclic torques, stresses, current, and power swings that occur with respect to the electrical center, somewhere out in the system. A generator can be severely stressed due to these violent swings that impact the shaft, mechanical systems, and also the foundations of the machine.

ANSI/IEEE device 78, out-of-step protection, will trip out the generator on the first swing after it falls out of step. This can prevent much damage and stresses to the electrical and mechanical systems and foundations. The out-of-step protective relays are impedance-single blinder-type, and a plot of the impedance locus is required for appropriate settings; the stable and unstable swings must be distinguished.

Figure 12-22 is the impedance plot in *R*-*X* plane of a generator as it continues to swing after it pulls out of step.

Example 12-5 A three-bus system, as shown in Fig. 12-23, is considered for transient stability. As a first step, the bus voltages, loads, and load flow data at $t = 0^-$ are superimposed in Fig. 12-23. Then, it can be converted to impedances and admittances (Fig. 12-24). The loads are replaced by constant impedances. In stability study, it is essential to have dynamic models of the rotating loads, that is, induction and synchronous motors, as these impact the results significantly. Here we consider all loads of constant impedance type, for simplicity. The impedance data of the transmission lines are shown in Table 12-1.

Prior load flow results are inputted; for example, generator 1 voltage can be calculated from:

$$V_{g1} = V_1 + j(r_g + X_d)\left(\frac{P_1 + jQ_1}{V_1}\right)^*$$

Based on Fig. 12-24, the system equations can be written as:

$$Y_{11}V_1 + Y_{12}V_2 + Y_{13}V_3 + Y_{14}V_{g1} + Y_{15}V_\infty = 0$$

$$Y_{22}V_2 + Y_{21}V_1 + Y_{23}V_3 + Y_{26}V_{g2} = 0$$

$$Y_{33}V_3 + Y_{31}V_1 + Y_{32}V_2 = 0$$

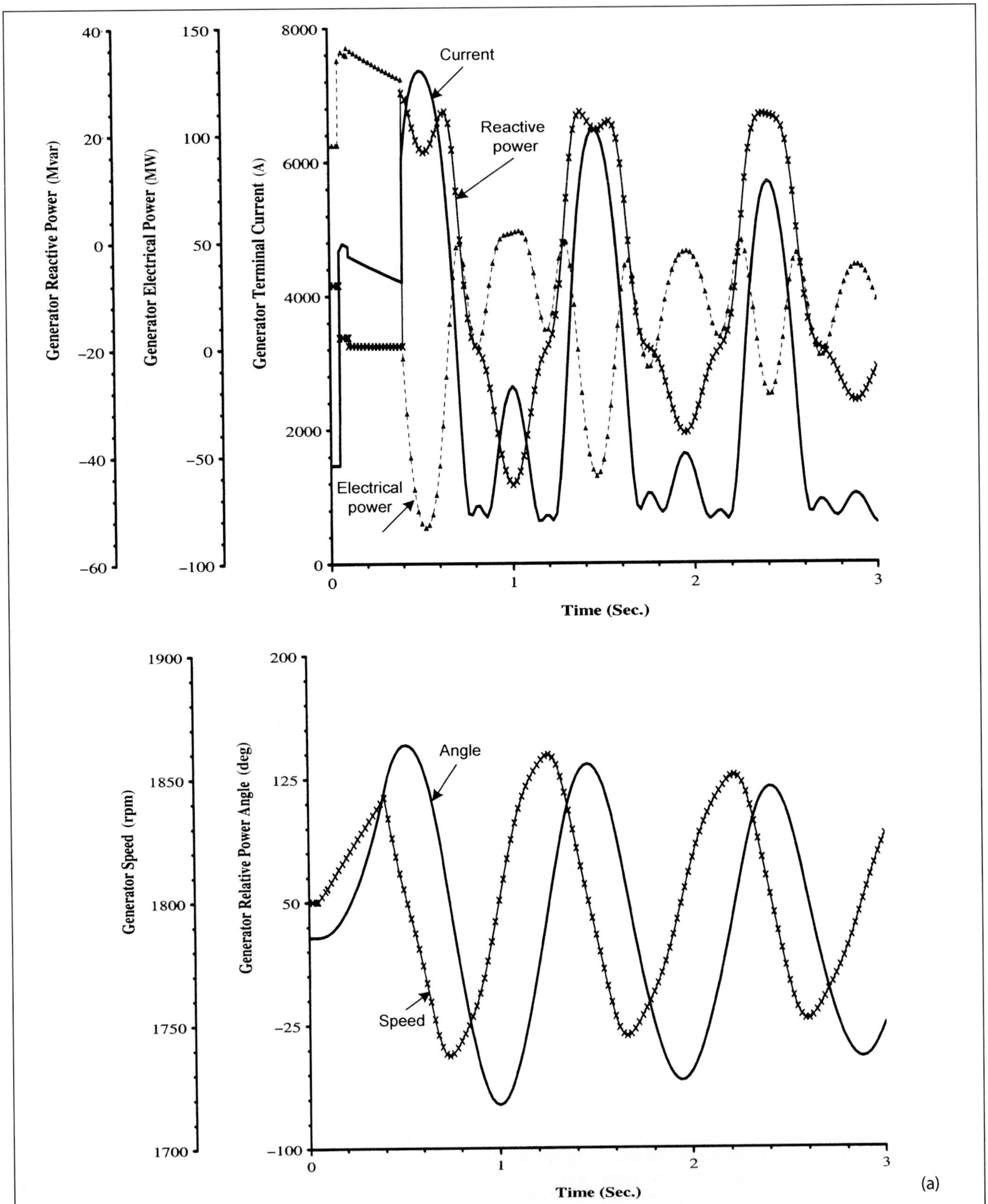

FIGURE 12-21 Transients for simulation of fault condition, Example 12-4, when fault is cleared in 0.2 s, but 37.75 MW of load shed at 0.1 s. (*a*) Transients in generator G1. (*b*) Transient of the bus recovery voltage. The transients show stability of the system, with fast load shedding. (*Continued*)

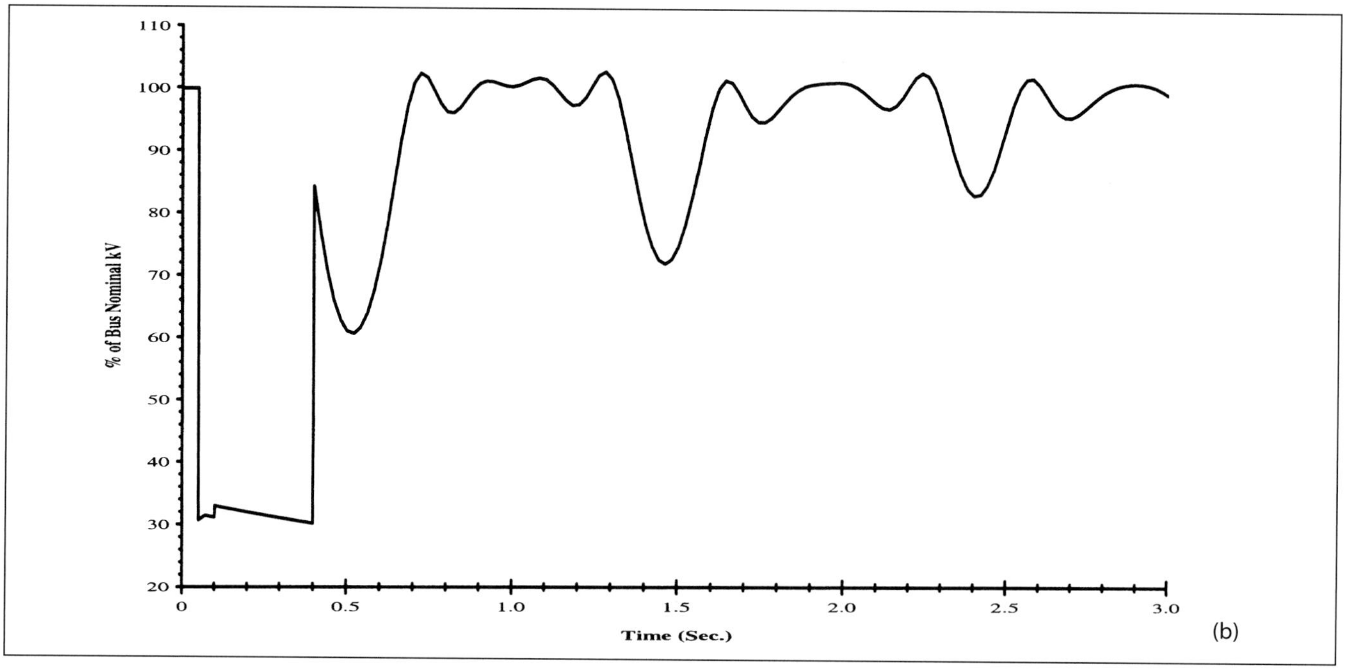

FIGURE 12-21 (*Continued*)

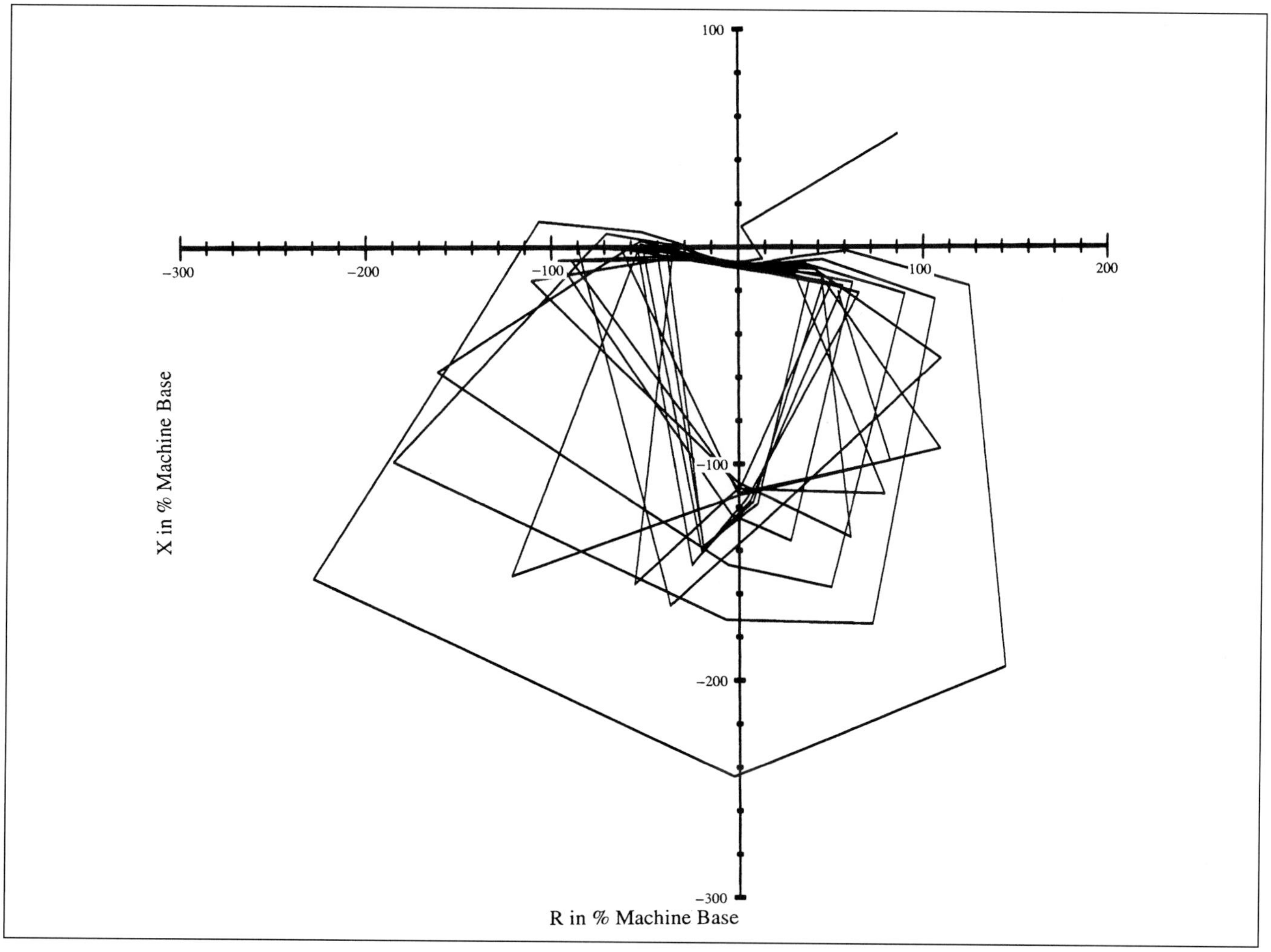

FIGURE 12-22 Impedance locus of a generator in *R*-*X* plane, as it goes on swinging after pulling out of step.

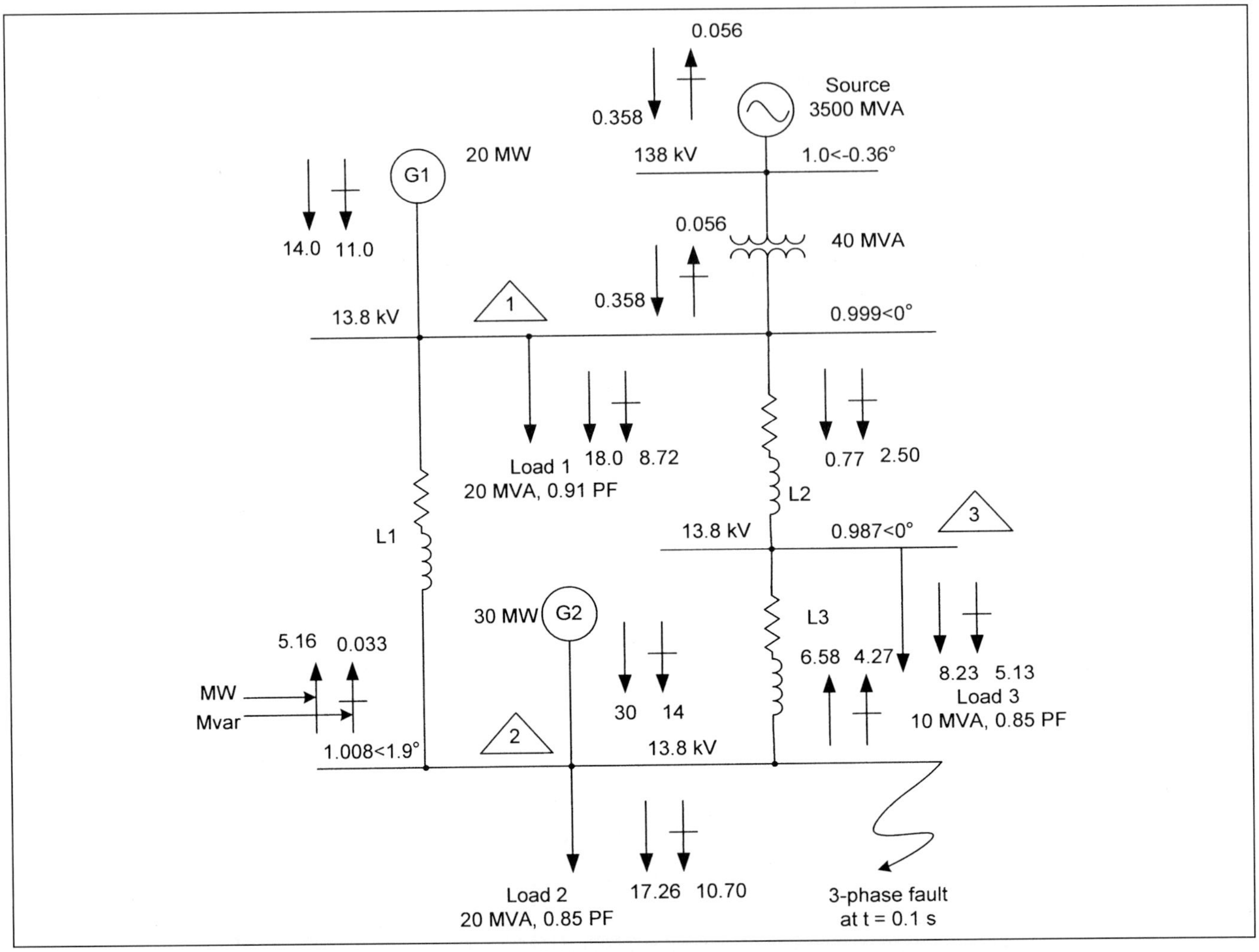

FIGURE 12-23 A power system configuration for transient stability study, Example 12-5.

And Y-matrix can be written as:

	1	2	3	4	5	6
1	$5.405 - j10.887$	$-0.289 + j1.512$	$-0.387 + j2.013$	$-0.021 + j1.023$	$-0.146 + j4.003$	0
2	$-0.289 + j1.512$	$4.945 - j7.636$	$-0.387 + j2.013$	0	0	$-0.033 + j1.47$
3	$-0.387 + j2.013$	$-0.387 + j2.013$	$9.426 - j9.447$	0	0	0
4	$-0.021 + j1.023$	0	0	$0.021 - j1.023$	0	0
5	$-0.146 + j4.003$	0	0	0	$0.146 - j4.003$	0
6	0	$-0.033 + j1.47$	0	0	0	$0.033 - j1.47$

The source impedance is combined with the 40-MVA transformer impedance. The results of the load flow pertain to $t = 0^-$. The loads, voltages, generation, and angles will be calculated in the specified interval at each incremental K.

The generator swing equations can be written in the state form and solved with the modified Euler's method for transient stability calculations. A flowchart is given in Fig. 12-25.

Case 1: A three-phase fault is simulated at bus 2 at $t = 0.1$ s and cleared at 0.3 s, that is, the fault is sustained for 0.2 s.

The transients are shown in Fig. 12-26. It is obvious that generator 2 is more susceptible to fall out of step as compared to generator 1. Because it is closer to the fault location, it undergoes much larger swings, though it has a larger rating compared to generator 1. Though under normal operating conditions, little power is drawn from the utility source through 40-MVA transformer, but under a disturbance, the utility source tends to support the system from falling out of step. It is, however, interesting to note that though swings of generator 1 are much lower than that of generator 2, these show diverging trend, giving rise to dynamic stability problems.

Case 2: The fault is now cleared at 0.4 s, that is, it is sustained for 0.3 s. Generator 2 falls out of step; transients depicted in

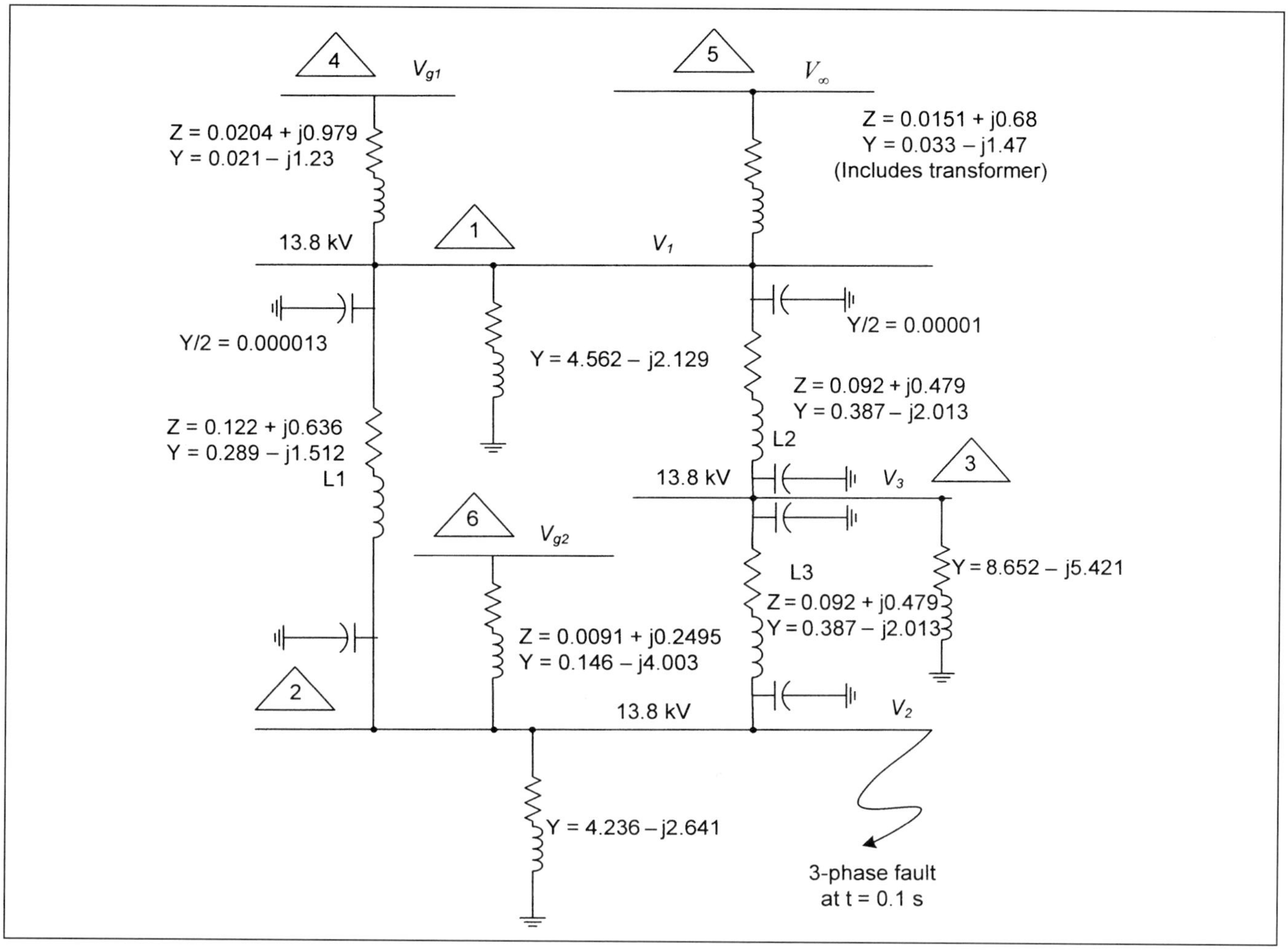

FIGURE 12-24 The system in Fig. 12-23 converted to admittances.

TABLE 12-1 System Data for Example 12-5

ITEM DESIGNATION	RAW IMPEDANCE DATA
Power grid	3500 MVA, three-phase symmetrical short circuit, $X/R = 15$
G1	23.53 MVA, 13.8 kV, 0.85 power factor, 2-pole $X'_d = 23\%$, $X_d = 200\%$, $X_q = 190\%$, $X_l = 11.0\%$, $X/R = 48$, total H including coupling and prime mover = 5.5 ($WR^2 = 43226$ lb-ft^2), $K_1 = 5\%$, saturation modeled, $S_{100} = 1.070$, $S_{120} = 1.12$, $S_{break} = 0.80$, resistance grounded. Excitation and speed governing not modeled
G2	35.3 MVA, 13.8 kV, 0.85 power factor, 2-pole, $X'_d = 23\%$, $X_d = 200\%$, $X_q = 190\%$, $X_l = 11.0\%$, $X/R = 48$, total H including coupling and prime mover = 5.5 (WR^2 43226 lb-ft^2), $K_1 = 5\%$, saturation modeled, $S_{100} = 1.070$, $S_{120} = 1.12$, $S_{break} = 0.80$, resistance grounded. Excitation and speed governing not modeled
Transformer T1	40.0 MVA, 138–13.8 kV, $Z = 9\%$, $X/R = 27.30$
Line L1	1033.50 KCMIL, ACSR, 54 strands, 5' flat spacing, GMD = 6.3 ft, GMR = 0.042 ft, outside diameter = 1.246", one ground wire, placed 6 ft above phase conductors, height of conductors 30 ft, soil resistivity = 100 Ω/m, untransposed conductors, calculated impedance per mile: positive and negative sequence = $0.099 + j0.608\ \Omega$, $Y = 7.11281\ \mu S$, zero-sequence per mile: $0.06007 + j2.68652\ \Omega$, line length = 10560 ft
Lines L2 and L3	Same data as line L1, except length = 7920 ft each
Fault	Three-phase bolted fault occurs at bus at 0.1 s
Fault clearance	Study conducted for two cases, fault cleared at 0.3 s, i.e., lasting for 0.2 s and cleared at 0.4 s, i.e., fault lasting for 0.3 s

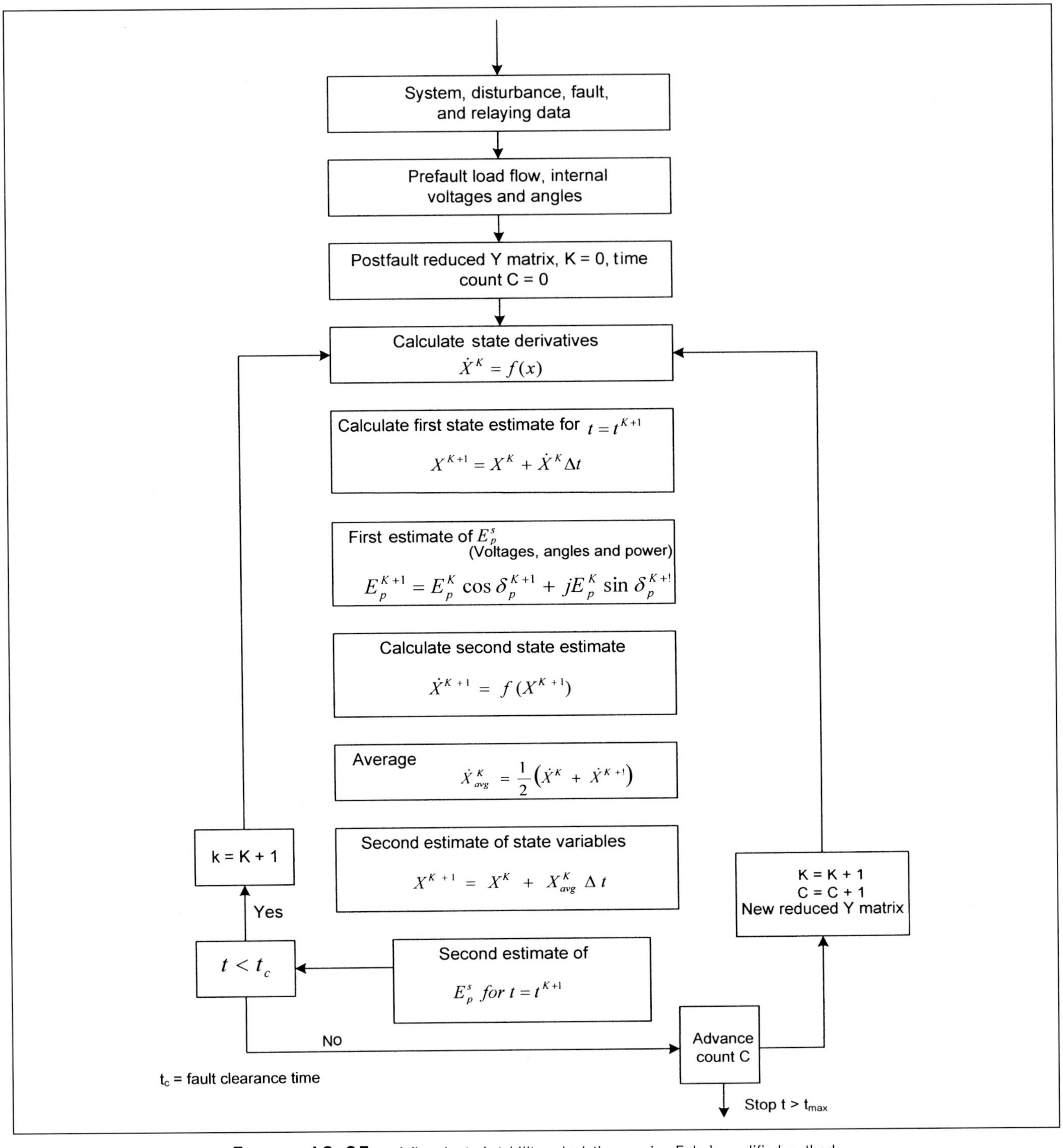

FIGURE 12-25 A flowchart of stability calculations, using Euler's modified method.

Fig. 12-27. The situation is similar to that depicted in Fig. 12-21a, Example 12-3. Its torque angle has increased to approximately 180° at the fault clearance time and then keeps on oscillating violently. There are violent swings in the speed, currents, and power. Generator 1 remains stable and rides through the disturbance, though its torque angle is settling very slowly—thus, it keeps swinging for a long time after the perturbation is removed. The voltage of bus 1 slowly recovers, while bus 2 has large swings, in synchronism with the generator power angle swings as it draws power from and supplies power into the electrical system. Practically, the protective relays will take the generator off-line on the first swing, device 78, as discussed earlier.

Chapter 13 continues with this example and shows that in the same configuration and same fault condition lasting for 0.3 s, stability can be achieved by addition of a fast-response exciter.

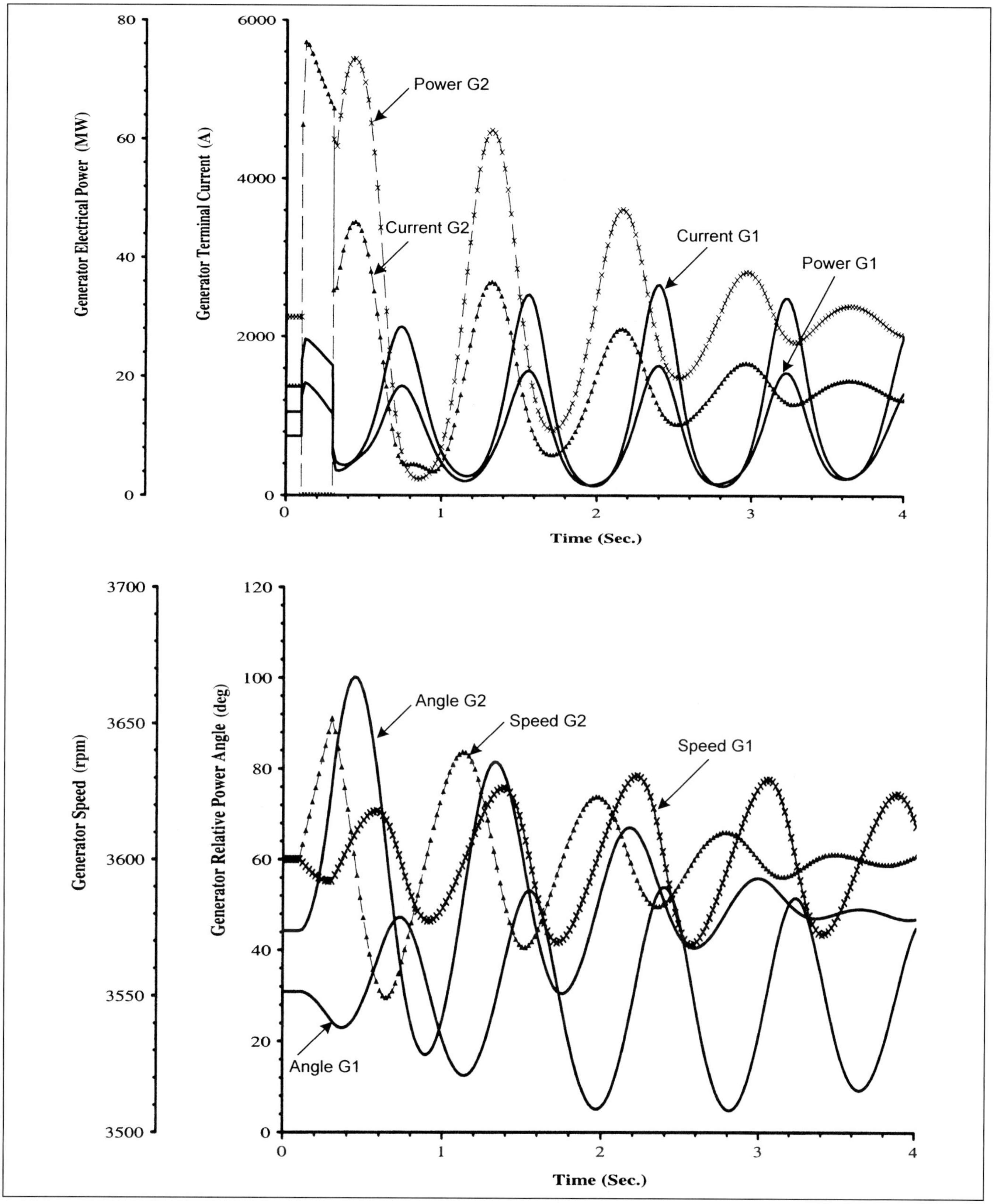

FIGURE 12-26 Transients in generators G1 and G2 for simulation of fault condition, Example 12-5; three-phase fault on bus 2 is cleared in 0.2 s. The transients show stability of both the generators G1 and G2.

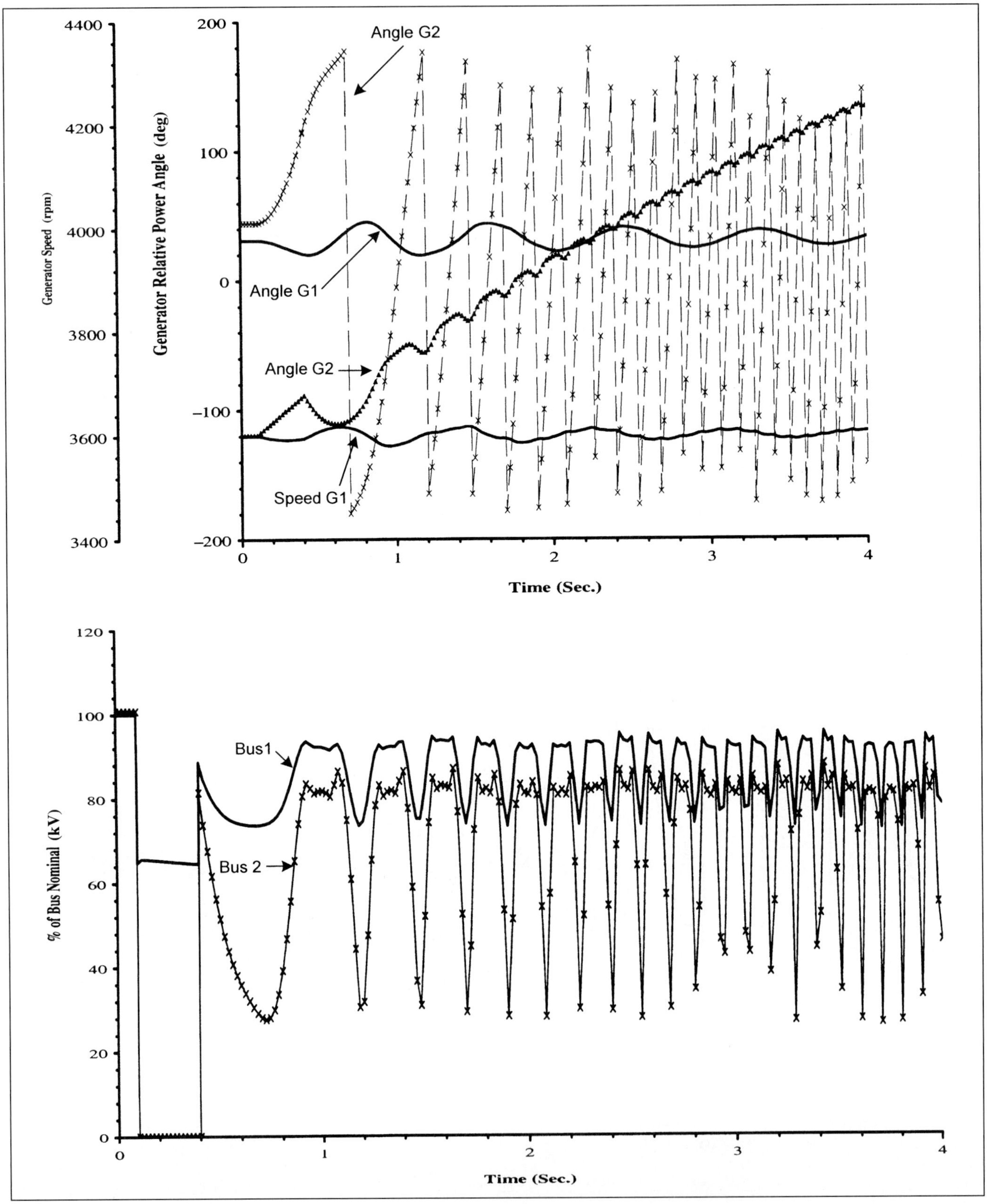

FIGURE 12-27 Transients in generators G1 and G2 for simulation of fault condition, Example 12-5; three-phase fault on bus 2 is cleared in 0.3 s. The transients show instability of generator G2, while generator G1 is stable. Bus voltage transients are also depicted.

12-10 SMALL-SIGNAL STABILITY

The following types of subsynchronous frequency oscillations are recognized:

- Local plant mode oscillations
- Interarea mode oscillations
- Torsional mode oscillations
- Control mode oscillations

12-10-1 Local Mode Oscillations

Local mode oscillations occur when units at a generating station oscillate with respect to the rest of the power system. The typical frequency of oscillation is 1 to 2 Hz, depending mainly on the impedance of the transmission system. Stronger transmission systems tend to have a higher local mode frequency. The oscillations are termed local because behavior is mainly localized at one point, with the rest of the system experiencing much less effects. Spontaneous local mode oscillations tend to occur when a very weak transmission link exists between a machine and its load center. An example can be an isolated power station (wind power generation), sending power along a long transmission line. The systems can be usually modeled accurately for the study. Consider the following parameters:

$H = 4$ kW-s/kVA

$\delta_0 = 45°$

$X'_d = 0.25, \quad X_e = 0.45$ pu

$E'_q = 1.05$ pu, and $V = 1.0$ pu

Then from Eq. (12-20), $K_1 = 1.0605$ and from Eq. (12-21), ω_n = rad/s = 1.125 Hz.

If oscillations of this frequency are experienced at the generating plant, these will most likely be local mode oscillations. An external reactance of approximately 0.50 per unit or less, including impedance of the step-up transformer, means that the undamped local mode oscillations will not be a problem.

12-10-2 Interarea Oscillations

As the name suggests, interarea oscillations are associated with machine in one part of the system oscillating against machines in the other part of the system. These are more complex and mean that a combination of the machine in one part of the system is swinging against machines in another part of the system. The simplest form of representation is interlinking two separate power systems or a group of generators through transmission tie lines.

The frequency of oscillations is in the range of 0.1 to 1 Hz. This is lower than local mode oscillations because of the high inertia of a large group of machines.

The modeling for interarea oscillations can be complex. A power system containing n generators has $(n - 1)$ normal modes of oscillations, with their own natural frequency and profiles of oscillations. Separating these and finding which one is easily excitable and has potential of undamped oscillations can be a difficult task. Frequency domain techniques using state–space analysis have been used. Most oscillation modes in generators are positively damped, and thus attention is focused on those modes that contain enough negative damping to give rise to persistent rotor oscillation or the oscillations increasing with time.

Even when the system is unstable on large signal perturbations, the natural damping of the system, represented by positive K_D term in Eq. (12-17), prevents any sustained oscillations.

It is recognized that normal feedback control action of voltage regulators and speed governors has the potential of contributing to negative damping, which can cause undamped modes of dynamic oscillations.

12-10-3 Torsional Oscillations

Torsional oscillations are due to interactions of the generating units' excitation systems with prime mover controls, that is, speed-governing systems (see Chap. 16).

12-10-4 Control Mode Oscillations

Control mode oscillations are due to poorly tuned controls of excitation systems, SVCs, and prime movers. These are the usual causes of instability of control modes.

12-11 EIGENVALUES AND STABILITY

An introduction to matrix theory is first provided to proceed with this section. Consider a square matrix $\bar{A}$. Then the matrix $\bar{A} - \lambda\bar{I}$ is called the *characteristic matrix*. Λ is a scalar and $\bar{I}$ is a unit matrix. The determinant $|\bar{A} - \lambda\bar{I}|$, when expanded, gives a polynomial which is called the *characteristic polynomial* of $\bar{A}$, and the expression $|\bar{A} - \lambda\bar{I}| = 0$ is called the *characteristic equation* of matrix $\bar{A}$. The roots of the characteristic equation are called the *characteristic roots* or *eigenvalues*.

- Any square matrix and its transpose have the same eigenvalues.
- The sum of the eigenvalues of a matrix is equal to the trace of the matrix. The sum of the elements on the principal diagonal is called the *trace* of the matrix.
- The product of the eigenvalues of a matrix is equal to the determinant of the matrix. If $\lambda_1, \lambda_2, \ldots, \lambda_n$ are the eigenvalues of $\bar{A}$, then the eigenvalues of:

$$
\begin{aligned}
&k\bar{A} \text{ are } k\lambda_1, k\lambda_2, \ldots, k\lambda_n \\
&\bar{A}^m \text{ are } \lambda_1^m, \lambda_2^m, \ldots, \lambda_n^m \\
&\bar{A}^{-1} \text{ are } 1/\lambda_1, 1/\lambda_2, \ldots, 1/\lambda_n
\end{aligned}
\tag{12-56}
$$

- Zero is a characteristic root of the matrix only if matrix is singular.
- The characteristic roots of the triangular matrix are diagonal elements of the matrix.
- The characteristic roots of a real symmetric matrix are all real.

12-11-1 Characteristic Vectors

Each characteristic root λ has a corresponding nonzero vector $\bar{x}$, which satisfies the equation:

$$|\bar{A} - \lambda\bar{I}|\bar{x} = 0 \tag{12-57}$$

The nonzero vector is called the *characteristic vector* or eigenvector. The eigenvectors are therefore not unique.

Now consider the linearization of dynamic system state equations (see Chap. 2). We can write:

$$\dot{\bar{x}} = \bar{A}\bar{x} + \bar{B}\bar{r}$$

or

$$\Delta\dot{x} = \bar{A}\Delta x + \bar{B}\Delta r \tag{12-58}$$

Taking Laplace transform into consideration:

$$s\Delta x(s) - \Delta x(0) = \bar{A}\Delta x(s) + \bar{B}\Delta r(s)$$

or

$$\Delta x(s) = (s\bar{I} - \bar{A})^{-1}[\Delta x(0) + \bar{B}\Delta r(s)] \tag{12-59}$$

Then the characteristic equation is:

$$\det(s\bar{I} - \bar{A}) = 0 \tag{12-60}$$

- A real eigenvalue corresponds to a nonoscillatory mode.
- A negative eigenvalue corresponds to a decaying mode.
- A positive eigenvalue represents aperiodic instability.
- Complex eigenvalues occur in conjugate pairs, and each pair corresponds to an oscillatory mode. Also see Chap. 3.

We can write the two swing equations in the matrix form:

$$\begin{vmatrix} \Delta\dot{\omega}_r \\ \Delta\dot{\delta} \end{vmatrix} = \begin{vmatrix} -\dfrac{K_D}{2H} & -\dfrac{K_1}{2H} \\ \omega_0 & 0 \end{vmatrix} \begin{vmatrix} \Delta\omega_r \\ \Delta\delta \end{vmatrix} + \begin{vmatrix} \dfrac{1}{2H} \\ 0 \end{vmatrix} \Delta T_m \tag{12-61}$$

This is of the form:

$$\dot{x} = \bar{A}\bar{x} + \bar{B}\bar{r} \tag{12-62}$$

The block diagram representation is illustrated in Fig. 12-28, from which:

$$\Delta\delta = \frac{\omega_0}{s}\left[\frac{1}{2Hs}\left(-K_1\Delta\delta - K_D s\frac{\Delta\delta}{\omega_0} + \Delta T_m\right)\right]$$

Or

$$s^2\Delta\delta + \frac{K_D}{2H}s\Delta\delta + \frac{K_1}{2H}\omega_0\Delta\delta = \frac{\omega_0}{2H}\Delta T_m \tag{12-63}$$

The characteristic equation is:

$$s^2 + \frac{K_D}{2H}s + \frac{K_1\omega_0}{2H} = 0 \tag{12-64}$$

Referring to Chap. 3 and comparing with Eq. (3-1), it is reproduced as follows:

$$s^2 + 2\varsigma\omega_n s + \omega_n^2 = 0 \tag{12-65}$$

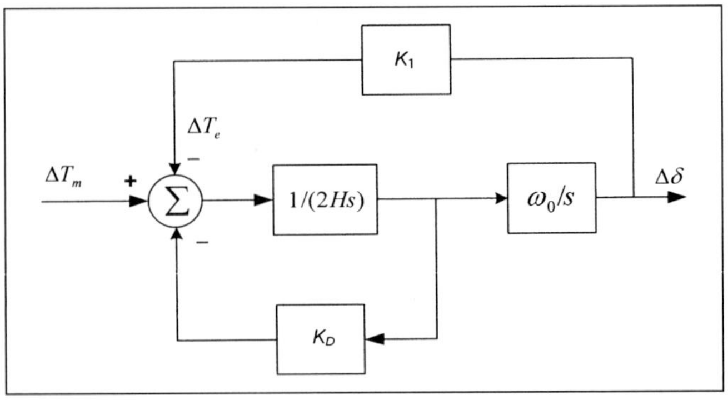

FIGURE 12-28 A control block circuit diagram of the swing equation of a generator.

The undamped natural frequency is given by ω_n, and the damping ratio is:

$$\varsigma = \frac{1}{2}\frac{K_D}{2H\omega_n} = \frac{1}{2}\frac{K_D}{\sqrt{K_1 2H\omega_0}} \tag{12-66}$$

Example 12-6 A 18-kV, 100-MVA generator connected through a step-up transformer of 100 MVA to two parallel 230-kV lines is considered, (Fig. 12-29*a*). The system data are in Table 12-2, and the results of load flow with calculated bus voltages are superimposed in this figure. The 230-kV lines have a charging requirement of 78.26 Mvar at no load; also see Chap. 4. The small-signal stability characteristics of the system are analyzed when line 2 is lost by opening breakers at either end at 0.1 s; there is no fault in the system. Then, only one line remains in service. The bus voltages at load flow after L2 is taken out of service are shown in Fig. 12-29 *b*. Therefore:

$$\delta = 29.76°$$
$$E_g = 1.11$$
$$E_t = 1.0$$

E_g is the voltage behind generator transient reactance and E_t is the bus voltage. For simplicity of hand calculation, we will ignore resistance component; then $X_e = 0.39$ per unit.

$$K_1 = \frac{1.11 \times 1}{0.39}\cos 29.76° = 2.447\,\text{pu torque/rad}$$

From Eq. (12-61):

$$\begin{vmatrix} \Delta\dot{\omega}_r \\ \Delta\dot{\delta} \end{vmatrix} = \begin{vmatrix} -0.5 & -0.2447 \\ 377 & 0 \end{vmatrix} \begin{vmatrix} \Delta\omega_r \\ \Delta\delta \end{vmatrix} + \begin{vmatrix} 0.1 \\ 0 \end{vmatrix} \Delta T_m$$

TABLE 12-2 System Data for Example 12-6

DESIGNATION	DATA
Power grid	An infinite bus
G1	18 kV, 60 Hz, 2-pole, 100 MVA, $X'_d = 25\%$, $X_d = X_q = 200\%$, $X/R = 60$, $H = 5.0$, $K_D = 5$ (machine MVA base)
Transformer T1	100.0 MVA, 18–230 kV, delta, wye, solidly grounded, $Z = 10\%$ (100-MVA base), $X/R = 34$
Line L1, 200 mi	1351 ACSR, 2 ground wires, soil resistivity = 100 Ω/m. $Z^+ = 0.07003$ Ω (25°C) + $j0.75996$ Ω/mi and $Y = 5.64688$ μS/mi. Only positive is sequence impedance considered in calculations
Lines L2, 100 mi	Data same as for L1

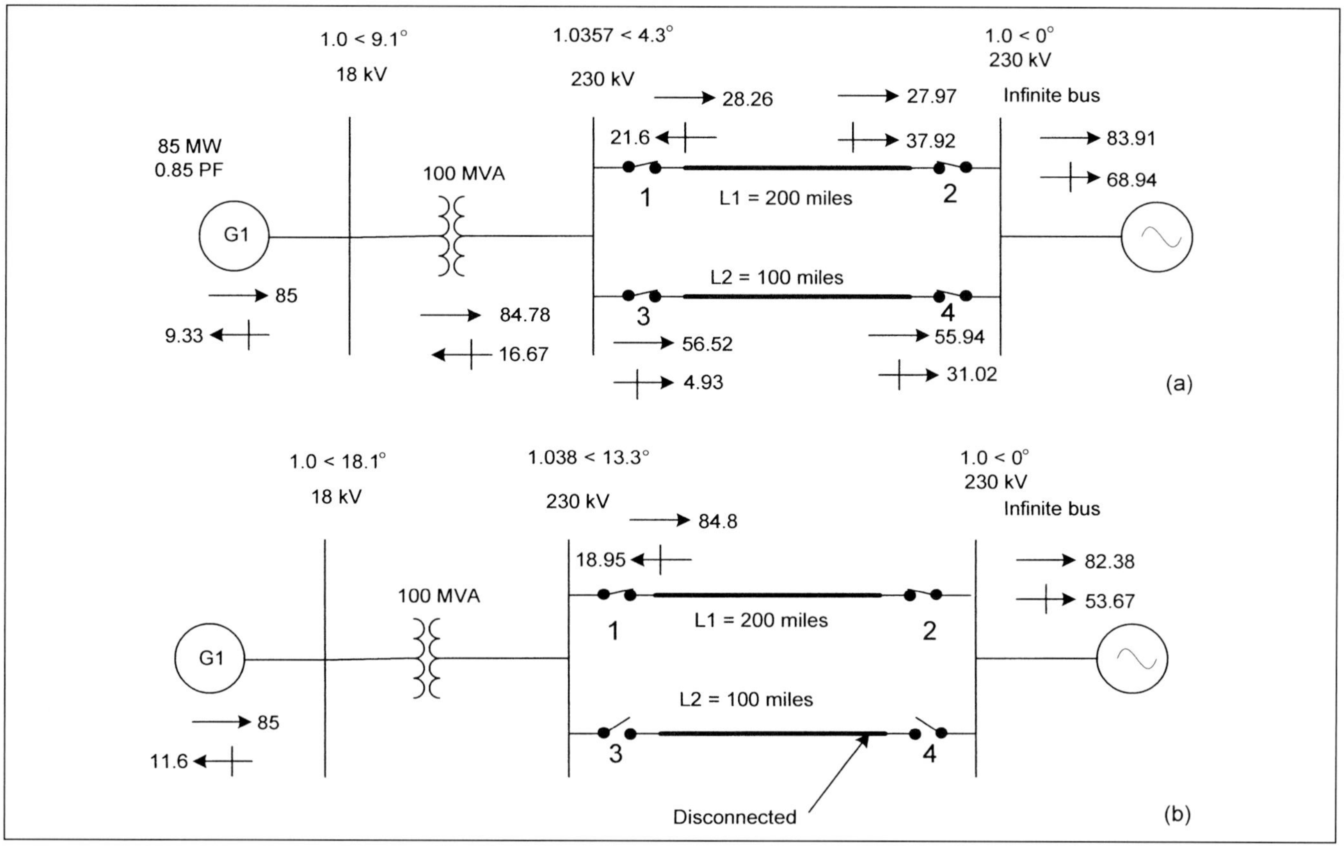

FIGURE 12-29 (*a*) Power system configuration of a generator connected to an infinite bus through duplicate tie lines. (*b*) One tie line opened suddenly, without a fault, Example 12-6.

The eigenvalues of the state matrix are:

$$\begin{vmatrix} -0.5-\lambda & -0.2447 \\ 377 & -\lambda \end{vmatrix} = 0$$

$$\lambda^2 + 0.5\lambda + 92.25 = 0$$

Therefore:

$$\omega_n = \sqrt{92.25} = 9.60 \text{ rad/s} = 1.513 \text{ Hz}$$

$$\varsigma = \frac{0.5}{2 \times 9.60} = 0.02604$$

The damped frequency is:

$$\omega_d = \omega_n \sqrt{1 - 0.02604^2} = 1.509 \text{ Hz}$$

$$\lambda_1, \lambda_2 = -0.250 \pm j9.596$$

Write the state equation:

$$(\bar{A} - \lambda \bar{I})\bar{\phi} = 0$$

Thus:

$$\begin{vmatrix} -0.5-\lambda_1 & -0.2447 \\ 377 & -\lambda_1 \end{vmatrix} \begin{vmatrix} \phi_{11} \\ \phi_{21} \end{vmatrix} = 0$$

For:

$$\lambda = -0.250 - j9.596$$

$$377.0\phi_{11} + (0.250 + j9.596)\phi_{21} = 0$$

We know that eigenvectors corresponding to eigenvalues are not unique; assume:

$$\phi_{21} = 1.0$$

$$\phi_{11} = -0.000663 + j0.0254$$

Similarly, for:

$$\lambda = -0.250 - j9.596$$

$$\phi_{22} = 1$$

$$\phi_{12} = -0.000663 - j0.0254$$

The right eigenvector model matrix is:

$$\bar{\phi} = \begin{vmatrix} -0.000663 + j0.0254 & -0.000663 - j0.0254 \\ 1 & 1 \end{vmatrix}$$

The left model matrix is:

$$\bar{\phi}^{-1} = \begin{vmatrix} -j19.841 & 0.5 - j0.013 \\ j19.841 & 0.5 + j0.013 \end{vmatrix}$$

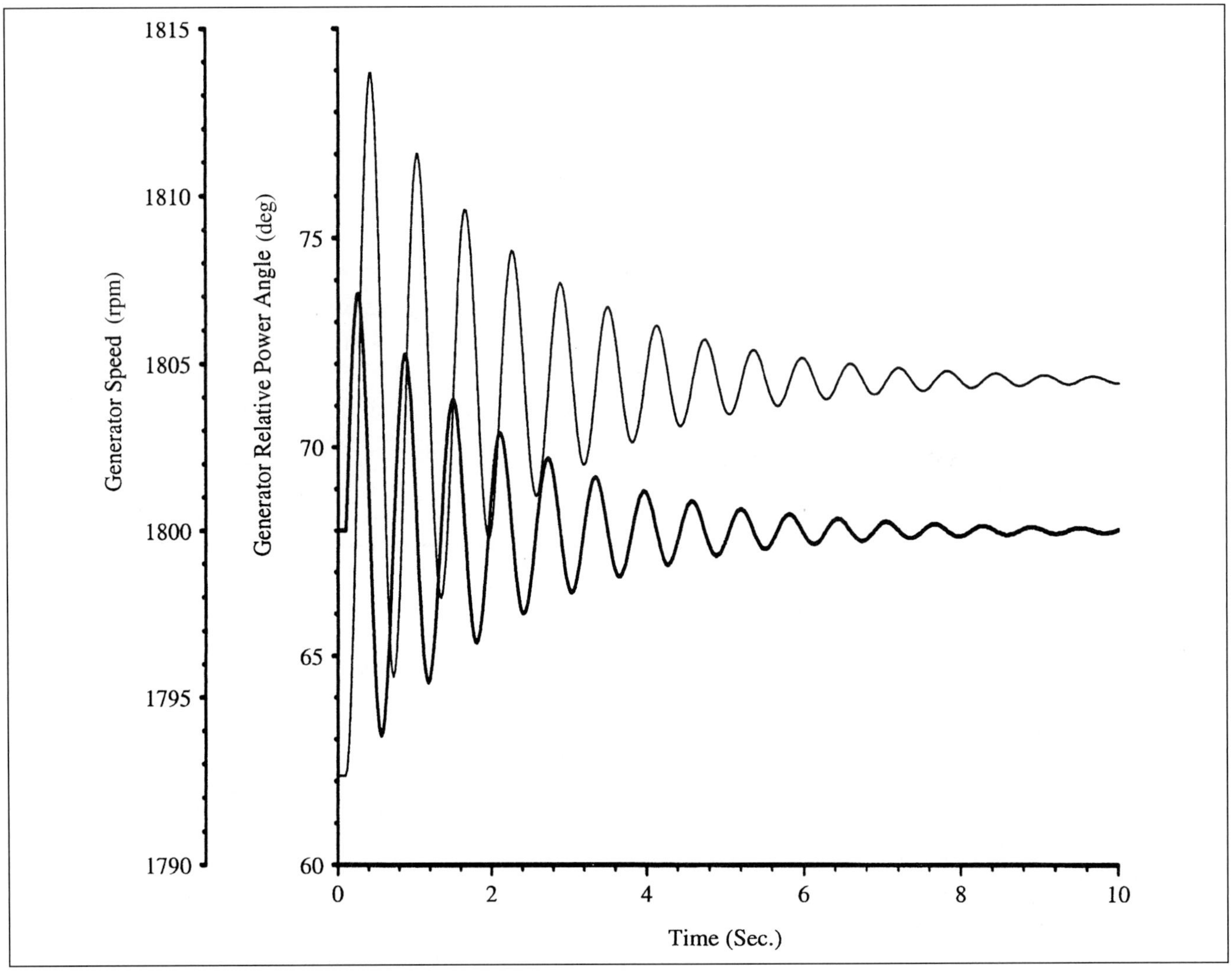

FIGURE 12-30 Simulation of transients, Example 12-6; opening of a parallel tie line.

The time response is given by:

$$\begin{vmatrix}\Delta\omega_r(t)\\ \Delta\delta t\end{vmatrix} = \bar{\phi}\begin{vmatrix}c_1e^{\lambda_1 t}\\ c_2e^{\lambda_2 t}\end{vmatrix}$$

where:

$$\begin{vmatrix}c_1\\ c_2\end{vmatrix} = \bar{\phi}^{-1}\begin{vmatrix}\Delta\omega_r(0)\\ \Delta\delta(0)\end{vmatrix}$$

Consider that $\Delta\omega_r = 0, \Delta\delta = 0.04$ rad, at $t = 0$. Then:

$$\begin{vmatrix}c_1\\ c_2\end{vmatrix} = \begin{vmatrix}-j19.841 & 0.5-j0.013\\ j19.841 & 0.5+j0.013\end{vmatrix}\begin{vmatrix}0\\ 0.04\end{vmatrix}$$
$$= \begin{vmatrix}0.02-j0.00052\\ 0.02+j0.00052\end{vmatrix}$$

Then:

$$\Delta\omega_r(t) = \phi_{11}c_1e^{\lambda_1 t} + \phi_{12}c_2e^{\lambda_2 t}$$
$$\Delta\delta(t) = \phi_{21}c_1e^{\lambda_1 t} + \phi_{22}c_2e^{\lambda_2 t}$$

Substituting the values:

$$\Delta\omega_r(t) = (-0.000663 + j0.0254)(0.02 - j0.000052)e^{(-0.250+j9.596)t}$$
$$+(-0.000663 - j0.0254)(0.02 + j0.000052)e^{(-0.250-j9.596)t}$$
$$= -0.0001e^{-0.250t}\sin 9.596t$$

Similar expression can be written for δ. This is a second-order system with oscillatory response, having a damped frequency of 9.596 rad/s = 1.527 Hz. The oscillations decay with a time constant of 1/0.25 s.

Figure 12-30 shows simulated transients. The decay of the transients is much faster compared to the calculations because of simplifying assumptions; that is, we are neglecting resistance, capacitance, and saturation in the calculations.

The small-signal stability simulations, when applied to large multi-machine systems, become fairly involved due to the requirements of simultaneous solution of differential and algebraic equations representing:

1. Synchronous machines
2. Excitation system and stabilizers
3. Interconnecting transmission networks

4. Other devices such as SVCs, HVDC converters, and so on

5. Prime mover speed governors

6. Static and dynamic loads such as induction and synchronous motors

These equations must be solved simultaneously. The state variables required for simulation may run into thousands; this is well outside the range of conventional eigenvalue analysis. Techniques like selected subset of eigenvalues associated with the complete system response, such as AESOPS (Analysis of Essentially Spontaneous Oscillations in Power Systems),[8] have been developed. PEALS (Program for Eigenvalue Analysis of Large Systems) uses AESOPS and modified Arnoldi method.[9,10]

AESOPS computes the eigenvalues associated with rotor angle modes, one complex conjugate pair at a time. This is done by applying to the generator an external sinusoidal torque, expressed in terms of initial estimate of the eigenvalue. Then, the complex frequency response is calculated and a revised estimate of eigenvalue arrived at by steady-state response. The process is repeated until following estimates converge with a specified tolerance. The resulting analysis indicates eigenvalues of the mode of oscillations in which the selected generator participates significantly.

12-12 VOLTAGE STABILITY

A power system must be able to maintain its voltage profile under various operating conditions. ANSI Std. C84.1-1989 specifies the preferred nominal voltages and associated nonstandard nominal system voltages in the United States. The electrical equipment operate within a certain narrow range of voltage variations, and utilities strive to maintain the voltages at the consumer premises within the specified limits and according to statuary regulations. Thus, the voltage and power flow should be controlled simultaneously. The voltage stability here excludes large voltage dips that may occur on faults and considers the instability due to slow variations in network and load characteristics. The operation of load tap changing equipment on transformers, voltage regulators, automatic generation control, load shedding, and switchable reactive power devices (capacitors) are rather slow acting. This implies that the dynamics of load variation are slower than the dynamics occurring in the state vector of stability model.

Other important factors are the contingencies, event disturbances, and dynamic load response to the voltage variations. We can, therefore, analyze the problem in two possible manners:

1. Dynamic framework, using dynamic models of the loads and voltage-correcting devices

2. Steady-state framework, that is, load flow

The voltage instability/voltage collapse can be generally analyzed using a time-domain stability program, though it requires extensive computational effort. We may say that with respect to original stable operating point, the system is voltage stable relative to slow or dynamic problem if the postdisturbance system trajectory approaches stable steady states where voltage magnitudes are acceptable, as well as the transient voltage magnitudes are acceptable. The task of accessing voltage stability is, therefore, to find that s.e.p that addresses both—the static and dynamic aspects.

12-12-1 P-Q Characteristics, Critical Reactance

Let us consider a simple load flow situation first. We talk about *P-V* and *Q-V* relations. The basic concept can be gathered by considering load flow over a transmission line, ignoring the capacitance effects (Fig. 12-31*a*).

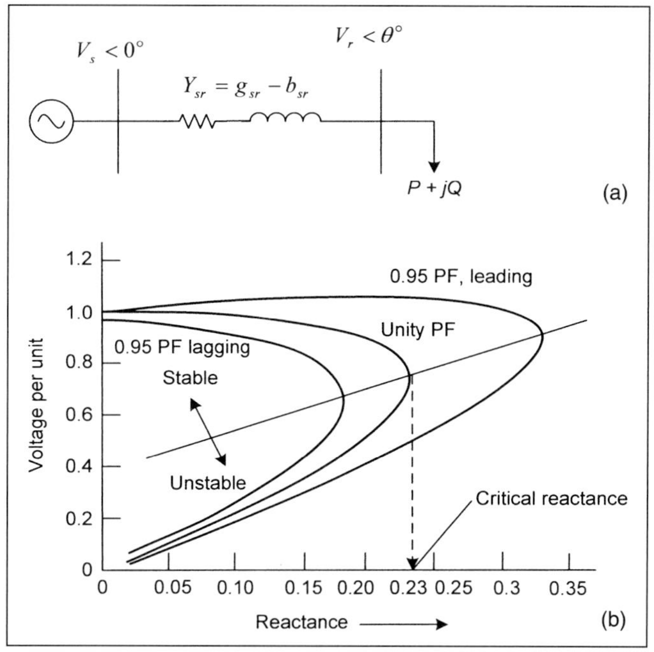

FIGURE 12-31 (*a*) Power flow over a tie line. (*b*) Voltage instability, power factor, and critical reactance.

The load demand is shown as $P + jQ$, the series admittance $Y_{sr} = g_{sr} + b_{sr}$, and $Z = R_{sr} + jX_{sr}$. The power flow from the source bus is given by:

$$\begin{aligned} P + jQ &= V_r e^{-j\theta}[(V_s - V_r e^{j\theta})(g_{sr} + jb_{sr})] \\ &= [(V_s V_r \cos\theta - V_r^2)g_{sr} + V_s V_r\, b_{sr} \sin\theta] \\ &\quad + j[(V_s V_r \cos\theta - V_r^2)b_{sr} - V_s V_r\, g_{sr} \sin\theta] \end{aligned} \tag{12-67}$$

If we neglect resistance

$$\begin{aligned} P &\doteq V_s V_r b_{sr} \sin\theta \\ Q &\doteq (V_s V_r \cos\theta - V_r^2) b_{sr} \end{aligned} \tag{12-68}$$

We could replace θ with δ, the difference between the sending-end and receiving-end voltage angle. If the receiving-end load changes by a factor $\Delta P + \Delta Q$

$$\begin{aligned} \Delta P &= (V_s b_{sr} \sin\theta)\Delta V + (V_s V_r b_{sr} \cos\theta)\Delta\theta \\ \Delta Q &= (V_s \cos\theta - 2V_r) b_{sr} \Delta V - (V_s V_r b_{sr} \sin\theta)\Delta\theta \end{aligned} \tag{12-69}$$

where V_r is the scalar change in voltage V_r and $\Delta\theta$ change in angular displacement. If θ is eliminated from Eq. (12-69) and resistance is neglected, a dynamic voltage equation of the system is obtained as follows:

$$V_r^4 + V_r^2(2QX_{sr} - V_s^2) + X_{sr}^2(P^2 + Q^2) = 0 \tag{12-70}$$

The positive real roots of this equation are:

$$V_r = \left[\frac{-2QX_{sr} + V_s^2}{2} \pm \frac{1}{2}\sqrt{(2QX_{sr} - V_s^2)^2 - 4X_{sr}^2(P^2 + Q^2)}\right]^{1/2} \tag{12-71}$$

This equation shows that in a lossless line, the receiving-end voltage V_r is a function of sending-end voltage V_s, series reactance

X_{sr}, and receiving-end real and reactive power. Voltage problems are compounded when reactive power flows over heavily loaded active power circuits. If reactive power is considered 0 and the sending-end voltage is 1 per unit, then Eq. (12-71) reduces to:

$$V_r = \left[\frac{1}{2} \pm \frac{\sqrt{1-4X_{sr}^2P^2}}{2}\right]^{1/2} \quad (12\text{-}72)$$

For two equal values of V_r:

$$(1-4X_{sr}^2P^2)^{1/2} = 0$$

That is:

$$X_{sr} = 1/2P \quad (12\text{-}73)$$

This value of X_{sr} may be called a *critical reactance*. Figure 12-31*b* shows characteristics of receiving-end voltage against system reactance for constant power flow at lagging and leading power factors, with sending-end voltage maintained constant. This figure shows, for example, that at a power factor of unity, voltage instability occurs for any reactance value exceeding 0.23 per unit. This reactance is the critical reactance. For a system reactance less than the critical reactance, there are two values of voltages, one higher and the other lower. The lower voltage represents unstable operation, requiring large amount of source current. For a system reactance close to the critical reactance, voltage instability can occur for a small positive excursion in the power demand. As the power factor improves, a higher system reactance is permissible for the power transfer. The *voltage instability* can be defined as the limiting stage beyond which the reactive power injection does not elevate the system voltage to normal. The system voltage can only be adjusted by reactive power injection until the system voltage stability is maintained. From Eq. (12-71), the critical system reactance at voltage stability limit is obtained as follows:

$$\begin{aligned}(2QX_{sr} - V_s^2) &= 4X_{sr}^2(P^2+Q^2)\\ 4X_{sr}^2P + 4X_{sr}QV_s^2 - V_s^4 &= 0\end{aligned} \quad (12\text{-}74)$$

The solution of this quadratic equation gives:

$$\begin{aligned}X_{sr,critical} &= \frac{V_s^2}{2}\left[\frac{Q \pm \sqrt{P^2+Q^2}}{P^2}\right]\\ &= \frac{V_s^2}{2P}(-\tan\phi + \sec\phi)\end{aligned} \quad (12\text{-}75)$$

where ϕ is the power factor angle.

Enhancing the thermal capacity of radial lines by use of shunt capacitors, SVC, and synchronous condensers is increasingly common. However, there is a limit to which capacitors can extend the load-carrying capability.

In practice, the phenomena of collapse of voltage is more complex. Constant loads are assumed in the previous scenario; however, the load dynamics are a very important factor. At lower voltages, the loads may be reduced, though this is not always true, that is, an induction motor may not stall until the voltage has dropped more than 25 percent, and even then, the magnetic contactors in the motor starter supplying power to the motor may not drop out. This lockout of the motors and loss of some loads may result in voltage recovery, which may start the process of load interruption afresh, as the motors try to reaccelerate on the return voltage.

From Eq. (12-69), as θ is normally small:

$$\frac{\Delta Q}{\Delta V} = \frac{V_s - 2V_r}{X_{sr}} \quad (12\text{-}76)$$

If the three phases of the line connector are short-circuited at the receiving end, the receiving-end short-circuit current is:

$$I_{rsc} = \frac{V_s}{X_{sr}} \quad (12\text{-}77)$$

This assumes that the resistance is << reactance. At no load, $V_r = V_s$; therefore:

$$\frac{\partial Q}{\partial V} = -\frac{V_r}{X_{sr}} = -\frac{V_s}{X_{sr}} \quad (12\text{-}78)$$

Thus:

$$\left|\frac{\partial Q}{\partial V}\right| = \text{current} \quad (12\text{-}79)$$

Alternatively, we could say that:

$$\frac{\Delta V_s}{V} \approx \frac{\Delta V_r}{V} = \frac{\Delta Q}{S_{sc}} \quad (12\text{-}80)$$

where S_{sc} is the short-circuit level of the system. This means that the voltage regulation is equal to the ratio of the reactive power change to the short-circuit level. This gives the obvious result that receiving-end voltage falls with the decrease in system short-circuit capacity or increase in system reactance. A stiffer system tends to uphold the receiving-end voltage.

The voltage-reactive power stability problem is more involved than portrayed earlier. The power flow has to be considered from the generation level to the transmission, subtransmission level, and finally to the consumer level. As reactive power does not travel well over long distances, it becomes a local problem. The utility companies operate with an agreed voltage level at intertie connections, and provide for their own reactive power compensation to meet this requirement. It can be said that there are no true hierarchical structures in terms of reactive power flow.

The voltage instability is not a single phenomenon and is closely coupled to the electromagnetic stability and shares all aspects of active power stability, though there are differences.

Consider the equations of active and reactive power flow, and Eq. (12-67) for the short line. Let the voltages be fixed. The angle (phasor difference between the sending-end and receiving-end voltage vectors) varies with receiving bus power, as shown in (Fig. 12-32*a*). Beyond the maximum power drawn, there is no equilibrium point. Below the maximum power drawn, there are two equilibriums, one in the stable state and the other in the unstable state. A load flow below the maximum point is considered statically stable.

A similar curve can be drawn for reactive power flow (Fig. 12-32*b*). The angles are fixed, and the bus voltage magnitude changes. The character of this curve is the same as that of active power flow curve. A reactive power demand above the maximum reactive power results in a nonexistence of a load flow situation. The point ΔV^U is statically unstable, corresponding to $\Delta\delta^U$. If we define:

$$\begin{aligned}|P_{(max)} - P| &< \varepsilon\\ |Q_{(max)} - Q| &< \varepsilon\end{aligned} \quad (12\text{-}81)$$

Then, howsoever small ε may be, there will always be two equilibrium points. Such an equilibrium point may be called a *bifurcation point*.

The derivatives $\partial p/\partial\delta, \partial Q/\partial V$ are zero at the static stability limit. The goal of the local bifurcation theory is to investigate static solutions, and stability property of static solutions may alter the dynamic behavior.

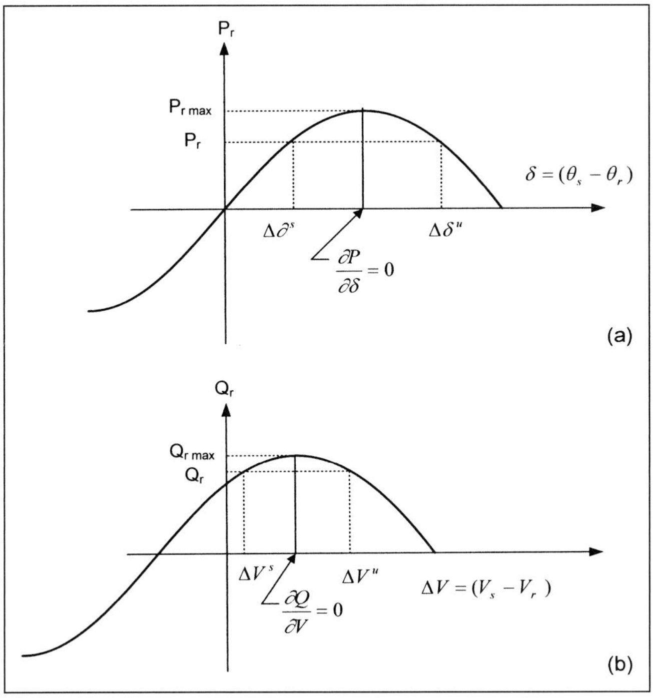

FIGURE 12-32 (*a*) and (*b*) Real power flow and reactive power flow instability, respectively.

Power systems are normally operated near s.e.p. When the system parameters are away from the bifurcation values, but change slowly and continuously, it is likely that s.e.p changes position but remains a s.e.p or the old s.e.p lies in the stability boundary of new s.e.p. The ways in which the system may lose stability are: (1) It may happen that one s.e.p may coalesce into another s.e.p and disappear in a saddle-node bifurcation. (2) A s.e.p and unstable limit cycle coalesce and disappear, and an unstable equilibrium point (u.e.p) emerges. (3) The s.e.p bifurcates into an u.e.p surrounded by a stable limit cycle.

12-12-2 Proximity to Instability

For dynamic analysis, the general structure of the equations is:

$$\begin{aligned} \dot{x} &= f(x,\bar{V}) \quad \text{differential equations} \\ \bar{I}(x,\bar{V}) &= \bar{Y}_N \bar{V} \quad \text{algebraic equations} \end{aligned} \tag{12-82}$$

where $\bar{Y}_N$ is node admittance matrix, and $\bar{V}, \bar{I}$ are voltage vectors. From Newton–Raphson decoupled load flow:[11]

$$\begin{vmatrix} \Delta\bar{P} \\ \Delta\bar{Q} \end{vmatrix} = \begin{vmatrix} \bar{J}_{P\theta} & \bar{J}_{PV} \\ \bar{J}_{Q\theta} & \bar{J}_{QV} \end{vmatrix} \begin{vmatrix} \Delta\bar{\theta} \\ \Delta\bar{V} \end{vmatrix} \tag{12-83}$$

where J ($\bar{J}_{P\theta}, \bar{J}_{PV}, \bar{J}_{Q\theta}, \bar{J}_{QV}$) are Jacobian matrices.

For $\Delta P = 0$, the reduced Jacobian matrix J_R is:

$$\bar{J}_R = \left[\bar{J}_{QV} - \bar{J}_{Q\theta}^{-1}\bar{J}_{PV}\right] \tag{12-84}$$

And

$$\Delta\bar{V} = \bar{J}_R^{-1}\Delta\bar{Q} \tag{12-85}$$

The voltage stability characteristics can be ascertained by computing eigenvalues and eigenvectors of the reduced Jacobian matrix. We can write:

$$\bar{J}_R = \bar{\varepsilon}_r \bar{\Lambda} \bar{\varepsilon}_l \tag{12-86}$$

where $\bar{\varepsilon}_r, \bar{\varepsilon}_l$, and $\bar{\Lambda}$ are the right eigenvector matrix, the left eigenvector matrix, and diagonal eigenvector matrix, respectively. Note that $\bar{\varepsilon}_r^{-1} = \bar{\varepsilon}_l$.

From these equations:

$$\Delta\bar{V} = \sum_i \frac{\bar{\varepsilon}_{ri}\bar{\varepsilon}_{li}}{\lambda_i}\Delta\bar{Q} \tag{12-87}$$

where each eigenvalue λ_i and the corresponding right and left eigenvectors, $\varepsilon_{ri}, \varepsilon_{li}$, define i-th mode of Q-V response. Also:

$$v = \Lambda^{-1} q \tag{12-88}$$

where:

$$\bar{v} = \bar{\varepsilon}_l \Delta\bar{V} \text{ is the vector of modal voltage variations} \tag{12-89}$$

$$\bar{q} = \bar{\varepsilon}_r \Delta\bar{Q} \text{ is the vector of modal reactive power variations} \tag{12-90}$$

The V-Q sensitivity at bus k is:

$$\frac{\partial \bar{V}_k}{\partial \bar{Q}_k} = \sum_i \frac{\bar{\varepsilon}_{rki}\bar{\varepsilon}_{lki}}{\lambda_i} \tag{12-91}$$

Equation (12-91) shows that the V-Q sensitivities cannot identify individual voltage collapse modes $\bar{J}_R, \bar{Y}_N$ as symmetrical if resistances are neglected; then, eigenvalues and eigenvectors *of* $\bar{J}_R$ are real, and right and left eigenvectors of an eigenvalue of $\bar{J}_R$ are equal. The magnitude of eigenvalues can provide an estimate of proximity to instability. The application of modal analysis determines how stable the system is and how much extra load can be added. At the stability critical point, areas and elements that participate in each mode can be identified.

12-12-3 Shortest Distance to Instability

This involves finding smallest load in MW and Mvar, which imposed on the initial conditions, cause the power flow Jacobian to be singular. The power flow equations can be organized as:

$$f(\mathrm{x},\rho) = g\begin{vmatrix} V \\ \theta \end{vmatrix} - \begin{vmatrix} P \\ Q \end{vmatrix} = 0 \tag{12-92}$$

where x is the system state vector and ρ is the parameter vector, given by:

$$\mathrm{x} = \begin{vmatrix} V \\ \theta \end{vmatrix}, \quad \rho = \begin{vmatrix} P \\ Q \end{vmatrix} \tag{12-93}$$

Let $\bar{J}_x, \bar{J}_\rho$ be the Jacobian matrices of function f. For a given parameter ρ_i, a system state vector x_i can be obtained by solving Eq. (12-92). The system reaches its voltage stability critical limit if parameter vector ρ^* and the corresponding state vector $\mathbf{x}^*$ are such that Jacobian $\bar{J}_x$ is singular. This is done for a given operating point ($\mathbf{x}_0$, ρ_0) by finding the parameter vector ρ^* on surface S so that distance between ρ_0 and ρ^*, $k = (\rho^* - \rho_0)$, is a minimum. The Jacobian becomes singular when:

$$\rho^* = \rho_0 + k\varepsilon_1 \tag{12-94}$$

Thus, starting from an initial estimate of ε, stress the system by increasing ρ in direction of ε_l, and iterate until ε_l converges to critical ε*. Then $\rho^* = \rho_0 + k^*\varepsilon^*$ is the equilibrium condition.

12-13 LOAD MODELS

Load modeling has a profound impact on stability. Figure 12-33 shows the effect of change of operating voltage with respect to constant current and constant MVA for three load types: (1) constant KVA, (2) constant current, and (3) constant impedance. Heavy industrial motor loads are approximately constant MVA loads, while commercial and residential loads are mainly constant impedance loads. Classification into commercial, residential, and industrial is rarely adequate, and one approach has been to divide the loads into individual load components. The other approach is based on measurements. Thus, the two approaches are:

- Component-based models
- Models based on measurements

A component-based model is a *bottom-up* approach in the sense that different load components comprising the loads are identified. Each load component is tested to determine the relations between real and reactive power requirements versus voltage and frequency. A load model in exponential or polynomial form can then be developed from the test data.

The measurement approach is a *top-down* approach in the sense that the model is based on the actual measurement. The effect of the variation of voltage on active and reactive power consumption is recorded and, based on this load model, is developed.

A composite load, that is, a consumer load consisting of heating, air-conditioning, lighting, computers, and television is approximated by combining load models in certain proportions based on load surveys. This is referred to as *load window*. Construction of a load window requires certain data, that is, load saturation, composition, and diversity data. Any number of load windows can be defined.

The load models are normalized to rated voltage, rated power, and rated frequency, and are expressed in per unit. The exponential

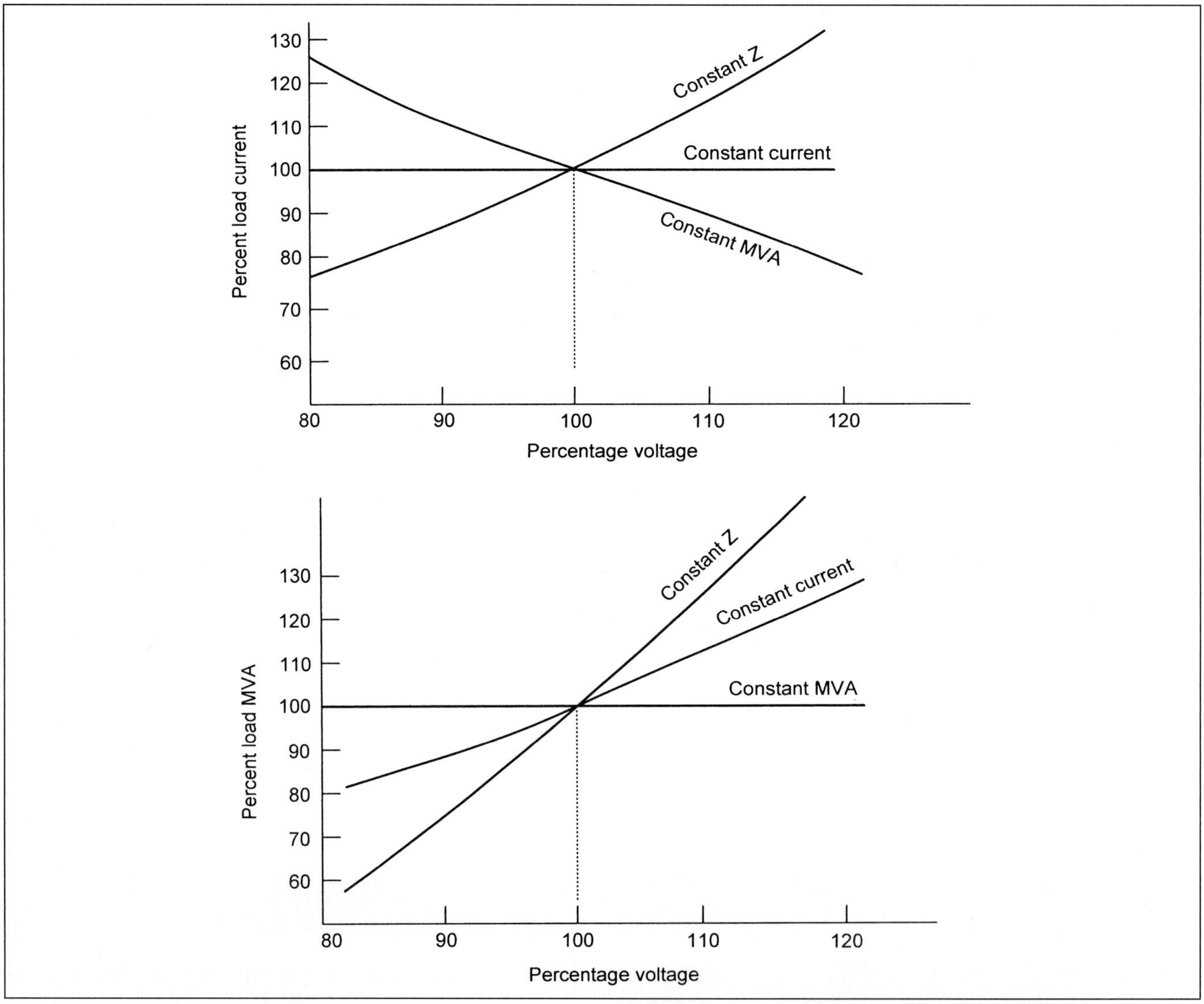

FIGURE 12-33 Behavior of loads on voltage dips, with respect to percent load current and MVA, depending on the load type.

load models are:

$$\frac{P}{P_n}=\left|\frac{V}{V_n}\right|^{\alpha_v}\left|\frac{f}{f_n}\right|^{\alpha_f} \tag{12-95}$$

$$\frac{Q}{P_n}=\frac{Q_0}{P_0}\left|\frac{V}{V_n}\right|^{\beta_v}\left|\frac{f}{f_n}\right|^{\beta_f} \tag{12-96}$$

where V is the initial value of voltage and P_0 is the initial value of power. V_n is the adjusted voltage and P_n is the power corresponding to this adjusted voltage. The exponential factors depend on the load type. The frequency dependence of loads is ignored for load flow studies, but can be important for transient and dynamic stability studies. Another form of load model is polynomial model.

Consider the equation:

$$P=V^n \tag{12-97}$$

For all values of n, $P=0$ when $V=0$, and $P=1$ when $V=1$, as it should be. Differentiating:

$$\frac{dP}{dV}=nV^{n-1} \tag{12-98}$$

For $V=1$, $dP/dV=n$. The value of n can be found by experimentation if dP/dV, that is, change in active power with change in voltage, is obtained. Also, by differentiating $P=VI$:

$$\frac{dP}{dV}=I+V\frac{dI}{dV} \tag{12-99}$$

For V and n equal to unity, the exponential n from Eqs. (12-98) and (12-99) is:

$$n=1+\frac{\Delta I}{\Delta V} \tag{12-100}$$

The exponential n for a composite load can be found by experimentation if the change in current for a change in voltage can be established. For a constant power load, $n=0$; for a constant current load, $n=1$; and for a constant MVA load, $n=2$. The composite loads are a mixture of these three load types and can be simulated with values of n between 0 and 2. Following are some quadratic expressions for the various load types.[12,13]

Air conditioning:

$$\begin{aligned}P&=2.18+0.268V-1.45V^{-1}\\Q&=6.31-15.6V+10.3V^2\end{aligned} \tag{12-101}$$

Fluorescent lighting:

$$\begin{aligned}P&=2.97-4.00V+2.0V^2\\Q&=12.9-26.8V+14.0V^2\end{aligned} \tag{12-102}$$

Induction motor loads:

$$\begin{aligned}P&=0.720+0.109V+0.172V^{-1}\\Q&=2.08+1.63V-7.60V^2+4.08V^3\end{aligned} \tag{12-103}$$

In sequential load method of analysis, the stability situation is examined in certain time windows: (1) In the first second, motors slow down and generator voltage regulators operate, (2) 1 to 20 s, when the excitation limits occur, (3) 20 to 60 s when generator overexcitation limits operate, and (4) 1 to 10 min or longer when the ULTC (under load tap changing), AGC (automatic generation contsol), and phase angle regulators operate.

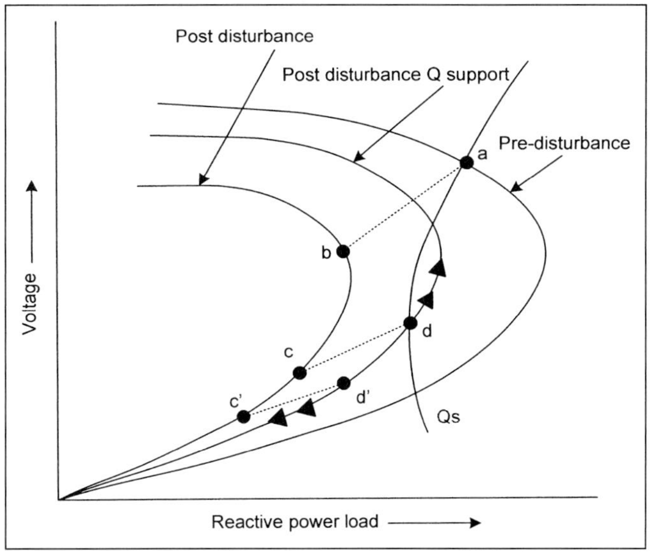

FIGURE 12-34 Voltage stability characteristics, as a function of pre- and postdisturbance *V-Q* characteristics and reactive power support.

In Fig. 12-34, consider a system operating at point *a*, prior to disturbance, intersection of *V-Q* characteristics, and dynamic load curve Q_s. A disturbance reduces *V-Q* operating point to *b*. However, at *b* the reactive power load demand is greater than what can be supplied. This load demand can only be met by lowering the operating voltage, and the operating point is now at *c*.

Now consider that postdisturbance reactive power support is added, as shown in the figure. The operating point will shift to *d* at higher voltage. This depends on how fast the postdisturbance reactive power support is restored. The operating point can as well be *d′* for slower restoration of the reactive power support.

At point *d*, if the reactive load demand is lower than the supply system capability, the voltage will recover. Conversely at point *d′*, if the load demand exceeds the supply system, the voltage will collapse monotonically.

We highlighted the importance of reactive power flow demand as an important contributing factor to voltage instability. Usually, the voltage collapse involves highly overloaded lines or loss of a line. Other initiating causes are loss of a generation, sudden increase in load due to contingency load flow, outage of an SVC or large reactive power compensating unit, and failure of ULTC on transformers or voltage regulators.

The voltage instability can be prevented by the proper application of reactive power compensating devices, control of generator reactive power output capability, ULTC, undervoltage and underfrequency load shedding, and coordination and protection.

Example 12-7 A system configuration of Fig. 12-35 is considered, with system data, as shown in Table 12-3. Generators G1 and G2, each rated at 200 MW, step up their power output to 400 kV, and are interconnected through a 400-kV line, 150 mi long. The load flow balance and calculated bus voltages are shown. The load consists of a 20000-hp motor at 13.8 kV and a 50-Mvar capacitor bank, in addition to 350 MVA of load at 0.85 power factor connected to 400-kV system, bus 3. The generators are modeled with type AC2 excitation systems and speed-governing models type ST (see Chap. 13).

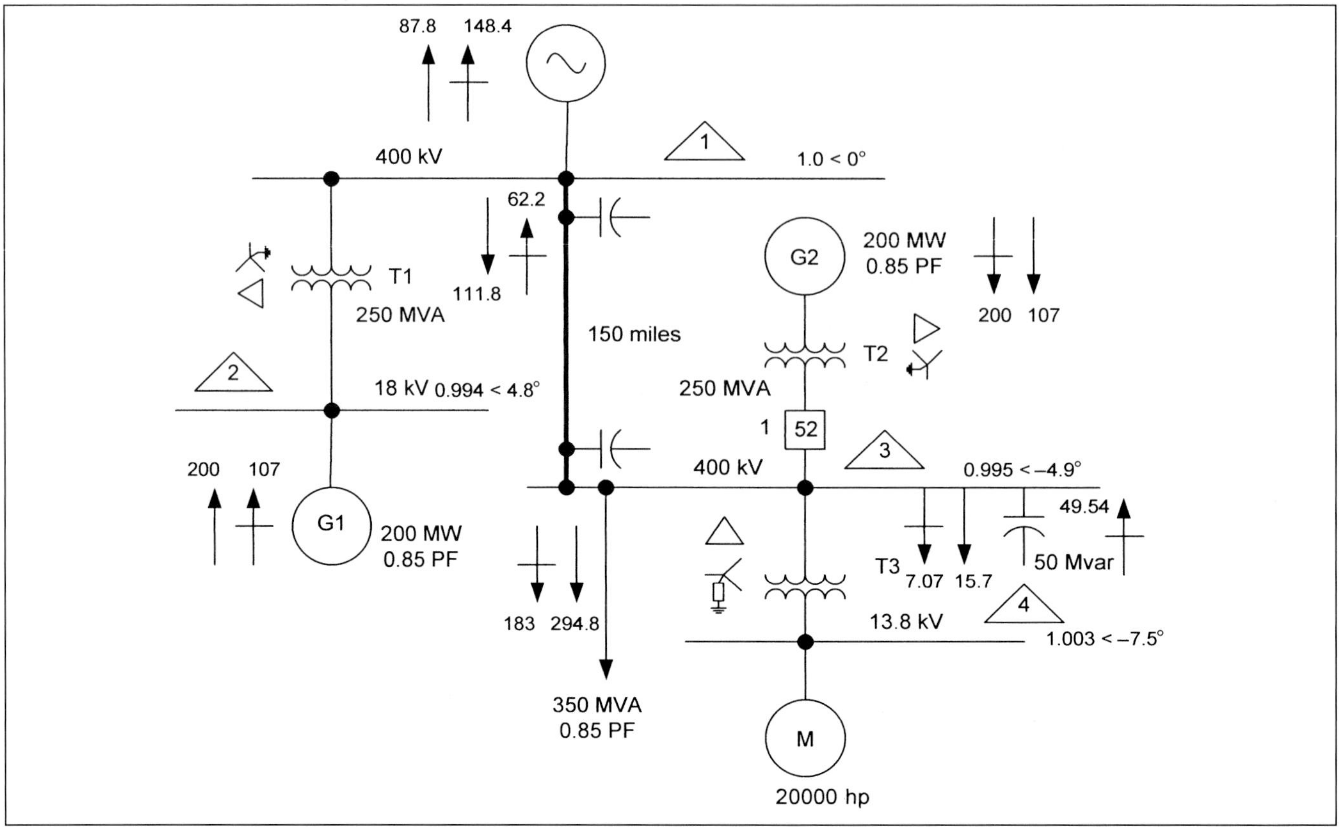

FIGURE 12-35 A three-bus system for dynamic stability study. Breaker 52 suddenly opened to isolate generator G2, Example 12-7.

TABLE 12-3 Data for Example 12-7

DESIGNATION	DATA
400-kV utility source	$Z_1 = Z_2 = 0.0995 + j0.995$ (pu 100-MVA base) $Z_0 = 0.100 + j1.9975$ (pu 100-MVA base)
G1 and G2	200 MW, 0.85 pF, $X_d = 110\%$, $X_q = 108\%$, $T'_{do} = 5.6$ s, $X''_d = 12\%$, $X'_d = 23\%$, $X_l = 11\%$, model saturation, model exciter type AC2A, model governing system type ST; see Chap. 13 $H = 2.5$ (reactances on machine MVA base), high-resistance grounded through a distribution transformer, ground current = 8 A
Transformers T1 and T2	250 MVA, two-winding, low voltage delta, 400 kV wye-grounded, 30° standard phase shift, set taps at 400 kV at +5% (420 kV), $Z = 10\%$ (on transformer MVA base), $X/R = 50$
Transformer T3	50 MVA, two-winding, 400-kV delta-connected, 13.8-kV wye-connected, low-resistance grounded, $Z = 14.5\%$, on transformer MVA base, set tap on high side = −2.5% (390 kV)
Capacitor bank	Provides 50 Mvar at rated voltage of 400 kV
20000-hp motor	Full-load current = 738 A, efficiency = 95.12%, full-load power factor = 94%, full-load speed = 1768 rpm, locked rotor current = 637% of full-load current, locked rotor pf = 6.76%, double age, $X/R = 89.152$, $T'_d = 1.225$ s, $H = 1.925$, polynomial load model = 100 ω^2

Consider now that breaker 1 is suddenly opened at 0.1 s, with no fault in the system. This isolates generator G2 and the load demand must be met by the utility source through 150-mi transmission line and remaining generator G1 in service.

Figure 12-36 shows transients in generator G1 which gradually diverge over a period of time. This generator pulls out of step at around 8 to 9 s and keeps oscillating. Figure 12-37 depicts transients in the 20000-hp motor, which escalate at the time the generator G1 pulls out of step; the motor ultimately locks out. Figure 12-38 illustrates bus voltage transients. The field voltage and current transients, not shown, are of similar patterns.

Though not illustrated graphically, even if generator G1 is disconnected from the system by protective relaying, the 20000-hp motor still locks out and does not recover. In any interconnected power system, such dynamic behavior is unacceptable.

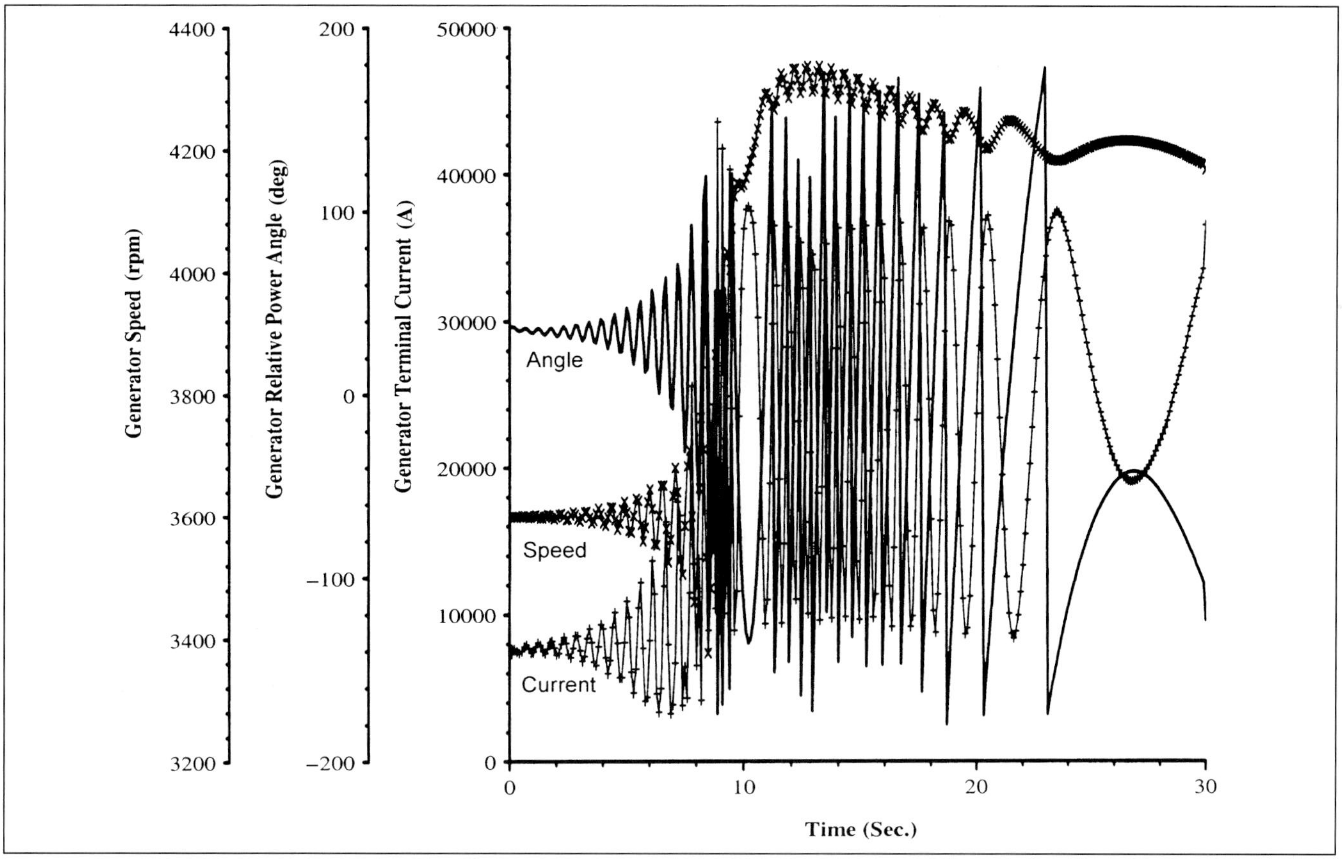

FIGURE 12-36 Diverging transients in generator G1, simulated for 30 s, Example 12-7.

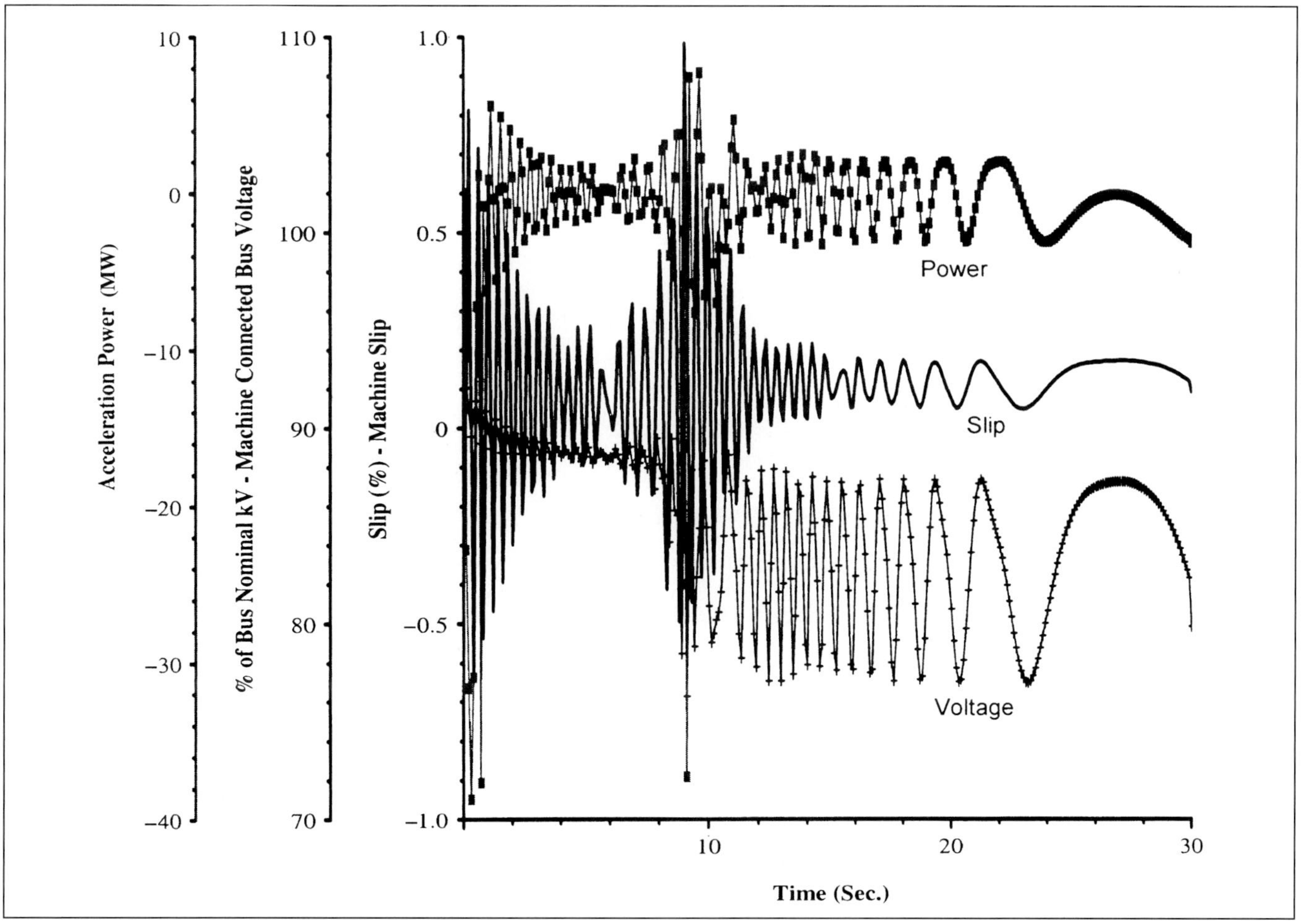

FIGURE 12-37 Transients in 20000-hp induction motor, Example 12-7.

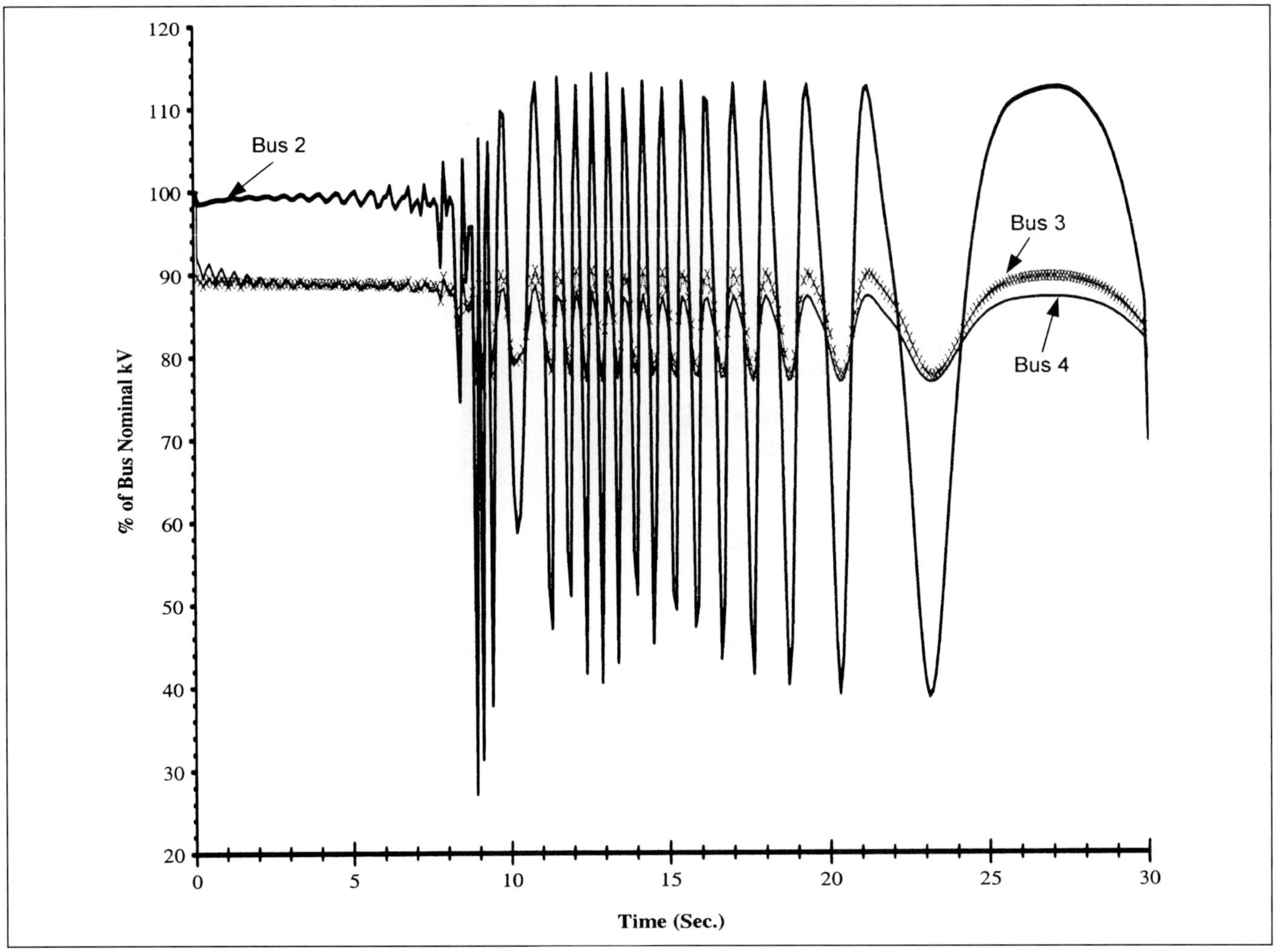

FIGURE 12-38 Bus voltage transients, Example 12-7.

12-14 DIRECT STABILITY METHODS

The time-domain methods described earlier for the analysis of stability problems may involve solving thousands of algebraic and nonlinear differential equations, complexity depending on the models used and the power system studied. This may be fairly time consuming and may slow down the computing speed.

The time-domain methods analyze the prefault, faulted, and the postfault systems. The direct methods integrate the faulted system only and determine, without examining the postfault system, whether the system will be stable after fault clearance. The system energy, when the fault is cleared, is compared to a critical energy value.

The reader may study Appendix E before proceeding with this section.

The differential equations of stability problems have a general structure given by:

$$\dot{x}(t) = f[\dot{x}(t)] \tag{12-104}$$

The state vector x(*t*) is in the Euclidean space R^n and function *f*: $\mathrm{R}^n \rightarrow \mathrm{R}^n$ satisfies the condition of existence and uniqueness of solutions.

A state vector $\hat{\mathrm{x}}$ is called an equilibrium point of Eq. (12-104) if:

$$f(\hat{\mathrm{x}}) = 0 \tag{12-105}$$

An equilibrium point is hyperbolic if the Jacobian of $f(\cdot)$ at $\hat{\mathrm{x}}$, denoted by $J_f(\hat{\mathrm{x}})$, has no eigenvalues with a zero real part. It is an asymptotically stable equilibrium point if all eigenvalues of its corresponding Jacobian have negative real parts; otherwise it is an unstable equilibrium point, u.e.p.

If the Jacobian of the equilibrium point $\hat{\mathrm{x}}$ has exactly one eigenvalue with a positive real part, it is called *type-one equilibrium point*. For a type *k* equilibrium point, Jacobian has exactly *k* eigenvalues with positive real parts.

The stable and unstable manifolds are described as:

$$\begin{aligned} &\mathrm{W}^s(\hat{\mathrm{x}}) := \{x \in \mathbf{R}^n : \Phi_t(x) \rightarrow \hat{\mathrm{x}} \quad t \rightarrow \infty\} \\ &\mathrm{W}^u(\hat{\mathrm{x}}) := \{x \in \mathbf{R}^n : \Phi_t(x) \rightarrow \hat{\mathrm{x}} \quad t \rightarrow -\infty\} \end{aligned} \tag{12-106}$$

Every trajectory in stable manifold converges to $\hat{\mathrm{x}}$ as $t \rightarrow +\infty$, and every trajectory in the unstable manifold also converges to $\hat{\mathrm{x}}$ as $t \rightarrow -\infty$. For a s.e.p, there exists a number $\delta > 0$ so that:

$$\begin{aligned} &\| x_0 - \hat{\mathrm{x}} \| < \delta \\ &\Phi_t(x_0) \rightarrow \hat{\mathrm{x}}, \quad \text{as } t \rightarrow \infty \end{aligned} \tag{12-107}$$

If δ is arbitrarily large, then $\hat{\mathrm{x}}$ is called a *global stable equilibrium point*. There can be many physical systems containing stable equilibrium points but not the global stable equilibrium points.

A stability region for these systems is defined as:

$$A(x_s) := \left[x \in R^n : \lim_{t\to\infty} \Phi_t(x) = x_s\right] \tag{12-108}$$

The stability region, also called the region of attraction, $A(x_s)$, is an open invariant and connected set. Its boundary is called the stability boundary of x_s and denoted as $\partial A(x_s)$.

12-14-1 Stability Boundary

If an energy function exists that has an asymptotically stable equilibrium point x_s, but not globally asymptotically stable, then the stability boundary $\partial A(x_s)$ is contained in the set which is a union of the stable manifolds of the u.e.p's on the stability boundary $\partial A(x_s)$.

$$\partial A(x_s) \subseteq \bigcup_{x_i \in [E \cap \partial A(x_s)]} W^s(x_i) \tag{12-109}$$

Transient Stability Models for Direct Analysis The two prevalent models are:

1. Network reduction models, where all loads are constant impedance loads and the entire system is reduced to generator internal buses (Example 12-4)

2. Network preserving models to overcome shortcomings of network reduction models[14]

We will discuss only network reduction models. The transient stability model can be written in the compact form as:[15]

$$\begin{aligned} \bar{T}\dot{x} &= -\frac{\partial U}{\partial x}(x,y) + g_1(x,y) \\ \dot{y} &= z \\ \bar{M}\dot{z} &= -\bar{D}z - \frac{\partial U}{\partial y}(x,y) + g_2(x,y) \end{aligned} \tag{12-110}$$

where $x \in R^n, y$ and $z \in R^m$. $\bar{T}, \bar{M},$ and $\bar{D}$ are positive diagonal matrices, g_1 (x, y) and g_2 (x, y) are transfer conductance of the reduced network, and $U(x, y)$ is a smooth function.

A condition for existence of the energy function for the system with zero-transfer conductance ($g_{ij} = 0$) is:

$$\begin{aligned} W(x,y,z) &= K(z) + U(x,y) \\ &= \frac{1}{2}\bar{z}^t \bar{M}\, z + U(x,y) \end{aligned} \tag{12-111}$$

If along every nontrivial trajectory $x(t)$, $y(t)$, $z(t)$ with bounded function value **W**, $x(t)$ is also bounded, then $\mathbf{W}(x, y, z)$ is an energy function of Eq. (12-110).

12-14-2 Controlling u.e.p Method

The controlling u.e.p method has been around since 1970s and is the most viable method for direct analysis of practical power systems.[16–18] The four basic steps are:

- An energy function is constructed for the postfault system.
- Energy is computed immediately after the fault clearing time is reached.
- The energy is computed for the fault on trajectory.
- The energy of the state, when the fault is cleared, is compared to a critical energy for stability assessments.

Consider a power system with:

$$\begin{aligned} &\text{Prefault SEP} = X_s^{\text{pre}} \\ &\text{Fault trajectory } X_f(t) \\ &\text{Postfault SEP} = X_s^{\text{post}} \end{aligned} \tag{12-112}$$

Let there be an energy function for the postfault system, and X_s^{pre} lies inside the stability region of X_s^{post}. Then, the controlling u.e.p with respect to the fault on trajectory $X_{f(t)}$ is the u.e.p of the postfault system, whose stable manifold contains the exit point of $X_{f(t)}$. A sustained fault on the trajectory must exit the stability boundary $\partial A(x_s)$ of a postfault system. The exit point must lie on the stable manifold of the u.e.p. This u.e.p is the controlling u.e.p of the fault on the trajectory.

The controlling u.e.p method is applied in the following steps:

1. The controlling u.e.p, X_{co}, is found for a given fault on the trajectory.

2. The critical energy is the value of energy function at controlling u.e.p:

3. $v_{cr} = V(X_{co})$

4. The stability boundary is established for the fault on the trajectory by using connected energy surface of the energy function passing through controlling u.e.p, X_{co}. This boundary contains the s.e.p, Xs.

5. Calculate the value of energy function at the fault clearance time, using fault on the trajectory.

6. $v_f = V(X_f t_f)$

7. If $v_f < v_{cr}$, then the point $X_{f(cl)}$ is located within the stability boundary and the postfault system is stable; otherwise it may be unstable. Thus, an approximation of the stability boundary of the postfault system is carried out to which the fault trajectory is heading. It uses the connected controlling energy surface passing through the controlling u.e.p to establish the stability boundary. If, when the fault is cleared, the system state lies inside the stability boundary, then the postfault system is stable.

Figure 12-39 illustrates this. In this figure, X_{co} is the controlling u.e.p with respect to the fault on the trajectory $X_{f(t)}$. X_{cl} is the closest u.e.p on the postfault system S^{post}. X_e is the exist point of the sustained fault on trajectory $X_{f(t)}$, which is contained in the stable manifold $W^s(X_{co})$ of the controlling u.e.p. $X_{co}, X_1, \ldots \partial S[V(X_{co})]$ and X_{co}. $\partial S[V(X_{cl})]$ are the constant energy surfaces passing through X_{co} and X_{cl}, respectively. The sustained fault on the trajectory must first hit $\partial S[V(X_{co})]$, then $\partial S[V(X_{cl})]$, and finally the stability boundary at exit point X_e.

The task of calculating the exit point, and controlling u.e.p in Fig. 12-39, is complex and requires iterative procedure in time-domain approach.

The energy function consists of kinetic energy and potential energy terms. Even when the system is stable, some amount of kinetic energy may not be absorbed. Some of the kinetic energy is responsible for intermachine motion between the generators and does not contribute to the separation of the severely disturbed generators from the system. The potential energy terms in the energy function have a path-dependent term, which is approximated.

12-14-3 BCU Method

The difficulty of calculating the controlling u.e.p is overcome in the BCU method, which does not calculate the u.e.p of the original

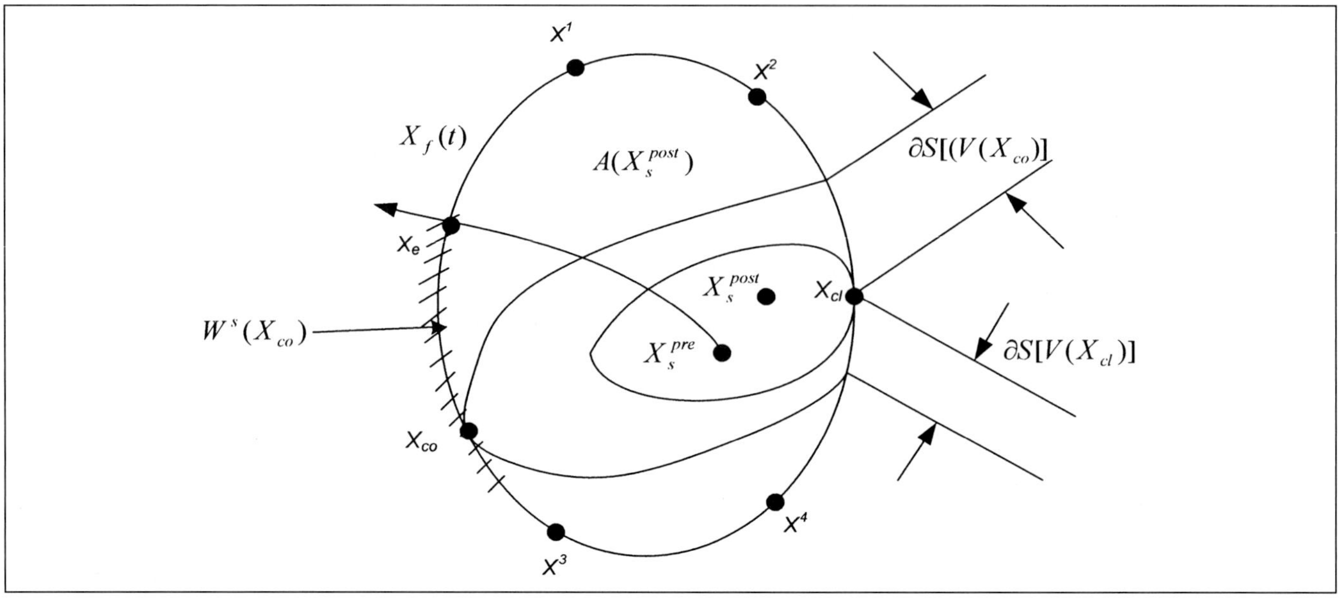

FIGURE 12-39 Illustration of controlling u.e.p method, direct stability analysis method.

model, but of a reduced state model. BCU stands for *b*oundary of *c*ontrolling *u*.e.p.[19–21]

In applying this method for a power system, an associated reduced-state model is first developed, so that:

1. The location of equilibrium points in the reduced-state model correspond to that in the original system.

2. The type of equilibrium points are the same as that in the original system.

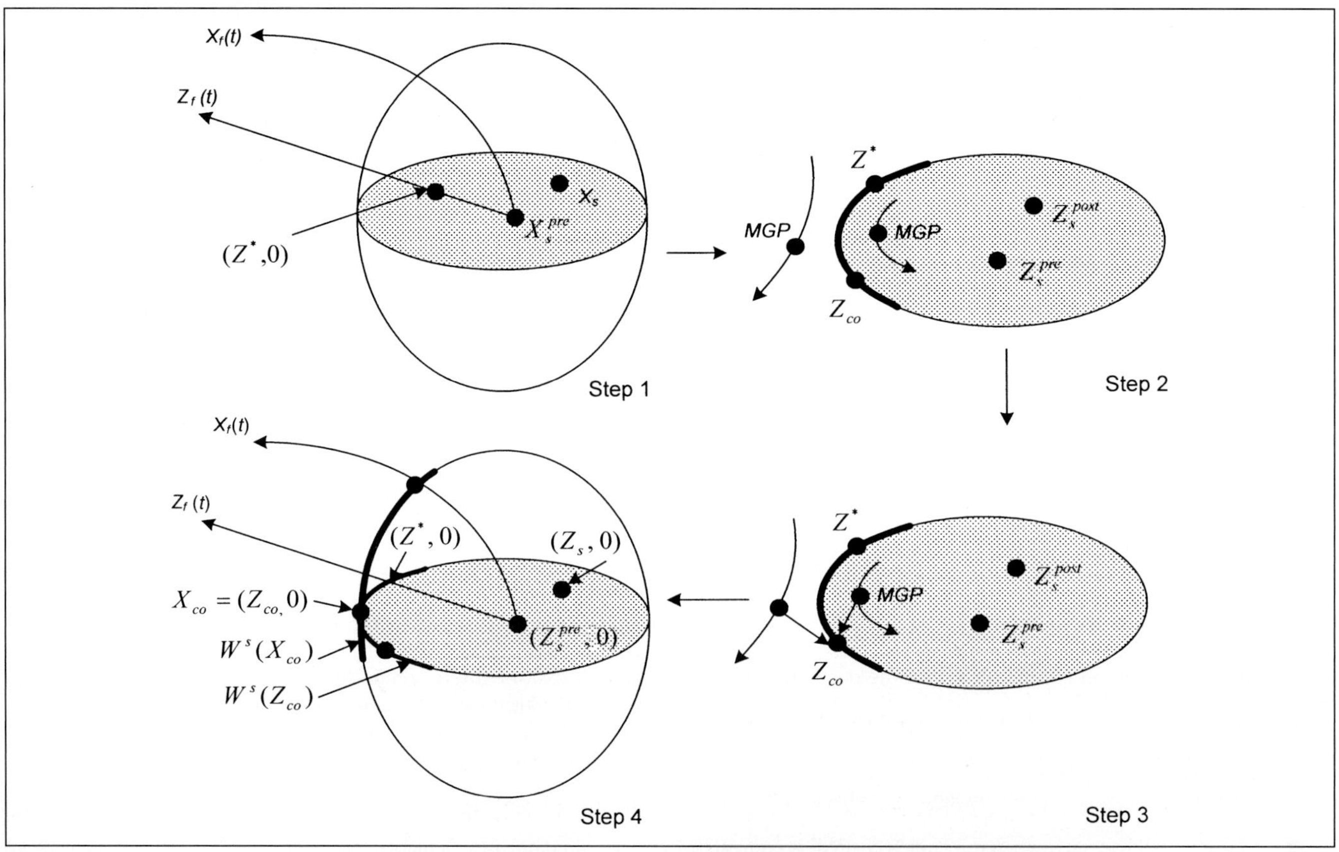

FIGURE 12-40 Four steps in the numerical BCU method of direct stability analysis method.

3. The reduced system has an energy function.

4. An equilibrium point on the reduced model lies on the stability boundary of the reduced system only if the equilibrium point is on the stability boundary of the original system.

5. Without resorting to the time domain iterative approach, it is possible to detect when a projected fault on the trajectory hits the stability boundary.

The various steps of calculation in a numerical network preserving BCU method are shown in Fig. 12-40. Steps 1 through 3 calculate the controlling u.e.p of the reduced-state system and step 4 relates the controlling u.e.p of the reduced state to the original system. In step 1, the exit point of the projected fault trajectory is calculated by first local maxima of the potential energy along the sustained fault trajectory. In step 3, minimum gradient point (MGP) is used to search for controlling u.e.p. MGP can be used as an initial guess, and if it is close to controlling u.e.p, it will converge to the controlling u.e.p, otherwise it may converge to another equilibrium point or even diverge. A robust nonlinear solution is required.

This provides a brief overview of the direct methods. The cited references provide further reading. This chapter is an overview of the power system stability, and the concepts are continued further in Chap. 13. The power system stability remains one of the subjects of great interest to electrical engineers, especially in the utility industry. Many industrial facilities have their own generation. These machines and embedded generation run in synchronism with the utility systems and have their own problems of stability (Example 13-3).

PROBLEMS

1. A generator delivers 1.0 pu power to an infinite bus, which reduces to 0.6 under a fault. The prefault maximum power is 2 pu, and postfault (after fault clearance) it is 1.25 pu. Neglect saliency and saturation. Determine critical clearing angle for $H = 5$ and critical clearing time of the breakers for stability. The breakers have an interrupting time of 3 cycles. Plot the swing curve.

2. A 150 MW, 0.85 power factor, $H = 5.5$ generator runs at 20 MW, 0.8 pF. The power load suddenly changes to 100 MW. Determine acceleration and deacceleration of the rotor.

3. A 100 mi transmission line has the following parameters:

$R = 0.1094\ \Omega$, $X = 0.7654\ \Omega$, $Y = 5.456\ \mu\text{S}$ (Siemens).

If the sending-end and receiving-end voltages are kept the same, what is the steady-state stability limit of the line?

4. Obtain a Y matrix and a reduced Y matrix for the system shown in Fig. 12-19 and calculate the generator internal voltage.

5. A generator with 0.3 pu transient reactance supplies power to an infinite bus through a line of 0.4 pu reactance. The generator no-load voltage is 1.1 pu, and that of infinite bus is 1.0 pu. $H = 3.5$. If the generator is loaded to 80 percent, what is the natural frequency of oscillation and maximum power transfer capability for small perturbations?

6. A two-pole 200-MVA generator has moment of inertia of the rotor = 150,000 kg-m^2. Convert it into H.

7. 150 MW of power is transmitted over a line which has a static stability limit of 300 MW. Calculate the maximum load that can be switched without loss of stability.

8. What are the limitations of "equal area criterion" of stability?

9. Based on Fig. 12-17*a*, construct block diagrams of synchronous generator, using data as shown in Table 10-2. Also construct phasor diagram.

10. Does system neutral grounding impact stability? Illustrate with an example.

11. A transmission line has the following parameters:

length = 200 km, voltage = 400 kV three-phase line-to-line, reactance = 0.4 Ω/km, $B_c = 5.0\ \mu$S (Siemens)/km, $\alpha = 0.00006$ Np/km, $\beta = 0.0013$ rad/km.

Plot a curve showing power transmitted on x-axis versus the ratio of receiving-end voltage to sending-end voltage at unity power factor and 0.8 power factor lagging.

12. Consider a system described by:

$P_m = 2.5$ pu, during fault $r_1P_m = 0.80$ pu, after fault clearance $r_2P_m = 2.0$ pu, $H = 3.0$, $f = 60$ Hz, $\Delta t = 0.03$ s, $P_e = 1.25$ per unit.

Calculate rotor angle and angular velocity at the end of 0.03 s using Runge–Kutta and Euler's modified method.

13. Drive the system equations for E'' model, step by step (Eqs. 12-53 to 12-55).

14. In Example 12-5, a simple opening of a generator breaker results in instability of a generator 150 mi away. Is it an example of bad system design? Consider that all system components are designed for emergency load flow conditions and have enough load carrying capability. What could be done to improve stability?

15. Write a one page description of V-Q sensitivity analysis without mathematical equations. What is implied by "distance-to-voltage instability?"

16. Write a one page comparison of controlling u.e.p and BCU direct stability methods without mathematical equations.

REFERENCES

1. A. J. Gonzales, G. C. Kung, C. Raczkowski, C. W. Taylor, and D. Thonn, "Effects of Single- and Three-Pole Switching and High Speed Reclosing on Turbine Generator Shafts and Blades," *IEEE Trans. PAS-103*, pp. 3218–3228, Nov. 1984.
2. CIGRE SC38-WG02 Report, "State of the Art in Non-Classical Means to Improve Power System Stability," *Electra*, no. 118, pp. 88–113, May 1988.
3. IEEE Working Group Report, "Turbine Fast Valving to Aid System Stability: Benefits and Other Considerations," *IEEE Trans. PWRS-1*, pp. 143–153, Feb. 1986.
4. *Electrical Transmission and Distribution Handbook*, 4th ed., Westinghouse Electric Corporation, PA, 1964.
5. P. M. Anderson and A. A. Fouad, *Power System Control and Stability*, IEEE Press, NJ, 1994.
6. P. Kundar, *Power System Stability and Control*, McGraw Hill, New York, 1994.
7. IEEE Std. 1110, IEEE Guide for Synchronous Generator Modeling Practices and Applications in Power System Stability Analyses, 2002.

8. R. T. Byerly, R. J Bennon, and D. E. Sherman, "Eigenvlaue Analysis of Synchronizing Power Flow Oscillations in Large Power Systems," *IEEE Trans. PAS-101*, pp. 235–243, Jan. 1982.

9. N. Martins, "Efficient Eigenvalue and Frequency Response Methods Applied to Power System Small-Signal Stability Studies," *IEEE Trans. PWRS-1*, pp. 217–225, Feb. 1986.

10. P. Kundur, G. J. Rogers, D. Y. Wong, L. Wang, and M. G. Lauby, "A Comprehensive Computer Program for Small Signal Stability Analysis of Power Systems," *IEEE Trans. PWRS-5*, pp. 1076–1083, Nov. 1990.

11. J. C. Das, *Power System Analysis*, Chapter 12, "Load Flow Methods," Marcel Dekker, New York, 2002.

12. D. J. Hill, "Nonlinear Dynamic Load Models with Recovery for Voltage Stability Study," *IEEE Trans. Power Systems*, vol. 8, no. 1, pp. 166–176, 1993.

13. EPRI, Load Modeling for Power Flow and Transient Stability Computer Studies, Report EL-5003, 1987.

14. A. R. Bergen and D. J. Hill, "A Structure Preserving Model for Power Stability Analysis," *IEEE Trans. PAS-100*, pp. 25–35, 1981.

15. H. D. Chiang, C. C. Chu, and G. Cauley, "Direct Stability Analysis of Electrical Power Systems Using Energy Functions: Theory, Applications, and Perspective," *Proc. IEEE* vol. 83, no. 11, pp. 1497–1529, 1995.

16. A. A. Fouad and V. Vittal, *Power System Transient Stability Analysis Using the Transient Energy Function Method,* Prentice Hall, Englewood Cliffs, NJ, 1991.

17. T. Athay, R. Podmore, and S. Virmani, "A Practical Method for Direct Analysis of Transient Stability," *IEEE Trans. PAS-98*, no. 2, pp. 573–584, 1979.

18. M. A. Pai, *Energy Function Analysis for Power System Stability,* Kluwer Academic Publishers, Boston, MA, 1989.

19. H. D. Chiang, F. F. Wu, and P. P. Varaiya, "A BCU Method for Direct Analysis of Power System Transient Stability," *IEEE Trans. Power Syst.*, vol. 8, no. 3, pp. 1194–1208, 1994.

20. F. A. Rehimi, M. G. Lauby, J. N. Wrubel, K. L. Lee, "Evaluation of Transient Energy Function Method for On-Line Dynamic Security Assessment," *IEEE Trans. PS*, vol. 8, no. 2, pp. 497–507, 1993.

21. EPRI, *User Manual for Direct Version 4.0 EPRI TR-105886s*, Palo Alto, CA, December 1995.

FURTHER READING

V. Ajjarapu and B. Lee, "Voltage Stability and Long Term Stability Working Group, Bibliography on Voltage Stability," *IEEE Trans. Power Systems*, vol. 13, no. 1, pp. 115–125, 1998.

W. T. Carson, *Power System Voltage Stability*, McGraw Hill, New York, 1994.

H. D. Chaing, J. S. Thorp, I. Dobson, R. J. Thomas, and L. Fekih-Ahmed, "On the Voltage Collapse in the Power Systems," *IEEE Trans. Power Systems*, vol. 5, no. 2, pp. 601–611, 1990.

H. D. Chiang, F. F. Wu, and P. P. Varaiya, "Foundations of Direct Methods for Power System Stability Analysis," *IEEE Trans. Circuits Syst., CAS-34*(2), pp. 160–173, 1987.

CIGRE Task Force 38.02.05, Load Modeling and Dynamics, *Electra*, May 1990.

S. B. Crary, *Power System Stability*, vol. 2, Wiley, New York, 1947.

P. L. Dandeno, R. L. Hauth, and R. P. Schulz, "Effects of Synchronous Machine Modeling in Large Scale System Studies," *IEEE Trans. PAS-92*, pp. 926–933, 1973.

EPRI Report EL-5798, The Small Signal Stability Program Package, vol. 1. Final report of Project 2447-1, Nov. 1990.

F. P. de Mello, J. W. Feltes, T.F. Laskowski, L. J. Oppel, "Simulating Fast and Slow Dynamic Effects in Power Systems," *IEEE Computer Applications, Power*, vol. 5, no. 3, pp. 33–38, 1992.

J. Deuse and M. Stubbe, "Dynamic Simulation of Voltage Collapses," *IEEE Trans. Power Systems*, vol. 8, no. 3, pp. 894–900, 1993.

IEEE Committee Report, "Dynamic Stability Assessment Practices in North America," *IEEE Trans. Power Systems*, vol. 3, no. 3, pp. 1310–1321, 1988.

IEEE Committee Report, "Supplementary Definitions and Associated Tests for Obtaining Parameters for Synchronous Machines Stability Simulations," *IEEE Trans. PS, PAS-99*, pp. 1625–1633, 1980.

E. W. Kimbark, *Power System Stability*, vol. 3, Wiley, New York, 1956.

K. Ogata, *State-Space Analysis of Control Systems*, Prentice Hall, Englewood Cliffs, NJ, 1967.

R. J. O'Keefe, R. P. Schulz, and N. B. Bhatt, "Improved Representation of Generator and Load Dynamics on Study of Voltage Limited Power System Operations," *IEEE Trans. Power Systems,* vol. 12, no. 1, pp. 304–312, 1997.

W. W. Price, K. A. Wirgau, A. Murdoch, J. V. Mitsche, E. Vaahedi, and M. A. El-Kady, "Load Modeling for Power Flow and Transient Stability Computer Studies," *IEEE Trans. PS*, vol. 3, no. 1, pp. 180–187, 1988.

V. A. Venikov, *Transient Phenomena in Electrical Power Systems*, Pergamon Press, Macmillan, New York, 1964.

C. C. Young, "Equipment and System Modeling for Large Scale Stability Studies," *IEEE Trans. PAS-91*, vol. 1, pp. 99–109, 1972.

CHAPTER 13

EXCITATION SYSTEMS AND POWER SYSTEM STABILIZERS

The correct modeling and tuning of the excitation systems have a profound effect on power system stability—both transient and dynamic. Consider a synchronous machine operating on an infinite bus of constant voltage and frequency. There is no external impedance added between the generator and the infinite bus, and the saliency is neglected. We will study the effect of variation of generator excitation in such a system. By definition of the infinite bus, its voltage V is fixed and is independent of the generator operation. (A large generator in close proximity of the system under study does impact the voltage and frequency.) The synchronous impedance of generator is Z_s, where:

$$Z_s = r + X_l + X_{ad} = r + X_d \tag{13-1}$$

X_{ad} is the reactance due to armature reaction, and X_l is the leakage reactance (Chap. 10). Z_s is assumed to be constant for the following discussions.

If the generator is running at no load, and its emf is adjusted by field excitation, exactly equal to the terminal voltage, then neglecting no-load and core losses, no current flows in or out of the generator (Fig. 13-1*a*).

If generator is underexcited then E is less than the terminal voltage V; a leading current I_r flows out of the generator which adds to the field ampere turns by direct armature reaction. This current is entirely reactive as no power is being supplied to or taken out of the generator (Fig. 13-1*b*).

If the excitation is increased, then E is greater than the terminal voltage V, and a lagging demagnetizing current circulates in the stator, which reduces the net excitation, so that again there is no power flow into or out of the machine (Fig. 13-1*c*).

Consider now that the generator supplies current into the infinite bus. Figure 13-1*d* shows the operation when the generator supplies current exactly at unity power factor, Fig. 13-1*e* when it is underexcited, and, finally, Fig. 13-1*f* when it is overexcited. When the generator is underexcited, current leads the voltage, and when overexcited, the current lags the voltage. In overexcited and underexcited conditions, a component of current I_r adds to the active component I_a, so that the voltage triangle E, V, and the drop I_aZ_s, and $I_r Z_s$ satisfy the required conditions.

For a constant power output, the I_a and I_aZ_s are constant. The variation of I_r changes component I_rZ_s, that is, the operating power factor of the generator. Changes in load or excitation change the power angle or torque angle δ; it reverses when the generator acts like a motor, and when δ increases excessively, the machine becomes unstable.

13-1 REACTIVE CAPABILITY CURVE (OPERATING CHART) OF A SYNCHRONOUS GENERATOR

With these basic concepts the so-called reactive capability curve of the generator can be constructed. Consider Fig. 13-2, where I is the generator load current at a power factor of ϕ, and V is the terminal voltage which is equal to the infinite bus voltage. E and δ are as defined before.

For constant current and MVA, IX_d is constant (here we neglect resistance to demonstrate the simplicity of construction), and therefore its locus is a circle with center at V. If excitation is held constant, then its center is at point O. Then we can write:

$$\begin{aligned} Vq &\equiv \text{MVA} \\ qp &\equiv \text{Mvar} \\ Vp &\equiv \text{MW} \end{aligned} \tag{13-2}$$

For zero excitation $E = 0$, $IX_s = V$, and I is purely reactive, leading, and corresponding to VI vars per phase. For static stability limit $\delta = 90°$, and therefore the horizontal line through O gives the stability limit.

Using these observations, an operating chart (commonly called reactive capability curve) can be constructed, as shown in Fig. 13-3. Consider a synchronous reactance of 167 percent. For zero excitation current 100/167 = 60 percent of full-load value, which fixes point O in Fig. 13-3. This also represents 60 percent of the full-load MVA, in the form of leading Mvar, which fixes all the MVA and Mvar scales. A horizontal line from V gives MW. Circles drawn with V as center give MWs, and those drawn with center O give excitation levels. Hundred percent excitation corresponds to fixed terminal voltage OV.

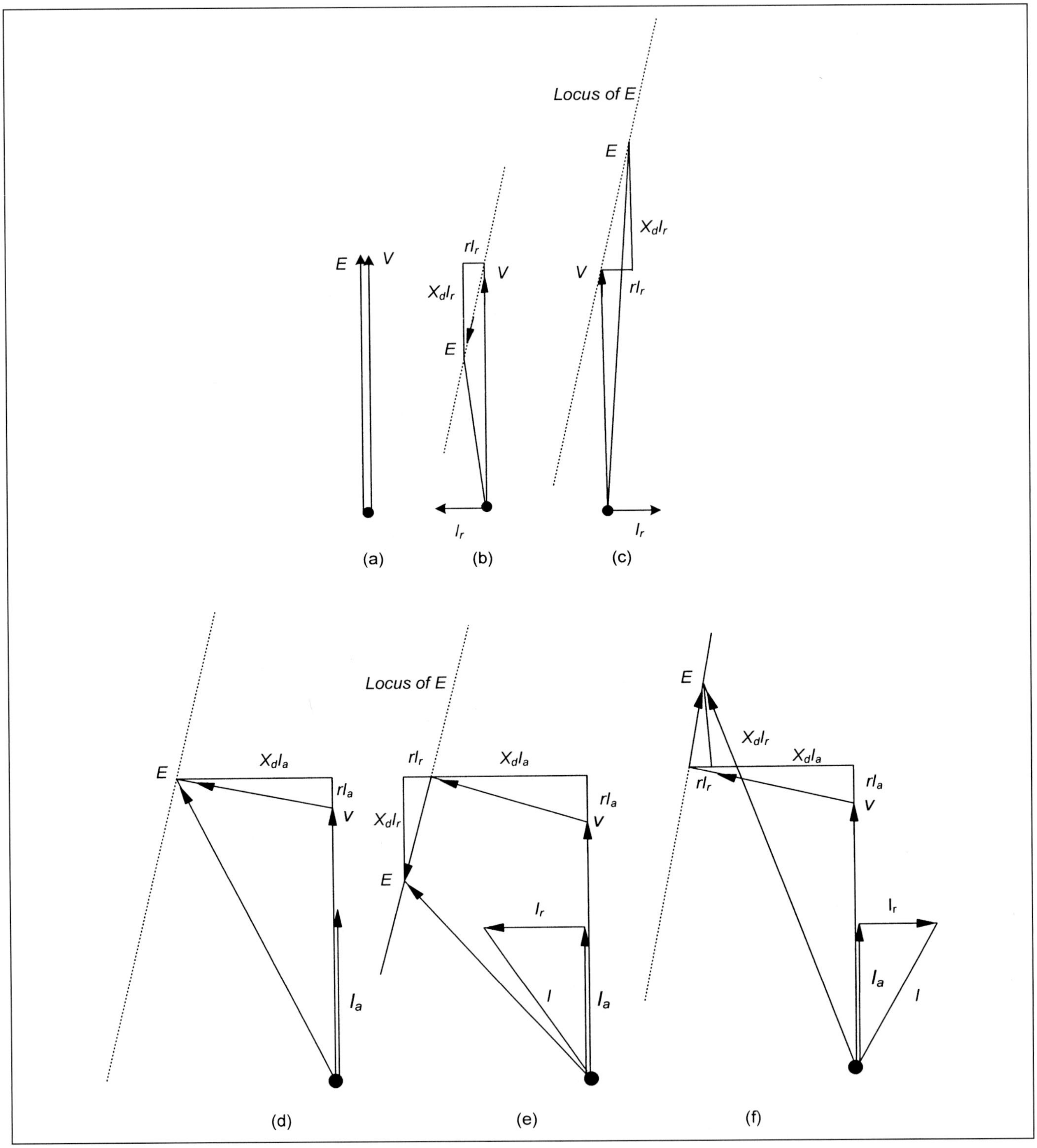

FIGURE 13-1 A synchronous machine on infinite bus. Generator at no load, (*a*) normal, (*b*) underexcited, and (*c*) overexcited. Generator at full load, (*d*) unity power factor, (*e*) underexcited, and (*f*) overexcited.

The working area of the generator can now be marked, with known generator specifications. Consider that the generator provides rated output at 0.9 power factor. This fixes the part of curve *pq*. If we assume that the excitation limit is 260 percent, then arc *mn* can be described, and the limitation of the stator current corresponding to maximum MW gives arc *nq*, with center *V*.

The line *pq* cannot be continued to the stability limit, as a small excursion in load will cause instability. The other consideration is the core heating when the generator absorbs reactive power from the system and acts like a reactor. Point *v* on the stability limit shows a MW of 60 percent at 100 percent excitation. Reduce it by 10 percent to give point *w* and point *u* on the operating area.

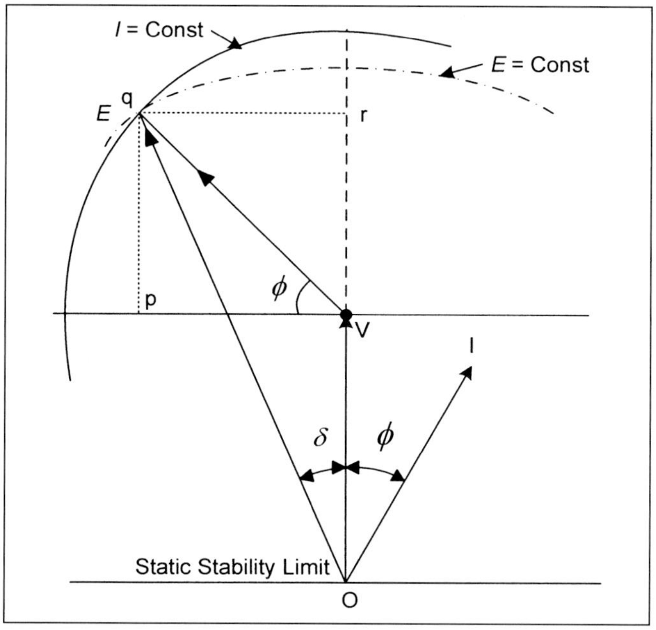

FIGURE 13-2 Generator on infinite bus, showing generator operation, constant current, and constant excitation loci.

This procedure can be followed at all points where the excitation circles are crossing the stability line, giving 10 percent MW in hand before the instability is reached.

A practical reactive capability curve is as shown in Fig. 13-4; the underexcitation limiter (UEL) is an auxiliary control to limit the underexcited reactive power current. The UEL uses synchronous generator voltage and current inputs to determine the limit start point and provide necessary feedback. This limit rises well above the steady-state limit, curve *A*, given by the construction as shown in the Fig. 13-4. The term UEL is synonymous with underexcited reactive volt-ampere limit (URAL) and manufacturer's excitation limit (MEL). The excitation systems may have an overexcitation limiter also, to prevent overheating of the generator rotor. According to ANSI/IEEE standards, a generator shall operate successfully at its rated kVA, frequency, and power factor at any voltage not more than 5 percent above or below the rated voltage, but not necessarily in accordance with the standards established for rated voltage operation.

The generator capability curves for voltages other than the rated voltage will differ. For hydrogen-cooled generators, two or three curves are supplied, depending upon the hydrogen pressure. The dotted curve *C* in Fig. 13-4 shows operation at higher than the rated voltage. On a short-time basis, it may be permissible to increase reactive power capability within the thermal limits of the generator stator, rotor, and the exciter. Curve *C* shows a 20-min capability. This comes handy as the generators are the primary source of

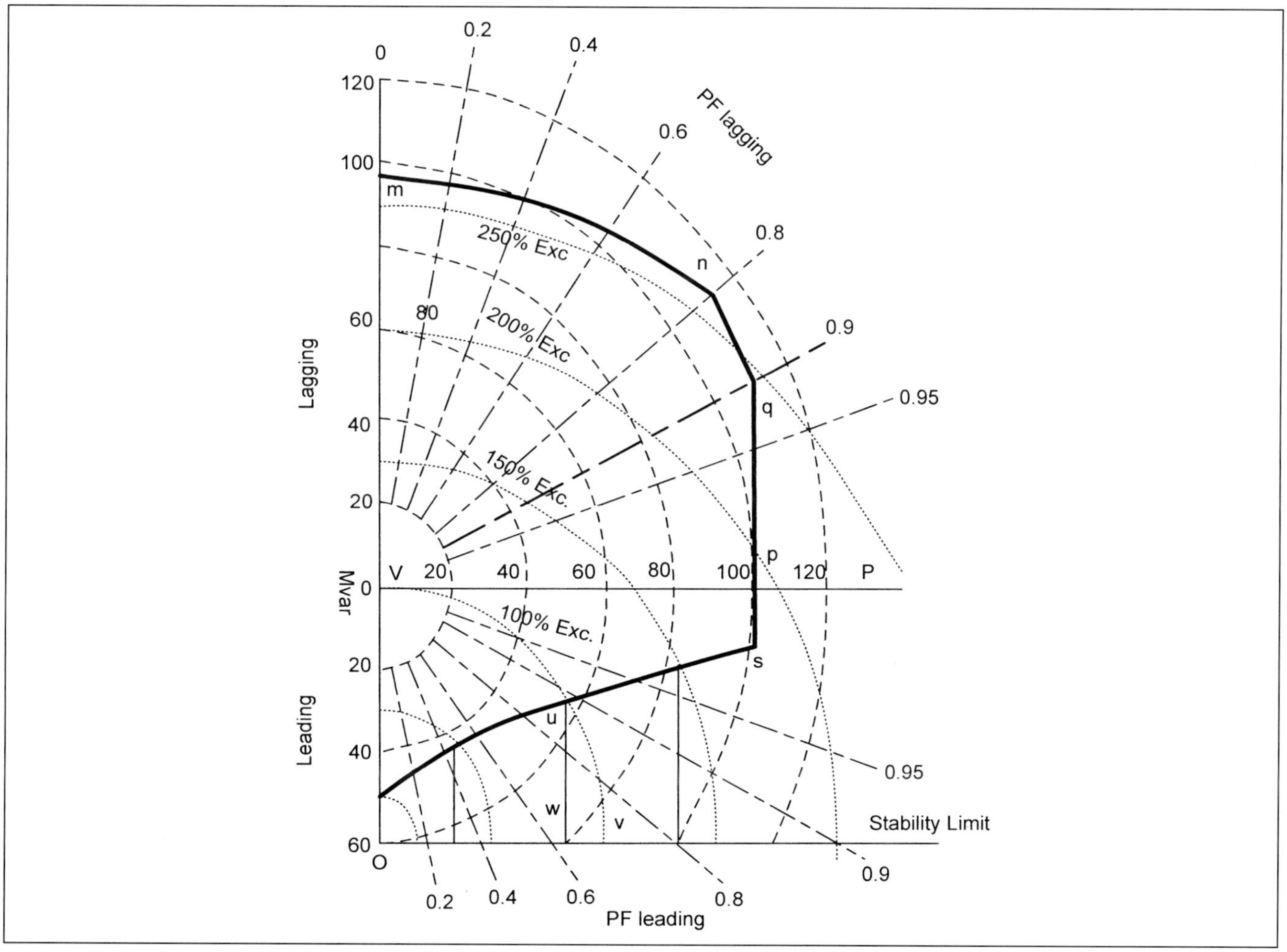

FIGURE 13-3 Operating chart of a turbogenerator.

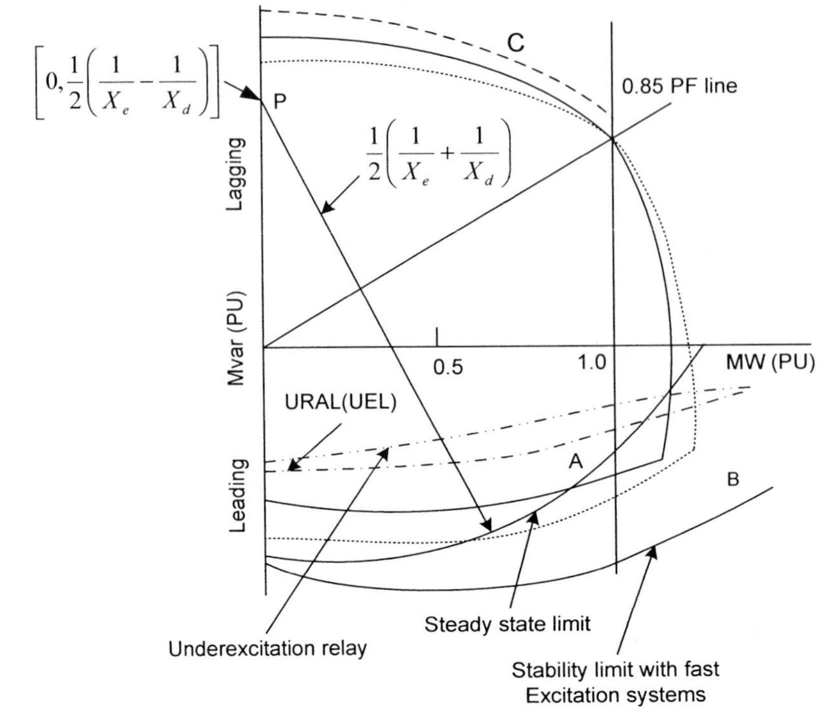

FIGURE 13-4 Reactive capability (*P-Q*) curve of a generator, showing steady-state stability limit, URAL, and stability limit with fast excitation system. Curve *C* shows emergency, 20-min reactive capability.

voltage regulation and may be called upon to supply greater reactive power under contingency conditions. The UEL limit can be circular type, straight-line type, or of multiple segments. The other limits imposed on the excitation system are field current or overexcitation limit and volts per hertz limit (V/Hz).

13-2 STEADY-STATE STABILITY CURVES

A generator has $X_d = 1.78$ pu and operates through an external reactance $X_e = 0.4$ pu, then the steady-state stability curve is given by:

$$\text{Center } P=0 \quad Q=\frac{V^2}{2}\left[\frac{1}{X_e}-\frac{1}{X_d}\right]=j0.97\text{pu}$$
$$\text{Radius}=\frac{V^2}{2}\left[\frac{1}{X_e}+\frac{1}{X_d}\right]=1.53\text{pu} \tag{13-3}$$

The permissible vars for any active power output can be calculated from:

$$P^2+\left[Q-\frac{V^2}{2}\frac{(X_d-X_e)}{X_dX_e}\right]^2=\left[V^2\left(\frac{X_d+X_e}{2X_dX_e}\right)\right]^2 \tag{13-4}$$

This is the equation of the circle of steady state stability illustrated in Fig. 13-4. At $P = 0$, the permissible vars, the generator operating at rated voltage = 0.969 pu.

These equations can be derived from the equivalent circuit of a generator, with a synchronous reactance of X_d, connected to an infinite bus through an external reactance of X_e.

13-3 SHORT-CIRCUIT RATIO

The short-circuit characteristics of a generator relate the armature reaction to the field excitation. The field excitation produces a small flux that generates an emf to circulate the short-circuit current through the resistance and leakage reactance. At high values of short-circuit currents, nonlinearity can arise, which is ignored here.

Referring to Fig. 13-5*a*, the short-circuit ratio is defined as:

$$\text{SCR}=\frac{F_v}{F_{sc}} \tag{13-5}$$

where F_v = field excitation to generate rated voltage on open-circuit characteristics, and F_{sc} = field excitation to circulate short-circuit current equal to rated current of the generator on terminal short circuit.

The SCR is, therefore, the ratio of the pu excitation for rated voltage on open circuit and pu excitation for rated armature current on short circuit. Short-circuit ratio is approximately reciprocal of the synchronous reactance, defined in per unit value for normal voltage and current. However, synchronous reactance for a given load is affected by the saturation, while SCR is unique to the machine.

The significance of higher SCR can be examined with reference to Fig. 13-5*b* and *c*. Figure 13-5*b* shows a generator with stronger field (larger field flux) which requires a larger short-core. F_a is the armature mmf, and the field mmf F_t must partly oppose F_a, and the resultant F_e develops the gap flux on which internal voltage *E* depends. With fewer armature turns F_a is reduced, so that gap length is increased to absorb F_e. Figure 13-5*c* shows the opposite, a larger F_a, that is, larger armature turns. The comparison of Fig. 13-5*b* and *c* shows that a stronger field and larger air gap gives a smaller torque angle δ, that is, a stiffer machine. Thus, SCR is a measure of relative stability.

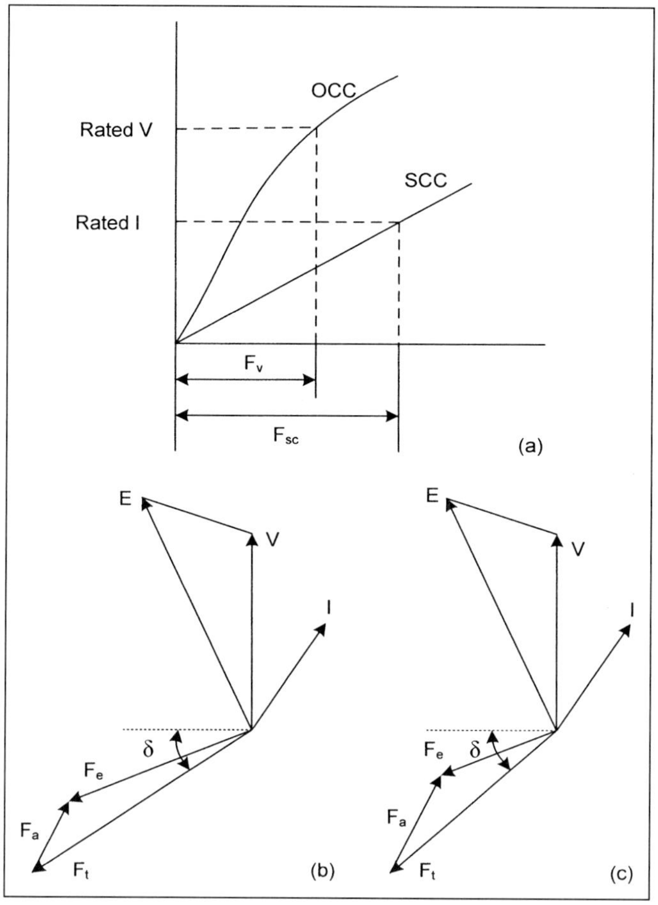

FIGURE 13-5 (*a*) Short-circuit ratio of a generator, (*b*) effect of relative construction: large flux, and small long-core, and (*c*) effect of small field flux and large short-core on torque angle δ.

Over the course of years the SCR of synchronous generators has reduced to around 0.5 or sometimes, lower. The enhancement in stability limit is obtained by modern fast-response excitation systems (Fig. 13-4).

13-4 PER UNIT SYSTEMS

In constructing the above reactive capability chart of the generator, we used an excitation voltage limit, that is, ceiling excitation voltage of 2.6 times the rated voltage. This requires some explanation. In defining rated voltage of the excitation system, there are some choices: (1) the rated voltage of the exciter, (2) the rated load field voltage, and (3) the voltage required to circulate the field current on the air-gap line. Anderson[1] recommends using (2), but IEEE standard 421.5[2] embraces all these definitions and even enlarges them. The ceiling voltage under load is determined with excitation-supplying ceiling current, which is the maximum current the excitation system can supply for a specified time. For excitation systems whose supply depends upon synchronous machine voltage and current, the nature of disturbance and specific machine parameters should be considered. For rotating exciter, the ceiling voltage is determined at rated speed.

Synchronous machine currents and voltages are defined in per unit. Rated stator current and rated terminal voltage of the machine are one per unit quantities. The rated field current is the current required to produce rated terminal voltage on air-gap line, and one per unit field voltage is the corresponding voltage. The excitation system models must interface with the machine models. Signals with per unit synchronous machine terminal voltage must be normalized to the same base, the exciter output current in per unit on the field current base, and the exciter output voltage in per unit on field voltage base.

The earlier reciprocal system is not in use. The non-reciprocal system is much in use, and the base voltage, v_{Fb}, is the voltage on the air-gap line to produce a no-load voltage equal to the generator terminal voltage.

Consider that a field current $I_F = 1000$ A at field voltage $v_F = 200$ V is required to produce rated generator voltage at no-load on the air-gap line; then in the nonreciprocal pu system, the base values of field voltage and current by definition are $V_{Fb} = 200$ V and $I_{Fb} = 1000$ A. In the reciprocal system if L_{ad} (unsaturated) = 1.8 pu, and $r_F = 0.0014$ pu, then:

$$i_{fb(recip)} = L_{ad}\, I_{Fb} = 1.8 \times 1000 = 1800 \text{ A}$$

$$u_{fb(recip)} = \frac{L_{ad}}{r_F} V_{Fb} = \left(\frac{1.8}{0.0014}\right) 200 = 257.1 \text{ kV}$$

13-5 NOMINAL RESPONSE OF THE EXCITATION SYSTEM

The excitation system response is determined from the curve shown in Fig. 13-6[3]. In this figure the rate of increase or decrease of the excitation output voltage is given by line *ac*, so that area *acd* is equal to area *abd*. This means that the rate of rise, if maintained constant, will develop the same voltage-time area as obtained from the curve for a specified period, 0.5 s shown in the figure. The *response ratio* is (*ce-ao*)/[(ao)0.5] pu V/s. Nominal response is used as a figure of merit for comparing different types of excitation systems; misleading results can occur if different types of limiters or different values of inductances are permitted.

The response ratio is an approximate measure of the rate of rise of the exciter open-circuit voltage in 0.5 s, if the excitation control is adjusted suddenly in the maximum increase position. This can be considered more like a step input of sufficient magnitude to drive the exciter to its ceiling voltage, when the exciter is operating under no-load conditions.

13-5-1 Fast-Response Systems

The time of 0.5 s in IEEE standards was chosen because it approximates to one-half period of natural electromechanical oscillation of average power system. Modern fast systems may reach ceiling voltage in much smaller time. Fast-response systems are defined as the

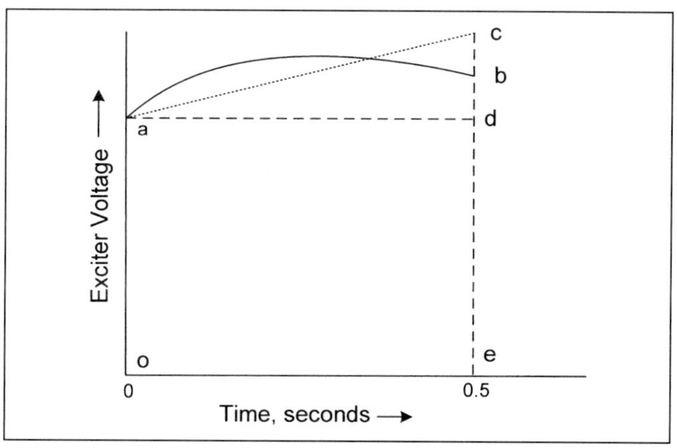

FIGURE 13-6 Determination of excitation system response, according to Ref 2.

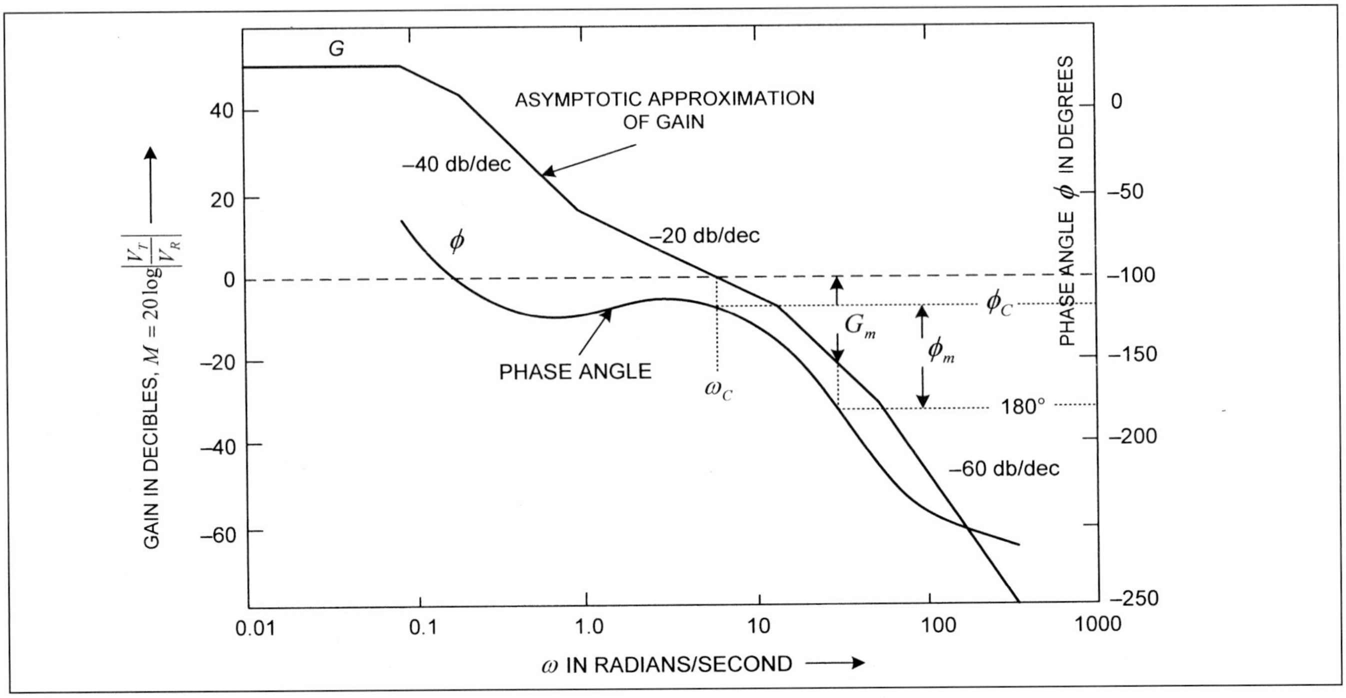

FIGURE 13-7 Open-loop frequency response, stability of excitation systems.

ones which reach ceiling voltage in a time of 0.1 s or less, and *oe* in Fig. 13-6 is replaced by 0.1 s.

13-5-2 Stability of the Excitation Systems

The excitation system stability refers to the ability of the system to control the field voltage of the generator, so that transient changes and oscillations in the regulated voltage are suppressed. This leads from one stable operation to another. Some factors that impact stability are the speed of operation, the nature of response, whether over damped or under damped, rise time, and overshoots and settling time. Concepts of stability of control systems discussed in Chap. 3 are applicable. The small-signal criteria are used to evaluate the performance of closed-loop excitation control systems.

Typical open-loop frequency response, in the form of Bode plot of the excitation system with synchronous machine open circuited, is shown in Fig. 13-7.[3] The closed-loop frequency response is shown in Fig. 13-8. In general, a gain margin G_m of 6 dB or more and a

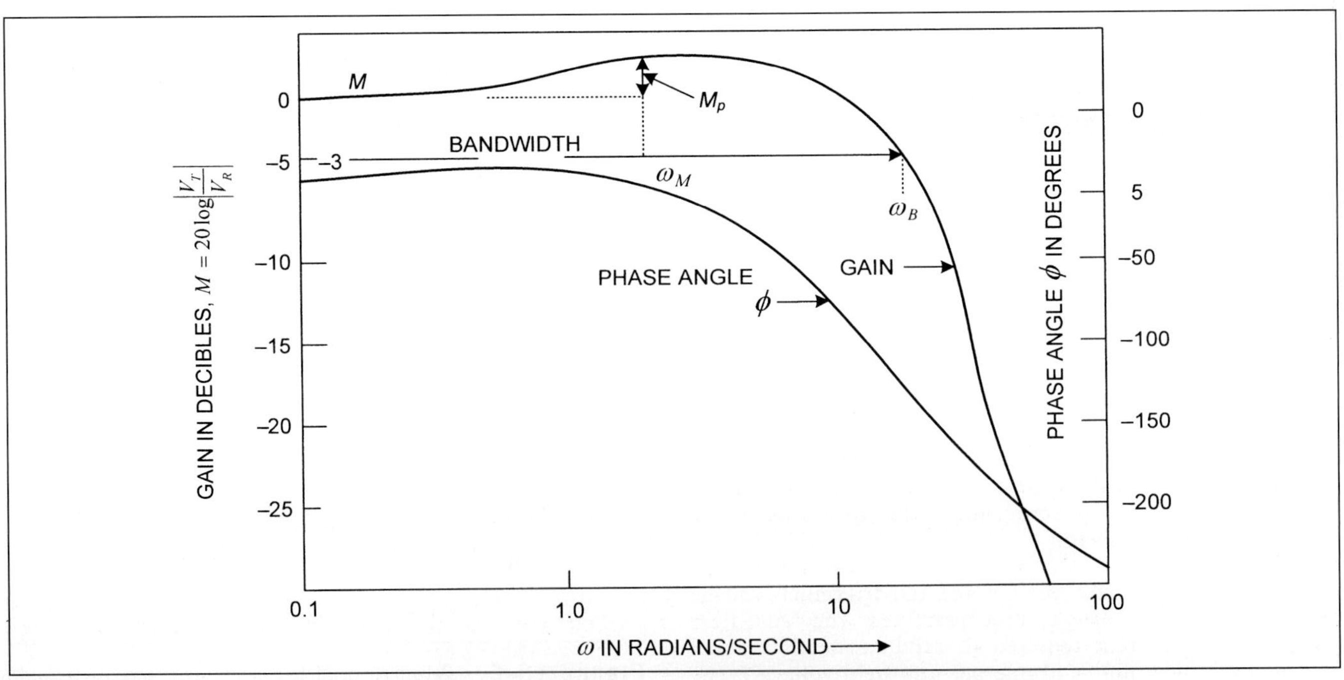

FIGURE 13-8 Closed-loop frequency response, stability of excitation systems.

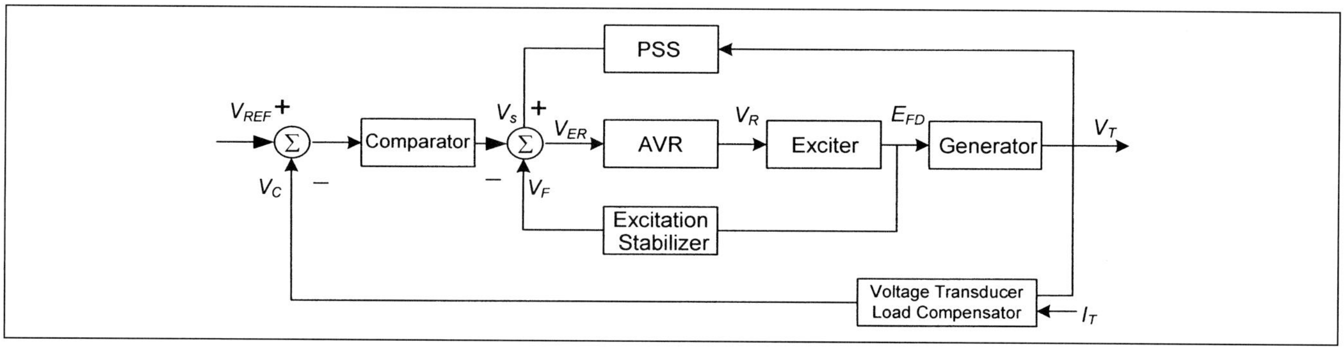

FIGURE 13-9 Basic control circuit block diagram of excitation systems.

phase margin of ϕ_m 40° or more is recommended for most feedback control systems (also see Chap. 3).

Relative stability of the closed-loop control system can be determined from properties of the open-loop transfer function, provided open-loop transfer function does not have poles and zeros in the right half s-plane.

With respect to closed-loop frequency response, peak value M_P in decibels of the amplitude response is also a measure of stability. A high value of $M_P > 1.6$ dB is indicative of oscillatory system exhibiting large overshoot. In general, 1.1 dB $\leq M_P \leq$ 1.6 dB is an acceptable design.

13-6 BUILDING BLOCKS OF EXCITATION SYSTEMS

Figure 13-9 shows basic block diagram of the excitation systems. The explanation of each block function is as follows (Fig. 13-10):

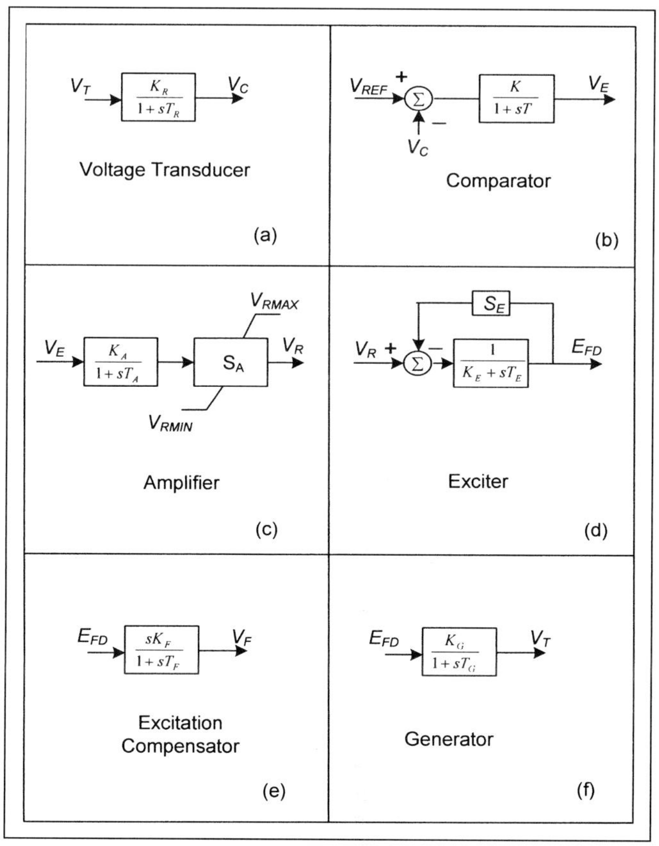

FIGURE 13-10 Building blocks of an excitation system.

1. *Voltage transducer.* The generator voltage is sensed by a PT, which is rectified, and these blocks can be represented by Fig. 13-10*a*:

$$V_c = \frac{K_R \times V_T}{1+sT_R} \tag{13-6}$$

where V_T is the terminal rms voltage V_c is the output dc voltage, K_R is the proportionality constant, and T_R is the time constant of PT and rectifier assembly (this is small).

Figure 13-11 shows a variation, where load compensator is present. Note that only one time constant is recognized in the model shown, though load sensing may have different time constants. When load sensing is not there, $R_c = X_c = 0$, and the system returns to that of Fig. 13-10*a*. The values of R_c and X_c can take positive or negative values. In most cases, the value of R_c is negligible. The input variables of synchronous machine voltage and current must be in the phasor form for the comparator calculations.

2. *Comparator.* The comparator compares the V_c against a fixed reference V_{REF}, and generates an output voltage V_E, the error voltage. It could be an electronic difference amplifier with negligible time constant or a nonlinear bridge circuit with practically zero time constant; the control circuit block diagram is shown in Fig. 13-10*b*.

$$V_E = \frac{K}{1+sT}(V_{REF} - V_c) \tag{13-7}$$

where K is the gain in the regulator, V_{REF} is the reference voltage, and T is approximately equal to 0 for a passive circuit.

3. *Amplifier.* The amplifier takes the error signal. Also the summation point receives a signal from the stabilizer, which is added, and also a signal from the excitation stabilizer, which is a negative feedback and is subtracted (Fig. 13-9). The amplifier may be a rotating amplifier, a magnetic amplifier, or an electronic amplifier. Considering linear voltage amplification, K_A is the gain with time constant T_A. The saturation value of the amplifier is given within limits, $V_{RMIN} < V_R < V_{RMAX}$ (Fig. 13-10*c*).

$$V_R = \frac{K_A}{1+sT_A} V_E \tag{13-8}$$

where V_R is the regulator output voltage, K_A is the linear voltage gain, and T_A is the time constant.

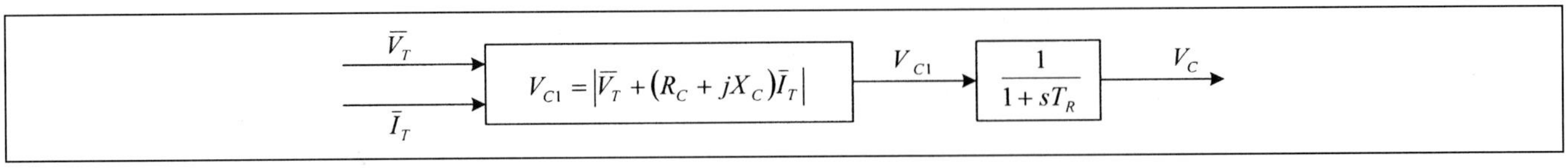

FIGURE 13-11 Terminal voltage transducer and optional load compensation elements.

4. *Exciter.* The exciter output voltage is a function of the regulator voltage. The block diagram in Fig. 13-10*d* represents this operation. The exciter is a boost-buck system:

$$E_{FD} = \frac{V_R - E_{FD}S_E}{K_E + sT_E} \tag{13-9}$$

where E_{FD} is the exciter output voltage, K_E is the exciter proportionality constant, T_E is the exciter time constant, and S_E is the exciter saturation function.

The constant T_E is explained further in Sec. 13-7.

5. *Excitation compensator.* The compensation circuit in the control system adds to the stability. It can be a feedback or lead/lag compensation. The block diagram representing the comparator is shown in Fig. 13-10*e*.

$$V_F = \frac{sK_F}{1+sT_F}E_{FD} \tag{13-10}$$

where V_F is the exciter feedback signal, K_F is the rate feedback constant, and T_F is the time constant of feedback system.

6. *Generator.* The final element in the control system is the generator itself, whose terminal voltage is a function of the exciter output voltage. The generator is at load. When the generator is open-circuited, the open-circuit time constant, T_{do}, applies. When the generator is short-circuited, T_d' applies. The time constant dependent upon load is designated by T_G. The block diagram is shown in Fig. 13-10*f*.

$$V_T = \frac{K_G}{1+sT_G}E_{FD} \tag{13-11}$$

where K_G is the generator gain from exciter voltage and T_G is the generator time constant, which varies between two extremes of no-load and short-circuit time constant.

13-7 SATURATION CHARACTERISTICS OF EXCITER

The open-circuit characteristics (OCC) relate terminal voltage at open circuit and normal speed to field excitation. Under these conditions, the terminal voltage measured is induced emf, which depends upon the total flux linking the armature. It is a measure of the saturation in the magnetic circuit. At low levels of excitation, the OCC is linear, and the main reluctance in the circuit is that of air gap. The part of the excitation devoted to the gap is defined as air line.

As the excitation is increased, the iron parts suffer a considerable decline in permeability, and the OCC for upper range of field excitation has a smaller slope. There is definite knee point linking the two curves. The saturation increases as the excitation is increased.

13-7-1 DC Exciters

For dc commutator exciters' two saturation functions, S_E (E_{FD}) is defined as a multiplier of per unit exciter output voltage to represent the increase in exciter excitation requirements due to saturation, illustrated in Fig. 13-12. At a given exciter output voltage, quantities, *A*, *B*, and *C* are defined as the excitation to produce that voltage on the constant resistance load saturation curve, on air-gap line, and on no-load saturation open circuit curve, respectively.[2] The constant resistance load saturation is used for all dc commutator exciters and ac exciters type AC5A (see Sec. 13-9).

$$S_E(E_{FD}) = \frac{A-B}{B} \tag{13-12}$$

Two values of dc commutator exciter voltage are applicable: E_{FD1} and E_{FD2}. Voltage E_{FD1}, for which $S_E(E_{FD1})$ is specified, is near the exciter ceiling voltage and voltage E_{FD2}, for which $S_E(E_{FD2})$ is specified at a lower value, commonly near 75 percent of E_{FD1}.

Consider saturation characteristics, (shown in Fig. 13-12). Define saturation as $S_e(E_x)$. Then, change in field current I_{ef} is function of E_x and the saturation factor is given by:

$$\Delta I_{ef} = E_x S_e(E_x) \tag{13-13}$$

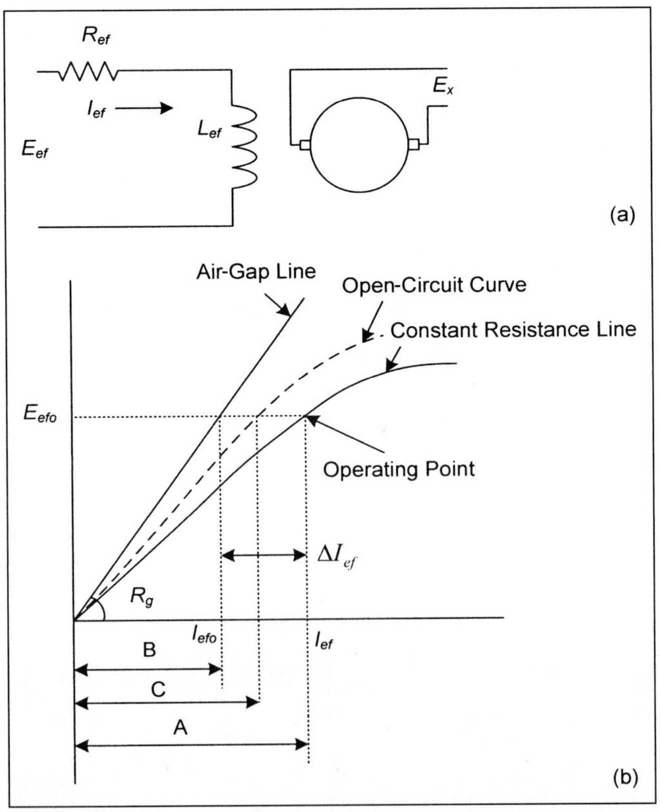

FIGURE 13-12 (*a*) Separately excited dc exciter. (*b*) Exciter load saturation curves.

Also in the field circuit, we can write the equation:

$$E_{ef} = R_{ef}I_{ef} + L_{ef}\frac{dI_{ef}}{dt} \tag{13-14}$$

Also:

$$I_{ef} = \frac{E_x}{R_g} + \Delta I_{ef} = \frac{E_x}{R_g} + E_x S_e(E_x) \tag{13-15}$$

where R_g is the slope of the air-gap line. Thus, combining the above relations, we can write:

$$E_{ef} = \left(\frac{R_{ef}}{R_g} + R_{ef}S_e(E_x)\right)E_x + \frac{1}{K_E}\frac{dE_x}{dt} \tag{13-16}$$

Let generator base field voltage be E_{fb}:

$$E_{xb} = E_{fb}$$
$$I_{efb} = \frac{E_{fb}}{R_g} \qquad R_{gb} = R_g \tag{13-17}$$

In per unit the saturation factor is:

$$S_e(E_x) = R_g S_e(E_x) \tag{13-18}$$

Equation (13-16) is reduced to:

$$E_{ef} = \frac{R_{ef}}{R_g}E_x[1 + S_e(E_x)] + \frac{1}{K_e}\frac{dE_x}{dt} \tag{13-19}$$

$1/K_e$ is inverse of a time constant:

$$\frac{1}{K_e} = \frac{L_{ef}}{R_g}\frac{I_{ef0}}{E_{x0}} = \frac{L_{ef}}{R_g} = T_E \tag{13-20}$$

I_{ef0} and E_{x0} correspond to an operating point in pu. Then, from Eq. (13-9):

$$E_{ef} = E_x[K_E + S_E(E_x)] + T_E\frac{dE_x}{dt} \tag{13-21}$$

where:

$$K_E = \frac{R_{ef}}{R_g}$$
$$S_E(E_x) = S_e(E_x)\frac{R_{ef}}{R_g} \tag{13-22}$$

Block circuit diagram in Fig. 13-13*a* shows the dc exciter representation. For small-signal analysis, it can be simplified to Fig. 13-13*b* and then *c*.

$$\Delta E_{ef} = \Delta E_f \frac{(1 + sT)}{K} \tag{13-23}$$

$$K = \frac{1}{S_E(E_{x0}) + K_E} \qquad T = KT_E \tag{13-24}$$

Thus, for small signal deviation, the dc exciter is represented by a single time constant. Both K and T vary with the operating point. For self-excited dc exciter, the model is similar but $K_E = R_{ef}/R_g - 1$.

13-7-2 AC Exciters

The ac exciters generate the dc power required for generator field windings, through an ac alternator and either stationary or rotating rectifiers (brushless systems). The rectifier system is generally a three-phase bridge circuit; these sources have an internal impedance which is predominantly inductive. The $V_f(I_f)$ curve of the rectifier is non-linear and depends upon the diode commutation; the overlapping angle and the alternator reactance plays a major role. The rectifier regulation characteristics are determined from the equations in Fig. 13-14.[2] Three operation modes are recognized. For small values of K_C, only mode 1 needs be modeled, that is, for type ST1A exciter model, I_N should not be greater than 1.

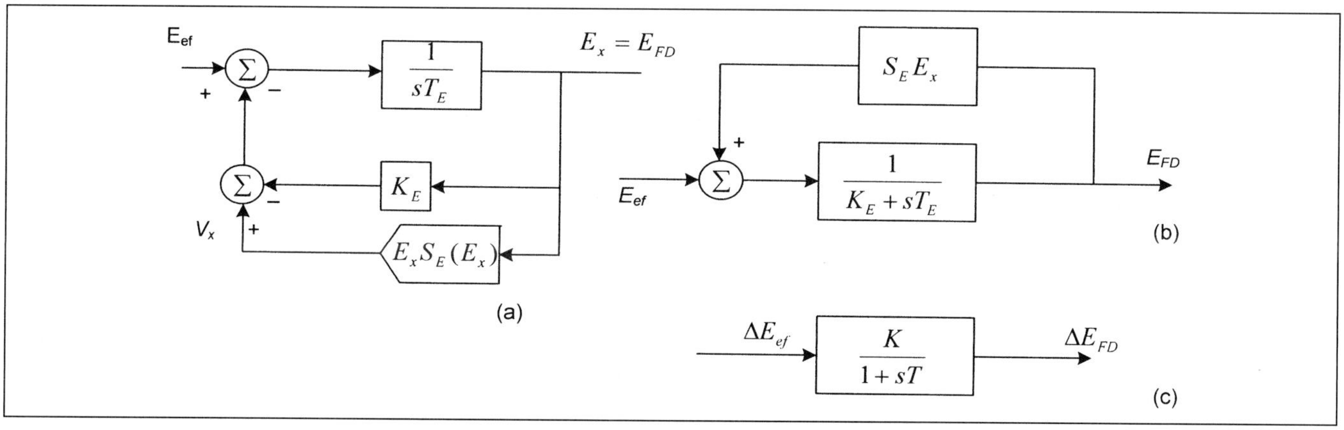

FIGURE 13-13 (*a*) Block circuit diagram of a dc exciter, (*b*) and (*c*) simplification of the block circuit diagram for small signal deviations.

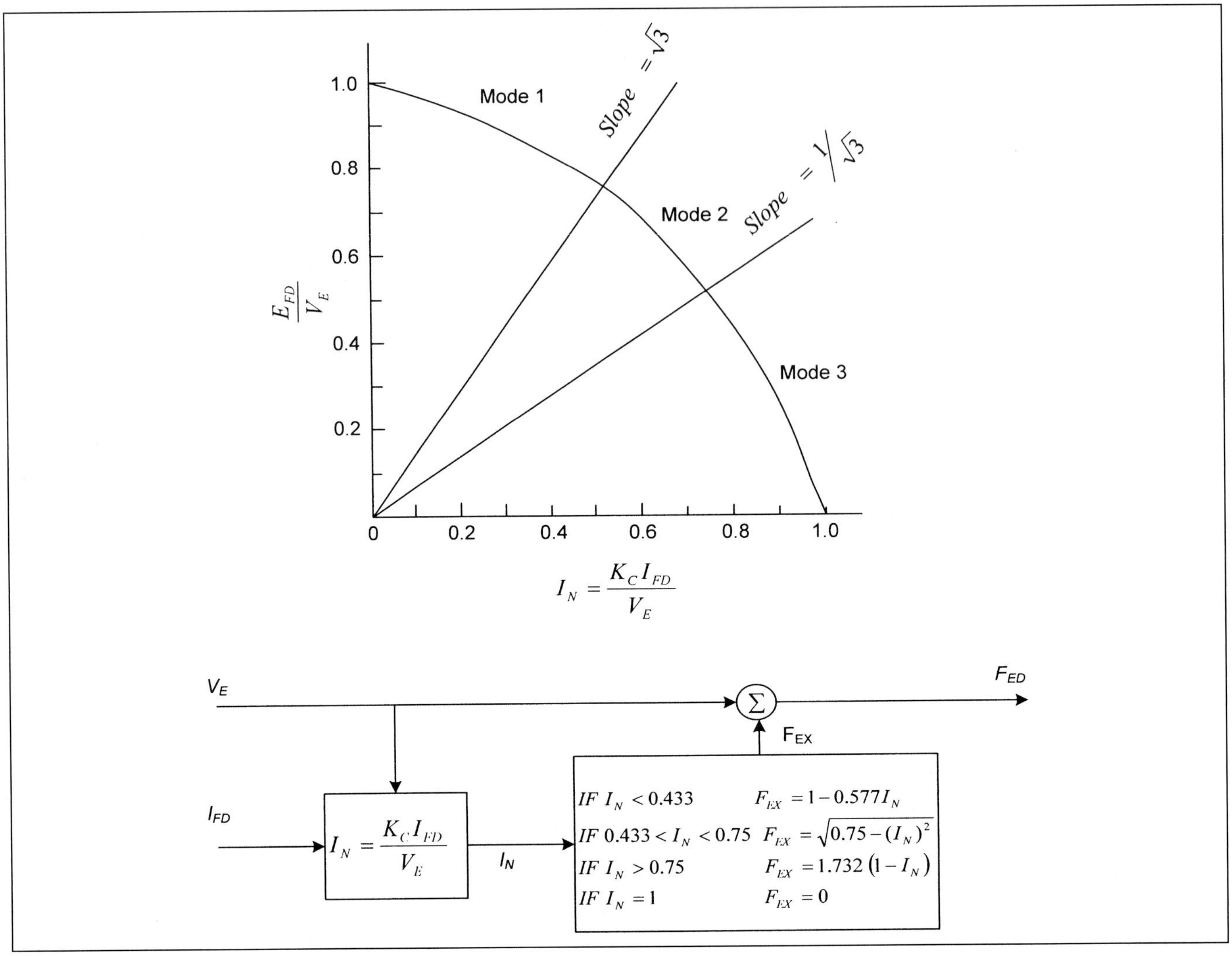

FIGURE 13-14 Rectifier regulation characteristics.[2]

The quantities E_{FD}, I_{FD}, V_E, and K_C are all in per unit on a synchronous machine field base.

For ac exciters (except for exciter type AC5A), the saturation factor S_E (V_f) is calculated from no-load saturation curve and the air-gap line of the ac exciter. From Fig. 13-12,

$$S_E(V_E) = \frac{C - B}{B} \tag{13-25}$$

Again two voltages V_{E1} and V_{E2} are defined for the alternator-rectifier excitation voltage. V_{E1} voltage, for which S_E (V_{E1}) is specified, is close to the exciter open-circuit ceiling voltage and the voltage V_{E2}, for which $S_E(V_{E2})$ is defined, is at a lower value, nearly 75 percent of V_{E1}.

13-7-3 Windup and Nonwindup Limits

Figure 13-15*a* and *b* shows the windup and nonwindup limiters applied to a single-time block constant. For the windup limit, the system equation is:

$$\frac{dv}{dt} = \frac{u - v}{T} \tag{13-26}$$

and the following limiting action applies:

$$\begin{aligned} &L_N < v < L_X && \text{then } y = v \\ &v \geq L_X && \text{then } y = L_X \\ &v \leq L_N && \text{then } y = L_N \end{aligned} \tag{13-27}$$

For the nonwindup limit, the system equation is:

$$f = \frac{u - v}{T} \tag{13-28}$$

and the following limiting action applies:

$$\begin{aligned} &L_N < y < L_X && && \text{then } \frac{dy}{dt} = f && \\ &y \geq L_X && \text{and } f > 0 && \text{then } \frac{dy}{dt} = 0 && y = L_X \\ &y \leq L_N && \text{and } f < 0 && \text{then } \frac{dy}{dt} = 0 && y = L_N \end{aligned} \tag{13-29}$$

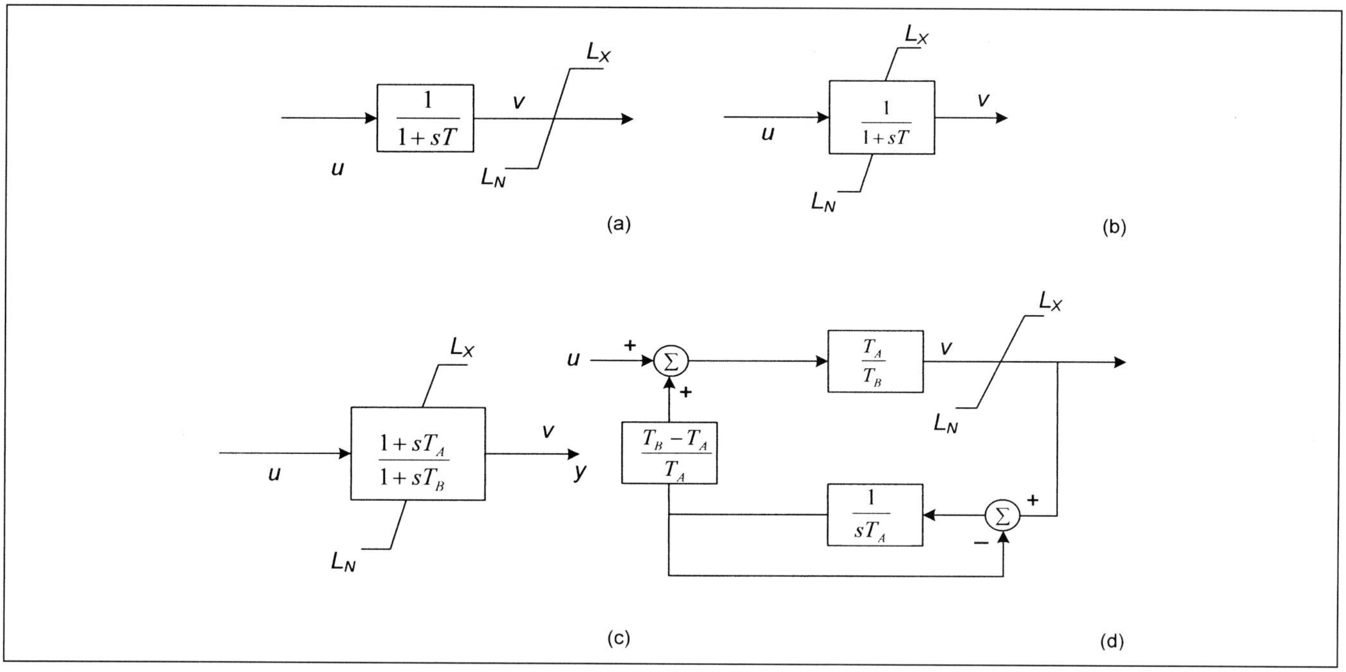

FIGURE 13-15 (*a*) and (*b*) Windup and nonwindup limiters applied to a single-time block constant. (*c*) Lead-lag function with nonwindup limits. (*d*) Physical realization of (*c*).

Lead-lag function with nonwindup limits is shown in Fig. 13-15*c*, and its physical realization is shown in Fig. 13-15*d*. Following relations apply in these cases:

$$\begin{aligned} &L_N \le v \le L_x && \text{then } y = v \\ &v > L_X && \text{then } y = L_X \\ &v < L_N && \text{then } y = L_N \end{aligned} \qquad (13\text{-}30)$$

13-7-4 Gating Functions

Gating functions are used to allow control of one of the two input signals, depending upon their relative size with respect to each other. In Fig. 13-16*a*. For the LV gate:

$$\begin{aligned} &u \le v && y = u \\ &u > v && y = v \end{aligned} \qquad (13\text{-}31)$$

For the HV gate in Fig. 13-16*b*:

$$\begin{aligned} &u \ge v && y = u \\ &u < v && y = v \end{aligned} \qquad (13\text{-}32)$$

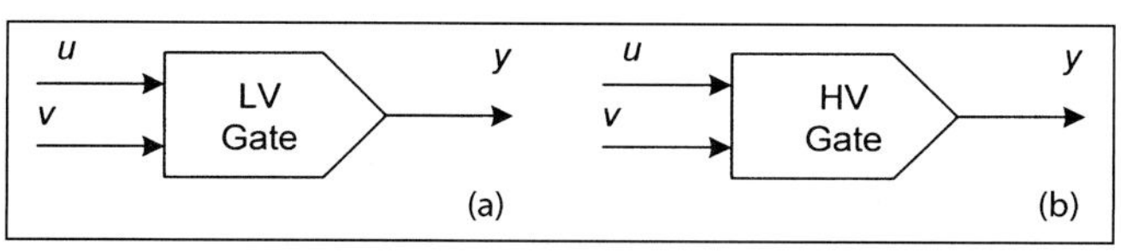

FIGURE 13-16 (*a*) LV and (*b*) HV gates.

13-8 TYPES OF EXCITATION SYSTEMS

An earlier IEEE report[4] classified the excitation systems into four types: (1) type 1 excitation system represented continuously acting regulator and exciter, (2) type 2 system represented rotating rectifier (brushless) system, (3) type 3 represented static system with terminal potential and current supplies, and (4) type 4 represented noncontinuously acting regulator. These models are still in use, and their control circuit diagrams are not shown. Currently the types of excitation systems and AVR models can be described according to IEEE standard,[2] which classifies these into three major categories, based upon the excitation power source:

- DC excitation systems utilize a dc current generator with a commutator for the excitation power source (e.g., IEEE types DC1A, DC2A, and DC3A).
- AC excitation systems use an alternator and either stationary or rotating rectifiers (brushless) to produce the dc power (e.g., IEEE types AC1A, AC2A, AC3A, AC4A, AC5A, and AC6A).
- ST type excitation systems use static controlled rectifiers for the dc power supply (e.g., IEEE types ST1A and ST2A).

The type designations in parenthesis are from IEEE standard,[3] which describes excitation system models for the stability studies. This reference gives the block circuit diagrams of each of the exciter types and also typical time constants. Manufacturers specify the time constants and data based upon the IEEE models, and this data may differ from the specimen data included in IEEE standards. For any rigorous stability study, the manufacturer's data should be used. Further, many models may be manufacturer-specific and may not exactly confirm to IEEE types, especially for equipment of European origin.

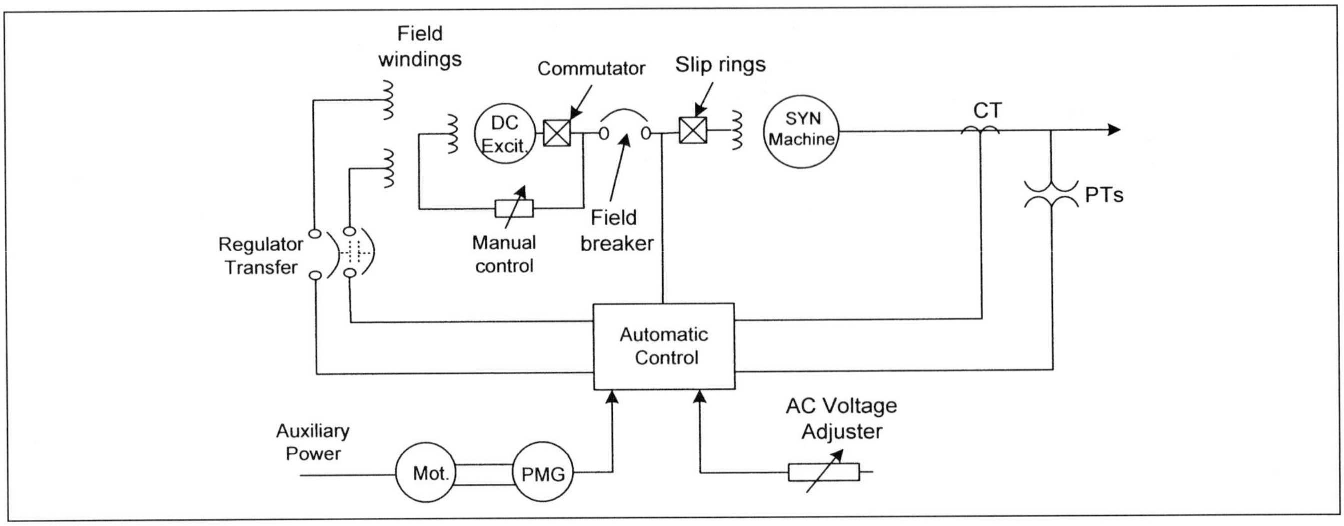

FIGURE 13-17 A dc commutator exciter system with manual control and regulator transfer, represented by DC1A.

13-8-1 The DC Exciter

The dc exciters are being replaced with ac exciters; earlier dc exciters were extensively used for machines in the range of 100 MVA and below. The commutators and brushes in the rotating dc exciters require much maintenance. A typical exciter is shown in Fig. 13-17; The shaft is coupled with the main generator, and the excitation is supplied to the generator field windings through slip rings. The dc excitation systems are not being built currently, though many old systems are still in use. The dc exciters use a main dc exciter, which may be separately mounted (motor generator set) or shaft mounted (with static amplifier). A rotating amplifier (amplydine) was earlier used. These have a relatively slow response. The exciters may be noncontinuously acting type, that is, the control action is not continuous and changes in steps—in other words "raise-lower" mode, with a change in terminal voltage of the machine large enough to exceed the threshold. These exciters have basically two different rates, depending upon the magnitude of the voltage error. For small changes, the adjustments are made with a motor-operated rheostat. Large voltage deviation results in a resistor being shorted.

Type DC1A Excitation System Figure 13-17 shows a dc generator exciter with static magnetic amplifier deriving its power from a permanent magnet generator (PMG) set. Often the frequency is increased to 420 Hz to increase the amplifier response. The exciter has two fields, one for buck and another for boost. A third field provides for the manual operation.

The control circuit block diagram of a continuously acting dc commutator exciter type DC1A is shown in Fig. 13-18. This model represents many commercial excitation systems that have been in use, namely, Westinghouse types (Mag-A-Stat™, Rototrol™, Silverstat™, and TRA™), GE types (GDA and amplidyne regulators), and Brown Boveri types (AB and KC regulators).

As shown in this diagram, the principal input V_C, from the terminal voltage transducer and load compensator at the junction, is subtracted from V_{REF}; then the stabilizing voltage V_S is added and the feedback voltage V_F is subtracted to form an error signal, which is amplified. K_A and T_A are the major time constants associated with amplifier; T_C and T_B are small and can be neglected. In steady state V_S and V_F are zero. The voltage regulator utilizes power sources which are essentially unaffected by brief transients of the synchronous machine and auxiliary buses.

The exciter may be separately excited or self-excited. When self-excited, the value of K_E is selected so that V_R is initially equal to zero. This represents operator action for tracking the voltage regulator. The saturation function is as discussed before. A common expression for *Vx* is:

$$V_x = A_{EX}e^{B_{EX}E_X} \tag{13-33}$$

The block diagram of Fig. 13-13*a* can be reduced as shown in Fig. 13-13*b* and *c*, by considering small signal response. As the operators track the voltage regulator by manual adjustments, it means that K_E is not fixed, but varies with operating condition.

Example 13-1 A generator is operating with E_{FD} = 1.7 pu and V_t = 1.0 pu. Consider A_{EX} = 0.01 and B_{EX} = 1.5. Find K_E and $V_{REF} = E_{ef}$.

K_E takes a value so that V_{REF} = 0. Thus:

$$\begin{aligned} K_E E_{FD} &= -V_x \\ &= -A_{EX}e^{B_{EX}E_X} = -0.01e^{1.5\times1.7} \\ K_E &= -\frac{1.281}{1.7} = -0.75 \end{aligned}$$

Under steady-state $V_{REF} = V_C = V_T$ = 1.0 pu. Typical data with self-excited exciter is:

$$K_A = 400 \qquad T_A = 0.89 \qquad T_E = 1.15$$

13-8-2 The Static Exciters

A static exciter schematic is shown in Fig. 13-19. The potential source for the excitation power is derived from the generator bus itself. The ac voltage through the power potential transformer (PPT) is rectified through controlled rectifiers and is proportional to the generator bus voltage. Due to static controls and components, the response is fast. These systems are modeled with IEEE model type ST1. Some examples of this model are Westinghouse-type PS static excitation with WTA or WHS regulators, and Canadian General Electric Silcomatic exciters. The model ST2 is for a compound source static exciter, where the current support is added in addition to voltage through a short-circuit CT (SCT). A phasor combination of the terminal voltage and current is formed. One example of this excitation system is the GE SCT-PPT type excitation system, which has been popular. On a short circuit close to the generator terminals, the voltage will be much reduced, and as PPT is connected to the same generator bus, its output is much reduced. On the other hand, short-circuit current increases, and the current support

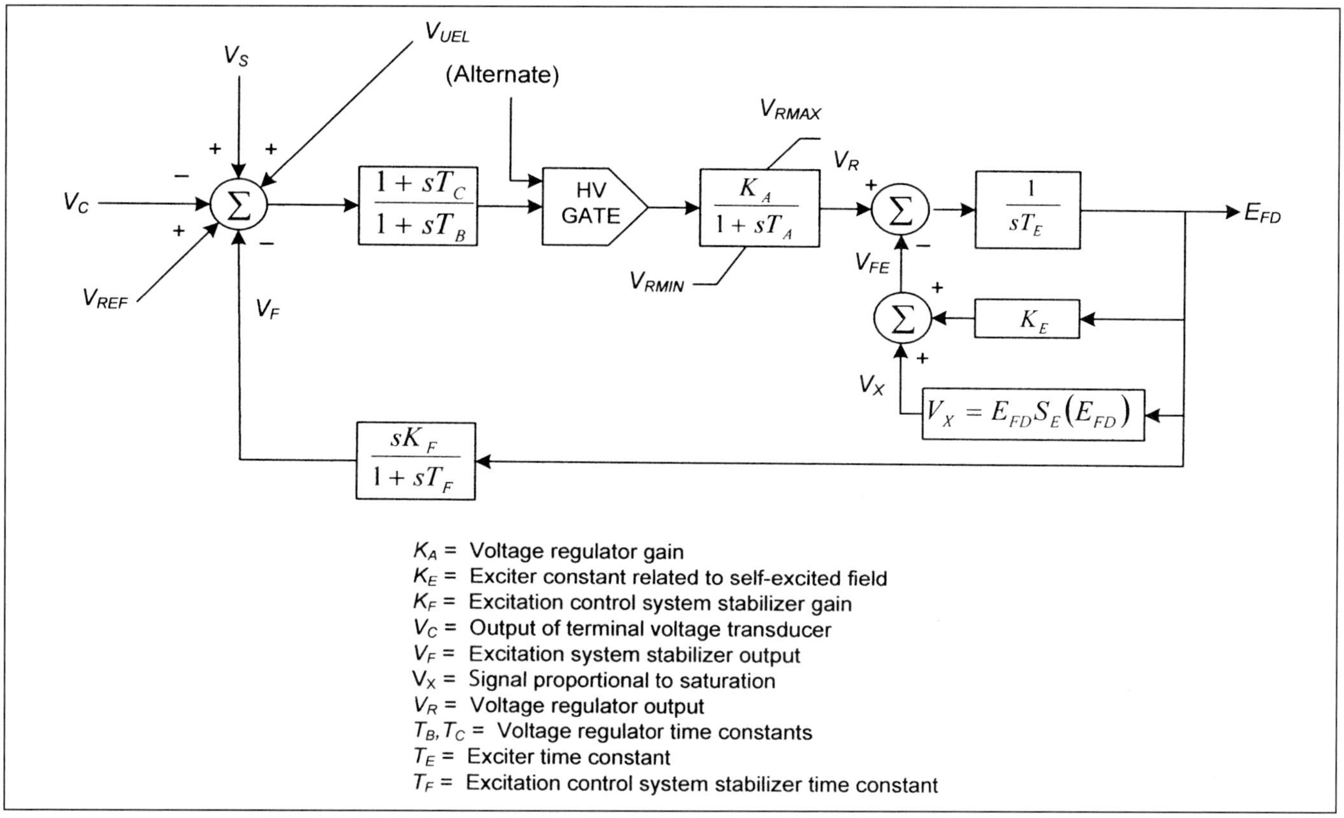

FIGURE 13-18 A dc commutator exciter, type DC1A.

through an SCT can hold the excitation voltage better, enhancing the stability limit. We will concentrate on ST1 type exciters, and there are many commercial exciter types which meet this model.

The computer model for stability analysis is shown in Fig. 13-20. The inherent time constants are small. The transient gain reduction can be implemented in the forward path via time constants T_B and T_C (in which case K_F is set to zero) or through feedback path by suitable choice of K_F and T_F. Time constants T_{C1} and T_{B1} allow for possibility of representing transient gain increase, and T_{C1} is normally > T_{B1}.

The input-output relation is assumed to be linear by proper choice of K_A. In some systems, the bridge relationship is not linearized, though a linearization of the characteristics is generally satisfactory.

In many cases, the internal limits on V_1 can be neglected. The field voltage limits, which are a function of both terminal voltage and synchronous machine current, should be modeled. As a result of the high forcing capability of the system, a field current limiter is sometimes employed to protect the generator rotor and exciter. The limit start setting is defined as I_{LR} and the gain as K_{LR}. To ignore this

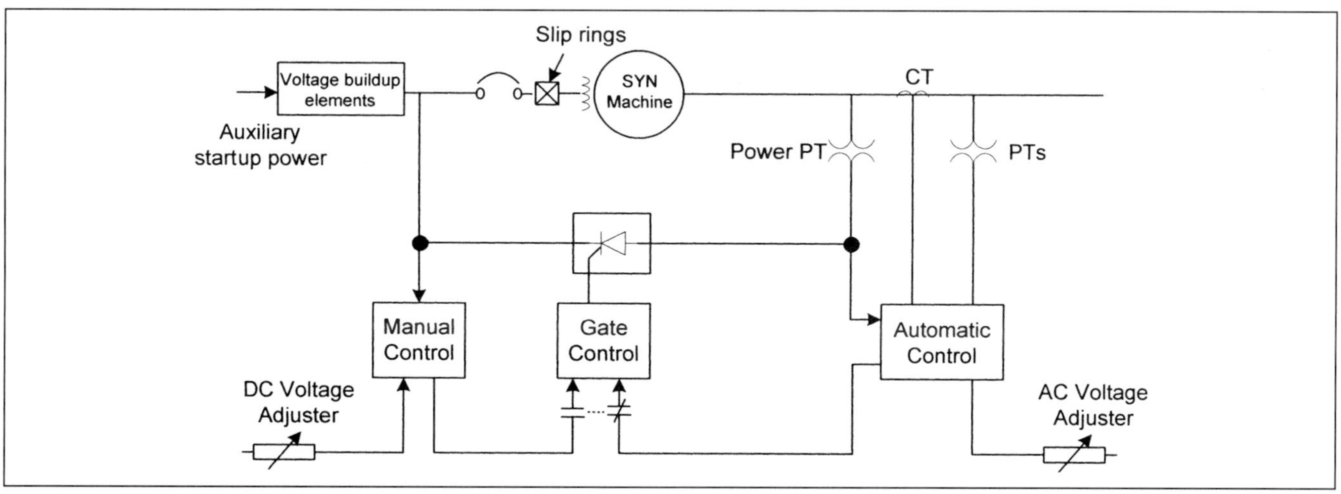

FIGURE 13-19 A static excitation system with the potential source employing controlled rectifiers, where the excitation power is supplied through slip rings, represented by ST1A.

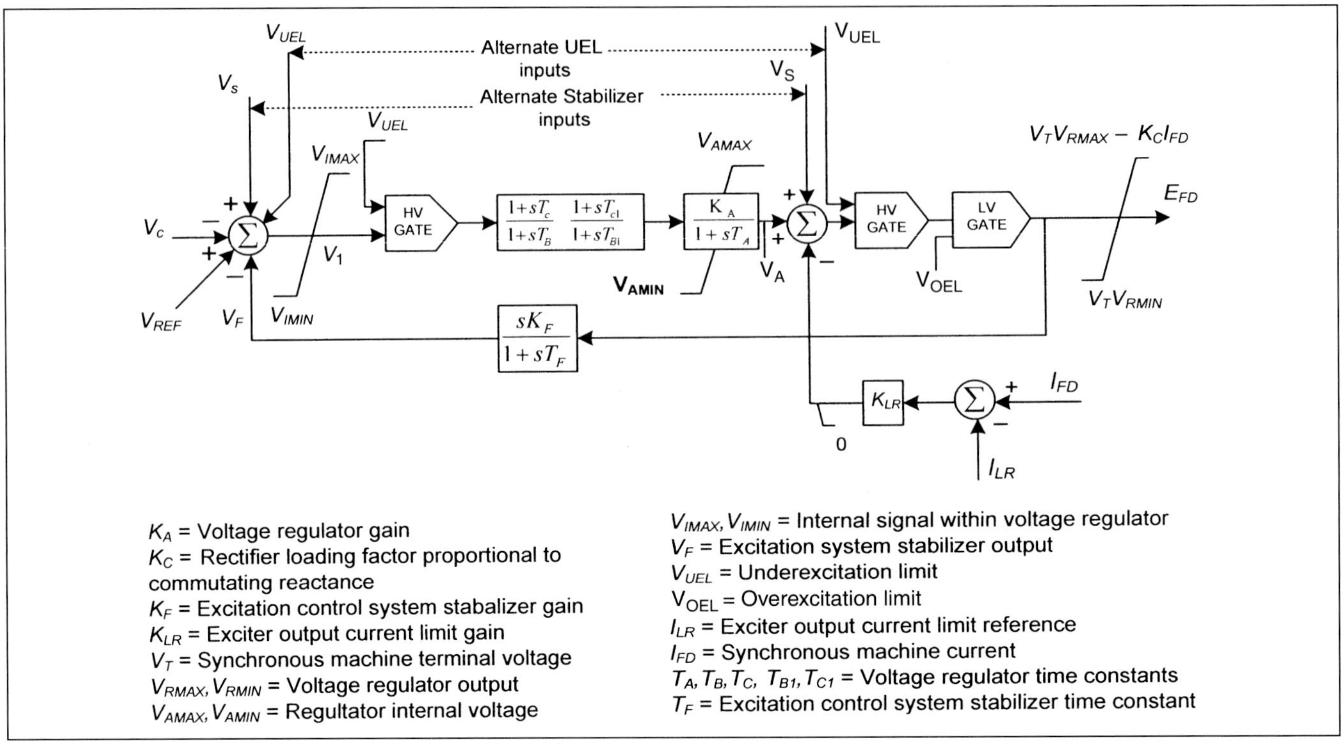

FIGURE 13-20 A static exciter system, type ST1A.

limit set $K_{LR} = 0$. The model is also applicable to systems with half-controlled bridge, in which case $VR_{MIN} = 0$.

Figure 13-19 shows a manual regulator in addition to the automatic control regulator. In case of failure of the automatic regulator, a transfer can be automatically made to manual mode. It is possible to have a step-less autotransfer to manual, in case the automatic regulator fails. The automatic regulators can also be duplicated in redundant mode for large machines. For large generators above, say, 80 MW, it is usual to have redundant fully automatic regulators in automatic transfer mode. There is no hard and fast limit with respect to the rating of the machine, and the power supply reliability, depending upon the usage of the generator, dictates the choice. The PT supplies to regulators are invariably arranged through redundant PTs with autotransfer, in case a PT fuse operates or the output power of one PT is lost.

13-8-3 The AC Exciter Type AC2A

The brushless ac exciters, using a PMG, require low auxiliary power to control the field of the rotating ac pilot exciter, which has a rotating armature and stationery field (Fig. 13-21). This provides

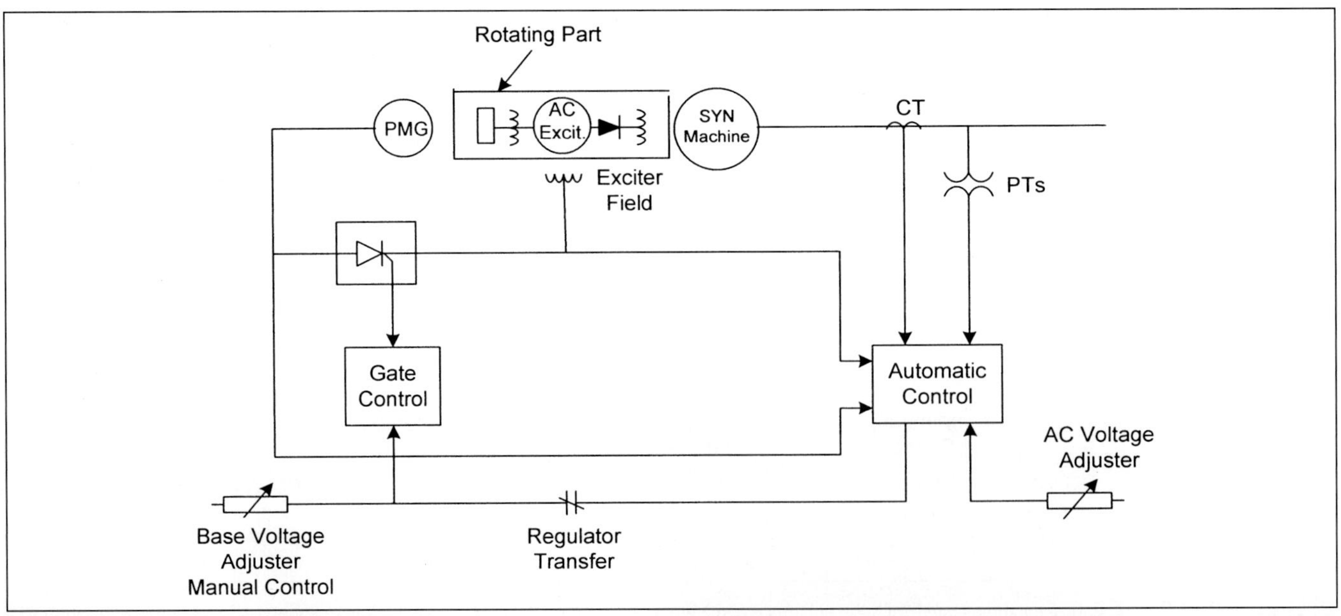

FIGURE 13-21 A brushless rotating rectifier exciter system, represented by AC1A or AC2A.

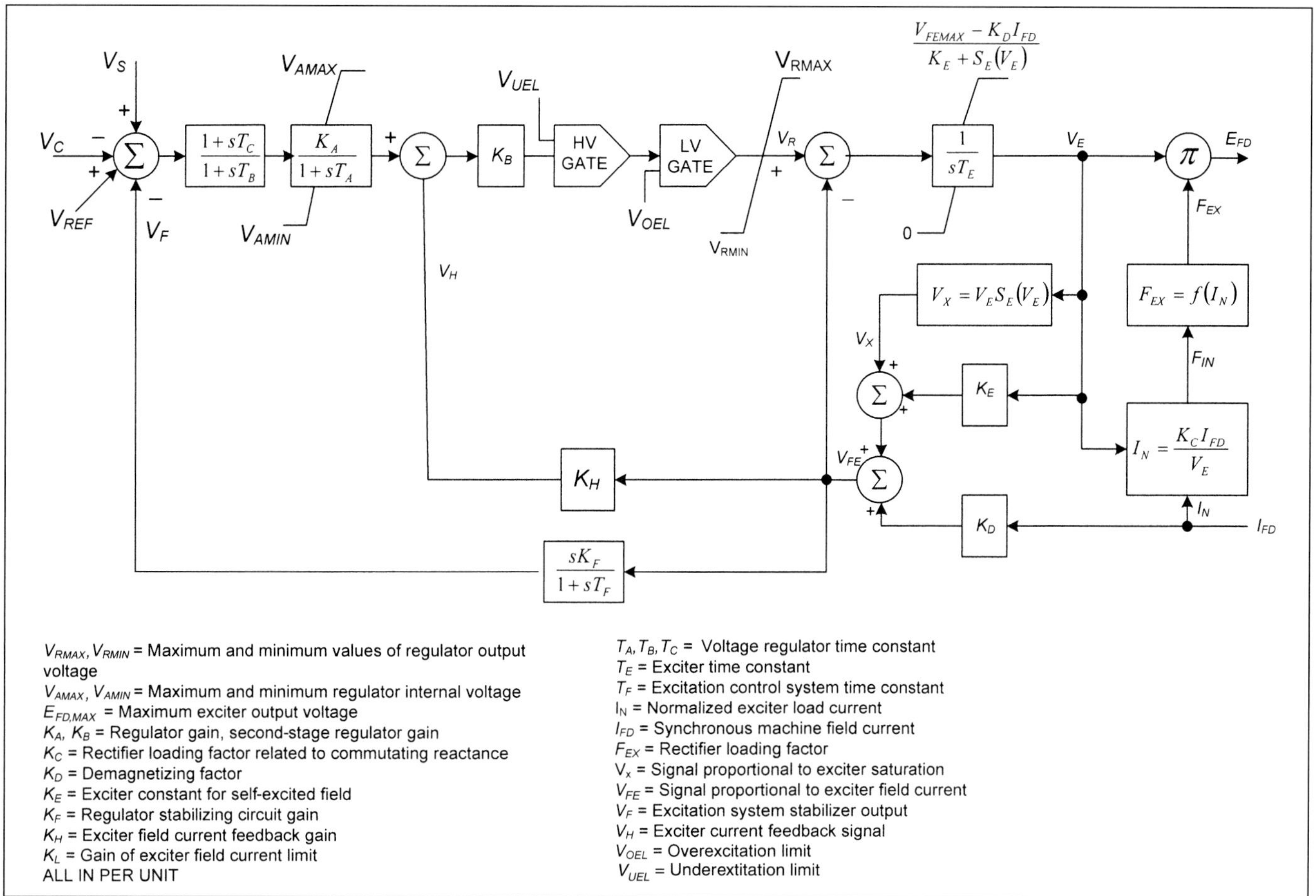

FIGURE 13-22 Control circuit exciter type AC2A.

the black start capability. These systems can be represented with IEEE type AC1A or AC2A models. The brush type ac exciters have a shaft-connected ac pilot exciter. These may have controlled or noncontrolled stationary rectifier assemblies. These are modeled with IEEE type AC3 or AC4A models.

The AC2 model represents a high initial response system, and its block circuit diagram is shown in Fig. 13-22. The alternator main exciter is used with noncontrolled rectifiers. The model is similar to AC1A, but has two additional field current feedback loops. An example of this exciter is Westinghouse high initial response brushless excitation system. The demagnetizing effect of load current I_{FD} on exciter alternator output voltage V_E is accounted for in the feedback path that includes the constant K_D. This is a function of the generator's synchronous and transient reactances. The exciter voltage drop due to excitation is simulated by inclusion of constant K_C, which is a function of commutating reactance and the rectifier regulation curve (Fig. 13-14).

V_{FE} proportional to the exciter field current is derived from the summation of signals from the exciter output voltage V_E, multiplied by $K_E + S_E[V_E]$ which represents saturation and I_{FD} multiplied by the demagnetization term K_D. The exciter field current signal V_{FE} is used as the input to the excitation system stabilizing block with output V_F.

The exciter time constant compensation consists of a direct feedback V_H, around the exciter field time constant, reducing its effective value, thereby increasing the small signal response bandwidth of the excitation system. The time constant is reduced proportional to the product of gains K_B and K_H of the compensation loop and is normally more than an order of magnitude lower than the time constant without compensation. To obtain a high response, a very high forcing voltage V_{RMAX} is applied to the exciter field. A limit-sensing exciter field current serves to allow high forcing but limits the current.

Other exciter models, their time constants, and their circuit diagrams are in Ref. 2. Manufacturers will usually include a control circuit diagram of their excitation system and AVR at the time of equipment purchase with all the relevant time constants and data applicable to the specific machine.

Example 13-2 We will continue with Example 12-5. Referring to Fig.12-27, generator $G2$ is unstable for a three-phase fault on Bus 2, fault duration 0.3 s. The generators were modeled without excitation systems. Now add a type AC2 exciter only on generator $G2$ which lost stability. The AC2 exciter parameters are the same as shown in Table 13-2.

Figure 13-23*a* and *b* depicts transients in generator $G2$ and also field circuits. $G2$ power angle swings to 152° in the first swing, but recovers in the subsequent swings which show stability. Also Bus 2 voltage, which had collapsed to zero during the fault, recovers on fault removal (Fig. 13-23*c*). These transients can be compared with those shown in Fig. 12-27.

Example 13-3 Figure 13-24 depicts a power system with load flow data and the system data is tabulated in Table 13-1. It shows the normal load flow and bus voltages with both generators in service.

A three-phase fault occurs on Bus 4. The critical fault clearance time is calculated with four excitation system models: AC1A, AC2A, DC3A, and ST1A, and the results are given in Table 13-2, which also shows typical data of the excitation system models.

The plots of transients are not included. The maximum difference in the stability limit is seen to be 0.029 s (1.74 cycles), exciter model AC2 giving maximum critical clearance time. The results of this example should not be generalized, much will depend upon the power system being studied and the nature of disturbance.

Example 13-4 This example is a study of transient stability of two small generators of 3.125 MVA and 5.0 MVA connected in a power system of Fig. 13-25.[5] The generators pull out of step for faults and voltage dips in the 44-kV weak utility system (three-phase short circuit at 44 kV = 3 kA rms symmetrical). The system

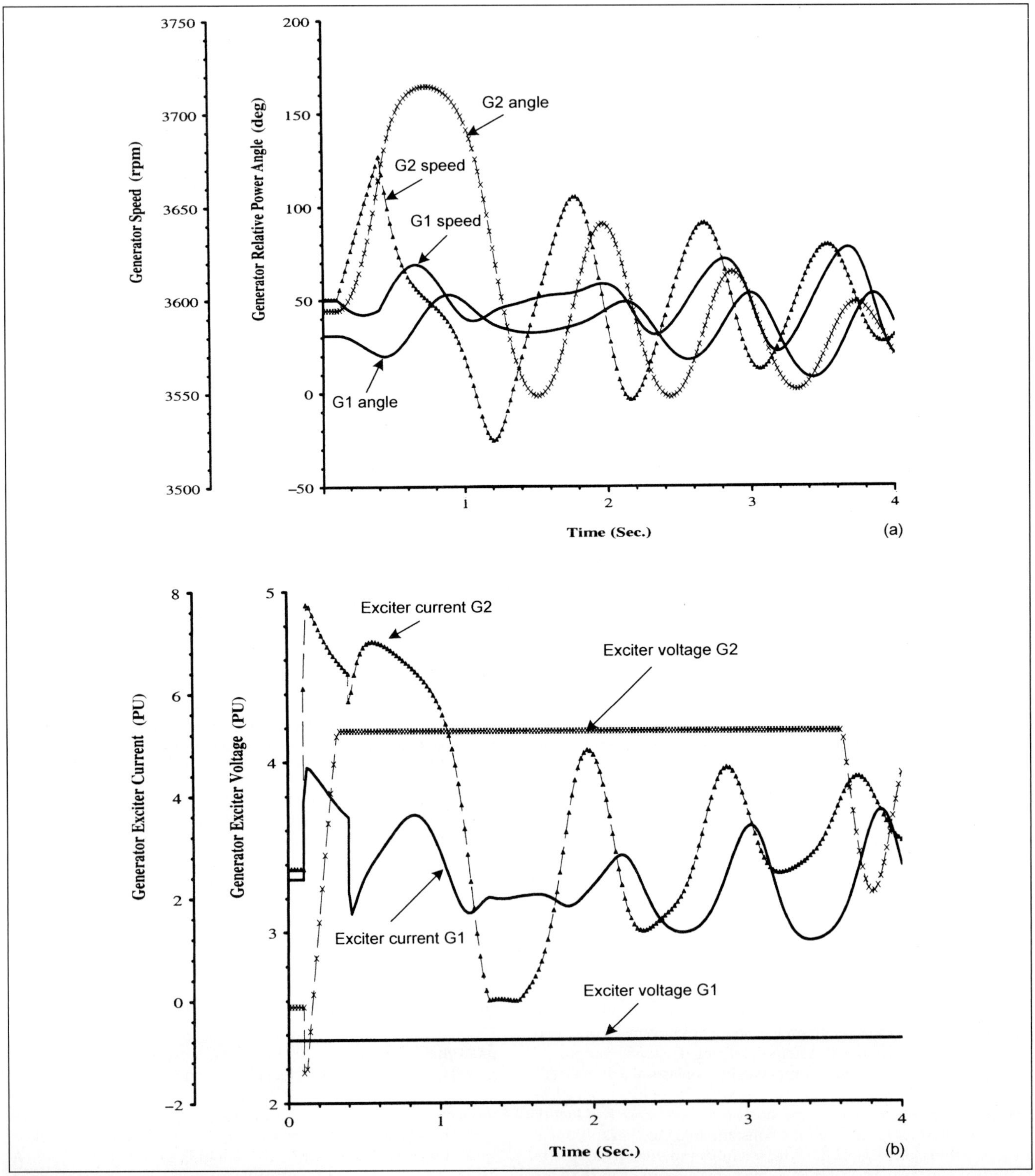

FIGURE 13-23 (*a*) *G*1 and *G*2 transients and field circuit transients, with an exciter type AC2A applied to *G*2. (*b*) Recovery transients of the bus voltage.

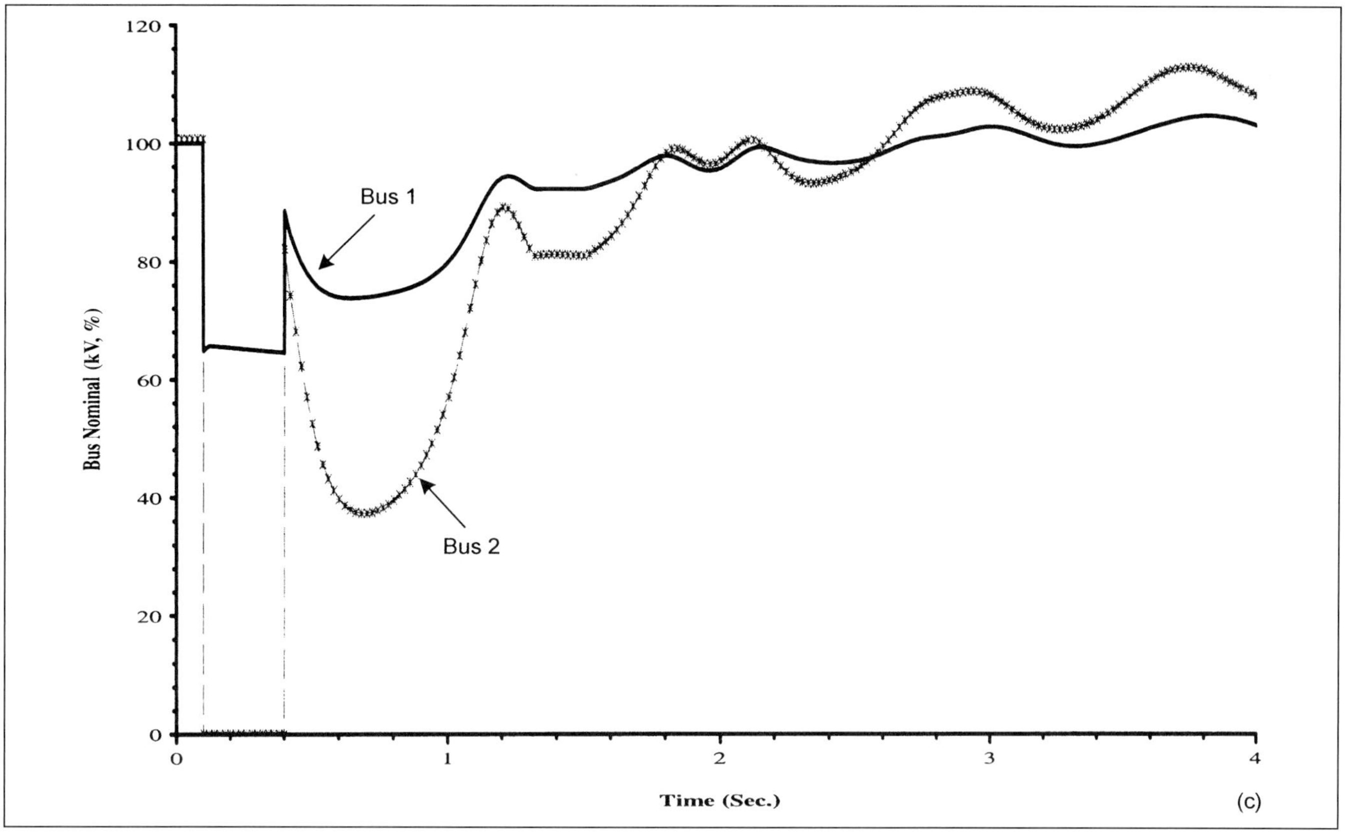

FIGURE 13-23 *(Continued)*

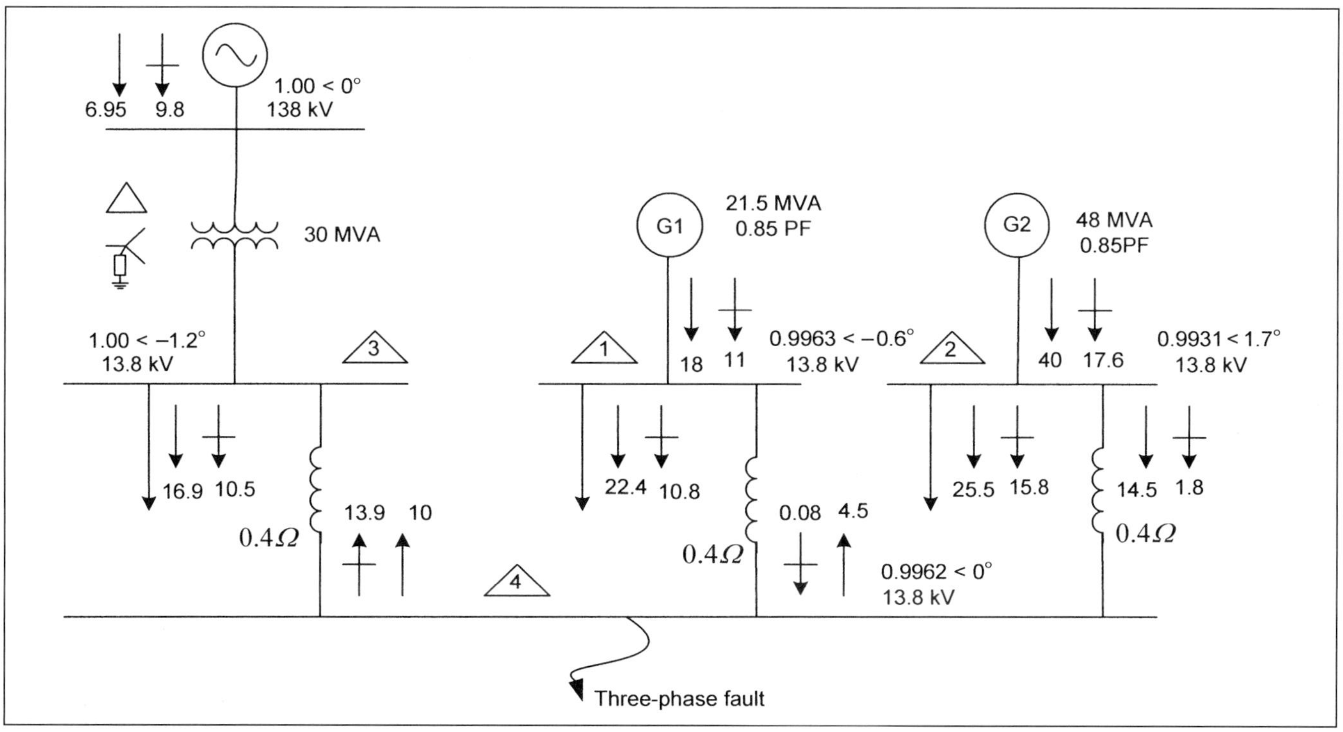

FIGURE 13-24 A power system configuration for study of critical clearing time of exciter types DC3A, ST1A, AC1A, and AC2A.

TABLE 13-1 System Data for Transient Stability Study

DESCRIPTION	DATA
Utility source, 138 kV	$Z_1 = Z_2 = 0.0995 + j1.989$, $Z_0 = 0.374 + j5.9885$, 100-MVA base
Transformer $T1$	30-MVA, 138-kV winding = delta-connected, 13.8-kV winding wye connected and resistance grounded through 400-A resistor, standard phase shift = –30°, Z = 10% on 30-MVA base, X/R = 23.7, set primary tap at –2.5 (134.55 kV)
Loads	Bus 1-25 MVA, 0.9 PF, Bus 2-20 MVA, 0.85 PF, Bus 3 = 30 MVA, 0.85 PF, constant Z type
Current-limiting reactors $R1$, $R2$, and $R3$	2000 A, 0.4 Ω, Q = 60
$G1$	21.5 MVA, 2 pole, 60 Hz, 0.85 PF, Model 3, $X''_d = 14.4$, $X''_q = 13.9$, $X_2 = 13.5$, $X_0 = 12$, $X_d = 227$, $X_q = 213$, $X_{dv} = 242$, $X_{qv} = 226$, $X'_d = 21.9$, $X'_q = 53.6$, $X_l = 15.7$ all on generator base MVA of 21.5 $T'_{do} = 4.8$ s, $T''_{do} = 0.023$ s, $T'_{qo} = 0.404$ s, $T''_{qo} = 0.055$ s, damping = 5% $H = 4.6$
$G2$	48 MVA, 2 pole, 60 Hz, 0.85 PF, Model 3, $X''_d = 14.4$, $X''_q = 13.9$, $X_2 = 13.5$, $X_0 = 12$, $X_d = 227$, $X_q = 213$, $X_{dv} = 242$, $X_{qv} = 226$, $X'_d = 21.9$, $X'_q = 53.6$, $X_l = 15.7$ all on generator base MVA of 21.5 $T'_{do} = 4.8$ s, $T''_{do} = 0.023$ s, $T'_{qo} = 0.404$ s, $T''_{qo} = 0.055$ s, damping = 5% $H = 4.7$

TABLE 13-2 Calculated Critical Clearing Time and Exciter Data

EXCITER TYPE	CRITICAL FAULT CLEARING TIME*	EXCITER PARAMETERS†
AC1A	0.232 s	$V_{RMAX} = 14.5$, $V_{RMIN} = -14.5$, $S_{EMAX} = 0.1$, $S_{E0.75} = 0.03$, $E_{FD} = 4.18$, $K_A = 400$, $K_C = 0.2$, $K_D = 0.38$, $K_E = 1$, $K_F = 0.03$, $T_A = 0.02$, $T_B = T_C = 0$, $T_E = 0.8$, $T_F = 1$, $T_R = 0$
AC2A	0.234 s	$V_{RMAX} = 105$, $V_{RMIN} = -95$, $S_{EMAX} = 0.04$, $S_{E0.75} = 0.01$, $E_{FD} = 4.4$, $V_{AMAX} = 8$, $V_{AMIN} = -8$, $K_A = 400$, $K_B = 25$, $K_C = 0.28$, $K_D = 0.35$, $K_E = -1$, $K_F = 0.03$, $K_H = 1$, $K_L = 1$, $T_A = 0.01$, $T_B = T_C = 0$, $T_E = 0.6$, $T_F = 1$, $T_R = 0$, $V_{OEL} = 11$ (S_{EMAX} and $S_{E0.75}$ = the value of excitation fuction at E_{FDMAX} and $E_{FD0.75}$ to simulate saturation)
DC3A	0.220 s	$V_{RMAX} = 1$, $V_{RMIN} = 0$, $S_{EMAX} = 0.28$, $S_{E0.75} = 0.07$, $E_{FD} = 3.03$, $K_E = 0.05$, $K_V = 0.05$, $T_E = 0.5$, $T_R = 0$, $T_{RH} = 20$
ST1A	0.205 s	$V_{RMAX} = 4.6$, $V_{RMIN} = 0$, $V_{MAX} = 2.4$, $V_{MIN} = -2.4$, $K_A = 52$, $K_C = 0.05$, $K_F = 0.114$, $T_A = 0.01$, $T_B = 0.92$, $T_{C1} = 0$, $T_{B1} = 0$ $T_C = 0$, $T_F = 0$, $T_R = 0$

*Three-phase fault on Bus 4 (Fig. 13-24).
†Control circuit diagrams: AC2A, Fig. 13-22; ST1A, Fig. 13-20. For control circuit diagrams of AC1A and DC3A, see Ref. 2.

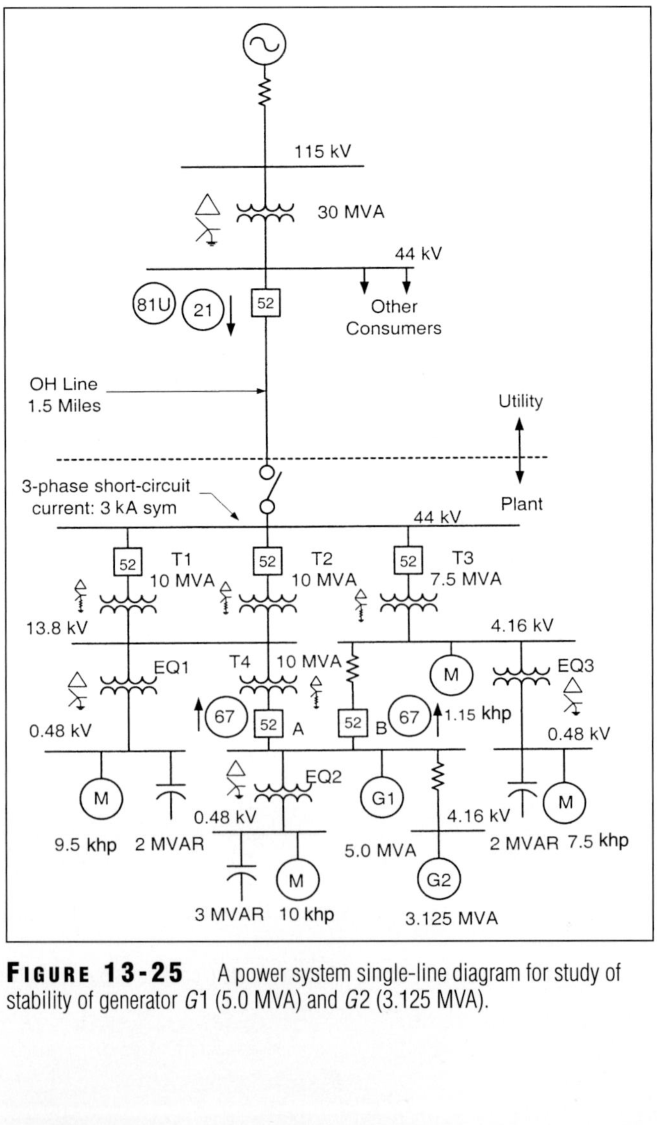

FIGURE 13-25 A power system single-line diagram for study of stability of generator $G1$ (5.0 MVA) and $G2$ (3.125 MVA).

data, generator, turbine, and exciter models are not shown. The following relaying data is applicable:

- Eighty percent of 44-kV line is protected, in first zone of distance protection, where utility breaker interrupting time is three cycles, relaying time is one cycle, and fault is cleared in four cycles.

- The second zone of distance protection extends to remaining 20 percent of line; partly looking into primary windings of transformers *T*1, *T*2, and *T*3, fault clearance time is 18 cycles, including interrupting time of breaker, second zone distance timer, and relay operating time.

- Breakers A and B of 4.16 kV have reverse current protection (device 67) to isolate generators from the utility source when reverse flow of current is fed into the utility system on a fault—these are proven to be too slow to affect separation before pullout conditions.

The analysis shows that stability cannot be achieved by any means for three-phase faults in the utility—fast load shedding (practical limit is set by breaker interrupting time, fault sensing, and some operational delay, say minimum of five cycles), replacement with modern fast excitation systems, and fast intertripping in about four cycles, that is, simultaneous fault clearance. After three-phase faults, in order of decreasing severity, the fault types are: (1) double line-to-ground fault, (2) phase-to-phase fault, (3) single line-to-ground fault.

Figure 13-26*a* and *b* depict the behavior of voltage and frequency, respectively, for a phase-to-phase fault on the generator bus. These transients are plotted on the basis that phase-to-phase fault is cleared in the first zone and second zone of distance protection. Generation is separated in 18 cycles maximum time from the instance of fault occurrence, and simultaneously 15.65 khp of loads are shed, leaving 5 khp essential loads (Fig. 13-25).

Note that for a fault cleared in 18 cycles in the second zone of distance protection, the frequency does not dip below 59.40 Hz, while for a fault cleared in 4 cycles it dips to 55.75 Hz. The voltage behaves opposite to frequency; it dips by 34 percent for the fault cleared in

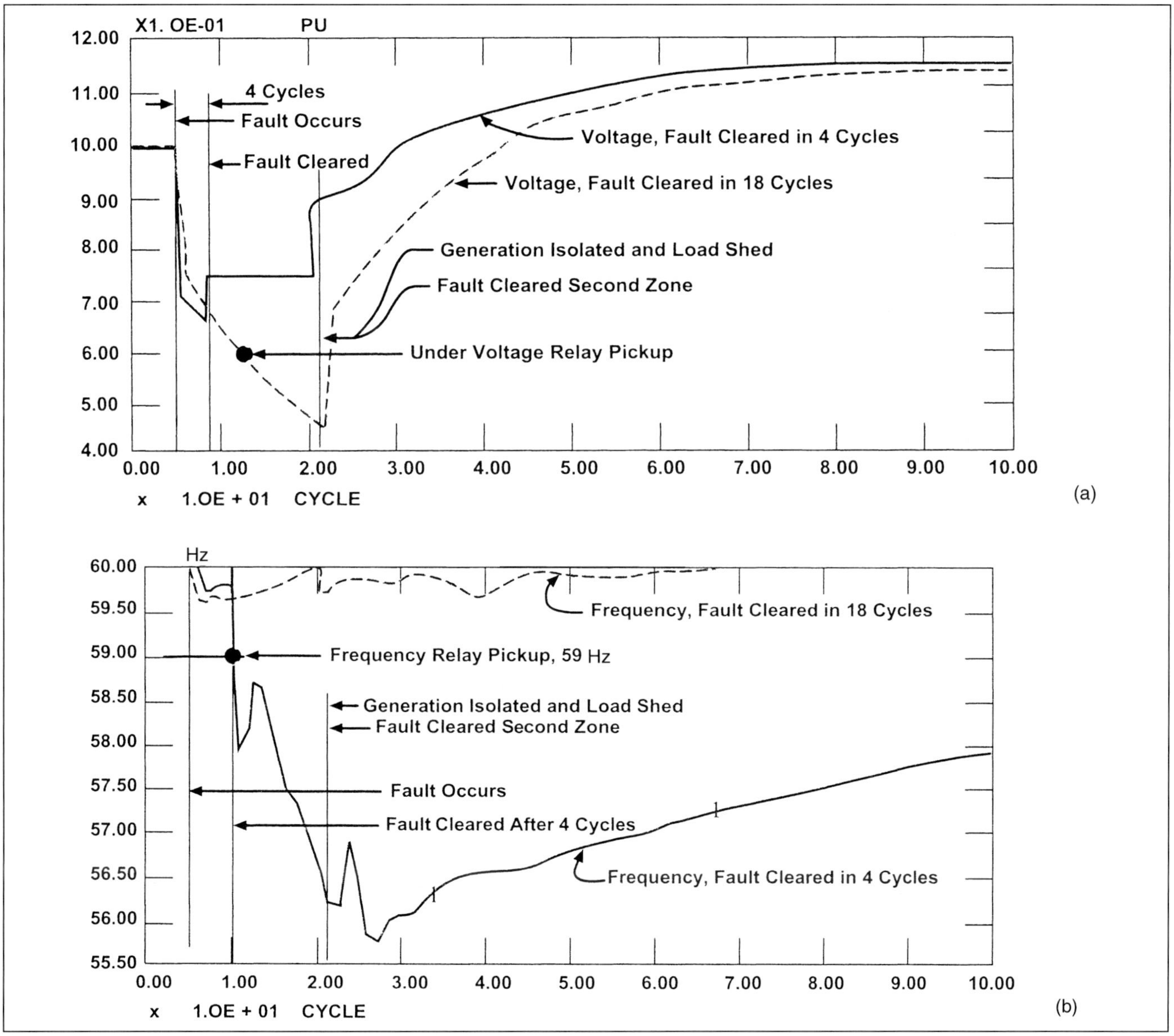

FIGURE 13-26 (*a*) Voltage and (*b*) frequency transients.

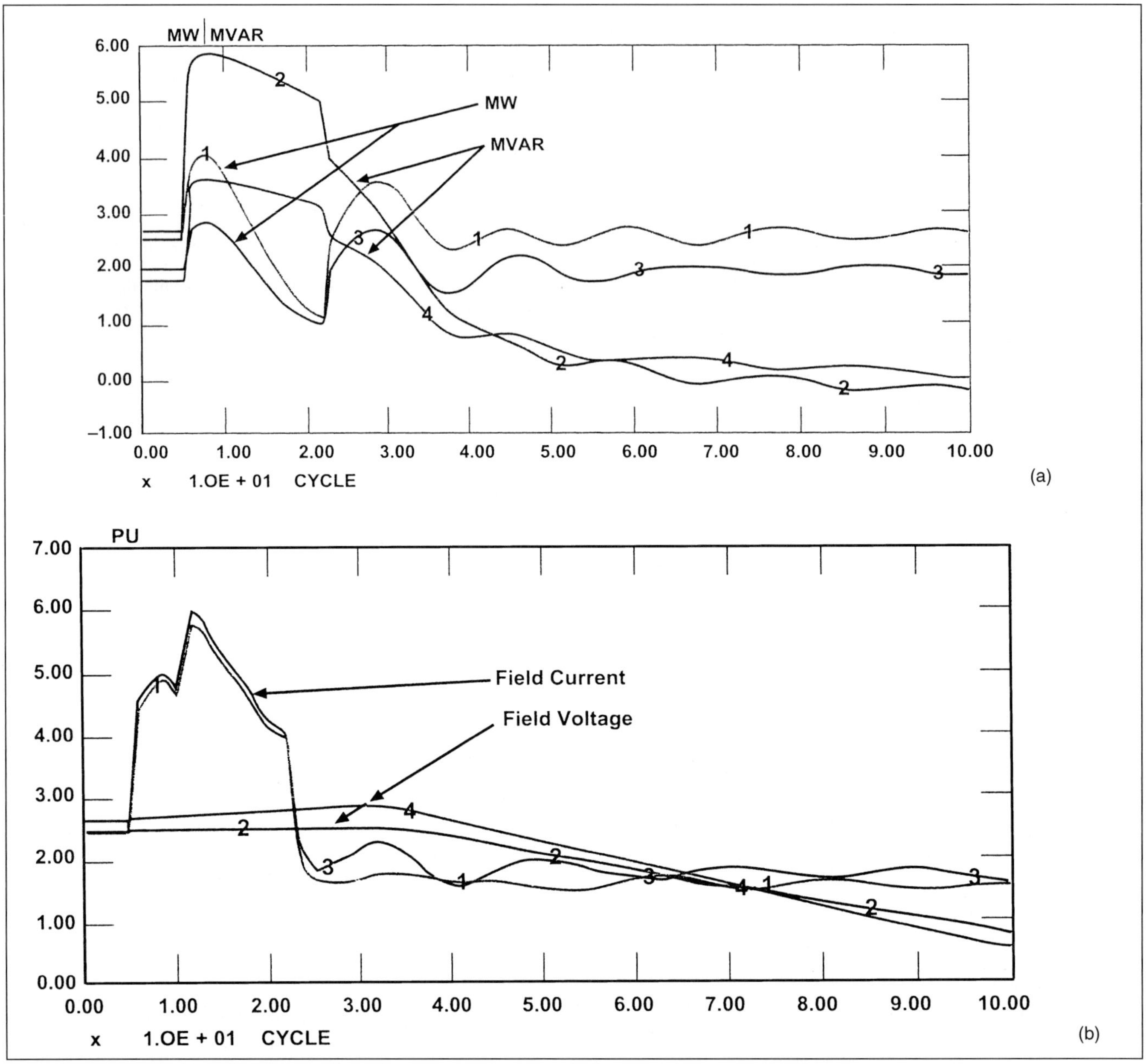

Figure 13-27 (*a*) Active and reactive power and (*b*) field current and voltage transients.

four cycles and dips by 55 percent for a fault cleared in 18 cycles. This provides the appropriate settings on voltage and frequency relays:

- An underfrequency relay is set at 59 Hz and time delay of two cycles.
- An instantaneous three-phase undervoltage relay is set at 0.6-pu voltage in conjunction with timer with a delay of four cycles.

These time delays are purposeful, in order to ride through fast system transients. Figure 13-27*a* and *b* illustrate active and reactive power transients and field current and voltage transients, respectively. Fig. 13-28*a* and *b* show transients in the equivalent 5-khp motor load that is retained in service.

This example illustrates that it is not always practical to design for and achieve stability for the worst fault type. The underfrequency and undervoltage load shedding schemes should be considered in combination. Some other strategies of enhancement of stability, such as speed of relaying or intertripping, may not always be feasible.

13-9 POWER SYSTEM STABILIZERS

PSS acts through the automatic voltage regulator of an excitation system and provides positive damping to generator rotor angle swings. These swings are in broad range of frequencies:

- Intertie mode. 0.1 to 1.0 Hz
- Local mode. 1 to 2 Hz
- Interplant mode. 2 to 3 Hz

Typically, a PSS tuning study is performed to optimize damping for a broad range of frequencies.

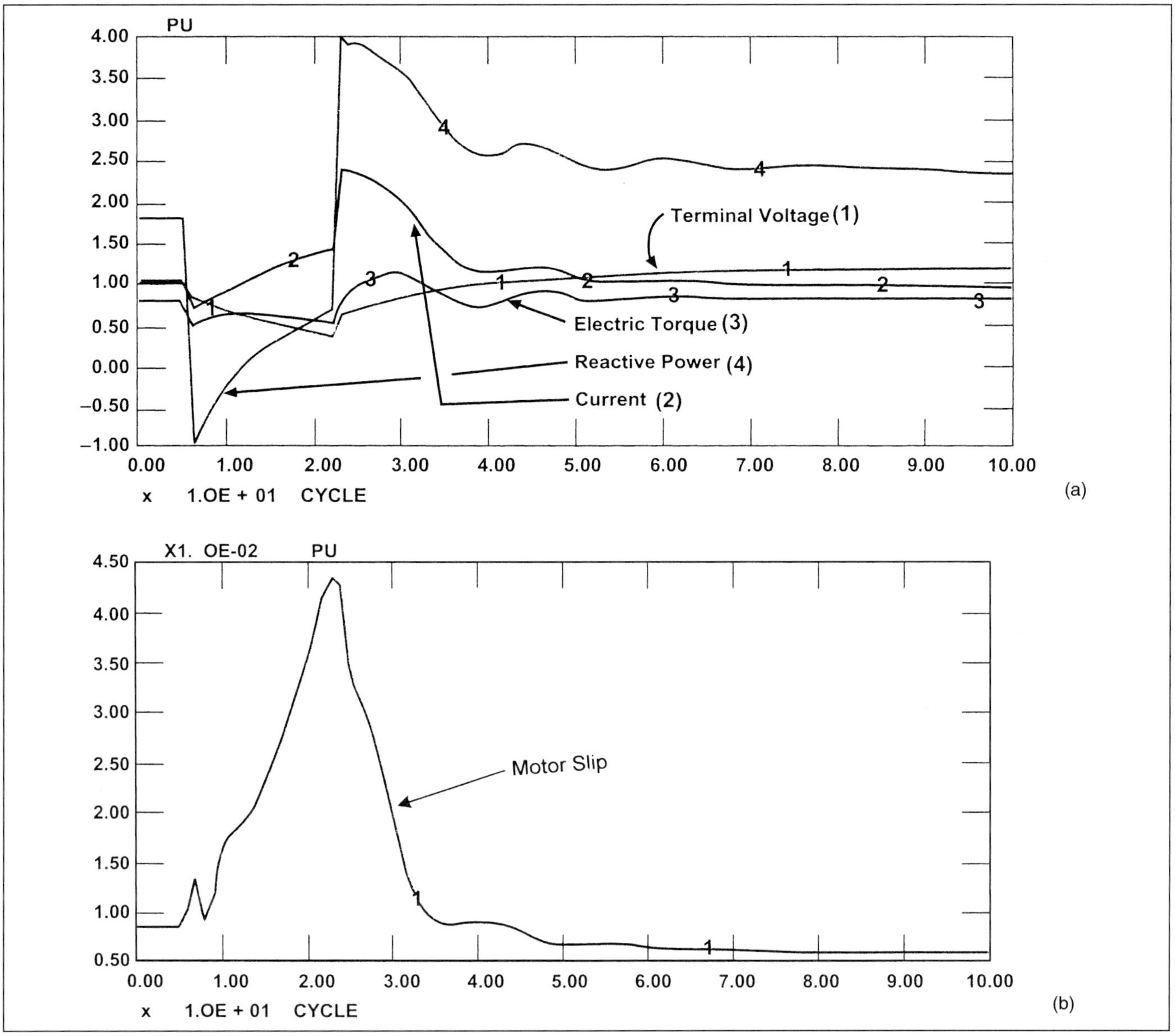

Figure 13-28 Transients in the 5-khp motor load retained in service.

Historically, the measurements of rotor speed, bus frequency, and electrical power signals have been used as input signals. Some considerations are gain, phase compensation, susceptibility to other interactions, and noise in the transducers. Consider a generator under so-called PV control. The voltage regulator adjusts the generator excitation in response to the terminal voltage. A decrease in generator terminal voltage results in a boost of the excitation and more reactive power is produced in the system to control the voltage, and, conversely, for a rise in the voltage, a generator may act even as a reactor to absorb reactive power from the system. While this is the most common mode of operation, the generators are sometimes operated under constant reactive power mode also.

The reactance of generator field windings delays rapid build up of generator terminal voltage. The swing equation can be modified to show changes in the machine flux $\Delta E'_q$:

$$\frac{2H}{\omega}\frac{d^2\delta}{dt^2}+\frac{K_D}{\omega}\frac{d\delta}{dt}+K_1\Delta\delta+K_2\Delta E'_q=0 \tag{13-34}$$

The term $K_2\Delta E'_q$ is mainly determined by the changes in the excitation level. Consider the phasor diagram of Fig. 13-29. The vertical axis represents the synchronizing axis. The components of restoring power along this axis, in the positive direction, tend to increase the frequency of dynamic oscillations. The x axis is the damping axis. Components of restoring power which are in phase with the machine speed or frequency provide damping to these oscillations. Components which are out of phase with machine speed and frequency and lie in the first quadrant tend to cancel the natural damping. In Fig. 13-29, the dynamic oscillations will be damped, as the resultant forcing action for oscillations points to left of vertical axis. The angle $\Delta\delta$ lags the speed deviation by $\Delta w = 90°$ and the terminal voltage deviation ΔV by 180°. Figure 13-29 is for excitation system under manual control.

Under automatic voltage regulator control, the generator terminal voltage deviation ΔV creates the error signal Δe, which initiates the control action to correct the terminal voltage with certain time constants. If the machine is connected to a weak transmission system, the phase lags between the error signal and

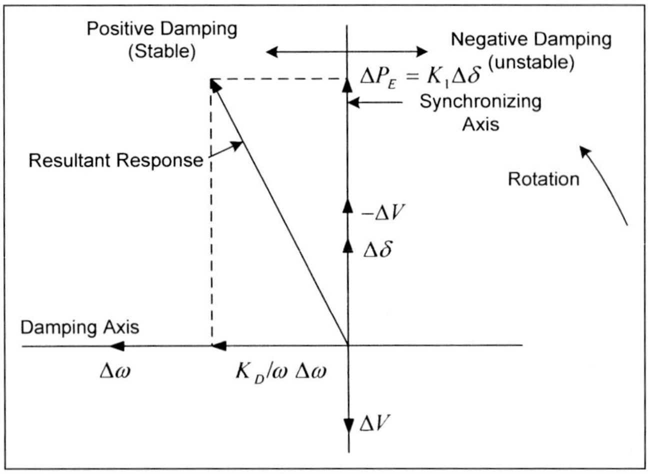

FIGURE 13-29 Phasor relationships of signals and torque components, fixed excitation.

generator flux may result in a forcing action which points in the negative region, giving rise to undamped oscillations; these may grow in magnitude.

A fast response excitation system and a high effective gain will act to magnify the negative damping contribution. This results in oscillatory instability. When the power angle is above 70°, between the generator internal voltage and infinite bus voltage, the unit tends to have local mode stability problems under voltage control.

In Chap. 10, we have derived a generator model on infinite bus with constants *K*1, *K*2, *K*3, *K*4, *K*5, and *K*6. See Fig. 10-22. We showed that:

$$\Delta T_e = K_1\Delta\delta + K_2\Delta E'_q$$
$$\Delta E'_q = \frac{K_3}{1+sT_3}(\Delta E_{FD} - K_4\Delta\delta) \tag{13-35}$$

where T_3 is written in place of $T'_{do}k_3$. Combining these two equations:

$$\Delta T_e = K_1\Delta\delta + \frac{K_3K_2}{1+sT_3}\Delta E_{FD} - \frac{K_2K_3K_4}{1+sT_3}\Delta\delta \tag{13-36}$$

The last term is variation in E'_q, transients caused by electromagnetic torque due to power angle variation. At steady state, or low oscillating frequency, and for $\omega < \frac{1}{T_3}$:

$$\Delta T_e \approx K_1\Delta\delta - K_2K_3K_4\Delta\delta \tag{13-37}$$

Thus, the field flux variations produce a negative synchronizing torque component if:

$$K_1 \le K_2K_3K_4 \tag{13-38}$$

The system becomes monotonically unstable.

13-9-1 PSS and Oscillatory Stability

It seems pertinent, therefore, that a supplementary signal can be provided which can increase damping by sensing some measurable quantity. Not only the undamping effect of voltage regulator is cancelled, but the steady-state stability can also be enhanced. The supplementary signals are changes in the shaft speed, frequency, and electrical power: ($\Delta\omega$, Δf, ΔP_e).

Consider the Eq. (13-35); at higher oscillating frequencies, the last term becomes:

$$\Delta T_e = -\frac{K_2K_3K_4}{1+sT_3}\Delta\delta = \frac{K_2K_3K_4}{\omega T_3}(j\Delta\delta) \tag{13-39}$$

The air-gap torque deviations caused by E'_q lead power angle deviation by 90° and thus are in phase with speed variations, $\Delta\omega_r$. The field winding flux linkage variation produces a positive damping torque component.

A block circuit diagram of PSS is shown in Fig. 13-30. The transducer converts the sensing signal into a voltage signal. This output is phase shifted by an adjustable lead-lag network, which compensates the time delays in generator field and excitation system. The resulting signal is amplified and sent through a washout module which continuously balances the PSS output and prevents it from biasing the generator voltage for prolonged frequency or power excursions. The signal limiter causes the output signal to limit on-load rejection and retain beneficial effect of forcing during disturbances.

Figure 13-31 shows a typical phase relationship. The change $\Delta\omega$ is phase shifted by ϕ_1 in the lead-lag network. This is the PSS output signal which is applied to the voltage regulator. The time delays in the generator and excitation result in a forcing function which has a large damping component. By increasing PSS amplifier gain, this forcing action can be increased subject to the limiting action of the signal limiter.

13-9-2 IEEE-Type PSS1A Model

This model is shown in Fig. 13-32 and has a single input (V_{S1}), power frequency or speed. T_6 may be used to represent a transducer time constant; stabilizer gain is set by K_S, and signal washout by T_5. Next, A_1 and A_2 allow some of the low-frequency effects of high-frequency torsional filters, used in some stabilizers, to be accounted for. This block can also be used for shaping the gain and characteristics of the stabilizer. The next two blocks allow for

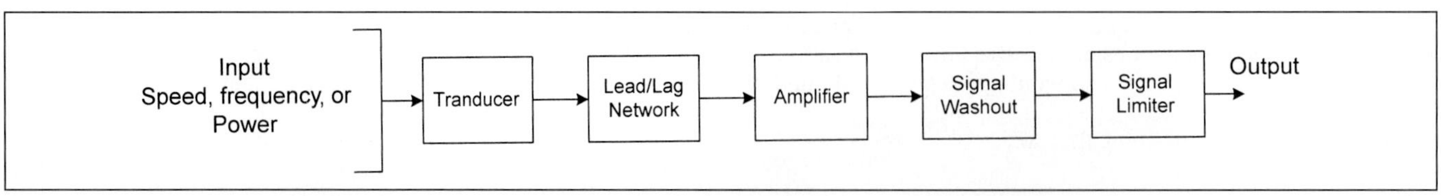

FIGURE 13-30 Major elements of a power system stabilizer.

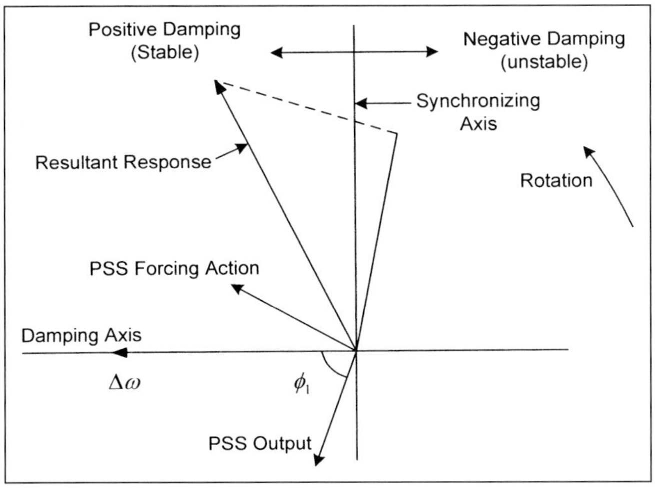

FIGURE 13-31 Phasor relationships of signals, generator under-voltage regulator control with PSS.

lead-lag compensation, as set by T_1 to T_4. Stabilizer output can be limited in many ways, and the model shows simple output limits: V_{STMAX} and V_{STMIN}.

13-9-3 IEEE-Type PSS2A Model

The block diagram is shown in Fig. 13-33. The design represents stabilizers with dual inputs, and in particular the model can be used for:

1. Stabilizers, which in the frequency range of system oscillations, act as electrical power input stabilizers. These use the speed or frequency input for generation of an equivalent mechanical signal, to make the total signal insensitive to mechanical power change.

2. Stabilizers that use a combination of speed (or frequency) and electrical power. These systems usually use the speed directly, without phase-lead compensation and add a signal proportional to the electrical power.

The parameters used for these two types vary considerably. For each input, two washouts can be represented: T_{W1} and T_{W4}, along with transducer or integrator time constant: T_6, T_7. For the first type of stabilizer, K_{S3} will normally = 1, and K_2 will be $T_7/2H$, where H is the inertia constant. Input V_{SI1} would normally represent speed or frequency, and V_{SI2} would be a power signal. The indices N (an integer up to 4) and M (an integer up to 2) allow a ramp-tracking or simpler filter characteristics to be represented. Phase compensation is provided by two lag-lead or lead-lag blocks T_1 and T_4.

13-10 TUNING A PSS

This has been a subject of much research over the past 20 years.[6,7–10] The following is a brief description:

Phase compensation method. The stabilizer is adjusted for phase lags through the generator and excitation system, so that the torque changes are in phase with the speed changes. This is the simplest approach.

Root locus method. This involves shifting the eigenvalues associated with the power system mode of oscillation by adjusting the poles and zeros in the *s* plane. The approach works directly with the closed-loop characteristics of the system as opposed to open-loop nature of the phase compensation method.

State-space approach. It is a powerful mathematical tool. A gain matrix of the multiple variable control system can be calculated and an objective function, such as integral of the squared error over time, can be minimized.

Basic concepts. To provide damping, the PSS must produce a component of electrical torque on the rotor which is in phase with speed variations. In other words, the transfer function of PSS must compensate for the gain and phase characteristics of the exciter, generator, and the power system.

Figure 13-34, which shows a linearized small signal model of a single machine, is from a paper by Concordia and deMello.[11] (This is a slight modification of Fig. 10-22). This figure shows an

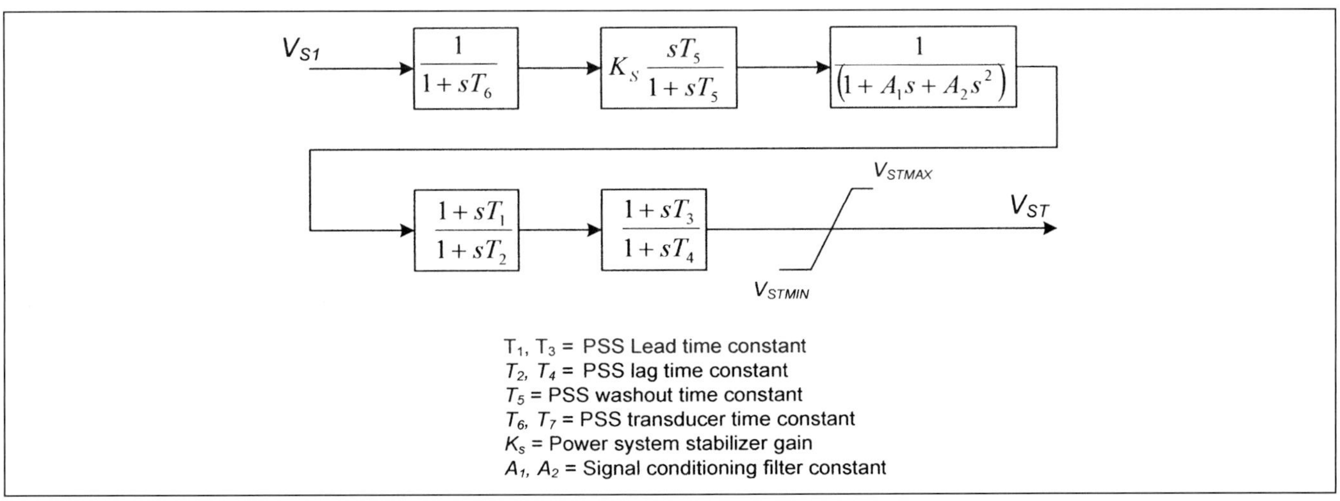

FIGURE 13-32 Single-input power system stabilizer PSS1A.

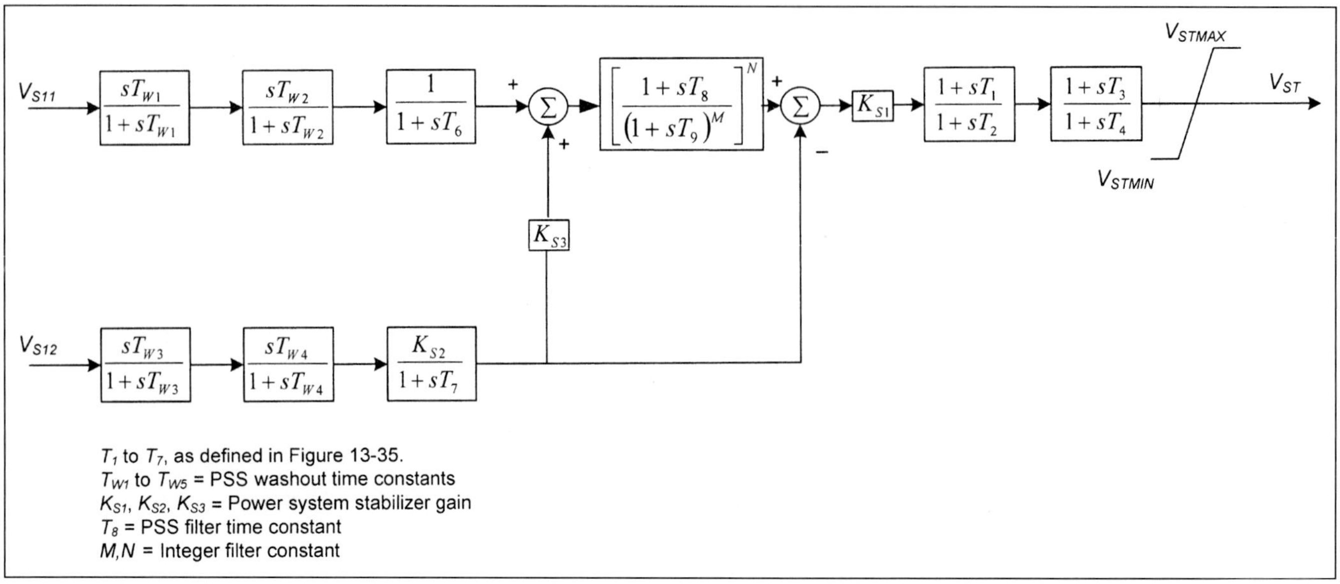

FIGURE 13-33 Dual-input power system stabilizer PSS2A.

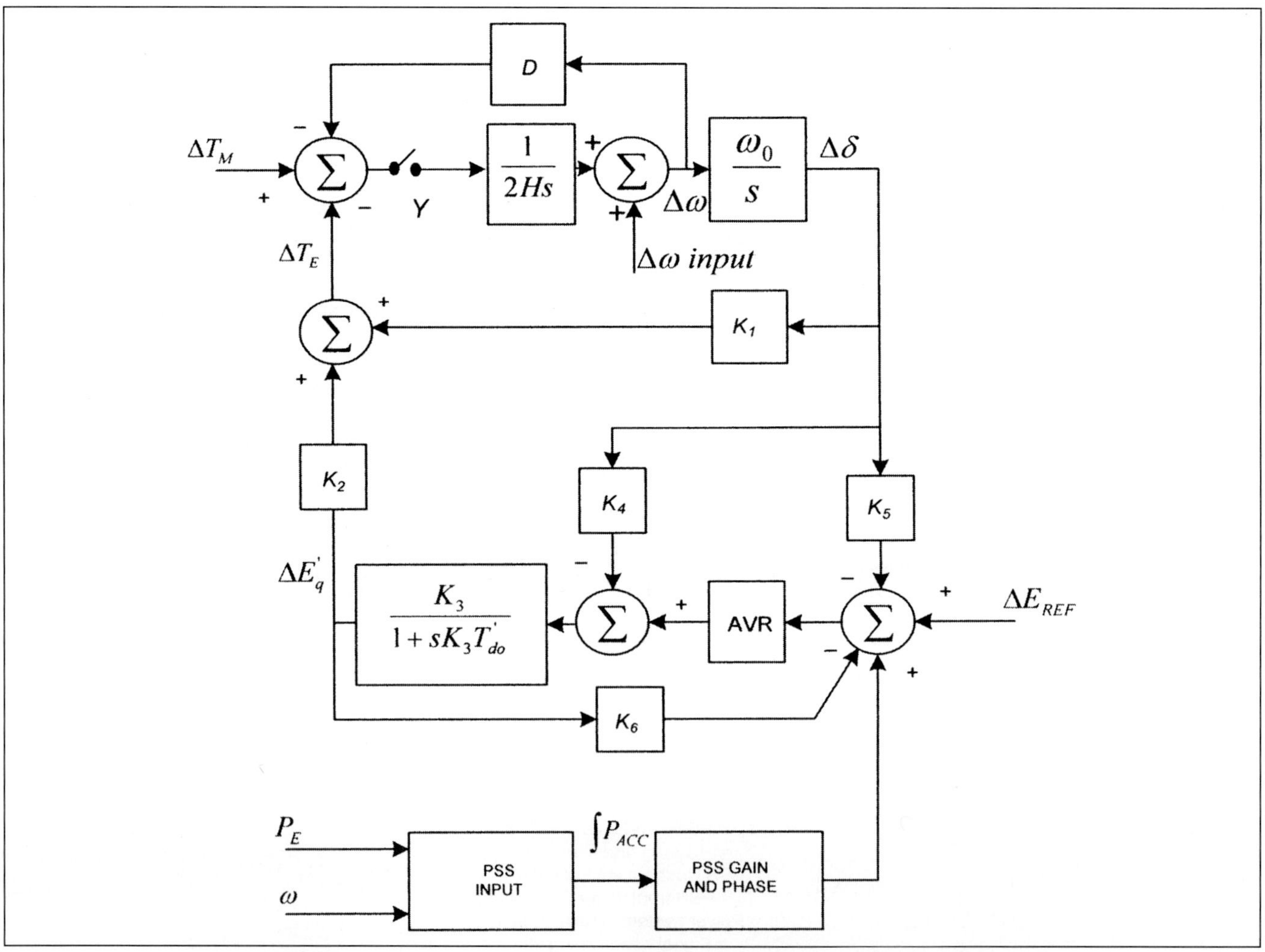

FIGURE 13-34 Small signal model of a generator with PSS and AVR inputs.

integral of accelerating power type PSS, which is synthesized from measurements of speed and electrical power:

$$\frac{d\omega}{dt}=\frac{1}{2H}(T_M-T_E)=\frac{1}{2H}(T_{ACC}) \tag{13-40}$$

We can divide the control circuit into two parts. The upper part shows the relation between the accelerating torque and speed and angle deviations. D is the damping associated with dampers, amortisseurs, and system loads. K_1 is inherent synchronizing coefficient. The lower part of the figure is a representation of the generator flux dynamics, T'_{do}, K_3, AVR, and PSS. Ignoring PSS for a moment, the lower part involving flux dynamics and AVR forms a loop, which can be thought to be in parallel with K_1 and D. This loop has a component which is in phase with damping D and also K_1 and it adds or subtracts from these terms. An uncompensated system can be considered with switch Y open, and for changes in speed, $\Delta\omega$, the changes in electrical torque are considered. If changes in torque are in phase with speed, we have a pure damping term D. In practice, a detailed model for transient stability study is used to calculate frequency response.

Figure 13-35 shows the frequency and gain plots of the uncompensated system, shown in solid lines. This can be divided into two components. The first term is from AVR input to electrical torque, $\Delta T_E/\Delta E_{REF}$, and the second part is transfer function from speed to PSS input signal, $\Delta\int P_{ACC}/\Delta\omega$. We see that the major component in the phase lag is the AVR loop. The PSS input does not alter the phase lag curve significantly.

The objective is choosing the phase compensation and lead-lag functions so that the total compensated phase between speed and torque is zero or slightly lagging. We concentrate on frequency range of interest where power system oscillations exist, that is, are between 0.1 and 3.0 Hz. This is shown in Fig. 13-36.

The PSS gain settings can be done using root-locus method. The closed-loop poles will migrate from open-loop system poles to open-loop system zeros as the gain is increased from zero to infinity (Chap. 3). This means that instability will develop at some value of the gain. Thus, an optimum condition can be defined as the one at which the damping is the maximum. A sequence of eigenvalues, roots of the closed-loop system including PSS, are plotted on s plane. The root-loci vary with:

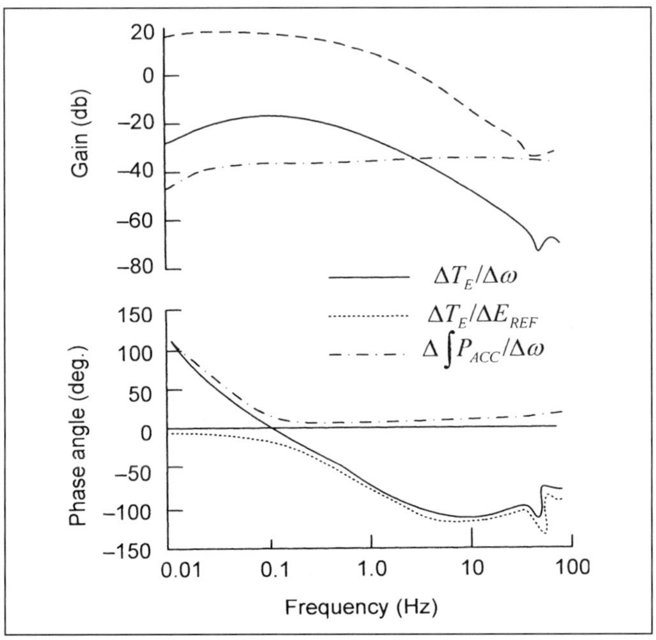

FIGURE 13-35 Bode plot of tuning of a PSS for small angle stability and phase and gain plots of an uncompensated system.

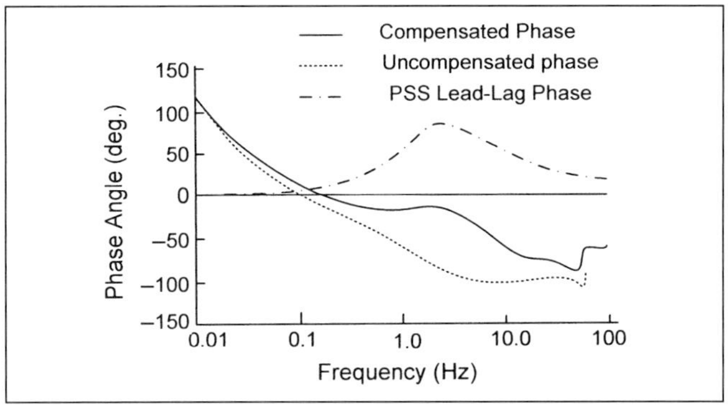

FIGURE 13-36 Phase angle plot of a compensated system.

- *External reactance.* The local mode frequency is higher for the strong system and lower for the weak system.
- *AVR settings K_A and K_F.* For large values of regulator K_A, the eigenvalues will be toward the right of the j-ω axis and will move toward the left as feedback gain is increased.
- *Excite mode.* As the local mode damping is increased, sometimes another mode called excite mode is brought into play, which decreases the damping. If PSS gain is increased, and it crosses over to the right half plane, instability will occur.
- *Torsional oscillations.* The high frequency gain can affect the turbine-generator torsional oscillations.

Example 13-5 The example considers the application of a PSS to a small signal oscillation problem. Consider a generator of 235.3 MVA, 18 kV, with data as shown in Table 13-3. The generator is connected through a step-up transformer of 200 MVA, 18 to 230 kV, and system configuration as shown in Fig. 13-37. It supplies a load of 100 MVA at 0.85 power factor from 230-kV Bus 1, and the rest of the load is supplied into the utility system through a system impedance Xe = 40 percent. Bus 2 of 230 kV is created halfway

TABLE 13-3 System Data for Example 13-5

DESCRIPTION	DATA
*G*1	235.3 MVA, 2 pole, 18 kV, 60 Hz, 0.85 PF, HR grounded through a distribution transformer, ground fault current limited to 10 A. $X''_d=15$, $X_2=12$, $X_0=12$, $r_a=0.15$, $r_2=r_0=0.12$, $X_d=211$, $X_q=201$, $X'_d=23$, $X'_q=46$, $X_l=13$, $X''_q=12$, $T'_{do}=6.8$, $T''_{do}=0.04$ s, $T''_{qo}=0.08$, $T'_{qo}=0.59$, $H=5$, damping = 5%. All reactances on generator MVA base.
PSS1A	V_{SI} = speed, $K_S=3.15$, $V_{STMAX}=0.9$, $V_{STMin}=-0.9$, $A_1=A_2=0$, $T_1=0.76$, $T_2=0.1$, $T_3=0.76$, $T_4=0.1$, $T_5=1$, $T_6=0.1$
Exciter	AC2A, data as in Table 13-2

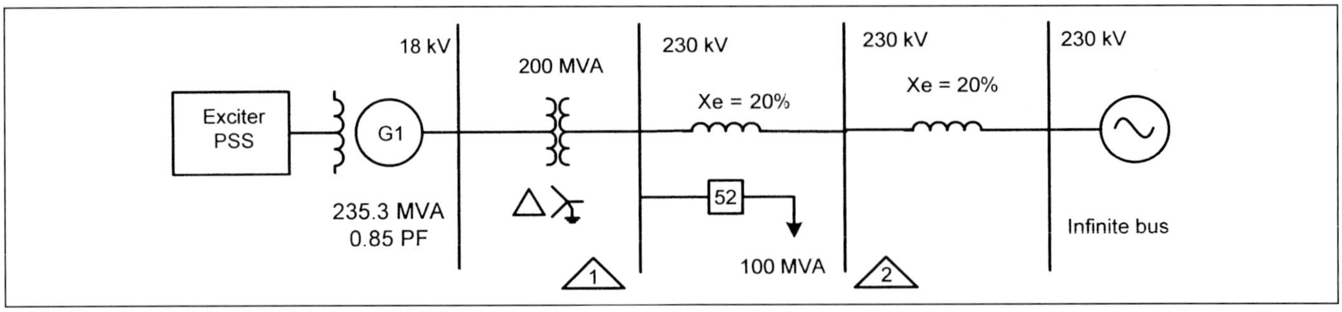

FIGURE 13-37 A 235.3-MVA generator connected to an infinite bus through a reactance and supplying load.

between the supply system impedance of 40 percent to plot the nature of transients.

Large utility generators are connected through a step-up transformer, and the system impedance X_e partly represents the impedance of the step-up transformer and partly the utility high-voltage system source impedance.

First, consider that the generator is provided with IEEE exciter type AC2A, and no PSS. Breaker 52 in Fig. 13-34 is suddenly opened to remove the load. The resulting transients are shown in Fig. 13-38*a* and *b*. The terminal current, generator-relative power angle, exciter current, and Bus 1 and Bus 2 voltages decay slowly. For example, even after 10 s, the current has decayed to only 13 percent of its original value and so also the other transients.

Next, apply PSS parameters as shown in Table 13-3. This is IEEE-type PSS1A, control circuit diagram as shown in Fig. 13-32. The transients are replotted with PSS in service (Fig. 13-39*a* and *b*), and this shows that transients practically vanish within 3 s.

13-11 MODELS OF PRIME MOVERS

In the above discussions, we referred to models of prime movers and speed governors. Example 12-7 uses a simple steam turbine model. A steam turbine during start-up can initially pick up about 10 percent of its rated output quickly, and then the load can only be increased slowly, approximately at the rate of 2 to 3 percent per minute. The thermal shock to the turbine is the limiting factor. Same is true about a boiler. An increase in steam flow demanded by the turbine results in pressure drop, and an increase in fuel to the boiler is required to restore pressure. The speed-governing systems have a response time delay of 3 to 5 s. The response of hydraulic turbines will be even slower.

Therefore, the models of prime movers and speed-governing systems may not be required for large rotor angle study, period of study couple of seconds, but these are required for small signal stability and long-term dynamic studies for operational procedures. In some studies which involve the loss of generation or load, the response of prime movers has been demonstrated to have an important effect on tie-line loading.

The prime mover models,[12–16] depending upon the prime mover (steam, gas, or hydraulic turbines or diesel engines), are even more varied than the excitation system models. Manufacturers put forward control circuit models specific to their product. The models vary, for example, single reheat or double reheat, simple or compound steam turbine, number of turbine sections in tandem, that is, high pressure (HP), intermediate pressure (IP), and low pressure (LP), time constants of crossover piping, and inlet volumes. In addition, models of surge tanks and boilers are also incorporated in the studies. Figure 13-40 shows a compound double reheat steam turbine model, with some typical time constants.

13-12 AUTOMATIC GENERATION CONTROL

The objective of automatic generation control (AGC) is to regulate frequency to the nominal value and to maintain interchange of power between the control areas by adjusting the output of the selected generators. This control is commonly referred to as load frequency control (LFC). Associated with this is the unit commitment and economical dispatch of power, discussed in many power system texts.[16–18]

The frequency in the power systems is closely regulated and does not vary much. Based solely on the frequency alone, a utility company may island a cogeneration facility if the frequency falls, say, below 59.5 Hz for about 12 cycles. This varies, and the numbers are provided only as a typical example.

The composite power frequency characteristics of a power system depend upon the combined effects of the droop characteristics of speed governors (which allow the units to share the load demand). The steady-state frequency deviation following a load change may be written as:

$$\Delta f_{ss} = \frac{-\Delta P_L}{(1/R_1 + 1/R_2) + (D_1 + D_2)} = \frac{-\Delta P_L}{1/R_{eq} + D} \tag{13-41}$$

where D is the load damping, and R_{eq} is defined as the ratio (in percent) of the percentage speed or frequency change and percent power output change, that is, $R = \Delta f/\Delta P$. The characteristics of a composite load may be defined as:

$$\Delta P_e = \Delta P_L + D\Delta\omega_r \tag{13-42}$$

where ΔP_L is the nonfrequency-sensitive load change, $D\Delta\omega_r$ is the frequency-sensitive load change, and D is the load damping factor.

From Eq. (13-41), the frequency response characteristics are:

$$\beta = \frac{-\Delta P_L}{\Delta f_{ss}} = \frac{1}{R_{eq}} + D \tag{13-43}$$

β is expressed in megawatts per hertz and is a measure of the stiffness of the system. Consider two interconnected areas. The change of load in Area 1, results in frequency reduction in both the areas and a tie line flow of power between Areas 1 and 2. This flow is due to the regulation characteristics of the areas. For change in Area 1 load, we write:

$$\Delta P_{12} = \frac{\Delta P_{L1}\beta_2}{\beta_1 + \beta_2} \qquad \Delta f = \frac{-\Delta P_{L1}}{\beta_1 + \beta_2} \tag{13-44}$$

Similarly, a change in Area 2 is negative of the change in load in Area 1:

$$\Delta P_{12} = -\Delta P_{21} = \frac{\Delta P_{L2}\beta_2}{\beta_1 + \beta_2} \qquad \Delta f = \frac{-\Delta P_{L2}}{\beta_1 + \beta_2} \tag{13-45}$$

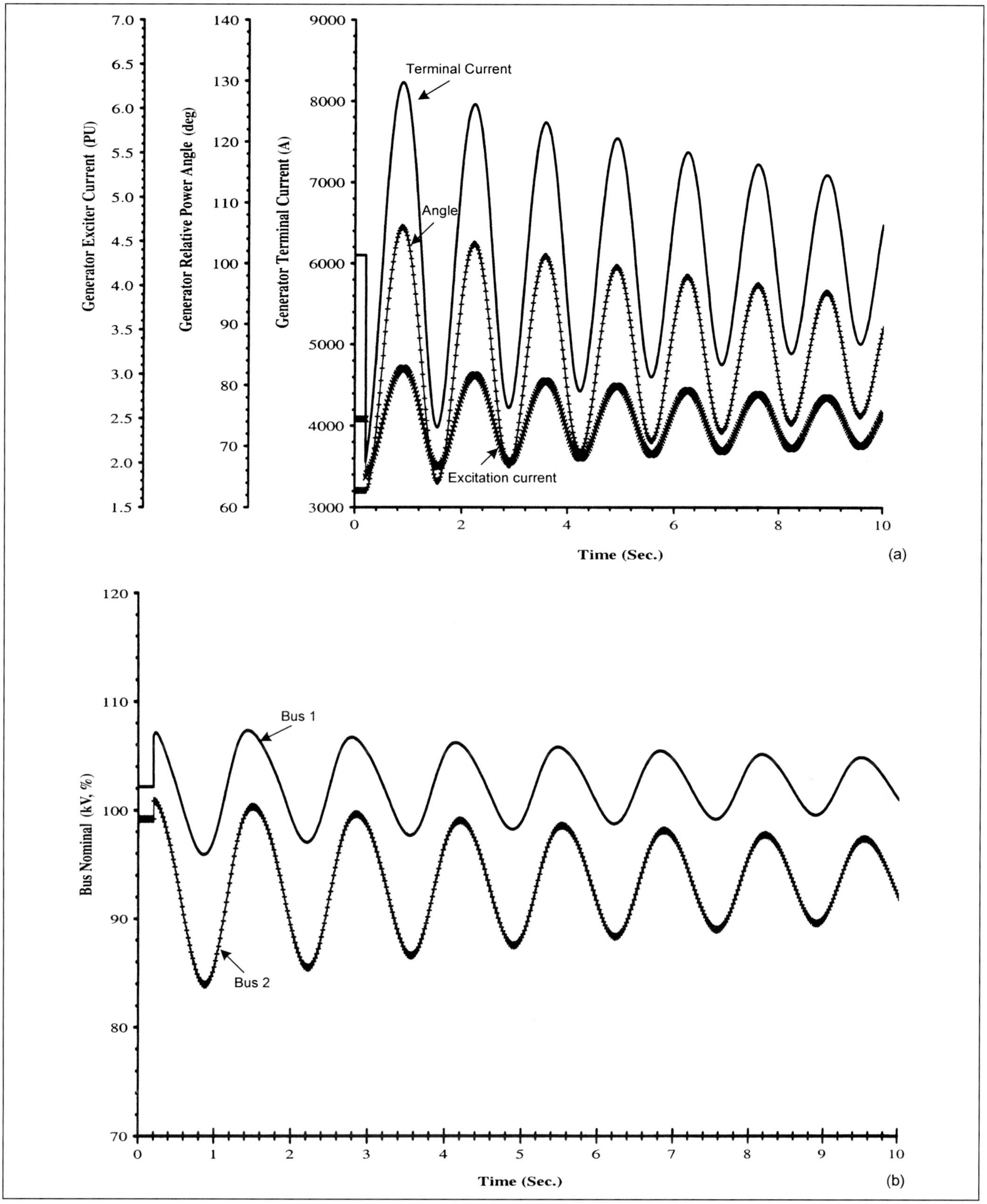

FIGURE 13-38 (*a*) Generator terminal current, relative power angle (deg.) and exciter current (pu). (*b*) Bus 1 and Bus 2 voltages. Sudden disconnection of 100-MVA load in Fig. 13-37. No PSS is provided.

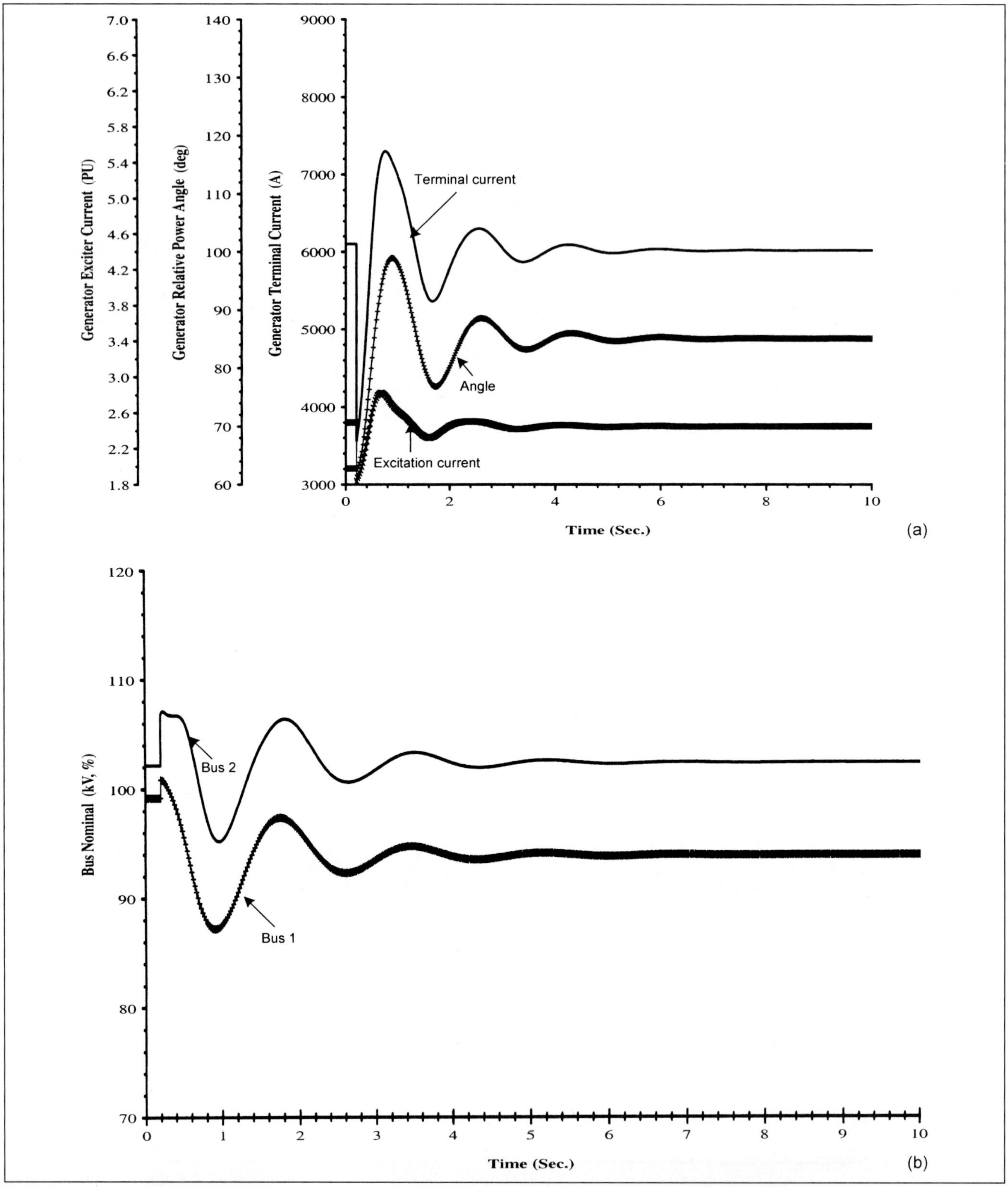

FIGURE 13-39 Fast decay of generator and bus transients in Fig. 13-38, with PSSIA, data as in Table 13-3.

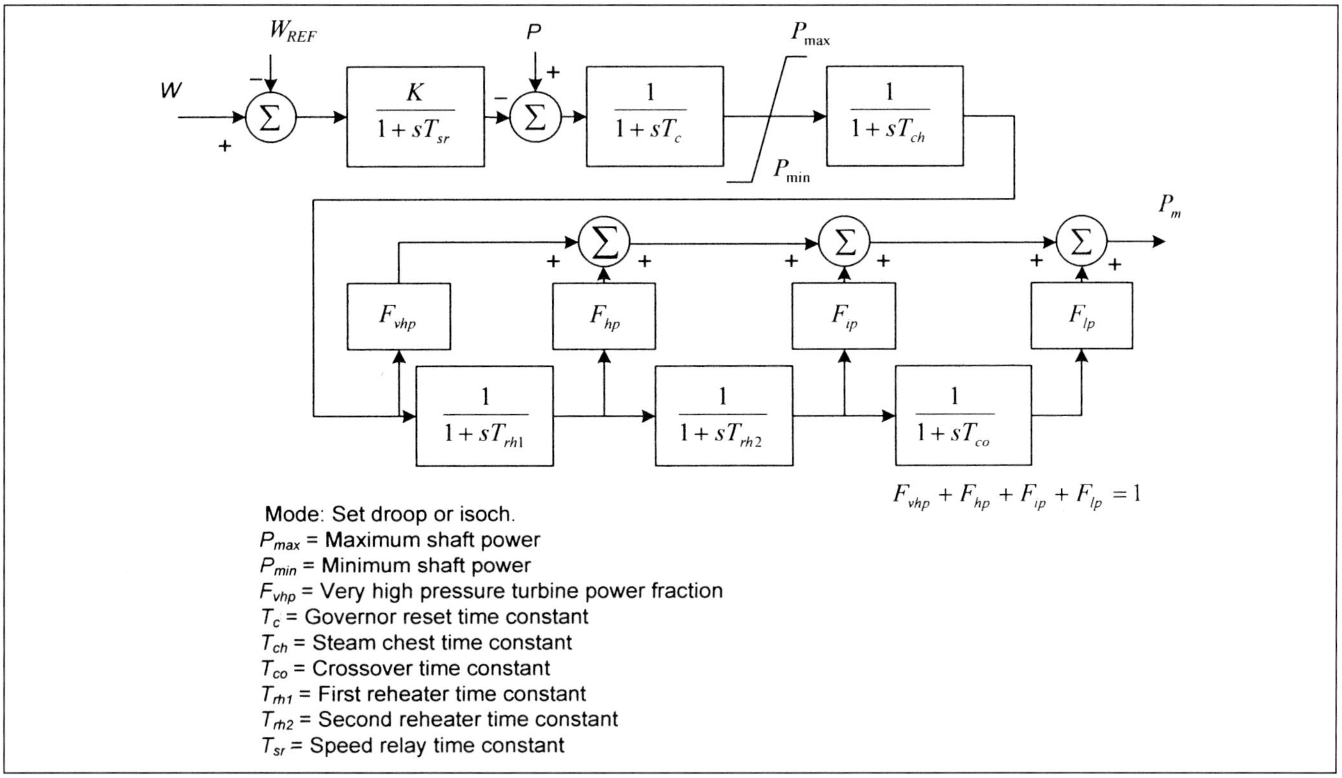

FIGURE 13-40 Control circuit diagram of a compound double reheat steam turbine.

From a control strategy point of view, the above situation is not desirable. We should control each area so that the controls correct for changes in that area only. This can be done by adding a control component of tie-line load flow deviation to frequency deviation, weighted by a bias factor. This control signal is called area control error (ACE). For an area we can write:

$$\mathrm{ACE}_1 = A_1 \Delta P_{12} + \beta_1 \Delta f = 0 \tag{13-46}$$

For a change in frequency due to load change in Area 1, the following expression holds:

$$\Delta f_R = \frac{-\Delta P_{L1}}{\beta_1 + \beta_2} \tag{13-47}$$

Equation (13-46) will give $ACE_2 = 0$ for Area 2, which signifies, that load change in Area 1 does not impact the supplementary control in Area 2.

When more than two areas are interconnected, the power flows could split over parallel paths, and are not necessarily confined to tie lines. Under an abnormal condition, it may happen that the generation-load mismatch is not corrected in all the areas, say due to insufficient generation. The other areas may take over and power transfers above the scheduled values may occur, and frequency may deviate. The controls are limited by the amount of stored energy in the generating units and how fast the generation can be changed.[17–19]

A factor that needs to be considered is the dead band of the speed governor. IEEE standards specify a maximum dead band of 0.06 percent (0.036 Hz) for governors of large steam turbines,[17] and maximum speed dead band of 0.02 percent and maximum blade control dead band of 1.0 percent for hydraulic turbines.[18] If the frequency deviation is small, it may remain in the dead band, though the response of individual generating units will be random.

13-12-1 Underfrequency Load Shedding

When the areas are islanded, to prevent prolonged operation on underfrequency, which may have a detrimental effect on the turbines (Chap. 16), underfrequency load shedding is resorted to. The two parameters that determine the rate of frequency decay are:

- Frequency-sensitive characteristics of the loads
- Inertia constant H

The frequency decay is given by:

$$\Delta f = -\Delta L(1 - e^{-t/T})\,(1/D) \tag{13-48}$$

where D is the load damping constant and $T = 2H/D$.

The procedural details for frequency load shedding are described in Ref. 20. The load may be shed based upon frequency alone, or for larger imbalances, it is shed in increments as the rate of frequency decay increases.

13-13 ON-LINE SECURITY ASSESSMENTS

Power system security refers to the capability of the system to withstand sudden disturbances, faults, loss of generation, tie lines, transformers, and other power equipment. Due to the very nature of power systems, the time frame of change varies from very short intervals (under fault conditions) to even days (unit outage, load management forecast, and dispatch).

We talk of online static and dynamic contingency security assessments (SSA and DSA), though these are somewhat related.[21–24] In the static security assessments, a set of credible contingencies which are likely to occur are gathered and the system response ascertained. The objective is to ensure that the system operates in a secure state. Software packages of varying complexity have been implemented in the energy management systems (EMS) to meet this objective. However these operate on steady-state analysis.

In modern power systems, SSA is evidently not enough and DSA must be integrated. DSA is concerned with power system stability as system conditions and operations change due to contingencies. This requires handling of a large set of nonlinear differential equations in addition to the nonlinear algebraic equations involved in SSA. This could be a stupendous task. Thus, potential contingencies which may give rise to instabilities should be categorized, in a two-step process:

1. Screening to ascertain that the contingency is definitely stable. A detailed assessment of these can then be performed. Screening a large number of contingencies and filtering out the ones that need further analysis is a fundamental step that makes DSA feasible. Reliability, speed, efficiency, and online computation are some of the essential requirements for classifying. There are two approaches: artificial intelligence (AI) approach and energy function approach. AI techniques, like artificial neural networks, pattern recognition, and decision tree techniques, first perform off-line computations and try to capture the essential stability features of the system—this may not always be applicable to online simulations. Sequences of BCU classifiers for dynamic security assessment have been developed.

2. Contingencies which are undecided or identified as unstable are then sent to a time-domain stability program.

Thus, the architecture of DSA consists of state monitors, estimators, and powerful contingency screening to send the undecided or unstable cases to online time-domain stability analysis. The outputs will be corrective or preventive actions.

PROBLEMS

1. Draw the reactive capability curve of 200-MW generator to scale, following the construction of Fig. 13-4. Consider $X_d = 200$ percent (on machine MVA base), ceiling excitation voltage = 255 percent, and the rated power factor of the machine = 0.85.

2. The generator in Prob. 1 is connected to an infinite bus through an impedance of 0.4 pu on 100-MVA base. Draw the static stability limit curve.

3. Why do the manufactures today offer synchronous generators with lower SCR compared to their predecessors? Comment on SCR with respect to stability.

4. Figure 13-P1 shows the control circuit block diagram of an excitation system, Type 1.[4] Write state-space equations.

5. A 200-MW, 0.85-power factor, 18-kV generator has: $L_{ad} = L_{aq} = L_{fd} = 1.56$, $l_d = 0.12$, $r_F = 0.0005$, and $r = 0.002$, all in pu on generator MVA base. The field current required to generate rated voltage on air-gap line is 600 A, and the corresponding field voltage = 80 V. What are the base values for E_{FD} and i_F in per unit nonreciprocal and reciprocal systems?

6. What is the kinetic energy stored by a 50-MW, 60-Hz, two-pole generator with $H = 4$ kW/kVA? If the generator electrical load suddenly changes to 20 MW, will the rotor accelerate or decelerate? What will be the generator speed after 10 cycles?

7. In an existing system consisting of a generator, step-up transformer, and given load, what measures will be the most practical to implement to enhance stability?

8. Define a high-response excitation system according to IEEE. Describe why high-response excitation system can give rise to negative damping and contribute to instability?

9. Explain why it is necessary to use undervoltage settings along with underfrequency settings for effective load shedding for enhancing stability.

10. Study two control circuit block diagrams of steam turbines, and write a one-page explanation of each circuit. Define reheat time constant and time constant of a steam vessel.

REFERENCES

1. P. M. Anderson and A. A. Fouad, *Power System Control and Stability,* IEEE Press, NJ, 1994.
2. IEEE Std. 421.5, IEEE Recommended Practice for Excitation System Models for Power System Stability Studies, 1992.
3. IEEE Std. 421.2, IEEE Guide for Identification, Testing, and Evaluation of Dynamic Performance of Excitation Control Systems, 1990.
4. IEEE Committee Report, "Computer Representation of Excitation System," *IEEE Trans.* PAS, vol. 87, pp. 1460–1464, June 1968.
5. J. C. Das, "Transient Stability of Small Generators Connected to a Weak Utility System, A Case Study," *IEEE Trans. Industry Applications,* vol. 41, no. 1, pp 155–162, Jan./Feb. 2005.
6. K. E. Bollinger, A. Laha, R. Hamilton, and T. Harras, "Power System Stabilizer Design Using Root Locus Methods," *IEEE Trans. PAS,* vol.-94, pp. 1484–1488, Sept./Oct. 1975.

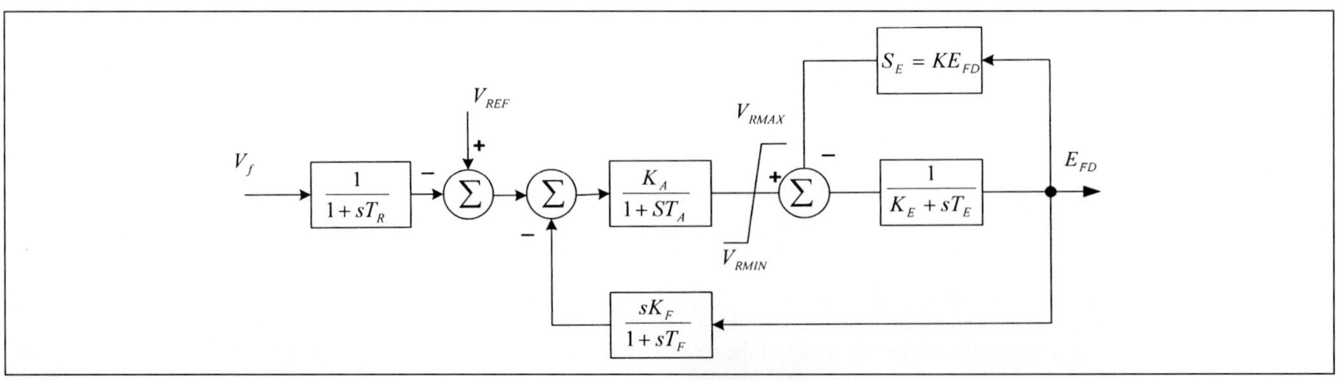

FIGURE 13-P1 Excitation system Type 1.

7. E. V. Larsen and D. A. Swann, "Applying Power System Stabilizers, Part 1: General Concepts," Paper 80, SM 558-7, in Conference Record, IEEE PES Summer Meeting, Minneapolis, July 1980.
8. Ibid, "Part II: Performance Objectives and Tuning Concepts," Paper 80, SM 559-5.
9. Ibid, "Part III: Practical Considerations," Paper 80, SM, 560-3.
10. F. P. DeMello, J. S. Czuba, P. A. Rusche, and J.Wills, "Developments in Application of Stabilizing Measures Through Excitation Control," CIGRE Paper 38–05, 1986.
11. F. P. De Mello and C. Concordia, "Concepts of Synchronous Machine Stability as Affected by Excitation Control," *IEEE Trans.PAS,* vol. 88, pp. 316–329, April 1969.
12. IEEE Committee Report, "Dynamic Models of Steam and Hydro Turbines in Power System Studies," *IEEE Trans.*, vol. PAS-92, pp. 1904–1951, Nov./Dec. 1973.
13. IEEE Working Group Report, "Hydraulic Turbine and Turbine Control Models for System Dynamic Studies," *IEEE Trans.*, vol. PWRS-7, no. 1, pp. 167–179, Feb. 1992.
14. IEEE Working Group Report, "Dynamic Models for Fossil Fueled Steam Units in Power System Studies," *IEEE Trans.*, vol. PWRS-6, no. 2, pp. 753–761, May 1991.
15. P. L. Dandeno, P. Kundur, and J. P. Bayne, "Hydraulic Unit Dynamic Performance Under Normal and Island Conditions—Analysis and Validation," *IEEE Trans.*, vol. PAS-97, pp. 2134–2143, Nov./Dec. 1978.
16. EPRI Report EL-6627, Long Term Dynamic Simulation: Modeling Requirements, Dec. 1989.
17. IEEE Standard 122, Recommended Practice for Functional and Performance Characteristics of Control Systems for Steam Turbine-Generator Units, 2003.
18. IEEE Standard 125, Recommended Practice for Preparation of Equipment Specifications for Speed Governing for Hydraulic Turbines Intended to Drive Electrical Generators, 2007.
19. IEEE AGC Task Force Report, "Understanding Automatic Generation Control," *IEEE Trans.*, vol. PWRS-7, no. 3, pp. 1106–1122, Aug. 1992.
20. H. E. Lokay and V. Burtnyk, "Application of Underfrequency Relays for Automatic Load Shedding," *IEEE Trans.*, vol. PAS-87, pp. 776–783, March 1968.
21. D. H. Berry, R. D. Brown, J. J. Redmond, and W. Watson, "Underfrequency Protection of Ontario Hydro System," CIGRE paper 32-14, Aug./Sept. 1970.
22. S. C. Savulescu, *Real-Time Stability Assessment in Modern Power System Control Centers*, IEEE Press, Series on Power Engineering, NJ, February 2009.
23. B. Scott, O. Alsac, and A. J. Monticelli, "Security Analysis and Optimization," *Proc. IEEE*, vol. 75, no. 12, pp. 1623–1644, 1987.
24. Y. Mansour, E. Vaahedi, A. Y. Chang, B. R. Corns, B. W. Garrett, K. Demaree, T. Athay, and K. Cheung, "BC Hydro's on Line Transient Stability Assessment (TSA) Model Development, Analysis, and Post Processing," *IEEE Trans.*, vol. PS10, pp. 241–253, 1995.

FURTHER READING

ANSI/IEEE Std. 421.1, IEEE Standard Definitions of Excitation Systems for Synchronous Machines, 1986.

W. T. Carson, *Power System Voltage Stability,* McGraw Hill, New York, 1994.

H. D. Chiang and R. Jean-Jumeau, "Towards a Practical Performance Index for Predicting Voltage Collapse in Power Systems," *IEEE Trans.*, vol. PS10, no. 2, pp. 584–592, 1995.

J. C. Das, Power System Relaying Wiley Encyclopedia of Electrical and Electronic Engineers, vol. 17, pp. 71–86, 2002.

J. Deuse and M. Stubbe, "Dynamic Simulation of Voltage Collapses," *IEEE Trans.*, vol. PS8, no. 3, pp. 894–900, 1993.

IEEE Tutorial Course, Power System Stabilization via Excitation Control, 81-EHO 175-0 PWR, Sep. 1980.

P. Kundur, *Power System Stability and Control*, McGraw Hill, New York, 1993.

J. A. Momoh, *Electrical Power System Applications of Optimization*, Marcel Dekker, New, York, 2001.

Voltage Stability and Long Term Stability Working Group, Bibliography on Voltage Stability, *IEEE Trans. PS*, vol. 13, no. 1, pp. 115–125, 1998.

CHAPTER 14

TRANSIENT BEHAVIOR OF TRANSFORMERS

Transients in the transformers originate due to saturation, switching and lightning, high-flux densities, energy storage between inductance and capacitance of the windings, part-winding resonances, and winding connections.

A power transformer is an important component of the power system. The transformation of voltages is carried out right from generating voltage to transmission, subtransmission, and distribution, and also consumer level. The installed capacity of the transformers in a power system may be seven to eight times the generating capacity. The special class of transformers includes furnace, converter, regulating, rectifier, phase shifting, traction, welding, and instrument (current and potential transformers). Saturation of current transformers is important for relaying applications, which is not discussed here.

14-1 FREQUENCY-DEPENDENT MODELS

Transformer models for transients must consider nonlinearity and frequency-dependent behavior. Table 14-1 from CIGRE modeling guidelines[1] shows the importance of modeling of various parameters. Further, the modeling and ascertaining of the test results for modeling are also dependent upon the transformer winding connections and core construction. Ascertaining the modeling parameters for a particular model based on the nameplate details is not adequate and special tests may be required.[2]

The concept of leakage flux, and total flux, mutual and self reactances in a circuit of two magnetically coupled coils is described in Chap. 10, and a matrix model can be written as:

$$\begin{vmatrix} V_1 \\ V_2 \\ \cdot \\ V_n \end{vmatrix} = \begin{vmatrix} r_{11} & r_{12} & \cdot & r_{1n} \\ r_{21} & r_{22} & \cdot & r_{2n} \\ \cdot & \cdot & \cdot & \cdot \\ r_{n1} & r_{n2} & \cdot & r_{nn} \end{vmatrix} \begin{vmatrix} i_1 \\ i_2 \\ \cdot \\ i_n \end{vmatrix} + \begin{vmatrix} L_{11} & L_{12} & \cdot & L_{1n} \\ L_{21} & L_{22} & \cdot & L_{2n} \\ \cdot & \cdot & \cdot & \cdot \\ L_{n1} & L_{n2} & \cdot & L_{nn} \end{vmatrix} \frac{d}{dt} \begin{vmatrix} i_1 \\ i_2 \\ \cdot \\ i_n \end{vmatrix} \tag{14-1}$$

For transient calculations, we can write:

$$\bar{V} = \bar{R}\,\bar{i} + \bar{L}[d\bar{i}/dt] \tag{14-2}$$

where $\bar{R}$, and $\omega\bar{L}$ are real and imaginary parts of the impedance matrix. In case of low magnetization currents, the transformer should be described by an admittance matrix:

$$\bar{I} = \bar{Y}\bar{V} \tag{14-3}$$

For transient simulation, this expression becomes:

$$[d\bar{i}/dt] = \bar{L}^{-1}\bar{V} + \bar{L}^{-1}\bar{R}\,\bar{i} \tag{14-4}$$

These models are linear, neglect saturation, and have limited application with respect to transients.

14-2 MODEL OF A TWO-WINDING TRANSFORMER

A two-winding transformer model can be derived from the circuit diagram shown in Fig. 14-1 and the corresponding phasor diagram shown in Fig. 14-2. The transformer supplies a load current I_2 at a terminal voltage V_2 and lagging power factor angle ϕ_2. Exciting the primary winding with voltage V_1 produces changing flux linkages. Though, the coils in a transformer are tightly coupled by interleaving the windings and are wound on a magnetic material of high permeability, all the flux produced by primary windings does not link the secondary. The winding leakage flux gives rise to leakage reactance. In Fig. 14-2, Φ_m is the main or mutual flux, assumed constant. The EMF induced in the primary winding is E_1, which lags Φ_m by 90°. In the secondary winding, the ideal transformer produces an EMF E_2 due to mutual flux linkages. There has to be a primary magnetizing current, even at no load, in time phase with its associated flux to excite the core. The pulsation of flux in the core produces losses. Considering that the no-load current is sinusoidal, which is not true under magnetic saturation, it must have a core loss component due to hysteresis and eddy currents:

$$I_0 = \sqrt{I_m^2 + I_e^2} \tag{14-5}$$

where I_m is the magnetizing current, I_e is the core loss component of the current, and I_0 is the no-load current. I_m and I_e are in phase quadrature. The generated EMF because of flux Φ_m is given by:

$$E_2 = 4.44 f n_2 \Phi_m \tag{14-6}$$

TABLE 14-1 Modeling Guidelines for Transformers*

TRANSFORMER	GROUP 1 0.1 Hz TO ≈ 3 kHz	GROUP 2 60 Hz TO ≈ 3 kHz	GROUP 3 10 kHz TO ≈ 3MHz	GROUP 4 100 kHz TO ≈ 50 MHz
Short-circuit impedance	Very important	Very important	Important only for surge transfer	Negligible
Saturation	Very important	See note below	Negligible	Negligible
Frequency-dependent series losses	Very important	Important	Negligible	Negligible
Hysteresis and iron losses	Important only for resonance phenomena	Important only for transformer energizing	Negligible	Negligible
Capacitance coupling	Negligible	Important for surge transfer	Very important for surge transfer	Very important for surge transfer

Note: Very important for transformer energizing and load rejection with high-voltage increases, otherwise negligible.

*These guidelines are based on CIGRE recommendations.[1]

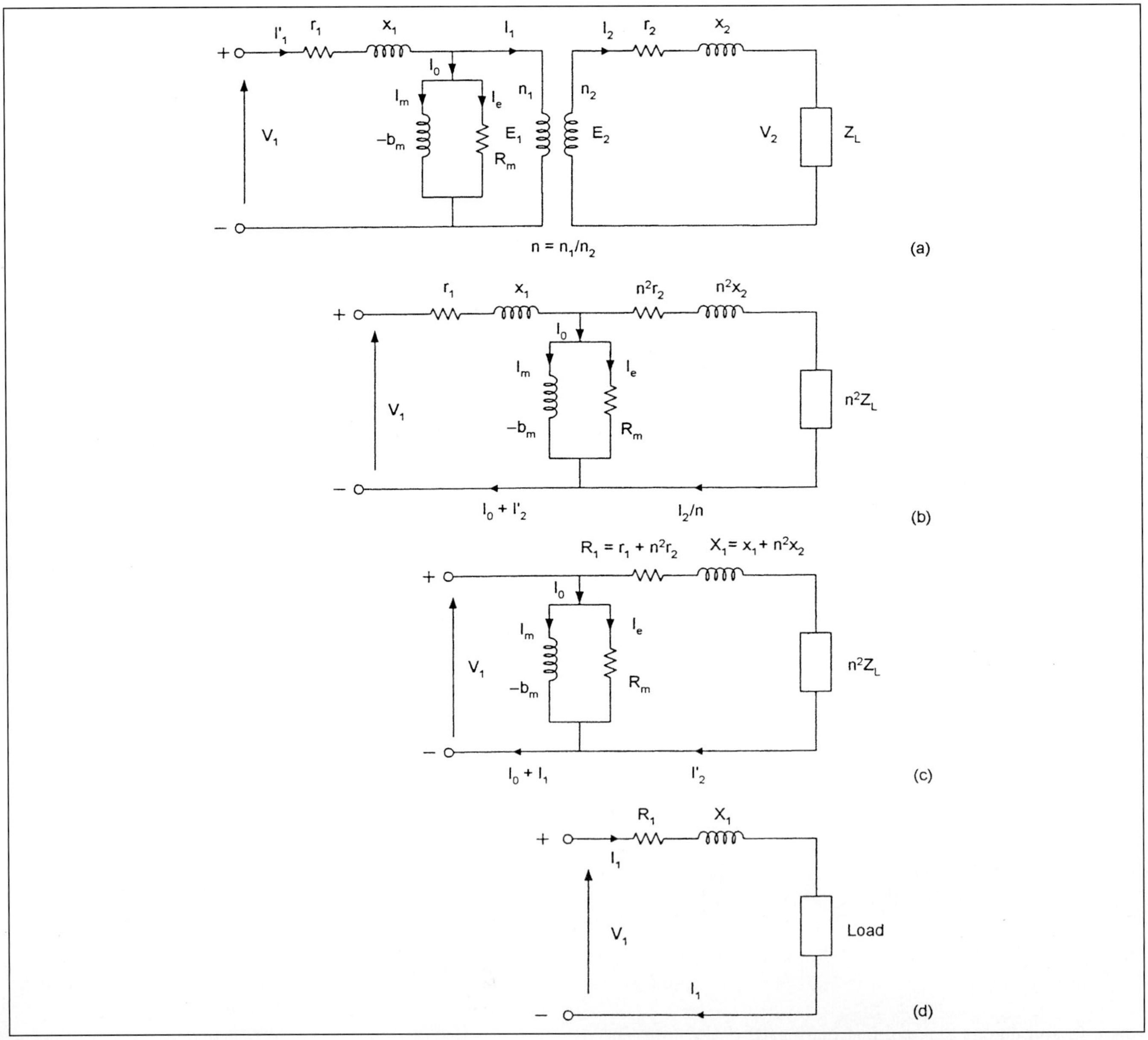

FIGURE 14-1 (*a*) Equivalent circuit of a two-winding transformer. (*b*), (*c*), and (*d*) Simplifications to the equivalent circuit.

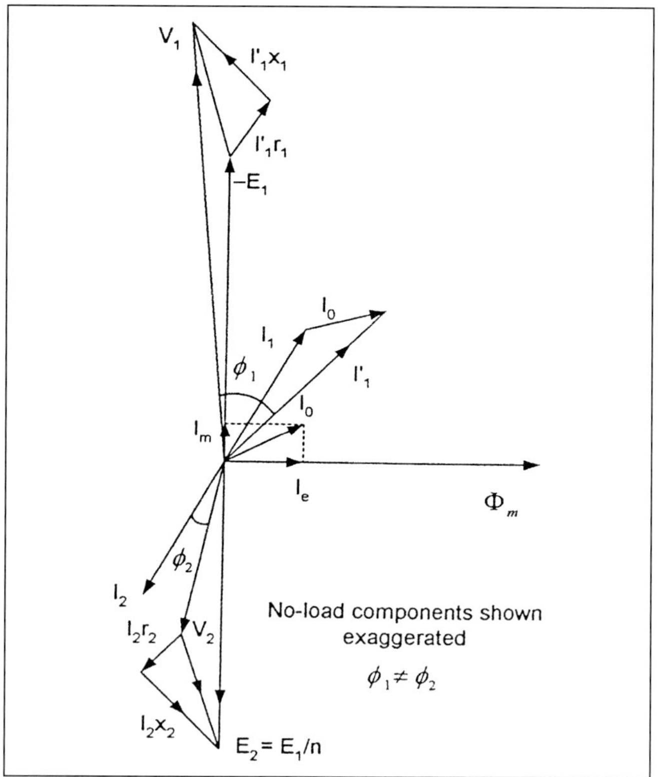

FIGURE 14-2 Phasor diagram of a two-winding transformer.

where E_2 is in volts when Φ_m is in weber per square meter, n_2 is the number of secondary turns, and f is the frequency. As primary ampere turns must be equal to the secondary ampere turns, that is, $E_1 I_1 = E_2 I_2$ we can write:

$$\frac{E_1}{E_2} = \frac{n_1}{n_2} = n$$
$$\text{and} \quad \frac{I_1}{I_2} \approx \frac{n_2}{n_1} = \frac{1}{n} \tag{14-7}$$

where n_1 is the number of primary turns and n is the transformation ratio. The current relation holds because the no-load current is small. The terminal relations can now be derived. On the primary side, the current is compounded to consider the no-load component of the current, and the primary voltage is equal to $-E_1$ (to neutralize the EMF of induction) and I_1r_1 and I_1x_1 drop in the primary windings. On the secondary side, the terminal voltage is given by the induced EMF $E_2 - I_2r_2$, and I_2x_2 drops in the secondary windings. The equivalent circuit is therefore as shown in Fig. 14-1*a*. The transformer is an ideal lossless transformer of turn ratio n.

The secondary resistance and reactance can be referred to the primary side. The secondary windings of n_2 turns can be replaced with an equivalent winding referred to primary, where the copper loss in the windings and the voltage drop in the reactance are the same as in the actual winding. The resistance and reactance of the equivalent windings are denoted as r_2' and x_2':

$$I_1^2 r_2' = I_2^2 r_2$$
$$r_2' = r_2\left(\frac{I_2^2}{I_1^2}\right) \approx r_2\left(\frac{n_1}{n_2}\right)^2 = n^2 r_2 \tag{14-8}$$
$$x_2' = x_2\left(\frac{I_2^2}{I_1^2}\right) \approx x_2\left(\frac{n_1}{n_2}\right)^2 = n^2 x_2$$

Now the transformer is an ideal transformer with no losses and has a turn's ratio of unity and no secondary resistance or reactance. By transferring the load impedance also to the primary side, the unity ratio ideal transformer can be eliminated and the magnetizing circuit is pulled out to the primary terminals without appreciable error (Fig. 14-1*b* and *c*). In Fig.14-1*d*, the magnetizing and core loss circuit is altogether omitted. The equivalent resistance and reactance are:

$$R_1 = r_1 + n^2 r_2$$
$$X_1 = x_1 + n^2 x_2 \tag{14-9}$$

Thus, on a simplified basis, the transformer positive or negative sequence model is given by its percentage reactance specified by the manufacturer, generally, on the transformer natural cooled MVA rating base. This reactance remains fairly constant and is obtained by a short-circuit test on the transformer. The magnetizing circuit components are obtained by an open-circuit test.

The expression for hysteresis loss is:

$$P_h = K_h f B_m^s \tag{14-10}$$

where K_h is a constant, and s is the Steinmetz exponent, which varies from 1.5 to 2.5, depending upon the transformer core material; generally it is = 1.6. The eddy current loss is:

$$P_e = K_e f^2 B_m^2 \tag{14-11}$$

where K_e is a constant. Eddy current loss occurs in core laminations, conductors, tank, and clamping plates. The core loss is the sum of eddy current and hysteresis loss. In Fig. 14-2, the primary power factor angle ϕ_1 is > ϕ_2.

We discussed the sequence impedances of transformers in Chap. 9 (Figs. 9-4 and 9-5). We know that the transformer winding connections have a profound effect on the zero-sequence impedance which can be an open circuit. A phase shift occurs between the primary and secondary voltages of a three-phase transformer, depending upon the winding connections, as discussed in Chap. 9.

The model thus derived is adequate for power frequency operation. For 60-Hz load flow and short-circuit calculations, the transformer is represented with equivalent impedance as derived above. This model cannot be used for any transient studies. Note that no capacitance between the windings or to the core and ground, no saturation, and no core losses are considered. The transfer of power takes place through inductive couplings and capacitances are neglected. The tap adjustments can be taken care of by the equivalent Π or T circuit, as discussed below.

14-3 EQUIVALENT CIRCUITS FOR TAP CHANGING

Figure 14-3*a* shows an ideal transformer, with a series impedance as derived above. Then:

$$V_s = (V_r - ZI_r)/n \tag{14-12}$$

As the power through the transformer is an invariant:

$$V_s I_s^* + (V_r - ZI_r) I_r^* = 0 \tag{14-13}$$

Then substituting $I_s/(-I_r) = n$, $Y = \dfrac{1}{Z}$

$$I_s = [n(n-1)Y]V_s + nY(V_s - V_r)$$
$$I_r = nY(V_r - V_s) + [(1-n)Y]V_r \tag{14-14}$$

These equations give the equivalent circuits illustrated in Fig. 14-3*b* and *c* for the equivalent T and Π circuits.

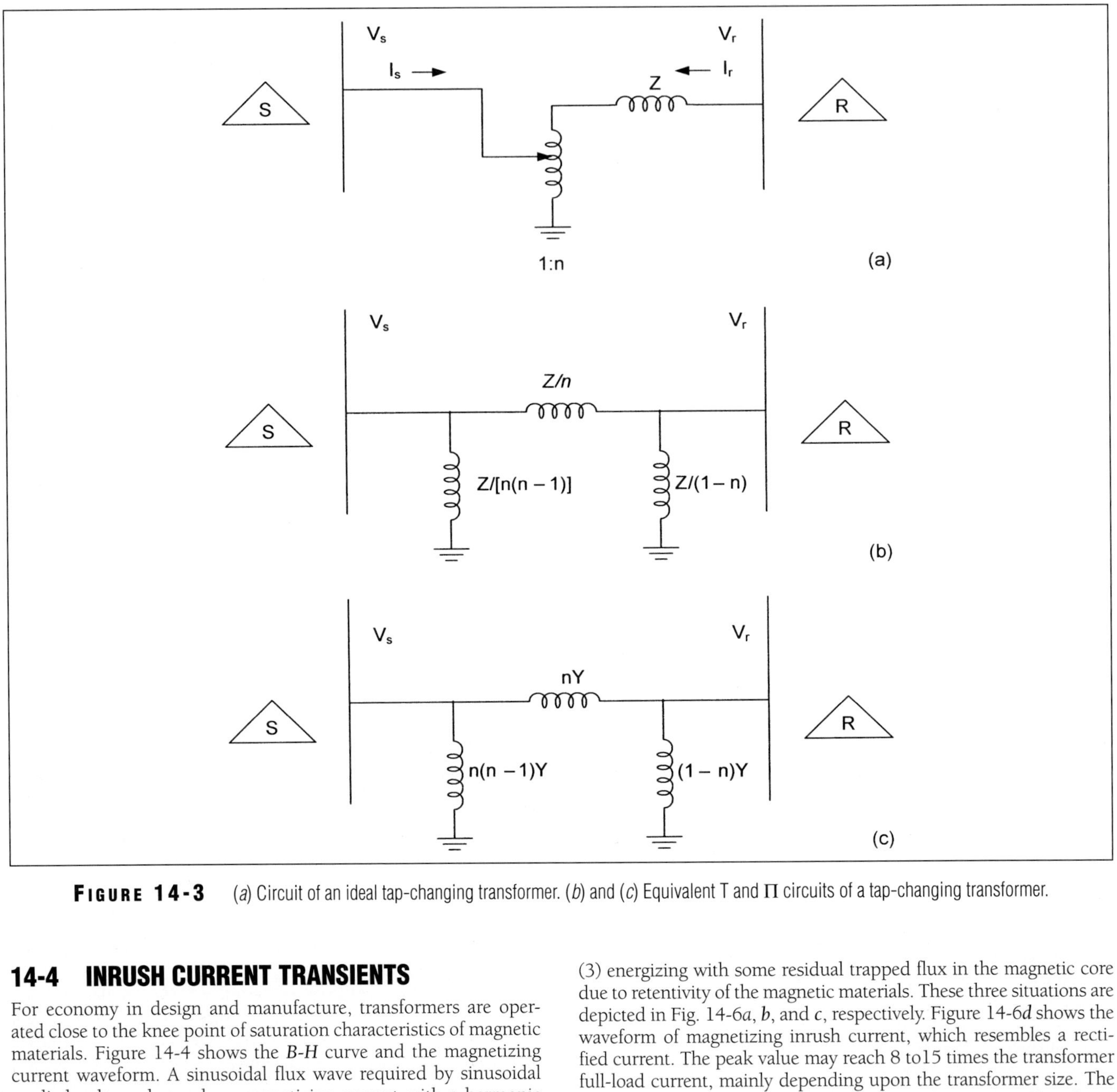

FIGURE 14-3 (*a*) Circuit of an ideal tap-changing transformer. (*b*) and (*c*) Equivalent T and Π circuits of a tap-changing transformer.

14-4 INRUSH CURRENT TRANSIENTS

For economy in design and manufacture, transformers are operated close to the knee point of saturation characteristics of magnetic materials. Figure 14-4 shows the *B-H* curve and the magnetizing current waveform. A sinusoidal flux wave required by sinusoidal applied voltage demands a magnetizing current with a harmonic content. Conversely, with sinusoidal magnetizing current, the induced EMF is peaky and the flux is flat topped.

An explanation of the generation of the peaky magnetizing current considering third harmonic is provided in Fig. 14-5. A sinusoidal EMF E_a generates a sinusoidal current flow, I_a in lagging phase-quadrature with E_a. This sets up a flat topped flux wave, ϕ_1—which can be resolved into two components, ϕ_a, the fundamental flux wave, and ϕ_3, the third harmonic flux wave, (higher harmonics are neglected). The third harmonic flux can be supposed to produce a third harmonic EMF E_3 and corresponding third harmonic current I_3, which when summed up with I_a makes the total current drawn peaky.

On energizing a power transformer, a large magnitude of inrush current will flow, which can be resolved into harmonics including a dc component. Figure 14-6 depicts three conditions of energizing a power transformer: (1) the switch closed at the peak value of the voltage, (2) the switch closed at zero value of the voltage, and (3) energizing with some residual trapped flux in the magnetic core due to retentivity of the magnetic materials. These three situations are depicted in Fig. 14-6*a*, *b*, and *c*, respectively. Figure 14-6*d* shows the waveform of magnetizing inrush current, which resembles a rectified current. The peak value may reach 8 to15 times the transformer full-load current, mainly depending upon the transformer size. The asymmetrical loss due to conductor and core heating rapidly reduces the flux wave to symmetry about the time axis, and typically, the inrush currents last for a short duration of approximately 0.1 s.

The inrush current is rich in the second harmonic; also third, fourth, and fifth harmonics are generated in addition to a dc bias. Presence of power capacitors increases the transients and their decay time. Differential relays used for protection of transformers inhibit switching current harmonics to prevent operation of the switching devices under inrush conditions.

14-5 TRANSIENT VOLTAGES IMPACTS ON TRANSFORMERS

Transformers normally operate under steady-state excitation and voltages are controlled to specific limits. The generated fundamental frequency EMF is given by:

$$V = 4.44 f T_{\mathrm{ph}} B_m A_c \tag{14-15}$$

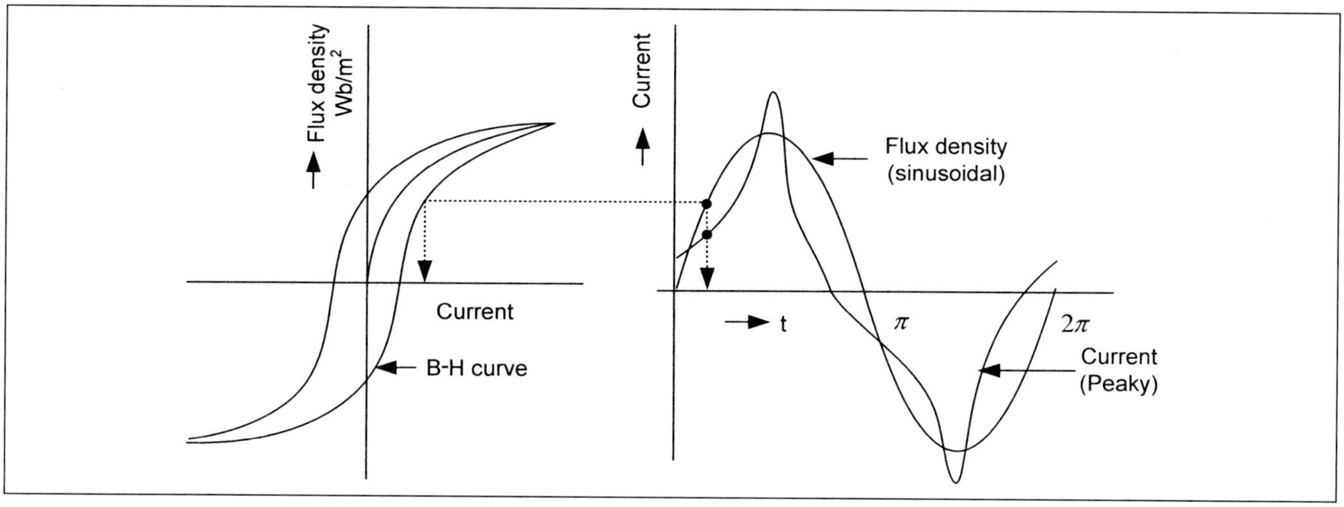

FIGURE 14-4 *B-H* curve of a magnetic material and peaky transformer magnetizing current.

where T_{ph} is the number of turns in a phase, B_m is the flux density (consisting of fundamental and higher-order harmonics), and A_c is the area of core. Thus, the factor V/f is a measure of the overexcitation, though these currents do not normally cause a wave distortion of any significance. Exciting currents increase rapidly with voltage and ANSI/IEEE standard C57.12.00 requires that transformers are capable of operating continuously at 10 percent above rated secondary voltage at no load without exceeding limiting temperature rise.[3] Under certain system upset conditions, the transformers may be subjected to even higher voltages and overexcitation. An example is a transformer connected to a generator in step-up configuration (GSU—generator step-up transformer). Overexcitation of transformers in steady state can produce harmonics. Also consider that:

1. A total load rejection of a generator will subject the transformer to higher voltages and overexcitaion. The protection is provided by V/f relays (device 24).

2. The transformers supplying nonsinusoidal loads have increased eddy current, pulsation, and core losses and are derated according to ANSI/IEEE standard[4] and UL Standard.[5] This is not discussed in this book.

3. Dynamic voltages to which the transformers may be subjected refer to low-frequency voltages 0 to 1500 Hz, which are damped oscillatory voltages.

4. Excitation by switching and lightning surges and fault voltages may give rise to chopped voltage waveforms and higher stresses.

5. The term "very fast transients" encompasses excitation with voltages of rise time in the range of 50 to100 ns (Chaps. 1 and 18).

6. Transformers often have high internal winding ringing frequencies. Certain switching operations can excite these frequencies so that voltage escalation occurs deep inside

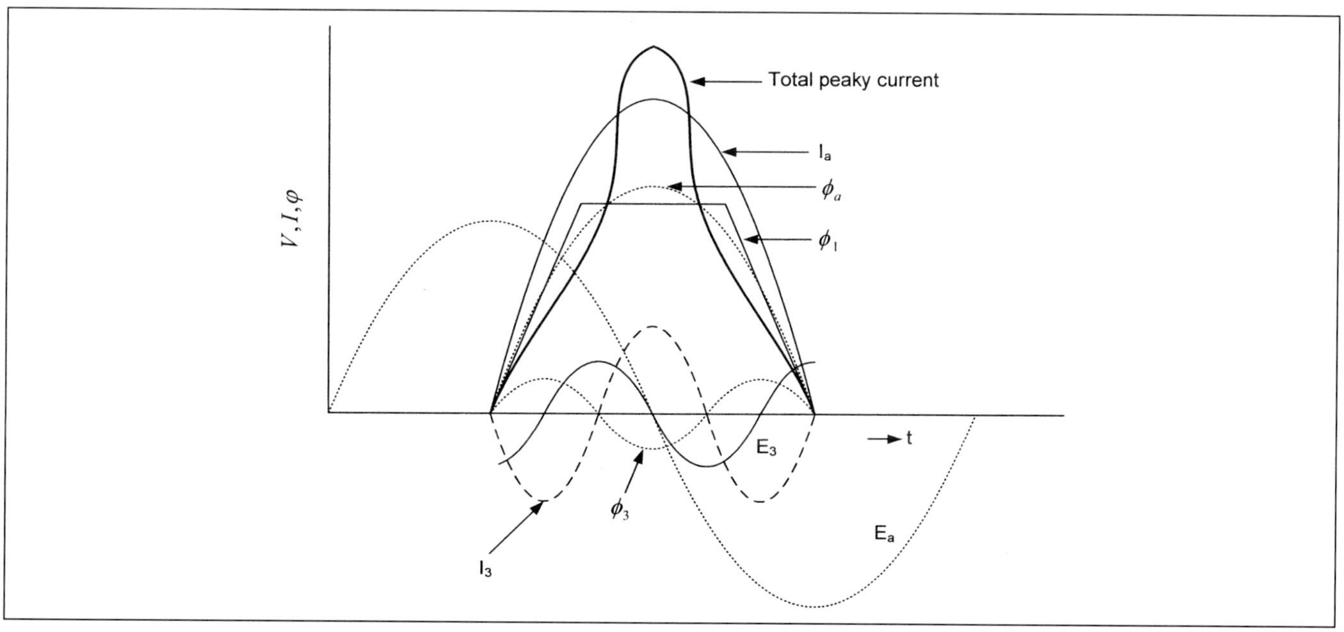

FIGURE 14-5 Origin of a flat-topped flux wave in a transformer and peaky magnetization current.

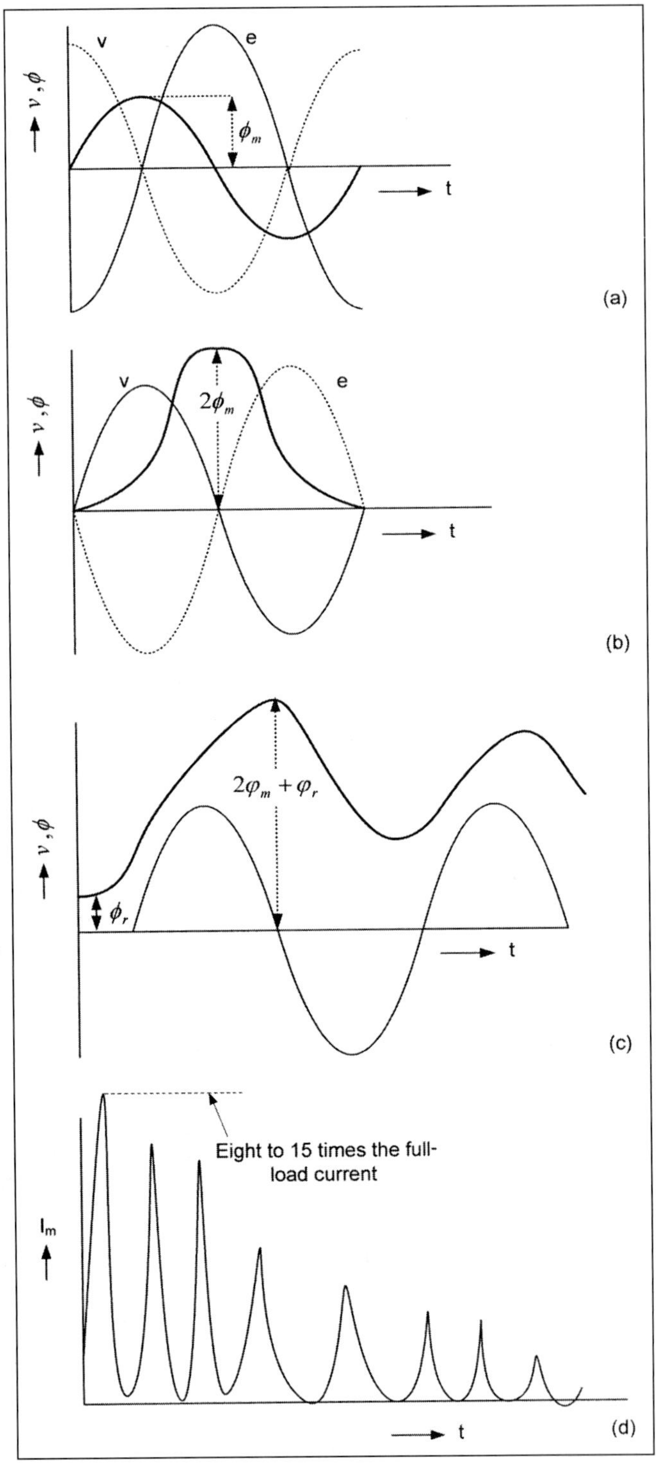

FIGURE 14-6 Inrush current transients: (*a*) switch closed at the peak of the applied voltage, (*b*) switch closed at voltage zero crossing, (*c*) with prior trapped magnetic flux in the core, and (*d*) profile of transient inrush current.

transformer windings, creating excessive intera-winding stresses. Traditional surge arresters applied at transformer terminals are ineffective in stopping this phenomena, and resistance and capacitance snubber circuits have been shown to be effective. Also see Example 20-3.

7. Power systems, transformers, and circuit breakers can form a dynamic system during switching operation, and on rare occasions this may require action to mitigate transient frequencies produced during switching. Circuit breakers may impose frequencies on transformers that may be resonant with or in power system elements.

8. The surge voltages across the windings have unequal distribution and tend to concentrate on the first few turns.

9. A phenomenon of *oscillating neutrals* occurs in ungrounded wye-wye connected transformers. In delta-connected windings, the impedance to third harmonics is negligible, and a very small third harmonic EMF is required to circulate the third harmonic current additive to the fundamental frequency. This maintains sinusoidal flux. In wye-wye connected transformers with isolated neutrals, all the third harmonics are either directed inward or outward; these cancel between the lines, no third harmonic currents flow, and the flux wave is flat topped. The effect on the ungrounded wye neutral is to make it oscillate at three times the fundamental frequency, giving rise to distortion of the phase voltages. Practically, wye-wye ungrounded connection is not used. Even when the windings are grounded, a tertiary delta winding is provided to circulate the third harmonic currents. Sometimes this tertiary winding can be designed to serve load at a different voltage.

The power frequency, lightning impulse, switching surge, and chopped-wave test voltages according to ANSI/IEEE standards are described in Chap. 17.

14-5-1 Electromagnetic Forces

In a power transformer, the vector sum of MMFs is zero at each instant, neglecting the small amount required to magnetize the core. Under short-circuit conditions, the force can be expressed by the equation:

$$F_t = F_{\max}\left(\frac{1}{2} + e^{-2t/\tau} - 2e^{-t/\tau}\cos\omega t + \frac{1}{2}\cos 2\omega t\right) \qquad (14\text{-}16)$$

The total force F_t consists of:

- Two unidirectional components: one constant and the other decreasing with time.
- Two alternating components: one of fundamental frequency, decreasing with time, and the other of double frequency with a smaller but constant magnitude.

It is necessary to calculate flux patterns to determine the forces. Figure 14-7*a* and *b* illustrates flux and force patterns in core-type and shell-type transformers, respectively. The forces are repulsive in pairs of windings having MMFs in opposite directions, that is, carrying currents in opposite directions. In the concentric windings, core-type transformers, forces act mostly in the radial direction. There is no significant axial contribution over approximately two-thirds of the total winding length. In the shell-type transformers, forces act mostly perpendicularly to pancake coil surfaces. At the end of the coils, where the magnetic field lines bend, forces exhibit significant components that are axial in core-type transformers and parallel to pancake surface in shell-type transformers.

The forces tend to move the windings apart due to opposing MMFs. Figure 14-7*c* and *d* illustrates the short-circuit current and the corresponding electromagnetic force transients, respectively. Short-circuit stresses under through fault conditions can result in winding failures. The transformer manufacturers in the United States should meet the through-fault withstand capability of transformers specified

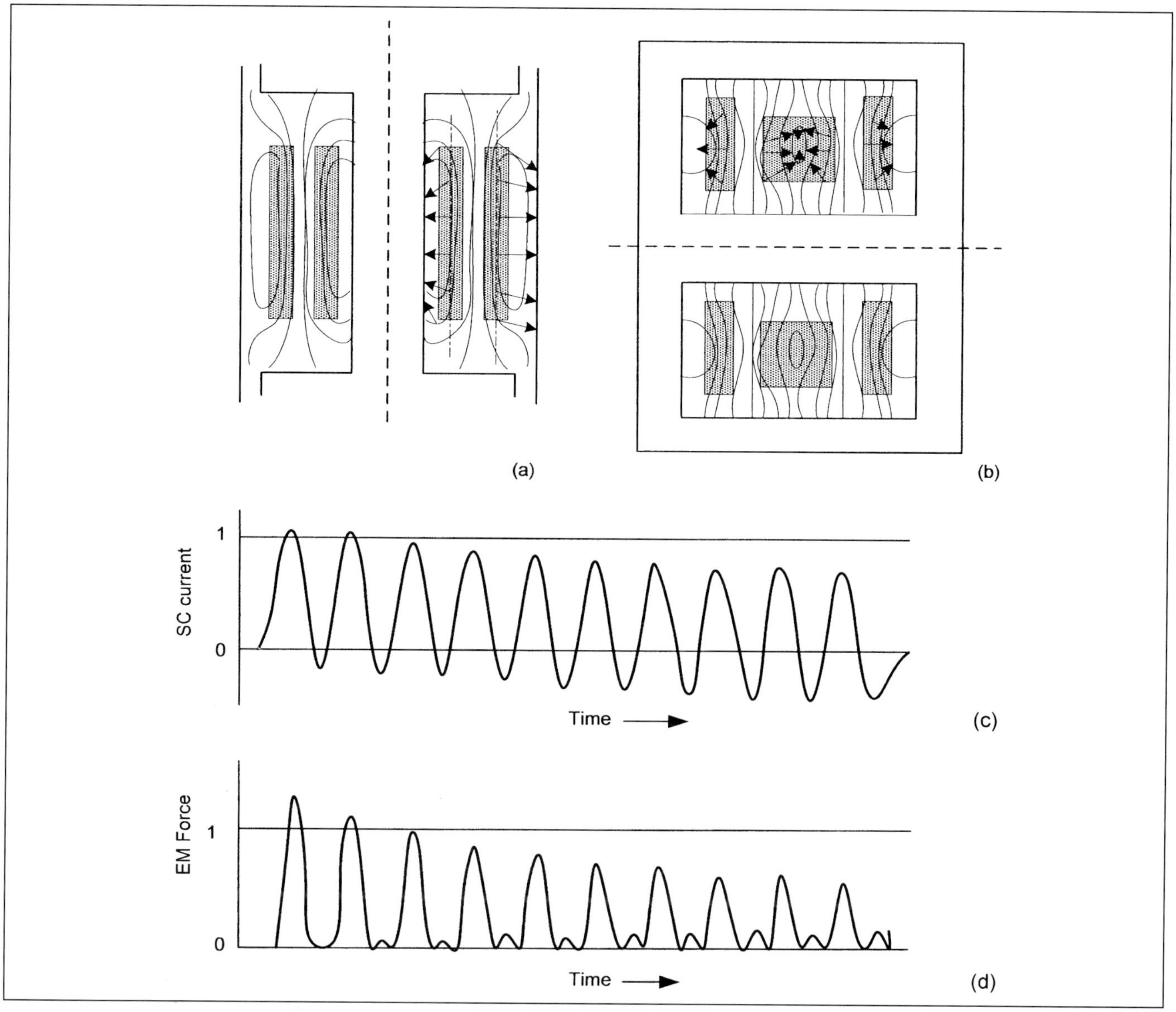

FIGURE 14-7 (*a*) Magnetic flux and (*b*) force patterns in a core-type and shell-type transformer. (*c*) Transient short-circuit current and (*d*) electromagnetic force waveforms.

in ANSI/IEEE Standard C37.91[6] for liquid-immersed transformers and the ANSI/IEEE "Guide for Dry-Type Transformers"[7] for dry-type transformers. There are four categories, depending upon the transformer kVA rating, as follows:

- *Category I.* 5 to 500 kVA single phase and 15 to 500 kVA three phase
- *Category II.* 501 to 1667 kVA single phase and 501 to 5000 kVA three phase
- *Category III.* 1668 to 10,000 kVA single phase and 5001 to 30,000 KVA three phase
- *Category IV.* Above 10,000 kVA single phase and above 30,000 kVA three phase

Figure 14-8 shows the through-fault curves for Category III transformers. Curve (a) is applicable for faults that will occur frequently, typically more than five in transformer lifetime (transformer connected to overhead systems), and curve (b) is applicable for faults that will occur less frequently, typically less than five in transformer lifetime. The transformers in the industrial distribution systems are connected through cables. The curves consider both thermal and mechanical damage. Curve (b) shows that the transformer must withstand a maximum through-fault current of 25 times the rated base current (full-load current based upon rated KVA), for 2 s. In curve (a), the through-fault withstand for 2 s is a function of the transformer impedance as shown in Fig. 14-8. The distribution of currents in the windings depends upon their connections. A three-phase fault is considered. In some applications, for example, UAT (unit auxiliary transformers in generating stations) higher through-fault withstand capability as compared to standard capability may be required.

14-6 MATRIX REPRESENTATIONS

The matrix representation of transformers can be written based on short-circuit and excitation test data. Consider a single-phase

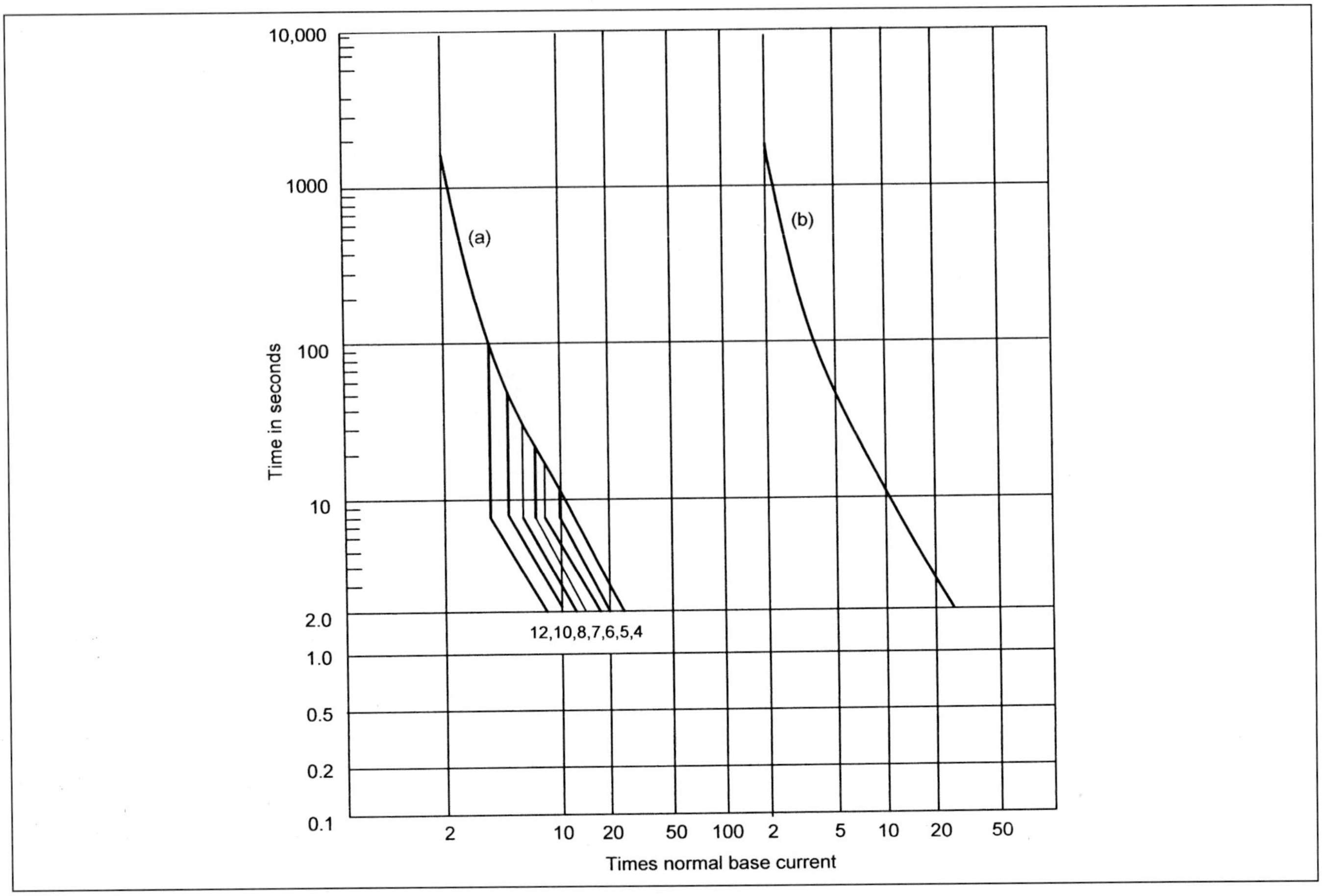

FIGURE 14-8 Through-fault withstand capability of Category III transformers, according to ANSI/IEEE standard.[6] Curve (*a*) shows frequent faults and curve (*b*) shows infrequent faults in the lifetime of the transformer.

two-winding transformer. If magnetizing reactance is not considered:

$$\bar{R} = \begin{vmatrix} R_1 & 0 \\ 0 & R_2 \end{vmatrix} \quad \text{and} \quad \omega\bar{L}^{-1} = \begin{vmatrix} 1/X & -1/X \\ -1/X & 1/X \end{vmatrix} \tag{14-17}$$

Equation (14-17) is written in pu. If the short-circuit impedance of the transformer is 0.005 + j0.10 pu, then:

$$\bar{R} = \begin{vmatrix} 0.0025 & 0 \\ 0 & 0.0025 \end{vmatrix} \quad \omega\bar{L}^{-1} = \begin{vmatrix} 10 & -10 \\ -10 & 10 \end{vmatrix} \tag{14-18}$$

Here R_1 pu is assumed equal to R_2 pu. This can be converted to actual values.

The positive and negative sequence circuits of three-winding transformers are shown in Fig. 9-5. Consider a wye-grounded *H*, delta *L*, and delta *M* connection. The positive sequence connection is a wye connection. The manufacturers specify short-circuit data, for example, as:

$$X_{HL} = 0.117 \qquad X_{LM} = 0.241 \qquad X_{HM} = 0.115 \text{ pu}$$

X_H, X_L, X_M can be found from the following expressions:

$$\begin{aligned} X_H &= \frac{1}{2}(X_{HL} + X_{HM} - X_{LM}) \\ X_L &= \frac{1}{2}(X_{LM} + X_{HL} - X_{HT}) \\ X_M &= \frac{1}{2}(X_{HM} + X_{LM} - X_{HL}) \end{aligned} \tag{14-19}$$

Using these equations:

$$X_H = -000045 \qquad X_L = 0.001215 \qquad X_M = 0.001195$$

Convert these to delta equivalent.

$$D_{HL} = 889.484 \qquad D_{HM} = 904.371 \qquad D_{LM} = -33.495$$

We can write:

$$\begin{aligned} \omega\bar{L}^{-1} &= \begin{vmatrix} D_{HL} + D_{HM} & -D_{HL} & -D_{HM} \\ -D_{HL} & D_{HL} + D_{LM} & -D_{LM} \\ -D_{HM} & -D_{LM} & D_{HM} + D_{LM} \end{vmatrix} \\ &= \begin{vmatrix} 17.9385 & -8.8948 & -9.0437 \\ -8.8948 & 8.5599 & 0.3349 \\ -9.0437 & 0.3349 & 8.7088 \end{vmatrix} \end{aligned} \tag{14-20}$$

In steady state, Eq. (14-1) can be written as:

$$\begin{vmatrix} V_1 \\ V_2 \\ \cdot \\ V_n \end{vmatrix} = \begin{vmatrix} Z_{11} & Z_{12} & \cdot & Z_{1n} \\ Z_{21} & Z_{22} & \cdot & Z_{2n} \\ \cdot & \cdot & \cdot & \cdot \\ Z_{n1} & Z_{n2} & \cdot & Z_{nn} \end{vmatrix} \begin{vmatrix} I_1 \\ I_2 \\ \cdot \\ I_n \end{vmatrix} \tag{14-21}$$

As per matrix theory, if all coils are open circuited and coil k is energized, the measured values of I_k and $V_i\,(V_1, V_2 \ldots V_n)$ give:

$$Z_{ik} = V_i / I_k \tag{14-22}$$

Continuing with Eq. (14-18), short-circuit impedance of 0.005 + j0.10 pu can be divided into two equal parts, and then, the magnetizing impedance which is relatively high, say 99.0 pu, can be connected in the middle. This gives a T circuit of the transformer. Then, the input impedance = 99.5 pu, and the matrix equation in the steady state is:

$$\begin{vmatrix} V_1 \\ V_2 \end{vmatrix} = \left\{ \begin{vmatrix} 0.0025 & 0 \\ 0 & 0.0025 \end{vmatrix} + j \begin{vmatrix} 99.5 & 99 \\ 99 & 99.5 \end{vmatrix} \right\} \begin{vmatrix} I_1 \\ I_2 \end{vmatrix} \tag{14-23}$$

The difference between the elements is small, and it could lead to ill-conditioned network, Z becoming infinity for zero exciting current. This problem is solved by constructing Y-matrix.

$$\begin{vmatrix} I_1 \\ I_2 \\ \cdot \\ I_n \end{vmatrix} = \begin{vmatrix} Y_{11} & Y_{12} & \cdot & Y_{1n} \\ Y_{21} & Y_{22} & \cdot & Y_{2n} \\ \cdot & \cdot & \cdot & \cdot \\ Y_{n1} & Y_{n2} & \cdot & Y_{nn} \end{vmatrix} \tag{14-24}$$

Then:

$$\overline{L}^{-1} = j\omega \overline{Y} \tag{14-25}$$

The Y matrix can be built as follows from reduced matrices, short-circuit data. For transient analysis, R and X are separated.

$$\begin{aligned} Y_{in} &= Y_{ni} = -\sum_{k=1}^{n-1} Y_{ik(\text{reduced})} \quad \text{for } i \neq n \\ Y_{nn} &= -\sum_{i=1}^{n-1} Y_{in} \end{aligned} \tag{14-26}$$

The equation is in pu. This can be illustrated by recalculating the matrix arrived in Eq. (14-20):

$$X_{\text{reduced}} = j \begin{vmatrix} X_{\text{HM}} & X_{\text{M}} \\ X_{\text{M}} & X_{\text{LM}} \end{vmatrix} = j \begin{vmatrix} 0.1150 & 0.1195 \\ 0.1195 & 0.2410 \end{vmatrix} \tag{14-27}$$

Find inverse:

$$Y_{\text{reduced}} = \frac{1}{j} \begin{vmatrix} 17.9385 & -8.89484 \\ -8.8948 & 8.5599 \end{vmatrix} \tag{14-28}$$

This gives:

$$\begin{aligned} \overline{L}^{-1} &= \omega \begin{vmatrix} 17.9385 & -8.89484 & -(17.93856 - 8.8948) = -9.04372 \\ -8.89484 & 8.55989 & -(-8.89484 + 8.55989) = 0.33495 \\ -9.04372 & 0.33495 & -(-9.04372 + 0.33495) = 8.70877 \end{vmatrix} \\ &= \omega \begin{vmatrix} 17.9385 & -8.89484 & -9.04372 \\ -8.89484 & 8.55989 & 0.33495 \\ -9.04372 & 0.33495 & 8.70877 \end{vmatrix} \end{aligned} \tag{14-29}$$

This is the same matrix as in Eq. (14-20). For three-phase n-coil transformers, we can write:

$$\begin{vmatrix} \overline{Z}_s & \overline{Z}_m & \overline{Z}_m \\ \overline{Z}_m & \overline{Z}_s & \overline{Z}_m \\ \overline{Z}_m & \overline{Z}_m & \overline{Z}_s \end{vmatrix} \tag{14-30}$$

where $\overline{Z}_s$ is the self-impedance of the coil on one leg, and $\overline{Z}_m$ is its mutual impedance to the coils on the other two legs. As in transmission lines:

$$\begin{aligned} Z_s &= \frac{1}{3}(Z_0 + 2Z_+) \\ Z_m &= \frac{1}{3}(Z_0 - Z_+) \end{aligned} \tag{14-31}$$

For transient analysis, the resistance and inductance parts are separated. This can be done from the short circuit test data. The winding resistance forms a diagonal matrix and $\overline{L}^{-1} = j\omega\overline{Y}$. The matrix $\overline{Y}$ is built from reactance values $j\omega L$.

14-7 EXTENDED MODELS OF TRANSFORMERS

The transient behavior of transformers has been a subject of much study and research over the last 100 years, and the literature is rich on this subject. An exact representation may be very complex. A transient transformer model should address saturation, hysteresis, eddy current, and stray losses. Saturation plays an important role in determining the transient behavior of the transformer. Extended transformer models can be very involved, and these are not required in every type of study (Table 14-1). The models in Fig. 14-9 for each category in Table 14-1 are based upon CIGRE guidelines with surge transfer.[1] A model not suited to this type of transient study may be prone to errors. For lightning and switching transients' studies, it is necessary to include capacitance of the transformers as high-frequency surges will be transferred more through electrostatic couplings rather than the electromagnetic couplings. There are many approaches to the models, some of which are briefly discussed.

14-7-1 EMTP Model Satura

Figure 14-10 shows a single-phase model. R_{MAG} remains constant, and it is calculated from excitation losses. The nonlinear inductor is modeled from transformer excitation data and its nonlinear V-I characteristics. The transformer's core saturates sharply and there is a definite knee point. Often a two-slope piece-wise linear inductor is adequate to model such curves, though complete analytical functions for representation of saturable elements have been proposed.[1] The saturation data is not supplied as a flux-current relation, but as a rms-voltage and rms-current curve. The Satura routine in EMTP converts this voltage-current relation to flux-current data. The input data is presented in per unit values with regard to winding connections and base current and voltage:

$$\begin{aligned} I_{\text{mag}} &= \left[I_{\text{ex}}^2 - (P_{\text{ex}} / V_{\text{ex}})^2 \right]^{1/2} \\ R_m &= \frac{V_{\text{ex}}^2}{P_{\text{ex}} - RI_{\text{ex}}^2} \end{aligned} \tag{14-32}$$

where R = ac resistance considered in excitation test, P_{ex} = excitation in watts, I_{ex} = excitation current, V_{ex} = excitation voltage, and R_m = magnetizing resistance representative of core losses, considered constant. There is a linear interpolation between the assumed values and finite-difference approximation to sinusoidal excitation. The hysteresis is ignored.

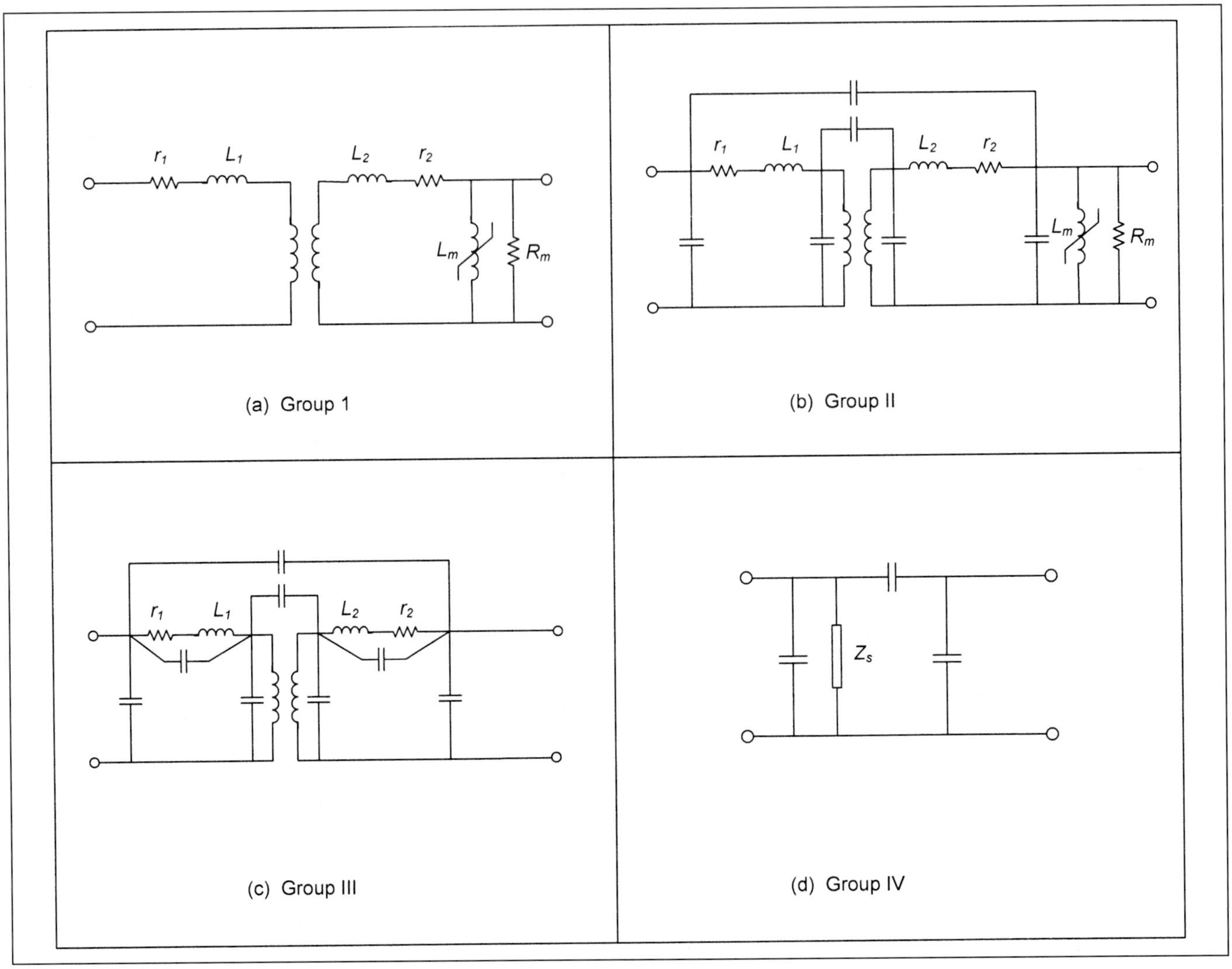

FIGURE 14-9 Transformer models based upon CIGRE.[1] Figures (*a*), (*b*), (*c*), and (*d*) for Group I, II, III, and IV transients, respectively.

Example 14-1 Consider a 50-kVA, 2400/240-volt, two-winding, single-phase, 60-Hz transformer. Its saturation test data is given in Table 14-2. The following relations apply:

$$V_{\text{base}} = 240 \text{ V} \quad S_{\text{base}} = 50 \text{ kVA} \quad I_{\text{base}} = 208.33 \text{ A} \quad V_{\text{mag}} = \frac{V_{\text{ex}}}{V_{\text{base}}} \tag{14-33}$$

Using Eq. (14-33) the input data for Satura is shown in Table 14-3. The calculated I-λ data in EMTP subroutine, nonlinear inductor is shown in Table 14-4 and Fig. 14-11. This is a two-slope curve and is modeled as a sequence of points, with linear interpolation between assumed values and finite-difference approximation to sinusoidal excitation. This representation is considered adequate except that for ferroresonance phenomena, a more detailed representation of the saturation characteristics may be required.

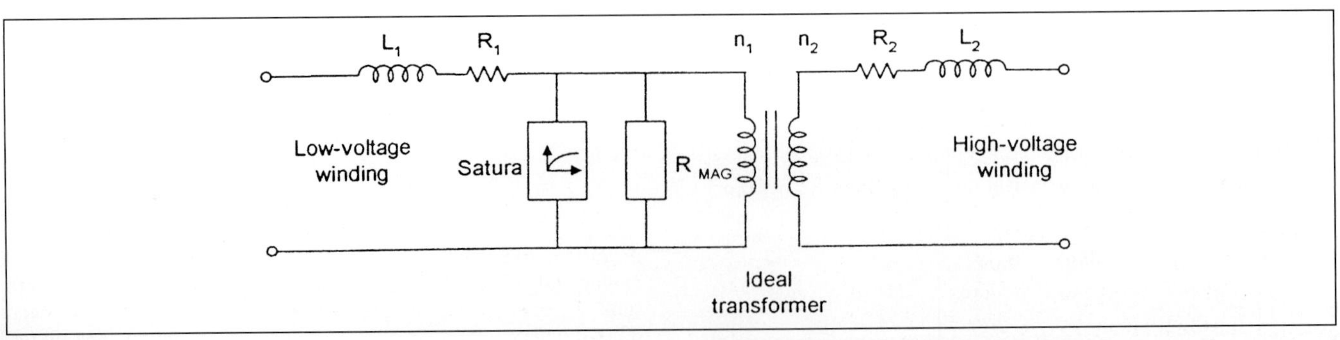

FIGURE 14-10 Single-phase transformer, EMTP model Satura.

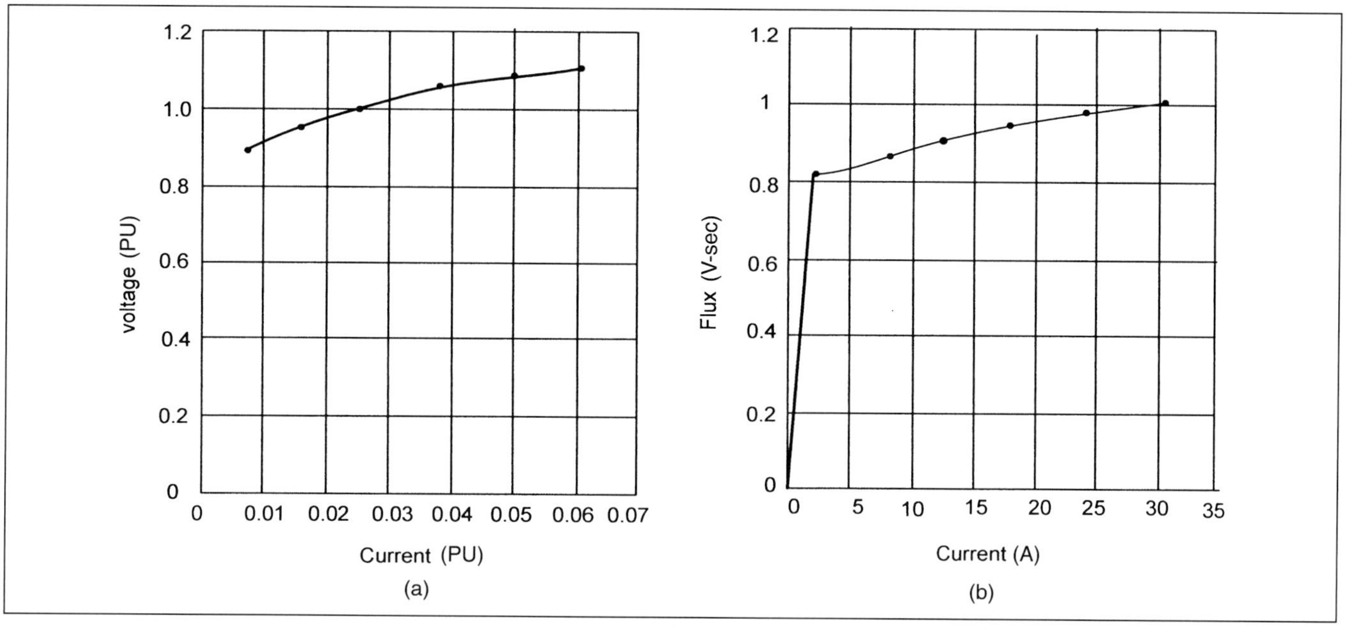

FIGURE 14-11 (*a*) Given RMS excitation data, voltage verses current. (*b*) Conversion of the excitation data into two-piece I-λ curve (saturation curve).

TABLE 14-2 Saturation Test Data for a Single Phase Transformer

EXCITATION VOLTAGE, V_{ex} (V, rms)	EXCITATION CURRENT, I_{ex} (A, rms)	EXCITATION POWER, P(W)
216	1.75	150.66
229	3.45	169.63
240	5.41	186.00
252	8.07	205.07
259	10.48	216.95
264	12.72	225.06

TABLE 14-4 Results of Calculation (Current versus Flux)

CURRENT (A)	FLUX (V-s)
2.27128591E+00	8.10284685E–01
8.49859062E+00	8.59802082E–01
1.24164919E+01	9.00316316E–01
1.80862562E+01	9.45332132E–01
2.44179342E+01	9.72341621E–01
3.05139341E+01	9.90347948E–01

TABLE 14-3 Calculated Magnetization Voltage and Current for Input to Satura, EMTP

V_{MAG} (PU)	I_{MAG} (PU)
0.9	0.007709
0.955	0.01619
1.0	0.02570
1.05	0.03855
1.08	0.05012
1.1	0.06091

The transformer is connected to a 2400-V single-phase ac source and serves a load of 50 kVA at 0.9 power factor (Fig. 14-12). It is energized by closing switch S_1 at 0.0125 s and then the load is applied by closing switch S_2 at 0.833 s.

Figure 14-13*a* depicts the primary current flow from the 2400-V source transients, while Fig. 14-13*b* shows excitation current and secondary load current transients. The excitation current resembles a rectified wave-shape.

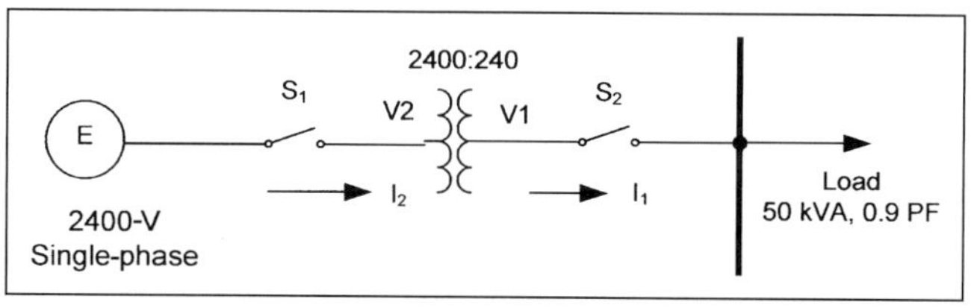

FIGURE 14-12 A circuit for study of switching transients in a single-phase, 2400 to 240-V transformer.

14-7-2 EMTP Model Hysdat

In this model, the hysteresis loop, represented in 4 to 5 points to 20 to 25 points, can be modeled:

- *Level 1.* 4 to 5 data points
- *Level 2.* 10 data points
- *Level 3.* 15 data points
- *Level 4.* 20 to 50 data points.

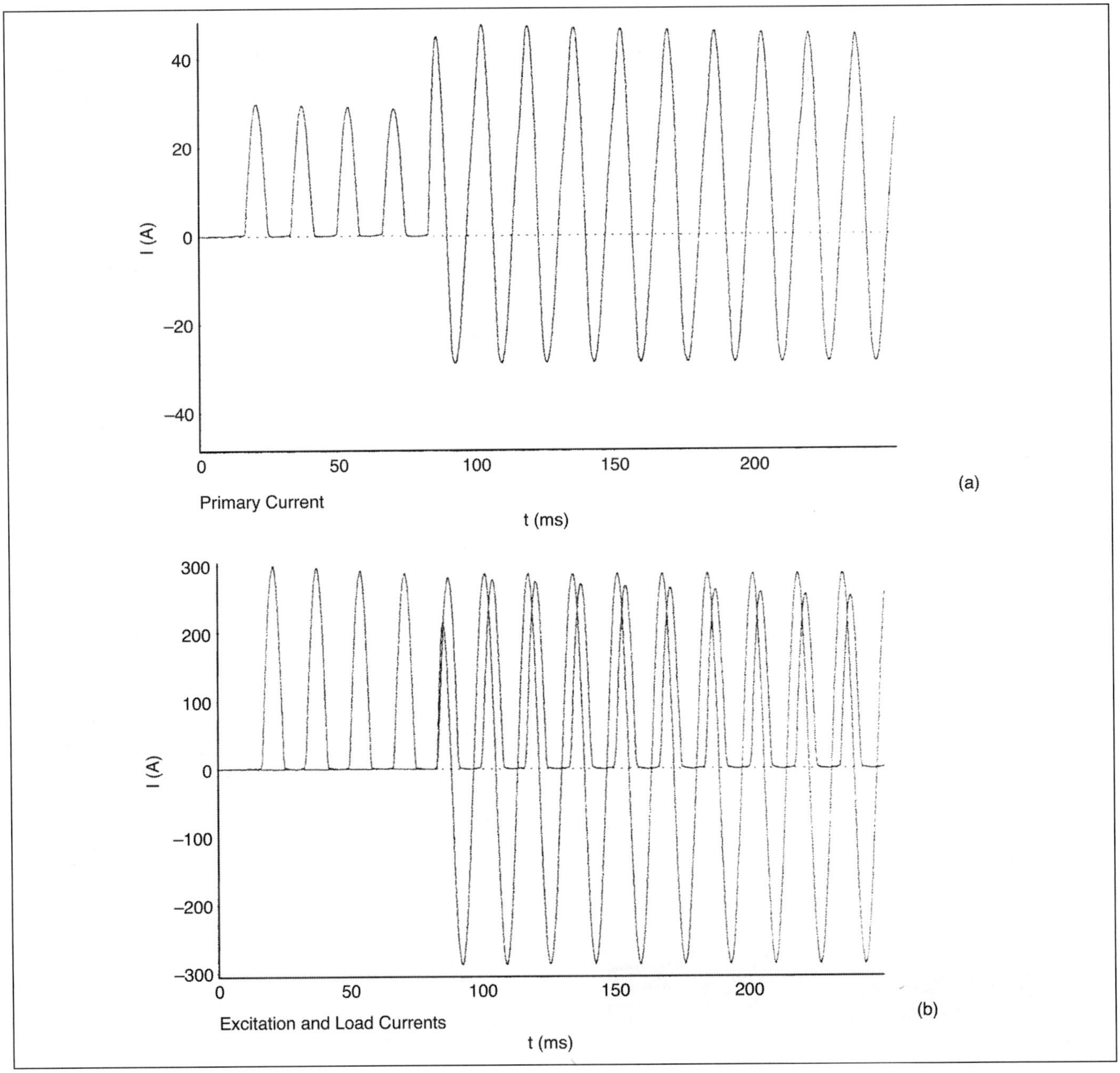

FIGURE 14-13 (*a*) Primary transformer current, and (*b*) excitation and load current transients, EMTP routine Satura.

The saturation current and flux in volt-seconds (V-s) is required to be specified. The scaling of the loop depends upon the geometry, the number of turns, and other constructional factors of the transformer. This scaling can be fully specified by location of the positive saturation point. This is the point on the first quadrant where hysteresis loop changes from being multivalued to single valued (Fig. 14-14). One way of determining these coordinates from the dc magnetizing curve is as follows.

Beginning at the right, that is, the linear part on the normal magnetization curve, a straight edge is used to extrapolate this line back to the left. The point where this line and the actual curve first begin to diverge is taken as the positive saturation point. Similarly the negative saturation point can be plotted; the positive and negative saturation points are shown with dark circular dots. A dc magnetizing curve is more readily available from a transformer manufacturer than the hysteresis loop.

Example14-2 The transients in Example 14-1 are recalculated with hysteresis data shown in Table 14-5. The switching sequence, the loading data, and the transformer data remains the same. The primary excitation current is now shown in Fig. 14-15.

14-7-3 Another Model of the Hysteresis Loop

The locus of the midpoints of the loop is obtained by measurements at four points and its displacement by a *consuming function*, whose maximum value is ob.[8] The segment ef changes periodically by half-wave symmetry (Fig. 14-16). The consuming function can be written as:

$$f(x) = -\text{ob}\sin\omega t \tag{14-34}$$

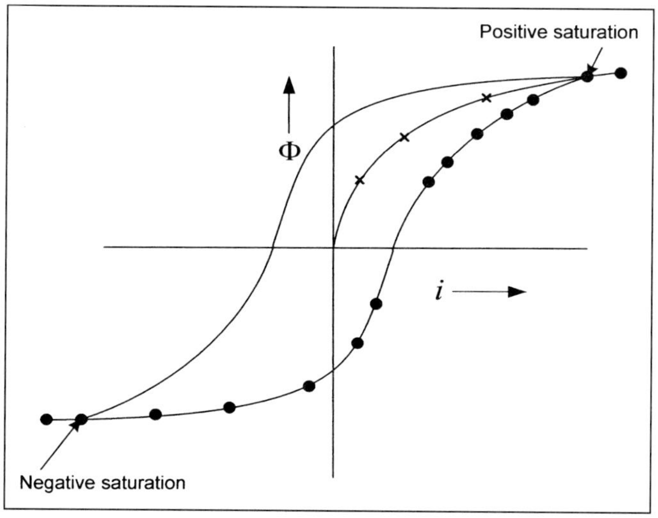

FIGURE 14-14 Hysteresis loop and EMTP model Hysdat.

TABLE 14-5 Computation of Hysteresis Curve, Using EMTP Hysteresis Routine*†

CURRENT (A)	FLUX (V-s)
−6.09855000E+00	−9.43743706E−01
−7.62318750E−01	−8.93982674E−01
4.57391250E−01	−8.23630871E−01
1.06724625E+00	−6.4061788E−01
2.05826063E+00	5.54806906E−01
3.20173875E+00	7.32116329E−01
5.48869500E+00	8.40789847E−01
1.01388394E+01	9.15145412E−01
2.43942000E+01	9.72342000E−01
3.35420250E+01	9.78061659E−01

*ARMCO silicon steel stampings
†Level 2—10 data points

The periphery can be represented by 16 line segments:

$$\begin{aligned} i &= (i_k - m_k\phi_k) + m_k\phi - \text{ob}\sin\omega t \\ \phi_{k-1} &< |\phi| \le \phi_k \qquad k = 1, 2, \ldots, 16 \end{aligned} \tag{14-35}$$

14-7-4 EMTP Model Becatran

Another popular transformer model is Becatran. It can be used to derive a linear $[A]$-$[R]$ or $[R]$-$[\omega L]$ representation of a single-phase or three-phase, core-type or shell-type, two-winding, three-winding, or multiple-winding transformers. (The matrix $\overline{A}$ is an inverse of $\overline{L}$.)

Test data from excitation test and short-circuit test is required for the model. For three-phase transformers, having one or more delta-connected windings, the core construction, shell or core type, is important. A closed delta acts like a short-circuited circuit to zero sequence currents. It is, therefore, assumed that all delta connections are open during the zero sequence excitation tests. Excitation losses can be taken into account or neglected. For low-reluctance transformers, that is, (i) transformers banks of single phase, (ii) three-phase, shell-type transformers, and (iii) three-phase, four-limb

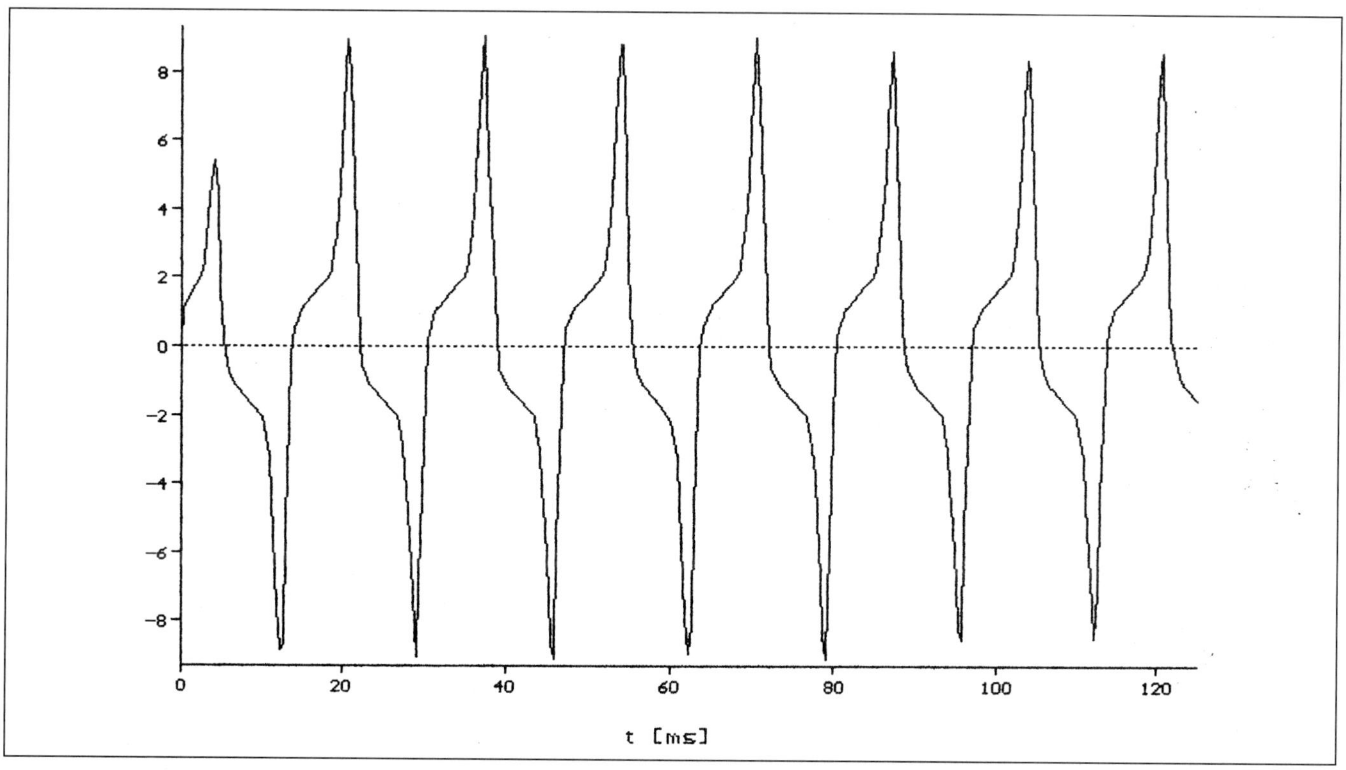

FIGURE 14-15 Excitation current transient, EMTP model Hysdat.

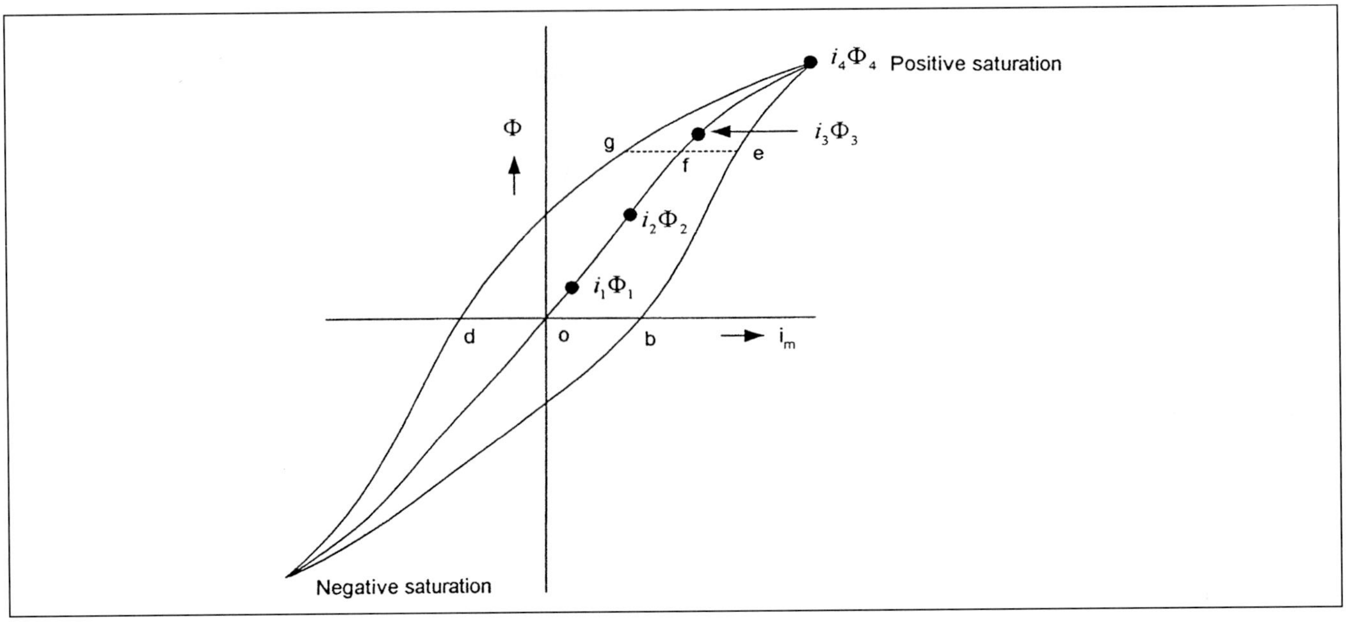

FIGURE 14-16 Piecewise hysteresis loop curve fitting.

or five-limb transformers, flux closes its path over core material and the reluctance is low. The zero sequence excitation current is small. In three-limb core-type transformers, the homopolar flux will close its path through the air and the tank, and the reluctance of the path is high. The zero sequence excitation current is important and the resulting excitation losses cannot be ignored. This is illustrated in Fig. 14-17. In a core-type transformer (Fig. 14-17*a*), the sum of the fluxes in each phase in a given direction along the cores is zero; however, the flux going up in one limb must return through the other two, that is, the magnetic circuit is completed through the other two phases in parallel. Now consider the zero sequence flux, which will be directed in one direction only in each of the three limbs. The return path lies not through the core, but through the insulating medium and tank. In three separate single-phase transformers connected in three-phase or shell-types transformers, the magnetic circuits of each phase are complete in itself and do not interact, Fig. 14-17(*b*).

Stray capacitances are again ignored in this representation. Thus, the model is applicable to a few kHz. The inductive and resistive parts of the short-circuit impedance are treated separately by this model, and therefore the model holds for low frequencies (a few kHZ).

The nonlinear behavior cannot be included in the Becatran model properly, but can be taken into account by elements, in Satura or Hysdat, connected to the windings close to the core. The saturable transformer model of EMTP (not discussed here) works well for two-winding transformers, but can become numerically unstable for three-winding transformers.

For a low value of the magnetizing current, the admittance matrix becomes nearly singular and the output option $[R]$-$[\omega L]$ cannot be used.

Example 14-3 Consider a 40-MVA, 138 to 13.8-kV three-phase, wye-wye-connected transformer, both wye-connected windings are grounded. Following test data is available: Excitation loss = 19 kW, excitation current = 3 A, excitation voltage = 13.2 kV, short-circuit losses = 195 kW, short-circuit current = 160 A, short-circuit voltage = 36 kV. (Excitation test conducted from the low-voltage winding side.)

Homopolar measurements are: excitation losses = 120 kW, excitation current = 500 A, excitation voltage = 1.5 kV, short-circuit losses = 9 kW, short-circuit current = 80 A, short-circuit voltage = 3 kV.

It is required to calculate the transformer parameters for input into Becatran routine. The calculation is straightforward; based upon the given data:

$$I_{\text{excpositive}} = 3\left(\frac{13.8}{13.2}\right) = 3.136A$$

$$I_{\text{excpositive}}(\%) = 3.136\times10^{-3}\times\frac{\sqrt{3}\times13.8}{40}\times100 = 0.1874 \text{ percent}$$

and:

$$\text{loss}_{\text{excpositive}} = 19\left(\frac{13.8}{13.2}\right)^2 = 20.77 \text{ kW}$$

Similarly:

$$I_{\text{exczero}} = \frac{500}{3}\times\frac{13.8}{\sqrt{3}\times1.5} = 885 \text{ A}$$

$$I_{\text{exczero}}(\%) = 885\times10^{-3}\times\frac{\sqrt{3}\times13.8}{40}\times100 = 52.88 \text{ percent}$$

and:

$$\text{loss}_{\text{exzero}} = 120\left(\frac{13.8}{\sqrt{3}\times1.5}\right)^2 = 3385 \text{ kW}$$

$$Z_{\text{positive}} = \frac{36}{0.160}\times\frac{40}{\sqrt{3}\times138^2}\times100 = 27.28 \text{ percent}$$

$$Z_{\text{zero}} = 3\times\frac{3\times10^3}{80}\times\frac{40}{138^2}\times100 = 23.63 \text{ percent}$$

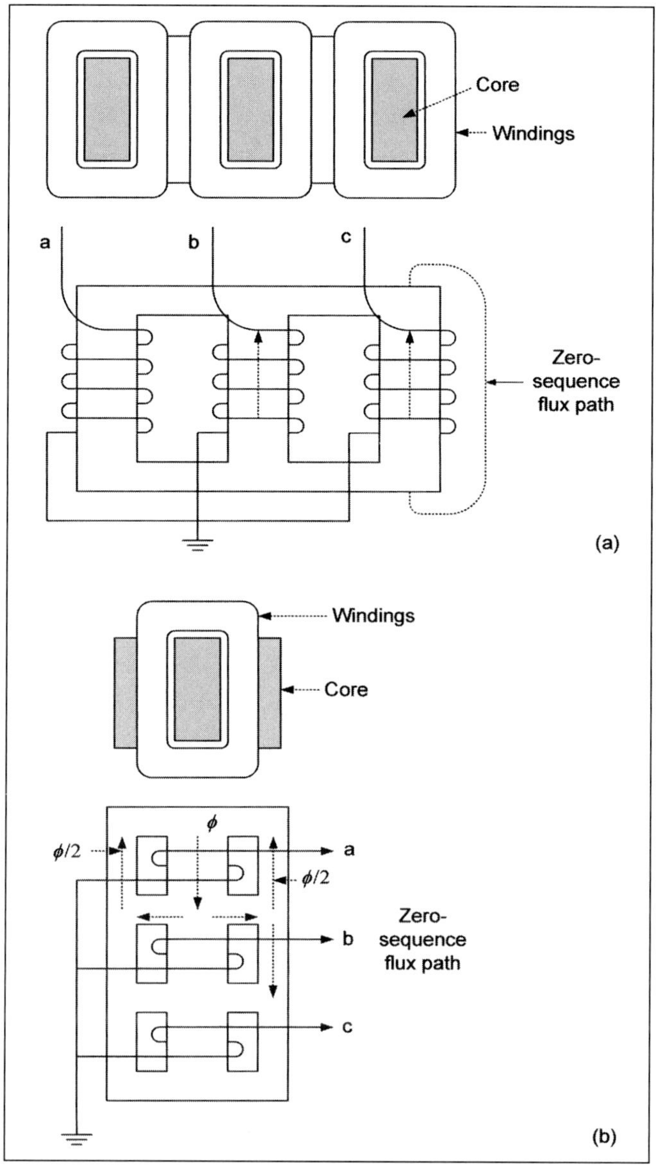

FIGURE 14-17 (*a*) Core-type three-phase transformer, flux paths for zero sequence currents. (*b*) Shell-type of transformer, zero sequence flux paths.

From the short-circuit test and dividing the resistance in 50 percent ratio on the primary and secondary side, the winding resistances are:

$$R_1 = 0.5\left[\frac{1}{3}\times\frac{195000}{160^2}\right] = 1.269\ \Omega$$

$$R_2 = 1.269\left(\frac{13.8}{138}\right)^2 = 0.0127$$

This data can be inputted to Becatran routine in EMTP. The calculated inductance matrix is shown below, each element truncated to five digits only.

$$\bar{L}^{-1} = \begin{vmatrix} 0.00102 & -0.01018 & 0.00005 & -0.0005 & 0.00005 & -0.0005 \\ -0.01018 & 0.10638 & -0.0005 & 0.00956 & -0.0005 & 0.00956 \\ 0.00005 & -0.0005 & 0.00102 & -0.01018 & 0.00005 & -0.0005 \\ -0.0005 & 0.00956 & -0.01018 & 0.10638 & -0.00005 & 0.00956 \\ 0.00005 & -0.0005 & 0.00005 & -0.00005 & 0.00108 & -0.01017 \\ -0.0005 & 0.00956 & -0.0005 & 0.00956 & -0.01017 & 0.10638 \end{vmatrix} \tag{14-36}$$

Note that this matrix is symmetrical.

Nonlinearity in Core Losses Figure 14-18 depicts a frequency-domain approach and considers that winding resistance and leakage reactance remain constant and the nonlinearity is confined to the core characteristics.[9] The core loss is modeled as a superimposition of losses occurring in fictitious harmonic and eddy current resistors. The magnetizing characteristics of the transformer are defined by a polynomial expressing the magnetizing current in terms of flux linkages:

$$i_M = A_0 + A_1\lambda + A_2\lambda^2 + A_3\lambda^3 + \cdots \tag{14-37}$$

Only a specific order of harmonic currents flow in to the appropriate G_h resistors. From Eqs. (14-10) and (14-11), the core loss equation is:

$$\begin{aligned} P_{fe} &= P_h + P_e \\ &= K_h B^s f + K_e B^2 f^2 \end{aligned} \tag{14-38}$$

For a sinusoidal voltage, this can be written as:

$$P_{fe} = k_h f^{1-s} E^s + k_e E^2 \quad (k_h \neq K_h \quad \text{and} \quad k_e \neq K_e) \tag{14-39}$$

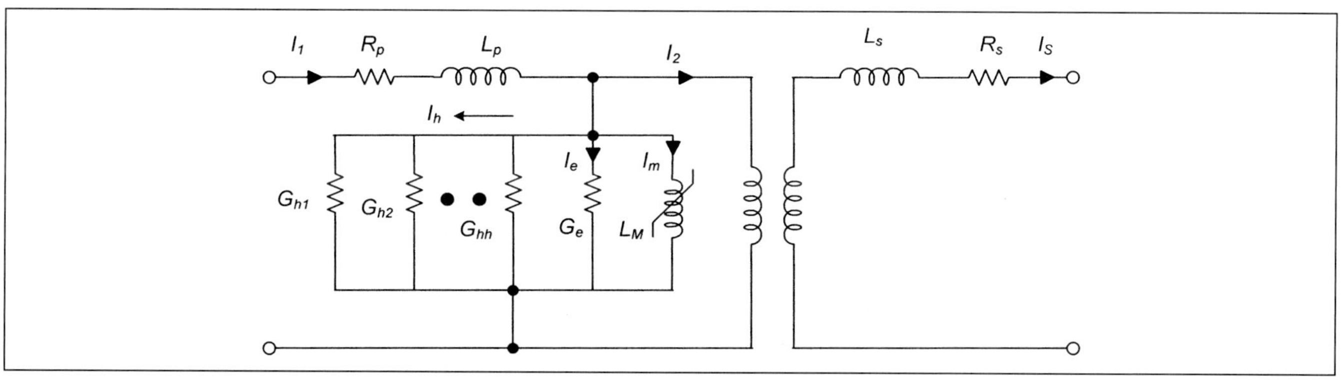

FIGURE 14-18 Transformer nonlinear shunt model with superimposition of harmonic currents in resistors.

This defines two conductances: G_h for hysteresis loss and G_e for eddy current loss, given by:

$$G_h = k_h f^{1-s} E^{s-2} \qquad G_e = k_e \tag{14-40}$$

14-8 EMTP MODEL FDBIT

A frequency-dependent transformer model is shown in Fig. 14-19*a* and *b*. The component building blocks are:

Power frequency module (BCTRAN, TOPMEG, and TRELEG)

Hysteresis/saturation module

Eddy current module

High-frequency transformer (HFT) module

It represents the behavior of the transformer over a wide range of frequency. The problem is generating the data and fitting it in the transformer model. Raw input data can be in the form of field measurements of $Y(\omega)$—the frequency scan measurements of nodal admittance matrix of the transformer. Then it has to be approximated with rational functions to be represented as FDB branches, see Fig. 14-19*b*. As an example, a two-winding, three-phase transformer can be represented by a two-port three-phase pi circuit, with nodal admittance

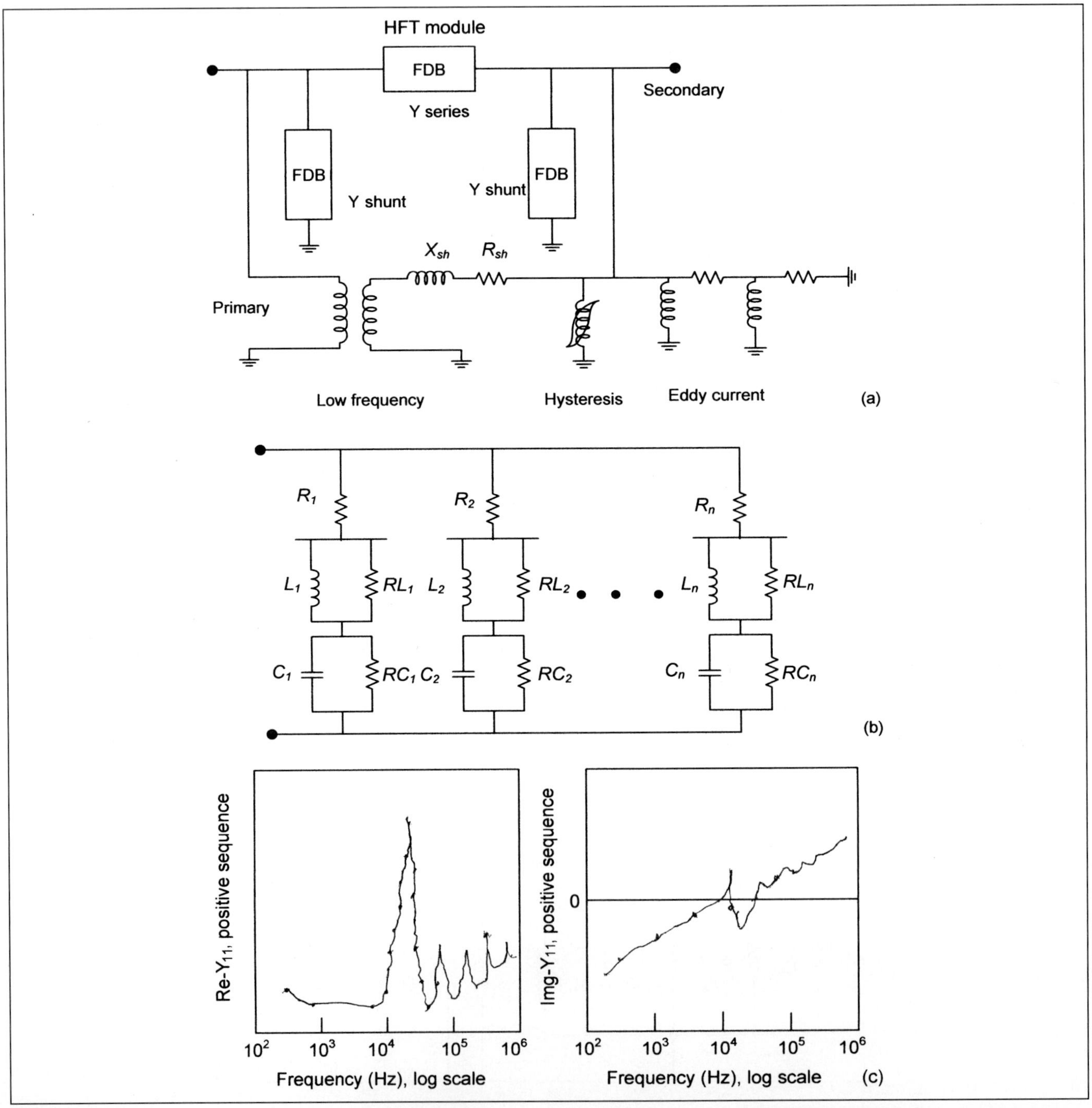

FIGURE 14-19 (*a*) Transformer model FDBIT in EMTP. (*b*) The high-frequency module realized with RLC network. (*c*) Real and imaginary parts of $\overline{Y}_{11}$ (positive sequence). Small dots show measured data.

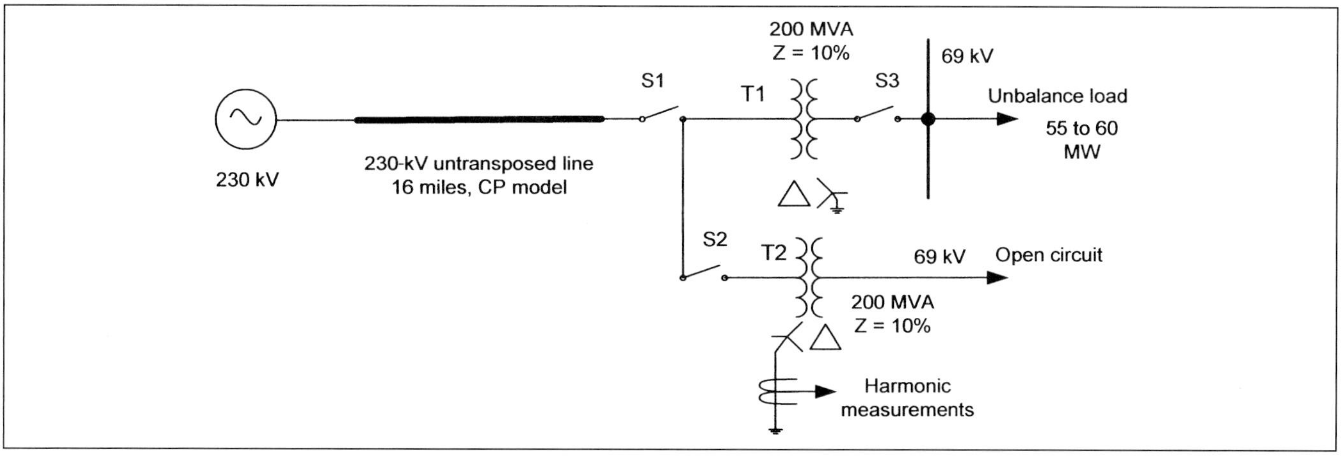

FIGURE 14-20 A power system model for study of transformer inrush current transients and sympathetic inrush.

matrix portioned in 3 × 3 blocks.

$$\bar{Y} = \begin{vmatrix} \bar{Y}_{11} & \bar{Y}_{12} \\ \bar{Y}_{21} & \bar{Y}_{22} \end{vmatrix} \tag{14-41}$$

where $\bar{Y}_{ij}$ is a 3 × 3 matrix of the form shown below and is assumed to be balanced,

$$\bar{Y}_{ij} = \begin{vmatrix} y_s & y_m & y_m \\ y_m & y_s & y_m \\ y_m & y_m & y_s \end{vmatrix} \tag{14-42}$$

Elements of pi circuit are described by balanced matrices; these can be modeled using positive and zero sequence parameters. The FDB branches represent approximations by rational functions of Y_{shunt} and Y_{series} of the pi circuit in zero and positive sequence.

The elements of Y are approximated with rational functions which contain real as well as complex conjugate poles and zeros. These are then realized with RLC networks, combined to produce the parameters of an equivalent Π circuit (Fig. 14-19*b*). Specimen measured data points of real and imaginary parts of $\bar{Y}_{11}$ are shown in this figure with small dots. The actual curves are fitted based upon measured data. Similar curves are obtained for $\bar{Y}_{11}$-zero sequence. The model provides:

- The inductive behavior at low frequencies which includes frequency-dependent effects due to skin effects and iron core eddy current losses.
- Series and parallel resonances from mid to high frequencies caused by winding-to-winding and winding-to-ground capacitances.
- Predominantly, capacitive behavior at high frequencies.

Example 14-4 A power system configuration is shown in Fig. 14-20. Two 200-MVA, 230–69 kV transformers of single-phase banks, winding connections as shown in this figure are connected at the end of a 16-mi, 230-kV untransposed line for which CP model is used. There are three simulation switches: *S*1, *S*2, and *S*3. Transformer *T*1 serves an unbalanced load if switches *S*1 and *S*3 are closed. The transformer models consider excitation characteristics.

First consider that all switches are open and the inrush transients of transformer *T*1 at no load on closing switch *S*1 at 8.33 ms are simulated. Figure 14-21 shows the switching current transients for time duration of 200 ms, in three phases.

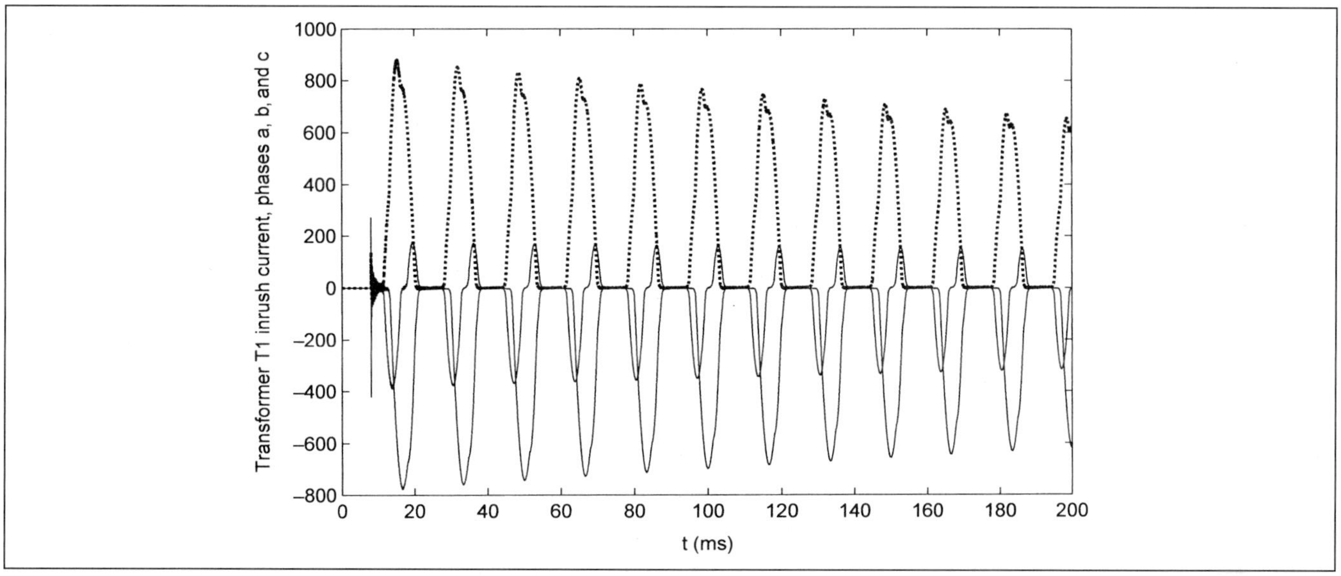

FIGURE 14-21 Transformer *T*1 inrush current transients, phases a, b, and c.

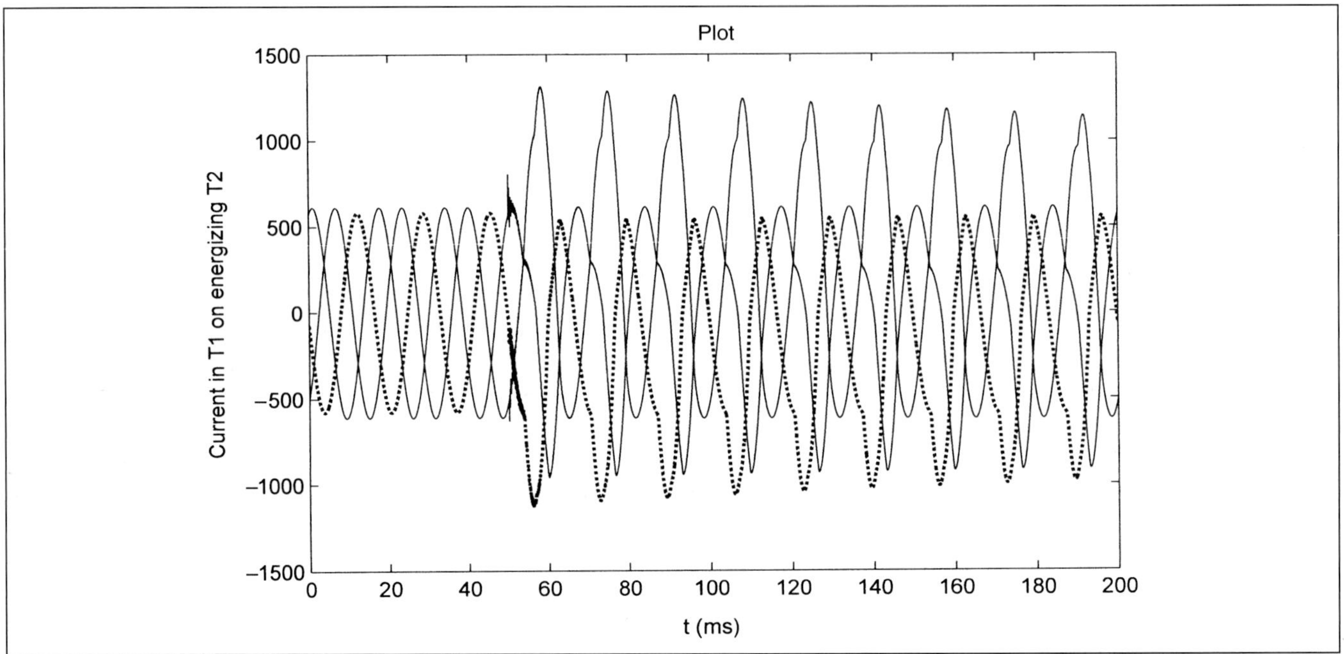

FIGURE 14-22 Sympathetic inrush current in transformer *T*1 on switching transformer *T*2 of Fig. 14-20.

14-9 SYMPATHETIC INRUSH

When a transformer is energized on the same bus to which an existing operating transformer is connected, the operating transformer experiences a surge in current on switching the transformer being energized. This phenomenon is called *sympathetic inrush* and is illustrated in the following example.

Example 14-5 The power system configuration is as in Fig. 14-20. The switches *S*1 and *S*3 remain closed for a long time, that is, transformer *T*1 is in a steady state and supplies unbalance load with an unbalance = 12 percent maximum. Transformer *T*2 does not serve any load and is energized by closing switch *S*2 at 50 ms. The impact of energizing *T*2 on the current transients in *T*1 is illustrated in Fig. 14-22, which shows that the already energized transformer *T*1 on the same bus experiences a transient inrush which increases its current by a factor of 2.6.

The harmonic current measurements of the transformer *T*2, in its primary neutral circuit, are shown in Fig. 14-23 which is a plot of second, third, fourth, and fifth harmonics.

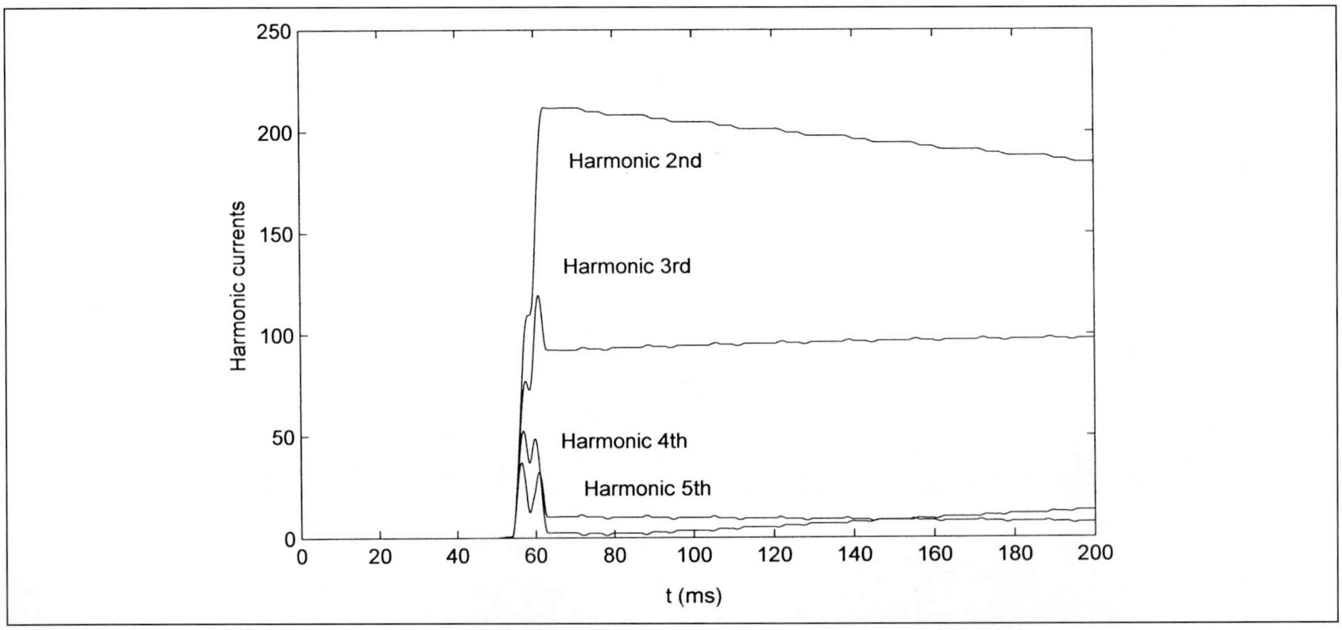

FIGURE 14-23 Harmonic currents in transformer *T*2 neutral on switching.

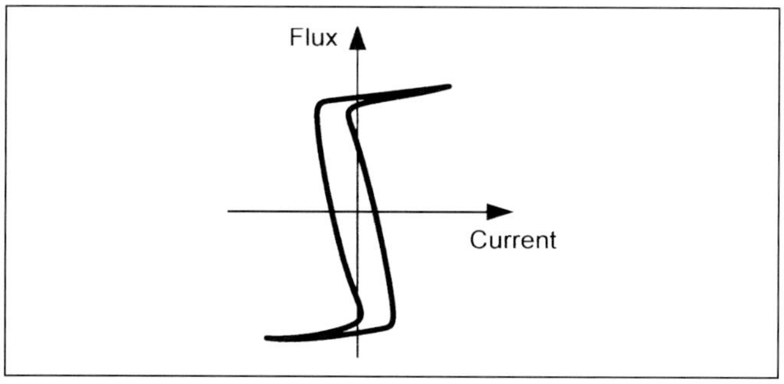

FIGURE 14-24 Cobra-type hysteresis characteristics with capacitance effects included.

14-10 HIGH-FREQUENCY MODELS

The initial surge transfer through a transformer is a function of the transformer and secondary-side external stray capacitance. Thus, knowledge of the transformer capacitances is required. Most of the time this data may not be readily available, and it is a function of transformer winding construction, end-turn reinforcement, and whether the transformer is of shell- or core-type construction. For an accurate analysis, this data should be obtained from the manufacturer and can also be estimated from constructional details.[10,11] Table 14-1 shows that modeling of transformer capacitances is important for some slow transient phenomena, like ferroresonance. These capacitances can confound measurements of transformer excitation characteristics, causing a "cobra" flux curve (Fig. 14-24). The capacitive transfer of voltages to low-voltage winding may be described as a capacitive voltage division.

At higher frequencies, the simplest circuit as seen from the low-voltage winding side consists of an EMF source in series with a transfer capacitance. The equivalent EMF on the secondary side is reduced by a factor, which is not so well defined. The loading of the secondary cables and switchgear bus, connected to the secondary side of the transformer adds lumped capacitance to the circuit, and the circuit acts like a voltage divider, reducing the transferred overvoltage peak (Figure 14-25).

With the transformer isolated from the distribution system, the lumped capacitance is dependent upon the resonant frequency of the transformer and will change in ratio f/f_0, where f is the frequency of the impressed oscillation and f_0 is the natural resonant frequency of the transformer windings.

A very detailed model may include each winding turn and turn-to-turn inductances and capacitances. A disk-layer winding or pancake sections are shown in Fig. 14-26*a*.[12] Each numbered rectangular block represents cross section of a turn. Winding line terminal is at *A* and winding continues beyond *E*. Each section can be represented by a series of inductance elements with series and shunt capacitances as shown in Fig. 14-26*b*. Though the model looks complex, the mutual inductances are not shown, resistances are not represented, and all interturn capacitances are omitted. This circuit will be formidable in terms of implementation. For most applications, representation of each turn is not justified, and by successive lumping, a much simpler model is obtained (Fig. 14-26*c*). Figure 14-27 shows a further simplified model of the capacitances associated with a two-winding transformer. The symbols in this figure are defined as follows:

C_{bh}. High-voltage bushing capacitance

C_b. Secondary bushing capacitance

C_{hg}. Distributed capacitance of the high-voltage windings

C_{lg}. Distributed capacitance of the low-voltage winding

C_{hl}. Distributed capacitance between the high- and low-voltage winding

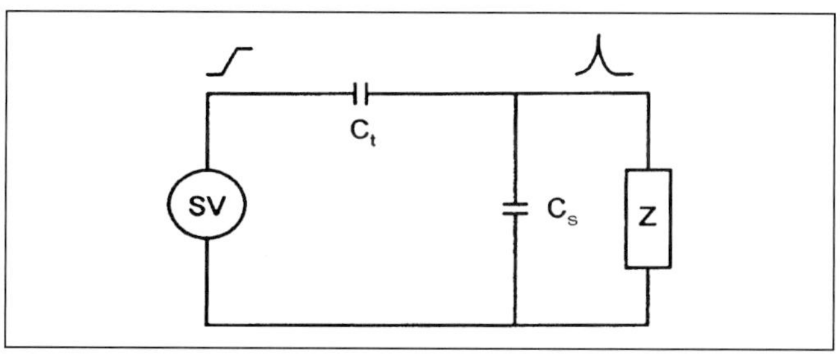

FIGURE 14-25 An elementary model for surge transference through the transformer.

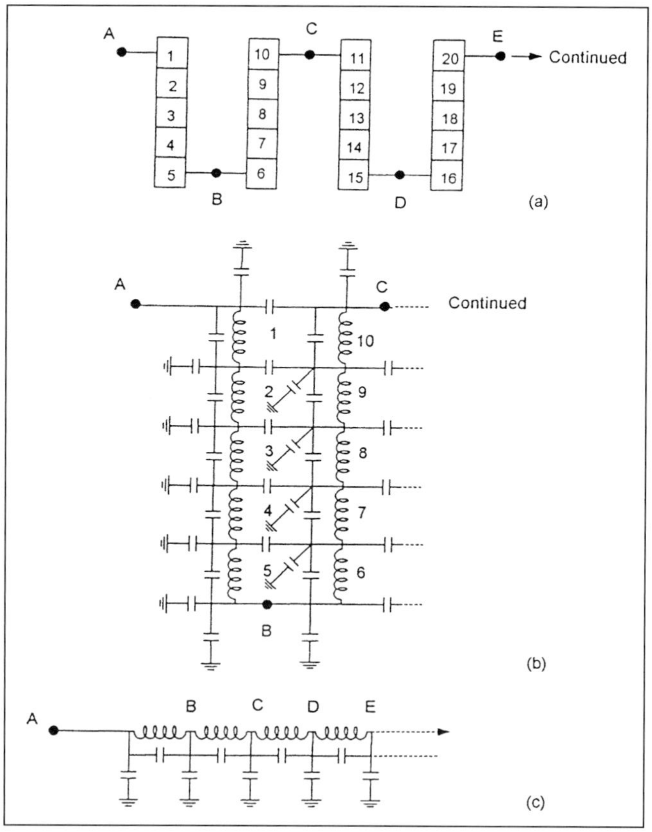

FIGURE 14-26 (*a*) Winding turns in a pancake coil. (*b*) Circuit of winding inductances and capacitances. (*c*) Simplified circuit model.

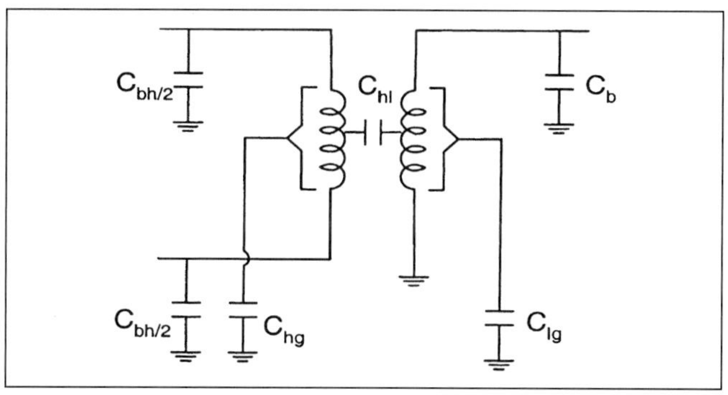

FIGURE 14-27 Circuit model of a two-winding transformer with capacitances.

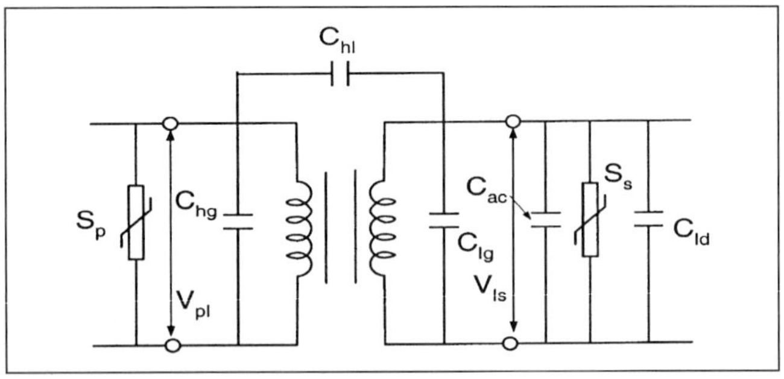

FIGURE 14-28 Circuit model of a transformer with transformer capacitances, external load capacitance, and primary and secondary surge arresters.

By combining the bushing capacitance into the winding capacitances, the model is further simplified as shown in Fig. 14-28, which considers:

C_{ac}. Surge capacitors on the secondary side of the transformer, if provided, to limit the transferred surge voltage

C_{ld}. Capacitance of the load, (cables and distribution system), connected to the secondary of the transformer.

Sp and S_S. primary and secondary surge arresters, respectively

This simplified model can be used for capacitance surge transfer. Table 14-6 shows typical capacitances of core- and shell-type transformers. It is seen that these capacitances vary over a wide range.

The transformer model described in Ref. 12 determines how a voltage applied to one set of terminals is transferred to the other set of terminals. This high-frequency model duplicates transformer behavior over a wide range of frequencies and acts like a filter suppressing some harmonics and passing others. For example, the use of frequency-dependent model is important when determining the surge that will appear on a generator bus (secondary side of the transformer), and when a steep-fronted surge impinges on the high-voltage terminals of the generator step-up transformer.

14-11 SURGE TRANSFERENCE THROUGH TRANSFORMERS

The lightning and switching surges can be transferred from one voltage level to another through transformer couplings.[13] A distribution system, which may not be directly exposed to the overvoltages of atmospheric origin, but is connected to a utility system through a transformer of high turns ratio will be exposed to overvoltages on the secondary side due to overvoltages on the primary windings. The resulting stresses on the distribution system may exceed the BIL levels, unless surge arresters are provided to limit the transferred voltages.

The most common primary distribution voltage in industrial systems is 13.8 kV. However, for large power demands, the utility system voltage may be as high as 230 kV. The surge transfer through the transformers depends upon the voltage, turn ratio, and electrostatic and electromagnetic couplings of the windings.

The lightning and steep-fronted waves are partially transferred through the electromagnetic coupling, which is the mechanism that governs the transformer operation at power frequencies and depends upon the turn ratio. The transfer of surges of this nature is less than 40 percent of the magnitude by using conventional

TABLE 14-6 Typical Capacitance of Transformers (Capacitance in nF)

TRANSFORMER	CORE TYPE			SHELL TYPE			AUTOTRANSFORMERS		
MVA	C_{hg}	C_{hl}	C_{lg}	C_{hg}	C_{hl}	C_{lg}	C_{hg}	C_{hl}	C_{lg}
1	1.2–14	1.2–17	3.1–16						
2	1.4–16	1–18	3–16						
5	1.2–14	1.1–20	5.5–17						
7	2.7–11	3.5–17	8–16						
10	4–7	4–11	8–18						
25	2.8–4.2	2.5–18	5.2–20	4–7	10–17	4–8	8–18	3.5–8	3–8
50	4–6.8	3.4–11	3–24	6–9	8–16	4–15	5.5–10	5.3–13	6–17
75	3.5–7	5.5–13	2.8–30	7–13	7–17	4–25	7–11	6–20	11–18
100	3.3–7	5–13	4–40	6–14	7–19	5–30	8.5–12	9–20	12–17
200				4.5–18	8–25	5–27	8–40	7–22	16–26
300				5.5–21	11–22	20–40	6–40	5–24	12–32
400				8–24	14–21	17–45	6–25	5–20	11–23
500				6.5–30	17–24	18–46	6–22	5–20	10–24
600				5–40	16–29	16–46	6–20	4.5–20	14–26
700				4–39	15–30	14–43	6–18	5–20	12–28
1000				4–41	11–30	6–54	6–12	7–23	16–30

turn ratio, and hence electrostatic effects dominate the coupling of transients from the primary to the secondary windings. For slower switching surges the electromagnetic coupling effect predominates, and the surges due to electromagnetic couplings can be estimated:

The switching surge transferred through the inductive coupling is given by Ref. 14:

$$V_{ss} = \left[\frac{2nZ_2}{Z_1 + n^2 Z_2}\right] \times V_{ps}[e^{-(Z_1 n^2 Z_2/L_m(Z_1+n^2Z_2))t} - e^{[-(Z_1+n^2Z_2)/L_s]t}] \tag{14-43}$$

where Z_1 is the primary surge impedance, Z_2 is the secondary surge impedance, n is the turn ratio, L_m is the transformer magnetizing inductance, L_s is the transformer leakage inductance, V_{ss} is the switching surge transferred to the secondary, and V_{ps} is the peak value of the switching surge on the primary, limited by the primary surge arrester.

There are other factors, that is, the surge arrester lead length, the location of arrester with respect to the protected transformer, and the effectiveness of grounding, which also impact the surge transference (Chap. 20). The selection of an appropriate surge arrester is an important consideration. In a multiplegrounded system, a part of the lightening impulse to the primary system could travel to the ground through the secondary neutral conductor. This ground path could induce current in the secondary conductors that could enter the low-voltage side and cause failure of the high-voltage windings. In industrial systems, the secondary neutral is commonly grounded through 200- to 400-A resistor and, thus, the secondary windings can be considered isolated.

First, a simplified approach to the estimation of the initial voltage spike transferred to secondary windings, due to lightning strokes, is discussed without detailed system modeling. The transformer is provided with primary (high-voltage) surge arresters, a circuit model as shown in Fig. 14-28. The initial voltage spike can be calculated from the following expression (see Chap. 20 for the surge arrester characteristics):

$$V_{ls} = SpV_{pl} \tag{14-44}$$

where V_{ls} is the lightning surge transferred to the secondary of the transformer, S is the factor depending upon the transformer winding capacitance and capacitance of the low-voltage side connections, p is the factor which allows superimposition of lightening surge voltage over power frequency voltage. p is 1.05 for wye-wye and delta-delta connected transformers, and p is 1.15 for delta-wye connected transformers. V_{pl} is the primary surge protective level of the surge arresters. With the transformer energized at no load, the factor S is given by:

$$S = \frac{C_{hl}}{(C_{hl} + C_{lg} + C_{ext})} \tag{14-45}$$

where C_{ext} is the secondary-side external capacitance, that is, of cables connected to the secondary of the transformer.

Equations (14-44) and (14-45) show that the initial voltage spike is independent of the transformer turn ratio, and it is simply a function of C_{hl}, C_{lg}, and the external capacitance. This results in a simplified transformer model, neglects all saturation, but it is not accurate.

Example 14-6 A 40-MVA, 230 to 13.8-kV transformer has primary windings in delta connection and the secondary windings are wye-connected. The primary windings have a BIL of 900-kV and the secondary winding BIL is 110-kV. The 230-kV system is solidly grounded and the 13.8-kV system is low-resistance grounded

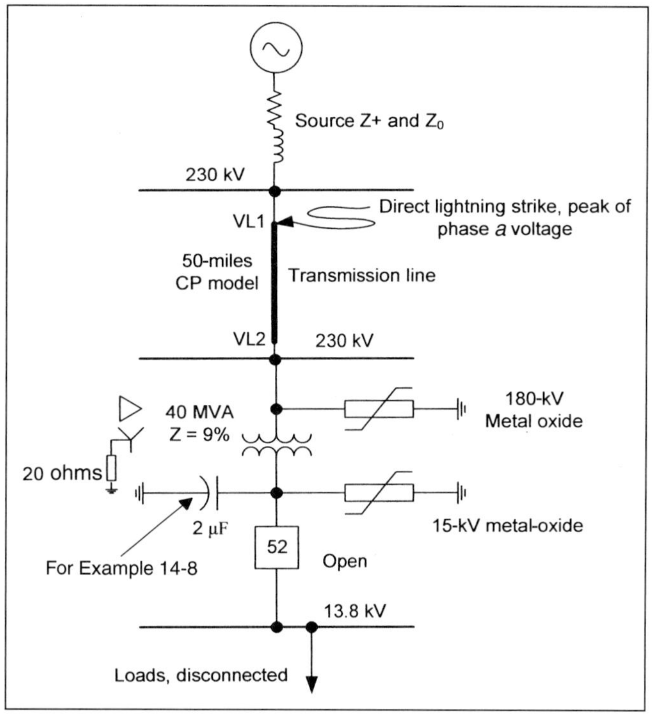

FIGURE 14-29 Power system connection diagram for study of switching transients.

through a 400-A resistor. Let the primary winding be protected with a metal oxide, station-class surge arrester of 180-kV rated voltage. Consider that C_{hg} = 10 nF and C_{lg} = 20 nF. The primary surge protective level of 180 kV surge arrester = 468 kV (See Chap. 20). Then from Eqs. (14-44) and (14-45), the secondary voltage spike = 179 kV, which is superimposed upon power frequency voltage, that is, a peak of 198.5 kV. This is much above the secondary BIL of 110 kV, commonly applied for a 13.8-kV system.

Example 14-7 This example illustrates EMTP simulations of a lightning surge transferred through a transformer when: (a) no primary or secondary surge arresters are provided, (b) only primary surge arresters are provided, and (c) both primary and secondary surge arresters are provided. The configuration for the study is shown in Fig. 14-29; the transformer secondary is open circuited. A direct lightning stroke occurs on phase a conductor of the transmission line. The following models are used:

- *Transformer.* Capacitance model as shown in Fig. 14-28, percentage impedance = 9 percent on 40-MVA base, capacitance values are: C_{hg} = 10 nF, C_{lg} = 20 nF, and C_{hl} = 10 nF.
- *230-kV transmission line.* Flat conductor layout with one overhead ground wire. A constant parameter traveling wave model (see Chap. 4).
- *Source impedance.* Lumped impedance behind a voltage source.
- *Direct lightning stroke.* The stroke on phase a on 230-kV side at the bus labeled VL1 is modeled using EMTP single exponential model for the transient. Surge current = 20 kA with 8/20 μs wave at t = 0.4667 ms, at the peak of the phase a voltage (See Chap. 20).
- *Surge arrester models.* These are discussed in Chap. 20.

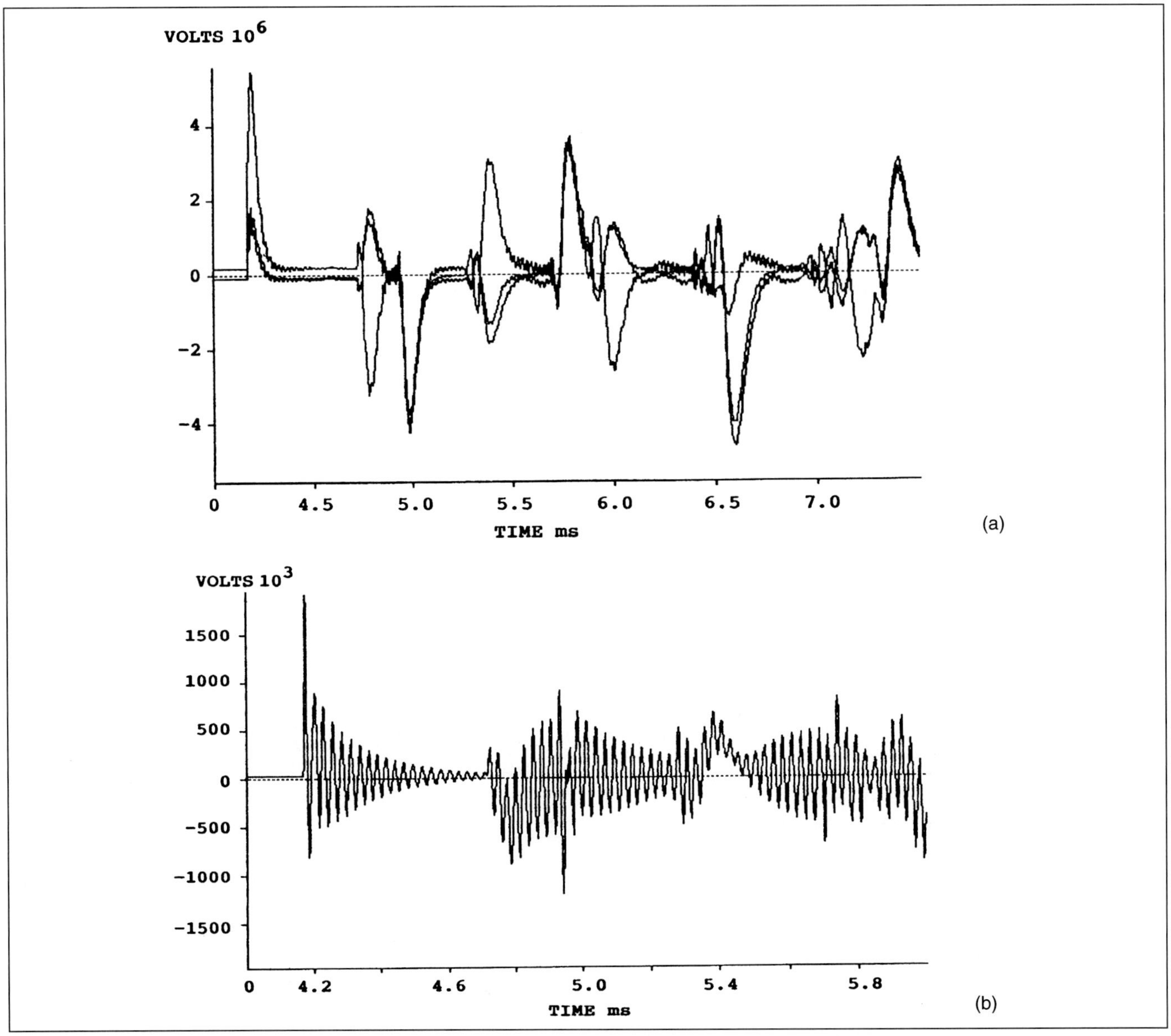

FIGURE 14-30 (*a*) Voltage transients on transformer primary windings in phases *a*, *b*, and *c*, and (*b*) voltage transient in phase *a* of the secondary windings, no surge arresters.

(a) *Model with no primary or secondary surge arresters.*

The voltage profiles on 230-kV and 13.8-kV sides are shown in Fig. 14-30*a* and *b*, respectively. Figure 14-30*a* shows the surge voltages for all three phases on the transformer primary windings, while Fig. 14-30*b* shows the secondary surge voltage for phase *a* only, for clarity. It is seen that multiple peaks occur due to traveling wave phenomenon and reflections from the source end of the 230-kV line. The negative maximum of the primary voltage in phase *c* occurs after approximately 0.25 ms of the positive maxima in phase *a*.

The maxima of phase *a* voltage, directly struck shows a voltage of 5400-kV peak (phase-to-neutral). The phases *b* and *c* soon get coupled and the negative maximum is 4060-kV. The BIL level of the primary insulation is 900-kV peak. The voltage transferred to secondary reaches a peak of 2000-kV (phase-to-neutral), a very high transferred voltage indeed for a 13.8-kV (rms) secondary winding, which has a BIL of 110-kV. Thus, without surge arresters on the primary of the transformer a very high surge voltage impacts the windings, which can result in immediate failure. The simulation demonstrates the importance of primary surge arrester, provisions of which are a standard practice in the industry.

(b) *Model with only primary surge arresters provided.*

A 180-kV metal oxide gapless surge arrester, characteristics as discussed in Chap. 20, is provided on the primary terminals of the transformer, directly mounted on the transformer,

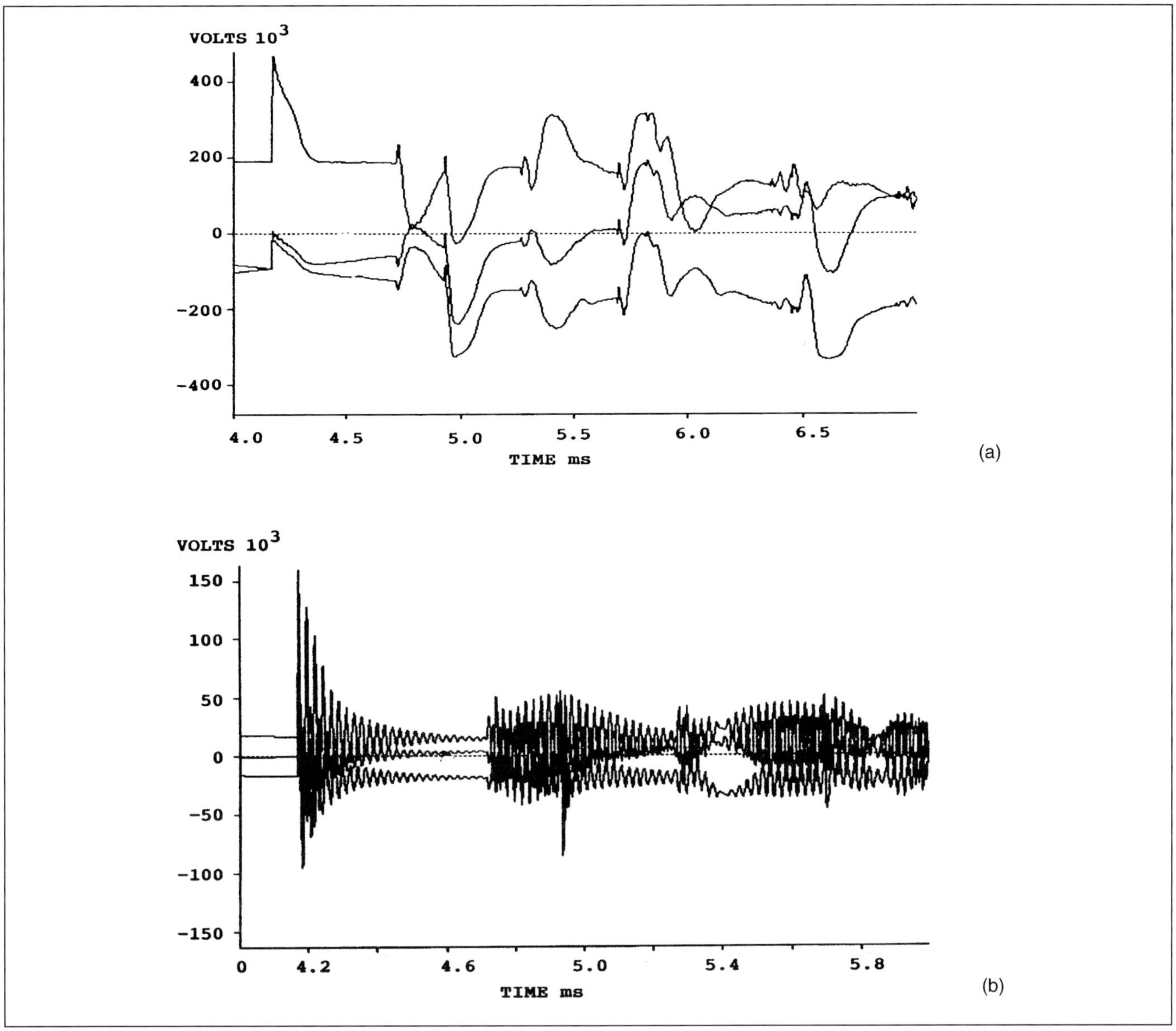

FIGURE 14-31 Voltage transients on transformer (*a*) primary and (*b*) secondary windings in phases *a*, *b*, and *c*; primary surge arresters only.

which is a usual installation. This ignores the arrester lead lengths. Figure 14-31*a* and *b* depicts the transients on the primary (230-kV) windings and the surge voltage transferred to the secondary (13.8-kV) windings, respectively, in all the three phases. The model of the surge arrester is discussed in Chap. 20; the manufacturer's data is used. The surge voltages are reduced:

- *230-kV windings*. 465 kV peak to ground
- *13.8-kV windings*. 160 kV peak to ground

The primary surge arresters have drastically reduced the surge voltage on the primary windings and also the voltage transferred to secondary windings. Though the primary insulation is protected, the secondary is not. It is exposed to a stress equal to 1.45 times the BIL of 110 kV.

(c) *Model with primary and secondary arresters provided.*

In addition to the primary arrester, a secondary arrester of 15-kV rated voltage, metal oxide, gapless type is mounted directly on the transformer terminals. The simulations of transients in this case are shown in Fig. 14-32*a* and *b*. The primary surge voltage pattern remains unaltered as in case (b), but the secondary surge voltage is reduced to a peak of 34.5 kV.

Example 14-8 This example is a continuation of Example 14-7 and is a simulation of the surge transferred to the secondary when in addition to the primary and secondary surge arresters a secondary surge capacitor of 2 μF is also provided.

The simulation in Fig. 14-33 shows that the secondary voltage is further reduced to 27 kV peak, phase-to-ground. Example 14-7 illustrates that the transformers must be protected with primary

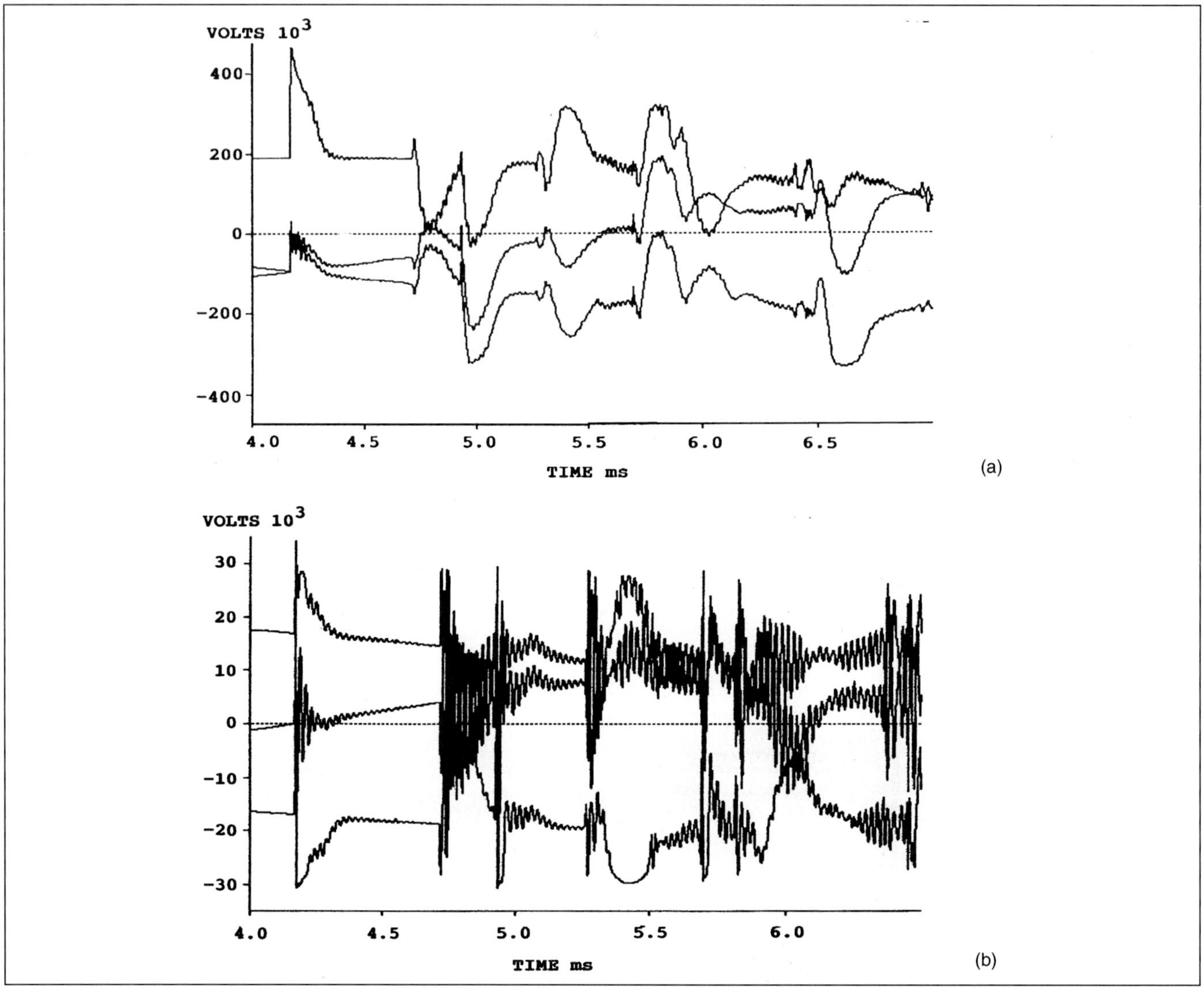

FIGURE 14-32 Voltage transients on transformer (*a*) primary and (*b*) secondary windings in phases *a*, *b*, and *c*; primary and secondary surge arresters.

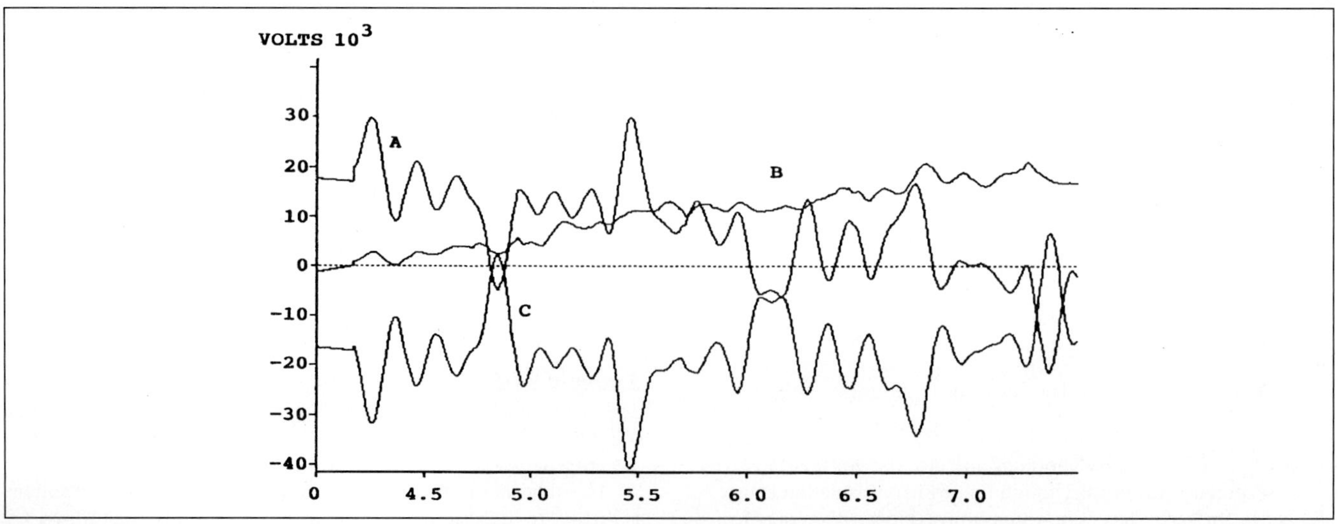

FIGURE 14-33 Voltage transients on transformer secondary windings in phases *a*, *b*, and *c*; primary and secondary surge arresters, and 2-μF secondary surge capacitor.

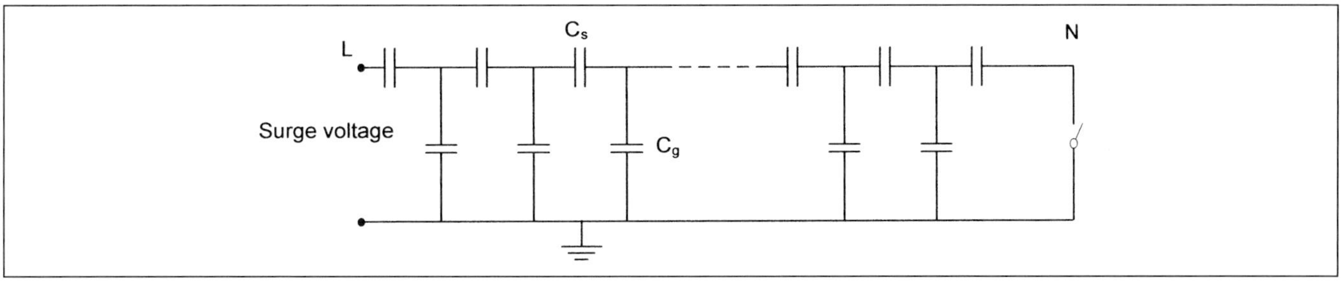

FIGURE 14-34 A capacitance ladder network for distribution of surge voltage across transformer windings.

and secondary arresters, even if a rigorous study is not conducted. In fact the selection of appropriate surge protection may lower the required BIL for the transformer windings, resulting in cost savings.

14-12 SURGE VOLTAGE DISTRIBUTION ACROSS WINDINGS

The surge distribution across a transformer winding from the line to neutral terminal is nonlinear, and the line-side turns are exposed to a much steeper voltage gradient. For study of such surge distribution, we could derive the circuit of Fig. 14-34 from Fig. 14-26*c* by eliminating reactance portion of windings altogether, and retaining only the capacitances. For a distributed parameter line, Eq. (4-11) is reproduced below:

$$V = V_1 e^{\gamma x} + V_2 e^{-\gamma x} \tag{14-46}$$

By analogy, we can write a similar equation for the circuit of Fig. 14-34:

$$E = A_1 e^{px} + A_2 e^{-px} \tag{14-47}$$

where:

$$p = \frac{1}{l}\sqrt{\frac{C_g}{C_s}} \tag{14-48}$$

and A_1 and A_2 are the constants depending upon the terminal conditions. Consider that the neutral terminal is grounded in Fig. 14-34, then at $x = 0$, $E = 0$, and at $x = l$, $E = V$.

This gives:

$$A_1 = -A_2 = \frac{V}{e^{pl} - e^{-pl}} = \frac{V}{2\sinh\alpha} \tag{14-49}$$

where:

$$\alpha = pl = \sqrt{\frac{C_g}{C_s}} \tag{14-50}$$

Substituting in Eq. (14-47) gives:

$$\begin{aligned} E &= \frac{V}{2\sinh\alpha}(e^{px} - e^{-px}) \\ &= \frac{V\sinh(\alpha x/l)}{\sinh\alpha} \end{aligned} \tag{14-51}$$

If the neutral is isolated, then at $x = l$, $E = V$ and at $x = 0$, $Is = 0$ or $dE/dx = 0$

This will give:

$$A_1 = A_2 = \frac{V}{2\cosh\alpha}$$

and

$$E = \frac{V\cosh(\alpha l/x)}{\cosh\alpha} \tag{14-52}$$

Curves of distribution from line end to neutral can be plotted. The distribution will not be uniform, and the voltage will be much higher on the winding turns toward the line side; the steepness depends upon α.

Example 14-9 Referring to Fig. 14-34, there are four sections, with series capacitance of 1 nF and shunt capacitance to ground of 5 nF. A surge voltage ramp of 100 kV, approximately 4 μs duration is applied to the line terminal, while the neutral is grounded. The results of the EMTP simulation, with the surge voltage distribution at the beginning of each section are shown in Fig. 14-35. The surge voltage distribution is highly nonlinear.

Now model the four sections each with a series reactance of 0.01 mH (Fig. 14-26*c*). The surge voltage distribution across each section is illustrated in Fig. 14-36. This shows an oscillatory response and unequal surge distribution; sections 2 and 3 have exactly the same oscillatory pattern.

In this model, the capacitances and reactances are identical in each of the four sections. The capacitance of the windings is a function of the winding geometry. Practically, the winding capacitance will vary, and so also the surge distribution.

Earlier the end turns of the transformers were reinforced and strengthened to protect them from high insulation stresses. This did not prove to be effective in most cases. Currently, metallic shields in strategic positions are placed adjacent to the coils. These shields are so placed and connected that the capacitance current flowing through them and the transformer windings tends to compensate for the ground capacitance current, thereby making the current through the series capacitance more uniform. Another method is to adjust capacitance distribution depending upon how the windings are interconnected.

14-13 DUALITY MODELS

Duality-based models, also called topology-based models, can be used to represent transformers.[15] These models are based upon core topology and utilize the correspondence between electric and magnetic circuits, as expressed by the principle of duality. Voltage, current, and inductance in electrical circuits correspond to flux, MMF, and reluctance, respectively, in the magnetic circuit.

$$\begin{aligned} I &= \frac{V}{R} \\ \Phi &= \frac{\text{MMF}}{l/\mu\mu_0 a} = \frac{\text{MMF}}{S} \end{aligned} \tag{14-53}$$

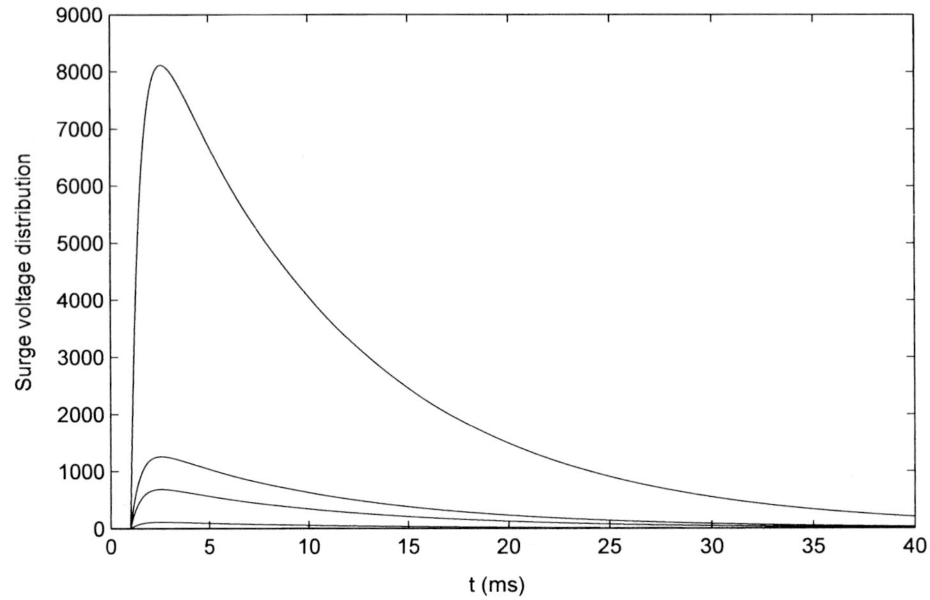

FIGURE 14-35 Simulation of surge voltage distribution, four capacitance, sections with ground capacitances.

where l is the length of the magnetic path, a is the cross-sectional area, and S is the reluctance, which is analogous to resistance in electrical circuit and determines the MMF necessary to produce a given magnetic flux. Permeance is reverse of reluctance.

Figure 14-37*a* shows electrical equivalent circuit of a three-winding core-type transformer, portraying magnetic coupling in three-limbed and five-limbed transformers.[15] Nonlinear inductances correspond to iron flux paths in the magnetic circuit, permitting each core limb to be modeled separately. Each L_k represents top and lower yokes and each L_b represents a wound limb. L_0 represents flux path through the air, outside the core, and around the windings. Finally, the ladder network between linear inductances, L_0 and L_b, represents winding leakages through the air. Inductances L_h and L_y represent unequal flux linkages between turns due to finite winding

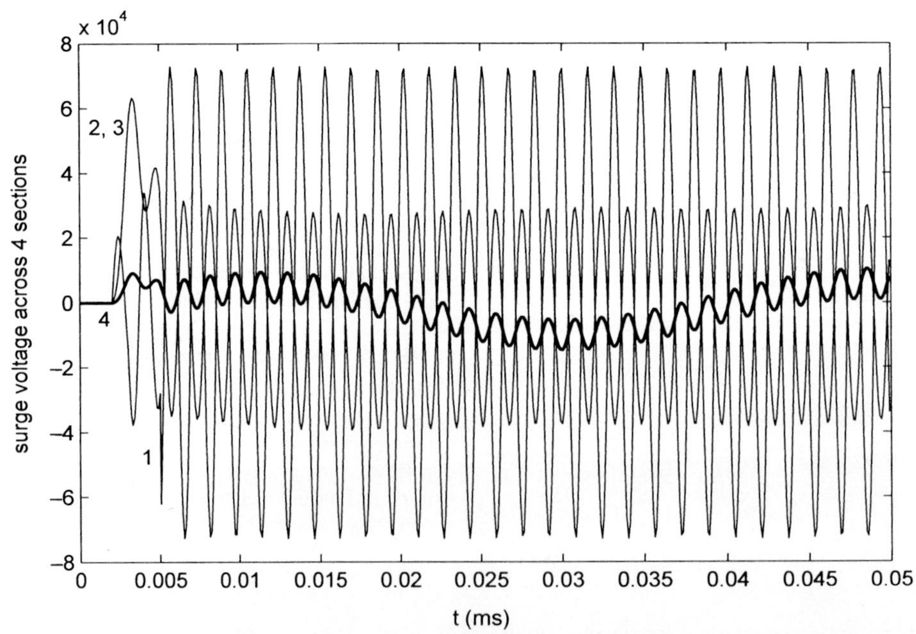

FIGURE 14-36 Simulation of surge voltage distribution; four capacitance and inductance sections with ground capacitances.

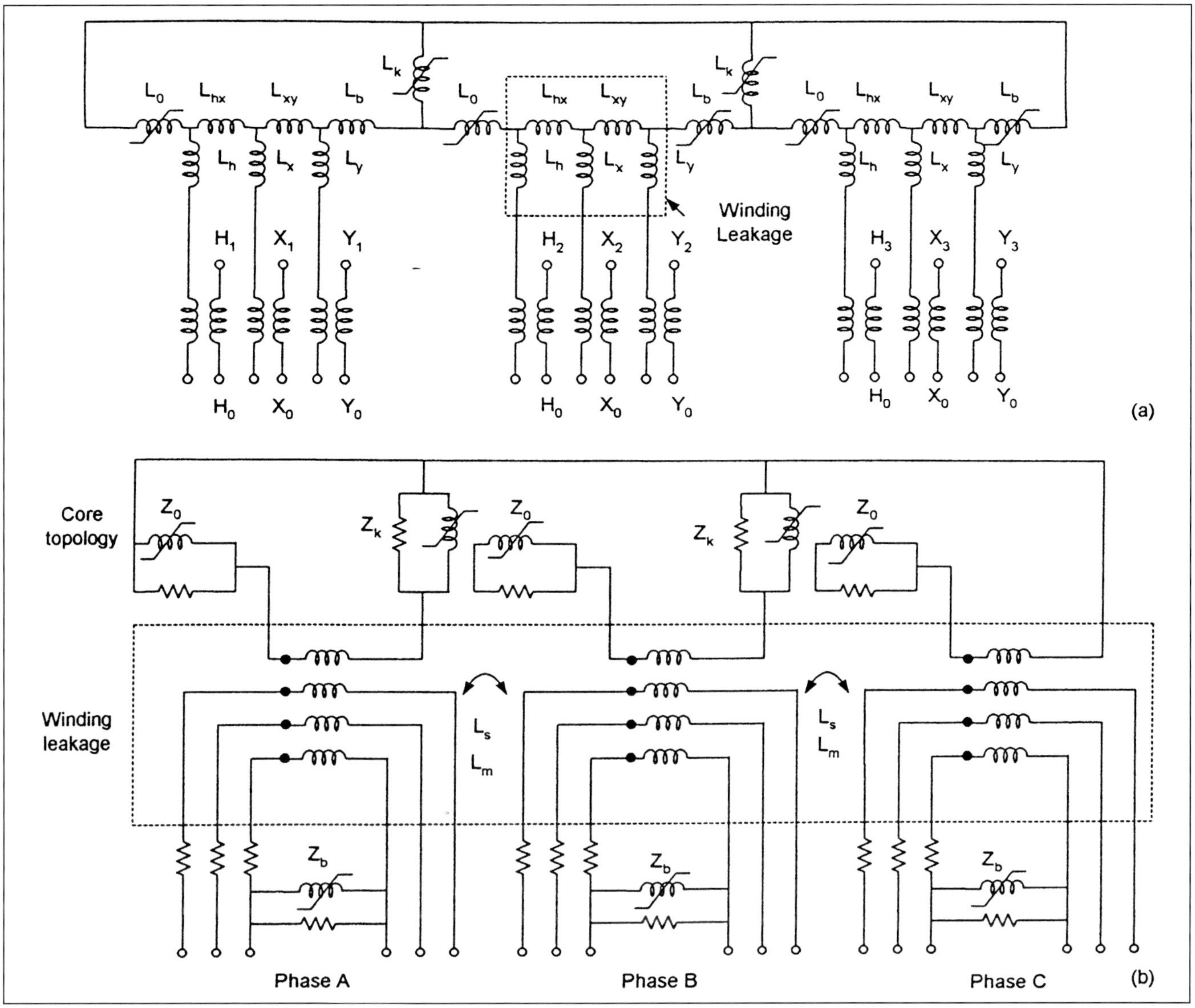

FIGURE 14-37 (*a*) Duality-based model for a core-type, three-winding transformer. (*b*) Simplified circuit derived from Fig. (*a*).

radial build and these are small compared to L_0 and L_b. This model is simplified as shown in Fig. 14-37*b*. The various inductances are calculated based upon short-circuit tests.

Duality models can be used for low-frequency transient studies, such as short circuits, inrush currents, ferroresonance, and harmonics. The model can be implemented in EMTP TOPMAG data calculation function. For three-limbed and five-limbed units, each limb is modeled individually and interfaced to an admittance matrix, reproducing the exact magnetic coupling between the windings. An additional three-phase fictitious winding is required for this interface, since core limbs are electrically isolated from the windings. $N\,(N{-}1)/2$ positive sequence short-circuit test values are required for an N-winding five-limbed or single-phase transformers. For three-limbed transformers zero sequence short-circuit test impedance or zero sequence excitation current is required.[15]

14-14 GIC MODELS

Geomagnetically induced currents (GIC) flow on the earth's surface due to solar magnetic disturbance (SMD), and these are typically of 0.001 to 0.1 Hz and can reach peak values of 200 A. They can enter transformer windings through grounded neutrals (Fig. 14-38), and bias the transformer core to half-cycle saturation. As a result, the transformer magnetizing current is greatly increased. Harmonics increase and these could cause reactive power consumption, capacitor overload, and false operation of protective relays.

A GIC model is shown in Fig.14-39;[16] four major flux paths are included. All R elements represent reluctances in different branches. Subscripts c, a, and t stand for core, air, and tank, respectively, and 1, 2, 3, and 4 represent major branches of flux paths. Branch 1 represents sum of core and air fluxes within the excitation windings, branch 2 represents flux path in yoke, branch 3 represents sum of fluxes entering the side leg, part of which leaves the side leg and enters the tank, and branch 4 represents flux leaving the tank from the center leg. An iterative program is used to solve the circuit of Fig. 14-39 so that nonlinearity is considered.

14-15 FERRORESONANCE

Ferroresonance can occur when a nonlinear inductance (iron-core reactance of a transformer) is excited in parallel with capacitance through high impedance (Chap. 7). Overvoltages up to three to four

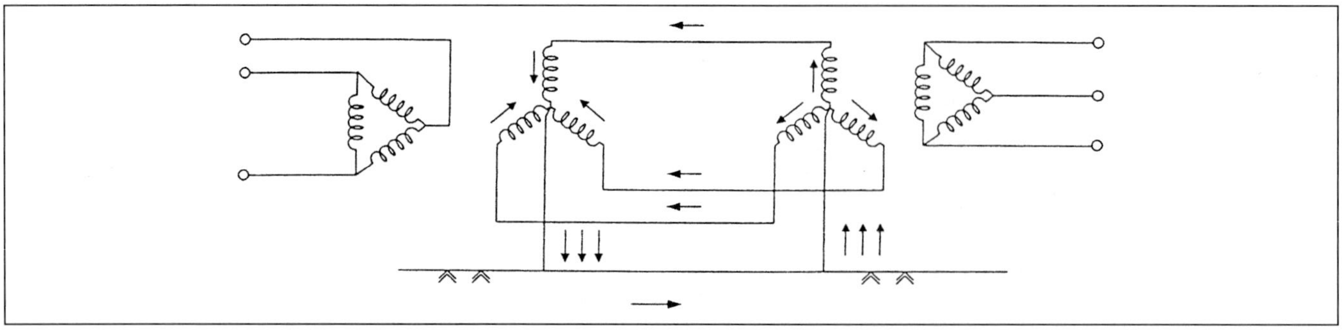

FIGURE 14-38 GIC entering the grounded neutrals of wye-connected transformers.

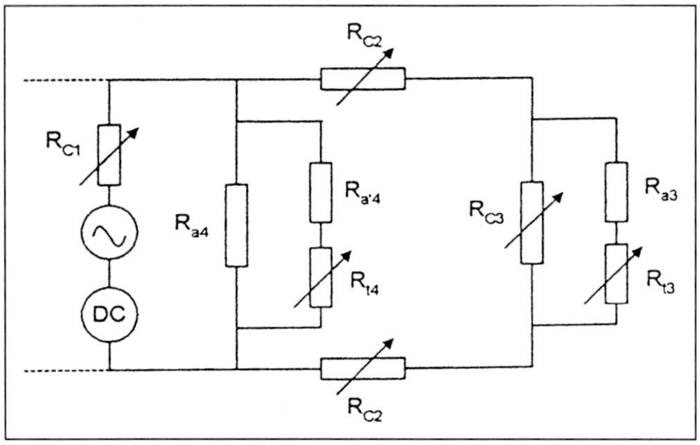

FIGURE 14-39 Transformer model for GIC simulation.

times the nominal voltage can occur. Ferroresonance occurs because of reactance of a distribution transformer X_m, and the capacitance of the bushings and underground cables form a resonant circuit. Highly distorted voltage waveforms are produced which include a 60-Hz component. Considerable harmonics are generated. We may postulate, based upon the current literature, that:

- High overvoltages of the order of 4.5 pu can be created.
- Resonance can occur over a wide range of X_c/X_m,[17] possibly in the range $0.1 < X_c/X_m < 40$.
- Resonance occurs only when the transformer is unloaded, or very lightly loaded. Transformers loaded to more than 10 percent of their rating are not susceptible to ferroresonance.

The capacitance of cables varies between 40 to 100 nF per 1000 ft, depending upon conductor size. However, the

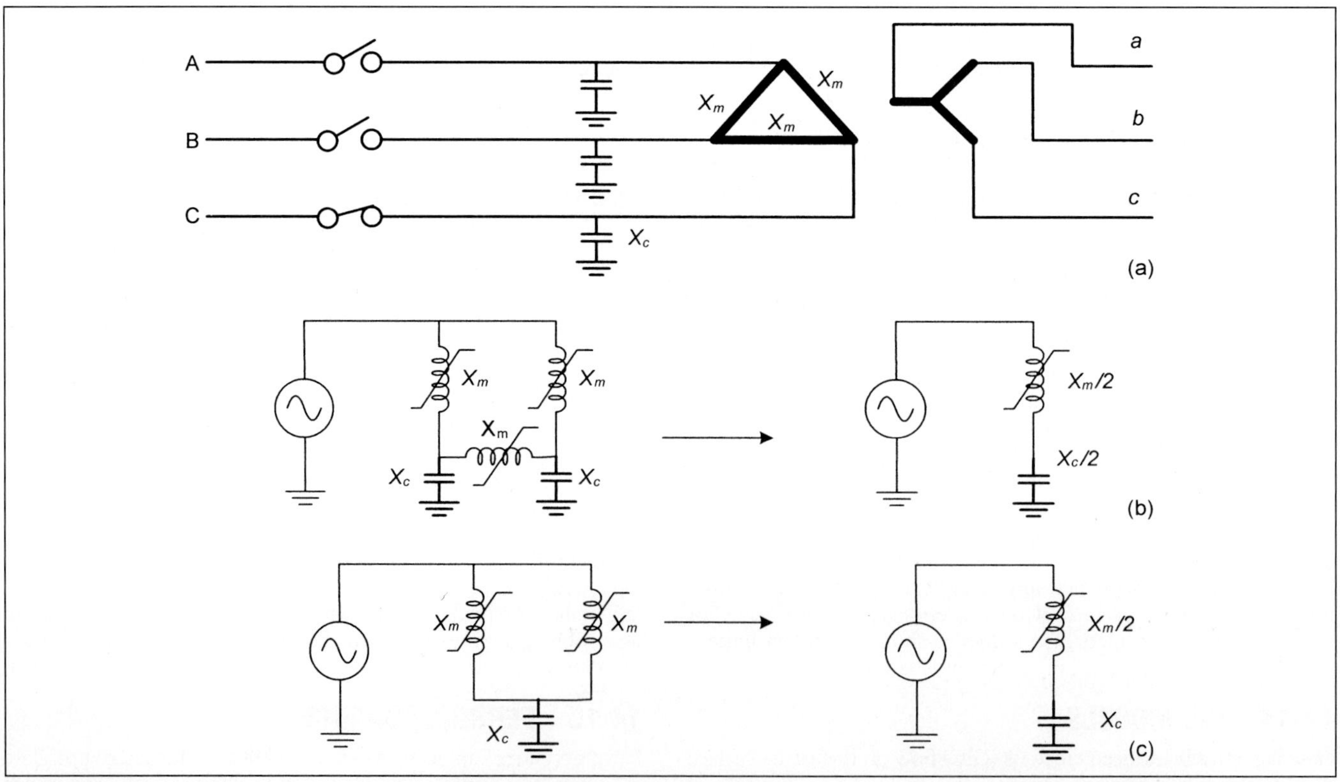

FIGURE 14-40 (*a*) A circuit for possible ferroresonance. Equivalent circuit derived with (*b*) one phase closed and (*c*) two phases closed.

magnetizing reactance of a 35-kV transformer is several times higher than that of a 15-kV transformer; the ferroresonance can be more damaging at higher voltages. For delta-connected transformers, the ferroresonance can occur for less than 100 ft of cable. Therefore, the grounded wye-wye transformer connection has become the most popular in underground distribution system in North America. It is more resistant, though not totally immune to ferroresonance.

Series Ferroresonance Figure 14-40*a* shows a basic circuit of ferroresonance. Consider that current-limiting fuses in one or two lines operate. X_c is the capacitance of the cables and transformer bushings to ground. Also the switch may not close simultaneously in all the three phases, or while opening, the phases may not open simultaneously. We can draw the equivalent circuits with one phase closed and also with two phases closed as shown in Fig. 14-40*b* and *c*, respectively.

Similar equivalent circuit can be drawn for wye-wye ungrounded transformer. The minimum capacitance to produce ferroresonance can be calculated from $X_c/X_m = 40$, say:

$$C_{\text{mres}} = \frac{2.21\times 10^{-7}\,\text{MVA}_{\text{transf}} I_m}{V_n^2} \tag{14-54}$$

where I_m is the magnetizing current of the transformer as a percentage of the full-load current. Typically, magnetizing current of a distribution transformer is 1 percent to 3 percent of the transformer full-load current. $\text{MVA}_{\text{trasf}}$ is the rating of transformer in megavolt-amperes, C_{mres} is the minimum capacitance for resonance in picofarads, and V_n is the line-to-neutral voltage in kilovolts. Table 14-7 gives the approximate values of C_{mres}.

Parallel Ferroresonance A less common condition of ferroresonance is illustrated in Fig. 14-41*a* and *b*.[18] This involves the mutual X_m formed between the windings at the same voltage level in four- or five-leg core-type transformers. These may resonate even when connected in grounded wye. However, the overvoltages are limited to 1.5 pu, because equivalent circuit is a parallel LC combination, and X_m limits the voltage by saturation.

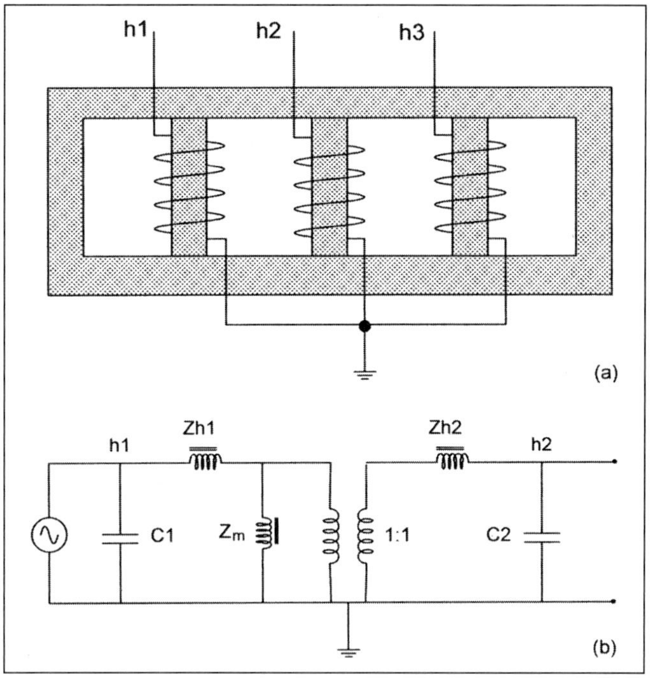

FIGURE 14-41 (*a*) Mutual coupling ferroresonance. (*b*) Equivalent circuit for mutual coupling ferroresonance.

Example 14-10 The curve of Fig. 7-20*c* can be represented by two-piece saturation characteristics of a nonlinear reactor

TABLE 14-7 Capacitance Limits for Ferroresonance (C_{mres} in Picofarads)

TRANSFORMER KVA		SYSTEM VOLTAGE (KV)		
		8.32/4.8	12.5/7.2, 13.8/7.99	25/14.4, 27.8/16
SINGLE PHASE	THREE PHASE	C_{mf}	C_{mf}	C_{mf}
15	5	72	26	16
25	10	119	43	11
50		339	86	21
75	25	358	129	32
100		477	172	43
150	50	719	258	64
225	75	1070	387	97
300	100	1432	516	129
500	167	2390	859	215
750	250	3580	1290	322
1000		4770	1720	430
1500		7190	2580	645
2000		9550	3440	859

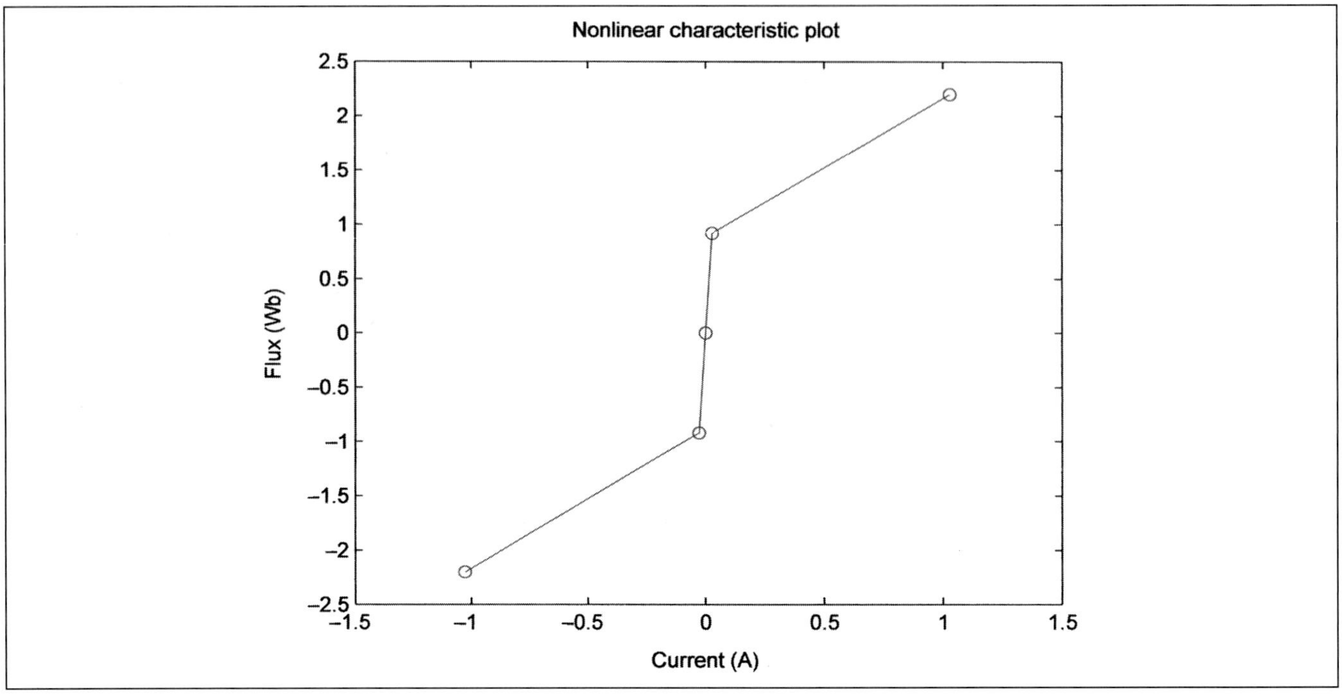

FIGURE 14-42 A piecewise curve of an iron core (nonlinear) reactor.

(Fig. 14-42). The reactor is energized through a 60-Hz voltage of 240 V and a parallel capacitor of 1.7 μF in two cases: (a) with no prior magnetic flux, and (b) 0.07 Wb of initial flux. The simulation results of the voltage across the reactor and the current through it are shown in Fig. 14-43*a* and *b*, respectively. Without a prior flux, there are no current or voltage transients. With initial flux, the voltage rises to 4 pu of the applied voltage and the current waveshape can be compared with that of Fig. 14-15.

14-16 TRANSFORMER RELIABILITY

When properly specified and applied for the specific applications, a transformer is a fairly reliable piece of equipment. IEEE standard[19] gives a failure rate $\lambda = 0.0030$ per year. However, the downtime associated with this failure rate is high, r (downtime in hours per failure) = 342 for repairs of the faulty unit and $r = 130.0$ if the unit is replaced. These data are applicable for transformers of 601 to

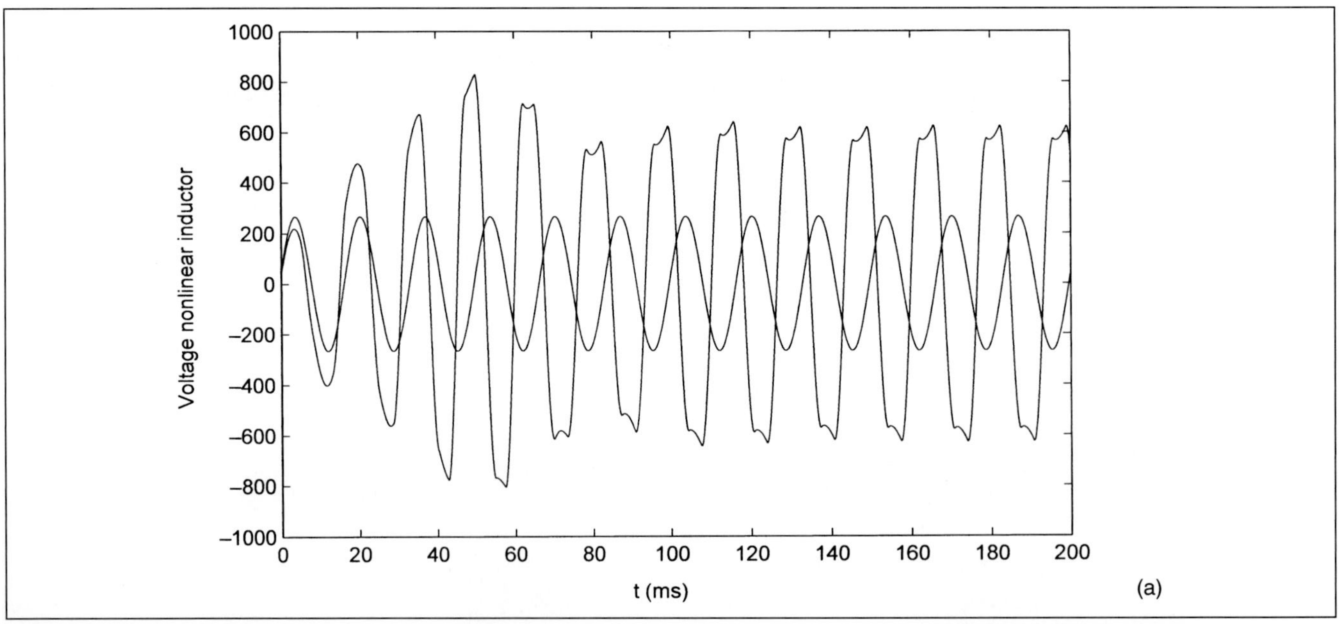

FIGURE 14-43 Simulation of transients with a parallel capacitor of 1.7 μF, (*a*) non-linear reactor and (*b*) voltage and current transients respectively with no prior magnetic flux and with 0.07 Wb of magnetic flux. (*Continued*)

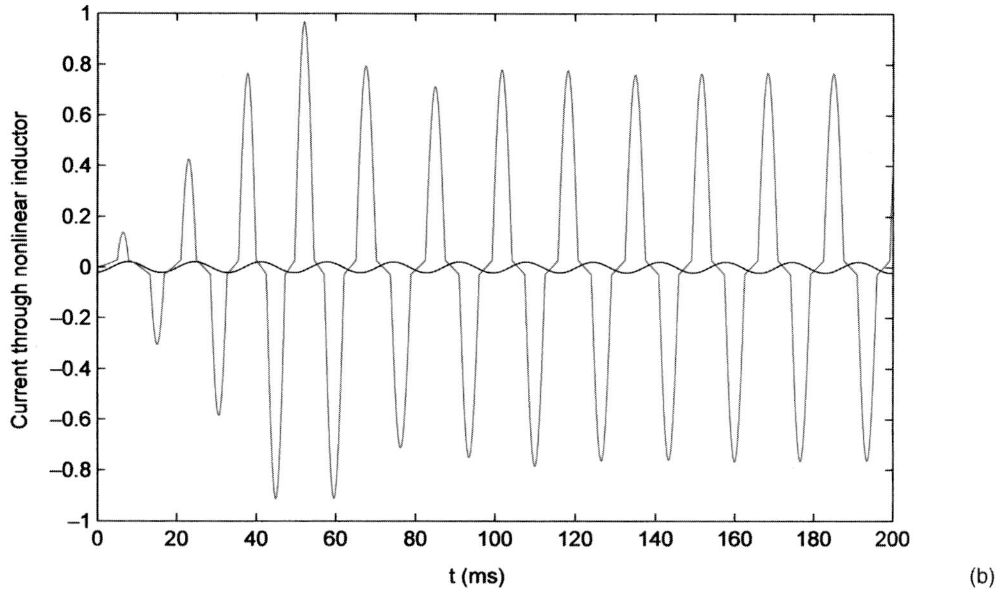

FIGURE 14-43 (*Continued*)

15 kV. For reliable power supply systems, it is common practice to design the systems so that failure of one unit does not precipitate in complete loss of power, and it is customary to provide standby or redundant transformers. Transformers may be operated in parallel and sized, so that on loss of one of the parallel running unit entire load can be supplied. The protective relaying can ensure that the faulty unit is automatically taken out of service, with least interruption to the load, for example, bus transfer systems discussed in Chap. 16.

PROBLEMS

1. A 30-MVA, 138 to 13.8-kV, delta-wye-connected transformer has a percentage impedance of 9 percent. What is the magnitude and duration of short-circuit currents that it should withstand according to ANSI/IEEE standards, both for frequently occurring faults and less frequently occurring faults?

2. In Example14-3, consider that a transformer of the same specifications is connected in delta-wye grounded configuration. How will it impact the test data?

3. Explain the zero sequence flux patterns and impedances in three-limbs, core-type, five-limb core-type and shell-type three-phase transformers. Compare core-type construction with shell-type construction.

4. The transformer in Prob. 1 is provided with ±2.5 and ±5 percent off-load taps on the high-voltage winding. Construct equivalent T and Π impedance circuits in ohms when the transformer primary windings are at plus 5 percent taps.

5. The tests on a 3000-kVA, 13.8 to 4.16-kV, delta-wye-connected transformer give a copper loss of 20 kW on the delta side and 18 kW on the wye side. Find winding resistances, and refer secondary winding resistances to primary windings. If the total reactance is 5.5 percent, find primary and secondary reactances, assuming that these are divided in the same ratio as the resistances.

6. The transformer of Prob. 5 is tested on open circuit on the 4.16-kV side. Rated primary voltage at power input = 12 kW, and no load current = 3 A. Find magnetizing circuit parameters.

7. What is the predominant harmonic in the inrush current of a transformer?

8. The following points relate to the excitation test data supplied by a manufacturer for a transformer of 750 MVA, with a wye-connected, high-voltage winding of 400 kV, and a delta-connected, low-voltage winding of 18 kV:

$V_{ex} = 22.76$, $I_{ex} = 9.0$, $P_{ex} = 210.5$

$V_{ex} = 35.2$, $I_{ex} = 80.75$, $P_{ex} = 580$

Convert these values to V_{rms} and I_{rms} per unit base, similar to conversion of Tables 14-2 and 14-3.

9. Two 10-MVA transformers of the same impedance of 5.5 percent, but with voltage ratios of 138 to 13.8 kV and 138 to 13.2 kV, are suddenly paralleled by closing a switch at the crest of secondary voltage. Derive a time-current equation of the circulating current.

10. Consider a transformer with five winding sections and the ratio α in Eq. (14-50) = 3. Calculate the voltage across each section of the winding for a surge of 250-kV peak applied to the line terminals, neutral grounded and ungrounded.

11. Explain sympathetic inrush.

12. Draw circuit diagrams of single-phase transformers: (1) ignore all saturation and losses, (2) include saturation, (3) include hysteresis, (4) include saturation, hysteresis, and core loss, (5) include winding capcitances to ground, and (6) include all other capacitances.

13. A transformer is represented with a capacitance of 15 nF, where the secondary is an open circuit, the surge impedance = 300 ohms, and the capacitance to ground = 10 nF. What is the

magnitude of the surge transferred to the secondary for (1) delta-wye-connected transformer, and (2) wye-wye transformer?

14. The transformer of Prob. 1 is operated at 50 Hz. What can be expected with respect to its continuous operation at this frequency?

15. Without conducting a rigorous study, recommend installation of surge arresters for surge protection of the transformer windings. What is the role of a secondary surge capacitor?

16. Draw a sketch showing the distortion in phase voltages on phenomenon of oscillating neutrals in ungrounded wye-wye-connected transformers.

17. Why is an admittance matrix, rather than an impedance matrix, used in formation of the transformer winding data for transient analysis?

REFERENCES

1. CIGRE Working Group 33.02, Guidelines for Representation of Network Elements When Calculating Transients, CIGRE Brochure 39, 1990.
2. IEEE Modeling and Analysis of System Transients Using Digital Programs, Document TP-133-0, 1988.
3. ANSI/IEEE Std. C57.12, General Requirements for Liquid Immersed Distribution, Power and Regulating Transformers, 2006.
4. ANSI/IEEE Std. C57.110, IEEE Recommended Practice for Establishing Transformer Capability When Supplying Non-Sinusoidal Loads, 1998.
5. UL Standard 1561, General Purpose and Power Transformers, 1990.
6. ANSI/IEEE Std. C37.91, IEEE Guide for Protecting Power Transformers, 2008.
7. ANSI/IEEE, IEEE Guide for Dry-Type Transformer Through Fault Current Duration, 2006.
8. C. E. Lin, J. B. Wei, C. L. Huang, and C. J. Huang, "A New Method for Representation of Hysteresis Loops," *IEEE Trans. PD.*, vol. 4, pp. 413–419, 1989.
9. J. D. Green and C. A. Gross, "Non-Linear Modeling of Transformers," *IEEE Trans. Industry Applications*, vol. 24, pp. 434–438, 1988.
10. T. Adielson, A. Carlson, H. B. Margolis, J. A. Halladay. "Resonant Overvoltages in EHV Transformers: Modeling and Applications," *IEEE Trans. PAS*, vol. PAS-100, no. 7, pp. 3563–3572, 1981.
11. P. S. Maruvada and N. H Cavallius, "Capacitance Calculations for Some Basic High Voltage Electrode Configurations," *IEEE Trans. PAS*, vol. 94, no. 5, pp. 1708–1713, Sept./Oct. 1975.
12. W. J. McNutt, T. J. Blalock, R. A. Hinton, "Response of Transformer Windings to System Transient Voltages," *IEEE Trans. PAS*, vol. PAS-93, pp 457–467, 1974.
13. J. C. Das, "Surge Transference Through Transformers," *IEEE Industry Magazine*, vol. 9, no. 5, pp. 24–32, Oct. 2003.
14. A. Greenwood, *Electrical Transients in Power Systems*, John Wiley, New York, 1991.
15. A. Narang and R, Brierley, "Topology-Based Magnetic Model for Steady State and Transient Studies for Three-Phase Core-Type Transformers," *IEEE Trans. PS*, vol. 9, no. 3, pp. 1337–1349, Aug. 1994.
16. S. Lu, Y. Liu, and J. D. R. Ree, "Harmonics Generated from a DC Biased Transformer," *IEEE Trans. PD*, vol. 8, pp. 725–731, 1993.
17. R. H. Hopkinson, "Ferroresonance During Single-Phase Switching of Three-Phase Distribution Transformer Banks," *IEEE Trans. PAS*, vol. 84, pp. 289–293, 1965.
18. D. R. Smith, S. R. Swanson, and J. D. Borst, "Overvoltages with Remotely Switched Cable Fed Grounded Wye-Wye Transformers," *IEEE Trans. PAS*, vol. PAS-94, pp. 1843–1853, 1975.
19. ANSI/IEEE Std. 493, IEEE Recommended Practice for Design of Reliable Industrial and Commercial Power Systems, 2007.

FURTHER READING

Canadian/American EMTP User Group, ATP Rule Book, Portland, OR, 1987–1992.

X. Chen and S. S. Venkta, "A Three-Phase Three-Winding Core-Type Transformer Model for Low-Frequency Transient Studies," *IEEE Trans. PD*, vol. 12, pp. 775–782, 1997.

F. de Leon and A. Semlyen, "Complete Transformer Model for Electromagnetic Transients," *IEEE Trans. PD*, vol. 9, no. 1, pp. 231–239, 1994.

F. de Leon and A. Semlyen, "Efficient Calculations of Elementary Parameters of Transformers," *IEEE Trans. PD*, vol. 7, no. 1, pp. 376–383, Jan. 1992.

A. Morched, L. Marti, and J. Ottenvangers, "A High Frequency Transformer Model for EMTP," *IEEE Trans. PD*, vol. 8, no. 3, pp. 1615–1626, 1993.

G. C. Paap, A. A. Alkema, and L. van der Sluis, "Overvoltages in Power Transformers Caused by No-Load Switching," *IEEE Trans. PD*, vol. 10, no. 1, pp. 301–307, Jan. 1995.

C. W. Plummer, G. L. Goedde, E. L. Pettit Jr, J. S. Godbee, M. G. Hennessey, and W. I. Pawaukee. "Reduction in Distribution Transformer Failure Rates and Nuisance Outages Using Improved Lightning Protection Concepts," *IEEE Trans. PWRD*, vol. 10, no. 2, pp. 768–777, 1995.

E. J. Tarasiewicz, A. S. Morched, A. Narang, and E. P. Dick, "Frequency Dependent Eddy Current Models for Non-linear Iron Cores," *IEEE Trans. PS*, vol. 8, no. 2, pp. 588–597, May 1993.

P. M. Vaseen, "Transformer Model for High Frequencies," *IEEE Trans. PWRD*, vol. 3, no. 4, pp. 1761–1768, 1988.

CHAPTER 15

POWER ELECTRONIC EQUIPMENT AND FACTS

Nonlinear loads are on the increase, and it is estimated that in the next 7 to 8 years 60 percent of the loads served from the utility systems will be of nonlinear nature. These loads are a source of electrical noise and harmonics, and while polluting the power supply, are themselves less tolerant to the poor power quality. Power electronics are already impacting the electrical utility industry in the form of flexible ac transmission systems (FACTS), which can mitigate some of the inherent problems of conventional ac transmission systems. Some examples of nonlinear loads are:

- Adjustable speed drive systems
- Cycloconverters
- Arc furnaces, which are further discussed in Chap. 16
- Switch mode power supplies (SMPs) (Chap. 19)
- Copy machines, television sets, computers, and peripherals
- Static var compensators
- HVDC transmission
- Electric traction
- Wind and solar power generation
- Battery charging, fuel cells, and UPS systems
- Slip frequency recovery schemes of large induction motors
- Fluorescent lighting and electronic ballasts

It is not the intention to go into discussions of the operation of these equipments. From the power quality point of view, the harmonic generation from these loads is of concern. From the application point of view, the FACTS controllers are used to enhance the performance of electrical power transmission, mitigate oscillations, and improve transient stability.

15-1 THE THREE-PHASE BRIDGE CIRCUITS

Figure 15-1 shows a three-phase fully controlled six-pulse current source converter circuit and current and voltage waveforms. The sequential firing of the thyristors is shown in Table 15-1. At any one time, two thyristors conduct. The firing frequency is six times the fundamental frequency, and the firing angle α can be measured from the point O shown in Fig. 15-1. With a large reactor the output current can be assumed to be continuous and the input current is rectangular of pulse width $2\pi/3$ duration and amplitude I_d. A Fourier analysis of the input current waveform gives:

$$i_a = \frac{2\sqrt{3}}{\pi} I_d \left[\cos\omega t - \frac{1}{5}\cos 5\omega t + \frac{1}{7}\cos 7\omega t - \frac{1}{11}\cos 11\omega t + \frac{1}{13}\cos 13\omega t - \ldots \right] \tag{15-1}$$

In general, the theoretical order of harmonics is given by:

$$h = pm \pm 1 \quad m = 1, 2, \ldots \tag{15-2}$$

where h is the order of the harmonic, and p is the pulse number, which is defined as the total number of successive nonsimultaneous commutations occurring within the converter during each cycle when operating without phase control. The harmonics given by Eq. (15-2) are an integer of the fundamental frequency and are called characteristic harmonics. Certain type of pulsed loads and integral cycle controllers produce distorted waveforms having a submultiple of power frequency. Some loads like arc furnaces, arcing devices, fluorescent, and mercury and sodium vapor lighting produce waveforms which are aperiodic. If the operations are kept constant over a length of time, the waveform may be almost periodic. Cycloconverters and arcing loads produce "interharmonics," which are defined as "between the harmonics of the power frequency voltage and current, further frequencies can be observed, which are not integers of the fundamental frequency. These appear as discrete frequencies or wide-band spectrum." This has attracted much attention in recent times.

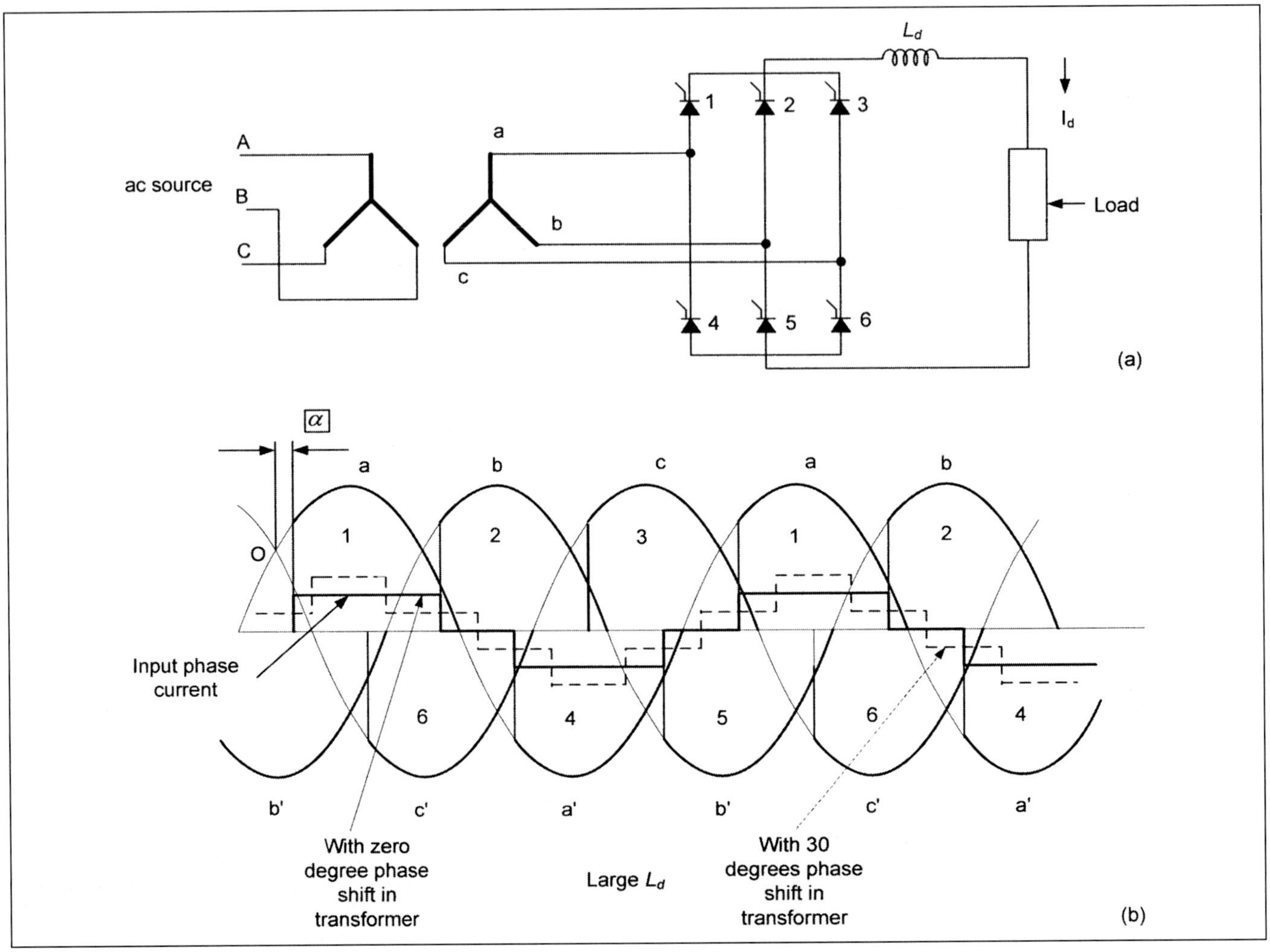

FIGURE 15-1 (*a*) Circuit diagram of three-phase, fully controlled six-pulse current source converter (*b*) Voltage and line-current waveforms for a certain delay angle α.

TABLE 15-1 Firing Sequence of Thyristors in Six-Pulse Converter

CONDUCTING THYRISTOR	5,3	1,5	6,1	2,6	4,2	3,4
Thyristor to be fired	1	6	2	4	3	5
Thyristor turning off	3	5	1	6	2	4

Also from Eq. (15-1) the theoretical magnitude of a harmonic is given by:

$$I_h = \frac{I_1}{h} \tag{15-3}$$

where h is the order of the harmonic, and I_1 is the fundamental frequency current. Thus, a six-pulse three-phase converter produces harmonics of the order of: 5th = 20 percent, 7th = 14.28 percent, 11th = 9.09 percent and so on. Practically, the harmonics will be much lower, as discussed further. The second-order and third-order harmonics are ideally absent, but practically, converters do produce a small percentage of these harmonics due to some asymmetry in firing angles and three-phase circuits. These are called noncharacteristic harmonics.

The ripple content in the output voltage wave is shown in Fig. 15-2. The form factor is the measure of the shape of the output voltage or current and is defined as:

$$\mathrm{FF} = \frac{I_{\mathrm{rms}}}{I_{\mathrm{dc}}} \tag{15-4}$$

The ripple factor, which is a measure of the ripple content of the output current or voltage, is defined as the rms value of the output voltage or current, including all harmonics, divided by average value:

$$\mathrm{RF} = \sqrt{\left(\frac{I_{\mathrm{rms}}}{I_{\mathrm{dc}}}\right)^2 - 1} = \sqrt{\mathrm{FF}^2 - 1} \tag{15-5}$$

The ripple factor of a six-pulse converter, with zero firing angle is 0.076, and the lowest harmonic in the dc output of the converter is sixth. Thus, a converter may be considered a splitter of harmonics, odd harmonics going into the line side and even

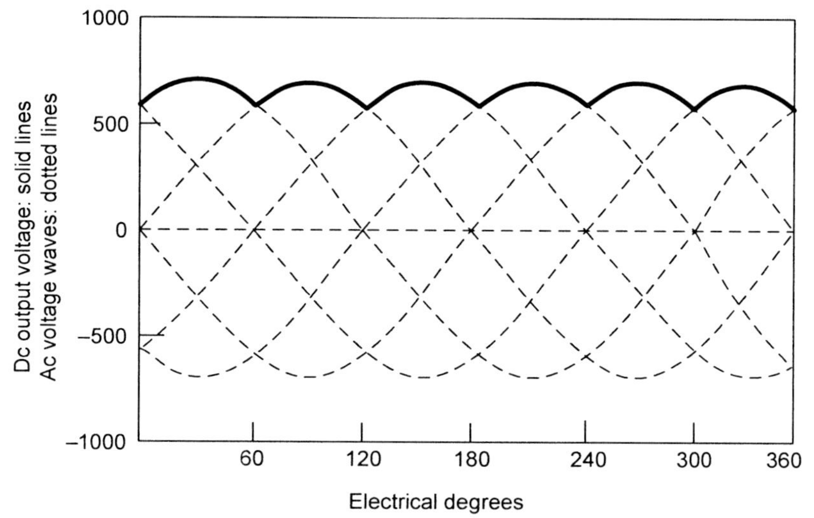

FIGURE 15-2 Ripple content in the output dc voltage of a six-pulse converter.

harmonics into the load side. As the pulse number increases, the ripple in the dc output voltage and the harmonic content in the input are reduced.

15-1-1 Phase Multiplication

The winding connections of the input transformer impact the input current waveform. The harmonic spectrum in Eq. (15-1) is obtained when there is 0° phase shift between the voltage vectors of the primary and secondary windings of the input transformer Fig. 15-1*a*. The angular displacement in windings of polyphase transformers was discussed in Chaps. 9 and 14.

Consider now a 30° phase shift in the transformer winding connections, that is, a delta-wye or wye-delta connected transformer. For example, Fig. 15-1*b* shows this alternate waveform in dotted lines, with this phase shift. A Fourier analysis of this waveform gives:

$$i_a = \frac{2\sqrt{3}}{\pi} I_d \left[\cos\omega t + \frac{1}{5}\cos 5\omega t - \frac{1}{7}\cos 7\omega t - \frac{1}{11}\cos 11\omega t + \frac{1}{13}\cos 13\omega t + \ldots \right] \tag{15-6}$$

If we combine the two waveforms of the input currents, the harmonics of the order of 5th, 7th, 17th, and 19th cancel out. Practically, approximately 75% cancellation is achieved. Figure 15-3 shows the combination of these two waveforms; and a closer approximation to sinusoidal wave shape is achieved. The combination acts like a 12-pulse system.

The concept is widely used to reduce the harmonic distortion in the input current waveforms, say in HVDC and ASD systems. A 24-pulse operation can be obtained with four transformers having a 15° phase shift. This will eliminate, though not totally, all lower-order harmonics of the order of below 23rd.

Some other technologies for harmonic mitigations, for example, multilevel converters, interphase reactors, and active current shaping are not discussed.

15-1-2 Commutation and Overlap Angle

Figure 15-1 assumes that the conduction of current from one SCR to another is instantaneous. The commutation is delayed by an angle μ due to source inductance, and during this period short circuit occurs through the conducting devices, the circulating current being limited by the source impedance. The angle μ is called the overlap angle. When $\alpha = 0$, the short-circuit conditions correspond to the maximum asymmetry and μ is large, that is, slow initial rise. At 90° the conditions are that of zero asymmetry with its fast rate of rise of current. Figure 15-4 shows the effect of overlap angle. Note that the slow rise of current reduces the harmonics.

Commutation produces two primary notches per cycle and four secondary notches of lesser magnitude, which are due to notch reflection from the other legs of the bridge. Figure 15-5 shows the voltage notching in the fully controlled bridge with large output reactor. The notch area must be limited according to standards[1] by adding additional source reactors, if required.

The depth of the notch is given by the commutation angle:

$$\mu = \cos^{-1}[\cos\alpha - (X_s + X_t)I_d] - \alpha \tag{15-7}$$

where X_s and X_t are the system and input transformer reactances, respectively, and I_d is the dc current in per unit on converter base. Figure 15-6 illustrates four waveforms:

1. Rectangular current waveform or ideal textbook waveform (Fig. 15-6*a*)

2. Waveform modified by commutation angle (Fig. 15-6*b*)

3. Waveform with ripple content (Fig. 15-6*c*)

4. A discontinuous waveform due to large firing angle, giving rise to much higher harmonics in the input supply system. (Fig. 15-6*d*)

These line harmonics can be estimated using the methodology described in Refs 1, 2 and 3. The dc voltage is given by:

$$E_d = 2\left[\frac{3}{2\pi}\int_{-\pi/3+\alpha}^{\pi/3+\alpha} E_m \cos\omega t\, d(\omega t)\right] = \frac{3\sqrt{3}}{\pi} E_m \cos\alpha \tag{15-8}$$

The output voltage can be reduced to zero and with increasing control angle, the converter will act as an inverter. A bidirectional

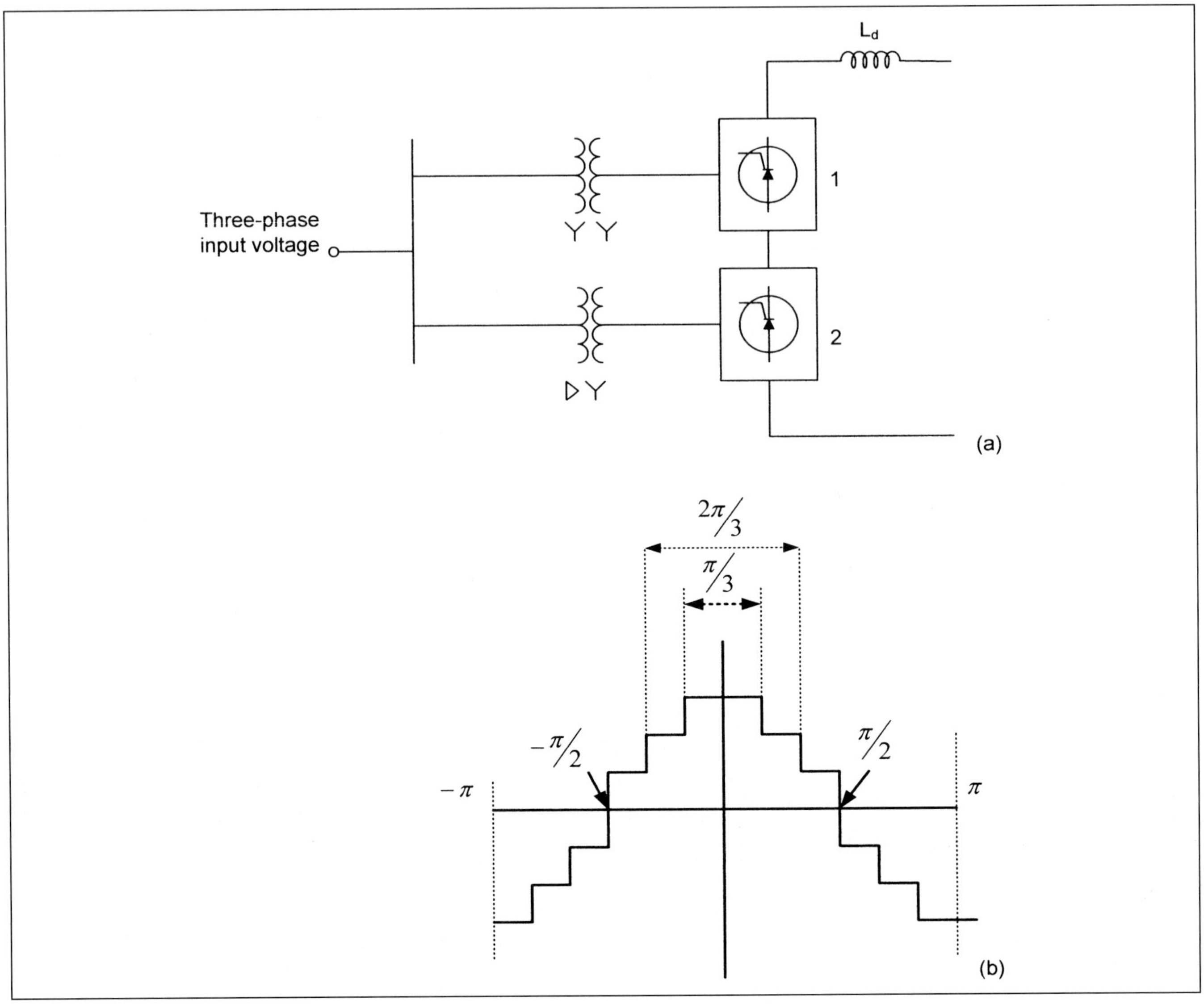

FIGURE 15-3 (*a*) Circuit diagram of harmonic elimination with phase multiplication. (*b*) Input current waveform with ideal commutation.

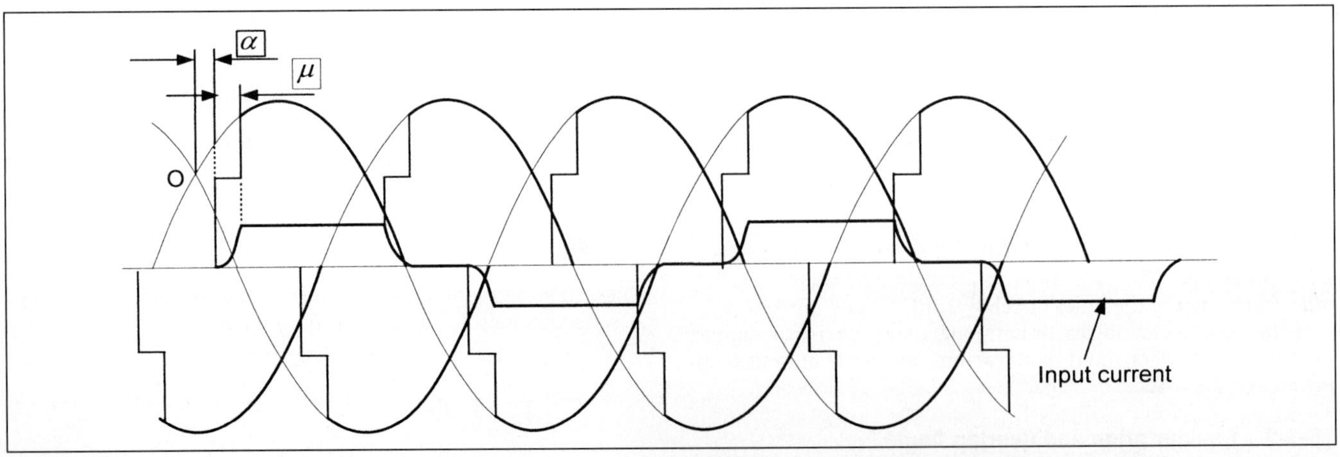

FIGURE 15-4 Effect of commutation angle μ on the input current waveform.

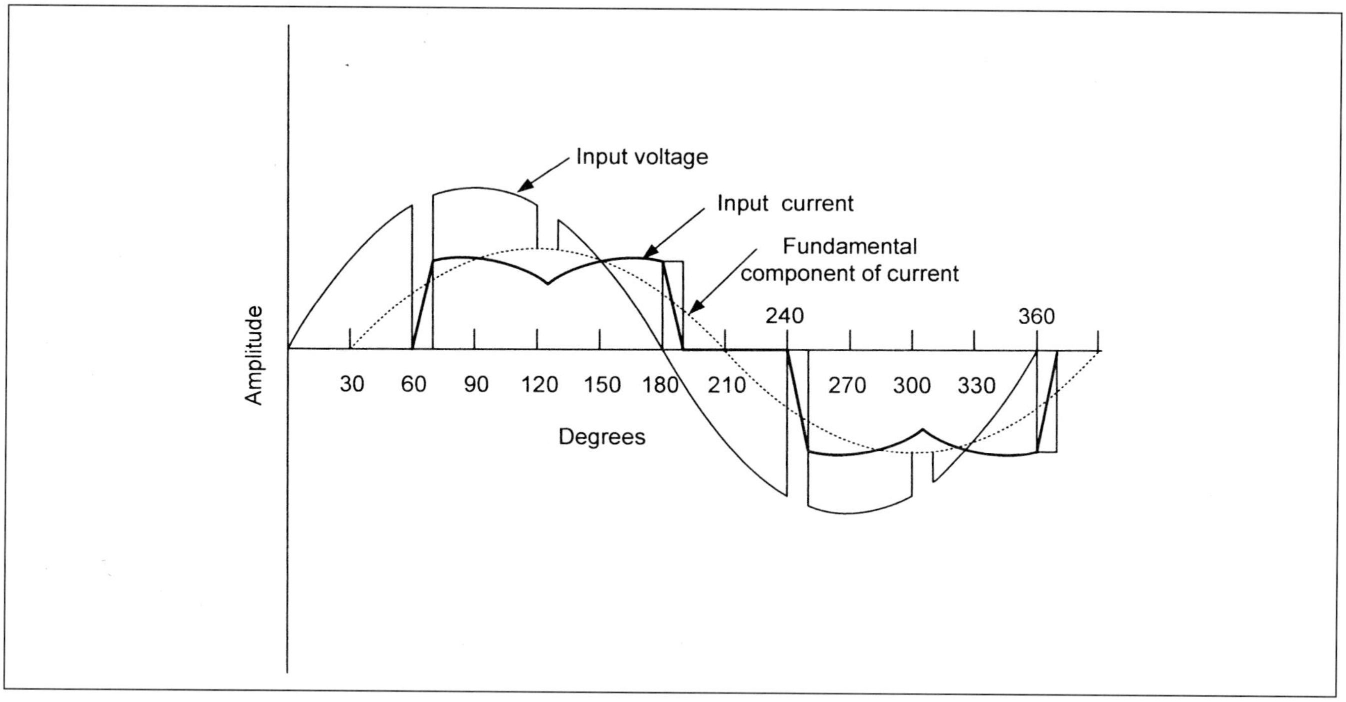

FIGURE 15-5 Voltage notching due to commutation in a six-pulse current source converter and the input current waveform.

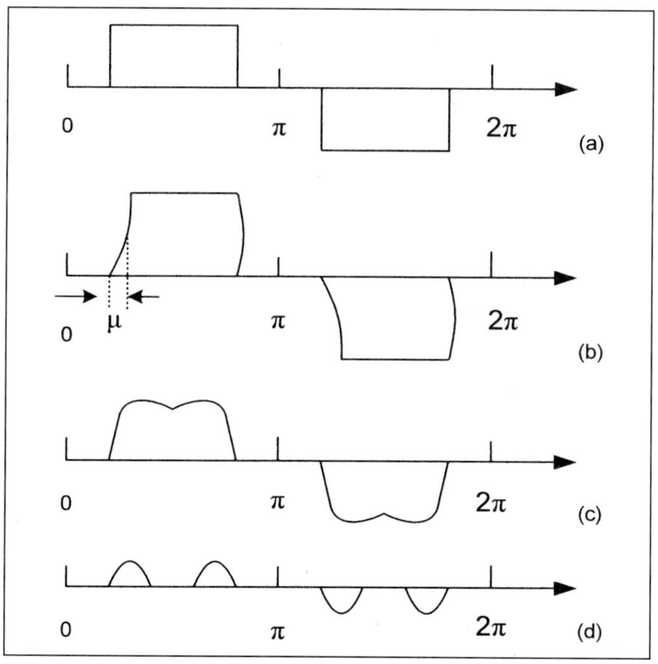

FIGURE 15-6 (*a*) A rectangular ideal input current waveform, (*b*) a waveform with commutation angle, (*c*) a waveform with ripple content, and (*d*) a discontinuous waveform due to large delay angle control.

flow of active power is possible from ac to dc and dc to ac, but a line-commutated current source converter, with thyristors (turn-on control devices only), draws reactive power from the supply system (Fig. 15-7). This reactive power is due to phase shift between fundamental of ac current and voltage, and is strongly dependent upon the firing angle α. The commutation process also introduces further displacement of the ac current. Figure 15-7 shows that the converter is consuming reactive power both in the rectifier and inverter mode. In addition, a certain amount of reactive power is lost in the rectifier transformers too. The reactive power demand, for a single converter operation, is usually 50 to 60 percent of the transmitted active power.

For HVDC transmission, which requires conversion of ac to dc and dc to ac, these converters have been extensively used, though these may lead to commutation failures. The reactive power is managed through switched capacitor filters; it can be drawn from the system and can even be controllable, and load-dependent controls can be provided. An advantage of thyristors is that these can handle two to three times the power that gate turn-off (GTO), integrated gate-commutated thyristors (IGCTs), or MOS turn-off thyristors (MTO) can handle. In the late 1990s, HVDC transmission using voltage source converters with pulse width modulation (PWM) was introduced and commercially called HVDC-*light*. Figure 15-8 shows solid-state converter developments, thyristors and IGBTs over the years.

15-2 VOLTAGE SOURCE THREE-PHASE BRIDGE

The FACTS use voltage source bridge, rather than current source configuration. The requirements of FACTS controllers are:

- The converter should be able to act as an inverter or rectifier with leading or lagging reactive power, that is, four-quadrant operation is required, as compared to the current source line commutated converter which has two quadrant operations.

- The active and reactive power should be independently controllable with control of phase angle.

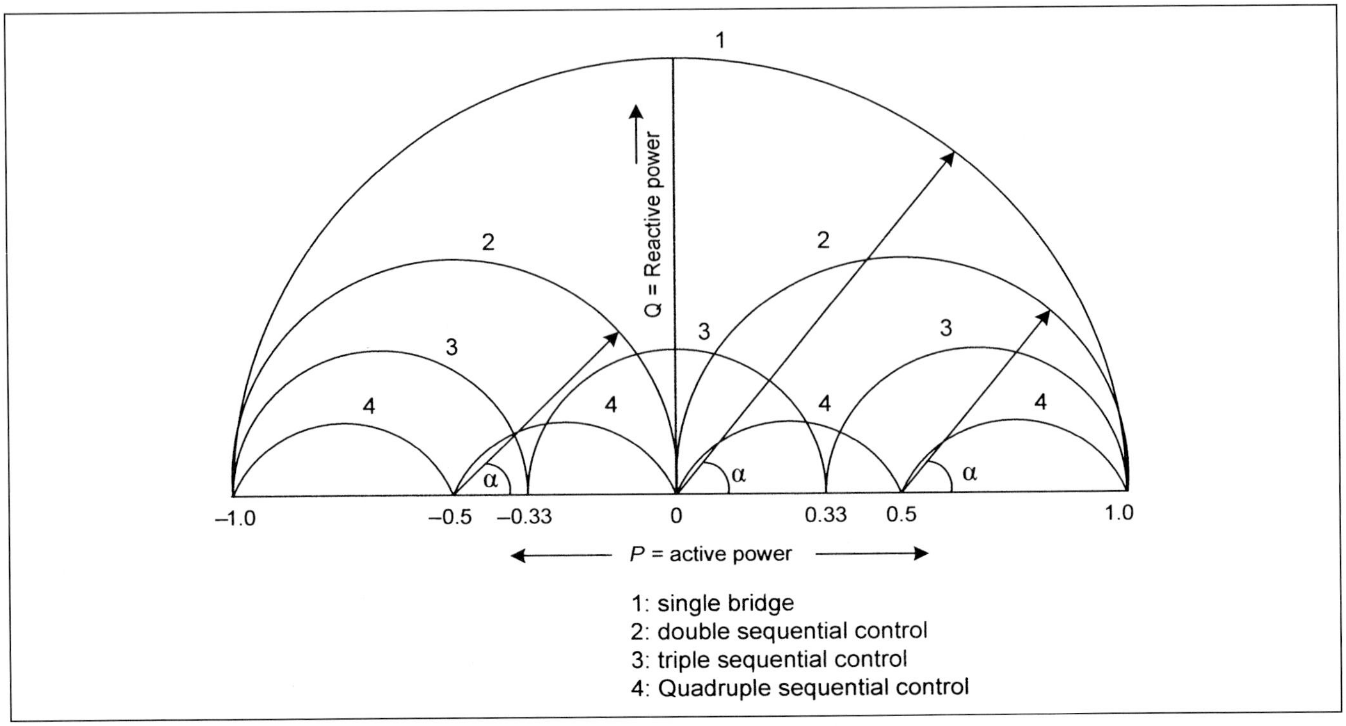

FIGURE 15-7 Reactive power requirements of current source converters.

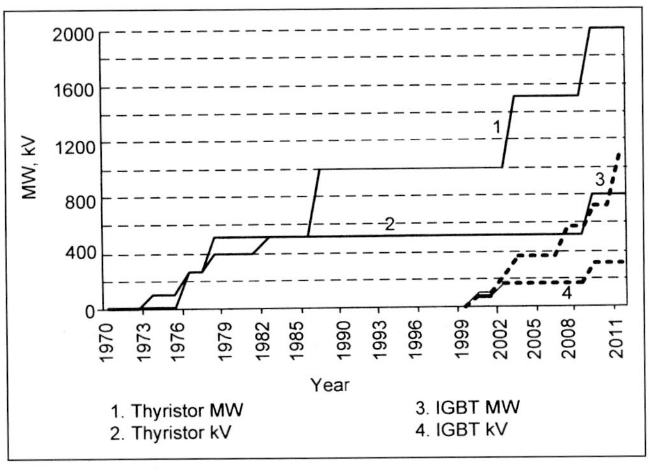

FIGURE 15-8 Developments of solid-state technology: thyristors and IGBTs.

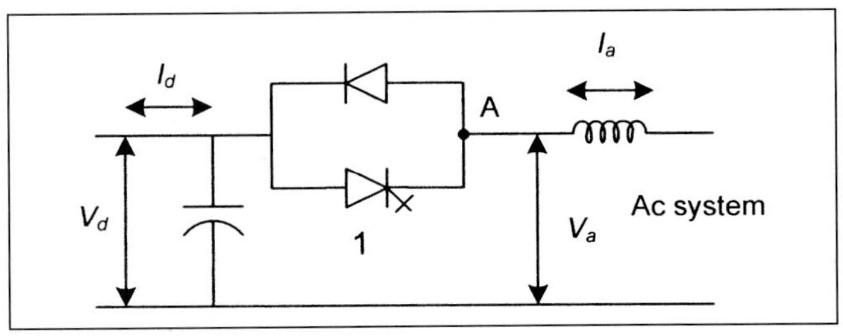

FIGURE 15-9 A single-valve voltage source converter.

The principle is illustrated with respect to single-valve operation (Fig. 15-9). Consider that dc voltage remains constant and the turn-off device is turned on by gate control. Then the positive of dc voltage is applied to terminal *A*, and the current flows from $+V_d$ to *A*, that is, inverter action. If the current flows from *A* to $+V_d$, even when device 1 is turned on, it will flow through the parallel diode, rectifier action. Thus, the power can flow in either direction. A valve with a combination of turn-off device and diode can handle power in either direction.

Figure 15-10*a* extends this concept for a fully controlled single-phase bridge and shows the conduction of the devices. With 1 and 2 turned on, the output voltage $V_{ab} = +V_d$, and with 3 and 4 turned on it is $-V_d$. This voltage waveform occurs Fig. 15-10*b*, irrespective of the ac current flow magnitude and phase angle. The respective conduction of devices is shown and when the current reverses, it flows through diodes 1′, 2′. The ac current flow is an interaction of the square wave generated by the converter, the ac voltage, and the impedance. Figure 15-10*c* shows that the current flow from the ac system is sinusoidal, I_{ab} leading V_{ab} by angle θ. Figure 15-10*d* shows the current waveform of I_d and Fig. 15-10*e* shows the voltage across the valve elements.

15-3 THREE-LEVEL CONVERTER

Figure 15-11 shows the circuit of a three-level converter and associated waveforms. In Fig. 15-11*a*, each half of the phase leg is split into two series-connected circuits, midpoint connected through diodes, which ensure better voltage sharing between the two sections. Waveforms in Figs. 15-11*b* through *e* are obtained corresponding to one three-phase leg. Waveform Fig. 15-11*b* is obtained with 180°

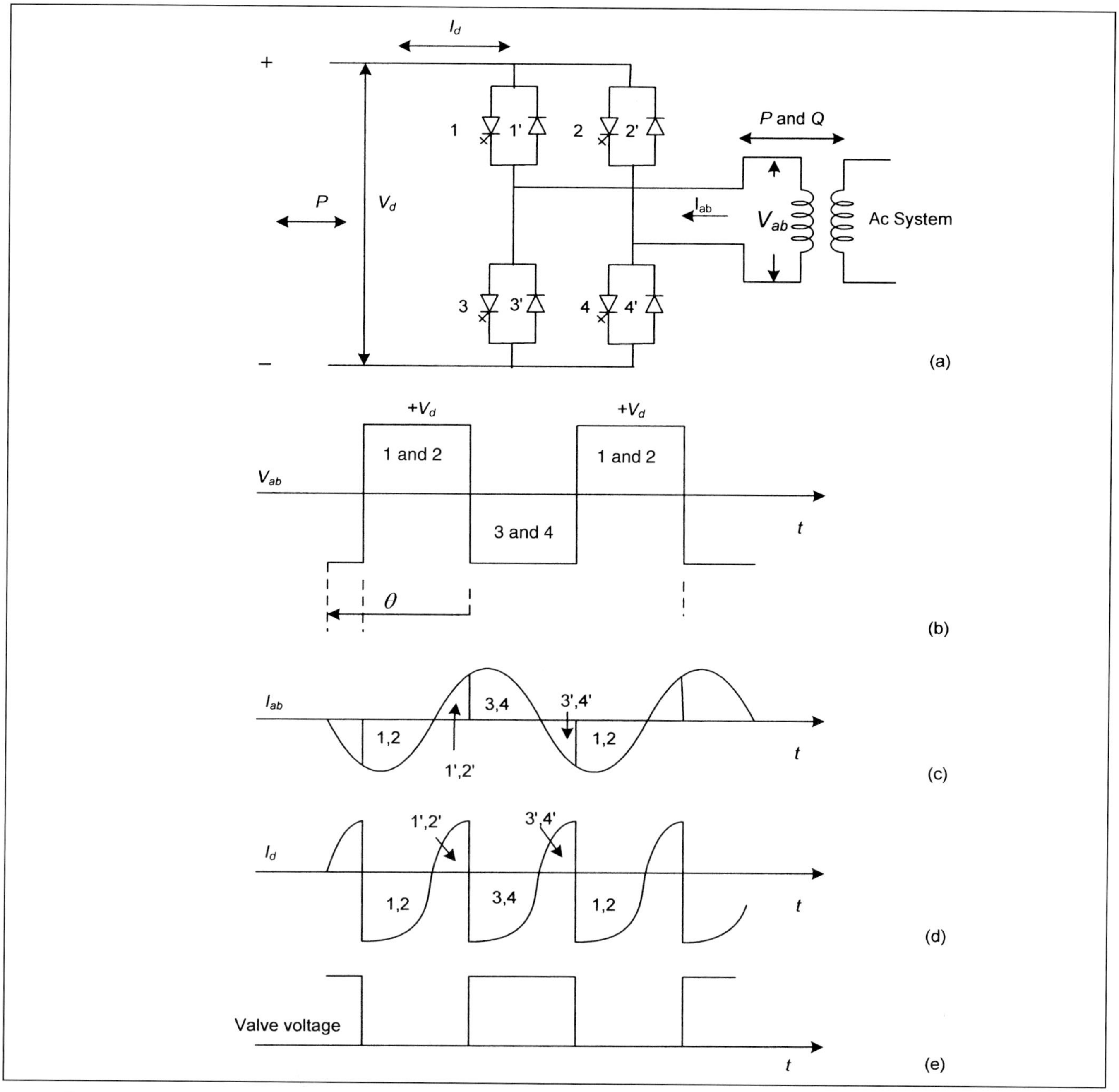

FIGURE 15-10 (*a*) Single-phase full-wave voltage source converter connection diagram. (*b*), (*c*), (*d*), and (*e*) Operational waveforms.

conduction of the devices. Waveform Fig. 15-11*c* is obtained if 1 is turned off and 2A is turned on at an angle α *earlier* than for 180° conduction. The ac voltage V_a is clamped to zero with respect to mid point N of the two capacitors. This occurs because devices 1A and 2A conduct and, in combination with diodes, clamp the voltage to zero. This continues for a period of 2α, until 1A is turned off and 2 is turned on and voltage is now $-V_d/2$, with both the 2 and 2A turned off and 1 and 1A turned on. The angle α is variable and output voltage Va is square waves $\sigma = 180° - 2\alpha°$. The converter is called a three-level converter as dc voltage has three levels, $V_d/2$, 0, and $-V_d/2$. The magnitude of the ac voltage can be varied without changing the magnitude of the dc voltage by varying angle α.

Fig. 15-11*d* shows the voltage V_b, and Fig. 15-11*e*, the phase-to-phase voltage V_{ab}.

The harmonic and fundamental rms voltages are given by:

$$V_h = \frac{2\sqrt{2}}{\pi}\left(\frac{V_d}{2}\right)\frac{1}{2}\sin\frac{h\alpha}{2}$$

$$V_f = \frac{2\sqrt{2}}{\pi}\left(\frac{V_d}{2}\right)\sin\frac{\alpha}{2} \qquad (15\text{-}9)$$

$V_f = 0$ at $\alpha = 0$ and maximum at $\alpha = 180°$.

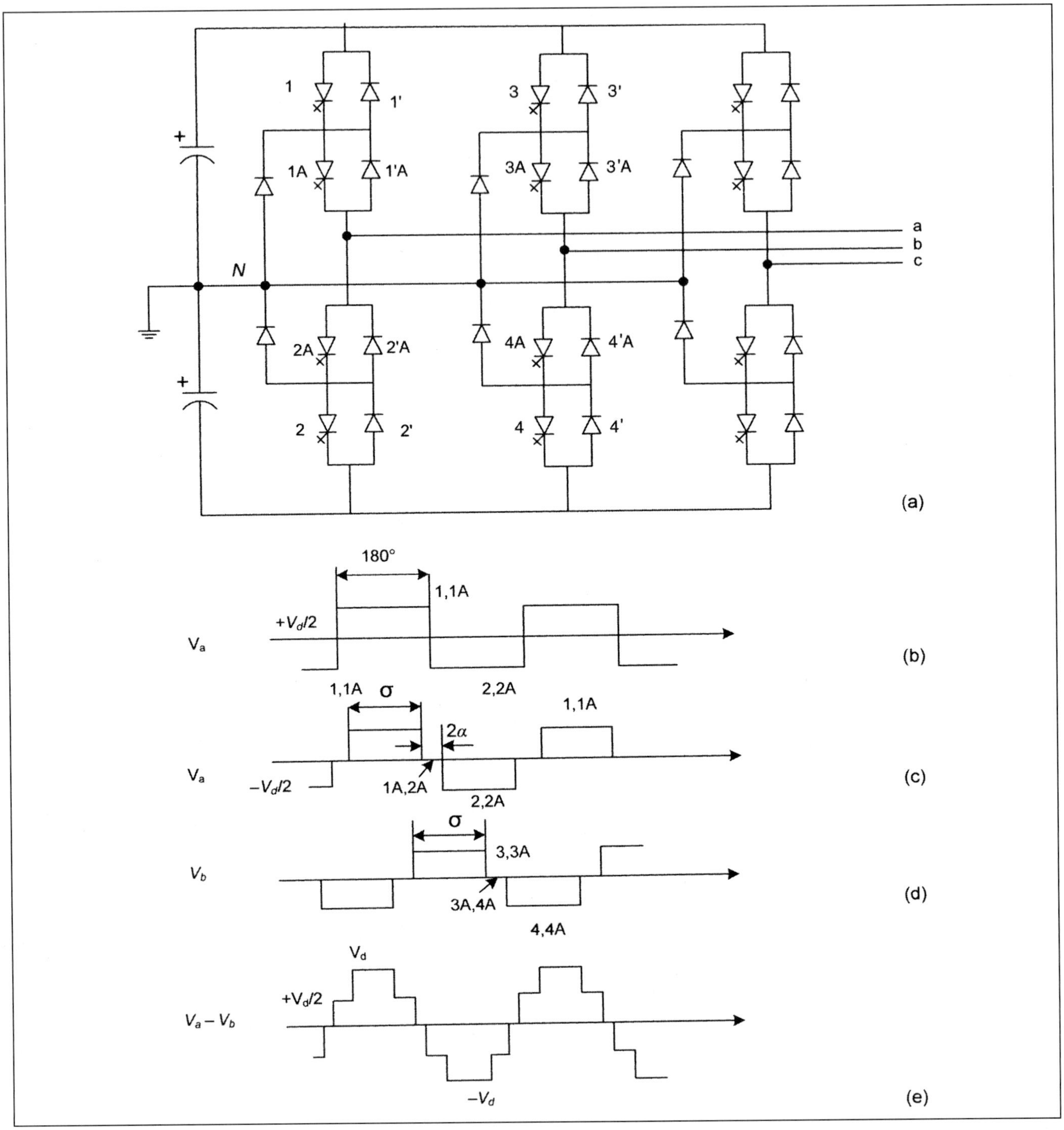

FIGURE 15-11 (*a*) A three-level, three-phase voltage source converter. (*b*), (*c*), (*d*), and (*e*) Operational waveforms.

15-3-1 Pulse Width Modulation

Consider one-phase leg of the three-level converter in Fig. 15-11*a*. Figure 15-12 shows the PWM signals, which are produced using sawtooth wave (triangular) shown at nine times the fundamental frequency. Turn-off and turn-on pulses to the conducting devices correspond to the crossing of the sawtooth wave with the sine wave of the corresponding phase. The negative slope of the sawtooth wave crossing sine wave of phase *a* results in turn-on pulse for device 1 and turn-off for device 2, and vice versa for the positive slope of the sawtooth wave. The resulting ac voltage with respect to midpoint *N* is shown in shaded blocks.

Any even multiple of sawtooth frequency will create asymmetry about zero crossing. The ac voltage can be changed by increasing or decreasing the amplitude of the sine wave, considering sawtooth voltage wave of constant magnitude. Increasing amplitude of sine wave will increase conduction of device 1 and decrease conduction

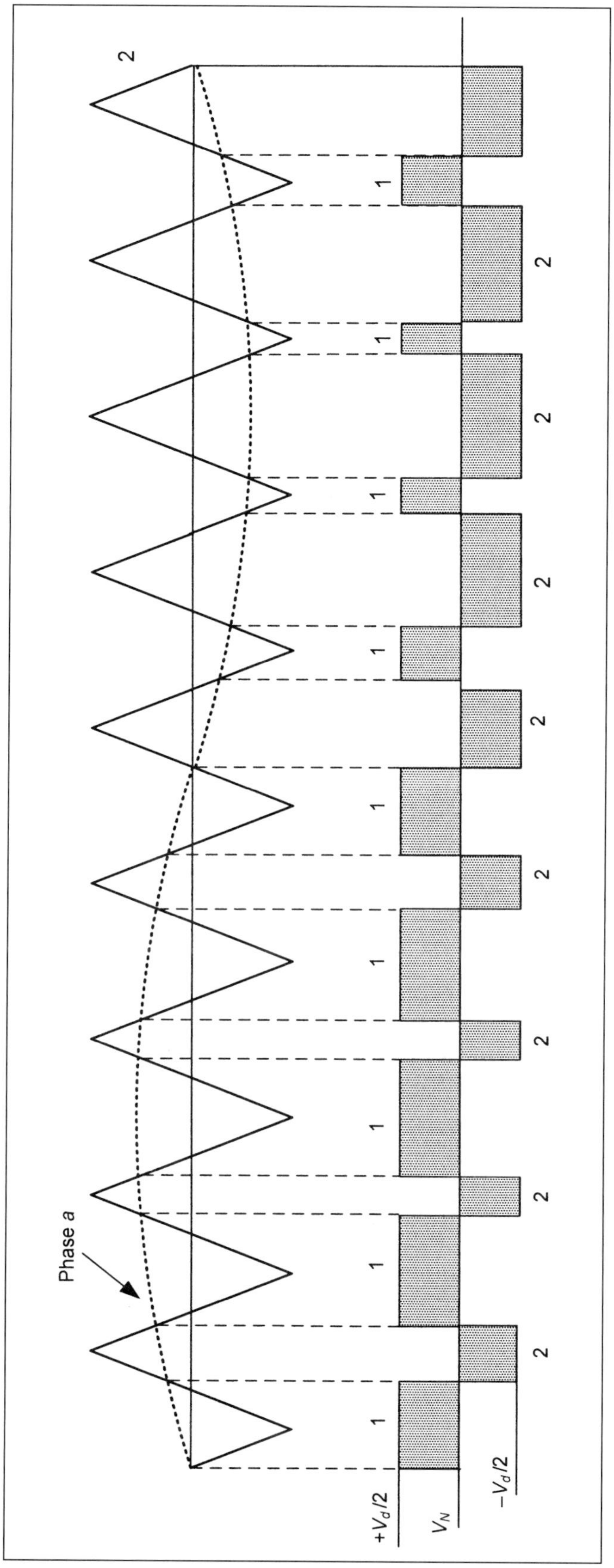

FIGURE 15-12 Pulse width modulation (PWM), triangular wave frequency nine times the fundamental frequency. One phase leg of three-level converter in Fig. 15-11*a* is shown.

time of device 2. The ac voltage varies linearly with variation of ac control voltage from zero to maximum. If the control ac voltage peak and the sawtooth voltage peak are the same, the middle notch will disappear. The output voltage can become a rectangular wave per cycle. Harmonics are produced, though even harmonics and triplen harmonics are absent due to symmetry. Developments of five-level converters are described in Refs. 4 and 5.

15-4 STATIC VAR COMPENSATOR (SVC)

The var requirements in transmission lines swing from lagging to leading, depending upon the load. Shunt compensation by capacitors and reactors is one way. However, it is slow, and power circuit breakers have to be derated for frequent switching duties. This does not provide variable control of the reactive power requirements, and the compensation can be varied only in some discrete steps. A compensator can not only provide variable fast reactive power compensation, but it can also regulate the bus voltage and improve transient stability. In Chap. 7, we discussed that for a transmission line the best location for the var compensation is at the midpoint of the line. More recently gate turn-off devices have been used to supply and absorb reactive power without the use of capacitors and inductors.

The IEEE and CIGRE definition of a static var generator (SVG) embraces different semiconductor power circuits with their internal control enabling them to produce var output proportional to an input reference. A SVG becomes an SVC when equipped with external or system controls which derive its reference from power system operating requirements and variables. SVCs can be classified into the following categories:

- Thyristor-controlled reactor(TCR) (Fig. 15-13*a*)
- Thyristor-switched capacitor (TSC) (Fig. 15-13*b*)
- Fixed capacitor and thyristor-controlled reactor (FC-TCR) (Fig. 15-13*c*)
- Thyristor-switched capacitor and thyristor-controlled reactor (Fig. 15-13*d*)

15-4-1 FC-TCR

An FC-TCR (Fig. 15-13*c*) provides discrete leading vars from the capacitors and continuously lagging vars from thyristor-controlled reactors. The capacitors are used as tuned filters, as considerable harmonics are generated by thyristor control. The design parameters consider the following:

- Stabilization of the system voltage, within certain limits, around the rated voltage under normal operating conditions
- Reduction of voltage fluctuations following a system disturbance
- Control of the reactive load flow between the network and FC-TCR
- Damping power swings in the system
- Compensation or reduction of reactive power unbalance
- Improvements in transient stability

The steady-state characteristics of an FC-TCR are shown in Fig. 15-14. The control range is *AB* with a positive slope, determined by the firing angle control.

$$Q_a = |b_c - b_{1(\alpha)}| V^2 \tag{15-10}$$

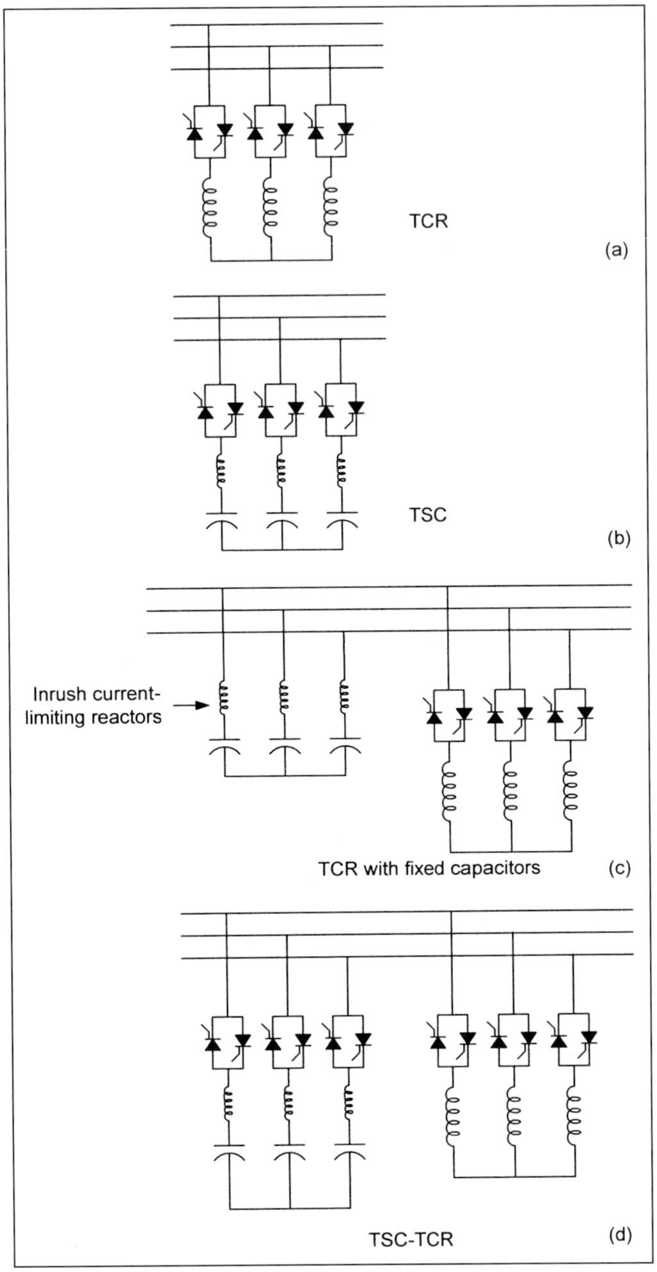

FIGURE 15-13 Conceptual circuit diagrams: (*a*) Thyristor-controlled reactor (TCR). (*b*) Thyristor-switched capacitor (TSC). (*c*) Thyristor-controlled reactor with fixed capacitors. (*d*) Thyristor-controlled reactor and thyristor-switched capacitor (TSC-TCR).

where b_c is the susceptance of the capacitor, and $b_{1(\alpha)}$ the susceptance of the inductor at firing angle α. As the inductance is varied, the susceptance varies over a large range. The voltage varies within limits $V \pm \Delta V$. Outside the control interval *AB*, the FC-TCR acts like an inductor in the high-voltage range and like a capacitor in the low-voltage range. The response time is of the order of one to two cycles. The compensator is designed to provide emergency reactive and capacitive loading beyond its continuous steady-state rating.

Consider a TCR, controlled by two thyristors, as shown in Fig. 15-15*a*. If both thyristors are gated at maximum voltage, the reactor is directly connected across the supply system voltage, producing a 90° lagging current (ignoring the resistance). If the gating is delayed, the waveforms in Fig. 15-15*b* will result. The instantaneous current through the reactor can be written as:

$$i = \sqrt{2}\frac{V}{X}(\cos\alpha - \cos\omega t) \quad \text{for } \alpha < \omega t < \alpha + \beta \qquad (15\text{-}11)$$
$$= 0 \ for\ \alpha + \beta < \omega t < \alpha + \pi$$

where V is the line-to-line fundamental rms voltage, α is the gating angle, and β is the conduction angle. The fundamental component can be written as:

$$I_f = \frac{\beta - \sin\beta}{\pi X}V \qquad (15\text{-}12)$$

Assuming balanced gating angles, only odd harmonics are produced; their rms value is given by:

$$I_h = \frac{4V}{\pi X}\left[\frac{\sin(h+1)\alpha}{2(h+1)} + \frac{\sin(h-1)\alpha}{2(h-1)} - \cos\alpha\frac{\sinh\alpha}{h}\right] \qquad (15\text{-}13)$$

Harmonic filters are provided to mitigate the harmonics.

15-4-2 TSC-TCR

A TSC-TCR provides thyristors control for both the reactive power control elements, the capacitors and reactors. Improved performance under large system disturbances and lower power loss are obtained. Figure 15-16 shows the *V-I* characteristics. A certain short-time overload capability is provided both in the maximum inductive and capacitive regions (shown for the inductive region in Fig. 15-16). Voltage regulation with a given slope can be achieved in the normal operating range. The maximum capacitive current decreases linearly with the system voltage, and the SVC becomes a fixed capacitor when the maximum capacitive output is reached. The voltage support capability decreases with a decrease in system voltage.

The SVCs are ideally suited to control the varying reactive power demand of large fluctuating loads, that is, rolling mills and arc furnaces, dynamic overvoltages due to load rejection, and in HVDC converter stations for fast control of reactive power flow.

Compensation by sectionalizing is based upon a midpoint dynamic shunt compensator. With a dynamic compensator at the midpoint, the line tends to behave like a symmetrical line. The power transfer equation for compensated lines was discussed in Chap. 7. The advantage of static compensators is apparent. The midpoint voltage will vary with the load, and an adjustable midpoint susceptance is required to maintain constant voltage magnitude. With rapidly varying loads, the reactive power demand should be able to be corrected fast, with least overshoot and voltage rise. Figure 15-17 shows transient power angle curves for uncompensated line, with phase angle shift, shunt compensation, and series compensation. With the midpoint voltage held constant, the angles between the two systems can each approach 90°, for a total static stability limit angle of 180°.

The power system oscillation damping can be obtained by rapidly changing the output of the SVC from capacitive to inductive, so as to counteract the acceleration and deceleration of interconnected machines. The transmitted electrical power can be increased by capacitive vars when the machines accelerate, and it can be decreased by reactive vars when the machines decelerate.

The SVCs do not have capability to control the active power flow. The first application of the SVC to voltage control was demonstrated on the Tri-State G&T system in 1977 by General Electric Co. Another SVC designed by Westinghouse and EPRI was operational in 1978.

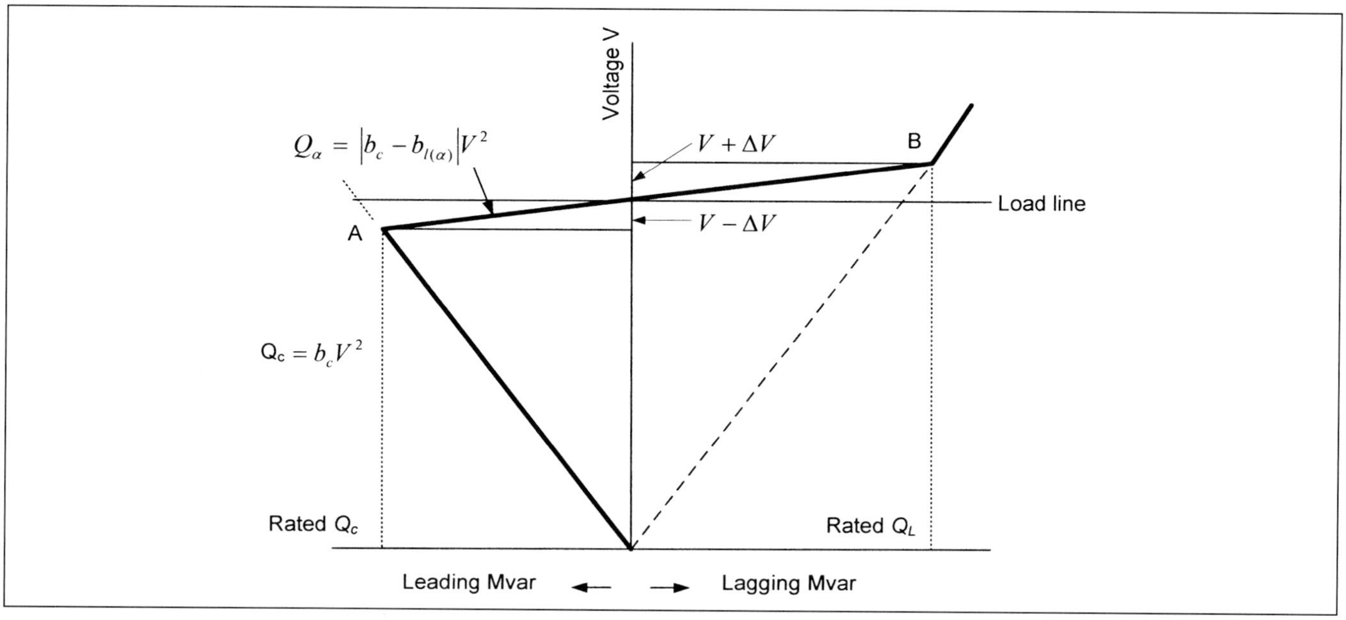

FIGURE 15-14 Steady-state *V-Q* characteristics of a fixed capacitor and thyristor-controlled reactor (FC-TCR).

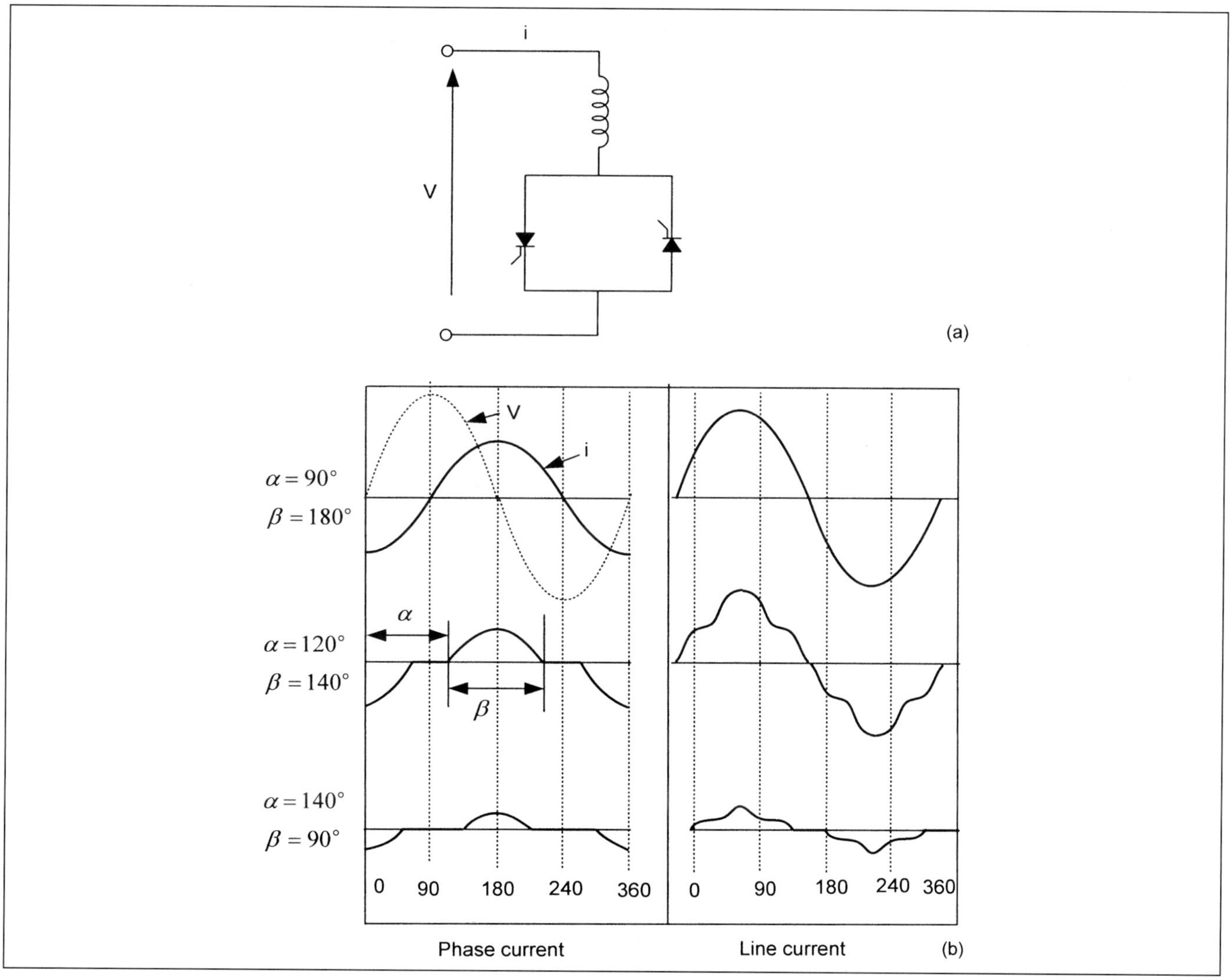

FIGURE 15-15 (*a*) Basic circuit of a thyristor-controlled reactor. (*b*) Current waveforms due to varying conduction and firing angles.

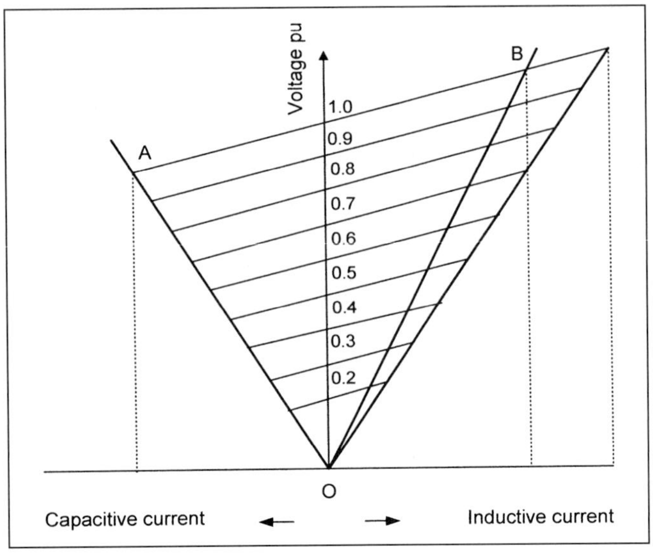

FIGURE 15-16 *V-I* characteristics of a SVC (TSC-TCR) showing reduced output at lower system voltages.

Example 15-1 Figure 15-18 is a simplified system configuration connection for application of an SVC. Consider that a two-phase short circuit occurs at $t = 0.1$ s by closing switch SW and subsequently opening this switch at $t = 0.2$ s, that is, the short-circuit duration is 0.1 s. The SVC is rated for ± 50 Mvar and is controlled in the line voltage control mode.

Figure 15-19 shows the associated transients and EMTP simulation of the system. Figure 15-19*a* shows voltages on 230-kV bus 1, Fig. 15-19*b* shows the phase voltages at the load in per unit, Fig. 15-19*c* shows the active and reactive power output of SVC and load, and Fig. 15-19*d* shows the two-phase fault current. It is seen that the SVC supplies 40 Mvar reactive transiently, yet the voltages under fault conditions dip severely.

15-5 SERIES CAPACITORS

Series compensation is used for (1) voltage stability, as it reduces the series reactive impedance to minimize the receiving-end voltage variations and the possibility of voltage collapse, (2) improvement of transient stability by increasing the power transmission by maintaining the midpoint voltage during swings of the machines, and (3) power oscillation damping by varying the applied compensation so as to counteract the accelerating and decelerating

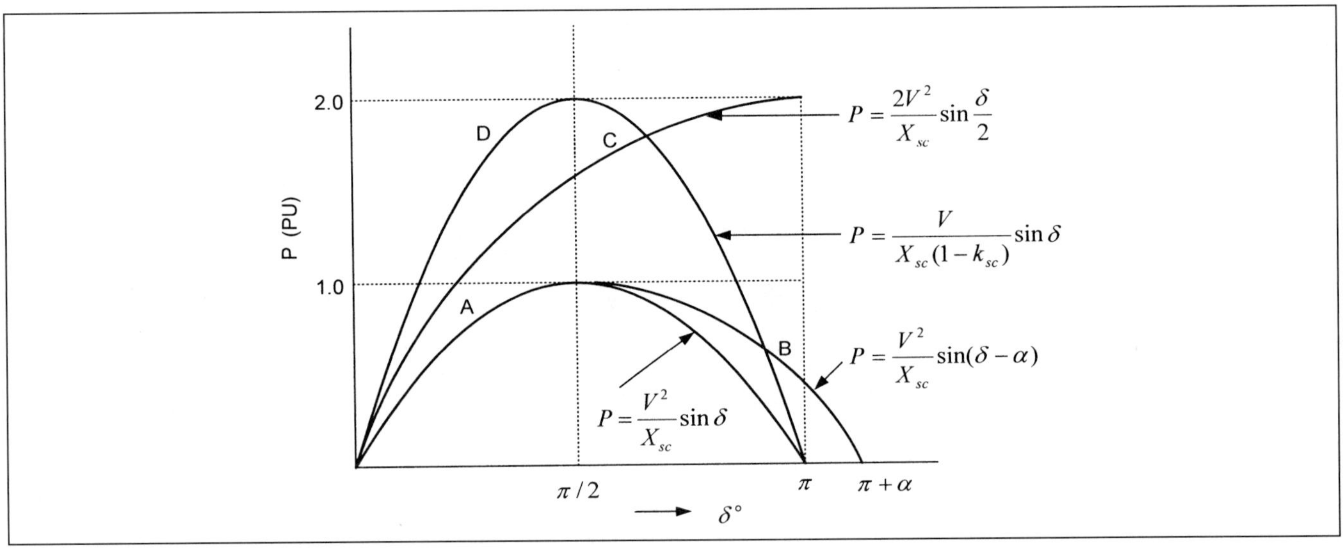

FIGURE 15-17 *P-δ* characteristics. A: uncompensated line; B: with phase-angle regulator; C: with midpoint shunt compensation; and D: with series compensation.

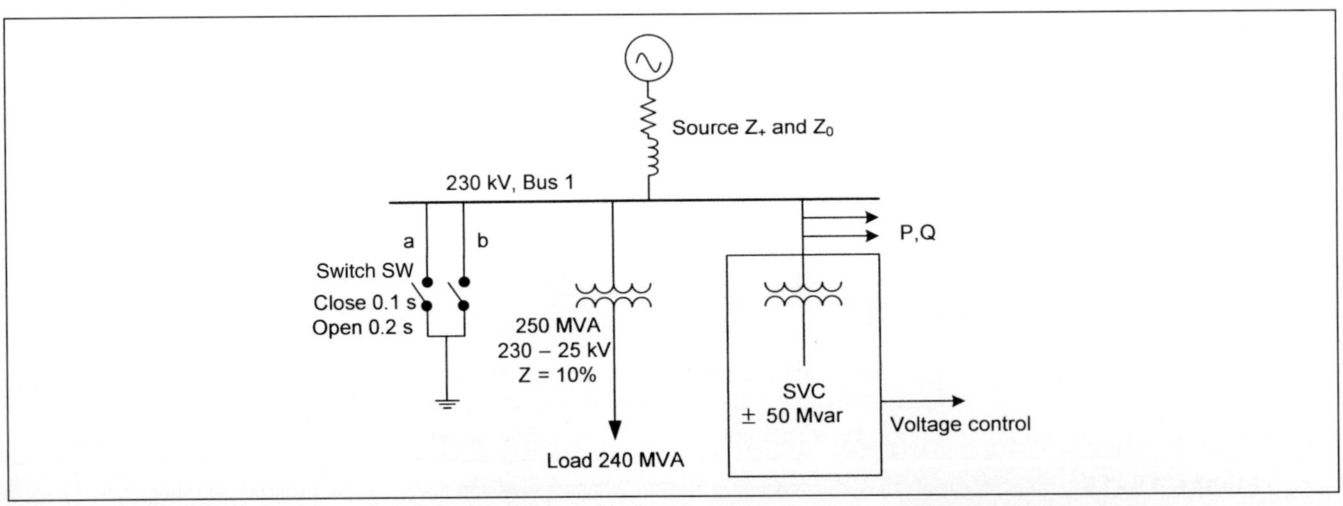

FIGURE 15-18 Application of a ±50 Mvar SVC for study of transients under two-phase–to–ground fault conditions.

swings of the machines. A fixed type of series compensation can, however, give rise to subsynchronous oscillations.

An implementation schematic of the series capacitor installation is shown in Fig. 15-20. The performance under normal and fault conditions should be considered. Under fault conditions, voltage across capacitor rises, and unlike a shunt capacitor, a series capacitor sees many times its rated voltage due to fault currents. A zinc oxide varistor in parallel with the capacitor may be adequate to limit this voltage. In some applications, the varistor will reinsert the bank immediately on termination of a fault. For locations with high fault currents, a parallel fast-acting triggered gap is introduced which operates for more severe faults. When the spark gap triggers, it is followed by closure of bypass breaker. Immediately after the fault is cleared, to realize the beneficial effect of series capacitor on stability, it should be reinserted quickly, and the main gap is made self-extinguishing. A high-speed reinsertion scheme can reinsert the series capacitors in a few cycles. The bypass switch must close at voltages in excess of nominal, but not at levels too low to initiate main gap spark-over. The varistor should be properly chosen, based on:

- Maximum voltage appearing across capacitor at specified current, normally determined by system studies.
- Energy-handling capability, which is the thermal capability—the maximum temperature at which the varistors can continue to operate and withstand capability, that is, maximum short-time energy that can be withstood.

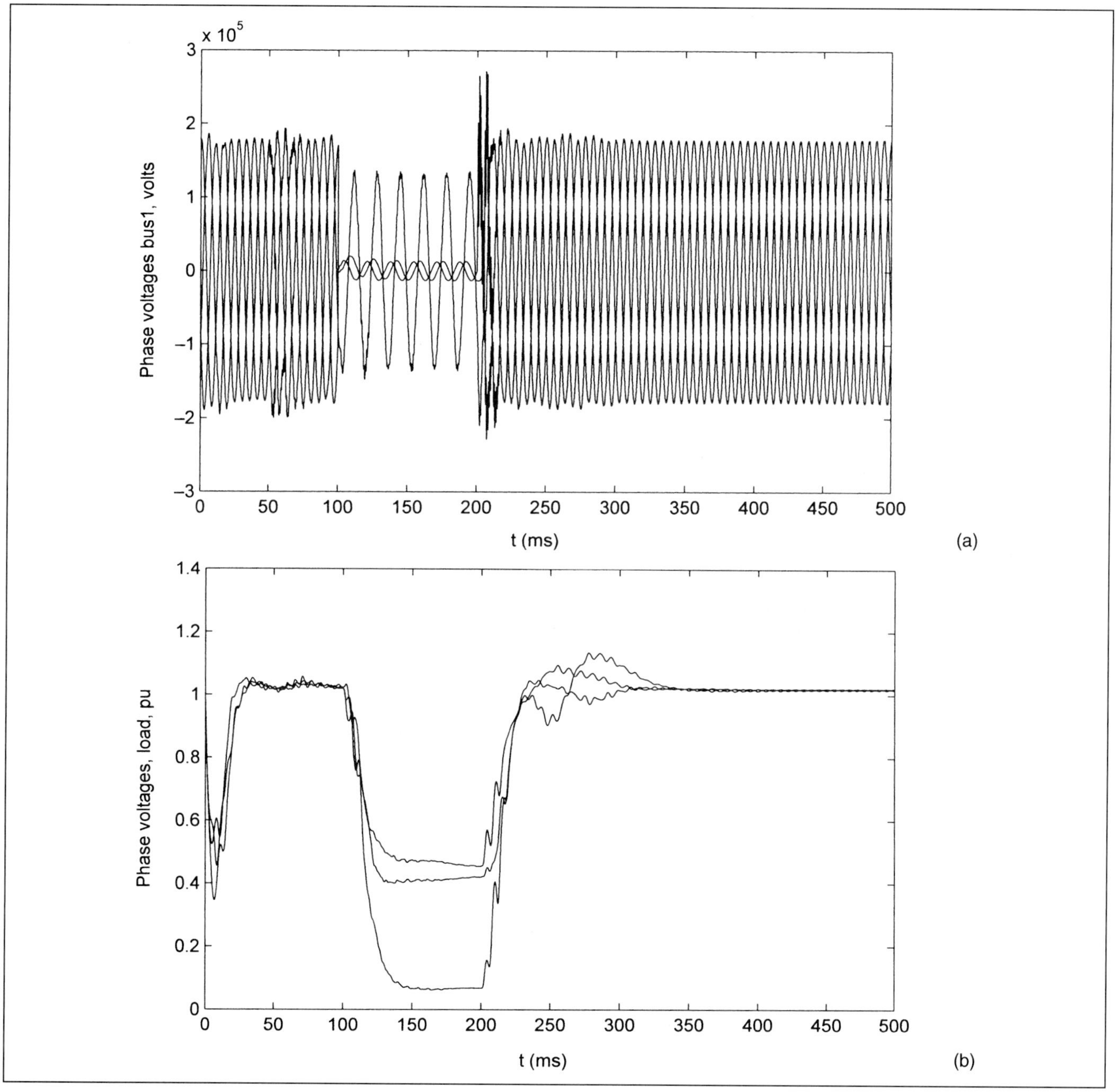

FIGURE 15-19 (a) Phase voltages on 230-kV bus 1 of Fig. 15-18. (*b*) Phase voltages at load. (*c*) SVC *P* and *Q* and load *Q* transients. (*d*) Two-phase fault currents, fault cleared in 100 ms. (*Continued*)

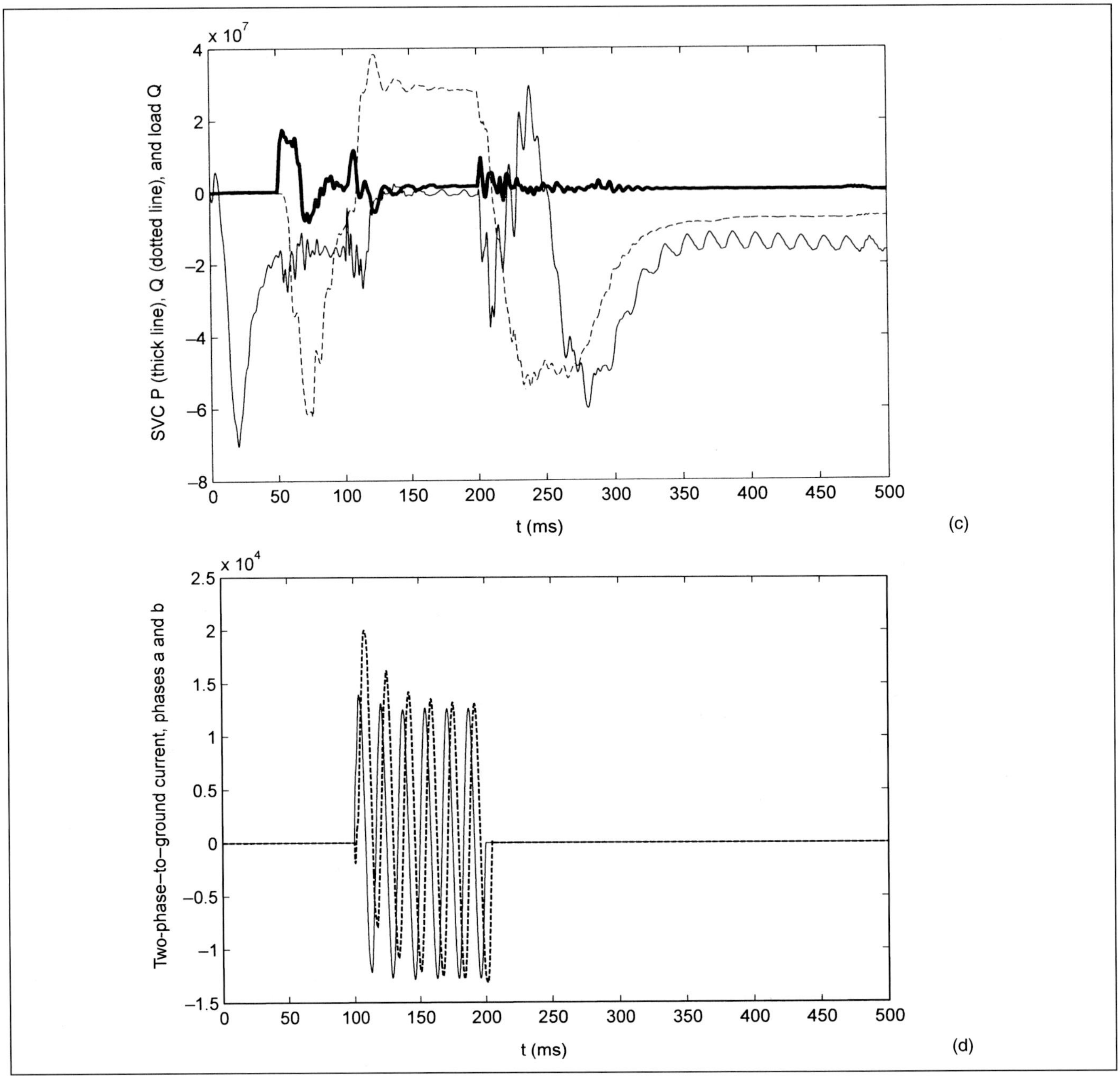

FIGURE 15-19 *(Continued)*

- Extremely large currents at high frequency will be present, and these may produce a large overpressure within the enclosure. The varistors are a special design for the series capacitor application.

The discharge reactor limits the magnitude and frequency of the current through the capacitor when the gap sparks-over. This prevents damage to the capacitors and fuses. A series capacitor must be capable of carrying the full line current. Its reactive power rating is:

$$I^2X_c \text{ per phase} \quad (15\text{-}14)$$

Thus, the reactive power output varies with the load current. Figure 15-21 shows the impact of series versus shunt compensation at midpoint of a transmission line. Both systems are, say, designed to maintain 95 percent midpoint voltage. The midpoint voltage does not vary much when an SVC is applied, but with series compensation, it varies with load. However, for a transfer of power higher than the SVC control limit, the voltage falls rapidly as the SVC hits its ceiling limit, while series compensation holds the midpoint voltage better.

A series capacitor has a natural resonant frequency given by:

$$f_n = \frac{1}{2\pi\sqrt{LC}} \quad (15\text{-}15)$$

where f_n is usually less than the power system frequency. At this frequency, the electrical system may reinforce one of the frequencies of

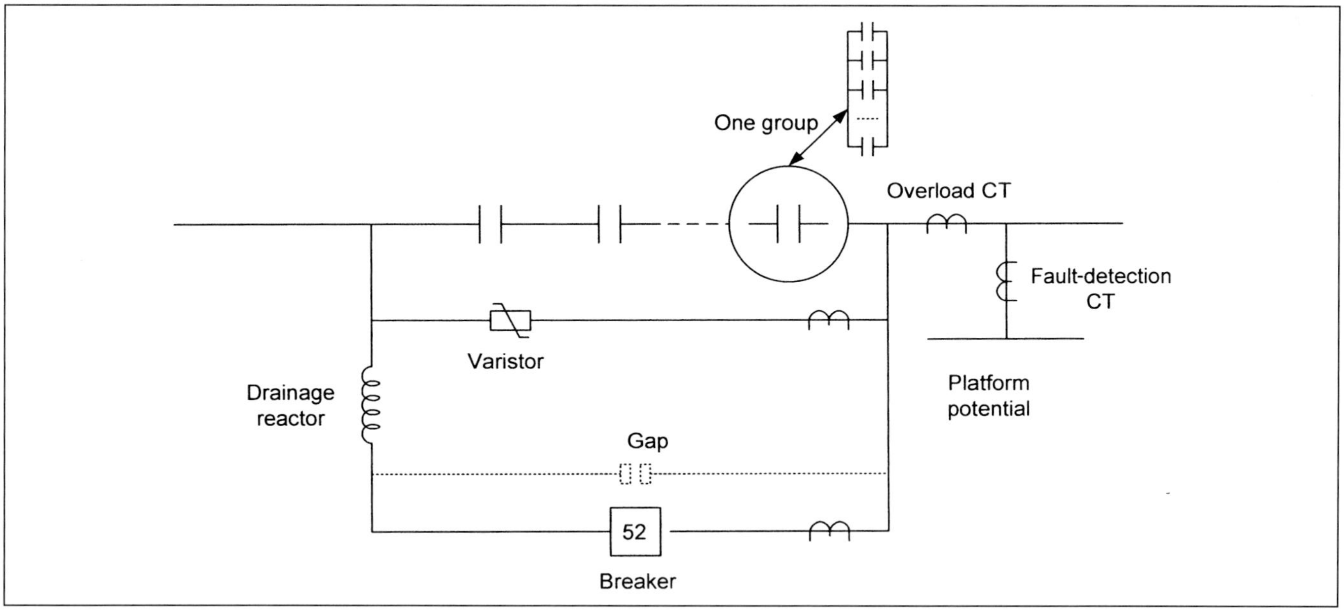

FIGURE 15-20 Schematic diagram of a series capacitor installation.

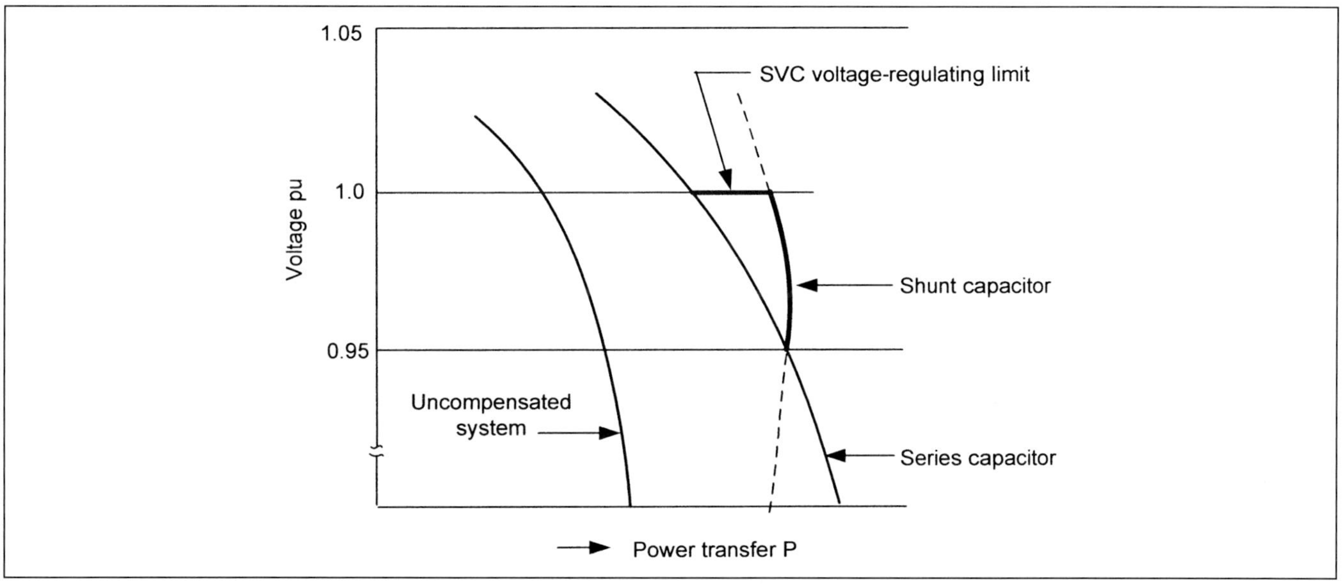

FIGURE 15-21 Series versus shunt compensation; power transfer characteristics for a midpoint compensated line.

the mechanical resonance, causing *subsynchronous resonance* (SSR). If f_r is the subsynchronous resonance frequency of the compensated line, then at resonance

$$2\pi f_r L = \frac{1}{2\pi f_r C}$$
$$f_r = f\sqrt{K_{sc}} \quad (15\text{-}16)$$

This shows that the subsynchronous resonance occurs at frequency f_r, which is equal to normal frequency multiplied by the square root of the degree of compensation, typically, between 15 to 30 Hz. As the compensation is in 25 to 75 percent range, f_r is lower than f. The transient currents at subharmonic frequency are superimposed upon power frequency component and may be damped out within a few cycles by the resistance of the line. Under certain conditions, subharmonic currents can have a destabilizing effect on rotating machines. If the electrical circuit oscillates, then the subharmonic component of the current results in a corresponding subharmonic field in the generator. This field rotates backward with respect to the main field and produces an alternating torque on the rotor at the difference frequency f-f_r. If the mechanical resonance frequency of the shaft of the generator coincides with this frequency, damage to the generator shaft can occur. A dramatic voltage rise can occur if the generator quadrature axis reactance and the system capacitive reactance are in resonance. There is no field winding or voltage regulator to control quadrature axis flux in a generator. Magnetic circuits of transformers can be driven to saturation and surge arresters can fail. The inherent dominant subsynchronous frequency characteristics of the series capacitor can be modified by a parallel connected TCR. If the series capacitor is thyristor- or GTO-controlled, then the whole operation changes. It can be modulated to damp out any subsynchronous as well as low-frequency oscillations.

Example 15-2 Consider a 300-MVA, 22-kV generator, connected to a step-up transformer of 350 MVA, which is delta-wye connected, 22–400 kV, wye windings solidly grounded, which feeds into a 400-mi long line. A CP model of the transmission line is modeled. A series capacitor compensation of 50 percent at the midpoint of the transmission line is provided. For subsynchronous oscillations, the shaft mass system of steam turbine generators is modeled with a number of masses of certain inertia constants connected together through spring constants. External torques can be applied to each of the masses, for example, turbine, generator, and exciter masses (see Chap. 16). An EMTP simulation of the frequency scan at the 400-KV side of the step-up transformer is shown in Fig. 15-22*a*. This shows resonances at 19 Hz and close to the fundamental frequency. A three-phase fault occurs at the secondary of the transformer at 1 s, and cleared at 1.1 s, fault duration of 6 cycles. The resulting torque transients in the 300-MVA synchronous machine masses are shown in Fig. 15-22*b*, total simulation time 5 s. It is seen that these transients do not decay even after 4 s of fault clearances and depending upon the system parameters, can even

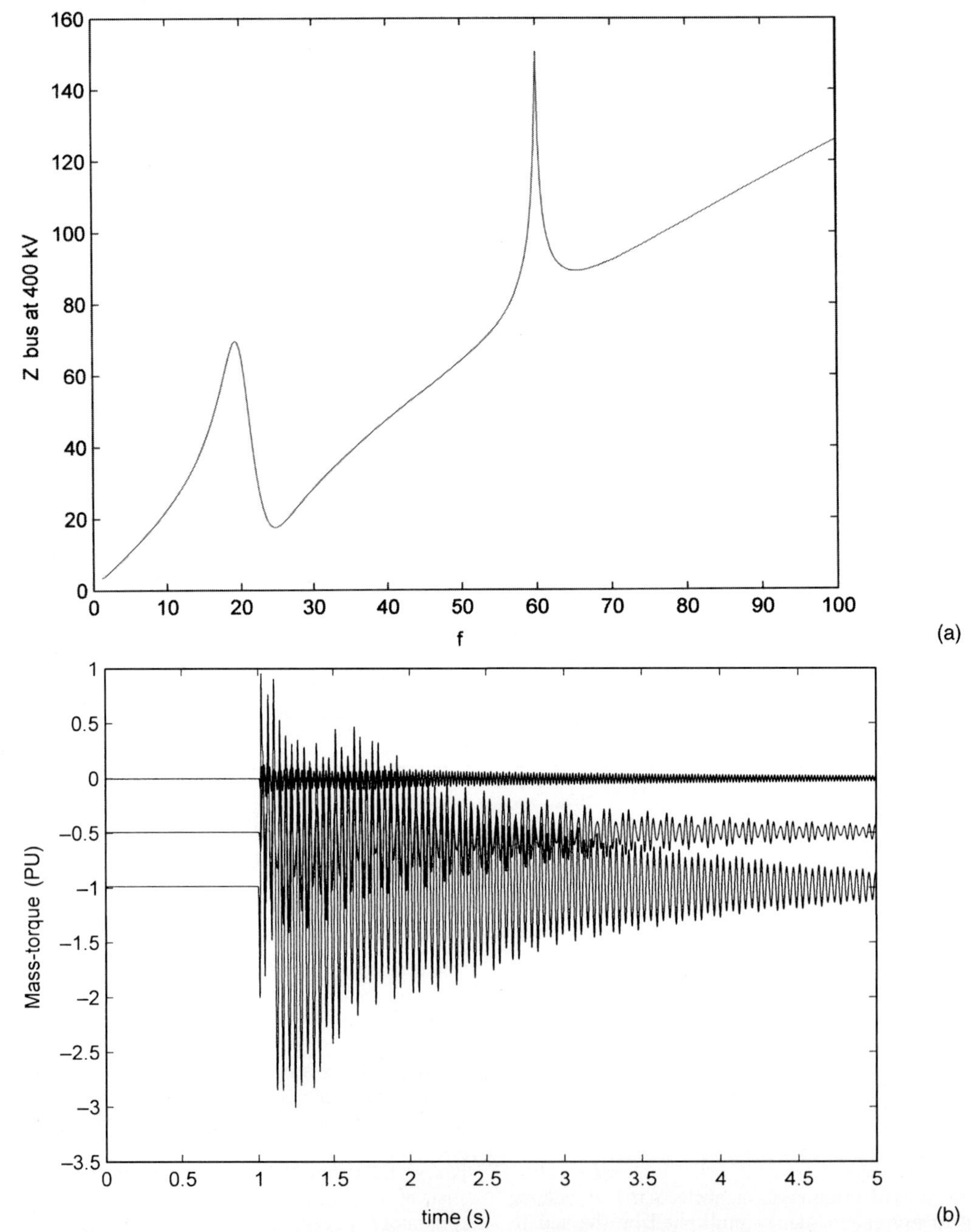

FIGURE 15-22 (*a*) Frequency scan of 400-kV line. (*b*) Shaft transients for three-mass model. (*c*) Angular speed transient mass 1, with no external torque.

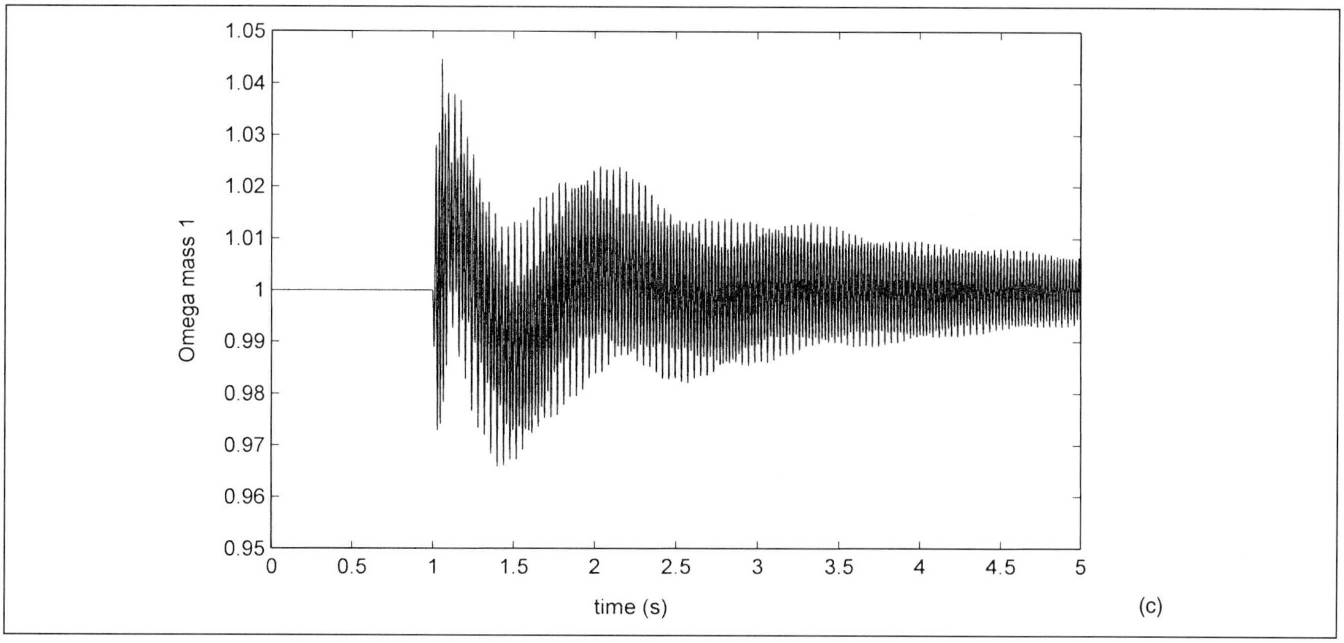

FIGURE 15-22 (*Continued*)

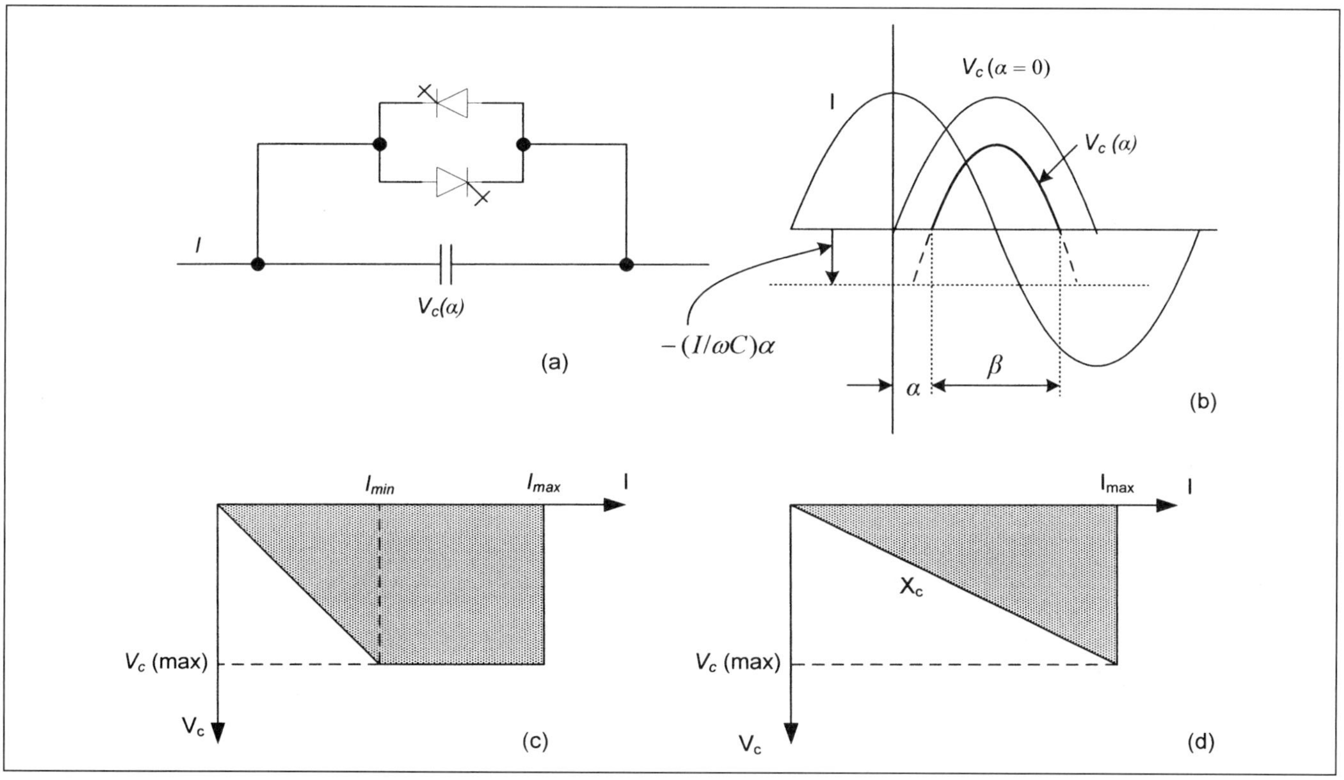

FIGURE 15-23 (*a*) GTO-controlled series capacitor. (*b*) Turn-off delay control. (*c*) Compensating voltage versus current characteristics with voltage control mode. (*d*) Reactance control mode.

diverge, imposing stresses on the generator shaft and mechanical systems. The angular frequency of mass 1 (zero external torque which will give maximum swings) is plotted in Fig. 15-22*c*. This shows violent speed variations. The frequency relays or vibrations probes may isolate the generator from the system.

15-5-1 GTO-Controlled Series Capacitor

A series capacitor can be controlled by parallel thyristors/GTOs. Figure 15-23*a* shows a GTO-controlled series capacitor (GCSC) scheme, which changes the impedance of the circuit. The objective is to control the voltage V_c across the capacitor at a given line current I. When the GTO is conducting, then the voltage across the capacitor is zero, and when nonconducting it is at the maximum. For controlling the voltage, the closing and opening of the GTO can be controlled in synchronism with the power frequency; a positive half cycle control is shown in Fig. 15-23*b*. The GTO valve is controlled to close whenever the capacitor voltage crosses zero. The turnoff in each half cycle is controlled by turn-off delay angle α $(0 \le \alpha \le \pi/2)$. Capacitor voltage at $\alpha = 0$ and at a finite angle α is clearly shown. The capacitor voltage can be expressed as:

$$V_c = \frac{1}{C}\int_{\alpha}^{\omega t} i(t)dt = \frac{I}{\omega C}(\sin\omega t - \sin\alpha) \tag{15-17}$$

The voltage can, therefore, be continuously controlled. Varying the fundamental capacitor voltage at a fixed line current can be considered variable capacitive impedance. We can write that the fundamental capacitor voltage is equal the capacitive reactance as a function of delay angle α.

$$V_c(\alpha) = X_c(\alpha) = \frac{I}{\omega C}\left(1 - \frac{2}{\pi}\alpha - \frac{1}{\pi}2\alpha\right) \tag{15-18}$$

The two modes of operations are voltage control mode and the impedance compensation mode, which are shown in Fig. 15-23*c* and *d*, respectively. In the voltage compensation mode, X_c is selected to maintain rated compensating voltage at I_{min}. As the current increases, the delay angle is increased to reduce the capacitor injection, thereby maintaining the compensating voltage with increased current. In the impedance compensation mode, maximum X_c is maintained at any line current up to the maximum. X_c is chosen to provide maximum compensation at rated current, that is, $X_c = V_{cmax}/I_{max}$. For zero compensating impedance, the capacitor is bypassed by the GTO, and for the maximum compensation, GTO is open and the capacitor is fully inserted. Harmonics analogous to TCR are produced:

$$V_h = \frac{I}{\omega C}\frac{4}{\pi}\left[\frac{\sin\alpha\cos(h\alpha) - n\cos\alpha\sin(h\alpha)}{h(h^2-1)}\right], \tag{15-19}$$

where $n = 2k+1$

$k = 1, 2, 3.....$

15-6 FACTS

The concept of flexible ac transmission systems was developed by EPRI and many FACTS operating systems are already implemented.[6] The world's first thyristor-controlled series capacitor was put in service on Bonneville Power Authority's 500-KV line in 1993. FACTS are made possible by modern high-power electronics and the fast speed of operation. We refer to the devices used in the transmission systems, voltage levels 121 kV and upward, as FACTS, and the devices applied at 3 to 34 kV, Distribution custom power.

We have examined the problem of power flow over transmission lines and role of SVCs series and shunt compensation to change the impedance of the line or its transmission angle. Advances in recent years in power electronics, software, microcomputers, and fiber optic transmitters that permit signals to be sent to and fro from high-voltage levels make possible the design and use of fast FACTS.

Another thrust leading to FACTS is the use of electronic devices in processes, industry, and the home which has created an entirely new demand for power quality that the energy providers must meet. Power quality problems include voltage sags and swells, high-frequency line-to-line surges, steep wave fronts or transients caused by switching of loads, harmonic distortions, and outright interruptions, which may extend over prolonged period (see Chap. 19). The tolerance of processes to power quality problems are being investigated more thoroughly, however, the new processes and electronic controls are more sensitive to power quality problems.

TABLE 15-2 Applications of FACTS Devices

Application	STATCOM	SSSC	UPFC	NGR-SSR Damper
Voltage control	X	X	X	
Var compensation	X		X	
Series impedance control		X	X	X
MW flow control		X	X	
Transient stability	X	X	X	X
Dynamic stability	X	X	X	X
System isolation		X	X	
Damping of oscillations	X	X	X	X

The basic electronic devices giving rise to this thrust in the power quality are tabulated in Table 15-2, with their applications. FACTS devices control the flow of ac power by changing the impedance of the transmission line or the phase angle between the ends of a specific line. FACTS controllers can provide the required fast control by increasing or decreasing the power flow on a specific line and responding almost instantaneously to stability problems. FACTS devices can be used to dampen the subsynchronous oscillations which can be damaging to rotating equipment, that is, generators. These devices and their capabilities are briefly discussed in this chapter.

15-7 SYNCHRONOUS VOLTAGE SOURCE

A solid-state synchronous voltage source (SS) can be described analogous to a synchronous machine. A rotating synchronous condenser has a number of desirable characteristics, that is, high capacitive output current at low voltages and source impedance that does not cause harmonic resonance with the transmission network. It has a number of shortcomings too, that is, slow response, rotational instability, and high maintenance.

An SS can be implemented with voltage source inverter using GTOs. An elementary six-pulse voltage source inverter with a dc voltage source can produce a balanced set of three quasi- square waveforms of a given frequency. The output of the six-pulse inverter will contain harmonics of unacceptable level for transmission-line application, and multipulse inverters can be implemented by a variety of circuit arrangements.

The reactive power exchange between the inverter and the ac system can be controlled by varying the amplitude of the three-phase voltage produced by an SS. Similarly, the real power exchange between the inverter and the ac system can be controlled by phase-shifting the output voltage of the inverter with respect to the ac system. Figure 15-24*a* shows the coupling transformer, the inverter, and an energy source which can be dc capacitor, battery, or superconducting magnet.

The reactive and real power generated by the SS can be controlled independently and any combination of real-power generation/absorption with var generation and absorption is possible as shown in Fig. 15-24*b*. The real power supplied/absorbed must be supplied by the storage device, while the reactive power exchanged is internally generated in the SS. The reactive power exchange is controlled by varying the amplitude of three-phase voltage. For a voltage greater than the system voltage, the current flows through the reactance from the inverter into the system, that is, the capacitive power is generated. If the amplitude of output voltage is reduced below

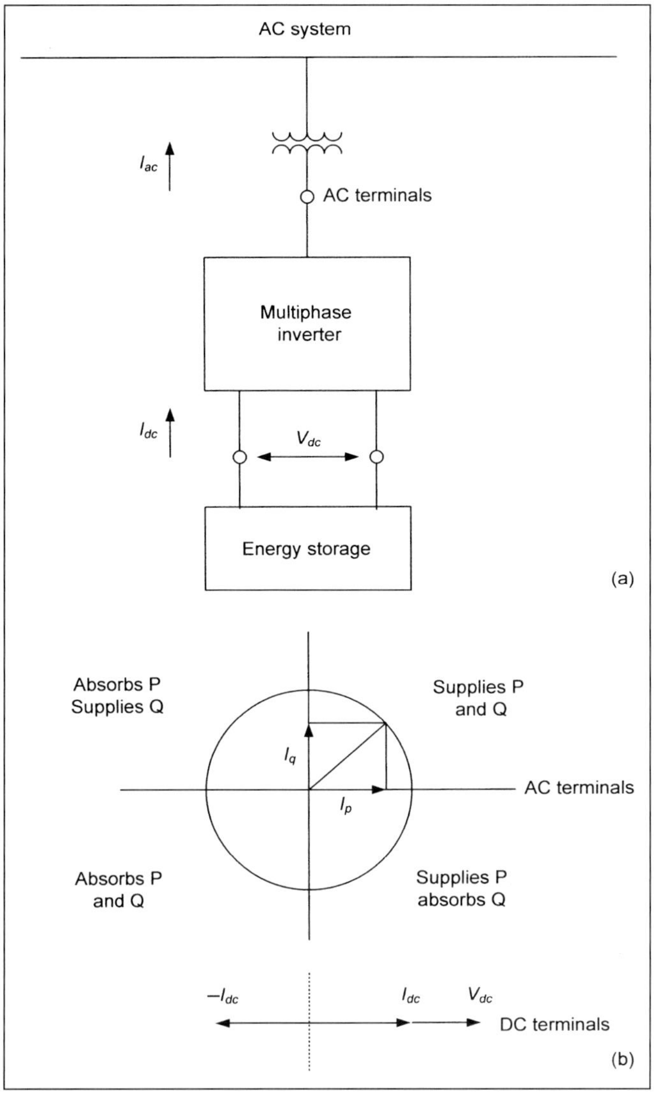

FIGURE 15-24 (*a*) Basic circuit of a shunt-connected static synchronous source (SS). (*b*) Possible modes of active and reactive power compensation.

that of system voltage, the inverter absorbs reactive power. The reactive power is, thus, exchanged between the system and the inverter, and the real power input from the dc source is zero. In other words, the inverter simply interconnects the output terminals in such a way that the reactive power currents can freely flow through them.

The real power exchange is controlled by phase shifting the output voltage of the inverter with respect to the system voltage. If the inverter voltage leads the system voltage, it provides real power to the system from its storage battery. This results in a real component of the current through tie reactance, that is, in phase opposition to the ac voltage. Conversely, the inverter will absorb real power from the system to its storage device if the voltage is made to lag the system voltage. This bidirectional power exchange capability of the SS makes possible the complete temporary support of the ac system.

15-8 STATIC SYNCHRONOUS COMPENSATOR

The three-phase 12-pulse bridge (Fig. 15-11) can be used in the schematic of an inverter-based shunt static synchronous compensator (STATCOM), sometimes called static condenser (STATCON). Its power and control circuit is essentially shown in Fig. 15-25. It is a shunt reactive power compensating device and can be considered as an SS with storage device as a dc capacitor. A GTO-based power converter produces an ac voltage in-phase with the transmission line voltage. When the voltage source is greater than the line voltage ($V < V_0$), leading vars are drawn from the line and the equipment appears as a capacitor; when voltage source is less than the line voltage ($V > V_0$), lagging reactive current is drawn. The basic building block is a voltage-source inverter which converts dc voltage at its terminals to three-phase ac voltage. A STATCON may use many six-pulse inverters, output phase-shifted and combined magnetically to give a pulse number of 24 or 48 for the transmission systems. Using the principle of harmonic neutralization, the output of n inverters, with relative phase displacements, can be combined to produce an overall multiphase system. The output waveform is nearly sinusoidal, and the harmonics present in the output voltage and input current are small. This ensures waveform quality without passive filters. Figure 15-26 shows the output voltage and current waveform of a 48-pulse STATCON, generating reactive power.

The *V-I* characteristics are shown in Fig. 15-27. As compared to the characteristics of an SVC, a STATCON is able to provide rated reactive current under reduced voltage conditions. It also has transient overload capacity, both in the inductive and capacitive region; the limit is set by the junction temperature of the semiconductors. By contrast a SVC can only supply diminishing output current with decreasing system voltage. The ability to produce full capacitive current at low voltages makes it ideally suitable for improving the first swing (transient) stability. Dynamic performance capability far exceeds that of a conventional SVC. It has been shown that the current system can transition from full-rated capacitive to full-rated inductive vars in approximately a *quarter cycle*.

A STATCON, just like an SVC behaves like an ideal midpoint compensator, until the maximum capacitive output current is reached. The reactive power output of a STATCON varies linearly with the system voltage, while that of an SVC varies with the square of the voltage. In an SVC, thyristor-controlled reactors produce high harmonic content, as the current waveform is chopped off in the phase-controlled rectifiers and passive filters are required. A STATCON uses phase multiplication or pulse-width modulation and the harmonic generation is a minimum. Figure 15-28 shows PWM, the reference voltage, the triangular carrier, and the desired output voltages. The PWM signals are produced using two-carrier triangular signals; two different offsets of opposite sign are added to the triangular waveform to give the required carriers. These are compared with the reference voltage waveform, and the output and its inverse is used for the gating signals—four gating signals, one for each IGBT in a leg. Compare this with Fig. 15-12.

The gating signals for the GTOs are generated by the internal converter control in response to the demand for reactive power and/or real power reference signals. The reference signals are provided by the external or system control, from operator intervention, and system variables, which determine the functional operation of STATCOM. If it is operated as an SVC, the reference input to the internal control is the reactive current. The magnitude of the ac voltage is directly proportional to the dc capacitor voltage. Thus, the internal control should establish the required dc voltage on the capacitor.

The internal controls must establish the capability to produce a synchronous output voltage waveform that forces the real and reactive power exchange with the system. As one option, the reactive output current can be controlled *indirectly* by controlling the dc capacitor voltage (as magnitude of ac voltage is directly proportional to the dc capacitor voltage),which is controlled by the angle of the output voltage. Alternatively, it can be *directly* controlled by internal voltage control mechanism, that is, PWM of the converter, in which case, the dc voltage is kept constant by control of the angle.

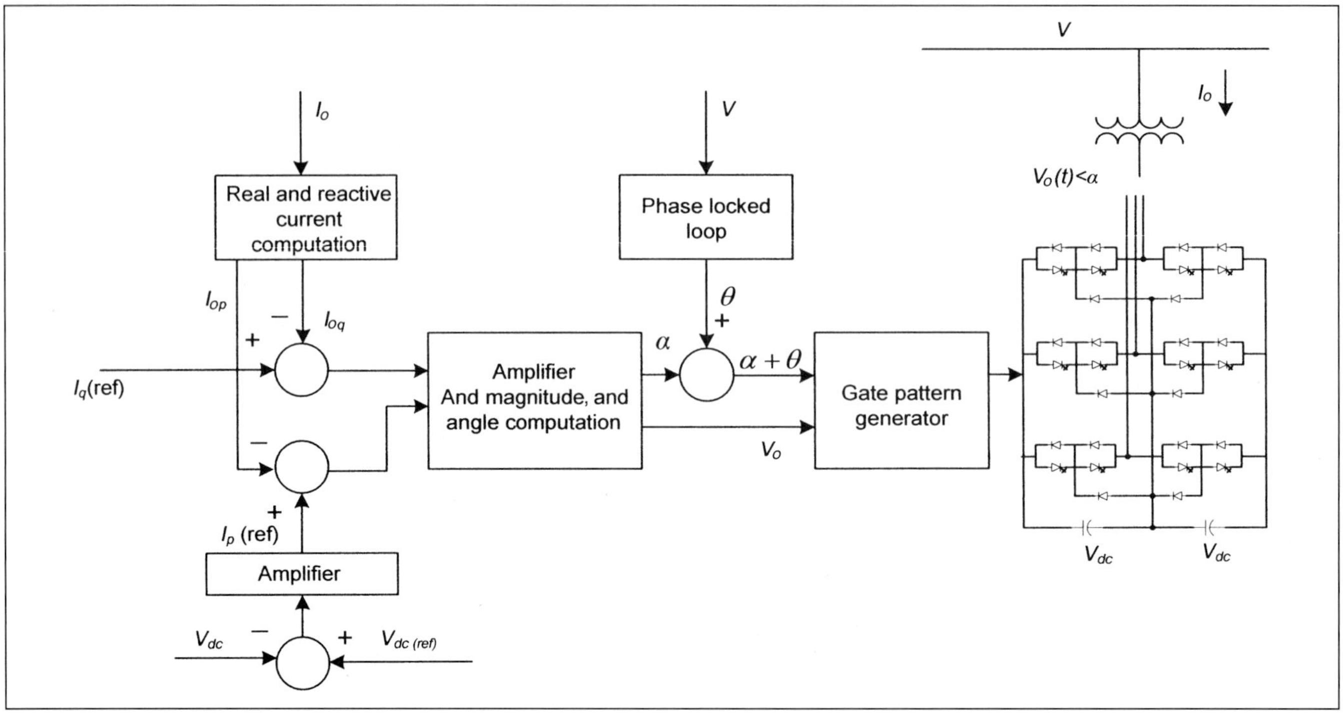

FIGURE 15-25 Schematic diagram of a STATCON (or STATCOM) with internal control circuit block diagram.

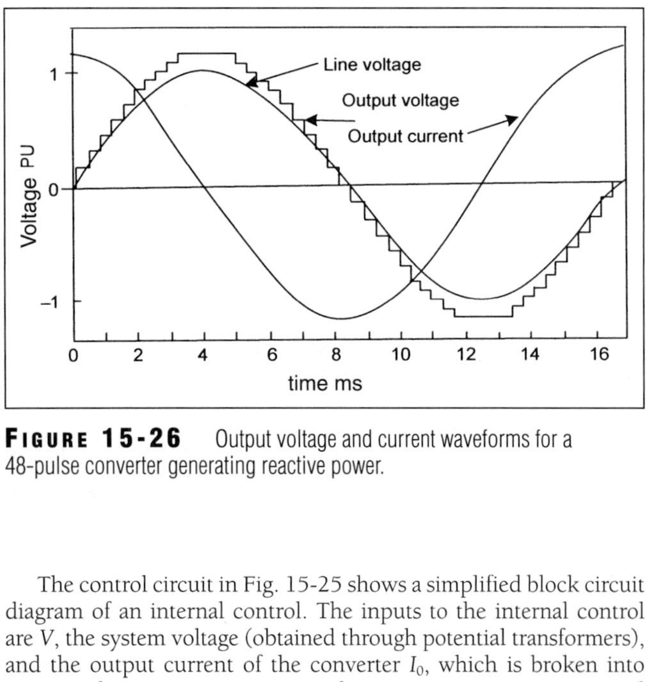

FIGURE 15-26 Output voltage and current waveforms for a 48-pulse converter generating reactive power.

The control circuit in Fig. 15-25 shows a simplified block circuit diagram of an internal control. The inputs to the internal control are V, the system voltage (obtained through potential transformers), and the output current of the converter I_0, which is broken into active and reactive components. These components are compared with an external reactive current reference, determined from the compensation requirements, and the internal real current reference, derived from dc voltage regulation loop. Voltage V operates a phase-locked loop that produces the synchronizing signal angle θ. The reactive current error amplifier produces control angle α. After amplification, the real and reactive current signals are converted into magnitude and angle of the required converter output voltage. The dc voltage reference determines the real power the converter must absorb from the system to supply losses. Summarizing, the following advantages are obtained:

- Interface real power sources
- Higher response to system changes
- Mitigation of harmonics as compared to an FC-TCR
- Superior low-voltage performance

A STATCON of ±100 Mvar is installed at TVA system at Sullivan.[6]

15-9 STATIC SERIES SYNCHRONOUS COMPENSATOR

A static series synchronous compensator (SSSC) may also be called a series power flow controller (SPFC). The basic circuit is that of an SS which is in *series* with the transmission line (Fig. 15-29). We have observed that conventional series compensation can be considered as reactive impedance in series with the line, and the voltage across it is proportional to the line current. A series capacitor increases the transmitted power by a certain percentage, depending upon the series compensation for a given load. In contrast, a SSSC injects a compensating voltage, V_q, in series with the line *irrespective of the line current*. The transmitted power is:

$$P=\frac{V^2}{X_{sc}}\sin\delta+\frac{V}{X_{sc}}V_q\cos\left(\frac{\delta}{2}\right) \tag{15-20}$$

The transmitted power is increased by a fixed percentage over the power capability of an uncompensated line, in the range $0\le\delta\le 90°$. While a capacitor can only increase the transmitted power, the SSSC can also decrease it by simply reversing the

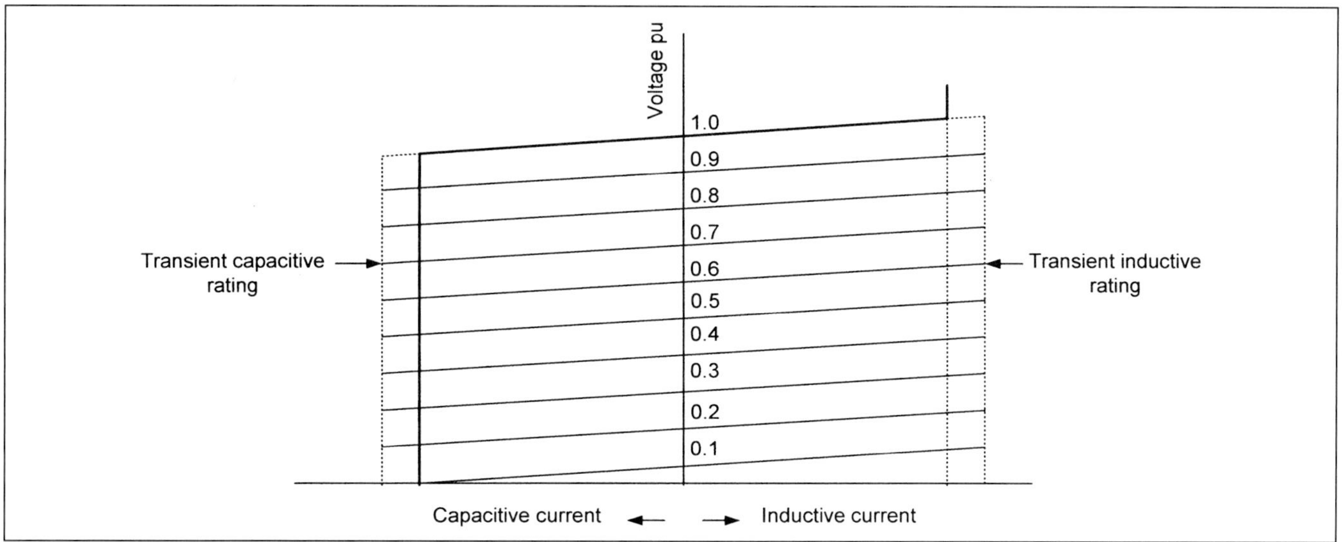

FIGURE 15-27 *V-I* characteristics of a STATCON.

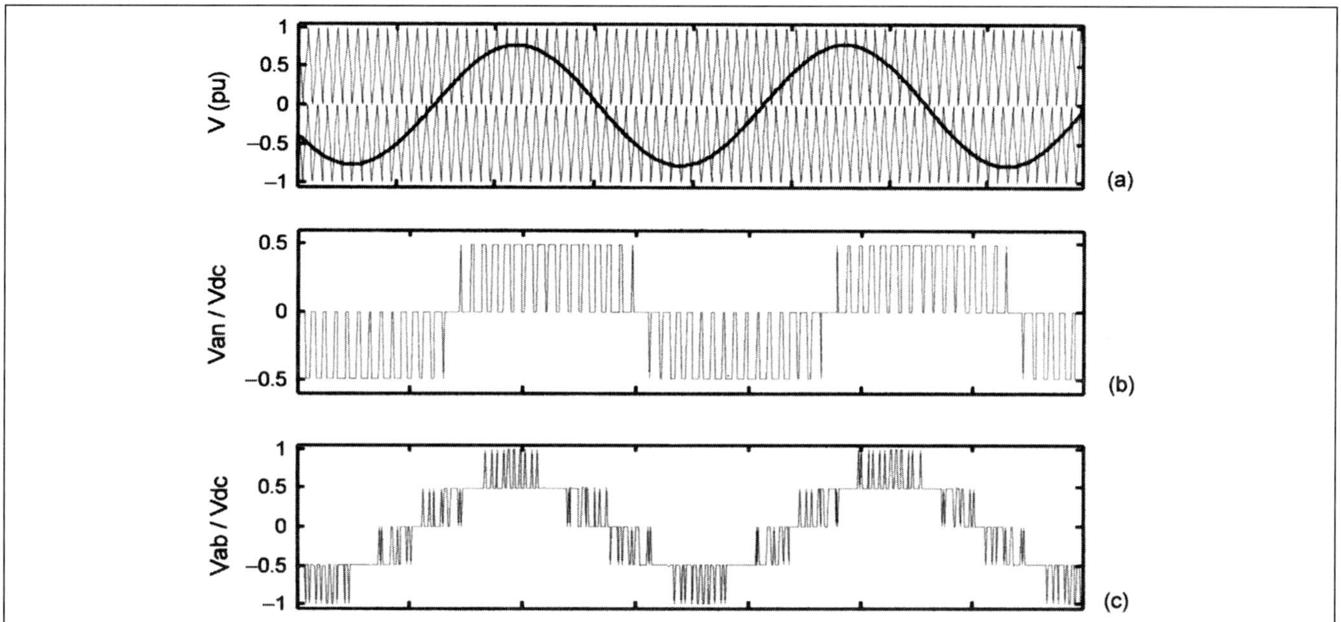

FIGURE 15-28 (a) Offset triangular carrier waves and reference voltage waveform. (*b*) Phase and (*c*) line voltages.

polarity of the injected voltage. The reversed voltage adds directly to the reactive power drop in the line, as if the reactive line impedance is increased. If this reversed polarity voltage is larger than the voltage impressed across the line, by sending and receiving end systems, that is:

$$|V_q| > |V_s - V| + IX_L \tag{15-21}$$

the power flow will reverse.

Thus, stable operation of the system is possible with positive and negative power flows, and due to response time being less than a cycle, the transition from positive to negative power flow is smooth and continuous. Figure 15-30*a* and *b* compares the P/δ curves of a series capacitor and SSSC. Figure 15-30*b* shows reversal of power with the SSSC.

Therefore, the SSSC can negotiate both the reactive and active power with ac system, simply by controlling the angular position of the injected voltage with respect to the line current. One important application is simultaneous compensation of both reactive and resistive elements of the line impedance. By applying series compensation, the X/R ratio decreases. As R remains constant the ratio is now $(X_L - X_c)/R$. As a result, the reactive component of the current supplied by the receiving-end system progressively increases, while the real component of the current transmitted to the receiving end progressively decreases. An SSSC can inject a component of voltage in antiphase to that developed by the line resistance

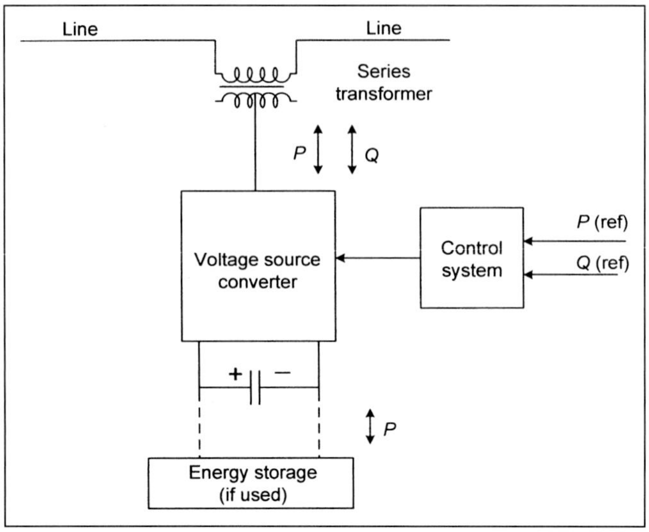

FIGURE 15-29 Functional block diagram of static synchronous series compensator (SSSC), with voltage source converter.

drop to counteract the effect of resistive voltage drop on the power transmission.

The dynamic stability can be improved, as the reactive line compensation with simultaneous active power exchange can dampen power system oscillations. During periods of angular acceleration (increase of torque angle of the machines), an SSSC with a suitable energy dc supply (which can be from a bus or any other source) can provide maximum capacitive line compensation to increase the active power transfer and also absorb active power, acting like a damping resistor in series with the line.

The problems of subsynchronous resonance stated in Section 15-5 are avoided. SSSC is essentially an ac voltage source which operates only at the fundamental frequency, and its output impedance at other frequencies, theoretically, will be zero, though SSSC does have a small inductive impedance of the coupling transformer. An SSSC does not form a series resonant circuit with the line inductance, rather it can damp out the subsynchronous oscillations that may be present due to existing series capacitor installations.

Figure 15-31 shows the characteristics of an SSSC. The VA rating is simply the product of maximum line current and the maximum series compensating voltage. Beyond the maximum rated current, the voltage falls to zero. The maximum current rating is practically the maximum steady-state line current. In many practical applications, only capacitive series line compensation is required, and an SSSC can be combined with a fixed series capacitor. If the device is connected to a short line with infinite buses, unity voltages at sending and receiving ends, and constant phase angle difference, the characteristic can be represented by a circle[7] in *P*-*Q* plane with:

$$\text{Center} = \frac{S_0 Z^*}{2R} \qquad \text{Radius} = \left| \frac{S_0 Z^*}{2R} \right| \tag{15-22}$$

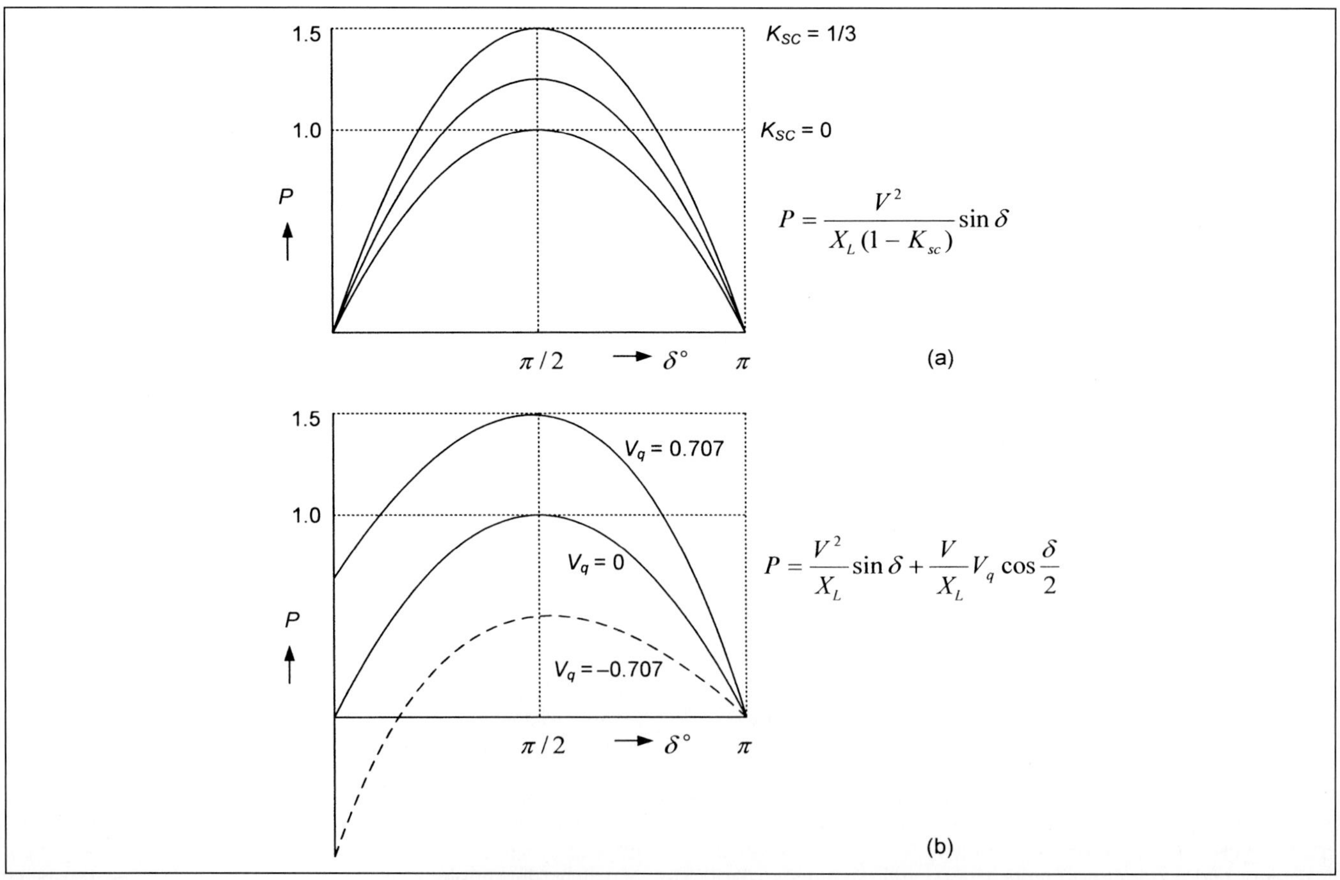

FIGURE 15-30 (*a*) *P*-*δ* characteristics of series capacitor compensation. (b) *P*-*δ* characteristics of SSSC as a function of the compensating voltage V_q.

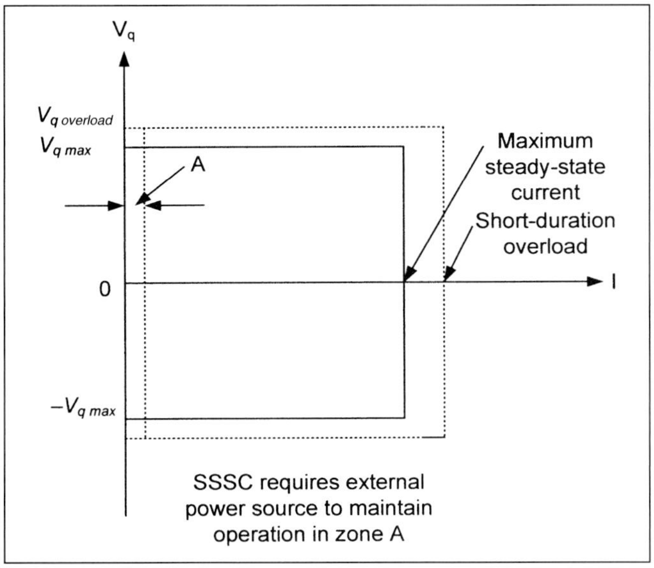

FIGURE 15-31 SSSC *V-I* characteristics with overload capability shown in dotted lines.

where Z* is the complex conjugate of Z, the line series impedance = $R + j\,X_L$ and $S_0 = P_0 + jQ_0$ is the uncompensated power flow. The operating characteristics are defined by the *edge* of the circle only. From Eq. (15-22) the maximum theoretical power amplification is defined as the ration:

$$|S/S_0| = \frac{Z^*}{2R} \tag{15-23}$$

This is approximately 50 percent of the *X/R* ratio. A comparison can be made with series variable compensation, with respect to Fig. 15-32, which shows the characteristics for series compensation

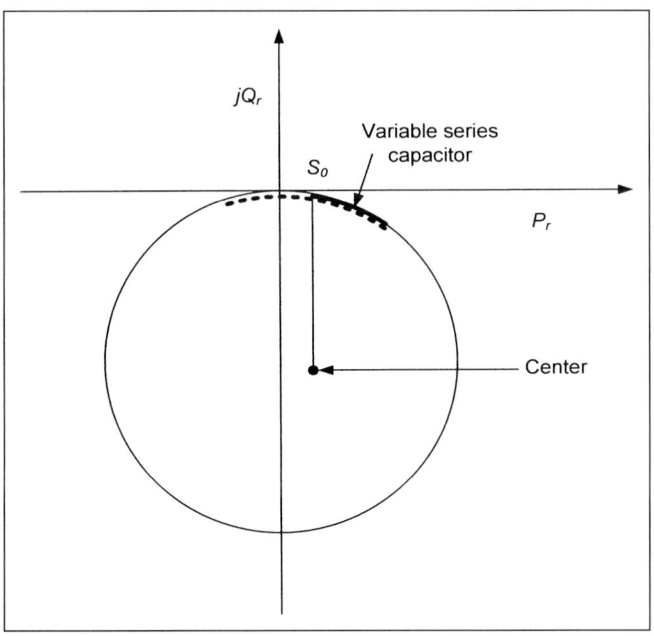

FIGURE 15-32 Operating characteristics of SSSC in *P-Q* plane showing comparison with variable series capacitor compensation.

versus SSSC. As SSSC can decrease the power transfer by behaving inductively, it acts akin to phase angle regulator (PAR). The portion of the characteristic to the left of origin shows power reversal capability of an SSSC.

15-10 UNIFIED POWER FLOW CONTROLLER

A unified power controller consists of two voltage source-switching converters, a series and shunt converter, and a dc link capacitor (Fig. 15-33). This can be considered a combination of an SSSC and a STATCON. The arrangement functions as an ideal ac-to-ac power converter in which real power can flow in either direction between ac terminals of the two converters and each converter can independently generate or absorb reactive power at its own terminals. Converter 2 injects an ac voltage V_{pq} with controllable magnitude and phase angle (0 to 360°) at the power frequency in series with the line. This injected voltage can be considered as an SS. The current *I* and V_{pq} result in an active and reactive power exchange between it and the ac system. The real power exchanged at the ac terminal is converted into dc power which appears as dc link voltage.

The basic function of Converter 1 is to absorb the real power demanded by Converter 2 at the dc link. This dc link power is converted back to ac and coupled to a transmission line through a shunt transformer. Converter 1 can also generate or absorb controllable reactive power, and thereby provide independent shunt reactive compensation of the line. The real power supplied to the system by the series converter must be taken from the system by Converter 1.

Consider the power transfer characteristics in a two-machine system through a reactance with a UPFC connecting two infinite buses, unity voltage. If the injected voltage V_{pq} is zero, the system is like a conventional system. The control characteristics can be illustrated as follows, with respect to Fig. 15-34. If we make $V_{pq} = \pm\Delta V$ (angle $\theta = 0$), that is, change only the magnitude with constantly variable in-phase and antiphase injection, terminal voltage regulation is obtained (Fig. 15-34*c*). This is like an under load tap changing (ULTC) transformer with an infinite number of taps in the control region.

Series reactive compensation (Fig. 15-34*d*) is obtained by injecting $V_{pq} = V_c$ in quadrature with current *I*. This is similar to series compensation obtained by series capacitive and inductive line compensation attained by the SSSC. The injected series voltage can be kept constant, independent of the current variation. Alternatively, it can be varied in proportion to the line current to obtain compensation, much akin to a series capacitor.

Phase angle control is shown in Fig. 15-34*e*. Here $V_{pq} = V_\beta$ is injected with an angular relationship with respect to V_s. This achieves the desired phase shift β (advance or retard), without change in the magnitude. Thus, UPFC can act as a perfect phase angle regulator, which can supply reactive power with internal var generation.

Finally, Fig. 15-34*f* shows an operation unique to UPFC. Multifunction power flow control based upon terminal voltage regulation (Fig. 15-34*c*), series capacitive compensation (Fig. 15-34*d*), and phase shifting Fig. 15-34*e* can be combined:

$$V_{pq} = \Delta V + V_C + V_\beta \tag{15-24}$$

With no injection of V_{pq}, and in terms of normalized maximum power, we can write:

$$\begin{aligned} P &= \frac{V^2}{X}\sin\delta = \sin\delta \quad (V^2/X) = 1 \\ Q &= \frac{V^2}{X}(1-\cos\delta) = 1-\cos\delta \end{aligned} \tag{15-25}$$

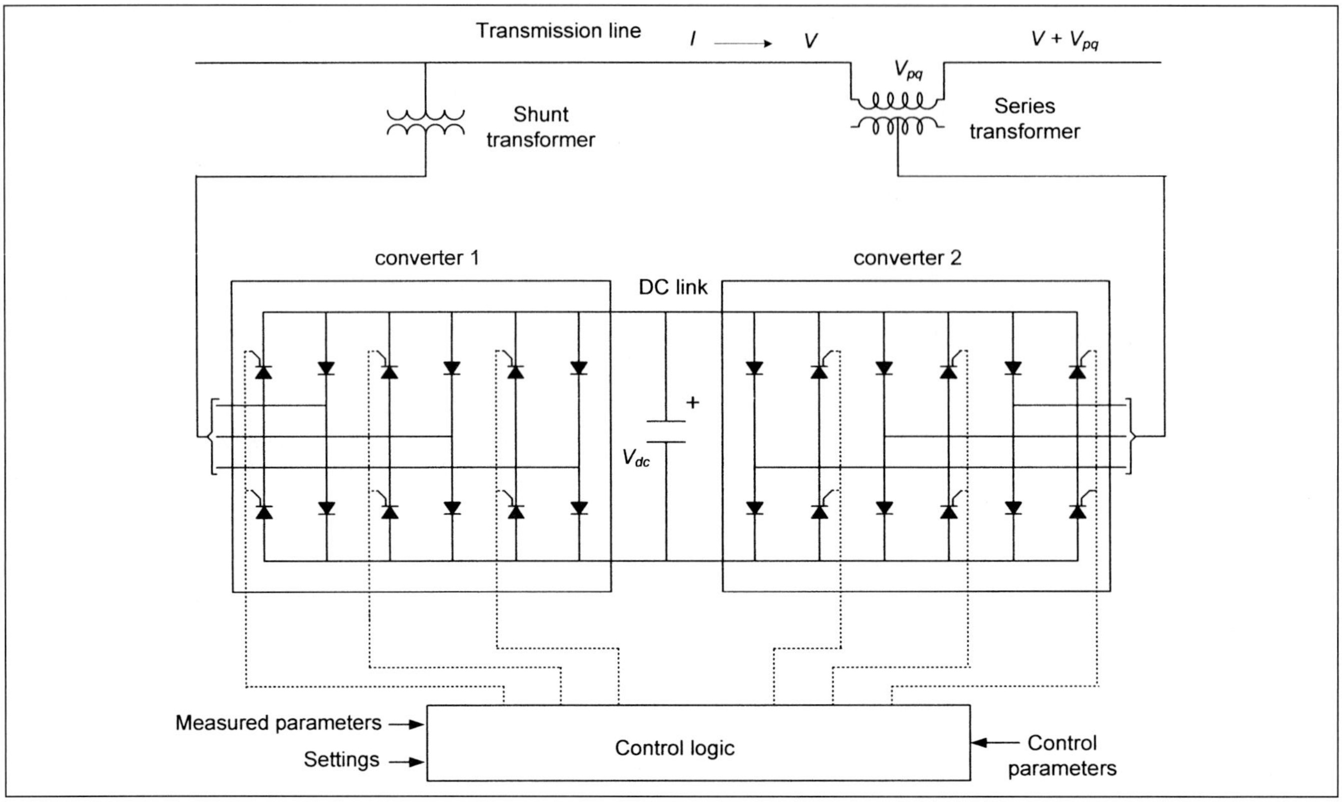

FIGURE 15-33 Schematic of UPFC controller with two voltage source converters and dc link.

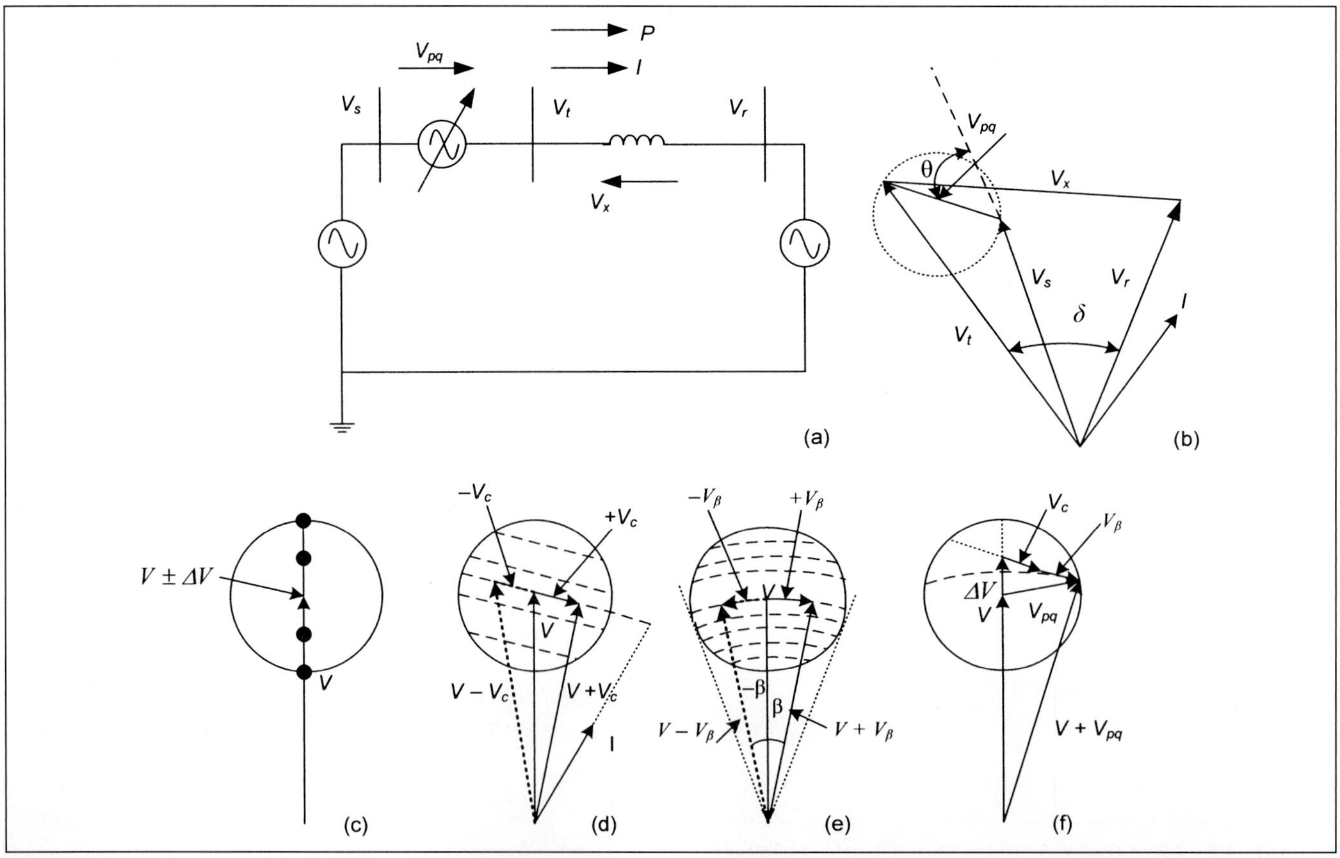

FIGURE 15-34 (*a*) Basic circuit to illustrate UPFC action and injected voltage V_{pq} in a two-bus system. (*b*) Phasor diagram and relevant angles. (*c*) Voltage regulation only. (*d*) Line impedance compensation. (*e*) PAR (phase angle regulator). (*f*) Control of voltage, impedance, and angle.

Therefore:

$$[Q(\delta)+1]^2 + [P(\delta)]^2 = 1 \tag{15-26}$$

which is the equation of a circle (Fig. 15-35*a*). With compensation through UPFC, V_{pq}, the transmitted power P and the reactive power $-jQ$, supplied by receiving end is given by:

$$P - jQ = V_r\left(\frac{V_s + V_{pq} - V_r}{jX}\right)^* \tag{15-27}$$

The asterisk means conjugate of the complex number. The power transfer characteristics of a UPFC for two-machine system connected through a short line of reactance X can be represented by a circle on P-Q plane, described by:

$$(P_R - P_0)^2 + (Q_R - Q_0)^2 = \left(\frac{V \times V_{pq\,\max}}{X}\right)^2 \tag{15-28}$$

where P_0, and Q_0 are the line uncompensated real and reactive power, respectively. The center is at the uncompensated power level S_0 and radius is Vpq/X.

The operation of P-Q control is shown with respect to transmission angles δ of 0°, 60°, and 90°, in Fig. 15-35*b*, *c*, and *d*, respectively. Consider $V_{pq\,\max} = 0.5$ pu, X = 1.0 pu, $V = 1$ pu, that is, the UPFC circle = 0.5 pu, which can be placed at any transmission angle. At $\delta = 0°$, no active and reactive power can be transmitted, and UPFC will improve active power flow by 0.5 pu in either direction, without impacting reactive power flow. This assumes that the sending- and receiving-end sources should be able to absorb the active power without change of angle, though UPFC can force the system to supply or absorb reactive power from the other end. At $\delta = 60°$, the control range of active and reactive power is given by $P_1 - P_2$ and $0 - Q_1$. Figure 15-35*d* shows the controllable region for $\delta = 90°$.

The allowable operating range with UPFC is anywhere *inside* the circle, while the SPFC operating range is the circle itself (Fig. 15-36). A comparison with phase angle regulator and series capacitors is superimposed in this figure. These devices provide one-dimensional

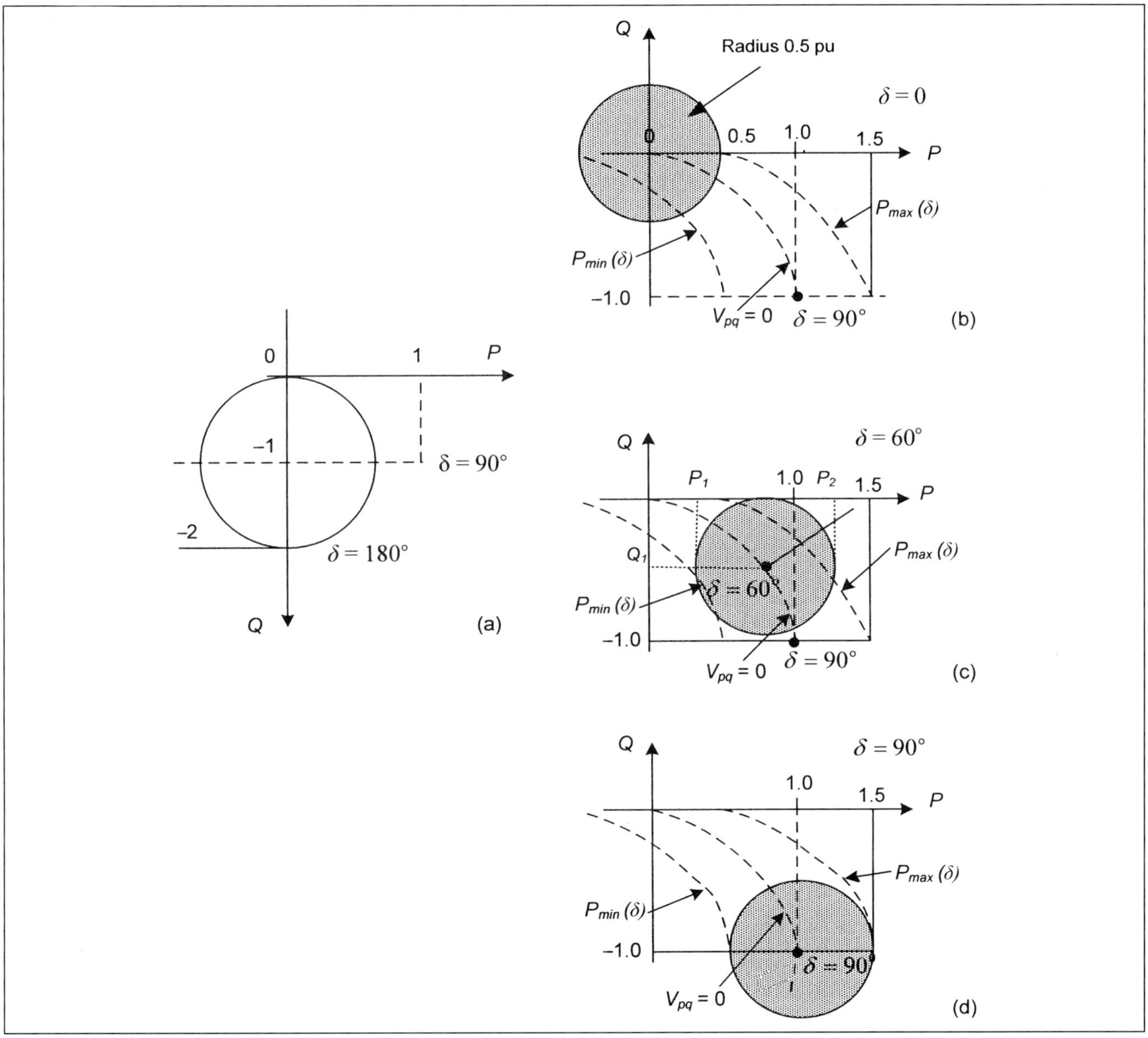

FIGURE 15-35 (*a*) *P-Q* plot of an uncompensated line. (*b*) UPFC compensation placed at $\delta = 0°$, (*c*) 60°, and (*d*) 90°.

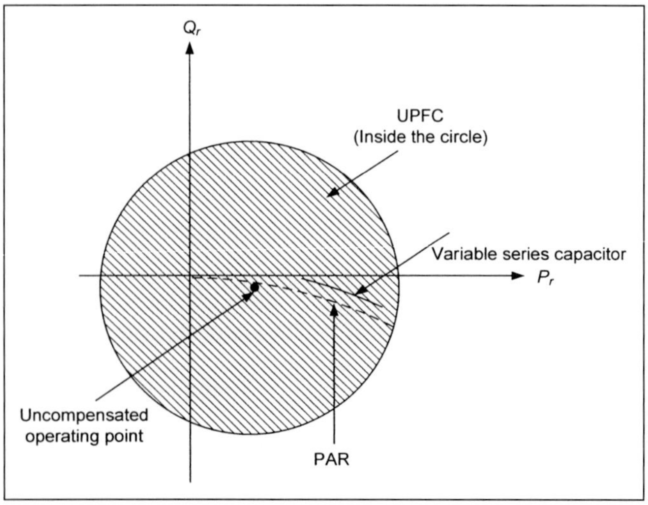

FIGURE 15-36 Relative control characteristics in *P-Q* plane of UPFC: variable series capacitor and PAR.

control, while UPFC provides simultaneous and independent *P* and *Q* control over a wide range. This figure is drawn for:

- UPFC with $V_{pq} = 0.25$ pu
- Fixed plus variable capacitor with 30 to 50 percent compensation
- Phase angle regulator with a symmetrical no-load shift of 14.4°.[7]

15-11 NGH-SSR DAMPER

Subsynchronous resonance can occur with series compensation Section 15-5. Many countries around the world including the United States and Canada use series compensation for the transmission lines, but SSR and possible higher stresses on machine shafts is of consideration. NGH-SSR (Narain Hingorani—the inventor) subsynchronous resonance suppressor counteracts subsynchronous resonance. The principle of operation can be explained with reference to Fig. 15-37*a*. The subsynchronous voltage V_C, when combined with 60-Hz voltage wave V_f, produces a voltage V_α, shown in dotted lines in this figure. Note that some half-cycles of V_α become longer than the normal half-cycle period of 8.333 ms. These distorted waveshapes represent the voltage across the capacitor. This unbalance charge on the capacitor interacts with system inductance to produce oscillations. If this unbalance charge is eliminated, the system will be detuned to any other frequency, other than the fundamental.

The basic scheme (Fig. 15-37*b*) consists of an impedance in series with thyristor switch is connected across the capacitor. The impedance is a small resistance determined from the peak current capability of the thyristor switch in series with a inductor, the value of which is dictated by *di/dt* limit of the thyristors. The thyristor switch is controlled so that when a current zero is detected, the following half-cycle period is timed. If the half-cycle exceeds the set time of 8.33 ms, the corresponding thyristor is turned on to discharge the capacitor and bring about a current zero sooner. The switch stops conducting when the current zero is obtained. At each capacitor voltage zero, a new count starts.

With advancements in modern power electronics and control, FACTS are becoming more attractive for improving the operation of power systems at all levels of electrical energy. Table 15-2 shows application of FACTS controllers, while Table 15-3 shows custom power solutions that can be applied at utility end and customer

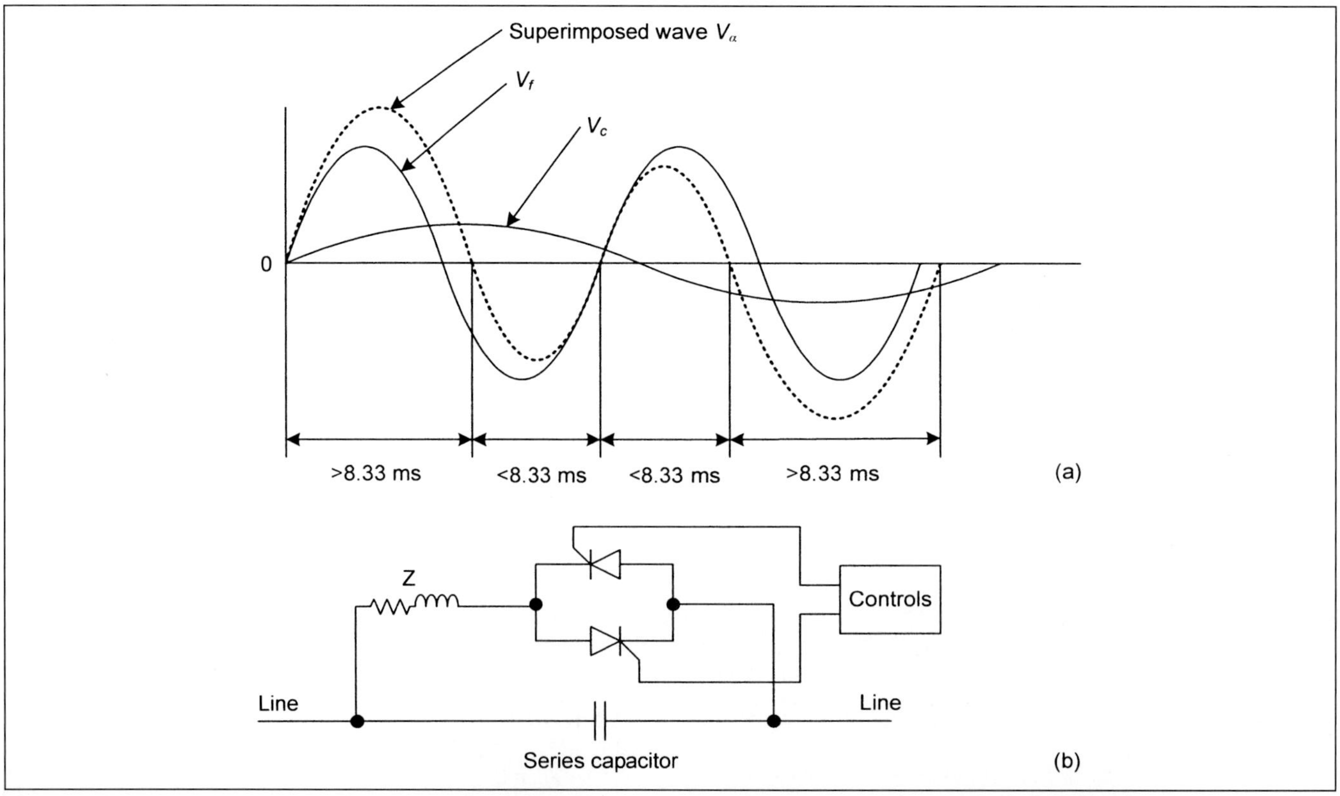

FIGURE 15-37 (*a*) Superimposition of a 60-Hz wave on a subsynchronous oscillation. (*b*) Principle of NGH-SSR damper.

TABLE 15-3 Custom Power Solutions of Power Quality Problems

DISTURBANCE	UTILITY-SIDE SOLUTIONS	CONSUMER-SIDE SOLUTIONS
Voltage sags	Dynamic voltage restorer (DVR) Static condenser DSTATCON	Line conditioner UPS systems
Voltage swells	DVR Fault current limiter High-energy surge arrester	Line conditioner UPS Voltage regulator
Interruption	Solid-state circuit breaker DSTATCON	UPS M-G set
Harmonic distortion	Active and passive filters DVR DSTATCON	Harmonic filters Line conditioner
Noise	-	Grounding and shielding Filters Line conditioner
Transients	Surge arresters	Surge suppressors Line conditioner

end. We see that the DVR and DSTATCOM are synonymous with FACTS-UPFC and STATCOM controllers.

15-12 DISPLACEMENT POWER FACTOR

The fundamental power relations, when the current and voltage waveforms are distorted, due to any reason—applications of nonlinear loads, power electronics, and harmonic resonance—need to be redefined. Nonlinearity also occurs in conventional power systems, for example, saturation, tooth ripples, and harmonic torques in motors, switching of transformers, which produce distortions in the current and voltage waveforms.

For sinusoidal waveforms, by definition:

$$\begin{aligned} S &= \sqrt{P^2+Q^2} = \sqrt{P^2+(P\tan\phi)^2} = P\sqrt{(1+\tan\phi^2)} \\ &= P\sec\phi = P/\cos\phi \end{aligned} \tag{15-29}$$

where S is the apparent power, P is the active power, and Q is the reactive power. The power factor angle under these conditions is defined as:

$$\phi = \tan^{-1}\left(\frac{Q}{P}\right) = \cos^{-1}\left(\frac{S}{P}\right) \tag{15-30}$$

In terms of fundamental currents and voltages (no harmonics and distortions):

$$\begin{aligned} P &= V_f I_f \cos(\phi-\delta) \\ Q &= V_f I_f \sin(\phi-\delta) \\ S &= V_f I_f \end{aligned} \tag{15-31}$$

where V_f, and I_f are the fundamental rms currents and voltages, respectively. In presence of harmonics, change definition of reactive power as:

$$S_h = V_f\sqrt{I_f^2+\sum_{h=2}^{\infty} I_h^2} = \sqrt{P^2+Q^2+D^2} = \sqrt{S_f^2+D^2} \tag{15-32}$$

Based upon Eq. (15-32), the power factor is:

$$\begin{aligned} \mathrm{PF}_T &= \frac{P_f}{\sqrt{S_f^2+D^2}} = \frac{1}{\sqrt{1+\mathrm{THD}_i}}\cos\phi \\ &= \mathrm{PF}_f \times \mathrm{PF}_{\text{displacement}} \end{aligned} \tag{15-33}$$

This means total power factor is conventional power factor multiplied by displacement power factor. The current harmonic distortion THD_i is defined in Chap. 6. The other theories with respect to nonsinusoidal power relations are Fryze theory[8] and Kusters and Moore theory.[9] The maximum theoretical power factor of a converter is given by the expression:

$$\mathrm{PF}_t = \frac{q}{\pi}\sin\left(\frac{\pi}{q}\right) \tag{15-34}$$

where q is the number of converter pulses and π/q is the angle in radians, $q \neq 1$. This ignores commutation overlap and no phase retards and neglects transformer magnetizing current. For a six-pulse converter it will be 0.955. For a 12-pulse converter, the theoretical power factor will be 0.988. With commutation overlap and phase retard, the power factor is:[10]

$$\mathrm{PF}_t = \frac{E_d' I_d}{\sqrt{3}E_L I_L} = \frac{3}{\pi}\frac{1}{\sqrt{3f(\mu,\alpha)}}\left(\cos\alpha - \frac{E_x}{E_{\text{do}}}\right) \tag{15-35}$$

where

$$E_d' = E_d + E_t + E_f$$

E_d is the average direct voltage under load, E_t is the resistance drop, E_f is the total forward drop per circuit element, I_d is the dc load current in average amperes, E_L is the primary line-to-line ac voltage, I_L is the ac primary line current in amperes, α is the phase retard angle, μ is the angle of overlap or commutation angle, E_{do} is the theoretical dc voltage, and E_x is the direct voltage drop due to commutation reactance, and

$$f(\mu,\alpha)=\frac{\sin\mu[2+\cos(\mu+2\alpha)-\mu[1+2\cos\alpha\cos(\mu+\alpha)}{[2\pi\cos\alpha-\cos(\mu+\alpha)]^2} \tag{15-36}$$

The displacement power factor is:

$$\cos\phi_1'=\frac{\sin^2\mu}{\sqrt{\mu^2+\sin^2\mu-2\mu\sin\mu\cos\mu}} \tag{15-37}$$

The above equations neglect the effect of transformer magnetizing currents. The correction for transformer magnetizing current is approximately given by:

$$\cos\phi_1=\cos[\text{arc}\cos\phi_1'+\text{arc}\tan(1_{\text{mag}}/I_1)] \tag{15-38}$$

where $\cos\phi_1'$ is the displacement power factor, not including transformer magnetization current. See Ref. 1 for further details.

15-13 INSTANTANEOUS POWER THEORY

The instantaneous reactive power theory as propounded by Nabae and Akagi[11,12] is of much interest in control algorithms of switching compensators. It uses Clark's transformation (Chap. 4) to transform three-phase quantities to two orthogonal axes, called here α and β. The zero sequence components in the matrix are omitted. The description of power properties of circuits, using instantaneous voltage and current values, without use of Fourier transforms and harmonic analysis, has drawn much interest and simplified the mathematical operations. In the two coordinate system:

$$\begin{vmatrix} e_\alpha \\ e_\beta \end{vmatrix}=\sqrt{\frac{2}{3}}\begin{vmatrix} 1 & -\frac{1}{2} & -\frac{1}{2} \\ 0 & \frac{\sqrt{3}}{2} & -\frac{\sqrt{3}}{2} \end{vmatrix}\begin{vmatrix} e_a \\ e_b \\ e_c \end{vmatrix} \tag{15-39}$$

and:

$$\begin{vmatrix} i_\alpha \\ i_\beta \end{vmatrix}=\sqrt{\frac{2}{3}}\begin{vmatrix} 1 & -\frac{1}{2} & -\frac{1}{2} \\ 0 & \frac{\sqrt{3}}{2} & -\frac{\sqrt{3}}{2} \end{vmatrix}\begin{vmatrix} i_a \\ i_b \\ i_c \end{vmatrix} \tag{15-40}$$

Define instantaneous real and reactive powers as p and q, respectively, then:

$$\begin{vmatrix} p \\ q \end{vmatrix}=\begin{vmatrix} e_\alpha & e_\beta \\ -e_\beta & e_\alpha \end{vmatrix}\begin{vmatrix} i_\alpha \\ i_\beta \end{vmatrix} \tag{15-41}$$

Here p and q are not conventional watts and vars. These are defined by the instantaneous voltages in one phase and the instantaneous current in the other phase.

$$\begin{aligned} p&=e_\alpha i_\alpha+e_\beta i_\beta \\ &=e_a i_a+e_b i_b+e_c i_c=\frac{dW}{dt} \end{aligned} \tag{15-42}$$

To define instantaneous reactive power, the space vector of imaginary power is defined as:

$$\begin{aligned} q&=e_\alpha i_\beta+e_\beta i_\alpha \\ &=\frac{1}{\sqrt{3}}[i_a(e_c-e_b)+i_b(e_a-e_c)+i_c(e_b-e_a)] \end{aligned} \tag{15-43}$$

Equation (15-43) can be written as:

$$\begin{vmatrix} i_\alpha \\ i_\beta \end{vmatrix}=\begin{vmatrix} e_\alpha & e_\beta \\ -e_\beta & e_\alpha \end{vmatrix}^{-1}\begin{vmatrix} p \\ q \end{vmatrix} \tag{15-44}$$

This can be divided into two kinds of currents:

$$\begin{vmatrix} i_\alpha \\ i_\beta \end{vmatrix}=\begin{vmatrix} e_\alpha & e_\beta \\ -e_\beta & e_\alpha \end{vmatrix}^{-1}\begin{vmatrix} p \\ 0 \end{vmatrix}+\begin{vmatrix} e_\alpha & e_\beta \\ -e_\beta & e_\alpha \end{vmatrix}^{-1}\begin{vmatrix} 0 \\ q \end{vmatrix} \tag{15-45}$$

This can be written as:

$$\begin{vmatrix} i_\alpha \\ i_\beta \end{vmatrix}=\begin{vmatrix} i_{\alpha p} \\ i_{\beta p} \end{vmatrix}+\begin{vmatrix} i_{\alpha q} \\ i_{\beta q} \end{vmatrix} \tag{15-46}$$

where $i_{\alpha p}$ is the α-axis instantaneous active current:

$$i_{\alpha p}=\frac{e_\alpha}{e_\alpha^2+e_\beta^2}p \tag{15-47}$$

and $i_{\alpha p}$ is the α-axis instantaneous reactive current:

$$i_{\alpha q}=\frac{e_\alpha}{e_\alpha^2+e_\beta^2}q \tag{15-48}$$

also $i_{\beta p}$ is the β-axis instantaneous active current:

$$i_{\beta p}=\frac{e_\alpha}{e_\alpha^2+e_\beta^2}p \tag{15-49}$$

and $i_{\beta p}$ is the β-axis instantaneous reactive current:

$$i_{\beta q}=\frac{e_\alpha}{e_\alpha^2+e_\beta^2}q \tag{15-50}$$

The following equations exist:

$$\begin{aligned} p&=e_\alpha i_{\alpha p}+e_\beta i_{\beta p}\equiv P_{\alpha p}+P_{\beta p} \\ 0&=e_\alpha i_{\alpha q}+e_\beta i_{\beta q}\equiv P_{\alpha q}+P_{\beta q} \end{aligned} \tag{15-51}$$

where α-axis instantaneous active and reactive powers are:

$$P_{\alpha p}=\frac{e_\alpha^2}{e_\alpha^2+e_\beta^2}p \quad P_{\alpha q}=\frac{-e_\alpha e_\beta}{e_\alpha^2+e_\beta^2}q \tag{15-52}$$

The β-axis instantaneous active and reactive powers are:

$$P_{\beta p}=\frac{e_\beta^2}{e_\alpha^2+e_\beta^2}p \quad P_{\beta q}=\frac{e_\alpha e_\beta}{e_\alpha^2+e_\beta^2}q \tag{15-53}$$

The sums of instantaneous active powers in the two axes coincides with the instantaneous real power in the three-phase circuit, and the instantaneous reactive powers cancel each other and make no contributions from the source to the load.

This needs some explanation. In a converter, the instantaneous reactive power is the power circulating between the source and converter, while the instantaneous reactive power on the output is the instantaneous reactive power between the converter and the load. Therefore, there is no relation between the reactive powers on the input and the output sides, and instantaneous imaginary power on the input is not equal to the instantaneous imaginary power on the output side. Assuming zero active power loss in the converter, the active power on the input is equal to the active power output.

15-14 ACTIVE FILTERS

Mitigation of harmonics has been discussed using passive filters, phase multiplication, or higher pulse number converters. Active current shaping and active filters are yet other alternatives.

In active current shaping, the input current of the converters can be forced to follow a sinusoid in phase with voltage, addressing the need for reactive power compensation as well as harmonic elimination.[13]

Active filters inject harmonic distortion into the coupling point of the nonlinear load which is of opposite polarity to the distortion caused by the nonlinear load. These can be classified according to their connections in the circuit:

- In series connection
- In parallel shunt connection
- Hybrid connection of active and passive filters

An obvious advantage of active filters is that the harmonics are eliminated at the source. The cost and available size are of consideration.[14]

The series and shunt connections are shown in Fig. 15-38*a* and *b*. A harmonic current source can be represented by a Norton equivalent circuit, and it can be implemented with a voltage-fed PWM inverter to inject a harmonic current of the same magnitude as the load, but of harmonics of opposite polarity. In the shunt connection (Fig. 15-38*a*), the load current will be sinusoidal so long as the load impedance is much higher than the source impedance. This means that the injected harmonic cancellation current I_{fl} flows only in the source impedance Z_s and none in the load impedance I_L, that is, the load impedance Z_L is infinite. Practically, this will not be achieved. A shunt connection with voltage source cancellation inverter will be more effective where the output reactor resists the change of current.

In a series connection, a voltage V_f is injected in the series with the line and compensates the voltage distortion produced by the nonlinear load. A series active filter is more suitable for harmonic compensation of diode rectifiers where the dc voltage of the inverter is derived from a capacitor, which opposes the change of voltage.

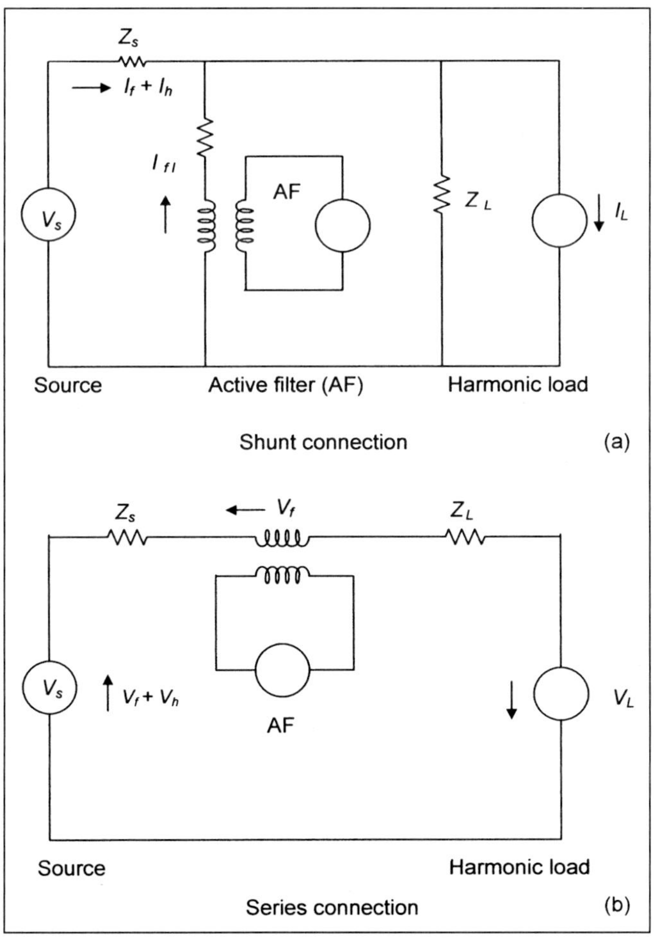

FIGURE 15-38 (*a*) Shunt connection of an active filter. (*b*) Series connection of an active filter.

PROBLEMS

1. Mathematically derive Eqs. (15-1), (15-6), and (15-8) of the text.

2. The voltage underload flow at certain bus in an interconnected system dips by 10 percent, while the voltages on adjacent buses are held constant. The available short-circuit current at this bus is 21 kA, rms symmetrical, and voltage is 230 kV. Find the reactive power injection to restore voltage to its rated value of 230 kV.

3. Draw the circuit of a current-source six-pulse converter with input transformers of 0° phase shift. Plot current profiles in the input transformer windings—primary and secondary, line currents, currents in each leg of the bridge circuit, and the output current. Calculate their relative magnitudes. Repeat with input transformer of −30° phase shift in the input transformer windings.

4. A 100-MVA load at 0.8 power factor and at 13.8 kV is supplied through a short transmission line of 0.1 pu reactance (100-MVA base). Calculate the reactive power loss and the load voltage. Size a capacitor bank at load terminals to limit the voltage drop to 2 percent at full load. What are the capacitor sizes for three-step switching to maintain the load voltage no more than 2 percent below the rated voltage as the load varies from zero to full load?

5. A 200-MW, 18-kV, 0.85-power factor generator is connected through a 200-MVA step-up transformer of 18 to 500 kV and of 10 percent reactance to 500 kV system. The transformer primary windings are provided with a total of five taps, two below the rated voltage of 2.5 and 5 percent, and two above rated voltage of 2.5 and 5 percent and one at the rated voltage. Assuming that initially the taps are set at rated voltage of 18 to 500 kV, what is the generation voltage to take full-rated reactive power output from the generator? If the operating voltage of the generator is to be limited to the rated voltage, find the tap setting on the transformer. Find the reactive power loss through the transformer in each case. Neglect resistance.

6. A 100-mi long (160-km) line has series $R = 0.3\ \Omega$, $L = 3$ mH, and shunt capacitive reactance = 0.2 MΩ/mi. It delivers a load of 100 MVA at 0.85 power factor at 138 kV. If the sending-end voltage is maintained at 145 kV, find the Mvar rating of the synchronous condenser at receiving end at no load and at full load.

7. Why have current source converters been in use for HVDC transmission in the past? Explain the recent shift to voltage source converters.

8. Plot P-δ curves of 230-kV uncompensated line, with midpoint shunt compensation of 0.6 and also with series compensation of 0.6. The line is 200-mi long and has a series reactance of 0.8 Ω/mi with a susceptance $y = 5.4 \times 10^{-6}$ S/mi.

9. Compare the performance of a conventional SVC device with STATCON. Draw their relative *V-I* curves.

10. Compare UPFC with STATCON and SCCC. Can a STATCON be used in a transmission line for active power control?

11. Make a table of all the electronic and conventional devices for the reactive power generation and their relative characteristics.

12. In Fig. 15-33, (a), Can converter 1 supply active power to the system? (b) Can converter 2 supplies active power to the system? (c) Can both converters supply active power to the system. Choose the correct option and explain the reasoning.

REFERENCES

1. IEEE Std. 519, IEEE Recommended Practices and Requirements for Harmonic Control in Electrical Power Systems, 1992.
2. J. C. Das, "Estimating Line Harmonics-Appendix G," *Power System Analysis: Short-Circuit, Load Flow, and Harmonics*, Marcel Dekker, 2002.
3. A. D. Graham and E. T. Schonholzer, "Line Harmonics of Converters with DC-Motor Loads," *IEEE Trans. Industry Applications*, vol. 19, pp. 84–93, 1983.
4. N. Hatti, Y. Kondo, and H. Akagi, "Five Level Diode-Clamped PWM Converters Connected Back-to-back for Motor Drives," *IEEE Trans. Industry Applications*, vol. 44, no. 4, pp. 1268–1276, July/Aug. 2008.
5. H. Akagi, H. Fujita, S. Yonetani, and Y. Kondo, " A 6.6 kV Transformer-Less STATCOM Based on a Five Level Diode-Clamped PWM Converter: System Design and Experimentation of 200-V, 10 kVA Laboratory Model," *IEEE Trans. Industry Applications*, vol. 44, no. 2, pp. 672–680, March/ April 2008.
6. C. Schauder, M. Gernhard, E. Stacey, T. Lemak, L. Gyugi, T.W. Cease, and A. Edris, "Development of a 100 Mvar Static Condenser for Voltage Control of Transmission Systems," in *Conf. Record, IEEE/PES 1994 Summer Meeting*, San Francisco, CA, 1994.
7. R. J. Nelson, J. Bian, and S. L. Williams, "Transmission Series Power Flow Control," *IEEE Trans. PD*, vol. 10, no. 1, pp. 504–510, Jan. 1995.
8. IEEE Std. 1459, IEEE Trial-Use Standard Definitions for the Measurement of Electrical Power Quantities Under Sinusoidal, Non-Sinusoidal, Balanced or Unbalanced Conditions, 2000.
9. N. L. Kusters and W. J. M. Moore, "On the Definition of Reactive Power Under Non-Sinusoidal Conditions," *IEEE Trans.*, vol. PAS-99, pp. 1845–1854, 1980.
10. IEEE Std. 519, IEEE Recommended Practices and Requirements for Harmonic Control in Electrical Power Systems, 1992.
11. H. Akagi, Y. Kanazawa, and A. Nabae, "Generalized Theory of Instantaneous Reactive Power in Three-Phase Circuits," *Proc. International Power Electronics Conference*, pp. 1375–1386, Tokyo.
12. H. Akagi, E. H. Watanbe, and M. Aredes, *Instantaneous Power Theory and Applications to Power Conditioning*, Wiley–Intersciences, 2007.
13. H. Akagi, "Trends in Active Filters for Power Conditioning," *IEEE Trans. Industry Applications*, vol. 32, pp. 1312–1322, 1996.
14. J. C. Das, "Passive Filters—Potentialities and Limitations," *IEEE Trans. Industry Applications*, vol. 40, no. 1, pp. 232–241, Jan./ Feb. 2004.

FURTHER READING

H. Akagi, "New Trends in Active Filters for Power Conditioning," *IEEE Trans. on Industry Applications*, vol. 32, no. 6, pp. 1312–1322, Nov./Dec. 1996.

J. Arrillaga, *High Voltage Direct Current Transmission, IEEE Press*, London, 1983.

M. H. Baker, "An Assessment of FACTS Controllers for Transmission System Enhancement," *CIGRE-SC14, International Colloquium on HVDC and FACTS*, Montréal, Sep. 1995.

B. J. Baliga, "Power Semiconductor Devices for Variable Frequency Drives," Chapter 1, *Power Electronics and Variable Frequency Drives*, B. K. Bose, ed, IEEE Press, Piscataway, NJ, 1996.

B. K. Bose, "Evaluation of Modern Power Semiconductor Devices and Their Impact for Power Transmission," *IEEE Trans. Industry Applications*, vol. 28, no. 2, pp. 403–413, March/April 1992.

C. A. Canizares, "Analysis of SVC and TCSC in Voltage Collapse," in *Conf. Record, IEEE PES 1998 Summer Meeting*, San Diego, CA, July 1998.

CIGRE WG 30-01, Task Force No. 2, Static Var Compensators, I. A. Erinmez, ed., 1986.

R. Doncker, "High Power Semiconductor Development for FACTS and Custom Power Applications," *EPRI Conf. on Future Power Delivery*, Washington, DC, April 1996.

J. B. Ekanayake and N. Jenkins, "A Three Level Advanced Static var Compensator," *IEEE Trans. on Power Delivery*, vol. 10, no. 2, pp. 540–545, April 1996.

L. Gyugyi, R. A. Otto, T. H. Putman, "Principles and Applications of Static, TC Shunt Compensators," *IEEE Trans. PAS*, vol. PAS-97, no. 5, pp. 1935–1945, Sept./Oct. 1978.

L. Gyugyi, C. D. Schauder, and K. K. Sen. "Static Synchronous Series Compensator: A Solid State Approach to the Series Compensation of Transmission Lines," FACT Technical Papers, Westinghouse Electric Corporation, pp. 1–8, Orlando, FL, August 1996.

L. Gyugyi, C. D. Schauder, and K. K. Sen, "Static Synchronous Series Compensator: A Solid State Approach to the Series Compensation of the Transmission Lines," *IEEE Trans. PD*, vol. 12, no. 1, pp. 406–417, Jan. 1997.

C. J, Hatziadoniu and F. E. Chalkiadakis, "A 12-Pulse Static Synchronous Compensator for the Distribution System Employing the Three Level GTO Inverter," in *Conf. Record, IEEE 1997 PES Meeting*, Paper No. PE-542-PWRD-0-01, 1997.

N. G. Hingorani and L. Gyugyi, *Understanding FACTS*, IEEE Press, New York, 2000.

N. G. Hingorani, "A New Scheme for Subsynchronous Resonance Damping of Torsional Oscillations and Transient Torque, Part 1," *IEEE Trans.*, vol. PAS-100, no. 4, pp. 1852–1855, April 1981.

N. G. Hingorani, "Flexible AC Transmission," *IEEE Spectrum*, pp. 40–45, April 1993.

IEEE Committee Report, "VAR Management–Problem Recognition and Control," *IEEE Trans.*, vol. PAS-103, pp. 2108–2116, Aug. 1984.

T. W. Kay, P. W. Sauer, R. D. Shultz, and R. A. Smith, "EHV & UHV Line Loadability Dependence on Var Supply Capability," *IEEE Trans.*, vol. PAS-101, no. 9, pp. 3568–3575, 1982.

P. Kessel and H. Glavitsch, "Estimating the Voltage Stability of a Power System," *IEEE Trans.*, vol. PWRD-1, no. 3, pp. 346–354, July 1986.

T. J. E. Miller, *Reactive Power Control in Electrical Power Systems*, John Wiley, 1982.

S. Mori and K. Matsuno, "Development of Large Static var Generator Using Self Commutated Inverters for Improving Power System Stability," *IEEE Trans. PS*, vol. 8, no. 1, pp. 371–377, Feb. 1993.

K. K. Sen., "SSSC-Static Synchronous Series Compensator: Theory, Modeling and Applications," *IEEE Trans. PD*, vol. 13, no. 1, pp. 241–246, Jan. 1998.

CHAPTER 16

FLICKER, BUS TRANSFER, TORSIONAL DYNAMICS, AND OTHER TRANSIENTS

In this chapter we will study:

- Flicker
- Auto bus transfer schemes
- Cogging and crawling of induction motors
- Tooth ripples in machines
- Reciprocating compressors
- Torsional vibrations
- Out-of-phase closing and synchronizing

16-1 FLICKER

Voltage flicker occurs due to operation of rapidly varying loads, which affect the system voltage. This can cause annoyance by causing visible light flicker on tungsten filament lamps. The human eye is most sensitive to light variations in the frequency range of 5 to 10 Hz, and voltage variations of less than 0.5 percent in this frequency range can cause annoying flicker on tungsten lamps. Percentage pulsation of voltage related to frequency, at which it is most perceptible, from various sources is shown in Fig. 16-1.[1] In this figure, solid lines are composite curves of voltage flicker by General Electric Company (General Electric Review, August 1925); Kansas Power and Light Company, Electrical World May 19, 1934; T&D Committee EEI, October 14, 1934, Chicago; Detroit Edison Company; West Pennsylvania Power Company; and Public Service Company of North Illinois. Dotted lines show voltage flicker allowed by two utilities, reference Electrical World November 3, 1958 and June 1961.

Though Fig. 16-1 has been in use for a long time, it was superseded in IEEE Std. 1453.[2] The solid-state compensators and loads may produce modulation of the voltage magnitude that is more complex than what was envisaged in the original flicker curves. This standard adopts IEC standard 61000-4-15[3] in total. Define:

$$\text{Plt} = \sqrt[3]{\frac{1}{12} \times \sum_{j=1}^{12} Pst_j^3} \tag{16-1}$$

where Plt is a measure of long-term perception of flicker obtained for a 2-h period. This is made up of 12 consecutive Pst values, where Pst is a measure of short-term perception of flicker for 10-min interval. This value is the standard output of IEC flicker meter. Further qualification is that IEC flicker meter is suited to events that occur once per hour or more often. The curves in Fig. 16-1 are still useful for infrequent events like a motor start, once per day, or even as frequent as some residential air conditioning equipment. Figure 16-2 depicts comparison of IEEE and IEC for flicker irritation.

For acceptance of flicker causing loads, IEC Standards (Refs. 4, 5, and 6) are recommended. The application of shape factors allows the effect of loads with voltage fluctuations other than the rectangular to be evaluated in terms of Pst values. Further research is needed in issues related to the effect of interharmonics on flicker and flicker transfer coefficients from HV to LV electrical power systems.[2,7]

Two levels, planning level and compatibility levels, are defined. *Compatibility level* is the specified disturbance level in a specified environment for coordination in setting the emission and immunity limits. *Planning level*, in a particular environment, is adopted as a reference value for limits to be set for the emissions from large loads and installations, in order to coordinate those limits with all the limits adopted for equipment intended to be connected to the power supply system.

As an example, planning levels for Pst and Plt in MV (voltages >1 kV and <35 kV), HV (voltages >35 kV and < 230 kV) and EHV (voltages >230 kV) are shown in Table 16-1, and compatibility levels for low voltage and medium voltage power systems are shown in Table 16-2.

Arc furnaces cause flicker because the current drawn during melting and refining periods is erratic and fluctuates widely, and the power factor is low. There are other loads that can generate

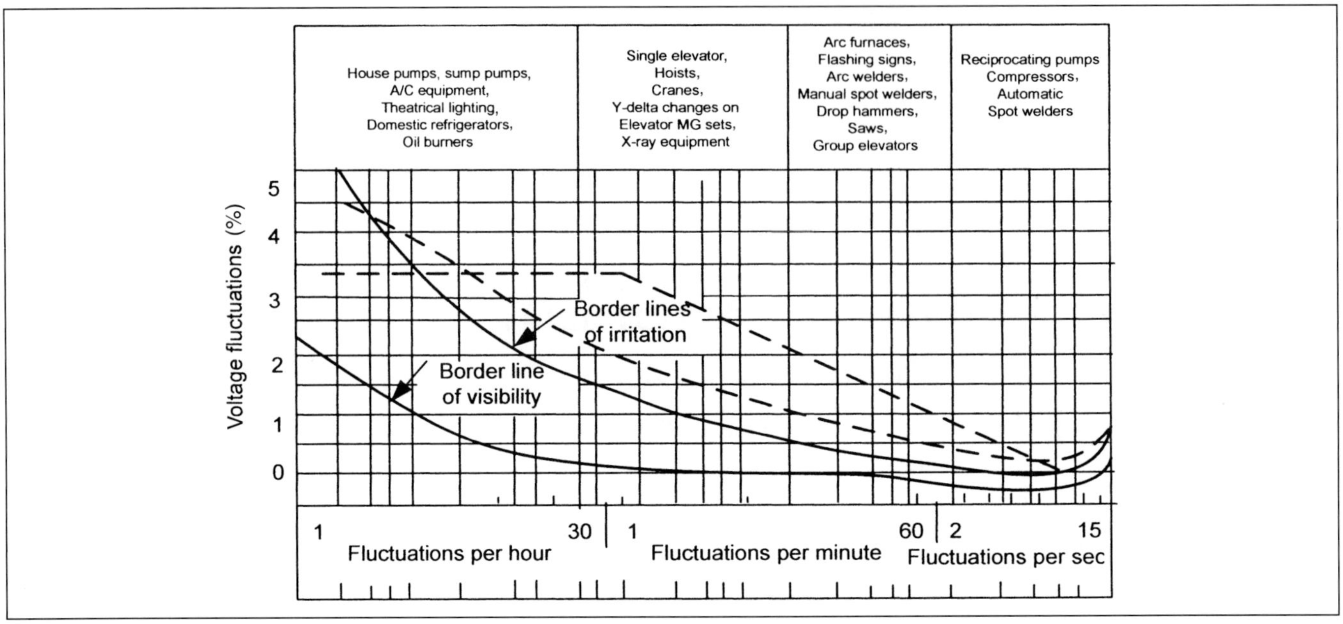

FIGURE 16-1 Tolerable limits of flicker. *Source: IEEE Standard 519.*[1]

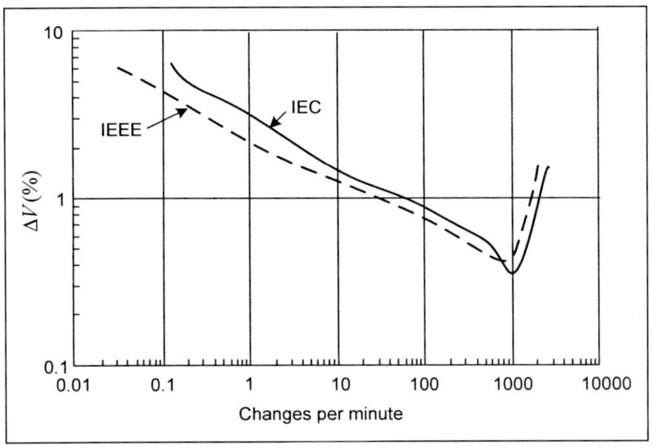

FIGURE 16-2 Comparison of IEC and IEEE standards with respect to flicker tolerance. *Source: IEEE Standard 1453.*[2]

TABLE 16-1 Planning Levels for Pst and Plt in MV, HV, and EHV Power Systems

	Planning Levels	
	MV	HV-EHV
Pst	0.9	0.8
Plt	0.7	0.6

TABLE 16-2 Compatibility Levels for Pst and Plt in LV and MV Systems

	Compatibility Level
Pst	1.0
Plt	0.8

flicker, for example, large spot welding machines often operate close to the flicker perception limits. Industrial processes may comprise a number of motors having rapidly varying loads or starting at regular intervals, and even domestic appliances, such as cookers and washing machines, can cause flicker on weak systems. However, the harshest load for flicker is an arc furnace. During the melting cycle of a furnace, the reactive power demand is high. Figure 16-3 gives typical performance curves of an arc furnace, and Fig. 16-4 is the frequency spectrums of the currents during melting and refining. Though the signature of an arc furnace current is random and no periodicity can be assigned (thus Fourier analysis cannot be used), some harmonic spectrums have been established, and Table 16-3 shows typical harmonics during melting and refining stage. Note that even harmonics are produced during melting stage. The high reactive power demand and poor power factor causes cyclic voltage drops in the supply system. Reactive power flow in an inductive element requires voltage differential between sending end and receiving ends and there is reactive power loss in the element itself (Chap. 12). When the reactive power demand is erratic, it causes corresponding swings in the voltage dips, much depending upon the stiffness of the system behind the application of the erratic load. This voltage drop is proportional to the short-circuit MVA of the supply system and the arc furnace load.

For a furnace installation, the short-circuit voltage depression (SCVD) is defined as:

$$\mathrm{SCVD} = \frac{2\,\mathrm{MW_{furnace}}}{\mathrm{MVA_{sc}}} \tag{16-2}$$

where $\mathrm{MW_{furnance}}$ is the installed load of furnace, and $\mathrm{MVA_{sc}}$ is the short-circuit level of the utility's supply system. This gives an idea whether potential problems with flicker can be expected. An SCVD of 0.02 to 0.025 may be in the acceptable zone, between 0.03 to 0.035 in the borderline zone, and above 0.035 objectionable. When there are multiple furnaces, these can be grouped into one equivalent MW load. The worst flicker occurs during the first 5 to 10 min of each heating cycle and decreases as the ratio of the solid metal to liquid metal decreases.

The significance of $\Delta V/V$ and number of voltage changes are illustrated in Fig. 16-5.[3] This shows a 50-Hz waveform, having a

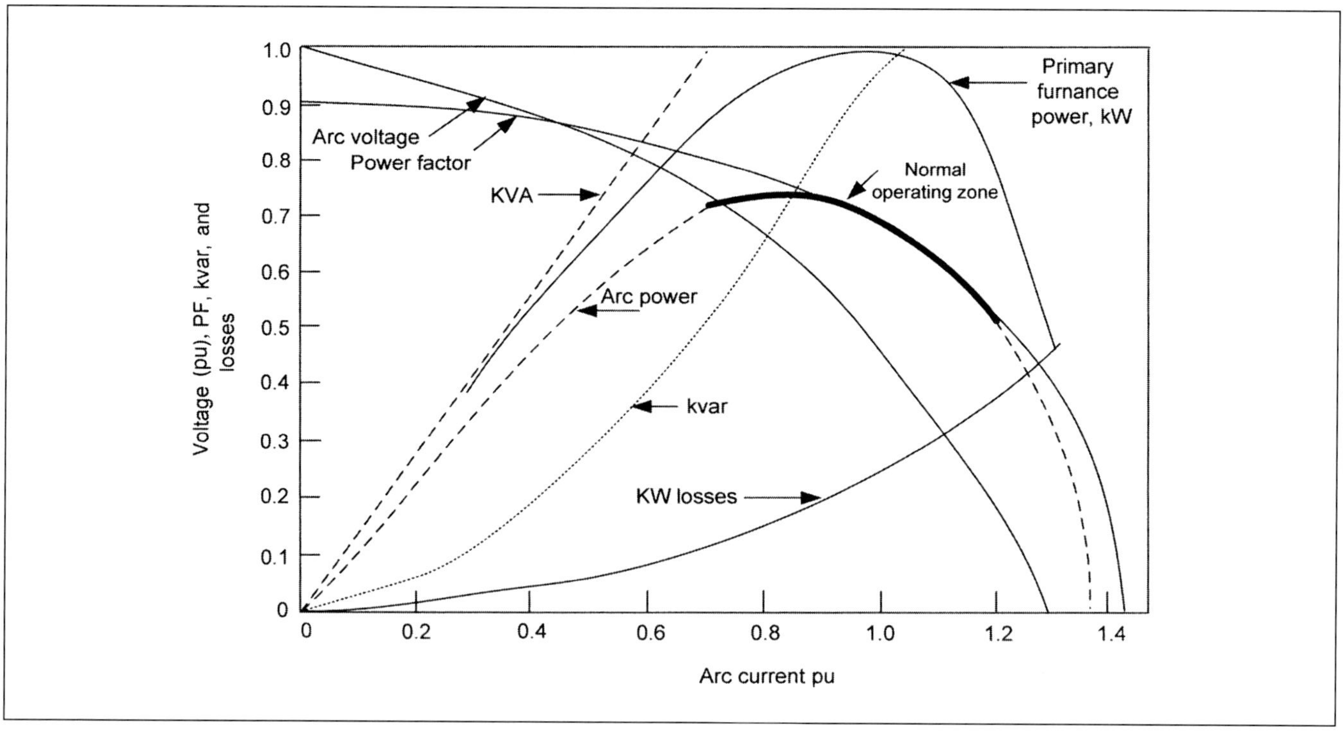

FIGURE 16-3 Operating characteristics of an arc furnace.

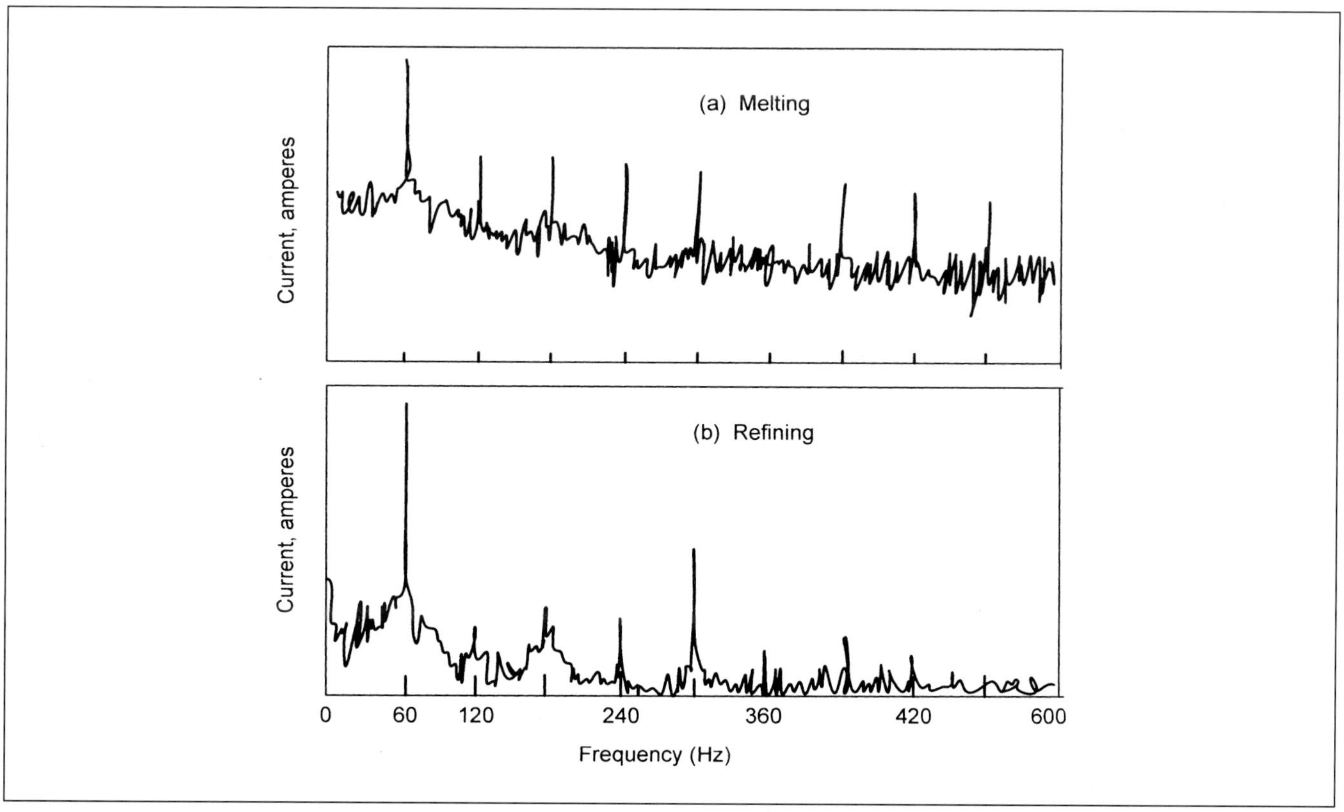

FIGURE 16-4 Frequency spectrums of currents during (*a*) melting and (*b*) refining operation of an arc furnace.

TABLE 16-3 Harmonic Content of Arc Furnace Current

	Harmonic Current, % of Fundamental				
	Harmonic Order				
Operating Condition	**2**	**3**	**4**	**5**	**7**
Melting (active arc)	7.7	5.8	2.5	4.2	3.1
Refining (stable arc)	0	2.0	0	2.1	0

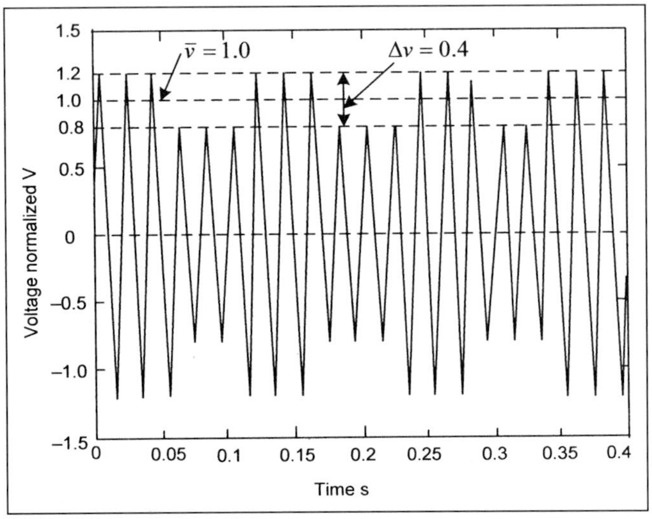

FIGURE 16-5 Modulation with rectangular voltage change $\Delta V/V =$ 40 percent, 8.8 Hz, 17.6 changes per second.

1.0 average voltage with a relative voltage change $\Delta v/\bar{v} = 40$ percent and with 8.8-Hz rectangular modulation. It can be written as:

$$v(t) = 1 \times \sin(2\pi \times 50t) \times \left[1 + \frac{40}{100} \times \frac{1}{2} \times \text{signum}\ (2\pi \times 8.8 \times t)\right] \quad (16\text{-}3)$$

Each full period produces two distinct changes: one with increasing magnitude and one with decreasing magnitude. Two changes per period with a frequency of 8.8 Hz give rise to 17.6 changes per second.

16-1-1 Control of Flicker

During the past years, high-power thyristor valves have been used to connect capacitors and reactors in shunt with power lines. When it is essential to compensate load fluctuations within a few ms, SVCs have been applied. Large TCR flicker compensators of 200 MW have been installed for arc furnace installations. Closed-loop control is necessary due to randomness of load variations, and complex circuitry is required to achieve response times less than one cycle. Significant harmonic distortion may be generated. Harmonic filters are required. TSCs have also been installed and these have inherently one cycle delay as the capacitors can only be switched when their terminal voltage matches the system voltage. Thus, the response time is slower. SVCs employing TSCs do not generate harmonics, but the resonance with system and transformers needs to be checked.

We discussed STATCOM in Chap. 15. It has been long recognized that reactive power can be generated without use of bulk capacitors and reactors and STATCOM makes it possible. As discussed in Chap. 15, it is capable of operating with leading or lagging power factors. With design of high bandwidth control capability, STATCOM can be used to force three-phase currents of arbitrary waveshape through the line reactance. This means that it can be made to supply nonsinusoidal, unbalanced, randomly fluctuating currents demanded by the arc furnace. With a suitable choice of dc capacitor, it can also supply the fluctuating real power requirements, which cannot be achieved with SVCs. Harmonics are not of concern (Fig. 15-26).

The instantaneous reactive power on the source side is the reactive power circulating between the electrical system and the device, while reactive power on the output side is the instantaneous reactive power between the device and its load. There is no relation between the instantaneous reactive powers on the load and source side, and the instantaneous imaginary power on the input is not equal to the instantaneous reactive power on the output (Chap. 15). The STATCOM for furnace compensation may use vector control based on the concepts of instantaneous active and reactive power, i_α and i_β (Chap. 15).

Figure 16-6 shows flicker reduction factor as a function of flicker frequency STATCOM versus SVC.[8] Flicker mitigation with a fixed reactive power compensator and an active compensator—a hybrid solution for welding processes—is described in Ref. 9.

16-2 AUTOTRANSFER OF LOADS

Autotransfer of loads is adopted to restore power to running loads by switching to an alternate source of power. Figure 16-7*a* shows that two step-down transformers, connected to different sources of power on the primary side, serve loads connected to a single bus. One source is labeled as standby. On loss of the normal source, the power is restored automatically by switching to the standby source.

An extension of this concept is to bring the standby source of power, say, a diesel generator, on line automatically to serve the load. It implies that on loss of normal power, the standby generator will be started, and the power is restored after a time delay, during which all rotating motor loads connected to the bus will shutdown and have to be restarted back in a certain sequence. This does not give rise to transients and is not discussed further.

A more common configuration for load transfer is shown in Fig. 16-7*b*. Here the sources are labeled 1 and 2, and the bus section breaker is open under normal operation, when both the sources are available. Each source supplies the loads connected to the sectionalized bus to which it is connected. On failure of any one source, the breaker controlling that source is tripped, and the bus section breaker is closed. Each transformer should be sized, not only for the total bus load, but also to limit voltage drops when the connected loads accelerate.

The transients that may occur on such operation depends on how fast the power is restored. Many fast bus transfer schemes have been implemented, especially for generating stations, where shutdown of auxiliary station service loads will precipitate a shutdown of the entire generating station.

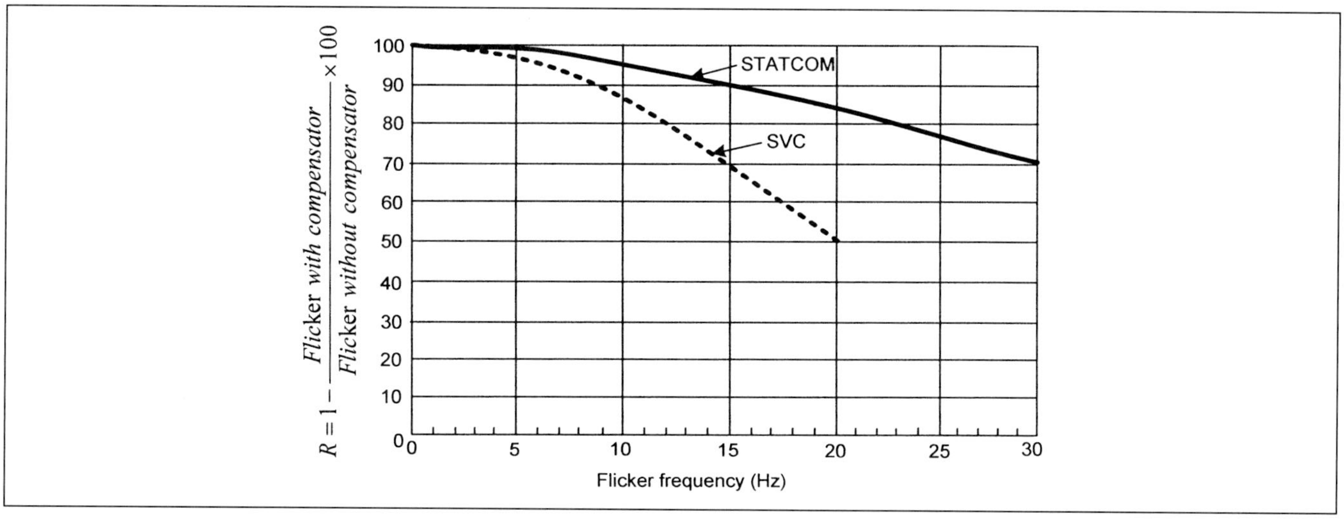

FIGURE 16-6 Flicker factor R with STATCOM and SVC.

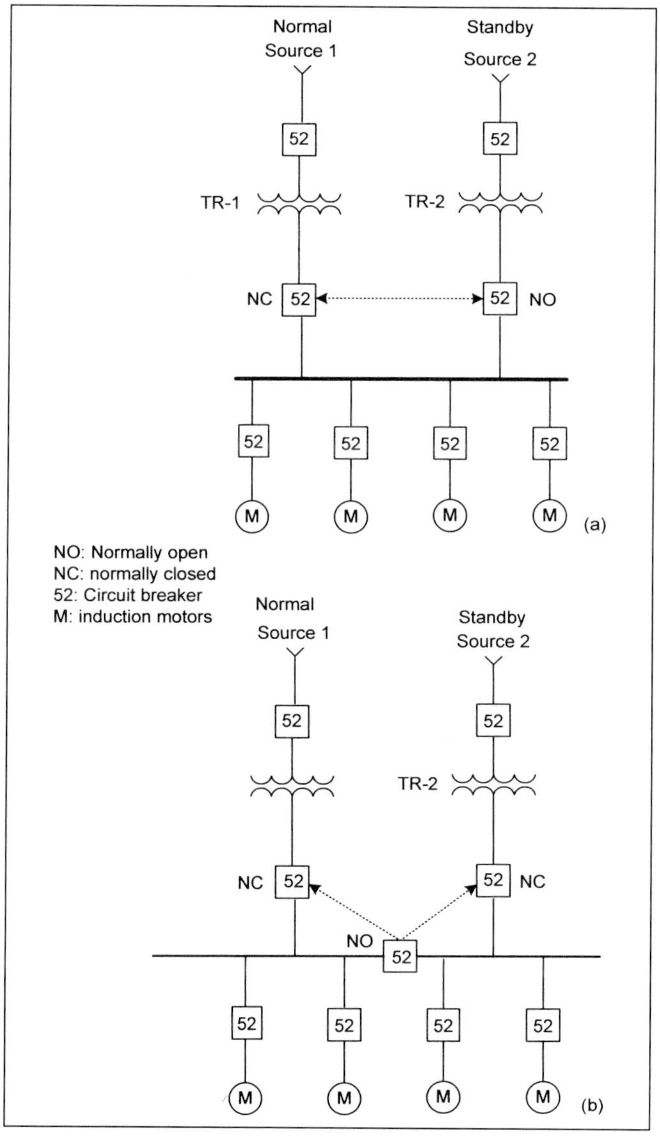

FIGURE 16-7 (*a*) Load transfer scheme without bus section breaker. (*b*) Load transfer scheme with bus section breaker.

16-2-1 Induction Motor Behavior on Loss of Power Supply

Reconnection of motors to a power supply after an interruption, while they are still in rotation, may give rise to large transient torques and current surges. After disconnection from the supply lines, the motor internal voltage will decay at a rate determined by the inertia, and motor and load characteristics. A phase difference will therefore exist between the incoming line (alternate source to which the motors are being switched) and the motor decaying residual voltage. Negative transient torques of seven times per unit torque may be produced, and the maximum may not occur exactly at 180° phase opposition of the decaying motor voltage and the incoming supply voltage. The motor leakage reactance and resistance will have an impact on the decay of transients, and high damping coefficients are desirable. There are successive time delays at which the transient torques will be zero on reconnection. Figure 16-8[10] shows the transient negative torques versus reconnection time for a 2000-hp motor, open circuit time constant, 2 s, and breakdown torque, 150 percent. The high negative torques can be avoided with fast switching.

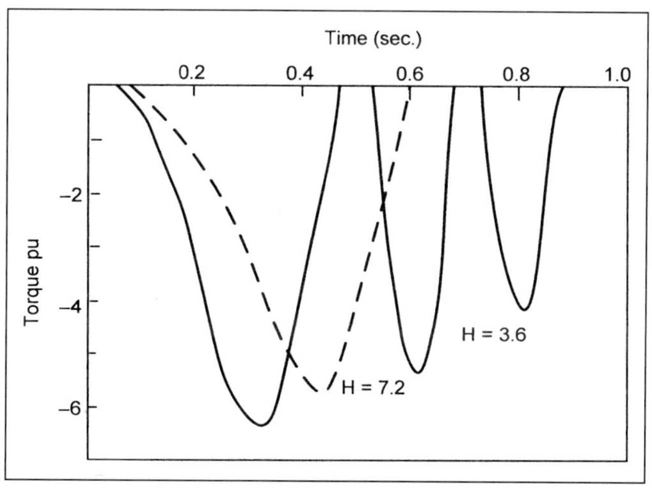

FIGURE 16-8 Negative torques on reconnection of a 2000-hp motor to the supply system with $H = 3.6$ and 7.2.

This time delay is dependent upon the inertia, motor size, and preloading. Therefore, there are three parameters to be considered:

- Motor decaying residual voltage
- Phase angle
- The oscillating shaft torque and transient electrical air gap torque

From Chap.11, for the short-circuited rotor of a squirrel cage induction motor, v_{dr} and v_{qr} are zero, and when the stator is disconnected from the supply system, i_{ds} and i_{qs} are also zero. Therefore, we can write:

$$\begin{vmatrix} v_{ds} \\ v_{qs} \end{vmatrix} = \begin{vmatrix} Mp & 0 \\ 0 & Mp \end{vmatrix} \begin{vmatrix} i_{dr} \\ i_{qr} \end{vmatrix} \tag{16-4}$$

and

$$\begin{vmatrix} 0 \\ 0 \end{vmatrix} = \begin{vmatrix} r_r + L_r p & L_r\omega_r \\ -L_r\omega_r & r_r + L_r p \end{vmatrix}$$

Since the speed of the motor varies, Eq. (16-4) is nonlinear and can be solved by numerical integration. It is, however, necessary to know i_{dr} and i_{qr} as the starting points. The stator current will fall to zero in a very short time. As the flux in the machine cannot change suddenly, the change in stator currents must be accompanied by a change in the rotor currents to maintain constant flux linkages after the motor is disconnected from the supply and before disconnection from the supply. These boundary conditions can then be used to determine the decay of rotor currents and stator voltages after disconnection.

For calculating the voltage decay, it is permissible to lump the motor loads into an equivalent motor horsepower, having average electrical characteristics and inertia constant; the slope of the decay line is related to the *X/R* of the motor. Analytical expressions for deceleration of the motor are given in Chap. 11.

Example 16-1 The residual voltage of a 6.6-kV, four-pole, 10000-hp, double-cage induction motor is simulated using EMTP. The motor parameters are: load plus motor inertia 12000 lb-ft^2, L_s (stator reactance) = 0.32, R_s (stator resistance) = 0.023, R_{r1} (cage 1 resistance) = 0.12, R_{r2} (cage 2 resistance) = 0.05, L_{r1} (cage 1 reactance) = 0.27, L_{r2} (cage 2 reactance) = 0.294, all values in ohms. The motor is running at no load when disconnected from the supply system. Figure 16-9 shows the decay in residual voltage. Figure 16-10[11] shows residual voltage and phase shift in degrees for various motor sizes, after disconnection from the supply system. Commonly the decays are represented in a polar diagram (Fig. 16-11).

16-2-2 Currents and Torques on Reswitching

When the stator and rotor currents at the instant of switching are known, the transients currents and, therefore, the transient torque using Eq. (16-4) can be calculated.

Consider that the autotransfer occurs at a time t_m, when the residual emf of the motor is E_m. Then the resultant voltage vector is:

$$E_r = \sqrt{E_s^2 + E_m^2 - 2E_m E_s \cos\theta} \tag{16-5}$$

where E_s is the supply system voltage. The voltage should be limited to a value of 1.33 puV/Hz, to prevent overfluxing and damaging the motor. Phase angle of small low-inertia motors drops quickly and falls rapidly in and out of phase with the supply voltage (Fig. 16-10).

16-2-3 Bus Transfer Strategies

1. *Fast bus transfer*

Figure 16-11 suggests that large transients can be avoided by fast reclosing, so that the motor internal voltage is not much displaced, and the incoming supply voltage is almost in phase with the decaying emf.

2. *Residual voltage transfer*

An alternative will be to let the motor speed fall, and the internal voltage decay to a low value, say 25 percent of the supply

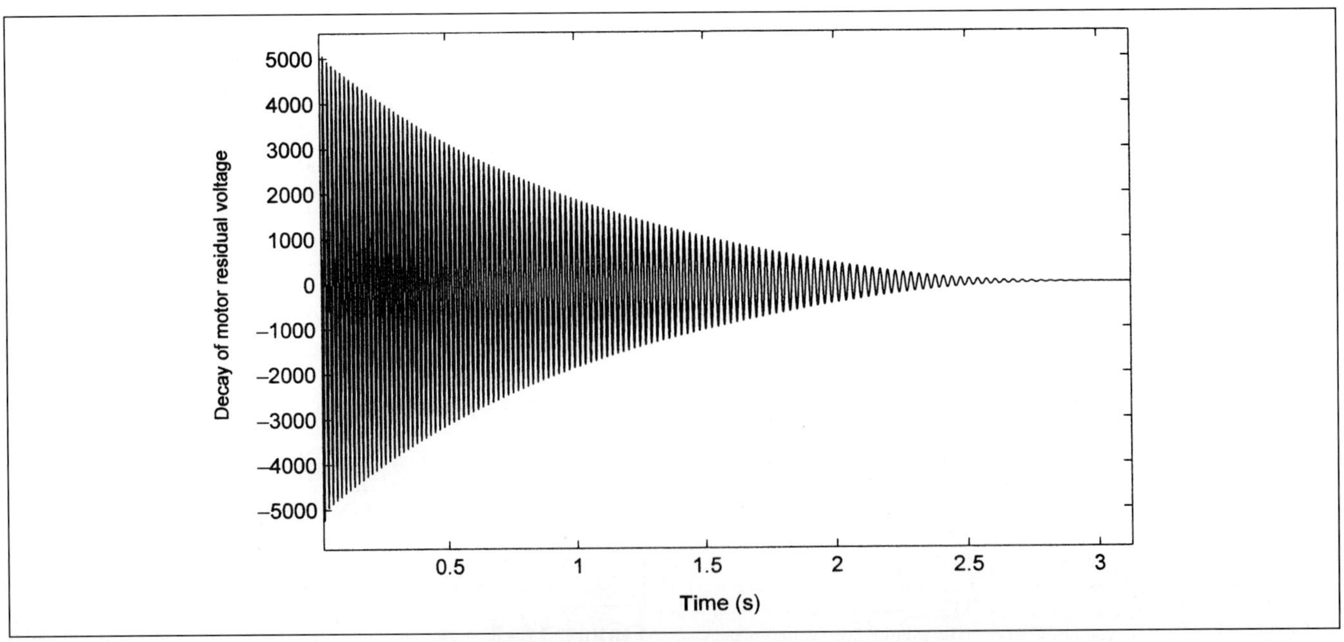

FIGURE 16-9 EMTP simulation of the decay of a 6.6-kV, 10000-hp induction motor residual voltage on disconnection from the supply system.

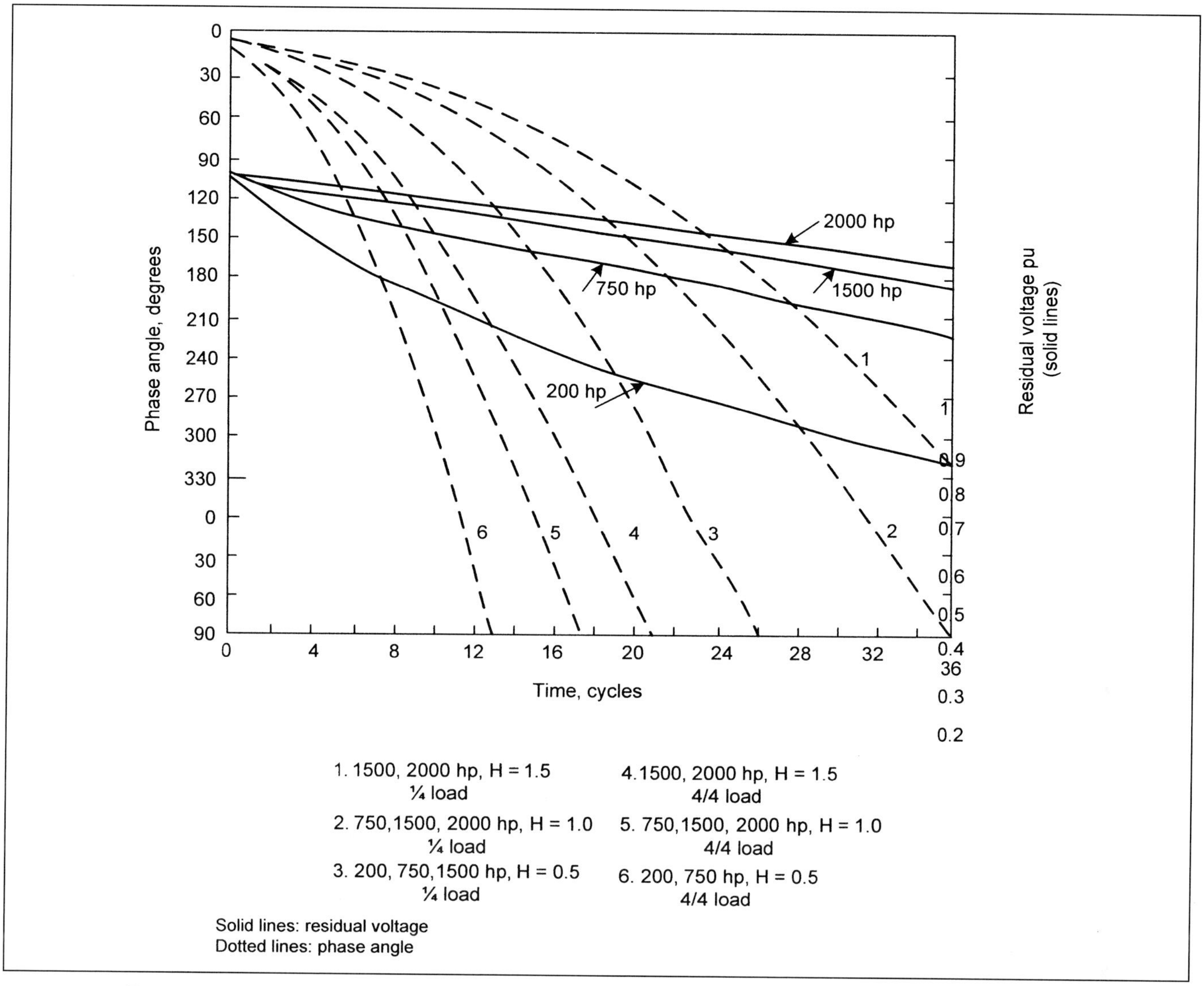

FIGURE 16-10 Residual voltage and phase angle of induction motors on disconnection from power supply system.[11]

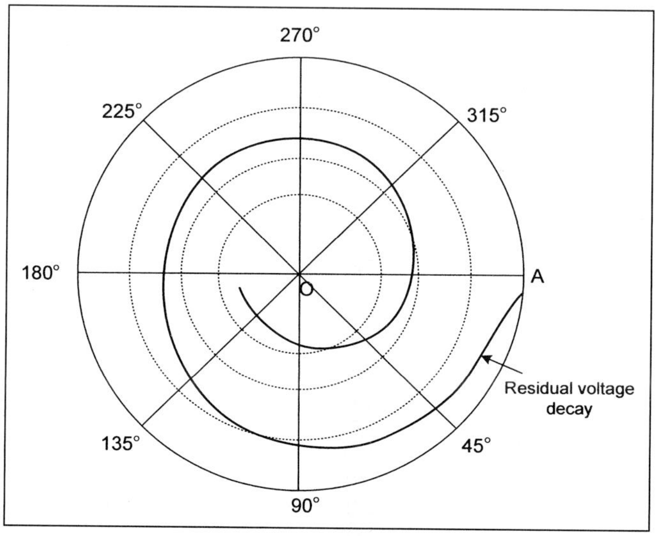

FIGURE 16-11 Polar representation of the decay of an induction motor on disconnection from the power supply system.

voltage to avoid transient torques. Tests made on motors of 100 to 1500 hp establish that the critical time for maximum inrush currents vary from 15 to 30 cycles. The voltage decay in a group of motors will be faster than a single large motor of an equivalent size. However, this mode of transfer defeats the purpose of maintaining continuity of operation of connected loads. The motors may decelerate to a level where the reacceleration may not be possible on reconnection.

3. *In-phase transfer*

When the motor is disconnected from the supply lines, its voltage and frequency falls, as illustrated in points 1 and 2. The supply frequency, however, remains constant. The residual voltage falls in and out of phase with the supply voltage, at an increasing rate, as the motors slow down. If the alternate supply breaker is closed exactly when the voltages are in phase, bumpless transfer can be achieved. Care has to be exercised, as the two sources may not be exactly in-phase when the transfer is started, and the alternate supply phase may change further when the normal supply breaker is opened due to changes in the system interconnections.

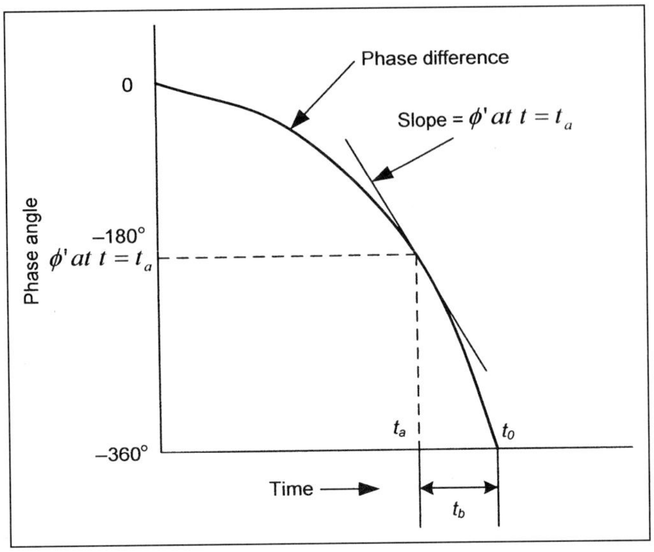

FIGURE 16-12 To illustrate in-phase bus transfer.

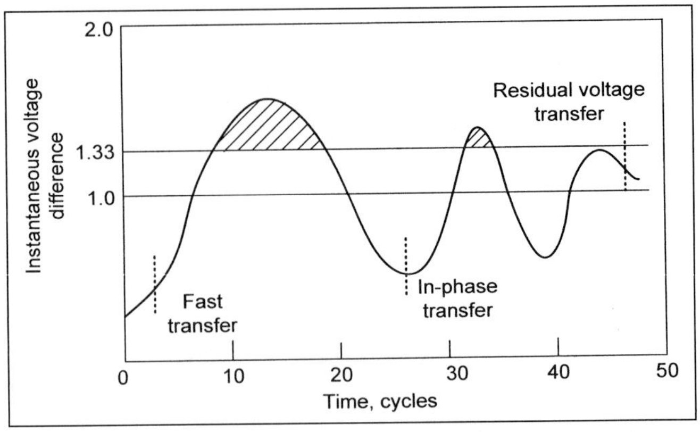

FIGURE 16-13 Motor voltage stresses with respect to transfer mode: instantaneous voltages at the second breaker must be a minimum when the breaker is closed to connect to alternate source of power.

The in-phase transfer must consider the circuit breaker closing time and close the circuit breaker ahead of zero phase by breaker closing time (Fig. 16-12). Assume that the breaker closing time is t_b seconds. Then the relay must initiate closing of the breaker t_b seconds before the phase angle reaches 360°, which means at time t_a. In order to predict when the phase angle will be zero, Taylor series can be used.[12]

The phase-time curve can be estimated by Taylor series, considering the first three terms:

$$\phi(t)=\phi(t_1)+\phi'(t_1)(t-t_1)+\frac{\phi''}{2}(t-t_1)^2 \tag{16-6}$$

where $\phi(t_1)$, $\phi'(t_1)$, and $\phi''(t_1)$ are the value of phase at t_1, the derivative at t_1, and second derivative at t_1, respectively. For $\phi(t) = 0$:

$$\phi(t_1+t_b)=\phi(t_1)+\phi'(t_1)t_b+\frac{\phi''(t_1)}{2}t_b^2 \tag{16-7}$$

Equation (16-22) gives prospective phase value at t_b when $\phi(t_1 + t_b) = 0$, by predetermination of ϕ and ϕ' and knowing the value of t_b. The in-phase transfer will be ineffective if at the time of transfer there is not enough bus voltage to support motor reacceleration.

A transfer logic controller is available that performs all the three modes of transfers, user selectable. Figure 16-13 shows the three modes of transfer with respect to the voltage *difference profile* across the bus section breaker—the fast transfer and in-phase transfer are better compared to residual voltage transfer with respect to motor stresses. If we consider the residual voltage of the motor to completely decay to zero, the voltage across the bus section breaker is 1 pu, while for the fast bus transfer and in-phase transfer, it will be much lower and less stressful to the motor. The motor transients will be much reduced.

16-2-4 Momentary Paralleling

During planned load transfer, momentary paralleling of two sources is widely used, that is, bus section breaker in Fig. 16-7*b* is closed before any of the two source breakers are opened. (Synchronism check features are incorporated.) A question arises that when the two sources can be paralleled in the shutdown or planned load transfer modes, why can these not continuously run in parallel? Appropriate protective relaying schemes are available for two continuously running parallel circuits, which will selectively trip the faulty circuit and the loads will experience lesser transients, as compared to switching on failure of a source. While continuous parallel operation is much desired from the operational point of view, paralleling increases the short-circuit duties on the circuit breakers. During momentary paralleling, a calculated risk is taken, that for the time duration involved to transfer loads the probability of a fault is low. If a fault does occur, it can destroy the equipment and can be hazardous to human life. This practice should be carefully reviewed.

16-2-5 Fault Conditions

The fault conditions should be distinguished from simple loss of source voltage, say due to inadvertent switching. The fault voltage dips will be more severe, depending upon the location and nature of the fault. Obviously for any fault in the bus itself or in the transformers (Fig. 16-7), the autotransfer is blocked. A fault voltage dip will give rise to transients in the connected motor loads, and these may not survive operation when the alternate source of power is restored.

Sometimes simultaneous opening and closing of the breakers is implemented. The control signals to open and close are applied at the same time. The dead time including sensing of the sources may be one to three cycles. There is some risk involved that a breaker may fail to open. Alternatively, the normal breaker is first opened and interlocked, so that the standby breaker cannot be closed, unless the normal breaker has opened. The dead times may be 10 cycles or more. The stability of loads will depend how fast the power is restored; the stability of motors on voltage dips is discussed in Chap. 11.

16-2-6 Drop Out of Motor Contactors

Another parameter of practical consideration is that magnetic contactors for motor starting will drop out quickly—for low-voltage motor starters, the contactors may drop out in the very first cycle of voltage dip exceeding 30 percent. NEMA E2 starters are used for starting and control of medium voltages (in a certain range), and these have dc contactors, which may take a little longer to drop out, depending upon the trapped residual magnetism in the coils. It is possible that the motors are tripped out from service because of drop out of contactors, while these may still be able to ride through the voltage dip.

Latched type of contactors and stabilizing the contactors for undervoltage dips for a certain time duration are possible options. This has to be attempted carefully, as on voltage restoration all the connected motor loads, which lost speed during the voltage dip, will take high inrush currents to accelerate. This large current may cause further voltage dips in the system impedance, a cumulative effect, to precipitate a shutdown.

16-2-7 Autotransfer of Synchronous Motors

Autotransfer of power on synchronous motors is not attempted. The phase angle between the motor generated voltage and supply connection will vary from 0° to 360° per slip cycle. An autoclosing operation may produce short-circuit current equal to 2.5 times the normal short-circuit current. This will subject the windings to stresses approaching 6.5 times the normal short-circuit currents. The transient torques produced by the fault currents have oscillatory components and the peak torque may be as high as 30 to 40 times the normal full-load torque. On a voltage dip, the excitation to the motor may be removed and reapplied shortly thereafter, if the system conditions are favorable. (See resynchronizing in Chap. 11.)

Example 16-2 The transients on reconnection of induction motor loads are studied with EMTP modeling. Consider that five motors of the specifications as in Example 16-1 are connected to a 6.6-kV bus, served from two alternate sources in a system configuration as in Fig. 16-7*a*. The normal source of power is tripped, normal breaker is opened, and the standby breaker is closed; the total duration of loss of power (including relaying and closing time of breakers) is 500 ms (arbitrary here, for the example). Each of the two transformers in Fig. 16-7*a* are rated at 80 MVA, transformer impedance = 9 percent on 80-MVA base. Figure 16-14*a* shows that in approximately 500 ms the bus voltage in phase *a* has decayed by 43 percent. The restoration of power at 500 ms does not give rise

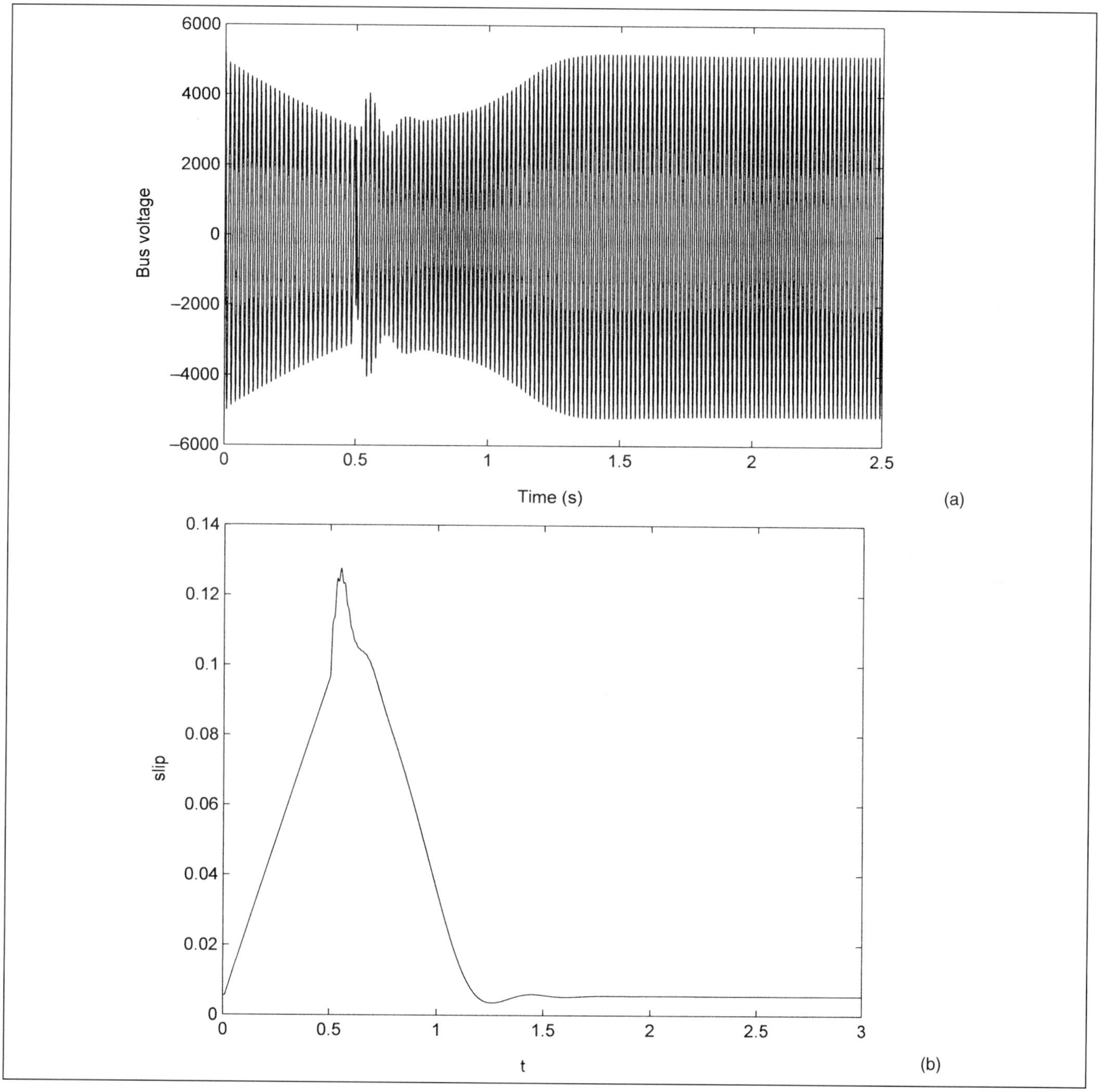

Figure 16-14 EMTP simulation of transients in five, 10000-hp, 6.6-kV, bus-connected motors, dead time = 500ms: (*a*) Bus voltage, (*b*) slip of a 10000-hp motor, (*c*) transient current in phase *a*, and (*d*) transient torque.

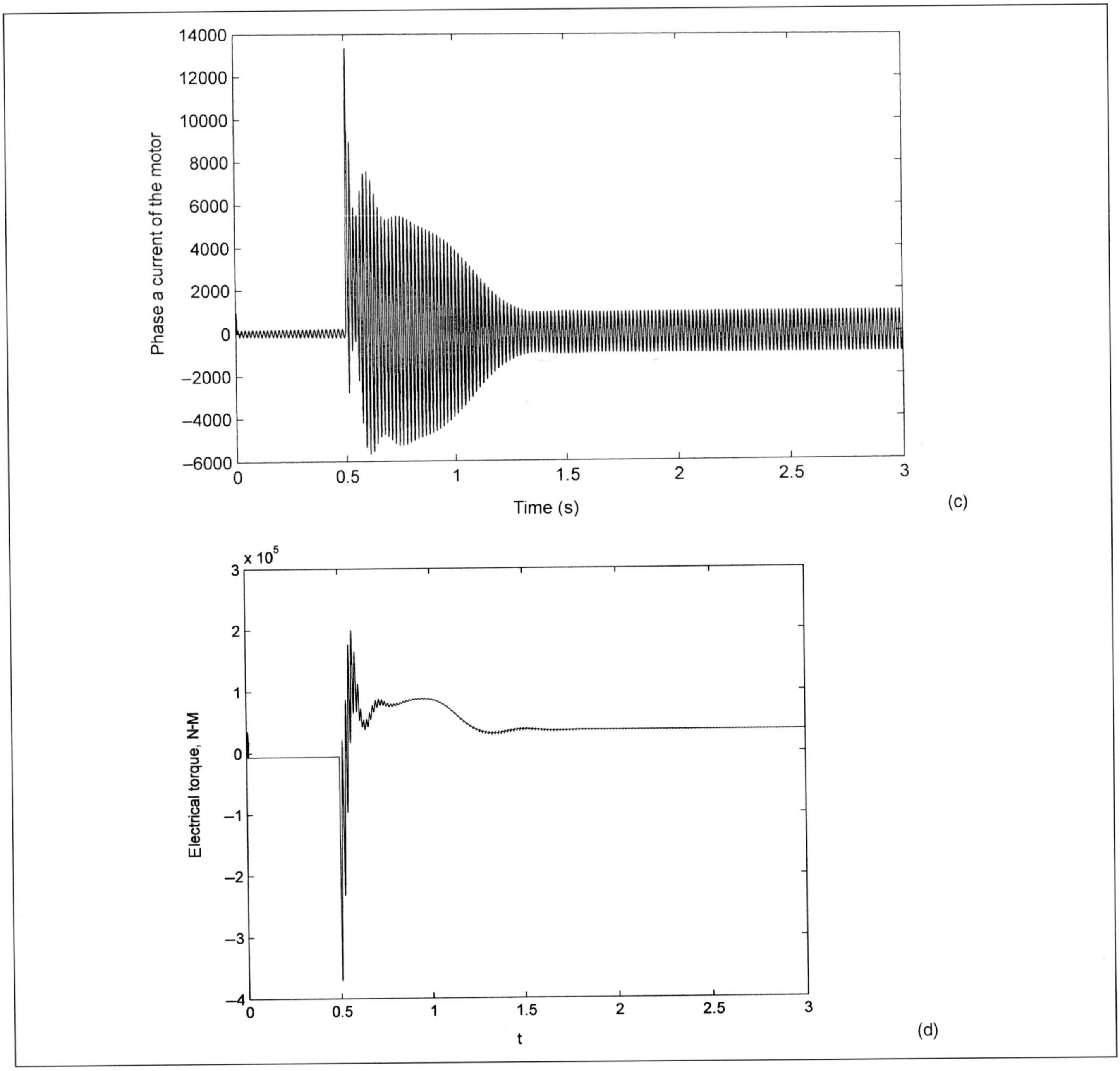

FIGURE 16-14 (*Continued*)

to much voltage escalation. The motors accelerate and remain stable; the slip which had increased to approximately 13 percent rapidly decreases (Fig. 16-14*b*). The current in phase *a* and the transient electromagnetic torque (newton-meters) of one of the motors are shown in Fig. 16-14*c* and *d*, respectively. The current on reconnection jumps to 13800 A, approximately 15.7 times the full-load current of the motor. The 80-MVA transformer will therefore experience a sudden loading of 13.8 × 5 = 69 kA. The ANSI/IEEE short-time withstand capability of the transformer for through-fault is 77.76 kA for 2 s. The transient load for a couple of cycles can be safely allowed.

16-3 STATIC TRANSFER SWITCHES AND SOLID-STATE BREAKERS

The static transfer switches (STS) can affect a transfer in approximately one-fourth of a cycle, depending upon the topology of the circuit. STSs are commonly used in UPS systems (Chaps. 19), and are commercially available for medium-voltage applications. Thyristor-based STS systems have the advantage of low cost and low conduction losses, compared to GTO- or IGBT- based technologies. The natural commutation of thyristors in a STS system is dependent upon load power factor and depth of voltage sag, which prolongs the transfer time to more than one-fourth of the cycle.

Figure 16-15*a* depicts the power circuit (only single phase shown), with maintenance and load bypass mechanical switches. Two sets of thyristors are connected in opposite directions to pass ac positive and negative half-cycles. Figure 16-15*b* is a diagram of transfer logic. The phase voltages V_a, V_b, and V_c are transformed to V_d, V_q, and V_0, using Park's transformation. The voltage $V_p = \sqrt{V_d^2 + V_q^2}$. This is compared with a reference and the error signal is passed on to an LP filter, which attenuates voltage transients.

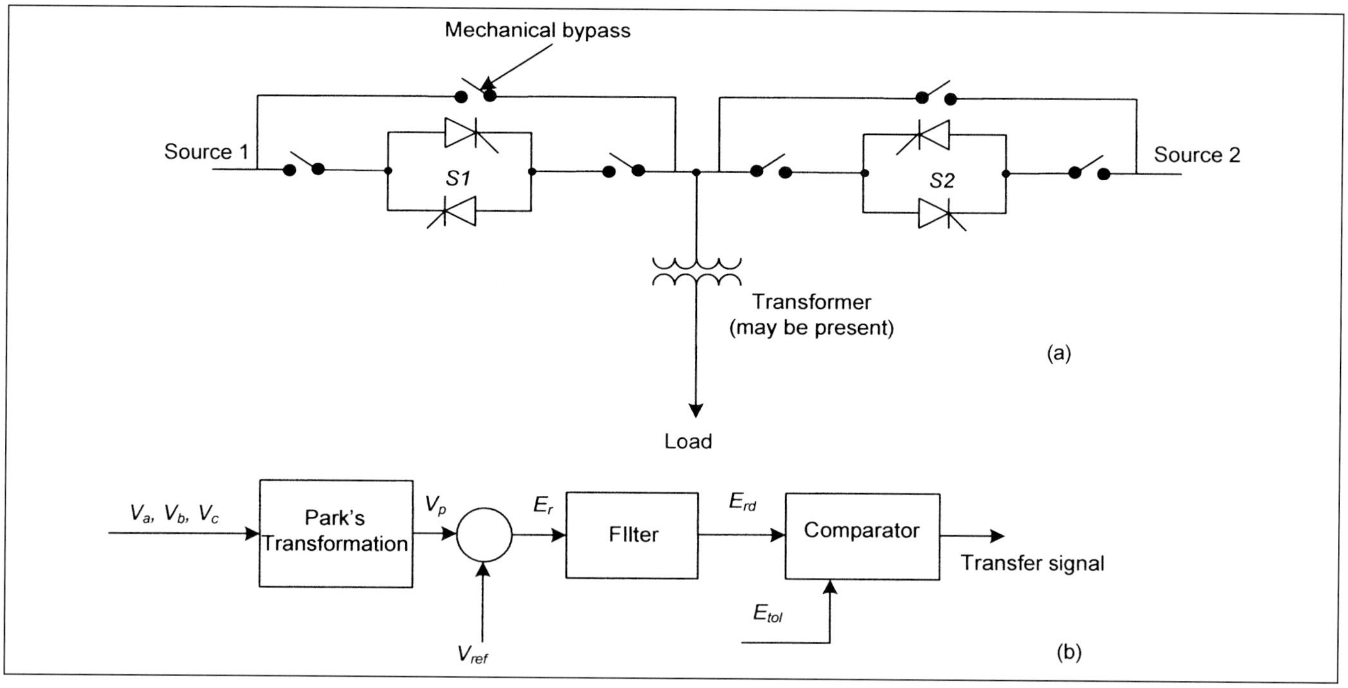

FIGURE 16-15 (*a*) Circuit of a thyristor-based static switch (only one phase shown). (*b*) Block circuit diagram of a transfer logic.

Filter output is compared to tolerance limit E_{tol}, and the output of the comparator is the transfer signal, which initiates a transfer.[13] The regenerative load transfer is discussed in Ref. 14; the maximum transfer time is determined by the zero crossing of the system voltage. The total transfer time varies from 6.5 to 10.5 ms, depending upon the nature of the disturbance; under voltage or fault conditions. If there is a transformer connected to the load side (Fig. 16-15*a*), it can be subjected to large transients due to fast transfer.

Solid-state circuit breakers (SSBs) offer considerable advantage over mechanical breakers. The short-circuit fault current is reduced. The voltage dips due to three-phase fault clearance (lasting for about 100 ms or more) can be reduced to 100 μs. The impediments in the development have been material costs and on-state losses.

At the 15-kV level, the SSBs can be built using GTOs or SCRs. The SCRs have better blocking voltages, lesser losses, and higher current ratings. GTOs can interrupt current with negligible delay and can turn off the fault current during the first half cycle when the overload condition is detected. This interruption must take place before exceeding the current interrupting rating of GTOs. Figure 16-16*a* shows a hybrid solution applied to a 15-kV breaker. It consists of parallel branches, composed of GTOs and SCRs. The GTO section conducts load current in the steady state. It is rated for maximum line currents but not for fault currents. It opens rapidly when the preset level of the current is exceeded, say 3000 A. A number of GTO switches are connected in series with snubber circuits and metal-oxide arresters. The SCR section which incorporates pairs of antiparallel connected thyristor devices in series is normally kept open and has no continuous current rating. It conducts fault currents (say 15 kA) for a period of 10 to 15 cycles, so that the downstream devices on the load side may coordinate. Further topologies of SSBs are described in Ref. 15.

Solid-state circuit breakers can be used as a bus section switch or static transfer switches. Figure 16-16*b* shows the load-transfer characteristics of two SSBs connected in a system configuration of Fig. 16-7*a*, normal breakers replaced with SSBs. The load current is basically unaffected due to transfer.

16-4 COGGING AND CRAWLING OF INDUCTION MOTORS

Parasitic magnetic fields are produced in an induction motor due to harmonics in the mmf originating from:

- Windings
- Certain combination of rotor and stator slotting
- Saturation
- Air gap irregularity
- Unbalance and harmonics in the supply system voltage

IEEE standard 519[1] lays down the permissible limits of harmonic current injections into the supply system. A Fourier analysis of the waveshape of MMFs in a three-phase winding reveals that it has a constant fundamental and harmonics of the order of 5, 7, 11, 13, . . . or $6m + 1$, and $6m - 1$, where m is any integer. All harmonics of the order of third and their multiples (triplen harmonics) are absent, because a spread of 120° in windings of a three-phase machine eliminates these, though not perfectly. The effective phase spread for a third harmonic will be therefore 120° × 3 = 360°. The harmonics move with a speed reciprocal to their order, either with or against the fundamental. Harmonics of the order of $6m + 1$ move in the same direction as the fundamental field, while those of $6m - 1$ move in the opposite direction.

These harmonics can be considered to produce, by an additional set of rotating poles, rotor emfs, currents, and harmonic torques akin to the fundamental frequency, but synchronous speeds, depending upon the order of the harmonics. The resultant speed torque curve will then be a combination of the fundamental and harmonic torques. This produces a saddle in the torque-speed characteristics, and the motor can crawl at the lower speed of one-seventh of the fundamental (Fig. 16-17).

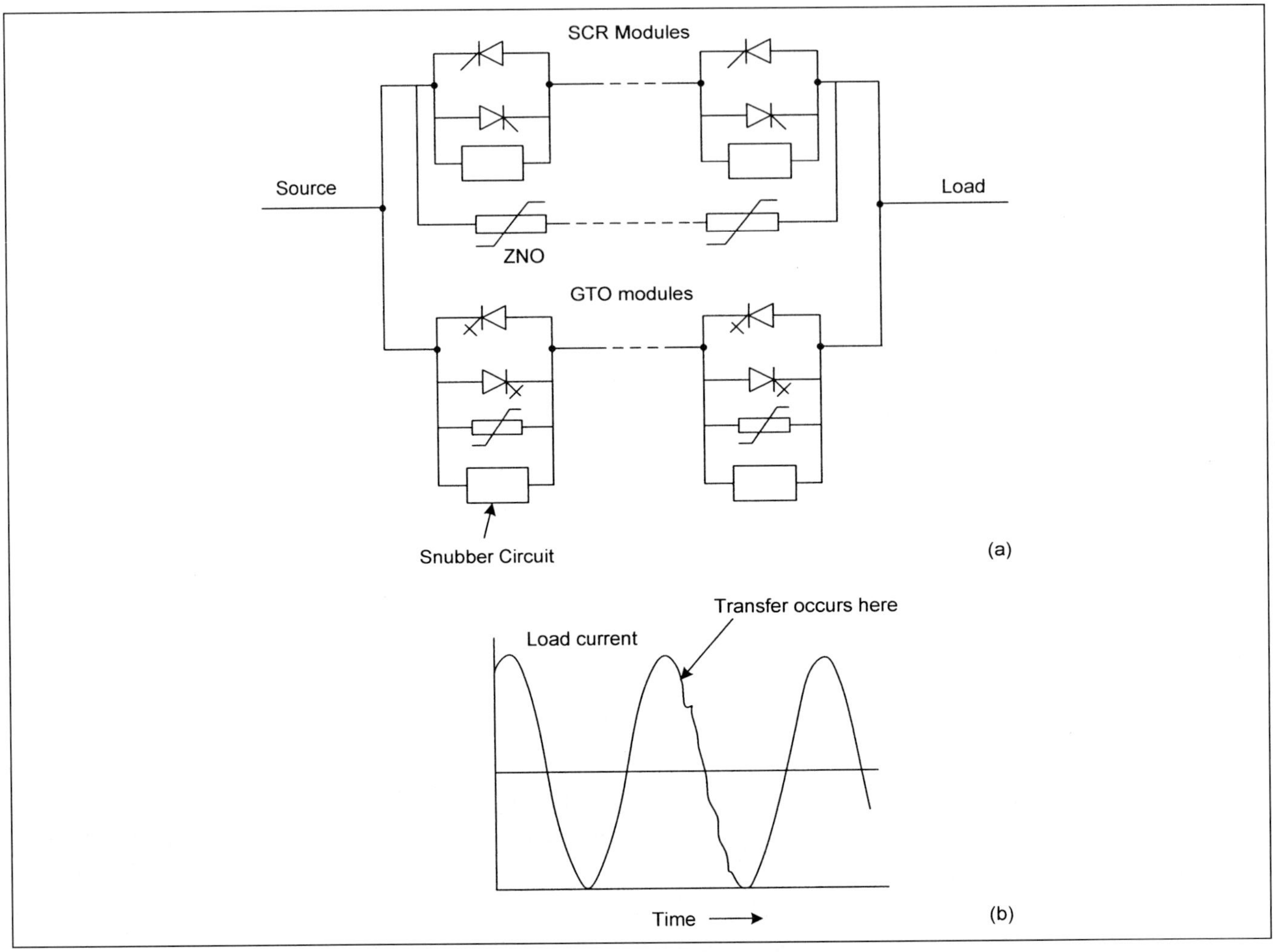

FIGURE 16-16 (*a*) A hybrid static circuit breaker: GTO and SCR switches in parallel. (*b*) Load transfer transient with static breakers.

This harmonic torque can be augmented by stator and rotor slotting. In *n*-phase winding, with g slots per pole per phase, emf distribution factors of the harmonics are:

$$n = 6Ag \pm 1 \tag{16-8}$$

where *A* is any integer, 0, 1, 2, 3, ... The harmonics of the order $6Ag + 1$ rotate in the same direction as the fundamental, while those of order $6Ag - 1$ rotate in the opposite direction.

Consider 24 slots in the stator of a 4-pole machine. Then $g = 2$ and 11th and 13th harmonics will be produced strongly. The harmonic induction torque thus produced can be augmented by the rotor slotting. Consider a rotor with 44 slots. The 11th harmonic has 44 half-waves in the air gap each corresponding to a rotor bar. This will accentuate 11th harmonic torque and produce strong vibrations. If the numbers of stator slots are equal to the number of rotor slots, the motor may not start at all, a phenomenon called *cogging*.

16-4-1 Harmonic Synchronous Torques

The rotor mmf has a harmonic content and with certain combination of the stator and rotor slots, it is possible to get a stator and rotor harmonic torque producing a harmonic *synchronizing* torque. There will be a tendency to develop sharp synchronizing torque at some subsynchronous speed (Fig. 16-17*b*).

The rotor slotting will produce harmonics of the order of:

$$n = \frac{S_2}{p} \pm 1 \tag{16-9}$$

where S_2 is the number of rotor slots. Here, the plus sign means rotation with the machine. Consider a motor with $S_1 = 24$ and $S_2 = 28$. The stator produces reversed 11th harmonic (reverse going) and 13th harmonic (forward going). The rotor develops a reversed 13th and forward 15th harmonic. The 13th harmonic is produced both by stator and rotor, but of opposite rotation. The synchronous speed of the 13th harmonic is 1/13 of the fundamental synchronous speed. Relative to rotor, it becomes:

$$-\frac{(n_s - n_r)}{13} \tag{16-10}$$

where n_s is the synchronous speed and n_r is the rotor speed. The rotor, therefore rotates its own 13th harmonic at a speed of:

$$-\frac{(n_s - n_r)}{13} + n_r \tag{16-11}$$

relative to the stator. The stator and the rotor 13th harmonic fall into step when:

$$+\frac{n_s}{13} = -\frac{(n_s - n_r)}{13} + n_r \tag{16-12}$$

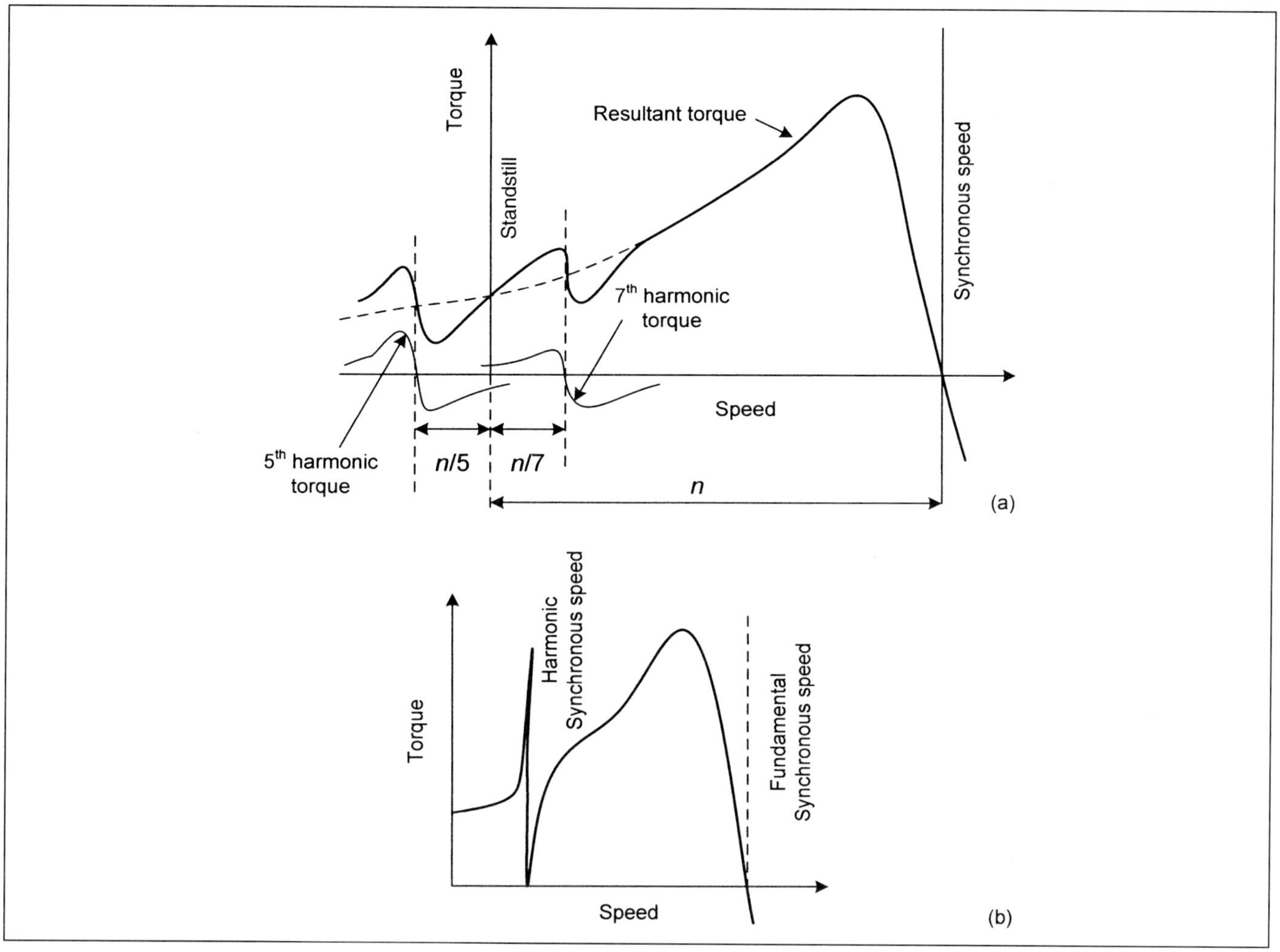

FIGURE 16-17 (*a*) Harmonic induction torques in an induction motor. (b) Harmonic synchronous torques.

This gives $n_r = n_s/7$, that is, torque discontinuity is produced not by 7th harmonic but by 13th harmonic in the stator and rotor rotating in opposite directions.

The harmonic torques are avoided in the design of machines by proper selection of the rotor and stator slotting.

16-4-2 Tooth Ripples

Figure 16-18 shows the tooth ripples in the air-gap flux distribution of an induction motor, somewhat exaggerated. These are produced because of variation of air-gap permeance. The frequency of the flux pulsations corresponds to the rate at which slots cross the pole face, given by $2gf$, where g is the number of slots per pole (as defined before) and f is the system frequency. This stationary pulsation may be considered as two waves of fundamental space distribution rotating at angular velocity $2g\omega$ in the forward and backward direction. The component fields will have velocities of $(2g \pm 1)\omega$ relative to armature windings and will generate harmonic emfs of frequencies $(2g \pm 1)f$ cycles per second. However, this is not the main source of tooth ripples. Since the ripples are due to slotting, these do not move with respect to the stator conductors, and these cannot generate an emf of pulsation. With respect to rotor, the flux wave has a relative velocity of $2g\omega$ and generates emfs of $2gf$ frequency. Such currents superimpose an MMF variation of $2gf$ on the resultant pole MMF. These can be again resolved into forward- and backward-moving components, with respect to the *rotor*, and $(2g \pm 1)\omega$, with respect to the stator.

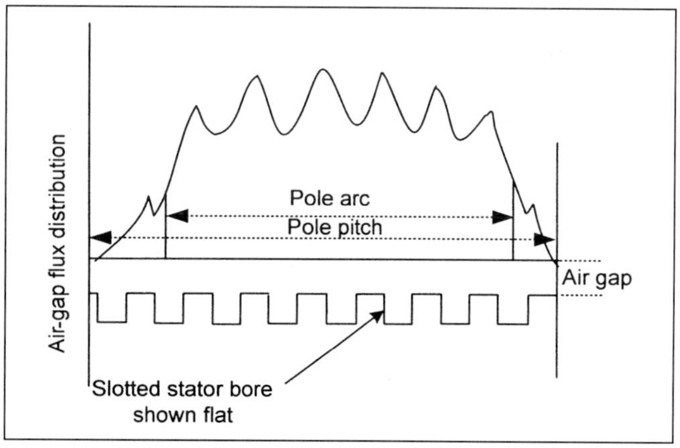

FIGURE 16-18 Tooth ripples in induction motors.

Thus, stator emfs of $(2g \pm 1)f$ are generated which are the main tooth ripples.

16-5 SYNCHRONOUS MOTOR-DRIVEN RECIPROCATING COMPRESSORS

Synchronous motors driving reciprocating compressors give rise to current pulsations in the stator current due to the nature of the

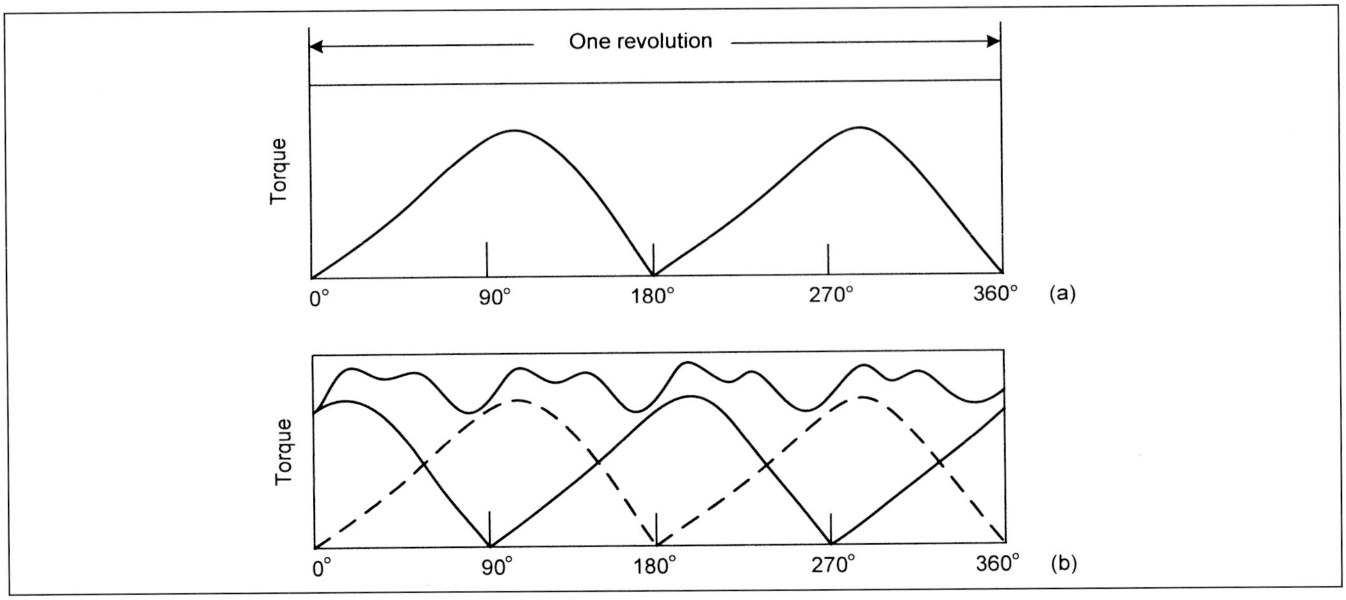

FIGURE 16-19 (*a*) Crank effort diagram for one-cylinder, double-acting compressor or two-cylinder, single-acting compressor. (*b*) Crank effort diagram for two-cylinder, double-acting compressor or four-cylinders, single-acting compressor.

reciprocating compressor torque variations. Figure 16-19*a* shows a crank effort diagram for a one-cylinder, double acting compressor or a two-cylinder, single-acting compressor, and Fig. 16-19*b* shows a crank effort diagram for a two-cylinder, double-acting compressor or a four-cylinder, single-acting compressor. The torque varies per revolution and this gives rise to variation in stator current, oscillogram in Fig.16-20. In some cases, step unloading of compressors is used, which introduces additional irregularity in the crank effort diagram and increases the current pulsation. Figure 16-21 shows resolution of a compressor to equivalent rotating masses, also useful for torsional vibrations discussed in Sec. 16-6.

NEMA standards limit current pulsation to 66 percent of rated full-load current, corresponding to an angular deviation of approximately 5 percent from the uniform speed. The required flywheel effect to limit current pulsation is proportional to compressor factor "*X*" given by:

$$\mathrm{WK}^2 = 1.34 X f P_r \left(\frac{100}{\mathrm{rpm}}\right)^4 \tag{16-13}$$

where X is the compressor factor, f is the frequency, P_r is the synchronizing power, and WK^2 is the total inertia in lb. ft^2.

A system having a mass in equilibrium, and a force tending to return this mass to its initial position if displaced, has a natural frequency of oscillation. For a synchronous motor, it is given by:

$$f_n = \frac{35200}{n_s}\sqrt{\frac{P_r f}{\mathrm{WK}^2}} \tag{16-14}$$

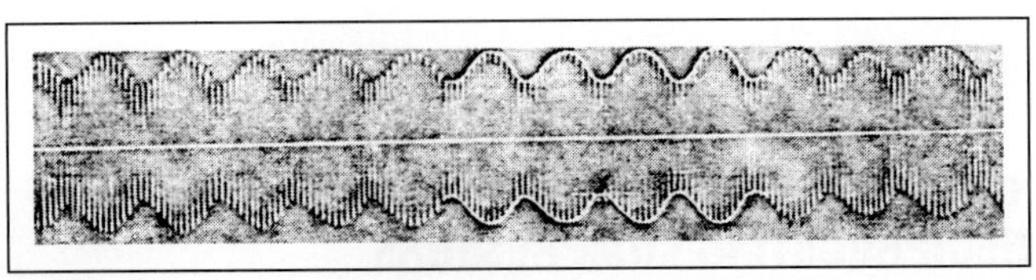

FIGURE 16-20 Oscillogram of current input to motor, driving a vertical two-cylinder, single-acting compressor.

where f_n is the natural frequency of oscillation, and n_s is the synchronous speed. This assumes that there is no damping and the motor is connected to an infinite system. The forcing frequency, due to crank effort diagram, should differ from the natural frequency by at least 20 percent.

We defined synchronizing power for the generator in Chap. 12. The definition for the synchronous motor is analogous. Similar to a synchronous generator (Chap. 10), the phasor diagrams of a synchronous motor are depicted in Fig. 16-22*a*, *b*, and *c* at leading power factor, unity power factor, and lagging power factor, respectively. The resistance is ignored in these figures. The power output can be expressed as:

$$P = V(I_q \cos\delta + I_d \sin\delta) \tag{16-15}$$

This can be shown to be equal to:

$$P = \frac{VE_t}{x_d}\sin\delta + \frac{V^2(x_d - x_q)}{2x_d x_q}\sin 2\delta \tag{16-16}$$

A disturbance, in relative position of angle δ, results in a synchronizing power flow in the machine. A synchronizing torque is developed which opposes the torque angle change due to the disturbance, that is, a load change. The final torque angle may be reached only after a series of oscillations, which determine the transient stability of the machine, depending upon whether these converge or diverge, akin to stability of a synchronous generator.

The synchronizing power in per unit on a machine can be written as:

$$P_r = \frac{dP}{d\delta} = \frac{VE_t}{x_d}\cos\delta + \frac{V^2(x_d - x_q)}{x_d x_q}\cos 2\delta \tag{16-17}$$

Figure 16-23 shows the synchronizing power with respect to torque angle. For any change in P, we can write the equation:

$$\Delta P = P_r \Delta\delta \tag{16-18}$$

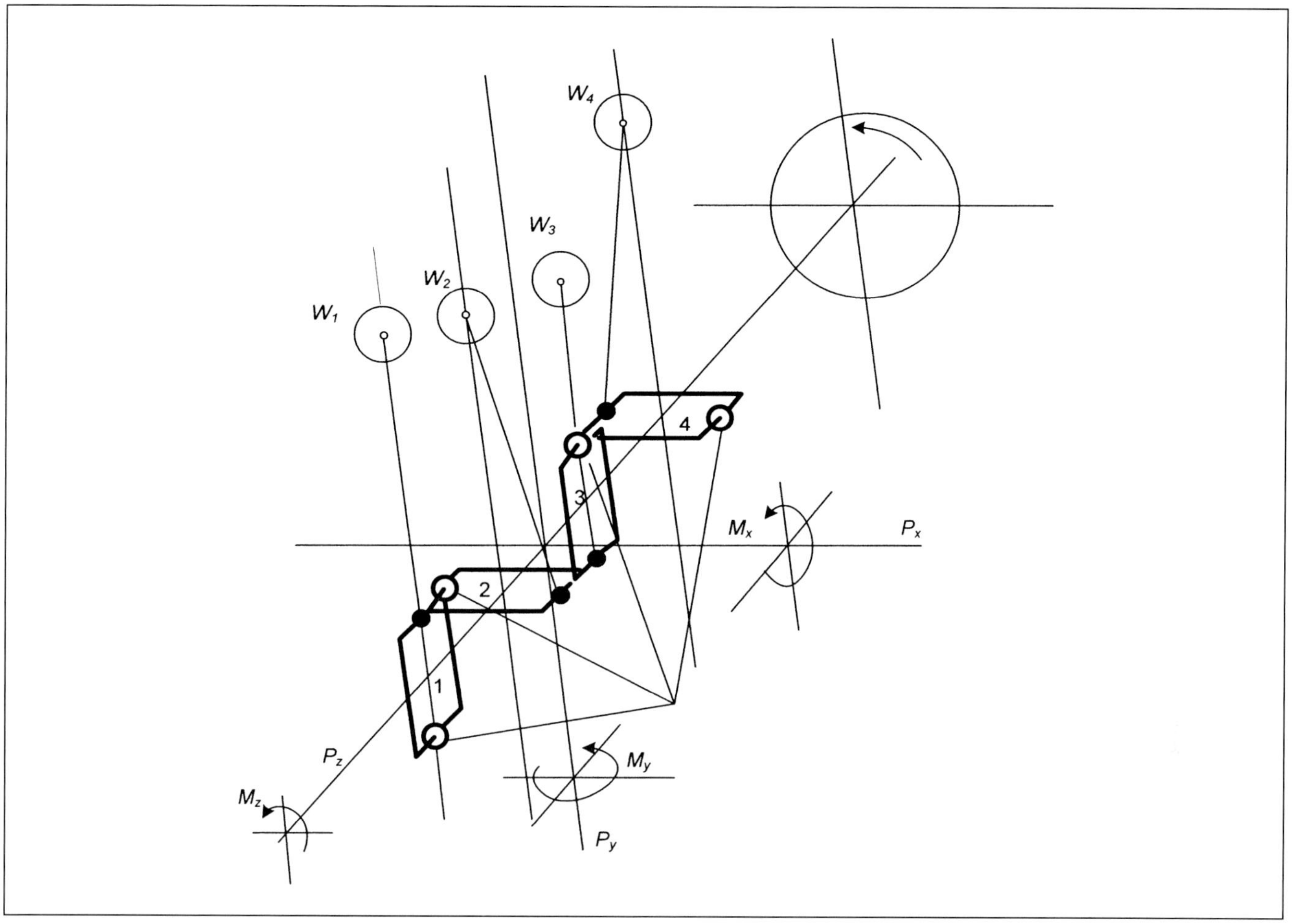

FIGURE 16-21 Resolution of a compressor to equivalent rotating masses.

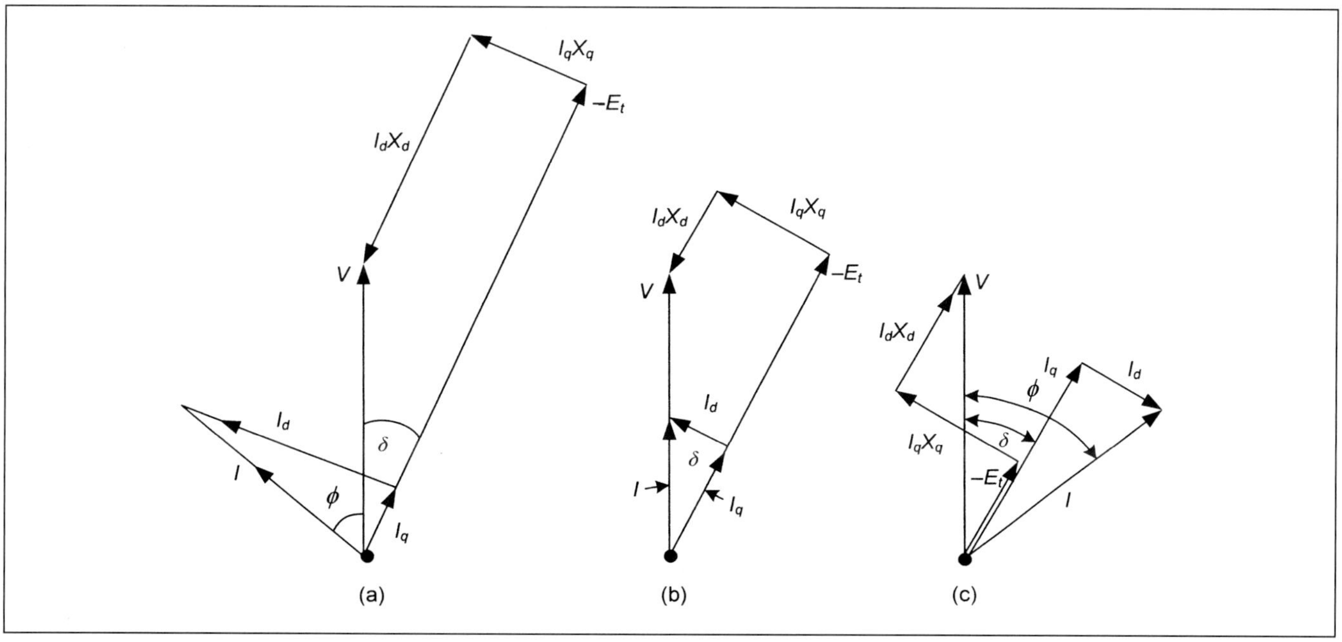

FIGURE 16-22 Phasor diagrams of a synchronous motor operating at (*a*) leading, (*b*) unity and, (*c*) lagging power factor.

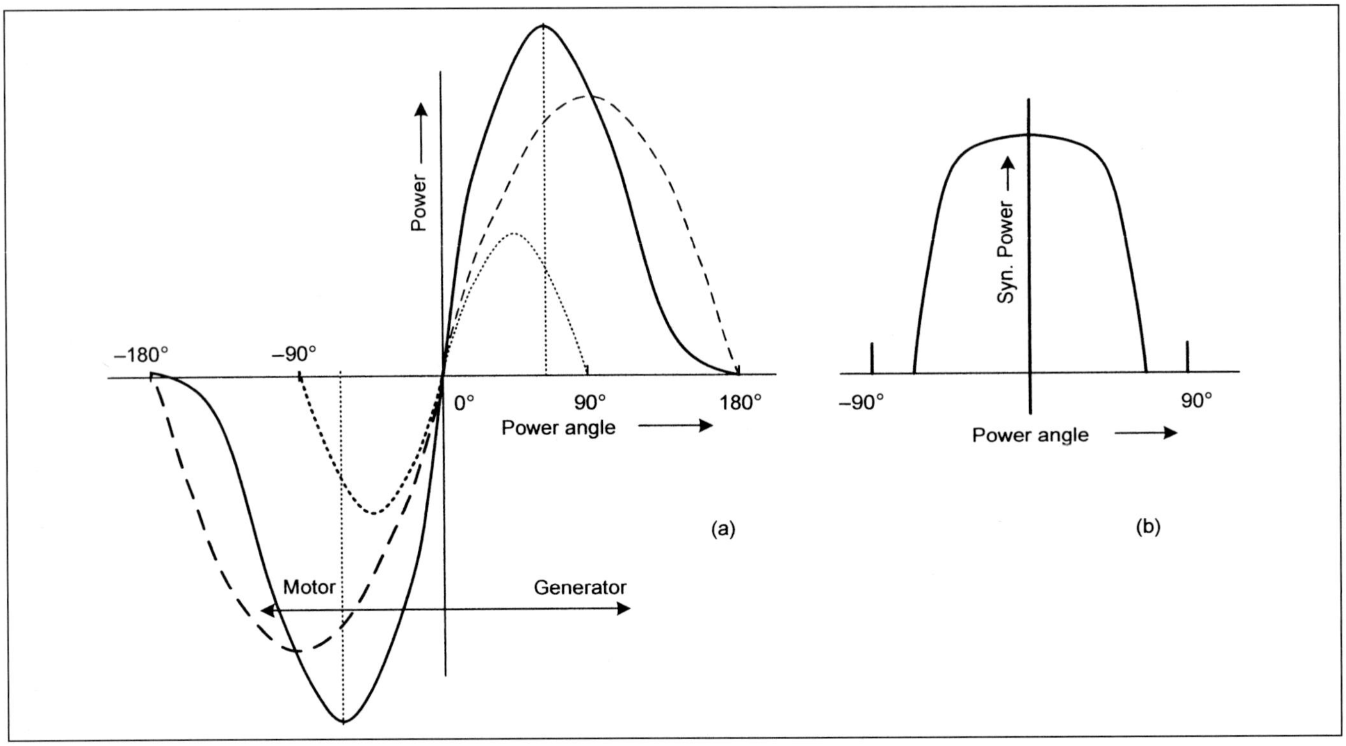

FIGURE 16-23 (*a*) Power angle diagram of a synchronous generator/motor. (*b*) Synchronizing power.

The damping torque occurs due to induced currents in the damper windings, mainly because of quadrature axis flux. The damping power P_{ds} at any slip s can be written as:

$$P_{ds} = s\frac{P_0}{s_0} \tag{16-19}$$

where s_0 is the slip, which will produce *asynchronously* the rated power P_0 of the machine. We can define the slip as the change in the torque angle with time:

$$s = \frac{1}{\omega}\frac{d\delta}{dt} \tag{16-20}$$

Then the acceleration or retardation of the rotor is given by:

$$a = \frac{ds}{dt} = \frac{1}{\omega}\frac{d^2\delta}{dt^2} \tag{16-21}$$

From above relations, we can write:

$$P_{ds} = \frac{P_0}{s_0\omega}\frac{d\delta}{dt} \tag{16-22}$$

Lastly, the power of inertia is:

$$P_j = Jg\omega_0^2 a = \frac{T_a P_0}{\omega}\frac{d^2\delta}{dt^2} \tag{16-23}$$

where T_a is the a accelerating time constant, J is the moment of inertia, and g is the gravitational constant.

The power balance equation becomes:

$$\frac{T_a P_0}{\omega}\frac{d^2\delta}{dt^2} + \frac{P_0}{s_0\omega}\frac{d\delta}{dt} + P_s\delta = P_a \tag{16-24}$$

where P_a is the net power of acceleration or retardation. This differential equation without forcing function is:

$$\frac{d^2\delta}{dt^2} + \rho\frac{d\delta}{dt} + \nu^2\delta = 0 \tag{16-25}$$

where the natural frequency is:

$$\nu = \sqrt{\frac{\omega P_s}{T_a P_0}} = \sqrt{\frac{\omega P_s}{Jg\omega_0^2}} \tag{16-26}$$

and:

$$\rho = \frac{1}{\nu T_a} \tag{16-27}$$

Example 16-3 A machine with T_a = 10 s, and synchronizing power two times the rated power has a natural frequency of:

$$\nu = \sqrt{\frac{2\pi \times 60 \times 2}{10}} = 8.8$$

The oscillations in cycles per second (cps) is $\nu/2\pi = 1.41$ cps. This is fairly low. Figure 16-24 shows current oscillations of a salient-pole machine, fitted with damper cage, on sudden loading. The solution of Eq. (16-25) is of the form:

$$\delta = Ae^{-\frac{\rho}{2}t}\cos(\nu t + B) \tag{16-28}$$

where A and B are the constants of integration. From Eqs. (16-20) and (16-21), the rotor slip is given by:

$$s = -\frac{\nu}{\omega}Ae^{-\frac{\rho}{2}}\sin(\nu t + B) \tag{16-29}$$

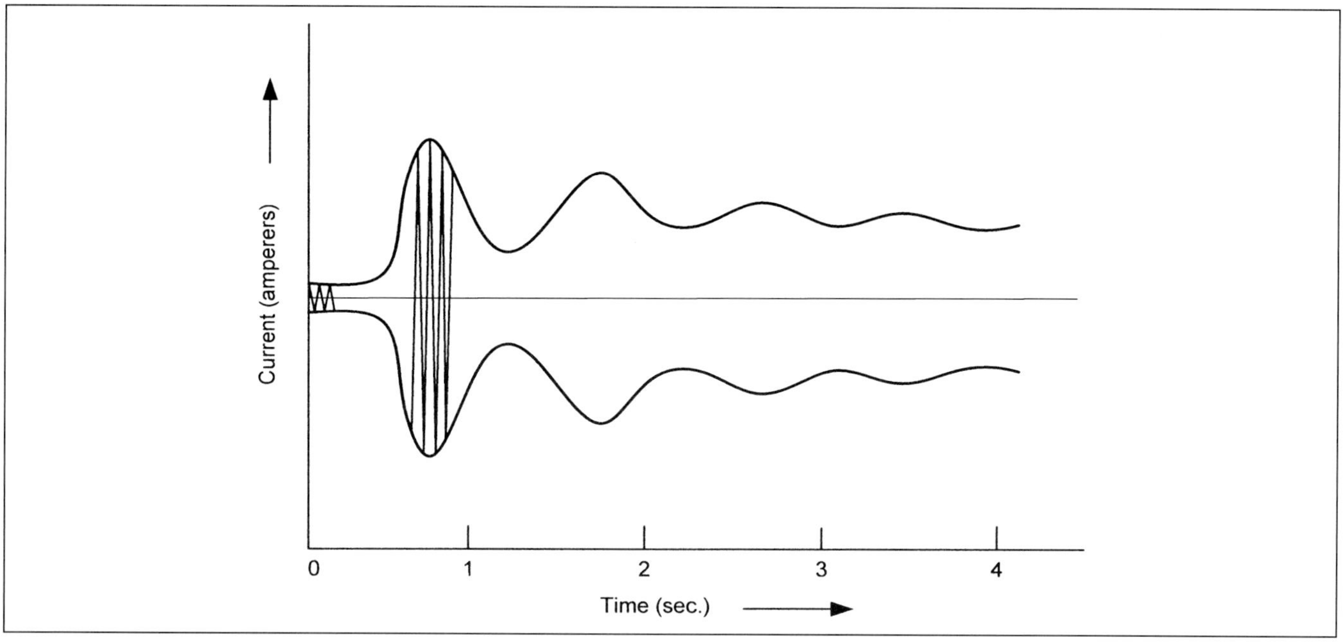

FIGURE 16-24 Current oscillations on sudden loading a salient-pole synchronous motor.

16-5-1 Forced Oscillations

If the driving power is not constant and fluctuates (due to fluctuating load) and P_m is the mean, P_a is the amplitude of power oscillation at frequency λ, then from Eq. (16-24)

$$\frac{d^2\delta}{dt^2}+\rho\frac{d\delta}{dt}+\nu^2\delta=\nu^2\Lambda e^{j\lambda t} \tag{16-30}$$

where Λ is given by:

$$\Lambda=\frac{\omega}{T_a P_0}\frac{P_a}{\nu^2}=\frac{P_a}{P_s} \tag{16-31}$$

Equation (16-46) can be solved for the forced oscillation and gives the magnitude of the oscillatory component as:

$$|A|=\frac{\Lambda}{\sqrt{\left[1-\left(\frac{\lambda}{\nu}\right)^2\right]^2+\left(\frac{\rho\lambda}{\nu^2}\right)^2}} \tag{16-32}$$

This is the resonance formula. Then, the forced amplitude of the synchronous power oscillation is:

$$P'=P_s|A| \tag{16-33}$$

The resonance curves are shown in Fig. 16-25.[16] The curves peak around $\lambda/\nu=1$. Further apart is the disturbing frequency with respect to natural frequency of the motor, lesser will be its effect.

A NEMA *X-Y* curve for determining compressor factor is in Fig. 16-26. This plots compressor factor against current pulsation for a given type of compressor. The total exchange of power with the network is:

$$\sqrt{\frac{1+\left(\frac{\rho\lambda}{\nu^2}\right)^2}{\left(1-\left(\frac{\lambda}{\nu}\right)^2\right)^2+\left(\frac{\rho\lambda}{\nu^2}\right)^2}}P_a=V_mP_a \tag{16-34}$$

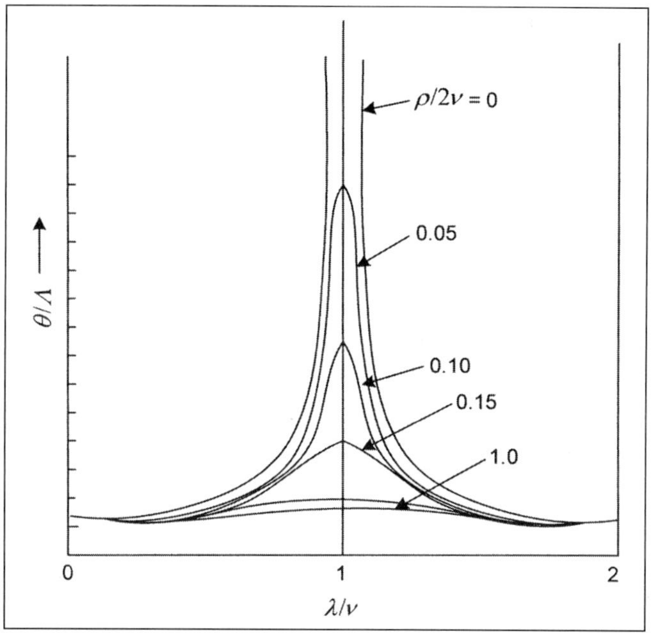

FIGURE 16-25 Resonance and damping characteristics.

For constant excitation, the current fluctuations are:

$$\Delta I=V_mP_a \tag{16-35}$$

The voltage fluctuations can be calculated as:

$$\Delta V=\Delta I\times Z \tag{16-36}$$

Consider a 900-hp synchronous motor with motor plus load inertia = 660 kg-m^2, operating at a torque angle δ = 28°. T_a, the accelerating time constant = 1.64 s. Following the above calculation method, the swing in power at the ratio υ/λ can be calculated. The motor compressor drive may pass through one or more

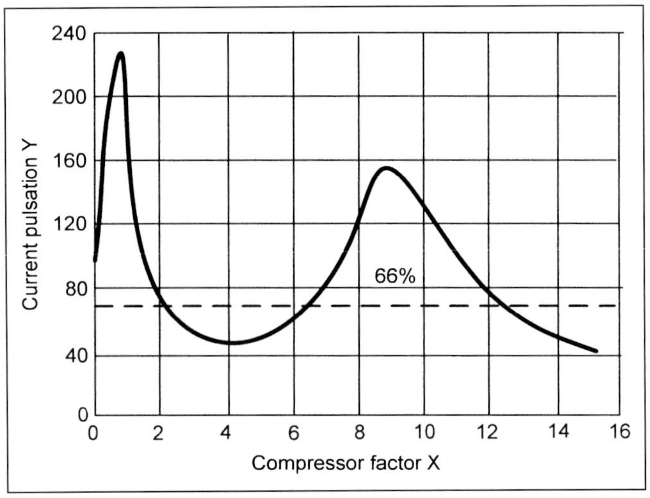

FIGURE 16-26 NEMA compressor factor curve.

natural torsional frequencies, and amplification of these may result (Sec.16-6).

16-5-2 Several Synchronous Motors on the Same Bus

In case several synchronous motors are connected to the same bus and are driving oscillatory loads, the oscillating voltage drop and input power of one machine may be superimposed on the other machines on the same bus. A situation may arise where the motor may alternately draw and pass energy with disturbing frequency of the mechanical load. The electric power supplied by one motor may be fed into other motors and accentuate the oscillations. The amplitude of the oscillating load of a compressor motor is passed on to the other motors with different proportionality ratios. The natural period of oscillations can be changed with:

- Changing the inertia of the drive system
- Excitation system
- Varying resistance of the damper windings
- Changing the air gap
- Varying the source reactance of the power supply system through reactors or transformers

A large change in the flywheel will be necessary to make an appreciable change in the natural frequency. When the reciprocating loads are driven by induction motors, the cranks are displaced from each other, due to different load slips of the motors so that overall fluctuations are reduced. The induction motors require smaller flywheel effects.

16-6 TORSIONAL DYNAMICS

Torsional vibrations are responsible for failure of drive system components and can stress or shear the turbine blades in generating units. Figure 16-27*a* shows a simple torsional model in steady-state torqued condition at rest or at constant speed. The electrical torque and the load torque are constant and in balance. There is no relative motion between the masses, but there is a twist in the shaft, with a spring constant of *K*. Note the relative positions of the angles of twist.

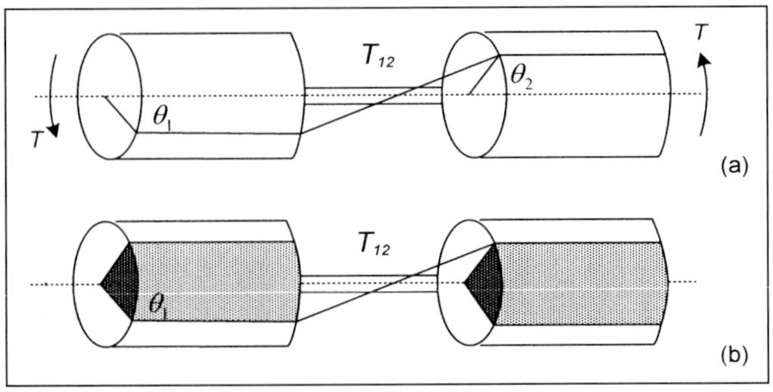

FIGURE 16-27 (*a*) Torsional model with two shaft-coupled rotating masses under steady-state, constant torque. (*b*) Free oscillations with the torque removed.

If the steady-state torques were removed, the two inertias would vibrate about the zero torque axes (Fig. 16-27*b*). In the absence of any damping, these vibrations will continue with peak torque and twist equal to initial steady-state values. The resonant frequency will be given by:

$$f_0 = \sqrt{\frac{K(J_1+J_2)}{J_1J_2}} \tag{16-37}$$

The stored energy in the system is converted to kinetic energy two times per cycle, and the two inertias oscillate in opposition to each other (Fig. 16-28). If one of the torques is removed, an oscillation with smaller amplitude will occur.

16-6-1 Steady-State Excitation

Consider that a steady-state excitation of frequency f_0 is applied to the system shown in Figure 16-27*a*. A torsional vibration will be excited, and it may continue to grow in magnitude, until the energy loss per cycle is equal to the energy that the small disturbance adds to the system during a cycle. The amplification curves of the resonant system are in Fig. 16-29. This can be compared with the curves in Fig. 16-25. If the excitation frequency varies at a certain rate, the torsional vibrations will be amplified as the system passes through the resonant point.

The driven load may have a positive slope, that is, the load increases with the speed. This occurs for fans and blowers. The load may have a negative slope, that is, conveyers and crushers. If the motor torque is removed, the negative load slope tends to give increasing torque pulsations.

An induction motor produces transient torques during starting (Chap. 11), however, at the instant of switching, the excitation is at power frequency and decays fairly rapidly as the motor accelerates. As discussed in Chap. 11, a synchronous motor, in addition to the initial fixed frequency excitation like an induction motor, produces a slip-frequency excitation which varies from 120 Hz at starting to 0 Hz at synchronism. When a synchronous machine pulls out, it will produce a sinusoidal excitation at the pull-out frequency, till it is disconnected or resynchronized. Figure 16-30 gives basic torque-frequency–pulsation relation of a synchronous motor. The critical speed, with twice the slip frequency during starting, cannot be avoided. The critical speed, with the slip frequency excitation following pull out, can be avoided, if the machine is deenergized in time.

16-6-2 Excitation from the Mechanical System

There can be excitations from the mechanical system too, which are proportional to speed and may occur at a frequency of multiple of shaft revolution or integral multiple of gear tooth passing frequency.

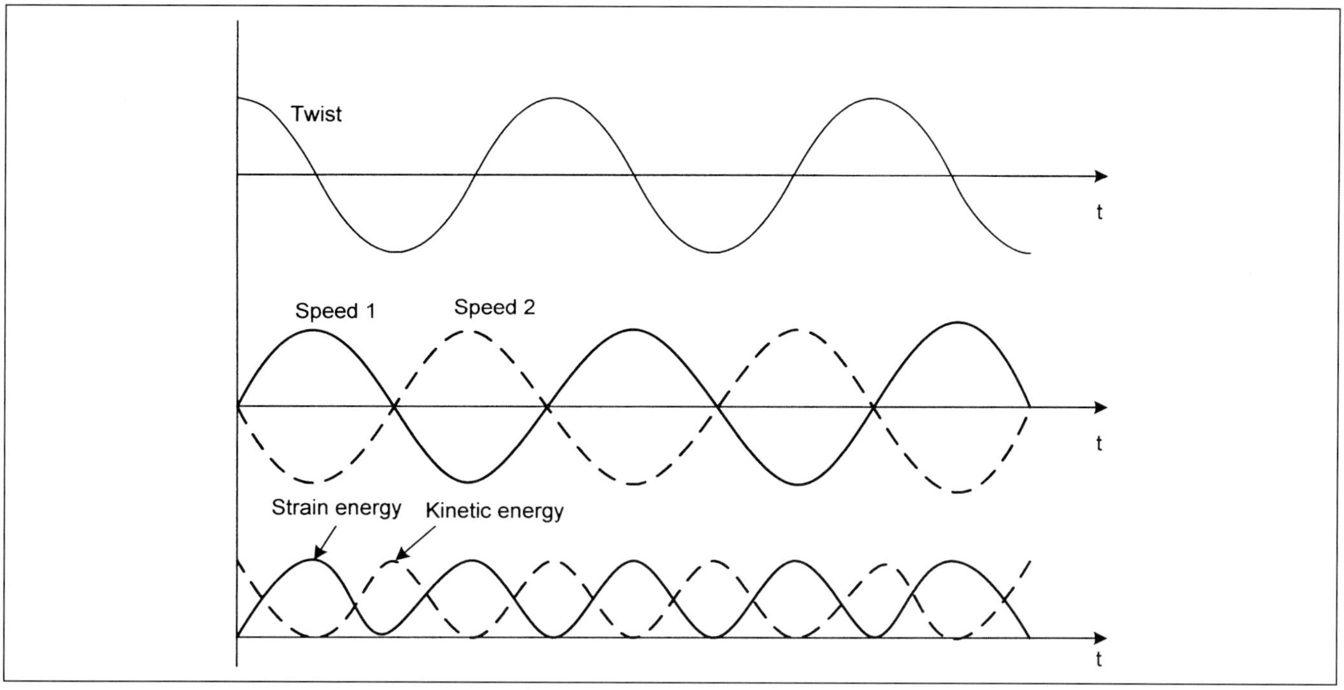

FIGURE 16-28 Vibrating torsional system showing shaft twist, inertia speeds, and energies versus time.

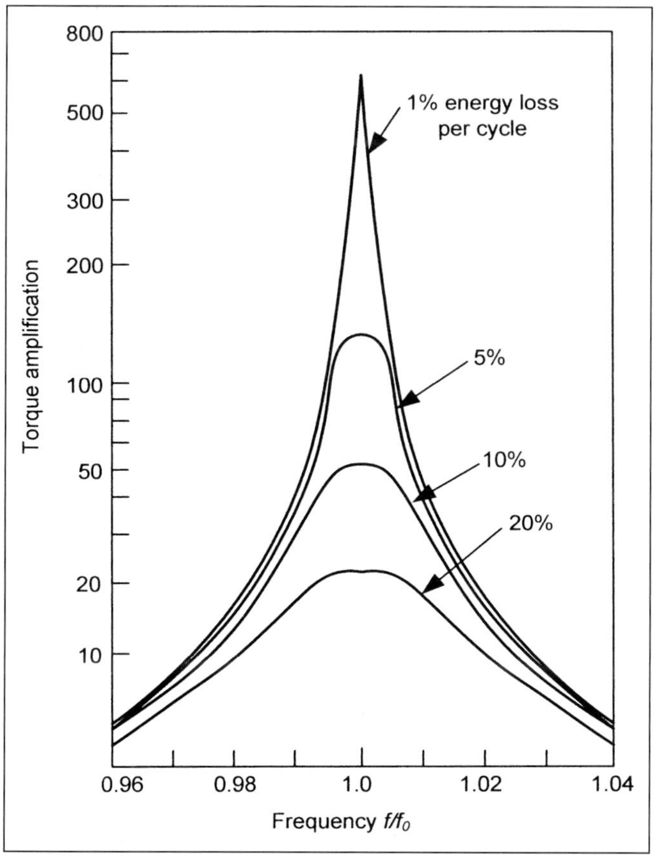

FIGURE 16-29 Amplification curves of the resonant system.

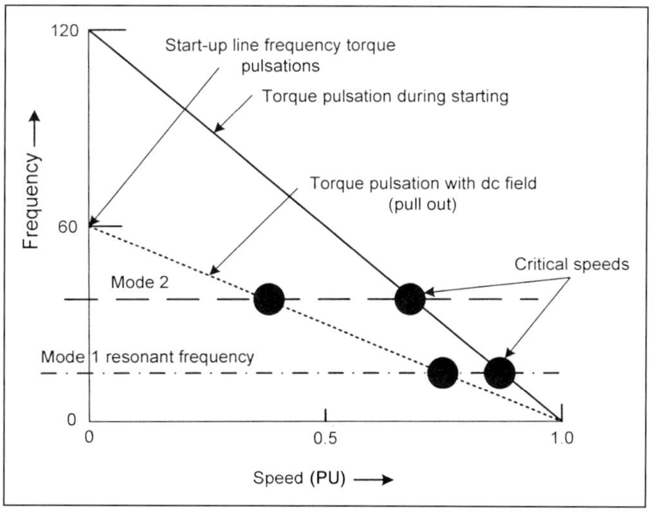

FIGURE 16-30 Torque pulsation frequency characteristics of synchronous motor during starting and following overload (with dc field).

These are due to imperfections in the mechanical system and generally of smaller magnitude. Excitations can also occur from the load system. A mechanical jam may produce severe dynamic torques.

It may be difficult to totally avoid some amplification of the torques during starting, however, it should show damping as the drive train quickly passes through the vibration mode. Figure 16-31 shows the results of a torsional analysis of a salient-pole synchronous refiner motor of 30000-hp, 13.8-kV, 1.0-pf motor during

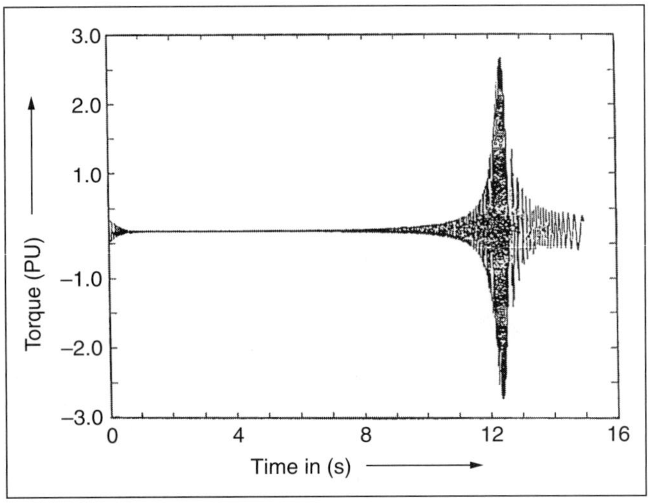

FIGURE 16-31 Torsional vibration analysis results of a 30000-hp synchronous motor driving a pulp mill refiner.

start-up—a system model with seven inertias and six spring constants is used in the analysis. Maximum shaft torque and minimum shaft torque during starting are 2.8 and −2.5 pu, respectively. A fatigue analysis of the shaft can also be made.[17,18,19]

The torsional analysis requires a host of motor and load data, starting characteristics, inertias, and spring constants. These are summarized in Table 16-4. A torsional analysis may discover many natural frequencies of the system.

16-6-3 Steam Turbines of Generating Plant

Full- or partial-load rejection or overloading of generators gives rise to overfrequency and underfrequency operation. For example, on clearing a system fault, the complete load may be shed, and this will cause the turbine to overspeed and frequency to rise. Assuming 5 percent governor droop characteristics, a 50 percent load rejection will result in approximately 2.5 percent rise in the frequency. This condition does not last long as the control action will quickly restore speed. On the other hand, overloading can be caused due to many reasons, and enough loads may not be shed, giving rise to a prolonged operating at underfrequency.

Twofold effects occur on the generator and turbine, and turbine is considered to be more restrictive. The rating of a generator is reduced when operating at a lower frequency, and prolonged operation on overload at a reduced frequency may result in generator damage if its short-time thermal capability is exceeded.

The turbine is more restrictive because of possible mechanical resonance that we have discussed above, due to the mechanical resonance frequency coinciding or nearing the reduced system frequency. These exciting frequencies may dramatically increase the vibratory stresses, resulting in damaging or cracking of some part of the turbine blade structure. Turbine manufacturers provide time limits for abnormal frequency operation, and Fig. 16-32 is a composite curve for five manufacturers[20] for general guidelines, though specific information for a unit is readily available with the manufacturer. The abnormal frequency operation varies considerably with the manufacturers. The speed limits are rather tight, and the resonant frequencies lie fairly close to the supply system frequency. The protection is provided by coordinating settings on frequency relays with appropriate time delays to trip the generating units.

16-6-4 Analysis

The *n*-spring-connected rotating masses can be described by the equation:

$$J\frac{d\bar{\omega}_m}{dt}+\bar{D}\bar{\omega}_m+\bar{H}\bar{\theta}_m=\bar{T}_{\text{turbine}}-\bar{T}_{\text{generator}} \tag{16-38}$$

$\bar{H}$ is the tri-diagonal matrix of stiffness coefficients, $\bar{\theta}_m$ is the vector of angular positions, $\bar{\omega}_m$ is the vector of mechanical speeds, $\bar{T}_{\text{turbine}}$ is the vector of torques applied to turbine stages, and $\bar{D}$ is the tri-diagonal matrix of damping coefficients. The moment of inertia and damping coefficients are available from design data. The spring action creates a torque proportional to the angle twist.

Figure 16-33*a* shows a torsional system model for the steam turbine generator. The masses will rotate at one or more of the turbine mechanical natural frequencies called *torsional mode frequencies*. When the mechanical system oscillates under such steady state at one or more natural frequencies, the relative amplitude and phase

TABLE 16-4 Data Required for Torsional Vibration Analysis

PARAMETER	DESCRIPTION
M_m (kg-m)	Maximum transient shaft torque during starting
M_s (kg-m)	Breakaway torque refiner
F_1 (kg)	Transferred thrust load to motor at zero-end gap in thrust bearings, both directions
P_1 (kW)	Power loss in the refiner during idling
Critical damping (%)	Critical damping in the shaft system
Fatigue analysis	Data includes shaft diameter, speed ratio, material, shear stress, and stress concentration factor due to step change in shaft diameter
J_1, J_2, J_3, J_4 (kg-m²/lb-ft²)	Rotating inertias
K_1, K_2, K_3 (N-m/rad/lb-in/rad)	Spring constants
Motor	Starting speed torque characteristics, average and oscillating torques, effect of variation of system voltage and starting conditions
Load	Starting torque speed characteristics

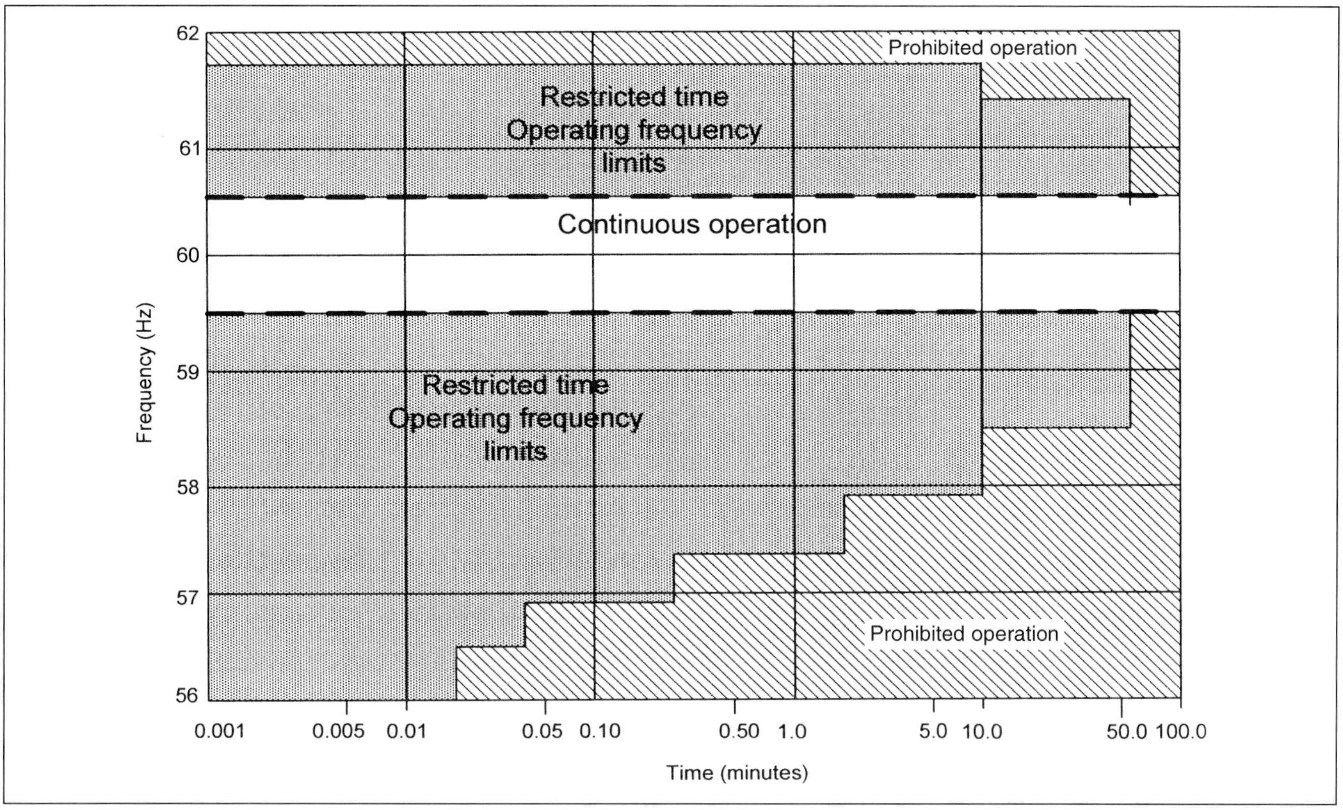

FIGURE 16-32 Abnormal frequency tolerance limits of steam-turbine generators, consolidated curves of five manufacturers. *Source: ANSI/IEEE standard C37.106.*[20]

of individual turbine-rotor elements are fixed and are called *mode shapes* of torsional motion (Fig. 16-33*b*).[21]

Torsional mode damping quantifies the decay of torsional oscillations. The ratio of natural log of the successive peaks of oscillation is called *logarithmic decrement*. The decrement factor is defined as the time in seconds to decay from the original point to $1/e$ of its value.

The modal spring-mass model is a mathematical representation of Fig. 16-33*a* for oscillation in mode n, given by:

$$\begin{vmatrix} J_1 & & & \\ & J_2 & & \\ & & \cdot & \\ & & & J_n \end{vmatrix} \begin{vmatrix} \ddot{\theta}_1 \\ \ddot{\theta}_2 \\ \\ \ddot{\theta}_n \end{vmatrix} + \begin{vmatrix} K_{12} & -K_{12} & & & \\ -K_{12} & K_{12} \; K_{23} & -K_{23} & & \\ & -K_{23} & \cdot & \cdot & -K_{n1,n} \\ & & \cdot & \cdot & \\ & & & -K_{n1,n} & K_{n1,n} \end{vmatrix} \begin{vmatrix} \theta_1 \\ \theta_2 \\ \cdot \\ \theta_n \end{vmatrix} = \begin{vmatrix} T_1 \\ T_2 \\ \cdot \\ T_n \end{vmatrix} \tag{16-39}$$

The derivation follows from eigenvectors and frequencies of the spring-mass model. It is seen there are n second-order differential equations of motion for an n mass model and coupled to one another by spring elements.

Diagonalization of the stiffness term would yield n decoupled equations called the *modal-spring mass models*. This diagonalization can be accomplished by coordinate transformation from a reference frame in the rotors to a reference frame of the eigenvectors. (See Example 15-2.)

16-7 OUT-OF-PHASE SYNCHRONIZATION

Out-of-phase synchronization can occur in power stations due to human error and failure of autosynchronizing and synchronizing check relays. We discussed some aspects of it in Chap. 10. Here, it can be shown that resulting fault current due to out-of-phase synchronization will have delayed current zeros. These should not be confused with the delayed current zeros discussed in Chap. 10, due to high X/R ratio of large generators and certain combination of generator reactances and time constants.

The problem can be significant for generators with a high subtransient current component, which means high X'_d/X_d and small T_d and a large dc time constant T_a. During an out-of-phase synchronizing, currents in the armature windings may exhibit delayed zeros. The inertia constant of the generator influences these phenomena. Whether it will lead to a problem of generator circuit breaker operation depends upon the protective relay settings—in some plants circuit breaker tripping may be blocked on out-of-phase synchronizing.

Example 16-4 EMTP analysis of 18-kV, 200-MVA generator of the following specifications, on out-of-phase synchronizing is demonstrated:

$$X_d = 1.78 \quad X'_d = 0.215 \quad X''_d = 0.145 \quad T'_d = 0.69\text{s} \quad T''_d = 0.015$$

$$X_q = 1.75 \quad X'_q = 0.35 \quad X''_q = 0.145 \quad T'_q = 0.148\text{s} \quad T''_q = 0.015\text{s}$$

$$r_a = 0.0012 \quad X_a = 0.12 \quad H = 5.62$$

All reactance and resistance values are in per unit. The generator is provided with a voltage regulator, a PSS, and a governor. The generator is connected through a step-up transformer to a 230-kV system. The transients on out-of-phase closing of the high-side circuit breaker are depicted in Fig. 16-34, phase currents for an out-of-phase closing angle of 120°. The current zero is not obtained in phase a.

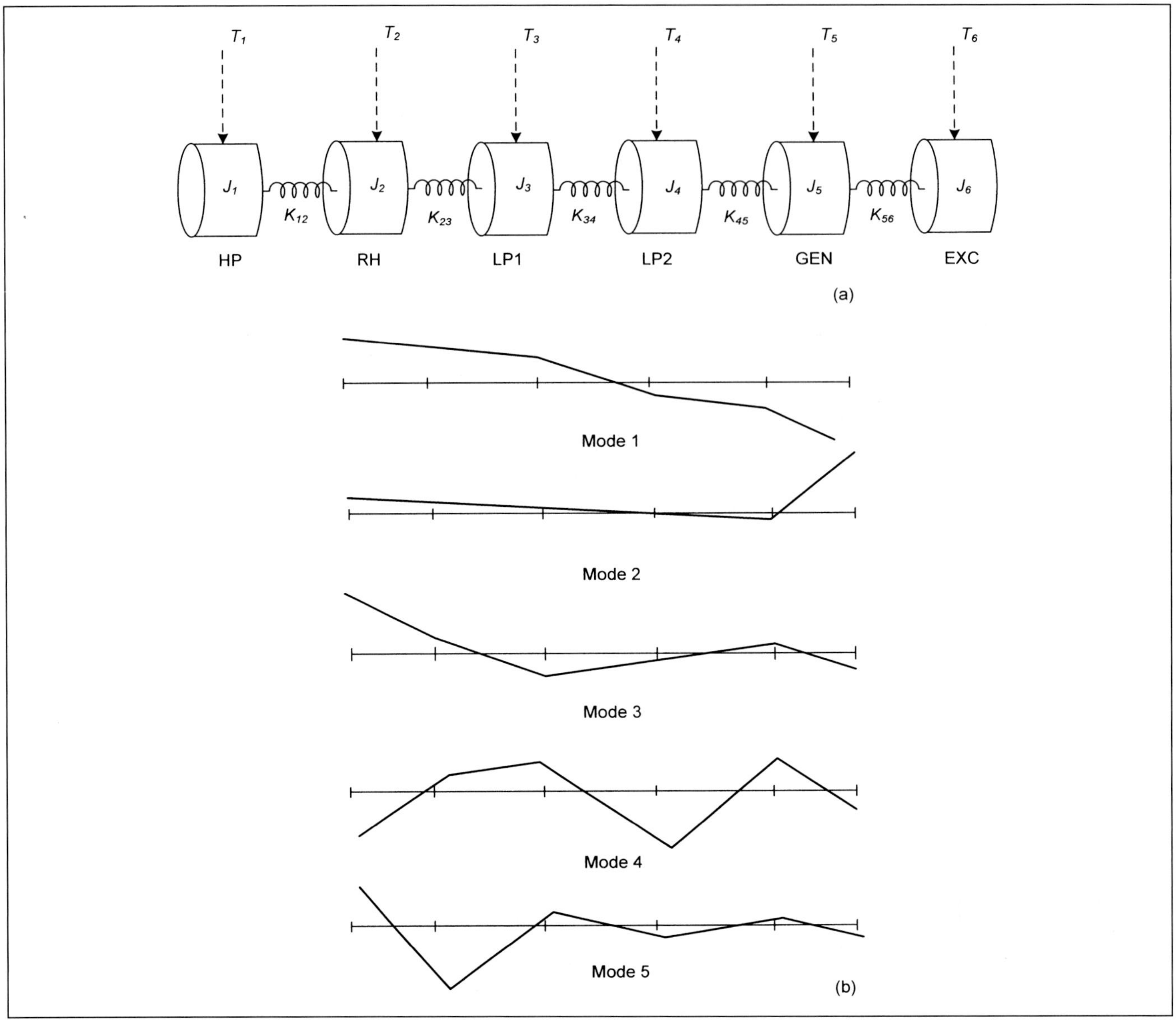

FIGURE 16-33 (*a*) Rotating mass model of steam turbine generator. (*b*) Oscillation modes.

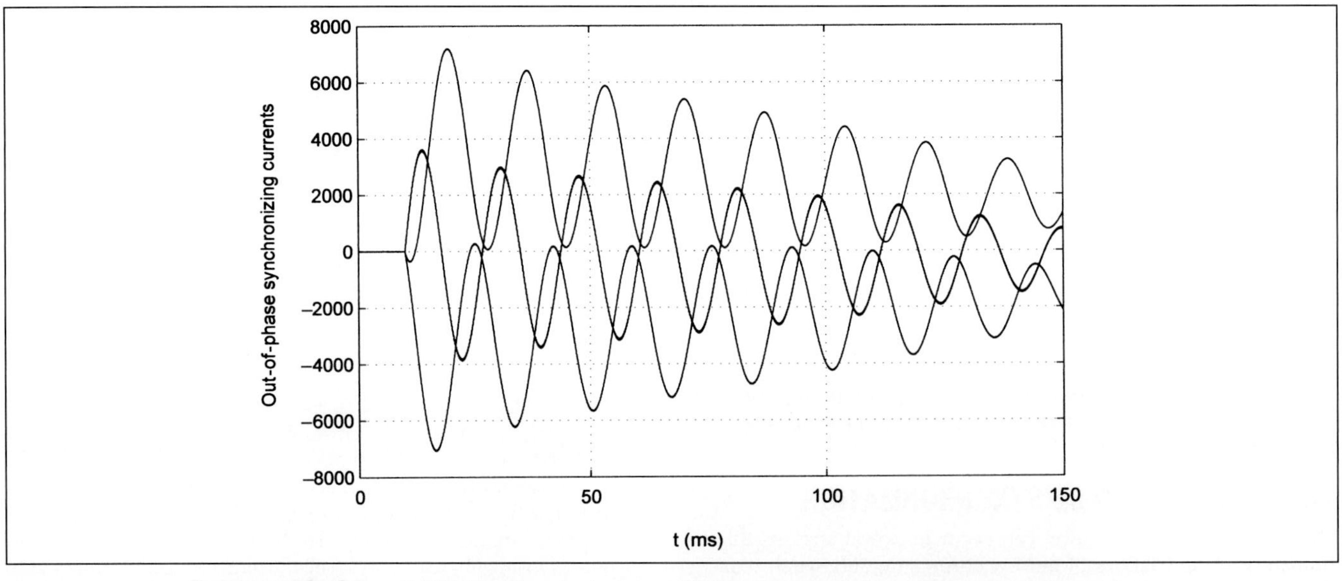

FIGURE 16-34 EMTP simulation of out-of-phase closing. Current zero is not obtained in phase *a* current.

PROBLEMS

1. A 2.3-kV motor load of 5 MW is connected to 13.8-2.4 kV, 10-MVA transformer of percentage impedance of 9.5 percent on a 10-MVA base. The cyclic starting and stopping of the motors gives rise to unacceptable flicker. List all the means of reducing the flicker, with their relative advantages and disadvantages and cost effectiveness. Consider that the starting current of the composite induction motor loads is six times the full-load current. Consider a combined operating power factor of 0.92 and motor efficiency of 94 percent.

2. Explain why a STATCOM is more effective in controlling flicker as compared to an equivalent SVC.

3. A total furnace load of 50 MVA is connected to a system having a short-circuit impedance of $0.01 + j0.1$ pu on 100-MVA base. Can flicker be a problem.

4. Calculate the phase angle at the time of switching of an induction motor on restoration of the supply voltage, given that supply system voltage = 1.0 pu, the residual motor voltage at the time of reswitching = 0.8 pu, and the resultant voltage vector = 1.5 pu. Is the statement that an induction motor residual voltage and fluxes decay approximately at the same rate correct? Explain the reasons.

5. An auxiliary distribution system at medium voltage, having two sources of power in a system configuration as shown in Fig. 16-7*a*, serves generator auxiliary loads, mainly consisting of rotating induction motors and some static loads. What is the appropriate choice for an autobus transfer scheme?

6. Why are the torsional vibration frequencies in steam turbines close to the 60-Hz operating frequency? Can these frequencies be placed much away from the operating frequency (see Fig. 16-32)?

7. Describe the forced excitation types for torsional vibrations. Which of these modes can give rise to a greater possibility of torsional vibration resonance?

8. An induction motor has four poles and 24 slots. Which harmonics will predominate for harmonic induction torques?

9. Distinguish between harmonic induction torques and harmonic synchronous torques of an induction motor.

10. The three-phase power supply to an induction motor contains predominant fifth and seventh harmonics. What can be expected with respect to performance and starting of the induction motor?

11. Why it is necessary that S_2, the rotor slots in an induction motor, should not exceed approximately 50 to 60 percent of the stator slots?

12. Consider a 1000-hp, four-pole, 4-kV synchronous motor in which P_r = two times the motor rating. The motor and load inertia is 5000 lb-ft^2. Calculate:

- The compressor factor X
- The natural frequency of oscillation
- Accelerating time constant

13. A three-phase synchronous motor of 2000 hp, 4.0 kV, and 60 Hz is connected to a stiff electrical system. It has a moment of inertia of 9000 kg-m^2, synchronizing coefficient of 500 kW/electrical rad, and damping torque of 2000 N-m/electrical rad/s. The motor is driving a reciprocating compressor, with a harmonic torque of 1500 N-m at angular frequency of 2 rad/s. Calculate maximum deviation in load angle and pulsation of synchronizing power.

14. Calculate the natural frequency of oscillation of a 1500-hp, six-pole, synchronous motor operating at a power angle of 30° at full load. Total moment of inertia of rotating parts is 2000 kg-m^2.

REFERENCES

1. IEEE Std. 519, Recommended Practice and Requirements for Harmonic Control in Electrical Systems, 1992.
2. IEEE Std. 1453, Recommended Practice for Measurements and Limits of Voltage Fluctuations and Associated Light Flicker on AC Power Systems, 2004.
3. IEC 61000-4-15, Electromagnetic Compatibility (EMC)—Part 4: Testing and Measurement Techniques—Section 15: Flicker Meter: Functional and Design Specifications, 2003.
4. IEC 61000-3-3, Electromagnetic Compatibility (EMC)—Part 3: Limits—Section 3: Limitation of Voltage Changes, Voltage Fluctuations and Flicker, in Public Low-Voltage Supply Systems, for Equipment with Rated Current ≤ 16A per Phase and Not Subject to Conditional Connection, 2008.
5. IEC-61000-3-11, Electromagnetic Compatibility (EMC)—Part 3–11, Limits: Limitation of Voltage Fluctuations and Flicker in Low-Voltage Power Supply Systems for Equipment with Rated Current < = 75A, 2000.
6. IEC 61000-3-8, Electromagnetic Compatibility (EMC)—Part 3: Limits—Section 8: Signalling on Low-Voltage Electrical Installations—Emission Levels, Frequency Bands and Electromechanical Disturbance Levels, 1997.
7. S. M. Halpin and V. Singhvi, "Limits for Interharmonics in the 1-100 Hz Range Based Upon Lamp Flicker Considerations," *IEEE Trans. Power Delivery*, vol. 22, no. 1, pp. 270–276, Jan. 2007.
8. C. D. Schauder and L. Gyugyi, "STATCOM for Electric Arc Furnace Compensation," EPRI Workshop, Palo Alto, CA, 1995.
9. M. Routimo, A. Makinen, M. Salo, R. Seesvuori, J. Kiviranta, and H. Tuusa, "Flicker Mitigation with Hybrid Compensator," *IEEE Trans. Industry Applications*, vol. 44, no. 4, pp. 1227–1238, Jul./Aug. 2008.
10. J. C. Das, "Effects of Momentary Voltage Dips on the Operation of Induction and Synchronous Motors," *IEEE Trans. Industry Applications*, vol. 26, no. 4, pp. 771–781, Jul./Aug. 1990.
11. *Westinghouse Electrical Transmission and Distribution Reference Book*, Chapter 22: Flicker, Westinghouse Electric Corporation, East Pittsburg, PA, 1964.
12. D. L. Hornak and D. W. Zipse, "Automated Bus Transfer Control for Critical Industrial Processes," *IEEE Trans. Industry Applications*, vol. 27, no. 5, pp. 862–871, Sept./Oct. 1991.
13. H. Mokhtari, S. B. Dewan, and M. R. Irvani, "Performance Evaluation of Thyristor Based Static Transfer Switch," *IEEE Trans. Power Delivery*, vol. 15, no. 3, pp. 960–966, Jul. 2000.
14. H. Mokhtari, S. B. Dewan, and M. R. Irvani, "Effect of Regenerative Load on Static Transfer Switch Performance," *IEEE Trans. Power Delivery*, vol. 16, no. 4, pp. 619–624, Oct. 2001.

15. C. Meyer and R. W. De Doncker, "Solid-State Circuit Breaker Based on Active Thyristor Topologies," *IEEE Trans. Power Electron.*, vol. 21, no. 2, pp. 450–458, Mar. 2006.
16. J. C. Das, "Oscillations Due to Synchronous Motors Driving Reciprocating Compressors," *Journal of Institution of Engineers*, India, vol. 63, EL 4, pp. 185–189, Feb. 1983.
17. E. L. Owen, "Torsional Coordination of High Speed Synchronous Motors, Part 1," *IEEE Trans. Industry Applications*, vol. 17, pp. 567–571, Nov./Dec. 1981.
18. E. L. Owen, H. D. Snively, and T. A. Lipo, "Torsional Coordination of High Speed Synchronous Motors, Part 2," *IEEE Trans. Industry Applications*, vol. 17, pp. 572–580, Nov./Dec. 1981.
19. C. B. Meyers, "Torsional Vibration Problems and Analysis of Cement Industry Drives," *IEEE Trans. Industry Applications*, vol. 17, no. 1, pp. 81–89, Jan./Feb. 1981.
20. ANSI/IEEE C37.106, IEEE Guide for Abnormal Frequency Protection for Power Generating Plants, 1987.
21. IEEE Subsynchronous Resonance WG, "Terms, Definitions and Symbols for Subsynchronous Oscillations," *IEEE Trans. PAS*, vol. PAS-104, no. 6, pp. 1326–1334, Jun. 1985.

FURTHER READING

H. Akagi and A. Nabe, "The p-q Theory in Three-Phase Systems Under Non-Sinusoidal Conditions," *ETEP*, vol. 3, pp. 27–30, 1993.

P. Ballensperger and H. Meyer, "Overvoltages Resulting from Disconnection of High-Voltage Motors," *BBC Review*, vol. 40, no. 9, pp. 342–349, 1953.

I. M. Canay, D. Braun, and G. S. Koppl, "Delayed Current Zeros Due to Out-of-Phase Synchronizing," *IEEE Trans. EC*, vol. 13, no. 2, pp. 124–132, Jun. 1998.

P. T. Cheng and Y. H. Chen, "Design of an Impulse Commutated Bridge for the Solid-State Transfer Switch," *IEEE Trans. Industry Applications*, vol. 44, no. 4, pp. 1249–1258, Jul./Aug. 2008.

J. M. Daly, "Load Transfer Strategies for Machine and Other Inrush Loads," *IEEE Trans. Industry Applications*, vol. 34, no. 6, pp. 1404–1410, Nov./Dec. 1998.

J. C. Das, "Applications of Synchronous Motors in Pulp and Paper Industry," TAPPI Joint Conference Process Control, Electrical and Information, pp. 165–190, Vancouver, BC, Canada, Mar. 1998.

S. Kreitzer, J. Obermeyer, and R. Mistry, "The Effects of Structural and Localized Resonances on Induction Motor Performance," *IEEE Trans. Industry Applications*, vol. 44, no. 5, pp. 1367–1375, Sept./Oct. 2008.

D. G. Lewis and W. D. Marsh, "Transfer of Steam-Electric Generating Station Auxiliary Bus," *Trans. AIEEE*, pp. 322–331, Jun. 1955.

S. R. Mendis, M. T. Bishop, and J. F. Witte, "Investigations of Voltage Flicker in Electric Arc Furnace Systems," *IEEE Industry Magazine*, no. 2, pp. 34–38, 1986.

C. D. Schauder and L. Gyugyi, "STATCOM for Electric Arc Furnace Compensation," EPRI Workshop, 1995.

E. S. Thomas, "Forgoing Flicker," *IEEE Industry Magazine*, vol. 13, pp. 34–41, Jul./Aug. 2007.

W. T. Thomson, *Theory of Vibration with Applications*, 3rd ed., Prentice Hall, Englewood Cliffs, NJ, 1988.

UIE (International Union for Electricity Applications) WG Disturbances, *Flicker Measurements and Evaluation*, Revised, 2d ed., 1991, www.uie.org.

K. G. Williams and A. Khayer, "Auto-Transfer of Induction Motors in Industrial Systems," *Electrical Review*, pp. 395–397, Mar. 1968.

C. C. Young and J. Dunki-Jacobs, "The Concept of In-Phase Transfer Applied to Industrial Systems Serving Essential Service Motors" *AIEEE Trans.*, pp. 508–516, Jan. 1961.

CHAPTER 17

INSULATION COORDINATION

The electrical insulation is an important element of the electrical apparatuses, and its long-term integrity is directly related to the performance integrity. Probable life expectancy is assigned to electrical equipment, based on the deterioration and aging of the insulation and field experience. As an example, the life expectancy of a power transformer is 20 to 25 years, though, practically much older transformers are in service. Much has been written about the breakdown phenomena in insulation, and the research continues as better insulation materials are developed.

17-1 INSULATING MATERIALS

Development of solid dielectric cables up to 500 kV is discussed in Chap. 4. The advent of thermosetting resins (which tend to set and harden on application of heat) compared to thermoplastic resins (which melt on application of heat and solidify on removal of heat, e.g., polyvinyl chloride, PVC) raised the temperature of solid insulation materials in usage. Usage of epoxy resins in liquid and solid form was a further development which revolutionized the insulation systems of the rotating machines, switchgear, and enclosed busbars. Polytetrafluoroethelene (PTFE), commonly known as Teflon, is chemically inert, better than gold and platinum, and can be used at high temperature. It can be molded into various intricate shapes as an insulating material, does not require lubrication, and can be machined.

The types of insulation systems vary widely depending upon their usage and application—gases dielectrics for circuit breakers, mineral oil and solid insulation systems for cables, mineral oil and less flammable liquids for transformer insulations, porcelain and glass-epoxy materials for insulators, and SF_6 as an insulating medium, as discussed in Chap. 8. Air itself is an insulating medium. The mechanisms of breakdown in each of these mediums and physical and chemical properties vary. Two types of insulations are:

- *Self-restoring insulation.* The electrical power systems which use air as an insulating medium for external insulation are called the self-restoring insulation systems. The breakdown in air is strongly dependent on gap configuration, shape, and polarity of the surge and the ambient conditions. In an outdoor environment, the effects of rain, fog, humidity, and pollution become important. For GIS systems, the effects of temperature, pressure, and internal irregularities play an important role. Failure of air insulation is self-healing, as removal of overvoltage will restore the insulation, though not in every case of overvoltage. It is therefore, acceptable to tolerate a small probability of insulation failure to minimize the cost. Probabilistic methods are, therefore, applied for this type of insulation. We can say that the methodology tends to optimize the cost versus failure rate, though it is not so easy to accomplish this objective due to incomplete data of the systems and extensive modeling that may be required.

- *Non-selfrestoring insulation.* An insulation failure can cause a permanent fault and damage. This is true of solid dielectric materials. The mechanical stresses also impact the insulation strength. Degradation of insulation tends to increase with time. Any flashover is undesirable and unacceptable. The insulation characteristics must be selected with zero percent chance of flashover with some margin of safety. Practically, the cost of equipment and voltage is an important factor. An 800-MVA transformer will receive a more thorough analysis compared to a 500-kVA distribution transformer in meeting these criteria.

17-2 ATMOSPHERIC EFFECTS AND POLLUTION

Pollution and atmospheric conditions have a profound impact on external insulation.[1] Rain together with pollution can drastically reduce the insulation strength. Fog, dew formation, and light rain with contamination are the worst conditions. The design of line or substation insulation is based on wet conditions. For substation insulation, BSL is defined for wet conditions, while BIL is tested under dry conditions. It is assumed that the rain will not severely degrade the insulation strength and wet BIL = dry BIL. This may not be entirely true. For line insulation CFO, dry CFO is obtained from parametric tests and wet CFO is obtained by multiplying dry CFO with an experimentally determined correction factors.

For self-restoring insulation, the probability of flashover can be defined by an insulation strength characteristic curve modeled by a cumulative Gaussian distribution function (App. F), considered valid at least four standard deviations below CFO. The *statistical* BIL or BSL, defined as the 10 percent probability of flashover, is given by equation:

$$\text{BIL or BSL} = \text{CFO}\left(1 - 1.28\frac{\sigma}{\text{CFO}}\right) \tag{17-1}$$

where σ/CFO is the coefficient of variation. The truncation point is usually neglected. For lightning impulses, this coefficient is about 3 percent, and for switching impulses for line and tower insulation, this is assumed to be between 6 and 7 percent.

TABLE 17-1 Contamination Site Severity[2,3]

Site Severity	ESSD (mg/cm²) CIGRE	IEEE
None	0.0075–0.015	
Very light	0.015–0.03	0.03
Light	0.03–0.06	0.03–0.06
Average/moderate	0.06–0.12	0.06–0.10
Heavy	0.12–0.24	>0.10
Very heavy	0.24–0.48	
Exceptional	>0.48	

17-2-1 Contamination Severity

This is divided into two general classes:

- *Industrial*. The industrial contamination is caused by particles driven by wind and deposited on insulators surfaces. The contamination may be dust, cement, fly ash, limestone, and the like. These materials contain salt and form a conducting layer when wetted. The contamination is specified in terms of milligrams per square centimeter. To standardize industrial contamination, the equivalent salt deposit density (ESSD) is used. The *surface conductivity* is defined as the ratio of the power frequency current over a sample insulator to the applied voltage.

- *Sea*. Saltwater spray driven by wind is specified in terms of grams per liter or in kilograms per cubic meter of water.

Table 17-1 lists ESSD according to IEEE[3] and CIGRE[2]

17-2-2 Insulation Strength of Insulators

The design of air-insulated substation insulation is universally based on rain conditions. According to IEEE recommendations,[3] the insulation strength is defined as ratio of withstand voltage and connecting length. The withstand voltage is determined by clean fog test method. Table 17-2 gives the standard insulators in a string for voltages from 138 to 765 kV. The standard suspension insulator has a leakage distance of 292 mm (11.5 in).

IEC recommendations[4] describe the creepage distance for ceramic or glass insulators for different contamination severities. The specific leakage distance L_c is determined by different contamination test methods. The recommendation does not cover some environmental situations like ice in heavy pollution, heavy rain, arid areas, and the like. Table 17-3[4] shows the recommendations of minimum specific creepage distance in millimeteres per kilovolt.

TABLE 17-3 IEC-Recommended Creepage Distances[4]

Pollution Level	Minimum Specific Creepage Distance (mm/kV)
I: Light	27.7
II: Medium	34.6
III: Heavy	43.3
IV: Very heavy	53.7

Methods to determine contamination are:

English salt fog method

The contamination and wetting are applied simultaneously by spraying saltwater, and the degree of contamination is defined by the salt in the sprayed solution. The insulator surface is purely resistive during the test period of 1 hr:

$$L_C = 2.34 S_a^{0.224} = 7.09\ \mathrm{ESSD}^{0.224} \qquad (17\text{-}2)$$

where S_a is the amount of salt in water used for test in kilograms per cubic meter.

Wet contamination test

The test may be done with "light mixture" (contaminant defined by the amount of dry contaminant and the surface has a resistive-capacitive impedance) or "heavy mixture" (the contamination defined by surface conductivity).

German Kieselghur method

$$L_C = 1.42\sigma_w^{0.387} = 8.15\ \mathrm{ESSD}^{0.387} \qquad (17\text{-}3)$$

where σ_w is the insulatsor surface conductivity in μs.

Clean fog test

The test uses a slow or quick wetting process, and the insulator surface changes from dry to wet. In case of slow wetting, the surface is capacitive-resistive.

TABLE 17-2 Number of Standard Insulator Units (5.75 in × 10 in) (146 mm × 254 mm)[3]

System Voltage	Number of Standard Units Required for Contamination Severity I-Strings/V-Strings: Very Light	Light	Moderate	Heavy
138	6/6	8/7	9/7	11/8
161	7/7	10/8	11/9	13/10
230	11/10	14/12	16/13	19/15
345	16/15	21/17	24/19	29/22
500	25/22	32/27	37/29	44/33
765	36/22	47/39	53/42	64/48

Leakage distance = 11.5 in (292 mm) for the standard insulator unit

Japanese fog withstand method

$$L_C = 7.14\,\mathrm{ESSD}^{0.246} \qquad (17\text{-}4)$$

Some contamination countermeasures are:

- To increase the number of insulators, as discussed above, is the most common method.
- To use fog-type insulators, which may have a leakage distance equal to 150 percent of the standard insulators; though increasing the leakage distance is not the only criterion for satisfactory performance. Leakage distance to length in ratios of 2.9 to 4.5 are available.
- Nonceramic insulators have hydrophobic surfaces, which generally improve performance. The withstand voltage per connected length improves by 40 to 100 percent.
- The insulator's surface at room temperature may be coated with vulcanized silicon rubber (RTV coatings), which produces a hydrophobic surface and improves performance.
- Application of silicon greases improves performance, but it has to be removed and re-applied periodically. The technique can be applied to sea-salt contaminations. Silicon greases, because of their temperature stability, can be used at higher temperatures. The greases envelope the pollution and prevent occurrence of a discharge.
- Washing and cleaning of deenergized insulators has been used. Washing techniques can be applied to energized insulators too.

Monitoring contamination is not so well established. Leakage current monitoring may be unreliable because it is dependent on the water present and the contamination. A contamination monitor was developed, which used ellipsometry to determine optical parameters of the contamination, and its thickness. However, without wavelength scanning, it lacked the ability to distinguish different types of contaminations. The alternative of ESSD measurement is laborious. Spectrophotometry to measure the thickness and refractive index has been suggested.

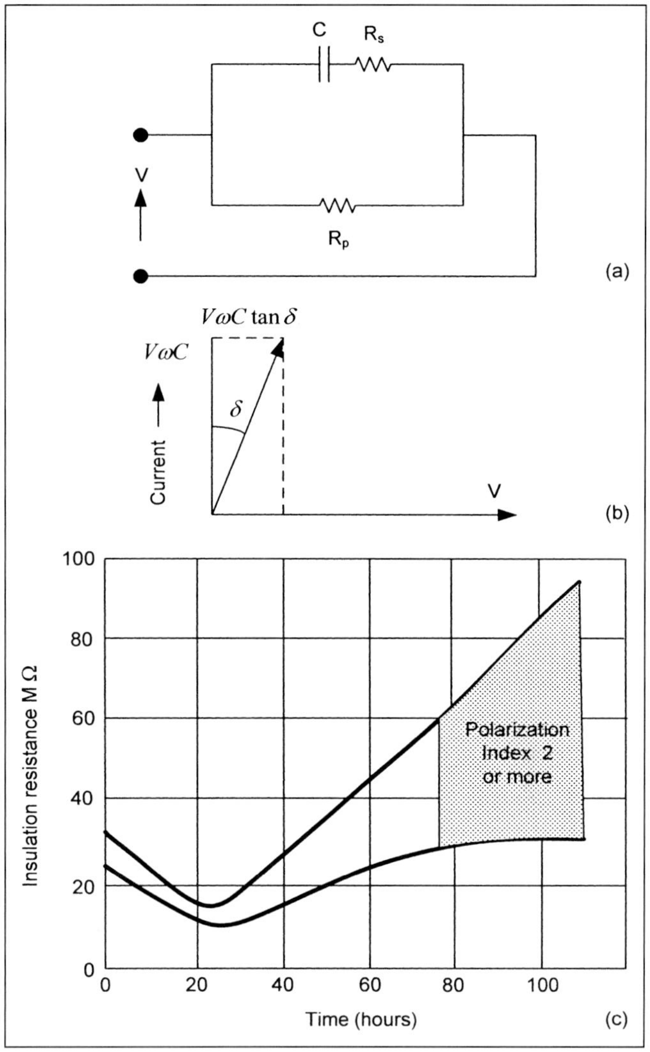

FIGURE 17-1 (*a*) Equivalent circuit of a leaky dielectric. (*b*) Phasor diagram. (*c*) Change in 1 minute and 10 minutes insulation resistance during drying of class B insulated armature winding; initial winding temperature 25° C, and final winding temperature 75° C.

17-3 DIELECTRICS

The term *dielectric* is synonymous with insulation. A dielectric or electrical insulation, when placed between two electrodes at different potentials, permits only a small negligible current to flow. The electrical resistivity may be of the order of 10^{20} to 10^{6} Ω-cm. Ideally, no current should flow through a perfect dielectric, but practically, an equivalent circuit can be drawn as shown in Fig. 17-1*a* and accompanying phasor diagram (Fig. 17-1*b*). For a perfect dielectric, the phase angle δ, called the *loss angle or dissipation factor*, should be zero. As the insulation deteriorates this angle increases, and measurements of the loss angle have been used to compare the deterioration over a period of time. We can write:

$$\begin{aligned} \delta &= \tan^{-1}(2\pi f C_s R_s) \quad &&\text{when } R_p = 0 \\ &= \tan^{-1}(1/2\pi f C_p R_p) \quad &&\text{when } R_s = 0 \end{aligned} \qquad (17\text{-}5)$$

On application of dc voltage, there is an inrush current (absorption current) which decays, and the conduction current remains, which will flow indefinitely. When the voltage is removed, a recovery current flows, much akin to a discharge of a capacitor. This model of a leaky capacitor is a simplified model of a dielectric.

Practically, the insulation resistance measurement at dc voltages of 500 to 5000 V is carried out, and the value increases as the duration of applied voltage increases. If the insulation is dirty or moist, the steady value will be reached quicker. The polarization index is the ratio of the 10-min resistance value to 1-min resistance value and is a rough indication of the condition of insulation, Fig. 17-1*c*.

For an insulating material, a host of mechanical, chemical, and aging characteristics must be considered, depending upon the insulation usage. These physical parameters, like heat resistance, thermal aging, ozone and corona resistance, moisture resistance, flame resistance and toxicity (for cable insulation systems), and surface contamination are as important as the electrical characteristics. Only the thermal aspect is touched upon.

17-3-1 Heat Resistance

The electrical and mechanical properties of the electrical insulating materials (EIM) are temperature dependent, thus, the standards specify the maximum temperature rise to which a specific EIM can be continuously subjected. The term electrical insulated systems (EIS) denotes one or more EIMs, with conducting parts. The limiting temperatures are based on consideration of reliability and useful life in service. As an example, the permissible temperature rises,

TABLE 17-4 Temperature Rise Limits for Machines with a 1.0 Service Factor[5]

Item	Machine Part	Method of Temperature Determination	Temperature Rise (°C)* Class of Insulation System A	B	F	H
a	Insulated windings					
	1. All horsepower (kW) ratings	Resistance	60	80	105	125
	2. 1500 hp or less	Embedded detector†	70	90	115	140
	3. Over 1500 hp					
	a. 7000 V or less	Embedded detector†	65	85	110	135
	b. Over 7000 V	Embedded detector†	60	80	105	125
b	The temperature attained by cores, squirrel cage windings, collector rings, and miscellaneous parts, such as brush holders, brushes, and the like, shall not injure the insulation in any respect.					

*For machines which operate under prevailing barometric pressure and which are designed not to exceed the specified temperature rise at altitudes from 1000 to 4000 m, the temperature rise, as checked by tests at low altitudes, shall be less than those listed by 1 percent of the specified temperature rise for each 100 m of altitude in excess of 1000 m.

†Embedded detectors are located in the slots of the machines and can be resistance elements or thermocouples.

for insulation of large induction motors, from NEMA[5] are shown in Table 17-4. The temperature rises are based on 40° C ambient. There is a certain hot-spot temperature of insulation, and the specified temperature rises in the standards are below the hottest-spot temperature by 10° to 15° C. The allowance between the allowable temperature rise and maximum hottest-spot temperature is determined by thermal analysis or calculations, substantiated by testing on prototype equipment. The concept of the hottest spot temperature is that the temperature rise throughout a machine or electrical apparatus will not be uniform, and the hottest-spot temperature may be embedded in a certain section of insulation, for example, in the interlayer slot windings of an induction motor or in the windings of a transformer. The electrical and mechanical properties will be influenced in different ways and to different degrees as a function of temperature and thermal aging. Some mechanisms influencing thermal aging are:

- Oxidation that can lead to molecular cross-linking
- Continuous molecular polymerization, which may increase physical and electrical strength first, but may subsequently lead to decreased flexibility, embitterment, and early failure under mechanical stress
- Hydrolytic degradation in which moisture reacts with insulation under the influence of heat
- Chemical breakdown of constituents
- Loss of volatile constituents

It can be safely said that much analysis, industry practice, and experience is behind the application of an insulation system to an electrical apparatus. Yet, the electrical insulations do break down due to electrical stresses, contamination, inappropriate application, surface contamination, tracking, treeing, and aging. When this happens, short circuit can occur, and it may involve phases and ground; the single-line-to-ground fault being the most common.

17-4 INSULATION BREAKDOWN

The insulation capability defines the maximum electrical stresses that it can withstand before there is a flashover, which may be of temporary nature, that is, overhead lines. The insulation capabilities are defined based on rigorous testing with impulse test waveforms. The breakdown of insulation is a function of the nature of dielectric, for example, air, oil, SF_6, and vacuums. The breakdown phenomena in various mediums have been widely studied, and it is not the intention to go into much details of it. A breakdown between two electrodes in any medium will occur if sufficiently high voltage is applied, which will result in an avalanche of charged particles between the electrodes. The *average* applied field, E_a, is simply the voltage divided by the gap distance between the electrodes, but the electrical field is rarely uniform. As the electrical field is increased, electrons emitted by cathode accelerate toward anode and collide with gas molecules to release more electrons, and the critical factor in establishing electron avalanche is the *gap distance*. For practical purposes, the developments in insulation systems are centered on decreasing the gap for a given applied voltage for the compactness and cost reductions of the electrical equipment.

Practically, the breakdown process is more complex. It is accelerated by formation of streamers, created by localized distortion of electrical field by space charge. These streamers appear at a certain number of free electrons in the cloud (10^8), and the increased electrical field in the space charge could cause further increase in energy and collisions. The critical point occurs when photons are emitted from the collisions in the steamers. The electrical resistance across the gap is reduced, and the current through this reduced resistance causes *more* heating, creating more ions. At this point an arc is formed between the electrodes, with very high temperatures and thermal agitation. Thus, to define insulation coordination it can be said that:

All voltages (switching, lightning, power frequency) that occur in practice and usage of an insulation system do *not initiate a gap breakdown*. The various factors impacting insulation breakdown are:

- Electrical field intensity and uniformity
- Polarity of asymmetrical gaps
- The type of insulating medium, that is, restorable gas insulation and nonrestorable gas insulation characteristics
- The type of electrical stress, that is, the time variation and duration of electrical field

Table 17-5 gives relative breakdown voltages of gaseous insulating mediums.

TABLE 17-5 Relative Breakdown Voltages for Different Gases

GAS TYPE	RELATIVE BREAKDOWN VOLTAGE
Air	1
Sulphur Hexafluoride, SF_6	2.2
Freon, CCl_2F_2, CF_3SF_3	2.4–3
Carbon tetrachloride, CCl_4	6

Some important factors in the insulation breakdown process can be itemized:

Effect of Voltage Polarity The effect of polarity is that a positive space charge is created by bombardment of molecules by electrons, and it alters the local electrical field density. The enhancement depends upon pre-existing nonuniformities. In a point-plane gap, the field near the plane is more uniform than near the point, and the space charge causes a negative-point plane gap to break at higher voltages than a positive point-plane gap; Fig. 17-2*a* and *b* clearly illustrate this.

When an electrode in air is positively charged, the breakdown voltage is *lower* than if the electrode is negatively charged. Practically, in the high-voltage gaps, the high-voltage conductor is more stressed and is more irregularly shaped. For the air-porcelain insulation, for example, tower or line insulation, positively charged impulses produce the lower insulation strength. For self-restoring insulation, the standards specify that the polarity which gives the lower insulation strength must be used to establish BIL and BSL. *Though not so explicitly stated, this polarity is universally positive.*

Effect of Gas Density The effect of gas density is shown in Fig. 17-3; the Paschen's curve is obtained by varying the nitrogen pressure at a constant temperature. As the pressure is reduced, the breakdown voltage decreases to a minimum. This can be explained that with decrease of pressure, the molecular separation increases and the electrons gain more energy before collision. From the Paschen minimum, further reduction of pressure causes an increase in breakdown voltage, and toward the vacuum, the gas itself becomes less important.[6] This very low-pressure region may be classified as breakdown in vacuum.

Breakdown in Vacuum Vacuum interruption is extensively used in circuit breakers and is briefly discussed here. Pressures bellow 10^{-3} mm of mercury are considered high vacuum. The charged particles are unlikely to cause collision with a residual gas molecule, thus ionization by collision is minimal. Electrons can, however,

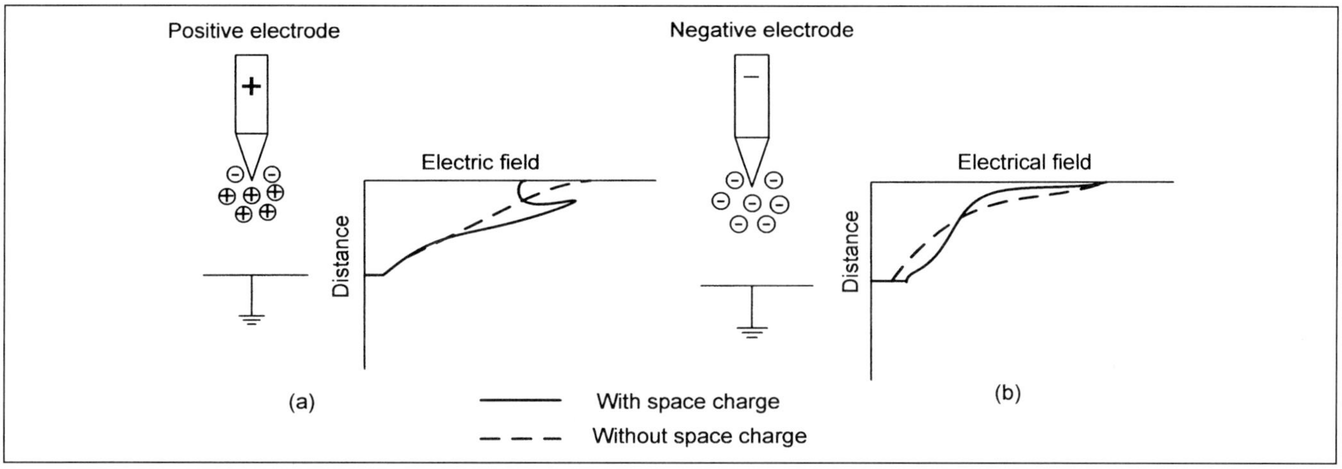

FIGURE 17-2 (*a*) Space charge builds up in a positive point-plane gap and its electric field strength. (*b*) Space charge builds up in a negative point-plane gap and its electric field strength.

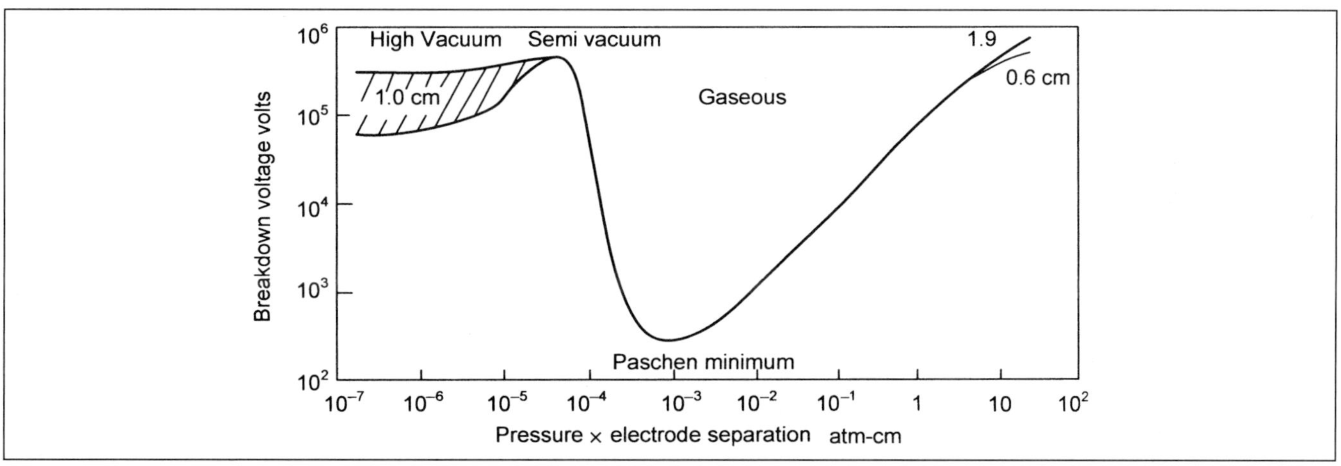

FIGURE 17-3 Paschens curve of pressure dependence of breakdown voltage of nitrogen.

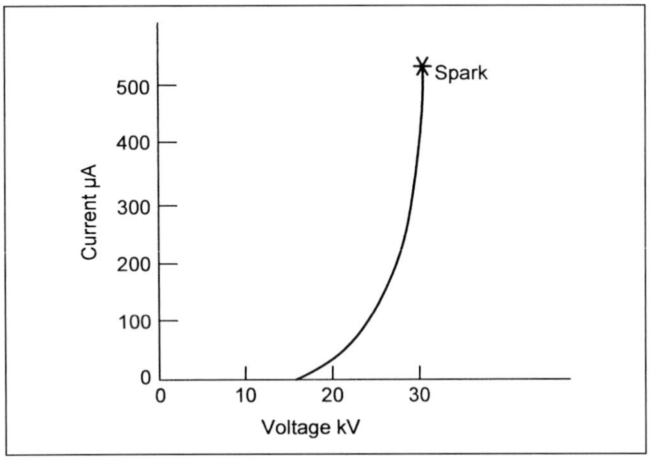

FIGURE 17-4 Breakdown in vacuum.

be emitted from metal surfaces by high electric field intensity, a phenomenon called field emission:

$$I = Ce^{-\frac{B}{E}} \tag{17-6}$$

where I is the emission current, E is the, the electrical field intensity, and B and C are the constants. As the contacts in vacuum separate, initially the gap is small and electrical intensity is high to cause electron emission, giving rise to a breakdown spark. The voltage not changing much, the current escalates. This phenomena is called *vacuum breakdown* or *vacuum spark* (Fig. 17-4). This shows the characteristics for 0.5-mm rough steel gaps. The nature of this characteristic depends upon surface conditions and material of electrodes.

Secondary emission takes place by bombardment of high-energy particles on surface of electrodes. The current leaves the electrodes from a few hot spots, and the current densities are high at these spots. The arc core temperatures are of the order of 6000 to 15000 K. At such high temperatures, emission takes place from electrode surfaces, and is called thermal emission. The creep of materials, occluded gases in the material, and the vacuum chamber material create special design problems in vacuum circuit breakers. The vacuum gap regains its dielectric strength at a rate of approximately 5 to 10 kV/μs. An arc cannot persist in ideal vacuum, and the separation of contacts carrying current causes vapor to be released from the contact material giving rise to plasma. The restrikes were common in the earlier designs of the vacuum breakers and much development has taken place in the contact materials to limit this phenomena.

Time Factor in Breakdown Process It is evident that the mechanism of breakdown involves movement of electrons and their ionization to form an avalanche action. Thus, the time is an important factor. The time to begin a breakdown is called the *statistical time lag* and is the time interval before the first electron becomes free to start an avalanche. Then there is *formative time lag*, which is the time taken by discharge to propagate through the gap, depending upon the electrical field intensity. The overall time to breakdown decreases as the voltage increases, which increases the field intensity for a given gap. This gives rise to *volt-time breakdown curve* discussed further.

Breakdown in Liquid Dielectrics The breakdown voltages in liquid dielectrics are, generally, higher than that in gases. This is due to the higher density of molecules in liquids. The breakdown in SF_6 is a special phenomena described in Chap. 8. The impurities in liquids, for example, dissolved gas bubbles, cause local field concentrations and partial bridging of gaps. Therefore, there are stringent requirements for purity of insulating fluids like transformer mineral oil, which also serves a cooling medium. In practical designs, the liquid volume in an apparatus, say a transformer, is determined by the cooling requirements and electrical clearances. The electrical stress is low, 5 percent of the breakdown value.

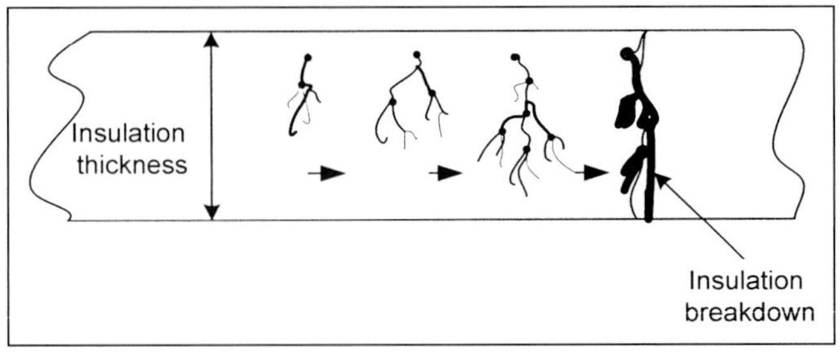

FIGURE 17-5 Progressive treeing phenomena in solid non-self-restoring insulation, leading to ultimate breakdown.

Breakdown in Solid Dielectrics In a uniform dielectric, the breakdown mechanism will be similar to that of a gas. The molecular spacing is much smaller, and critical level of acceleration between collisions can occur only over a narrow range of conditions. The electrical field stress causing the breakdown is defined as the intrinsic electrical strength. A probability of thermal runaway occurs because of increase in dielectric losses due to temperature.

The solid dielectrics are rarely uniform and perfectly homogeneous. An important mechanism of breakdown is the partial discharges that occur in the voids and cavities left in the dielectric during manufacture. These voids create localized areas of higher electrical stress. The progressive elongation of the voids, due to enlargement of voids, is the prominent cause of breakdown in polyethylene (PE) and cross-linked polyethylene (XLPE) cables, a phenomena known as *treeing* (Fig. 17-5). This figure shows progressive treeing, and a flashover will occur when the gap is bridged.

Surface tracking can cause long-term degradation through chemical changes, carbonization, and erosion. The electrical field strength is higher than average at the electrode/dielectric interface. The surface abrasions, contamination, and moisture can eventually lead to a complete breakdown.

17-4-1 Consequences of Insulation Breakdown

The consequences can be evaluated depending on the type of insulation. Important criteria are whether the insulation is self-healing or restorable or nonrestorable. A breakdown of solid dielectrics is unlikely to be restorable, as it will do permanent damage, that is, puncture of the insulation. Surface tracking (e.g., insulators) may be restorable. The liquid insulation may be contaminated after an electrical discharge (e.g., bulk oil and minimum oil-content circuit breakers), and full dielectric strength may not be recoverable. The number of fault current interruptions in a circuit breaker is limited, after which the medium must be tested and replenished. In essence, transmission line breakdowns can be qualified as restorable. All other insulations, regardless of type, are considered nonrestorable. The insulation system in an electrical apparatus may not be exclusively restorable or nonrestorable—the bushings on the top of a transformer tank and the transformer windings inside the tank.

Thus, the various insulating mediums have much different characteristics with respect to withstand characteristics of voltage stresses. The test standards standardize the lightning, switching, and power frequency overvoltages that the insulations need to withstand. The designs of an electrical apparatus must cater to the specified test requirements.

17-5 INSULATION CHARACTERISTICS—BIL AND BSL

An impulse test wave as shown in Fig. 5-18 is supposed to mimic the lightning impulse impinged on the electrical apparatus and is North American practice of testing for BIL. It is extensively used as a measure of insulation characteristics. The BIL level may be considered equal to the peak value of 1.2/50-μs wave that will *not* cause breakdown.

The BIL specified in the standards for various equipments are arrived at by industry practices, consensus between manufacturers and utilities, and application of surge arresters.

Data in Table 7-1 is specific to the liquid-immersed transformers. Similar data is applicable to other electrical equipment, for example, circuit breakers, outdoor substations and bushings, and gas-insulated substations. These BSL and BIL data are not reproduced here, but are required for practical insulation coordination in a power system. Values similar to Table 7-1 are shown in the relevant standards. For the same rated voltage, it is possible to select lower or upper values of BIL, which also relate to the switching surge levels. This selection will depend upon a number of factors, for example, the Keraunic lightning level of a place, the atmospheric pollution, (an installation close to a sea shore with salt-laden sprays versus a clean atmospheric conditions in a temperate climate), application of surge arresters, the overvoltage studies, and the like. As the system voltage rises, the high cost of equipment requires more careful evaluation of overvoltages of various origins. This requires extensive computational requirements for overvoltage predictions using TNAs or digital computers.

IEC[7] divides the electrical systems into Range I (1 kV < V_m ≤ 245 kV) and Range II (V_m > 245 kV) with respect to the standard withstand voltages and testing procedures. IEEE draft standard Proposal PC62.82/D2 for insulation coordination[8] does the same. The standard withstand voltages for Class I (15 kV < V_m ≤ 242 kV) and Class II (V_m > 242 kV) are shown in Tables 17-6 and 17-7, respectively.

Both ANSI/IEEE and IEC specify the standard values of BSL and BIL, and generally a choice can be made out of these standard values depending upon the availability of standard manufactured equipment. To select an insulation level for a certain voltage without regard to the standard equipment availability is not prudent. Equipment goes through rigorous testing according to standards to meet the requirements of BIL and BSL.

BSL for circuit breakers is only given for voltages above 362 kV. Also for circuit breakers, for each system voltage, two BSL values are given. BSL for a 362-kV circuit breaker increases from 825 (closed position) to 900 kV in the open position. BSLs/BILs for GIS are given in Table 17-8.[9]

Phase-to-Phase and Phase-to-Ground Withstand Internationally an extensive study has been made whether the phase-to-ground insulation levels can also be applied to phase-to-phase conditions. The results show that the values of phase-to-ground power frequency, temporary overvoltages, and switching overvoltages are inadequate for phase-to-phase conditions. IEC specifies phase-to-phase BSL, as shown in Table 17-9.[7] The test for phase-to-phase BSL consists of applying equal positive and negative switching impulses to each electrode. The phase-to-phase BSL is 1.5 to 1.7 times the phase-to-ground BSL.

In ANSI standards, BSLs are only specified for maximum system voltages at and above 300 kV. *Phase-to-phase BSLs are not standardized in the United States.*

The various combinations of the withstand voltages specified in the standards must be selected after appropriate studies. The equipment withstand voltages are standardized within the ratings in the standards. For example, users can lay down specifications for any of the combinations for their maximum system operating voltage.

TABLE 17-6 Standard Withstand Voltages for Class I[8]

Highest Voltage for Equipment V_m (Phase-to-Phase) (kV, rms)	Low-Frequency, Short-Duration Withstand (Phase-to-Ground) (kV, rms)	Lightning Impulse Withstand Voltage (kV) Crest
15	34	95
		110
26.2	40	125
	50	150
36.2	50	150
	70	200
48.3	95	120
72.5	95	250
	140	350
121	140	350
	185	450
	230	550
145	185	450
	230	550
	275	650
169	230	550
	275	650
	325	750
242	275	650
	325	750
	360	825
	395	900
	480	975
		1050

TABLE 17-7 Standard Withstand Voltages for Class II[8]

Highest Voltage for Equipment V_m (Phase-to-Phase) (kV, rms)	BIL (Phase-to-Ground) (kV, Peak)	BSL (Phase-to-Ground) (kV Peak)
362	900	650
	975	750
	1050	825
	1175	900
	1300	975
		1050
420	1050	850
	1175	950
	1300	1050
	1425	
550	1300	1175
	1425	1300
	1550	1425
	1675	1550
	1800	
800	1800	1300
	1925	1425
	2050	1550
		1675
		1800

TABLE 17-8 Voltage Ratings for Gas-Insulated Substations[9]

System Voltage	Type Test Voltages		
Rated Maximum Voltage (Phase-to-Phase) (kV, rms)	Rated BIL (kV) Crest	Low-Frequency (Phase-to-Ground) Withstand (kV, rms)	Switching Surge Withstand (kV), Crest
72.5	350	160	
121	550	215	
145	650	310	
169	750	365	
242	900	425	
362	1050	500	825
550	1550	740	1175
800	2100	960	1550

Note: Disconnect switch open-gap withstand shall be 10 percent higher than the substation type test values.

TABLE 17-9 IEC BIL and BSL[7]

Maximum System Voltage (kV)	Phase-to-Ground BSL_g (kV, Peak)	Ratio BSL_p/BSL_g	BIL (kV, Peak)*
300	750	1.5	850, 950
	850	1.5	950, 1050
362	850	1.5	950, 1050
	950	1.5	1050, 1175
420	850	1.6	1050, 1175
	950	1.5	1175, 1300
	1050	1.5	1300, 1425
550	950	1.70	1175, 1300
	1050	1.60	1300, 1425
	1175	1.50	1425, 1550
800	1300	1.70	1675, 1800
	1425	1.70	1800, 1950
	1550	1.60	1950, 2100

*For each of the two BILs in this column.

17-5-1 Tail-Chopped Wave

This simulates the stresses on the system insulation, when it is subjected to a lightning impulse of 1.2/50 μs, and the voltage suddenly collapses on the tail of the wave shape due to an insulation breakdown elsewhere in the system. The time to chop varies with the peak voltage (Fig. 17-6).

17-5-2 Front-Chopped Wave

Flashover occurs on the rising edge of the 1.2/50-μs wave. The time to chop t_c is determined as shown in Fig. 17-7. This impulse is used when the rate of rise of voltage is a critical parameter. It mimics the behavior in the first few microseconds of a high-magnitude lightning surge prior to insulation flashover.

17-5-3 Standard Switching Impulse

The standard switching impulse is 100/1000-μs wave (Fig. 19-11). Table 7-1 shows BIL levels of transformers and also switching surge levels. While statistical methods are used for self-restoring insulation, for non-self-restoring insulation, these procedures cannot be used. The insulation strength is specified by conventional BSL.

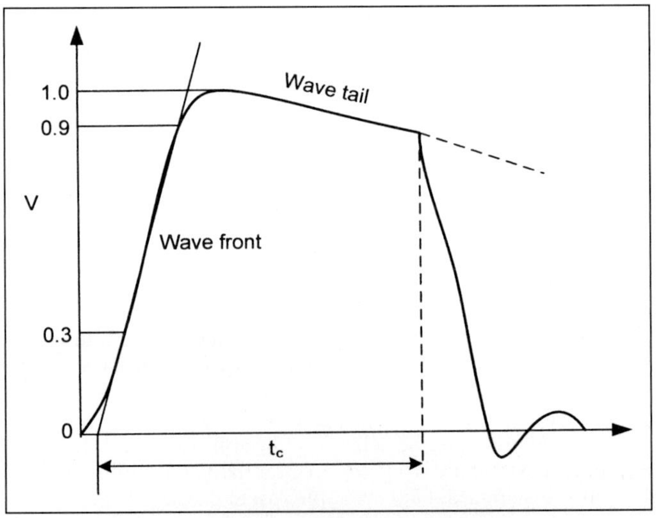

FIGURE 17-6 Tail-chopped 1.2/50-μs wave.

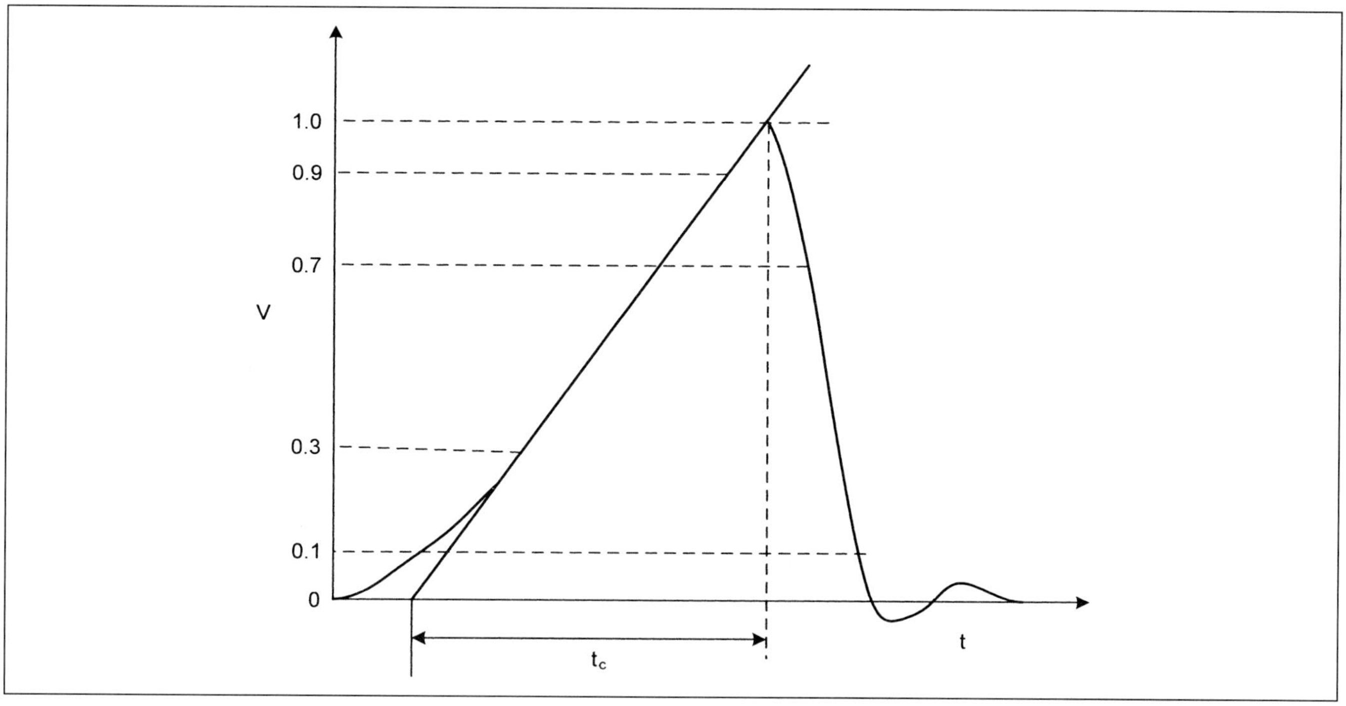

FIGURE 17-7 Front-chopped 1.2/50-μs wave.

The required BSL is the arrester switching impulse discharge voltage multiplied by protective ratio (Chap. 20).

The switching impulse strength is dependent upon components of the phase-to-phase switching impulse, and both positive and negative SOVs must be known. The collection of SOV's may be:

- The positive and negative polarity SOVs
- The positive and phase-to-phase SOVs
- The negative and phase-to-phase SOVs

In the collection of data for positive and negative SOVs, the time constants considered are: (1) the time for which maximum positive SOV occurs and (2) the time for which the maximum phase-to-phase SOVs occur. For air clearances, in general, the maximum positive SOV is more severe, and for non-self-restoring insulation, the maximum phase-to-phase SOV is more severe.

17-6 VOLT-TIME CHARACTERISTICS

As the time duration of application of surge voltage to insulation decreases, it can withstand higher stresses. This is an important factor in the insulation coordination with surge arrester characteristics.

For restorable insulation systems, the volt-time characteristics can be plotted by applying a series of standard test impulses with progressively higher peak values and noting down the time to breakdown in each case, which gives volt-time (*V-t*) characteristics (Fig. 17-8). The breakdown occurs in a smaller time as the voltage peak is raised. Note that the breakdown is possible even after the peak of the wave, as shown for time t_1 in this figure. There is a time delay between the initiation of discharge and its breakdown. In this case, the breakdown voltage is taken as the peak of the wave, though breakdown occurs on the tail. The time to breakdown of insulation is an important parameter, when two or more components are subjected simultaneously to the same impulse stress.

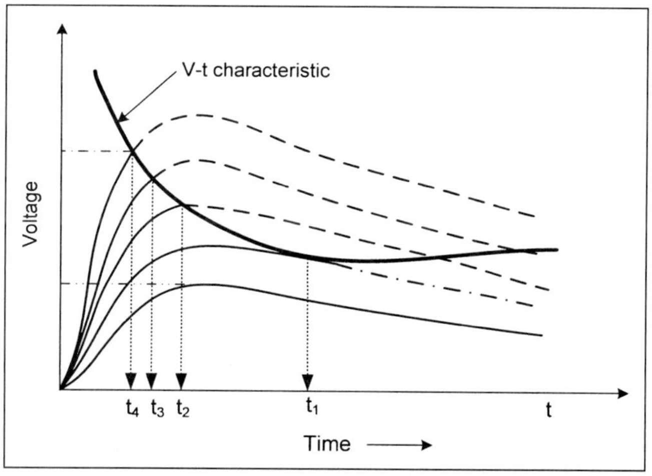

FIGURE 17-8 Construction of volt-time breakdown characteristics.

17-7 NONSTANDARD WAVE FORMS

A question arises, that what is the impact of irregular wave shapes that occur at various points in a system? These irregular wave shapes are unidirectional or oscillatory. The insulation strength is based on standard BIL tests. A correlation between the test wave shapes and the irregular wave shapes is analyzed in Ref. 10. It relates a nonstandard wave to the transformer volt-time curve as shown in Fig. 17-9. This curve is a composite of three test points, *A* the standard full wave, *B* the chopped wave, and *C* the front of wave. The equivalence considers that:

1. The nonstandard wave remains below the transformer withstand curve at all points.

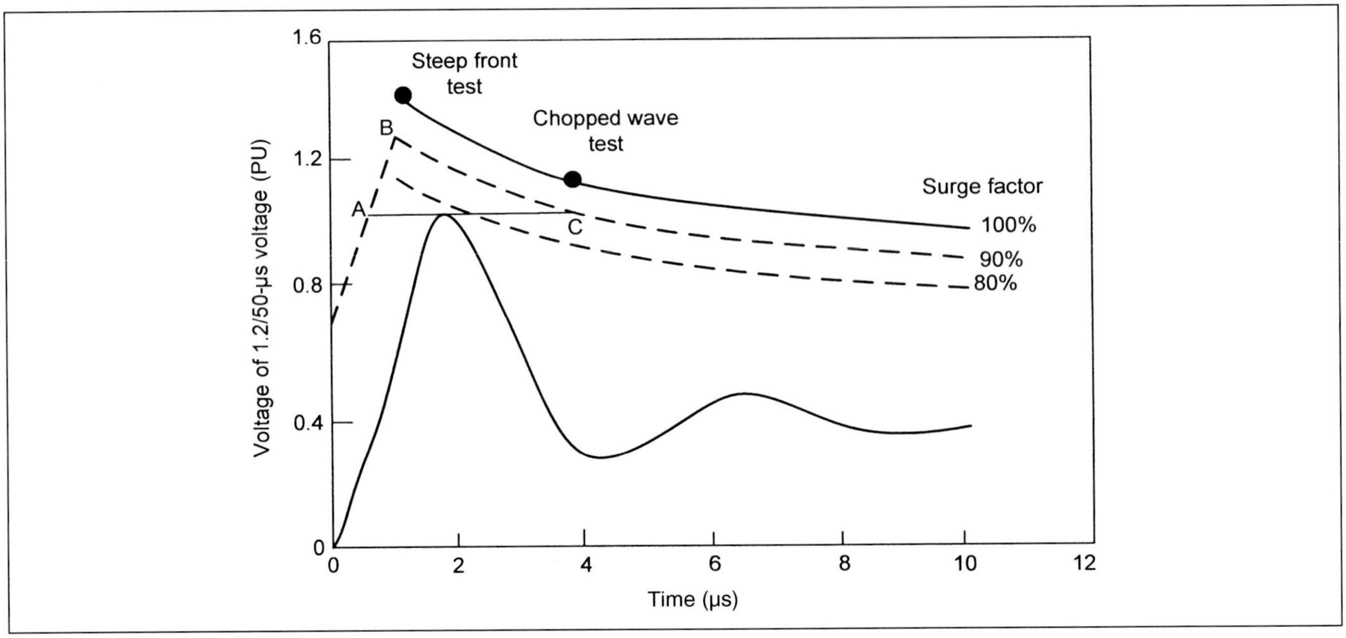

FIGURE 17-9 Voltage breakdown characteristics for nonstandard impulse wave forms.

2. The surge must not be above the line AC for longer than half the time of AC. This second condition is imposed as the nonstandard wave may impose more stresses on the insulation than the standard wave shapes.

A number of graphs at various percentages of ordinates of the volt-time curve are drawn. A percentage surge factor is assigned to the nonstandard wave shape, and a margin of 10 percent is required, for example, a 90 percent surge factor for nonstandard wave is acceptable for the construction shown in Fig. 17-9.

IEEE standard[3] recommends a minimum protective margin of 1.15 on BIL for non-self-restoring insulation. Industry uses higher margins. For self-restoring insulation, the protective margin is not recommended, and the recommended method is to raise mean time between failures (MTBF) by increasing its value. The higher MTBF indirectly results in the increase in the severity of incoming surge.

17-8 PROBABILISTIC CONCEPTS

Breakdown in insulations is a random phenomenon and involves complex breakdown mechanisms. It can be said that there is no single impulse level above which the breakdown always occurs and below which the breakdown never occurs. Some impulse applications may produce flashover, while others with the same amplitude may be withstood by the insulation. This is particularly true for large gaps. There is a band of impulse magnitudes over which breakdown is likely to occur.

The probability distribution of the likelihood of breakdown across this band of impulse waves is statistically called the *normal distribution* or *Gaussian distribution* (App. F). Figure 17-10 shows that the band is characterized by a mean value at which there is a 50-50 chance of a breakdown, and a band outside at which the breakdown has essentially 100% and 0% probability (when the curve is extended to infinity). The mean value of the breakdown voltage for the standard impulse voltage wave shape is also known as the critical breakdown (flashover) voltage (CFO), which has significance in defining insulation withstand levels. The time to breakdown at this voltage is typically 3 to 20 μs. This means that the breakdown occurs on the tail of the test wave as shown in Fig. 17-6. The width of the normal distribution is specified in terms of σ, the standard deviation of breakdown voltage. The distribution indicates that:

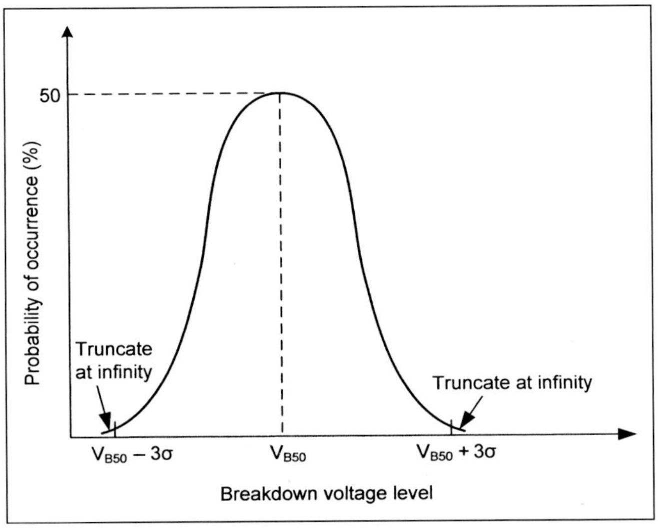

FIGURE 17-10 Gaussian distribution.

$$V_{B0} \approx V_{B0.13} = V_{B50} - 3\sigma$$
$$V_{B100} \approx V_{B99.87} = V_{B50} + 3\sigma \tag{17-7}$$

where $V_{B0.13}$ is the impulse level at which breakdown occurs in 0.13 percent cases of full-impulse applications, for example, the breakdown probability is low. $V_{B99.87}$ is the impulse level at which the breakdown occurs in 99.87 percent cases of full-wave applications, for example, the breakdown probability is high. For a rod gap, this approximates phase-to-phase impulse behavior of overhead line insulation:

$$\sigma = 0.03\, V_{B50} \tag{17-8}$$

Therefore:

$$V_{B0} = 0.91\, V_{B50}$$
$$V_{B100} = 1.09\, V_{B50} \qquad (17\text{-}9)$$

For air-insulated equipment, BIL rating is normally assigned so that it represents a probability of breakdown of 9.7 percent or less, typically:

$$\text{BIL} \le V_{B50} - 1.3\sigma = 0.96\, V_{B50} \qquad (17\text{-}10)$$

Nonrestorable Insulation As the breakdown causes permanent damage to the insulation, the volt-time characteristics cannot be established by repeated applications of test impulse voltage. A BIL is dictated by an appropriate standard, and testing is performed to ensure that there is no breakdown for the specified levels. The magnitude of various types of test impulses is determined by multiplying BIL with appropriate factors. For most oil-immersed apparatus, these factors are shown in Table 17-10.[11]

We can write an expression for the breakdown probability function in Gaussian distribution as:

$$P(V) = \frac{1}{\sigma\sqrt{2\pi}} \int_{-\infty}^{V} \exp\left[-\frac{1}{2}\left(\frac{V - V_{50}}{\sigma}\right)^2\right] \qquad (17\text{-}11)$$

This can be written as:

$$P(V) = \frac{1}{\sqrt{2\pi}} \int_{-\infty}^{z} \exp\left(-\frac{z^2}{2}\right) \qquad (17\text{-}12)$$

where:

$$z = \frac{V - V_{50}}{\sigma} = \frac{V - CFO}{\sigma} \qquad (17\text{-}13)$$

This cumulative Gaussian distribution function is shown in Fig. 17-11*a*. If this function is plotted with $(V - V_{50})/\sigma$ as the y axis, the breakdown function becomes a straight line (Fig. 7-11*b*). We can write σ as:

$$\sigma = \left[\sum (V - V_{50})^2/n\right]^{1/2} \qquad (17\text{-}14)$$

where n is the number of tests. Table 17-11 can be constructed (also see App. F). Extensive tables for Gaussian distribution are available in many texts (App. F).

If the voltage is twice the standard deviation below the critical voltage ($z = -2$), the probability of disruptive discharge is 2.3 percent, and the probability of withstand, therefore, $= (1 - P) = 0.977$(97.7 percent). While σ is a measure of the scatter or dispersion of observations about the mean or CFO, σ/CFO is called the coefficient of variation, as the standard deviation is normally given in per unit or percentage of CFO. It varies with the wavefront, wet/dry conditions.[12] Therefore, the line insulation voltage denoted by V_3 can be written as:

$$V_3 = \text{CFO}\left(1 - 3\frac{\sigma}{\text{CFO}}\right) \qquad (17\text{-}15)$$

Based on the system studies like EMTP, V_3 can be substituted with calculated maximum voltage. This leads to the *deterministic design* of transmission lines provided V_3 can be accurately calculated. A relation between CFO and strike distance, S, here meaning clearances, is given by:[13]

$$\text{CFO} = k_g \frac{3400}{1 + 8/S} \qquad (17\text{-}16)$$

where CFO is in kilovolts. The variable k_g is called the gap factor, and at the center phase of a tower, it is given by:

$$k_g = 1.25 + 0.005\left(\frac{h}{S} - 6\right) + 0.25(e^{-8W/S} - 0.20) \qquad (17\text{-}17)$$

where h is the conductor height and W is the tower width. Usually kg = 1.20 for lattice type towers and 1.25 for steel poles where tower width is small. For lightning impulse, the CFO may be determined from:

$$\text{CFO}^+ = 560S \quad \text{(positive polarity)}$$
$$\text{CFO}^- = 605S \quad \text{(negative polarity)} \qquad (17\text{-}18)$$

For wet conditions, multiply Eq. (17-16) by 0.96, (4 percent decrease). As outer phase has a higher CFO, multiply Eq. (17-16) by 1.08; V_3 is increased by 10% for fronts of 1000 μs or longer. For V-strings the insulator string length is a minimum of 1.05 times the strike distance to structure. For I-string insulators, multiply Eq. (17-16) by 1.10. The distance S is minimum of the following three distances:

1. Strike distance to tower side
2. Strike distance to upper truss
3. Insulator string length divided by 1.05

TABLE 17-10 Factors for Estimating Withstand Voltages of Mineral Oil–Immersed Equipment[11]

IMPULSE DURATION	BIL MULTIPLIER	EQUIPMENT TYPE
Front of wave (0.5 μs)	1.30–1.50	Transformers and reactors
Chopped wave (2 μs)	1.29	Breakers 15.5 kV and above
Chopped wave (3 μs)	1.10–1.15	Transformers and reactors
Chopped wave (3 μs)	1.15	Breakers 15.5 kV and above
Full wave (1.2/50 μs)	1.00	Transformer and reactor windings
Switching surge, 250/2500 wave	0.83	Transformer and reactor windings
Switching surge, 250/2500 wave	0.63–0.69	Bushings
Switching surge, 250/2500 wave	0.63–0.69	Breakers 362–800 kV*

*Includes air blast and SF6 breakers.

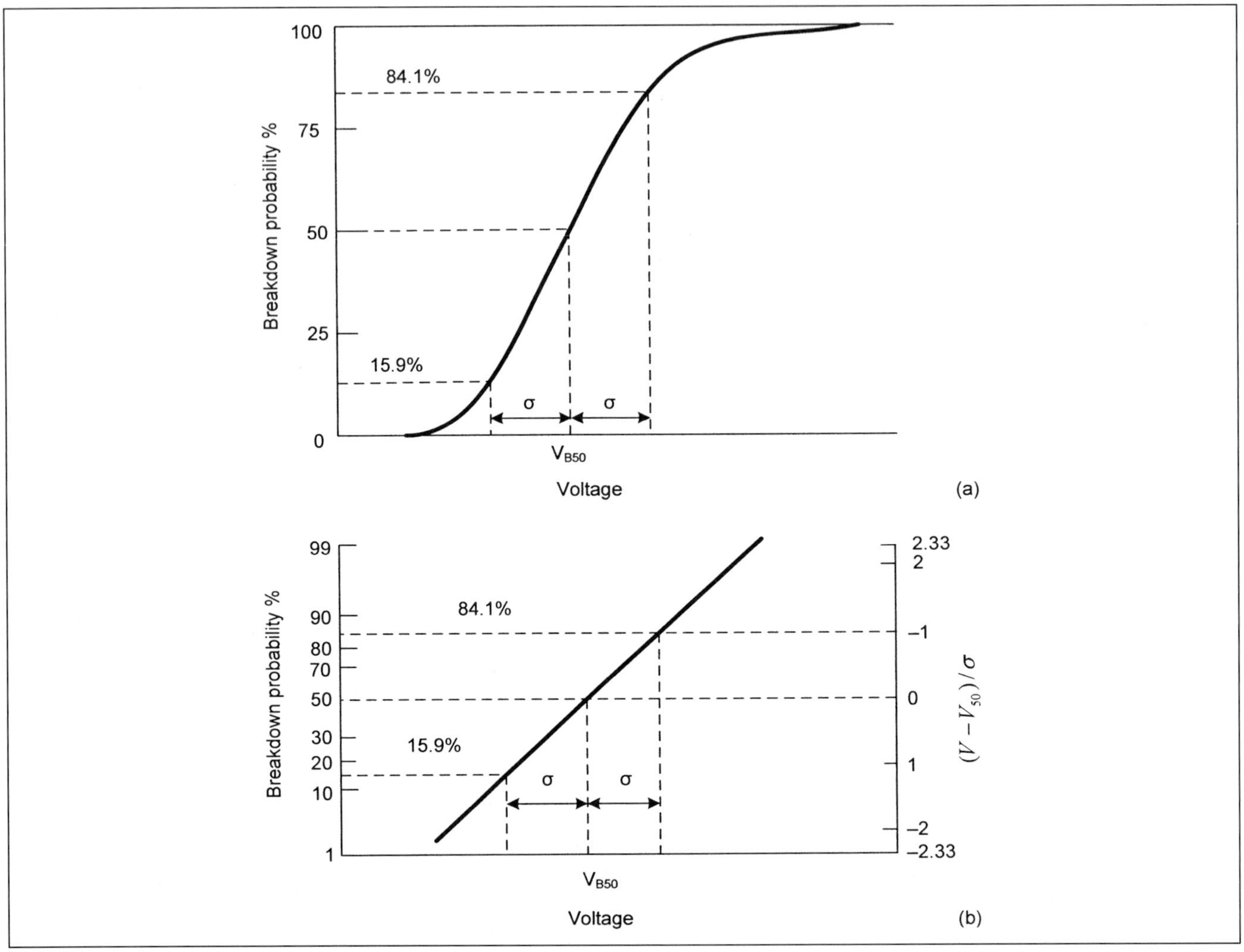

FIGURE 17-11 (*a*) Breakdown probability of external insulation, linear scale Gaussian distribution (*b*) Breakdown probability of external insulation, Gaussian scale, σ, standard deviation.

TABLE 17-11 Normal (Gaussian) Distribution

z	−3.0	−2.0	−1.28	−1.0	0	1.0	1.28	2.0	3.0
P(%)	0.13	2.3	10	15.9	50	84.1	90	97.7	99.87

Example 17-1 An insulator has a V_{50} = 1000 V with σ = 5 percent. A switching voltage of 900 kV is applied to the insulator. Then:

$$z = \frac{900-1000}{0.05\times 1000} = -2$$

This means −2 standard deviations below the mean, and referring to Table 17-11, P = 2.3 percent. There is a 2.3 percent chance of flashover every time the insulator is subjected to a 900-V switching surge.

Example 17-2 Calculate the strike distance for a 400-kV line, with standard atmospheric condition. Wet coefficient of variation = 5 percent and maximum switching surge = 2.5 pu. Width of tower W = 1.5 m and height of tower h = 20 m. Maximum system voltage from Table 17-7 = 420 kV. Switching surge voltage = 1050 kV. From Eq. (17-16) CFO = 1050/0.85 = 1235 kV.

Note that S appears in both Eqs. (17-16) and (17-17). Therefore, iteration is required to calculate S. A value of 3.1 m satisfies both the equations by hit and trial. This gives center-phase clearance of 3.1 m. The strength of the outer phases will be greater.

Example 17-3 Calculate the recommended number of standard insulators for the line in Example 17-2. From Table 17-2, the standard insulators are 5.75 × 10 in. Increase the calculated 3.1 m by 5 percent, say = 3.26 m, then the number of insulators required are 24 (I-Strings). These have a leakage distance of 7008 mm. According to Table 17-3, IEC recommendations, for lightly polluted conditions, a specific creepage of 420 × 27.7 = 11634 mm is required for light pollution conditions. It is prudent to consider both the clearances and the creepage, depending upon the severity of contamination.

Nonstandard Atmospheric Conditions The following expressions for nonstandard atmospheric conditions are from Ref. 13

$$V_{ns} = \delta^m H_c^w V_s \tag{17-19}$$

where V_{ns} is the voltage for nonstandard conditions, V_s is the voltage for standard conditions, δ is the relative air density, H_c is the humidity correction factor, and m and w are constants depending upon G_0 given by:

$$G_0 = \frac{\mathrm{CFO}_s}{500S} \qquad (17\text{-}20)$$

$$m = 1.25G_0(G_0 - 0.2)$$

where S is the striking distance in meters and CFOs under standard atmospheric conditions. Further explanations of the factors are in Ref. 3.

Flashover of Gaps in Parallel The number of insulation components in parallel between phases to ground, say in a transmission line, may run into thousands, depending upon the station configuration. For example, in a transmission line, thousands of insulator strings connect a phase conductor to ground. With two gaps in parallel, assume that: (1) only one of the two gaps first flashes over, (2) a total flashover will be initiated by the gap that has shortest time to breakdown. The probability of flashover of the first gap will be:

$$P_1' = P_1(1-P_2) + (P_1P_2)P_{1,2} \qquad (17\text{-}21)$$

where P_1 and P_2 are the probabilities of first and second gaps flashing first, respectively, P_1P_2 is the probability of simultaneous flashover, and $P_{1,2}$ is the probability that first gap is faster than the second gap 2 to break. A similar expression for second gap will be:

$$P_2' = P_2(1-P_1) + (P_1P_2)P_{2,1} \qquad (17\text{-}22)$$

Also:

$$P_{1,2} + P_{2,1} = 1 \qquad (17\text{-}23)$$

$P_{1,2}$ and $P_{2,1}$ are the functions of the probability density distributions $f_1(t_1)$ and $f_2(t_2)$. Thus:

$$P_{1,2} = \int_0^\infty \int_t^\infty f_1(t_1) f_2(t_2)\,dt_1 dt_2 \qquad (17\text{-}24)$$

If we consider Gaussian distribution, mean values T_1 and T_2 and corresponding deviations:

$$P_{1,2} = \frac{1}{2} + \frac{1}{\sqrt{2}} \int_0^{(T_2-T_1)/\sqrt{\sigma_1^2-\sigma_2^2}} \exp\left(\frac{-t^2}{2}\right) dt \qquad (17\text{-}25)$$

The overall probability of breakdown of two gaps is:

$$P_o = P_1' + P_2' = P_1 + P_2 - P_1P_2 \qquad (17\text{-}26)$$

Identical Gaps in Parallel If the probability of breakdown of a single gap is P, then for n identical gaps in parallel, the probability of breakdown P_n is given by:

$$P_n = 1 - (1-P)^n \qquad (17\text{-}27)$$

If the distribution of the breakdown probability of a single gap, P, is accepted to be cumulative Gaussian distribution, the distribution of P_n is not Gaussian. If we consider insulators in parallel on a transmission system, the probability of withstand decreases as the numbers of insulators in parallel increase. From Eq. (17-27), if the probability of withstand of a single insulator is considered 99 percent, and the insulators on successive towers are considered in parallel, then 20 insulators in parallel give a withstand probability of 82 percent. The attenuation of the traveling waves should be considered in evaluating the line's flashover probability.

17-9 MINIMUM TIME TO BREAKDOWN

The time to breakdown (TBD) of a system of parallel gaps or insulators is also a random quantity. The distribution of TBD is important for insulation coordination and surge protection. The probability distribution of *collective* TBD is composed of TBD probability densities of individual components, each being weighted for a corresponding probability of occurrence of breakdown. Let us consider two gaps in parallel with breakdown probabilities of P_1 and P_2 and TBD distributions of $f_1(t)$ and $f_2(t)$ under a given impulse. Then the TBD distribution of the combination is:

$$f_c = \frac{\phi[P_1, P_2, f_1(t), f_2(t)]}{\int_0^\infty \phi[P_1, P_2, f_1(t), f_2(t)]\,dt} \qquad (17\text{-}28)$$

where, in view of Eq. (17-21):

$$\varphi = P_1(1-P_2)f_1(t) + P_2(1-P_1)f_2(t) + P_1P_2f_{12}(t) \qquad (17\text{-}29)$$

and the denominator must fulfill the constraint:

$$\int_0^\infty f_c(t)\,dt = 1 \qquad (17\text{-}30)$$

The overall distribution of TBD of a n-gap system can be shown to be:

$$f_c(t) = \frac{1}{P_c}\sum_{r=1}^{n} \frac{n!}{(n-r)!} P^r (1-P)^{n-r} f_{(r)}(t) \qquad (17\text{-}31)$$

where r considers different combinations of gaps, that is, singles, twos, threes, and so on out of n gaps. If individual breakdown probabilities are small, all combinational contributions to $f_c(t)$ can be neglected and the combination probability approaches equal to that of a single gap $f(t)$.

17-10 WEIBULL PROBABILITY DISTRIBUTION

A fundamental problem of Gaussian distribution is the truncation value. Physically, no discharge can occur below the minimum value of V. The function is, therefore, truncated at $V_{B0} = V_{B50} - 3\sigma$. Yet, with Gaussian distribution it is not, so IEC[4] adopts Weibull distribution. See App. F for further details and the differences of distribution functions, Gaussian, and Weibull. Figure F-5*a* and *b* diagrammatically represent the comparison.

17-11 AIR CLEARANCES

In installations like electrical substations, which cannot be tested as a whole, it is necessary to ensure that the dielectric strength is adequate. The switching and lightning withstand voltages in air should be greater than the withstand voltages specified in the standards at the worst atmospheric conditions, and again a choice of the appropriate withstand levels can be made through simulations and simplified calculations.

IEC[4] provides tables for *minimum* clearances with a conservative approach based on BIL and different electrode configurations (rod or conductor to structure). Also tables for *minimum* clearances are provided for switching impulse withstand voltage (750 kV or higher) for phase-to-ground and phase-to-phase. These are not reproduced here.

The equivalent IEEE standard, Draft Guide for Recommended Electrical Clearances and Insulation Levels in Air Insulated Electrical Power Substations is yet in the draft form.[13] NESC,[14] specifies safety clearances, which override the calculations and all other standard specifications.

The draft standard[13] considers many factors like altitude, contamination, reduced air gaps due to corona rings, arcing due to air break switches and expulsion fuses, and areas of high keraunic level.

17-11-1 Clearances Based on BIL

Phase-to-ground clearances based upon BIL are given by the equation:

$$S = \frac{V_{\text{crest ph-g}}}{CFO_{\text{gradient}}} \qquad (17\text{-}32)$$

where CFO gradient has been taken as 605 kV/m, a value found satisfactory for typical geometry of air-insulated substations, and S is the metal-to-metal strike distance in meters. With respect to selection of BIL and clearances for the *transformer*:

1. If the time to crest t_T is > 3μs, then BIL = $1.15V_t$.

2. If the time to crest t_T is < 3μs, and $V_t/V_d \leq 1.10$ then BIL = $1.15(V_t/1.10)$. V_d is the calculated arrester discharge voltage and V_t is the voltage at the transformer terminals.

3. In addition to above criteria, if the tail time constant of the incoming surge is > 30 μs, then BSL = $(1.15/0.83)V_d$.

Therefore Eq. (17-32) can be written as:

$$S = \frac{1.15\text{BIL}}{605} = \frac{\text{BIL}}{526} \qquad (17\text{-}33)$$

The draft standard[13] postulates that at the struck point, the phase-to-phase voltage is less than the phase-to-ground impulse voltage, and phase-to-phase impulse voltages seldom exceed the phase-to-ground impulse voltages. As a result, the phase-to-phase clearances for lighting impulse can be equal to phase-to-ground clearances.

17-11-2 Phase-to-Ground Clearances Based on Switching Surges

The statistical probability, Gaussian distribution, reflecting 10 percent probability of flashover is given by Eq. (17-1), repeated here:

$$\begin{aligned}\text{BSL} &= \text{CFO} - 1.28\sigma \\ &= \text{CFO}\left(1 - 1.28\frac{\sigma}{\text{CFO}}\right)\end{aligned} \qquad (17\text{-}34)$$

For self-restoring insulation, the coefficient of variation is smaller for lightning impulses than for the switching impulses. For lightning impulses, this coefficient is 3 percent, and for switching impulses on line or tower insulation, it increases to an average of 5 percent. For station insulation, it is generally 0.06 or 0.07. With a value of 0.07 (= σ/CFO), Eq. (17-34) is:

$$\text{BSL}_{\text{PH-G}} = 0.9104\,\text{CFO} \qquad (17\text{-}35)$$

Above 240 kV, switching surge conditions usually dominate. The following relation holds for flashover of air gap and strike distance for various air-gap configurations up to at least 30 m.

$$S = \frac{8}{\dfrac{(3400k_g)\delta^m}{\text{CFO}} - 1} \qquad (17\text{-}36)$$

where k_g is the gap factor depending upon the gap configuration. It is one for rod-plane gap and a maximum of 1.35 for rod-rod (horizontal) gap. For substation air clearances, $k_g = 1.3$ is suggested. δ^m is the altitude correction factor which is 1.0 at sea level. Substituting in Eq. (17-36)

$$S = \frac{8}{\dfrac{(3400\times 1.3)}{\text{BSL}/0.9104} - 1} = \frac{8}{(4024/\text{BSL} - 1)} \qquad (17\text{-}37)$$

The factor δ, relative air density, is:

$$\delta = 0.997 - 0.106(A) \qquad (17\text{-}38)$$

where A is altitude in kilometers, and m is a constant defined by:

$$m = 1.25G_0(G_0 - 0.2) \qquad (17\text{-}39)$$

where:

$$G_0 = \frac{\text{CFO}}{500S} = \frac{1.07\text{BSL}}{500S} = \frac{\text{BSL}}{467S} \qquad (17\text{-}40)$$

This requires an iterative solution.

17-11-3 Phase-to-Phase Clearances Based on Switching Surges

As stated before, phase-to-phase clearances are not yet established in ANSI/IEEE. With respect to Table 17-8, which gives IEC factors of 1.5 to 1.7, Hileman found ratios as high as 1.25 to 1.40.[15] Table 7 in the draft standard[13] gives gap factor and coefficient of variation for phase-to-phase switching impulses, from which S can be calculated.

17-12 INSULATION COORDINATION

The objective of insulation coordination, a sort of overvoltage coordination, is to minimize the number of insulation failures and, consequently, the number of interruptions. Thus, the insulation coordination is selection of the electrical strength of the electrical equipment and its application in relation to the overvoltages that appear in the electrical system, taking into account the characteristics of the surge protective devices, so that an economical and operationally acceptable level of *probability* is achieved.

The words "probability" and "surge arresters" are noteworthy. The probability should be at an acceptable level to limit damage to the equipment or effect continuity of service. Also use of surge protective devices to achieve the desired end of reducing the expense of installations by reducing the overvoltages is an important consideration. Without surge protection devices, an electrical system will be phenomenally expensive to design, and proper applications of surge protection devices become of paramount importance as the system voltage rises. A reader may like to refer to Chap. 20 on surge arresters concurrently with this chapter, as practically, the insulation coordination at all voltage levels must consider the application of surge arresters. The electrical insulation is selected to withstand dielectric stresses, classified as follows:

- *Power-frequency voltage under normal operation.* This voltage for the purpose of insulation coordination is considered to be the highest system voltage. The system operating voltage is regulated by the utilities and a 5 percent overvoltage limit seems appropriate. This may be exceeded in some cases, that is, close to a generating station. ANSI standard C84.1-1989 specifies the nominal voltages and the operating voltage ranges for equipment operating from 120 V to 34.5 kV in the United States. See Chap. 7 for further discussions.

- *Temporary overvoltages*. These arise from ground faults, load rejection, and ferroresonance, and we have discussed these in the previous chapters.
- Switching overvoltages, discussed in previous chapter.
- Lightning overvoltages, discussed in previous chapter.

In other words, all the transients categorized in Table 1.3 must be considered. Figure 17-12 shows a flowchart for the selection of the insulation characteristics. This is rather an oversimplification of a complex process. The electrical system characteristics, calculations of overvoltages for the particular system, and application of surge arresters are the key elements, which have varying parameters, not standardized by equations.

Experience shows that developments of generalized equations for expected overvoltages are difficult due to the large number of parameters affecting the results. Further, high levels of skills are required for such calculations. Apart from calculations, experience and practice play an important role.

In each category, the overvoltages at the equipment terminals need to be calculated and the application of surge arresters accounted for. Both phase-to-phase and phase-to-ground overvoltages need to be calculated. Specific considerations apply with respect to the systems protected. As an example, all lightning events within a certain distance from substation cause higher overvoltages at the protected equipment than an assumed value, and all events outside the distance cause lower values. It has been documented that backflashovers do not occur at a tower close to the substation due to lower impedance of substation ground. Though the previous chapters provide practical data, simulations, and other considerations that are applicable, standards need to be consulted for further guidelines.

17-12-1 Types of Electrical Systems

For the purpose of insulation coordination, the electrical systems can be subdivided into four major categories: transmission lines, substations, industrial and commercial power systems, and GIS installations.

1. *Transmission lines*. We have discussed overvoltages on OH lines in the previous chapters, particularly Chaps. 4, 5, and 7. Flashovers across insulator strings or phase conductors to ground wires do not lead to permanent failures but transient overvoltages. These momentary overvoltages should be minimized at an optimum cost, and statistical procedures are used, as stated before. The basic parameters involved are:

- Shielding designs for direct strokes
- Type of grounding systems; utility high voltage systems are invariably solidly grounded (Chap. 21)
- Estimation of the number of lightning strokes, shielding failures, probability of backflashovers, as discussed in the previous chapters
- Pollution and atmospheric impacts
- Tower footing impedance, voltage on tower top, and cross-arms due to a lightning stroke

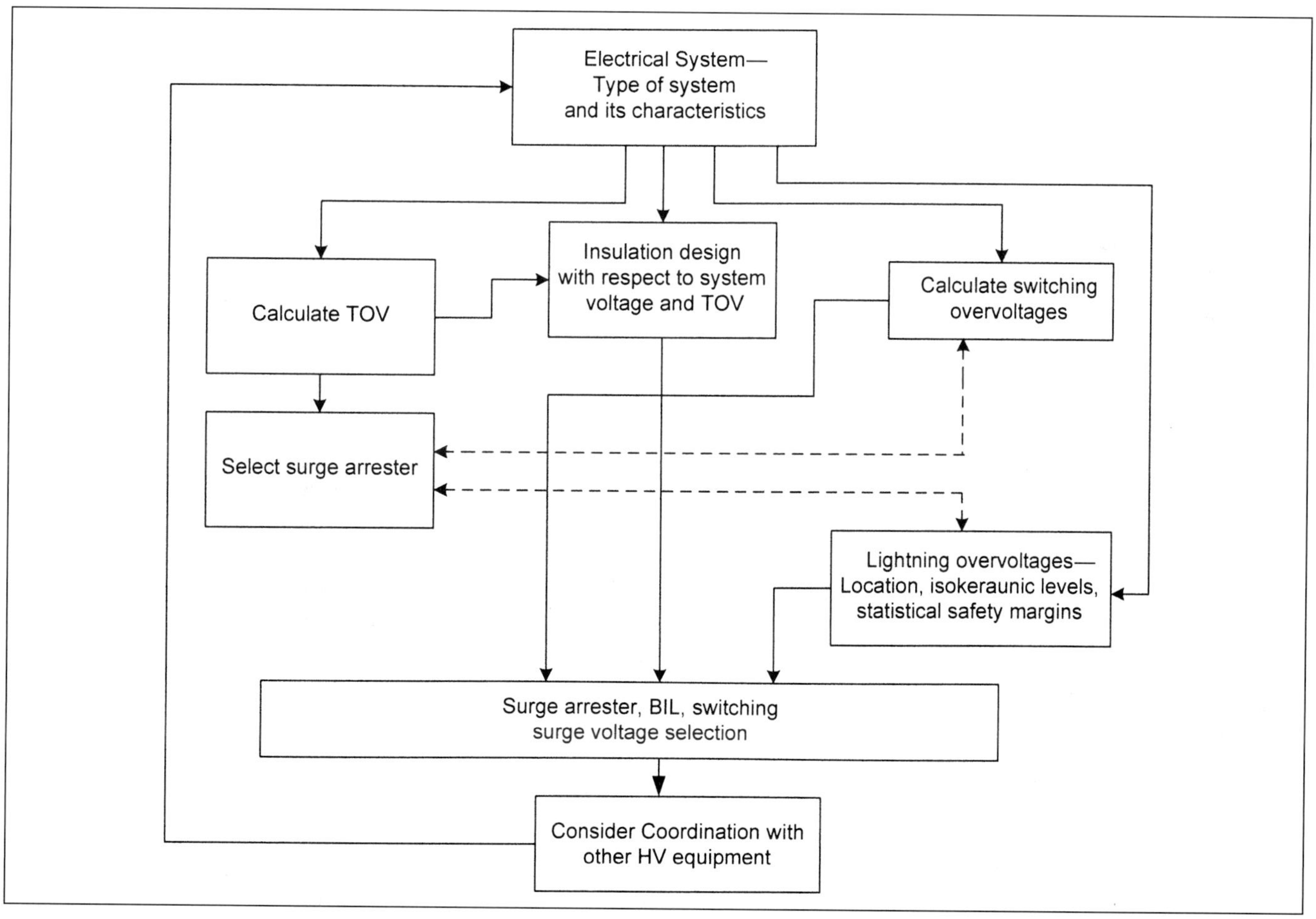

Figure 17-12 Simplified flowchart for insulation coordination.

2. *Substations.* Substations contain transformers and other valuable equipment with non-self-restoring insulation which must be guarded against internal breakdowns. Repairs/replacement of a large transformer will be expensive and will result in much downtime. This cannot be tolerated for continuity of power supply. Even a risk of flashover in air with consequent disturbances should be kept to a minimum. Substations are invariably shielded from atmospheric lightning by suitable arrangement of ground wires on high masts, rising above the substation outdoor structures. These ground wires are continuous with the incoming/outgoing transmission lines/distribution lines/sky wires, and are grounded at each mast and also at substation outdoor structures. The effectiveness of grounding plays an important role in surge diversion. The equipment to be protected varies: large transformers, circuit breakers, reactors, capacitor banks, bus work, and the like. It is standard practice to provide surge arresters on the primary and secondary side of each power transformer on the incoming/outgoing lines from the substations, and at the junctions of the cables/overhead lines. The lightning/switching overvoltages occur due to:

- Direct lightning strokes due to shielding failures
- Lightning/switching surges from the lines, entering and leaving the substation
- Backflashovers
- Lightning strikes in the vicinity of the substation, giving rise to induced lightning overvoltages

The rolling sphere model and other lightning protection models for structures are discussed in Chap. 22.

The mechanism of induced voltage on the line for a nearby lightning stroke is illustrated in Fig. 17-13, and several models are reported in literature[16,17] (see also Chap. 5). A simplified equation is:

$$v = Z_0 Ih\left[\left(\frac{1}{2y}\right)\frac{1+(x+\beta y)}{\sqrt{x^2+2y^2+2\beta xy-\beta^2y^2}}+\frac{2\beta x+y}{y^2+(2\beta x+y)^2} + \frac{1+2\beta^2x+\beta y-\beta x}{\sqrt{x^2+2y^2+2\beta xy-\beta^2y^2}}\right] \tag{17-41}$$

where:

$$\begin{aligned}\beta &= 0.004I^{0.64}+0.068 \text{ (first stroke)}\\ &= 0.004I^{0.86}+0.18 \text{ (subsequent strokes)}\end{aligned} \tag{17-42}$$

Z_0 is the surge impedance and I is the stroke current and other parameters are shown in Fig. 17-13.

The lighting-induced voltages in transmission lines and substations for strokes to ground in the vicinity of the installations are generally limited to approximately 300 kV and can be ignored for system voltages above approximately 145 kV. Also the substations are generally considered perfectly shielded and a direct lighting stroke on the substation structures is excluded. Therefore, the surges entering the substation are a result of strokes on lines close to the substation, backflashover, or shielding failure. The backflashover gives rise to overvoltages of steep wavefront as shown in the EMTP simulation in Chap. 5.

3. *Industrial systems.* When the industrial systems are connected to utility overhead systems, they will be exposed to the overvoltages of atmospheric origin through the OH lines. The industrial systems vary in complexity and power usage, and some large industrial systems may receive large chunks of power, even of the order of 80 to 200 MW across high-voltage substations located within the industrial premises. Cogeneration modes are also common, and an industrial plant may generate excess power than what it consumes, depending upon the processes. Thus, lightning and switching surges are of consideration, the system grounding playing an important role—most industrial systems have 22, 13.8, 4.16, or 2.4 kV as the primary distribution voltages, and these medium-voltage distribution systems are invariably resistance grounded (Chap. 21). Sudden load rejection, temporary overvoltages due to faults, switching surges, impingement of lightning surges through utility connections, and induced lightning voltages due to nearby strokes—all enter into an informed surge protection strategy.

4. *GIS.* See Chap. 18.

17-12-2 Capacitance to Ground of Substation Equipment

Table 17-12 gives the stray capacitance to ground of various substation equipments. Figure 17-14 shows the circuit breaker diagrams and minimum capacitance values used in the lightning studies. A capacitor acts like a short circuit to a surge when uncharged and an open circuit when it is charged. When two equipments are located close to each other, the capacitances can be combined. Comparative simulations show that support structures do not impact the simulation results and can be neglected. However, the capacitance

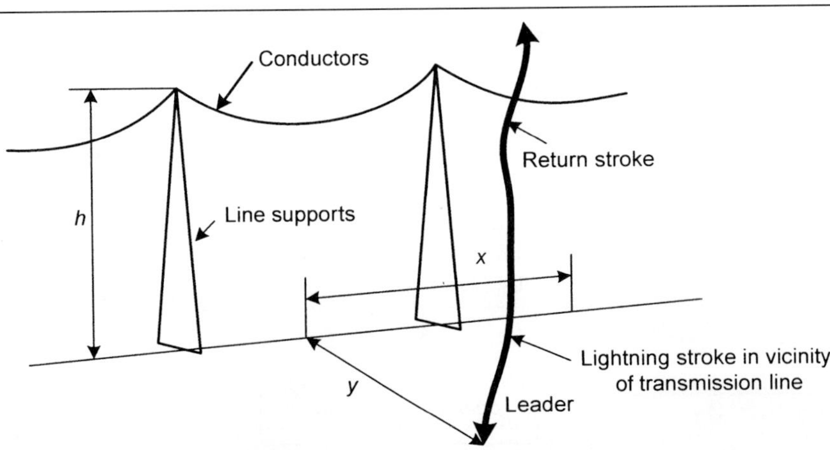

FIGURE 17-13 Induced voltages due to nearby lightning flash.

TABLE 17-12 Capacitance to Ground Data

EQUIPMENT	CAPACITANCES TO GROUND (PF)		
	115 KV	400 KV	765 KV
Disconnect switch	100	200	160
Circuit breaker (dead tank, outdoor)	100	150	600
Bus support insulator	80	120	150
CVT	8000	5000	4000
Magnetic PT	500	550	600
CT	250	680	800
Autotransformer (capacitances depend upon MVA)	3500	2700	5000

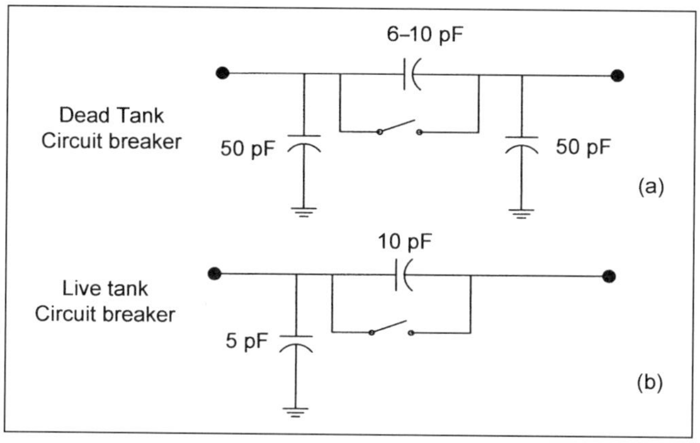

FIGURE 17-14 Capacitances of circuit breakers.

to ground of the insulators should be represented. To evaluate the stresses on the transformer, the conservative approach is to consider that the transformer is open-circuited on the secondary with no load connected to it.

Lumped capacitance near the substation entrance can reduce the slope of the steep-fronted incoming wave fronts; this is akin to the phenomena discussed in Chap. 20 for surge protection of the rotating machinery. The shapes of reflected and transmitted waves will be the same at lossless discontinuities, and presence of capacitance or inductance will alter this picture. A lumped shunt capacitance can be modeled by an equivalent line stub. A capacitive voltage transformer (CVT) has a capacitance of 2000 to 8000 pF. Suspension insulator has a capacitance of 80 pF/unit (divide by the number of units in the string). Busbar surge impedance can be calculated like overhead lines, and conductor capacitances can be increased by the capacitance of insulators.

17-13 REPRESENTATION OF SLOW FRONT OVERVOLTAGES (SFOV)

As an example, consider the energization and reenergization of transmission lines, which give slow front overvoltages of transmission lines, discussed in Chap. 7. The study of Fig. 17-15 from Ref. 4 is interesting; it uses the same symbols as in Ref 4. This figure gives a simplified overvoltage circle, given by values of phase-to-earth overvoltages and phase-to-phase overvoltages for the considered probability of 2 percent. The term "earth" in IEC standards and European practice is synonomous with "ground" in ANSI/IEEE Standards, USA. The area R gives the most critical stress. The symbols in this figure are defined as follows. U_p is the amplitude of phase-to-phase overvoltage, U^- is the negative switching component in phase-to-phase insulation test, U^+ is the positive switching component in phase-to-phase insulation test, U_e^+, and U_e^- are the positive and negative switching component in phase-to-earth insulation test, U_c^+, and U_c^- are the positive and negative components defining the center of the circle which describes phase-phase-earth slow-front overvoltages, and U_0^+ is the equivalent phase-to-earth component used to represent most critical phase-to-phase overvoltage.

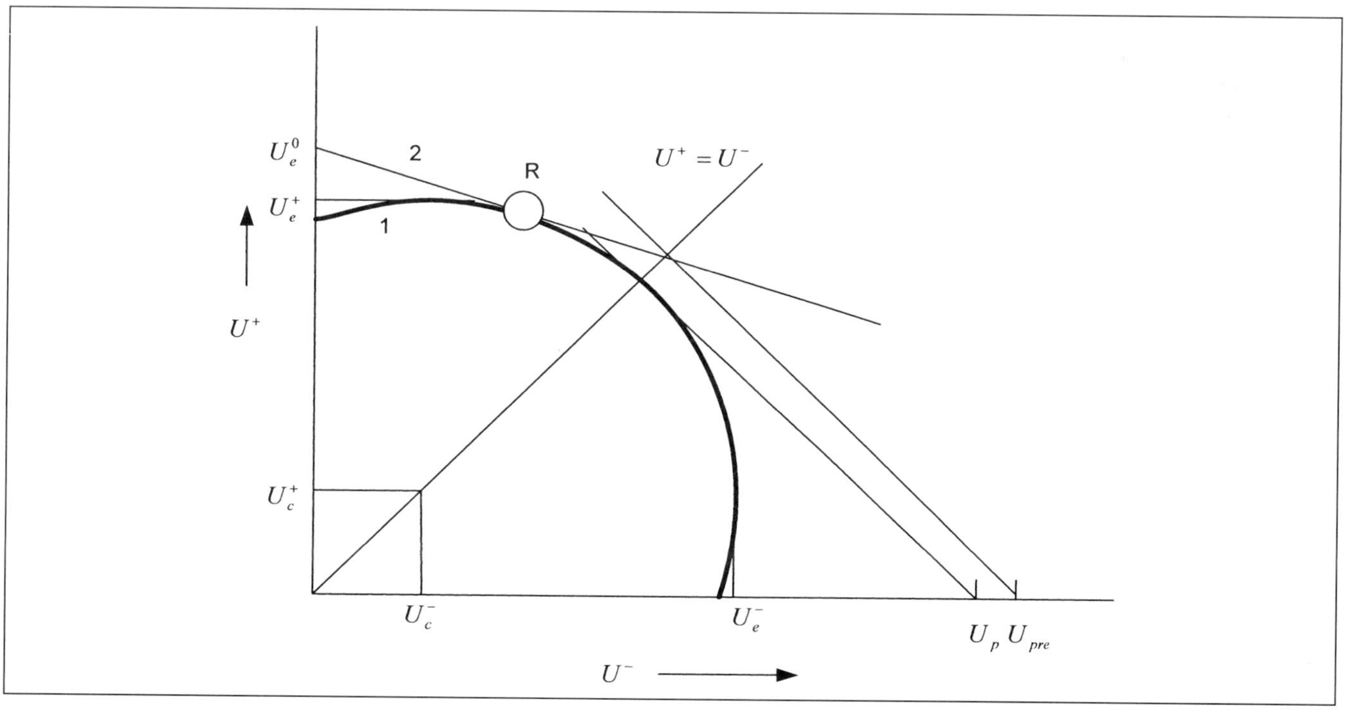

FIGURE 17-15 Curve 1, simplified overvoltage circle, SFOV, phase-to-earth and phase-to-phase. Curve 2, 50 percent flashover characteristics of the insulation, R, maximum stress.[4]

IEC 71-1[7] defines the representative voltages between phases as consisting of two components with equal amplitude and opposite polarity, U^+, and U^- and by an equivalent positive voltage:

$$U_0^+ = U^+ + \beta U^- \tag{17-43}$$

This overvoltage is situated on line $U^+ = U^-$ or $\alpha = 0.5$. The most critical stress depends upon the insulation characteristics and inclination β. The factor β depends on the electrode geometry and, in particular on the gap spacing D between phases versus height H between phase-to-earth.

$$\begin{aligned} H >> D \quad & \beta \to 1 \\ H << D \quad & \beta \to 0 \end{aligned} \tag{17-44}$$

Between these two extremes, β assumes values between 0 and 1. The center and radius of the circle in Fig. 17-15 are given by:

$$\begin{aligned} U_c^+ = U_c^- &= \frac{U_p - \sqrt{2U_e}}{2-\sqrt{2}} \\ r &= \frac{2U_e - U_p}{2-\sqrt{2}} \end{aligned} \tag{17-45}$$

17-14 RISK OF FAILURE

Overvoltages that stress the system insulation can be expressed by statistical distribution of magnitude and also time. The breakdown probability can also be expressed as a function of the same two quantities. Therefore, overvoltages, which fall within interval dV_m and with times to crest within dt, give risk of failure, written as:

$$R = \int_0^\infty \int_0^\infty P(V_m, t) f(V_m, t) dV_m \, dt \tag{17-46}$$

It is customary to simplify Eq. (17-46) and express the risk of failure as:

$$R = \int_0^\infty P(V_m) f(V_m) dV_m \tag{17-47}$$

Figure 17-16 shows a graphical representation of this equation and the general method by which the probability of failure can be assessed. It assumes that $P(V_m)$ and $f(V_m)$ are not correlated. The equation applies for a single piece of insulation. If number of pieces are connected together on the same plane and are subjected to the same overvoltage, it may be assumed that the overall risk is equal to that of the risk of single insulation multiplied by the number of insulations in parallel. For a three-phase system, the risk R can be multiplied by three or, alternatively, an overvoltages probability density can be established by considering only the highest value of overvoltages on the three phases. The mathematical model of R in Eq. (17-47) is based on the following assumptions:

1. Peaks other than the highest in the wave shape of overvoltages are discarded. This gives the calculated risk of failure lower than the actual risk.

2. The wave shape of the highest peak is assumed to be equal to the standard switching or lightning impulse test wave shape. This gives a calculated risk which may be higher than the actual risk.

3. The highest overvoltages peaks are assumed to be of the same polarity. This gives a risk factor higher than the actual risk.

The overall risk factor as calculated by Eq. (17-46) is normally conservative. The accuracy of calculation depends upon accuracy in determination of the overvoltages in the system. A direct implication is that it is possible to coordinate the security levels of the various parts of the system according to the consequences of a breakdown. The insulation strength is selected to obtain a probability of failure to the required safety level. Thus, it is necessary to define acceptable safety factors as follows:

1. *Statistical withstand voltage (SWV)*. It is already defined in Eq. (17-1) as the peak value of a *switching or lightning* impulse test voltage at which insulation exhibits, under specified conditions, a probability of withstand equal to 90 percent. In many cases, a larger margin between CFO and statistical withstand voltage is adopted: $z = -2$ to $z = -3$.

2. *Statistical overvoltage (SOV)*. The switching or lightning overvoltages are applied to the measurement as a result of

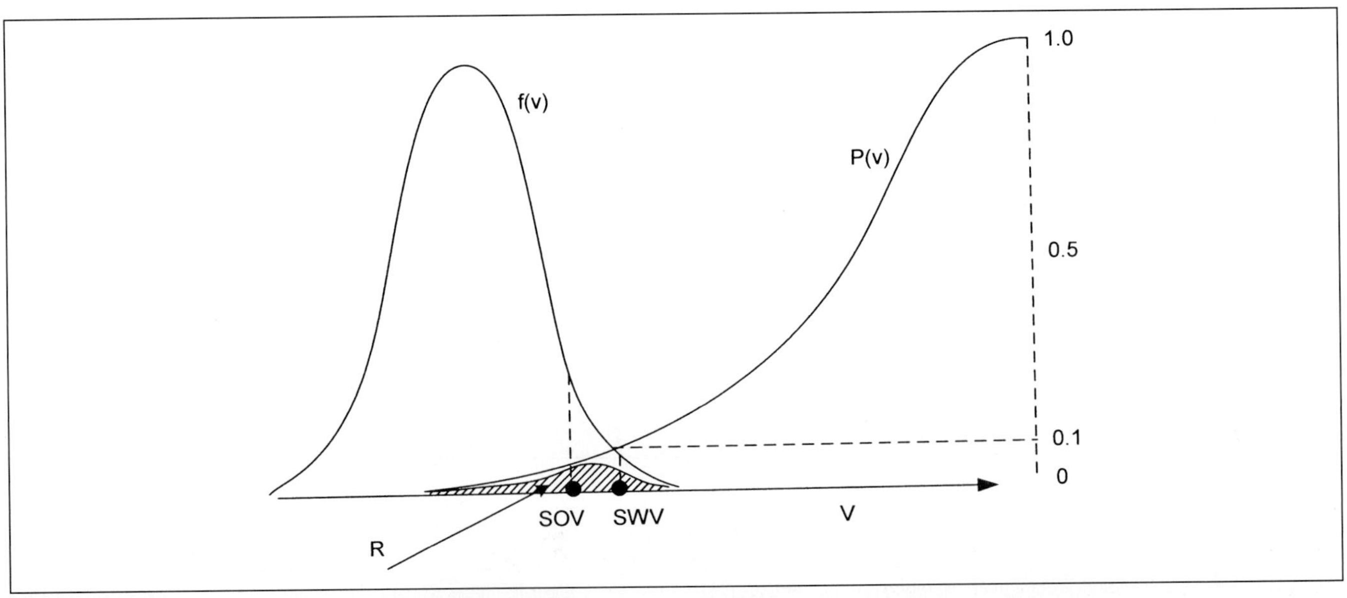

FIGURE 17-16 Computation of risk of insulation failure.

an event of one specific type on the system, the peak value of which has a probability of being exceeded by 2 percent. For Gaussian distribution:

$$E_2 = \mu_0 + 2.054\sigma_0 \qquad (17\text{-}48)$$

CIGRE working group[18] provides some guidelines of σ_0. It can be taken as 2.8 pu for high-speed reclosing without closing resistors and 1.8 pu with closing resistors. Also see Ref. 3.

3. *Statistical safety factor*. A simplified procedure to evaluate risk of failure was given in IEC 71-Insulation Coordination, 1971, now revised as Ref. 4. It designated a safety factor.

For a given type of event, it is a ratio of the appropriate statistical impulse withstand voltage and statistical overvoltages:

$$\text{SSF} = \frac{\text{SWV}}{\text{SOV}} \qquad (17\text{-}49)$$

If SSF is increased, R is decreased. Sensitivity analysis and evaluations of the risk failure can be made on this basis. The complete distribution of overvoltages and the discharge probability is defined by one point only, that is, SWV and SOV. Figure 17-17*a* and *b* shows frequency distributions of the overvoltages and insulation strengths, where V_s and V_w denote SWV and SOV, respectively. In Fig. 17-17*c*, *d*, and *e*, superimposition of the characteristics shows risk factors by darkened areas: R_1, R_2, and R_3, where $R_1 > R_2 > R_3$. As SOV is increased, the risk factor decreases. In IEC,[4] the SSF factor is denoted by an equivalent term as statistical coordination factor K_{cs}:

$$K_{cs} = \frac{U_{cw}}{U_{e2}} \qquad (17\text{-}50)$$

The simplified statistical method for slow-front overvoltages in Ref. 4 states that correlation between risk of failure and K_{cs} appears to be only slightly effected by the parameters of overvoltage distribution. The standard describes two representations of the probability distribution:

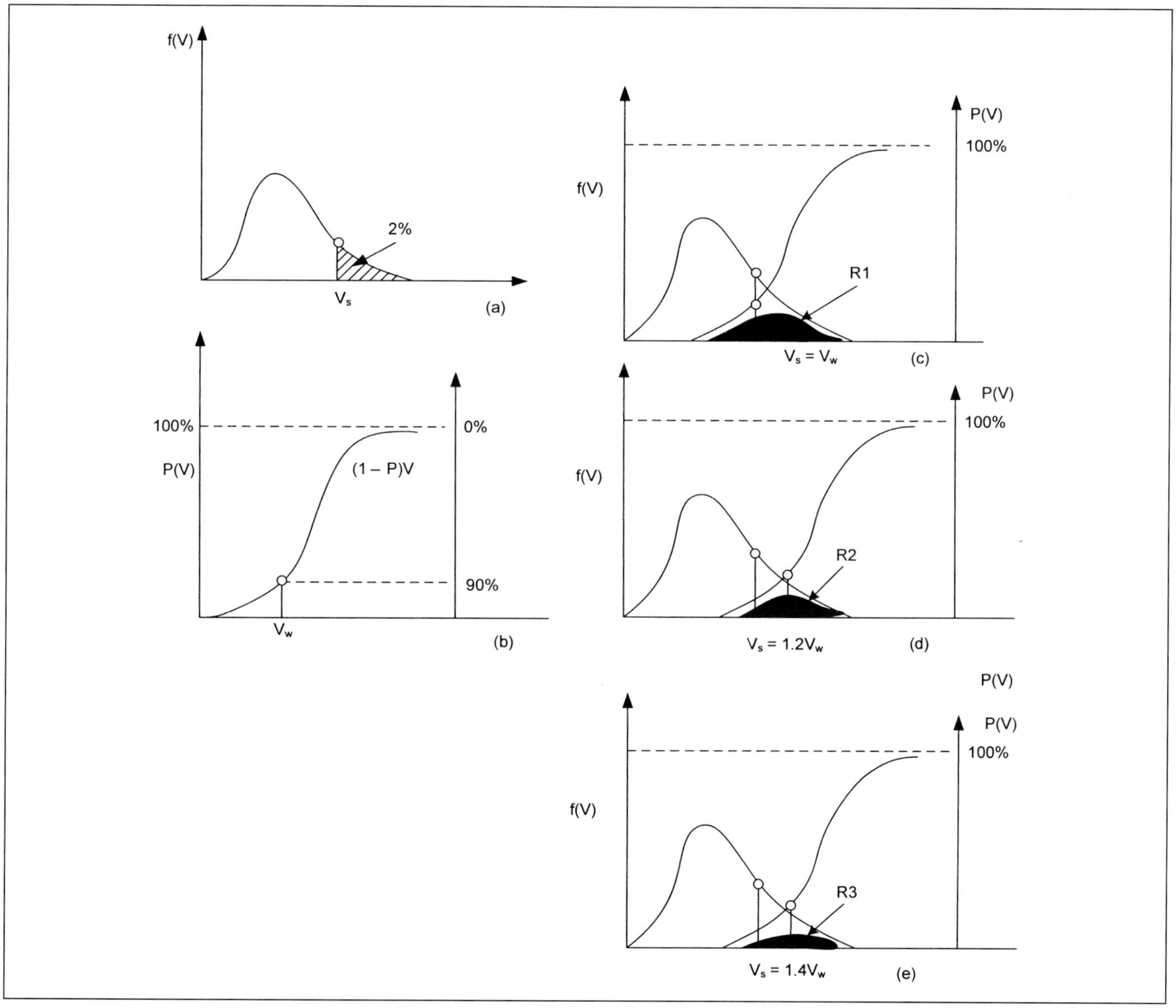

FIGURE 17-17 Simplified statistical method for calculation and reduction of insulation failure rate. (*a*) Overvoltage and (*b*) insulation strength probability functions. (*c*), (*d*), and (*e*) Reduction of risk of failure factor by increasing insulation strength.

Phase-Peak Method From each switching operation, the highest peak value of overvoltage on each phase-to-earth, or between each combination of phases, is included in the voltage probability distribution. This distribution is then assumed to be equal for each of the three insulations: phase-to-earth, phase-to-phase, or longitudinal. (Longitudinal insulation is an insulation configuration between terminals belonging to the same phase, but which are separated into two independently energized parts, e.g., open switching device.) For 2 percent value:

$$\text{Deviation } \sigma_e = 0.25(U_{e2} - 1) \tag{17-51}$$

$$\text{Truncation value } U_{et} = 1.25U_{e2} - 0.25 \tag{17-52}$$

Case-Peak Method From each switching operation, the highest peak value of overvoltage of all three phase-to-earth or between all the three phases is included in the voltage probability distribution.

One energization yields one data point, the highest peak, phase-to-earth, and phase-to-phase.

For 2 percent value:

$$\text{Deviation } \sigma_e = 0.17(U_{e2} - 1) \tag{17-53}$$

$$\text{Truncation value } U_{et} = 1.13U_{e2} - 0.13 \tag{17-54}$$

where U_{e2} is phase-to-earth overvoltage having a 2 percent probability of being exceeded. Figure 17-18 from IEC[4] depicts the risk factors for both of these methods.

Example 17-4 Consider a 765-kV system, phase-to-earth voltage = 625-kV peak and an SOV of 1255 kV peak = 2.0 pu. Let SWV = 1300 kV = 2.08 pu. Then from Eq. (17-49), SSF = 1.04, and from Figure 17-18, $R = 10^{-2}$. In this case, R is the same from both the curves in Fig. 17-18, the phase-peak and case-peak methods.

Example 17-5 Consider a string of 10 insulators at 500 kV. Desired probability of 10 insulator string is 0.5 percent. A switching surge of 2.5 times the system voltage occurs. Standard mean deviation is 6 percent for the switching surge. Calculate the required CFO for the stated conditions.

The probability of 10 insulator gaps is given by:

$$P_{10} = 0.005 = 1 - (1 - P)^{10}$$

Therefore:

$$P = 1 - (1 - 0.005)^{1/10} = 0.000501$$

This corresponds to standard deviation below mean, z = –3.29. Therefore:

$$500 \times 2.5 = \text{CFO}(1 - 3.29 \times 0.06)$$

$$\text{CFO} = 1557\text{V}$$

17-15 COORDINATION FOR FAST-FRONT SURGES

17-15-1 Deterministic Method

For fast-front lightning overvoltages, a deterministic coordination factor, K_{cd}, of 1 is applied.[4] This is because for the lightning overvoltages, probability effects are included. For fast-front switching overvoltages, the same relations apply as for slow-front switching overvoltages.

Figure 17-19 from this standard shows the factor K_{cd} as a function of ratio U_{ps}/U_{e2}. Here, U_{ps} is the switching surge protective level of the surge arrester. Curve (a) is applicable to phase-to-earth overvoltages and also applies to longitudinal overvoltages. Curve (b) is applicable to phase-to-phase overvoltages.

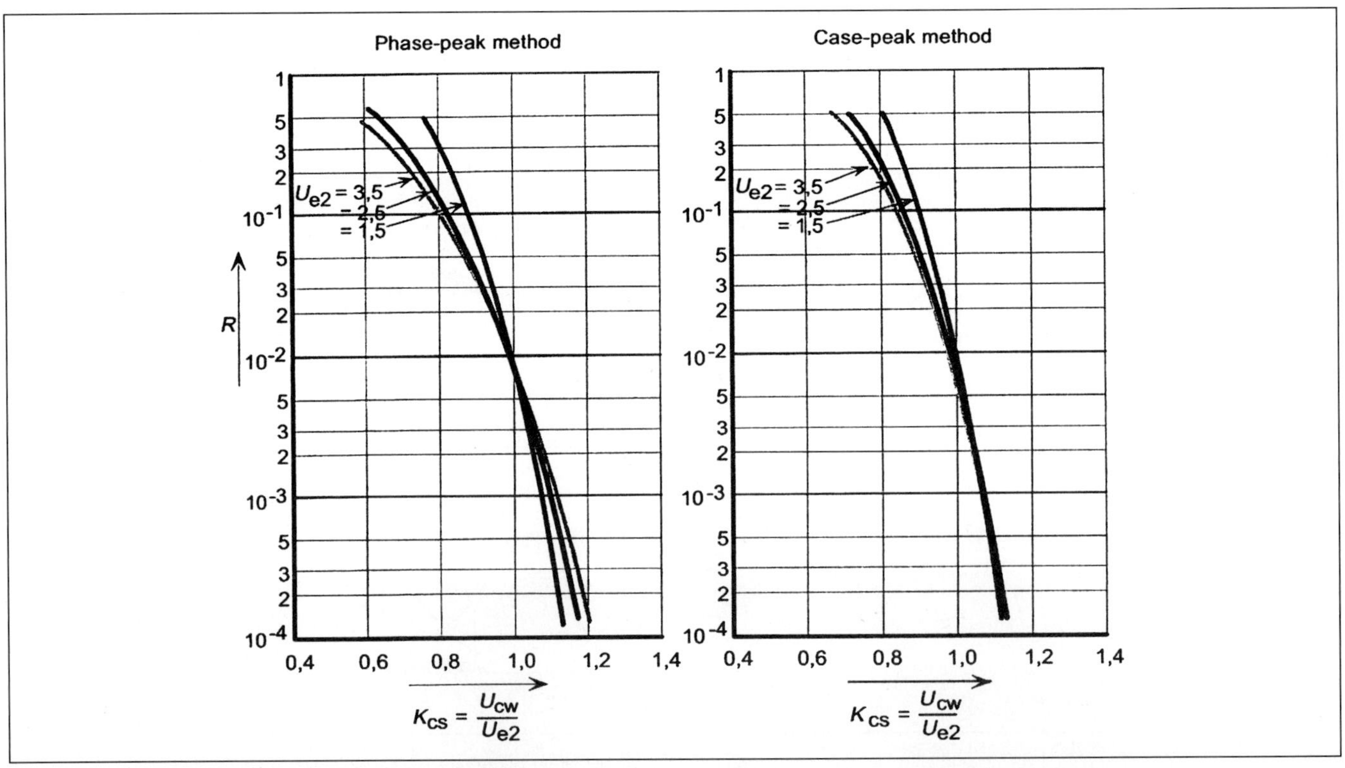

FIGURE 17-18 Risk of failure, external insulation SFOV, as a function of statistical coordination factor.[4]

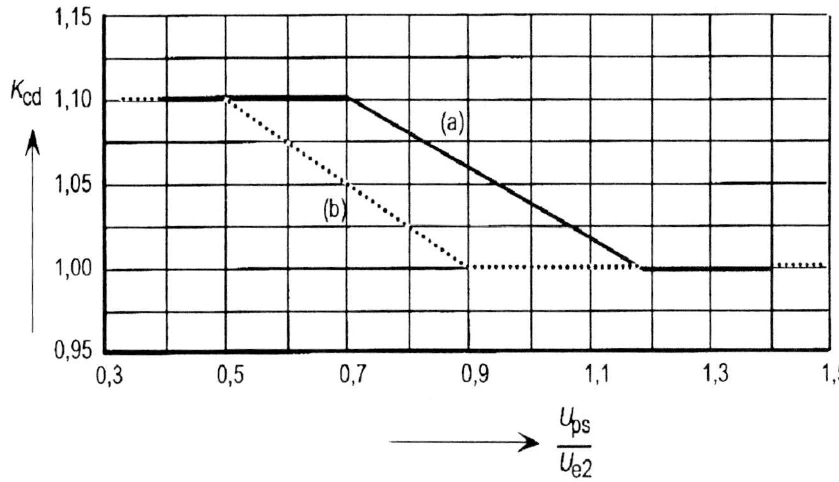

FIGURE 17-19 Evaluation of deterministic coordination factor.[4]

17-15-2 Statistical Method

The statistical method in Ref. 4 for fast-front overvoltages considers recommended probability distributions representative of lightning overvoltages (Chap. 5).

17-16 SWITCHING SURGE FLASHOVER RATE (SSFOR)

For general design, it is recommended that the design value of SSFOR is set at 0.5 flashover per 100 km-yr. Either Brown-Whitehead or IEC equations can be employed for the calculations of SSFOR.[4,22] From Eq. (17-24), the incremental probability of a flashover can be expressed as:

$$dP = f_s(V) f_{st}(V) dV = p f_s(V) dV \tag{17-55}$$

where function $f_{st}(V)$ denotes strength, and $f_s(V)$ stress, and $f_s(V) = p$. The strength/stress ratio and SOV profile along the line are important factors. Then:

$$\text{SSFOR} = \frac{1}{2}\int_{E}^{E_m} p f_s(V) dV \tag{17-56}$$

Factor 1/2 means that negative SOV distribution is neglected.

The integration in Eq. (17-56) is carried out from system line-to-neutral voltage, E, and maximum switching overvoltage, E_m. By definition, $E = 1.0$ pu. For n-towers:

$$\text{SSFOR} = \frac{1}{2}\int [1 - q^n] f_s(V) dV \tag{17-57}$$

where:

$$P_n = 1 - (1-P)^n = 1 - q^n \tag{17-58}$$

This is a function of:

- SSF
- change of SOV along the line
- The type of probability distribution
- The number of towers n

The Brown's Method If the strength is represented by a single-value function, CFO_n, which denotes the strength of n towers given by:

$$\text{CFO}_n = \text{CFO}\left(1 - z_f \frac{\sigma_f}{\text{CFO}}\right) \qquad z_f = \frac{\text{CFO}_n - \text{CFO}}{\sigma_f} \tag{17-59}$$

where we may call z_f as the reduced variate, then SSFOR is:

$$\text{SSFOR} = \frac{1}{2}\int_{\text{CFO}_n}^{E_m} f_s(V) dV \tag{17-60}$$

Equation (17-60) can be estimated for *any* SOV *distribution*. For Gaussian distribution:

$$\text{SSFOR} = \frac{1}{2}\left[1 - F\left(\frac{\text{CFO}_n - \mu_0}{\sigma_0}\right)\right] \qquad z = \frac{V - \mu_0}{\sigma_0} \tag{17-61}$$

Here μ_0 replaces V_{50} in the definition of z. The CFO_n of n towers is defined at probability of 0.5:

$$0.5 = 1 - (1-P)^n \qquad P = 1 - \sqrt[n]{0.5} \tag{17-62}$$

Example 17-6 Considering a probability = 0.023 percent, percentage deviation = 5 percent, CFO = 1000 V, Gaussian distribution, $\sigma_0/E_2 = 0.12$, no arrester (from Ref. 3), and $\sigma_f/\text{CFO} = 5$ percent, find SSFOR.

From Table 17-11, $z = -2.0$. From Eq. (17-59)

$$-2.0 = \frac{\text{CFO}_n - \text{CFO}}{\sigma} = \frac{(\text{CFO}_n/\text{CFO}) - 1}{0.05}$$

$$\text{CFO}_n/\text{CFO} = 0.865$$

For Gaussian distribution, $E_2 = 900$ kV and given that $\sigma_0/E_2 = 0.12$, from Eq. (17-48), $\mu_0 = 678$ kV. Then SSFOR from Eq. (17-61) is:

$$\frac{1}{2}\left[1 - F\left(\frac{865 - 678}{104}\right)\right] = \frac{1}{2}[1 - F(1.79)] = \frac{0.014}{100}$$

$F(Z)$ is obtained from the Gaussian distribution tables.The SFFOR will vary with the type of distribution, that is, skew distribution (App. F).

17-17 OPEN BREAKER POSITION

A breaker when opened for a long time will have its associated disconnects also open. A lightning flash is composed of one or more subsequent strokes, and a stroke may occur while the breaker is open and unprotected by the surge arresters in the substation. A remedy will be to use a surge arrester at the line side of the breaker.

17-18 MONTE CARLO METHOD

The parameters impacting the lightning performance are subject to large variations. The important ones are:

- Current magnitudes of strokes to lines
- Weather conditions and insulator swings
- Tower footing resistance
- Lightning strike locations
- Shielding failures and backflashovers
- Values of power frequency voltages at the time of lightning stroke
- Soil resistivity

Because lightning parameters vary, the effect of lightning must be evaluated on statistical basis. The randomly selected parameters for each lightning incident should determine the outcome—whether a flashover will occur. The analysis must be repeated for the number of variants to obtain a long-term flashover rate. The numbers of incidents normally correspond to the total number of strokes hitting the line over the period of study, 15 to 20 years.

When many statistical distributions have to be combined, the Monte Carlo computer program, based on random selection of relevant parameters including weathering effects, can be used. Anderson was the first to apply this technique.[19,20] Anderson and Barthold[21] refined the approach by a program, Metifor, which creates a digital model of the climate.

17-19 SIMPLIFIED APPROACH

According to IEC[4], the voltage at a point in a substation can be calculated from the following equation:

$$E_t = E_d + 2\frac{S\tau}{n} \tag{17-63}$$

where E_t is the voltage at a certain equipment, E_d is the discharge voltage of the surge arrester, which is the protection level of the surge arrester, S is the steepness of the incoming surge, n is the number of lines, and τ is the transit time between the arrester and the protected equipment, considering the separation distance and the lead length. The discharge voltage of the arrester is not a constant parameter and depends upon the current that it conducts.

The steepness of the incoming surge is given by:

$$S = \frac{K_c}{d + S_l} \tag{17-64}$$

where K_c is the corona constant given in Ref. 15. The suggested values in kilovolts-kilometers per microsecond are 700, 1000, 1700, and 2000 for single conductor, two conductor bundle, 3 or 4 conductor bundle, and 6 or 8 conductor bundle, respectively. S_l is the span length in meters, and d is calculated from the equation:

$$d = \frac{1}{n(\text{BFR})(\text{MTBF})} \tag{17-65}$$

A substation configuration is shown in Fig. 17-20.[3] n is the number of lines terminating at the substation, for calculation of the voltages at transformer and transformer bus, and for all other equipment n is equal to 1, regardless of the number of lines. BFR is the backflashover rate. It can be estimated using IEEE and CIGRE methods (Chap. 5). BFR varies significantly with the system voltage; the BFR for 345- and 500-kV lines is 0.3 to 0.6 flashovers per 100 km-year, and for 138- to 230-kV lines, it ranges from 0.6 to 2. The incoming surge to the station is caused by a backflash or shielding failure and is selected for the desired reliability of the station. Typical values of MTBF range from about 50 to 200 years for air-insulated substations and 200 to 800 years for gas-insulated substations.

As an example, consider MTBF = 100 years, BFR = 2.0 flashovers per 100 years, then distance to flashover $d = 0.5$ km. V_B, V_J, and V_T are the surge voltages at the breaker, at arrester connections, and at the transformer terminals, respectively. Let the lowercase subscripts V_b, V_j, V_t, denote the *total* voltages to ground, including superimposition of power frequency voltages. T_A, T_T, and T_B are the travel times. Two examples in Ref. 3 compare the results of simplified calculations with actual simulations.

Example 17-7 Consider a 345-kV system, protected with a gapless surge arrester of 276-kV rated voltage. The maximum discharge voltage according to vendor's data for 40-kA discharge current is 808 kV. Span length = 350 m, BFR = 0.8/100 km-yr,

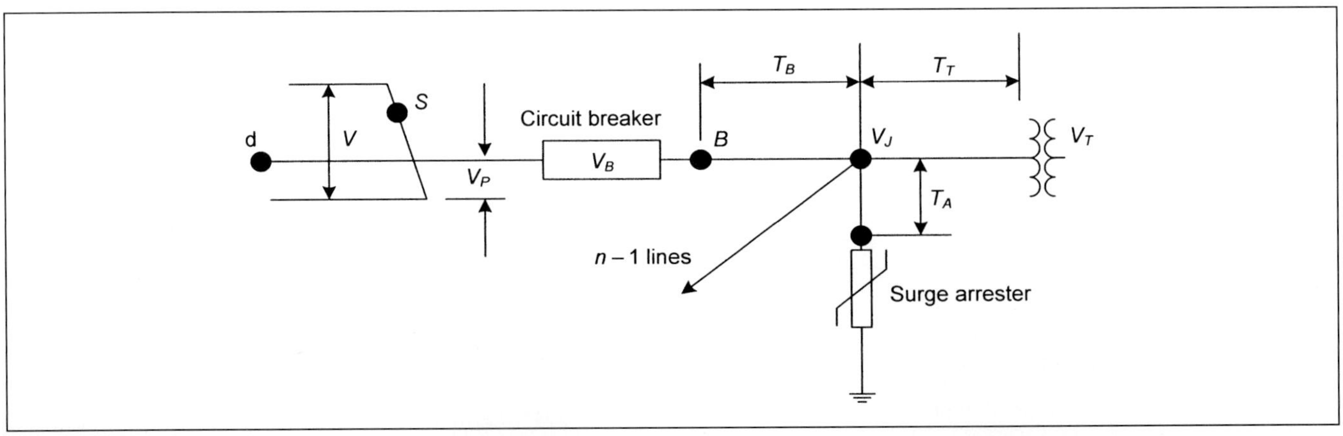

FIGURE 17-20 Simplified representation of a substation with multiple lines for calculations of insulation coordination according to IEEE method.[3]

MTBF = 400 years, K_c for single conductor = 675 kV-km/μs, $n = 3$, and $\tau = 0.1$ μs. Then, from Eq. (17-65)

$$d = \frac{100}{400 \times 0.8} = 0.31 \text{ km}$$

Steepness of incoming surge is:

$$S = \frac{675}{0.31 + 0.35} = 1022 \text{ kV/μs}$$

Therefore from Eq. (17-63)

$$E_t = 808 + 2\left(\frac{1022}{3}\right)(0.1) = 876.1 \text{ kV}$$

Larger the number of lines connected, smaller is the E_t. We can select a BIL equal to 1.2 times E_t, for example, 1051 kV. Standard BIL is 1050 kV, which is the minimum acceptable value.

17-20 SUMMARY OF STEPS IN INSULATION COORDINATION

Primarily, the maximum voltages occurring in the system, overhead lines, substations, or industrial systems need to be ascertained. This is rather a complex situation due to many parameters and statistical nature of the problem. Simplified methods are of academic interest and simulations using EMTP are popular, the results depending upon the accuracy of the models. The steps in insulation coordination can be summarized as:

1. The surge arrester location and selection of appropriate ratings is important (Chap. 20). To design a system without surge arresters will be prohibitively expensive (Example 6-8 of an EMTP simulation).

2. The incoming surge steepness and magnitude is selected. The magnitude is approximately 1.25 times the CFO. The wavefront is calculated. Select station clearances considering weather conditions, ground flash density, and location of the substation, altitude above sea level and pollution level. Select preliminary BIL and BSL levels and calculate CFO. The maximum steepness of surge occurs for a backflashover. Standards provide guidelines for selection of BIL and BSL for various equipments, that is, insulators, transformers, circuit breakers, CTs, PTs, and the like.

3. Consider that some lines may be out of service. The steepness of the wavefront increases as the number of lines is reduced.

4. Consider appropriate safety factors on BIL, BSL, and CFO. These are interrelated as shown in above equations. Depending upon MTBF, the severity of incoming surge will change.

5. A capacitance at the entrance of a station, for example, a CVT, will reduce the steepness of the wavefront.

6. Simulate the system and calculate maximum voltages at the points of interest, that is, circuit breakers, transformers, surge arresters. Revisit the selected BIL and BSL, and apply appropriate safety factors.

PROBLEMS

1. What is the difference between thermosetting thermoplastic and epoxy resins for use as insulating materials?

2. Give five reasons why standards specify different test BIL and BSL levels for the same system voltage.

3. During impulse testing of a restorable insulation, a flashover occurs on the tail of test wave, while the insulation has successfully withstood the peak. Explain this phenomenon with respect to breakdown mechanism in insulations.

4. As the system voltage rises, the insulation coordination with respect to switching surges becomes important, shadowing the insulation coordination with respect to lightning surges. Explain.

5. What is the loss angle of a dielectric? Explain its significance. Will it increase or decrease as the insulation ages?

6. The breakdown voltage for insulation will be higher for (a) positive polarity tests, (b) negative polarity tests, (c) polarity does not make a difference. Provide explanations for the alternative selected.

7. Why the breakdown voltage decreases from high vacuum to mild vacuum and then again increases, Pachen's curve?

8. Distinguish between field emission and secondary emission in a breakdown process.

9. Calculate the leakage distance of an insulator string according to IEEE and IEC methods for a service voltage of 500 kV, medium polluted atmosphere.

10 An insulation under test gave a flashover voltage probability of 5 percent, 150 kV. Probability was 60 percent as the voltage was increased to 180 kV. Estimate CFO and corresponding standard deviation.

11. In Problem 10, five insulators are connected in parallel. What is the overall probability for 150 kV?

12. Consider two dissimilar insulations:
   a. Breakdown probability = 10 percent and time to flashover = 100 to 120 μs
   b. Breakdown probability = 15 percent and time to flashover = 110 to 130 μs

Estimate the overall probability of the two systems in parallel.

13. Determine the clearances and insulator length for a 500-kV line, where switching surge = 2 pu, maximum voltage = 550 kV (pu), variation = 5 percent, wet conditions, decrease CFO by 5 percent, tower width = 2 m, and tower height = 30 m.

14. In Problem 13, calculate the probability of at least one flashover for switching surge of 2.0 and 2.5 pu.

15. Calculate the steepness of an incoming surge for a 500-KV system, where MTBF = 400 years, number of lines = 3, assume appropriate BFR, S_l, and K_c for a single-conductor transmission line.

16. In Figure 17-1*c* why does the insulation resistance dips initially when the insulation is drying up?

17. The lightning performance can be improved by counterpoises and lowering tower footing resistance close to a substation. Explain why. What is the surge impedance of a counterpoise?

18. Calculate the strike distance and insulator string length for a 400-kV tower for design of SSFOR = 1.0/100. Altitude = 1.5 km, $E_2 = 690$, $\sigma_0/E_2 = 0.10$, σ_f/CFO = 0.05, $n = 200$, tower width = 1.2 m, and conductor height = 13 m.

19. Estimate the BIL of a transformer and switchgear in a 230-kV substation, given the following data: Line insulation = 15 discs

of standard insulators, conductor diameter = 2.54 cm, average height = 13 m, protected zone = 1.6 km, surge impedance = 400 Ω, transformer capacitance = 1000 pF, earthing factor = 1.35, temporary overvoltage = 1.10 pu for 1 s, distance of surge arrester to transformer = 10 m, distance of surge arrester to line entrance = 60 m, two lines leading in different directions are connected to the substation. Choose appropriate surge arrester rating. (Peruse Chap. 20, and then attempt this problem.)

REFERENCES

1. K. Feser and A. Pigini, "Influence of Atmospheric Conditions on the Dielectric Strength of External Insulation, *Electra*, no. 112, p. 83, 1987.
2. CIGRE Technical Bulletin 63, Guide to Procedures for Estimating the Lightning Performance of Transmission Lines, 1991.
3. IEEE Standard 1313.2, IEEE Guide for Application of Insulation Coordination, 1999.
4. IEC 60071-2, Insulation Coordination-Application Guide, 1996.
5. NEMA MG-1, Motors and Generators, Large Machines, Parts 20 and 21, 1987.
6. C. M. Cooke and A. H. Cookson, "The Nature and Practice of Gases and Electrical Insulators," *IEEE Trans. EI*, pp. 239–248, 1978.
7. IEC 60071-1, Insulation Coordination, Part 1: Terms, Definitions, Principles and Rules, 1993.
8. IEEE PC62.82.1/D2, Draft Standard for Insulation Coordination—Definitions, Principles, and Rules, 2008.
9. ANSI/IEEE C37.122.1 IEEE Guide for Gas Insulated Substations, 1993(R2002).
10. G. D. Breuer, R. H. Hopkinson, I. B. Johnson, and A. J. Shultz, "Arrester Protection of High Voltage Stations Against Lightning," *IEEE Trans. Power Apparatus and Systems*, vol. 79, pp. 414–423, 1960.
11. IEEE Standard C62.22, Guide for Application of Metal Oxide Surge Arresters for Alternating Current Systems, 1997.
12. G. W. Alexander and H. R. Armstrong, " Electrical Design of 345 kV Double Circuit Transmission Line, Including the Influence of Contamination," *IEEE Trans. PAS*, no. 85, pp. 656–665, 1966.
13. IEEE P1427, Draft 13, Draft Guide for Recommended Electrical Clearances and Insulation Levels in Air Insulated Electrical Power Substations, 2004.
14. NESC, ANSI C-2, National Electric Safety Code, 1993. (Published by IEEE).
15. A. R. Hileman, *Insulation Coordination of Power Systems*, Taylor and Francis Group, Boca Raton, FL, 1999.
16. A. J. Eriksson, M. F. Stringefellow, and D. V. Meal, "Lightning Induced Overvoltages on Overhead Distribution Lines," *IEEE Trans.*, PAS 101, pp. 960–968, 1982.
17. P. Chowdhuri, *Electromagnetic Transients in Power Systems*, John Wiley, New York, 1996.
18. CIGRE Working Group, "Switching Overvoltages in EHV and UHV Systems with Special Reference to Closing and Reclosing Transmission Lines," *Electra*, pp. 70-122, Oct. 1973.
19. J. G. Anderson, "Monte Carlo Computer Calculation of Transmission Line Lightning Performance," *AIEEE Trans, Power Apparatus and Systems*, no. 80, pp. 414–419, 1961.
20. M. A. Sargent, "Monte Carlo Simulation of Lightning Performance of Overhead Shielding Networks of High Voltage Stations," *IEEE Trans. PAS*, Part III, no. 91, pp. 1651–1656, 1972.
21. J. G. Anderson and L. O. Barthold, "Metifor, a Statistical Method of Insulation Coordination Design of EHV Lines," *IEEE Trans. Power Apparatus and Systems*, no. 83, pp. 271–280, 1964.
22. G. W. Brown, "Designing EHV Lines to A Given Outage Rate—Simplified Techniques," *IEEE Trans. PAS*, pp. 379–383, 1978.

FURTHER READING

EHV Transmission Line Reference Book, Edison Electric Institute, New York, 1968.

G. Gallet, G. LeRoy, R. Lacey, and I. Kromel, "General Expressions for Positive Switching Impulse Strength up to Extra Long Air Gaps," *IEEE Trans. PAS*, pp. 1989–1993, Nov./Dec. 1975.

T. Harada, Y. Aihara, and Y. Aoshima, "Influence of Humidity on Lightning and Switching Impulse Flashover Voltages," *IEEE Trans.*, PAS-90, no. 4, pp. 1433–1442, 1971.

IEEE Standard 1243, IEEE Guide for Improving the Performance of Transmission Lines, 1997.

IEEE Standard. 1313.1, IEEE Standard for Insulation Coordination—Definitions, Principles, and Rules, 1996.

A. M. Mousa and R. J. Wehling, "A Survey of Industry Practices Regarding Shielding of Substations Against Direct Lightning Strokes," Paper 92, WM 224-6, PWRD, 1992.

Transmission Line Reference Book, 345 kV and Above, 1st ed., EPRI, Palo Alto, CA, 1975.

Transmission Line Reference Book, 345 kV and Above, 2d ed., EPRI, Palo Alto, CA, 1982.

M. A. Uman, *Lightning*, McGraw-Hill, New York, 1969.

CHAPTER 18

Gas-Insulated Substations—Very Fast Transients

GIS utilize the superior insulating properties of SF_6 gas, and the assembly consists of grounded metal equipment modules, that is, circuit breakers, current and voltage transformers, disconnects and grounding switches, interconnecting bus, and connections to the electrical power system. The modules are assembled together using bolted flanges with an O-ring seal system for the enclosure and sliding plug-in contact for the conductor. Three-bus enclosure or single-bus enclosure configuration is used, depending on the application voltage and other considerations. Figure 18-1 shows typical assembly of GIS.

The assembly can be located in enclosed buildings, and the space requirement may be only 3 percent of a conventional outdoor air-insulated substation. Thus, the aesthetics, community considerations, and location in crowded areas are some of the obvious advantages; where the real estate is at a premium cost, the onsite cost of GIS may come out to be comparative with air insulated substation.

18-1 CATEGORIZATION OF VFT

In this chapter, we will not be so much concerned about assembly and mechanical arrangements, but the very fast transients (VTF) that occur in GIS. These can be divided into two categories:

- *Internal transients*. These produce overvoltages between the inner conductor and the encapsulation and traveling waves inside the enclosure. These create insulation stresses in GIS.

- *External transients*. These are due to traveling waves and radiation outside the GIS. These include transient enclosure voltage (TEV), transient electromagnetic fields (TEMF), and overvoltages on overhead lines and equipment. These can cause stress on secondary and adjacent high-voltage equipment.

Internal transients are generated in GIS during normal operation by switching of disconnects and breakers and by dielectric breakdown. The collapse of voltage across the contacts of a switching device or to ground in case of a dielectric breakdown occurs in 3 to 5 ns. This is rapid enough to excite resonances within GIS at a frequency up to about 100 MHz. Traveling waves of very short rise time occur which propagate in either direction from the breakdown location. The propagation throughout the substation can be analyzed by representing GIS sections as low-loss distributed parameter transmission line, each section having a certain characteristic impedance and transit time. Akin to transmission line traveling wave theory (Chap. 4), the traveling waves are reflected and refracted at every point where these meet an impedance discontinuity. The generated transients depend on the GIS configuration, and superimposition of reflected and refracted waves at discontinuities like "T" junctions, bushings, or breakers. Thus, transient voltages can increase above the original values and very high-frequency oscillations can occur.

18-2 DISCONNECTOR-INDUCED TRANSIENTS

Isolators (also called disconnects/disconnectors) are installed in several locations in a GIS and are referred to as busbar isolators or line isolators. Two situations can be examined during switching action of an isolator.

1. Two networks are to be isolated, for example, on the right and left of the GIS; each network has its distributed inductance and capacitance. Under unfavorable conditions, there can be a phase shift between the voltage of the two networks, and a voltage of twice the system voltage can arise across the poles of the isolator (Fig. 18-2).

2. The second more frequent use is to connect or disconnect unloaded portions of the GIS. The second network on the right in Fig. 18-2, system 2, is absent and the isolator disconnects a part of GIS or overhead line from the source.

Disconnector operation typically involves slow-moving contacts that result in numerous discharges during operation. A restrike occurs every time the voltage between contacts exceeds the dielectric strength of the gaseous medium. Each restrike generates a spark, which equalizes the potential between the contacts. Following spark extinction, the source and load-side voltages again deviate and another spark occurs when the voltage across the contacts reaches new dielectric strength. Now consider the opening of disconnect with flashover. Assume that the load side of the interrupter

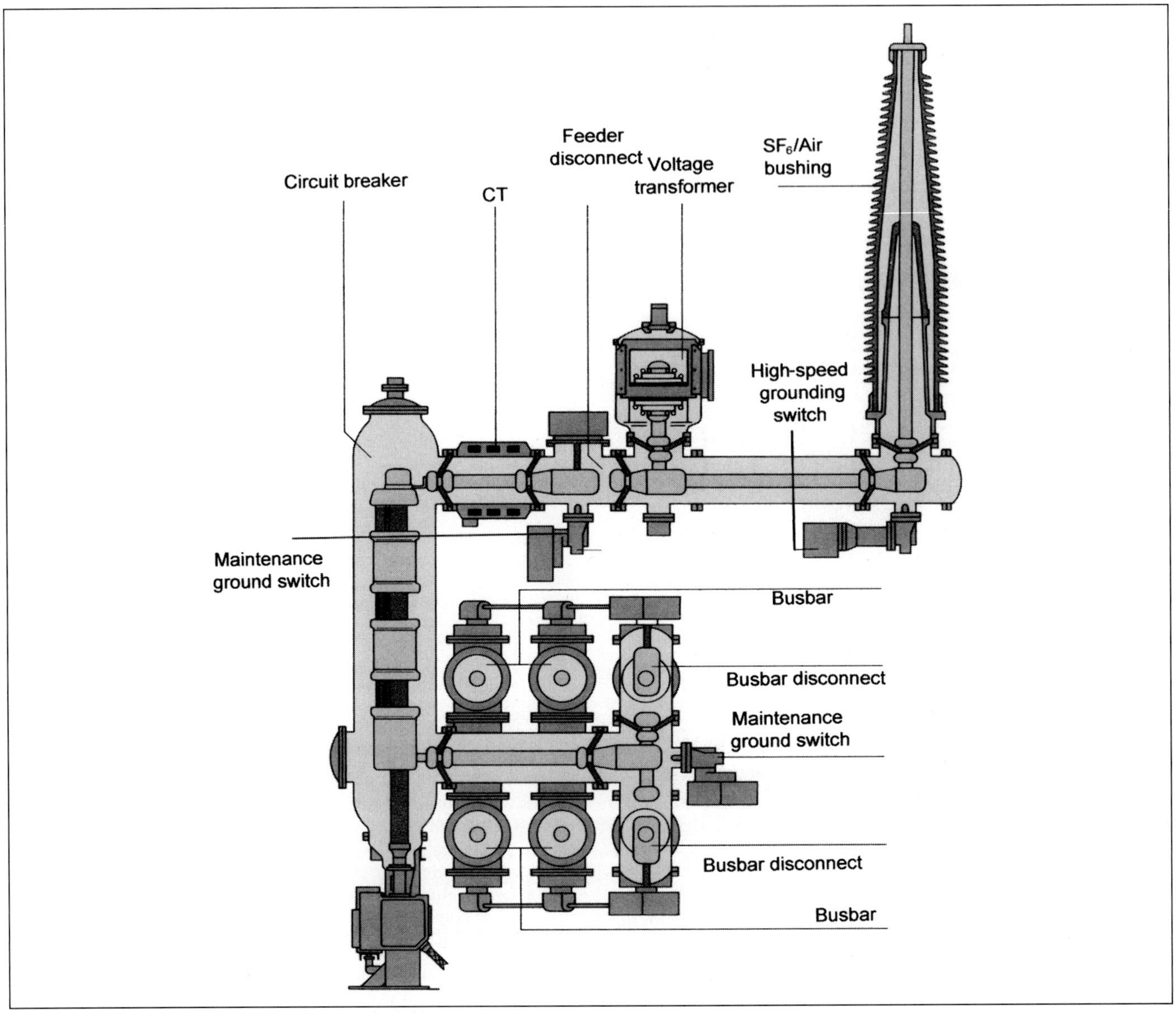

Figure 18-1 Cross section through a section of GIS assembly.

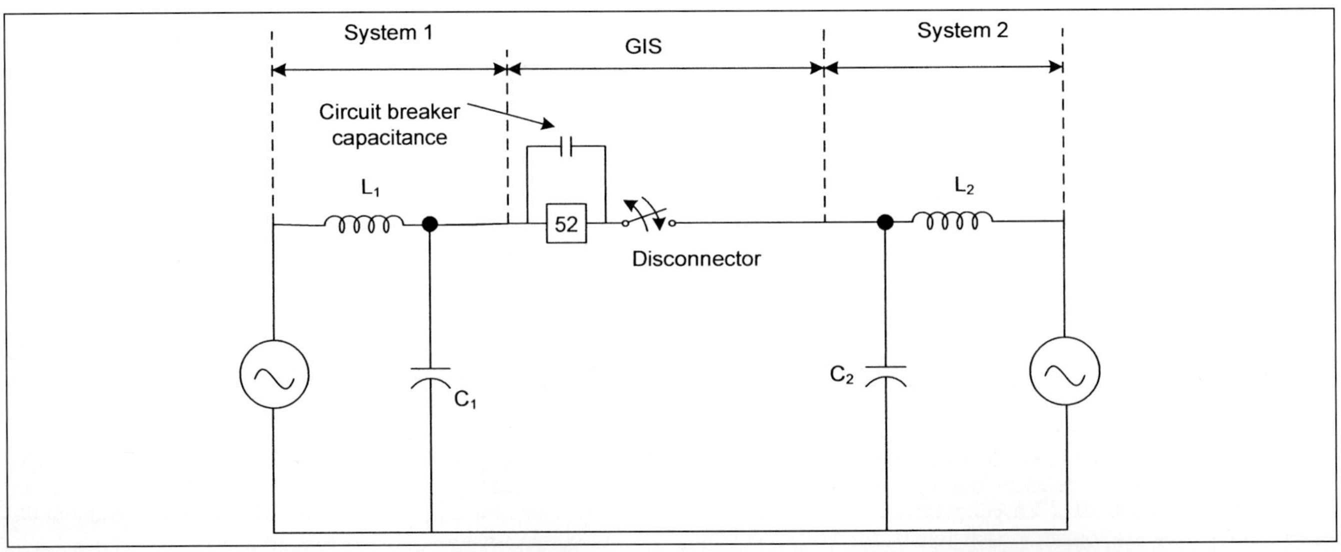

Figure 18-2 A disconnector in GIS, interconnecting two systems, possible switching in phase opposition.

has trapped charge V_1 and surge impedance Z_1, and the source side has a trapped charge V_s and surge impedance Z_s. At the time of breakdown (sparking occurs as soon as the voltage between the source and load exceeds the dielectric strength across the contacts), the voltage on the load side goes from V_1 to:

$$V_1 + (V_s - V_1)\frac{Z_1}{Z_1 + Z_s} \tag{18-1}$$

And the voltage on the source side goes from V_s to:

$$V_s - (V_s - V_1)\frac{Z_s}{Z_1 + Z_s} \tag{18-2}$$

After restrike, a high-frequency current will flow through the spark and equalize the capacitive load voltage to the source voltage. The potential difference across the contacts will fall and the spark will extinguish. A subsequent restrike occurs when the voltage between the contacts reaches the new dielectric strength level, determined by the speed of the parting of contacts and the disconnect characteristics.

In the simplest case, consider $Z_s = Z_1$, $V_s = 1$ pu, and $V_1 = 0$, that is, a disconnector connecting a previously grounded load and the first prestrike occurring at peak of the ac source voltage. V_s and V_1 go to 0.5 pu in about 3 to 5 ns. For a GIS operating at 420 kV, this implies a change of 210 kV in, say, 4 ns or rate of change of approximately 50 MV/μs. In terms of direction of propagation, the voltage changes by 210 kV over a distance of 130 cm or a change of 1.6 kV/cm.

The transient magnitude of the voltage on one side of disconnect can exceed if the surge impedances differ on the two sides of the disconnector, that is, if the disconnector is near a "T" or a capacitive voltage transformer is located close to the disconnector.

Figure 18-3 illustrates these phenomena. If we assume that the breakdown voltage of the gap increases with increasing separation of the contacts as they part, then at the start of the contact separation, voltages on V_1 and V_2 are equal. As the voltage $V_2 - V_1$ across the gap changes and exceeds the dielectric strength of the gap, a breakdown occurs and first restrike occurs. Both poles of the disconnecting switch are electrically connected by the conducting spark, and V_2 is rapidly changed to V_1. At this instant, the transient current through the gap interrupts. As the supply-side voltage changes, the second restrike occurs with an increased breakdown voltage, V_B, due to larger gap. The voltage V_2 follows voltage V_1 in steps until at the end of the switching process, the gap can no longer be broken down due to increased spacing.

The discharge during each individual restrike begins as the voltage across the contacts collapses, and in SF_6, it occurs in 10^{-2} μs and is directly related to the formation of spark channel. With a typical voltage decrease of 10 kV in 1 ns, it gives rise to a traveling wave, which propagates away from the gap into the installation. After a certain travel time, the wave reaches the open end of the GIS, is reflected back, and crosses the gap, which may be still bridged by the arc. At a further discontinuity, for example, connection to an OH line, it is split into reflected and transmitted components. The reflected wave travels a second time toward the open end of GIS and is again reflected. Thus, the discharge transient shows a double periodicity of the traveling wave. If the circuit has T-connections, then the discharge transient will have components of different frequency. Very different magnitudes can occur, depending on the installation.

Example 18-1 An open-ended GIS has $Z_g = 75\ \Omega$, which is to be disconnected from an OH line of $Z_1 = 320\ \Omega$. A restrike occurs at $V_b = 500$ kV, at a contact gap of 3 cm. Find the initial amplitude of the voltage and current entering into GIS and how it ends.

The initial amplitude V_{ing} entering GIS can be easily calculated as no reflection is present yet, and the breakdown voltage V_b distributes itself corresponding to surge impedances of the lines. The voltage wave entering the GIS is:

$$V_{ing} = \frac{V_b Z_g}{Z_g + Z_1 + R_s} \tag{18-3}$$

where R_s is the spark resistance and is neglected, initially. This gives 95 kV. Thus, the voltage wave into the transmission line is 405 kV and into GIS 95 kV. The initial value of current wave is given by:

$$I_{in} = \frac{V_{ing}}{Z_g} = 1.26 \text{ kA} \tag{18-4}$$

The spark resistance, which was neglected earlier, can be found from Toepler equation:

$$R_s = \frac{K_T l}{\int_0^t i\,dt} \tag{18-5}$$

where R_s is spark resistance in ohms, K_T is Toepler coefficient = $(0.4 - 0.8)10^{-4}$ Vs/cm for SF_6 at 1 to 20 bar, and l is spark length in cm. This shows that the resistance falls below 10 Ω in 20 ns.

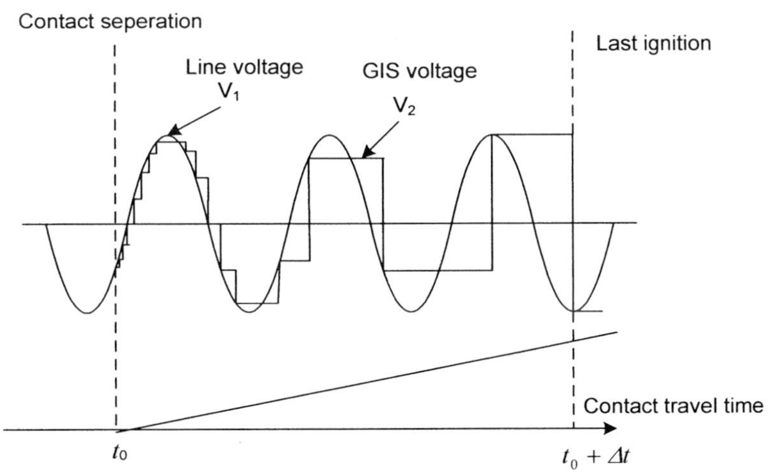

FIGURE 18-3 Voltage on the open-ended GIS side of the disconnector without loss of charge.

The power carried by the wave into the transmission line is:

$$P = \frac{V_{\text{inl}}^2}{Z_l} \tag{18-6}$$

where V_{inl} voltage entering transmission line = 405 kV.

This gives a power of 512 MW into the line. In comparison, the power loss in the spark resistance and GIS, for practical purposes, is so small that it can be ignored. Thus, the transient discharge oscillations in GIS are mainly damped by wave transmitted into the transmission line; in other words, in a loss-free GIS open at both ends, if a surge voltage V_o enters GIS, the discharge oscillation will continue indefinitely. Considering reflection now, the reflection factor is (Chap. 4):

$$\rho = \frac{Z_1 - Z_g}{Z_1 + Z_g} \tag{18-7}$$

The transmitted wave of $(1+P)V_0$ runs into the line. At the end of reflection process, the voltage on line 2 is reduced by reflection constant and the energy stored in the wave in GIS goes down by ρ^2. Though the waves may be more complex, the fundamental concept that the wave energy is removed by transmitted waves holds. In isolator models, the arc extinction takes place when the current has fallen to 1 to 5 A. Though this current to be interrupted is small, high *di/dt* values occur due to high frequency. As the current reduces, it can be interrupted at a high-frequency zero crossing.

The total duration of transients has been estimated from optical observations of the spark, with the use of so-called streak photographs on a high-speed moving film. The duration of luminous period is determined by smearing of the spark picture. These pictures can also show sequential restrike with time interval between them.

The behavior on a closing operation is similar and the load-side voltage will follow the supply-side voltages until the contacts are made.

18-3 BREAKDOWN IN GIS—FREE PARTICLES

If a breakdown occurs within GIS, the transient magnitude is generally much greater than for a switching operation. The voltage goes from initial value of *V* (1 per unit) to zero, again in 3 to 5 ns. Though the in-service breakdowns are rare, and when these occur, they are induced by switching-related transient or by areas of high-voltage stress within GIS. The electrical field in a GIS is approximately uniform due to circular enclosures. During manufacture, extreme care is taken to eliminate any irregularities and free particles, which can be a potential problem in breakdown in SF_6. This becomes important as the operating voltage is raised. Extreme care is required in handling and field assembly to eliminate free conducting particles. Most dielectric failures will generate a transient larger at least by a factor of two than the switching transient. During testing, the voltage is raised to 2.5 times the normal operating voltage, and a breakdown under such a condition will generate transients three to four times greater.

The VHF transients thus generated due to breakdown or across switch contacts propagate away from their source, reflecting and refracting at impedance mismatches. This results in a complex pattern of traveling waves within the substation that may superimpose to generate greater voltages in some locations as compared to the others. The switching overvoltages to which a substation is subjected is a function of station configuration and the characteristics of the switching devices.

18-3-1 Trapped Charges

When a disconnector operates on a floating section of switchgear, a trapped charge may be left on the floating section due to capacitance. Left to themselves, these trapped charges will decay very slowly, from hours to days, as a result of leakage through spacers. Large trapped charges are undesirable because they will not only result in overvoltages, but may also levitate particles. These particles may be scattered across the insulating surfaces, reducing their breakdown strength. A trapped charge of 1pu implies that the first breakdown on closing the disconnector will occur at 2 pu across the switch contacts and may lead to conductor-to-ground overvoltage up to 2.5 pu.

The disconnector operation with +1 pu voltage on one side and −1 pu voltage on the other side is the worst-case scenario. In theory, this can give rise to peak waveform amplitude of 3 pu. Numerous studies indicate theoretical worst-case transient-induced overvoltages in the range of 2.8 pu relative to power frequency peak voltage. Typically, a magnitude little over 2 pu is more practical. The probability density of trapped charges is discussed in (Ref. 1), which shows that only 5.9 percent trapped charges in pu, normalized by the peak voltage of the applied ac voltage, are of 1.0 per unit. 65.3 percent are below 0.5 per unit. The probability density of trapped charge left on a capacitive load during opening of a disconnector with arcing time of 0.6 s and 0, 15, 30 percent and contact breakdown asymmetry as a result of 1000 Monte Carlo simulations is shown in (Ref. 2). This shows that the trapped charge left on the load side is less than 0.3 pu.

18-3-2 Speed of Operation

Overvoltages are dependent on the speed of the operation of the disconnector. The voltage drop at the disconnector before striking and the trapped charge that remains on the load side are of concern. For a slow-speed interrupter, the maximum trapped charge reaches 0.5 pu and a voltage collapse of 1.5 pu. For these cases, the overvoltages may reach up to 2 pu. The trapped charge increases with a high-speed interrupter, and can be a maximum of 1 pu. The highest overvoltages reach 2.5 pu. The frequencies depend on the length of the GIS section and are in the range of 1 to 50 MHz. Larger disconnector-induced transients occur during opening as a result of the statistical variation in nature of disconnector operation, documented in (Ref. 3). In case of a high-speed disconnector, the probability of an occurrence of prestrike by interpolar breakdown voltage of 1.8 to 2.0 per unit is less than 1 percent.[1]

During closing, a reversal of the previous description takes place. The first strike invariably occurs at the peak of the power frequency voltage. When the GIS is fed through a transformer, oscillations of the whole system can occur. The high-frequency transients may couple through PTs and CTs, interfere with the secondary equipment, and cause electromagnetic coupling problems.

18-3-3 Operating Voltage of GIS

The magnitude of switching-induced transients is proportional to the operating voltage, while the dimensioning of GIS say up to 550 kV, is proportional to the basic impulse levels impulse (BIL). Thus, the severity of the switching-induced transients increases with the increasing voltage class. Consider a BIL of 1550 kV for 550-kV GIS and a BIL of 1050 kV for 230-kV GIS. The ratio of the BIL with respect to operating voltage is 3.45 for 550-kV GIS and 5.6 for 230-kV GIS.

In the past, 25 percent failures in 550-kV GIS occurred due to switching-induced transients, though the transients should not have produced a voltage of 3.45 pu, IEEE Std. C37.122.1.[4] (This standard contains a bibliography of 1518 technical papers and references on GIS.) When a defect is present in the conductor inside GIS, the withstand voltage can decrease by a factor of 2 or more, and a decrease in the surge waveform rise time is from 100 to 10 μs.[4] This points to the necessity of appropriate field testing.

18-3-4 Maximum VFT Stresses

The maximum VFT stresses under normal operating conditions occur due to disconnector operation. The maximum voltage is a function of the trapped charge on the load side, the geometry of GIS, and the disconnect voltage at the time of breakdown. The following conclusions can be drawn based on the published literature:

- The trapped charge is mainly dependent on disconnector characteristics. The faster the switch, the greater is the mean value of the voltage.
- The capacitive-graded GIS-air or GIS-oil bushings impact the rise time of the transient, which is further discussed in Sec. 18-8-2.
- For slow switches, the probability of restrikes and prestrikes in the voltage range of 1.8 to 2 per unit is small, though it cannot always be ignored.
- The breakdown voltage of the disconnector gap may vary for positive and negative values, which effects the trapped charge distribution.

18-4 EXTERNAL TRANSIENTS

Internally generated transients propagate throughout GIS to reach external connections and bushings, where they cause transient enclosure voltages (TEV) and traveling waves that propagate along overhead transmission lines.

18-4-1 Transient Enclosure Voltages

Transient enclosure voltages are also called *transient ground potential rises* (TGPR). These are short-duration high-voltage transients that appear on the enclosure of GIS because internal transients couple to the enclosure at its discontinuities. Figure 18-4 illustrates this. The propagation of an internally generated transient to an air termination and its refraction to outside transmission line/cable is the dominant mechanism for damping out the switching transient. At the air termination, we can decipher three transmission lines as follows:

1. The GIS enclosure is at ground for the GIS conductor, but it acts as a conductor with respect to the station ground and ground mat. At higher frequencies, GIS enclosure, generally about 4 ft from the ground, is really *two distinct conducting surfaces*, interior and exterior.[5] Thus, part of the energy couples to the transmission line formed by outside of the GIS enclosure relative to earth and station ground.
2. The GIS conductor to GIS enclosure.
3. Overhead line to ground.

The reflection coefficient at the transition:

$$\rho_{tr} = \frac{(Z_{enc} + Z_{oh})Z_{gis}}{Z_{enc} + Z_{oh} + Z_{gis}} \tag{18-8}$$

Then, refraction coefficient is $1 + \rho_{tr}$.
The peak transient ground wave into outside enclosure is:

$$(1+\rho_{tr})\frac{-Z_{enc}}{Z_{enc} + Z_{oh}} \tag{18-9}$$

This simplifies to:

$$V_{gr} = -2V_i\left[\frac{Z_{enc}}{Z_{enc} + Z_{oh} + Z_{gis}}\right] \tag{18-10}$$

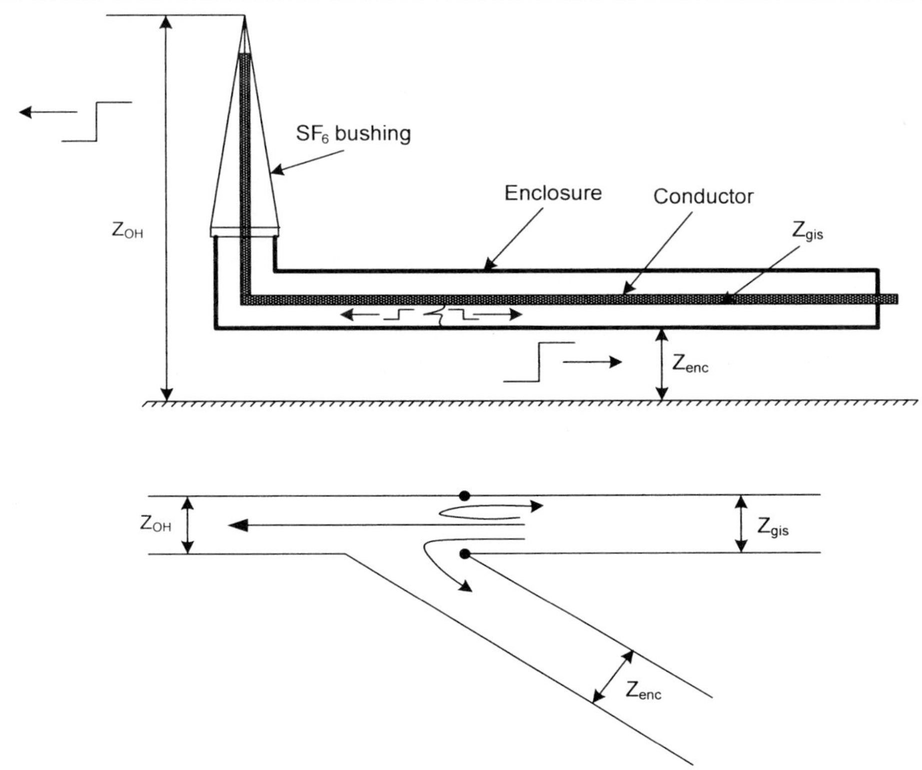

FIGURE 18-4 Voltage transmission on an insulation breakdown within GIS.

where V_i is the voltage within GIS incident on termination; Z_{gis} is the GIS bus duct impedance, typically 60 Ω; Z_{oh} is the overhead line surge impedance, typically 300 to 400 Ω; and Z_{enc} is the impedance of the GIS enclosure relative to station ground, typically 150 Ω.

The negative sign in Eq. (18-10) implies reversal of the waveform with respect to the internal transient. This assumes perfect junction between the lines. The effect of vertical bushings is ignored. The real SF_6 systems are not so simple as a configuration of a central conductor, enclosure, and ground plane, yet the magnitude of the *initial* step transient that is coupled to the enclosure-to-ground plane can be estimated with reasonable accuracy from Eq. (18-10). The surge impedance Z_{enc} can be calculated from the following equation:

$$Z_{enc} \approx 60 \cos h^{-1}(h/r) \tag{18-11}$$

where r is the radius of the enclosure and h is the height. For a 500-kV GIS, consider r of the order of 25 to 30 cm, and height 1.5 m. This gives a surge impedance of approximately 150 Ω.

Although the initial TGPR response may be like a step function, reflections from external ground leads and internal discontinuities will cause overall response to be oscillatory. These oscillations are largely dependent on the physical length of the bus and are in the range of 5 to 50 MHz.

18-4-2 Effect of the Ground Straps

In Fig. 18-4, one way to reduce TGPR is to reduce Z_{enc} and add ground straps from the enclosure to the ground plane. The ground straps have significant effect on the waveshape that propagates away from the terminations, though most ground straps may be too long and too inductive for effective grounding of TGPR. The ground lead can be considered as a vertical transmission line with surge impedance that varies with the height. The attachment of the ground strap to the enclosure constitutes a T-connection. When the transient reaches a ground strap, a division takes place, which reduces the magnitude of the transmitted wave. This can be expressed by the coupling coefficient, as follows:

$$k = \frac{2Z_{strap}}{2Z_{strap} + Z_{enc}} \tag{18-12}$$

where Z_{strap} is the surge impedance of the ground strap. This is normally > Z_{enc}, the attenuation is normally small, $k = 0.8$. Practically, the surge impedance of the ground strap varies along its length, but may be considered constant for simplification.

Now consider the effect of the grounding grid. The wave traveling down the ground strap meets the low resistance of the grounding grid. At this point, the wave is reflected back and reduced along the grounding lead to the enclosure, where it will tend to cancel the original waveform, the phenomenon similar to the one illustrated in Example 5-3, in connection with a lattice diagram of a struck tower. The traveling wave from the air/SF_6 bushing will be altered by the reflected ground wave after a time interval equal to twice the transit time along the ground strap. Depending on this time, the incident wave may be reduced in magnitude or attain its peak before the reflected wave from the ground reduces its magnitude. The end result may be a peaked wave, as shown in Fig. 18-5.

The maximum TGPR will be approximately 60 percent of the voltage incident on GIS-to-air termination and of opposite polarity. The maximum transient likely to be incident on the air termination is 1.3 pu which will generate 0.8 pu on the enclosure at the base of the termination. This voltage will last only for nanoseconds. TGPR due to dielectric breakdown of GIS, especially during testing, is of much larger magnitude than that caused by switching. Note the following points:

- The ground connection has no effect on the transient generated by the breakdown, the transient magnitude of GIS-to-air termination, or transient ground rise. It will have only an effect when the transient ground rise reaches the ground connections many nanoseconds later.

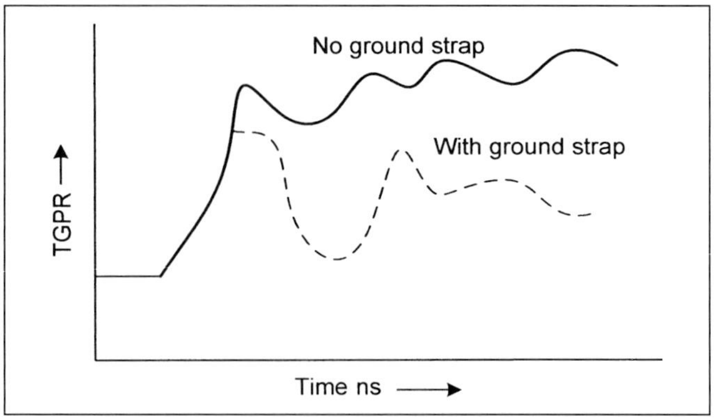

FIGURE 18-5 TGPR with and without ground straps.

- The rise time of the wave which emerges from GIS-to-air termination will be of the order of 10 ns.

- The wave takes about 4 ns to propagate from GIS enclosure to ground where it gets inverted and reflected back to the enclosure, canceling the voltage on the enclosure.

- Transient ground rise can cause sparks across electrical discontinuities in the electrical enclosure and can induce transients on poorly configured control cables that may cause incorrect operation of electrical and electronic equipment.

- Transient ground rise in indoor GIS is reduced through the use of a metal building with a coaxial connection between the metal wall and the enclosure. The interior of the building acts like a Faraday cage and reflects transients that propagate from external GIS to air terminations back toward the building.

- When a GIS is connected to a high-pressure fluid-filled cable, the enclosures of two systems are isolated and bypassed with a polarization cell to apply cathodic protection to the cable pipe.[6] The transient can cause a failure of the insulation between the cable pipe and GIS. The problem can be eliminated by extending the insulator to the point so that it does not flash over.

18-5 EFFECT OF LUMPED CAPACITANCE AT ENTRANCE TO GIS

When a capacitance, for example, a CVT is installed at the entrance of a GIS, under fault conditions the capacitor will discharge, the frequency determined by inductance and capacitance of the discharge loop (Fig. 18-6). For a loop inductance of 20 μH and capacitance of 1000 pF, the discharge frequency will be about 1 MHz and, therefore, skin effect considerations apply. The refracted waveforms of higher frequency are superimposed upon the lower-frequency waveform due to discharge loop. Induced potentials may be generated in the nearby metallic loops. A discharge waveform is shown in the EMTP simulation in an example later.

The waveform can be approximated by the following expression:[7]

$$V = V_l e^{-\alpha_l t} \cos \omega_l t + V_h e^{-\alpha_h t} \sin \omega_h t \tag{18-13}$$

where V_l is peak value of low-frequency component of enclosure transient voltage; V_h is high-frequency component of enclosure

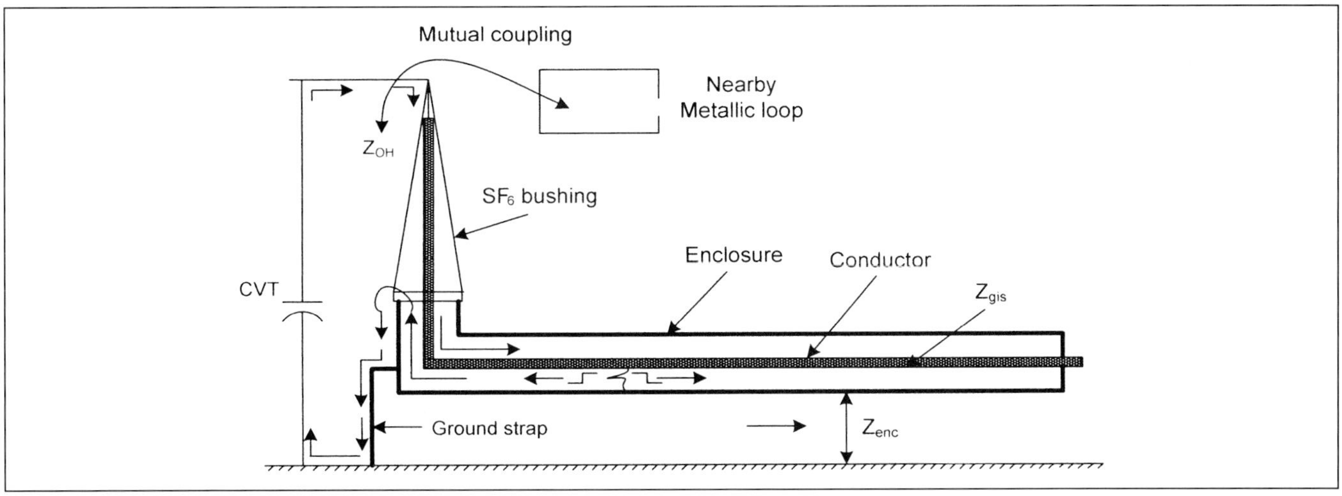

FIGURE 18-6 Impact of external capacitance discharge on TGPR.

transient voltage; α_l and α_h are attenuation factors of low-frequency and high-frequency components, respectively; and ω_l and ω_h are angular frequency of low-frequency and high-frequency components, respectively.

For shock hazard considerations, the body energy can be calculated from the following equation:

$$E = \frac{1}{4R_B}\left[\frac{V_l^2}{\alpha_l} + \frac{V_h^2}{\alpha_h}\right] \tag{18-14}$$

where the body resistance R_B varies from 50 Ω for good contact to 100000 Ω for dry contact, typical value being 500 Ω.

18-6 TRANSIENT ELECTROMAGNETIC FIELDS

The electromagnetic fields are radiated from the enclosure and will stress the secondary equipment, especially the electronic controls and the computer equipment. Measurements of these fields have been made on GIS.[8] At a distance of 9 cm, an electrical field of 60 kV/m with main oscillating frequency of 10 MHz was observed, and at a distance of 5 m from the enclosure, an electrical field of 5 kV/m but higher frequency was observed in an 800-kV GIS. While the electrical field strength rapidly reduces with distance, its frequency increases. Reflections and refractions from internal structures are observed.

18-7 BREAKDOWN IN SF_6

We discussed some superior insulating properties of SF_6 as an electronegative gas and breakdown mechanism in Chap. 8. With respect to insulation coordination in GIS, it is an important criterion. Here, the insulation has to be considered nonrestoring, and in the event of a fault, long downtimes have to be accepted. To ensure reliability, this failure rate should be kept low. Due to almost perfect metal casing, atmospheric and environmental effects are minimum, yet for the steep-fronted waves, the insulation properties are quite different. Due to small gap and high field strengths, the voltage-time characteristics of SF_6 are comparatively flatter as compared to that of air insulation, and this is of much importance. It becomes all the more necessary that the overvoltages are limited by surge arresters.

We will not go into the breakdown theories in detail, except to show breakdown time lags in SF_6 on the application of a surge voltage, (Fig. 18-7).

The two criteria for breakdown are that:

1. There must be enough electrical fields of sufficient strength and adequate distribution to produce avalanches according to streamer criteria. For a given electrode configuration, this condition is met by voltage U_0 at time t_0.

2. There must be one primary electron in a suitable location to initiate first avalanche. Such electrons are produced due to cosmic, ultraviolet, and radioactive radiations. Electrons are emitted at the cathode by electrical and thermal excitations and photon emissions. As these occur randomly, the time lag t_s (statistical time lag) indicates availability of an initiatory electron.

According to streamer criteria a first avalanche occurs followed by condensing avalanches bridging the gap.

There is a third time delay t_f (formative time lag) which is the sum of two time delays—one time lag for the formation of streamer (negligible) and second for the formation of spark.

The total time lag after the application of voltage impulse is t_c, which is calculated as follows:

$$t_c = t_0 + t_s + t_f \tag{18-15}$$

Figure 18-7 shows the difference between a step voltage and lightning impulse.[9]

18-7-1 Voltage-Time Characteristics

Figure 18-8*a* shows the VI characteristics when U_0 is exceeded. For zero-percent probability, only t_f is considered. Then t_s is added later on for 95 percent probability. The voltage-time characteristics obtained by this method are shown in Fig. 18-8*b*. Further research shows that the breakdown voltages are higher if the impulse is chopped, of the order of 10 percent or more, and the influence of the wavefront is much smaller as compared to the wave-tail on the breakdown voltage and the breakdown probability.[10]

The SF_6 pressure, stressed electrode area, impurities, and electrode surface roughness are some factors that influence the breakdown on switching and lightning surges.

When applied to actual assembly of GIS, the field strength is approximately uniform and BIL is approximately equal to chopped wave test. Table 18-1 shows the BIL and BSL levels of GIS from IEEE Std. C37.122.[11] The importance of BSL under disconnector operation has been highlighted earlier. It may be prudent to raise the BSL level when specifying GIS over the values specified in the standards for better security. The safety margins considered are normally 20 percent above the maximum crest of the voltage calculated, in GIS.

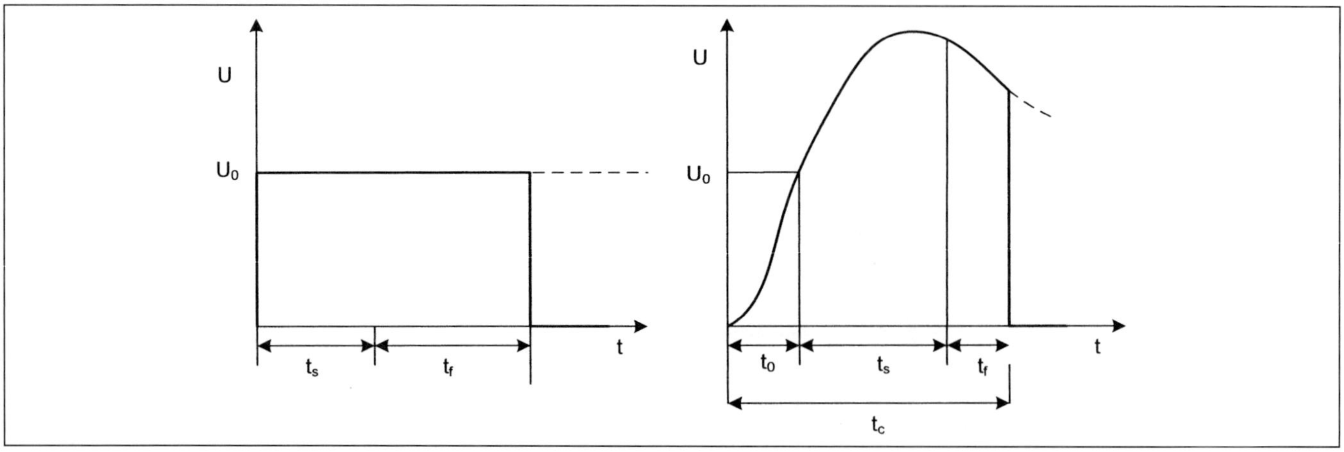

FIGURE 18-7 Breakdown time lags in SF_6 for step voltage change and lightning impulse.

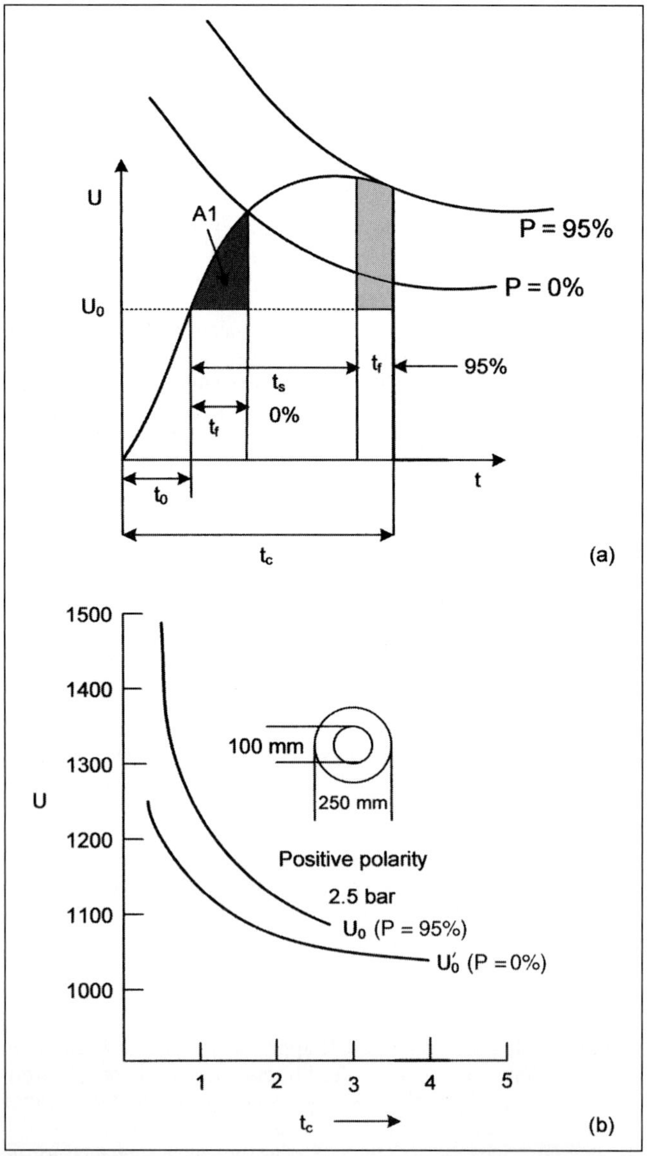

FIGURE 18-8 (*a*) Application of volume-time and voltage-time criteria for development of voltage-time characteristics of SF_6. (*b*) Calculated voltage-time characteristics of SF_6, based on (*a*).

18-8 MODELING OF TRANSIENTS IN GIS

The modeling of transients in GIS is not straightforward. The internal damping is determined by a number of factors, including spark resistance, and refraction to lines and cables connected to the substation. The accurate modeling of these components and the relationship between the enclosure and ground are necessary.[3,12] The computing of an accurate waveform at an interval of a meter in the substation would be impractical. For a rise time of the order of 10 ns, any connection larger than 1 ft should be thought of as a transmission line with surge impedance and a propagation time. An approximate approach is acceptable so long as it is verified against measurements.

IEEE reports in (Refs. 13 and 14) provide modeling guidelines for VFT. The modeling can be single-phase or three-phase, depending on the study to be performed. An electrical equivalent circuit composed of lumped elements and distributed parameter line models is used. The skin effect at high frequencies can produce appreciable attenuation; however, often this effect is ignored for conservatism. The calculations of internal transients may be performed using distributed parameter models for which conductor enclosure is taken into account, assuming that external enclosure is perfectly grounded. For TEV, a second mode of enclosure to ground is also considered.

18-8-1 Bus Work, Conductors, and Cables

For frequencies up to 100 MHz, the bus duct can be represented as a lossless transmission line. The inductance and capacitance of a coaxial single-phase cylindrical configuration with conductor radius of r and internal enclosure radius of R is given by:

$$L = \frac{\mu_0}{2\pi}\ln\frac{R}{r}$$
$$C = \frac{2\pi\varepsilon}{\ln(R/r)} \tag{18-16}$$

Therefore, the surge impedance is:

$$Z = \sqrt{L/C} = \frac{\sqrt{\mu_0\varepsilon}}{2\pi}\ln\frac{R}{r} \approx 60\ln\frac{R}{r} \quad (\varepsilon = \varepsilon_0) \tag{18-17}$$

Experimental results show that propagation velocity in GIS ducts is approximately 0.95 to 0.96 of the speed of light.

The bus work and conductors between discontinuity points and connections between substation equipments can be represented by line sections, which can be modeled by untransposed distributed parameter sections with surge impedances. The minimum

TABLE 18-1 BIL and BSL of GIS

System Voltage	Type Test Voltages		
Rated Maximum Voltage Phase-to-Phase	Rated BIL, kV Crest	Low-Frequency Phase-to-Ground Withstand kV, rms	Switching Impulse Withstand kV Crest
72.5	350	160	
121	550	215	
145	650	310	
169	750	365	
242	900	425	
362	1050	500	825
550	1550	740	1175
800	2100	960	1550

ANSI/IEEE Std. C37.122.1, IEEE Guide for Gas Insulated Substations, 1993 (R2002).[4]

conductor length with distributed parameter representation dictates the simulation step in EMTP which may become of the order of 1 ns (typically 5 ns). The surge impedance of GIS (60 to 75 Ω) is considerably smaller than that of air-insulated line or bus work (300 to 400 Ω), and considerable overvoltages can develop at open disconnect positions. The surge impedance of cables is between 30 and 60 Ω and velocity of propagation 1/3 to 1/2 of speed of light.

18-8-2 Bushings

A detailed model consists of several sections of transmission lines in a series that can be used, as the bushing gradually changes the surge impedance of the GIS to that of the line. Rigorously, the graded bushings are modeled with a group of coaxial transmission line sections created by concentric grading foils. A further detailed model considers the coupling between the conductors and shielding electrodes. The incident wave divides between the concentric coaxial transmission lines formed by the foils. On a simplified basis, the bushing can be represented by equivalent single-section surge impedance and an equivalent surge capacitance. A lumped resistor represents the losses.[15] Figure 18-9 shows the simulation results with these models. The solid line indicates the response with concentric transmission lines, the dashed line is the response when the bushing is represented with single-surge impedance, and dotted lines are the response with lumped capacitance. The response is for a step input in all the cases.[15]

18-8-3 Power Transformer

At high frequencies, the transformer behaves like a series and shunt capacitance network. See Chap. 14.

18-8-4 Surge Arresters

Surge arrester models are discussed in Chap. 20. If the switching operations do not produce voltages high enough for the metal-oxide arresters to conduct, these can be modeled as capacitance to ground. A detailed model represents each internal shield and block individually, and includes travel times along shield sections, and capacitance between these sections, capacitance between blocks and shields, and the blocks (Chap. 20). Because the GIS conductors are inside a grounded metal enclosure, the lightning impulses can only enter through the connection of GIS to the rest of the system. Cables and direct transformer connections are not subject to lightning strikes; therefore, only air-to-SF_6 bushing connections are a lightning concern. The conventional metal-oxide arresters may not be effective on high-rise switching surges.

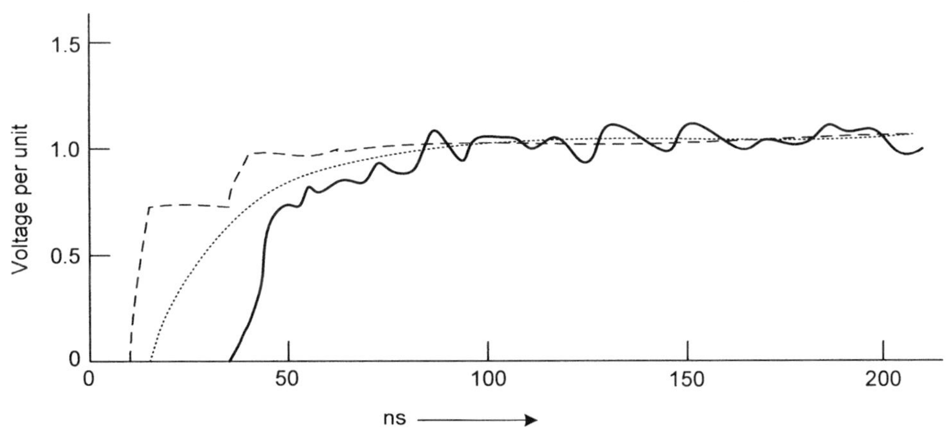

FIGURE 18-9 Response of bushing models—solid line represents a number of transmission line elements; dashed line, bushing represented by a single transmission line; dotted line, bushing represented by equivalent lumped capacitance.[15]

18-8-5 Circuit Breakers

A closed circuit breaker can be represented by a lossless line, length equal to the physical length of the breaker, with propagation velocity reduced to 95 percent that of light. An open-circuit breaker is not so simple to model. Circuit breakers with several chambers may contain grading capacitors, which are not arranged symmetrically.

The accuracy of simulation depends on the details of the modeling of each GIS component. To achieve reasonable results, highly accurate models are required. The arrangement of spacers, shielding electrodes, and varying diameters can be accurately simulated from physical parameters derived from manufacturer's drawings.

Figure 18-10, adapted from (Ref. 14), shows a detailed model of a line feeder connected to an overhead line with length of each section in meter, surge impedance, and capacitances. It shows 420-kV GIS model with duplicate busbars, disconnectors, circuit breaker, current, and potential transformers. Note each small section is

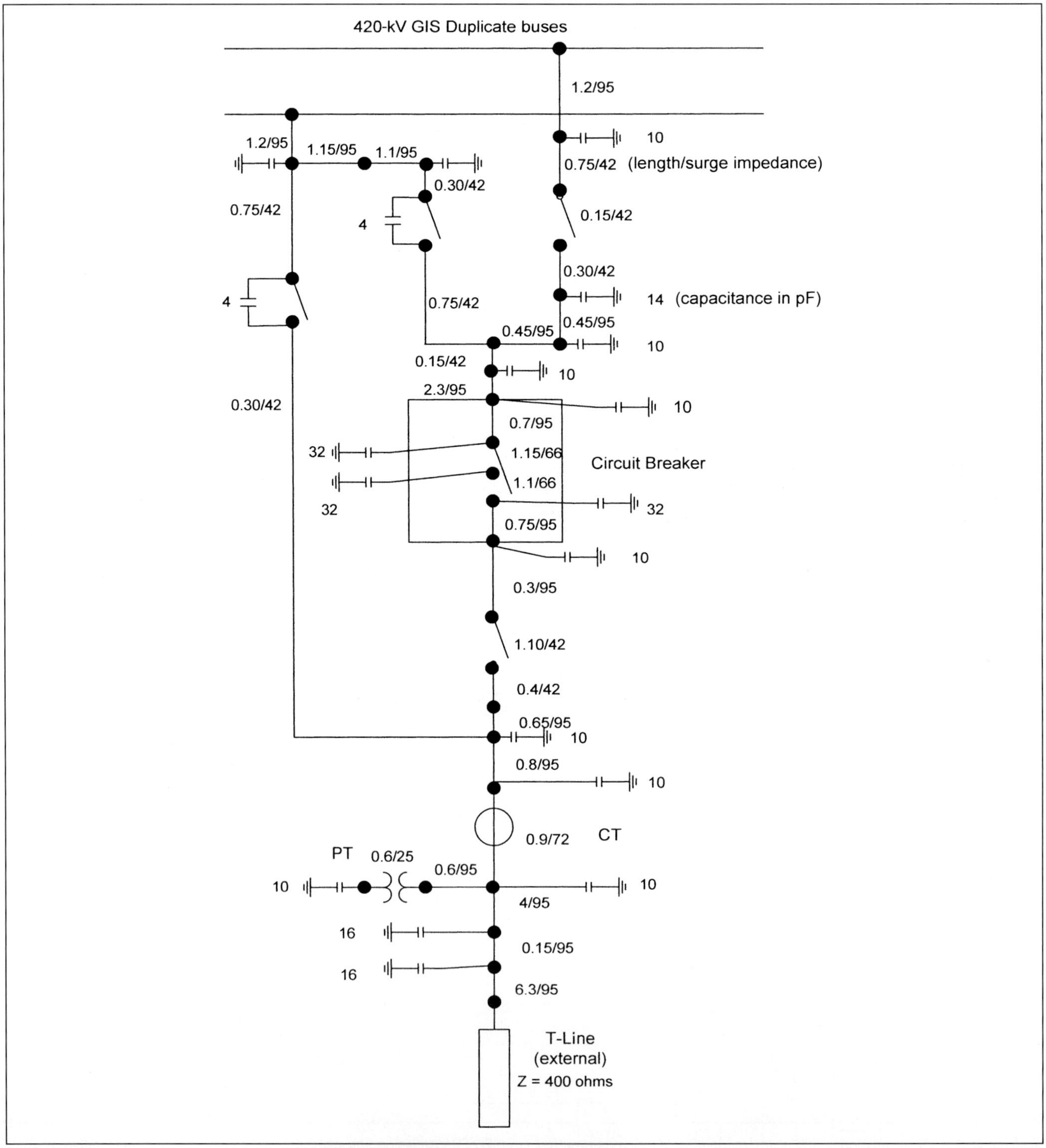

FIGURE 18-10 Detailed model of a feeder, based on Ref. 14.

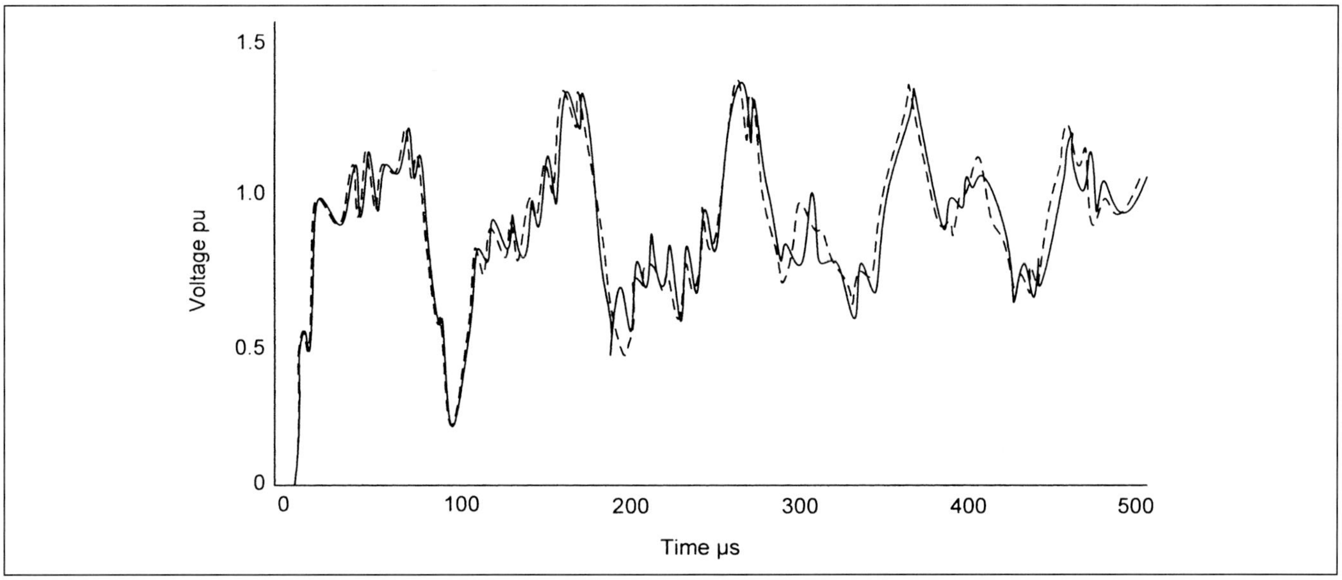

FIGURE 18-11 Simulated versus actual response based on model, Fig. 18-10.[14]

modeled with a line length and its surge impedance. Figure 18-11 shows comparison of computer simulation and actual measurements of disconnector-induced overvoltages in a 420-kV GIS. The effect of spacers, elbows, corona shields, and other hardware is considered, and the simulation closely follows the actual results.

18-9 INSULATION COORDINATION

The configurations of GIS vary. In fact, all the bus arrangements popular in outdoor air-insulted substations can be implemented in GIS, for example: (1) single-bus arrangement, (2) main and transfer bus, (3) ring bus, (4) double-bus single breaker, (5) double-bus double breaker, and (6) breaker and one-half schemes. These are not shown here. One difference is the use of maintenance ground switches; Fig. 18-12 shows 500-kV GIS single-line diagram of connections, with two incoming circuits, two outgoing circuits, and a bus section circuit-breaker for coupling the duplicate busbars. The assembly, thus, consists of disconnect switches, circuit breakers, surge arresters, voltage and current transformers, gas–oil bushings, air bushings, and gas barriers. Note the use of grounding switches with fault closing capability for maintenance. The power transformers (not shown in Fig. 18-12) may be throat-connected to GIS bus,

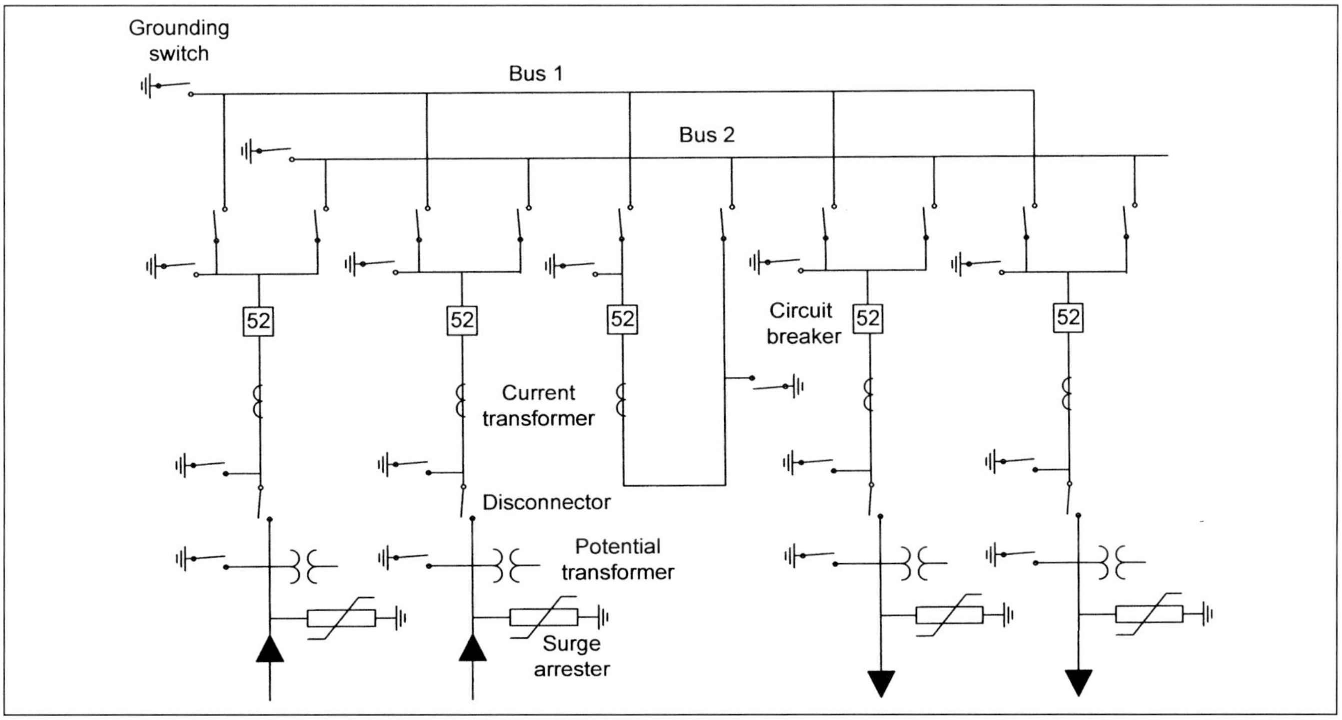

FIGURE 18-12 Implementation of a double bus bar with bus section power system configuration in GIS; disconnectors and grounding switches are shown.

or the transformers may be connected through an open air bus. The surge arresters may be connected at each line entrance and exit, and they can be conventional air-insulated arresters outside the GIS enclosure; alternatively, gas-insulated arresters may be accommodated inside the GIS enclosure. The latter have an advantage that the surge arrestee lead lengths will be short.

Thus, the modeling of transients in GIS on EMTP with all the connections, terminations, surge arresters, potential transformers, taps, and bushings becomes rather complex. On a simplistic basis, assuming a certain incoming surge and modeling the entire GIS as a bus of certain surge impedance, a hand calculation of the voltage, say at the end of an open GIS terminal can be made based on the lattice diagrams discussed in Chap. 4.

The considerations for the insulation coordination are similar as discussed in Chap. 17, for air-insulated substations, modified by the short transit time of the incoming surges in the GIS.

Again we consider:

- Back flashovers and shielding failures
- Switching and lightning overvoltages
- Application of surge arresters and their locations
- Tower-footing resistance, especially close to GIS
- Presence of capacitive voltage transformer (outside GIS) at the entrance of GIS
- Impact of bushing capacitance, entrance, and outgoing
- Influence of T-connections and branched connections

It is noted that the high overvoltages in GIS are caused by back flashovers, especially if these occur close to GIS, which may result in insulation failure. For simulation, only a few spans of overhead line close to GIS need to be considered. The tower footing resistance plays an important role. The greater the tower footing resistance, the greater the distance from GIS that should be considered. Back flashovers generate overvoltages of high rate of rise at the point of flashover. Thus, a back flashover close to the GIS means overvoltages of high rise time and magnitude.

Only the lightning strokes close to GIS contribute to the risk of insulation failure. Voltage within GIS will depend on the steepness of the incoming voltage, the relative surge impedance of GIS and overhead line/cable, and the length of the GIS, as it determines the wave travel time. Initially, the overvoltages will increase with the length of GIS bus; however, when the $2T$, where T is the travel time in the GIS, reaches the duration of the entering voltage, the superimposition of the reflected waves is less pronounced and the overvoltages decrease.

When a coupling capacitor is connected (say a CVT outside the GIS) at the entrance, it reduces the overvoltages in GIS as the steepness of the incoming voltage is reduced. This will delay the travel time of the surge, say to an open disconnector, and there will be short delay in voltage built up on GIS. The capacitor decreases the MTBF.

Shielding failures, when the lightning stroke lands on the phase conductor rather than on ground wire, should not normally cause flashover in a well-designed overhead line. Even with flashovers, the overvoltages entering the substation are unlikely to have a front time <1 μs.

18-10 SURGE ARRESTERS FOR GIS

As discussed in an earlier section, the voltage-time characteristics of breakdown in SF_6 are flatter compared to that of air insulation. It means that an incoming surge is likely to cause breakdown in the GIS before breakdown occurs in air. A high incoming overvoltage tends to increase the rate of voltage buildup in the entire GIS with the same arrester characteristics. The arrester will conduct earlier, but the reflection time of the negative wave from the arrester to reach any point in the GIS remains the same. Increasing the protective level for fast front transients, as compared to air-insulated substations, is one option, and a higher margin of 20 percent can be applied. An arrester on the incoming, which may be located outside GIS, is normally provided and its insulation level should be chosen consistent with GIS (Fig. 18-13). A CVT on the incoming junction of OH line and GIS will slow down the fast front of incoming surges. For large substations, an arrester inside GIS at location 2 on the source side of the first disconnector can be provided, though it has been debated whether such a location is helpful at all for the VFT generated by the disconnector operation. Though metal-oxide arresters have been demonstrated to respond to VFT, it is doubtful

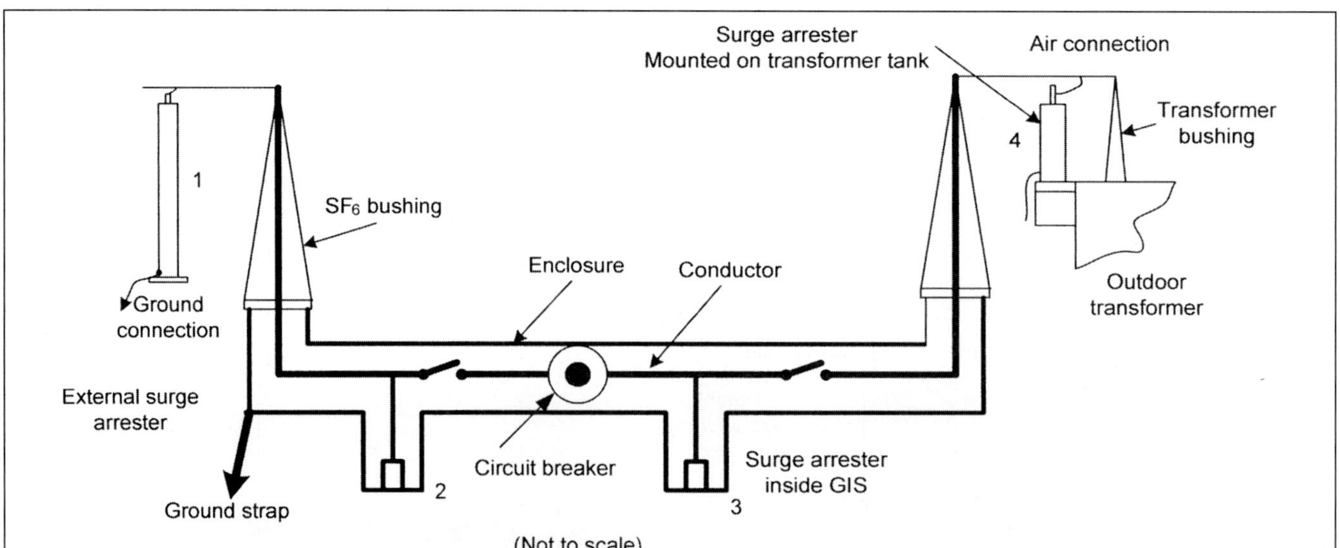

FIGURE 18-13 Location of surge arresters in GIS and external to GIS.

whether the dimensions involved will allow control of VFT surges of nanoseconds. An arrester at location 4, on the primary side of the transformer, is invariably provided. Again, the transformer windings will be subjected to high frequencies. With one arrester only at the incoming, the maximum overvoltage will occur at the far end of GIS at transformer connection. With two arresters, it will occur somewhere in the middle of GIS.[16,17]

When a GIS has direct connection to a transformer, much impedance discontinuities are avoided and only degradation is caused by the slight delay in the rise time due to bushings. At the transformer terminals, the surge will appear with no reduction in magnitude and slight degradation in the rise time until it is attenuated by reflections and surge arresters. The surge capacitance of the power transformer has rather a small impact on the overvoltages within GIS.

The influence of a number of connected branches on maximum overvoltages is shown in Fig. 18-14. In this figure, V_d is the discharge voltage of the arrester. The figure represents equal length of branches and is drawn for a system voltage of 420 kV and arrester discharge voltage of 830 kV. The ratio of the line-to-GIS surge impedances is 5 and L is the distance to the disconnector.

Special arresters, indoor or outdoor types, for use with GIS have been developed. These have high energy absorbing capability. The active part is installed in metal enclosure filled with SF_6 gas under pressure and forms an impervious pressure system according to IEC Std. 60694,[18] Chap. 20.

Example18-2 A system configuration of 750-kV GIS section is shown in Fig. 18-15. The bushing 1 is connected to a short spur of 750-kV line, and bushing 2 is connected to a power transformer, the secondary of the transformer is open circuited. The length of the three sections and bushings in GIS are modeled with surge impedances, bushing capacitances, and so on. A lightning stroke, 200 kA, 3/100 μs wave front, occurs on the top of tower, which results in a backflashover of phase *c*. The resulting voltage transients at points *VT*, V_m, and V_a in Fig. 18-15 are shown in Figure 18-16*a* to *c*, respectively. These show similar patterns. The phase *c* voltage rises to approximately 1750 kV peak (= 2.86 times the normal) at the transformer terminals. The surge arrester currents SA1 and SA2 are depicted in Figure 18-17*a* and *b*, respectively. The discharge current through phase *c*, arrester SA1, has a peak of 6000 A.

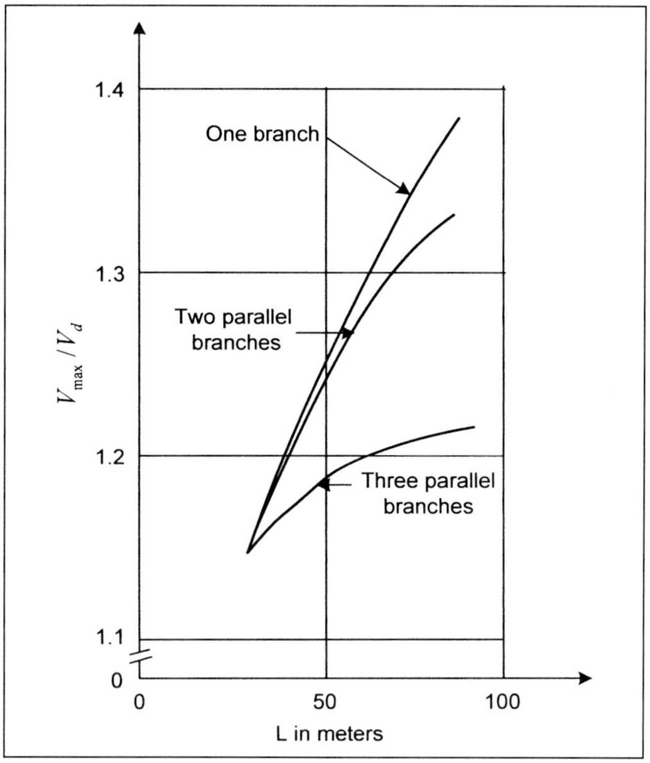

FIGURE 18-14 Influence of parallel connected branches of equal length in a 420-kV GIS.

Example 18-3 Now consider that phase c flashes over to ground at 1000 kV. The resulting voltage transient patterns at *VT*, V_a, and V_m are simulated in Fig. 18-18. Figure 18-19*a* and *b* illustrates the flashover voltage and current transients to ground, respectively.

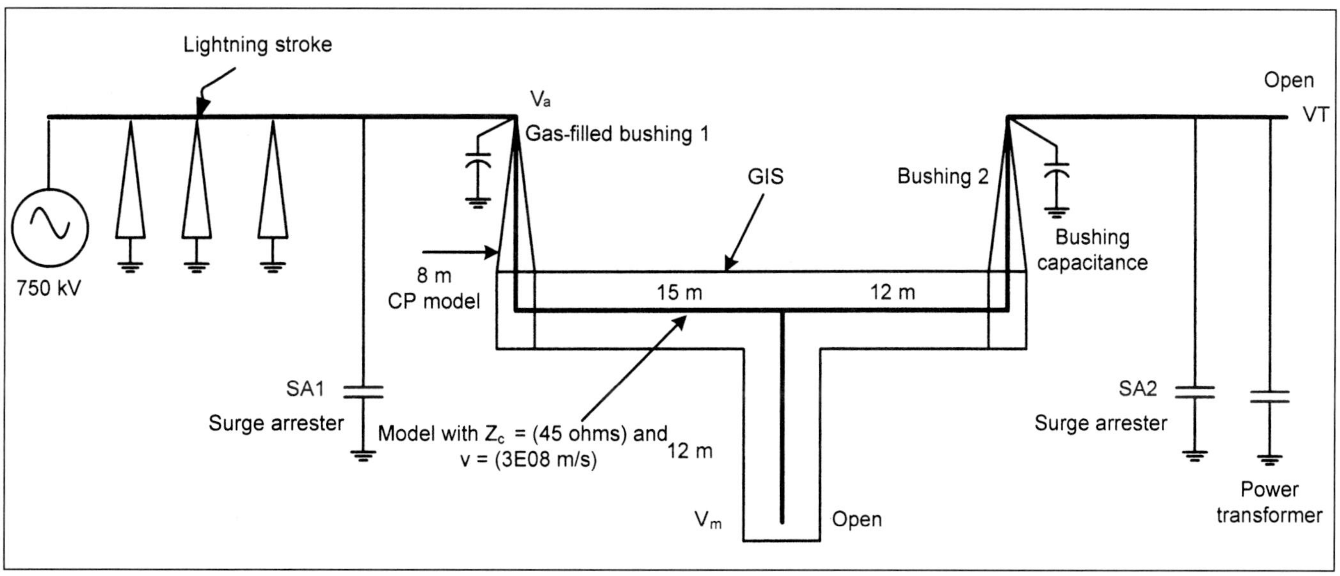

FIGURE 18-15 A GIS interconnected with a spur of 750-kV transmission line for study of transients on backflash, phase *c*; Examples 18-1 and 18-2.

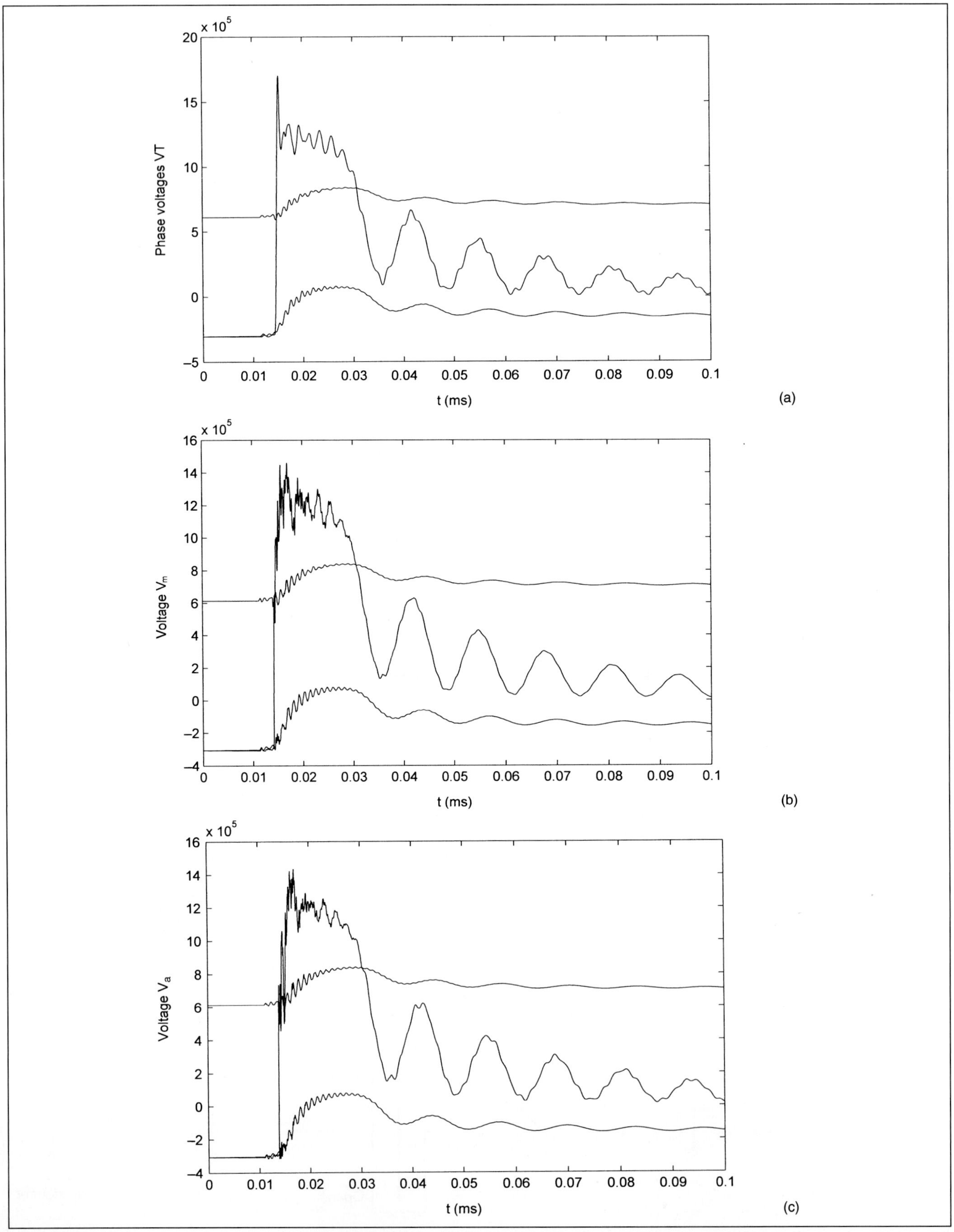

FIGURE 18-16 *a, b, c,* Voltages *VT*, V_m, and V_a, respectively, on backflash conductor phase *c*; see text.

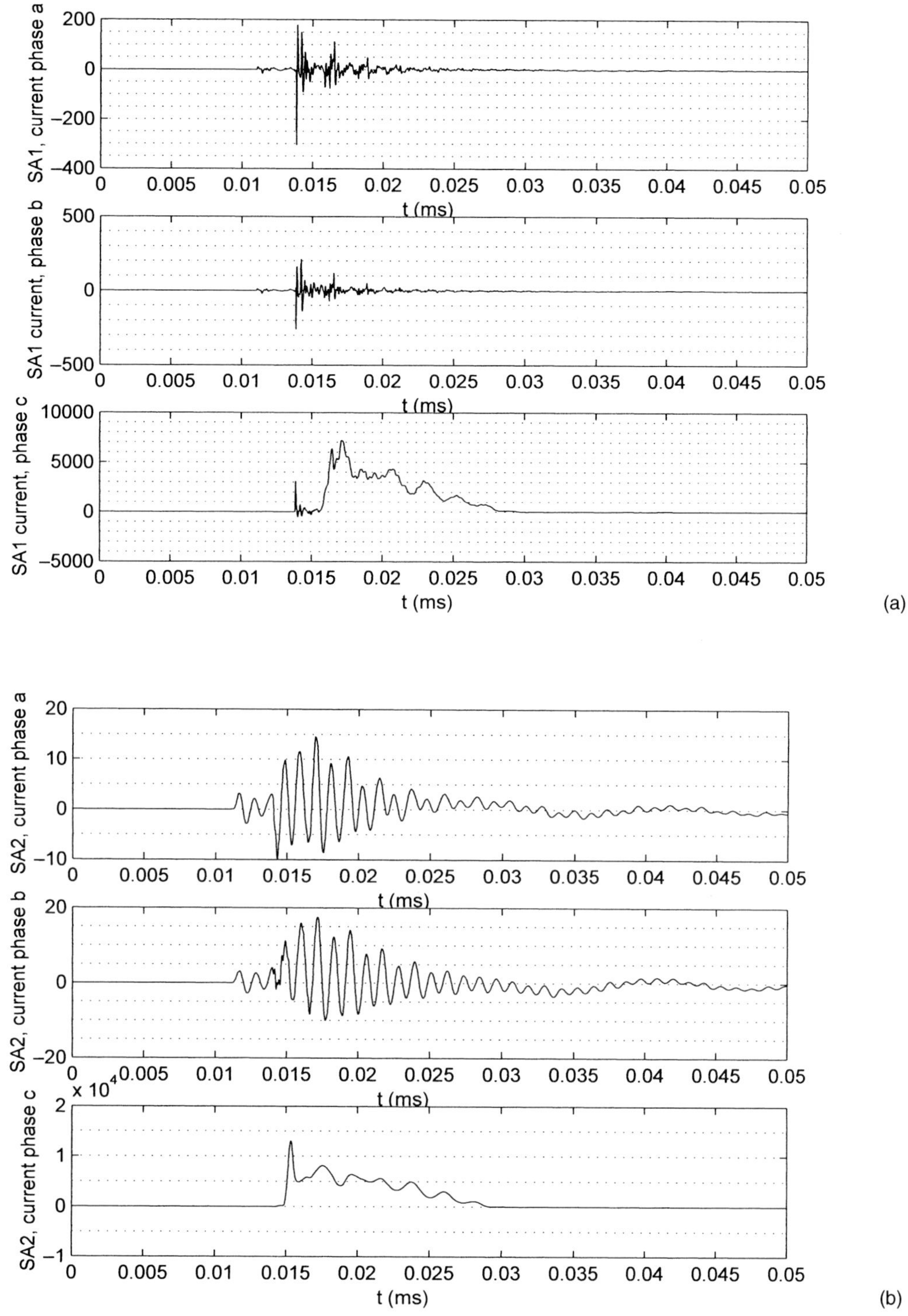

FIGURE 18-17 *a* and *b*, Surge arrester currents, SA1 and SA2, respectively.

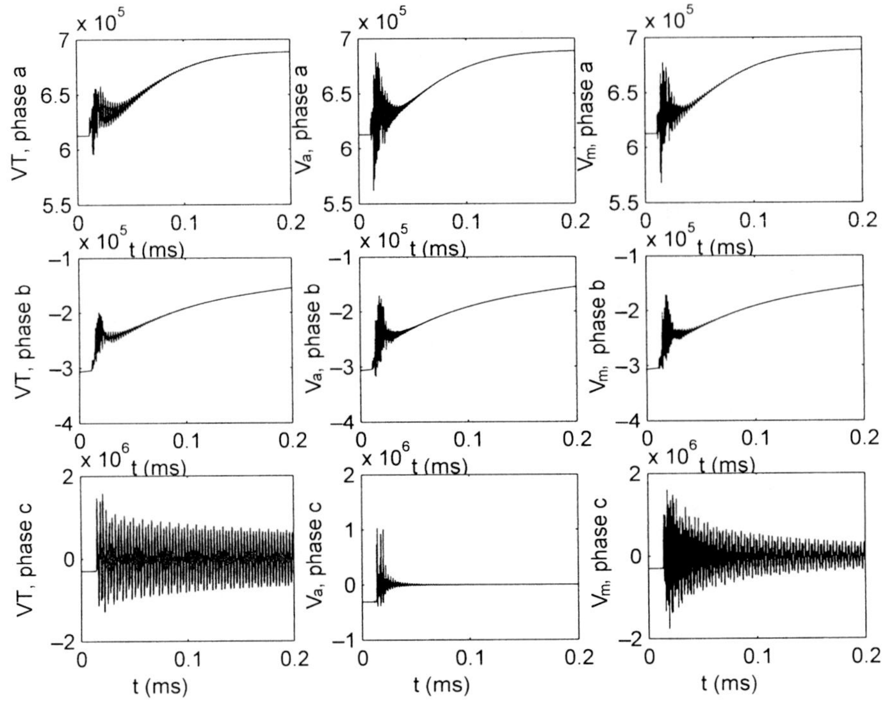

FIGURE 18-18 Voltages VT, V_a, and V_m on ground flash at 1000 kV in phase c; Example 18-2.

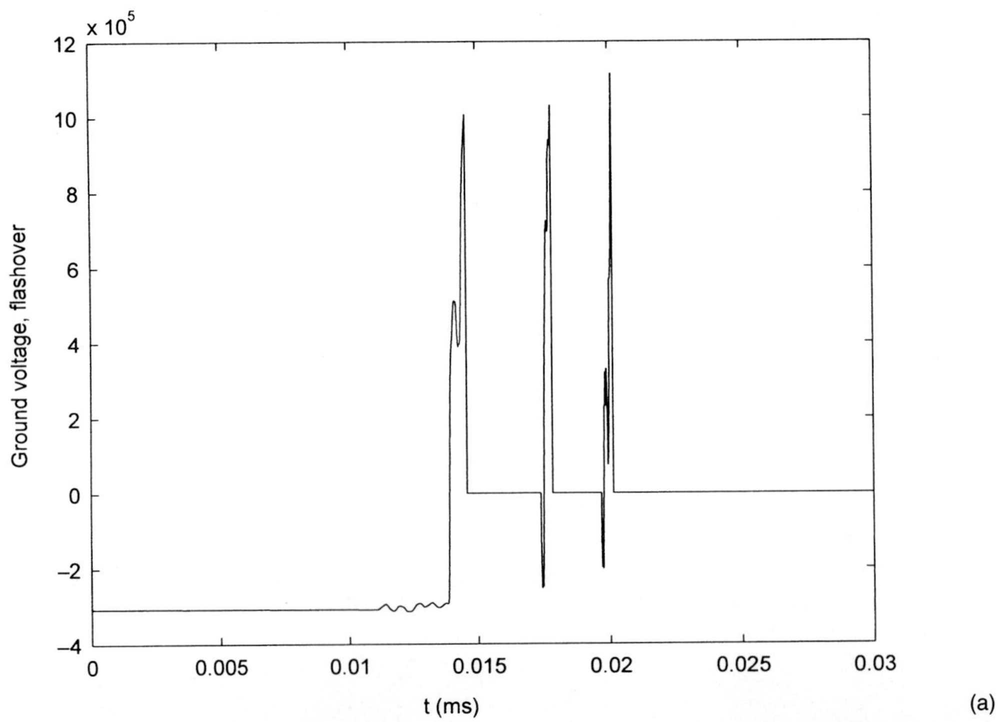

FIGURE 18-19 *a* and *b*, Voltage and current, flashover to ground, phase *c*; Example 18-2.

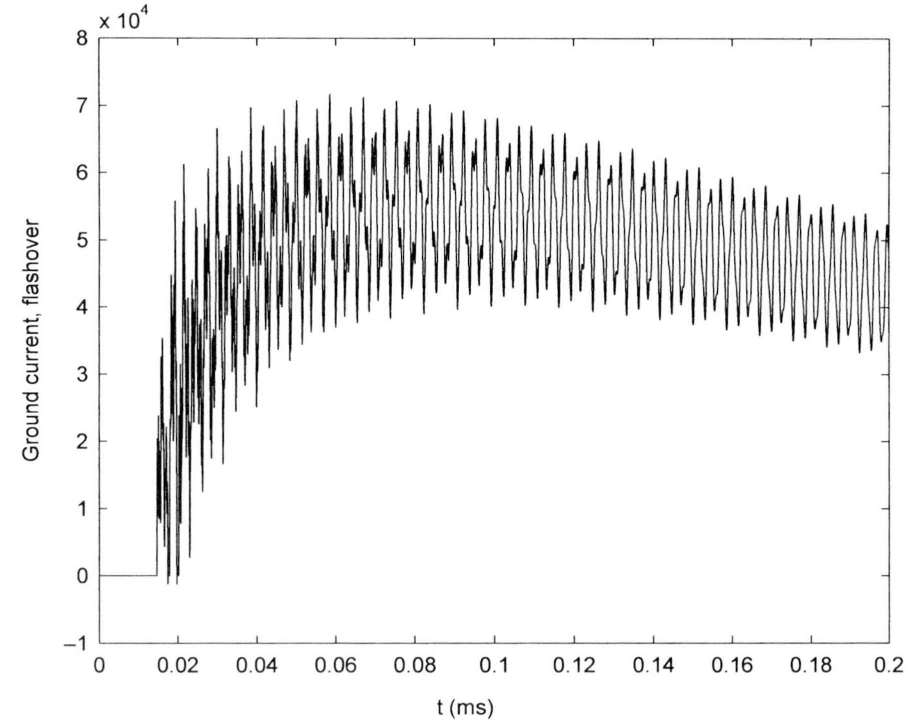

FIGURE 18-19 (*Continued*)

PROBLEMS

1. During operation of a disconnector, the voltage on the source side is 1.2 per unit, and the load side is 0.5 per unit. Source side $Z = 100\ \Omega$, the load side $Z = 75\ \Omega$. Assuming that a sparkover takes place at the peak of the voltage, calculate the source and load-side voltages after sparkover.

2. A breakdown occurs at 750 kV in the middle of a 20-m section of GIS, connected to an OH line of $Z = 400\ \Omega$, and open at the other end. Considering a surge impedance of 75 Ω at either end of GIS, and a velocity of propagation of 3E08 (m/s), calculate (a) voltage at open end at breakdown and after first reflection and (b) voltage at junction of OH line and GIS on breakdown and after first reflection. How will these voltages behave in subsequent reflections? Will they increase or decrease?

3. A GIS is connected to a transmission line at one end and terminated in a transformer at the other end. A surge arrester is installed at the transformer end only. Discuss the limitations of this scheme. At what point in the GIS will the transient voltage peak? If a surge arrester is installed on the incoming line, how will the location of maximum surge voltage in GIS change?

4. Draw a freehand sketch of the waveshape of a disconnector-induced transient on opening the disconnector, superimposed upon the system voltage.

5. A GIS is connected to an OH line of surge impedance = 400 Ω. Consider a surge impedance of 100 Ω between the GIS and its enclosure, and a surge impedance of 75 Ω for the GIS, open at the other end. Calculate the peak transient voltage on the enclosure in per unit. Also calculate the voltage at the open end of the GIS and junction of OH line and GIS. Explain the polarity of the enclosure voltage.

6. Calculate surge impedance of a bus, with enclosure diameter 30 cm and conductor diameter 4 in. What is the estimated capacitance?

7. How does the insulation coordination of GIS differ from an air-insulated substation?

8. Write a half-page note on the voltage-time characteristics of SF_6, without any mathematical equations.

REFERENCES

1. S. Yanabu, H. Murase, H. Aoyagi, H. Okubo, and Y. Kawaguchi, "Estimation of Fast Transient Overvoltages in Gas Insulated Substations," *IEEE Trans. PD*, vol. 5, no. 4, pp. 1875–1882, Nov. 1990.
2. S. A. Boggs, A. Krenicky, A. Plisse, and D. Schlicht, "Disconnect Switch Induced Transients and Trapped Charges in Gas-Insulated Substations," *IEEE Trans. PAS*, vol. 101, no. 10, pp. 3593–3602, Oct. 1982.
3. S. A. Boggs, N. Fujimoto, M. Collod, and E. Thuries, "The Modeling of Statistical Operating Parameters and the Computation of Operation Induced Waveforms for GIS Disconnectors," CIGRE Paper 13-15, 1984.
4. IEEE Std. C37.122.1, IEEE Guide for Gas Insulated Substations, 1993. (This standard contains a bibliography of 1518 technical papers and references on GIS.)
5. N. Fujimoto, E. P. Dick, S. A. Boggs, and G. L. Ford, "Transient Ground Potential Rise in Gas Insulated Substation: Experimental Studies," *IEEE Trans. PAS*, vol. 101, pp. 3603–3619, Oct. 1982.

6. N. Fujimoto, S. J. Croall, and S. M. Forty, "Techniques for the Protection of Gas-Insulated Substation to Cable Interface," *IEEE Trans. PD*, vol. 3, no. 4, pp. 1650–1655, Oct. 1988.
7. G. L. Ford and L. A. Geddes, "Transient Ground Potential Rise in Gas Insulted Substations—Assessment of Shock Hazard," *IEEE Trans. PAS*, vol. 101, no. 10, pp. 3620–3629, Oct. 1982.
8. J. Meppelink, K. Diederich, K. Feser, and W. Pfaff, "Very Fast Transients in GIS," *IEEE Trans. PD*, vol. 4, pp. 223–233, Jan. 1989.
9. I. M. Bortnik, C. M. Cooke, "Electrical Breakdown and Similarity Low in SF_6 at Extra High Voltages," *IEEE Trans. PAS*, vol. 91, no. 5, pp. 2196–2203, 1972.
10. T. H. Teich and W. S. Zaengi, Dielectric Strength of an SF_6 Gap, In *Current Interruption in High-Voltage Network*, K. Ragaller, ed., Plenum Press, New York, 1978.
11. IEEE Std. C37.122, IEEE Standard for Gas Insulated Substations, 1993.
12. N. Fujimoto, H. A. Stuckless, and S. A. Boggs, Calculation of Disconnector Induced Overvoltages in GIS, In *Gaseous Dielectrics*, vol. IV, L. Christophorou, ed., p. 473, Pergamon Press, New York, 1984.
13. IEEE Report, "Modeling Guidelines for Fast Front Transients," *IEEE Trans. PD*, vol. 11, no. 1., pp. 493–506, Jan. 1996.
14. IEEE Report, "Modeling and Analysis Guidelines for Very Fast Transients," *IEEE Trans. PD*, vol. 11, no. 4, pp. 2029–2035, Oct. 1996.
15. N. Fujimoto and S. A. Boggs, "Characteristics of Disconnector-Induced Short Rise Time Transients Incident on Externally Connected Power System Components," *IEEE Trans. PD*, vol. 3, no. 3, pp. 961–970, Jul. 1988.
16. M. E. Potter and T. O. Sokoly, "Development of Metal-Oxide Varistors for Gas Insulated Surge Arresters," IEEE Trans. PAS, vol. PAS-101, no. 7, pp. 2217–2220, July 1982.
17. M. Mitani, "Analysis of Lightning Surge Phenomena and Examination of Insulation Coordination in a Three-Core Type Gas Insulated Bus," In *Conference Proceedings, IEEE PES Summer Meeting, Paper A 79 401-1,* 8 pages, July 1979.
18. IEC Std. 60099—Part 4, Metal Oxide Surge Arresters Without Gaps for AC Systems, 2009.

FURTHER READING

H. W. Anderi, C. L. Wagner, and T. H. Dodds, "Insulation Coordination for Gas Insulated Substations," *IEEE Trans. PAS*, vol. 92, no. 5, pp. 1922–1930, Sept./Oct. 1973.

W. Boeck and K. Pettereson, "Fundamentals and Specific Data of Metal-Enclosed Substations for the Insulation Coordination," CIGRE Paper 23-03, 14 pages, Aug. 1978.

P. F. Coventry and A. Wilson, "Fast Transients in Gas Insulated Substations—An Experimental and Theoretical Investigation," *Gaseous Dielectrics*, vol. VI, pp. 503–508, 1991.

R. Erikson and H. Holmborn, "Insulation Coordination of Gas-Insulted Substations," *Proceedings of the International Symposium on Gaseous Dielectrics*, Knoxville, TN, March 6–8, 1978.

IEC 60071-2, Insulation Coordination, Part 2: Application Guide, 1996.

T. Kobayashi, S. Mori, H. Koshiishi, K. Niromija, M. Mitsuki, H. Yokoyama, T. Hara, "Development of Compact 500 kV, 8000 A Gas Insulated Transmission Line—Study on Insulation Design," *IEEE Trans. PAS*, vol. 103, no. 11, pp. 3154–3164, Nov. 1984.

J. M. Meek and J. G. Craggs, *Electrical Breakdown of Gases*, John Wiley and Sons, New York, 1978.

D. B. Miller and P. M. Harrison, "Measurements and Modeling of Switching Transients in Gas-Insulated Transmission Lines," *Conference Proceedings, IEEE SOUTHEASCON' 87*, Tampa, FL, vol. 2, pp. 456–460, Apr. 1987.

H. Raether, *Electron Avalanches and Breakdown*, Butterworth, London, 1964.

Klaus Ragller, ed., *Surges in High-Voltage Networks*, Plenum Press, London, 1980.

T. H. Teich and W. S. Zanegi, The Dielectric Strength of an SF_6 Gap, In *Current Interruptions in High Voltage Networks*, K. Ragaller ed., pp. 269–297, Plenum Press, New York, 1978.

B. Wahlstron, H. Holmborn, and A. Schei. "Overvoltage Protection of Metal Enclosed SF_6 Substations: Insulation Coordination Philosophy and Surge Arrester Characteristics," CIGRE Paper 33–03, 15 pages, Aug. 1976.

J. M. Wetzer, M. A. Van Houten, and P. C. T. Ven der Laan, "Prevention of Breakdown Due to Overvoltages Across Interruptions of GIS Enclosure," *Gaseous Dielectrics*, vol. VI, pp. 531–536, 1991.

CHAPTER 19

TRANSIENTS AND SURGE PROTECTION IN LOW-VOLTAGE SYSTEMS

The mechanism of generation of surge voltages on low-voltage power systems is somewhat akin to the switching transients and direct or indirect effects of lightning in the high-voltage systems. In low-voltage systems, the surges may appear in any combination of line, neutral, or grounding conductors. These may be of sufficient magnitude, duration, rate of change, periodic or random in nature to cause damage or operational upset. The effectiveness of grounding plays a significant role. Telephone and data/communication lines also provide a means of conducting surges into a facility, and need to be protected. *Electrical line noise* generally implies any voltage less than two times the nominal peak voltage of the system, and it can cause malfunctions in computer software. To distinguish between the electrical noise and transient voltage surges, the latter generally implies transient voltage amplitude which exceeds two times the nominal peak voltage of the electrical system.

19-1 MODES OF PROTECTION

Figure 19-1 shows *seven modes* of protection in three-phase, wye-grounded systems. These are phase to ground (L-G, three modes), phase to neutral (L-N, three modes), and, finally, line to ground (L-G mode). The L-L and L-N modes are called *normal modes*, while L-G and N-G are called *common modes*; the term "common" comes from an old term used in the telecommunication industry meaning from ground.

Most commercial facilities are fed from a utility transformer that transforms the utility primary distribution voltage, normally 12.47 kV to a common voltage of 480 or 208 V. Three-phase conductors and the neutral conductor feed the loads, while the fifth ground wire provides the ground reference. When an external disturbance enters the building, it is coupled between one or more phase wires and ground or neutral. Most transient phenomena, that is, switching of capacitors, utility load shedding, accidents involving power loss, and transfer switches (which transfer power form one source to another) may produce a transient at the service entrance. Thus, all modes shown in the Fig. 19-1 should be protected. The ground conductor is grounded at the distribution pole, and even if a ground grid is utilized to provide a zero reference point, there will be transient voltage differential developed when the ground current is shunted to ground.

19-2 MULTIPLE-GROUNDED DISTRIBUTION SYSTEMS

Figure 19-2 shows typical grounding practice for wye service entrance served by a wye multiple-grounded medium voltage system in North America. Note the multiple grounds of the neutral conductor protected neutral (PEN). The practice of grounding of commercial and residential facilities in the United States requires that the neutral conductor is bonded to the ground conductor at the service entrance, and both are bonded to the building ground. There cannot be N-G surge at the service entrance. However L-N surges within the building can produce N-G surges at the end of a branch circuit.

Further implications of multiple-grounded distribution systems are shown in Fig. 19-3. National Electric Safety Code (NESC)[1] requires that the neutral on multiple-grounded wye distribution systems have a minimum of four earth connections per mile. This also applies to direct buried underground cables. The voltage between neutral and earth can originate from a variety of sources. A 60-Hz voltage can exist between objects connected to neutral and earth. A short-duration transient can exist, when the lightning current is dissipated into the earth. A differential voltage between the neutral and ground is more likely to occur when the same service transformer feeds two or more consumers.

Figure 19-4 shows the grounding practice for industrial establishments. Here the neutral is grounded only at one point, at the source. This figure shows that the neutral from the utility transformer is not required to be run for industrial plant medium-voltage, three-phase loads. In case the industrial plant needs some loads like lighting and controls to be served from low-voltage grounded systems, these lower voltages are served from a separate transformer with artificially derived neutral. In case the service is at low voltage, a neutral may be run to supply phase-to-neutral

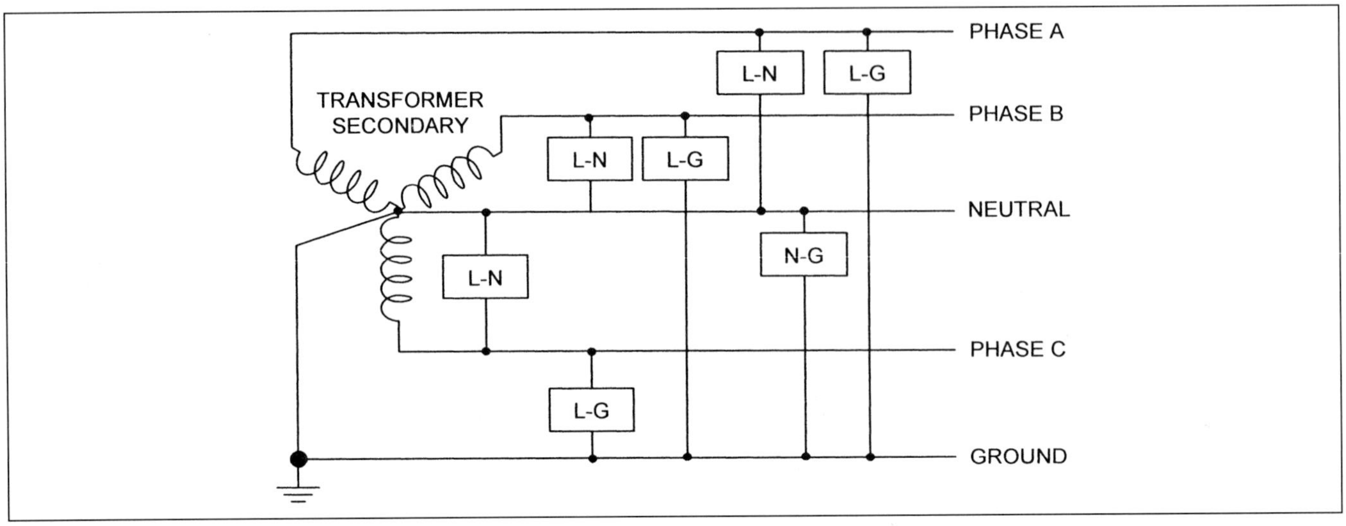

FIGURE 19-1 Seven modes of protection, common and normal modes, in a three-phase, four-wire, wye-connected, low-voltage system.

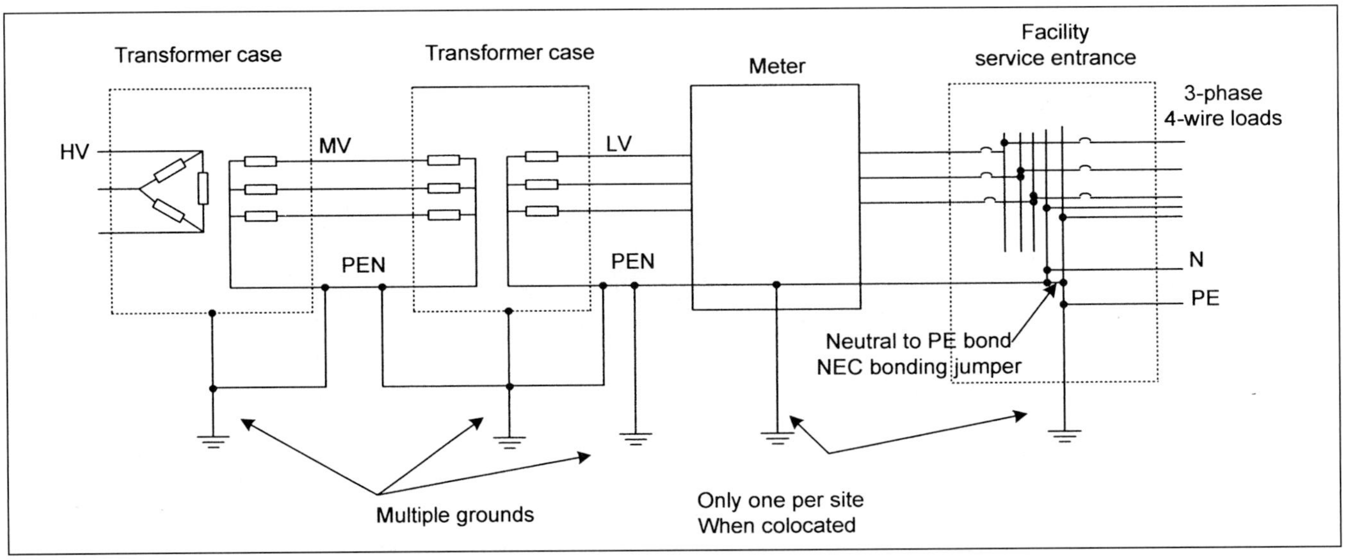

FIGURE 19-2 Typical grounding practice for wye-service entrance served by multiple-grounded, medium-voltage system in North American systems.

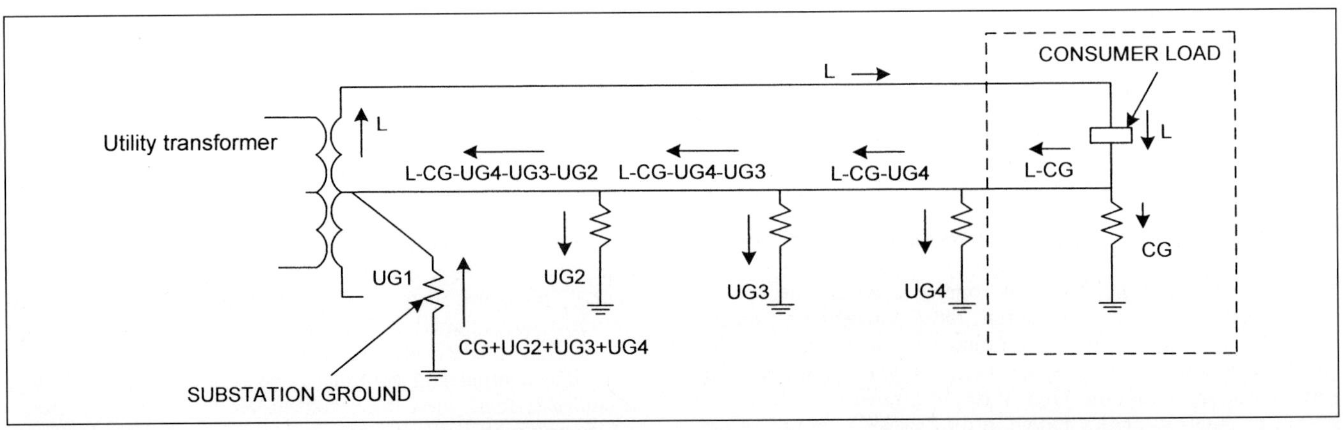

FIGURE 19-3 Distribution of load currents in phase and neutral/ground conductor in a multiple-grounded system; the neutral/ground conductor develops differential voltages.

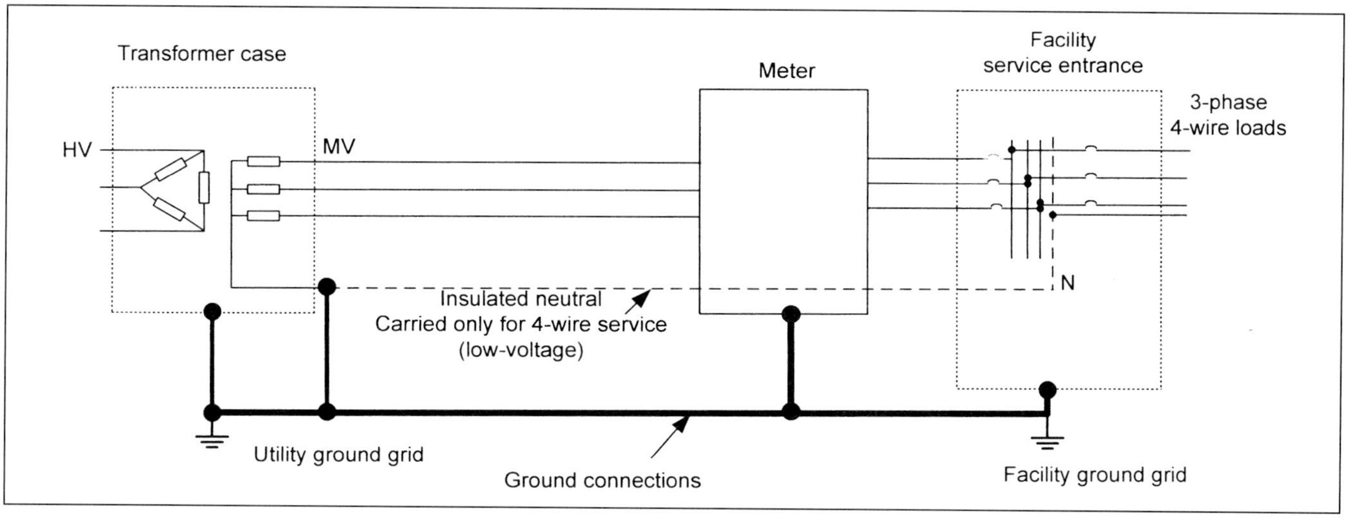

FIGURE 19-4 Typical grounding practice for industrial distribution systems, the transformer neutral is grounded only at the source in North American systems.

loads, *but it is* not grounded anywhere in the plant except at the service transformer. There is no bonding of neutral conductor with the ground at the service entrance, a practice which is invariably followed for industrial medium- or high-voltage grounded systems or separately derived industrial systems.

19-2-1 Equivalent Circuit of Multiple-Grounded Systems

Figure 19-3 of a multiple-grounded system shows that the grounds at various points cannot be at the same potential. This figure shows the current flow in the multiple-grounded neutral under normal operation. The load current flows through line to neutral, but as the neutral is grounded at the consumer premises (ground CG) and also at multiple points, the neutral current returns to the utility transformer through multiple paths; the sharing of current depends upon the relative impedances of the grounding circuit. An equivalent impedance diagram is shown in Fig. 19-5. The system may be analyzed using symmetrical components or equivalent circuit concepts. The line may pass through a region of high soil resistivity; each grounding point can be modeled individually along with the section of the feeder separating it from the adjacent grounds.

National Electric Code (NEC)[2] article 250.106 requires that lightning protection system ground terminals shall be bonded to the building or structure electrode system (so that the potential differential between these is minimized). Further NEC articles mandate that electrical equipment is bonded and connected in a manner to establish a path of sufficiently low impedance. Where the ground resistance of a single ground electrode exceeds 25 Ω, additional ground rods are mandated to lower the resistance to 25 Ω. This value of 25 Ω in NEC is too high; an ideal ground should provide a near-zero resistance between bonded components and the ground electrode of the facility to limit ground potential rise on a surge current. With the grounding resistance of 25 Ω allowed by NEC, high surge currents will produce high ground potential rise (GPR), which can damage the surge protection devices.

Consider the dissipation of a lightning surge near the distribution system (Fig. 19-6). If the lightning current is not effectively dissipated through the arrester, the result can be flashover of the

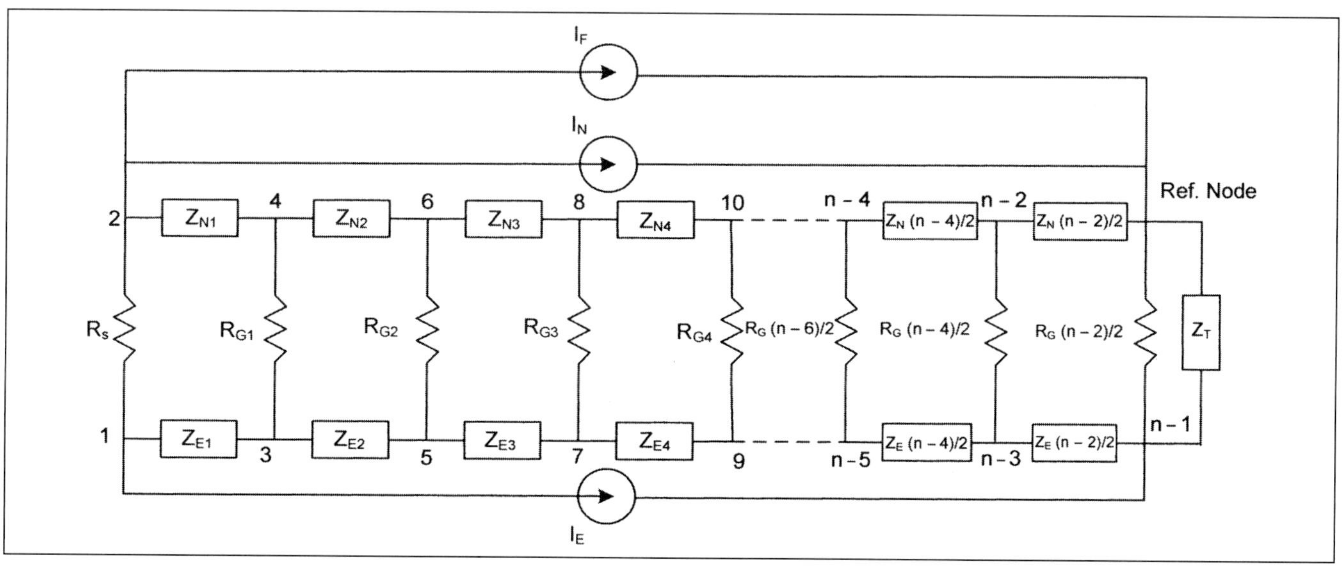

FIGURE 19-5 An equivalent circuit diagram of a multiple-grounded system for a line-to-ground fault at a remote node.

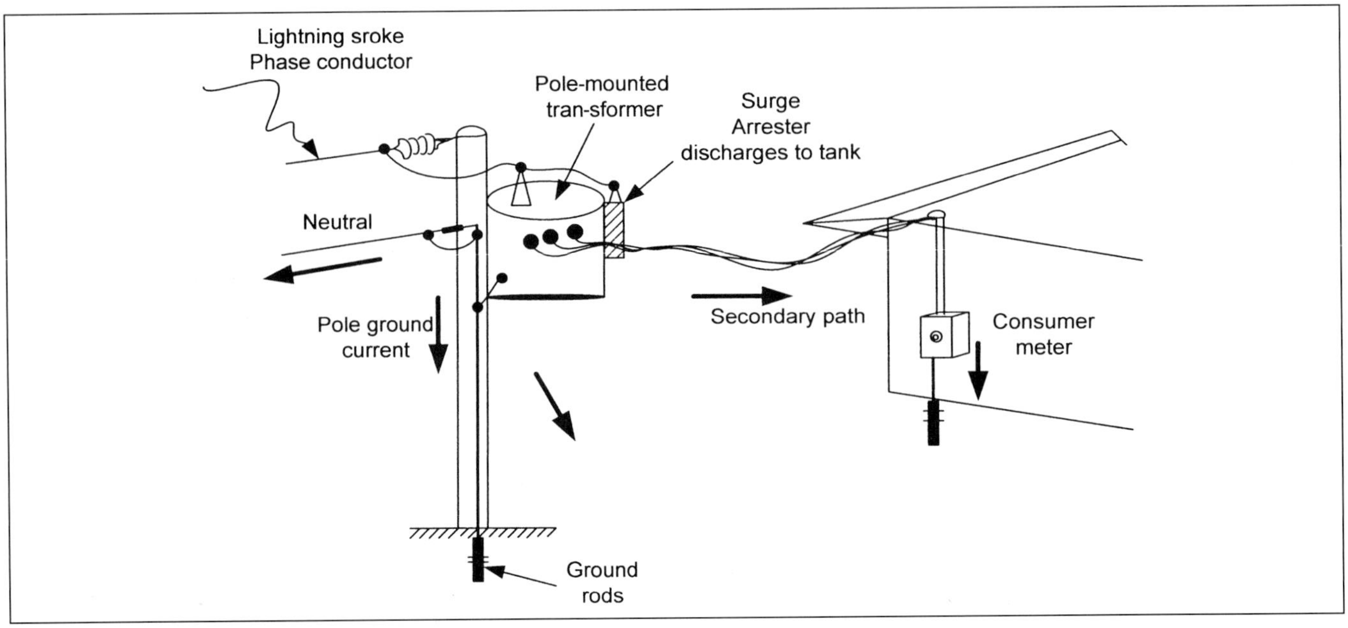

FIGURE 19-6 Distribution of lightning stroke current in a pole-mounted transformer service to a premises.

insulation and impingement of the surge on the consumer apparatus. This figure shows a surge arrester mounted on the transformer tank and the surge current will pass through the tank to the neutral that acts as ground conductor. The division of the surge current is shown in thick arrows. A surge voltage will appear on the neutral conductor and the consumer premises as the neutral is bonded to the consumer ground. This surge voltage will depend upon a number of factors—the downward lead length, the resistance of the grounding electrodes, and the surge impedances of the various paths. Generally, it may not be detrimental to the premises, especially when the recommendations of surge protection of low-voltage systems and the category of installations are followed, however, a possibility of flashover cannot be ruled out. The grounding of underground cable distribution system is more important, in the sense that lightning surges and wave fronts may double on cables. The voltage is reduced by close connection of the surge arrester to the cable terminations. The parameters of surge performance of grounding systems under high impulse currents of short rise time have been discussed earlier in Chap. 5, and will be further discussed in Chap. 22. It may be necessary to state here the grounding and bonding requirements laid down in NESC[1] and NEC.[2]

- NESC requires grounded items on joint poles (e.g., for power and communication) to be bonded together using either single grounding conductor or bonding the supply grounding conductor to the communication grounding conductor, except where a certain separation is maintained (Rule 97A), in which case there should be insulation between the grounding conductors. A hazardous potential difference can exist between the two conductors. The Rule 215C3 requires bonding between messengers at typically four times per mile.

- It is required that a common ground electrode system should be created by the two utilities for the communication and power supply systems. If separate electrode systems are used, these should be bonded together with a minimum #6 AWG conductor. This is to ensure that dangerous potentials do not exist between the two grounding systems. A user of computer modems, fax machines, answering machines, and other communication equipment could be exposed to an electrical shock apart from damage to the equipment.

- NEC (250.24(A)(5)) prohibits bonding of the equipment grounding conductor and neutral inside the premises. Note that the neutral will carry a current, and a difference of potential exists between the neutral and the ground conductor. Again, bonding is required to metal water pipes, which limits the possibility of a potential difference between the water system and other noncurrent-carrying parts within the building. [A coupling can occur through soil resistivity (Chap. 22)]. It is prohibited using *interior metal water piping system* located more than 5 ft from the entrance to the building from being used to interconnect grounding electrodes within the building.

19-3 HIGH-FREQUENCY CROSS INTERFERENCE

Consider electronic devices that are connected together through a data cable. The digital equipment radiates high-frequency energy back on to the power line and must meet FCC specifications. Every device containing any digital clock frequencies above 10 kHz has to be registered and approved to meet FCC specifications. In a building hundreds of such devices may be housed throughout, and electrical noise can sum up to major pulses, causing errors or upsets. Every computer or electronics circuit board contains ground paths, which are eventually connected to the building ground. This ground path passes over all the sensitive chips, CPUs, controllers memories, and the like. One input goes to ground and a voltage comparison of data to ground is made. Data signals are very small. In Fig. 19-7, the devices are connected through a data cable and are possibly served from different ac sources. The grounds are not uniformly the same at two locations, especially at high frequencies, and even a few feet of distance can create a voltage difference. Because of higher resistance of the computer circuit boards, the difference in ground voltages tend to concentrate at logic chips, which can cause errors.

Strategic installations of proper power-conditioning equipment can mitigate potential cross-reference hazards, though improper use of surge suppressor can worsen the problem by diverting a significant amount of energy into the ground circuit near the computer or its terminals. Consider a 5000-V surge and an impedance of 1 Ω between ground connections. This will generate a potential rise of 5000 V. Lowering of grounding resistance becomes important and

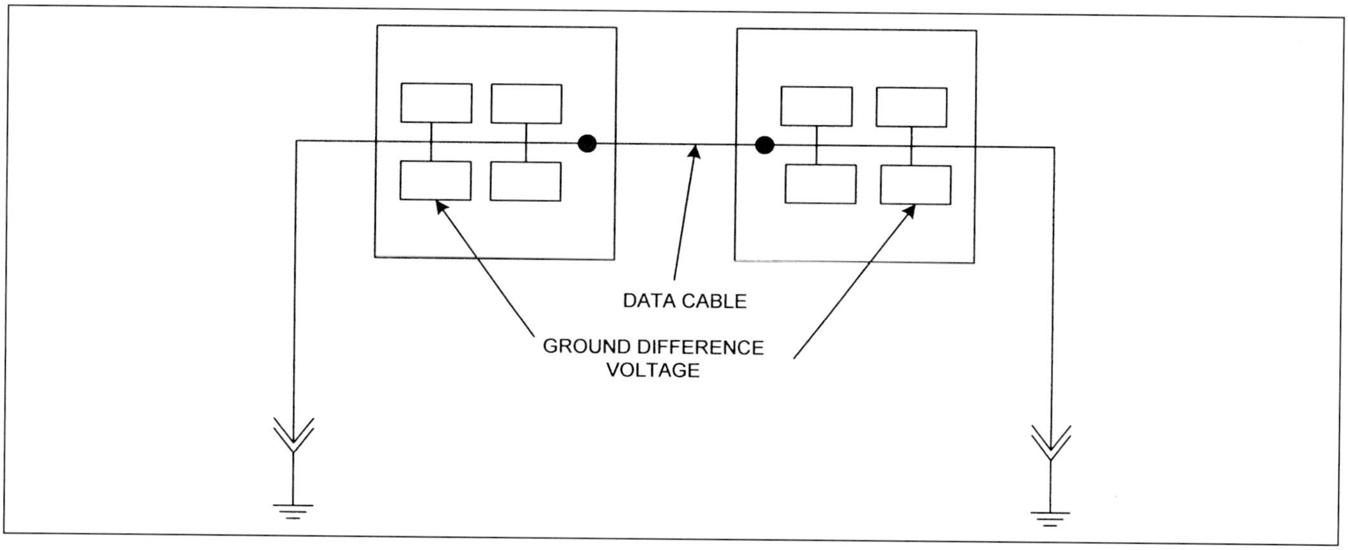

FIGURE 19-7 To illustrate ground differential voltage on two computer systems connected through a data cable.

larger size ground conductors can be run to the source or entrance ground. All data communication equipment should be referenced to a single ground network, properly designed to eliminate differential voltages. Embedding of ground mats under the electronic equipment, which are then connected to the main grounding system to establish an equipotential surface, are common means. TVSS should not randomly shunt surge impulses to ground. Further discussions and designs of grounding systems for electronic and data processing equipment are not the scope of this book. To quote from NEC:

"To protect the sensitive electronic equipment, knowledge of transient voltages and their waveforms, frequency of occurrence, energy levels, and failure threshold of the protected equipment is required." An interested reader may see Refs. 3, 4, and 5. IEEE standard[3] has 603 pages and provides extensive bibliography.

19-4 SURGE VOLTAGES

We have amply discussed the lighting and switching overvoltages in previous chapters. In low-voltage systems there are more coupling modes for these overvoltages.

19-4-1 Lightning

The coupling methods are: (1) direct coupling, associated with lightning energy in the incoming conductors, (2) inductive couplings produced by magnetic fields during a lightning strike, and (3) capacitive couplings derived from the charged ions passing over conductors. Shielding is often used to bleed off these charges to negate capacitive coupling effects. The lightning produces overvoltages through the following coupling mechanisms:

1. A direct stroke on primary circuits, that is, on a phase conductor, will inject high currents into primary circuits; the coupling mode can be through the phase and ground conductors and transformers. A direct stroke can also occur on the secondary conductors.

2. A nearby stroke that sets up electromagnetic fields, which in turn gives rise to overvoltages.

3. Surge transference through the transformers (Chap. 14).

4. Couplings through the common ground impedance paths when a large amount of ground current flows, say due to cloud-to-ground discharges. This couples through the common grounding impedance paths and gives rise to transient voltage differentials.

19-4-2 Switching

The switching transients originate due to normal and abnormal conditions, as discussed earlier:

1. Capacitor switching, and fault clearing (Chaps. 6 and 8)

2. Load switching and load turn-off (Chap. 5)

3. Operation of current limiting fuses (Chap. 8)

4. Multiple reignitions or restrikes within the switching devices themselves (Chap. 8)

5. Resonating circuits associated with switching devices (Chaps. 2 and 6)

6. Commutation notches in electronic power converters (Chap. 15)

19-5 EXPOSURE LEVELS

The rate of occurrence of surges varies over large limits and is not easily predictable in a particular power system. IEEE[6] classifies the exposure levels as shown in Fig. 19-8:

Low exposure. Low exposure occurs in systems in geographical areas known for low lightning activity, with little load or capacitor switching activity.

Medium exposure. Medium exposure occurs in systems in geographical areas known for medium to high lightning activity, or with significant switching transients. Both or only one of these causes may be present.

High exposure. The more severe conditions result from extensive exposure to lightning or unusually severe switching surges.

Figure 19-8 shows that the surges can be the result of original surge or the remnant resulting from the sparkover of the clearances.

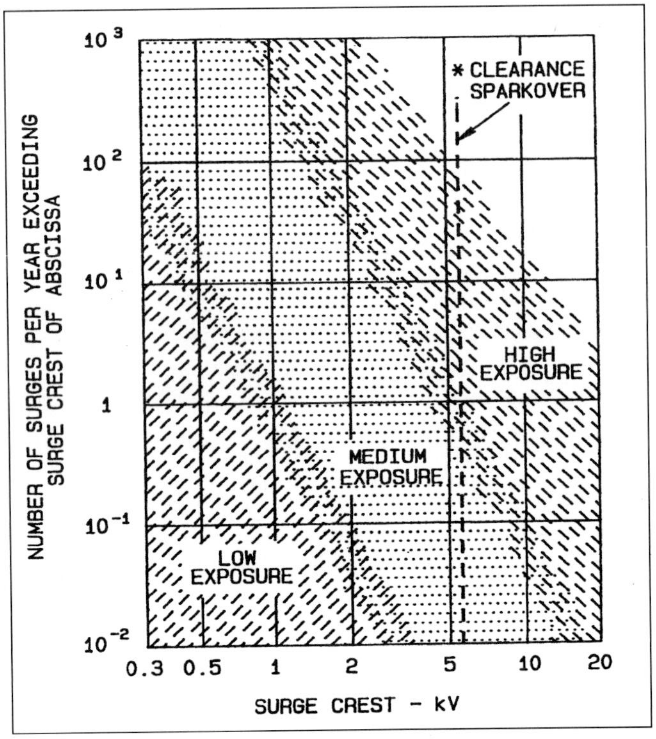

FIGURE 19-8 Rate of surge occurrence versus voltage level at unprotected locations. *Source: ANSI/IEEE Standard C62.41.*[6]

For outdoors, the clearances will be higher, a sparkover of 10 kV (1.2/50-μs wave) is typical, though 20 kV is possible. For indoor wiring devices used in 120-, 240-, or 480-V distributions, a sparkover of 6 kV is adequate between the phase and the ground.

19-6 TEST WAVE SHAPES

This section should be read in conjunction with Sec. 19-7 on location categories as there are cross-references between these two sections. A total of five test waveforms are specified, which are discussed in the following sections.

19-6-1 Combination Wave

The lightning-impulse (1.2/50 μs and discharge-current 8/20 μs) wave shapes are shown in Chap. 5. The 1.2/50-μs and 8/20-μs waves taken together are called *combination waves*; open circuit voltage and short-circuit current, respectively. The terminology according to IEC is "impulse" waves. See Chap. 5, Eq. (5-13) through Eq. (5-18) for these wave shapes.

19-6-2 100-kHz Ring Wave

The 0.5/100-kHz ring wave shape, shown in Fig. 19-9,[6] is a reasonable representation of surge voltages in 120- and 240-V ac systems. A surge, even if originally unidirectional, excites natural oscillation frequencies,[7] and the oscillatory surge may have different frequencies. The wave shape is characterized by:

$$\text{Rise time} = 0.5\,\mu\text{s} \pm 0.15\,\mu\text{s} \quad \text{and}$$

$$\text{ringing frequency} = 100\text{ kHz} \pm 20\text{ kHz} \tag{19-1}$$

The amplitude decays, so that the amplitude ratio of the second peak to first peak is between 40 and 110 percent. The ratio of the third peak to the second peak and of the fourth peak to third peak is between 40 to 80 percent. The equations are:

$$V(t) = AV_p(1-e^{-t/\tau_1})e^{(t/\tau_2)\cos\omega t} \tag{19-2}$$

where $\tau_1 = 0.533$ μs, $\tau_2 = 9.788$ μs, $A = 1.59$, and $\omega = 2\pi 10^5$ *rad-s*$^{-1}$. The rise time is defined as the time between the 10 and 90 percent amplitude points on the leading edge of waveform. The frequency is calculated from the first and third zero crossing after the initial peak.

The nominal amplitude of the first peak of either open-circuit voltage V_p or the short-circuit current I_p is selected according to the severity involved. The ratio V_p/I_p is specified as $12\,\Omega \pm 3\,\Omega$ for category B requirements and $30\,\Omega \pm 8\,\Omega$ for category A requirements. No short-circuit waveform is specified for 100-kHz ring wave.

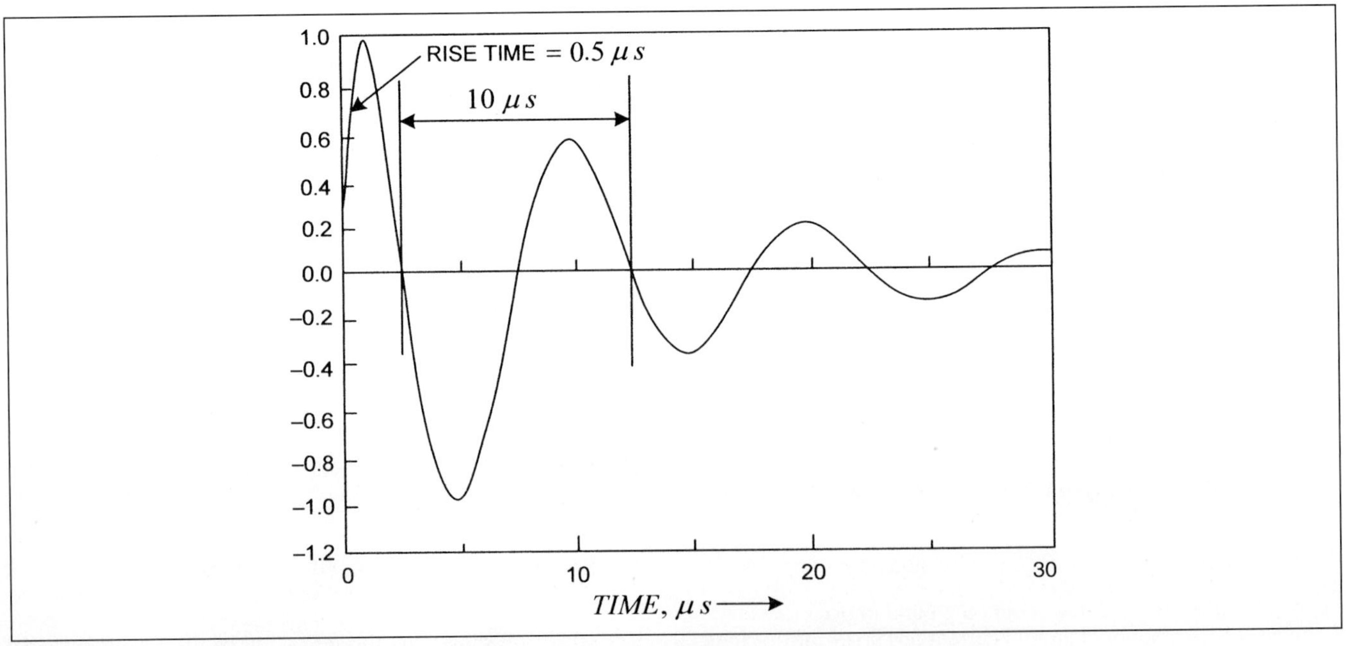

FIGURE 19-9 100-kHz ring wave. *Source: ANSI/IEEE Standard C62.41.*[6]

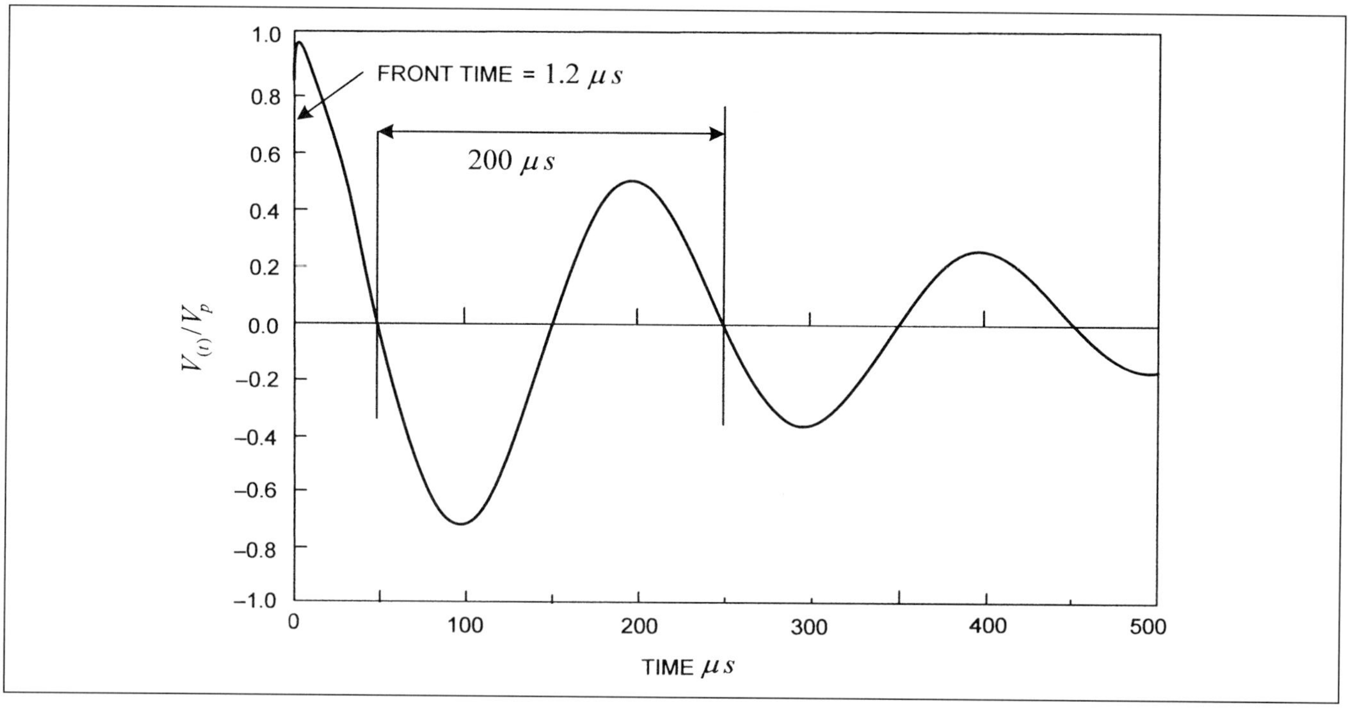

FIGURE 19-10 Wave form of 5-kHz ring wave. *Source: ANSI/IEEE Standard C62.41.*[6]

19-6-3 5-kHz Ring Wave

The waveform is shown in Fig. 19-10[6] and is defined by its open-circuit voltage parameters:

Rise time: $1.5\,\mu s + 0.5\,\mu s$ and

ringing frequency: $5\text{ kHz} \pm 1\text{ kHz}$

Ratio of adjacent peaks of opposite polarity = 60 to 80 percent (19-3)

The equation is same as for 100-kHz ring wave with the following constants:

$$\tau_1 = 0.7356\,\mu s, \tau_2 = 280.4\,\mu s, A = 1.027, \omega = \pi 10^4\,rad\text{-}s^{-1} \qquad (19\text{-}4)$$

The location categories in Figure 19-13 are based on the limiting effects of inductance of the branch circuits at frequencies associated with two standard test waves, which cause amplitude of current to decrease as distance from the source of surge increases. The 5-kHz ring wave is applicable to all categories of location.

19-6-4 10/1000-μs Wave

This waveform is shown in Fig. 19-11. The characteristics are:
Open circuit voltage:

Front time: $10\,\mu s(+0,-5)\,\mu s$

Duration: $1000\,\mu s(+1000,-0)\,\mu s$ (19-5)

Short-circuit current:

Front time: $10\,\mu s(+0,-5)\,\mu s$

Duration: $1000\,\mu s(+1000,-0)\,\mu s$ (19-6)

The equation is:

$$I(t) = AI_p(1-e^{-t/\tau_1})e^{-t/\tau_2} \qquad (19\text{-}7)$$

where: $\tau_1 = 3.827\,\mu s, \tau_2 = 1404\,\mu s, A = 1.019$. The long-term waveform reduces the effect of inductance and is applicable to all location categories.

19-6-5 Electrical Fast Transient

Circuit opening by air-gap switches has been recognized as producing a succession of clearing and reignitions, which generate bursts of fast ringing surges. The transients have been associated with arcing phenomena under the label, "showering arc".[3] IEC[8] requires a test involving bursts of surges of 5-ns rise time and 50-ns duration. IEEE C37.90.1[9] intended for protective relays and relay systems also includes a fast transient specification of rise time 10 ns and duration of 150 ns. The individual EFT in a burst is defined as:

Rise time: $5\text{ ns} \pm 1.5\text{ ns}$

Duration: $50\text{ ns} \pm 1.5\text{ ns}$ (19-8)

The duration is defined as the full width at half-maximum (FWHM), that is, the time difference between the 50 percent amplitude points at the leading edge and trailing edge of each pulse. Individual pulses occur in bursts with duration of 15 ms ± 3ms. Within each burst the repetition rate is specified as a function of peak open-circuit voltage.

For peaks $\leq 2\text{kV}$: $5\text{ kHz} \pm 1\text{ kHz}$

For peaks $> 2\text{kV}$: $2.5\text{ kHz} \pm 0.5\text{ kHz}$ (19-9)

The equation is:

$$V(t) = AV_p(1-e^{-t/\tau_1})e^{(t/\tau_2)} \qquad (19\text{-}10)$$

where $\tau_1 = 3.5$ ns, $\tau_2 = 55.6$ ns, and $A = 1.270$. Figure 19-12 shows the waveform.

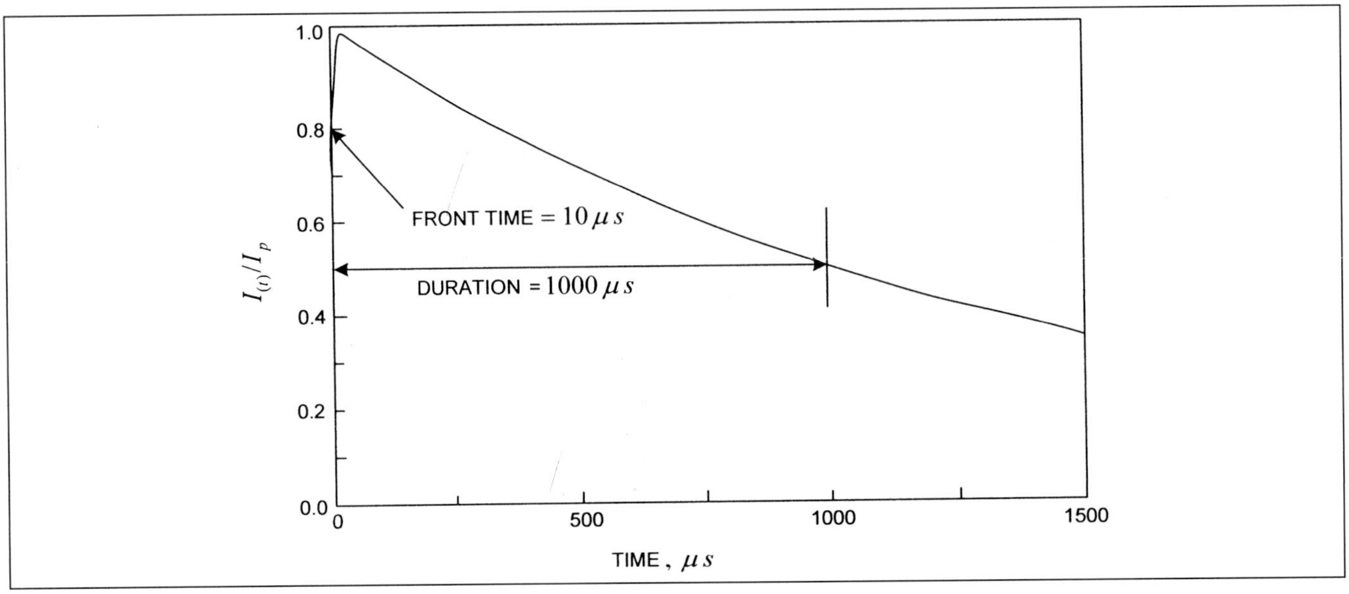

FIGURE 19-11 Waveform of 10/1000-μs current surge. *Source: ANSI/IEEE Standard C62.41.*[6]

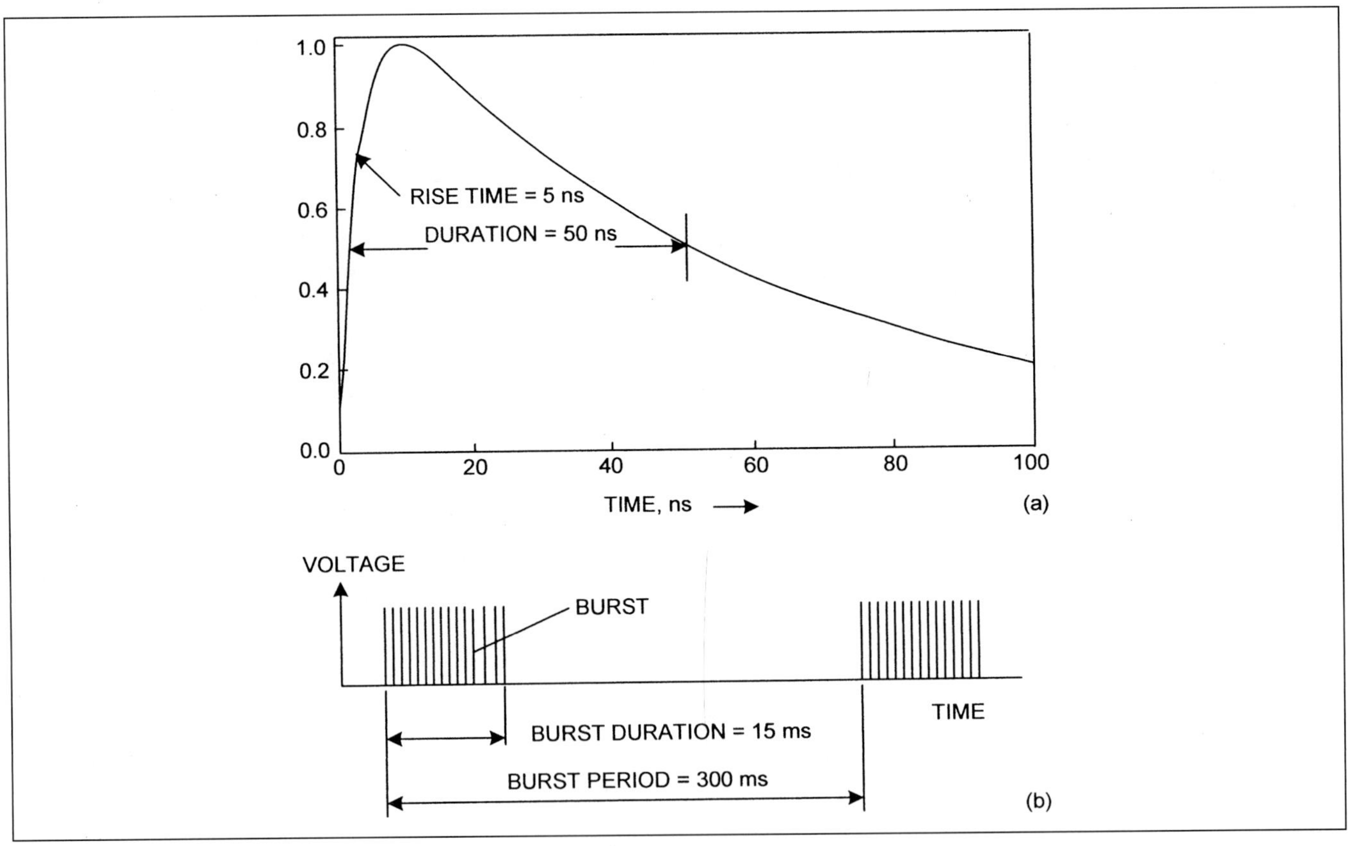

FIGURE 19-12 (*a*) Waveform of EFT pulse. (*b*) Pattern of EFT bursts. *Source: ANSI/IEEE Standard C62.41.*[6]

19-7 LOCATION CATEGORIES

IEEE C62.41[6] divides locations into three broad categories, as shown in Fig. 19-13*a*. The range of possible surge impedances and difficulty of selecting a specific value is addressed in these categories. Location category C is likely to be exposed to substantially higher voltages than the location category B, and higher exposure rates of Fig. 19-8 could apply with open-circuit voltages of 10 kV and discharge currents of 10 kA or more. The demarcation between category B and C is taken at the meter or at the main disconnect of the utility service. If the service is at high voltage, the demarcation is at the secondary of the supply service transformer. For surges originating in the utility supply, the source impedance may be considered constant, while the series impedance of the mains increases from outside the location to within the building. Open-circuit voltages for

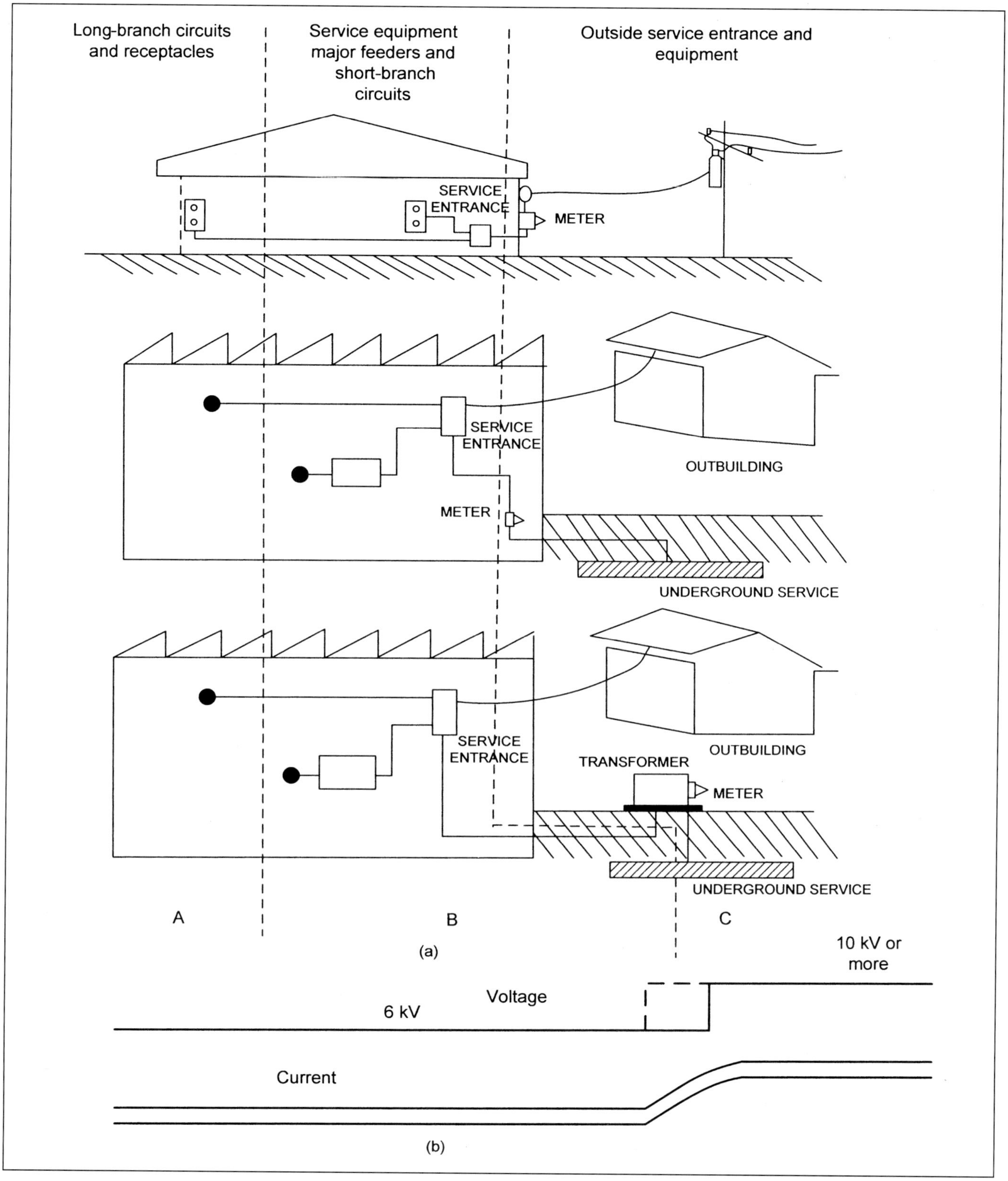

FIGURE 19-13 (*a*) Location categories. (*b*) Voltage staircases and current down-slopes according to location categories. *Source: ANSI/IEEE Standard C62.41.*[6]

surges other than the fast transients show little attenuation within the building.[7] Secondary arresters having 10-kA, 4/10-μs ratings have been successfully employed, effectively diverting associated surge currents. Direct lightning strikes will produce higher currents.[10] A location-C category surge protection device could be used in locations A and B, but this is not necessary or economical.

The location categories are described in qualitative terms, as an attempt to describe the scenarios of surges impinging at the service entrance or those generated inside the building. The propagation of surges is a phenomenon that does not recognize arbitrary boundaries, but is influenced by the components inside the installation. As an example, a surge protection device on the primary of a transformer does impact the secondary distribution. The concept resets on the propagation and dispersion of surge currents. The surge currents at the service meet the high surge impedance of the wiring system, and with or without flashovers, reduce the surge current diverted into the branch circuits. In contrast, a voltage surge below the point of flashover of clearances will run unimpeded into the subcircuit wiring. This concept based on IEEE Std[11] is shown in Fig. 19-13*b*.

Consider now a direct flash to the structure. The current will seek dispersion into local and remote earth electrodes and in the neutral conductor in multiple grounded systems. IEEE Std[11] recognizes that coupling of surge voltages will occur within the building power system:

1. Shift in the potential of conductors by rise of ground potential due to flow of surge current may be represented by a unidirectional pulse.

2. Induction of voltage or current in the circuit loops by the electromagnetic field of the lightning may be a ring-shape wave.

These two types of surges involve less energy-delivery capability than those associated with direct dispersion of the lightning current. In the absence of published measurements made during these rare occurrences, it is difficult to assign quantitative recommendations.[11] The application of test waveforms in categories A, B, and C is summarized in Table 19-1.

A reference to IEC[12] can be made which is more specific with respect to cloud-to-ground lightning strike on a facilitie's electrical service. It discusses the effect of direct stroke current (practically a rare occurrence on low-voltage incoming service because the conductors are strung at a much lower level from the ground) and assumes that 50 percent of the total lightning energy enters the earth grounding conductors, and the remaining 50 percent is divided by the numbers of conductors effected. It defines a worst peak current of 200 kA with a 10 × 350 μs waveform. After 50 percent of the current enters the grounding system, the remaining current is divided into two paths (L and N) giving 25 kA per conductor entering a facility (Fig. 19-14). A structure as defined in this standard can be anything from a high-rise building to an outdoor shelter containing SCADA or telecommunication equipment. Similar to ANSI/IEEE, lightning protection zones (LPZ) are introduced, see Chap. 22 for further discussions.

Example 19-1 We have simulated capacitor switching and resonant circuits in Chaps. 2 and 6, and these clearly show the nature of the transients. This example shows an EMTP simulation of propagation of transients in a commercial 480-V wiring system. The simplified system configuration is in Fig. 19-15. This shows a 600-kVA, 12.47- to 0.48-kV, three-phase transformer, with multiple grounded service connection to premises. (Practically, one service transformer may serve a number of residential/small consumer loads.) Consider that the service is overhead, 1-m long, 4/0 Penguin (211.6 KCMIL, 6 strand) with a neutral/ground conductor of 115.6 KCMIL (7 strands, #8). The neutral is multiple grounded at four points, equal distance from the service transformer.

A lightning stroke of 1.2/50-μs wave and 10-kV peak occurs at 1/4 of a mile from the service transformer on phase *a* to ground at 0.02 ms. The three-phase bus serves mixed loads, a 100-hp induction motor, a 208-V load of 150 kVA through a 200-kVA transformer,

TABLE 19-1 Summary of Test Requirements for Location Categories

LOCATION CATEGORY	A	B	C
100-kHz ring wave	All coupling modes except N-G: A1—low, 2 kV, 0.07 kA A2—medium, 4kV, 0.13 kA A3—high, 6kV, 0.2 kA Effective impedance = 30 Ω in all cases	All coupling modes except N-G: B1—low, 2 kV, 0.17 kA B2—medium, 4 kV, 0.33 kA B3—high, 6 kV, 0.5 kA Effective impedance = 12 Ω*	No requirement
Combination wave, 1.2/50 μs and 8/20 μs	No provision	All coupling modes except N-G: B1—low, 2 kV, 1.0 kA B2—medium, 4 kV, 2.0 kA B3—high, 6 kV, 3.0 kA Effective impedance = 2 Ω*	All coupling modes except N-G: C1—low, 6 kV, 3.0 kA C2—medium, 10 kV, 5.0 kA C3—high, 20 kV, 10 kA Effective impedance = 2 Ω*
5/50-ns bust	Test severity 1, 1 kV Test severity II, 2 kV Test severity III, 4 kV	Test severity 1, 1 kV Test severity II, 2 kV Test severity III, 4 kV	No provisions
10/1000 μs	Low (residential): none Medium (commercial)—1.0 U_{pk},† source $Z = 1.0\ \Omega$ High (Industrial)—1.3 U_{pk},† source $Z = 0.25\ \Omega$	As for category A	As for category A
5-kHz ring wave	Low—(far from switched capacitor banks) = none Medium—1.0 U_{pk},† $Z = 1–5\ \Omega$ High—(near large switched banks) 1.8 U_{pk},† source impedance = 0.5 to 1 Ω	As for category A	As for category A

*The effective impedance of surge source, emulated by surge generator is defined as the ratio of peak voltage to peak current, but it is not pure resistance.

†U_{pk} to be added to the mains voltage for the phase angle at which the surge is applied.

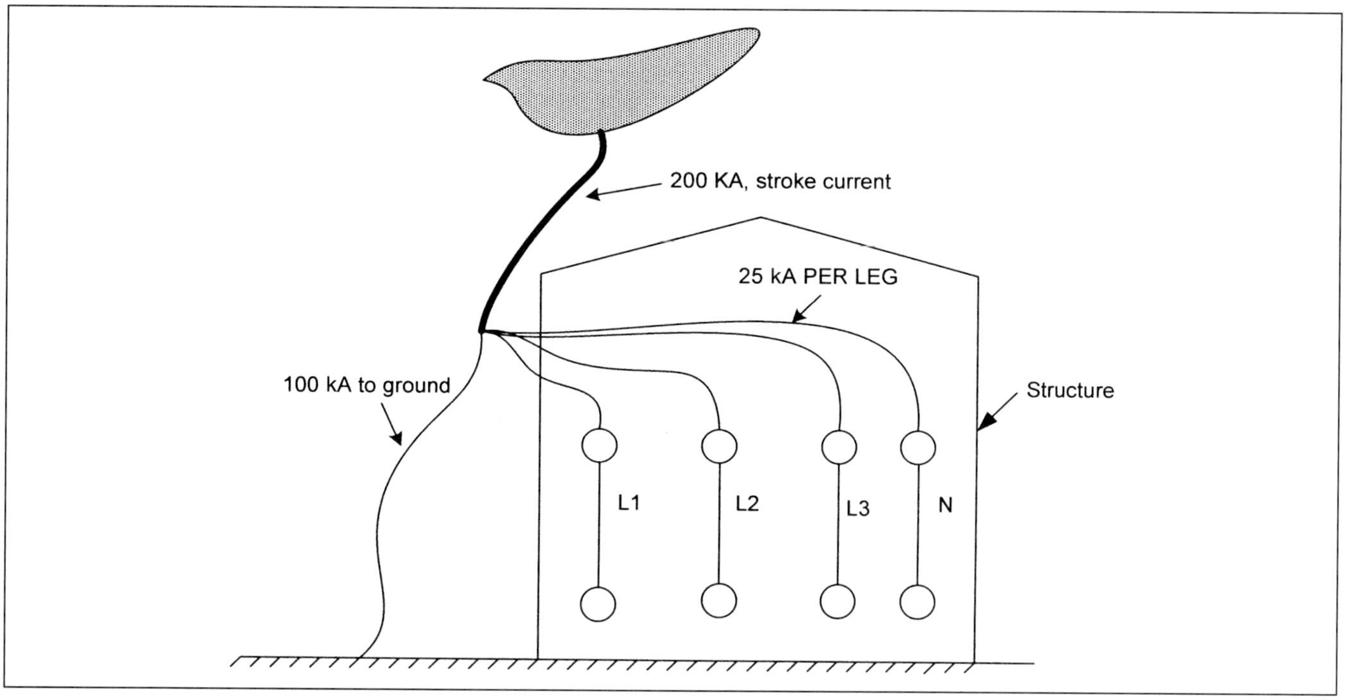

Figure 19-14 Distribution of direct lightning stroke current, IEC.[12]

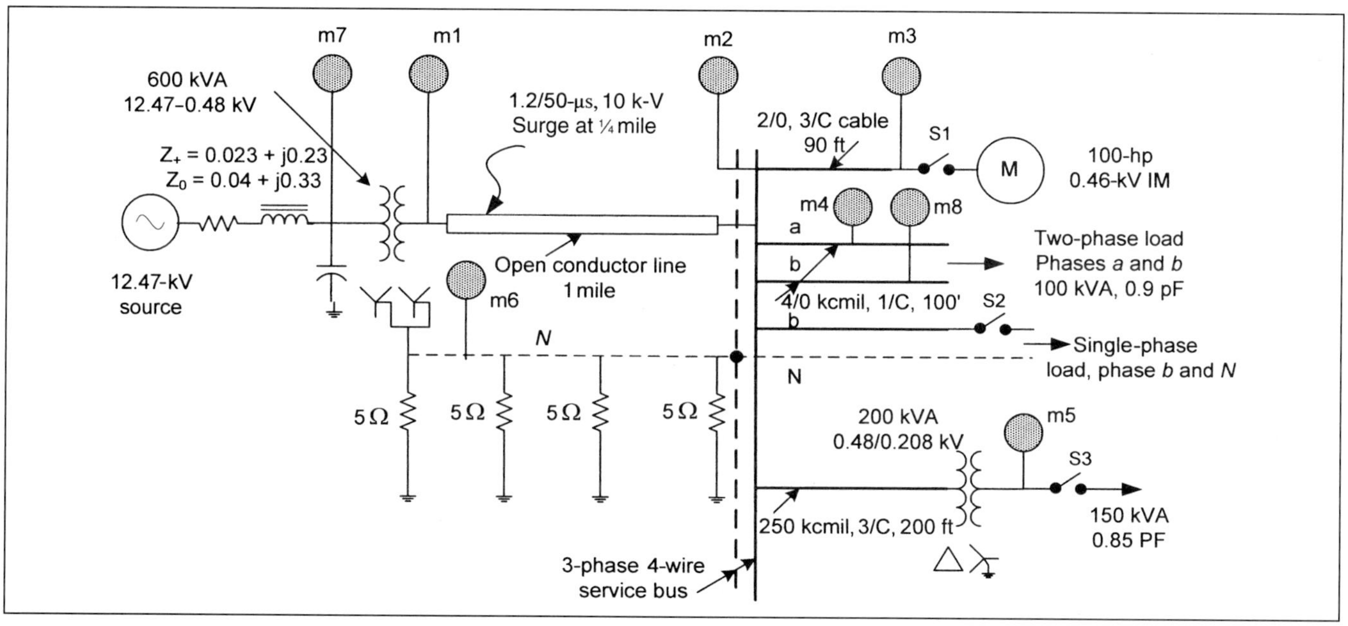

Figure 19-15 A typical commercial distribution system with incoming utility service connections for EMTP simulation of surges in the wiring system.

a single-phase load, and also a two-phase load of 100 kVA between phases *a* and *b*. Except for the two-phase load all other loads are switched off.

The voltages at the metering points *m*1 through *m*8 are shown in Fig. 19-16*a* and *b*. This shows oscillatory nature of response. The voltage couples to other phases, and high-frequency superimposed oscillations can be observed on the input voltage to the 600-kVA transformer. The neutral-to-ground voltage (*m*6) shows oscillations of small magnitude. There is voltage escalation in phase *a* (*m*2) at the service bus for a short duration. If we assume that the dielectric strength and clearances would limit the voltage at the service bus to 6 kV, the amplitudes throughout the distribution system would be reduced.

19-8 SURGE PROTECTION DEVICES

We may define SPDs (surge protection devices) as an assembly of one or more components intended to limit or divert surges. The device contains at least one nonlinear component. For applications to low-voltage systems, we may subdivide SPDs into two distinct categories:

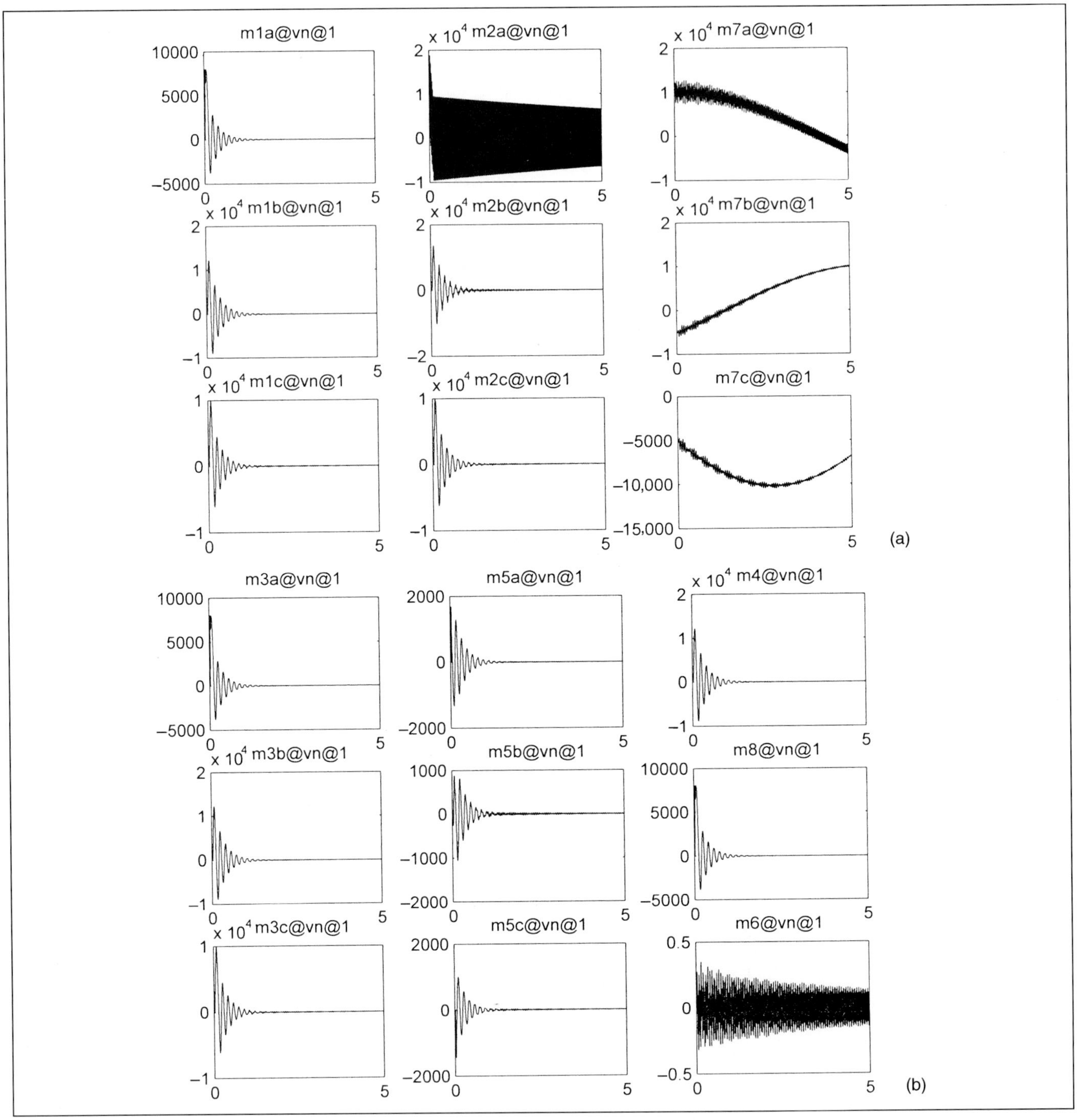

FIGURES 19-16 Simulated surge patterns.

1. *Secondary surge arresters.* Secondary surge arresters are rarely exposed to direct lightning strikes and are rated 1000 V or less. These meet the category C test requirements and are, therefore, installed between the transformer and the service entrance main overcurrent protective device. It appears that the application of secondary surge arresters directly across the transformer LV terminals can greatly reduce the possibility of transformer failures. In Fig. 19-6, if the lightning strikes the primary line, the current is not only discharged through the primary arrester to the pole-ground lead, but also through the consumer-load ground. Besides, the surges can couple to the primary side and vice versa, through the common ground. Some utilities that did not use secondary surge protection on their transformers are reconsidering this practice. The voltage ratings are 175 and 650 V for metal-oxide and valve-type arresters. As yet, no MCOV ratings have been assigned for metal-oxide arresters. Duty cycle discharge current is 1.5 kA, with 8 × 20-μs impulses and 22 discharges; high-current short duration test is 10 kA, 4 × 10 μs, and two discharges. The performance characteristics of metal-oxide and valve-type arresters are shown in Tables 19-2 and 19-3, respectively.

TABLE 19-2 Performance Characteristics of Secondary Metal-Oxide Surge Arresters

Arrester Rating (V)	Equivalent FOW Protective Level (kV)	Discharge Voltage for 8 × 20 μs Current Wave				Discharge Capability for 8 × 20 μs Current Wave (kA)
		At 1.5 kA (kV)	At 5 kA (kV)	At 10 kA (kV)	At 20 kA (kV)	
175	1.2	0.78	0.94	1.1	1.4	20
650	5.0	2.2	2.6	2.9	3.5	20

FOW: Front-of-wave.

TABLE 19-3 Performance Characteristics of Secondary Valve-Type Surge Arresters

	Impulse Sparkover Voltages			Discharge Voltage for 8 × 20 μs Current Wave	
	Front of Wave				
Arrester Rating (V)	Rate of Rise (kV/μs)	Sparkover Level (kV)	1.2 × 50 μs Sparkover Level (kV)	At 5 kA (kV)	At 10 kA (kV)
175	10	2.5–3.0	2.1–2.5	1.0–1.5	1.4–1.8
650	10	2.9–3.8	2.5–3.5	2.2–3.8	2.9–5.0

2. *Transient voltage surge suppressors.* The term TVSS is unique to UL 1449 standard.[13] TVSS is installed on the load side of a service entrance and provides protection of the equipment against transient surge voltages. These are rated 600 V or less and are required to have category B surge ratings.

19-8-1 Application of SPDs

There are many considerations for the proper application of SPDs. The effect of lead length becomes important when SPDs are separately mounted, away from the equipment that they protect. UL test data for integral panel board devices shows that 1ft of #8 stranded connecting wire adds 165 V. An externally mounted device will require several feet of connections, adding hundreds of volts to the suppressor let-through performance. Thus, better surge protection is achieved by integrating the SPDs within panel boards, switchgear, load centers, that is, within the equipment these SPDs devices protect. Table 19-4 shows the effect of cable connections versus surge current and increased system voltage level. We will revert to the separation distance in Chap. 20, however, the effect on surge protection is much pronounced in low-voltage systems. Some other considerations for the application are:

- SPDs should have required protection modes at the point of connection to the power system.
- SPD should be selected for the appropriate location category and should have been tested for that category.
- The lead lengths for connections to the required modes of suppression should be the shortest possible. Integrated surge protection devices should be considered.
- SPDs must be rated for the short-circuit currents at the point of application
- The SPDs may consist of a combination of individual components; however, it is important to consider ratings as applied to the power system to be protected.
- The system grounding is an important consideration; the NEC limit of 25 Ω grounding resistance is too high for practical purposes.
- Application of SPDs for electronic, data processing, and communication circuits requires special considerations. The restrikes, current chopping, and lightning surges originating in the power systems can be detrimental to these circuits.[3,4,14]
- Environmental constraints, that is, ambient temperature, humidity, and altitude, are a consideration, and an appropriate enclosure type should be selected.

19-8-2 Characteristics of SPDs

Some characteristics of SPDs are:

1. *Surge current capability.* It is defined as the maximum level of surge current that an SPD can withstand for a single surge event. An SPD in a geographical location like Florida will be highly susceptible to lightning. A high level of surge current capability is desirable in such areas. Note that C62.41[6] does not recommend maximum peak ampere ratings when defining location categories A to C. The manufacturers specify the peak ampere rating based on the summation of capabilities of individual suppression components associated with particular suppression mode.

TABLE 19-4 Effect of Cable Connections on Suppressor Voltage Rating

Description	Surge Current 3 kA			Surge Current 10 kA		
Suppressor voltage rating (V)	500	500	500	650	650	650
10 ft of open conductor voltage	1650			5010		
5-ft twisted and 10-ft total conductor voltage		500			1500	
1-ft twisted and 2-ft total conductor voltage			200			600
Increased system voltage, suppressor voltage plus conductor voltage	2170	1000	700	5660	2150	1250

TABLE 19-5 UL 1449 Suppression Voltage Rating for TVSS*

SERVICE VOLTAGE	L-N	L-G	N-G	L-L	MCOV
120/240 V, single phase	330	330	330	700	150
208 Y/120 V, three phase, four wire	330	330	330	700	150
240/120 V, three phase, high leg delta	700/330	330	330	1200/700	275/150
480Y/277 V, three phase , four wire	600	600	600	1200	320
600Y/347 V, three phase, four wire	1000	1000	900	1800	420

Note: All values are shown in volts

*Input surge voltage is 1.2 × 50 μs at 6 kV, and current is 8 × 20 μs at 500 to 3000 A. These are high exposure ratings.

- *Service entrance equipment*. Surge current capability 100 to 250 kA
- *Downstream distribution*. Surge current capability 80 to 150 kA

2. *Clamping voltage*. It is the measure of the SPD to limit the voltage a surge suppressor will pass to the load that it protects. It is the performance measure of a surge arrester to attenuate a transient. The UL suppression voltage ratings of TVSS are shown in Table 19-5.

3. *Response time*. TVSS response time is measured in ns. The response time of 1 ns is common for TVSS in the industry, while some manufacturers are claiming response time of even 1 ps. The speed of light is 0.3 mm/ps and electrical signals through conductors travel at a smaller speed. The measurements of response time in picoseconds are much dependent upon the test setup.

4. *Sparkover voltage*. This voltage for the low-voltage systems is defined as 6 kV. Surge voltage in the low-voltage systems is limited by the circuit's dielectric strength. The decrease in the peak surge current between category C and category A is a direct correlation of various influencing factors such as circuit impedance and current capability.

5. *Joule ratings or energy delivery capability*. An SPD may contain many components, like silicon avalanche diodes (SAD), metal-oxide varistors (MOVs), and gas tubes. MOVs and SADs typically fail due to excessive energy, and these components are rated in terms of joules, which these are capable of handling. The energy absorption will vary with the type of pulse. The significant parameter is not "the energy contained in the surge," but the actual energy that can be deposited in a surge-absorbing device.[15] In a purely resistive load, the surge energy can be expressed as V^2/R, however, for a nonlinear surge protection device this relation is not so simple. Thus, the energy-delivery capability of a surge is difficult to calculate when surge is diverted or shunted by a nonlinear device. A surge of high amplitude and short duration has the potential of upsetting the equipment operation but has little energy. A surge of high amplitude and long duration has the potential of high energy deposition.[6] It appears that the maximum surge current and clamping voltage seem to be reliable parameters for comparison between the SPDs performance, rather than the joules.

In Chap. 20, the energy-handling capability of the surge arresters applied at medium and high voltages is an important parameter of selection of the arrester. However, the arrester resistance and the surge voltage at the application of the surge arrester are calculated using the published data of the arrester for the switching surge test wave shapes. For SPDs at low voltage, there is no such test standard for specifying the switching surge tests, Some attempts have been made to specify in terms of specific energy as $\int (V^2/R)dt$, where R is arbitrarily taken as 50 Ω. However, varistor impedance is not constant and varies with the current and voltage in a nonlinear manner, and the energy deposited in the varistor varies with the source impedance, amplitude, and waveform.[15]

It is important that SPDs are capable of suppressing short-duration surges resembling and including that of lightning activity; it is also equally important that SPDs suppress long-duration surges. Suppressors intended to protect branch panels or outlets will be required to suppress long-duration surges more frequently than the surges induced by lightning activity.

19-9 SPD COMPONENTS

The component surge protection devices may be defined as a discrete SPD involving a specific technology and installed as a component within a surge protector, for example gas tubes, MOVs, and avalanche junction semiconductor devices. Rarely, are these used in isolation as examined below.

SPDs can be categorized as *voltage-limiting type* and *voltage-switching type*. The voltage-limiting types have high impedance when no surge is present and limit voltage progressively and smoothly by reducing its impedance. On the other hand, a voltage-switching device ("crowbar") can have a sudden voltage collapse to a low impedance state, that is, air gaps and gas tubes. These create short circuit in the system in response to a surge. The power-flow current is a concern, and the surge energy has to be dissipated elsewhere in the circuit. These do not provide an interruption-free service, unless used in conjunction with other components to avoid short circuit being applied to the line. The published impulse breakdown voltage is defined for specific rate of voltage rise.

The component varistors and avalanche junction semiconductors are voltage-limiting SPDs that behave like nonlinear impedances, and no short circuit occurs when responding to a surge. These are available in a variety of rated parameter values. A comparison of the parameters is shown in Table 19-6.

19-9-1 Gas Tubes

Gas tubes are cold cathode discharge tubes. There are two, sometimes three, electrodes sealed inside an enclosure that is filled with gas. As the gap is subjected to increasing field intensity due to a voltage surge, it breaks down at some voltage determined by the rate of rise of the surge and the design of the gas tube. The dc breakdown potential for a certain gas or gas mixture follows Paschen's law (Chap. 17). The line is, therefore, crowbarred or shorted until the voltage across the tube drops below its extinguishing level. One common method to interrupt the follow-current is to install a fuse ahead of the gas tube.

TABLE 19-6 Comparison of Different Device Parameters for SPDs

RATED DEVICE PARAMETER	GAS TUBES/AIR GAPS	MOVS	AVALANCHE JUNCTION SEMICONDUCTORS
DC breakdown voltage	75 to >1200 V		
Nominal varistor voltage, V_n		3.5 to >1200 V	
Breakdown voltage, V_{BR}			3 to > 850 V
Capacitance	0.5–5 pF	3–30000 pF	5–10000 pF
Maximum single-impulse discharge current	5000–100000 A (8/20) 10–2000 A (10/1000)		
Rated single-pulse transient current, I_{tm}		8–80000 A (8/20)	
Rated peak impulse current, I_{PPM}			2.5–40 A (10/1000)
Impulse breakdown voltage	250 to > 1500 V (100 V/μs)		
Clamping voltage		17–4000 V (8/20)	7–540 V (10/1000)
Rated RMS voltage, V_m		3–1500 V	
Rated stand-off voltage, V_{wm}			3–700 V
DC standby current, Id		0.5–200 μA	0.01–1000 μA
Insulation resistance	1–10 GΩ		

Another method is to use a varistor in series with a spark gap. A finite time is needed to start the electron avalanche action and breakdown; therefore, the gas tubes are slow to operate. Also the faster the rate of rise, the higher the overvoltage required to operate the tube. Damage to the protected equipment can occur on fast rising surges.

Telephone companies, usually, install a gas tube across the telephone wiring at the point of entry to a home or commercial facility. Communication interfaces RS-422 and signal circuits are similarly protected. The grounding and mounting is governed by provisions of NEC[2] or ANSI C2.[1]

The three-electrode tube provides both normal- and common-mode protection in a single device. The two outer electrodes are connected to the signal lines being protected, and the center electrode is connected to the ground (Fig. 19-17*a* and *b*).

19-9-2 Metal Oxide Varistors

Figure 19-18 shows the volt-ampere characteristics of a metal-oxide varistor, a contraction of variable resistor. This shows three regions: (1) leakage region, (2) normal varistor operation, (3) upturn region.

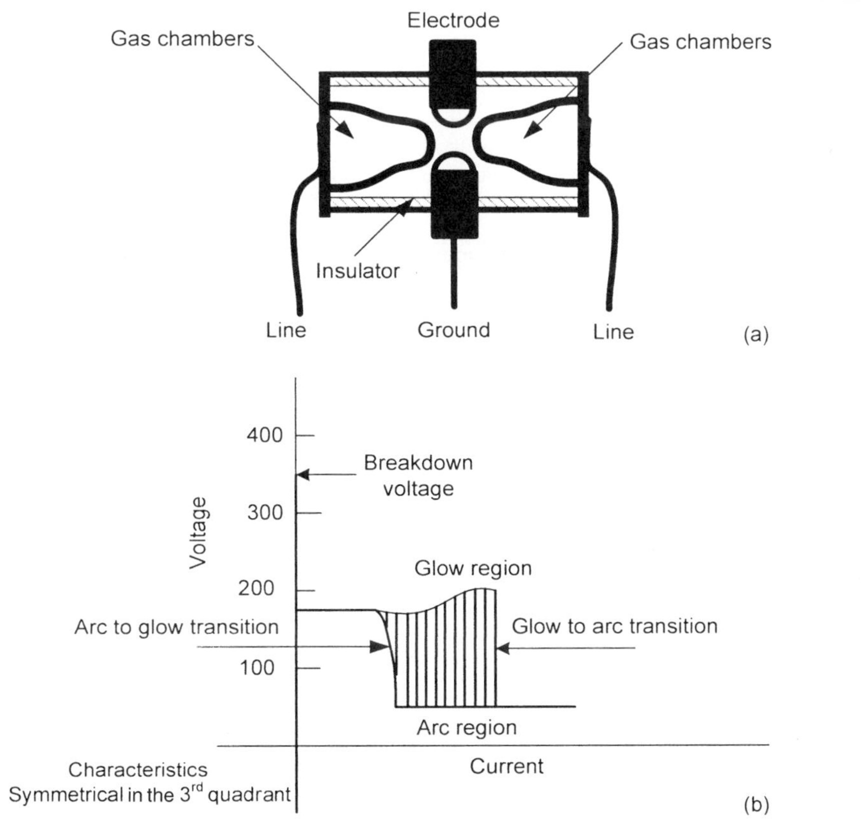

FIGURE 19-17 (*a*) A three-electrode gas tube. (*b*) Operating characteristics of the gas tube.

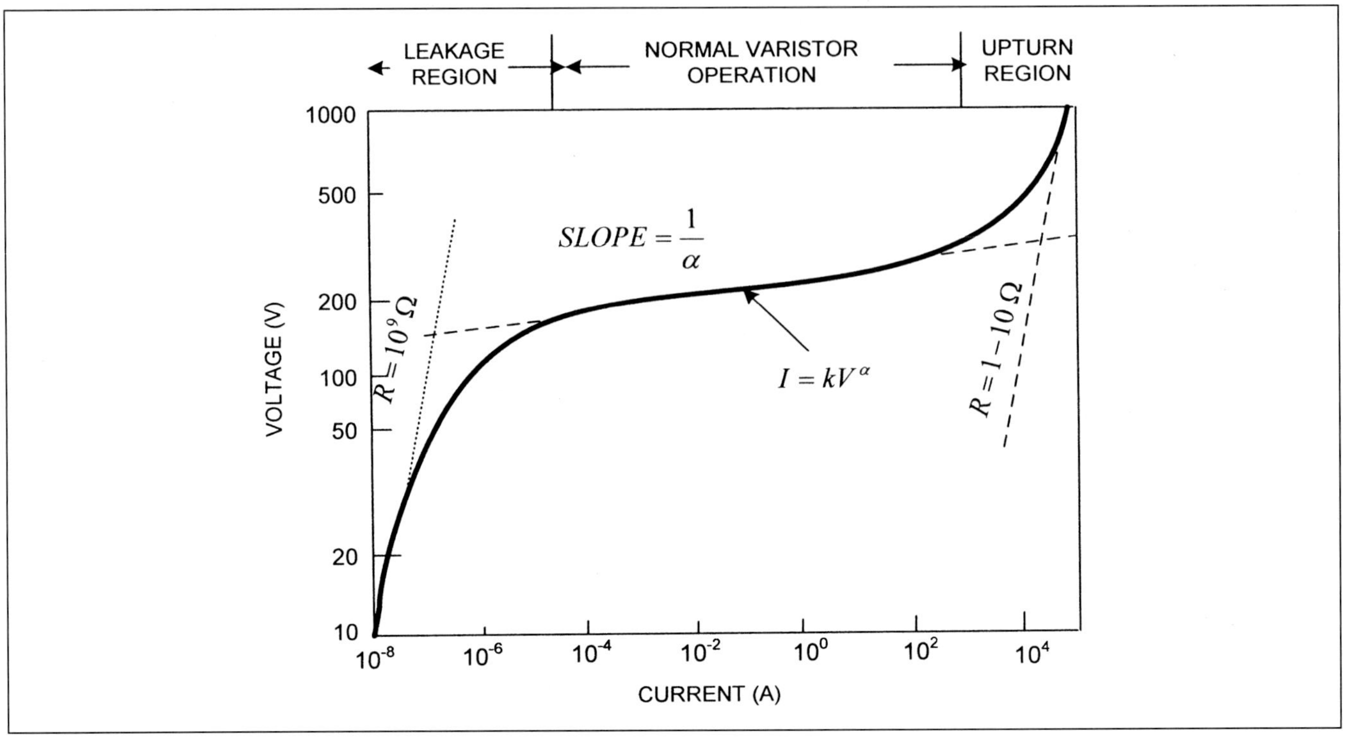

FIGURE 19-18 *V-I* characteristics of a MOV.

The metal-oxide particles are compressed under very high pressure to form various size discs, electrical leads are bonded to the discs, and the finished component is covered with an insulating material. The conduction mechanism of MOVs results from semiconductor junctions at the boundary of zinc-oxide grains formed during sintering process. A varistor may be considered a multijunction device with many grains in series parallel combination. The equivalent circuit model is shown in Fig. 19-19. When exposed to surges, large amount of currents are conducted, and there will be energy deposited in the varistor during its surge diversion function. At low levels of current, R_{OFF}, which is of high value, is the prevailing component. In the conduction region, resistor R_x takes progressively decreasing values. At large currents in the upturn region, R_{ON} becomes a significant part of the total device resistance, causing an upturn with value of R_{ON} as asymptote. Parallel capacitor can pass a current which may be higher than the dc leakage current. Lead reactance significantly affects the behavior of the varistor. Considerations are applied in the thermal design of an application, considering ambient temperature and temperature rise when passing a surge.

Clamping voltage has been defined previously. The maximum surge current I_{tm} is the current 8/20-μs waveshape, which a varistor is rated to withstand for a single surge. Rated rms voltage is the maximum voltage that can be applied continuously. The dc standby current is the current conducted by the varistor from a power source. The self-heating of the varistor should not be significant compared to its rated transient average power dissipation. Rated recurrent peak voltage is the maximum designated value of a repetitive nonsinusoidal power source of specified wave shape that can be applied continuously between terminals of the device. The nonlinear exponent α shown in Fig. 19-18 is the measure of varistor nonlinearity according to

$$I = kV^{\alpha} \tag{19-11}$$

where k is a device constant.

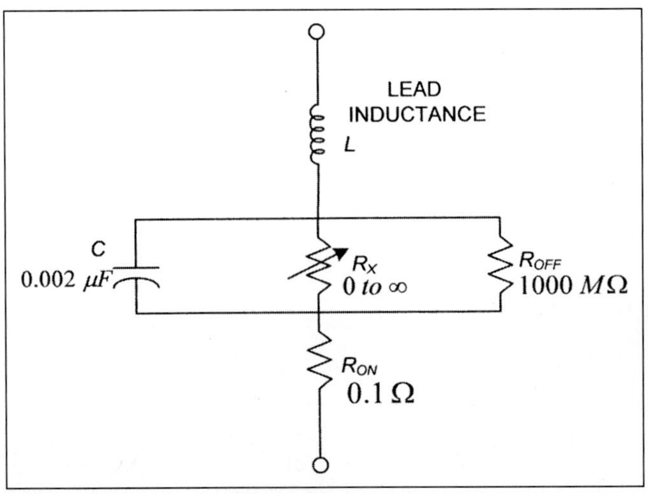

FIGURE 19-19 Equivalent circuit of a MOV.

The voltage overshoot shown in Fig. 19-20 occurs due to lead inductance. The following definitions are applicable: V_c is the clamp voltage for 8/20-μs wave, V_2 is equal to $(V_1 + V_2)/2$, $V_{os} = V_1 - V_c$ is the voltage overshoot, $t_2 - t_c$ is the overshoot duration, t_c is the time for the device to reach its clamp voltage, t_2 is the time for device to decay to 50 percent of its overshoot voltage, t_1 is the time for device to reach its peak value, and $t_1 - t_c$ is the response time.

Measurements of the clamping voltage across a lead-mounted varistor can indicate values exceeding levels observed with standard 8 × 20 μs waveform. This higher voltage is referred as voltage overshoot, V_{os}. For fast response solid-state devices, the voltage overshoot, response time, and overshoot duration are the phenomena primarily caused by the lead inductance.

A procedure for evaluating the cumulative effect of surges on varistors has been devised.[16] The method constructs a model of the surge environment and derives a set of surges of specific magnitude. This enables calculation of an annual rate of consumption of the rated pulse life of varistors, which can then be selected so that

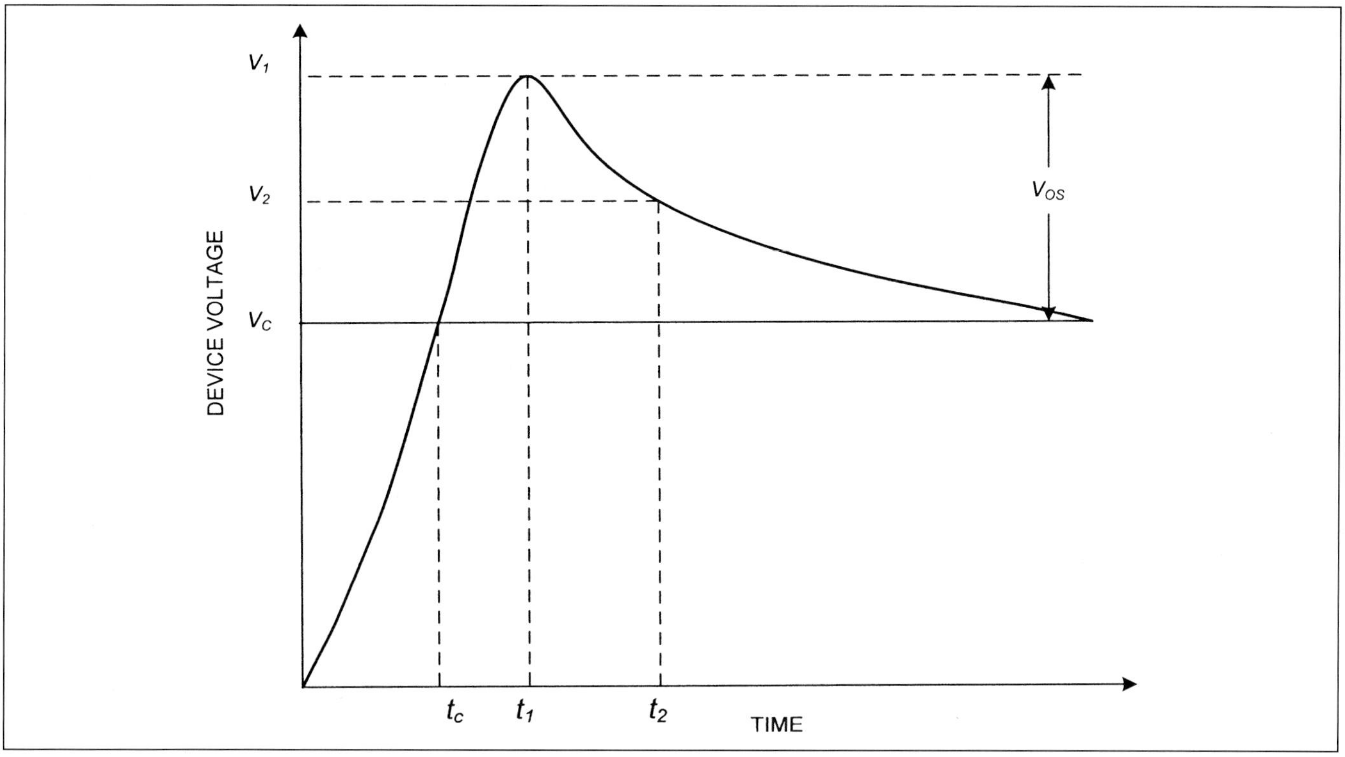

FIGURE 19-20 To illustrate voltage overshoot, response time, and voltage overshoot duration of a MOV.

the rated life will not be consumed within some the time duration, chosen as the design goal.

The response time of a *leadless* MOV can be less than 0.5 ns. Practically, with attached leads it is much higher. Figure 19-21 shows the derating curves for 40-mm diameter MOVs. The numbers across the curve can be taken as the relative value of the varistor in service for the specified surge environment. A MOV subjected to low-duration, low-current surges can last for a very long time, while it can fail quickly if subjected to high-current, long-duration surges.

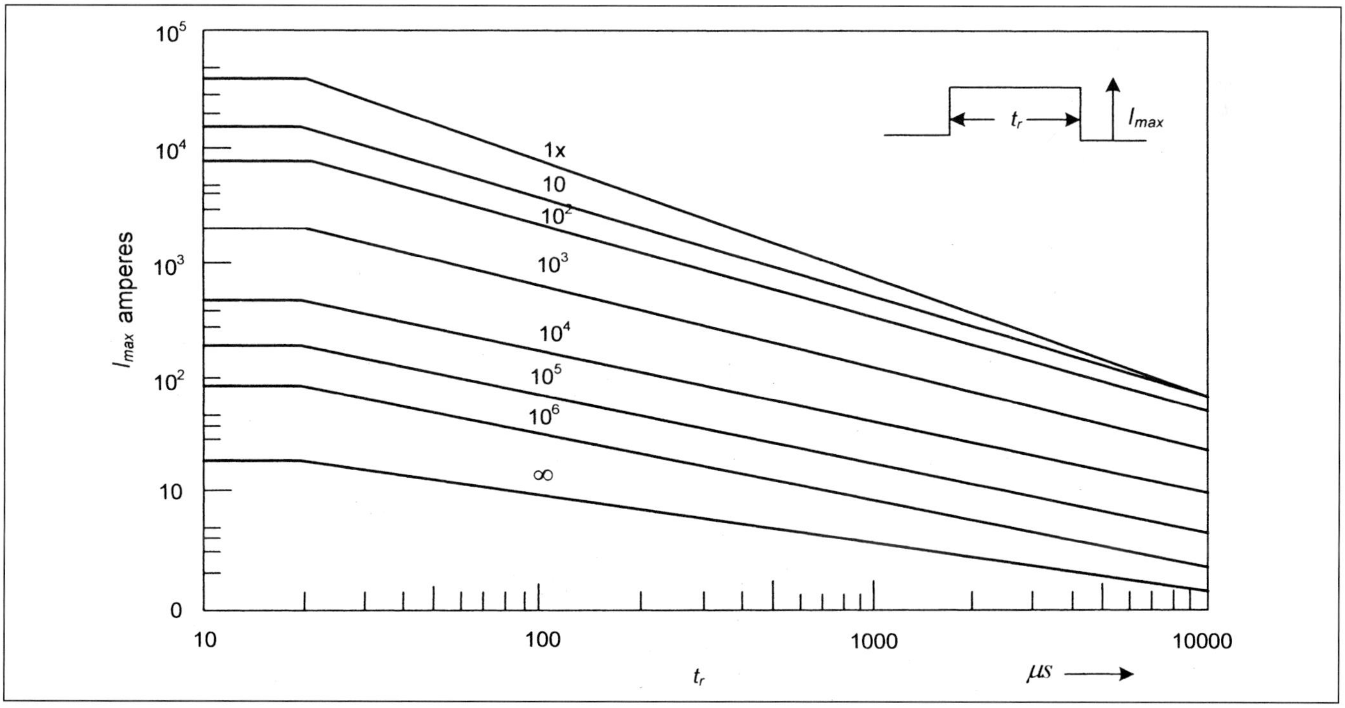

FIGURE 19-21 Derating curves of 40-mm diameter MOVs. *Reproduced with permission from Phoenix Contact (www. phoenixcontact.com). All rights reserved.*

NEC[2] requires category 3 or category 5 cables for modern data circuits. The specifications for these cables call for a very low capacitance because adding capacitance to the data signal can attenuate the normal data flow. Installing parallel MOVs on a data circuit has the same *parasitic* capacitance effects. Combined with the inductance of the cables, a low-pass filter is created that result in considerable signal attenuation in circuits with frequencies above 30 kHz. Therefore, MOVs are not well suited for high-frequency applications.

To summarize the advantages and disadvantages, MOVs can handle relatively large current levels, very specific turn-on time, a wide range of protection levels, but the performance decays with time, as the crystalline structure is etched away every time a surge passes through it. Also, they are not suitable for high-frequency applications.

19-9-3 Avalanche Junction Semiconductors

Silicon avalanche diodes (SADs) are two-terminal devices having symmetrical or asymmetrical characteristics, and their series and parallel combinations are used to enhance the voltage and power-handling capability. Figure 19-22 shows avalanche diode *V-I* characteristics for a *P-N* junction.

Clamping voltage V_c is the crest (peak) voltage across the device, assuming that external leads do not adversely add to this voltage. A waveform of 10/1000 μs is used as the standard test current impulse. The clamping voltage varies as a function of the applied peak impulse current. Rated peak impulse current I_{PPM} is the maximum value of peak impulse current for a minimum of 10 pulses, using 10/1000-μs waveform and maximum duty cycle of 0.01 percent without causing device failure. Rated stand-off or working rms voltage V_{WM} is assigned by the manufacturers, and is the maximum continuous operating voltage of the avalanche diode that is determined by multiplying the minimum breakdown voltage by 0.9 or 0.95. These voltage ratings are considered the maximum circuit, system, or line voltage when operating on device full temperature range. Rated peak surge current I_{PP} is the same as defined for MOVs. The breakdown voltage V_{BR} is measured at 1 mA, which may be stated as a minimum or nominal value and is a characteristic of the avalanche point for a *P-N* junction. In Fig. 19-22, V_{FS} is the forward voltage and I_{FS} is the forward current; I_{FSM} is the maximum forward current.

Silicon avalanche suppressor diodes were developed to protect sensitive electronics in aerospace industry. They exhibit a response time of less than 5 ns and limit transients to a low voltage level, typically less than 300 V (for a 120-V ac system). These have small leakage current. The disadvantages are low energy-handling capability and capacitance concerns for high-frequency applications.

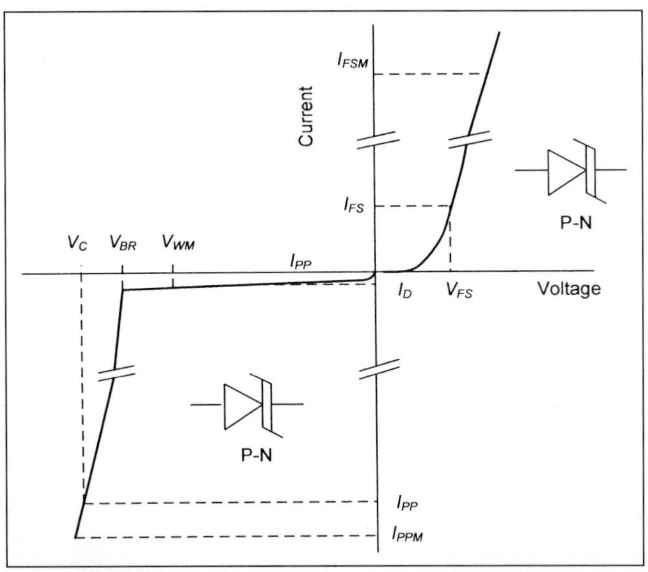

FIGURE 19-22 Bipolar characteristics of a SAD.

Diodes used in the reverse breakdown region are called *zener diodes*. Strictly only devices of breakdown voltage less than 5 V use zener mechanism, and it is not a bipolar device. SADs for transient voltage suppression are manufactured as bipolar devices. The zener diode is designed primarily for voltage regulation and is slower to respond than SADs.

19-9-4 Failure Modes of SPD Components

Gas tube arresters may fail because of mechanical shock, corrosion, hermetic seal failure, or operation on excessive large surges. They may or may not interfere with the system operation. Failure due to short circuits, low breakdown voltage, and low insulation resistance can be detected by the user of the protected equipment and are preferable where protection of property, people, or terminal equipment is of importance. The high breakdown failure mode is not normally noticeable, and may be used where uninterrupted system operation is more important.

MOVs can fail in the short-circuit mode with violent rupture and shattering of the material, which may pose a danger of the flying debris. This can be mitigated by use of overcurrent protection devices, generally, current-limiting fuses, which can isolate the faulty MOV or the entire circuit depending upon the location, as shown in Figure 19-23*a* and *b*. The varistor can have suitable packaging to contain overheating and shattering. Encapsulation may not be the answer to this problem. Initially, the encapsulating material acts as a barrier to expansion, but it has limited effectiveness in terms of confining explosion and may increase the magnitude and violence of an event. On the other hand, nonencapsulated devices will fail in a graceful, noneventful manner. The MOVs exhibit thermal runaway when the limiting voltage is set too close to the peak of the sine wave. Higher leakage current passing through the device leads to overheating.

An avalanche diode can fail due to short circuit, which has two modes. In a shorted junction, the surface or bulk material has failed, evidenced by melting of silicon material. The second failure mode is melting of internal materials due to arc across semiconductor material caused by a long-duration surge. The device can fail in open-circuit mode when breakdown voltage exceeds 150 percent of the pretested value at a value used to obtain V_{BR} or lower current. A device is considered a failure when the clamping voltage exceeds 120 percent of the pretested voltage.

MOVs have some disadvantages with respect to silicon avalanche diodes. A MOV cannot maintain a stable protection level V_{PL} as the current conducted increases (Fig. 19-18). A MOV may be considered failed if its initial V_{PL} has shifted by more than ±10 percent from the original value. Degradation with use is another consideration. With repeated conduction paths, the zinc-oxide particles do not perform consistently. There is evidence that once the degradation starts, it will continue in that direction. A comparison of the characteristics of single-SPD devices is shown in Table 19-7. This shows that all the desirable features cannot be obtained with one technology.

19-10 CONNECTION OF SPD DEVICES

19-10-1 Single-stage Parallel Connection

This connection is shown in Fig. 19-24 and is the most popular. The devices may consist of high-energy, tight-clamping, and fast-response characteristics. This configuration has two limitations, namely, the clamping level is a function of transient rise time, and the lead length is a consideration when installing parallel units. The limitation occurs because of use of one protective component, which may have high surge capability, but slow response, that is, gas tubes.

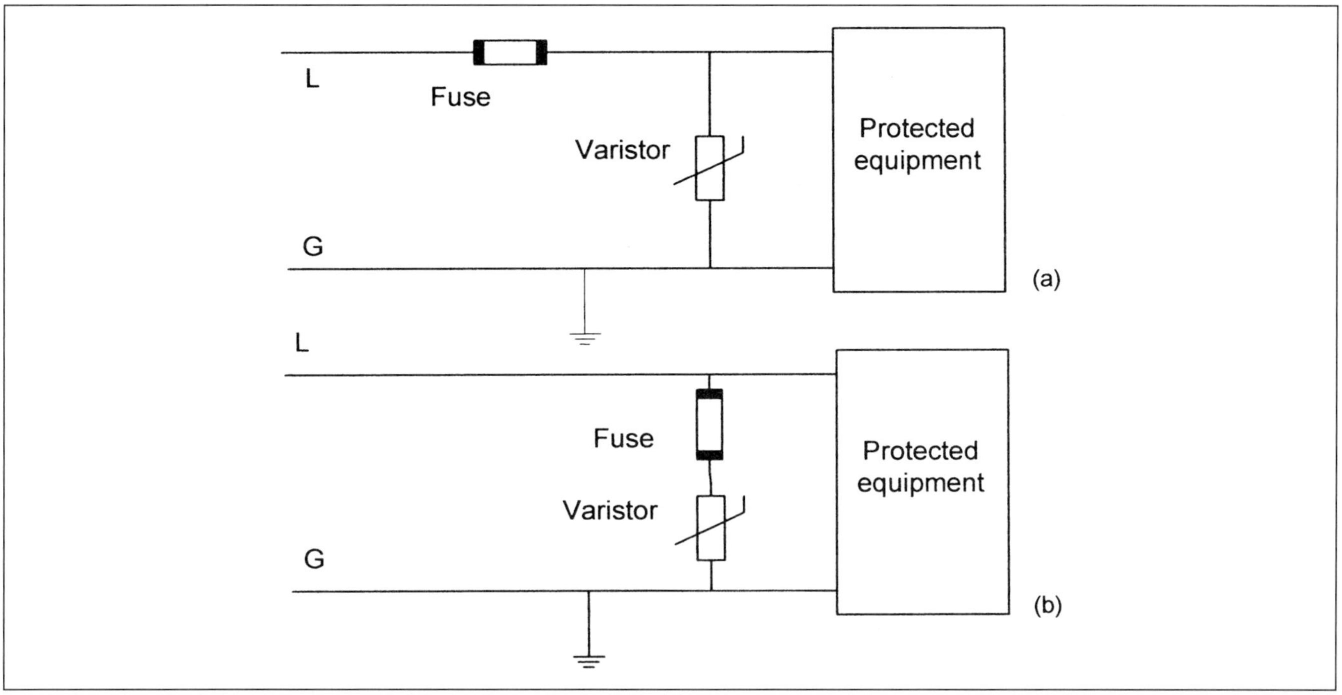

FIGURE 19-23 Alternative methods of protection of MOVs through current-limiting fuses.

TABLE 19-7 Characteristics of SPDs

TYPE OF SPD	RESPONSE TIME < 5 ns	SURGE CURRENT CAPABILITY	HIGH-ENERGY CAPABILITY	SAFE CLAMP VOLTAGE	LONG LIFE	CONNECTION METHOD
Gas tube	No	High	High	No	Yes	Only L-G
MOV	Yes	Medium	Medium	Yes	Limitations*	L-L, L-N, L-G**
Avalanche diode	Yes	Low	Low	Yes	*	L-L, L-N, L-G**
Selenium rectifiers	No	Medium	Medium	No	Medium	L-L, L-N

Note: Service interruption occurs with gas tubes.

*May fail explosively under category C surge conditions.

**For L-G connection, life expectancy will be short.

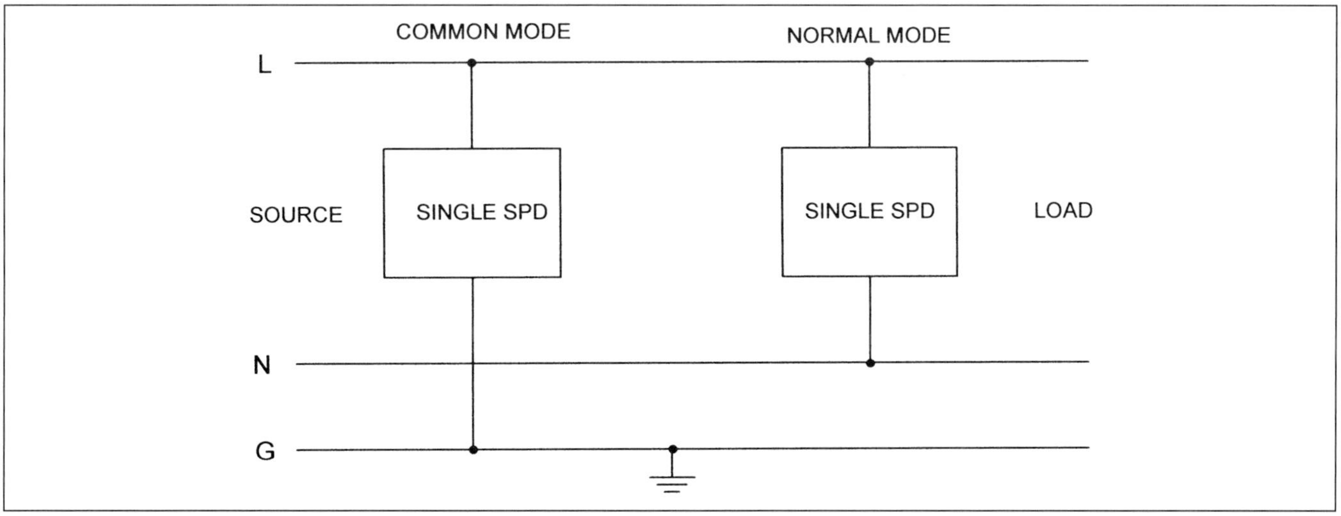

FIGURE 19-24 Shunt connections of a single SPD.

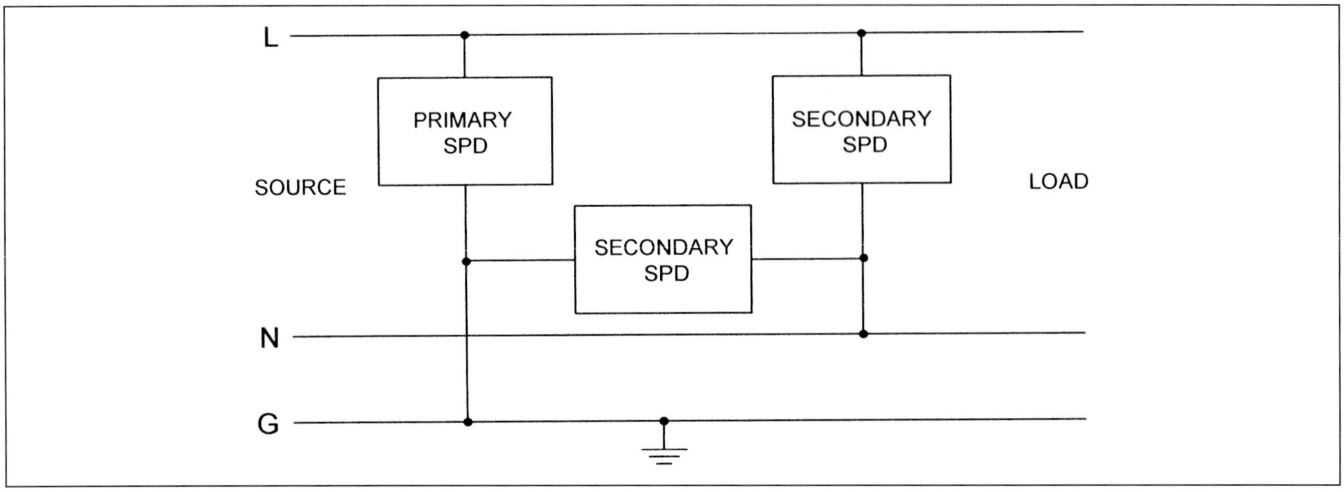

Figure 19-25 Multistage parallel connection of SPDs.

19-10-2 Multistage Parallel Connections

Parallel protectors can be improved by adding more stages of the different types of protective components. This enhances reliability and response for both common-mode and normal-mode protection. The primary surge suppression component provides a path for the surge current, while the secondary surge suppression component provides tight clamping (Fig. 19-25).

19-10-3 Series Hybrid Design

A series inductor or control element provides protection by delaying the rapidly rising waveform and protecting the secondary components. The series element with parallel capacitance attenuates electrical noise. The primary surge component is connected in common mode and diverts a larger amount of energy, simultaneously protecting the secondary surge components which may be MOVs or avalanche diodes. These components are usually installed line-to-neutral or line-to-line. These respond within 5 ns. Unlike parallel units, series hybrid systems maintain uniform lead-length distances between protection modules and phase conductors.

Figure 19-26 shows a three-stage protection solution and corresponding attenuation of the voltage surge. Each component shares in the overall reduction of the transient and is applied within its surge-handling capability. Some problems identified with series hybrid circuits are:

- In case of bidirectional transients, the power dissipation through primary and secondary devices is of consideration. The circuit is designed for transients from source to load side.
- In case of harmonics and load regulation, there is load-dependent voltage drop across the inductor. Possibility of resonances with inductor and capacitances exist.

Figure 19-27 shows a circuit accommodating filters for higher-frequency noise attenuation. Digital equipment radiates energy back into the power system. Any device containing any digital clock frequencies or pulses above 10 kHz must be registered and approved by FCC specifications. This limits the radiation from a single device. However, in a single facility, hundreds of such devices may contribute to sum up the noise in major pulses, which can cause system errors. Strategic installations of power-conditioning equipment with high-frequency filtering can obviate this cross-reference hazard. The filtering must be capable of broad range filtering, for example, from 100 kHz to 100 MHz.

19-10-4 Lattice Circuit

Figure 19-28 shows a lattice circuit using SADs. The first stage responds in 5 ns, and if the transient is of high magnitude, the first stage triggers the second stage to provide additional dissipation capability. The SCR2 acts merely as a switch to trigger the stage on. The clamping voltage V_2 is lower than V_1. When the first stage is turned off, the equipment downstream is protected at a lower clamp voltage of V_2.

19-10-5 Snubber Circuits

A sudden reverse voltage is applied to semiconductors in many applications after it has been turned off. With high inductive loads it can be a problem, and the high *dv/dt* may turn on the device or even result in its failure. A snubber circuit consists of a capacitor and resistor in series which is connected across the semiconducting device. Consider that a SCR is connected to a source impedance of R_s and L_s. Let R and C be the values of resistance and capacitance in the snubber circuit that need to be determined. When the circuit is energized and assuming that the capacitor is initially uncharged, the maximum *dv/dt* across SCR is:

$$\frac{dv}{dt} = \frac{ER}{L_s} \qquad (19\text{-}12)$$

$$R + R_s = 2\sqrt{L_s/C}$$

The minimum capacitance required is given by the expression:

$$C = \frac{1}{2L_s}\left(\frac{0.564E_m}{dv/dt}\right)^2 \qquad (19\text{-}13)$$

where E_m is the rms applied voltage to the circuit and *dv/dt* is the maximum safe tolerable value for the SCR. The resistance can be calculated from:

$$R = 2\sigma\sqrt{L_s/C} \qquad (19\text{-}14)$$

where σ, the damping factor, may be taken approximately as 0.65.

Table 3-1 in Ref. 3 defines surge parameters affecting equipment failure modes. For example, for semiconductors, peak amplitude, maximum rate of rise, and I^2t (energy let through) in the devices are surge parameters of concern. In addition, for triacs, source impedance is of concern.

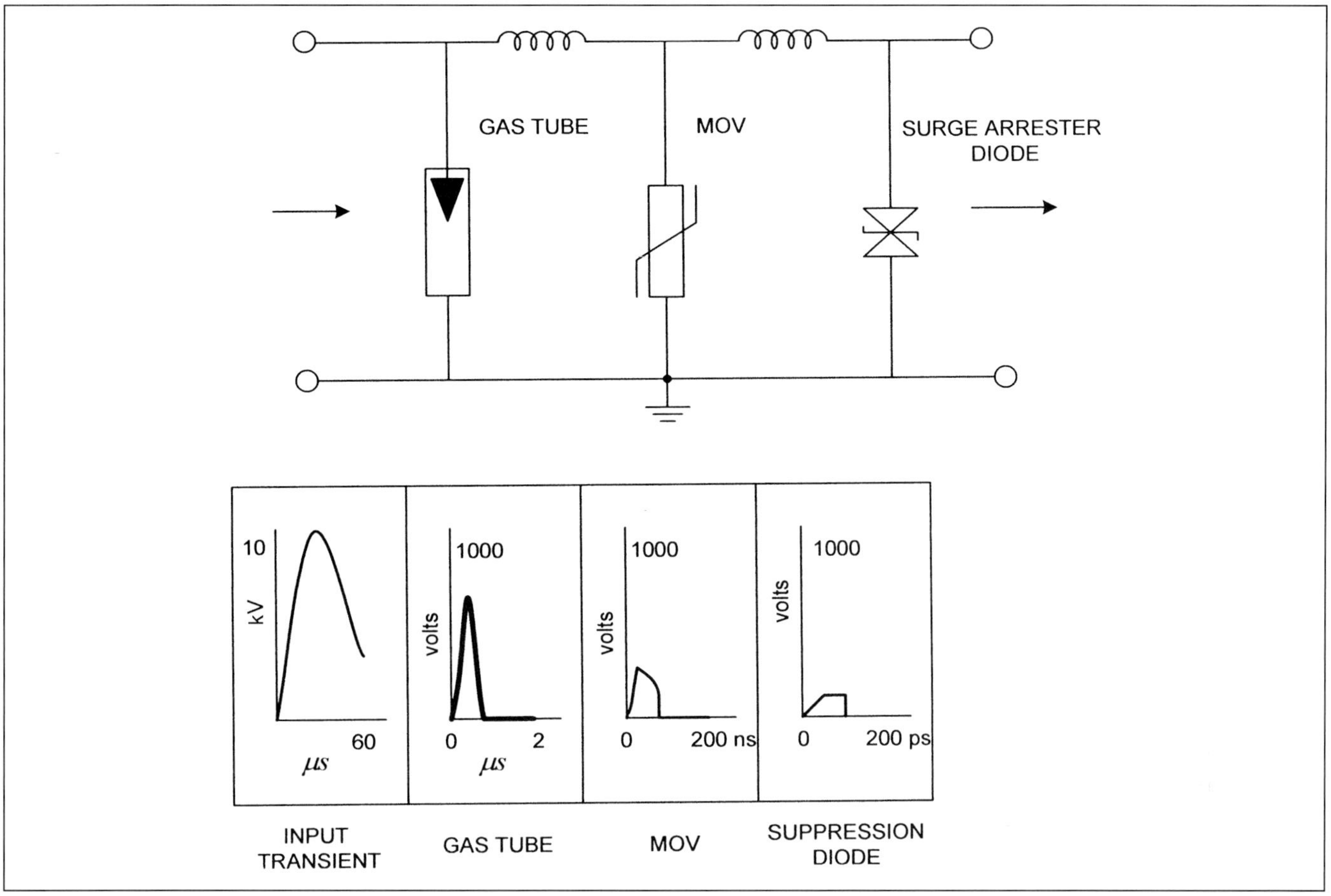

FIGURE 19-26 Three-stage surge protection with coupling inductors. *Reproduced with permission from Phoenix Contact (www. phoenixcontact. com).*

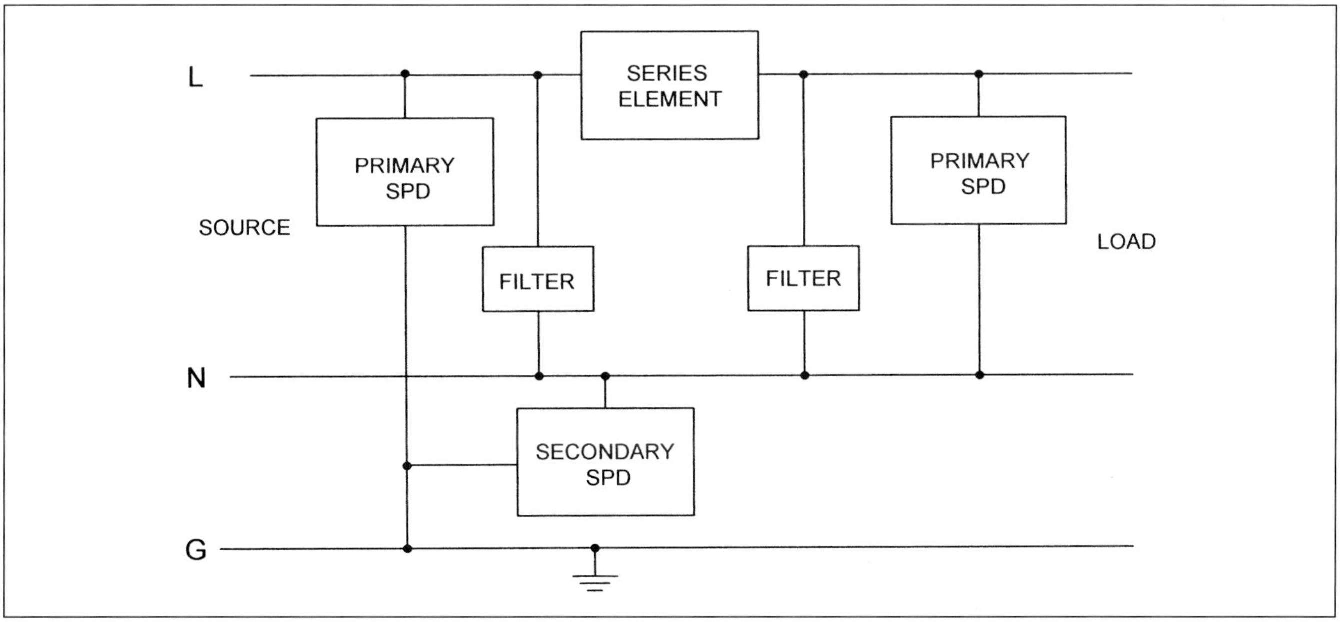

FIGURE 19-27 A hybrid connection of SPDs with filters.

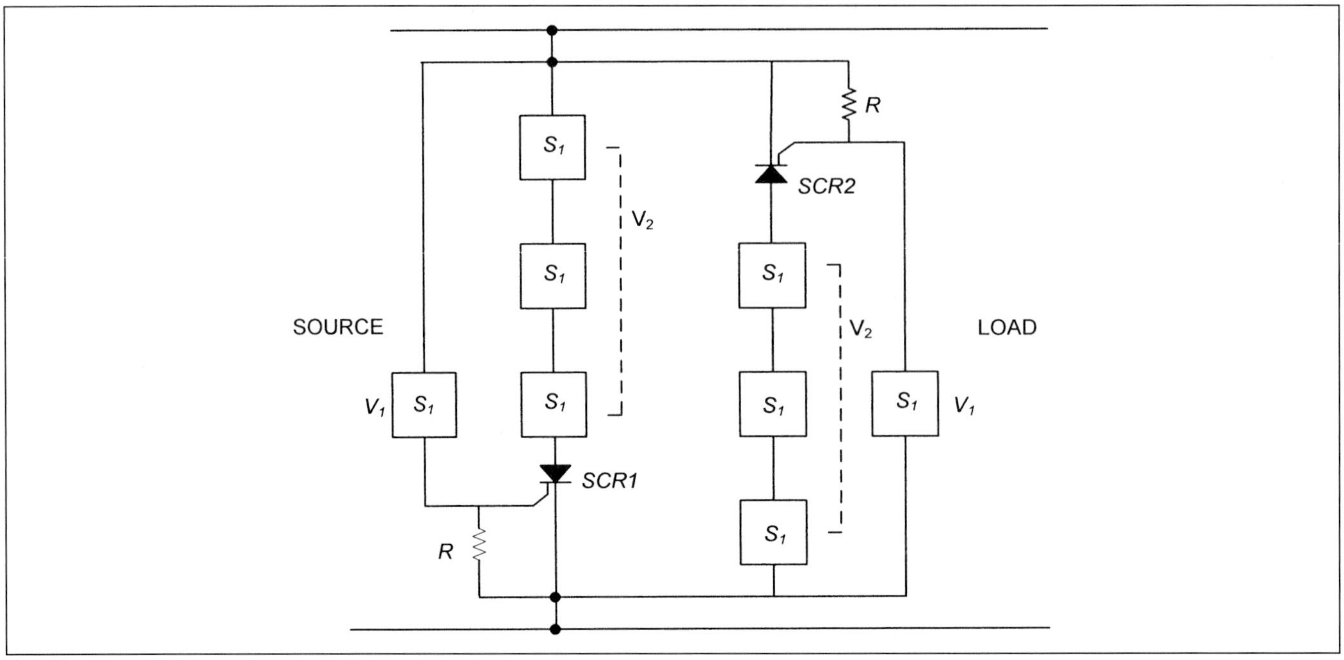

FIGURE 19-28 A lattice circuit for surge protection using SADs.

19-11 POWER QUALITY PROBLEMS

Transients and noise are two important concerns of power quality. Voltage sag is reduction of nominal system voltage for more than 0.01s and less than 2.5 s. A swell is increase in the nominal voltage over the same time limits. Low and high voltages are increases or decreases in voltages for more than 2.5 s. There can be outright interruptions, frequency variations, transients caused by load interruption, and harmonic pollution, as shown in Fig. 19-29. Frequency variations are more of a concern for engine-generator-based distribution systems. Voltage waveform distortions (harmonics) occur due to nonlinear loads and dissymmetry, that is, unbalanced loads cause voltage unbalance. Power quality problems cost the U.S. industry some billions of dollars per year, though these can be effectively mitigated in a system with proper designs and selection of equipment

The nonlinear loads are increasing at a fast rate, and the power quality problems in the distribution systems are compounding. The nonlinear loads degrade the power quality, and these loads are much sensitive to the pollution they produce. We discussed these in Chap. 16. Here, another example of switch mode power supplies (SMPs) can be stated, which creates severe line pollution. The switch-mode power supplies have replaced the conventional power supplies for computers, copiers, TV sets, and other household appliances. These have some advantages, for example, greater tolerance to rms voltage fluctuations, fundamental frequency variations, smaller compact size, greater efficiency, and lower manufacturing costs, yet these create new power quality problems.

Figure 19-30*a* and *b* show conventional- and switch-mode power supply systems, respectively. The conventional power supply has a main input transformer, a full-wave bridge rectifier, and the main ripple frequency is at 120 Hz. The current drawn from the supply system is relatively linear; capacitors C_1 and C_2 and inductor act as a passive filter. In SMP, the incoming voltage is rectified at line voltage and high dc voltage is stored in capacitor C_1. The switcher and controls switch dc from C_1 at high rate (10 to 100 kHz). These high-frequency pulses are stepped-down in a transformer and rectified. The switcher eliminates series regulator and losses in conventional supplies, and the end product is more efficient with lighter components. Four common configurations used with the SMPs are: flyback, push-pull, half bridge, and full bridge.

The current drawn from the supply system is discontinuous and flows in pulses (Fig. 19-31). This current spectrum has high harmonic content. The third harmonic is 81.0 percent, fifth harmonic is 60.6 percent, and seventh is 37 percent. There are considerable percentages of higher harmonics. In a three-phase, four-wire system, if the loads are not fairly balanced, the neutral can be overloaded with third harmonics and a neutral conductor size equal to phase conductors may be necessary.[2] Third harmonic traps and filters can be provided. Harmonic filters, active or passive, are the most effective in mitigating the harmonic problems (Chap. 6).

Switch-mode power supplies produce 180- to 100-kHz impulses through the rectification process. Due to lack of main input transformer mass and greater internal capacitive coupling, SMPs have much greater susceptibility to transient surge voltages. The fast-rise, high-frequency transients enter the SMPs through the low-impedance paths and are dissipated within the SMPs or logic circuits and hardware. A failure can occur.

Figure 19-29 shows the interaction of various technologies and their effectiveness for a particular power supply problem. It is not the intention to discuss the various mitigation technologies shown in Fig. 19-29. Let us consider, as an example, uninterruptible power supply (UPS) systems.[5] A conceptual circuit is shown in Fig. 19-32*a*. The ac power is rectified, a battery system floated on the dc output, and the dc voltage is reconverted to ac. A static bypass switch can reduce the mean time between failures (MTBF) and switches over the load to another standby ac source in case of a component failure in the main rectifier-battery-inverter lineup. The static switch transfers fast, practically, in-phase transfer, with little voltage drop on transfer. In Fig.19-32*b*, the interruption in the voltage is hardly noticeable. Two lineups of rectifiers, battery systems, and inverters can be connected in redundant or load-sharing mode, again with a bypass, depending upon the required service reliability of the load.

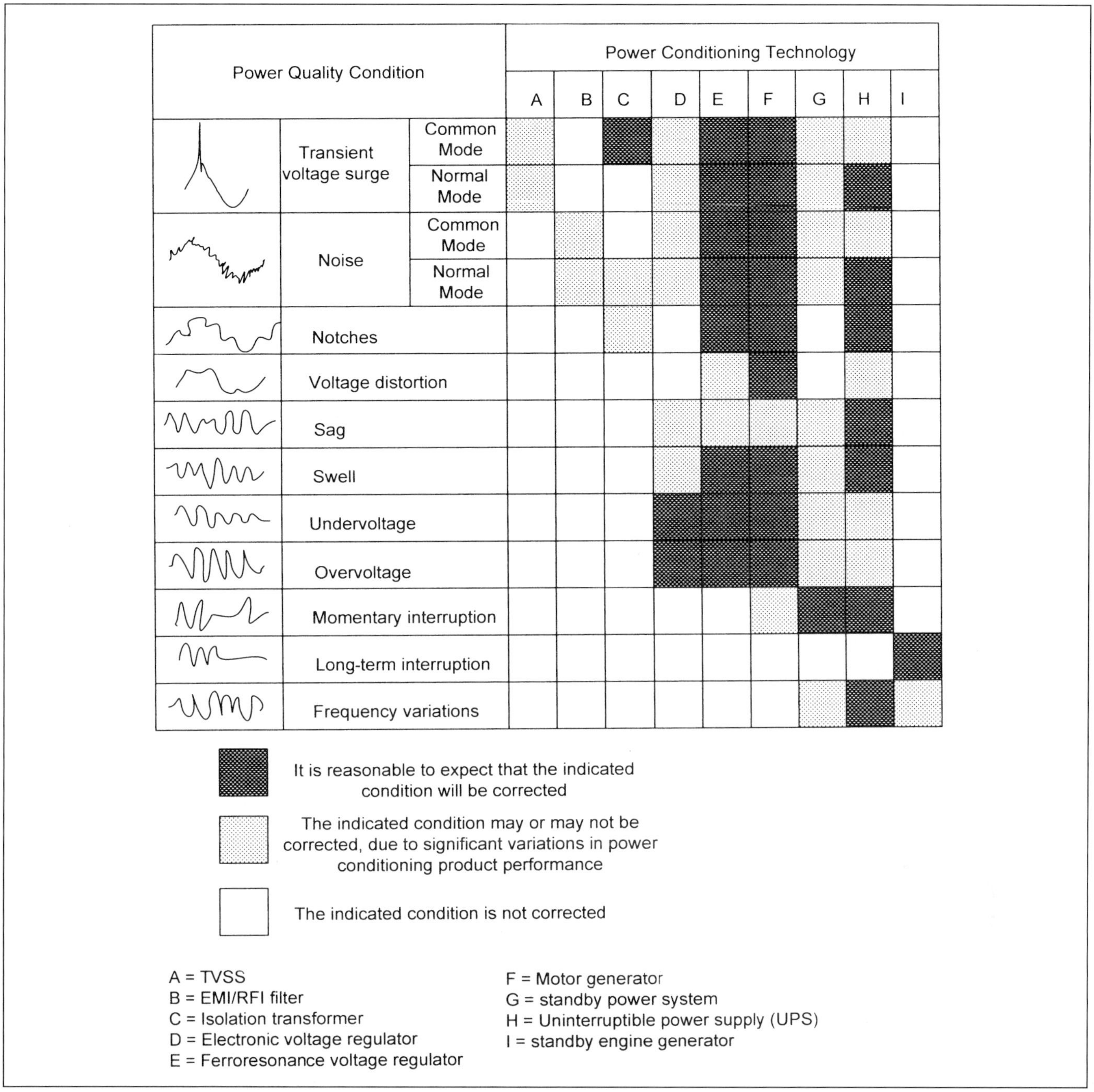

FIGURE 19-29 Power quality conditions and conditioning technologies to overcome power quality problems.

The first step in identifying the power quality problems and their impact on the processes is to distinguish which operating characteristics of the concerned equipment is likely to cause an equipment malfunction or shutdown. Methods of characterizing rms disturbances for calculations of performance indices have been developed. An important parameter is low-voltage sag and its time duration. A recommended method of summarizing expected voltage sag is shown in Fig. 19-33, which presents performance as constant supply sag contours, similar to a topographical contour map. This can be directly compared with the equipment sensitivity. The problem lies in forming the voltage contour map at a location. The equipment sensitivity is sometimes specified by the manufacturers.

19-12 SURGE PROTECTION OF COMPUTERS

With respect to protection of computers and electronics, over the course of years, the component density on an IC has increased (it is now possible to accommodate more than couple of million transistors on a chip the size of a postage stamp). This has increased the vulnerability to transients. The memory circuit of a chip can be upset with as little as 10^{-9} J of energy (Fig. 19-34), and the chip can be destroyed by 10^{-5} J. The higher energy enters through (1) customers ac/dc power supply systems and (2) data lines. Both must have surge protection. It is estimated that 88.5 percent of power-line disturbances are of transient type:

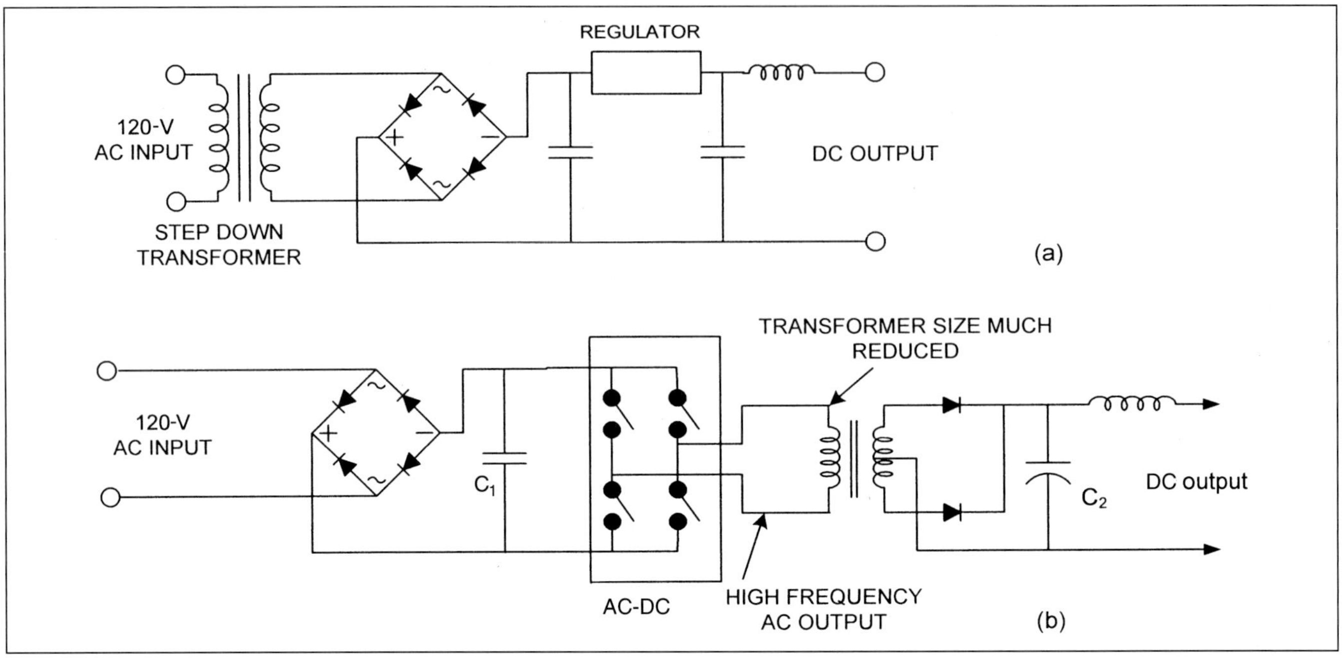

FIGURE 19-30 (*a*) A conventional power supply, no longer in use. (*b*) Switch-mode power supply circuit diagram.

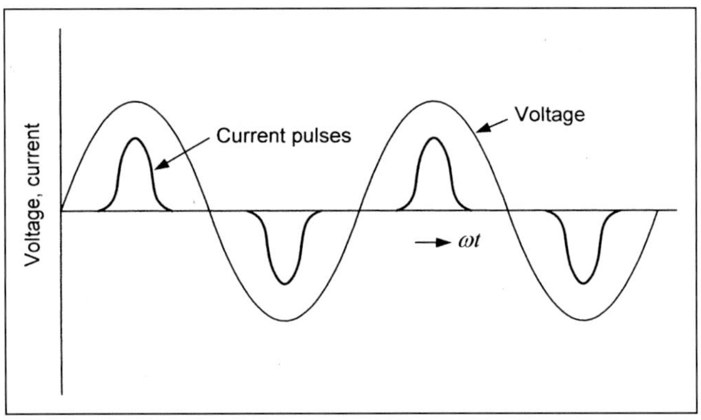

FIGURE 19-31 Discontinuous line-current input demanded by a SMP, containing high percentage of harmonics.

- Oscillatory decaying transients: 49 percent
- Voltage impulse transients: 39.5 percent
- Undervoltage and overvoltages: 11.0 percent
- Voltage outages: 0.5 percent
- Total is 100 percent

A voltage transient creates the following types of stresses on sensitive electronic equipment:

Upset errors. There can be a data error or an altered memory. This upset is caused by a low-level transient of voltage amplitude sufficiently intense to cause a component to change state.

Damage. There can be immediate component damage. If the transient is of large magnitude/time duration, a burnout can occur.

Erosion—long-term failure. The effect may not become apparent for sometime, but the component erodes and degrades, which may result in failure in days or weeks.

19-12-1 Causes of Semiconductor Failure

The causes of semiconductor failure are:

1. Overstress due to transients, lightning, and electrostatic discharge (ESD).[17] ESD phenomena generate electromagnetic fields from dc to low gigahertz. The ESD event includes not only the discharge current, but also the electromagnetic fields and corona effects before and during discharge. Sudden transfer of charge between bodies of differing electrostatic potentials occurs. Electrostatic induction is defined as a means by which portions of humans (hands, fingers, or hips) or other items can become differentially charged. Physical charge transfer occurs when charged particles are physically transferred to an object that is itself not creating the charge. Most common methods of charging is the act of walking. The human body in the process of walking builds up electrical charge with respect to its surroundings. Walking on a floor with foot wear of an insulating material leads to charge buildup on the underside of shoe sole by triboelectrification (involves electron or ion transfer upon contact due to frictional localized heating of microscopic contact areas on solid surfaces). A transient voltage of 35 kV can occur to microprocessor-based equipment from a simple walk across a synthetic carpet (Table 19-8). (It is common experience that on a dry winter day, one can get a mild shock by touching the metal door knob of a door or see blue sparks on friction of nylon garments in the dark.) The voltage levels caused by ESD can

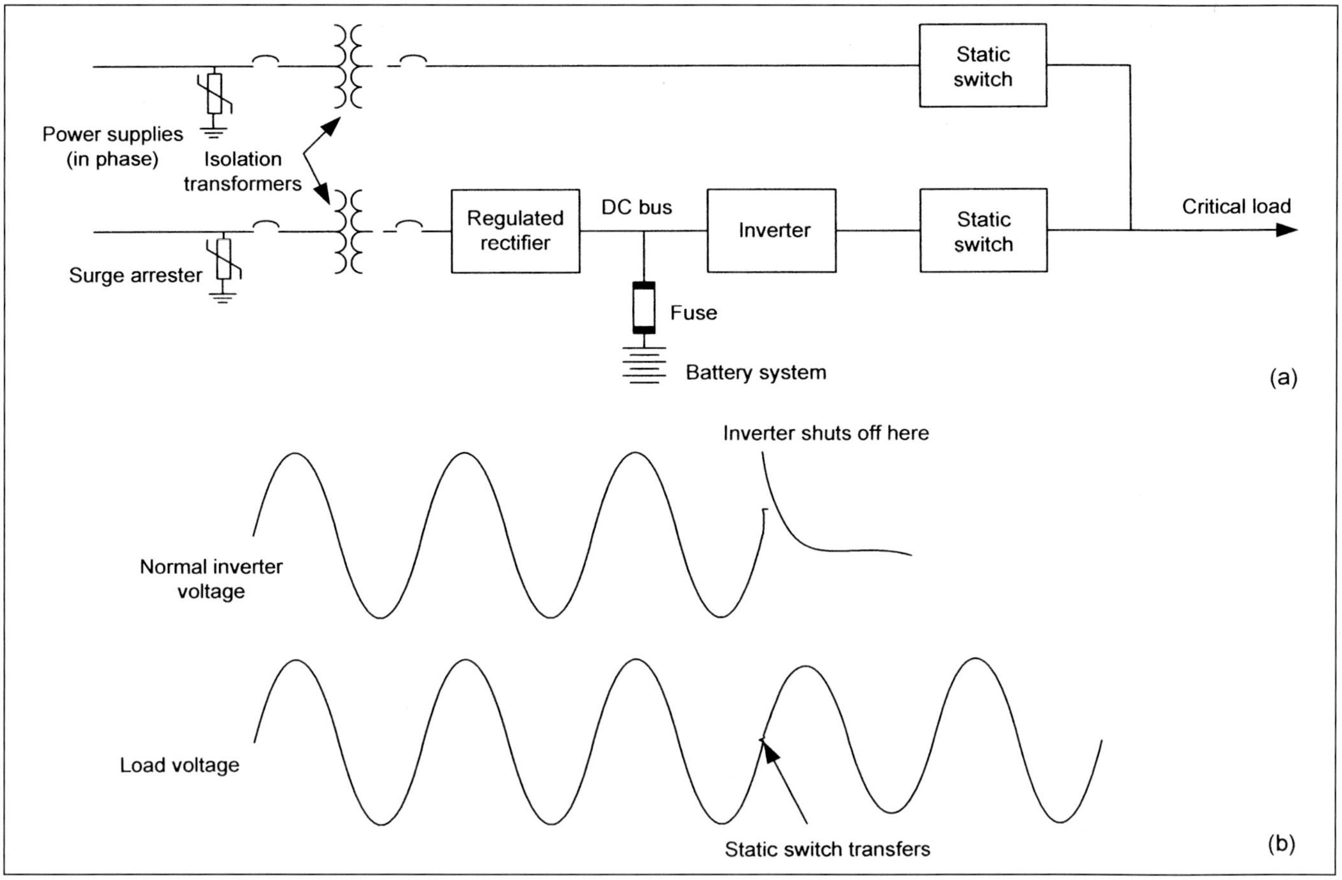

FIGURE 19-32 (*a*) Circuit diagram of an UPS system. (*b*) Critical load transfer waveforms through a static switch.

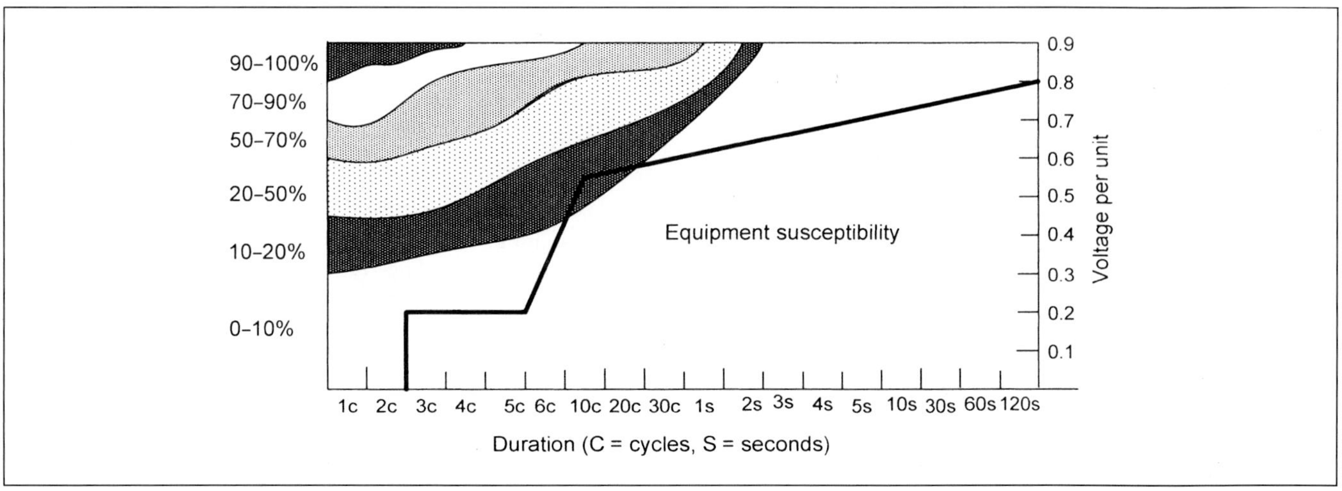

FIGURE 19-33 Topographical concepts applied to voltage sags and equipment susceptibility to ride through voltage sags.

be quite high. This type of overstresses account for 65 percent of the failures.

2. Thirty percent of the failures occur because of environmental conditions, that is, high temperature, high moisture, and high humidity.

3. Five percent of the failures occur due to defects in the manufacturing, usually within 39 days of operation.

This shows that transients and ESD account for the bulk of the failures.

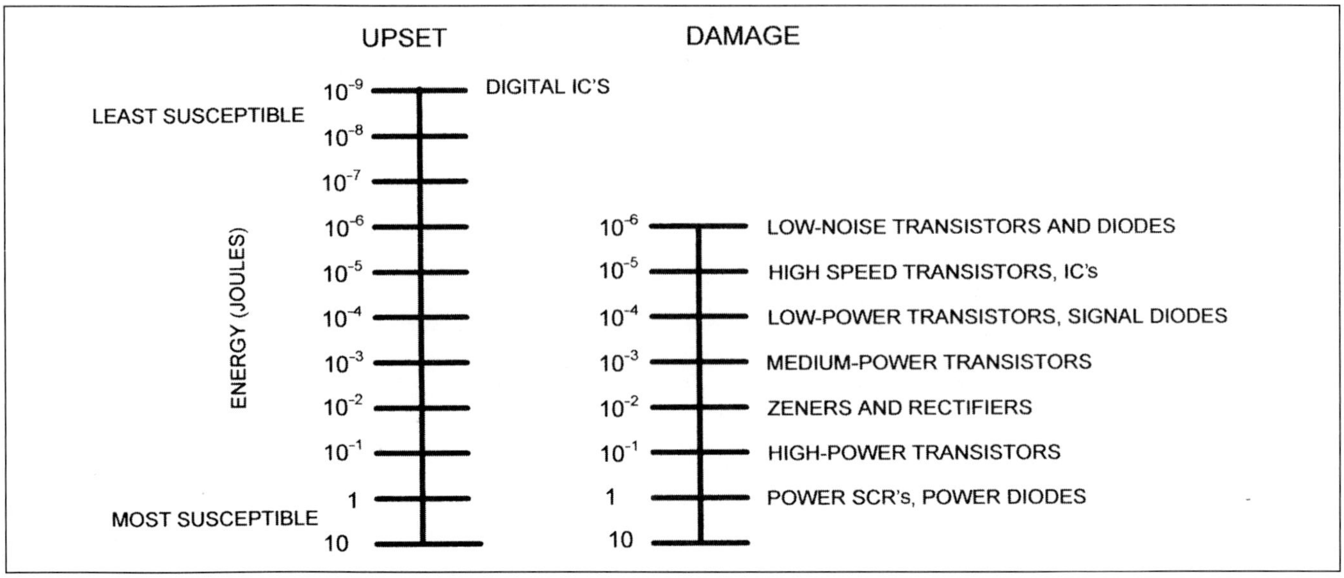

FIGURE 19-34 Upset and damage energy levels of digital ICs.

TABLE 19-8 Typical Generated Static Voltages

		Static Voltage Developed (kV)	
No.	**Event**	**Low Humidity (20%)**	**High Humidity (80%)**
1	Walking across synthetic carpet	35	1.5
2	Handling polyethylene bag	20	0.6
3	Sitting on foam cushion	18	1.5
4	Sliding plastic box on carpet	18	1.5
5	Heat shrinkable film on PC board	16	3
6	Pulling Mylar tape from PC board	12	1.5
7	Walking across vinyl floor	12	0.25

19-13 POWER QUALITY FOR COMPUTERS

Table 19-9 shows typical range of power quality and load parameters of a major computer manufacturer.[5] Computer manufacturers specify maximum momentary voltage deviations for their equipment to operate without errors and without sustaining damage. Transient conditions are specified in terms of amplitude and time. Some manufacturers may specify total loss of power from 1 ms to 1 cycle. A figure of 8.3 ms was common for the older equipment. Impulse tolerances are of much shorter duration.

19-13-1 CBEMA Curves

Figure 19-35*a* shows an envelope of voltage tolerances that was put forward by Computer and Business Equipment Manufacturer's Association (CBEMA) working group and included in IEEE standard.[5] This has been superseded with a new curve in Fig. 19-35*b* in 2000 and is called, "New Information Technology Industry Council (ITI, formerly, CBEMA) curve (2000)." It attempts to define everything from specification criteria for electronic equipment to the basis of power quality performance contracts between electrical utilities and large industrial consumers. Yet it is a partial picture of the immunity limits in modern office electronic equipment. CBEMA curve does not address noise immunity. This criterion is addressed as energy delivery criterion. The operation of data links, which should operate successfully without noise-related interference, is a data-transfer criterion. Also reference grounds should operate at equal potentials and free of transient voltage shifts, as discussed in Sec. 19-3.

Another industry standard that defines voltage sag ride-through capability for semiconductor processing, and automated test equipment is SEMI F47-0200-2000 (Semiconductor Equipment and Materials International, Inc.).[18] The tolerable limits defined in this curve are: 50 percent of nominal voltage for a time period of 0.05 to $\leq$ 0.2 s; 70 percent for 0.2 to $\leq$ 0.5 s; and 80 percent for 0.5 to $\leq$ 1.0 s.

19-14 TYPICAL APPLICATION OF SPDs

Figure 19-36 shows application of SPDs in a typical distribution system. The characteristics and selection of the SPD at each location follows the guidelines discussed above.

TABLE 19-9 Typical Range of Input Power Quality and Load Parameters of a Major Computer Manufacturer

PARAMETER	RANGE
Voltage limit, steady-state all phases*	+6%, −13%
Voltage disturbances, all phases*	Surge +15% for 0.5 s maximum Sag −18% for 0.5 s maximum Transient voltage 150-200% for 0.2 ms
Harmonic content†	5% maximum with equipment operating
Electromagnetic compatibility†	1 V/m maximum
Frequency limits*	60 Hz ± 0.5
Frequency rate of change*	1 Hz/s (slew rate)
Three-phase voltage unbalance†	2.5% of arithmetic average
Three-phase load unbalance‡	5-20% maximum for any one phase
Power factor‡	0.8-0.9
Load demand‡	0.75-0.85

*These parameters depend on power source.

†These parameters are a function of interaction of the source and the equipment load.

‡These parameters are a function of the equipment. The harmonic content of the voltage is computed as the sum of all harmonic voltages added vectorially.

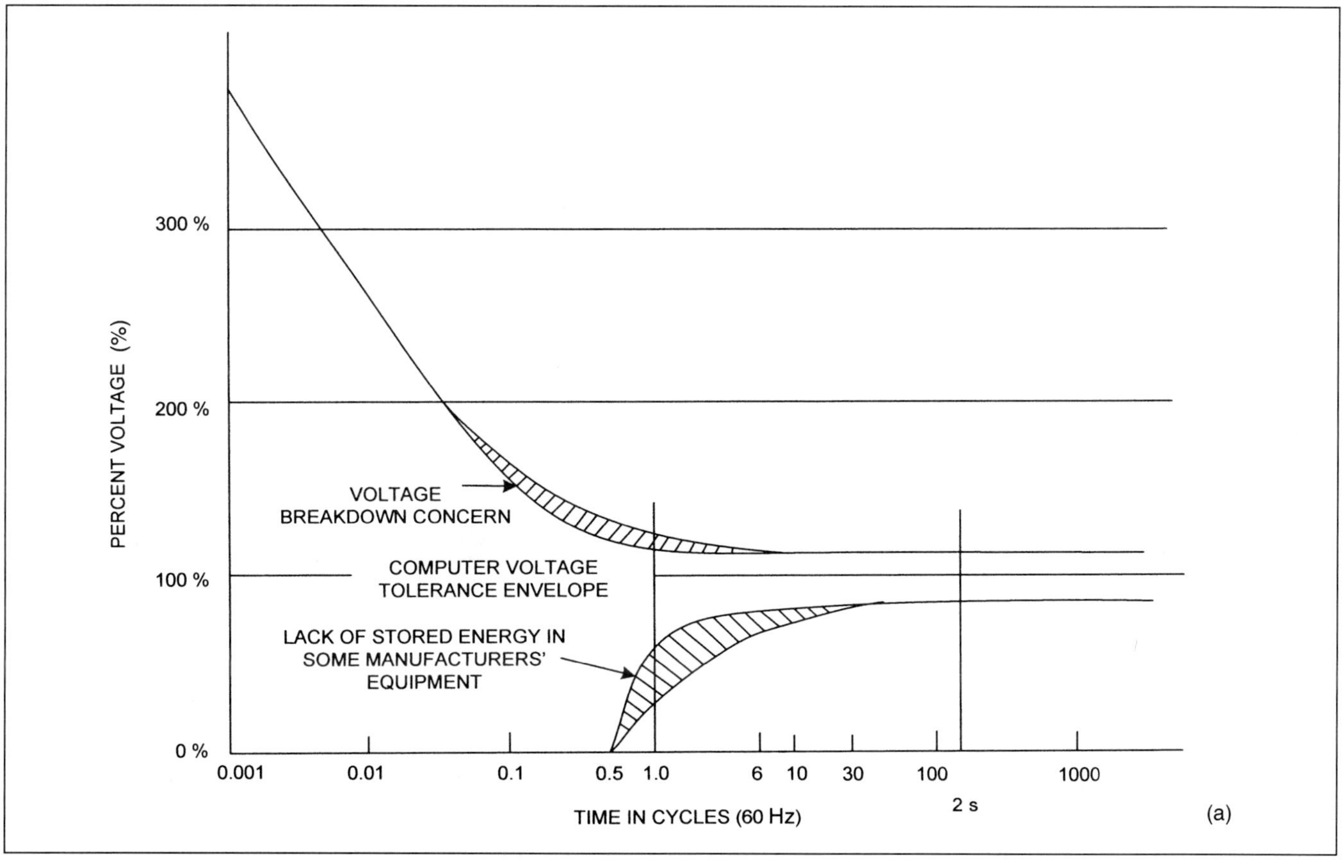

FIGURE 19-35 (*a*) CBEMA curve (old), now superseded by (*b*) new ITI (CBEMA) curve.

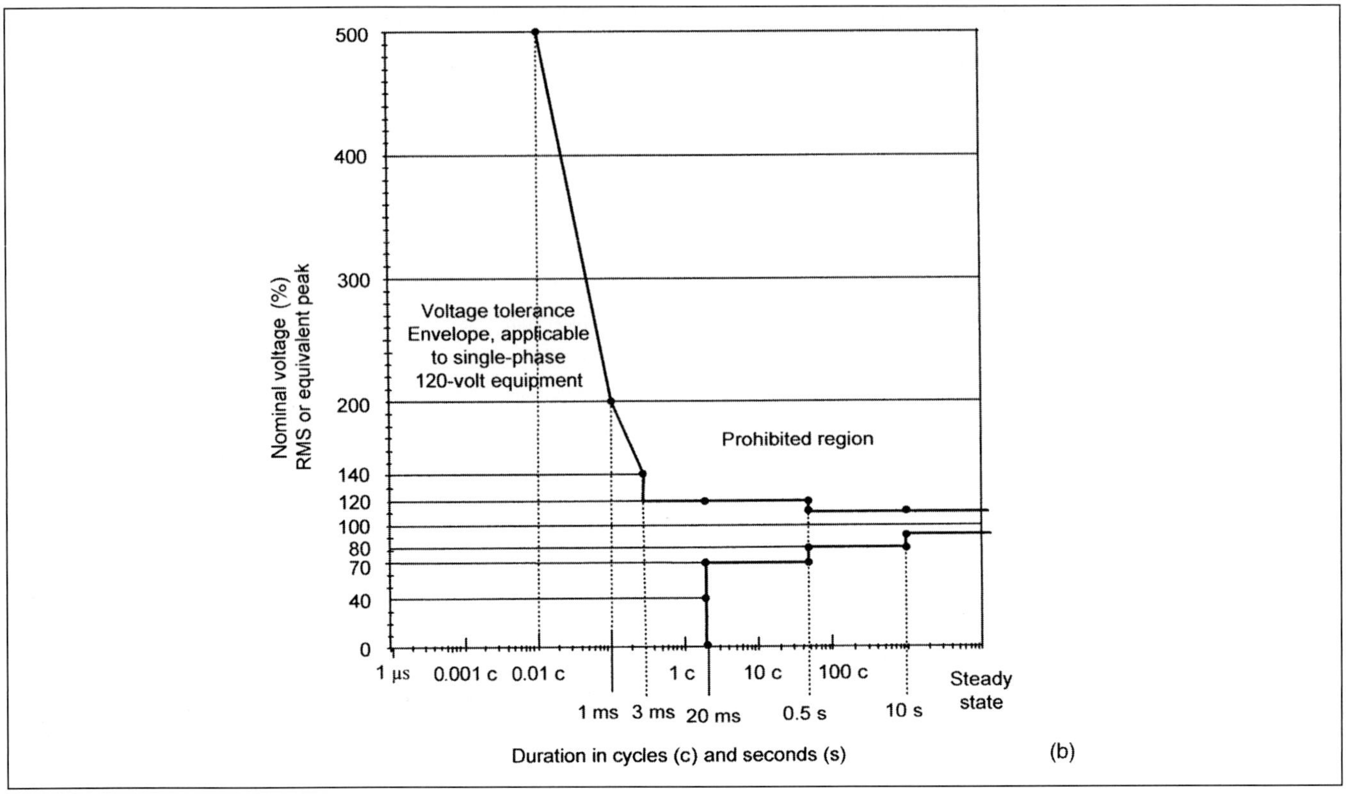

FIGURE 19-35 (*Continued*)

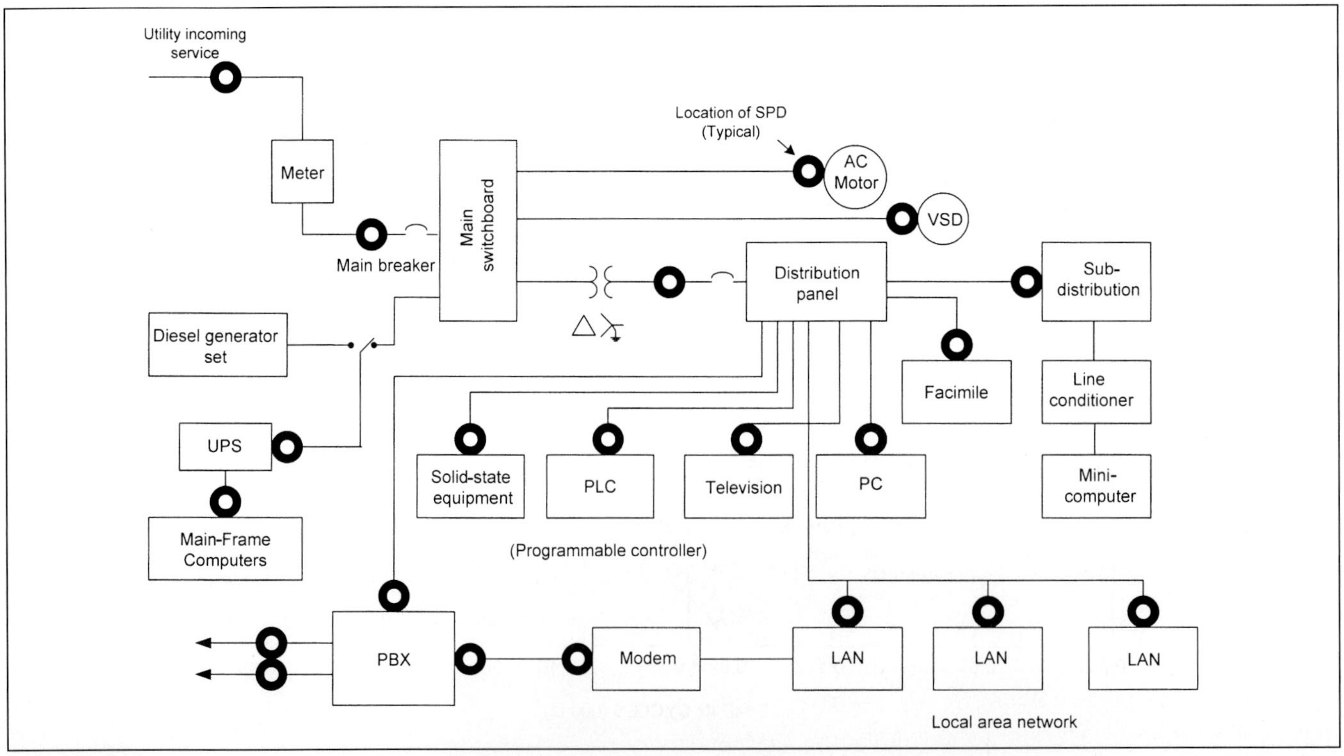

FIGURE 19-36 Typical applications of SPDs in a distribution system.

PROBLEMS

1. Distinguish between electrical noise and transients. Draw a freehand sketch illustrating the two.

2. Describe normal mode and common mode of surge protection for a three-phase, four-wire, low-voltage system.

3. The multiple-grounding systems of utility services to commercial and residential buildings give rise to differential voltages between neutral and ground. What are the relative advantages and disadvantages? Why is this not applied to industrial and HV systems? Comment on the statement: "A person taking a shower in his home gets an electrical shock." Can it be true?

4. How are the lightning surges coupled to low-voltage systems?

5. Explain why standards specify that the maximum lightning surge voltage will get limited to 6 kV in locations of medium exposures.

6. Compare ANSI/IEEE location categories with IEC. What are the major conceptual differences?

7. Recommend a SPD specification for application in category C location, without any rigorous study.

8. Distinguish between transient voltage surge suppressors (TVSS), surge protection devices (SPDs), and SSAs (secondary surge arresters). Can a TVSS device be used in category C location?

9. Write eight considerations that are applicable when selecting a SPD for the low-voltage systems.

10. Write a note on the controversy surrounding surge-current capability versus energy-handling capability of a TVSS device. How can data on surge-handling energy capability be misleading?

11. What are the UL limits of clamp voltages for 120-V, 240-V, and 480-V, low-voltage systems?

12. What are major differences between gas tube, MOV, and selenium rectifiers with respect to speed of response to a surge? Which of these devices has the highest life?

13. MOVs have been shattered on handling surges in some installations. What are the common remedial measures to mitigate this problem?

14. Compare varistors with selenium rectifiers. Write a note on the failure mode of these two devices.

15. Describe the effect of lead length on the operation of an MOV. An MOV is connected with a lead length of 10 ft. By what percentage will its clamp voltage change?

16. The overshoot of voltage in a varistor occurs because of (1) high surge voltage, (2) varistor characteristics,(3) rate of rise of surge voltage, (4) lead length, (5) all of these. Which is correct?

17. Give reasons why multistage parallel designs of TVSS are necessary. What is the limitation of a single-stage shunt device? What are some objections to the use of series devices in multistage parallel arrangements?

18. A 120-V, single-phase ac power supply has a voltage swell of 10 percent for 3 s for a computer application. Is it acceptable? What is the maximum time duration for safe tolerance of the overvoltage, keeping the same magnitude?

19. What is the major cause of failure of semiconductors?

20. Write the category classification (A, B, or C) on each of the surge protective devices shown in Fig. 19-36.

21. Based on this chapter and a manufacturer's catalogue, write down specifications of all SPDs shown in Fig. 19-36.

22. A ring test wave is specified in the standards. What type of switching surge can produce a ring wave in a low-voltage system?

23. Design a snubber circuit for an SCR; its maximum dv/dt = 100 V/μs, source inductance = 0.2 mH, applied voltage = 480 V rms.

24. In Example 19-1 (Fig. 19-15), single-phase loads and loads served by 200 kVA are shown disconnected in the simulation results obtained in Fig. 19-16*a* and *b*. What will be the effect of connecting these loads in service on the nature of transients? Will the amplitudes and decay rate increase or decrease? Consider that the 100-hp motor in Fig. 19-15 is operating at steady state when the system is impacted with the surge. What can be expected in the current, terminal voltage, direct axis, and quadrature axis transients of the motor?

REFERENCES

1. ANSI C2, Electric Safety Code, 2002.
2. NFPA70, National Electric Code, 2008.
3. IEEE Std. 1100, IEEE Recommended Practice for Powering and Grounding Electronic Equipment (Emerald Book), 1999.
4. IEEE Std. 518, IEEE Guide for Installation of Electrical Equipment to Minimize Noise Inputs to Controllers from External Sources, 1982.
5. IEEE Std. 446, IEEE Recommended Practice for Emergency and Standby Power Systems for Industrial and Commercial Applications, 1995.
6. ANSI/IEEE Std. C62.41, IEEE Recommended Practice on Surge Voltages in Low Voltage AC Power Circuits, 1991.
7. F. D. Martzloff, "Coupling, Propagation and Side Effects of Surges in an Industrial Building Wiring System," *IEEE Trans.*, IA-26, no. 2, pp. 193–203, Mar./Apr. 1990.
8. IEC 62305-1, Protection Against Lightning, Part-1, General Principles, 2010.
9. IEEE Std. C37.90.1, Surge Withstand Capability (SWC) Tests for Protection Relays and Relay Systems, 1989.
10. P. Chowdhri, "Estimation of Flashover Rates of Overhead Power Distribution Lines by Lightning Strokes to Nearby Ground," *IEEE Trans.*, PWRD-4, no. 3, pp. 1982–1987, Jul. 1989.
11. IEEE Std. C62.41.1, IEEE Guide on Surge Environment in Low-Voltage (1000 V or Less) AC Power Circuits, 2002.
12. IEC 62305-4, Protection Against Lightning, Part 4, Failure of Electrical and Electronic Systems Within Structures, 2006. (This is a Revision of Earlier IEC 61312-1, Protection Against LEMP, 1995, Now Withdrawn).
13. UL 1449, 2nd ed., Standard for Transient Voltage Surge Suppressors, July 1987.
14. IEEE Std. C62.36, IEEE Standard Test Methods for Surge Protectors Used in Low Voltage Data, Communication, and Signaling Circuits, 1994.

15. T. Key, A. Mansoor and F. Martzloff, "No Joules for Surges: Relevant and Realistic Assessment of Surge Stress Threats," *International Conference on Harmonics and Quality of Power,* Las Vegas, Sep. 1996.

16. F. D. Martzloff, "Matching Surge Protective Devices to Their Environment," *IEEE Trans. IA,* vol. IA-21, no. 1, Jan. /Feb. 1985.

17. IEEE Std. C62.47, IEEE Guide on Electrostatic Discharge: Characteristics of the ESD Environment, 1997. (This standard was archived in 2003).

18. SEMI F47-0200, Specifications for Semiconductor Processing Equipment Voltage Sag Immunity, 2000.

FURTHER READING

ANSI/IEEE Std. C62.11, IEEE Standard for Metal Oxide Surge Arresters for AC Power Circuits (> 1 kV), 1999.

IEC 60664, Insulation Coordination for Equipment Within Low Voltage Systems, Part 1, 1980.

IEEE Std. C62.33, IEEE Standard Test Specifications for Varistor Surge-Protective Devices, 1982.

IEEE Std. C62.35, IEEE Standard for Test Specifications for Avalanche Junction Semiconductor Surge Protective Devices, 1987.

IEEE Std. C62.42, IEEE Guide for the Application of Component Surge Protection Devices for Use in Low-Voltage (Equal to or Less than 1000V AC or 1200 V DC) Circuits, 2005.

IEEE Std. C62.45, Guide on Surge Testing for Equipment Connected to Low-Voltage AC Power Circuits, 2002.

IEEE Std. C62.1 IEEE Standard for Gapped Silicon Carbide Surge Arresters for AC Power Circuits, 1989.

NEMA LS-1, Low Voltage Surge Protection Devices, 1992.

F. D. Martzloff and T. F. Leedy, "Electrical Fast Transients: Application and Limitations," *IEEE Trans.*, IA-26, no. 1, pp. 151–159, Jan./Feb. 1990.

C. R. Paul and K. B. Hardin, "Diagnosis and Reduction of Conducted Noise Emissions," *IEEE Trans. EMC,* pp. 553–560, Nov. 1988.

R. B. Standler, "Protection of Small Computers from Disturbances on the Mains," in *Conf. Record, IEEE IAS, Annual Meeting,* pp. 1482–1487, Oct. 1988.

E. S. Thomas, J. B. Dagenhart, R. A. Barber, and A. L. Clapp, "Distribution System Grounding Fundamentals," *IEEE Industry Applications Magazine*, vol. 11, no. 5, Sept./Oct. 2005.

CHAPTER 20

SURGE ARRESTERS

Previous chapters have examples of applications of surge arresters. The applications of surge arresters are based on interrelated aspects, like (1) knowledge of system temporary overvoltages, switching, and lightning surges, (2) system grounding and ground resistance of the grid to which the arrester is connected, Chap. 21, (3) equipment being protected, and (4) characteristics of the surge arresters. The surge arresters may be applied at:

- Transmission line terminations and origins
- OH line supports—transmission and distribution line towers and poles—at certain intervals, for insulation protection
- An open circuit breaker
- Protection of substation (transmission, distribution, industrial, or generation)
- Junctions of overhead lines and cables, in OH distribution
- Shunt and series capacitor banks
- Circuit breakers—TRV control and medium-voltage vacuum contactors
- Rotating machinery, motors, and generators
- Gas-insulated substations
- Power transformers of all types—generating, step-up, step-down, distribution, pad mounted, substation, and a variety of winding connections; auto-transformers, step voltage regulators, Chap. 14
- Series reactors, Chap. 15
- Protection of drive systems, Chap. 6
- Protection of low-voltage systems, Chap. 19

Each of these applications has some generalities and some specifics. General concepts are discussed and applied to some situations, without an attempt to be comprehensive about *all* applications of surge arresters.

20-1 IDEAL SURGE ARRESTER

An ideal surge protection device should clamp the system voltage to its normal value, when an overvoltage condition arises. Normally it should have an infinite resistance, which should change precisely under an overvoltage condition so that enough current is conducted to clamp the voltage. This transition should occur instantaneously without causing disturbance to the power system, and the device itself should be immune to failure, including overloading and high-current conduction. Further, the performance should be consistent with *any type* of surge or applied voltage—of different wave shapes, rise times, and frequencies. A surge outside the protected zone should not cause a flashover inside the protected zone, and strokes within a protected zone should not cause flashovers inside or outside the protected zone.

The practical surge protection tends to simulate this ideal. Thus, the basic function of the arrester is to limit transient voltage to some lower level, depending on the arrester characteristics. This level may be called the *protective level*. It will be somewhat higher than the system voltage and become important in insulation coordination. During operation, the arresters dissipate energy and should not be damaged on this transient energy surge through them.

20-2 ROD GAPS

Rod gaps or air gap devices were the earliest and are no longer in popular use. Rod gaps were installed across transformer bushings and insulators. Once the air gap breaks down under an overvoltage condition, that is, the lightning surge, it draws the arc away from the equipment being protected but will maintain the arc even after the lightning surge has been discharged. The arc may extinguish at a current zero; however, due to prolonged time of arcing, the system ground fault relays may operate, resulting in interruption of service and the very reliability of the service may be jeopardized. The electrodes, which may be rounded or pointed, across which the arc is drawn will rapidly deteriorate. Though simple and rugged, the rod gaps have serious limitation of application as surge protective devices.

The rod gaps are designed to withstand 150 percent of system voltage under the worst atmospheric conditions, humidity, and relative air density. The 60-Hz withstand voltage of the gap, $V_{W,60}$:

$$V_{W,60} = \sqrt{2} \times 1.5 \times V_{TOV} \tag{20-1}$$

where V_{TOV} is the temporary overvoltage in kV rms.

V_B, the mean breakdown voltage of the gap, is related to 60-Hz breakdown voltage:

$$\begin{aligned} V_B &= V_{W,60} + 3\sigma \\ &= 1.06 V_{W,60} \end{aligned} \tag{20-2}$$

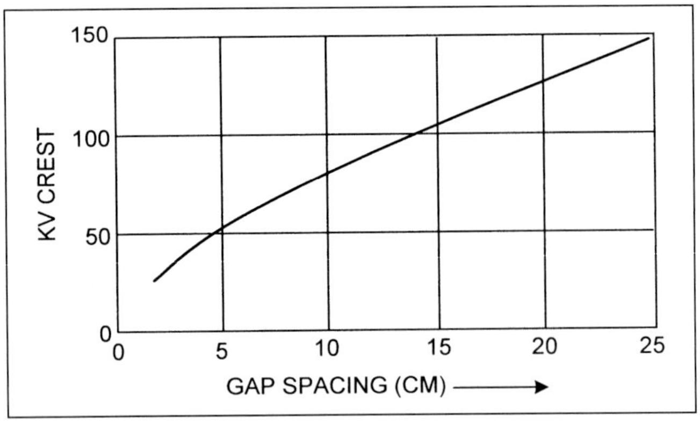

FIGURE 20-1 Standard mean breakdown voltage for rod gaps, 60 Hz.

where σ is the standard deviation of the breakdown voltage, taken as 2 percent for 60-Hz voltage $V_{W,60}$. This voltage can be corrected for atmospheric conditions:

$$V_M = V_B \frac{K_h}{R_{ad}} \tag{20-3}$$

where V_M is the maximum system voltage, and K_h and R_{ad} are the humidity and relative air density correction factors, respectively. Figure 20-1 illustrates the gap spacing in centimeters with respect to 60-Hz standard mean breakdown voltage.

20-3 EXPULSION-TYPE ARRESTERS

Expulsion-type arresters are stated here as of historical interest. These were the next development in the surge protection and were intended to break up the follow current, so that the continuity of services could be maintained without operation of ground fault protective devices. Two air gaps, one upper and one lower, were provided in series. The breakdown of the upper gap led to the breakdown of the lower gap, which had a plug of fibrous material to cool the arc, so that the current was interrupted at a current zero. These arresters had limited useful life because the fibrous material was consumed and had different operating characteristics, depending on the voltage wave shape. These are no longer in use.

20-4 VALVE-TYPE SILICON CARBIDE ARRESTERS

Valve-type arresters are basically an air gap in series with a nonlinear resistor of the silicon carbide type. Though replaced with metal-oxide gapless arresters, these are still in wide use.

The breakdown characteristic of the air gap in the arrester governs the initial voltage at which surge-limiting action begins. The silicon carbide blocks have a nonlinear characteristic, so that their resistance decreases as the current increases. Figure 20-2*a* shows a schematic representation of the arrester and Fig. 20-2*b* shows a cross section through the arrester, functional only.

The silicon carbide crystals are bonded with an inert material by sintering in a kiln at 1000°C. The outer surface of a valve block has an epoxy or ceramic collar to prevent flashovers when conducting high currents. The VI characteristics of silicon carbide valve block is controlled by grains of silicon carbide within the block. The heating

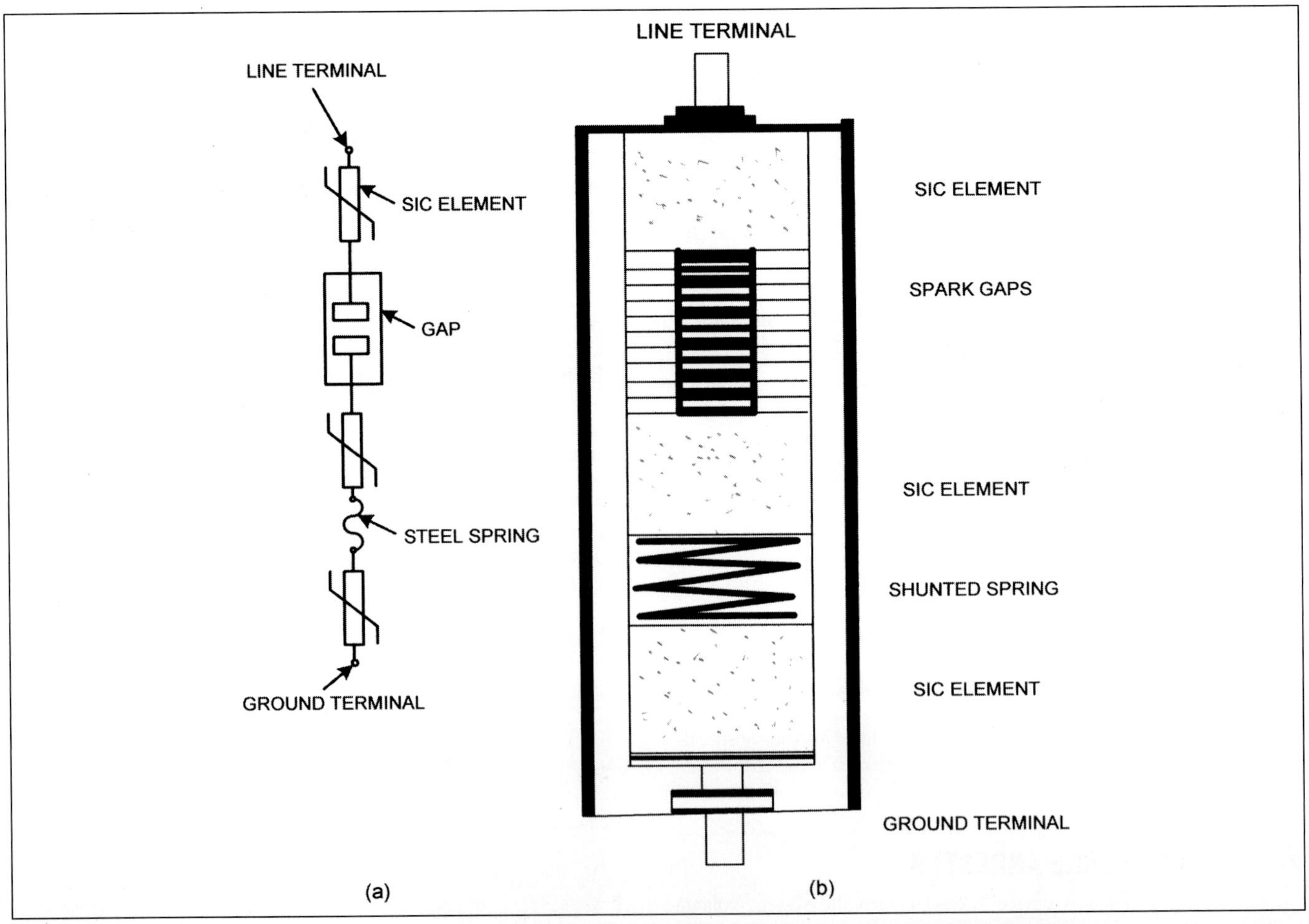

FIGURE 20-2 (*a*) Circuit representation of a valve-type surge arrester. (*b*) Cross-sectional view.

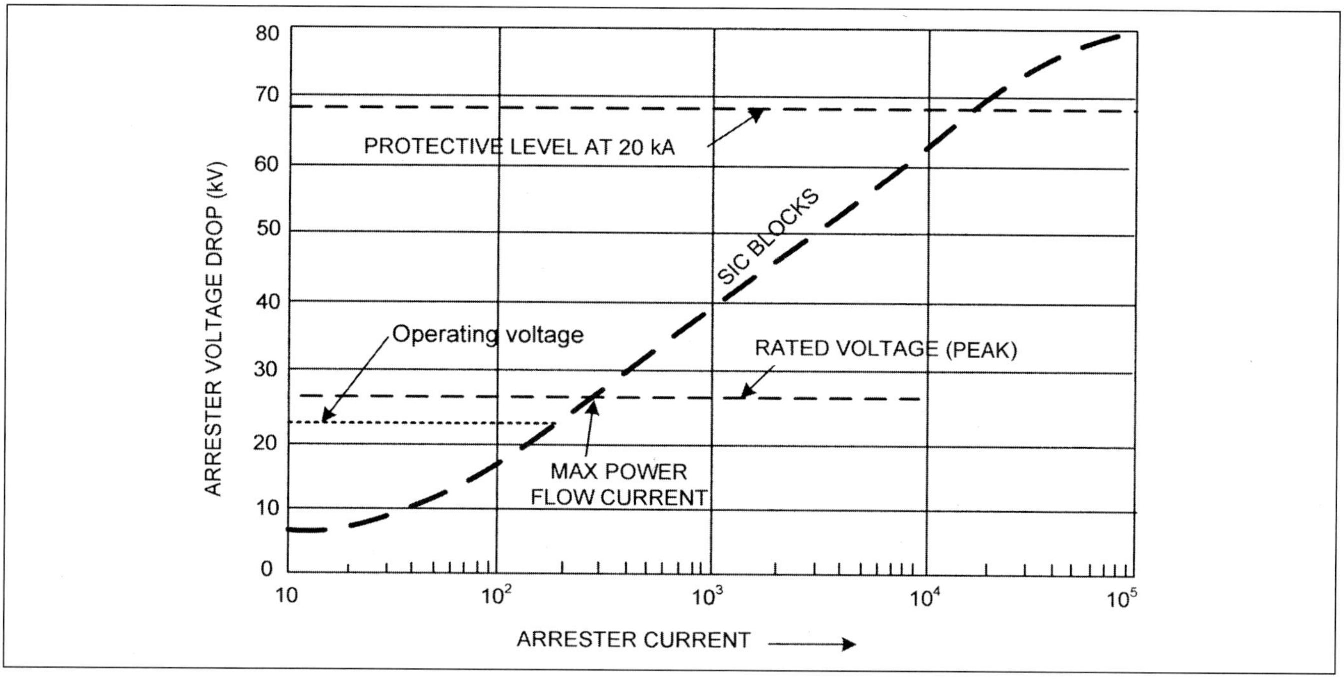

FIGURE 20-3 VI characteristics of a distribution class surge arrester.

occurs due to contact points within the block, and thus, there is a distinct relationship between the current magnitude, wave shape, and the temperature at the point of contact. The silicon carbide has a negative coefficient of resistance, which decreases with increasing temperature and increasing current. The characteristics are tailored to provide voltage limiting with lightning discharge current wave shapes, keeping the power flow currents low, which are interrupted by the air gap.

The blocks have little overload capability and can be destroyed if subjected to energy discharge beyond their design capabilities.

The VI characteristics are measured using standard 8/20 μs test wave shape for peak magnitudes of 1.5, 3, 5, 10, and 20 kA. A VI characteristic of distribution class surge arrester is shown in Fig. 20-3. The silicon elements are chosen to give a certain protective level at a specified lightning discharge current, 20 kA in this figure. The power flow current after discharge of a surge is interrupted by air gap at a current zero.

There are a number of series gaps, each rated at 1 to 1.5 kV. Each gap may consist of a contoured brass, copper, or stainless steel plate separated from the adjacent plate by a ceramic spacer. The use of modular gaps makes it feasible to have different ratings by assembling varying number of gaps, so that

- Ability to interrupt power flow current is increased.
- Increase in total arc voltage drop is obtained, thus enhancing power flow current-interrupting capability of the arrester.
- Response to steep-fronted waves is improved.
- Response to different voltage wave shapes can be modified by control of voltage between individual gaps.

As the heating is dictated by current, short-duration surges create higher voltage drops than the long-duration surges. Figure 20-4 shows the discharge voltage as a function of rise time of the impulse current. Manufacturers publish data for the switching surge capabilities.

Surge protection levels are not standardized and the manufacturers are required to publish maximum values for each design; these data are required for proper application.

The arrester continues to conduct until the arc in the air gap is extinguished at current zero. Due to nonlinearity, the current through the surge arrester is nonsinusoidal. Practically, the gap design is more complex than what has been described earlier.

Preinsertion Gap For the purpose of surge protection, it is desirable to have gap breakdown at a fixed voltage level regardless of the voltage wave shape. However, the gap reacts to instantaneous voltage and voltage rises at different rates, depending on the maximum surge level and its rate of rise. The *statistical time lag*, which is the time taken by the initial electrons to start the avalanche process, can be minimized by incorporating a pre-ionization gap in addition to the main arc gap. A smaller gap is placed into the circuit and coupled into it by large capacitive or resistive impedance.

Grading Components In case of even slight nonuniform distribution of voltage across the gaps, the gaps with higher voltage applied across it will break down first and the breakdown will cascade throughout the assembly. Consequently, the arresters are fitted with some form of intentional grading components, that is, grading resistors. This provides a brief description of the constructional features; the characteristics for surge suppression are of more interest.

20-4-1 Arrester Ratings

Arrester Class The arresters' ratings depend on the class of the arresters, which are as follows:

- Station class arresters, voltage ratings 3 to 682 kV
- Intermediate class arresters, voltage ratings 3 to 120 kV
- Distribution class arresters, voltage ratings 1 to 30 kV

The distribution class arresters are further divided into *heavy-duty distribution class arresters* and *normal-duty distribution class arresters*.

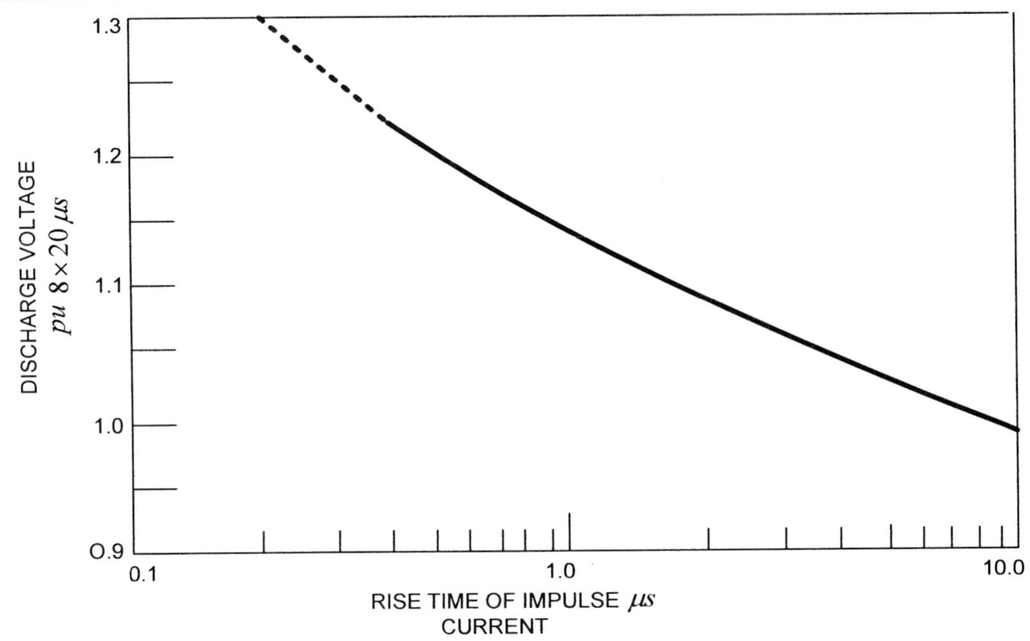

FIGURE 20-4 Correction factors for nonstandard current impulses.

Voltage Rating A valve-type arrester must be able to withstand rated voltage for 24 min, while discharging 5 to 10 kA of standard lightning currents at 1-min interval. Successful operation means that the power flow current after each discharge should be interrupted. This rated voltage is, therefore, defined by the duty cycle tests. The standards publish the rated voltages. An arrester will not be able to continuously withstand its rated voltage. The selection of the rated voltage of the arrester for a particular application is described further.

Overvoltage Rating The valve-type arresters are applied so that their rated voltage is greater than the system overvoltage produced by single line-to-ground faults. The definition of COG and its calculations have been described in Chap. 9. Considerations should also be applied to the overvoltages and their duration, which may be generated at the location of the arrester due to contingency operations, that is, severe load dropping (Chap. 7). Overvoltage characteristics of arresters are an important parameter.

The ability to withstand overvoltage demands that the arrester gaps should not spark over when nonconducting. When conducting, the arrester should be able to withstand increased stresses and have the capability to interrupt higher power follow currents. The distribution class surge arresters do not have a well-defined overvoltage capability, while intermediate class and station class surge arresters have better overvoltage withstand capabilities. The performance of distribution class arresters can, therefore, only be expressed in terms of increasing probability of failure with higher overvoltages.

Manufacturers publish the overvoltage capability curve of the arresters. Figure 20-5 shows such a curve for station and intermediate class arresters.

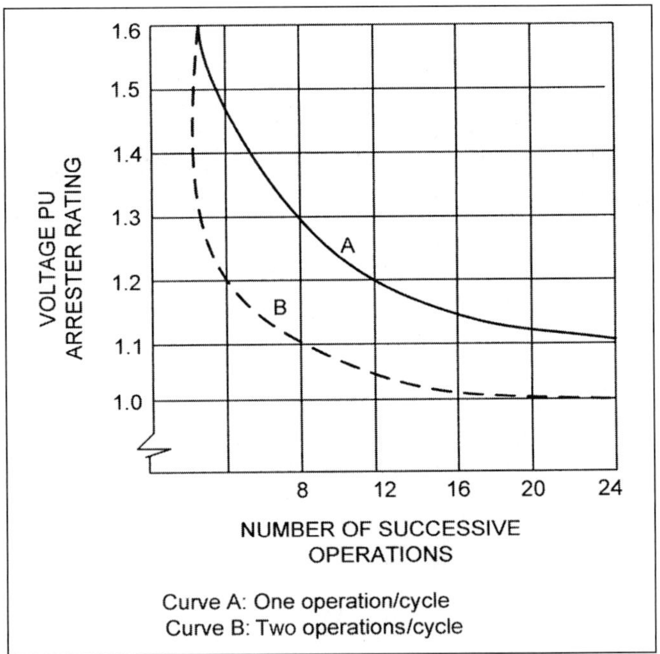

FIGURE 20-5 Overvoltage operating capability of a station class and intermediate class surge arresters.

60-Hz Sparkover Voltage The 60-Hz sparkover voltage level is specified by the manufacturers. This does not mean that the arrester can withstand this voltage continuously. The limiting factor is probability of the failure of grading resistors in the arrester assembly.

Lightning Impulse Sparkover Voltage The front of wave sparkover is tested with 1.2 × 50-μs standard wave, with a peak magnitude selected to give a rise time of 25 kV/μs for each 3 kV of arrester rating; that is, an arrester of 24 kV should be tested for 200 kV/μs, with a test impulse peak of 225 kV. Ten impulses, five of each polarity, are applied, and maximum sparkover voltage is taken as the published value.

A 1.2 × 50-μs impulse sparkover test determines the highest 1.2 × 50-μs impulse for which there is a 50 percent probability of sparkover occurring within 3 μs. A series of impulses is applied for the test calculations. The published value is the mean of all the applied impulses.

Lightning Surge Discharge Performance A lightning discharge interacts with the system, and the magnitude and shape of the discharge current can vary. The tests determine the capability of the arrester to discharge standard current surges and to interrupt the succeeding power flow current. Lightning current test wave shape is 8 × 20 μs. Discharge voltage current characteristics are obtained for currents of 1.5-, 3.0-, 5.0-, 10.0-, 20-, and 40-kA crest.

High Current Short-Duration Test Two current impulses are applied to an arrester or an arrester-prorated section. The magnitude of the impulse is 10 kA for distribution class arresters, and 65 kA for intermediate and station class arresters, using 4/10-μs wave, with a tolerance of –0 percent to 50 percent to accommodate the characteristics of the test equipment. The test specimen should withstand these impulses without damage and major deterioration.

Switching Surge Discharge Performance Switching surges are of longer duration and, therefore, may contain more energy than lightning surges. Distribution class surge arresters will not be normally subjected to switching surges and, therefore, their switching surge capabilities are not specified. Also, switching surge sparkover voltage is not specified for intermediate and station class surge arresters of voltage rating less than 60 kV. The switching surge sparkover test includes three sequences, each identical, except that a different standard test wave is used:

- 30/60 μs
- 150/300 μs
- 1000/2000 μs

Transmission Line Discharge Test (TLD) The test standards also specify that station and intermediate class surge arresters be subjected to current discharges generated by a test circuit that approximates an OH line.[1,2] This is called a TLD. The parameters of TLD test are specified in Refs. 1 and 2.

In addition, high current short-duration tests are conducted (HCSD, 4/10-μs wave). The nonstandard tests indicate that 90 percent of randomly selected arresters will survive one 20-kA discharge and 70 percent will survive one HCSD discharge.

Table 20-1 shows protective characteristics of valve-type station class arresters.[3] Similar characteristics for distribution class arresters and intermediate class arresters are available in Ref. 3. Switching surge absorption capabilities of valve-type arresters are specified in Table 20-2, which seems to be conservative. The ability of a surge arrester of a given rating to absorb energy depends on the shape of the switching surge.

20-5 METAL-OXIDE SURGE ARRESTERS

20-5-1 Construction and Operation

A metal-oxide gapless arrester consists of several metal-oxide elements connected line to ground. The resistance of the zinc oxide changes with voltage, but the change is more dramatic as compared to that of silicon carbide. Figure 20-6 shows a cross section through a typical arrester, and Fig. 20-7 shows the VI characteristics of the zinc elements (the material consists of approximately 90 percent of zinc oxide; 10 percent are oxides of other metals, like aluminum, antimony, barium, bismuth, cobalt, and so on) versus that of silicon carbide. The extreme nonlinearity allows zinc-oxide arrester to be designed so that these conduct only a small current under normal system voltage conditions, and a much larger current at slightly higher voltage. Figure 20-7 shows that low-current zone is affected by temperature and frequency. For 100°C, three curves are shown: dc VI, ac resistance VI, and ac total VI. For currents above 1 A, the VI characteristic is highly nonlinear, and the voltage drop remains constant over a wide range of currents. In high-current-density region, the device loses its nonlinearity due to the linear resistance of the zinc-oxide grains.

Considerable power dissipation occurs due to passage of leakage current. Metal-oxide surge arresters are susceptible to thermal runaway.

The construction of the assembly is very similar to silicon carbide gap-type arresters, except the gap. There are gapped metal-oxide surge arresters, which are not discussed.

20-5-2 Equivalent Circuit

An equivalent circuit of the metal-oxide surge arrester is shown in Fig. 20-8.[4] The leakage current is mostly capacitive, and the capacitance of the arrester C_p varies with its class, from 400 to 2000 pF. The inductance L_E is 10 to 20 nH and can be ignored at 60 Hz. C_1 and C_2 are constants and I is the arrester current. The capacitance of metal-oxide surge arresters is important in modeling and is shown in Table 20-3.

With normal voltage applied, the current is, therefore, mostly leading and has a resistive component, which is distorted. As the nonlinearity increases, the distortion in the resistive component of the current is increased. This change in current at different voltage levels means that there is no constant relation between the voltage, current, and power dissipation. An equation that models the VI characteristics has been developed,[4] and is given by:

$$V = \frac{C_1}{C_2 - \ln I} + RI \qquad (20\text{-}4)$$

where V is resistive voltage drop, C_1 and C_2 are constants, I is the arrester current, and R is the equivalent valve element resistance. Constants C_1 and C_2 vary somewhat with the manufacturers and the dimensions of the arrester.

A so-called "alpha equation" to model the VI characteristics is much in use and is given by:

$$V = KI^{1/\alpha} \qquad (20\text{-}5)$$

where V is resistive voltage drop, I is arrester current, K is proportionality constant, and α is nonlinearity constant. The alpha equation is less accurate compared to Eq. (20-4).

The nonlinearity constants K and α can be determined from solutions of two simultaneous nonlinear equations, using high and low levels of arrester voltages and currents. Alternatively, it can be assumed 4.5 for valve-type and 25 for metal-oxide arresters. K is given by the following equation:

$$\begin{aligned} K &= 0.96\, V_{ds} \text{ for intermediate or station class arresters} \\ &= 1.73\, V_{ow} \text{ for riser pole-type units} \\ &= 1.96\, V_{ow} \text{ for distribution class arresters} \end{aligned} \qquad (20\text{-}6)$$

V_{ds}, V_{OW} are defined in the following section.

20-5-3 Energy Absorption Capability

An advantage of metal-oxide arresters is that they can absorb more energy compared to gapped surge arresters. The energy handling capability is published by the manufacturers and is dependent on thermal stability limit and thermal shock limit of the arrester. The following list shows the typical range (manufacturer's literature should be consulted for accuracy):

- Distribution class: 1 to 1.9 kJ/kV (MCOV)
- Intermediate class: 2.7 to 3.4 kJ/kV (MCOV)
- Station class (2.7 to 48 kV): 5 kJ/kV (MCOV)
- Station class (54 to 36 kV): 9 kJ/kV (MCOV)

The kV here pertains to the arrester MCOV rating, discussed further in Sec. 20-5-4.

Pressure Relief Current A criterion in the selection of the arrester class is pressure relief current limits, which should not be exceeded at the arrester location above the system short-circuit currents

TABLE 20-1 Protective Characteristics of Gapped Silicon-Carbide Station Arresters

Voltage Range of the Arrester (kV rms)	Impulse Sparkover Voltage: Front-of-Wave Rate of Rise of Test Voltage (kV/μs)	Impulse Sparkover Voltage: Front-of-Wave Rate of Rise kV Crest (Range of Maxima)	Impulse Sparkover Voltage: 1.2/50-μs kV Crest (Range of Maxima)	Switching Surge Sparkover Voltage kV Crest (Range of Maxima)	Discharge Voltage for 8/20-μs Discharge Current Wave: kV Crest for 1500 A (Range of Maxima)	kV Crest for 3000 A (Range of Maxima)	kV Crest for 5000 A (Range of Maxima)	kV Crest for 10000 A (Range of Maxima)	kV Crest for 20000 A (Range of Maxima)	kV Crest for 40000 A (Range of Maxima)
3	25	10–18	10–14	—	4.7–6	5.3–6.5	6–7	6.7–7.5	7.7–8.3	–9.2
9	75	28.5–38	24–32	—	13.9–17	16–18	17.8–19	20–21.5	22.9–24.3	–28
15	125	45–47	40–51	—	23.1–27.5	26.6–30	29.5–32	33.4–36	38.2–40	–46
24	200	71–86	62–77	—	36.9–44	42.5–48	47–51	53.4–57	61–63.5	–74
36	300	107–118	92–108	—	55.3–66	63.7–72	70.5–76	80–85	91.5–94.5	–111
48	400	143–148	122–132	—	73.8–88	84.9–96	94–102	106–114	122–126	–148
72	600	204–226	169–190	163–178	114–131	130–144	141–155	159–170	180–189	–222
96	800	270–295	218–245	218–225	151–174	173–192	188–218	212–227	240–253	–296
120	1000	338–360	272–300	272–275	188–218	216–240	235–272	265–285	300–319	–370
168	1400	460–525	380–404	380–381	263–305	303–336	329–362	371–399	420–442	–517
192	1600	520–600	427–460	426–435	300–348	346–384	376–414	424–455	480–505	–591
258	2000	760–790	575–620	573–585	402–438	465–474	505–515	569–575	650–666	–795
294	2000	875–885	653–675	653–675	458–472	528–532	576–595	635–653	735–758	–906
372	2000	1078–1100	870–890	790–830	562–610	655–680	726–738	809–826	932–955	1136–1145
420	2000	1200–1250	980–1005	890–940	634–713	739–770	819–830	913–930	1050–1070	1176–1294
468	2000	1326–1390	1090–1110	992–1045	707–794	823–860	913–930	1018–1040	1170–1200	1310–1441
540	2000	1515–1555	1274–1280	1145–1200	814–890	949–990	1052–1070	1173–1195	1350–1390	1646–1663
612	2000	1700–1765	1440–1480	1300–1370	924–1010	1076–1130	1193–1220	1330–1360	1531–1580	1865–1885
684	2000	1880–1960	1610–1680	1455–1525	1031–1130	1153–1260	1331–1360	1489–1520	1709–1765	2063–2107

All the standard voltage ranges are not shown in this table. See Ref. 3 for further details.

TABLE 20-2 Switching Surge Absorption Capability of Typical Valve-Type Arresters

ARRESTER CLASS	NOMINAL ENERGY CAPABILITY PER KV OF ARRESTER RATING (J/KV)	ESTIMATED ONE-SHOT ENERGY CAPABILITY, PER KV OF ARRESTER RATING (J/KV)
Station	2000	3000
Intermediate	1000	1500
Distribution	500	750

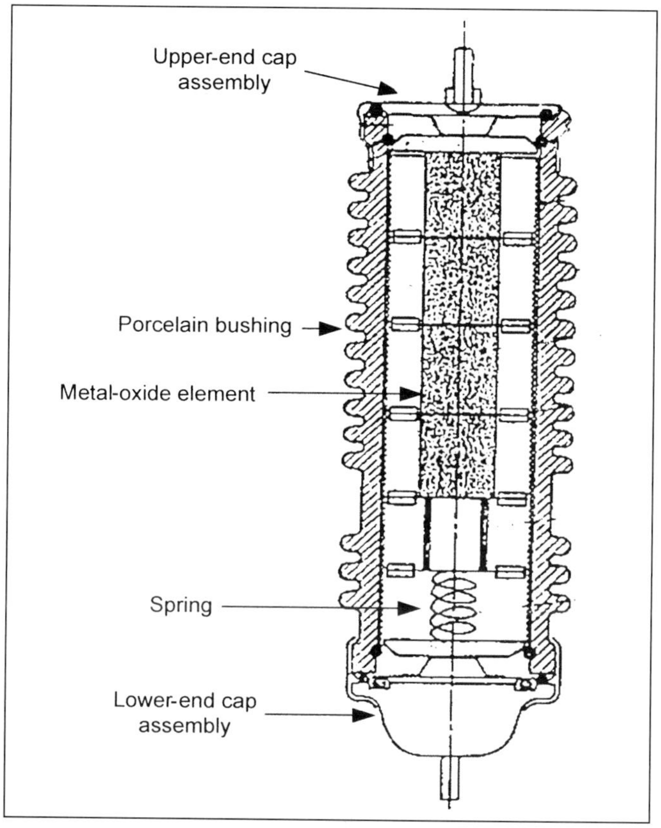

FIGURE 20-6 Cross section through a gapless metal-oxide surge arrester.

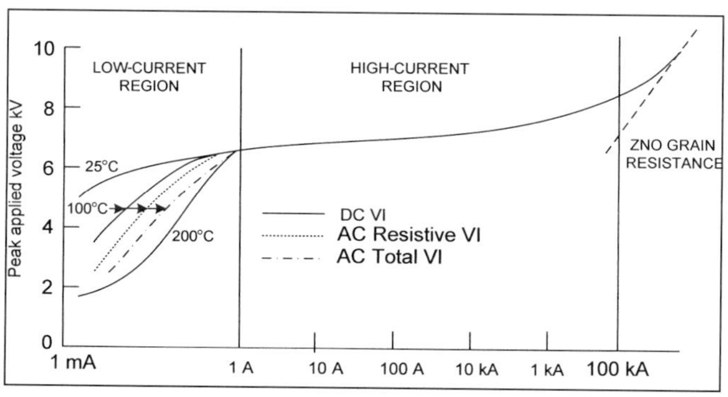

FIGURE 20-7 VI characteristics of a station class metal-oxide surge arrester.

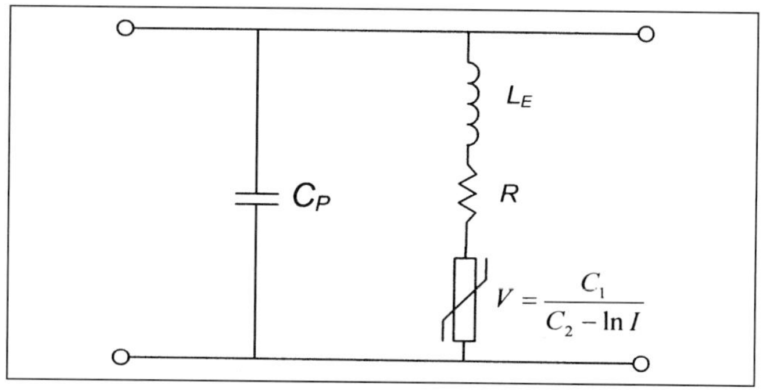

FIGURE 20-8 Equivalent circuit of a metal-oxide element.

TABLE 20-3 Capacitance of Metal-Oxide Surge Arresters

ARRESTER CLASS	SUBCLASS	CAPACITANCE (PF)
Distribution	Normal	400
	Heavy duty	700
	Riser pole	1200
Intermediate	—	1200
Station		2000

and duration. The pressure relief currents are specified in Ref. 5 and vary from 40 to 80 kA rms symmetrical. If the pressure relief capability of a surge arrester is exceeded, the discs may crack or rupture. This will reduce the arrester internal discharge resistance. It may not jeopardize the insulation protection characteristics. In case of complete failure, line-to-ground arc will develop and pressure will build inside the arrester. The pressure will be safely vented to the outside atmosphere and an external arc will be established, provided the fault current is within the pressure relief capability of the arrester. Standards specify 65 kA rms symmetrical (169-kA peak, first crest), though arresters with higher pressure relief capability are available. Many applications require high energy handling capability; for example, for the application of surge arresters to capacitor banks, energy capability becomes an important consideration.

Metal-oxide gapless arresters change smoothly from conducting a small leakage current to kiloamperes of surge current as the voltage is increased. Manufacturers do not specify a conduction threshold above which the dissipation of switching surge energy may be of concern. The conduction threshold voltage is taken as the instantaneous terminal voltage or discharge voltage at which the surge arrester current is 10 A. The operation on switching surge is not of concern if:

$$V_{ds} \geq 1.25 V_s \tag{20-7}$$

and

$$V_{ow} \geq \frac{V_s}{1.1\sqrt{2}} \tag{20-8}$$

where V_s is switching surge overvoltage in peak, V_{ow} is power frequency voltage withstand capability of the surge arrester for 10 s in kV rms, and V_{ds} is maximum switching surge discharge voltage in kV peak at 500 A for intermediate class and 1000 A for station class arresters. Equation (20-7) is used for station class and intermediate class surge arresters, while Eq. (20-8) is for distribution class surge arresters.

20-5-4 Maximum Continuous Operating Voltage (MCOV)

The MCOV is defined as the maximum continuous operating voltage to which the unit may be subjected continuously, and remain thermally stable after being subjected to standard impulse duty tests. It is implied that the arrester is not capable of withstanding voltages above its MCOV rating on a continuous basis. It is the maximum voltage at which the arrester can be permanently energized; MCOV $\geq K_T V_{\text{lng}}$, where V_{lng} is the line-to-ground voltage and K_T is voltage tolerance factor, which considers allowances for variations due to voltage regulation. It may be typically taken as 1.05, but it can be higher as well. The arresters are normally connected from phase to ground. An arrester on a corner-grounded system or a delta-connected system with fault on one phase will be exposed to phase-to-phase voltage and must withstand phase-to-phase voltages. Therefore, the circuit connections, for example, delta, single-phase wye, as well as arrester connections, must be considered.

System grounding is another consideration. In high-resistance grounded systems, especially for industrial process plants, an immediate shutdown is avoided on occurrence of first line-to-ground fault, and the fault may be allowed to persist for hours. In such applications, the arrester MCOV must be chosen to withstand the higher phase-to-ground voltages that may even exceed line-to-line voltages, continuously.

20-5-5 Duty Cycle Voltage Rating

The *duty cycle voltage* is defined as the maximum permissible voltage between its terminals at which the arrester is designed to perform its duty cycle. Duty cycle voltage rating is above MCOV. The arrester will operate at duty cycle voltage rating for a limited duration. Test procedures are described in Ref. 1.

20-5-6 Temporary Overvoltage (TOV)

Figure 20-9 shows TOV capability of station class Tranquell™ surge arresters of General Electric Company. Low-magnitude TOVs are those that cause a rise in the element temperature, and the largest surge voltages will cause a surge arrester to reach its thermal stability limit of 200°C in 1 to 200 s. The high-magnitude TOVs are those that cause leakage currents in excess of 1 A, so that the system voltage waveform begins to distort due to voltage drop in the system impedance. It is, therefore, not possible to generalize the overvoltage capabilities of the surge arresters. An EMTP program can be used for case-by-case studies. Failure can occur due to rise in temperature caused by TOV, which may drive the arrester beyond its thermal stability limit.

The TOV is also a function of the arrester type (Fig. 20-10). This figure shows that the distribution class surge arresters have a higher TOV. The stability limit voltage also depends on the arrester design. This figure also depicts that the stability limit voltage for intermediate and distribution class arresters is above the duty cycle voltage. Generally, for distribution class arresters, TOV is not of major concern.

We discussed TOV in Chap. 7. A TOV can be defined as a voltage condition that lasts at least for a cycle and is limited in duration. Figure 20-11 provides a prospective of voltages across the arrester and their time duration with reference to TOV, switching, and lightning surges based on MCOV rating of the arrester. If a rigorous study is not available, the overvoltage on single line-to-ground fault can be considered (Chap. 9). The TOV capability must exceed the magnitude and duration of expected temporary overvoltages, considering operating time of the primary and backup protective devices. The curves in Fig. 20-9 illustrate the duration and magnitude of the TOV that can be applied to the arrester before the arrester voltage is reduced to MCOV. Also, the permissible durations are a function of the energy that may have been discharged through the arrester just prior to the application of TOV. This prior energy is mostly the result of discharge of a switching surge.

Manufacturers sometimes publish curves of prospective switching surge versus arrester discharge energy applicable to their arresters that can be applied for a general estimate of the TOV capability. Figure 20-12 shows such a curve for 336-kV and 300-kV metal-oxide arrester for 400-kV system for Tranquell station class arresters. If the switching surge is 2.5 per unit, $[400\times(\sqrt{2/3})\times 2.5\text{ kV}]$ and arrester rating is 336 kV, then, for a 100-mi-long line, the energy read from this figure is 1.1 kJ/kV. If the station class arrester has an energy capability of 7.2 kJ/kV, six surges can be discharged within 1 min period. After 1 min rest period, the discharge can be repeated. Note that the energy-handling capability of arresters is specified for a certain magnitude of switching current and surge impedances.

Some other system considerations and connections of equipment being protected must be considered. A delta-wye step-down transformer that has its only source of ground at the substation may become ungrounded on operation of line breaker and back fed from the wye-side, exposing the surge arrester connected line-to-ground to line-to-line voltage.

For line-to-ground–connected surge arresters, we may summarize that:

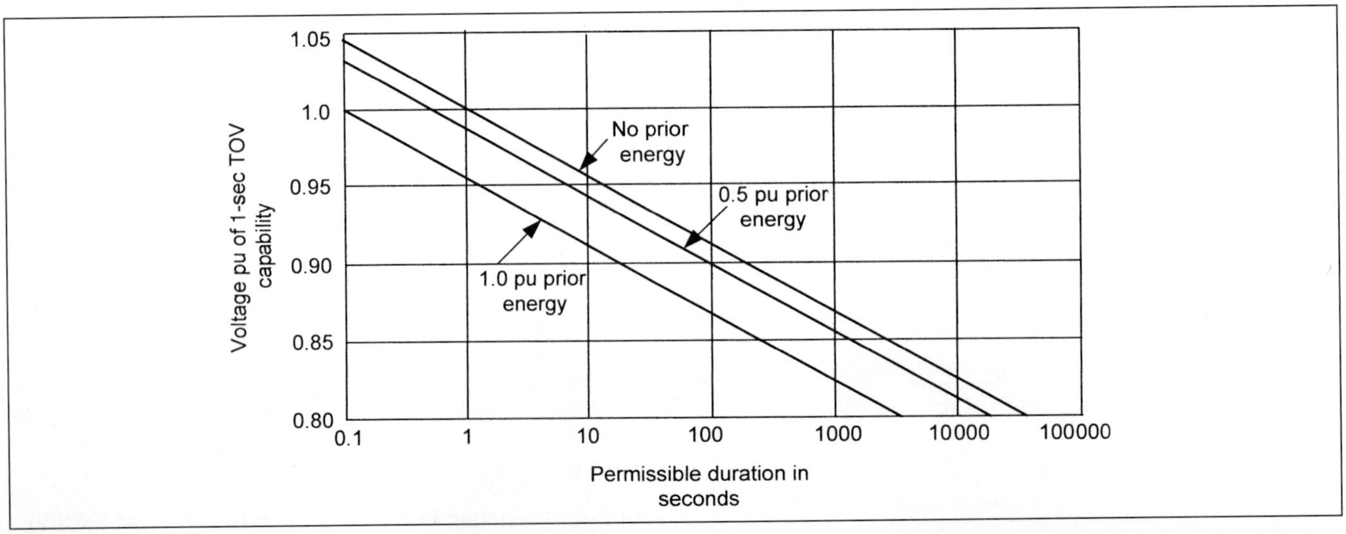

FIGURE 20-9 Overvoltage capability of station class, General Electric Company Tranquell™ station class surge arresters. *Reproduced with permission.*

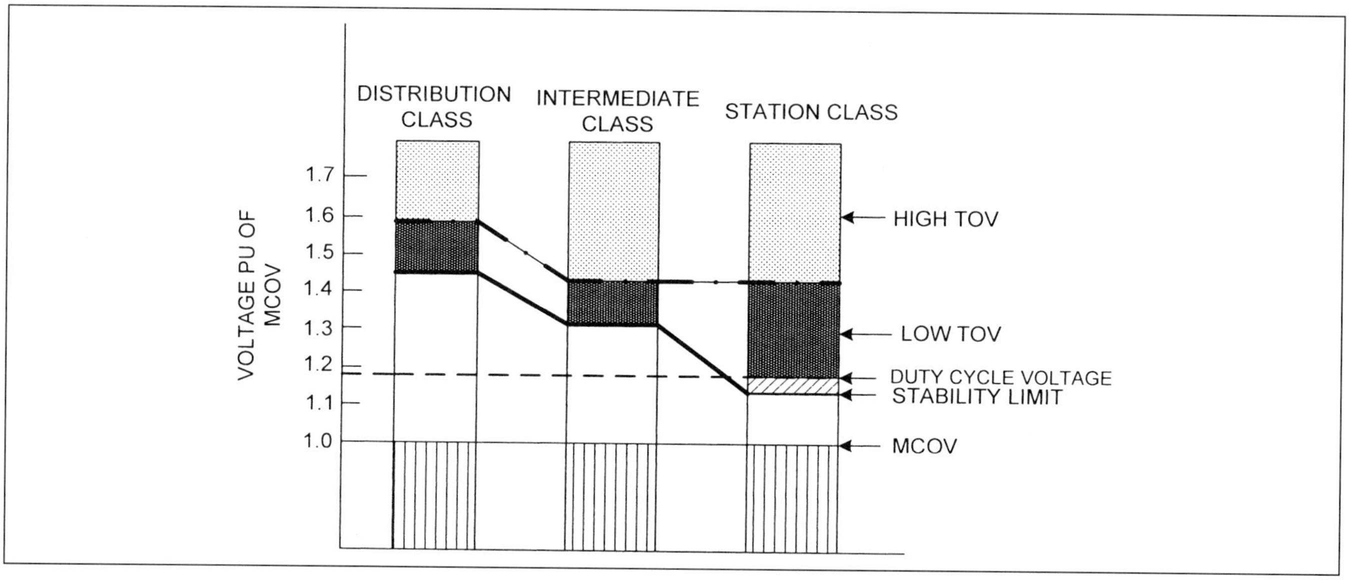

FIGURE 20-10 Relationship between four operating regions, different classes of surge arresters.

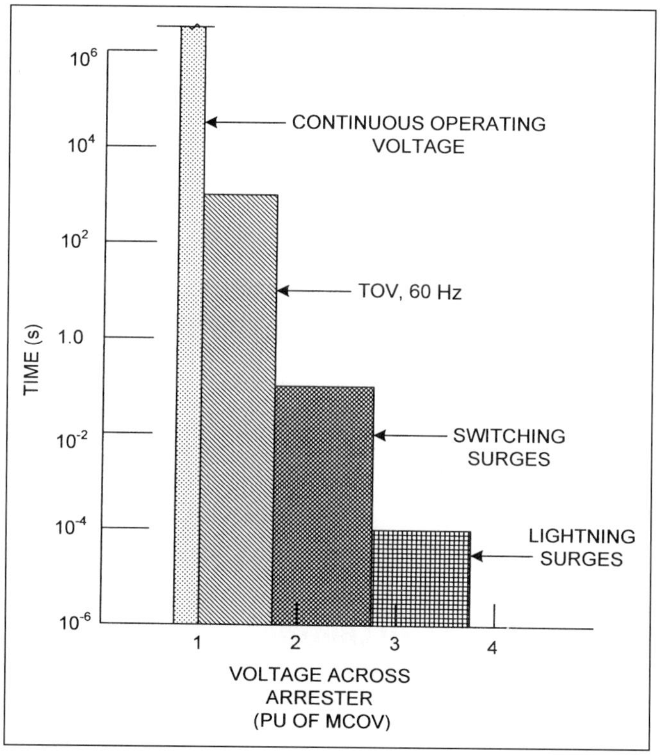

FIGURE 20-11 Voltage across surge arrester, generalization, for various overvoltage conditions.

- For solidly grounded systems, an arrester rating of 80 percent of maximum phase-to-phase voltage with no fault on the system can be used.
- For multigrounded wye open-wire circuits in distribution systems, the nominal phase-to-ground operating voltage is multiplied by 1.25 and for cable spacer circuits by 1.50.
- For ungrounded and impedance grounded systems, MCOV is selected equal to line-to-line voltage, or more, where the faults can be removed in a few seconds.

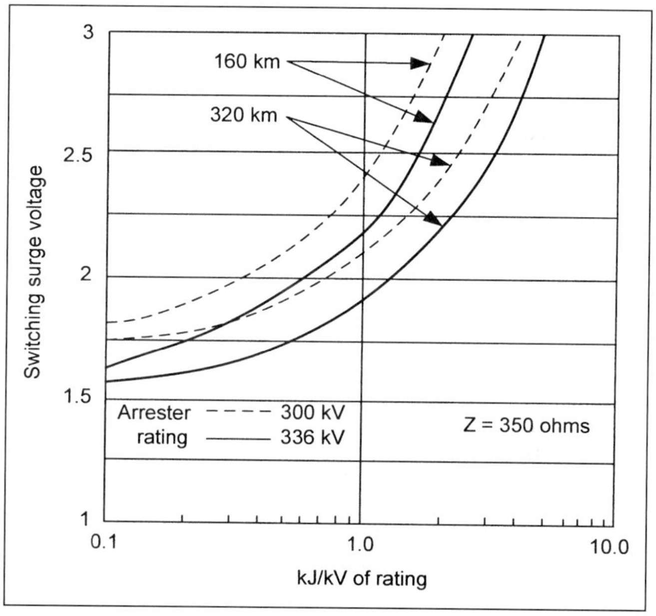

FIGURE 20-12 Prospective switching surge versus arrester discharge energy, 400-kV system, General Electric Company Tranquell™ station class surge arresters; 300 kV and 326 kV rated voltages. *Reproduced with permission. All rights reserved.*

- Ungrounded or resistance-grounded systems may require MCOV = 1.25 times or higher than line-to-line voltage.

Table 20-4 shows the typical arrester-rated voltages for grounded neutral circuits and for high-impedance grounded, ungrounded, or temporarily ungrounded systems. High-voltage systems are not impedance grounded. The surge arrester data shown for high-voltage impedance grounded/ungrounded system may be applicable for special situations.

20-5-7 Lightning and Switching Surge Discharge Currents

The tests determine the ability of arrester to operate successfully under application of sequence of current impulses while energized at MCOV.

TABLE 20-4 Selection of Arrester-Rated Voltage With Respect to System Grounding

NOMINAL SYSTEM VOLTAGE	GROUNDED NEUTRALS	HIGH IMPEDANCE-GROUNDED, UNGROUNDED, OR TEMPORARILY UNGROUNDED SYSTEMS
4.16	3.0	
	4.5	4.5
		5.1
13.8	10	
	12	
		15
		18
34.5	27	
	30	
		36
		39
138	108	
	120	
		132
		144
230	172	
	180	
	182	
		228
		240
400	300	At such high voltages, there are no impedance-grounded or ungrounded systems
	312	
	336	
	360	

Table 20-5 shows lightning impulse classifying currents 8/20-μs wave shape for discharge voltage characteristics,[1] and Table 20-6 shows switching surge classifying currents to obtain the discharge voltage characteristics with a time to crest of 45 to 60 μs. We will further discuss these lightning and switching currents in Sec. 20-6.

Table 20-7 shows the characteristics of Tranquill metal-oxide station class arresters of General Electric Company, reproduced with permission. Tables 20-8 and 20-9 depict the VI characteristics of arresters for front of wave (FOW), 8/20 μs, 36/90 μs, and also for 1-ms wavefront, courtesy of General Electric Company. The data shown in these tables may not be readily available from a manufacturer. The data shown in these tables are used throughout the following examples.

Table 20-10 provides the arrester discharge voltage data for various arrester classes. The factors shown are generalized and are not of a particular manufacturer.

TABLE 20-5 Lightning Impulse Classifying Currents

CLASSIFICATION OF THE ARRESTER	IMPULSE VALUE (PEAK A)
Station (800 kV, maximum system voltage)	20000
Station (550 kV, maximum system voltage)	15000
Station (below 550 kV, maximum system voltage)	10000
Intermediate	5000
Distribution, heavy duty	10000
Distribution, normal duty	5000
Distribution, light duty	5000

TABLE 20-6 Switching Surge Classifying Current

SYSTEM VOLTAGE, MAXIMUM	STATION CLASS (CREST A)	INTERMEDIATE CLASS
3–150	500	500
151–325	1000	—
326–900	2000	—

20-6 RESPONSE TO LIGHTNING SURGES

The steepness of the lightning surge voltage varies from the standard wave shape and so does the discharge voltage. Statistical data may be available for a certain location of rate of rise of the surge voltages at arrester locations, which will differ for direct strokes and induced voltages. The system configuration also plays a role. A surge will be reflected at any impedance discontinuity, such as line end, fault, line tap, or junction between OH and underground systems. A metal-oxide surge arrester will begin conducting a lightning surge current when the voltage reaches approximately 80 percent of the FOW peak discharge voltage. The test procedure for FOW sparkover results in a wave front which is determined by the test apparatus and not by the surge arrester. Thus, practically the FOW response is determined more by the system and characteristics of the originating surge rather than the surge arrester.

The voltage response of the arrester to a lightning surge after it begins its surge-limiting action is termed *voltage response*. This depends on the peak current discharged, which is termed coordinating current. *The coordinating current* is one key parameter in the insulation

TABLE 20-7 Gapless Metal-Oxide Surge Arrester Data

Arrester Rating (kV rms)	MCOV (kV rms)	TOV (1s, kV rms)	FOW (kV Crest)	Maximum Discharge Voltage (kV Crest) at Indicated Current for 8/20-μs Current Wave: 1.5 kA	3.0 kA	5.0 kA	10 kA	15 kA	20 kA	40 kA	Max. Switching Protective Level (kV Crest) at Indicated Current: kV	kA
2.7	2.2	3.1	7.8	5.9	6.2	6.5	6.9	7.4	7.8	8.9	5.4	0.5
5.1	4.2	6.0	14.8	11.2	11.8	12.3	13.1	14.0	14.7	16.9	10.3	0.5
7.5	6.1	8.8	21.4	16.2	17.0	17.7	18.9	20.2	21.2	24.3	14.8	0.5
9.0	7.65	11.0	26.6	20.2	21.1	22.0	23.5	25.1	26.4	30.2	18.4	0.5
15.0	12.7	18.3	44.2	33.5	35.1	36.6	39.1	41.8	43.9	50.3	30.6	0.5
18.0	15.3	22.0	53.3	40.4	42.3	44.1	47.1	50.3	52.8	60.6	36.8	0.5
27	22.0	31.7	76.5	58.0	60.8	63.3	67.7	72.3	75.9	87.0	52.9	0.5
30	24.4	35.2	84.9	64.3	67.4	70.3	75.1	80.2	84.2	96.5	58.7	0.5
45	36.5	52.6	128	96.8	102	106	113	121	127	146	88.3	0.5
54	44	63	144	111	116	120	127	135	141	159	102	0.5
72	58	85	191	148	154	160	169	179	188	212	136	0.5
96	78	113	255	197	206	213	225	239	250	282	181	0.5
108	87	127	287	222	232	240	254	270	282	318	204	0.5
120	98	142	321	249	259	269	284	301	315	355	235	1.0
144	117	170	382	296	309	320	338	359	375	423	280	1.0
168	136	198	446	345	360	373	394	418	437	493	326	1.0
192	156	226	509	394	411	426	450	477	499	563	372	1.0
240	194	283	635	491	513	531	562	596	623	703	465	1.0
258	209	304	683	528	551	571	604	641	670	755	518	2.0
276	224	325	730	565	589	611	646	685	716	808	554	2.0
294	238	347	778	602	628	650	688	730	763	860	590	2.0
300	243	354	795	615	641	665	703	745	779	879	603	2.0
360	292	424	953	737	769	797	843	894	934	1054	723	2.0

(All standard arrester ratings are not shown.)

Courtesy of General Electric Company; reproduced with permission.

TABLE 20-8 Arrester VI Characteristics for Ratings 2.7 to 48 kV

	Arrester Voltage in Per Unit of the 10 kA, 8/20-µs Discharge Voltage							
	Test Wave							
	FOW		8/20 µs		36/90 µs		1-ms Wavefront	
Current (A)	**Min**	**Max**	**Min**	**Max**	**Min**	**Max**	**Min**	**Max**
1			0.608	0.663	0.599	0.653	0.596	0.650
10			0.645	0.696	0.635	0.686	0.631	0.681
100			0.695	0.743	0.685	0.732	0.676	0.722
500			0.754	0.794	0.743	0.782	0.738	0.777
1000	0.891	0.933	0.787	0.824	0.776	0.812	0.769	0.805
2000	0.939	0.977	0.829	0.863	0.817	0.850	0.807	0.841
5000	1.013	1.061	0.895	0.937				
10000	1.110	1.132	0.981	1.000				
15000	1.168	1.210	1.031	1.069				
20000	1.210	1.271	1.069	1.123				
40000	1.329	1.458	1.174	1.288				

TABLE 20-9 Arrester VI characteristics for Ratings 54 to 360 kV

	Arrester Voltage in Per Unit of the 10-kA, 8/20-µs Discharge Voltage							
	Test Wave							
	FOW		8/20 µs		36/90 µs		1-ms Wavefront	
Current (A)	**Min**	**Max**	**Min**	**Max**	**Min**	**Max**	**Min**	**Max**
1			0.647	0.691	0.645	0.689	0.640	0.684
10			0.682	0.725	0.674	0.717	0.671	0.713
100			0.734	0.769	0.722	0.756	0.716	0.750
500			0.790	0.819	0.775	0.803	0.762	0.790
1000	0.927	0.958	0.820	0.847	0.802	0.828	0.787	0.813
2000	0.972	0.996	0.860	0.881	0.839	0.859	0.839	0.859
5000	1.044	1.070	0.923	0.946				
10000	1.117	1.131	0.988	1.000				
15000	1.167	1.200	1.032	1.061				
20000	1.209	1.254	1.069	1.109				
40000	1.318	1.414	1.166	1.251				

TABLE 20-10 Arrester Discharge Voltage With Respect to Its Class at 20-kA Coordinating Current

Arrester Class	**Discharge Voltage in pu Based on Distribution Class Normal Duty Arrester as 1 pu**
Distribution—normal duty	1.0
Distribution—heavy duty	0.87
Distribution—riser pole	0.78
Intermediate	0.71
Station	0.68

coordination. The appropriate coordinating current for lightning surges depends on the effectiveness of shielding. Table 20-11 contains coordinating currents that are found satisfactory in most situations.[5]

Manufacturers publish data of the discharge voltage of the surge arresters for 8/20 µs current surges (Table 20-7). As the discharge voltage will depend on the wave shape and duration of associated discharge current, it may be necessary to correct the published data to account for wave shapes encountered in the actual surge applications, using data from Tables 20-8 and 20-9.

For insulation coordination and application of surge arresters, it may be necessary to reduce the discharge voltage for a better protective margin.

Increasing the arrester class is one way to reduce the discharge voltage, as shown in Table 20-10. The other advantages of choosing a higher arrester classification are:

TABLE 20-11 Currents for Determining Discharge Voltages in Shielded Substations with Shielded Incoming Lines

MAXIMUM SYSTEM VOLTAGE	DISCHARGE CURRENT (A)
15	*
36.5	*
72.5	5000
121	5000
145	5000
242	10000
362	10000
550	15000
800	20000

*Generally unshielded lines.

- Failures are less likely.
- Higher energy-handling capability.
- Discharge voltage does not increase rapidly with discharge current.

Metal-Oxide Surge Arresters in Parallel Another alternative to cater for higher discharge currents is to connect metal-oxide gapless arresters in parallel. The discharge current for each arrester is reduced and, therefore, the discharge voltage is reduced. This option can be exercised with metal-oxide arresters *only* as exact matching of discharge characteristics is required. The valve-type arresters with air gap have a broad tolerance of characteristics, and the current sharing between parallel units will be unpredictable. The gapped-type surge arresters should, therefore, never be paralleled. Table 20-12 shows the effect of parallel connection of distribution class arresters. Most beneficial improvement is obtained with two arresters in parallel.

A configuration that can be used for resistance-grounded medium-voltage systems is shown in Fig. 20-13*b*. Again, only metal-oxide surge arresters should be used in this configuration. This is the application of surge protection to a 13.8-kV, impedance-grounded system. Normally, say, 18-kV arresters are required for phase-to-ground connection, as shown in Fig. 20-13*a*. If the system *was* solidly grounded, 12-kV arresters would be adequate for phase-to-ground connection, as the TOV is limited. For configuration shown in Fig. 20-13*b*, 12-kV arresters are chosen for connection from phase to a *common wye point,* which is then connected to ground through a 6-kV arrester. This behaves like a 18-kV arrester for phase-to-ground surges, same as in Fig. 20-13*a*; however, for phase-to-phase overvoltages, the effective rating of surge arresters in Fig. 20-13*b* is 24 kV versus 36 kV in Fig. 20-13*a*.

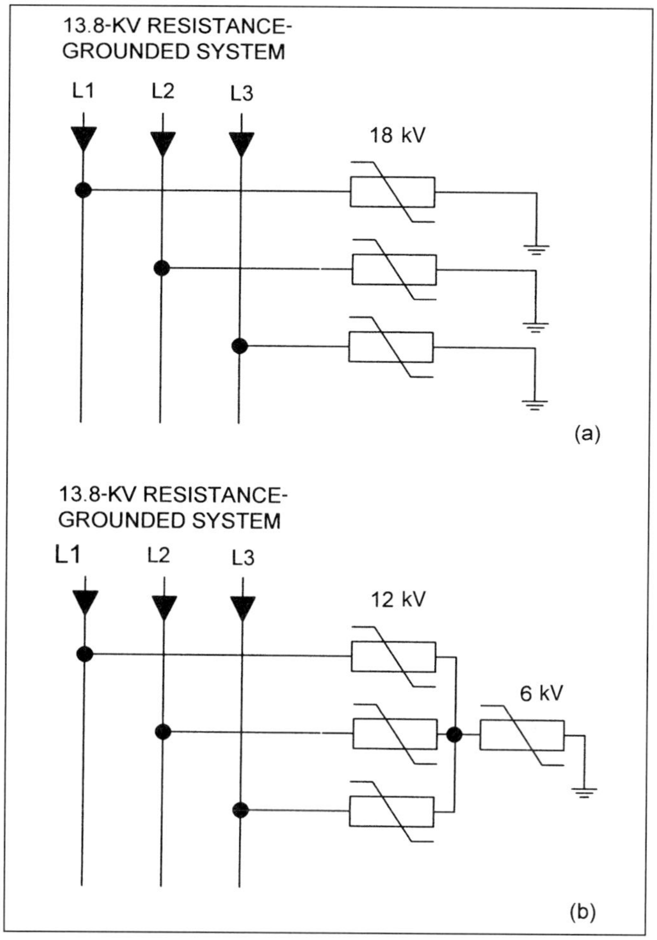

FIGURE 20-13 (*a*) Conventional phase-to-ground connection of three surge arresters, rated voltage 18 kV, for 13.8-kV resistance-grounded system. (*b*) Configuration with four arresters of 12 kV and 6 kV for better control of phase-to-phase surges; see text.

TABLE 20-12 Effect of Parallel Distribution Class Arresters on Discharge Voltage

		DISCHARGE VOLTAGE REDUCTION	
NUMBER OF UNITS	COORDINATING CURRENT IN EACH UNIT	NORMAL DUTY ARRESTER PU	HEAVY DUTY ARRESTER PU
1	20	1.00	1.00
2	10	0.87	0.86
3	6.7	0.83	0.82
4	5	0.80	0.80

20-7 SWITCHING SURGE DURABILITY

For insulation coordination, an estimate of the energy through the arrester for a switching surge is required. The energy dissipated is related to prospective switching surge magnitude, its wave shape, and system impedance. This can be best determined through detailed system studies with TNA or EMTP. In the absence of this analysis, the energy absorbed can be estimated from the following equations:[5]

$$J = D_L E_A I_A / v$$
$$I_A = (E_s - E_A)/Z \qquad (20\text{-}9)$$

where J is the discharge energy in kJ, E_A is the arrester discharge voltage for the switching current I_A in kA, D_L is line length in km, v is speed of light, E_s is prospective switching surge voltage in kV, and Z is single-phase surge impedance of the line in ohms.

Manufacturer's data are required for the discharge energy capability, generally expressed as kJ/kV of rating voltage rating or kJ/MCOV.

Example 20-1 A surge arrester of 258 kV rating has a maximum switching protective level of 518-kV, 2-kA switching impulse current—data taken from Table 20-7. The line length is 300 km of surge impedance 350 Ω and the switching surge level is 900 kV. Then from Eq. (20-9), I_A is 1.09 kA. However, from Eq. (20-9), we cannot calculate the discharge current correctly as the discharge voltage of the arrester is not constant and varies with the current. The internal resistance of the arrester should be calculated which can be taken as a straight line with constant slope. Table 20-9 shows the arrester voltage for a surge of 36/90 μs and 2000 A discharge current is 0.859 (maximum value) of the arrester voltage for 10-kA, 8/20-μs wave. From Table 20-7, the arrester discharge voltage for 10-kA, 8/20-μs wave is 604 kV. Therefore, 604 × 0.859 = 518 kV, which is the same as shown under maximum switching protection level for 2 kA current in Table 20-7. If the first estimate of discharge current of 1.09 kA is used, then again referring to Table 20-9, this scaling factor is reduced to 0.828, which gives a discharge voltage of 604 × 0.828 = 500 kV. Now, recalculate the discharge current, which increases to 1.148 kA. From Eq. (20-9), the energy is 1033 kJ or 4.0 kJ/kV. This is well within the capability of the station class surge arrester.

Approximately, the energy can be calculated from $J = 0.5CE^2$, where C is the total capacitance of the line. This gives 1157 kJ.

Example 20-2 In Example 5-6, a metal-oxide surge arrester of 258 kV is applied on the transformer primary; see also Fig. 5-32. The EMTP simulation of power through the arrester is shown in Fig. 20-14*a* and its histogram in Fig. 20-14*b*. The arrester has an energy-handling capability of 7.2 kJ/kV, a total of 1857.6 kJ. The

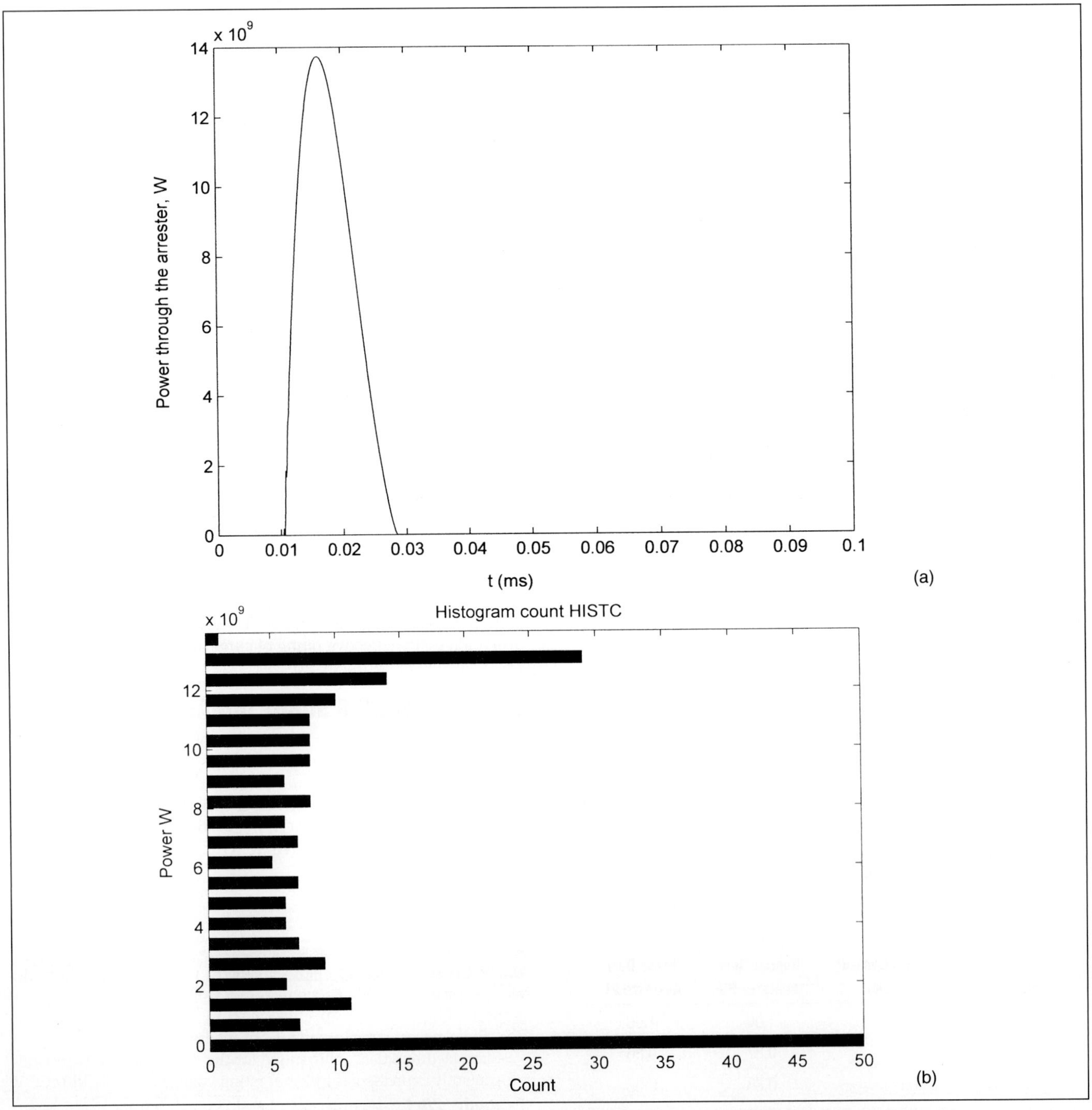

FIGURE 20-14 (*a*) Power through the surge arrester, Example 20-2. (*b*) Histogram of the power through the arrester.

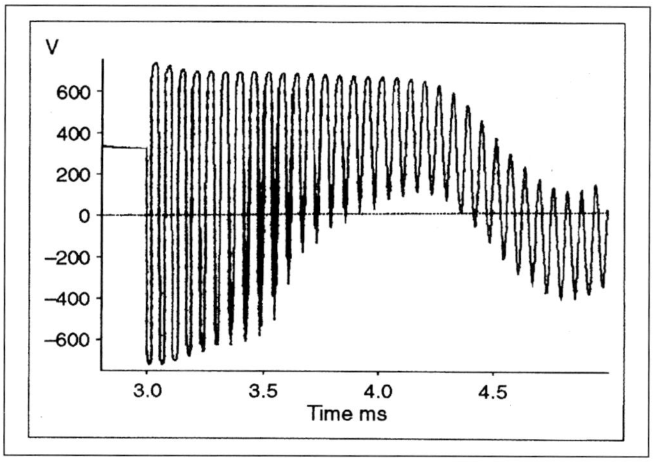

Figure 20-15 Reduction of secondary voltage magnitude and oscillation frequency, Example 20-3. Compare this with Fig. 6-28; see text.

calculated energy absorbed by the arrester is 210.8 kJ, which is a safe application. The arrester can discharge approximately six surges of this nature.

Example 20-3 This is a continuation of Example 6-8. Figure 6-28 shows the high-frequency oscillations of secondary voltage of transformer T2, and furthermore this voltage is 7.65 times the rated secondary voltage of 480 V. A secondary surge arrester in conjunction with a 2-μF surge capacitor reduces this voltage magnitude to 1.71 per unit. Comparing Fig. 6-28 with Fig. 20-15, the oscillation frequency is also reduced.

Example 20-4 In Example 18-1, the study of transients in a GIS, the surge arrester currents SA1 and SA2 are shown in Fig. 18-17*a* and 18-17*b*, respectively. The power through the surge arrester is depicted in Fig. 20-16. It hits an instantaneous peak of 1.9 GW, and note that *two* surge arresters have been used in parallel. The surge arresters for GIS are special, high-energy arresters (Chap. 19).

20-8 ARRESTER LEAD LENGTH AND SEPARATION DISTANCE

The surge arresters are mostly connected line to ground. The lead length is totaled on both sides of the arrester. The lead length may be defined as the conductor length totaled on both sides of the arrester housing and the ends of the leads of the protected equipment. This must be considered with respect to the point of surge impact (Fig. 20-17). This figure clearly shows how the separation distance becomes lead length and vice a versa with respect to the point of surge impact. In Fig. 20-17*a*, the separation distance is *c* and the lead length is $a+b$. In Fig. 20-17*b*, the lead length is $a+b+c$. Note that the dimensions (length of the arrester itself) are not included in the lead length. Even a minor amount of stray inductance in the lead length can create a significant voltage drop in the lead length due to rapid rise in increase of current. The bends in the lead length add to this inductance, and therefore, the leads to the arrester on the equipment side and on ground side are run as straight and direct as possible. This voltage drop affects the protection level provided by the surge arrester. The voltage drop in the lead length is given by:

$$V_{\text{le}} = L_{\text{le}} S_{\text{ld}} \tag{20-10}$$

where V_{le} is the voltage drop across the lead in kV, L_{le} is the inductance of the lead length in μH, and S_{ld} is the slope of the rise of discharge current in kA/μs. Adding the voltage drop across the lead length to ascertain a protective level with the arrester as installed should modify the surge arrester discharge voltage.

20-8-1 Separation Distance

The *separation distance* is the length of connection separating the line terminal of the arrester from the line terminal of the protected equipment. It should not be confused with the lead length (Fig. 20-17).

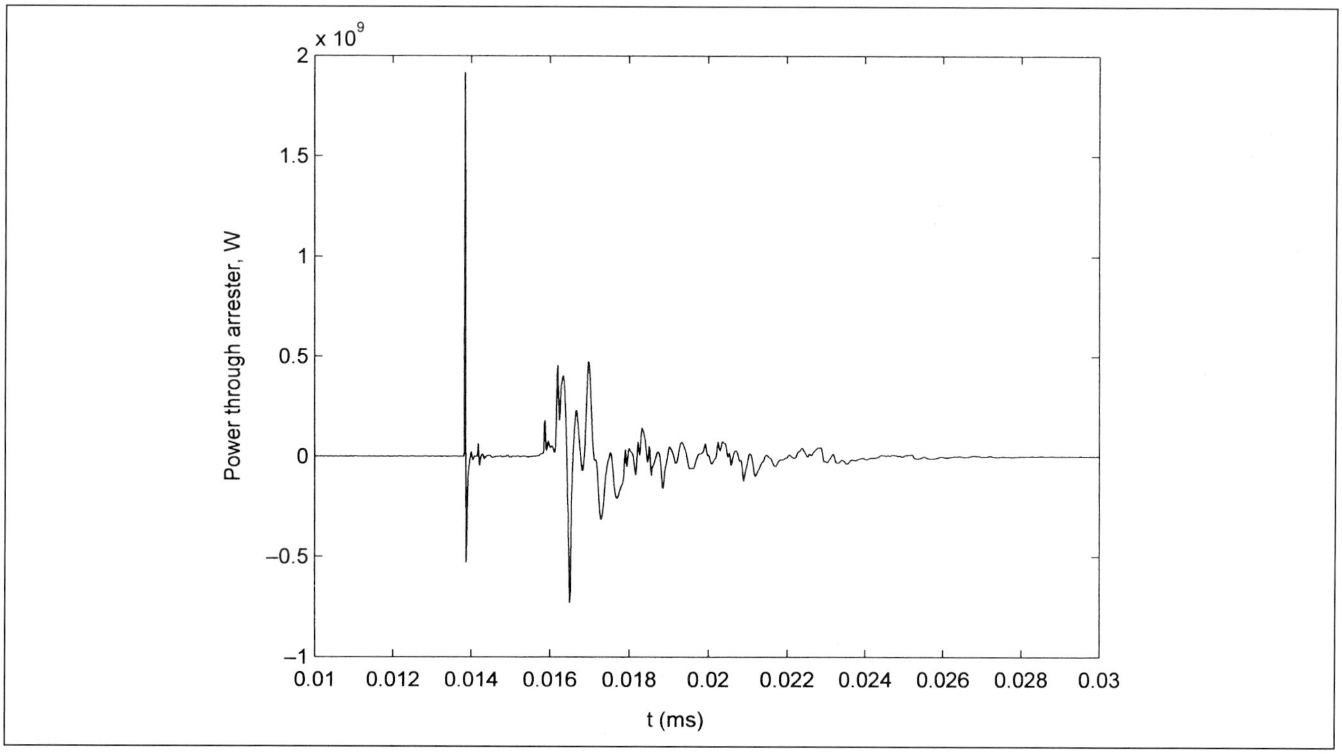

Figure 20-16 Transient power through the arrester, Example 20-4.

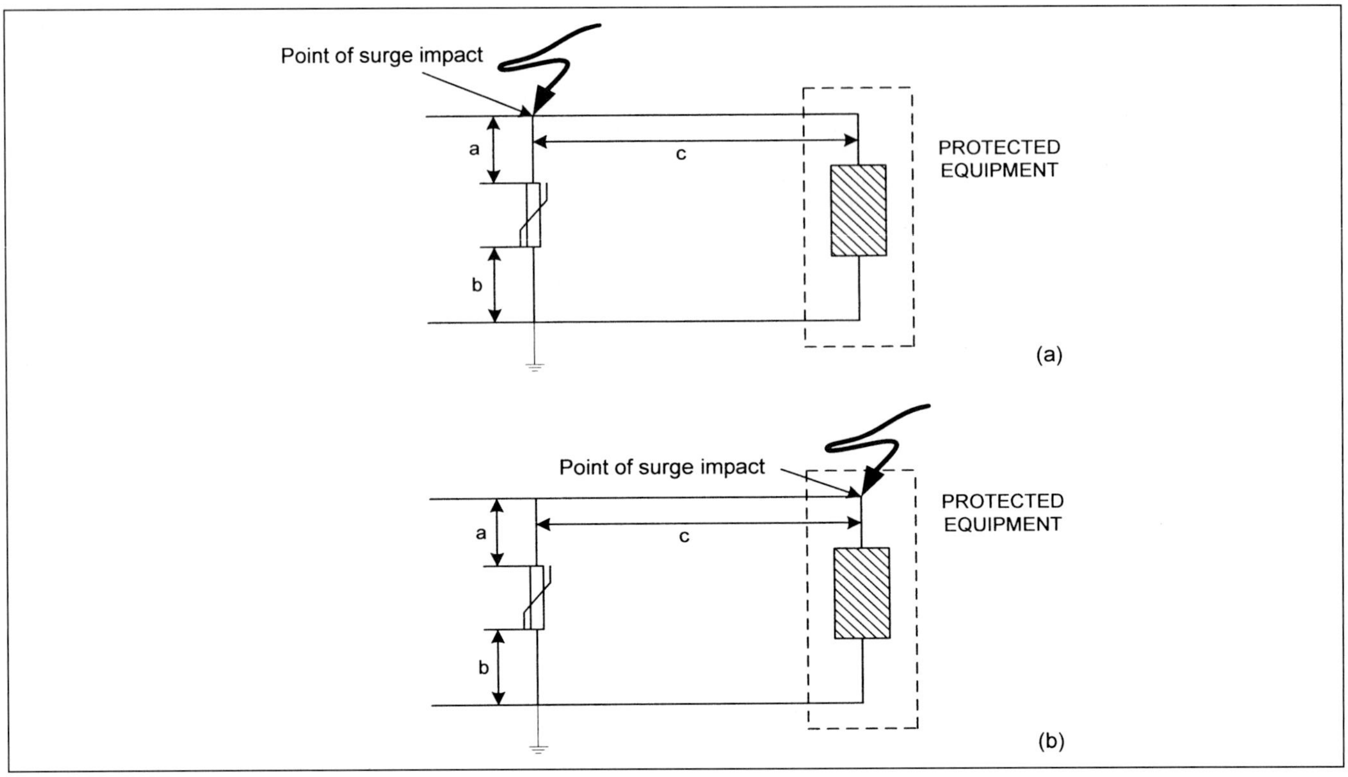

FIGURE 20-17 (*a*) and (*b*) To explain arrester lead length and separation distance with respect to the point of impact of the surge.

Separation distance becomes important when it is not possible to locate the surge arrester directly at the terminals of the protected equipment. It typically occurs in the substations and at transition points between the OH and underground distribution.

On a lightning surge, the surge arresters operate in a few microseconds or less to affect voltage changes. These changes propagate approximately at the speed of light, and the separation creates a delay of some microseconds. The instantaneous surge voltage at equipment terminals will therefore be different from that at the arrester terminals. Therefore, the surge arrester effectiveness is reduced with increasing separation.

If the separation distance exceeds a certain critical distance, its effect on the equipment protective level must be investigated and accounted for.

Example 20-5 The protective level of an arrester is 160 kV, and separation plus lead length is 60 m from the equipment. Considering a propagation velocity of 300 m/μs, the travel time of the surge for a distance of 60 m is 0.2 μs. Incident surge wave is of 200 kV/μs. The incident wave will travel past the arrester unchanged, as it is below the conduction threshold of the arrester. At $t = 0.2$ μs, it arrives at the equipment, which can be considered an open circuit of line for illustrative purposes, and is reflected back positively toward the arrester. The rate of rise of voltage at the equipment is effectively doubled. At $t = 0.4$ μs, the voltage wave arrives at the arrester terminals and the rate of rise is doubled. At $t = 0.6$ μs, the voltage reaches conduction level of the arrester and a negative wave of 400 kV/μs rate of rise is reflected toward the equipment. The negative wave arrived at $t = 0.8$ μs is doubled and reflected back toward the arrester. Because of negative polarity, the voltage at the equipment terminals starts decreasing. Further reflections may cause a positive sawtooth voltage, but the highest voltage at the equipment occurs at 0.8 μs. Construction of a lattice diagram will facilitate the construction of sawtooth waveform shown in Fig. 20-18.

For a simple configuration, as shown in Fig. 20-19, with arrester ahead of the equipment to be protected, the voltage at the equipment terminals can be determined from the following equation:

$$V_{\mathrm{pe}} = V_{\mathrm{pa}} + 2S_f \frac{d_{\mathrm{es}} + d_a}{\lambda} \tag{20-11}$$

where V_{pe} is the voltage at the protected equipment in kV, V_{pa} is the initial protective level of the surge arrester in kV, S_f is the rate of rise of the front in kV/μs, d_{es} is the separation distance in m, d_a is the lead length in m, and λ is the propagation velocity in m/μs.

20-8-2 Reflections from Other Equipment

Reflections from other equipment can be accounted for by IEEE method.[5] This is particularly suitable for analyzing the effect of separation in transformer substations. The following explanations specify how a two-transformer and three-line system configuration, Fig. 20-20*a* and *b*, can be reduced to the simple equivalent circuit shown in Fig. 20-19. The separation distance and the lead length should be properly calculated. Refer to Fig. 20-20*b* for the following steps:

1. Remove the transformer that is not being considered for surge protection. Assuming that surge protection of transformer T1 is of interest, transformer T2 is removed.

2. Identify junction between d_{es}, separation distance, and d_a, the lead length, and the line that has an incoming surge. This junction is "*c*" in Fig. 20-20.

3. Remove all lines connected to the lead length d_a, between junction *c* and the arrester terminal. Thus, line 3 is removed.

4. Multiply rate of rise of incoming surge by $3/(n + 2)$, where n is the number of lines remaining in service after the lines in rule 3 are removed. Referring to Fig. 20-20*b*, $n = 2$.

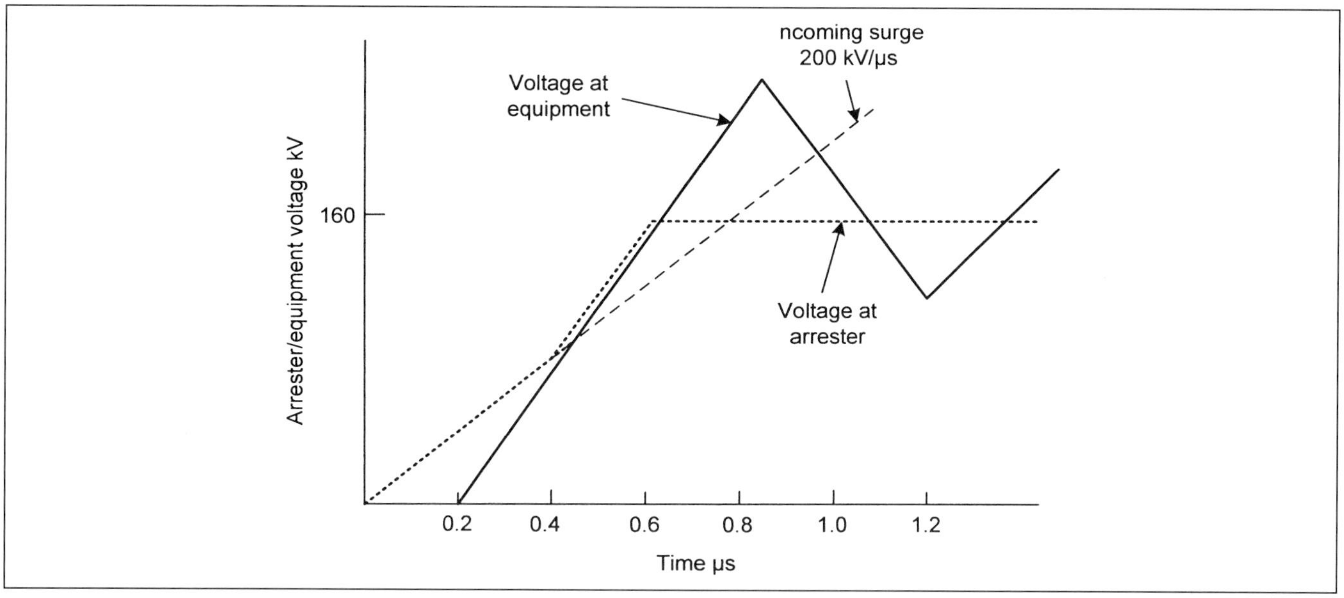

FIGURE 20-18 Sawtooth voltage at the protected equipment, surge arrester located some distance away from the protected equipment, Example 20-5.

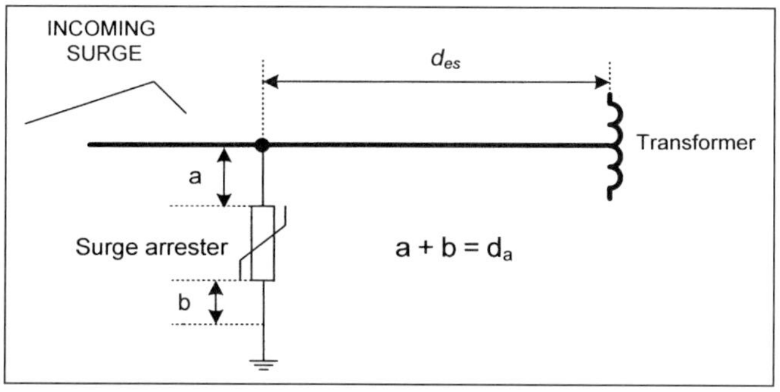

FIGURE 20-19 Example of surge protection with line terminated at a single transformer (protected equipment), surge arrester located some distance away from the protected equipment.

To calculate the maximum permissible separation, follow the procedure shown in Example 20-6.

Example 20-6 In Fig. 20-20, lead length d_a is 71 ft and separation distance d_{es} is 29 ft. Arrester has a protective level of 254 kV at a rate of rise of 900 kV/μs. Following rules 1, 2, and 3, transformer T2 in Fig. 20-20 is removed, line L3 is removed, and according to rule 4, the modified rate of rise is:

$$S_f = 900\times\frac{3}{n+2} = 900\times\frac{3}{4} = 675 \text{ kV}/\mu\text{s}$$

The separation distance within which the equipment is protected can be said to be dependent on the steepness of the incoming wave. Let the surge impedance be 450 Ω, velocity of propagation v be 984 ft/μs, and BIL be 450 kV. Then calculate chopped-wave withstand (CWW) by using the following equation:

$$\begin{aligned}
\text{CWW} &= 1.1\ \text{BIL} = 495 \text{ kV}\\
di/dt &= 2S_f/Z = 3.0 \text{ kA}/\mu\text{s}\\
L &= d_a \times 0.4 = 28\ \mu\text{H}\\
V_{sa} &= V_{pa} + L(di/dt) = 254 + 28\times 3 = 338 \text{ kV}\\
V_t &= \text{CWW}/1.15 = 430 \text{ kV}\\
V_t/V_{sa} &= 1.27
\end{aligned} \tag{20-12}$$

where V_t is the maximum stress allowed at the transformer and V_{sa} is the voltage at the surge arrester from function point c to ground. For calculation of inductance of lead length, a value of 4 μH/ft is assumed. The separation distance is calculated from the curve in Ref. 5 which gives 32.5 ft. This curve is approximated by the following analytical equation:

$$D = \left[\frac{0.385(vV_{sa})}{d_{sa}}\right]\times\left[\frac{0.957\text{BIL} - V_{sa}}{2.92V_{sa} - 0.957\text{BIL}}\right] \tag{20-13}$$

20-9 APPLICATION CONSIDERATIONS

Figure 20-21 shows a general flowchart for application of surge arresters. The importance of correct system data, study results of overvoltages, equipment to be protected, and location of surge arrester are the starting points. Tables 20-1 and 20-7 show that all the characteristics of the arrester are based on the rated voltage of the arrester. The flowchart in Fig. 20-21 shows the interrelatedness of the activities and arrester parameters. The objective is to select an arrester which will:

- Provide required protective levels in the standards with respect to the insulation strength.
- Not be damaged when discharging the surges at the required location.
- Not operate under normal system operations. As discussed earlier, the basic considerations are selection of MCOV, TOV, switching surge energy discharged, the pressure relief current, and effects of ambient temperature and altitudes.
- Apply special considerations for use with capacitor banks, GIS, electronic equipment, rotating machines, and even transformers; that is, dry-type transformers may require closer coordination of the surge protection devices due to, generally, lower winding surge withstand capabilities.

20-9-1 Protective Levels

The *protective level* is the maximum crest voltage that appears across arrester terminals under specified conditions. For gapless arresters,

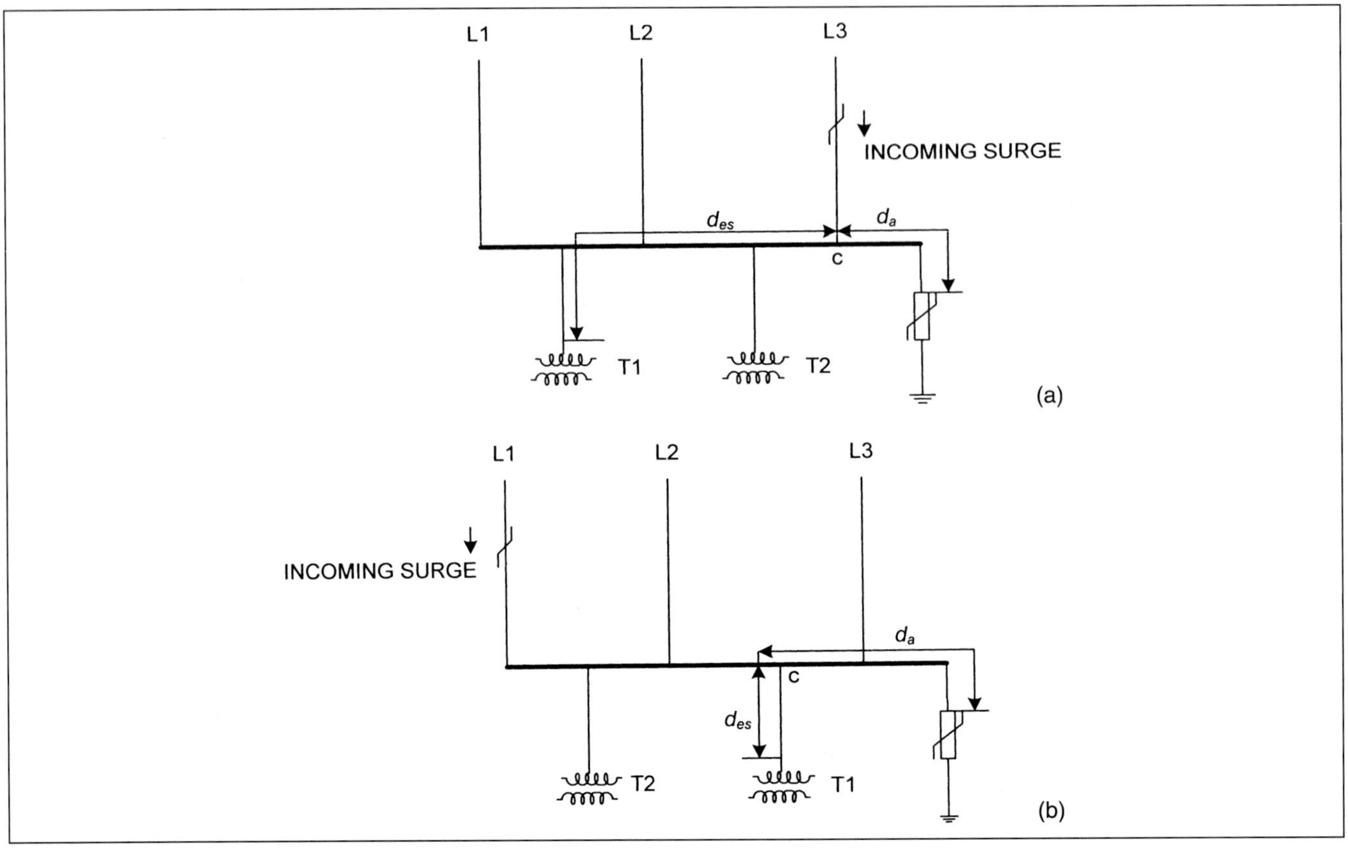

FIGURE 20-20 (*a*) and (*b*) Two-transformer and three-line substation surge protection, calculation of maximum separation.

it is the arrester discharge voltage for a specified discharge current. For gapped arresters, the protective level is the higher of the gap sparkover voltage or discharge voltage. The following protective levels are specified in ANSI/IEEE standards.[3,5]

FOW (Front-of-Wave Protective Level) This level is defined as the higher of the FOW sparkover voltage for specified wave shapes or arrester discharge voltage resulting with the application of lightning *impulse classifying current* magnitude cresting in 0.5 μs.

The lightning impulse classifying currents are given in Table 20-11. The currents are dependent on shielding. The table is for completely shielded lines for some distance from the substations, which means that predominant mechanism will be backflashovers. The arrester current can be calculated from the following equation:

$$I = I_c = \frac{3.84(E_{\text{CFO}} - E_c)}{Z_0} \tag{20-14}$$

where I is arrester discharge current in kA, I_c is arrester coordinating current in kA, E_{CFO} is the positive CFO of insulation in kV, E_c is arrester discharge voltage in kV, Z_0 is single-phase surge impedance of line, and factor 3.84 is the correction factor based on studies.[5]

The discharge currents in stations with nonshielded lines should not be less than 20 kA. In severe thunderstorm areas, keraunic level of 40 or more, higher levels may be desirable.

LPL (Lightning Impulse Protective Level) The higher value of the lightning impulse sparkover for 1.2/50-μs lightning impulse or arrester discharge voltage that results from an 8/20-μs wave. The selection of the discharge current depends on the following:

- The importance of the installation
- The line insulation and grounding effects
- The probability of occurrence of higher currents

SPL (Switching Impulse Protective Level) The higher value of the switching impulse sparkover or arrester discharge voltage that results from a current wave with a time to crest of 45 to 60 μs or gap sparkover from similar waves. The arrester coordinating current for the switching surges is a complex function of the surge arrester characteristics and the details of the system. The effective impedance seen by the arrester during a switching surge can vary from several hundreds ohms for arresters connected to OH lines to tens of ohms for arresters connected near cables and capacitor banks.

20-9-2 Insulation Withstand

BIL, BSL, and CWW values are required for the insulation coordination. These may be taken from the relevant equipment standards. Transmission and distribution line strength is described by CFO, and a standard deviation of 5 percent of CFO. The BIL and BSL may be conventional or statistical, CFO – 1.28 σ (Chap. 17). Factors for estimating the withstand voltage of mineral-oil-immersed equipment are provided in Ref. 5. The negative polarity lightning impulse CFO is approximately 600 kV/m, and for the positive polarity it is 560 kV/m. Bus and line support insulators' breakdown voltage is about 1.3 to 1.4 times CFO.

20-9-3 Insulation Coordination with Surge Arresters

Two methods of insulation coordination are described:[3,5]

1. Tabulation of protective ratios (PRs) or margins (PMs)
2. Graphical representation

Three-Point Method In this method, CWW, BIL, and switching surge withstand (BSL) are compared with the corresponding defined protective levels. The basic assumption is made that the

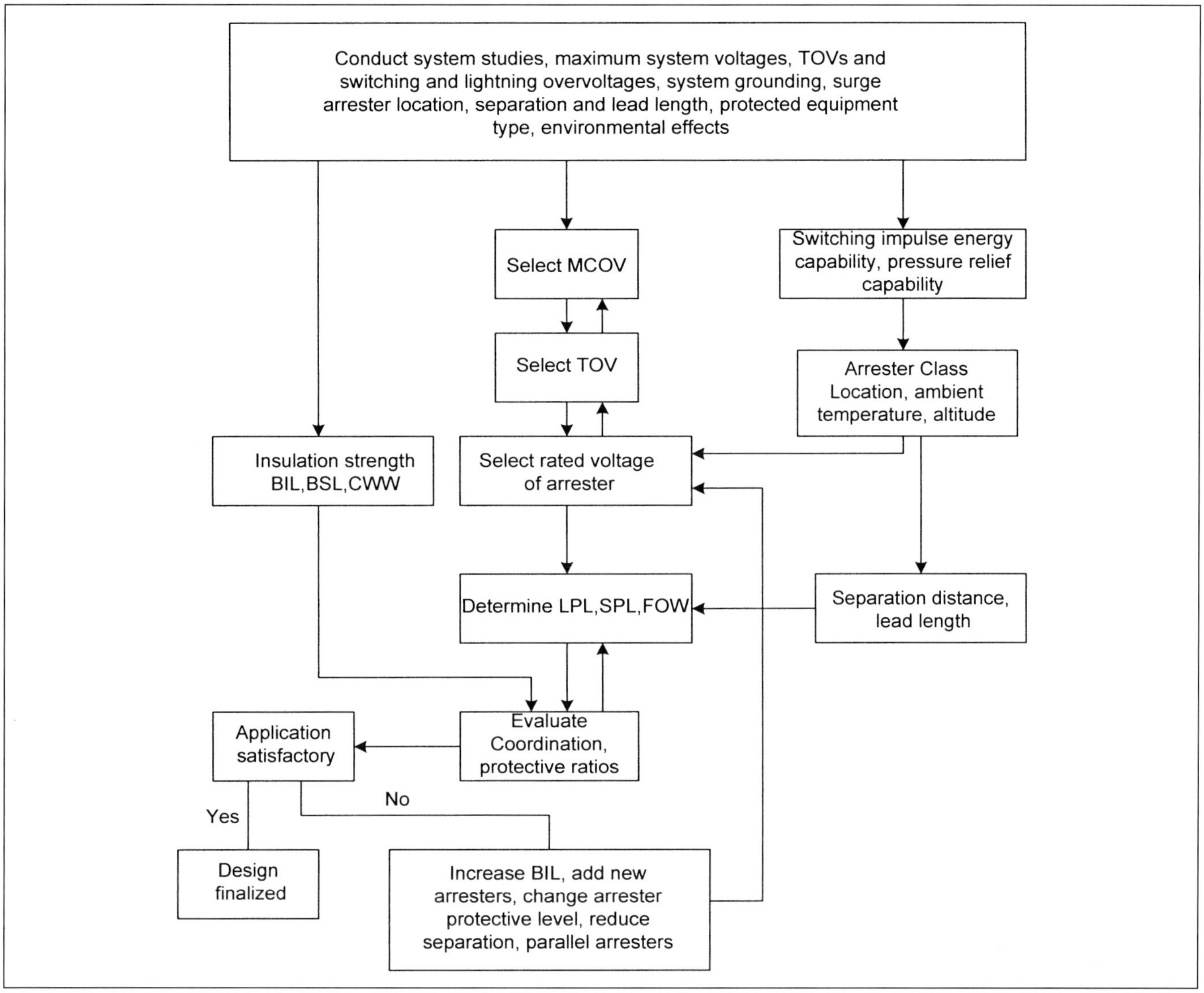

FIGURE 20-21 Flowchart for application of surge arresters.

insulation will be protected over the entire range of lightning and switching impulses, provided that the margin is adequate. Allowances for separation effects are included.

The following PRs are defined

$$PR_{L1S} = \frac{CWW}{FOW} = PM_{L1S} = [PR_{L1S} - 1]\,100$$

$$PR_{L2S} = \frac{BIL}{LPL} = PM_{L2S} = [PR_{L2S} - 1]\,100 \qquad (20\text{-}15)$$

$$PR_s = \frac{BSL}{SPL} = PM_s = [PR_s - 1]\,100$$

The PR limits for coordination are:

$$PR_{L1} \geq 1.2 \quad PR_{L1S} \geq 1.15$$

$$PR_{L2} \geq 1.2 \quad PR_{L2S} \geq 1.15 \qquad (20\text{-}16)$$

$$PR_s \geq 1.15 \quad PR_s \geq 1.15$$

PR_{L1S} refers when there is significant separation. If time to crest is equal to 2μs or less, $PRW_{L1S} = \frac{CWW}{V_T} \geq 1.1$ where V_T is determined as in Example 20-6; otherwise, $PR_{L2s} = \frac{BIL}{V_T} \geq 1.15$. For switching impulse, PR_s is same, $\frac{BSL}{SPL} \geq 1.15$. The PM limits for coordination are:

$$PM_{L1} \geq 20 \quad PM_{L1S} \geq 15$$

$$PM_{L2} \geq 20 \quad PM_{L2S} \geq 15 \qquad (20\text{-}17)$$

$$PM_s \geq 15 \quad PM_s \geq 15$$

For distribution system insulation coordination, the following PMs are acceptable:

$$PM(1) = \left(\frac{CWW}{(FOW + L\,di/dt)} - 1\right)100 \qquad (20\text{-}18)$$

$$PM(2) = (BIL/LPL - 1)100$$

where *di/dt* is calculated by dividing the crest current by the time to crest. Inductive voltage drop varies as a function of the magnitude of the current impulse as well as its rate of rise. A 10 kA current with rise time of 8 μs gives a voltage drop of 0.5 kV/ft, but with 1 μs rise time, it becomes 4 kV/ft.

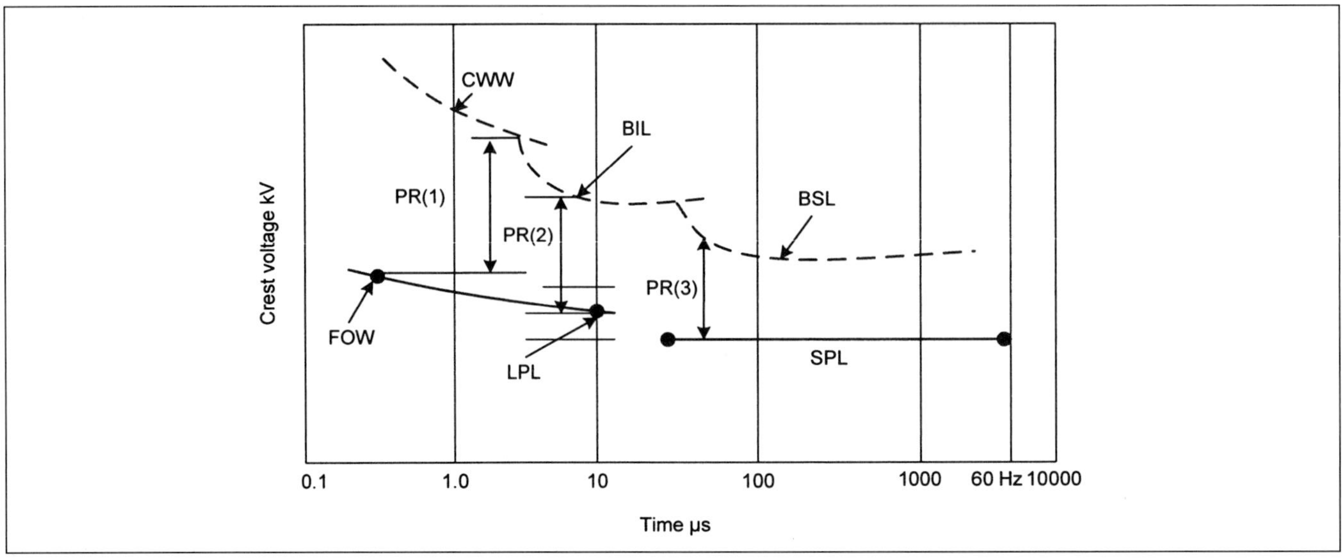

FIGURE 20-22 Typical volt-time curves for coordination of surge arrester protective levels with insulation withstand strength of liquid-filled transformers.

In general, PM(1) and PM(2) should both be at least 20 percent. For oil-filled, solid, or inorganic insulation, CWW can be assumed 1.15 times BIL, and for dry-type organic insulation, CWW can be assumed equal to BIL.

The discharge voltage of an arrester is greater for less frequent high crest lightning discharge surges and increases with higher rates of rise of lightning current. Currents higher than 20 kA can be considered.

Graphical Coordination Curve Another method of insulation coordination is the graphical coordination curve comparison method. Plot three points for the published voltages for the specific arrester under consideration: (1) FOW, (2) LPL at the coordinating current, and (3) switching surge protective level as a straight line. Connect the points with a curve of approximately the shape in Fig. 20-22, for oil-filled transformers.[5] If a manufacturer voltage-time sparkover curve is available, it can be used instead of approximation. Draw a ladder of lines each extending from 5 to 10 μs at levels corresponding to 5, 10, and 20 kA. At 8/20-μs test wave, plot CWW, BIL, and SPL curves. This is illustrated in the following example.

Example 20-7 For a 345-kV system nominal voltage, the data for application are: maximum voltage = 362 kV, rated switching impulse voltage = 2.5 per unit peak, temporary overvoltage = 1.5 per unit for 1.4 s, line length = 300 km, surge impedance = 350 Ω, available short-circuit current = 40 kA rms symmetrical. The power transformer to be protected has a BIL = 1050 kV, CWW = 1155 kV, and BSL = 870 kV. Average ambient temperature = 40°C.

As the system is solidly grounded, an 80 percent arrester is selected. This gives an arrester rated voltage of 276 kV. From Table 20-7, following are the arrester parameters:

MCOV = 224 kV rms

TOV = 325 kV rms

FOW = 730 kV crest

Switching surge protective level = 554 kV at 2-kA switching discharge current. To ascertain LPL, read from this table, the discharge voltage kV crest for 8/20 μs wave at the qualifying current of 10 kA for 345 kV voltage. This gives 646 kV.

Also, the arrester has an energy-handling capability of 8.9 kJ/kV, pressure relief capability of 65 kA rms symmetrical, and can be located in an average ambient temperature of 45°C without any derating.

Establish the base voltage in terms of arrester rated voltage = $(362\times\sqrt{2})/\sqrt{3} = 295$ kV crest. Then, the switching surge is 740 kV and temporary overvoltage is 0.96 per unit. Referring to Fig. 20-9, the temporary overvoltage can be withstood for 9, 6, and 0.9 s with no prior energy, 0.5 per unit prior energy, and 1.0 per unit prior energy, respectively. This application is acceptable. Now, calculate the switching energy for a surge of 745 kV, similar to Example 20-1. This gives a switching current of 0.67 kA, and J = 1.80 kJ/kV. The arrester has a capability of 8.9 kJ/kV.

The PRs can now be calculated:

$$PR(1) = 1155/730 = 1.58$$

$$PR(2) = 1050/646 = 1.62$$

$$PR(3) = 870/554 = 1.57$$

These exceed the minimum required values, and the application is safe.

20-10 SURGE ARRESTER MODELS

Silicon carbide, gapless metal-oxide, and gapped metal-oxide models are supported in EMTP type 99; pseudo-nonlinear resistance and type 92 piecewise linear resistance models can be used; however, type 92 ZnO model allows single or multiple exponential segments, gapped or gapless arresters and is the preferred model for all arresters. Nonlinear resistance segments can be modeled by the following equation:

$$i = p\left(\frac{V}{V_{\text{ref}}}\right)^q \tag{20-19}$$

where p and q are the characteristics of the arrester and V_{ref} is arbitrary. It can be rated voltage, or maximum switching surge protective level (for switching surge model of the arrester). This is depicted in Fig. 20-23.

Station Class Silicon-Carbide Arresters Switching surges can be modeled with single exponential as follows:

$$\begin{aligned} &V_{\text{ref}} = \sqrt{2}V_{\text{rated}} \\ &q = 14 \\ &p = 500[V_{\text{rated}}/V_{\text{sparkover}}]^{14} \qquad V_{\text{rated}} < 48\text{ kV} \\ &p = 3000[V_{\text{rated}}/V_{\text{sparkover}}]^{14} \qquad V_{\text{rated}} > 48\text{ kV} \end{aligned} \tag{20-20}$$

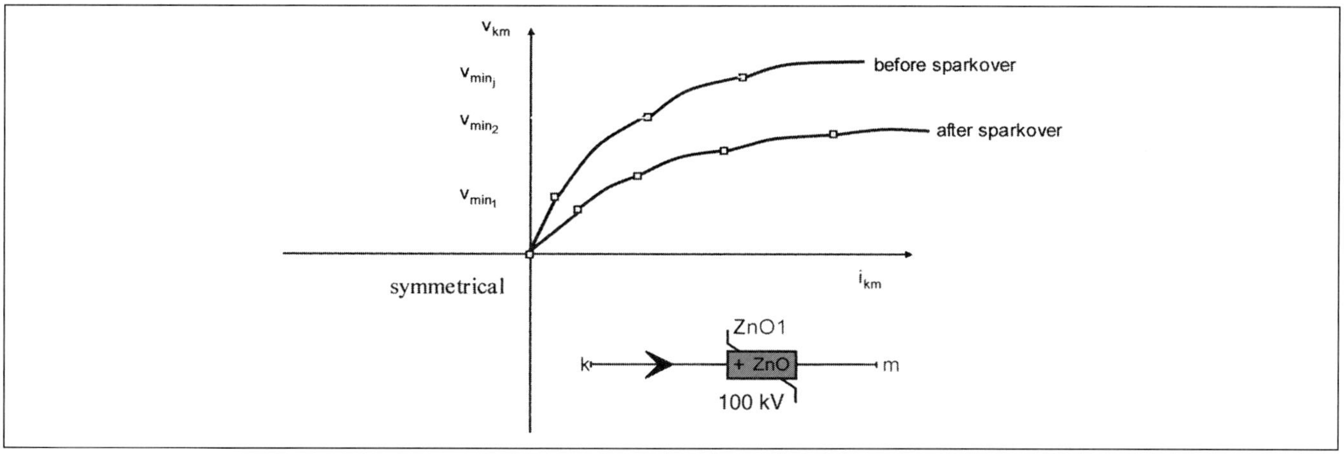

Figure 20-23 Modeling of metal-oxide surge arrester with discrete segments of different slopes.

Lightning surges can be modeled as follows:

$$V_{ref} = \sqrt{2}V_{rated}$$
$$q = 14 \qquad (20\text{-}21)$$
$$p = 1000[0.7962/S^{0.1}]^{14}$$

where S is steepness of the discharge current in kA/ms.

Station Class Metal-Oxide Arresters

$$V_{ref} = \sqrt{2}V_{rated}$$
$$p = 1000[1.0/c]^q$$
$$q = 17.2, c = 1.292 \qquad V_{rated} = 54 \text{ to } 360 \text{ kV} \qquad (20.22)$$
$$q = 21, c = 1.306 \qquad V_{rated} = 396 \text{ to } 444 \text{ kV}$$

Lightning surges can be modeled as follows:

$$V_{ref} = \sqrt{2}V_{rated}$$
$$p = 1000[1.0/cS^{1/\beta}]^q \qquad (20.23)$$

where c, q, and β are shown in Table 20-13.

Use of supporting program called ARRDAT from known VI characteristics is the most popular model.

Lightning Surge Models Double exponential model is given by the following equation:

$$f(t) = V_m[e^{\alpha t} - e^{\beta t}] \qquad (20\text{-}24)$$

The analytical equations for standard test wave shapes are given in Chap. 5 and 19. Some disadvantages of the double exponential model are that (1) it is only an approximation of measured lightning currents, (2) amplitude does not correspond to peak value of the impulse, and (3) numerical problems, that is, subtraction of two exponentials.

A single exponential model is given by:

$$i(t) = \frac{I_0}{n}\frac{K^n}{1+K^n}e^{-t/\tau} \qquad K = \frac{t}{\tau_1} \qquad (20\text{-}25)$$

where τ_1 is the wavefront time constant proportional to front duration, time interval between $t = 0$ and wave peak; t is proportional to stroke duration, time interval between $t = 0$ and point on tail where amplitude falls to 50 percent of the peak value; I_0 is the peak value; and n is factor influencing rate of rise, typically $n = 5$ to 10.

For a 8/20-μs current wave, amplitude 10 kA:

$$I_0 = 10 \text{ kA} \qquad \tau_1 = 8E6 \qquad \tau = 20E6 \qquad (20\text{-}26)$$

Any lightning, switching or test wave shape, and ramp or impulse can be modeled using analytical or graphical models, for example, point-to-point wave model.

A more accurate model of metal-oxide surge arrester can be generated for switching or lightning surges, using the manufacturer's VI characteristics from the data in Tables 20-8 and 20-9. As an example of generating this model for a 90-kV metal-oxide surge arrester for the lightning impulses, scaling factors for 8/20-μs wave are specified in Table 20-9, from 1 to 40 kA, for the voltage based on coordinating current of 10 kA. Two columns, giving maximum and minimum scaling factors, are provided. For conservatism, the maximum values are used. The result of EMTP simulation of exponents and various segments is shown in Fig. 20-24. For switching surge model, 36/90-μs data are used. Therefore, an appropriate surge arrester model for the type of surge involved should be applied. For fast transients,[6,7] the model may consist of parallel capacitance and series inductance shunted by a resistor.

Table 20-13 Constants c, q, and β for Metal-Oxide Surge Arrester Models, EMTP Empirical Parameters

Arrester-Rated Voltage (kV)	Discharge Current Range (kA)	c	q	β
60–360	3–10	1.454	31.1	17.7
	10–40	1.182	8.2	17.7
396–588	3–10	1.500	52.0	17.3
	10–40	1.350	16.9	17.3

20-11 SURGE PROTECTION OF AC MOTORS

The surge withstand capability of motors is not so well defined. A reasonable value of the impulse strength can be 125 percent of the crest value of the factory high potential test.[8] As the high voltage

Multiplier	Exponent	V_{min}
1.1861205468829285E+03	4.7938732313226595E+01	7.4695410866304812E-01
4.9074760345085809E+02	3.9080310194276308E+01	9.0517751479289887E-01
2.8291790968182090E+02	2.5549425982617397E+01	9.6011242603550218E-01
3.3416328571273164E+02	1.8958115539679803E+01	1.0255780755319184E+00
5.9797442815944339E+02	1.2710771468726582E+01	1.0976225257806505E+00
2.6233287974672799E+03	6.1496816510229406E+00	1.2527820600433313E+00
9999		

FIGURE 20-24 An EMTP model derived based on 90-kV station class surge arrester, data as shown in Tables 20-7 and 20-9.

test is conducted for 1 min with a voltage equal to twice the rated voltage of the machine + 1000 V, the BIL of motor windings is:

$$\text{BIL} = 1.25\sqrt{2}[V_{\text{nameplate}} + 1000]\text{V} \tag{20-27}$$

Now compare the BIL of a 4000-V rated motor with that of a transformer and enclosed switchgear:

- Motor BIL = 15.9 kV
- Oil-filled transformer = 75 kV
- Enclosed switchgear = 60 kV

This shows that compared to other electrical apparatus, the motor BIL is low, and furthermore, the motor windings have a complex structure. Another factor that needs to be considered is that medium-and-high voltage motors are normally controlled through vacuum contactors, which may restrike.

20-11-1 Surge Protection of Vacuum Contactors

In Chap. 8, it was mentioned that the chopping current of vacuum circuit breakers had been reduced, and the restrikes, common in the initial development of vacuum technology, were prevented by using better contact materials. However, we must distinguish between the vacuum circuit breakers and contactors, as the latter are more prone to restrikes. An area of possible multiple ignitions is shown in Fig. 20-25 with respect to locked rotor current of the motor (for application of motors up to 6.6 kV).[9] The frequency shown in this figure is given by:

$$f_n = \frac{1}{2\pi\sqrt{LC}} \tag{20-28}$$

where L is ungrounded motor inductance at locked rotor condition, l is length of the cable connection from the contactor to the motor, and C_v is capacitance of the cable per unit length ($C = lC_v$). Essentially Eq. (20-28) is a resonance formula of cable capacitance with motor locked rotor reactance. Therefore, a certain length of cable, if exceeded, can lead to reignition. This shows that (1) motor insulation is more vulnerable to stresses on account of lower BIL and (2) there is a possibility of restrikes in the switching devices. As an example, for a 500-hp motor, overvoltage protection device is needed if #4 shielded cable length, used to connect the motor to the switching vacuum contactor, exceeds approximately 470 ft. This is based on the cable data.

20-11-2 Motor Insulation

The motor insulation has two different functions: (1) there is insulation between turns, and then (2) a group of turns comprise a coil, which has to be insulated from ground. It is impractical to employ the same insulation between turns as the ground insulation of a coil. The windings present a complex system of distributed inductance and capacitance. The response is entirely different to impulse surges at high frequency, as compared to 60-Hz response. The surge phenomenon can be viewed as a circuit problem based on transmission line theory. The propagation rate is much lower through closely coupled winding in the slots of the magnetic material.

Akin to transformers, the voltage across the windings will not be uniformly distributed and the first few turns will be severely stressed. With respect to propagation velocity in the turns embedded in the slots (magnetic material), the propagation velocity is of the order of 10 to 20 m/μs, and the velocity in the end windings that overhangs in air can be ignored. This brings about the concept of electrical length of a winding expressed as the time required for the wave to transit from one end to another end of the windings.

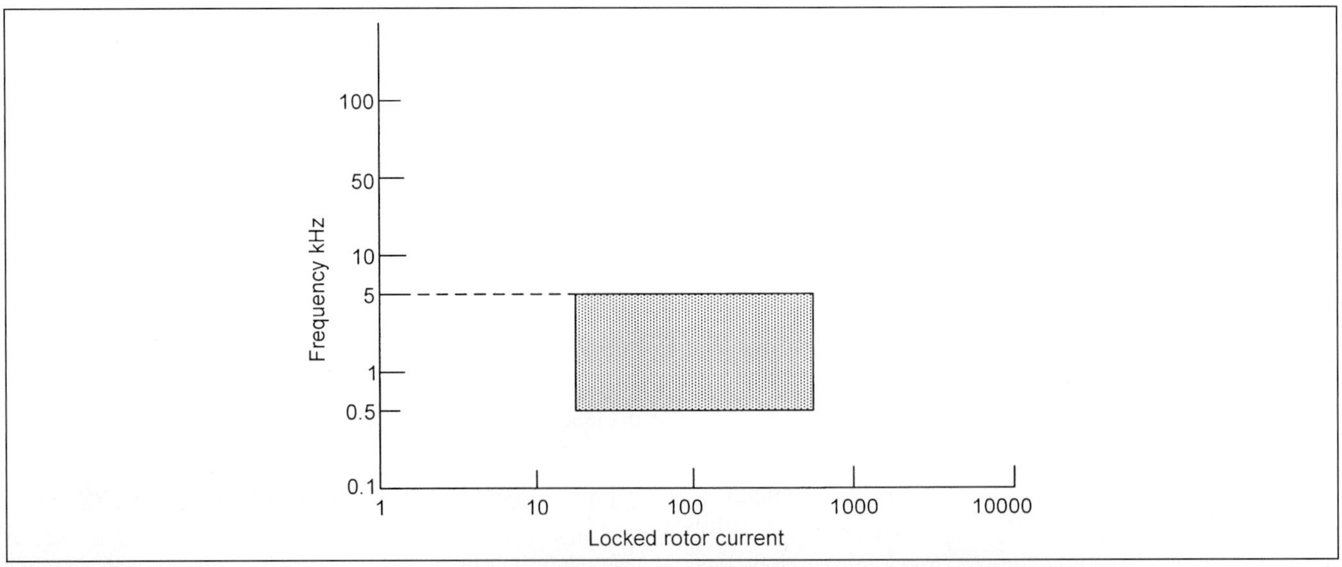

FIGURE 20-25 Zone of probable restrikes in vacuum contactors, based on the motor locked rotor current and frequency of oscillation; see text.

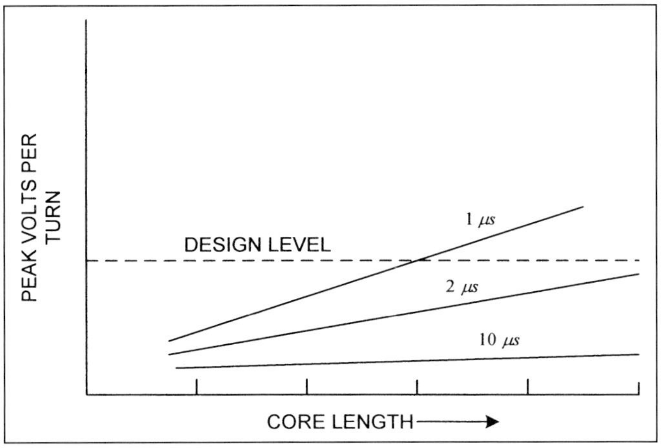

FIGURE 20-26 Effect of rise of surge voltage on the inter-turn voltage stress in motor windings versus motor core length.

Assume that a medium voltage motor has a core length of 0.5 m and a propagation velocity of 10 m/μs. A crest of approximately twice the rated motor voltage from line to ground, say 6800 V for a motor of 4.16 kV, with a rise time to crest of 1 μs impacts the motor. Then, the surge voltage between the first and second turn of the line terminal is 680 V. The steep-fronted waves concentrate on the first few turns, and will be attenuated as they progress through the winding. If the rise time of the wave is reduced, so does the voltage stresses on the turn insulation (Fig. 20-26). The standard practice has been to slope the impulse wave so that the time to reach the maximum voltage is 10 μs. The desired limitation is achieved by the application of a surge capacitor at the motor terminals. A surge capacitor of size 0.5 μF has been commonly applied (Fig. 20-27).

Example 20-8 A 0.5-μF capacitor is applied at the motor terminals and incoming surge is 6800 V, with rise time 1 μs. Most connections between the motor starter and motor will have surge impedance of the order of 20 to 50 Ω. Considering a surge impedance of 50 Ω, the voltage builds up according to the following equation:

$$E_m = E_s(1 - e^{-t/t'}) \qquad (20\text{-}29)$$

where t' is the time constant = RC = $25(10^{-6})$ s.

At $t = t_0$, when the surge arrives at the capacitor terminals, the voltage E_m is zero and the current through the capacitor is 6800/50 = 136 A peak. Therefore, the rate of change, dE_m/dt (maximum) = $136/(0.5 \times 10^{-6}) = 272$ V/μs.

For surge duration of 5 μs, and from Eq. (20-29), the voltage across the capacitor is 1233 V peak.

The effectiveness of a capacitor connected at a point other than the motor terminals will not be optimum. An incoming surge, before the capacitor becomes effective, will be reflected back from the motor terminals and will produce almost twice the rate of rise. The following design basis parameters for the surges can be adopted:[10]

1. 10 kA, 4/10 μs current wave to simulate direct lightning stroke
2. 60 kV, 1.2/50 μs voltage wave to simulate indirect lightning or induced lightning stroke
3. 1.5 kA, 8/20 μs current wave to simulate switching surge

The capacitors will impact the initial rate of rise, while the surge arresters *do not*. These are provided in parallel with capacitors to limit the surge magnitude. A LC filter is suggested in Ref. 10. With appropriate choice of components, the output from the filter, for any of three types of surges, is reduced below the motor BIL for motors of 480 to 5000 V. The inductor rating is of the order of 0.5 to 1 mH, and capacitor rating is 0.5 to 1 μF.

20-11-3 Machines Directly Connected to OH Lines

Figure 20-28 shows a protection scheme for the rotating machines directly connected to the OH lines. The surge protection capacitors and surge arresters are shown, both for the line and neutral side of the motor. The neutral-side surge arrester and capacitor connected between the common wye-connected neutral point of the motor to ground may be required in some cases. In addition, a surge arrester is shown on the line side. The arrester A_L, along with arrester A_M at machine terminals and capacitor C_S, reduces the voltage rate of rise at machine terminals to a value so that the machine turn insulation is protected. The arrester A_L discharges much of surge current and limits the voltage that is applied to the transmission line inductance. The separation distance D is of consideration, and is a function of the machine voltage class, system grounding, and ground resistance.

20-12 SURGE PROTECTION OF GENERATORS

Generators operating in synchronism with utility sources may island due to a fault or disconnection, giving rise to overvoltages.

Conventionally, surge arresters paralleled with surge capacitors are provided for the surge protection of the generators, located in the generator line terminal compartment itself. A generator, whether connected in a step-up transformer configuration, or bus connected and operated in synchronism with the utility, can be subjected to the following possible overvoltages:[11]

- *Power frequency overvoltages*. Load rejection may leave the generator connected to an OH line, cable, or capacitor, and this situation can lead to self-excitation (due to energy absorbed in

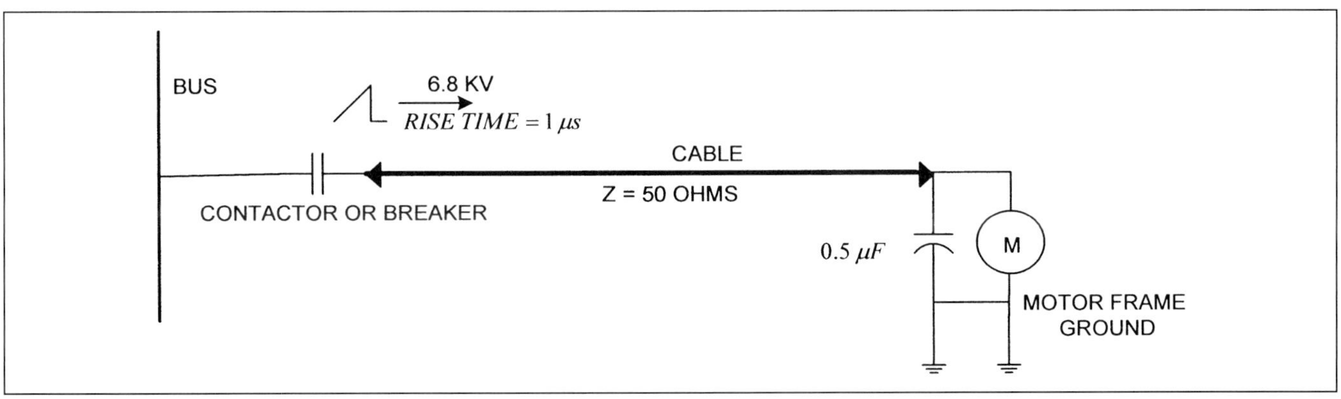

FIGURE 20-27 Surge protection of a motor with a surge capacitor alone.

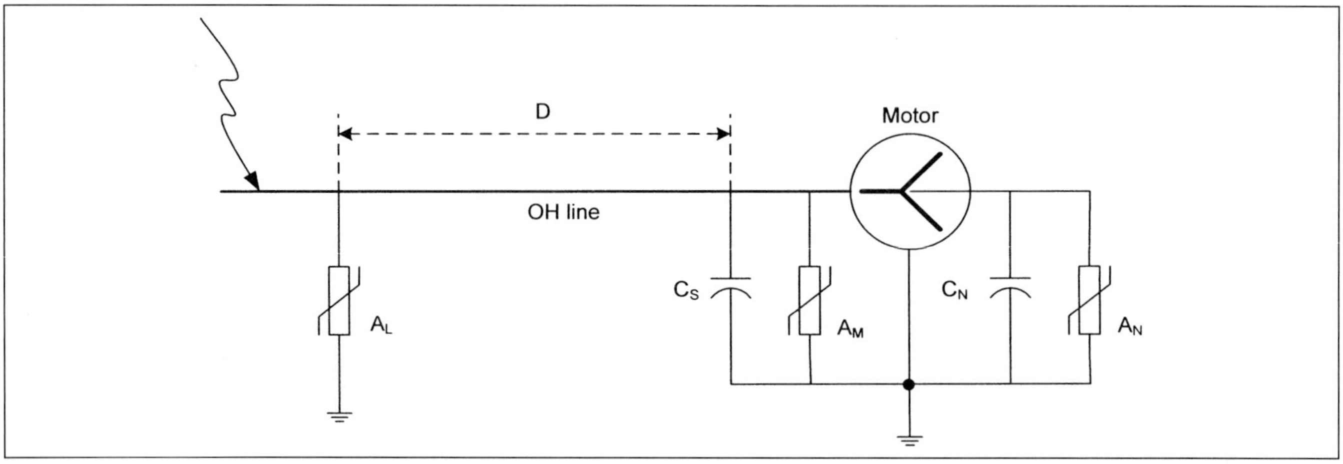

FIGURE 20-28 Surge protection of motors directly connected to an overhead line.

the system capacitance) and harmonic resonance may occur. The overvoltage will depend on the exciter response and operating reactive power. Modern exciters may limit the voltage to 1.05 per unit; however, the arrester on the high side of the transformer will operate on these overvoltages, which may reach two times or more.

- A neutral resistor left out of service can result in ferroresonance overvoltages, involving the capacitance of generators and potential transformers. These overvoltages may reach 2.4 per unit.

- An HV to LV fault in the step-up transformer can present large stresses in the generator.

20-13 SURGE PROTECTION OF CAPACITOR BANKS

Certain considerations for application of surge arresters to capacitor banks are enumerated in Chaps. 6 and 8. Surge arresters can be used to provide protection for lightning transients, switching transients, interruption of capacitor currents, recovery voltages, restrikes, and capacitor switching.

Lightning surges Capacitor banks with neutrals grounded will be charged by the lightning stroke current, and when protected by a metal-oxide surge arrester, the arrester will limit this voltage to its protective level. At the termination of stroke, the surge arrester will cease to conduct, leaving some charge on the capacitor. The discharge energy is of consideration. For an ungrounded capacitor bank, relatively little energy is added to the capacitors and high-discharge energy surge arresters are not required.

Depending on the capacitor bank size and location, these may be self-protecting, especially if surge arresters for lightning protection are provided to protect other equipment close to the capacitor banks, and substation shielding is in place. In any case, a simulation is required considering substation layout, bus connections, transformers, surge arresters, and capacitor banks. The energy on backflashovers and shielding failures and its impact on overvoltages on capacitors can be estimated by detailed simulation.

An important factor is the origin of surge on the transmission lines. Capacitor banks located in shielded substations will see a surge propagated from a remote location, and the surge energy will be small. The limiting magnitude of the surge current is:

$$I = \frac{\text{CFO}}{Z} \tag{20-30}$$

If the time to half value is 400 μs, CFO is 2500 kV for 500-kV system, and surge impedance is 350 Ω, then surge current is 7.14 kA, and the energy in the surge is:

$$Q = I\tau/2 = 1.47\text{C} \tag{20-31}$$

A capacitor bank reduces the transient voltage caused by a lightning surge. An ungrounded bank ties the three phases together, reducing the equivalent surge impedance and transient voltage caused by the surge. A grounded wye-bank provides low impedance path, slowing the surge considerably. The energy dissipated in the arrester can be estimated from:

$$E = \int vi\,dt \tag{20-32}$$

If the discharge voltage of the arrester is assumed constant with current, then approximately:

$$E = v\int i\,dt = vQ \tag{20-33}$$

For 1.47 C, as calculated in Eq. (20-31), and a surge arrester of, say 276 kV, discharge voltage at 10 kA = 646 kV crest, the energy in kJ/kV is 3.44, which is well within the capability of the arrester. The arrester has a rating of 7 kJ/kV and is safely applied for a surge having a charge of 3.5 C. The charge in a direct lightning stroke on the conductor close to the location of capacitor bank can be higher. Ungrounded wye-connected capacitor banks have very little effect on the surge arrester.

Switching Surges In Chap. 6, some switching transients and the overvoltages produced due to connection and disconnection of the capacitor banks were discussed. In Chap. 8, current interruption, recovery voltages, and overvoltages due to breaker operation in presence of a ground fault were discussed. The location of arresters to control these transients were also covered in these chapters. Here, it is highlighted that for switching transients, irrespective of neutral grounding and circuit configuration, the energy requirements of the arrester must be checked for proper application with proper system modeling. The following equation may be used for estimate purposes:

$$E = \frac{1}{2}C(V_s^2 - V_p^2)\ \text{J} \tag{20-34}$$

where V_s is the surge voltage and V_p is the protective level of the arrester. There may be a need for multiple-column arresters in certain applications.

With some reiteration of the discussions in earlier chapters, the following areas indicate specific considerations:

1. The capability of primary and backup switchgear to limit TRV and possibility of restrikes. The breaker technology has come a long way to produce restrike-free breakers, but this possibility of a restrike cannot be entirely eliminated. Switching transient of various origins can occur, but restrikes result in the highest arrester duty. The arresters minimize the risk of restrikes and are an insurance against unforeseen resonant conditions. The worst-case scenario is a two-phase restrike, with full charge on the capacitors due to a previous operation.

2. Higher phase-to-phase voltages on transformer windings can occur when a capacitor is energized, and a transformer remains connected to the capacitor on a radial line.[12]

3. To control voltages due to resonance when switching capacitors in series or parallel with transformers or other capacitors at lower voltages, surge arresters are required; see Chap. 6.

4. To control overvoltage on inductively coupled lower voltage systems, surge arresters are required; see Chaps. 6 and 14.

The transient voltage on capacitor bank and the recovery voltage across the switching device can be reduced during a restrike by installing arresters on the capacitor side of the switching device. The two-phase restrike can be considered as a conservative approach. Connecting arresters line to ground on ungrounded capacitor banks may not limit the overvoltages. Figure 20-29 shows that the arresters may be connected phase to ground, phase to neutral, or phase to phase. For a particular application, in each of these cases, the arrester characteristics, protective level, and durability on switching surges can be calculated. In this figure, V_g is the phase-to-neutral voltage of the system, Z_g can be zero to infinity, Z_s is the source impedance, the switches show two-phase restrike, L is the series inductance, and C is the capacitor bank, which in all three cases is shown ungrounded.

Figure 20-29*a* depicts arresters connected phase to ground, and these do not limit the trapped charges on the capacitors, which can be higher than 2.0 per unit. Figure 20-29*b* illustrates arresters connected phase to neutral, and these can reduce the trapped charge

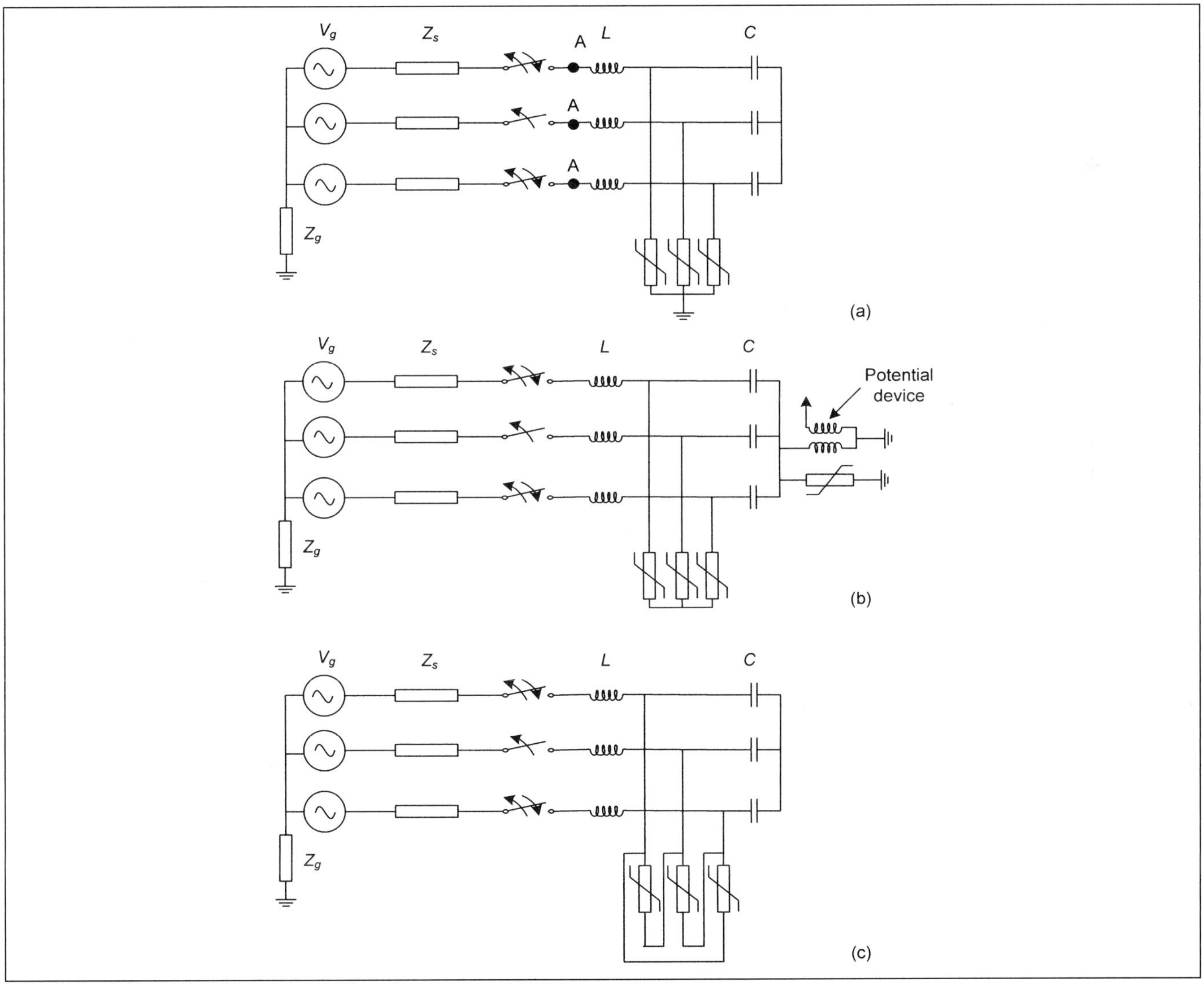

Figure 20-29 Surge protection of ungrounded wye-connected capacitor banks. (*a*) Surge arresters connected phase to ground. (*b*) Surge arresters connected to an ungrounded neutral point and a surge arrester connected neutral to ground across neutral potential device. (*c*) Delta-connected surge arresters for protection of phase-to-phase surges.

to a lower value, but the connection does not limit the neutral high-frequency overvoltages. A potential device is normally connected to monitor the fuse failure or phase unbalance in capacitor banks, and a surge arrester connected neutral to ground can be added (Fig. 20-29*b*). This will reduce the recovery voltage across the switching device and the possibility of a restrike. Figure 20-29*c* shows delta-connected surge arresters. When a series reactor is provided, either to limit inrush current or to form a single-tuned filter at some harmonic of interest, location of the surge arrester at source side of the reactor, point A in Fig. 20-29*a*, will not protect the capacitors. The voltage may be as high as two times of that at the reactor.

The MCOV selection should be based on rms voltage in the presence of harmonics. Also on switching of capacitors, the overvoltages can reach even 10 percent above system voltage for short duration.

Example 20-9 13.8 kV, 12 Mvar three-phase ungrounded wye, capacitor bank, short-circuit level at 13.8 kV = 40 kA symmetrical is simulated with two-pole restrikes of the switching device using EMTP. Figure 20-30*a* shows phase-to-phase voltages, which escalate to 2.97 per unit, and the neutral voltage in Fig. 20-30*b* rises to a peak of approximately 8 kV.

With 12.0-kV-rated metal-oxide surge arresters connected phase to neutral and a 6-kV arrester between neutral and ground, as in Fig. 20-29*b*, phase-to-phase voltages are reduced to maximum of 1.84 per unit (Fig. 20-31*a*). The power through the 12-kV surge arrester in phase *c* is shown in Fig. 20-31*b*. The energy handled is 5.04 kJ/kV. This exceeds the energy-handling capability of the arrester which is 4 kJ/kV.

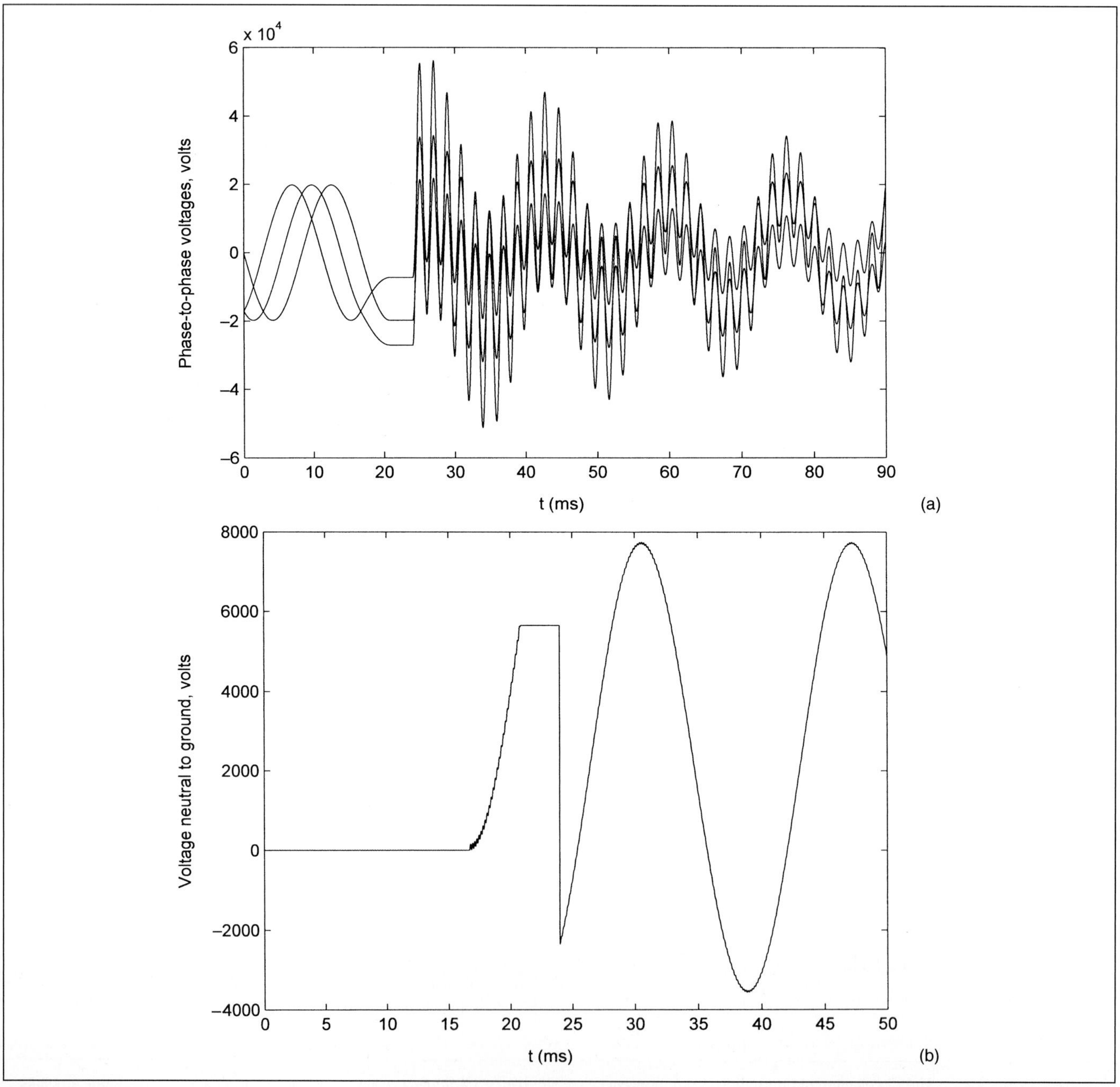

FIGURE 20-30 (*a*) Two-phase restrike (without arcing), phase-to-phase overvoltages; wye-connected ungrounded capacitor banks, Example 20-9. (*b*) Neutral-to-ground overvoltage.

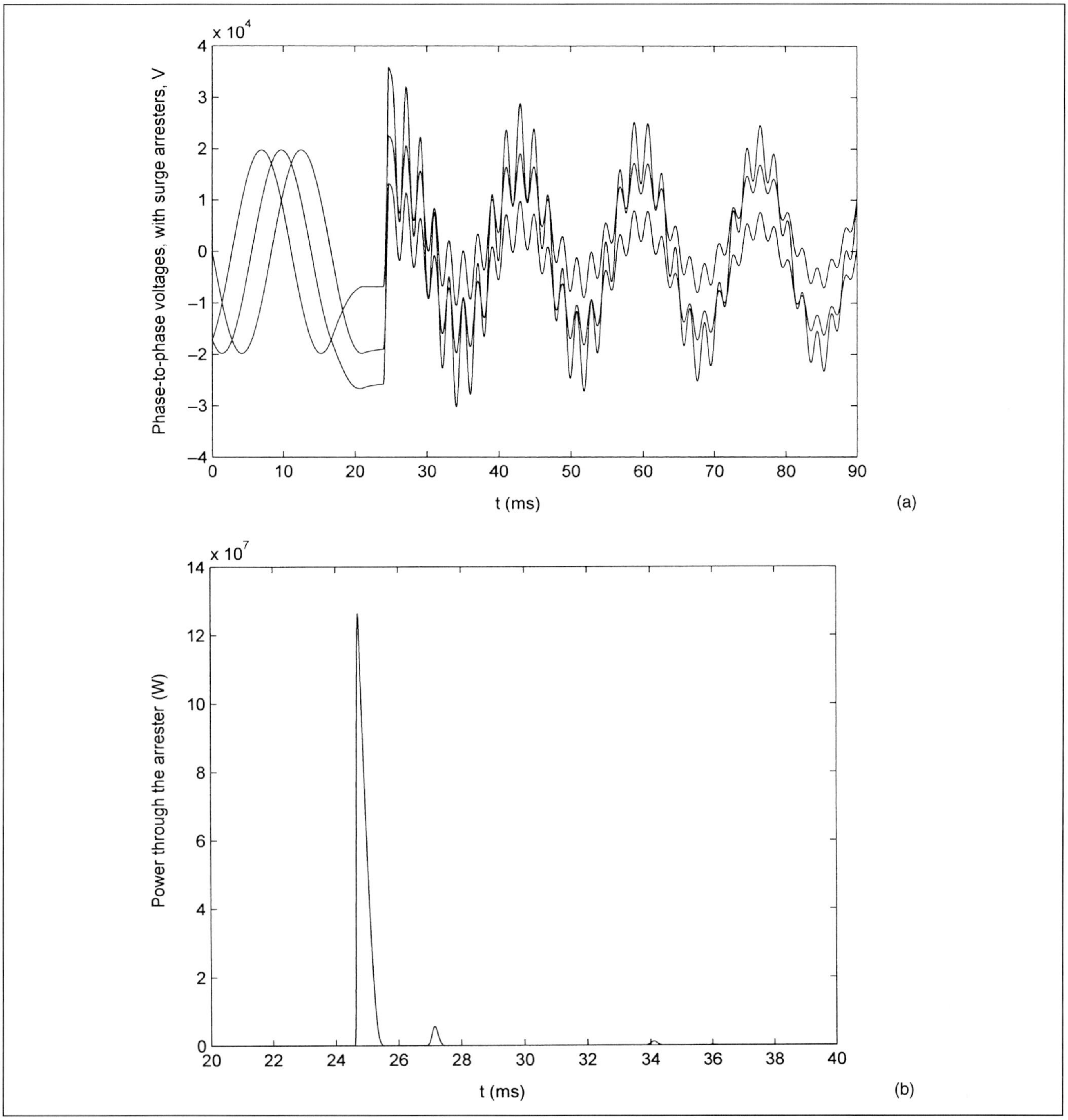

FIGURE 20-31 (*a*) Two-phase restrike (without arcing), phase-to-phase overvoltages; wye-connected ungrounded capacitor banks, with surge arresters, Example 20-9. (*b*) Power through 12-kV arresters.

20-14 CURRENT-LIMITING FUSES

A current-limiting fuse (CLF) is designed to reduce equipment damage by interrupting the rising fault current before it reaches its peak value. Within its current-limiting range, the fuse operates within 1/4 to 1/2 cycle. The total interrupting time consists of melting time (sometimes called the prearcing time) and the arcing time. The let-through current can be lower than the prospective fault current and rms symmetrical available current can be lower than the let-through peak current. Figure 20-32 shows the let-through characteristics of CLFs. Note that the threshold current, or the critical current at which the current-limiting action starts, point A in Fig. 20-32, depends on the current rating of the fuse. A noteworthy feature is that the let-through energy is much reduced, because of current limitation and fast operating time of 1/4 to 1/2 cycle.

When a CLF operates in the current-limiting range, it abruptly introduces a high resistance to reduce the fault current magnitude and duration. An arc voltage, much higher than the system voltage, is generated, which forces the current to zero. The fusible

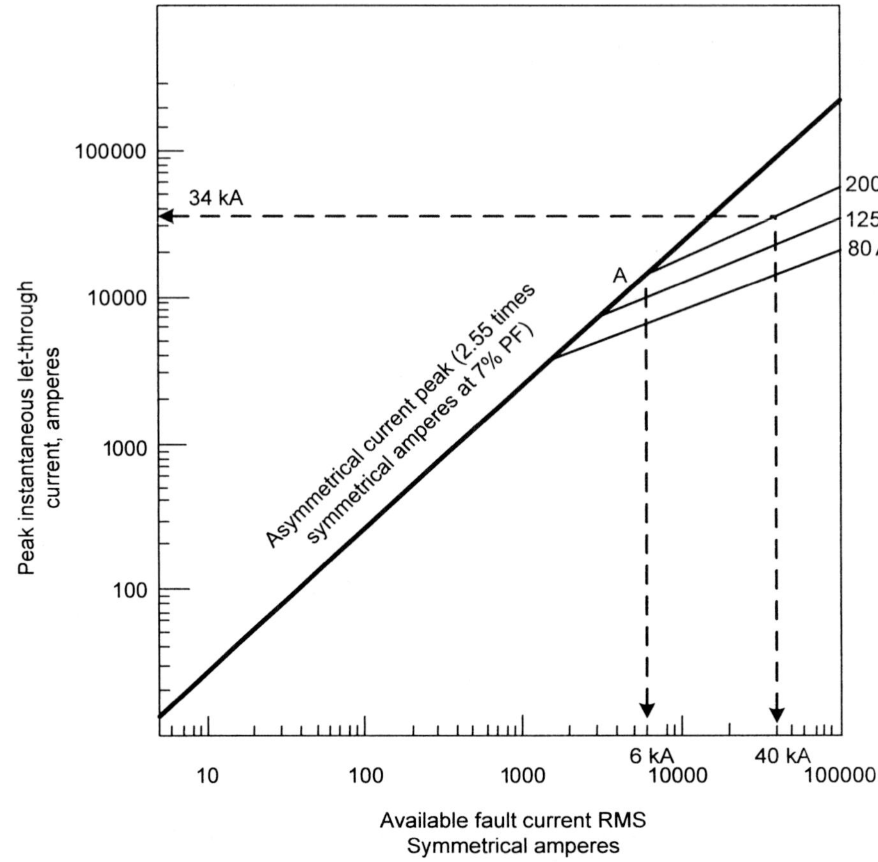

FIGURE 20-32 Let-through characteristics of current-limiting fuses.

element of nonhomogeneous cross section may be perforated or notched and while operating, it first melts at the notches, because of reduced cross-sectional area at the notch. Each melted notch forms an arc, which lengthens and disperses the element material into the surrounding medium. Controlling the size and shape of notches controls the arcs in series, and hence the rate of rise of the arc voltage and its magnitude. The voltage crests in a few hundred microseconds. The peak voltage occurs approximately at the peak let-through current. This is illustrated in Fig. 20-33. This let-through peak voltage must be limited in the design of the CLF to meet ANSI/IEEE standard requirements[13] (Table 20-14). For example, for a 15-5 kV maximum rated voltage, the voltage peak should not exceed 70 kV for CLF of rating less than or equal to 12 A, and 49 kV for CLF ≥ 12 A.

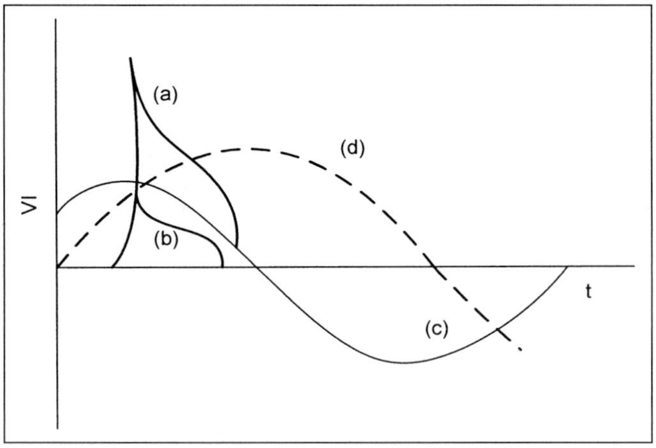

FIGURE 20-33 Arc voltage generated on operation of a current-limiting fuse. Curve (*a*) surge voltage, (*b*) peak surge current, (*c*) circuit voltage, (*d*) short-circuit current.

20-14-1 CLFs with Surge Arresters

Typically, for primary protection of substation transformers in industrial environment up to a certain kVA rating, CLFs are used. The UL listing of less inflammable liquid-immersed transformers for indoor installations is based on limiting the let-through energy that will be released into a fault in the transformer. This limitation of energy prevents explosion and damage to personnel and property.

At the same time, in these applications of CLFs, a surge arrester may be present upstream of the transformer for protection of transformer feeder or close to the transformer. It is possible that the surge arrester may be damaged due to operation of the fuse, because of the high arc voltage that is generated. Figure 20-34 is an equivalent circuit diagram for calculation of switching surge generated by the operation of a CLF and energy diverted to surge arrester.

To calculate the energy absorbed by the surge arrester, a difficulty arises as both the CLFs and the surge arrester have nonlinear voltage-current characteristics. Two solutions are possible, graphical approach and EMTP simulation. For illustrative purposes, a graphical method of calculation, followed by EMTP simulation results, is illustrated.

TABLE 20-14 Maximum Permissible Arc Voltage of Current Limiting Fuses

RATED MAXIMUM VOLTAGE	MAXIMUM PEAK ARC VOLTAGE (KV)	
	THROUGH 12 A	>12 A
2.8	13	9
5.5	25	18
8.3	38	26
15.5	70	49
22.0	117	70
27.0	123	84

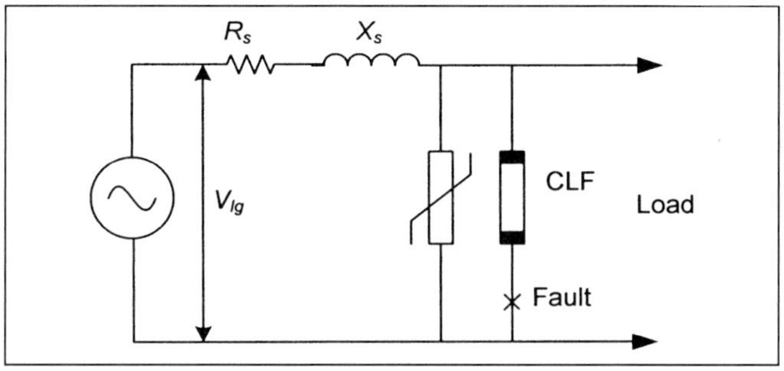

FIGURE 20-34 Circuit configuration of a surge arrester due to operation of a current-limiting fuse.

The peak arc voltage generated by a fuse should be corrected for system operating voltage and short-circuit current.[14] It is given by the following expression:

$$V_{fa} = K_v . K_i . V_{fm} \quad (20\text{-}35)$$

where V_{fa} is actual fuse arc voltage in crest; V_{fm} is maximum fuse arc voltage, from Table 20-14; K_v is voltage adjustment factor due to nonhomogeneous nature of CLF; and K_i is current adjustment factor based on the ratio of short-circuit current and critical current of the fuse.

The maximum energy will be stored in the system inductance at the instant the CLF begins to interrupt the fault current. When the arc voltage and the system voltage are coincidental, the current begins to be limited. This energy is given by:

$$E_s = \frac{1}{2} L_s I_{lt}^2 \quad (20\text{-}36)$$

TABLE 20-15 Division of Current Between Surge Arrester and Current-Limiting Fuse

VOLTAGE	I_{clf}	I_a
0	0	34000
5	1113	32887
10	3744	30256
20	12591	21409
25	18607	15393
30	25600	8400
35.28	34000	0

where E_s is energy stored in the system inductance in J, L_s is system reactance in H, and I_{lt} is peak let-through current of CLF in A for the available short-circuit current.

The system impedance can be approximately calculated from the following equation:

$$Z_s = \frac{K_0 V_{lg}}{I_f} \quad (20\text{-}37)$$

where K_0 is overvoltage factor, V_{lg} is line-to-neutral voltage in kV rms, I_f is fault current in kA, and Z_s is system impedance.

The nonlinear characteristics of the fuse can be represented by:

$$I_{clf} = K_F V_{clf}^{1.75} \quad (20\text{-}38)$$

where I_{clf} is instantaneous fuse current in A, V_{clf} is instantaneous fuse arc voltage in kV, and K_F is fuse arc voltage constant. This constant is given by:

$$K_F = \frac{I_{lt}}{V_{fa}^{1.75}} \quad (20\text{-}39)$$

where V_{fa} and I_{lt} have been defined earlier.

The current division between the surge arrester and fuse is calculated by establishing the operating point on their VI curves. Then, the arrester current is given by:

$$I_a = I_{lt} - I_{clf} = I_{lt} - K_F V_{clf}^{1.75} \quad (20\text{-}40)$$

Example 20-10 A 200E CLF, rated voltage = 15.5 kV, is applied for the primary (13.8 kV) protection of a 2500 kVA, 13.8 to 2.4-kV transformer, and the 13.8-kV system is low resistance-grounded. Overvoltage factor = 1.06, maximum arc voltage = 49 kV peak, as per ANSI/IEEE standards, critical current from the fuse let-through characteristics = 6.5 kA rms (Fig. 20-32), and maximum system line-to-ground voltage V_{lg} = 8.45 kV. The available short-circuit current at 13.8-kV system is 40 kA rms. The adjusted arc voltage V_{fa} = 35.28 crest, according to Eq. (20-35), all adjustment factors not shown.

As the 13.8-kV system is resistance grounded, an arrester of rated voltage of 15 kV is selected, characteristics as in Table 20-7.

The let-through current for a fault of 40 kA sym is 34 kA peak (Fig. 20-32). From Eq. (20-36), the energy stored in the system is 305 kJ. The 15-kV station class surge arrester has a single-shot energy capability of 60 kJ. As the energy stored in the system is greater than the energy-handling capability of the arrester, the division of energy should be calculated.[15]

From Eq. (20-39), $K_F = 34000/(35.28)^{1.75} = 66.57$. The division of current is shown in Table 20-15. Now, the VI characteristics of the surge arrester based on data in Table 20-7 and arc voltage versus I_{clf} of the fuse calculated in Table 20-15 can be plotted together, and the crossing of these two characteristics gives the operating point. This gives a voltage of 30 kV and current of 6 kA. Voltage in terms of fuse arc generated voltage = 31/35.28 = 0.88, and current in terms of fuse let-through = 6/34 = 0.176.

Another graph is needed to calculate the energy through the arrester. This is plotted in Fig. 20-35. Enter the curve on the *x*-axis at calculated value of 0.176, go to peak source voltage curve (per unit of operating point voltage, maximum curve is for 0.7, while the calculated value here is 0.88), and read off from the vertical axis the arrester energy as a percentage of the total energy. Approximately 20 percent of the energy passes through the arrester, that is 61 kJ, which is close to its maximum single-shot capability. An EMTP simulation using Fig. 20-34 gives 66 kJ.

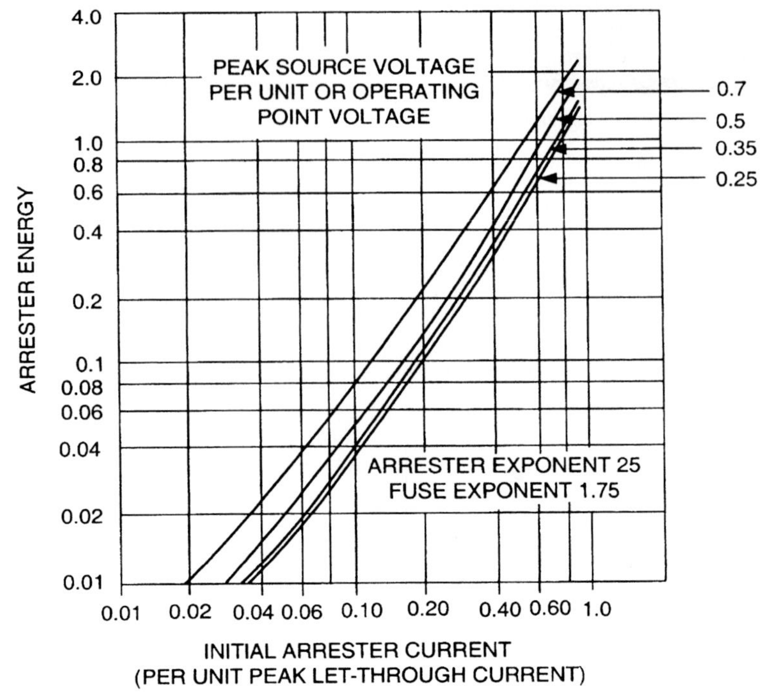

FIGURE 20-35 Arrester energy versus initial current for metal-oxide surge arresters.

PROBLEMS

1. How does the surge protection of motors connected to OH lines differ from those connected through small cable lengths?

2. Explain the limitations of gap-type arresters for protection of capacitor banks.

3. Draw a freehand sketch of the waveform through a metal-oxide surge arrester in (1) low-current region, less than 1 A and (2) high-current region.

4. A gapless arrester is selected for a certain application. It meets all the requirements of insulation coordination and PR ratios, except the energy-handling capability. What can be done? Discuss all the possible alternatives and their ramifications, say, Examples 20-9 and 20-10.

5. The following system data are provided for selection of a surge arrester for a 138-kV system:

- The system is solidly grounded.
- The temporary overvoltages are 1.2 times lasting for 10 s.
- The COG is 1.1 on a phase-to-ground fault, ground fault clearance time = 1 s.
- On load rejection, a temporary overvoltage of 120 percent occurs for 5 s.
- For selection of TOV, the prior energy is 0.5 per unit.

Using the tables in this chapter, select rated voltage, MCOV, and TOV of a gapless metal-oxide arrester.

6. How is the maximum stability voltage related to TOV? Which class of arrester, distribution, intermediate, or station is better?

7. To obtain a better PM, the discharge voltage of a gapless metal-oxide arrester is required to be reduced. What can be done?

8. In Fig. 20-20, $d_a = 40$ m, $d_{es} = 100$ m, switching surge-1000 kV/μs. Surge arrester discharge voltage at 10 kA = 284 kV crest, line length = 300 km. What is the voltage at the protected equipment (transformer T1)?

9. In Prob. 8, BIL of power transformer is 550 kV, system voltage is 138 kV. Using the tables in this chapter, find CWW and the gapless arrester-rated voltage, MCOV, TOV, FOW, and calculate the three PRs. Are these within the acceptable limits specified in IEEE standards?

10. In Prob. 9, plot the values similar to Fig. 20-22.

11. What is the limitation of a two-exponential model of a surge arrester in EMTP?

12. Select a station class, gapless zinc-oxide surge arrester for a 400-kV system based on the following data:

System solidly grounded, power frequency overvoltages = 1.07 per unit, TOV with 0.5 per unit prior energy, and a temporary overvoltage of 1.4 per unit for 1s. Maximum switching surge level = 2 per unit, line length = 600 km, surge impedance = 400 Ω. Use the tables and figures in this chapter.

13. How do the phase-to-phase overvoltages at a remote location, connected through a transmission line, arise in capacitor bank switching? What is the nature of phase-to-ground overvoltages? Will the voltages be higher or lower in a system with high short-circuit power?

14. A single-tuned capacitor filter bank, wye-connected, has ungrounded neutral. Draw a sketch showing the optimum locations of the surge arresters. What is the purpose of a surge arrester connected from the capacitor bank neutral to ground?

15. Plot the CLF and surge arrester VI curve in Example 20-10 and verify the operating point calculated in this example.

16. In Example 20-9, the size of the capacitor bank is increased to 20 Mvar. Will it impact the phase-to-phase voltage profiles?

REFERENCES

1. ANSI/IEEE Standard C62.11, IEEE Standard for Metal Oxide Surge Arresters for AC Power Circuits >1 kV, 2005 (IEEE Standard C62.11a, 2008).
2. ANSI/IEEE Standard C62.1, Standard for Gapped Silicon Carbide Arresters for AC Power Systems, 1989.
3. ANSI/IEEE Standard C62.2, Guide for the Application of Gapped Silicon-Carbide Surge Arresters for Alternating Current Systems, 1997.
4. M. V. Lat, "A Method for Performance Prediction of Metal Oxide Arresters," *IEEE Trans. PAS*, vol. 104, no. 10, pp. 2665–2675, Oct. 1985.
5. IEEE Standard C62.22, Guide for Application of Metal Oxide Surge Arresters for Alternating Current Systems, 1997.
6. D. W. Durbak, "Zinc Oxide Arrester Model for Fast Front Surges," *EMTP Newsletter*, vol. 5, no. 1, Jan. 1985.
7. W. Schmidt, J. Meppelink, B. Richter, K. Feser, L. E. Kehl, and D. Qiu, "Behavior of Metal Oxide Surge Arrester Blocks to Fast Transients," *IEEE Trans. PD*, pp. 292–300, Jan. 1989.
8. IEEE Committee Report, "Impulse Voltage Strength of Rotating Machines," *IEEE Trans. PAS*, vol. 100, no. 8, pp. 4041–4053, Aug. 1981.
9. S. F. Frag and R. G. Bartheld, "Guidelines for Application of Vacuum Contactors," *IEEE Trans. IA*, vol. 22, no. 1, pp. 102–108, Jan./Feb. 1986.
10. S. M. Dillard and T. D. Greiner, "Transient Voltage Protection for Induction Motors Including Electrical Submersible Pumps," *IEEE Trans. IA*, vol. IA-23, no. 2, pp. 365–370, Mar./Apr. 1987.
11. E. P. Dick, B. K. Gupta, J. W. Porter, and A. Greenwood, "Practical Design of Generator Surge Protection," *IEEE Trans. PD*, vol. 6, no. 2, pp. 736–743, Apr. 1991.
12. P. Kirby, C. C. Erven, and O. Nigol, "Discharge Capacity of Metal Oxide Valve Elements," *IEEE Trans. PD*, pp. 1656–1665, Oct. 1988.
13. ANSI Standard C37.46, ANSI Specifications for Power Fuses and Disconnect Switches, 1981.
14. M. V. Lat, Application Guide for Surge Arresters on Distribution Systems, Research Report H3Z 2p9, Canadian Electrical Association, 1988.
15. J. C. Das, "Coordination of Lightning Arresters and Current Limiting Fuses," *IEEE Trans. IA*, vol. 38, no. 3, pp. 744–751, May/Jun. 2002.

FURTHER READING

J. E. Harder, A. E. Hughes, and J. Vosicky, "Analytical Method for Coordination of Surge Arresters with Current Limiting Fuses," *IEEE Trans. IA*, vol. IA-17, no. 5, pp. 445–453, Sep./Oct. 1981.

A. R. Hileman and K. H. Weck, "Protection Performance of Metal Oxide Surge Arresters," *Electra*, pp. 133–146, Dec. 1990.

IEC Standard 60099, Part 4, Metal Oxide Surge Arresters Without Gaps for AC Systems, 2006–07.

IEC Standard 60099, Part 5, Selection and Application Recommendations, 2000–03.

IEEE Switchgear Committee and Surge Protective Devices Working Group, *IEEE Trans. PAS*, vol. PAS-91, pp. 1075–1078, May/Jun. 1972.

IEEE Standard C62.23, IEEE Application Guide for Surge Protection of Electric Generating Plants, 1995.

IEEE Standard 1299/C62.22.1, IEEE Guide for the Connections of Surge Arresters to Protect Insulated Shielded Electrical Power Cable Systems, R2003.

IEEE Working Group 3.4.11, "Modeling of Metal Oxide Surge Arrester," *IEEE Trans. PD*, pp. 302–309, Jan. 1992.

IEEE Working Group 3.4.17, "Impact of Shunt Capacitor Banks on Substation Surge Environment, and Surge Arrester Applications," *IEEE Trans. PD*, vol. 11, no. 4, pp. 1798–1807, Oct. 1996.

D. W. Jackson, "Analysis of Surge Capacitor Lead Connections for the Protection of Motors," *IEEE Trans. PAS*, vol. 103, no. 9, pp. 2605–2611, 1984.

R. A. Jones and H. S. Fortson, Jr., "Considerations of Phase-to-Phase Surges in the Application of Capacitor Banks," *IEEE Trans. PD*, vol. 1, no. 3, pp. 240–245, Jul. 1986.

M. P. McGranaghan, W. E. Reid, S. W. Law, and D. W. Gresham, "Overvoltage Protection of Capacitor Banks," *IEEE Trans. PAS*, vol. 103, no. 8, pp. 2326–2336, Aug. 1984.

NEMA LA-1, Surge Arresters, 1999.

J. Osterhout, "Comparison of IEC and U.S. Standards for Metal Oxide Surge Arresters," *IEEE Trans. PD*, pp. 2002–2006, Oct. 1992.

CHAPTER 21

TRANSIENTS IN GROUNDING SYSTEMS

The grounding systems can be studied under two classifications: (1) system grounding and (2) equipment grounding. System grounding refers to the electrical connection between the phase conductors and ground and dictates the manner in which the neutral points of wye-connected transformers and generators or of artificially derived neutral systems through delta-wye or zig-zag transformers are grounded. The equipment grounding refers to the grounding of the exposed metallic parts of the electrical equipment, which can become energized and create a potential to ground—say due to a breakdown of insulation or fault—and can be a potential safety hazard. The safety of the personnel and human life is of importance. Also in the previous chapters we noted that the grounding of lightning arresters, low-voltage surge protection devices, and tower footing resistance impact the performance, backflashovers, and control of overvoltages. Thus, low resistance of grounding systems is important on both the counts—personal safety and system considerations. Ninety percent of the soil in the United States has a resistivity of approximately 90 Ω-m . Low resistance in soils of high resistivity can be achieved by special devices and soil treatment, as is demonstrated in Example 21-3. The methods of system grounding are:

- Solidly grounded systems
- Low-resistance grounded systems
- High-resistance grounded systems
- Reactance grounded systems
- Ungrounded systems

We can add resonant grounding systems (reactance fault neutralizers) to this categorization, and we will briefly study the characteristics of each of these grounding systems. Figure 21-1 shows basic connections of various grounding methods. Table 21-1 summarizes the grounding methods with respect to system voltage levels, though variations do exist. The system grounding also varies in various parts of the world, according to the established engineering practices in a country.

21-1 SOLID GROUNDING

In a solidly grounded system, there is no intentional impedance between the system neutral and ground. A power system is solidly grounded when the generator, power transformer, or grounding transformer neutral is directly connected to the ground. Note that a solidly grounded system is not a zero impedance circuit due to the sequence impedances of the grounded equipment itself. These systems, in general, meet the requirements of an "effectively grounded" system in which ratio X_0/X_1 is positive and less than 3.0 and ratio R_0/X_0 is less than 1, where X_1, X_0, and R_0 are the positive-sequence reactance, zero-sequence reactance, and zero-sequence resistance, respectively. The coefficient of grounding was defined in Chap. 9. To recapitulate, COG is the ratio of E_{Lg}/E_{LL} in percentage, where E_{Lg} is the highest rms voltage on an unfaulted phase, at a selected location, during a fault effecting one or more phases to ground and E_{LL} is the rms phase-to-phase power frequency voltage obtained at that location with the fault removed. These systems are, generally, characterized by COG of 80 percent. Approximately, a surge arrester with its rated voltage calculated on the basis of the system voltage multiplied by 0.8 can be applied. We will consider solidly grounded systems in two distinct situations in the following sections.

21-1-1 Grounding of Utility EHV, HV, and Distribution Systems

The utility systems at transmission, subtransmission, and distribution levels are solidly grounded. The main reason for this is that on occurrence of a ground fault, enough ground fault current should be available to selectively trip the faulty circuit. The utility generators, connected in step-up configuration to a generator transformer, are invariably high resistance grounded (Fig. 21-1). If a generator neutral is left ungrounded, there is a possibility of generating high voltages through inductive-capacitive couplings (see Sec. 21-3). Ferroresonance can also occur due to the presence of generator PTs.

The utility substations serving large chunks of power at high voltages for industrial plants through delta-wye transformers have wye windings low-resistance grounded. The most common secondary voltages of distributions for the industrial plants are 13.8 kV, 4.16 kV, and 2.4 kV.

21-1-2 Low-Voltage Industrial Distribution Systems

The low-voltage systems in industrial power distribution systems used to be solidly grounded. However, this trend is changing and high resistance grounding is being adopted. The solidly grounded

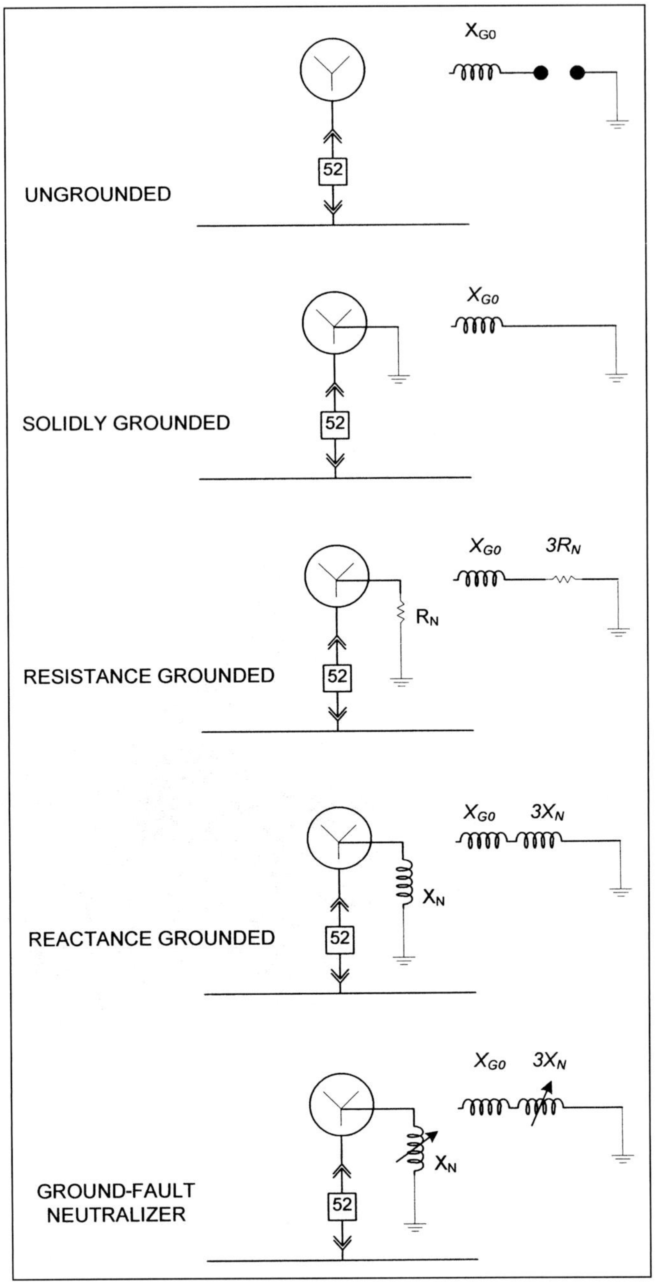

Figure 21-1 Methods of system grounding.

systems have an advantage of providing effective control of overvoltages, which become impressed or are self-generated in the power system by insulation breakdowns and restriking faults. Yet, these give the highest arc fault current and consequent damage and require immediate isolation of the faulty section. Single-line-to-ground fault currents can be higher than the three-phase fault currents. These high magnitudes of fault currents have twofold effect:

- Higher burning or equipment damage
- Interruption of the processes, as the faulty section must be selectively isolated without escalation of the fault to unfaulted sections

The arcing faults are caused by insulation failure, loose connections, or accidents. The arc behavior is difficult to predict and model because of spasmodic nature of the arc fault. This is due to elongation and blowout effects and arc reignition. Arc travels from point to point and physical flexing of cables and structures can occur. Arcing faults can exhibit low levels because of the impedance of the arc fault circuit itself. Arcing faults can be discontinuous requiring a certain minimum voltage for reignition. The limits of the acceptable damage to material for arc fault currents of 3000 to 26000 A in 480-V systems have been established by testing[1,2] and are given by the following equation:

$$\text{Fault damage} \propto (I)^{1.5}t \tag{21-1}$$

where I is the arc fault current and t is the duration is seconds.

$$V_D = K_s(I)^{1.5}t(\text{in})^3 \tag{21-2}$$

where K_s is the burning rate of material in in^3 /As$^{1.5}$, V_D is acceptable damage to material in in^3, I is the arc fault current, t is the duration of flow of fault current, and K_s depends upon type of material and is given by:

$$\begin{aligned} K_s &= 0.72 \times 10^{-6} \text{ for copper} \\ &= 1.52 \times 10^{-6} \text{ for aluminum} \\ &= 0.66 \times 10^{-6} \text{ for steel} \end{aligned} \tag{21-3}$$

NEMA[1] assumes a practical limit for the ground fault protective devices, so that:

$$(I)^{1.5}t \lhd 250I_r \tag{21-4}$$

I_r is the rated current of the conductor, bus, disconnect, or circuit breaker to be protected.

Combining these equations, we can write:

$$V_D = 250K_sI_r \tag{21-5}$$

As an example, consider a circuit of 4000 A. Then the NEMA practical limit is 1.0×10^6 (A)$^{1.5}$ s and the permissible damage to copper, from Eq. (21-5) is 0.72 in^3. To limit the arc fault damage to this value, the maximum fault clearing time can be calculated. Consider that the arc fault current is 20 kA. Then, the maximum fault clearing time, including the relay operating time and breaker interrupting time, is 0.35 s. It is obvious that vaporizing 0.72 in^3 of copper on a ground fault, which is cleared according to established standards, is still determinant to the operation of the equipment. Shutdown and repairs will be needed after the fault incidence.

The arc fault current is not of the same magnitude as the three-phase fault current, due to voltage drop in the arc. In the low-voltage 480-V systems, it may be 50 to 60 percent of the bolted three-phase current, while for medium-voltage systems it will approach three-phase bolted fault current, but somewhat lower. An arcing fault releases incident energy measured in cal/cm^2, and this can cause fatal body burns to a worker who may happen to be maintaining the electrical apparatus. In recent times, there has been much emphasis on limiting the arc flash hazard, calculating it precisely, and using appropriate personal protective equipment (PPE) for personal safety and avoiding fatal arc-flash burns. This subject is not discussed here and an interested reader may see Refs. 3 and 4.

Due to high arc fault damage and interruption of processes, the solidly grounded systems are not in much use in the industrial distribution systems.

However, ac circuits of less than 50 V, circuits of 50 to 1000 V for supplying premises wiring systems, and single-phase 120/240-V control circuits must be solidly grounded according to NEC.[5]

Figure 21-2 shows a sustained arc fault current in a 3/16-in gap in a 480-V three phase system.[2] Experimentally, an arc is established

TABLE 21-1 System Grounding with Respect to Voltage Levels

System Voltage	System Grounding Methods				
		Impedance			
	Solid	Low *R*	High *R*	Reactance	Resonant
Extra high voltage, >345 kV	X				
High voltage, 115 to 230 kV	X				
Medium voltage, 2.4 to 69 kV, utility systems	X				X*
Medium voltages, 2.4, 4.16, 13.8, 23.0 kV, industrial systems		X			
Medium voltages, 2.4, 4.16 kV, industrial systems			X		
Low voltages <1000 V	X		X		
Circuits <50 V, control circuits 120/240 V (single phase) and circuits of 50–1000 V for premises wiring, as per NEC	X				
Utility generators connected through step-up transformers			X		
Industrial bus-connected generators	X†	X‡	X‡	X§	

*Not in common use in the United States; some European countries and Russia use this method. Generally applied at 15 kV and above for distribution systems.

†For small single-phase machines and three-phase machines on low-voltage and medium-voltage systems, say up to 1000 kVA.

‡Grounding of bus-connected industrial generators through low resistance typically to limit the ground current to 200 or 400 A has been a common practice. Recent trends are toward hybrid grounding—a combination of HR and low-resistance grounding.

§Not in common use in the United States.

The ungrounded systems are not shown in this Table, as these are not in use. In a power system, tripping of a grounded source may result in a section of the system becoming ungrounded. Appropriate protective measures are required, for example, protection through neutral displacement tuned voltage relays to 60 Hz. Possibility of ferroresonance exists in this scenario.

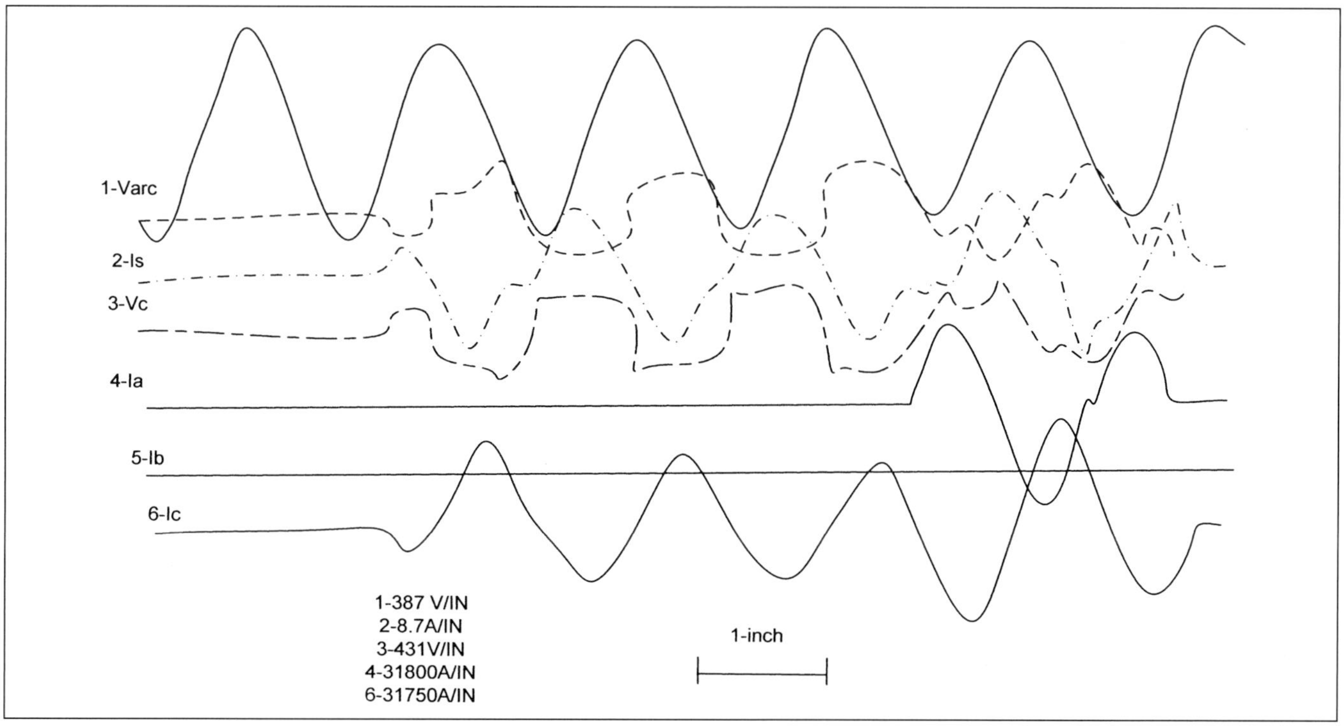

FIGURE 21-2 Arc fault in a 3/16-in gap, 480-V system.

between phase *c* of the bus and ground, and a current of 1100 A flows. After three-cycle, phase *a* is involved and the arc current for two-line to enclosure is 18,000 A. Arc energy equals 7790 kW-cycles.

21-2 RESISTANCE GROUNDING

An impedance grounded system has a resistance or reactance connected in the neutral circuit to ground, as shown in Fig. 21-1. In a low-resistance grounded system, the resistance in the neutral circuit is so chosen that the ground fault is limited to approximately full-load current or even lower, typically 200 to 400 A. The arc fault damage is reduced, and these systems provide effective control of the overvoltages generated in the system by resonant capacitive-inductive couplings and restriking ground faults (see Sec. 21-3). Though the ground fault current is much reduced, it cannot be allowed to be sustained, and selective tripping must be provided to isolate the faulty section. For a ground fault current limited to 400 A, the pickup sensitivity of modern ground fault devices can be even lower than 5 A. Considering an available fault current of 400 A and the relay pickup of 5 A, approximately 98.75 percent of the transformer or generator windings from the line terminal to neutral are protected. This assumes a linear distribution of voltage across the winding. (Practically, the pickup will be higher than the low set point of 5 A.) The incidence of ground fault occurrence toward the neutral decreases as square of the winding turns. Medium-voltage distribution systems in industrial distributions are commonly low-resistance grounded.

The low-resistance grounded systems are adopted at medium voltages, 13.8 kV, 4.16 kV, and 2.4 kV for industrial distribution systems. Also industrial bus-connected generators are commonly low-resistance grounded. An industry practice has been, generally, to limit the fault current to 400 A, or lower in some cases. A recent trend in industrial bus-connected medium-voltage generator grounding is hybrid grounding system.[6]

21-2-1 High-Resistance Grounded Systems

High-resistance grounded systems limit the ground fault current to a low value, so that an immediate disconnection on occurrence of a ground fault is not required. It is well documented that to control over voltages in the high-resistance grounded systems, the grounding resistor should be so chosen that

$$R_n = \frac{V_{\text{ln}}}{3I_c} \tag{21-6}$$

where V_{ln} is the line-to-neutral voltage and I_c is the stray capacitance current of each phase conductor. Figure 21-3 shows transient voltage in percent of normal line-to-ground crest voltage versus the resistor kW/charging capacitive kVA. The transients are a minimum when this ratio is unity. This leads to the requirement of accurately calculating the stray capacitance currents in the system.[7] Cables, motors, transformers, surge arresters generators all contribute to the stray capacitance current. Surge capacitors connected line to ground must be considered in the calculations. Once the system stray capacitance is determined, then, the charging current per phase, I_c, is given by:

$$I_c = \frac{V_{\text{ln}}}{X_{c0}} \tag{21-7}$$

where X_{c0} is the capacitive reactance of each phase, stray capacitance considered lumped together.

This can be illustrated with an example. A high-resistance grounding system for a wye-connected neutral of a 13.8–0.48 kV transformer is shown in Fig. 21-4*a*. This shows that the stray capacitance current of all the distribution system connected to the secondary of the transformer is 0.21 A per phase, assumed to be

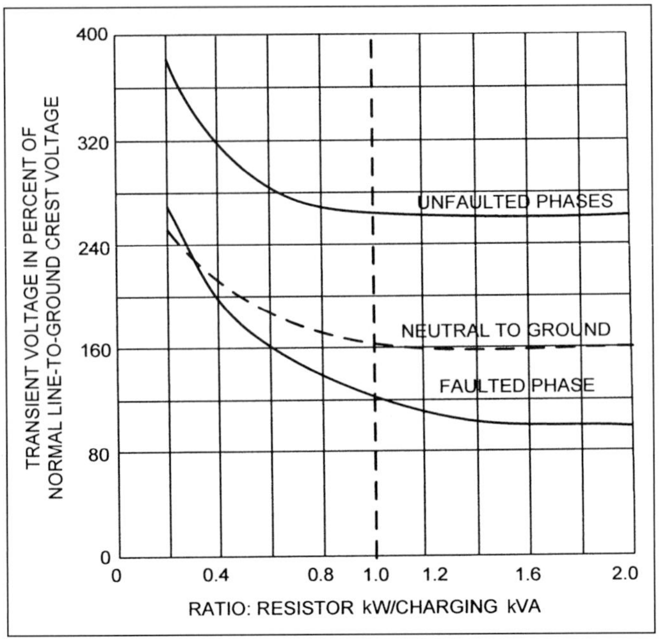

FIGURE 21-3 Overvoltages versus ratio of resistor kW/charging kVA.

balanced in each phase. Generally, for the low-voltage distribution systems, a stray capacitance current of 0.1 A per MVA of transformer load can be taken, though this rule of thumb is no substitute for accurate calculations of stray capacitance currents. Figure 21-4*a* shows that under no fault condition, the vector sum of three capacitance currents is zero, as these are 90° displaced with respect to each voltage vector, and therefore, 120° displaced with respect to each other. Thus, the grounded neutral does not carry any current and the neutral of the system is held at the ground potential, Fig. 21-4*b*. As:

$$I_{c1} + I_{c2} + I_{c3} = 0 \tag{21-8}$$

no capacitance current flows into the ground. On occurrence of a ground fault, say in phase *a*, the situation is depicted in Fig. 21-4*c* and *d*. The capacitance of faulted phase *a* is short-circuited to ground. The faulted phase, assuming zero fault resistance is at the ground potential (Fig. 21-4*d*), and the other two phases have line-to-line voltages with respect to the ground. Therefore, the capacitance current of the unfaulted phases *b* and *c* increases proportional to the voltage increase, that is, $\sqrt{3} \times 0.21 = 0.365\,\text{A}$. Moreover, this current in phases *b* and *c* reverses and flows through the transformer windings and sums up in the transformer winding of phase *a*. Figure 21-4*e* shows that this vector sum = 0.63 A.

Now consider that the ground current through the grounding resistor is limited to 1 A only. This is acceptable according to Eq. (21-6) as the stray capacitance current is 0.63 A. This resistor ground current also flows through transformer phase winding *a* to the fault and the total ground fault current is $I_g = \sqrt{1^2 + 0.63^2} = 1.182\,\text{A}$ (Fig. 21-4*e*).

The above analysis assumes a full neutral shift, ignores the fault impedance itself, and assumes that the ground grid resistance and the system zero-sequence impedances are zero. Practically, the neutral shift will vary (Fig. 21-5).

Fault Detection, Alarms, and Isolation As the ground fault currents are low, special means of fault detection and isolation are required. Figure 21-6*a* and *b* show this. In Fig. 21-6*a*, alarm/trip can be provided through neutral-connected voltage relays, device 59, connected across a part of the neutral grounding resistor.

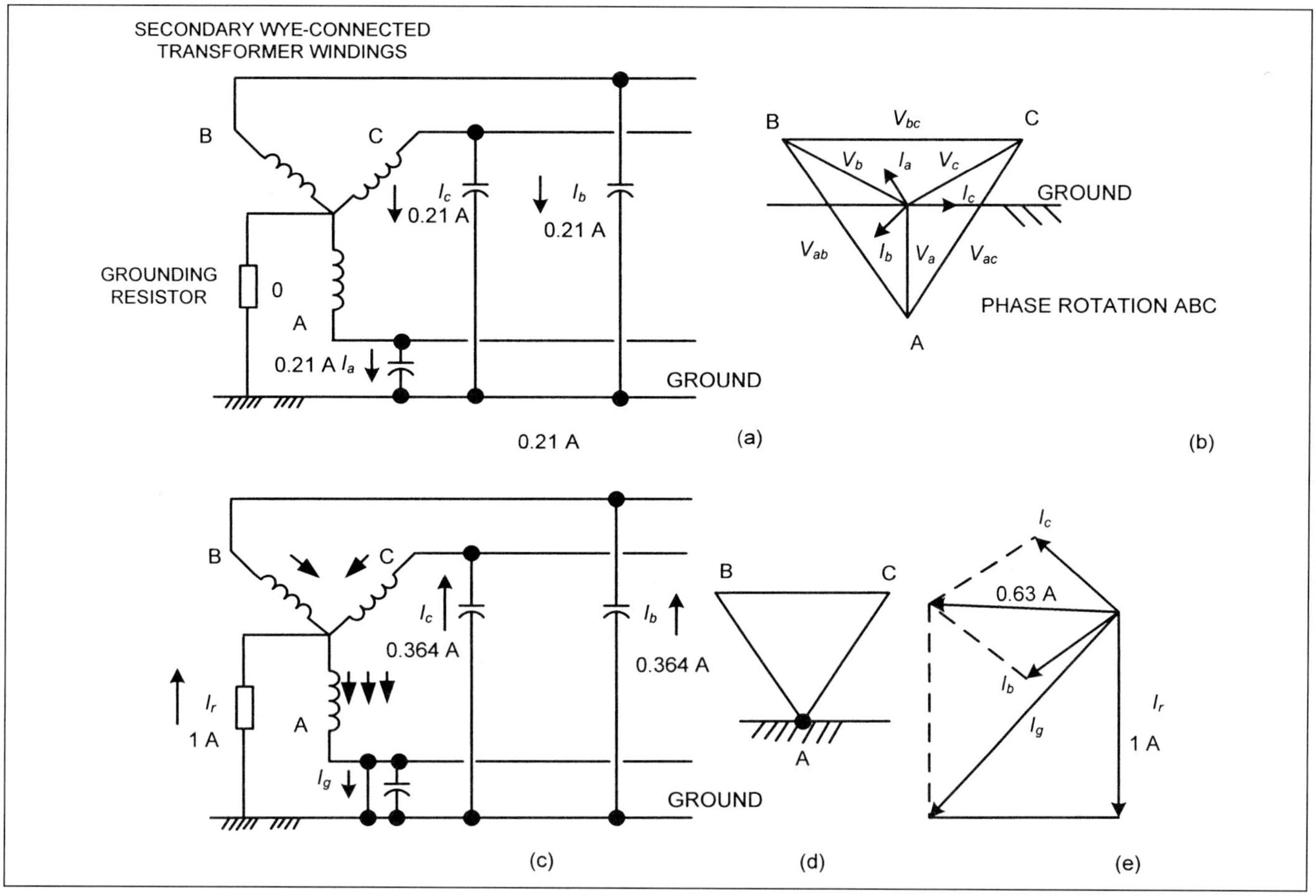

FIGURE 21-4 (*a*) and (*b*). The stray capacitance currents and voltages in a low-voltage wye-connected high-resistance grounding system under no-fault conditions. (*c*) The flow of capacitance and ground currents, phase *a* faulted to ground. (*d*) Voltages to ground for phase *a* grounded. (*e*) Phasor diagram of summation of capacitance and resistor currents.

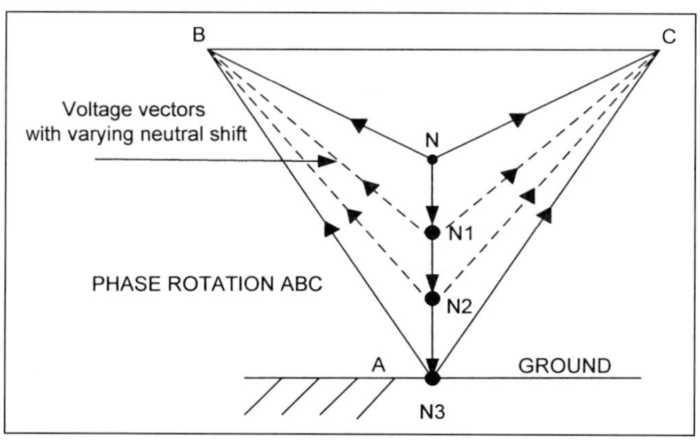

FIGURE 21-5 Phase-to-phase and phase-to-ground voltages on a line-to-ground fault, with varying degree of neutral shifts.

This relay should preferably be a rectifier type of relay to sense harmonic currents that may flow through the ground resistor. Consider a feeder ground fault; the flow of capacitive and resistive components of the current are shown in Fig. 21-6*a*. The sensitive ground fault sensor and relay on each unfaulted feeder see the capacitive current related to that feeder only, but the sensor and relay on the faulted circuit see total ground fault current less the feeder's own capacitive current. The sensitive ammeter can monitor the capacitive current in each feeder and indicate the state of the system by monitoring the capacitance current.

Figure 21-6*b* shows a ground fault localization scheme, popular in the industry and called a "pulsing-type high-resistance grounding system." The pulses are created by a current sensing relay and cyclic timer, 62, at a frequency of approximately 20/min by alternatively shorting and opening a part of the grounding resistor through a contactor. These can be traced to the faulty circuit with a clip-on ammeter.

Advantages of HR Systems Some obvious advantages are:

- The resistance limits the ground fault current and, therefore, reduces burning and arcing effects in switchgear, transformers, cables, and rotating equipment.
- It reduces mechanical stresses in circuits and apparatus carrying fault current.
- Reduces arc blast or flash hazard to personnel who happen to be in close proximity of ground fault.
- Reduces line-to-line voltage dips due to ground fault and three-phase loads can be served.
- Control of transient overvoltages is secured by proper selection of the resistor.

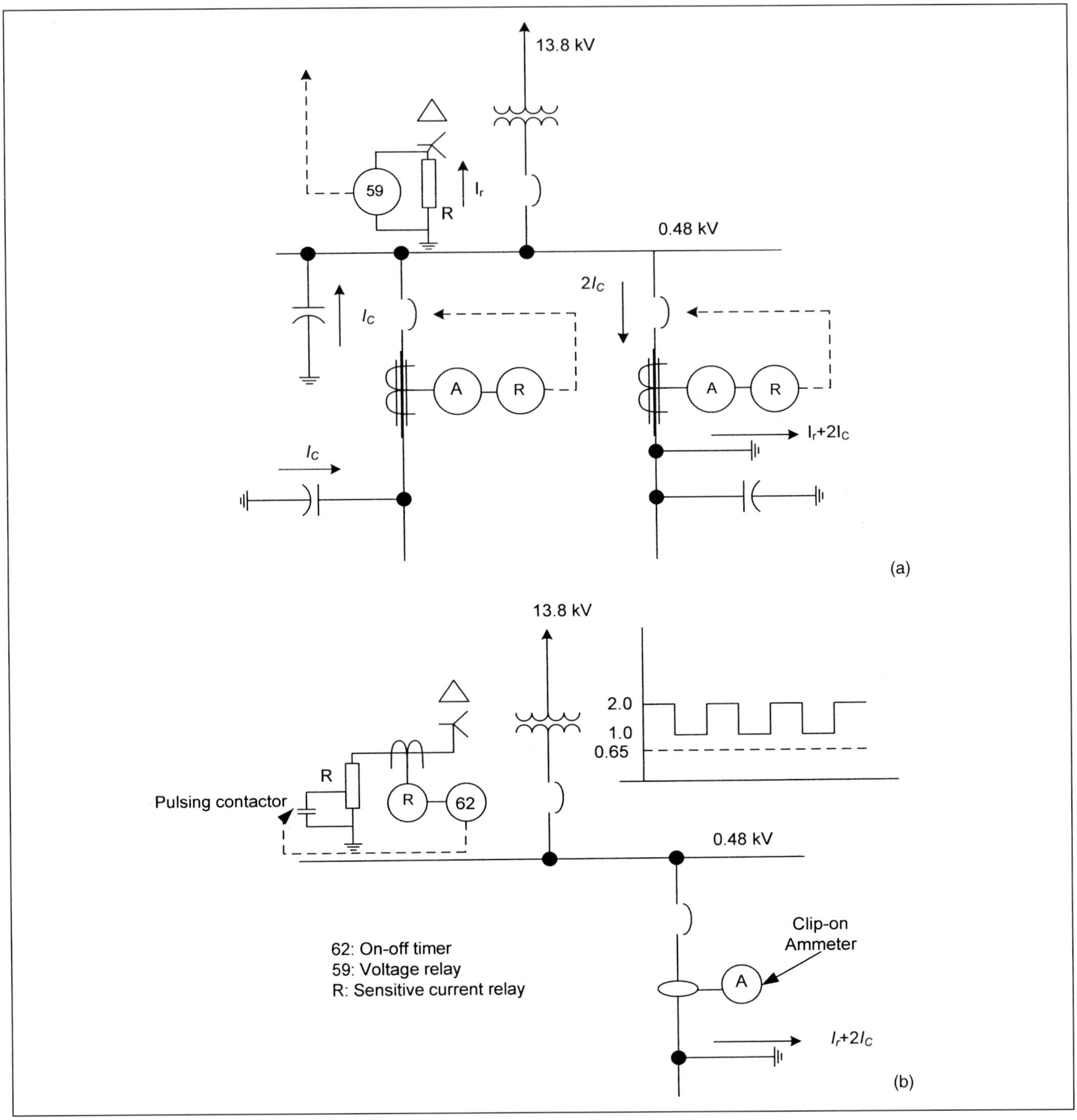

FIGURE 21-6 (*a*) Alarm and fault detection system in a high-resistance grounded system. (*b*) Pulsing-type high-resistance grounding system and fault detection.

Limitation of HR Systems The limitation of the system is that the capacitance current should not exceed approximately 10 A to prevent immediate shutdowns. As the system voltage increases, so does the capacitance currents. This limits the applications of HR system to systems of rated voltages of 4.16 kV and below.

Though immediate shutdown is prevented, the fault situation should not be prolonged, the fault should be localized and removed. There are three reasons for this:

1. Figure 21-4*d* shows that the unfaulted phases have voltage rise by a factor of $\sqrt{3}$ to ground. This increases the normal insulation stresses between phase-to-ground. This may be of special concern for low-voltage cables. If the time required to deenergize the system is indefinite, 173 percent insulation level for the cables must be selected.[8] However, NEC does not specify 173 percent insulation level, and for 600-V cables, insulation levels correspond to 100 and 133 percent. Also Ref. 8 specifies that the actual operating voltage on cables should not exceed 5 percent during continuous operation and 10 percent during emergencies. This is of importance when 600-V nominal three-phase systems are used for power distributions. The dc loads served through six-pulse converter

systems will have a dc voltage of 648 and 810 V, respectively, for 480 and 600-V rms ac systems.

2. Low levels of fault currents if sustained for long time may cause irreparable damage. Though the burning rate is slow, the heat energy released over the course of time can damage cores and windings of rotating machines even for ground currents as low as 3 to 4 A. This has been demonstrated in test conditions.[9]

3. A first ground fault left in the system increases the probability of a second ground fault on other phase. If this happens, then it amounts to a two-phase-to-ground fault with some interconnecting impedance, depending upon the fault location. The potentiality of equipment damage and burnout increases.

21-3 UNGROUNDED SYSTEMS

In an ungrounded system, there is no intentional connection to ground except through potential transformers or metering devices of high impedance. In reality, an ungrounded system is coupled to ground through distributed phase capacitances. It is difficult to assign X_0/X_1 and R_0/X_0 values for ungrounded systems. The ratio X_0/X_1 is negative and may vary from low to high values and COG may approach 120 percent. These systems provide no effective control of transient and steady-state voltages above ground. A possibility of resonance with high-voltage generation, approaching 5 times or more of the system voltage exists for values of X_0/X_1 between 0 and −40. For the first phase-to-ground fault, the continuity of operations can be sustained, though unfaulted phases have $\sqrt{3}$ times the normal line-to-ground voltage. All unremoved faults, thus, put greater than normal voltage on system insulation and increased level of conductor and motor insulation may be required. The grounding practices in the industry are withdrawing from this method of grounding.

If an inductor of certain size gets connected to ground, then the possibility of high voltages exist, as the overvoltage is given by:

$$V_{ov} = \frac{X_L}{X_{co}/3 - X_L} V_t \tag{21-9}$$

where V_t is the applied terminal voltage, X_L is the inductance of the grounded inductor. Figure 21-7 shows three possible cases of a high inductance that may get connected to a phase and give rise to resonant voltages.

1. The coil of a motor starter may be inadvertently connected between phase and ground due to a ground fault.

2. A fuse can operate in one phase due to a ground fault, this connects the reactances of the other two phases in parallel between phases and ground.

3. A broken grounded conductor on the load side of the transformer connects reactances of two phases in parallel between phase and ground.

The phenomena of arcing grounds and resulting overvoltages which may escalate to 5 to 6 times the normal rated voltage can be explained with reference to Fig. 21-8. It is somewhat similar to restrikes discussed in Chap. 8. Intermittent ground faults can give rise to these phenomena.

Figure 21-8*a* shows normal voltage vectors rotating counterclockwise. Consider that phase *a* is grounded, as shown in Figure 21-8*b*. The current in a capacitor is zero when the voltage is at its peak, and therefore at the instant shown in Fig. 21-8*b* the capacitance is charged to the line voltage and the current is zero, thus the arc tends to extinguish. During the next half cycle, as the voltage vectors rotate, the phase *a* voltage changes from zero to twice the line voltage, dotted lines. This value of line-to-ground potential of phase *a* may be sufficient to break down the gap in the ground fault circuit, which got extinguished half cycle before. Thus, a pulse current flows and the phase *a* voltage may swing between plus and minus 2 times the rated voltage at a frequency of 20 to 100 times the fundamental, due to presence of reactance in the circuit. If it was a solid metallic connection between phase *a* and ground, it would leave the phase *a* conductor at ground potential. Associated with the transitory oscillation of the voltage, there will be a corresponding oscillatory charge current. This transient charging current or restrike current will again reach zero when the system voltage triangle is at its maximum excursion in the negative direction (lower

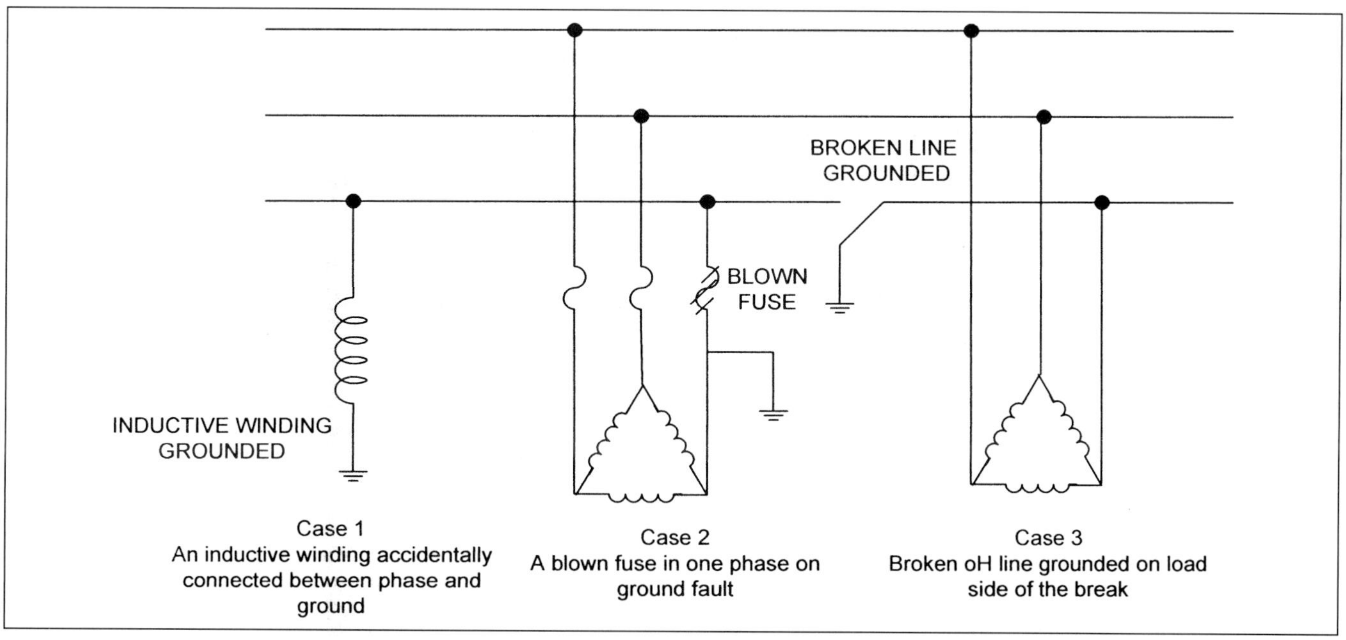

FIGURE 21-7 Connections of an inductance to ground for faults in ungrounded systems.

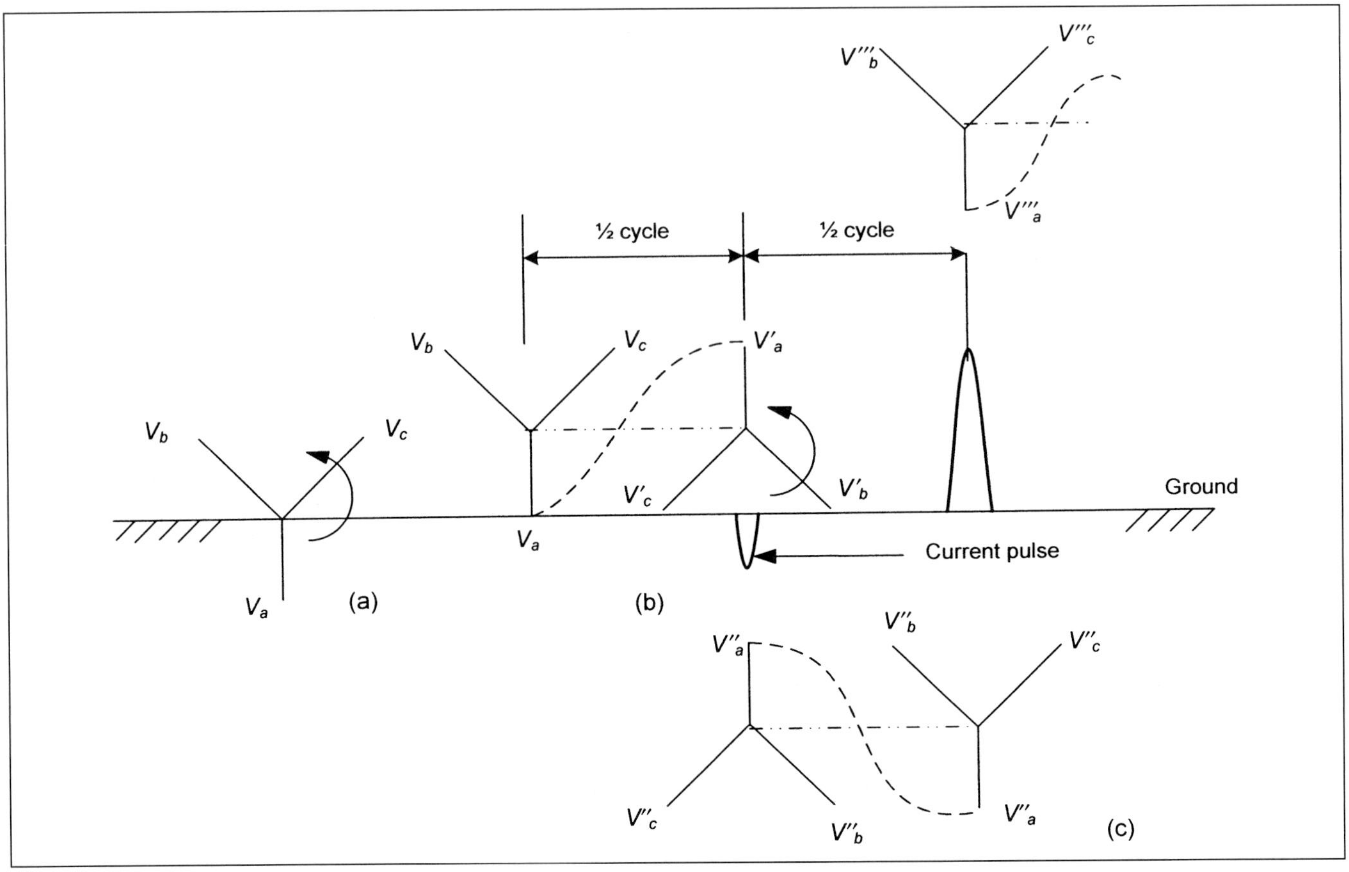

FIGURE 21-8 (*a*), (*b*), and (*c*). Illustration of arcing grounds in ungrounded systems and consequent development of high voltages.

part of Fig. 21-8*c*). In the next half cycle as the voltage vectors rotate further, the phase *a* voltage will escalate from –2 to –4, as indicated in lower part of Fig. 21-8*c*. This increases voltage across the gap, which may again result in a restrike. Also see capacitor restrikes discussed in Chap. 8.

21-4 REACTANCE GROUNDING

In reactance grounding, a reactor is connected between the system neutral and ground, the magnitude of the ground fault current that will flow depends upon the size of the reactor. The ground fault current should be at least 25 percent and preferably 60 percent of the three-phase fault current to prevent serious transient overvoltages (X_0 < or equal to $1.0X_1$). This current is considerably higher than in a resistance grounded system, and the reactance grounding is not an alternative to resistance grounding. The system is generally used for grounding of small generators, so that the generator ground fault current does not exceed three-phase fault current and three-phase four-wire loads could be served. Reactance grounded systems are not common in the United States.

21-4-1 Resonant Grounding

Figure 21-1 also shows a resonant grounding system, ground-fault neutralizer. A reactor can be connected between the neutral of a system and ground and tuned to the system charging current so that the resulting current is resistive and of low magnitude. This current is in phase with the line-to-neutral voltage, so that the current and voltage zeros occur simultaneously. The system is used for voltages above 15 kV, consisting of overhead transmission or distribution lines. The system is rarely used for industrial and commercial establishments. A disadvantage of the system is that the resonant tuning can change due to switching conditions, that is, when a part of the system may be out of service or when the system expansion takes place. This grounding method is not common in the United States, though sometimes used in Europe (Sweden) and Russia.

21-4-2 Artificially Derived Neutrals

Many times it is required to ground delta-connected transformer windings and other ungrounded separately derived systems. A neutral can be artificially derived in a delta-connected system with zigzag transformer or a wye-delta-connected grounding transformer. Figure 21-9*a* shows a zigzag transformer. Windings a_1 and a_2 are on the same limb and have the same number of turns, but are wound in opposite directions. The zero-sequence currents in the two windings on the same limb have, therefore, canceling ampere turn effect. The impedance to the zero sequence currents is that due to leakage flux of the windings. For the positive and negative sequence currents, neglecting magnetizing currents, the connection has infinite impedance. Figure 21-9*b* shows distribution of zero-sequence currents on a ground fault ahead of a zigzag transformer. A wye-delta grounding transformer circuit can be similarly drawn. The neutrals of large utility generators directly connected to the step-up transformers (GSUs) are not directly connected to ground through a high resistance but through a distribution transformer which is loaded on the secondary windings with a grounding resistor. An example will clarify the calculations.

Example 21-1 The example illustrates deriving an artificial neutral through a wye-delta grounding transformer in a delta-connected system. Figure 21-10*a* shows a 5-MVA, 34.5 to 2.4-kV, delta-delta

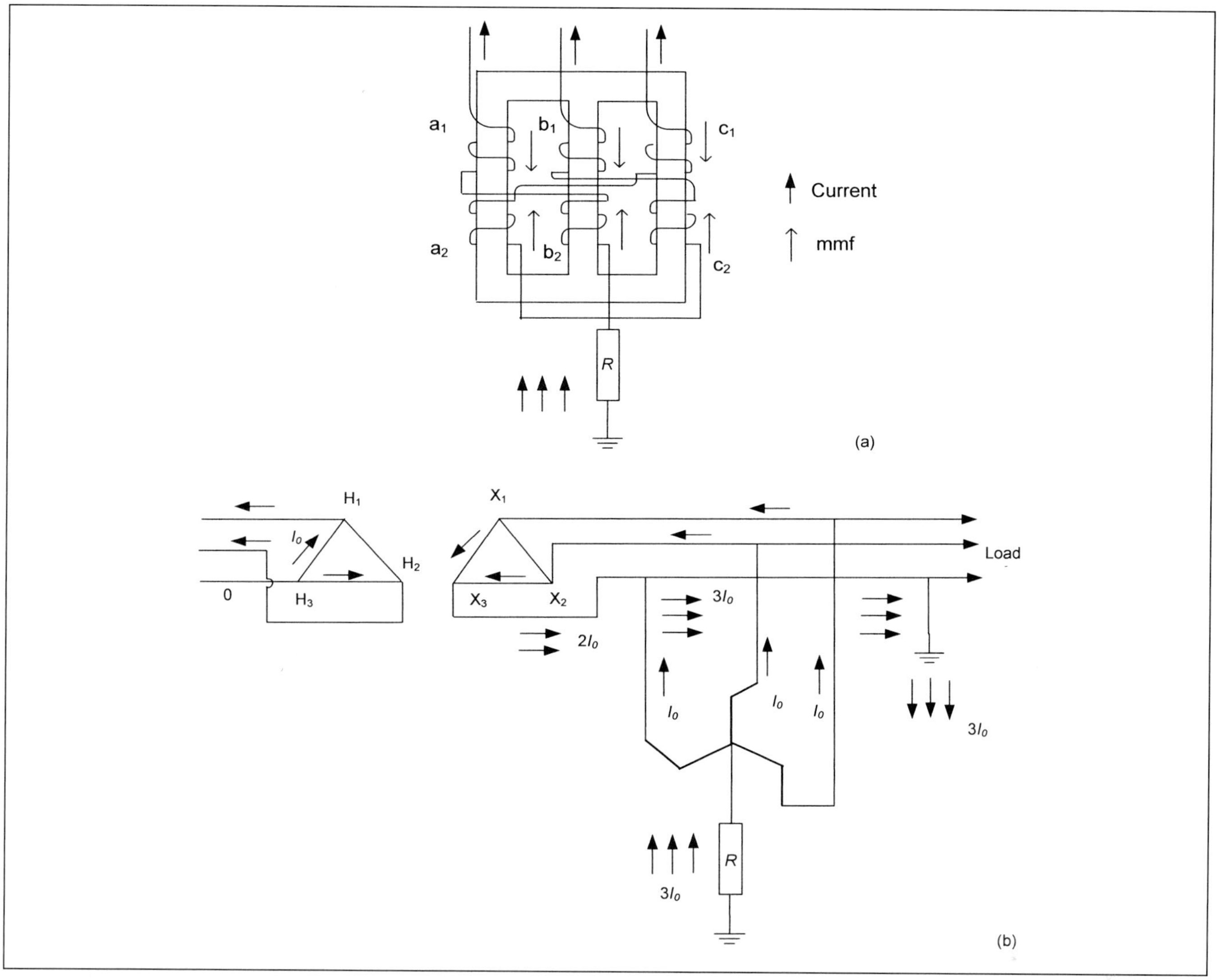

FIGURE 21-9 (*a*) Flow of ground fault currents in the windings of a zig-zag transformer. (*b*) Application of a zig-zag transformer to derive an artificial neutral and current flows for a downstream ground fault.

transformer serving industrial motor loads. The wye-delta distribution grounding transformer is loaded with a secondary resistor.

Given that the capacitance charging current of the system is 8 A, and the system and equipment data is as shown in Table 21-2, it is required to calculate the value of the resistor to limit the ground fault current to 10 A. Neglect transformer resistance and source resistance. The ground current has to be higher than the capacitance current Eq. (21-6).

The connection of sequence networks is shown in Fig. 21-10*b* and the given impedance data is reduced to a common 100-MVA base, as shown in Table 21-2. Note that the zero sequence impedance of the motors is infinite as the motor wye-connected neutrals are left ungrounded. This is the common practice in the United States. (This contrasts with grounding practices in some European countries, where the motor neutrals are grounded.) The source zero-sequence impedance can be calculated based on the assumption of equal positive and negative sequence reactances. The motor voltage is 2.3 kV, and therefore its per unit reactance on 100-MVA base in per unit is given by:

$$\left(\frac{16.7}{1.64}\right)\left(\frac{2.3}{2.4}\right)^2 = 9.35$$

Similarly, the grounding transformer per unit calculations should be adjusted for correct voltages:

$$X_0 = \left(\frac{1.5}{0.06}\right)\left(\frac{2.4}{2.4/\sqrt{3}}\right)^2 = 75$$

The equivalent positive and negative sequence reactances are: 1.41 per unit (pu) each. The zero-sequence impedance of the grounding transformer is $50 + j75$ per unit. The total fault current should be limited to $10 - j8 = 12.80$ A. Thus, the required impedance is:

$$Z_t = \left(\frac{2400/\sqrt{3}}{12.8/3}\right) = 324.8\,\Omega$$

The base ohms (100-MVA base) = 0.0576. The required Z_t = 324.8/ base ohms = 5638.9 per unit. This shows that the system positive and negative sequence impedances are low compared to the desired total impedance in the neutral circuit. The system positive and negative sequence impedances can, therefore, be neglected.

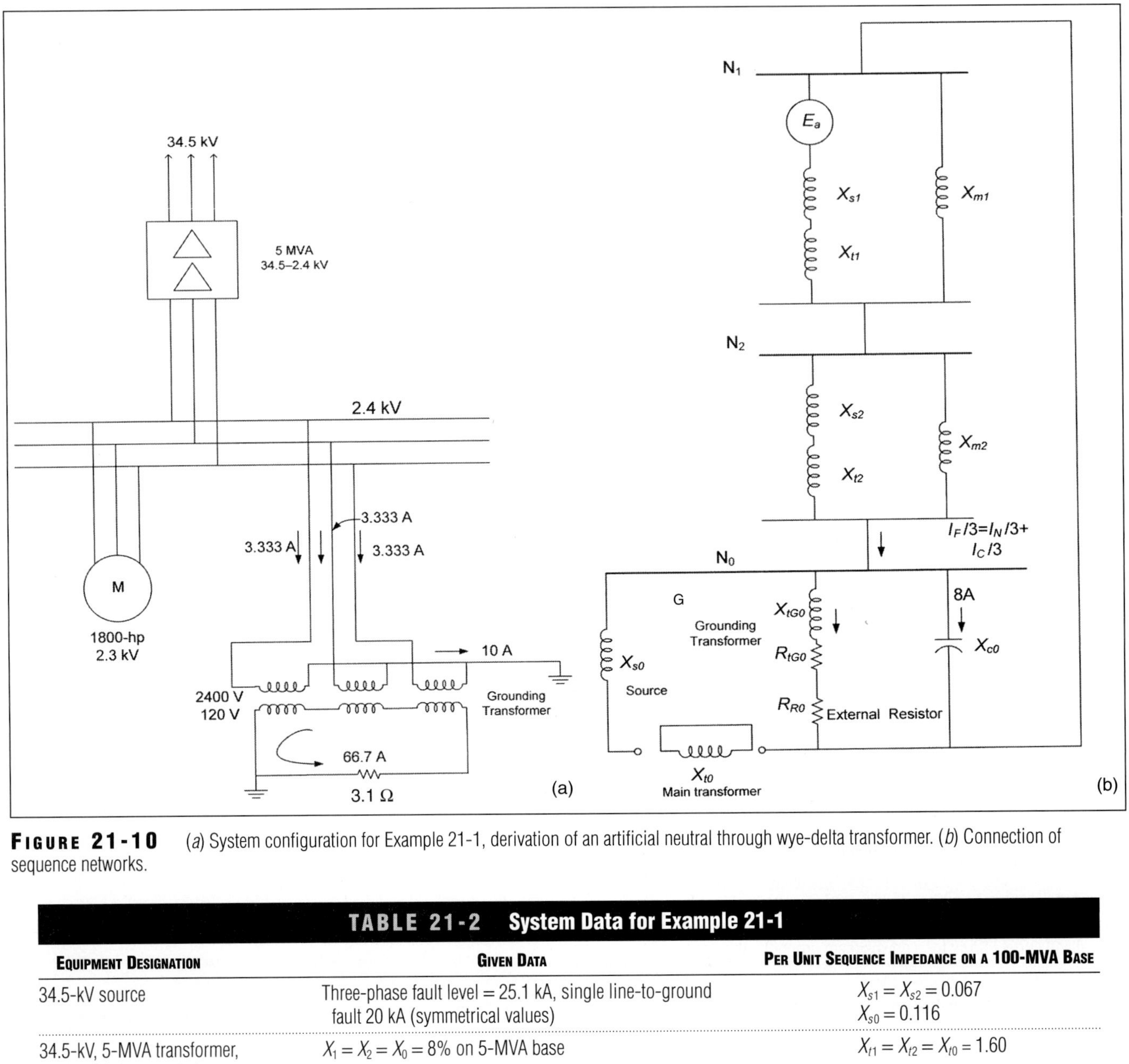

FIGURE 21-10 (*a*) System configuration for Example 21-1, derivation of an artificial neutral through wye-delta transformer. (*b*) Connection of sequence networks.

TABLE 21-2 System Data for Example 21-1

EQUIPMENT DESIGNATION	GIVEN DATA	PER UNIT SEQUENCE IMPEDANCE ON A 100-MVA BASE
34.5-kV source	Three-phase fault level = 25.1 kA, single line-to-ground fault 20 kA (symmetrical values)	$X_{s1} = X_{s2} = 0.067$ $X_{s0} = 0.116$
34.5-kV, 5-MVA transformer, delta-delta connected	$X_1 = X_2 = X_0 = 8\%$ on 5-MVA base	$X_{t1} = X_{t2} = X_{t0} = 1.60$
2.3-kV 1800-hp (1640-kVA) motor load	Locked rotor reactance = 16.7%	$X_{m1} = X_{m2} = 9.35$ (Note 1) X_{m0} = infinite
Grounding transformer, 60-kVA, wye-delta connected 2400:120V	$X_0 = 1.5\%$ $R_0 = 1.0\%$	$X_0 = 75$ $R_0 = 50$

Note: Negative sequence impedance of a rotating machine is not the same as positive sequence impedance. Here an assumption is made for simplicity of calculations; subscripts 1, 2, 0 refer to positive, negative, and zero sequence impedances.

$I_{R0} = 10/3 = 3.33$ A. Therefore, $Z_{R0} = (2400/\sqrt{3})/3.33 = 416.09$ $\Omega = 416.09/$ base $\Omega = 7223.9$ per unit. The additional resistor to be inserted is:

$$R_{R0} = \sqrt{Z_{R0}^2 - X_{tG0}^2} - R_{tG0}$$

$$= \sqrt{7223.9^2 - 75^2} - 50$$

$$= 7173.5 \text{ pu}$$

where R_{R0} is added resistor, Z_{R0} is total impedance in ground circuit, X_{tG0} and R_{tG0} are reactance and resistance of the grounding transformer, Fig. 21–10*b*.

Multiplying by base ohms, the required resistance = 413.2 Ω. These values are in symmetrical component equivalents. In actual values, referred to 120-V secondary, the resistance value is:

$$R_R = \left(\frac{120}{2400}\right)^2 413.2 \times 3 = 3.1\,\Omega$$

If we had ignored all the sequence impedances, including that of the grounding transformer, the calculated value is 3.12 Ω. This is often done in the calculations for grounding resistance for high-resistance grounded systems, and all impedances, including that of the grounding transformer, can be ignored without appreciable error in the final results. The grounding transformer should be rated to permit continuous operation, with a ground fault on the system. The per phase grounding transformer requirement is 2.4 (kV) × 3.33 (A) = 7.92 kVA, that is, a total of 7.92 × 3 = 23.76 kVA. The grounding transformer of the example is therefore adequately rated.

21-5 GROUNDING OF VARIABLE-SPEED DRIVE SYSTEMS

A brief reference can be made to the grounding of variable-speed drive systems (VSD) systems, where special considerations apply. Consider a three-phase, six-pulse bridge rectifier circuit (Fig. 21-11). Only two phases conduct at a time, and the dc plus and negative voltages to mid-point are shown. These voltages do not add to zero and the mid-point oscillates at thrice the ac supply system frequency. The dc positive and negative buses have common mode voltages and its magnitude changes with the firing angle. The peak of the voltage is approximately 0.5 V_{ln}, where V_{ln} is the peak line-to-neutral point input voltage.

Figure 21-12 shows common mode voltages on the motor isolated neutral-to-ground of a 4160-V drive system GTO inverter. The peak line-to-ground $V_{l\text{-}g}$ voltage is 4100 V and the waveform has a frequency of 60-Hz. The peak neutral-to-ground voltage is 2500 V and has a frequency of 180 Hz. The operation of the output bridge creates a common mode voltage by exactly the same mechanism as the input bridge does, where the back emf of the motor is analogous to the line voltage.

Thus, the worst-case condition for the common mode voltage is no-load, full speed operation, as the phase-back angle is 90° for both the converters, and the motor voltage is essentially equal to the line voltage. The sum of both common mode voltages is approximately V_{ln} at six times the input frequency. Since the input and output frequencies are generally different, the motor experiences a waveform with beat frequencies of both input and output frequencies, and there will be instances when twice the rated voltage is experienced, (Fig. 21-12).[10]

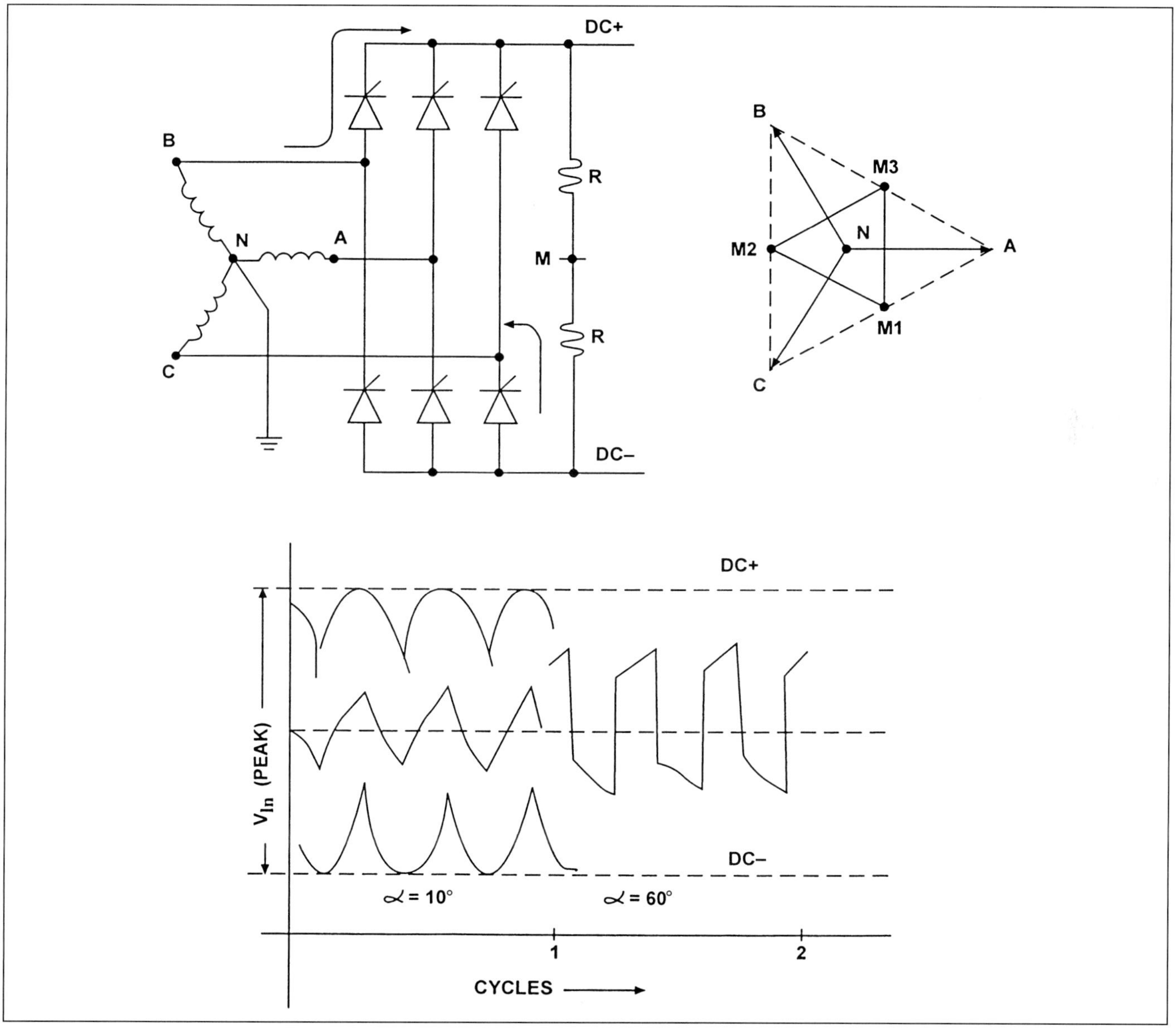

FIGURE 21-11 Generation of common mode voltage in a six-pulse converter.

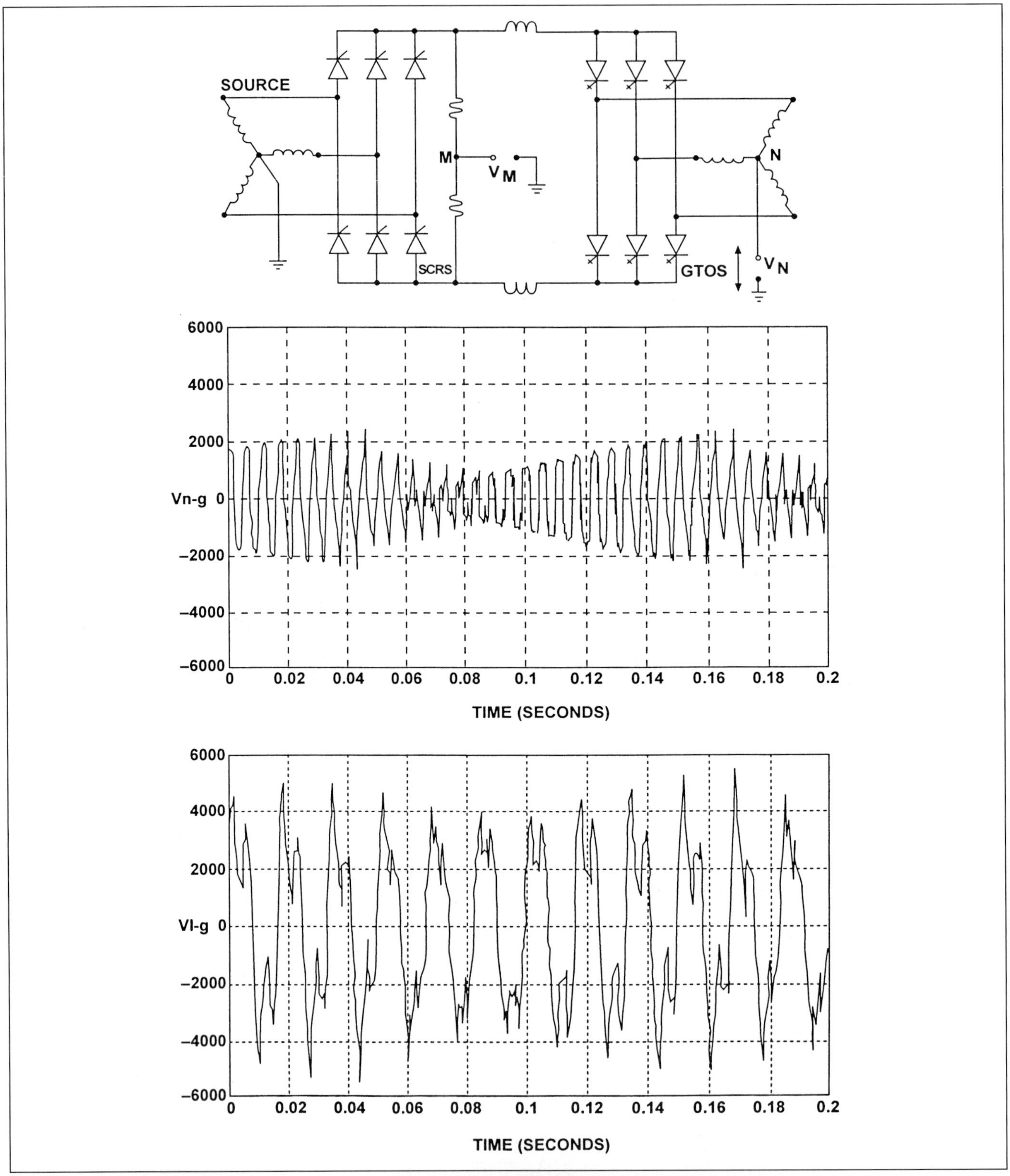

Figure 21-12 Common mode neutral-to-ground and line-to-ground voltages in a CSI for a drive system.[10]

The grounding must ensure that the motor insulation system is not stressed beyond its design level. Figure 21-13 shows three possible methods. In Fig. 21-13*a* output drive isolation transformer is provided with its secondary grounded. The transformer primary winding insulation can tolerate the common mode voltage swings better than the motor insulation. This option is tricky as the drive isolation transformer is subjected to residual dc offset in the drive and high harmonic content passing through the transformer windings. A better location for the drive isolation transformer would be at the input line end (Fig. 21-13*b*), and the secondary winding

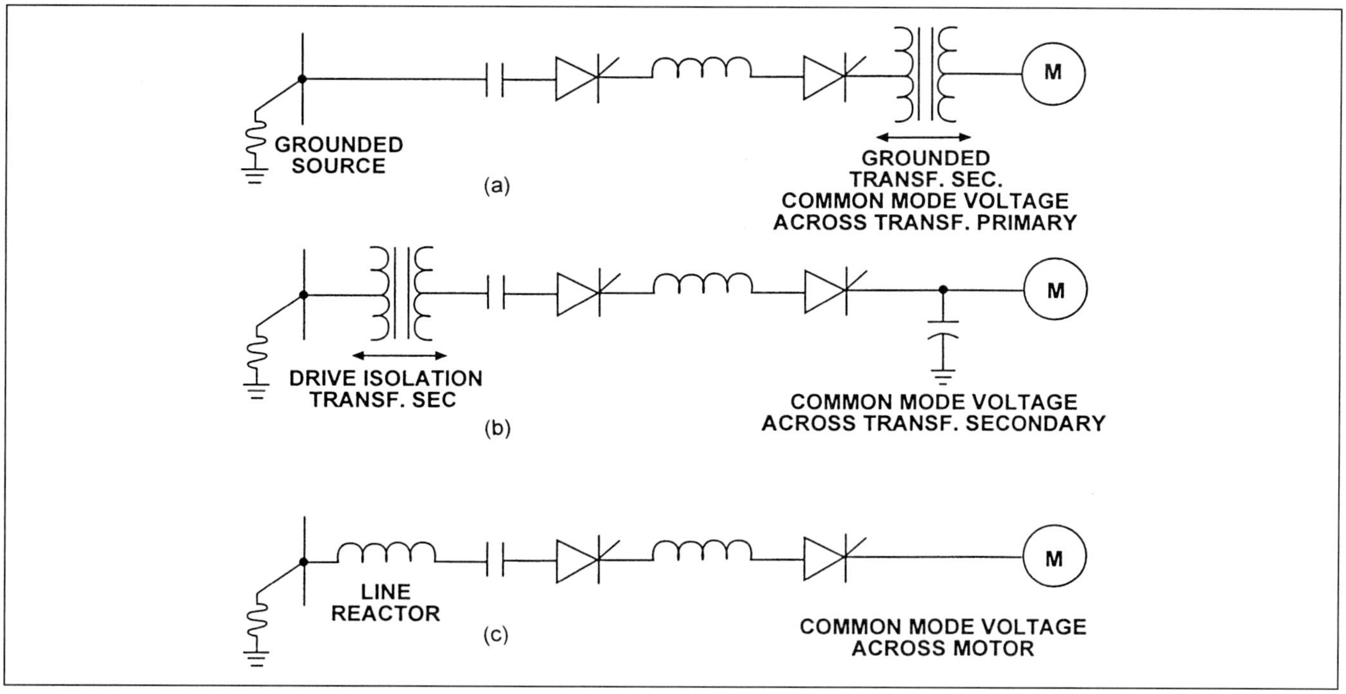

FIGURE 21-13 (*a*) Isolation transformer with grounded secondary to withstand common mode voltage. (*b*) Drive isolation transformer at input, common mode voltage across transformer secondary. (*c*) Input line reactor, common mode voltage across drive motor windings.

is left ungrounded. The neutral of the filter capacitors on the output of the inverter provides a convenient place to ground the load side of the inverter and a small resistor rated 1 A or so is enough to do it. The line-to-neutral voltages must, then, be taken into account for the insulation of the secondary of the transformer. Also the cables from the transformer secondary to the input rectifier bridge must be rated for higher voltage to ground; for example, for a 4.16-kV drive system, cables with 173 percent insulation level will be required; alternatively, 8-kV cables can be used.

An input or output drive transformer increases the cost and reduces the efficiency of the drive system. In a transformer-less GTO drive system, a line reactor may be used on the input side to affect commutation notches, but the motor insulation must withstand twice the normal voltage stress to ground (Fig. 21-13*c*).

21-6 GROUNDING FOR ELECTRICAL SAFETY

One of the major objectives of grounding is the safety of the personnel and to ensure freedom from dangerous electrical shock voltage exposure. This becomes of greater importance in utility high-voltage substations, as the high-voltage systems are solidly grounded and the fault currents are high. We can categorize as follows:

1. Utility substations in isolation, that is, located far away from the loads served

2. Utility substations located within the premises of industrial distribution systems

3. Industrial systems in isolation, that is, located away from the utility substations

Let us first concentrate on Situation 1 above. IEEE Standard 80-2000, "Guide for Safety in AC Substation Grounding"[11] is the main document addressing this subject and the methodology described in this standard is widely used in the United States; though it has some limitations which we will discuss in the sections to follow. Due to complex and iterative nature of the grounding system design, the computer programs are invariably used for calculations using IEEE method and other methods like finite element methods. The criteria for safety are described in the following section.

21-6-1 Criteria for Safety

The shock situation is illustrated in Fig. 21-14. On occurrence of a ground fault, the potential of the local earth rises with respect to the remote earth. This is shown as ground potential rise (GPR) in Figure 21-14, and is simply given by GPR = $I_G R_G$, where I_G is the current returning through grid to earth and R_G is the grid resistance. The calculation of I_G is not so simple. It is not the bolted fault current at the location. All the available ground fault current does not give rise to GPR, and a split factor is applicable depending upon the number of sky wires, tower footing resistance, soil resistivity, connections to grounding conductors, and remote and local grounds, which is discussed further in Section 21-6-2.

Figure 21-14 shows that the safety criteria are to establish safe *step and touch* voltages in the grounding grid area during a fault, when someone may be walking in the substation or touching a grounded metallic frame or structure. Also transferred voltage must be considered. (Note the profile of the potential with respect to grounding grid conductors buried at a certain depth below the soil.) All these voltages are shown in this figure and are defined as step voltage, touch voltage, mesh voltage, and transfer voltage. These are discussed in the following sections.

Step Voltage The *step voltage* is defined as the difference in surface potential experienced by a person bridging a distance of 1m with his feet, without contacting any other grounded object. This is given by:

$$E_{\text{step},50} = (1000 + 6C_s(h_s, K)\rho_s)\,0.116/\sqrt{t_s}$$
$$E_{\text{step},70} = (1000 + 6C_s(h_s, K)\rho_s)\,0.157/\sqrt{t_s} \qquad (21\text{-}10)$$

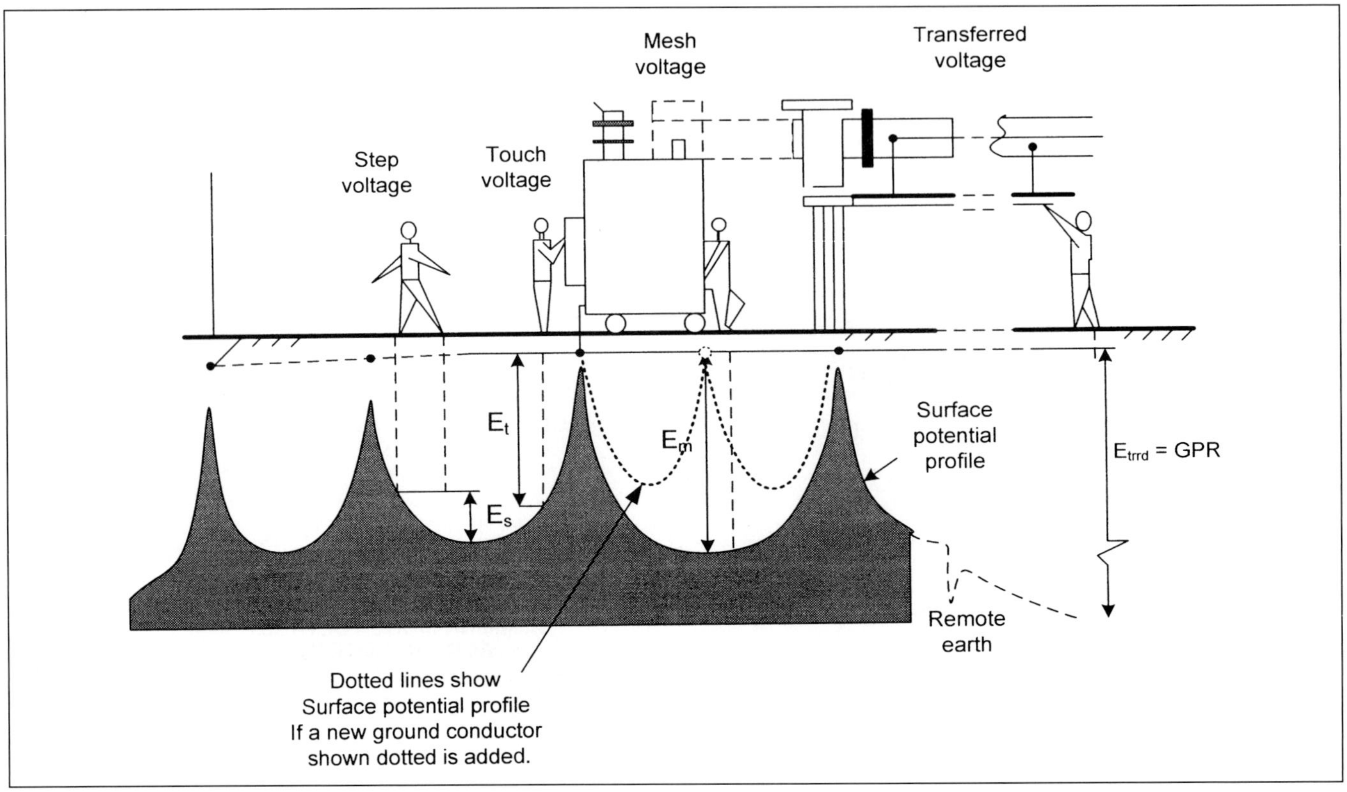

FIGURE 21-14 Shock hazard situation, step, touch, mesh, and transfer voltages, and GPR. *The figure is adapted from ANSI/IEEE Std. 80.*[11]

where $E_{step,50}$ and $E_{step,70}$ signify the tolerable limits for a body weight of 50 and 70 kg, respectively.

Touch Voltage The touch potential is defined as the potential difference between the ground potential rise and the surface potential at the point where a person is standing while at the same time having a hand in contact with a grounded structure. This is shown as E_t in Fig. 21-14 and is given by:

$$E_{touch,50} = (1000 + 1.5C_s(h_s, K)\rho_s)\,0.116/\sqrt{t_s}$$
$$E_{touch,70} = (1000 + 1.5C_s(h_s, K)\rho_s)\,0.157/\sqrt{t_s} \qquad (21\text{-}11)$$

where $E_{touch,50}$ and $E_{touch,70}$ signify the tolerable limits for a body weight of 50 and 70 kg, respectively. An explanation of Eqs. (21-10) and (21-11) follows.

These equations are derived from the tolerable limits of shock hazard current that can pass through the human body without a shock hazard. (The threshold of fibrillation is 107 mA rms for a body weight of 70 kg, and the equations use a lower value of 91 mA rms.)

The surface resistivity of the soil, denoted by ρ_s, is an important parameter in the equations. The higher the surface resistivity, the greater will be the tolerable step and touch potentials and more economical the grounding grid design. In all grounding grid designs, it is a standard practice to increase the surface resistivity by installing crushed rock to a depth of 4 to 6 in (h_s = depth of the crushed rock) throughout the substation area. Table 21-3 gives the resitivity of some materials used in the substations. Consider, for example, that the natural soil has a resistivity of 90 Ω-m. If the surface material #3 in Table 21-3 is used, the surface resistivity improves to 3000 Ω-m. Therefore, we observe that:

- To have low ground grid resistance, a lower soil resistivity is desirable.
- To have increased step and touch voltages for personal safety, a higher surface soil resistivity is desirable.
- Crushed rock treatment of higher soil resistivity on top of low-resistivity soil meets these two conflicting requirements.

Factor $C_s = 1$ for no protective surface layer. When a surface layer of crushed rock of higher resistivity is installed compared to the bare soil, then $C_s < 1$. The factor C_s considers derating the nominal value of the surface resistivity material, based upon thickness of the rock layer that is laid in the substation. An expression for C_s from Ref. 11 is:

$$C_s = \frac{1}{0.96}\left[1 + 2\sum_{n=1}^{n=\infty} \frac{K^n}{\sqrt{1 + (2nh_s/0.08)^2}}\right] \qquad (21\text{-}12)$$

where K, the reflection factor is given by:

$$K = \frac{\rho - \rho_s}{\rho + \rho_s} \qquad (21\text{-}13)$$

where ρ is the resistivity of natural soil, *without surface treatment*. Values of C_s are plotted in a graph form in Ref. 11, not reproduced here.

The Shock Duration The higher the time for which the ground fault is sustained, the lower will be the tolerable step and touch potentials and more difficult the grid design. Thus, high-speed fault clearance is an important factor. In Eqs. (21-10) and (21-11), t_s is the shock duration in s. The shock duration is conservatively chosen considering the ground fault clearance time (relay operating time plus breaker interrupting time). For safety and conservatism, it is assumed that the primary ground fault protection fails to operate and the fault is cleared in backup protection zone with additional time delay. Fault clearance times used in the calculations generally

TABLE 21-3 Resistivity of Substation Surface Materials

No.	Surface Material	Resistivity (Ω-m) Dry	Wet
1	Crusher run granite with fines	140×10^6	1300
2	#57 washed granite similar to 3/4 in gravel	190×10^6	8000
3	Clean limestone slightly coarser than #2	7×10^6	2000–3000
4	Washed granite similar to 3/4 in gravel	2×10^6	10000
5	Washed granite similar to pea granite	40×10^3	5000
6	Crushed aggregate base granite (with fines)	Varies	500–1000
7	Concrete	2000–10000	50–100
8	Concrete	1200–280000	21–63
9	Asphalt	Varies	10000
10	Asphalt	2×10^6–30×10^6	10000–6×10^6

vary from 0.25 to 0.5 s. As stated in Ref. 11 this time may be much higher for smaller substation and may approach 3 s. This defines all the terms in Eqs. (21-10) and (21-11).

Mesh Voltage Again referring to Fig. 21-14, the mesh voltage E_m is defined as the maximum touch voltage to be found within a mesh of the grounding system. The maximum touch voltage is equal to the mesh voltage.

Transfer Voltage The transferred voltage E_{trrd} is a special case of the touch voltage, where a voltage is transferred into or out of the substation due to metallic connections. These include buried pipe lines close to the substation, cable racks, grounded shields of cables, and cable trays, entering and leaving the substation. On a first thought, isolation of local and remote grounds seems to be an alternative to prevent problems of transferred potentials. Practically, this is not attempted. Sometimes, insulting barriers are installed in the substation fences, but to isolate all metallic connections coming in and getting out of a substation is not practical. The couplings can take place through the soil also, as discussed in Section 21-9.

Again, referring to Fig. 21-14, consider that the module of placement of grid conductors is reduced by installing another grid conductor in between the existing two grid conductors. Then the consequent change in the surface potential profile is as shown in the dotted lines. Obviously the step, touch, and mesh potentials are all reduced. Thus, on a simplistic basis, an effective grounding grid design optimizes the grid conductors and their placement modules.

21-6-2 Design Parameters of Grounding Grids

Soil Resistivity Measurements The soil resistivity at the site of location is an important criterion for design of the grounding grid. The GPR is given by the ground fault current that flows through the earth and the resistance of the grounding grid plays an important role. Depending upon the soil resistivity, special electrodes and soil treatment may become necessary to have an acceptable grid resistance. The soil resistivity over a given area will be rarely uniform and will vary with the depth, moisture content, type of earth, that is, bedrock, wet organic soil, sandy loom, etc. The soil resistivity is measured using Wenner's Four-Pin method or by laboratory testing of soil borings at different depth; the former method is commonly used.

These measurements at different depths will form a sort of scatter plot and a least square line can be fitted by computer programs (App. F). A two-layer soil model results, which breaks soil resistivity into two numbers, with respect to the depth of the soil. Figure 21-15*a* and *b* shows the equivalent soil resistivity for two-layer soil models, commonly used for substation designs. Note that the IEEE calculations are based on a single-layer model.

The Worst Ground Fault Current The maximum of the single-line-to-ground or double-line-to-ground fault bolted current and their *X/R* ratios are required. In Chap. 9 we studied the calculations of unsymmetrical fault currents. The available ground fault current in a system depends upon the method of system neutral grounding. In the calculations of worst ground fault currents, the resistance of the ground grid is often ignored. A single line-to-ground fault is given by the expression:

$$I = \frac{3E}{\sqrt{(R_1 + R_2 + R_0 + 3R_f + 3R_g)^2 + (X_1 + X_2 + X_0)^2}} \tag{21-14}$$

where R_g is the ground grid resistance and R_f is the fault resistance. The expressions for double line-to-ground fault are in Chap. 9. A worst grounding fault is that which gives the maximum flow of zero-sequence current. This depends upon the relative values of the sequence impedances. For $Z_0Z_1 > Z_2^2$, the single line-to-ground fault current will be higher, for $Z_0Z_1 < Z_2^2$, double line-to-ground fault will be the worst type of fault, Figure 9-1.

The ground grid resistance and fault resistances are ignored in the fault calculations for the grounding grid designs.

The Split Factor Consider the following scenarios:

1. There is no ground connection of the neutral at the local substation (e.g., a delta transformer high-voltage primary windings). For a ground fault on the high-voltage side, all the ground fault current must flow to the remote-grounded neutral.

2. There is no remote ground, and the high-voltage primary winding neutral is grounded locally only. The entire fault current will circulate locally and none flows to the remote ground.

3. For a situation where both the neutrals at local and remote substations are grounded, the fault current will split, partly flowing through the local ground and partly returning to the remote ground, the split factor depending upon the relative ground resistances.

Practically, the situation with respect to split factor is more complex. The ground wires of the transmission lines are normally connected to station ground and these divert a considerable amount

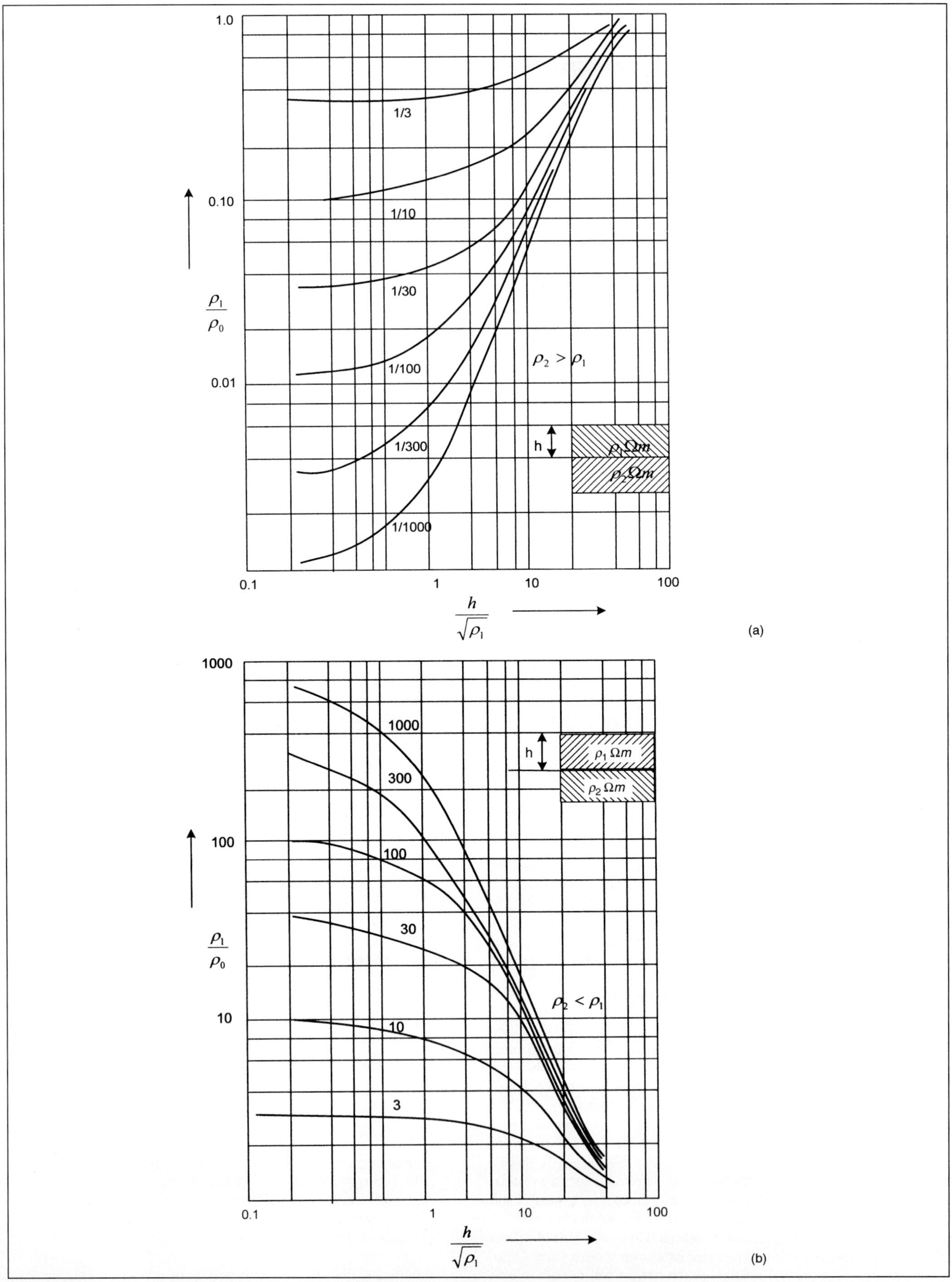

FIGURE 21-15 (*a*) Resistivity of two-layer soil model, lower layer ρ_2 of higher resistivity than the top layer ρ_1. (*b*) Lower layer ρ_2 of lower resistivity than the top layer ρ_1.

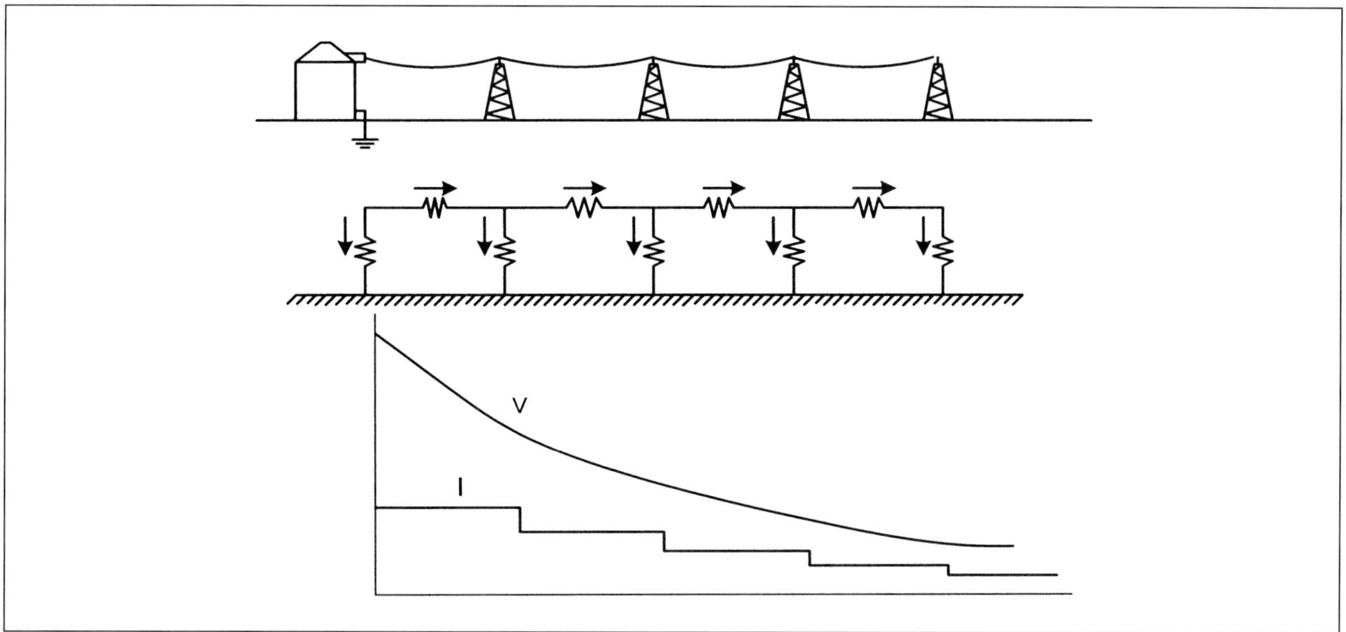

FIGURE 21-16 Flow of earth current through an earth wire.

of current from the ground. A similar effect occurs when pipes and cables are buried and their sheaths and armor are in effective contact with the ground. Figure 21-16 shows flow of current through a ground wire and this forms a sort of ladder network. If the line is sufficiently long it behaves with the source like an equivalent impedance independent of length. Also see the distribution of ground current in multiple grounded distribution systems, (Chap. 19).

Endrenyi[12] reduces the tower impedances and ground wires in identical spans to an equivalent impedance. In the matrix method,[13] an impedance matrix is derived for each span of the line. Dawalibi[14] provides algorithms for deriving simple equations to solve for current in each tower. Meliopoulos[15] introduces an equivalent conductor to represent effect of earth using Carson's formula (App. D). Thus, the techniques that model the phase conductors, ground wires, tower footing resistance in detail will give the best determination of the split factor S_f. Apart from EMTP, other computer programs are available for the calculation of the split factor. These are, for example, SMECC (EPRI report EL-2682/2[16] developed by EPRI and PATHS, developed for EPRI). PATHS was developed to simulate faults on the transmission lines, but has been used to find the grounding grid current for simple substations. The IEEE Std. 80[11] includes simplified curves for calculating the split factor, based on the calculated grid resistance, number of transmission and distribution lines with ground wires, and whether the current is contributed solely from the remote sources or local/remote sources and also tower footing resistance. These curves are not reproduced here.

Sizing the Grid Conductor and Connectors The short-time temperature rise of the conductor can be obtained from the following equation by Sverak:[17]

$$I = 5.0671.10^{-6} A\left[\left(\frac{\text{TCAP}}{t_c \alpha_r \rho_r}\right) \ln\left(\frac{K_0 + T_m}{K_0 + T_a}\right)\right]^{1/2} \tag{21-15}$$

where I is the rms current in kA, A is the conductor cross section in circular mils, T_m is the maximum allowable conductor temperature in °C, T_a is the ambient temperature in °C, T_r is the reference temperature of material constants in °C, α_0 is the thermal coefficient of resistivity at 0°C, α_r is the thermal coefficient of resistivity at reference temperature—α_r = 0.00393 for annealed soft copper wire, conductivity 100 percent, ρ_r is the resitivity of the ground conductor at reference temperature T_a in μΩ-cm = 1.7241 at 20° C for annealed soft copper wire, $K_0 = 1/\alpha_0$ = 234 for soft annealed copper wire at 0° C, t_c = time of current flow in s, TCAP = thermal capacity factor = 3.422 J/cm³/°C for soft annealed copper.

TCAP for various description of grounding conductors is given in Ref. 18.

For mechanical reasons, a conductor size less than 2/0 AWG is not used, and generally the minimum size of conductor in high-voltage substations is 4/0 AWG. The connectors must meet the same criteria as the conductors, that is, corrosion resistance, electrical conductivity, current-carrying capability, and mechanical strength. Standards define the test performance requirements. Connectors may be brazed type, using a brazing method, the filler material melts above 450°C; exothermic welded type; and pressure type. The maximum temperature rise limit T_m is 450°C for brazed connections and 350 to 250°C for bolted connectors.

The Grid Resistance The grid resistance is given by:

$$R_g = \frac{\rho}{4}\sqrt{\frac{\pi}{A}} + \frac{\rho}{L} \tag{21-16}$$

where L is the total length of the buried conductor and A is the area of the grid in square meters.

This shows that there is an upper limit to the reduction of the grid resistance in a certain given area by increasing the length of the buried conductor. For grid depths between 0.25 and 2.5 m, the following expression can be used:

$$R_g = \rho\left[\frac{1}{L} + \frac{1}{\sqrt{20A}}\left(1 + \frac{1}{1 + h\sqrt{20/A}}\right)\right] \tag{21-17}$$

Total resistance of the grid with conductors and ground rods is given by:

$$R_g = \frac{R_1R_2 - R_{12}^2}{R_1 + R_2 - 2R_{12}} \tag{21-18}$$

where R_1 is the resistance of grid conductors,

$$R_1 = (\rho_1/\pi l_1)[\ln(2l_1/h' + K_1(l_1/\sqrt{A}) - K_2] \tag{21-19}$$

R_2 is the resistance of ground rods,

$$R_2 = (\rho_a/2n\pi l_2)[\ln(8l_2/d_2) - 1 + 2K_1(l_2/\sqrt{A})(\sqrt{n}-1)^2] \tag{21-20}$$

R_{12} is the mutual resistance between the group of grid conductors and group of ground rods,

$$R_{12} = (\rho_a/\pi l_1)(\ln(2l_1/l_2 + K_1(l_1/\sqrt{A}) - K_2 + 1) \tag{21-21}$$

The terms used in the above equations are defined below:

ρ_1 is the soil resistivity encountered by grid conductors buried at depth h Ω-m, ρ_a is the apparent soil resistivity seen by ground rod in Ω-m, H is the thickness of upper soil layer in m, ρ_2 is the soil resistivity from depth h downward in Ω-m, l_1 is the length of grid conductors in m, l_2 is the length of ground rods in m, h is the depth of grid burial in m, $h' = \sqrt{d_1 h}$ for conductors buried at depth h, or $0.5d_l$ for conductors at earth surface, A is the area of grid in m^2, n is the number of ground rods, K_1 and K_2 are the constants related to the geometry of the system,[11] d_1 is the diameter of grid conductor in m, d_2 is the diameter of ground rods in m, a is the short-side of grid in m, and b is the long side of grid in m.

When the tops of the rods are at the same depth as the grid conductors

$$\rho_a = l_2(\rho_1\rho_2)/\rho_2(H-h) + \rho_1(l_2 + h - H) \tag{21-22}$$

The Maximum Grid Current The maximum grid current is given by:

$$I_G = C_p D_f S_f I_g \tag{21-23}$$

where D_f is the decrement factor for the entire duration of the fault current, to account for the asymmetry in the ground fault current:

$$D_f = \left[1 + \frac{T_a}{t_f}(1 - e^{-2t_f/T_a})\right]^{1/2} \tag{21-24}$$

where T_a is the effective X/R ratio of the system subtransient impedance, t_f is duration of fault, s_f is the split factor, I_g is the ground fault current.

The factor C_p considers the future changes in the system, which may give rise to higher ground fault currents. The split factor can be defined as the ratio of grid current divided by $3I_0$

Maximum Step and Touch Potentials The mesh and step voltages are given by:

$$E_m = \rho K_m K_i I_G/L$$
$$E_s = \rho K_s K_i I_G/L \tag{21-25}$$

The geometric factor K_m is given by:

$$K_m = \frac{1}{2\pi}\left[\ln\left(\frac{D^2}{16hd} + \frac{(D+2h)^2}{8Dd} - \frac{h}{4d}\right) + \frac{K_{ii}}{K_h}\ln\frac{8}{\pi(2n-1)}\right] \tag{21-26}$$

$K_{ii} = 1$ for grids with ground rods along the perimeter, or for grids with ground rods in the corners, as well as both along the perimeter and throughout the grid area.

$$K_{ii} = \frac{1}{(2n)^{2/n}} \tag{21-27}$$

for grids with no ground rods or for grids with only a few ground rods, none located in the corners or on the perimeter.

$$K_h = \sqrt{1 + h/h_0} \tag{21-28}$$

where h_0 is the 1-m, reference depth of the grid, D is the spacing between parallel conductors in m, h is the depth of burial of the grid conductors in m, n is the number of parallel conductors in one direction, and d is the diameter of the grid conductor in m.

K_i is the corrective factor for the mathematical grid model:

$$K_i = 0.656 + 0.172n \tag{21-29}$$

For grids with ground rods, E_m can be written as:

$$E_m = \frac{\rho I_G K_m K_i}{L_c + 1.15L_r} \tag{21-30}$$

where L_c is the total length of the conductor and L_r is the total length of the ground rods. For grids with only a few ground rods, change the denominator to $L_c + L_r$.

Step Voltage In Eq. (21-25) the $L = L_c + 1.15\,L_r$ or $L = L_c + L_r$. Similar qualifications apply for E_{mesh}. For burial depths 0.25 m < h < 2.5 m:

$$K_s = \frac{1}{\pi}\left[\frac{1}{2h} + \frac{1}{D+h} + \frac{1}{D}(1 - 0.5^{n-2})\right] \tag{21-31}$$

The step voltage can now be calculated by Eq. (21-25).

21-6-3 Limitations of the Simplified Equations

There are certain assumptions in deriving the above equations as discussed in Ref. 11. For square or rectangular grids:

$$n \le 25$$
$$0.25\text{ m} \le h \le 2.5\text{m}$$
$$d < 0.25h$$
$$D > 2.5\text{ m} \tag{21-32}$$

For equally spaced rectangular grids, the value of n for determining factor K_i and K_m for mesh voltage E_m should be:

$$n = \sqrt{n_a n_b} \tag{21-33}$$

where n_a and n_b are the number of conductors in each direction. For calculating the step voltage, the value of n for K_i and K_s is the maximum of n_a and n_b.

Example 21-2 This example is a hand calculation of a grid design to illustrate the iterative process and complexity of these calculations. Consider the design with the following given parameters:

Total single line-to-ground fault current $3I_0 = 12$ kA.

$X/R = 15$.

Area of the rectangular grid = 84 × 57 m.

Soil resistivity = 270 Ω-m. (Practically, it will rarely be uniform and the two-layer soil model is derived from the measurements.) Here we will consider a uniform soil resistivity. This can be an arithmetic average of the measurements for illustrative purposes.

Surface material is clean limestone with, resistivity = 3000 Ω-m.

Depth of the surface layer = 0.152 m (approximately 6 in).

Depth of grid conductor below natural soil $h = 0.5$m.

Ambient temperature $T_a = 40°$C.

Duration of current flow $t_f = 0.5$ s.

Duration of current flow for sizing grid conductor = 0.5 s.

The substation has a transformer with high-voltage, delta primary, and wye-connected secondary, solidly grounded. The remote source is solidly grounded, 100 mi away. The substation has two incoming lines with ground wires and two outgoing lines with ground wires.

1. Calculate the conductor size from Eq. (21-15).

$$I = 5.671\times10^{-6}A\sqrt{\left(\frac{TCAP}{t_c\alpha_r\rho_r}\right)\ln\left(\frac{K_0+T_m}{K_0+T_a}\right)}$$

T_m is the maximum allowable temperature that varies depending on the type of connections. For brazed joints (thermo weld), it is considered 450°C, but for pressure-type connections, the limit is 250 to 350°C.

Consider the lower limit of 250°C and soft annealed copper conductors, the constants of which are given in Eq. (21-15). Though the current giving rise to GPR will be lower, the *conductors are sized for the maximum fault current available.* The factor C_p considers future increase in the system short-circuit level. For this example, let us consider $C_p = 1$.

Substituting these values:

$$1 = 5.0671\times10^{-6}A\sqrt{\left(\frac{3.422}{0.5\times0.00393\times1.7241}\right)\ln\left(\frac{234+250}{234+40}\right)}$$

This gives an area A in circular mils of 99089. Thus, a conductor size of 2/0 AWG (133100 CM) is acceptable. However, we use 4/0 AWG (211600 CM) based on mechanical considerations. Welded-type connections can be used at the perimeter conductors and every forth module lengthwise and breadth wise, while compression-type connectors can be used elsewhere for ease of installation. The diameter of 4/0 conductor is 1.167 cm.

2. Calculate tolerable step and touch potentials.

The touch and step potentials can be calculated based on the soil resistivity $\rho = 270\ \Omega/\text{m}$, top layer of limestone $\rho_s = 3000\ \Omega/\text{m}$:

$$K = \frac{\rho-\rho_s}{\rho+\rho_s} = -0.834$$

Then from Ref. 11, $C_s = 0.790$

Consider a body weight of 50 kg. The step potential is:

$$E_{\text{step50}} = (1000+6C_s(h_s,K)\rho_s)0.116/\sqrt{t_s}$$

$$= (1000+6\times0.790\times3000)0.116/\sqrt{0.5} = 1275.0\text{ V}$$

The touch potential is:

$$E_{\text{touch50}} = (1000+1.5C_s(h_s,K)\rho_s)0.116/\sqrt{t_s}$$

$$= (1000+1.5\times0.790\times3000)0.116/\sqrt{0.5} = 747.2\text{ V}$$

3. Estimate the length of required buried conductor.

Next estimate the length of the buried conductor. The following expression can be used:

$$\frac{K_mK_i\rho I_G}{L} < [1000+1.5C_s(h_s,K)\rho_s]0.116/\sqrt{t_s} \qquad (21\text{-}34)$$

Equation (21-34) signifies that $E_{m\,(\text{calculated})}$ is less than $E_{\text{touch (tolerable)}}$. We calculated the RHS of the above equation as 742.2 V. We do not know K_m, K_i, and S_f to calculate the grid current, and as a first step, some assumptions can be made on the values of these parameters (based upon experience). Consider arbitrarily, $K_m = 0.7$, $K_i = 0$ 6, split factor of the current cannot be found unless the grid resistance is known and that cannot be calculated unless the length of the buried conductor and ground rods to be provided are known. Assume a split factor of 0.7, which gives grid current of $I_G = 8.4$ kA. Ignoring D_f for this preliminary calculation and substituting these values gives an initial estimate of conductor length = 1275 m. Consider 40 10 ft long copper ground rods of 1 in diameter provided at the perimeter of the grid.

Proceed with the preliminary design on the basis of the assumption. Calculate grid resistance. Estimate split factor from the graphs in Ref. 11, calculate D_f of current and the maximum I_G giving rise to GPR according to Eq. (21-23). Calculate the maximum step and touch potentials according to equations previously described.

4. Check and reiterate the calculations.

These calculations reveal that the calculated touch voltage exceeds the permissible safe touch voltage of 747.2 V, while the step voltage is within the safe limits. Thus, the design should be modified by manual iteration of the involved parameters, until the desired results are obtained.

All these iterative calculation steps are not shown. The design satisfying the requirements is described as follows:

- The grid consists of 20 conductors lengthwise, that is, parallel to x axis and 29 conductors parallel to y axis. This gives a uniform spacing of conductors throughout the rectangular grid area of 87 × 60 m, giving total length of conductors equal to 3723 m.

- Note that according to IEEE 80 calculation procedures, the unequal spacing of the conductors is not permissible, and these should be spaced at uniform interval, here 3 m. The minimum spacing is limited to no less than 2.5 m.

- Retain 40 ground rods of 1 in diameter, and 10 ft long at the grid perimeter.

5. Run final calculations.
Calculate the grid resistance:

$$R_g = 270\left[\frac{1}{3843} + \frac{1}{\sqrt{20\times 84\times 57}}\left(1+\frac{1}{1+0.5\sqrt{20/84\times 57}}\right)\right]$$
$$=1.716\ \Omega$$

Note that the length of the ground rods is added in the length of the total grid conductors. This resistance value agrees closely with the computer-based calculations, using finite element method (see further discussions below), which gives 1.649 Ω versus 1.716 Ω with hand calculations.

Calculate the D_f from Eq. (21-24):

$$D_f = \sqrt{1+\frac{15}{2\pi f\times 0.5}(1-e^{-2\times 0.5\times 2\pi f/15})} = 1.039$$

Estimate current split factor. Consider 100 percent remote contribution, that is, all the current must return to remote grounded source as the incoming lines from remote are connected to a step-down transformer with delta primary windings, which block the flow of zero-sequence currents for a high-voltage side fault. Consider two transmission lines and two feeder lines entering and leaving the substation, and a tower footing resistance for transmission lines of 100 Ω and that of feeder lines equal to 200 Ω. This gives a split factor of 0.5, graphically determined from IEEE 80.[11]

Thus, the current giving rise to GPR:

$$I_g = (3I_0)S_f D_f C_p = 12\times 0.5\times 1.039\times 1 = 6.234 \text{ kA}$$

Therefore, the GPR = 10,700 V.
Calculate K_m from Eq. (21-24)

$$K_m = \frac{1}{2\pi}\left[\ln\left(\frac{3^2}{16\times 0.5\times 0.01167}+\frac{(3+2\times 0.5)^2}{8\times 3\times 0.5}-\frac{0.5}{4\times 0.01167}\right) + \frac{K_{ii}}{K_h}\ln\frac{8}{\pi(2n-1)}\right]$$

In the above equation, for calculation of mesh voltage factor K_m, irregularity factor K_i is calculated as follows:

$$n = \sqrt{29\times 20} = 24 \quad \text{for} \quad E_m$$

$$K_{ii} = 1$$

$$K_h = \sqrt{1+0.5/1} = 1.2247$$

Substituting these values:

$K_m = 0.30$

Also:

$$K_i = 0.656+0.172\times 24 = 4.784$$

Thus, E_m is calculated from Eq. (21-25)

$$E_m = \frac{\rho I_G K_m K_i}{L_c+1.15L_r} = \frac{270\times 6234\times 0.32\times 4.784}{3723+1.15\times 120} = 667 \text{ V}$$

This is acceptable.
Calculate the factors for the step voltage from Eq. (21-31):

$$K_s = \frac{1}{\pi}\left[\frac{1}{2\times 0.5}+\frac{1}{3+0.5}+\frac{1}{3.}(1-0.5^{18})\right] = 0.515$$

And:

$$K_{ii} = 0.656+20\times 0.172 = 4.096$$

Therefore from Eq. (21-25):

$$E_s = \frac{270\times 6234\times 4.096\times 0.515}{3723+1.15\times 120} = 919.50 \text{ V}$$

6. Is the calculated design safe?
Note that the maximum touch voltage cannot be greater than E_m. Thus, we have completed a design with step and touch voltages below the safe limits. The calculated touch and step potentials are approximately 89 and 72 percent, respectively, of the safe tolerable limits.

Based on these calculation results, should we conclude that the grid design is safe?

This cannot be said with certainty because of the limitations of IEEE method applied in the above calculations. The calculation procedure does not account for the areas of high risk near the corners of the grid.

7. Check with computer based calculations.
The computer based calculations of *exactly* the same system give the following results:

- Maximum touch potential = 698.07 V versus 667 V calculated above by hand calculations.

- Maximum GPR = 9856 V versus 10700 V by hand calculation.

While these numbers are comparable, the profile of voltages through a cross section of the grid based on computer calculations can reveal areas of high touch and step voltages. This cannot be easily hand calculated, pointing to limitation of such calculation. Though, such calculations are academically instructive. With computer-based calculations the voltage at any point in the grid can be calculated, and it is not the same in every mesh. It tends to rise toward the periphery and in the grid corners.

Figure 21-17 shows a profile of the touch potential through the middle of the grid. It is seen that touch potential everywhere is acceptable. Figure 21-18 shows a similar profile at the outermost conductor. This shows that the touch potential exceeds the tolerable limits by 143 V, approximately 2 m from the grid parameter. It is necessary to provide another run of grid conductor or add more ground rods around the periphery of the grid at 2 m level. Generally, the grid conductor module is reduced to 50 percent at the corners.

Figure 21-19 shows a three-dimensional sketch of the touch voltage contours. Note that the potential tends to rise toward the corners, though the actual values cannot be read.

The calculation procedure in ANSI/IEEE standard does not permit modeling of grid conductors at unequal spacing and gives no idea with respect to the voltage contours in the grid. Some qualifications are:

1. The geometric parameters should not exceed the limits specified in Eq. (21-32). However, in the foregoing example, these did not, yet we have areas of high potential risk.

2. A two-layer model is not supported. In the foregoing example, we considered only a single-layer model. A model with a top-layer resistivity to a certain depth and lower-layer resistivity below that depth, based on test measurements, can be derived by currently available computer programs.

3. An unsymmetrical grid, that is, L-shaped with projections, cannot be analyzed with respect to danger points in the

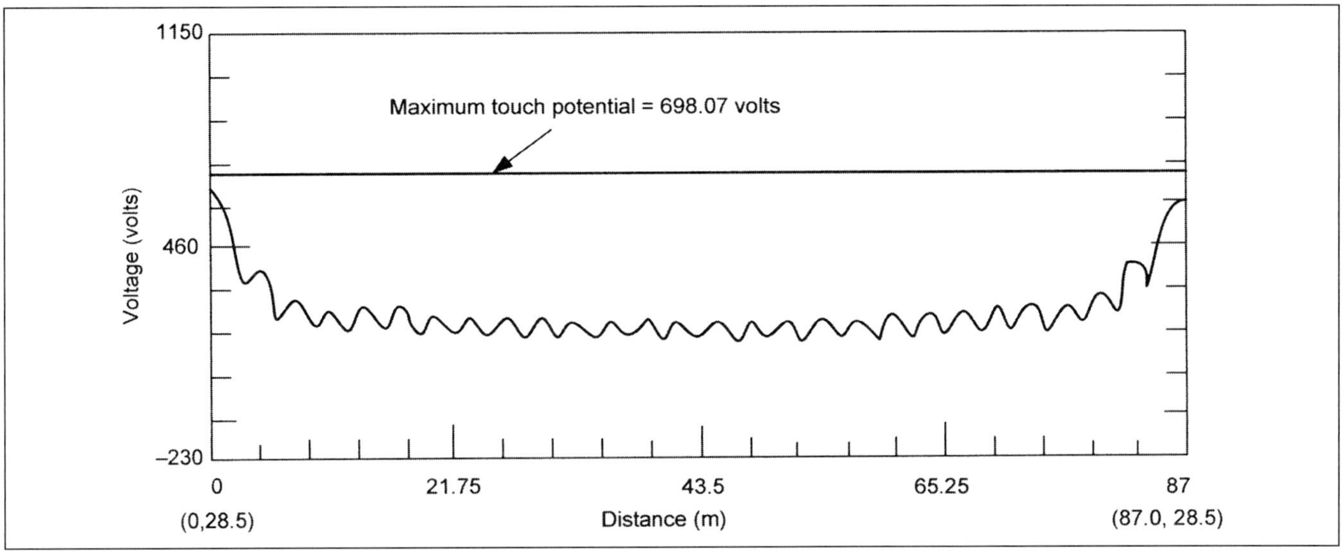

FIGURE 21-17 Profile of touch voltage through the middle cross section of the grid. This shows touch voltage below maximum safe touch voltage of 698.07 V.

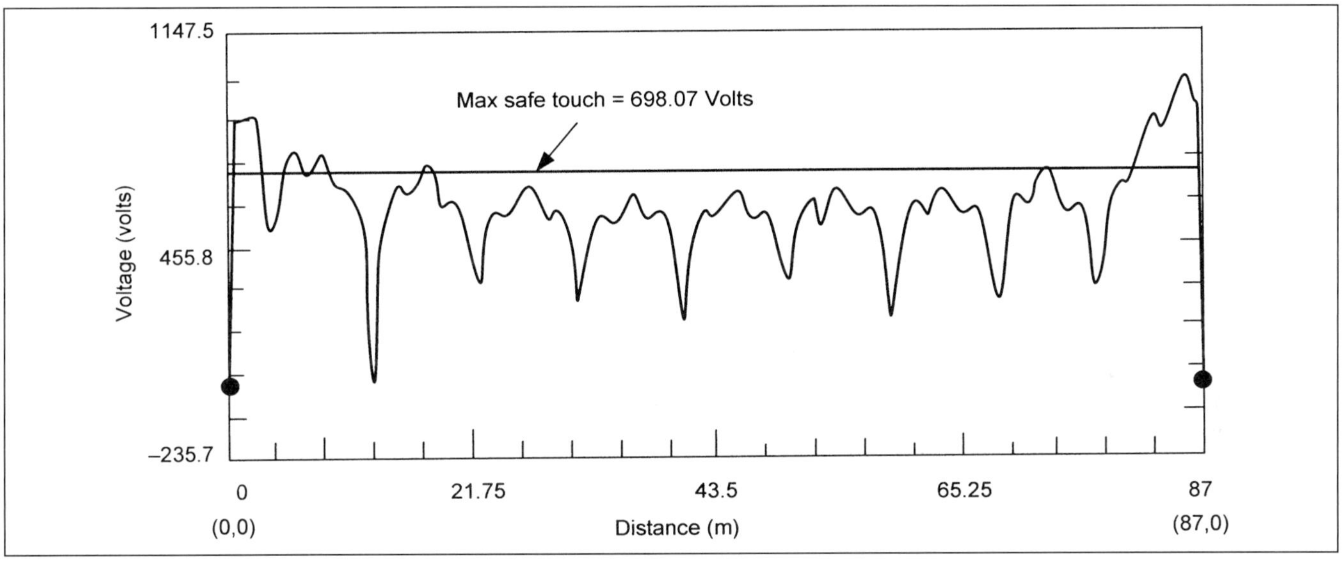

FIGURE 21-18 Profile of touch voltage at the periphery conductor of the grid. This shows higher than the safe touch voltage at the corners.

corners and bends. It is difficult to control the potentials at the corners in such a grid shape.

4. Some utilities adopt a grid design where the ground conductors are buried at two levels. The lower depth has conductors of larger size, while the top conductors are of smaller cross section, connected to the lower grid conductors at certain intervals. Though the design is not so common and somewhat controversial, an economy in conductor material is an advantage. Such a design cannot be analyzed using IEEE methods.

5. EPRI has developed grounding grid analytical programs, which calculate the split factors and can handle unequal spacing of conductors, two-layer designs with unequal conductor spacing, and flexibility in determining local danger points.

21-7 FINITE ELEMENT METHODS

Most computer algorithms use finite element methods[18] in addition to the IEEE method and have the graphical capabilities to plot three-dimensional profiles as well as surface potentials through any cross section of the grid. An effective design using unequal spacing and economizing the cost of the materials can be implemented. In the finite element methods, the current density need not be uniform.

The current flux at any point A, with respect to an elemental section of the grid conductor, as shown in Fig. 21-20, is:

$$\xi = \int_{y1}^{y2} \frac{\sigma_1}{4\pi} \frac{dy}{R^2} r \tag{21-35}$$

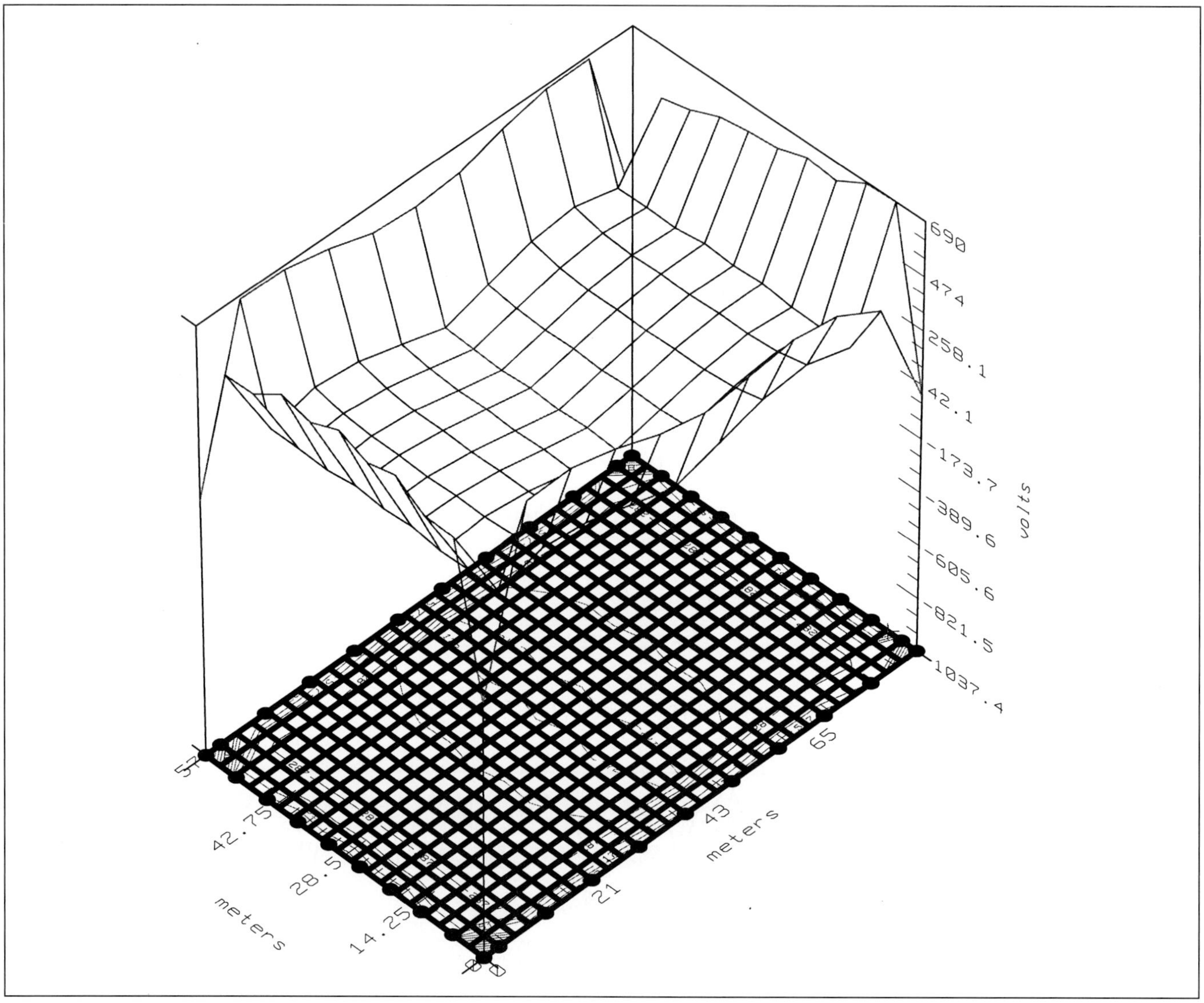

FIGURE 21-19 A three-dimensional plot of the touch voltages.

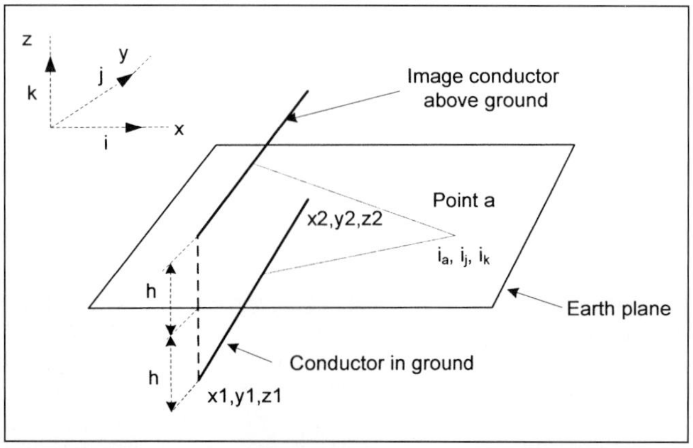

FIGURE 21-20 Illustration of the concept of finite element method for grounding grid design.

where, from Fig. 21-20, ξ is the current per unit area at any point, σ_1 is the current flowing to ground per unit length of conductor (current density) and:

$$R=\sqrt{(i-x_1)^2+(j-y_1)^2+(k-z_1)^2}$$
$$r=\frac{(i-x_1)i+(i-y_1)j+(i-z_1)k}{R} \tag{21-36}$$

The electrical field at any point is then:

$$E=\rho\xi \tag{21-37}$$

where ρ is the soil resistivity. Then the voltage at point a is:

$$v=\int E.dl \tag{21-38}$$

The conductor length can be divided into small finite elements. The current density will not be uniform. In practice, variation of the current density along the length of an element will have little effect

Dimensioning of primary electrodes

Element	X1	Y1	Z1	X2	Y2	Z2	Length	Radius	Current
0001	0.0	0.0	0.0	24.4	0.0	0.0	24.38	0.0067	22.1752
0002	24.4	0.0	0.0	48.8	0.0	0.0	24.38	0.0067	18.8225
0003	48.8	0.0	0.0	73.1	0.0	0.0	24.38	0.0067	17.5993
0004	73.1	0.0	0.0	97.5	0.0	0.0	24.38	0.0067	16.8991
0005	97.5	0.0	0.0	121.9	0.0	0.0	24.38	0.0067	16.5769
0006	121.9	0.0	0.0	146.3	0.0	0.0	24.38	0.0067	16.5769
0007	146.3	0.0	0.0	170.7	0.0	0.0	24.38	0.0067	16.8991
0008	170.7	0.0	0.0	195.1	0.0	0.0	24.38	0.0067	17.5993
0009	195.1	0.0	0.0	219.4	0.0	0.0	24.38	0.0067	18.8225
0010	219.4	0.0	0.0	243.8	0.0	0.0	24.38	0.0067	22.1752
0011	0.0	3.0	0.0	24.4	3.0	0.0	24.38	0.0067	14.1232
0012	24.4	3.0	0.0	48.8	3.0	0.0	24.38	0.0067	12.9482
0013	48.8	3.0	0.0	73.1	3.0	0.0	24.38	0.0067	12.0818
0014	73.1	3.0	0.0	97.5	3.0	0.0	24.38	0.0067	11.5965
0015	97.5	3.0	0.0	121.9	3.0	0.0	24.38	0.0067	11.3711
0016	121.9	3.0	0.0	146.3	3.0	0.0	24.38	0.0067	11.3711
0017	146.3	3.0	0.0	170.7	3.0	0.0	24.38	0.0067	11.5966
0018	170.7	3.0	0.0	195.1	3.0	0.0	24.38	0.0067	12.0818
0019	195.1	3.0	0.0	219.4	3.0	0.0	24.38	0.0067	12.9482
0020	219.4	3.0	0.0	243.8	3.0	0.0	24.38	0.0067	14.1232
0021	0.0	6.1	0.0	24.4	6.1	0.0	24.38	0.0067	11.0370
0022	24.4	6.1	0.0	48.8	6.1	0.0	24.38	0.0067	10.8395
023	48.8	6.1	0.0	73.1	6.1	0.0	24.38	0.0067	10.0823
4	73.1	6.1	0.0	97.5	6.1	0.0	24.38	0.0067	9.6671
	97.5	6.1	0.0	[illegible]	6.1	0.0	24.38	0.0067	[illegible]
	[illegible]	6.1	[illegible]		[illegible]	0.0	24.38	[illegible]	
	[illegible]	6.1							

FIGURE 21-21 A section of a computer printout showing the finite element approach in dimensioning the grounding electrodes.

upon the calculated voltages. The variation between elements will be significant. The variation in current density can be accounted for by modeling the elements in sections. Then the total voltage at a point is the summation of all the elemental sections in the grid.

Figure 21-21 shows a section of the computer printout. Note the difference in current in the elements. A user can increase the number of sections of a conductor for analysis for greater accuracy.

As an extension, multiple images of each conductor are used to perform analysis in two-layer soil models. The advantage is that the technique accounts for the finite length of the element and can handle grid designs with large asymmetry. Special danger points, that is, exterior to the fence along the perimeter, and the grid corners can be ascertained and the grid design modified for the optimum results. Figure 21-22 shows the special areas in L-shaped grid where the hazards potentials should be investigated.

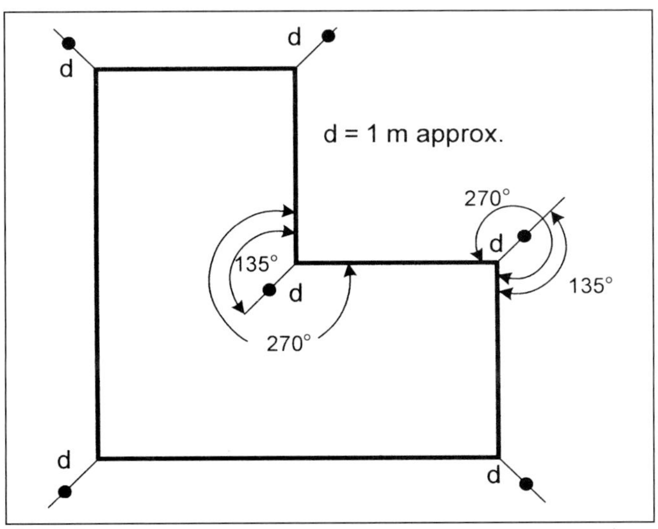

FIGURE 21-22 Danger points, where the voltages can be high, at the corners of an L-shaped grid.

It is practical to extend crushed rock layer some distance outside the fence, while the fence itself is included in the grid area, about 1 to 1.5 m inside the grid conductors. A grid conductor run about 0.8 m away outside the fence will help to reduce the shock hazard. Note the high resistivity of asphalt shown in Table 21-3. Outside the grid, from the grid conductors on the boundary, the voltage will fall gradually over a distance.

Example 21-3 This example is a computer simulation of a relatively difficult grid design as the soil resistivity measurements are 762 Ω-m. The grid is 242 × 121 m, rectangular. The computer calculations optimize the design with unequal conductor spacing, 26 conductors along *x*-axis, and 72 conductors along *y* axis, shown in the grid layout in Fig. 21-23. Note the closer spacing of the conductors toward the perimeter. The calculated GPR is 12556 V, with a grid current of 6.77 kA. The grid resistance is 1.8531 Ω. The tolerable step and touch voltages are 550.45 Ω and 1709 V, respectively.

Due to rocky soil, Chemrods® are used,[19] and each ground rod is installed horizontally at a shallow depth of 0.5 m due to the rocky nature of the soil and difficulty in drilling. Each horizontally installed rod of 8 ft length in artificial soil fill is treated like a 40 ft 1.5 in rod driven vertically in the soil. The three-dimensional conductor profile of the grid is shown in Fig. 21-24. A three-dimensional profile of voltages is shown in Fig. 21-25. Compare the voltage counter profile with Fig. 21-19. By closer spacing of the ground grid conductors toward the periphery, the touch voltages are even lower compared to the voltages in the middle grid meshes.

21-8 GROUNDING AND BONDING

Consider the following scenarios:

1. The lightning currents that are to be conducted to ground can be very large. Low-impedance grounding is required to dissipate the electrostatic charge and prevent a strike and to provide a path for lightning surge currents. The efficacy of lightning protection system can be marred by an ineffective grounding system. In the lightning protection system of structures, the metallic portions of windows and frames, are bonded to the down conductors to prevent side flashes.

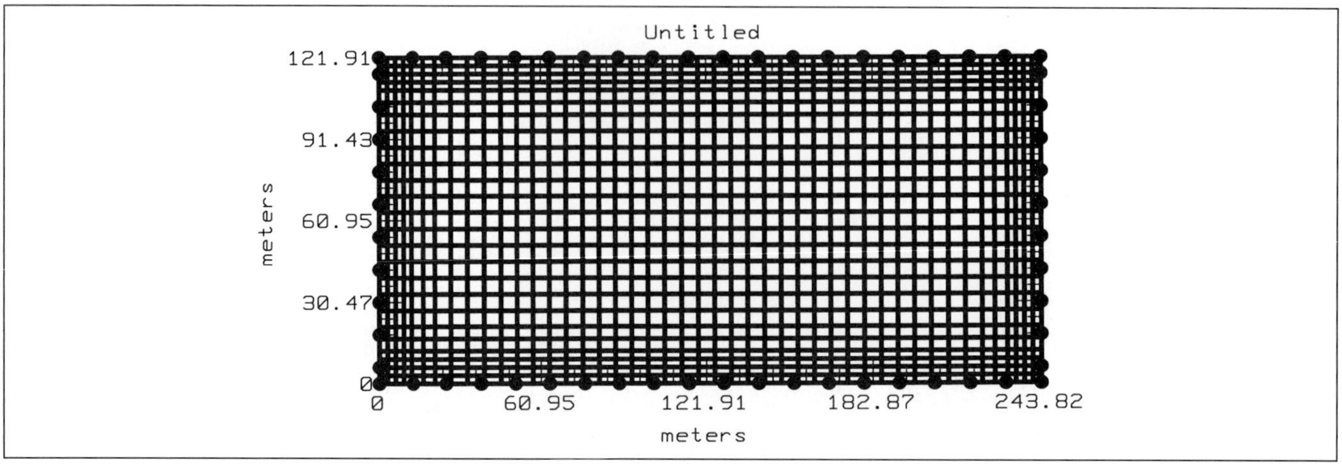

Figure 21-23 Plan view of grounding conductors and rods, unequal spacing.

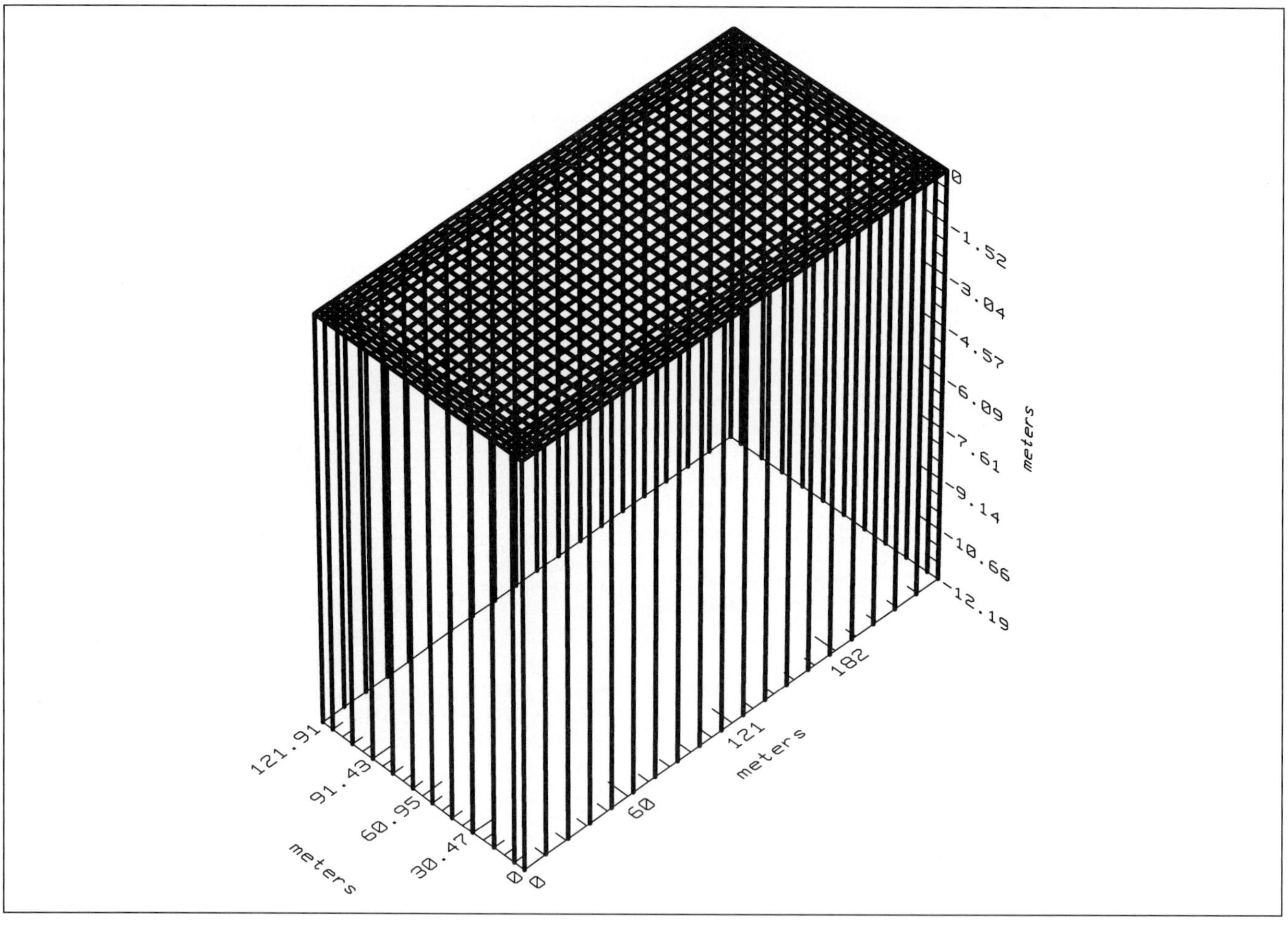

Figure 21-24 A three-dimensional view of the grounding grid electrodes with simulated ground rods 40 ft deep.

2. Potential hazards can occur if the tanks and vessels and floating roof tanks storing inflammable liquids are not grounded according to appropriate standard requirements.

3. The data control and computer system grounding has special requirements.

4. The electrical noise and shielding can pose problems in data processing and control systems in case of ineffective or inappropriate grounding.

5. The protection of human life and risk of fire hazard and explosion is directly related to proper system and equipment grounding.

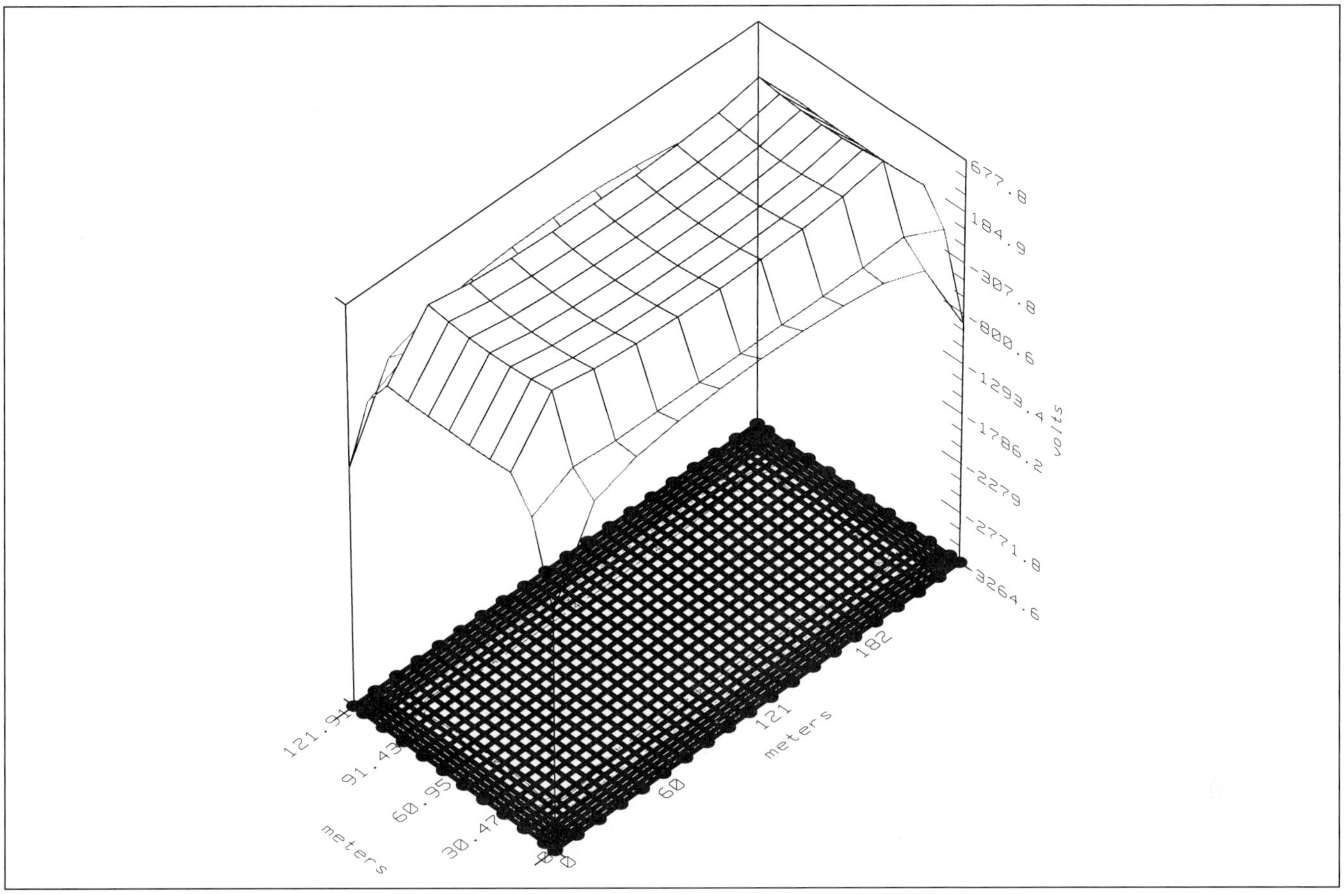

FIGURE 21-25 A three-dimensional plot of the touch voltage profile of the grid.

Lack of bonding can give rise to differential voltages with respect to the grounded equipment and pose shock hazards. In an industrial plant, the plant grid is essentially composed of steel reinforcing bars in the concrete footings, foundations, and floor slabs. These electrodes are supplemented at each building and structure by bare copper conductors connecting together individual footings, foundations, water pipes, and driven rods. At each building or structure, the grounding system is bonded to structural steel, large metallic tanks, and the frames of process and electrical equipment. Metallic pipes and conduits are bonded to the grounding electrode system by virtue of being supported on or from structural steel. The neutrals of separately derived systems or their secondary side neutral impedances are also bonded to the building grounding electrode system. This creates an equipotential surface and provides the required safety for personnel and operation of the protective relays.

Let us consider the grounding grid designs in a situation where a high voltage utility substation is located next to or in the same premises as a large industrial facility. It is not uncommon to see 230-kV high voltage step-down substations consisting of a number of transformers and industrial plant loads of the order of 150 to 200 MW. The distribution voltage may be at multivoltage level, say, at 13.8, 4.16, 2.4, and low voltage. These distribution systems will, invariably be high resistance or low resistance grounded; thus limiting the magnitude of ground fault currents. The maximum ground fault current giving rise to GPR will, therefore, occur for a high-voltage fault in the utility substation. As discussed earlier, there is no effective way to isolate all the metallic connections, cable shields, cable racks, and ground connections between the utility substation and the industrial distribution systems. In fact, an industrial plant grid will have much lower ground resistance as compared to the utility substation grid. With all the ground systems bonded together, the transfer potentials (Fig. 21-14) should not be a problem. Here the ground current will split between (1) the remote source, and (2) industrial distribution grounding system and utility grounding grids. This leads to fairly complex calculations to analytically calculate the split factor.

21-9 FALL OF POTENTIAL OUTSIDE THE GRID

Figure 21-26[20] shows that the potential will fall gradually outside the grid area. The earth potential distribution from the edge of a power station grid with respect to remote earth point is shown for various grid sizes.

First consider an analytical calculation of the fall of grid potential. Consider a grid of area 31416 m^2, 2000 m of buried ground conductor, and a soil resistivity of 200 Ω-m (sand and clay). This gives a grid resistance of 0.6 Ω; the current flowing in the earth is 11.25 kA. Applying split factor and decrement factor, the calculated GPR = 6.75 kV.

The voltage at a distance x from the grid can be calculated from the following equation:

$$\frac{V(x)}{\text{GPR}} = \frac{0.64}{(1+4r/L)}\sin^{-1}(r/x)\ \text{rad} \tag{21-39}$$

where $V(x)$ is the voltage at distance x measured from the center of the grid, and r is the radius of a circle with an area equal to the grid area. This is independent of the soil resistivity and is merely a function of grid geometry, and distance x should be $\geq 3r$.

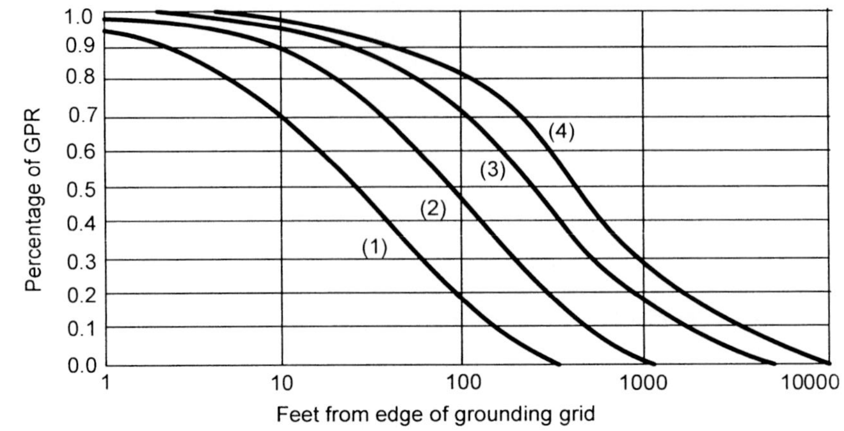

FIGURE 21-26 Fall of potential outside a grid. (1) A small grid area 1600 ft^2 (150 m^2), (2) a medium-small grid area, approximately 35000 ft^2 (3350 m^2), (3) a medium-large grid area, approximately 290000 ft^2 (27000 m^2), (4) a large grid area, approximately 935000 ft^2 (87000 m^2). *Source: ANSI/IEEE Std.* 367.[20]

Assuming a totally isolated grid, the voltage at 300 m from the grid will be 1.19 kV. Thus, the zone of the fall of potential outside the grid extends over a considerable distance. These calculations can be correlated with Fig. 21-26. A large separation distance is required to reduce the coupled voltage to 100 V. The impact of GPR and its zone of influence can be summarized as follows:

- The potential can couple to other nearby grids. Isolation of two grids in close proximity is not possible and these should be interconnected.
- The potential can be transferred to a metal part in the ground in the vicinity of the grid. Consider a buried telecommunication cable. The potential can enter the telecommunication circuits through cable sheath or telecommunication grounding electrodes.
- Figure 21-27 shows that the contours of potential around a grid area are not uniform and get distorted due to presence of buried pipes and metal objects. In this figure, A is the area of the power station grid. Bs are points of equal values of ground

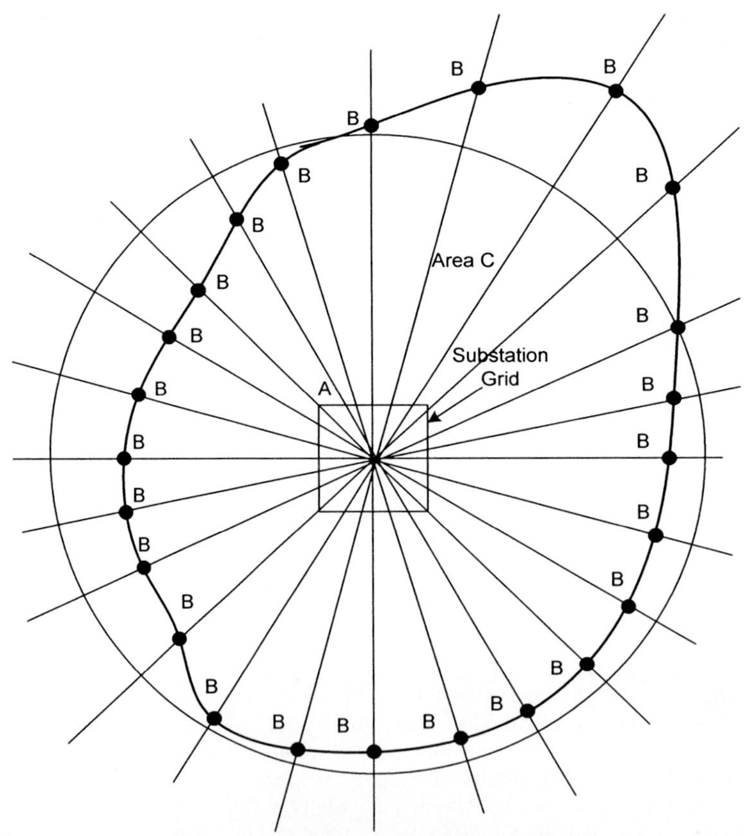

FIGURE 21-27 Zone of influence of fall of potential outside a power station grid to safe levels.

resistance from station ground grid where the GPR is reduced to an acceptable level. Note the departure from ideal circular locus. Area C is the zone of influence of GPR.

21-10 INFLUENCE ON BURIED PIPELINES

From the above discussions, it is amply clear that soil itself acts like a coupling medium for a potential generated somewhere else in the power system, and the effect of resultant GPR is present at a considerable distance from the original site of GPR. Often pipe lines and power lines run in parallel (at least for some distance), and due to the high cost of acquiring land and right-of-way, it is not practical to avoid such parallelism. In the pipeline work, the following equation is used to ascertain the distance at which the effect of parallelism is considered negligible:

$$d \approx 100\sqrt{\rho} \tag{21-40}$$

This means that for a high soil resistivity say 500 Ω-m, a separation of 2.24 km is required. High voltages can be induced in the pipelines and currents can flow due to such parallelism. These voltages and currents can have the following effects:

- Damage the pipeline insulation and cause flashovers.
- Swap the normal cathodic protection currents provided for the pipelines and render the cathodic protection inoperative.
- These can be a source of danger and shock hazard to operating personnel.

The two conditions to be studied are (1) ground faults, which will raise the tower footing and surrounding ground voltage, as discussed in earlier chapters, and (2) voltages induced under normal operating conditions. This occurs because the magnetic flux around the conductors and supports is not perfectly balanced. Mutual inductances exist between the overhead line and the pipeline, which can be estimated using Carson's equations, (App. D). The pipeline acts like a parallel transmission line of certain characteristics' impedance. The voltage and current induced in the pipeline is given by the expressions:

$$V_s = \frac{-jM\omega J}{2\gamma_p}[e^{-\gamma_p(L-x)} - e^{-\gamma_p x}]$$
$$I_s = \frac{-jM\omega J}{2Z}[2 - e^{-\gamma_p(L-x)} - e^{-\gamma_p x}] \tag{21-41}$$

where V_s and I_s are the current and voltages at any distance x, respectively, L is the length of parallelism, γ_p = propagation constant, Z is the characteristic impedance, M is the mutual inductance, and J is the current in the power line. Figure 21-28 shows the current and voltage profiles. Equation (21-41) is applicable when pipe line is terminated in its characteristics impedance.

Table 21-4 is an actual calculation of the induced voltages on a 30-in pipeline sections paralleled with 400-kV line carrying a current of 100 A (line configuration not described here). The induced voltage is a function of line-coating resistance (screening effects not shown). The remedial measures to reduce induced voltages on parallelism include lowering tower footing resistance, earth wires on towers, metallic screen (buried metallic conductor), and increasing spacing between the power line and pipeline.[21]

21-11 BEHAVIOR UNDER LIGHTNING IMPULSE CURRENT

In Chap. 5, there is a brief discussion of the tower footing resistance under lightning impulse currents. The lightning current is at high frequencies, from some hundreds of kHz to 1 MHz, and the soil resitivity also varies over a wide range. The transient behavior of the ground electrode system is determined more by inductive phenomena than the resistive effects. The rate of rise of the front of current impulse is of major importance, as it increases the inductive voltage drop.[22,23,24]

The high values of lightning currents, with very short front durations, can result in high current density in the layers of soil, adjacent to the grounding conductors and electrodes. IEEE standard[25] has a table which provides mathematical expressions for calculating the ground resistance of grounding electrodes of various shapes. For a single ground rod of length L and radius r, the ground resistance is given by:

$$R = \frac{\rho}{2\pi L}\left(\ln\frac{4L}{r} - 1\right) \tag{21-42}$$

The current density J at a distance a from the rod for an injected current I, is $J = I/2\pi aL$ and the voltage gradient E is ρJ.

For high currents, when the gradient E exceeds a certain level, soil moisture is evaporated, and arcs are produced within the soil around the electrode in a certain area surrounded by streamers. The streamers and arc zones can be modeled as an area of zero resistivity.

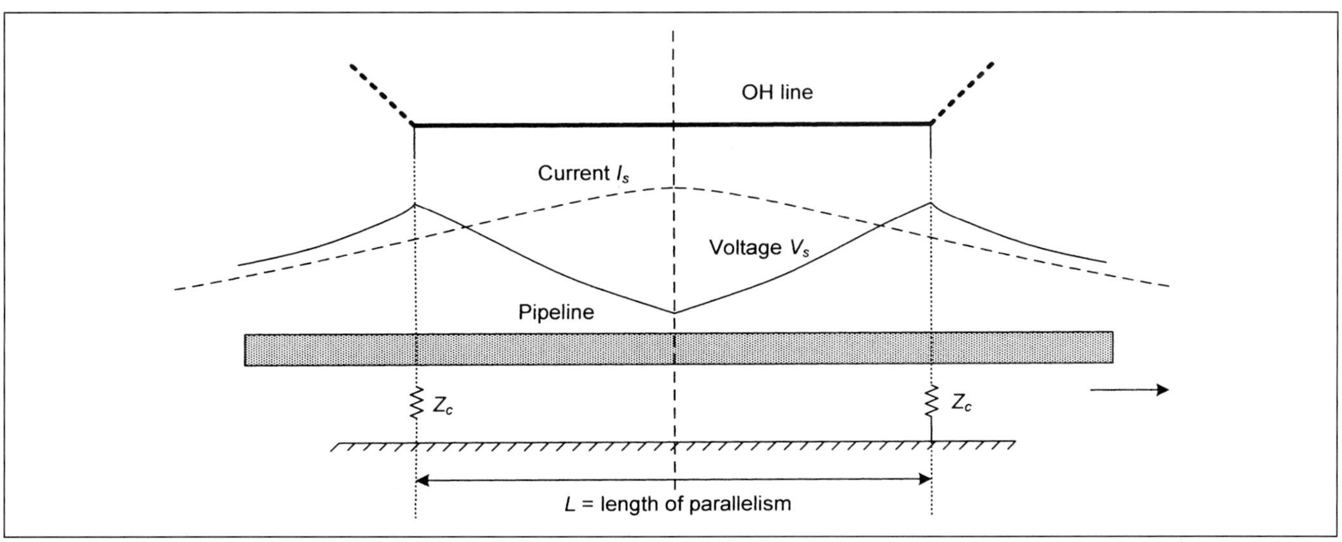

FIGURE 21-28 Voltage and currents induced in a pipeline in parallel with a power line; the pipeline terminated in its characteristic impedance.

TABLE 21-4 Induced Voltages in a 30 in Pipeline Paralleled Sections with a 400-kV Line

SOIL RESISTIVITY (Ω-m)	SEPARATION (m)	LENGTH OF PARALLELISM (km)	INDUCED VOLTAGE (kV)
10	160	16.7	20
25	150	30.7	25
22	100	27.4	23
60	460	2.4	13.3

Note: The transmission line carries a current of 100 A. The pipeline has a coating resistance of 10 Ω/km, linear resistance of 0.2 Ω/km, and linear inductance of 0.5 Ω/km.

Thus, for certain critical field strength, the diameter of the grounding rod is artificially increased. This means that a ground rod can be modeled as hemisphere electrode.

For a hemisphere of radius r[11]:

$$R = \frac{\rho}{2\pi r} \tag{21-43}$$

Similar expressions for J and E are: $J = I/2\pi ra^2$ and $E = \rho J$. Now expand the radius r to $\rightarrow$ a new radius $= a$, and eliminating a from the above relations, the impulse resistance is:

$$R_i = \sqrt{\frac{E_0 \rho}{2\pi I}} \tag{21-44}$$

where E_0 is the critical gradient. A critical gradient of 1000 kV/m is suggested in Ref. 22, though 400 kV/m is more common. There has to be a sufficient current I_g to produce the critical gradient. Then, from the current density and voltage gradient for the hemisphere electrode:

$$I_g = \frac{2\pi a^2 E_0}{\rho} = \frac{1}{2\pi}\frac{\rho E_0}{R^2} \tag{21-45}$$

Here, we have set $a = r$. Then in terms of I_g and from Eq. (21-44)

$$R_i = R\sqrt{\frac{I_g}{I}} \tag{21-46}$$

It can be shown that R_i is a function of log of I and a further modification to this equation for ground rods is:

$$R_i = \frac{R}{\sqrt{1+(I/I_g)}} \tag{21-47}$$

We applied this equation in Chap. 5. An equation to estimate the tower footing inductance is:[23]

$$L_f = 60T_t \ln\frac{t_f}{T_t} \tag{21-48}$$

where T_t is the tower travel time, and t_f is the time to crest of the stroke current.

21-11-1 Impulse Response of Grounding Grids

When analyzing a grid system for impulse currents, consider Fig. 21-29*a*, where the impulse current is injected into a certain point in the system. An elemental section can be considered as shown in Fig. 21-29*b* and it acts like a distributed parameter transmission line. Based on the current theoretical studies and experimental work, the following conclusions are summarized:

1. The impulse current dispersion is dependent upon the electrode length, soil resisitvity, peak value of the surge current, and the time to crest.

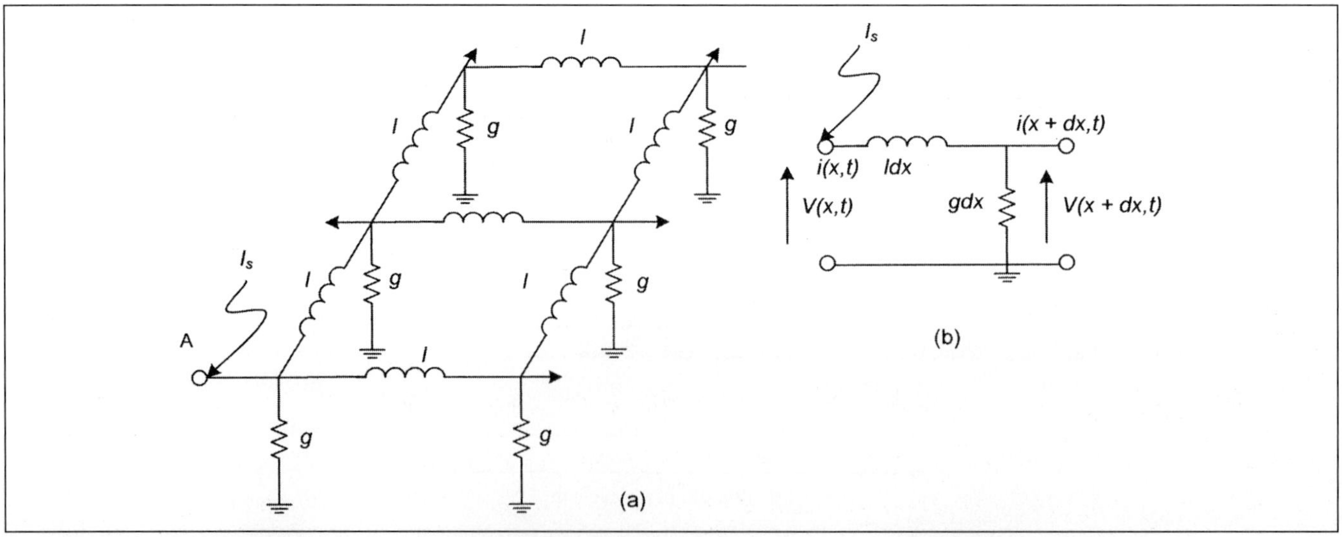

FIGURE 21-29 (*a*) Injection of impulse current in a grounding grid. (*b*) Simulation like a traveling wave circuit.

2. An analytical expression to calculate the effective length of earth electrode is:

$$l_e = K_0\sqrt{\rho\tau} \tag{21-49}$$

where l_e is the effective length in m, and τ is the time to crest in μs, and ρ is the resistivity in Ω-m. The factor K_0 depends upon the geometric construction. It can be taken as 1.40 for the single conductor energized at one end, 1.55 for the single conductor energized in the middle, and 1.65 for conductors arranged in mesh formation, energized at the center. The effective length signifies the distance at which the conventional ground resistance does not undergo any significant reduction.

3. The localized earth electrodes will present lower resistance if lightning impulse current is sufficiently high to cause soil ionization.

Selection and applications of grounding systems for low-voltage to high-voltage levels and their behavior under power frequency and surge currents is rather an extensive field. The chapter provides an overview of the system and equipment grounding and points to further investigations and reading.

PROBLEMS

1. A 480-V system is solidly grounded. Calculate the permissible damage in cubic inches to copper, steel, and aluminum for a rated current of 2000 A.

2. In Prob. 1, calculate the maximum fault clearing time of a ground fault relay to limit the damage to the permissible NEMA limits. The arcing current is 30 kA, and the operating time of the circuit breaker is half a cycle.

3. Comment on the statement that arcing ground phenomena occurs only in the ungrounded systems.

4. What are the typical X_0/X_1, R_0/X_0 ratios for an effectively grounded system? What is typical COG for solidly grounded systems?

5. The utilities solidly ground their transmission systems and also distribution and subdistribution systems, except the generators, which are high resistance grounded. Why? If a generator neutral is left ungrounded, what harmful effects can occur?

6. An 18-kV, three-phase generator is to be high resistance grounded through a distribution transformer of 240-V secondary windings. The calculated capacitance current of the generator circuit is 8 A. Calculate (1) voltage ratio and kVA rating of the transformer, and (2) the secondary loading resistor value across 240-V secondary windings of the distribution transformer and its current rating. What should be the primary voltage rating of the distribution transformer? Could there be a problem if $18/\sqrt{3}$ kV is selected, as the generator windings are wye-connected and rated for 18 kV?

7. Why are 13.8-kV industrial distribution systems not high resistance grounded? State two technical reasons.

8. Comment on the statement: "On a cloudy rainy day, a deer was grazing outside the fence of a high-voltage substation. Suddenly it fell unconscious to ground, as if subjected to an electrical shock. Then it got on its feet and fled."

9. Describe the importance of providing crushed rock in the design of substation grounding grids, especially in areas of low soil resistivity.

10. Consult IEEE Std. 80 to find out the split factor of the ground fault current given that:

Single line-to-ground fault current = 20 kA

Ground mat resistance = 1 Ω

Tower footing resistance = 20 Ω

Number of transmission lines leaving/entering the substation = 4

Number of sky wires = 4

Remote and local contribution are 25% and 75%, respectively.

Calculate the split factor based on the calculation procedure given in this standard.

11. Design a grounding grid, given the following parameters:

Ground fault current = 11 kA rms symmetrical, $X/R = 10$

Split factor = 0.6

Capacity factor = 1.1

Ambient temperature = 40°C

Maximum allowable temperature for compression joints: 350°C

Grid area = 70 m × 80 m rectangular

Soil resistivity = 200 Ω-m

Crushed rock laid to a depth of 0.1 m, resistivity = 3000 Ω-m

Grid conductor buried to a depth of 0.6 m

Fault duration = 0.5 s = shock duration.

Number of ground rods = 30, along the periphery and grid corners. Each rod is 3/4 in diameter and 10 ft long.

Calculate:

- Size of grid conductor
- Tolerable step and touch potentials
- Decrement factor
- Grid resistance
- GPR
- Factors K, K_m, K_i, E_m etc.

Iterate and finalize the grid design. Prepare a layout of the final design. Does the design need to be modified? Where will the additional conductors/ground rods be added without a rigorous calculation to ascertain local danger points? Modify the calculated design and prepare a final sketch of the layout.

12. Calculate the fall of potential outside the grid area at a distance of 100 and 200 m in Prob. 11. At what distance will the voltage fall to 100 V?

REFERENCES

1. NEMA PB1-2, Application Guide for Ground Fault Protective Devices for Equipment, 1977.
2. H. L. Stanback, "Predicting Damage from 277 Volt Single-Phase-to-Ground Arcing Faults," *IEEE Trans. IA*, vol. 13, no. 4, pp. 307–314, Jul./Aug. 1977.

3. NFPA 70E, Electrical Safety Requirements for Employee Workplaces, 2004.
4. IEEE Std. 1584, IEEE Guide for Performing Arc-Flash Calculations, 2002.
5. ANSI/NFPA 70, National Electric Code, 2008.
6. J. C. Das, "Ground Fault Protection of Bus-connected Generators in an Interconnected 13.8 kV System," *IEEE Trans. IA*, vol. 43, no. 2, pp. 453–461, Mar./Apr. 2007.
7. D. S. Baker, "Charging Current Data for Guess Work-free Design of High Resistance Grounded Sytems," *IEEE Trans. IA*, vol. IA-15, no. 2, pp. 136–140, Mar./Apr. 1979.
8. ICEA Pub. S-61-40, NEMA WCS, Thermoplastic Insulated Wire for Transmission and Distribution of Electrical Energy, 1979.
9. J. R. Dunki-Jacobs, "The Reality of High Resistance Grounding," *IEEE Trans.* IA, vol. IA-13, pp. 469–475, Sept./Oct. 1977.
10. J. C. Das and H. Osman, "Grounding of AC and DC Low-Voltage and Medium-Voltage Drive Systems," *IEEE Trans. IA*, vol. 34, no. 1., pp. 205–216, Jan./Feb. 1998.
11. ANSI/IEEE Std. 80, IEEE Guide for Safety in Substation Grounding, 2000.
12. J. Endrenyi, "Analysis of Transmission Tower Potentials During Ground Faults," *IEEE Trans. PAS*, vol. 86, pp. 1274–1283, Oct. 1967.
13. S. A. Sebo, "Zero Sequence Current Distribution Along Transmission Lines," *IEEE Trans. PAS*, vol. 88, pp. 910–919, Jun. 1969.
14. F. Dawalibi, "Ground Fault Distribution Between Soil and Neutral Conductors," *IEEE Trans. PAS*, vol. 99, no. 2, pp. 452–461, Mar./Apr. 1980.
15. A. P. Meliopoulos, A. Papalexopoulos, and R. P. Webb, "Current Division in Substation Grounding System," *Proc. of the 1982 Protective Relaying Conf.*, Georgia Institute of Technology, Atlanta, GA, 1982.
16. EPRI Final Report, EL-2682, Vol. 1, "Analysis Techniques for Power Substation Grounding Systems," Oct. 1982.
17. J. G. Sverak, "Sizing of Ground Conductors Against Fusing," *IEEE Trans. PAS*, vol. 100, no. 1, pp. 51–59, Jan. 1981.
18. P. P. Silvester and R. L. Ferrari, *Finite Elements for Electrical Engineers*, 3rd ed., Cambridge University Press, New York, 1996.
19. Lightning Eliminators and Consultants, Chem Rod and GAF (Ground Augmentation Fill) Product Data, www.lecglobal.com.
20. ANSI/IEEE Std. 367, IEEE Recommended Practice for Determining the Electrical Power Station Ground Potential Rise and Induced Voltages from a Power Fault, 1996.
21. J. C. Das, "Influence of HV Line Parallelism on Buried Pipe Lines," *Electrical India*, vol. XX, no. 1, pp. 27–35, Jan. 1980.
22. E. E. Oette, "A New General Estimation for Predicting the Impulse Resistance of Concentrated Earth Electrodes," *IEEE Trans.* PD, vol. 4, no. 2, pp. 1329–1337, 1989.
23. W. A. Chisholm and E. Janischewskyj, "Lightning Surge Response of Grounding Electrodes," *IEEE Trans. PD*, vol. 4, no. 1, pp. 1329–1337, Apr. 1989.
24. CIGRE Working Group 33.01, Guide to Procedures for Estimating the Lightning Performance of Transmission Lines, Brochure 63, Oct. 1991.
25. IEEE Std. 142, IEEE Recommended Practice for Grounding of Industrial and Commercial Power Systems, 1991.

FURTHER READING

R. R. Conrad and D. Dalasta, "A New Ground Fault Protective System for Electrical Distribution Circuits," *IEEE Trans.*, vol. IGA-3, no. 3, May/Jun. 1967.

F. Dawalibi and D. Mukhedkar, "Parametric Analysis of Grounding Grids," IEEE PES Winter Meeting, Paper No. F79243-7, 1979.

M. J. Frazier and J. Dabkowski, "Magnetic Coupled Longitudinal Electrical Filed Measurements on Two Transmission Lines," *IEEE Trans. PAS*, vol. 104, no. 4, pp. 933–940, Apr. 1985.

General Electric Industrial Power System Data Book, 1954.

IEEE Std. 399, Brown Book, Power System Analysis, 1997.

E. B. Joy, A. P. Meliopoulos, and R. P. Webb, "Touch and Step Calculation for Substation Systems," IEEE PES Winter Meeting, Paper No. A79052-2, 1979.

R. H. Kaufman and J. C. Page, "Arcing Fault Protection for Low Voltage Power Distribution Systems—Nature of the Problem," *IEEE Trans. IA*, vol. 79, pp. 160–167, Jun. 1960.

R. C. Quist, "Voltages to Ground in Load Commutated Inverters," *IEEE Trans. IA*, vol. 24, pp. 526–530, May/Jun. 1988.

W. F. Robertson and J. C. Das, "Grounding Medium Voltage Mobile or Portable Equipment," *Industry Applications Magazine*, vol. 6, no. 3, pp. 33–42, May/Jun. 2000.

R. Ruenburg, *Transient Performance of Electrical Power Systems*, McGraw Hill, New York, 1954.

R. Verma and D. Mukhedkar, "Ground Fault Current Distribution in Substation Towers and Ground Wires," *IEEE Trans. PAS*, vol. 98, pp. 724–730, 1979.

Chapter 22

Lightning Protection of Structures

Surge protection of the incoming services, electrical supply lines, and communication lines was discussed in the previous chapters. In a likewise manner, the direct and nearby cloud-to-ground discharges can be hazardous to the structures, human life, and the contents in the structures. To safeguard against lightning hazards, risk management approach is adopted. In this approach, the lightning risks are identified, frequencies of events and their consequences estimated, and if these are above a tolerable limit, lightning protection is provided. The protection provided should reduce the risk below an acceptable level. In this approach, a comparison can be drawn with the statistical methods used for insulation coordination. The type of structure, its occupancy, risk of lightning damage, and cost are simultaneously considered in arriving at an engineered solution.

Variations exist with respect to practices in Europe according to IEC standards and NFPA 780 (National Fire Protection Association) used in the United States. Both these practices are briefly discussed.

22-1 PARAMETERS OF LIGHTNING CURRENT

The lightning parameters considered for the protection in IEC standards[1–4] are:

- *Peak value of the first stroke*. The lowest value of the downward flashes is important for the choice of air terminal systems to prevent direct flashes to the structures and has the highest statistical values for protection against electrodynamic effects.
- *Maximum rate of rise*. This is important for inductive effects and overvoltages.
- *Flash duration and total charge in the flash*. The highest statistical values are important for limiting the thermal effect at the impact point of lightning flash.
- *Specific energy in flash*. The highest value of statistical distribution is important for selection of conductor sizes to prevent damage due to thermal effects and implementing suitable grounding systems.

22-2 TYPES OF STRUCTURES

The structures according to IEC 62305-3[3] are classified as:

1. Common structures.
2. Structures with risks of explosion; that is, zone 0 as classified in IEC 60079-10[5] for storing solid explosive materials. IEC does not consider hazardous zones types 1 and 2 at risk of explosion due to low probability of simultaneous presence of lightning and inflammable vapors.
3. Structures with electronic systems in which a large amount of electronic installations exist, including telecommunication equipment.
4. Structures dangerous to environment, which may cause biological, chemical, or radioactive emissions and hazard as a consequence of lightning; for example, chemical, petrochemical, and nuclear plants.

22-2-1 Risk of Fire

For risk of fire, the structures are classified into three categories based on the specific fire load of the structure, which is calculated as the ratio of the total amount of combustible material in the structure divided by the overall surface area of the structure. It is specified in kg/m^2.

1. *Structures with high risk of fire*. Structures that are constructed of combustible materials or have roofs of combustible materials, and structures with a specific fire load of 45 kg/m^2.
2. *Structures with ordinary risk of fire*. Structures with specific fire load of 20 to 45 kg/m^2.
3. *Structures with low risk of fire*. Structures with a specific fire load of <20 kg/m^2 and structures that contain combustible materials occasionally.

22-3 RISK ASSESSMENT ACCORDING TO IEC

The following risks are to be taken into account according to the type of loss:

R_1—loss of human life

R_2—loss of service to public

R_3—loss of cultural heritage

R_4—economic loss—loss of structural value, contents, and normal activity in the structure

Figure 22-1, based on IEC 62305-3,[3] shows these risks from different types of damages. Risk components are specified.

1. Lightning flashes direct to the structure may generate:

   R_A—risk of shock due to touch and step potentials

   R_B—risk component due to fire, explosion, mechanical, and chemical effects inside the structure

   R_C—risk component due to failure of electrical and electronic systems

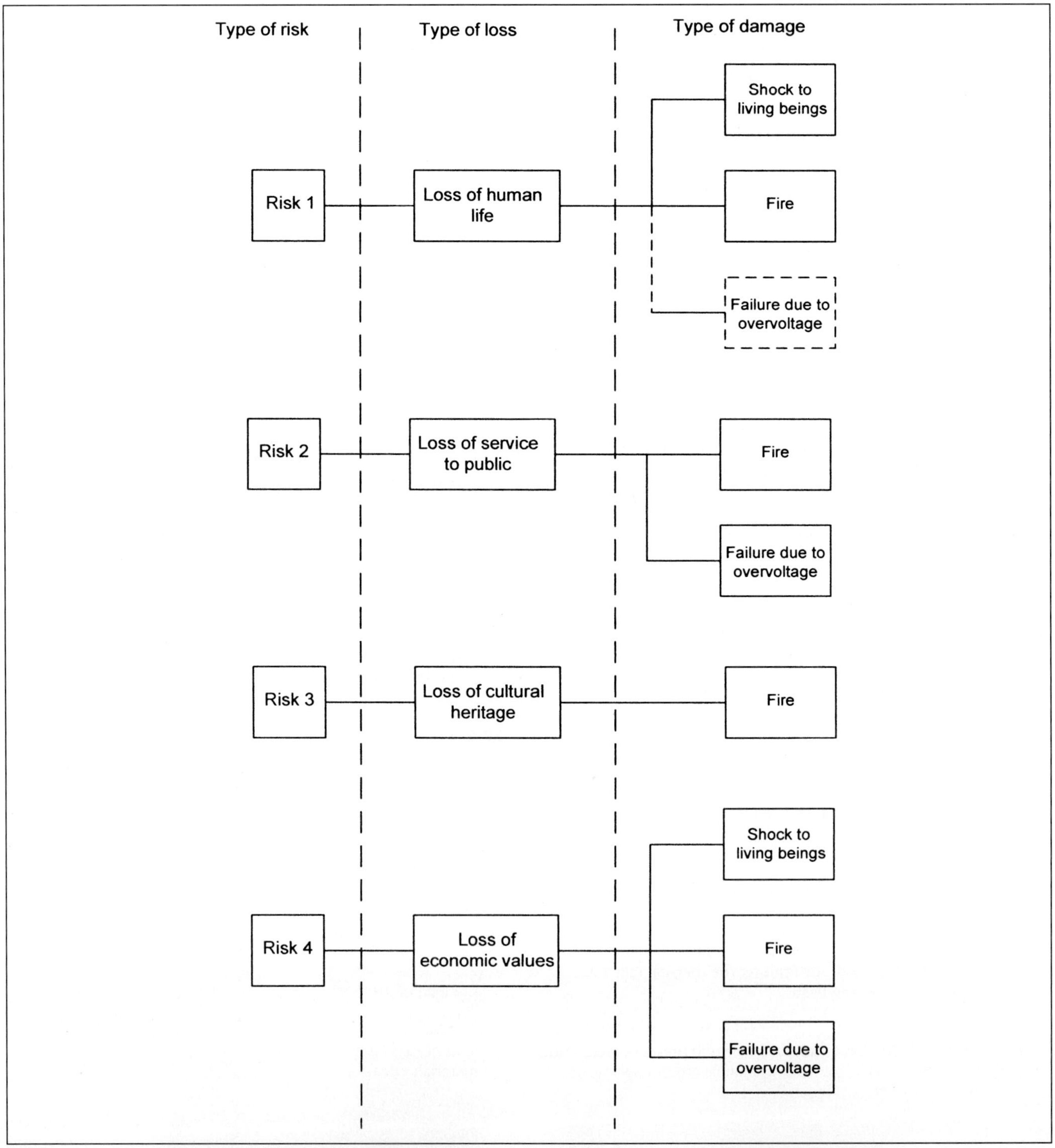

FIGURE 22-1 Risk categories, type of loss, and type of damage according to IEC.

2. Lightning flashes to ground near the structure generate:

   R_M—risk due to failure of electrical and electronic components inside the structure due to overvoltages

3. Lightning flashes direct to incoming lines generate:

   R_U—risk due to shock of living beings

   R_V—risk related to fire, mechanical, and chemical effects

   R_W—component due to failure of electrical and electronic equipment

4. Lightning flashes to ground near the incoming line will have:

   R_Z—risk due to failure of electrical and electronic equipment

Note that the type of damage may be ultimately the same, but the various components are labeled with respect to the type of lightning activity and loss.

IEC 62305-1[1] and 62305-2[2] define the risk as probable annual loss in a structure due to lightning and suggests the following expression:

$$R = NPL \tag{22-1}$$

where N is the number of flashes, P is the probability of damage, and L is the value of consequent loss. The probability P is not so simple to calculate. The probability that a spark could trigger a fire is given by:

$$P = p_s p_f \tag{22-2}$$

where p_s is the probability of a spark and p_f is the probability of a fire. The probability of failure of electronic system due to overvoltages caused by a direct flash to structure is given by:

$$P = [1-(1-p_r)(1-p_i)] \tag{22-3}$$

where p_r is failure due to overvolatge by resistive coupling and p_i is the failure due to inductive coupling.

The factor L is dependent on the use of the structure, the attendance time of the personnel in the structure, the type of service provided, and the value of goods affected by the damage. The factor R consists of the sum of its components, with reference to point of strike, as follows:

$$R = R_D + R_I \tag{22-4}$$

where R_D is the risk due to direct flashes to the structure and R_I is the risk due to indirect flashes to the structure. These are given by:

$$R_D = R_A + R_B + R_C \tag{22-5}$$

$$R_I = R_M + R_U + R_V + R_W + R_Z \tag{22-6}$$

For various types of damages:

$$R = R_s + R_F + R_O \tag{22-7}$$

where risk of shock to living beings is:

$$R_s = R_A + R_U \tag{22-8}$$

The risk related to physical damage is:

$$R_F = R_B + R_V \tag{22-9}$$

The risk related to failure of electrical and electronic systems is:

$$R_O = R_C + R_M + R_W + R_Z \tag{22-10}$$

TABLE 22-1 Typical Values of Tolerable Risk Factor R_T (IEC Standards)

TYPE OF LOSS	R_T
Loss of human life	10^{-5}
Loss of service to public	10^{-3}
Loss of cultural heritage	10^{-3}

IEC 62305-2[2] provides details of the calculations of risk factors. Table 22-1 shows typical values of tolerable risk factors. The objective is to reduce R to a maximum level R_T. If $R \le R_T$, lightning protection is not necessary; if $R > R_T$, protection measures shall be adopted to reduce $R \le R_T$.

22-4 CRITERIA FOR PROTECTION

According to IEC 62305-1,[1] the lightning protection system (LPS) consists of both the external lightning protection system and the internal lightning protection system inside the structure. The external LPS should be designed to achieve the following:

1. Intercept a direct lightning strike to the structure
2. Conduct the lightning current safely to ground
3. Disperse it into the earth

Protection measures are required to avoid hazard of touch-and-step potential for persons outside the structure or in the vicinity of down conductor system. This can also be achieved by increasing the surface resistivity of the soil outside the structure and of the floors inside the structure; see Table 21-3.

Four sets of LPS, types I, II, III, and IV, are defined, based on the corresponding lightning protection level. For each lightning protection level, a set of maximum and minimum current parameters have been stipulated in IEC 62305-1[1]. These values are shown in Table 22-2. These four levels may be considered construction rules for external LPS. For each level, the maximum and minimum lightning current parameters are fixed.

As an example, for protection level 1, the specified fixed maximum values do not exceed with a probability of 99 percent. Ten percent positive and 90 percent negative flashes are considered. Values taken from positive flashes must have probabilities below 10 percent and those from negative flashes below 1 percent. The minimum values of the lightning current amplitude influence the positioning of air termination system of an LPS in order to intercept the lightning flashes direct to the structure.

The minimum values of the lightning current parameters fixed in IEC 62305-1[1] and the rolling sphere radius, according to protection level, are shown in Table 22-3. Flashes with peak values less than the specified minimum value of current may still strike the structure. An explanation is provided in the next section.

22-4-1 Lightning Flashes to Earth

IEC standards consider two basic types of lightning flashes: (1) downward flashes initiated by a downward leader from cloud to earth and (2) upward flashes initiated by an upward leader from an earthed structure to cloud. In a flat territory, mostly downward flashes occur (Fig. 22-2). For exposed and higher structures, upward flashes become dominant (Fig. 22-3).

The profiles of short strokes and long strokes according to IEC 62305-1[1] and 62305-4[4] are shown in Fig. 22-4*a* and *b*, respectively. Short strokes are typically below 2 ms, and long stroke typically 2 ms < T_{long} < 1 s. The components in the upward stroke are the first

TABLE 22-2 Lightning Current Parameters

Occurrence	Parameter	Symbol	Unit	Protection Level I	Protection Level II	Protection Level III–IV
First short stroke	Peak current	I	kA	200	150	100
	Short-stroke charge	Q_{short}	C	100	75	50
	Specific energy	W/R	MJ/Ω	10	5.6	2.5
	Time parameters	T_1/T_2	μs/μs		10/350	
Subsequent short stroke	Peak current	I	kA	50	37.5	25
	Average steepness	di/dt	kA/μs	200	150	100
	Time parameters	T_1/T_2	μs/μs		0.25/100	
Long stroke	Long-stroke charge	Q_{long}	C	200	150	100
	Time parameter	T_{long}	s		0.5	
Flash	Flash charge	Q_{flash}	C	300	225	150

Source: From IEC 62305-1, 2006[1], and IEC 62305-3, 2006.[3]

TABLE 22-3 Minimum Value of Lightning Current and Related Rolling Sphere Radius

Protection Level	Rolling Sphere Radius, R (m)	Minimum Peak Current, I (kA)
I	20	3
II	30	5
III	45	10
IV	60	16

Source: From IEC 62305-3, 2006.[3]

long stroke without or with up to 10 superimposed short strokes. All short stroke parameters of the upward strokes are less than that of downward flashes. For lightning protection, the lightning parameters for upward strokes are considered to be taken care of by maximum values of the downward flashes.

22-4-2 Dimensioning LPS

For dimensioning of the LPS, and with reference to Table 22-2, consider the following parameters:

1. Effects of lightning are related to peak value of the current and specific energy, W/R.

2. The thermal effects are related to specific energy, W/R, when resistive coupling is involved and to charge Q when arc develops to the installations.

3. The sparking caused by inductive coupling is related to di/dt of the lightning current front.

4. Values I, Q, and W/R related to mechanical effects are determined from the positive flashes.

Based on the probability concepts and analyses of the lightning surges, a weighted probability can be determined. The lightning current falls between the minimum and maximum values, as defined in Table 22-4.

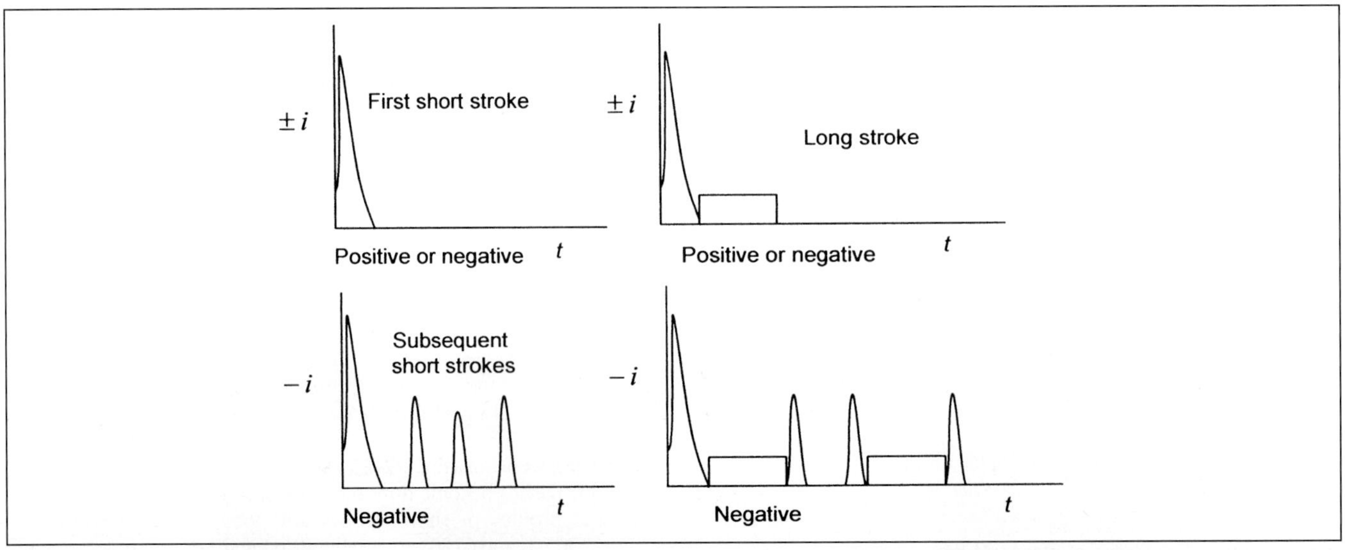

Figure 22-2 Components of downward flashes typical in flat territory and to lower structures, IEC.

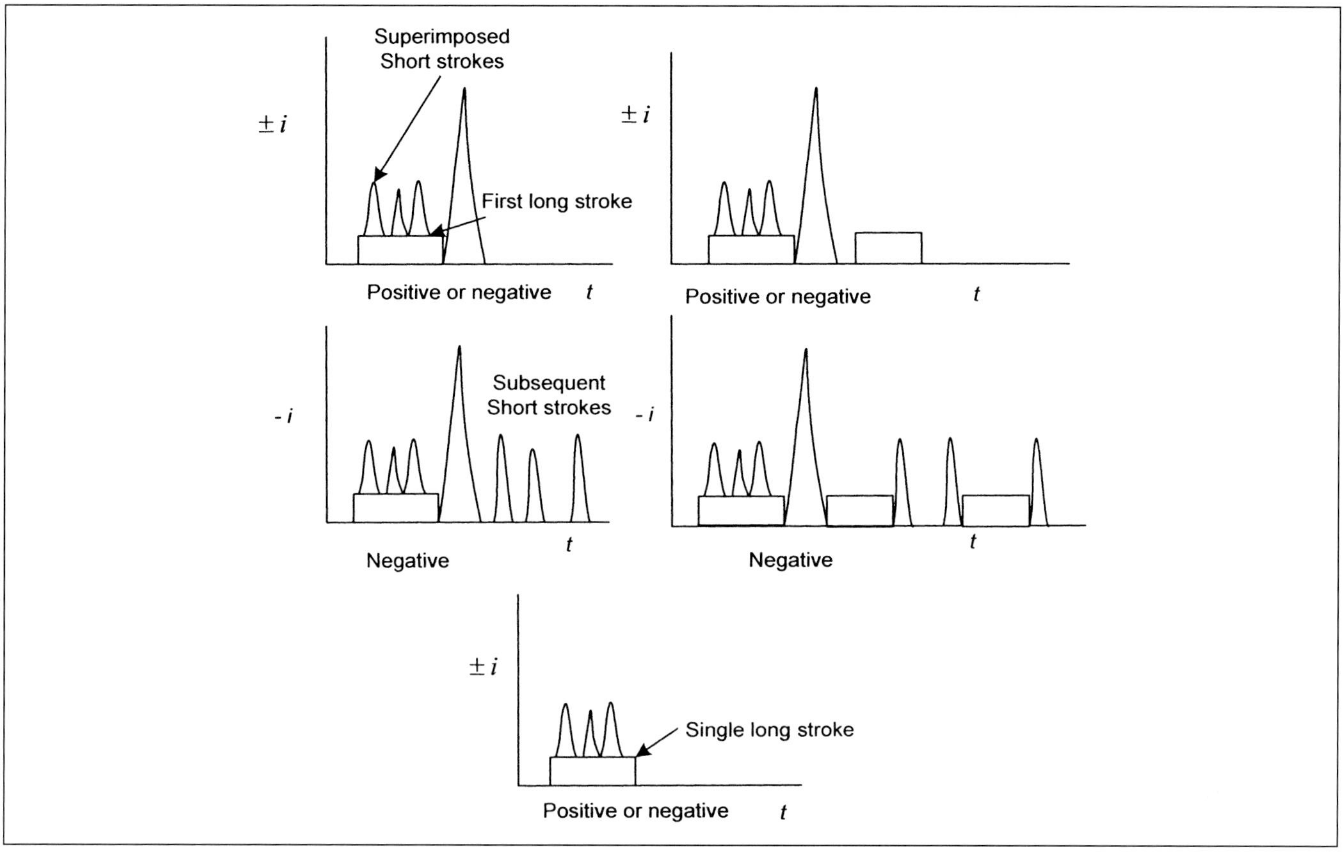

FIGURE 22-3 Components of upward flashes typical to higher structures, IEC.

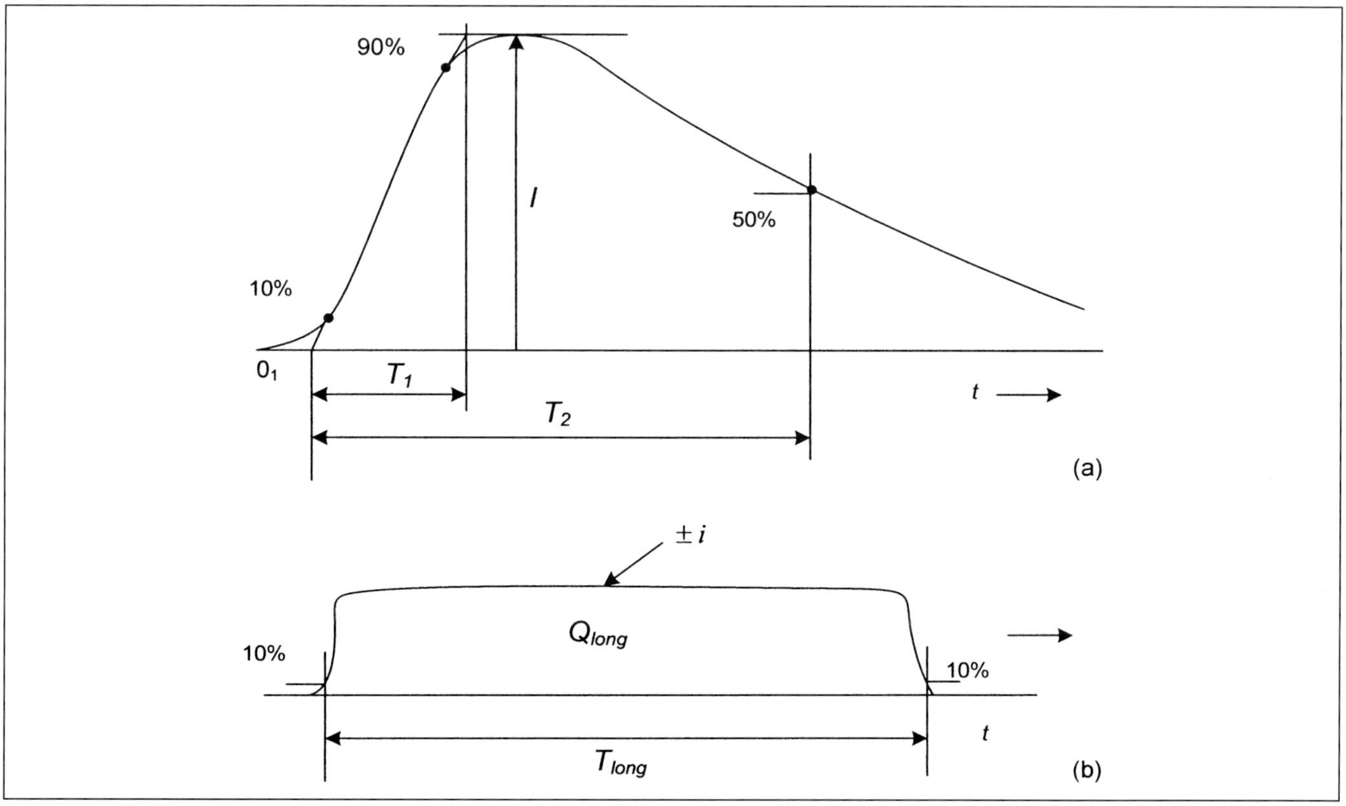

FIGURE 22-4 (*a*) Short-stroke parameters, IEC. (*b*) Long-stroke parameters, IEC.

TABLE 22-4 Probability for the Limits of the Lightning Current Parameters

PROBABILITY	LIGHTNING PROTECTION LEVEL			
PROBABILITY VALUES ↓	I	II	III	IV
Higher than minimum, Table 22-2	0.99	0.97	0.91	0.84
Lower than maximum, Table 22-3	0.99	0.98	0.97	0.97

Source: From IEC 62305-3, 2006.[3]

22-5 PROTECTION MEASURES

The protection measures based on the contents of Tables 22-2, 22-3, and 22-4 are shown in Table 22-5.

22-5-1 Positioning of Air Terminals

Three methods are suggested in IEC 62305-3,[3] and, as shown in Table 22-5, they are as follows:

1. The protection angle method
2. The rolling sphere method
3. The mesh method or the Faraday cage Method

The rolling sphere method is based on electrogeometric model. It was first proposed by Lee.[6] The position of the downward leader approaching the grounded structure defines a distance from the top of the structure called the *striking distance*. It is related to the charge in the downward leader and the peak value of the lightning current. Different relations have been proposed between the striking distance and the peak value of lightning current (also see Chap. 5).

IEEE Working Group's[7] proposal is accepted by IEC, and the relation is (also see Chap. 5):

$$D_s = 10I^{0.65} \tag{22-11}$$

where D_s is the striking distance and I is the surge current. This does not consider the height of the structure. The model in Ref. 8 provides a rational basis for taking account of the structure height. According to this model, the *attractive radius*, defined as the maximum distance from the structure for which a downward leader having a definite charge is captured by the structure itself, may be evaluated as a function of the downward leader charge. The impact of the lightning to structure is only possible when the tip of the downward leader reaches the volume above the structure defined by the attractive radius, given by:

$$R = I^{\alpha}0.84h^{0.6} \tag{22-12}$$

where:

$$\alpha = 0.7h^{0.02} \tag{22-13}$$

where h is the height of the structure in m. For structures ranging from 10 to 100 m, and the average amplitude current of 35 kA, the simplified formula is:

$$R = 14h^{0.6} \tag{22-14}$$

Another formula[9] for a free-standing structure of up to 60 m and lightning current of 31 kA is:

$$R = 24.6h^{0.4} \tag{22-15}$$

Figure 22-5 shows that as the height of the structure increases, the number of strokes per annum increases. Empire Estate Building in Manhattan, New York, is struck by lightning on an average of 20 times per year.

Without sifting through various expressions, the IEC 62305-3[3] makes the applications easy by specifying the rolling sphere radius for various lightning protection levels, as shown in Table 22-5. The protection with rolling sphere method is shown in Fig. 22-6. The air terminal system is adequate if no point of the structure to be protected comes in contact with a sphere with radius R, as shown in Table 22-5, rolling around and on top of the structure in all possible positions. IEC recognizes that with structure heights greater than the radius of rolling sphere, side flashes may occur, though the probability is small.

Figure 22-7 shows the protection angle method. Note how the angle varies with the height of the structure. The termination point of the curve, in black dots, shows the height limit of the protection angle method on the abscissa. Only mesh methods or rolling

TABLE 22-5 Minimum Values of Rolling Sphere Radius, Mesh Size, and Protection Angle Corresponding to Lightning Protection Levels

	PROTECTION METHOD		
LPL	ROLLING SPHERE RADIUS, R (m)	FARADAY CAGE, MESH, m × m	PROTECTION ANGLE
I	20	5 × 5	See Fig. 22-7
II	30	10 × 10	
III	45	15 × 15	
IV	60	20 × 20	

Source: From IEC 62305-3, 2006.[3]

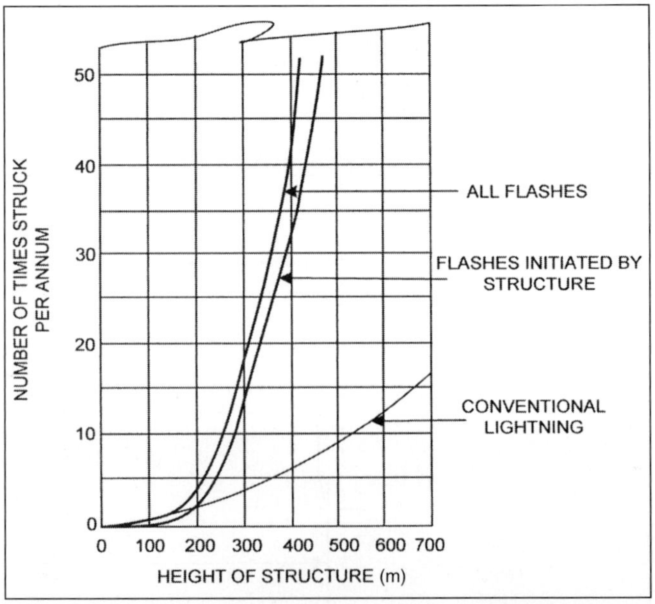

FIGURE 22-5 Increase in incidence of lightning with the structure height.

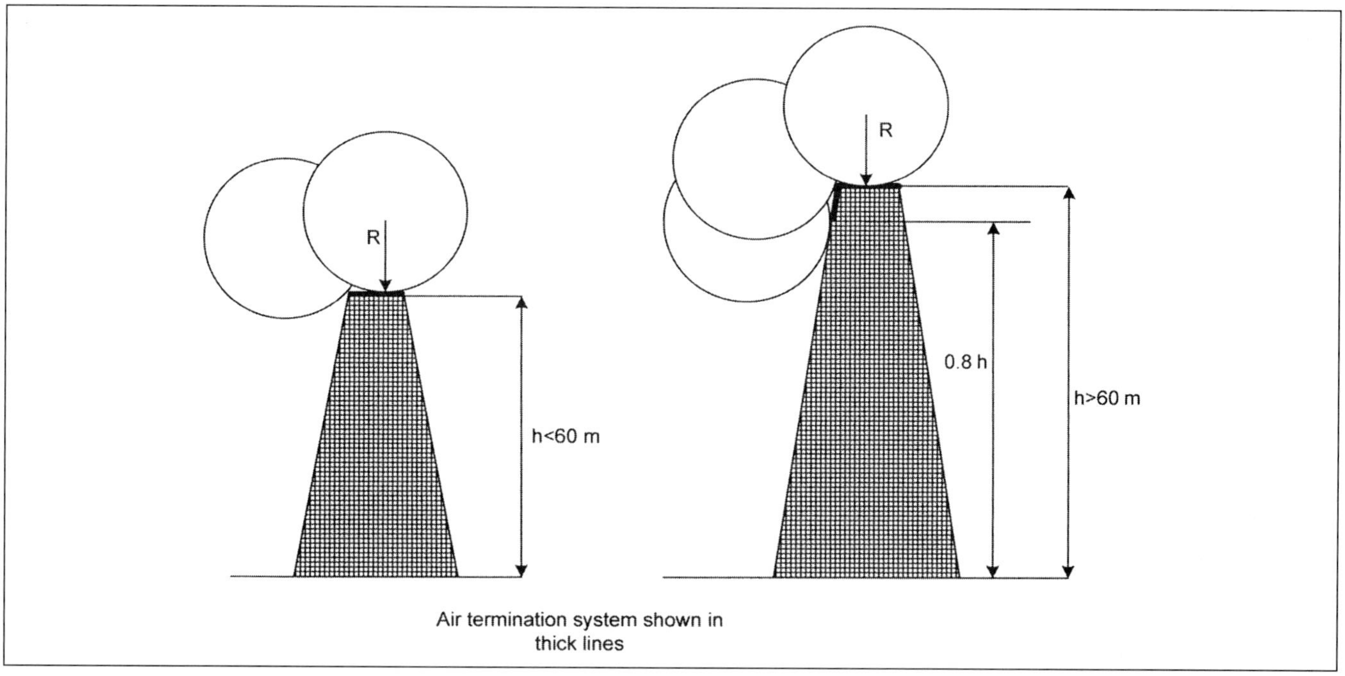

FIGURE 22-6 Design of LPS air termination system according to rolling sphere radius, IEC.

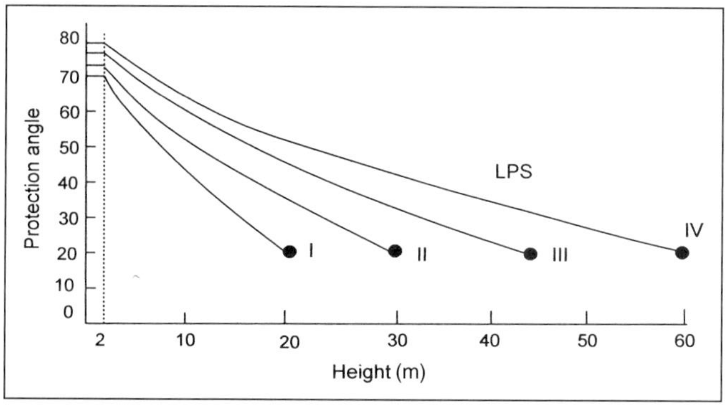

FIGURE 22-7 Lightning protection by protection angle method. Termination in black dots shows the maximum height on x-axis to which protection angle method can be applied.

sphere method apply for greater heights. The height h is the height of the air terminal above the area to be protected; this means add height of the highest point of the structure to the length of the air terminal. This is illustrated in Fig. 22-8. The angle α_1 corresponds to the height of the air terminal h_1, while the angle α_2 corresponds to the height h_2 above ground = $h_1 + h$. The angle does not change for heights below 2 m.

The following types of air terminations are recognized:

1. Rods
2. Catenary wires
3. Meshed conductors

The radioactive terminals or terminals (see section 22-12-2) with intensified ionization, if used, are to be positioned only as conventional terminals.

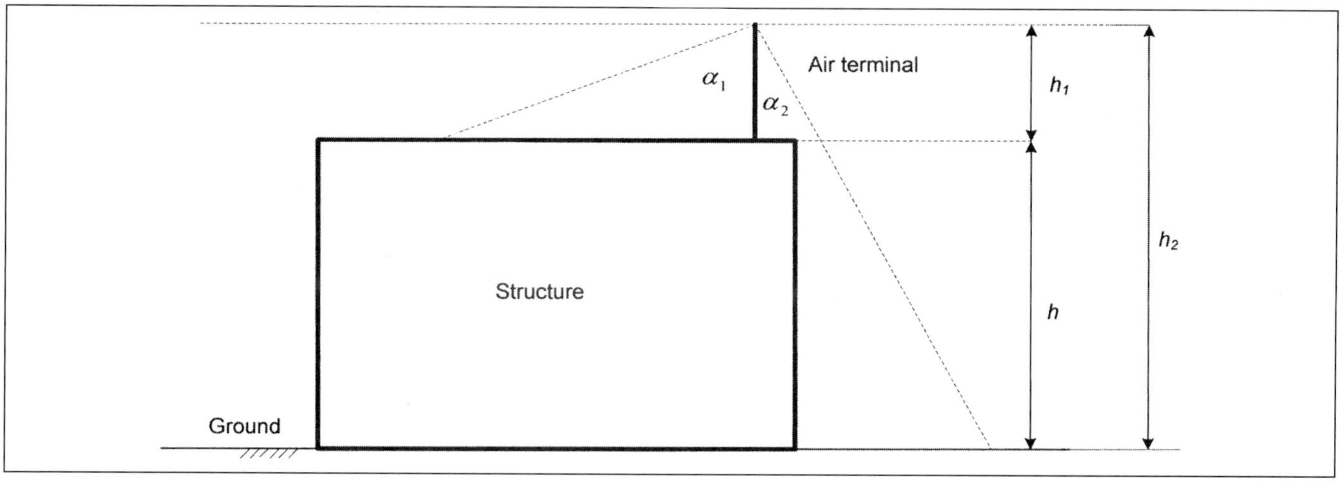

FIGURE 22-8 Structure volume protected by a vertical air termination system, IEC concept.

22-5-2 The Step-and-Touch Voltages

IEC 62305-3[3] specifies that risk for persons for the step-and-touch voltages is negligible if *one* of the following conditions is fulfilled:

1. The probability of persons approaching or the time of their presence outside the structures and close to down conductors is very low.

2. Insulation over the exposed conductor is provided to withstand 100 kV, 1.2/50-μs BIL, which requires at least 3-mm cross-linked polyethylene. (The insulation thickness varies according to the conductor size and voltage rating. According to NEC, the insulation thickness for 601-2000 V nonshielded types RHH and RHW cables conductor size 501-1000 KCMIL is 3.05 mm. Cables rated for higher voltages and of smaller conductor size are required.)

3. The resistivity of surface layer of the soil for a distance of 3 m from the conductor is no less than 5000 Ω/m. Asphalt in a layer of 5 cm thick will meet this requirement.

4. The natural down conductors consist of several columns of extensive metalwork of the structure, or several pillars of interconnected steel, electrically continuous.

5. Equipotential surface by means of meshed grounding system is provided for controlling the step voltage; see Chap. 21.

22-6 TRANSIENT BEHAVIOR OF GROUNDING SYSTEM

The behavior of grounding electrode systems under impulse currents is discussed in Sec. 21-11. Equation (21-48) is repeated here:

$$L_e = K_0(\rho T_1)^{1/2} \qquad (22\text{-}16)$$

To disperse the lightning current into the soil through the ground electrodes, the maximum energy that a human body can tolerate in transient conditions and the risks involved if this level is exceeded are considered in IEC 62305-3.[3] The maximum energy is 20 W-s. The standard lays down the minimum dimensions of the ground electrode system based on the probabilistic lightning parameters, the earth resistivity, and the LPS level. There are two basic types of earth electrode arrangements that need to be considered.

- Type A comprises horizontal and vertical ground electrodes (i.e., conductors and ground rods) connected to each down conductor.

- Type B comprises a ring conductor external to the structure and is in contact with soil for at least 80 percent of its total length or foundation electrodes, that is, interconnected reinforcing steel of the foundations or other suitable metal structures.

For type B electrode system, the mean radius r of the area enclosed by the ring earth electrode or foundation earth electrode shall not be less than l_1 from Fig. 22-9.[3]

$$r > l_1 \qquad (22\text{-}17)$$

where l_1 is greater than r, additional radial or vertical or inclined electrodes shall be added whose individual lengths, l_r horizontal and l_v vertical, are given by:

$$l_r = l_1 - r$$
$$l_v = (l_1 - r)/2 \qquad (22\text{-}18)$$

The number of additional electrodes shall not be less than the number of down conductors with a minimum of two and will be connected to the ring bus with equal spacing, as far as practical.

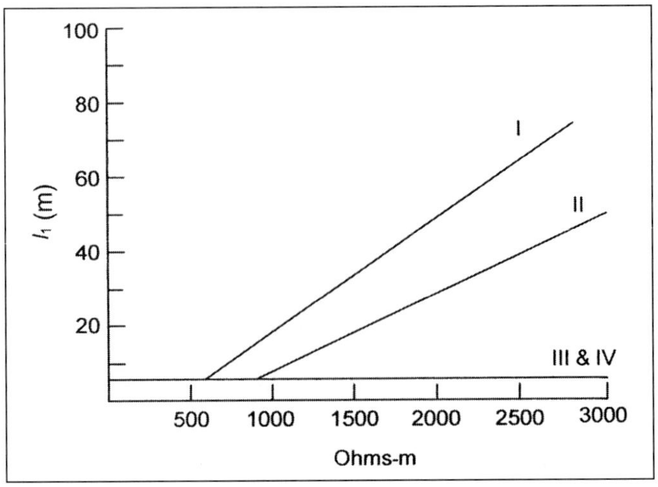

FIGURE 22-9 Minimum length l_1 versus soil resistivity required according to LPS system type, IEC classifications. Types III and IV are independent of soil resistivity.

22-7 INTERNAL LPS SYSTEMS ACCORDING TO IEC

We discussed categories A, B, and C for the lightning protection according to ANSI/IEEE standards in Chap. 19. Figure 22-10 from IEC 62305-4[4] defines the zones of lightning electromagnetic impulse (LEMP) severity. Each individual zone is characterized by different changes in electromagnetic conditions at its boundaries. The protection should be completed with proper shielding, bonding, and grounding. Services entering the structure should avoid overhead connections to avoid flashes occurring directly on overhead services, reduce level of overvoltages by shielding, divert lightning current, and consider increased BILs where needed. Figure 22-10 shows the following locations of SPDs according to IEC standards:

1. At the boundary LPZ 0/X, SPDs should meet class I requirements, test impulse current = 10/350 μs.

2. At the boundary LPZ X/Y (X>0, Y>1), SPDs should meet class II requirements, test impulse current = 8/20 μs.

See IEC 62305-4[4] for further discussions and applications.

22-8 LIGHTNING PROTECTION ACCORDING TO NFPA STANDARD 780

NFPA 780[10] is the guiding document for the lightning protection of structures in the United States. There are more high-rise structures in the United States than in the rest of the world. NFPA first adopted specifications for protection of buildings against lightning in 1904, and subsequently, these specifications have been revised at least 24 times, and 2008 is the current edition. This standard has specific requirements for lightning protection of:

- Ordinary structures
- Special occupancies
- Fixed and floating roof tanks
- Heavy-duty stacks
- Structures containing flammable vapors
- Watercraft

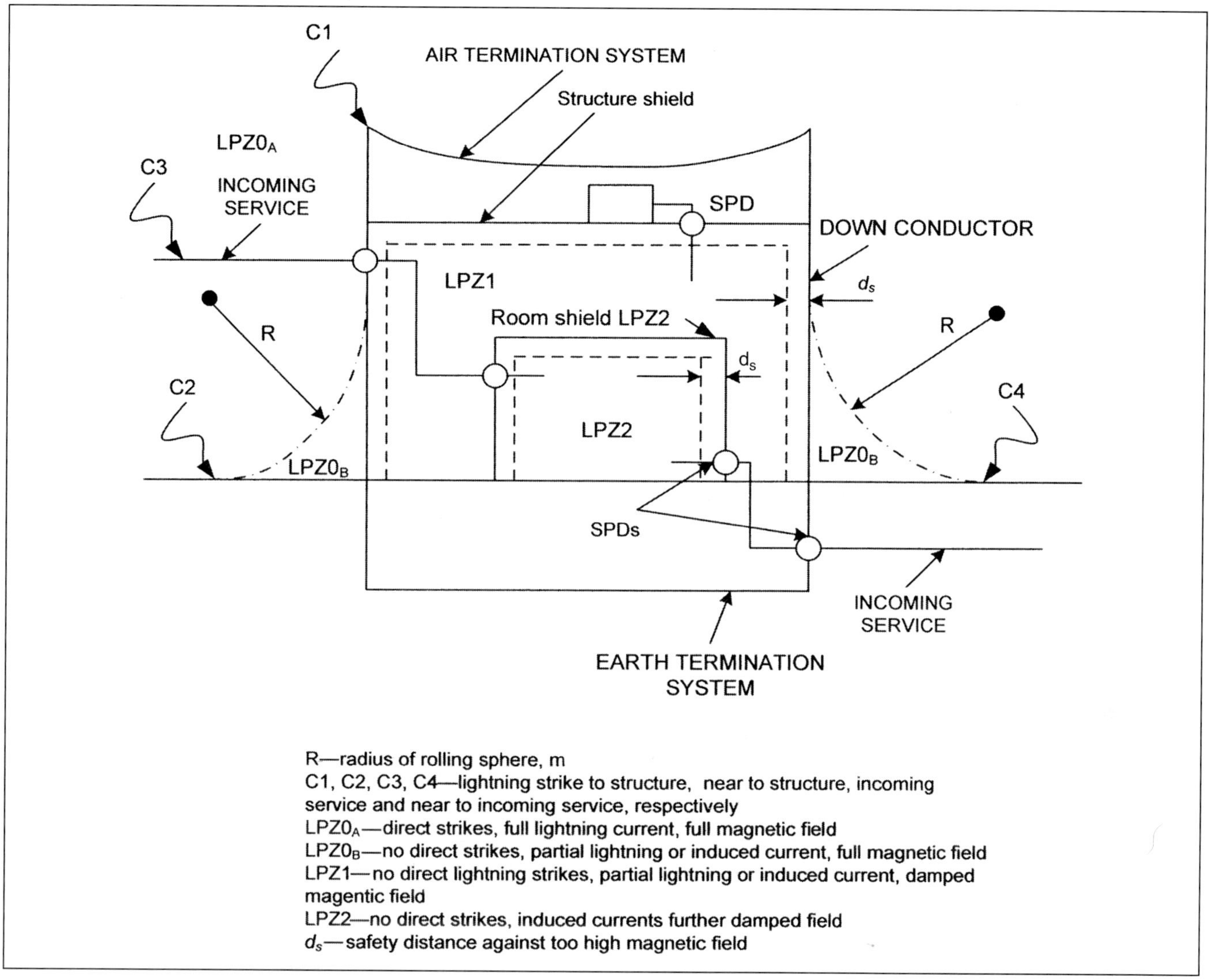

FIGURE 22-10 LEMP zones according to IEC; see text.

- Wind-turbines
- Protection of livestock in fields
- Protection of trees
- Picnic grounds, playgrounds, ball parks, and other open spaces
- Parked aircraft, and so on

These specifics are not discussed. The standard excludes protection of electrical generating, transmission and distribution systems, and also explosive manufacturing buildings and magazines. The basic tenats of this standard and lightning protection philosophy are based on Faraday cage, also called integral system or Franklin system.

22-9 LIGHTNING RISK ASSESSMENT ACCORDING TO NFPA 780

In addition to direct losses, such as destruction of the building, fire, and killing of livestock, indirect losses arise due to loss of business and essential properties. A whole community may depend on the integrity of a single structure for their safety and comfort and the examples cited are a police station, a fire station, water-pumping plant, telecommunication facility, and so on. The damage to museums and cultural sites will be an irreplaceable loss of cultural heritage. In most cases, the need for lightning protection may be obvious, for example:

- Large crowds
- Service continuity
- Very high lightning flash frequency
- Tall isolated structures
- Buildings containing explosives or flammable materials
- Buildings containing irreplaceable cultural heritage

Apart from lightning flash density, the risk assessment accounts for the following factors:

1. Building environment
2. Type of construction

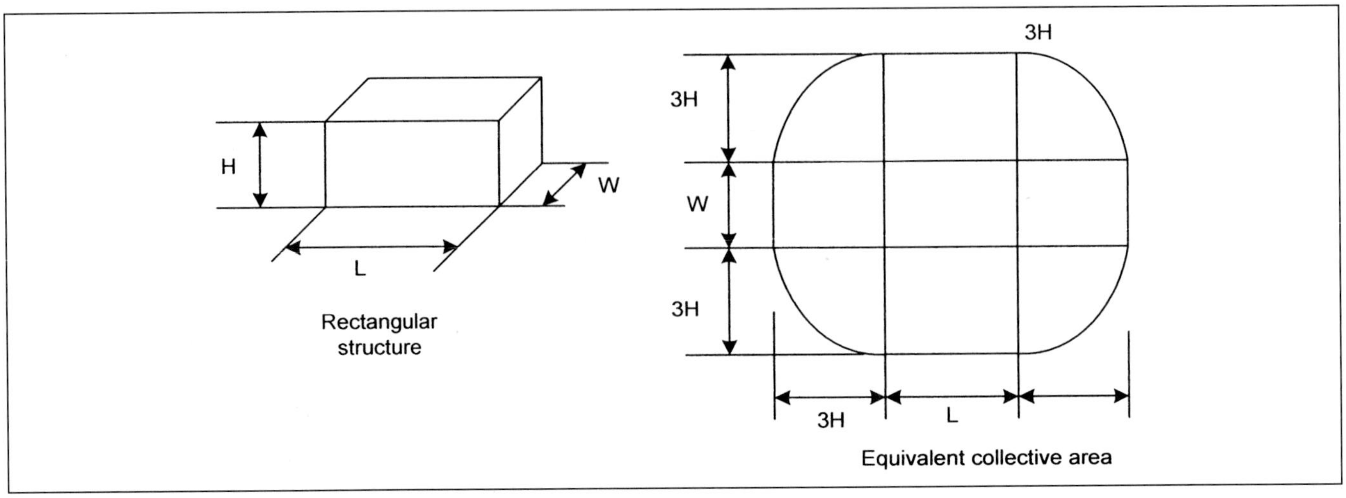

FIGURE 22-11 Calculation of equivalent collective area for a rectangular structure according to NFPA 780.

3. Structure occupancy

4. Lightning stroke consequences

The following expression for the yearly lightning strike frequency is specified:

$$N_d = N_g \times A_e \times C_1 \times 10^{-6} \qquad (22\text{-}19)$$

where N_d is yearly stroke frequency, and N_g is the yearly flash density in the region where the structure is located. Reference 10 provides a map of the United States, data collected by U.S. National Lightning Detection Network (NLDN), showing the contours of N_g. A_e is equivalent collective area of the structure, and C_1 is environmental coefficient.

The *equivalent collective area* of a structure is defined as the area obtained by extending a line with a slope of 1 to 3 from the top of the structure to ground, completely around the structure. The calculation of equivalent area for various structure shapes is explained with accompanying figures. Figure 22-11, for example, shows this calculation for a rectangular structure of length *L*, width *W*, and height *H*. The equivalent area in m^2 is:

$$A_e = LW + 6H(L+W) + \pi 9H^2 \qquad (22\text{-}20)$$

The environmental coefficient accounts for the topography of site of the structure and any objects located within a distance of 3*H* from the structure that can effect the collective area. These factors are tabulated in the standard. As an example, for a structure located within a space containing structures or trees of the same height or taller within a distance 3*H*, $C_1 = 0.25$.

The tolerable lightning frequency, N_c, is considered as:

$$N_c = \frac{1.5\times 10^{-3}}{C} \qquad (22\text{-}21)$$

where:

$$C = C_2 \times C_3 \times C_4 \times C_5 \qquad (22\text{-}22)$$

where the coefficients C_2, C_3, C_4, and C_5 take into consideration the construction, occupancy, contents, and so on. Briefly, these constants are tabulated in Table 22-6.

If $N_d > N_c$, the LPS should be installed. If $N_d \le N_c$, the LPS is optional. Note that the standard does not use the words "not required."

When a LPS is installed, the following additional measures should be considered:

- Limiting step-and-touch potentials
- Restricting fire propagation
- Limiting induced overvoltages
- Reduceing effects of lightning-induced voltages to sensitive electronic equipment

Compare these requirements with IEC standards discussed earlier.

Example 22-1 A rectangular structure, $L = 30$ m, $W = 20$ m, $H = 20$ m, is considered. The structure is located in Texas with $N_g = 3$. The structure is surrounded by smaller structures within a distance 3*H*, has metal construction with metal roof, stores irreplaceable cultural items of exceptional value, is normally occupied, continuity of services is required, and there are no environmental effects. The risk assessment is as follows:

From Eq. (22-20), calculate the equivalent collective area, for the given dimensions. This gives $A_e = 17909.9$ m^2.

Calculate expected lightning strike frequency to the structure from Eq. (22-21). Here $N_g = 3$, and from the Table in this standard with respect to location topology, $C_1 = 0.5$. Substituting in Eq. (22-19), $N_d = 0.026$—yearly lightning strike frequency to the structure.

Read factors C_2, C_3, C_4, and C_5 from Table 22-6, based on the given conditions. These factors are $C_2 = 0.5$, $C_3 = 4.0$, $C_4 = 1.0$, $C_5 = 5.0$, $C = 10$.

Calculate tolerable lightning frequency to the structure, from Eq. (22-21). This gives $N_c = 0.15 \times 10^{-3}$ yearly lightning strike frequency to the structure.

As $N_d > N_c$, the lightning protection is required. Now consider that the same structure, exactly under the same physical environment, construction, occupancy, and usage, is located in Florida, with $N_g = 12$. Then the yearly frequency N_d becomes 0.104, and more care must be taken in designing the lightning protection for such a structure.

22-10 PROTECTION OF ORDINARY STRUCTURES

An *ordinary structure* is defined as any structure that is used for ordinary purposes, whether commercial, industrial, farm, institutional, or residential.

TABLE 22-6 Risk Assessment According to NFPA 780[10] (Determination of Coefficients C_1, C_2, C_3, C_4, C_5)

ENVIRONMENTAL COEFFICIENT C_1

Relative structure location	C_1
Structure located within a space containing structures or trees of the same height or taller within a distance of 3*H*	0.25
Structure surrounded by smaller structures within a distance 3*H*	0.5
Isolated structure, no other structure located within 3*H*	1
Isolated structure on a hilltop	2

STRUCTURE COEFFICIENT C_2

Structure	Metal Roof	Nonmetallic Roof	Flammable Roof
Metal	0.5	1.0	2.0
Nonmetallic	1.0	1.0	2.5
Flammable	2.0	1.0	3.0

STRUCTURE CONTENTS COEFFICIENT C_3

Structure contents	C_3
Low value and inflammable	0.5
Standard value and nonflammable	1.0
High value, moderate flammability	2.0
Exceptional value, flammable, computer and electronics	3.0
Exceptional value, irreplaceable cultural items	4.0

STRUCTURE OCCUPANCY COEFFICIENT C_4

Structure occupancy	C_4
Unoccupied	0.5
Normally occupied	1.0
Difficult to evacuate or risk of panic	3.0

LIGHTNING CONSEQUENCE COEFFICIENT C_5

Lightning consequence	C_5
Continuity of facility services, not required, no environmental effect	1.0
Continuity of facility services required, no environmental effect	5.0
Consequences to the environment	10.0

The construction materials used for lightning protection are divided into two classifications:

1. Class I materials, consisting of lightning conductors, air terminals, ground terminals, and associated fittings, are required for lightning protection of structures not exceeding 75 ft (23 m).

2. Class II materials, of similar nature as class I materials, are used for structures exceeding 75 ft (23 m).

The Tables in NFPA 780[10] provide specifications and sizing of air terminals, main conductor, bonding conductor, and so on. Class II materials are of heavier size and cross-sectional area.

Roof types of structures with respect to shape, (e.g., flat, mansard, hip, gable, and so on), construction materials, slope, multilevels, conductor placements, air terminal height, and support systems are specified. The zone of protection is dependent on the geometry of the structure. Here we limit our discussions to the rolling sphere method and strike termination devices on a flat roof.

22-11 NFPA ROLLING SPHERE MODEL

The zone of protection depicting rolling sphere model is shown in Fig. 22-12. It has a radius of 46 m (150 ft). The zone of protection includes the space not intruded by a rolling sphere.

- Where the sphere is tangent to the earth and resting against the strike termination device, all space in the vertical plane between the two points of contact and under the sphere is considered to be in the zone of protection.
- A zone of protection is also formed when such a sphere is resting on two or more strike termination devices and includes the space in the vertical plane under the sphere and between those devices, Fig. 22-13. Note that the zone of protection is limited to the space above the horizontal plane of the lowest structure. For structures exceeding 46 m height above earth or "above a lower strike termination device," the protection zone is the vertical plane between the points of contact and also the point where the sphere is resting against a vertical surface of the structure and the lower termination device to earth.

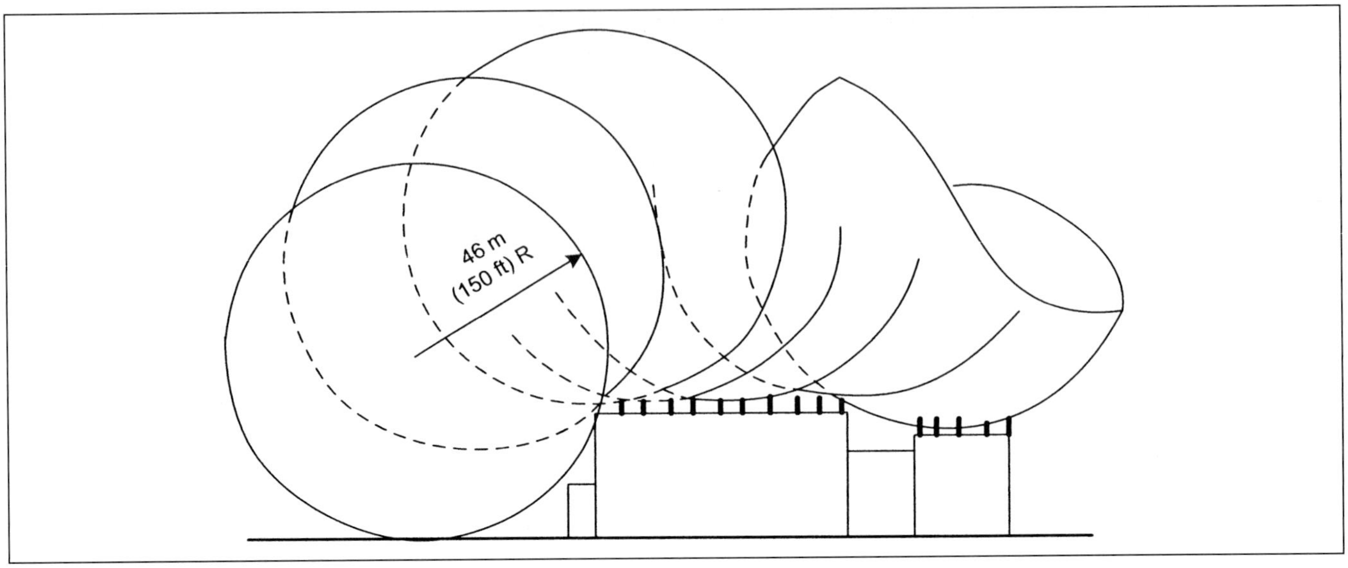

FIGURE 22-12 Protection by rolling sphere method, NFPA 780.

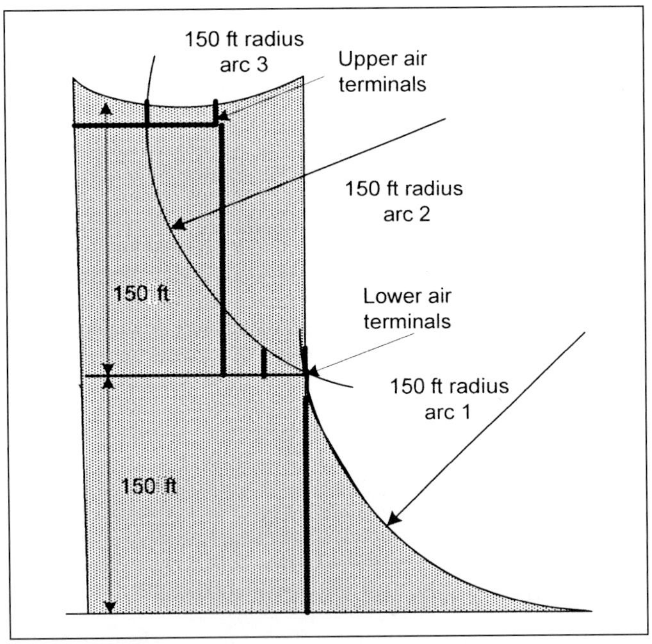

FIGURE 22-13 Rolling sphere method applied to structures higher than 150 ft.

Figure 22-14 depicts 46-m geometric model for structures up to height of 46 m. Based on the height of the strike termination device for the protected structure (7.6, 15, 23, 30, and 46 m) above ground, reference to the appropriate curve shows anticipated zone of protection for objects and roofs at lower elevations. The horizontal protective distance found from the curve can also be calculated from:

$$d = \sqrt{h_1(300-h_1)} - \sqrt{h_2(300-h_2)} \tag{22-23}$$

where d is horizontal distance in ft, h_1 is height of the higher roof, and h_2 is height of the lower roof.

Equation (22-23) is based on a striking distance of 46 m. For Eq. (22-23) to be valid, the sphere shall be tangent to lower roof or in contact with the earth, and in contact with vertical portion of the higher portion of the structure.

Figure 22-13 shows the principle applied to structures above 150 ft height. The shaded area shows the protection zone. The difference in heights between the upper and lower roofs and earth shall be 46 m or less.

Figure 22-15 shows a conduction LPS for flat roof, with their major components and spacing.

Example 22-2 Design an air termination system according to NFPA 780 for a rectangular flat roof, with dimensions 100 × 80 m.

Based on Fig. 22-15, the system requires:

- 94 air terminals
- 7 longitudinal conductors, spaced approximately 13.33 m apart
- 7 cross conductors
- 10 down conductors to grounding grid at the base of the structure

Material specifications are not discussed.

22-12 ALTERNATE LIGHTNING PROTECTION TECHNOLOGIES

This section discusses some alternate lightning protection technologies. The research continues in this important field. However, it is important to state that the manufacturers presented their cases for incorporation of alternate lightning protection technologies while the technical committee of NFPA 780 was considering the 2004 revision. The committee did not accept incorporation of any of these technologies in the standard. Also, the 2008 revision does not incorporate any of these technologies.

22-12-1 Franklin Rods—Sharp versus Blunt Tips

There is some controversy ranging with respect to sharp versus blunt Franklin rods. The air terminal is meant to attract and direct the stroke leader, and its effectiveness is assessed on the basis of

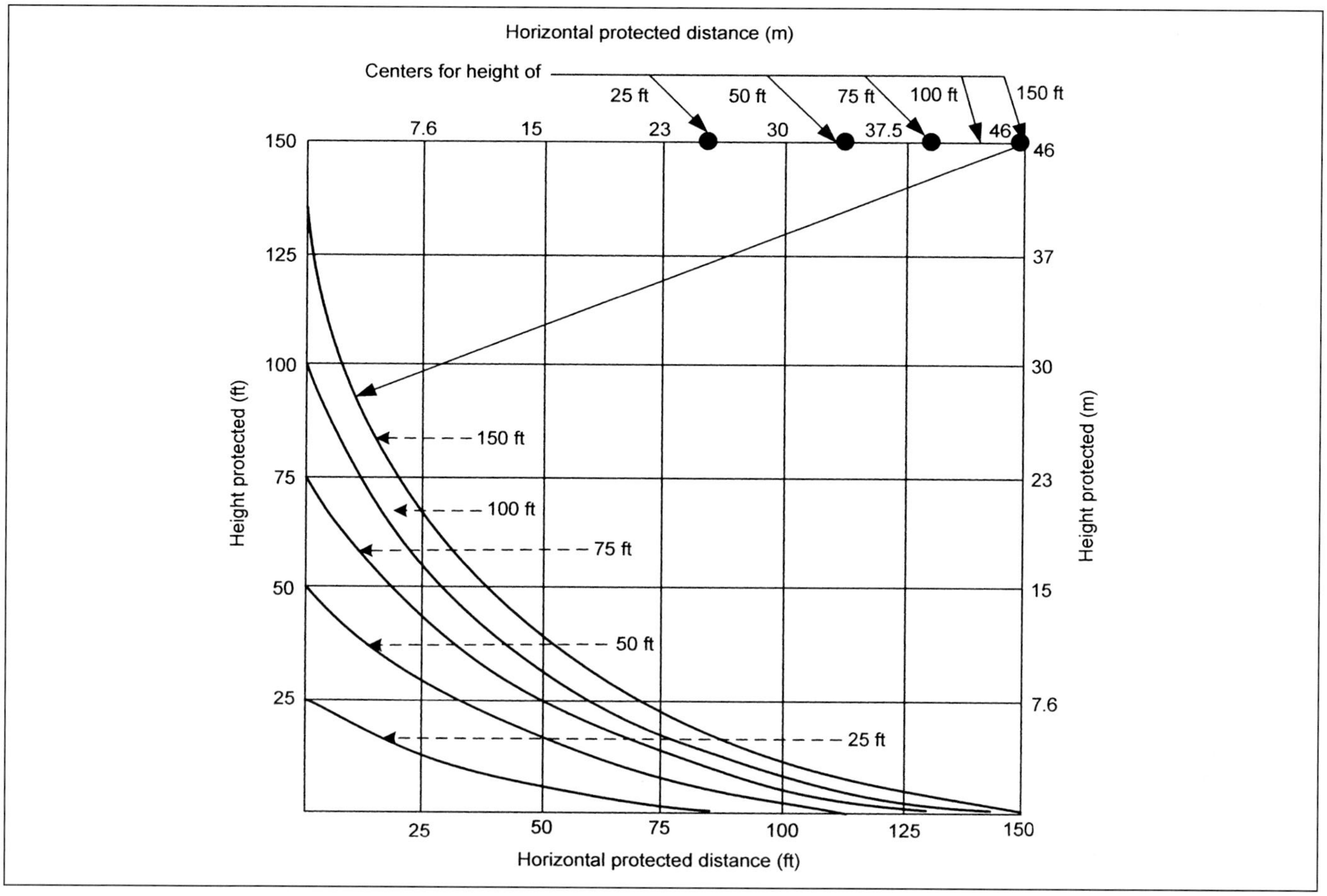

FIGURE 22-14 Zone of protection utilizing rolling sphere method.

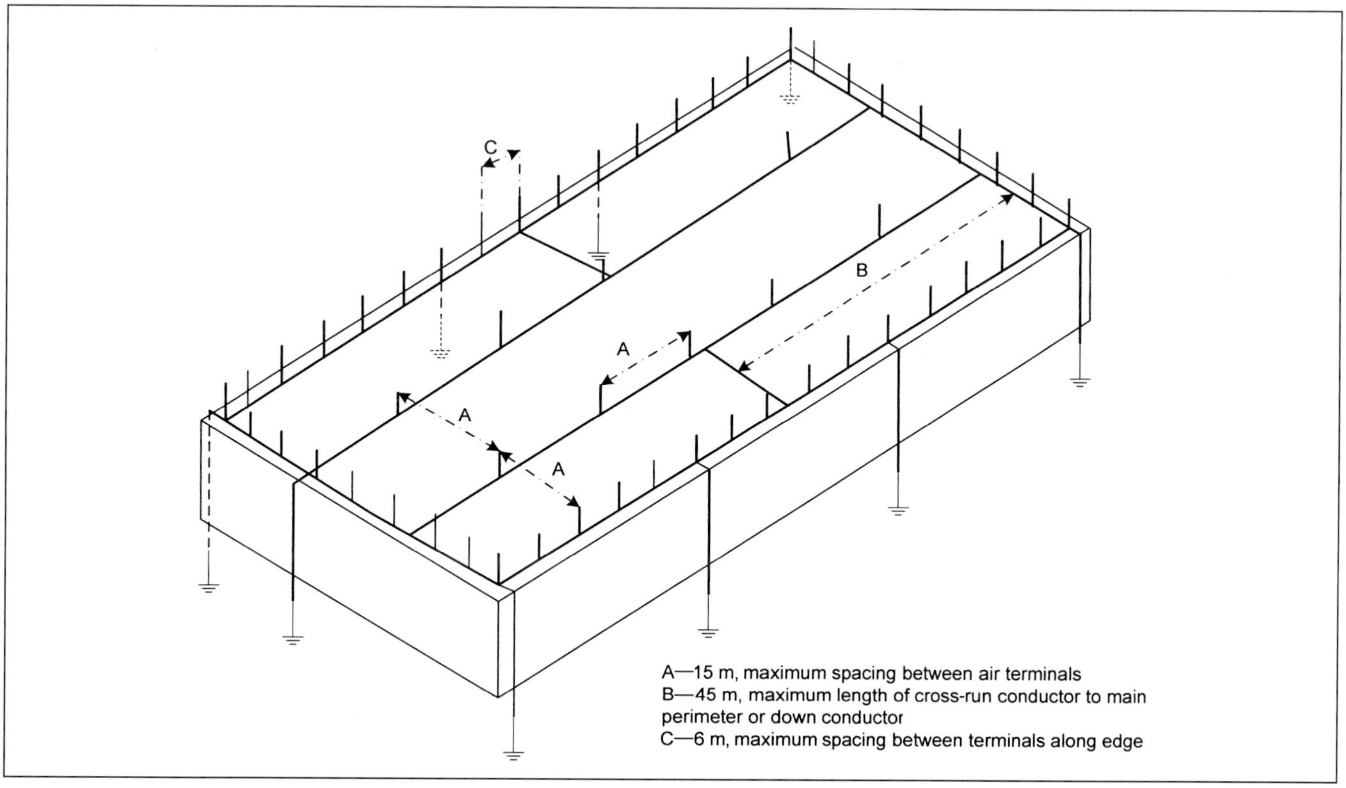

FIGURE 22-15 Air terminals on a flat or gently sloping roof, NFPA 780.

probability of capture of charged particles for a given configuration. Langmuir researchers (Langmuir Labs, New Mexico, United States[11]), experimented with sharp-pointed lightning rods on the top of a mountain that did not work, and the lightning instead struck the nearby trees because of the buildup of corona discharge around the rod's tip. This shifted the positive charged wave away from the rod, seeking a lesser disruptive path. Further research is reported in Ref. 12, and the following relations are presented.

High probability of lightning striking the Franklin rod:

$$H+D>\Delta L+D \qquad H>\Delta L \tag{22-24}$$

High probability of lightning striking the ground:

$$H+D<\Delta L+D \qquad H<\Delta L \tag{22-25}$$

Equal probability of lightning striking the ground and rod:

$$H+D=\Delta L+D \qquad H=\Delta L \tag{22-26}$$

where ΔL is gain in the length H as an upward connecting leader is formed by the approaching downward leader, and D is an empirical relation between the leader current and striking distance (Whitehead's relation).

22-12-2 Ionizing Terminals (Radioactive Lightning Terminals)

These are not so new in the market and have been available for quite some time. These are based on the use of radio isotopes in the metal of the air terminal. It is usually of the pointed design, and the number of elements are supposed to control the reach of the collector. However, a controversy has prevailed. Figure 22-16 shows the results of two test programs conducted by independent researchers in Sweden and Switzerland. A radioactive terminal and a sharp Franklin rod were installed at the same height and

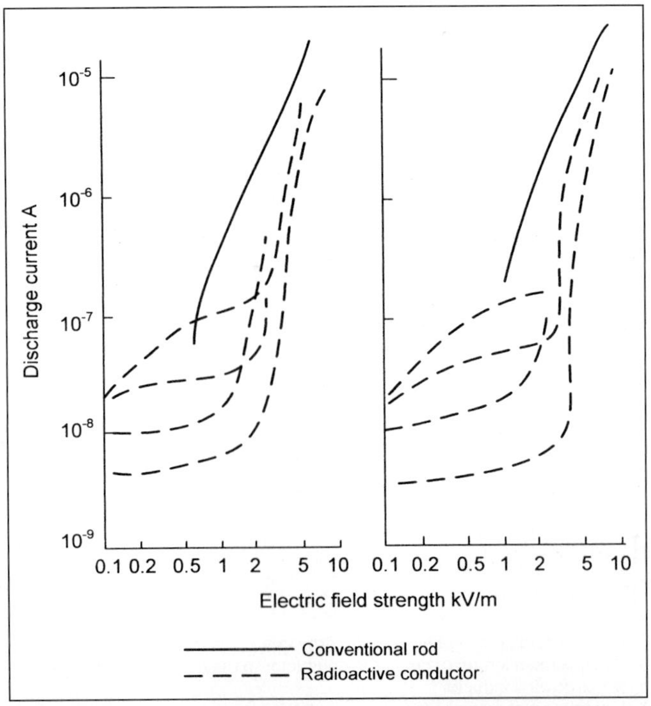

FIGURE 22-16 Conventional rod versus radioactive terminals, discharge patterns; two test results by independent authorities.

discharge currents were measured before and during the thunderstorm. Figure 22-16 shows that when the field strength exceeded 1000 V/m, both the radioactive terminal and the sharp-pointed rod emitted the same ionization current. Average ΔL for enhanced air terminals (radioactive terminals) under thunderstorm conditions is longer than ΔL for Franklin rod[12]. But under fair weather conditions, with low absolute humidity, reverse is true. The authors conclude that sharp air terminals are more efficient than blunt ones in attracting flashovers, and that ionized terminals are more efficient than nonionized terminals.

22-12-3 Laser Beam Systems

The laser beam system was developed by Leonard Ball.[13] The laser was termed "mode-locked" system that produced large-scale multiphoton ionization (MPI) in the right place at the right time. It claims to divert the downward leader and the resulting channel from distances of several kilometers. The major considerations are cost, state of development, and the problem of diverting the stroke energy to earth without damaging the laser itself.

22-12-4 Lightning Prevention (Dissipation Systems)

The dissipation array system (DAS, U.S. Patent No. 4180698) is supposed to prevent a lightning stroke to both the protected area and the array itself. Figure 22-17 shows the component parts—an ionizer, a ground current collector, and interconnections to ground grid. The ionizer consists of an array of multiple sharp points, designed for the specific applications. It provides ionization to transfer the collected charge through it and the surrounding air molecules. The earth charge collector (ECC) are the conventional grounding systems. These are augmented, so that the grounding resistance is lowered. It collects the charge induced in the area and equipment to be protected. And finally, the charge conducting system provides a low impedance path between EEC and ionizer. It claims to:

- Prevent direct stroke to the protected facility
- Reduce electrostatic field within the sphere of influence
- Provide an area free of lightning-related electromagnetic phenomena

A lightning stroke neutralizes the charge differential between the base of a storm cloud and its image charge on the earth below. By making the intervening air space a leaky dielectric, the same function can be achieved slowly. DAS encourages the trickle charge concept through the use of point discharge.

An electrostatic field near a pointed conductor tends to concentrate on the point, enhancing the electrical field. Under fair weather conditions, an electrical field exists at the ground. At an altitude increase, the air has high conductivity, and the surface of earth, which is also a good conductor, constitutes a sort of capacitor, two plates charged to different potentials. The positive field at ground is about 130 V/m.

The point discharge current will start to flow from the tip of an earthed vertical conductor as soon as the electrical field exceeds a critical value, which is required to initiate ionization by collision.

Under a thunderstorm, the base of the cloud has mostly negative charge, and upper part of cloud is positively charged (Chap. 5). The negatively charged base will attract positive charges from the ground. Thus, under fair weather conditions, while the cloud is negatively charged, it becomes positively charged under a thunderstorm. The point discharge current, from the array of thousands of sharpened points, will carry away positively charged ions. These will move away from the array, and depending on the wind velocity, these will form a space charge that reduces electrical field around the points, since lines of force will terminate on these ions rather than

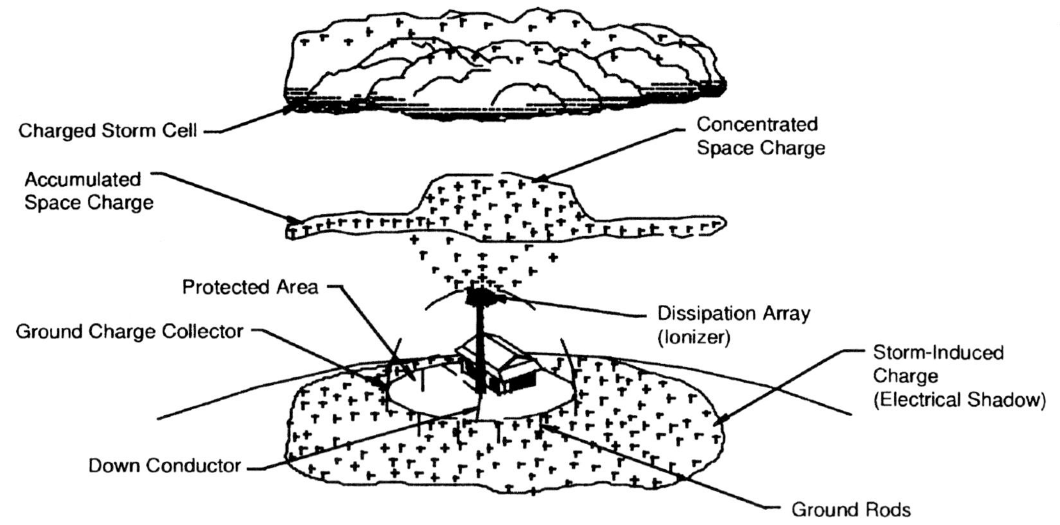

FIGURE 22-17 Principals of DAS. (*Copyright Lightning Eliminators and Consultants, Inc. Reproduced with permission.*)

on the sharpened points. This will progress till the electrical field is reduced below corona level. As the space charge is carried away by the wind, the electrical fields near the points increase again, and the new corona discharge starts. Thus, point discharge currents have a pulsed nature at some interval of time. As a result, the initiation of the upward leader would not take place, and the lightning strike will be eliminated.

Figure 22-18 shows a hemispherical array, consisting of multiple sharp points—this array is called an *ionizer*. Depending on the structure to be protected, the arrays of different configurations, dimensions, and shapes have been fabricated.

Two extensive investigations of the system were conducted. J. Hughes organized the first investigation, "Review of Lightning Protection Technology for Tall Structures," Lyndon B. Johnson Space Center, Clear Lake City, Houston, Texas. Twelve analysts presented different views about the efficacy of multipoint systems.[14] The conclusions drawn were:

- Single point corona currents exceeded multipoint corona currents.
- A single point at a height of 50 ft always gave more corona than DAS at the same height.
- The DAS did not eliminate lightning.
- The lightning stroke is photographed, striking an array many times with measured currents of the order of 30 to 50 kA.

In another investigation, the DAS was installed on an air traffic control tower. Examination of the multipoint array after lightning

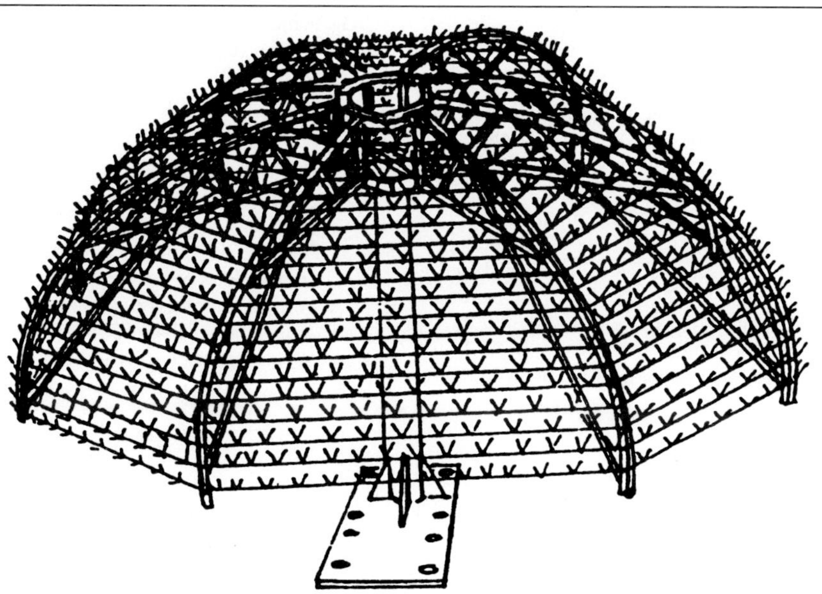

FIGURE 22-18 A hemisphere multipoint array, typical DAS system. (*Copyright Lightning Eliminators and Consultants, Inc. Reproduced with permission.*)

stroke showed missing spikes and mechanical expansion, and bulging of down conductors was noted, showing that these passed large amounts of lightning current.

Yet, in many practical installations covering substations, tall structures, communication facilities, power stations, and the petroleum and paper industries, the system is reported to be effective.[15] Thus, it may be inferred that the system does not entirely eliminate lighting strikes, but is still effective in most applications.

22-12-5 Coaxial Cable as a Down Conductor

A low impedance path from the terminals to ground is imperative. Tests demonstrate that with a steep voltage, wave front, such as created by lightning surge impedance of conventional conductors, increases to a point where side flashes can occur. Specially manufactured large bare copper conductors have been used, and bonding to metallic parts of the structures is a common practice while the conductors are run as straight as possible, without kinks and bends to the ground terminations. A coaxial cable is a recent development and has a surge impedance of 16 to 126 Ω versus 500 to 5000 Ω for conventional open wire conductors.

22-13 IS EMF HARMFUL TO HUMANS?

Associations have been made between various cancers and leukemia and electromagnetic fields in some epidemiological studies, which attracted a lot of public concern and exposure in the media. However, these studies cannot be called conclusive. Some studies were based on indirect assessment of exposure, in geographical areas close to the power lines, rather than actual measurement of flux density. Some studies, which suggested associations between health and magnetic fields, indicated exposures that are less than the theoretical threshold level for creation of an electrical current density in head or trunk, comparable in magnitude to the current density levels that occur in normal body processes. Other studies indicated that weak magnetic fields are associated with an adverse response while strong ones are not. Data on some other studies will even suggest a potential beneficial effect. In June 2001, an expert scientific group of IARC (International Agency for Research on Cancer) reviewed studies and extremely low magnetic fields (ELFs) were classified as "possibly carcinogenic." To put this classification in proper prospective, note that "coffee" and "gasoline engines" are also classified as possibly carcinogenic.

22-13-1 EMF Sources

ELFs occur in every office environment and wherever electricity is used. Electrical fields arise from electrical charges, measured in V/m, and magnetic fields arise from motion of electrical charges, that is, current, and expressed in Telsa (T) or sometimes Gauss (10000 G = IT). These fields are not shielded by most common materials. Their frequencies are 50 or 60 Hz, with higher harmonics up to 300 Hz and lower harmonics up to 5 Hz. The magnetic field values in homes in North America average 0.11 μT, and electrical fields average tens of V/m. Readings in the field taken under a 380-kV line at the point of greatest sag showed a magnetic flux density of 15 to 20 μT, and electrical field can be several hundred V/m. Close to appliances, the magnetic fields may be much higher.

22-13-2 Levels

According to a study by WHO, interaction between current and muscle cells occurs above a body current density of 1000 mA/m^2. The lowest level for detection of biological effects is 10 mA/m^2. Below 1 mA/m^2, there are no biological effects. A current density of 1.2 mA/m^2 corresponds to 5 kV/m for the electric field, and 100 μT for the magnetic field. These are the values adopted in some European countries with short-term higher values.

22-13-3 Monitoring and Survey

Monitoring can be performed using sensitive and accurate gauss meters. The meter should be able to measure a flux of 0.01 μT. A large number of readings may be taken over a course of time. Evaluating a personal exposure involves determining weighted average and mean density at waist level.

22-13-4 Screening

If the levels are found to be high, possibly close to load centers, switchgear, cables, and metal trays, low content carbon steel or nickel content alloys are used for shielding. A magnetic shield should have low reluctance and remain unsaturated.

WHO published a new monograph reviewing the scientific literature on health effects of ELF in 2007.[16] This evaluated all the available evidence accumulated over three decades of research. A technical guide for measurements of low-frequency electrical and magnetic fields near overhead power lines is discussed in Ref. 17. Characteristics of ELF magnetic fields are discussed in Ref. 18.

This chapter provides a glimpse into the fascinating art and science of protection of structures against lightning. Research continues in this field. Approximately 1000 persons are hit with lightning every year in the United States. EPRI organized a mock power distribution system in the state of Florida, the area of highest lightning activity in the United States, and with RTL studies investigated the effect of lightning on buried cables. It was found that the ground flashes of lightning created channels of intense heat, which fused the sand and clay together into molten glass-like crystals, and these channels ran across the concrete walls of the trenches to seek the buried cables.[19]

PROBLEMS

1. Calculate the equivalent length of the grounding conductor for a lightning surge of rise time = 200 kA/μs, surge current = 100 kA, soil resistivity = 500 Ω-m. Consider that the point of impact is the center of a mesh-grounding electrode system.

2. A building has a ground conductor run along its parameter, total length = 400 m. Considering a soil resistivity of 500 Ω-m, and according to IEC standards, is it adequate to prevent shock hazard?

3. What is the recommended value of the resistivity of soil around a structure within a range of 3 m from the conductor according to IEC to prevent hazard due to step-and-touch potential? How can this resistivity be obtained in a soil of natural resistivity of 100 Ω-m?

4. Compare IEC LEMP zones of protection with ANSI/IEEE categories in Chap. 19.

5. Describe the differences between NFPA 780 and IEC rolling sphere methods of lightning protection. IEC standard does not recommend any intermediate lightning termination system for structures higher than 60 m, and accepts the risk of side flashes. NFPA rolling sphere method requires a radius of 46 m (150 ft) tangent to earth of lower level of the structure. Comment.

6. A structure is 95 m tall. How will the lightning protection be approached with respect to IEC and NFPA standards?

7. Write an expression for (1) fire risk and (2) shock hazard risk based on the component risk factors according to IEC discussed in this chapter.

8. Conduct a risk analysis, according to NFPA 780, to ascertain if a LPS is required for the following building installation:

Building dimensions = 80 × 100 m, 60 m high factor, C_1 = 0.25, N_g = 4, nonmetal construction with nonmetallic roof, high value and moderate flammability, normally occupied, continuity of service required, with no environmental effects.

9. In Prob. 8, design a roof lightning termination system according to NFPA 780. Calculate the number of roof conductors, spacing of roof conductors, number of air terminals, and number of down conductors.

REFERENCES

1. IEC 62305-1, Protection Against Lightning, Part 1—General Principles, 2006.
2. IEC 62305-2, Protection Against Lightning, Part 2—Risk Management, 2006.
3. IEC 62305-3, Protection Against lightning, Part 3—Physical Damage to Structures and Life Hazard, 2006.
4. IEC 62305-4, Protection Against Lightning, Part 4—Electrical and Electronic Systems Within Structures, 2006.
5. IEC 60079-10, ed. 4.0, Electrical Apparatus for Explosive Gas Atmospheres, Part 10—Classification of Hazardous Areas, 2009.
6. R. H. Lee, "Lightning Protection of Buildings," *IEEE Trans. IA*, vol. 15, pp. 237–240, 1979.
7. IEEE Working Group Report, "A Simplified Method of Estimating Lightning Performance of Transmission Lines," *IEEE Trans. PAS*, vol. PAS 104, no. 4, pp. 912–932, 1985.
8. A. J. Eriksson, "The Incident of Lightning Strikes to Power Lines," *IEEE Trans. Power Delivery*, PWRD-2, pp. 859–870, 1987.
9. F. A. M. Rizk, "Modeling of Transmission Line Exposure to Direct Lightning Strokes," *IEEE Trans. Power Delivery*, vol. 5, no. 4, pp. 2009–2029, 1990.
10. NFPA 780, Standard for Installation of Lightning Protection Systems, 2008.
11. D. Eskow, "Striking Back at Lightning," *Popular Mechanics*, Aug. 1983.
12. K. P. Heary, A. Z. Chaberski, S. Gumley, J. R. Gumley, F. R. Richens, and J. H. Moran, "An Experimental Study of Ionization Air Terminal Performance," in Conf. Record, *IEEE Power Engineering Society Summer Meeting*, Portland, Oregon, July 1998.
13. L. M. Ball, "The Laser Lightning Rod System: Thunderstorm Domestication," *Applied Optics*, vol. 13, pp. 2292–2295, Oct. 1974.
14. R. M. Bent and S. K. Llewellyn, "An Investigation of the Lightning Elimination and Strike Reduction Properties of Dissipation Arrays," in Conf. Record, *Review of Lightning Protection Technology for Tall Structures*, Lyndon B. Johnson Space Flight Center, Clear Lake City, Houston, TX, Nov. 1976.
15. R. B. Carpenter, Jr. and M. M. Drabkin, Lightning Strike Protection, LEC, Inc. Colorado, March 2005. Available at: http://www.lightningeliminators.com
16. WHO Environmental Health Criteria No. 238, Geneva 2007. Available at: http:/www.who.int/peh-emf/publications/elf_ehc/en/index.html
17. CIGRE Work Group, GT C4.203, "Technical Guide for Measurement of Low Frequency Electrical and Magnetic Fields Near Overhead Power Lines," *Electra*, no. 243, pp. 32–39, April 2009.
18. CIGRE Work Group, TF C4.205, "Characteristics of ELF Magnetic Fields," *Electra*, no. 231, pp. 62–69, April 2007.
19. NOVA-Lightning, Videocassette, WGBH, P.O. Box 2284, S. Burlington, VT 05407-2284, USA.

FURTHER READING

I. Cotton and N. Jenkins, "Lightning Protection of Wind Turbines," *International Conference and Exhibition on Lightening protection*, Paper No. 6.1, Solihull, West Midlands, UK, 1998.

J. C. Das, "Lightning Protection of Industrial and Commercial Facilities—What is New?" *Construction Journal of India*, vol. II, Issue 1, pp. 32–38, Jan./Feb. 1999.

IEEE Std. C95.6, IEEE Standard for Safety Levels with Respect to Human Exposure to Electromagnetic Fields, 0–3 kHz, 2002.

M. E. Morris, R. J. Fisher, G. H. Schnetzer, K. E. Merewether, and R. E. Jorgenson, "Rocket Triggered Lightning Studies for Protection of Critical Assets," in Conf. Record, *IEEE Ind. and Commercial Power System Conference*, 1993, St. Petersburg, Florida.

NFPA 115, Standard for Laser Fire protection, 2003.

B. Richardson, "Lightning Protection of Hi-Tech Buildings," *EC&M*, Aug. 1989.

WHO Media Center. Fact Sheet No. 322, June 2007.

CHAPTER 23

DC Systems, Short Circuits, Distributions, and HVDC

AC and dc systems are generally interconnected at some interface and the transient behavior of one impacts the other, though there can be stand-alone dc systems. The short-circuit transients, dc current interruption, trends in residential and commercial dc distributions, and HVDC transmission systems are discussed in this chapter. From the transient analysis point of view HVDC systems provide considerable challenges.

23-1 SHORT-CIRCUIT TRANSIENTS

IEC standard[1] is a comprehensive document on the subject; though the dc short-circuits currents are covered in some other publications too, which are referred in proper context in this chapter. This IEC standard addresses calculation of short-circuit currents in dc auxiliary installations in power plants and substations and does not include other dc systems, for example, electric railway traction and transit systems.

The dc sources analyzed for the short-circuit transients are:

- Lead acid batteries
- DC motors and generators
- Converters in three-phase bridge configuration
- Smoothing capacitors.

Figure 23-1 shows typical time current profiles of these sources and Fig. 23-2 shows approximation function assumed in Ref. 1. The function is described by:

$$i_1(t) = i_p \frac{1-e^{-t/\tau_1}}{1-e^{-t_p/\tau_1}} \tag{23-1}$$

$$i_2(t) = i_p[(1-\alpha)e^{-(t-t_p)/\tau_2} + \alpha]\ t \geq t_p \tag{23-2}$$

$$\alpha = \frac{I_k}{i_p} \tag{23-3}$$

where I_k is the quasi-steady-state short-circuit current, I_p is the peak short-circuit current, t_p is the time to peak, τ_1 is the rise time constant, and τ_2 is the decay time constant.

The quasi-steady-state current I_k is conveniently assumed as the value at 1 s after beginning of the short circuit. If no definite maximum is present, as shown in Fig. 23-1*a* for the converter current, then the function is given by Eq. (23-1).

23-1-1 Short Circuit of a Lead Acid Battery

We studied the transient in an RL circuit in Chap. 2. The short circuit of a battery is analogous:

$$i = \frac{E_B}{R}(1-e^{-(R/L)t}) \tag{23-4}$$

where E_B is internal battery voltage, R is the total resistance in the circuit, which is given by:

$$R = R_B + R_C + R_L \tag{23-5}$$

where R_B is the internal cell resistance, R_C is the resistance of cell interconnections, and R_L is the lead resistance. These resistances must be accurately calculated. Similarly the inductance L is:

$$L = L_{CC} + L_{BC} + L_L \tag{23-6}$$

where L_{CC} is the inductance of the cell structure, L_{BC} is the inductance of the battery cells considered as bus bars, and L_L is the inductance of the battery leads. The internal inductance of the cells can be considered to be zero. Again these inductances are small and should be accurately calculated.

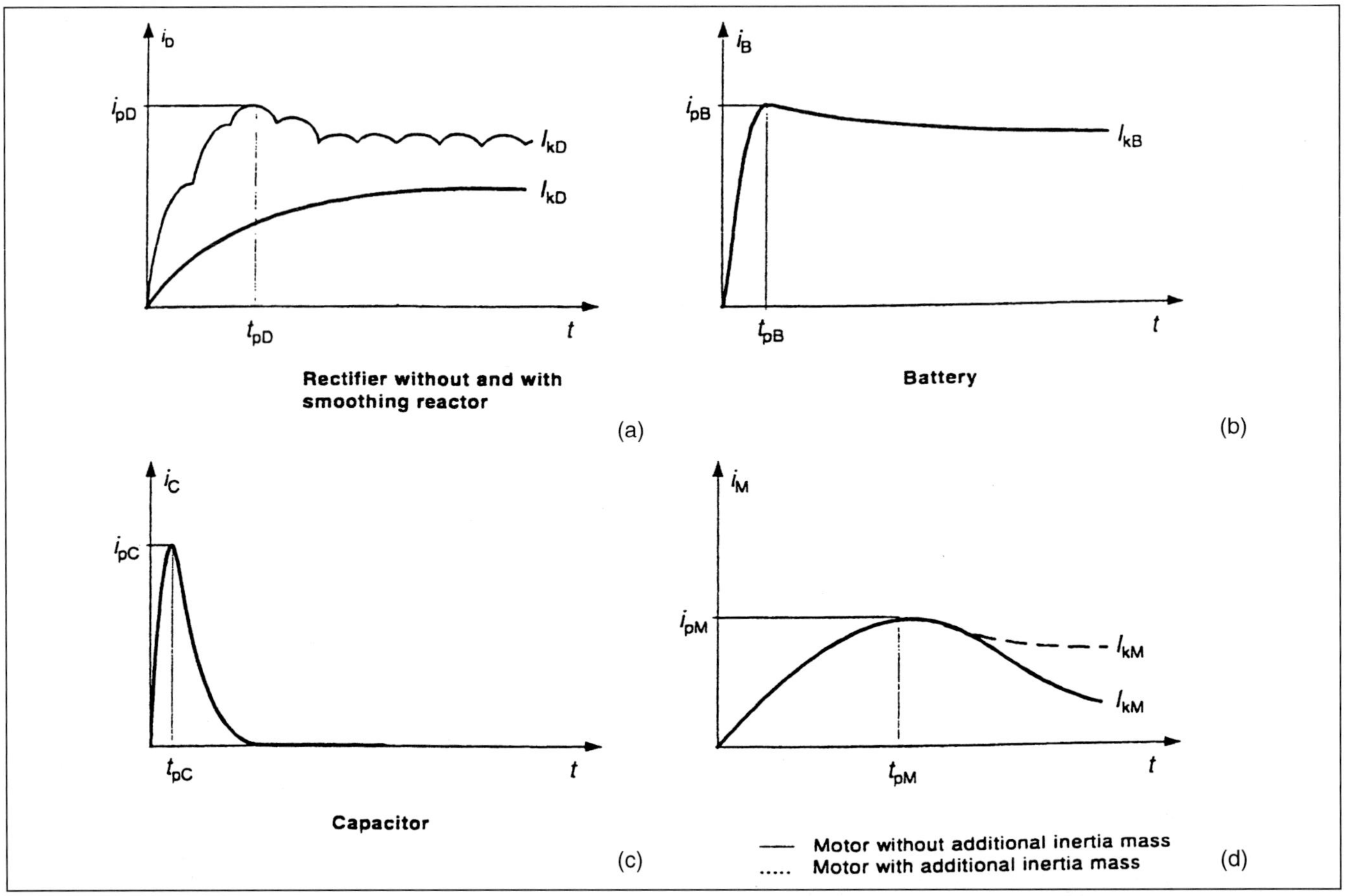

FIGURE 23-1 Short-circuit current profiles of dc sources. (*a*) Rectifier with and without smoothing reactor. (*b*) Lead acid battery. (*c*) Capacitor. (*d*) Dc motor with and without additional inertia mass.[1]

From Eq. (23-4) the maximum battery short-circuit current is:

$$I_{Bsc} = \frac{E_B}{R} \tag{23-7}$$

and the initial rate of rise is:

$$\frac{di_B}{dt} = \frac{E_B}{L} \tag{23-8}$$

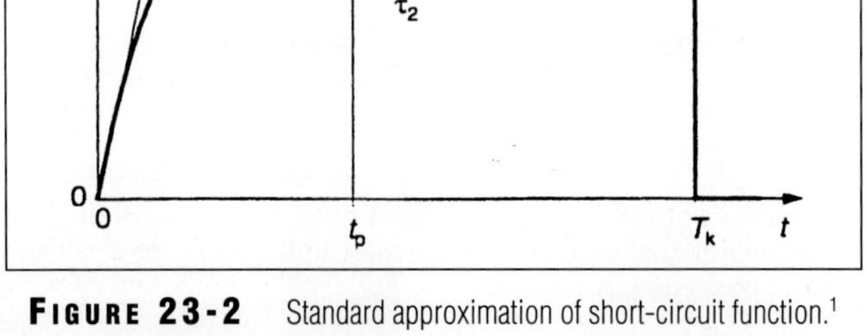

FIGURE 23-2 Standard approximation of short-circuit function.[1]

The battery internal resistance is normally specified by the battery manufacturer. An expression from Ref. 2 is:

$$R_B = R_{cell} N = \frac{R_p}{N_p} \tag{23-9}$$

where R_{cell} is the resistance in ohms per cell, N is the number of cells, R_p is the resistance in ohms per positive plate, and N_p is the number of positive plates in a cell. R_p is given by:

$$R_p = \frac{V_1 - V_2}{I_2 - I_1}\ \Omega/\text{positive plate} \tag{23-10}$$

where V_1 is the cell voltage and I_1 is the corresponding rated discharge current per plate. Similarly, V_2 is the cell voltage and I_2 is the discharge current at V_2. Another expression from Ref. 3 is:

$$R_B = \frac{E_B}{100 \times I_{8hr}}\ \Omega \tag{23-11}$$

where I_{8hr} is the 8-h ampere rating of the battery to 1.75 V per cell at 25°C. R_B is not a constant quantity and depends on the charge state of the battery. A discharged battery has a much higher cell resistance.

Example 23-1 Consider a 120-V battery of 200 A-h rating, 8-hr rate of discharge to 1.75 V, each cell having dimensions of height 7.9 in, length is 10.7 in, width is 6.8 in. The battery is rack mounted, 30 cells per row as shown in Fig. 23-3; cell connectors are 250 KCMIL, diameter is 0.575 in. The battery is connected through a cable of approximately 100 ft, cable resistance 5 mΩ, and inductance 14 μH to a circuit breaker.

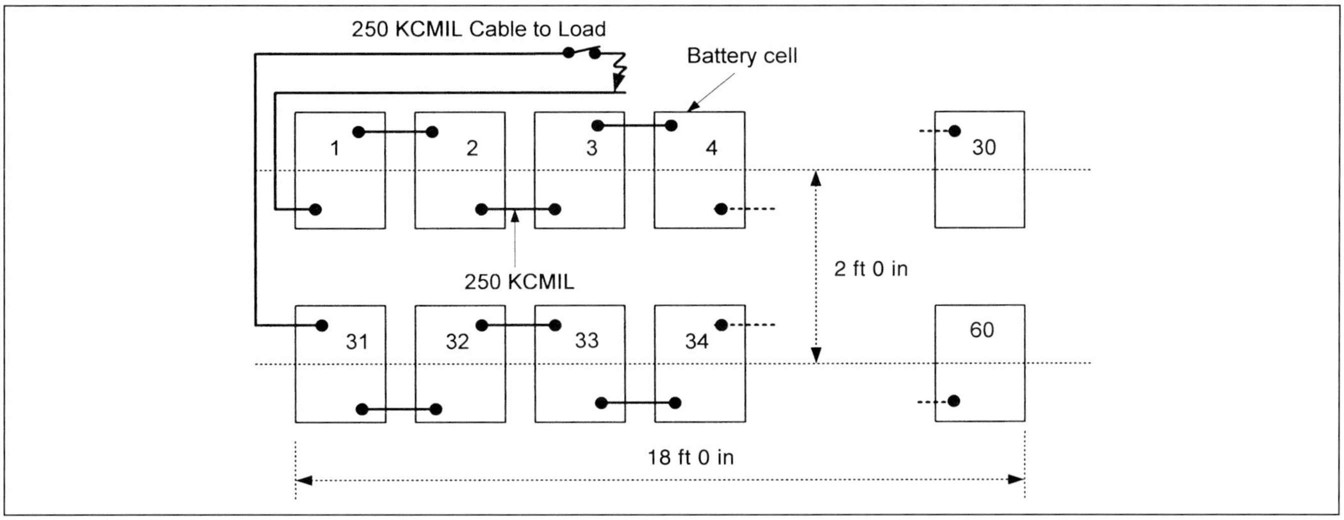

FIGURE 23-3 Lead acid battery system layout for calculation of battery short-circuit current.

Calculations According to IEEE Equations The equation for the battery short-circuit current can be written based on Eq. (23-4) by calculating all the resistance and inductance in the circuit.

The manufacturer specifies the following expression for internal resistance of this battery:

$$R_B = \frac{31E_B}{I_{8\text{hr}}}\ \text{m}\Omega$$

Substituting the values, this gives a battery resistance of 18.6 mΩ. If Eq. (23-11) is used, then R_B = 6 mΩ. To use Eqs. (23-9) and (23-10), a battery discharge curve published by the manufacturer is required. This shows initial cell voltage versus the amperes per plate and also the various discharge rate curves.

For purpose of this example, let us consider R_B = 18.6 mΩ. From Fig. 23-3, the total length of battery connectors is 28 ft. The resistance at 25°C can be read from appropriate tables in electrical handbooks or analytically calculated; it is equal to 1.498 mΩ. Thus, the total resistance in the battery circuit including lead length is 25.1 mΩ. The maximum short-circuit current is therefore:

$$\frac{120}{25.1}\times 10^3 = 4781\ \text{A}$$

The inductance of two round conductors of radius r, spaced at distance d is given by:

$$L = \frac{\mu_0}{\pi}\left(0.25 + \ln\frac{d}{r}\right) \tag{23-12}$$

where L is in henry per meter, μ_0 is the permeability in vacuum = $4\pi\times 10^{-7}$ H/m. From Fig. 23-3, the distance d = 24 in., and r, the radius of 250 KCMIL conductor, is 0.2875 in. Substituting these values gives L_{CC} = 1.87 μH/m. For an 18-ft loop length, L_{CC} = 10.25 μH

The inductance of the battery cells can be determined by considering each row of cells like a bus bar at spacing d = 24 in, height of bus bar h = height of cell = 7.95 in, width of bus bar w = width of cell = 6.8 in. The inductance is given by:

$$L_{BC} = \frac{\mu_0}{\pi}\left(\frac{3}{2} + \ln\frac{d}{h+w}\right)\text{H/m} \tag{23-13}$$

Substituting the values gives: L_{BC} = 4.36 μH. The total inductance in the system = 10.25 + 4.36 + 14 = 28.61 μH. The initial rate of rise of current is:

$$\frac{120}{28.61\times 10^{-6}} = 4.19\times 10^6\ \text{A/s}$$

and the time constant is:

$$\frac{28.61\times 10^{-6}}{25.089\times 10^{-3}} = 1.14\ \text{ms}$$

The current, therefore, reaches 0.63 × 4781 = 3012 A in 1.14 ms, and in 2.28 ms it will be 0.87 × 4781 = 4159 A.

Calculations as per IEC Standards In the IEC method of calculation[1] the battery resistance R_B is multiplied by a factor of 0.9, while all other resistances remain unchanged. Also, if the open-circuit voltage of the battery is not known, then use $E_B = 1.05U_{nB}$, where U_{nB} = 2.0 V/cell for the lead acid battery. The peak current is given by:

$$i_{pB} = \frac{E_B}{R_{BBr}} \tag{23-14}$$

where i_{pB} is the peak current and R_{BBr} is the total equivalent resistance.

The time to peak and the rise time are read from the curves in IEC,[1] which are not reproduced, based on $1/\delta$, which is defined as follows:

$$\frac{1}{\delta} = \frac{2}{(R_{BBr}/L_{BBr}) + (1/T_B)} \tag{23-15}$$

The time constant T_B = 30 ms and L_{BBr} is the total equivalent inductance to the fault point. The decay time constant τ_{2B} = 100 ms. The quasi-steady-state current is given by:

$$I_{kB} = \frac{0.95E_B}{R_{BBr} + 0.1R_B} \tag{23-16}$$

This considers that battery voltage falls and the internal cell resistance increases after a short circuit. All equations in IEC are in MKS units. The curves in IEC for calculations of short-circuit currents in this example and other examples to follow are not reproduced. An interested reader may like to get a copy of this standard.[4]

Example 23-2 Example 23-1 is repeated with IEC method of calculation. The total resistance and inductance in the battery circuit are:

$$R_{\text{BBr}} = 0.9 \times 18.6 + 1.498 + 5 = 23.238 \text{ m}\Omega$$

$$L_{\text{BBr}} = 28.61 \ \mu\text{H}$$

Therefore, the maximum short-circuit current is:

$$i_{\text{pB}} = \frac{1.05 \times 120}{23.238 \times 10^{-3}} = 5.42 \text{ kA}$$

Calculate $1/\delta$ from Eq. (23-15)

$$\frac{1}{\delta} = \frac{2}{\dfrac{23.238 \times 10^{-3}}{28.61 \times 10^{-6}} + \dfrac{1}{30 \times 10^{-3}}} = 2.37 \text{ ms}$$

Then from the curve in Ref. 1, time to peak is approximately 5.4 ms, and the rise time is 1.2 ms. The quasi-steady-state current is:

$$I_{\text{kB}} = \frac{0.95 \times 1.05 \times 120}{23.238 + 0.1(18.6)} = 4.77 \text{ kA}$$

Short-circuit current plots according to this calculation are shown in Fig. 23-4(a).

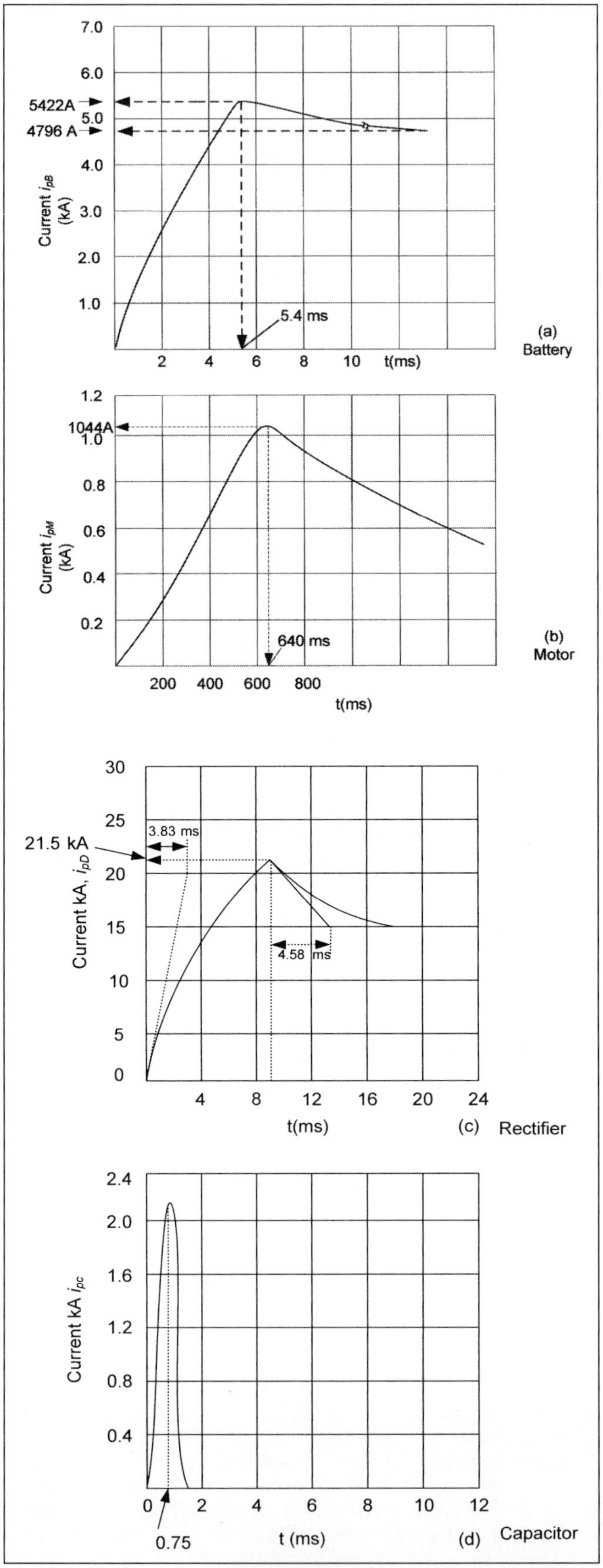

FIGURE 23-4 Calculated short-circuit current-time plots. (*a*) Lead acid battery. (*b*) DC motor. (*c*) Rectifier. (*d*) Capacitor.

23-1-2 Short Circuit of DC Motors and Generators

ANSI/IEEE Method of Calculation An expression for the short-circuit currents of dc motors and generators is:[3]

$$i_a = \frac{e_0}{r'_d}(1 - e^{-\sigma_a t}) - \left(\frac{e_0}{r'_d} - \frac{e_0}{r_d}\right)(1 - e^{-\sigma_f t}) \tag{23-17}$$

where i_a is the per unit current, e_0 is the internal emf prior to short circuit in per unit, r_d is the steady-state effective resistance of the machine in per unit, r'_d is the transient effective resistance of the machine in per unit, σ_a is the armature circuit decrement factor, and σ_f is the field circuit decrement factor.

The first part of the equation has armature time constant, which is relatively short and controls the buildup and peak of the short circuit current, the second part is determined by the shunt field excitation and it controls the decay of the peak value. The problem with the calculation is that the time constants in this equation are not time-invariant. Saturation causes the armature circuit decrement factor to increase as the motor saturates. Approximate values suggested for saturated conditions are 1.5 to 3.0 times the unsaturated value and conservatively a value of 3.0 can be used. The unsaturated value is applicable at the start of the short-circuit current and the saturated value at the maximum current. Between these two extreme values, the decrement is changing from one value to another. Figure 23-5 shows the approximate curve of the short-circuit current and its equivalent circuit. For the first two-thirds of the curve, the circuit is represented by machine unsaturated inductance L'_a, and for last one-third part of the short-circuit curve L'_a is reduced to 1/3 with series transient resistance. The peaks short-circuit current in per unit is given by:

$$i'_a = \frac{e_0}{r'_d} \tag{23-18}$$

The transient resistance r'_d in per unit requires some explanation. It is the effective internal resistance:

$$r'_d = r_w + r'_b + r'_x \tag{23-19}$$

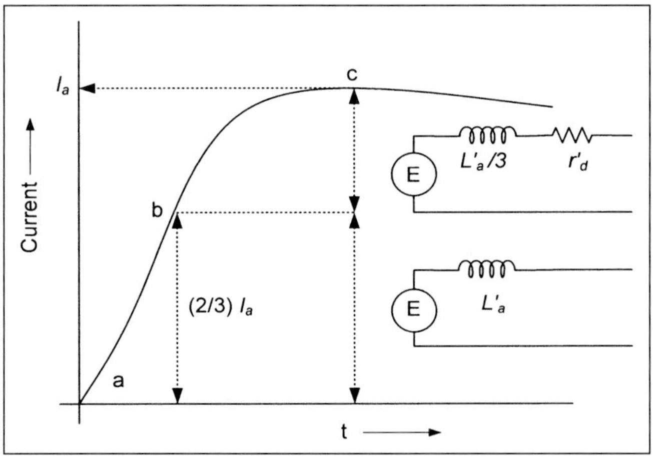

FIGURE 23-5 Short-circuit time curve for a dc motor or generator, showing two distinct time constants.

where r_w is the total resistance of the windings in the armature circuit, r'_x is equivalent to flux reduction in per unit, and r'_b is the transient resistance equal to reactance voltage and brush contact resistance in per unit. The flux reduction and distortion are treated as ohmic resistance. The values of transient resistance r'_d in per unit are given graphically in the AIEE Committee report[5] of vintage 1950, depending on the machine rating, voltage, and speed. The transient resistance is not constant and there is a variation band. The machine load may also affect the transient resistance.[6] Similarly, the steady-state resistance is defined as:

$$r_d = r_w + r_b + r_x \tag{23-20}$$

where r_b is the steady-state resistance equivalent to reactance voltage and brush contact in per unit, and r_x is steady-state resistance equivalent to flux reduction in per unit. The maximum rate of rise of the current is dependent on armature unsaturated inductance. The unit inductance is defined as:

$$L_{a1} = \frac{V_1}{I_a}\frac{2\times 60}{2\pi P N_1} \tag{23-21}$$

Per unit inductance is the machine inductance L'_a divided by the unit inductance:

$$C_x = \frac{L'_a}{L_{a1}} = \frac{PN_1 L'_a}{19.1}\frac{I_a}{V_1} \tag{23-22}$$

This can be written as:

$$L'_a = \frac{19.1 C_x V_1}{PN_1 I_a} \tag{23-23}$$

where P is the number of poles, N_1 is the base speed, V_1 is the rated voltage, I_a is the rated machine current, and C_x varies with the type of machine. Charts of initial inductance plotted against unit inductance show a linear relation for a certain group of machines. For this purpose, the machines are divided into four broad categories as follows:

Motors: $C_x = 0.4$ for motors without pole face windings.

Motors: $C_x = 0.1$ for motors with pole face windings.

Generators: $C_x = 0.6$ for generators without pole face windings.

Generators: $C_x = 0.2$ for generators with pole face windings.

The armature circuit decrement factor is:

$$\sigma_a = \frac{r'_d 2\pi f}{C_x} \tag{23-24}$$

The maximum rate of rise of current in amperes per second is given by:

$$\frac{di_a}{dt} = \frac{V_1 e_0}{L'_a} \tag{23-25}$$

Rate of rise of current can also be expressed in terms of per unit rated current:

$$\frac{di_a}{dt} = \frac{PN_1 e_0}{19.1 C_x} \tag{23-26}$$

In Eqs. (23-25) and (23-26), e_0 can be taken equal to unity without appreciable error. More accurately e_0 can be taken as 0.97 per unit for motors and 1.03 for generators. The inductance L'_a is given in tabular and graphical forms in an AIEE publication.[7] Figure 23-6 shows L'_a values in mH for certain motor sizes, without pole face windings. The recent IEEE standard[8] reiterates this calculation method.

Example 23-3 Calculate the terminal short-circuit current of a 230-V, 150-hp, 1150-rpm motor. The motor armature current is 541 A.

From graphical data in Ref. 3, the transient resistance = 0.068 per unit. The inductance L'_a from graphical data in Fig. 23-6 is 1.0 mH. Then, the peak short-circuit current is:

$$I'_a = \frac{I_a}{r'_d} = \frac{541}{0.068} = 7956 \text{ A}$$

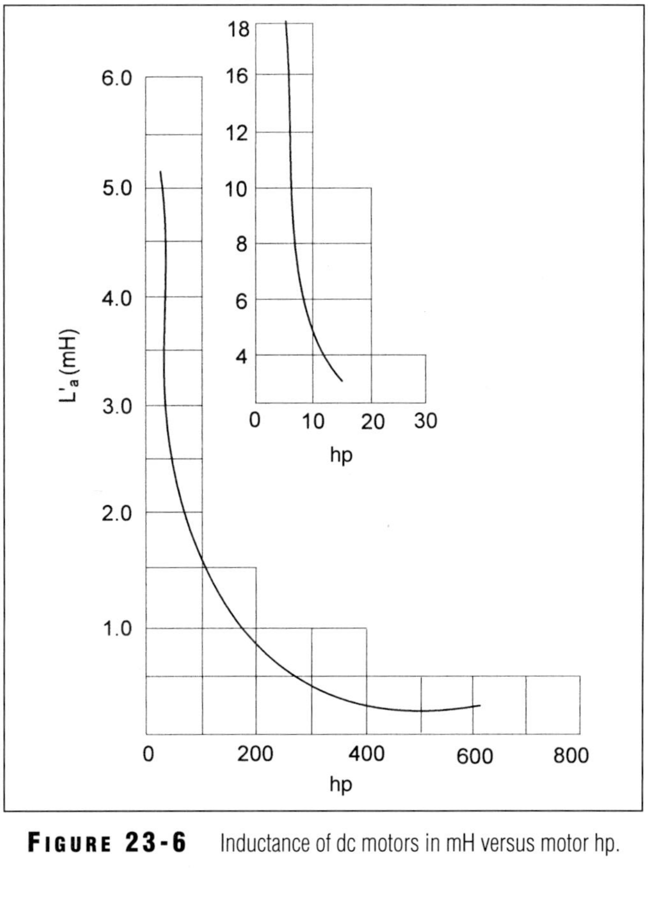

FIGURE 23-6 Inductance of dc motors in mH versus motor hp.

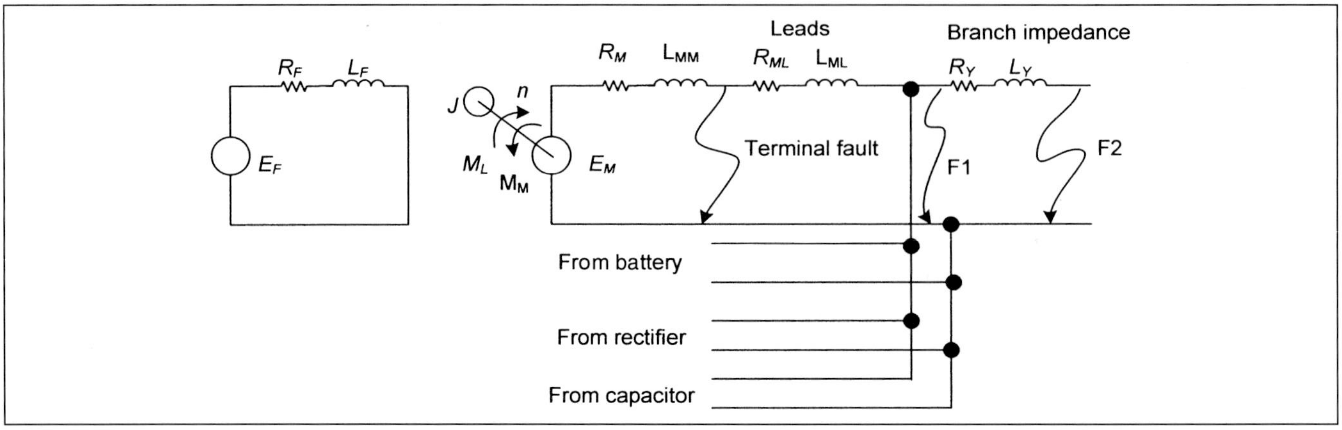

FIGURE 23-7 Equivalent circuit: short circuit of a dc motor. Total short circuit current-summation of partial short-circuit currents from battery, rectifier, and capacitor.

and the initial rate of rise of the current is:

$$\frac{di_a}{dt} = \frac{V_1}{L'_a} = \frac{230}{1\times10^{-3}} = 230 \text{ kA/s}$$

As shown in Fig. 23-5, time constant changes at point b.

$$\frac{L'_a}{3r'_d}$$

The base ohms are V_a/I_a = 230/541 = 0.425. Therefore, r'_d = (0.425) (0.068) = 0.0289 Ω. This gives a time constant of 11.52 ms.

IEC Method of Calculation The resistance and inductance network for short circuit of a dc machine with a separately excited field is shown in Fig. 23-7. The equivalent resistance and inductance are:

$$R_{\text{MB}} = R_M + R_{\text{ML}} + R_y$$
$$L_{\text{MB}} = L_M + L_{\text{ML}} + L_y \qquad (23\text{-}27)$$

where R_M and L_M are the resistance of the armature circuit, including brushes, R_{ML} and L_{ML} are the resistance and inductance of the conductor in the motor circuit, respectively, and R_y and L_y are the resistance and inductance of the common branch, respectively, if present. The time constant of the armature circuit to the point of short circuit is given by:

$$\tau_M = \frac{L_{\text{MB}}}{R_{\text{MB}}} \qquad (23\text{-}28)$$

The quasi-steady-state short-circuit current is given by:

$$I_{\text{KM}} = \frac{L_F}{L_{\text{0F}}} \frac{U_{\text{rM}} - I_{\text{rM}} R_M}{R_{\text{MB}}} \quad I_{\text{KM}} = 0 \quad \text{when } n \to 0 \qquad (23\text{-}29)$$

where L_F is the equivalent saturated inductance of the field circuit on short circuit, L_{0F} is the equivalent unsaturated inductance of the field circuit at no load, U_{rM} is the rated voltage of the motor, I_{rM} is the rated current of the motor, n is the motor speed, and n_n is the rated motor speed.

The peak short-circuit current of the motor is given by:

$$i_{\text{pM}} = \kappa_M \frac{U_{\text{rM}} - I_{\text{rM}} R_M}{R_{\text{MB}}} \qquad (23\text{-}30)$$

At normal speed or decreasing speed with $\tau_{\text{mec}} \geq 10\tau_F$, the factor $\kappa_M = 1$, where τ_{mec} is mechanical time constant given by:

$$\tau_{\text{mec}} = \frac{2\pi J n_0 R_{\text{MB}} I_{\text{rM}}}{M_r U_{\text{rM}}} \qquad (23\text{-}31)$$

where J is the moment of inertia and M_r is the rated torque of the motor.

The field time constant is given by:

$$\tau_F = \frac{L_F}{R_F} \qquad (23\text{-}32)$$

for $\tau_{\text{mec}} \geq 10\tau_F$ the time to peak and time constant are given by:

$$t_{\text{pM}} = \kappa_{\text{1M}} \tau_{\text{M}}$$
$$\tau_{\text{1M}} = \kappa_{\text{2M}} \tau_{\text{M}} \qquad (23\text{-}33)$$

The factors κ_{1M}, κ_{2M} are taken from the curves in IEC[1] and are dependent on τ_F/τ_M and L_F/L_{0F}. For decreasing speed with $\tau_{\text{mec}} < 10\tau_F$, the factor κ_M is dependent upon $1/\delta = 2\tau_M$ and ω_0.

$$\omega_0 = \sqrt{\frac{1}{\tau_{\text{mec}} \tau_M} \left(1 - \frac{I_{\text{rM}} R_M}{U_{\text{rM}}}\right)} \qquad (23\text{-}34)$$

where ω_0 is undamped natural angular velocity and δ is the decay coefficient; κ_M is derived from the curves in the IEC standard.[1]

For decreasing speed with $\tau_{\text{mec}} < 10\tau_F$, the time to peak τ_M is read from the curve in IEC standard, and the rise time constant is given by:

$$\tau_{\text{1M}} = \kappa_{\text{3M}} \tau_M \qquad (23\text{-}35)$$

where κ_{3M} is read from the curves in Ref. 1, which are not reproduced here.

The decay time constant τ_{2M} for nominal speed or decreasing speed with $\tau_{\text{mec}} \geq 10\tau_F$ is given by:

$$\tau_{\text{2M}} = \tau_F \qquad \text{when } n = n_n = \text{const}$$
$$\tau_{\text{2M}} = \frac{L_{\text{0F}}}{L_F} \kappa_{\text{4M}} \tau_{\text{mec}} \qquad \text{when } n \to 0 \text{ with } \tau_{\text{mec}} \geq 10\tau_F \qquad (23\text{-}36)$$

For decreasing speed with $\tau_{\text{mec}} < 10\tau_F$:

$$\tau_{2M} = \kappa_{4M}\tau_{\text{mec}} \tag{23-37}$$

where κ_{4M} is again read from curve in Ref. 1, not reproduced here. Thus, IEC calculations require extensive motor data and application of number of graphical data in the standard.

Example 23-4 Calculate the short-circuit current of a 15-hp, 1150-rpm, six-pole, 115-V, dc motor for a terminal fault. The armature current = 106 A, the armature and brush circuit resistance = 0.1 Ω, and inductance in the motor circuit = 8 mH; τ_F = 0.8 s, $\tau_{\text{mec}} > 10\tau_F, L_{0F}/L_F = 0.5$, and $\tau_{\text{mec}} = 20$ s.

There is no external resistance or inductance in the motor circuit, therefore $R_{MBr} = R_M = 0.10\ \Omega$. IEC is not specific about calculation of motor circuit resistance. The time constant is:

$$\tau_M = \frac{L_M}{L_R} = \frac{8\times 10^{-3}}{0.10} = 80 \text{ ms}$$

The quasi-steady-state current from Eq. (23-29) is:

$$0.5\left(\frac{115-(0.10)(106)}{0.10}\right) = 522 \text{ A}$$

From Eq. (23-30), the peak current = 1044 A, because for $\tau_{\text{mec}} > 10\tau_F$, factor κ_M in Eq. (23-30) is equal to 1. The time to peak and time constant are given by Eq. (23-33). From Curve in IEC[1] and $\tau_F/\tau_M = 10$ and $L_F/L_{0F} = 0.5$, factor $\kappa_{1M} = 8.3$ and $\kappa_{2M} = 3.7$. Therefore, time to peak = 640 ms and time constant $\tau_{1M} = 296$ ms. The short-circuit profile is plotted in Fig. 23-4*b*.

23-1-3 Short-Circuit Current of a Rectifier

ANSI/IEEE Method The typical current-time curve for short circuit of a rectifier is shown in Fig. 23-8. The maximum current is reached at half cycle after the fault occurs. The peak at half cycle is caused by the same phenomena that creates dc offset in ac short-circuit calculations. The magnitude of this peak is dependent upon *X*/*R* ratio, the ac system source reactance, rectifier transformer impedance, and also the resistance and reactance through which the current flows in the dc system. The addition of resistance or inductance in the dc system reduces this peak and, depending upon the magnitude of these components, the peak may be entirely eliminated, with smoothing dc reactor as shown in Fig. 23-1. Figure 23-8 shows three regions of the rectifier short-circuit current. The region A covers the initial rise of current, the peak current occurs in region B, and region C covers the time after one cycle till the current is interrupted.

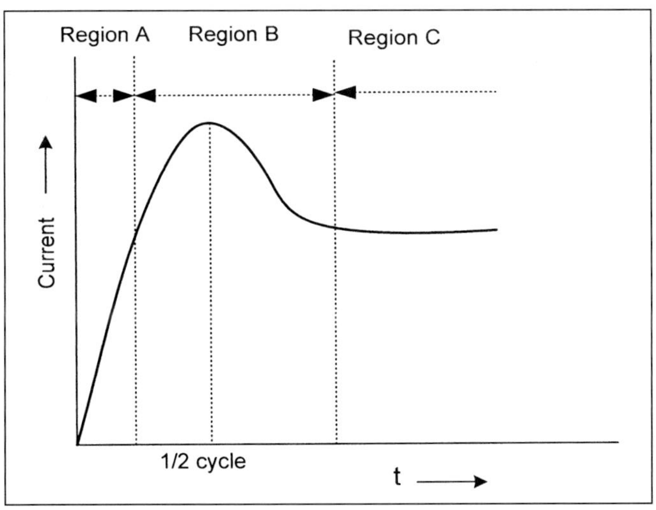

FIGURE 23-8 Short-circuit current profile of a rectifier.

The initial rate of rise of dc short-circuit current for bolted fault varies with the magnitude of the sustained short-circuit current. The addition of inductance in the dc circuit tends to decrease the rate of rise.

An equivalent circuit of the rectifier short-circuit current is developed with a voltage source and equivalent resistance and inductance. The equivalent resistance varies with rectifier terminal voltage, which is a function of the short-circuit current. The equivalent resistance is determined from rectifier regulation curve by an iterative process, which can admirably lend itself to iterative computer solution. The equivalent inductance is determined from sustained short-circuit current for a bolted fault and rated system voltage. The magnitude of the peak current is determined from ac and dc system impedance characteristics.[3]

The following step-by-step procedure can be used:

1. Calculate total ac system impedance $Z_C = R_C + X_C$ in ohms. Convert to per unit impedance z_C, which may be called the commutating impedance in per unit on rectifier transformer kVA base. This is dependent on the type of rectifier circuit. For double-wye, six-phase circuit, the conversion is given by:

$$Z_C = z_C \times 0.6 \times \frac{E_D}{I_D}\Omega \tag{23-38}$$

where E_D and I_D are rectifier rated dc voltage and current.

2. Assume a value of rectifier terminal voltage e_{da} under faulted condition and obtain factor K_2 from Fig. 23-9. Then the *preliminary* calculated value of the sustained short-circuit current is given by:

$$I_{da} = \frac{K_2}{z_C} I_D \tag{23-39}$$

The equivalent rectifier resistance is then given by:

$$R_R = \frac{(E_D - E_{da})}{I_{da}}\Omega \tag{23-40}$$

E_{da} is the assumed rectifier terminal voltage in pu under fault conditions.

$$E_{da} = e_{da} \times E_D$$

3. The sustained value of the fault current is:

$$I_{dc} = \frac{E_D}{R_R + R_D} \tag{23-41}$$

where R_D is the resistance external to the rectifier. The rectifier terminal voltage in volts is:

$$E_{dc} = E_D - I_{dc}R_R \tag{23-42}$$

4. The value of $E_{da} = e_{da} \times E_D$ should be within 10 percent of the calculated value E_{dc}, the rectifier terminal voltage under sustained short-circuit current. The iterative process is repeated until the desired tolerance is achieved.

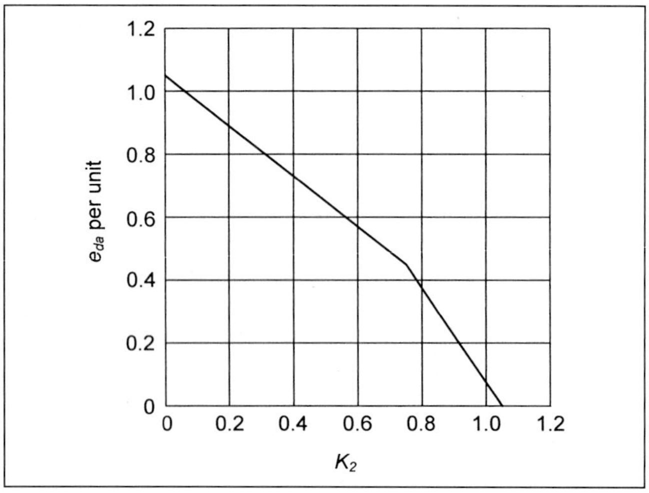

FIGURE 23-9 Sustained fault current factor versus rectifier terminal voltage.

Example 23-5 Consider a 100-kW source at 125 V dc. The dc resistance of the feeder cable is 0.004 Ω. Let the ac source and rectifier transformer impedance z_C = 0.05 pu and I_D = 800 A. Calculate the rectifier resistance for a fault at the end of the cable. Assume e_{da} = 0.5 per unit, that is, E_{da} = 62.5 V. K_2 from Fig. 23-9 = 0.70. Therefore:

$$I_{da} = \frac{K_2}{z_C} I_D = \frac{0.70}{0.05} \times 800 = 11200 \text{ A}$$

Then R_R is:

$$R_R = \frac{E_D - E_{da}}{I_{da}} = \frac{62.5}{11200} = 0.00558\ \Omega$$

$$I_{dc} = \frac{E_D}{R_R + R_D} = \frac{125}{0.00558 + 0.004} = 13048 \text{ A}$$

$$E_{dc} = 125 - (13048)(0.00558) = 72.18 \text{ V}$$

This does not satisfy $E_{da} = e_{da} \times E_D$ We can iterate once more for closer estimate of R_R:

E_{dc} = 72.18 V
e_{da} = 49 V (0.392 pu)
K = 0.77
I_{da} = 12320 A
R_R = 0.006169 Ω
I_{dc} = 12292.5 A
E_{dc} = 49.17 V

Now, E_{da} = 0.392 × 125 V = 49 V which is acceptable. To calculate rate of rise of the current, the rectifier inductance L_R is required. This is calculated based upon terminal short-circuit of the rectifier.

On a terminal short-circuit, voltage is zero and from Fig. 23-9, K_2 = 1.02. This gives a terminal short-circuit current of 16320 A. Then from Ref. 3, rectifier inductance is given by:

$$L_R = \frac{E_D}{360 \times I_{da}}$$

This gives 0.0213 mH. Considering that the cable has an inductive reactance of 0.003 Ω, (= 0.00796 mH), the rate of rise of current is (125)/(0.00796 + 0.0213) = 4.27 kA/s. The peak current can be 1.4 to 1.6 times the sustained current depending upon ac and dc system parameters.[3]

Thus the short-circuit currents of converters are high and the semiconductor devices are protected with fast-acting, current-limiting fuses, specially designed to reduce fuse let-through energy.

IEC Method of Calculation The equivalent circuit diagram of the short-circuit of a rectifier is shown in Fig. 23-10. The maximum short-circuit current is given by minimum impedance $Z_{Q\min}$, which is obtained from the maximum short-circuit current $I''_{kQ\max}$ of the ac system:

$$Z_{Q\min} = \frac{cU_n}{\sqrt{3}I''_{kQ\max}} \tag{23-43}$$

The minimum short-circuit current is given by:

$$Z_{Q\max} = \frac{cU_n}{\sqrt{3}I''_{kQ\min}} \tag{23-44}$$

The resistance and inductance on the ac side are:

$$\begin{aligned} R_N &= R_Q + R_P + R_T + R_R \\ X_N &= X_Q + X_P + X_T + X_R \end{aligned} \tag{23-45}$$

where R_Q and X_Q are the short-circuit resistance and reactance of the ac source referred to the secondary of the rectifier transformer, R_P and X_P are the short-circuit resistance and reactance of the power supply cable referred to the secondary side of the transformer, R_T and X_T are the short-circuit resistance and reactance of the transformer referred to the secondary side of the transformer, and R_R and X_R are the short-circuit resistance and reactance of the commutating reactor if present.

Similarly on the dc side:

$$\begin{aligned} R_{DBr} &= R_S + R_{DL} + R_y \\ L_{DBr} &= L_S + L_{DL} + L_y \end{aligned} \tag{23-46}$$

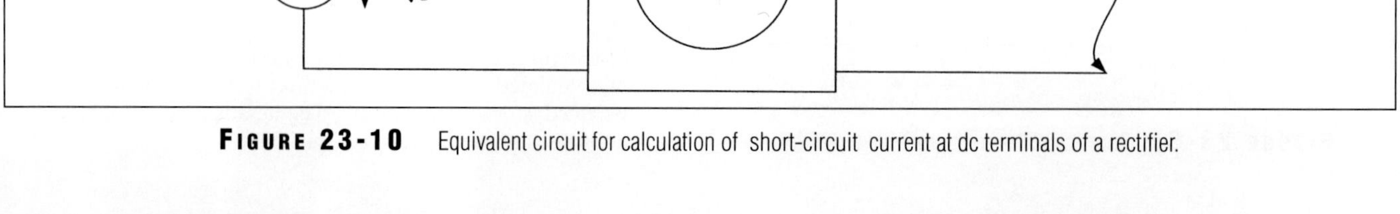

FIGURE 23-10 Equivalent circuit for calculation of short-circuit current at dc terminals of a rectifier.

where R_S, R_{DL}, and R_y are the resistances of dc saturated smoothing reactor, the conductor in the rectifier circuit, and the common branch, respectively, and L_S, L_{DL}, and L_y are the corresponding inductances. The quasi-short-circuit current is given by:

$$i_{kD} = \lambda_D \frac{3\sqrt{2}}{\pi} \frac{cU_n}{\sqrt{3}Z_n} \frac{U_{rTLV}}{U_{rTHV}} \tag{23-47}$$

where Z_n is the impedance on the ac side. The factor λ_D as a function of R_N/X_N and R_{DBr}/R_N is estimated from the curves in Ref. 1. Analytically, it is given by:

$$\lambda_D = \sqrt{\frac{1+(R_N/X_N)^2}{1+(R_N/X_N)^2[1+0.667(R_{DBr}/R_N)]^2}} \tag{23-48}$$

The peak short-circuit current is given by:

$$i_{pD} = \kappa_D I_{kD} \tag{23-49}$$

where the factor κ_D is dependent on:

$$\frac{R_N}{X_N}\left[1+\frac{2R_{DBr}}{3R_N}\right] \quad \text{and} \quad \frac{L_{DBr}}{L_N} \tag{23-50}$$

It can be estimated from the curves in Ref. 1 or analytically:

$$\kappa_D = \frac{i_{pD}}{I_{kD}} = 1+\frac{2}{\pi}e^{-\left(\frac{\pi}{3}+\phi_D\right)\cot\phi_D}\sin\phi_D\left(\frac{\pi}{2}-\arctan\frac{L_{DBr}}{L_N}\right) \tag{23-51}$$

where

$$\phi_D = \arctan\frac{1}{\frac{R_N}{X_N}\left(1+\frac{2}{3}\frac{R_{DBr}}{R_N}\right)} \tag{23-52}$$

Time to peak t_{pD}, when $\kappa_D \geq 1.05$ is given by:

$$t_{pD} = (3\kappa_D+6)\ \text{ms} \quad \text{when} \frac{L_{DBr}}{L_N} \leq 1$$

$$t_{pD} = \left[(3\kappa_D+6)=4\left(\frac{L_{DBr}}{L_N}-1\right)\right]\ \text{ms} \quad \text{when} \frac{L_{DBr}}{L_N} > 1 \tag{23-53}$$

If $\kappa_D < 1.05$, the maximum current, compared with the quasi-steady-state short-circuit current, is neglected, and $t_{pD} = T_k$ is used. The rise time constant for 50 Hz given by:

$$\tau_{1D} = \left[2+(\kappa_D-0.9)\left(2.5+9\frac{L_{DBr}}{L_N}\right)\right]\ \text{ms} \quad \text{when } \kappa_D \geq 1.05$$

$$\tau_{1D} = \left[0.7+\left[7-\frac{R_N}{X_N}\left(1+\frac{2}{3}\frac{L_{DBr}}{L_N}\right)\right]\right. \times \left.\left(0.1+0.2\frac{L_{DBr}}{L_N}\right)\right]\ \text{ms} \quad \text{when } \kappa_D < 1.05 \tag{23-54}$$

For simplification:

$$\tau_{1D} = \frac{1}{3}t_{pD} \tag{23-55}$$

The decay time constant τ_{2D} for 50 Hz is given by:

$$\tau_{2D} = \frac{2}{\frac{R_N}{X_N}\left(0.6+0.9\frac{R_{DBr}}{R_N}\right)}\ \text{ms} \tag{23-56}$$

The time constants for 60-Hz systems are not given in Ref. 1.

Example 23-6 A three-phase rectifier is connected to a three-phase 480-V system through a 100-kVA transformer, 480 to 120 V, of 3 percent impedance, transformer $X/R = 4$. The available short circuit on 480-V side is 36 kA, and the fault point $X/R = 6$. The dc smoothing reactor has an inductance of 5 μH and the resistance of the cable connection is 0.002 Ω. It is required to calculate the short-circuit current and plot its profile.

The ac side source impedance and transformer impedance reflected on the secondary side are:

$$R_Q + jX_Q = 0.00008 + j0.00048\ \Omega$$
$$R_T + jX_T = 0.001 + j0.00419\ \Omega$$

Therefore:

$$R_N + jX_N = 0.0011 + j0.004671\ \Omega$$

On the dc side:

$$R_{DBr} = 0.002\ \Omega \quad \text{and} \quad L_{DBr} = 5\ \mu\text{H}$$

Therefore:

$$\frac{R_N}{X_N} = 0.24 \quad \text{and} \quad \frac{R_{DBr}}{R_N} = 2$$

Then from Eq. (23-48)

$$\lambda_D = \sqrt{\frac{1+(0.24)^2}{1+(0.24)^2(1+0.667)(2.0)^2}} = 0.897$$

The quasi-steady-state current from Eq. (23-47) is:

$$I_{kD} = (0.897)\left(\frac{3\sqrt{2}}{\pi}\right)\left(\frac{1.05\times 480}{\sqrt{3}\times 0.0048}\right)\left(\frac{120}{480}\right) = 18.36\ \text{kA}$$

To calculate the peak current, calculate the ratios:

$$\frac{R_N}{X_N}\left(1+\frac{2}{3}\frac{R_{DBr}}{R_N}\right) = (0.24)(1+0.667\times 2) = 0.56$$

$$\frac{L_{DBr}}{L_N} = \frac{5\times 10^{-6}}{0.0128\times 10^{-3}} = 0.392$$

Thus from Eq. (23-52):

$$\phi_D = \tan^{-1}\left(\frac{1}{1+0.667(2)}\right) = 60.75°$$

From Eq. (23-51), $\kappa_D = 1.204$. Therefore, the peak short-circuit current is:

$$i_{pD} = \kappa_D I_{kD} = 1.204 \times 18.36 = 22.10 \text{ kA}$$

The time to peak is given by Eq. (23-53)

$$t_{pD} = (3\kappa_D + 6) \text{ ms} = (3 \times 1.204 + 6) = 9.62 \text{ ms}$$

The rise time is given by Eq. (23-54) and is equal to 3.83 ms, and the decay time constant from Eq. (23-56) is 4.58 ms.

The short-circuit current is plotted in Fig. 23-4*c*. The intermediate values can be plotted using Eqs. (23-1) and (23-2). The calculations are for a 50-Hz system. For a 60-Hz system the peak will occur in approximately 8.3 ms.

23-1-4 Short Circuit of a Charged Capacitor

IEC Method of Calculation The resistances and inductances in the capacitor circuit are:

$$R_{CBr} = R_C + R_{CL} + R_y \qquad L_{CBr} = L_{CL} + L_y \tag{23-57}$$

where R_C is the equivalent dc resistance of the capacitor, and R_{CL} and L_{CL} are the resistances and inductances of the conductor in the capacitor circuit, respectively. The steady short-circuit current of the capacitor is zero, and the peak current is given by:

$$i_{pC} = \kappa_C \frac{E_C}{R_{CBr}} \tag{23-58}$$

where E_C is the capacitor voltage before the short circuit, and κ_C is read from the curves in Ref. 1 based on:

$$\frac{1}{\delta} = \frac{2L_{CBr}}{R_{CBr}} \qquad \omega_0 = \frac{1}{\sqrt{L_{CBr}C}} \tag{23-59}$$

If $L_{CBr} = 0$, then $\kappa_C = 1$.

The time to peak is read from curves in Ref. 1. If $L_{CBr} = 0$, then the time to peak $t_{PC} = 0$. The rise time constant is:

$$\tau_{1C} = \kappa_{1C} t_{pC} \tag{23-60}$$

where κ_{1C} is read from the curves in Ref. 1. The decay time constant is:

$$\tau_{2C} = \kappa_{2C} R_{CBr} C \tag{23-61}$$

where κ_{2C} is again read from the curves in Ref. 1. These curves are not reproduced.

Example 23-7 A 120-V, 100-μF capacitor has $R_{CBr} = 0.05\ \Omega$ and L_{CBr} = 10 mH. The terminal short-circuit profile is required to be calculated.

$$\frac{1}{\delta} = \frac{2 \times 10 \times 10^{-3}}{0.05} = 0.4$$

Also:

$$\omega_0 = \frac{1}{\sqrt{10 \times 10^{-3} \times 100 \times 10^{-6}}} = 1000$$

From curves in Ref. 1, $\kappa_C = 0.92$. The peak current from Eq. (23-58) is (0.92) × (120/0.05) = 2208 A. The time to peak from curves in Ref. 1 = 0.75 ms and $\kappa_{1C} = 0.58$. From Eq. (23-60) the rise time constant is $0.58 \times 0.75 = 0.4335$ ms. Also $\kappa_{2C} = 1$ and from Eq. (23-61) the decay time constant is 5 μs. The short-circuit current profile is plotted in Fig. 23-4*d*.

23-1-5 Total Short-Circuit Current

In Fig. 23-7 the total short-circuit current at the fault point F1 is the sum of partial short-circuit currents from various sources. For a fault at F2, the total short-circuit current is calculated by adding resistance and reactance of the common branch to all the partial current calculations, and then a correction factor is applied.

For a short circuit at F2, after the common branch (Fig. 23-7), the partial short-circuit currents are calculated by adding the resistance and inductance of the common branch to the equivalent circuit. A correction factor is then applied. These correction factors from every source are obtained from following equations (as described in Ref. 1):

$$i_{pcorj} = \sigma_j i_{pj} \qquad i_{kcorj} = \sigma_j i_{kj} \tag{23-62}$$

Example 23-8 The total short-circuit current profile at F1 is plotted. From the previous examples, the time to peak, the magnitude, and the decay time are different for the partial short-circuit currents as plotted in Fig. 23-4. Thus, a graphical approach is taken to summate the currents, as shown in Fig. 23-11. The maximum peak is approximately 27.3 kA and it occurs at 9 ms after the fault.

The short-circuit current from the rectifier predominates. The short-circuit current from the capacitor is a high-rise pulse which quickly decays to zero. The dc motor short-circuit current rises slowly. Smaller dc motors have high armature inductance (Fig. 23-6). The relative magnitudes of the short-circuit currents contributed by

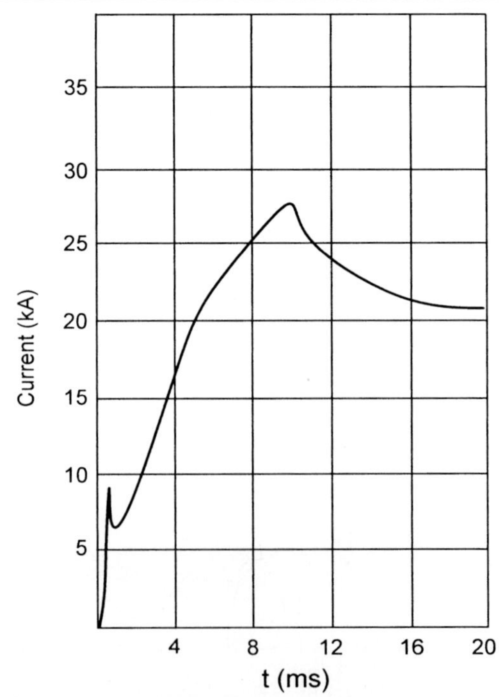

FIGURE 23-11 Total short-circuit current—summation of partial short-circuit currents.

the sources can vary depending on the ratings: this and connections can give varying profiles of the total short-circuit current and time to peak. The total profile is important for properly applying dc circuit breakers.

23-1-6 Matrix Methods

The above calculations are based on the theorem of superimposition. Matrix techniques, as for ac systems, are equally applicable to dc systems too. In an example,[3] three sources of current, a generator, a rectifier, and a battery, are considered in parallel. The resistances and inductances of the system components are calculated, and separate resistance and reactance networks are constructed, much akin to ac short-circuit calculation. These networks are reduced to a single resistance and reactance. Then the maximum short-circuit current is simply given by the source voltage divided by the resistance and its rate of rise by equivalent time constant. This is rather an oversimplification and assumes that all sources have the same voltage. When voltages differ, then the partial short-circuit calculations as described above can be made. For calculations of current from a rectifier, an iterative procedure has been demonstrated.

23-2 CURRENT INTERRUPTION IN DC CIRCUITS

Current interruption in ac circuits (Chap. 8) concentrates on the circuit breakers operating when the current is passing through zero. We alluded to high-resistance current interruption or rheostatic breaker. There are no current zeros in dc circuits. Thus, a current zero must be forced. There are two methods—the current can be forced to zero by increasing the arc voltage to a level that is higher than the system voltage (rheostatic breaker) or by injecting into the circuit a voltage which is of opposite polarity to that of driving voltage. This means that a reverse current flows into the source.

23-2-1 Rheostatic Breaker

An ideal rheostatic circuit breaker inserts a constantly increasing resistance in the circuit until the current to be interrupted drops to zero. The arc gets extinguished when the system voltage can no longer maintain the arc because of high voltage drops. The arc length is increased and the arc resistance acquires high value. The energy stored in the system is gradually dissipated in the arc. The volt-ampere characteristic of a *steady* arc is given by:

V_{arc} = anode voltage + cathode voltage + voltage drop across the arc

$$A+\frac{C}{I_{arc}}+\left(B+\frac{D}{I_{arc}}\right)d \tag{23-63}$$

where I_{arc} is the arc current, V_{arc} is the voltage across the arc, d is the length of the arc, and A, B, C, and D are constants. For small arc lengths, the voltage across arc length can be neglected:

$$V_{arc}=A+\frac{C}{I_{arc}} \tag{23-64}$$

Voltage across the arc reduces as the current increases. Energy dissipated in the arc is:

$$E_{arc}=\int_0^t ivdt \tag{23-65}$$

Equation (23-65) can be written in the general form as:

$$E_{arc}=\int_0^t i_m^2 r(\sin\omega t)^2 dt \tag{23-66}$$

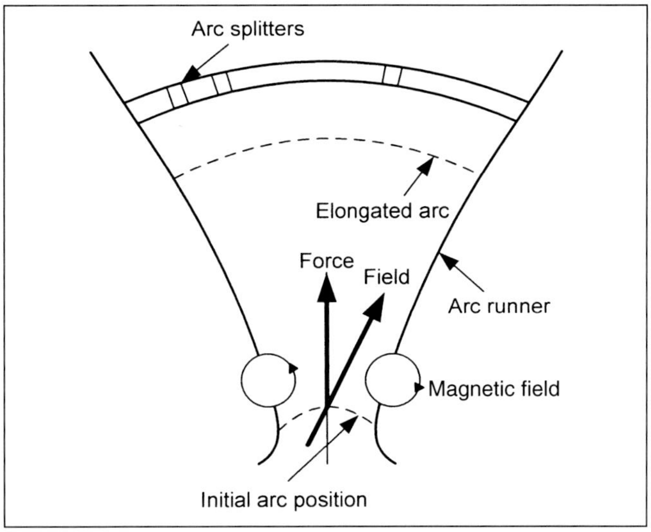

FIGURE 23-12 A rheostatic breaker, lengthening of arc. Splitter plates schematically shown.

For a steady arc, that is, current not varying (dc arc), we can write:

$$E_{arc}=V_{arc}I_{arc}t \tag{23-67}$$

The approximate variation of arc resistance r with time t is obtained for different parameters of the arc by experimentation and theoretical analysis.

In a rheostatic breaker, if the arc current is assumed constant, the arc resistance can be increased by increasing the arc voltage. Therefore, the arc voltage and the arc resistance can be increased by increasing the arc length. The arc voltage increases until it is greater than the voltage across the contacts. At this point, the arc is extinguished. If arc voltage remains lower, the arc will continue to burn until the contacts are destroyed.

Figure 23-12 shows a practical design of the arc lengthening principle. The arc originates at the bottom of the arc chutes and is blown upward by the magnetic force. It is split by arc splitters, which may consist of resin-bonded plates of high-temperature fiber glass, placed perpendicular to the arc path. Blow-out coils, in some breaker designs, subject the arc to a strong magnetic field forcing it upward in the arc chutes.

Figure 23-13 shows the dc arc characteristics for different arc lengths. The resistance characteristic is shown by a straight line. For arc length given by curve a, the arc resistance characteristics shown by straight line intersects at point A. The arc voltage is less than the supply voltage and the arc will continue to burn. As the arc length is increased, the arc voltage increases above the supply voltage and the arc is extinguished. During the arcing time, the supply source continues to give up energy (energy stored in the inductance of the system); the longer the arc burns the greater is the energy. Thus, the entire interruption process is a question of energy balance.

This principle has been successfully employed in some commercial designs of medium-voltage ac breakers also. In these breaker designs, the current can be easily extinguished at current zero, there is practically no current chopping, and the breaker mechanism is very light, except for a large arc chute with splitters. In the case of dc, some chopping can be expected as there is no current zero.

Consider a dc current flowing in a parallel LC circuit. If this current is suddenly interrupted (instantaneously), the recovery voltage situation can be simulated by forcing an equal and opposite current

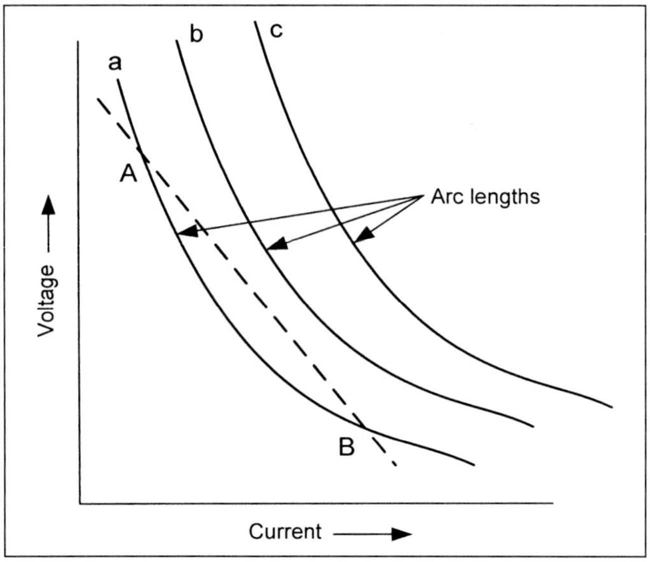

FIGURE 23-13 Relation of arc and system voltage during dc current interruption.

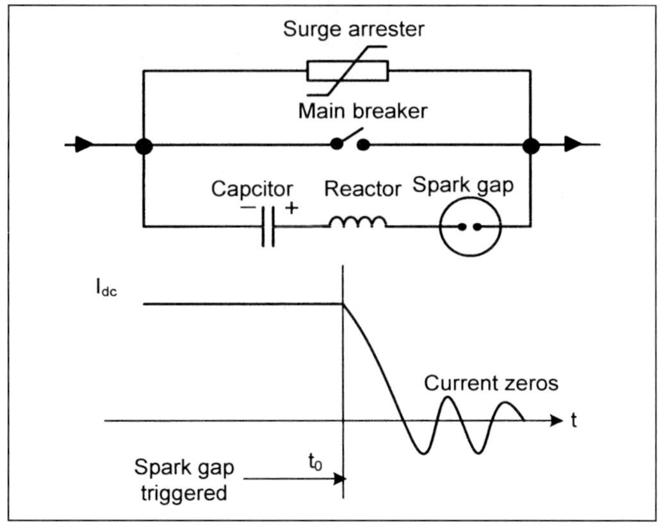

FIGURE 23-14 Interruption of high-energy dc currents, commutating circuit, parallel surge arrester and current zeros brought by reverse current flow.

through the interrupting breaker. Referring to Chap. 2, the recovery voltage in this case will be:

$$V_r(s) = \frac{i}{C}\left(\frac{1}{s^2 + \frac{1}{LC}}\right) \tag{23-68}$$

and its solution is:

$$V_r(t) = i\sqrt{\frac{L}{C}}\sin\frac{t}{\sqrt{LC}} \tag{23-69}$$

where i is the current to be interrupted. In case a ramp is used, that is, the current is gradually brought to zero, the rate of rise of the recovery voltage can be slowed, that is, the slower the chopping rate, the slower will be the recovery voltage. A ramp signal in Laplace transform can be written as:

$$i = -\left[\frac{a}{s^2} - \frac{a}{s^2}e^{-t_c s}\right] \tag{23-70}$$

where $a = \frac{i_0}{t_c}$ = slope of the ramp, and the solution with this signal is:

$$V_r(t) = \frac{-iL}{t_c}\left[\left(t - \cos\frac{t}{\sqrt{LC}}\right) - \left(1 - \cos\frac{t-t_c}{\sqrt{LC}}\right)u(t - t_c)\right] \tag{23-71}$$

23-2-2 HVDC Breakers

A reader may pursue this section after Sec. 23-4, HVDC Transmission. HVDC circuit breakers are not necessary and are not used. Thyristor control takes care of the overcurrent, short circuits, and abnormal conditions in HVDC converters, a metallic return transfer breaker (MRTB) is, however, required.

In high-voltage dc circuit breakers, the energy stored in the system can be high $0.5\ LI_d^2$ = 10 to 25 MJ. Artificial current zeros are produced by a reverse flow of current. Figure 23-14 shows main breaker contacts in parallel with a charged capacitor, reactor, and vacuum gap. This circuit is called a commutating circuit. When the main breaker contacts open, the vacuum gap is triggered at t_0. The precharged capacitor discharges violently through main circuit causing discharge current to flow in opposition to the main current. The main current drops down and oscillates with current zeros. The current is interrupted by the main circuit breaker at one of the current zeros. Vacuum gap seals in and the commutating circuit is opened.

The energy in the main dc pole is partly absorbed by the commutating circuit and partly by the zinc-oxide surge arrester which can absorb up to 20 kJ/kV. The arc in the main circuit is quenched by interrupting medium, air, or SF6. The stresses are much higher because of higher energy associated with dc current interruption and higher TRV appears across the breaker pole.

The commutation circuits can vary. In the passive commutating circuit, a parallel LC resonant circuit is used. For HVDC application, the main pole breakers are not used; however, MRTB is used in bipolar HVDC transmission, and the bipolar mode is changed to monopolar mode during a fault on one of the poles.[9,10]

The HVDC breaker types identified by CIGRE working group are classified on the following basis:

- Switching time
- Current to be interrupted
- TRV
- Switching energy

The current to be interrupted is a function of switching time, converter control, and short-circuit ratio of HVDC system with respect to the ac system. The short circuit rises to a maximum value within 30 to 50 ms; thereafter, converter control brings it down to 5 percent of full-load current in 120 ms:

At $t = 0$, $I_s = I_d$

At $t = 15$ ms, $I_s = 1.25\ I_d$

At $t = 30$ ms, $I_s = 3\ I_d$

At $t = 150$ ms, $I_s < 0.25\ I_d$

TABLE 23-1 Classification of HVDC Breakers

Type	Complexity	Time	Dependence Upon Converter Control	Current Capability	TRV
A	Most complex	< 15 ms	No	Up to 1.25 I_d	Highest
B1	Less complex	60–90 ms	Yes	Low	High
B2	Least complex	90–120 ms	Yes	Low	Low

The classification according to CIGRE is:

- *Type A:* It is fast breaker, less than 15 ms, and does not wait for converter action to reduce dc voltage.

- *Type B:* It is slow and waits for converter action to reduce dc line current to low value. It is classified into two types B1 and B2. Table 23-1 shows the characteristics of these breaker types.

23-3 DC INDUSTRIAL AND COMMERCIAL DISTRIBUTION SYSTEMS

The dc distribution systems are being considered for commercial facilities. A greater percentage of office and commercial building loads are electronic in nature, which have dc as the internal operating voltage. Fuel and solar cells and batteries can be directly connected to a dc system, and the double conversion of power, first from dc to ac and then from ac to dc, can be avoided. Figure 23-15 shows the

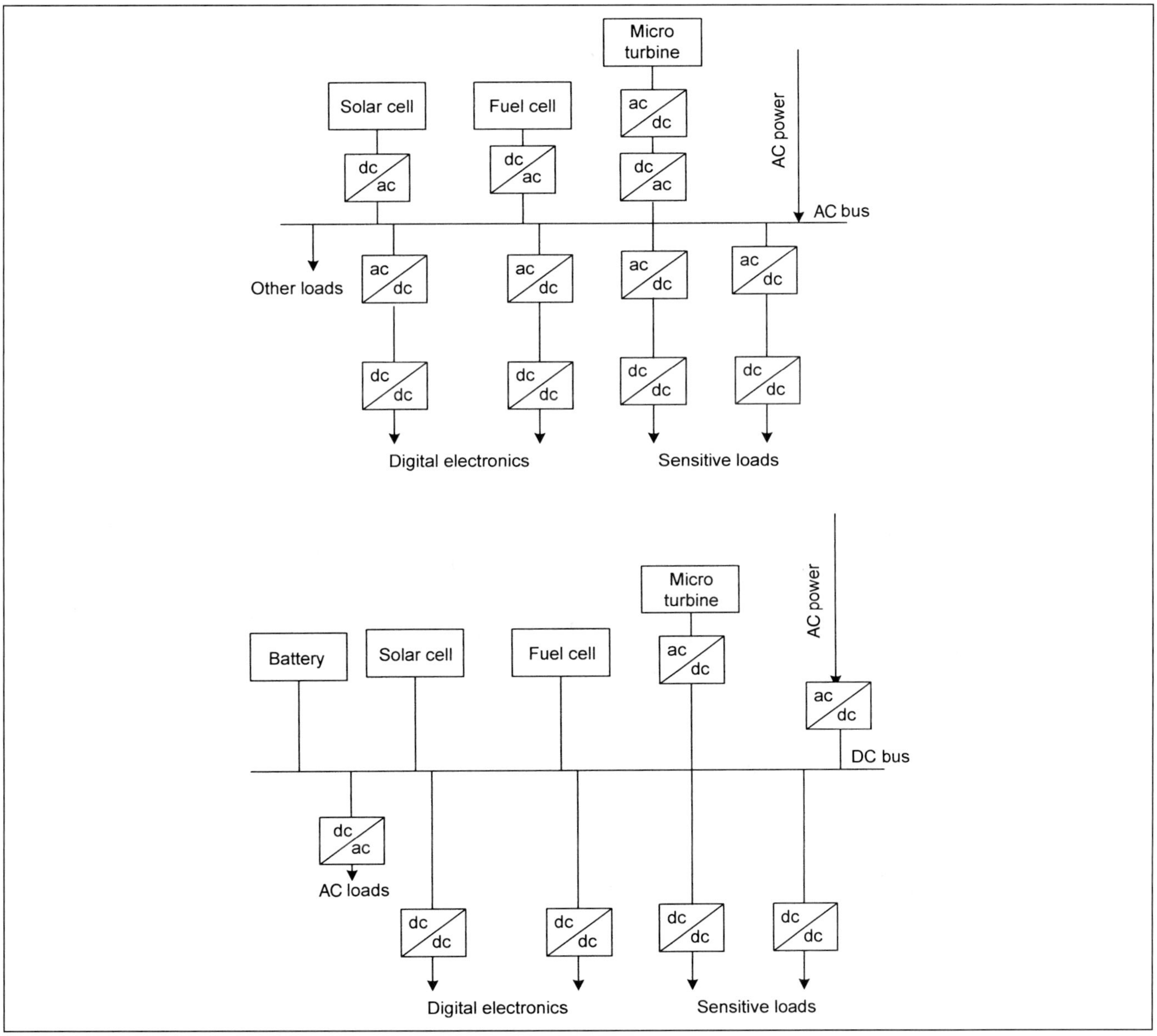

FIGURE 23-15 Ac versus dc distribution systems for commercial facilities.

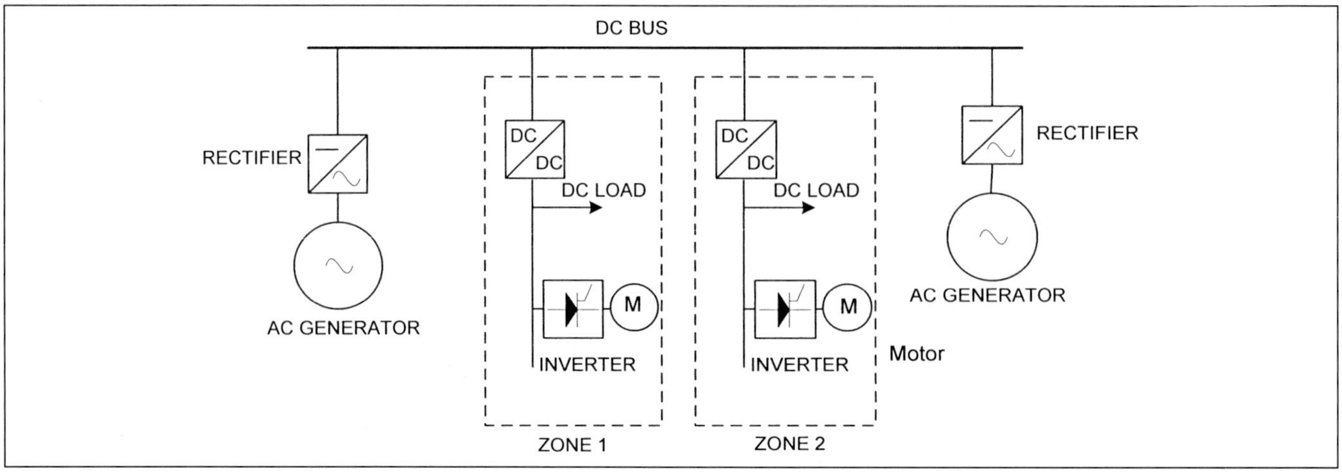

Figure 23-16 A dc distribution system for U.S. Navy, based on Ref. 13.

conceptual base. A case study is presented in Ref. 11 which compares reliability, voltage drops, cable sizing, grounding, and safety, ac versus dc distribution system, which is a part of Department of Electrical Power Engineering, Chalmers University of Technology, Gothenburg, Sweden.

For high reliability UPS systems are used for computers and other sensitive loads. Distributed generation units, like solar cells and fuel cells, generate power at dc, and natural gas microturbines, which generate power at high frequency, need double conversion of power (Fig. 23-15).

With dc distribution, the electronic loads can be supplied more effectively, and by choosing a proper voltage level, one conversion stage is avoided. This results in savings of energy and equipment costs, and so dc distribution becomes attractive. Solar and fuel cells can be directly connected to the dc bus, while for microturbines, a rectifier is needed. A battery block can be directly connected.

The prospective dc distribution opens up an entirely new set of issues—system studies, short-circuit calculations, application of circuit breakers, load flow, contingency analysis, voltage levels, starting impact of loads, and conversion technologies. The transient response of many household appliances, incandescent and fluorescent lamps, coffee makers, computers, and monitors are shown in Ref. 12.

An example of industrial distribution is shown in Fig. 23-16, dc shipboard distribution system,[13] envisaged by U.S. Navy. Two steam turbine generators are connected through rectifiers to a 7000-V dc bus. The dc loads in a zone are served through dc-dc converters and the ac loads through an inverter. The dc-dc converters isolate the loads from the rest of the system, and any fault in that zone does not escalate to other areas. Similarly, the rectifiers isolate the generators and synchronizing and frequency control requirements of the generators are relaxed. The dc bus becomes immune to ground faults through proper system grounding. Thus, ease of maintenance, reliability, and confinement of a system to a limited area are the obvious advantages. More power can be handled on a cable circuit with dc than with ac. Simulation transients with PSCAD/EMTD are shown in Ref. 14. The rectifiers, inverters, and converters use 18-kHz PWM schemes for switching and have controllers to regulate the output voltages.

23-4 HVDC TRANSMISSION

The cumulative megawatts of HVDC systems around the world approach 100 GW. HVDC has a long history, but a transition point occurred when thyristor valves took over mercury arc rectifiers in the late 1970s. The major technology leap was in Brazil with the 3150 MW ± 600 kV Itaipu project commissioned from 1984 to 1987. The overhead line is 800 km long and each 12-pulse converter is rated 790 MW, 300 kV. HVDC is finding major applications in countries like India and China, and a large number of thyristor valve–based systems are planned—the power levels and distances are such that ± 800 kV may be needed. An HVDC technology review paper is at weblink.[15] Another Web site of interest is that of CIGRE Study Committee B4, HVDC, and Power Electronic Equipment.[16]

HVDC Light The IGBTs for motor drives have begun to find applications in HVDC systems at the lower-end of power usage. These operate using PWM techniques; there is little or no need for reactive power compensation as the converters can generate active and reactive power. The systems have found applications in offshore wind farms and short-distance XLPE-type cable systems. The largest system to date is the 330-MW cross-sound dc link between Connecticut and Long Island.

23-4-1 HVDC Advantages

An HVDC interconnection properly planned and designed may achieve the following advantages:

1. An interlinking of two ac systems through interconnections which may have become overloaded during load development poses problems of trip out, system separation, and load/frequency management. There are limitations of load transfer with respect to deviations of frequency in the interconnected systems. A dc link on the other hand is immune to such fluctuations and power transfer remains steady.

2. It is possible to improve the stability of the network as a whole by introducing control parameters from the ac network, and the disturbances in the ac network can be damped out by modulating power flow through the dc link.

3. An ac interconnection is a synchronous link, and with increased generating capability, it requires equipment of increasing short-circuit ratings. HVDC link is an *asynchronous* connection, and the short-circuit levels of interconnected systems remain unchanged. Also frequency disturbances in one system do not impact the interconnected system. HVDC links are in operation, interconnecting 40- or 50-Hz systems with 60-Hz systems.

4. A *synchronous* HVDC link is a combination of HVDC link in parallel with ac tie lines. Power transfer through parallel dc line can be controlled to have a stabilizing effect on parallel ac tie line. The stabilizing controls are initiated after first opening the ac line, continue during the dead time and subsequent return of the ac system to the equilibrium point.

5. The power through an HVDC link can be regulated more precisely and rapidly, change of the order of 30 MW/min or more. The limitations are imposed by the ac system reserve capacity and generation.

6. An HVDC link does not transmit reactive power (though converters require reactive power to operate). Thus, there is no loss of reactive power in the HVDC line itself. Continuous charging currents are absent. Thus, transmission losses are reduced.

7. The skin effect is absent in dc currents, and the conductors can have a uniform current density through their cross section.

8. The phase-to-phase clearances and, phase-to-ground clearances are smaller for dc transmission. HVDC towers are physically smaller for the same ac voltage level, and right of way (ROW) is reduced.

9. With respect to line loading, the ac line remains loaded below its thermal limit due to limits of transient stability, and conductors are not fully utilized. The voltage across the line varies due to absorption of reactive power, which is load power factor dependent. The voltage fluctuates with the load. An ac line cannot be loaded more than approximately 0.8 times the surge impedance loading. Such limitations do not exist in HVDC transmission. Line can be loaded up to the thermal limits of line or thyristor valves.

10. Intermediate substations are generally required in ac transmission for compensation located approximately 300 km apart—no such substations are required for HVDC transmission. It is point-to-point long-distance transmission, though multi-terminal operation is possible.

11. Ac transmission requires several three-phase conductors, while HVDC transmission requires two-pole conductors. Therefore, HVDC becomes more economical over ac transmission due to reduced tower size, conductor cost, reduced ROW, and reduced transmission losses. HVDC can utilize even earth return and may not need a double circuit.

12. An HVDC line can be operated with constant current or voltage regulation by suitable control of the thyristor valves and tap changer control.

23-4-2 HVDC Configurations and Operating Modes

Figure 23-17 shows common system configurations:

1. Monopolar systems are the simplest and most economical for moderate power transfer. Only two converters and one high-voltage connection is required. These have been used with low-voltage electrode lines and sea electrodes to carry return currents (Fig. 23-17*a*).

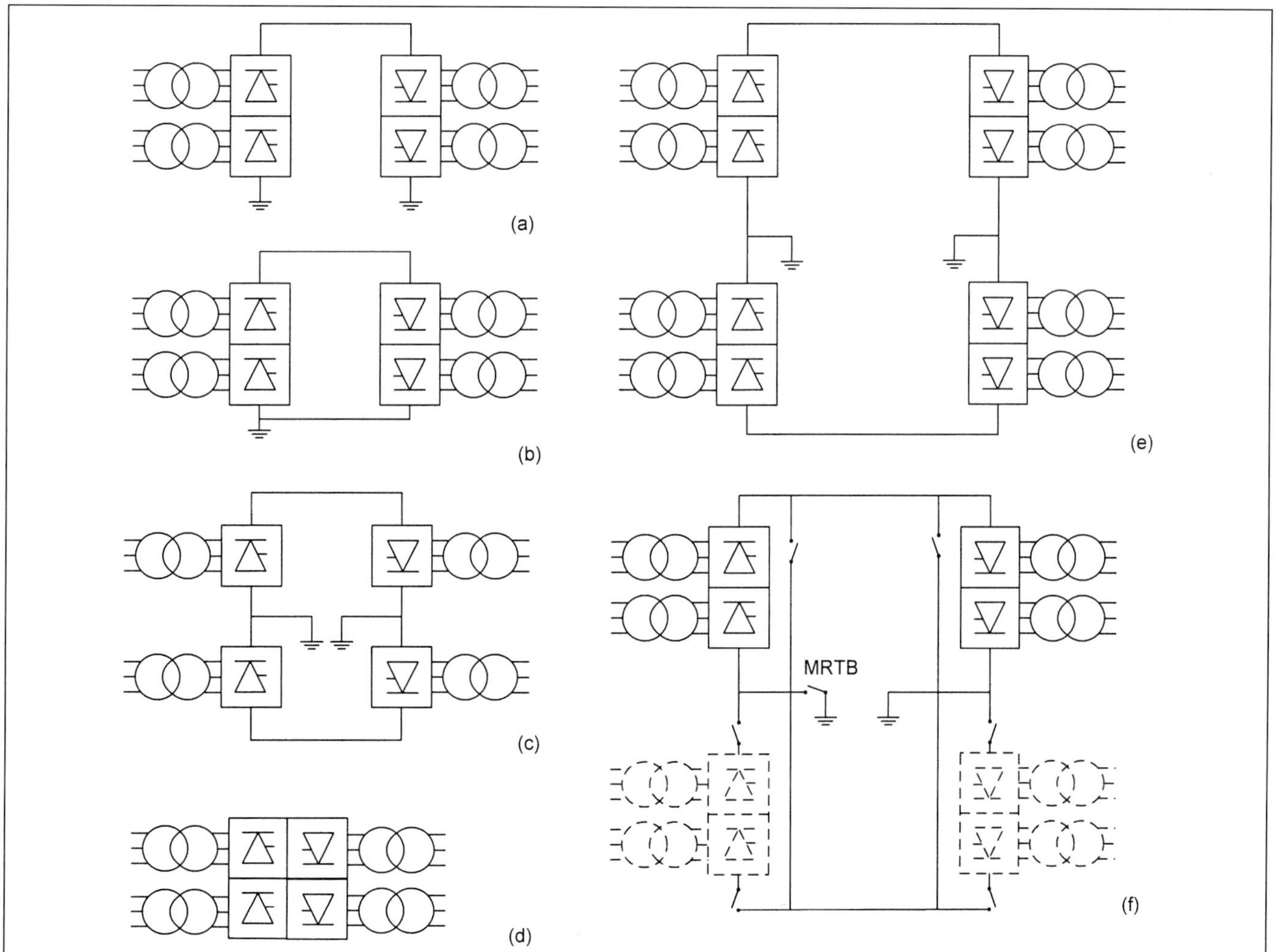

FIGURE 23-17 HVDC system configurations.

2. In congested areas, or soils of high resistivity, conditions may not be conducive to monopolar systems. In such cases, a low-voltage cable is used for the return path, and the dc circuit uses local ground connection for potential reference (Fig. 23-17*b*).

3. An alternative of monopolar systems with metallic return is that the midpoint of a 12-pulse converter can be connected to earth and two half voltage cables or line conductors can be used. The converter is operated only in 12-pulse mode, so that there are no stray currents (Fig. 23-17*c*).

4. Back-to-back systems are used for interconnection of synchronous networks and use ac lines to connect on either side. The power transfer characteristics are limited by the relative capabilities of adjacent ac systems. There are no dc lines. The purpose is to provide bidirectional exchange of power, easily and quickly. An ac link will have limitations in control over direction and amount of power flow. Twelve-pulse bridges are used. It is preferable to connect two back-to-back systems in parallel between the same ac buses (Fig. 23-17*d*).

5. The most common configuration is 12-pulse bipolar converter for each pole at the terminal. This gives two independent circuits each of 50 percent capacity. For normal balanced operation, there is no earth current. Monopolar earth return operation can be used during outage of the opposite pole (Fig. 23-17*e*).

6. The earth return option can be minimized during monopolar operation by using opposite pole line for metallic return through pole/converter bypass switches at each end. This requires a metallic return transfer breaker in the ground electrode line at one of the dc terminals to commutate the current from relatively low resistance of earth into that of dc line conductor. This metallic return facility is provided for most dc transmission systems (Fig. 23-17*f*).

For voltages above ± 500 kV, series-connected converters are used to reduce energy unavailability for individual converter outage or partial line insulation failure. By using two-series-connected converters per pole in a bipolar system, only 25 percent of the line capability is lost for a converter outage or if the line insulation is degraded and it can support only 50 percent of the rated line voltage.

23-4-3 Earth Return

The flow of direct current through earth can give rise to the following problems:

- Corrosion of buried metallic pipes, cable sheaths, and fences due to electrolytic action
- Disturbance in communication signals through metallic rails in the path of earth current
- Rapid consumption of electrode material
- Shocks to fish, sea life, and shore swimmers

The earth electrodes are installed far away from the main terminal substation earth of HVDC system; the location is selected such that there are no buried pipe lines, cables, and structure foundations. This may be at a distance of 5 to 10 km, and the electrodes are connected to the neutral point of the converter through an electrode line lightly insulated. The touch and step potentials, (Chap. 21) should be within prescribed limits.

For submarine cable HVDC transmission, the earth electrodes are in the form of sea electrode or shore electrode. The sea electrode is held in the sea water close to the shore and is installed a few kilometers away from the shore, while the shore electrode is buried on the seashore near the sea. The cathodic protection of buried metallic objects is one way of preventing corrosion.[17]

23-4-4 Terminal Layout

Figure 23-18 shows a practical layout of a terminal. While reactive power compensation and harmonic filters have been previously discussed, dc filters are required to limit interference with communication circuits, which may be inductively coupled to the dc line. The parameters to be considered are separation between the dc and communication lines, their shielding, the presence of ground wires, and the soil resistivity. This criterion is expressed as equivalent disturbing current. Disturbance effects are lower in bipolar designs. The filter design must account for all operating modes and harmonic sources. In Chap. 15 we noted that while the converter throws odd harmonics on to the ac supply system, the even harmonics of input frequency are transmitted to the load. A Fourier expression for the output voltage is:

$$e_0 = e_d + e_2 \sin 2\omega t + e_2' \cos 2\omega t + e_4 \sin 4\omega t + e_4' \cos 4\omega t + \cdots \tag{23-72}$$

where e_m and e_m' ($m = 2, 4, 6, \ldots$) are given by:

$$\begin{aligned} e_m &= \frac{2E_m}{\pi}\left(\frac{\sin(m+1)\alpha}{m+1} - \frac{\sin(m-1)\alpha}{m-1}\right) \\ e_m' &= \frac{2E_m}{\pi}\left(\frac{\cos(m+1)\alpha}{m+1} - \frac{\cos(m-1)\alpha}{m-1}\right) \end{aligned} \tag{23-73}$$

Generally 12th harmonic band pass filter with active filtering of higher order harmonics are provided.

23-4-5 Line-Commutated Current Source Converters

Line-commutated current source converters form the basis of high-power and high-voltage HVDC transmission. We discussed the control of dc voltage output of the current source converters, delay and commutation angle, harmonic generation, and mitigation through phase multiplication and the waveforms in Chap. 15. A fully controlled six-pulse converter will act as an inverter, if the firing angle is adjusted above 90°. At $\alpha = 90°$ the output voltage is zero, and the positive half and negative half waves are equal. As the angle is increased above 90°, inverter action is obtained. The firing angle is, however, limited and cannot be 180°, as explained further.

Figure 23-19 shows waveforms for rectifier and inverter operation. For inverter operation the output power factor is leading. The reactive power increases as α approaches 90° (Chap. 15).

For a six-pulse bridge, internal no-load voltage is:

$$V_{do} = \frac{3\sqrt{3}E_m}{\pi}\cos\alpha \tag{23-74}$$

Then, considering the dc voltage drop in the source inductance (commutating inductance):

$$\begin{aligned} V_d &= \frac{3\sqrt{3}E_m}{\pi}\cos\alpha - \frac{3\omega L_s I_d}{\pi} \\ &= \frac{3\sqrt{3}E_m}{\pi}\cos(\alpha+\mu) + \frac{3\omega L_s I_d}{\pi} \end{aligned} \tag{23-75}$$

We can write:

$$\frac{3}{\pi}X_c I_d = R_c I_d \tag{23-76}$$

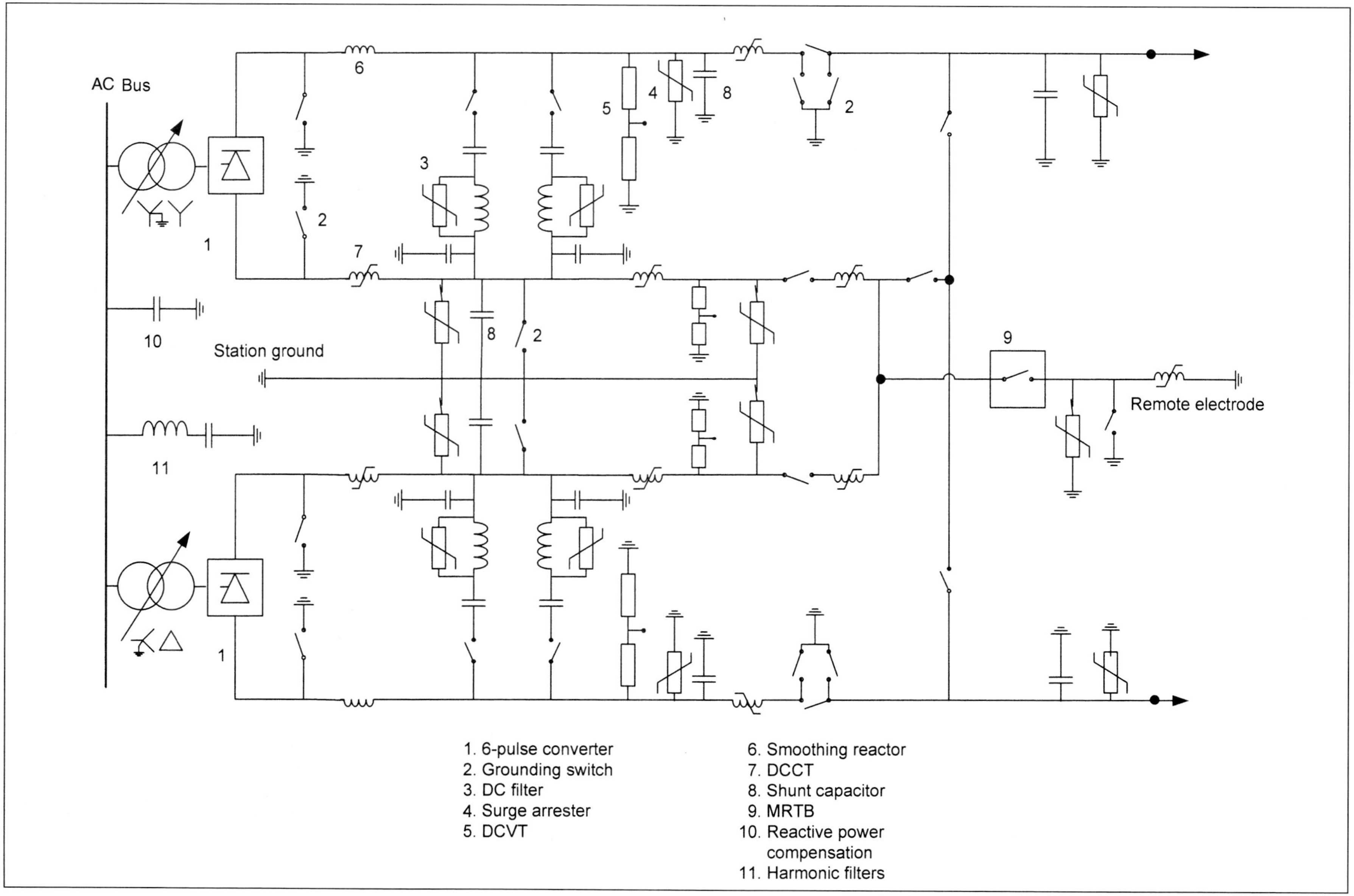

FIGURE 23-18 Details of dc terminal setup: bipolar HVDC system.

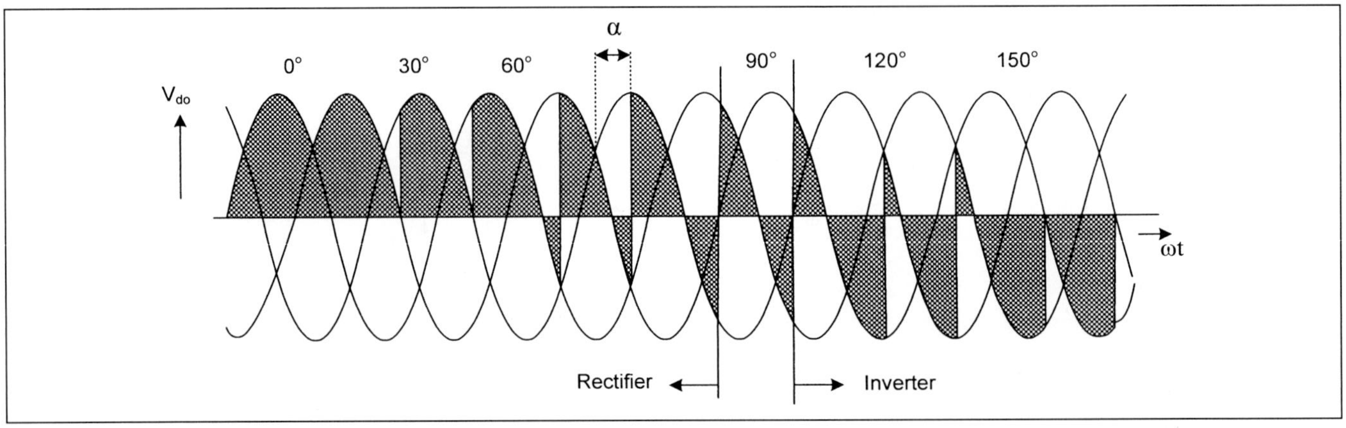

FIGURE 23-19 Effect of delay angle on voltage waveform—rectifier and inverter operation modes of a six-pulse converter.

where R_c is the equivalent commutating resistance, which does not represent a real resistance and does not consume any real power.

For rectifier *or inverter* operation a thyristor experiences a reverse voltage for $(\pi - \alpha - \mu)$ after the commutation has taken place. However, for inverter operation, angle $\alpha > 90°$. This means that for inverter operation the thyristors are reverse biased for a shorter duration as compared to the rectifier operation. Therefore, the firing angle should be limited so that angle $\gamma = (\pi - \alpha - \mu)$ is enough for proper commutation. Angle $(\pi - \gamma)$ is called the extinction angle.

Figure 23-20 shows the operating modes of converter as a rectifier and inverter. The angle α can be varied so that I_d is constant; this may be called mode 1 of operation. As source voltage V_s increases, the internal voltage V_{do} also increases so that difference between these two is a constant value. To increase V_d, the firing angle is advanced. The duration for which the SCR is reversed biased will reach a minimum value required at some level of internal voltage. Any further increase in α for increasing I_d will result in failure of commutation. This sets the limit for constant current mode. Any further increase in I_d is possible only by reducing α, so that margin γ remains constant; this is the other mode of operation. Here the dc voltage will decrease with an increase in current.

Figure 23-21 shows the inverter waveforms for $\alpha = 5\pi/6$. The voltage across one of the thyristors is also shown. Unless proper control is exercised to maintain margin angle, commutation failure can occur. Equation (23-75) can also be written as:

$$\frac{\pi}{3\sqrt{3}}\left(V_d + \frac{3\omega L_s}{\pi} I_d\right) = E_m \cos\alpha \tag{23-77}$$

To determine the instant of firing, replace α with ωt and $(\alpha + \mu)$ with $(\pi - \gamma_{min})$:

$$\frac{2\omega L_s}{\sqrt{3}} I_d - E_m \cos\gamma_{min} = E_m \cos\omega t \tag{23-78}$$

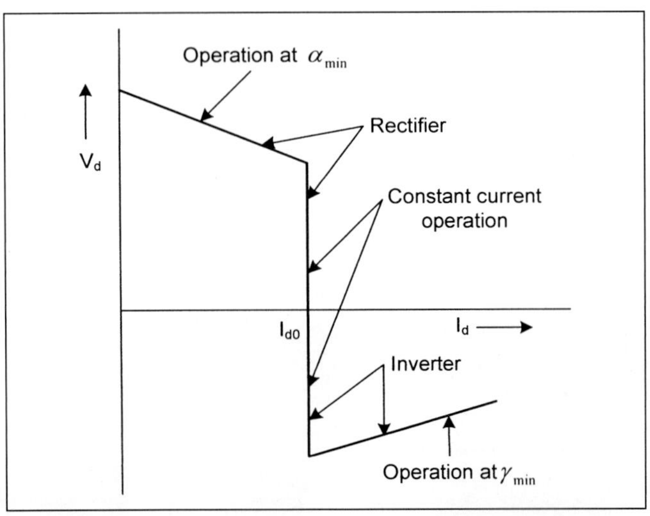

FIGURE 23-20 Rectifier/inverter characteristics of a converter.

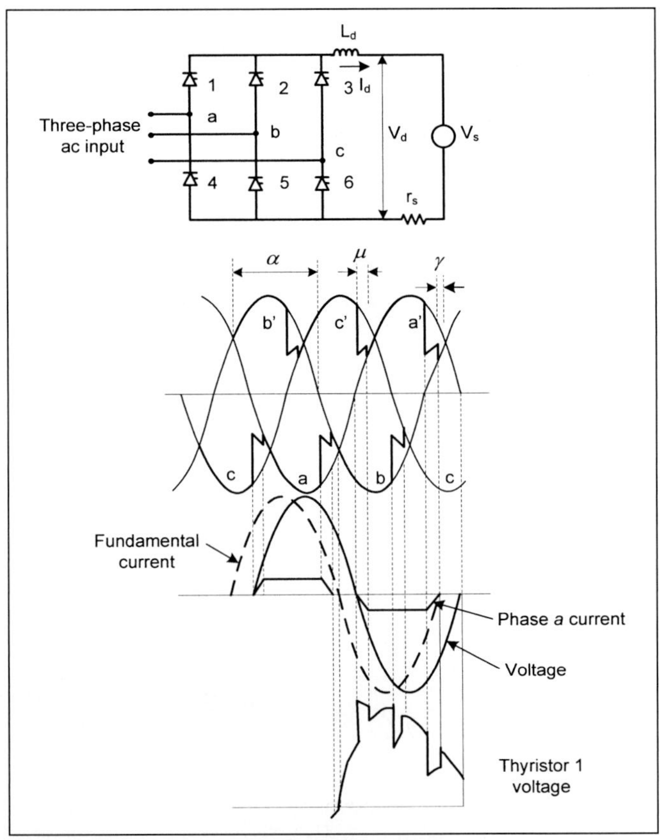

FIGURE 23-21 A six-pulse inverter: circuit diagram and current and voltage waveforms.

where γ_{min} is a constant. Therefore, the left side of Eq. (23-78) depends upon I_d. The right-hand side is the negative value of ac phase c voltage for firing thyristor 2. Whereas the right side is the same for all thyristors, the control voltage will be different for each thyristor and separate firing circuits are required.

Example 23-9 A six-pulse circuit operates in inverting mode with constant margin angle $\gamma_{min} = 20°$. Input voltage is 440-V dc, source resistance = 1 Ω, input current = 20 A. The ac voltage is 480 V, 60 Hz. Calculate source inductance. What is the output power factor?

Converter dc voltage is given by:

$$V_{do} = 440 + I_d r_s = 460 \text{ V}$$

Therefore:

$$\frac{3\omega L_s}{\pi} \times 20 = -460 + 3\sqrt{3}\cos\gamma_{min} = 70 \text{ V}$$

Then:

$$L_s = \frac{70 \times \pi}{20 \times 3 \times 377} \times 1000 = 9.70 \text{ mH/phase}$$

From Eq. (23-77):

$$\frac{3\sqrt{3}E_m}{\pi}\cos\alpha = -460 + 70 = -390$$

Therefore:

$$\cos\alpha = -\frac{390 \times \pi}{3\sqrt{3}} \times \frac{\sqrt{3}}{\sqrt{2} \times 480} = 0.60 \text{ leading}$$

$$\cos\gamma_{min} = 0.98 \text{ leading}$$

The output power factor = (0.60 + 0.98)/2 = 0.79 leading approximately.

Figure 23-22 shows definitions of the angles related to rectifier and inverter operation.

- α is the ignition delay angle.
- μ is the overlap angle.
- δ is the extinction delay angle = $\alpha + \mu$.

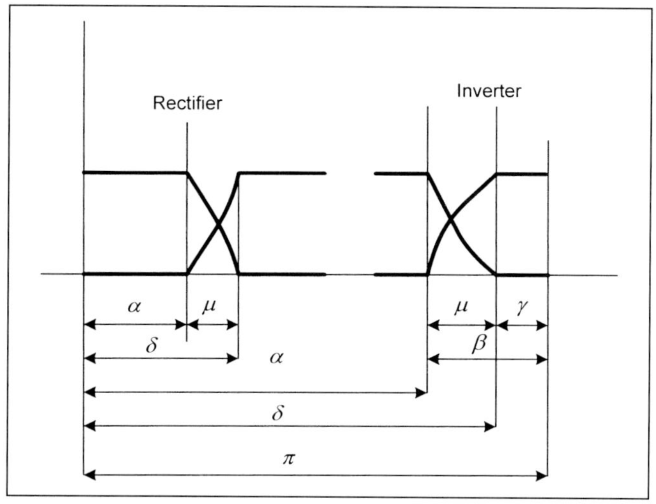

FIGURE 23-22 Various angles in rectifier and inverter operation.

- $\beta = \pi - \alpha$ is the ignition advance angle.
- $\gamma = \pi - \alpha$ is the extinction advance angle.
- $\mu = \delta - \alpha = \beta - \gamma$ is the overlap angle.

More completely, the V_d equations for rectifier and inverter mode are written as:

$$\begin{aligned} V_d &= V_{do}\cos\alpha - R_c I_d - R_{eq} I_d - V_{drop} \quad \text{rectifier} \\ V_d &= V_{do}\cos\gamma - R_c I_d + R_{eq} I_d + V_{drop} \quad \text{Inverter} \end{aligned} \tag{23-79}$$

where R_{eq} is the equivalent resistance representing losses in valves and auxiliaries, and V_{drop} is the voltage drop in the valve itself. The equivalent circuits of rectifier and inverter reduce to the simple circuits shown in Fig. 23-23.

The current flowing in the link between rectifier and inverter is:

$$I_d = \frac{V_{dor}\cos\alpha - V_{doi}\cos\gamma}{R_{cr} + R_L + R_{ci}} \tag{23-80}$$

where R_L is the dc line resistance, and the commutating resistances are as defined in Eq. (23-76).

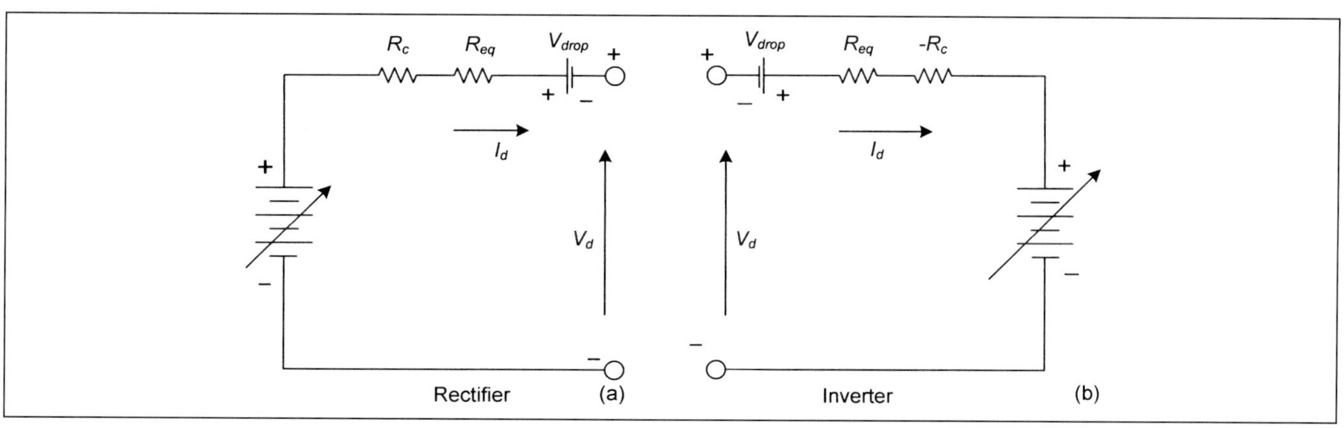

FIGURE 23-23 Equivalent circuits of (*a*) rectifier and (*b*) inverter.

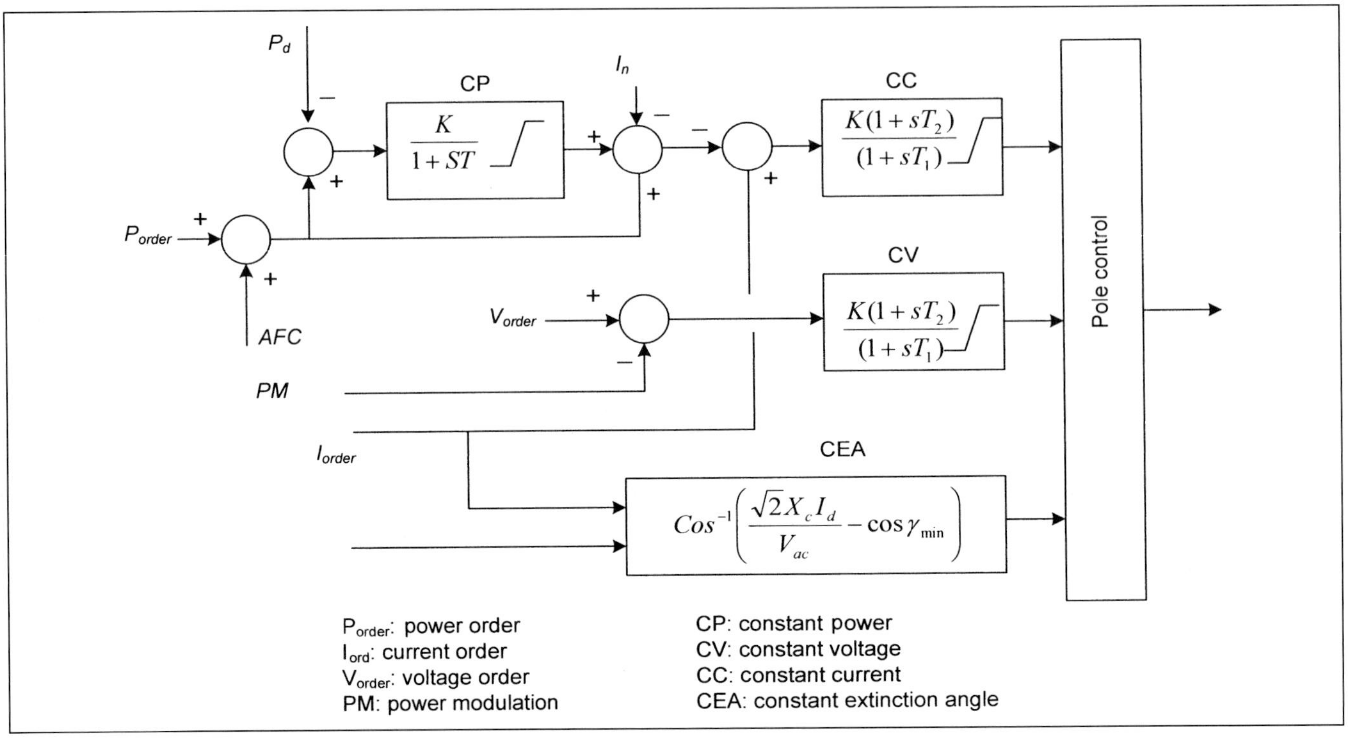

FIGURE 23-24 Control circuit diagram of a pole control, HVDC, LCC source converters.[16]

23-4-6 Control and Operation of HVDC

A load commutated converter (LCC) HVDC system is operated so as to maintain constant dc current and voltage. The control circuit of a phase is shown in block circuit diagram of Fig. 23-24.[16] This produces the operating characteristics shown in Fig. 23-25 by selecting the values amongst the outputs of constant current (CC), constant voltage (CV), and constant extinction angle (CEA). Extinction angle control is needed to avoid commutation failures. The two characteristics of the inverter and rectifier are symmetrical with respect to horizontal axis of the dc current. The interaction indicates an operating point in the steady state, points A or B. Usually the rectifier sets the dc current and the inverter the dc voltage. The actual operating points of the converters differ because of the voltage drop in the dc line. Current control gives fast response, while voltage control is comparatively slow to avoid unstable control action.

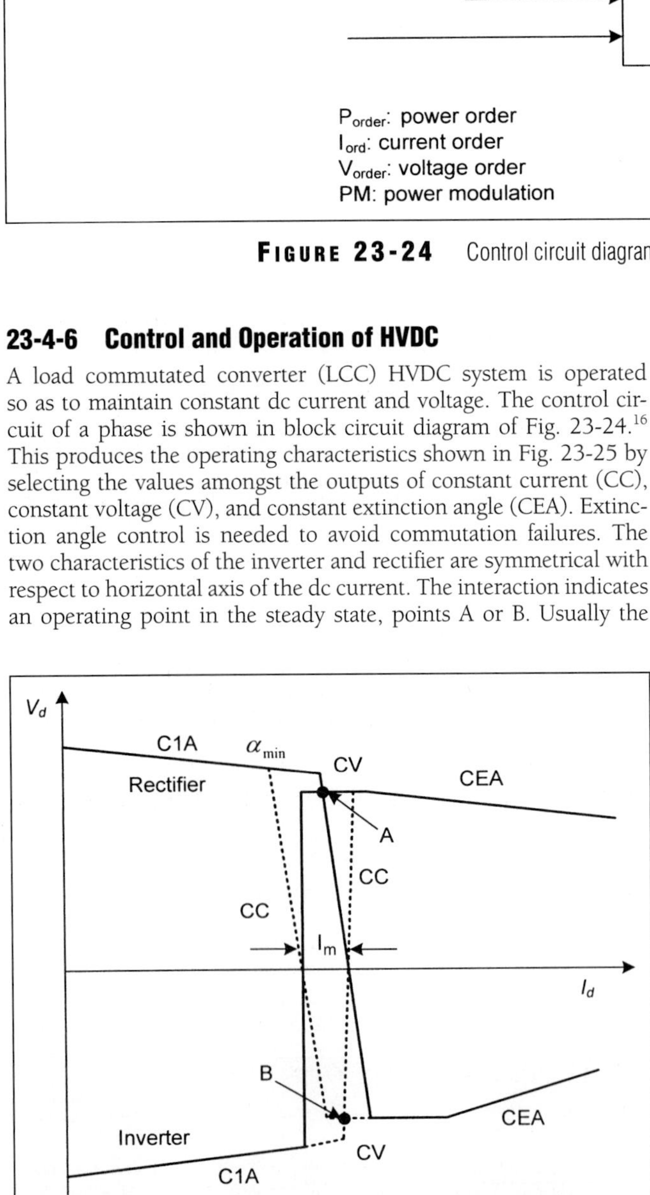

FIGURE 23-25 HVDC operation with each converter having rectifier and inverter characteristics.

The dc voltage can be controlled by varying the delay angle, which has a faster response, or by controlling the converter ac voltage by load tap changers on transformers, which gives a slow response. Load tap changing (LTC) is used to keep angles α and γ within a desired range. Also by increasing the angle α, the reactive power requirements increase.

At any time, only one terminal can control the dc power in a two-terminal system. The power control is integrated in parallel with the dc control loop and corrects the current order setting so as to adjust the actual dc power to the setting value.

Power inversion or an interchange of rectifier and converter operation can be done by changing the converter that subtracts the current margin I_m in Fig. 23-25. A new operating point B with reversed dc power is the intersection of two dotted lines. The power reversal action changes the dc polarity and the current direction remains as before.

There are ac/dc system interactions which must be considered. The protection and control functions of dc system can normally prevent spread of disturbances, but an improperly controlled terminal can even accentuate the disturbances. Trip of generator can bring about a block of HVDC, resulting in possible power flow, voltage, instability, operating reserve, and frequency response issues.

Consider a synchronous dc link. Figure 23-26 shows the power-angle characteristics and the impact of critical damping control on the stability. Curve A is for ac lines tripping, curve B for a parallel dc link, but with P_d control only and curves 3 and 4 with damping controls.

HVDC offers the flexibility of adding auxiliary controls to stabilize the system for transient and dynamic stability swings. An example for improving transient stability is to block HVDC and then slowly ramp it up to full output. Dynamic loops can be added to provide damping of oscillatory response. Other slower-acting controls can utilize an HVDC to share operating reserves across synchronous systems through frequency response loops.[18,19,20]

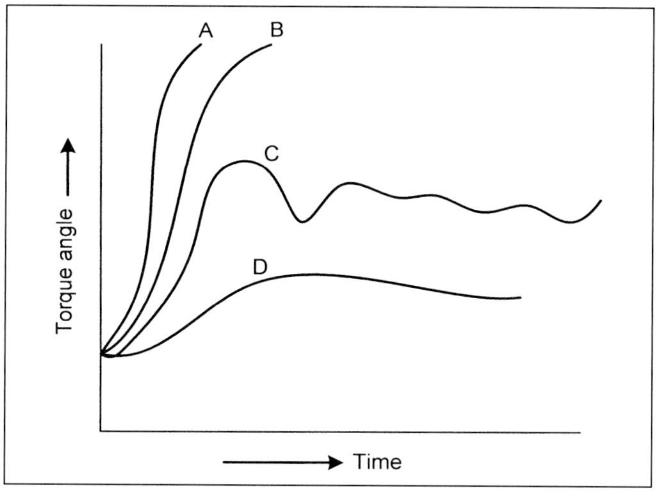

FIGURE 23-26 Torque angle stability characteristics, with and without HVDC synchronous link. Curve A for ac line tripping. Curve B—with parallel HVDC link and only P_d control. Curves C and D—with damping HVDC control.

23-4-7 Reactive Power Requirements

The converters consume reactive power both in the rectifier and inverter mode (Fig. 15-7). The reactive power demand is usually 50 to 60 percent of the active power demand.

Equation (23-76) can be written as:

$$V_d = V_{do}\cos\alpha - \Delta V_d = V_{do}\frac{\cos\alpha + \cos\delta}{2} \tag{23-81}$$

The power factor is given by:

$$\begin{aligned} \cos\phi &= 0.5(\cos\alpha + \cos(\alpha+\mu)) \quad \text{rectifier} \\ \cos\phi &= 0.5(\cos\gamma + \cos(\gamma+\mu)) \quad \text{inverter} \end{aligned} \tag{23-82}$$

To achieve high power, factor α and γ should be kept low. From Eq. (23-82)

$$\cos\phi \approx \frac{V_d}{V_{do}} \tag{23-83}$$

The reactive power can be calculated from the following expressions:

$$Q = V_d I_d \sqrt{\left(\frac{V_{d0}}{V_d}\right)^2 - 1} \tag{23-84}$$

As an example, if $V_{do} = 280$ kV, $V_d = 250$ kV, $I_d = 2$ kA, the reactive power requirement is 252 Mvar. At inverter end, the inverter feeds the ac load, therefore, the reactive power requirements of the ac load must be added.

The reactive power sources are discussed in Chap. 15. The reactive power can be supplied by SVC's, rotating condensers, and shunt capacitors. The shunt capacitors will supply a fixed amount of reactive power and will overcompensate at light loads.[21] The ac filters can serve the dual purpose of supplying reactive power requirements as well as mitigating harmonics (Chap. 6).

23-4-8 Short-Circuit Ratio

The definition of SCR here should not be confused with SCR for synchronous generators described in Chap. 10. Here SCR signifies how strong the ac system is with respect to the HVDC link. It is defined as:

$$\text{SCR} = \frac{S_{\text{MVA,min}}}{P_{d,\max}} \tag{23-85}$$

where $S_{\text{MVA,min}}$ is the minimum short-circuit MVA of the ac bus at the HVDC link, and $P_{d,\max}$ is the maximum rated bipolar power of the HVDC link. The minimum short-circuit level can be calculated based on the minimum three-phase short-circuit current of the ac system at the point of interconnection (Chap. 9).

The equivalent SCR (ESCR) considers the reactive power of shunt capacitor filters and other shunt capacitors if any.

$$\text{ESCR} = \frac{S_{\text{MVA,min}} - Q_c}{P_{d,\max}} \tag{23-86}$$

An ac system with SCR < 2 is considered weak, 2 to 4 as intermediate, and > 4 as strong. The impacts of low SCR (weak electrical system) are:[22]

- High dynamic overvoltages
- Voltage instability
- Harmonic resonance and objectionable flicker

With low SCR, the P_d/I_d curve has a stability limit above which increase in I_d *causes* a decrease in P_d. Control of voltage and recovery from disturbances becomes difficult. The dc system response may contribute to the collapse of the ac system. Synchronous condensers can be installed to increase SCR, as these contribute to the short-circuit currents.

The ac bus voltage depends on SCR, active power, and reactive power supplied by the ac filters. For a given power flow, ac bus voltage can be raised by increasing the shunt compensation.

On sudden load rejection, there will be transient rise in ac voltage, which should be limited to 1.1 pu. This is controlled by rapidly reducing reactive power provided by shunt capacitors. SVCs have an obvious advantage (Chap. 15).

The harmonic resonance problems are due to parallel resonance between ac capacitors, filters, and ac system at lower harmonics (Chap. 6).

The problems of flicker can arise as the switching of shunt capacitors and reactors may cause a large voltage swing in the vicinity of compensating equipment due to frequently switched compensating devices.

23-4-9 Typical Thyristor Current and Voltage Waveforms

Figure 23-27 shows typical waveforms of a thyristor voltage and current, under normal operation. The system represented is 230-kV dc link, 12-pulse converters, one converter served from wye-wye transformer, and the other through delta-wye transformer. This figure depicts the waveforms of one of the thyristors in six-pulse bridge connected to wye-wye transformer. Figure 23-28 shows the impact of a dc pole-to-ground fault on one of the lines that occurs at 200 ms and is cleared in 100 ms EMTP simulations.

23-4-10 VSC-Based HVDC: HVDC-Light

Voltage converter-based HVDC, also called HVDC-light, consists of a bipolar two-wire HVDC system with converters connected pole-to-pole (Fig. 23-29). DC capacitors are used to provide a stiff dc voltage source. There is no earth return operation. The converters are coupled to the ac system through ac phase reactors and converter transformers. Harmonic filters are located between the phase reactors and converter transformers. This avoids the harmonic loading

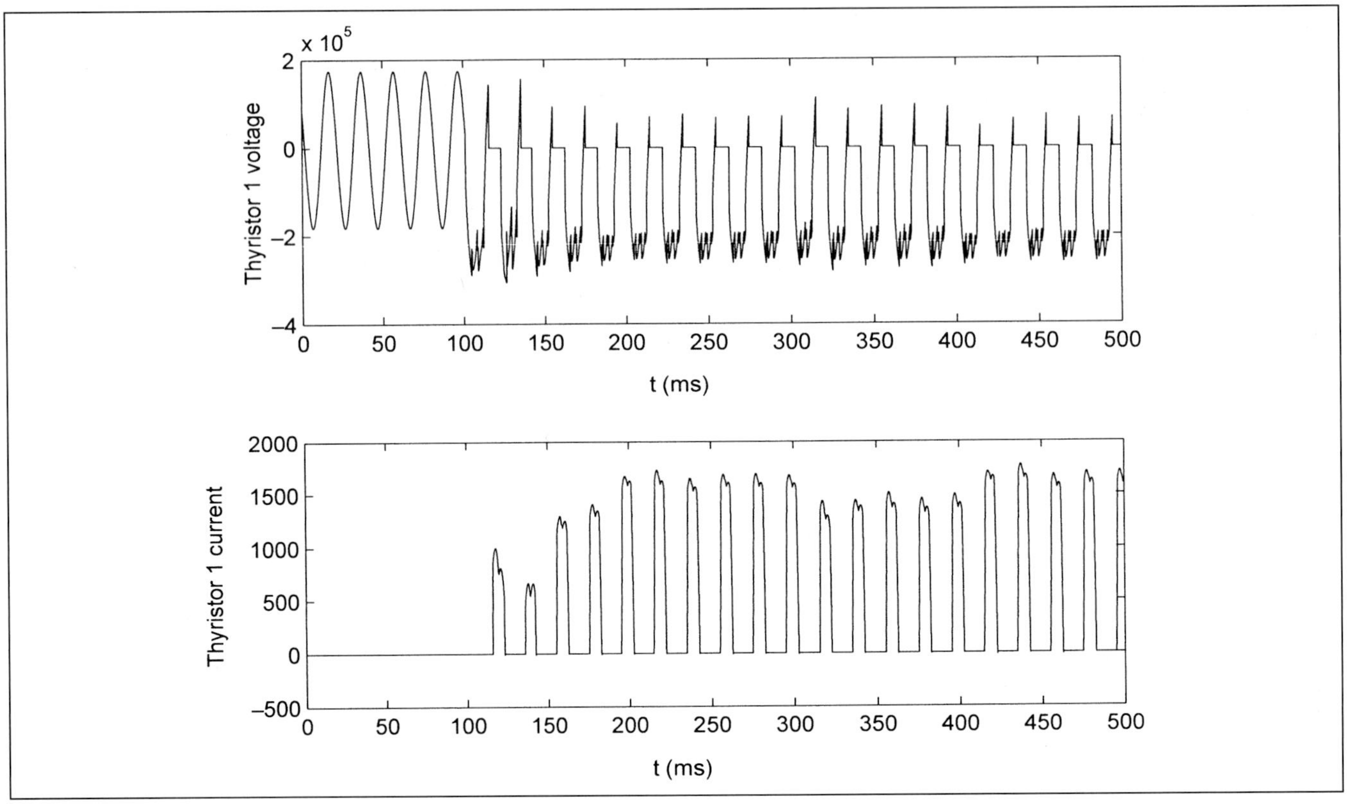

FIGURE 23-27 Typical waveforms, normal operation, voltage, and current in one of the thyristors of a six-pulse converter, 230-kV HVDC transmission.

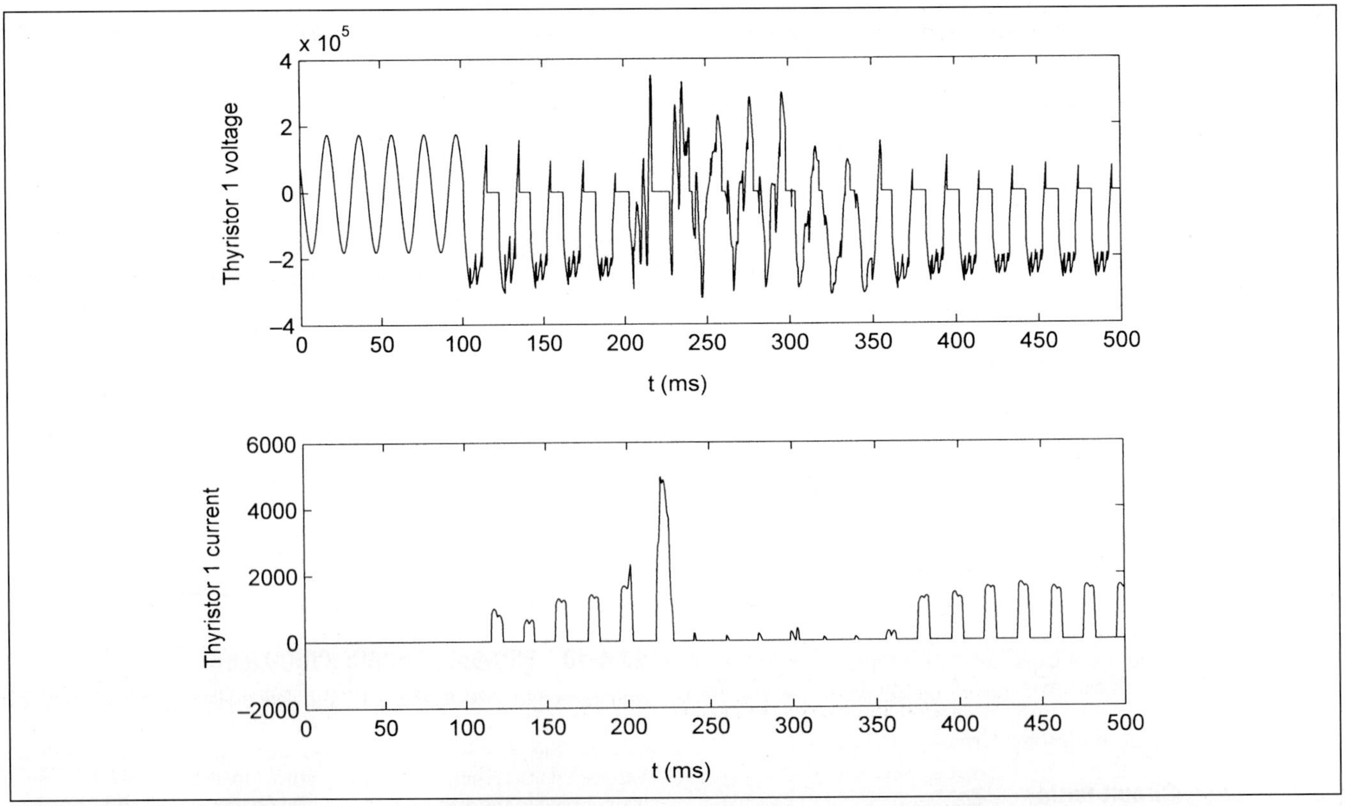

FIGURE 23-28 As in Fig. 23-27, but a dc line-to-ground fault occurs at 200 ms, cleared in 100 ms. Fault resistance = 0.1 Ω.

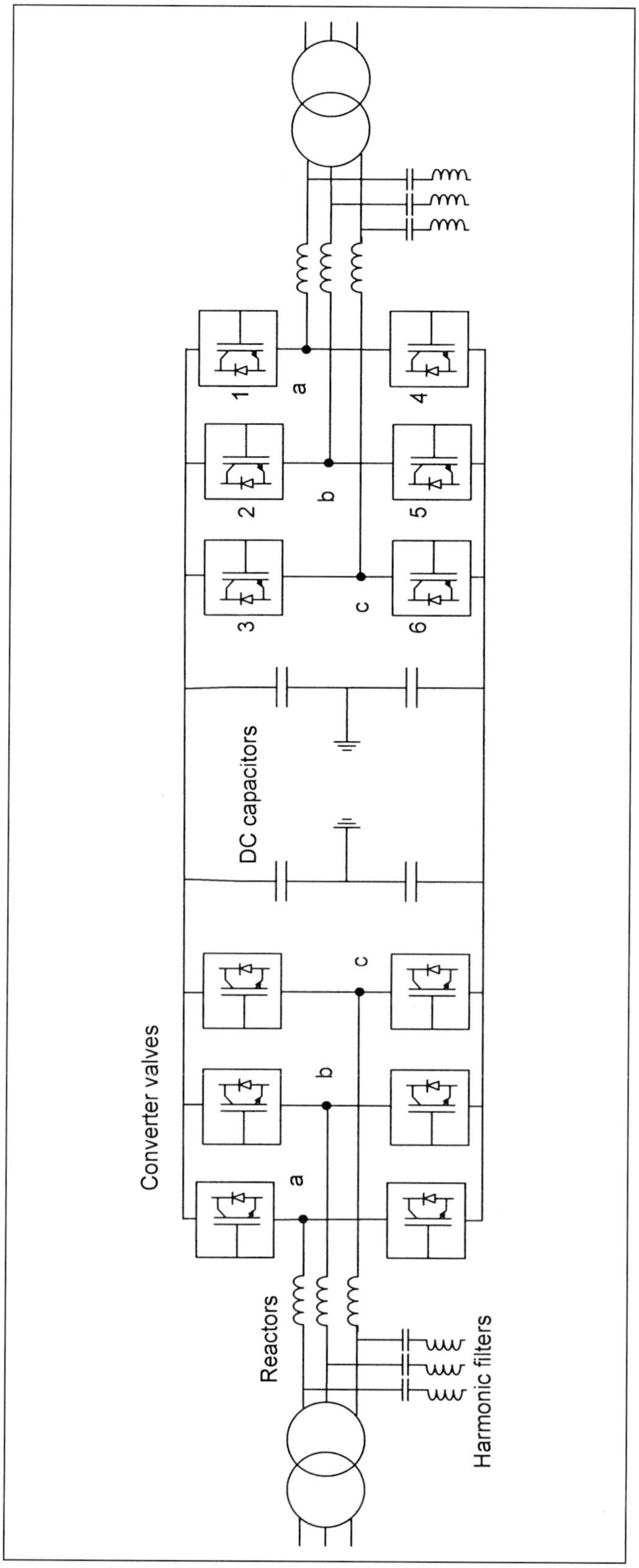

FIGURE 23-29 VSC-based HVDC.

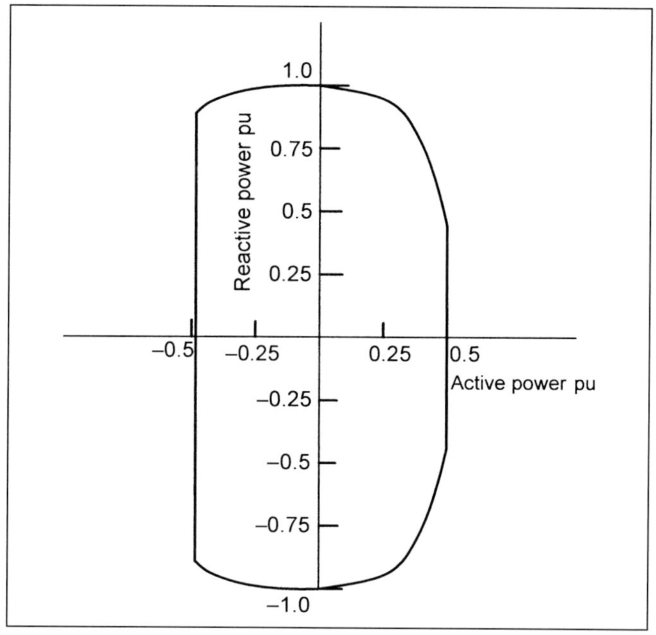

FIGURE 23-30 *P-Q* characteristics, VSC-based HVDC, and practical operating range.

and dc stresses on the converter transformers. The harmonic filters are much smaller due to PWM modulation (Chap. 15).[23,24]

The IGBT valves comprise series-connected IGBT positions. An IGBT exhibits low forward voltage drop and has a voltage-controlled capacitive gate. The complete IGBT position consists of an IGBT, an antiparallel diode, a voltage divider, and a water-cooled heat sink. IGBTs may be connected in series to switch higher voltages, much alike thyristors in conventional HVDC.

Active power can be controlled by changing the phase angle of the converter ac voltage with respect to filter bus voltage, whereas the reactive power can be controlled by changing the magnitude of the phase voltage with respect to filter bus voltage (Chap. 15). This allows separate active and reactive power control loops for HVDC system regulation. The active power loop can be set to control the dc voltage or the active power. In a dc link, one terminal may be set to control the active power, while the other terminal to control the dc side voltage. Either of these two modes can be selected independently. The operating range of VSC converter is shown in Fig. 23-30.

PROBLEMS

1. Plot the short-circuit profile of a lead acid battery: 240 V, 120 cells, 500 A-h at 8-hour rate of 1.75 V per cell at 25°C. Each cell is 15 in long, 7 in wide, and 10 in high. The cells are arranged in two-tier configuration, 30 cells per row, total 4 rows. Intercell connectors are 1 in × 1/2 in cross section, resistance 0.0321 m Ω/ft. Battery is connected through a cable of 0.002-Ω resistance and 15-μH inductance. The fault occurs at the end of the battery cable.

2. Calculate and plot short-circuit current profile of a dc motor of 100 hp, 230 V, 690 rpm, armature current = 256 A, and transient resistance = 0.12 Ω.

3. Calculate and plot the short-circuit current profile for a fault on the dc side of a rectifier, 480-V, three-phase ac power

supply system, short-circuit level of the ac system = 30 kA rms symmetrical, $X/R = 7.0$. The rectifier is supplied through a three-phase transformer of 400 kVA of percentage impedance = 4 percent, $X/R = 4$, the dc side equivalent resistance and reactance are 0.001 Ω and 4 μH. (The IEC methods of calculation illustrated in the text will require a copy of IEC standard.[1] However, methodology other than IEC procedures can be used for the solution of above problems.)

4. Based on the partial currents calculated in Probs. 1, 2, and 3, plot the total current. What is its peak value and time to peak? Select a suitable circuit breaker with respect to the short-circuit ratings.

5. A six-pulse converter is fed from a transformer of 230–110 kV 9% impedance. Determine dc voltage for $\alpha = 25°$ and $\mu = 15°$. If the dc current is 2000 A, what is the effective commutating reactance, rms fundamental component of ac current, power factor, and the reactive power requirement?

6. Repeat Prob. 5 for a 12-pulse converter.

7. A six-pulse converter is connected to a 480-V, 60-Hz, three-phase power source and operates at $\alpha = 30°$. The load current is maintained constant at 100 A, and load voltage = 360 V. What is the load resistance, source inductance, and angle μ?

8. Explain the impact of extinction angle on commutation failures. Describe all modes of operation of an HVDC link.

9. The ideal no load dc voltage of one pole at the rectifier end of dc link is 550 kV. The direct current = 1000 A and $V_d = 500$ kV. Calculate the reactive power requirements. Explain why the reactive power requirements at inverter end are higher.

10. Explain impact of short-circuit ratio on the stability limit.

11. A dc line-to-ground fault occurs on a 500-kV line, fault resistance = 5 Ω. Reactance of the smoothing reactor in series with rectifier = 0.5 H. What is the energy requirement of a dc circuit breaker?

12. Study an ac distribution system of a commercial facility and convert it into a dc distribution system. Include all single-line diagrams of distribution main switchboard and panels. Show details of all converters and cables, and calculate dc voltage drops in all main and subcircuits and make a comparative analysis.

REFERENCES

1. IEC Standard 61660-1, Short-Circuit Currents in DC Auxiliary Installations in Power Plants and Substations, 1997.
2. IEEE Standard 946, DC Auxiliary Power Systems for Generating Stations, 1992.
3. General Electric Company, *GE Industrial System Data Book*, Schenectady, New York, 1978.
4. www.iec.ch, IEC Standards on line.
5. AIEEE Committee Report. "Maximum Short-Circuit Current of DC Motors and Generators, Transient Characteristics of DC Motors and Generators," *AIEEE Trans.*, vol. 69, pp. 146–149, 1950.
6. A. T. McClinton, E. L. Brancato, and R. Panoff, "Maximum Short-Circuit Currents of DC Motors and Generators, Transient Characteristics of DC Motors and Generators," *AIEEE Trans.*, vol. 68, pp. 1100–1106, 1949.
7. A. G. Darling and T. M. Linville, "Rate of Rise of Short-Circuit Current of DC Motors and Generators," *AIEE Trans.*, vol. 71, pp. 314–325, 1952.
8. IEEE Standard 399, Power System Analysis, 1997.
9. J. J. Vithayathil, A. L. Courts, W. G. Peterson, N. G. Hingorani, S. Nilsson, and J. W. Porter, "HVDC Circuit Breaker Development and Field Tests," *IEEE Trans.*, vol. PAS-104, pp. 2693–2705, Oct. 1985.
10. CIGRE Joint Working Group 13/14-08, "Circuit Breakers for Meshed Multi-terminal HVDC Systems, Part 1: DC Side Substation Switching Under Normal and Fault Conditions," *Electra*, no. 163, pp. 98–122, Dec. 1995.
11. A. Sannino, G. Postiglione, and M. H. J. Bollen, "Feasibility of a DC Network for Commercial Facilities," *IEEE Trans. Industry Applications*, vol. 39, no. 5, pp. 1499–1507, Sep./Oct. 2003.
12. D. Nilsson and A. Sannino, "Load Modeling for Steady-State and Transient Analysis of Low-Voltage DC Systems," In Conference Record, IEEE I&CPS, Paper 0-7803-3/04, 2004.
13. M. E. Baran and N. R. Mahajan, "DC Distribution for Industrial Systems: Opportunities and Challenges," *IEEE Trans. Industry Applications*, vol. 39, no. 6, pp. 1596–1601, Nov./Dec. 2003.
14. J. G. Ciezki and R. W. Ashton, "Selection and Stability Issues Associated with a Navy Shipboard DC Zonal Electrical Distribution System," *IEEE Trans. Power Delivery*, vol. 15, pp. 665–669, Apr. 2000.
15. http://www.internetcad.com/pub/energy/technology_abb.pdf
16. www.cigre-b4.org: CIGRE Study Committee B4, HVDC, and Power Electronic Equipment.
17. M. Parker and E. G. Peattie, *Pipe Line Corrosion and Cathodic Protection, 3rd* ed., Gulf Professional Publishing, Houston, TX. 1995.
18. V. K. Sood, *HVDC and FACTS Controllers*, Kulwer, Norwell, MA, 2004.
19. A. Ekstrom and G. Liss, "A Refined HVDC Control System," *IEEE Trans. PS*, vol. PAS 89, pp. 723–732, May-Jun. 1970.
20. IEEE Committee Report, "HVDC Controls for System Dynamic Performance," *IEEE Trans.*, vol. PWRS-6, no. 2, pp. 743–752, May 1991.
21. IEEE Standard 1031, IEEE Guide for the Functional Specifications of Transmission Line Static VAR Compensators, 2000.
22. CIGRE and IEEE Joint Task Force Report, "Guide for Planning DC Links Terminating at AC Locations having Low Short-Circuit Capabilities, Part 1: AC/DC Interaction Phenomena," CIGRE Publication 68, Jun. 1992.
23. B. Jacobson, Y. Jiang-Hafner, P. Rey, and G. Asplund, "HVDC with Voltage Source Converters, and Extruded Cables for up to ±300 kV and 1000 MW," In Proc., CIGRE, pp. 84–105, 2006.
24. G. Asplund, K. Eriksson, H. Jiang, J. Lindberg, R. Palsson, and K. Stevenson, "DC Transmission Based on Voltage Source Converters," in Proceedings CIGRE, Paris, 1998.

FURTHER READING

J. Arrillaga, *High Voltage Direct Current Transmission*, 2d. ed. IEEE Press, Piscataway, NJ, 1998.

EPRI HVDC Electrode Design, EPRI Research Project 1467-1, Report, Palo Alto, EL-2020.

EPRI HVDC Transmission Line Reference Book, EPRI Report TR-102764, 1993.

A. Greenwood, *Electrical Transients in Power Systems*, John Wiley and Sons, New York, 1991.

E. W. Kimbark, *Direct Current Transmission*, Vol. 1, Wiley Interscience, New York, 1971.

N. Knudsen and F. Iliceto, "Contributions to the Electrical Design of HVDC Overhead Lines," *IEEE Trans.*, vol. PAS93, no. 1, pp. 233–239, 1974.

P. Kundur, *Power System Stability and Control*, Chapter 10: HVDC Transmission, EPRI, Palo Alto, 1993.

K. R. Padiyar, *Power Transmission Systems*, John Wiley, New York, 1990.

CHAPTER 24

SMART GRIDS AND WIND POWER GENERATION

In the years to come, the power generation, transmission, and distribution will undergo profound changes like improved environmental compatibility, reliability, and operational efficiency, integration of renewable energy technologies like wind, solar power, and distributed generation. The modern grid systems are being controlled and will be controlled and operated so that the dynamic state of the grid is known in the terms of:

1. Rotor angle stability and voltage stability
2. Increase/decrease in transmission capability that can take place in real-time over transmission systems
3. Control and regulation of power flow to maintain grid parameters
4. Remedial action schemes (RAS) and system integrated protection systems (SIPS)
5. Identification of the remedial measures that should be taken to avoid an extreme contingency, that is, cascading and blackouts
6. Physical implementation of corrective actions

The technologies driving the self-healing smart grid are: wide-area measurement systems (WAMSs), system integrity protection schemes (SIPSs), phasor measurement units (PMUs), energy management systems (EMSs), flexible AC transmission (FACTS) (Chap. 16), and communication systems, dynamic contingency analysis (DCA); all somewhat related.

It is amply clear from Chap. 12 that stability of a system is not a fixed identity and varies with the operating and switching conditions. Some not so common contingencies in a system can cascade and bring about a shutdown of a vital section. Historically, the great Northeast Blackout of November 9–10, 1965, and more recently the 2003 East Coast Blackout can be mentioned.

24-1 WAMS AND PHASOR MEASUREMENT DEVICES

Wide area network measurements have been around for the last 60 years and have been used in economic dispatch, generation control, and real-time measurements of power flows. Supervisory control and data acquisition systems (SCADAs) are of late-1960s origin and provided real-time state estimates of power systems. The measurement comprised a data window of several seconds, without regard to the instant at which the precise measurement was made, while the system may have drifted meanwhile from the instant of measurement. Developments in microprocessor-based relays got an impetus in 1970s, based on the requirements that symmetrical components of currents and voltages at relay locations be estimated from synchronized sampled data on a system-wide basis. In the 1980s global positioning systems (GPSs) were deployed and the prospect of synchronizing sampled data on a system-wide basis became a reality.

This requirement led to the concept of phasor measurement unit. The concept is simple; a sinusoidal waveform can be represented by a magnitude and phase angle (Fig. 24-1). The magnitude is the peak or rms value of the sinusoid. The phase angle is given by the frequency and the time reference. The synchrophasor representation $\boldsymbol{X}$ of a signal $x(t)$ is the complex value given by:

$$\begin{aligned} \boldsymbol{X} &= X_r + jX_i \\ &= (X_m/\sqrt{2})e^{j\phi} \\ &= \frac{X_m}{\sqrt{2}}(\cos\phi + j\sin\phi) \end{aligned} \tag{24-1}$$

where $X_m/\sqrt{2}$ is the rms value of signal $x(t)$ and ϕ is the phase angle relative to cosine function at normal system frequency synchronized to universal time coordinated (UTC). This angle is 0° when maximum of $x(t)$ occurs at the UTC second rollover [1 pulse per second (PPS) time signal] and −90° when the positive zero crossing occurs at the UTC second rollover. Synchrophasors are, thus, phasor values that represent power system sinusoidal waveforms referenced to nominal system frequency and coordinated with universal time. The phase angle is uniquely determined by the time of measurement, waveform, and system frequency. If a sinusoid is observed at intervals $(0, T_0, 2T_0, \ldots, nT_0, \ldots)$ leading to phasor representations $(X_0, X_1, X_2, \ldots)$ and observation time interval T_0 is an integer multiple of sinusoid $T = 1/f$, then a constant phasor is obtained at each observation. If observation time T_0 is not an integer multiple of T, the observed phasor has a constant

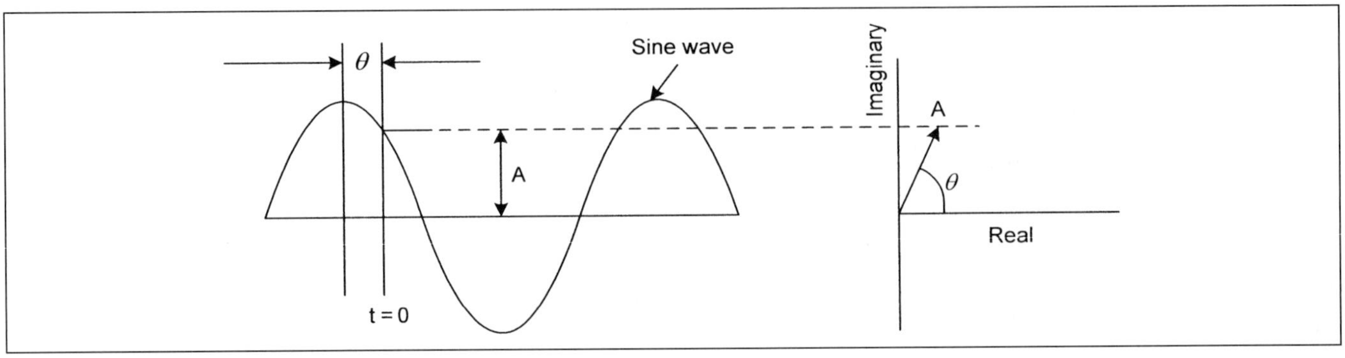

FIGURE 24-1 Concept of phasor representation of a sinusoidal waveform.

magnitude, but angles of the phasors (X_0, X_1, X_2,...) will change uniformly at a rate $2\pi(f - f_0)T_0$, where $f_0 = 1/T_0$.

System frequencies are not rock steady and can vary. An interconnected system runs at the same frequency and all phase angles rotate together, one way or the other. Because of this rotation, the phase angle measurements should be made exactly at the same rate. As an example, state estimators run at intervals ranging from a few seconds to 10s to minutes, and a phasor system running at 6+ samples/s cannot directly feed into the slower system. A solution would be to use synchronized samples drawn (periodically) from the full data set. This led to the development of IEEE standard C37.118[1] revised in 2005. The basic measurement requirements, including angle-time relationship, are detailed in this standard. The accuracy of phasor estimate is compared with a mathematically predicted value using a total vector error (TVE). This can be defined as root square difference of the values, and compliance with the standard[1] requires a difference within 1 percent under various conditions:

$$\text{TVE} = \sqrt{\frac{(X_r(n) - X_r)^2 + (X_i(n) - X_i)^2}{X_i^2 + X_i^2}} \qquad (24\text{-}2)$$

where $X_r(n)$ and $X_i(n)$ are the measured values and X_r and X_i are the theoretical values of the input signal at instant of time of measurement.

Obtaining a phasor equivalent of an arbitrary sinusoidal signal requires a sample of waveform taken at appropriate frequency—the quality of phase estimate has to be ensured. Discrete Fourier transform (DFT) is the most commonly used method of phase estimation. This technique uses the standard Fourier estimate applied over one or more cycles at nominal system frequency. At a sufficient sample rate and accurate synchronization with UTC, it produces an accurate and usable phasor value for most system conditions. Problems with DFT response like roll-off can occur with varying frequency and must be corrected, for example, by centering the measurement window. GPS is universally used for the UTC time reference. Other PMUs may rely on time signal, such as IRIG-B from an external GPS receiver.

The PMU functions are built into microprocessor-based multifunction relays (MMPR) and digital fault recorders (DFRs). These may have variable capabilities. Based on analogue inputs; three-phase quantities; and positive, negative, and zero sequence phasors can be outputted. The sensing elements, that is, the accuracy of potential and current transformers, become a question mark, and so far ANSI/IEEE relaying class accuracies are found to be adequate. Further work is being done by the North American Synchrophasor Initiative (NASPI) project.

Figure 24-2 shows a typical hierarchal system; the PMUs feed into the phasor data concentration (PDC) at a control center. PDCs are produced which interface with other products, such as monitor/control platforms and a data historian. PDCs connect to multiple PMUs and receive, parse, and sort incoming data. Due to sheer amount of data, it is an overwhelming computer processing task. IEEE standard[1] establishes PMU data protocols. Over the course of years, CPU processing power has increased. The number of incoming PDC devices that may be deployed is limited by CPU processing power.

Originally WAMSs were limited to single utilities. The interutility data exchange enables wide-area visibility. When interfacing with SCADA, the data must be reduced to match SCADA data rates and interface with protocols used by SCADA. (Most estimators draw data from SCADA). The widely used IEEE common format for transient data exchange (COMTRADE), IEEE Standard C37.111[2] developed for time sequence data, supports binary and floating points formats. Most control center applications use a data historian for analysis and trending. And generally, these will accept data at full rate.

The phasor estimation, primarily developed for steady-state signals, will be applied to system dynamics as the next step. Most power system dynamics are slower compared with the speed of phasor systems. An IEEE working group is formed to revise C37.118 to include dynamic performance requirements. DCA will make stability assessment and issue real-time control signals. The stability functional requirements will dictate the system performance.

24-2 SYSTEM INTEGRITY PROTECTION SCHEMES

System integrity protection schemes are automated systems that protect the grid against system contingencies and minimize the potential for wide outages. Without SIPS it may not be possible to provide for many contingencies, address transmission paths, alternate routes, corrective measures, and take prior warnings. A SIPS design is based on system studies of predefined contingencies for a variety of conditions. According to the IEEE Power System Relaying Committee (PSRC), the following is the list of SIPS measures:

- Generator and load rejection
- Underfrequency and undervoltage load shedding
- Adaptive load mitigation
- Out-of-step tripping
- Voltage and angular instability and advance warning schemes
- Overload and congestion mitigation

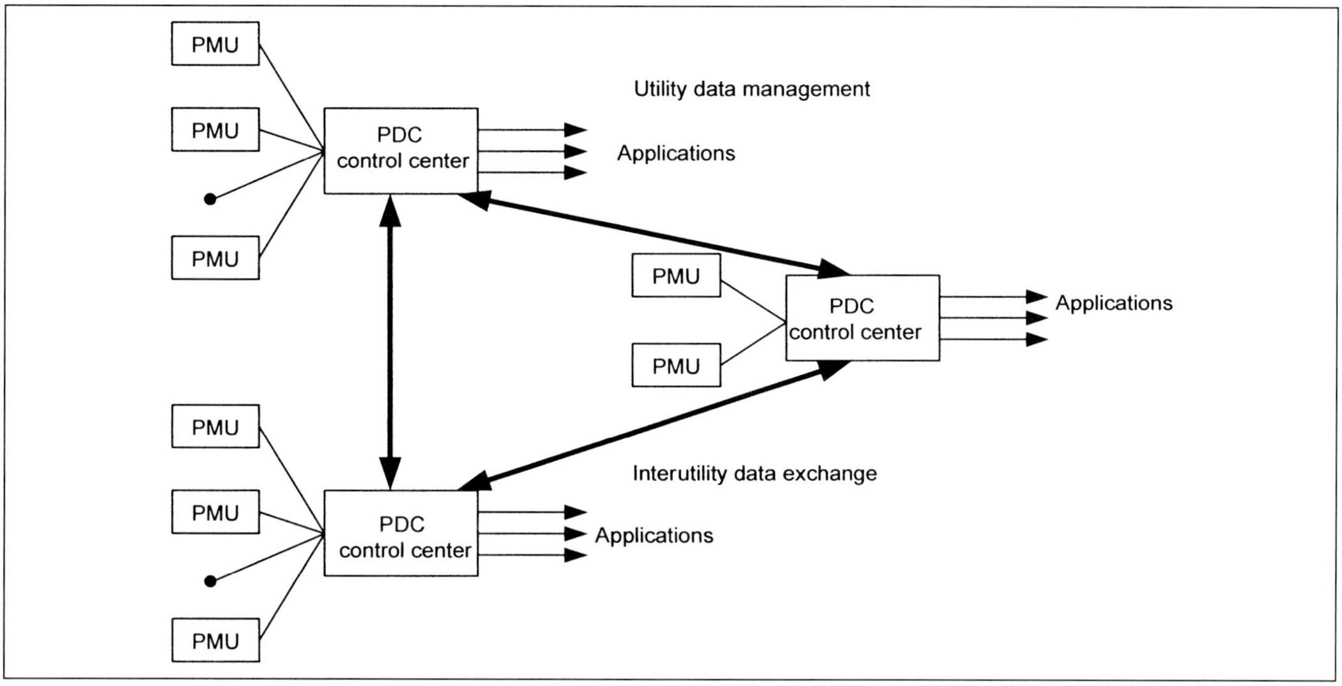

FIGURE 24-2 Typical star hierarchal system of PMUs feeding into PDCs that feed local and remote applications. These can extend to share data between utilities.

- System separation
- Shunt capacitor switching
- Tap changer control
- SVC/STATCOM control
- Turbine valve control
- HVDC controls
- Power system stabilizer control
- Discrete excitation
- Dynamic breaking
- Generator runback
- Bypassing series capacitor
- Black-start or gas turbine start-up
- ASGC actions
- Bus bar splitting

We have discussed the basic concepts of many of these items in this book. Applied to complex grid systems, SIPS is the last line of defense to protect the integrity of the power system and propagation of disturbances for severe system emergencies caused by unplanned operating conditions.

24-3 ADAPTIVE PROTECTION

The objective of adaptive relaying is to adjust relay performance or settings as per changing system conditions. It can be defined as, a protection philosophy which permits and seeks to make adjustments automatically in various protection functions in order to make them more attuned to prevailing system conditions. This is being achieved by phenomenal advancements in the microprocessor-based technology applied to protective relaying. For example, current differential schemes with high-speed communications can be applied to transmission lines.

Dependability and security are measures of reliability, which mutually oppose.[3] To be dependable, the protection must always trip, even if there are nuisance trips. Consider that a system is robust and a dependable protection is applied to it—if the system changes, whether due to planned or unplanned outages, the strength of the system to withstand same amount of trips becomes questionable. With WAMS and digital devices, it is possible to reorganize so as to reorient the relay performance from dependable to secure. Figure 24-3 shows a scheme, where three protective schemes can trip independently, without supervision to security, while tripping decisions are connected so that out of three at least two schemes should operate correctly. After the 2003 East Coast Blackout, National Electric Reliability Council (NERC) recommended removing all unnecessary zone 3 distance tripping to avoid, "over-tripping" by these elements.

Other smart grid issues can be itemized as follows:

- Requirements for renewable portfolio standards (RPS), limits on greenhouse gases (GHS), and demand response (DR).
- Advanced metering structures at consumer loads.
- Integration of solar, wind, nuclear, and geothermal facilities which pose their own challenges. For example, the large-scale solar plants or wind generation may be located in areas distant from existing transmission facilities. New protection and control strategies, interconnection standards—for example, low-voltage ride-through (LVRT) capabilities—forecasting, and scheduling are required.

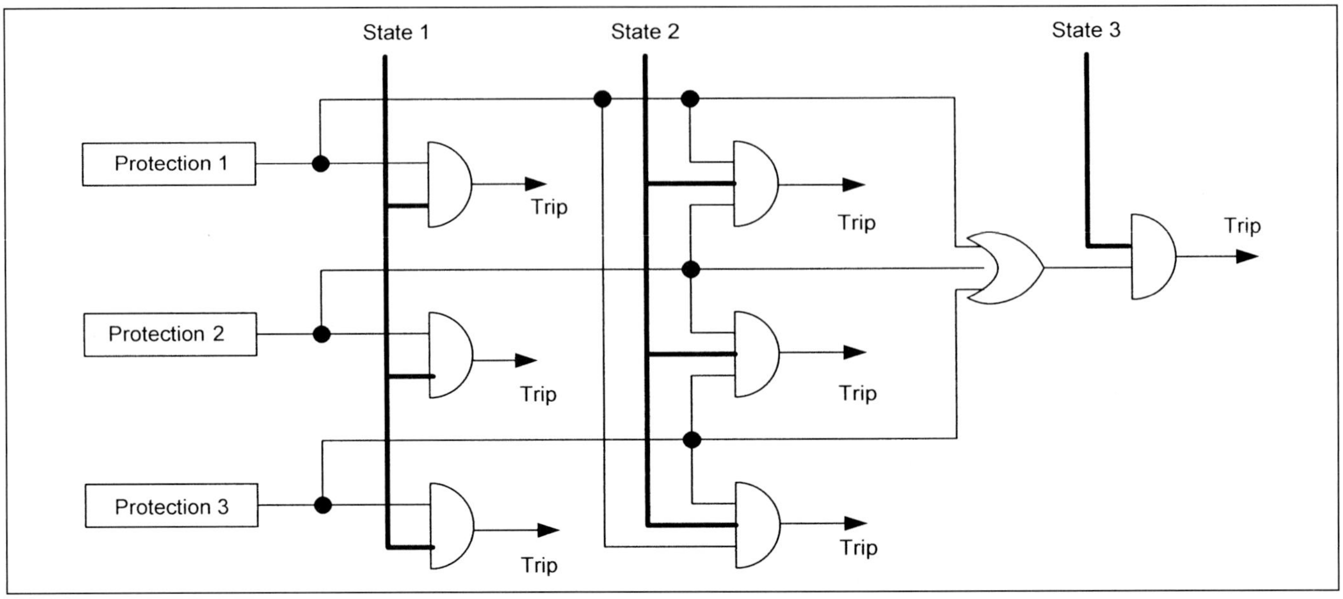

FIGURE 24-3 Schematic adaptive protection and redundancy.

- Managing circuit congestion, managing distribution system overloads.
- Role of information and automation technologies.

Figure 24-4 shows a wide-area control framework. Although it may not be entirely possible to avoid multiple blackouts, the probability, size, and impact of widespread outages can be reduced.

24-4 WIND-POWER STATIONS

A synopsis of wind-power generation is included. This is an important subject in view of rising energy costs and growing concerns of global climate changes and environmental effects. With respect to transient and stability analyses, interconnections of wind-power stations with grid systems pose even larger problems, and require more thorough analysis of the stability of the interconnection and system isolation under disturbances.

More than 75000 MW of wind-power generation has been added worldwide, out of which over 12000 MW is in the United States. Looking at energy penetration levels (ratio of wind power delivered by total energy delivered), Denmark leads, reaching a level of 20 percent or more, followed by Germany. Sometiimes the wind energy penetration exceeds 100 percent, with excess sold to Germany and NordPool. Nineteen off-shore projects operate in Europe producing 900 MW. United States off-shore wind energy resources are abundant.

In the United States, wind-power generation accounts for approximately 0.6 percent of the total, and Renewable Energy Laboratory (DOE/NREL) did an investigation of what 20 percent

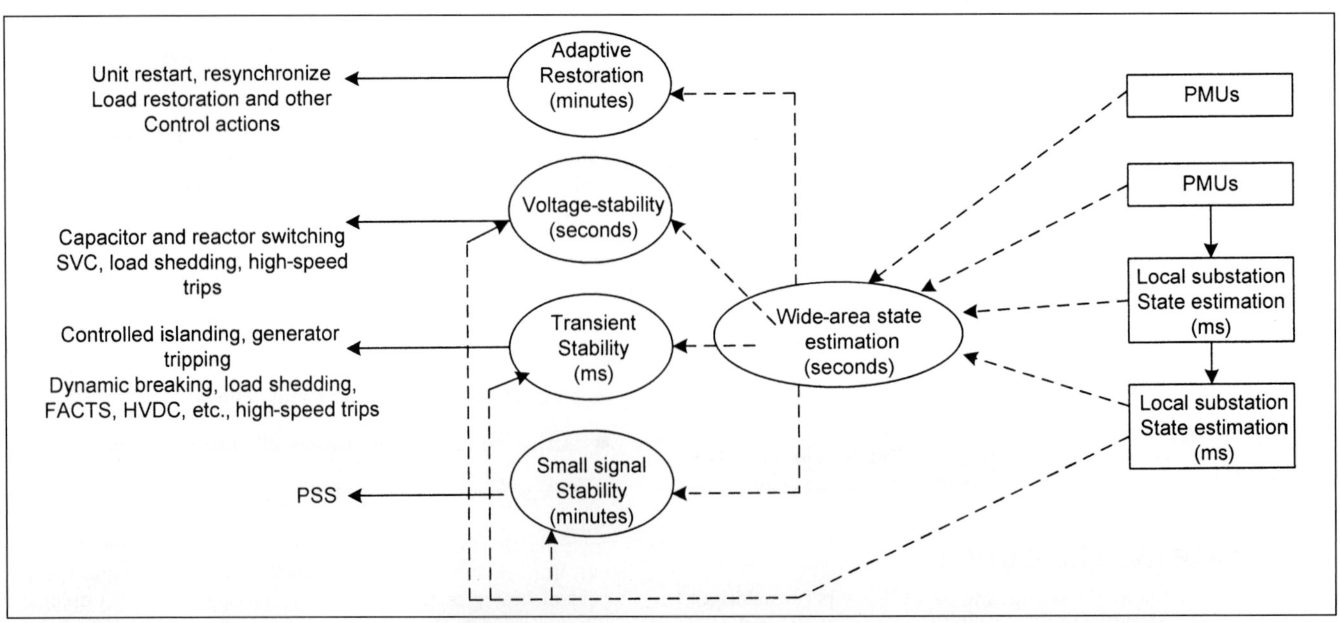

FIGURE 24-4 Wide-area control framework.

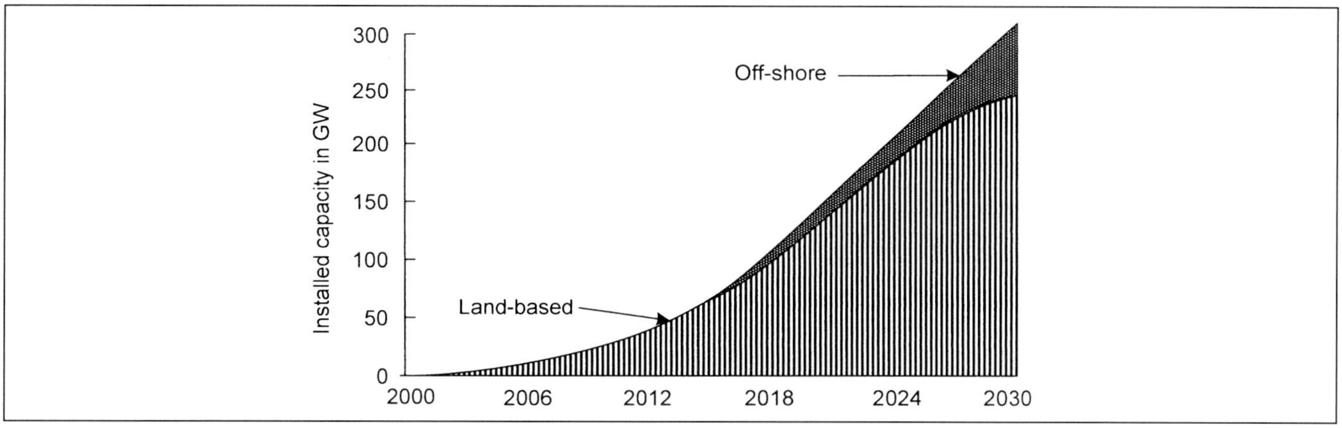

FIGURE 24-5 Twenty percent penetration of installed wind-power capacity: land-based and off-shore.

of energy from wind would look like in 2030 (Fig. 24-5). American Electric Power (AEP) produced a white paper that included a 765-kV network overlay for a U.S. power system that would increase reliability and allow for 400-GW of wind or other generation to be added. The political climate change in Washington can impact these decisions and accelerate exploitation of wind power. The modern wind-power plants can be as large as 300-MW and are often located within a short distance of each other.

24-5 WIND-ENERGY CONVERSION

Today's wind turbines all over the world have three-blade rotors, diameters ranging from 70 to 80 m and mounted atop 60 to 80 m or higher towers. The tower heights up to 160 m are a technical feasibility. The typical turbine installed in the United States in 2006 can produce about 1.5 MW of power. Higher-rated units and off-shore wind base plants may see a unit size of 5 MW or more by 2010. Figure 24-6 shows the developments of single-unit wind-power turbines in the United States.

Figure 24-7 is a schematic representation of electrical and mechanical features of a wind converter unit, with *upwind* rotor. The rotor speed which is of the order of 8 to 22 rpm is the input to gear box, and on the output side the speed is 1500 to 1800 rpm. The drive train dimensions are large, increasing the horizontal dimension of nacelle.

24-5-1 Drive Train

Several designs are under development to reduce the drive train weight and cost. One approach is to build direct drive permanent magnet generators that eliminate the complexity of the gear box. The slowly rotating generator will be larger in diameter, 4 to 10 m, and quite heavy. The decrease in cost and availability of rare-earth permanent magnets is expected to significantly affect size and weight. The generator designs tend to be compact and lightweight and reduce electrical loses in windings, as compared to wound rotor machines. Prototypes have been built—a 1.5-MW design with 56 poles is only 4 m in diameter versus 10 m for a wound rotor design. It is undergoing testing at the National Wind Technology Center.

A hybrid of direct drive approach uses a low-speed generator. The WindPACT drive train project has developed a single-stage planetary

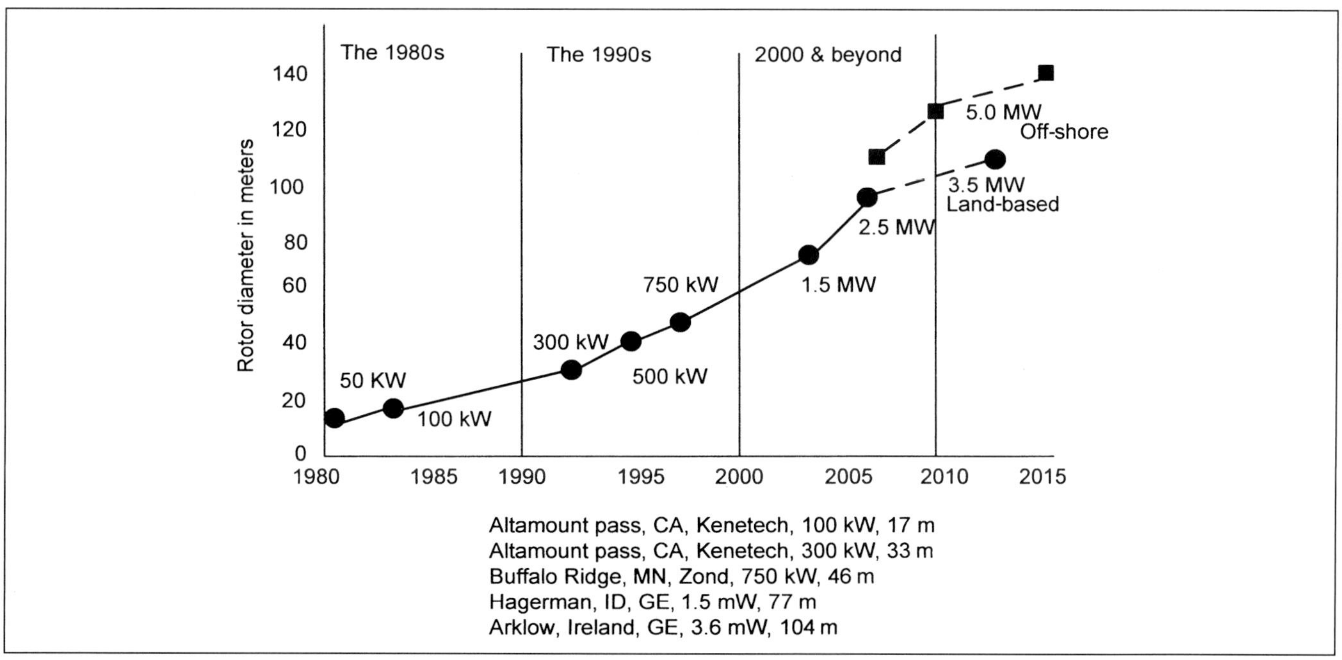

FIGURE 24-6 Development of single-unit wind turbines in the United States.

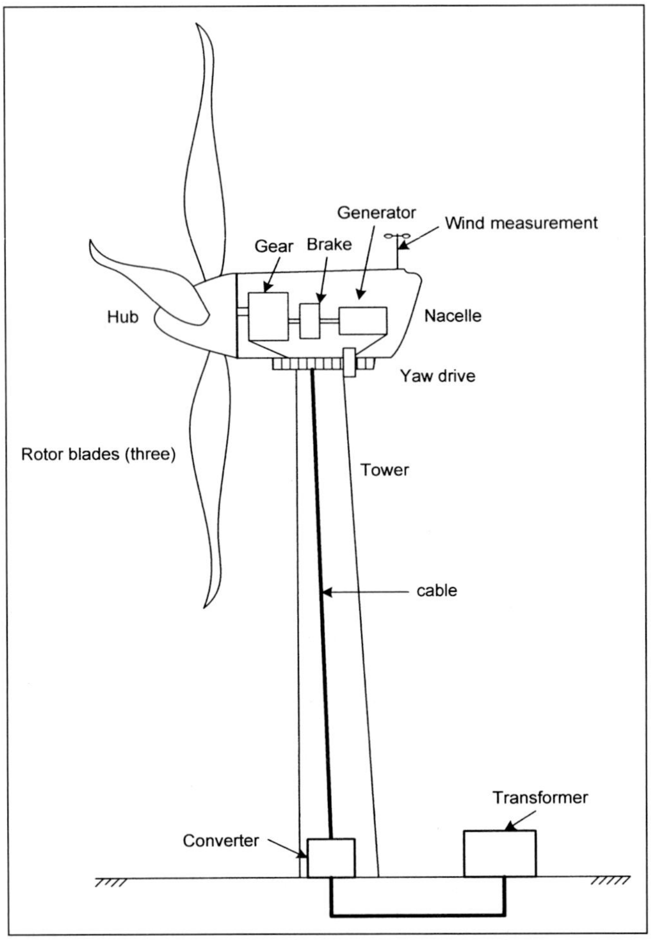

FIGURE 24-7 A schematic of a wind-power conversion unit, showing major mechanical and electrical components.

drive operating at a gear box ratio of 9.16:1. The gear box drives a 72-pole permanent magnet (PM) generator, and reduces the diameter of a 1.5-MW generator to 2 m.[4] Another development is the distributed drive train.

24-5-2 Towers

Depending upon the nacelle weight, the towers are constructed from steel or steel/concrete. The tower heights are up to 100 m. Wind speeds are height dependent, and lattice towers of up to 160 m height can be constructed. A relationship between tower mass and tower height shows sharp increase in the tower mass per meter height as turbine output increases, approximately 2000 kg/m for turbines in the output range of 1.5 MW. Large plants tend to have significantly greater mass per meter as the mast height increases. There is an ongoing effort to develop advanced tower designs that are easily transported and installed and are cost effective.

The wind speeds are not uniform over the area of the rotor. The rotor profiles result in different wind speeds at the blades nearest to the ground level compared to the top of the blade travel. Gusts and changes in wind speed impact the rotor unequally. Influences due to tower shadow or windbreak effects cause fluctuations in power or torque.

24-5-3 The Rotor Blades

As the wind turbines increase in output so do their blades. During the 1980's, the blade length was only 8 m. It has increased to 70 m for many land-based wind units. Improved blade designs have kept weight growth much lower than the geometric escalation of blade lengths. Work continues in the application of lighter and stronger carbon fiber in highly stressed areas to stiffen the blades, improve fatigue resistance, and simultaneously reduce the weight. Research continues in the development of lighter blades, such as carbon fiber and fiberglass.

24-6 THE CUBE LAW

Wind originates from a difference in temperature and pressure in air mass. A change in these parameters alters the air density ρ. The force exerted on volume of air V is given by:

$$F = Vg\Delta\rho \tag{24-3}$$

where g is the gravitational constant. This produces kinetic energy:

$$E = \frac{1}{2}mv^2 \tag{24-4}$$

where m is the mass of the air, v is the velocity which is assumed constant, and E is the wind power. This is rather an oversimplification. If we consider an air volume of a certain cross section and a swirl-free speed, upstream of the turbine and downstream of the turbine, it will result in a reduction in speed, with a corresponding broadening of the cross sectional area (wake decay). Thus, practically v is not constant. However, with this simplification of constant speed, the basic tenets of turbine output are still valid.

The following equation can be written for the air mass:

$$m = \rho A v \tag{24-5}$$

where A is the rotor area. Substituting Eq. (24-5) in Eq. (24-4), the theoretical power output is:

$$P = \frac{1}{2}\rho A v^3 c_p \tag{24-6}$$

where the wind-speed-dependent coefficient c_p describes the amount of energy converted by the wind turbine. It is in the range of 0.4 to 0.5.

All the energy in a moving stream of air cannot be captured. A block wall cannot be constructed because some air must remain in motion after extraction. On the other hand, a device which does not slow the air will not extract any energy. The optimal blockage is called *Betz limit*, which is around 59 percent. The aerodynamic performance of blades has improved dramatically, and it is possible to capture about 80 percent of the theoretical limit. The new aerodynamic designs also minimize fouling due to dirt and bugs that accumulate at the leading edge and can reduce efficiency.

According to Betz, the maximum wind power turbine output is:

$$P = \frac{16}{27} A_R \frac{\rho}{2} v_1^3 \tag{24-7}$$

where A_R is the air flow in the rotor area and v_1 is the wind velocity far upstream of the turbine. The maximum is obtained when:

$$v_2 = \frac{2}{3}v_1 \quad \text{and} \quad v_3 = \frac{1}{3}v_1 \tag{24-8}$$

where v_3 is the reduced velocity after broadening of the air stream past the rotor, and v_2 is the velocity in the rotor area. The ratio of the power absorbed by turbine to that of moving air mass is:

$$P_0 = A_R \frac{\rho}{2} v_1^3 \tag{24-9}$$

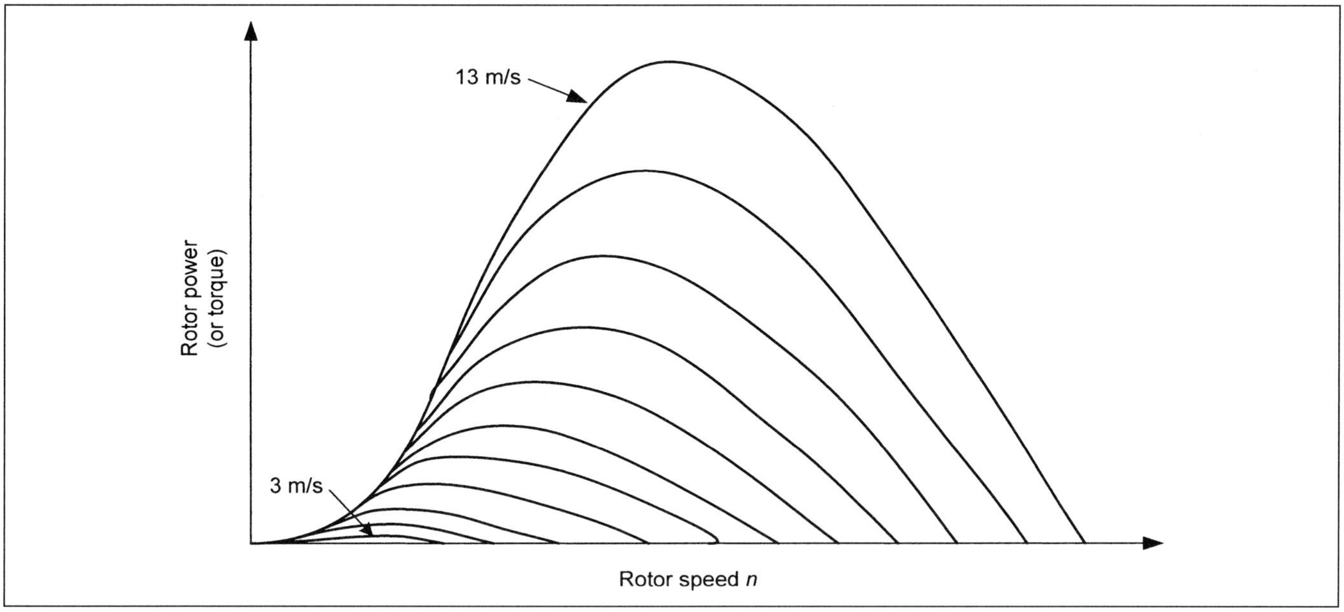

FIGURE 24-8 Rotor power of a wind generating unit, based on rotor speed and wind velocity.

and

$$c_p = \frac{P}{P_0} \tag{24-10}$$

Another way of defining c_p is with respect to λ, which is defined as:

$$\lambda = \frac{v_{TS}}{v_{RP}} = \frac{2\pi nr}{v_{RP}} \tag{24-11}$$

where v_{TS} is the blade tip speed and v_{RP} is the speed of the rotor plane, and v_{TS} can be written as:

$$v_{TS} = 2\pi nr \tag{24-12}$$

where r is the rotor radius, m, and n is the speed, s^{-1}. Typical values of λ are 8 to 10, and the tip-speed ratio influences the power coefficient c_p. Also c_p is dependent on wind speed. Combining these relations, Fig. 24-8 shows power or torque versus the wind speed. (Torque is simply power divided by angular velocity ω.) Figure 24-9

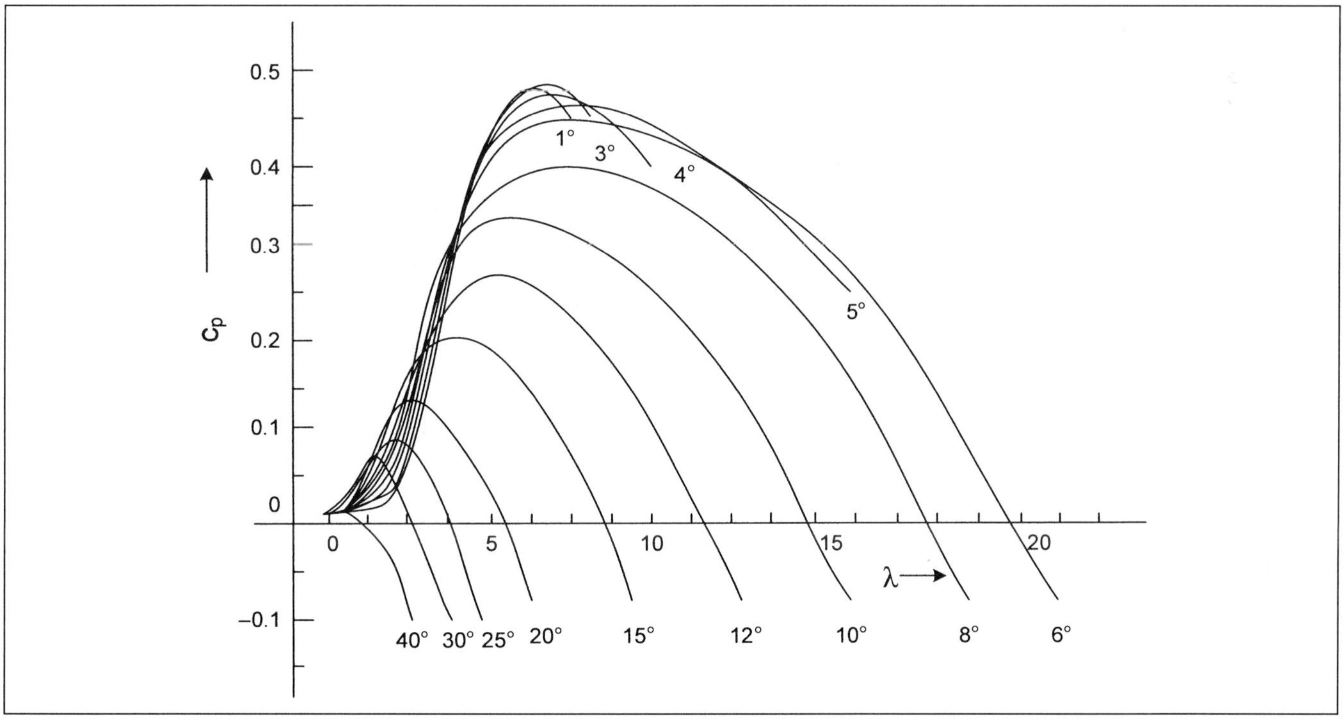

FIGURE 24-9 c_p-λ characteristics of a wind turbine unit with blade pitch angle.

shows performance coefficient as a function of tip-speed ratio with blade pitch angle as a parameter (see Sec. 24-7).

24-7 OPERATION

Figure 24-10 shows power curves for typical modern turbine. The turbine output is controlled by rotating the blades about their long axis to change angle of attack; this process is called "controlling the blade pitch." The turbine is pointed into the wind by rotating the nacelle about the tower, which is called the "yaw control." Modern turbines operate with rotor positioned on the windward side of tower, which is referred as an "upward rotor." A turbine generally starts producing in winds of about 12 mph, reach a maximum power at about 28 to 30 mph and shutdown, or "feather the blades," at about 50 mph. The amount of energy in the wind available for extraction by turbine increases with cube of speed, thus a 10 percent increase in speed means a 33 percent increase in energy. While the output increases proportional to rotor-swept area, the volume of material, and thus the cost, increases as cube of the diameter. Controllers integrate signals from dozens of sensors to control rotor speed, blade pitch angle, generator torque, power conversion voltage, and phase angle.

Wind-speed changes may occur over long periods of time or suddenly within a matter of seconds. The system has to be protected from the sudden gusts of wind. The mechanical loads are determined by dynamic forces, and knowledge of dynamic wind behavior at a location is necessary for proper component ratings, and the mechanical power acting on the turbine should be limited.

There are three methods to achieve it:

1. Stall
2. Active stall
3. Pitch-control

The stall control exercises adjustable clutches on rotating blade tips to shut down. Under normal conditions, laminar air flow occurs on the blades. The lift values corresponding to angle of attack are achieved at low drag components. With wind speeds exceeding nominal values at which the generator-rated output occurs, higher angles of attack and stalling occurs. This is achived by properly profiling the blades. The lift forces and lift coefficient are reduced and the drag forces and coefficients increased (Fig. 24-11*a*). Stall-regulated machines are often designed with asynchronous generators of higher nominal output, and rigid coupling with the grid is obtained. An active stall control is achived with turnable rotor blades (Fig. 24-11*b*).

Variable blade pitch allows direct control of turbine. By varying the blade pitch it is possible to control the torque of the turbine and by further adjustments bring about a stall condition (Fig. 24-11*c*). The control and regulation system is complex. In high power applications, pitch control is used. For the design and control of pitch adjustments, a host of mechanical moments and forces must be considered. These inclule forces and moments due to blade deflection, lift on blades, moments on propeller due to teetering, and frictional moments.[5]

24-7-1 Speed Control

The curves of fixed- and variable-speed generators are marked in Fig. 24-12. This shows that when the turbine is driven by a synchronous generator, varying the generator frequency at a certain wind speed will give operation at $n/n_1 = 1$, where n corresponds to the grid frequency. The turbine is constrained to follow the grid frequency. Sufficient turbine torque to drive the generator is available for wind speeds above approximately 3.6 m/s for synchronous generators to about 3.8 m/s for asynchronous generators. Under variable-frequency generator operation, the speed of rotation can be freely set within the given limits. The turbine utilization of the available wind power is optimized. A ramp control of the active power output is possible.

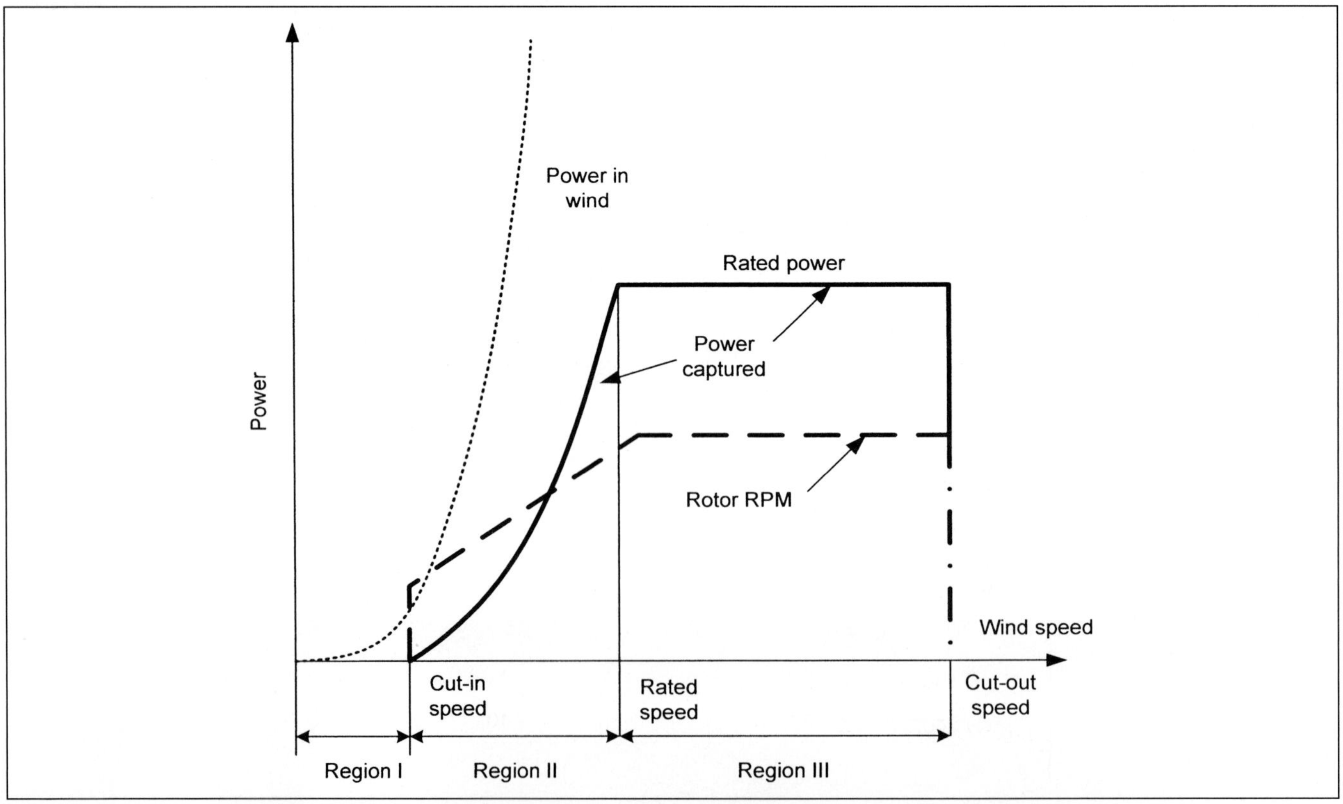

FIGURE 24-10 Typical wind generating unit operating curve, power verses wind speed.

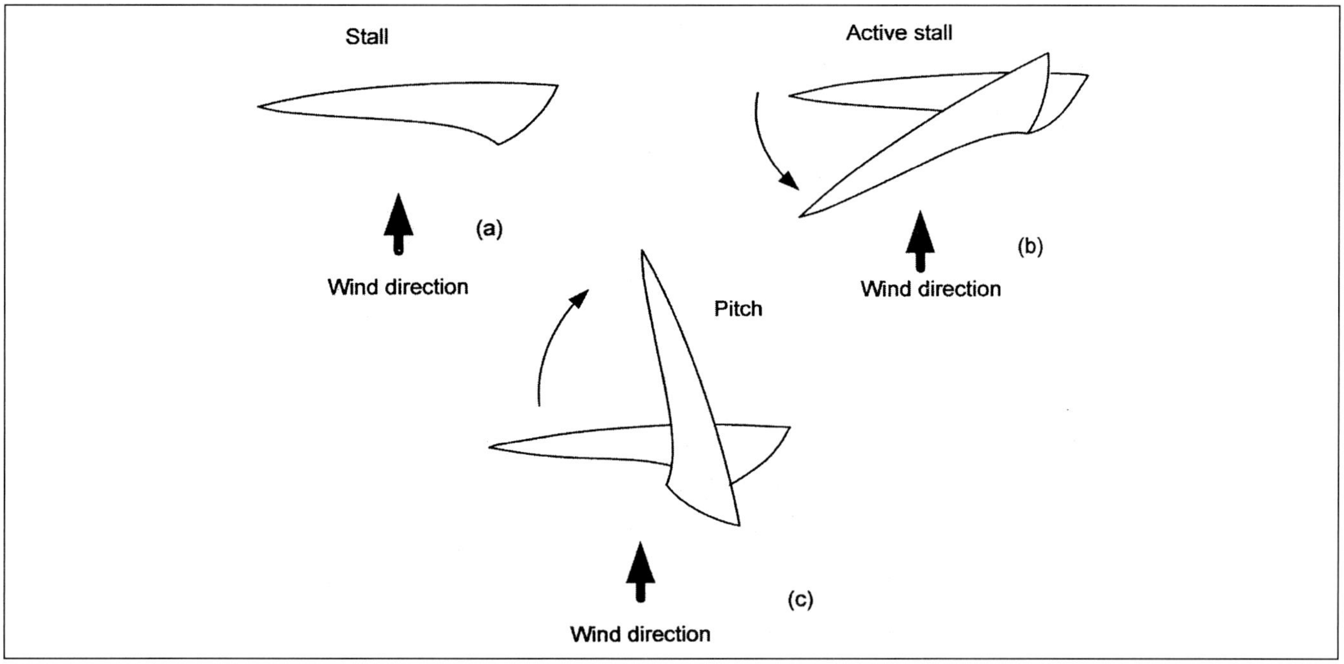

FIGURE 24-11 Limiting the mechanical power in a wind turbine. (*a*) Stall control. (*b*) Active stall control. (*c*) Pitch control.

24-7-2 Behavior Under Faults and Low-Voltage Ride-Through

Figure 24-13 shows the recommendation of Western Electricity Coordinating Council (WECC) wind generation task force (WGTF) with respect to proposed voltage ride-through requirements for all wind generators. A three-phase fault is cleared in 9 cycles, and the post-fault voltage recovery dictates whether the wind-power-generating plant can remain online. The requirement does not apply to faults that will occur between the wind generator terminals and the high side of GSU (generator step-up unit transformer), and the wind plants connected to transmission network via a radial line will not be required to ride-through the fault on that line.

Figure 24-13 also addresses the high and low voltage profile that must be maintained. Thus, the voltage should remain within a certain operating region defined in Fig. 24-13. Induction generators, degrade power system voltage performance as these require excitation reactive power. The Federal Energy Regulatory Commission (FERC) orders 661 and 661A require new wind generators to have capability to control their reactive power within 0.95 leading to 0.95 lagging range. As this requirement can be expensive to comply, FERC requires it only if the interconnection study shows that it is needed. Modern wind generators provide this capability from power electronics that control the real power operation of the machine.

24-8 WIND GENERATORS

Induction generators are either squirrel-cage or wound-rotor induction types, and their operation, circuit diagrams, and modeling is discussed in Chap. 11. For wind-power generation, the generators are rotating-field-type. The coupling with the grid, directly or through inverters, is of significance. Mostly induction generators are used.

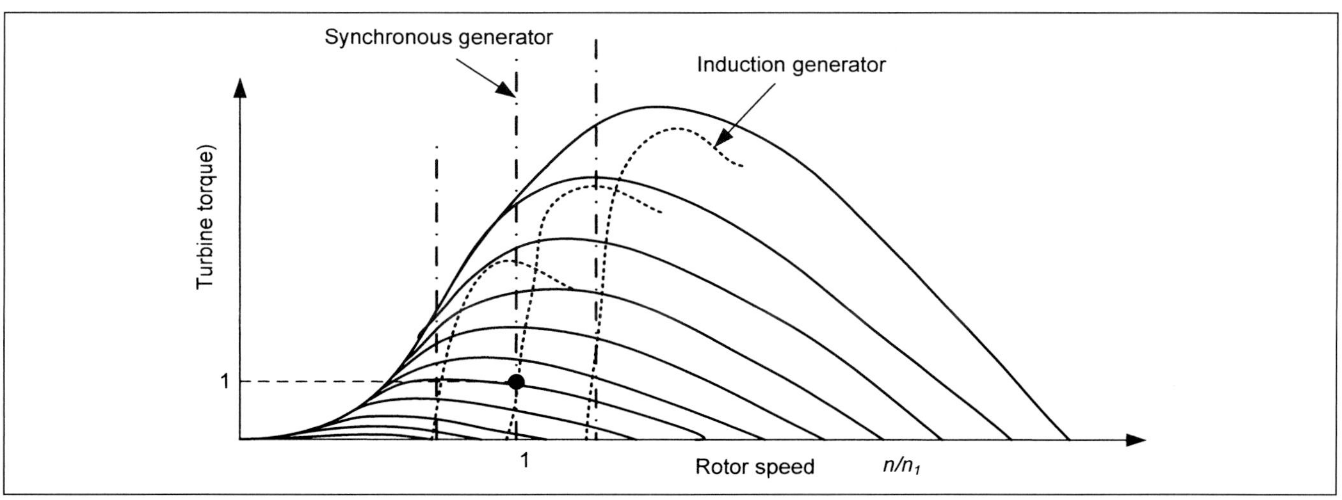

FIGURE 24-12 Wind turbine torque-speed characteristics by variation of generator frequency, superimposed on wind-power curves.

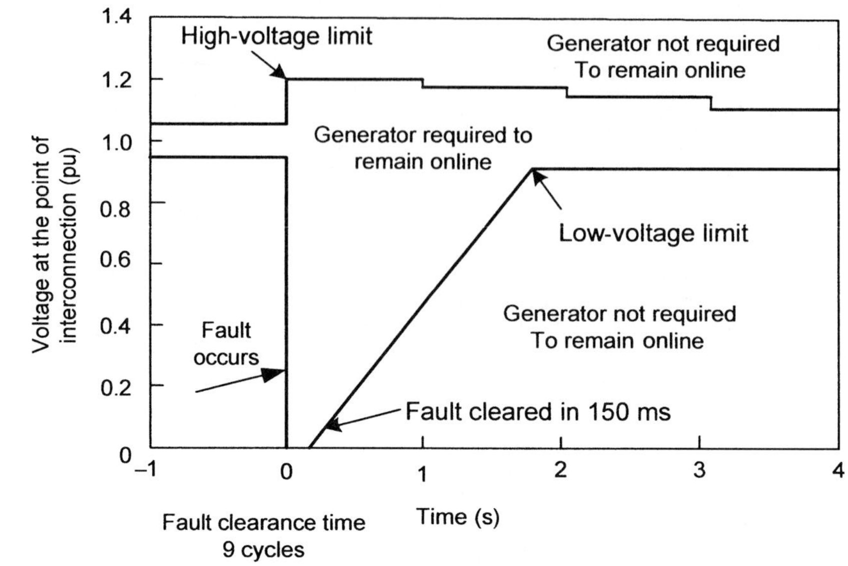

FIGURE 24-13 Proposed WECC voltage ride-through requirements of wind generators.

The induction generator must draw its reactive power requirement from the grid source. When capacitors, SVCs, and rotary phase shifters are connected, the operational capabilities can be equivalent with synchronous machines, though resonance with grid inductance is a possibility. Induction generators produce harmonic and synchronous pulsating torques, akin to induction motors (Chap. 11). A synchronous machine provides control of operating conditions, leading or lagging by excitation control. With respect to interconnection with the grid, the schemes exist that are discussed in Secs. 24-8-1 through 24-8-3.

24-8-1 Direct Coupled Induction Generator

The direct coupled induction machine is generally of four-pole type; a gear box transforms the rotor speed to a higher speed for generator operation above synchronous speed. It requires reactive power from grid or ancillary sources, and starting after a blackout may be a problem. Wind dependent power surges produce voltage drops and flicker. The connection to the grid is made through thyristor switches which are bypassed after start. A wound-rotor machine has the capability of adjusting the slip and torque characteristics by inserting resistors in the rotor circuit, and the slip can be increased at an expense of more losses and heavier weight (Fig. 24-14*a*). The system will not meet the current regulations of connection to grid and may be acceptable for isolated systems.

24-8-2 Induction Generator Connected to Grid through Full-Size Converter

The induction generator is connected to the grid through two back-to-back voltage source converters. Because of the full-power rating of the inverter, the cost of electronics is high. The wind-dependent power spikes are damped by the dc link. The grid-side inverter need not be switched in and out so frequently and harmonic pollution occurs.

24-8-3 Doubly Fed Induction Machine (DFIM)

The stator of the induction machine is directly connected to the grid, while the rotor is connected through voltage source converter (Fig. 24-14*b*). The energy flow over the converter in the rotor circuit is bidirectional. In subsynchronous mode, the energy flows to the rotor, and in supersynchronous mode, it flows from rotor to the grid. The ratings of the converter are much reduced, generally one-third of the full power, and depend on the speed range of the turbine. The power rating is:

$$P = P_s \pm P_r \tag{24-13}$$

where P_s and P_r are the stator and rotor powers. But the rotor has only the slip frequency induced in its windings, therefore, we can write:

$$P_r = P_a \times s \tag{24-14}$$

where s is the slip and P_a is rotor air gap power. For a speed range of ± 30 percent, the slip is ± 0.3, and a third of converter power is required. Also we can write:

$$n_s = \frac{f_r \pm f}{p} 120 \tag{24-15}$$

where p is the number of pairs of poles.

Synchronous generators can be brush type or brushless type of permanent magnet excitation systems. These are also connected to the grid much like asynchronous machines. The excitation power has to be drawn from the source, unless the generator is of permanent magnet type. Figures 24-14*c* and *d* show typical connections.

24-9 POWER ELECTRONICS

The entire output of variable voltage, variable frequency output of a wind generator, sometimes called the wild ac, is converted to direct current, which is then converted to utility quality ac power. The frequency converters condition the electrical energy from the wind generators and dampen the influence on the grid connections. With respect to the topology of the electronic devices, we have:

- Current-controlled rectifier
- Voltage source rectifier
- Voltage source inverters
- Current source inverters

The rectifiers may be (1) uncontrolled, (2) bridges with dc/dc regulators, and (3) controlled rectifiers. Again, pulse-controlled

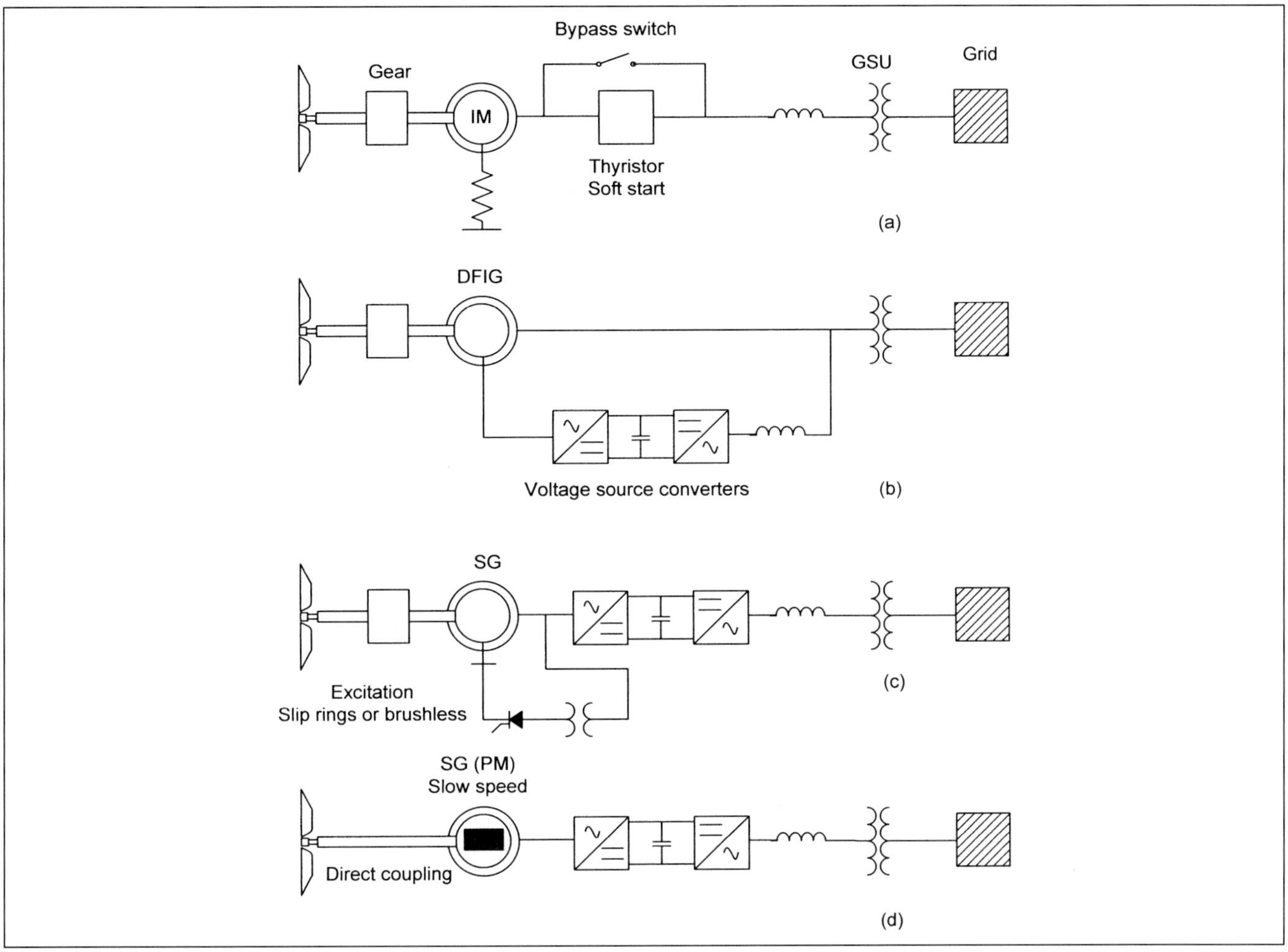

FIGURE 24-14 Grid connections of wind generators. (*a*) Direct connection of an induction generator, stall regulated. (*b*) Connections of a DFIG, variable-speed generator with pitch regulation. (*c*) Synchronous generator, brush-type or brushless with voltage source converters, pitch regulated. (*d*) Gearless connection of a low-speed permanent magnet generator.

IGBT topology with phase multiplication to reduce harmonic generation is the preferred choice, as active and reactive control of power can be exercised (Chap. 15).

Consider a synchronous generator grid connection. The constant voltage dc link is supplied by a controlled rectifier bridge, and the rectifier is current controlled so that magnitude and phase angle of the generator current is controlled by triggering of the rectifier. By phase shifting the generator current in underexcited and overexcited regions, the generator voltage can be controlled and matched to the dc link.

Short-circuit current calculations according to machines integrated in industrial systems may not always be valid. For example, in a DFIM, the stator current for a nearby fault may be limited to nearly rated current if rotor power converter remains active. It may, however, be disabled by crowbar circuit for protection during fault, in which case the fault current will be several times the rated curremt for a few cycles.

Switching of devices in converters gives rise to interference emission over a wide spectrum. The electromagnetic compatibility should be ensured according to relevant standards.

24-9-1 Reactive Power Control

Apart from the reactive power need of the induction machine itself, the passive connecting element consumes reactive power. These elements are transformers and cable connections to the point of grid. Then the reactive power is required by loads. Problems of transient low voltages can occur when the wind power generation is connected to relatively weak grid systems.

Some wind-generating plants have added SVCs, DSTATCOMs, and STATCOMs (Chap. 15), which control the power factor to unity at the point of interconnection. The Argonne Mesa wind plant in New Mexico has a DSTATCOM, which controls the power factor to utility at the point of interconnection at Guadalupe 345-kV station bus. Four mechanically switched capacitor banks are located in the collector substation some 2 mi away from the interconnect substation. The DSTATCOM controls determine the required reactive power output based on voltage and current measurements at 345-kV collector bus.

Figure 24-15 shows the impact of high winds on the 1.5-MW generators, connected to a 230-kV transmission line. Line-drop compensation algorithms are utilized to synthesize voltage at the point of interconnection, located approximately 75 km from the wind plant. The flicker index of the voltage is less than 2 percent at the point of interconnection. In spite of considerable variation in the wind speed, the plant output is relatively stable.

Figure 24-16 shows field tests results on an active power regulator and power rate limiter on an operating 30-MW wind plant. Initially, the output is curtailed to 10 MW, and during the tests the active power command is raised in four 5-MW increments. The transition between each step ramp-rate is controlled to 2 MW/min.

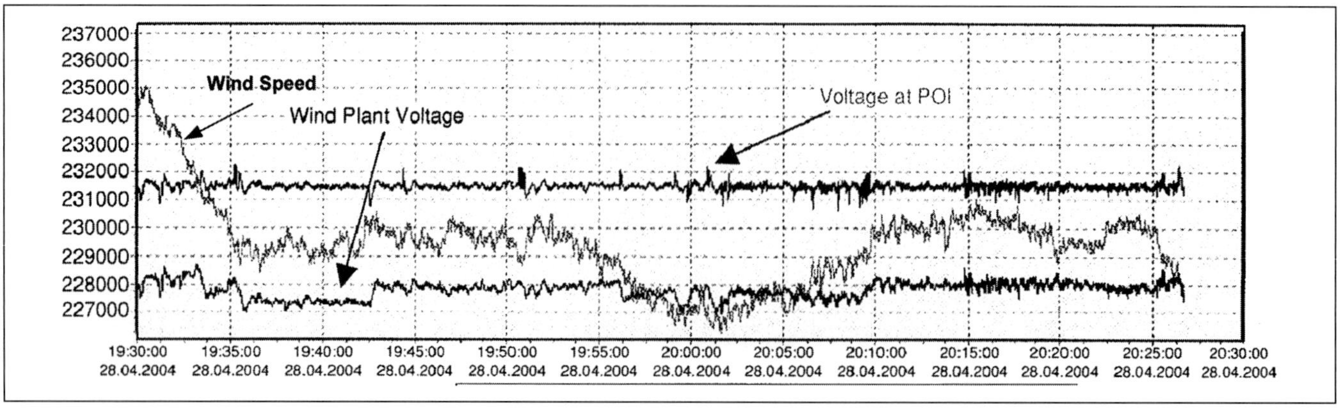

FIGURE 24-15 Wind-plant voltage response and regulation at the point of interconnection.

24-9-2 Harmonics

Generation of harmonics and mitigation is discussed in Chaps. 6 and 15. Even harmonics in wind generation can arise due to unsymmetrical half waves and may appear at fast load changes. Subharmonics can be produced due to periodical switching with variable frequency. Interharmonics can be generated when the frequency is not synchronized to the fundamental frequency, which may happen at low- and high-frequency switching.

The interharmonics due to back-to-back configuration of two converters can be calculated according to IEC.[6]

$$f_{n,m} = [(p_1 k_1) \pm 1] f_1 \pm (p_2 k_2) F \tag{24-16}$$

where $f_{n,m}$ is the interharmonics frequency, f_1 input frequency, F is output frequency, p_1 and p_2 are pulse numbers of the two converters, respectively. Interharmonics are also generated due to speed-dependent frequency conversion between rotor and stator of DFIM and as side bands of characteristic harmonics of PWM converters. Noncharacteristic harmonics can be generated due to grid unbalance.

24-10 COMPUTER MODELING

Modeling is required in the planning stage to conduct interconnection studies, to establish grid reliability, and to simulate energy capture for hybrid plants. A forecast is required for the wind power generation energy capability. This requires wind density probability curves and history of wind samples at the location of wind generation. Also simulation is required for aerodynamics, mechanical dynamics, and structures—this gives ideas of static and dynamic loads, predicted power curves, vibration modes, and control system response. Electrical transients in generators and power electronics need to be simulated. Excess starting currents, behavior under short-circuit and under voltage transients, and isolation from grid can be studied in a time frame varying from a few microseconds to several seconds. GE PSLF/PSDS and Siemens PTI/PSSE programs are designed for study of large-scale interconnected systems, yet there is not much sharing

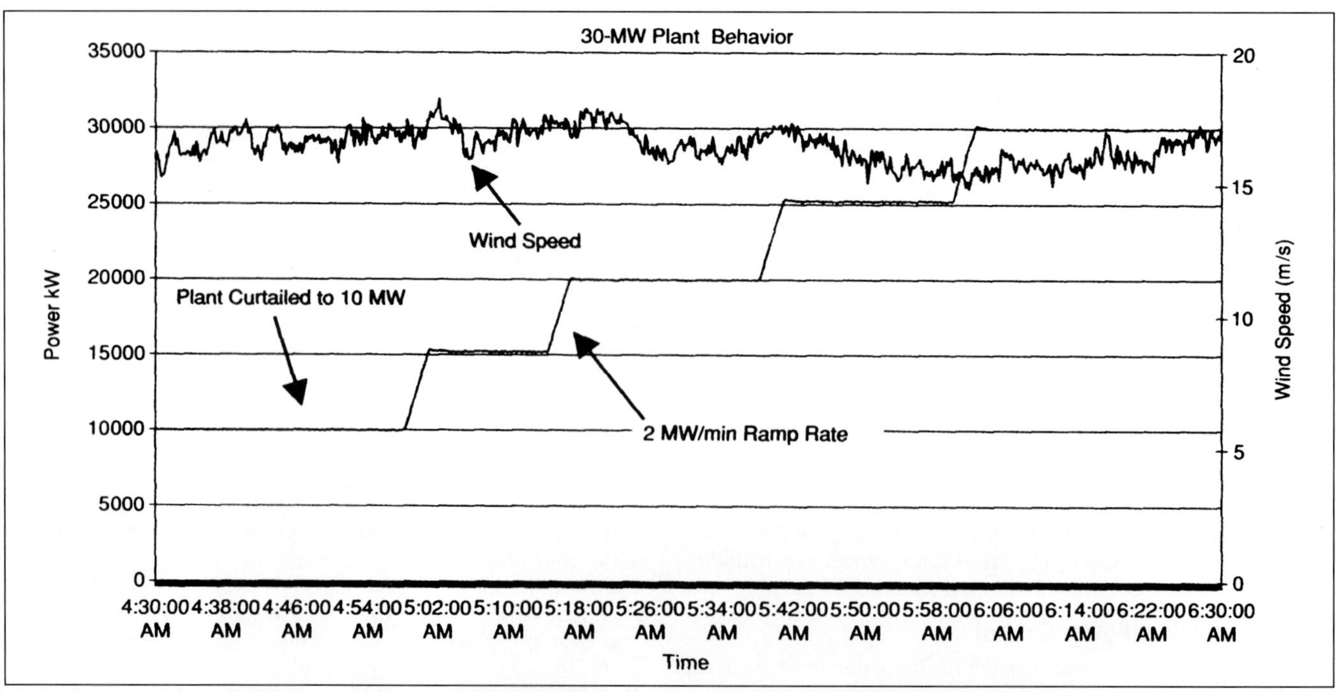

FIGURE 24-16 Active power response of a wind plant with ramp control.

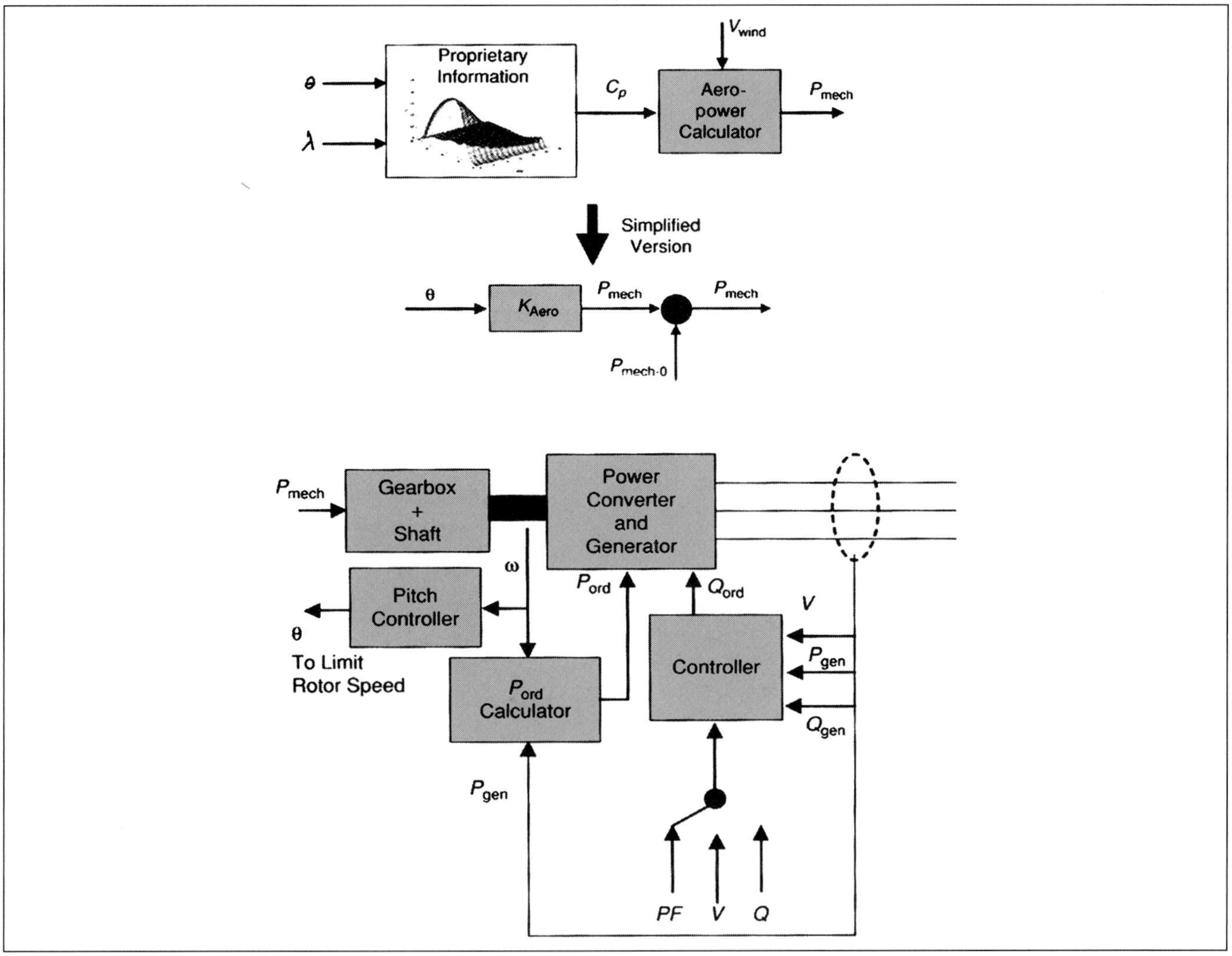

FIGURE 24-17 Simplified block circuit diagram of aerodynamic and mechanical systems for modeling.

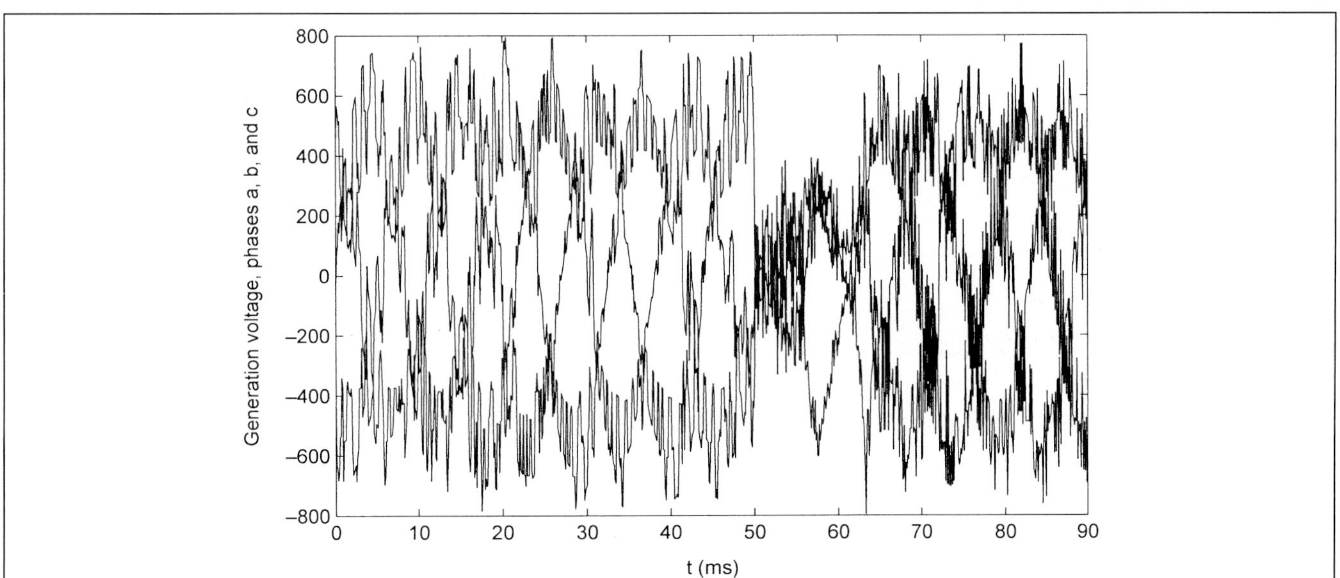

FIGURE 24-18 Output generated voltage of a 2-MW wind-generating DFIG unit, three-phase fault at 50 ms, cleared in 6 cycles, EMTP simulation.

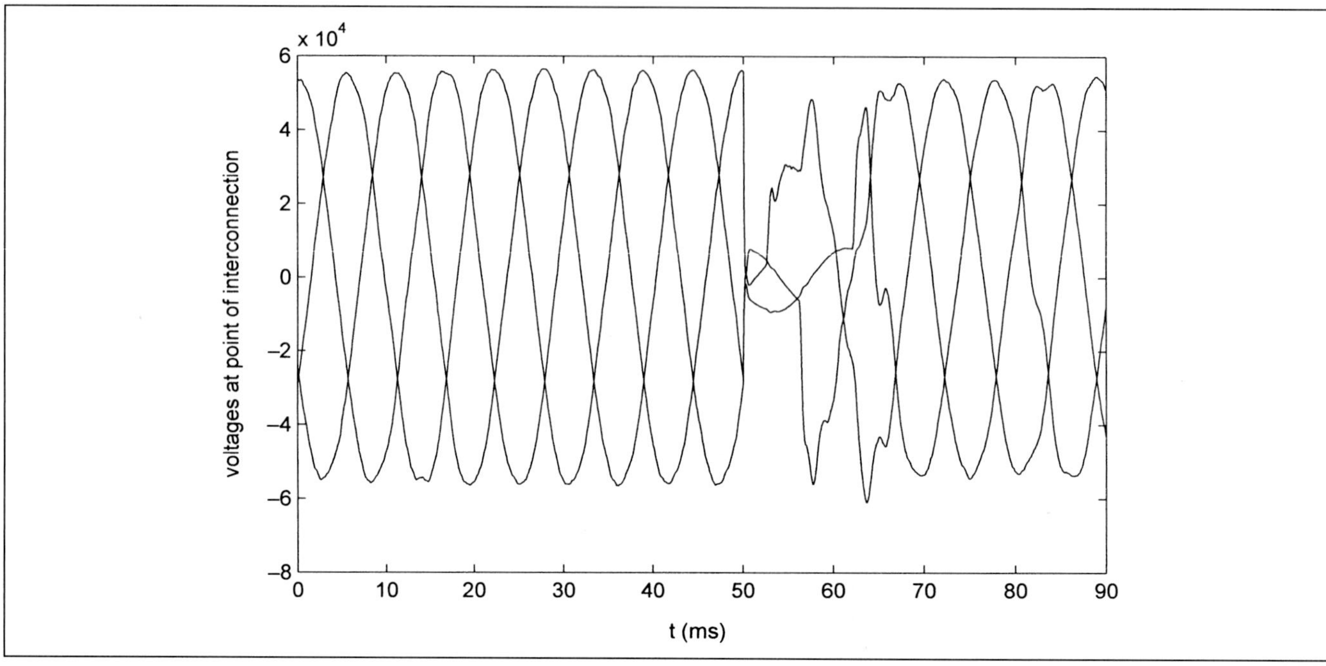

FIGURE 24-19 Voltages at the point of interconnection.

of the data between consultants, utilities, and power planners. Engineering design models are implemented in three-phase simulation programs like EMTP and PSCAD. A simplification of turbine aerodynamic and mechanical systems models is depicted in Fig. 24-17.

Example 24-1 This example is a simulation of transients using EMTP for a wind-generation plant of 22 MW, 2-MW DFIM units, connected to utility system at 69 kV, through a GSU of 26 MVA. A three-phase fault occurs on the 69-KV side of the GSU transformer at 50 ms, and cleared at 51 ms, fault duration 1 ms (= 6 cycles).

Figure 24-18 shows the generated voltage. Figure 24-19 depicts voltage at the point of interconnection. Figure 24-20 shows the stator and rotor currents in a DFIM, and also electrical power.

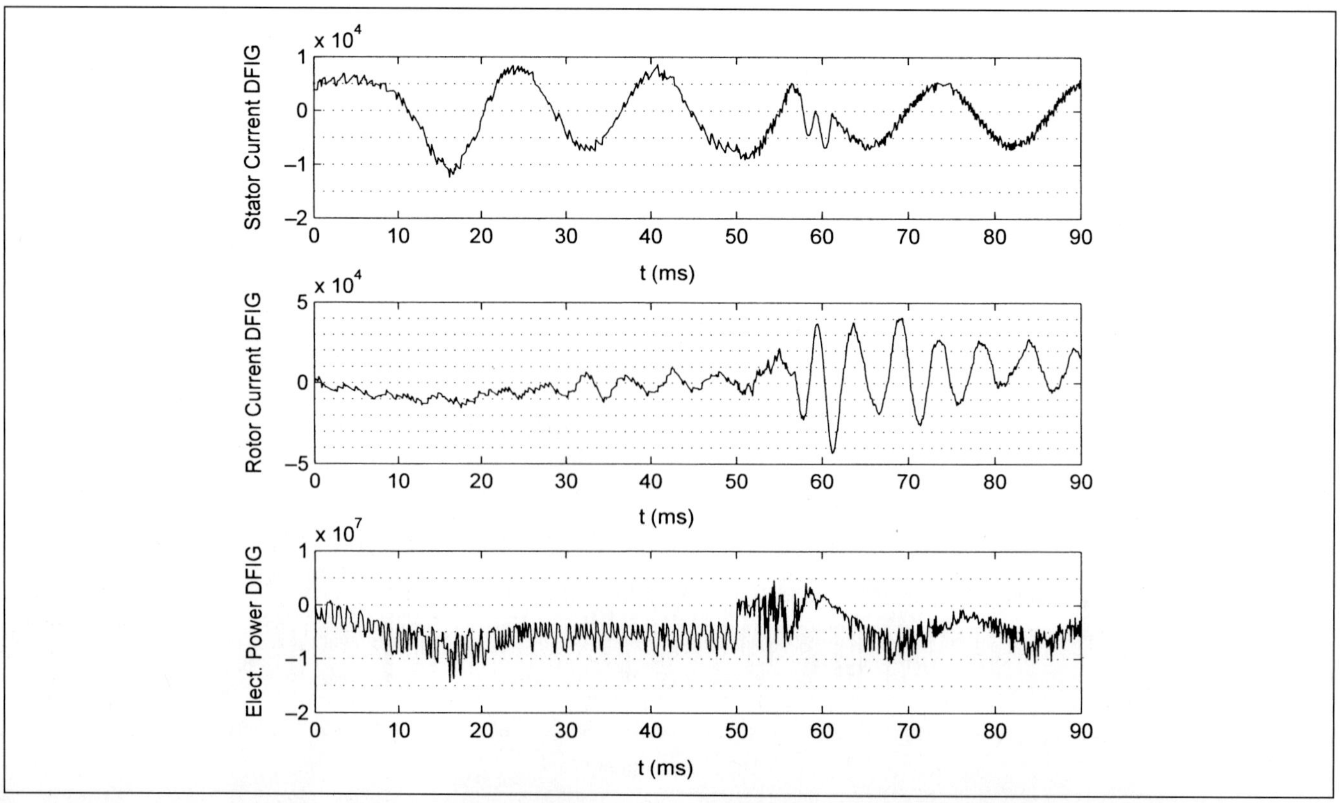

FIGURE 24-20 DFIG stator and rotor currents and electrical power output.

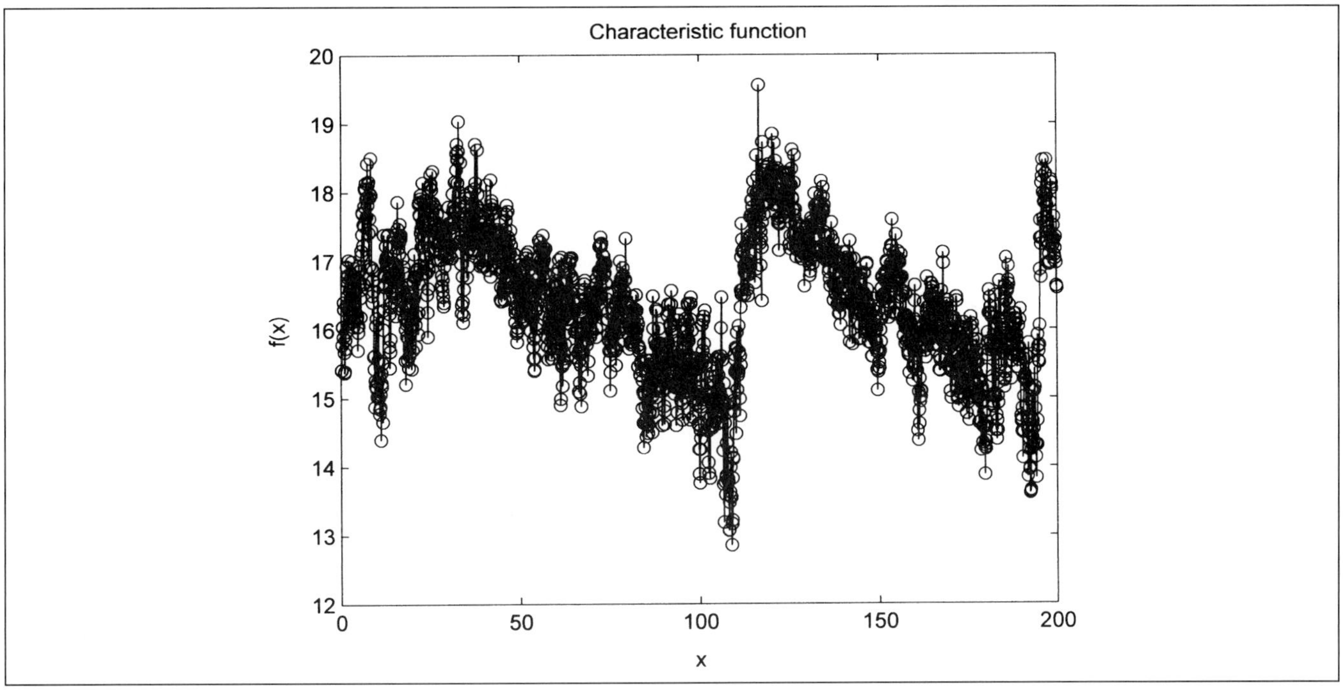

FIGURE 24-21 Real wind speed function, 2000 points, 200 intervals.

Figure 24-21 illustrates representation of wind speed function, 2000 sample points, interval 0 to 200.

24-11 FLOATING WIND TURBINES

The wind patterns change with height above ground. At a height of approximately 1000 ft above ground, the winds are comparatively steady and power can be produced 24 hours a day. It is not practical to construct 1000-ft high towers. A video of an innovative idea of helium-filled, lighter-than-air, floating wind turbines and development of a successful prototype is at Ref. 7. The construction of a successful prototype describes a number of hurdles, such as design of blades, testing in wind tunnels, stability as the turbine climbs through the wind turbulence around 300-ft above ground, the innovations in design of a tether with copper-core cable to transmit generated power to ground, the strength of tether to support tons of weight to withstand forces which the floating turbine will experience during the climb-up. Some millions of floating turbines can meet all the energy requirements of the world, and set back the clock on global warming. It may not be a commercial viability, but is indicative of necessity-based ingenuity in the process of development of new technologies for better life on earth.

REFERENCES

1. IEEE Std. C37.118, IEEE Standard for Synchrophasors for Power Systems, 2005.
2. IEEE Std. C37.111, IEEE Standard for Common Format for Transient Data Exchange (COMTRADE) for Power Systems, 1999.
3. J. C. Das, "Power System Relaying" in *Wiley Encyclopedia of Electrical and Electronic Engineers*, vol. 17, pp. 71–86, 2002.
4. The following reports can be found at: www.nrel.gov/publications/ D. A. Griffen, WindPACT Turbine Design Scaling Studies Technical Area 1—Composite Blades for 80–120 m Rotor, 21 Mar. 2000–15 Mar. 2001; NREL Rep. SR-500-29492; G. Bywaters, V. John, J. Lynch, P. Mattila, G. Norton, J. Stowell, M. Salata, O. Labath, A. Chertok, and D. Hablanian, Northern Power Systems WindPact Drive Train Alternative Design Study Report: Period of Performance, Apr. 12, 2001–Jan. 31, 2005; NREL Rep. SR-500-35524, 2004; M. W. LaNier, LWST Phase 1 Conceptual Design Study: Evaluation of Design and Construction Approaches for Economical Hybrid Steel/ Concrete Wind Turbine Towers, Jun. 28, 2002–Jul. 31, 2004; NREL Rep. SR-500-36777.
5. Dynamic Models for Wind Farms for Power System Studies, http://www.energy.sintef.no/wind/IEA.asp.
6. IEC 61000-2-4, Electromagnetic Compatibility, Part 2. Environmental Section 4: Compatibility Levels in Industrial Plants for Low-Frequency Conducted Disturbances.
7. Discovery Broadcast Channel—Green Earth-Infinite Winds, www.discovery.com.

FURTHER READING

E. A. DeMeo, W. Grant, M. R. Milligan, and M. J. Schuerger, "Wind Power Integration," *IEEE Power Energy Magazine*, vol. 3, no. 6, pp. 38–46, 2005.

EPA, Renewable Portfolio Standards Fact Sheet, http://www.epa.gov/chp/state-policy/renewable_fs.html.

GE Energy and AWS Truewind, Ontario Wind Integration Study, http://www.ieso.ca/imoweb/pubs/marketreports/OPA-Report-200610-1.pdf.

S. Heier, *Grid Integration of Wind Energy Conversion Systems*, 2d ed., John Wiley, New York, 2009.

IEC-61400-21, Wind Turbine Generator Systems. Part 21: Measurement and Assessment of Power Quality Characteristics of Grid Connected Wind Turbines, 2001.

IEEE Std. 1001-1988, Guide for Interfacing Dispersed Storage and Generation Facilities with Electric Facility Systems, 1988.

North American SynchroPhasors Initiative (NASPI), http://www.naspi.org.

A. G. Phadke, J. S. Throp, and M. G Adamiak, "A New Measurement Technique of Tracking Voltage Phasors, Local System Frequency, and Rate of Change of Frequency," *IEEE Trans. PAS*, vol. 102, no. 5, pp. 1025–1038, May 1983.

R. Strzelecki and G. Benysek, eds., *Power Electronics in Smart Electrical Energy Networks*, Springer-verlag, London, 2008.

UWIG Modeling User Group, Dynamic Model Validation for the GE Wind Turbine, www.uwig.org.

Western Electricity Coordinating Council Disturbance Monitoring Reports, www.wecc.biz.

APPENDIX A

DIFFERENTIAL EQUATIONS

Differential equations form a powerful analytical tool in electrical engineering. A fundamental knowledge of differential equations is a prerequisite for understanding transients. This appendix provides an overview.

An equation involving differential coefficients is called a differential equation. An example is:

$$L\frac{d^2q}{dt^2}+R\frac{dq}{dt}+\frac{q}{c}=E\sin\omega t \tag{A-1}$$

This is an equation of the second order and first degree. The order of the differential equation is the same as the order of the highest differential coefficient present in the equation. The degree pertains to the exponential of the highest differential. For example, the following equation has an order of two and a degree of three:

$$\left[\frac{d^2y}{dx^2}\right]^3+y\left[\frac{dy}{dx}\right]^4+y^4=0 \tag{A-2}$$

In general, the following equation:

$$a_0\left[\frac{d^nv}{dt^n}\right]^n+a_1\left[\frac{d^{n-1}v}{dt^{n-1}}\right]^{n+1}+\cdots+a_{n-1}\left[\frac{dv}{dt}\right]+a_nv=y(t) \tag{A-3}$$

is an equation of the nth order and nth degree. Note that the exponential of the lower derivative can be higher than the exponential of the nth derivative.

In this equation $a_0, a_1, \ldots, a_n$ are the constants, $v(t)$ is the dependent variable, t is the time and is an independent variable, and $y(t)$ is the forcing function, sometimes called the excitation. Differentiating the ordinary equations and eliminating the arbitrary constants can form the differential equation. Consider an equation of the form:

$$y=A\cos x+B\sin x \tag{A-4}$$

Differentiating it twice gives:

$$\frac{dy}{dx}=-A\sin x+B\cos x$$
$$\frac{d^2y}{dx^2}=-(A\cos x+B\sin x) \tag{A-5}$$

Thus:

$$\frac{d^2y}{dx^2}=-y \quad \text{or} \quad \frac{d^2y}{dx^2}+y=0 \tag{A-6}$$

This is a second-order equation obtained by eliminating the two constants A and B in the original equation. The original equation forms a *solution* of the differential equation arrived in Eq. (A-6).

■ It can be generalized that the order of the differential equation obtained will be equal to the constants in the original equation.

A-1 HOMOGENEOUS DIFFERENTIAL EQUATIONS

A differential equation of the form:

$$\frac{dy}{dx}=\frac{h(x,y)}{f(x,y)} \tag{A-7}$$

is a homogeneous equation if $h(x, y)$ and $f(x, y)$ are of the same degree. Therefore, the following equation is a homogeneous equation:

$$(y^2-xy)\,dx+x^2dy=0 \tag{A-8}$$

Because, by rearranging, Eq. (A-8) can be written as

$$\frac{dy}{dx}=\frac{xy-y^2}{x^2} \tag{A-9}$$

The criterion in Eq. (A-7) is satisfied. Its solution can be found by substituting as follows:

$$y=vx$$
$$\frac{dy}{dx}=v+x\frac{dv}{dx} \tag{A-10}$$

By equating and simplifying, the original equation becomes:

$$x\frac{dv}{dx}=-v^2 \tag{A-11}$$

or

$$-\frac{dv}{v^2}=\frac{dx}{x} \tag{A-12}$$

Integrating and reverse substituting, the solution is:

$$\frac{x}{y}=\log x+C \tag{A-13}$$

where C is a constant.

A-2 LINEAR DIFFERENTIAL EQUATIONS

Consider a nonhomogeneous equation written as:

$$\frac{dy}{dx}+Py=Q \qquad \text{(A-14)}$$

where P and Q are functions of x or are constants. P and Q cannot be the functions of y. The Eq. (A-14) is called a *linear differential equation*. Its solution can be found by multiplying both sides by:

$$e^{\int Pdx} \qquad \text{(A-15)}$$

Thus, Eq. (A-14) becomes:

$$e^{\int Pdx}\left(\frac{dy}{dx}+Py\right)=Qe^{\int Pdx}$$

Therefore:

$$\frac{d}{dx}\left[ye^{\int Pdx}\right]=Qe^{\int Pdx}$$

Integrating both sides

$$ye^{\int Pdx}=\int(Qe^{\int Pdx})dx+C \qquad \text{(A-16)}$$

This is the required solution. The factor $e^{\int Pdx}$ in Eq. (A-16) is called the *integrating factor*, abbreviated as IF. The solution can therefore be written in the following form:

$$y\text{IF}=\int Q\text{IF}\,dx+C \qquad \text{(A-17)}$$

Example A-1 Solve the following equation:

$$(x^2+5)\frac{dy}{dx}+4xy=x$$

Here IF is:

$$e^{\int[(4x/x^2+5)dx]}=e^{2\log(x^2+5)}=e^{\log(x^2+5)^2}=(x^2+5)^2$$

Thus, the solution can be written as:

$$y(x^2+5)^2=\int\left(\frac{x}{x^2+5}\right)(x^2+5)^2dx+C$$

$$=\frac{x^4}{4}+\frac{5x^2}{2}+C$$

A-3 BERNOULLI'S EQUATION

Sometimes it is possible to reduce a differential equation to the form in Eq. (A-14) by simple substitution. Consider the equation of the form:

$$\frac{dy}{dx}+Py+Qy^n \qquad \text{(A-18)}$$

where P and Q are constants or function of x, but not of y as before. This can be reduced to linear form by dividing by y^n and then substituting the following expression:

$$\frac{1}{y^{n-1}}=z$$

Therefore:

$$\frac{1-n}{y^n}\frac{dy}{dx}=\frac{dz}{dx}$$

With this transformation Eq. (A-18) becomes:

$$\frac{dz}{dy}+P(1-n)z=Q(1-n) \qquad \text{(A-19)}$$

This is a linear equation and can be solved as before.

Example A-2 Solve:

$$\tan y\frac{dy}{dx}+\tan x=\cos y\cos^2 x$$

The equation can be written as:

$$\sec y\tan y\frac{dy}{dx}+\sec y\tan y=\cos^2 x$$

By substituting $z=\sec y$, and differentiating:

$$\frac{dz}{dx}=\sec y\tan y\frac{dy}{dx}$$

Thus, the original equation reduces to:

$$\frac{dz}{dx}+z\tan x=\cos^2 x$$

Its solution is:

$$\sec y=(\sin x+C)\cos x$$

A-4 EXACT DIFFERENTIAL EQUATIONS

An equation of the form:

$$Mdx+Ndy=0 \qquad \text{(A-20)}$$

is said to be an exact differential equation, if:

$$\frac{\partial M}{\partial y}=\frac{\partial N}{\partial x} \qquad \text{(A-21)}$$

where:

$$\frac{\partial M}{\partial y}$$

is the partial differential coefficient of M with respect to y, keeping x constant; similarly:

$$\frac{\partial N}{\partial x}$$

is the partial differential coefficient of N with respect to x keeping y constant. The exact differential equations can be solved in the following three steps:

1. Integrate M with respect to x keeping y constant.
2. Integrate with respect to y, only the terms in N which are devoid of x.
3. The solution is given by the sum of the results of Steps 1 and 2.

Example A-3 Test for exactness and solve the following equation:

$$(x^4-2xy^2+y^4)dx-(2x^2y-4xy^3+\cos y)dy=0$$

Here:

$$M=(x^4-2xy^2+y^4)$$

$$N=(2x^2y-4xy^3+\cos y)$$

This satisfies the criteria in Eq. (A-21) and is therefore an exact equation. The solution is given by:

$$\int (x^4 - 2xy^2 + y^4)dx + \int -\cos y dy = C$$

Thus, the solution is:

$$\frac{x^5}{5} - x^2y^2 + xy^4 - \sin y = C$$

A-5 CLAIRAUT'S EQUATION

The equation:

$$y = px + f(p) \tag{A-22}$$

where:

$$p = \frac{dy}{dx}$$

is called the *Clairaut's equation*. Differentiating with respect to x:

$$\frac{dy}{dx} = p + x\frac{dp}{dx} + f(p)\frac{dp}{dx}$$

However, $dy/dx = p$. Substituting and rearranging:

$$[x + f'(p)]\frac{dp}{dx} = 0 \qquad \text{or} \qquad \frac{dp}{dx} = 0 \qquad \text{Thus, } p = a$$

Then:

$$y = ax + f(a) \tag{A-23}$$

is the required solution.

A-6 COMPLEMENTARY FUNCTION AND PARTICULAR INTEGRAL

The solution of the differential equation in Eq. (A-14) can be written as summation of complementary function and particular integral. Let us revisit the solution of Eq. (A-14) derived in Eq. (A-16). Equation (A-14) can be written as:

$$ye^{\int Pdx} = \int (Qe^{\int Pdx})\,dx + C$$

or:

$$y = Ce^{-\int Pdx} + e^{-\int Pdx}\int (Qe^{\int Pdx})\,dx$$

$$y = Cu + v$$

where:

$$u = e^{-\int Pdx} \qquad \text{and} \qquad v = e^{-\int Pdx}\int Qe^{\int Pdx}\,dx$$

Differentiate:

$$u = e^{-\int Pdx}$$

with respect to x. Then:

$$\frac{du}{dx} = -Pe^{-\int Pdx} = -Pu$$

$$\frac{du}{dx} + Pu = 0$$

or

$$\frac{d(Cu)}{dx} + P(Cu) = 0$$

This shows that $y = Cu$ is the solution of:

$$\frac{dy}{dx} + Py = Q \tag{A-24}$$

■ In general, the complementary function is the solution of the differential equation with Q set to zero. The solution will have constants which are equal to the degree of the equation, and these constants are evaluated from the initial conditions. The solution presents a *steady state* when the transients have decayed to zero. This conclusion will be amplified with examples to follow.

Differentiate:

$$v = e^{-\int Pdx}\int (Qe^{\int dx})dx$$

with respect to x:

$$\frac{dv}{dx} = -Pe^{-\int Pdx}\int (Qe^{\int Pdx})\,dx + e^{-\int Pdx}Qe^{\int Pdx}$$

or:

$$\frac{dv}{dx} = -Pv + Q$$

$$\frac{dv}{dx} + Pv = Q \tag{A-25}$$

This shows that $y = v$ is the solution of the original differential equation [Eq. (A-14)].

■ In general, the particular integral does not contain a constant and represents the *transient response* of the system represented by the differential equation. The complete solution is the sum of complementary function and particular integral. We will apply these concepts to the solution of second-order differential equations, which occur very frequently in electrical systems.

A-7 FORCED AND FREE RESPONSE

Let us introduce here the ideas of free and forced response. Consider a differential equation of the form:

$$\sum_{i=0}^{n} a_i \frac{d^i y}{dt^i} = \sum_{i=0}^{m} b_i \frac{d^i u}{dt^i} \tag{A-26}$$

where $u = u(t)$ is the known input and $y = y(t)$ the unknown output. The free response of a differential equation is the solution of the differential equation when the input $u(t)$ is identically zero. Then the differential equation reduces to:

$$\sum_{i=0}^{n} a_i \frac{d^i y}{dt^i} = 0 \tag{A-27}$$

The forced response of a differential equation is the solution of the differential equation when all the initial conditions are identically zero; that is:

$$y(o), \left.\frac{dy}{dt}\right|_{t=0}, \ldots, \left.\frac{d^{n-1}y}{dt^{n-1}}\right|_{t=0} \tag{A-28}$$

This means that the forced response depends only on the input $u(t)$. For a constant coefficient ordinary differential equation the forced response can be written in terms of a convolution integral:

$$y_b(t)=\int_0^t w(t-\tau)\left[\sum_{i=0}^{m} b_i \frac{d^i u(\tau)}{d\tau^i}\right] d\tau \tag{A-29}$$

here $w(t-\tau)$ is the weighting function, also called the *Kernel* of the differential equation. This form of convolution integral describes a casual system. Note that $w(t)$ is a free response of the differential equation and therefore requires n initial conditions for complete specifications. These initial conditions fix the values of the constants $C_1, C_2 \ldots, C_n$.

■ The total response is the sum of the free response and forced response. We also use the terms steady-state response and transient response. The *steady-state response* is that part of the total response which does not approach zero as the time approaches infinity. The *transient response* is that part of the total response which approaches zero as the time approaches infinity.

A-8 LINEAR DIFFERENTIAL EQUATIONS OF THE SECOND ORDER (WITH CONSTANT COEFFICIENTS)

The general form of the equation is:

$$\frac{d^2y}{dx^2}+P\frac{dy}{dx}+Qy=R \tag{A-30}$$

where P and Q are constants and R is a function of x or constant. Define a differential operator D, so that:

$$Dy=\frac{dy}{dx} \qquad D^2y=\frac{d^2y}{dx} \tag{A-31}$$

Then the complementary function is found by putting $R = 0$ in Eq. (A-25):

$$D^2y+PDy+Qy=0$$

Assume that:

$$y=C_1e^{mx}$$

is the complementary function of Eq. (A-25). Then:

$$C_1e^{mx}(m^2+Pm+Q)=0$$

The equation:

$$m^2+Pm+Q=0 \tag{A-32}$$

is called the *auxiliary equation*.

A-9 CALCULATION OF COMPLEMENTARY FUNCTION

Let us solve the Eq. (A-32). Three distinct cases arise:

Case 1: The equation has real and different roots. The complementary function is:

$$y=C_1e^{m_1x}+C_2e^{m_2x} \tag{A-33}$$

Case 2: The equation has real and equal roots. In this case, we can write:

$$(D-m)(D-m)\,y=0$$

Substitute:

$$(D-m)y=v$$

then:

$$(D-m)v=0 \quad \text{or} \quad v=C_1e^{mx}$$

Therefore, we can write:

$$(D-m)y=C_1e^{mx}$$

This can be written as:

$$\frac{dy}{dx}-my=C_1e^{mx}$$

The solution is:

$$\begin{aligned} ye^{-mx} &= \int (C_1e^{mx})e^{-mx}dx+C_2 \\ &=C_1x+C_2 \end{aligned} \tag{A-34}$$

Thus, the complementary function is:

$$y=(C_1+C_2x)e^{mx} \tag{A-35}$$

Case 3: The equation has imaginary roots. Let the roots be $\alpha \pm i\beta$. Then the solution will be:

$$\begin{aligned} y&=C_1e^{(\alpha+i\beta)x}+C_2e^{(\alpha-i\beta)x} \\ &=e^{\alpha x}[C_1e^{i\beta x}+C_2e^{-i\beta x}] \\ &=e^{\alpha x}[C_1(\cos\beta x+i\sin\beta x)+C_2(\cos\beta x-i\sin\beta x)] \\ &=e^{\alpha x}[(C_1+C_2)\cos\beta x+i(C_1-C_2)\sin\beta x] \end{aligned} \tag{A-36}$$

Equation (A-36) can be written as:

$$y=e^{\alpha x}[A\cos\beta x+B\sin\beta x] \tag{A-37}$$

Example A-4 Solve the differential equation:

$$\frac{d^2y}{dx^2}-6\frac{dy}{dx}+9y=0$$

The auxiliary equation is:

$$D^2y-6Dy+9=0$$

The roots are real and equal to 3,3. Therefore, the solution is:

$$y=(C_1+C_2x)e^{3x} \tag{A-38}$$

The constants C_1 and C_2 can be found from the initial conditions. If the given initial conditions are: $y(0) = 0$ and $y'(0) = 3$, then substituting $y = 0$ and $x = 0$ in Eq. (A-38):

$$C_1=0 \quad \text{and} \quad y=C_2xe^{3x}$$

On differentiating:

$$y'=C_2e^{3x}+3C_2xe^{3x}$$

Substituting $x = 0$ and $y' = 3$ in the above equation, $C_2 = 3$. Therefore, the solution satisfying the given conditions is:

$$y=3xe^{3x}$$

Example A-5 Solve the differential equation:

$$\frac{d^2y}{dx^2}+6y+10=0$$

for $y(0)=2$ and $y'(0)=y''(0)$

The roots of the auxiliary equation are $3 \pm i$, therefore, the complementary function is:

$$y=e^{-3x}(A\cos x+B\sin x)$$

$y = 2$, at $x = 0$, therefore, substituting, $A = 2$

$$y=e^{-3x}(2\cos x+B\sin x)$$

Differentiating:

$$\frac{dy}{dx}=e^{-3x}[-(3B+2)\sin x+(B-6)\cos x]$$

Differentiating again:

$$\frac{d^2y}{dx^2}=e^{-3x}[(-6B+16)\cos x+(8B+12)\sin x]$$

Equating $y' = y''$ at $x = 0$, we get $B = 3.33$. Thus, the solution is:

$$y=e^{-3x}(2\cos x+3.33\sin x)$$

A-10 HIGHER-ORDER EQUATIONS

The complementary function of higher-order equations can be similarly found. Consider nth-order auxiliary equation:

$$a_0D^n+a_1D^{n-1}+\cdots+a_{n-1}D+a_n=0 \tag{A-39}$$

It has n roots, namely $m_1, m_2, m_3 \ldots, m_n$; some of them can be equal, imaginary, or real:

$$a_0(D-m_1)(D-m_2)\cdots(D-m_n)=0 \tag{A-40}$$

Then, its complementary function can be written based on the relations already derived.

Example A-6 Find the solution of the differential equation:

$$\frac{d^4y}{dx^4}+8\frac{d^3y}{dx^3}+25\frac{d^2y}{dx^2}+36\frac{dy}{dx}+20=0$$

The auxiliary equation is:

$$D^4+8D^3+25D^2+36D+20=0$$

This can be factored as follows:

$$(D^2+4D+4)(D^2+4D+5)=(D+2)(D+2)(D^2+4D+5)$$

Thus, the roots are 2, 2, $-2 \pm i$. The solution is:

$$y=e^{2x}(C_1+C_2x)+e^{-2x}(C_3\cos x+C_4\sin x)$$

A-11 CALCULATIONS OF PARTICULAR INTEGRALS

The expressions for the calculation of particular integrals are shown in Table A-1. We will prove some of these expressions.

A. *Prove that:*

$$\frac{1}{f(D)}e^{ax}=\frac{1}{f(a)}e^{ax} \tag{A-41}$$

We know that:

$$De^{ax}=ae^{ax}$$

$$D^2e^{ax}=a^2e^{ax}$$

$$D^ne^{ax}=a^ne^{ax}$$

if

$$f(D)e^{ax}=(D^n+C_1D^{n-1}+\cdots+C_{n-1}D+C_n)e^{ax}$$

Then

$$f(D)e^{ax}=(a^n+C_1a^{n-1}+\cdots+C_{n-1}a+C_n)e^{ax}$$

or

$$e^{ax}=f(a)\frac{1}{f(D)}e^{ax}$$

$$\frac{1}{f(D)}e^{ax}=\frac{1}{f(a)}e^{ax}$$

If $f(a)$ is zero, then this rule fails. In this case:

$$\frac{1}{f(D)}e^{ax}=x\frac{1}{f'(D)}e^{ax}=x\frac{1}{f'(a)}e^{ax} \tag{A-42}$$

If $f'(a)$ is also zero, then:

$$\frac{1}{f(D)}e^{ax}=x^2\frac{1}{f''(a)}e^{ax} \tag{A-43}$$

B. *Prove that:*

$$\sin ax=f(-a^2)\frac{1}{f(D^2)}\sin ax$$

Proceed as follows:

$$D(\sin ax)=a\cos ax$$

$$D^2(\sin ax)=D(a\cos ax)=-a^2\sin ax$$

$$D^4(\sin ax)=D^2D^2\sin(ax)=D^2(-a^2\sin ax)=(-a^2)^2\sin ax$$

In general:

$$(D^2)^n\sin ax=(-a^2)^n\sin ax$$

$$\frac{1}{f(D^2)}\sin ax=\frac{\sin ax}{f(-a^2)} \tag{A-44}$$

Thus, we can write:

$$f(D^2)\sin ax=f(-a^2)\sin ax$$

$$\frac{1}{f(D^2)}f(D^2)\sin ax=\frac{1}{f(D^2)}f(-a^2)\sin ax$$

or

$$\sin ax=f(-a^2)\frac{1}{f(D^2)}\sin ax$$

This proves the identity.

TABLE A-1 Rules for Finding Particular Integrals in Differential Equations

A.	$\frac{1}{f(D)}e^{ax}=\frac{1}{f(a)}e^{ax}$ if $f(a)=0$ $\frac{1}{f(D)}e^{ax}=x\frac{1}{f'(a)}e^{ax}$ if $f'(a)=0$ $\frac{1}{f(D)}e^{ax}=x^2\frac{1}{f''(a)}e^{ax}$
B.	$\frac{1}{f(D)}x^n=[f(D)]^{-1}x^n$
C.	$\frac{1}{f(D^2)}\sin ax=\frac{\sin ax}{f(-a^2)}$ and $\frac{1}{f(D^2)}\cos ax=\frac{\cos ax}{f(-a^2)}$
D.	$\frac{1}{f(D)}e^{ax}\phi(x)=e^{ax}\frac{1}{f(D+a)}\phi(x)$
E.	$\frac{1}{f(D)}x^n\sin ax=\mathrm{Im}\left[e^{iax}\frac{1}{f(D+ia)}x^n\right]$ and $\frac{1}{f(D)}x^n\cos ax=\mathrm{Ir}\left[e^{iax}\frac{1}{f(D+ia)}x^n\right]$ Im stands for the imaginary part and Ir for the real part.
F.	The general equation for finding the particular integral of any function $\phi(x)$: $\frac{1}{D-a}\phi(x)=e^{ax}\int e^{-ax}\phi(x)dx$

C. *Prove that:*

$$\frac{1}{f(D)}e^{ax}\phi(x)=e^{ax}\frac{1}{f(D+a)}\phi(x) \tag{A-45}$$

This equation is sometimes called *heavyside shifting theorem.*

$$D[e^{ax}\phi(x)]=e^{ax}D\phi(x)+ae^{ax}\phi(x)=e^{ax}(D+a)\phi(x)$$

$$\begin{aligned}D^2[e^{ax}\phi(x)]&=D[e^{ax}(D+a)\phi(x)]\\&=e^{ax}(D^2+aD)\phi(x)+ae^{ax}(D+a)\phi(x)\\&=e^{ax}(D+a)^2\phi(x)\end{aligned}$$

In general:

$$D^n[e^{ax}\phi(x)]=e^{ax}(D+a)^n\phi(x)$$

$$f(D)[e^{ax}\phi(x)]=[e^{ax}f(D+a)\phi(x)]$$

$$e^{ax}\phi(x)=\frac{1}{f(D)}[e^{ax}f(D+a)\phi(x)] \tag{A-46}$$

Substitute:

$$f(D+a)\phi(x)=X$$

$$e^{ax}\frac{1}{f(D+a)}X=\frac{1}{f(D)}[e^{ax}X]$$

or:

$$\frac{1}{f(D)}[e^{ax}\phi(x)]=e^{ax}\frac{1}{f(D+a)}\phi(x)$$

D. *Prove the following general equation in Table A-1:*

$$\frac{1}{D-a}\phi(x)=e^{ax}\int e^{-ax}\phi(x)dx \tag{A-47}$$

Assume that the particular integral is:

$$\frac{1}{D-a}\phi(x)=y$$

Then:

$$(D-a)\frac{1}{(D-a)}\phi(x)=(D-a)y$$

or:

$$\phi(x)=Dy-ay$$

This is a linear differential equation. Its solution is:

$$ye^{-\int adx}=\int e^{-\int adx}\phi(x)dx$$

or:

$$ye^{-ax}=\int e^{-ax}\phi(x)$$

Which proves Eq. (A-47).

E. *General solution with multiple roots:*

$$\mathrm{PI}=\frac{1}{(D-s_1)(D-s_2)}x \qquad \text{(A-48)}$$

Consider that s_1 and s_2 are the roots of the characteristic equation. Then the particular integral can be found by successive integrations

$$\mathrm{PI}=\frac{1}{D-s_1}\left[\frac{1}{D-s_2}x\right]$$

but:

$$\frac{1}{D-s_2}x=e^{s_2x}\int e^{-s_2x}x\,dx$$

Substituting:

$$\mathrm{PI}=\frac{1}{D-s_1}[e^{s_2x}\int e^{-s_2x}x\,dx]$$

$$=e^{s_1x}\int e^{(s_2-s_1)x}\int e^{-s_2x}x(dx)^2$$

In general if $s_1, s_2, s_3, \ldots, s_n$ are the roots of the auxiliary equation, the particular integral is:

$$e^{s_1x}\int e^{(s_2-s_1)x}\int e^{(s_3-s_2)x}\int\cdots\int e^{(s_n-s_{n-1})x}\int e^{-s_nx}x(dx)^n \qquad \text{(A-49)}$$

A-12 SOLVED EXAMPLES

Example A-7 Solve:

$$\frac{d^2y}{dt^2}-4\frac{dy}{dt}=4y=6xe^{2x}+5e^{-2x}+\log 3$$

The roots of the auxiliary equation are 2, 2, and therefore the complementary function is:

$$(C_1+C_2x)e^{2x}$$

The particular integral is:

$$\frac{1}{D^2-4D+4}6xe^{2x}+\frac{1}{D^2-4D+4}5e^{-2x}+\frac{1}{D^2-4D+4}\log 3$$

$$=6e^{2x}\frac{1}{D^2}x+5e^{-2x}\frac{1}{4+8+4}+\log 3\frac{1}{D^2-4D+4}e^{0x}$$

$$=x^3e^{2x}+\frac{5}{16}e^{-2x}+\frac{1}{4}\log 3$$

The complete solution is:

$$(C_1+C_2x+x^3)e^{2x}+\frac{5}{16}e^{-2x}+\frac{1}{4}\log 3$$

Example A-8 Solve:

$$\frac{d^2y}{dx^2}+4\frac{dy}{dx}+4=x^3$$

The roots of the auxiliary equation are –2, –2 and therefore the complementary function is:

$$(C_1+C_2x)\,e^{-2x}$$

The particular integral is:

$$\frac{1}{(D+2)^2}x^3=\frac{1}{4}\left(1+\frac{D}{2}\right)^{-2}x^3$$

Expand by binomial theorem; particular integral is:

$$\frac{1}{4}\left[1-2\frac{D}{2}+\frac{(-2)(-3)}{2!}\frac{D^2}{4}+\frac{(-2)(-3)(-4)}{3!}\frac{D^3}{8}+\cdots\right]x^3$$

$$=\frac{1}{4}\left[x^3-Dx^3+\frac{3}{4}D^2x^3-\frac{1}{2}D^3x^3\right]$$

$$=\frac{1}{4}\left[x^3-3x^2+\frac{9}{2}x-3\right]$$

Thus, the complete solution is:

$$(C_1+C_2x)\,e^{-2x}+\frac{1}{4}\left(x^3-3x^2+\frac{9}{2}x-3\right)$$

Example A-9 Solve:

$$\frac{d^2y}{dx^2}+2\frac{dy}{dx}+5y=2\sin 3x$$

The roots of the auxiliary equation are $1 \pm 2i$. Thus, the complementary function is:

$$(A\cos 2x+B\sin 2x)\,e^{-x}$$

The particular integral is:

$$\frac{1}{D^2+2D+5}2\sin 3x=2\frac{1}{(-3^2)+2D+5}\sin 3x$$

$$=\frac{2}{2D-4}\left[\frac{2D+4}{2D+4}\right]\sin 3x=\frac{4D+8}{4D^2-16}\sin 3x$$

$$=-\frac{1}{13}(D+2)\sin 3x=-\frac{1}{13}(3\cos 3x+2\sin 3x)$$

The complete general solution is:

$$y=(A\cos 2x+B\sin 2x)\,e^{-x}-\frac{1}{13}(3\cos 3x-2\sin 3x)$$

Example A-10 Solve:

$$\frac{d^2y}{dx^2}-3\frac{dy}{dx}+2=3xe^{2x}+2e^{x}\cos x$$

The roots of the auxiliary equation are 1 and 2. Therefore, the complementary function is:

$$C_1e^{2x}+C_2e^x$$

The particular integral is:

$$\frac{1}{D^2-3D+2}3xe^{2x}+\frac{1}{D^2-3D+2}2e^x\cos x$$

$$=3e^{2x}\frac{1}{(D+2)^2-3(D+2)+2}x+2e^x\frac{1}{(D+1)^2-3(D+1)+2}\cos x$$

$$=3e^{2x}\frac{1}{D(1+D)}x+2e^x\frac{1}{D^2-D}\cos x$$

$$=3e^{2x}\frac{1}{D}(1+D)^{-1}x+2e^x\frac{1}{-1-D}\cos x$$

$$=3e^{2x}\frac{1}{D}[1-D+D^2+\cdots]x-2e^x\frac{(1-D)}{(1+D)(1-D)}\cos x$$

$$=3e^{2x}\left[\frac{1}{D}-1+D+\cdots\right]x-e^x(1-D)\cos x$$

$$=3e^{2x}\left(\frac{x^2}{2}-x+1\right)-e^x(\cos x+\sin x)$$

Example A-11 Find particular integral of:

$$\frac{d^2y}{dx^2}+y=x^2\sin x$$

The particular integral is given by:

$$\frac{1}{D^2+1}x^2\sin x=\text{Im}\left[\frac{1}{D^2+1}x^2e^{ix}\right]$$

$$=\text{Im}\left[e^{ix}\frac{1}{(D+i)^2+1}x^2\right]$$

$$=\text{Im}\left[e^{ix}\frac{1}{2iD}\left(1+\frac{D}{2i}\right)^{-1}x^2\right]$$

The solution is given by:

$$=\text{Im}\frac{e^{ix}}{2i}\left[\frac{x^3}{3}+\frac{ix^2}{2}-\frac{x}{2}\right]$$

$$\text{PI}=-\frac{1}{2}\left[\frac{x^3}{3}\cos x-\frac{x}{2}\cos x-\frac{x^2}{2}\sin x\right]$$

Example A-12 Solve:

$$\frac{d^2y}{dx^2}+4y=\sec 2x$$

The complementary function is:

$$C_1\cos 2x+C_2\sin 2x$$

The particular integral is:

$$\frac{1}{D^2+4}\sec 2x=\frac{1}{(D+2i)(D-2i)}\sec 2x$$

$$=\frac{1}{4i}\left[\frac{1}{D-2i}-\frac{1}{D+2i}\right]\sec 2x$$

$$=\frac{1}{4i}\frac{1}{D-2i}\sec 2x-\frac{1}{4i}\frac{1}{D+2i}\sec 2x$$

Now:

$$\frac{1}{D-2i}\sec 2x=e^{2ix}\int e^{-2ix}\sec 2x dx$$

$$=e^{2ix}\int\frac{\cos 2x-i\sin 2x}{\cos 2x}dx$$

$$=e^{2ix}\int(1-i\tan 2x)dx$$

$$=e^{2ix}\left[x+\frac{i}{2}\log\cos 2x\right]$$

Also:

$$\frac{1}{D+2i}\sec 2x=e^{-2ix}\left[x-\frac{i}{2}\log\cos 2x\right]$$

The particular integral is therefore:

$$\frac{1}{4i}\left[e^{ix}\left(x+\frac{i}{2}\log\cos 2x\right)-e^{-2ix}\left(x-\frac{i}{2}\log\cos 2x\right)\right]$$

This can be further simplified to:

$$\frac{x}{2}\sin 2x+\frac{1}{4}\cos 2x\log\cos 2x$$

Example A-13 Find the particular integral of:

$$\frac{d^3y}{dx^2}+4\frac{d^2y}{dx}+6\frac{dy}{dx}=x$$

The roots of the auxiliary equation are 0, (s_1), (2 – 2i) (s_2), and –(2 + 2i) (s_3). The particular integral is:

$$\text{PI}=\frac{1}{D(D+2+2i)(D+2-2i)}x$$

$$=\int e^{-(2+2i)x}\int e^{4ix}\int e^{(2-2i)x}x(dx)^3$$

A-13 HOMOGENEOUS LINEAR DIFFERENTIAL EQUATIONS

An equation of the form:

$$a_nx^n\frac{d^ny}{dx^n}+a_{n-1}x^{n-1}\frac{d^{n-1}y}{dx^{n-1}}+\cdots+a_0y=\phi(x) \tag{A-50}$$

where $a_0, a_1, a_2, \ldots, a_n$ are constants is called a homogeneous linear differential equation. By substituting:

$$x=e^z \qquad z=\log_e Z \qquad \frac{d}{dz}=D$$

The equation is reduced to a differential equation with constant coefficients:

$$\frac{dy}{dx}=\frac{dy}{dz}\frac{dz}{dx}=\frac{1}{x}\frac{dy}{dz}$$

or:

$$x\frac{dy}{dx}=Dy \tag{A-51}$$

Similarly, it can be proved that:

$$x^2\frac{d^2y}{dx^2}=D(D-1)y$$
$$x^3\frac{d^3y}{dx^3}=D(D-1)(D-2)y \tag{A-52}$$

Example A-14 Solve:

$$x^2\frac{d^2y}{dt^2}-2x\frac{dy}{dx}=x+8$$

Using Eq. (A-51) the equation is reduced to:

$$D(D-1)y-2Dy=e^z+8$$
$$(D^2-3D)y=e^z+8$$

The roots of the auxiliary equation are 0 and 3. Therefore the complementary function is:

$$C_1+C_2e^{3z}$$

The particular integral is:

$$\frac{1}{D^2-3D}(e^z+8)=\frac{1}{D^2-3D}e^z+\frac{8}{D^2-3D}e^{0z}$$
$$=\frac{1}{1-3}e^z+8z\frac{1}{2D-3}e^{0z}=-\frac{e^z}{2}-\frac{8z}{3}$$

The complete solution is:

$$y=C_1+C_2e^{3z}-\frac{e^z}{2}-\frac{8z}{3}$$
$$=C_1+C_2x^3-\frac{x}{3}-\frac{8}{3}\log x$$

A-14 SIMULTANEOUS DIFFERENTIAL EQUATIONS

When two or more dependent variables are functions of a single independent variable, the equations involving their derivates form simultaneous differential equations. In the transient solutions of electrical networks, many simultaneous differential equations are required to be solved. Consider the equations:

$$\frac{dx}{dy}+ay=t$$
$$\frac{dy}{dt}+bx=\sin t \tag{A-53}$$

Here x and y are both functions of t. The procedure of solving simultaneous differential equations is somewhat similar to solution of simultaneous algebraic equations. Matrix techniques are shown in Chap. 2. Here we will illustrate the concept with a simple example.

Example A-15 Solve the following simultaneous equations:

$$\frac{dx}{dt}+\frac{dy}{dt}+3x+2y=0$$
$$\frac{dy}{dt}+4x+2y=0$$

We can write the equations as:

$$(D+3)x+(D+2)y=0$$
$$4x+(D+2)y=0$$

From these equations x can be eliminated by multiplying the first equation by 4 and the second equation by $(D+3)$ and then subtracting. This gives:

$$(D+2)(D+3)y-4(D+2)y=0$$
$$(D^2+D-2)y=0$$

This can be solved for y, roots are 1 and –2, and thus the solution is:

$$y=C_1e^{-2t}+C_2e^t$$
$$Dy=-2C_1e^{-2t}+C_2e^t$$

Substituting the solution thus found and solving for x, we get:

$$x=-\frac{3}{4}C_2e^t$$

A-15 PARTIAL DIFFERENTIAL EQUATIONS

Partial differential equations are encountered in transmission line models, wave equations, heat flow, and other engineering applications. A brief description and methods of solution are provided. The equations contain partial differential coefficients, independent variables, and dependent variables. Let us denote the independent variables with x and y and the dependent variable by z; then the partial differential coefficients are:

$$\frac{\partial z}{\partial x}=p \qquad \frac{\partial z}{\partial y}=q$$
$$\frac{\partial^2 z}{\partial x^2}=r \qquad \frac{\partial^2 z}{\partial x\partial y}=s \tag{A-54}$$

The selection of p, q, r, and s is arbitrary. Consider an equation:

$$P_p+Q_q=R \tag{A-55}$$

This is called *Lagrange's* linear equation, where P, Q, and R are functions of x, y, and z. Such an equation can be solved by following steps:

- *Step 1*: Write the auxiliary equation:

$$\frac{dx}{P}=\frac{dy}{Q}=\frac{dz}{R} \tag{A-56}$$

(The proof of this relation is not provided.)

- *Step 2*: Let $u=c_1$ and $v=c_2$ be the solutions. (A-57)
- *Step 3*: Then $f(u,v)=0$ or $u=\phi(v)$ is the solution. (A-58)

Example A-16 Solve $xp+yq=z$, where

$$p=\frac{\partial z}{\partial x} \qquad q=\frac{\partial z}{\partial y}$$

The auxiliary equation is:

$$\frac{dx}{x}=\frac{dy}{y}=\frac{\partial z}{z}$$

Thus, from first two equations

$$\log x=\log y+\log c_1$$

and from the last two equations

$$\log y=\log z+\log c_2$$
$$y/z=c_2,\ x/y=c_1$$

and the solution is:

$$f\left(\frac{x}{y},\frac{y}{z}\right)=0$$

A-15-1 *N*th-Order Linear Partial Differential Equations

An equation of the type,

$$a_0\frac{\partial^n z}{\partial x^n}+a_1\frac{\partial^n z}{\partial x^{n-1}\partial y}+\cdots+a_n\frac{\partial^n z}{\partial y^n}=F(x,y) \quad \text{(A-59)}$$

is a homogeneous linear partial differential equation of nth order. It is homogeneous because all the terms contain derivatives of the same order. The solution is somewhat akin to ordinary differential equations, complementary functions, and particular integrals.

A-15-2 Complementary Function

Consider an equation:

$$a_0\frac{\partial^2 z}{\partial x^2}+a_1\frac{\partial^2 z}{\partial x\partial y}+a_2\frac{\partial^2 z}{\partial y^2}=0 \quad \text{(A-60)}$$

Write this equation as:

$$(a_0D^2+a_1DD'+a_2D'^2)=0$$

Put $D = m$, $D' = 1$, then the equation reduces to:

$$a_0m^2+a_1m+a_2=0$$

If the roots are real and unequal, then:

$$\text{CF}=f_1(y+m_1x)+f_2(y+m_2x) \quad \text{(A-61)}$$

If the roots are real and equal, then:

$$\text{CF}=f_1(y+mx)+xf_2(y+mx) \quad \text{(A-62)}$$

Nonhomogeneous Linear Differential Equations Consider the equation of the form:

$$p-mq=az \quad \text{(A-63)}$$

The Lagrange equation is:

$$\frac{dx}{1}=\frac{dy}{-m}=\frac{dz}{az} \quad \text{(A-64)}$$

Thus, from first two relations:

$$y+mx=c$$

and from first and third relation:

$$x=\frac{1}{a}\log z+c$$

$$ax=\log z+ac$$

$$z=e^{ax-ac}=c_2e^{ax}$$

Therefore:

$$Z=e^{ax}f(y+mx) \quad \text{(A-65)}$$

A-15-3 Particular Integrals

The rules for finding particular integrals in partial differential equation are tabulated in Table A-2.

Case 1:

$$\text{PI}\ \frac{1}{f(D,D')}e^{ax+by}=\frac{e^{ax+by}}{f(a,b)} \quad \text{(A-66)}$$

(Put $D = a$, $D' = b$)

Case 2:

$$\text{PI}\ \frac{1}{f(D^2,DD',D'^2)}\sin(ax+by)=\frac{\sin(ax+by)}{f(-a^2,-ab,-b^2)} \quad \text{(A-67)}$$

$$D^2=-a^2 \qquad DD'=-ab \qquad D'^2=-b^2$$

TABLE A-2 Rules for Finding Particular Integrals in Partial Differential Equations

A.	$\text{PI}\frac{1}{f(D,D')}e^{ax+by}=\frac{1}{f(a,b)}e^{ax+by}$
B.	$\text{PI}\frac{1}{f(D^2,DD',D'^2)}\sin(ax+by)=\frac{\sin(ax+by)}{f(-a^2-ab-b^2)}$
C.	$\text{PI}\frac{1}{f(D^2,DD',D'^2)}\cos(ax+by)=\frac{\cos(ax+by)}{f(-a^2-ab-b^2)}$
D.	$\text{PI}\frac{1}{f(D,D')}x^my^m=[f(D,D')]^{-1}x^my^m$
E.	$\text{PI}\frac{1}{f(D,D')}\phi(x,y)$ Resolve $\frac{1}{f(D,D')}$ in partial fractions, considering $f(D,D')$ as function of D only. $\text{PI}=\frac{1}{D-mD'}\phi(x,y)=\int\phi(x,c-mx)dx$

Case 3:

$$\text{PI } \frac{1}{f(D^2, DD', D'^2)}\cos(ax+by) = \frac{\cos(ax+by)}{f(-a^2, -ab, -b^2)} \quad \text{(A-68)}$$

Case 4:

$$\text{PI } \frac{1}{f(D, D')}x^m y^n = [f(D, D')]^{-1}x^m y^n \quad \text{(A-69)}$$

Expand $f[(D, D')]^{-1}$ in ascending powers of D or D' and operate on $x^m y^n$ term by term.

Case 5:

Any function, $F(x, y)$

$$\text{PI } \frac{1}{f(D, D')}F(x, y)$$

Resolve $1/f(D, D')$ into partial fractions. Considering that $f(D, D')$ is a function of D alone:

$$\text{PI} = \frac{1}{D - mD'}F(x, y) = \int F(x, c - mx)dx \quad \text{(A-70)}$$

where c is replaced with $y + mx$ after integration.

Example A-17 Solve:

$$\frac{\partial^3 z}{\partial x^3} - 2\frac{\partial^3 z}{\partial x \partial y} = 2x^2 y$$

The auxiliary equation is:

$$(D^3 - 2D^2 D') = 0$$

$$m^2(m-2) = 0$$

Therefore, the complimentary function is:

$$\text{CF} = f_1(y) + xf_2(y) + f_3(y + 2x)$$

$$\text{PI} = \frac{1}{D^3 - 2D^2 D'}2x^2 y$$

$$= 2\frac{1}{D^3\left(1 - \frac{2D'}{D}\right)}x^2 y$$

$$= \frac{2}{D^3}\left(1 - \frac{2D'}{D}\right)^{-1} x^2 y = \frac{2}{D^3}\left(1 + \frac{2D'}{D} \cdots\right)x^2 y$$

$$= \frac{2}{D^3}\left(x^2 y + \frac{2}{D}x^2\right) = \frac{2}{D^3}\left(x^2 y + \frac{2}{3}x^3\right)$$

$$= 2y\frac{x^5}{3\times4\times5} + \frac{4x^6}{3\times4\times5\times6}$$

Thus, the complete solution is CF + PI.

Example A-18 Solve the wave equation:

$$\frac{\partial^2 y}{\partial t^2} = c^2\frac{\partial^2 y}{\partial x^2}$$

Let $y = XT$, where X is a function of x only and T is a function of t only. Then we can write:

$$\frac{\partial y}{\partial t} = X\frac{dT}{dt} \qquad \frac{\partial^2 y}{\partial t^2} = X\frac{d^2 T}{dt^2}$$

Note that T and X are functions of a single variable only. Similarly

$$\frac{\partial y}{\partial x} = T\frac{dX}{dt} \qquad \frac{\partial^2 y}{\partial x^2} = T\frac{d^2 X}{dt^2}$$

Substituting these results in the original equation:

$$X\frac{d^2 T}{dt^2} = C^2 T\frac{d^2 X}{dx^2}$$

Thus by separating the variables, we have:

$$\frac{\left(\frac{d^2 T}{dt^2}\right)}{C^2 T} \quad \text{and} \quad \frac{\left(\frac{d^2 X}{dx^2}\right)}{X} = k(say)$$

$$\frac{d^2 T}{dt^2} - kC^2 T = 0$$

$$\frac{d^2 X}{dx^2} - kX = 0$$

The auxiliary equations are:

$$m^2 - kC^2 = 0 \qquad m = \pm C\sqrt{k}$$

$$m^2 - k = 0 \qquad m = \pm\sqrt{k}$$

Three cases arise:

Case 1: $ki > 0$

$$T = C_1 e^{C\sqrt{k}t} + C_2 e^{-C\sqrt{k}t} \qquad X = C_3 e^{\sqrt{k}x} + C_4 e^{-\sqrt{k}x} \quad \text{(A-71)}$$

Thus the solution is:

$$y = (C_1 e^{C\sqrt{k}t} + C_2 e^{-C\sqrt{k}t}) \times (C_3 e^{\sqrt{k}x} + C_4 e^{-\sqrt{k}x})$$

Case 2: $k < 0$ (wave motion)

$$y = (C_5 \cos C\sqrt{k}t + C_6 \sin C\sqrt{k}t) \times (C_7 \cos\sqrt{k}x + C_8 \sin\sqrt{k}x) \quad \text{(A-72)}$$

Case 3: $k = 0$

$$y = (C_9 t + C_{10}) \times (C_{11}x + C_{12}) \quad \text{(A-73)}$$

The constants can be evaluated from the boundary conditions, as illustrated before.

Partial differential equations appear in the wave propagation on transmission lines (Chap. 4). The nonlinear differential equations and their solutions are discussed in App. G. A nonlinear differential equation is nonhomogeneous and has variable coefficients. An important nonlinear differential equation in electrical engineering is the swing equation of a synchronous machine (Chap. 12). Numerical methods are used for solution of nonlinear differential equations (App. G).

FURTHER READING

R. Bronson and G. Costa, *Differential Equations*, 3d ed., *Schaum's Outline Series*, McGraw Hill, New York, 2006.

J. Cronin, *Ordinary Differential Equations: Introduction to Qualitative Theory*, Chapman and Hall/CRC, Boca Raton, FL, 2008.

P. DuChateau and D. W. Zachmann, *Partial Differential Equations*, *Schaum's Outline Series*, McGraw Hill, New York, 1986.

H. Edwards and D. Penny, *Differential Equations and Boundary Value Problems—Computing and Modeling*, Pearson Education, Upper Saddle River, NJ, 2004.

S. J. Farlow, *Partial Differential Equations for Scientists and Engineers*, Dover Publications, New York, 1993.

E. D. Rainville and P. E. Bedient, *Short Course in Differential Equations*, 4th ed., Macmillan, New York, 1969.

R. G. Watts, *Essentials of Applied Mathematics for Scientists and Engineers*, Morgan and Claypool Publishers, San Rafael, CA, 2007.

APPENDIX B

LAPLACE TRANSFORM

The classical approach to the solution of differential equations is described in App. A, and the methodology can be used for a variety of excitation functions. Laplace transformation transforms the exponential and transcendental functions and their combinations into algebraic equations. Differentiation and integration are transformed into multiplication and division. Effective use of step and impulse responses is made. The boundary values are considered in the formation of the transform. After a solution of these geometric equations is found, the inverse Laplace transform is applied to obtain the solution in the time domain. This three-step process of solving the differential equations may simplify the calculations.

If $f(t)$ is a function defined for all positive values of t, then,

$$F(s)=\int_0^\infty e^{-st}f(t)dt \tag{B-1}$$

is called the Laplace transform of $f(t)$, provided the integral exists. We write it the form:

$$\mathcal{L}[f(t)]=F(s) \tag{B-2}$$

If $F(s)$ is the Laplace transform of $f(t)$ and denoted by above equation, then,

$$f(t)=\mathcal{L}^{-1}F(s) \tag{B-3}$$

is called the inverse Laplace transform of $F(s)$.

Tables B-1 and B-2 provide Laplace transforms of some functions and their inverses. We will prove one result shown in Table B-1.

Example B-1 Prove that:

$$\mathcal{L}(t^n)=\frac{n!}{s^{n+!}} \tag{B-4}$$

By definition:

$$\mathcal{L}(t^n)=\int_0^\infty e^{-st}t^n\,dt$$

Substitute $x = st$

$$\mathcal{L}(t^n)=\int_0^\infty e^{-x}\left(\frac{x}{s}\right)^n\frac{dx}{s}$$

$$=\frac{1}{s^{n+1}}\int_0^\infty e^{-x}x^n\,dx$$

$$=\frac{n!}{s^{n+1}}$$

Some properties of the transform are:

$$\text{(a)}\ \mathcal{L}[af_1(t)+bf_2(t)]=a\mathcal{L}[f_1(t)]+b\mathcal{L}[f_2(t)] \tag{B-5}$$

where a and b are not functions of t.

$$\text{(b)}\ \mathcal{L}[e^{at}f(t)]=F(s-a) \tag{B-6}$$

This is called *first shifting theorem*. Proofs of Eqs. (B-5) and (B-6) are not provided.

Example B-2 Find the Laplace transform of $t\sin at$

$$\mathcal{L}(t\sin at)=\mathcal{L}\left(t\frac{e^{iat}-e^{-iat}}{2i}\right)$$

$$=\frac{1}{2i}[\mathcal{L}(te^{iat})-\mathcal{L}(te^{-iat})]$$

$$=\frac{1}{2i}\left[\frac{1}{(s-ia)^2}-\frac{1}{(s+ia)^2}\right]$$

$$=\frac{2as}{(s^2+a^2)}$$

B-1 METHOD OF PARTIAL FRACTIONS

The resolution of the transformed geometric equations into partial fractions is key to the solution of differential equations using Laplace transform. A function can be converted into its partial fractions:

$$\frac{N(x)}{D(x)}=\frac{N(x)}{G(x)H(x)L(x)}=\frac{A(x)}{G(x)}+\frac{B(x)}{H(x)}+\frac{C(x)}{L(x)} \tag{B-7}$$

The numerator $N(x)$ must be of lower order than the denominator $D(x)$. Mathematical texts describe the resolution of a function into partial fractions—here an overview is provided through examples.

Example B-3 Find inverse Laplace transform of:

$$\frac{2s+1}{s(s+1)(s+2)}$$

TABLE B-1 Some Important Laplace Transform Pairs

1.	$\mathcal{L}\, u(t)=\frac{1}{s}$
2.	$\mathcal{L}(t^n)=\frac{n!}{s^{n+1}} \quad n=0,1,2,3,\ldots$
3.	$\mathcal{L}(e^{at})=\frac{1}{s-a} s>a$
4.	$\mathcal{L}(\sin at)=\frac{a}{s^2+a^2}$
5.	$\mathcal{L}(\cos at)=\frac{s}{s^2+a^2}$
6.	$\mathcal{L}(\cosh at)=\frac{s}{s^2-a^2}$
7.	$\mathcal{L}(\sinh at)=\frac{a}{s^2-a^2}$
8.	$\mathcal{L}(e^{-at}\sin \omega t)=\frac{\omega}{(s+a)^2+\omega^2}$
9.	$\mathcal{L}(e^{-at}\cos \omega t)=\frac{s+a}{(s+a)^2+\omega^2}$
10.	$\mathcal{L}(\sin \omega t+\theta)=\frac{s\sin\theta+\omega\cos\theta}{s^2+\omega^2}$
11.	$\mathcal{L}(\cos \omega t+\theta)=\frac{s\cos\theta-\omega\sin\theta}{s^2+\omega^2}$
12.	$\mathcal{L}(te^{at})=\frac{1}{(s-a)^2}$
13.	$\mathcal{L}\left[\frac{t^{n-1}}{(n-1)!}\right]=\frac{1}{s^n} \quad n=\text{integer}$
14.	$\mathcal{L}\left[\frac{1}{(n-1)!}t^{n-1}e^{at}\right]=\frac{1}{(s-a)^n}$

This can be written as:

$$\frac{2s+1}{s(s+1)(s+2)}=\frac{A}{s}+\frac{B}{s+1}+\frac{C}{s+2}$$

$$2s+1=A(s+1)(s+2)+Bs(s+2)+Cs(s+1)$$

A, B, and C can be evaluated by:

$$A=\left.\frac{2s+1}{(s+1)(s+2)}\right|_{s=0}=\frac{1}{2} \qquad (s=0)$$

$$B=\left.\frac{2s+1}{s(s+2)}\right|_{s=-1}=1 \qquad (s=-1)$$

$$C=\left.\frac{2s+1}{s(s+1)}\right|_{s=-2}=\frac{-3}{2} \qquad (s=-2)$$

TABLE B-2 Some Important Inverse Laplace Transform Pairs

1.	$\mathcal{L}^{-1}\left(\frac{1}{s}\right)=1$
2.	$\mathcal{L}^{-1}\left(\frac{1}{s^n}\right)=\frac{t^{n-1}}{(n-1)!}$
3.	$\mathcal{L}^{-1}\left(\frac{1}{s-a}\right)=e^{at}$
4.	$\mathcal{L}^{-1}\left(\frac{a}{s^2+a^2}\right)=\sin at$
5.	$\mathcal{L}^{-1}\left(\frac{s}{s^2+a^2}\right)=\cos at$
6.	$\mathcal{L}^{-1}\left(\frac{a}{s^2-a^2}\right)=\sinh at$
7.	$\mathcal{L}^{-1}\left(\frac{s}{s^2-a^2}\right)=\cosh at$
8.	$\mathcal{L}^{-1}\left(\frac{1}{(s-a)^2+b^2}\right)=\frac{1}{b}e^{at}\sin bt$
9.	$\mathcal{L}^{-1}\left(\frac{s-a}{(s-a)^2+b^2}\right)=e^{at}\cos bt$
10.	$\mathcal{L}^{-1}\left(\frac{1}{(s-a)^2-b^2}\right)=\frac{1}{b}e^{at}\sinh bt$
11.	$\mathcal{L}^{-1}\left(\frac{s-a}{(s-a)^2-b^2}\right)=e^{at}\cosh bt$
12.	$\mathcal{L}^{-1}\left(\frac{s^2-a^2}{(s^2+a^2)^2}\right)=t\cos at$

Therefore, the inverse Laplace transform is:

$$\frac{1}{2}(1)+e^{-t}-\frac{3}{2}e^{-2t}$$

B-1-1 Repeated Linear Factors

A general equation for resolution into partial fractions is of the form:

$$\frac{N(x)}{X^m G(x)}=\frac{A_0}{X^m}+\frac{A_1}{X^{m-1}}+\cdots+\frac{A_{m-1}}{X}+\frac{F(x)}{G(x)} \qquad \text{(B-8)}$$

where

$$F(x)=f_0+f_1x+f_2x^2+\cdots$$
$$G(x)=g_0+g_1x+g_2x^2+\cdots \qquad \text{(B-9)}$$
$$N(x)=n_0+n_1x+n_2x^2+\cdots$$

This gives the following relations:

$$A_0=\frac{n_0}{g_0} \qquad A_1=\frac{n_1-A_0g_1}{g_0} \qquad A_2=\frac{n_2-A_0g_2-A_1g_1}{g_0} \qquad \text{(B-10)}$$

In general:

$$A_k = \frac{1}{g_0}\left[n_k - \sum_{i=0}^{k-1} A_i g_k - 1\right] \tag{B-11}$$

The method of application of these equations to obtain partial fractions is illustrated in the following example.

Example B-4 Resolve into partial fractions:

$$\frac{s^2+2}{s^3(s^2+2s-4)}$$

We can write the following fractions:

$$\frac{s^2+2}{s^3(s^2+2s-4)} = \frac{A_0}{s^3}+\frac{A_1}{s^2}+\frac{A_2}{s}+\frac{Ds+E}{s^2+2s-4}$$

We could write $A_0 = A$, $A_1 = B$, $A_2 = C$. Then

$$A_0 = \frac{n_0}{g_0} = \frac{2}{-4} = -\frac{1}{2}$$

n_0 is the numerator with $s = 0$, and g_0 is the denominator with $s = 0$. $n_1 = 0$ because the numerator does not have an s term, g_1 is 2, as the denominator has $2s$ term. $n_2 = 1$ as numerator has s^2 term, etc. Substituting these values:

$$A_1 = \frac{n_1 - A_0 g_1}{g_0} = \frac{0+\frac{1}{2}(2)}{-4} = -\frac{1}{4}$$

$$A_2 = \frac{n_2 - A_0 g_2 - A_1 g_1}{g_0} = \frac{1-\left(-\frac{1}{2}\right)(1)-\left(-\frac{1}{4}\right)(2)}{-4} = -\frac{1}{2}$$

Calculate D and E as follows:

$$D = n_3 - A_0 g_4 - A_1 g_3 - A_2 g_2 = \frac{1}{2}$$

$$E = n_3 - A_0 g_3 - A_1 g_2 - A_2 g_1 = \frac{5}{4}$$

The partial fractions are:

$$-\frac{1}{2}\left[\frac{1}{s^3}+\frac{1}{2s^2}+\frac{1}{s}-\frac{2s+5}{2(s^2+2s-4)}\right]$$

Example B-5 Find inverse Laplace transform of:

$$\frac{s^2}{(s^2+a^2)(s^2+b^2)}$$

Resolve into partial fractions:

$$\frac{s^2}{(s^2+a^2)(s^2+b^2)} = \frac{a^2}{a^2-b^2}\frac{1}{s^2+a^2} - \frac{b^2}{a^2-b^2}\frac{1}{s^2+b^2}$$

$$= \frac{1}{a^2-b^2}\left[\frac{a^2}{s^2+a^2} - \frac{b^2}{s^2+b^2}\right]$$

Thus, we can write:

$$\frac{1}{a^2-b^2}\left[a^2\left(\frac{1}{a}\sin at\right) - b^2\left(\frac{1}{b}\sin bt\right)\right]$$

Therefore, the inverse transform is:

$$\frac{1}{a^2-b^2}[a\sin at - b\sin bt]$$

B-2 LAPLACE TRANSFORM OF A DERIVATIVE OF *f*(*t*)

$$\mathcal{L}[f'(t)] = s\mathcal{L}[f(t) - f(0)]$$

$$\mathcal{L}f'(t) = \int_0^\infty e^{-st} f'(t)\,dt = \left[e^{-st} f(t)\right]_0^\infty - \int_0^\infty (-se^{st}) f(t)\,dt$$

$$= -f(0^+) + s\mathcal{L}f(t)$$

$$= s\mathcal{L}f(t) - f(0^+) \tag{B-12}$$

■ The Laplace transform of a derivative of *f*(*t*) corresponds approximately to the multiplication of Laplace transform of *f*(*t*) by *s*.

In general:

$$\mathcal{L}f^n(t) = s^n \mathcal{L}[f(t) - s^{n-1} f(0^+) - s^{n-2} f'(0^+) - \cdots - f^{n-1}] \tag{B-13}$$

B-3 LAPLACE TRANSFORM OF AN INTEGRAL

The Laplace transform of integral of *f*(*t*) is given by:

$$\mathcal{L}\int_0^t f(t)\,dt = \frac{1}{s}F(s) \tag{B-14}$$

We can also write:

$$\int_0^t f(t)\,dt = \mathcal{L}^{-1}\left[\frac{1}{s}F(s)\right] \tag{B-15}$$

■ The Laplace transform of an integral of *f*(*t*) corresponds to the division of Laplace transform of *f*(*t*) by *s*.

Example B-6 Find inverse Laplace transform of:

$$\frac{1}{s(s^2+1)}$$

Inverse Laplace transform of:

$$\mathcal{L}^{-1}\left(\frac{1}{s^2+1}\right) = \sin t$$

Therefore:

$$\mathcal{L}^{-1}\left(\frac{1}{s(s^2+1)}\right) = \int_0^t \sin t\,dt = [-\cos t]_0^t = -\cos t + 1$$

Example B-7 Find inverse Laplace transform of:

$$\frac{8}{s^4-4s^2}$$

$$\mathcal{L}^{-1}\frac{8}{(s^2-4)} = 4\sinh 2t$$

$$\mathcal{L}^{-1}\frac{8}{s(s^2-4)} = 4\int_0^t \sinh 2t\,dt = 2[\cosh 2t]_0^t = 2(\cosh 2t - 1)$$

$$\mathcal{L}^{-1}\frac{8}{s^2(s^2-4)} = \int_0^t 2(\cosh 2t - 1)\,dt = \sinh 2t - 2t$$

B-4 LAPLACE TRANSFORM OF *tf*(*t*)

If:

$$\mathcal{L}[f(t)] = F(s)$$

then:

$$\mathcal{L}[t^n f(t)] = (-1)^{n-1} \frac{d^n}{ds^n}[F(s)] \tag{B-16}$$

The proof is not provided.

Example B-8 Find the Laplace transform of:

$$te^t \sin 3t$$

We know that:

$$\mathcal{L} \sin 3t = \frac{3}{s^2+9}$$

$$\mathcal{L}[e^t \sin 3t] = \frac{3}{(s-1)^2+9}$$

Then from Eq. (B-16):

$$\mathcal{L}(te^t \sin 3t) = \frac{d}{ds}\left(\frac{3}{s^2-2s+10}\right) = \frac{3(2s-2)}{(s^2-2s+10)^2}$$

B-5 LAPLACE TRANSFORM OF (1/*t*) *f*(*t*)

If:

$$\mathcal{L}[f(t)] = F(s)$$

then:

$$\mathcal{L}\left[\frac{1}{t} f(t)\right] = \int_s^\infty F(s)ds \tag{B-17}$$

A proof is not provided.

Example B-9 Find the Laplace transform of:

$$\frac{\sin 2t}{t}$$

$$\mathcal{L} \sin 2t = \frac{2}{s^2+4}$$

$$\mathcal{L}\frac{\sin 2t}{t} = \int_s^\infty \frac{2}{s^2+4} ds = 2 \times \frac{1}{2}\left[\tan^{-1}\frac{s}{2}\right]_s^\infty$$

$$= \frac{\pi}{2} - \tan^{-1}\frac{s}{2} = \cot^{-1}\frac{s}{2}$$

B-6 INITIAL-VALUE THEOREM

When the Laplace transform is known, the corresponding time function can be evaluated at $t = 0$, by the initial-value theorem:

$$f(0^+) = \lim_{s\to\infty} sF(s) \tag{B-18}$$

Example B-10 Find the value of function below at $t = 0$

$$\frac{2a(s^2-a)}{(s^2-2as+2a^2)}$$

Write the function as:

$$\frac{2a - \dfrac{2a^2}{s^2}}{\left(1 - \dfrac{2a}{s} + \dfrac{2a^2}{s^2}\right)}$$

Therefore the value at $t = 0$ is $2a$.

B-7 FINAL-VALUE THEOREM

When the Laplace transform is known, the corresponding function can be found at $t = \infty$, by the final-value theorem:

$$f(\infty) = \lim_{s\to 0} sF(s) \tag{B-19}$$

Example B-11 The final value of the function in Example B-10 is -1, because as $s \to 0$, function $= \dfrac{-2a^2}{2a^2} = -1$.

B-8 SOLUTION OF DIFFERENTIAL EQUATIONS

Ordinary linear differential equations with constant coefficients can be easily solved by the Laplace transform method, without finding the general solution.

Example B-12 Solve the following differential equation with Laplace transform:

$$\frac{d^2y}{dt^2} - 4\frac{dy}{dt} + 4y = 64 \sin 2t$$

The initial conditions are $Y(0) = 0$, $Y'(0) = 1$.

Step 1. Taking Laplace transform of both sides and considering the boundary conditions specified, we can write:

$$[s^2 y(s) - 4sy(0) - Y'(0)] - 4[sy(s) - Y(0)] + 4y(s) = \frac{128}{s^2+4}$$

$$s^2 y(s) - 1 - 4sy(s) + 4y(s) = \frac{128}{s^2+4}$$

$$(s-2)^2 y(s) = 1 + \frac{128}{s^2+4}$$

Step 2. Resolving into partial fractions:

$$y(s) = \frac{1}{(s-2)^2} - \frac{8}{s-2} + \frac{16}{(s-2)^2} + \frac{8s}{s^2+4}$$

Step 3. Apply inverse transform:

$$y(t) = \mathcal{L}^{-1}\left[-\frac{8}{s-2} + \frac{17}{(s-2)^2} + \frac{8s}{s^2+4}\right]$$

$$= -8e^{2t} + 17te^{2t} + 8\cos 2t$$

This example illustrates the general method of solution of differential equations using Laplace transform.

B-9 SOLUTION OF SIMULTANEOUS DIFFERENTIAL EQUATIONS

Solutions of two simultaneous differential equations can be obtained using Laplace transform. We take the transform of the differential equations, solve them like algebraic equations, eliminate variables by substitution, and take inverse transform.

Example B-13 Solve the following differential equations:

$$\frac{dx}{dt}+4y=0$$

$$\frac{dy}{dt}-9x=0$$

For $X(0) = 2$ and $Y(0) = 1$. Taking Laplace transform:

$$sx(s)-X(0)+4y(s)=0$$

$$sy(s)-Y(0)-9x(s)=0$$

Substituting the values:

$$sx(s)-2+4y(s)=0$$

$$sy(s)-1-9x(s)=0$$

Therefore:

$$y(s)=\frac{s+18}{s^2+36} \qquad x(s)=\frac{2s-4}{s^2+36}$$

Taking inverse Laplace transforms:

$$x=-\frac{2}{3}\sin 6t+2\cos 6t$$

$$y=\cos 6t+3\sin t$$

B-10 UNIT-STEP FUNCTION

The unit-step function is defined as:

$$u(t-a)=0 \qquad t<0$$
$$u(t-a)=1 \qquad t\geq a \qquad a\geq 0 \qquad \text{(B-20)}$$

See Fig. B-1*a*. Laplace transform of the unit function is given by:

$$\mathscr{L}[u(t-a)]=\int_0^\infty e^{-st}u(t-a)dt=\frac{e^{-as}}{s} \qquad \text{(B-21)}$$

B-11 IMPULSE FUNCTION

An impulse function is shown in Fig. B-1*b*. A large force acts for a short time duration. The unit impulse function is the limiting function:

$$\delta(t-1)=\frac{1}{\varepsilon} \qquad a<t<a+\varepsilon$$
$$= 0 \text{ otherwise} \qquad \text{(B-22)}$$

The value of the function becomes infinite as $\varepsilon \to 0$, that is, the area of the rectangle shown in Fig. B-1*b* is unity. The unit impulse function is, therefore, defined as:

$$\delta(t-a)=\infty \qquad t=a \qquad \delta(t-a)=0 \qquad t\neq a \qquad \text{(B-23)}$$

and $\int_0^\infty \delta(t-a)dt = 1$

Laplace transform of unit-step function is given by:

$$\int_0^\infty f(t)\delta(t-a)dt=\int_a^{a+\varepsilon} f(t)\frac{1}{\varepsilon}dt$$

$$=(a+\varepsilon-a)f(\lambda)\frac{1}{\varepsilon}$$

$$=f(\lambda)$$

where:

$$\lambda>a \qquad \text{and} \qquad <a+\varepsilon \qquad \text{(B-24)}$$

$$\int_0^\infty f(t)\delta(t-a)dt=f(a) \text{ as } \varepsilon\to 0 \qquad \text{(B-25)}$$

B-12 GATE FUNCTION

A gate function can be created by two unit-step functions (Chap. 2) and written as:

$$g_t(T)=u(t-t_0)-u(t-t_0-T)$$

The function is shown in Fig. B-1*c*. It is rectangular pulse starting at t_0 and lasting for T. The function is useful in finding Laplace transform of other functions.

B-13 SECOND SHIFTING THEOREM

If:

$$\mathscr{L}[f(t)]=F(s)$$

then:

$$\mathscr{L}[f(t-a)u(t-a)]=e^{-as}F(s) \qquad \text{(B-26)}$$

Consider the following functions:

$$f(t-t_0) \qquad f(t-t_0)u(t) \qquad f(t)u(t-t_0) \qquad f(t-t_0)u(t-t_0)$$

These are all different and the second shifting theorem applies only to the last function.

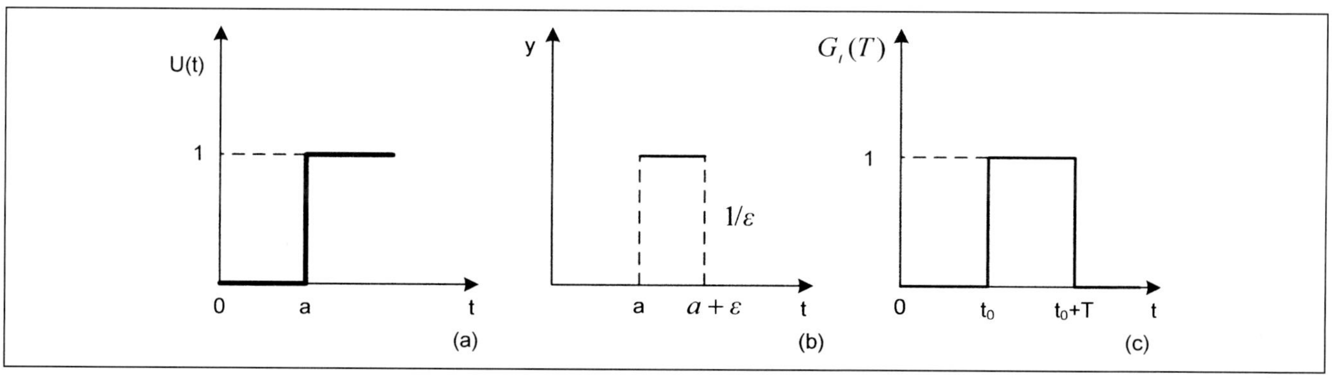

FIGURE B-1 (*a*) Step function, (*b*) impulse function, (*c*) gate function.

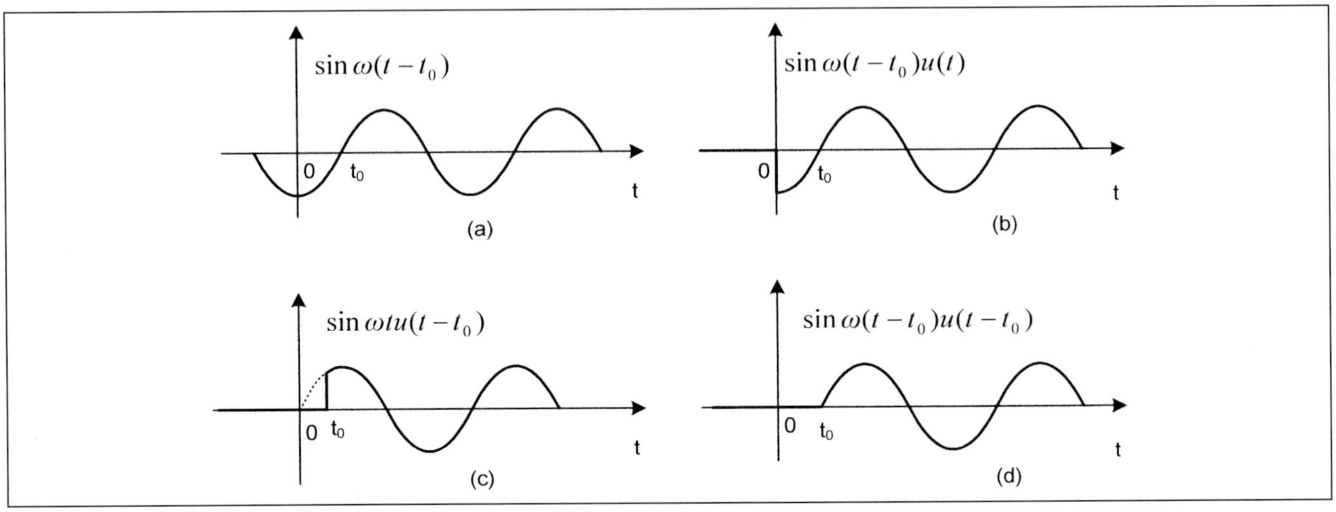

FIGURE B-2 Four different functions as marked as (*a*), (*b*), (*c*), and (*d*).

Example B-14 Find the Laplace transform of the following functions:

1. $f(t-t_0)=\sin\omega(t-t_0)$
2. $f(t-t_0)u(t)=\sin\omega(t-t_0)u(t)$
3. $f(t)u(t-t_0)=\sin\omega t\, u(t-t_0)$
4. $f(t-t_0)u(t-t_0)=\sin\omega(t-t_0)\,u(t-t_0)$

These four functions are shown graphically in Fig. B-2*a*, *b*, *c*, and *d*, respectively. The first and the second functions are different but these have the same Laplace transform. A reader may decipher why it is so. The Laplace transform is given by:

$$\mathcal{L}[\sin\omega(t-t_0)]=\mathcal{L}[\sin\omega t\cos\omega t_0-\cos\omega t\sin\omega t_0]$$
$$=\frac{\omega\cos\omega t_0-s\sin\omega t_0}{s^2+\omega^2}$$

The Laplace transform of (*c*) is:

$$\mathcal{L}[\sin\omega t\, u(t-t_0)]=\int_{t_0}^{\infty}(\sin\omega t)e^{-st}dt$$
$$=e^{-t_0 s}\left[\frac{\omega\cos\omega t_0-s\sin\omega t_0}{s^2+\omega^2}\right]$$
$$=\frac{1}{2j}\int_{t_0}^{\infty}[e^{(-s+j\omega)t}-e^{(-s-j\omega)t}]dt$$

To find Laplace transform of the fourth function shown in Fig. B-2*d*, apply second shifting theorem:

$$\mathcal{L}[\sin\omega(t-t_0)u(t-t_0)]=e^{-t_0 s}\mathcal{L}(\sin\omega t)$$
$$=e^{-t_0 s}\left(\frac{\omega}{s^2+\omega^2}\right)$$

Example B-15 Find the inverse transform of:

$$\frac{e^{-2s}}{s^2}$$

This can be solved by second shifting theorem:

$$\mathcal{L}^{-1}\left(\frac{1}{s^2}\right)=t$$
$$\mathcal{L}^{-1}\left(\frac{e^{-2s}}{s^2}\right)=(t-2)u(t-2)$$

This is zero for $0<t<2$ and is $t-2$ for $t>2$, a step function.

A theorem of importance, which can be helpful in transient solution when the excitation is a step function, is:

$$\mathcal{L}f(t)u(t-a)=e^{-as}\mathcal{L}[f(t+a)] \tag{B-27}$$

The proof of this theorem is not provided. Its application is illustrated in the following example.

Example B-16 Consider that an LC circuit is excited by a step voltage given by:

$$v(t)=1 \quad 0<t<1$$
$$=0 \text{ otherwise}$$

It is required to find the current in the circuit. The differential equation is:

$$\mathcal{L}\frac{d^2q}{dt^2}+\frac{q}{c}=v(t)$$

Consider $\mathcal{L}=1\ H$ and $C=1F$, then taking Laplace transform:

$$s^2q(s)-sQ(0)-Q'(0)+q(s)=\int_0^{\infty}v(t)e^{-st}\,dt$$

Consider the initial conditions that $Q(0)=Q'(0)=0$, We can write:

$$q(s)(s^2+1)=\int_0^1 te^{-st}\,dt+\int_1^{\infty}0e^{-st}\,dt$$
$$=\left[t\frac{e^{-st}}{-s}\right]_0^1-\int_0^1\frac{e^{-st}}{-s}\,dt$$

This gives:

$$q(s) = \frac{1}{s^2+1}\left[-\frac{e^{-s}}{s} - \frac{e^{-s}}{s^2} + \frac{1}{s^2}\right]$$

Taking inverse Laplace transforms:

$$q(t) = \mathcal{L}^{-1}\left(\frac{-e^{-s}}{s(s^2+1)}\right) - \mathcal{L}^{-1}\left(\frac{e^{-s}}{s^2(s^2+1)}\right) + \mathcal{L}^{-1}\left(\frac{1}{s^2(s^2+1)}\right)$$

From Table B-2, we can write:

$$\mathcal{L}^{-1}\left(\frac{1}{s(s^2+1)}\right) = \int_0^t \sin t\,dt = 1 - \cos t$$

$$\mathcal{L}^{-1}\left(\frac{1}{s^2(s^2+1)}\right) = \int_0^t (1-\cos t)dt = t - \sin t$$

Now from Eq. (B-27)

$$\mathcal{L}^{-1}\left[\frac{-e^{-s}}{s(s^2+1)}\right] = -[1-\cos(t-1)]u(t-1)$$

$$\mathcal{L}^{-1}\left[\frac{e^{-s}}{s^2(s^2+1)}\right] = [(t-1)-\sin(t-1)]u(t-1)$$

Thus, $q(t)$ is given by:

$$q = -[1-\cos(t-1)]u(t-1) - [(t-1)-\sin(t-1)]u(t-1)$$

Example B-17 Write the Laplace transform of the ramp function shown in Fig. B-3*a*.

The function can be considered to consist of three components, as shown in Fig. B-3*b*, *c*, and *d*. Then:

$$f(t) = f_1(t) + f_2(t) + f_3(t)$$

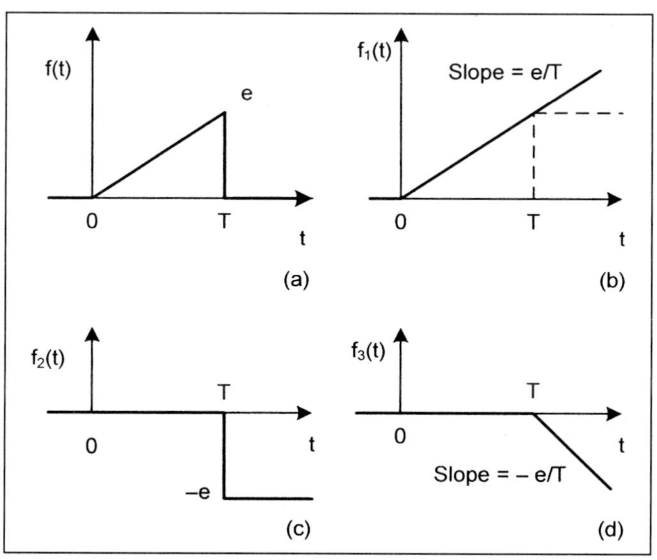

FIGURE B-3 (*a*) Ramp function. (*b*), (*c*), and (*d*) Ramp function decomposed into three functions.

From Fig. B-3, we can write:

$$f_1(t) = \frac{e}{T}tu(t) \qquad F_1(s) = \frac{e}{Ts^2}$$

$$f_2(t) = -eu(t-T) \qquad F_2(s) = -\frac{e}{s}e^{-Ts}$$

$$f_3(t) = -\frac{e}{T}(t-T)u(t-T) \qquad F_3(s) = -\frac{e}{Ts^2}e^{-Ts}$$

Thus, the Laplace transform is:

$$F(s) = F_1(s) + F_2(s) + F_3(s)$$

$$= \frac{e}{Ts^2}[1-(Ts+1)e^{-Ts}]$$

We could arrive at the same result by using the function in Fig. B-1*c*. The multiplication of a function with a gate function means that it will have nonzero values only in the duration of the gate. Thus for the ramp, we can write:

$$f(t) = \frac{e}{T}tg(t) = \frac{e}{T}t[u(t) - u(t-T)]$$

$$F(s) = \frac{e}{T}[\mathcal{L}(tu(t)) - \mathcal{L}(tu(t-T))]$$

This gives the same result as before.

B-14 PERIODIC FUNCTIONS

If $f(t)$ is a periodic function with period T, then the Laplace transform of $f(t)$ is given by:

$$\mathcal{L}[f(t)] = \frac{\int_0^T e^{-st} f(t)dt}{1-e^{-sT}} \qquad \text{(B-28)}$$

The proof of Eq. (B-28) is not provided. We can also postulate that:

- The Laplace transform of a periodic function with period T is equal to $1/(1 - e^{-Ts})$ times the Laplace transform of the first cycle.

Example B-18 Find the Laplace transform of the waveform given by:

$$f(t) = 2t \qquad 0 \le t \le 3$$

$$\mathcal{L}[f(t)] = \frac{1}{1-e^{-3s}}\int_0^3 e^{-st}t\,dt$$

$$= \frac{1}{1-e^{-3s}}2\left[\frac{te^{-st}}{-s} - (1)\frac{e^{-st}}{s^2}\right]_0^3$$

$$= \frac{2}{1-e^{-3s}}\left[\frac{3e^{-st}}{-s} + \frac{1-e^{-3s}}{s^2}\right]$$

Periodicity occurs in dc rectifier circuits.

Example B-19 Find the Laplace transform of rectifier half-wave shape, given by:

$$f(t) = \sin \omega t \qquad 0 < t < \pi/\omega$$

$$= 0 \qquad \pi/\omega < t < 2\pi/\omega$$

From Eq. (B-28)

$$\mathcal{L}[f(t)] = \frac{1}{1-e^{-2\pi s/\omega}} \int_0^{2\pi/\omega} e^{-st} f(t)\,dt$$

$$= \frac{1}{1-e^{-2\pi s/\omega}} \left(\int_0^{\pi/\omega} e^{-st} \sin dt + \int_{\pi/\omega}^{2\pi/\omega} e^{-st} \times 0\,dt \right)$$

Integral $e^{ax} \sin bx$ is given by:

$$\int e^{ax} \sin bx\,dx = e^{ax} \frac{(a \sin bx - b \cos bx)}{a^2+b^2}$$

Therefore:

$$\mathcal{L}[f(t)] = \frac{1}{1-e^{-2\pi s/\omega}} \left[\frac{e^{-st}(-s \sin \omega t - \omega \cos \omega t)}{s^2+\omega^2} \right]_0^{\pi/\omega}$$

Simplification of these yields:

$$\mathcal{L}[f(t)] = \frac{\omega}{(s^2+\omega^2)[1-e^{-\pi s/\omega}]}$$

We could arrive the same result by using gate function. The Laplace transform of the single half-sine cycle is first found.

$$f(t) = (\sin \omega t) G(\pi/\omega) = (\sin \omega t)[u(t) - u(t-\pi/\omega)]$$

Therefore:

$$\mathcal{L}[f(t)] = \mathcal{L}[\sin \omega t u(t)] - \mathcal{L}[\sin \omega t\, u(t-\pi/\omega]$$

$$= F_1(s) + F_2(s)$$

Solving, this gives:

$$F_1(s) = \frac{\omega}{s^2+\omega^2}$$

$$F_2(s) = \frac{\omega}{s^2+\omega^2} e^{-\pi s/\omega}$$

Combining for the half cycle, we can write:

$$F(s) = \frac{\omega}{s^2+\omega^2}(1+e^{-\pi s/\omega})$$

Then from Eq. (B-28) Laplace transform of periodic half-wave rectifier is:

$$F(s) = \frac{1+e^{-\pi s/\omega}}{1-e^{-2\pi s/\omega}} \frac{\omega}{s^2+\omega^2}$$

$$= \frac{1}{1-e^{-\pi s/\omega}} \frac{\omega}{s^2+\omega^2}$$

The same result as arrived at before.

B-15 CONVOLUTION THEOREM

Let:

$$\mathcal{L}[f_1(t)] = F_1(s) \quad \text{and} \quad \mathcal{L}[f_2(t)] = F_2(s)$$

Then:

$$\mathcal{L}\left[\int_0^t f_1(x) f_2(t-x)\,dx \right] = F_1(s) F_2(S) \tag{B-29}$$

or:

$$F_1(s) F_2(s) = \mathcal{L}^{-1} \int_0^t f_1(x) f_2(t-x)\,dx$$

This can be proved as follows:

$$\mathcal{L}\left[\int_0^t f_1(x) f_2(t-x)\,dx \right]$$

$$= \int_0^\infty e^{-st} \int_0^t f_1(x) f_2(t-x)\,dx\,dt$$

$$= \int_0^\infty \int_0^t e^{-st} f_1(x) f_2(t-x)\,dx\,dt = \int_0^\infty \int_x^\infty e^{-st} f_1(x) f_2(t-x)\,dt\,dx$$

$$= \int_0^\infty e^{-sx} f_1(x)\,dx \int_x^\infty e^{-s(t-x)} f_2(t-x)\,dt$$

$$= \int_0^\infty e^{-sx} f_1(x) F_2(s)\,dx$$

$$= \left[\int_0^\infty e^{-sx} f_1(x)\,dx \right] F_2(s) = F_1(s) F_2(s)$$

Example B-20 Find the inverse Laplace transform of:

$$\mathcal{L}^{-1}\left[\frac{s^2}{(s^2+a^2)(s^2+b^2)} \right] \qquad a \neq b$$

We know that:

$$\mathcal{L}[\cos at] = \frac{s}{s^2+a^2} \qquad \mathcal{L}[\cos bt] = \frac{s}{s^2+b^2}$$

Therefore from convolution theorem:

$$\mathcal{L}^{-1}\left[\frac{s^2}{(s^2+a^2)(s^2+b^2)} \right]$$

$$= \int_0^t \cos ax \cos b(t-x)\,dx$$

$$= \frac{1}{2} \int_0^t \cos[(a-b)x+bt]\,dx + \frac{1}{2} \int_0^t \cos[(a+b)x-bt]\,dx$$

$$= \left[\frac{\sin[(a-b)+bt]}{2(a-b)} \right]_0^t + \left[\frac{\sin[(a+b)x-bt]}{2(a+b)} \right]_0^t$$

$$= \frac{a \sin at - b \sin bt}{a^2-b^2}$$

B-16 INVERSE LAPLACE TRANSFORM BY RESIDUE METHOD

$f(x)$ is the sum of residues of $e^{sx}F(s)$ at poles of $F(s)$. This will be illustrated with an example:

Example B-21 Find the inverse Laplace transform of:

$$\frac{1}{(s+1)(s^2+1)}$$

We can write:

$$\mathcal{L}^{-1}\left[\frac{1}{(s+1)(s^2+1)}\right] = \text{sum of residues of } e^{st}F(s) \text{ at the poles}$$

The poles are $s=-1$, $s=+i$, $s=-i$. Residue of $e^{st}F(s)$ at $s=-1$

$$\lim_{s\to -1}(s+1)\frac{e^{st}}{(s+1)(s^2+1)} = \frac{e^{-t}}{2}$$

Similarly, residue of $e^{st}F(s)$ at $s=i$:

$$\lim_{s\to i}(s-i)\frac{e^{st}}{(s+1)(s^2+1)} = \lim_{s\to i}\frac{e^{st}}{(s+1)(s+i)} = -\frac{e^{it}(1+i)}{4}$$

and

$$\lim_{s\to -i}(s+i)\frac{e^{st}}{(s+1)(s^2+1)} = \frac{e^{-it}(i-1)}{4}$$

Therefore, the required inverse is:

$$\frac{e^{-t}}{2} - \frac{e^{it}(1+i)}{4} + \frac{e^{-it}(i-1)}{4} = \frac{e^{-t}}{2} - \frac{e^{it}+e^{-it}}{4} - \frac{i(e^{it}-e^{-it})}{4}$$

$$= \frac{e^{-t}}{2} - \frac{1}{2}\cos t + \frac{1}{2}\sin t$$

We could have obtained the same results by partial fraction method, as before.

B-17 CORRESPONDENCE WITH FOURIER TRANSFORM

We can arrive at some correspondence between Fourier transform and Laplace transform. Many texts will show the Fourier transform pair as:

$$X(\omega) = \int_{\infty}^{-\infty} x(t)e^{-j\omega t}\,dt$$

$$x(t) = \frac{1}{2\pi}\int_{-\infty}^{\infty} X(\omega)e^{j\omega t}\,d\omega \qquad \text{(B-30)}$$

If a factor $e^{-\sigma t}$, which may be called a convergence factor, is applied to unilateral Fourier transform, we have:

$$X(\omega) = \int_0^{\infty}[e^{-\sigma t}f(t)]e^{-j\omega t}\,dt = \int_0^{\infty} f(t)e^{-(\sigma+j\omega)t}\,dt = \int_0^{\infty} f(t)e^{-st}\,dt$$

(B-31)

where $s = \sigma + j\omega$. Then we can write Eq. (B-32) as:

$$F(s) = \int_0^{\infty} f(t)e^{-st}\,dt = \mathcal{L}f(t) \qquad \text{(B-32)}$$

Again by definition, the inverse Fourier transform of Eq. (B-31) is:

$$e^{-\sigma t}f(t) = \frac{1}{2\pi}\int_{-\infty}^{\infty} X(\omega)e^{j\omega t}\,d\omega \qquad \text{(B-33)}$$

or:

$$f(t) = \frac{1}{2\pi j}\int_{\sigma-j\omega}^{\sigma+j\omega} F(s)e^{st}\,ds \qquad \text{(B-34)}$$

The real part σ should be large enough to make the integral converge. When the integral exists we say that the function is Laplace transformable.

The Laplace transform method is extensively covered in the mathematical texts and its applications to the electrical engineering transients. Yet, the limitations are that it is not suitable for computer-based applications of analysis of large power systems. Also, nonlinearity and saturation cannot be taken into account; see App. G for further discussions.

FURTHER READING

D. K. Cheng, *Analysis of Linear Systems*, Addison-Wesley Publishing Company, Reading, MA, 1959.

R. A. DeCarlo and P. M Lin, *Linear Circuit Analysis-Time Domain, Phasor and Laplace Transform Approaches*, 2d ed., Oxford University Press, New York, 2001.

S. Goldman, *Laplace Transform Theory and Electrical Transients*, Dover Publications, New York, 1966.

P. K. F. Kuhjittig, *Introduction to Laplace Transform*, Plenum Press, New York, 1978.

M. B. Reed and G. B. Reed, *Electrical Network Theory, Laplace Transform Technique*, Scranton International Textbook Company, 1968.

M. G. Smith, *Laplace Transform Theory*, Van Nostrand, London/New York, 1966.

APPENDIX C

Z-TRANSFORM

The role of z-transform in discrete system analysis is the same as that of Laplace and Fourier transforms in continuous systems. Difference equations are based on discrete systems (Chaps. 2 and 3), and their analysis is carried out by using z-transform. A signal in continuous-time system can be sampled at small intervals (Fig. C-1). Then we can write:

$$f^*(t) = f(t)\delta_T(t) \tag{C-1}$$

where the superscript * means that $f(t)$ is a sampled discrete function. We can write it as $f(k)$ and $\delta_T(t)$ represents a periodic train of unit impulses spaced T seconds apart. Thus for $f(t) = 0$, $t < 0$,

$$f(k) = \sum_{n=0}^{\infty} f(nT)\delta(t - nT) \tag{C-2}$$

If we take the Laplace transform of Eq. (C-2):

$$\mathscr{L}[f(k)] = \mathscr{L}\left[\sum_{n=0}^{\infty} f(nT)\delta(t-nT)\right] = \sum_{n=0}^{\infty} f(nT)e^{-nTs} \tag{C-3}$$

Write:

$$z = e^{Ts} \tag{C-4}$$

Then z-transform of a sequence {f(k)} is defined as:

$$Z[\{f(k)\}] = F(z) = \sum_{k=-\infty}^{k=\infty} f(k)z^{-k} \tag{C-5}$$

$F(z)$ is a polynomial in z. The z-transform exists only if the sum given in Eq. (C-5) converges. z is an operator of z-transform, z is a complex variable defined by $\mu + j\upsilon$ (where μ and υ are real variables) and $F(z)$ is the z-transform of $\{f(k)\}$. Consider a sequence:

$$f(k) = \{10, 7, 4, \hat{1}, -1, 0, 3\}$$

then

$$F(z) = 10z^3 + 7z^2 + 4z + 1 - \frac{1}{z} + 0 + \frac{3}{z^3}$$

(The ^ as superscript of 1 denotes zero of the sequence.) Often, $\{f(k)\}$ is defined over equally spaced intervals of time 0, T, $2T$, . . . , kT. The dependence on T is usually suppressed and the variable arguments T and kT are used interchangeably where there is no ambiguity.

z-transform is sometimes defined as the transformation:

$$z \equiv e^{sT} \tag{C-6}$$

which amounts to exponential change of variables between the complex variable $z = \mu + j\upsilon$ and the complex variable $s = \sigma + j\omega$ in the Laplace transform domain, where T is the sample period of the discrete-time system. This definition implies a sequence $\{f(k)\}$ *or* $\{f(kT)\}$, obtained by ideal sampling of a continuous signal $f(t)$ at uniformly spaced times kT, where $k = 1, 2, \ldots$ The z-transform of a two-sided noncasual sequence is:

$$\{f(k)\} = \{\ldots, f(-2), f(-1), f(0), f(1), f(2), \ldots\} \tag{C-7}$$

The inverse Laplace transform:

$$\{f(k)\} = Z^{-1}[F(z)] \tag{C-8}$$

Now, consider the single-sided geometric sequence, denoted by its k-th term as:

$$\begin{aligned} f(k) &= 0 \qquad k < 0 \\ &= a^k \qquad k \geq 0 \end{aligned} \tag{C-9}$$

Then from the definition of z-transform in Eq. (C-5):

$$F(z) = \sum_{k=0}^{\infty} a^k z^{-k} \tag{C-10}$$

This converges only if:

$$|az^{-1}| < 1 \qquad |z| > |a|$$

$$F(z) = \frac{1}{1 - az^{-1}} \qquad |z| > |a| \tag{C-11}$$

The region of convergence in the complex plane is shown in Fig. C-2*a*. Now consider the sequence:

$$\begin{aligned} f(k) &= -b^k \qquad k < 0 \\ &= 0 \qquad k \geq 0 \end{aligned}$$

Then:

$$\begin{aligned} F(z) &= \sum_{k=-\infty}^{k=-1} -b^k z^{-k} = \sum_{k=1}^{\infty} -b^k z^{-k} \\ &= \frac{1}{1 - bz^{-1}} \qquad |z| < |b| \end{aligned} \tag{C-12}$$

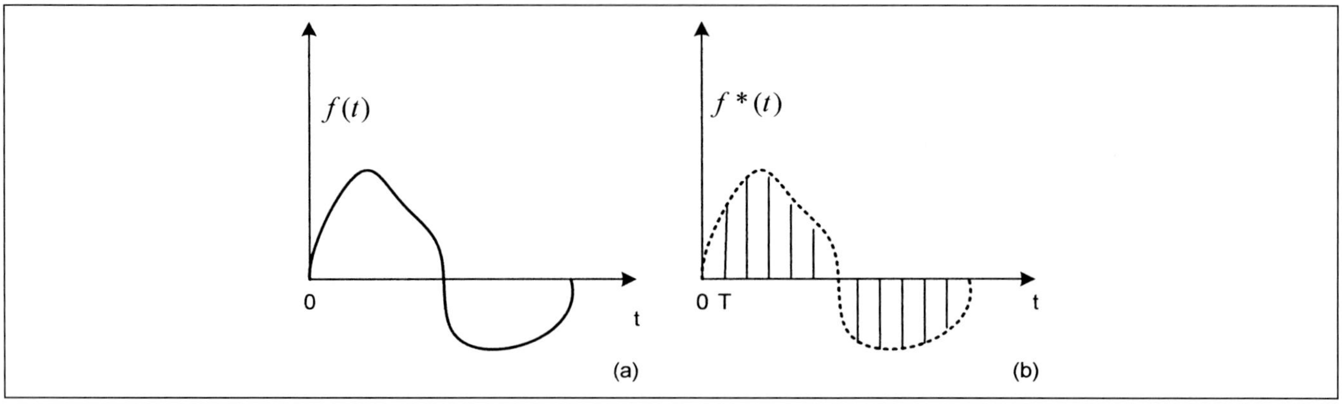

FIGURE C-1 (*a*) Function in time domain. (*b*) Function considered as sequence of impulses at sampled rate.

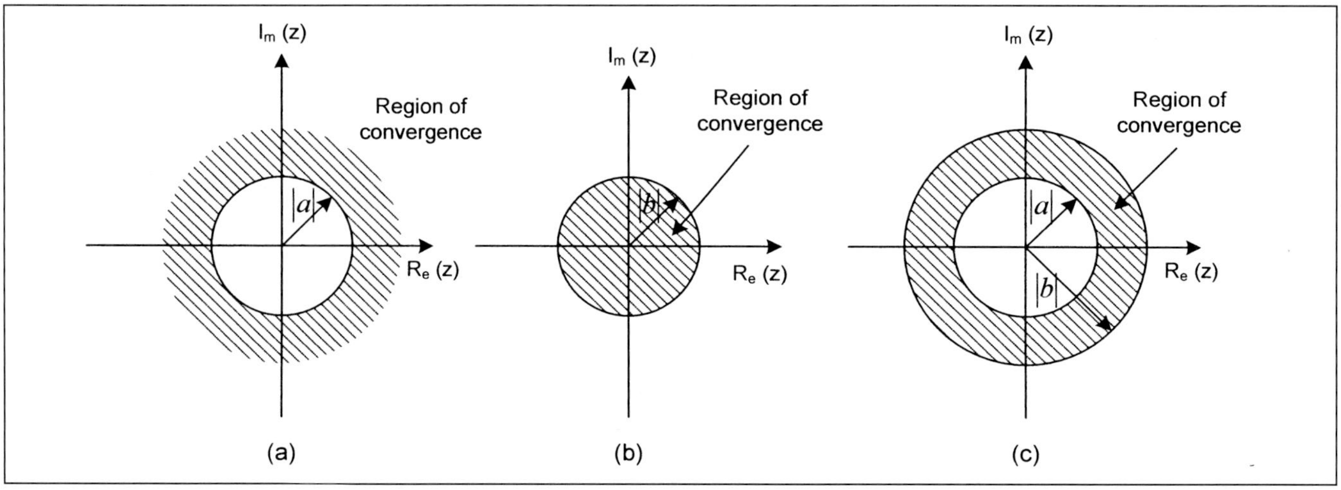

FIGURE C-2 Region of convergences and z-transforms.

This is shown in Fig. C-2*b*. If $b = a$, the two sequences have the same z-transform, that is, these are not unique. It follows that if we are dealing with single-sided sequences $k \geq 0$, their z-transforms are unique.

Consider now a two-sided sequence, if we consider that $|a|<|z|<|b|$, then the region of convergence is shown in Fig. C-2*c*. This region exists only if $|b|>|a|$. The transform does not exist if $|b|=|a|$.

Discrete unit impulse. Consider a discrete unit impulse:

$$\delta(k)=1 \quad k=0$$
$$=0 \quad k\neq 0 \tag{C-13}$$

Then:

$$Z[\delta(k)]=\sum_{k=-\infty}^{\infty}\delta(k)z^{-k}=1 \tag{C-14}$$

Discrete unit step.

$$u(k)=0 \quad k<0$$
$$=1 \quad k\geq 0 \tag{C-15}$$

Then:

$$Z[u(k)]=\frac{1}{1-z^{-1}} \tag{C-16}$$

C-1 PROPERTIES OF z-TRANSFORM

C-1-1 Linearity

If $f(k)$ and $g(k)$ can be added and a and b are constants, then:

$$Z\{af(k)+bg(k)\}=aZ[\{f(k)\}]+bZ[\{f(g)\}] \tag{C-17}$$

Example C-1 The z-transform of the $\{f(k)\}$ where:

$$f(k)=\begin{vmatrix}4^k & k<0\\ 2^k & k\geq 0\end{vmatrix}$$

can be written as:

$$Z[\{f(k)\}]=\sum_{k=-\infty}^{k=-1}4^k z^{-k}+\sum_{k=0}^{k=\infty}2^k z^{-k}$$

Example C-2 Find z-transform of:

$$\sin(k+1)$$

$$F(z)=\sum_{k=0}^{k=\infty}\sin(k+1)z^{-k}=\sum_{k=0}^{k=\infty}\frac{e^{i(k+1)}-e^{-(k+1)}}{2i}z^{-k}$$

$$=\frac{1}{2i}\sum_{k=0}^{k=\infty}e^{i(k+1)}z^{-k}-\frac{1}{2i}\sum_{k=0}^{k=\infty}e^{-i(k+1)}z^{-k}$$

$$=\frac{1}{2i}e^{i}\sum_{k=0}^{k=\infty}(e^{i}z^{-1})^{k}-\frac{1}{2i}e^{-i}\sum_{k=0}^{k=\infty}(e^{-i}z^{-1})^{k}$$

$$=\frac{1}{2i}e^{i}[1+(e^{i}z^{-1})+(e^{i}z^{-1})^{2}+\cdots]$$

$$-\frac{1}{2i}e^{-i}[1+(e^{-i}z^{-1})+(e^{-i}z^{-1})^{2}+\cdots]$$

$$=\frac{e^{i}}{2i}\frac{1}{1-e^{i}z^{-1}}-\frac{e^{-i}}{2i}\frac{1}{1-e^{-i}z^{-1}}$$

This on further simplification gives:

$$F(z)=\frac{z^{2}\sin 1-z\sin 2}{z^{2}-2z\cos 1+1}$$

C-1-2 Theorem

If {f(k)} = F(z) and {g(k)} = G(z) and a and b are constants then:

$$Z^{-1}[aF(z)+bG(z)]=aZ^{-1}[F(z)]+bZ^{-1}[G(z)] \qquad \text{(C-18)}$$

The proof is not provided.

C-1-3 Change of Scale

If Z[{$f(k)$}] = $F(z)$ then:

$$Z[\{a^{k}f(k)\}]=F\left(\frac{z}{a}\right) \qquad \text{(C-19)}$$

Proof of Eq. (C-19) is not provided.

Example C-3 Find z- transform of:

$$a^{k}\sin bk \qquad k\geq 0$$

We know that:

$$Z(\sin bk)=\frac{z\sin b}{z^{2}-2z\cos b+1}$$

Then, from Eq. (C-19)

$$Z[a^{k}\sin bk]=\frac{(Z/a)\sin b}{(z/a)^{2}-[2(z/a)]\cos b+1}$$

$$=\frac{az\sin b}{z^{2}-2az\cos a+a^{2}}$$

C-1-4 Shifting Property

If:

$$Z[\{f(k)\}]=F(z),$$

then:

$$Z[\{f(k\pm n)\}]=Z^{\pm n}F(z) \qquad \text{(C-20)}$$

This is applicable to two-sided sequence. The proof is not provided. For one-sided sequences:

$$f(k+n)u(k)\leftrightarrow z^{n}[F(z)-f(0)-z^{-1}f(1)\cdots-z^{-(n-1)}f(n-1)]$$

$$f(k-n)u(k-n)\leftrightarrow z^{-n}F(z) \qquad \text{(C-21)}$$

It follows that for the casual sequence:

$$\begin{aligned}&Z[\{f(k-1)\}]=z^{-1}F(z) \quad \text{as} \quad f(-1)=0\\&Z[\{f(k+1)\}]=zF(z)-zf(0)\\&Z[\{f(k+2)\}]=z^{2}F(z)-z^{2}f(0)-zf(1)\end{aligned} \qquad \text{(C-22)}$$

C-1-5 Multiplication by *k*

If:

$$Z[\{f(k)\}]=F(z)$$

then:

$$Z[\{kf(k)\}]=-z\frac{d}{dz}F(z) \qquad \text{(C-23)}$$

In general:

$$Z[\{k^{n}f(k)\}]=\left(-z\frac{d}{dz}\right)^{n}F(z) \qquad \text{(C-24)}$$

Thus, if $f(k) = 1$ then we know that:

$$Z[\{1\}]=\frac{1}{1-z^{-1}}$$

From Eq. (C-23):

$$Z[\{k\}]=-z\frac{d}{dz}(1-z^{-1})^{-1}$$

$$=z(1-z^{-1})^{2}\left(\frac{1}{z^{2}}\right)=z^{-1}(1-z^{-1})^{-2} \qquad \text{(C-25)}$$

C-1-6 Division by *k*

If:

$$Z[\{f(k)\}]=F(z)$$

then:

$$Z\left[\left\{\frac{f(k)}{k}\right\}\right]=-\int^{z}z^{-1}F(z)dz \qquad \text{(C-26)}$$

C-2 INITIAL-VALUE THEOREM

For one-sided sequences, if:

$$Z[\{f(k)\}]=F(z) \qquad k\geq 0$$

then:

$$f(0)=Lim_{z\to\infty}F(z) \qquad \text{(C-27)}$$

The proof of Eq. (C-27) is not provided.

Example C-4 The z- transform of

$$\left\{\left(\frac{1}{2}\right)^{k}\right\}=\frac{z}{z-1/2}$$

Then from the initial value theorem in Eq. (C-27)

$$\lim_{z\to\infty}\left(\frac{z}{z-1/2}\right)=1$$

C-3 FINAL-VALUE THEOREM

The final-value theorem is given by:

$$\lim_{k\to\infty} f(k) = \lim_{z\to 1}(z-1)F(z) \qquad \text{(C-28)}$$

Provided the limit exits; that is, the sequence has a final value. For the final value theorem to be applicable $F(z)$ should not have poles outside the unit circle. The proof of Eq. (C-28) is not provided.

Example C-5 Find the final value of the sequence:

$$[1-(1/4)^k]$$

The z-transform of the sequence is:

$$\frac{(1/4)z}{z^2-(5z/4)+(1/4)}$$

Thus, from Eq. (C-28):

$$\lim_{z\to\infty}\left(1-\left(\frac{1}{4}\right)^k\right) = \lim_{z\to 1}(z-1)\frac{(1/4)z}{z^2-(5z/4)+(1/4)}$$

$$= \lim_{z\to 1}\frac{z(z-1)}{4z^2-5z+1}$$

$$= \lim_{z\to 1}\frac{z(z-1)}{(z-1)(z-1/4)} = 4/3$$

C-4 PARTIAL SUM

If:

$$Z[\{f(k)\}] = F(z),$$

then:

$$Z\left[\left\{\sum_{n\to-\infty}^{k} f(n)\right\}\right] = \frac{F(z)}{1-z^{-1}} \qquad \text{(C-29)}$$

The proof of Eq. (C-29) is not provided.

C-5 CONVOLUTION

If there are two sequences,

$$\{f(k)\} \text{ and } \{g(k)\},$$

then their convolution:

$$\{h(k)\} = \{f(k)\}^*\{g(k)\} \qquad \text{(C-30)}$$

Consider casual sequences:

$$F(z) = \{f(0)+f(1)z^{-1}+f(2)z^{-2}+\cdots\}$$

$$G(z) = \{g(0)+g(1)z^{-1} = g(2)z^{-2}+\cdots\}$$

then:

$$F(z)G(z) = f(0)g(0)+\{f(1)g(0)+f(0)g(1)\}z^{-1}$$

$$+\{f(0)g(2)+f(1)g(1)+f(2)g(0)\}z^{-2}+\cdots$$

$$= h(0)+h(1)z^{-1}+h(2)z^{-2}+\cdots$$

$$= Z\{h(k)\} = z\{f(k)\}^*\{g(k)\}$$

Example C-6 Find the z-transform of the sequence:

$$\{f(k)\} = \sum_{k=0}^{k=\infty} 3^k \sum_{k=0}^{k=\infty} 4^k$$

We know that:

$$Z[\{3^k\}] = [1+3z^{-1}+3^2z^{-2}+3^3z^{-3}+\cdots] = \frac{1}{1-3Z^{-1}}$$

$$Z[\{4^k\}] = [1+4z^{-1}+4^2z^{-2}+4^3z^{-3}+\cdots] = \frac{1}{1-4Z^{-1}}$$

Then from Eq. (C-30):

$$Z[\{f(k)\}] = Z[\{3^k\}\{4^k\}] = \frac{1}{(1-3z^{-1})(1-4z^{-1})}$$

Table C-1 provides z transforms of some important sequences.

C-6 INVERSE z-TRANSFORM

The sequence $\{f(k)\}$ can be found from $F(z)$ and is defined as the inverse z-transform:

$$Z^{-1}F(z) = \{f(k)\} \qquad \text{(C-31)}$$

The inverse z-transform can only be settled if the region of convergence is given. The inverse transform can be found by:

- Direct division
- Binomial expansion and partial fractions
- Residue method

Example C-7 Find inverse z-transform of:

$$\frac{2z}{z-a} \qquad |z|>|a| \qquad \text{and} \qquad |z|<|a|$$

We will use binomial expansion:

$$\frac{2z}{z-a} = \frac{2z}{z}\frac{1}{1-\dfrac{a}{z}} = 2\left(1-\frac{a}{z}\right)^{-1}$$

$$= 2\left[1+\frac{a}{z}+\frac{a^2}{z^2}+\cdots\right]$$

$$= 2+2az^{-1}+2a^2z^{-2}+\cdots+2a^kz^{-k}$$

$$= \{2a^k\}z^{-k}$$

Thus:

$$Z^{-1}\left(\frac{2z}{z-a}\right) = [2a^k]$$

For $|z|<|a|$:

$$\frac{2z}{z-a} = -\frac{2z}{a-z} = -\frac{1}{a}\frac{2z}{[1-(z/a)]} = \frac{-2z}{a}[1-(z/a)]^{-1}$$

$$= -\frac{2z}{a}\left[1+\frac{z}{a}+\frac{z^2}{a^2}+\cdots\right]$$

$$= -\frac{2z}{a}-\frac{2z^2}{a^2}-\frac{2z^3}{a^3}-\cdots$$

Therefore:

$$\{f(k)\} = \left[\cdots,-\frac{1}{a^3},-\frac{1}{a^2},-\frac{1}{a}\right]$$

TABLE C-1 z-Transform of Some Important Sequences

SEQUENCE $k \geq 0$	z-TRANSFORM	CONDITIONS
$[f(k)]$	$F(z)$	
$\delta(k)$	1	
$u(k)$ or 1	$(1-z^{-1})^{-1}$	
k	$-z\dfrac{d}{dz}(1-z^{-1})^{-1}$	$\lvert z\rvert>1$
k^n	$\left(-z\dfrac{d}{dz}\right)^n(1-z^{-1})^{-1}$	$\lvert z\rvert>1$
a^k	$(1-az^{-1})^{-1}$	$\lvert z\rvert>\lvert a\rvert$
$\dfrac{(k+1)(k+2)\cdots(k+n-1)}{(n-1)!}a^k$	$\dfrac{z^n}{(z-a)^n}$	$\lvert z\rvert>a, k\geq 0, n\geq 2$
${}^kC_n$	$z^{-n}(1-z^{-1})^{-(n+1)}$	$\lvert z\rvert>1$
${}^{k+n}C_n a^k$	$(1-az^{-1})^{-(n+1)}$	$\lvert z\rvert>\lvert z\rvert$
${}^nC_k$	$(1-z^{-1})^n$	$0\leq k\leq n,\ \lvert z\rvert>0$
$k^n a^k$	$\left(-z\dfrac{d}{dz}\right)^n(1-az^{-1})^{-1}$	$\lvert z\rvert<\lvert a\rvert$
$a^{\lvert k\rvert}$	$(1-a^2)(1-az)(1-az^{-1})^{-1}$	$\lvert a\rvert<\lvert z\rvert<\dfrac{1}{\lvert a\rvert}$
$\sin\alpha k$	$\dfrac{z\sin\alpha}{z^2-2z\cos\alpha+1}$	$k\geq 0$
$c^k\sin\alpha k$	$\dfrac{cz\sin\alpha}{z^2-2cz\cos\alpha+c^2}$	$k\geq 0$
$\cos\alpha k$	$\dfrac{z^2-z\cos\alpha}{z^2-2z\cos\alpha+1}$	$k\geq 0$
$c^k\cos\alpha k$	$\dfrac{z^2-cz\cos\alpha}{z^2-2cz\cos\alpha+c^2}$	$k\geq 0$
$\cosh\alpha k$	$\dfrac{z^2-z\cosh\alpha}{z^2-2z\cosh\alpha+1}$	$k\geq 0$
$c^k\cosh\alpha k$	$\dfrac{z^2-cz\cosh\alpha}{z^2-2cz\cosh\alpha+c^2}$	$k\geq 0$
$\sinh\alpha k$	$\dfrac{z\sinh\alpha}{z^2-2z\cosh\alpha+1}$	$k\geq 0$
$c^k\sinh\alpha k$	$\dfrac{cz\sinh\alpha}{z^2-2cz\cosh\alpha+c}$	$k\geq 0$

C-7 INVERSION BY PARTIAL FRACTIONS

Assuming that $F(z)$ is a rational function in z:

$$F(z)=\frac{b_m z^m+b_{m-1}z^{m-1}+\cdots+b_0}{z^n+a_{n-1}z^{n-1}+\cdots+a_0} \qquad m\le n \tag{C-32}$$

The inverse z-transformer can be obtained by partial fractions. For $m = n$, divide numerator by denominator to separate out the constant term:

$$F(z)=a+\frac{N(z)}{D(z)} \tag{C-33}$$

Now, the polynomial $N(z)$ is one order less than $D(z)$. Thus, for $m < n$

$$\begin{aligned}F(z)&=\frac{N(z)}{(z-p_1)(z-p_2)\cdots(z-p_n)}\\&=\frac{C_1}{z-p_1}+\frac{C_2}{z-p_2}+\cdots+\frac{C_n}{z-p_n}\end{aligned} \tag{C-34}$$

where $p_1, p_2, \ldots, p_n$ are the poles of $F(z)$.

1. *Linear nonrepeated factor.*

$$Z^{-1}\left(\frac{z}{z-a}\right)=Z^{-1}\frac{1}{1-az^{-1}}=\{a^k\} \qquad |z|>|a| \tag{C-35}$$

$$Z^{-1}\left(\frac{z}{z-a}\right)=-Z^{-1}\frac{z/a}{1-z/a}=\{-a^k\} \qquad |z|<|a| \tag{C-36}$$

2. *Linear repeated factor.* Let the repeated factor be:

$$\frac{z}{(z-b)^r} \qquad r\ge 2 \qquad |z|>b$$

$$\begin{aligned}Z^{-1}\left(\frac{z}{(z-b)r}\right)&=Z^{-1}\left[z^{-(r-1)}\frac{z^r}{(z-b)^r}\right]\\&=\frac{(k+2-r)(k+3-r)\cdots ku(k-r+1)b^{(k-r+1)}}{(r-1)!} \qquad r\ge 2 \quad |z|>|b|\end{aligned} \tag{C-37}$$

$$Z^{-1}\left(\frac{z}{(z-b)r}\right)=-\frac{(k+2-r)(k+3-r)\cdots kb^{(k-r+1)}}{(r-1)!}u(-k-r-1) \qquad |z|<|b| \tag{C-38}$$

3. *Quadratic nonrepeated factor.* Let the factor be:

$$\frac{Mz+N}{z^2+pz+q} \tag{C-39}$$

Compare Eq. (C-39) with:

$$Z[(c^k\cos\alpha k)]=\frac{z^2-cz\cos\alpha}{z^2-2cz\cos\alpha+c^2} \tag{C-40}$$

Or compare with:

$$Z[(c^k\cosh\alpha k)]=\frac{z^2-cz\cosh\alpha}{z^2-2cz\cosh\alpha+c^2} \tag{C-41}$$

where:

$$c^2=q \qquad p=-2c\cos\alpha \qquad \text{or} \qquad =2c\cosh\alpha$$

$$\frac{p}{-2c}=\cos\alpha \qquad \left|\frac{p}{-2c}\right|=<1 \qquad \text{or} \qquad >1$$

For $\left|\frac{p}{2c}\right|<1$:

$$\frac{Mz^2+Nz}{z^2+pz+q}=\frac{Mz(z-c\cos\alpha)+[(Mc\cos\alpha+N)/(c\sin\alpha)](cz\sin\alpha)}{z^2-2cz\cos\alpha+c^2} \tag{C-42}$$

$$Z^{-1}\left(\frac{Mz^2+Nz}{z^2+pz+q}\right)=M[c^k\cos\alpha]+\left(\frac{Mc\cos\alpha+N}{c\sin\alpha}\right)[c^k\sin\alpha k] \tag{C-43}$$

Similarly for $\left|\frac{p}{2c}\right|>1$:

$$Z^{-1}\left(\frac{Mz^2+Nz}{z^2+pz+q}\right)=M[c^k\cosh\alpha]+\left(\frac{Mc\cosh\alpha+N}{c\sinh\alpha}\right)[c^k\sinh\alpha k] \tag{C-44}$$

where:

$$c^2=q \qquad p=2c\cosh\alpha \qquad \left|\frac{p}{2c}\right|=\cosh\alpha>1$$

Example C-8 Find inverse z-transform of:

$$\frac{z^2+3z}{z^2+z+1/9} \qquad z>1/3$$

Here:

$$q=c^2=\frac{1}{9} \qquad c=\pm 1/3 \qquad p=1$$

For $c=-1/3$:

$$\left|\frac{p}{2c}\right|=3/2>1$$

$$1=2c\cosh\alpha$$

$$\cosh\alpha=3/2$$

$$\sinh\alpha=\sqrt{5}/2 \qquad (\therefore \cosh^2\alpha-\sinh^2\alpha=1)$$

From Eq. (C-42):

$$\frac{z^2+3z}{z^2+z+1/9}=\frac{z(z-c\cosh\alpha)+[(c\cosh\alpha+3)/(c\sinh\alpha)](cz\sinh\alpha)}{z^2-2cz\cosh\alpha+c^2}$$

Substituting the numerical values and simplifying:

$$\frac{z^2+3z}{z^2+z+1/9}=\frac{z(z-c\cosh\alpha)}{z^2-2cz\cosh\alpha+c^2}-\frac{15}{\sqrt{5}}\frac{cz\sinh\alpha}{z^2-2cz\cosh\alpha+c}$$

Therefore:

$$Z^{-1}\frac{z^2+3z}{z^2+z+1/9}=\left(-\frac{1}{3}\right)^k\cosh\alpha k-\frac{15}{\sqrt{5}}\left(-\frac{1}{3}\right)^k\sinh\alpha \qquad k\ge 0$$

C-8 INVERSION BY RESIDUE METHOD

The residue theorem states that:

$$f(k)=\sum \text{residues of } z^{k-1}F(z) \quad \text{at its poles} \tag{C-45}$$

where the residue of a simple pole z_i is:

$$[(z-z_i)z^{k-1}F(z)]_{z=z_i} \tag{C-46}$$

TABLE C-2 Inverse z-Transforms

Partial Fraction	Inverse z-Transform: $\lvert z\rvert > \lvert a\rvert, k>0$	Inverse z-Transform: $\lvert z\rvert < \lvert a\rvert, k<0$
$\frac{z}{z-a}$	$a^k u(k)$	$-a^k u(k)$
$\frac{z^2}{(z-a)^2}$	$(k+1)a^k$	$-(k+1)a^k$
$\frac{z^3}{(z-a)^3}$	$\frac{1}{2!}(k+1)(k+2)a^k u(k)$	$\frac{1}{2!}(k+1)(k+2)a^k u(k)(-k+2)$
$\frac{z^n}{(z-a)^n}$	$\frac{1}{(n-1)!}(k+1)....(k+n-1)a^k u(k)$	$-\frac{1}{(n-1)!}(k+1)....(k+n-1)a^k$
$\frac{1}{z-a}$	$a^{k-1}u(k-1)$	$-a^{k-1}u(-k)$
$\frac{1}{(z-a)^3}$	$\frac{1}{2}(k-2)(k-1)a^{k-3}u(k-3)$	$-\frac{1}{2}(k-2)(k-1)a^{k-3}u(-k+2)$

The residue of a pole is its coefficient in the partial fraction expansion. For repeated poles, order r, at pole $z = z_i$ is:

$$\left[\frac{1}{(r-1)!}\frac{d^{r-1}}{dz^{r-1}}(z-z_i)^r z^{k-1}F(z)\right]_{z=z_i} \quad \text{(C-47)}$$

Example C-9 Find inverse z-transform of:

$$Z^{-1}\left(\frac{z}{(z-2)(z-3)}\right)$$

Residue at $z = 2$ is given by:

$$\left[(z-2)z^{k-1}\frac{z}{(z-2)(z-3)}\right]_{z=2} = -2^k$$

Residue at $z = 3$:

$$\left[(z-3)z^{k-1}\frac{z}{(z-2)(z-3)}\right]_{z=3} = 3^k$$

Thus, the required inverse is sum of the residues:

$$3^k - 2^k$$

Table C-2 provide inverse z-transforms.

C-9 SOLUTION OF DIFFERENCE EQUATIONS

Difference equations arise in electrical engineering when there is a recurrence in system configurations. As an example, an electrical network consisting of a chain of identical components—a combination of series and shunt impedances as in transmission lines—is best analyzed by difference equations. While the differential equations contain derivatives of the dependent variable, difference equations contain differences of values of the dependent variables at discrete equally spaced values of the independent variable. Following is an example of second-order linear difference equation:

$$x(t+2T)+a_1x(t+T)+a_0x(t)=e(t) \quad \text{(C-48)}$$

Coefficients, which may be negative are assumed to be constants. The functions may be defined only at $t = nT$, where $n = 0, 1, 2, \ldots$. It is of second-order as it contains an ordinate $x(t + 2T)$ and is linear as it does not contain powers, except first powers. If $T = 1$, the equation becomes:

$$x(n+2)+a_1x(n+1)+a_0x(n)=e(n) \quad \text{(C-49)}$$

The classical method of solving difference equations is akin to differential equations, the complementary function, and particular integral.

In Chap. 1, we stated that the origin of electromagnetic transient analysis program (EMTP), which has become the most important analytical tool for simulation of electrical system transients all over the world, started with a paper by H.W. Dommel published in 1969.[1] The paper proposed digital solution of electromagnetic transients based on difference equations.

Example C-10 Solve the difference equation:

$$y_{k+3}-3y_{k+2}+3y_{k+1}-y_k=u(k)$$

Take z-transform of both the sides:

$$Z[y_{k+3}]-3Z[y_{k+2}]+3Z[y_{k+1}]-Z[y_k]=Z[u(k)]=[z^3Y(z)-z^3Y(0)$$
$$-z^2Y(1)-zY(2)]-3[z^2Y(z)-z^2Y(0)-zY(1)]+3[3Y(z)$$
$$-zY(0)]-Y(z)=\frac{1}{1-z^{-1}}$$

Consider the initial conditions that $Y(0) = Y(1) = Y(2) = 0$. Then substituting and simplifying:

$$[z^3-3z^2+3z-1]Y(z)=\frac{1}{1-z^{-1}}$$

$$(z-1)^3Y(z)=\frac{1}{1-z^{-1}}$$

$$Y(z)=z^{-3}(1-z^{-1})^4$$

Therefore:

$$y_k = \text{Coefficient of } z^{-k} \quad \text{in } z^{-3} = z^{-3}(1-z^{-1})^{-4}$$
$$= \text{Coefficient of } z^{-k} \quad \text{in } z^{-k-3} = (1-z^{-1})^{-4}$$
$$y_k = \frac{(k-2)(k-1)k}{6} \qquad k \geq 3$$

C-10 STATE VARIABLE FORM

Akin to differential equations, it is often desirable to describe a system with a set of first-order difference equations, rather than by one or more nth-order difference equations. Consider an nth-order, single-input, linear, and constant-coefficient difference equation:

$$\sum_{i=0}^{n} a_i y(k+i) = u(k) \tag{C-50}$$

This can be replaced with:

$$\begin{aligned} x_1(k+1) &= x_2(k) \\ x_2(k+1) &= x_3(k) \\ &\vdots \\ x_n(k+1) &= -\frac{1}{a_n}\left[\sum_{i=0}^{n-1} a_i x_{i+1}(k)\right] + \frac{1}{a_n}u(k) \end{aligned} \tag{C-51}$$

or in the matrix form:

$$\begin{vmatrix} x_1(k+1) \\ x_2(k+1) \\ . \\ x_n(k+1) \end{vmatrix} = \begin{vmatrix} 0 & 1 & 0 & . & 0 \\ 0 & 0 & 1 & . & 0 \\ . & . & . & . & 1 \\ -a_0/a_n & -a_1/a_n & . & . & -a_{n-1}/a_n \end{vmatrix} \begin{vmatrix} x_1(k) \\ x_2(k) \\ . \\ x_n(k) \end{vmatrix} + \begin{vmatrix} 0 \\ 0 \\ . \\ 1/a_n \end{vmatrix} u \tag{C-52}$$

or:

$$\mathbf{x}(k+1) = \bar{\mathbf{A}}\mathbf{x}(k) = \bar{\mathbf{b}}u \tag{C-53}$$

Consider the solution of the difference equation:

$$x(k+2) + \frac{5}{6}x(k+1) + \frac{1}{6}x(k) = u(k) \tag{C-54}$$

In the state variable form, the equation can be written as:

$$\begin{aligned} x_1(k+1) &= x_2(k) \\ x_2(k+1) &= -\frac{5}{6}x_2(k) - \frac{1}{6}x_1(k) + u(k) \end{aligned} \tag{C-55}$$

Consider the state form of the difference equation for multi-input, multi-output systems, a modification of Eq. (C-53). See also Chap. 2.

$$\mathbf{x}(k+1) = \bar{\mathbf{A}}\mathbf{x}(k) + \bar{\mathbf{B}}\mathbf{r}(k) \qquad \text{(state equation)} \tag{C-56}$$

Consider $\mathbf{r}(k) = 1$. The z-transform of the vector-matrix form of the equation is:

$$z\mathbf{X}(z) - z\bar{\mathbf{x}}(0) = \bar{\mathbf{A}}\mathbf{X}(z) + \frac{z}{z-1}\bar{\mathbf{B}} \tag{C-57}$$

or:

$$\mathbf{X}(z) = z(z\mathbf{I} - \bar{\mathbf{A}})^{-1}\bar{\mathbf{x}}(0) + \frac{z}{z-1}(z\mathbf{I} - \bar{\mathbf{A}})^{-1}\bar{\mathbf{B}} \tag{C-58}$$

Here, we have:

$$\bar{\mathbf{A}} = \begin{vmatrix} 0 & 1 \\ -\frac{1}{6} & -\frac{5}{6} \end{vmatrix} \quad \bar{\mathbf{B}} = \begin{vmatrix} 0 \\ 1 \end{vmatrix} \quad \bar{\mathbf{x}}(k) = \begin{vmatrix} x_1(k) \\ x_2(k) \end{vmatrix} \quad \bar{\mathbf{x}}(0) = \begin{vmatrix} 0 \\ 1 \end{vmatrix} \tag{C-59}$$

Therefore:

$$\begin{aligned} z\mathbf{I} - \bar{\mathbf{A}} &= \begin{vmatrix} z & -1 \\ \frac{1}{6} & z+\frac{5}{6} \end{vmatrix} \\ (z\mathbf{I} - \bar{\mathbf{A}})^{-1} &= \frac{1}{z^2 + \frac{5}{6}z + \frac{1}{6}} \begin{vmatrix} z+\frac{5}{6} & 1 \\ -\frac{1}{6} & z \end{vmatrix} \end{aligned} \tag{C-60}$$

This gives:

$$\mathbf{X}(z) = \begin{vmatrix} \dfrac{z}{z^2+\frac{5}{6}z+\frac{1}{6}} \\ \dfrac{z}{z^2+\frac{5}{6}z+\frac{1}{6}} \end{vmatrix} + \begin{vmatrix} \dfrac{z}{(z-1)\left(z^2+\frac{5}{6}z+\frac{1}{6}\right)} \\ \dfrac{z^2}{(z-1)\left(z^2+\frac{5}{6}z+\frac{1}{6}\right)} \end{vmatrix} \tag{C-61}$$

The first term is the free response and the second term is the forced response:

$$\mathbf{x}(k) = \begin{vmatrix} \frac{1}{2} - 2\left(-\frac{1}{2}\right)^k + \frac{3}{2}\left(-\frac{1}{3}\right)^k \\ \frac{1}{2} - 2\left(-\frac{1}{2}\right)^k + \frac{3}{2}\left(-\frac{1}{3}\right)^k \end{vmatrix} \qquad k = 0, 1, 2, \ldots \tag{C-62}$$

REFERENCES

J. A. Cadzow and H. R. Martens, *Discrete Time Control Systems*, Prentice Hall, Englewood Cliffs, NJ, 1970.

H. W. Dommel, "Digital Computer Solution of Electromagnetic Transients in Single- and Multiphase Networks," *IEEE Trans. PAS*, vol. PAS-88, no. 4, pp. 388–399, Apr. 1969.

S. Goldberg, *Difference Equations*, Wiley, New York, 1958.

W. D. Humpage, *z-Transform Electromagnetic Transient Analysis in High Voltage Networks*, Peter Peregrinus (IEE-UK), 1982.

E. I. Jury, *Theory and Application of z-transform Method*, Wiley, New York, 1964.

B. C. Kuo, *Discrete Data Control Systems, Science Tech.*, Champaign, IL, 1974.

E. J. Muth, *Transform Methods with Applications to Engineering and Operations Research*, Prentice Hall, Englewood Cliffs, NJ, 1977.

I. J. Nagrath and M. Gopal, *Systems Modeling and Analysis*, Tata McGraw Hill, New Delhi.

R. Vich, *z-Transform Theory and Application*, Kluwar Academic, New York, 1987.

APPENDIX D

SEQUENCE IMPEDANCES OF TRANSMISSION LINES AND CABLES

The concept of symmetrical components is basic to the understanding of transformation of sequence impedances. It is assumed that a reader has some knowledge of this subject, which is covered in many texts.[1–4] Also, a fair knowledge of matrix manipulation is assumed.

Practically, the transmission or cable systems constants will be calculated using computer-based subroutines in EMTP-like programs which provide evaluation routines for low-to-high frequency models. Some data is available in tabular form in the texts.[5–8] The basis of these calculations and required transformations is of interest for system modeling, fault studies, and transients analyses, and to formulate a theoretical base.

D-1 AC RESISTANCE OF CONDUCTORS

The conductor ac resistance is dependent on frequency and proximity effects, temperature, and bundle conductor effects. Spiraling of conductors with a certain pitch increases the length of wound conductor. The resistance increases linearly with temperature and is given by the following equation:

$$R_2 = R_1\left(\frac{T+t_2}{T+t_1}\right) \qquad \text{(D-1)}$$

where R_2 is the resistance at temperature t_2 and R_1 is the resistance at temperature t_1. T is the temperature coefficient which depends on the conductor material. It is 234.5 for annealed copper, 241.5 for hard drawn copper, and 228.1 for aluminum. The resistance is read from manufacturer's data, databases in computer programs, or generalized tables.

The skin effect considers that as the frequency increases, the current distribution in the cross section of the conductor will not be uniform. The current tends to flow more densely near the outer surface of the conductor than toward the center. This is because an ac flux results in induced emfs which are greater at the center than at the circumference, so that potential difference tends to establish currents that oppose the main current at the center and assist the current at the circumference. The result is that the current is forced to the outside, reducing the effective area of the conductor. The effect is utilized in high-ampacity hollow conductors and tubular bus bars, to save material costs. The skin effect is given by:

$$Y_{cs} = F(x_s) \qquad \text{(D-2)}$$

where Y_{cs} are the skin effect losses in the conductor and $F(x_s)$ is a skin effect function:

$$x_s = 0.875\sqrt{f\frac{k_s}{R_{dc}}} \qquad \text{(D-3)}$$

where k_s depends on conductor construction, and f is the frequency.

Proximity effect is also considered in cables. This occurs because of distortion of current between two conductors in close proximity. The concentration of current occurs in parts of bus bars or conductors closest to each other (currents flowing in forward and return paths). The expressions and graphs for calculating proximity effect are given in Ref. 9.

Thus, we can write the following equation describing the ac and dc resistance of the conductor, considering both—the skin effect as well as proximity effect.

$$R_{ac}/R_{dc} = 1 + Y_{cs} + Y_{cp} \qquad \text{(D-4)}$$

The internal inductance and resistance ac to dc ratios using Bessel's functions is provided in Ref. 10. The internal impedance of conductors is comparatively small. Its frequency-dependent behavior can be simulated by dividing the conductor into n hollow cylinders with an internal reactance L and conductivity $G_r = 1/R_r$. This model can be replicated by an equivalent circuit consisting of a series of connections of hollow cylinders (Fig. D-1).[11] The values of L_r and G_r for each cylinder can be calculated, depending upon radius r_r.

$$r_r = r_n\left(\frac{r}{n}\right)^m \qquad \text{(D-5)}$$

A good approximation of real $R(f)$ and $X(f)$ curves can be achieved by varying the exponent m. For example, with $n = 5$ and $m = 0.15$,

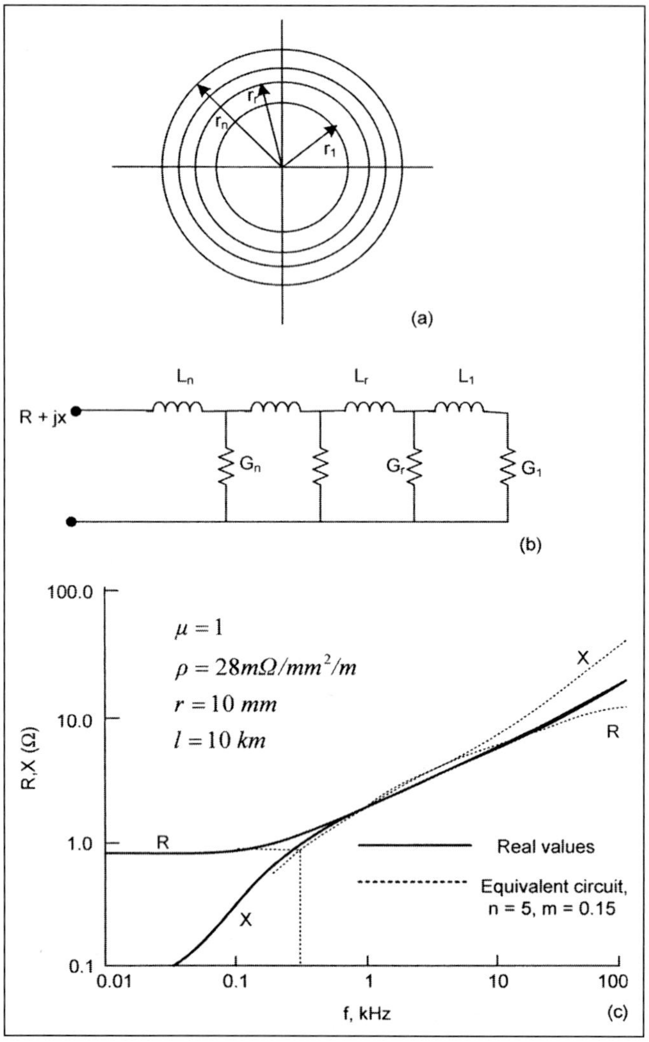

FIGURE D-1 (*a*) Conductor divided into concentric cylinders. (*b*) Equivalent circuit. (*c*) Simulation results of internal resistance and reactance, actual versus equivalent circuit.

a good agreement of the equivalent circuit with real performance of conductor up to 100 kHz is achieved.

$$R = R_{\text{int}} = \frac{\rho l}{\pi r_n^2} \qquad X = 2\pi f L_{\text{int}} = 2\pi f \frac{\mu l}{8\pi} \qquad \text{for } f << f_x$$

$$R = R_{\text{int}}\sqrt{\frac{f}{f_x}} \qquad X = 2\pi f L_{\text{int}}\sqrt{\frac{f}{f_x}} \qquad \text{for } f >> f_x \tag{D-6}$$

D-2 INDUCTANCE OF TRANSMISSION LINES

The *internal* inductance of a solid, smooth, round metallic cylinder of infinite length is due to its internal magnetic field when carrying an alternating current and is given by:

$$L_{\text{int}} = \frac{\mu_0}{8\pi}\ \text{H/m} \quad \text{(henry per meter)} \tag{D-7}$$

where μ_0 is the permeability $= 4\pi 10^{-7}$(H/m). Its *external* inductance is due to flux outside the conductor and is given by:

$$L_{\text{ext}} = \frac{\mu_0}{2\pi}\ln\left(\frac{D}{r}\right)\ \text{H/m} \tag{D-8}$$

where D is any point at a distance D from the surface of the conductor and r is the conductor radius. In most inductance tables, $D = 1$ ft and adjustment factors are tabulated for higher conductor spacing. The total reactance is:

$$L = \frac{\mu_0}{2\pi}\left[\frac{1}{4} + \ln\frac{D}{r}\right] = \frac{\mu_0}{2\pi}\left[\ln\frac{D}{e^{-1/4}r}\right] = \frac{\mu_0}{2\pi}\left[\ln\frac{D}{\text{GMR}}\right]\ \text{H/m} \tag{D-9}$$

where GMR is called the geometric mean radius and is equal to $0.7788r$. It can be defined as the radius of a tubular conductor with an infinitesimally thin wall that has the same external flux out to a radius of 1 ft as the external and internal flux of a solid conductor to the same distance.

D-2-1 Inductance of a Three-Phase Line

We can write the inductance matrix of a three-phase line in terms of flux linkages λ_a, λ_b, and λ_c:

$$\begin{vmatrix} \lambda_a \\ \lambda_b \\ \lambda_c \end{vmatrix} = \begin{vmatrix} L_{aa} & L_{ab} & L_{ac} \\ L_{ba} & L_{bb} & L_{bc} \\ L_{ca} & L_{cb} & L_{cc} \end{vmatrix} \begin{vmatrix} I_a \\ I_b \\ I_c \end{vmatrix} \tag{D-10}$$

The flux linkages λ_a, λ_b, and λ_c are given by:

$$\lambda_a = \frac{\mu_0}{2\pi}\left[I_a \ln\left(\frac{1}{\text{GMR}_a}\right) + I_b \ln\left(\frac{1}{D_{ab}}\right) + I_c \ln\left(\frac{1}{D_{ac}}\right)\right]$$

$$\lambda_b = \frac{\mu_0}{2\pi}\left[I_a \ln\left(\frac{1}{D_{ba}}\right) + I_b \ln\left(\frac{1}{\text{GMR}_b}\right) + I_c \ln\left(\frac{1}{D_{bc}}\right)\right] \tag{D-11}$$

$$\lambda_c = \frac{\mu_0}{2\pi}\left[I_a \ln\left(\frac{1}{D_{ca}}\right) + I_b \ln\left(\frac{1}{D_{cb}}\right) + I_c \ln\left(\frac{1}{\text{GMR}_c}\right)\right]$$

where D_{ab}, D_{ac}, . . . are the distances between conductor of a phase with respect to conductors of b and c phases. L_{aa}, L_{bb}, and L_{cc} are the self-inductance of the conductors and L_{ab}, L_{ac}, . . . are the mutual inductances. If we assume a symmetrical line, that is, GMR of all three conductors is equal and also the spacing between the conductors is equal, the equivalent inductance per phase is:

$$L = \frac{\mu_0}{2\pi}\ln\left(\frac{D}{\text{GMR}}\right)\ \text{H/m} \tag{D-12}$$

The phase-to-neutral inductance of a tree-phase symmetrical line is the same as the inductance/conductor of a two-phase line.

D-3 TRANSPOSED LINE

A transposed line is shown in Fig. D-2. Each phase conductor occupies the position of other two phase conductors for one-third length of the length. The purpose is to equalize the phase inductances and reduce unbalance. The inductance derived for symmetrical line is still valid and the distance D in Eq. (D-8) is substituted by the geometric mean distance (GMD). It is given by:

$$\text{GMD} = (D_{ab}D_{bc}D_{ca})^{1/3} \tag{D-13}$$

For a mathematical relation, consider a rotation matrix $\bar{R}$, given by:

$$\bar{R} = \begin{vmatrix} 0 & 0 & 1 \\ 1 & 0 & 0 \\ 0 & 1 & 0 \end{vmatrix} \tag{D-14}$$

It can be demonstrated that *premultiplying* by $\bar{R}$ will move the third row in a 3 × 3 impedance matrix to row 1; postmultiplying

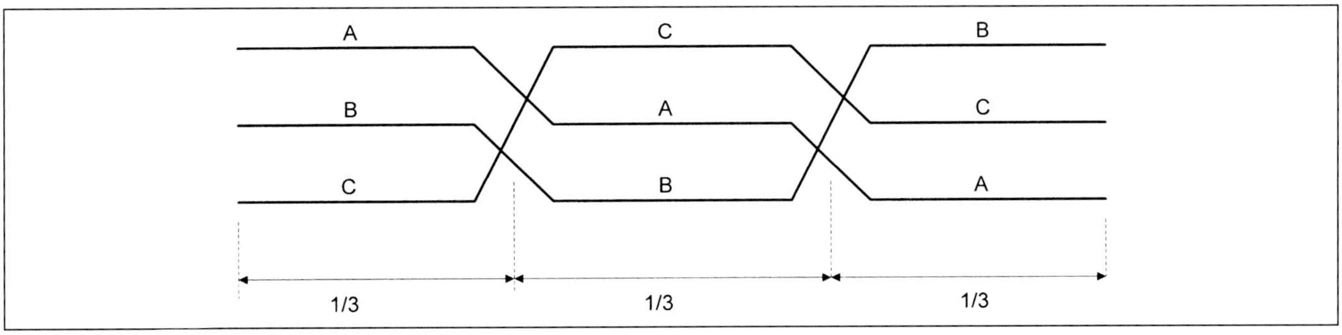

FIGURE D-2 Transposed transmission line.

by $\bar{R}^{-1}$ will move the third column to position 1; premultiplying by $\bar{R}^{-1} = \bar{R}^2$ will move the first row to the position of the third row; postmultiplying by $\bar{R}$ will move the first column to position 3. These manipulations can be used to compute the overall impedance of a transposed line. Figure D-2 shows a transposed line in three sections. In the first section, we can write:

$$\begin{vmatrix} V_{a1} \\ V_{b1} \\ V_{c1} \end{vmatrix} = \begin{vmatrix} Z_{11-1} & Z_{12-1} & Z_{13-1} \\ Z_{21-1} & Z_{22-1} & Z_{23-1} \\ Z_{31-1} & Z_{32-2} & Z_{33-1} \end{vmatrix} \begin{vmatrix} I_{a1} \\ I_{b1} \\ I_{c1} \end{vmatrix} \tag{D-15}$$

or in compact form:

$$\bar{V}_{abc-1} = \bar{Z}_{123-1}\bar{I}_{abc-1} \tag{D-16}$$

In the second section, phases *c-a-b* correspond to positions *a-b-c* of the first section. Therefore, the following transformation applies:

$$\bar{R}^{-1}\bar{V}_{123} = \left[\bar{R}^{-1}Z_{123-1}\bar{R}\right]\bar{R}^{-1}I_{123}$$

which is the same as:

$$V_{231} = Z_{231-2}I_{231} \tag{D-17}$$

In the third section, phases *b-c-a*, correspond to the phases *a-b-c* in the first section. Thus, the following transformation applies:

$$\bar{R}\bar{V}_{123} = \left[\bar{R}Z_{123-1}\bar{R}^{-1}\right]\bar{R}I_{123} \tag{D-18}$$

which is the same as:

$$V_{312} = Z_{312-3}I_{312}$$

Then the overall impedance matrix of the line is:

$$\bar{Z} = \begin{vmatrix} (Z_{11-1}+Z_{22-2}+Z_{33-3}) & (Z_{12-1}+Z_{23-2}+Z_{31-3}) & (Z_{13-1}+Z_{21-2}+Z_{32-3}) \\ (Z_{21-1}+Z_{32-2}+Z_{13-3}) & (Z_{22-1}+Z_{33-2}+Z_{11-3}) & (Z_{23-1}+Z_{31-2}+Z_{12-3}) \\ (Z_{31-1}+Z_{12-2}+Z_{23-3}) & (Z_{32-1}+Z_{13-2}+Z_{21-3}) & (Z_{33-1}+Z_{11-2}+Z_{22-3}) \end{vmatrix} \tag{D-19}$$

Example D-1 Consider a transposed line with three *equal* sections, though practically, it will not be possible to transpose exactly in three sections. Moreover, the utilities are getting away from transposition of the high-voltage lines.

Consider:

$$Z_{11} = Z_{22} = Z_{33} = 0.2 + j2.0$$

$$Z_{12} = Z_{21} = Z_{23} = Z_{32} = 0.1 + j1.0$$

$$Z_{13} = Z_{31} = 0.1 + j1.2$$

Then for the first section:

$$\bar{Z}_{123} = \begin{vmatrix} 0.2+j2.0 & 0.1+j1.0 & 0.1+j1.2 \\ 0.1+j1.0 & 0.2+j2.0 & 0.1+j1.0 \\ 0.1+j1.2 & 0.1+j1.0 & 0.2+j2.0 \end{vmatrix}$$

Then, for the second section,

$$\bar{Z}_{231} = \bar{R}^{-1}\bar{Z}_{123}\bar{R} = \begin{vmatrix} 0.2+j2.0 & 0.1+j1.0 & 0.1+j1.0 \\ 0.1+j1.0 & 0.2+j2.0 & 0.1+j1.2 \\ 0.1+j1.0 & 0.1+j1.2 & 0.2+j1.2 \end{vmatrix}$$

Similarly,

$$\bar{Z}_{312} = \bar{R}\bar{Z}_{123}\bar{R}^{-1} = \begin{vmatrix} 0.2+j2.0 & 0.1+j1.2 & 0.1+j1.0 \\ 0.1+j1.2 & 0.2+j2.0 & 0.1+j1.0 \\ 0.1+j1.0 & 0.1+j1.0 & 0.2+j2.0 \end{vmatrix}$$

The sum of the three sections [Eq. (D.19)] is:

$$\bar{Z} = \begin{vmatrix} 0.6+j6.0 & 0.3+j3.2 & 0.3+j3.2 \\ 0.3+j3.2 & 0.6+j6.0 & 0.3+j3.2 \\ 0.3+j3.2 & 0.3+j3.2 & 0.6+j6.0 \end{vmatrix}$$

All the mutual impedances are exactly balanced. This can be decoupled to a diagonal matrix with symmetrical component transformation.

D-4 COMPOSITE CONDUCTORS

A transmission line with composite conductors is shown in Fig. D-3. Consider that group X is composed of n conductors in parallel, each of which carries $1/n$ of the line current, and the group Y is composed of m parallel conductors, each of which carries $-1/m$ of the return current. Then L_x, the inductance of conductor group X is:

$$L_x = 2\times10^{-7}\ln\frac{\sqrt[nm]{(D_{aa}D_{ab}D_{ac}\cdots D_{am}),\ldots,(D_{na}D_{nb}D_{nc}\cdots D_{nm})}}{\sqrt[n^2]{(D_{aa}D_{ab}D_{ac}\cdots D_{an}),\ldots,(D_{na}D_{nb}D_{nc}\cdots D_{nn})}} \tag{D-20}$$

We write Eq. (D-20) as:

$$L_x = 2\times10^{-7}\ln\left(\frac{D_m}{D_{sx}}\right)\ \text{H/m} \tag{D-21}$$

The term D_m abbreviated for numerator of Eq. (D-20) can be called mutual GMD and denominator GMR.

Similarly:

$$L_y = 2\times10^{-7}\ln\left(\frac{D_m}{D_{sy}}\right)\ \text{H/m} \tag{D-22}$$

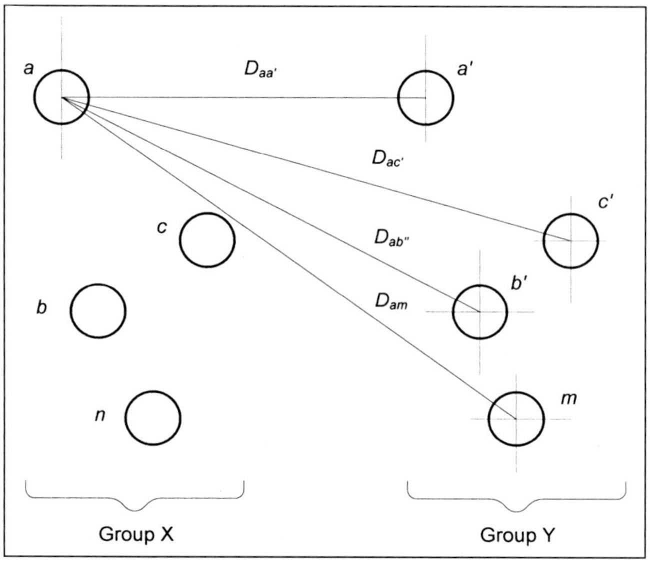

FIGURE D-3 Inductance of composite conductors.

The total inductance is:

$$L=(L_x+L_y)\text{ H/m} \tag{D-23}$$

D-5 IMPEDANCE MATRIX

Consider a symmetrical three-phase line with 3 × 3 matrix having equal self-impedances and mutual impedances:

$$\bar{Z}_{abc}=\begin{vmatrix} Z & M & M \\ M & Z & M \\ M & M & Z \end{vmatrix} \tag{D-24}$$

It can be diagonalized (decoupled) with symmetrical components:

$$\begin{aligned}\bar{Z}_{012}&=\bar{T}_s^{-1}\bar{Z}_{abc}\bar{T}_s\\ &=\frac{1}{3}\begin{vmatrix} 1 & 1 & 1 \\ 1 & a & a^2 \\ 1 & a^2 & a \end{vmatrix}\bar{Z}_{abc}\begin{vmatrix} 1 & 1 & 1 \\ 1 & a^2 & a \\ 1 & a & a^2 \end{vmatrix}\\ &=\frac{1}{3}\begin{vmatrix} Z+2M & 0 & 0 \\ 0 & Z-M & 0 \\ 0 & 0 & Z-M \end{vmatrix}\end{aligned} \tag{D-25}$$

But if we start with an original three-phase system, not completely balanced, we will have different results. Ignoring mutual impedances, let:

$$\bar{Z}_{abc}=\begin{vmatrix} Z_1 & 0 & 0 \\ 0 & Z_2 & 0 \\ 0 & 0 & Z_3 \end{vmatrix} \tag{D-26}$$

Then attempting to decouple it will give:

$$\bar{Z}_{012}=\frac{1}{3}\begin{vmatrix} Z_1+Z_2+Z_3 & Z_1+a^2Z_2+aZ_3 & Z_1+aZ_2+aZ_3 \\ Z_1+aZ_2+aZ_3 & Z_1+Z_2+Z_3 & Z_1+a^2Z_2+Z_3 \\ Z_1+a^2Z_2+aZ_3 & Z_1+aZ_2+aZ_3 & Z_1+Z_2+Z_3 \end{vmatrix} \tag{D-27}$$

If we start with equal self-impedances and unequal mutual impedances, the results will be the same. The resulting matrix is nonsymmetrical. In the application of symmetrical component theory, it is assumed that the system is perfectly balanced before an unbalance condition occurs, for example, the study of a line-to-ground fault.

Generally, for transmission lines, the off-diagonal elements of the sequence impedance matrix are zero. In high-voltage transmission lines which are transposed, this is generally true and the mutual couplings between phases are almost equal. However, the same cannot be said of distribution lines, and these may have unequal off-diagonal terms. In many cases the off-diagonal terms are smaller than the diagonal terms, and the errors introduced in ignoring these will be small. Sometimes equivalence can be drawn by the equations:

$$\begin{aligned}Z_s&=\frac{Z_{aa}+Z_{bb}+Z_{cc}}{3}\\ Z_m&=\frac{Z_{ab}+Z_{bc}+Z_{ca}}{3}\end{aligned} \tag{D-28}$$

that is, an average of the self-and mutual impedances can be taken. The sequence impedance matrix then has only diagonal terms. See Example D-2.

D-6 THREE-PHASE LINE WITH GROUND CONDUCTORS

A three-phase transmission line has couplings between phase-to-phase conductors and also between phase-to-ground conductors. Consider a three-phase line with two ground conductors, as shown in Fig. D-4. The voltage V_a can be written as:

$$\begin{aligned}V_a&=R_aI_a+j\omega L_aI_a+j\omega L_{ab}I_b+j\omega L_{ac}I_c+j\omega L_{aw}I_w+j\omega L_{av}I_v\\ &\quad-j\omega L_{an}I_n+V_a'+R_nI_n+j\omega L_nI_n-j\omega L_{an}I_a-j\omega L_{bn}I_b\\ &\quad-j\omega L_{cn}I_c-j\omega L_{wn}I_w-j\omega L_{vn}I_v\end{aligned} \tag{D-29}$$

where $R_a, R_b, \ldots, R_n$ are the resistances of phases $a, b, \ldots, n$, respectively, $L_a, L_b, \ldots, L_n$ are the self-inductances, and $L_{ab}, L_{ac}, \ldots, L_{an}$ are the mutual inductances.

This can be written as:

$$\begin{aligned}V_a&=(R_a+R_n)I_a+R_nI_b+R_nI_c+j\omega(L_a+L_n-2L_{an})I_a\\ &\quad+j\omega(L_{ab}+L_n-L_{an}-L_{bn})I_b+j\omega(L_{ac}+L_n-L_{an}-L_{cn})I_c+R_nI_w\\ &\quad+j\omega(L_{aw}+L_n-L_{an}-L_{wn})I_w+R_nI_v+j\omega(L_{av}+L_n-L_{an}-L_{vn})I_v+V_a'\\ &=Z_{aa-g}I_a+Z_{ab-g}I_b+Z_{ac-g}I_c+Z_{aw-g}I_w+Z_{av-g}I_v\end{aligned} \tag{D-30}$$

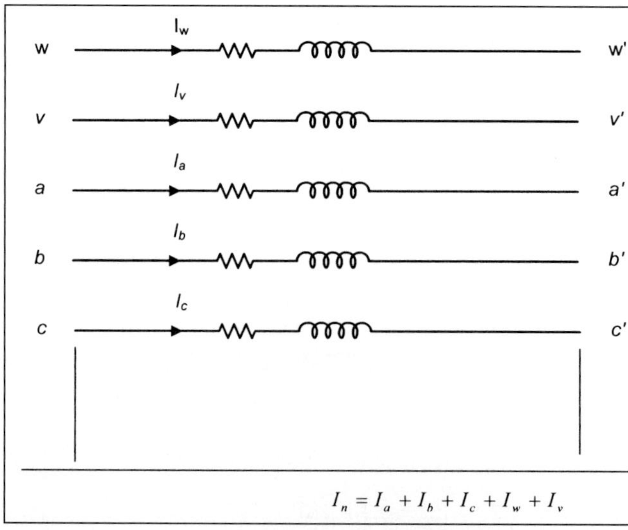

FIGURE D-4 Transmission line section with two ground conductors.

where $Z_{aa\text{-}g}$ and $Z_{bb\text{-}g}$ are the self-impedance of a conductor with ground return, $Z_{ab\text{-}g}$ and $Z_{ac\text{-}g}$ are the mutual impedances between two conductors with common earth return.

Similar equations apply to voltages of other phases and ground wires. Then the following matrix holds for the voltage differentials between terminals marked w, v, a, b, c and w', v', a', b' and c':

$$\begin{vmatrix} \Delta V_a \\ \Delta V_b \\ \Delta V_c \\ \Delta V_w \\ \Delta V_v \end{vmatrix} = \begin{vmatrix} Z_{aa-g} & Z_{ab-g} & Z_{ac-g} & Z_{aw-g} & Z_{av-g} \\ Z_{ba-g} & Z_{bb-g} & Z_{bc-g} & Z_{bw-g} & Z_{bv-g} \\ Z_{ca-g} & Z_{cb-g} & Z_{cc-g} & Z_{cw-g} & Z_{cv-g} \\ Z_{wa-g} & Z_{wb-g} & Z_{wc-g} & Z_{ww-g} & Z_{wv-g} \\ Z_{va-g} & Z_{vb-g} & Z_{vc-g} & Z_{vw-g} & Z_{vv-g} \end{vmatrix} \begin{vmatrix} I_a \\ I_b \\ I_c \\ I_w \\ I_v \end{vmatrix} \tag{D-31}$$

In the partitioned form this matrix can be written as:

$$\begin{vmatrix} \Delta \bar{V}_{abc} \\ \Delta \bar{V}_{wv} \end{vmatrix} = \begin{vmatrix} \bar{Z}_A & \bar{Z}_B \\ \bar{Z}_C & \bar{Z}_D \end{vmatrix} \begin{vmatrix} \bar{I}_{abc} \\ \bar{I}_{wv} \end{vmatrix} \tag{D-32}$$

Considering that the ground wire voltages are zero:

$$\begin{aligned} \Delta \bar{V}_{abc} &= \bar{Z}_A \bar{I}_{abc} + \bar{Z}_B \bar{I}_{vw} \\ 0 &= \bar{Z}_C \bar{I}_{abc} + \bar{Z}_D \bar{I}_{vw} \end{aligned} \tag{D-33}$$

Thus:

$$\begin{aligned} \bar{I}_{vw} &= -\bar{Z}_D^{-1} \bar{Z}_C \bar{I}_{abc} \\ \Delta \bar{V}_{abc} &= \left(\bar{Z}_A - \bar{Z}_B \bar{Z}_D^{-1} \bar{Z}_C\right) \bar{I}_{abc} \end{aligned} \tag{D-34}$$

This can be written as:

$$\Delta \bar{V}_{abc} = \bar{Z}_{abc} \bar{I}_{abc}$$

$$\bar{Z}_{abc} = \bar{Z}_A - \bar{Z}_B \bar{Z}_D^{-1} \bar{Z}_C = \begin{vmatrix} Z_{aa'-g} & Z_{ab'-g} & Z_{ac'-g} \\ Z_{ba'-g} & Z_{bb'-g} & Z_{bc'-g} \\ Z_{ca'-g} & Z_{cb'-g} & Z_{cc'-g} \end{vmatrix} \tag{D-35}$$

The five-conductor circuit is reduced to *an*-equivalent three-conductor circuit. The technique is applicable to circuits with any number of ground wires provided the voltages are zero in the lower portion of the voltage vector.

D-7 BUNDLE CONDUCTORS

Consider bundle conductors, consisting of two conductors per phase (Fig. D-5). The original circuit of conductors a, b, c, and a', b', c' can be transformed into an equivalent conductor system of a'', b'', *and* c''.

Each conductor in the bundle carries a different current and has different self-and mutual impedance because of its specific location. Let the currents in the conductors be I_a, I_b, I_c, and I'_a, I'_b, and I'_c, respectively. The following primitive matrix equation can be written as:

$$\begin{vmatrix} V_a \\ V_b \\ V_c \\ \hline V'_a \\ V'_b \\ V'_c \end{vmatrix} = \begin{vmatrix} Z_{aa} & Z_{ab} & Z_{ac} & | & Z_{aa'} & Z_{ab'} & Z_{ac'} \\ Z_{ba} & Z_{bb} & Z_{bc} & | & Z_{ba'} & Z_{bb'} & Z_{bc'} \\ Z_{ca} & Z_{cb} & Z_{cc} & | & Z_{ca'} & Z_{cb'} & Z_{cc'} \\ - & - & - & - & - & - & - \\ Z_{a'a} & Z_{a'b} & Z_{a'c} & | & Z_{a'a'} & Z_{a'b'} & Z_{a'c'} \\ Z_{b'a} & Z_{b'b} & Z_{b'c} & | & Z_{b'a'} & Z_{b'b'} & Z_{b'c'} \\ Z_{c'a} & Z_{c'b} & Z_{c'c} & | & Z_{c'a'} & Z_{c'b'} & Z_{c'c'} \end{vmatrix} \begin{vmatrix} I_a \\ I_b \\ I_c \\ \hline I'_a \\ I'_b \\ I'_c \end{vmatrix} \tag{D-36}$$

This can be partitioned so that:

$$\begin{vmatrix} \bar{V}_{abc} \\ \bar{V}_{a'b'c'} \end{vmatrix} = \begin{vmatrix} \bar{Z}_1 & \bar{Z}_2 \\ \bar{Z}_3 & \bar{Z}_4 \end{vmatrix} \begin{vmatrix} \bar{I}_{abc} \\ \bar{I}_{a'b'c'} \end{vmatrix} \tag{D-37}$$

For symmetrical arrangement of bundle conductors $\bar{Z}_1 = \bar{Z}_4$. Modify so that lower portion of the vector goes to zero. Assume that:

$$\begin{aligned} V_a &= V'_a = V''_a \\ V_b &= V'_b = V''_b \\ V_c &= V'_c = V''_c \end{aligned} \tag{D-38}$$

Then upper part of the matrix can be subtracted from the lower part:

$$\begin{vmatrix} V_a \\ V_b \\ V_c \\ \hline 0 \\ 0 \\ 0 \end{vmatrix} = \begin{vmatrix} Z_{aa} & Z_{ab} & Z_{ac} & | & Z_{aa'} & Z_{ab'} & Z_{ac'} \\ Z_{ba} & Z_{bb} & Z_{bc} & | & Z_{ba'} & Z_{bb'} & Z_{bc'} \\ Z_{ca} & Z_{cb} & Z_{cc} & | & Z_{ca'} & Z_{cb'} & Z_{cc'} \\ -- & -- & -- & | & -- & -- & -- \\ Z_{a'a}-Z_{aa} & Z_{a'b}-Z_{ab} & Z_{a'c}-Z_{ac} & | & Z_{a'a'}-Z_{aa'} & Z_{a'b'}-Z_{ab'} & Z_{a'c'}-Z_{ac'} \\ Z_{b'a}-Z_{ba} & Z_{b'b}-Z_{bb} & Z_{b'c}-Z_{bc} & | & Z_{b'a'}-Z_{ba'} & Z_{b'b'}-Z_{bb'} & Z_{b'c'}-Z_{bc'} \\ Z_{c'a}-Z_{ca} & Z_{c'b}-Z_{cb} & Z_{c'c}-Z_{cc} & | & Z_{c'a'}-Z_{ca'} & Z_{c'b'}-Z_{cb'} & Z_{c'c'}-Z_{cc'} \end{vmatrix} \begin{vmatrix} I_a \\ I_b \\ I_c \\ \hline I'_a \\ I'_b \\ I'_c \end{vmatrix} \tag{D-39}$$

We can write it in the partitioned form as:

$$\begin{vmatrix} \bar{V}_{abc} \\ 0 \end{vmatrix} = \begin{vmatrix} \bar{Z}_1 & \bar{Z}_2 \\ \bar{Z}_2^t - \bar{Z}_1 & \bar{Z}_4 - \bar{Z}_2 \end{vmatrix} = \begin{vmatrix} \bar{I}_{abc} \\ \bar{I}_{a'b'c'} \end{vmatrix} \tag{D-40}$$

Write:

$$\begin{aligned} I''_a &= I_a + I'_a \\ I''_b &= I_b + I'_b \\ I''_c &= I_c + I'_c \end{aligned} \tag{D-41}$$

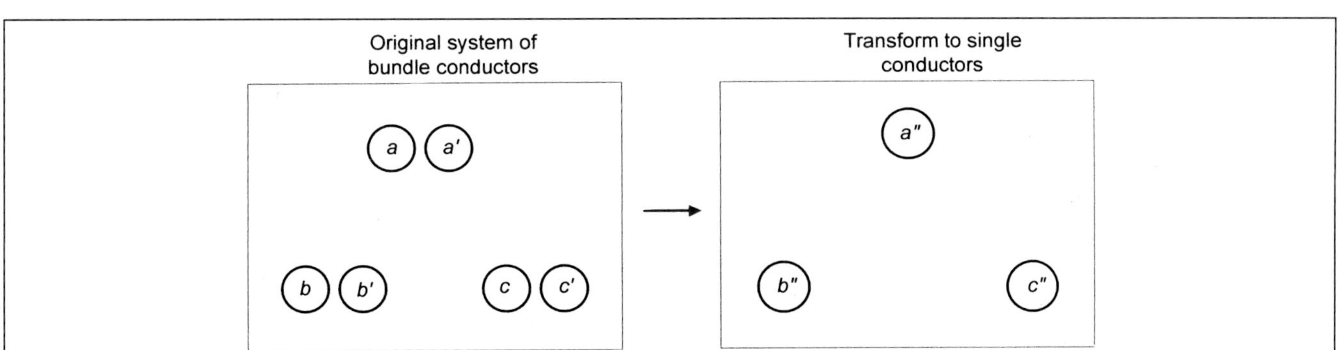

FIGURE D-5 Transformation of bundle conductors to equivalent single-conductor system.

We can write the following matrix in the partitioned form:

$$\begin{vmatrix} \bar{V}_{abc} \\ 0 \end{vmatrix} = \begin{vmatrix} \bar{Z}_1 & \bar{Z}_2 - \bar{Z}_1 \\ \bar{Z}_2^t - \bar{Z}_1 & (\bar{Z}_4 - \bar{Z}_2) - (\bar{Z}_2^t - \bar{Z}_1) \end{vmatrix} = \begin{vmatrix} \bar{I}''_{abc} \\ \bar{I}_{a'b'c'} \end{vmatrix} \qquad \text{(D-42)}$$

This can now be reduced to the following 3 × 3 matrix as before:

$$\begin{vmatrix} V''_a \\ V''_b \\ V''_c \end{vmatrix} = \begin{vmatrix} Z''_{aa} & Z''_{ab} & Z''_{ac} \\ Z''_{ba} & Z''_{bb} & Z''_{bc} \\ Z''_{ca} & Z''_{cb} & Z''_{cc} \end{vmatrix} \begin{vmatrix} I''_a \\ I''_b \\ I''_c \end{vmatrix} \qquad \text{(D-43)}$$

See Example D-3 for an application.

D-8 CARSON'S FORMULA

The theoretical value of $Z_{abc\text{-}g}$ can be calculated by Carson's formula (circa 1926). This is of importance even today in calculations of line constants. For an n-conductor configuration, the earth is assumed an infinite uniform solid with a constant resistivity. Figure D-6 shows image conductors in the ground at a distance equal to the height of the conductors above ground and exactly in the same formation, with same spacing between the conductors. A flat conductor formation is shown in Fig. D-6.

$$Z_{ii} = R_i + 4\omega P_{ii} G + j\left[X_i + 2\omega G \ln\frac{S_{ii}}{r_i} + 4\omega Q_{ii} G\right] \ \Omega/\text{mi}$$

$$Z_{ij} = 4\omega P_{ii} G + j\left[2\omega G \ln\frac{S_{ij}}{D_{ij}} + 4\omega Q_{ij} G\right] \ \Omega/\text{mi} \qquad \text{(D-44)}$$

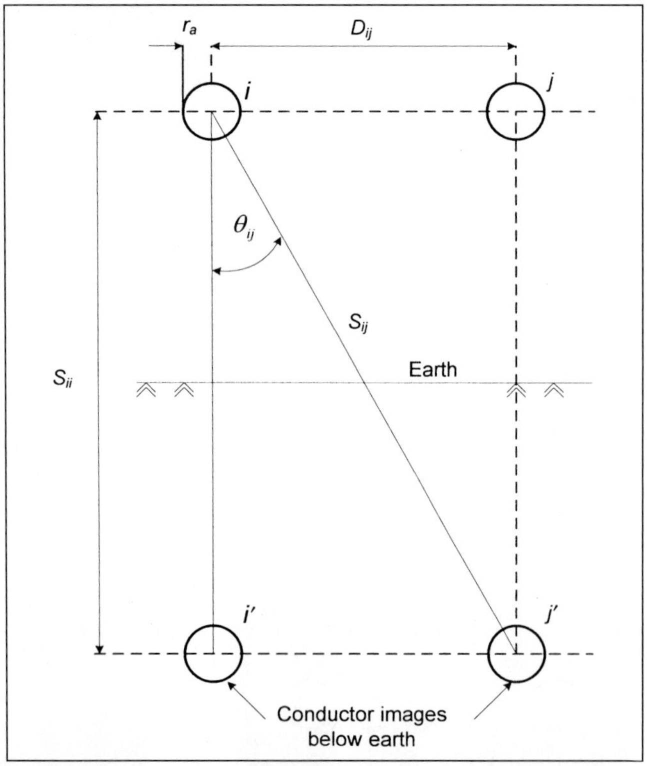

FIGURE D-6 Line conductors and their images at equal depth in the ground, Carson's formula.

where Z_{ii} is the self-impedance of conductor i with earth return, Ω/mi, Z_{ij} is the mutual impedance between conductors i and j, Ω/mi, R_i is the resistance of conductor, Ω/mi, S_{ii} is the conductor to image distance of the ith conductor to its own image, D_{ij} is the distance between conductors i and j, r_i is the radius of conductor, ft, S_{ij} is the conductor-to-image distance of the ith conductor to the image of jth conductor, ω is the angular frequency, G is 0.1609347 × 10^{-7} Ω-cm/abohm-mi, GMR$_i$ equal to geometric mean radius of conductor i, ρ is the soil resistivity, and θ_{ij} is the angle as shown in Fig. D-6.

Expressions for P and Q are:

$$P = \frac{\pi}{8} - \frac{1}{3\sqrt{2}}k\cos\theta + \frac{k^2}{16}\cos 2\theta\left(0.6728 + \ln\frac{2}{k}\right)$$

$$+\frac{k^2}{16}\theta\sin\theta + \frac{k^3\cos 3\theta}{45\sqrt{2}} - \frac{\pi k^4 \cos 4\theta}{1536}$$

$$Q = -0.0386 + \frac{1}{2}\ln\frac{2}{k} + \frac{1}{3\sqrt{2}}\cos\theta - \frac{k^2\cos 2\theta}{64} + \frac{k^3\cos 3\theta}{45\sqrt{2}}$$

$$-\frac{k^4\sin 4\theta}{384} - \frac{k^4\cos 4\theta}{384}\left(\ln\frac{2}{k} + 1.0895\right) \qquad \text{(D-45)}$$

where $k = 8.565\times10^4 S_{ij}\sqrt{f/\rho}$, S_{ij} is in feet, ρ is soil resistivity in ohms-meter, and f is the system frequency. This shows dependence on frequency as well as soil resistivity.

D-8-1 Approximations to Carson's Equations

These approximations involve P and Q and the expressions are given by:

$$P_{ij} = \frac{\pi}{8}$$

$$Q_{ij} = -0.03860 + \frac{1}{2}\ln\frac{2}{k_{ij}} \qquad \text{(D-46)}$$

Using these assumptions, f = 60 Hz and soil resistivity = 100 Ω-m, the equations reduce to:

$$Z_{ii} = R_i + 0.0953 + j0.12134\left(\ln\frac{1}{\text{GMR}_i} + 7.93402\right) \ \Omega/\text{mi}$$

$$Z_{ij} = 0.0953 + j0.12134\left(\ln\frac{1}{D_{ij}} + 7.93402\right) \ \Omega/\text{mi} \qquad \text{(D-47)}$$

Equation (D-47) is of practical significance for calculations of line impedances.

Example D-2 Consider an unsymmetrical overhead line configuration, as shown in Fig. D-7. The phase conductors consist of 556.5 KCMIL (556500 circular mils) of ACSR conductor consisting of 26 strands of aluminum, 2 layers, and 7 strands of steel. From the properties of ACSR conductor tables, the conductor has a resistance of 0.1807 Ω at 60 Hz and its GMR = 0.0313 ft at 60 Hz. Conductor diameter = 0.927 in.

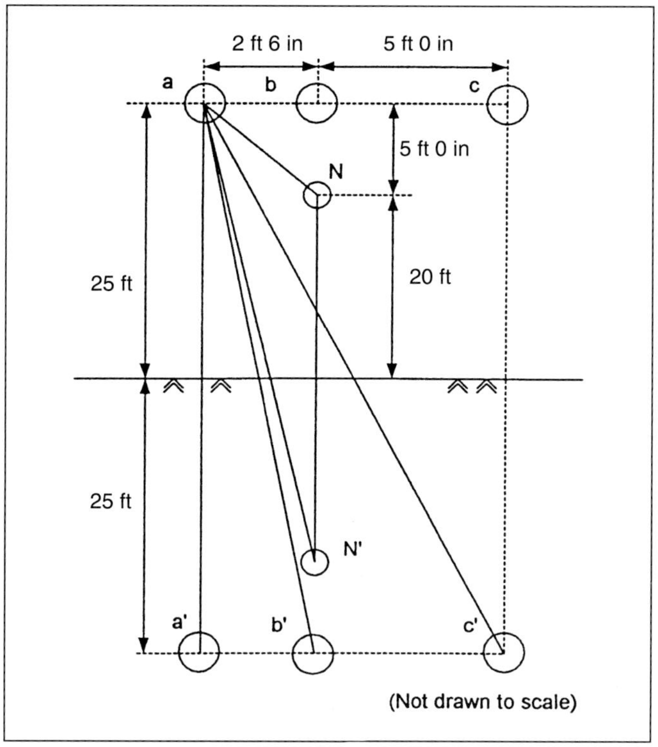

FIGURE D-7 Line configuration for calculations of line constants.

The neutral consists of 336.4 KCMIL, ACSR conductor resistance = 0.259 Ω/mi at 60 Hz and 50°C and GMR = 0.0278 ft, and conductor diameter is 0.806 in.

Using Eq. (D-47) (all in ohms):

$$Z_{aa} = Z_{bb} = Z_{cc} = 0.1859 + j1.3831$$

$$Z_{nn} = 0.3543 + j1.3974$$

$$Z_{ab} = Z_{ba} = 0.0953 + j0.8515$$

$$Z_{bc} = Z_{cb} = 0.0953 + j0.7674$$

$$Z_{ca} = Z_{ac} = 0.0953 + j0.7182$$

$$Z_{an} = Z_{na} = 0.0953 + j0.7539$$

$$Z_{bn} = Z_{nb} = 0.0953 + j0.7674$$

$$Z_{cn} = Z_{nc} = 0.0953 + j0.7237$$

It is required to form a primitive Z matrix, convert it to 3 × 3 Z_{abc} matrix and then to develop sequence impedance matrix Z_{012}.

Therefore the primitive impedance matrix is:

$$\bar{Z}_{\text{prim}} = \begin{vmatrix} 0.1859+j1.3831 & 0.0953+j0.08515 & 0.0953+j0.7182 & 0.0953+j0.7539 \\ 0.0953+j0.08515 & 0.1859+j1.3831 & 0.0953+j0.7624 & 0.0953+j0.7674 \\ 0.0953+j0.7182 & 0.0953+j0.7624 & 0.1859+j1.3831 & 0.0953+j0.7237 \\ 0.0953+j0.7539 & 0.0953+j0.7674 & 0.0953+j0.7237 & 0.3543+j1.3974 \end{vmatrix}$$

Eliminate the last row and column:

$$\bar{Z}_{abc} = \begin{vmatrix} 0.1846+j0.9825 & 0.0949+j0.4439 & 0.0921+j0.3334 \\ 0.0949+j0.4439 & 0.1864+j0.9683 & 0.0929+j0.3709 \\ 0.0921+j0.3334 & 0.0929+j0.3709 & 0.1809+j1.0135 \end{vmatrix}$$

Convert to Z_{012} by using symmetrical component transformation

$$\bar{Z}_{012} = \bar{T}_s^{-1}\bar{Z}_{abc}\bar{T}_s$$

$$\bar{Z}_{012} = \begin{vmatrix} 0.3705+j1.7536 & 0.0194+j0.0007 & -0.0183+j0.0055 \\ -0.0183+j0.0055 & 0.0907+j0.6054 & -0.0769-j0.0146 \\ 0.0194+j0.0007 & 0.0767+j0.0147 & 0.0907+j0.6054 \end{vmatrix} \Omega/\text{mi}$$

This shows mutual coupling between sequence impedances. We could average out the self and mutual impedances according to Eq. (D-28):

$$Z_s = \frac{Z_{aa} + Z_{bb} + Z_{cc}}{3} = 0.184 + j0.9973$$

$$Z_m = \frac{Z_{ab} + Z_{bc} + Z_{ca}}{3} = 0.0933 + j0.38271$$

Then the matrix Z_{abc} becomes:

$$\bar{Z}_{012} = \begin{vmatrix} 0.3706+j1.7627 & 0 & 0 \\ 0 & 0.0907+j0.6146 & 0 \\ 0 & 0 & 0.0907+j0.6146 \end{vmatrix}$$

which is a decoupled diagonal matrix.

Example D-3 Figure D-8 shows a high-voltage line with bundle ACSR conductors. Each conductor has a cross sectional area of 636000 (3 layers of 54 strands of aluminium, 0.1085 inch diameter and 7 strands of steel, 0.1085 inch diameter) circular mils. Conductor GMR = 0.0329 ft, resistance = 0.1688 Ω/mi, diameter = 0.977 in, and the spacings are as shown in Fig. D-8. Calculate the primitive impedance matrix and reduce it to 3 × 3 matrix, Then convert it to sequence component matrix. From Eq. (D-47) and the specified spacings in Fig. D-8, matrix Z_1 is:

$$\bar{Z}_1 = \begin{vmatrix} 0.164+j1.3770 & 0.0953+j0.5500 & 0.0953+j0.4659 \\ 0.953+j0.5500 & 0.164+j1.3770 & 0.0953+j0.5500 \\ 0.0953+j0.4659 & 0.0953+j0.5500 & 0.164+j1.3770 \end{vmatrix}$$

This is also equal to Z_4, as the bundle conductors are identical and symmetrically spaced. Matrix $\bar{Z}_2$ is:

$$\bar{Z}_2 = \begin{vmatrix} 0.0953+j0.8786 & 0.0953+j0.5348 & 0.0953+j0.4581 \\ 0.0953+j0.5674 & 0.0953+j0.8786 & 0.0953+j0.5348 \\ 0.0953+j0.4743 & 0.0953+j0.08786 & 0.0953+j0.8786 \end{vmatrix}$$

The primitive matrix is 6 × 6 given by Eq. (D-39) formed by partitioned matrices according to Eq. (D-40). From these two matrices, we will calculate matrix equation [Eq. (D-42)].

$$\bar{Z}_1 - \bar{Z}_2 = \begin{vmatrix} 0.069+j0.498 & j0.0150 & j0.0079 \\ -j0.0171 & 0.069+j0.498 & j0.0150 \\ -j0.00841 & -j0.0170 & 0.069+j0.498 \end{vmatrix}$$

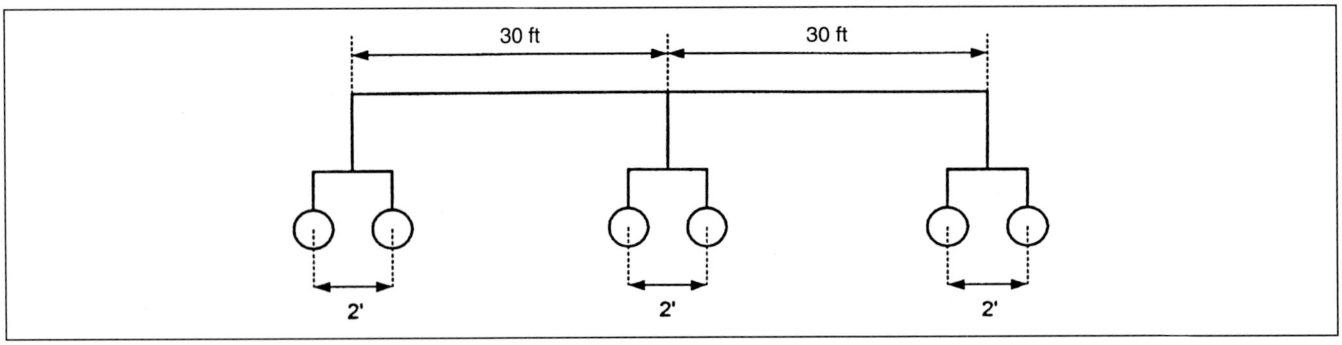

FIGURE D-8 Bundle conductors matrix transformation.

and

$$\bar{Z}_k = (\bar{Z}_1 - \bar{Z}_2) - (\bar{Z}_2^t - \bar{Z}_1) = \begin{vmatrix} 0.138 + j0.997 & -j0.0022 & -j0.0005 \\ -j0.0022 & 0.138 + j0.997 & -j0.0022 \\ -j0.0005 & -j0.0022 & 0.138 + j0.997 \end{vmatrix}$$

The inverse is:

$$\bar{Z}_k^{-1} = \begin{vmatrix} 0.136 - j0.984 & 0.000589 - j0.002092 & 0.0001357 - j0.0004797 \\ 0.0005891 - j0.002092 & 0.136 - j0.981 & 0.0005891 - j0.002092 \\ 0.0001357 - j0.0004797 & 0.0005891 - j0.002092 & 0.136 - j0.981 \end{vmatrix}$$

Then, the matrix $(\bar{Z}_2 - \bar{Z}_1)\bar{Z}_k^{-1}(\bar{Z}_2^t - \bar{Z}_1)$ is:

$$\begin{vmatrix} 0.034 + j0.2500 & -0.000018 - j0.000419 & 0.0000363 - j0.0003871 \\ -0.000018 - j0.000419 & 0.034 + j0.2500 & -0.000018 - j0.000419 \\ 0.0000363 - j0.000387 & -0.000018 - j0.000419 & 0.034 + j0.2500 \end{vmatrix}$$

Note that the off-diagonal elements are comparatively smaller as compared to the diagonal elements. The required 3 × 3 transformed matrix is then Z_1 minus above matrix:

$$\bar{Z}_{\text{transformed}} = \begin{vmatrix} 0.13 + j1.127 & 0.095 + j0.55 & 0.095 + j0.466 \\ 0.095 + j0.55 & 0.13 + j1.127 & 0.095 + j0.55 \\ 0.095 + j0.466 & 0.095 + j0.55 & 0.13 + j1.127 \end{vmatrix} \ \Omega/\text{mi}$$

The sequence impedance matrix is ($\bar{Z}_{012} = \bar{T}_s^{-1}\bar{Z}_{abc}\bar{T}_s$):

$$\bar{Z}_{012} = \begin{vmatrix} 0.32 + j2.171 & 0.024 - j0.014 & -0.024 - j0.014 \\ -0.024 - j0.014 & 0.035 + j0.605 & -0.048 + j0.028 \\ 0.024 - j0.014 & 0.048 + j0.028 & 0.035 + j0.605 \end{vmatrix} \ \Omega/\text{mi}$$

D-9 CAPACITANCE OF LINES

The shunt capacitance per unit length of a two-wire, single-phase transmission line is:

$$C = \frac{\pi\varepsilon_0}{\ln(D/r)} \ \text{F/m(farads per meter)} \tag{D-48}$$

where ε_0 is the permittivity of free space = 8.854×10^{-12} and other symbols are as defined before. For a three-phase line with equilaterally spaced conductors, the line to neutral capacitance is:

$$C = \frac{2\pi\varepsilon_0}{\ln(D/r)} \ \text{F/m} \tag{D-49}$$

For unequal spacing, D is replaced with GMD from Eq. (D-8). The capacitance is affected by ground and the effect is simulated by a mirror image of the conductors exactly at the same depth as the height above the ground. These mirror-image conductors carry charges which are of opposite polarity to those of above ground conductors (Fig. D-9). From this figure, the capacitance to ground is:

$$C_n = \frac{2\pi\varepsilon_0}{\ln(\text{GMD}/r) - \ln\left(\sqrt[3]{S_{ab'}S_{bc'}S_{ca'}} / \sqrt[3]{S_{aa'}S_{bb'}S_{cc'}}\right)} \tag{D-50}$$

Using the notations in Eqs. (D-21) and (D-22), this can be written as:

$$C_n = \frac{2\pi\varepsilon_0}{\ln(D_m/D_s)} = \frac{10^{-9}}{18\ln(D_m/D_s)} \tag{D-51}$$

D-9-1 Capacitance Matrix

The capacitance matrix of a three-phase line is:

$$\bar{C}_{abc} = \begin{vmatrix} C_{aa} & -C_{ab} & -C_{ac} \\ -C_{ba} & C_{bb} & -C_{bc} \\ -C_{ca} & -C_{cb} & C_{cc} \end{vmatrix} \tag{D-52}$$

This is diagrammatically shown in Fig. D-10*a*.The capacitance between the phase conductor *a* and *b* is C_{ab} and capacitance between phase- *a* conductor and ground is: $C_{aa}-C_{ab}-C_{ac}$. If the line is perfectly

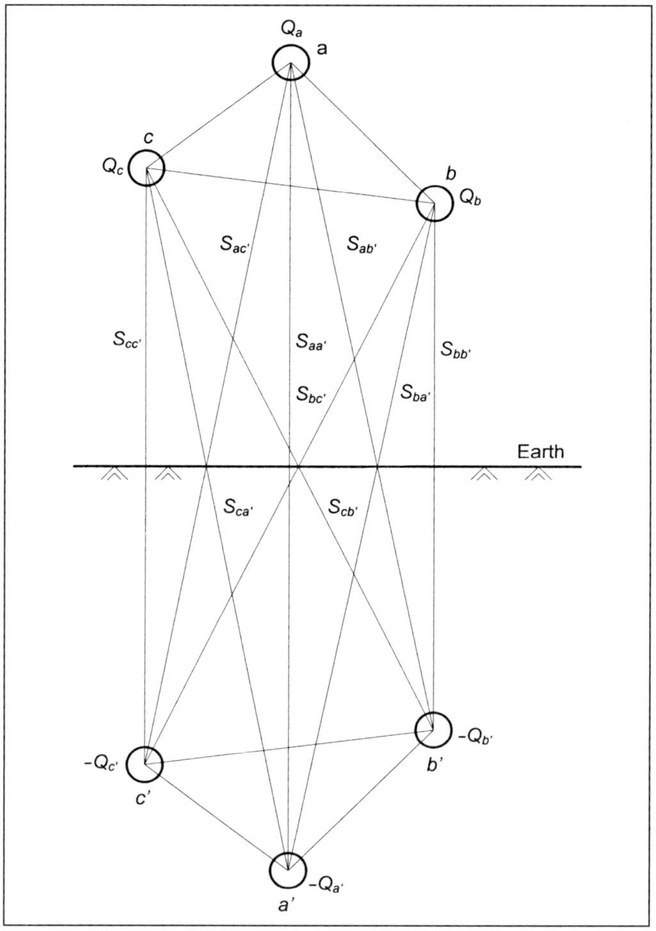

FIGURE D-9 Mirror-image conductors for calculation of line capacitances, conductors, spacing, mirror images, and electrical charges.

symmetrical, all diagonal elements are the same and all off-diagonal elements of the capacitance matrix are identical:

$$\bar{C}_{abc} = \begin{vmatrix} C & -C' & -C' \\ -C' & C & -C' \\ -C' & -C' & C \end{vmatrix} \quad \text{(D-53)}$$

Symmetrical component transformation is used to diagonalize the matrix:

$$\bar{C}_{012} = \bar{T}_s^{-1}\bar{C}_{abc}\bar{T}_s = \begin{vmatrix} C-2C' & 0 & 0 \\ 0 & C+C' & 0 \\ 0 & 0 & C+C' \end{vmatrix} \quad \text{(D-54)}$$

The zero, positive, and negative sequence networks of capacitance of a symmetrical transmission line are shown in Fig. D-10*b*. The eigenvalues are $C - 2C'$, $C + C'$, and $C + C'$, and. $C + C'$ can be written as $3C' - C - 2C'$; that is, it is equivalent to line capacitance of a three-phase system minus line-to-ground capacitance of a three-conductor system.

In a capacitor $V = Q/C$. The capacitance matrix can be written as:

$$\bar{V}_{abc} = \bar{P}_{abc}\bar{Q}_{abc} = \bar{C}_{abc}^{-1}\bar{Q}_{abc} \quad \text{(D-55)}$$

where $\bar{P}$ is called the potential coefficient matrix; that is,

$$\begin{vmatrix} V_a \\ V_b \\ V_c \end{vmatrix} = \begin{vmatrix} P_{aa} & P_{ab} & P_{ac} \\ P_{ba} & P_{bb} & P_{bc} \\ P_{ca} & P_{cb} & P_{cc} \end{vmatrix} \quad \text{(D-56)}$$

where

$$P_{ii} = \frac{1}{2\pi\varepsilon_0}\ln\frac{S_{ii}}{r_i} = 11.17689\ln\frac{S_{ii}}{r_i}$$
$$P_{ij} = \frac{1}{2\pi\varepsilon_0}\ln\frac{S_{ij}}{D_{ij}} = 11.17689\ln\frac{S_{ij}}{D_{ij}} \quad \text{(D-57)}$$

where S_{ij} is the conductor to image distance below ground, ft, D_{ij} is the conductor-to-conductor distance, ft, r_i is the radius of the conductor, ft, and ε_0 is the permitivity of the medium surrounding the conductor = 1.424×10^{-8} F/mi for air.

For sine wave voltage and charge, the equation can be expressed as:

$$\begin{vmatrix} I_a \\ I_b \\ I_c \end{vmatrix} = j\omega \begin{vmatrix} C_{aa} & -C_{ab} & -C_{ac} \\ -C_{ba} & C_{bb} & -C_{bc} \\ -C_{ca} & -C_{cb} & C_{cc} \end{vmatrix} \quad \text{(D-58)}$$

Capacitance of three-phase lines with ground wires and with bundle conductors can be addressed as in the calculations of inductances. The primitive *P* matrix can be partitioned and reduces to 3 × 3 matrix.

Example D-4 Calculate the matrix *P* and *C* for the Example D-2. Neutral is 30 ft above ground and configuration of Fig. D-7 is applicable.

The mirror images of the conductors are drawn in Fig. D-7. This facilitates calculation of spacings required in Eq. (D-57) for *P* matrix. Based on the geometric distances and conductor diameter the primitive *P* matrix is:

$$\bar{P} = \begin{vmatrix} P_{aa} & P_{ab} & P_{ac} & P_{an} \\ P_{ba} & P_{bb} & P_{bc} & P_{bn} \\ P_{ca} & P_{cb} & P_{cc} & P_{cn} \\ P_{na} & P_{nb} & P_{nc} & P_{nn} \end{vmatrix}$$

$$= \begin{vmatrix} 80.0922 & 33.5387 & 21.4230 & 23.3288 \\ 33.5387 & 80.0922 & 25.7913 & 24.5581 \\ 21.4230 & 25.7913 & 80.0922 & 20.7547 \\ 23.3228 & 24.5581 & 20.7547 & 79.1615 \end{vmatrix}$$

This is reduced to 3 × 3 matrix:

$$\bar{P} = \begin{vmatrix} 73.2172 & 26.3015 & 15.3066 \\ 26.3015 & 72.4736 & 19.3526 \\ 15.3066 & 19.3526 & 74.6507 \end{vmatrix}$$

Therefore, the required $\bar{C}$ matrix is inverse of $\bar{P}$ and $\bar{Y}_{abc}$ is:

$$\bar{Y}_{abc} = j\omega\bar{P}^{-1} = \begin{vmatrix} j6.0141 & -j1.9911 & -j0.7170 \\ -j1.9911 & j6.2479 & -j1.2114 \\ -j0.7170 & -j1.2114 & j5.5111 \end{vmatrix} \mu\text{S/mi}$$

D-10 CABLE CONSTANTS

Construction of cables varies widely, mainly a function of insulation type, method of laying and voltage of application. For high-voltage applications above 230 kV or 325 kV, oil-filled, paper-insulated cables are used, though recent trends see development of solid dielectric cables. A three-phase solid dialectic cable has three conductors enclosed within a sheath and because the conductors are much closer to each other than those in an overhead line and permittivity of insulating medium is much higher than that of air, the shunt capacitive reactance is much smaller as compared to a overhead line. Thus, use of a *T* or Π model is required even for shorter cable lengths, see Chap. 4.

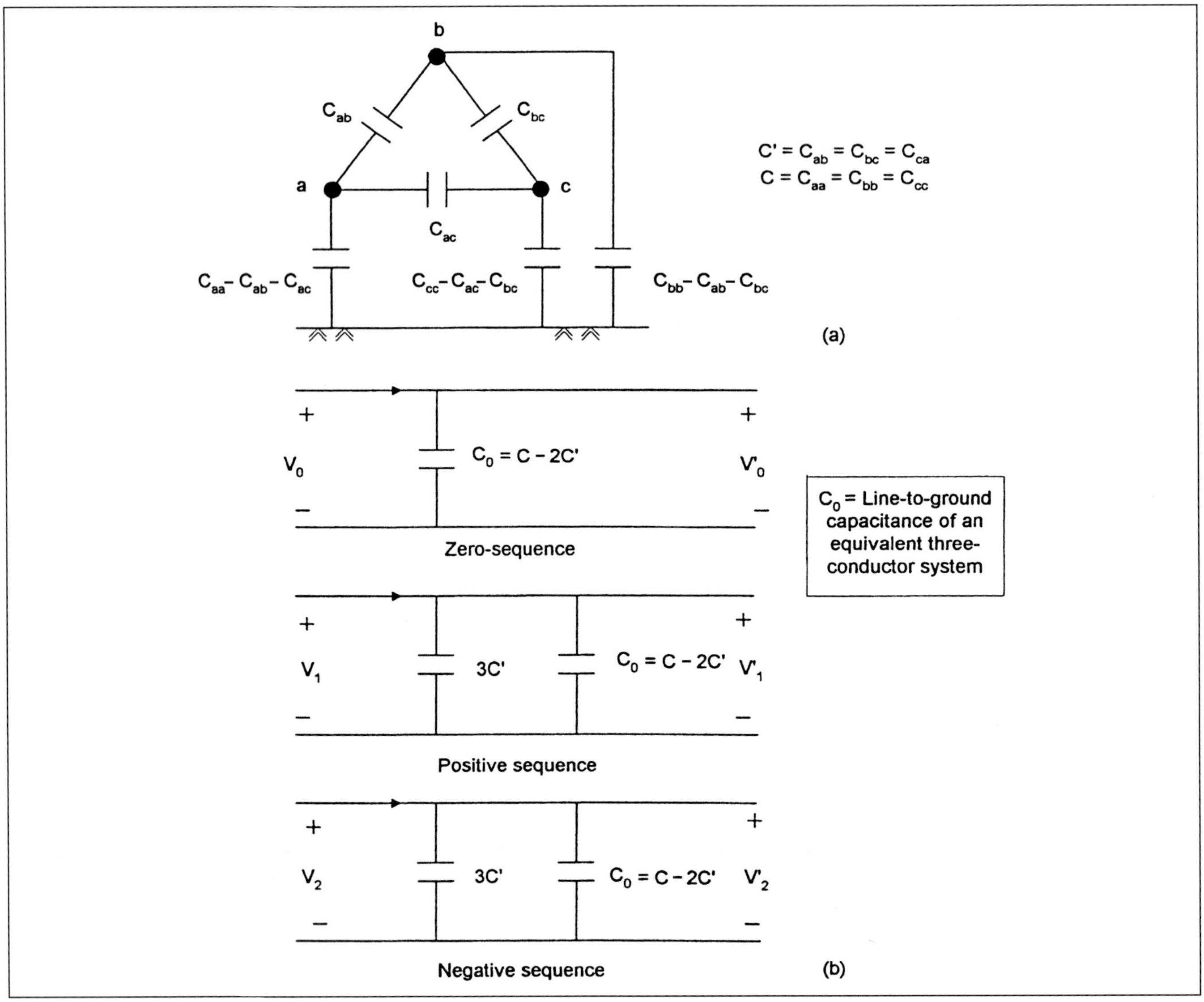

FIGURE D-10 (*a*) Capacitance of a three-phase line. (*b*) Equivalent positive, negative, and zero sequence network of capacitances.

The inductance per unit length of a single conductor cable is given by:

$$L = \frac{\mu_0}{2\pi} \ln \frac{r_1}{r_2} \tag{D-59}$$

where r_1 is the radius of the conductor and r_2 is the radius of the sheath; that is, cable outside diameter divided by 2.

When single-conductor cables are installed in magnetic conduits, the reactance may increase by a factor of 1.5. Reactance is also dependent on conductor shape; that is, circular or sector-shaped, and on magnetic binders in three-conductor cables. Manufacturers' data is normally used. The flow of currents in the magnetic grounded sheaths increases the effective conductor resistance.

D-10-1 Shunt Admittance of Cables

In a single-conductor cable, the capacitance per unit length is given by:

$$C = \frac{2\pi\varepsilon\varepsilon_0}{\ln(r_1/r_2)} \text{ F/m} \tag{D-60}$$

By change of units, this can be written as:

$$C = \frac{7.35\varepsilon}{\log(r_1/r_2)} \text{ pF/ft}$$

Note that ε is the permittivity of the dielectric medium and not free space. The capacitances in a three-conductor cable are shown in Fig. D-11. This assumes a symmetrical construction and the capacitances between conductors and from conductors to the sheath are equal. The circuit of Fig. D-11*a* is successively transformed and Fig. D-11*d*, which shows that the net capacitance per phase = $C_1 + 3\,C_2$. Table D-1 gives typical values of the dielectric constants of the cables.

D-10-2 Zero-Sequence Impedance of the OH Lines and Cables

The zero-sequence impedance of the lines and cables is dependent on the current flow through a conductor and return through the ground or sheaths, and encounters the impedance of these paths. The zero-sequence current flowing in one phase also encounters

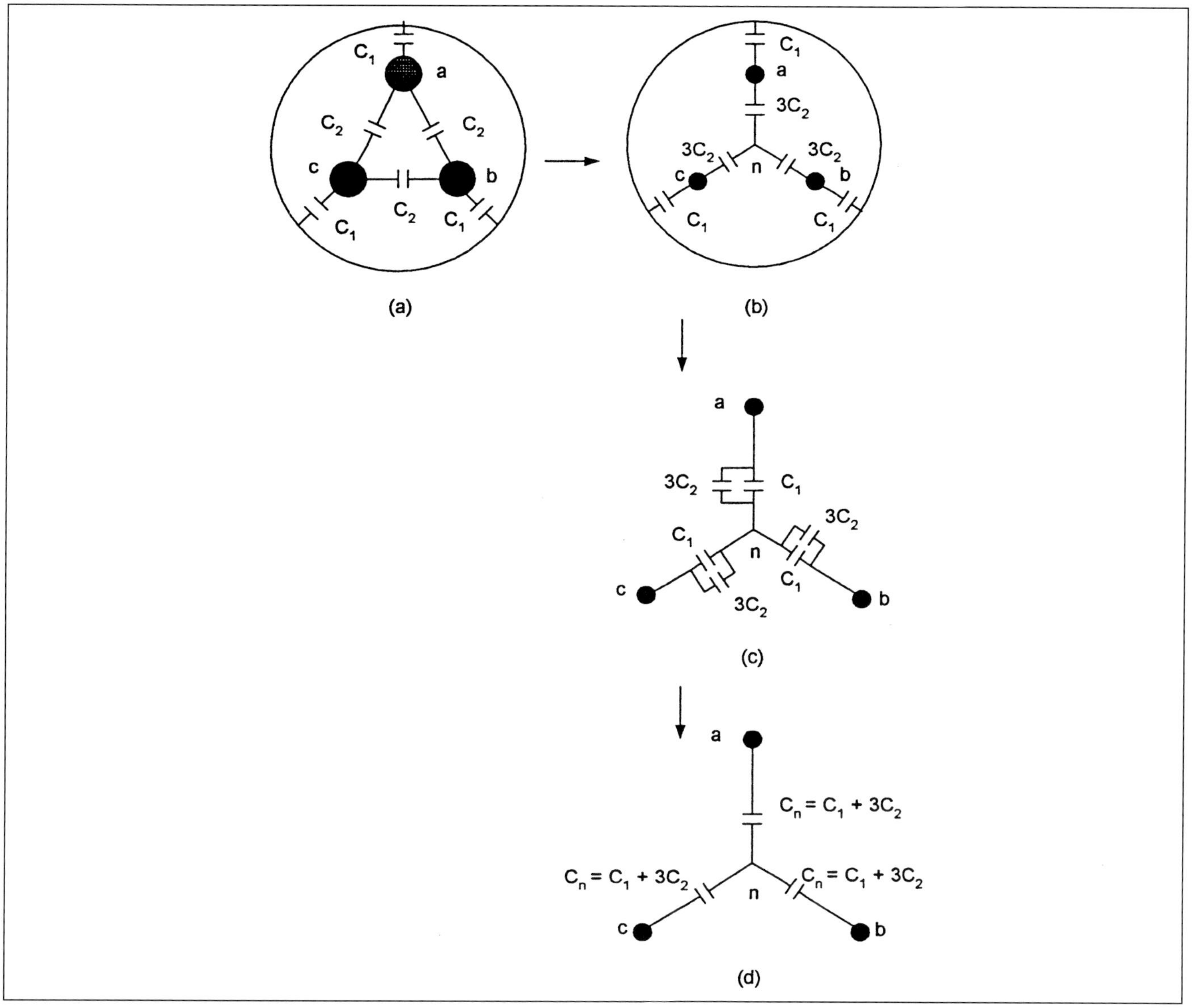

FIGURE D-11 (*a*) Capacitances in a three-phase cable. (*b*) and (*c*) Equivalent circuits. (*d*) Final capacitance circuit.

TABLE D-1 Relative Permittivity of Insulating Materials

TYPE OF INSULATION	PERMITTIVITY (ε)
Polyvinyl chloride (PVC)	3.5–8.0
Ethylene-propylene insulation	2.8–3.5
Polyethylene insulation	2.3
Cross-linked polyethylene (XLPE)	2.3–6.0
Impregnated paper	3.3–3.7
Mass-impregnated	4.2
Fluid filled	3.5

the currents arising out of that from conductor self-inductance, from mutual inductance to other two phase conductors, from the mutual inductance to the ground and sheath return paths, and from the self inductance of the return paths. Tables and analytical expressions are provided in Ref. 7. As an example, the zero-sequence impedance of a three-conductor cable with a solidly bonded and grounded sheath is given by:

$$z_0 = r_c + r_e + j0.8382\frac{f}{60}\log_{10}\frac{D_e}{\text{GMR}_{3c}} \tag{D-61}$$

where r_c is the ac resistance of one conductor Ω/mi, r_e is the ac resistance of earth return (depending on equivalent depth of earth return, soil resistivity, taken as 0.286 Ω/mi), D_e is the distance to equivalent earth path (see Ref. 7), and GMR_{3c} is the geometric mean radius of conducting path made up of three actual conductors taken as a group:

$$\text{GMR}_{3c} = \sqrt[3]{\text{GMR}_{1c}S^2} \tag{D-62}$$

where GMR_{1c} is the geometric mean radius of individual conductor and $S = (d + 2t)$, where d = diameter of the conductor and t is the thickness of the insulation.

D-10-3 Cable Constant Calculation Routines

Cable constant calculation routines are included in EMTP-type programs. The electrical representation is a series impedance matrix Z and a shunt matrix Y. The basic inputs required for these cable constant routines (See Chap. 7 for a calculation) require material properties apart from the cable construction, cable geometry, method of laying, burial depth, and the like.[12,13] These routines take account of skin effect and can include proximity effects.[14,15] Tables D-2 and D-3 show the resistivity of conductive materials and parameters of semiconductive layers, respectively.

The resistivity of the surrounding medium depends strongly on soil characteristics. Submarine cables are designed with a magnetic steel armor and the permeability depends on wire diameter, laying angle, and intensity of circumferential magnetic field. Curves for the permeability of round steel wire armor due to magnetic field in circumferential direction is included in Ref. 15.

Most extruded insulations, including XLPE and PE, are practically lossless up to 1 MHz, whereas paper insulation exhibits significant losses at lower frequencies.

The losses are associated with complex frequency dependant permittivity.

$$\varepsilon_r = \varepsilon_r'(\omega) - j\varepsilon_r''(\omega) \tag{D-63}$$

$$\tan\delta(\omega) = \frac{\varepsilon_r''}{\varepsilon_r'} \tag{D-64}$$

Brenien and Johansen fitted a Debye model to measured frequency response of insulation samples of low-pressure, fluid-filled cable, in the frequency range of 10 to 100 MHz.[16] The permittivity is given by:

$$\varepsilon_r = 2.5 + \frac{0.94}{1+(j\omega.6.10^{-9})^{0.315}} \tag{D.65}$$

High-frequency transients in cable systems essentially propagate as decoupled coaxial waves between cores and sheath and transient behavior is sensitive to modeling of cable parameters:

1. Increasing core resistivity increases attenuation and slightly decreases propagation velocity.

2. Increasing sheath resistivity or decreasing sheath thickness increases attenuation.

3. Increasing insulation permittivity increases cable capacitance. This decreases velocity and surge impedance.

4. With fixed insulation thickness, adding semiconducting screens increases inductance of the core-sheath loop without changing the capacitance. This decreases velocity and surge impedance.[17]

TABLE D-2 Resistivity of Conductive Materials

Material	Resistivity (ρ), Ω-m
Copper	1.728E-8
Aluminum	2.83E-8
Lead	22E-8
Steel	18E-8

TABLE D-3 Parameters of Semiconductive Layers (Extruded Insulation)

Resistivity (Ω-m)	<1E-3
Permittivity	>1000

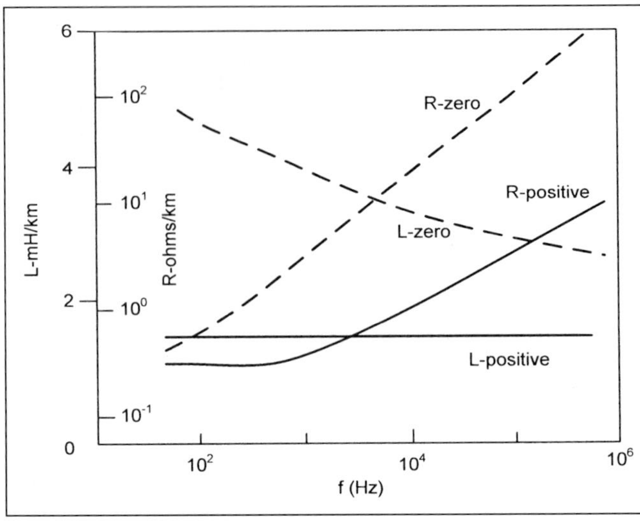

Figure D-12 Variation of transmission line parameters due to skin effects. Conductors in flat formation, spacing 2 m apart, height above ground = 12 m, ground resistivity = 100 Ω-m.

D-11 FREQUENCY-DEPENDENT TRANSMISSION LINE MODELS

The transmission line models for transient studies should be appropriate for the involved frequency range (Chap. 4). The parameters vary with frequency depending on:

- Soil resistivity
- Skin effect
- Conductor height
- Proximity effects

Some curves of variations with frequency are provided in Ref. 18. Figure D-12 shows variations of positive- and zero-sequence resistances and inductances due to skin effects.

REFERENCES

1. J. L. Blackburn, *Symmetrical Components for Power System Engineering*, Marcel Dekker, New York, 1993.
2. L. J. Myatt, *Symmetrical Components,* Pergamon Press, Oxford, UK, 1968.
3. C. F. Wagner and R. D. Evans, *Symmetrical Components,* McGraw Hill, New York, 1933.
4. P. M. Anderson, *Analysis of Faulted Systems*. Iowa State University Press, Ames, IA, 1973.
5. D. G. Fink, Ed. *Standard Handbook for Electrical Engineers*, 10th ed., McGraw Hill, New York, 1969.
6. T. Croft, W. Summers, F. Hastwells, *American Electrician's Handbook*, 15th ed., McGraw Hill, New York, 1970.

7. Central Station Engineers. *Electrical Transmission and Distribution Reference Book,* 4th ed., East Pittsburg, PA, Westinghouse Corp, 1964.
8. The Aluminum Association. *Aluminum Conductor Handbook,* 2nd ed., Washington, DC, 1982.
9. J. H. Neher and M. H. McGrath, "The Calculation of the Temperature Rise and Load Capability of Cable Systems," *AIEEE Trans.*, Part III, vol. 76, Oct. 1957.
10. W. D. Stevenson, *Elements of Power System Analysis,* 2d ed., McGraw Hill, New York. 1962.
11. CIGRE, Working Group 33.02 (Internal Voltages), Guidelines for Representation of Network Elements When Calculating Transients, Paris.
12. L. M. Wedephol and D. J. Wilcox, "Transient Analysis of Underground Power-Transmission Systems. System-Model and Wave-Propagation Characteristics," *Proc. IEE,* vol. 120, no. 2, pp. 253–260, Feb. 1973.
13. A. Amentani, "A General Formation of Impedance and Admittance of Cables," *IEEE Trans. PAS,* vol. 99, no. 3, pp. 902–909, May/Jun. 1980.
14. Y. Yin and H. W. Dommel, "Calculations of Frequency Dependent Impedances of Underground Power Cables with Finite Element Method," *IEEE Trans. On Magnetics*, vol. 25, no. 4, pp. 3025–3027, Jul. 1989.
15. G. Bianchi and G. Luoni, "Induced Current Losses in Single Core Submarine Cables," *IEEE Trans. PAS*, vol. 95, pp. 49–58, Jan./Feb. 1976.
16. O. Brenien and I. Johansen, "Attenuation of Traveling Waves in Single-Phase High Voltage Cables," *Proc. IEE*, vol. 118, no. 6, pp. 787–793, Jun. 1971.
17. IEEE PES TF, Parameter Determination for Modeling System Transients, Part II: Insulated Cables, IEEE WG.
18. IEEE PES TF, Parameter Determination for Modeling System Transients, Part I: Overhead Lines, IEEE WG.

APPENDIX E

ENERGY FUNCTIONS AND STABILITY

Generally, we seek explicit solutions to circuit problems, and circuit analysis programs give accurate solutions based on some specific initial conditions. However, many times we seek the qualitative behavior of the system; that is, is the system stable for all initial conditions after a disturbance? Will it keep oscillating? Thus a new solution must be found for each initial condition and examined, which may not be even practical. A useful technique of analyzing can be to study a quantity derived from the system itself—for example, energy function—and use it to draw conclusions about the system.

E-1 DYNAMIC ELEMENTS

In Chap. 2, we defined a dynamic system as the one whose behavior changes with time. Neglecting wave propagation and assuming that the voltages and currents external to the system are bounded functions of time, a broad definition can be that a dynamic circuit is a network of n-terminal resistors, capacitors, and inductors, which can be nonlinear. We call capacitors and inductors dynamic elements. We can define the nonlinear resistors, inductors, and capacitors by the following equations:

$$R(v,i)=0 \qquad C(q,v)=0 \qquad L(\phi,i)=0 \tag{E-1}$$

These relations are constituency relations. A nonlinear resistor is described by the current and voltage and a nonlinear capacitor by the charge q and voltage v across it. If an element is time varying we write:

$$R(v,i,t)=0 \qquad C(q,v,t)=0 \qquad L(\phi,i,t)=0 \tag{E-2}$$

A resistor is voltage controlled if the current i is a function of the voltage across it. The constituency relation is written as $R(i, v) = 0$ and is expressed as:

$$i=\hat{i}v \tag{E-3}$$

and it is current controlled if we write:

$$v=\hat{v}i \tag{E-4}$$

Same notations apply to capacitors and inductors. For a capacitor we write:

$$C(q,v)=0 \quad \text{or} \quad q=\hat{q}v \text{ (voltage controlled)} \qquad v=\hat{v}q \text{ (charge controlled)} \tag{E-5}$$

and for an inductor:

$$i=\hat{i}(\phi) \qquad \phi=\hat{\phi}(i) \tag{E-6}$$

E-2 PASSIVITY

Passivity implies that the nonlinear elements dissipate energy; that is, these do not have a power source. A nonlinear resistor with constituency relationship $i=g(v)$ is passive if $vg(v)\geq 0$ for all v. It is strictly passive if $vg(v)>0$ for all $v\neq 0$. It is eventually passive if there exists a $k>0$ such that $vg(v)\geq 0$, where $|v|>k$. It is eventually strictly passive if $vg(v)>0$, where $|v|>k$.

E-3 EQUILIBRIUM POINTS

A dynamic circuit is called autonomous or time-independent if it does not contain any time-varying elements or sources. A stationary or equilibrium point of a dynamic circuit is defined as that state of the circuit which does not change with time. In a network of resistors, capacitors, and inductors, this means that all branch currents of capacitors are zero and all voltages on inductors are also zero; that is, a capacitor is open circuited and an inductor is short circuited, reducing the network to a resistive circuit. To find the equilibrium points of a dynamic system, find operating points of the resistive circuit. Then for each capacitor, the branch voltage at the operating point is found to ascertain all the charges which satisfy $C_i(q_i,v_i)=0$ and similarly for inductors to find all the fluxes. Each operating point of the resistive circuit can correspond to many equilibrium points of the dynamic circuit.

E-4 STATE EQUATIONS

If we eliminate variables v and i, the state equations can be written as:

$$\frac{dq_c}{dt} = \mathbf{f_c}(q_c, \phi_l, t)$$
$$\frac{d\phi_l}{dt} = \mathbf{f_l}(q_c, \phi_l, t) \qquad \text{(E-7)}$$

where q_c and ϕ_l are the vectors corresponding to charges on capacitors and fluxes in the inductors. If all the capacitors are charge-controlled and all the inductors are flux controlled—that is, $v = \hat{v}(q)$ and $i = \hat{i}(\phi)$—then the state equations can be written in terms of capacitor voltage v_c and inductor currents i_l:

$$v_c = g_c(q_c)$$
$$\frac{dv_c}{dt} = \nabla g_c(q_c) \times \frac{dq_c}{dt} = \nabla g_c(x)\Big|_{x=g_c^{-1}(v_c)} \times \mathbf{f_c}[g_c^{-1}(v_c), g_l^{-1}(i_l), t]$$
$$= \mathbf{h_c}(v_c, i_l, t)$$
$$i_l = g_l(\phi_l)$$
$$\frac{di_l}{dt} = \nabla g_l(\phi_l) \times \frac{d\phi_l}{dt} = \nabla g_c(x)\Big|_{x=g_l^{-1}(i_l)} \times \mathbf{f_l}[g_c^{-1}(v_c), g_l^{-1}(i_l), t]$$
$$= \mathbf{h_l}(v_c, i_l, t) \qquad \text{(E-8)}$$

This assumes that $\nabla g_c, g_c^{-1}, \nabla g_i$, and g_i^{-1} exist.

Even if the state equations exist, there is no guarantee that a solution exits. In Chap. 2, we described that the first-order state equations are equal to the order of the system. We assume that solutions to state equations [Eq. (E-9)] exist and are unique for all time, and f is continuous. A dynamic circuit may have solutions which escape to infinity in finite time. This will not happen if:

- There are no loops or cut sets consisting only of capacitors and/or inductors.
- All capacitors and inductors are eventually strongly locally passive.
- All resistors are eventually passive.

E-5 STABILITY OF EQUILIBRIUM POINTS

The state equation in Eq. (E-7) can be written as:

$$\dot{\mathbf{x}} = f(x) \qquad \text{(E-9)}$$

Equilibrium point is a state $\mathbf{x}^*$ so that:

$$\dot{\mathbf{x}} = 0 \qquad \text{(E-10)}$$

that is,

$$f(x^*) = 0$$

If the state of a circuit is at equilibrium point, then it remains there for all time. To be more precise, we should consider noise. A sphere balanced on a pin will become unstable on the slightest disturbance. Equilibrium point is asymptotically stable, if all initial conditions nearby do not leave a neighborhood, and converge to the equilibrium point as the time passes.

E-6 HARTMAN-GROBMAN LINEARIZATION THEOREM

This theorem[1] allows to look at the stability of the equilibrium point by examining the eigenvalues of the linearization. If x^* is a an equilibrium point, then $f(x^*) = 0$. Consider that $Jf(x^*)$, the Jacobian matrix of f at x^*, does not have purely imaginary eigenvalues. If all the eigenvalues have negative real parts, then x^* is asymptotically stable; otherwise it is unstable.

E-7 LYAPUNOV FUNCTION

The Lyapunov function maps the state of the system $\mathbf{x}$ into simple, usually scalar quantity $V(x)$. Then we observe $V(x)$ evolve with time and draw conclusions about the system. Energy, like quantity, is sometimes, used for the Lyapunov function. Let a and b be strictly increasing functions, so that $a(0) = b(0) = 0$. A function:

$$V : \mathbf{R}^n \to \mathbf{R} \qquad \text{(E-11)}$$

is a Lyapunov function if:

$$b(|x|) \le V(x) \le a(|x|) \quad \text{for all values of } x \qquad \text{(E-12)}$$

Consider a system:

$$\dot{\mathbf{x}} = f(x) \qquad \text{(E-13)}$$

If V is a differentiable Lyapunov function, so that:

$$\nabla V \cdot f(x) \le -c(|x|) \qquad \text{(E-14)}$$

for some strictly increasing function c so that $c(0) = 0$, then the origin is globally asymptotically stable and all trajectories converge toward the origin. The construction of V to show stability in this manner is called the Lyapunov direct method.

E-8 LASALLE'S INVARIANT PRINCIPLE

For an autonomous system, consider Eq. (E-9) and that Eq. (E-11) is a differentiable function so that $V(x) \ge C$ for all x and some constant C and:

$$\frac{dV}{dt} \equiv \nabla V \cdot f(x) \le -W(x) \le 0 \qquad \text{(E-15)}$$

for all $t > 0$ and for some continuous W. Define:

$$E = [x : W(x) = 0] \qquad \text{(E-16)}$$

Then each solution of $x(t)$ of Eq. (E-9) approaches $E \cup \{\infty\}$. The V is still called a Lyapunov function.

E-9 ASYMPTOTIC BEHAVIOR

A dynamic circuit is completely stable if all trajectories converge toward a stable equilibrium point. In Chap. 2, we could divide the response of linear systems into a transient or zero-input response and the steady-state solution. The transient response goes to zero if the system is stable, through some decaying oscillations or asymptotically. This cannot apply to nonlinear systems, but convergence to a steady-state solution irrespective of initial conditions can be defined. Consider that the state equations exist. If all solutions of state equations, $\dot{\mathbf{x}} = f(x,t)$, are bounded and any two solutions converge toward each other asymptotically, say $[x(t)$ and $y(t)]$ converge toward each other asymptotically:

$$\lim_{t \to \infty} \|x(t) - y(t)\| = 0 \qquad \text{(E-17)}$$

then the system has a unique steady-state solution. A dynamic circuit has a unique steady state if:[2]

- Capacitors, inductors, and/or voltage sources do not form any exclusive loop.
- Capacitors, inductors, and/or current sources do not form any exclusive cut set.
- All capacitors and inductors are weakly nonlinear and passive.
- All resistors are strongly locally passive.

E-10 PERIODIC INPUTS

Even simple circuits, when driven by periodic inputs, exhibit chaotic behavior. The state trajectories contain frequencies which are not integral combinations of driving frequencies. A dynamic circuit will have steady-state waveforms with frequency components only at integral combination of driving frequencies, if certain conditions similar to the ones described in Sec. E-9 are met.[2]

The nonlinear circuits can generate beat frequencies. In Chap. 2, Example 2-15 shows that beat frequencies are produced in linear circuits when excited by frequencies close to the resonant frequencies. The same phenomena occur in nonlinear circuits. Given two sinusoidal inputs, the beat frequency is:

$$\omega_{a,b} = a\omega_1 + b\omega_2 \tag{E-18}$$

The output can be written as:

$$y(t) = f(\sin \omega_1 t + \sin \omega_2 t) \tag{E-19}$$

This can be expanded using Taylor's series around zero. The Manley-Rowe equations describe the relative power in various beat frequency components. These equations are:

$$\begin{aligned} \sum_{a=0}^{\infty}\sum_{b=0}^{\infty} \frac{aP_{a,b}}{\omega_{a,b}} &= 0 \\ \sum_{a=0}^{\infty}\sum_{b=0}^{\infty} \frac{bP_{a,b}}{\omega_{a,b}} &= 0 \end{aligned} \tag{E-20}$$

The energy concepts described in this appendix, though not exhaustive, are relevant to direct stability methods in Chap. 12.

REFERENCES

1. P. Hartman, *Ordinary Differential Equations*, Wiley, New York, 1964.
2. L. O. Chua, *Introduction to Nonlinear Network Theory*, McGraw Hill, New York, 1964.

FURTHER READING

L. O. Chua, "Nonlinear Circuits," *IEEE Trans. Circuits Syst.*, vol. 31, pp. 69–87, 1984.

D. W. Jordan and P. Smith, *Nonlinear Ordinary Differential Equations*, Oxford, London, 1977.

J. P. LaSalle, "An Invariance Principal in Theory of Stability," in *Differential Equations and Dynamical Systems*, J. K. Hale and J. P. LaSalle, Ed., pp. 277–286, Academic Press, New York, 1967.

J. M. Manley and H. E. Rowe, "Some General Properties of Nonlinear Elements—Part1. General Energy Relations," *Proc. IEE*, vol. 44, pp. 904–913, Jul. 1956.

APPENDIX F

STATISTICS AND PROBABILITY

The statistical and probabilistic concepts are used in the insulation coordination. A brief background is included here.

F-1 MEAN, MODE, AND MEDIAN

The frequency distribution data can be represented by a histogram, frequency polygon, frequency curve, bar chart, and pie diagrams. The arithmetic mean for n numbers, $x_1, x_2, \ldots x_n$, is given by:

$$x_m = \frac{\sum x}{n} \tag{F-1}$$

If x_1 occurs f_1 times, x_2 occurs f_2 times, ... and f_n occurs n times, then the arithmetic mean is:

$$x_m = \frac{\sum fx}{\sum f} \tag{F-2}$$

If a is the assumed arithmetic mean and d is the deviation of the variate x from a, we can write:

$$\frac{\sum fd}{\sum f} = \frac{\sum f(x-a)}{\sum f} = x_m - \frac{a\sum f}{\sum f} = x_{m-a}$$

$$x_m = a + \frac{\sum fd}{\sum f} \tag{F-3}$$

Median is the measure of the central item of the distribution when it is arranged in ascending or descending order. For an odd n, it is given by:

$$M_d = \frac{n+1}{2} \text{ th item} \tag{F-4}$$

For even frequency, there are two middle terms, and the mean of $n/2$ and $(n/2+1)$ terms gives the median. Mode is defined as the size of the variable in a population that occurs most frequently:

$$\text{Mean} - \text{mode} = 3(\text{mean} - \text{median}) \tag{F-5}$$

Geometric mean is defined as:

$$G = (x_1 \times x_2 \times \ldots\ldots \times x_n)^{1/n} \tag{F-6}$$

F-2 MEAN AND STANDARD DEVIATION

Average deviation or mean deviation is defined as the mean of the absolute values of the deviations of a given set of numbers from their arithmetic mean.

$$\text{Mean deviation} = \frac{\sum_{n=1}^{n} f_n |x_n - x_m|}{\sum f} \tag{F-7}$$

where $x_1, x_2, \ldots x_n$ is a set of numbers with frequencies $f_1, f_2, \ldots f_n$. Standard deviation is defined as square root of the mean of the square of the deviation from the arithmetic mean:

$$\text{SD} = \sigma = \sqrt{\frac{\sum_{n=1}^{n} f_n (x_n - x_m)^2}{\sum f}} \tag{F-8}$$

The square of the SD, σ^2, is called variance. It is also called the second moment about the mean and is denoted by μ_2. The coefficient of variance is given by:

$$\frac{\sigma}{x_m} \times 100 \tag{F-9}$$

The standard deviation can be found from the following expression:

$$\sigma = \sqrt{\frac{\sum fd^2}{\sum f} - \left(\frac{\sum fd}{\sum f}\right)^2} \tag{F-10}$$

Example F-1 Consider the following sample data:

(6, 4) (7, 6) (8, 10) (9, 11) (10, 7) 11, 4) (12, 3)

The first number in the parenthesis is the size, the second its frequency. It is required to find the mean and standard deviation.

Assume a = 10. To calculate mean and standard deviation, Table F-1 is constructed. From this table:

$$\sum f = 45 \qquad \sum fd = -53 \qquad \sum fd^2 = 239$$

TABLE F-1 Calculation of Mean and Standard Deviation

x	f	$d=x-a$	fd	fd^2
6	4	−4	−16	64
7	6	−3	−18	108
8	10	−2	−20	40
9	11	−1	−11	11
10	7	0	0	0
11	4	1	4	4
12	3	2	8	12
	$\sum f=45$		$\sum fd=-53$	$\sum fd^2=239$

Assumed mean $a=10$

Therefore from Eq. (F-3)

$$x_m = 10 + \frac{\sum fd}{\sum f} = 8.82$$

$$\sigma = \sqrt{\frac{239}{45} - \left(\frac{-53}{45}\right)^2} = 1.98$$

F-3 SKEWNESS AND KURTOSIS

Skewness is opposite of symmetry. In a symmetrical series, the mode, the median, and x_m are the same. Figure F-1*a* and *b* shows positive and negative skewness. Define coefficient of skewness by:

$$\frac{\text{Mean} - \text{mode}}{\sigma} \quad \text{(F-11)}$$

Measure of kurtosis is given by:

$$\beta_2 = \frac{\mu_4}{\mu_2^2} \quad \text{(F-12)}$$

where:

$$\mu_2 = \frac{\sum (x - x_{\text{am}})^2}{\sum f}$$

$$\mu_4 = \frac{\sum (x - x_{\text{am}})^4}{\sum f}$$

If $\beta_2 = 3$, the curve is normal, or mesokurtic. If $\beta_2 > 3$, the curve is peaked or leptokurtic; if $\beta_2 < 3$, the curve is flat-topped or platykuric (Fig. F-1*c*).

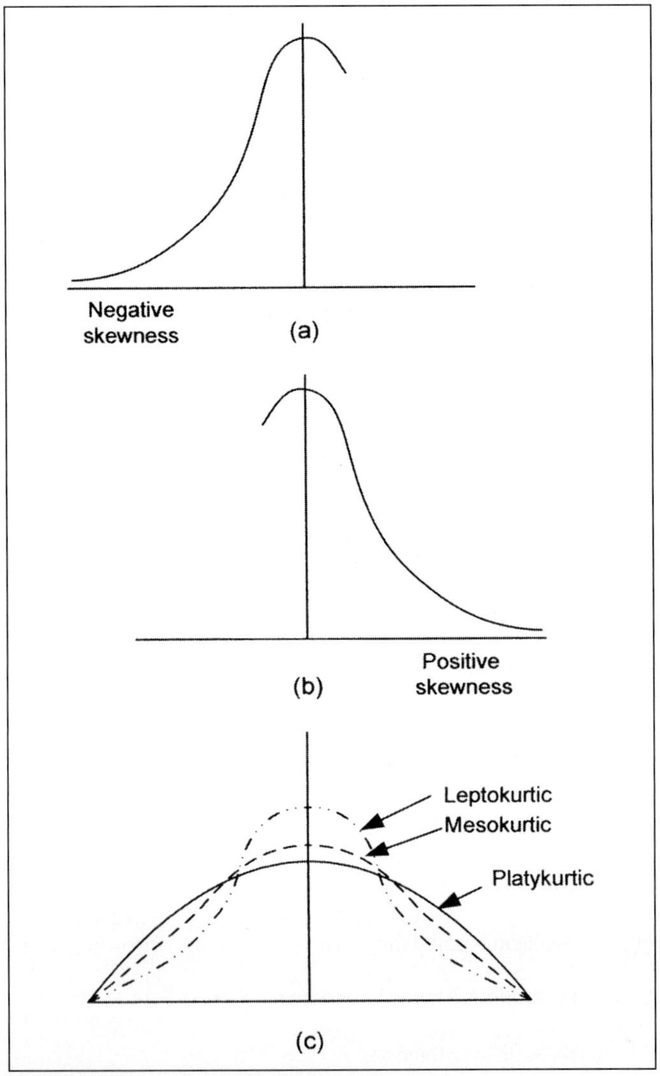

FIGURE F-1 Illustration of (*a*) positive and (*b*) negative skewness. (*c*) Illustration of kurtosis.

F-4 CURVE FITTING AND REGRESSION

The purpose of regression is to estimate one of the variables (dependent variable) from the other (independent variable). If y is estimated from x by some equation, it is referred to as regression. In other words, if the scatter diagram of two variables indicates some relation between these variables will be concentrated around a

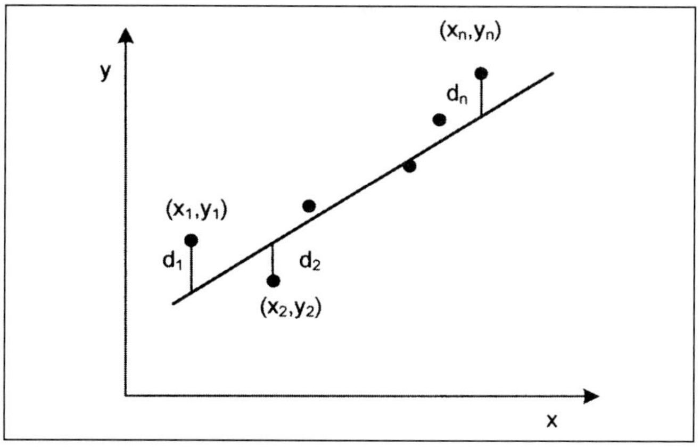

FIGURE F-2 Criteria of fitting a least-square line in a scatter plot.

curve. This curve is called the curve of regression. When the curve is a straight line it is called a line of regression.

For some given data points, more than one curve may seem to fit. Intuitively, it will be hard to fit an appropriate curve in a scatter diagram and variation will exist. Referring to Fig. F-2, a measure of goodness for the appropriate fit can be described as:

$$d_1^2 + d_2^2 + \cdots + d_n^2 = a \quad \text{minimum} \tag{F-13}$$

A curve meeting these criteria is said to fit the data in the *least-square sense* and is called a least-square regression curve, or simply a least-square curve—straight line or parabola. The least-square line imitating the points $(x_1, y_1),..(x_n, y_n)$ has the equation:

$$y = a + bx \tag{F-14}$$

The constants a and b are determined from solving simultaneous equations, which are called the normal equations for the least-square line:

$$\sum y = an + b\sum x$$
$$\sum xy = a\sum x + b\sum x^2 \tag{F-15}$$

This gives:

$$a = \frac{\left(\sum y\right)\left(\sum x^2\right) - \left(\sum x\right)\left(\sum xy\right)}{n\sum x^2 - \left(\sum x\right)^2}$$
$$b = \frac{n\sum xy - \left(\sum x\right)\left(\sum y\right)}{n\sum x^2 - \left(\sum x\right)^2} \tag{F-16}$$

The constant b can be written as:

$$b = \frac{\sum (x - x_{\text{mean}})(y - y_{\text{mean}})}{\sum (x - x_{\text{mean}}^2)} \tag{F-17}$$

This yields:

$$y_{\text{mean}} = a + bx_{\text{mean}} \tag{F-18}$$

From above we can also write the least-square line as:

$$y - y_{\text{mean}} = b(x - x_{\text{mean}})$$
$$= \frac{\sum (x - x_{\text{mean}})(y - y_{\text{mean}})}{\sum (x - x_{\text{mean}})^2}(x - x_{\text{mean}}) \tag{F-19}$$

Constant b is the slope of the line Eq. (F-14). The least-square line passes through $(x_{\text{mean}}, y_{\text{mean}})$ which is called the centroid of the data. Slope b is independent of origin of coordinates; that is, for translation of axis given by: $x = x' + h$, $y = y' + k$, where h and k are constants. In Eq. (F-19), x and y can be replaced with x' and y'.

Similarly, for least-square line of x on y:

$$x - x_{\text{mean}} = \frac{\sum (x - x_{\text{mean}})(y - y_{\text{mean}})}{\sum (y - y_{\text{mean}})^2}(y - y_{\text{mean}}) \tag{F-20}$$

The least-square line can be written in terms of variance and covariance. The sample variance and covariance are given by:

$$S_x^2 = \frac{\sum (x - x_{\text{mean}})^2}{n}$$
$$S_y^2 = \frac{\sum (y - y_{\text{mean}})^2}{n} \tag{F-21}$$
$$S_{xy} = \frac{\sum (x - x_{\text{mean}})(y - y_{\text{mean}})}{n}$$

In terms of these the least-square lines of y on x and x on y are:

$$y - y_{\text{mean}} = \frac{S_{xy}}{S_x^2}(x - x_{\text{mean}})$$
$$x - x_{\text{mean}} = \frac{S_{xy}}{S_y^2}(y - y_{\text{mean}}) \tag{F-22}$$

A sample correlation coefficient can be defined as:

$$r = \frac{S_{xy}}{S_x S_y} \tag{F-23}$$

Example F-2 Given the data points (x, y) as:

(1, 1), (3, 2), (4, 5), (6, 7), (7, 6), (9, 8), (12, 10), (15, 16)

fit a least-square line with x as independent variable, y as dependent variable.

Table F-2 shows the various steps pf calculations. Substituting the values arrived in this Table in Eq. (F-15), we have:

$$8a + 57b = 55$$

$$57a + 561b = 543$$

where n is the number of samples = 8. Solving these equations, $a = -0.719$, $b = 0.975$. Therefore the least-square line is:

$$y = -0.719 + 0.975x$$

If x is considered as the dependent variable and y as the independent variable, then:

$$8c + 55d = 57$$

$$55c + 535d = 543$$

Solution of which gives, $c = 0.507$ and $d = 0.963$.

An application of fitting the least-square line is the scatter plot of soil resistivity measurements over a given area for grounding grid designs.

TABLE F-2 Fitting the Least-Square Line

x	y	x^2	xy	y^2
1	1	1	1	1
3	2	9	6	4
4	5	16	20	25
6	7	36	42	49
7	6	49	42	36
9	8	81	72	64
12	10	144	120	100
15	16	225	240	256
$\sum x=57$	$\sum y=55$	$\sum x^2=561$	$\sum xy=543$	$\sum y^2=535$

F-4-1 Least-Square Parabola

As an extension of the above least-square example, the equation for a least-square parabola to fit a set of sample points is given by:

$$y=a+bx+c^2x \qquad \text{(F-24)}$$

where:

$$\begin{aligned}\sum y &= na+b\sum x+c\sum x^2\\ \sum xy &= a\sum x+b\sum x^2+c\sum x^3\\ \sum x^2y &= a\sum x^2+b\sum x^3+c\sum x^4\end{aligned} \qquad \text{(F-25)}$$

F-4-2 Multiple Regression

If there is linear relation between a dependent variable z and two linear variables x and y, we seek an equation of the form:

$$z=a+bx+cy \qquad \text{(F-26)}$$

This is called a *regression equation* of z on x and y. Again, a, b, c can be determined from the following equations:

$$\begin{aligned}\sum z &= na+b\sum x+c\sum y\\ \sum xz &= a\sum x+b\sum x^2+c\sum xy\\ \sum yz &= a\sum y+b\sum xy+c\sum y^2\end{aligned} \qquad \text{(F-27)}$$

F-5 PROBABILITY

If the outcome of an event A is m successes and n failures, all these likely to occur equally (coin toss falling on heads or tails), then the probability of success is:

$$p=\frac{m}{m+n} \qquad \text{(F-28)}$$

and of failure is:

$$q=\frac{n}{m+n} \qquad \text{(F-29)}$$

Thus $p + q = 1$, equal to 100 percent.

If $p_1, p_2, p_3, \ldots, p_n$ are separate probabilities of *mutually exclusive* events, then the probability P that any of these events will happen is given by:

$$P=p_1+p_2+\cdots+p_n \qquad \text{(F-30)}$$

This is the addition law of probability.

If two events A and B are not mutually exclusive, then the probability of the event that either A or B or both will occur is given by:

$$P(A\cup B)=P(A)+P(B)-P(A\cap B) \qquad \text{(F-31)}$$

The probability of two events happening together is the product of their probabilities:

$$P(AB)=P(A)\times P(B) \qquad \text{(F-32)}$$

This is the multiplication law of probability.

The probability of an event happening r times in n trials is given by:

$$P(r)={}^nC_r p^r q^{n-r} \qquad \text{(F-33)}$$

This is $(r + 1)$th term of $(q + p)^n$.

$$(q+p)^n=q^n+{}^nC_1q^{n-1}p+{}^nC_2q^{n-2}p^2+\cdots+{}^nC_rq^{n-r}p^r+\cdots+p^n \qquad \text{(F-34)}$$

This is the binomial distribution $P(r)$.

Example F-3 Consider that 10 percent of manufactured cars have defective seat belts. Determine the probability that out of 10 cars sold, 2 will have defective seat belts. Here $p = 0.1$, $q = 0.9$, $n = 10$

Probability of zero defective seat belts is:

$$P(0)={}^{10}C_0(0.1)^0(0.9)^{10}=0.3487$$

Probability that one seat belt is defective is:

$$P(1)={}^{10}C_1(0.1)^1(0.9)^9=0.3874$$

Probability that two seat belts are defective is:

$$P(2)={}^{10}C_2(0.1)^2(0.9)^8=0.1937$$

Probability of at most two defective seat belts is:

$$P(0)+P(1)+P(2)=0.9298$$

F-6 BINOMIAL DISTRIBUTION

The binomial distribution is given by Eq. (F-34).

$$\text{Mean of the distribution} = np \quad \text{(F-35)}$$

$$\text{Standard deviation SD} = \sigma = \sqrt{npq} \quad \text{(F-36)}$$

F-7 POISSON DISTRIBUTION

The Poisson distribution is a particular limiting form of the binomial distribution, where p or q is very small and n is very large.

Poisson distribution is:

$$P(r)=\frac{m^r e^{-m}}{r!} \quad \text{(F-37)}$$

where m is the mean of the distribution.

$$\text{The SD} = \sqrt{m}=\sigma \qquad \text{variance} = \mu_2=\sigma^2=m \quad \text{(F-38)}$$

F-8 NORMAL OR GAUSSIAN DISTRIBUTION

This is a continuous distribution and plays an important role in measurement statistics. It is applied in insulation coordination and lightning statistics (Chaps. 5 and 17). It is derived as the limiting form of binomial distribution for large values of n and p, while q is not very small.

As the measurements increase, the curve most commonly fitted is a bell-shaped curve, called the normal or Gaussian distribution. The density function for this distribution is given by:

$$p(x)=\frac{1}{\sigma\sqrt{2\pi}}e^{-\frac{1}{2}\left(\frac{x-\mu}{\sigma}\right)^2} \qquad -\infty<x<\infty \quad \text{(F-39)}$$

where μ is the mean and σ is the SD.

The distribution is completely defined if the parameters σ and μ are known. The corresponding distribution function is given by:

$$P(x)=P(X\le x)=\frac{1}{\sigma\sqrt{2\pi}}\int_{-\infty}^{x} e^{-\frac{1}{2}\left(\frac{x-\mu}{\sigma}\right)^2}dx \quad \text{(F-40)}$$

The probability that x takes between values x_1 and x_2 is given by:

$$P(x_1<x<x_2)=\int_{x_1}^{x_2}\frac{1}{\sigma\sqrt{2\pi}}e^{-\frac{(x-\mu)^2}{2\sigma^2}}dx \quad \text{(F-41)}$$

If we substitute:

$$z=\frac{x-\mu}{\sigma} \quad \text{(F-42)}$$

where z is the standardized variable corresponding to x, then the mean of z is 0 and the variance is 1. In such a case the density function becomes:

$$f(z)=\frac{1}{\sqrt{2\pi}}e^{-\frac{z^2}{2}} \quad \text{(F-43)}$$

This is referred as *standard normal density function*. The corresponding distribution function is:

$$F(z)=P(Z\le z)=\frac{1}{\sqrt{2\pi}}\int_{-\infty}^{z}e^{-\frac{z^2}{2}}dz \quad \text{(F-44)}$$

A graph of the density function is shown in Fig. F-3.

- The curve is symmetrical about the y-axis. The mean, median, and mode coincide at the origin.
- The area of the curve is equal to the total number of observations.
- $f(z)$ decreases rapidly as z increases numerically. The curve extends on either side of the origin.

In Fig. F-3, the areas within 1, 2, and 3 standard deviations of the mean are indicated.

$$P(-1\le Z\le 1)=0.6827$$

About 68.27 percent of the values lie between $(\mu - \sigma)$ and $(\mu + \sigma)$

$$P(-2\le Z\le 2)=0.9545$$

About 95.45 percent of the values lie between $(\mu - 2\sigma)$ and $(\mu + 2\sigma)$

$$P(-3\le Z\le 3)=0.9973 \text{ percent}$$

About 99.73 percent of the values lie between $(\mu - 3\sigma)$ and $(\mu + 3\sigma)$

The total area under the curve is 1, divided into two equal parts. The Table F-3 gives the area shown from 0 to z [$= (x - \mu)/\sigma$].

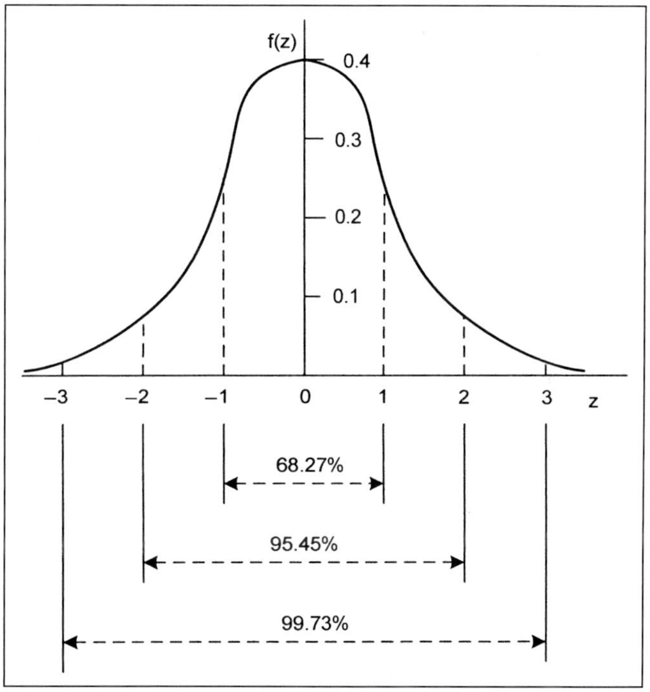

FIGURE F-3 Gaussian or normal distribution graph of the density function.

TABLE F-3 Area Under the Standard Normal Curve from 0 to z

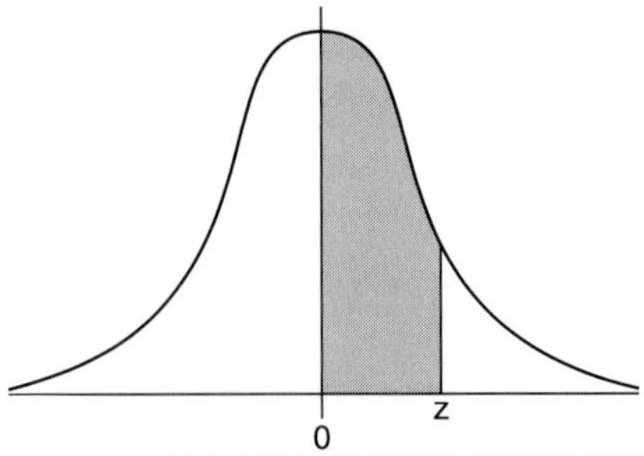

z	0	1	2	3	4	5	6	7	8	9
0.0	0.0000	0.0040	0.0080	0.0120	0.0160	0.0199	0.0239	0.0279	0.0319	0.0359
0.1	0.0398	0.0438	0.0478	0.0517	0.0557	0.0596	0.0636	0.0675	0.0714	0.0754
0.2	0.0793	0.0832	0.0871	0.0910	0.0948	0.0987	0.1026	0.1064	0.1103	0.1141
0.3	0.1179	0.1217	0.1255	0.1293	0.1331	0.1368	0.1406	0.1443	0.1480	0.1517
0.4	0.1554	0.1591	0.1628	0.1664	0.1700	0.1736	0.1772	0.1808	0.1844	0.1879
0.5	0.1915	0.1951	0.1985	0.2019	0.2054	0.2088	0.2123	0.2157	0.2190	0.2224
0.6	0.2258	0.2291	0.2324	0.2357	0.2389	0.2422	0.2454	0.2486	0.2518	0.2549
0.7	0.2580	0.2612	0.2642	0.2673	0.2704	0.2734	0.2764	0.2794	0.2823	0.2852
0.8	0.2881	0.2910	0.2939	0.2967	0.2996	0.3023	0.3051	0.3078	0.3106	0.3133
0.9	0.3159	0.3186	0.3212	0.3238	0.3264	0.3289	0.3315	0.3340	0.3365	0.3389
1.0	0.3413	0.3438	0.3461	0.3485	0.3508	0.3531	0.3554	0.3577	0.3599	0.3621
1.1	0.3643	0.3665	0.3686	0.3708	0.3729	0.3749	0.3770	0.3790	0.3810	0.3830
1.2	0.3849	0.3869	0.3888	0.3907	0.3925	0.3944	0.3962	0.3980	0.3997	0.4015
1.3	0.4032	0.4049	0.4066	0.4082	0.4099	0.4115	0.4131	0.4147	0.4162	0.4177
1.4	0.4192	0.4207	0.4222	0.4236	0.4251	0.4265	0.4279	0.4292	0.4306	0.4319
1.5	0.4332	0.4345	0.4357	0.4370	0.4382	0.4394	0.4406	0.4418	0.4429	0.4441
1.6	0.4452	0.4463	0.4474	0.4484	0.4495	0.4505	0.4515	0.4525	0.4535	0.4545
1.7	0.4554	0.4564	0.4573	0.4582	0.4591	0.4599	0.4608	0.4616	0.4625	0.4633
1.8	0.4641	0.4649	0.4656	0.4664	0.4671	0.4678	0.4686	0.4693	0.4699	0.4706
1.9	0.4713	0.4719	0.4726	0.4732	0.4738	0.4744	0.4750	0.4756	0.4761	0.4767
2.0	0.4772	0.4778	0.4783	0.4788	0.4793	0.4798	0.4803	0.4808	0.4812	0.4817
2.1	0.4821	0.4826	0.4830	0.4834	0.4838	0.4842	0.4846	0.4850	0.4854	0.4857
2.2	0.4861	0.4864	0.4868	0.4871	0.4875	0.4878	0.4881	0.4884	0.4887	0.4890
2.3	0.4893	0.4896	0.4898	0.4901	0.4904	0.4906	0.4909	0.4911	0.4913	0.4916
2.4	0.4918	0.4920	0.4922	0.4925	0.4927	0.4929	0.4931	0.4932	0.4934	0.4936
2.5	0.4938	0.4940	0.4941	0.4943	0.4945	0.4946	0.4948	0.4949	0.4951	0.4952
2.6	0.4953	0.4955	0.4956	0.4957	0.4959	0.4960	0.4961	0.4962	0.4963	0.4964
2.7	0.4965	0.4966	0.4967	0.4968	0.4969	0.4970	0.4971	0.4972	0.4973	0.4974
2.8	0.4974	0.4975	0.4976	0.4977	0.4977	0.4978	0.4979	0.4979	0.4980	0.4981
2.9	0.4981	0.4982	0.4982	0.4983	0.4984	0.4984	0.4985	0.4985	0.4986	0.4986
3.0	0.4987	0.4987	0.4987	0.4988	0.4988	0.4989	0.4989	0.4989	0.4990	0.4990
3.1	0.4990	0.4991	0.4991	0.4991	0.4992	0.4992	0.4992	0.4992	0.4993	0.4993
3.2	0.4993	0.4993	0.4994	0.4994	0.4994	0.4994	0.4994	0.4995	0.4995	0.4995
3.3	0.4995	0.4995	0.4995	0.4996	0.4996	0.4996	0.4996	0.4996	0.4996	0.4997
3.4	0.4997	0.4997	0.4997	0.4997	0.4997	0.4997	0.4997	0.4997	0.4997	0.4998
3.5	0.4998	0.4998	0.4998	0.4998	0.4998	0.4998	0.4998	0.4998	0.4998	0.4998
3.6	0.4998	0.4998	0.4999	0.4999	0.4999	0.4999	0.4999	0.4999	0.4999	0.4999
3.7	0.4999	0.4999	0.4999	0.4999	0.4999	0.4999	0.4999	0.4999	0.4999	0.4999
3.8	0.4999	0.4999	0.4999	0.4999	0.4999	0.4999	0.4999	0.4999	0.4999	0.4999
3.9	0.5000	0.5000	0.5000	0.5000	0.5000	0.5000	0.5000	0.5000	0.5000	0.5000

Tables of distribution to higher decimal accuracies are available. Note that the area to the right and left becomes 0.5 only at plus and minus infinity.

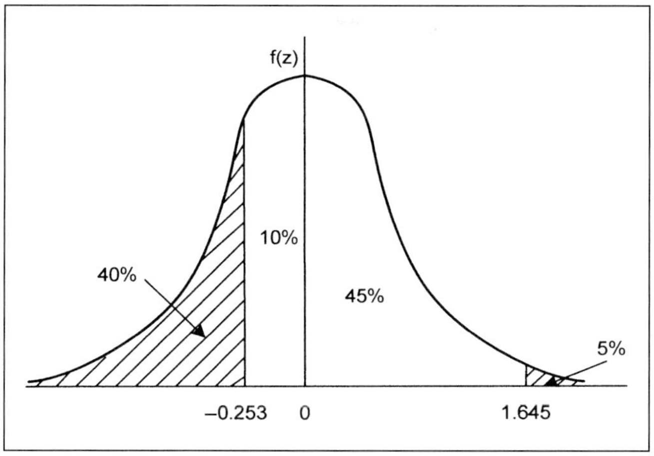

FIGURE F-4 Gaussian distribution of data in Example F-4.

Example F-4 Consider that in a normal distribution, 40 percent of the data is under 50 and 5 percent is over 80. What is the mean and SD.

The graphical construction from the given data is shown in Fig. F-4. This gives:

$$z_1 = \frac{50-\mu}{\sigma} \quad \text{and} \quad z_2 = \frac{80-\mu}{\sigma}$$

From the percentage distribution shown, area between 0 and $z_1 = 0.5 - 0.4 = 0.1$. From Table F-3, the value of $z_1 = -0.253$ (interpolate)

$$\frac{50-\mu}{\sigma} = -0.253$$

For the percent distribution shown, area between 0 and $z_2 = 0.5 - 0.05 = 0.45$. From Table F-3, the value of $z_2 = 1.645$ (interpolate)

$$\frac{80-\mu}{\sigma} = 1.645$$

Thus $\mu = 53.9989$ and SD $= \sigma = 15.8061$.

Example F-5 Consider production of a product where $\mu = 110$, and the standard deviation is 1.0. What percentage of the product deviation falls between 109 and 111?

$$z_1 = \frac{109-110}{1} = -1$$

$$z_2 = \frac{111-110}{1} = 1$$

The area between 0 and z_1 = area between 0 and z_2 = twice the area between 0 and z_2 = 0.683 from the Table F-3. Thus, 68.26 percent of the product will fall between 109 and 111.

F-9 WEIBULL DISTRIBUTION

IEC standard[1] recommends use of the Weibull distribution to represent the discharge capability of the external insulation. Note that the Gaussian distribution is unbounded to the right and left and is defined between plus and minus infinity. From the electrical application point of view there is a probability of flashover even with minus infinity; that is, when the voltage is zero.

The general expression for Weibull distribution is:

$$P(V) = 1 - e^{-\left(\frac{V-\delta}{\beta}\right)^{\gamma}} \tag{F-45}$$

where δ is truncation value, β is the scale parameter, and γ is the shape parameter.

Equation (F-45) can be modified to depict the discharge probability of insulation with a truncated discharge probability by substituting the truncation value δ and the scale factor β.

$$\begin{aligned} \delta &= V_{50} - N\sigma \\ \beta &= N\sigma(\ln 2)^{-\frac{1}{\gamma}} \end{aligned} \tag{F-46}$$

where N is the number of conventional deviations between V_{50} and V_0 of a self-restoring insulation, and σ is the conventional deviations of the discharge probability function $P(V)$ of a self-restoring insulation.

This leads to the modified Weibull function:

$$P(V) = 1 - 0.5^{\left(1+\frac{V-V_{50}}{N\sigma}\right)^{\gamma}} \tag{F-47}$$

The exponent is determined by the condition that:

$$P(V_{50} - \sigma) = 0.16 \tag{F-48}$$

This gives:

$$\gamma = \frac{\ln\left[\frac{\ln(1-0.16)}{\ln 0.5}\right]}{\ln(1-(1/N))} \tag{F-49}$$

For external insulation, it is assumed that no discharge is possible (withstand probability = 100 percent) at a truncation value:

$$V_0 = V_{50} - 4\sigma \tag{F-50}$$

Substituting $N = 4$ in Eq. (F-49) gives $\gamma = 4.85$, approximated as percentage. Introduce the normalized variable:

$$x = (V - V_{50})/\sigma \tag{F-51}$$

The Weibull flashover probability distribution is:

$$P(V) = 1 - 0.5^{\left(1+\frac{x}{4}\right)^{5}} \tag{F-52}$$

The comparison with Gaussian distribution is shown in Fig. F-5*a* and *b*.

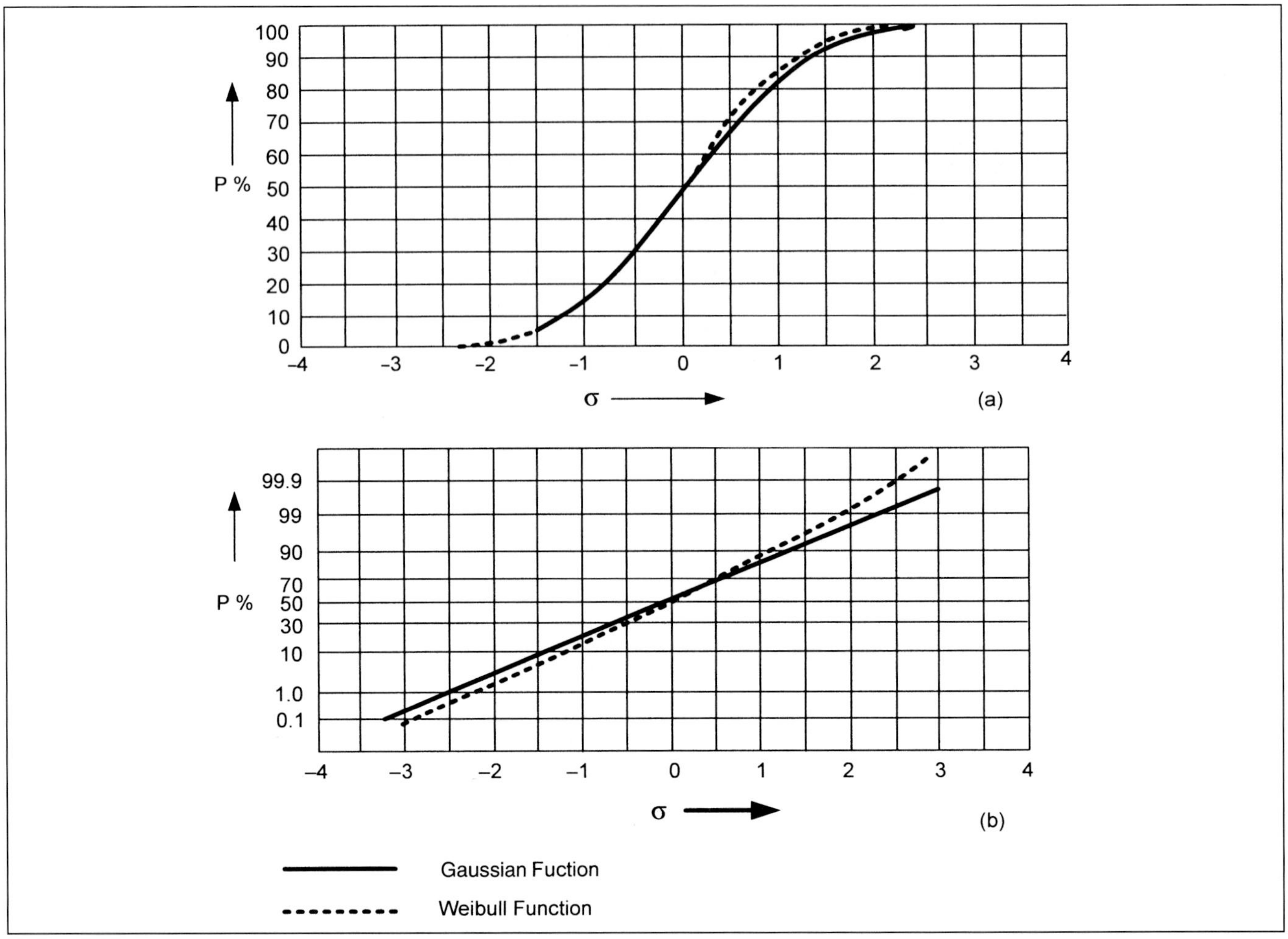

Figures F-5 Comparison of Gauss and Weibull functions of disruptive discharge probability of self-restoring insulation. (*a*) Described on a linear scale. (*b*) Described on a Gaussian scale.[1]

REFERENCE

1. IEC Std. 60071-2, Insulation Coordination, Part-2: Application Guide, 1996.

FURTHER READING

F. M. Dekking, C. Kraaikamp, H. P. Loupaä, L. E. Meester, *A Modern Introduction to Probability and Statistics (Springer Texts in Statistics)*, Springer Verlag, London, 2005.

J. L. Devore, *Probability and Statistics for Engineering and Sciences*, Brooks/Cole (Duxbury Press), 2007.

M. R. Spiegel, J. J. Schiller, R. A. Srinivasan, *Theory and Problems of Probability and Statistics*, 2d ed., *Schaum's Outline Series*, McGraw Hill, New York, 2000.

APPENDIX G

NUMERICAL TECHNIQUES

We used Laplace transform method in Chap. 2 for solution of simple circuits, and toward the end of this chapter concluded that the theoretical analysis methods using Laplace transforms are not suited to large systems; nonlinearity and saturation cannot be accounted for. Computerized methods to take account of nonlinearity may use:

- Piecewise linearization, example of saturation in transformers (Chap. 14)
- Exponential segments of arrester models (Chap. 19)
- One-time step-delay methods—pseudo-nonlinear devices
- Iterative Newton methods

G-1 NETWORK EQUATIONS

Consider a network of connections as shown in Fig. G-1. For the inductance branch, nodes 1 and 3, we can write:

$$v = L\frac{di}{dt} \tag{G-1}$$

In terms of difference equation:

$$\frac{v(t)+v(t-\Delta t)}{2} = L\frac{i(t)-i(t-\Delta t)}{\Delta t} \tag{G-2}$$

This can be written as:

$$i_{13}(t) = \frac{\Delta t}{2L}(v_1(t)-v_3(t))+\text{hist}_{13}(t-\Delta t) \tag{G-3}$$

where hist_{13} term is known from the preceding time step:

$$\text{hist}_{13}(t-\Delta t) = i_{13}(t-\Delta t)+\frac{\Delta t}{2L}[v_1(t-\Delta t)-v_3(t-\Delta t)] \tag{G-4}$$

For the capacitance circuit, we can similarly write:

$$\text{hist}_{14}(t-\Delta t) = -i_{14}(t-\Delta t)-\frac{2C}{\Delta t}(v_1(t-\Delta t)-v_4(t-\Delta t)) \tag{G-5}$$

For the transmission line, ignoring losses:

$$\begin{aligned} i_{15}(t) &= \frac{1}{Z}v_1(t)+\text{hist}_{15}(t-\tau) \\ \text{hist}_{15}(t-\tau) &= -\frac{1}{Z}v_5(t-\tau)-i_{51}(t-\tau) \end{aligned} \tag{G-6}$$

where z is the surge impedance and τ is the line length/velocity of propagation. Therefore for node 1, we can write:

$$\left(\frac{1}{R}+\frac{\Delta t}{2L}+\frac{2C}{\Delta t}+\frac{1}{Z}\right)v_1(t)-\frac{1}{R}v_2(t)-\frac{\Delta t}{2L}v_3(t)-\frac{2C}{\Delta t}v_4(t)$$
$$= i_1(t)-\text{hist}_{13}(t-\Delta t)-\text{hist}_{14}(t-\Delta t)-\text{hist}_{15}(t-\tau) \tag{G-7}$$

For any type of network with n nodes, we can write the general equation:

$$\bar{G}\bar{v}_t = \bar{i}_t - \text{his}\bar{t} \tag{G-8}$$

where $\bar{G}$ is the nxn symmetrical nodal conductance matrix, $\bar{v}_t$ is the vector of n node voltages, $\bar{i}_t$ is the vector of current sources, and $\text{his}\bar{t}$ is the vector of n known history terms.

Some nodes will have known voltages or may be grounded. Equation (G-8) can be partitioned into a set of nodes A with known voltages and a set B with unknown voltages, and the unknown voltages can be found by solving for $\bar{v}_{At}$:

$$\bar{G}_{AA}\bar{v}_{At} = \bar{i}_{At} - \text{his}\bar{t}_A - \bar{G}_{AB}\bar{v}_{Bt} \tag{G-9}$$

G-2 COMPENSATION METHODS

In compensation methods, the nonlinear elements are simulated as current injections, which are superimposed on linear network after a solution without the nonlinear elements has been just found.

Figure G-2 shows one nonlinear element across linear network. The current i_{km} must satisfy the following two equations:

$$v_{km} = v_{km-0} - R_{\text{Thev}}i_{km} \tag{G-10}$$

$$v_{km} = f(i_{km}, di_{km}/dt, t, \ldots) \tag{G-11}$$

where $v_{km\text{-}0}$ is the solution without nonlinear branch connected. R_{Thev} is the instantaneous equivalent circuit between nodes k and m.

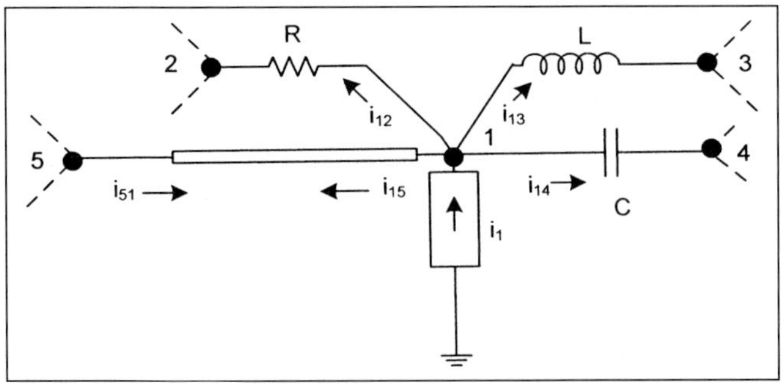

FIGURE G-1 A network around node 1 (for writing node equations).

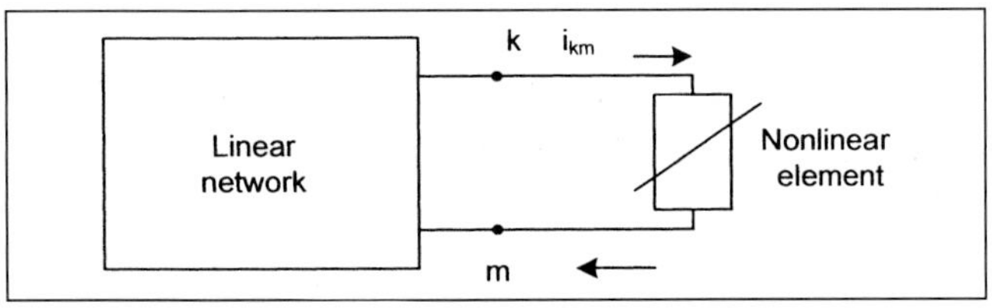

FIGURE G-2 One nonlinear element connected to a linear network.

Equation (G-11) is the relationship of the nonlinear branch. To find Thévenin resistance, a current of 1A is injected into the node k and taken out from node m. The network equation (G-9) can be written as:

$$\bar{G}_{AA}\bar{V}_A = \bar{k}_A \tag{G-12}$$

where $\bar{k}_A$ is known from right-hand side of Eq. (G-9). It is replaced with a vector whose components are all zero, except for + 1.0 in row k and −1.0 in row m. Then one repeat solution is performed with right-hand side vector which produced a vector $\bar{r}_{\text{Thev}}$. This vector is the difference of the kth and mth columns of the inverse V matrix $\bar{G}_{AA}^{-1}$. Then:

$$\bar{R}_{\text{Thev}} = \bar{r}_{\text{Thev}-k} - \bar{r}_{\text{Thev}-m} \tag{G-13}$$

Thus the solution proceeds as follows:

- Compute node voltages without a nonlinear element connected, with a repeat solution of Eq. (G-12).
- From this vector and other known voltages, extract open-circuit voltage:

$$v_{km-0} = v_{k-0} - v_{m-0} \tag{G-14}$$

- Solve two scalar equations [Eqs. (G-10) and (G-11)] simultaneously to find i_{km}. Newton-Raphson method can be used to solve Eq. (G-11). If Eq. (G-11) is given as a point-by-point, piecewise linear curve, then intersection of two curves is found by a search process (Fig. G-3).
- The final solution is found by superimposition, which is permissible as the rest of the system is linear.

The method can be used to simulate M nonlinear branches with current sources. M vectors, $\bar{r}_{\text{Thev}-1},\ldots,\bar{r}_{\text{Thev}-M}$ must be precomputed and recomputed, whenever switch changes position. The Thévenin equivalent resistance becomes a $M \times M$ matrix, $\bar{R}_{\text{Thev}}$.

$$\bar{v}_{km} = \bar{v}_{km-0} - \bar{R}_{\text{Thev}}\bar{i}_{km} \tag{G-15}$$

and:

$$\bar{v}_A = \bar{v}_{A-0} - (\bar{r}_{\text{Thev}-1},\ldots,\bar{r}_{\text{Thev}-M})\bar{i}_{km} \tag{G-16}$$

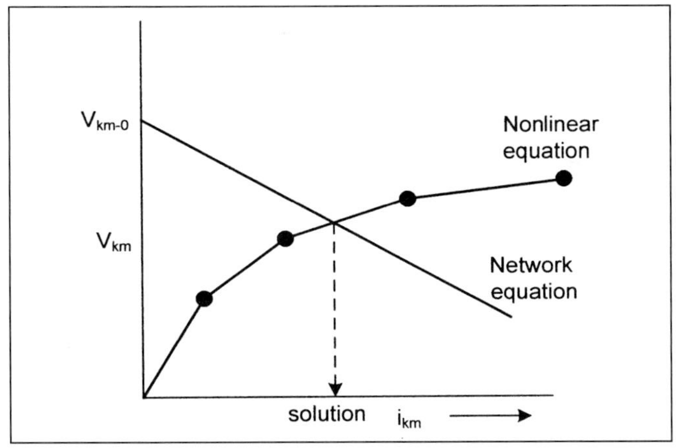

FIGURE G-3 Solution of network equation and nonlinear element.

G-3 NONLINEAR INDUCTANCE

The simulation shown in Fig. G-3 is straightforward for a nonlinear resistor defined by $v_{km} = f(i_{km})$. For nonlinear inductance the solution is not so direct as the nonlinear characteristics are of the form:

$$\lambda = f(i) \tag{G-17}$$

The flux λ being integral over the voltage $v = v_k - v_m$

$$\lambda(t) = \lambda(t-\Delta t) + \int_{t-\Delta t}^{t} v(u)\,du \tag{G-18}$$

This is solved by trapezoidal rule of integration, which converts flux $\lambda(t)$ into linear function of $v(t)$:

$$\lambda(t) = \frac{\Delta t}{2} v(t) + \text{hist}(t-\Delta t) \tag{G-19}$$

With a known history term:

$$\text{hist}(t-\Delta t) = \lambda(t-\Delta t) + \frac{\Delta t}{2} v(t-\Delta t) \tag{G-20}$$

Substituting Eq. (G-19) into Eq. (G-17) produces a resistance relationship by first shifting the origin by $+\text{hist}(t-\Delta t)$ and rescaling λ-axis to v-axis with a multiplication factor of $2/\Delta t$. The v-i characteristics are solved with network equation in a similar way to nonlinear resistance. Figure G-4a and b shows representation of a linear and nonlinear inductance. A numerical problem can arise with nonlinear elements if Δt is too large. Artificial damping or hysteresis can occur.

G-4 PIECEWISE LINEAR INDUCTANCE

As discussed in Chap. 16, the saturation characteristics of transformers can be represented by piecewise linear inductance with two slopes (Fig. G-5a). Such an inductance can be simulated by two inductances L_1 and L_p in parallel (Fig. G-5b). The flux in L_p is always computed by integrating voltage $v_k - v_m$ independent of the switch position. The switch is closed whenever $|\lambda| \geq \lambda_{\text{saturation}}$ and opened again when $|\lambda| < \lambda_{\text{saturation}}$. In EMTP, there is an option to start the simulation from a user-specified residual flux. A piecewise linear resistance can be similarly represented (Fig. G-6).

G-5 NEWTON-RAPHSON METHOD

Newton-Raphson method is extensively used in power-system load flow and is best suited for system of nonlinear equations. To illustrate,

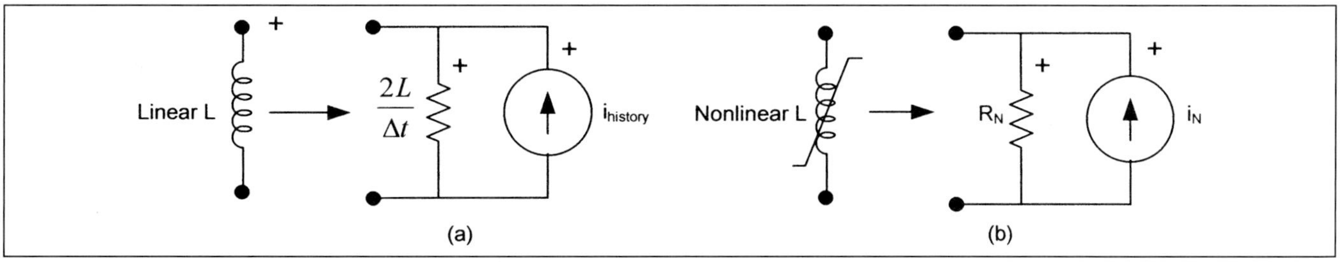

FIGURE G-4 (*a*) Representation of a linear inductor and (*b*) nonlinear inductor.

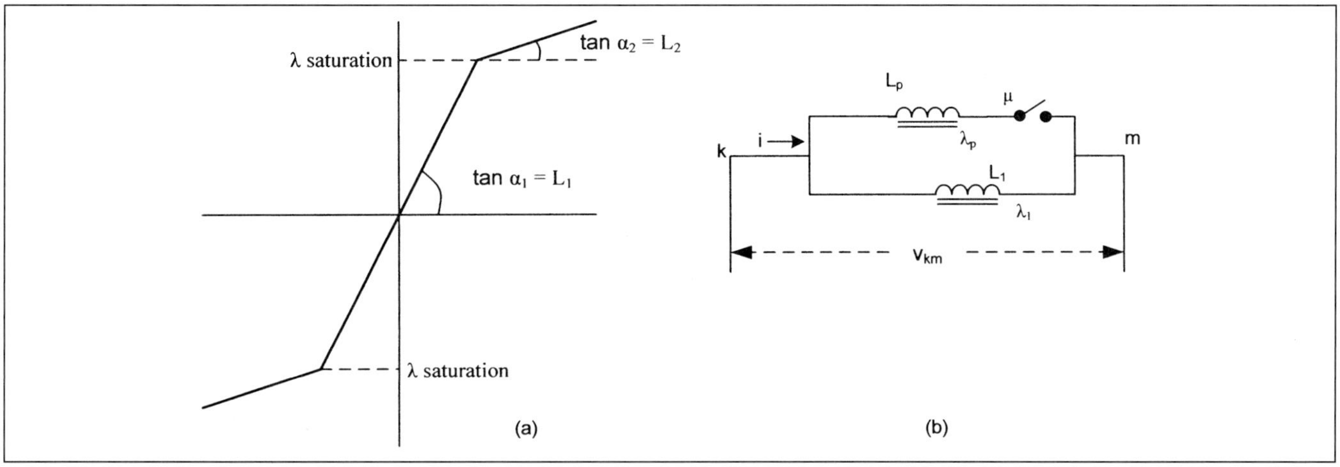

FIGURE G-5 (*a*) Two-piece representation of a nonlinear inductance. (*b*) Circuit implementation of nonlinear inductance.

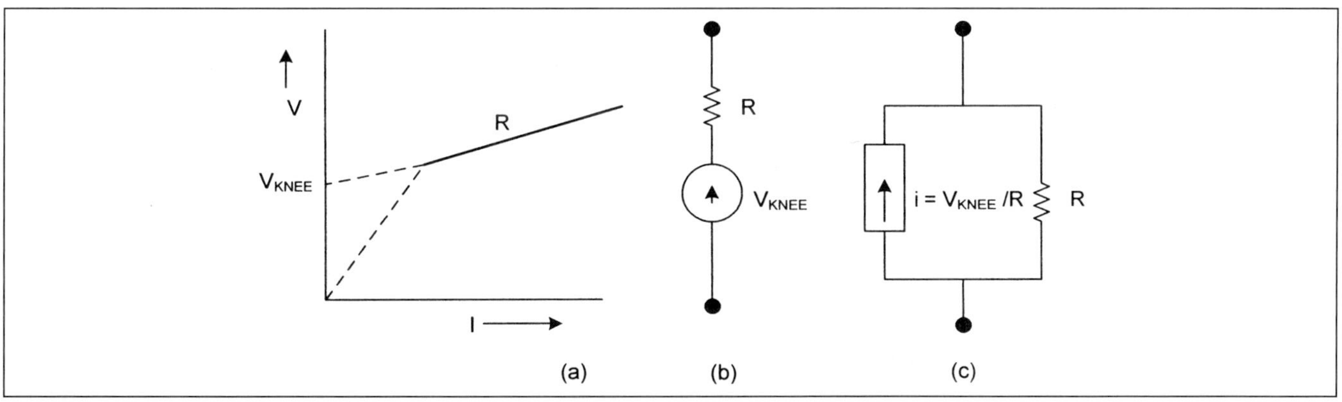

FIGURE G-6 (*a*) Piecewise resistor. (*b*) Voltage source representation. (*c*) Current source representation.

consider that the elements are nonlinear resistances. Equation (G-15) is rewritten as:

$$\bar{v}_{km} - \bar{v}_{km\text{-}0} + \bar{R}_{\text{Thev}}\bar{i}_{km} = 0 \qquad \text{(G-21)}$$

$\bar{i}_{km}$ can be replaced with a diagonal matrix $f(\bar{v}_{km})$ whose elements are *i-v* characteristics of *M* nonlinear resistances, the nonlinear resistance represented by a power function of the form:

$$i = p\left(\frac{v}{v_{\text{ref}}}\right)^{q} \qquad \text{(G-22)}$$

where *p*, v_{ref}, and *q* are constants and the voltage region is divided into segments. Thus, we write:

$$\bar{i}_{km} = f(\bar{v}_{km}) \qquad \text{(G-23)}$$

Applying Newton-Raphson method to Eq. (G-21)

$$\bar{R}_{\text{Thev}}\left[\frac{df_{km}}{dv_{km}}\right]^{-} + \bar{U}\Delta\bar{v}_{km} = \bar{v}_{km\text{-}0} - \bar{v}_{km} - \bar{R}_{\text{Thev}} f(\bar{v}_{km}) \qquad \text{(G-24)}$$

where the matrix on the left-hand side (Jacobian matrix), and the matrix on the right-hand side are solved with appropriate answers from the last iteration step $h-1$. The improved solution is found by solving the system of linear equations:

$$\overline{v}_{km}^{h} = \overline{v}_{km}^{h-1} + \Delta\overline{v}_{km} \qquad \text{(G-25)}$$

In Eq. (G-24), $[df_{km}/dv_{km}]^{-}$ is diagonal matrix of i-v characteristics, and it destroys the symmetry of the Jacobian. To maintain symmetry the Jacobian can be multiplied by inverse matrix $[df_{km}/dv_{km}^{-}]^{-1}$ and solve the equation for variable $\Delta\overline{x}$.

$$\left[\overline{R}_{\text{Thev}} + \left[\frac{df_{km}}{dv_{km}}\right]^{-1}\right]\Delta\overline{x} = \overline{v}_{km-0} - \overline{v}_{km} - \overline{R}_{\text{Thev}}\overline{f(v_{km})} \qquad \text{(G-26)}$$

The Jacobian is now symmetric. The diagonal elements of $[df_{km}/dv_{km}]^{-1}$ are reciprocals of $[df_{km}/dv_{km}]$. After $\Delta\overline{x}$ has been found, the voltage corrections are:

$$\Delta v_{km} = \frac{\Delta x_{km}}{df_{km}/dv_{km}} \qquad \text{(G-27)}$$

G-6 NUMERICAL SOLUTION OF LINEAR DIFFERENTIAL EQUATIONS

In Chap. 2, we showed that a linear differential equation can be put in the state variable form:

$$\dot{\mathbf{x}} = \overline{A}\overline{x} + \overline{g}(t) \qquad \text{(G-28)}$$

The closed-form solution of Eq. (G-28), which carries from the state of the system at $t-\Delta t$ to t, is:

$$x(t) = e^{\overline{A}\Delta t}[x(t+\Delta t)] + \int_{t-\Delta t}^{t} e^{\overline{A}(t-u)}g(u)\,du \qquad \text{(G-29)}$$

where the matrix $e^{\overline{A}\Delta t}$ is called the transition matrix. It is transformed into a diagonal matrix, whose elements are evaluated using eigenvalues λ_i of matrix $\overline{A}$ and matrix of eigenvectors (modal matrix $\overline{M}$) of $\overline{A}$. An efficient method of finding eigenvalues is QR transformation,[1] and for finding eigenvectors the "inverse iteration scheme",[2] which has been modified by J. E. Van Ness.[3] The diagonalized matrix is:

$$\overline{M}^{-1}e^{\overline{A}\Delta t}\overline{M} = e^{\overline{\Lambda}\Delta t} \qquad \text{(G-30)}$$

where $\overline{\Lambda}$ is the diagonal matrix of eigenvalues λ_i. Once the diagonal elements are found these can be converted back to:

$$e^{\overline{A}\Delta t} = \overline{M}e^{\overline{\Lambda}\Delta t}\overline{M}^{-1} \qquad \text{(G-31)}$$

Equation (G-31) becomes:

$$x(\overline{t}) = \overline{M}e^{\overline{\Lambda}\Delta t}\overline{M}^{-1}[x(t-\Delta t)] + \int_{t-\Delta t}^{t} \overline{M}e^{\Lambda(t-u)}\overline{M}^{-1}g(u)\,du \qquad \text{(G-32)}$$

The convolution integral can be evaluated in closed form for many types of functions $g(t)$.

G-7 LAPLACE TRANSFORM

Laplace transform involves ratio of polynomials and the poles and zeros thereof. The task of computing the coefficients of polynomials in a network function of $P(s)/Q(s)$ is not only time consuming but also prone to serious numerical inaccuracies, especially when the polynomials are of higher degree. The so-called topological formula approach[4] to computing these network functions involves finding the trees of a network and then computing the sum of corresponding tree-admittance branches. The number of trees may run into millions for a network of only 20 nodes and 40 branches. The computation of the roots of polynomials $P(s)$ and $Q(s)$ is "hazardous" because these roots may be extremely sensitive to errors in coefficients. The polynomial approach is not matched to the network analysis tasks which the computer is called upon to handle. The eigenvalue approach is much better suited and gives the theoretical information that Laplace transform methods are designated to provide.[5]

G-8 TAYLOR SERIES

The matrix exponential $e^{\overline{A}\Delta t}$ can be approximated by Taylor series:

$$e^{\overline{A}\Delta t} = \overline{U} + \Delta t\overline{A} + \frac{\Delta t^2}{2!}\overline{A}^2 + \frac{\Delta t^3}{3!}\overline{A}^3 + \cdots\cdots \qquad \text{(G-33)}$$

The method runs into eigenvalue problem: When matrix $\overline{A}$ has a large eigenvalue, which means a small time constant, the time step Δt must be kept small in order to permit rapid convergence of Eq. (G-33).[5] In stiff systems, there are large differences in the magnitudes of eigenvalues and the largest eigenvalues produce ripples, which are not of interest. The method of using Eq. (G-33) becomes numerically unstable for a given number of terms if Δt is not sufficiently small to trace uninteresting small ripples. The method becomes identical with fourth-order Runge-Kutta method if fifth and higher-order terms are neglected in Eq. (G-33).

A rational approximation of the transition matrix is provided by E. J. Davison:[6]

$$e^{\overline{A}\Delta t} = \left(\overline{U} - \frac{\Delta t}{2}\overline{A} + \frac{\Delta t^2}{4}\overline{A}^2 - \frac{\Delta t^3}{12}\overline{A}^3\right)^{-1}\left(\overline{U} + \frac{\Delta t}{2}A + \frac{\Delta t^2}{4}\overline{A}^2 + \frac{\Delta t^3}{12}\overline{A}^3\right) \qquad \text{(G-34)}$$

A lower-order approximation, which is numerically stable for all Δt, neglects second- and higher-order terms of Eq. (G-34)

$$e^{\overline{A}\Delta t} = \left(\overline{U} - \frac{\Delta t}{2}\overline{A}\right)^{-1}\left(\overline{U} + \frac{\Delta t}{2}\overline{A}\right) \qquad \text{(G-35)}$$

This is identical with trapezoidal rule of integration discussed next.

G-9 TRAPEZOIDAL RULE OF INTEGRATION

Write Eq. (G-28) as an integral equation:

$$\overline{x}(t) - \overline{x}(t-\Delta t) + \int_{t-\Delta t}^{t} (\overline{A}\overline{x}(u) + \overline{g}(u))\,du \qquad \text{(G-36)}$$

By using linear interpolation on x and g between $t-\Delta t$ and t, assuming x is known at t:

$$\overline{x}(t) = \overline{x}(t-\Delta t) + \frac{\Delta t}{2}\overline{A}(\overline{x}(t-\Delta t) + \overline{x}(t))$$

$$+\frac{\Delta t}{2}(\overline{g}(t-\Delta t) + \overline{g}(t)) \qquad \text{(G-37)}$$

Linear interpolations imply that the area under the integral in Eq. (G-36) are approximated by trapezoidals (Fig. G-7), and therefore the name "trapezoidal rule of integration." The method is identical with central difference quotients:

$$\frac{\overline{x}(t) - \overline{x}(t-\Delta t)}{\Delta t} = \overline{A}\frac{\overline{x}(t-\Delta t) + \overline{x}(t)}{2} + \frac{\overline{g}(t-\Delta t) + \overline{g}(t)}{2} \qquad \text{(G-38)}$$

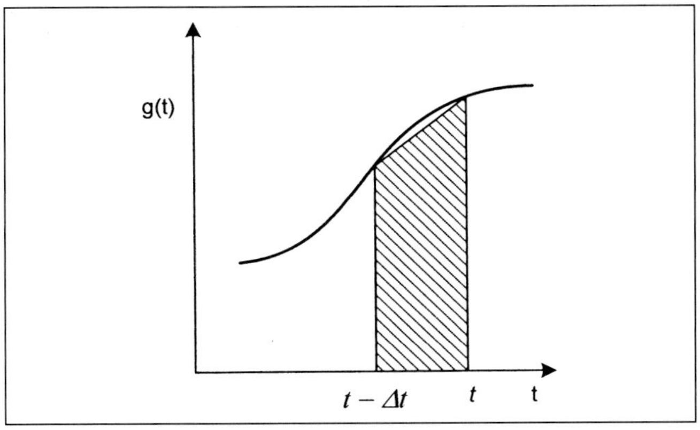

FIGURE G-7 Trapezoidal rule of integration.

This can be written as:

$$\left(\bar{U}-\frac{\Delta t}{2}\bar{A}\right)\bar{x}(t)=\left(\bar{U}+\frac{\Delta t}{2}\bar{A}\right)\bar{x}(t-\Delta t)+\frac{\Delta t}{2}(\bar{g}(t-\Delta t)+\bar{g}(t)) \tag{G-39}$$

This equation when premultiplied by $[(\bar{U}-\Delta t)(\bar{A}/2)]^{-1}$ gives the approximate transition matrix of Eq. (G-35). Working with trapezoidal rule of integration requires solution of linear, algebraic equations in each time step. If Δt is not changed, and so long the network modifications occur because of switching and nonlinear effects, the matrix $[(\bar{U}-\Delta t)(\bar{A}/2)]^{-1}$ for the system of equations remains unchanged. It is therefore best to triangularize this matrix once at the beginning and again whenever network changes occur.

The trapezoidal rule applied to the state equations results in the same answers as the trapezoidal rule applied to individual branch equations, which are then assembled into node equations [Eq. (G-7)]. The trapezoidal rule is of lower-order accuracy than many other methods, however, it is numerically stable, which is more important in power system transient analysis.

G-10 RUNGE-KUTTA METHODS

The method can be used for the solution of ordinary differential equations.

$$\dot{x}=f(x,t) \tag{G-40}$$

The fourth-order method seems to be widely used. Starting from the known value $x(t-\Delta t)$, the slope is calculated at point (Fig. G-8*a*):

$$\frac{\Delta x^{(1)}}{\Delta t}=f(x(t-\Delta t)) \qquad t-\Delta t \tag{G-41}$$

This is used to obtain an approximate value of $x^{(1)}$ at the midpoint 1.

$$x^{(1)}=(x(t-\Delta t))+\frac{1}{2}\Delta x^{(1)} \tag{G-42}$$

Now the slope is recalculated at mid-point 1 (Fig. G-8*b*):

$$\frac{\Delta x^{(2)}}{\Delta t}=(f(x^{(1)})) \qquad t-\frac{\Delta t}{2} \tag{G-43}$$

This is used to obtain a second approximate value $x^{(2)}$ at mid-point 2.

$$x^{(2)}=\left[(x(t-\Delta t))+\frac{1}{2}\Delta x^{(2)}\right] \tag{G-44}$$

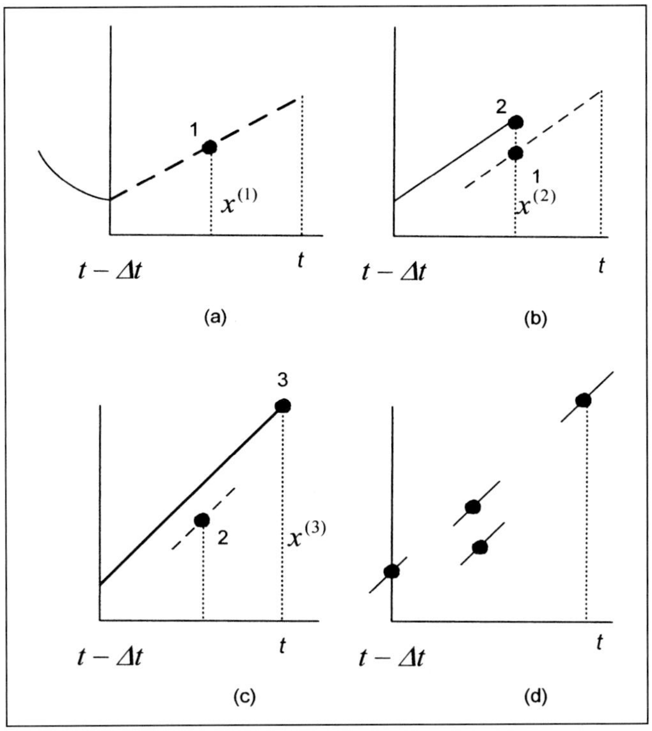

FIGURE G-8 Fourth-order Runge-Kutta method.

Then the slope is evaluated third time, now at mid-point 2 (Fig. G-8*c*):

$$\frac{\Delta x^{(3)}}{\Delta t}=\left[(f(x^{(2)})),t-\frac{\Delta t}{2}\right] \tag{G-45}$$

This is used to get an approximate solution at point 3:

$$x^{(3)}=\left[(x(t-\Delta t))+\frac{1}{2}\Delta x^{(3)}\right] \tag{G-46}$$

Finally, the slope is evaluated for a fourth time at point 3 (Fig. G-8(*d*)):

$$\frac{\Delta x^{(4)}}{\Delta t}=[(f(x^{(3)})),t] \tag{G-47}$$

From these four slopes the final value at t is obtained by using their weighted averages:

$$x(t)=x(t-\Delta t)+\frac{\Delta t}{6}\frac{\Delta x^{(1)}}{\Delta t}+2\frac{\Delta x^{(2)}}{\Delta t}+2\frac{\Delta x^{(3)}}{\Delta t}+\frac{\Delta x^{(4)}}{\Delta t} \tag{G-48}$$

The fourth-order Runge-Kutta formula is identical with fourth-order Taylor series expansion of of the transition matrix, if the differential equations are linear, at least for the autonomous systems with $g(t)=0$ in Eq. (G-8), which becomes:

$$\frac{\Delta x^{(1)}}{\Delta t}=\bar{A}(x(t-\Delta t)) \qquad x^{(1)}=\left(\bar{U}+\frac{\Delta t}{2}\bar{A}\right)(x(t-\Delta t)) \tag{G-49}$$

The second slope is:

$$\frac{\Delta x^{(2)}}{\Delta t}=\left(\bar{A}+\frac{\Delta t}{2}\bar{A}^2\right)(x(t-\Delta t)) \tag{G-50}$$

and

$$x^{(2)}=\left(\bar{U}+\frac{\Delta t}{2}\bar{A}+\frac{\Delta t^2}{4}\bar{A}^2\right)(x(t-\Delta t)) \tag{G-51}$$

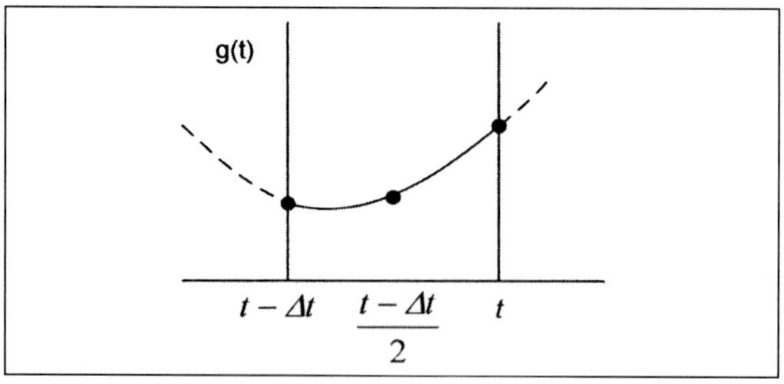

FIGURE G-9 Simpson's rule.

Then the third slope becomes:

$$\frac{\Delta x^{(3)}}{\Delta t}=\left(\bar{A}+\frac{\Delta t}{2}\bar{A}^2+\frac{\Delta t^2}{4}\bar{A}^3\right)(x(t-\Delta t)) \tag{G-52}$$

and

$$x^{(3)}=\left(\bar{U}+\Delta t\bar{A}+\frac{\Delta t^2}{2}\bar{A}^2+\frac{\Delta t^3}{4}\bar{A}^3\right)(x(t-\Delta t)) \tag{G-53}$$

The fourth slope is calculated as:

$$\frac{\Delta x^{(4)}}{\Delta t}=\left(\bar{A}+\Delta t\bar{A}^2+\frac{\Delta t^2}{2}\bar{A}^3+\frac{\Delta t^3}{4}\bar{A}^4\right)(x(t-\Delta t)) \tag{G-54}$$

Finally, the new value is obtained as:

$$x(t)=\left(\bar{U}+\Delta t\bar{A}+\frac{\Delta t^2}{2}\bar{A}^2+\frac{\Delta t^3}{6}\bar{A}^3+\frac{\Delta t^4}{24}\bar{A}^4\right)(x(t-\Delta t)) \tag{G-55}$$

which is the Taylor series approximation of the transition matrix in Eq. (G-33). If $\bar{A} = 0$, that is, if x is simply the integral over the known function $g(t)$, the fourth-order Runge-Kutta method is identical with Simpson's rule of integration, in which the curve is approximated as a parabola going through three known points in $t-\Delta t, t-\Delta t/2$, and t (Fig. G-9). The Runge-Kutta method is subject to instability if Δt is not chosen small enough.

G-11 PREDICTOR-CORRECTOR METHODS

These methods can again be used for ordinary differential equations. Apply trapezoidal rule to state equation [Eq. (G-40)]:

$$x^{(h)}=x(t-\Delta t)+\frac{\Delta t}{2}(f[x(t-\Delta t),t-\Delta t])+f(x^{(h-1)},t) \tag{G-56}$$

In the general time-varying or nonlinear case, the direct solution is not possible. Here h indicates iteration step. The iteration works as follows:

1. A predicator formula is used to obtain a predicted guess for solution at time t.

2. The corrected solution is found in iteration step $h = 1, 2, \ldots$ by inserting the approximate solution.

3. If the difference $x^{(h)}-x^{(h-1)}$ is small, then integration from $t-\Delta t$ to t is completed, otherwise return to step 2.

Equation (G-56) is a second-order corrector formula. To start an iteration process, a predicator formula is needed. This can be obtained by midpoint rule:

$$x^{(0)}=x(t-2\Delta t)+2\Delta t(f[x(t-\Delta t),t-\Delta t) \tag{G-57}$$

The difference in step 3 is used to (a) decide whether step size Δt should be reduced, and (b) improve prediction of the next step. It is generally better to shorten the time step Δt, then to use corrector formula repeatedly in step 2. It is assumed that the difference between the predicted and corrected values changes slowly over the time step.

In addition to second-order methods, there are higher-order methods, and fourth-order predictor-corrector methods seem to be used more often. Amongst these methods are Meline's method and Hamming's method; the latter is usually more stable numerically. The convergence and numerical stability of corrector formulas are more important than those of the predicator formulas. These two should be of the same order in error terms. There are different classes of predictors: Adams-Bashforth predictors, obtained by integrating Newton backward interpolation formulas, Meline-type formulas, obtained by an open Newton-Cotes forward integrating formula, and others. Formulas requiring values at $t - 2\Delta t$ are not self-starting; Runge-Kutta methods are sometimes used to build enough history points.

G-12 RICHARDSON EXTRAPOLATION AND ROMBERG INTEGRATION

Instead of using higher-order methods, the second-order trapezoidal rule is used more than once in the interval between $t - \Delta t$ and t to improve the accuracy. Assume that normal step size Δt is used to find $x^{(1)}$ at t from Fig. G-10. Now repeat the integration with the step size $\Delta t/2$, and perform two integration steps to obtain $x^{(2)}$. With these two values an intelligent guess can be made as to where the solution will end up if the step size were decreased more and more. This extrapolation toward $\Delta t = 0$, is called Richardson extrapolation and gives a better answer.

$$x(t)=x^{(2)}+\frac{1}{3}(x^{(2)}-x^{(1)}) \tag{G-58}$$

The accuracy can be further improved by repeating the integration between $t - \Delta t$ and t with 4, 8, 1, ... intervals. The corresponding extrapolation formula for $\Delta t \rightarrow 0$ is known as *Romberg integration*.

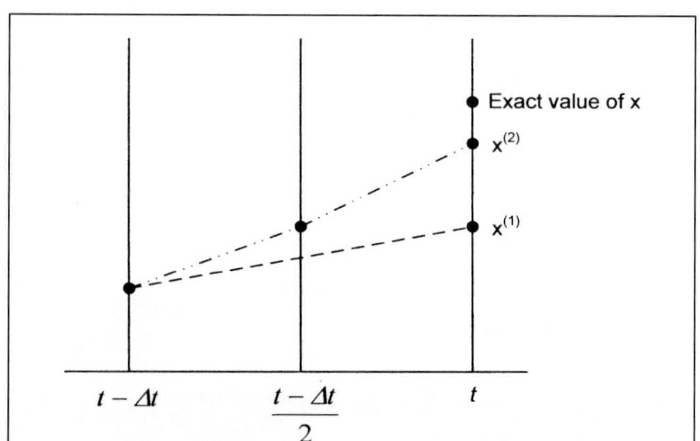

FIGURE G-10 Richardson's extrapolation.

REFERENCES

1. J. G. F. Francis, "The QR Transformation," *Computer Journal*, vol. 4, pp. 332–345, 1961.
2. J. H. Wilkinson, *The Algebraic Eigenvalue Problem*, Oxford University Press, London, 1965.
3. J. E. Van Ness, "The Inverse Iteration Method for Finding Eigenvectors," *IEEE Trans. Automatic Control*, vol. AC-14, pp. 63–66, Feb. 1969.
4. S. Seshu and M. B. Reed, *Linear Graphs and Electrical Networks*, Addison-Wesley, Reading, MA, 1961.
5. F. H. Branin, "Computer Methods of Network Analysis," *Proc. IEEE*, vol. 55, pp. 1787–1801, Nov. 1967.
6. E. J. Davidson, "A High Order Crank-Nicholson Technique for Solving Differential Equations," *Computer Journal*, vol. 10, pp. 195–197, Aug. 1967.

FURTHER READING

EMTP Theory Book, EMTP user Group, www.emtp.org.

A. Ralston, A *First Course in Numerical Analysis*, McGraw Hill, New York, 1965.

Index

T

U

V

W